Errata Sheet

for

ENCYCLOPEDIA OF MINERALS

Willard Lincoln Roberts

George Robert Rapp, Jr.

Julius Weber


Color, page 10   Transpose illustrations for AXINITE,
Hyuga, Japan (#2, column 1) and
AZURITE, Concepcion del Oro, Zacate-
cas, Mexico (#3, column 2).

Color, page 108   Transpose illustrations for SELLAITE,
Vesuvius, Italy (#2, column 2) and
SEMSEYITE, Julcani mine, Huancavelica,
Peru (#3, column 2).

Color, page 117   Transpose illustrations for TELLURITE,
Moctezuma, Sonora, Mexico (#2, column
1) and TENNANTITE, Dresser quarry,
Walton, Nova Scotia, Canada (#3, column 2).

 Van Nostrand Reinhold Company

New York/Cincinnati/Toronto/London/Melbourne

# Encyclopedia of
# Minerals

# Encyclopedia of Minerals

**WILLARD LINCOLN ROBERTS**
Curator of Mineralogy, Museum of Geology
South Dakota School of Mines & Technology
Rapid City, South Dakota

**GEORGE ROBERT RAPP, JR.**
Associate Professor of Mineralogy
Department of Geology and Geophysics, University of Minnesota
Minneapolis, Minnesota

**JULIUS WEBER**
Associate, Department of Mineralogy
American Museum of Natural History, New York, New York
Research Associate, Department of Mineralogy
Royal Ontario Museum, Toronto, Canada

**VNR** VAN NOSTRAND REINHOLD COMPANY
NEW YORK  CINCINNATI  TORONTO  LONDON  MELBOURNE

Van Nostrand Reinhold Company Regional Offices:
New York Cincinnati Chicago Millbrae Dallas

Van Nostrand Reinhold Company International Offices:
London Toronto Melbourne

Library of Congress Catalog Card Number: 74-1155
ISBN: 0-442-26820-3

Manufactured in the United States of America

Published by Van Nostrand Reinhold Company
450 West 33rd Street, New York, N.Y. 10001

Published simultaneously in Canada by Van Nostrand Reinhold Ltd.

15 14 13 12 11 10 9 8 7 6 5 4 3 2 1

**Library of Congress Cataloging in Publication Data**

Roberts, Willard Lincoln.
    Encyclopedia of minerals.

    Includes bibliographical references.
    1. Mineralogy—Dictionaries.    I. Rapp, George
Robert, 1931–      joint author.    II. Weber, Julius,
joint author.    III. Title.
QE355.R6          549'.03          74-1155
ISBN 0-442-26820-3

*This book is respectfully dedicated to the Friends of Mineralogy*

# FOREWORD

In its descriptive aspects, the science of mineralogy requires the acquisition and the permanent preservation of mineral specimens. The reasons for this are varied. They include the need to provide documentation for published descriptions, especially of new minerals, and the provision of described specimens on which new research can be brought into context with earlier studies. More importantly, minerals often are quite variable in their characters, making it difficult to attain a complete written description, or to capture in words the subtle variations in color and other outward characters by which very similar minerals can be distinguished and, by the trained eye, brought to sight identification. Instruction in mineralogy thus necessarily involves the direct examination of hand specimens, in classroom study trays and in systematic exhibit collections, to supplement the textbook accounts.

Colored drawings of specimens were employed in numerous mineralogical works of the nineteenth century and before, notably in the "British Mineralogy, or Coloured Figures Intended to Elucidate the Mineralogy of Great Britain" by James Sowerby, published in London in 1804–17. Photography later became widely used although not on the comprehensive scale of the present work. This book describes approximately 2,200 minerals. It also provides the numerical data afforded by X-ray powder diffraction and optical methods of study, that are needed for a rigorous characterization, together with color photographs of representative specimens of many of the described minerals. Here, an advantage of photography appears in that deliberate selection can be made of specimens that best reveal the features to be depicted. For those to whom recourse to mineral collections is inconvenient or, in the case of rare minerals, unavailing, or for those who wish only a quick reference or overview, the color plates of this useful encyclopedia are invaluable.

CLIFFORD FRONDEL

# INTRODUCTION

A book, whether textbook, intellectual history, compilation, or whatever, must be written with some audience in mind. We have written for a very broad audience indeed (perhaps too broad). University professors and students, research institute scientists, museum curators, knowledgeable collectors, (and surely others!) find a regular need to check up on some detail, a forgotten or new mineral name, a misremembered formula. To date, a handy desk volume containing the most frequently needed items has not been available in English. Excellent volumes and sets abound, but each is too specialized, too limited in coverage, too technical, or too out-of-date to serve as a handy quick-reference volume.

To satisfy this apparent gap, Willard Roberts and George Rapp, Jr. set about in 1967 to compile such a reference encyclopedia. Because the world of minerals is (or can be) dramatically visual, Julius Weber joined the effort in 1972 to bring this added dimension to the written text.

We feel the need, in a work of this sort, to explain to the user the rationale used in the selection of specific mineral data from the abundant, often conflicting, mineralogical and crystallographic literature. Nearly one-third of the mineral species have seriously conflicting (i.e., incompatible) data published in major recognized sources. This has often prompted us to use a formula from one source, X-ray data from a second, optical data from a third, and so forth. Although we take full responsibility for our choice of the data for inclusion, we accept no responsibility for ultimate accuracy of these data. We have chosen what we believe to be the best information available and we have used the symbol (?) to indicate our lack of full acceptance of many data.

Indeed, it would take at least a small laboratory research project to resolve each of the conflicts and questions. Unfortunately, we have not found it possible to remove all the conflicts. Occasionally, we have included cell data not compatible with X-ray or chemical formula data, etc. To "remove" these conflicts would have unnecessarily delayed publication. Anyone seeking to identify interesting mineralogical problems needs only to attempt a complete description of a small mineral group to uncover the unsatisfactory state of our present knowledge. For many species that are included (for the sake of completeness) we have indicated that the species status is in doubt or uncertain. In retrospect, possibly we should have used this disclaimer more frequently.

We also have a category generally listed as "inadequately described mineral." One can only be appalled at the frequency with which mineral names are proposed (and now regularly cited) for which we have little but a tentative formula and some optical data. It was only our concern for the harrowed user, who wants to know what information is available, that caused us to include many of these interlopers among the authenticated.

The actual abstracting and recording of data were done on a standard worksheet which we designed at the onset of this work. Thus, we were firmly chained to our original inclinations and biases. Our only regrets along this line relate to our unrealized expectations that our rigid format would lead to a high degree of internal consistency. Even at the corrections stage of the final manuscript we found that another "pass" through the 5,000 worksheets to insure another "consistency" took days if not weeks. We trust that those inconsistencies (in the form of the data) will not unduly hamper our readers.

The mineral species are presented in alphabetical order because a more refined (e.g., Dana system) order might well render the volume difficult to use by a large group of potential readers not familiar with structural/chemical classification. It remains our hope that this volume belongs as much on the encyclopedia shelf of the small town library as on the desk of the established mineral collector. As mineralogists, we are acquainted with the inherent problems of end-member names vs. group names and related aspects of nomenclature. Rather than define precisely our "system" in this forward, we believe a few minutes of browsing through your favorite minerals will acquaint you with our treatment of the naming problem. We have retained common usage of the prefixes *meta*, *para* and so forth as part of mineral names, even though we believe that certain of these usages are ill-advised.

We have not attempted to include all synonyms, varietal names, gem names, spelling variants, descriptive names, mixtures, group names, chemical names, etc., found in the literature. Only those names most frequently encountered in contemporary use have been included. With rare exception, we have not listed discredited species. Nearly every issue of the American Mineralogist contains a section by Michael Fleischer on new minerals and discredited species. Readers interested in the more esoteric nomenclature or in historical usage and development are referred to Dana's "System of Mineralogy," sixth and seventh editions, and Hey's "Chemical Index of Minerals" (see below for more complete references to these works). Abbreviations, symbols, and technical words used are defined later in this introduction or included in the glossary.

Although the background of one of the authors (Rapp) is in mineral (phase) equilibria studies, it was felt that the abundance of recent data in this field precludes any fair treatment of it in a compilation of this sort. Unlike the excellent work now being published in phase equilibria and crystal-structure analyses, the work on systematic optical constants of minerals, particularly their relations to newly established structural variations, has lagged regrettably.

We have not included artificial "minerals" except for a very few such as *austenite* which find their way very frequently into general discussions in the common mineralogical literature. We *have* included data (particularly cell constants and X-ray data) taken from studies on the artificial analogues of minerals when data on natural species were unavailable, highly suspect, or out-of-date. When data on artificial phases have been included, they are normally so indicated.

The mineral properties we have chosen to include are those we believe most likely to be required from a quick-reference encyclopedia. Where a standard property is omitted, it can be assumed that we were unable to find an acceptable entry. Symbols used are generally those in common usage. Sometimes physicists or metallurgists use different symbols than mineralogists. Because of expected usage, we have chosen those familiar to mineralogists and collecters.

Following is an item by item introduction to each of the properties.

*Chemical Formula:* Until the crystal structure is known or at least until an accurate cell (size and symmetry) is determined, the chemical formula may present more of a problem than is generally realized. For some mineral species, the amount of material available for analysis may have been too small to allow accurate determination of $H_2O$ or OH. Fairly frequently, we had the "choice" of three or even four non-equivalent formulas. We have indicated the ionic charge on iron and manganese.

*Hardness:* Mineralogists have long used *Mohs hardness*, the scale of ability of one material to scratch another. This scale is not linear as the numbers 1–10 might indicate but rather, as is often the case in science, is nearly a logarithmic scale if the indentation hardness is accepted as the quantitative measure. Mohs was exceedingly perceptive in his choice of the ten minerals to represent these hardness intervals, as is attested by the longevity of this scale. Mohs' scale reads as follows:

| | | | |
|---|---|---|---|
| 1 | Talc | 6 | Orthoclase |
| 2 | Gypsum | 7 | Quartz |
| 3 | Calcite | 8 | Topaz |
| 4 | Fluorite | 9 | Corundum |
| 5 | Apatite | 10 | Diamond |

In a few descriptions we have entered a Vickers or Talmage micro-indentation hardness without converting such data to the Mohs scale.

*Density:* By definition, density equals mass per unit volume and is therefore dimensional. The density entries are given with the *grams/cm³* assumed. The density observed in most natural species varies appreciably with chemical solid solution. Although density ranges are often entered, they are not necessarily the widest ranges possible. Densities *calculated* from cell constants and formula data are distinguished from those *measured* on natural specimens.

*Cleavage:* Most crystalline substances contain certain rational planes in their structure (usually with simple Miller indices) that are potential crystal faces or, when the bonds across the plane are few in number or weak, *cleavage planes*. The description of a cleavage must contain its crystallographic orientation, as given by its Miller index, and the quality and ease of production of the cleavage surface. Such quality is usually described as *perfect, good, distinct, indistinct*, and so forth. Cleavage data have not been determined for some species with a fine-grained or massive habit.

A special note on symbols representing crystal planes:
( )    Miller indices included in parentheses denote a single crystal plane.
{ }    Miller indices included in braces denote all the planes of the form generated by the symmetry of the class.
[ ]    Miller indices included in square brackets denote all the planes in the given zone.

*Habit:* The shape imparted to a mineral by the relative development (or lack) of crystal faces is referred to as habit. Common habits are described as fibrous, massive, columnar, lamellar, and so forth. Habit is controlled mainly by crystal structure although both the physical and chemical environment during growth may determine the habit.

*Color-Luster:* Many, if not most mineral species are found to exhibit a range of colors depending on chemical solid solution or finely dispersed extraneous colorants. (See "Color of Minerals" by George Rapp, Jr. Houghton Mifflin, 1971, 30 pp.) Most of the basic color in minerals is imparted by transition elements, especially iron, manganese, titanium, chromium, and copper. Despite its many variations and complexities, mineral color may provide important clues to identification. A more powerful tool for identification is the color of the *streak* of a mineral, the mineral in powdered form, most often produced by rubbing the mineral on a sheet of unglazed porcelain. *Luster* is the nature and degree of light reflectance from the surface of a mineral. Luster may be *metallic* or *nonmetallic*. Several varieties of nonmetallic luster are normally described as *vitreous, greasy, resinous, adamantine*, etc. All faces or other surfaces on a given mineral do not necessarily exhibit the same luster.

*Mode of Occurrence:* To give even a marginally adequate petrologic description desired by geologists or to provide locality summaries desired by collectors is far beyond the scope of this volume. We have attempted to present only the barest information to give the reader unfamiliar with a given species some idea of the "nature of the beast." We have also included in this section some new and unreported occurrences.

*Best Reference in English:* Here, we have been obliged frequently to list two or more references where all references are inadequate or do not contain good bibliographies or are in conflict in a manner that we cannot easily resolve. The best reference is not necessarily the one that contains the best *or* most up-to-date *or* more used (by us) data. Rather, the best reference was chosen on the basis of being the most complete or handiest reference to consult for additional information.

The crystal data entries presented special difficulties. Many kinds of scientists (physicists, metallurgists, biochemists, physical chemists, mathematicians, etc.) have long contributed important mineral data. Various schemes of geometric representation and crystallographic nomenclature have grown up with accompanying use, misuse and misunderstanding. The assumed mathematical rigor of crystallography seems to melt away when it is not known, for example, which of the six orthorhombic orientations were used in many references when the one claimed is manifestly impossible. Rules, conventions, and physical "constants" (for example in X-ray measurements) have all evolved through time, thus adding to the problem. Although at first it seemed that if enough time was available nearly all questions and inconsistencies could be resolved, we are now prepared for a long series (welcome for the next edition) of letters pointing out that for some entries our space group, Z, and lattice constants are *mutually exclusive*. Time and possibly wit were insufficient for the task.

*Crystal System:* Data are lacking for an exceptionally large (it seems to us) number of commonly accepted species. Also, we haven't done any better than most in arriving at and transforming all rhombohedral/hexagonal mineral data into one coherent pattern. Occasionally we have followed the common pattern of "having it both ways" where rhombohedral lattices may be referred to both rhombohedral and hexagonal axes.

*Class:* Classes are designated by their point group symmetry. Note that as mineralogists we use the *second setting* for monoclinic crystals. The monoclinic holohedral point group symmetry is 2/m and the one symmetry direction is defined as *b*. Note also that the rhombohedral R cell is a triple cell when indexed on hexagonal axes.

*Space Group:* Space group designations are given exclusively in the Hermann-Mauguin notation. Readers desiring an explanation of space group (and other crystallographic and crystal chemical) nomenclature and symbols are referred to:

"Elementary Crystallography," by Martin Buerger, (John Wiley & Sons, 1956).

"International Tables for X-Ray Crystallography," edited by N. F. M. Henry and K. Lonsdale, Kynoch Press (Birmingham, England) three volumes, 1952–63.

"Crystallography and Crystal Chemistry," by F. Donald Bloss, (Holt, Rinehart and Winston, 1971).

Z, the number of chemical formula units per unit cell, is governed by the stated formula and the lattice constants. We have recalculated a large number of Z's where the literature data are incorrect.

*Lattice Constants:* For many minerals more than one set of cell data was available (polymorphic, rhombohedral/hexagonal, or otherwise nonequivalent). We have tried to include these and to remove any ambiguity by attaching space group and Z data to each set of lattice constants. All, or nearly all, X-ray measurements reported before 1949 were expressed in kX units. At first we converted a few of these to angstrom units (1 kX = 1.00202Å). Later we let all reported kX measurements stand but indicated them in the text. Our procedure was changed, in part, because we do not want to represent this book as the best reference for persons seeking third decimal place information.

*Strongest Diffraction Lines:* Readers are referred to the Powder Diffraction File of the Joint Committee on Powder Diffraction Standards for the latest powder diffraction information. However, not all mineral species are yet represented in the File; many of our entries are from the literature and a few are from our own investigations. Some older work and some recent work on poor quality material use a semi-quantitative description of relative intensity where VS = very strong, S = strong, W = weak and VW = very weak. We have converted these to the normal scale using the following conversions; VS = 100, S = 60, W = 40, and VW = 20.

*Optical Constants:* When the wavelength is not specified, we believe it is safest to assume Na light. The optical constants of most minerals tend to vary over wide ranges because of chemical substitution. Although we have tried to include a statement of the range, we have not attempted to be comprehensive or inclusive.

We have chosen, as is often the case in condensed summaries of scientific data, to omit referencing each entry. Indeed, we have taken so many liberties with some of the data in our selection, synthesis, and cross-correlation process, that we feel solely responsible for many entries. Instead, the books and journals we *most frequently* consulted for data used in this compilation are listed below. This list is not exhaustive, but can serve as a guide for those seeking additional information on minerals.

*Acta Crystallographica*, Section B: Structural Crystallography and Crystal Chemistry, published for the International Union of Crystallography.

*American Mineralogist*, Journal of the Mineralogical Society of America.

"ASTM Powder Diffraction File," published by the Joint Committee on Powder Diffraction Standards.

*Canadian Mineralogist*, Journal of the Mineralogical Association of Canada.

"Calculated X-ray Powder Patterns for Silicate Minerals," by I. Y. Borg and D. K. Smith, Geological Society of America, Memoir 122.

"Clay Mineralogy," by R. E. Grim (McGraw-Hill, 1968).

"Crystal Data: Determinative Tables," Third Edition, Vol. II, by Donnay *et al.*, American Crystallographic Association, 1973.

"Crystallographic Data for the Calcium Silicates," by L. Heller and H. F. W. Taylor, Her Majesty's Stationery Office, 1956.

"Crystal Chemical Classification of Minerals," by A. S. Povarennykh, Plenum Press, 1972.

"Elements of Optical Mineralogy," Fourth Edition, Part II, Descriptions of Minerals, by A. N. Winchell and H. Winchell (John Wiley & Sons, Inc., 1951).

"Feldspars," by T. F. W. Barth (Wiley-Interscience, 1969).

"Glossary of Mineral Species 1971," by M. Fleischer, published by Mineralogical Record, 1971.

"Index of Mineral Species and Varieties Arranged Chemically (and Supplement)," by M. H. Hey, British Museum (Natural History), 1955.

*Mineralogical Magazine*, Journal of The Mineralogical Society.

"Mineralogische Tabellen," Third Edition, by H. Strunz, 1957.

"Mineralogy of Rare Elements," Volume II of Geochemistry and Mineralogy of Rare Elements and Genetic Types of Their Deposits, K. A. Vlasov, Editor, Academy of Sciences of the USSR, in translation, 1966.

"Mineralogy and Types of Deposits of Selenium and Tellurium," by N. D. Sindeeva, in translation (Wiley-Interscience, 1964).

*Mineralogical Abstracts*, published jointly by The Mineralogical Society of Great Britain and The Mineralogical Society of America.

*Mineralogical Record*, published by the Mineralogical Record, Inc.

"The Ore Minerals and Their Intergrowths," by Paul Ramdohr, Pergamon Press, 1969.

*Pyroxenes and Amphiboles: Crystal Chemistry and Phase Petrology*, edited by J. J. Papike *et al.*, Special Paper Number Two of the Mineralogical Society of America, 1969.

"Rock-Forming Minerals" (in five volumes), by W. A. Deer, R. A. Howie, and J. Zussman (John Wiley, 1962).

*Structure Reports*, published for the International Union of Crystallography.

"Systematic Mineralogy of Uranium and Thorium," by Clifford Frondel, *USGS Bulletin* 1064.

"System of Mineralogy (of Dana)," Seventh Edition, Volumes I–III, by C. Palache, H. Berman and C. Frondel (John Wiley).

The authors owe a debt not only to the above sources and the original researchers, but to others who have contributed directly to this volume. First and foremost, Jean Roberts deserves the equivalent of co-authorship for her almost daily contribution to the tedious task of search, review, critique and compilation. Larry Johnson and Alta Walker also contributed to this task. Paul B. Moore, David Garske, and Tibor Zoltai have aided the authors with information and wise counsel. Finally, without the patient assistance of Van Nostrand Reinhold Managing Editor, Mrs. Alberta Gordon, this volume could not have gone to press. To the aforementioned, to others who have kindly assisted with special problems, and to those who, in the future, will call our attention to oversights and inaccuracies, our deep gratitude.

WLR
GRR

# PREFACE

The last twenty-five years has been an era of extraordinary progress in practical and theoretical photographic technology. Despite the vast outpouring of new optics and both physical and chemical material, the extent of application of these new tools in the sciences has been uneven. This is explained in part by the needs of each discipline and the difficulties, especially economic, encountered in meeting these needs.

I believe that all of us in mineralogy have much to gain from a greater use of photography in our science. I note with gratification the increased interest in macro and micro imaging systems and imaging and recording techniques. Mineral lectures are now more profusely illustrated; the photographs are better and impart more information. The story of "beautiful minerals" has stimulated the publication of many illustrated books of varying photographic quality. Finding the proper specimens to illustrate exceptional mineral and esthetic quality requires that the great museums and private collections be culled for their finest examples. When such minerals are photographed by careful, knowledgeable and dedicated technicians, the result is a good alternative to a personal view of the original.

In this work the photography of the minerals presented a very special problem. For the same reason the choice of specimens offered a unique opportunity. In keeping with the high purpose of my colleagues in the text of the book, I suggested the use of the mineral specimen with micro crystals as the basis for the photographic documentation. In almost every instance, specimens of similar or equal quality are available to all who collect and study minerals.

The rationale for the production of the photographs is a very personal one. Its origin and subsequent growth began a long time ago. Let me explain.

My life has been a journey through two absorbing scenes—minerals and microscopy. It is easy to fall in love with minerals; the microscope is another matter: I do not quite know how my love for the microscope began, but I do know about the minerals. We were very fortunate—when I was a child I lived a few miles from the Brooklyn Children's Museum. This became my second home. This little museum was staffed by able, devoted, naturalist-teachers, and directed by an extraordinary woman—Anna Billings Gallup. Most of us spent most of our weekends in the museum park and some even rushed through school homework to spend an exhilarating hour or two there during the week. The mineralogy teachers were outstanding. This was especially true of Mr. John Claudius Boyle, a man whom we all loved and who, in later life, honored me with his friendship. Jack Boyle was a born teacher, a dedicated mineralogist with a natural instinct of child psychology. We followed him about with devotion—we devoured his information as a rabbit devours lettuce. All micromineralogists should think of him with gratitude for he enriched our scene with Neal Yedlin and Lou Perloff and, through them, thousands more.

My interest in microscopy was also born when I was quite young. Mrs. Gallup lent me a monocular, double objective Leitz microscope and thereby changed my whole life. Most of my adult years were spent working with transmitted light microscopy in biological and medical research. During that time I have been fortunate to enjoy the fruits of the continuing changes brought about by new glass, computers, and imaginative, practical application of revolutionary theory in optical design and function. The microscope has been turned from a tool for visual study—mainly qualitative, with simple photographic accessories—to a superb instrument for quantitative analysis and precise photographic recording.

While these advances were mainly in the transmission instruments, the microscope used in studying specimens in reflected light has also been greatly improved and fitted with the necessary components for photomicrography, polarization and interference. Radiant sources with optically designed condenser systems have been standardized, improved and built-into the transmission microscope; similar units are now available for reflected light use. This insures greater versatility and control in illumination and desired image formation.

Color film is to me still a modern miracle. I had used it when it was made up of starch granules, and it is difficult for young people now to imagine the excitement of Kodachrome, the phenomenon of Polaroid. In teaching and research in the sciences, color films have been of inestimable value. The royal color of hemoglobin bathes the cells of living things, and the histological and cytological stains are brilliant in serving to differentiate the various components. Such equally brilliant colors await us in the micromineral specimens.

It was natural for me to turn to the microscope to examine specimens with small crystals. The splendor of the scene, the wealth of varied colors and form, the intricacies of structure combined to complete a breath-taking view. I was an immediate captive. I wasted no time in setting up a microscope specifically to study minerals. Now the hidden problems began to show themselves. The conventional "thin slide," biological transmission microscope is not designed optically or mechanically to meet the needs of the increased dimension of even the small mineral specimen. We must turn to the stereo microscope, and we are indeed fortunate that the use of this instrument in industrial, microcircuitry has promoted and speeded many useful improvements such as modular design, and Zoom optics with extended magnification range and accessories for photographic recording. Present stereo microscopes provide sharp images with improved color corrected lenses whose greater working distance enables us to study most specimens with ease. Where then is the problem?

All of the advantages described above apply to the production of the visual image of the specimen; it is the recording of this image that presents the problem. To properly photograph the crystal specimen we must meet a number of criteria. The specimen must be oriented—this is best accomplished if it has been previously mounted. In this way the specimen is secure, protected and readily available for observation or storage. Hence the name "micromount."

These specimens with small crystals offer us an almost inexhaustible cornucopia of mineral and crystal detail, crystal structure, diverse habit, association, paragenesis and metamorphosis; all are accessible for study and record at a resolution beyond the ability of the naked eye and simple lens. The magnification range of the reflecting microscope and the resolution attained with this instrument enhances the value of the photographic record of the micro-crystal. In this way we can obtain close studies of the subjects with much greater definition than the conventional photographic or "macro" lens affords.

It is useful to define macro and micro. The distinction is not one of magnification, but rather the optical system used. A photomacrograph is obtained with the use of a single optical imaging system with or without bellows or tubes. A photomicrograph is made by imaging and magnifying with one system and then continuing the imaging and magnification with another, (i.e., objective x ocular). In this system (the microscope) the light gathering power and resolution is considerably greater. Almost all the reproductions in this book are photomicrographs. Space precludes a description of the apparatus, the special accessories and radiants and the techniques used in this work. These will be reported separately.

I would like to emphasize that a photograph is only a compendium of the consideration and regard for many parameters present in the subject and in the image forming and image transfer apparatus. If the picture is made carelessly, it will equal nothing more than the record of a scant squint. It can, however, be a stimulating, fact-divulging documentation worthy of prolonged study and open to the extraction of considerable information. It will be a pleasure to look at and will represent the best permanent record, especially where inexorable, natural changes alter or destroy the original subject.

I conclude with this observation: In these specimens, nature speaks truly of herself—she tells us a many faceted story with marvelous continuity that provides us with a bridge of understanding from the far back unknown, mysterious, maphitic beginnings to the fantastic format of the barely understood, provocative present.

Nature sets up the conditions, we proscribe the problems. In the sub-visual mineral world, the range of colors, shapes, designs is astonishing. It is all so familiar and yet all so new. The surprise lies in the symmetry, the mystery, and delight of each "micro acre."

There is an architecture above classic, above romantic, above modern; a geometry, an attitude, a formality and an esthetic unto itself; simple and sumptuous, controlled and wild, random and related, exotic and quiet; never drab, never unapproachable, never imitative, always original.

I hope that the reader will find my efforts useful, and that he will be encouraged to photograph his own fine mineral specimens.

I congratulate all of you who share my interest. I envy all of you who have more beautiful, more instructive specimens than I have. I fervently hope our numbers grow.

JULIUS WEBER

# ACKNOWLEDGMENTS

I welcome the opportunity to record my appreciation to all my friends who have helped me in many ways, especially the following:

Louis Perloff—for decades this shy, retiring savant has served mineralogists with advice, demonstrations, lectures and cheerful willingness to tackle tricky problems of identification. His friendship has been the greatest asset in my mineral work. Long before this Encyclopedia, at a time when my own minerals were meager, Mr. Perloff brought a thousand of his finest specimens to my laboratory to enable me to determine the program for optimum photographic reproduction of the varied surfaces, color and magnification requirements confronting me. His zeal for mineralogy is unexcelled; his mineralogical memory phenomenal. During the work on this book his unerring advice has been a beacon that illuminated my way through the panorama of problems. His collection of superb micromounts, second only in size, but unexcelled for excellence, is the source of many illustrations. He was chiefly responsible for the identification, classification and initial choice of photographs from the thousands of transparencies made.

Neal Yedlin—the towering figure in micromineralogy whose cognomen "Mr. Micromounter" is so well deserved. He is responsible for the resurgence in the art of micromounting since 1948. His informative columns in *Rocks and Minerals* and the *Mineralogical Record* stimulated the great increase in the number of collectors. His lectures, demonstrations and aid in identification is a well-spring of continuing support to micro mineralogists everywhere. His collection of micromounts, the largest in the world, is the source of many of the photographs. He also helped with the checking of the identification and with classifying and choosing the photographs.

Clifford Frondel—Dana Prof. of Mineralogy of Harvard University, renowned mineralogist, who placed the entire Bement collection at my disposal. Many of the fine specimens were picked from the more than 2,500 specimens assembled in this collection. His encouragement and advice is a constant source of strength.

Joseph Mandarino—Curator of Dept. of Mineralogy, Royal Ontario Museum, Toronto, Canada, for his guidance and encouragement and his help with beautiful specimens.

Paul Ney—Prof. of Mineralogy, University of Cologne, West Germany; Curator of The Mineralogical Museum. A dear friend who has been of the greatest help to me from the very beginning of the work. I am indebted to him not only for the many choice specimens which he carefully picked for photomicrography and for his advice, but especially for his early recognition of the substance of my efforts and his constant encouragement.

Brian Mason—prestigious investigator of meteorites, Curator of Meteoritics, Dept. of Mineral Sciences, Smithsonian Institution, who has broadened our horizons of the cosmos and introduced me to our planetary building blocks.

Willard Roberts—for his help in obtaining rare species and phosphates, and for his continuing help and guidance.

Marvin Deshler—mineralogist and magnificent micromounter who devoted long hours to carefully mounting hundreds of selected species for photography. His help in this way increased measurably the number of illustrations, for his fine work freed me from specimen preparation and permitted me to spend more time on photography.

Pierre Bariand—Curator of The Mineralogical Museum, University of Paris, for his gifts of specially chosen specimens.

Philip Goodell—Asst. Prof., City College of the City University, New York, for his help with many beautiful and instructive specimens, especially the rarer sulfo salts of Peru.

John van Itallie—lover of nature and scholar of shells. He is president of the diamond company which bears his name. He patiently scrutinized the shipments of diamonds passing through his firm selecting the best for color, shape and surface features. More than 500 diamond photographs were made during the course of the work.

John Anthony—Prof. of Mineralogy, University of Arizona in Tucson, for his gift of fine mineral specimens.

David Cook—Dept. of Mineral Sciences, Harvard University, who helped me greatly with specimens from his personal collection.

Harry Schwartzmann—Pres. of the National Instrument Co. of Mamaroneck—without his engineering, imaginativeness, inventiveness and patience most of the necessary adaptations and instrumentation changes for the microscope would not have been available.

Joseph G. Blum, Esq.—enthusiastic worker in macro and micro stereo photography, for his patience and constant help and advice.

Andrew Azan—Pres., Aristo Grid Lamp Co.; pioneer in the development of the cold cathode lamp. He fabricated to specifications the special tubular radiants with synchronized electronic flash and the necessary power sources which proved so valuable in the photography of metals.

Sidney Braginsky—Manager of Precision Instruments, Olympus Corp. of America—for his continuing help and cooperation with the Nomarski Interference Microscope. This instrument was used in the special surface photomicrographs of the diamonds and many other crystals.

Mel Zane—my assistant, for his patience and forebearance, who helped me during many long hours of exacting chemical processing, mounting transparencies and divers other laboratory labors.

Herbert Ohlmeyer—for his great help in the electron micrography comparison studies.

Pat Daley—for his advice and help on the special color film used in the photography.

Jerrine Anthony—for her kind help in labeling transparencies.

Eliot Kahn—for his help in the preparation and storage of the gross and micro specimens.

Although my personal collection of more than 9,000 micro crystal specimens has been an important source for the photographs, the following collectors were helpful with specific species or individual mounts.

Violet Anderson
Joel Arem
Henry Barwood
Douglas Berndt
Jules Bernhardt
George Bideaux
Robert Boa
Roger L. Bostard
James G. Camilleri
Albert H. Chapman
Ben Chromy
Herbert and Geneva Corbett
Philip Cosminsky
Myer Crumb
Rock Currier
Susie Davis
Ray DeMark
Paul Desautels
Virginia Deshler
Mary Dodds
Adolph F. Dosse
Steve G. Dulla
Peter J. Dunn

Glenn Elsfelder
Peter Embrey
Richard Gaines
Robert I. Gait
Joey Galt
Barnett Goff
Ronald Gooley
Hatfield Goudey
Richard Green
Robert Griffis
Jean Hall
Richard and Elna Hauck
William A. Henderson
Jean Hiebert
William Hunt
George Jellenik Jr.
Ole Johnsen
Russ Kenaga Jr.
Vandall King
Alex Kipfer
Steven Kiss
M. Z. Kissileff
Rustam Z. Kothavala

William Kurtz
Ervan F. Kushner
Robert Lambert
William Larsen
Wayne and Donna Leicht
Charles Lewis
Roger and Peg Marble
John H. Marshall Jr.
A. L. McGuinness
Frank Melanson
Robert Mudra
William Mueller
Lennart Narlund
Manuel Ontiveros
Cynthia Peat
Ole Petersen
William Pinch
Frederick Pough
Julian Reasenberg
Mike Ridding
Ivan Robinson
Art Rocker
W. Jack Rodekohr

Leo Rosenhahn
Abe Rosenzweig
Argimiro Santos
Gene Schlepp
Bruno Scortecci
Curt Segeler
John C. Seguin
Milton and Hilda Sklar
Phyllis Sonnenberg
Francesco Spertini
Hans Stalder
Muriel Starke
Terence Szenics
Al Valenti
Frances Villemagne
William Wall
Charles and Marcelle Weber
David Wilber
Betty Williams
Carroll E. Withers
Anthony Worth
Jack R. Young
Julius and Miriam Zweibel

especially also—

| | | |
|---|---|---|
| Vern Brooks | Peter Larson | Peter Hurrell |
| Milo Olmstead | James Honert | David H. and Susan Garske |
| Frank Tinsley | E. P. Bottley | Archie Le Croy |

Help from the following institutions and museums is also gratefully acknowledged:

American Museum of Natural History, New York, N.Y.
Harvard University, Boston, Mass.
Smithsonian Institution, Washington, D.C.
British Museum of Natural History, London, England
Royal Ontario Museum, Toronto, Canada
University of Arizona, Tucson, Ariz.
Mineralogical Museum, University of Cologne, Germany
Mineralogical Museum, University of Berne, Switzerland
Mineralogical Museum, University of Copenhagen, Denmark
Washington State University
City College of the City University of New York
Mineralogical Museum, University of Paris

*Dedicated to Mary Weber*

*Whose patience, understanding and help, shortened the downcast hours and lengthened the sunny fruitful days.*

**JW**

# GLOSSARY

**acicular** Needle-shaped, spiny.

**amorphous** Lacking a crystalline structure, i.e., a solid phase without a symmetrical internal arrangement of atoms.

**amygdaloidal** Containing vesicles normally filled with secondary minerals such as zeolites.

**angstrom unit (Å)** A measure of distance equal to $10^{-8}$ centimeters.

**anisotropic** In crystal optics, the characteristic whereby physical properties vary with crystallographic direction; all crystals except those in the cubic system are anisotropic.

**basal** Referring to planes or faces that form parallel to the base (bottom) of a crystal.

**biaxial** Having two optic axes and three indices of refraction.

**birefringence** An optical property of crystals, whereby a beam of light upon entering the crystal is split into two beams which travel with unequal velocities.

**bonding** *See* ionic bond, covalent bond, metallic bond.

**Bravais lattice** Synonym for crystal lattice; named after the French mathematician who determined that there are only 14 possible crystal lattices; *see* crystal lattice.

**cell** *See* unit cell.

**chatoyancy** An optical property of some minerals, whereby narrow bands of wavy sheen are seen in reflected light; the bands move as the mineral is turned.

**clay** (1) A mineral fragment having a diameter less than 1/256 mm; (2) a sediment having a predominance of clay-sized particles; (3) a mineral group of phyllosilicates.

**class** *See* crystal class.

**cleavage** The property of a mineral whereby it is constrained to break along regular crystallographic planes (indicative of the internal crystal structure of the mineral); cleavage is named by combining the Miller index of the crystallographic plane with an evaluation of the perfection of the cleavage.

**color** Result of interaction of light with composition and structure (of a mineral); mineral color may be radically changed by trace impurities of transition elements.

**concretion** A compact, rounded sedimentary rock mass formed by localized cementation around a nucleus; concretions often have odd outlines that suggest a spurious relation to similar-looking objects such as turtle shells, eggs and bones. (*See* septarian nodule.)

**covalent bond** The crystalline bond formed between two chemical elements by sharing electron orbitals; a shared-electron bond.

**contact metamorphism** Thermal metamorphism related to the intrusion of molten rock whereby pre-existing rocks near the contact with the hot molten rock are metamorphosed.

**cryptocrystalline** Rock textural term signifying that the grain size of the minerals contained in the rock is so small that the minerals cannot be distinguished even with the aid of a microscope.

**crystal axes** Three directions through the crystal that often parallel prominent edges between crystallographic planes; these three directions provide the geometric framework to which all crystallographic properties can be referred.

**crystal class** One of 32 possible combinations of symmetry elements; mathematically synonymous with point group; each of the 32 point groups are characterized by a symbol denoting the combination of symmetry elements unique to the point group (e.g., 2/m, $\bar{3}$, 4/m, $\bar{4}$3m).

**crystal face** The external planar surfaces that bound a crystal; crystal faces are parallel to planar arrangements of atoms in the internal crystal structure and are designated by Miller indices.

**crystal form** The assemblage of identical (symmetrically equivalent) faces on a crystal (e.g., all six faces of a cube).

**crystal lattice** The regular, symmetrical arrangement of hypothetical points (positions) in a crystal whereby each point has identical surroundings (i.e., direction and distance to other identical points in the crystal).

**crystallization** The process whereby crystals are formed from a gaseous, liquid, molten, or dispersed state.

**crystallographic plane** Any plane within or bounding a crystal which can be mathematically related to the crystal axes.

**crystallography** The study of crystals and the crystalline state.

**crystal structure** The orderly geometric arrangement of atoms in a crystal; each crystal structure conforms to the geometry of one of the 14 crystal (Bravais) lattices and one of the 230 space groups.

**crystal system** One of the six groups of crystal classes; based on the internal symmetry and a specified arrangement of reference crystal axes; the crystal systems are: cubic, hexagonal, tetragonal, orthorhombic, monoclinic, and triclinic.

**cubic system** Synonymous with isometric system, the crystal system with the highest symmetry.

**dendritic** Crystallizing with an external form having a branching pattern.

**density** The mass per unit volume of a substance; usually given in grams per cubic centimeter; water has a density of 1 g/cm$^3$, gold 19.3 g/cm$^3$.

**diffraction** The process whereby the direction of a beam of electromagnetic energy, such as X-rays, is altered by interaction with a crystal structure; the precise change in direction is governed by the spacing of atomic planes in the crystal.

**diffraction lines (X-ray)** Sometimes called diffraction peaks, these lines (or peaks) represent the diffracted X-ray beam and occur at specific angles to the incident beam; the angles are a function of atomic spacings in the crystal; each diffraction line of a crystal occurs at its unique angle and with a distinct intensity; the sum total of all the diffraction lines (angle plus intensity) is unique to a given crystal (mineral) and can be used as an identification fingerprint.

**double refraction** Synonymous with birefringence.

**drusy** Surface encrusted with small projecting crystals.

**ductile** Property of being able to be stretched, drawn or hammered without breaking.

**efflorescence** A process which forms a fluffy crystalline powder on a rock surface by evaporation in arid regions.

**elastic** A property of solid materials whereby strains are totally recoverable and deformation is independent of time.

**enantiomorphism** The property of some crystals whereby they can crystallize with right- or left-handed forms (e.g., quartz).

**epitaxy** Overgrowth of one crystal on another whereby the shared plane of atoms is consistent (or nearly so) with the atomic arrangement in both crystals.

**etch figure** A pattern of pitting produced on a face by solution; the pattern reflects the internal symmetry.

**evaporite** A sedimentary rock or mineral formed by precipitation from saline waters.

**exsolution** A process whereby a homogeneous crystal phase separates into two or more crystal phases (minerals) without the addition or removal of new chemical constituents to the system.

**extinction** The complete darkness observed in a birefringent mineral at two orientations between crossed nicols in a petrographic microscope.

**fluorescence** The type of luminescence whereby the light emission ceases when the external excitation ceases.

**formula (chemical formula)** An expression indicating the relative proportion of chemical elements in the composition of a mineral.

**fracture** Property of breaking along directions other than crystallographic planes; described by the resultant shape such as fibrous, conchoidal, etc.

**gangue** The portion of an ore which is not economically valuable.

**gem** A mineral that possesses intrinsic value because of its beauty, hardness and rarity, making it desirable as jewelry; also the cut-and-polished form of the mineral.

**geode** A small globular cavity found in some sedimentary rocks, normally partly filled by inward projecting crystals.

**glass** A highly viscous state of matter exhibiting an arrangement of atoms intermediate between the liquid and crystalline states; very few liquids will form glasses; most natural molten silicates are glass formers.

**gossan** The weathered and oxidized zone overlying a sulfide deposit; contains a concentration of hydrated iron oxide.

**habit** The characteristic crystal form of a mineral.

**hackly** The mineral fracture that produces a jagged surface.

**hemimorphism** The property of a crystal whereby the two ends of the crystal have different forms.

**hexagonal** Having six-sided forms; *see* also crystal system.

**hardness** The property of resistance of a mineral to being scratched.

**hydrothermal** The process and product of mineral formation by precipitation from hot aqueous solution.

**hydrous** Containing water.

**hygroscopic** Property of attracting or absorbing moisture from the air.

**I (intensity)** A measure of the X-ray quanta refracted from a given set of parallel crystallographic planes.

**igneous** Having crystallized from a molten material (magma).

**inosilicate** A type of silicate crystal structure characterized by the linkage of $SiO_4$ tetrahedra into linear chains (pyroxenes have single chains, amphiboles have double chains).

**ionic bond** The crystalline bond formed between two oppositely charged ions (e.g., in halite between $Na^+$ and $Cl^-$).

**iridescence** A spectral display of color in a mineral caused by the interference of light passing between layers of different refractive index.

**isometric** *See* cubic system.

**isomorphism** The property of two or more distinct but chemically similar substances to crystallize in the same crystal class; such substances may form a continuous isomorphous series.

**isotropic** In crystal optics the characteristic whereby physical properties are the same (equal) in all directions; cubic crystals are isotropic.

**lamellar** A sheaf-like texture similar to leaves in a book.

**lattice** *See* crystal lattice.

**lattice constants** The dimensions of the unit cell in a crystal structure; also called cell constants.

**locality** The place of origin of a mineral or rock specimen.

**lode** A primary mineral deposit in a vein or rock mass as opposed to a secondary (e.g., placer) deposit.

**luminescence** The emission of light of a different wavelength than that of the excitation radiation.

**luster** The quality and intensity of the reflection of light from the surface of a mineral, described in such terms as metallic, vitreous, resinous, etc.

**malleable** The property of a substance whereby it can be plastically deformed by hammering.

**metallic bond** A weak covalent bond found in metals; each metal atom forms an electron-pair bond with each adjacent atom in turn.

**metamict** The property of a mineral resulting from internal radioactive bombardment whereby enough disruption of the crystal structure occurs to make it amorphous to X-rays.

**metamorphism** Mineralogical and textural adjustments of solid rocks to changes in the physical and chemical environment.

**metastable** Condition of being relatively stable under given physical and chemical conditions but capable of being transformed to a more thermodynamically stable form if sufficient energy is available for the transformation.

**meteoric** (1) Pertaining to water of atmospheric origin; (2) having an origin related to meteorites or meteors.

**micaceous** Composed of thin sheets resembling mica.

**microcrystalline** Rock texture consisting of crystals small enough to be visible only under a microscope.

**micron** A measure of distance equal to $10^{-3}$ mm.

**Miller indices** A set of three or four symbols (e.g., $\{100\}$ or $\{122\}$) used to specify the orientation of a crystallographic plane in relation to the crystal axes.

**Mohs hardness scale** A scale of ten minerals by which mineral hardness may be rated; from softest (1) to hardest (10). The scale is: (1) talc, (2) gypsum, (3) calcite, (4) fluorite, (5) apatite, (6) orthoclase, (7) quartz, (8) topaz, (9) corundum, (10) diamond.

**monochromatic light** Electromagnetic radiation composed of a single wavelength.

**monoclinic** One of the six crystal systems, characterized by three unequal axes, two inclined to one another and the third perpendicular to the plane of the first two.

**morphology** The study of form and the processes and structures that determine form.

**native element** A chemical element existing in an uncombined state in nature (e.g., carbon as diamond or graphite).

**nesosilicate** A type of silicate crystal structure characterized by the linking of $SiO_4$ tetrahedra through other cations rather than the sharing of oxygens among $SiO_4$ tetrahedra.

**nodule** (1) A small, compact, rounded rock lump exhibiting greater hardness or chemical resistance than the surrounding rock mass; (2) a concretionary lump of manganese and iron oxides found on ocean bottoms.

**occurrence** The physical and genetic origin of a mineral, its mineral and rock associations.

**ocherous** Containing or resembling ocher, a yellowish red variety of earthy hematite.

**octahedral** Having eight-sided forms or pertaining to an octahedron (eight faces).

**oolitic** A rock or mineral texture composed of small, round accretions resembling fish roe.

**opalescence** Somewhat pearly luster such as shown by some opal.

**opaque** Property of being impervious to visible light.

**optic axis** The direction in an anisotropic crystal along which there is no birefringence; the included angle between the two optic axes in biaxial crystals is called 2V.

**optical constants** See refractive index, optic axes, and optic sign.

**optic sign (+)** or (−) a quantitative relation among refractive indices, e.g., in uniaxial minerals if $\epsilon > \omega$ the sign is defined as positive.

**ore** Naturally occurring concentrations of minerals that can be mined and sold at a profit.

**orthorhombic** One of the six crystal systems, characterized by three mutually perpendicular axes of unequal length.

**parting** Breaking of a mineral along special planes of weakness caused by deformation or twinning.

**pegmatite** Very coarse-grained granitic rock representing the last stages of crystallization of a magma; pegmatites often contain concentrations of lithium, boron, beryllium, and rare earth minerals.

**phosphorescence** The type of luminescence whereby the substance continues to emit light after the external excitation has ceased.

**phyllosilicate** A type of silicate structure characterized by the sharing of three of the four $SiO_4$ tetrahedral oxygens to form sheets.

**piezoelectric** An effect in certain crystals whereby an electric potential is generated when mechanical stress is applied.

**pinacoid** A crystal form composed of two parallel faces.

**placer** A mineral deposit formed by the mechanical (density) concentration of mineral grains by moving water.

**pleochroism** The property of anisotropic crystals to differentially absorb various wavelengths of light in different crystallographic directions giving rise to different colors in different directions.

**plutonic** Pertaining to igneous rocks formed at great depths.

**polarized light** Light that is constrained to vibrate in a single plane; used in the optical analyses of rocks and minerals with the polarizing microscope.

**polymorphism** The property of a chemical substance to crystallize in more than one form (e.g., carbon as diamond and graphite, $FeS_2$ as pyrite and marcasite).

**polytype** A type of polymorph due to the internal atomic arrangements whereby identical atomic planes can be packed or stacked in more than one manner.

**powder diffraction** X-ray diffraction by a crystalline powder; common method used in mineral identification.

**pseudomorph** A secondary mineral whose crystal shape is inherited from the pre-existing mineral it replaced; this shape does not reflect the internal crystal structure of the mineral.

**pyramidal** Having the symmetry or shape of a pyramid.

**pycnometer** A vessel used to measure the density of minerals.

**pyroelectric** An effect in certain crystals whereby an electric potential is generated by certain changes in temperature.

**rare earth (R.E.)** One of a group of somewhat rare metallic elements with consecutive atomic numbers 57 to 71.

**refractive index** In crystal optics the refractive index is the ratio of the velocity of light in a vacuum to the velocity of light in the crystal. Isotropic substances have one index, conventionally labeled N (or n), uniaxial crystals have two indices labeled $\epsilon$ and $\omega$, biaxial crystals have three indices labeled $\alpha, \beta, \gamma$ (or X, Y, Z).

**reniform** Kidney-shaped.

**rhombohedral** A set of trigonal crystal axes related to the rhombohedron, a parallelepiped with six identical rhomb faces; calcite has rhombohedral cleavage; also the name of a crystal class and a general name used when speaking about $\bar{3}$ symmetry.

**rotation axes** See symmetry axes.

**sectile** Property of a mineral whereby it is capable of being cut with a knife.

**secondary (mineral)** Formed later than the primary minerals associated with it or enclosing it and usually formed at the expense of some primary mineral.

**sedimentary** Pertaining to the process of sedimentation or the rocks formed from this process.

**semiprecious stone** A gemstone of less value than a precious stone, ordinarily one with a lower hardness and brilliance and/or greater abundance.

**septarian nodule** A roughly spheroidal concretion, from a few inches to several feet in diameter, usually composed of calcium carbonate or clay-ironstone; often hollow or transected by veins of a cementing material such as calcite.

**solid solution** A single crystal (mineral) phase that can vary in composition over a definite range (e.g., the plagioclase feldspars).

**specific gravity** The ratio of the density of the mineral to the density of water; specific gravity is a dimensionless number.

**space group** One of 230 possible symmetrical arrangements of atoms in space; each crystal structure will have the symmetry of one of the space groups.

**space group symbols** Each of the 230 space groups are characterized by (1) a Bravais lattice (such as P, I, F, C, etc.) followed by a point group (crystal class) designation, the latter may contain complex symmetry operations such as glide planes (instead of mirror planes) or screw axes (instead of simple rotation axes) (e.g., $P2_1/c$, $Ia3d$ instead of $P2/m$, $Im3m$).

**streak** The color of a mineral in its powdered form, usually obtained by drawing the mineral across unglazed porcelain.

**symmetry** The angular or linear repeat pattern of crystals formed by the ordered internal arrangement of the atoms.

**symmetry axes** Lines through the center of a crystal about which the crystal may be rotated whereby there will be 2, 3, 4, or 6 repetitions of its crystal elements (e.g., faces).

**symmetry planes** Planes of mirror symmetry in crystals.

**tabular** A crystal habit having two prominent parallel faces giving the mineral a tablet-shaped appearance.

**tarnish** A surface alteration of color and luster; found especially in copper minerals.

**tektosilicate** Type of silicate crystal structure characterized by the sharing of all $SiO_4$ tetrahedral oxygens resulting in three-dimensional framework structures.

**tetragonal** Crystal forms or operations based on the number four; *see* also crystal system.

**transition element (transition metal)** A series of elements occupying certain positions in the periodic table; iron and manganese are transition elements whose electronic structure results in their major contribution to the color of minerals.

**translucent** Capable of transmitting a diffuse light but not transparent.

**triclinic** The lowest symmetry system of the six crystal systems.

**trigonal** Crystal forms or operations based on the number 3; *see* also crystal system.

**twinning** An intergrowth of two or more crystals of the same mineral along or related to definite crystallographic planes or axes.

**uniaxial** Having only one optic axis; hexagonal, tetragonal and trigonal crystals are uniaxial.

**unit cell** The fundamental parallelipiped of a crystal structure that contains its full symmetry and chemical composition.

**variety** A recognized variation in a given mineral species; a variety is normally distinguished by color, habit or similar external characteristics.

**vein** A mineral seam filling a fracture in a host rock.

**vitreous** A luster resembling that of glass.

**vug** A small cavity in a rock, usually lined with crystals.

**wavelength ($\lambda$)** The distance between corresponding points on two successive waves; a distinguishing characteristic of electromagnetic radiation (X-rays, light, etc.).

**weathering** Destructive chemical and physical processes that alter and disintegrate minerals at or near the earth's surface.

**X-ray diffraction** Diffraction of an X-ray beam by crystalline substances. *See* diffraction lines.

**Z** The number of chemical formula multiples per unit cell.

**zone** A set of planes (e.g., crystal faces) with mutually parallel intersections, this common direction being termed the zone axis.

**zoning** Systematic variation in the composition of a crystal, usually from core to outside edge.

# A

## ABERNATHYITE
$KUO_2AsO_4 \cdot 3H_2O$

CRYSTAL SYSTEM: Tetragonal
CLASS: 4/mmm
SPACE GROUP: P4/ncc; P4/nmm (Pseudo)
Z: 2
LATTICE CONSTANTS:
  $a = 7.17$
  $c = 9.08$
3 STRONGEST DIFFRACTION LINES:
  9.14 (100)
  3.84 ( 80)
  3.34 ( 80)
OPTICAL CONSTANTS:
  $e = 1.570$
  $\omega = 1.597$
  HARDNESS: ~2½
  DENSITY: >3.32 (Meas.)
              3.572 (Calc.)
  CLEAVAGE: {001} perfect
  HABIT: Crystals thin to thick tabular and occur singly or in groups. Prominent forms are {001} and {110}. As crystalline coatings. Largest crystals about 0.5 mm on an edge.
  COLOR-LUSTER: Yellow; weakly vitreous. Crystals clear and transparent. Streak pale yellow. Fluoresces yellow-green under long- and short-wave ultraviolet light.
  MODE OF OCCURRENCE: Occurs very rarely as a secondary mineral, associated with yellow-brown earthy scorodite, coating fractures in sandstone at the Fuemrol No. 2 mine, Temple Mountain, Emery County, Utah; also sparsely disseminated in uraniferous lignite in the lower Ludlow member of the Fort Union formation in the Cave Hills and Slim Buttes areas, Harding County, South Dakota.
  BEST REF. IN ENGLISH: Thompson, M. E., Ingram, Blanche, and Gross, E. B., *Am. Min.*, **41**: 82-90, (1956). Ross, Malcolm, and Evans, Howard T., Jr., *Am. Min.*, **49**: 1578-1602 (1964).


## ABSITE = Brannerite


## ABUKUMALITE = Britholite–(Y)


## ACANTHITE
$Ag_2S$
Dimorphous with argentite

CRYSTAL SYSTEM: Monoclinic
CLASS: 2/m
SPACE GROUP: $P2_1/n$
Z: 4
LATTICE CONSTANTS:
  $a = 7.87$
  $b = 6.91$
  $c = 4.23$
  $\beta = 99°35'$
3 STRONGEST DIFFRACTION LINES:
  2.606 (100)
  2.440 ( 80)
  2.383 ( 75)
  HARDNESS: 2-2½
  DENSITY: 7.22 (Meas.)
              7.27 (Calc.)
  CLEAVAGE: Not determined. Fracture uneven. Sectile.
  HABIT: Crystals short to long prismatic, up to 2.2 cm long.
  COLOR-LUSTER: Iron black. Opaque. Metallic, sometimes brilliant.
  MODE OF OCCURRENCE: Occurs in Colorado at the Pelican Dives mine, Silver Plume district, and at the Little Emma mine, Clear Creek County; at the Enterprise mine, Dolores County; and at the Double Header lode, Summit County. It is also found in Canada, Mexico, Czechoslovakia, and Germany. Acanthite is stable below 179°C. (Argentite is the stable form between 179°C and 586°C.)
  BEST REF. IN ENGLISH: Palache, et al., "Dana's System of Mineralogy," 7th Ed., v. I, p. 191-192, New York, Wiley, 1944.


## ACHAVALITE
FeSe

CRYSTAL SYSTEM: Tetragonal, hexagonal
CLASS: 4/m 2/m 2/m; 6/m 2/m 2/m
SPACE GROUP: P4/nmm, $P6_3/mmc$
Z: 2

LATTICE CONSTANTS:
(tetragonal)   (hexagonal)
$a = 3.765$    $a = 3.61$
$c = 5.518$    $c = 5.87$

3 STRONGEST DIFFRACTION LINES:
3.125 (100)
1.923 (100)
1.889 ( 80)

HARDNESS: ~ 2½
DENSITY: Not determined.
CLEAVAGE: Not determined.
HABIT: Massive.
COLOR-LUSTER: Dark gray. Opaque. Metallic.

Magnetic.
MODE OF OCCURRENCE: Occurs intergrown with clausthalite in a calcite matrix at Cerro de Cacheuta, Mendoza, Argentina.
BEST REF. IN ENGLISH: Olsacher, Juan, *Min. Abs.*, 12: 236 (1953).

**ACHROITE** = Colorless variety of tourmaline

**ACMITE** = Variety of **aegirine** occurring as long prismatic, vertically striated crystals with acute terminations

## ACTINOLITE
$Ca_2(Mg, Fe)_5 Si_8 O_{22}(OH)_2 -$
20–100 mol. % $Ca_2 Fe_5 Si_8 O_{22}(OH)_2$
Amphibole Group
Var. Ferroactinolite
    Nephrite
    Byssolite

CRYSTAL SYSTEM: Monoclinic
CLASS: 2/m
SPACE GROUP: C2/m
Z: 2
LATTICE CONSTANTS:
$a \simeq 9.85$
$b \simeq 18.1$
$c \simeq 5.3$
$\beta \simeq 104°50'$

3 STRONGEST DIFFRACTION LINES:
8.38 (100)
3.12 (100)
2.71 ( 90)

OPTICAL CONSTANTS:
$\alpha$ : 1.599–1.688
$\beta$ : 1.612–1.697
$\gamma$ : 1.622–1.705
$2V_\alpha$ : 86°–65°

HARDNESS: 5-6
DENSITY: 3.0-3.44 (Meas.)
CLEAVAGE: {110} good
      {100} parting

Fracture uneven to subconchoidal. Brittle; compact varieties often tough.
HABIT: Crystals commonly long-bladed; less frequently short and stout. Usually in fibrous or thin columnar ag-

gregates, often radiated. Also massive, fibrous or granular. Twinning on {100} common, simple, lamellar.
COLOR-LUSTER: Light green to blackish green or black. Transparent to nearly opaque. Vitreous; sometimes dull.
MODE OF OCCURRENCE: Occurs widespread, mainly in contact and regionally metamorphosed dolomites, magnesian limestones, and low-grade ultrabasic rocks. Typical occurrences are found in Alaska (nephrite), California, Idaho (ferroactinolite), Colorado, Arizona, Wyoming (nephrite), South Dakota, Minnesota, Pennsylvania (byssolite), Vermont, New Jersey, Massachusetts, and Virginia (byssolite). Also in Canada, Scotland, England, Sweden, Switzerland, Japan, and New Zealand (nephrite).
BEST REF. IN ENGLISH: Deer, Howie, and Zussman, "Rock Forming Minerals," v. 2, p. 249–262, New York, Wiley, 1963.

## ADAMITE
$Zn_2 AsO_4 OH$
Dimorphous with Paradamite

CRYSTAL SYSTEM: Orthorhombic
CLASS: 2/m 2/m 2/m
SPACE GROUP: Pnnm
Z: 4
LATTICE CONSTANTS:
$a = 8.30$
$b = 8.51$
$c = 6.04$

3 STRONGEST DIFFRACTION LINES:
2.45 (100)
4.90 ( 90)
2.97 ( 90)

OPTICAL CONSTANTS:
$\alpha = 1.722$
$\beta = 1.742$
$\gamma = 1.763$
$2V(+) = 88°$

Considerable variation in optical properties of natural material.
HARDNESS: 3½
DENSITY: 4.32-4.48 (Meas.)
      4.435 (Calc.)
CLEAVAGE: {101} good
      {010} poor

Fracture subconchoidal to uneven. Brittle.
HABIT: Crystals varied in habit; commonly elongated along *b*-axis, tabular, or equant. Often as druses of numerous interlocked crystals with wedge-shaped terminations or as radial aggregates or spheroids implanted on matrix.
COLOR-LUSTER: Usually vivid yellowish green; also various shades of yellow and green, colorless, white, bluish green, violet to rose. Transparent. Vitreous. Sometimes fluoresces yellowish green under ultraviolet light.
MODE OF OCCURRENCE: Occurs as a secondary mineral in the oxidation zone of ore deposits, often associated with limonite, calcite, hemimorphite, smithsonite, azurite, and malachite. Found at the Iron Blossom mine, Juab County, and with austinite at the Gold Hill mine, Tooele County, Utah; in Inyo, San Bernardino, and Santa Cruz counties, California; and at the Simon mine, Mineral

County, Nevada. Magnificent specimens occur at the Ojuela mine, Mapimi, Durango, Mexico; at Tsumeb, South West Africa; and at Laurium, Greece. Other occurrences are found in Chile, Italy, France, Germany, Turkey, and Algeria.

BEST REF. IN ENGLISH: Palache, et al., "Dana's System of Mineralogy," 7th Ed., v. II, p. 864-866, New York, Wiley, 1951.

## ADELITE
$CaMgAsO_4OH$

CRYSTAL SYSTEM: Orthorhombic
CLASS: 2 2 2
SPACE GROUP: $P2_12_12_1$
Z: 4
LATTICE CONSTANTS:
  $a = 7.47$
  $b = 8.94$
  $c = 5.88$
OPTICAL CONSTANTS:
  $\alpha = 1.712$
  $\beta = 1.721$
  $\gamma = 1.731$
  $(+)2V \sim 90°, 68°36'$
HARDNESS: 5
DENSITY: 3.73 ± 0.03 (Meas.)
        3.79 (Calc.)
CLEAVAGE: Fracture conchoidal to uneven.
HABIT: Crystals rare, elongated [100]. Usually massive.
COLOR-LUSTER: Colorless, gray to bluish gray, yellow to yellowish gray, pale green. Transparent. Resinous.
MODE OF OCCURRENCE: Occurs in manganese ore deposits associated with braunite, hausmannite, and other minerals. Found in Sweden at Långban; and in Nordmark at Jacobsberg and at the Kittel and Moss mines.
BEST REF. IN ENGLISH: Palache, et al., "Dana's System of Mineralogy," 7th Ed., v. II, p. 804-806, New York, Wiley, 1951.

## ADULARIA = A low-temperature form of K-feldspar, may be either monoclinic or triclinic

## AEGIRINE (Acmite)
$NaFe^{3+}Si_2O_6$
Pyroxene Group
Var. Blanfordite

CRYSTAL SYSTEM: Monoclinic
CLASS: 2/m
SPACE GROUP: C2/c
Z: 4
LATTICE CONSTANTS:
  $a = 9.65$
  $b = 8.79$
  $c = 5.29$
  $\beta = 107.4°$

3 STRONGEST DIFFRACTION LINES:
  2.900 (100)
  6.369 ( 90)
  4.416 ( 80)
OPTICAL CONSTANTS:
  $\alpha = 1.750-1.776$
  $\beta = 1.780-1.820$
  $\gamma = 1.800-1.836$
  $2V_\alpha(-) = 60°-70°$
HARDNESS: 6
DENSITY: 3.55-3.60 (Meas.)
        3.576 (Calc.)
CLEAVAGE: {110} good
         {100} parting
Fracture uneven. Brittle.
HABIT: Crystals long prismatic, vertically striated or furrowed; terminations blunt (aegirine) or acute (acmite). Also in groups or tufts of acicular to capillary crystals, or in felted aggregates of minute fibers. Twinning on {100} common, simple or lamellar.
COLOR-LUSTER: Dark green to greenish black (aegirine); reddish brown, dark green to black (acmite). Translucent to opaque. Vitreous to somewhat resinous. Streak pale yellowish gray.
MODE OF OCCURRENCE: Occurs widespread as a characteristic component of alkaline rocks such as syenites and carbonatites, also in certain metamorphic rocks; authigenic acmite is found in rocks of the Green River formation in Utah, Wyoming, and Colorado. Typical occurrences are found in California, Montana, South Dakota, Arkansas, Minnesota, and Massachusetts. Also at Mont St. Hilaire, Quebec, Canada, and at localities in Greenland, Scotland, Norway, USSR, Nigeria, Kenya, India, and elsewhere.
BEST REF. IN ENGLISH: Deer, Howie, and Zussman, "Rock Forming Minerals," v. 2, p. 79-91, New York, Wiley, 1963.

## AENIGMATITE
$Na_2Fe_5TiSi_6O_{20}$

CRYSTAL SYSTEM: Triclinic
CLASS: $\bar{1}$
SPACE GROUP: $P\bar{1}$
Z: 2
LATTICE CONSTANTS:
  $a = 10.41$
  $b = 10.81$
  $c = 8.93$
3 STRONGEST DIFFRACTION LINES:
  8.11 (100)
  3.14 (100)
  2.705 ( 80)
OPTICAL CONSTANTS:
  $\alpha = 1.81$
  $\beta = 1.82$
  $\gamma = 1.88$
  $(+)2V = 32°$
HARDNESS: 5½
DENSITY: 3.74-3.85 (Meas.)
        3.843 (Calc.)

CLEAVAGE: {010} perfect
{100} perfect
Brittle.

HABIT: Crystals long prismatic. Twinning on {1$\bar{1}$0} common, sometimes lamellar.

COLOR-LUSTER: Black. Nearly opaque. Streak reddish brown.

MODE OF OCCURRENCE: Occurs as minute phenocrysts in lavas in Sonoma County, California; in sodalite-syenite in the Julianehaab district, Greenland; in the liparite lavas of the island Pantelleria, Mediterranean Sea; and widespread in the rocks of East Africa.

BEST REF. IN ENGLISH: Winchell and Winchell, "Elements of Optical Mineralogy," 4th Ed., Pt. 2, p. 477–478, New York, Wiley, 1951. Cannillo, E., et al., *Am. Min.*, **56**: 427–446 (1971).


# AËRINITE = A mixture of aluminosilicates of Fe, Mg, and Ca


# AERUGITE
$Ni_9As_3O_{16}$

CRYSTAL SYSTEM: Monoclinic
CLASS: 2/m or 2 or m
SPACE GROUP: C2/m or C2 or Cm
Z: 2
LATTICE CONSTANTS:
$a$ = 10.29
$b$ = 5.95
$c$ = 9.79
$\beta$ = 110°19′
3 STRONGEST DIFFRACTION LINES:
2.060 (100)
3.76 ( 80)
2.329 ( 60)
HARDNESS: 4
DENSITY: 5.85 ± 0.07 (Meas. Synthetic)
5.95 (Calc.)
CLEAVAGE: Not determined.
HABIT: Massive; finely crystalline.
COLOR-LUSTER: Green.
MODE OF OCCURRENCE: Occurs with xanthiosite, various nickel-cobalt-arsenic minerals, traces of decomposing pitchblende and various alteration products, in comby quartz vein-material at the South Terras mine, St. Stephen-in-Brannel, Cornwall, England; also found at Johanngeorgenstadt, Saxony, Germany.

BEST REF. IN ENGLISH: Davis, R. J., Hey, M. H., and Kingsbury, A. W. G., *Min. Mag.*, **35**: 72–83 (1965).


# AESCHYNITE (Eschynite)
(Ce, Ca, Fe, Th)(Ti, Nb)$_2$(O, OH)$_6$

CRYSTAL SYSTEM: Orthorhombic
CLASS: 2/m 2/m 2/m
SPACE GROUP: Pmnb
Z: 4

LATTICE CONSTANTS:
$a$ = 7.55
$b$ = 10.97
$c$ = 5.42
3 STRONGEST DIFFRACTION LINES:
2.975 (100)
3.024 ( 80)
3.106 ( 35)
OPTICAL CONSTANT:
$N$ = 2.26
HARDNESS: 5–6
DENSITY: 5.19 ± 0.05 (Meas.)
CLEAVAGE: None.
{010} parting, distinct.
Fracture conchoidal. Brittle.

HABIT: Crystals prismatic parallel to [001], up to 10 cm long; also tabular {010}; rarely acicular. Also massive, compact.

COLOR-LUSTER: Black, brownish black, brown, brownish yellow. Subtranslucent; sometimes transparent in thin fragments. Adamantine, submetallic; resinous on fracture. Streak dark brown to black.

MODE OF OCCURRENCE: The mineral occurs chiefly in feldspathoid and nepheline-feldspathoid pegmatites in association with feldspar, biotite, muscovite, zircon, corundum, and sphene. Found at Hitterö, Norway; in the Vishnevye and Ilmenskie Mountains, Urals, USSR; and at Bayun-Obo, China.

BEST REF. IN ENGLISH: Palache, et al., "Dana's System of Mineralogy," 7th Ed., v. I, p. 793–797, New York, Wiley, 1944.


# AESCHYNITE-(Y) (Syn. Priorite)
(Y, Ca, Fe, Th)(Ti, Nb)$_2$(O, OH)$_6$

CRYSTAL SYSTEM: Orthorhombic
CLASS: 2/m 2/m 2/m
SPACE GROUP: Pmnb
Z: 4
LATTICE CONSTANTS:
$a:b:c$ = 0.4746 : 1 : 0.6673
OPTICAL CONSTANTS:
Isotropic (metamict).
$N$ = 2.142
HARDNESS: 5–6
DENSITY: 4.95 ± 0.10 (Meas.)
CLEAVAGE: None
{010} parting, distinct.
Fracture conchoidal. Brittle.

HABIT: Crystals prismatic parallel to {001} or {100}; also tabular {010}. Also massive, compact.

COLOR-LUSTER: Black, brownish black, brown, brownish yellow, yellowish. Subtranslucent; sometimes transparent in thin fragments. Adamantine, submetallic; resinous on fracture. Streak reddish yellow.

MODE OF OCCURRENCE: Occurs chiefly in feldspathoid and nepheline-feldspathoid pegmatites in association with feldspar, biotite, muscovite, zircon, corundum, and sphene; also as a detrital mineral in placer deposits. Found at Hitterö and elsewhere in Norway; in the Ilmenskie

Mountains, Urals, USSR; and at localities in Madagascar and Swaziland, Africa.

BEST REF. IN ENGLISH: Palache, et al., "Dana's System of Mineralogy," 7th Ed., v. I, p. 793–797, New York, Wiley, 1944.

## AFGHANITE
$(Na, Ca, K)_{12}(Si, Al)_{16}O_{34}(Cl, SO_4, CO_3)_4 \cdot H_2O$
Cancrinite Group

CRYSTAL SYSTEM: Hexagonal
CLASS: 6/m 2/m 2/m
SPACE GROUP: $P6_3/mmc$
Z: 3
LATTICE CONSTANTS:
$a = 12.77$
$c = 21.35$
3 STRONGEST DIFFRACTION LINES:
3.688 (100)
3.298 (100)
4.82 ( 80)
OPTICAL CONSTANTS:
$\beta = 1.523$
$\gamma = 1.529$
+2V = ?
HARDNESS: 5½–6
DENSITY: 2.55 (Meas.)
2.65 (Calc.)
CLEAVAGE: {100} perfect
HABIT: Massive
COLOR-LUSTER: Bluish, transparent.
MODE OF OCCURRENCE: Occurs in the lapis lazuli mine at Sar-e-Sang, Badakhshan Province, Afghanistan. It forms the core of a crystal of lazurite, associated with sodalite, nepheline, phlogopite, olivine, and pyrite.
BEST REF. IN ENGLISH: Bariand, P., Cesbron, F., and Giraud, R., Am. Min., 53: 2105 (1968).

## AFWILLITE
$Ca_3Si_2O_4(OH)_6$

CRYSTAL SYSTEM: Monoclinic
CLASS: 2 (m)
SPACE GROUP: $P2_1$ (Ia)
Z: 4
LATTICE CONSTANTS:
| ($P2_1$) | (Ia) |
|---|---|
| $a = 11.39$ | $a = 13.23$ |
| $b = 5.47$ | $b = 5.632$ |
| $c = 13.09$ | $c = 11.68$ |
| $\beta = 98°26'$ | $\beta = 98°42'$ |

3 STRONGEST DIFFRACTION LINES:
| | |
|---|---|
| 2.83 (100) | 3.19 (100) |
| 6.61 ( 90) | 2.84 (100) |
| 3.18 ( 90) | 2.74 (100) |

OPTICAL CONSTANTS:
($P2_1$)
$\alpha = 1.6169$
$\beta = 1.6204$
$\gamma = 1.6336$
(+)2V = 54°40'

HARDNESS: 3
DENSITY: 2.62 ± 0.01 (Meas.)
CLEAVAGE: {001} perfect
{100} good
HABIT: Crystals elongated parallel to [010]; usually tabular parallel to {101}. Also massive.
COLOR-LUSTER: Colorless or white; transparent to translucent. Vitreous.
MODE OF OCCURRENCE: Occurs as small crystals up to 3 mm in length along cracks in blocks of contact rock on the floor of the 910-foot level of the Commercial Quarry, Crestmore, California, associated with quartz, thaumasite, merwinite, gehlenite, and calcite. Also in a dolerite inclusion in kimberlite at the Dutoitspan diamond mine, Kimberley, South Africa; and as small crystals with calcite in a spurrite rock at Scawt Hill, Antrim County, Ireland.
BEST REF. IN ENGLISH: Switzer, George and Bailey, Edgar H., Am. Min., 38: 629–633 (1953).

## AGARDITE
$(Y, Ca)_2Cu_{12}(AsO_4)_6(OH)_{12} \cdot 6H_2O$
Mixite Group
General formula for mixite group:
$A_2Cu_{12}(AsO_4)_6(OH)_{12} \cdot 6H_2O$, where
A = Bi (mixite), Ca, Y.

CRYSTAL SYSTEM: Hexagonal
LATTICE CONSTANTS:
$a = 13.55$
$c = 5.87$
3 STRONGEST DIFFRACTION LINES:
10.73 (100)
2.938 ( 80)
2.451 ( 80)
OPTICAL CONSTANTS:
$\epsilon = 1.782$
$\omega = 1.701$
HARDNESS: 3–4
DENSITY: 3.72 ± 0.05 (meas.)
3.66 ± 0.04 (calc. from x-ray)
HABIT: Crystals acicular, elongated [0001].
COLOR-LUSTER: Blue-green.
MODE OF OCCURRENCE: Occurs as acicular crystals up to a few millimeters in length in the oxidation zone of the copper deposit of Bou-Skour, Morocco, associated with azurite, malachite, cuprite, native copper, quartz, and unidentified minerals. It has also been found at four other localities, including Tintic, Utah.
BEST REF. IN ENGLISH: Am. Min., 55: 1447–1448 (1970).

## AGATE = Variety of quartz

## AGRINIERITE
$K_2CaU_6O_{20} \cdot 9H_2O$

CRYSTAL SYSTEM: Orthorhombic
CLASS: 2/m 2/m 2/m
SPACE GROUP: Cmmm
Z: 8

LATTICE CONSTANTS:
$a = 14.04$
$b = 24.07$
$c = 14.13$
$\beta = 90°$
3 STRONGEST DIFFRACTION LINES:
7.08 (100)
3.128 (100)
3.485 ( 80)
OPTICAL CONSTANTS:
$\propto \| [001]$
$\gamma \| [010]$
2V = 55°
HARDNESS: Not determined.
DENSITY: 5.7 (Meas.)
　　　　　5.62 (Calc.)
CLEAVAGE: {001} good
HABIT: Crystals tabular on {001} with pseudohexagonal section bounded by the trace of (010). Sector twinning on {110}.
COLOR-LUSTER: Orange
MODE OF OCCURRENCE: The mineral occurs in association with uranophane in small cavities in "gummite" at Margnac, in the Massif Central, France.
BEST REF. IN ENGLISH: Cesbron, F., Brown, W. L., Bariand, P., and Geffroy, J., *Min. Mag.*, **38**: 781–789 (1972).

## AGUILARITE
$\beta$-$Ag_4SSe$
Argentite Group

CRYSTAL SYSTEM: Cubic at high temperatures.
3 STRONGEST DIFFRACTION LINES:
2.42 (100)
2.19 ( 60)
4.09 ( 50)
HARDNESS: 2½
Talmadge hardness A.
DENSITY: 7.40–7.53 (Meas.)
CLEAVAGE: None.
Fracture hackly. Sectile.
HABIT: Intergrown rudely dedecahedral crystals with pitted surfaces and mostly rounded edges; also massive.
COLOR-LUSTER: Bright lead gray on fresh surfaces; dull iron gray to iron black on exposed surfaces. Opaque; metallic, brilliant.
MODE OF OCCURRENCE: Occurs associated with pearceite, argentite, silver, and calcite at the San Carlos mine, Guanajuato, Mexico; also reported from the Comstock Lode, Virginia City, Nevada.
BEST REF. IN ENGLISH: Palache, et al., "Dana's System of Mineralogy," 7th Ed., v. I, p. 178–179, New York, Wiley, 1944. Early, J. W., *Am. Min.*, **35**: 337–364 (1950).

## AHLFELDITE
$(Ni, Co)SeO_3 \cdot 2H_2O$

CRYSTAL SYSTEM: Monoclinic
CLASS: 2/m

SPACE GROUP: $P2_1/n$
Z: 4
LATTICE CONSTANTS:
$a = 7.53$
$b = 8.76$
$c = 6.43$
$\beta = 99°05'$
3 STRONGEST DIFFRACTION LINES:
5.69 (100)
3.426 ( 80)
2.992 ( 75)
OPTICAL CONSTANTS:
$\alpha = 1.709$
$\beta = 1.752$
$\gamma = 1.787$
$(-)2V = 85°$
HARDNESS: 2–2½
DENSITY: 3.37 (Meas.)
　　　　　3.51 (Calc.)
CLEAVAGE: {110} fair
　　　　　　{103} fair
Fracture conchoidal. Brittle.
HABIT: Crystalline crusts; rarely well-formed crystals ranging from 0.25 to 1.0 mm in length, elongated along c-axis or tabular flattened on {110}. Forms: {110}, {011}, {032}, {103}, and {$\bar{1}$01}.
COLOR-LUSTER: Brownish to reddish; transparent; vitreous.
MODE OF OCCURRENCE: Occurs as an alteration product of penroseite in the Pacajake silver mine near Hiaco, about 24 km east northeast of Colquechaca on the eastern slope of the Central Cordillera, Bolivia.
BEST REF. IN ENGLISH: Aristarain, L. F., and Hurlbut, C. S., Jr., *Am. Min.*, **54**: 448–456 (1969). Goni, J., and Guillemin, C., *Am. Min.*, **39**: 850 (1954).

## AIKINITE
$CuPbBiS_3$

CRYSTAL SYSTEM: Orthorhombic
CLASS: 2/m 2/m 2/m
SPACE GROUP: Pbnm
Z: 4
LATTICE CONSTANTS:
$a = 11.30$
$b = 11.65$
$c = 4.00$
3 STRONGEST DIFFRACTION LINES:
3.68 (100)
3.19 ( 80)
2.88 ( 70)
HARDNESS: 2–2½
DENSITY: 7.07 ± 0.01 (Meas.)
　　　　　7.22 (Calc.)
CLEAVAGE: {010} (?) indistinct
Fracture uneven.
HABIT: Crystals prismatic to acicular; striated parallel to elongation. Usually massive.
COLOR-LUSTER: Blackish gray; tarnishes brown or copper-red. Opaque. Metallic. Streak grayish black.

MODE OF OCCURRENCE: Occurs in veins with gold, galena, and white quartz in the Beresovsk district, Ural Mountains, USSR. Reported from several widely scattered mining districts in Colorado; at the Sells and South Hecla mine, Salt Lake County, and at the Cyclone mine, Tooele County, Utah; in Adams and Butte counties, Idaho; at Gold Hill, Rowan County, North Carolina; and at deposits in Mexico, France, and Tasmania.

BEST REF. IN ENGLISH: Palache, et al., "Dana's System of Mineralogy," 7th Ed., v. I, p. 412–413, New York, Wiley, 1944.

## AINALITE = Variety of cassiterite containing up to 9% $Ta_2O_5$

## AJOITE
$Cu_6Al_2Si_{10}O_{29} \cdot 5H_2O$

CRYSTAL SYSTEM: Monoclinic
3 STRONGEST DIFFRACTION LINES:
  12.4  (100)
  3.34 ( 25)
  6.19 (  9)
OPTICAL CONSTANTS:
  $\alpha$ = 1.565
  $\beta$ = 1.590
  $\gamma$ = 1.650
  (+)2V = 68°
HARDNESS: Not determined.
DENSITY: 2.96 (Meas.)
HABIT: Usually massive; rarely as platy crystals or laths elongated parallel to c and flattened parallel to {010}. Crystals less than 1 mm in length.
COLOR-LUSTER: Near Ridgeway's Venice green. (Bluish green.)
MODE OF OCCURRENCE: Occurs associated with dark blue shattuckite at Ajo, Pima County, Arizona.
BEST REF. IN ENGLISH: Schaller, W. T., and Vlisidis, Angelina C., Am. Min., 43: 1107–1111 (1958).

## AKAGANÉITE
$\beta$-FeOOH

CRYSTAL SYSTEM: Tetragonal
CLASS: 4/m
SPACE GROUP: I4/m
Z: 8
LATTICE CONSTANTS:
  a = 10.48
  c = 3.023
3 STRONGEST DIFFRACTION LINES:
  7.40  (100)
  3.311 (100)
  1.635 (100)
DENSITY: 3.555 (Calc.)
HABIT: Resembles "limonite."
MODE OF OCCURRENCE: Occurs at the Akagané mine, Iwate Prefecture, Japan, where it appears to have formed from pyrrhotite.

BEST REF. IN ENGLISH: MacKay, A. L., Min. Mag., 33: 270–280 (1962).

## AKATOREITE
$Mn_9(Si, Al)_{10}O_{23}(OH)_9$

CRYSTAL SYSTEM: Triclinic
CLASS: $\bar{1}$ or 1
SPACE GROUP: $P\bar{1}$ or P1
Z: 1
LATTICE CONSTANTS:
  a =  8.344    $\alpha$ = 104°29'
  b = 10.358    $\beta$ =  93°38'
  c =  7.627    $\gamma$ = 103°57'
3 STRONGEST DIFFRACTION LINES:
  4.665 (100)
  3.310 ( 90)
  2.214 ( 80)
OPTICAL CONSTANTS:
  $\alpha$ = 1.698
  $\beta$ = 1.704
  $\gamma$ = 1.720
  (+)2V = 65.5°
HARDNESS: 6
DENSITY: 3.48 ± 0.01 (Meas.)
          3.47 (Calc.)
CLEAVAGE: {010} excellent
          {0$\bar{1}$2} imperfect
HABIT: Crystals elongate prismatic and striated parallel to a-axis. As massive to radiating twinned columnar aggregates. Twinning on {0$\bar{2}$1}.
COLOR-LUSTER: Yellow-orange to orange-brown. Vitreous.
MODE OF OCCURRENCE: Occurs associated with rhodochrosite and pyroxmangite in a manganiferous metachert and carbonate lens on the southeastern margin of the Haast Schist Group, near the mouth of Akatore Creek, Eastern Otago, New Zealand.
BEST REF. IN ENGLISH: Read, Peter B., and Reay, Anthony, Am. Min., 56: 416–426 (1971).

## AKDALAITE
$4Al_2O_3 \cdot H_2O$

CRYSTAL SYSTEM: Hexagonal
CLASS: 622 or 6
SPACE GROUP: $P6_122$ or $P6_1$
Z: 18 (?)
LATTICE CONSTANTS:
  a = 12.87
  c = 14.97
3 STRONGEST DIFFRACTION LINES:
  2.11  (100)
  1.418 (100)
  1.393 (100)
OPTICAL CONSTANTS:
  $\omega$ = 1.747
  $\epsilon$ = 1.741
HARDNESS: Microhardness 1085 kg/mm$^2$.

DENSITY: 3.68 ± 0.02 (Meas.)
3.673 (Calc.)
CLEAVAGE: None.
Fracture irregular. Brittle.
HABIT: Crystals tabular and elongated tabular up to
0.8 mm long and 0.1 mm wide, showing hexagonal outline.
As lens-like or nest-like aggregates.
COLOR-LUSTER: White. Translucent. Vitreous to
porcelain-like.
MODE OF OCCURRENCE: Occurs in veinlets cutting
amesite-fluorite-muscovite rock and fluorite-magnetite-
diopside-vesuvianite-andradite skarn, in the Solvech fluorite
deposit, Karagandin region, Kazakhstan.
BEST REF. IN ENGLISH: Shpanov, E. P., Sidorenko,
G. A., and Stolyarova, T. I., *Am. Min.*, **56**: 635 (1971).

# ÅKERMANITE

$MgCa_2Si_2O_7$
Isomorphous with gehlenite
Melilite Group

CRYSTAL SYSTEM: Tetragonal
CLASS: $\bar{4}2m$
SPACE GROUP: $P\bar{4}2_1m$
Z: 2
LATTICE CONSTANTS:
$a = 7.84$
$c = 5.01$
3 STRONGEST DIFFRACTION LINES:
2.87 (100)
3.09 ( 30)
1.76 ( 30)
OPTICAL CONSTANTS:
$\epsilon = 1.640$
$\omega = 1.632$
HARDNESS: 5–6
DENSITY: 2.944 (Meas.)
2.922 (Calc.)
CLEAVAGE: {001} distinct
{110} indistinct
Fracture uneven to conchoidal. Brittle.
HABIT: Crystals usually short prismatic. Commonly
massive, granular. Twinning on {100}, {001}.
COLOR-LUSTER: Colorless, grayish, green, brown.
Transparent to translucent. Vitreous to resinous.
MODE OF OCCURRENCE: Occurs in calcium-rich basic
eruptive rocks; in thermally metamorphosed impure lime-
stones; and in furnace slags and artificial melts. Occurrences
are essentially the same as for gehlenite (q.v.).
BEST REF. IN ENGLISH: Deer, Howie, and Zussman,
"Rock Forming Minerals," v. I, p. 236–255, New York,
Wiley, 1962.

# AKROCHORDITE

$Mn_4Mg(AsO_4)_2(OH)_4 \cdot 4H_2O$

CRYSTAL SYSTEM: Monoclinic
CLASS: 2/m
SPACE GROUP: $P2_1/c$
Z: 2

LATTICE CONSTANTS:
$a = 5.70$ Å
$b = 17.60$
$c = 6.752$
$\beta = 99°48'$
3 STRONGEST DIFFRACTION LINES:
4.40 (100)
8.79 ( 80)
2.75 ( 50)
OPTICAL CONSTANTS:
$\alpha = 1.672$
$\beta = 1.676$
$\gamma = 1.683$
HARDNESS: 3½
DENSITY: 3.26 (Meas.)
3.29 (Calc.)
CLEAVAGE: Two, in mutually perpendicular directions.
HABIT: As spherical or wart-like aggregates of minute
crystals.
COLOR-LUSTER: Reddish-brown with yellow tint.
Translucent.
MODE OF OCCURRENCE: Occurs sparingly in haus-
mannite ore associated with barite and pyrochroite at
Långban, Sweden.
BEST REF. IN ENGLISH: Moore, Paul B., *Am. Min.*, **53**:
1779 (1968). Palache, et al., "Dana's System of Mineral-
ogy," 7th Ed., v. II, p. 927, New York, Wiley, 1951.

# AKSAITE

$MgB_6O_{10} \cdot 5H_2O$

CRYSTAL SYSTEM: Orthorhombic
CLASS: 2/m 2/m 2/m
SPACE GROUP: Pbca
Z: 8
LATTICE CONSTANTS:
$a = 12.54$
$b = 24.35$
$c = 7.484$
3 STRONGEST DIFFRACTION LINES:
6.4 (100)
6.1 ( 50)
4.72 ( 50)
OPTICAL CONSTANTS:
$\alpha = 1.472$
$\beta = 1.503$
$\gamma = 1.526$
$(-)2V = 80°$
HARDNESS: ~2.5
DENSITY: 1.99 ± 0.01 (Meas.)
1.972 (Calc.)
CLEAVAGE: {100} probable
{010} probable
HABIT: Crystals elongated on [001], flattened on (100).
Faces of prismatic zone striated parallel to [100]. Charac-
teristic forms are {100}, {010}, and {021}. Crystals up to
7 mm in length.
COLOR-LUSTER: Light gray, colorless in small crystals.
MODE OF OCCURRENCE: Occurs in fine-grained rock
salt containing kieserite, anhydrite, and preobrazhenskite at
Ak-saĭ, Kazakhstan.

BEST REF. IN ENGLISH: Blazko, L. N., Kondrat′eva, V. V., Yarzhemskii, Ya. Ya., *Am. Min.*, **48**: 209–210 (1963). Clark, Joan R., and Erd, Richard C., *Am. Min.*, **48**: 930–935 (1963).

## AKTASHITE
$Cu_6Hg_3As_5S_{12}$

CRYSTAL SYSTEM: Hexagonal
CLASS: $3, \bar{3}$
SPACE GROUP: R3 or $R\bar{3}$
Z: 3
LATTICE CONSTANTS:
 $a$ = 13.72 Å
 $c$ = 9.32
3 STRONGEST DIFFRACTION LINES
 3.12 (100)
 1.909 ( 54)
 1.627 ( 40)
HARDNESS: Microhardness with 50 g load 300–346, av. 313 kg/mm$^2$.
DENSITY: Not determined.
CLEAVAGE: None.
Brittle.
HABIT: As xenomorphic grains, rarely in crystals resembling trigonal pyramids.
COLOR-LUSTER: White in reflected light. Opaque.
MODE OF OCCURRENCE: Aktashite occurs in the Gal Khaya deposit, Yakutia, and in the Aktash mercury deposit, Gornyi Altai, Soviet Union. Associated minerals include quartz, pyrite, calcite, sphalerite, stibnite, luzonite, chalcopyrite, chalcostibite, mercurian tetrahedrite, tennantite, enargite, dickite, cinnabar and orpiment.
BEST REF. IN ENGLISH: Vasil'ev, V. I., *Am. Min.*, **56**: 358–359 (1971). Gruzdev, V. S.; Chernitsova, N. M.; and Shumkova, N. G., *Am. Min.*, **58**: 562 (1973).

## ALABANDITE
MnS

CRYSTAL SYSTEM: Cubic
CLASS: m3m
SPACE GROUP: Fm3m
Z: 4
LATTICE CONSTANT:
 $a$ = 5.214
3 STRONGEST DIFFRACTION LINES:
 2.612 (100)
 1.847 ( 50)
 1.509 ( 20)
OPTICAL CONSTANT:
 $N$ = 2.70
HARDNESS: 3½–4
DENSITY: 3.95–4.04 (Meas.)
         4.05 (Calc. x-ray)
CLEAVAGE: {001} perfect
Fracture uneven. Brittle.
HABIT: Crystals cubic or octahedral. Usually massive, granular. Twinning on {111}.

COLOR-LUSTER: Iron black; tarnishes brown. Opaque. Submetallic. Streak green.
MODE OF OCCURRENCE: Occurs chiefly in metallic sulfide vein deposits, often associated with sphalerite, pyrite, galena, rhodochrosite, rhodonite, calcite, and quartz. Found sparingly in Alum Rock Park, Santa Clara County, California; at Bisbee and abundantly at Tombstone, Arizona; in White Pine County, Nevada; at Wickes, Montana; and in Boulder, Mineral, Park, San Juan, and Summit counties, Colorado. It also occurs in Mexico, Peru, France, Germany, Roumania, Japan, and at Alabanda, Asia Minor.
BEST REF. IN ENGLISH: Palache, et al., "Dana's System of Mineralogy," 7th Ed., v. I, p. 207–208, New York, Wiley, 1944.

## ALAMOSITE
$PbSiO_3$

CRYSTAL SYSTEM: Monoclinic
CLASS: 2/m
Z: 12
LATTICE CONSTANTS:
 $a$ = 12.254 Å
 $b$ = 7.055
 $c$ = 11.243
 $\beta$ = 113°9′
3 STRONGEST DIFFRACTION LINES:
 3.34 (100)
 2.30 (100)
 3.56 ( 80)
OPTICAL CONSTANTS:
 $\alpha$ = 1.947
 $\beta$ = 1.961
 $\gamma$ = 1.968
 $(-)2V = 65°$
HARDNESS: 4½
DENSITY: 6.49 (Meas.)
CLEAVAGE: {010} perfect
HABIT: Crystals fibrous; in radiating aggregates.
COLOR-LUSTER: Colorless, white. Transparent to translucent. Adamantine.
MODE OF OCCURRENCE: Occurs in association with cerussite, wulfenite, quartz, and other minerals, at Alamos, Sonora, Mexico.
BEST REF. IN ENGLISH: Winchell and Winchell, "Elements of Optical Mineralogy," Pt. 2, 4th Ed., p. 455, New York, Wiley, 1951.

## ALASKAITE = A mixture

## ALBITE
$mNaAlSi_3O_8$ with $nCaAl_2Si_2O_8$
$Ab_{100}An_0$ to $Ab_{90}An_{10}$
Plagioclase Feldspar Group
Var. Cleavelandite
  Pericline
  Peristerite

CRYSTAL SYSTEM: Triclinic
CLASS: $\bar{1}$
SPACE GROUP: C$\bar{1}$
Z: 4
LATTICE CONSTANTS:

| (high temp.) | (low temp.) |
|---|---|
| $a = 8.15$ | $a = 8.14$ |
| $b = 12.88$ | $b = 12.79$ |
| $c = 7.11$ | $c = 7.16$ |
| $\alpha = 93°22'$ | $\alpha = 94°20'$ |
| $\beta = 116°18'$ | $\beta = 116°34'$ |
| $\gamma = 90°17'$ | $\gamma = 87°39'$ |

3 STRONGEST DIFFRACTION LINES:

| (high temp.) | (low temp.) |
|---|---|
| 3.176 (100) | 3.196 (100) |
| 3.752 ( 30) | 3.780 ( 25) |
| 3.211 ( 30) | 6.39 ( 20) |

OPTICAL CONSTANTS:

| (high temp.) | (low temp.) |
|---|---|
| $\alpha = 1.527$ | $\alpha = 1.527$ |
| $\beta = 1.532$ | $\beta = 1.531$ |
| $\gamma = 1.534$ | $\gamma = 1.538$ |
| $(-)2V = 45°$ | $(+)2V = 77°$ |

HARDNESS: 6-6½
DENSITY: 2.60-2.63 (Meas.)
2.619 (Calc.)
CLEAVAGE: {001} perfect
{010} nearly perfect
{110} imperfect
Fracture uneven to conchoidal. Brittle.

HABIT: Crystals commonly tabular, flattened along $b$-axis, often platy. Usually massive, lamellar or granular; laminae often curved, sometimes in large divergent aggregates. Twinning very common; simple, multiple, and repeated. Principal twin laws: Pericline, Albite, Carlsbad, Baveno, Manebach.

COLOR-LUSTER: White to colorless; occasionally bluish, gray, reddish, greenish. Transparent to subtranslucent. Vitreous; sometimes pearly. Streak white.

MODE OF OCCURRENCE: Occurs widespread as a very common constituent of pegmatites, granites, rhyolites, andesite, syenite, and other alkaline igneous rocks; also in gneiss, crystalline schists, granular limestone and marble; in hydrothermal veins; and as an authigenic mineral in sedimentary deposits. Typical occurrences are found in California, Arizona, New Mexico, Colorado, Wyoming, South Dakota (cleavelandite), Maine, New Hampshire, Massachusetts, Connecticut, New York, and at Amelia, Virginia (cleavelandite). Also in Canada (peristerite), Brazil (cleavelandite), England, Norway, Sweden, France, Italy, Switzerland (pericline), Austria, Germany, USSR, India, Japan, and elsewhere.

BEST REF. IN ENGLISH: Deer, Howie and Zussman, "Rock Forming Minerals," v. 4, p. 94-165, New York, Wiley, 1963. T. F. W. Barth, "Feldspars," (1969).

**ALDANITE** = Variety of thorianite containing lead and uranium

## ALDZHANITE
Ca, $B_2O_3$, Cl, Mg, Mn (?)
(Inadequately described mineral)

CRYSTAL SYSTEM: Orthorhombic
LATTICE CONSTANTS:
$a = 12.76$
$b = 14.59$
$c = 8.19$
3 STRONGEST DIFFRACTION LINES:
7.46 (100)
3.07 ( 90)
3.49 ( 80)
OPTICAL CONSTANTS:
$\alpha = 1.600$
$\gamma = 1.620$
DENSITY: 2.21 (Meas.)
HABIT: Crystals dipyramidal.
COLOR-LUSTER: Colorless to pale rose.
MODE OF OCCURRENCE: Found in the insoluble residues of carnallite-bischofite rock, with boracite, hilgardite, anhydrite, and carbonates presumably in Kazakhstan.
BEST REF. IN ENGLISH: Avrova, N. P., Bocharov, V. M., Khalturina, I. I., and Yunosova, Z. R., *Am. Min.*, **56**: 1122 (1971).

## ALGODONITE
[$Cu_6As$] $Cu_3As$ proposed

CRYSTAL SYSTEM: Hexagonal
CLASS: 6/m 2/m 2/m
SPACE GROUP: P$6_3$/mmc
Z: 2
LATTICE CONSTANTS:
$a = 2.586$
$c = 4.228$
3 STRONGEST DIFFRACTION LINES:
1.989 (100)
2.11 ( 40)
2.25 ( 20)
HARDNESS: 4
DENSITY: 8.38 (Meas.)
8.71-8.72 (Calc.)
CLEAVAGE: None.
Fracture subconchoidal.
HABIT: Rarely as sharp, distorted crystals showing forms {0001}, {10$\bar{1}$0}, {11$\bar{2}$0}, {30$\bar{3}$1}, and {22$\bar{4}$1}. Usually massive, granular.
COLOR-LUSTER: Silver white to steel gray; tarnishes rapidly and becomes dull brown on exposure. Opaque. Metallic.
MODE OF OCCURRENCE: Occurs in Michigan as distinct crystals at Painsdale, Houghton County, and as crystalline masses at the Mohawk mine and others in Keweenaw County, and also from Baraga County. In Chile it is found in the Cerro de Los Seguas, Rancagua, and at the Algodones mine, Coquimbo.
BEST REF. IN ENGLISH: Palache, et al., "Dana's System of Mineralogy," 7th Ed., v. I, p. 171, New York, Wiley, 1944.

## ALLACTITE
$Mn_7(AsO_4)_2(OH)_8$

CRYSTAL SYSTEM: Monoclinic

CLASS: 2/m
SPACE GROUP: $P2_1/a$
Z: 2
LATTICE CONSTANTS:
  $a = 11.03$
  $b = 12.12$
  $c = 5.51$
  $\beta = 114°04'$
3 STRONGEST DIFFRACTION LINES:
  3.06 (100)
  3.71 ( 65)
  3.28 ( 55)
OPTICAL CONSTANTS:
  $\alpha = 1.755$
  $\beta = 1.772$
  $\gamma = 1.774$
  $(-)2V \simeq 0$
HARDNESS: 4½
DENSITY: 3.83 (Meas.)
         3.94 (Calc.)
CLEAVAGE: {001} distinct
Fracture uneven. Brittle.
  HABIT: Crystals slender prismatic or tabular. Sometimes as rosette-like aggregates.
  COLOR-LUSTER: Light to dark brownish red or purplish red. Translucent. Vitreous, oily on fracture.
  MODE OF OCCURRENCE: Occurs associated with calcite, franklinite, willemite, and fluorite at Franklin and Sterling Hill, Sussex County, New Jersey. It also is found in Sweden at Långban, and at the Brattfors and Moss mines in the Nordmark district.
  BEST REF. IN ENGLISH: Palache, et al., "Dana's System of Mineralogy," 7th Ed., v. II, p. 785–787, New York, Wiley, 1951.  Moore, Paul B., *Am. Min.*, **53**: 733–741, (1968).

## ALLANITE (Orthite)
$(Ca, Fe^{2+})_2(R, Al, Fe^{3+})_3Si_3O_{12}OH$
Epidote Group

CRYSTAL SYSTEM: Monoclinic
CLASS: 2/m
SPACE GROUP: $P2_1/m$
Z: 2
LATTICE CONSTANTS:
  $a = 10.22$
  $b = 5.75$
  $c = 8.95$
  $\beta = 115°$
3 STRONGEST DIFFRACTION LINES:
  (metamict-heated)    (non-metamict)
  2.96 (100)           2.91 (100)
  3.50 ( 80)           2.92 ( 90)
  2.67 ( 80)           2.86 ( 50)
OPTICAL CONSTANTS:
  $\alpha = 1.64-1.78$
  $\beta = 1.65-1.80$
  $\gamma = 1.66-1.81$
  $(-)2V$ = large (50°–80°) occasionally (+)
HARDNESS: 5½–6

DENSITY: 3.9–4.0
CLEAVAGE: None.
Fracture conchoidal to uneven. Brittle.
  HABIT: Crystals usually tabular; also long prismatic to acicular.  Commonly compact massive, bladed, or as embedded grains. Twinning on {100} common, polysynthetic.
  COLOR-LUSTER: Brown to black. Resinous or pitchy submetallic. Translucent to opaque.
  MODE OF OCCURRENCE: Occurs widespread in deposits of various genetic types, especially in granites and granite pegmatites or in gneiss and other metamorphic rocks. Found in the United States in California, Wyoming, Colorado, Texas, Arkansas, Pennsylvania, Massachusetts, Connecticut, New York, New Jersey, North Carolina, Virginia, and in ash beds of Miocene, Oligocene, and Pleistocene age in several western states.  Other notable occurrences are found in Canada, Greenland, Norway, Sweden, Germany, USSR, China, Japan, and Madagascar.
  BEST REF. IN ENGLISH: Vlasov, K. A., "Mineralogy of Rare Elements," v. II, p. 302–308, Israel Program for Scientific Translations, 1966.

## ALLARGENTUM
$Ag_{1-x}Sb_x$
ε-phase of Ag-Sb system
(usually, but not necessarily, containing some Hg)

CRYSTAL SYSTEM: Hexagonal
Z: 2
LATTICE CONSTANTS:
  $a = 2.95$
  $c = 4.77$
3 STRONGEST DIFFRACTION LINES:
            (synthetic)
  2.370 (100)    2.245 (100)
  2.252 ( 60)    2.545 ( 70)
  2.548 ( 40)    2.396 ( 70)
HARDNESS: Microhardness 189 kg/mm²
DENSITY: 10.12 (Calc.)
         10.0 (Meas. synthetic)
CLEAVAGE: Not determined.
  HABIT: Massive, in complex intergrowths; as small grains.
  COLOR-LUSTER: Silver; metallic; opaque.
  MODE OF OCCURRENCE: Occurs in the silver ores from Cobalt, Ontario; in ore from the Red Lake area of Ontario; and has also been reported in ore from the Consols Mine, Broken Hill, Australia.
  BEST REF. IN ENGLISH: Petruk, W., Cabri, L. J., Harris, D. C., Stewart, J. M., and Clark, L. A., *Can. Min.*, **10** (pt. 2): 163–172 (1970).

## ALLCHARITE = Goethite

## ALLEGHANYITE
$Mn_5Si_2O_8(OH)_2$
Manganese analogue of chondrodite

CRYSTAL SYSTEM: Monoclinic
CLASS: 2/m

SPACE GROUP: P2₁/a

Z: 2

LATTICE CONSTANTS:

$a = 10.46$

$b = 4.86$

$c = 8.3$

$\beta = 109°8'$

3 STRONGEST DIFFRACTION LINES:

1.799 (100)

2.860 ( 80)

2.598 ( 60)

OPTICAL CONSTANTS:

$\alpha = 1.756$

$\beta = 1.780$

$\gamma = 1.792$

$(-)2V = 72°$

HARDNESS: 5.5

DENSITY: 4.02 (Meas.)

CLEAVAGE: None.
Brittle.

HABIT: Crystals slender plate-like, deeply striated by twinning; also as unstriated stout platy crystals; sometimes in fan-shaped aggregates. Usually massive or as irregular grains. Twinning in {001}, common.

COLOR-LUSTER: Brown to pinkish brown to deep pink. Transparent to translucent. Vitreous to dull.

MODE OF OCCURRENCE: Occurs in association with franklinite-willemite-zincite ore and in veinlets with leuco-phoenicite at Franklin, and with manganoan calcite, serpentine, magnesium-chlorophoenicite, hetaerolite, tephroite, or zincite at Sterling Hill, Sussex County, New Jersey; in veins with tephroite, spessartine, rhodonite, galaxite, and other minerals near Bald Knob, Alleghany County, North Carolina; with tephroite and other manganese minerals at the Germolis prospect near Fiddletown, Amador County, and elsewhere in California; and at the Benallt mine, Wales.

BEST REF. IN ENGLISH: Winchell and Winchell, "Elements of Optical Mineralogy," 4th Ed., Pt. 2, p. 516, New York, Wiley, 1951.  Cook, David, *Am. Min.*, **54**: 1392–1398 (1969).

## ALLEMONTITE
AsSb

CRYSTAL SYSTEM: Hexagonal

CLASS: 3̄m

SPACE GROUP: R3̄m

Z: 3

LATTICE CONSTANTS:

$a = 4.044$

$c = 10.961$

3 STRONGEST DIFFRACTION LINES:

2.92 (100)

2.01 ( 70)

2.13 ( 60)

HARDNESS: 3–4

DENSITY: 5.8–6.2 (Meas.)

6.277 (Calc.)

CLEAVAGE: Perfect, one direction.

HABIT: As indistinct crystals; usually massive; reniform, mammillary, lamellar; also fine granular.

COLOR-LUSTER: Tin white, reddish gray; tarnishes gray or brownish black. Opaque. Metallic. Streak gray.

MODE OF OCCURRENCE: Occurs at the Ophir mine, Virginia City, Nevada; at Atlin and Alder Island, British Columbia, Canada; near Allemont, Isère, France; and in Italy, Germany, Sweden, Czechoslovakia, and Roumania.

BEST REF. IN ENGLISH: Palache, et al., "Dana's System of Mineralogy," 7th Ed., v. I, p. 130–132, New York, Wiley, 1944.

## ALLEVARDITE = Rectorite

## ALLOCLASITE
(Co, Fe) AsS
Dimorphous with Cobaltite

CRYSTAL SYSTEM: Orthorhombic

CLASS: 2 2 2

SPACE GROUP: P22₁2₁

Z: 2

LATTICE CONSTANTS:

$a = 4.641, 4.631, 4.662$

$b = 5.606, 5.605, 5.606$

$c = 3.415, 3.430, 3.415$

3 STRONGEST DIFFRACTION LINES:

2.750 (100)

2.469 ( 90)

1.817 ( 70)

HARDNESS: 5

DENSITY: 6.166 (Meas.)

CLEAVAGE: {101} perfect

{010} distinct
Fracture uneven. Brittle.

HABIT: Crystals short prismatic. Usually massive, commonly in columnar to hemispherical aggregates.

COLOR-LUSTER: Steel gray. Opaque. Metallic. Streak nearly black.

MODE OF OCCURRENCE: Occurs in association with bismuthinite in calcite at the Elizabeth mine, Oravicza, Roumania, also found in Westphalia, Germany, and at the Dogatani mine, Japan.

BEST REF. IN ENGLISH: Kingston, P. W., *Can. Min.*, **10**: 838–846 (1971).

## ALLOPALLADIUM
Pd

CRYSTAL SYSTEM: Hexagonal

HARDNESS: Not determined.

DENSITY: Not determined.

CLEAVAGE: {0001} good.

HABIT: As small hexagonal tablets.

COLOR-LUSTER: Silver white to pale steel gray. Opaque. Metallic.

MODE OF OCCURRENCE: Occurs associated with native gold and clausthalite at Tilkerode, and elsewhere in the Harz Mountains, Germany, and in hortonolite-dunite from Mooihoek, Transvaal, Africa.

BEST REF. IN ENGLISH: Palache, et al., "Dana's System of Mineralogy," 7th Ed., v. I, p. 113-114, New York, Wiley, 1944.

## ALLOPHANE
$Al_2SiO_5 \cdot nH_2O$
Kaolinite Group (Kandite)

CRYSTAL SYSTEM: Not determined.
3 STRONGEST DIFFRACTION LINES:
  11.0 (100   ) (Usually amorphous)
   3.3 (  80bb)
   2.2 (  60b )
OPTICAL CONSTANTS:
  $N$ = 1.47-1.49
HARDNESS: 2-3
DENSITY: 1.85-1.89 (Meas.)
CLEAVAGE: None
Fracture conchoidal to earthy. Very brittle.
  HABIT: Massive; as crusts resembling hyalite opal; also stalactitic or as powdery aggregates.
  COLOR-LUSTER: Colorless, white, green, bluish, yellow, brown. Transparent to translucent. Vitreous, resinous or waxy. Streak uncolored.
  MODE OF OCCURRENCE: Occurs mainly in fissures and cavities in ore veins; also in coal deposits. Typical occurrences are found throughout the western United States, especially in California, Nevada, Utah, Colorado, and New Mexico. Also in Lebanon and Lehigh counties, Pennsylvania; near Richmond, Massachusetts; and at localities in Sardinia and Germany. Crystalline allophane is reported from Wheal Hamblyn, Bridestone, Devonshire, England.
  BEST REF. IN ENGLISH: Winchell and Winchell, "Elements of Optical Mineralogy," 4th Ed., Pt. 2, p. 531, New York, Wiley, 1951.

## ALLUAUDITE
$(Na, Fe^{3+}, Mn^{2+})PO_4$

CRYSTAL SYSTEM: Monoclinic
CLASS: 2/m
SPACE GROUP: C2/c
Z: 12
LATTICE CONSTANTS:
  $a$ = 12.004
  $b$ = 12.533
  $c$ = 6.404
  $\beta$ = 114.4°
3 STRONGEST DIFFRACTION LINES:
  2.73 (100)
  6.30 (  80)
  3.07 (  65)
OPTICAL CONSTANTS:
  $\alpha$ = 1.760
  $\beta$ = 1.765
  $\gamma$ = 1.775
  (+)2V = moderate
HARDNESS: 5-5½
DENSITY: 3.52 (Meas.—Pringle, S. D.)
         3.584 (Meas.—Chanteloube, France)

CLEAVAGE: {100} good
          {010} good
HABIT: Massive, compact granular, radiating fibrous, and globular aggregates.
  COLOR-LUSTER: Dull greenish black; dirty yellow, brownish yellow to brownish black. Subtranslucent to opaque. Streak and powder dirty yellow to brownish.
  MODE OF OCCURRENCE: Occurs in large masses in pegmatite at several mines in the Custer-Pringle area, Custer County, and at the Etta mine, Pennington County, South Dakota. Also found in pegmatite at Chanteloube, France; Buranga, Ruanda; Varuträsk, Sweden; Sukula and Lemnäs, Finland; and in several other pegmatite deposits.
  BEST REF. IN ENGLISH: Fisher, D. Jerome, *Am. Min.*, **40**: 1100-1109 (1955). *Am. Min.*, **50**: 1647-1669 (1965). Moore, *Am. Min.*, **56**: 1955-1975 (1971).

## ALMANDINE (Almandite)
$Fe_3^{2+}Al_2Si_3O_{12}$
Garnet Group

CRYSTAL SYSTEM: Cubic
CLASS: 4/m $\bar{3}$ 2/m
SPACE GROUP: Ia3d
Z: 8
LATTICE CONSTANT:
  $a$ = 11.526
3 STRONGEST DIFFRACTION LINES:
  2.569 (100)
  1.540 (  50)
  2.873 (  40)
OPTICAL CONSTANT:
  $N$ = 1.830
HARDNESS: 7-7½
DENSITY: 4.1-4.3 (Meas.)
         4.313 (Calc.)
CLEAVAGE: None.
          {110} parting sometimes distinct.
Fracture uneven to conchoidal. Brittle.
  HABIT: Crystals usually dodecahedrons or trapezohedrons; also in combination, or with hexoctahedron. Also massive, compact; fine or coarse granular; or as embedded grains.
  COLOR-LUSTER: Deep red, brownish red, brownish black. Transparent to translucent. Vitreous to resinous. Streak white.
  MODE OF OCCURRENCE: Occurs widespread, chiefly in schists, gneiss, and other metamorphic rocks; in contact zones and in some igneous rocks; and as a detrital mineral in sedimentary deposits. Important occurrences are found at Wrangell, Alaska; at many places in California; in the Emerald Creek district, Benewah County, Idaho; in the Black Hills, South Dakota; near Salida, Chaffee County, Colorado; at Michigamme, Michigan; at Avondale, Delaware County, and elsewhere in Pennsylvania; at Roxbury, Connecticut; as giant crystals at the Barton mine, North Creek, New York, and at many other localities in the New England states. Other notable occurrences are found in Canada, Brazil, Uruguay, Greenland, Norway, Sweden, Austria, Ceylon, India, Japan, Madagascar, Tanzania, and Australia.

BEST REF. IN ENGLISH: Deer, Howie, and Zussman, "Rock Forming Minerals," v. 1, p. 77–112, New York, Wiley, 1962.

## AL-NONTRONITE = Variety of Chloropal
$Fe_2^{3+}[AlSi_3O_{10}](OH)_2 \cdot nH_2O$

## AL-SAPONITE = Variety of saponite containing aluminum

## ALSTONITE
$CaBa(CO_3)_2$
Dimorphous with Barytocalcite.

CRYSTAL SYSTEM: Orthorhombic
CLASS: 2/m 2/m 2/m
Z: 2
LATTICE CONSTANTS:
  $a = 4.99$
  $b = 8.77$
  $c = 6.11$
3 STRONGEST DIFFRACTION LINES:
  3.68 (100)
  3.12 ( 62)
  2.13 ( 37)
OPTICAL CONSTANTS:
  $\alpha = 1.5261$
  $\beta = 1.6710$
  $\gamma = 1.6717$
  $(-)2V = 6°$
HARDNESS: 4–4½
DENSITY: 3.67 (Meas.)
         3.67 (Calc.)
CLEAVAGE: {110} imperfect
Fracture uneven.
  HABIT: Crystals commonly twinned; as pseudohexagonal prisms, pseudodihexagonal dipyramids, and as steep-sided dipyramids. Apparent dipyramidal faces striated horizontally. Twinning on {110} and {130}.
  COLOR-LUSTER: Colorless, white, grayish, pinkish. Transparent to translucent. Vitreous. Streak white. Fluoresces weak yellow under long-wave ultraviolet light.
  MODE OF OCCURRENCE: Occurs as a low-temperature hydrothermal vein mineral at the Brownley Hill mine near Alston, Cumberland; at New Brancepeth, Durham; and in the Fallowfield mine near Hexham, Northumberland, England.
  BEST REF. IN ENGLISH: Palache, et al., "Dana's System of Mineralogy," 7th Ed., v. II, p. 218–219, New York, Wiley, 1951.

## ALTAITE
PbTe

CRYSTAL SYSTEM: Cubic
CLASS: m3m
SPACE GROUP: Fm3m
Z: 4
LATTICE CONSTANT:
  $a = 6.452 \pm 6$
3 STRONGEST DIFFRACTION LINES:
  3.23 (100)
  2.28 ( 80)
  1.442 ( 50)
HARDNESS: 2–3
DENSITY: 8.19 (Meas.)
         8.27 (Calc. x-ray)
CLEAVAGE: {001} perfect
Fracture uneven to conchoidal. Brittle.
  HABIT: Crystals rare; cubic or octahedral. Commonly massive.
  COLOR-LUSTER: Tin white with yellowish tint; tarnishes bronze. Opaque. Metallic. Streak black.
  MODE OF OCCURRENCE: Occurs in vein deposits often associated with native gold, silver, antimony, tellurium, sulfides, and tellurides. Found in Gaston County, North Carolina; in small amounts in many telluride-bearing ores in Boulder, Clear Creek, Lake, and Saguache counties, Colorado; in Dona Ana County, New Mexico; in Calaveras, Madera, Nevada, and Tuolumne counties, California; and sparingly in other western states. It also occurs at Kirkland Lake, Ontario, and at other places in Canada; in Chile at Coquimbo; at Kalgoorlie, Western Australia; and at several localities in the USSR, especially in the Urals and Altai Mountains.
  BEST REF. IN ENGLISH: Palache, et al., "Dana's System of Mineralogy," 7th Ed., v. I, p. 205–207, New York, Wiley, 1944.

## ALUM = Group Name
See: Potash Alum
    Ammonia Alum
    Soda Alum

## ALUMINITE
$Al_2SO_4(OH)_4 \cdot 7H_2O$

CRYSTAL SYSTEM: Monoclinic or orthorhombic
Z: 1
3 STRONGEST DIFFRACTION LINES:
  9.0 (100)
  7.8 (100)
  3.72 (100)
OPTICAL CONSTANTS:
  $\alpha = 1.459$
  $\beta = 1.464$
  $\gamma = 1.470$
  $(+)2V = large \sim 90°$
HARDNESS: 1–2
DENSITY: 1.66–1.82 (Meas.)
CLEAVAGE: Fracture of aggregates earthy. Friable.
  HABIT: As nodular masses composed of minute fibers; spherulitic; as veinlets.
  COLOR-LUSTER: White to grayish white. Opaque. Dull, earthy.
  MODE OF OCCURRENCE: Occurs in Mount Vernon Canyon, Jefferson County, and probably at the Silver Ledge mine, San Juan County, Colorado; at Green River,

Utah; and coating limestone at Joplin, Missouri. It also is found at Newhaven, Sussex, England; in the vicinity of Halle, Saxony, Germany; and in Czechoslovakia, France, Italy, USSR, and in the Salt Range, Punjab, India.

BEST REF. IN ENGLISH: Palache, et al., "Dana's System of Mineralogy," 7th Ed., v. II, p. 600–601, New York, Wiley, 1951.

## ALUMINOCOPIAPITE
$(Mg, Fe^{2+})(Fe^{3+}, Al)_4 (SO_4)_6 (OH)_2 \cdot 20H_2O$

CRYSTAL SYSTEM: Triclinic
CLASS: $\bar{1}$
SPACE GROUP: $P\bar{1}$
Z: 1
LATTICE CONSTANTS:
$a = 7.251$ $\alpha = 93°59'$
$b = 18.161$ $\beta = 102°17'$
$c = 7.267$ $\gamma = 97°58'$
3 STRONGEST DIFFRACTION LINES:
9.2 (100)
18.1 ( 80)
5.58 ( 80)
OPTICAL CONSTANTS:
$\beta = 1.535$
$\gamma = 1.585$
DENSITY: 2.163 (Calc.)
HABIT: Massive; as thin crusts and coatings.
COLOR-LUSTER: Yellowish.
MODE OF OCCURRENCE: Occurs as a thin local efflorescence in sheared and crushed bedrock along the north bank of Mosquito Fork of the Forty Mile River, Alaska.
BEST REF. IN ENGLISH: Jelly, James H., and Foster, Helen L., *Am. Min.*, **52**: 1220–1223 (1967).

## ALUMOFERROASCHARITE = A mixture of szaibelyite and hydrotalcite

## ALUMOHYDROCALCITE
$CaAl_2 (CO_3)_2 (OH)_4 \cdot 3H_2O$

CRYSTAL SYSTEM: Monoclinic
Z: 1
3 STRONGEST DIFFRACTION LINES:
6.25 (100)
6.50 ( 70)
3.23 ( 60)
OPTICAL CONSTANTS:
$\alpha = 1.500$
$\beta = 1.560$
$\gamma = 1.584$
$(-)2V = 64°$
HARDNESS: 2½
DENSITY: 2.231–2.4 (Meas.)
CLEAVAGE: {100} perfect
{010} imperfect
Brittle.
HABIT: As chalky masses of fibrous crystals.
COLOR-LUSTER: White to pale blue, also yellowish, gray, violet.

MODE OF OCCURRENCE: Occurs as an alteration of allophane in the Khakassy district, Siberia, USSR, associated with allophane, volborthite, malachite, and other minerals. Also found at a dolomite quarry near Bergisch-Gladbach, Germany.
BEST REF. IN ENGLISH: Palache, et al., "Dana's System of Mineralogy," 7th Ed., v. II, p. 280–281, New York, Wiley, 1951.

## ALUMOTUNGSTITE (Species status uncertain)
$(W, Al)_{16} (O, OH)_{48} \cdot xH_2O$

## ALUNITE
$KAl_3 (SO_4)_2 (OH)_6$

CRYSTAL SYSTEM: Hexagonal
CLASS: 3m or $\bar{3}$m
SPACE GROUP: R3m or R$\bar{3}$m
Z: 3
LATTICE CONSTANTS:
$a = 6.96$
$c = 17.35$
3 STRONGEST DIFFRACTION LINES:
2.99 (100)
2.89 (100)?
2.293 ( 80)
OPTICAL CONSTANTS:
$\omega = 1.572$
$\epsilon = 1.592$
HARDNESS: 3½–4
DENSITY: 2.6–2.9 (Meas.)
2.82 (Calc.)
CLEAVAGE: {0001} distinct
{01$\bar{1}$2} in traces
Fracture conchoidal. Brittle.
HABIT: Crystals rhombohedral, often pseudocubic; also tabular or lenticular; as druses or aggregates. Usually massive, granular to dense compact; also columnar or fibrous.
COLOR-LUSTER: White; often discolored grayish, yellowish, reddish, brownish. Transparent to nearly opaque. Vitreous, pearly on base. Streak white.
MODE OF OCCURRENCE: Occurs widespread chiefly as a mineral formed by sulfotaric processes on volcanic rocks, and less commonly as a vein mineral in ore deposits. Found at many localities in the western United States, especially at Marysvale, Utah, and in Nevada, Colorado, and California. Other well-known deposits occur in Newfoundland, Italy, Spain, Hungary, Czechoslovakia, China, and Australia.
BEST REF. IN ENGLISH: Palache, et al., "Dana's System of Mineralogy," 7th Ed., v. II, p. 556–560, New York, Wiley, 1951.

## ALUNOGEN
$Al_2 (SO_4)_3 \cdot 16H_2O$

CRYSTAL SYSTEM: Triclinic
CLASS: $\bar{1}$
Z: 1

LATTICE CONSTANTS:

$a:b:c = .8355:1:.6752$    $\alpha = 89°58'$
$\beta = 97°26'$
$\gamma = 91°52'$

3 STRONGEST DIFFRACTION LINES:

4.48 (100)
13.34 ( 60)
4.39 ( 30)

OPTICAL CONSTANTS:

$\alpha = 1.475$
$\beta = 1.478$    (Indices vary with water content.)
$\gamma = 1.485$
$(+)2V = 31°, 69°$

HARDNESS: 1½–2

DENSITY: 1.77 (Meas.)
1.78 (Meas.-Synthetic)

CLEAVAGE: {010} perfect

HABIT: Crystals prismatic with a six-sided outline; small and rare. Usually as an efflorescence; as fibrous crusts or crystalline masses. Twinning on {010}.

COLOR-LUSTER: Crystals colorless; transparent; vitreous. Aggregates white, yellowish, reddish; translucent; vitreous to silky.

Soluble in water. Taste acid, biting.

MODE OF OCCURRENCE: Occurs chiefly as an efflorescence or cavity filling in slates, shales, or coal formations containing iron sulfides; in the oxidation zone of pyritic ore deposits; and in areas of volcanic activity. Found in the United States as large crystalline plates at the Dexter No. 7 mine, Calf Mesa, San Rafael Swell, Utah; at The Geysers, Sonoma County, and at many other places in California; in the Alum Mountain district, Grant County, New Mexico; and in Arizona, Colorado, North Carolina, New York, and other states. It also occurs at Paracutin, Mexico; in Canada, Peru, Chile, France, Italy, Czechoslovakia; and in New South Wales, Australia.

BEST REF. IN ENGLISH: Palache, et al., "Dana's System of Mineralogy," 7th Ed., v. II, p. 537–540, New York, Wiley, 1951.

## ALURGITE = Variety of muscovite

## ALVANITE
$Al_3VO_4(OH)_6 \cdot 2½H_2O$

CRYSTAL SYSTEM: Monoclinic

LATTICE CONSTANT:

$\beta = 115°$

3 STRONGEST DIFFRACTION LINES:

4.477 (100)
1.484 ( 90)
1.982 ( 80)

OPTICAL CONSTANTS:

$\alpha = 1.658$
$\gamma = 1.714$
$(-)2V = 80°-85°$

HARDNESS: 3–3½

DENSITY: 2.41 (Meas.)

CLEAVAGE: {010} perfect

HABIT: Mica-like platelets of hexagonal form. Faces: a{100}, b{010}, c{001}, and d{101}. Polysynthetic twinning plane parallel to cleavage plane.

COLOR-LUSTER: Light bluish green to bluish black; vitreous, pearly on cleavage. Streak white.

MODE OF OCCURRENCE: Occurs in the oxidation zone of the vanadiferous clay-anthraxolite horizon at several mines in the Kurumsak and Balasauskandyk ore fields, northwestern Kara-Tau.

BEST REF. IN ENGLISH: Ankinovich, E. A., *Am. Min.*, **44**: 1325–1326 (1959).

## ALVAROLITE = Mangantantalite

## AMAKINITE
(Fe, Mg)(OH)$_2$

CRYSTAL SYSTEM: Trigonal

CLASS: 3m

SPACE GROUP: R3m

Z: 12

LATTICE CONSTANTS:

$a = 6.917$
$c = 14.52$

3 STRONGEST DIFFRACTION LINES:

2.30 (100)
1.728 ( 90)
2.80 ( 80)

OPTICAL CONSTANTS:

uniaxial (+)
$\omega = 1.707$
$\epsilon = 1.722$

HARDNESS: 3½–4

DENSITY: 2.98 ± 0.01

CLEAVAGE: Fracture irregular.

HABIT: Irregular grains, rarely as well-formed rhombohedra up to 1 cm in size. Weakly magnetic.

COLOR-LUSTER: Pale green to yellow-green.

MODE OF OCCURRENCE: Found as thin veins (up to 3 cm) and pockets in fissured kimberlite in a drill core at a depth of 300 m in the "Lucky Eastern" kimberlite pipe, presumably in the Yakut A.S.S.R.

BEST REF. IN ENGLISH: Kozlov, I. T. and Levshov, P. P., *Am. Min.*, **47**: 1218 (1962).

## AMALGAM = Mercurian silver (arquerite)
Ag, Hg

## AMARANTITE
$Fe^{3+}SO_4OH \cdot 3H_2O$

CRYSTAL SYSTEM: Triclinic

CLASS: $\bar{1}$

Z: 4

LATTICE CONSTANTS:

$a = 8.90$    $\alpha = 95°38½'$
$b = 11.56$    $\beta = 90°23½'$
$c = 6.64$    $\gamma = 97°13'$

3 STRONGEST DIFFRACTION LINES:
  11.3 (100)
  8.69 (100)
  3.57 ( 80)
OPTICAL CONSTANTS:
  $\alpha = 1.516$
  $\beta = 1.598$
  $\gamma = 1.621$
  $(-)2V = 30°$
HARDNESS: 2½
DENSITY: 2.189–2.286 (Meas.)
  2.197 (Calc.)
CLEAVAGE: {010} perfect
  {100} perfect
Brittle.

HABIT: Crystals short to long prismatic, somewhat flattened to nearly square in cross-section; striated [001]. Usually as columnar or bladed masses; also as matted or radiating aggregates of needle-like crystals.

COLOR-LUSTER: Red, orangish red to brownish red. Transparent. Vitreous. Streak yellow.

Decomposed by water.

MODE OF OCCURRENCE: Occurs associated with magnesium copiapite in the Santa Maria Mountains, near Blythe, Riverside County, California; minutely crystallized with melanterite in old mine tunnels at the Broken Hill mine, near Hayward, Pennington County, South Dakota; and in Chile at Chuquicamata, Quetena, Alcaparrosa, Paposa, Tierra Amarilla, and near Sierra Gorda.

BEST REF. IN ENGLISH: Palache, et al., "Dana's System of Mineralogy," 7th Ed., v. II, p. 611–612, New York, Wiley, 1951.

# AMARILLITE
$NaFe^{3+}(SO_4)_2 \cdot 6H_2O$

CRYSTAL SYSTEM: Monoclinic
CLASS: 2/m
Z: 1
LATTICE CONSTANTS:
  $a : b : c = 0.7757 : 1 : 1.1482$
  $\beta = 95°37'$
OPTICAL CONSTANTS:
  $\alpha = 1.532$
  $\beta = 1.555$
  $\gamma = 1.591$
  $(+)2V = large$
HARDNESS: 2½–3
DENSITY: 2.19 (Meas.)
CLEAVAGE: {110} good, but difficult
Fracture conchoidal.
HABIT: Crystals thick tabular; equant; rarely prismatic.
COLOR-LUSTER: Pale yellow with faint greenish tinge. Transparent. Brilliant, vitreous to adamantine.

Soluble in water. Taste astringent.

MODE OF OCCURRENCE: Occurs at Tierra Amarilla, near Copiapo, Chile, as veinlets in massive coquimbite.

BEST REF. IN ENGLISH: Palache, et al., "Dana's System of Mineralogy," 7th Ed., v. II, p. 468–469, New York, Wiley, 1951.

# AMAZONITE = Green variety of microcline

# AMBATOARINITE
## (inadequately described mineral)
$Sr_5(La, Ce, etc.)_{10}(CO_3)_{17}O_3$ (?)

CRYSTAL SYSTEM: Orthorhombic (?)
OPTICAL CONSTANTS:
  $\alpha > 1.658$
  optically (–)
HARDNESS: Not determined.
HABIT: As microscopic crystals, often in skeletal groups.
COLOR-LUSTER: Pink to reddish or black (due to impurities.)
MODE OF OCCURRENCE: Occurs enclosed in plates of celestite in veins of manganese-bearing calcite and quartz in metamorphosed limestone at Ambatoarina, Madagascar.

BEST REF. IN ENGLISH: Palache, et al., "Dana's System of Mineralogy," 7th Ed., v. II, p. 293, New York, Wiley, 1951.

# AMBER (Hydrocarbon)
Variable C : H : O ratios

# AMBLYGONITE
$(Li, Na)AlPO_4(F, OH)$
Amblygonite series

CRYSTAL SYSTEM: Triclinic
CLASS: $\bar{1}$
SPACE GROUP: $P\bar{1}$
Z: 2
LATTICE CONSTANTS:
  $a = 5.18$    $\alpha = 109°29'$
  $b = 7.03$    $\beta = 97°46½'$
  $c = 5.03$    $\gamma = 106°37'$
3 STRONGEST DIFFRACTION LINES:
  4.64  (100)
  3.151 (100)
  2.925 (100)
OPTICAL CONSTANTS:
  $\alpha = 1.578$
  $\beta = 1.593$    (vary with % F)
  $\gamma = 1.598$
  $(-)2V \simeq 50°$
HARDNESS: 5½–6
DENSITY: 3.08 (Meas.)
  3.065 (Calc.)
CLEAVAGE: {100} perfect
  {110} good
  {0$\bar{1}$1} distinct
  {001} imperfect
Fracture conchoidal to uneven. Brittle.
HABIT: Crystals equant to short prismatic; commonly with rough faces. Usually as large cleavable masses. Twinning on {$\bar{1}$11} common.
COLOR-LUSTER: Usually white to grayish white; also colorless, yellowish, pinkish, tan, greenish, bluish. Transparent to translucent. Vitreous to greasy; pearly on cleavages.

MODE OF OCCURRENCE: Occurs in granite pegmatites often in masses and crystals of very large size. Found at many places in the Black Hills, South Dakota, especially at the Beecher Lode, Custer County, as masses up to 200 tons in weight; as rounded masses as much as 20 feet long and 4-8 feet in thickness at the Giant-Volney mine at Tinton, Lawrence County; and as masses 20-40 feet long at the Hugo, Ingersoll, and Peerless mines, Keystone, Pennington County. It also is found in Arizona, New Mexico, California, New Hampshire, and at several localities in Maine, notably as fine transparent crystals at Newry, Oxford County. Among other occurrences, it is found in France, Germany, Czechoslovakia, Sweden, and Brazil.

BEST REF. IN ENGLISH: Palache, et al., "Dana's System of Mineralogy," 7th Ed., v. II, p. 823-827, New York, Wiley, 1951.

## AMEGHINITE
$NaB_3O_5 \cdot 2H_2O$

CRYSTAL SYSTEM: Monoclinic
CLASS: 2/m
SPACE GROUP: C2/c
Z: 4
LATTICE CONSTANTS:
  $a$ = 18.454
  $b$ = 9.895
  $c$ = 6.322
  $\beta$ = 104°20′
3 STRONGEST DIFFRACTION LINES:
  3.065 (100)
  3.147 ( 76)
  2.548 ( 29)
OPTICAL CONSTANTS:
  $\alpha$ = 1.429
  $\beta$ = 1.528
  $\gamma$ = 1.538
  (−)2V = 33°
HARDNESS: 2½
DENSITY: 2.030 ± 0.006 (Meas.)
         2.037 (Calc.)
CLEAVAGE:  {100} good
           {010} poor
           {001} poor
Brittle; conchoidal fracture.
HABIT: Crystals elongated on [010], with a maximum length of 5 mm, frequently bent; flattened on {001}.
COLOR-LUSTER: Colorless, shows strong pale blue fluorescence and phosphorescence under ultraviolet light.
MODE OF OCCURRENCE: Found at the Tincalayu borax deposit, Salta, Argentina. It occurs in small nodules embedded in borax associated with ezcurrite and rivadavite.
BEST REF. IN ENGLISH: Aristarain, L. F., and Hurlbut, C. S., Jr., *Am. Min.*, 52: 935-945 (1967).

## AMESITE
$(Mg, Fe^{2+})_4 Al_4 Si_2 O_{10} (OH)_8$
Chlorite Group

CRYSTAL SYSTEM: Hexagonal
CLASS: 6
SPACE GROUP: P6
Z: 4
LATTICE CONSTANTS:
  $a$ = 5.31
  $c$ = 14.04
3 STRONGEST DIFFRACTION LINES:
  7.06 (100)
  3.52 (100)
  1.925 ( 70)
OPTICAL CONSTANTS:
  $\alpha$ = 1.597
  $\beta$ = 1.600
  $\gamma$ = 1.615
  $2V_\gamma$ = 10°-14°
HARDNESS: 2½-3
DENSITY: 2.77 (Meas.)
CLEAVAGE: {0001} perfect
HABIT: Crystals hexagonal plates, in foliated aggregates.
COLOR-LUSTER: Pale green. Translucent. Pearly on cleavage.
MODE OF OCCURRENCE: Occurs in association with corundophilite, diaspore, and magnetite at Chester, Massachusetts; also at the Saranovskoye chromite deposit, north Urals, USSR.
BEST REF. IN ENGLISH: Deer, Howie, and Zussman, "Rock Forming Minerals," v. 3, p. 166-167, New York, Wiley, 1962.

## AMETHYST = Variety of quartz.

## AMINOFFITE
$Ca_2 (Be, Al)Si_2 O_7 (OH) \cdot H_2O$

CRYSTAL SYSTEM: Tetragonal
CLASS: 4/m 2/m 2/m
SPACE GROUP: I4/mmm
Z: 12
LATTICE CONSTANTS:
  $a$ = 13.8
  $c$ = 9.8
3 STRONGEST DIFFRACTION LINES:
  2.614 (100)
  2.840 ( 90)
  4.02 ( 80)
OPTICAL CONSTANTS:
  $\omega$ = 1.647
  $\epsilon$ = 1.637
  (−)
HARDNESS: 5½
DENSITY: 2.94 (Meas.)
         3.124 (Calc.)
CLEAVAGE: {001} imperfect
Fracture conchoidal. Brittle.
HABIT: Crystals pyramidal with only two forms—p{111} and c{001}.
COLOR-LUSTER: Colorless. Transparent. Vitreous.
MODE OF OCCURRENCE: Occurs as small (0.5-1.0 mm) euhedral crystals in veins and in cavities in massive magnetite and goethite at Långban, Sweden.

BEST REF. IN ENGLISH: Vlasov, K. A., "Mineralogy of Rare Elements," v. II, p. 129–134, Israel Program for Scientific Translations, 1966.    Moore, Paul B., *Am. Min.,* **53:** 1418-1420 (1968).

## AMMONIA ALUM (Tschermigite)
$NH_4 Al(SO_4)_2 \cdot 12H_2O$

CRYSTAL SYSTEM:  Isometric
CLASS:  $2/m \bar{3}$
SPACE GROUP:  Pa3
Z:  4
LATTICE CONSTANT:
  $a = 12.215$
3 STRONGEST DIFFRACTION LINES:
  4.327 (100)
  4.079 ( 80)
  3.273 ( 75)
OPTICAL CONSTANT:
  $N = 1.457$
  HARDNESS:  1½
  DENSITY:  1.645 (Meas.)
             1.642 (Calc.)
  CLEAVAGE:  Fracture conchoidal.
  HABIT:  Crystals octahedral.  As an efflorescence or as columnar or fibrous masses.
  COLOR-LUSTER:  Colorless, white.  Transparent.  Crystals brilliant, vitreous; fibrous masses silky.

Soluble in water.  Taste sweetish and astringent.
  MODE OF OCCURRENCE:  Occurs in California as an efflorescence at Sulphur Bank, Lake County, and as crystals, crusts, and crystalline masses associated with other sulfates at numerous places in upper Geyser Creek Canyon, The Geysers, Sonoma County.  Found in abundance in bituminous shale near Wamsutter, Wyoming; in lignitic shale in southern Utah; and in Quay County, New Mexico.  It also occurs in brown coal at Tschermig and at other localities in Czechoslovakia; in Hungary; and as a fumarolic deposit in Italy, Sicily, and Zaire.
  BEST REF. IN ENGLISH:  Palache, et al., "Dana's System of Mineralogy," 7th Ed., v. II, p. 475–476, New York, Wiley, 1951.

## AMMONIA-NITER
$NH_4 NO_3$

CRYSTAL SYSTEM:  Orthorhombic
CLASS:  2/m 2/m 2/m
SPACE GROUP:  Pmmn
Z:  2
LATTICE CONSTANTS:
  $a = 4.93$
  $b = 5.44$
  $c = 5.73$
3 STRONGEST DIFFRACTION LINES:
  3.08  (100)
  2.722 ( 75)  (Synthetic)
  3.96  ( 65)

OPTICAL CONSTANTS:
  $\alpha = 1.413$
  $\beta = 1.611$
  $\gamma = 1.637$
  $(-)2V = 35°$
  DENSITY:  1.72 (Meas. synthetic)
             1.722 (Calc.)
  MODE OF OCCURRENCE:  Reported to occur in the earth of Nicojack Cave, Tennessee.  No verification of its occurrence in nature.
  BEST REF. IN ENGLISH:  Palache, et al., "Dana's System of Mineralogy," 7th Ed., v. II, p. 305, New York, Wiley, 1951.

## AMMONIOBORITE
$(NH_4)_3 B_{15} O_{20} (OH)_8 \cdot 4H_2O$

CRYSTAL SYSTEM:  Monoclinic
CLASS:  2/m
SPACE GROUP:  C2/c
Z:  4
LATTICE CONSTANTS:
  $a = 25.27$
  $b = 9.65$
  $c = 11.56$
  $\beta = 94°17'$
3 STRONGEST DIFFRACTION LINES:
  3.16 (100)
  3.09 (100)
  5.70 ( 60)
  HARDNESS:  Not determined.
  DENSITY:  1.765 (Meas.)
             1.758 (Calc.)
  CLEAVAGE:  No apparent cleavage.
  HABIT:  As microscopic platy crystals grouped in parallel position; compact, fine-grained granular masses.
  COLOR-LUSTER:  White.
  Soluble in water.
  MODE OF OCCURRENCE:  Occurs admixed with larderellite and sassolite in the boric acid lagoon at Larderello, Tuscany, Italy.
  BEST REF. IN ENGLISH:  Merlino, Stefano, and Sartori, Franco, *Science,* **171** (3969): 377–379 (1971). *Am. Min.,* **44:** 1150-1158 (1959).

## AMMONIOJAROSITE
$NH_4 Fe_3^{3+} (SO_4)_2 (OH)_6$
Alunite Group

CRYSTAL SYSTEM:  Hexagonal
CLASS:  3m
SPACE GROUP:  R3m
Z:  3
LATTICE CONSTANTS:
  $a = 7.20$
  $c = 17.00$
3 STRONGEST DIFFRACTION LINES:
  3.10 (100)
  5.10 ( 60)
  1.99 ( 40)

OPTICAL CONSTANTS:
  $\omega = 1.800$
  $\epsilon = 1.750$
  $(-)$
HARDNESS: Not determined.
DENSITY: 3.028 (Calc.)
CLEAVAGE: Not determined.
HABIT: As irregular flattened nodules composed of microscopic tabular grains, some with hexagonal outline.
COLOR-LUSTER: Light yellow. Dull and waxy to earthy.
MODE OF OCCURRENCE: Occurs with buddingtonite and other minerals at the Sulphur Bank mercury deposits, Lake County, California; near Wamsutter, Wyoming; on the west side of the Kaibab fault in southern Utah; and in pyritic shales at Valachov, Czechoslovakia.
BEST REF. IN ENGLISH: Palache, et al., "Dana's System of Mineralogy," 7th Ed., v. II, p. 562–563, New York, Wiley, 1951.


## AMOSITE = An asbestiform variety of cummingtonite


## AMPANGABEITE = Samarskite


## AMPHIBOLE = Mineral group of general formula:
$A_2 B_5 (Si, Al)_8 O_{22} (OH)_2$

See:
| | |
|---|---|
| Actinolite | Hastingsite |
| Anthophyllite | Holmquistite |
| Arfvedsonite | Hornblende |
| Barkevikite | Kaersutite |
| Clinoholmquistite | Katophorite |
| Crossite | Kôzulite |
| Cummingtonite | Magnesiokatophorite |
| Dashkesanite | Magnesioriebeckite |
| Eckermannite | Mboziite |
| Edenite . | Pargasite |
| Ferririchterite | Richterite |
| Fluor-richterite | Riebeckite |
| Gedrite | Tirodite |
| Glaucophane | Tremolite |
| Grunerite | Tschermakite |


## ANALCIME (Analcite)
$NaAlSi_2O_6 \cdot H_2O$
Zeolite Group

CRYSTAL SYSTEM: Cubic
CLASS: $4/m \bar{3} 2/m$
SPACE GROUP: Ia3d
Z: 16
LATTICE CONSTANT:
  $a \simeq 13.7$
3 STRONGEST DIFFRACTION LINES:
  3.43  (100)
  5.61  ( 80)
  2.925 ( 80)

OPTICAL CONSTANTS:
  $N = 1.479-1.493$
HARDNESS: 5-5½
DENSITY: 2.22-2.29 (Meas.)
CLEAVAGE: {001} in traces
Fracture subconchoidal. Brittle.
HABIT: Crystals usually well-formed trapezohedrons or modified cubes. Also massive granular; compact with concentric structure. Twinning on {001}, {110}, lamellar.
COLOR-LUSTER: Colorless, white, gray, yellowish, pink, greenish. Transparent to translucent. Vitreous.
MODE OF OCCURRENCE: Occurs widespread chiefly in basalts and other igneous rocks; as an alteration product of nepheline and sodalite; and in siltstones, sandstones, and other sedimentary rocks. Found as fine crystals in Cowlitz County, Washington; at Ritter Hot Springs, Grant County, Oregon; at several places in California; at Table Mountain, Jefferson County, Colorado; in Houghton County, Michigan; and in the trap rock of northern New Jersey. Other notable occurrences are found at Mont St. Hilaire, Quebec, and in the Bay of Fundy district, Nova Scotia; at many places in Scotland; and at localities in Iceland, Norway, Ireland, Italy, Germany, Czechoslovakia, Australia, and elsewhere.
BEST REF. IN ENGLISH: Deer, Howie, and Zussman, "Rock Forming Minerals," v. 4, p. 338–350, New York, Wiley, 1963. *Am. Min.*, **44**: 300-313 (1959).


## ANANDITE
$(Ba, K)(Fe, Mg)_3 (Si, Al, Fe)_4 O_{10} (O, OH)_2$
Mica Group

CRYSTAL SYSTEM: Monoclinic
CLASS: 2/m, m
SPACE GROUP: C2/c, Cc
Z: 2
LATTICE CONSTANTS:
  $a = 5.412$
  $b = 9.434$
  $c = 19.953$
  $\beta = 94°52'$
3 STRONGEST DIFFRACTION LINES:
  3.320 (100)
  4.995 ( 85)
  2.490 ( 80)
OPTICAL CONSTANTS:
  $\alpha = 1.85$
  $\gamma = 1.88$ } (on cleavage flakes)
  $(+)2V = ?$
HARDNESS: 3-4
DENSITY: 3.94 (Meas.)
        3.94 (Calc.)
CLEAVAGE: {001} perfect
HABIT: Massive; as bands and lenses from ½ inch to 2 inches or more in thickness. Flakes sometimes exhibit hexagonal outlines indicating poorly developed prism faces.
COLOR-LUSTER: Black. Nearly opaque.
MODE OF OCCURRENCE: Occurs associated with magnetite, chalcopyrite, pyrite, and pyrrhotite in the iron-bearing zones of the Wilagedera iron ore prospect in the Northwestern Province of Ceylon.

BEST REF. IN ENGLISH: Pattiaratchi, D. B., Saari, Esko, and Sahama, Th. G., *Am. Min.*, **52**: 1586 (1967).

## ANAPAITE
$Ca_2 Fe^{2+}(PO_4)_2 \cdot 2H_2O$

CRYSTAL SYSTEM: Triclinic
CLASS: $\bar{1}$
SPACE GROUP: $P\bar{1}$
Z: 1
LATTICE CONSTANTS:
  $a = 6.41$    $\alpha = 101°34\frac{1}{2}'$
  $b = 6.88$    $\beta = 104°05\frac{1}{2}'$
  $c = 5.86$    $\gamma = 71°03\frac{1}{2}'$
3 STRONGEST DIFFRACTION LINES:
  3.135 (100b)
  2.866 ( 70 )
  3.72  ( 60b)
OPTICAL CONSTANTS:
  $\alpha = 1.602$
  $\beta = 1.613$
  $\gamma = 1.649$
  $(+)2V = 54°$
HARDNESS: $3\frac{1}{2}$
DENSITY: 2.81 (Meas.)
         2.80 (Calc.)
CLEAVAGE: {001} perfect
          {010} distinct
HABIT: Crystals tabular; as crusts of subparallel crystals and as rosette-like aggregates.
COLOR-LUSTER: Green to greenish white. Transparent. Vitreous. Streak white.
MODE OF OCCURRENCE: Found as layers of pale green crystals at a depth of 500 feet in a core from the Lewis well, Kings County, California. It also occurs in phosphatic geodes in Miocene clay at Bellaver de Cerdena, Gerona Province, Spain; in a bituminous clay rock at Messel, Hesse, Germany; and in oolitic iron ore near Anapa, Taman peninsula, USSR.
BEST REF. IN ENGLISH: Palache, et al., "Dana's System of Mineralogy," 7th Ed., v. II, p. 731–732, New York, Wiley, 1951.

## ANATASE (Octahedrite)
$TiO_2$
Trimorphous with rutile and brookite

CRYSTAL SYSTEM: Tetragonal
CLASS: 4/m 2/m 2/m
SPACE GROUP: I4/amd
Z: 4
LATTICE CONSTANTS:
  $a = 3.73$
  $c = 9.37$
3 STRONGEST DIFFRACTION LINES:
  3.51  (100)
  1.891 ( 33)
  2.379 ( 22)
OPTICAL CONSTANTS: Vary markedly with wavelength and temperature.

HARDNESS: 5½–6
DENSITY: 3.82–3.97 (Meas.)
         3.899 (Calc.)
CLEAVAGE: {001} perfect
          {011} perfect
Fracture subconchoidal. Brittle.
HABIT: Crystals steep pyramidal, often striated at right angles to c-axis; also blunt pyramidal, tabular, or prismatic. Often highly modified. Twinning on {112}, uncommon.
COLOR-LUSTER: Various shades of brown to deep blue or black; also nearly colorless, grayish, greenish, bluish green, pale lavender. Transparent to nearly opaque. Adamantine or metallic-adamantine. Streak colorless, white, to pale yellow.
MODE OF OCCURRENCE: Occurs chiefly in gneiss, schist, detrital deposits, and as an accessory mineral in many types of igneous and metamorphic rocks. Found in weathered quartz diorites in the Coast Range counties and elsewhere in California; in Boulder and Gunnison counties, Colorado, and at localities in Arkansas, Massachusetts, Virginia, and North Carolina. It also occurs in Canada, Brazil, England, Wales, Norway, France, Italy, Switzerland, Austria, Czechoslovakia, and USSR.
BEST REF. IN ENGLISH: Palache, et al., "Dana's System of Mineralogy," 7th Ed., v. I, p. 583–588, New York, Wiley, 1944.

## ANAUXITE = Kaolinite

## ANCYLITE
$(Ce, La)_4 (Sr, Ca)_3 (CO_3)_7 (OH)_4 \cdot 3H_2O$

CRYSTAL SYSTEM: Orthorhombic
CLASS: 2/m 2/m 2/m
SPACE GROUP: Pmmm
Z: 1
LATTICE CONSTANTS:
  $a = 4.97$
  $b = 7.24$ (Calcian)
  $c = 8.50$
  $a : b : c = 0.686 : 1 : 1.175$
3 STRONGEST DIFFRACTION LINES:
  4.34 (100)
  3.71 (100)
  2.96 (100)
OPTICAL CONSTANTS:        (Calcian)
  $\alpha = 1.625$         $\alpha = 1.654$
  $\beta = 1.700$          $\beta = 1.733$
  $\gamma = 1.735$         $\gamma = 1.772$
  $(-)2V = 66°$            $(-)2V = 60°-70°$
HARDNESS: 4–4½
DENSITY: 3.95 (Meas.)
CLEAVAGE: None.
Fracture splintery. Brittle.
HABIT: Crystals stout to long prismatic, elongated along c-axis; also as minute pseudo-octahedral crystals with curved faces, and as tiny disseminated grains.
COLOR-LUSTER: Colorless, gray, pale pink, pale yellow to brown or brownish red. Transparent to subtranslucent. Vitreous on faces, greasy on fracture surfaces.

MODE OF OCCURRENCE: The mineral occurs in Montana in hydrothermal veins associated with mafic and felsic alkalic igneous rocks in the Bearpaw Mountains, and as very fine grains intimately intergrown with fine-grained quartz, calcite, and some barite, in masses several feet across, in the Sheep Creek Vein, Ravalli County. It is also found in carbonatite at Mont St. Hilaire, Quebec, Canada; in pegmatitic veinlets in nepheline-syenite at Narsarsuk, Greenland, in the Langesundfjord district, Norway, and in the Khibina tundra, Kola Peninsula, USSR; and in cavities in granite at Baveno, Italy. Nd-rich calcian ancylite occurs as minute crystals, typically perched on calcite or embedded in cellular calcite and quartz masses, at the Keystone Trappe Rock quarry, Cornog, Chester County, Pennsylvania.

BEST REF. IN ENGLISH: Palache, et al., "Dana's System of Mineralogy," 7th Ed., v. II, p. 291-293, New York, Wiley, 1951. Keidel, F. A., Montgomery, A. Wolfe, C. W., and Christian, R. P., *Min. Rec.*, 2: 18-25, 36 (1971).

# ANDALUSITE

$Al_2SiO_5$

CRYSTAL SYSTEM: Orthorhombic
CLASS: 2/m 2/m 2/m
SPACE GROUP: Pnnm
Z: 4
LATTICE CONSTANTS:
  $a = 7.78$
  $b = 7.92$
  $c = 5.57$
3 STRONGEST DIFFRACTION LINES:
  5.54 (100)
  4.53 ( 90)
  2.77 ( 90)
OPTICAL CONSTANTS:
  $\alpha = 1.629\text{-}1.640$
  $\beta = 1.633\text{-}1.644$
  $\gamma = 1.638\text{-}1.650$
  $(-)2V_\alpha = 73°\text{-}86°$
HARDNESS: 6½-7½
DENSITY: 3.13-3.16 (Meas.)
CLEAVAGE: {110} distinct—{110}:{1$\bar{1}$0} = 89°
  {100} indistinct
Fracture uneven to subconchoidal. Brittle.

HABIT: Crystals prismatic, nearly square in cross section; usually coarse. Also massive, compact; in columnar or fibrous aggregates. Twinning on {101}, rare.

COLOR-LUSTER: Usually pink, reddish brown, rose red, or whitish; also grayish, yellowish, violet, greenish. Variety chiastolite exhibits cruciform pattern of carbonaceous impurities when viewed in cross section. Transparent to nearly opaque. Vitreous to subvitreous. Streak uncolored.

MODE OF OCCURRENCE: Occurs chiefly in slates and argillaceous schists as a contact mineral; also in mica schist, gneiss, and related rocks; as a detrital mineral; and rarely in granites or granitic pegmatites. Common associated minerals include sillimanite, kyanite, cordierite, and corundum. Found in large quantities in the Inyo Range, Mono County, and at many other places in California; in the Black Hills, South Dakota; and at localities in Idaho, Colorado, New Mexico, Pennsylvania, Maine, and Massachusetts. Also in Brazil, Scotland, Sweden, France, Spain, Austria, Germany, Czechoslovakia, USSR, South Africa, Ceylon, Korea, and Australia.

BEST REF. IN ENGLISH: Deer, Howie, and Zussman, "Rock Forming Minerals," v. 1, p. 129-136, New York, Wiley, 1962.

# ANDERSONITE

$Na_2CaUO_2(CO_3)_3 \cdot 6H_2O$

CRYSTAL SYSTEM: Hexagonal
CLASS: $\bar{3}$, 3
SPACE GROUP: R$\bar{3}$, R3
Z: 16
LATTICE CONSTANTS:
  $a = 18.009$ Å
  $c = 23.838$
3 STRONGEST DIFFRACTION LINES:
  13.0 (100)
  7.93 (100)
  5.67 (100)
HARDNESS: 2½
DENSITY: 2.8 (Meas.)
  2.875 (Calc.)
CLEAVAGE: Not determined.
HABIT: Crystals rhombohedral, up to 1 cm or more in size; also as clusters of minute pseudocubic crystals. As thick crusts and veinlets.

COLOR-LUSTER: Bright yellow-green of various shades; transparent to translucent; vitreous to pearly. Fluoresces bright yellow-green in ultraviolet light.

MODE OF OCCURRENCE: Occurs as a secondary mineral associated with bayleyite, swartzite, schroeckingerite, and gypsum as an efflorescence on mine walls at the Hillside mine near Bagdad, Yavapai County, Arizona; in Utah as thick crusts, veinlets, and crystals up to one-half inch in size at the Atomic King No. 2 and nearby mines in Cane Wash near Moab, San Juan County, and as an efflorescence on mine walls at the Parco No. 23 mine and at the Skinny No. 1 mine southeast of Thompsons, Grand County; in New Mexico as superb complex crystals over one-half inch in size encrusting sandstone in the Ambrosia Lakes district, McKinley County; and as an efflorescence with liebigite at the carnotite locality at Jim Thorpe, Pennsylvania.

BEST REF. IN ENGLISH: Frondel, Clifford, *USGS Bull.*, **1064**: 115-117 (1958).

# ANDESINE

mNaAlSi$_3$O$_8$ with nCaAl$_2$Si$_2$O$_8$
Ab$_{70}$An$_{30}$ to Ab$_{50}$An$_{50}$
Plagioclase Feldspar Group

CRYSTAL SYSTEM: Triclinic
CLASS: $\bar{1}$
SPACE GROUP: C$\bar{1}$
Z: 4
LATTICE CONSTANTS:
  $a = 8.14$      $\alpha = 93°26'$
  $b = 12.86$     $\beta = 116°28'$
  $c = 7.17$      $\gamma = 89°59'$

## 3 STRONGEST DIFFRACTION LINES:

3.21 (100)
3.18 ( 90)
4.04 ( 80)

## OPTICAL CONSTANTS:

$\alpha = 1.543$
$\beta = 1.476-1.5480$
$\gamma = 1.551$
$(+)2V = 76°-86°$ and $83°(-)$

HARDNESS: 6-6½
DENSITY: 2.66-2.68 (Meas.)
CLEAVAGE: {001} perfect
{010} nearly perfect
{110} imperfect
Fracture uneven to conchoidal. Brittle.

HABIT: Crystals tabular, flattened along b-axis; not common. Usually massive, cleavable, granular, or compact. Twinning common. Principal twin laws: Carlsbad, albite, and pericline.

COLOR-LUSTER: Colorless, white, gray. Transparent to translucent. Vitreous. Streak white.

MODE OF OCCURRENCE: Occurs widespread as a common rock-forming mineral in igneous rocks with intermediate silica content, such as andesite; also in amphibolites, charnockites, and other metamorphic rocks. Typical occurrences are found in California, Utah, Colorado, South Dakota, Minnesota, New York, and North Carolina. Also in Colombia, Argentina, Greenland, Norway, France, Italy, Germany, South Africa, India, and Japan.

BEST REF. IN ENGLISH: Deer, Howie, and Zussman, "Rock Forming Minerals," v. 4, p. 94-165, New York, Wiley, 1963. Barth, T. F. W., "Feldspars," New York, Wiley, 1969.

## ANDORITE (Sundtite; Webnerite)

$AgPbSb_3S_6$

CRYSTAL SYSTEM: Orthorhombic
CLASS: 2/m 2/m 2/m
SPACE GROUP: Pmma
Z: 4

## LATTICE CONSTANTS:

$a = 13.01$
$b = 19.19$
$c = 4.27$ (true cell may have $c = 4 \times 4.27$ for Andorite IV or $c = 6 \times 4.27$ for Andorite VI)

## 3 STRONGEST DIFFRACTION LINES:

3.30 (100)
2.90 ( 80)
3.45 ( 40)

HARDNESS: 3-3½
DENSITY: 5.38 (Meas.)
5.44 (Calc.)
CLEAVAGE: None.
Fracture conchoidal. Brittle.

HABIT: Crystals stout prismatic, striated parallel to elongation; also thick to thin tabular. Commonly massive. Twinning on {110}.

COLOR-LUSTER: Dark steel gray, sometimes tarnished iridescent or yellowish. Opaque. Metallic. Streak black.

MODE OF OCCURRENCE: Occurs as thin tabular crystals at the Thompson mine, Darwin district, Inyo County, California; at the Keyser mine, Nye County, Nevada; at Takla Lake, British Columbia, Canada; with cassiterite, sulfides, and sulfosalts at Oruro and elsewhere in Bolivia; and at Felsobanya, Roumania.

BEST REF. IN ENGLISH: Palache, et al., "Dana's System of Mineralogy," 7th Ed., v. I, p. 457-459, New York, Wiley, 1944.

## ANDRADITE

$Ca_3 Fe_2^{3+} Si_3 O_{12}$
Garnet Group
var. Topazolite
Demantoid
Melanite
Schorlomite

CRYSTAL SYSTEM: Cubic
CLASS: 4/m $\bar{3}$ 2/m
SPACE GROUP: Ia3d
Z: 8

## LATTICE CONSTANT:

$a = 12.048$

## 3 STRONGEST DIFFRACTION LINES:

2.696 (100)
3.015 ( 60)
1.6112 ( 60)

## OPTICAL CONSTANT:

$N = 1.887$

HARDNESS: 6½-7
DENSITY: 3.7-4.1 (Meas.)
3.849 (Calc.)
CLEAVAGE: None.
{110} parting sometimes distinct
Fracture uneven to conchoidal. Brittle.

HABIT: Crystals usually dodecahedrons or trapezohedrons; also in combination, or with hexoctahedron. Also massive, compact; fine or coarse granular; or as embedded grains.

COLOR-LUSTER: Various shades of yellowish green, green, greenish brown, brown, reddish brown, grayish black, black. Transparent to nearly opaque. Vitreous to resinous.

MODE OF OCCURRENCE: Occurs chiefly in chlorite schist and serpentinite (demantoid, topazolite); in alkaline igneous rocks (melanite, schorlomite); and in metamorphosed limestones or contact zones. Notable occurrences are found at Garnet Hill, Calaveras County, and near the benitoite locality (melanite) San Benito County, California; near Stanley Butte, Graham County, Arizona; on Iron Hill, Gunnison County, Colorado (melanite); at Magnet Cove, Hot Spring County, Arkansas (schorlomite); at Cornwall, Lebanon County, and French Creek Mines, Chester County, Pennsylvania; at Franklin and Sterling Hill, Sussex County, New Jersey; in the Ala Valley, Piedmont (demantoid), and elsewhere in Italy; and at localities in Greenland, Norway, Sweden, Switzerland, Czechoslovakia, Roumania, USSR, and Uganda.

BEST REF. IN ENGLISH: Deer, Howie, and Zussman, "Rock Forming Minerals," v. 1, p. 77-112, New York, Wiley, 1962.

# ANDREWSITE
$(Cu, Fe^{2+}) Fe_3^{3+}(PO_4)_3(OH)_2$

CRYSTAL SYSTEM: Orthorhombic
CLASS: 2 2 2
SPACE GROUP: $B2 2_1 2$
Z: 4
LATTICE CONSTANTS:
 $a = 14.16$
 $b = 16.83$
 $c = 5.18$
3 STRONGEST DIFFRACTION LINES:
 3.22 (100)
 2.12 ( 80)
 5.01 ( 50)
OPTICAL CONSTANTS:
 $\alpha = 1.813$
 $\beta = 1.820$
 $\gamma = 1.830$
 (+)2V = large
HARDNESS: 4
DENSITY: 3.475 (Meas.)
CLEAVAGE: In two directions parallel to fiber length.
HABIT: Botryoidal aggregates with radial-fibrous structure.
COLOR-LUSTER: Green to bluish green. Silky.
MODE OF OCCURRENCE: Occurs associated with limonite, chalcosiderite, and other minerals at the West Phoenix mine, near Liskeard, Cornwall, England.
BEST REF. IN ENGLISH: Palache, et al., "Dana's System of Mineralogy," 7th Ed., v. II, p. 802, New York, Wiley, 1951.

# ANGARALITE = A chlorite

# ANGELELLITE
$Fe_4As_2O_{11}$

CRYSTAL SYSTEM: Triclinic
CLASS: $\bar{1}$
SPACE GROUP: $P\bar{1}$
Z: 1
LATTICE CONSTANTS:
 $a = 5.03$     $\alpha = 114°$
 $b = 6.49$     $\beta = 116°$
 $c = 7.11$     $\gamma = 81°$
3 STRONGEST DIFFRACTION LINES:
 3.152 (100)
 2.997 ( 70)
 2.856 ( 50)
OPTICAL CONSTANTS:
 $\alpha = 2.13$
 $\beta = 2.2$
 $\gamma = 2.40$
 (+)2V = large
HARDNESS: 5½
DENSITY: 4.867 (Meas.)
      4.862 (Calc.)
CLEAVAGE: {001}
Fracture conchoidal. Brittle.

HABIT: Crystals usually tabular with {001} and {00$\bar{1}$}, dominant. As globular and crystalline incrustations.
COLOR-LUSTER: Blackish brown; adamantine to semimetallic. Streak reddish brown.
MODE OF OCCURRENCE: Occurs on andesite, probably deposited from fumarolic vapors, at the Cerro Pululus tin mine, northwestern Argentina.
BEST REF. IN ENGLISH: Ramdohr, Paul, Ahlfeld, F., and Berndt, F., *Am. Min.*, **44**: 1322–1323 (1959).

# ANGLESITE
$PbSO_4$

CRYSTAL SYSTEM: Orthorhombic
CLASS: 2/m 2/m 2/m
SPACE GROUP: Pnma
Z: 4
LATTICE CONSTANTS:
 $a = 8.45$
 $b = 5.38$
 $c = 6.93$
3 STRONGEST DIFFRACTION LINES:
 3.001 (100)
 4.26 ( 87)
 3.333 ( 86)
OPTICAL CONSTANTS: (Na)
 $\alpha = 1.8771$
 $\beta = 1.8826$
 $\gamma = 1.8937$
 (+)2V = 75°24'
HARDNESS: 2½–3
DENSITY: 6.38 ± 0.01 (Meas.)
      6.36 (Calc.)
CLEAVAGE: {001} good
      {210} distinct
      {010} indistinct
Fracture conchoidal. Brittle.
HABIT: Crystals widely varied in habit; elongate parallel to $c$, $b$, or $a$-axis. Usually thin to thick tabular; also prismatic or equant. Massive, coarse to fine granular; nodular, stalactitic; sometimes as massive, fine-grained concentric alteration layers enclosing unaltered galena cores.
COLOR-LUSTER: Colorless, white, yellowish, gray, and pale shades of green or blue. Transparent to opaque. Adamantine; also resinous or vitreous. Streak colorless. Sometimes fluoresces yellow under ultraviolet light.
MODE OF OCCURRENCE: Occurs as a characteristic secondary mineral in lead deposits, usually formed by the oxidation of galena. Found in the United States as superb colorless crystals up to five inches in length at the Wheatley mines, Chester County, Pennsylvania; in the Tintic and Park City districts, Utah; abundantly as fine crystals in the Coeur d'Alene district, Shoshone County, Idaho; and at many localities in Arizona, New Mexico, Colorado, California, and other western states. It also occurs as fine doubly terminated crystals in sulfur at Los Lamentos, Chihuahua, and at other places in Mexico; as good crystallized specimens at Matlock and Cromford, Derbyshire, England; at Leadhills and Wanlockhead, Scotland; on the island of Anglesey, Wales; and as excellent crystals at localities in Sardinia, Germany, USSR, Tunisia; at Tsumeb, South West Africa; at Broken Hill, New South Wales, Australia; and at Dundas, Tasmania.

BEST REF. IN ENGLISH: Palache, et al., "Dana's System of Mineralogy," 7th Ed., v. II, p. 420–424, New York, Wiley, 1951.

## ANHYDRITE
$CaSO_4$

CRYSTAL SYSTEM: Orthorhombic
CLASS: 2/m 2/m 2/m
SPACE GROUP: Amma
Z: 4
LATTICE CONSTANTS:
 $a = 6.94$
 $b = 6.97$
 $c = 6.20$
3 STRONGEST DIFFRACTION LINES:
 3.49 (100)
 2.849 ( 35)
 2.328 ( 20)
OPTICAL CONSTANTS: (Na)
 $\alpha = 1.5698$
 $\beta = 1.5754$
 $\gamma = 1.6136$
 $(+)2V = 43°41'$
HARDNESS: 3½
DENSITY: 2.98 (Meas.)
 2.963 (Calc.)
CLEAVAGE: {010} perfect
 {100} nearly perfect
 {001} good to imperfect
Fracture splintery to uneven. Brittle.
 HABIT: Crystals equant; thick tabular, or prismatic; rare. Usually massive, fine to coarse granular, sometimes fibrous. Twinning on {011}; rarely on {120}.
 COLOR-LUSTER: Colorless, white, gray, bluish, violet, pinkish, reddish, brownish. Transparent to translucent, vitreous to greasy to pearly. Streak white or grayish white.
 MODE OF OCCURRENCE: Occurs widespread as an important rock-forming mineral, often associated with gypsum, salt beds, dolomite, or limestone; as a hypogenic mineral in hydrothermal vein deposits; in cavities in igneous trap rock; and rarely as a sublimation product. It is found as extensive strata in many parts of western South Dakota; in the vicinity of Carlsbad, New Mexico and adjacent parts of Texas; with zeolites in cavities in igneous flow-rocks in New Jersey and Massachusetts; in New York, at Ajo, Arizona, and at many other places in the United States. It also occurs as fine specimens at the Faraday uranium mine, Bancroft, Ontario, and as bedded deposits in Nova Scotia. Among many other occurrences, it is found in the salt deposits of Germany, Austria, Poland, France, and India.
 BEST REF. IN ENGLISH: Palache, et al., "Dana's System of Mineralogy," 7th Ed., v. II, p. 424–428, New York, Wiley, 1951.

## ANILITE
$Cu_7S_4$

CRYSTAL SYSTEM: Orthorhombic
CLASS: 2/m 2/m 2/m

SPACE GROUP: Pnma
Z: 4
LATTICE CONSTANTS:
 $a = 7.89$
 $b = 7.84$
 $c = 11.01$
3 STRONGEST DIFFRACTION LINES:
 1.96 (100)
 2.77 ( 65)
 3.20 ( 57)
HARDNESS: ~3
DENSITY: 5.68 (Calc.)
CLEAVAGE: None. Sectile.
 HABIT: Crystals prismatic or platy (up to 5 mm in size) similar to chalcocite in appearance.
 COLOR-LUSTER: Bluish gray. Opaque. Metallic. Streak black.
 MODE OF OCCURRENCE: Occurs associated with djurleite in drusy parts of the quartz vein at the Ani Mine, Akita, Japan.
 BEST REF. IN ENGLISH: Morimoto, Nobuo; Koto, Kichiro; and Shimazaki, Yoshiko, *Am. Min.*, **54**: 1256–1268 (1959)

## ANKERITE
$Ca(Fe, Mg)(CO_3)_2$
Manganoan = $Mn^{2+}$ in substitution for $Fe^{2+}$

CRYSTAL SYSTEM: Hexagonal
 (Trigonal)
CLASS: $\bar{3}$
SPACE GROUP: $R\bar{3}$
Z: 3
LATTICE CONSTANTS:
 Fe : Mg = 1:1.1
 $a = 4.822$
 $c = 16.11$ } Vary with ratio of Fe to Mg.
3 STRONGEST DIFFRACTION LINES:
 2.899 (100)
 2.199 ( 6)
 1.812 ( 6)
OPTICAL CONSTANTS:
 Fe : Mg = 1 : 1.1
 $\omega = 1.728$
 $\epsilon = 1.531$ } Vary with ratio of Fe to Mg.
uniaxial (–)
HARDNESS: 3½–4
DENSITY: 2.97 (Meas.)
CLEAVAGE: $\{10\bar{1}1\}$ perfect
Fracture subconchoidal. Brittle.
 HABIT: Commonly simple rhombohedrons. Massive, fine to coarse granular. Twinning on {0001}, $\{10\bar{1}0\}$, and $\{11\bar{2}0\}$ common.
 COLOR-LUSTER: White, gray, yellowish brown to brown. Translucent to subtranslucent, vitreous to pearly.
 MODE OF OCCURRENCE: Occurs as a gangue mineral at the Homestake gold mine, Lead, South Dakota; in sulfide veins in the Coeur d'Alene region, Idaho; at the Antwerp iron mine, Jefferson County, New York; at Oldham, Lancashire, England; at Erzberg, Styria, Austria; and at localities in Czechoslovakia, Hungary, and Algeria.

BEST REF. IN ENGLISH: Palache, et al., "Dana's System of Mineralogy," 7th Ed., v. II, p. 208-217, New York, Wiley, 1951.

## ANNABERGITE (Cabrerite)
$Ni_3(AsO_4)_2 \cdot 8H_2O$
Erythrite-Annabergite Series

CRYSTAL SYSTEM: Monoclinic
CLASS: 2/m
SPACE GROUP: I2/m
Z: 2
LATTICE CONSTANTS:
  $a = 10.122$
  $b = 13.284$
  $c = 4.698$
  $\beta = 104°45'$
3 STRONGEST DIFFRACTION LINES:
  6.58 (100)
  2.98 ( 30)
  3.18 ( 26)
OPTICAL CONSTANTS:
  $\alpha = 1.622$
  $\beta = 1.658$
  $\gamma = 1.687$
  2V = 84° (+), but also (−)
HARDNESS: 1½–2½
DENSITY: 3.07 (Meas.)
       3.23 (Calc.)
CLEAVAGE: {010} perfect
          {100} indistinct
          {$\bar{1}02$} indistinct
Flexible in thin laminae. Sectile.
  HABIT: Crystals prismatic, flattened on {010}, commonly striated. Usually as crystalline crusts, earthy masses, or powders.
  COLOR-LUSTER: White, gray, pale green to intense yellow-green. Transparent to translucent. Weakly adamantine, pearly on {010}; also dull. Streak lighter shade than color.
  MODE OF OCCURRENCE: Occurs as a secondary mineral in the oxidation zone of nickel-bearing ore deposits. Found associated with retgersite at the Lovelock mine, Humboldt County, Nevada; at localities in Inyo, Los Angeles, Santa Cruz, and Tulare Counties, California; in Custer and Fremont Counties, Colorado; and in Houghton County, Michigan. It also occurs in the Cobalt district, Ontario, Canada; at Schneeberg and at other places in Germany; in the Sierra Cabrera, Almeria Province, Spain; as fine crystals (cabrerite) at Laurium, Greece; and in Austria and Sardinia.
  BEST REF. IN ENGLISH: Palache, et al., "Dana's System of Mineralogy," 7th Ed., v. II, p. 746-750, New York, Wiley, 1951.

## ANOPHORITE = Variety of arfvedsonite
$Na_3Mg_3Fe^{2+}, Fe^{3+}(Ti, Si)_8O_{22}(OH)_2$

## ANORTHITE
$mCaAl_2Si_2O_8$ with $nNaAlSi_3O_8$
$Ab_{10}An_{90}$ to $Ab_0An_{100}$
Plagioclase Feldspar Group

CRYSTAL SYSTEM: Triclinic
CLASS: $\bar{1}$
SPACE GROUP: P$\bar{1}$
Z: 8
LATTICE CONSTANTS:
  $a = 8.18$      $\alpha = 93°10'$
  $b = 12.88$     $\beta = 115°51'$
  $c = 14.16$     $\gamma = 91°13'$
3 STRONGEST DIFFRACTION LINES:
  3.20 (100)
  3.18 ( 75)
  4.04 ( 60)
OPTICAL CONSTANTS:
  $\alpha = 1.577$
  $\beta = 1.585$
  $\gamma = 1.590$
  (−)2V = 78°
HARDNESS: 6–6½
DENSITY: 2.74–2.76 (Meas.)
CLEAVAGE: {001} perfect
          {010} nearly perfect
          {110} imperfect
Fracture uneven to conchoidal. Brittle.
  HABIT: Crystals usually short prismatic. Also massive, cleavable, with coarse lamellar or granular structure. Twinning common. Principle twin laws: Carlsbad, Manebach, Baveno, albite, and pericline.
  COLOR-LUSTER: Colorless, white, grayish, reddish. Transparent to translucent. Vitreous. Streak white.
  MODE OF OCCURRENCE: Occurs as a rock-forming mineral in basic plutonic and volcanic rocks such as anorthosite, norite, and lavas; in certain metamorphic rocks; and in meteorites. Typical occurrences are found in California at Pala Mountain in San Diego County, at Grass Valley, Nevada County, and near Middletown in Lake County; on Italian Mountain, Gunnison County, Colorado; and at localities in Greenland, England, Sweden, Italy, Sicily, Finland, and India. Also as large lava-coated crystals on the island of Miyake, Japan.
  BEST REF. IN ENGLISH: Deer, Howie, and Zussman, "Rock Forming Minerals," v. 4, p. 94-165, New York, Wiley, 1963. Barth, T. F. W., "Feldspars," New York, Wiley, 1969.

## ANORTHOCLASE
(Na, K)AlSi$_3$O$_8$
Alkali Feldspar Group

CRYSTAL SYSTEM: Triclinic
CLASS: $\bar{1}$
SPACE GROUP: C$\bar{1}$
Z: 4
LATTICE CONSTANTS:
  $a \simeq 8.2$
  $b \simeq 12.8$
  $c \simeq 7.1$
  $\beta \simeq 116°$
3 STRONGEST DIFFRACTION LINES:
  3.211 (100)
  3.243 ( 90)
  4.106 ( 16)

OPTICAL CONSTANTS:
$\alpha = 1.518-1.527$
$\beta = 1.522-1.532$
$\gamma = 1.522-1.534$
$(-)2V = 18°-54°$
HARDNESS: 6-6½
DENSITY: 2.56-2.62 (Meas.)
2.57 (Calc.)
CLEAVAGE: {001} perfect
{010} perfect
{100}, {110}, {$\bar{1}$10}, {$\bar{2}$01} partings.
Fracture uneven. Brittle.

HABIT: Crystals usually short prismatic, blocky, sometimes orthorhombic or tetragonal in aspect; also tabular, flattened along $b$ axis. Commonly massive; cleavable to granular, lamellar, or cryptocrystalline. Twinning very common; simple, multiple, and repeated. Principal twin laws: Carlsbad, Baveno, Manebach.

COLOR-LUSTER: Colorless, white, gray, yellowish, reddish, greenish. Transparent to translucent. Vitreous; somewhat pearly on cleavages. Streak white.

MODE OF OCCURRENCE: Occurs mainly in volcanic rocks such as andesite, phonolite, and trachyte. Typical occurrences are found at Cripple Creek and elsewhere in Colorado; at Franklin, Sussex County, New Jersey; and at localities in Scotland, Norway, Sicily, Italy, Germany, Nigeria, Kenya, Mongolia, Australia, and Antarctica.

BEST REF. IN ENGLISH: Deer, Howie, and Zussman, "Rock Forming Minerals," v. 4, p. 6-93, New York, Wiley, 1963. Barth, T. F. W., "Feldspars," New York, Wiley, 1969.

## ANTARCTICITE
$CaCl_2 \cdot 6H_2O$

CRYSTAL SYSTEM: Hexagonal
CLASS: 3 2
SPACE GROUP: P32
Z: 1
LATTICE CONSTANTS:
$a = 7.89$
$c = 3.95$
3 STRONGEST DIFFRACTION LINES:
2.16 (100)
3.93 ( 75)
2.78 ( 63)
OPTICAL CONSTANTS:
$\omega$ - 1.550
$\epsilon = 1.495$
uniaxial (-)
HARDNESS: 2-3
DENSITY: 1.715 ± 0.010 (Meas.)
1.700 (Calc.)
CLEAVAGE: Basal-perfect
Prismatic-good to perfect.
Brittle.
HABIT: Acicular aggregates up to 15 cm long.
COLOR-LUSTER: Colorless; vitreous.
MODE OF OCCURRENCE: Found crystallizing from the brine of Don Juan Pond, Victoria Land, Antarctica, and at Bristol Dry Lake, San Bernardino County, California.

BEST REF. IN ENGLISH: Torii, Tetsuya, and Ossaka, Joyo, *Science*, **149** (3687): 975-977 (1965). Dunning, G. E., and Cooper, J. F., Jr., *Am. Min.*, **54**: 1018-1025 (1969).

## ANTHOINITE
$Al_2W_2O_9 \cdot 3H_2O$

CRYSTAL SYSTEM: Monoclinic or Triclinic
Z: 10 (?)
LATTICE CONSTANTS:
(monoclinic but not entirely satisfactory)
$a = 9.33$
$b = 8.17$
$c = 13.68$
$\beta = 95°40'$
3 STRONGEST DIFFRACTION LINES:
4.195 (100)
3.070 (100)
3.052 (100)
OPTICAL CONSTANT:
$n = 1.81-1.82$
HARDNESS: 1
DENSITY: ~4.6
4.96 (Calc.)
CLEAVAGE: Not determined.
HABIT: Tabular crystals up to 3 $\mu$ in size. As chalk-like masses.
COLOR-LUSTER: White.
MODE OF OCCURRENCE: Occurs in placer concentrates and in quartz veins associated with ferberite at Mt. Misobo, Kalima district, Zaire, and also at Ruanda in the Kifuruwe region.

BEST REF. IN ENGLISH: Palache, et al., "Dana's System of Mineralogy," 7th Ed., v. II, p. 1097-1098, New York, Wiley, 1951. Niggli, Ernst, and Jäger, Emilie, *Am. Min.*, **43**: 384 (1958).

## ANTHONYITE
$Cu(OH, Cl)_2 \cdot 3H_2O$

CRYSTAL SYSTEM: Monoclinic
CLASS: 2/m
LATTICE CONSTANTS:
$a : b : c = 0.6898 : 1 : 0.4271$
$\beta = 112°38'$
3 STRONGEST DIFFRACTION LINES:
5.84 (100)
4.14 ( 70)
3.99 ( 60)
OPTICAL CONSTANTS:
$\alpha = 1.526$
$\beta = 1.602$
$\gamma = 1.602$
2V = 3°
HARDNESS: 2
CLEAVAGE: {100} good
Sectile.
HABIT: Prismatic crystals up to 0.5 mm in length; often curved along $c$-axis; plane of curvature is {010}; bending occurs readily on {100}.

COLOR-LUSTER: Lavender.
MODE OF OCCURRENCE: Occurs in pockets and as encrustations on basalt at the Centennial mine near Calumet, Michigan.
BEST REF. IN ENGLISH: Williams, Sidney A., *Am. Min.*, **48**: 614–619 (1963).

## ANTHOPHYLLITE
$(Mg, Fe)_7Si_8O_{22}(OH)_2$
Amphibole Group

CRYSTAL SYSTEM: Orthorhombic
CLASS: 2/m 2/m 2/m
SPACE GROUP: Pnma
Z: 4
LATTICE CONSTANTS:
 $a$ = 18.5–18.6
 $b$ = 17.7–18.1
 $c$ = 5.27–5.32
3 STRONGEST DIFFRACTION LINES:
 3.05 (100)
 3.24 ( 60)
 8.26 ( 55)
OPTICAL CONSTANTS:
 $\alpha$ = 1.596–1.694
 $\beta$ = 1.605–1.710
 $\gamma$ = 1.615–1.722
 $(\pm)2V_\gamma$ = 78°–111°
HARDNESS: 5½–6
DENSITY: 2.85–3.57 (Meas.)
 3.385 (Calc. Mg = Fe)
CLEAVAGE: {110} perfect
 {010} imperfect
 {100} imperfect
HABIT: Crystals prismatic, rare. Usually massive, fibrous or lamellar.
COLOR-LUSTER: White, gray, greenish, brownish green, clove-brown, yellowish brown, dark brown. Transparent to nearly opaque. Vitreous to silky, somewhat pearly on cleavage. Streak uncolored or grayish.
MODE OF OCCURRENCE: Occurs only in metamorphic rocks, such as schists and gneisses, or in metasomatic rocks. Found relatively widespread, at times as the principal constituent of the rock. Typical occurrences are found in California, Arizona, Montana, Colorado, Pennsylvania, New York, and North Carolina. Also in Canada, Greenland, Norway, Italy, Austria, Czechoslovakia, and elsewhere.
BEST REF. IN ENGLISH: Deer, Howie, and Zussman, "Rock Forming Minerals," v. 2, p. 211-229, New York, Wiley, 1963.

## ANTIGORITE
$Mg_3Si_2O_5(OH)_4$
Serpentine Group

CRYSTAL SYSTEM: Monoclinic
CLASS: m
SPACE GROUP: Cm
Z: 16
LATTICE CONSTANTS:
 $a$ = 43.53
 $b$ = 9.259

 $c$ = 7.263
 $\beta$ = 91°8'
3 STRONGEST DIFFRACTION LINES:
 7.29 (100)
 2.53 (100)
 3.61 ( 80)
OPTICAL CONSTANTS:
 $\alpha$ = 1.557
 $\beta$ = 1.566
 $\gamma$ = 1.571
 $(-)2V$ = 61°
HARDNESS: 2½–3½
Feel smooth, sometimes greasy.
DENSITY: ~2.61 (Meas.)
 2.58 (Calc.)
CLEAVAGE: {001} perfect
 {010} distinct
 {100} distinct
Fracture usually conchoidal or splintery.
HABIT: Crystals flaky or lath-like, very minute. Usually massive, very fine grained, compact; also lamellar, foliated, columnar, or fibrous. Twinning not common.
COLOR-LUSTER: White, yellowish, various shades of green and yellowish to brownish green, bluish white to bluish green, brownish red. Translucent to nearly opaque. Resinous, greasy, pearly, waxy, earthy. Streak white.
MODE OF OCCURRENCE: Occurs widespread, often admixed with chrysotile, as the principal constituent of serpentines—a mineral group derived from the alteration of ultrabasic rocks. Typical occurrences are found in Washington, California, Utah, Arizona, Colorado, Wyoming, South Dakota, Pennsylvania, New York, New Jersey, Massachusetts, Rhode Island, Maryland, and North Carolina. Also in Canada, Venezuela, England, Norway, Italy, Switzerland, Austria, Germany, Poland, Finland, USSR, South Africa, China, New Zealand, and Australia. Chrysotile-free antigorite occurs in Val Antigorio, Novara Province, Piedmont region, Italy.
BEST REF. IN ENGLISH: Deer, Howie, and Zussman, "Rock Forming Minerals," v. 3, p. 170-190, New York, Wiley, 1962.

## ANTIMONITE = Stibnite

## ANTIMONPEARCEITE
$(Ag, Cu)_{16}(Sb, As)_2S_{11}$
Dimorphous with polybasite
Pearceite-Antimonpearceite Series

CRYSTAL SYSTEM: Monoclinic
CLASS: 2/m
SPACE GROUP: C2/m
Z: 2
LATTICE CONSTANTS:
 $a$ = 12.8
 $b$ = 7.4
 $c$ = 11.9
 $\beta$ = 90°
3 STRONGEST DIFFRACTION LINES:
 3.00 (100)
 2.84 ( 90)
 3.11 ( 50)

HARDNESS: 3
DENSITY: 6.33–6.35 (Meas.)
CLEAVAGE: None.
Fracture irregular to conchoidal. Brittle.
HABIT: Crystals thin tabular, pseudohexagonal, up to 2 cm across. In subparallel and rosette-like groups. Twinning on {110}.
COLOR-LUSTER: Black. Opaque. Submetallic. Streak black.
MODE OF OCCURRENCE: Occurs in Mexico at an unspecified locality in Sonora; also in association with argentite and amethyst at Guanajuato.
BEST REF. IN ENGLISH: Frondel, Clifford, *Am. Min.*, **48**: 565–572 (1963).


# ANTIMONY
Sb

CRYSTAL SYSTEM: Hexagonal
CLASS: $\bar{3}$m
SPACE GROUP: R$\bar{3}$m
Z: 6
LATTICE CONSTANTS:
  $a$ = 4.299
  $c$ = 11.25
3 STRONGEST DIFFRACTION LINES:
  3.109 (100)
  2.248 ( 70)
  1.368 ( 67)
HARDNESS: 3–3½
DENSITY: 6.688 (Calc.)
CLEAVAGE: {0001} perfect, easy.
         {10$\bar{1}$1} distinct
         {10$\bar{1}$4} imperfect
         {11$\bar{2}$0} indistinct
Fracture uneven. Very brittle.
HABIT: Crystals pseudocubic or thick tabular. Commonly massive, lamellar or radiated; also as rounded masses with granular texture. Twinning on {10$\bar{1}$4}, common.
COLOR-LUSTER: Tin white. Opaque. Brilliant metallic. Streak gray.
MODE OF OCCURRENCE: Occurs associated with stibnite at a number of localities in the Havilah and Kernville areas, Kern County, California, and has been reported from Butte, El Dorado, and Riverside counties. It also is found at Nuevo Tepache, Sonora, Mexico; in Quebec and New Brunswick, Canada; at Huasco, Atacama, Chile; in France, Sweden, Germany, Sardinia, Czechoslovakia, Borneo, and at several places in Australia.
BEST REF. IN ENGLISH: Palache, et al., "Dana's System of Mineralogy," 7th Ed., v. I, p. 132–133, New York, Wiley, 1944.


# ANTLERITE
$Cu_3 SO_4 (OH)_4$

CRYSTAL SYSTEM: Orthorhombic
CLASS: 2/m 2/m 2/m
SPACE GROUP: Pnam
Z: 4

LATTICE CONSTANTS:
  $a$ = 8.22  kX
  $b$ = 11.97
  $c$ = 6.02
3 STRONGEST DIFFRACTION LINES:
  4.86  (100)
  2.566 ( 85)
  3.60  ( 75)
OPTICAL CONSTANTS:
  $\alpha$ = 1.726
  $\beta$ = 1.738
  $\gamma$ = 1.789
  (+)2V = 53°
HARDNESS: 3½
DENSITY: 3.88 (Meas.)
         3.936 (Calc.)
CLEAVAGE: {010} perfect
          {100} indistinct
Fracture conchoidal to uneven. Brittle.
HABIT: Crystals thick tabular, short prismatic, or equant. As aggregates of fibrous or acicular crystals; as cross-fiber veinlets; granular; as coatings.
COLOR-LUSTER: Emerald green to blackish green, light green. Translucent. Vitreous. Streak pale green.
MODE OF OCCURRENCE: Occurs as a secondary mineral in the oxidation zone of copper deposits in arid regions. Found at the Antler mine, Mohave County, and at Bisbee, Cochise County, Arizona; in the Darwin district, Inyo County, and in Madera and Shasta counties, California; at the Northern Light mine, near Black Mountain, Nevada; and at Kennecott, Alaska. It is also found in Coahuila, Mexico; in Kazakhstan, USSR; and as the principal copper ore mineral at Chuquicamata, Chile.
BEST REF. IN ENGLISH: Palache, et al., "Dana's System of Mineralogy," 7th Ed., v. II, p. 544–546, New York, Wiley, 1951.


# APATITE = Mineral group of general formula:
$A_5 (PO_4)_3 (F, OH, Cl)_3$

See: Belovite
     Britholite
     Britholite (Y)
     Carbonate-apatite
     Chlorapatite
     Ellestadite
     Fermorite
     Fluorapatite
     Hedyphane

Hydroxylapatite
Hydroxylellestadite
Mimetite
Pyromorphite
Strontium-apatite
Svabite
Vanadinite
Wilkeite


# APHTHITALITE (Glaserite)
$K_3 Na(SO_4)_2$ to $KNa_3 (SO_4)_2$

CRYSTAL SYSTEM: Hexagonal
CLASS: $\bar{3}$2/m
SPACE GROUP: P$\bar{3}$m
Z: 1
LATTICE CONSTANTS:
  $a$ = 5.65 kX
  $c$ = 7.29

3 STRONGEST DIFFRACTION LINES:
2.85 (100)
2.97 ( 90)
2.09 ( 90)
OPTICAL CONSTANTS:
$\omega$ = 1.487–1.491
$\epsilon$ = 1.492–1.499
uniaxial (+)
HARDNESS: 3
DENSITY: 2.71 (Meas.)
2.72 (Calc.)
CLEAVAGE: $\{10\bar{1}0\}$ fair
$\{0001\}$ poor
Fracture conchoidal to uneven. Brittle.

HABIT: Crystals thin to thick tabular; also as pseudo-hexagonal twins resembling aragonite; massive; as crusts; and as bladed aggregates. Twinning on $\{0001\}$ or $\{11\bar{2}0\}$.

COLOR-LUSTER: Colorless, white; also gray, greenish, blue, or reddish due to inclusions. Transparent, translucent, opaque. Vitreous to resinous. Taste bitter.

MODE OF OCCURRENCE: Occurs associated with trona, borax, and hanksite at Searles Lake, San Bernardino County, and as crusts and efflorescences from Deep Spring Lake, Inyo County, California. It is also found in fumaroles of volcanoes as at Mount Vesuvius, Italy, and Mount Etna, Sicily; and in potash deposits near Carlsbad, New Mexico, and near Strassfurt and at other places in Germany.

BEST REF. IN ENGLISH: Palache, et al., "Dana's System of Mineralogy," 7th Ed., v. II, p. 400–403, New York, Wiley, 1951.

# APJOHNITE
$Mn^{2+}Al_2(SO_4)_4 \cdot 24$ (or 22) $H_2O$

CRYSTAL SYSTEM: Monoclinic
CLASS: 2/m
SPACE GROUP: P2/m
Z: 1
OPTICAL CONSTANTS:
$\alpha$ = 1.478
$\beta$ = 1.482
$\gamma$ = 1.482
(–)2V small
HARDNESS: 1½
DENSITY: 1.78 (Meas.)
CLEAVAGE: Not determined.
HABIT: As masses and crusts of fibrous or acicular crystals.
COLOR-LUSTER: Colorless, white, and pale shades of yellow, green, and pink. Silky.

Soluble in water.

MODE OF OCCURRENCE: Occurs in large masses under an overhanging cliff at the headwaters of Little Pigeon Creek, Alum Cave, Sevier County, Tennessee; at Delagoa Bay, Mozambique; and at Szomolnok, Czechoslovakia.

BEST REF. IN ENGLISH: Palache, et al., "Dana's System of Mineralogy," 7th Ed., v. II, p. 527–528, New York, Wiley, 1951.

# APLOWITE
$CoSO_4 \cdot 4H_2O$

CRYSTAL SYSTEM: Monoclinic
CLASS: 2/m
SPACE GROUP: P2$_1$/n
Z: 4
LATTICE CONSTANTS:
$a$ = 5.94
$b$ = 13.56
$c$ = 7.90
$\beta$ = 90°30′
3 STRONGEST DIFFRACTION LINES:
4.46 (100)
5.44 ( 90)
3.95 ( 80)
OPTICAL CONSTANTS:
1.528 (min.)
1.536 (max.)
HARDNESS: 3
DENSITY: 2.33 (Meas.)
2.359 (Calc.)
CLEAVAGE: Not determined.
HABIT: Fine-grained; as a water-soluble efflorescence.
COLOR-LUSTER: Pink; vitreous. Streak white.
MODE OF OCCURRENCE: Occurs associated with moorhouseite on a 2 × 4 × 8 inch specimen of sulfides in a barite-siderite matrix from the Magnet Cove Barium Corporation mine about 2½ miles southwest of Walton, Nova Scotia.

BEST REF. IN ENGLISH: Jambor, J. L., and Boyle, R. W., *Can. Min.*, **8**: 166–171 (1965).

# APOPHYLLITE
$KCa_4Si_8O_{20}(F, OH) \cdot 8H_2O$

CRYSTAL SYSTEM: Tetragonal or pseudo-tetragonal
CLASS: 4/m 2/m 2/m
SPACE GROUP: P4/mnc
Z: 2
LATTICE CONSTANTS:
$a$ = 9.00
$c$ = 15.8
3 STRONGEST DIFFRACTION LINES:
3.943 (100)
2.976 ( 70)
1.578 ( 55)
OPTICAL CONSTANTS:
$\omega \simeq 1.53$–1.54
$\epsilon \simeq 1.53$–1.54 (variable)
HARDNESS: 4½–5
DENSITY: 2.3–2.4 (Meas.)
2.378 (Calc.)
CLEAVAGE: $\{001\}$ perfect
$\{110\}$ imperfect
Fracture uneven. Brittle.
HABIT: Crystals commonly pseudocubic, often modified by $\{111\}$; also tabular, pyramidal, or long and square prismatic. Prism faces vertically striated, often brilliant; basal

pinacoid dull or rough; bipyramidal faces often uneven. Also massive, lamellar or granular. Twinning on {111}, rare.

COLOR-LUSTER: Colorless, white, grayish; also pale yellowish, greenish, reddish. Vitreous to pearly. Transparent to translucent. Streak white.

MODE OF OCCURRENCE: Occurs chiefly as a secondary mineral in cavities in basalt and related rocks in association with prehnite, calcite, analcime, stilbite and other minerals; less commonly in cavities in granite, gneiss, and limestone; sometimes as a low-temperature hydrothermal mineral in sulfide ore veins. Notable occurrences are found in Brush Canyon, Los Angeles County, and at other places in California; in Lane County, Oregon; at Table Mountain, Jefferson County, Colorado; in the Lake Superior copper district, Michigan; at the French Creek mines, Chester County, Pennsylvania; in the trap rocks of northeastern New Jersey; and as fine crystals near Centreville, Fairfax County, Virginia. Also as exceptional specimens in Mexico, Brazil, Canada, Iceland, Ireland, Scotland, Sweden, Germany, Finland, Czechoslovakia, and near Poona, India.

BEST REF. IN ENGLISH: Winchell and Winchell, "Elements of Optical Mineralogy," 4th Ed., Pt. 2, p. 394–395, New York, Wiley, 1951.

# ARAGONITE
$CaCO_3$
Trimorphous with Calcite and Vaterite

CRYSTAL SYSTEM: Orthorhombic
CLASS: 2/m 2/m 2/m
SPACE GROUP: Pmcn
Z: 4
LATTICE CONSTANTS:
  $a = 4.94$
  $b = 7.94$
  $c = 5.72$
3 STRONGEST DIFFRACTION LINES:
  3.396 (100)
  1.977 ( 65)
  3.273 ( 52)
OPTICAL CONSTANTS:
    $\alpha = 1.530$
    $\beta = 1.681$
    $\gamma = 1.685$
  $(-)2V = 18°$
HARDNESS: 3½–4
DENSITY: 2.947 ± 0.002 (Meas.)
         2.944 (Calc.)
CLEAVAGE: {010} distinct
           {110} indistinct
           {011} indistinct
Fracture subconchoidal. Brittle.

HABIT: Crystals acicular or chisel-shaped, elongated along c-axis; rarely thick tabular or pyramidal. Crystals frequently twinned {110} forming sixling prisms, nearly hexagonal in cross section, commonly striated. Also microcrystalline to coarsely crystalline, columnar, fibrous, stalactitic, coralloidal, and pisolitic.

COLOR-LUSTER: Colorless, white, yellowish, gray, green to bluish green, blue, pale to deep lavender, reddish, brown. Transparent to translucent. Vitreous to resinous.

Often fluoresces greenish white, green, yellowish, pink, or bluish under ultraviolet light. Occasionally phosphoresces greenish white under short-wave ultraviolet light.

MODE OF OCCURRENCE: Occurs widespread in a variety of low-temperature deposits formed near the surface, especially in limestone caverns, around hot springs and geysers, in the oxidized zone of ore deposits, and in many sedimentary and metamorphic rocks. It is found in a great variety of forms in Wind Cave, Custer County, South Dakota; as fine crystals in mines of the Magdelena district, Socorro County, New Mexico; and as pseudohexagonal tablets as much as 3 inches in diameter in Larimer County, Colorado, and at several localities in New Mexico. Found in Europe as excellent sixling twins at Molina de Aragon, Spain; as fine coralloidal specimens (flos-ferri) at several places in Carinthia, Austria; as delicate aggregates of acicular crystals in Cumberland, England; as superb crystals with celestite and native sulfur in Sicily; and as superb transparent crystals up to 3 inches long near Horschenz, Czechoslovakia. Plumbian and zincian varieties occur at Tsumeb, South West Africa. Many marine organisms secrete aragonite as their skeletal material.

BEST REF. IN ENGLISH: Palache, et al., "Dana's System of Mineralogy," 7th Ed., v. II, p. 182–193, New York, Wiley, 1951.

# ARAKAWAITE = Veszelyite

# ARAMAYOITE
$Ag(Bi, Sb)S_2$

CRYSTAL SYSTEM: Triclinic
CLASS: $\bar{1}$
SPACE GROUP: $P\bar{1}$
Z: 6
LATTICE CONSTANTS:
  $a = 8.32$     $\alpha = 103°54'$
  $b = 8.83$     $\beta = 90°$
  $c = 7.73$     $\gamma = 100°23'$
3 STRONGEST DIFFRACTION LINES:
  2.82 (100)
  3.22 ( 40)
  1.94 ( 30)
HARDNESS: 2½
DENSITY: 5.602 (Meas.)
        5.624 (Calc.)
CLEAVAGE: {010} perfect
          {100} fair
          {001} poor
Flexible; inelastic. Sectile.

HABIT: As aggregates of thin broad plates. Striated {001} and {100}. Twinning on {$\bar{1}$01}, common.

COLOR-LUSTER: Iron black. Nearly opaque; thin fragments translucent and deep red in color. Metallic.

MODE OF OCCURRENCE: Occurs associated with tetrahedrite, stannite, and pyrite at the Animas mine, Chocoya, Potosi, Bolivia.

BEST REF. IN ENGLISH: Palache, et al., "Dana's System of Mineralogy," 7th Ed., v. I, p. 427–429, New York, Wiley, 1944.

## ARCANITE
$K_2SO_4$

CRYSTAL SYSTEM: Orthorhombic
CLASS: 2/m 2/m 2/m
SPACE GROUP: Pmcn
Z: 4
LATTICE CONSTANTS:
  $a$ = 5.76
  $b$ = 10.05
  $c$ = 7.46
3 STRONGEST DIFFRACTION LINES:
  2.903 (100)
  3.001 ( 71)
  2.886 ( 53)
OPTICAL CONSTANTS:
  $\alpha$ = 1.4935
  $\beta$ = 1.4947
  $\gamma$ = 1.4973
  (+)2V = 67°
HARDNESS: 2
DENSITY: 2.663 (Meas. Synthetic)
        2.660 (Calc. Synthetic)
CLEAVAGE: {010} good
          {001} good
HABIT: As thin plates, pseudohexagonal due to cyclic twinning.
COLOR-LUSTER: Colorless, yellowish. Transparent. Vitreous.
MODE OF OCCURRENCE: Found as yellow crystals in a pine railroad tie in Tunnel No. 1 of the Santa Ana Tin Mining Company in Trabuco Canyon, Orange County, California.
BEST REF. IN ENGLISH: Palache, et al., "Dana's System of Mineralogy," 7th Ed., v. II, p. 399–400, New York, Wiley, 1951.

## ARDEALITE
$Ca_2HPO_4SO_4 \cdot 4H_2O$

CRYSTAL SYSTEM: Monoclinic
Z: 2
LATTICE CONSTANTS:
  $a$ = 5.67 kX
  $b$ = 14.64
  $c$ = 6.28
  $\beta$ = 113°50′
HARDNESS: Not determined.
DENSITY: 2.30 (Meas.)
        2.38 (Calc.)
CLEAVAGE: Not determined.
HABIT: As very fine grained powdery masses.
COLOR-LUSTER: Pale yellow.
MODE OF OCCURRENCE: Occurs associated with gypsum and brushite in a phosphate deposit in a limestone cavern at Cioclovina, Transylvania, Roumania.

BEST REF. IN ENGLISH: Palache, et al., "Dana's System of Mineralogy," 7th Ed., v. II, p. 1010–1011, New York, Wiley, 1951.

## ARDENNITE
$Mn_5Al_5(As, V)O_4Si_5O_{20}(OH)_2 \cdot 2H_2O$

CRYSTAL SYSTEM: Orthorhombic
CLASS: 2/m 2/m 2/m
SPACE GROUP: Pnmm
Z: 2
LATTICE CONSTANTS:
  $a$ = 8.72
  $b$ = 18.56
  $c$ = 5.83
3 STRONGEST DIFFRACTION LINES:
  2.574 (100)
  2.911 ( 70)
  4.21 ( 60)
OPTICAL CONSTANTS:
  $\beta$ = 1.74–2.0
  (+)2V = 0–50°, 68°–70°
HARDNESS: 6–7
DENSITY: 3.620 (Meas.)
CLEAVAGE: {010} perfect
          {110} distinct
Very brittle.
HABIT: Prismatic; as crystalline aggregates and small rosettes of tapered acicular radiating crystals.
COLOR-LUSTER: Yellow to yellowish brown, black. Subadamantine.
MODE OF OCCURRENCE: Occurs, associated with calcite and cuprosklodowskite, as rosettes of crystals that coat a joint surface in Jurassic Todilto limestone, about 20 miles northwest of Grants, New Mexico. Also found in the Ala valley, Piedmont, Italy; abundantly as fine crystalline aggregates at Salm-chateau in the Ardennes, Belgium; and at the Kajlidongri manganese mine, Jhabua District, Madhya Pradesh, India.
BEST REF. IN ENGLISH: Winchell and Winchell, "Elements of Optical Mineralogy," 4th Ed., Pt. II, p. 529, New York, Wiley, 1951.

## ARFVEDSONITE
$Na_{2-3}(Fe, Mg, Al)_5Si_8O_{22}$
Amphibole Group

CRYSTAL SYSTEM: Monoclinic
CLASS: 2/m
SPACE GROUP: C2/m
Z: 2
LATTICE CONSTANTS:
  $a \simeq 9.9$
  $b \simeq 18.0$
  $c \simeq 5.3$
  $\beta \simeq 104°$
3 STRONGEST DIFFRACTION LINES:
  3.161 (100)
  2.732 ( 80)
  8.51 ( 70)

OPTICAL CONSTANTS:
$\alpha = 1.674-1.700$
$\beta = 1.679-1.709$
$\gamma = 1.686-1.710$
$(-)2V_\alpha = 0°-50°$ (data uncertain)
HARDNESS: 5-6
DENSITY: 3.37-3.50 (Meas.)
CLEAVAGE: {110} perfect
{010} parting
Fracture uneven. Brittle.

HABIT: Crystals long prismatic, often tabular {010}, but rarely distinctly terminated. Also in prismatic aggregates. Twinning on {110} simple, lamellar.

COLOR-LUSTER: Greenish black, black. Nearly opaque. Vitreous. Streak dark bluish gray.

MODE OF OCCURRENCE: Occurs as a characteristic constituent of plutonic alkali igneous rocks such as nepheline-syenite. Typical occurrences are found in the Jamestown district, Boulder County, Colorado; at Red Hill near Moultonboro, Carroll County, New Hampshire; in the Julianehaab district, Greenland; and at localities in Norway, Finland, and USSR.

BEST REF. IN ENGLISH: Deer, Howie, and Zussman, "Rock Forming Minerals," v. 2, p. 364-374, New York, Wiley, 1963.

# ARGENTITE
$Ag_2S$
Dimorphous with acanthite

CRYSTAL SYSTEM: Cubic
CLASS: $4/m\,\bar{3}\,2/m$
SPACE GROUP: Im3m
Z: 2
LATTICE CONSTANTS:
$a = 4.91$
3 STRONGEST DIFFRACTION LINES:
3.17 (100)
2.24 (100)
1.819 (100)
HARDNESS: 2-2½
DENSITY: 7.2-7.34 (Meas.)
7.05 (Calc.)
CLEAVAGE: {001} poor
{011} poor
Fracture subconchoidal. Very sectile.

HABIT: Crystals usually cubic or octahedral; also dodecahedral; commonly modified, and often distorted. Usually massive; branching; reticulated; fine-granular; and as a coating. Twinning on {111}, as penetration twins, common.

COLOR-LUSTER: Lead gray to blackish lead gray. Opaque. Metallic.

MODE OF OCCURRENCE: Occurs in moderately low-temperature sulfide ore deposits, commonly associated with native silver, silver sulfosalts, and galena; often as microscopic inclusions in the latter. Found at numerous localities in the western United States, especially in California, Nevada, Colorado, and Montana. It also occurs in British Columbia, Ontario, and elsewhere in Canada; as superb crystals and as a major ore in the silver districts of Mexico; and,

among others, at localities in Honduras, Chile, Bolivia, England, Germany, Czechoslovakia, Sardinia, and Norway.

BEST REF. IN ENGLISH: Palache, et al., "Dana's System of Mineralogy," 7th Ed., v. I, p. 176-178, New York, Wiley, 1944.

# ARGENTOJAROSITE
$AgFe_3(SO_4)_2(OH)_6$

CRYSTAL SYSTEM: Hexagonal
CLASS: 3m
SPACE GROUP: R3m
Z: 3
LATTICE CONSTANTS:
$a = 7.22$ kX
$c = 16.40$
OPTICAL CONSTANTS:
$\omega = 1.882$
$\epsilon = 1.785$
uniaxial (-)
HARDNESS: Not determined.
DENSITY: 3.66 (Meas.)
3.81 (Calc.)
CLEAVAGE: {0001}
HABIT: Crystals very minute, micaceous, flattened on {0001} with hexagonal outline. As fine-grained masses and coatings.

COLOR-LUSTER: Yellow to brown. Brilliant.

MODE OF OCCURRENCE: Occurs as a secondary mineral at the Tintic Standard mine, Juab County, associated with anglesite, barite, and quartz, and at the Buffalo mine and Chloride Point, Tooele County, Utah.

BEST REF. IN ENGLISH: Palache, et al. "Dana's System of Mineralogy," 7th Ed., v. II, p. 565, New York, Wiley, 1951.

# ARGENTOPYRITE
$AgFe_2S_3$
Dimorphous with sternbergite

CRYSTAL SYSTEM: Orthorhombic
CLASS: 2/m 2/m 2/m
SPACE GROUP: Pmmn
Z: 4
LATTICE CONSTANTS:
$a = 6.64$
$b = 11.47$
$c = 6.45$
3 STRONGEST DIFFRACTION LINES:
3.341 (100)
3.318 (100)
1.808 ( 70)
HARDNESS: 3½-4
DENSITY: 4.25 (Meas.)
4.27 (Calc.)
CLEAVAGE: None.
Fracture uneven. Brittle.

HABIT: Minute pseudohexagonal crystals with forms $b${010}, $m${110}, $n${120}, $x${011}, and $c${001}; termination faces usually rough. Interpenetration and lamellar twinning present.

COLOR-LUSTER: Steel gray to tin white; often tarnished shades of bronze brown, green, blue, yellow, and purple. Metallic.

MODE OF OCCURRENCE: Occurs associated with arsenic, some proustite and occasional grains of pyrite at Freiberg, Saxony, and with chloanthite, proustite, and siderite at Joachimstal, Czechoslovakia.

BEST REF. IN ENGLISH: Murdoch, J., and Berry, L. G., *Am. Min.*, **39**: 475–485 (1954).

## ARGYRODITE (Canfieldite)
$Ag_8GeS_6$
Argyrodite Series

CRYSTAL SYSTEM: Cubic
CLASS: $4/m \bar{3} 2/m$
SPACE GROUP: Im3m
Z: 32
LATTICE CONSTANT:
  $a = 21.11$
3 STRONGEST DIFFRACTION LINES:
  3.02 (100)
  1.863 ( 50)
  2.66 ( 40)
HARDNESS: 2½
DENSITY: 6.26–6.29 (Meas.)
         6.32 (Calc.)
CLEAVAGE: Fracture conchoidal to uneven. Brittle.
HABIT: Crystals octahedral, dodecahedral (up to 6 cm in diameter), or as combinations of these forms. As crystal aggregates, botryoidal crusts, or massive. Twinning on {111} common, as spinel twins.
COLOR-LUSTER: Steel gray with reddish tint; rapidly tarnishes black with purple or bluish tint. Opaque. Metallic. Streak grayish black.
MODE OF OCCURRENCE: Occurs in vein deposits commonly associated with argentite, other silver minerals, and sulfides. Found at several deposits in the vicinity of Colquechaca, Bolivia, and Freiberg, Saxony, Germany.
BEST REF. IN ENGLISH: Palache, et al., "Dana's System of Mineralogy," 7th Ed., v. I, p. 356–358, New York, Wiley, 1944.

**ARITE** = Variety of niccolite containing up to 6% Sb

**ARIZONITE** = A mixture of hematite, ilmenite, rutile, and anatase

## ARMALCOLITE
$(Mg, Fe)Ti_2O_5$
Pseudobrookite Group

CRYSTAL SYSTEM: Orthorhombic
CLASS: 2/m 2/m 2/m
LATTICE CONSTANTS:
  $a = 9.743$
  $b = 10.024$
  $c = 3.738$

4 STRONGEST DIFFRACTION LINES:
  3.468 (100)
  1.958 ( 80)
  2.763 ( 25)
  2.454 ( 25)
DENSITY: 4.94 (Calc.)
HABIT: As embedded grains, usually rectangular in outline, with longest dimension 100–300 $\mu m$. Commonly with ilmenite overgrowth.
COLOR-LUSTER: Gray. Opaque.
MODE OF OCCURRENCE: Occurs in rocks collected by the Apollo XI mission, at Tranquillity Base, Moon.
BEST REF. IN ENGLISH: *Proc. Apollo XI Lunar Sci. Conf.*, **1**: 55–63 (1970).

## ARMANGITE
$Mn_3(AsO_3)_2$

CRYSTAL SYSTEM: Hexagonal
CLASS: $\bar{3} 2/m$, 3m or 3 2
SPACE GROUP: R3m, R32 or R$\bar{3}$m
Z: 9
LATTICE CONSTANTS:
  $a = 13.44$ kX
  $c = 8.72$
3 STRONGEST DIFFRACTION LINES:
  2.76 (100)
  2.94 ( 70)
  2.43 ( 70)
OPTICAL CONSTANTS:
  $\omega = 2.01$
  $\epsilon = 1.99$
  uniaxial (–)
HARDNESS: ~4
DENSITY: 4.43 (Meas.)
         4.47 (Calc.)
CLEAVAGE: {0001} fair
HABIT: Crystals short prismatic. Twinning on {02$\bar{2}$1}, lamellar.
COLOR-LUSTER: Black. Transparent in thin splinters. Streak brown.
MODE OF OCCURRENCE: Occurs associated with hematite in calcite-barite veinlets at Långban, Sweden.
BEST REF. IN ENGLISH: Palache, et al., "Dana's System of Mineralogy," 7th Ed., v. II, p. 1031–1032, New York, Wiley, 1951.

## ARMENITE
$BaCa_2Al_6Si_9O_{30} \cdot 2H_2O$

CRYSTAL SYSTEM: Hexagonal
CLASS: 6mm
SPACE GROUP: P6cc or P6$_3$mc
Z: 2
LATTICE CONSTANTS:
  $a = 10.690$
  $c = 13.898$
3 STRONGEST DIFFRACTION LINES:
  3.86 (100)
  3.41 ( 90)
  2.91 ( 90)

OPTICAL CONSTANTS:
$\alpha$ = 1.551
$\beta$ = 1.559
$\gamma$ = 1.562
$(-)2V = 60°$
HARDNESS: 7½
DENSITY: 2.76 (Meas.)
2.787 (Calc.)
CLEAVAGE: {010} perfect
{110} distinct
Brittle.
HABIT: Crystals prismatic.
COLOR-LUSTER: Grayish green. Translucent. Vitreous.
MODE OF OCCURRENCE: Occurs in association with axinite, pyrrhotite, and quartz on a specimen (British Museum No. Bm 1947, 290) collected in 1877 from the silver-bearing calcite veins of the Armen mine near Kongsberg, Norway.
BEST REF. IN ENGLISH: *Am. Min.*, **26**: 235 (1941).

## ARNIMITE = Antlerite

## ARROJADITE
$(Na, Ca)_2 (Fe^{2+}, Mn^{2+})_5 (PO_4)_4$ (?)

CRYSTAL SYSTEM: Monoclinic
CLASS: 2/m
SPACE GROUP: C2/m
Z: 12
LATTICE CONSTANTS:
$a$ = 16.60
$b$ = 10.02
$c$ = 23.99
$\beta$ = 93°37'
3 STRONGEST DIFFRACTION LINES:
3.04 (100)
2.72 ( 80)
3.22 ( 60)
OPTICAL CONSTANTS:
$\alpha$ = 1.664
$\beta$ = 1.670
$\gamma$ = 1.675
$2V = 86°$
HARDNESS: 5
DENSITY: 3.55 (Meas.)
CLEAVAGE: {001} distinct
{201} indistinct
Fracture subconchoidal to uneven.
HABIT: Large cleavable masses.
COLOR-LUSTER: Dark green; vitreous to greasy. Translucent.
MODE OF OCCURRENCE: Occurs as large cleavable masses in granitic pegmatite associated with quartz, graftonite, sphalerite, cassiterite, beryl, muscovite, and spodumene at the Nickel Plate mine, Keystone, Pennington County, South Dakota; also as small cleavage masses in pegmatite at the Smith mine, Chandlers Mill, Newport, New Hampshire; and at the Serra Blanca pegmatite, Picuhy, Brazil.

BEST REF. IN ENGLISH: Lindberg, Mary Louise, *Am. Min.*, **35**: 59–76 (1950). "Dana's System of Mineralogy," 7th Ed., v. II, p. 679–681, New York, Wiley, 1951.

## ARSENATE-BELOVITE = Synonym for talmessite.
Belovite of Nefedov.
$Ca_2 Mg(AsO_4)_2 \cdot 2H_2O$

## ARSENIC
As

CRYSTAL SYSTEM: Hexagonal
CLASS: $\bar{3}$ 2/m
SPACE GROUP: R$\bar{3}$m
Z: 6
LATTICE CONSTANTS:
$a$ = 3.77
$c$ = 10.57
3 STRONGEST DIFFRACTION LINES:
2.771 (100)
3.52 ( 26)
1.879 ( 26)
HARDNESS: 3½
DENSITY: 5.72–5.73 (Meas.)
5.75 (Calc.)
CLEAVAGE: {0001} perfect
Fracture uneven. Brittle.
HABIT: As small rhombohedral or acicular crystals. Usually granular massive; in concentric layers; also stalactitic, reniform, columnar, reticulated. Twinning on {10$\bar{1}$4}, rare.
COLOR-LUSTER: Tin white, tarnishes to dark gray. Opaque. Somewhat metallic on fresh fracture.
MODE OF OCCURRENCE: Occurs chiefly in hydrothermal veins, dolomitic limestone, and in the anhydrite cap rock of salt domes. Found in Inyo, Monterey, and Nevada counties, California; at the Winnfield salt dome, Louisiana; and at Washington Camp, Santa Cruz County, Arizona. It also occurs as fine specimens on Alder Island, British Columbia, and elsewhere in Canada; at Copiapó, Chile; in Italy, France, Germany, Czechoslovakia, Roumania, Borneo, Western Australia, and as small crystals at Akadani, Japan.
BEST REF. IN ENGLISH: Palache, et al., "Dana's System of Mineralogy," 7th Ed., v. I, p. 128–130, New York, Wiley, 1944.

## ARSENIOPLEITE = Caryinite

## ARSENIOSIDERITE
$Ca_3 Fe_4^{3+}(OH)_6 (H_2O)_3 [AsO_4]_4$

CRYSTAL SYSTEM: Monoclinic
CLASS: 2/m
SPACE GROUP: A2/a
Z: 8
LATTICE CONSTANTS:
$a$ = 17.76(4) Å
$b$ = 19.53
$c$ = 11.30
$\beta$ = 96.0°

3 STRONGEST DIFFRACTION LINES:
8.84 (100)
2.772 ( 80)
5.62 ( 50)
OPTICAL CONSTANTS:
$\alpha$ = 1.815
$\beta$ = 1.898
$\gamma$ = 1.898
Biaxial (−), small 2V
HARDNESS: 4½ (granular)
       to 1½ (fibrous)
DENSITY: 3.58, 3.60 (Meas.)
       3.59 (Calc.)
CLEAVAGE: {100} good
HABIT: Massive, as radial-fibrous aggregates; also as granular pseudomorphs, or concretionary.
COLOR-LUSTER: Yellow to dark brown. Opaque except in small grains. Submetallic to silky. Streak ochreyellow.
MODE OF OCCURRENCE: Arseniosiderite occurs at the Eureka mines, Tintic district, Juab County, and at the Gold Hill (Western Utah) mine, Tooele County, Utah; at Franklin, Sussex County, New Jersey; at Tagish Lake, Yukon, Canada (yukonite); at Mapimi and Mazapil, Mexico (mazapilite); and at localities in France, Austria, and Germany.
BEST REF. IN ENGLISH: Palache, et al., "Dana's System of Mineralogy," 7th Ed., v. II, p. 953–955, New York, Wiley, 1951. Moore, Paul Brian, *Am. Min.*, **59**:48–59 (1974).

## ARSENOBISMITE
Near $Bi_2AsO_4(OH)_3$

CRYSTAL SYSTEM: Cubic
3 STRONGEST DIFFRACTION LINES:
3.11 (100)
6.06 ( 90)
1.843 ( 80)
OPTICAL CONSTANT:
$N$ > 1.86
HARDNESS: Not determined.
DENSITY: ~5.7
CLEAVAGE: Not determined.
HABIT: As friable or ocherous microcrystalline masses.
COLOR-LUSTER: Yellowish brown to yellowish green.
MODE OF OCCURRENCE: Occurs abundantly in oxidized ore at the Mammoth mine, Tintic district, Juab County, Utah, and with bismutite and bindheimite at Tazna, Bolivia.
BEST REF. IN ENGLISH: Palache, et al., "Dana's System of Mineralogy," 7th Ed., v. II, p. 907, New York, Wiley, 1951.

## ARSENOCLASITE
$Mn_5(OH)_4(AsO_4)_2$

CRYSTAL SYSTEM: Orthorhombic
CLASS: 2 2 2
SPACE GROUP: $P2_12_12_1$
Z: 4

LATTICE CONSTANTS:
$a$ = 9.31 Å
$b$ = 5.75
$c$ = 18.84
3 STRONGEST DIFFRACTION LINES:
2.933 (100)
2.739 ( 75)
4.55 ( 70)
OPTICAL CONSTANTS:
   $\alpha$ = 1.787
   $\beta$ = 1.810
   $\gamma$ = 1.816
(−)2V = 53°26′
HARDNESS: 5-6
DENSITY: 4.16 (Meas.)
       4.21 (Calc.)
CLEAVAGE: {010} perfect
HABIT: Massive, granular.
COLOR-LUSTER: Red
MODE OF OCCURRENCE: Occurs associated with adelite and sarkinite along fissures in hausmannite-bearing dolomite at Långban, Sweden.
BEST REF. IN ENGLISH: Moore, Paul B., *Am. Min.*, **53**: 1779 (1968), Palache, et al., "Dana's System of Mineralogy," 7th Ed., v. II, p. 801–802, New York, Wiley, 1951.

## ARSENOLAMPRITE
As
Polymorph of arsenic

CRYSTAL SYSTEM: Orthorhombic
Z: 8
LATTICE CONSTANTS:
$a$ = 3.63
$b$ = 4.45
$c$ = 10.96
3 STRONGEST DIFFRACTION LINES:
5.44 (100)
2.72 (100)
1.12 (100)
HARDNESS: 2
DENSITY: 5.3–5.5 (Meas.)
       5.577 (Calc.)
CLEAVAGE: Perfect, one direction.
HABIT: Plates and veinlets; fibrous foliated structure.
COLOR-LUSTER: Dark gray. Opaque. Brilliant metallic. Streak black.
MODE OF OCCURRENCE: Occurs associated with loellingite, native silver, and a chlorite-like mineral in carbonate at Cěrny Dul, Czechoslovakia; at Marienberg, Saxony, Germany; at Ste. Marie aux Mines, Alsace, France; and in the vicinity of Copiapó, Chile.
BEST REF. IN ENGLISH: Johan, Zdenek, *Am. Min.*, **45**: 479–480 (1960).

## ARSENOLITE
$As_2O_3$
Dimorph of claudetite

CRYSTAL SYSTEM: Cubic

CLASS: $4/m\bar{3}2/m$
SPACE GROUP: Fd3m
Z; 16
LATTICE CONSTANT:
 $a = 11.0457$
3 STRONGEST DIFFRACTION LINES:
 3.195 (100)
 6.394 ( 63)
 2.541 ( 38)
OPTICAL CONSTANT:
 $N = 1.755$
 HARDNESS: 1½
 DENSITY: $3.87 \pm 0.01$
  3.868 (Calc.)
 CLEAVAGE: {111} distinct
Fracture conchoidal.
 HABIT: Crystals octahedral or capillary. Also as crusts, stalactitic, botryoidal, earthy to powdery.
 COLOR-LUSTER: White; also tinted yellowish, bluish, or reddish. Transparent. Vitreous to silky.

Taste biting, sweet.
 MODE OF OCCURRENCE: Occurs as a secondary mineral formed by the alteration of arsenopyrite or other arsenic-bearing minerals; less commonly as a sublimation product resulting from mine fires or burning coal deposits. Found in California, Nevada, South Dakota, and Kansas. Also in Canada, Peru, France, Italy, Germany, Hungary, Czechoslovakia, and elsewhere.
 BEST REF. IN ENGLISH: Palache, et al., "Dana's System of Mineralogy," 7th Ed., v. I, p. 543–544, New York, Wiley, 1944.

## ARSENOPALLADINITE
= Name proposed for naturally occurring $Pd_3As$

## ARSENOPYRITE (Cobaltian-Danaite)
FeAsS

CRYSTAL SYSTEM: Monoclinic
CLASS: 2/m
SPACE GROUP: $P2_1/c$
Z: 4
LATTICE CONSTANTS:
 $a = 5.74$
 $b = 5.68$
 $c = 5.79$
 $\beta = 112.17°$
3 STRONGEST DIFFRACTION LINES:
 2.677 (100)
 2.662 (100)
 2.418 ( 95)
 HARDNESS: 5½–6
 DENSITY: $6.07 \pm 0.15$ (Meas.)
  6.18 (Calc.)
 CLEAVAGE: {101} distinct
  {010} in traces
Fracture uneven. Brittle.
 HABIT: Crystals usually short prismatic, somewhat elongated parallel to c-axis, or less commonly to b-axis. Often striated parallel to c-axis. Also massive; compact granular,

or columnar. Twinning on {100}, {001}, {101}, and {012} common; the last producing star-like trillings or cruciform twins.
 COLOR-LUSTER: Silver white to steel gray. Opaque. Metallic; sometimes tarnished iridescent, yellowish, or dull gray. Streak black.
 MODE OF OCCURRENCE: Occurs widespread as the most abundant arsenic mineral in many different types of deposit. Common in medium- to high-temperature veins, in contact metasomatic deposits and metamorphic rocks, and sparsely in basic rocks and pegmatites. Notable localities occur throughout the United States, Canada, Mexico, Bolivia, Brazil, England, Norway, Sweden, France, Portugal, Italy, Switzerland, Austria, Germany, Czechoslovakia, Hungary, Zaire, and Japan.
 BEST REF. IN ENGLISH: Palache, et al., "Dana's System of Mineralogy," 7th Ed., v. I, p. 316–322, New York, Wiley, 1944.

## ARSENOSULVANITE (Lazarevićite)
$Cu_3(As, V)S_4$

CRYSTAL SYSTEM: Cubic
CLASS: $\bar{4}3m$
SPACE GROUP: $P\bar{4}3m$
Z: 1
LATTICE CONSTANT:
 $a = 5.257$
 HARDNESS: 3½
 DENSITY: 4.01, 4.2
  4.39 (Calc.)
 HABIT: Tiny grains.
 COLOR-LUSTER: Bronze yellow.
 MODE OF OCCURRENCE: Occurs in quartz-calcite veins cutting bituminous limestone in Mongolia; also as microscopic grains in copper ore, associated with enargite, luzonite, covellite and pyrite, at the Tilva Mika deposit, Bor, eastern Serbia.
 BEST REF. IN ENGLISH: Betekhtin, A. G., *Am. Min.*, **40**: 368–369 (1955). Sclar, C. B., and Drovenik, Matija, *Bull. Geol. Soc. Am.*, **71**: 1970 (1960) [abstract].

## ARSENPOLYBASITE
$(Ag, Cu)_{16}(As, Sb)_2S_{11}$
Dimorphous with pearceite
Polybasite-Arsenpolybasite Series

CRYSTAL SYSTEM: Monoclinic
CLASS: 2/m
SPACE GROUP: C2/m
LATTICE CONSTANTS:
 $a = 26.08$
 $b = 15.04$
 $c = 23.84–23.95$
 $\beta = 90°$
3 STRONGEST DIFFRACTION LINES:
 3.00 (100)
 3.19 ( 90)
 2.88 ( 80)
 2.69–2.64 ( 60) - Best diagnostic line. Absent in Pearceite Series.

HARDNESS: 3
DENSITY: 6.18-6.23 (Meas.)
CLEAVAGE: {001} imperfect
Fracture uneven. Brittle.
HABIT: Crystals thin tabular, pseudohexagonal, up to 6 cm across. Twinning on {110}, repeated.
COLOR-LUSTER: Black. Opaque. Metallic. Streak black.
MODE OF OCCURRENCE: Occurs as large crystals in association with argentite and chalcopyrite at the Neuer Morgenstern mine, Freiberg, Saxony, Germany; also found at Quespisiza, Chile.
BEST REF. IN ENGLISH: Frondel, Clifford, *Am. Min.*, **48**: 565-572 (1963).

## ARSENSULFURITE = Amorphous mixture As + S
HARDNESS: 2½
HABIT: Massive; as isotropic crusts.
COLOR-LUSTER: Red-brown.
MODE OF OCCURRENCE: Occurs in the crater of Papandagan volcano, Java; also from the solfotara near Naples, Italy.
BEST REF. IN ENGLISH: Palache, et al., "Dana's System of Mineralogy," v. I, p. 144, New York, Wiley, 1944.

## ARSENURANYLITE
$Ca(UO_2)_4(AsO_4)_2(OH)_4 \cdot 6H_2O$

CRYSTAL SYSTEM: Orthorhombic
LATTICE CONSTANTS:
  $a = 15.40$
  $b = 17.40$
  $c = 13.768$
3 STRONGEST DIFFRACTION LINES:
  7.72 (100)
  3.85 (100)
  8.41 ( 80)
OPTICAL CONSTANTS:
  $\alpha = 1.737$
  $\gamma = 1.766$
HARDNESS: ~2½
DENSITY: 4.25 (Calc.)
CLEAVAGE: {001} perfect
HABIT: Fine scales; lichen-like.
COLOR-LUSTER: More orange than phosphuranylite.
MODE OF OCCURRENCE: Occurs associated with meta-zeunerite, uranospinite, novacekite, schoepite, and para-schoepite in the oxidation zone of a deposit containing arsenic-bearing sulfides at an unspecified locality in USSR.
BEST REF. IN ENGLISH: Belova, L. N., *Am. Min.*, **44**: 208 (1959).

## ARTHURITE
$Cu_2Fe_4(AsO_4, PO_4, SO_4)_4(O, OH)_4 \cdot 8H_2O$

CRYSTAL SYSTEM: Monoclinic
CLASS: 2/m
SPACE GROUP: $P2_1/c$
Z: 1

LATTICE CONSTANTS:
  $a = 10.09$
  $b = 9.62$
  $c = 5.55$
  $\beta = 92.2°$
3 STRONGEST DIFFRACTION LINES:
  4.28 (100)
  4.81 (100)
  6.97 (100)
DENSITY: 3.02 (Meas.)
CLEAVAGE: Brittle.
HABIT: Crystals long prismatic, usually as hemispherical or globular aggregates with a radial fibrous or stellate structure. Also as crusts.
COLOR-LUSTER: Apple green to emerald green; vitreous.
MODE OF OCCURRENCE: Occurs as thin crusts usually intimately mixed with pharmacosiderite or an unidentified mineral of the alunite-beudantite family or both at the Hingston Down Consols mine, Calstock, Cornwall, England. Also as hemispherical crystal aggregates up to 2 mm in diameter in cracks and vugs in rock from Majuba Hill, Pershing County, Nevada; and at Potrerillos, Atacama Province, Chile.
BEST REF. IN ENGLISH: Davis, R. J., and Hey, M. H., *Min. Mag.*, **33**: 937-941 (1964).    Clark, A. H., and Sillitoe, R. H., *Min. Mag.*, **37**: 519-520 (1969).    Davis, R. J., and Hey, M. H., *Min. Mag.*, **37**: 520-521 (1969).

## ARTINITE
$Mg_2CO_3(OH)_2 \cdot 3H_2O$

CRYSTAL SYSTEM: Monoclinic
CLASS: 2
SPACE GROUP: C2
Z: 2
LATTICE CONSTANTS:
  $a = 16.56$
  $b = 3.15$
  $c = 6.22$
  $\beta = 99°9'$
3 STRONGEST DIFFRACTION LINES:
  2.736 (100)
  5.34 ( 65)
  3.69 ( 50)
OPTICAL CONSTANTS:
  $\alpha = 1.488$
  $\beta = 1.534$
  $\gamma = 1.556$
  $(-)2V = 70°$
HARDNESS: 2½
DENSITY: 2.02 ± 0.01 (Meas.)
         2.047 (Calc.)
CLEAVAGE: {100} perfect
          {001} good
Very brittle.
HABIT: As sprays and crusts of acicular crystals; as spherical or flattened aggregates of radiating fibers; and as fibrous veinlets.
COLOR-LUSTER: White. Transparent. Crystals vitreous; fibrous aggregates, silky.

MODE OF OCCURRENCE: Occurs associated with hydromagnesite or other magnesium-bearing minerals coating fracture surfaces in serpentinized ultrabasic rocks. Found in the United States as excellent radiating fibrous aggregates on serpentine near the Gem mine and at other places in San Benito County, and along White Creek, Fresno County, California; at Luning, Nevada; as superb sprays of acicular crystals associated with hydromagnesite on Long Island, New York; at Hoboken, New Jersey; and at Eden Mills, Vermont. Fine specimens also are found in the Aosta Valley and at other places in Italy.

BEST REF. IN ENGLISH: Palache, et al., "Dana's System of Mineralogy," 7th Ed., v. II, p. 263–264, New York, Wiley, 1951.

## ARZRUNITE (inadequately described mineral)
Near $Cu_4Pb_2SO_4Cl_6(OH)_4 \cdot 2H_2O$

CRYSTAL SYSTEM: Orthorhombic (?)
HARDNESS: Not determined.
DENSITY: Not determined.
CLEAVAGE: Not determined.
HABIT: As druses of minute prismatic crystals.
COLOR-LUSTER: Blue, bluish green.
MODE OF OCCURRENCE: Found at the Buena Esperanza mine, Challacolla, Tarapacá province, Chile.
BEST REF. IN ENGLISH: Palache, et al., "Dana's System of Mineralogy," 7th Ed., v. II, p. 130–131, New York, Wiley, 1951.

## ASBECASITE
$Ca_3(Ti,Sn)(As_6Si_2Be_2O_{20})$

CRYSTAL SYSTEM: Hexagonal
CLASS: 3m
SPACE GROUP: P3c
Z: 2
LATTICE CONSTANTS:
$a = 8.36$
$c = 15.30$
3 STRONGEST DIFFRACTION LINES:
3.23 (100)
1.570 ( 70)
1.153 ( 70)
OPTICAL CONSTANTS:
$\omega = 1.86$
$\epsilon = 1.83$
(–)
HARDNESS: 6½–7
DENSITY: 3.70 (Meas.)
3.71 (Calc.)
CLEAVAGE: Rhombohedral.
HABIT: Rhombohedral crystals up to 5 mm in size.
COLOR-LUSTER: Lemon yellow; transparent. Vitreous.
MODE OF OCCURRENCE: Occurs on cleft faces in orthogneisses of the Monte Leone nappe, southern Binnatal, Switzerland.
BEST REF. IN ENGLISH: St. Graeser, *Am. Min.*, **52**: 1583–1584 (1967). Cannillo, Elio; Giuseppetti, Giuseppe; and Tadini, Carla, *Am. Min.*, **55**: 1818 (1970).

## ASCHARITE = Szaibelyite

## ASHCROFTINE
$KNaCaY_2Si_6O_{12}(OH)_{10} \cdot 4H_2O$

CRYSTAL SYSTEM: Tetragonal
CLASS: 4/m 2/m 2/m, 4mm, 422, $\bar{4}$2m
SPACE GROUP: I4/mmm, I4mm, I422, I$\bar{4}$m2
Z: 16
LATTICE CONSTANTS:
$a = 24.044$
$c = 17.553$
3 STRONGEST DIFFRACTION LINES:
17.1 (100)
12.0 ( 80)
3.10 ( 60)
HARDNESS: Not determined.
DENSITY: 2.61 (Meas.)
2.53 (Calc.)
CLEAVAGE: {100} perfect
{001} distinct
HABIT: Crystals fibrous prismatic. As a powder.
COLOR-LUSTER: Pink.
MODE OF OCCURRENCE: Occurs in pegmatitic pockets in augite syenite at Narsarsuk, Greenland.
BEST REF. IN ENGLISH: Deer, Howie, and Zussman, "Rock Forming Minerals," v. 4, p. 401–407, New York, Wiley, 1963. Moore, P. B., Bennett, J. M., and Louisnathan, S. J., *Min. Mag.*, **37**: 515–517 (1969).

## ASHTONITE = Mordenite

## ASTRAKHANITE = Bloedite

## ASTROLITE = Muscovite

## ASTROPHYLLITE
$(K,Na)_3(Fe,Mn)_7Ti_2Si_8O_{24}(O,OH)_7$

CRYSTAL SYSTEM: Triclinic
CLASS: $\bar{1}$
SPACE GROUP: P$\bar{1}$
Z: 2
LATTICE CONSTANTS:
$a = 11.72$   $\alpha = 90°$
$b = 5.41$   $\beta = 94°$
$c = 21.14$   $\gamma = 103°$
3 STRONGEST DIFFRACTION LINES:
10.6 (100)
3.51 ( 80)
2.77 ( 60)
OPTICAL CONSTANTS:
$\alpha = 1.678$
$\beta = 1.703$ variable
$\gamma = 1.733$
(+)2V = 70°–80° (occasionally –)
HARDNESS: 3
DENSITY: 3.3–3.4 (Meas.)

CLEAVAGE: {010} perfect
{100} poor
Laminae brittle.

HABIT: Crystals bladed, up to 15 cm long; often in stellate groups.

COLOR-LUSTER: Bronze yellow to golden yellow. Translucent in thin laminae. Submetallic, pearly.

MODE OF OCCURRENCE: Occurs in association with quartz, feldspar, riebeckite and zircon at St. Peters Dome, El Paso County, Colorado. Also found in nepheline-syenites at Mont St. Hilaire, Quebec, Canada; in the Julianehaab district, Greenland; on the small islands in the Langesundfjord, Norway; and on the Iles de Los, Guinea.

BEST REF. IN ENGLISH: Winchell and Winchell, "Elements of Optical Mineralogy," 4th Ed., Pt. 2, p. 480-481, Wiley, 1951.

# ATACAMITE
$Cu_2Cl(OH)_3$
Dimorph of Paratacamite

CRYSTAL SYSTEM: Orthorhombic
CLASS: 2/m 2/m 2/m
SPACE GROUP: Pmcn
Z: 4
LATTICE CONSTANTS:
$a = 6.84$
$b = 9.13$
$c = 6.01$
3 STRONGEST DIFFRACTION LINES:
5.40 (100)
5.00 (100)
2.82 (100)
OPTICAL CONSTANTS:
$\alpha = 1.831$
$\beta = 1.861$
$\gamma = 1.880$
$(-)2V = 74°56'$ (Calc.)
HARDNESS: 3-3½
DENSITY: 3.76 (Meas.)
3.78 (Calc.)
CLEAVAGE: {010} perfect
{101} fair
Fracture conchoidal. Brittle.

HABIT: Crystals slender prismatic, elongated along c-axis. Also tabular {010}; rarely pseudo-octahedral. Often striated. As crusts of subparallel crystals; also as coarsely crystalline aggregates; massive, fibrous, granular to compact; as sand. Twinning, both contact and penetration, common.

COLOR-LUSTER: Bright green to blackish green. Transparent to translucent. Vitreous to adamantine. Streak apple green

MODE OF OCCURRENCE: Occurs as a secondary mineral in the oxidation zone of copper deposits. Found in the United States at Goffs, San Bernardino County, California; in the Tintic district, Juab County, Utah; at the Iron Hill mine, Carbonate, Lawrence County, South Dakota; and at Bisbee and other localities in Arizona. It is also found at Boleo and El Toro, Baja California, Mexico; abundantly at many places in the provinces of Atacama, Antofagasta, and Tarapacá, Chile; in Peru; as fine crystals in the Burra district and at other places in South Australia; in Cornwall, England; and in the Bogoslowsk district, USSR.

BEST REF. IN ENGLISH: Palache, et al., "Dana's System of Mineralogy," 7th Ed., v. II, p. 69-73, New York, Wiley, 1951.

# ATELESTITE
$Bi_8(AsO_4)_3O_5(OH)_5$

CRYSTAL SYSTEM: Monoclinic
CLASS: 2/m
SPACE GROUP: $P2_1/a$
Z: 1
LATTICE CONSTANTS:
$a = 10.88$
$b = 7.42$
$c = 6.98$
$\beta = 107°13'$
$a:b:c = 0.933:1:1.505$
3 STRONGEST DIFFRACTION LINES:
3.239 (100 b)
3.116 ( 40 )
2.725 ( 30 )
OPTICAL CONSTANTS:
$\alpha = 2.14$
$\beta = 2.15$
$\gamma = 2.18$
$(+)2V = 44°$
HARDNESS: 4½-5
DENSITY: 6.82 (Meas.)
6.95 (Calc.)
CLEAVAGE: {001} indistinct
Fracture subconchoidal.

HABIT: Crystals minute, tabular. Also as mammillary or spherical crystalline aggregates with smooth surface.

COLOR-LUSTER: Bright yellow to yellowish green or wax yellow. Transparent to translucent. Resinous to adamantine.

MODE OF OCCURRENCE: Occurs associated with bismutite and other minerals at the Neuhilfe and Weisser Hirsch mines in the Schneeberg district, Saxony, Germany.

BEST REF. IN ENGLISH: Palache, et al., "Dana's System of Mineralogy," 7th Ed., v. II, p. 792-793, New York, Wiley, 1951.

# ATHABASCAITE
$Cu_5Se_4$

CRYSTAL SYSTEM: Orthorhombic
Z: 4
LATTICE CONSTANTS:
$a = 8.227$
$b = 11.982$
$c = 6.441$
3 STRONGEST DIFFRACTION LINES:
3.235 (100)
1.997 ( 80)
3.015 ( 60)
HARDNESS: Microhardness 68.8-93.5 kg/mm². Average 78.1 kg/mm².

DENSITY: 6.59 (Calc.)
HABIT: As elongated lath-shaped grains which range up to 100 $\mu$ in length and 40 $\mu$ in width.
COLOR-LUSTER: Strongly bireflecting in reflected light with colors in air ranging from light gray to bluish gray.
MODE OF OCCURRENCE: Occurs as lath-shaped inclusions in umangite and as stringers and veinlets in hematite-stained carbonate vein material in basalt at the Martin Lake Mine, Lake Athabasca area, northern Saskatchewan, Canada.
BEST REF. IN ENGLISH: Harris, D. C., Cabri, L. J., and Kaiman, S., *Can. Min.*, **10** (pt. 2): 207–215 (1970).

## ATTAKOLITE (Attacolite)
$(Ca, Mn, Sr)_3 Al_6 (PO_4, SiO_4)_7 \cdot 3H_2O$

CRYSTAL SYSTEM: Orthorhombic
Z: 4
LATTICE CONSTANTS:
  $a$ = 11.38
  $b$ = 13.22
  $c$ = 14.08
3 STRONGEST DIFFRACTION LINES:
  3.09 (100)
  3.13 ( 80)
  4.34 ( 70)
OPTICAL CONSTANTS:
  $\alpha$ = 1.655
  $\beta$ = 1.664
  $\gamma$ = 1.675
  (+)2V = 84°
HARDNESS: Not determined.
DENSITY: 3.229 (Meas.)
HABIT: Massive, indistinctly crystalline.
COLOR-LUSTER: Pale red.
MODE OF OCCURRENCE: The mineral occurs associated with berlinite and lazulite in the iron mine at Västanå, southernmost Sweden.
BEST REF. IN ENGLISH: Palache, et al., "Dana's System of Mineralogy," 7th Ed., v. II, p. 845, New York, Wiley, 1951. Geijer, Per, and Gabrielson, O., *Am. Min.*, **51**: 534 (1966).

## ATTAPULGITE = Palygorskite

## AUERLITE = Variety of thorite high in $PO_4$

## AUGELITE
$Al_2 PO_4 (OH)_3$

CRYSTAL SYSTEM: Monoclinic
CLASS: 2/m
SPACE GROUP: C2/m
Z: 4
LATTICE CONSTANTS:
  $a$ = 13.124
  $b$ = 7.988
  $c$ = 5.066
  $\beta$ = 112°15'

3 STRONGEST DIFFRACTION LINES:
  3.338 (100)
  3.506 ( 90)
  4.00 ( 80)
OPTICAL CONSTANTS: (Na)
  $\alpha$ = 1.5736
  $\beta$ = 1.5759
  $\gamma$ = 1.5877
  (+)2V = 50°49'
HARDNESS: 4½–5
DENSITY: 2.696 (Meas.)
         2.704 (Calc.)
CLEAVAGE: {110} perfect
          {$\bar{2}$01} good
          {001} imperfect
          {$\bar{1}$01} imperfect
Fracture uneven. Brittle.
HABIT: Crystals thick tabular, also prismatic to acicular or as thin triangular plates. Also massive.
COLOR-LUSTER: Colorless, white, yellowish, pale aquamarine-blue, pale rose. Transparent to translucent. Vitreous; pearly on perfect cleavage surface. Streak white.
MODE OF OCCURRENCE: Occurs as transparent colorless crystals up to one inch in size at the andalusite ore body at White Mountain, Mono County, California; as large cleavable masses and small crystals associated with morinite, wardite, apatite, and montebrasite at the Hugo mine, Keystone, South Dakota; as small crystals associated with lazulite, albite, and quartz at the Smith mine near Newport, and as small crystals with whitlockite at the Palermo mine, near North Groton, New Hampshire; and at localities in Bolivia and Sweden.
BEST REF. IN ENGLISH: Palache, et al., "Dana's System of Mineralogy," 7th Ed., v. II, p. 871–872, New York, Wiley, 1951. Araki, T., Finney, J. J., and Zoltai, T., *Am. Min.*, **53**: 1096–1103 (1968).

## AUGITE
$(Ca, Na)(Mg, Fe, Al)(Si, Al)_2 O_6$
Pyroxene Group
var. Titanaugite

CRYSTAL SYSTEM: Monoclinic
CLASS: 2/m
SPACE GROUP: C2/c
Z: 4
LATTICE CONSTANTS:
  $a \simeq 9.8$
  $b \simeq 9.0$
  $c \simeq 5.25$
  $\beta \simeq 105°$
3 STRONGEST DIFFRACTION LINES:
  2.99 (100)
  1.62 (100)
  1.43 (100)
OPTICAL CONSTANTS:
  $\alpha$ = 1.671–1.735
  $\beta$ = 1.672–1.741
  $\gamma$ = 1.703–1.761
  (+)2V$_\gamma$ = 25°–60°
HARDNESS: 5½–6

DENSITY: 3.23-3.52 (Meas.)
CLEAVAGE: {110} good
{100} parting
{010} parting
Fracture uneven to conchoidal. Brittle.

HABIT: Crystals usually short and stout prismatic. Also massive, compact, and as disseminated grains and granular aggregates; rarely fibrous. Twinning on {100} common, simple and multiple; also multiple on {001}.

COLOR-LUSTER: Pale brown to dark brown or purplish brown, greenish, black. Translucent to nearly opaque. Vitreous to dull. Streak grayish green.

MODE OF OCCURRENCE: Occurs widespread as a characteristic component of basalts, dolerites and gabbros; less frequently in ultrabasic rocks, intermediate rocks, and high-grade metamorphic rocks. Typical occurrences are found in California, Oregon, Montana, Colorado, Arizona, South Dakota, Minnesota, New York, New Hampshire, Vermont, and Virginia. Also in Canada, Greenland, Scotland, Norway, Sweden, Germany, Italy, Sicily, Czechoslovakia, Finland, USSR, South West Africa, India, Japan, and elsewhere.

BEST REF. IN ENGLISH: Deer, Howie, and Zussman, "Rock Forming Minerals," v. 2, p. 109-142, New York, Wiley, 1963.

# AURICHALCITE
$(Zn, Cu)_5 (CO_3)_2 (OH)_6$

CRYSTAL SYSTEM: Orthorhombic
CLASS: 222
SPACE GROUP: $B22_1 2$
Z: 4
LATTICE CONSTANTS:
$a$ = 27.78
$b$ = 6.40
$c$ = 5.25
3 STRONGEST DIFFRACTION LINES:
6.78 (100)
2.61 ( 80)
3.68 ( 70)
OPTICAL CONSTANTS:
$\alpha$ = 1.654-1.661
$\beta$ = 1.740-1.749
$\gamma$ = 1.743-1.756
$(-)2V$ = very small
HARDNESS: 1-2
DENSITY: 3.96 (Meas.)
4.23 (Calc. for Zn:Cu = 2.65:1)
CLEAVAGE: {010} perfect
Very brittle.

HABIT: Crystals acicular or slender laths, elongated along $c$-axis. Commonly as delicate tufted aggregates or incrustations. Rarely granular, columnar, or laminated.

COLOR-LUSTER: Pale green, greenish blue, sky blue. Transparent. Silky to pearly.

MODE OF OCCURRENCE: Occurs widespread as a secondary mineral in oxidized zinc-copper deposits, and rarely in pegmatites. Found in the United States associated with smithsonite at the 79 Mine, Banner district, Gila County, and at Bisbee, Cochise County, Arizona; in Cottonwood Canyon, Salt Lake County, and at other places in Utah; in the Darwin District, Inyo County, California; at Kelly, Socorro County, New Mexico; and in pegmatite associated with smithsonite, hemimorphite, hydrozincite, columbite, and cassiterite, at the Tin Mountain mine, Custer County, South Dakota. It is also found abundantly as exceptional specimens at Mapimi, Durango, Mexico, and occurs in England, Scotland, Spain, France, Italy, Sardinia, Greece, Roumania, USSR, Japan, Zaire, and at Tsumeb, South West Africa.

BEST REF. IN ENGLISH: Palache, et al, "Dana's System of Mineralogy," 7th Ed., v. II, p. 249-250, New York, Wiley, 1951.

# AURICUPRIDE (Cuproauride) = Name proposed for naturally occurring intermetallic compound $Cu_3 Au$

# AURORITE
$(Mn, Ag, Ca)Mn_3O_7 \cdot 3H_2O$

CRYSTAL SYSTEM: Triclinic
CLASS: $\bar{1}$
3 STRONGEST DIFFRACTION LINES:
6.94 (100)
3.46 ( 70)
4.06 ( 50)
HARDNESS: <3
HABIT: As irregular masses and platy or scaly grains. Largest grains less than $8 \mu$ in size.

COLOR-LUSTER: In reflected light strongly birefringent and anisotropic, showing color changes from cream white to medium gray.

MODE OF OCCURRENCE: Occurs in veinlets distributed through black calcite, associated with argentian todorokite, cryptomelane, pyrolusite, and quartz at the Aurora mine (Treasure Hill), Hamilton, Nevada.

BEST REF. IN ENGLISH: Radtke, A. S., Taylor, C. M., and Hewett, D. F., Econ. Geol., 62: 186-206 (1967).

# AUROSMIRIDIUM = Mixture of osmiridium and gold(?)

# AUROSTIBITE
$AuSb_2$

CRYSTAL SYSTEM: Cubic
CLASS: m3
SPACE GROUP: Pa3
Z: 4
LATTICE CONSTANT:
$a$ = 6.66
3 STRONGEST DIFFRACTION LINES:
2.01 (100)
2.98 ( 75)
3.33 ( 70)
HARDNESS: ~3
DENSITY: 9.98 (Meas.)
9.91 (Calc.)

HABIT: As rounded anhedral grains less than 350 $\mu$ in diameter.

COLOR-LUSTER: Galena-like with a slight pinkish tinge in polished section. Opaque. Metallic. Bornite-like tarnish on minute grains in hand specimen.

MODE OF OCCURRENCE: Occurs in dolomitic carbonate and quartz with gold, freibergite, stibnite, jamesonite, chalcostibite, bournonite, arsenopyrite, pyrite, chalcopyrite, and sphalerite in gold ores from the Giant Yellowknife mine, Northwest Territories; and with gold, freibergite, galena, tennantite, bournonite, chalcopyrite, sphalerite, arsenopyrite, gersdorffite, and pyrite at the Chesterville mine, Larder Lake, Ontario, Canada.

BEST REF. IN ENGLISH: Graham, A. R., and Kaiman, S., *Am. Min.*, **37**: 461–469 (1952).


# AUSTENITE (artificial)
(Fe, C)


# AUSTINITE
$CaZnAsO_4OH$

CRYSTAL SYSTEM: Orthorhombic
CLASS: 2 2 2
SPACE GROUP: $P2_12_12_1$
Z: 4
LATTICE CONSTANT:
 $a$ = 7.43 kX
 $b$ = 9.00
 $c$ = 5.90
3 STRONGEST DIFFRACTION LINES:
 3.171 (100)
 2.801 (100)
 2.637 (100)
OPTICAL CONSTANTS:
 $\alpha$ = 1.759
 $\beta$ = 1.763
 $\gamma$ = 1.783
 (+)2V ~ 47°
 HARDNESS: 4–4½
 DENSITY: 4.13 (Meas.)
      4.37 (Calc.)
 CLEAVAGE: {011} good
Brittle.

HABIT: As minute prismatic or acicular crystals; as drusy crusts, and radial-fibrous nodules and incrustations.

COLOR-LUSTER: Colorless, white, pale yellow, bright green. Transparent to translucent. Subadamantine (crystals); silky when fibrous.

MODE OF OCCURRENCE:  Occurs associated with adamite on limonite at the Gold Hill mine, Tooele County, Utah. It also is found in the Ojuela mine, Mapimi, Durango, Mexico; as fibrous veinlets in the Lilli mine, near Lomitos, Bolivia; and at Tsumeb, South West Africa.

BEST REF. IN ENGLISH: Palache, et al,, "Dana's System of Mineralogy," 7th Ed., v. II, p. 809–810, New York, Wiley, 1951.    Radcliffe, Dennis, and Simmons, W. Bruce Jr., *Am. Min.*, **56**: 1359–1365 (1971).


# AUTUNITE
$Ca(UO_2)_2(PO_4)_2 \cdot 10-12H_2O$

CRYSTAL SYSTEM: Tetragonal
CLASS: 4/m 2/m 2/m
SPACE GROUP: I4/mmm
Z: 2
LATTICE CONSTANTS:
 $a$ = 6.989
 $c$ = 20.63
3 STRONGEST DIFFRACTION LINES:
 10.3 (100)
 4.96 ( 80)
 3.59 ( 70)
OPTICAL CONSTANTS:
 $\omega$ = 1.577–1.578
 $\epsilon$ = 1.553–1.555
 Usually biaxial (–), sometimes uniaxial (–) 2V variable ranging up to 53°.
 HARDNESS: 2–2½
 DENSITY: 3.05–3.2 (Meas.)
      3.14 (Calc. for 10½$H_2O$)
 CLEAVAGE: {001} perfect
      {100} indistinct
Flexible in thin plates.

HABIT: Crystals very thin to thick tabular on {001} with rectangular or octagonal shape. Commonly in fan-like aggregates, as crusts, and as foliated or scaly aggregates. Also as small disseminated grains and as earthy masses.

COLOR-LUSTER: Bright to pale sulfur yellow, lemon yellow, greenish yellow, pale green to dark green. Transparent to translucent. Vitreous, pearly on {001}. Earthy masses dull. Streak pale yellow. Fluoresces strong yellowish green in ultraviolet light.

MODE OF OCCURRENCE: Occurs as a secondary mineral formed by the alteration of uraninite or other uranium-containing minerals in pegmatites, granites, hydrothermal veins, and sedimentary deposits. Found in the United States as magnificent specimens composed of thick platy crystals more than an inch on edge from seams in granitic rock at the Daybreak mine, Mt. Spokane, Washington; also found widespread in small amounts in pegmatites in the Keystone and Custer districts, South Dakota; at numerous localities in Colorado, Utah, California, New Mexico, and other western states; and in pegmatites in North Carolina and in many pegmatites in the New England area. In Australia it is found in the Rum Jungle area, Northern Territory, and in thick masses at Mt. Painter in the Flinders Range. In Europe it occurs at several localities near Autun, Saône-et-Loire, France; at several mines in Saxony, Germany; at Sabugal and other places in Portugal; and in fine specimens at St. Austel and Redruth, Cornwall, England.

BEST REF. IN ENGLISH: Frondel, Clifford, *USGA Bull.* 1064, p. 160–170 (1958).    Takano, Yukio, *Am. Min.*, **46**: 812–822 (1961).


# AVELINOITE = Cyrilovite

## AVICENNITE
$Fe_2O_3 \cdot 7Tl_2O_3$

CRYSTAL SYSTEM: Cubic
CLASS: m3
SPACE GROUP: Ia3
LATTICE CONSTANT:
  $a = 9.12$
3 STRONGEST DIFFRACTION LINES:
  3.03 (100)
  1.859 (100)
  1.584 ( 90)
DENSITY: 9.574 (Calc.)
CLEAVAGE: Indistinct. Fracture uneven. Very brittle.
HABIT: Crystals less than 1 mm in size, somewhat resembling perovskite.
COLOR-LUSTER: Grayish black; metallic. Streak grayish black.
MODE OF OCCURRENCE: Occurs in a hematite-calcite vein cutting banded marmorized and silicified limestones near their contact with granite-gneisses of the Ketmenchinsk intrusive near the village of Dzhuzumli, Mt. Zirabulaksk region, Bukhara.
BEST REF. IN ENGLISH: Karpova, Kh. N., Kon'kova, E. A., Larkin, E. D., and Savel'ev, V. F., *Am. Min.*, **44**: 1324-1325 (1959).

## AVOGADRITE
$(K,Cs)BF_4$

CRYSTAL SYSTEM: Orthorhombic
CLASS: 2/m 2/m 2/m
SPACE GROUP: Pnma
Z: 4
LATTICE CONSTANTS:
  a = 8.664
  b = 5.480
  c = 7.028
3 STRONGEST DIFFRACTION LINES:
  3.41 (100)
  3.26 ( 80)
  3.06 ( 75)
OPTICAL CONSTANTS:
    $\alpha = 1.3239$
    $\beta = 1.3245$
    $\gamma = 1.3247$
$(-)2V$ = very large
HARDNESS: Not determined.
DENSITY: 3.00 (Calc.)
CLEAVAGE: Not determined.
HABIT: Crystals minute, tabular to platy {001}, sometimes elongated. As dense crusts.
COLOR-LUSTER: Colorless, white; yellowish to reddish when impure.

Taste bitter.
MODE OF OCCURRENCE: Occurs admixed with sassolite and other salts as a sublimation product around fumaroles on Mt. Vesuvius, Italy.
BEST REF. IN ENGLISH: Palache, et al., "Dana's System of Mineralogy," 7th Ed., v. II, p. 97-98, New York, Wiley, 1951.

## AWARUITE = Nickel-iron

## AXINITE GROUP
$(Ca, Mn, Fe, Mg)_3 Al_2 BSi_4 O_{15} (OH)$
  Ferroaxinite - Ca > 1.5; Fe > Mn
  Manganaxinite - Ca > 1.5; Mn > Fe
  Tinzenite - Ca < 1.5; Mn > Fe

CRYSTAL SYSTEM: Triclinic
CLASS: $\bar{1}$
SPACE GROUP: $P\bar{1}$
Z: 2
LATTICE CONSTANTS:
  a = 7.15    $\alpha = 88°04'$
  b = 9.16    $\beta = 81°36'$
  c = 8.96    $\gamma = 77°42'$
3 STRONGEST DIFFRACTION LINES:
  2.812 (100)
  3.16 ( 90)
  3.46 ( 80)
OPTICAL CONSTANTS:
    $\alpha = 1.674–1.693$
    $\beta = 1.681–1.701$
    $\gamma = 1.684–1.704$
  $(-)2V_\alpha = 63°-80°$
HARDNESS: 6½-7
DENSITY:    3.26-3.36 (Meas.)
        3.316 (Calc. Mn = Fe)
CLEAVAGE:  {100} good
          {001} poor
          {110} poor
          {011} poor
Fracture uneven to conchoidal. Brittle.
HABIT: Crystals usually tabular, wedge-shaped. Often in bladed aggregates. Also massive, lamellar; sometimes granular.
COLOR-LUSTER: Usually violet-brown; also colorless, yellowish, or pale violet to reddish. Transparent to translucent. Vitreous. Streak uncolored.
MODE OF OCCURRENCE: Occurs chiefly as a mineral of contact metamorphism and metasomatism, frequently associated with calcite, quartz, prehnite, zoisite, actinolite, hedenbergite, andradite, and other minerals. Found at numerous localities in California, especially as large gem-quality violet-colored crystals near Coarse Gold, Madera County; as large-bladed masses near Luning, Mineral County, Nevada; at Bethlehem, Northampton County, Pennsylvania; and as fine yellow micro crystals at Franklin, Sussex County, New Jersey. It also occurs as fine crystals at localities in Baja California; at Bourg d'Oisans, Isère, France; in Cornwall, England; and in Switzerland (tinzenite), Norway, Germany, Finland, USSR, Tasmania, and at the Toroku mine, Miyazaki Prefecture, Japan.
BEST REF. IN ENGLISH: Deer, Howie, and Zussman, "Rock Forming Minerals," v. 1, p. 320-327, New York, Wiley, 1962. Sanero, E., and Gottardi, G., *Am. Min.*, **53**: 1407-1411 (1968).

## AZOPROITE
$(Mg, Fe^{2+})_2 (Fe^{3+}, Ti, Mg)BO_5$
Ludwigite Group

CRYSTAL SYSTEM: Orthorhombic
CLASS: 2/m 2/m 2/m
SPACE GROUP: Pbam
LATTICE CONSTANTS:
$a = 9.26$
$b = 12.25$
$c = 3.01$
3 STRONGEST DIFFRACTION LINES:
2.57 (100)
5.07 ( 80)
2.16 ( 60)
OPTICAL CONSTANTS:
$\alpha = 1.799$
$\beta = 1.822$
$\gamma = 1.855$
$(+)2V > 70°$
HARDNESS: ~ 5½
DENSITY: 3.63 ± 0.02 (Meas.)
CLEAVAGE: {010} good
{001} less good
Fracture conchoidal. Brittle.

HABIT: Crystals prismatic, 1–20 mm long and 0.1–5 mm wide.

COLOR-LUSTER: Black. Adamantine. Transparent in section.

MODE OF OCCURRENCE: Occurs in association with calcite, ludwigite, clinohumite, baddeleyite, tazheranite, perovskite, and geikielite, in magnesian skarns in the contact aureole of the Tazheran alkalic massif, west of Lake Baikal, Siberia, USSR.

BEST REF. IN ENGLISH: Konev, A. A., Lebedeva, V. S., Kashaev, A. A., and Ushchapovskaya, Z. F., *Am. Min.*, **56**: 360 (1971).


# AZURITE
$Cu_3(CO_3)_2(OH)_2$

CRYSTAL SYSTEM: Monoclinic
CLASS: 2/m
SPACE GROUP: $P2_1/a$
Z: 2
LATTICE CONSTANTS:
$a = 10.35$
$b = 5.85$
$c = 5.00$
$\beta = 92°20'$
3 STRONGEST DIFFRACTION LINES:
3.516 (100)
2.224 ( 70)
5.15 ( 55)
OPTICAL CONSTANTS: (Na)
$\alpha = 1.730$
$\beta = 1.758$
$\gamma = 1.836$
$(+)2V = 67°$
HARDNESS: 3½–4
DENSITY: 3.773 ± 0.003 (Meas.)
3.78 (Calc.)
CLEAVAGE: {011} slightly imperfect
{100} fair
{110} in traces
Fracture conchoidal. Brittle.

HABIT: Crystals widely varied in habit and commonly highly modified. Usually tabular or short prismatic, also rhombohedral or equant. Also massive; as nodular concretions; stalactitic; earthy; and as films and stains. Twinning rare.

COLOR-LUSTER: Light blue to very dark blue; usually azure blue. Transparent to nearly opaque. Vitreous; massive varieties sometimes dull. Streak blue, lighter than color.

MODE OF OCCURRENCE: Occurs widespread as a secondary mineral in the oxidation zone of copper deposits, closely associated with malachite and other secondary minerals. Found widespread in the mining districts of the western United States, especially at Bisbee, Cochise County, and at Morenci, Greenlee County, Arizona, where it occurs as splendid crystal groups; and at many places in New Mexico, Nevada, Utah, and California. It also occurs as fine specimens at the San Carlos mine, Mazapil, Zacatecas, and at other localities in Mexico; at Chessy, France; in Sardinia, Italy; at Laurium, Greece; in the Altai and Ural Mountains, USSR; at Broken Hill and other places in Australia; and as magnificent crystals and crystal groups, sometimes completely or partly replaced by malachite, at Tsumeb, South West Africa.

BEST REF. IN ENGLISH: Palache, et al., "Dana's System of Mineralogy," 7th Ed., v. II, p. 264–269, New York, Wiley, 1951.

# B

**BABABUDANITE** = Variety of Amphibole

## BABEFPHITE
$BaBe(PO_4)(O, F)$

CRYSTAL SYSTEM: Tetragonal
CLASS: 4/m 2/m 2/m
SPACE GROUP: I4/amd
Z: 1
LATTICE CONSTANTS:
$a =$ 4.89
$c =$ 16.74
3 STRONGEST DIFFRACTION LINES:
3.190 (100)
2.163 (100)
1.516 (100)
OPTICAL CONSTANTS:
$\omega =$ 1.629
$\epsilon =$ 1.632
(+)
HARDNESS: Microhardness approx. 140–200 kg/mm²
DENSITY: 4.31 (Meas.)
4.44 (Calc.)
CLEAVAGE: None observed. Very brittle.
HABIT: Grains of isometric and rarely of flattened tabular form up to 1 mm X 1.5 mm in size.
COLOR-LUSTER: White; vitreous to greasy.
MODE OF OCCURRENCE: Found in heavy concentrates along with zircon, ilmenorutile, fluorite, phenakite, and scheelite in "a rare-metal fluorite deposit in Siberia" genetically associated with subalkalic syenites, in the eluvial deposits located directly above the ore body.
BEST REF. IN ENGLISH: Nazarova, A. S., Kuznetsova, N. N., and Shaskin, D. P., *Am. Min.*, **51**: 1547 (1966).

## BABINGTONITE
$Ca_2 Fe^{2+} Fe^{3+} Si_5 O_{14} OH$

CRYSTAL SYSTEM: Triclinic
CLASS: $\bar{1}$
SPACE GROUP: $P\bar{1}$
Z: 2

LATTICE CONSTANTS:
$a =$ 7.36    $\alpha = 91°31'$
$b =$ 11.52    $\beta = 93°51'$
$c =$ 6.58    $\gamma = 104°4'$
3 STRONGEST DIFFRACTION LINES:
2.75 (100)
2.87 ( 80)
3.12 ( 70)
OPTICAL CONSTANTS:
$\alpha =$ 1.720
$\beta =$ 1.731
$\gamma =$ 1.753
(+)2V = 76°
HARDNESS: 5½–6
DENSITY: 3.36 (Meas.)
3.26 (Calc.)
CLEAVAGE: {001} perfect
{1$\bar{1}$0} imperfect
Fracture uneven. Brittle.
HABIT: Crystals short prismatic or platy.
COLOR-LUSTER: Greenish black to brownish black. Vitreous.
MODE OF OCCURRENCE: Occurs chiefly in cavities and coating fractures in granite, trap rock, and gneiss. Found as fine crystals at the Lane trap rock quarry near Westfield, Hampden County, and at Blueberry Mountain, Middlesex County, Massachusetts; in Passaic County, New Jersey; in Devon, England; at Arendal, Norway; at Baveno, Piedmont, Italy; and near Herborn, Hessen, Germany.
BEST REF. IN ENGLISH: Winchell and Winchell, "Elements of Optical Mineralogy," 4th Ed., Pt. 2, p. 462, New York, Wiley, 1951.

**BÄCKSTRÖMITE** = Mixture of feitknechtite and hausmannite

## BADDELEYITE
$ZrO_2$

CRYSTAL SYSTEM: Monoclinic
CLASS: 2/m
SPACE GROUP: $P2_1/c$
Z: 4

LATTICE CONSTANTS:
$a = 5.21$
$b = 5.26$
$c = 5.375$
$\beta = 99°15'$

3 STRONGEST DIFFRACTION LINES:
3.16 (100)
2.826 ( 80)
2.611 ( 60)

OPTICAL CONSTANTS:
$\alpha = 2.13$
$\beta = 2.19$
$\gamma = 2.20$
$(-)2V = 30°$

HARDNESS: 6½

DENSITY: 5.739 ± 0.005 (Meas.)
5.825 (Calc.)

CLEAVAGE: {001} nearly perfect
{010} imperfect
{110} imperfect
Fracture subconchoidal to uneven. Brittle.

HABIT: Crystals short to long prismatic, also tabular and somewhat elongated; rarely equant. Also botryoidal masses, radially fibrous with concentric banding. Untwinned crystals rare. Polysynthetic twinning and twinning on {100} and {110} common; on {201} rare.

COLOR-LUSTER: Colorless to yellow, green, reddish or greenish brown, brown to nearly black. Greasy to vitreous. Dark crystals submetallic. Streak white to brownish white.

MODE OF OCCURRENCE: Occurs in a corundum-syenite near Bozeman, Montana; in situ and as rolled pebbles at several localities in Minas Geraes, Mato Grosso, and Sao Paulo, Brazil; as a minor constituent of carbonatite rocks at Phalaborwa in the Eastern Transvaal; on Mount Somme, Italy; and as a detrital material in placer concentrates from Zaire, and in the gem gravels in Ceylon.

BEST REF. IN ENGLISH: Palache, et al., "Dana's System of Mineralogy," 7th Ed., v. I, p. 608–610, New York, Wiley, 1944. *Am. Min.*, **40**: 275–282 (1955).

**BADENITE** = Variety of smaltite or safflorite

# BAFERTISITE
$BaFe_2TiSi_2O_9$

CRYSTAL SYSTEM: Monoclinic
CLASS: 2/m
SPACE GROUP: $P2_1/m$
Z: 2
LATTICE CONSTANTS:
$a = 10.98$
$b = 6.80$
$c = 5.36$
$\beta = 94°$

3 STRONGEST DIFFRACTION LINES:
2.65 (100)
2.11 ( 40)
1.72 ( 40)

OPTICAL CONSTANTS:
$\alpha = 1.808$
$\gamma = 1.860$
$(-)2V = 54°$
HARDNESS: ~ 5
DENSITY: 3.96–4.25 (Meas.)
3.8 (Calc.)
CLEAVAGE: One distinct.
One poor.
HABIT: Aggregates of acicular crystals 1.5 cm long.
COLOR-LUSTER: Bright red, yellowish red to light brown.

MODE OF OCCURRENCE: Occurs associated with aegirine, fluorite, barite, and bastnaesite in hydrothermal veins in the Baiyun-Obo (Bayoune-Obo) deposit, Inner Mongolia.

BEST REF. IN ENGLISH: Ch'i-Jui, Peng, *Am. Min.*, **45**: 754 (1960).

# BAKERITE
$Ca_4B_4(BO_4)(SiO_4)_3(OH)_3 \cdot H_2O$

CRYSTAL SYSTEM: Monoclinic
CLASS: 2/m
SPACE GROUP: $P2_1/c$
Z: 1
LATTICE CONSTANTS:
$a = 4.82Å$
$b = 7.60$
$c = 9.60$
$\beta = 90°12'$

3 STRONGEST DIFFRACTION LINES:
3.11 (100)
2.85 ( 60)
2.236 ( 60)

OPTICAL CONSTANTS:
$\alpha = 1.624$
$\beta = 1.635$
$\gamma = 1.654$
$(-)2V = 85°$
HARDNESS: 4½
DENSITY: 2.88 (Meas.)
CLEAVAGE: Not determined.

HABIT: Crystals stout prismatic with {001} and {$\bar{1}$11} only; also thin tabular with {001} dominant and {$\bar{1}$11} and {012} as narrow modifying faces. Crystals do not exceed 0.2 mm in longest dimension. Also as nodules and veins, dense and microcrystalline.

COLOR-LUSTER: Crystals transparent, colorless; vitreous. Massive material white.

MODE OF OCCURRENCE: Crystals of bakerite occur as a coating on the surface of massive bakerite or of a mixture of pellets of bakerite in a clay-like material at the Sterling Borax Mine, Tick Canyon, Los Angeles County, California. Also found as veinlets and lenses in an altered volcanic rock in Upper Baker Canyon, Death Valley, Inyo County, California, and associated with howlite near Yermo, San Bernadino County, California.

BEST REF. IN ENGLISH: Murdoch, Joseph, *Am. Min.*, **47**: 919–923 (1962).

# BALAVINSKITE (Species status uncertain)
$Sr_2B_6O_{11} \cdot 4H_2O$

**BALDAUFITE** = Hureaulite

## BALKANITE
$Cu_9 Ag_5 HgS_9$

CRYSTAL SYSTEM: Orthorhombic
CLASS: 222, mm2, or 2/m 2/m 2/m
SPACE GROUP: P222, Pmm2, or Pmmm
Z: 1
LATTICE CONSTANTS:
$a = 10.62$
$b = 9.42$
$c = 3.92$
3 STRONGEST DIFFRACTION LINES:
2.98 (100)
2.55 (100)
1.955 ( 90)
OPTICAL CONSTANTS: Strongly anisotropic; white-gray in polished sections.
HARDNESS: 3½
DENSITY: 6.318 (Meas. Synthetic)
CLEAVAGE: None
HABIT: Crystals prismatic, rod-like, elongated along c-axis, almost isometric in cross section, striated parallel to elongation, from 0.01 to 0.2 mm. Usually as grains up to 2–3 mm. Twinning polysynthetic.
COLOR-LUSTER: Steel gray. Opaque. Metallic.
MODE OF OCCURRENCE: The mineral occurs in association with bornite, chalcocite, djurleite, digenite, tennantite, stromeyerite, mercurian silver, and cinnabar, in high-grade copper ores at the Sedmochislenitsi mine, Vratsa district, in the western part of the Stara Planina, Bulgaria.
BEST REF. IN ENGLISH: Atanassov, Vasil A., and Kirov, Georgi N., *Am. Min.*, **58**: 11–15 (1973).

## BAMBOLLAITE
$Cu(Se, Te)_2$

CRYSTAL SYSTEM: Tetragonal
CLASS: 4/m
SPACE GROUP: $P4_2/n$
LATTICE CONSTANTS:
$a = 5.46$
$c = 5.63$
3 STRONGEST DIFFRACTION LINES:
3.19 (100)
1.961 ( 70)
1.653 ( 50)
HARDNESS: 3
DENSITY: 5.64
CLEAVAGE: Not determined.
HABIT: Massive; microcrystalline.
COLOR-LUSTER: Slate gray. Opaque. Metallic.
MODE OF OCCURRENCE: Occurs in association with klockmannite, native selenium, native tellurium, chalcomenite, tellurite, paratellurite, illite, quartz, and calcite at Mina Moctezuma, Sonora, Mexico.
BEST REF. IN ENGLISH: Harris, D. C., and Nuffield, E. W., *Can. Min.*, **8**: 397 (1965).

## BANALSITE
$Na_2 BaAl_4 Si_4 O_{16}$

CRYSTAL SYSTEM: Orthorhombic
CLASS: mm2 or 2/m 2/m 2/m
SPACE GROUP: Ibc2 or Ibcm
Z: 4
LATTICE CONSTANTS:
$a = 8.50$
$b = 9.97$
$c = 16.73$
3 STRONGEST DIFFRACTION LINES:
3.50 (100)
3.19 (100)
2.07 (100)
OPTICAL CONSTANTS:
$\alpha = 1.5695$
$\beta = 1.571$
$\gamma = 1.5775$
$(+)2V = 41°$
HARDNESS: 6
DENSITY: 3.065 (Meas.)
CLEAVAGE: {110} distinct
{001} distinct
HABIT: Massive.
COLOR-LUSTER: White. Translucent. Vitreous; pearly on cleavage.
MODE OF OCCURRENCE: Occurs in a vein in manganese ore at Rhiw, Lleyn Peninsula, Caernarvonshire, Wales.
BEST REF. IN ENGLISH: Winchell and Winchell, "Elements of Optical Mineralogy," 4th Ed., Pt. 2, p. 260, New York, Wiley, 1951.

## BANDYLITE
$CuB(OH)_4 Cl$

CRYSTAL SYSTEM: Tetragonal
CLASS: 4/m
SPACE GROUP: P4/n
Z: 2
LATTICE CONSTANTS:
$a = 6.13$ kX
$c = 5.54$
3 STRONGEST DIFFRACTION LINES:
5.59 (100)
3.08 ( 80)
2.54 ( 80)
OPTICAL CONSTANTS:
$\omega = 1.691$
$\epsilon = 1.641$ (Na)
(–)
HARDNESS: 2½
DENSITY: 2.810 (Meas.)
2.81 (Calc.)
CLEAVAGE: {001} perfect.
Flexible.
HABIT: Crystals tabular {001} or equant. As small flattened oval or rounded aggregates randomly scattered on matrix.
COLOR-LUSTER: Deep blue with greenish tint. Translucent. Vitreous; pearly on cleavage. Streak pale blue.

MODE OF OCCURRENCE: Occurs as a secondary mineral associated with atacamite and eriochalcite at Mina Quetena, near Calama, Chile.

BEST REF. IN ENGLISH: Palache, et al., "Dana's System of Mineralogy," 7th Ed., v. II, p. 373–374, New York, Wiley, 1951. *Am. Min.*, **23**: 85 (1938).

# BANNISTERITE
$(K, Na, Ca)Mn_{10}Al_2Si_{15}O_{44} \cdot 10H_2O$
Dimorph of ganophyllite

CRYSTAL SYSTEM: Monoclinic
CLASS: 2/m
SPACE GROUP: A2/a
Z: 20
LATTICE CONSTANTS:
 $a = 22.20$
 $b = 16.32$
 $c = 24.70$
 $\beta = 94°20'$
3 STRONGEST DIFFRACTION LINES:
 12.33 (100)
 3.44 ( 20)
 4.10 ( 16)
OPTICAL CONSTANTS:
 $\alpha = 1.544$
 $\beta = 1.586$
 $\gamma = 1.588$
 2V = small
HARDNESS: 4
DENSITY: 2.92 (Meas.)
     2.936 (Calc.)
CLEAVAGE: {001} perfect
HABIT: As aggregates of anhedral plates up to 5 cm across cleavage surfaces.
COLOR-LUSTER: Dark brown. Translucent.
MODE OF OCCURRENCE: Occurs in association with greenish black manganoan amphibole, calcite, and barite at Franklin, Sussex County, New Jersey. Also found at the Benallt mine, Caernarvonshire, Wales, and at the Ananai mine, Kochi, Japan.
BEST REF. IN ENGLISH: Smith, Marie Lindberg, and Frondel, Clifford, *Min. Mag.*, **36**: 893–913 (1968).

# BAOTITE
$Ba_4(Ti, Nb)_8Si_4O_{28}Cl$

CRYSTAL SYSTEM: Tetragonal
CLASS: 4/m
SPACE GROUP: $I4_1/a$
Z: 4
LATTICE CONSTANTS:
 $a = 20.02$
 $c = 6.006$
3 STRONGEST DIFFRACTION LINES:
 1.3346 (100)
 1.4221 ( 25)
 3.52 ( 10)

OPTICAL CONSTANTS:
 $\omega = 1.944$
 $\epsilon > 2.00$
 (+)
HARDNESS: 6
DENSITY: 4.71 ± 0.01 (Meas.)
CLEAVAGE: {110} two directions at 90°
Hackley fracture. Brittle.
HABIT: Grains and partly faced crystals 0.8–4 mm in size (Montana); isometric masses, often with rectangular outline, up to 10 cm in size (China).
COLOR-LUSTER: Light brown to black; vitreous.
MODE OF OCCURRENCE: Found as small brilliant crystals associated with aeschynite in a carbonatic vein in the upper reaches of Sheep Creek, Ravalli County, Montana. Discovered near the town of Baotou, Inner Mongolia, Chinese People's Republic.
BEST REF. IN ENGLISH: Heinrich, E. Wm., Boyer, Wm. H., and Crowley, F. A., *Am. Min.*, 47:987–993 (1962).

# BARARITE
$(NH_4)_2SiF_6$
Dimorphous with Cryptohalite

CRYSTAL SYSTEM: Hexagonal
CLASS: $\bar{3}$ 2/m
SPACE GROUP: $R\bar{3}m$
Z: 1
LATTICE CONSTANTS:
 $a = 5.76$
 $c = 4.77$
OPTICAL CONSTANTS:
 $\omega = 1.406$
 $\epsilon = 1.391$
 (−)
HARDNESS: ~ 2½
DENSITY: 2.152 (Meas.)
     2.144 (Calc.)
CLEAVAGE: {0001} perfect
HABIT: As crusts. Crystals tabular {0001}, sometimes twinned with twin-plane inclined to {0001}. Also mammillary or arborescent.
COLOR-LUSTER: White, vitreous.

Taste saline.
MODE OF OCCURRENCE: Occurs as a sublimate over a burning coal seam at Barari, Jhari coal field, India, associated with cryptohalite and sulfur. It is also found as a sublimation product at Mt. Vesuvius, Italy.
BEST REF. IN ENGLISH: Palache, et al., "Dana's System of Mineralogy," 7th Ed., v. II, p. 106–107, New York, Wiley, 1951.

# BARBERTONITE
$Mg_6Cr_2CO_3(OH)_{16} \cdot 4H_2O$
Dimorphous with Stichtite

CRYSTAL SYSTEM: Hexagonal
Z: 1

LATTICE CONSTANTS:
$a = 6.17$
$c = 15.52$
OPTICAL CONSTANTS:
$\omega = 1.557$
$\epsilon = 1.529$
(–)
HARDNESS: 1½–2
DENSITY: 2.10 (Meas.)
2.11 (Calc.)
CLEAVAGE: {0001} perfect
Laminae flexible, inelastic. Greasy feel.

HABIT: Massive; in matted or twisted masses of fibers or plates flattened {0001}; also as cross-fiber veinlets.

COLOR-LUSTER: Deep lilac to rose pink. Translucent to transparent. Pearly to waxy. Streak white to pale lilac.

MODE OF OCCURRENCE: Occurs in serpentine rock associated with stichtite and chromite. Found at Cunnigsburgh, Scotland; Dundas, Tasmania; and Kaapsche Hoop, Barberton district, Transvaal, South Africa.

BEST REF. IN ENGLISH: Palache, et al., "Dana's System of Mineralogy," 7th Ed., v. I, p. 659, New York, Wiley, 1944.

## BARBIERITE = Variety of microcline

## BARBOSALITE
$Fe^{2+}Fe^{3+}_2(PO_4)_2(OH)_2$

CRYSTAL SYSTEM: Monoclinic
CLASS: 2/m
SPACE GROUP: $P2_1/c$
Z: 2
LATTICE CONSTANTS:
$a = 7.25$
$b = 7.46$
$c = 7.49$
$\beta = 120°15'$
3 STRONGEST DIFFRACTION LINES:
3.361 (100)
3.313 ( 80)
4.84 ( 60)
OPTICAL CONSTANTS:
$\alpha = 1.77$
$\gamma = 1.835$
HARDNESS: 5½–6
DENSITY: 3.60 (Meas.)
CLEAVAGE: Not observed.
HABIT: Crystals prismatic, short, small. Massive; also as powdery crusts.
COLOR-LUSTER: Greenish blue to almost black. Vitreous to dull and earthy.
MODE OF OCCURRENCE: Occurs in pegmatite in masses up to several hundred pounds in weight associated with pyrite, and often containing small vugs lined with small prismatic crystals of barbosalite, large crystals of metastrengite, strengite, and other phosphates at the Bull Moose mine, Custer County, South Dakota. Originally found as a very fine grained layer between heterosite and porous triphylite at the Sapucaia pegmatite mine, Minas

Geraes, Brazil. Subsequently identified from several additional pegmatites in the Black Hills, South Dakota and other pegmatite districts in many parts of the world.

BEST REF. IN ENGLISH: Lindberg, M. L., and Pecora, W. T., Am. Min., 40: 952–966 (1955).

## BARIANDITE
$V_2O_4 \cdot 4V_2O_5 \cdot 12H_2O$

CRYSTAL SYSTEM: Monoclinic
CLASS: m or 2/m
SPACE GROUP: Cc or C 2/c
Z: 2
LATTICE CONSTANTS:
$a = 11.70$
$b = 3.63$
$c = 29.06$
$\beta = 101°30'$
3 STRONGEST DIFFRACTION LINES:
14.20 (100)
3.48 ( 60)
3.43 ( 60)
OPTICAL CONSTANTS: Pleochroism greenish brown on [010], bottle-green perpendicular to [010]. Extinction parallel, ns above 1.85. In reflected light, shows green internal reflections, strongly pleochroic in brownish gray.
HARDNESS: Not determined.
DENSITY: 2.7 (Meas.)
CLEAVAGE: {001} perfect
HABIT: As fibers up to 3 mm long, platy on {001}.
COLOR-LUSTER: Black. Nearly opaque.
MODE OF OCCURRENCE: The mineral occurs associated with duttonite and lenoblite in the oxidation zone of the uranium deposit at Mounana, Gabon.
BEST REF. IN ENGLISH: Cesbron, Fabien, and Vachey, Helene, Am. Min., 57: 1555 (1972).

## BARITE
$BaSO_4$

CRYSTAL SYSTEM: Orthorhombic
CLASS: 2/m 2/m 2/m
SPACE GROUP: Pnma
Z: 4
LATTICE CONSTANTS:
$a = 8.85$ kX
$b = 5.43$
$c = 7.13$
3 STRONGEST DIFFRACTION LINES:
3.442 (100)
3.101 ( 97)
2.120 ( 80)
OPTICAL CONSTANTS:
$\alpha = 1.6362$
$\beta = 1.6373$ (Na)
$\gamma = 1.6482$
(+)2V = 37°
HARDNESS: 3–3½
DENSITY: 4.50 (Meas.)
4.480 (Calc.)

CLEAVAGE: {001} perfect
{210} distinct
{010} imperfect

Fracture uneven. Brittle.

HABIT: Crystals commonly thin to thick tabular, often large; also short to long prismatic; equant. As aggregates of crystals, as rosettes (desert roses) with sand inclusions. Also massive, cryptocrystalline to granular; lamellar; stalactitic; concretionary; columnar to fibrous; and earthy.

COLOR-LUSTER: Colorless, white to grayish, yellowish to brown, bluish, greenish, reddish. Transparent to subtranslucent. Vitreous to resinous; sometimes pearly. Sometimes fluorescent and phosphorescent under ultraviolet light.

MODE OF OCCURRENCE: Occurs widespread chiefly in medium- to low-temperature hydrothermal vein deposits; in sedimentary rocks as veins, lenses, cavity fillings, and in concretions; in residual clay deposits; in cavities in igneous rocks; and as a hot springs deposit. Found as exceptional amber-brown crystals in cavities in large concretions in Fall River, Meade, and Pennington Counties, South Dakota; as fine blue tabular crystals near Stoneham, Weld County, and at other places in Colorado; as "desert roses" near Norman, Oklahoma; at many localities in Missouri; as attractive crystal aggregates at Palos Verdes, Los Angeles County, California; as fine white to blue crystal groups in the Rosiclare district, Illinois; and as excellent crystals at Cheshire, Connecticut. It occurs in England as outstanding crystals and crystal groups in Cumberland, Northumberland, and Westmoreland; and as fine stalactitic masses in Derbyshire. Fine specimens also are found in Germany, Norway, Roumania, Czechoslovakia, and France.

BEST REF. IN ENGLISH: Palache, et al., "Dana's System of Mineralogy," 7th Ed., v. II, p. 408–415, New York, Wiley, 1951.

## BARKEVIKITE

$(Na, K)Ca_2(Fe, Mg, Mn)_5(Si, Al)_8O_{22}(OH, F)_2$
Amphibole Group

CRYSTAL SYSTEM: Monoclinic
CLASS: 2/m
SPACE GROUP: C2/m
Z: 2
LATTICE CONSTANTS:
$a \simeq 9.9$
$b = 18.34$
$c = 5.34$
$\beta \simeq 106°$
OPTICAL CONSTANTS:
$\alpha = 1.685-1.691$
$\beta = 1.696-1.700$
$\gamma = 1.701-1.707$
$(-)2V_\alpha \simeq 40°-50°$
HARDNESS: 5–6
DENSITY: 3.35–3.44 (Meas.)
CLEAVAGE: {110} perfect
{001} parting
{100} parting

Fracture uneven. Brittle.

HABIT: Crystals long prismatic, commonly well developed and showing terminal faces. Also as prismatic aggregates. Twinning on {100} simple.

COLOR-LUSTER: Black. Nearly opaque. Vitreous.

MODE OF OCCURRENCE: Occurs in alkaline rocks such as trachytes, phonolites, and nepheline and sodalite syenites. Found in Fresno, Los Angeles, and San Benito counties, California; in Custer County, Colorado; and in camptonite dikes in Lawrence County, South Dakota. Also found in the Sherbrooke district, Quebec, Canada; in Greenland; as large crystals at Lugar, Ayrshire, Scotland; in the Langesundfjord district, southern Norway; on Alnö Island, Sweden; and elsewhere.

BEST REF. IN ENGLISH: Deer, Howie, and Zussman, "Rock Forming Minerals," v. 2, p. 328–332, New York, Wiley, 1963.

## BARNESITE

$Na_2V_6O_{16} \cdot 3H_2O$

CRYSTAL SYSTEM: Monoclinic
CLASS: 2/m
SPACE GROUP: P2/m
Z: 1
LATTICE CONSTANTS:
$a = 12.17$
$b = 3.602$
$c = 7.78$
$\beta = 95°2'$
3 STRONGEST DIFFRACTION LINES:
7.90 (100)
3.12 ( 70)
3.45 ( 35)
OPTICAL CONSTANTS:
$\alpha = 1.797$
$\beta > 2.0$
$\gamma > 2.0$
(−)
HARDNESS: Not determined.
DENSITY: 3.15 (Meas.)
3.21 (Calc.)
CLEAVAGE: Not determined.
HABIT: Microscopic bladed to fibrous crystals in loose aggregates or radiating clusters. In continuous coatings, velvety with botryoidal appearance.

COLOR-LUSTER: Brilliant dark red; adamantine. On exposure slightly brownish red.

MODE OF OCCURRENCE: Found in small cavities and fractures and as interstitial filling in sandstone in the oxidized zone of a vanadiferous uranium deposit in the Cactus Rat group of mines, 15 miles east southeast of Thompson, Grand County, Utah.

BEST REF. IN ENGLISH: Weeks, Alice D., Ross, Daphne R., and Marvin, Richard F., *Am. Min.*, 48; 1187–1195 (1963).

## BARRANDITE = Aluminian strengite

## BARRINGERITE

$(Fe, Ni)_2P$

CRYSTAL SYSTEM: Hexagonal

CLASS: $\bar{6}m2$
SPACE GROUP: $P\bar{6}2m$
LATTICE CONSTANTS:
  $a = 5.87$Å
  $c = 3.44$
HARDNESS: Not determined.
DENSITY: 6.92 (Calc.)
CLEAVAGE: Not determined.
HABIT: Grains less than 1 $\mu$m in diameter; as bands 10–15 $\mu$m wide and several hundred microns long.
COLOR-LUSTER: White—similar to kamacite, bluish compared to schreibersite.
MODE OF OCCURRENCE: Occurs along the contacts between schreibersite and troilite in the Ollague pallasite meteorite.
BEST REF. IN ENGLISH: Buseck, P. R., *Science*, **165**: 169–171 (1969).

## BARRINGTONITE
$MgCO_3 \cdot 2H_2O$ (?)

CRYSTAL SYSTEM: Triclinic
Z: 4
LATTICE CONSTANTS:
  $a = 9.155$     $\alpha = 94°00'$
  $b = 6.202$     $\beta = 95°32'$
  $c = 6.092$     $\gamma = 108°72'$
3 STRONGEST DIFFRACTION LINES:
  8.682 (100)
  3.093 (100)
  2.936 (100)
OPTICAL CONSTANTS:
  $\alpha = 1.458$
  $\beta = 1.473$
  $\gamma = 1.501$
  $(+)2V = 73°44'$ (Calc.)
HARDNESS: Not determined.
DENSITY: 2.825 (Calc.)
CLEAVAGE: {001} distinct
          {100} distinct
          {010} distinct
HABIT: Minute nodular with radiating fibrous or acicular structure.
COLOR-LUSTER: Colorless.
MODE OF OCCURRENCE: Occurs in association with nesquehonite as encrustations on the surface of Tertiary olivine basalt under Rainbow Falls, Sempill Creek, Barrington Tops, New South Wales, Australia.
BEST REF. IN ENGLISH: Nashar, Beryl, *Min. Mag.*, **34**: 370–372 (1965).

## BARSANOVITE = Eudialyte

## BARYLITE
$BaBe_2Si_2O_7$

CRYSTAL SYSTEM: Orthorhombic
CLASS: mm2
Z: 4

SPACE GROUP: $Pn2_1a$
LATTICE CONSTANTS:
  $a = 9.8$Å
  $b = 11.65$
  $c = 4.63$
3 STRONGEST DIFFRACTION LINES:
  3.38 (100)
  2.92 (100)
  3.02 ( 80)
OPTICAL CONSTANTS:
  $\alpha = 1.69$
  $\beta = 1.69$–1.70
  $\gamma = 1.70$
  $(-)2V = 40°27'$
HARDNESS: 7
DENSITY: 4.046 ± 0.020
          4.038 (Calc.)
CLEAVAGE: {001} perfect
          {100} perfect
Brittle.
HABIT: Crystals thin tabular; also prismatic with square cross section, sometimes massive, as veinlets or disseminated grains.
COLOR-LUSTER: Colorless, white, bluish. Transparent to translucent. Vitreous.
MODE OF OCCURRENCE: Occurs as lamellar segregations in association with hedyphane and willemite in calcite veins at Franklin, New Jersey; as veinlets and disseminated grains in fenitic gneisses near Seal Lake, Labrador; as crystals in association with hedyphane, barite, garnet, and calcite in hematite, at Långban, Sweden; and in calcite veins that cut fenites in the Vishnevye Mountains, USSR. Recently found as superb crystals up to two inches long in an amazonite pocket in Park County, Colorado.
BEST REF. IN ENGLISH: Vlasov, K. A., "Mineralogy of Rare Elements," v. II, p. 87–89, Israel Program for Scientific Translations, 1966. *Am. Min.*, **41**: 512–513 (1956).

## BARYSILITE
$MnPb_8(Si_2O_7)_3$

CRYSTAL SYSTEM: Hexagonal
CLASS: $\bar{3}\,2/m$
SPACE GROUP: $R\bar{3}c$
Z: 6
LATTICE CONSTANTS:
  $a = 9.82$
  $c = 38.3$
3 STRONGEST DIFFRACTION LINES:
  3.198 (100)
  3.307 ( 80)
  2.960 ( 80)
HARDNESS: 3
DENSITY: 6.55–6.706 (Meas.)
          6.70 (Calc.)
CLEAVAGE: {0001} distinct
HABIT: Massive; curved lamellar structure.
COLOR-LUSTER: White, pinkish white; tarnishes on exposure.

MODE OF OCCURRENCE: Occurs as thin films or veinlets in zinc ore, associated with willemite, garnet, hardystonite, and axinite, at Franklin, Sussex County, New Jersey. Also at Långban and at the Harstig mine at Pajsberg, Vermland, Sweden.

BEST REF. IN ENGLISH: Glasser, F. P., *Am. Min.*, **49**: 1485–1488 (1964).

## BARYTOANGLESITE = Variety of anglesite

## BARYTOCALCITE
$BaCa(CO_3)_2$
Dimorphous with alstonite

CRYSTAL SYSTEM: Monoclinic
CLASS: 2/m, 2
SPACE GROUP: $P2_1$, $P2_1/m$
Z: 2
LATTICE CONSTANTS:
  $a = 8.15$
  $b = 5.22$
  $c = 6.58$
  $\beta = 106°18'$
3 STRONGEST DIFFRACTION LINES:
  3.125 (100)
  3.140 ( 90)
  4.018 ( 40)
OPTICAL CONSTANTS:
  $\alpha = 1.525$
  $\beta = 1.684$
  $\gamma = 1.686$
  $(-)2V = 15°$
HARDNESS: 4
DENSITY: 3.66–3.71 (Meas.)
         3.65 (Calc.)
CLEAVAGE: {210} perfect
          {001} imperfect
Fracture subconchoidal to uneven. Brittle.
  HABIT: Crystals short to long prismatic or equant. Striated. Also massive.
  COLOR-LUSTER: White, gray, yellowish, greenish. Transparent to translucent. Vitreous, resinous. Fluoresces pale yellowish under ultraviolet light.
  MODE OF OCCURRENCE: Occurs as cleavable masses and as large crystals associated with calcite, barite, and fluorite in veins in limestone, at Alston Moor, Cumberland, England. It is also found at Långban, Sweden, and at the Himmelsfürst mine, Freiberg, Saxony, Germany.
  BEST REF. IN ENGLISH: Palache, et al., "Dana's System of Mineralogy," 7th Ed., v. II, p. 220–221, New York, Wiley, 1951.

## BARYTOCELESTINE = Variety of celestite

## BARYTOLAMPROPHYLLITE
$Na_2(Ba, Sr)_2Ti_3(SiO_4)_4(OH, F)_2$
Ba analogue of lamprophyllite

CRYSTAL SYSTEM: Monoclinic
CLASS: 2/m, 2
SPACE GROUP: C2/m, C2
Z: 1
LATTICE CONSTANTS:
  $a = 19.96$
  $b = 7.07$
  $c = 5.43$
  $\beta = 96°30'$
3 STRONGEST DIFFRACTION LINES:
  2.85 (100)
  2.19 ( 90)
  1.51 ( 90)
OPTICAL CONSTANTS:
  $\alpha = 1.742$–1.743
  $\beta = 1.754$ (Calc.)
  $\gamma = 1.776$–1.778
  $(+)2V = 29°$–30°
HARDNESS: 2–3
DENSITY: 3.62–3.66 (Meas.)
         3.61 (Calc.)
CLEAVAGE: {100} perfect
          {011} good
          {010} imperfect
Brittle.
  HABIT: Foliated aggregates of cleavage rhombs with angles of 128° and 104°.
  COLOR-LUSTER: Dark brown; vitreous.
  MODE OF OCCURRENCE: Occurs associated with aegirine, nepheline, K feldspar, cancrinite, and apatite in ijolite, Lovozero intrusive, Kola Peninsula, USSR.
  BEST REF. IN ENGLISH: Peng, Tze-chung, and Chang, Chien-hung, *Am. Min.*, **51**: 1549 (1966).

## BASALUMINITE
$Al_4SO_4(OH)_{10} \cdot 5H_2O$

CRYSTAL SYSTEM: Hexagonal (?)
Z: 24
LATTICE CONSTANTS:
  $a = 22.56$
  $c = 18.72$
3 STRONGEST DIFFRACTION LINES:
  9.4  (100)
  4.68 ( 80)
  3.68 ( 70)
OPTICAL CONSTANT:
  $N = 1.515$–1.519 (Na)
HARDNESS: Not determined.
DENSITY: 2.12 (Meas.)
         2.24 (Calc.)
CLEAVAGE: Not determined. Fracture conchoidal when compact.
  HABIT: Massive, microcrystalline; compact to powdery.
  COLOR-LUSTER: White; often stained yellow, orange, or brown by impurities. Dull.
  MODE OF OCCURRENCE: Occurs at Temple Mountain, Utah; in Crawford County, Kansas; in septarian concretions at the Jude coal mine in Liberty Township, Marion County, Iowa; at Shoals, Indiana; in Pulaski and Marshall counties, Tennessee; at Fort Foote, Maryland; and in Roanoke and

Washington counties, Virginia. Also in England associated with hydrobasaluminite, gypsum, allophane, and aragonite as veinlets and coatings on joint surfaces at the Lodge Pit siderite deposit, Irchester, Northamptonshire, and in a deposit at Clifton Hill, Brighton, Sussex; and at Epernay, Marne, France.

BEST REF. IN ENGLISH: Palache, et al., "Dana's System of Mineralogy," 7th Ed., v. II, p. 586, New York, Wiley, 1951.

## BASSANITE
$2CaSO_4 \cdot H_2O$

CRYSTAL SYSTEM: Monoclinic
CLASS: 2
SPACE GROUP: A2
Z: 6
LATTICE CONSTANTS:
  $a = 12.70$
  $b = 6.83$
  $c = 11.94$
  $\beta = 90°36'$
3 STRONGEST DIFFRACTION LINES:
  3.00 (100)
  6.01 ( 95)
  2.80 ( 50)
OPTICAL CONSTANTS:
  $\omega = 1.55$
  $\epsilon = 1.57$
HARDNESS: Not determined.
DENSITY: 2.55 (Meas.)
         2.71 (Calc.)
HABIT: Microscopic needles in parallel arrangement; as pseudomorphs after gypsum.
COLOR-LUSTER: Colorless, white.
MODE OF OCCURRENCE: Occurs in cavities of leucite-tephrite rocks, and with gibbsite in fumaroles, on Mt. Vesuvius, Italy; also found in well borings (vibertite) at Nappan, Cumberland County, Nova Scotia.
BEST REF. IN ENGLISH: Palache, et al., "Dana's System of Mineralogy," 7th Ed., v. II, p. 476, New York, Wiley, 1951.

## BASSETITE
$Fe^{2+}(UO_2)_2(PO_4)_2 \cdot 8H_2O$

CRYSTAL SYSTEM: Monoclinic
Z: 2
LATTICE CONSTANTS:
  $a = 7.01$
  $b = 17.07$
  $c = 6.98$
  $\beta = 90°32'$
3 STRONGEST DIFFRACTION LINES:
  4.89 (100)
  3.46 (100)
  8.59 ( 60)
OPTICAL CONSTANTS:
  $\alpha = 1.603$
  $\beta = 1.610$ (Na)
  $\gamma = 1.617$
  $2V \sim 90°$

HARDNESS: $\sim 2\frac{1}{2}$
DENSITY: 3.4 (Meas.)
         3.6 (Calc.)
CLEAVAGE: {010} perfect
HABIT: Crystals thin lozenge-shaped plates flattened on {010}. Forms {010}, {111}, {101}, {110}, and {001}. Often occurs as parallel growths with torbernite and uranospathite.
COLOR-LUSTER: Olive green, greenish brown, yellowish brown, bronze-yellow, yellow. Transparent. Vitreous, bronze on {010}. Not fluorescent.
MODE OF OCCURRENCE: Occurs in Utah at the Fuemrol mine, Temple Mountain, and at the Denise No. 1 mine on the Bowknot of the Green River, Emery County, Utah. It is also found as small crystals in oxidized vein material carrying pyrite and uraninite at the Wheal Basset, Redruth, Cornwall, England.
BEST REF. IN ENGLISH: Frondel, Clifford, USGS Bull. 1064, p. 200–204 (1958).

## BASTINITE = Lithium hureaulite

## BASTNAESITE (Bastnasite)
$(Ce, La)(CO_3)F$

CRYSTAL SYSTEM: Hexagonal
CLASS: $\bar{6}m2$
SPACE GROUP: $C\bar{6}2c$
Z: 6
LATTICE CONSTANTS:
  $a = 7.16$
  $c = 9.79$
3 STRONGEST DIFFRACTION LINES:
  2.879 (100)
  3.564 ( 70)
  4.88 ( 40)
OPTICAL CONSTANTS:
  $\omega = 1.72$ (average)
  $\epsilon = 1.82$
  (+)
HARDNESS: 4–4½
DENSITY: 4.78–5.2 (Meas.)
         5.02 (Calc.)
CLEAVAGE: {0001} parting, distinct to perfect
          {10$\bar{1}$0} indistinct
Fracture uneven. Brittle.
HABIT: Usually tabular on $c$ {0001}; rarely short prismatic with $a$ {2$\bar{1}\bar{1}$0}, $m$ {10$\bar{1}$0}, $q$ {10$\bar{1}$1}, $p$ {10$\bar{1}$2}, and $t$ {10$\bar{1}$3}. Also as large anhedral masses; granular.
COLOR-LUSTER: Wax yellow to reddish brown; transparent to translucent. Vitreous to greasy.
MODE OF OCCURRENCE: Occurs in great abundance associated with other rare earth minerals and barite in dolomite breccias associated with syenitic intrusives at Mountain Pass, San Bernardino County, California; also found in pegmatite in the Pikes Peak region, El Paso County, Colorado; in a contact metamorphic amphibole-skarn at Bastnäs, Västmanland, Sweden; and at many other localities throughout the world.
BEST REF. IN ENGLISH: Palache, et al., "Dana's System of Mineralogy," 7th Ed., v. II, p. 289–291, New York, Wiley, 1951. Donnay, Gabrielle, and Donnay, J. D. H., *Am. Min.*, **38**: 932–963 (1953).

# BASTNAESITE-(Y)
$(Y, Ce)(CO_3)F$

CRYSTAL SYSTEM: Hexagonal
CLASS: $\bar{6}m2$
SPACE GROUP: $C\bar{6}2c$
Z: 6
LATTICE CONSTANTS:
  $a = 6.57$
  $c = 9.48$
  $c/a = 1.441$
3 STRONGEST DIFFRACTION LINES:
  2.78  (100)
  1.948 (100)
  3.43  ( 75)
OPTICAL CONSTANTS:  (+) with high birefringence
  HARDNESS: 4-4½
  DENSITY: 3.9-4.0 (Meas.)
  CLEAVAGE: Not determined.
  HABIT: Massive; fine-grained.
  COLOR-LUSTER: Brick red.
  MODE OF OCCURRENCE: Occurs as pseudomorphs up to 8cm long after gagarinite in microcline-quartz pegmatite, Kazakhstan.
  BEST REF. IN ENGLISH: Mineev, D. A., Lavrishcheva, T. I., and Bykova, A. V., *Am. Min.*, 57: 594 (1972).

# BATISITE
$Na_2BaTi_2(Si_2O_7)_2$

CRYSTAL SYSTEM: Orthorhombic
CLASS: mm2
SPACE GROUP: Ima2
Z: 4
LATTICE CONSTANTS:
  $a = 10.41$
  $b = 13.85$
  $c = 8.06$
3 STRONGEST DIFFRACTION LINES:
  2.91 (100)
  3.39 ( 50)
  2.16 ( 50)
OPTICAL CONSTANTS:
  $\alpha = 1.727$
  $\beta = 1.732$
  $\gamma = 1.789$
  $(+)2V = 7°-40°$
  HARDNESS: 5.9
Microhardness 764kg/mm².
  DENSITY: 3.432 (Meas.)
  CLEAVAGE: {100} good
  HABIT: As elongated crystals showing {010}, {001}, {110}, {031}, {150}, {011}, and {310}; up to 10 cm long.
  COLOR-LUSTER: Dark brown. Powder rosy.
  MODE OF OCCURRENCE: Occurs associated with nepheline (zeolitized), aegirine, arfvedsonite, uranothorite, ramsayite, eudialyte, apatite, and adularia-like orthoclase in the Inaglina nepheline syenite pegmatite, Central Aldan.
  BEST REF. IN ENGLISH: Kravchenko, S. M., and Vlasova, E. V., *Am. Min.*, 45: 908-909 (1960).

# BAUMHAUERITE
$Pb_3As_4S_9$ or $Pb_4As_6S_{13}$

CRYSTAL SYSTEM: Monoclinic
CLASS: 2/m
SPACE GROUP: $P2_1/m$
Z: 3
LATTICE CONSTANTS:
  $a = 22.68$
  $b =  8.32$
  $c =  7.92$
  $\beta = 97°17'$
3 STRONGEST DIFFRACTION LINES:
  4.149 (100)
  2.957 ( 90)
  3.594 ( 80)
  HARDNESS: 3
  DENSITY: 5.33  (Meas.)
          5.449 (Calc.)
  CLEAVAGE: {100} perfect
Fracture conchoidal.
  HABIT: Crystals short prismatic or tabular, commonly striated and sometimes rounded.  Twinning on {100}, polysynthetic.
  COLOR-LUSTER: Steel gray to lead gray, sometimes with iridescent tarnish. Opaque. Metallic. Streak chocolate brown.
  MODE OF OCCURRENCE: Occurs in crystalline dolomite associated with sartorite, rathite, and other rare sulfosalts, at the Lengenbach quarry in the Binnental, Valais, Switzerland.
  BEST REF. IN ENGLISH: Palache, et al., "Dana's System of Mineralogy," 7th Ed., v. I, p. 460-462, New York, Wiley, 1944.

**BAUXITE** = A rock name; various oxides and hydroxides of aluminum or of aluminum and iron, often with clay.

**BAVALITE** = Variety of daphnite

# BAVENITE
$Ca_4(Be,Al)_4Si_9(O,OH)_{28}$

CRYSTAL SYSTEM: Orthorhombic
CLASS: 2/m 2/m 2/m or mm2
SPACE GROUP: Amma or Am2a
Z: 4
LATTICE CONSTANTS:
  $a = 19.34$
  $b = 22.90$
  $c = 4.96$
3 STRONGEST DIFFRACTION LINES:
  3.71 (100)
  3.35 ( 90)
  3.22 ( 80)
OPTICAL CONSTANTS:
  $\alpha = 1.578-1.586$
  $\beta = 1.579-1.588$
  $\gamma = 1.583-1.593$
  $(+)2V = 22°-60°$
  HARDNESS: 5½

DENSITY: 2.71–2.74 (Meas.)
  2.81 (Calc.)
CLEAVAGE: {100} perfect
  {001} indistinct
Brittle.
HABIT: Crystals prismatic; also as fibrous or radial-lamellar aggregates. Twinning on {100}.
COLOR-LUSTER: White, greenish white, pinkish, pale brown. Vitreous. Silky. Streak white.
MODE OF OCCURRENCE: Occurs chiefly in granite pegmatites in association with beryl; less commonly in skarns as an alteration product of helvite. Found at Pala, San Diego County, California; at the Rutherford mine, Amelia, Virginia; in miarolitic cavities of granite pegmatites at Baveno, Italy; and at localities in Switzerland, Poland, Czechoslovakia, USSR, and Western Australia.
BEST REF. IN ENGLISH: Fleischer, Michael, and Switzer, George, *Am. Min.*, **38**: 988–993 (1953). Vlasov, K. A., "Mineralogy of Rare Elements," v. II, p. 143–147, Israel Program for Scientific Translations, 1966.

# BAYERITE
$Al(OH)_3$

CRYSTAL SYSTEM: Hexagonal
CLASS: $\bar{3}$
SPACE GROUP: $P\bar{3}$
Z: 2
LATTICE CONSTANTS:
  $a \simeq 5.02$–$5.04$
  $c \simeq 4.73$–$4.77$
3 STRONGEST DIFFRACTION LINES:
  4.72 (100)
  4.36 ( 70)
  2.21 ( 65)
HARDNESS: Not determined.
DENSITY: 2.53 (Meas.)
  2.54 (Calc.)
CLEAVAGE: Not determined
HABIT: Fibrous, very fine.
COLOR-LUSTER: Colorless.
MODE OF OCCURRENCE: Occurs in association with calcite and gypsum in a sedimentary rock containing portlandite and ettringite in Hartrurim, Israel.
BEST REF. IN ENGLISH: Gross, S., and Heller, L., *Min. Mag.*, **33**: 723–724 (1963).

# BAYLDONITE
$(Pb,Cu)_3(AsO_4)_2(OH)_2$

CRYSTAL SYSTEM: Monoclinic
3 STRONGEST DIFFRACTION LINES:
  3.14 (100)
  3.21 ( 70)
  2.93 ( 60)
OPTICAL CONSTANTS:
  $\alpha = 1.95$
  $\beta = 1.97$
  $\gamma = 1.99$
  (+)2V = large

HARDNESS: 4½
DENSITY: 5.5 (Meas.)
CLEAVAGE: Not determined.
HABIT: Usually massive, fine-granular to powdery. Also as crusts and minute mammillary concretions with fibrous structure.
COLOR-LUSTER: Various shades of yellowish green. Subtranslucent. Resinous.
MODE OF OCCURRENCE: Occurs as a secondary mineral in the oxidation zone of copper-bearing ore deposits. Found abundantly associated with azurite, mimetite, olivenite, and other minerals at Tsumeb, South West Africa; at St. Day, Cornwall, England; and at Diou, Allier, France.
BEST REF. IN ENGLISH: Palache, et al., "Dana's System of Mineralogy," 7th Ed., v. II, p. 929–930, New York, Wiley, 1951.

# BAYLEYITE
$Mg_2UO_2(CO_3)_3 \cdot 18H_2O$

CRYSTAL SYSTEM: Monoclinic
CLASS: 2/m
SPACE GROUP: $P2_1/a$
Z: 4
LATTICE CONSTANTS:
  $a = 26.65$
  $b = 15.31$
  $c = 6.53$
  $\beta = 93°4'$
3 STRONGEST DIFFRACTION LINES:
  7.66 (100)
  13.1 ( 90)
  3.83 ( 60)
OPTICAL CONSTANTS:
  $\alpha = 1.453$–$1.455$
  $\beta = 1.490$–$1.492$
  $\gamma = 1.498$–$1.502$
  (–)2V = 30° (Calc.)
HARDNESS: Not determined.
DENSITY: 2.05 (Meas.)
  2.06 (Calc.)
CLEAVAGE: Not determined. Fracture conchoidal.
HABIT: Crystals short prismatic along {001}; as crusts.
COLOR-LUSTER: Yellow; becomes pale yellow to yellowish white on exposure. Transparent. Vitreous; dull on dehydration. Fluoresces weak yellowish green.
MODE OF OCCURRENCE: Occurs as a water-soluble secondary mineral at the Hillside mine near Bagdad, Yavapai County, Arizona; with andersonite as crusts on sandstone in the Ambrosia Lakes district, McKinley County, New Mexico; in the Pumpkin Buttes area of the Powder River Basin, Wyoming; and widespread at many deposits in the Colorado Plateau region. It is also found as an efflorescence with epsomite and gypsum in a mine tunnel at Azegour, Morocco.
BEST REF. IN ENGLISH: Frondel, Clifford, USGS Bull. 1064, p. 112–115 (1958).

# BAZZITE
$Be_3(Sc,Al)_2Si_6O_{18}$
Scandium analogue of beryl.

CRYSTAL SYSTEM: Hexagonal
CLASS: 6/m 2/m 2/m
SPACE GROUP: P6/mcc
Z: 2
LATTICE CONSTANTS:
  $a = 9.521$
  $c = 9.165$
3 STRONGEST DIFFRACTION LINES:
  3.29 (100)
  2.94 (100)
  1.65 ( 80)
OPTICAL CONSTANTS:
  $\omega = 1.627$
  $\epsilon = 1.607$
  HARDNESS: Microhardness 851–897 $kg/mm^2$.
  DENSITY: 2.77 ± 0.01 (Meas.)
  CLEAVAGE: Not determined. Brittle.
  HABIT: Crystals acicular.
  COLOR-LUSTER: Intense blue. Transparent. Vitreous.
  MODE OF OCCURRENCE: Occurs in small amounts in drusy cavities and as inclusions in quartz and fluorite in pegmatites of the Kentsk granitic massif of Kazakhstan, USSR.
  BEST REF. IN ENGLISH: Christyakova, M. B., Moleva, V. A., and Razmanova, Z. P., *Min. Abs.*, **18**: 115 (1967).


## BEARSITE
$Be_2(AsO_4)(OH)\cdot 4H_2O$
Arsenic analogue of moraesite.

CRYSTAL SYSTEM: Monoclinic
Z: 12
LATTICE CONSTANTS:
  $a = 8.55$
  $b = 36.90$
  $c = 7.13$
  $\beta = 97°49'$
3 STRONGEST DIFFRACTION LINES:
  6.95 (100)
  3.31 ( 80)
  4.23 ( 60)
OPTICAL CONSTANTS:
  $\alpha = 1.490$
  $\beta \simeq \gamma$
  $\gamma = 1.502$
  (−)2V
  HARDNESS: Not determined.
  DENSITY: 1.8–2.0 (Meas.)
          2.199 (Calc.)
  CLEAVAGE: Not determined.
  HABIT: As fine incrustations and tangled fibrous aggregates of prismatic crystals with longitudinal striations, tenths to hundredths of a millimeter in size.
  COLOR-LUSTER: White.
  MODE OF OCCURRENCE: Occurs coating pharmacosiderite and arseniosiderite associated with tyrolite, scorodite-mansfieldite, conichalcite, sodium uranospinite, and metazeunerite in the zone of oxidation of a polymetallic deposit at an unspecified locality in Kazakhstan.
  BEST REF. IN ENGLISH: Kopchenova, E. V., and Sidorenko, G. A., *Am. Min.*, **48**: 210–211 (1963).


## BEAVERITE
$Pb(Cu,Fe,Al)_3(SO_4)_2(OH)_6$

CRYSTAL SYSTEM: Hexagonal
CLASS: 3m
SPACE GROUP: R3m
Z: 3
LATTICE CONSTANTS:
  $a = 7.203$
  $c = 16.94$
3 STRONGEST DIFFRACTION LINES:
  5.85  (100)
  3.03  ( 95)
  2.276 ( 45)
OPTICAL CONSTANTS:
  $\omega = 1.85$ (variable)
  uniaxial (−)
  HARDNESS: Not determined.
  DENSITY: 4.36 (Meas.)
          4.31 (Calc.)
  CLEAVAGE: Not determined.
  HABIT: As earthly and friable masses composed of microscopic hexagonal plates.
  COLOR-LUSTER: Yellow.
  MODE OF OCCURRENCE: Occurs as a secondary mineral in the oxidized zone of lead-copper deposits. Found in Utah, associated with plumbojarosite, at the Horn Silver mine, Beaver County; at the Centennial Alta mine, Salt Lake County; and at the Hidden Treasure mine, Tooele County. It also occurs admixed with plumbojarosite at the Boss mine, Yellow Pine district, Nevada.
  BEST REF. IN ENGLISH: Palache, et al., "Dana's System of Mineralogy," 7th Ed., v. II, p. 568, New York, Wiley, 1951.


## BECKELITE = Synonym of britholite


## BECQUERELITE
$CaU_6O_{19}\cdot 11H_2O$

CRYSTAL SYSTEM: Orthorhombic
CLASS: 2/m 2/m 2/m or mm2
SPACE GROUP: Pnma or Pn2$_1$a
Z: 4
LATTICE CONSTANTS:
  $a = 13.86$
  $b = 12.38$
  $c = 14.96$
3 STRONGEST DIFFRACTION LINES:
  7.44 ± .05  (100)
  3.20 ± .01  ( 35)
  3.73 ± .015 ( 30)
OPTICAL CONSTANTS:
  $\alpha = 1.730$
  $\beta = 1.825$
  $\gamma = 1.830$
  (−)2V ∼ 30°
  HARDNESS: 2½
  DENSITY: 5.14 ± 0.06 (Meas.)
          5.10 (Calc.)

CLEAVAGE: {001} perfect
{101} imperfect
{010} imperfect
{110} imperfect

HABIT: Crystals usually prismatic, elongated [010]; occasionally tabular on {001}. Also as fine-grained aggregates and coatings.

COLOR-LUSTER: Golden to lemon yellow, less commonly orange-yellow. Adamantine to greasy. Transparent. Streak yellow.

MODE OF OCCURRENCE: Occurs as a secondary mineral usually closely associated with uraninite, from which it has been derived. Found as crystals up to 3cm in length associated with fourmarierite, soddyite, dewindtite, schoepite, and curite, at Shinkolobwe, Katanga, Zaire; at several localities in the Colorado Plateau uranium province; sparingly as an alteration product of uraninite in pegmatites; and in other types of uraninite deposits.

BEST REF. IN ENGLISH: Christ, C. L., and Clark, Joan R., *Am. Min.*, **45**: 1026–1061 (1960).

## BEDENITE = Variety of hornblende

## BEEGERITE (inadequately described mineral)
$Pb_6Bi_2S_9$

CRYSTAL SYSTEM: Cubic (?)
HARDNESS: Not determined
DENSITY: 7.27 (Meas.)
CLEAVAGE: {001} ? perfect
HABIT: Crystals tetrahedrons; usually indistinctly crystallized. Also massive, fine granular to dense.

COLOR-LUSTER: Light to dark gray. Opaque. Metallic.

MODE OF OCCURRENCE: Occurs in Colorado at the Baltic lode on Revenue Mountain, Park County, and as small particles and patches as much as 1 cm in diameter disseminated through quartz and associated with pyrite and chalcopyrite at the Treasury Vault mine, Summit County. Found also in Switzerland and in the Minusinsk district, Yeniseisk, Siberia, USSR.

BEST REF. IN ENGLISH: Palache, et al., "Dana's System of Mineralogy," 7th Ed., v. I, p. 392–393, New York, Wiley, 1944.

## BEFANAMITE = Variety of thortveitite

## BEHIERITE
(Ta,Nb)BO$_4$

CRYSTAL SYSTEM: Tetragonal
CLASS: 4/m 2/m 2/m
SPACE GROUP: I4$_1$/amd
Z: 4
LATTICE CONSTANTS:
$a$ = 6.206
$c$ = 5.472
3 STRONGEST DIFFRACTION LINES:
4.10 (100)
3.10 ( 71)
2.327 ( 71)

HARDNESS: 7–7.5
DENSITY: 7.86 ± 0.05 (Meas.)
7.91 (Calc.)
CLEAVAGE: {110} distinct
{010} distinct
Fracture subconchoidal.

HABIT: As octahedral crystals, 0.5–7.0 mm in size.

COLOR-LUSTER: Grayish pink; adamantine. Streak white.

MODE OF OCCURRENCE: Occurs very sparingly associated with rubellite, pollucite, manganoan apatite, lepidolite, quartz, and albite in pegmatite at Manjaka, Madagascar.

BEST REF. IN ENGLISH: Mrose, Mary E., and Rose, H. J. Jr., *Geol. Soc. Am.*, Abstracts 1961 Ann. Meetings, p. 235, (1961).

## BEHOITE
β-Be(OH)$_2$

CRYSTAL SYSTEM: Orthorhombic
CLASS: 222
SPACE GROUP: P2$_1$2$_1$2$_1$
Z: 4
LATTICE CONSTANTS:
$a$ = 4.64
$b$ = 7.05
$c$ = 4.55
3 STRONGEST DIFFRACTION LINES:
2.38 (100)
3.93 ( 90) synthetic
3.80 ( 80)
OPTICAL CONSTANTS:
α = 1.533
β = 1.544
γ = 1.548
(−)2V = 82°
HARDNESS: ~4
DENSITY: 1.92 ± 0.01 (Meas.)
1.924 (Meas. Synthetic)
CLEAVAGE: Breaks with conchoidal fracture; no obvious cleavage.

HABIT: Euhedral to subhedral crystals of pseudo-octahedral habit, up to 1mm diameter.

COLOR-LUSTER: Colorless; matrix materials discolor some crystals to pastel shades of brown. Vitreous luster.

MODE OF OCCURRENCE: Behoite crystals and aggregates occur in altered zones or rinds around vitreous masses of gadolinite at the Rode Ranch pegmatite, Llano County, Texas. A typical alteration rind is approximately 2cm thick on an ovaloid mass of gadolinite 5 × 8cm. (Ehlmann et al., *Econ. Geol.*, **59**: 1348–1360 [1964].) Also occurs sparsely disseminated in altered volcanic tuff at Honeycomb Hill, Juab County, Utah.

BEST REF. IN ENGLISH: Ehlmann, Arthur J., and Mitchell, Richard S., *Am. Min.*, **55**: 1–9 (1970).

## BEIDELLITE
$(Na,Ca/2)_{0.33}Al_2(Al,Si)_4O_{10}(OH)_2 \cdot nH_2O$
Montmorillonite (Smectite) Group

CRYSTAL SYSTEM: Monoclinic
CLASS: 2/m
SPACE GROUP: C2/m
Z: 1
3 STRONGEST DIFFRACTION LINES:
  4.52 (100)
  2.61 (100)
  2.55 (100)
OPTICAL CONSTANTS:
  (−)2V = small
  HARDNESS: 1–2
  DENSITY: Variable, 2–3.
  CLEAVAGE: {001} perfect.
  HABIT: Crystals thin plates, extremely minute; in clay-like masses.
  COLOR-LUSTER: White, reddish, brownish gray. Waxy to vitreous, also dull.
  MODE OF OCCURRENCE: Occurs mainly as a constituent of bentonitic clay deposits or as a hydrothermal alteration product in ore veins. Typical occurrences are found in Los Angeles and Sierra counties, California; at the Castle Dome mine, Gila County, Arizona; at Beidell, Saguache County, and at other places in Colorado; in Tooele and Utah counties; Utah; at the Black Jack mine, Owyhee County, Idaho; near Chust, Carpathian Ukraine; and elsewhere.
  BEST REF. IN ENGLISH: Weir, A. H., and Greene-Kelly, R., *Am. Min.*, **47**: 137–146 (1962). Deer, Howie, and Zussman, "Rock Forming Minerals," v. 3, p. 226–245, New York, Wiley, 1962.

# BELLINGERITE
$3Cu(IO_3)_2 \cdot 2H_2O$

CRYSTAL SYSTEM: Triclinic
CLASS: $\bar{1}$
SPACE GROUP: $P\bar{1}$
Z: 1
LATTICE CONSTANTS:
  $a = 7.22$ kX     $\alpha = 105°06'$
  $b = 7.82$        $\beta = 96°57\frac{1}{2}'$
  $c = 7.92$        $\gamma = 92°55'$
3 STRONGEST DIFFRACTION LINES:
  3.72 (100)
  3.35 ( 90)
  3.17 ( 90)
OPTICAL CONSTANTS:
  $\alpha = 1.890$
  $\beta = 1.90$
  $\gamma = 1.99$
  (+)2V = medium
  HARDNESS: 4
  DENSITY: 4.89 ± 0.01 (Meas.)
           4.932 (Calc.)
  CLEAVAGE: Fracture subconchoidal. Brittle.
  HABIT: As tiny isolated crystals and as veinlets. Crystals prismatic [001] and somewhat tabular {100}, {100} striated parallel [001]. Twinning on {$\bar{1}$01}.
  COLOR-LUSTER: Light green. Transparent to translucent. Vitreous. Streak very pale green.

  MODE OF OCCURRENCE: Occurs as a secondary mineral, associated with gypsum and leightonite, at Chuquicamata, Chile.
  BEST REF. IN ENGLISH: Palache, et al., "Dana's System of Mineralogy," 7th Ed., v. II, p. 313–315, New York, Wiley, 1951.

# BELOVITE (of Borodin and Kazakova)
$(Sr, Ce, Na, Ca)(PO_4)_3 (O, OH)$
Strontium analogue of hydroxylapatite

CRYSTAL SYSTEM: Hexagonal
CLASS: 6/m
Z: 2
LATTICE CONSTANTS:
  $a = 9.62$
  $c = 7.12$
3 STRONGEST DIFFRACTION LINES:
  2.87  (100)
  1.998 ( 80)
  1.900 ( 80)
OPTICAL CONSTANTS:
  $\omega = 1.660$
  $\epsilon = 1.640$
    (−)
  HARDNESS: 5
  DENSITY: 4.19
  CLEAVAGE: Prismatic and pinacoidal, imperfect. Fracture irregular. Brittle.
  HABIT: Prismatic crystals up to 2cm in size with ($10\bar{1}0$) and (0001) prominent.
  COLOR-LUSTER: Honey yellow; vitreous, greasy on fracture.
  MODE OF OCCURRENCE: Occurs in ussingite, formed by late-stage replacement of microcline, in the central part of a pegmatite in nepheline syenite at an unspecified locality, presumably on the Kola Peninsula.
  BEST REF. IN ENGLISH: Borodin, L. S., and Kazakova, M. E., *Am. Min.*, **40**: 367–368 (1955).

# BELOVITE (of Nefedov) = Talmessite, arsenate-belovite

# BELYANKINITE
$Ca(Ti,Zr,Nb)_6O_{13} \cdot 14H_2O$

CRYSTAL SYSTEM: Orthorhombic or monoclinic
OPTICAL CONSTANTS:
  $\alpha = 1.740$
  $\beta = 1.775 – 1.780$
  $\gamma \approx 1.778$ ($\gamma - \beta = 0.002 – 0.003$)
  (−)2V
  HARDNESS: 2–3
  DENSITY: 2.32–2.40
  CLEAVAGE: One perfect.
Fracture uneven.
  HABIT: As cleavable masses up to 20 × 12 × 0.5 cm in size.
  COLOR-LUSTER: Light yellow to brownish yellow; vitreous to greasy, pearly on cleavage.

MODE OF OCCURRENCE: Occurs in nepheline syenite pegmatite composed mainly of microcline, nepheline, and aegirine at an unspecified locality, presumably on the Kola Peninsula.

BEST REF. IN ENGLISH: Gerasimovsky, V. I., and Kazakova, M. E., *Am. Min.*, 37: 882 (1952).

## BEMENTITE
$Mn_8Si_6O_{15}(OH)_{10}$

CRYSTAL SYSTEM: Monoclinic
Z: 1
LATTICE CONSTANTS:
$a = 7.5$
$b = 9.8$
$c = 5.65$
3 STRONGEST DIFFRACTION LINES:
7.30 (100)
2.51 ( 90)
2.81 ( 70)
OPTICAL CONSTANTS:
$\alpha = 1.602-1.624$
$\gamma = \beta = 1.632-1.650$
HARDNESS: 6
DENSITY: 2.9-3.1 (Meas.)
CLEAVAGE: {001} perfect, micaceous
{010} perfect
{100} perfect
HABIT: Crystals fibrous or lamellar; usually massive, granular or horn-like; also in radiate or stellate masses; rarely minute stalactitic.

COLOR-LUSTER: Brown, yellow, gray; darkening on exposure. Translucent. Pearly on cleavage; usually waxy.

MODE OF OCCURRENCE: Occurs as one of the chief minerals in primary manganese ore deposits at numerous localities in California; in the Olympic Mountains, Washington; at Franklin, Sussex County, New Jersey; and at Långban, Sweden.

BEST REF. IN ENGLISH: Palache, Charles, USGS Prof. Paper 180, p. 117-118 (1935).

## BENITOITE
$BaTiSi_3O_9$

CRYSTAL SYSTEM: Hexagonal
CLASS: $\bar{6}m2$
SPACE GROUP: $P\bar{6}c2$
Z: 2
LATTICE CONSTANTS:
$a = 6.60$
$c = 9.71$
3 STRONGEST DIFFRACTION LINES:
3.72 (100)
2.742 ( 75)
3.32 ( 40)
OPTICAL CONSTANTS:
$\omega = 1.757$
$\epsilon = 1.804$
(+)

HARDNESS: 6-6½
DENSITY: 3.64-3.68 (Meas.)
3.688 (Calc.)
CLEAVAGE: {10$\bar{1}$1} indistinct
Fracture conchoidal to uneven. Brittle.

HABIT: Crystals pyramidal or tabular, usually flattened along c-axis, and somewhat triangular in shape.

COLOR-LUSTER: Blue, purple, pink, white, colorless; often varicolored in a single crystal. Transparent to translucent. Vitreous. Streak uncolored. Fluoresces bluish under short-wave ultraviolet light.

MODE OF OCCURRENCE: Occurs in California in a section of drill core taken near Rush Creek, Fresno County; as small grains from a drill hole in the Lazard area, Lost Hills, Kern County; and as superb crystals in association with neptunite, joaquinite, and natrolite in serpentine at two localities near the headwaters of the San Benito River in San Benito County. It occurs also in Eocene sands in southwest Texas, and in sands of the Owithe Valley, Belgium.

BEST REF. IN ENGLISH: Winchell and Winchell, "Elements of Optical Mineralogy," 4th Ed., Pt. 2, p. 453, New York, Wiley, 1951.

**BENJAMINITE** = A mixture of berryite, matildite, and lindströmite.

## BENSTONITE
$MgCa_6(Ba,Sr)_6(CO_3)_{13}$

CRYSTAL SYSTEM: Hexagonal
CLASS: $\bar{3}$ (?)
SPACE GROUP: R3(?)
Z: 3
LATTICE CONSTANTS:
$a = 18.28$
$c = 8.67$
3 STRONGEST DIFFRACTION LINES:
3.08 (100)
3.92 ( 40)
2.536 ( 30)
OPTICAL CONSTANTS:
$\omega = 1.690$
$\epsilon = 1.527$
(-)
HARDNESS: 3-4
DENSITY: 3.596-3.66 (Meas.)
3.648 (Calc.)
CLEAVAGE: Rhombohedral, good; angles near those of calcite.

HABIT: As flat, unmodified rhombohedral crystals up to ½ inch across; also as crusts of interlocking saddle-shaped crystals, and as cleavable masses with cleavage faces up to 1 cm across.

COLOR-LUSTER: Snow white to ivory, pale yellow to yellowish brown. Translucent. Vitreous. Fluoresces weak yellow or weak to strong red under short and long wave ultraviolet light.

MODE OF OCCURRENCE: Occurs in veins associated with milky quartz, barite, and calcite at a barite mine about

2½ miles east northeast of Magnet Cove, Hot Spring County, Arkansas; as crystals arranged epitaxially around spine-like crystals of calcite, associated with fluorite and sphalerite, at the Minerva mine, Cave in Rock, Hardin County, Illinois; and with calcite in hausmannite ore at Långban, Sweden.

BEST REF. IN ENGLISH: Lippmann, Friedrich, *Am. Min.*, 47: 585-598 (1962). White, John S., and Jarosewich, Eugene, *Min. Rec.*, 1 (4): 140-141 (1970).

# BERAUNITE
$Fe^{2+}Fe_5^{3+}(OH)_5(H_2O)_4(PO_4)_4 \cdot 2H_2O$

CRYSTAL SYSTEM: Monoclinic
CLASS: 2/m
SPACE GROUP: C2/c
Z: 4
LATTICE CONSTANTS:
 $a = 20.80-20.646$
 $b = 5.156-5.129$
 $c = 19.22-19.213$
 $\beta = 93.34°$
3 STRONGEST DIFFRACTION LINES:
 10.37 (100)
 4.825 ( 60)
 3.082 ( 60)
OPTICAL CONSTANTS:
 $\alpha = 1.775$
 $\beta = 1.786$ (variable)
 $\gamma = 1.815$
usually (+)2V ~ med. large
HARDNESS: 3½-4
DENSITY: 3.01 (Meas.)
 2.962 (Calc.)
CLEAVAGE: {100} good
HABIT: Crystals small, rare, tabular and somewhat elongated [010]; vertically striated. Usually as radial-fibrous aggregates, foliated crusts and globules, or disk-like concretions. Twinning on {100} as interpenetration twins.
COLOR-LUSTER: Reddish brown to dark red; also dull grayish green to dark greenish brown. Translucent. Vitreous to dull; pearly on cleavage. Streak yellow to greenish brown.
MODE OF OCCURRENCE: Occurs as a secondary mineral in iron ore deposits, and as an alteration product of primary phosphate minerals in granite pegmatites. Found as an alteration product of triphylite at the Big Chief and Hesnard mines, Keystone, Pennington County, South Dakota, and at the Palermo mine, North Groton, New Hampshire. In the United States it also is found in Arkansas, New Jersey, and Pennsylvania; other occurrences are found in Ireland, Germany, Czechoslovakia, and USSR.
BEST REF. IN ENGLISH: Palache, et al., "Dana's System of Mineralogy," 7th Ed., v. II, p. 959-961, New York, Wiley, 1951. Moore, P. B., *Am. Min.*, 55: 135-169 (1970).

# BERBORITE
$Be_2(BO_3)(OH,F) \cdot H_2O$

CRYSTAL SYSTEM: Hexagonal

LATTICE CONSTANTS:
 $a = 4.43$
 $c = 5.33$
3 STRONGEST DIFFRACTION LINES:
 5.3 (100)
 3.11 (100)
 2.656 (100)
OPTICAL CONSTANTS:
 $\omega = 1.580$
 $\epsilon = 1.485$
 (-)
HARDNESS: 3
DENSITY: 2.200 ± 0.003 (Meas.)
CLEAVAGE: {0001} perfect
Fracture irregular.
HABIT: Crystals 0.1-0.5 mm in diameter. Forms $c\{0001\}$, $l'\{11\bar{2}2\}$, $q\{10\bar{1}1\}$ (most common), also $m\{10\bar{1}0\}$, $a'\{11\bar{2}0\}$, $f'\{33\bar{6}4\}$, $g\{11\bar{2}1\}$, $p'\{2\bar{1}\bar{1}8\}$, and $f\{\bar{6}33\bar{4}\}$. Twinning noted.
COLOR-LUSTER: Colorless, vitreous.
MODE OF OCCURRENCE: Found in dumps of "old workings of one of the skarn deposits in the northwestern part of the (Soviet) Union," associated with hambergite, magnetite, vesuvianite, sphalerite, fluorite, helvite, and apatite.
BEST REF. IN ENGLISH: Nefedov, E. I., *Am. Min.*, 53: 348-349 (1968).

# BERESOVITE = Phoenicochroite

# BERGENITE
$Ba(UO_2)_4(PO_4)_2(OH)_4 \cdot 8H_2O$

CRYSTAL SYSTEM: Orthorhombic
CLASS: 2/m 2/m 2/m
SPACE GROUP: Bmmb
Z: 4
LATTICE CONSTANTS:
 $a = 16.05$
 $b = 17.76$
 $c = 13.86$
3 STRONGEST DIFFRACTION LINES:
 7.78 (100)
 3.88 (100)
 3.08 (100)
OPTICAL CONSTANTS:
 $\alpha = 1.660$
 $\beta = 1.690$
 $\gamma = 1.695$
(-)2V > 45°
HARDNESS: Not determined.
DENSITY: > 4.1 (Meas.)
 2.72 (Calc.)
CLEAVAGE: Not determined.
HABIT: Crystals thin tabular.
COLOR-LUSTER: Yellow. Fluorescent weak orange-brown under short- and long-wave ultraviolet light.
MODE OF OCCURRENCE: Found associated with much uranocircite, some torbernite, renardite, autunite, barium uranophane, and unidentified uranium minerals on a mine

dump at Streuberg near Bergen on the Treib, Vogtland, Saxony, Germany.

BEST REF. IN ENGLISH: Bültemann, Hans W., and Moh, Günter H., *Am. Min.*, **45**: 909 (1960).

# BERLINITE
$AlPO_4$

CRYSTAL SYSTEM: Hexagonal
CLASS: 3 2
SPACE GROUP: $P3_1 21$ or $P3_2 21$
Z: 3
LATTICE CONSTANTS:
  $a = 4.92$
  $c = 10.91$
3 STRONGEST DIFFRACTION LINES:
  3.369 (100)
  4.28 ( 25)
  1.835 ( 16)
OPTICAL CONSTANTS:
  $\omega = 1.5235$
  $\epsilon = 1.529$
  (+)
HARDNESS: $\sim 6\frac{1}{2}$
DENSITY: 2.64 (Meas.)
        2.618 (Calc.)
CLEAVAGE: None. Fracture conchoidal.
HABIT: Massive, granular. Synthetic crystals similar in habit to quartz.
COLOR-LUSTER: Colorless, grayish, pale rose. Transparent to translucent. Vitreous.
MODE OF OCCURRENCE: Occurs at the Westanå iron mine, near Näsum, Sweden, associated with augelite and other phosphates.
BEST REF. IN ENGLISH: Palache, et al., "Dana's System of Mineralogy," 7th Ed., v. II, p. 696–697, New York, Wiley, 1951.

# BERMANITE
$Mn^{2+}Mn_2^{3+}(PO_4)_2(OH)_2 \cdot 4H_2O$

CRYSTAL SYSTEM: Monoclinic (pseudo-orthorhombic)
CLASS: 2
SPACE GROUP: $P2_1$
Z: 2
LATTICE CONSTANTS:
  $a = 5.425$
  $b = 19.210$
  $c = 5.425$
  $\beta = 110°24'$
3 STRONGEST DIFFRACTION LINES:
  9.68 (100)
  5.08 ( 65)
  4.81 ( 65)
OPTICAL CONSTANTS:
  $\alpha = 1.687$
  $\beta = 1.725$
  $\gamma = 1.748$
  (−)2V = 74°
HARDNESS: $3\frac{1}{2}$

DENSITY: 2.84 (Meas.)
        2.840 (Calc.)
CLEAVAGE: {001} perfect
         {110} imperfect
Brittle.
HABIT: Tabular crystals, often twinned, up to 0.5 mm in size. In subparallel, fan-shaped, or rosette-like aggregates of crystals, and as lamellar masses.
COLOR-LUSTER: Pale red to dark reddish brown; vitreous to slightly resinous.
MODE OF OCCURRENCE: Occurs in pegmatite, associated with metastrengite, hureaulite, and other phosphates, coating seams in triplite, on the 7-U-7 Ranch, near Hillside, Arizona; in South Dakota as well-formed crystals in vugs in altered triphylite at the Tip-Top mine, Custer, and at several other pegmatite mines in the Custer and Keystone districts; also found at the Stewart Mine, Pala, California; Fletcher quarry, North Groton, New Hampshire; Williams prospect, Coosa County, Alabama; near Tanti, Córdoba, Argentina; Mangualde, Portugal; Sapucaia pegmatite, Minas Geraes, Brazil; and at an unspecified locality in Madagascar.
BEST REF. IN ENGLISH: Hurlbut, C. S. Jr., and Aristarain, L. F., *Am. Min.*, **53**: 416–431 (1968).

# BERNDTITE
$SnS_2$

CRYSTAL SYSTEM: Hexagonal
CLASS: $\bar{3} 2/m$
SPACE GROUP: $P\bar{3}m1$
Z: 1
3 STRONGEST DIFFRACTION LINES:
  2.78 (100)
  5.9 ( 50)
  2.14 ( 50)
HARDNESS: Very soft.
DENSITY: 4.5 (Synthetic)
CLEAVAGE: Not determined.
HABIT: Crystals tabular, minute.
COLOR-LUSTER: Gray in reflected light with very intense brownish to orange-yellow internal reflections. Transparent to translucent. Streak golden yellow.
MODE OF OCCURRENCE: Occurs as very fine inclusions in pyrite which has replaced stannite in tin sulfide ore, from a zone of secondary enrichment or oxidation at Cerro de Potosi, Bolivia.
BEST REF. IN ENGLISH: Moh, Gunter H., and Berndt, Fritz, *Am. Min.*, **50**: 2107 (1965).

# BERRYITE
$Pb_2(Cu,Ag)_3Bi_5S_{11}$

CRYSTAL SYSTEM: Monoclinic
CLASS: 2/m
SPACE GROUP: $P2_1/m$
Z: 6

LATTICE CONSTANTS:
$a = 12.72$
$b = 4.02$
$c = 58.07$
$\beta = 102\frac{1}{2}°$
3 STRONGEST DIFFRACTION LINES:
3.47 (100)
2.89 ( 80)
2.80 ( 70)
DENSITY: 6.7 (Meas.)
          7.11 (Calc.)
HABIT: Crystals lath-like; up to 1mm in length.
COLOR-LUSTER: Lead gray. Opaque. Metallic.
MODE OF OCCURRENCE: Occurs in the lower part of the siderite-rich cryolite with fluorite, topaz, weberite, ivigtite, quartz, cosalite, galena, aikinite, chalocopyrite, sphalerite, pyrrhotite, marcasite, pyrite, native bismuth, gustavite and phase X, at the cryolite mine at Ivigtut, Greenland. Also found on the type specimen of cupro-bismuthite from Park County, Colorado, and on specimens from the Nordmark mines, Sweden.
BEST REF. IN ENGLISH: Nuffield, E. W., and Harris, D. C., *Can. Min.*, 8: 400 (1965) (abstr.). Nuffield, E. W., and Harris, D. C., *Can. Min.*, 8: 407–413 (1966). Karup-Møller, *Can. Min.*, 8: 414–423 (1966).

## BERTHIERINE = Variety of chamosite

## BERTHIERITE
$FeSb_2S_4$

CRYSTAL SYSTEM: Orthorhombic
CLASS: 2/m 2/m 2/m
SPACE GROUP: Pnam
Z: 4
LATTICE CONSTANTS:
$a = 11.44$
$b = 14.12$
$c = 3.76$
3 STRONGEST DIFFRACTION LINES:
3.68 (100)
2.63 ( 80)
4.37 ( 60)
HARDNESS: 2–3
DENSITY: 4.64 (Meas.)
          4.645 (Calc. x-ray)
CLEAVAGE: {010} indistinct. Brittle.
HABIT: Crystals prismatic, vertically striated. Massive, prismatic to fibrous, radial or plumose. Also granular.
COLOR-LUSTER: Dark steel gray. Commonly tarnished brownish or iridescent. Opaque. Metallic. Streak dark brownish gray.
MODE OF OCCURRENCE: Occurs chiefly in low-temperature vein deposits, commonly associated with stibnite and quartz. Found associated with colemanite at Boron, Kern County, California; at the Good Hope mine, Gunnison County, Colorado; at the Cochenor-Willans mine, Red Lake, Ontario, and near Lake George, New Brunswick, Canada; and at deposits in Mexico, Bolivia, Chile, Peru, England, France, Germany, Czechoslovakia, Hungary, Australia, and Japan.

BEST REF. IN ENGLISH: Palache, et al., "Dana's System of Mineralogy," 7th Ed., v. I, p. 481–482, New York, Wiley, 1944. *Am. Min.*, 40: 226–238 (1955).

## BERTHONITE = Bournonite

## BERTOSSAITE
$(Li, Na)_2(Ca, Fe, Mn)Al_4(PO_4)_4(OH, F)_4$

CRYSTAL SYSTEM: Orthorhombic
CLASS: 2/m 2/m 2/m or mm2
SPACE GROUP: Imaa or Iaa2
Z: 4
LATTICE CONSTANTS:
$a = 11.48$
$b = 15.73$
$c = 7.23$
3 STRONGEST DIFFRACTION LINES:
3.056 (100)
3.286 ( 70)
3.104 ( 70)
OPTICAL CONSTANTS:
$\alpha = 1.624$
$\beta = 1.636$
$\gamma = 1.642$
(−)2V moderately large
HARDNESS: 6
DENSITY: 3.10 (Meas.)
          3.10 (Calc.)  $Li_2CaAl_4(PO_4)_4(OH)_4$
CLEAVAGE: {100} good
Fracture uneven to subconchoidal.
HABIT: Massive.
COLOR-LUSTER: Pale pink; vitreous.
MODE OF OCCURRENCE: Occurs associated with amblygonite, lazulite-scorzalite, augelite, brazilianite, apatite, crandallite, and quartz in the Buranga lithium pegmatite in Rwanda, Africa.
BEST REF. IN ENGLISH: Von Knorring, O., and Mrose, M. E., *Can. Min.*, 8: 668 (1966) (abstr.).

## BERTRANDITE
$Be_4Si_2O_7(OH)_2$
Var: Gelbertrandite (colloidal)
     Sphaerobertrandite (spherulitic)

CRYSTAL SYSTEM: Orthorhombic
CLASS: mm2
SPACE GROUP: $Ccm2_1$
Z: 4
LATTICE CONSTANTS:
$a = 15.22$
$b = 8.69$
$c = 4.54$
3 STRONGEST DIFFRACTION LINES:
4.38 (100)
3.19 ( 90)
2.54 ( 80)

OPTICAL CONSTANTS:
$\alpha = 1.584-1.591$
$\beta = 1.603-1.605$
$\gamma = 1.611-1.614$
$(-)2V = 73°-81°$
HARDNESS: 6-7
DENSITY: 2.60 (Meas.)
2.61 (Calc.)
CLEAVAGE: {001} perfect
{110} distinct
{010} distinct
{100} distinct
Brittle.

HABIT: Crystals thin tabular or prismatic, up to 3.0 cm in size. Often pseudomorphous after beryl. Twinning on {011} or {021} common.

COLOR-LUSTER: Colorless, pale yellow. Transparent. vitreous, pearly on {001}.

MODE OF OCCURRENCE: Occurs chiefly in granite pegmatites, greisens, aplites, and pneumatolytic-hydrothermal veins, in association with beryl, feldspar, quartz, phenacite, and tourmaline. Found in Colorado in Chaffee, Clear Creek, Jefferson, and Larimer counties, and abundantly at the Boomer and nearby mines and prospects in Park County; in the Pala pegmatites, San Diego County, California; at several localities in Maine; at South Acworth, New Hampshire; and at Amelia, Virginia. Also at localities in England, Norway, France, Switzerland, Germany, Czechoslovakia, USSR, and at Mica Creek, Mt. Isa, Queensland, Australia.

BEST REF. IN ENGLISH: Vlasov, K. A., "Mineralogy of Rare Elements," v. II, p. 89–96, Israel Program for Scientific Translations, 1966.

# BERYL
$Be_3 Al_2 Si_6 O_{18}$
Var. Emerald
Aquamarine
Morganite (Vorobyevite)
Goshenite
Heliodor

CRYSTAL SYSTEM: Hexagonal
CLASS: 6/m 2/m 2/m
SPACE GROUP: P6/mmc
Z: 2
LATTICE CONSTANTS:
$a = 9.21$
$c = 9.17$
Values depend on chemical composition.
3 STRONGEST DIFFRACTION LINES:
2.867 (100)
3.254 ( 95)
7.98 ( 90)
OPTICAL CONSTANTS:
$\omega = 1.566-1.602$
$\epsilon = 1.562-1.594$
$(-)$
HARDNESS: 7½-8
DENSITY: 2.6-2.9 (Meas.)
2.640 (Calc.)

CLEAVAGE: {0001} indistinct
Fracture uneven to conchoidal. Brittle.

HABIT: Crystals usually short to long prismatic. Combining hexagonal prism {1010} and pinacoid {0001}: terminations sometimes modified by small hexagonal pyramid or dihexagonal pyramid faces. Often striated vertically or severely etched. Forms isolated crystals measuring from fractions of an inch to 6 feet in diameter and 18 feet long, and crystal aggregates up to 100 tons in weight. Also coarse columnar to compact.

COLOR-LUSTER: Colorless, white, light green, bluish green, green, greenish yellow, yellow, pink, pinkish orange, red, pale blue, blue. Transparent to translucent. Vitreous.

MODE OF OCCURRENCE: Occurs chiefly in granite pegmatites, biotite schists, greisens, and pneumatolytic hydrothermal veins. Found widespread in the United States, notably in Riverside and San Diego counties, California (morganite); as fine red crystals in rhyolite in the Thomas Mountains, Utah; abundantly as large crystals in Custer and Pennington counties, South Dakota; as beautiful crystals (aquamarine) at Mt. Antero, Colorado; in Alexander County North Carolina (emerald); and at localities in Idaho, Montana, New Mexico, Pennsylvania, Massachusetts, Connecticut, New Hampshire and numerous places in Maine. Other notable occurrences are found in Canada, Mexico, Colombia (emerald), Brazil (emerald, aquamarine, morganite, heliodor), Ireland (aquamarine), Elba (morganite), Austria (emerald), USSR (emerald, morganite, aquamarine), Madagascar (aquamarine, morganite), and at several localities in Africa.

BEST REF. IN ENGLISH: Vlasov, K. A., "Mineralogy of Rare Elements," v. II, p. 98–108, Israel Program for Scientific Translations, 1966.

# BERYLLITE (Berillite)
$Be_3 SiO_4 (OH)_2 \cdot H_2 O$

CRYSTAL SYSTEM: Orthorhombic or monoclinic
3 STRONGEST DIFFRACTION LINES:
4.01 (100)
2.34 (100)
3.64 ( 90)
OPTICAL CONSTANTS:
$\alpha = 1.541$
$\beta = $ ?
$\gamma = 1.560$
$(-)2V$
HARDNESS: 1
DENSITY: 2.196 (Meas.)
CLEAVAGE: Not determined.

HABIT: Spherulites 2-3 mm in diameter, or as fibrous crusts up to 2mm thick.

COLOR-LUSTER: White. Transparent. Silky.

MODE OF OCCURRENCE: Occurs rarely and in small amounts in cavities and coating albite and epididymite in the center of a zoned pegmatite that cuts aegirine lujavrite in the Lovozero Massif, Kola Peninsula, USSR.

BEST REF. IN ENGLISH: Kuzmenko, M. V., *Am. Min.*, **40**: 787-788 (1955). Vlasov, K. A., "Mineralogy of Rare Elements," v. II, p. 96-98, Israel Program for Scientific Translations, 1966.

# BERYLLONITE
NaBePO$_4$

CRYSTAL SYSTEM: Monoclinic
CLASS: 2/m
SPACE GROUP: P2$_1$/n
Z: 12
LATTICE CONSTANTS:
  $a$ = 8.13 kX
  $b$ = 7.76
  $c$ = 14.17
  $\beta$ = 90°00′
3 STRONGEST DIFFRACTION LINES:
  2.84 (100)
  3.65 ( 90)
  2.28 ( 70)
OPTICAL CONSTANTS:
  $\alpha$ = 1.5520
  $\beta$ = 1.5579 (Na)
  $\gamma$ = 1.561
  (−)2V = 67°56′
HARDNESS: 5½–6
DENSITY: 2.84 (Meas.)
         2.794 (Calc.)
CLEAVAGE: {010} perfect
          {100} good, interrupted
          {101} indistinct
Fracture conchoidal. Brittle.
  HABIT: Crystals tabular to short prismatic; often highly modified. Crystal faces frequently etched or rough. Twinning on {101} common; also polysynthetic.
  COLOR-LUSTER: Colorless, white, pale yellow. Transparent to translucent. Vitreous; pearly on perfect cleavage.
  MODE OF OCCURRENCE: Occurs sparingly in granite pegmatite at two localities in Maine. Found at Newry associated with albite, tourmaline, herderite, and eosphorite; and in a disintegrated pegmatite outcrop at McKean Mountain, near Stoneham, with muscovite, albite, smoky quartz crystals, and other minerals.
  BEST REF. IN ENGLISH: Palache, et al., "Dana's System of Mineralogy, 7th Ed., v. II, p. 677–679, New York, Wiley, 1951. Wehrenberg, John P., *Am. Min.*, **39**: 397 (1954).

# BERYLLOSODALITE
Na$_4$BeAlSi$_4$O$_{12}$Cl

CRYSTAL SYSTEM: Tetragonal, cubic
LATTICE CONSTANTS:
  (tetragonal)   (cubic)
  $a$ = 8.583    $a$ = 8.72
  $c$ = 8.817
3 STRONGEST DIFFRACTION LINES:
  3.95 (100)
  2.53 ( 80)
  2.05 ( 80) broad
OPTICAL CONSTANTS:
  (tetragonal)   (cubic)
  $\omega$ = 1.496   $N$ = 1.495
  $\epsilon$ = 1.502
     (+)

HARDNESS: ~ 4
DENSITY: 2.28
CLEAVAGE: Bipyramidal-distinct. Fracture conchoidal.
HABIT: Massive; cryptocrystalline.
COLOR-LUSTER: Rose, bluish, greenish, white (changing to light pink in strong sunlight). Translucent. Vitreous. Fluoresces strongly rose-colored under ultraviolet light.
  MODE OF OCCURRENCE: Occurs as a hydrothermal alteration product of chkalovite in pegmatites of Mt. Sengischorr and Mt. Punkaruaiv, Lovozero massif, Kola Peninsula, USSR, and as veins cutting chkalovite in an albite-analcime vein in nepheline syenite pegmatite, Tugtup agtakôrfia, Ilimaussaq, Greenland.
  BEST REF. IN ENGLISH: Semenov, E. I., and Bykova, A. V., *Am. Min.*, **46**: 241 (1961).

# BERZELIANITE
Cu$_2$Se

CRYSTAL SYSTEM: Cubic
CLASS: 4/m $\bar{3}$ 2/m
SPACE GROUP: Fm3m
Z: 4
LATTICE CONSTANTS:
  $a$ = 5.739
3 STRONGEST DIFFRACTION LINES:
  2.02 (100)
  3.32 ( 90)
  1.726 ( 80)
HARDNESS: 2
Talmadge hardness B.
DENSITY: 6.65 (Meas.)
         6.90 (Calc.)
CLEAVAGE: None. Brittle.
HABIT: Thin veinlets and dendritic crusts; disseminated.
COLOR-LUSTER: Shiny lead gray with faint bluish tint; old surfaces dusty metallic black; opaque; metallic.
  MODE OF OCCURRENCE: Occurs disseminated in a highly weathered siliceous gangue associated with native gold and lead at Aurora, Nevada; as blebs and veinlets in grayish pink calcite gangue associated with athabascaite, umangite, and chalcocite at the Martin Lake mine, Lake Athabasca area, northern Saskatchewan, Canada. Also found associated with other selenides at Zorge and in the iron ores at Lehrbach, Harz Mountains, Germany; finely disseminated in calcite veins in serpentine at the copper mine of Skrikerum, Kalmar, Sweden; and at Cerro de Cacheuta, Mendoza, Argentina.
  BEST REF. IN ENGLISH: Palache, et al., "Dana's System of Mineralogy," 7th Ed., v. I, p. 182–183, New York, Wiley, 1944. Early, J. W., *Am. Min.*, **35**: 337–364 (1950).

# BERZELIITE
(Ca, Na)$_3$(Mg, Mn)$_2$(AsO$_4$)$_3$

CRYSTAL SYSTEM: Cubic
CLASS: 4/m $\bar{3}$ 2/m
SPACE GROUP: Ia3d
Z: 8

LATTICE CONSTANT:
$a$ = 12.35–12.46
3 STRONGEST DIFFRACTION LINES:
2.75 (100)
1.71 ( 70)
1.65 ( 70)
OPTICAL CONSTANT:
$N$ = 1.71
HARDNESS: 4½–5
DENSITY: ~ 4.08 (Mg end-member)
CLEAVAGE: None.  Fracture uneven to subconchoidal.
Brittle.
HABIT: Crystals rare; as minute trapezohedrons with small modifying faces.  Usually massive or as rounded grains.
COLOR-LUSTER: Sulfur to orange-yellow. Transparent to translucent. Resinous.
MODE OF OCCURRENCE: Occurs in limestone skarn associated with caryinite, hausmannite, tephroite, rhodonite, and other minerals.  Found in Sweden at Långban; at the Sjö mine, Grythytte parish; and at the Moss mine, Nordmark.
BEST REF. IN ENGLISH: Palache, et al., "Dana's System of Mineralogy," 7th Ed., v. II p. 681–683, New York, Wiley, 1951.

# BETAFITE
(Ca, Fe, U)$_{2-x}$(Nb, Ti, Ta)$_2$O$_6$(OH, F)$_{1-z}$
Pyrochlore Group

CRYSTAL SYSTEM: Cubic
CLASS: 4/m $\bar{3}$ 2/m
SPACE GROUP: Fd3m
Z: 8
LATTICE CONSTANT:
$a$ = 10.31
3 STRONGEST DIFFRACTION LINES:
2.96  (100)
1.814 ( 45)
1.546 ( 40)
Metamict
OPTICAL CONSTANT:
$N \simeq 1.92$ (variable)
HARDNESS: 3–5½
DENSITY: 4.15 (Meas.)
CLEAVAGE: None.
Fracture conchoidal to uneven. Brittle.
HABIT: Crystals octahedral, often with modifying faces of {011}; occasionally distorted by flattening parallel to a pair of faces of {011}.  Crystals occur up to 6 inches or more in diameter.
COLOR-LUSTER: Black, brown, greenish brown, yellowish brown, yellow. Translucent to opaque.  Submetallic, waxy, greasy, vitreous. Crystal exteriors often altered and dull. Not fluorescent.
MODE OF OCCURRENCE: Occurs as a primary mineral in pegmatites with zircon, fergusonite, thorite, columbite, euxenite, allanite, pyroxene, biotite, sphene, beryl, and many other minerals. In the United States found in small amounts associated with monazite, gahnite, and columbite-tantalite at the Brown Derby No. 1 mine, Gunnison County, Colorado; as small crystals with cyrtolite in the Cady Mountains north of Hector, San Bernardino County, California;

and in the Pidlite pegmatite, Mora County, New Mexico. In Canada found as superb crystals up to 3 or more inches in diameter associated with zircon crystals at the Silver Crater mine, Bancroft, Ontario.  Occurs in relative abundance as excellent large crystals and crystal groups at many places in Madagascar; and it is also found in Norway, USSR, Spain, Manchuria, and Peru.
BEST REF. IN ENGLISH: Frondel, Clifford, USGS Bull. 1064, p. 320–325 (1958).

# BETA-LOMONOSOVITE
Na$_2$Ti$_2$Si$_2$O$_9$·NaH$_2$PO$_4$ (?)

CRYSTAL SYSTEM: Triclinic
LATTICE CONSTANTS:
$a$ =  5.28      $\alpha$ = 102°24′
$b$ =  7.05      $\beta$ =  96°51′
$c$ = 14.50      $\gamma$ =  90°
3 STRONGEST DIFFRACTION LINES:
2.77 (100)
3.45 ( 90)
13.53 ( 80)
OPTICAL CONSTANTS:
$\alpha$ = 1.670
$\beta$ = 1.770
$\gamma$ = 1.779
(–)2V = 10°–20°
HARDNESS: ~ 4
DENSITY: 2.95–2.98
CLEAVAGE: One perfect.
Fracture uneven. Brittle.
HABIT: Tabular to platy masses up to 5 × 4 × 0.3 cm in size.
COLOR-LUSTER: Pale yellow-brown, brown, rose.  Vitreous to pearly on cleavage. Vitreous to greasy on fracture.
MODE OF OCCURRENCE: Occurs associated with microcline, aegirine, arfvedsonite, eudialyite, sodalite, and nepheline in alkalic pegmatites of the Lovozero massif, Kola Peninsula.
BEST REF. IN ENGLISH: Gerasimovskii, V. I., and Kazakova, M. E., *Am. Min.*, 48: 1413–1414 (1963).

# BETA-ROSELITE
Ca$_2$(Co, Mg)(AsO$_4$)$_2$·2H$_2$O
Isostructural with fairfieldite

CRYSTAL SYSTEM: Triclinic
CLASS: $\bar{1}$
SPACE GROUP: P$\bar{1}$
Z: 1
LATTICE CONSTANTS:
$a$ = 5.88      $\alpha$ = 112°19′
$b$ = 7.67      $\beta$ =  71°12′
$c$ = 5.58      $\gamma$ = 119°41′
3 STRONGEST DIFFRACTION LINES:
2.77 (100)
3.20 ( 80)
3.07 ( 80)

OPTICAL CONSTANTS:

$\alpha = 1.723$
$\beta = 1.737$
$\gamma = 1.756$
$(-)2V = 80°\text{-}90°$
HARDNESS: 3½–4
DENSITY: 3.71 (Meas.)
CLEAVAGE: {010} perfect
HABIT: As granular masses; distinct crystals not common.
COLOR-LUSTER: Dark rose red. Transparent to translucent. Vitreous.
MODE OF OCCURRENCE: Occurs as distinct crystals with erythrite and roselite at Bou Azzer, Morocco, and associated with pale pink cobaltian calcite and dark rose-colored roselite in vein material consisting principally of quartz and pyrite at Schneeberg, Saxony, Germany.
BEST REF. IN ENGLISH: Frondel, Clifford, *Am. Min.*, **40**: 828–833 (1955).

# BETA-URANOPHANE (β- uranotile)
$Ca(UO_2)_2(SiO_3)_2(OH)_2 \cdot 5H_2O$

CRYSTAL SYSTEM: Monoclinic
CLASS: 2/m
SPACE GROUP: $P2_1/c$
Z: 4
LATTICE CONSTANTS:
$a = 6.64$
$b = 15.55$
$c = 14.01$
$\beta = 91°$
3 STRONGEST DIFFRACTION LINES:
7.83 (100)
3.90 ( 90)
3.51 ( 60)
OPTICAL CONSTANTS:
$\alpha = 1.660\text{-}1.678$
$\beta = 1.682\text{-}1.723$
$\gamma = 1.689\text{-}1.730$
2V = small to 71°
HARDNESS: ~ 2½–3
DENSITY: 3.90 (Meas.)
3.93 (Calc.)
CLEAVAGE: {010} perfect
{100} reported
Fracture conchoidal. Brittle.
HABIT: Crystals prismatic, up to 1 mm long, elongated along c-axis, with square or rectangular cross section; sometimes flattened on {010}. Crystals often terminated by large face of {001}. As felt-like coatings or radial to fan-shaped aggregates of acicular crystals; and as dense aggregates, often pseudomorphous after uraninite.
COLOR-LUSTER: Yellowish green to yellow. Transparent to translucent. Crystals vitreous; fibrous aggregates silky; dense massive aggregates greasy to waxy. Fluoresces weak green in ultraviolet light.
MODE OF OCCURRENCE: Occurs as a secondary mineral in the same manner and association as its dimorph, uranophane. Found in the United States in pegmatites in Mitchell County, North Carolina; Bedford, New York; Newry, Maine; Ruggles and Palermo mines in New Hampshire; and in the Black Hills, South Dakota. It is also found in Pennsylvania, Utah, Montana, and New Mexico. In Canada it is found as excellent micro crystals in vugs at the Faraday Uranium mine, Bancroft, Ontario, and at other deposits in Ontario and Quebec. It occurs as crusts of tiny needles on altered uraninite at Joachimsthal, Bohemia, and in uranium deposits in Wölsendorf, Bavaria; in France; Portugal; and Argentina.
BEST REF. IN ENGLISH: Frondel, Clifford, USGS Bull. 1064, p. 307–311 (1958).

# β - URANOTILE = Beta-uranophane

# BETEKHTINITE (Betechtinite)
$Cu_{20}FePb_2S_{15}$

CRYSTAL SYSTEM: Orthorhombic
CLASS: 2/m 2/m 2/m
SPACE GROUP: Immm
Z: 2
LATTICE CONSTANTS:
$a = 3.85$
$b = 14.67$
$c = 22.8$
3 STRONGEST DIFFRACTION LINES:
1.832 (100)
2.93 ( 90)
3.08 ( 80)
HARDNESS: ~ 3
DENSITY: 5.96–6.05 (Meas.)
HABIT: Irregular masses; needle-like from a few tenths of a millimeter in diameter to 2 cm in length and 5 mm in diameter. Some needles have overgrowths of native silver.
MODE OF OCCURRENCE: Occurs in masses cut by bornite, chalcopyrite, and galena, and as needles associated with calcite in veins cutting the Mansfeld copper shale, Germany; also found at Djeskagan, Ud. S. S. R.
BEST REF. IN ENGLISH: Schüller, Arno, and Wohlmann, Erika, *Am. Min.*, **41**: 371–372 (1956). *Acta. Cryst.* **12**: 646 (1959).

# BETPAKDALITE
$CaFe_2H_8(AsO_4)_2(MoO_4)_5 \cdot 10H_2O$

CRYSTAL SYSTEM: Probably Monoclinic
3 STRONGEST DIFFRACTION LINES:
8.75 (100)
3.63 ( 90)
1.53 ( 80)
OPTICAL CONSTANTS:
$\alpha = 1.809$
$\beta = 1.821$
$\gamma = 1.857$
(+)2V = 60° (Calc.)
HARDNESS: ~ 3
DENSITY: 2.98–3.05 (Meas.)
CLEAVAGE: Not determined.
HABIT: Crystals short prismatic with {hkO} and {hOl} faces strongly developed; most are oriented intergrowths of

2 or 3 individuals on {010}; individual crystals 0.005–0.025 mm in size. As powdery, finely crystalline aggregates.

COLOR-LUSTER: Bright lemon yellow with weak greenish or brownish tint. Aggregates dull to waxy; powder vitreous.

MODE OF OCCURRENCE: Occurs associated with quartz, wolframite, huebnerite, pyrite, arsenopyrite, jarosite, ferrimolybdite, opal, hydromica, gypsum, and limonite in the oxidation zone of the Karaoba wolframite deposit, Bet-Pak-Dal Desert, Central Kazakhstan.

BEST REF. IN ENGLISH: Ermilova, L. P., and Senderova, V. M., *Am. Min.*, **47**: 172–173 (1962).

# BEUDANTITE
$PbFe_3 AsO_4 SO_4 (OH)_6$

CRYSTAL SYSTEM: Trigonal
CLASS: $\bar{3} 2/m$
SPACE GROUP: $R\bar{3} 2/m$
Z: 3
LATTICE CONSTANTS:
  $c:a = 1.184$
3 STRONGEST DIFFRACTION LINES:
  3.08 (100)
  5.99 ( 80)
  3.67 ( 70)
OPTICAL CONSTANTS:
  $\omega = 1.96$
  $\epsilon = $    ?
  uniaxial (–) occasionally (–)2V medium
HARDNESS: 3½–4½
DENSITY: 4–4.3 (Meas.)
CLEAVAGE: {0001} easy
HABIT: Crystals rhombohedral, often pseudocubic.
COLOR-LUSTER: Dark green, brown, black. Transparent to translucent. Vitreous to resinous. Streak greenish or grayish yellow.
MODE OF OCCURRENCE: Occurs as a secondary mineral in the oxidation zone of ore deposits. Found at the Mammoth mine, Tiger, Pinal County, Arizona; at Mina San Felix, near Caborca, Sonora, Mexico; and at deposits in France, Germany, Greece, and Western Australia.
BEST REF. IN ENGLISH: Palache, et al., "Dana's System of Mineralogy," 7th Ed., v. II, p. 1001–1002, New York, Wiley, 1951.

# BEUSITE
$(Mn, Fe, Ca, Mg)_3 (PO_4)_2$

CRYSTAL SYSTEM: Monoclinic
CLASS: 2/m
SPACE GROUP: $P2_1/c$
Z: 4
LATTICE CONSTANTS:
  $a = 8.78$
  $b = 11.52$
  $c = 6.15$
  $\beta = 99°25'$

3 STRONGEST DIFFRACTION LINES:
  3.49  (100)
  2.863 (100)
  2.708 ( 60)
OPTICAL CONSTANTS:
  $\alpha = 1.702$
  $\beta = 1.703$
  $\gamma = 1.722$
  (+)2V = 25°
HARDNESS: 5
DENSITY: 3.702 (Meas.)
         3.715 (Calc.)
CLEAVAGE: {010} good
          {100} fair
HABIT: Crystals prismatic, rough, up to 30 cm long; interlaminated with lithiophilite.
COLOR-LUSTER: Reddish brown. Translucent. Vitreous. Streak pale pink.
MODE OF OCCURRENCE: Found in granite pegmatites at Los Aleros, Amanda, and San Salvador, in San Luis Province, Argentina.
BEST REF. IN ENGLISH: Hurlbut, C. S., Jr., and Aristarain, L. F., *Am. Min.*, **53**: 1799–1814 (1968).

# BEYERITE
$Ca(BiO)_2 (CO_3)_2$

CRYSTAL SYSTEM: Tetragonal
CLASS: 4/m 2/m 2/m
SPACE GROUP: I4/mmm
Z: 2
LATTICE CONSTANTS:
  $a = 3.767$
  $c = 21.690$
3 STRONGEST DIFFRACTION LINES:
  2.84 (100)
  2.14 ( 80)
  1.75 ( 80)
OPTICAL CONSTANTS:
  $\omega = 2.13$
  $\epsilon = 1.97$
   (–)
HARDNESS: 2–3
DENSITY: 6.56 (Meas.)
         6.58 (Calc.)
CLEAVAGE: Fracture conchoidal.
HABIT: Crystals thin flattened rectangular plates, also as compact very fine grained earthy masses.
COLOR-LUSTER: White, yellowish white to bright yellow; also grayish green and gray. Crystals vitreous; massive material earthy.
MODE OF OCCURRENCE: Occurs as a secondary mineral commonly associated with bismutite. Found as minute crystals in small cavities in bismutite near Wickenburg, Maricopa County, Arizona; as compact masses in pegmatite at the Stewart mine, Pala, San Diego County, California; at the Mica Lode and School Section pegmatites in Eight Mile Park, Fremont County, and at the Meyers Ranch pegmatite, Park County, Colorado; and at Schneeberg, Saxony, Germany.

BEST REF. IN ENGLISH: Palache, et al., "Dana's System of Mineralogy," 7th Ed., v. II, p. 281–282, New York, Wiley, 1951.

## BIANCHITE
$(Zn, Fe)SO_4 \cdot 6H_2O$

CRYSTAL SYSTEM: Monoclinic
CLASS: 2/m
SPACE GROUP: C2/c
Z: 8
LATTICE CONSTANTS:
  $a{:}b{:}c = 1.3847{:}1{:}3.3516$  Zn only
  $a{:}b{:}c = 1.3788{:}1{:}3.3324$  Zn:Fe = 2:1
    $\beta = 98°30'$
3 STRONGEST DIFFRACTION LINES:
  4.42  (100)
  4.03  ( 90)
  2.965 ( 80)
OPTICAL CONSTANTS:
    $\alpha = 1.465$
    $\beta = 1.494$
    $\gamma = 1.495$
  $(-)2V = 10°$
HARDNESS: $\sim 2\frac{1}{2}$
DENSITY: 2.07 (Meas. for $ZnSO_4 \cdot 6H_2O$)
         2.031 (Meas. for $ZnFeSO_4 \cdot 6H_2O$ with
              Zn:Fe = 2:1)
CLEAVAGE: Not determined.
HABIT: As crusts of indistinct crystals. Synthetic crystals tabular.
COLOR-LUSTER: Colorless, white, yellowish. Transparent. Vitreous.

Soluble in water.
MODE OF OCCURRENCE: Occurs at Boleslaw, near Olkusz, Poland, and on the walls of a mine at Raibl, near Predil, in the Julian Alps, Italy, associated with gypsum, melanterite, goslarite, and hydrozincite.
BEST REF. IN ENGLISH: Palache, et al., "Dana's System of Mineralogy," 7th Ed., v. II, p. 495–496, New York, Wiley, 1951.

## BIDEAUXITE
$Pb_2 AgCl_3 (F, OH)_2$

CRYSTAL SYSTEM: Isometric
CLASS: $4/m \, \overline{3} \, 2/m$
SPACE GROUP: Fd3m
Z: 16
LATTICE CONSTANT:
  $a = 14.132$
3 STRONGEST DIFFRACTION LINES:
  2.718 (100)
  3.530 ( 90)
  2.497 ( 90)
OPTICAL CONSTANT:
  $N = 2.192$
  HARDNESS: 3

DENSITY: 6.274 (Meas.)
         6.256 (Calc.)
CLEAVAGE: None.
Fracture conchoidal. Brittle; verges on sectile.
HABIT: Crystals small (2–7 mm) showing forms {100}, {111}, {011}, {113}, {114}, {112}, {116}, and {029}. Also massive.
COLOR-LUSTER: Colorless. Adamantine. Pale lavender and dull on exposure to strong light. Streak white.
MODE OF OCCURRENCE: Occurs enveloping and replacing boleite, associated with leadhillite, matlockite, anglesite, cerussite, covellite, and quartz, at the Mammoth-St. Anthony mine, Tiger, Pinal County, Arizona.
BEST REF. IN ENGLISH: Williams, S. A., *Min. Mag.*, 37: 637–640 (1970).

## BIEBERITE
$CoSO_4 \cdot 7H_2O$

CRYSTAL SYSTEM: Monoclinic
CLASS: 2/m
SPACE GROUP: $P2_1/c$
Z: 4
LATTICE CONSTANTS:
  $a = 14.13$
  $b = 6.55$
  $c = 11.00$
  $\beta = 105°05'$
3 STRONGEST DIFFRACTION LINES:
  4.87 (100)
  3.76 ( 75)
  4.82 ( 55)
OPTICAL CONSTANTS:
    $\alpha = 1.4748$
    $\beta = 1.4820$  (Na)(artif.)
    $\gamma = 1.4885$
  $(+)2V = 88°$
HARDNESS: $\sim 2$
DENSITY: 1.948 (Meas.)
         1.942 (Calc.)
CLEAVAGE: {001} perfect
          {110} fair
HABIT: As stalactites and crusts.
COLOR-LUSTER: Rose red, flesh red. Translucent. Vitreous. Dehydrates readily becoming opaque and mealy.

Soluble in water.
MODE OF OCCURRENCE: Occurs as a secondary mineral on pyrrhotite at the Island Mountain deposit, Trinity County, and as an alteration of linnaeite in the Klau quicksilver mine, Santa Lucia Range, San Luis Obispo County, California. It is also found near Copiapó, Chile; at Bieber in Hesse, and at Siegen, Westphalia, Germany; at Leogang, Salzburg, Austria; at Chalanches, Isère, France; and from Lomagundi, Rhodesia.
BEST REF. IN ENGLISH: Palache, et al., "Dana's System of Mineralogy," 7th Ed., v. II, p. 505–507, New York, Wiley, 1951.

# BIKITAITE
LiAlSi$_2$O$_6$ · H$_2$O

CRYSTAL SYSTEM: Monoclinic
CLASS: 2/m
SPACE GROUP: P2$_1$/m
Z: 2
LATTICE CONSTANTS:
  $a$ = 8.63
  $b$ = 4.95
  $c$ = 7.64
  $\beta$ = 114°34′
3 STRONGEST DIFFRACTION LINES:
  3.46 (100)
  3.37 (100)
  4.20 ( 90)
OPTICAL CONSTANTS:
  $\alpha$ = 1.510
  $\beta$ = 0.521
  $\gamma$ = 1.523
  (−)2V = 45°
HARDNESS: 6
DENSITY: 2.29 (Meas.)
        2.29 (Calc.)
CLEAVAGE: {100} perfect
          {001} good
Conchoidal fracture.
HABIT: Crystals prismatic, pseudo-orthorhombic in appearance, up to 6 cm in length and 1 cm across. Also massive.
COLOR-LUSTER: Colorless, transparent; vitreous. Streak white.
MODE OF OCCURRENCE: Occurs associated with eucryptite, quartz, stilbite, calcite, and allophane in a lithium-rich pegmatite at Bikita, Southern Rhodesia.
BEST REF. IN ENGLISH: Hurlbut, Cornelius S. Jr., *Am. Min.*, **42**: 792–797 (1957); **43**: 768–770 (1958).

# BILINITE
Fe$^{2+}$Fe$_2^{3+}$(SO$_4$)$_4$ · 22H$_2$O
Halotrichite Group

CRYSTAL SYSTEM: Monoclinic (?)
OPTICAL CONSTANT:
  $\beta$ = 1.500
HARDNESS: ~2
DENSITY: 1.875 (Meas.)
CLEAVAGE: Not determined.
HABIT: As radial-fibrous aggregates.
COLOR-LUSTER: White to yellowish.
MODE OF OCCURRENCE: Occurs at Schwaz near Bilina, Czechoslovakia, as an alteration of iron sulfide in lignite.
BEST REF. IN ENGLISH: Palache, et al., "Dana's System of Mineralogy," 7th Ed., v. II, p. 529, New York, Wiley, 1951.

# BILLIETITE
BaU$_6$O$_{19}$ · 11H$_2$O

CRYSTAL SYSTEM: Orthorhombic

CLASS: 2/m 2/m 2/m or mm2
SPACE GROUP: Pnmn or Pn2n
Z: 2
LATTICE CONSTANTS:
  $a$ = 7.14
  $b$ = 12.08
  $c$ = 15.10
3 STRONGEST DIFFRACTION LINES:
  7.56 (100)
  3.183 ( 35)
  3.78 ( 25)
OPTICAL CONSTANTS:
  $\alpha$ = 1.76 (Calc.)
  $\beta$ = 1.800
  $\gamma$ = 1.805
  2V$_\alpha$ ≃ 37°
HARDNESS: Not determined.
DENSITY: 5.28 ± 0.01 (Meas.)
        5.27 (Calc.)
CLEAVAGE: {001} perfect
          {110} imperfect
          {010} imperfect
HABIT: Crystals tabular on {001}, pseudohexagonal. Twinning common on {110}, crystals range in size from microscopic up to 2mm.
COLOR-LUSTER: Yellowish amber, yellowish brown. Transparent to translucent. Adamantine. Streak yellow.
MODE OF OCCURRENCE: Occurs sparingly as an alteration product of uraninite, associated with fourmarierite, torbernite, and uranophane, at Shinkolobwe, Katanga, Zaire.
BEST REF. IN ENGLISH: Christ, C. L., and Clark, Joan R., *Am. Min.*, **45**: 1026–1061 (1960).

# BILLINGSLEYITE
Ag$_7$AsS$_6$

CRYSTAL SYSTEM: Orthorhombic
CLASS: 222
SPACE GROUP: C222$_1$
Z: 8
LATTICE CONSTANTS:
  $a$ = 14.96
  $b$ = 14.99  (synthetic)
  $c$ = 10.56
3 STRONGEST DIFFRACTION LINES:
  3.03 (100)
  2.80 ( 70)
  2.47 ( 70)
HARDNESS: 2½
DENSITY: 5.92 ± 0.02 (Meas.)
        5.90 (Calc.)
CLEAVAGE: Slightly sectile; not malleable.
HABIT: Massive; as fine-grained aggregates.
COLOR-LUSTER: Dark lead gray. Opaque. Metallic.
MODE OF OCCURRENCE: Found at the North Lily mine, East Tintic district, Utah, associated with argentite, tennantite, bismuthinite, galena, and pyrite. Believed to have occurred in a body of high-grade silver ore that was stoped in the Ophir shale above the 1200 level in 1930.
BEST REF. IN ENGLISH: Frondel, Clifford, and Honea, Russell M., *Am. Min.*, **53**: 1791–1798 (1968).

# BINDHEIMITE

$Pb_{2-y}Sb_{2-x}(O, OH, H_2O)_{6-7}$ or $Pb_2Sb_2O_6(O, OH)$

CRYSTAL SYSTEM: Cubic
CLASS: $4/m \bar{3} 2/m$
SPACE GROUP: Fd3m
Z: 8
LATTICE CONSTANT:
  $a = 10.41$
3 STRONGEST DIFFRACTION LINES:
  3.01 (100)
  1.85 ( 25) (synthetic)
  2.61 ( 18)
OPTICAL CONSTANT:
  $N = 1.84-1.93$
  HARDNESS: 4-4½
  DENSITY: 4.6-7.32 (Meas.)
  CLEAVAGE: Fracture conchoidal to earthy.
  HABIT: Massive, cryptocrystalline; as crusts, nodular or reniform masses with concentric layering; pseudomorphous; rarely opaline.
  COLOR-LUSTER: Yellow to brown or reddish brown; also greenish, gray, white. Translucent to opaque. Resinous, dull, or earthy.
  MODE OF OCCURRENCE: Occurs widespread as a secondary mineral in the oxidation zone of lead-antimony ore deposits. Found in Fresno, Inyo, San Bernardino, and Santa Cruz counties, California; at numerous deposits in Nevada; in Beaver, Box Elder, Juab, Summit, and Tooele counties, Utah; in the Black Hills, South Dakota; and in the Coeur d'Alene district, Shoshone County, Idaho. It also is found in Mexico, Bolivia, England, Italy, Austria, USSR, Algeria, Australia, and at many other localities.
  BEST REF. IN ENGLISH: Palache, et al., "Dana's System of Mineralogy," 7th Ed., v. II, p. 1018-1020, New York, Wiley, 1951.

# BINNITE = Tennantite

# BIOTITE

$K(Mg, Fe)_3(Al, Fe)Si_3O_{10}(OH, F)_2$
Mica Group
Var: Lepidomelane
     Manganophyllite
     Siderophyllite

CRYSTAL SYSTEM: Monoclinic (1M), monoclinic (2M), trigonal (3T)
CLASS: m (1M polymorph), 2/m (2M polymorph), 3 2 (3T polymorph)
SPACE GROUP: Cm, C2/c, $C3_1 12$ or $C3_2 12$
Z: 2
LATTICE CONSTANTS:

| | 1M | 2M | 3T |
|---|---|---|---|
| $a =$ | 5.3 | 5.3 | 5.3 |
| $b =$ | 9.2 | 9.2 | |
| $c =$ | 10.2 | 20.2 | 30.0 |
| $\beta =$ | 100° | 95° | |

3 STRONGEST DIFFRACTION LINES:
  10.1 (100)
  3.37 (100) (2M)
  2.66 ( 80)

OPTICAL CONSTANTS:
  $\alpha = 1.565-1.625$
  $\beta = 1.605-1.696$
  $\gamma = 1.605-1.696$
  $(-)2V_\alpha = 0°-25°$
  HARDNESS: 2½-3
  DENSITY: 2.7-3.4 (Meas.)
  CLEAVAGE: {001} perfect
Thin laminae flexible to brittle.
  HABIT: Crystals tabular or short prismatic, often hexagonal in cross section. Usually as massive aggregates of cleavable scales, or disseminated. Twinning on {001}, twin axis [310].
  COLOR-LUSTER: Black, dark shades of brown, reddish brown or green, rarely white. Transparent to nearly opaque. Splendent, almost submetallic; also vitreous, often pearly on cleavage. Streak uncolored.
  MODE OF OCCURRENCE: Occurs very widespread chiefly in granites, pegmatites, gabbros, norites, diorites, schists, phyllites, and gneisses. Typical occurrences are found in Alaska, California, Idaho, Colorado (siderophyllite), Arizona, New Mexico, South Dakota, Pennsylvania, Maine, New Hampshire, Massachusetts, New York, Connecticut, and North Carolina. Also in Canada, Ireland, Scotland, Norway (lepidomelane), Sweden (manganophyllite), Italy, Austria, Poland, USSR, Nigeria, Japan, and New Zealand.
  BEST REF. IN ENGLISH: Deer, Howie, and Zussman, "Rock Forming Minerals," v. 3, p. 55-84, New York, Wiley, 1962.

# BIPHOSPHAMMITE

$NH_4H_2PO_4$

CRYSTAL SYSTEM: Tetragonal
CLASS: $\bar{4} 2m$
SPACE GROUP: $I\bar{4} 2d$
Z: 1
LATTICE CONSTANTS:
  $a = 7.4935$
  $c = 7.340$
3 STRONGEST DIFFRACTION LINES:
  3.75 (100)
  5.24 ( 90)
  3.02 ( 90)
OPTICAL CONSTANTS:
  $\omega = 1.525$
  $\epsilon = 1.480$
  HARDNESS: Very soft.
  DENSITY: 2.04
  CLEAVAGE: None.
  HABIT: Crystals prismatic, tapered, with pyramidal terminations. As rounded crusts with radiating groups of crystals (up to $2 \times 0.2$ mm).
  COLOR-LUSTER: Colorless to light buff. Transparent to translucent.

  Soluble in water.
  MODE OF OCCURRENCE: Occurs, apparently as a crystalline product of the liquid fraction of bat guano, in Murra El Elevyn cave, Western Australia. Also reported to occur in the guano of Guañape Island.

BEST REF. IN ENGLISH: Pryce, M. W., *Min. Mag.*, 38: 965-967 (1972).

## BIRINGUCCITE (Hoeferite)
$Na_4B_{10}O_{17} \cdot 4H_2O$

CRYSTAL SYSTEM: Monoclinic (?)
3 STRONGEST DIFFRACTION LINES:
  10.32 (100)
  3.453 ( 80)
  5.17 ( 60)
OPTICAL CONSTANTS:
  $\alpha = 1.496$
  $\beta = 1.539$
  $\gamma = 1.557$
  $(-)2V = 62.7°$
HARDNESS: Not given
DENSITY: Not given.
CLEAVAGE: [001] good
         [100] good
HABIT: Tiny laminae or needles with hexagonal cross section.
COLOR-LUSTER: Not given.
MODE OF OCCURRENCE: Occurs as incrustations of a mechanically inseparable mixture of biringuccite, nasinite, thenardite, orpiment, and minor silicates on tubing of the "Hole of the Storehouse" drilled in 1927 at Lardarello, Tuscany, Italy.
BEST REF. IN ENGLISH: Cipriani, Curzio, and Vannuccini, Piero, *Am. Min.*, 48: 709-711 (1963).

## BIRNESSITE
$(Na, Ca)Mn_7O_{14} \cdot 3H_2O$

CRYSTAL SYSTEM: Hexagonal
LATTICE CONSTANTS:
  $a = 5.82$
  $c = 14.62$
3 STRONGEST DIFFRACTION LINES:
            (synthetic)
  7.27 (100)    2.44 (100)
  2.44 ( 70)    4.75 ( 90)
  1.41 ( 70)    7.2 ( 80)
OPTICAL CONSTANTS:
  $\omega = 1.73$ (variable)
  $\epsilon = 1.69$
    $(-)$
HARDNESS: 1½
DENSITY: 3.0 (Meas.)
CLEAVAGE: Not determined.
HABIT: Minute grains, some exhibiting well-defined faces.
COLOR-LUSTER: Black; dark brown in transmitted light. Nearly opaque.
MODE OF OCCURRENCE: Occurs cementing gravel in a pan 0.5-1.5 inches thick at a depth of about 12 feet at Birness, Aberdeenshire, Scotland; also found at Cummington, Massachusetts; at Sterling Hill, New Jersey; and on colemanite at Boron, California.
BEST REF. IN ENGLISH: Jones, L. H. P., and Milne, Angela A., *Min. Mag.*, 31: 283-288 (1956).

## BIRUNITE (Inadequately described mineral.)
$8.5CaSiO_3 \cdot 5.5CaCO_3 \cdot CaSO_4 \cdot 15H_2O$
(Formula: Near thaumasite.)

CRYSTAL SYSTEM: Orthorhombic (?)
3 STRONGEST DIFFRACTION LINES:
  2.595 (100)
  1.781 (100)
  1.939 ( 80)
OPTICAL CONSTANTS:
  $\alpha = 1.527$
  $\beta = ?$
  $\gamma = 1.531$
  $(+)2V = ?$
HARDNESS: 2
DENSITY: 2.36
CLEAVAGE: One perfect.
HABIT: Fibrous incrustations 2.0-3.0 mm thick.
COLOR-LUSTER: White; dull.
MODE OF OCCURRENCE: Occurs bordering veinlets of thaumasite that fills fractures in enstatitic rock in one part of the Kurgashinkan deposit, Almalyk ore field, Uzbekistan S.S.R.
BEST REF. IN ENGLISH: Badalov, S. T., and Golovanov, I. M., *Am. Min.*, 44: 907 (1959).

## BISBEEITE = Chrysocolla plus plancheite

## BISCHOFITE
$MgCl_2 \cdot 6H_2O$

CRYSTAL SYSTEM: Monoclinic
CLASS: 2/m
SPACE GROUP: C2/m
Z: 2
LATTICE CONSTANTS:
  $a = 9.90$
  $b = 7.15$
  $c = 6.10$
  $\beta = 93°42'$
3 STRONGEST DIFFRACTION LINES:
  4.10 (100)
  2.65 ( 75)
  2.88 ( 50)
OPTICAL CONSTANTS:
  $\alpha = 1.495$
  $\beta = 1.507$
  $\gamma = 1.528$
  $(+)2V = 79°24'$
HARDNESS: 1-2
DENSITY: 1.56 (Meas.)
CLEAVAGE: None.
Fracture conchoidal to uneven.
HABIT: Crystals short prismatic. Usually granular and foliated; also fibrous.
COLOR-LUSTER: Colorless, white. Transparent to translucent. Vitreous to dull.

Deliquescent. Taste bitter.
MODE OF OCCURRENCE: Occurs in salt deposits at Strassfurt, Vienenburg, and Leopoldshall in northern Germany.

BEST REF. IN ENGLISH: Palache, et al., "Dana's System of Mineralogy," 7th Ed., v. II, p. 46–47, New York, Wiley, 1951.

# BISMITE

$\alpha$-$Bi_2O_3$

Dimorphous with sillenite

CRYSTAL SYSTEM: Monoclinic, pseudo-orthorhombic
CLASS: 2/m
SPACE GROUP: $P2_1/c$
Z: 4
LATTICE CONSTANTS:
  $a = 5.83$
  $b = 8.14$
  $c = 7.48$
  $\beta = 67°04'$
3 STRONGEST DIFFRACTION LINES:
  3.253 (100)
  2.693 ( 40)
  2.708 ( 40)
OPTICAL CONSTANTS:
  biaxial indices >2.42.
  HARDNESS: 4½, decreasing in earthy material.
  DENSITY: 8.64–9.22 (Meas.)
           9.370 (Calc.)
  CLEAVAGE: Not determined.   Fracture of massive material uneven to earthy.
  HABIT: Massive, compact, very fine grained.   Also earthy and powdery.
  COLOR-LUSTER: Yellow, greenish yellow, grayish green. Translucent; small fragments transparent. Subadamantine to dull. Streak yellow to grayish.
  MODE OF OCCURRENCE: Occurs as an alteration product of native bismuth and other bismuth minerals. Reported from California, Utah, New Mexico, Colorado, South Dakota, and North Carolina.   Also from Mexico, Bolivia, England, France, Germany, Czechoslovakia, USSR, and Tasmania.
  BEST REF. IN ENGLISH: Palache, et al., "Dana's System of Mineralogy," 7th Ed., v. I, p. 599–600, New York, Wiley, 1944.

# BISMOCLITE

BiOCl

CRYSTAL SYSTEM: Tetragonal
CLASS: 4/m 2/m 2/m
SPACE GROUP: P4/nmm
Z: 2
LATTICE CONSTANTS:
  $a = 3.883$
  $c = 7.347$
3 STRONGEST DIFFRACTION LINES:
  3.44  (100)
  2.677 ( 95)
  2.753 ( 75)
OPTICAL CONSTANTS:
  $\omega = 2.15$
  $\epsilon = ?$
  uniaxial (−)

HARDNESS: 2-2½
DENSITY: 7.717 (Meas.)
         7.76 (Calc.)
CLEAVAGE: {001} perfect
Plastic.
  HABIT: Massive; compact, earthy, columnar to fibrous; as tiny scale-like crystals.
  COLOR-LUSTER: Whitish, grayish, yellowish brown. Transparent to translucent.   Greasy, silky; pearly on cleavage. Massive material earthy to dull.
  MODE OF OCCURRENCE: Occurs as a secondary mineral formed by the alteration of native bismuth or bismuthinite.   In the United States it is found associated with iodyrite at Goldfield, Nevada; and associated with bismutite, cerussite, jarosite, and alunite at the Blue Bell and Eagle mines, Tintic district, Juab County, Utah.   It is also found near Jackals Water, Namaqualand, South Africa; and at Bygoo, New South Wales, Australia.
  BEST REF. IN ENGLISH: Palache, et al., "Dana's System of Mineralogy," 7th Ed., v. II, p. 60–62, New York, Wiley, 1951.

# BISMUTH

Bi

CRYSTAL SYSTEM: Trigonal
CLASS: $\bar{3}$ 2/m
SPACE GROUP: R$\bar{3}$m
Z: 6
LATTICE CONSTANTS:
  $a = 4.537$
  $c = 11.838$
3 STRONGEST DIFFRACTION LINES:
  3.28  (100)
  2.273 ( 41)
  2.39  ( 40)
HARDNESS: 2-2½
DENSITY: 9.70–9.83 (Meas.)
         9.798 (Calc.)
CLEAVAGE: {0001} perfect, easy
          {10$\bar{1}$1} good
          {10$\bar{1}$4} imperfect
Brittle; sectile.
  HABIT: As indistinct crystals; usually massive, cleavable; also in reticulated and branching shapes; granular; foliated. Twinning on {10$\bar{1}$4}, common.
  COLOR-LUSTER: Silver white with pinkish tint; tarnishes iridescent. Opaque. Metallic. Streak silver white.
  MODE OF OCCURRENCE: Occurs chiefly in hydrothermal veins associated with silver, cobalt, nickel, uranium, and tin minerals; less commonly in pegmatites, quartz veins, and placer deposits.   Found at several places in California, especially in the pegmatites of San Diego County; in high-temperature vein deposits and pegmatites in Custer and Pennington counties, South Dakota; in placers and vein deposits at numerous localities in Colorado; in the Chesterfield district, South Carolina; and at Monroe, Connecticut. It also occurs as excellent specimens in the Cobalt district, Ontario, and at Great Bear Lake, Canada; in abundance at many deposits in Bolivia; and at localities in England, France, Germany, Czechoslovakia, Sweden, South Africa, Madagascar, and Australia.

BEST REF. IN ENGLISH: Palache, et al., "Dana's System of Mineralogy," 7th Ed., v. I, p. 134–135, New York, Wiley, 1944.

## BISMUTHINITE
$Bi_2S_3$

CRYSTAL SYSTEM: Orthorhombic
CLASS: 2/m 2/m 2/m
SPACE GROUP: Pbnm
Z: 4
LATTICE CONSTANTS:
 $a = 11.13$
 $b = 11.27$
 $c = 3.97$
3 STRONGEST DIFFRACTION LINES:
 3.569 (100)
 3.118 ( 80)
 3.530 ( 60)
HARDNESS: 2
DENSITY: 6.78 (Meas.)
 6.81 (Calc.)
CLEAVAGE: {010} perfect, easy
 {100} imperfect
 {110} imperfect
Flexible. Sectile.
HABIT: Crystals prismatic, stout to slender or acicular; vertically striated; less commonly dipyramidal. Usually massive, foliated or fibrous.
COLOR-LUSTER: Lead gray to tin white; often tarnished iridescent or yellowish. Opaque. Metallic. Streak lead gray.
MODE OF OCCURRENCE: Occurs chiefly in low- to high-temperature hydrothermal vein deposits and in granite pegmatites, commonly associated with other sulfides and native bismuth. Found at numerous localities in California; in Beaver County, Utah; at Wickes, Jefferson County, Montana; in pegmatites in the Keystone and Custer districts, Black Hills, South Dakota; in Delaware County, Pennsylvania; and at Haddam, Connecticut. It also occurs in Canada and Mexico; as significant ore deposits in Bolivia; and in numerous European and Australian mining districts.
BEST REF. IN ENGLISH: Palache, et al., "Dana's System of Mineralogy," 7th Ed., v. I, p. 275–278, New York, Wiley, 1944.

## BISMUTITE
$(BiO)_2CO_3$

CRYSTAL SYSTEM: Tetragonal (also reported as
 orthorhombic)
CLASS: 4/m 2/m 2/m
SPACE GROUP: I4/mmm
Z: 2
LATTICE CONSTANTS:
 $a = 3.859$ kX
 $c = 13.658$
3 STRONGEST DIFFRACTION LINES:
 2.95 (100)
 2.14 (100)
 1.62 (100)

OPTICAL CONSTANTS: Mean index ranges from ~2.12 to ~2.30 due to variable water content.
HARDNESS: Variable, usually 2½–3½.
DENSITY: Variable; 6.1–7.7 (Meas.)
 8.15 (Meas. synthetic)
 8.28 (Calc.)
CLEAVAGE: {001} distinct
HABIT: Massive, dense to earthy. As fibrous or chalcedonic crusts, and as scaly to lamellar aggregates.
COLOR-LUSTER: Yellowish white to brownish yellow to brown; also greenish, gray to black. Sometimes discolored bright green or blue by copper minerals. Transparent to subtranslucent. Vitreous to pearly to dull.
MODE OF OCCURRENCE: Occurs widespread as a secondary mineral in the oxidized zone of veins and pegmatites containing primary bismuth minerals. Found in the United States especially in Colorado, Arizona, New Mexico, South Dakota, Utah, California, Connecticut, and North Carolina. Among many other places, it occurs in Mexico, Peru, Bolivia, Brazil, England, France, Germany, USSR, Australia, and Madagascar.
BEST REF. IN ENGLISH: Palache, et al., "Dana's System of Mineralogy," 7th Ed., v. II, p. 259–262, New York, Wiley, 1951.

## BISMUTOFERRITE
$BiFe_2(SiO_4)_2(OH)$

CRYSTAL SYSTEM: Orthorhombic
3 STRONGEST DIFFRACTION LINES:
 7.63 (100)
 3.87 (100)
 2.90 ( 70)
OPTICAL CONSTANTS:
 $\alpha = 1.93$
 $\beta = 1.97$
 $\gamma = 2.01$
DENSITY: 4.47 (Meas.)
HABIT: Massive, compact. Also powdery.
COLOR-LUSTER: Bright yellow, green.
MODE OF OCCURRENCE: Occurs associated with quartz, chalcedony, bismuth, cobaltite, and arsenopyrite in veins in shale at Schneeberg, and associated with quartz at Ullersreuth and Gersdorf, Saxony, Germany.
BEST REF. IN ENGLISH: Milton, Charles, Axelrod, Joseph M., and Ingram, Blanche, *Am. Min.*, **43**: 656–670 (1958).

## BISMUTONIOBITE
Hypothetical end member $BiNbO_4$

## BISMUTOTANTALITE
$(Bi, Sb)(Ta, Nb)O_4$

CRYSTAL SYSTEM: Orthorhombic
CLASS: 2/m 2/m 2/m or mm2
SPACE GROUP: Pcmn or Pcn2₁
Z: 4

LATTICE CONSTANTS:
$a = 4.97$
$b = 11.80$
$c = 5.66$
3 STRONGEST DIFFRACTION LINES:
3.148 (100)
2.945 ( 60)
3.555 ( 30)
OPTICAL CONSTANTS:
$\alpha = 2.388$
$\beta = 2.403$
$\gamma = 2.428$ (Li)
(+)2V = 80°
HARDNESS: 5
DENSITY: 8.51 (Meas. after heating–Uganda)
8.73 (Calc.–Uganda)
8.84 (Meas.–Brazil)
8.98 (Calc.–Brazil)
CLEAVAGE: {010} perfect
{101} distinct
Fracture subconchoidal.
HABIT: Crystals stout prismatic, large, often irregular. Also massive, as stream pebbles.
COLOR-LUSTER: Light brown to black; adamantine to submetallic. Streak yellow-brown to black.
MODE OF OCCURRENCE: Found in pegmatite associated with muscovite, black tourmaline, and small amounts of cassiterite at Gamba Hill, southwest Uganda, Africa, and in light brown, rounded stream pebbles, up to two inches in diameter, at Acari, Campina Grande area, Brazil.
BEST REF. IN ENGLISH: Hurlbut, Cornelius S. Jr., *Am. Min.*, **42**: 178–183 (1957).

# BITYITE
$Ca(Al, Li)_2 [(Al, Be)_2 Si_2 (O, OH)_{10}] \cdot H_2O$

CRYSTAL SYSTEM: Monoclinic
CLASS: 2/m
SPACE GROUP: C2/c
Z: 4
LATTICE CONSTANTS:
$a = 4.98$
$b = 8.67$
$c = 18.74$
$\beta \simeq 90°$
3 STRONGEST DIFFRACTION LINES:
2.480 (100)
1.45 (100)
2.043 ( 90)
OPTICAL CONSTANTS:
$\alpha = 1.651$
$\beta = 1.659$
$\gamma = 1.661$
(-)2V = 35°-52°
HARDNESS: 5½
DENSITY: 3.02-3.07 (Meas.)
3.14 (Calc.)
CLEAVAGE: {001} perfect
HABIT: Crystals thin tabular, pseudohexagonal, up to 2mm in diameter. As rosettes and dense micaceous aggregates.

COLOR-LUSTER: White, yellow, brownish white. Transparent.

Easily fusible.
MODE OF OCCURRENCE: Occurs sparingly in lithium-rich pegmatites. Found associated with pink tourmaline, albite, and lepidolite at Maharitra, on Mt. Bity, Madagascar, and in association with albite, beryl, bavenite, columbite, and cassiterite, at Londonderry, Western Australia.
BEST REF. IN ENGLISH: Vlasov, K. A., "Mineralogy of Rare Elements," v. II, p. 114–116, Israel Program for Scientific Translations, 1966.

# BIXBYITE
$(Mn, Fe)_2 O_3$

CRYSTAL SYSTEM: Cubic
CLASS: 2/m $\bar{3}$
SPACE GROUP: Ia 3
Z: 16
LATTICE CONSTANT:
$a = 9.365$
3 STRONGEST DIFFRACTION LINES:
2.72 (100)
1.657 ( 90)
1.421 ( 80)
HARDNESS: 6-6½
DENSITY: 4.945 (Meas.)
5.068 (Calc.)
CLEAVAGE: {111} in traces
Fracture subconchoidal to uneven.
HABIT: Crystals cubic, sometimes highly modified. Twinning on {111} as penetration twins.
COLOR-LUSTER: Black. Opaque. Brilliant metallic to submetallic. Streak black.
MODE OF OCCURRENCE: Occurs associated with topaz in cavities in rhyolite in the Thomas Mountains, Juab County, Utah; as fine highly modified crystals in rhyolite from San Luis Potosi, Mexico; and at localities in Sweden, Spain, India, and South Africa.
BEST REF. IN ENGLISH: Palache, et al., "Dana's System of Mineralogy," 7th Ed., v. I, p. 550-551, New York, Wiley, 1944.

# BJAREBYITE
$(Ba, Sr)(Mn, Fe, Mg)_2 Al_2 (OH)_3 (PO_4)_3$

CRYSTAL SYSTEM: Monoclinic
CLASS: 2/m
SPACE GROUP: $P2_1/m$
Z: 2
LATTICE CONSTANTS:
$a = 8.930$
$b = 12.073$
$c = 4.917$
$\beta = 100.15°$
3 STRONGEST DIFFRACTION LINES:
3.090 (100)
8.81 ( 70)
2.681 ( 70)

OPTICAL CONSTANTS:
  $\alpha = 1.692$
  $\beta = 1.695$
  $\gamma = 1.710$
  Biaxial (+), $2V \sim 35°$
HARDNESS: $4^+$
DENSITY: 4.02 (Calc.)
CLEAVAGE: $\{010\}$ perfect
                $\{100\}$ perfect
HABIT: Crystals very complex, steeply terminated, up to 3 mm in greatest dimension.
COLOR-LUSTER: Emerald green with faint bluish tinge. Subadamantine. Streak white.
MODE OF OCCURRENCE: Bjarebyite occurs as a rare secondary mineral derived from metasomatized ambly-gonite-scorzalite (in probable close association with triphylite) at the Palermo No. 1 pegmatite, near North Groton, New Hampshire. It is found in open cavities associated with amblygonite, augelite, childrenite, siderite, scorzalite, quartz, minor sulfides, Fe-Mn oxides, and palermoite.
  BEST REF. IN ENGLISH: Moore, Paul B., Lund, Dennis H., and Keester, Kenneth L., *Min. Rec.* (in press).

## BLAKEITE
$Fe_2(TeO_3)_3$

CRYSTAL SYSTEM: Unknown
3 STRONGEST DIFFRACTION LINES:
  3.00 (100)
  2.54 ( 90)
  1.72 ( 80)
OPTICAL CONSTANTS: Nearly isotropic.
  $N = 2.16$ (Li)
HARDNESS: $>2$
DENSITY: $>3.1$ (Meas.)
CLEAVAGE: Friable.
HABIT: Massive, as microcrystalline crusts.
COLOR-LUSTER: Dark brown to dark reddish brown. Dull. Streak yellowish brown.
MODE OF OCCURRENCE: Occurs as an uncommon associate of emmonsite and mackayite in the oxidized portion of a vein containing pyrite and native tellurium at Goldfield, Nevada.
  BEST REF. IN ENGLISH: Palache, et al., "Dana's System of Mineralogy," 7th Ed., v. II, p. 643, New York, Wiley, 1951. Sindeeva, N. D., "Mineralogy and Types of Deposits of Selenium and Tellurium," Wiley, 1964.

## BLANFORDITE = Manganese-rich variety of aegirine

## BLENDE = Sphalerite

## BLIXITE
$Pb_2Cl(O, OH)_{2-x}$

CRYSTAL SYSTEM: Orthorhombic
Z: 8

LATTICE CONSTANTS:
  $a = 5.832$
  $b = 5.694$
  $c = 25.47$
3 STRONGEST DIFFRACTION LINES:
  2.93 (100)
  3.88 ( 80)
  1.660 ( 80)
OPTICAL CONSTANTS:
  $\alpha \sim 2.05$
  $\beta$ ?
  $\gamma \sim 2.20$
  (+)$2V = 80°$
HARDNESS: $\sim3$
DENSITY: 7.35
CLEAVAGE: One distinct.
HABIT: As thin crystalline coatings.
COLOR-LUSTER: Pale yellow; vitreous, sometimes dull. Streak pale yellow.
MODE OF OCCURRENCE: Occurs as coatings on fissures in dolomite impregnated with hausmannite and sometimes associated with native copper, also as a coating of a fissure in manganophyllite skarn in the "Amerika" stope, Långban, Sweden.
  BEST REF. IN ENGLISH: Gabrielson, Olof, Parwel, Alexander, and Wickman, Frans E., *Am. Min.*, **45**: 908 (1960).

## BLOCKITE = Penroseite

## BLOEDITE (Blödite)
$Na_2Mg(SO_4)_2 \cdot 4H_2O$

CRYSTAL SYSTEM: Monoclinic
CLASS: 2/m
SPACE GROUP: $P2_1/a$
Z: 2
LATTICE CONSTANTS:
  $a = 11.04$ kX
  $b = 8.15$
  $c = 5.49$
  $\beta = 100°41'$
3 STRONGEST DIFFRACTION LINES:
  3.25 (100)
  4.56 ( 95)
  3.29 ( 95)
OPTICAL CONSTANTS:
  $\alpha = 1.483$
  $\beta = 1.486$
  $\gamma = 1.487$
  (−)$2V = 71°$
HARDNESS: 2½–3
DENSITY: 2.25 (Meas.)
              2.27 (Calc.)
CLEAVAGE: None. Fracture conchoidal. Brittle.
HABIT: Crystals short prismatic, often rich in forms, and sometimes 5 or more centimeters in size. Also massive granular or compact.
COLOR-LUSTER: Colorless; also grayish, bluish green or reddish due to inclusions. Transparent.

Taste slightly saline.

MODE OF OCCURRENCE: Occurs in lacustrine and oceanic salt deposits, and also in the nitrate deposits of the Atacama Desert in Chile. Found in California as large well-developed crystals at Soda Lake, San Luis Obispo County; as a constituent of saline crusts in Deep Spring Lake, Inyo County; and as a 6–12 inch thick layer interstratified with thenardite and clay at the Bertram sodium sulfate deposits in Imperial County. It also occurs at the Laguna Salina, Torrance County, New Mexico, and at deposits in Germany, Poland, Austria, Sicily, USSR, and India.

BEST REF. IN ENGLISH: Palache, et al., "Dana's System of Mineralogy," 7th Ed., v. II, p. 447–450, New York, Wiley, 1951.

## BLOMSTRANDINE = Polycrase

## BLOMSTRANDITE = Variety of betafite

## BLOODSTONE = Variety of quartz

## BOBIERRITE
$Mg_3(PO_4)_2 \cdot 8H_2O$

CRYSTAL SYSTEM: Monoclinic
CLASS: 2/m
SPACE GROUP: $P2_1/c$
Z: 4
LATTICE CONSTANTS:
  $a$ = 9.946 kX
  $b$ = 27.654
  $c$ = 4.639
  $\beta$ = 104°01'
3 STRONGEST DIFFRACTION LINES:
  6.96 (100)
  2.94 ( 27)
  8.04 ( 18)
OPTICAL CONSTANTS:
  (Minn.)          (N.Z.)
    $\alpha$ = 1.510    $\alpha$ = 1.5468
    $\beta$ = 1.520    $\beta$ = 1.5533
    $\gamma$ = 1.543    $\gamma$ = 1.5820
  (+)2V = 71°, 53°
  HARDNESS: 2–2½
  DENSITY: 2.195 (Meas. Synthetic)
           2.17 (Calc.)
  CLEAVAGE: {010} perfect
HABIT: As minute acicular or fibrous crystals; also massive, as flattened aggregates, lamellar, and as minute crystalline aggregates.
COLOR-LUSTER: Colorless, white. Transparent. Weakly vitreous.
MODE OF OCCURRENCE: Found in cavities in a fossil elephant tusk near Edgerton, Pipestone County, Minnesota; with apatite near Bamle, Norway; in guano on Mejillones island, Chile; and in New Zealand.

BEST REF. IN ENGLISH: Palache, et al., "Dana's System of Mineralogy," 7th Ed., v. II, p. 753–754, New York, Wiley, 1951. Frazier, A. William, Lehr, James R., and Smith, James P., *Am. Min.*, **48**: 635–641 (1963).

## BOEHMITE
$AlO \cdot OH$
Dimorph of diaspore

CRYSTAL SYSTEM: Orthorhombic
CLASS: 2/m 2/m 2/m
SPACE GROUP: Amam
Z: 4
LATTICE CONSTANTS:
  $a$ = 3.78
  $b$ = 11.8
  $c$ = 2.85
3 STRONGEST DIFFRACTION LINES:
  6.11  (100)
  3.164 ( 65)
  2.346 ( 53)
OPTICAL CONSTANTS:
    $\alpha$ = 1.638, 1.646
    $\beta$ = 1.645          (somewhat uncertain)
    $\gamma$ = 1.651, 1.661
  (-)2V = moderate; (+)2V = 80°
  HARDNESS: 3
  DENSITY: 3.070 (Calc.)
  CLEAVAGE: {010} good
  HABIT: Crystals microscopic, tabular. Commonly in pisolitic aggregates or disseminated.
  COLOR-LUSTER: White, brown.
  MODE OF OCCURRENCE: Occurs as a major constituent of bauxite, associated with gibbsite, kaolinite, and other supergene minerals; also as a hydrothermal product in pegmatites related to alkaline igneous rocks. Found at the Alberhill clay pits, Riverside County, California; in the Linwood-Barton district, Georgia; and at deposits in Scotland, France, Italy, Germany, Hungary, and elsewhere.

BEST REF. IN ENGLISH: Palache, et al., "Dana's System of Mineralogy," 7th Ed., v. I, p. 645–646, New York, Wiley, 1944.

## BØGGILDITE
$Na_2 Sr_2 Al_2 PO_4 F_9$

CRYSTAL SYSTEM: Monoclinic
CLASS: 2/m
SPACE GROUP: $P2_1/c$
Z: 4
LATTICE CONSTANTS:
  $a$ = 5.24
  $b$ = 10.48
  $c$ = 17.66 (variable, up to 18.52 reported)
  $\beta$ = 107.35°
3 STRONGEST DIFFRACTION LINES:
  3.162 (100)
  3.893 ( 80)
  3.960 ( 65)
OPTICAL CONSTANTS:
    $\alpha$ = 1.462
    $\beta$ = 1.466
    $\gamma$ = 1.469
  (+)2V = 78°–80°
  HARDNESS: 4–5
  DENSITY: 3.66

CLEAVAGE: Not determined.
HABIT: Massive.
COLOR-LUSTER: Flesh red.
MODE OF OCCURRENCE: Occurs associated with siderite, fluorite, black cryolite, quartz, mica, and sphalerite at the contact of cryolite with greisen at Ivigtut, Greenland.
BEST REF. IN ENGLISH: Pauly, Hans, and Moeller, Chr. K., *Am. Min.*, **41**: 959 (1956). Bøgrad, Richard, *Am. Min.*, **39**: 848–849 (1954).

## BOHDANOWICZITE
$AgBiSe_2$
Selenium analogue of matildite

CRYSTAL SYSTEM: Unknown
HARDNESS: 3.2
Microhardness 96 kg/mm$^2$.
HABIT: As anhedral grains. Twinning polysynthetic.
COLOR-LUSTER: Lead gray. Opaque. Metallic.
MODE OF OCCURRENCE: Occurs associated with clausthalite and bornite at Julienehaab, Greenland, and with uraninite, clausthalite, chalcocite, and emplectite in magnetite-fluorite veins and in fluorite-quartz-sulfide veins at Kletna, Poland.
BEST REF. IN ENGLISH: Banas, Marian, and Ottemann, Joachim, *Am. Min.*, **55**: 2135 (1970). *Am. Min.*, **53**: 2103 (1968).

## BOKITE
$KAl_3Fe_6V_6^{4+}V_{20}^{5+}O_{76} \cdot 30H_2O$

CRYSTAL SYSTEM: Unknown
3 STRONGEST DIFFRACTION LINES:
10.20 (100)
3.44 ( 90)
2.61 ( 80)
OPTICAL CONSTANTS:
$\alpha' + \gamma' = 2.01$–2.06
HARDNESS: ~3
DENSITY: 2.97–3.10
CLEAVAGE: One perfect parallel to elongation. One fair perpendicular to plane of perfect cleavage.
HABIT: Grains (0.1–0.3mm long) platy to columnar or wedge-shaped; as reniform crusts with radiating fibrous structure; also as veinlets.
COLOR-LUSTER: Black; semimetallic to dull. Streak black, sometimes with brownish tint.
MODE OF OCCURRENCE: Occurs in carbonaceous vanadiferous shales, Balasauskandyk area, Kara-Tau, USSR.
BEST REF. IN ENGLISH: Ankinovich, E. A., *Am. Min.*, **48**: 1180–1181 (1963).

## BOLDYREVITE  Species status uncertain. Cf. ralstonite.

## BOLÉITE
$Pb_9Cu_8Ag_3Cl_{21}(OH)_{16} \cdot H_2O$

CRYSTAL SYSTEM: Tetragonal
CLASS: 4/m 2/m 2/m

SPACE GROUP: I4/mmm
Z: 12
LATTICE CONSTANTS:
$a = 15.27$
$c = 60.94$
3 STRONGEST DIFFRACTION LINES:
4.40 (100)
3.83 (100)
3.13 (100)
OPTICAL CONSTANTS:
$\omega = 2.05$
$\epsilon = 2.03$
(−)
HARDNESS: 3–3½
DENSITY: 5.054 (Meas.)
5.10 (Calc.)
CLEAVAGE: {001} perfect
{101} good
{100} poor
HABIT: Crystals pseudo-cubic, pseudo-cubo-octahedral and pseudo-dodecahedral. Crystals often overgrown in parallel position by cumengite and pseudoboleite.
COLOR-LUSTER: Indigo blue. Translucent. Weakly vitreous; pearly on cleavages. Streak blue, with greenish tinge.
MODE OF OCCURRENCE: Occurs as a secondary mineral associated with cumengite, pseudoboleite, and other secondary lead minerals, rarely as crystals up to 2 cm on an edge, at Boléo, Baja California, Mexico. It is also found in Chile at Challacollo and at the San Augustin mine near Huantajaya in Tarapacá; and at the South mine, Broken Hill, New South Wales, Australia.
BEST REF. IN ENGLISH: Palache, et al., "Dana's System of Mineralogy," 7th Ed., v. II, p. 78–79, New York, Wiley, 1951.

## BOLIVARITE
$Al_2(PO_4)(OH)_3 \cdot 4$-$5H_2O$

3 STRONGEST DIFFRACTION LINES:  Amorphous to 1050°C.
OPTICAL CONSTANTS: Biaxial to uniaxial (+).
$\eta = 1.50$–1.51
HARDNESS: 2½–3½
DENSITY: 1.97–2.05 (Meas.)
CLEAVAGE: None.
Fracture conchoidal. Brittle.
HABIT: Massive; as cryptocrystalline crusts, botryoidal masses, and veinlets.
COLOR-LUSTER: Bright yellowish green to greenish white. Vitreous. Fluoresces strong bright green under ultraviolet light.
MODE OF OCCURRENCE: Occurs in crevices in granite near Pontevedra, Spain; also found in the weathering zone and phosphate-rich zone of the pegmatite of Kobokobo, Kivu, Zaire. Both occurrences contain abnormal amounts of uranium.
BEST REF. IN ENGLISH: Van Wambeke, L., *Min. Mag.*, **38**: 418–423 (1971).

# BOLTWOODITE
$K_2(UO_2)_2(SiO_3)_2(OH)_2 \cdot 5H_2O$

CRYSTAL SYSTEM: Orthorhombic (may be monoclinic with $b$-axis elongation).
3 STRONGEST DIFFRACTION LINES:
  6.81 (100)
  3.40 ( 90)
  2.95 ( 80)
OPTICAL CONSTANTS:
  $\alpha$ = 1.668–1.670
  $\beta$ = 1.695–1.696
  $\gamma$ = 1.698–1.703
  (–)2V = large
HARDNESS: 3½–4
DENSITY: Near 3.6 (Meas.)
CLEAVAGE: {010} perfect
           {001} imperfect
HABIT: Radiating acicular to fibrous; also dense microcrystalline pseudomorphic aggregates.
COLOR-LUSTER: Pale yellow; luster pearly on cleavage surfaces; radial aggregates vitreous to silky; dull to earthy in microcrystalline pseudomorphs. Fluoresces dull green under both long- and short-wave ultraviolet light.
MODE OF OCCURRENCE: Occurs in the outer silicate zone of alteration surrounding the zone of hydrated uranyl oxides, which encrust primary uraninite. Also found filling fractures and interstitial openings some distance from primary uraninite. Found in widely scattered localities with different geological environments. Originally discovered at the Delta mine, Emery County, Utah.
BEST REF. IN ENGLISH: Honea, Russell M., *Am. Min.*, **46**: 12–25 (1961).

# BONATTITE
$CuSO_4 \cdot 3H_2O$

CRYSTAL SYSTEM: Monoclinic
CLASS: m
SPACE GROUP: Cc
Z: 4
LATTICE CONSTANTS:
  $a$ = 5.59
  $b$ = 13.03
  $c$ = 7.34
  $\beta$ = 97.1°
3 STRONGEST DIFFRACTION LINES:
  4.42 (100)
  5.11 ( 70)
  3.65 ( 60)
OPTICAL CONSTANTS:
  $\alpha$ = 1.554
  $\beta$ = 1.577
  $\gamma$ = 1.618
DENSITY: 2.68 (Calc.)
HABIT: Concretions composed of minute crystals.
COLOR-LUSTER: Blue.
MODE OF OCCURRENCE: Occurs as a secondary mineral in the deposits of Cape Calamita, Elba; and also in the Bonaparte River area, Lillooet district, British Columbia, Canada.

BEST REF. IN ENGLISH: Garavelli, Carlo L., *Am. Min.*, **43**: 180 (1958). *Am. Min.*, **51**: 276 (1966). Jambor, J. L., *Can. Min.*, 7(2): 245–252 (1962).

# BONCHEVITE (species status uncertain)
$Pb_3Bi_2S_6$ (?)

CRYSTAL SYSTEM: Orthorhombic
CLASS: 2/m 2/m 2/m
SPACE GROUP: Bbmm
Z: 4
LATTICE CONSTANTS:
  $a:b:c$ = 0.9004 : 1 : 0.3249
3 STRONGEST DIFFRACTION LINES:
  3.50  (100)
  3.08  ( 80)
  1.939 ( 80)
HARDNESS: 2½
DENSITY: 6.92 (Meas.)
CLEAVAGE: {100} perfect
Fracture uneven to subconchoidal. Very brittle.
HABIT: Crystals acicular or long prismatic up to 2.0 cm in length. Forms: $m$ {110} dominant, also $e$ {011}, $a$ {100}, and $b$ {010}. {100} faces commonly striated along $c$-axis.
COLOR-LUSTER: Lead gray to steel gray; metallic.
MODE OF OCCURRENCE: Occurs associated with scheelite, pyrite, calcite, sphalerite, and molybdenite in quartz veins cutting gneisses and amphibolites close to the so-called Yugovo granite, Central Rhodopian Mts., Bulgaria.
BEST REF. IN ENGLISH: Kostov, Ivan, *Min. Mag.*, **31**: 821–828 (1958). *Am. Min.*, **55**: 1449 (1970).

# BOOTHITE
$CuSO_4 \cdot 7H_2O$

CRYSTAL SYSTEM: Monoclinic
LATTICE CONSTANTS:
  $a:b:c$ = 1.622 : 1 : 1.5000
  $\beta$ = 105°36'
OPTICAL CONSTANTS:
  $\alpha$ = 1.47
  $\beta$ = 1.48
  $\gamma$ = 1.49
  2V = large
HARDNESS: 2–2½
DENSITY: ~ 2.1
CLEAVAGE: {001} imperfect
HABIT: Usually massive with crystalline structure or fibrous. Rarely in crystals.
COLOR-LUSTER: Light blue. Transparent to translucent. Vitreous. Fibrous material silky or pearly.

Soluble in water. Taste harsh metallic.
MODE OF OCCURRENCE: Occurs in California at the Alma mine, Leona Heights, Alameda County, associated with melanterite and other copper and iron sulfates; as crystals and crystalline masses at Campo Seco, Calaveras County; and in the veins of a small gold prospect near the Tunnel Ranch, Figueroa Mountain, Santa Barbara County. It is also found at a pyrite mine at Sain-Bel, Rhône, France.

BEST REF. IN ENGLISH: Palache, et al., "Dana's System of Mineralogy," 7th Ed., v. II, p. 504–505, New York, Wiley, 1951.

## α - BORACITE (Stable above 265°)
$Mg_3B_7O_{13}Cl$

CRYSTAL SYSTEM: Cubic
CLASS: $\bar{4}3$ m
SPACE GROUP: $F\bar{4}3c$
Z: 8
3 STRONGEST DIFFRACTION LINES:
  2.05 (100)
  3.01 ( 90)
  2.70 ( 80)
MODE OF OCCURRENCE: See β-boracite.

## β - BORACITE
$Mg_3B_7O_{13}Cl$

CRYSTAL SYSTEM: Orthorhombic; also (pseudotetragonal).
CLASS: mm 2
SPACE GROUP: $Pca2_1$
Z: 4
LATTICE CONSTANTS:
  $a = 8.54$
  $b = 8.54$
  $c = 12.10$
3 STRONGEST DIFFRACTION LINES:
  2.058 (100)
  3.044 ( 70)
  2.72 ( 50)
OPTICAL CONSTANTS:
  $\alpha = 1.658–1.6622$
  $\beta = 1.662–1.6670$
  $\gamma = 1.668–1.6730$
  $(+)2V = 82\frac{1}{2}°$
HARDNESS: 7–7½
DENSITY: 2.95 (Meas.)
        2.97 (Calc.)
CLEAVAGE: None.
Fracture conchoidal to uneven.
  HABIT: Crystals cubic, dodecahedral, tetrahedral, pseudooctahedral or cubo-octahedral. Usually isolated and embedded. Also fine-grained or fibrous.
  COLOR-LUSTER: Colorless, white, gray, yellow, pale green to dark green, bluish green. Transparent to translucent. Vitreous.
  MODE OF OCCURRENCE: Occurs in the water-insoluble residue of rock salt from the Choctaw salt dome, Iberville Parish, Louisiana; at Otis, San Bernardino County, California; at Aislaby, Yorkshire, England; and at Luneville, La-Meurthe, France. In Germany it is found at many localities as fine crystals in bedded sedimentary deposits of halite, anhydrite, and gypsum; and also in oceanic-type potash deposits, especially in the Strassfurt and Hanover districts.
  BEST REF. IN ENGLISH: Palache, et al., "Dana's System of Mineralogy," 7th Ed., v. II, p. 378–381, New York, Wiley, 1951.

## BORAX
$Na_2B_4O_7 \cdot 10H_2O$

CRYSTAL SYSTEM: Monoclinic
CLASS: 2/m
SPACE GROUP: A2/a
Z: 4
LATTICE CONSTANTS:
  $a = 12.197$
  $b = 10.674$
  $c = 11.858$
  $\beta = 106°41'$
3 STRONGEST DIFFRACTION LINES:
  4.82 (100)
  5.68 ( 90)
  2.56 ( 65)
OPTICAL CONSTANTS:
  $\alpha = 1.4466$
  $\beta = 1.4687$
  $\gamma = 1.4717$
  $(-)2V = 39°58'$ (Calc.)
HARDNESS: 2–2½
DENSITY: 1.715 ± 0.005 (Meas.)
        1.70 (Calc.)
CLEAVAGE: {100} perfect
          {110} slightly imperfect
          {010} in traces
Fracture conchoidal. Very brittle.
  HABIT: Crystals short prismatic and somewhat tabular on {100}. Often densely grouped on massive borax. Twinning on {100} rare.
  COLOR-LUSTER: Colorless; also white to grayish, greenish, or bluish. Transparent to opaque. Vitreous; sometimes earthy. Streak white.

Soluble in water. Taste sweetish.
  MODE OF OCCURRENCE: Occurs in evaporated saline lake and playa deposits, as a deposit from hot springs, and as an efflorescence on the soil in arid regions. Found widespread and abundant in California, especially at the Boron open pit, Kramer, Kern County, in layers up to 10 feet in thickness; as fine large crystals at Borax Lake, Lake County; as excellent large crystals up to 6 inches in diameter at Searles Lake, San Bernardino County; and at other deposits in the state. It is also found in Nevada and New Mexico; in Salta Province, Argentina; and in Tibet, Kashmir, Iran, India, and the USSR.
  BEST REF. IN ENGLISH: Palache, et al., "Dana's System of Mineralogy," 7th Ed., v. II, p. 339–341, New York, Wiley, 1951.

## BORCARITE
$Ca_4MgH_6(BO_3)_4(CO_3)_2$

CRYSTAL SYSTEM: Monoclinic
CLASS: 2/m
SPACE GROUP: P2/m (?)
Z: 2
LATTICE CONSTANTS:
  $a = 17.52$
  $b = 8.40$
  $c = 4.46$
  $\beta = 92°30'$

3 STRONGEST DIFFRACTION LINES:
   7.57  (100)
   2.67  (100)
   1.886 (100)
OPTICAL CONSTANTS:
   $\alpha$ = 1.590
   $\beta$ = 1.651
   $\gamma$ = 1.657
   $(-)2V \simeq 30°$
HARDNESS: 4
DENSITY: 2.77± 0.01 (Meas.)
             2.765 (Calc.)
CLEAVAGE: {100} perfect - smooth
                {110} perfect - striated parallel to $C$
Also cleavages several directions (hKl) and (hOl) with angle $\beta$ 50–70°.
HABIT: Dense masses up to 0.5 meters in diameter, and as veins.
COLOR-LUSTER: Greenish blue to bluish green, sometimes nearly colorless.  Vitreous to slightly pearly on cleavages.
MODE OF OCCURRENCE: Occurs in kotoite marble and in ludwigite-szaibelyite-magnetite rock at an unspecified deposit in Siberia.  It contains inclusions of szaibelyite, calcite, serpentine, magnetite, and spinel.
BEST REF. IN ENGLISH: Pertzev, N. N., Ostravskaya, I. V., and Nikitina, I. B., *Am. Min.*, **50**: 2097 (1965).

# BOŘICKITE
~$CaFe_5(PO_4)_2(OH)_{11} \cdot 3H_2O$

CRYSTAL SYSTEM: Unknown
OPTICAL CONSTANT:
   $N$ = 1.57–1.67
HARDNESS: 3½
DENSITY: ~ 2.70
CLEAVAGE: Not determined.
HABIT: As compact reniform masses.
COLOR-LUSTER: Reddish brown. Opaque. Weak waxy. Streak reddish brown.
MODE OF OCCURRENCE: Found at Nenačovik, south of Kladno, Czechoslovakia, and at Leoben, Styria, Austria. A similar poorly defined substance (foucherite) found at Fouchères, Champagne, France, and at localities in Czechoslovakia, may be bořickite.
BEST REF. IN ENGLISH: Palache, et al., "Dana's System of Mineralogy," 7th Ed., v. II, p. 915–916, New York, Wiley, 1951.

# BORNHARDTITE
$Co_3Se_4$
Pyrite Group

CRYSTAL SYSTEM: Cubic
CLASS: 4/m $\bar{3}$ 2/m
SPACE GROUP: Fd3m
Z: 8
LATTICE CONSTANT:
   $a$ ~ 10.2

3 STRONGEST DIFFRACTION LINES:
   2.7 (100)
   2.4 (100)
   2.3 (100)
   Not reliable
HARDNESS: ~ 4
DENSITY: Not determined.
CLEAVAGE: Not determined.
HABIT: Massive; as microscopic rims on trogtalite crystals.
COLOR-LUSTER: Rose red.
MODE OF OCCURRENCE: Occurs associated with trogtalite, hastite, and an unnamed cobalt selenide as intergrowths in clausthalite at Steinbruch Trogtal, near Lautenthal, Harz, Germany.
BEST REF. IN ENGLISH: Ramdohr, Paul, and Schmitt, Marg., *Am. Min.*, **41**: 164–165 (1956).

# BORNITE
$Cu_5FeS_4$

CRYSTAL SYSTEM: Tetragonal
CLASS: $\bar{4}$2 m
SPACE GROUP: P$\bar{4}2_1$c
Z: 16
LATTICE CONSTANTS:
   $a$ = 10.94
   $c$ = 21.88
3 STRONGEST DIFFRACTION LINES:
   1.937 (100)
   3.18  ( 60)
   2.74  ( 50)
HARDNESS: 3
DENSITY: 5.079 (Meas.)
             5.09  (Calc. x-ray)
CLEAVAGE: {111} in traces
Fracture uneven to conchoidal.  Brittle.
HABIT: Crystals rare; cubic, octahedral, or dodecahedral, usually with curved or rough faces.  Commonly massive; compact or granular.  Twinning on {111}, often as penetration twins.
COLOR-LUSTER: Copper red or bronze; rapidly tarnishes iridescent purplish. Opaque. Metallic. Streak light grayish black.
MODE OF OCCURRENCE: Occurs widespread chiefly in mesothermal or magmatic hypogene deposits, commonly associated with chalcopyrite, pyrite, and chalcocite. Found in Virginia, North Carolina, and as fine crystals at Bristol, Connecticut; in the western United States at Butte, Montana, and at many copper mines in Arizona, Colorado, Alaska, and California.  It also is found in Canada, Chile, Peru, England, Italy, Germany, Kazakhstan, Madagascar, Tasmania, and South Africa.
BEST REF. IN ENGLISH: Palache, et al., "Dana's System of Mineralogy," 7th Ed., v. I, p. 195–197, New York, Wiley, 1944.   Morimoto, N., *Acta. Cryst.* **17**: 351–360 (1964).

**BOSPHORITE** = Oxidized vivianite

# BOTALLACKITE
$Cu_2Cl(OH)_3$

CRYSTAL SYSTEM: Monoclinic
CLASS: 2/m
SPACE GROUP: $P2_1/m$
Z: 1
LATTICE CONSTANTS:
  $a$ = 5.715
  $b$ = 6.124
  $c$ = 5.632
  $\beta$ = 92°45'
3 STRONGEST DIFFRACTION LINES:
  5.66 (100)
  2.40 ( 80)
  2.57 ( 70)
OPTICAL CONSTANTS:
  $\alpha$ = 1.775
  $\beta$ = 1.800
  $\gamma$ = 1.846
  (+)2V = mod. large
HARDNESS: Not determined.
DENSITY: ~ 3.6 (Meas.)
CLEAVAGE: One direction.
HABIT: As crusts of minute intergrown or columnar crystals; also powdery.
COLOR-LUSTER: Pale green to green; also pale bluish green.
MODE OF OCCURRENCE: Occurs associated with atacamite and paratacamite as a secondary mineral at the Botallack mine, St. Just, Cornwall, England.
BEST REF. IN ENGLISH: Palache, et al., "Dana's System of Mineralogy," 7th Ed., v. II, p. 76-77, New York, Wiley, 1951.

# BOTRYOGEN
$MgFe^{3+}(SO_4)_2OH \cdot 7H_2O$

CRYSTAL SYSTEM: Monoclinic
CLASS: 2/m
SPACE GROUP: $P2_1/n$
Z: 4
LATTICE CONSTANTS:
  $a$ = 10.50
  $b$ = 17.84
  $c$ = 7.12
  $\beta$ = 100°15'
3 STRONGEST DIFFRACTION LINES:
  8.87  (100)
  2.998 ( 80)
  6.29  ( 60)
OPTICAL CONSTANTS:
  $\alpha$ = 1.523
  $\beta$ = 1.530
  $\gamma$ = 1.582
  (+)2V = 42°
HARDNESS: 2-2½
DENSITY: 2.14   (Meas.)
        2.113 (Calc.)

CLEAVAGE: {010} perfect
          {110} good
Fracture conchoidal. Brittle.
HABIT: Crystals short to long prismatic, often striated [001]. Usually as botryoidal or reniform masses with radiating structure.
COLOR-LUSTER: Pale to dark reddish orange. Transparent to translucent. Vitreous. Streak pale yellowish.
MODE OF OCCURRENCE: Occurs at the Palisades mine, near Calistoga, and with copiapite at the Redington mine, Napa County, California; and at Cornwall, Lebanon County, Pennsylvania. It also is found in abundance at Chuquicamata, and at other places in Chile; at the Santa Elena mine, San Juan province, Argentina; at Paracutin, Mexico; and at localities in France, Germany, Sweden, and Iran.
BEST REF. IN ENGLISH: Palache, et al., "Dana's System of Mineralogy," 7th Ed., v. II, p. 617-618, New York, Wiley, 1951.

# BOULANGERITE
$Pb_5Sb_4S_{11}$

CRYSTAL SYSTEM: Monoclinic
CLASS: 2/m
SPACE GROUP: $P2_1/a$
Z: 8
LATTICE CONSTANTS:
  $a$ = 21.56
  $b$ = 23.51
  $c$ =  8.09
  $\beta$ = 100°48'
3 STRONGEST DIFFRACTION LINES:
  3.731 (100)
  3.218 ( 45)
  3.025 ( 40)
HARDNESS: 2½-3
DENSITY: 6.23 (Meas.)
         6.21 (Calc.)
CLEAVAGE: {100} good
Brittle; fibers flexible.
HABIT: Crystals long prismatic to acicular; striated parallel to elongation. Usually in compact fibrous masses; also plumose or schistose.
COLOR-LUSTER: Lead gray to bluish lead gray. Opaque. Metallic. Streak brownish gray to brown.
MODE OF OCCURRENCE: Occurs in low- to medium-temperature hydrothermal vein deposits, commonly associated with jamesonite and other lead sulfosalts, sulfides, carbonates, and quartz. Found in the United States at Silver City, Pennington County, South Dakota; at the Domingo and other nearby mines in Gunnison County, Colorado; at Rocker Gulch and Superior, Montana; in Union County, Nevada; in the Coeur d'Alene and Wood River Districts in Idaho; in Inyo, Mono, and Santa Cruz counties, California; and at the Cleveland mine, Stevens County, Washington. It also occurs in Mexico, Peru, France, Italy, Germany, Czechoslovakia, Yugoslavia, Sweden, USSR, and Algeria.
BEST REF. IN ENGLISH: Palache, et al., "Dana's System of Mineralogy," 7th Ed., v. I, p. 420-423, New York, Wiley, 1944.

# BOURNONITE

$CuPbSbS_3$

CRYSTAL SYSTEM: Orthorhombic
CLASS: mm2
SPACE GROUP: $Pnm2_1$
Z: 4
LATTICE CONSTANTS:
  $a = 8.10$
  $b = 8.65$
  $c = 7.75$
3 STRONGEST DIFFRACTION LINES:
  2.740 (100)
  1.768 ( 50)
  2.685 ( 45)
OPTICAL CONSTANTS:
  $\alpha = 3.141$
  $\beta = 3.166$  ($\lambda = 8521$ Å)
  $\gamma = 3.280$
  (+)2V = 52°
HARDNESS: 2½–3
DENSITY: 5.83 (Meas.)
         5.93 (Calc.)
CLEAVAGE: {010} imperfect
          {100} less perfect
          {001} less perfect
Fracture subconchoidal to uneven. Brittle.
  HABIT: Crystals short prismatic or tabular, often in sub-parallel aggregates. Commonly striated. Also massive, granular to compact. Twinning on {110} common forming cross or cogwheel aggregates.
  COLOR-LUSTER: Steel gray to iron black. Opaque. Metallic, brilliant to dull. Streak steel gray to iron black.
  MODE OF OCCURRENCE: Occurs chiefly in medium-temperature hydrothermal veins, commonly associated with galena, sphalerite, stibnite, chalcopyrite, chalcocite, tetrahedrite, siderite, and quartz. Found in the United States in Arkansas, Colorado, Montana, Nevada, Arizona, California, and as fine crystals at Park City, Utah. Other notable occurrences are found in Canada, Mexico, Chile, Peru, Bolivia, England, France, Spain, Italy, Germany, Roumania, Hungary, Czechoslovakia, and Australia.
  BEST REF. IN ENGLISH: Palache, et al., "Dana's System of Mineralogy," 7th Ed., v. I, p. 406–410, New York, Wiley, 1944.

# BOUSSINGAULTITE

$(NH_4)_2Mg(SO_4)_2 \cdot 6H_2O$

CRYSTAL SYSTEM: Monoclinic
CLASS: 2/m
SPACE GROUP: $P2_1/a$
Z: 2
LATTICE CONSTANTS:
  $a = 9.28$ kX
  $b = 12.57$
  $c = 6.20$
  $\beta = 107°06'$
3 STRONGEST DIFFRACTION LINES:
  4.22  (100)
  2.085 ( 95)
  5.11  ( 85)

OPTICAL CONSTANTS:
  $\alpha = 1.4716$
  $\beta = 1.4730$  (Na)
  $\gamma = 1.4786$
(+)2V = 51°11'
HARDNESS: 2
DENSITY: 1.722 (Meas.)
         1.718 (Calc.)
CLEAVAGE: {$\overline{2}01$} perfect
  HABIT: Crystals short prismatic. Usually stalactitic or as crusts.
  COLOR-LUSTER: Colorless, yellowish pink. Transparent.
  Soluble in water.
  MODE OF OCCURRENCE: Occurs abundantly as crusts and stalactites, associated with other sulfate minerals, at The Geysers, near Cloverdale, Sonoma County, and coating crevices of sandstones and shales at South Mountain, Ventura County, California. It is also found near Mahanoy City, Schuylkill County, Pennsylvania, as crystals formed during the burning of waste heaps of an anthracite coal mine, and in the boric acid fumaroles of Travale, Tuscany, Italy.
  BEST REF. IN ENGLISH: Palache, et al., "Dana's System of Mineralogy," 7th Ed., v. II, p. 455–456, New York, Wiley, 1951.

# BRACEWELLITE

CrO(OH)
Trimorphous with grimaldiite and guyanaite. Isostructural with goethite.

CRYSTAL SYSTEM: Orthorhombic
SPACE GROUP: Pbnm
  HARDNESS: Not determined.
  DENSITY: Not determined.
  CLEAVAGE: Not determined.
  HABIT: Massive; microcrystalline.
  COLOR-LUSTER: Dark brown.
  MODE OF OCCURRENCE: Occurs in complex microscopic intergrowth with eskolaite, guyanaite, grimaldiite, and macconnellite in "merumite" from alluvial gravels in Guyana.
  BEST REF. IN ENGLISH: Milton, C., et al., Geol. Soc. Am., Program 1967 Annual Meetings, p. 151–152 (1967). (abstr.)

# BRACKEBUSCHITE

$Pb_4(Mn, Fe)(VO_4)_4 \cdot 2H_2O$

CRYSTAL SYSTEM: Monoclinic
CLASS: 2/m
SPACE GROUP: $P2_1/m$
Z: 1
LATTICE CONSTANTS:
  $a = 8.92$
  $b = 6.16$
  $c = 7.69$
  $\beta = 111°47'$
3 STRONGEST DIFFRACTION LINES:
  3.25 (100)
  4.95 ( 80)
  2.76 ( 80)

OPTICAL CONSTANTS:
$\alpha = 2.28$
$\beta = 2.36$ (Li)
$\gamma = 2.48$
(+)2V = large
HARDNESS: Not determined.
DENSITY: 6.05 (Meas.)
6.11 (Calc.)
CLEAVAGE: Not determined.
HABIT: As striated acicular crystals, sometimes flattened parallel to elongation, in tufts and groups. Also botryoidal or dendritic.
COLOR-LUSTER: Dark brown to black. Translucent to nearly opaque. Submetallic. Streak yellow.
MODE OF OCCURRENCE: Occurs in the oxidation zone of lead-zinc deposits, associated with descloizite and vanadinite, in the western part of the Sierra de Córdoba, Córdoba province, Argentina.
BEST REF. IN ENGLISH: Palache, et al., "Dana's System of Mineralogy," 7th Ed., v. II, p. 1052–1053, New York, Wiley, 1951. Donaldson, D. M., and Barnes, W. H., *Am. Min.*, **40**: 597–613 (1955).

## BRADLEYITE
$Na_3MgPO_4CO_3$

CRYSTAL SYSTEM: Monoclinic
CLASS: 2/m
SPACE GROUP: $P2_1/m$
Z: 2
LATTICE CONSTANTS:
$a = 8.85$
$b = 6.63$
$c = 5.16$
$\beta = 90°25'$
3 STRONGEST DIFFRACTION LINES:
3.32 (100)
2.66 (100)
2.57 ( 80)
OPTICAL CONSTANTS:
$\alpha = 1.49$
$\beta =$ ?
$\gamma = 1.56$
(–)2V = small
HARDNESS: Not determined.
DENSITY: 2.734 (Meas.)
CLEAVAGE: Not determined.
HABIT: As fine-grained masses.
COLOR-LUSTER: Light gray due to clay impurities. Probably colorless or white.
MODE OF OCCURRENCE: Occurs admixed with shortite and montmorillonite in a drill core from oil shale in the Green River formation, about 20 miles west of Green River, Sweetwater County, Wyoming.
BEST REF. IN ENGLISH: Fahey, Joseph J., and Tunell, George, *Am. Min.*, **26**: 646–650 (1941). Fahey, J. J., USGS Prof. Paper **405**, 34–35 (1962).

## BRAGGITE
(Pt, Pd, Ni) S

CRYSTAL SYSTEM: Tetragonal
CLASS: 4/m, 4

SPACE GROUP: $P4_2/m$, $P4_2$
Z: 8
LATTICE CONSTANTS:
$a = 6.37$
$c = 6.58$
3 STRONGEST DIFFRACTION LINES:
2.86 (100)
2.93 ( 30)
2.64 ( 30)
HARDNESS: Not determined.
DENSITY: $\approx$10 (Meas.)
8.9 (Calc.)
CLEAVAGE: Not determined.
HABIT: As minute prisms and rounded grains.
COLOR-LUSTER: Steel gray. Opaque. Metallic.
MODE OF OCCURRENCE: Occurs associated with native platinum, sperrylite, and other precious-metal minerals, in concentrates from the platinum-bearing norites of the Bushveld complex, Transvaal, South Africa.
BEST REF. IN ENGLISH: Palache, et al., "Dana System of Mineralogy," 7th Ed., v. I, p. 259, New York, Wiley, 1944.

## BRAITSCHITE
$(Ca, Na_2)_7(Ce, La)_2B_{22}O_{43} \cdot 7H_2O$

CRYSTAL SYSTEM: Hexagonal
LATTICE CONSTANTS:
$a = 12.156$
$c = 7.377$
3 STRONGEST DIFFRACTION LINES:
4.283 (100)
3.021 ( 92)
10.52 ( 54)
OPTICAL CONSTANTS:
$\omega = 1.646$
$\epsilon = 1.647$
(+)
HARDNESS: Not determined.
DENSITY: 2.903±0.002 (Meas.)
2.837 (Calc.)
CLEAVAGE: Not determined.
HABIT: Crystals simple hexagonal plates having only prism and pinacoid faces. 0.1–70 microns in diameter.
COLOR-LUSTER: Colorless to white. Vitreous.
MODE OF OCCURRENCE: Occurs in the Cane Creek potash mine of the Texas Gulf Sulphur Company and in a core hole one mile south of the mine, near Moab, Grand County, Utah. The braitschite is in white to reddish pink nodules which constitute a zone in anhydrite rock immediately overlying a potash bed, throughout the workings of the mine. The nodules contain approximately 65% braitschite and about 35% quartz, anhydrite, dolomite, halite, small amounts of hematite, and a few crystals of chalcopyrite.
BEST REF. IN ENGLISH: Raup, Omer B., Gude, Arthur J. 3rd, Dwornik, E. J., Cuttitta, Frank, and Rose, Harry J. Jr., *Am. Min.*, **53**: 1081–1095 (1968).

**BRAMMALITE** = Variety of illite in which sodium is the interlayer cation

MODE OF OCCURRENCE: Occurs mainly in shales associated with coal deposits, as at Llandebie, South Wales.

## BRANDTITE
$Ca_2 Mn(AsO_4)_2 \cdot 2H_2O$

CRYSTAL SYSTEM: Monoclinic
CLASS: 2/m
SPACE GROUP: $P2_1/c$
Z: 2
LATTICE CONSTANTS:
 $a = 5.65$ kX
 $b = 12.80$
 $c = 5.65$
 $\beta = 99°30'$
3 STRONGEST DIFFRACTION LINES:
 2.98 (100)
 2.785 (100)
 2.606 ( 80)
OPTICAL CONSTANTS:
  $\alpha = 1.709-1.707$
  $\beta = 1.711$
  $\gamma = 1.724-1.729$
 (+)2V = 23°
HARDNESS: 3½
DENSITY: 3.67 (Meas.)
     3.70 (Calc.)
CLEAVAGE: {010} perfect
     {001} good
HABIT: Crystals slender prismatic, up to 8.0 mm long, 1.0 mm wide, and 0.2 mm thick. Also stout prismatic [001]. As radial groups and rounded or reniform masses with radial fibrous structure. {100} striated [001]. Twinning on {100}, common.
COLOR-LUSTER: Colorless to white, transparent to translucent. Vitreous.
MODE OF OCCURRENCE: Found associated with rhodochrosite and chalcopyrite in a vug in a coarse intergrowth of calcite, franklinite, brown willemite, and sphalerite at the Sterling Hill mine, Sussex County, New Jersey. Also associated with calcite, barite, galena, sarkinite, flinkite, caryopilite, and native lead at the Harstig mine, Pajsberg, Vermland, Sweden.
BEST REF. IN ENGLISH: Palache, et al., "Dana's System of Mineralogy," 7th Ed., v. II, p. 725–727, New York, Wiley, 1951.

## BRANNERITE
$(U, Ca, Ce)(Ti, Fe)_2O_6$

CRYSTAL SYSTEM: Monoclinic
CLASS: 2/m
SPACE GROUP: I2/m
Z: 2
LATTICE CONSTANTS:
 $a = 8.90$
 $b = 3.80$
 $c = 6.99$
 $\beta = 104°45'$

3 STRONGEST DIFFRACTION LINES:
 3.42 (100)
 1.903 ( 80)
 2.455 ( 70) Metamict
OPTICAL CONSTANTS:
 $N = 2.26$ (Li) Metamict
 $N = 2.30$ (Na)
  2.23
HARDNESS: 4½-5½
DENSITY: 4.2-5.4 (Meas. natural)
     6.35 (Meas. synthetic)
CLEAVAGE: None.
Fracture conchoidal.
HABIT: As indistinct crystals, embedded grains and masses; and as rounded detrital grains and pebbles.
COLOR-LUSTER: Black when fresh, also brownish green, yellowish brown, or yellow. Translucent to opaque; sometimes transparent in thin splinters. Vitreous when fresh; often resinous to dull. Streak dark greenish brown to yellowish brown. Not fluorescent.
MODE OF OCCURRENCE: Occurs chiefly as a primary mineral in pegmatites or as a detrital mineral derived from pegmatites or vein deposits. Found in the United States in placer deposits in Custer County, Idaho; at the California mine, Chaffee County and in Gilpin County, Colorado; and in Mono, Plumas, and San Bernardino counties, California. It also occurs in the Blind River district, Ontario, Canada; in the Vosges, France; near Fuenteovejuna, Cordoba, Spain; at Bou-Azzer, Morocco; and at several places in South Australia.
BEST REF. IN ENGLISH: Frondel, Clifford, USGS Bull. 1064, p. 333–337 (1958).

## BRANNOCKITE
$KLi_3 Sn_2 Si_{12}O_{30}$
Lithium-tin analog of osumiilite

CRYSTAL SYSTEM: Hexagonal
CLASS: 6/m 2/m 2/m
SPACE GROUP: P6/mcc
Z: 2
LATTICE CONSTANTS:
 $a = 10.02$
 $c = 14.25$
3 STRONGEST DIFFRACTION LINES:
 4.109 (100)
 2.905 ( 90)
 7.141 ( 80)
OPTICAL CONSTANTS:
 $\omega = 1.567$
 $\epsilon = 1.566$
 Uniaxial (-)
HARDNESS: Not determined.
DENSITY: 2.980 (Meas.)
     3.08 (Calc.)
CLEAVAGE: None
Brittle.
HABIT: Crystals very thin plates less than 1mm in size. Forms: {001}, {100}, {110} and {114}, with the pinacoid {001} dominant.

COLOR-LUSTER: Colorless. Transparent. Vitreous. Fluoresces bright blue-white under short-wave ultraviolet light.

MODE OF OCCURRENCE: Brannockite occurs in minute amounts in vugs and on fracture surfaces in leached pegmatite associated with bavenite, tetrawickmanite, pyrite, stannian titanite, albite, and quartz at the Foote Mineral Company's spodumene mine at Kings Mountain, Cleveland County, North Carolina.

BEST REF. IN ENGLISH: John S. White, Jr., Joel E. Arem, Joseph A. Nelan, Peter B. Leavens, and Richard W. Thomssen, *Min Rec.*, **4**: 73–76 (1973).


# BRAUNITE
$Mn^{2+}Mn_6^{3+}SiO_{12}$

CRYSTAL SYSTEM: Tetragonal
CLASS: 4/m 2/m 2/m
SPACE GROUP: $I4_1/acd$
Z: 8
LATTICE CONSTANTS:
  $a = 9.44$
  $c = 18.76$
3 STRONGEST DIFFRACTION LINES:
  2.72 (100)
  1.656 ( 70)
  2.14 ( 50)
HARDNESS: 6–6½
DENSITY: 4.72–4.83 (Meas.)
       4.67 (Calc. for Mn:Si = 7:1)
CLEAVAGE: {112} perfect
Fracture uneven to conchoidal. Brittle.
HABIT: Crystals pyramidal, striated. Also massive, granular. Twinning on {112}, common.
COLOR-LUSTER: Black, brownish black, steel gray. Opaque. Submetallic.
MODE OF OCCURRENCE: Occurs chiefly in veins and lenses as a secondary mineral formed under weathering conditions, usually associated with other manganese minerals. Found in the United States in Humboldt, Plumas, Santa Clara, Stanislaus, and Trinity counties, California; in Pitkin County, Colorado; and at localities in Texas, Arkansas, and Georgia. It also occurs in Panama, Brazil, Norway, Sweden, Italy, Germany, and at many manganese deposits in India.
BEST REF. IN ENGLISH: Palache, et al., "Dana's System of Mineralogy," 7th Ed., v. I, p. 551–554, New York, Wiley, 1944. *Am. Min.*, **52**: 20–30 (1967).


# BRAVAISITE = Interstratified mixture of illite and montmorillonite


# BRAVOITE
$(Ni, Fe)S_2$
Pyrite Group

CRYSTAL SYSTEM: Cubic
CLASS: 2/m $\bar{3}$

SPACE GROUP: Pa3. Pyrite structure.
Z: 4
LATTICE CONSTANTS:
  $a = 5.57$ (variable depending on amount of substitution)
  $a = 5.74$ (artif.)
3 STRONGEST DIFFRACTION LINES:
  2.78 (100)
  1.68 ( 75)
  2.49 ( 50)
HARDNESS: 5½–6
DENSITY: 4.82 (Meas.)
       4.83 (Calc.)
CLEAVAGE: {001}
Fracture uneven to conchoidal. Brittle.
HABIT: Crystals cubic, octahedral, or pyritohedral; indistinct. Usually as nodular masses or crusts with columnar or radial-fibrous structure.
COLOR-LUSTER: Steel gray. Opaque. Metallic.
MODE OF OCCURRENCE: Occurs as a coating on pyrite crystals at Rico, Dolores County, and with niccolite and pyrite at the Copper King mine, Boulder County, Colorado; on Spirit Mountain, Lower Copper River Valley, Alaska; at the Mill Close mine, Derbyshire, England; in the Cármenes district, Leon Province, Spain; and in the Siegen district and elsewhere in Germany.
BEST REF. IN ENGLISH: Palache, et al., "Dana's System of Mineralogy," 7th Ed., v. I, p. 290–291, New York, Wiley, 1944.


# BRAZILIANITE
$NaAl_3(PO_4)_2(OH)_4$

CRYSTAL SYSTEM: Monoclinic
CLASS: 2/m
SPACE GROUP: $P2_1/n$
Z: 4
LATTICE CONSTANTS:
  $a = 11.19$
  $b = 10.08$
  $c = 7.06$
  $\beta = 97°22'$
3 STRONGEST DIFFRACTION LINES:
  5.05 (100)
  2.99 ( 80)
  2.74 ( 80)
OPTICAL CONSTANTS:
  $\alpha = 1.602$
  $\beta = 1.609$
  $\gamma = 1.621–1.623$
  (+)2V = 71°, 75° (Calc.)
HARDNESS: 5½
DENSITY: 2.983±0.005 (Meas.)
       3.025 (Calc.)
CLEAVAGE: {010} good
Fracture conchoidal. Brittle.
HABIT: Crystals nearly equant to short prismatic [001], up to 12cm long and 8cm in width; also elongated [100], spear-shaped. Prism zone striated [001]. Also globular with radial fibrous structure.
COLOR-LUSTER: Colorless, pale yellowish to yellowish green; transparent; vitreous. Streak colorless.

MODE OF OCCURRENCE: Occurs as a hydrothermal mineral in cavities in pegmatite at the Palermo mine, near North Grafton, and also in pegmatite at the Smith mine near Newport, New Hampshire. Found as superb large gemmy crystals associated with muscovite, albite, apatite, and tourmaline in cavities in pegmatite near Conselheira Pena, and as spear-shaped crystals at Mantena in Minas Geraes; also as small (1-2 mm) grains and crystals associated with massive amblygonite, wardite, apatite, and other phosphates in the pegmatite of the alto Patrimonio, at Piedras Lavrades, Paraíaba, Brazil.

BEST REF. IN ENGLISH: Palache, et al., "Dana's System of Mineralogy," 7th Ed., v. II, p. 841-843, New York, Wiley, 1951.

# BREDIGITE
α-$Ca_2SiO_4$
Dimorph of larnite
Unstable at atmospheric temperatures and pressures.

CRYSTAL SYSTEM: Orthorhombic
CLASS: 2/m 2/m 2/m
SPACE GROUP: Pmnn
Z: 16
LATTICE CONSTANTS:
 $a$ = 10.92
 $b$ = 18.41
 $c$ = 6.75
3 STRONGEST DIFFRACTION LINES:
 2.730 (100)
 2.663 (100)
 2.259 ( 80)
OPTICAL CONSTANTS:
 α = 1.712
 β = 1.716
 γ = 1.725
(+)2V = 30° (variable)
HARDNESS: Not determined.
DENSITY: 3.40 (Meas.)
 3.38 (Calc.)
CLEAVAGE: {130} distinct
HABIT: As rounded grains. Twinning on {110}.
COLOR-LUSTER: Colorless to gray. Vitreous.
MODE OF OCCURRENCE: Occurs in association with larnite and γ-$Ca_2SiO_4$ in a contact zone around a syenite-monzonite intrusion in Marble Canyon, Culberson County, Texas. Also in the contact zone of limestone and Tertiary dolerite at Scawt Hill, near Larne, County Antrim, Ireland.
BEST REF. IN ENGLISH: Tilley, C. E., and Vincent, H. C. G., *Min. Mag.*, **28**: 255-271 (1948)    Douglas, A. M. B., *Min. Mag.*, **29**: 875-884 (1952).    Bridge, Thomas E., *Am. Min.*, **51**: 1766-1774 (1966).

# BREITHAUPTITE
NiSb

CRYSTAL SYSTEM: Hexagonal
CLASS: 6/m 2/m 2/m
SPACE GROUP: P6$_3$/mmc
Z: 2

LATTICE CONSTANTS:
 $a$ = 3.942
 $c$ = 5.155
3 STRONGEST DIFFRACTION LINES:
 2.86 (100)
 2.05 ( 40)
 1.97 ( 40)
HARDNESS: 5½
DENSITY: 7.591-8.23 (Meas.)
 8.63-8.70 (Calc.)
CLEAVAGE: None.
Fracture subconchoidal to uneven. Brittle.
HABIT: Crystals rare, commonly thin tabular. Usually massive, compact; also disseminated or arborescent. Twinning on {1$\bar{1}$01}.
COLOR-LUSTER: Light copper red, often with violet tint. Opaque. Metallic. Streak reddish brown.
MODE OF OCCURRENCE: Occurs at South Lorrain and elsewhere in the Cobalt district, Ontario, Canada, associated with native silver, niccolite, and cobaltite in calcite veins; near Sarrabus, Sardinia; and at Andreasberg, Harz Mountains, Germany.
BEST REF. IN ENGLISH: Palache, et al., "Dana's System of Mineralogy," 7th Ed., v. I, p. 238-239, New York, Wiley, 1944.

# BREUNNERITE = Variety of magnesite containing about 9% FeO

# BREWSTERITE
(Sr, Ba, Ca) (AlSi$_3$O$_8$)$_2$ · 5H$_2$O
Zeolite Group

CRYSTAL SYSTEM: Monoclinic
CLASS: 2/m
SPACE GROUP: P2$_1$/m
Z: 2
LATTICE CONSTANTS:
 $a$ = 6.77
 $b$ = 17.41
 $c$ = 7.66
 β = 93°04′
3 STRONGEST DIFFRACTION LINES:
 4.66 (100)
 2.922 ( 80)
 3.268 ( 40)
OPTICAL CONSTANTS:
 α = 1.510
 β = 1.512
 γ = 1.523
(+)2V = 65°
HARDNESS: 5
DENSITY: 2.453 (Meas.)
CLEAVAGE: {010} perfect
Brittle.
HABIT: Crystals short prismatic. Also as radial-fibrous or granular aggregates. Twinning on {010}, common.
COLOR-LUSTER: White, colorless; sometimes with grayish or yellowish tint. Transparent. Vitreous.

MODE OF OCCURRENCE: Occurs as a hydrothermal mineral in cavities in basalts and schists frequently associated with calcite. Found on Ash Creek, on or near the Sonoma County line, Mendocino County, California, in association with edingtonite; as fine crystals at Strontian, Argyllshire, Scotland; and at localities in Ireland, France, and Germany.

BEST REF. IN ENGLISH: Vlasov, K. A., "Mineralogy of Rare Elements," v. II, p. 206–207, Israel Program for Scientific Translations, 1966.

## BREWSTERLINITE
Liquid $CO_2$

## BREZINAITE
$Cr_3S_4$

CRYSTAL SYSTEM: Monoclinic
Z: 2
LATTICE CONSTANTS:
$a = 5.96$
$b = 3.425$
$c = 11.27$
$\beta = 91°32'$
HARDNESS: Not determined.
DENSITY: 4.12 (Calc.)
CLEAVAGE: Not determined.
HABIT: Anhedral grains 5–80 $\mu$ across.
COLOR-LUSTER: Brownish gray in polished sections under reflected light. Synthetic brezinaite is dull gray. Opaque.
MODE OF OCCURRENCE: Occurs in the metal matrix and contiguous to silicate inclusions in the Tucson iron meteorite.
BEST REF. IN ENGLISH: Bunch, T. E., and Fuchs, Louis H., *Am. Min.*, **54**: 1509–1518 (1969).

## BRIANITE
$Na_2CaMg(PO_4)_2$

CRYSTAL SYSTEM: Orthorhombic
CLASS: Probably 222
SPACE GROUP: Probably P222
Z: 16
LATTICE CONSTANTS:
$a = 13.38$
$b = 10.50$
$c = 18.16$
3 STRONGEST DIFFRACTION LINES:
2.625 (100)
3.734 ( 90)
2.679 ( 90)
OPTICAL CONSTANTS:
$\alpha = 1.598$
$\beta = 1.605$
$\gamma = 1.608$
$(-)2V = 63°$–$65°$
HARDNESS: 4–5

DENSITY: 3.17 (Calc.)
          3.10 (Meas. Synthetic)
CLEAVAGE: Not determined.
HABIT: Grains up to 0.1–0.2 mm.
COLOR-LUSTER: Colorless, transparent.
MODE OF OCCURRENCE: Occurs associated with whitlockite, panethite, albite, and enstatite in small pockets in the metallic phase of the Dayton octahedrite meteorite.
BEST REF. IN ENGLISH: Fuchs, L. H., Olson, E., and Henderson, E. P., *Am. Min.*, **53**: 508–509 (1968).

## BRIARTITE
$Cu_2(Fe, Zn)GeS_4$

CRYSTAL SYSTEM: Tetragonal
CLASS: 4mm or $\bar{4}$m2
SPACE GROUP: $I4_1md$ or $I\bar{4}d2$
Z: 2
LATTICE CONSTANTS:
$a = 5.32$
$c = 10.51$
3 STRONGEST DIFFRACTION LINES:
3.06 (100)
1.89 ( 50)
1.87 ( 50)
HARDNESS: Polishing hardness same as for chalcopyrite.
DENSITY: 4.337 (Calc.)
HABIT: Embedded grains up to 1–2 mm in size. Commonly shows polysynthetic twinning.
COLOR-LUSTER: Gray to gray-blue.
MODE OF OCCURRENCE: Occurs as inclusions in chalcopyrite, tennantite, renierite, and sphalerite at the Prince Leopold mine, Kipushi, Katanga, Zaire; and at Tsumeb, South West Africa.
BEST REF. IN ENGLISH: Francotte, J., Moreau, J., Ottenburgs, R., and Levy, C., *Am. Min.*, **51**: 1816 (1966).

## BRITHOLITE
$(Ca, Ce)_5(SiO_4, PO_4)_3(OH, F)$
Apatite Group
Var. Fynchenite or Thorium-britholite, Alumobritholite, Hydrobritholite

CRYSTAL SYSTEM: Hexagonal
CLASS: 6/m
SPACE GROUP: $P6_3/m$
Z: 2
LATTICE CONSTANTS:
$a = 9.63$
$c = 7.03$
3 STRONGEST DIFFRACTION LINES:
2.836 (100) (Heated, partly metamict.)
3.48 ( 80)
2.809 ( 80)
OPTICAL CONSTANTS:
Uniaxial $(-)$ $n = 1.77$–$1.81$  $\omega$-$\epsilon$ = .005–.008
also (+) $\omega = 1.78$, $\epsilon = 1.82$
Biaxial
$\alpha = 1.772$
$\beta = 1.775$  2V small (up to 44°)
$\gamma = 1.777$

HARDNESS: 5
DENSITY: 3.86 (Meas.)  Vlasov:
        3.95 (Calc.)  4.2–4.69 (Meas.)
                  4.65 (Calc.)
CLEAVAGE: None. Fracture conchoidal. Brittle.
HABIT: Crystals prismatic, hexagonal in cross section. Usually compact massive or as disseminated grains.
COLOR-LUSTER: Yellow, brown, greenish brown, to nearly black. Translucent. Adamantine, resinous.
MODE OF OCCURRENCE: Occurs chiefly in nepheline-syenites and in contact metasomatic deposits related to alkali syenites and granites. Found at Susan Lode, Custer County, South Dakota (thorium britholite); at Oka, Quebec, Canada; in the Julianehaab district, Greenland; at several places in the USSR; in Tungpei, China (thorium britholite); and elsewhere.
BEST REF. IN ENGLISH: Vlasov, K. A., "Mineralogy of Rare Elements," v. II, p. 297–300, Israel Program for Scientific Translations, 1966.

## BRITHOLITE-(Y) syn. Abukumalite, Yttrobritholite

$(Ca, Y)_5 (SiO_4, PO_4)_3 (OH,F)$
Apatite Group

CRYSTAL SYSTEM: Hexagonal
SPACE GROUP: $P6_3/m$
Z: 2
LATTICE CONSTANTS:
  $a = 9.43$
  $c = 6.81$
3 STRONGEST DIFFRACTION LINES:
  2.81 (100)
  2.75 ( 90)
  2.73 ( 80)
OPTICAL CONSTANTS:
  $\omega = 1.732$
  $\epsilon = 1.728$
  $(-)2V$ up to $44°$
HARDNESS: 5
DENSITY: 4.25 (Meas.)
CLEAVAGE: None. Fracture conchoidal. Brittle.
HABIT: Crystals hexagonal, flattened. Also in oval forms up to 2.5 cm in diameter.
COLOR-LUSTER: Black. Adamantine.
MODE OF OCCURRENCE: Occurs in pegmatite in the Abukuma Range, Iisaka, Fukushima Prefecture, Japan. Also found in alkali granite pegmatites of the European USSR.
BEST REF. IN ENGLISH: Vlasov, K. A., "Mineralogy of Rare Elements," v. II, p. 297–300, Israel Program for Scientific Translations, 1966.

## BROCHANTITE

$Cu_4 SO_4 (OH)_6$

CRYSTAL SYSTEM: Monoclinic
CLASS: 2/m
SPACE GROUP: $P2_1/a$
Z: 4

LATTICE CONSTANTS:
  $a = 13.05$ kX
  $b = 9.83$
  $c = 6.01$
  $\beta = 103°22'$
3 STRONGEST DIFFRACTION LINES:
  2.521 (100)
  3.90 ( 85)
  2.678 ( 50)
OPTICAL CONSTANTS:
  $\alpha = 1.728$
  $\beta = 1.771$ (Na)
  $\gamma = 1.800$
$(-)2V \simeq 77°$
HARDNESS: 3½–4
DENSITY: 3.97 (Meas.)
        4.09 (Calc.)
CLEAVAGE: {100} perfect
Fracture conchoidal to uneven. Brittle.
HABIT: Crystals stout prismatic to acicular; also tabular. Commonly as drusy crusts; also in groups and aggregates of crystals; massive, granular. Twinning on {100} common, providing pseudo-orthorhombic symmetry.
COLOR-LUSTER: Emerald green to blackish green; pale green. Transparent to translucent. Vitreous, pearly on cleavage. Streak pale green.
MODE OF OCCURRENCE: Occurs widespread as a secondary mineral in the oxidation zone of copper deposits. Found in the United States at the Blanchard mine, Socorro County, and in the Organ district, Dona Ana County, New Mexico; at many places in Arizona, especially at Bisbee, Cochise County, at Tiger, Pinal County, and in the Clifton-Morenci district, Greenlee County; and in California, Colorado, Idaho, Utah, and other western states. Many other occurrences are found in Mexico, northern Chile, England, France, Germany, Spain, Italy, USSR, Zaire, South West Africa, and New South Wales, Australia.
BEST REF. IN ENGLISH: Palache, et al., "Dana's System of Mineralogy," 7th Ed., v. II, p. 541–544, New York, Wiley, 1951.

## BROCKITE

$(Ca, Th, Ce) \{ (PO_4), (CO_3) \} \cdot H_2O$

CRYSTAL SYSTEM: Hexagonal
CLASS: 622
SPACE GROUP: P622
Z: 3
LATTICE CONSTANTS:
  $a = 6.98$
  $c = 6.40$
3 STRONGEST DIFFRACTION LINES:
  3.03 (100)
  4.37 ( 70)
  2.83 ( 70)
OPTICAL CONSTANTS:
  $\omega = 1.680$
  $\epsilon = 1.695$
  (+)
HARDNESS: Not determined.

DENSITY: 3.9±0.2 (Meas.)
4.0 (Calc.)
CLEAVAGE: Conchoidal fracture.
HABIT: Radial aggregates of fibroid grains. Also stubby, imperfect hexagonal prisms, usually with granulated margins. Average grain size 20 $\mu$ in length; rarely up to 50 $\mu$ long.
COLOR-LUSTER: Deep red-brown; pale yellow on thin edges; translucent. Greasy to vitreous.
MODE OF OCCURRENCE: Occurs as very fine grained massive aggregates in nodules up to 35mm in diameter associated with abundant inclusions of pyrite, and as earthy coatings in veins and altered granitic rocks near the Bassick mine, Querida, Custer County, Colorado. Also as yellow earthy coatings from the Hardwick mine and Nightingale shaft, Custer County, Colorado.
BEST REF. IN ENGLISH: Fisher, Frances G., and Meyrowitz, Robert, *Am. Min.*, 47: 1346–1355 (1962).

**BRÖGGERITE** = Variety of uraninite containing thorium

**BROMARGYRITE** = Bromyrite

**BROMELLITE**
BeO

CRYSTAL SYSTEM: Hexagonal
CLASS: 6mm
SPACE GROUP: P6₃mc
Z: 2
LATTICE CONSTANTS:
$a$ = 2.68
$c$ = 4.36
3 STRONGEST DIFFRACTION LINES:
2.061 (100)
2.337 ( 91)
2.189 ( 61)
OPTICAL CONSTANTS:
$\omega$ = 1.719
$\epsilon$ = 1.733
(+)
HARDNESS: ~9
DENSITY: 3.017 (Meas.)
3.008 (Calc.)
CLEAVAGE: $\{10\bar{1}0\}$ distinct
HABIT: Crystals prismatic, with well-developed forms $h\{0001\}$ and $m\{10\bar{1}0\}$, and imperfect $r\{10\bar{1}1\}$.
COLOR-LUSTER: White. Transparent. Vitreous.
MODE OF OCCURRENCE: Occurs as tiny crystals in calcite veinlets in hematite skarns and skarnitized limestones, associated with manganophyllite, richterite, and swedenborgite, at Långban, Sweden.
BEST REF. IN ENGLISH: Palache, et al., "Dana's System of Mineralogy," 7th Ed., v. I, p. 506–507, New York, Wiley, 1944.

**BROMLITE** = Alstonite

**BROMYRITE (Bromargyrite)**
AgBr

CRYSTAL SYSTEM: Cubic
CLASS: 4/m $\bar{3}$ 2/m
SPACE GROUP: Fm3m
Z: 4
LATTICE CONSTANT:
$a$ = 5.755 kX
3 STRONGEST DIFFRACTION LINES:
2.886 (100)
2.041 ( 55)
1.667 ( 16)
OPTICAL CONSTANT:
$N$ = 2.253
HARDNESS: 2½
DENSITY: 6.474 (Meas.)
6.50 (Calc.)
CLEAVAGE: None
Fracture uneven to subconchoidal. Sectile and ductile.
HABIT: Crystals usually cubic, often with modifying faces. Commonly massive, often in crusts and waxy coatings; rarely columnar or fibrous. Twinning on $\{111\}$, not common.
COLOR-LUSTER: Gray, greenish gray, greenish brown, yellowish. Transparent to translucent. Resinous to adamantine, waxy.
MODE OF OCCURRENCE: Occurs as a secondary mineral in the oxidation zone of silver deposits, often associated with native silver, iodyrite, and iron and manganese oxides. In the United States it is found in the Hechman mine, Gila County, and at Tombstone, Arizona; it also has been reported from Lake, Mineral, Park, and Summit counties, Colorado, and from several deposits in Idaho and Kern County, California. It is found at several localities in Mexico including the San Onofre and Plateros districts, Zacatecas; found abundantly at Chanarcillo, Chile; at Broken Hill, New South Wales, Australia; also in France, Germany, and the USSR.
BEST REF. IN ENGLISH: Palache, et al., "Dana's System of Mineralogy," 7th Ed., v. II, p. 11–14, New York, Wiley, 1951.

**BRONZITE** = Variety of enstatite

**BROOKITE**
TiO₂
Trimorphous with rutile and anatase

CRYSTAL SYSTEM: Orthorhombic
CLASS: 2/m 2/m 2/m
SPACE GROUP: Pcab
Z: 8
LATTICE CONSTANTS:
$a$ = 5.436
$b$ = 9.166
$c$ = 5.135
3 STRONGEST DIFFRACTION LINES:
3.51 (100)
2.90 ( 90)
3.47 ( 80)

OPTICAL CONSTANTS:
$\alpha = 2.5831$
$\beta = 2.5843$
$\gamma = 2.7004$
HARDNESS: 5½–6
DENSITY: 4.14 ± 0.06
           4.119 (Calc.)
CLEAVAGE: {120} indistinct
           {001} in traces
Fracture subconchoidal to uneven. Brittle.

HABIT: Found only as crystals of varied habit; commonly tabular and vertically striated; also prismatic, pseudohexagonal, or pyramidal.

COLOR-LUSTER: Light brown to dark brown or dark reddish brown; also black. Transparent to opaque. Adamantine to submetallic. Streak white to gray to yellowish.

MODE OF OCCURRENCE: Occurs chiefly in veins in gneiss and schist, as an accessory mineral in certain igneous and metamorphic rocks, and as a detrital mineral. Found as fine black lustrous crystals of varied habit at Magnet Cove, Arkansas, and at localities in California, Montana, North Carolina, New Jersey, New York, Massachusetts, and Maine. Other notable occurrences are found in Brazil, England, Wales, France, Switzerland, Italy, Austria, Czechoslovakia, and USSR.

BEST REF. IN ENGLISH: Palache, et al., "Dana's System of Mineralogy," 7th Ed., v. I, p. 588–593, New York, Wiley, 1944.

# BROWNMILLERITE
$Ca_2(Al, Fe)_2O_5$

CRYSTAL SYSTEM: Orthorhombic
CLASS: 2/m 2/m 2/m
SPACE GROUP: Ibmm
Z: 2
LATTICE CONSTANTS:
$a = 5.52$
$b = 14.44$
$c = 5.34$
3 STRONGEST DIFFRACTION LINES:
2.65 (100)
7.19 ( 50)
2.78 ( 50)
OPTICAL CONSTANTS:
$\alpha < 2.02$
$\beta$ and $\gamma > 2.02$
(−)2V
HARDNESS: Not determined.
DENSITY: 3.76 (Meas. av.)
          3.73 (Calc.)
CLEAVAGE: Not determined.
HABIT: As platelets, tetragonal in outline; usually 60 $\mu$ or less along an edge and 15 $\mu$ thick.
COLOR-LUSTER: Reddish brown.
MODE OF OCCURRENCE: Occurs associated with ettringite, calcite, wollastonite, mayenite, gehlenite, larnite, diopside, grossular, pyrrhotite, spinel, afwillite, hydrocalumite, and portlandite. It is found in several thermally metamorphosed marly limestone inclusions of Tertiary(?) age in effusive volcanic rocks of the Ettringer Bellerberg

(leucite-tephrite), related to the Quaternary extrusives of the Laacher See area, near Mayen, Eifel district, (Rhineland-Palatinate), West Germany.

BEST REF. IN ENGLISH: Hentschel, Gerhard, *Am. Min.*, **50**: 2106 (1965).

# BRUCITE
$Mg(OH)_2$

CRYSTAL SYSTEM: Hexagonal
CLASS: $\bar{3}$ 2/m
SPACE GROUP: P$\bar{3}$m
Z: 1
LATTICE CONSTANTS:
$a = 3.125$
$c = 4.75$
3 STRONGEST DIFFRACTION LINES:
2.365 (100)
4.77 ( 90)
1.794 ( 55)
OPTICAL CONSTANTS:
$\omega = 1.559–1.59$
$\epsilon = 1.580–1.60$
   (+)
HARDNESS: 2½
DENSITY: 2.39 ± 0.01
          2.40 (Calc.)
CLEAVAGE: {0001} perfect
Plates separable and flexible. Sectile.

HABIT: Crystals usually broad tabular; rarely acicular (manganoan). Usually foliated massive; also fibrous, scaly, or fine granular.

COLOR-LUSTER: White, pale green, gray, gray-blue, blue. Manganoan varieties yellowish to dark brown. Transparent. Pearly, waxy, or vitreous. Streak, white.

MODE OF OCCURRENCE: Occurs in serpentine, metamorphic limestones, and chloritic and dolomitic schists, commonly associated with hydromagnesite, magnesite, talc, calcite, aragonite, or chrysotile. Found in Fresno, Riverside, San Benito, San Bernardino, and San Francisco counties, California; in the Lodi district, Nevada; at the Mountain Lake mine, Salt Lake County, Utah; at the Tilly Foster iron mine, Brewster, Putnam County, New York; as exceptional crystals and broad plates at Wood's and Low's chromite mines, Texas, Lancaster County, and in dolomite in Berks County, Pennsylvania; and in serpentine at Hoboken, New Jersey. It also occurs at Asbestos, Quebec, Canada, as fibrous masses up to several feet in length; and at localities in Scotland, Sweden, Italy, Austria, and USSR.

BEST REF. IN ENGLISH: Palache, et al., "Dana's System of Mineralogy," 7th Ed., v. I, p. 636–639, New York, Wiley, 1944.

# BRÜGGENITE
$Ca(IO_3)_2 \cdot H_2O$

CRYSTAL SYSTEM: Monoclinic
CLASS: 2/m
SPACE GROUP: P$2_1$/c
Z: 4

LATTICE CONSTANTS:
$a = 8.505$
$b = 10.000$
$c = 7.498$
$\beta = 95°15'$
3 STRONGEST DIFFRACTION LINES:
3.048 (100)
3.232 ( 71)
4.235 ( 50)
OPTICAL CONSTANTS:
$\alpha = 1.773$
$\beta = 1.797$
$\gamma = 1.814$
$(-)2V = 88°$
HARDNESS: 3½
DENSITY: 4.24 (Meas.)
4.267 (Calc.)
CLEAVAGE: Fracture conchoidal. Brittle.
HABIT: Crystals anhedral; short to long columnar. Also as crystalline crusts.
COLOR-LUSTER: Colorless to bright yellow. Transparent to translucent. Vitreous.
MODE OF OCCURRENCE: Occurs in association with lauterite and soda niter at Pampa Pique III, Oficina Lauters, Chile.
BEST REF. IN ENGLISH: Mrose, M. E., Ericksen, G. E., and Marinenko, J. W., *Am. Min.*, **57**: 1911 (1972).

## BRUGNATELLITE
$Mg_6FeCO_3(OH)_{13} \cdot 4H_2O$

CRYSTAL SYSTEM: Hexagonal
CLASS: $\bar{3}$ or 3
SPACE GROUP: $P\bar{3}$ or P3
Z: 1
LATTICE CONSTANTS:
$a = 5.47$
$c = 15.97$
3 STRONGEST DIFFRACTION LINES:
7.93 (100)
3.96 ( 90)
2.00 ( 70)
OPTICAL CONSTANTS:
$\omega = 1.540$
$\epsilon = 1.510$
$(-)$
HARDNESS: ~2
DENSITY: 2.14 (Meas.)
2.21 (Calc.)
CLEAVAGE: {0001} perfect
HABIT: Massive; foliated or lamellar masses of small flakes flattened {0001}.
COLOR-LUSTER: Flesh pink to yellowish or brownish white. Translucent to transparent. Pearly. Streak white.
MODE OF OCCURRENCE: Occurs chiefly as coatings and crusts in hydrothermally altered serpentinic rocks. Found as an alteration product of melilite in the uncompahgrite rock of Iron Hill, Gunnison County, Colorado; and at localities in Liguria, Lombardy, and Piedmont, Italy.
BEST REF. IN ENGLISH: Palache, et al., "Dana's System of Mineralogy," 7th Ed., v. I, p. 660–661, New York, Wiley, 1944.

## BRUNOGEIERITE
$GeFe_2O_4$

CRYSTAL SYSTEM: Cubic
CLASS: $4/m\,\bar{3}\,2/m$
SPACE GROUP: Fd3m
Z: 8
LATTICE CONSTANT:
$a = 8.409$
3 STRONGEST DIFFRACTION LINES:
2.540 (100)
1.484 ( 90)
1.615 ( 70)
HARDNESS: >4½
DENSITY: 5.51 (Calc.)
CLEAVAGE: Not determined.
HABIT: Massive; as thin rims encrusting tennantite which itself encloses renierite.
COLOR-LUSTER: Gray. Opaque.
MODE OF OCCURRENCE: The mineral occurs in the lower, partly oxidized zones of the Tsumeb orebody, South West Africa.
BEST REF. IN ENGLISH: Ottemann, J., and Nuber, B., *Min. Abs.*, **24**: 73–805, p. 76 (1973).

## BRUNSVIGITE
$(Fe^{2+}, Mg, Al)_6(Si, Al)_4O_{10}(OH)_8$
Chlorite Group

CRYSTAL SYSTEM: Monoclinic
CLASS: 2/m
SPACE GROUP: C2/m
Z: 1
LATTICE CONSTANTS:
$a \approx 5.3$
$b \approx 9.2$
$c \approx 14.3$
$\beta \simeq 97°$
OPTICAL CONSTANTS:
$\alpha = 1.64$
$\beta = 1.66$
$\gamma = 1.67$
$2V \simeq 0°$
HARDNESS: ~2
DENSITY: 2.99–3.08 (Meas.)
CLEAVAGE: {001} perfect
HABIT: Crystals tabular, hexagonal in outline, microscopic. Usually massive, foliated or cryptocrystalline; sometimes in spherical radiated aggregates. Twinning on {001}, twin plane [310].
COLOR-LUSTER: Olive green, yellow green. Translucent. Pearly.
MODE OF OCCURRENCE: Occurs in veins in granite at the Silent Valley quarry, east Mourne Mountains, Northern Ireland; in cavities in the gabbro of the Radau Valley, Harz Mountains, Germany; and in spilite on Great Island, Three Kings Group, New Zealand.
BEST REF. IN ENGLISH: Deer, Howie, and Zussman, "Rock Forming Minerals," v. 3, p. 131–163, New York, Wiley, 1962.

## BRUSHITE
$CaH PO_4 \cdot 2H_2O$

CRYSTAL SYSTEM: Monoclinic
CLASS: 2/m
SPACE GROUP: I2/a
Z: 4
LATTICE CONSTANTS:
  $a = 5.88$
  $b = 15.15$
  $c = 6.37$
  $\beta = 117°28'$
3 STRONGEST DIFFRACTION LINES:
  7.62 (100)
  3.80 ( 30)
  1.90 ( 10)
OPTICAL CONSTANTS:
  $\alpha = 1.539$ (Na)
  $\beta = 1.546$
  $\gamma = 1.551$
  (+)2V = 86°
HARDNESS: 2½
DENSITY: 2.328 (Meas.)
         2.257 (Calc.)
CLEAVAGE: {010} perfect
          {001} perfect
HABIT: Crystals needle-like or prismatic to tabular {010}. Also earthy or powdery; foliated.
COLOR-LUSTER: Colorless to ivory yellow. Transparent to translucent. Vitreous; pearly on cleavage.
MODE OF OCCURRENCE: Occurs as nodular masses of platy crystals in the lower part of a bat guano and hair deposit in Pig Hole Cave, Giles County, Virginia. Also found as crystals up to 2 cm long in cavities in phosphorite at Quercy, near Limoges, France; in commercial quantities in a cave near Oran, Algeria; in guano from Aves Island in the Caribbean Sea; and widespread in small amounts in many other deposits throughout the world.
BEST REF. IN ENGLISH: Palache, et al., "Dana's System of Mineralogy," 7th Ed., v. II, p. 704–706, New York, Wiley, 1951.

## BUDDINGTONITE
$NH_4 AlSi_3 O_8 \cdot nH_2O$

CRYSTAL SYSTEM: Monoclinic
CLASS: 2 or 2/m
SPACE GROUP: $P2_1$ or $P2_1/m$
Z: 4
LATTICE CONSTANTS:
  $a = 8.571$
  $b = 13.032$
  $c = 7.187$
  $\beta = 112°44'$
3 STRONGEST DIFFRACTION LINES:
  3.81  (100)
  6.52  ( 96)
  3.381 ( 72)
OPTICAL CONSTANTS:
  $\alpha = 1.530$
  $\beta = 1.531$
  $\gamma = 1.534$

HARDNESS: 5½
DENSITY: 2.32 (Meas.)
         2.388 (Calc.)
CLEAVAGE: {001} good
          {010} distinct
Brittle; subconchoidal fracture.
HABIT: Commonly anhedral to cryptocrystalline; rarely as euhedral crystals up to 0.05 mm in size. Crystals correspond closely to orthoclase in habit and in interfacial angles.
COLOR-LUSTER: Crystals colorless, transparent, vitreous. Compact material is translucent, cloudy, and has an earthy luster. Streak of compact material light gray to darker yellowish gray.
MODE OF OCCURRENCE: Occurs as compact masses pseudomorphous after plagioclase, and as minute crystals lining cavities in Quaternary andesite and older rocks hydrothermally altered by ammonia-bearing hot-spring waters below the water table at the Sulphur Bank quicksilver mine, Lake County, California.
BEST REF. IN ENGLISH: Erd, Richard C., White, Donald E., Fahey, Joseph J., and Lee, Donald E., *Am. Min.*, **49**: 831–850 (1964).

## BUERGERITE
$NaFe_3^{3+}Al_6Si_6B_3O_{30}F$
Tourmaline Group

CRYSTAL SYSTEM: Hexagonal
CLASS: 3m
SPACE GROUP: R3m
Z: 3
LATTICE CONSTANTS:
  $a = 15.873$
  $c = 7.187$
3 STRONGEST DIFFRACTION LINES:
  2.58 (100)
  3.99 ( 85)
  2.96 ( 85)
OPTICAL CONSTANTS:
  $\omega = 1.735$
  $\epsilon = 1.655$
  (–)
HARDNESS: 7
DENSITY: 3.31 ± 0.01 (Meas.)
CLEAVAGE: Prismatic, distinct.
HABIT: Crystals short prismatic. Forms: {11$\bar{2}$0}, {30$\bar{3}$0}, {10$\bar{1}$1}, and {02$\bar{2}$1}.
COLOR-LUSTER: Dark brown, almost black, with a bronze schiller. Streak yellow-brown.
MODE OF OCCURRENCE: Occurs in rhyolite at Mexquitic, San Luis Potosí, Mexico.
BEST REF. IN ENGLISH: Donnay, Gabrielle, Ingamells, C. O., and Mason, Brian, *Am. Min.*, **51**: 198–199 (1966).

## BUETSCHLIITE
$K_2 Ca(CO_3)_2$

CRYSTAL SYSTEM: Hexagonal
CLASS: $\bar{3}$

SPACE GROUP: $R\bar{3}$
Z: 3
LATTICE CONSTANTS:
  $a = 5.38$
  $c = 18.12$
3 STRONGEST DIFFRACTION LINES:
  2.86 (100)
  3.03 ( 80)
  2.69 ( 80)
OPTICAL CONSTANTS:
  $\omega = 1.595$
  $\epsilon = 1.445$
   (–)
HARDNESS: Not determined.
DENSITY: Not determined.
CLEAVAGE: Not determined.
HABIT: As microscopic barrel-shaped crystals.
MODE OF OCCURRENCE: Occurs at many localities in the western United States as clinkers formed by the fusion of wood ash in partly burned fir, hemlock, and other trees.
BEST REF. IN ENGLISH: Milton, Charles, and Axelrod, Joseph M., *Am. Min.*, 32: 607 (1947).   Mrose, Mary E., Rose, H. J. Jr., and Marinkenko, J. W., *Geol. Soc. Amer. Spec. Paper* **101**, p. 146 (1966).

# BUKOVITE
$Cu_{3+x}Tl_2FeSe_{4-x}$

CRYSTAL SYSTEM: Tetragonal
CLASS: $\bar{4}2m$, 4mm, 4/m 2/m 2/m
SPACE GROUP: Possibly $I\bar{4}2m$, I4mm, I4/mmm
Z: 1
LATTICE CONSTANTS:
  $a = 3.976$
  $c = 13.70$
3 STRONGEST DIFFRACTION LINES:
  2.995 (100)
  2.600 ( 90)
  1.771 ( 80)
OPTICAL CONSTANTS: In reflected light pale gray, weakly pleochroic from creamy gray to gray. Anisotropy medium.
HARDNESS: 2
DENSITY: 7.40 (Calc.)
CLEAVAGE:  {001} good
                {100} imperfect
HABIT: Massive.
COLOR-LUSTER: Grayish brown. Opaque. Metallic.
MODE OF OCCURRENCE: Occurs in association with clausthalite, eskebornite, eucairite, umangite, and other selenides, in calcite veins at Bukov and Petrovice, western Bohemia, and at Předbořice, central Bohemia, Czechoslovakia.
BEST REF. IN ENGLISH: Johan, Zdenek, and Kvacek, Milan, *Am. Min.*, 57: 1910 (1972).

# BUKOVSKÝITE
$Fe_2^{3+}(AsO_4)(SO_4)(OH)\cdot 7H_2O$

CRYSTAL SYSTEM: Probably monoclinic

3 STRONGEST DIFFRACTION LINES:
  ~9.6 (100)
  ~9.2 (100)
   3.90 ( 70)
OPTICAL CONSTANTS:
  $\beta' = 1.570$-$1.582$
  $\gamma' = 1.626$-$1.631$
HARDNESS: Soft, nearly powdery on the surface.
DENSITY: 2.334 (Meas.)
CLEAVAGE: Fracture uneven, earthy.
HABIT: Microcrystalline; in transmitted light appears as colorless to yellowish minute crystals with average size $0.05 \times 0.007$ mm and maximum size $0.11 \times 0.015$ mm, often in radiating aggregates.
COLOR-LUSTER: Pale yellowish green to grayish green. Streak pale yellowish white to dirty yellow.
MODE OF OCCURRENCE: Occurs in nodules up to 60 cm in diameter as a weathering product of sulfides (mainly arsenopyrite and pyrite) in medieval dumps at Kaňk, near Kutná Hora, Bohemia, Czechoslovakia.
BEST REF. IN ENGLISH: Novák, F., Povondra, P., and Vtělenský, J., *Am. Min.*, 54: 991-992 (1969).

# BULTFONTEINITE
$Ca_2SiO_2(OH, F)_4$

CRYSTAL SYSTEM: Triclinic
CLASS: $\bar{1}$
SPACE GROUP: $P\bar{1}$
Z: 4
LATTICE CONSTANTS:
  $a = 8.34$       $\alpha = 91°59'$
  $b = 11.18$      $\beta = 94°17'$
  $c = 5.68$       $\gamma = 90°44'$
3 STRONGEST DIFFRACTION LINES:
  1.93 (100)
  8.12 ( 60)
  2.92 ( 60)
OPTICAL CONSTANTS:
  $\alpha = 1.587$
  $\beta = 1.590$ (Calc.)
  $\gamma = 1.597$
   (+)
HARDNESS: 4½
DENSITY: 2.73 (Meas.)
                2.75 (Calc.)
CLEAVAGE:  {010} distinct
                {100} distinct
HABIT: Crystals acicular, radiated; in spherulitic masses. Twinning on {010}, {100}, common.
COLOR-LUSTER: Colorless to pink.
MODE OF OCCURRENCE: Occurs in association with calcite, apophyllite, and natrolite at the Bultfontein, Dutoitspan, and Jagersfontein mines in the Kimberley district, South Africa.
BEST REF. IN ENGLISH: Parry, John, Williams, Alpheus F., and Wright, F. E., *Min. Mag.*, 23: 145-162 (1932). Murdoch, Joseph, *Am. Min.*, 40: 900-904 (1955).

# BUNSENITE

NiO

CRYSTAL SYSTEM: Cubic
CLASS: $4/m\ \bar{3}\ 2/m$
SPACE GROUP: Fm3m
Z: 4
LATTICE CONSTANT:
  $a = 4.1768$
3 STRONGEST DIFFRACTION LINES:
  2.088 (100)
  2.410 ( 91)
  1.476 ( 57)
OPTICAL CONSTANTS:
  $N = 2.37$ (Li)
  $N = 2.23$ (Na)
  HARDNESS: 5½
  DENSITY: 6.7–6.9 (Meas.)
         6.806 (Calc.)
  CLEAVAGE: Not determined.
  HABIT: Crystals octahedral, sometimes with cube or dodecahedron modifications.
  COLOR-LUSTER: Dark yellowish green. Transparent. Vitreous. Streak brownish black.
  MODE OF OCCURRENCE: Occurs in the oxidation zone of a nickel-uranium vein at Johanngeorgenstadt, Saxony, Germany, associated with cobalt and nickel arsenates and native bismuth.
  BEST REF. IN ENGLISH: Palache, et al., "Dana's System of Mineralogy," 7th Ed., v. I, p. 500–501, New York, Wiley, 1944.

# BURBANKITE

(Na, Ca, Sr, Ba, Ce)$_6$ (CO$_3$)$_5$

CRYSTAL SYSTEM: Hexagonal
CLASS: $6/m\ 2/m\ 2/m$
SPACE GROUP: P6$_3$/mmc
Z: 2
LATTICE CONSTANTS:
  $a = 10.53$
  $c = 6.47$
3 STRONGEST DIFFRACTION LINES:
  2.639 (100)
  3.041 ( 80)
  5.28 ( 70)
OPTICAL CONSTANTS:
  $\omega = 1.627$
  $\epsilon = 1.615$
    (–)
  HARDNESS: ~ 3½
  DENSITY: 3.50 (Meas.)
  CLEAVAGE: Prismatic {100}, distinct to imperfect.
  HABIT: Crystals anhedral, commonly less than 0.02 mm in largest dimension; also as cleavable masses up to 3 cm across.
  COLOR-LUSTER: Grayish yellow.
  MODE OF OCCURRENCE: Occurs with intimately intergrown ancylite as irregular masses or veinlets younger than calcite, and as separate crystals in calcite, in hydrothermal deposits composed essentially of silicates, carbonates, and

sulfides at the head of Big Sandy Creek, southeastern Hill County, Montana.
  BEST REF. IN ENGLISH: Pecora, W. T., and Kerr, Joe H., *Am. Min.*, 38: 1169–1183 (1953).

# BURKEITE

Na$_6$CO$_3$(SO$_4$)$_2$

CRYSTAL SYSTEM: Orthorhombic
Z: 12
LATTICE CONSTANTS:
  $a = 21.15$
  $b = 27.63$
  $c = 5.16$
3 STRONGEST DIFFRACTION LINES:
  2.78 (100)
  3.78 ( 80)
  2.58 ( 80)
OPTICAL CONSTANTS:
  $\alpha = 1.448$
  $\beta = 1.489$
  $\gamma = 1.493$
  $(-)2V = 34°$
  HARDNESS: 3½
  DENSITY: 2.57 (Meas.)
         2.56 (Calc.)
  CLEAVAGE: Fracture conchoidal. Brittle.
  HABIT: Crystals tabular with uneven and rough faces. Usually massive; also as spherical to irregular nodules, and as reticulated aggregates of platy crystals. Twinning on {110} common, cross-shaped.
  COLOR-LUSTER: White, pale tan, grayish. Transparent. Vitreous to greasy.

  Soluble in water.
  MODE OF OCCURRENCE: Occurs in layers up to one foot in thickness associated with trona at Searles Lake, San Bernardino County, and in muds, as efflorescences and crusts, at Deep Spring Lake, Inyo County, California. It also is found at Carbonate Lake, Grant County, Washington.
  BEST REF. IN ENGLISH: Palache, et al., "Dana's System of Mineralogy," 7th Ed., v. II, p. 633–634, New York, Wiley, 1951.

# BURSAITE

Pb$_5$Bi$_4$S$_{11}$

CRYSTAL SYSTEM: Probably Monoclinic
3 STRONGEST DIFFRACTION LINES:
  2.88 (100)
  3.38 ( 90)
  2.05 ( 90)
  HARDNESS: Not determined.
  DENSITY: > 6.2
  CLEAVAGE: "Tabular."
  HABIT: Small cleavable grains.
  COLOR-LUSTER: Gray; metallic. Opaque.
  MODE OF OCCURRENCE: Occurs associated with sphalerite, tremolite, pyrite, calcite, garnet, quartz, scheelite, chalcopyrite, and native bismuth in a large contact-metamorphic scheelite deposit, Uludağ, Turkey.

BEST REF. IN ENGLISH: Tolun, Rasit, *Bull. Min. Research and Exploration Inst. Turkey, Foreign Ed.*, No. 46-47, p. 106-127 (1954-55) (in English).

## BUSTAMITE

$(Ca, Mn)Si_2O_6$
Pyroxene Group

CRYSTAL SYSTEM: Triclinic
CLASS: $\bar{1}$
SPACE GROUP: $P\bar{1}$
Z: 12
LATTICE CONSTANTS:
$a = 15.46$     $\alpha = 89°34'$
$b = 7.18$     $\beta = 94°53'$
$c = 13.84$     $\gamma = 102°47'$
3 STRONGEST DIFFRACTION LINES:
2.872 (100)
3.195 ( 70)
2.982 ( 60)
OPTICAL CONSTANTS:
$\alpha = 1.662-1.692$
$\beta = 1.674-1.705$
$\gamma = 1.676-1.707$
$(-)2V_\alpha = 30°-44°$
HARDNESS: 5½-6½
DENSITY: 3.32-3.43 (Meas.)
CLEAVAGE: {100} perfect
{110} good
{1$\bar{1}$0} good
{010} poor
HABIT: Crystals tabular parallel to {001}, usually rough with rounded edges. Commonly massive, cleavable to compact, often with fibrous character.
COLOR-LUSTER: Pale flesh pink to brownish red. Transparent to translucent. Vitreous.
MODE OF OCCURRENCE: Occurs in manganese-bearing ore bodies, usually formed by metasomatic processes, or by metamorphism of sedimentary deposits. Found at Franklin and Sterling Hill, New Jersey, at the Treburland mine, Cornwall, England; at Långban, Sweden; in Iwate and Yamagata Prefectures, Japan; and at Broken Hill, New South Wales, Australia.
BEST REF. IN ENGLISH: Deer, Howie, and Zussman, "Rock Forming Minerals," v. II, p. 191-195, New York, Wiley, 1963.

## BUTLERITE

$Fe^{3+}SO_4OH \cdot 2H_2O$
Dimorphous with parabutlerite

CRYSTAL SYSTEM: Monoclinic
CLASS: 2/m
SPACE GROUP: $P2_1/m$
Z: 2
LATTICE CONSTANTS:
$a = 6.44$
$b = 7.31$
$c = 5.87$
$\beta = 108°35'$

3 STRONGEST DIFFRACTION LINES:
4.97 (100)
3.15 ( 80)
3.59 ( 60)
OPTICAL CONSTANTS:
$\alpha = 1.593-1.604$
$\beta = 1.665-1.674$
$\gamma = 1.741-1.731$
$(+)2V$ = large; sometimes $(-)$
HARDNESS: 2½
DENSITY: 2.55 (Meas.)
2.596 (Calc.)
CLEAVAGE: {100} perfect
HABIT: Crystals usually thick tabular; also octahedral, Twinning on {$\bar{1}$05} common.
COLOR-LUSTER: Orange to dark orange. Vitreous. Transparent. Streak pale yellow.
MODE OF OCCURRENCE: Occurs associated with parabutlerite, copiapite, and melanterite, at the Dexter No. 7 mine, Calf Mesa, San Rafael Swell, Utah; with other sulfates in the fire zone at the United Verde mine, Jerome, Arizona; at Chuquicamata, Chile; and with parabutlerite and other sulfates at the Santa Elena mine, La Alcaparrosa, San Juan province, Argentina.
BEST REF. IN ENGLISH: Palache, et al., "Dana's System of Mineralogy," 7th Ed., v. II, p. 608-609, New York, Wiley, 1951.

## BUTTGENBACHITE

$Cu_{19}Cl_4(NO_3)_2(OH)_{32} \cdot 2(or 3)H_2O$
Isomorphous with connellite

CRYSTAL SYSTEM: Hexagonal
CLASS: 6/m 2/m 2/m
SPACE GROUP: $P6_3/mmc$
Z: 2
LATTICE CONSTANTS:
$a = 15.82$
$c = 9.13$
3 STRONGEST DIFFRACTION LINES:
7.95 (100)
13.70 ( 95)
2.30 ( 95)
OPTICAL CONSTANTS:
$\omega = 1.738$
$\epsilon = 1.752$
(+)
HARDNESS: 3
DENSITY: 3.42 (Meas.)
3.442 (Calc.)
CLEAVAGE: Not determined.
HABIT: Crystals acicular, striated in direction of elongation. As radiated sprays of crystals and as felt-like aggregates.
COLOR-LUSTER: Azure blue. Translucent. Vitreous.
MODE OF OCCURRENCE: Occurs as a secondary mineral at Likasi, Katanga, Zaire.
BEST REF. IN ENGLISH: Palache, et al., "Dana's System of Mineralogy," 7th Ed., v. II, p. 572-573, New York, Wiley, 1951.

# BYSTROMITE

$MgSb_2O_6$

CRYSTAL SYSTEM: Tetragonal
CLASS: 4/m 2/m 2/m
SPACE GROUP: $P4_2/mnm$
Z: 2
LATTICE CONSTANTS:
  $a = 4.68$
  $c = 9.21$
3 STRONGEST DIFFRACTION LINES:
  3.32 (100)
  2.57 ( 90)
  1.73 ( 90)
OPTICAL CONSTANTS: Mean index varies from 1.86 to 1.91.
  HARDNESS: $\sim 7$
  DENSITY: 5.7 (Meas.)
             5.80 (Calc.)
  CLEAVAGE: None.
  HABIT: Massive; very fine grained.
  COLOR-LUSTER: Blue-gray. Streak light gray.
  MODE OF OCCURRENCE: Occurs intimately admixed with stibiconite at the La Fortuna and San Jose mines, El Antimonio district, Sonora, Mexico.
  BEST REF. IN ENGLISH: Mason, Brian, and Vitaliano, Charles J., *Am. Min.*, **37**: 53–57 (1952).

# BYTOWNITE

$mCaAl_2Si_2O_8$ with $nNaAlSi_3O_8$
$Ab_{30}An_{70}$ to $Ab_{10}An_{90}$
Plagioclase Feldspar Group

CRYSTAL SYSTEM: Triclinic
CLASS: $\bar{1}$
SPACE GROUP: $C\bar{1}$
Z: 4
LATTICE CONSTANTS:
  $a \simeq 8.17$    $\alpha = 93°20'$
  $b \simeq 12.86$   $\beta = 116°$
  $c \simeq 7.10$    $\gamma = 91°$
3 STRONGEST DIFFRACTION LINES:
  3.20 (100)
  4.03 ( 80)
  3.75 ( 80)
OPTICAL CONSTANTS:
    $\alpha = 1.561$
    $\beta = 1.565$
    $\gamma = 1.570$
    $(-)2V = 86°$
  HARDNESS: 6–6½
  DENSITY: 2.72–2.74 (Meas.)
             2.718 (Calc.)
  CLEAVAGE: {001} perfect
           {010} nearly perfect
           {110} imperfect
Fracture uneven to conchoidal. Brittle.
  HABIT: Crystals tabular, flattened along *b*-axis; not common. Usually massive, cleavable, granular, or compact. Twinning common. Principal twin laws: Carlsbad, albite, and pericline.
  COLOR-LUSTER: Colorless, white, gray. Transparent to translucent. Vitreous. Streak white.
  MODE OF OCCURRENCE: Occurs as a rock-forming mineral in basic plutonic and volcanic rocks such as anorthosite and norite; in certain metamorphic rocks; and in meteorites. Typical occurrences are found in Montana, South Dakota, Oklahoma, Minnesota, and Wisconsin. Also in Scotland, England, Sweden, Japan, and Transvaal, South Africa.
  BEST REF. IN ENGLISH: Deer, Howie, and Zussman, "Rock Forming Minerals," v. 4, p. 94–165, New York, Wiley, 1963. Barth, T. F. W., "Feldspars," New York, Wiley, 1969.

## CACOXENITE
$Fe_4^{3+}(PO_4)_3(OH)_3 \cdot 12H_2O$

CRYSTAL SYSTEM: Hexagonal
CLASS: 6/m 2/m 2/m
SPACE GROUP: P6/mmm
Z: 12
LATTICE CONSTANTS:
 $a$ = 27.669
 $c$ = 10.655
3 STRONGEST DIFFRACTION LINES:
 23.1 (100)
 11.9 (100)
 9.1 ( 14)
OPTICAL CONSTANTS:
 $\omega$ = 1.580-1.585, 1.600
 $\epsilon$ = 1.640-1.656, 1.680
 (+)
HARDNESS: 3-4
DENSITY: 2.26  (Meas.)
 2.252 (Calc.)
CLEAVAGE: Not observed.
 HABIT: Crystals minute, rare, acicular, sometimes with hexagonal cross section. Usually as radial or tufted aggregates or fibrous crusts; spherulitic.
 COLOR-LUSTER: Golden yellow, yellow, brownish yellow, reddish yellow; rarely greenish.
 MODE OF OCCURRENCE: Occurs as a secondary mineral commonly associated with other phosphates and with limonite. Found at Shady, Polk County, Arkansas; in Cherokee County, Alabama; at the Vanleer mine, Iron City, Tennessee; at Hellertown and elsewhere in Pennsylvania; on hematite at Antwerp, New York; and at Tonapah, Nevada. It also occurs in France, Germany, Czechoslovakia, and Sweden.
 BEST REF. IN ENGLISH: Palache, et al., "Dana's System of Mineralogy," 7th Ed., v. II, p. 997-999, New York, Wiley, 1951. *Am. Min.*, **51**: 1811-1814 (1966).

## CADMIUM OXIDE (Monteponite)
CdO

CRYSTAL SYSTEM: Cubic

CLASS: 4/m $\overline{3}$ 2/m
SPACE GROUP: Fm3m
Z: 4
LATTICE CONSTANT:
 $a$ = 4.689
3 STRONGEST DIFFRACTION LINES:
 2.712 (100)
 2.349 ( 88)
 1.661 ( 43)
OPTICAL CONSTANT:
 $N$ = 2.49  (Li)
 HARDNESS: 3
 DENSITY: 8.1-8.2 (Meas. Synthetic)
 8.238 (Calc.)
 CLEAVAGE: {111} reported-Synthetic crystals.
Sectile.
 HABIT: Crystals octahedral or cubo-octahedral. Also as a powder. Penetration twinning reported.
 COLOR-LUSTER: Brilliant black. Transparent.
 MODE OF OCCURRENCE: Occurs as a coating on hemimorphite at Genarutta, near Iglesias, Sardinia.
 BEST REF. IN ENGLISH: Palache, et al., "Dana's System of Mineralogy," 7th Ed., v. I, p. 502-503, New York, Wiley, 1944.

## CADMOSELITE
CdSe

CRYSTAL SYSTEM: Hexagonal
CLASS: 6mm
SPACE GROUP: $P6_3mc$
Z: 2
LATTICE CONSTANTS:
 $a$ = 4.271
 $c$ = 6.969
3 STRONGEST DIFFRACTION LINES:
 3.72 (100)
 2.15 ( 85)
 3.29 ( 75)
HARDNESS: Not determined.
DENSITY: 5.663 (Calc.)
CLEAVAGE: Perfect prismatic.

HABIT: Minute crystals showing hexagonal pyramid with horizontal striations, and base.

COLOR-LUSTER: Black; resinous to adamantine. Black streak.

MODE OF OCCURRENCE: Occurs associated with ferroselite, clausthalite, amorphous selenium, cadmian sphalerite, and pyrite, as fine xenomorphic disseminations cementing sandstone at an unspecified locality, presumably in the USSR.

BEST REF. IN ENGLISH: Bur'yanova, E. Z., Kovalev, G. A., and Komkov, A. I., *Am. Min.*, **43**: 623 (1958).

# CADWALADERITE
$Al(OH)_2Cl \cdot 4H_2O$

CRYSTAL SYSTEM: Cubic (?)
OPTICAL CONSTANT:
  $N = 1.513$
  HARDNESS: Not determined.
  DENSITY: 1.66 (Meas.)
  CLEAVAGE: None.
Fracture conchoidal.
  HABIT: As amorphous grains and small masses.
  COLOR-LUSTER: Lemon yellow; transparent. Vitreous.
  MODE OF OCCURRENCE: Occurs in a sulfate deposit associated with halite in Cerro Pintados, 80 km southeast of Iquique, Tarapacá province, Chile.
  BEST REF. IN ENGLISH: Palache, et al., "Dana's System of Mineralogy," 7th Ed., v. II, p. 77, New York, Wiley, 1951.

# CAFARSITE
$Ca_3(Fe, Ti)_3Mn(AsO_4)_6 \cdot 2H_2O$

CRYSTAL SYSTEM: Cubic
CLASS: $2/m\,\overline{3}$
SPACE GROUP: Pn3
Z: 8
LATTICE CONSTANT:
  $a = 16.01$
3 STRONGEST DIFFRACTION LINES:
  2.83 (100)
  2.75 ( 80)
  3.15 ( 70)
  HARDNESS: 5½–6
  DENSITY: 3.90 (Meas.)
          3.82 (Calc.)
  CLEAVAGE: None.
Fracture conchoidal.
  HABIT: Well-formed crystals with rough faces up to 3cm in size.
  COLOR-LUSTER: Dark brown; in thin splinters translucent red. Streak yellowish brown.
  MODE OF OCCURRENCE: Found on cleft faces in orthogneisses of the Monte Leone nappe, southern Binnatal, Canton Wallis, Switzerland; also from Italy.
  BEST REF. IN ENGLISH: St. Graeser, *Am. Min.*, **52**: 1584 (1967).

# CAFETITE
$CaFe_2Ti_4O_{12} \cdot 4H_2O$

CRYSTAL SYSTEM: Orthorhombic or monoclinic
Z: 6
LATTICE CONSTANTS:
  $a = 31.34$
  $b = 12.12$
  $c = 4.96$
  $\beta = 90°$
3 STRONGEST DIFFRACTION LINES:
  7.86 (100)
  2.56 ( 80)
  1.91 ( 70)
OPTICAL CONSTANTS:
  $\alpha = 1.95$
  $\beta = 2.08$
  $\gamma = 2.11$
  $(-)2V = 38°$
HARDNESS: 4–5
DENSITY: 3.28 (Meas.)
        3.19 (Calc.)
CLEAVAGE: Prismatic, two directions. Brittle.
HABIT: Crystals columnar to accicular, elongated (001), striated on prism faces. Dominant forms: {100}, {520}, and {210}. Tangled fibrous aggregates and radial-fibrous aggregates of acicular crystals.
COLOR-LUSTER: Pale yellow to colorless; adamantine.
MODE OF OCCURRENCE: Occurs with titanian magnetite, phlogopite, ilmenite, sphene, clinochlore, dysanalyte, baddeleyite, anatase, and an unidentified mineral in miarolitic cavities in pegmatites cutting the Afrikanda pyroxenite massif, Kola Peninsula, USSR.
BEST REF. IN ENGLISH: Kukharenko, A. A., Kondrat'eva, V. V., and Kovyazina, V. M., *Am. Min.*, **45**: 476 (1960).

# CAHNITE
$Ca_2BAsO_4(OH)_4$

CRYSTAL SYSTEM: Tetragonal
CLASS: $\overline{4}$
SPACE GROUP: I$\overline{4}$
Z: 2
LATTICE CONSTANTS:
  $a = 7.095$
  $c = 6.190$
3 STRONGEST DIFFRACTION LINES:
  3.56  (100)
  1.818 ( 55)
  2.640 ( 50)
OPTICAL CONSTANTS:
  $\omega = 1.662$
  $\epsilon = 1.663$
   (+)
HARDNESS: 3
DENSITY: 3.156 (Meas.)
        3.177 (Calc.)
CLEAVAGE: {110} perfect
Brittle.
HABIT: Single crystals pseudotetrahedral; commonly twinned on {110} presenting characteristic cross-like appearance.

COLOR-LUSTER: Colorless to white; transparent; vitreous.

MODE OF OCCURRENCE: Occurs rarely in cavities in axinite veinlets, or implanted on garnet crystals lining drusy cavities in franklinite ore, and also associated with rhodonite, at Franklin, New Jersey; also found at Capo di Bove, near Rome, Italy, and in a skarn iron ore deposit at the Klodeborg mine, Arendal, Norway.

BEST REF. IN ENGLISH: Palache, et al., "Dana's System of Mineralogy," v. II, 7th Ed., p. 386-387, New York, Wiley, 1951.    Prewitt, Charles T., and Buerger, M. J., *Am. Min.*, 46: 1077-1085 (1961).

## CAIRNGORM = Smokey quartz

## CALAMINE = Hemimorphite

## CALAVERITE
$AuTe_2$

CRYSTAL SYSTEM: Monoclinic
CLASS: 2/m or 2
SPACE GROUP: C2/m or C2
Z: 2
LATTICE CONSTANTS:
  $a = 7.19$
  $b = 4.40$
  $c = 5.08$
  $\beta = 90°30'$
3 STRONGEST DIFFRACTION LINES:
  2.99 (100)
  2.09 ( 90)
  2.91 ( 70)
OPTICAL CONSTANTS: (unreliable)
  HARDNESS: 2½-3
  DENSITY: 9.10-9.40 (Meas.)
              9.31 (Calc.)
  CLEAVAGE: None
Fracture uneven to subconchoidal.  Very brittle.
  HABIT: Crystals lath-like or bladed and as short to slender prisms striated in direction of elongation.  Also massive, granular.  Twinning on {101}, {111}, or {310}.
  COLOR-LUSTER: Brass yellow to silver white. Opaque. Metallic. Streak greenish to yellowish gray.
  MODE OF OCCURRENCE: Occurs in low- to high temperature vein deposits, commonly associated with native gold, fluorite, quartz, sulfides, sulfosalts, and tellurides. Found as one of the principal gold ore minerals of the Cripple Creek district, Teller County, and in lesser amounts in Boulder, Costilla, LaPlata, and Montezuma counties, Colorado; at the Stanislaus, Melones, and Morgan mines on Carson Hill, Calaveras County, and in El Dorado and Tuolumne counties, California; in the Kirkland Lake district and elsewhere in Ontario, Canada; and at localities in Mexico, El Salvador, Roumania, USSR, Philippine Islands, and Australia.
  BEST REF. IN ENGLISH: Vlasov, K. A., Mineralogy of Rare Elements," v. II, p. 698-703, Israel Program for Scientific Translations, 1966.

## CALCIBORITE
$CaB_2O_4$

CRYSTAL SYSTEM: Monoclinic (?)
Z: 1
3 STRONGEST DIFFRACTION LINES:
  3.445 (100)
  1.867 (100)
  1.970 ( 90)
HARDNESS: ~ 3½
DENSITY: 2.878
CLEAVAGE: None.
Fracture uneven, on aggregates conchoidal.
  HABIT: Massive; as radial aggregates.
  COLOR-LUSTER: White.
  MODE OF OCCURRENCE: Occurs associated with calcite and dolomite, with minor amounts of garnet, magnetite, and pyroxene, in deep drill cores at the contact of Middle Devonian limestone with quartz diorite in a copper deposit of the skarn type in the Urals.
  BEST REF. IN ENGLISH: Petrova, E. S., *Am. Min.*, 41: 815 (1956).   Malinko, S. V., Kuznetsova, N. N., Pensionerova, V. M., and Rybakova, L. I., *Am. Min.*, 49: 820 (1964).

## CALCIOCELESTINE = Variety of celestine containing small amounts of calcium substituting for strontium.

## CALCIOCOPIAPITE (Tusiite)
$CaFe_4(SO_4)_6(OH)_2 \cdot 19H_2O$

3 STRONGEST DIFFRACTION LINES:
  3.108 (100)
  3.027 (100)
  2.809 (100)
HABIT: Powdery crusts.
COLOR-LUSTER: Grayish yellow to brownish yellow.
MODE OF OCCURRENCE: Occurs associated with hydrous iron oxides, chalcanthite, and malachite as a supergene product of weathering of magnetite ores containing pyrite and calcite at Dashkesan, Middle Caucasus, Azerbaidjan.
  BEST REF. IN ENGLISH: Kashkai, M. A., and Aliev, R. M., *Am. Min.*, 47: 807-808 (1962).

## CALCIOFERRITE
$Ca_2Fe_2(PO_4)_3(OH) \cdot 7H_2O$

CRYSTAL SYSTEM: Hexagonal (?)
OPTICAL CONSTANTS:
  $\omega = 1.57-1.58$ (variable)
    (-)
HARDNESS: 2½
DENSITY: 2.53
CLEAVAGE: Perfect parallel to foliation.  Brittle.
HABIT: As foliated reniform or nodular masses.
COLOR-LUSTER: Yellowish white to yellow; also greenish yellow to yellowish green.  Opaque; thin laminae translucent. Pearly.

MODE OF OCCURRENCE: Occurs as nodules in a clay bed at Battenberg, Germany.

BEST REF. IN ENGLISH: Palache, et al., "Dana's System of Mineralogy," 7th Ed., v. II, p. 976–977, New York, Wiley, 1951. Mead, Cynthia W., and Mrose, Mary E., U.S. Geol. Surv. Prof. Pap. 600-D, p. D204–D206 (1968).

# CALCIOTALC
$CaMg_2Si_4O_{10}(OH)_2$

CRYSTAL SYSTEM: Orthorhombic
LATTICE CONSTANTS:
  $a$ = 5.18
  $b$ = 9.25
  $c$ = 3.079
3 STRONGEST DIFFRACTION LINES:
  9.25  (90)
  3.079 (90)
  1.518 (60)
OPTICAL CONSTANTS:
  $\alpha$ = 1.565
  $\beta$ =  ?
  $\gamma$ = 1.583
  $(-)2V$ = small
HARDNESS: 2–3
HABIT: Partial or complete pseudomorphs after actinolite; columnar structure (preserved from actinolite). Dense.
COLOR-LUSTER: Greenish gray; silky to weak pearly.
MODE OF OCCURRENCE: Occurs with normal talc as a hydrothermal alteration of diopside-actinolite in Archean rocks in the basin of the Leglier River, Aldan Region, southern Yakutia.
BEST REF. IN ENGLISH: Serdyuchenko, D. P., and Belov, N. V., *Am. Min.*, 45: 476–477 (1960).

# CALCIOTHORITE = Calcian variety of thorite

# CALCIOVOLBORTHITE (Tangeite)
$CaCuVO_4(OH)$

CRYSTAL SYSTEM: Orthorhombic
CLASS: 2 2 2
SPACE GROUP: $P2_12_12_1$
Z: 4
LATTICE CONSTANTS:
  $a$ = 7.45
  $b$ = 9.26
  $c$ = 5.91
3 STRONGEST DIFFRACTION LINES:
  2.88 (100)
  2.61 (100)
  4.15 ( 75)
OPTICAL CONSTANTS:
  $\alpha$ = 2.01
  $\beta$ = 2.05
  $\gamma$ = 2.09
  $(+)2V$ = 83°
HARDNESS: 3½

DENSITY: 3.75 (Meas.)
         3.82 (Calc.)
CLEAVAGE: One perfect.
HABIT: As scaly aggregates; also fibrous to dense.
COLOR-LUSTER: Greenish yellow to olive green to dark green. Subtranslucent. Vitreous; pearly on cleavage.
MODE OF OCCURRENCE: Occurs as a secondary mineral in the oxidation zone of deposits containing vanadium minerals. Found at Camp Signal, near Goffs, San Bernardino County, California; in Custer, Huerfano, Montrose, and Park counties, Colorado; in Grand County, Utah; and near Tombstone, Cochise County, Arizona. It also occurs at Friedrichroda, Thuringia, Germany; at several localities in the USSR; at Tsumeb, South West Africa; and at Luiswishi and other places in Katanga, Zaire.
BEST REF. IN ENGLISH: Palache, et al., "Dana's System of Mineralogy," 7th Ed., v. II, p. 817–818, New York, Wiley, 1951.

# CALCITE
$CaCO_3$
Trimorphous with aragonite and vaterite

CRYSTAL SYSTEM: Hexagonal
CLASS: $\bar{3}\,2/m$
SPACE GROUP: $R\bar{3}c$
Z: 6
LATTICE CONSTANTS:

| (Hexagonal) | (Trigonal [rhombohedral]) | |
|---|---|---|
| $a$ = 4.98 | $a$ = 6.36 | Z = 2 |
| $c$ = 17.02 | $\alpha$ = 46°06′ | |

3 STRONGEST DIFFRACTION LINES:
  3.035 (100)
  2.285 ( 18)
  2.095 ( 18)
OPTICAL CONSTANTS:
  $\omega$ = 1.65835    $\lambda$ = 588.99
  $\epsilon$ = 1.48645       20°C
      $(-)$
HARDNESS: 3
DENSITY: 2.7102 (Meas.)
CLEAVAGE· $\{10\bar{1}1\}$ perfect
Parting on $\{01\bar{1}2\}$ and sometimes on $\{0001\}$. Fracture conchoidal. Brittle.
HABIT: Crystals extremely varied in appearance; scalenohedrons and rhombohedrons most common. Also massive, fibrous, granular, stalactitic, chalky. Twinning common.
COLOR-LUSTER: Colorless or white when pure. Often various shades of gray, yellow, brown, red, green, blue, and black from impurities. Transparent to translucent. Vitreous to pearly; also dull. Streak white to grayish. Often fluorescent and phosphorescent under ultraviolet light in shades of green, yellow, blue, and red.
MODE OF OCCURRENCE: Calcite, in its various forms, is one of the most widely distributed minerals in the world. It is the chief constituent in all limestones and marbles and is found as a component of other sedimentary and metamorphic rocks. In many ore deposits it is often an abundant gangue mineral and has been found sparingly in pegmatites. It also occurs as crystalline incrustations and as stalactites and stalagmites in numerous caverns; in

geodes and concretions; and as a petrifying material, replacing fossil animal and plant remains. Exceptional specimens occur in the Missouri-Kansas-Oklahoma and Wisconsin-Illinois lead-zinc regions; on the Keweenaw Peninsula, Michigan, often enclosing native copper; at Charcas, San Luis Potosi and other places in Mexico; at Eskifjord, Iceland (iceland-spar); at Kongsberg, Norway with native silver; in Cumberland, Durham, Lancashire, and Cornwall, England; and at Tsumeb and Grootfontein, South West Africa. Superb specimens of sand-calcite are found in Washabaugh County, South Dakota and at Fontainebleau, France. Many marine organisms secrete magnesian calcite as their skeletal material.

BEST REF. IN ENGLISH: Palache, et al., "Dana's System of Mineralogy," 7th Ed., v. II, p. 142–160, New York, Wiley, 1951.

## CALCIUM CATAPLEIITE
$CaZrSi_3O_9 \cdot 2H_2O$

CRYSTAL SYSTEM: Hexagonal
CLASS: $6/m \ 2/m \ 2/m$
SPACE GROUP: $P6_3/mmc$
Z: 2
LATTICE CONSTANTS:
$a = 7.32$
$c = 10.15$
3 STRONGEST DIFFRACTION LINES:
2.96 (100)
3.96 ( 80)
3.06 ( 80)
OPTICAL CONSTANTS:
$\omega = 1.603$
$\epsilon = 1.639$
   (+)
HARDNESS: 4½–5
DENSITY: 2.77 (Meas.)
            2.78 (Calc.)
HABIT: Massive.
COLOR-LUSTER: Pale yellow to cream; vitreous to dull. Opaque, translucent on edges.
MODE OF OCCURRENCE: Occurs in cavities between crystals of microcline associated with pyrophanite, pyrochlore, titanolavenite, loparite, kupletskite, rare-earth apatite, and hiortdahlite in syenite pegmatites of the alkalic massif of Burpala, northern Baikal, USSR.
BEST REF. IN ENGLISH: Portnov, A. M., *Am. Min.*, **49**: 1153 (1964).

## CALCIUM-LARSENITE = Esperite

## CALCIUM URSILITE = Ursilite

## CALCLACITE
$Ca(CH_3COO)Cl \cdot 5H_2O$

CRYSTAL SYSTEM: Monoclinic
CLASS: $2/m$

SPACE GROUP: $P2_1/n$
Z: 4
LATTICE CONSTANTS:
$a = 10.41$
$b = 13.72$
$c = 6.82$
$\beta = 99°6'$
3 STRONGEST DIFFRACTION LINES:
8.27 (100)
3.24 (100)
2.43 (100)
OPTICAL CONSTANTS:
   $\alpha = 1.468$
   $\beta = 1.484$
   $\gamma = 1.515$
(+)2V = 80°
HARDNESS: Not determined.
DENSITY: 1.5
CLEAVAGE: Not determined.
HABIT: As hair-like efflorescences.
COLOR-LUSTER: White.
MODE OF OCCURRENCE: Occurs as a recent formation on pottery and on specimens of calcareous rocks and fossils kept in oak museum cases.
BEST REF. IN ENGLISH: Palache, et al., "Dana's System of Mineralogy," 7th Ed., v. II, p. 1107, New York, Wiley, 1951.

## CALCURMOLITE
$Ca(UO_2)_3(MoO_4)_3(OH)_2 \cdot 11H_2O$

CRYSTAL SYSTEM: Unknown
3 STRONGEST DIFFRACTION LINES:
7.76  7.60  7.85  (100  )
8.41  8.28  8.34  (70–50)
3.90  3.90  3.89  (50–60)
OPTICAL CONSTANTS:
$\omega = 1.816–1.827$     $\alpha = 1.770$
$\epsilon = 1.856–1.863$     $\beta = 1.816$
   (−)               $\gamma = 1.856–1.863$
HABIT: Prismatic crystals; also as massive, platy aggregates.
COLOR-LUSTER: Honey yellow. Fluoresces strong yellowish green under ultraviolet light.
MODE OF OCCURRENCE: Occurs as pseudomorphs after uraninite and is associated with uranophane, uranospinite, halloysite, betpakdalite, jarosite, and ferrimolybdite in the zone of oxidation of an undisclosed uranium-molybdenum ore deposit.
BEST REF. IN ENGLISH: Federov, O. V., *Am. Min.*, **49**: 1152–1153 (1964).

## CALEDONITE
$Cu_2Pb_5(SO_4)_3CO_3(OH)_6$

CRYSTAL SYSTEM: Orthorhombic
CLASS: $2/m \ 2/m \ 2/m$
SPACE GROUP: Pnmm
Z: 2

ABERNATHYITE   Temple Mountain, San Rafael Swell, Utah.

ACANTHITE   Chihuahua, Mexico.

ADAMITE, Cuprian   Gold Hill, Tooele County, Utah.

ADAMITE, Cuprian   Laurium, Greece.

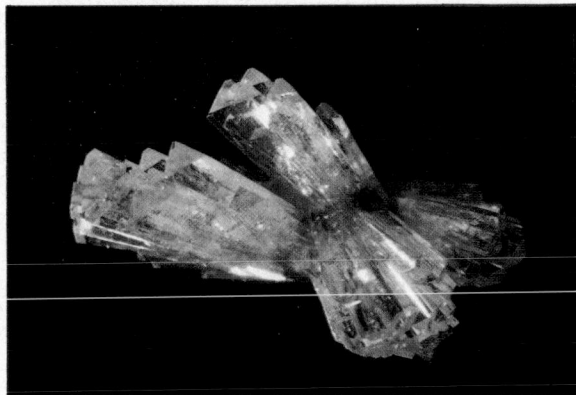

ADAMITE   Mapimi, Durango, Mexico.

AEGIRINE   Arch St. Pike, Little Rock, Arkansas.

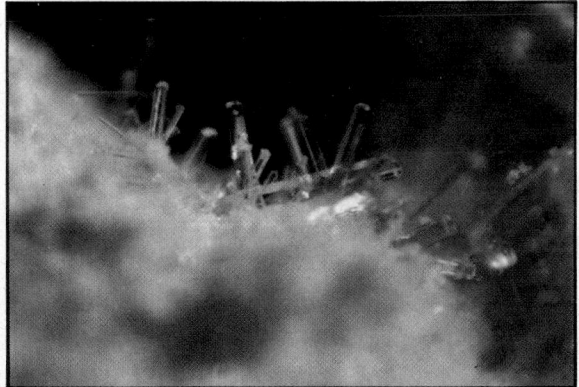

AFWILLITE   Boisse Jours, Auvergne, France.

AGARDITE   Clara mine, Wittichen, Germany.

1

AJOITE   New Cornelia mine, Pima County, Arizona.

ALBITE   Glendale, Pennington County, South Dakota.

ALLACTITE   Franklin, Sussex County, New Jersey.

ALLANITE (Orthite)   Laacher See, Eifel, Rhineland, Germany.

ALMANDINE in muscovite   Spruce Pine, Mitchell County, North Carolina.

ALMANDINE   Spruce Pine, Mitchell County, North Carolina.

AMBLYGONITE   Newry, Oxford County, Maine.

AMBLYGONITE   Newry, Oxford County, Maine.

ANALCIME   Cyclopean Islands, Italy.

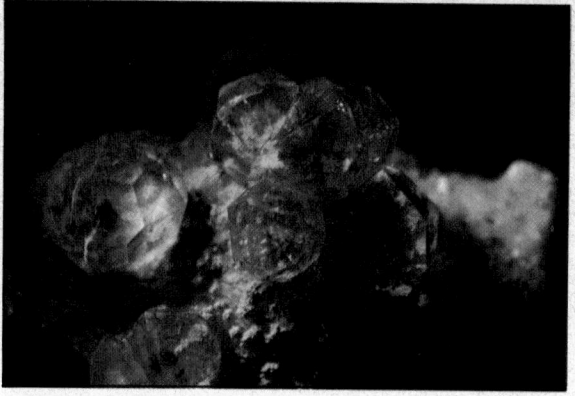

ANALCIME   Guam.

ANAPAITE   Prats Samsor, Lerida, Spain.

ANATASE on muscovite   Brook's Farm, Latimore Cleveland Co., North Carolina.

ANATASE   Binnental, Valais, Switzerland.

ANATASE with adularia   Homestake Mine, Lead, Lawrence Co., South Dakota.

ANATASE   Cavradi, Val Tavetsch, Grisons, Switzerland.

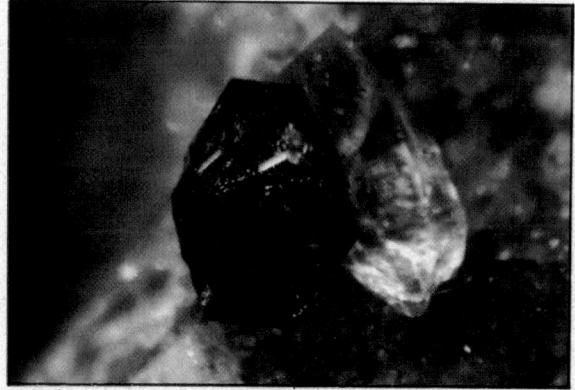

ANATASE   Binnental, Valais, Switzerland.

ANATASE    Binnental, Valais, Switzerland.

ANCYLITE, Calcian    Cornog, Chester County, Pennsylvania.

ANCYLITE with acmite    Mont St. Hilaire, Quebec, Canada.

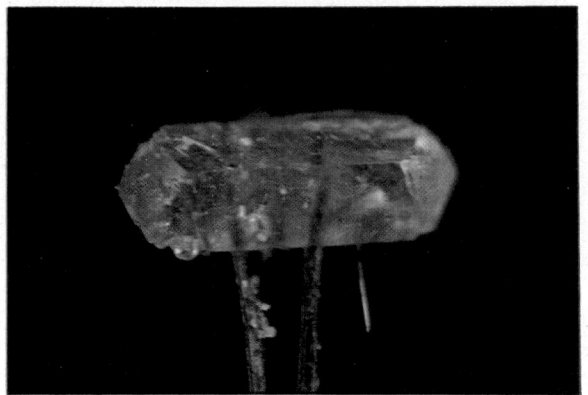

ANCYLITE    Mont St. Hilaire, Quebec, Canada.

ANDALUSITE    Near Oreville, Pennington County, South Dakota.

ANDERSONITE    Hillside mine, Yavapai County, Arizona.

ANGLESITE    Laurium, Greece.

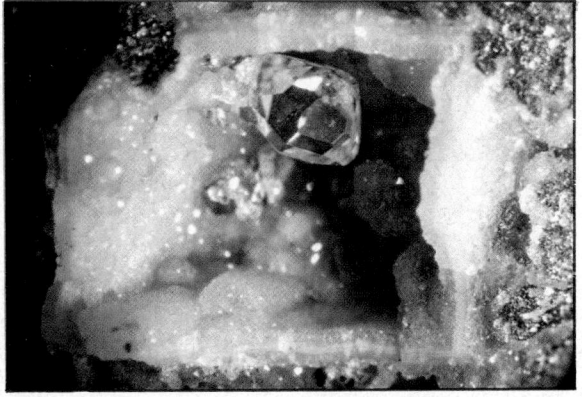

ANGLESITE    Eureka, Tintic district, Utah.

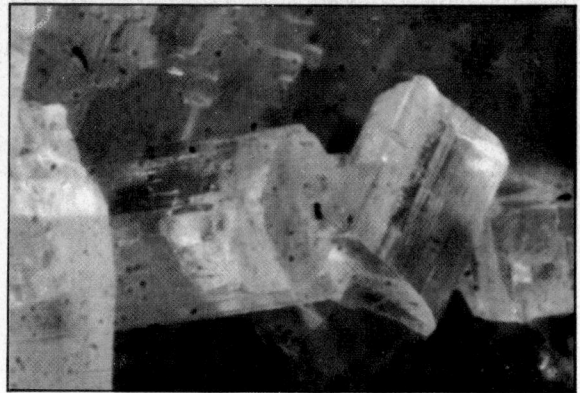

ANHYDRITE   Simpion Tunnel, Switzerland.

ANNABERGITE (Cabrerite)   Laurium, Greece.

ANNABERGITE (Cabrerite)   Laurium, Greece.

ANTLERITE   Chuquicamata, Chile.

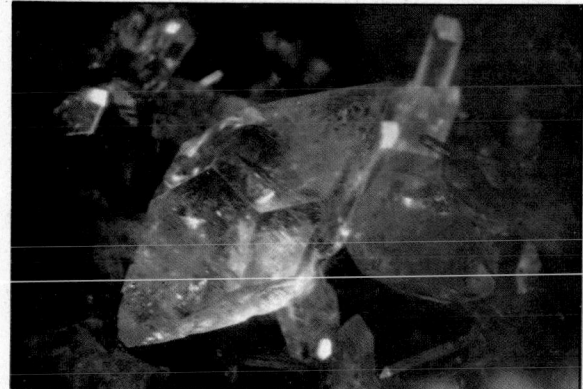

APOPHYLLITE   Guanajuato, Mexico.

APOPHYLLITE   Poona, India.

APOPHYLLITE   Snake Hill, Secaucus, New Jersey.

5

ARAGONITE with pyrite Zacatecas, Mexico.

ARGENTITE   Chanarcillo, Chile.

ARGENTITE   Freiberg, Saxony, Germany.

ARGENTITE with polybasite   Naica, Chihuahua, Mexico.

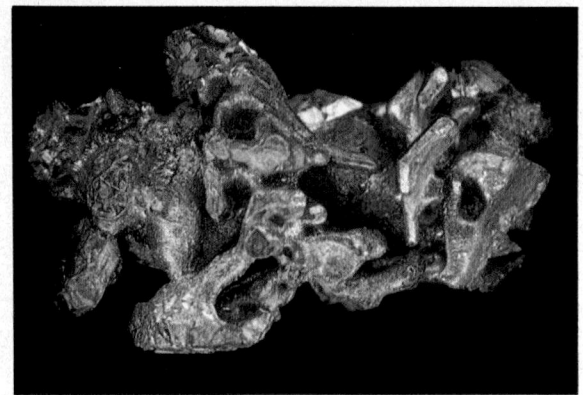

ARGENTITE   La Plata mine, Zacatecas, Mexico.

ARGENTOPYRITE with pyrargyrite   St. Andreasberg, Harz Mts., Germany.

ARGYRODITE   Colquechaca, Bolivia.

ARSENOLITE   Manhattan, Nye County, Nevada.

ARSENOPYRITE with sphalerite   Pribram, Bohemia, Czechoslovakia.

ARSENOPYRITE (cyclic twin)   Double Rainbow mine, Galena, South Dakota.

ARSENOPYRITE (twin)   Deloro, Ontario, Canada.

ARSENIOSIDERITE   Mapimi, Durango, Mexico.

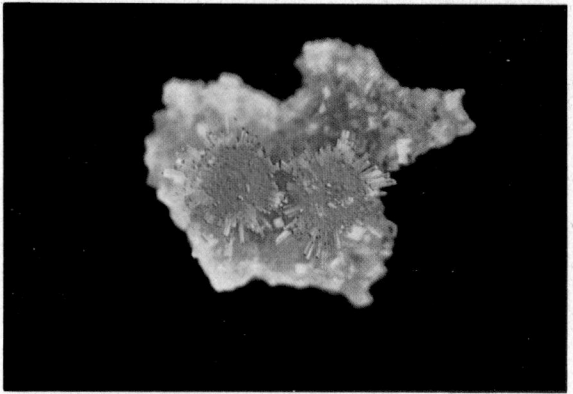

ARTHURITE   Majuba Hill, Pershing County, Nevada.

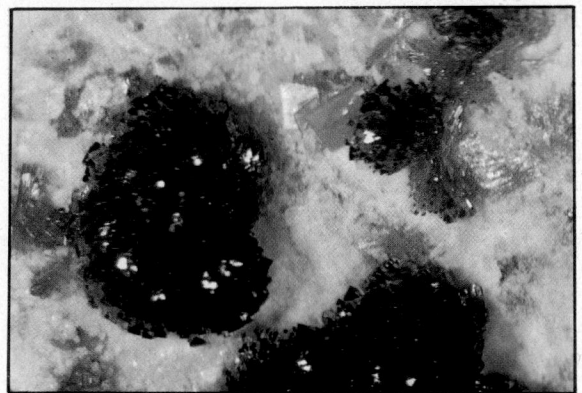

ARTHURITE with zeunerite   Majuba Hill, Pershing County, Nevada.

ARTINITE   Near Hernandez, San Benito County, California.

ARTINITE   Staten Island, New York.

ASTROPHYLLITE   Mont St. Hilaire, Quebec, Canada.

ASTROPHYLLITE   Mont St. Hilaire, Quebec, Canada.

ATACAMITE   Copiapo, Chile.

ATELESTITE   Schneeberg, Saxony, Germany.

AUGELITE   Oruro, Bolivia.

AUGITE   Vesuvius, Italy.

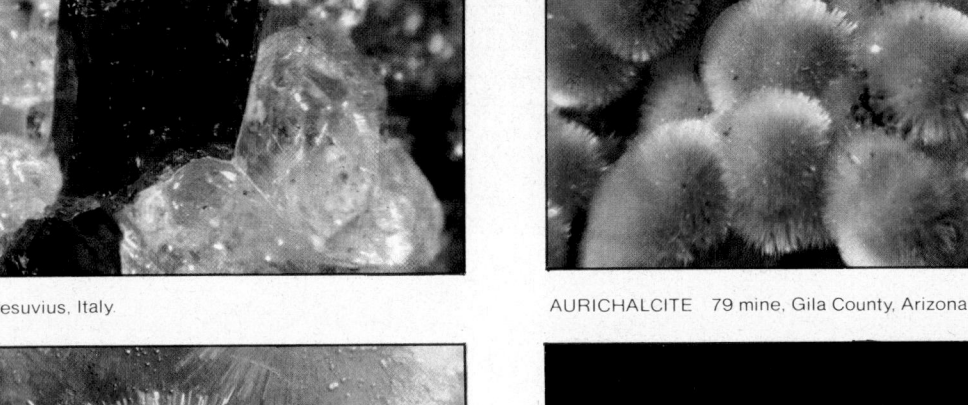

AURICHALCITE   79 mine, Gila County, Arizona.

AURICHALCITE   Jackpot mine, Lyon County, Nevada.

AURICHALCITE   79 mine, Gila County, Arizona.

AUSTINITE, Cuprian   Gold Hill, Tooele County, Utah.

AUSTINITE on conichalcite   Gold Hill, Tooele County, Utah.

AUSTINITE with conichalcite   Gold Hill, Tooele County, Utah.

9

AXINITE    Franklin, Sussex County, New Jersey.

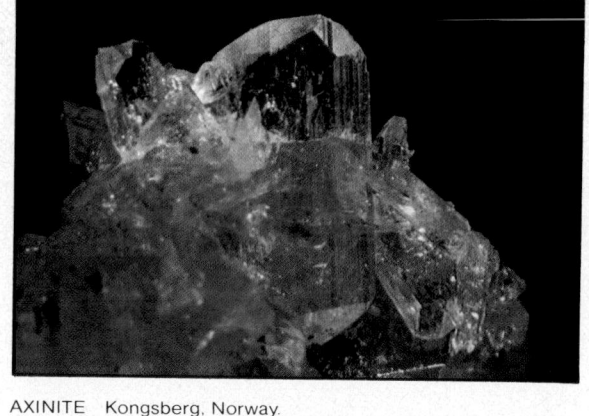

AXINITE    Kongsberg, Norway.

AXINITE    Hyuga, Japan.

AZURITE    Laurium, Greece.

AZURITE    Tsumeb, South West Africa.

AZURITE    Concepcion del Oro, Zacatecas, Mexico.

AZURITE    Concepcion del Oro, Zacatecas, Mexico.

AZURITE with malachite    Tsumeb, South West Africa.

BABINGTONITE   Durham, Middlesex County, Connecticut.

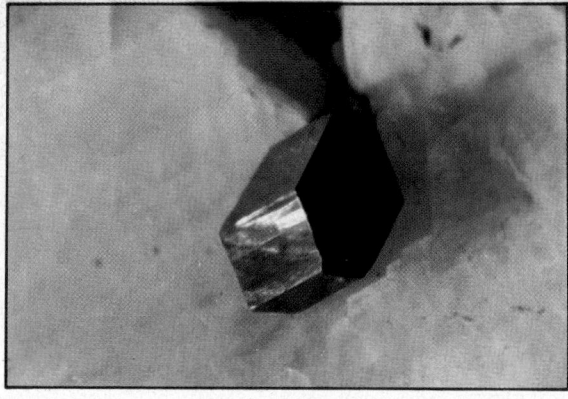

BABINGTONITE   Woburn, Middlesex County, Massachusetts.

BABINGTONITE on diopside   Manassas, Prince William County, Virginia.

BABINGTONITE   Durham, Middlesex County, Connecticut.

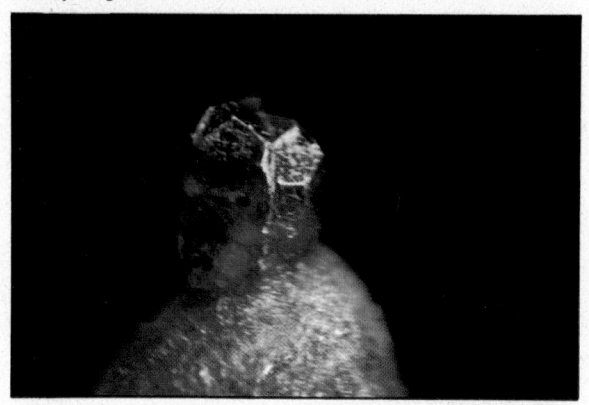

BADDELYITE   Jacupiranga, Sao Paulo, Brazil.

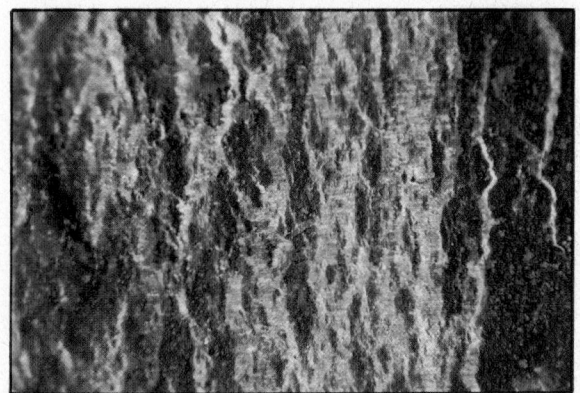

BARBERTONITE   New Amianthus mine, Kaapsche Hoop, Transvaal, South Africa.

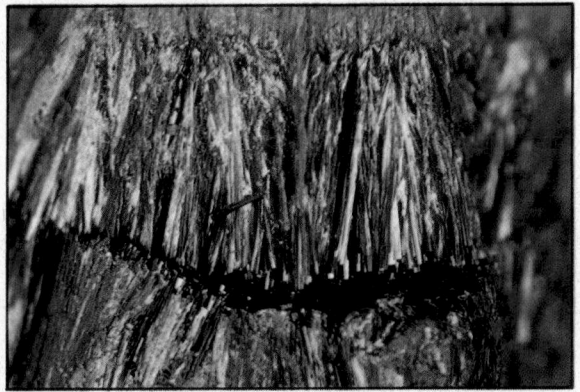

BARIANDITE   Mounana, Gabon.

11

BARITE with calcite    Elk Creek, Meade County, South Dakota.

BARITE with calcite    Elk Creek, Meade County, South Dakota.

BARITE    Cartersville, Bartow County, Georgia.

BARITE    Ward, Boulder County, Colorado.

BARITE    Magma mine, Superior, Pinal County, Arizona.

BARITE encrusted with pyrite    Zobes, Vogtland, Germany.

BAUMHAUERITE   Lengenbach quarry, Binnental, Switzerland.

BAUMHAUERITE   Lengenbach quarry, Binnental, Switzerland.

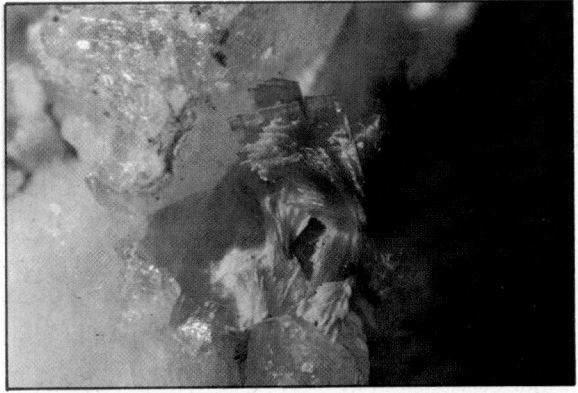

BAVENITE   Haddam, Middlesex County, Connecticut.

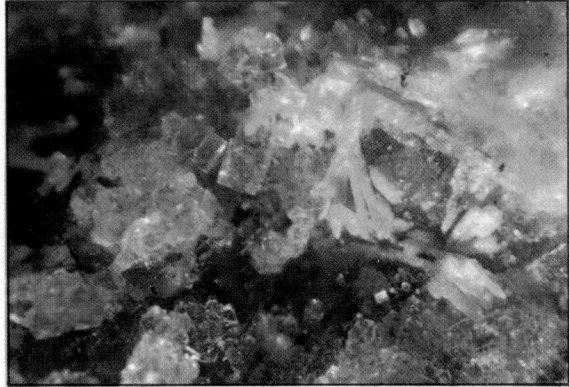

BAYLEYITE with andersonite   Ambrosia Lakes District, McKinley County, New Mexico.

BECQUERELITE   Shinkolobwe, Katanga, Zaire.

BENITOITE   San Benito County, California.

BENITOITE (pink)   San Benito County, California.

13

BERAUNITE with strunzite and strengite   Foote mine, Kings Mountain, North Carolina.

BERAUNITE   Indian Mountain, Cherokee County, Alabama.

BERGENITE   Bergen, Vogtland, Germany.

BERMANITE   Fletcher mine, North Groton, New Hampshire.

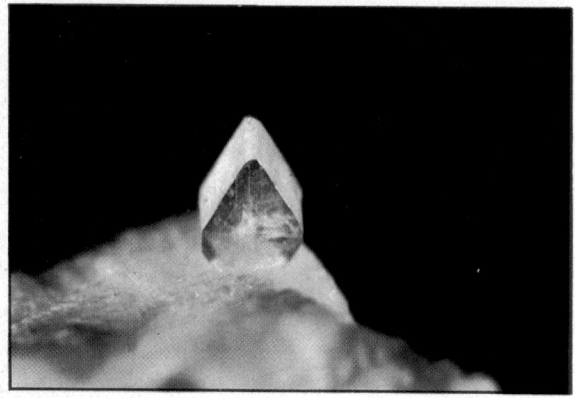

BERTRANDITE (twin)   Strickland quarry, Portland, Connecticut.

BERTRANDITE (twin)   West Haven, Connecticut.

BERYL   Muzo, Colombia.

BERYL (Aquamarine)   Mt. Wheeler mine, White Pine County, Nevada.

BERYL (Emerald)   Miask, Ural Mountains, USSR.

BERYL ("Trapiche" Emerald)   Muzo, Colombia.

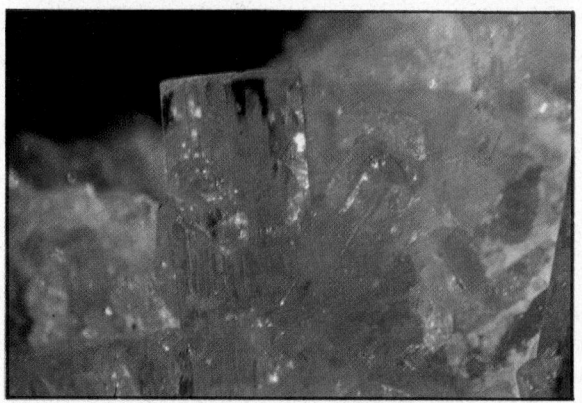

BERYL (Emerald)   Muzo, Colombia.

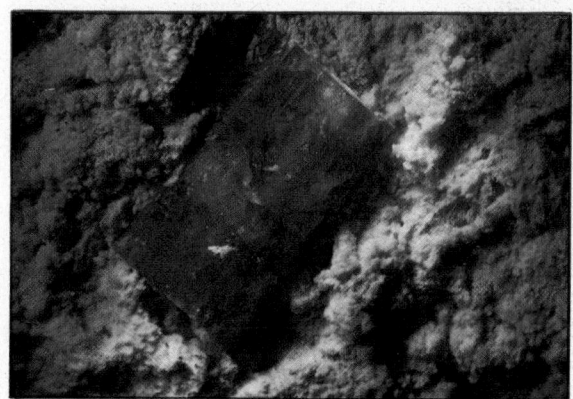

BERYL (Morganite)   Thomas Mountains, Juab County, Utah.

BETAFITE   Tete, Mozambique.

BETA-URANOPHANE   Black Forest, Germany.

BEUDANTITE   Tsumeb, South West Africa.

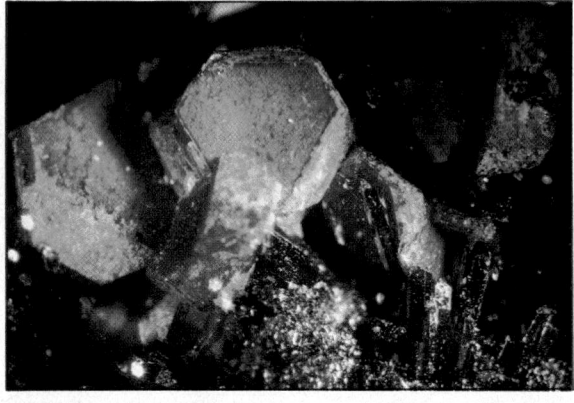

BEUDANTITE with carminite   Tsumeb, South West Africa.

BEUDANTITE   Laurium, Greece.

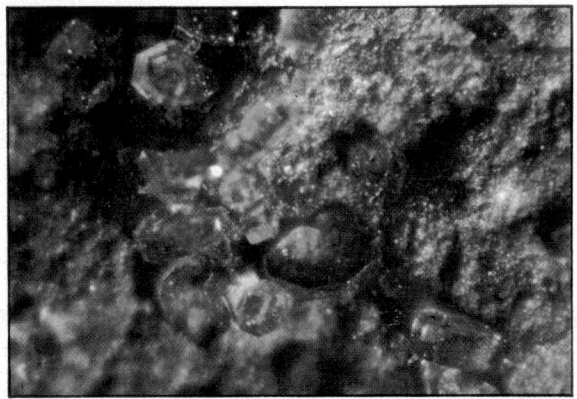

BILLIETITE   Menzenschwand, Germany.

BISMUTHINITE   Julcani mine, Huancavelica, Peru.

BISMUTHINITE inclusions in fluorite   Illogan, Cornwall, England.

BISMUTHINITE with chalcopyrite   Herdorf, Westphalia, Germany.

16

LATTICE CONSTANTS:
$a$ = 7.14 kX
$b$ = 20.06
$c$ = 6.55
3 STRONGEST DIFFRACTION LINES:
1.849 (100)
4.60 ( 80)
3.09 ( 80)
OPTICAL CONSTANTS:
$\alpha$ = 1.818
$\beta$ = 1.866
$\gamma$ = 1.909
(–)2V ~ 85°
HARDNESS: 2½–3
DENSITY: 5.6 (Meas.)
　　　　　5.68 (Calc.)
CLEAVAGE: {010} perfect
　　　　　{100} imperfect
　　　　　{101} imperfect
Fracture uneven. Brittle.

HABIT: As small elongated prismatic crystals, usually in divergent groups; also massive and as thin coatings. Commonly striated and with vicinal faces in the zone [001].

COLOR-LUSTER: Dark green, bluish green, blue. Transparent to translucent. Vitreous to resinous. Streak greenish white, bluish white.

MODE OF OCCURRENCE: Occurs widespread in small amounts in the oxidation zone of copper-lead deposits. Found in the United States, especially at the Mammoth mine, Tiger, Arizona, as superb blue crystals up to ½ inch in length; in California with linarite at the Wonder prospect, Darwin district, and at other places in Inyo County, and in Mono and San Bernardino counties; and at localities in New Mexico, Idaho, Montana, and Utah. It also is found in Canada, Chile, England, Scotland, France, Sardinia, USSR, and at Tsumeb, South West Africa.

BEST REF. IN ENGLISH: Palache, et al., "Dana's System of Mineralogy," 7th Ed., v. II, p. 630–632, New York, Wiley, 1951.

## CALKINSITE
$(La, Ce)_2 (CO_3)_3 \cdot 4H_2O$

CRYSTAL SYSTEM: Orthorhombic
CLASS: 222
SPACE GROUP: $P2_1 2 2_1$
Z: 4
LATTICE CONSTANTS:
$a$ = 9.57
$b$ = 12.65
$c$ = 8.94
3 STRONGEST DIFFRACTION LINES:
6.54 (100)
3.27 ( 50)
4.78 ( 40)
OPTICAL CONSTANTS:
$\alpha$ = 1.569
$\beta$ = 1.657
$\gamma$ = 1.686
(–)2V = 54°
HARDNESS: ~2½

DENSITY: 3.28 ± 0.01 (Meas.)
CLEAVAGE: {010} perfect
　　　　　{101} distinct
　　　　　{001} parting (?)
HABIT: Crystals platy parallel to {010}; plates commonly twinned, compound, and elongate parallel to $a$-axis. Single crystals do not exceed 1.0 mm in maximum dimension.

COLOR-LUSTER: Pale yellow.

MODE OF OCCURRENCE: Occurs with barite in vugs and associated with lanthanite, barite, and goethite in porous areas of weathered burbankite and ancylite in veins at the head of Big Sandy Creek, southeastern Hill County, Montana.

BEST REF. IN ENGLISH: Pecora, W. T., and Kerr, Joe H., *Am. Min.*, **38**: 1169–1183 (1953).

## CALLAGHANITE
$Cu_2 Mg_2 (CO_3)(OH)_6 \cdot 2H_2O$

CRYSTAL SYSTEM: Monoclinic
CLASS: 2/m
SPACE GROUP: C 2/c
Z: 4
LATTICE CONSTANTS:
$a$ = 10.06
$b$ = 11.80
$c$ = 8.24
$\beta$ = 107°18′
3 STRONGEST DIFFRACTION LINES:
7.45 (100)
6.17 (100)
3.87 ( 90)
OPTICAL CONSTANTS:
$\alpha$ = 1.559
$\beta$ = 1.653
$\gamma$ = 1.680
(–)2V = 55°
HARDNESS: 3–3½
DENSITY: 2.71 (Meas.)
　　　　　2.78 (Calc.)
CLEAVAGE: {111} perfect
　　　　　{$\bar{1}$11} perfect
Fracture irregular. Brittle.

HABIT: As minute pyramidal crystals; crystal forms are {111}, {$\bar{1}$11}, {221}, and {$\bar{2}$21}.

COLOR-LUSTER: Azure blue, transparent; vitreous. Streak white.

MODE OF OCCURRENCE: Occurs as tiny disseminated crystals, as encrustations, and as veinlets in magnesite and dolomite beds at the pits of Gabbs Refractories, Inc., near Gabbs, Nye County, Nevada.

BEST REF. IN ENGLISH: Beck, Carl W., and Burns, John H., *Am. Min.*, **39**: 630–635 (1954).

## CALOMEL
HgCl

CRYSTAL SYSTEM: Tetragonal
CLASS: 4/m 2/m 2/m

SPACE GROUP: I4/mmm
Z: 4
LATTICE CONSTANTS:
  *a* = 4.45 kX
  *c* = 10.89
3 STRONGEST DIFFRACTION LINES:
  3.164 (100)
  4.141 ( 97)
  1.970 ( 38)
OPTICAL CONSTANTS:
  $\omega$ = 1.973
  $\epsilon$ = 2.656
    (+)
HARDNESS: 1½
DENSITY: 7.16 (Meas.)
        7.176 (Calc.)
CLEAVAGE: {110} good
          {011} imperfect
Fracture conchoidal. Sectile; plastic.
HABIT: Crystals often complex; usually tabular {001},
pyramidal, prismatic [001], equant; as crusts of minute
crystals and as earthy masses and coatings.
COLOR-LUSTER: Colorless, white to yellowish gray,
gray, brown. Color darkens on exposure to light. Transparent to translucent. Adamantine. Fluoresces dark red in
ultraviolet light.
MODE OF OCCURRENCE: Occurs as an uncommon
secondary mineral formed by the alteration of minerals
containing mercury. In California it is found in small
amounts with native mercury, cinnabar, and eglestonite
near Palo Alto, and with cinnabar, native mercury, eglestonite, and montroydite near Redwood City, San Mateo
County; and with metacinnabar in the Redington mine,
Napa County. It occurs near Jackport, Pike County,
Arkansas; in the Denio district, Oregon; and as fine crystals
in the Terlingua district, Texas. It is also found at El
Doktor, Queretaro, Mexico, and in France, Spain, and
Germany.
BEST REF. IN ENGLISH: Palache, et al., "Dana's Sys-
tem of Mineralogy," 7th Ed., v. II, p. 25–28, New York,
Wiley, 1951.

## CALUMETITE
$Cu(OH, Cl)_2 \cdot 2H_2O$

CRYSTAL SYSTEM: Orthorhombic (?)
3 STRONGEST DIFFRACTION LINES:
  7.50  (100)
  2.481 ( 80)
  3.02  ( 60)
OPTICAL CONSTANTS:
  $\alpha$ = 1.666
  $\beta$ = 1.690
  $\gamma$ = 1.690
2V = 2°
HARDNESS: 2
CLEAVAGE: {001} good
Brittle.
HABIT: Spherules and sheaves of scaly crystals which
are subparallel on the perfect basal cleavage.
COLOR-LUSTER: Brilliant azure blue to powder blue;
basal cleavage pearly, bluish white.

MODE OF OCCURRENCE: Found with copper, cuprite,
malachite, atacamite, paratacamite, buttgenbachite, and
two unidentified copper chlorides in pockets and as encrustations on basalt at the Centennial mine near Calumet,
Michigan.
BEST REF. IN ENGLISH: Williams, Sidney A., *Am.
Min.*, 48: 614–619 (1963).

## CALZIRTITE
$CaZr_3TiO_9$

CRYSTAL SYSTEM: Tetragonal
CLASS: 4/mmm
SPACE GROUP: I4$_1$/acd
Z: 14
LATTICE CONSTANTS:
  *a* = 15.30
  *c* = 10.20
3 STRONGEST DIFFRACTION LINES:
  2.945 (100)
  1.801 (100)
  1.537 ( 90)
OPTICAL CONSTANTS:
  $\omega$ = 2.19–2.27
  $\epsilon$ = 2.30–2.36
    (+)
HARDNESS: 6–7
DENSITY: 5.01 ± 0.01 (Meas.)
CLEAVAGE: Brittle.
HABIT: Single crystals prismatic with faces {100},
{121}, {110}, {231}, {123}, and {112}. As small (not
more than 0.5 mm) isometric grains with many faces composed of twinned intergrowths of 2 or 3 crystals.
COLOR-LUSTER: Light brown to dark brown, light
green, brownish green; adamantine on crystal faces, greasy
on fractures. Transparent in thin sections.
MODE OF OCCURRENCE: Occurs in carbonatites and
alluvial deposits of the Meimecha-Kotui petrographical
province at the northern border of the Siberian platform;
in metasomatic calcite-forsterite-magnetite rocks of an
alkalic ultrabasic massif in Eastern Siberia; in carbonatites
of the Kola Peninsula; and at Tapira, Brazil.
BEST REF. IN ENGLISH: Zdorik, T. B., Sidorenko,
G. A., and Bykova, A. V., *Am. Min.*, 46: 1515 (1961).
Bulakh, A. G., Anastasenko, G. F., and Dakhiya, L. M.,
*Am. Min.*, 52: 1880–1885 (1967).

## CANASITE
$(Na, K)_5 (Ca, Mn, Mg)_4 (Si_2O_5)_5 (OH, F)_3$

CRYSTAL SYSTEM: Monoclinic
CLASS: 2/m
SPACE GROUP: C2/m
Z: 3
LATTICE CONSTANTS:
  *a* = 18.90
  *b* = 7.25
  *c* = 12.62
  $\beta$ = 112°

3 STRONGEST DIFFRACTION LINES:
  3.08  (100)
  2.907 ( 80)
  1.641 ( 80)
OPTICAL CONSTANTS:
  $\alpha = 1.534$
  $\beta = 1.538$
  $\gamma = 1.543$
  $(-)2V = 58°$ (?)
HARDNESS:  Not determined.
DENSITY:  2.707 (Meas.)
CLEAVAGE: One perfect and one very perfect at an angle of 118° to the first. Fracture splintery, breaks into long acute-angled or wedge-shaped pieces. Brittle. Yields felty mass when ground.
HABIT:  Grains up to 3.0 cm in diameter. Polysynthetic twinning; twinning plane at an angle of 8° to the less perfect cleavage.
COLOR-LUSTER: Greenish yellow, transparent to translucent; vitreous. Streak colorless.
MODE OF OCCURRENCE: Occurs associated with fenaksite and lamprophyllite in pegmatite in the Khibina Tundra, USSR.
BEST REF. IN ENGLISH: Dorfman, M. D., Rogachev, D. D., Goroshchenko, Z. I., and Uspenskaya, E. I., *Am. Min.*, **45**: 253–254 (1960).

# CANCRINITE
$(Na, Ca)_{7-8}Al_6Si_6O_{24}(CO_3, SO_4, Cl)_{1.5-2.0} \cdot 1-5H_2O$
Cancrinite Group

CRYSTAL SYSTEM:  Hexagonal
CLASS:  6
SPACE GROUP:  $P6_3$
Z:  1
LATTICE CONSTANTS:
  $a = 12.58-12.76$
  $c = 5.11-5.20$
3 STRONGEST DIFFRACTION LINES:
  3.19 (100)
  4.61 ( 67)
  3.61 ( 40)
OPTICAL CONSTANTS:
  $\omega = 1.507-1.528$
  $\epsilon = 1.495-1.503$
     $(-)$
HARDNESS:  5-6
DENSITY:  2.42-2.51 (Meas.)
CLEAVAGE:  $\{10\bar{1}0\}$ perfect
           $\{0001\}$ poor
Fracture uneven. Brittle.
HABIT:  Crystals prismatic, rare. Usually massive. Lamellar twinning rare.
COLOR-LUSTER: Colorless, white, yellow, orange, pink to reddish, pale blue to pale bluish gray. Transparent to translucent. Vitreous, pearly or greasy. Streak colorless.
MODE OF OCCURRENCE: Occurs as a primary mineral in certain alkali rocks and as a common alteration product of nepheline. Found at Iron Hill, Gunnison County, Colorado; at Litchfield, Kennebec County, Maine; in the Bancroft district, Ontario, and elsewhere in Canada; in the Fen area, southern Norway; and at localities in Roumania, Finland, USSR, Korea, China, India, Zaire, Uganda, and Kenya.
BEST REF. IN ENGLISH: Deer, Howie, and Zussman, "Rock Forming Minerals," v. 4, p. 310–320, New York, Wiley, 1963.

# CANFIELDITE
$Ag_8(Sn, Ge)S_6$
Argyrodite series

CRYSTAL SYSTEM:  Cubic
CLASS:  $4/m\,\bar{3}\,2/m$
SPACE GROUP:  Im3m
Z:  32
LATTICE CONSTANT:
  $a = 21.11$
3 STRONGEST DIFFRACTION LINES:
  3.02  (100)
  1.863 ( 50)
  2.66  ( 40)
HARDNESS:  2½
DENSITY:  6.276 (Meas.)
           6.49 (Calc. x-ray)
CLEAVAGE:  Fracture conchoidal to uneven. Brittle.
HABIT:  Crystals usually octahedral. Twinning on $\{111\}$ common, as spinel twins.
COLOR-LUSTER: Steel gray with reddish tinge; rapidly tarnishes black with purple or bluish tint. Opaque. Metallic. Streak grayish black.
MODE OF OCCURRENCE: Occurs in vein deposits commonly associated with argentite, other silver minerals, and sulfides. Found at several deposits in the vicinity of Colquechaca, Bolivia, and Freiberg, Saxony, Germany.
BEST REF. IN ENGLISH: Palache, et al., "Dana's System of Mineralogy," 7th Ed., v. I, p. 356–358, New York, Wiley, 1944.

# CANNIZZARITE
A lead bismuth sulfide

CRYSTAL SYSTEM:  Monoclinic
CLASS:  2/m
SPACE GROUP:  P2/m; C2/m
LATTICE CONSTANTS:

| (P2/m) | (C2/m) |
|---|---|
| $a = 4.13$ | $a = 7.07$ |
| $b = 4.10$ | $b = 4.10$ |
| $c = 15.5$ | $c = 15.5$ |
| $\beta = 99°$ | $\beta = 99°$ |

3 STRONGEST DIFFRACTION LINES:
  3.82 (100)
  3.01 ( 60)
  2.68 ( 60)
HARDNESS:  Not determined.
DENSITY:  6.7 (Meas.)
           6.81 (Calc.)
CLEAVAGE:  Slightly elastic; somewhat malleable.
HABIT:  As thin leafy blades; often striated parallel to elongation.

COLOR-LUSTER: Light gray; bluish to iridescent tarnish. Metallic.

MODE OF OCCURRENCE: Occurs very sparingly as a sublimation product at Vulcano, Lipari Island, Italy.

BEST REF. IN ENGLISH: Graham, A. R., Thompson, R. M., and Berry, L. G., *Am. Min.*, **38**: 536–544 (1953).

## CAPPELENITE

$(Ba, Ca, Ce, Na)_3 (Y, Ce, La)_6 (BO_3)_6 Si_3 O_9$

CRYSTAL SYSTEM: Hexagonal
3 STRONGEST DIFFRACTION LINES:
2.78 (100)
3.46 ( 50)
1.94 ( 50)
OPTICAL CONSTANTS:
$n \sim 1.76$
uniaxial (–)
HARDNESS: 6
DENSITY: 4.407 (Meas.)
CLEAVAGE: None.
Fracture conchoidal. Brittle.

HABIT: Crystals prismatic; main forms are hexagonal prism $\{10\bar{1}0\}$ and dipyramid $\{10\bar{1}3\}$.

COLOR-LUSTER: Greenish brown.

MODE OF OCCURRENCE: Occurs in nepheline-syenite pegmatites in association with woehlerite, rosenbuschite, and other minerals, on Lille-Arö, Langesundfjord, Norway.

BEST REF. IN ENGLISH: Vlasov, K. A., "Mineralogy of Rare Elements," v. II, p. 248, Israel Program for Scientific Translations, 1966.

## CARACOLITE

Near $Na_2 PbSO_4 ClOH$

CRYSTAL SYSTEM: Orthorhombic or hexagonal
Z: 1
LATTICE CONSTANTS:
$a:b:c$ = .584:1:0.422 (Orthorhombic)
OPTICAL CONSTANTS:
$\alpha$ = 1.743
$\beta$ = 1.754
$\gamma$ = 1.764
(–)2V = very large
HARDNESS: 4½
DENSITY: ~5.1
CLEAVAGE: Not determined.

HABIT: Crystals imperfect hexagonal prisms with hexagonal pyramid and large basal pinacoid. As a crystalline incrustation. Crystals may be pseudohexagonal trillings, analogous to aragonite.

COLOR-LUSTER: Colorless, grayish, greenish. Transparent. Vitreous.

MODE OF OCCURRENCE: Occurs as a crystalline incrustation composed of crystals up to one millimeter in size, associated with percylite, anglesite, and galena, at the Mina Beatriz, Atacama Province, Chile.

BEST REF. IN ENGLISH: Palache, et al., "Dana's System of Mineralogy," 7th Ed., v. II, p. 546–547, New York, Wiley, 1951.

## CARBOBORITE

$Ca_2 Mg(CO_3)(B_2 O_5) \cdot 10H_2 O$

CRYSTAL SYSTEM: Monoclinic
CLASS: 2/m
Z: 4
LATTICE CONSTANTS:
$a$ = 18.59
$b$ = 6.68
$c$ = 11.32
$\beta$ = 91°41'
3 STRONGEST DIFFRACTION LINES:
5.63 (100)
4.315 (100)
3.136 ( 80)
OPTICAL CONSTANTS:
$\alpha$ = 1.5069
$\beta$ = 1.5459
$\gamma$ = 1.5693
(–)2V = 75°
HARDNESS: 2
DENSITY: 2.12 (Meas.)
2.11 (Calc.)
CLEAVAGE: $\{100\}$ perfect
$\{\bar{1}11\}$ distinct
$\{001\}$ imperfect

HABIT: Crystals 0.5–3.0 mm long, resembling steep rhombohedrons. Dominant forms $a\{100\}$ and $\sigma\{\bar{1}11\}$; less common and small are $c\{011\}$, $m\{110\}$, $e\{111\}$, and $t\{\bar{1}01\}$.

COLOR-LUSTER: Colorless; vitreous. Fluoresces white under ultraviolet light, with pale green phosphorescence.

MODE OF OCCURRENCE: Occurs in a layer of hydroboracite containing seams of ulexite and gypsum, and apparently replacing ulexite, in a deposit of lacustrine origin at an undesignated locality in China.

BEST REF. IN ENGLISH: Hsien-Te, Hsieh, Tze-Chiang, Chien, and Lai-Pao, Liu, *Am. Min.*, **50**: 262–263 (1965).

## CARBOCERNAITE

$(Na, Ca, Sr, Ce) CO_3$

CRYSTAL SYSTEM: Orthorhombic
Z: 4
LATTICE CONSTANTS:
$a$ = 6.40
$b$ = 7.28
$c$ = 5.22
3 STRONGEST DIFFRACTION LINES:
3.00 (100)
2.015 ( 90)
1.813 ( 80)
OPTICAL CONSTANTS:
$\alpha$ = 1.569
$\beta$ = 1.679
$\gamma$ = 1.708
(–)2V = 52° (Calc.)
HARDNESS: 3
DENSITY: 3.53 (Meas.)
3.53 (Calc.)

CLEAVAGE: {100} poor
            {021} poor
            {010} poor
Brittle.

HABIT: Tiny grains and crystals. Forms: {100} most prominent, {010}, {001}, {021}, {540}, and {210} minor, {305} and {210} vicinal.

COLOR-LUSTER: Colorless, transparent. Turbid white, yellowish, rose, or brown when altered. Vitreous on crystal faces, greasy on fractures.

MODE OF OCCURRENCE: Occurs in dolomite-calcite carbonatite veins 0.5–1.0 meter in width in pyroxenites and ijolites of the Vuorjärvi massif, Kola Peninsula, as accessory grains and as crystals on walls of cavities, closely associated with chlorite and ankerite. Other associated minerals are sphalerite, galena, pyrite, and barite, and in the cavities alstonite, anatase, quartz, and zeolites.

BEST REF. IN ENGLISH: Bulakh, A. G., Kondrat'eva, V. V., and Baranova, E. N., *Am. Min.*, **46**: 1202 (1961).

# CARBONATE-APATITE (Dahllite, Francolite)
$Ca_5(PO_4, CO_3)_3(OH, F)$
Apatite Series

CRYSTAL SYSTEM: Hexagonal
CLASS: 6/m
SPACE GROUP: $P6_3/m$
Z: 2
LATTICE CONSTANTS:
 $a = 9.48$
 $c = 6.89$
3 STRONGEST DIFFRACTION LINES:
 2.822 (100)
 2.722 ( 90)
 3.451 ( 70)
OPTICAL CONSTANTS:
 $\omega = 1.603, 1.628$
 $\epsilon = 1.598, 1.619$
   (−)
HARDNESS: 5
DENSITY: 2.9–3.1 (Meas.)
CLEAVAGE: {0001} indistinct
            {10$\bar{1}$0} trace
Fracture uneven to conchoidal. Brittle.
HABIT: See fluorapatite.
COLOR-LUSTER: See fluorapatite.
MODE OF OCCURRENCE: Occurs in granite pegmatites in Maine, New Hampshire, and the Black Hills, South Dakota. Among many other localities it is also found in Wyoming (dahllite), Arkansas, Greenland, France, Norway, and USSR.

BEST REF. IN ENGLISH: Palache, et al., "Dana's System of Mineralogy, 7th Ed., v. II, p. 879–889, New York, Wiley, 1951.

# CARBONATE-CYANOTRICHITE
$Cu_4Al_2[(CO_3), (SO_4)](OH)_{13} \cdot 2H_2O$

CRYSTAL SYSTEM: Orthorhombic

3 STRONGEST DIFFRACTION LINES:
 4.21 (100)
 10.13 ( 93)
 5.03 ( 60)
OPTICAL CONSTANTS:
 $\alpha = 1.616$
 $\beta = ?$
 $\gamma = 1.677$
 $(+)2V = 55°-60°$
HARDNESS: ~2
DENSITY: 2.65–2.67
HABIT: Elongated platelets up to 4–5 mm long; fibrous aggregates
COLOR-LUSTER: Pale blue to azure blue; silky.
MODE OF OCCURRENCE: Occurs associated with alunite, jarosite, brochantite, azurite, malachite, spangolite, and phosphates and vandates of aluminum and iron, in the crust of weathering of Middle Cambrian shales at several localities in northwestern Kara-Tau, USSR. Also occurs at the Engle mine, Plumas County, California.

BEST REF. IN ENGLISH: Ankinovich, E. A., Gekht, I. I., and Zaitseva, R. I., *Am. Min.*, **49**: 441–442 (1964).

# CARBORUNDUM = Synthetic moissanite

# CARDENITE = Variety of saponite (?)
Near $(Mg, Fe^{2+}, Al)_3(Si, Al)_4O_{10}(OH)_2 \cdot nH_2O$

# CARLETONITE
$KNa_4Ca_4Si_8O_{18}(CO_3)_4(F,OH) \cdot H_2O$

CRYSTAL SYSTEM: Tetragonal
CLASS: 4/m 2/m 2/m
SPACE GROUP: P4/mbm
Z: 4
LATTICE CONSTANTS:
 $a = 13.178$
 $c = 16.695$
3 STRONGEST DIFFRACTION LINES:
 8.353 (100)
 4.171 (100)
 2.903 ( 90)
OPTICAL CONSTANTS:
 $\omega = 1.521$
 $\epsilon = 1.517$
   (−)
HARDNESS: 4–4½ on {001}
DENSITY: 2.45 (Meas.)
         2.426 (Calc.)
CLEAVAGE: {001} perfect
            {110} distinct
Fracture conchoidal.
HABIT: Massive.
COLOR-LUSTER: Pink to pale blue; small flakes colorless. Transparent to translucent. Vitreous to pearly, becoming slightly waxy upon prolonged exposure.
MODE OF OCCURRENCE: Occurs as masses up to 8 inches in diameter in association with pectolite, albite, arfvedsonite, calcite, fluorite, and apophyllite, in thermally

metamorphosed inclusions in the nepheline syenite at Mont St. Hilaire, Quebec, Canada.

BEST REF. IN ENGLISH: Chao, G. Y., *Am. Min.*, **56**: 1855–1866 (1971).

## CARLSBERGITE
CrN

CRYSTAL SYSTEM: Cubic
Z: 1
LATTICE CONSTANT:
 $a$ = 4.16
OPTICAL CONSTANTS: Isotropic.
 HARDNESS: High; probably greater than 1000 on Vickers scale
 DENSITY: Not determined.
 CLEAVAGE: Not determined.
 HABIT: As minute oriented platelets in kamacite, and as grain boundary precipitates a few microns in diameter.
 COLOR-LUSTER: Light gray with distinct rose-violet tint in reflected light.
 MODE OF OCCURRENCE: The mineral has been found in the Cape York meteorite and in more than 70 other iron meteorites.
 BEST REF. IN ENGLISH: Buchwald, V. F., and Scott, E. R. D., *Nat. Phys. Sci.*, **233**: 113–114 (1971).

## CARMINITE
$PbFe_2(AsO_4)_2(OH)_2$

CRYSTAL SYSTEM: Orthorhombic
CLASS: 2/m 2/m 2/m
SPACE GROUP: Amaa
Z: 8
LATTICE CONSTANTS:
 $a$ = 12.25
 $b$ = 16.52
 $c$ = 7.64
3 STRONGEST DIFFRACTION LINES:
 3.20 (100)
 2.580 ( 90)
 2.929 ( 80)
OPTICAL CONSTANTS:
 $\alpha$ = 2.070   2.05
 $\beta$ = 2.070   2.05
 $\gamma$ = 2.080   2.06
 (+)2V = medium (Na)
 HARDNESS: 3½
 DENSITY: 5.22 (Meas.)
     5.46 (Calc.)
 CLEAVAGE: {110} (?) distinct
Brittle.
 HABIT: Crystals minute, lath-like, elongated [001] and flattened {010}. As tufted aggregates, radial-fibrous spherical aggregates, or massive with radiated structure.
 COLOR-LUSTER: Carmine red to reddish brown. Translucent. Vitreous, pearly on cleavage. Streak reddish yellow.
 MODE OF OCCURRENCE: Occurs as fine microcrystals associated with scorodite, anglesite, cerussite, and other minerals at Mapimi, Durango, Mexico; at the Hingston Down

Consols mine, Calstock, Cornwall, England; at Bad Ems and Horhausen, Germany; and at Wyloo, Western Australia. It has also been reported from Nevada, Utah, and Colorado.
 BEST REF. IN ENGLISH: Palache, et al., "Dana's System of Mineralogy," 7th Ed., v. II, p. 912–913, New York, Wiley, 1951.    Finney, J. J., *Am. Min.*, **48**: 1–13 (1963).

## CARNALLITE
$KMgCl_3 \cdot 6H_2O$

CRYSTAL SYSTEM: Orthorhombic
CLASS: 2/m 2/m 2/m
SPACE GROUP: Pnna
Z: 12
LATTICE CONSTANTS:
 (synthetic)
 $a$ = 16.054
 $b$ = 22.508
 $c$ = 9.575
3 STRONGEST DIFFRACTION LINES:
 3.30 (100)
 2.92 ( 70)
 4.65 ( 50)
OPTICAL CONSTANTS:
 $\alpha$ = 1.4665
 $\beta$ = 1.4753
 $\gamma$ = 1.4937
 (+)2V = 70° 03′
 HARDNESS: 2½
 DENSITY: 1.602 (Meas.)
     1.598 (Calc.)
 CLEAVAGE: None.
Fracture conchoidal.
 HABIT: Crystals pseudohexagonal pyramidal; also thick tabular. Usually massive, granular.
 COLOR-LUSTER: Colorless, white; rarely yellow or blue; often reddish due to included hematite. Transparent to translucent. Shining, greasy.

Deliquescent in moist atmosphere. Taste bitter.
 MODE OF OCCURRENCE: Occurs chiefly as a component of extensive thick sedimentary saline deposits, often associated with kieserite, halite, sylvite, and polyhalite. In the United States it is found in large deposits in the Permian Basin of southeastern New Mexico, especially in the vicinity of Carlsbad, and also in adjacent parts of Texas; and it also occurs in Grand and San Juan counties, Utah. Among other localities, extensive deposits occur in Saskatchewan, Canada; in northern Germany; in Spain; and at Ozinki, Saratov, USSR.
 BEST REF. IN ENGLISH: Palache, et al., "Dana's System of Mineralogy," 7th Ed., v. II, p. 92–94, New York, Wiley, 1951.

## CARNASURTITE (Karnasurtite) = Rhabdophane ? (inadequately described mineral)
$CeTiAlSi_2O_9 \cdot 5H_2O$

CRYSTAL SYSTEM: Unknown

OPTICAL CONSTANTS:
$\omega = 1.617$
$\epsilon = 1.595$
$(-)$
HARDNESS: 2
DENSITY: 2.89-2.95 (Meas.)
CLEAVAGE: One perfect. One imperfect.
HABIT: Crystals tabular; also as hexahedral masses up to 1 cm in diameter.
COLOR-LUSTER: Yellow. Greasy.
MODE OF OCCURRENCE: Occurs in nepheline-sodalite syenite pegmatites at Mount Karnasurt and Mount Punkaruaiv, Kola Peninsula, USSR.
BEST REF. IN ENGLISH: Vlasov, K. A., "Mineralogy of Rare Elements," v. II, p. 324-325, Israel Program for Scientific Translations, 1966.

## CARNELIAN = Variety of quartz

## CARNOTITE
$K_2(UO_2)_2(VO_4)_2 \cdot 1\text{-}3H_2O$

CRYSTAL SYSTEM: Monoclinic
CLASS: 2/m
SPACE GROUP: $P2_1/a$
Z: 2
LATTICE CONSTANTS:
$a = 10.47$
$b = 8.41$
$c = 6.91$
$\beta = 103° \, 40'$
3 STRONGEST DIFFRACTION LINES:
6.56 (100)
3.12 ( 70)
3.53 ( 50)
OPTICAL CONSTANTS:
$\alpha = 1.750\text{-}1.78$
$\beta = 1.901\text{-}2.06$
$\gamma = 1.92\text{-}2.08$
$(-)2V$ = small, 43°-60°
HARDNESS: Not determined. Soft.
DENSITY: 4.70 (Meas.)
4.95 (Meas. Synthetic)
CLEAVAGE: {001} perfect, micaceous
Not brittle.
HABIT: Crystals microscopic in size with rhomboidal or diamond-shaped outline, elongated along $b$-axis and flattened on {001}. Disseminated or as coatings; as a powder or as microcrystalline aggregates; as firm, compact masses; and rarely as crusts or aggregates of tiny platy crystals.
COLOR-LUSTER: Bright yellow to golden yellow, also greenish yellow. Crystals pearly; fine-grained masses dull or earthy. Not fluorescent.
MODE OF OCCURRENCE: Occurs widespread in the Colorado Plateau region chiefly as disseminations or coatings in sandstones of Triassic or Jurassic age, often associated with tyuyamunite and metatyuyamunite. It is also found in sedimentary rocks at many localities in western South Dakota, especially near Edgemont, Fall River County; at several large deposits in Wyoming; and at Jim Thorpe,

Pennsylvania. It is found as an alteration product of davidite at Radium Hill near Olary, South Australia; at several localities in Katanga, Zaire, especially at Kambove where it occurs as excellent microcrystals; at El Borouj and Louis Gentil, Morocco; and at several deposits in northern Ferghana, Turkestan, USSR.
BEST REF. IN ENGLISH: Frondel, Clifford, USGS Bull. 1064, p. 243-247 (1958).

## CAROBBIITE
KF

CRYSTAL SYSTEM: Cubic
CLASS: $4/m\,\bar{3}\,2/m$
SPACE GROUP: Fm3m
Z: 4
LATTICE CONSTANT:
$a = 5.34$
3 STRONGEST DIFFRACTION LINES:
2.671 (100)
1.890 ( 63)
3.087 ( 29)
OPTICAL CONSTANT:
$N = 1.362$
DENSITY: 2.524 (Calc.)
CLEAVAGE: Cubic.
HABIT: Small cubic crystals.
COLOR-LUSTER: Colorless.
Very deliquescent.
MODE OF OCCURRENCE: Occurs in cavities in lavas from Vesuvius, Italy.
BEST REF. IN ENGLISH: Strunz, H., *Am. Min.*, **42**: 117 (1957).

## CARPATHITE (Pendletonite) (Karpatite)
$C_{24}H_{12}$

CRYSTAL SYSTEM: Monoclinic
CLASS: 2/m
SPACE GROUP: $P2_1/c$ or $P2/c$
Z: 2
LATTICE CONSTANTS:
$a = 10.035$    $a = 16.25$
$b = 4.695$    $b = 4.638$
$c = 16.014$    $c = 10.42$
$\beta = 69°$    $\beta = 111° \, 10'$
3 STRONGEST DIFFRACTION LINES:
9.50 (100)
7.43 (100)
3.96 ( 40)
OPTICAL CONSTANTS:
$\alpha = 1.780$    $\alpha = 1.76$
$\beta = 1.977\text{-}1.982$    $\beta = 1.78$
$\gamma = 2.05\text{-}2.15$    $\gamma > 1.85$
$(-)2V$    $(+)2V = 96°\text{-}115°$
HARDNESS: $> 1$. Flexible; almost plastic.
DENSITY: 1.35 (Meas.)
1.40 (Calc.)
CLEAVAGE: {100} very perfect
{001} very perfect
{$\bar{2}01$} very perfect

HABIT: Crystals usually thin tabular parallel to (001). Grains are bladed to acicular in habit.

COLOR-LUSTER: Pale yellow; vitreous; brilliant on fresh cleavages.

MODE OF OCCURRENCE: Found sparingly as bladed aggregates, occasionally as single or clustered crystals, intimately associated with quartz and cinnabar, in veins in a silicified matrix, at a small mercury deposit near the New Idria Mine, San Benito County, California.

BEST REF. IN ENGLISH: *Am. Min.*, **42**: 120 (1957). Murdoch, Joseph, and Geissman, Theodore A., *Am. Min.*, **52**: 611–616 (1967). *Am. Min.*, **54**: 329 (1969).

# CARPHOLITE (Karpholite)
$MnAl_2Si_2O_6(OH)_4$

CRYSTAL SYSTEM: Orthorhombic
CLASS: 2/m 2/m 2/m
SPACE GROUP: Ccca
Z: 8
LATTICE CONSTANTS:
$a = 13.86$
$b = 20.13$
$c = 5.12$
3 STRONGEST DIFFRACTION LINES:
5.73 (100)
5.08 ( 70)
2.62 ( 50)
OPTICAL CONSTANTS:
$\alpha = 1.62$
$\beta = 1.63$
$\gamma = 1.64$
$(+)2V = 67°$
HARDNESS: 5–5½
DENSITY: 2.9–3.04 (Meas.)
3.031 (Calc.)
CLEAVAGE: {010} perfect
HABIT: Fibrous; as radiated tufts.
COLOR-LUSTER: Straw yellow to wax yellow.
MODE OF OCCURRENCE: Occurs near Neuville, Ardennes, Belgium; at Schlaggenwald, Bohemia, Czechoslovakia; at Wippra, Harz Mountains, Germany; and at Radobilny, near Prilep, Macedonia, Yugoslavia.
BEST REF. IN ENGLISH: Dana, E.S., "System of Mineralogy," 6th Ed., p. 549, Wiley, 1892.

# CARPHOSIDERITE = Hydronium jarosite

# CARROLLITE
$CuCo_2S_4$
Linnaeite Group

CRYSTAL SYSTEM: Cubic
CLASS: 4/m 3̄ 2/m
SPACE GROUP: Fd3m
Z: 8
LATTICE CONSTANT:
$a = 9.477$

3 STRONGEST DIFFRACTION LINES:
2.86 (100)
1.674 ( 80)
1.825 ( 60)
HARDNESS: 4½–5½
DENSITY: 4.5–4.8 (Meas.)
4.83 (Calc.)
CLEAVAGE: {001} imperfect
Fracture subconchoidal to uneven.
HABIT: Crystals usually octahedral. Commonly massive; compact to granular. Twinning on {111}.
COLOR-LUSTER: Light gray to steel gray; rapidly tarnishes copper red or violet-gray. Opaque. Metallic.
MODE OF OCCURRENCE: Occurs in hydrothermal vein deposits commonly associated with other sulfide minerals. Found at the Patapsco mine, Finksburg, Carroll County, Maryland, and at Gladhammar, Sweden.
BEST REF. IN ENGLISH: Palache, et al., "Dana's System of Mineralogy," 7th Ed., v. I, p. 262–265, New York, Wiley, 1944.

# CARYINITE
$(Ca, Na, Pb, Mn)_3(Mn, Mg)_2(AsO_4)_{3-y}(OH)x$

CRYSTAL SYSTEM: Monoclinic
CLASS: 2/m
SPACE GROUP: $P2_1/c$
Z: 4
LATTICE CONSTANTS:
$a = 11.48$
$b = 13.17$
$c = 6.87$
$\beta = 99°$
3 STRONGEST DIFFRACTION LINES:
2.87 (100)
2.686 ( 50)
3.03 ( 30)
OPTICAL CONSTANTS:
$\alpha = 1.776$
$\beta = 1.780$
$\gamma = 1.805$
$(+)2V = 41°$
HARDNESS: 4
DENSITY: 4.29 (Meas.)
CLEAVAGE: {110} distinct
{010} distinct
Fracture subconchoidal to uneven. Brittle.
HABIT: Massive, cleavable; also fine-granular.
COLOR-LUSTER: yellowish brown to brown. Subtranslucent. Greasy.
MODE OF OCCURRENCE: Occurs in veinlets in limestone skarn associated with calcite, berzeliite, hausmannite, and other minerals, at Långban, Sweden.
BEST REF. IN ENGLISH: Palache, et al., "Dana's System of Mineralogy," 7th Ed., v. II, p. 683–684, New York, Wiley, 1951.

# CARYOCERITE = Thorium-rich variety of melanocerite

## CARYOPILITE (Ectropite, Ektropite)
$Mn_6(OH)_8Si_4O_{10}$
Dimorph of bementite

CRYSTAL SYSTEM: Monoclinic (?)
  HARDNESS: 3–3½
  DENSITY: 2.83–2.91 (Meas.)
  HABIT: Massive; in compact fibrous aggregates.
  COLOR-LUSTER: Brown.
  MODE OF OCCURRENCE: Occurs replacing rhodonite at Långban and at Harstigen, near Persberg, Sweden. Also found at the Nomura mine, Ehime Prefecture, Japan.
  BEST REF. IN ENGLISH: Kato, Toshio, *Am. Min.*, **49**: 446–447 (1964).

## CASSIDYITE
$Ca_2(Ni, Mg)(PO_4)_2 \cdot 2H_2O$
Isostructural with collinsite

CRYSTAL SYSTEM: Triclinic
Z: 1
LATTICE CONSTANTS:
  $a = 5.71$     $\alpha = 96° 50'$
  $b = 6.73$     $\beta = 6.73$
  $c = 5.41$     $\gamma = 104° 35'$
3 STRONGEST DIFFRACTION LINES:
  2.70 (100)
  3.03 ( 95)
  2.67 ( 79)
OPTICAL CONSTANTS:
  $\alpha = 1.64–1.65$
  $\beta = ?$
  $\gamma = 1.67–1.68$
HARDNESS: Not determined.
DENSITY: ~3.15
HABIT: Thin crusts and small spherules, finely fibrous.
COLOR-LUSTER: Pale green to bright green.
MODE OF OCCURRENCE: Occurs in cavities and cracks in weathered meteorites from the Wolf Creek crater in Western Australia.
  BEST REF. IN ENGLISH: White, John S. Jr., Henderson, E. P., and Mason, Brian, *Am. Min.*, **52**: 1190–1197 (1967).

## CASSITERITE
$SnO_2$

CRYSTAL SYSTEM: Tetragonal
CLASS: 4/m 2/m 2/m
SPACE GROUP: $P4_2/mnm$
Z: 2
LATTICE CONSTANTS:
  $a = 4.738$
  $c = 3.188$
3 STRONGEST DIFFRACTION LINES:
  3.351 (100)
  2.644 ( 81)
  1.765 ( 63)
OPTICAL CONSTANTS:
  $\omega = 2.006$ }
  $\epsilon = 2.0972$ } $\lambda = 585.1$
Uniaxial (+)

HARDNESS: 6–7
DENSITY: 6.99 (Meas.)
        6.995 (Calc.)
CLEAVAGE: {100} imperfect
         {110} indistinct
         {111} or {011} parting
Fracture subconchoidal to uneven. Brittle.
  HABIT: Crystals usually short prismatic; sometimes slender prismatic (needle-tin) or bipyramidal. Also coarse to fine granular, or as botryoidal and reniform masses with concentric and radial-fibrous structure (wood-tin). Twinning on {011}, common, as contact or penetration twins, often repeated.
  COLOR-LUSTER: Usually various shades of brown to brownish black or black; also colorless, gray, yellowish, greenish, red. Transparent to nearly opaque. Adamantine, splendent metallic adamantine, vitreous, often greasy on fracture. Streak white, grayish, brown.
  MODE OF OCCURRENCE: Cassiterite, the principal ore of tin, occurs chiefly in medium-to high-temperature hydrothermal veins or metasomatic deposits; also in granitic pegmatites, contact metamorphic deposits, rhyolite, and alluvial deposits. Found in the United States in Alaska, Washington, California, Nevada, South Dakota, Virginia, South Carolina, and in many granite pegmatites of the New England states. Notable deposits occur in Canada, Mexico, Bolivia, England, France, Portugal, Spain, Italy, Germany, Czechoslovakia, Finland, Malay Peninsula, Sumatra, China, Japan, Nigeria, Transvaal, South West Africa, Australia, and Tasmania.
  BEST REF. IN ENGLISH: Palache, et al., "Dana's System of Mineralogy," 7th Ed., v. I, p. 574–581, New York, Wiley, 1944.

## CASWELLITE = Altered biotite
Near $Ca_4(Mn^{3+}, Al, Fe)_4Si_6O_{20}(OH)_4$

## CATAPLEIITE
$(Na_2, Ca)ZrSi_3O_9 \cdot 2H_2O$

CRYSTAL SYSTEM: Hexagonal
CLASS: 6/m 2/m 2/m
SPACE GROUP: $P6_3/mmc$
Z: 2
LATTICE CONSTANTS:
  $a = 7.40$
  $c = 10.07$
3 STRONGEST DIFFRACTION LINES:
  3.94 (100)
  3.05 (100)
  2.96 (100)
OPTICAL CONSTANTS:
  $\omega = 1.596$
  $\epsilon = 1.624$
  (+)
HARDNESS: 5–6
DENSITY: 2.65–2.8 (Meas.)
        2.79 (Calc.)
CLEAVAGE: {10$\bar{1}$0} perfect
         {10$\bar{1}$1} imperfect
         {10$\bar{1}$2} imperfect
         {000$\bar{1}$} parting
Brittle.

HABIT: Crystals usually thin hexagonal plates; sometimes with small prism and pyramid faces. Also as lamellar masses. Twinning on $\{10\bar{1}0\}$ or $\{000\bar{1}\}$, common.

COLOR-LUSTER: Light yellow to yellowish brown to brown, salmon, yellowish red, rarely sky blue or colorless. Transparent to nearly opaque. Vitreous, greasy, or dull.

MODE OF OCUCRRENCE: Occurs in alkali rocks and pegmatites in association with feldspars, nepheline, sphene, mosandrite, eudialyte, zircon, and other minerals. Found at Magnet Cove, Hot Spring County, Arkansas; at Mont St. Hilaire, Quebec, Canada; in the Julianehaab district, Greenland; in the Langesundfjord district, Norway; in pegmatites of the Khibiny and Lovozero alkali massifs, and in alkali pegmatites of Tuva and Central Aldan, USSR; on the Islands of Los, Guinea; and in Madagascar.

BEST REF. IN ENGLISH: Vlasov, K. A., "Mineralogy of Rare Elements," v. II, p. 367–370, Israel Program for Scientific Translations, 1966.

## CATOPHORITE = Katophorite

## CATOPTRITE (Katoptrite)

$Mn_{13}Sb_2Al_4Si_2O_{28}$ or
$Mn_{14}Sb_2(Al, Fe)_4(SiO_4)_2O_{21}$

CRYSTAL SYSTEM: Monoclinic
CLASS: 2/m
SPACE GROUP: C2/m
Z: 2
LATTICE CONSTANTS:
  $a = 5.65$
  $b = 22.92$
  $c = 9.06$
  $\beta = 101°30'$
3 STRONGEST DIFFRACTION LINES:
  2.96 (100)
  8.88 ( 65)
  4.43 ( 45)
OPTICAL CONSTANTS:
  $\alpha = 1.92$
  $\beta = 1.95$
  $\gamma = 1.95$
  $(-)2V$ = small
HARDNESS: 5½
DENSITY: 4.56 (Meas.)
        4.65 (Calc.)
CLEAVAGE: $\{100\}$ perfect, micaceous
Brittle.

HABIT: Crystals usually tabular and somewhat elongated, also equant. As minute grains or masses up to 1 cm or more in size.

COLOR-LUSTER: Black. Opaque; thin flakes translucent and red in color. Metallic; brilliant.

MODE OF OCCURRENCE: Occurs associated with sonolite at the Brattfors mine near Nordmark; at the Sjö mine near Grythyttan; and at Langbån, Sweden.

BEST REF. IN ENGLISH: Palache, et al., "Dana's System of Mineralogy," 7th Ed., v. II, p. 1029–1030, New York, Wiley, 1951.    Moore, Paul B., *Am. Min.*, **51**: 1494–1500 (1966).

## CATTIERITE

$CoS_2$

CRYSTAL SYSTEM: Cubic
CLASS: 2/m $\bar{3}$
SPACE GROUP: Pa3
Z: 4
LATTICE CONSTANT:
  $a = 5.535$
3 STRONGEST DIFFRACTION LINES:
  2.75 (100)
  2.46 ( 60)
  1.66 ( 55)
HARDNESS: Not determined.
DENSITY: 4.80 (Calc.)
CLEAVAGE: $\{001\}$ perfect
HABIT: Massive; as granular intergrowths with other sulfides.

COLOR-LUSTER: Pinkish. Opaque. Metallic.

MODE OF OCCURRENCE: Occurs sparingly at the Shinkolobwe mine, Katanga, Zaire.

BEST REF. IN ENGLISH: Kerr, Paul F., *Am. Min.*, **30**: 483–497 (1945).

## CAVANSITE

$Ca(VO)(Si_4O_{10}) \cdot 4H_2O$

CRYSTAL SYSTEM: Orthorhombic
CLASS: 2/m 2/m 2/m
SPACE GROUP: Pcmn
Z: 4
LATTICE CONSTANTS:
  $a = 9.778$
  $b = 13.678$
  $c = 9.601$
3 STRONGEST DIFFRACTION LINES:
  7.964 (100)
  6.854 ( 50)
  6.132 ( 25)
OPTICAL CONSTANTS:
  $\alpha = 1.542$
  $\beta = 1.544$
  $\gamma = 1.551$
  Biaxial (+), $2V = 52 \pm 2°$
HARDNESS: $\sim$3–4
DENSITY: 2.21–2.31 (Meas.)
        2.33 (Calc.)
CLEAVAGE: $\{010\}$ good
Brittle.

HABIT: Crystals prismatic, elongated parallel to $c$, with dome terminations, usually $\sim$0.05 mm in diameter and 0.5–1.0 mm in length. Often as spherulitic rosettes up to 5 mm in diameter.

COLOR-LUSTER: Brilliant greenish blue. Transparent. Vitreous.

MODE OF OCCURRENCE: Cavansite is found associated with pentagonite, calcite, heulandite, apophyllite, and stilbite in cavities and veinlets in a brown tuff near Owyhee Dam in Malheur County, Oregon; also with calcite, heulandite, thomsonite, and occasionally copper, as cavity fillings, in amygdules, and in calcite veinlets in basalt and breccia at the Chapman quarry, near Goble, Columbia County.

BEST REF. IN ENGLISH: Lloyd W. Staples, Howard T. Evans, Jr., and James R. Lindsay, *Am. Min.*, **58**: 405–411, (1973).   Howard T. Evans, Jr., *Am. Min.*, **58**: 412–424 (1973).

# CEBOLLITE
$Ca_4Al_2Si_3O_{14}(OH)_2$

CRYSTAL SYSTEM: Orthorhombic (?)
3 STRONGEST DIFFRACTION LINES:
  2.73 (100)
  2.88 ( 90)
  2.59 ( 70)
OPTICAL CONSTANTS:
  $\alpha$ = 1.595
  $\beta$ = 1.60
  $\gamma$ = 1.628
  (+)2V = 58°
HARDNESS: 5
DENSITY: 2.96 (Meas.)
CLEAVAGE: Not determined.
HABIT: Fibrous.
COLOR-LUSTER: Colorless. Transparent. Vitreous.
MODE OF OCCURRENCE: Occurs as an alteration product of melilite near Cebolla Creek, Gunnison County, Colorado; also found at Scawt Hill, County Antrim, Ireland.
BEST REF. IN ENGLISH: Larsen, E. S., Jr., *Wash. Acad. Sci. J.*, **4**: 480–482 (1914).

# CELADONITE
$K(Mg, Fe^{2+})(Fe^{3+}, Al)Si_4O_{10}(OH)_2$
Mica Group

CRYSTAL SYSTEM: Monoclinic
CLASS: 2/m
SPACE GROUP: A2/m
Z: 2
LATTICE CONSTANTS:
  *a* = 10.27
  *b* = 9.02
  *c* = 5.21
  $\beta$ = 100°6′
3 STRONGEST DIFFRACTION LINES:
  2.580 (100)
  4.53 ( 85)
  3.64 ( 80)
OPTICAL CONSTANTS:
  $\alpha$ = 1.610–1.606
  $\gamma$ = 1.641–1.579
HARDNESS: ~2
DENSITY: 2.95–3.05 (Meas.)
CLEAVAGE: {001} perfect
HABIT: Earthy or as minute micaceous scales.
COLOR-LUSTER: Green, blue-green. Dull.
MODE OF OCCURRENCE: Occurs widespread in formations of altered volcanic rocks of intermediate to basaltic composition, commonly associated with montmorillonite, chalcedony, quartz, zeolites, and prehnite. Typical occur-

rences are found in the Wind River area, Washington; in the John Day formation, Oregon; and at localities in Scotland, Poland, USSR, Madagascar, Japan, and New Zealand.
BEST REF. IN ENGLISH: Wise, W. S., and Eugster, H. P., *Am. Min.*, **49**: 1031–1083 (1964).

# CELESTITE (Celestine)
$SrSO_4$

CRYSTAL SYSTEM: Orthorhombic
CLASS: 2/m 2/m 2/m
SPACE GROUP: Pnma
Z: 4
LATTICE CONSTANTS:
  *a* = 8.36
  *b* = 5.36
  *c* = 6.84
3 STRONGEST DIFFRACTION LINES:
  2.972 (100)
  3.295 ( 98)
  2.731 ( 63)
OPTICAL CONSTANTS:
  $\alpha$ = 1.6215
  $\beta$ = 1.6237
  $\gamma$ = 1,6308
  (+)2V $\approx$ 50°
HARDNESS: 3–3½
DENSITY: 3.97 ± 0.01 (Meas.)
         3.971 (Calc.)
CLEAVAGE: {001} perfect
          {210} good
          {010} indistinct
Fracture uneven. Brittle.
HABIT: Crystals usually thin to thick tabular, resembling barite, or elongated into lath-like shapes. Less commonly equant or pyramidal. Also as fibrous veinlets or nodules; in geodes; massive granular; lamellar; and earthy.
COLOR-LUSTER: Colorless and various shades of white, gray, blue, green, yellow, orange, red, and brown. Transparent to translucent. Vitreous, somewhat pearly on cleavages. Sometimes fluorescent under ultraviolet light.
MODE OF OCCURRENCE: Occurs widespread in sedimentary rocks, particularly in limestones; in hydrothermal vein deposits; and rarely in basic igneous rocks. Found in the United States as exceptional crystals at Clay Center, at Put-in-Bay on South Bass Island, and at other places in Ohio; at Maybee and other localities in Michigan; at Chittenango Falls, New York; at Lampasas and in the Mount Bonnell area in Texas; in geodes with colemanite at Borate in the Calico Hills, San Bernardino County, and at many other deposits in California. It also is found at numerous localities in Canada; as fine large blue crystals at Matehuala, San Luis Potosi, Mexico; abundantly in the vicinity of Bristol, Gloucestershire, England; as superb crystal groups at the sulfur deposits in Sicily; at numerous places in Germany, France, Austria, Switzerland, Italy, USSR, Egypt, and Tunisia; and as magnificent transparent blue crystals in geodes from Madagascar.
BEST REF. IN ENGLISH: Palache, et al., "Dana's System of Mineralogy," 7th Ed., v. II, p. 415–420, New York, Wiley, 1951.

# CELSIAN
$BaAl_2Si_2O_8$
Barium Feldspar Group
Dimorphous with paracelsian

CRYSTAL SYSTEM: Monoclinic
CLASS: 2/m
SPACE GROUP: $I2_1/c$
Z: 8
LATTICE CONSTANTS:
  $a = 8.627$
  $b = 13.045$
  $c = 14.408$
  $\beta = 115.2°$
3 STRONGEST DIFFRACTION LINES:
  3.47 (100)
  3.35 (100)
  3.02 ( 95)
OPTICAL CONSTANTS:
  $\alpha = 1.579-1.587$
  $\beta = 1.583-1.593$
  $\gamma = 1.588-1.600$
  $(+)2V_\gamma = 83°-92°$
HARDNESS: 6-6½
DENSITY: 3.10-3.45 (Meas.)
        3.420 (Calc.)
CLEAVAGE: {001} perfect
          {010} good
          {110} poor
Fracture uneven. Brittle.
HABIT: Crystals short prismatic with prominent prism faces. Also massive, cleavable. Carlsbad, Manebach, and Baveno twinning, common.
COLOR-LUSTER: Colorless, white, yellow. Transparent. Vitreous.
MODE OF OCCURRENCE: Occurs chiefly in the contact zones of manganese deposits. Found in California associated with quartz, diopside, witherite, and sanbornite in quartzite, near Rush Creek and Big Creek, Fresno County; with gillespite and sanbornite near Incline, Mariposa County; and at the Kalkar quarry at Santa Cruz, Santa Cruz County. It also occurs in the Alaska Range, Alaska; at the Benallt manganese mine at Rhiw, North Wales; at Jakobsberg, Sweden; and at localities in Italy, South West Africa, and Australia.
BEST REF. IN ENGLISH: Deer, Howie, and Zussman, "Rock Forming Minerals," v. 4, p. 166-178, New York, Wiley, 1963.

# CENOSITE (Kainosite)
$Ca_2(Ce, Y)_2(SiO_4)_3(CO_3) \cdot H_2O$

CRYSTAL SYSTEM: Orthorhombic
CLASS: 2/m 2/m 2/m
SPACE GROUP: Pmnb
Z: 4
LATTICE CONSTANTS:
  $a = 13.02$
  $b = 14.32$
  $c = 6.75$

3 STRONGEST DIFFRACTION LINES:
  6.52 (100)
  2.764 (100)
  3.29 ( 80)
OPTICAL CONSTANTS:
  $\alpha = 1.662$
  $\beta = 1.686$
  $\gamma = 1.692$
  $(-)2V = 40°$
HARDNESS: 5-6
DENSITY: 3.34-3.61 (Meas.)
        3.542 (Calc.)
CLEAVAGE: Two cleavages (at 90°). Fracture uneven. Brittle.
HABIT: Crystals short to long prismatic, up to 2 cm long; also pseudotetragonal.
COLOR-LUSTER: Light yellowish brown to dark chestnut brown; also colorless, yellow, light rose to light red. Transparent to translucent. Vitreous.
MODE OF OCCURRENCE: Occurs in the Henry granite pegmatite near Cotopaxi, Colorado, associated with albite, muscovite, fluorite, monazite, cyrtolite, euxenite, and doverite; at North Burgess and at the Bicroft uranium mine at Bancroft, Onatrio, Canada; in granite pegmatite at Igeltjern, Hitterö, Norway; in vugs in magnetite ore at the Ko mine, Nordmark, Sweden; in Alpine cleft veins in altered granite at Val Curnera and elsewhere in Switzerland; and in Tuva, USSR.
BEST REF. IN ENGLISH: Vlasov, K. A., "Mineralogy of Rare Elements," v. II, p. 246-247, Israel Program for Scientific Translations, 1966.   Heinrich, E. William, Borup, R. A., and Salotti, C. A., *Am. Min.*, 47: 328-336 (1962).

# CERARGYRITE (Chlorargyrite)
AgCl

CRYSTAL SYSTEM: Cubic
CLASS: 4/m $\bar{3}$ 2/m
SPACE GROUP: Fm3m
Z: 4
LATTICE CONSTANTS:
  $a = 5.556$
3 STRONGEST DIFFRACTION LINES:
  2.774 (100)
  1.962 ( 50)
  3.20 ( 50)
OPTICAL CONSTANT:
  $N = 2.071$
HARDNESS: 2½
DENSITY: 5.556 (Meas.)
        5.55 (Calc.)
CLEAVAGE: None. Fracture uneven to subconchoidal. Sectile and ductile.
HABIT: Crystals usually cubic with occasional modifying faces. Commonly massive, often in crusts and waxy coatings; rarely columnar or fibrous. Twinning on {111} common.
COLOR-LUSTER: Gray, greenish gray, yellowish. Colorless when fresh and pure. Often turns violet-brown or purple on exposure to light. Transparent to translucent. Resinous to adamantine, waxy.

MODE OF OCCURRENCE: Occurs as a secondary mineral in the oxidation zone of silver deposits, often associated with native silver, jarosite, iron and manganese oxides, cerussite, mimetite, pyromorphite, wulfenite, and malachite. Found widespread in the western United States, especially at Leadville and other silver districts in Colorado; at Carbonate, Lawrence County, South Dakota; as large masses and superb crystals at the Poorman mine, Owyhee County, and at other deposits in Nevada; at Tombstone, Arizona; and at many silver mines in Inyo, San Bernardino, and other counties in California. It also occurs in notable amounts in many parts of Mexico; in Bolivia, Peru, and Chile; in Saxony and the Harz Mountains, Germany; in England, France, Italy, Spain, and the USSR; and in large amounts at Broken Hill, New South Wales, Australia.

BEST REF. IN ENGLISH: Palache, et al., "Dana's System of Mineralogy," 7th Ed., v. II, p. 11–14, New York, Wiley, 1951.

# CERIANITE
$CeO_2$

CRYSTAL SYSTEM: Cubic
CLASS: $4/m \bar{3} 2/m$
SPACE GROUP: Fm3m
Z: 4
LATTICE CONSTANTS:
  $a \simeq 5.42$
3 STRONGEST DIFFRACTION LINES:
  3.124 (100)
  1.913 ( 51)
  1.632 ( 44)
OPTICAL CONSTANT:
  $N > 2$
HARDNESS: Not determined.
DENSITY: 7.216 (Calc.)
CLEAVAGE: No data.
HABIT: Minute octahedral crystals.
COLOR-LUSTER: Dark greenish amber, greenish yellow to buff. Translucent.
MODE OF OCCURRENCE: Occurs very sparingly in partly absorbed inclusions of wall-rock in a dike-like zone of carbonate rock cutting a nepheline syenite in Lackner Township, Sudbury district, Ontario, Canada, and as a secondary mineral from Morro do Ferro on the Poços de Caldaz plateau, Minas Geraes, Brazil.

BEST REF. IN ENGLISH: Graham, A. R., *Am. Min.*, **40**: p. 560–564 (1955). Frondel, Clifford, U.S.G.S. Bull. 1064, p. 53–55 (1958).

# CERITE
$(Ca, Mg)_2 (RE)_8 (SiO_4, FCO_3)_7 (OH, H_2O)_3$

CRYSTAL SYSTEM: Hexagonal
CLASS: $\bar{3} 2/m$ or 3m
SPACE GROUP: $R\bar{3}c$ or R3c
Z: 6
LATTICE CONSTANTS:
  $a \approx 10.82$
  $c \approx 37.71$

3 STRONGEST DIFFRACTION LINES:
  2.95 (100)
  1.954 ( 50)
  3.47 ( 42)
OPTICAL CONSTANTS:
  $\omega = 1.806-1.81$
  $\epsilon = 1.81-1.82$
  Uniaxial (+)
HARDNESS: ~5½
DENSITY: 4.78 (Meas.) (Mt. Pass, California)
         4.86 (Calc.)
CLEAVAGE: Not apparent. Fracture uneven.
HABIT: Crystals pseudo-octahedral, from 2.0 to 7.0 mm in size; commonly massive; granular. Crystal forms: $c\{0003\}$, $e\{01\bar{1}2\}$.
COLOR-LUSTER: Verona brown, clove brown, and cherry red to gray. Resinous.
MODE OF OCCURRENCE: Occurs as well-formed crystals associated with bastnaesite in one of the rare-earth-bearing veins of the Mountain Pass district, California; as an intimate intergrowth with allanite, epidote, tornebohmite, fluorite, monazite, uraninite, and quartz in the Jamestown district, Boulder County, Colorado. Also found in Papineau County, Quebec, Canada; at the Bastnaes mine near Riddarhyttan, Vastmanland, Sweden; in the Kyshtymsk (Kychtym) district, Ural Mountains, USSR; and at Mt. Tenbazan, Heiko Gun, Kôgen-dô, Korea.

BEST REF. IN ENGLISH: Glass, Jewell J., Evans, Howard T. Jr., Carron, M. K., and Hildebrand, F. A., *Am. Min.*, **43**: 460– 475 (1958).

**CEROLITE** = Mixture of serpentine and stevensite

# CEROTUNGSTITE
$(Ce, Nd)W_2O_6(OH)_3$

CRYSTAL SYSTEM: Monoclinic
CLASS: 2 or 2/m
SPACE GROUP: $P2_1$ or $P2_1/m$
Z: 2
LATTICE CONSTANTS:
  $a = 5.874$
  $b = 8.700$
  $c = 7.070$
  $\beta = 105°27'$
3 STRONGEST DIFFRACTION LINES:
  3.405 (100)
  2.273 ( 52)
  6.83 ( 36)
OPTICAL CONSTANTS: Under microscope yellow with slight greenish tint. ns $\alpha'$ 1.89m, $\beta'$ 1.95, $\gamma$ 2.02 (all ± 0.01), Z = b, Y nearly $\perp$ to (001), X nearly parallel to $a$.
HARDNESS: ~1
DENSITY: Not determined.
CLEAVAGE: {100} perfect
HABIT: Crystals bladed, up to 100 microns long; {001} dominant, also {101}, {100}, {011}, and {110}. As radiating groups. Twinning on {001}.
COLOR-LUSTER: Orange-yellow.
MODE OF OCCURRENCE: Found in association with

ferberite in the Kirwa and Nyamulilo mines, Kigezi district, Uganda.

BEST REF. IN ENGLISH: Sahama, Th. G., Von Knorring, Oleg, and Lehtinen, Martti, *Am. Min.*, **57**: 1558–1559 (1972).

## CERULÉITE (Inadequately described mineral)
$CuAl_4(AsO_4)_2(OH)_8 \cdot 4H_2O$

CRYSTAL SYSTEM: Unknown.
  HARDNESS: Not determined. Soft.
  DENSITY: 2.803 (Meas.)
  CLEAVAGE: Not determined.
  HABIT: Massive, compact, clay-like.
  COLOR-LUSTER: Turquoise blue.
  MODE OF OCCURRENCE: Found at the Emma Luisa gold mine near Huanaco, Taltal province, Chile, and at Wheal Gorland, St. Day, Cornwall, England.
  BEST REF. IN ENGLISH: Palache, et al., "Dana's System of Mineralogy," 7th Ed., v. II, p. 927–928, New York, Wiley, 1951.

## CERUSSITE
$PbCO_3$

CRYSTAL SYSTEM: Orthorhombic
CLASS: 2/m 2/m 2/m
SPACE GROUP: Pmcn
Z: 4
LATTICE CONSTANTS:
  $a = 5.1726$
  $b = 8.4800$
  $c = 6.1302$
3 STRONGEST DIFFRACTION LINES:
  3.593 (100)
  3.498 ( 50)
  2.487 ( 32)
OPTICAL CONSTANTS:
  $\alpha = 1.8036$
  $\beta = 2.0765$ (Na)
  $\gamma = 2.0786$
  $(-)2V = 9°0'$ (Calc.)
HARDNESS: 3–3½
DENSITY: 6.55 (Meas.)
          6.558 (Calc.)
CLEAVAGE: {110} distinct
          {021} distinct
          {010} trace
          {012} trace
Fracture conchoidal. Very brittle.
  HABIT: Crystals widely varied in habit; commonly twinned, often as sixlings. Single crystals usually tabular and elongated along *a*-axis; also equant; rarely acicular or thin tabular. Often striated. As clusters of crystals, reticulated masses, or "jackstraw" aggregates. Also massive, granular to dense and compact; rarely stalactitic, fibrous, or pulverulent.
  COLOR-LUSTER: Colorless, white, gray, smoky; sometimes blue, green, dark gray, or black due to inclusions. Transparent to subtranslucent. Adamantine, vitreous, resin-

ous, or pearly; sometimes submetallic. Streak colorless to white. Occasionally fluoresces yellowish in long-wave ultraviolet light.
  MODE OF OCCURRENCE: Occurs widespread as a secondary mineral in the oxidation zone of ore deposits, often associated with other secondary lead, zinc, or copper minerals. Found at numerous localities in the western United States, especially at the Mammoth mine, Pinal County, and at the Flux mine, Santa Cruz County, Arizona; in the silver-lead deposits of Idaho; and in Colorado, South Dakota, Utah, New Mexico, Montana, Nevada, and California. Fine specimens formerly were found at the Wheatley mines, Phoenixville, Pennsylvania. Notable foreign localities include Tsumeb, South West Africa; the Broken Hill mine, Zambia; Dundas, Tasmania; Broken Hill, New South Wales, Australia; Monte Poni and Monte Vecchio, Sardinia; and Leadhills, Lanarkshire, Scotland.
  BEST REF. IN ENGLISH: Palache, et al., "Dana's System of Mineralogy," 7th Ed., v. II, p. 200–207, New York, Wiley, 1951.

## CERVANTITE
$Sb_2O_4$

CRYSTAL SYSTEM: Orthorhombic
CLASS: mm2
SPACE GROUP: $Pbn2_1$
Z: 4
LATTICE CONSTANTS:
  $a = 4.79$
  $b = 5.43$
  $c = 11.73$
3 STRONGEST DIFFRACTION LINES:
              (synthetic)
  3.06 (100)    3.07 (100)
  2.91 ( 70)    2.94 ( 45)
  1.854 ( 70)   3.45 ( 35)
OPTICAL CONSTANTS:
  $N \simeq 1.67$–2.05
  $(-)2V =$ small
HARDNESS: 4–5
DENSITY: ~6.5 (Meas.)
          6.64 (Calc.)
CLEAVAGE: {001} excellent
          {100} distinct
HABIT: Massive; fine-grained.
COLOR-LUSTER: Yellow.
  MODE OF OCCURRENCE: Occurs with other antimony oxides at Brasina, near the Drina River, Zajača-Stolice district, western Serbia, Yugoslavia. Also as pseudomorphs after stibnite at Felsöbánya, Hungary (= Baia Sprie, Roumania), and at Pocca, Bolivia.
  BEST REF. IN ENGLISH: Gründer, W., Pätzold, H., and Strunz, H., *Am. Min.*, **47**: 1221 (1962).

## CESAROLITE (Inadequately described mineral)
$PbMn_3O_7 \cdot H_2O$ (?)

CRYSTAL SYSTEM: Unknown

**3 STRONGEST DIFFRACTION LINES:**
2.194 (100)
2.087 ( 80)
1.760 ( 50)
HARDNESS: 4½
DENSITY: 5.29
CLEAVAGE: Not determined.
HABIT: Massive; as botryoidal crusts and friable masses.
COLOR-LUSTER: Steel gray. Opaque. Submetallic to dull.
MODE OF OCCURRENCE: Occurs in cavities in galena at Sidi-amor-ben-Salem, Tunisia.
BEST REF. IN ENGLISH: Palache, et al., "Dana's System of Mineralogy," 7th Ed., v. I, p. 744, New York, Wiley, 1944.

## CESIUM-KUPLETSKITE = Kupletskite

## CHABAZITE
$CaAl_2Si_4O_{12} \cdot 6H_2O$
Zeolite Group

CRYSTAL SYSTEM: Hexagonal
CLASS: $\bar{3}\,2/m$
SPACE GROUP: $R\bar{3}m$
Z: 6
LATTICE CONSTANTS:
$a \simeq 13.8$
$c \simeq 15.0$
3 STRONGEST DIFFRACTION LINES:
9.46 (100)
5.56 (100)
5.03 (100)
OPTICAL CONSTANTS:
$\left.\begin{array}{c}\omega\\\epsilon\end{array}\right\}$ 1.470–1.494
(−)
HARDNESS: 4–5
DENSITY: 2.05–2.16 (Meas.)
CLEAVAGE: $\{10\bar{1}1\}$ distinct
Fracture uneven. Brittle.
HABIT: Crystals usually simple rhombohedrons resembling a cube; also complex, or tabular {0001}. Twinning on {0001} common, as penetration twins.
COLOR-LUSTER: Colorless, white, yellowish, pinkish, reddish white, greenish. Transparent to translucent. Vitreous. Streak uncolored.
MODE OF OCCURRENCE: Occurs chiefly in cavities in basalt, andesite, and other igneous rocks in association with other zeolites, calcite, and quartz; also in fractures in schists and crystalline limestones; and as a recent deposit of certain hot springs. Found relatively widespread. Notable occurrences are found in Mono, Nevada, Plumas, Riverside, Santa Clara, and Shasta counties, California; at Goble, Oregon; in Gunnison and Jefferson counties, Colorado; as superb crystals in the trap rocks of northeastern New Jersey; near Baltimore, Maryland; as fine crystals in the Bay of Fundy district, Nova Scotia; and at localities in Hawaii, Greenland, Ireland, Scotland, Italy, Sicily, Germany, Hungary, Czechoslovakia, USSR, India, Australia, and elsewhere.

BEST REF. IN ENGLISH: Deer, Howie, and Zussman, "Rock Forming Minerals," v. 4, p. 387–400, New York, Wiley, 1963.

## CHALCANTHITE
$CuSO_4 \cdot 5H_2O$

CRYSTAL SYSTEM: Triclinic
CLASS: $\bar{1}$
SPACE GROUP: $P\bar{1}$
Z: 2
LATTICE CONSTANTS:
$a = 6.12$ $\alpha = 97°35'$
$b = 10.7$ $\beta = 107°10'$
$c = 5.97$ $\gamma = 77°33'$
3 STRONGEST DIFFRACTION LINES:
4.73 (100)
3.71 ( 85)
3.99 ( 60)
OPTICAL CONSTANTS:
$\alpha = 1.516$
$\beta = 1.539$
$\gamma = 1.546$
(−)2V = 56°
HARDNESS: 2½
DENSITY: 2.286 (Meas.)
2.288 (Calc.)
CLEAVAGE: $\{1\bar{1}0\}$ imperfect
$\{110\}$ in traces
Fracture conchoidal.
HABIT: Crystals short prismatic; also thick tabular. Stalactitic; as fibrous veinlets; massive, granular.
COLOR-LUSTER: Pale blue to dark blue, greenish blue, greenish. Transparent to translucent. Vitreous to resinous. Streak colorless.

Soluble in water. Taste harsh metallic.
MODE OF OCCURRENCE: Occurs widespread as a secondary mineral in the oxidation zone of copper-bearing sulfide ore deposits. Found in the United States especially in California, Arizona, New Mexico, Colorado, Nevada, Montana, and Tennessee. It also occurs at Chuquicamata and at other places in Chile; in England, Ireland, Germany, Spain, and at many other localities.
BEST REF. IN ENGLISH: Palache, et al., "Dana's System of Mineralogy," 7th Ed., v. II, p. 488–491, New York, Wiley, 1951.

## CHALCEDONY = Variety of quartz

## CHALCOALUMITE
$CuAl_4(SO_4)(OH)_{12} \cdot 3H_2O$

CRYSTAL SYSTEM: Monoclinic
CLASS: 2
SPACE GROUP: $P2_1$    Z: 4
LATTICE CONSTANTS:
$a = 17.090,$ $b = 8.915,$ $c = 10.221,$ $\beta = 95°53'$

3 STRONGEST DIFFRACTION LINES:
  8.50 (100)
  4.25 ( 91)
  4.18 ( 23)
OPTICAL CONSTANTS:
  (+) 2V = 61° at 430 m$\mu$ to 46° at 620 m$\mu$.
HARDNESS: 2½
DENSITY: 2.29 (Meas.), 2.25 (Calc.)
CLEAVAGE: {100} perfect. Sectile.
  HABIT: Crystals thin triangular plates, commonly twinned; in spherules. Also microcrystalline to fibrous.
  COLOR-LUSTER: Light bluish green, pale blue, bluish gray. Transparent to translucent. Vitreous to dull.
  MODE OF OCCURRENCE: Occurs as a secondary mineral at the Copper Queen mine, Bisbee, Cochise County, Arizona, associated with azurite, cuprite, malachite, goethite and quartz.
  BEST REF. IN ENGLISH: Williams, S. A. and Khin, BaSaw, *Min. Rec.*, **2**: 126–127 (1971).

# CHALCOCITE
$Cu_2S$

CRYSTAL SYSTEM: Orthorhombic
CLASS: 2/m 2/m 2/m or mm2
SPACE GROUP: Cmmn or Cma2
Z: 96
LATTICE CONSTANTS:
  $a$ = 13.491
  $b$ = 27.323
  $c$ = 11.881
3 STRONGEST DIFFRACTION LINES:
  1.870 (100)
  1.969 ( 80)
  2.40 ( 70)
HARDNESS: 2½–3
DENSITY: 5.97 (Calc.)
        5.5–5.8 (Meas.)
  CLEAVAGE: {110} indistinct
Fracture conchoidal. Brittle; somewhat sectile.
  HABIT: Crystals usually well-formed pseudohexagonal prisms formed by twinning on {110}; also short prismatic or thick tabular single crystals, striated on {001}. Commonly massive. Twinning also on {112}, and {032}.
  COLOR-LUSTER: Blackish gray to black. Opaque. Metallic. Streak blackish gray.
  MODE OF OCCURRENCE: Occurs as an important ore of copper in the enriched zones of hydrothermal sulfide vein deposits, commonly associated with malachite, azurite, covellite, bornite, chalcopyrite, and pyrite. Found as excellent crystals and crystal groups at Butte, Montana, and at Bristol, Connecticut; as a major ore mineral in Tennessee, Montana, Utah, Nevada, New Mexico, Arizona, and Alaska. It also occurs as fine crystals in Cornwall, England; Messina, South West Africa; and at localities in Germany, Czechoslovakia, and the USSR. Other major deposits are found in Mexico, Peru, Chile, Spain, and numerous places in southern and central Africa.
  BEST REF. IN ENGLISH: Palache, et al., "Dana's System of Mineralogy," 7th Ed., v. I, p. 187–190, New York, Wiley, 1944.

# CHALCOCYANITE (Chalcokyanite)
$CuSO_4$

CRYSTAL SYSTEM: Orthorhombic
CLASS: 2/m 2/m 2/m
SPACE GROUP: Pmnb
Z: 4
LATTICE CONSTANTS:
  $a$ = 6.69
  $b$ = 8.39
  $c$ = 4.82
3 STRONGEST DIFFRACTION LINES:
  3.549 (100)
  2.62 ( 95)
  4.187 ( 75)
OPTICAL CONSTANTS:
  $\alpha$ = 1.724
  $\beta$ = 1.733 (Na)
  $\gamma$ = 1.739
  (−)2V = large
HARDNESS: 3½
DENSITY: 3.65 (Meas.)
        3.902 (Calc. synthetic)
CLEAVAGE: Not determined.
  HABIT: Usually as slightly elongated tabular crystals.
  COLOR-LUSTER: Pale green, sky blue, yellowish, brownish. Transparent to translucent.

  Very hygroscopic, readily soluble in water.
  MODE OF OCCURRENCE: Found as a sublimation product in fumaroles after various eruptions of Mount Vesuvius, Italy.
  BEST REF. IN ENGLISH: Palache, et al., "Dana's System of Mineralogy," 7th Ed., v. II, p. 429–430, New York, Wiley, 1951.    Strunz, H., *Am. Min.*, **46**: 758–759 (1961).

# CHALCOLITE = Torbernite

# CHALCOMENITE
$CuSeO_3 \cdot 2H_2O$

CRYSTAL SYSTEM: Orthorhombic
CLASS: 2 2 2
SPACE GROUP: $P2_12_12_1$
Z: 4
LATTICE CONSTANTS:
  $a$ = 6.671
  $b$ = 9.193
  $c$ = 7.384
3 STRONGEST DIFFRACTION LINES:
  5.39 (100)
  4.94 ( 90)
  3.35 ( 80)
OPTICAL CONSTANTS:
  $\alpha$ = 1.712
  $\beta$ = 1.732
  $\gamma$ = 1.732
  (−)2V ~ 0°
HARDNESS: 2–2½
DENSITY: 3.312 (Meas.)
        3.322 (Calc.)

CLEAVAGE: None.
HABIT: Crystals minute, prismatic, also acicular. Prism faces commonly rounded or striated.
COLOR-LUSTER: Bright blue. Transparent. Vitreous.
MODE OF OCCURRENCE: Occurs in Argentina as a secondary mineral formed by the oxidation of copper and lead selenides at Sierra Famatina and Sierra de Umango, La Rioja province, and in the Sierra de Cacheuta, Mendoza province. It also is found at the Hiaco mine, Pacajake, Bolivia.
BEST REF. IN ENGLISH: Palache, et al., "Dana's System of Mineralogy," 7th Ed., v. II, p. 638-639, New York, Wiley, 1951. *Am. Min.*, **49**: 1481-1485 (1964).


# CHALCONATRONITE
$Na_2Cu(CO_3)_2 \cdot 3H_2O$

CRYSTAL SYSTEM: Probably monoclinic
Z: 1
LATTICE CONSTANTS:
 $a = 13.72$
 $b = 6.12$
 $c = 9.70$
 $\beta = 91°18'$
3 STRONGEST DIFFRACTION LINES:
 6.92 (100)
 4.15 ( 80)
 3.68 ( 70)
OPTICAL CONSTANTS:
 $\alpha = 1.483$
 $\beta = 1.530$
 $\gamma = 1.576$
 2V = large. Probably (+)
HARDNESS: Low.
DENSITY: 2.27
HABIT: Fine-grained crusts. Crystals are small laths or pseudohexagonal plates.
COLOR-LUSTER: Greenish blue.
MODE OF OCCURRENCE: Found associated with cuprite and atacamite as a corrosion product of three ancient bronze objects from Egypt in the Fogg Art Museum, Harvard University.
BEST REF. IN ENGLISH: Frondel, Clifford, and Gettens, Rutherford, J., *Science*, **122** (3158): 75-76 (1955).


# CHALCOPHANITE
$ZnMn_3O_7 \cdot 3H_2O$

CRYSTAL SYSTEM: Triclinic
CLASS: $\bar{1}$
SPACE GROUP: $P\bar{1}$
Z: 2
LATTICE CONSTANTS:
 $a = 7.53$     $\alpha = 90°$
 $b = 7.53$     $\beta = 117°20'$
 $c = 8.20$     $\gamma = 120°$
3 STRONGEST DIFFRACTION LINES:
 6.96 (100)
 3.50 ( 60)
 4.08 ( 50)

OPTICAL CONSTANTS:
 $\omega > 2.72$
 $\epsilon \sim 2.72$
 (–)
HARDNESS: 2½
DENSITY: 3.98 (Meas.)
         3.827 (Calc.)
CLEAVAGE: {0001} perfect
Thin plates flexible.
HABIT: Crystals commonly tabular, minute; sometimes octahedral in aspect. Usually as drusy crusts. Also massive; dense, granular; platy; stalactitic or plumose.
COLOR-LUSTER: Bluish to iron black. Opaque, except in very thin fragments. Metallic. Streak brown, dull.
MODE OF OCCURRENCE: Occurs as a secondary mineral in the oxidation zone of zinc-bearing ore deposits. Found in the Leadville district, Lake County, Colorado, in sufficient quantity to constitute an ore mineral; at the Passaic mine, Sterling Hill, Sussex County, New Jersey; as fine specimens at Mapimi, Durango, Mexico; and at deposits in Palestine and Tasmania.
BEST REF. IN ENGLISH: Palache, et al., "Dana's System of Mineralogy," 7th Ed., v. I, p. 739-740, New York, Wiley, 1944.


# CHALCOPHYLLITE
$Cu_{18}Al_2(AsO_4)_3(SO_4)_3(OH)_{27} \cdot 36H_2O$

CRYSTAL SYSTEM: Hexagonal
CLASS: $\bar{3}\,2/m$
SPACE GROUP: $R\bar{3}m$
Z: 3
LATTICE CONSTANTS:
 $a = 10.77$
 $c = 57.5$
3 STRONGEST DIFFRACTION LINES:
 9.54 (100)
 4.79 (100)
 2.59 (100)
OPTICAL CONSTANTS:
 $\omega = 1.618$
 $\epsilon = 1.552$
 (–)
HARDNESS: 2
DENSITY: 2.67 (fully hydrated)
         2.64 (Calc.)
CLEAVAGE: {0001} perfect, micaceous
          {10$\bar{1}$1} in traces
Not brittle.
HABIT: Crystals thin tabular, six-sided, flattened on {0001}, sometimes striated. Also scaly, foliated massive, and as rosettes.
COLOR-LUSTER: Emerald green, grass green to bluish green. Transparent to translucent. Vitreous to adamantine or pearly. Streak lighter than color.
MODE OF OCCURRENCE: Occurs as a secondary mineral in the oxidation zone of copper-bearing ore deposits. Found associated with tyrolite, spangolite, cyanotrichite, and azurite at the Myler mine, Majuba Hill, Pershing County, and at Sodaville, Mineral County, Nevada; at the Eureka mines, Tintic district, Juab County, Utah; and at Bisbee,

Arizona. It also is found as excellent specimens at several localities in Cornwall, England; and in France, Germany, Austria, Hungary, USSR, and at the Teniente mine, near Rancagua, Chile.

BEST REF. IN ENGLISH: Palache, et al., "Dana's System of Mineralogy," 7th Ed., v. II, p. 1008–1010, New York, Wiley, 1951.

# CHALCOPYRITE
$CuFeS_2$

CRYSTAL SYSTEM: Tetragonal
CLASS: $\bar{4}$ 2m
SPACE GROUP: I$\bar{4}$2d
Z: 4
LATTICE CONSTANTS:
 $a$ = 5.24
 $c$ = 10.30
3 STRONGEST DIFFRACTION LINES:
 3.03  (100)
 1.854 ( 80)
 1.591 ( 60)
OPTICAL CONSTANTS:
 (–) uniaxial
HARDNESS: 3½–4
DENSITY: 4.35 (Meas.)
    4.40 (Calc.)
CLEAVAGE: {011} sometimes distinct
Fracture uneven. Brittle.
HABIT: Commonly as sphenoidal crystals resembling tetrahedrons; sphenoidal faces usually large, striated, and dull. Often as isolated crystals on matrix, in parallel growths, or as aggregates. Also massive, compact; reniform, botryoidal. Twinning on {112}, {012}, common.
COLOR-LUSTER: Brass yellow; commonly tarnishes iridescent. Opaque. Metallic. Streak greenish black.
MODE OF OCCURRENCE: Occurs widespread as one of the most important copper minerals, chiefly in medium- to high-temperature sulfide ore deposits. In the United States important deposits occur in Arizona, New Mexico, Utah, Montana, and Tennessee. Other major occurrences are found in Canada, Mexico, Chile, Peru, England, Spain, France, Italy, Germany, Austria, Czechoslovakia, Norway, Sweden, Japan, Tasmania, and Australia.
BEST REF. IN ENGLISH: Palache, et al., "Dana's System of Mineralogy," 7th Ed., v. I, p. 219–224, New York, Wiley, 1944.

# CHALCOPYRRHOTINE = Cubanite
$CuFe_4S_5$

# CHALCOSIDERITE
$CuFe_6(PO_4)_4(OH)_8 \cdot 4H_2O$
Turquoise Group

CRYSTAL SYSTEM: Triclinic
CLASS: $\bar{1}$
SPACE GROUP: P$\bar{1}$
Z: 1

LATTICE CONSTANTS:
 $a$ = 7.66 kx    $\alpha$ = 112°29'
 $b$ = 10.18     $\beta$ = 115°18'
 $c$ = 7.88      $\gamma$ = 69°00'
3 STRONGEST DIFFRACTION LINES:
 3.77 (100)
 3.39 ( 70)
 3.02 ( 60)
OPTICAL CONSTANTS:
 $\alpha$ = 1.775
 $\beta$ = 1.840
 $\gamma$ = 1.844
(–)2V ≃ 22°
HARDNESS: 4½
DENSITY: 3.22 (Meas.)
    3.26 (Calc.)
CLEAVAGE: {001} perfect
    {010} good
HABIT: Crystals short prismatic with large {001}, {010}, {1$\bar{1}$0}, and {$\bar{1}$11}. As crusts or sheaf-like groups of distinct crystals.
COLOR-LUSTER: Dark green. Transparent to translucent. Vitreous.
MODE OF OCCURRENCE: Occurs as a secondary mineral in the oxidation zone of copper-bearing ore deposits. Found at Bisbee, Arizona; as fine specimens at the Wheal Phoenix mine, Cornwall, England; and at Seigen, Westphalia, and Schneckenstein, Saxony, Germany.
BEST REF. IN ENGLISH: Palache, et al., "Dana's System of Mineralogy," 7th Ed., v. II, p. 947–951, New York, Wiley, 1951. *Am. Min.*, **50**: 227–231 (1965).

# CHALCOSTIBITE
$CuSbS_2$

CRYSTAL SYSTEM: Orthorhombic
CLASS: 2/m 2/m 2/m
SPACE GROUP: Pnam
Z: 4
LATTICE CONSTANTS:
 $a$ = 6.008
 $b$ = 14.456
 $c$ = 3.784
3 STRONGEST DIFFRACTION LINES:
 3.13  (100)
 3.00  ( 90)
 1.762 ( 50)
HARDNESS: 3–4
DENSITY: 4.8–5.0 (Meas.)
    5.010 (Calc.)
CLEAVAGE: {010} perfect
    {001} imperfect
    {100} imperfect
Fracture subconchoidal. Brittle.
HABIT: Crystals prismatic, flattened {010}; striated parallel to elongation. Also massive, granular. Twinning on {104}.
COLOR-LUSTER: Dark lead gray. Opaque. Metallic.
MODE OF OCCURRENCE: Occurs associated with various sulfosalts, sulfides, and quartz at Oruro and elsewhere in Bolivia; at Wolfsberg, Harz Mountains, Germany; in

Granada, Spain; and as large partially altered crystals at Rar el Anz, east of Casablanca, Morocco.

BEST REF. IN ENGLISH: Palache, et al., "Dana's System of Mineralogy," 7th Ed., v. I, p. 433–435, New York, Wiley, 1944.

## CHALCOTHALLITE (Inadequately described mineral)
$Cu_3 TIS_2$

CRYSTAL SYSTEM: Unknown
3 STRONGEST DIFFRACTION LINES:
  3.07 (100)
  2.48 ( 90)
  3.93 ( 40)
HARDNESS: 2–2½
DENSITY: 6.6 (Meas.)
CLEAVAGE: Cleavage in 3 mutually perpendicular directions; parallel to laminae perfect, the others good or fair. Splits easily into very fine flakes or rectangular plates.
HABIT: As lamellar aggregates up to 3 × 2 × 0.5 cm in size. Polysynthetic twinning noted under the microscope.
COLOR-LUSTER: Lead gray to iron black; metallic. Streak black. Tarnishes iridescent.
MODE OF OCCURRENCE: Occurs associated with chkalovite, epistolite, niobophyllite, analcime, natrolite, microcline, Li-mica, tugtupite, galena, sphalerite, and molybdenite in veins of ussingite cutting poikilitic sodalite syenite at Nakalaq, Ilimaussaq massif, Greenland.
BEST REF. IN ENGLISH: Semenov, E. I., Sørensen, H., Bessmertnaya, M. S., and Novorossova, L. E., *Am. Min.*, **53**: 1775 (1968).

## CHALCOTRICHITE = Capillary variety of cuprite

## CHAMBERSITE
$Mn_3 B_7 O_{13} Cl$

CRYSTAL SYSTEM: Orthorhombic
CLASS: mm2
SPACE GROUP: $Pca2_1$
Z: 4
LATTICE CONSTANTS:
  $a$ = 8.68
  $b$ = 8.68
  $c$ = 12.26
3 STRONGEST DIFFRACTION LINES:
  3.07 (100)
  2.74 ( 60)
  2.08 ( 60)
OPTICAL CONSTANTS:
  $\alpha$ = 1.732
  $\beta$ = 1.737
  $\gamma$ = 1.744
  (+)2V ≈ 83°
HARDNESS: 7
DENSITY: 3.49 (Meas.)
        3.48 (Calc.)
CLEAVAGE: None. Fracture subconchoidal to uneven.

HABIT: Subhedral to euhedral crystals ranging from less than 1 mm to 1.2 cm on an edge. Dominant form is the positive tetrahedron (111), commonly modified by faces of (001), (110), and (1$\bar{1}$1).
COLOR-LUSTER: Colorless to deep purple. Vitreous.
MODE OF OCCURRENCE: Found in brine returns from a gas storage well on Barber's Hill salt dome, Chambers County, Texas.
BEST REF. IN ENGLISH: Honea, Russell M., and Beck, Frank R., *Am. Min.*, **47**: 665-671 (1962).

## CHAMOSITE (a Septechlorite)
$(Mg, Fe^{2+})_3 Fe_3^{3+} (AlSi_3) O_{10} (OH)_8$
Chlorite Group

CRYSTAL SYSTEM: Monoclinic–Hexagonal
CLASS: 2/m (monoclinic)
        3m (hexagonal)
Z: 1 (monoclinic)
LATTICE CONSTANTS:
  (monoclinic)
  $a \sim 5.40$
  $b \sim 9.33$
  $c \sim 7.04$
  $\beta \approx 104.5°$
3 STRONGEST DIFFRACTION LINES:
  7.05 (100)
  3.52 (100)
  2.52 ( 90)
Probably a mixture of monoclinic and hexagonal.
HARDNESS: ~3
DENSITY: 3-3.4 (Meas.)
        3.031 (Calc. monoclinic)
CLEAVAGE: Not determined.
HABIT: Massive; compact or oölitic.
COLOR-LUSTER: Green, greenish gray to black.
MODE OF OCCURRENCE: Occurs in sedimentary iron-stones in association with siderite and kaolinite; also in lateritic clay deposits in association with clay minerals and iron oxides. Typical occurrences are found at Chamoson in the Rhone valley, Valais, and at other places in Switzerland; also at localities in England, France, Germany, Czechoslovakia, and elsewhere.
BEST REF. IN ENGLISH: Deer, Howie, and Zussman, "Rock Forming Minerals," v. 3, p. 164-169, New York, Wiley, 1962.

## CHAOITE
C
Polymorphous with diamond, graphite, lonsdaleite

CRYSTAL SYSTEM: Hexagonal
CLASS: 6/m 2/m 2/m
SPACE GROUP: P6/mmm
Z: 168
LATTICE CONSTANTS:
  $a$ = 8.948
  $c$ = 14.078

3 STRONGEST DIFFRACTION LINES:
4.47 (100)
4.26 (100)
4.12 ( 80)
HARDNESS: Not determined.
DENSITY: 3.43 (Calc.)
CLEAVAGE: Not determined.
HABIT: As thin lamellae (3–15 microns wide), alternating with graphite and perpendicular to the {0001} face of graphite.
MODE OF OCCURRENCE: Occurs in association with graphite, zircon, and rutile in shocked graphite gneisses from Mottingen in the Ries Crater, Germany. Also observed from the Goalpara and Dyalpur carbonaceous chondrites.
BEST REF. IN ENGLISH: El Goresy, A., and Donnay, G., *Science,* **161**: 363–364 (1968).

# CHAPMANITE
$Fe_2Sb(SiO_4)_2(OH)$
Antimony analogue of bismutoferrite

CRYSTAL SYSTEM: Orthorhombic
3 STRONGEST DIFFRACTION LINES:
3.56 (100)
3.17 (100)
7.64 ( 90)
HARDNESS: up to 2½
DENSITY: 3.69–3.75
HABIT: As lath-shaped crystals; compact masses; powdery.
COLOR-LUSTER: Olive green to deep yellow. Streak greenish yellow to deep yellow.
MODE OF OCCURRENCE: Occurs intimately associated with silver at the Keeley Mine, South Lorrain, near Cobalt, Ontario, Canada; also found in veinlets up to 3 cm wide cutting through gneisses containing various amounts of admixed graphite at Smilkov, near Votice, central Bohemia, Czechoslovakia; with sulfide ore from Velardeña, Durango, Mexico; and associated with quartz at Bräunsdorf, Saxony, Germany.
BEST REF. IN ENGLISH: Čech, F., and Povondra, P., *Am. Min.*, **49**: 1499–1500 (1964).

# CHATHAMITE = Variety of skutterudite containing up to 12% Fe

# CHAVESITE
Hydrated Ca, Mn Phosphate

CRYSTAL SYSTEM: Triclinic
LATTICE CONSTANTS:
$a = 5.79$     $\alpha = 99°44'$
$b = 13.07$    $\beta = 108°3'$
$c = 5.49$     $\gamma = 91°18.5'$
3 STRONGEST DIFFRACTION LINES:
3.35  (100)
2.945 ( 30)
2.23  ( 30)

OPTICAL CONSTANTS:
$\alpha = 1.60$
$\beta = 1.62$
$\gamma = 1.65$
(+)2V large
HARDNESS: ~3
CLEAVAGE: Two good, in prism zone, nearly perpendicular to each other.
HABIT: Crystalline coatings of tiny thin platy crystals.
COLOR-LUSTER: Colorless.
MODE OF OCCURRENCE: Occurs as a thin coating on lithiophilite, associated with hureaulite and tavorite, at the Boqueirão pegmatite near Parelhas, Rio Grande do Norte, Brazil.
BEST REF. IN ENGLISH: Murdoch, Joseph, *Am. Min.*, **43**: 1148–1156 (1958).

# CHELKARITE
$CaMgB_2O_4Cl_2 \cdot 7H_2O$

CRYSTAL SYSTEM: Orthorhombic
CLASS: 2/m 2/m 2/m
SPACE GROUP: Pbca
LATTICE CONSTANTS:
$a = 13.69$
$b = 20.84$
$c = 8.26$
3 STRONGEST DIFFRACTION LINES:
3.531 (100)
10.42 ( 90)
2.209 ( 80)
OPTICAL CONSTANTS:
$\alpha' = 1.520$
$\gamma = 1.558$
HARDNESS: Not determined.
DENSITY: Not determined.
CLEAVAGE: Perfect parallel to elongation. Breaks into fine needles.
HABIT: Crystals flattened long prismatic, less than 15 mm long.
COLOR-LUSTER: Colorless. Transparent.
MODE OF OCCURRENCE: Occurs in the insoluble residues of carnallite-bischofite rock, with boracite, hilgardite, anhydrite, and carbonates, presumably in Kazakhstan.
BEST REF. IN ENGLISH: Avrova, N. P., Bocharov, V. M., Khalturina, I. I., and Yunosova, Z. R., *Am. Min.*, **56**: 1122 (1971).

# CHENEVIXITE
$Cu_2Fe_2(AsO_4)_2(OH)_4 \cdot H_2O$

CRYSTAL SYSTEM: Orthorhombic
CLASS: 2/m 2/m 2/m
SPACE GROUP: Pman
Z: 8
LATTICE CONSTANTS:
$a = 12.3$
$b = 15.4$
$c = 10.7$

3 STRONGEST DIFFRACTION LINES:
  3.53 (100)
  2.56 ( 70)
  2.46 ( 70)
OPTICAL CONSTANTS:
  mean index ~ 1.88
  HARDNESS: 3½–4½
  DENSITY: 3.93 (Meas.)
            3.95 (Calc.)
  CLEAVAGE: Fracture subconchoidal to uneven. Brittle.
  HABIT: Massive; compact earthy to opaline.
  COLOR-LUSTER: Dark green; also olive green to greenish yellow. Subtranslucent. Greasy to dull. Streak greenish yellow.
  MODE OF OCCURRENCE: Occurs as thick masses associated with tyrolite, azurite, chrysocolla, and conichalcite, at the Eureka and Mammoth mines, Tintic district, Juab County, Utah; with olivenite at Chuquicamata, Chile; at Tsumeb and Klein Spitzkopje, South West Africa; at Tamanrasset, Hoggar, southern Algeria; and at Broken Hill, New South Wales, Australia.
  BEST REF. IN ENGLISH: Palache, et al., "Dana's System of Mineralogy," 7th Ed., v. II, p. 840–841, New York, Wiley, 1951.

## CHERALITE

(Th, Ca, Ce, La, U, Pb)(PO$_4$, SiO$_4$)
Monazite Group

CRYSTAL SYSTEM: Monoclinic
CLASS: 2/m
SPACE GROUP: P2$_1$/n
Z: 4
LATTICE CONSTANTS:
  $a$ = 6.717
  $b$ = 6.920
  $c$ = 6.434
  $\beta$ = 103°50'
3 STRONGEST DIFFRACTION LINES:
  3.07 (100)
  3.26 ( 90)
  2.86 ( 90)
OPTICAL CONSTANTS:
  $\alpha$ = 1.779
  $\beta$ = 1.780
  $\gamma$ = 1.816
  2V = 17.4–19°
  HARDNESS: 5
  DENSITY: 5.3 (Meas.)
            5.41 (Calc.)
  CLEAVAGE: {010} distinct
            {100} difficult } tentative
    Parting on {001} poor.
Fracture uneven. Brittle.
  HABIT: Masses up to 2 inches across.
  COLOR-LUSTER: Dark green to pale green; resinous to vitreous. Transparent.
  MODE OF OCCURRENCE: Occurs associated with black tourmaline, chrysoberyl, dark zircon, and smoky quartz in a kaolinized pegmatite dike at Kuttakuzhi in Halkulam taluk (parish), Travancore, southern India. It also occurs

sparsely in the adjacent wall-rock (kaolinized granite-gneiss) and in surface wash.
  BEST REF. IN ENGLISH: Bowie, S. H. U., and Horne, J. E. T., *Min. Mag.*, **30**: 93–99 (1953).    Finney, J. J., and Rao, N. Nagaraja, *Am. Min.*, **52**: 13–19 (1967).

## CHERNOVITE

YAsO$_4$
Xenotime Group

CRYSTAL SYSTEM: Tetragonal
CLASS: 4/m 2/m 2/m
SPACE GROUP: I4$_1$/amd
Z: 4
LATTICE CONSTANTS:
  $a$ = 7.039
  $c$ = 6.272
3 STRONGEST DIFFRACTION LINES:
  3.519 (100)
  2.644 (100)
  1.811 (100)
OPTICAL CONSTANTS:
  $\omega$ = 1.783
  $\epsilon$ = 1.879  (Na)
    (+)
  HARDNESS: 4.2–4.7, av. 4.5
  DENSITY: 4.866 (Calc.)
  CLEAVAGE: {010} perfect
Sometimes a gliding plane on {001}, rarely on {110}. Brittle.
  HABIT: Crystals prismatic, mostly about 0.25 mm long, maximum 0.65 mm long; forms {010}, {011}, and {001}, occasional narrow rough faces of {110}.
  COLOR-LUSTER: Colorless to pale yellow, with color zones. Vitreous.
  MODE OF OCCURRENCE: Occurs with molybdenian scheelite, albite, calcite, garnet, hastingsite, pyrolusite, and hematite in piemontite veinlets in liparite porphyry at the source of the Nyarta-syu-yu River, eastern Telpos-iz, near Polar Urals.
  BEST REF. IN ENGLISH: Goldin, B. A., Yushkin, N. P., and Fishman, M. V., *Am. Min.*, **53**: 1777 (1968).

## CHERT = Variety of quartz

## CHERVETITE

Pb$_2$V$_2$O$_7$

CRYSTAL SYSTEM: Monoclinic
CLASS: 2/m
SPACE GROUP: P2$_1$/a
Z: 4
LATTICE CONSTANTS:
  $a$ = 13.47
  $b$ = 7.32
  $c$ = 6.95
  $\beta$ = 107°25'
3 STRONGEST DIFFRACTION LINES:
  3.441 (100)
  3.428 (100)
  3.206 ( 80)

OPTICAL CONSTANTS:
$n \sim 2.2\text{-}2.6$
$(-)2V = 65°\text{-}75°$
HARDNESS: $< 3$
DENSITY: 6.30-6.49 (Meas.)
6.46 (Calc.)
CLEAVAGE: {100} and {010} doubtful.

HABIT: Crystals from less than 1 mm to several centimeters, always twinned on {100}. Also as pseudomorphs after francevillite. Prominent faces {100}, {221}, {2$\bar{2}$1}, {001}.

COLOR-LUSTER: Colorless to gray to brown; adamantine. Streak white.

MODE OF OCCURRENCE: Occurs associated with francevillite and more rarely with wulfenite in the zone of oxidation of the Mounana uranium mine, Dept. Haut-Ogooue, Republic of Gabon.

BEST REF. IN ENGLISH: Bariand, P., Chantret, F., Pouget, R., and Rimsky, A., *Am. Min.*, **48**: 1416 (1963).

## CHESSYLITE = Azurite

## CHEVKINITE
$Ce_4 Fe_2^{2+} Ti_3 (Si_2 O_7)_2 O_8$
Dimorph of perrierite

CRYSTAL SYSTEM: Monoclinic
CLASS: 2/m
SPACE GROUP: C2/m
Z: 2
LATTICE CONSTANTS:
$a = 13.56$
$b = 5.82$
$c = 11.21$
$\beta = 100°45'$
3 STRONGEST DIFFRACTION LINES:
3.17 (100)   2.74 (100)
3.14 (100)   3.20 ( 80)
2.71 (100)   4.67 ( 80)
OPTICAL CONSTANTS:
$\alpha = 1.967\text{-}1.973$
$\beta \simeq 2.02$
$\gamma \simeq 2.05$
$(-)2V = $ moderate
HARDNESS: 5-6
DENSITY: 4.3-4.67 (Meas.)
CLEAVAGE: None. Fracture conchoidal. Brittle.

HABIT: As slender pencil-shaped prisms, or as short stubby prisms to 0.5 mm long; occasionally twinned. Usually in irregular masses up to 10 cm in size.

COLOR-LUSTER: Dark reddish brown to velvety black. Nearly opaque. Resinous.

MODE OF OCCURRENCE: Occurs in pumice fragments and airfall ash associated with large ash-flow sheets from several areas of Cenezoic volcanism in the western United States, including Yellowstone Park, Wyoming; Jemez Mountains area, New Mexico; and southern and central Nevada. Also near Martin's Store, Nelson County, Virginia; in fayalite-quartz syenites at the Devil's Slide ring dike near Stark, New Hampshire; as an accessory mineral in alkali granite and syenite pegmatites on the Kola peninsula and elsewhere in the USSR; and in dolomite in contact with alkali granite in Mongolia.

BEST REF. IN ENGLISH: Vlasov, K. A., "Mineralogy of Rare Elements," v. II, p. 309-312, Israel Program for Scientific Translations, 1966.   Izett, G. A., and Wilcox, R. E., *Am. Min.*, **53**: 1558-1567 (1968).

## CHIASTOLITE = Variety of andalusite

## CHIKLITE = Ferririchterite

## CHILDRENITE
$(Fe, Mn)AlPO_4 (OH)_2 \cdot H_2O$
Childrenite Series

CRYSTAL SYSTEM: Orthorhombic
CLASS: mm2
SPACE GROUP: Bba2
Z: 8
LATTICE CONSTANTS:
$a = 10.38$
$b = 13.36$
$c = 6.911$
3 STRONGEST DIFFRACTION LINES:
2.81 (100)
5.27 ( 40)
2.42 ( 40)
OPTICAL CONSTANTS:
$\alpha = 1.63\text{-}1.645$
$\beta = 1.65\text{-}1.68$
$\gamma = 1.66\text{-}1.685$
$(-)2V = 40\text{-}45°$
HARDNESS: 5
DENSITY: 3.20 (Pure Fe end-member.)
3.186 (Calc.—Barnes, W., *Am. Min.*, **34**: 12-18 [1949] ).
CLEAVAGE: {100} poor
Fracture uneven to subconchoidal.

HABIT: Crystals equant or pyramidal to short prismatic [001] and thick tabular {010}; also platy {100}. Often doubly terminated. Twinning by reflection on {100} and {001}.

COLOR-LUSTER: Brown and yellowish brown. Transparent to translucent. Vitreous to resinous.

MODE OF OCCURRENCE: Occurs as fine crystals in hydrothermal vein deposits at several localities in Cornwall, England; in granite pegmatite at Greifenstein, Saxony, Germany and at the Hugo mine, Keystone and Helen Beryl mine, Custer, South Dakota; and as exceptional short prismatic crystals up to several centimeters in length in Minas Geraes, Brazil.

BEST REF. IN ENGLISH: "Dana's System of Mineralogy," 7th Ed., v. 2, p. 936-939, New York, Wiley, 1951. Hurlbut, Cornelius S. Jr., *Am. Min.*, **35**: 793-805 (1950).

## CHILENITE = Variety of silver containing bismuth

**CHILLAGITE** = Tungstenian wulfenite
$Pb(Mo, W)O_4$

**CHINGLUSUITE (Inadequately described mineral)**
$Na_4Mn_5Ti_3Si_{14}O_{41} \cdot 9H_2O$ (?)

CRYSTAL SYSTEM: Unknown
3 STRONGEST DIFFRACTION LINES: Metamict.
OPTICAL CONSTANT:
  $N = 1.582$
  HARDNESS: 2-3
  DENSITY: 2.151 (Meas.)
  CLEAVAGE: None.
Fracture uneven. Brittle.
  HABIT: Massive; as grains up to 5 mm in diameter.
  COLOR-LUSTER: Black. Opaque. Resinous. Streak brownish.
  MODE OF OCCURRENCE: Occurs in carbonatites in association with hackmanite, eudialyte, lamprophyllite, ramsayite, and other minerals in the Chinglusuai River valley, Lovozero Tundra, Kola peninsula, USSR.
  BEST REF. IN ENGLISH: Gerasimovsky, V. I., *Am. Min.*, **25**: 253 (1940).

**CHINOITE** = Libethenite

**CHIOLITE**
$Na_5Al_3F_{14}$

CRYSTAL SYSTEM: Tetragonal
CLASS: 4/m 2/m 2/m
SPACE GROUP: P4/mnc
Z: 2
LATTICE CONSTANTS:
  $a = 7.005$ kX
  $c = 10.39$
3 STRONGEST DIFFRACTION LINES:
  2.91 (100)
  5.18 ( 80)
  2.32 ( 70)
OPTICAL CONSTANTS:
  $\omega = 1.3486$
  $\epsilon = 1.3424$
  (−)
HARDNESS: 3½-4
DENSITY: 2.998 (Meas.)
        2.989 (Calc.)
CLEAVAGE: {001} perfect
        {011} distinct
  HABIT: Crystals dipyramidal, very minute; rare. Commonly as embedded masses.
  COLOR-LUSTER: Colorless, white. Transparent to translucent. Vitreous; pearly on basal cleavage.
  MODE OF OCCURRENCE: Occurs associated with cryolite in a large pegmatite at Ivigtut, Greenland, and in a cryolite-pegmatite associated with cryolithionite, thomsenolite, and other minerals, at Miask, Urals, USSR.
  BEST REF. IN ENGLISH: Palache, et al., "Dana's System of Mineralogy," 7th Ed., v. II, p. 123-124, New York, Wiley, 1951.

**CHIVIATITE (inadequately described mineral)**
$Pb_3Bi_8S_{15}$ (?)

CRYSTAL SYSTEM: Unknown
  HARDNESS: 2-3
  DENSITY: 6.92-7.15 (Meas.)
  CLEAVAGE: Three, parallel to elongation.
  HABIT: As foliated or columnar aggregates up to 4 cm long.
  COLOR-LUSTER: Lead gray. Opaque. Metallic.
  MODE OF OCCURRENCE: Occurs with pyrite and barite at Chiviato, and elsewhere in Peru; and with quartz in a cordierite-bearing amphibole rock at Fahlun, Sweden.
  BEST REF. IN ENGLISH: Palache, et al., "Dana's System of Mineralogy," 7th Ed., v. I, p. 474-475, New York, Wiley, 1944.

**CHKALOVITE**
$Na_2BeSi_2O_6$

CRYSTAL SYSTEM: Orthorhombic
CLASS: mm2
SPACE GROUP: Fdd2
Z: 24
LATTICE CONSTANTS:
  $a = 21.1$
  $b = 21.1$
  $c = 6.87$
3 STRONGEST DIFFRACTION LINES:
  4.02 (100 )
  2.48 (100 )
  2.76 ( 90b)
OPTICAL CONSTANTS:
  $\alpha = 1.544$
  $\beta = ?$
  $\gamma = 1.549$
  (+)2V = 78°
HARDNESS: ~6
DENSITY: 2.662 (Meas.)
        2.70 (Calc.)
CLEAVAGE: {100} imperfect
Fracture conchoidal. Brittle.
  HABIT: Crystals subhedral. Usually massive, in segregations up to 12 cm in diameter.
  COLOR-LUSTER: White. Transparent to translucent. Vitreous to somewhat greasy.
  MODE OF OCCURRENCE: Occurs sparingly in sodalite-syenite pegmatites in association with sodalite, ussingite, schizolite, steenstrupine, and natrolite. Found only in the Julianehaab district, Greenland, and on the Kola peninsula, USSR.
  BEST REF. IN ENGLISH: Vlasov, K. A., "Mineralogy of Rare Elements," v. II, p. 116-117, Israel Program for Scientific Translations, 1966.

**CHLOANTHITE** = Nickel-skutterudite

**CHLORALUMINITE**
$AlCl_3 \cdot 6H_2O$

CRYSTAL SYSTEM: Hexagonal
CLASS: $\bar{3}$ 2/m

SPACE GROUP: $R\bar{3}c$
OPTICAL CONSTANTS:
  $\omega = 1.560$
  $\epsilon = 1.507$
  $(-)$
HARDNESS: Not determined.
DENSITY: No data.
CLEAVAGE: No data.
HABIT: Crystals rhombohedral; as crystalline crusts and stalactites.
COLOR-LUSTER: Colorless, white, yellowish.
Deliquescent.
MODE OF OCCURRENCE: Found on Mt. Vesuvius, Italy in the eruptions of 1872 and 1906.
BEST REF. IN ENGLISH: Palache, et al., "Dana's System of Mineralogy," 7th Ed., v. II, p. 50, New York, Wiley, 1951.

# CHLORAPATITE
$Ca_5(PO_4)_3Cl$
Apatite Series

CRYSTAL SYSTEM: Hexagonal; Monoclinic
CLASS: 6/m (hex); 2/m (mono)
SPACE GROUP: $P6_3/m$ (hex); $P2_1/a$ (mono)
Z: 2 (hex), 4 (mono.)
LATTICE CONSTANTS:

| (hexagonal) | (monoclinic) |
|---|---|
| $a = 9.634$ | $a = 19.210$ |
| $c = 6.778$ | $b = 6.785$ |
| | $c = 9.605$ |
| | $\beta = 120°$ |

3 STRONGEST DIFFRACTION LINES:
  2.779 (100)
  2.861 ( 60)    Synthetic
  1.836 ( 20)
OPTICAL CONSTANTS:

| (hexagonal) | (monoclinic) |
|---|---|
| $\omega = 1.667$ | $\alpha = 1.665$ |
| $\epsilon = 1.664$ | $\gamma \sim \beta = 1.667$ |
| $(-)$ | $(-)2V \simeq 10°$ |

HARDNESS: 5
DENSITY: 3.1-3.2 (Meas.)
  3.174 (Calc.)
CLEAVAGE: {0001} indistinct
  {10$\bar{1}$0} trace
Fracture conchoidal to uneven. Brittle.
HABIT: Crystals short to long prismatic.
COLOR-LUSTER: Pinkish white, pale yellow. Transparent to translucent. Vitreous; sometimes chalky and dull due to included talc.
MODE OF OCCURRENCE: Occurs in calc-silicate marbles associated with actinolite, diopside, and lesser amounts of calcite, quartz, and talc at Bob's Lake, Oso Township, Frontenac County, Ontario, Canada. It also is found in veins in gabbroic rocks in southeastern Norway; at Kurokura, Kanagawa Prefecture, Japan; and as microgranules in meteorites.
BEST REF. IN ENGLISH: Hounslow, A. W., and Chao, G. Y., *Can. Min.*, **10**: 252-259 (1970).

# CHLORARGYRITE = Cerargyrite

# CHLORASTROLITE = Pumpellyite

# CHLORITE = Group name of general formula:
$A_6(AlSi_3)O_{10}(OH)_8$; A = Mg, $Fe^{2+}$, $Fe^{3+}$, Mn
See:

| | |
|---|---|
| Amesite | Greenalite |
| Brunsvigite | Kämmererite |
| Chamosite | Manandonite |
| Clinochlore | Nimite |
| Cookeite | Pennantite |
| Corundophilite | Penninite |
| Cronstedtite | Ripidolite |
| Daphnite | Sheridanite |
| Delessite | Sudoite |
| Diabantite | Thuringite |
| Gonyerite | |

# CHLORITOID
$(Mg, Fe^{2+})_2Al_4Si_2O_{10}(OH)_4$

CRYSTAL SYSTEM: Triclinic, monoclinic
CLASS: $\bar{1}$ and 2/m
SPACE GROUP: $P\bar{1}$ and C2/c
Z: 4 (triclinic)
  4 (monoclinic)
LATTICE CONSTANTS:
  (monoclinic)
  $a = 9.52$
  $b = 5.47$
  $c = 18.19$
  $\beta = 101°39'$
3 STRONGEST DIFFRACTION LINES:

| (monoclinic) | (triclinic) |
|---|---|
| 4.47 (100) | 4.47 (100) |
| 2.96 ( 90) | 2.46 ( 90) |
| 1.58 ( 80) | 2.97 ( 80) |

OPTICAL CONSTANTS:
  $\alpha = 1.713-1.730$
  $\beta = 1.719-1.734$
  $\gamma = 1.723-1.740$
  $2V_\gamma = 45°-68°$
HARDNESS: 6½
DENSITY: 3.61 (Meas.) 3.60 (Calc.) Monoclinic
  3.58 (Meas.) 3.56 (Calc.) Triclinic
CLEAVAGE: {001} perfect
  {110} distinct
  {010} parting
Laminae brittle.
HABIT: Crystals tabular, pseudohexagonal, rare. Usually massive, foliated, or as thin scales or plates. Twinning on {001}, common; often lamellar.
COLOR-LUSTER: Dark gray, greenish gray to greenish black. Translucent. Pearly on cleavage.
MODE OF OCCURRENCE: Occurs in regionally metamorphosed rocks such as mica schist, phyllite, and quartzite; as a hydrothermal alteration product in lavas and other rocks; and as a vein mineral. Common associated minerals are muscovite, chlorite, garnet, staurolite, and kyanite.

Typical occurrences are found in the Black Hills, South Dakota; Marquette County, Michigan; Lancaster County, Pennsylvania; Duchess County, New York; Kent County, Rhode Island; in the Deep River region of North Carolina; and elsewhere. Also in Ontario and Quebec, Canada; Scotland; Belgium; France; Italy; Austria; Switzerland; USSR; Japan; and Australia.

BEST REF. IN ENGLISH: Deer, Howie, and Zussman, "Rock Forming Minerals," v. 1, p. 161–170, New York, Wiley, 1962.


## CHLORMANGANOKALITE
$K_4MnCl_6$

CRYSTAL SYSTEM: Hexagonal
CLASS: $\bar{3}m$
SPACE GROUP: $R\bar{3}c$
Z: 6
LATTICE CONSTANTS:
  $a = 11.93$
  $c = 14.79$
3 STRONGEST DIFFRACTION LINES:
  2.55 (100)
  2.69 ( 80)
  5.90 ( 50)
OPTICAL CONSTANTS:
  $\omega = 1.59$
  $\epsilon - \omega$ very weak
  uniaxial (+)
  HARDNESS: 2½
  DENSITY: 2.31 (Meas.)
  CLEAVAGE: None.
Fracture conchoidal. Brittle.
  HABIT: Crystals rhombohedrons, sometimes in parallel groups.
  COLOR-LUSTER: Pale yellow to canary yellow. Transparent. Vitreous.
  MODE OF OCCURRENCE: Found lining cavities in blocks of scoria from Mt. Vesuvius, Italy, associated with halite, chlorocalcite, sylvite, and hematite crystals.
  BEST REF. IN ENGLISH: Palache, et al., "Dana's System of Mineralogy," 7th Ed., v. II, p. 109, New York, Wiley, 1951.


## CHLOROCALCITE
$KCaCl_3$

CRYSTAL SYSTEM: Orthorhombic
CLASS: 2/m 2/m 2/m
SPACE GROUP: Pnma
Z: 4
LATTICE CONSTANTS:
  $a = 7.551$
  $b = 10.442$
  $c = 7.251$
3 STRONGEST DIFFRACTION LINES:
  2.61 (100)
  3.14 ( 35)
  3.70 ( 30)
OPTICAL CONSTANTS:
  $\beta = 1.568$
  (–)
  HARDNESS: 2½–3
  DENSITY: 2.155 (Calc.)
  CLEAVAGE: Cube-like, with one direction better than the other two.
  HABIT: Crystals pseudocubic, sometimes with octahedral or dodecahedral-like modifications; also prismatic or tabular.
  COLOR-LUSTER: White, sometimes stained violet. Transparent.

  Deliquescent; taste bitter.
  MODE OF OCCURRENCE: Occurs as a sublimation product in fumaroles on Mt. Vesuvius, Italy.
  BEST REF. IN ENGLISH: Palache, et al., "Dana's System of Mineralogy," 7th Ed., v. II, p. 91–92, New York, Wiley, 1951.


## CHLOROMAGNESITE (Chlormagnesite)
$MgCl_2$

CRYSTAL SYSTEM: Hexagonal
CLASS: $\bar{3}$ 2/m
SPACE GROUP: $R\bar{3}m$
Z: 3
LATTICE CONSTANTS:
  $a = 3.58$ kX
  $c = 17.59$
3 STRONGEST DIFFRACTION LINES:
  2.56 (100)
  1.82 ( 63)
  2.96 ( 57)
OPTICAL CONSTANTS:
  $\omega = 1.675$
  $\epsilon = 1.59$
  (–)
  HARDNESS: Not determined.
  DENSITY: 2.325 (Meas. synthetic)
              2.44 (Calc.)
  CLEAVAGE: Not determined.
  HABIT: As deliquescent crusts composed of minute hexagonal plates.
  COLOR-LUSTER: Colorless, white.
  MODE OF OCCURRENCE: Occurs associated with sylvite and halite as a sublimate in fumaroles on Mt. Vesuvius, Italy.
  BEST REF. IN ENGLISH: Palache, et al., "Dana's System of Mineralogy," 7th Ed., v. II, p. 41, New York, Wiley, 1951.


## CHLOROPAL = Nontronite


## CHLOROPHOENICITE
$(Mn, Zn)_5 AsO_4 (OH)_7$

CRYSTAL SYSTEM: Monoclinic
CLASS: 2/m
SPACE GROUP: C2/m
Z: 2

LATTICE CONSTANTS:
  $a = 22.98$
  $b = 3.32$
  $c = 7.32$
  $\beta = 106°$
OPTICAL CONSTANTS:
  $a = 1.682$
  $\beta = 1.690$
  $\gamma = 1.697$
  $(-)2V \approx 83°$
HARDNESS: 3–3½
DENSITY: 3.46 (Meas.)
         3.47 (Calc.)
CLEAVAGE: {100} distinct
Brittle.
HABIT: Needles and matted fibers; crystals slender prismatic, elongated and striated [010].
COLOR-LUSTER: Light gray-green in daylight; purple-red in artificial light. Vitreous to pearly. Translucent.
MODE OF OCCURRENCE: Occurs associated with pyrochroite, gageite, willemite, calcite, tephroite, and leucophoenicite in zinc-manganese-iron oxide and silicate orebodies of the pyrometasomatic type in marbles, at Franklin and Sterling Hill, New Jersey.
BEST REF. IN ENGLISH: Palache, et al., "Dana's System of Mineralogy," 7th Ed., v. II, p. 778–780, New York, Wiley, 1951. Moore, Paul B., *Am. Min.*, **53**: 1110–1119 (1968).

# CHLOROTHIONITE
$K_2Cu(SO_4)Cl_2$

CRYSTAL SYSTEM: Orthorhombic
CLASS: 2/m 2/m 2/m
SPACE GROUP: Pmnb
Z: 4
LATTICE CONSTANTS:
  (synthetic)
  $a = 6.105$ kX
  $b = 7.697$
  $c = 16.132$
OPTICAL CONSTANTS:
  $(+)2V$ = moderately large
HARDNESS: 2½
DENSITY: 2.67 (Meas. Synthetic)
CLEAVAGE: {100}, {010}, {001} good
HABIT: As crystalline incrustations.
COLOR-LUSTER: Bright blue.

Soluble in water.
MODE OF OCCURRENCE: Found in fumaroles on Mount Vesuvius, Italy.
BEST REF. IN ENGLISH: Palache, et al., "Dana's System of Mineralogy," 7th Ed., v. II, p. 547, New York, Wiley, 1951.

# CHLOROTILE
$Cu_3(AsO_4)_2 \cdot 6H_2O$

CRYSTAL SYSTEM: Hexagonal
CLASS: 6/m or 6

SPACE GROUP: $P6_3/m$ or $P6_3$
Z: 4
LATTICE CONSTANTS:
  $a = 13.61$
  $b = 5.90$
3 STRONGEST DIFFRACTION LINES:
  12.1 (100)
  2.46 ( 90)
  4.48 ( 80)
OPTICAL CONSTANTS:
  $\omega = 1.725$–1.728
  $\epsilon > 1.81$
  $(-)$
HARDNESS: Soft.
DENSITY: 3.73 (Meas.)
         4.05 (Calc.)
CLEAVAGE: Not determined.
HABIT: Crystals prismatic, minute. Also fibrous and massive.
COLOR-LUSTER: Pale green, emerald green, bluish green. Transparent to translucent. Aggregates dull.
MODE OF OCCURRENCE: Occurs as a secondary mineral in the oxidation zone of ore deposits. Found in Germany at Schneeberg and Zinnwald in Saxony, and at the Clara mine, Black Forest. Also found at the Dome Rock copper mine, South Australia.
BEST REF. IN ENGLISH: Dana, E. S., "System of Mineralogy," 6th Ed., p. 814, (1892).

# CHLOROXIPHITE
$Pb_3CuCl_2O_2(OH)_2$

CRYSTAL SYSTEM: Monoclinic
CLASS: 2/m
SPACE GROUP: $P2_1/m$
Z: 2
LATTICE CONSTANTS:
  $a = 10.34$
  $b = 5.73$
  $c = 6.52$
  $\beta = 97°11'$
3 STRONGEST DIFFRACTION LINES:
  2.86 (100)
  10.3 ( 80)
  3.84 ( 80)
OPTICAL CONSTANTS:
  $\alpha = 2.16$
  $\beta = 2.24$
  $\gamma = 2.25$
  $(-)2V \sim 70°$
HARDNESS: 2½
DENSITY: 6.93 (Meas.)
         7.07 (Calc.)
CLEAVAGE: {$\bar{1}01$} perfect
          {100} distinct
Very brittle.
HABIT: As bladed crystals, up to 3.0 cm long, 1 cm across, and 1 mm thick, elongated along $c$-axis and flattened on {$\bar{1}01$}. In subparallel groupings.
COLOR-LUSTER: Dull olive green. Translucent. Resinous to adamantine. Streak pale yellowish green.

MODE OF OCCURRENCE: Occurs as a secondary mineral embedded in mendipite at Higher Pitts farm, Priddy, Mendip Hill, Somerset, England.

BEST REF. IN ENGLISH: Palache, et al., "Dana's System of Mineralogy," 7th Ed., v. II, p. 84-85, New York, Wiley, 1951.

## CHONDRODITE
$Mg_5Si_2O_8(F,OH)_2$
Humite Group

CRYSTAL SYSTEM: Monoclinic
CLASS: 2/m
SPACE GROUP: $P2_1/c$
Z: 2
LATTICE CONSTANTS:
$a = 7.87$
$b = 4.73$
$c = 10.27$
$\beta = 109°02'$
3 STRONGEST DIFFRACTION LINES:
2.272 (100)
2.252 (100)
1.737 ( 95)
OPTICAL CONSTANTS:
$\alpha = 1.592-1.615$
$\beta = 1.602-1.627$
$\gamma = 1.621-1.646$
$(+)2V_\gamma = 71°-85°$
HARDNESS: 6-6½
DENSITY: 3.16-3.26 (Meas.)
3.177 (Calc.)
CLEAVAGE: {100} indistinct
Fracture uneven to subconchoidal. Brittle.
HABIT: Crystals varied in habit, usually highly modified. Also massive. Twinning on {001} common, lamellar.
COLOR-LUSTER: Yellow, brown, red. Transparent to translucent. Vitreous.
MODE OF OCCURRENCE: Occurs chiefly in contact zones in limestone or dolomite; rarely in carbonatites. Found in Riverside and San Bernardino counties, California; as superb crystals at the Tilly Foster iron mine, Brewster, Putnam County, New York; at the Cardiff uranium mine, Wilberforce, Ontario, Canada; at Kafveltorp, Örebro, Sweden; near Pargas and elsewhere in Finland; at Mte. Somma, Italy; and in carbonatite at Loolekop, eastern Transvaal.
BEST REF. IN ENGLISH: Deer, Howie, and Zussman, "Rock Forming Minerals," v. 1, p. 47-58, New York, Wiley, 1962. Gibbs, G. V., Ribbe, P. H., and Anderson, C. P., *Am. Min.*, **55**: 1182-1194 (1970). *Am. Min.*, **54**: 391-411 (1969).

## CHRISTOPHITE = Variety of sphalerite containing about 26% Fe

## CHROMATITE
$CaCrO_4$

CRYSTAL SYSTEM: Tetragonal

CLASS: 4/m 2/m 2/m
SPACE GROUP: $I4_1/amd$
Z: 4
LATTICE CONSTANTS:
$a = 7.242$
$c = 6.290$
3 STRONGEST DIFFRACTION LINES:
3.62 (100)
2.68 ( 55)
1.85 ( 45)
OPTICAL CONSTANTS:
$\omega = 1.81-1.85$
$\epsilon = 1.84-1.88$
(+)
DENSITY: 3.142 (Calc.)
CLEAVAGE: None.
Fracture conchoidal.
HABIT: Finely crystalline; grain size 30-100 $\mu$.
COLOR-LUSTER: Yellow.
MODE OF OCCURRENCE: The mineral occurs with gypsum in Upper Cretaceous limestones and marls along the Jerusalem-Jericho highway in Jordan.
BEST REF. IN ENGLISH: Eckhardt, F. J., and Heimbach, W., *Am. Min.*, **49**: 439 (1964).

## CHROME-AUGITE =Variety of augite containing up to 3% $Cr_2O_3$

## CHROMINIUM = Phoenicochroite

## CHROMITE
$FeCr_2O_4$
Spinel Group

CRYSTAL SYSTEM: Cubic
CLASS: 4/m $\bar{3}$ 2/m
SPACE GROUP: Fd3m
Z: 8
LATTICE CONSTANT:
(synthetic)
$a = 8.344$
3 STRONGEST DIFFRACTION LINES:
2.52 (100)
1.60 ( 90)
1.46 ( 90)
OPTICAL CONSTANT:
$N = 2.08$
HARDNESS: 5½
DENSITY: 4.5-4.8 (Meas.)
5.09 (Calc. $FeCr_2O_4$)
CLEAVAGE: None.
Fracture uneven. Brittle.
HABIT: Crystals octahedral, sometimes modified by cube, ranging up to 1 cm along an edge. Usually massive, compact to fine granular.
COLOR-LUSTER: Black. Opaque. Metallic. Streak brown.
Sometimes weakly magnetic.

MODE OF OCCURRENCE: Occurs chiefly in olivine-rich (commonly serpentinized) igneous rocks, often associated with pyroxene, talc, magnetite, pyrrhotite, and uvarovite; also in placer deposits, and as a common component of meteorites. Found widespread in California, notably in the Coast Ranges from Santa Barbara County northward; also at localities in Oregon, Washington, Wyoming, Pennsylvania, Maryland, and North Carolina. Other notable deposits occur in Canada, Cuba, Norway, France, Bulgaria, Yugoslavia, USSR, Asia Minor, Rhodesia, India, Philippines, and New Caledonia.

BEST REF. IN ENGLISH: Palache, et al., "Dana's System of Mineralogy," 7th Ed., v. I, p. 709–712, New York, Wiley, 1944.

# CHRYSOBERYL
$BeAl_2O_4$

CRYSTAL SYSTEM: Orthorhombic
CLASS: 2/m 2/m 2/m
SPACE GROUP: Pmnb
Z: 4
LATTICE CONSTANTS:
  $a \simeq 5.47$
  $b \simeq 9.39$
  $c \simeq 4.42$
3 STRONGEST DIFFRACTION LINES:
  3.24 (100)
  2.091 ( 90)
  1.619 ( 80)
OPTICAL CONSTANTS:
  $\alpha = 1.746$
  $\beta = 1.748$
  $\gamma = 1.756$
  $(+)2V \simeq 70°$
HARDNESS: 8½
DENSITY: $3.75 \pm 0.10$
        3.69 (Calc.)
CLEAVAGE: {110} distinct
          {010} imperfect
          {001} indistinct
Fracture conchoidal to uneven. Brittle

HABIT: Crystals thin to thick tabular, flattened {001}; also short prismatic. Striated on {001} parallel to *a*-axis. Also granular massive. Twinning common on {130}, contact or interpenetrant, forming heart-shaped or pseudohexagonal twins.

COLOR-LUSTER: Various shades of yellowish green, yellow, gray, brown, blue-green, emerald green. Green varieties sometimes purplish red under artificial light. Occasionally chatoyant. Transparent to translucent. Vitreous.

MODE OF OCCURRENCE: Occurs chiefly in granite pegmatites, gneiss, mica schists, dolomitic marble, and in detrital sands and gravels. Transparent and chatoyant varieties are used extensively as gems. Found as large crystals and crystal aggregates at the Scott mine and elsewhere in Custer County, South Dakota; as large crystals near Golden, Colorado; and at a number of localities in Maine, New Hampshire, Connecticut, and New York. Other notable occurrences are found in Canada, Brazil, Ireland, Norway,

Italy, Switzerland, Germany, Finland, USSR, Zaire, Rhodesia, Madagascar, Ceylon, Burma, Japan, and Australia.

BEST REF. IN ENGLISH: Palache, et al., "Dana's System of Mineralogy," 7th Ed., v. I, p. 718–722, New York, Wiley, 1944.

# CHRYSOCOLLA
$Cu_2H_2Si_2O_5(OH)_4$

CRYSTAL SYSTEM: Orthorhombic (?)
3 STRONGEST DIFFRACTION LINES:
  1.494 (100)
  2.92 ( 80)
  8.3 ( 60)
OPTICAL CONSTANTS:
  $\alpha = 1.575–1.585$
  $\beta = 1.597$
  $\gamma = 1.598–1.635$
  $(-)2V$
HARDNESS: 2–4
DENSITY: 2.0–2.4 (Meas.)
CLEAVAGE: None.
Fracture uneven to conchoidal. Very brittle.

HABIT: Crystals acicular, microscopic, in radiating groups or close-packed aggregates. Usually cryptocrystalline; commonly opal-like; earthy. Sometimes botryoidal.

COLOR-LUSTER: Various shades of blue, blue-green, or green. Also brown to black when impure. Translucent to nearly opaque. Vitreous, earthy.

MODE OF OCCURRENCE: Occurs widespread as a common mineral in the oxidation zone of copper deposits. Typical occurrences are found in Pennsylvania, Michigan, and throughout the western United States, especially in Arizona, New Mexico, California, Nevada, Idaho, Utah, Montana, and Colorado. Other notable occurrences are found in Mexico, Chile, England, USSR, and Katanga, Zaire.

BEST REF. IN ENGLISH: Winchell and Winchell, "Elements of Optical Mineralogy," 4th Ed., Pt. 2, p. 420, New York, John Wiley & Sons, Inc., 1951.

# CHRYSOLITE
(Mg, Fe)$_2$SiO$_4$ (10–30 atomic percent Fe$^{2+}$)
Olivine Group

CRYSTAL SYSTEM: Orthorhombic
CLASS: 2/m 2/m 2/m
SPACE GROUP: Pbnm
Z: 4
LATTICE CONSTANTS:
  $a = 4.77$
  $b = 10.28$
  $c = 6.00$
3 STRONGEST DIFFRACTION LINES:
  2.51 (100)
  2.46 (100)
  3.88 ( 90)
HARDNESS: 7
DENSITY: 3.33–3.53 (Meas.)
CLEAVAGE: {010} imperfect
          {100} imperfect
Fracture conchoidal. Brittle.

HABIT: Crystals usually thick tabular, vertically striated, often with wedge-shaped terminations. Usually massive, compact or granular; in embedded grains, irregular or rounded. Twinning {100}, {011}, not common.

COLOR-LUSTER: Green, lemon yellow. Transparent to translucent. Vitreous to greasy. Streak uncolored.

MODE OF OCCURRENCE: Occurs as a common constituent of basic igneous rocks such as basalts, gabbros, and dolerites; in thermally metamorphosed impure limestones; and in meteorites. Typical occurrences are found in California, Arizona, New Mexico, Colorado, Minnesota, and North Carolina. Also in the Julianehaab district, Greenland; at localities in Scotland, Germany, Austria, Japan, Australia, South Africa; and as superb crystals on St. Johns Island in the Red Sea, Egypt.

BEST REF. IN ENGLISH: Deer, Howie, and Zussman, "Rock Forming Minerals," v. I, p. 1–33, New York, Wiley, 1962.

## CHRYSOPRASE = Variety of quartz

## CHRYSOTILE
$Mg_3Si_2O_5(OH)_4$
Serpentine group

CRYSTAL SYSTEM: Monoclinic
CLASS: 2/m
SPACE GROUP: Probably C2/m
Z: 4
LATTICE CONSTANTS:
$a \simeq 5.3$
$b \simeq 9.2$
$c \simeq 14.6$
$\beta \simeq 93°90'$
3 STRONGEST DIFFRACTION LINES:
7.36 (100)
3.66 (100)
2.456 ( 80)
OPTICAL CONSTANTS:
$\alpha = 1.532–1.549$
$\beta = ?$
$\gamma = 1.545–1.556$
$(-)2V$
HARDNESS: 2½
DENSITY: $\simeq 2.55$ (Meas.)
2.573 (Calc.)
CLEAVAGE: Fibrous parallel elongation. Fibers flexible, easily separable; rarely brittle.

HABIT: Massive; finely fibrous. Fibers tubular under electron microscope.

COLOR-LUSTER: White, gray, yellow, green, brownish. Translucent. Silky; sometimes greasy.

MODE OF OCCURRENCE: Occurs mainly as compact cross-fiber veinlets in serpentine. Typical occurrences are found in the Coast Ranges and on the west flank of the Sierra Nevada in California; near Globe, Gila County, Arizona; and at localities in New York and New Jersey. Major deposits are found in the metamorphic rocks of Quebec, especially at Thetford in Megantic County; also in Italy, Cyprus, USSR, Rhodesia, and Transvaal, South Africa.

BEST REF. IN ENGLISH: Deer, Howie, and Zussman, "Rock Forming Minerals," v. 3, p. 170–190, New York, Wiley, 1962.

## CHUDOBAITE
$(Na, K, Ca)(Mg, Zn, Mn)_2H(AsO_4)_2 \cdot 4H_2O$

CRYSTAL SYSTEM: Triclinic
CLASS: $\bar{1}$
SPACE GROUP: P$\bar{1}$
Z: 2
LATTICE CONSTANTS:
$a = 7.69$    $\alpha = 115°10'$
$b = 11.37$   $\beta = 95°54'$
$\gamma = 6.59$    $\gamma = 94°06'$
3 STRONGEST DIFFRACTION LINES:
10.163 (100)
2.979 ( 90)
3.440 ( 80)
OPTICAL CONSTANTS:
$\alpha = 1.583$
$\beta = 1.608$
$\gamma = 1.633$
$(-)2V = 79°$
HARDNESS: 2½–3
DENSITY: 2.94 (Meas.)
3.0 (Calc.)
CLEAVAGE: {010} very good
{100} good
HABIT: Crystals up to 0.5 cm in size. Forms: $a${100}, $b${010}, $c${001} dominant, also $m${110}, $n${120}, $L${180}, $d${101}, and $y${$\bar{1}$84}, and perhaps $x${186}.

COLOR-LUSTER: Pink, like kunzite.

MODE OF OCCURRENCE: Occurs associated with conichalcite, cuproadamite, and zincian olivenite at the 1000 meter level of the Tsumeb Mine, South West Africa.

BEST REF. IN ENGLISH: Strunz, H., *Am. Min.*, **45**: 1130 (1960).

## CHUKHROVITE
$Ca_3(Y, Ce)Al_2(SO_4)F_{13} \cdot 10H_2O$

CRYSTAL SYSTEM: Cubic
CLASS: 2/m$\bar{3}$
SPACE GROUP: Fd3
Z: 8
LATTICE CONSTANT:
$a = 16.80$
3 STRONGEST DIFFRACTION LINES:
2.193 (100)
1.834 (100)
3.261 ( 90)
OPTICAL CONSTANT:
$N = 1.440$
HARDNESS: ~3
DENSITY: 2.274–2.398, av. 2.353 (Meas.)
CLEAVAGE: Octahedral—indistinct. Fracture irregular. Brittle.

HABIT: Crystals range from less than 0.5 mm to 1.0 cm size. Forms: $a${100} and $o${111}.

COLOR-LUSTER: Colorless, white porcelain-like, sometimes with lilac tint; transparent to translucent. Vitreous when transparent, weakly pearly when white, greasy on fracture surfaces.

MODE OF OCCURRENCE: Occurs associated with gearksutite, creedite, halloysite, supergene fluorite, anglesite, and hydrous iron oxides in the secondary oxidation zone of the Kara-Oba molybdenum-tungsten deposit, Central Kazakhstan.

BEST REF. IN ENGLISH: Ermilova, L. P., Moleva, V. A., and Klevtsova, R. F., *Am. Min.*, **45**: 1132-1133 (1960).

# CHURCHITE = Weinschenkite

# CINNABAR
HgS
Dimorphous with metacinnabar

CRYSTAL SYSTEM: Hexagonal
CLASS: 32
SPACE GROUP: $P3_1 21$ or $P3_2 21$
Z: 3
LATTICE CONSTANTS:
$a = 4.160$
$c = 9.540$
3 STRONGEST DIFFRACTION LINES:
3.35 (100)
2.863 ( 95)
1.980 ( 35)
OPTICAL CONSTANTS:
$\omega = 2.905$
$\epsilon = 3.256$ $(\lambda = 598.5)$
(+)
HARDNESS: 2-2½
DENSITY: 8.090 (Meas.)
8.187 (Calc.)
CLEAVAGE: $\{10\bar{1}0\}$ perfect
Fracture conchoidal to uneven. Brittle; slightly sectile.

HABIT: Crystals rhombohedral to thick tabular; also stout to slender prismatic. Usually massive, fine-grained. Also as crystalline crusts or powdery coatings. Twinning on $\{0001\}$ common; often as penetration twins.

COLOR-LUSTER: Scarlet, brownish red, brown, black, lead gray. Transparent to translucent. Adamantine; also submetallic to dull. Streak scarlet to reddish brown.

MODE OF OCCURRENCE: Cinnabar, the chief ore of mercury, occurs in low-temperature deposits, commonly associated with native mercury, realgar, marcasite, pyrite, stibnite, calcite, quartz, and opal. Found in veins, sedimentary rocks, quartzites, trachytes, porphyries, and serpentine. In the United States it occurs widespread in California and Nevada, and is found in significant amounts in Douglas County, Oregon, at Terlingua, Texas, and in Pike County, Arkansas. Other major deposits occur in Mexico, Chile, Peru, Spain, Italy, Germany, USSR, and in the provinces of Hunan and Kweichow, China.

BEST REF. IN ENGLISH: Palache, et al., "Dana's System of Mineralogy," 7th Ed., v. I, p. 251-255, New York, Wiley, 1944.

# CITRINE = Variety of quartz

# CLARKEITE (Inadequately described mineral)
$(Na, K, Ca, Pb)_2 U_2 O_7 \cdot nH_2O$

CRYSTAL SYSTEM: Unknown
3 STRONGEST DIFFRACTION LINES:
3.17 (100)
3.34 ( 90)
5.77 ( 80)
OPTICAL CONSTANTS:
$\alpha = 1.997$
$\beta = 2.098$
$\gamma = 2.108$
$(-)2V = 30°-50°$
HARDNESS: 4-4½
DENSITY: 6.29-6.39 (Meas.)
CLEAVAGE: None observed.
Fracture conchoidal to splintery.

HABIT: Massive, microcrystalline, dense.

COLOR-LUSTER: Dark brown to chocolate brown, brownish yellow. Slightly waxy. Translucent. Streak yellowish brown.

MODE OF OCCURRENCE: Occurs as an alteration product of uraninite in pegmatites in the Spruce Pine district, North Carolina. Also found as pseudomorphs and partial pseudomorphs after uraninite crystals in fractured and partly kaolinized masses of white feldspar in the Ajmer district, Rajputane, India.

BEST REF. IN ENGLISH: Frondel, Clifford, USGS Bull. 1064, p. 95-98 (1958).

# CLAUDETITE
$As_2 O_3$
Dimorphous with arsenolite

CRYSTAL SYSTEM: Monoclinic
CLASS: 2/m
SPACE GROUP: $P2_1/n$
Z: 4
LATTICE CONSTANTS:
$a = 5.339$
$b = 12.984$
$c = 4.5405$
$\beta = 94°16.1'$
3 STRONGEST DIFFRACTION LINES:
3.245 (100)
3.454 ( 50)
2.771 ( 35)
OPTICAL CONSTANTS:
$\alpha = 1.871$
$\beta = 1.92$
$\gamma = 2.01$
$(+)2V = 58°$
HARDNESS: 2½
DENSITY: 4.15 (Meas.)
4.26 (Calc.)
4.185 (Calc. x-ray)
CLEAVAGE: $\{010\}$ perfect, micaceous
Very flexible.

HABIT: Crystals tabular, forming thin plates on {010}, elongated [001] with {111} and {$\bar{1}$11} prominent. Often as crusts.  Penetration and contact twins on {100} very common.

COLOR-LUSTER: Colorless to white, transparent. Vitreous; pearly on cleavage surfaces.

MODE OF OCCURRENCE: Occurs as a secondary mineral formed by the oxidation of arsenic minerals.  Found as crystals in a vein of kaolin, gypsum, halloysite, and sulfur at a sulfur prospect six miles north of the 4-S Ranch and 1½ miles west of the Colorado River, Imperial County, California; as a sublimation product in the fire zone of the United Verde mine, Jerome, Arizona; as well-formed crystals at the San Domingo mines, Portugal, and at Schmölnitz, Hungary; it also occurs at Calañas, Andalusia, Spain, and with orpiment and native sulfur at the LaSalle mine, Decazeville, Aveyron, France.

BEST REF. IN ENGLISH: Palache, et al., "Dana's System of Mineralogy," 7th Ed., v. I, p. 545–547, New York, Wiley, 1944.

## CLAUSTHALITE
PbSe

CRYSTAL SYSTEM:  Cubic
CLASS:  4/m $\bar{3}$ 2/m
SPACE GROUP:  Fm3m
Z:  4
LATTICE CONSTANT:
  $a$ = 6.147
3 STRONGEST DIFFRACTION LINES:
  3.06  (100)
  2.165 ( 70)
  3.54  ( 30)
  HARDNESS:  2½–3
Talmadge hardness B.
  DENSITY:  8.08–8.22 (Meas.)
          8.28 (Calc.)
  CLEAVAGE:  {001} good
Fracture granular. Brittle.
  HABIT:  Massive, fine-grained, sometimes foliated.
  COLOR-LUSTER:  Bright lead gray with faint bluish tint. Weathered surfaces dull grayish black with occasional reddish brown spots. Opaque; metallic.
  MODE OF OCCURRENCE:  Occurs associated with athabascaite, umangite, and other selenides in hematite-stained carbonate vein material in basalt at the Martin Lake mine, Lake Athabasca area, northern Saskatchewan, Canada.  Also found with penroseite at Colquechaca, Bolivia; at Cerro de Cacheuta, Mendoza, Argentina; at Fahlun, Sweden; at the Rio Tinto mines, Huelva Province, Spain; in Germany at Reinsberg, Saxony, and with other selenides at Clausthal, Tilkerode, Lehrbach, and Zorge in the Harz Mountains; and at Kweichow, China. In the United States it is abundant in ores of the Morrison formation and in Entrado sandstones in uranium-vanadium deposits of many regions of the Colorado Plateau.
  BEST REF. IN ENGLISH: Palache, et al., "Dana's System of Mineralogy," 7th Ed., v. I, p. 204–205, New York, Wiley, 1944.    Early, J. W., Am. Min., 35: 337–364 (1950).

**CLEAVELANDITE** = Platy variety of albite

**CLEIOPHANE** = Variety of sphalerite

**CLEVEITE** = Variety of uraninite

## CLIFFORDITE
UTe$_3$O$_8$

CRYSTAL SYSTEM:  Cubic
CLASS:  2/m $\bar{3}$
SPACE GROUP:  Pa3
Z:  8
LATTICE CONSTANT:
  $a$ = 11.371
3 STRONGEST DIFFRACTION LINES:
  3.273 (100)
  2.844 ( 80)
  2.007 ( 80)
OPTICAL CONSTANT:
  $N > 2.11$
  HARDNESS:  4
  DENSITY:  6.57 (Meas. synthetic)
          6.766 (Calc.)
  CLEAVAGE:  Not determined.
  HABIT:  As small octahedrons (up to 0.2 mm).
  COLOR-LUSTER:  Bright sulfur yellow; adamantine.
  MODE OF OCCURRENCE:  Occurs along joint surfaces in the oxidized zone of the San Miguel mine, a tellurium-gold-silver prospect near Moctezuma, Sonora, Mexico, associated with mackayite, barite, quartz, and limonite.  It has also been found in very small quantities at the Moctezuma mine in the same district, associated with native tellurium and paratellurite in small vugs.
  BEST REF. IN ENGLISH:  Gaines, Richard V., Am. Min.. 54: 697–701 (1969).

**CLIFTONITE** = Cubic pseudomorphs of graphite found in some meteorites

**CLINOBARRANDITE** = Variety of metastrengite

## CLINOCHLORE
(Mg, Fe$^{2+}$, Al)$_6$ (Si, Al)$_4$ O$_{10}$ (OH)$_8$
Chlorite Group

CRYSTAL SYSTEM:  Monoclinic
CLASS:  2/m
SPACE GROUP:  C2/m
Z:  2
LATTICE CONSTANTS:
  $a$ = 5.27
  $b$ = 9.21
  $c$ = 14.36
  $\beta$ = 96°58'

3 STRONGEST DIFFRACTION LINES:
7.12 (100)
3.56 ( 80)
2.55 ( 80)
OPTICAL CONSTANTS:
$\alpha \simeq 1.57$
$\beta \simeq 1.58$
$\gamma \simeq 1.59$
(+)
HARDNESS: 2-2½
DENSITY: 2.60-3.02 (Meas.)
2.668 (Calc.)
CLEAVAGE: {001} perfect
Laminae flexible, inelastic.
HABIT: Crystals commonly tabular, hexagonal in cross section. Usually massive, foliated, coarse scaly granular, fine granular, or earthy. Twinning on {001}, twin plane [310].
COLOR-LUSTER: Colorless, white, yellowish, pale green to deep grass green or olive green. Transparent to translucent. Pearly, greasy, or dull. Streak uncolored to greenish white.
MODE OF OCCURRENCE: Occurs widespread in a variety of geological environments; mainly in schists, serpentines, and other metamorphic rocks, and as a common product of the hydrothermal alteration of amphiboles, pyroxenes, and biotite in igneous rocks. Typical occurrences are found in California, Colorado, Montana, South Dakota, Pennsylvania, New York, and Vermont. Also in Italy, Switzerland, Austria, USSR, India, Japan, and New Zealand.
BEST REF. IN ENGLISH: Deer, Howie, and Zussman, "Rock Forming Minerals," v. 3, p. 131-163, New York, Wiley, 1962.

# CLINO-CHRYSOTILE = Structural variety of chrysotile
$Mg_3Si_2O_5(OH)_4$
Serpentine Group

CRYSTAL SYSTEM: Monoclinic
CLASS: 2/m
SPACE GROUP: A2/m
Z: 4
LATTICE CONSTANTS:
$a = 5.313$
$b = 9.120$
$c = 14.637$
$\beta = 93°10'$
3 STRONGEST DIFFRACTION LINES:
7.31 (100)
3.65 ( 70)
4.57 ( 50)
OPTICAL CONSTANTS:
$\alpha = 1.569$
$\gamma = 1.570$
$2V \sim 42°$
DENSITY: 2.53 (Meas.)
2.61 (Calc.)
HABIT: Massive; bladed.
COLOR-LUSTER: Dark green.
MODE OF OCCURRENCE: Occurs in a vein in serpentinite at the Butler Estate chrome mine, New Idria, San Benito County, California.

BEST REF. IN ENGLISH: Page and Coleman, USGS Prof. Paper 575-B, p. 103-107 (1967).

# CLINOCLASE
$Cu_3AsO_4(OH)_3$

CRYSTAL SYSTEM: Monoclinic
CLASS: 2/m
SPACE GROUP: $P2_1/a$
Z: 4
LATTICE CONSTANTS:
$a = 12.36$ kX
$b = 6.45$
$c = 7.23$
$\beta = 99°30'$
3 STRONGEST DIFFRACTION LINES:
3.587 (100)
3.139 ( 25)
7.21 ( 20)
OPTICAL CONSTANTS:
$\alpha = 1.756$
$\beta = 1.874$
$\gamma = 1.896$
$(-)2V = 50°$
HARDNESS: 2½-3
DENSITY: 4.33 (Meas.)
4.352 (Calc.)
CLEAVAGE: {001} perfect.
Fracture uneven. Brittle.
HABIT: Crystals elongated and tabular; also rhombohedral in aspect. As isolated crystals, druses, or grouped into rosettes.
COLOR-LUSTER: Dark greenish blue to greenish black. Transparent to translucent. Vitreous, pearly on cleavage. Streak bluish green.
MODE OF OCCURRENCE: Occurs as a secondary mineral commonly associated with olivenite. Found in the United States at the Eureka and Mammoth mines, Tintic district, Juab County, and at the Gold Hill and Western Utah Extension mines, Tooele County, Utah; and at the Myler mine, Majuba Hill, Pershing County, Nevada. It also is found at Collahuasi, Tarapacá, Chile; at several mines in Cornwall, and at the Bedford United mines near Tavistock, Devonshire, England; at Markirch and other places in Germany; and at the Kitabira mine, Yamaguchi Prefecture, Japan.
BEST REF. IN ENGLISH: Palache, et al., "Dana's System of Mineralogy," 7th Ed., v. II, p. 787-789, New York, Wiley, 1951.

# CLINOENSTATITE
$MgSiO_3$
Pyroxene Group
Dimorphous with enstatite

CRYSTAL SYSTEM: Monoclinic
CLASS: 2/m
SPACE GROUP: $P2_1/c$
Z: 8

LATTICE CONSTANTS:
 $a$ = 9.6065
 $b$ = 8.8146
 $c$ = 5.1688
 $\beta$ = 108.335°
3 STRONGEST DIFFRACTION LINES:
 2.88 (100)
 2.98 ( 95)
 3.17 ( 60)
OPTICAL CONSTANTS:
  $\alpha$ = 1.651
  $\beta$ = 1.654
  $\gamma$ = 1.660
 (+)2V = 53°
 HARDNESS: 5-6
 DENSITY: 3.19 (Calc.)
 CLEAVAGE: {110} distinct
Brittle.
 HABIT: Crystals short prismatic or tabular. Also massive, lamellar. Twinning on {100} common.
 COLOR-LUSTER: Colorless, yellow-brown, greenish. Transparent to translucent. Vitreous.
 MODE OF OCCURRENCE: Occurs rarely in igneous rocks and meteorites. Found as multiple-twinned phenocrysts in a porphyritic volcanic rock from the Cape Vogel area, Papua.
 BEST REF. IN ENGLISH: Deer, Howie, and Zussman, "Rock Forming Minerals," v. 2, p. 1–74, New York, Wiley, 1963.

# CLINOFERROSILITE
$FeSiO_3$
Pyroxene Group

CRYSTAL SYSTEM: Monoclinic
CLASS: 2/m
SPACE GROUP: $P2_1/c$
Z: 8
LATTICE CONSTANTS:
 $a$ = 9.53
 $b$ = 9.21
 $c$ = 5.15
 $\beta$ = 107°38′
3 STRONGEST DIFFRACTION LINES:
 3.03 (100)
 3.35 ( 80)
 3.23 ( 80)
OPTICAL CONSTANTS:
  $\alpha$ = 1.764
  $\beta$ = 1.767
  $\gamma$ = 1.792
 (+)2V = 25°
 HARDNESS: 5-6
 DENSITY: 4.068 (Calc.)
 CLEAVAGE: {110} distinct
Brittle.
 HABIT: Crystals needle-like, minute.
 COLOR-LUSTER: Colorless with amber tint. Transparent. Vitreous. Synthetic, greenish.
 MODE OF OCCURRENCE: Occurs as needles in lithophysae in obsidian from Coso Mountains, Inyo County, California; at Obsidian Cliff, Yellowstone National Park, Wyoming; and at Hrafntinn-uhryggur, Iceland.
 BEST REF. IN ENGLISH: Bowen, N. L., *Am. J. Sci.*, **30**(5): 481–494 (1935).

# CLINOHEDRITE
$CaZnSiO_3(OH)_2$

CRYSTAL SYSTEM: Monoclinic
CLASS: m
SPACE GROUP: Aa
Z: 4
LATTICE CONSTANTS:
 $a$ = 5.43
 $b$ = 15.94
 $c$ = 5.24
 $\beta$ = 103°56′
3 STRONGEST DIFFRACTION LINES:
 2.76 (100)
 3.23 ( 70)
 2.50 ( 60)
OPTICAL CONSTANTS:
  $\alpha$ = 1.662
  $\beta$ = 1.667
  $\gamma$ = 1.669
 (−)2V = large
 HARDNESS: 5½
 DENSITY: 3.28-3.335 (Meas.)
        3.25 (Calc.)
 CLEAVAGE: {010} perfect
Brittle.
 HABIT: Crystals usually prismatic or tabular, also wedge-shaped. Commonly massive, granular, or as lamellar aggregates.
 COLOR-LUSTER: Colorless to white or amethystine. Transparent. Vitreous; pearly on cleavage. Fluoresces orange under short-wave ultraviolet light.
 MODE OF OCCURRENCE: Occurs in association with willemite, axinite, hancockite, nasonite, calcite, franklinite, and other minerals, at Franklin, Sussex County, New Jersey.
 BEST REF. IN ENGLISH: Palache, Charles, USGS Prof. Paper 180, p. 106–108 (1935).

# CLINOHOLMQUISTITE = Structural variety of holmquistite
$(Na, Ca)(Al, Li, Mg, Fe)_7Si_8O_{22}(OH, F)_2$
Amphibole Group

CRYSTAL SYSTEM: Monoclinic
CLASS: 2/m
SPACE GROUP: P2/m
Z: 2
LATTICE CONSTANTS:
 $a$ = 9.80
 $b$ = 17.83
 $c$ = 5.30
 $\beta$ = 70°54′
3 STRONGEST DIFFRACTION LINES:
 7.93 (100)
 2.985 (100)
 2.70 (100)

OPTICAL CONSTANTS:
$\alpha = 1.610$
$\beta = 1.627$
$\gamma = 1.633$
$(-)2V = 55°-61°$
HARDNESS: 5-6
DENSITY: 3.00 (Meas.)
HABIT: Crystals long prismatic.
MODE OF OCCURRENCE: Occurs in crystals partly replaced by orthorhombic holmquistite from "Siberia."
BEST REF. IN ENGLISH: Ginzburg, I. V., *Am. Min.*, **52**: 1585-1586 (1967).

## CLINOHUMITE

$Mg_9Si_4O_{16}(F, OH)_2$
Humite Group

CRYSTAL SYSTEM: Monoclinic
CLASS: 2/m
SPACE GROUP: $P2_1/c$
Z: 2
LATTICE CONSTANTS:
$a = 13.71$
$b = 4.755$
$c = 10.29$
$\beta = 100°50'$
3 STRONGEST DIFFRACTION LINES:
1.74 (100)
5.02 ( 70)
3.70 ( 70)
OPTICAL CONSTANTS:
$\alpha = 1.629$
$\beta = 1.641$
$\gamma = 1.662$
$(+)2V = 73°$
HARDNESS: 6
DENSITY: 3.17-3.35 (Meas.)
3.279 (Calc.)
CLEAVAGE: {100} indistinct
Fracture uneven to subconchoidal. Brittle.
HABIT: Crystals varied in habit, usually highly modified. Twinning on {001} common, lamellar.
COLOR-LUSTER: Yellow, brown, white. Transparent to translucent. Vitreous.
MODE OF OCCURRENCE: Occurs in contact zones in dolomite, in veins, and in serpentine and talc schist. Found in Fresno, Monterey, and Riverside counties, California; at the Tilly Foster iron mine at Brewster, Putnam County, New York; on Mte. Somma and in the Ala Valley, Piedmont, Italy; in the Llanos de Juanar, Malaga, Spain; at Pargas and elsewhere in Finland; in the Lake Baikal district, Siberia, USSR; and in the Bhandara district, India.
BEST REF. IN ENGLISH: Deer, Howie, and Zussman, "Rock Forming Minerals," v. 1, p. 47-58, New York, Wiley, 1962. Jones, Norris W., Ribbe, P. H., and Gibbs, G. V., *Am. Min.*, **54**: 391-411, 1969.

## CLINOHYPERSTHENE

$(Mg, Fe)SiO_3$
Pyroxene Group

CRYSTAL SYSTEM: Monoclinic
CLASS: 2/m
SPACE GROUP: $P2_1/c$
Z: 8
LATTICE CONSTANTS:
$a \simeq 9.7$
$b \simeq 8.9$
$c \simeq 5.2$
$\beta = 108°$
OPTICAL CONSTANTS:
$\alpha = 1.680, 1.725$
$\beta = ?$
$\gamma = 1.700, 1.755$
$(+)2V = ?$
HARDNESS: 5-6
DENSITY: 3.2-3.55 (Meas.)
CLEAVAGE: {110} distinct
Brittle.
HABIT: Massive.
COLOR-LUSTER: Greenish to brownish. Translucent. Vitreous.
MODE OF OCCURRENCE: Occurs chiefly in meteorites. Also found at Broken Hill, New South Wales, Australia, and at the Bon Accord Quarry, near Pretoria, Transvaal, South Africa.
BEST REF. IN ENGLISH: Deer, Howie, and Zussman, "Rock Forming Minerals," v. 2, p. 8-41, New York, Wiley, 1963.

## CLINOPTILOLITE

$(Na, K, Ca)_{2-3}Al_3(Al, Si)_2Si_{13}O_{36} \cdot 12H_2O$
Zeolite Group

CRYSTAL SYSTEM: Monoclinic
CLASS: 2/m
SPACE GROUP: I2/m
Z: 4
LATTICE CONSTANTS:
$a = 7.41$
$b = 17.89$
$c = 15.85$
$\beta = 91°29'$
3 STRONGEST DIFFRACTION LINES:
8.92 (100)
2.974 ( 80)
3.897 ( 57)
OPTICAL CONSTANTS:
$\alpha = 1.478$ (variable)
$\beta = 1.480$
$\gamma = 1.481$
$2Vz = 32°-48°$
HARDNESS: 3½-4
DENSITY: 2.1-2.2 (Meas.)
CLEAVAGE: {010} perfect
HABIT: Crystals platy, minute, usually in clusters. Forms: (010), (001), (101), and (20$\bar{1}$).
COLOR-LUSTER: Colorless, white. Transparent to translucent. Vitreous.
MODE OF OCCURRENCE: Standard mineralogical constants such as specific gravity or hardness have not been determined with any degree of certainty. Can be clearly distinguished from heulandite by optical, x-ray, thermal,

and chemical means. Occurs disseminated in shale in Dewey, Stanley, and Lyman counties, South Dakota; in highly weathered amygdaloidal basalt in the Hoodoo Mountains, Wyoming; in altered vitreous tuff in San Luis Obispo County, and in fuller's earth in Kern County, California; at numerous bedded zeolite deposits in New Zealand; and in devitrified volcanic glass in New South Wales, Australia.

BEST REF. IN ENGLISH: Mason, Brian, and Sand, L. B., *Am. Min.*, **45**: 341–350 (1960). Mumpton, Frederick A., *Am. Min.*, **45**: 351–369 (1960). Deer, Howie, and Zussman, "Rock Forming Minerals," v. 4, p. 377–385, New York, Wiley, 1963. Wise, W. S., Nokleberg, W. J., and Kokinos, M., *Am. Min.*, **54**: 887–895 (1969).

# CLINOSAFFLORITE
(Co, Fe, Ni)As$_2$
Dimorphous with safflorite

CRYSTAL SYSTEM: Monoclinic
CLASS: 2/m
SPACE GROUP: P2$_1$/n
Z: 2
LATTICE CONSTANTS:
 $a$ = 5.040; 5.062; 5.121
 $b$ = 5.862; 5.851; 5.847
 $c$ = 3.139; 3.149; 3.094
 $\beta$ = 90°13'; 90°18'; 90°17'
3 STRONGEST DIFFRACTION LINES:
 2.531 (100)
 2.427 ( 80)
 2.422 ( 80)
 HARDNESS: 4½–5
 DENSITY: 7.46 (Calc.)
 CLEAVAGE: Not observed.
 HABIT: Massive; in intimate intergrowth with skutterudite.
 COLOR-LUSTER: Tin white. Opaque. Metallic.
 MODE OF OCCURRENCE: The mineral has been recognized in three specimens from the Cobalt area, Ontario, Canada.
 BEST REF. IN ENGLISH: Radcliffe, Dennis, and Berry, L. G., *Can. Min.*, **10**: 877–881 (1971).

# CLINO-UNGEMACHITE
Probably near K$_3$Na$_9$Fe(SO$_4$)$_6$(OH)$_3$ · 9H$_2$O

CRYSTAL SYSTEM: Monoclinic
CLASS: 2/m
LATTICE CONSTANTS:
 $a : b : c$ = 1.6327 : 1 : 1.7308
     $\beta$ = 110°40'
 HARDNESS: Not determined.
 DENSITY: Not determined.
 CLEAVAGE: Not determined.
 HABIT: Crystals thick tabular, flattened on {001}. Only six minute crystals found.
 COLOR-LUSTER: Colorless to pale yellow. Transparent. Vitreous.

MODE OF OCCURRENCE: Occurs at Chuquicamata, Chile, associated with jarosite, ungemachite, sideronatrite, and metasideronatrite.
BEST REF. IN ENGLISH: Palache, et al., "Dana's System of Mineralogy," 7th Ed., v. II, p. 597–598, New York, Wiley, 1951.

# CLINOVARISCITE = Metavariscite

# CLINOZOISITE
Ca$_2$Al$_3$Si$_3$O$_{12}$OH
Epidote Group
Dimorph of Zoisite

CRYSTAL SYSTEM: Monoclinic
CLASS: 2/m
SPACE GROUP: P2$_1$/m
Z: 2
LATTICE CONSTANTS:
 $a$ = 8.87–8.88
 $b$ = 5.59–5.61
 $c$ = 10.15–10.17
 $\beta$ = 115°27'
3 STRONGEST DIFFRACTION LINES:
 2.89 (100)
 2.79 ( 80)
 2.59 ( 70)
OPTICAL CONSTANTS:
 $\alpha$ = 1.670–1.715
 $\beta$ = 1.674–1.725
 $\gamma$ = 1.690–1.734
 (+)2V$_\gamma$ = 14°–90°
 HARDNESS: 6½
 DENSITY: 3.21–3.38 (Meas.)
     3.364 (Calc.)
 CLEAVAGE: {001} perfect
Fracture uneven. Brittle.
 HABIT: Crystals short to long prismatic, often deeply striated; also acicular. Commonly massive, coarse to fine granular; also fibrous, parallel, divergent, or radial. Twinning on {100}, lamellar; not common.
 COLOR-LUSTER: Colorless, pale yellow, gray, green, pink. Transparent to translucent. Vitreous. Streak uncolored, grayish.
 MODE OF OCCURRENCE: Occurs chiefly in regionally metamorphosed igenous and sedimentary rocks; in acid igneous rocks contaminated with calc-silicate material; as an alteration of plagioclase feldspars; and in contact zones. Typical occurrences are found in Los Angeles County and elsewhere in California; in the Nightingale district, Pershing County, Nevada; and at Allens Park, Boulder County, Colorado. Also at Pinos Altos in Baja California, and as radial aggregates in quartz at Alamos, Sonora, Mexico; at Timmons, Ontario, Canada; and at localities in Ireland, Italy, Switzerland, Austria, Czechoslovakia, India, and at many other places.
 BEST REF. IN ENGLISH: Deer, Howie, and Zussman, "Rock Forming Minerals," v. 1, p. 193–210, New York, Wiley, 1962. Dollase, W. A., *Am. Min.*, **53**: 1882–1898 (1968).

# CLINTONITE (Seybertite)
## (Xanthophyllite)
## (Brandisite)
$Ca(Mg, Al)_3 (Al_3 Si)O_{10} (OH)_2$
Mica Group

CRYSTAL SYSTEM: Monoclinic
CLASS: 2/m
SPACE GROUP: C2/m
Z: 2
LATTICE CONSTANTS:
 $a$ = 5.204
 $b$ = 9.026
 $c$ = 9.812
 $\beta$ = 100°21'
3 STRONGEST DIFFRACTION LINES:
 2.56 (100)
 3.21 ( 70)
 2.11 ( 70)
OPTICAL CONSTANTS:
 $\alpha$ = 1.643–1.648
 $\beta$ = 1.655–1.662
 $\gamma$ = 1.655–1.663
 (–)2V $\simeq$ 6°–32°
HARDNESS: 3½ on (001)
 6 perpendicular (001)
DENSITY: 3–3.1 (Meas.)
 3.096 (Calc.)
CLEAVAGE: {001} perfect
Laminae brittle.
HABIT: Crystals tabular, pseudohexagonal. Also massive, foliated or lamellar radiate. Twinning on {001}, twin axis [310].
COLOR-LUSTER: Colorless, yellowish, greenish, reddish brown, copper red. Transparent to translucent. Vitreous, pearly, pearly submetallic. Streak uncolored, yellowish or grayish.
MODE OF OCCURRENCE: Occurs chiefly in association with calcite, idocrase, grossular, spinel, clinopyroxene and phlogopite in crystalline limestones; less commonly in chloritic schists and contact zones. Typical occurrences are found at Crestmore, Riverside County, California (xanthophyllite); near Helena, Montana; at Amity, Orange County, New York (seybertite); on Mt. Monzoni, Val di Fassa, Trentino, Italy (brandisite); in the Pargas district, Finland; and in the Zlatoust district, southern Urals, USSR.
BEST REF. IN ENGLISH: Deer, Howie, and Zussman, "Rock Forming Minerals," v. 3, p. 99-102, New York, Wiley, 1962.

# COALINGITE
$Mg_{10} Fe_2 (OH)_{24} CO_3 \cdot 2H_2 O$

CRYSTAL SYSTEM: Hexagonal
LATTICE CONSTANTS:
 $a$ = 3.1
 $c \simeq$ 30 (?)
3 STRONGEST DIFFRACTION LINES:
 2.34 (100)
 4.20 ( 80)
 6.05 ( 50)

OPTICAL CONSTANTS:
 $\omega$ = 1.588–1.635
 $\epsilon$ = 1.560–1.590
 (–)2V small
HARDNESS: ~1–2
DENSITY: 2.33–2.42 (Meas.)
CLEAVAGE: Parallel to elongation; also prominent cleavage traces either normal or inclined at about 45° to elongation.
HABIT: Platelets 0.1–0.2 mm in size; aggregates of granular to elongate grains; bundles of fibers up to 5.0 mm in length.
COLOR-LUSTER: Reddish brown, pale brownish, straw-colored. Resinous.
MODE OF OCCURRENCE: Occurs in the surface weathering zone of the New Idria serpentinite about 35 miles northwest of Coalinga in Fresno and San Benito counties, California; also as fracture fillings in drill cores from the Muskox Intrusion, Northwest Territories.
BEST REF. IN ENGLISH: Mumpton, F. A., Jaffe, H. W., and Thompson, C. S., *Am. Min.*, **50**: 1893-1913 (1965). Jambor, J. L., *Am. Min.*, **54**: 437–447 (1969).

# COBALTITE
CoAsS

CRYSTAL SYSTEM: Cubic
CLASS: 2/m $\overline{3}$
SPACE GROUP: Pa3
Z: 4
LATTICE CONSTANTS:
 $a$ = 5.58
3 STRONGEST DIFFRACTION LINES:
 2.72 (100)
 1.800 (100)
 2.465 ( 90)
HARDNESS: 5½
DENSITY: 6.33 (Meas.)
 6.302 (Calc.)
CLEAVAGE: {001} perfect
Fracture uneven. Brittle.
HABIT: Crystals usually cubes, octahedrons, pyritohedrons, or combinations of cube and pyritohedron. Faces commonly striated like pyrite. Also massive, granular to compact. Twinning on {011} or {111}, rare.
COLOR-LUSTER: Tin white, steel gray with violet tint, and grayish black. Opaque. Metallic, brilliant to dull. Streak grayish black.
MODE OF OCCURRENCE: Occurs chiefly in high-temperature hydrothermal vein deposits or as disseminations in metamorphosed rocks. Found sparingly in the United States at the Copper King mine, Boulder City, Colorado; in Lemhi County, Idaho; and at a number of localities in Calaveras, Inyo, Madera, Mariposa, Mono, Nevada, and Placer counties, California. Important deposits occur at Cobalt and elsewhere in Ontario, Canada; in Sonora, Mexico; as superb crystals at Tunaberg, Hakansbö, Riddarhyttan, and at other places in Sweden; and at localities in England, Norway, Germany, USSR, India, and Australia.
BEST REF. IN ENGLISH: Palache, et al., "Dana's System of Mineralogy," 7th Ed., v. I, p. 296-298, New York, Wiley, 1944.

## COBALTOCALCITE (Spherocobaltite)
$CoCO_3$

CRYSTAL SYSTEM: Hexagonal
CLASS: $\bar{3}\,2/m$
SPACE GROUP: $R\bar{3}c$
Z: 6
LATTICE CONSTANTS:
  $a = 4.67$ kX
  $c = 15.13$
3 STRONGEST DIFFRACTION LINES:
  2.743 (100)
  3.551 ( 40)
  1.702 ( 30)
OPTICAL CONSTANTS:
  $\omega = 1.855$
  $\epsilon = 1.60$
  $(-)$
HARDNESS: 4
DENSITY: 4.13 (Meas.)
        4.11 (Calc.)
CLEAVAGE: $\{10\bar{1}1\}$ (presumed)
  HABIT: As small spherical masses with radiated structure; as crusts. Crystals rare.
  COLOR-LUSTER: Rose red. Often altered to gray, brown, or black on surface. Translucent. Vitreous to somewhat waxy.
  MODE OF OCCURRENCE: Occurs at Boleo, Baja California, Mexico; in the cobalt-nickel veins at Schneeberg, Saxony, Germany; in Liguria, Italy; and as exceptional specimens in the Katanga district, Zaire.
  BEST REF. IN ENGLISH: Palache, et al., "Dana's System of Mineralogy," 7th Ed., v. II, p. 175–176, New York, Wiley, 1951.

## COBALTOMENITE
$CoSeO_3 \cdot 2H_2O$

CRYSTAL SYSTEM: Monoclinic
CLASS: $2/m$
SPACE GROUP: $P2_1/n$
Z: 4
LATTICE CONSTANTS:
  $a = 7.58$
  $b = 8.73$
  $c = 6.59$
  $\beta = 98°30'$
HARDNESS: 2-2½
DENSITY: Not determined.
CLEAVAGE: Fracture conchoidal. Brittle.
  HABIT: As minute crystals, resembling erythrite; as crystalline crusts.
  COLOR-LUSTER: Rose red to dark red. Transparent. Vitreous.
  MODE OF OCCURRENCE: Occurs as coatings on sandstone at the A.E.C. Mine No. 8, Temple Mountain, Emery County, Utah; and in the Cerro de Cacheuta, Mendoza province, Argentina, with chalcomenite as an alteration product of selenides.
  BEST REF. IN ENGLISH: No good references in English.

## COBALT PENTLANDITE
$(Co, Fe, Ni)_9S_8$
Cobalt analogue of pentlandite

CRYSTAL SYSTEM: Cubic
CLASS: $4/m\,\bar{3}\,2/m$
SPACE GROUP: $Fm\,3m$
Z: 4
LATTICE CONSTANT:
  $a = 9.973$
3 STRONGEST DIFFRACTION LINES:
  3.008 (100)
  1.763 (100)
  1.918 ( 80)
  HARDNESS: Microhardness 245–310 kg/mm$^2$ (330–350 est. for pure $Co_9S_8$)
  DENSITY: 5.34 (Calc. for $Co_9S_8$)
  CLEAVAGE: $\{001\}$ distinct
Brittle.
  HABIT: As exsolved lamellae and minute crystals, rarely exceeding 4 mm in diameter. Also massive, fine-grained.
  COLOR-LUSTER: Bronze yellow, more lustrous than pentlandite. Opaque. Metallic.
  MODE OF OCCURRENCE: Occurs as intergrowths with linnaeite in association with chalcopyrite, sphalerite, and pyrrhotite at the Vauze mine, 15 miles north of Noranda, Quebec, Canada; also in association with other sulfides, notably pyrrhotite, at the Varislahti and Savonranta deposits, and at the Outokumpu mine, Finland.
  BEST REF. IN ENGLISH: Kouvo, Olavi; Huhma, Maija; and Vourelainen, Yrjö, *Am. Min.*, **44**: 897–900 (1959). Stumpfl, E. F., and Clark, A. M., *Am. Min.*, **50**: 2107–2108 (1965).

**COCCINITE** = Inadequately described mercuric iodide

**COCCOLITE** = Variety of diopside

## COCINERITE (Inadequately described species)
$Cu_4AgS$

CRYSTAL SYSTEM: Unknown
  HARDNESS: 2½
  DENSITY: 6.14 (Meas.)
  CLEAVAGE: Not determined.
  HABIT: Massive.
  COLOR-LUSTER: Silver gray; tarnishes black. Opaque. Metallic.
  MODE OF OCCURRENCE: Occurs sparingly in the oxidation zone at the Cocinera mine, Ramos, San Luis Potosi, Mexico.
  BEST REF. IN ENGLISH: Hough, Am. *J. Sci.*, **48**: 206 (1919).

## COCONINOITE
$Fe_2Al_2(UO_2)_2(PO_4)_4(SO_4)(OH)_2 \cdot 20H_2O$

CRYSTAL SYSTEM: Monoclinic

**3 STRONGEST DIFFRACTION LINES:**
 11.05 (100)
 5.52 ( 60)
 5.61 ( 40)
**OPTICAL CONSTANTS:**
 $\alpha = 1.550$
 $\beta = 1.588$
 $\gamma = 1.590$
 $(-)2V \approx 40°$
**HARDNESS:** Soft, hardness cannot be measured accurately.
**DENSITY:** 2.70 (Meas.)
 2.68 (Calc.)
**CLEAVAGE:** Not determined.
**HABIT:** As aggregates of microcrystalline grains.
**COLOR-LUSTER:** Light creamy yellow.
**MODE OF OCCURRENCE:** Occurs in the oxidized zone of uranium deposits in Utah and Arizona, and occurrences have been documented from Wyoming and New Hampshire.
 **BEST REF. IN ENGLISH:** Young, E. J., Weeks, A. D., and Meyrowitz, Robert, *Am. Min.*, **51**: 651–666 (1966).


# COERULEOLACTITE
$(Ca, Cu)Al_6 (PO_4)_4 (OH)_8 \cdot 4-5H_2O$
Calcium analogue of turquoise

**CRYSTAL SYSTEM:** Triclinic
**CLASS:** $\bar{1}$
**SPACE GROUP:** $P\bar{1}$
**3 STRONGEST DIFFRACTION LINES:**
 2.96 (100) broad
 3.70 ( 90)
 3.48 ( 50)
**OPTICAL CONSTANTS:**
 $\omega = 1.580$
 $\epsilon = 1.588$
 (+)
**HARDNESS:** 5
**DENSITY:** 2.57–2.696 (Meas.)
**CLEAVAGE:** Fracture conchoidal to uneven.
**HABIT:** As minutely crystalline or fibrous crusts, veinlets, and botryoidal aggregates.
**COLOR-LUSTER:** Milk white to light blue.
**MODE OF OCCURRENCE:** Occurs associated with wavellite at General Trimble's mine, East Whiteland, Chester County, Pennsylvania, and with limonite at the Rindsberg mine, Nassau, Germany.
 **BEST REF. IN ENGLISH:** Palache, et al., "Dana's System of Mineralogy," 7th Ed., v. II, p. 961, New York, Wiley, 1951.


# COESITE
$SiO_2$
Polymorphous with quartz, tridymite, cristobalite, stishovite

**CRYSTAL SYSTEM:** Monoclinic
**CLASS:** 2/m
**SPACE GROUP:** C2/c
**Z:** 16

**LATTICE CONSTANTS:**
 $a = 7.17$
 $b = 12.38$
 $c = 7.17$
 $\beta = 120°$
**3 STRONGEST DIFFRACTION LINES:**
 3.098 (100)
 3.432 ( 50)
 2.77 ( 15)
**OPTICAL CONSTANTS:**
 $\alpha = 1.593-1.599$
 $\beta = ?$
 $\gamma = 1.597-1.604$
 $(+)2V = 54°-64°$
**HARDNESS:** ~7½
**DENSITY:** 2.93 ± 0.02 (Meas.)
 2.93 (Calc.)
**CLEAVAGE:** None observed.
Fracture subconchoidal.
 **HABIT:** Irregular grains 5 to more than 50 $\mu$ in size. Synthetic crystals have simple gypsum-like habit, flattened on {010} and elongated on the *c*-axis.
 **COLOR-LUSTER:** Colorless, transparent; vitreous.
 **MODE OF OCCURRENCE:** Occurs abundantly in sheared and compressed areas of Coconino sandstone of Permian age, and as a subordinate constituent of sandstone that has largely been converted to glass at Meteor Crater, Arizona. Also found at the Wabar meteor crater, near Al Hadida, Arabia; at the Rieskessel meteoritic caldera, Bavaria, Germany; and in fossil meteorite craters at Kentland, Indiana, and near Sinking Springs, Ohio.
 **BEST REF. IN ENGLISH;** Frondel, C., "Dana's System of Mineralogy," 7th Ed., v. III, p. 310–316, New York, Wiley, 1962.


# COFFINITE
$U(OH)_{4x} (SiO_4)_{1-x}$

**CRYSTAL SYSTEM:** Tetragonal
**CLASS:** 4/m 2/m 2/m
**SPACE GROUP:** I4/amd
**Z:** 4
**LATTICE CONSTANTS:**
 $a = 6.92-6.94$
 $c = 6.22-6.31$
**3 STRONGEST DIFFRACTION LINES:**
 4.66 (100)
 3.47 (100)
 2.64 ( 70)
**OPTICAL CONSTANTS:**
 $\epsilon$ and $\omega \sim 1.73-1.75$ (+ and -)
**HARDNESS:** 5-6
**DENSITY:** 5.1 (Meas.)
**CLEAVAGE:** Pulverulent to friable or brittle. Fracture of aggregates earthy or irregular to subconchoidal.
 **HABIT:** As aggregates or disseminations of extremely fine particles; also well-crystallized, small, botryoidal masses.
 **COLOR-LUSTER:** Black. Dull to adamantine. Pale to dark brown under the microscope. Not fluorescent.
 **MODE OF OCCURRENCE:** Occurs at many localities in the Uravan mineral belt in southwestern Colorado and

southeastern Utah in the black vanadium-rich ores. Also at numerous localities in New Mexico, Oklahoma, Texas, Arizona, South Dakota, Wyoming, and in several European mining districts.

BEST REF. IN ENGLISH: Frondel, Clifford, USGS Bull. 1064, p. 285–289 (1958). Moench, Robert H., *Am. Min.*, 47: 26–33 (1962).


# COHENITE
$Fe_3C$

CRYSTAL SYSTEM: Orthorhombic
CLASS: 2/m 2/m 2/m
SPACE GROUP: Pnma
Z: 4
LATTICE CONSTANTS:
 $a = 5.06$
 $b = 6.73$
 $c = 4.51$
3 STRONGEST DIFFRACTION LINES:
 2.01 (100)
 2.06 ( 70)
 2.38 ( 65)
HARDNESS: 5½–6
DENSITY: 7.20–7.65 (Meas.)
 7.68 (Calc.)
CLEAVAGE: {100}
 {010}
 {001}
Very brittle.
HABIT: As elongated tabular crystals.
COLOR-LUSTER: Tin white; tarnishes light bronze to golden yellow on exposure. Opaque.

Strongly magnetic.
MODE OF OCCURRENCE: Occurs in the terrestrial irons at Ovifak and Niakornak, Greenland. It also is found in numerous iron meteorites, and as inclusions in diamonds from South Africa. (Called *cementite* when formed in steel making.)
BEST REF. IN ENGLISH: Palache, et al., "Dana's System of Mineralogy," 7th Ed., v. I, p. 122–123, New York, Wiley, 1944.


# COLEMANITE
$Ca_2B_6O_{11} \cdot 5H_2O$

CRYSTAL SYSTEM: Monoclinic
CLASS: 2/m
SPACE GROUP: $P2_1/a$
Z: 2
LATTICE CONSTANTS:
 $a = 8.743$
 $b = 11.264$
 $c = 6.102$
 $\beta = 110°7'$
3 STRONGEST DIFFRACTION LINES:
 3.13 (100)
 5.64 ( 50)
 3.85 ( 50)

OPTICAL CONSTANTS:
 $\alpha = 1.5863$
 $\beta = 1.5920$ (Na)
 $\gamma = 1.6140$
(+)2V = 55°
HARDNESS: 4½
DENSITY: 2.42 (Meas.)
 2.419 (Calc.)
CLEAVAGE: {010} perfect
 {001} distinct
Fracture subconchoidal to uneven. Brittle.
HABIT: Crystals commonly equant, short prismatic with complex terminations, and pseudorhombohedral. Also massive, cleavable, coarse to fine-granular; as rounded aggregates; and as geodes.
COLOR-LUSTER: Colorless, white, yellowish white, grayish. Transparent to translucent.
MODE OF OCCURRENCE: Occurs in playas and desiccated saline lake deposits in arid regions. Found widespread in California at localities in Inyo, Kern, Los Angeles, Riverside, San Bernardino, and Ventura counties; and in Nevada in Clark County. It also occurs in playa deposits in Jujuy Province, Argentina; abundantly in the Inder region, Kazakhstan, USSR; and as superb large crystals near Eskisehir, Turkey.
BEST REF. IN ENGLISH: Palache, et al., "Dana's System of Mineralogy," 7th Ed., v. II, p. 349–353, New York, Wiley, 1951.


# COLLINSITE
$Ca_2(Mg, Fe)(PO_4)_2 \cdot 2H_2O$

CRYSTAL SYSTEM: Triclinic
CLASS: $\bar{1}$
SPACE GROUP: $P\bar{1}$
Z: 1
LATTICE CONSTANTS:
 $a = 5.71$ $\alpha = 96°48.5'$
 $b = 6.73$ $\beta = 107°16.5'$
 $c = 5.39$ $\gamma = 104°32'$
3 STRONGEST DIFFRACTION LINES:
 2.69 (100)
 3.04 ( 80)
 1.669 ( 60)
OPTICAL CONSTANTS:
 $\alpha = 1.632$
 $\beta = 1.642$
 $\gamma = 1.657$
(+)2V = 80°
HARDNESS: 3½
DENSITY: 2.99 (Meas.)
 3.04 (Calc.)
CLEAVAGE: {001} perfect
 {010} perfect
Brittle.
HABIT: Crystals short prismatic to thin tabular; often in bundles, radiating outward in sheaf-like aggregates; sometimes doubly terminated. Usually as thick crusts with radial-fibrous structure.
COLOR-LUSTER: Colorless, white, light brown. Transparent to translucent. Vitreous or silky.

MODE OF OCCURRENCE: Occurs as fine crystals and radial aggregates lining cavities in altered phosphate nodules from granite pegmatite at the Tip Top mine, Custer County, South Dakota. Also found as alternate crusts with quercyite (carbonate-apatite), associated with small amounts of asphaltum, on fragments of andesite in a vein-like deposit at François Lake, British Columbia, Canada.

BEST REF. IN ENGLISH: Palache, et al., "Dana's System of Mineralogy," 7th Ed., v. II, p. 722–723, New York, Wiley, 1951.

## COLLOPHANE = Variety of apatite

## COLORADOITE
HgTe

CRYSTAL SYSTEM: Cubic
CLASS: $\bar{4}$3m
SPACE GROUP: F$\bar{4}$3m
Z: 4
LATTICE CONSTANT:
 $a = 6.448$
3 STRONGEST DIFFRACTION LINES:
 3.74 (100)
 2.29 ( 90)
 1.949 ( 70)
HARDNESS: 2½
DENSITY: 8.07–8.63 (Meas.)
 8.12 (Calc.)
CLEAVAGE: None.
Fracture subconchoidal to uneven.
Brittle; friable.
 HABIT: Massive, granular.
 COLOR-LUSTER: Iron black. Opaque. Metallic.
 MODE OF OCCURRENCE: Occurs in Colorado in the telluride ores of Boulder County, and in the Cripple Creek, La Plata, and Vulcan telluride districts; in California at the Stanislaus mine on Carson Hill, Calaveras County, and in the Norwegian mine near Tuttletown, Tuolumne County; and at Kalgoorlie, Western Australia.
 BEST REF. IN ENGLISH: Palache, et al., "Dana's System of Mineralogy," 7th Ed., v. I, p. 218–219, New York, Wiley, 1944.

## COLUMBITE
(Fe, Mn)(Nb, Ta)$_2$O$_6$
Columbite-Tantalite Series

CRYSTAL SYSTEM: Orthorhombic
CLASS: 2/m 2/m 2/m
SPACE GROUP: Pcan
Z: 4
LATTICE CONSTANTS:
 $a = 5.73$
 $b = 14.24$
 $c = 5.08$
3 STRONGEST DIFFRACTION LINES:
 2.96 (100)
 3.66 ( 50)
 1.721 ( 20)

OPTICAL CONSTANT:
 $\beta \approx 2.45$
HARDNESS: 6
DENSITY: 5.15 increasing linearly with increase in Ta$_2$O$_5$ content.
CLEAVAGE: {010} distinct
 {100} less distinct
Fracture uneven to subconchoidal. Brittle.
 HABIT: Crystals thin to thick tabular, short prismatic, equant; less commonly pyramidal. As large aggregates of parallel to divergent crystals; also massive, compact. Twinning on {201} common, often heart-shaped with pinnate striations parallel to {010}, or as pseudohexagonal trillings. Twinning on {203} or {501} uncommon.
 COLOR-LUSTER: Black to brownish black; manganoan varieties often reddish brown. Often tarnished iridescent. Opaque; except in very thin fragments. Submetallic to weakly vitreous. Streak black, brownish black to reddish brown.
 MODE OF OCCURRENCE: Occurs in granite pegmatites in association with albite, microcline, quartz, spodumene, beryl, lepidolite, montebrasite, tourmaline, muscovite, cassiterite, and other minerals; also in placer deposits in areas of pegmatites and granitic rocks. Found abundantly at numerous deposits in Custer, Lawrence, and Pennington counties, South Dakota; often as excellent crystals weighing up to 200 pounds or more, and as large masses ranging up to 18 tons in weight. Other notable occurrences in the United States are found in California, Idaho, Wyoming, Colorado, Maine, New Hampshire, Massachusetts, Connecticut, New York, Pennsylvania, North Carolina, South Carolina, Virginia, and Alabama. Other important deposits occur in Canada, Greenland, Brazil, Argentina, Norway, Sweden, France, Italy, Germany, USSR, India, Japan, Australia, Zaire, South West Africa, and Madagascar.
 BEST REF. IN ENGLISH: Palache, et al., "Dana's System of Mineralogy," 7th Ed., v. I, p. 780–787, New York, Wiley, 1944.

## COLUSITE
Cu$_3$(As, Sn, V)S$_4$

CRYSTAL SYSTEM: Cubic
CLASS: $\bar{4}$3m
SPACE GROUP: P$\bar{4}$3m
Z: 8
LATTICE CONSTANT:
 $a = 10.629$
3 STRONGEST DIFFRACTION LINES:
 3.075 (100)
 1.88 ( 80)
 1.60 ( 60)
HARDNESS: 3-4
DENSITY: 4.50 (Meas.)
 4.434 (Calc.)
CLEAVAGE: None.
 HABIT: Usually massive; crystals tetrahedral, modified by {012}.
 COLOR-LUSTER: Bronze; metallic. Opaque. Streak black.

MODE OF OCCURRENCE: Occurs associated with enargite, tetrahedrite, chalcocite, bornite, pyrite, and quartz at Butte, Montana, and associated with enargite in ore from the Red Mountain district, Ouray County, Colorado.

BEST REF. IN ENGLISH: "Dana's System of Mineralogy," 7th Ed., v. I, p. 386–387, New York, Wiley, 1944. Murdoch, Joseph, *Am. Min.*, **38**: 794–801 (1953). Dangel, Philip N., and Wuensch, B. J., *Am. Min.*, **55**: 1787–1791 (1970).

## COMBEITE
$Na_4Ca_3Si_6O_{16}(OH, F)_2$

CRYSTAL SYSTEM: Hexagonal
CLASS: 3m, 32, or $\bar{3}2/m$
SPACE GROUP: R3m, R32, or $R\bar{3}m$
Z: 3
LATTICE CONSTANTS:
  $a = 10.43$
  $c = 13.14$
3 STRONGEST DIFFRACTION LINES:
  2.657 (100)
  2.607 ( 80)
  3.304 ( 70)
OPTICAL CONSTANTS:
  $\epsilon = \omega = 1.598$
  (+)
HARDNESS: Not determined.
DENSITY: 2.844 (Meas.)
CLEAVAGE: None.
HABIT: As poorly developed stout hexagonal prisms a few tenths of a millimeter in length.
COLOR-LUSTER: Colorless.
MODE OF OCCURRENCE: Occurs associated with melilite, clinopyroxene, kalsilite, kirschsteinite, götzenite, sodalite, magnetite, perovskite, apatite, hornblende, biotite, and an unidentified mineral, in melilite-nephelinite lava from Mt. Shaheru, Zaire.
BEST REF. IN ENGLISH: Sahama, Th. G., and Hytönen, Kai, *Min. Mag.*, **31**: 503–510 (1957).

## COMPREIGNACITE
$K_2U_6O_{19} \cdot 11H_2O$

CRYSTAL SYSTEM: Orthorhombic
CLASS: 4/m 2/m 2/m
SPACE GROUP: Probably Pnmn
Z: 2
LATTICE CONSTANTS:
  $a = 7.16$
  $b = 12.14$
  $c = 14.88$
3 STRONGEST DIFFRACTION LINES:
  7.40 (100)
  3.53 ( 80)
  3.19 ( 80)
OPTICAL CONSTANTS:
  $\alpha < 1.790$
  $\beta = 1.798$
  $\gamma = 1.802$
  $(-)2V = 10°–15°$

HARDNESS: Not determined.
DENSITY: 5.03 (Meas.)
         5.13 (Calc.)
CLEAVAGE: {001} perfect
HABIT: Crystals minute, flattened on {001}. Twinning on {110} common.
COLOR-LUSTER: Yellow.
MODE OF OCCURRENCE: Occurs with other oxidation products of pitchblende ores of the Margnac deposit, Compreignac, France.
BEST REF. IN ENGLISH: Protas, J., *Am. Min.*, **50**: 807–808 (1965).

## CONGOLITE
$(Fe, Mg, Mn)_3ClB_7O_{13}$
Dimorphous with ericaite

CRYSTAL SYSTEM: Hexagonal
SPACE GROUP: R3c or $R\bar{3}c$
Z: 6
LATTICE CONSTANTS:
  $a$(hex.) = 8.6225
  $c$(hex.) = 21.054
  $\alpha_{rh}$ = 8.6042
  $\alpha = 60°10'$
3 STRONGEST DIFFRACTION LINES:
  2.725 (100)
  3.05  ( 80)
  2.061 ( 75)
OPTICAL CONSTANTS:
  Optically uniaxial, negative.
  $\epsilon = 1.731$
  $\omega = 1.755$
HARDNESS: Not determined.
DENSITY: Not measured, presumably close to that of ericaite (3.503).
CLEAVAGE: Not determined.
HABIT: Massive, fine-grained.
COLOR-LUSTER: Pale red. Transparent.
MODE OF OCCURRENCE: Occurs in the insoluble residue of a drill core from Brazzaville, Congo. Some so-called ericaite from Bischofferode, southern Harz, Germany, is congolite.
BEST REF. IN ENGLISH: Wendling, Emil, Hodenberg, Renate V., and Kühn, Robert, *Am. Min.*, **57**: 1315 (1972).

## CONICHALCITE
$CuCaAsO_4OH$

CRYSTAL SYSTEM: Orthorhombic
CLASS: 2 2 2
SPACE GROUP: $P2_12_12_1$
Z: 4
LATTICE CONSTANTS:
  $a = 7.42$ kX
  $b = 9.20$
  $c = 5.85$
3 STRONGEST DIFFRACTION LINES:
  2.842 (100)
  3.118 ( 90)
  2.594 ( 50)

OPTICAL CONSTANTS:
  $\alpha = 1.730-1.800$
  $\beta = 1.795-1.831$
  $\gamma = 1.771-1.846$
  $\pm 2V \sim 90°; (-) \sim 25°, (+) \sim 0°$
HARDNESS: 4½
DENSITY: 4.33 (Meas.)
            4.33 (Calc.)
CLEAVAGE: None.
Fracture uneven. Brittle.
HABIT: Crystals equant to short prismatic [010]. Commonly botryoidal to reniform crusts and masses with radial-fibrous structure.
COLOR-LUSTER: Yellowish green to emerald green. Subtranslucent. Vitreous to greasy.
MODE OF OCCURRENCE: Occurs as a secondary mineral in the oxidized zone of many copper deposits associated with limonite and a large variety of secondary copper minerals. It is found abundantly at many localities in Mexico, in South West Africa, Germany, Spain, Chile, USSR, and Poland. In the United States it is common in mining districts in Arizona, Utah, Nevada, and South Dakota.
BEST REF. IN ENGLISH: "Dana's System of Mineralogy," 7th Ed., v. II, p. 806-808, New York, Wiley, 1951. Berry, L. G., *Am. Min.*, **36**: 484-503 (1951).


# CONNELLITE
$Cu_{19}Cl_4SO_4(OH)_{32} \cdot 3H_2O$
Isomorphous with buttgenbachite

CRYSTAL SYSTEM: Hexagonal
CLASS: 6/m 2/m 2/m
SPACE GROUP: P6₃/mmc
Z: 2
LATTICE CONSTANTS:
  $a = 13.574$
  $c = 9.07$
3 STRONGEST DIFFRACTION LINES:
  8.00 (100)
  13.7 (100)
  2.29 (100)
OPTICAL CONSTANTS:
  $\omega = 1.724-1.738$
  $\epsilon = 1.746-1.758$
  (+)
HARDNESS: 3
DENSITY: 3.41 (Meas.)
            3.39 (Calc.)
CLEAVAGE: Not determined.
HABIT: As radiated groups of acicular crystals and as felt-like aggregates. Striated in direction of elongation.
COLOR-LUSTER: Azure blue, bluish green. Translucent. Vitreous. Streak pale greenish blue.
MODE OF OCCURRENCE: Occurs as a secondary mineral in the oxidation zone of copper-bearing ore deposits. Found as radiating groups up to 5 mm in diameter on gray schist near the Buchanan copper mine, near Daulton, Madera County, California; at the Grand Central mine, Juab County, Utah; at Ajo, Pima County, and in several mines at Bisbee, Cochise County, Arizona; as fine specimens at several mines

in Cornwall, England; and at localities in Sardinia, Algeria, and Namaqualand, South Africa.
BEST REF. IN ENGLISH: Palache, et al., "Dana's System of Mineralogy," 7th Ed., v. II, p. 572-573, New York, Wiley, 1951.


# COOKEITE (Chlorite-Ia)
$(Li, Al_4)Si_3AlO_{10}(OH)_8$
(Chlorite Group)

CRYSTAL SYSTEM: Monoclinic
CLASS: 2/m
SPACE GROUP: P2₁/a
Z: 4
LATTICE CONSTANTS:
  $a = 5.13$
  $b = 8.93$
  $c = 28.7$
  $\beta = 98°45'$
3 STRONGEST DIFFRACTION LINES:
  2.315 (100)
  4.70 ( 90)
  3.52 ( 90)
OPTICAL CONSTANTS:
  $\alpha = 1.572-1.576$
  $\beta = 1.579-1.584$
  $\gamma = 1.589-1.600$
Fluctuations due to degree of hydration.
(+)2V = 0-80°
HARDNESS: 2½-3½
DENSITY: 2.58-2.69 (Meas.)
CLEAVAGE: {001} perfect, micaceous
Flexible, inelastic.
HABIT: Crystals pseudohexagonal plates. As curved, radial arranged scales or spherulites measuring from a fraction of a millimeter to 3-4 mm.
COLOR-LUSTER: White, pink, greenish, yellowish, brown. Transparent to translucent. Pearly or silky.
MODE OF OCCURRENCE: Occurs in lithium-rich granite pegmatites in association with tourmaline, albite, microcline, quartz, lepidolite, and apatite. Found at Pala and Rincon, San Diego County, California; at the Bazooka prospect, Gunnison County, Colorado; at Haddam Neck, Connecticut; and at several places in Oxford County, Maine. It also occurs at Varutrask, Sweden; in various regions of the USSR; at Londonderry, Western Australia; and elsewhere.
BEST REF. IN ENGLISH: Vlasov, K. A., "Mineralogy of Rare Elements," v. II, p. 32-35, Israel Program for Scientific Translations, 1966.


# COOPERITE
PtS

CRYSTAL SYSTEM: Tetragonal
CLASS: 4/m 2/m 2/m
SPACE GROUP: P4/mmc
Z: 2
LATTICE CONSTANTS:
  $a = 3.47$
  $c = 6.10$

3 STRONGEST DIFFRACTION LINES:
 3.03  (100)
 1.510 ( 80)
 1.918 ( 70)
HARDNESS: 4–5
DENSITY: 9.5 (Meas.)
       10.0 (Calc.)
CLEAVAGE: {011}
Fracture conchoidal.
 HABIT: As irregular grains or distorted crystal fragments.
 COLOR-LUSTER: Steel gray. Opaque. Metallic.
 MODE OF OCCURRENCE: Occurs associated with native platinum, sperrylite, and other precious-metal minerals, in concentrates from the platinum-bearing norites of the Bushveld Complex, Transvaal, South Africa.
 BEST REF. IN ENGLISH: Palache, et al., "Dana's System of Mineralogy," 7th Ed., v. I, p. 258–259, New York, Wiley, 1944.

# COPIAPITE

$(Fe, Mg)Fe_4^{3+}(SO_4)_6(OH)_2 \cdot 20H_2O$

CRYSTAL SYSTEM: Triclinic
CLASS: $\bar{1}$
SPACE GROUP: $P\bar{1}$
Z: 1
LATTICE CONSTANTS:
 $a = 7.33$ kX    $\alpha = 93°51'$
 $b = 18.15$     $\beta = 101°30'$
 $c = 7.27$      $\gamma = 99°23'$
3 STRONGEST DIFFRACTION LINES:
 10.5 (100)
 3.43 ( 40)
 3.06 ( 40)
OPTICAL CONSTANTS:
  $\alpha = 1.509$
  $\beta = 1.532$
  $\gamma = 1.577$
 (+)2V = 73°
HARDNESS: 2½–3
DENSITY: 2.08–2.17 (Meas.)
       2.221 (Calc.)
CLEAVAGE: {010} perfect
       {$\bar{1}$01} imperfect
 HABIT: Crystals thin tabular, sometimes highly modified. Usually as loose aggregates composed of tiny scales; also as crusts, granular. Twinning on {010}, rare.
 COLOR-LUSTER: Golden yellow to yellow to orangish yellow; also greenish yellow to olive green. Transparent to translucent. Pearly.
 MODE OF OCCURRENCE: Occurs as a secondary mineral formed by the oxidation of pyrite or other sulfides, often associated with melanterite and other sulfates. Found as small brilliant yellow plates in the Dexter No. 7 mine, Calf Mesa, San Rafael Swell, Utah; in Alameda, Contra Costa, Kern, Lake, Napa, Riverside, San Bernardino, Shasta, and Trinity counties, California; in the Gilman district, Eagle County, Colorado; in Cochise and Yavapai counties, Arizona; and in Storey and Washoe counties, Nevada. It also is found at Chuquicamata and at other places in northern Chile; and at localities in France, Italy, Sweden, Germany, and Czechoslovakia.

BEST REF. IN ENGLISH: Palache, et al., "Dana's System of Mineralogy," 7th Ed., v. II, p. 623–627, New York, Wiley, 1951.

# COPPER

Cu

CRYSTAL SYSTEM: Cubic
CLASS: $4/m\,\bar{3}\,2/m$
SPACE GROUP: Fm3m
Z: 4
LATTICE CONSTANT:
 $a = 3.615$
3 STRONGEST DIFFRACTION LINES:
 2.088 (100)
 1.808 ( 46)
 1.278 ( 20)
HARDNESS: 2½–3
DENSITY: 8.94 (Meas.)
       8.929 (Calc.)
CLEAVAGE: None. Fracture hackly. Malleable and ductile.
 HABIT: Crystals cubic, octahedral, dodecahedral, tetrahexahedral. Commonly elongated, flattened, or distorted. Also arborescent, wire-like, massive, or powdery. Twinning on {111} common.
 COLOR-LUSTER: Pale rose; tarnishes rapidly to copper red, then to brown. Opaque. Metallic. Streak shining pale red.
 MODE OF OCCURRENCE: Occurs chiefly in the oxidation zone of copper-bearing sulfide ore deposits; in conglomerates and other sedimentary rocks near contacts with basic extrusive rocks; in cavities in basalts; and in sandstones, limestones, and other sedimentary deposits. Found as superb crystals up to one inch or more in size, and as masses up to 420 tons in weight in northern Michigan; as fine crystals at several places in Arizona, especially at Ajo and Bisbee; at Santa Rita, New Mexico; and in small amounts at many other localities in the United States. It also occurs in Canada, Mexico, Bolivia, Chile, England, Scotland, Italy, Germany, USSR, Australia, and Tsumeb, South West Africa.
 BEST REF. IN ENGLISH: Palache, et al., "Dana's System of Mineralogy," v. I, p. 99–102, New York, Wiley, 1944.

# COPPERAS = Melanterite

# COQUIMBITE

$Fe_{2-x}Al_x(SO_4)_3 \cdot 9H_2O$
Dimorphous with paracoquimbite

CRYSTAL SYSTEM: Hexagonal
CLASS: $\bar{3}\,2/m$
SPACE GROUP: $P\bar{3}c$
Z: 4
LATTICE CONSTANTS:
 $a = 10.922$
 $c = 17.084$

3 STRONGEST DIFFRACTION LINES:
8.26 (100)
2.759 ( 75)
5.45 ( 65)
OPTICAL CONSTANTS:
$\omega = 1.539$
$\epsilon = 1.548$
(+)
HARDNESS: 2½
DENSITY: 2.11 (Meas.)
2.109 (Calc.)
CLEAVAGE: $\{10\bar{1}1\}$ imperfect
$\{10\bar{1}0\}$ difficult
Fracture subconchoidal to uneven. Brittle.

HABIT: Crystals short prismatic, up to 15 mm long, with large basal pinacoids and small unequally developed pyramidal faces; pyramidal. Also massive, granular.

COLOR-LUSTER: Colorless to pale lavender to purple; also yellowish or greenish. Transparent. Vitreous.

Soluble in water. Taste astringent.

MODE OF OCCURRENCE: Occurs as superb large crystals associated with halotrichite, voltaite, roemerite, and other sulfates at the Dexter No. 7 mine, Calf Mesa, San Rafael Swell, Utah; at numerous localities in California; at Bisbee and Jerome, Arizona; and in Eagle County, Colorado. It is found abundantly at several places in northern Chile; at Szomolnok, Hungary; and in Germany, Spain, and Skouriotissa, Cyprus.

BEST REF. IN ENGLISH: Palache, et al., "Dana's System of Mineralogy," 7th Ed., v. II, p. 532–534, New York, Wiley, 1951. Fang, J. H., and Robinson, Paul D., *Am. Min.*, **55**: 1534–1540 (1970).


# CORDIERITE (Iolite, Dichroite)
$(Mg, Fe^{3+})_2 Al_4 Si_5 O_{18}$

CRYSTAL SYSTEM: Orthorhombic
CLASS: 2/m 2/m 2/m
SPACE GROUP: Cccm
Z: 4
LATTICE CONSTANTS:
$a \simeq 9.7$
$b \simeq 17.1$
$c \simeq 9.4$
3 STRONGEST DIFFRACTION LINES:
3.13 (100)
8.54 ( 80)
8.45 ( 80)
OPTICAL CONSTANTS:
$\alpha = 1.522\text{-}1.558$
$\beta = 1.524\text{-}1.574$
$\gamma = 1.527\text{-}1.578$
$(\pm)2V_\alpha = 65°\text{-}104°$
HARDNESS: 7-7½
DENSITY: 2.53-2.78 (Meas.)
CLEAVAGE: {010} distinct
{001} indistinct
{100} indistinct
Fracture conchoidal. Brittle.

HABIT: Crystals short prismatic, rectangular in cross section. Usually massive, compact, or as embedded grains.

Twinning on {110} or {130}, common; simple, cyclic, or lamellar. On {021} or {101}, rare.

COLOR-LUSTER: Various shades of blue or bluish violet, also smoky blue; rarely greenish, gray, yellowish, brown. Often exhibits strong pleochroism. Transparent to translucent. Vitreous. Streak uncolored.

MODE OF OCCURRENCE: Occurs in thermally altered aluminum-rich rocks; in gneisses and schists; in norites, granites, pegmatites, andesites, and other rocks, often formed by the assimilation of argillaceous material; and in alluvial gravels. Associated minerals include quartz, andalusite, sillimanite, spinel, and biotite. Found in California, Idaho, Colorado, Wyoming, South Dakota, and other western states; at Guilford and elsewhere in Connecticut; and in New Hampshire and New York. Notable occurrences also are found in Canada, Greenland, Scotland, England, Norway, Germany, Finland, Madagascar, Japan, Australia, and Antarctica.

BEST REF. IN ENGLISH: Deer, Howie, and Zussman, "Rock Forming Minerals," v. 1, p. 268–299, New York, Wiley, 1962.


# CORDYLITE
$(Ce, La)_2 Ba(CO_3)_3 F_2$

CRYSTAL SYSTEM: Hexagonal
CLASS: 6/m 2/m 2/m
SPACE GROUP: $P6_3/mmc$
Z: 2
LATTICE CONSTANTS:
$a = 5.1$
$c = 22.8$
OPTICAL CONSTANTS:
$\omega = 1.764$
$\epsilon = 1.577$
(–)
HARDNESS: 4½
DENSITY: 5.61 (Calc.)
CLEAVAGE: {0001} distinct (may be parting)
Fracture conchoidal. Brittle.

HABIT: Crystals minute, short prismatic with hexagonal dipyramidal terminations. Sometimes sceptre-shaped like quartz.

COLOR-LUSTER: Colorless to yellowish. Transparent when fresh; surfaces sometimes altered and dull. Greasy to adamantine; pearly on {0001}.

MODE OF OCCURRENCE: Occurs sparsely in pegmatitic veins in nepheline-syenite, associated with aegirine, synchisite, ancylite, and neptunite, at Narsarsuk, Greenland.

BEST REF. IN ENGLISH: Palache, et al., "Dana's System of Mineralogy," 7th Ed., v. II, p. 285–287, New York, Wiley, 1951.


# CORKITE
$PbFe_3 PO_4 SO_4 (OH)_6$

CRYSTAL SYSTEM: Hexagonal
CLASS: 3m
SPACE GROUP: R3m
Z: 3

LATTICE CONSTANTS:
$a = 7.22$
$c = 16.66$
3 STRONGEST DIFFRACTION LINES:
3.03  (100)
5.86  ( 65)
2.237 ( 55)
OPTICAL CONSTANTS:
$\omega = 1.93$
$\omega - \epsilon$ weak
uniaxial $(-)$
HARDNESS: 3½–4½
DENSITY: 4.295 (Meas.)
          4.423 (Calc.)
CLEAVAGE: {0001} perfect
HABIT: Crystals, rhombohedral, commonly pseudocubic. Also as loosely coherent fine-granular aggregates.
COLOR-LUSTER: Dark green, yellowish green to pale yellow. Vitreous to resinous; fine-granular aggregates shiny to dull.
MODE OF OCCURRENCE: Occurs in abundance associated with descloizite, vanadinite, pyromorphite, cerussite, and galena, at the Silver Queen mine, Galena, Lawrence County, South Dakota. It also is found at the Harrington-Hickory and Wild Bill mines, Beaver County, and at the Gold Hill and Honerine mines, Tooele County, Utah; in the Glendore iron mine, County Cork, Ireland; and at localities in Germany, Sardinia, Yugoslavia, and USSR.
BEST REF. IN ENGLISH: Palache, et al., "Dana's System of Mineralogy," 7th Ed., v. II, p. 1002–1003, New York, Wiley, 1951.

## CORNETITE
$Cu_3PO_4(OH)_3$

CRYSTAL SYSTEM: Orthorhombic
CLASS: 2/m 2/m 2/m
SPACE GROUP: Pbca
Z: 8
LATTICE CONSTANTS:
$a = 10.86$ kX
$b = 14.07$
$c = 7.10$
3 STRONGEST DIFFRACTION LINES:
3.04 (100)
4.29 ( 90)
3.17 ( 80)
OPTICAL CONSTANTS:
    $\alpha = 1.765$
    $\beta = 1.81$
    $\gamma = 1.82$
$(-)2V \sim 33°$
HARDNESS: ~4½
DENSITY: 4.10 (Meas.)
          4.10 (Calc.)
CLEAVAGE: None.
Brittle.
HABIT: Crystals short prismatic [001] to equant. {210} often rounded. As crusts of minute crystals.
COLOR-LUSTER: Dark blue to greenish blue. Transparent to translucent. Vitreous.

MODE OF OCCURRENCE: Occurs as fine dark blue single crystals and crusts at the Blue Jay and Empire-Nevada mines, Yerington, Nevada; in association with libethenite, brochanite, pseudomalachite, malachite, atacamite, and chrysocolla at Saginaw Hill, about seven miles southwest of Tucson, Arizona; as incrustations of dark blue crystals on compact, gray argillaceous sandstone at Bwana Mkubwa, Zambia; and at the Etoile du Congo mine, Katanga, Zaire, associated with pseudomalachite on fine-grained argillaceous sandstone.
BEST REF. IN ENGLISH: Berry, L. G., *Am. Min.,* **35**: 365–385 (1950). Palache, et al., "Dana's System of Mineralogy," 7th Ed., v. II, p. 789–791, New York, Wiley, 1951.

## CORNUBITE
$Cu_5(AsO_4)_2(OH)_4$

CRYSTAL SYSTEM: Triclinic
3 STRONGEST DIFFRACTION LINES:
4.72 (100)
2.562 (100)
2.489 (100)
HARDNESS: Not determined.
DENSITY: 4.64 (Meas.)
          4.8 (Calc.)
CLEAVAGE: Not determined.
HABIT: Fibrous; also massive, porcelaneous.
COLOR-LUSTER: Light green, apple green to dark green.
MODE OF OCCURRENCE: Occurs at five localities in Cornwall, one in Devon, and one in Cumberland, England. Associated minerals include cornwallite, olivenite, liroconite, and malachite.
BEST REF. IN ENGLISH: Claringbull, G. F., Hey, M. H., and Davis, R. J., *Min. Mag.,* **32**: 1–5 (1959).

## CORNWALLITE (Erinite)
$Cu_5(AsO_4)_2(OH)_4 \cdot H_2O$

CRYSTAL SYSTEM: Monoclinic
CLASS: 2/m
SPACE GROUP: $P2_1/a$
Z: 2
LATTICE CONSTANTS:
$a = 17.61$
$b = 5.81$
$c = 4.60$
$\beta = 92°15'$
3 STRONGEST DIFFRACTION LINES:
3.22 (100)
3.53 ( 90)
3.10 ( 90)
OPTICAL CONSTANTS:
    $\alpha = 1.81$
    $\beta = 1.815$
    $\gamma = 1.85$
$(+)2V =$ small
HARDNESS: 4½
DENSITY: 4.52 (Meas.)
          4.645 (Calc.)

CLEAVAGE: Fracture conchoidal. Not very brittle.

HABIT: As radiating botryoidal crusts up to 1 mm in thickness.

COLOR-LUSTER: Light green to very dark green. Sub-translucent. Dull, slightly resinous.

MODE OF OCCURRENCE: Occurs associated with olivenite, clinoclasite, azurite, malachite, and enargite at the Mammoth mine and Centennial Eureka mine, Tintic District, Utah; also at Eureka and Majuba Hill, Nevada; County Limerick, Ireland (erinite); and at several localities in Cornwall, England; in Scotland; and in Germany.

BEST REF. IN ENGLISH: Berry, L. G., *Am. Min.*, **36**: 484–503 (1951).

## CORONADITE
$Pb(Mn^{2+}, Mn^{4+})_8O_{16}$
Cryptomelane Group

CRYSTAL SYSTEM: Tetragonal
CLASS: 4/m
SPACE GROUP: I4/m
Z: 1
LATTICE CONSTANTS:
 $a = 9.84$
 $c = 2.85$
3 STRONGEST DIFFRACTION LINES:
 3.104 (100)
 3.466 ( 60)
 1.542 ( 50)
HARDNESS: 4½–5
DENSITY: 5.44 (Meas.)
     4.88 (Calc.)
CLEAVAGE: Not determined.
HABIT: Massive; as botryoidal crusts with fibrous structure.
COLOR-LUSTER: Black to dark gray. Opaque. Sub-metallic to dull. Streak brownish black.
MODE OF OCCURRENCE: Occurs in the oxidized zone at several mines in the Inyo-Argus Mountain area, Inyo County, California; in the upper levels of the Coronado vein, Clifton-Morenci district, Arizona; at Broken Hill, New South Wales, Australia; and in a manganese deposit at Bou Tazoult, Morocco.
BEST REF. IN ENGLISH: Palache, et al., "Dana's System of Mineralogy," 7th Ed., v. I, p. 742–743, New York, Wiley, 1944.

## CORRENSITE = Interlayered chlorite-vermiculite
Aluminosilicate of Mg and Fe

## CORUNDOPHILITE (Variety of chlorite)
$(Mg, Fe, Al)_6(Si, Al)_4O_{10}(OH)_8$

CRYSTAL SYSTEM: Triclinic, Monoclinic
CLASS: 2/m, 1
SPACE GROUP: C2/m, P1
Z: 1

LATTICE CONSTANTS:
 triclinic
 $a = 5.35$     $\alpha = 97°22'$
 $b = 14.36$    $\beta = 119°56'$
 $c = 5.34$     $\gamma = 86°20'$
OPTICAL CONSTANTS:
  $\alpha = 1.58$
  $\beta = 1.584$
  $\gamma = 1.596$
 $(+)2V \leqslant 30°$
HARDNESS: 2–3
DENSITY: 2.85 (Meas.)
CLEAVAGE: {001} perfect.
Thin laminae flexible, inelastic.
HABIT: Massive, foliated or granular. Twinning on {001}.
COLOR-LUSTER: Dark green. Translucent. Somewhat pearly on cleavage.
MODE OF OCCURRENCE: Occurs in association with margarite in the emery deposits at Chester, Hampden County, Massachusetts, and in the Black Lake area, Quebec, Canada.
BEST REF. IN ENGLISH: Deer, Howie, and Zussman, "Rock Forming Minerals," v. 3, p. 131–163, New York, Wiley, 1962.

## CORUNDUM
$\alpha - Al_2O_3$
Var.: Emery
   Ruby
   Sapphire

CRYSTAL SYSTEM: Hexagonal
CLASS: $\bar{3}$ 2/m
SPACE GROUP: R$\bar{3}$c
Z: 6
LATTICE CONSTANTS:
 $a = 4.751$
 $c = 12.97$
3 STRONGEST DIFFRACTION LINES:
 2.085 (100)
 2.552 ( 90)
 1.601 ( 80)
OPTICAL CONSTANTS:
 $\omega = 1.7653–1.7760$
 $\epsilon = 1.7573–1.7677$
 (–)
HARDNESS: 9
DENSITY: 4.0–4.1 (Meas.)
     3.987 (Calc.)
CLEAVAGE: None. {0001} parting, often perfect; interrupted. {10$\bar{1}$1} parting, often prominent. Fracture conchoidal to uneven. Brittle; tough when compact.
HABIT: Well-developed crystals common, often large. Steep-pyramidal, prismatic, tabular, rhombohedral; rough or rounded barrel-shaped; often deeply furrowed or striated. Also massive, compact, granular (emery), or lamellar. Twinning on {10$\bar{1}$1}, common, lamellar; also penetration or arrowhead twins. On {0001}, less common.
COLOR-LUSTER: Commonly gray, brown, bluish; also various shades of red, green, blue, yellow, orange, purple,

and colorless. Sometimes multicolored. Transparent to translucent. Vitreous to adamantine, sometimes pearly on base. Sometimes fluorescent under ultraviolet light.

MODE OF OCCURRENCE: Occurs widespread in crystalline limestone and dolomite, gneiss, chlorite schist, mica schist, granite, nepheline syenite, and other crystalline rocks. Gem varieties occur chiefly in placer deposits. Notable localities in the United States are found in California, Yogo Gulch and elsewhere in Montana, Colorado, Pennsylvania, Massachusetts, Connecticut, New York, New Jersey, North and South Carolina, Virginia, Alabama, and Georgia. Other important localities occur in Canada, Norway, Sweden, France, Italy, Switzerland, Greece, USSR, Ceylon, India, Burma, Thailand, Afghanistan, Japan, Transvaal, South Africa, Madagascar, and Australia.

BEST REF. IN ENGLISH: Palache, et al., "Dana's System of Mineralogy," 7th Ed., v. I, p. 520–527, New York, Wiley, 1944.

# CORVUSITE
$V_2O_4 \cdot 6V_2O_5 \cdot nH_2O$

CRYSTAL SYSTEM: Unknown
LATTICE CONSTANTS:
$a = 11.6$
$b = 3.65$
$c = ?$
$\gamma = 90°$
3 STRONGEST DIFFRACTION LINES:
1.796 (100)
2.28 ( 70)
3.11 ( 50)
HARDNESS: 2½–3
DENSITY: 2.82 (?)
CLEAVAGE: Not determined. Fracture conchoidal.
HABIT: Massive; irregular-shaped flakes less than 0.2 microns in size.
COLOR-LUSTER: Brown, bluish black, greenish black. Opaque. Streak same as color.
MODE OF OCCURRENCE: Occurs as an impregnation in sandstone at numerous mines in the Edgemont uranium-vanadium mining district, Fall River County, South Dakota; at the Jack Claim, La Sal Mountains, Grand County, Utah; at the Ponto No. 3 claim, San Miguel County, Colorado; and widespread at other localities in the carnotite region of the Colorado Plateau. Also found at Balasauskandyk, N. W. Kara-Tau, Kazakhstan, USSR.
BEST REF. IN ENGLISH: Palache, et al., "Dana's System of Mineralogy," 7th Ed., v. I, p. 602–603, New York, Wiley, 1944.

# CORYNITE = Arsenian ullmannite

# COSALITE
$Pb_2Bi_2S_5$

CRYSTAL SYSTEM: Orthorhombic
CLASS: 2/m 2/m 2/m
SPACE GROUP: Pbnm
Z: 8

LATTICE CONSTANTS:
$a = 19.09$
$b = 23.89$
$c = 4.058$
3 STRONGEST DIFFRACTION LINES:
3.44 (100)
2.81 ( 30)
3.37 ( 25)
HARDNESS: 2½–3
DENSITY: 6.86–6.99 (Meas.)
7.12 (Calc.)
CLEAVAGE: Not determined. Fracture uneven. Capillary fibers flexible.
HABIT: Crystals prismatic, commonly elongated to needle-like and capillary forms. Often massive, dense, or in radiating prismatic, fibrous, or feathery aggregates.
COLOR-LUSTER: Lead gray to steel gray to silver white. Opaque. Metallic. Streak black.
MODE OF OCCURRENCE: Occurs chiefly in medium-temperature hydrothermal deposits, in metamorphic deposits, and in pegmatites. Found in the United States at the Homestake mine, Lead, South Dakota, associated with pyrite, sphalerite, carbonates, and quartz; in Colorado at the Comstock mine, LaPlata County, the Gladiator mine, Ouray County, and at the Yankee Girl mine, San Juan County; and at Deer Park, Stevens County, Washington. It also occurs in Mexico, Canada, Sweden, Switzerland, Hungary, Roumania, Australia, Japan, and Madagascar.
BEST REF. IN ENGLISH: Palache, et al., "Dana's System of Mineralogy," 7th Ed., v. I, p. 445–447, New York, Wiley, 1944.

# COSTIBITE
CoSbS
Polymorph of paracostibite

CRYSTAL SYSTEM: Orthorhombic
CLASS: mm2
SPACE GROUP: $Pmn2_1$
Z: 2
LATTICE CONSTANTS:
$a = 3.603$
$b = 4.868$
$c = 5.838$
3 STRONGEST DIFFRACTION LINES:
2.596 (100)
2.503 ( 90)
1.908 ( 80)
OPTICAL CONSTANTS: Weakly bireflecting and weakly anisotropic.
HARDNESS: VHN microhardness 781 kg/mm² with 15 g load.
DENSITY: 6.89 (Calc.)
HABIT: Massive; as lamellae up to 0.2 X 1.4 mm in size.
COLOR-LUSTER: Grayish. Opaque. Metallic.
MODE OF OCCURRENCE: Occurs at the Consols Mine, Broken Hill, New South Wales, Australia, as inclusions in loellingite with associated ullmannite, willyamite, and pyrargyrite.
BEST REF. IN ENGLISH: Cabri, L. J., Harris, D. C., and Stewart, J. M., *Am. Min.*, 55: 10–17 (1970).

## COTUNNITE
$PbCl_2$

CRYSTAL SYSTEM: Orthorhombic
CLASS: 2/m 2/m 2/m
SPACE GROUP: Pmnb
Z: 4
LATTICE CONSTANTS:
$a$ = 7.67 kX
$b$ = 9.15
$c$ = 4.50
3 STRONGEST DIFFRACTION LINES:
3.581 (100)
3.890 ( 73)
2.778 ( 56)
OPTICAL CONSTANTS:
$\alpha$ = 2.199
$\beta$ = 2.217 (Na)
$\gamma$ = 2.260
(+)2V = 67°12′ (Calc.)
HARDNESS: 2½
DENSITY: 5.88 (Calc.)
CLEAVAGE: {010}, perfect
Fracture subconchoidal. Somewhat sectile.
HABIT: Usually massive, granular. Crystals often flattened {010} and elongated along $c$-axis.
COLOR-LUSTER: Colorless, white, yellowish, greenish. Transparent to translucent. Adamantine; also pearly or silky.

Soluble in water.
MODE OF OCCURRENCE: Occurs as an alteration product of galena in the Tintic district, Juab County, Utah; as thin veinlets in chalcocite in the Bentley district, Mohave County, Arizona; in Antofagasta and Tarapacá provinces, Chile; in the La Pampa district, Pallasca province, Peru; and as a sublimation product in fumaroles on Mt. Vesuvius, Italy.
BEST REF. IN ENGLISH: Palache, et al., "Dana's System of Mineralogy," 7th Ed., v. II, p. 42-44, New York, Wiley, 1951.

## COULSONITE
$FeV_2O_4$
Spinel Group

CRYSTAL SYSTEM: Cubic
CLASS: 4/m $\bar{3}$ 2/m
SPACE GROUP: Fd3m
Z: 8
LATTICE CONSTANT:
$a$ = 8.297
3 STRONGEST DIFFRACTION LINES:
2.50 (100)
2.07 ( 80)
2.93 ( 60)
HARDNESS: 4.5-5
DENSITY: 5.17-5.20 (Meas.)
5.15 (Calc.)
HABIT: As microscopic subhedral crystals.
COLOR-LUSTER: Bluish gray; metallic. Opaque. Powdered form dark brown to black.

MODE OF OCCURRENCE: Found closely associated with magnetite and chlorine-rich scapolite in the metamorphosed igneous rocks of the Buena Vista Hills located some 20 miles southeast of Lovelock, Nevada.
BEST REF. IN ENGLISH: Radtke, Arthur S., *Am. Min.*, 47: 1284-1291 (1962).

## COUSINITE (Species status uncertain.)
$MgU_2^{4+}Mo_2O_{11} \cdot 6H_2O$ (?)

CRYSTAL SYSTEM: Unknown
HABIT: Thin blades.
COLOR-LUSTER: Black. Vitreous.
MODE OF OCCURRENCE: Occurs as an alteration product of ore containing uraninite and molybdenite at the Shinkolobwe mine, Katanga, Zaire.
BEST REF. IN ENGLISH: Vaes, J. F., *Am. Min.*, 44: 910 (1959).

## COVELLITE (Covelline)
CuS

CRYSTAL SYSTEM: Hexagonal
CLASS: 6/m 2/m 2/m
SPACE GROUP: P6_3/mmc
Z: 6
LATTICE CONSTANTS:
$a$ = 3.802
$c$ = 16.43
3 STRONGEST DIFFRACTION LINES:
2.813 (100)
1.896 ( 75)
3.048 ( 65)
OPTICAL CONSTANTS:
$\omega$ = 1.45
$\epsilon$ = ? (Na)
(+)
HARDNESS: 1½-2
DENSITY: 4.681 (Meas.)
4.68 (Calc.)
CLEAVAGE: {0001} perfect, easy
Thin laminae flexible. Fracture uneven. Brittle.
HABIT: Crystals thin tabular hexagonal plates, flattened on {0001}, and frequently exhibiting hexagonal striations on basal pinacoid. Usually massive, foliated.
COLOR-LUSTER: Light to very dark indigo blue, often with bright purplish iridescence. Opaque. Submetallic to dull. Streak shining gray-black.
MODE OF OCCURRENCE: Occurs chiefly in the zone of secondary enrichment in copper sulfide deposits, commonly associated with chalcopyrite, pyrite, chalcocite, bornite, and enargite. Found in the United States as superb crystal groups and large foliated masses at Butte, Montana; and at localities in Wyoming, South Dakota, Colorado, Utah, California, and Alaska. It also occurs in Argentina, New Zealand, Philippines, Germany, Austria, Yugoslavia, and as fine crystals and crystal groups at the Calabona mine, Alghero, Sardinia, Italy.
BEST REF. IN ENGLISH: Palache, et al., "Dana's System of Mineralogy," 7th Ed., v. I, p. 248-251, New York, Wiley, 1944.

## CRANDALLITE (Pseudowavellite)
$CaAl_3(PO_4)_2(OH)_5 \cdot H_2O$

CRYSTAL SYSTEM: Triclinic, hexagonal
SPACE GROUP: R lattice
Z: 3
LATTICE CONSTANTS:
  (triclinic dimorph)
    $a = 7.010$    $\alpha = 103°10'$
    $b = 9.819$    $\beta = 91°14'$
    $c = 9.697$    $\gamma = 90°34'$
3 STRONGEST DIFFRACTION LINES:
  (hexagonal)    (triclinic dimorph)
  2.95 (100)     2.97 (100)
  2.98 ( 50)     2.18 ( 45)
  2.16 ( 40)     5.75 ( 35)
OPTICAL CONSTANTS:
  (hexagonal)    (triclinic)
  $\omega = 1.60-1.622$   $\alpha = 1.602$
  $\epsilon = 1.61-1.631$   $\beta = 1.608$
    (+)       $\gamma = 1.615$
         $(+)2V = 70°-75°$
HARDNESS: 5
DENSITY: 2.78-2.92 (Meas.)
CLEAVAGE: {0001} perfect
HABIT: Crystals minute trigonal prisms terminated by {0001}, or as rosettes of fibers. Commonly massive, as nodular masses or spherules with fibrous, fine-granular, or chalcedony-like structure.
COLOR-LUSTER: Yellow to yellowish white to white or gray. Transparent to subtranslucent. Vitreous; also dull and chalky.
MODE OF OCCURRENCE: Occurs abundantly associated with a wide variety of rare secondary phosphate minerals in variscite nodules near Fairfield, Utah County; with barite and quartz at the Brooklyn mine, Tintic District, Juab County; and at other places in Utah. It also is found in Esmeralda County, Nevada; at the Everly and Hugo mines, Keystone, Pennington County, South Dakota; in Lawrence County, Indiana; and in Brazil, Bolivia, Germany and Senegal. An unnamed apparently triclinic dimorph of crandallite occurs in the sediments of the Bajo de Santa Fé, El Patén, Guatamala.
BEST REF. IN ENGLISH: Palache, et al., "Dana's System of Mineralogy," 7th Ed., v. II, p. 835-837, New York, Wiley, 1951. *Am. Min.*, **48**: 1144-1153 (1963).

## CREDNERITE
$CuMnO_2$

CRYSTAL SYSTEM: Monoclinic
CLASS: 2/m
SPACE GROUP: A2/m
Z: 2
LATTICE CONSTANTS:
  $a = 5.898$
  $b = 2.884$  (synthetic)
  $c = 5.530$
  $\beta = 104°36'$

3 STRONGEST DIFFRACTION LINES:
  2.71 (100)
  2.42 (100)
  2.85 ( 80)
HARDNESS: 4
DENSITY: 5.34 (Meas.)
        5.46 (Calc.)
CLEAVAGE: {001} perfect
        {100} perfect
        {$\bar{1}$11} good
        {111} poor
        {212} very poor
HABIT: As thin six-sided plates in radiating hemispherical, or spherulitic groupings; also as earthy coatings.
COLOR-LUSTER: Iron black; opaque; bright metallic. Streak black with brownish tint.
MODE OF OCCURRENCE: Occurs intergrown with hausmannite and psilomelane as a secondary mineral at Friedrichroda, Thuringia, Germany; with cerussite, hydrocerussite, and malachite at Higher Pitts, Mendip Hills, Somerset, England; and massive near Calistoga, Napa County, California.
BEST REF. IN ENGLISH: McAndrew, John, *Am. Min.*, **41**: 276-287 (1956). "Dana's System of Mineralogy," 7th Ed., v. 1, p. 722-723, New York, Wiley, 1944. Gaudefroy, C., Dietrich, J., Permingeat, F., and Picot, P., *Am. Min.*, **51**: 1819 (1966).

## CREEDITE
$Ca_3Al_2SO_4(F, OH)_{10} \cdot 2H_2O$

CRYSTAL SYSTEM: Monoclinic
CLASS: 2/m
SPACE GROUP: C2/c
Z: 4
LATTICE CONSTANTS:
  $a = 13.88$ kX
  $b = 8.56$
  $c = 9.98$
  $\beta = 94°24'$
3 STRONGEST DIFFRACTION LINES:
  3.48 (100)
  7.3 ( 90)
  6.9 ( 90)
OPTICAL CONSTANTS:
    $\alpha = 1.461$
    $\beta = 1.478$ (Na)
    $\gamma = 1.485$
  $(-)2V = 64°22'$
HARDNESS: 4
DENSITY: 2.713 (Meas.)
CLEAVAGE: {100} perfect
Fracture conchoidal. Brittle.
HABIT: Crystals short prismatic to acicular; as sprays or clusters. Also as radiated aggregates, as drusy knobby masses, and as embedded crystals and grains.
COLOR-LUSTER: Colorless, white, amethystine. Transparent. Vitreous.
MODE OF OCCURRENCE: Occurs in cavities of banded fluorite, as embedded radial masses of crystals in fine-granular white barite, and as loose, doubly terminated

crystals embedded in halloysite in the fluorite-barite mine at Wagon Wheel Gap, near Creede, Mineral County, Colorado. In California clusters of crystals occur embedded in pyrite and lining vugs at Darwin, Inyo County; and in Nevada it was found in the oxidized zone of fluorite-quartz veins at Granite, Nye County. It also occurs as superb groups composed of crystals exceeding one inch in length, often associated with gypsum and galena, at Santa Eulalia, Chihuahua, Mexico; and as fine crystal aggregates in the tin veins of Colquiri, Bolivia.

BEST REF. IN ENGLISH: Palache, et al., "Dana's System of Mineralogy," 7th Ed., v. II, p. 129–130, New York, Wiley, 1951.

# CRICHTONITE = Ilmenite

# α-CRISTOBALITE
$SiO_2$
Polymorphous with quartz, tridymite, coesite, stishovite
Var. Lussatite

CRYSTAL SYSTEM: Tetragonal
CLASS: 4 2 2
SPACE GROUP: $P4_12_12$ or $P4_32_12$
Z: 4
LATTICE CONSTANTS:
  $a = 4.97$
  $c = 6.91–6.93$
3 STRONGEST DIFFRACTION LINES:
  4.05 (100)
  2.485 ( 20)
  2.841 ( 13)
OPTICAL CONSTANTS:
  $\omega = 1.484$
  $\epsilon = 1.487$ (white light)
    (–)
HARDNESS: 6½
DENSITY: 2.33 ± 0.01 (Meas.)
         2.32 (Calc.)
CLEAVAGE: None.
Brittle.
HABIT: Crystals usually octahedral, rarely cubical, under 4.0 mm in size. Also massive; stalactitic; spherulitic; as crusts and botryoidal aggregates. Massive varieties distinctly fibrous. Twinning on {111}common.
COLOR-LUSTER: White or milky white, gray, bluish gray, yellowish, brownish, etc. Translucent to opaque.
MODE OF OCCURRENCE: Occurs widespread chiefly in igneous rocks such as andesites, rhyolites, trachytes, and obsidian, or as opal or opaline silica formed at relatively low temperatures. Found at many localities in the western United States, notably as grayish spherulites in obsidian near Coso Hot Springs, Inyo County, California; at Crater Lake, Oregon; in Yellowstone Park, Wyoming; in the San Juan region of Colorado; as octahedra and spinel twins up to 4 mm in size (inverted from beta-cristobalite) in andesite at Cerro San Cristobal, Pachuca, Mexico; and at localities in France, Germany, Hungary, Czechoslovakia, USSR, India, Japan, New Zealand, and elsewhere. Localities for opaline varieties are listed under opal.

BEST REF. IN ENGLISH: Frondel, C., "Dana's System of Mineralogy," v. III, p. 273–286, New York, Wiley, 1962.

# CROCIDOLITE = Fibrous variety of riebeckite

# CROCOITE
$PbCrO_4$

CRYSTAL SYSTEM: Monoclinic
CLASS: 2/m
SPACE GROUP: $P2_1/n$
Z: 4
LATTICE CONSTANTS:
  $a = 7.10$ kX
  $b = 7.40$
  $c = 6.80$
  $\beta = 102°27'$
3 STRONGEST DIFFRACTION LINES:
  3.28 (100)
  3.03 ( 65)
  3.48 ( 55)
OPTICAL CONSTANTS:
  $\alpha = 2.29$
  $\beta = 2.36$ (Li)
  $\gamma = 2.66$
  (+)2V = 57°
HARDNESS: 2½–3
DENSITY: 5.99 ± 0.03 (Meas.)
         6.108 (Calc.)
CLEAVAGE: {110} distinct
          {001} indistinct
          {100} indistinct
Fracture conchoidal to uneven. Very brittle.
HABIT: Crystals usually slender prismatic, elongated along c-axis; nearly square in cross section, and commonly hollow; often striated parallel to c-axis. Rarely octahedral or acute rhombohedral. As "jackstraw" crystal aggregates; also massive.
COLOR-LUSTER: Brilliant red-orange or orange; less commonly red, yellow. Adamantine to vitreous. Transparent to translucent. Streak orange-yellow.
MODE OF OCCURRENCE: Occurs as a secondary mineral in the oxidation zone of ore deposits containing lead and chromium, often associated with cerussite, pyromorphite, wulfenite, vanadinite, and other secondary minerals. Found in California associated with wulfenite at the Darwin mines, Inyo County, and in the El Dorado mine, near Indio, Riverside County. It also occurs at several places in Arizona, especially at the Mammoth mine, Tiger, Pinal County. Superb crystals up to 3½ inches long, and matrix specimens several feet across, occur at mines in the Dundas district, Tasmania; excellent crystals also occur in mines of the Beresov district and at other places in the USSR; at Goyabeira, Minas Geraes, Brazil; and in Roumania, Rhodesia, and the Philippine Islands.

BEST REF. IN ENGLISH: Palache, et al., "Dana's System of Mineralogy," 7th Ed., v. II, p. 646–649, New York, Wiley, 1951.

# CRONSTEDTITE (a septachlorite)
$Fe_2^{2+}Fe_2^{3+}SiO_5(OH)_4$

CRYSTAL SYSTEM: Monoclinic
CLASS: m
SPACE GROUP: Cm
Z: 2
LATTICE CONSTANTS:
  $a = 5.49$
  $b = 9.52$
  $c = 7.32$
  $\beta = 104°31'$
3 STRONGEST DIFFRACTION LINES:
  7.09 (100)
  3.54 ( 85)
  2.72 ( 50)
OPTICAL CONSTANT:
  (–)
  HARDNESS: 3½
  DENSITY: 3.34–3.45 (Meas.)
           3.586 (Calc.)
  CLEAVAGE: {001} perfect
Thin cleavages somewhat elastic.
  HABIT: Crystals 3- or 6-sided pyramids; also fibrous.
  COLOR-LUSTER: Greenish black, brownish black to black; thin laminae emerald green by transmitted light. Translucent to nearly opaque. Streak dark olive green.
  MODE OF OCCURRENCE: Typical occurrences are found near Ouro Preto, Minas Geraes, Brazil; at Truro and elsewhere in Cornwall, England; at Kuttenberg and Pribřam, Bohemia, Czechoslovakia; and at Kisbánya, Roumania.
  BEST REF. IN ENGLISH: Deer, Howie, and Zussman, "Rock Forming Minerals," v. 3, p. 164–169, New York, Wiley, 1962.

# CROOKESITE
$(Cu, Tl, Ag)_2 Se$

CRYSTAL SYSTEM: Tetragonal
CLASS: 4/m 2/m 2/m
SPACE GROUP: I4/mmm
Z: 2
LATTICE CONSTANTS:
  $a = 10.40$
  $c = 3.93$
3 STRONGEST DIFFRACTION LINES:
  3.29 (100)
  2.59 (100)
  3.00 ( 80)
  HARDNESS: 2½–3
Talmadge hardness C.
  DENSITY: 6.90 (Meas.)
           7.71 (Calc.)
  CLEAVAGE: Two fairly well developed cleavages at right angles. Brittle.
  HABIT: Massive, compact.
  COLOR-LUSTER: Lead gray; opaque; metallic.
  MODE OF OCCURRENCE: Occurs as finely disseminated specks and small veinlets with other selenides in translucent calcite with minor quartz at the copper mine of Skrikerum, Kalmar, Sweden; and associated with clausthalite, umangite, berzelianite, and other minerals at the Pinkey Fault uranium mine, Saskatchewan, Canada.
  BEST REF. IN ENGLISH: Palache, et al., "Dana's System of Mineralogy," 7th Ed., v. I, p. 183, New York, Wiley, 1944. Early, J. W., *Am. Min.*, **35**: 337–364 (1950).

# CROSSITE
$Na_2 (Mg, Fe^{3+})_3 (Fe^{3+}, Al)_2 Si_8 O_{22} (OH)_2$
Amphibole Group

CRYSTAL SYSTEM: Monoclinic
CLASS: 2/m
SPACE GROUP: C2/m
Z: 2
LATTICE CONSTANTS:
  $a = 9.647$
  $b = 17.905$
  $c = 5.316$
  $\beta = 103.60°$
3 STRONGEST DIFFRACTION LINES:
  8.31 (100)
  3.08 ( 75)
  2.71 ( 40)
OPTICAL CONSTANTS:
  $\alpha = 1.659$
  $\beta = 1.670$
  $\gamma = 1.674$
  (+)2V = 59.8°
  HARDNESS: 6
  DENSITY: 3.11–3.211 (Meas.)
           3.223 (Calc.)
  CLEAVAGE: {110} perfect
Fracture uneven. Brittle.
  HABIT: Crystals prismatic, often lath-like. Commonly massive; fibrous, columnar, or granular. Twinning on {100} simple, lamellar.
  COLOR-LUSTER: Blue, grayish. Translucent. Vitreous to dull.
  MODE OF OCCURRENCE: Occurs as a constituent of crystalline schists in Contra Costa, Los Angeles, Mendocino, Orange, and San Diego counties, California; in Custer and Teller counties, Colorado; and at localities in Wales, Scotland, USSR, Japan, and elsewhere.
  BEST REF. IN ENGLISH: Deer, Howie, and Zussman, "Rock Forming Minerals," v. 2, p. 333–351, New York, Wiley, 1963.

# CRYOLITE
$Na_3 AlF_6$

CRYSTAL SYSTEM: Monoclinic
CLASS: 2/m
SPACE GROUP: P2₁/n
Z: 2
LATTICE CONSTANTS:
  $a = 5.40$
  $b = 5.60$
  $c = 7.78$
  $\beta = 90°11'$

3 STRONGEST DIFFRACTION LINES:
1.941 (100)
2.747 ( 95)
3.883 ( 60)
OPTICAL CONSTANTS:
$\alpha = 1.3376$
$\beta = 1.3377$ (Na)
$\gamma = 1.3387$
(+)2V = 43°
HARDNESS: 2½
DENSITY: 2.97 (Meas.)
2.963 (Calc.)
CLEAVAGE: None.
Parting on {001} and {110}. Fracture uneven. Brittle.
HABIT: Crystals cuboidal; also short prismatic. {110} faces commonly striated. Usually massive, coarse granular. Twinning very common.
COLOR-LUSTER: Colorless, white, brownish, reddish, rarely gray to black. Transparent to translucent. Vitreous to greasy. Streak white.
MODE OF OCCURRENCE: Occurs in Colorado as small masses in quartz-feldspar pegmatite and veins associated with astrophyllite, riebeckite, and zircon crystals at St. Peters Dome, El Paso County; and also in a drill core from the Green River formation of the Piceance Basin in the western part of the state. It occurs abundantly in a large pegmatite at Ivigtut, Greenland, associated with numerous aluminum fluorides, siderite, galena, sphalerite, and other minerals; and also is found in small amounts at Sallent, Huesca Province, Spain, and at Miask, Urals, USSR.
BEST REF. IN ENGLISH: Palache, et al., "Dana's System of Mineralogy," 7th Ed., v. II, p. 110-113, New York, Wiley, 1951.

# CRYOLITHIONITE
$Li_3 Na_3 Al_2 F_{12}$

CRYSTAL SYSTEM: Cubic
CLASS: $4/m\,\overline{3}\,2/m$
SPACE GROUP: Ia3d
Z: 8
LATTICE CONSTANT:
$a = 12.097$ kX
3 STRONGEST DIFFRACTION LINES:
1.96 (100)
4.29 ( 80)
2.21 ( 80)
OPTICAL CONSTANT:
$N = 1.3395$ (Na)
HARDNESS: 2½-3
DENSITY: 2.770 (Meas.)
2.772 (Calc.)
CLEAVAGE: {011} distinct
Fracture subconchoidal to uneven. Brittle.
HABIT: Crystals dodecahedral, up to 17 cm in size, sometimes modified by {112}.
COLOR-LUSTER: Colorless, white. Transparent. Vitreous.
MODE OF OCCURRENCE: Occurs in abundance closely associated with cryolite in pegmatite at Ivigtut, Greenland; and also near Miask, Ilmen Mountains, USSR.

BEST REF. IN ENGLISH: Palache, et al., "Dana's System of Mineralogy," 7th Ed., v. II, p. 99-100, New York, Wiley, 1951.

**CRYOPHYLLITE** = Ferroan lepidolite

**CRYPHIOLITE** = Mixture of apatite and sellaite

# CRYPTOHALITE
$(NH_4)_2 SiF_6$
Dimorphous with bararite

CRYSTAL SYSTEM: Cubic
CLASS: $4/m\,\overline{3}\,2/m$
SPACE GROUP: Fm3m
Z: 4
LATTICE CONSTANT:
$a = 8.337$ kX
3 STRONGEST DIFFRACTION LINES:
4.844 (100)
2.422 ( 45)
2.098 ( 35)
OPTICAL CONSTANT:
$N = 1.369$ (Na)
HARDNESS: 2½
DENSITY: 2.011 (Meas. Synthetic)
2.029 (Calc.)
CLEAVAGE: {111} perfect
HABIT: Massive, as crusts or masses with mammillary surface; arborescent. Synthetic crystals cubo-octahedral or octahedral.
COLOR-LUSTER: Colorless, white, gray. Transparent. Vitreous.

Taste saline.
MODE OF OCCURRENCE: Occurs as a sublimate over burning coal seams at Libusín in the Kladno coal basin, Czechoslovakia, and at Barari, Jharia coal field, India. It also occurs as a sublimation product on Mt. Vesuvius, Italy.
BEST REF. IN ENGLISH: Palache, et al., "Dana's System of Mineralogy," 7th Ed., v. II, p. 104-105, New York, Wiley, 1951.

# CRYPTOMELANE
$K(Mn^{2+}, Mn^{4+})_8 O_{16}$

CRYSTAL SYSTEM: Tetragonal
CLASS: 4/m
SPACE GROUP: I4/m
Z: 1
LATTICE CONSTANTS:
$a = 9.84$
$c = 2.85$
4 STRONGEST DIFFRACTION LINES:
2.39 (100)
6.90 ( 90)
3.10 ( 80)
4.90 ( 80)

HARDNESS: 6–6½ to as low as 1 for massive and fibrous varieties.

DENSITY: ~4.3

4.39 (Calc.)

CLEAVAGE: Not determined. Fracture conchoidal.

HABIT: Usually massive, fine-grained, often loosely aggregated or porous; less commonly compact and cleavable. Also botryoidal or radial-fibrous to distinct individual fibers.

COLOR-LUSTER: Steel gray to bluish gray to dull black; often tarnishes dull grayish black. Opaque. Metallic, submetallic, or dull. Streak brownish black.

MODE OF OCCURRENCE: Occurs as a secondary mineral formed under surface conditions of temperature and pressure, often in association with pyrolusite and other manganese oxides. Typical occurrences are found in the Lookout Mining District in Inyo County, and at the Logan mine, San Bernardino County, California; at Philipsburg, Montana; near Central City, Lawrence County, South Dakota; at Tombstone, Arizona; near Deming, New Mexico; and near Mena, Arkansas. Also at Nsuta, Ghana, and at many other places.

BEST REF. IN ENGLISH: Richmond, Wallace E., and Fleischer, Michael, *Am. Min.*, 27: 607–610 (1942).

## CSIKLOVAITE (Inadequately described mineral)
$Bi_2TeS_2$

CRYSTAL SYSTEM: Unknown

HARDNESS: Not determined.

DENSITY: Not determined.

CLEAVAGE: Not determined.

HABIT: Massive; only observed in polished sections fringing tetradymite.

MODE OF OCCURRENCE: Occurs in small amounts in association with chalcopyrite, tetradymite, and calcite at Csiklova, Rumania.

BEST REF. IN ENGLISH: Vlasov, K. A., "Mineralogy of Rare Elements," v. II, p. 736–737, Israel Program for Scientific Translations, 1966.

## CUBANITE
$CuFe_2S_3$

CRYSTAL SYSTEM: Orthorhombic

CLASS: 2/m 2/m 2/m

SPACE GROUP: Pcmn

Z: 4

LATTICE CONSTANTS:

$a$ = 6.46

$b$ = 11.12

$c$ = 6.23

3 STRONGEST DIFFRACTION LINES:

3.22 (100)

1.867 ( 80)

1.750 ( 70)

HARDNESS: 3½

DENSITY: 4.03–4.18 (Meas.)

4.076 (Calc.)

CLEAVAGE: None.

Parting on {110} and {1$\bar{3}$0}. Fracture conchoidal. Magnetic.

HABIT: Crystals thick tabular, elongated [001]. {001} striated. Usually massive. Twinning on {110} common.

COLOR-LUSTER: Bronze to brass yellow. Opaque. Metallic.

MODE OF OCCURRENCE: Occurs in deposits formed at relatively high temperatures, commonly associated with chalcopyrite, pyrrhotite, pyrite, magnetite, and sphalerite. Found in California at the Daulton mine, Madera County, and in the Walker mine, near Spring Garden, Plumas County; at Fierro, New Mexico; and on Prince William Sound, Alaska. It also occurs abundantly at Sudbury, Ontario, Canada; at Barracanao, Cuba; and at localities in Mexico, Brazil, Italy, Norway, and Sweden.

BEST REF. IN ENGLISH: Palache, et al., "Dana's System of Mineralogy," 7th Ed., v. I, p. 243–246, New York, Wiley, 1944.

## CUMENGÉITE
$Pb_4Cu_4Cl_8(OH)_8 \cdot H_2O$

CRYSTAL SYSTEM: Tetragonal

CLASS: 4/m 2/m 2/m

SPACE GROUP: I4/mmm

Z: 10

LATTICE CONSTANTS:

$a$ = 15.20

$c$ = 24.76

3 STRONGEST DIFFRACTION LINES:

2.38 (100)

4.86 ( 90)

3.73 ( 60)

OPTICAL CONSTANTS:

$\omega$ = 2.041

$\epsilon$ = 1.926

(−)

HARDNESS: 2½

DENSITY: 4.67 (Meas.)

4.60 (Calc.)

CLEAVAGE: {101} good

{110} distinct

{001} poor

HABIT: Crystals octahedral or cubo-octahedral. As parallel overgrowths on boleite and pseudoboleite crystals, sometimes giving regular groupings that simulate twins.

COLOR-LUSTER: Indigo blue. Translucent. Weakly vitreous. Streak sky blue.

MODE OF OCCURRENCE: Occurs as a secondary mineral associated with boleite and pseudoboleite at Boleo, Baja California, Mexico, occasionally as superb simulated twinned crystals as much as 2 cm on an edge.

BEST REF. IN ENGLISH: Palache, et al., "Dana's System of Mineralogy," 7th Ed., v. II, p. 79–80, New York, Wiley, 1951.

## CUMMINGTONITE
$(Mg, Fe)_7Si_8O_{22}(OH)_2$

Amphibole Group

Var. Dannemorite

Amosite

CRYSTAL SYSTEM: Monoclinic
CLASS: 2/m
SPACE GROUP: C2/m
Z: 2
LATTICE CONSTANTS:
  $a \simeq 9.6$
  $b \simeq 18.3$
  $c \simeq 5.3$
  $\beta \simeq 101°50'$
3 STRONGEST DIFFRACTION LINES:
  8.33 (100)
  3.06 ( 70)
  2.756 ( 70)
OPTICAL CONSTANTS:
  $\alpha = 1.635-1.665$
  $\beta = 1.644-1.675$
  $\gamma = 1.655-1.698$
  $(+)2V_\gamma = 65°-90°$
HARDNESS: 5-6
DENSITY: 3.10-3.47 (Meas.)
CLEAVAGE: {110} good
HABIT: Fibrous or fibro-lamellar, often radiated. Twinning on {100} very common, simple, lamellar.
COLOR-LUSTER: Dark green, grayish green, brown; sometimes white, light gray. Translucent to nearly opaque. Silky.
MODE OF OCCURRENCE: Occurs chiefly in contact or regionally metamorphosed rocks. Found in California, Colorado, Arizona, Massachusetts, and in abundance at the Homestake gold mine at Lead, Lawrence County, and elsewhere in the Black Hills, South Dakota. Other typical occurrences are found in Canada, Scotland, Sweden (dannemorite), Finland, USSR, Madagascar, Japan, and New Zealand.
BEST REF. IN ENGLISH: Deer, Howie, and Zussman, "Rock Forming Minerals," v. 2, p. 234-248, New York, Wiley, 1963.

# CUPRITE
$Cu_2O$

CRYSTAL SYSTEM: Cubic
CLASS: $4/m \bar{3} 2/m$
SPACE GROUP: Pn3m
Z: 2
LATTICE CONSTANT:
  $a = 4.252$
3 STRONGEST DIFFRACTION LINES:
  2.465 (100)
  2.135 ( 37)
  1.510 ( 27)
OPTICAL CONSTANT:
  $N_{red} = 2.849$
HARDNESS: 3½-4
DENSITY: 6.14
        6.15 (Calc.)
CLEAVAGE: {111} interrupted
          {001} rare
Fracture conchoidal to uneven. Brittle.
HABIT: Crystals cubic, octahedral, or dodecahedral, commonly highly modified. Also as cubes greatly elongated into hair-like shapes (chalcotrichite) and occurring as wads or mats. Also compact massive, granular or earthy.
COLOR-LUSTER: Various shades of brownish red, red, and purplish red to almost black. Translucent to transparent. Adamantine or submetallic to earthy. Streak brownish red, shining.
MODE OF OCCURRENCE: Occurs widespread, often as an important ore mineral, in the oxidized zone of copper deposits. Frequent associated minerals include native copper, azurite, malachite, chalcocite, tenorite, antlerite, and iron oxides. In the United States, notable deposits are found in Pennsylvania, Tennessee, Montana, Colorado, Utah, Idaho, Nevada, California, and especially in Arizona and New Mexico. Other notable deposits occur in Mexico, Bolivia, Chile, England, France, Germany, Hungary, USSR, Australia, Tasmania, Japan, Zaire, and Tsumeb, South West Africa.
BEST REF. IN ENGLISH: Palache, et al., "Dana's System of Mineralogy," 7th Ed., v. I, p. 491-494, New York, Wiley, 1944.

# CUPROAURIDE = Auricupride

# CUPROBISMUTITE
$Cu_{10}B_{12}S_{23}$
Dimorphous with emplectite

CRYSTAL SYSTEM: Monoclinic
CLASS: 2/m
SPACE GROUP: C2/m
Z: 12
LATTICE CONSTANTS:
  $a = 17.65$
  $b = 3.93$
  $c = 15.24$
  $\beta = 100°30'$
3 STRONGEST DIFFRACTION LINES:
  3.07 (100)
  2.73 ( 60)
  3.25 ( 40)
HARDNESS: 2 (?)
DENSITY: 6.47 (Meas.)
        6.44 (Calc.)
CLEAVAGE: Not determined.
HABIT: Very thin blades measuring about 1 mm in greatest dimension; characteristically show a slight twisting about an axis parallel to elongation.
COLOR-LUSTER: Dark bluish black. Opaque. Metallic. Streak black.
MODE OF OCCURRENCE: Occurs in association with quartz, chalcopyrite, and wolframite at the Missouri mine, Hall's Valley, Park County, Colorado.
BEST REF. IN ENGLISH: Dana, E. S., "System of Mineralogy," 6th Ed., p. 110-111. Nuffield, E. W., *Am. Min.*, 37: 447-452 (1952).

# CUPROCOPIAPITE
$CuFe_4(SO_4)_6(OH)_2 \cdot 20H_2O$
Copiapite Group

CRYSTAL SYSTEM: Triclinic
CLASS: $\bar{1}$
SPACE GROUP: P$\bar{1}$
Z: 1
LATTICE CONSTANTS:

$a = 7.34 \quad \alpha = 93°51'$
$b = 18.19 \quad \beta = 101°30'$
$c = 7.28 \quad \gamma = 99°23'$

3 STRONGEST DIFFRACTION LINES:

3.56 (100)
8.81 ( 80)
5.82 ( 50)

OPTICAL CONSTANTS:

$\alpha = 1.558$
$\beta = 1.575$
$\gamma = 1.620$
$(+)2V = 63°$

HARDNESS: 2½–3
DENSITY: 2.08–2.17 (Meas.)
2.23 (Calc.)
CLEAVAGE: {010} perfect
{$\bar{1}$01} imperfect
HABIT: Crystals tabular. Usually as loose scaly aggregates; also granular or as crusts.
COLOR-LUSTER: Various shades of yellow and greenish yellow. Transparent to translucent. Pearly.
MODE OF OCCURRENCE: Occurs as a secondary mineral formed by the oxidation of pyrite and other sulfides at Chuquicamata, Chile.
BEST REF. IN ENGLISH: Palache, et al., "Dana's System of Mineralogy," 7th Ed., v. II, p. 623–627, New York, Wiley, 1951.

## CUPRODESCLOIZITE = Mottramite

## CUPROMAGNESITE (Species status uncertain.)
(Cu, Mg)SO$_4$·7H$_2$O

## CUPRORIVAITE
CaCuSi$_4$O$_{10}$

CRYSTAL SYSTEM: Tetragonal
CLASS: 4/m 2/m 2/m
SPACE GROUP: P4/ncc
Z: 4
LATTICE CONSTANTS:

$a = 7.30$
$c = 15.12$

3 STRONGEST DIFFRACTION LINES:

3.29 (100)
3.78 ( 90)
3.00 ( 90)

OPTICAL CONSTANTS:

$\omega = 1.633$
$\epsilon = 1.590$
$(-)$

HARDNESS: ~ 5
DENSITY: 3.08 (Meas.)
3.09 (Calc.)

CLEAVAGE: {001} perfect
Brittle.
HABIT: Crystals tabular {001}, rarely also {102} and possibly {110}.
COLOR-LUSTER: Blue; vitreous.
MODE OF OCCURRENCE: Found as minute crystals associated with quartz at Vesuvius, Italy.
BEST REF. IN ENGLISH: Mazzi, Fiorenzo, and Pabst, A., Am. Min., 47: 409–411 (1962).

## CUPROSKLODOWSKITE (Jachimovite)
Cu(UO$_2$)$_2$(SiO$_3$)$_2$(OH)$_2$·5H$_2$O

CRYSTAL SYSTEM: Triclinic
CLASS: $\bar{1}$
SPACE GROUP: P$\bar{1}$
Z: 1
LATTICE CONSTANTS:

$a = 9.21 \quad \alpha = 90°$
$b = 6.63 \quad \beta = 110°$
$c = 7.06 \quad \gamma = 108°30'$

3 STRONGEST DIFFRACTION LINES:

8.16 (100)
4.82 ( 90)
6.06 ( 70)

OPTICAL CONSTANTS:

$\alpha = 1.654–1.655$
$\beta = \gamma = 1.664–1.667$
$(-)2V = $ very small or zero

HARDNESS: 4
DENSITY: 3.85 (Meas.)
3.83 (Calc.)
CLEAVAGE: {100} distinct
HABIT: Crystals acicular, up to 1 cm long, elongated along c-axis. Forms: {100}, {010}, and {110}. As stellate aggregates of acicular crystals, and as silky coatings and crusts with radial-fibrous or matted-fibrous structure.
COLOR-LUSTER: Pale green to emerald green, also olive green. Transparent to translucent. Subvitreous to dull or silky. Not fluorescent.
MODE OF OCCURRENCE: Occurs as a secondary mineral formed by the alteration of primary copper minerals and uraninite. Found in San Juan County, Utah at the Posey mine in Red Canyon associated with becquerelite and brochantite, and also at the Frey No. 4 mine in Frey Canyon. In New Mexico it is found with calcite and ardennite in Jurassic Todilto limestone near Grants. At the Nicholson mine, Lake Athabaska, Saskatchewan, Canada it occurs with uranophane and kasolite; in Katanga, Zaire as fine specimens associated with vandenbrandeite, kasolite, and sklodowskite, and at Luiswishi with vandenbrandeite; in Morocco found with dipotase at Amelal in the Argana-Bigoudine region; at Joachimsthal, Bohemia it occurs with liebigite, uranophane, and beta-uranophane; and also found at Johanngeorgenstadt, Saxony.
BEST REF. IN ENGLISH: Frondel, Clifford, USGS Bull. 1064, p. 304–307 (1958).

## CUPROSPINEL
CuFe$_2$O$_4$
Spinel Group

CRYSTAL SYSTEM: Cubic
CLASS: 4/m 3̄ 2/m
SPACE GROUP: Fd3m
Z: 8
LATTICE CONSTANT:
  $\alpha$ = 8.369
3 STRONGEST DIFFRACTION LINES:
  2.517 (100)
  1.479 ( 60)
  2.96 ( 50)
OPTICAL CONSTANTS: Isotropic
  HARDNESS: Microhardness (VHN$_{100}$) 920–1081.
  DENSITY: Not determined.
  CLEAVAGE: None
  HABIT: Massive; as irregular grains up to about 0.1 mm in diameter.
  COLOR-LUSTER: Black. Opaque. Streak black.
  MODE OF OCCURRENCE: Found in association with hematite in highly oxidized material from an ore dump on the property of Consolidated Rambler Mines Limited near Baie, Verte, Newfoundland.
  BEST REF. IN ENGLISH: Nickel, E. H., *Can. Min.*, **11**: 1003–1007 (1973).

## CUPROSTIBITE
Cu$_2$ (Sb, Tl)

CRYSTAL SYSTEM: Tetragonal
Z: 2
LATTICE CONSTANTS:
  *a* = 3.99
  *c* = 6.09
3 STRONGEST DIFFRACTION LINES:
  2.07 (100)
  2.56 ( 50)
  2.82 ( 40)
  HARDNESS: Microhardness 220 Kg/mm².
  DENSITY: 8.42 (Calc.)
  CLEAVAGE: One cleavage.
Fracture uneven.
  HABIT: Massive; as fine-grained aggregates up to 1.5 mm in diameter.
  COLOR-LUSTER: Steel gray with violet-red tint on fresh fracture; metallic. Alters to incrustations and ochers of green, orange, and bright red color.
  MODE OF OCCURRENCE: Occurs associated with loellingite and antimonian silver, with very little chalcopyrite and chalcothallite, in a vein of ussingite cutting sodalite syenites at Mt. Nakalak, Ilimaussaq alkalic massif, Greenland.
  BEST REF. IN ENGLISH: Sørensen, H., Semenov, E. I., Bezsmertnaya, M. S., and Khalezova, E. B., *Am. Min.*, **55**: 1810–1811 (1970).

## CUPROTUNGSTITE
Cu$_2$ WO$_4$ (OH)$_2$

CRYSTAL SYSTEM: Unknown

3 STRONGEST DIFFRACTION LINES:
  3.07 (100)
  1.24 ( 90)
  1.91 ( 80)
OPTICAL CONSTANTS:
  Mean index = 2.15
  HARDNESS: Not determined.
  DENSITY: 5.40 (Meas.)
  CLEAVAGE: Not determined.
  HABIT: Massive, microcrystalline, friable to compact. As crusts.
  COLOR-LUSTER: Brownish green or yellowish green to emerald green. Vitreous, waxy, dull. Streak greenish yellow to greenish gray.
  MODE OF OCCURRENCE: Occurs replacing the outer portions of molybdenian scheelite in a contact-metamorphic copper deposit at the South Peacock mine, Seven Devils district, Adams County, Idaho. It also is found in Inyo and Kern counties, California; in the San Andres Mountains, Socorro County, New Mexico; at Cave Creek, Maricopa County, Arizona; and in Mexico, Chile, Spain, Sardinia, Australia, Japan, and Transvaal, Africa.
  BEST REF. IN ENGLISH: Palache, et al., "Dana's System of Mineralogy," 7th Ed., v. II, p. 1091–1092, New York, Wiley, 1951.

## CUPROVUDYAVRITE = Variety of vudyavrite

## CURIENITE
Pb(UO$_2$)$_2$ (VO$_4$)$_2$ · 5H$_2$O

CRYSTAL SYSTEM: Orthorhombic
CLASS: 2/m 2/m 2/m
SPACE GROUP: Pcan
Z: 4
LATTICE CONSTANTS:
  *a* = 10.40
  *b* =  8.45
  *c* = 16.34
3 STRONGEST DIFFRACTION LINES:
  8.19  (100)
  3.005 (100)
  4.10 ( 80)
OPTICAL CONSTANTS:
  $\alpha, \beta, \gamma > 2$
  $(-)2V = 66°$
  HARDNESS: Not determined.
  DENSITY: 4.88 (Meas.)
          4.94 (Calc.)
  HABIT: Microcrystalline powder.
  COLOR-LUSTER: Canary yellow.
  MODE OF OCCURRENCE: Occurs on tabular crystals of francevillite in mineralized sandstones of the Mounana uranium-vanadium mine, Haut-Ogoue, Gabon, Africa.
  BEST REF. IN ENGLISH: Cesbron, Fabien, and Morin, Noel, *Am. Min.*, **54**: 1220 (1969).

## CURITE
Pb$_3$U$_8$O$_{27}$ · 4H$_2$O(?)

CRYSTAL SYSTEM: Orthorhombic

CLASS: mm2
SPACE GROUP: Pna2
Z: 2
LATTICE CONSTANTS:
 $a$ = 1.250
 $b$ = 13.01
 $c$ = 8.40
3 STRONGEST DIFFRACTION LINES:
 6.23 (100)
 3.14 ( 80)
 3.96 ( 70)
OPTICAL CONSTANTS:
 $\alpha$ = 2.05–2.06
 $\beta$ = 2.07–2.11
 $\gamma$ = 2.12–2.15
 (–)2V = large
HARDNESS: 4–5
DENSITY: 7.4 (Meas.)
 7.37 (Calc.)
CLEAVAGE: {100}
Brittle.
HABIT: Crystals long prismatic to acicular, elongated along $c$-axis. Commonly massive, fine granular or as porous masses of fine needles; also as opaline crusts. Crystals striated parallel $c$-axis.
COLOR-LUSTER: Deep orange-red to orange; transparent to translucent. Adamantine.
MODE OF OCCURRENCE: Occurs as excellent minute crystals and crystal aggregates associated with torbernite, soddyite, fourmarierite, kasolite, and other secondary uranium minerals at Shinkolobwe and elsewhere in the Katanga district, Zaire. Also found in the South Alligator district, N. T., Australia; with uranophane at Malakialina, Madagascar; and at La Crouzille, Puy-de-Dôme, France.
BEST REF. IN ENGLISH: Frondel, Clifford, USGS Bull. 1064, p. 92–95 (1958).

## CURTISITE = Idrialite

## CUSPIDINE (Custerite)
$Ca_4Si_2O_7(F,OH)_2$

CRYSTAL SYSTEM: Monoclinic
Z: 4
LATTICE CONSTANTS:
 $a$ = 10.93
 $b$ = 10.57
 $c$ = 7.57
 $\beta$ = 110°07′
3 STRONGEST DIFFRACTION LINES:
 3.062 (100)
 2.943 ( 35)
 3.259 ( 30)
OPTICAL CONSTANTS:
 $\alpha$ = 1.591
 $\beta$ = 1.596
 $\gamma$ = 1.602
 (+)2V = 76°
HARDNESS: 5–6

DENSITY: 2.8–2.99 (Meas.)
 2.964 (Calc.)
CLEAVAGE: {001} distinct
 {110} distinct
Brittle.
HABIT: Crystals spear-shaped, very small. Commonly massive, fine granular. Twinning on {001}, lamellar.
COLOR-LUSTER: Colorless, white, greenish gray, pale rose red. Transparent to translucent. Vitreous.
MODE OF OCCURRENCE: Occurs in metamorphic rock at Crestmore, Riverside County, California; in a limestone contact zone at the Empire mine, Custer County, Idaho; in very small amounts with nasonite at Franklin, Sussex County, New Jersey; and at Mte. Somma and elsewhere in Italy.
BEST REF. IN ENGLISH: Winchell and Winchell, "Elements of Optical Mineralogy," 4th Ed., Pt. 2, p. 480, New York, John Wiley & Sons, 1951.

## CUSTERITE = Cuspidine

## CYANITE = Kyanite

## CYANOCHROITE
$K_2Cu(SO_4)_2 \cdot 6H_2O$

CRYSTAL SYSTEM: Monoclinic
CLASS: 2/m
SPACE GROUP: P2$_1$/a
Z: 2
LATTICE CONSTANTS:
 $a$ = 9.09
 $b$ = 12.14
 $c$ = 6.18
 $\beta$ = 104°28′
3 STRONGEST DIFFRACTION LINES:
 3.70 (100)
 4.21 ( 80)
 4.09 ( 60)
OPTICAL CONSTANTS:
 (Na)
 $\alpha$ = 1.4836
 $\beta$ = 1.4864
 $\gamma$ = 1.5020
 (+)2V = 46°32′
HARDNESS: Not determined
DENSITY: 2.224 (Meas.)
 2.223 (Calc.)
CLEAVAGE: {$\bar{2}$01} perfect
HABIT: Crystals tabular; as crystalline crusts.
COLOR-LUSTER: Greenish blue. Transparent. Vitreous.
Soluble in water.
MODE OF OCCURRENCE: Occurs as a fumarolic deposit formed during various eruptions of Mt. Vesuvius, Italy.
BEST REF. IN ENGLISH: Palache, et al., "Dana's System of Mineralogy," 7th Ed., v. II, p. 454–455, New York, Wiley, 1951.

# CYANOTRICHITE
$Cu_4Al_2SO_4(OH)_{12} \cdot 2H_2O$

CRYSTAL SYSTEM: Probably orthorhombic
Z: 1
LATTICE CONSTANTS:
  $a = 10.16$
  $b = 12.61$
  $c = 2.90$
3 STRONGEST DIFFRACTION LINES:
  10.2 (100)
  3.88 ( 90)
  5.26 ( 80)
OPTICAL CONSTANTS:
  $\alpha = 1.588-1.591$
  $\beta = 1.617-1.620$
  $\gamma = 1.654-1.655$
  (+)2V = 82°
HARDNESS: Not determined.
DENSITY: 2.74-2.95 (Meas.)
         2.88 (Calc.)
HABIT: As velvety coatings and aggregates of minute acicular crystals; also in tufts and as radial-fibrous veinlets and aggregates.
COLOR-LUSTER: Pale sky blue to azure blue. Silky. Streak pale blue.
MODE OF OCCURRENCE: Occurs as a secondary mineral in the oxidized zone of copper-bearing ore deposits, often associated with brochantite and other copper minerals. Found as superb tufted aggregates with brochantite crystals at the Grandview mine, Coconino County, and in the Morenci district, Greenlee County, Arizona; at the Myler mine, Majuba Hill, Pershing County, and in the Mason Park area, Lyon County, Nevada; and at the American Eagle mine, Juab County, Utah. It also occurs in Scotland, France, Italy, Roumania, Greece, USSR, and Namaqualand, South Africa.
BEST REF. IN ENGLISH: Palache, et al., "Dana's System of Mineralogy," 7th Ed., v. II, p. 578-579, New York, Wiley, 1951.


# CYLINDRITE
$Pb_3Sn_4Sb_2S_{14}$

CRYSTAL SYSTEM: Orthorhombic
Z: 72
LATTICE CONSTANTS:
  $a = 17.1$
  $b = 11.6$
  $c = 70.0$
3 STRONGEST DIFFRACTION LINES:
  3.85 (100)
  2.88 (100)
  5.73 ( 50)
HARDNESS: 2½
DENSITY: 5.46 ± 0.03
CLEAVAGE: Good circular separation parallel to elongation of cylinders; poor separation across elongation. Somewhat malleable.
HABIT: Massive, in cylindrical forms; also as spherically grouped aggregates.

COLOR-LUSTER: Blackish lead gray. Opaque. Metallic. Streak black.
MODE OF OCCURRENCE: Occurs in tin-bearing veins, associated with franckeite, sphalerite, and pyrite, at Oruro, Poopó, Colquechaca, and Huanuni, Bolivia.
BEST REF. IN ENGLISH: Palache, et al., "Dana's System of Mineralogy," 7th Ed., v. I, p. 482-483, New York, Wiley, 1944.


# CYMOPHANE = Synonym for chrysoberyl


# CYMRITE
$BaAl_2Si_2O_8 \cdot H_2O$

CRYSTAL SYSTEM: Hexagonal
Z: 1 (pseudo-cell), 64 (true cell)
LATTICE CONSTANTS:
  (pseudo)      (true)
  $a = 5.334$   $a = 42.6$
  $c = 7.705$   $c = 7.67$
3 STRONGEST DIFFRACTION LINES:
  2.96 (100)
  3.96 ( 90)
  2.67 ( 70)
OPTICAL CONSTANTS:
  $\omega = 1.611-1.617$
  $\epsilon = 1.603-1.606$ (Na)
       (−)
HARDNESS: 2-3
DENSITY: 3.413 (Meas.)
         3.451 (Calc.)
CLEAVAGE: {0001} perfect
          {10$\bar{1}$0} imperfect
Fracture uneven. Brittle.
HABIT: Crystals hexagonal prisms up to 1mm long, or as crude hexagonal plates up to 7mm in diameter. Also massive, fibrous.
COLOR-LUSTER: Colorless; often dark green or brown due to inclusions or alteration. Transparent to translucent. Vitreous to satiny.
MODE OF OCCURRENCE: Occurs at the Benallt manganese mine, Rhiw, Carnarvonshire, Wales; in a copper deposit at Ruby Creek and at Bonanza Creek, Brooks Range, Alaska; and associated with calcite, albite, and lawsonite in a jadeite metagraywacke on the Pacheco Pass Road, San Benito County, California.
BEST REF. IN ENGLISH: Runnels, Donald D., *Am. Min.*, 49: 158-165 (1964). *Am. Min.*, 52: 1885 (1967).


# CYRILOVITE (Avelinoite)
$NaFe_3^{3+}(PO_4)_2(OH)_4 \cdot 2H_2O$

CRYSTAL SYSTEM: Tetragonal
CLASS: 4 2 2
SPACE GROUP: $P4_12_12$ or $P4_32_12$
Z: 4
LATTICE CONSTANTS:
  $a = 7.32$
  $c = 19.40$

3 STRONGEST DIFFRACTION LINES:
  4.85 (100)
  3.186 ( 80)
  2.658 ( 80)
OPTICAL CONSTANTS:
  $\omega$ = 1.803
  $\epsilon$ = 1.769
  (-)
HARDNESS: Not determined.
DENSITY: 3.081 (Meas.)
            3.09 (Calc.)
HABIT: Crystals squat; dominant forms {001} and {113}; {012} occasionally present. Individual crystals usually less than 0.1mm; often in intergrown aggregates.

COLOR-LUSTER: Orange to brownish yellow. Vitreous. Powder is yellow.
MODE OF OCCURRENCE: The mineral occurs associated with metastrengite and leucophosphite in an altered zone surrounding frondelite at the Sapucaia pegmatite mine, Minas Geraes, Brazil. Also found as minute crystals in pegmatite at Cyrilov, near Velké Mežiřiči, West Moravia, and in Madagascar.
BEST REF. IN ENGLISH: Lindberg, Mary Louise, *Am. Min.*, 42: 204–213 (1957).


**CYRTOLITE** = Variety of zircon

# D

## DACHIARDITE
Near $(K, Na, Ca)_5 Al_5 Si_{19} O_{48} \cdot 18H_2O$
Zeolite Group
Dimorph of mordenite

CRYSTAL SYSTEM: Monoclinic
CLASS: 2/m
SPACE GROUP: C2/m
Z: 1
LATTICE CONSTANTS:
  $a = 18.03$
  $b = 7.52$
  $c = 10.20$
  $\beta = 104°46'$
3 STRONGEST DIFFRACTION LINES:
  3.452 (100)
  3.204 (100)
  1.873 ( 75)
OPTICAL CONSTANTS:
    $\alpha = 1.494$
    $\beta = 1.496$
    $\gamma = 1.499$
  (+)2V = 70°
HARDNESS: 4-4½
DENSITY: 2.165-2.206 (Meas.)
            2.138 (Calc.)
CLEAVAGE: {001} perfect
            {100} perfect
Brittle.
HABIT: Crystals prismatic, always twinned on {110}, often cyclic tetragonal.
COLOR-LUSTER: Colorless. Transparent. Vitreous.
MODE OF OCCURRENCE: Occurs with other zeolites in the granitic pegmatite of San Piero, Campo, Elba.
BEST REF. IN ENGLISH: Gottardi, Glauco, *Am. Min.*, **46**: 769 (1961). Gottardi, G., and Meier, W. M., *Am. Min.*, **47**: 190-191 (1962). Galli, *Periodica Mineral.* (Roma), **34**: 129-136 (1965).

## DADSONITE
$Pb_{11} Sb_{12} S_{29}$

CRYSTAL SYSTEM: Monoclinic
CLASS: 2, m, or 2/m
SPACE GROUP: P2, Pm, or P2/m
Z: 1
LATTICE CONSTANTS:
  $a = 19.05$
  $b = 4.11$
  $c = 17.33$
  $\beta = 96°20'$
3 STRONGEST DIFFRACTION LINES:
  3.38 (100)
  3.78 ( 70)
  2.840 ( 70)
HARDNESS: 2½
DENSITY: 5.76 (Calc.)
CLEAVAGE: Not determined.
HABIT: Massive, in grains less than 0.5mm in diameter.
COLOR-LUSTER: Lead gray. Opaque. Metallic. Streak black.
MODE OF OCCURRENCE: Found very sparingly at Yellowknife, N.W. Territory; as minute grains associated with robinsonite in a small prospect pit in Precambrian marble at Madoc, Ontario, Canada; and at the Red Bird Mercury mine, Pershing County, Nevada.
BEST REF. IN ENGLISH: Jambor, J. L., *Min. Mag.*, **37**: 437-441 (1969).

## DAHLLITE = Carbonate-apatite

## DALYITE
$K_2 ZrSi_6 O_{15}$

CRYSTAL SYSTEM: Triclinic
CLASS: 1
SPACE GROUP: P$\bar{1}$
Z: 1
LATTICE CONSTANTS:
  $a = 7.51$     $\alpha = 106°$
  $b = 7.73$     $\beta = 113.5°$
  $c = 7.00$     $\gamma = 99.5°$

**3 STRONGEST DIFFRACTION LINES:**
4.20 (100)
3.58 (100)
3.08 (100)
**OPTICAL CONSTANTS:**
$\alpha$ = 1.575
$\beta$ = 1.590
$\gamma$ = 1.601
$(-)2V \approx 72°$
**HARDNESS:** 7½
**DENSITY:** 2.84 ± 0.02 (Meas.)
**CLEAVAGE:** {101} good
{010} good
{100} less distinct
Brittle.

**HABIT:** Short prismatic crystals, 0.05 to 0.5mm in size. Forms {$\bar{1}$01}, {100}, {1$\bar{1}$0}, {110}, and {$\bar{1}$11} most common. Twinning with (100) as composition plane.

**COLOR-LUSTER:** Colorless. Vitreous. Transparent.

**MODE OF OCCURRENCE:** Occurs as a rare accessory, about 0.2% of the rock, in medium-grained, pinkish gray alkali granites mainly composed of microperthite and quartz with aegirine and an amphibole on Ascension Island, Atlantic Ocean.

**BEST REF. IN ENGLISH:** Van Tassel, R., *Min. Mag.*, **29**: 850–857 (1952).

**DAMOURITE** = Variety of muscovite

**DANAITE** = Variety of arsenopyrite containing up to 9% cobalt

## DANALITE
$(Fe, Mn, Zn)_4 Be_3 Si_3 O_{12} S$
Helvite Group

**CRYSTAL SYSTEM:** Cubic
**CLASS:** $\bar{4}$3m
**SPACE GROUP:** P$\bar{4}$3n
**Z:** 2
**LATTICE CONSTANTS:**
$a$ = 8.20–8.15
**3 STRONGEST DIFFRACTION LINES:**
3.35 (100)
1.932 ( 70)
2.193 ( 50)
**OPTICAL CONSTANT:**
$N$ = 1.753–1.771
**HARDNESS:** 5½–6
**DENSITY:** 3.31–3.46 (Meas.)
3.37 (Calc.)
**CLEAVAGE:** {111} in traces
Fracture uneven. Brittle.

**HABIT:** Crystals octahedral or dodecahedral, up to 10cm along an edge. Also massive, granular.

**COLOR-LUSTER:** Gray, lemon yellow, flesh pink, red, reddish brown, brown. Becomes brownish black on weathering. Transparent to translucent. Vitreous to greasy or resinous.

**MODE OF OCCURRENCE:** Occurs chiefly in granite pegmatites; gneiss; contact metasomatic deposits; and hydrothermal veins. Found at Bartlett, Carroll County, New Hampshire; at Gloucester and Rockport, Essex County, Massachusetts; at Grants Mills, Rhode Island; at Needlepoint Mountain, McDame area, British Columbia, Canada; near Redruth, Cornwall, England; and at localities in Sweden and USSR.

**BEST REF. IN ENGLISH:** Vlasov, K. A., "Mineralogy of Rare Elements," v. II, p. 119–127, Israel Program for Scientific Translations, 1966.

## DANBURITE
$CaB_2 Si_2 O_8$

**CRYSTAL SYSTEM:** Orthorhombic
**CLASS:** 2/m 2/m 2/m
**SPACE GROUP:** Pnam
**Z:** 4
**LATTICE CONSTANTS:**
$a$ = 8.048
$b$ = 8.763
$c$ = 7.731
**3 STRONGEST DIFFRACTION LINES:**
3.564 (100)
2.654 ( 75)
2.961 ( 70)
**OPTICAL CONSTANTS:**
(Na)
$\alpha$ = 1.6303
$\beta$ = 1.6332
$\gamma$ = 1.6360
$(-)2V \simeq 88°$
**HARDNESS:** 7
**DENSITY:** 2.97–3.02 (Meas.)
2.995 (Calc.)
**CLEAVAGE:** {001} very indistinct
Fracture conchoidal. Brittle.

**HABIT:** Crystals prismatic, diamond-shaped in cross section, with wedge-like or pointed terminations. Vertically striated. Habit similar to topaz. Also massive, granular.

**COLOR-LUSTER:** Colorless, white, pale pink, light to dark yellow, yellowish brown, brown; sometimes greenish due to inclusions. Transparent to translucent. Vitreous to somewhat greasy. Streak uncolored.

**MODE OF OCCURRENCE:** Occurs with feldspar in dolomite at Danbury, Fairfield County, Connecticut; at Russell, St. Lawrence County, New York; as superb crystals up to 3 inches long in association with calcite, quartz, sphalerite, pyrite, chalcopyrite, and apophyllite, at Charcas, San Luis Potosi, Mexico; on the Piz Valatscha, Uri, Switzerland; at Maharita on Mt. Bity and elsewhere in Madagascar; as fine crystals associated with axinite at the Toroku mine, Miyazake Prefecture, and at other places in Kyushu, Japan; and as excellent etched yellow crystals in gravels of the Mogok district, Burma.

**BEST REF. IN ENGLISH:** Winchell and Winchell, "Elements of Optical Mineralogy," 4th Ed., Pt. 2, p. 258–259, New York, John Wiley & Sons, 1951.

**DANNEMORITE** = Manganoan cummingtonite
$(Fe,Mn,Mg)_7Si_8O_{22}(OH)_2$


**DAPHNITE**
$(Mg,Fe)_3(Fe,Al)_3(Si,Al)_4O_{10}(OH)_8$
Chlorite Group

CRYSTAL SYSTEM: Monoclinic
CLASS: 2/m
SPACE GROUP: C 2/m
Z: 2
LATTICE CONSTANTS:
  $a = 27.86$
  $b = 9.36$
  $c = 5.40$
  $\beta = 94°5'$
3 STRONGEST DIFFRACTION LINES:
  Similar to thuringite.
OPTICAL CONSTANTS:
    $\beta = 1.65-1.67$
  $\gamma - \alpha \simeq .009$
  $(-)2V =$ small
HARDNESS: 2-3
DENSITY: 3.20 (Meas.)
CLEAVAGE: {001} perfect
Laminae somewhat flexible, inelastic.
  HABIT: Massive; in spherical or botryoidal aggregates; foliated. Twinning on {001}, twin plane [310].
  COLOR-LUSTER: Dark green. Translucent. Pearly. Streak greenish.
  MODE OF OCCURRENCE: Occurs in association with quartz and arsenopyrite at Penzance, Cornwall, England; also found in Côtes-du-Nord, France.
  BEST REF. IN ENGLISH: Deer, Howie, and Zussman, "Rock Forming Minerals," v. 3, p. 131-163, New York, Wiley, 1962.


**DARAPSKITE**
$Na_3NO_3SO_4 \cdot H_2O$

CRYSTAL SYSTEM: Monoclinic
CLASS: 2/m
SPACE GROUP: $P2_1/m$
Z: 2
LATTICE CONSTANTS:
  $a = 10.564$
  $b = 6.913$
  $c = 5.1890$
3 STRONGEST DIFFRACTION LINES:
  10.29 (100)
  3.456 ( 35)
  2.865 ( 35)
OPTICAL CONSTANTS:
    $\alpha = 1.388$
    $\beta = 1.479$
    $\gamma = 1.486$
  $(-)2V =$ small to 27°
HARDNESS: ~2½

DENSITY: 2.202 (Meas.)
CLEAVAGE: {100} perfect
          {010} very good
Fracture uneven. Brittle.
  HABIT: Crystals long prismatic; also thin tabular (100), elongated $c$. Largest single crystals are 1-2 cm long and 3-4mm thick. As platy to granular material admixed with other saline materials, and as euhedral crystals in cavities. Polysynthetic twinning parallel to {100}.
  COLOR-LUSTER: Colorless, transparent to translucent; vitreous.
  MODE OF OCCURRENCE: Occurs widespread in the nitrate deposits in the Atacama Desert of northern Chile; in veins as much as 6 inches wide, associated with kroehnkite, bloedite, mirabilite, and epsomite in the copper deposit at Chuquicamata, Chile; with soda-niter and niter in the nitrate deposits of Death Valley, California; in saline material from caves in limestone, Funeral Mountains, California; and in saline arid soil on the Roberts Massif, Shackleton Glacier, Antarctica.
  BEST REF. IN ENGLISH: Ericksen, George E., and Mrose, Mary E., *Am. Min.*, 55: 1500-1517 (1970).


**DASHKESANITE** = Variety of hastingsite


**DATOLITE**
$CaBSiO_4OH$

CRYSTAL SYSTEM: Monoclinic
CLASS: 2/m
SPACE GROUP: $P2_1/c$
Z: 4
LATTICE CONSTANTS:
  $a = 4.84$
  $b = 7.60$
  $c = 9.62$
  $\beta = 90°09'$
3 STRONGEST DIFFRACTION LINES:
  3.114 (100)
  2.855 ( 65)
  2.189 ( 60)
OPTICAL CONSTANTS:
    $\alpha = 1.622-1.626$
    $\beta = 1.649-1.658$
    $\gamma = 1.666-1.670$
  $(-)2V_\alpha = 72°-75°$
HARDNESS: 5-5½
DENSITY: 2.8-3.0 (Meas.)
          3.05 (Calc.)
CLEAVAGE: None.
Fracture uneven to conchoidal. Brittle.
  HABIT: Crystals commonly short prismatic often with a variety of forms. Also granular or as porcelain-like compact masses.
  COLOR-LUSTER: Colorless, white, pale yellowish, pale greenish, or tinted pink, reddish or brownish by impurities. Transparent to translucent. Vitreous. Streak uncolored.
  MODE OF OCCURRENCE: Occurs chiefly as a secondary mineral in cavities and veins in basic igneous rocks, commonly associated with zeolites, prehnite, and calcite; also

in crevices in granite, gneiss, serpentine, and other rocks, and in metallic veins. Found in Colusa, Inyo, Riverside, San Bernardino, and San Francisco counties, California; in the copper district of Keweenaw County, Michigan; as superb large crystals at the Lane Quarry, Westfield, Massachusetts; in the trap rock of northern New Jersey; and at localities in Connecticut. Other notable occurrences are found in Canada, Mexico, England, Norway, Italy, Germany, Czechoslovakia, USSR, Japan, and Tasmania.

BEST REF. IN ENGLISH: Deer, Howie, and Zussman, "Rock Forming Minerals," v. 1, p. 171–175, New York, Wiley, 1962.

# DAUBRÉEITE
BiO(OH, Cl)

CRYSTAL SYSTEM: Tetragonal
CLASS: 4/m 2/m 2/m
SPACE GROUP: P4/nmm
Z: 2
LATTICE CONSTANTS:
 $a$ = 3.85  kX
 $c$ = 7.40
OPTICAL CONSTANTS:
  $\omega$ = 1.91
  $\epsilon$ = ?
  $\omega - \epsilon \sim$ .01
  (−)
 HARDNESS: 2–2½
 DENSITY: 7.56 (Calc. for OH:Cl ~ 1:1)
 CLEAVAGE: {001} perfect
Plastic.
 HABIT: Massive; earthy.
 COLOR-LUSTER: Yellowish. Dull.
 MODE OF OCCURRENCE: Occurs mixed with clay at the Constancia mine, Tazna, Bolivia.
 BEST REF. IN ENGLISH: Palache, et al., "Dana's System of Mineralogy," 7th Ed., v. II, p. 60–62, New York, Wiley, 1951.

# DAUBRÉELITE
FeCr$_2$S$_4$
Linnaeite Group

CRYSTAL SYSTEM: Cubic
CLASS: 4/m $\overline{3}$ 2/m
SPACE GROUP: Fd3m
Z: 8
LATTICE CONSTANT:
 $a$ = 9.966
3 STRONGEST DIFFRACTION LINES:
 1.77 (100)
 3.02 ( 80)
 2.51 ( 60)
 HARDNESS: Not determined.
 DENSITY: 3.81 (Meas.)
        3.842 (Calc. x-ray)
 CLEAVAGE: One distinct. Fracture uneven. Brittle.
 HABIT: Massive; scaly to platy.
 COLOR-LUSTER: Black. Opaque. Metallic, brilliant.

Not magnetic.
 MODE OF OCCURRENCE: Occurs in small amounts in many meteorites often intergrown with troilite.
 BEST REF. IN ENGLISH: Palache, et al., "Dana's System of Mineralogy," 7th Ed., v. I, p. 265, New York, Wiley, 1944.

# DAVIDITE
(Fe$^{2+}$,Ce,U)$_2$(Ti,Fe$^{3+}$,V,Cr)$_5$O$_{12}$

CRYSTAL SYSTEM: Hexagonal
CLASS: 3 or $\overline{3}$
SPACE GROUP: R3 or R$\overline{3}$
Z: 9
LATTICE CONSTANTS:
 $a$ = 10.37
 $c$ = 20.87
3 STRONGEST DIFFRACTION LINES:
 (metamict)
 2.895 (100)
 3.42 ( 72)
 2.850 ( 60)
OPTICAL CONSTANT:
 $n \sim$ 2.3 (metamict)
 HARDNESS: ~6
 DENSITY: 4.42 (Meas. Arizona)
        4.49 (Calc. Arizona)
 CLEAVAGE: Fracture subconchoidal to uneven. Brittle.
 HABIT: Massive, as rough cuboidal crystals, and as distinct crystals either tabular on {0001} or pyramidal with ditrigonal pyramids dominant.
 COLOR-LUSTER: Black to grayish black; oxidized surfaces brownish black, dark brown, reddish; opaque; vitreous to submetallic.
 MODE OF OCCURRENCE: Occurs as a primary mineral at widespread localities including the Pandora prospect, Quijotoa Mountains, Pima County, Arizona; in the Red River placer gravels, Idaho County, Idaho; at Iveland, Satesdal, Norway; at Radium Hill, Crocker's Well, and Billeroo, South Australia, and at Thackeringa, New South Wales; also as crystal fragments up to 12 inches long and 22 pounds in weight at Mavusi, in the Tete district, Mozambique.
 BEST REF. IN ENGLISH: Frondel, Clifford, USGS Bull. 1064, p. 337–341 (1958).  Pabst, A., *Am. Min.*, **46**: 700–718 (1961).

# DAVIESITE = Hemimorphite

# DAVISONITE (Dennisonite) (Inadequately described mineral)
Ca$_3$Al(PO$_4$)$_2$(OH)$_3$ · H$_2$O (?)

CRYSTAL SYSTEM: Hexagonal (?)
OPTICAL CONSTANTS:
 $\omega$ = 1.601
 $\epsilon$ = 1.591
 (−)
HARDNESS: 4½

DENSITY: 2.85 (Meas.)
CLEAVAGE: Basal, perfect.
HABIT: As botryoidal or spherulitic crusts composed of stout fibers.
COLOR-LUSTER: White.
MODE OF OCCURRENCE: Occurs lining cavities in crandallite in variscite nodules at Fairfield, Utah County, Utah. It also is found at Damasio, near Diamantina, Minas Geraes, Brazil.
BEST REF. IN ENGLISH: Palache, et al., "Dana's System of Mineralogy," 7th Ed., v. II, p. 939-940, New York, Wiley, 1951.

# DAVYNE (Microsommite)
$(Na, Ca, K)_8 Al_6 Si_6 O_{24} (Cl, SO_4, CO_3)_{2-3}$
Cancrinite Group

CRYSTAL SYSTEM: Hexagonal
CLASS: 6
SPACE GROUP: $P6_3$
Z: 1
LATTICE CONSTANTS:
  $a = 12.70$
  $c = 5.33$
3 STRONGEST DIFFRACTION LINES:
  4.80 (100)
  3.67 (100)
  3.28 (100)
OPTICAL CONSTANTS:
  $\omega \sim 1.519$
  $\epsilon \sim 1.520$
  (+)
HARDNESS: 6
DENSITY: 2.42-2.53 (Meas.)
CLEAVAGE: $\{10\bar{1}0\}$ perfect
                $\{0001\}$ poor
Fracture uneven to conchoidal. Brittle.
HABIT: Crystals prismatic, minute, vertically striated. Often in groups.
COLOR-LUSTER: Colorless to white. Transparent. Vitreous.
MODE OF OCCURRENCE: Occurs in ejected masses and in leucite-rich lava on Mte. Somma, Italy.
BEST REF. IN ENGLISH: Deer, Howie, and Zussman, Rock Forming Minerals," v. 4, p. 310-320, New York, Wiley, 1963.

# DAWSONITE
$NaAlCO_3(OH)_2$

CRYSTAL SYSTEM: Orthorhombic
CLASS: 2/m 2/m 2/m
SPACE GROUP: Imam
Z: 4
LATTICE CONSTANTS:
  $a = 6.73$
  $b = 10.36$
  $c = 5.575$

3 STRONGEST DIFFRACTION LINES:
  5.70 (100)
  3.385 ( 16)
  1.690 ( 14)
OPTICAL CONSTANTS:
  $\alpha = 1.462-1.466$
  $\beta = 1.537-1.542$
  $\gamma = 1.589-1.596$
  $(-)2V = 76°46'$ (Na)
HARDNESS: 3
DENSITY: 2.44 (Meas.)
            2.46 (Calc.)
CLEAVAGE: $\{110\}$ perfect
Brittle.
HABIT: Crystals acicular to blade-like; as radial incrustations and sprays of needle-like crystals.
COLOR-LUSTER: Colorless, white. Transparent. Crystals vitreous; aggregates silky.
MODE OF OCCURRENCE: Occurs as a low-temperature mineral disseminated in shale of the Green River formation, Piceance Basin, western Colorado; in Canada near McGill University, Montreal, and as fine microscopic crystals associated with weloganite at St. Michel, Montreal Island, Quebec; in the province of Siena, Italy; at Komana, Drin Valley, northern Albania; in the Permian Coal Measures of the Sydney Basin at Muswellbrook, New South Wales, Australia; near Tenès, Algeria; and altering from nepheline in arid, saline soils at Olduvai Gorge, Tanzania.
BEST REF. IN ENGLISH: Palache, et al., "Dana's System of Mineralogy," 7th Ed., v. II, p. 276-278, New York, Wiley, 1951.    Smith, J. W., and Milton, C., *Econ. Geol.*, **61**: 1029-1042 (1966).

# DEERITE
$(Fe, Mn)_{13} (Fe, Al)_7 Si_{13} O_{44} (OH)_{11}$

CRYSTAL SYSTEM: Monoclinic
CLASS: 2/m
SPACE GROUP: $P2_1/a$
Z: 2
LATTICE CONSTANTS
  $a = 10.755$
  $b = 18.874$
  $c = 9.568$
  $\beta = 107°12'$
3 STRONGEST DIFFRACTION LINES:
  9.03 (100)
  3.01 ( 70)
  2.64 ( 55)
OPTICAL CONSTANTS:
  $\alpha = 1.840$
  $\gamma = 1.870$
DENSITY: 3.837 (Meas.)
CLEAVAGE: $\{110\}$ good
HABIT: Acicular crystals, amphibole-like in cross section.
COLOR-LUSTER: Black, transparent on thin edges.
MODE OF OCCURRENCE: Occurs in some of the metamorphosed shales, siliceous ironstones, and impure limestones of the Franciscan formation in the Laytonville district, Mendocino County, California.

BISMUTHINITE with pyrite   Cornwall, England.

BISMUTH   Botallack mine, Cornwall, England.

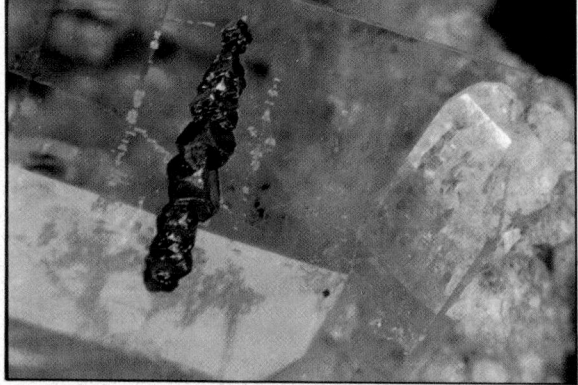

BIXBYITE oriented on topaz   Thomas Mountains, Juab County, Utah.

BIXBYITE   Thomas Mountains, Juab County, Utah.

BJAREBYITE   Palermo mine, North Groton, New Hampshire.

BJAREBYITE   Palermo mine, North Groton, New Hampshire.

BLOEDITE   Soda Lake, San Luis Obispo County, California.

17

BOLEITE   Mammoth-St. Anthony mine, Pinal County, Arizona.

BOLEITE   Mammoth-St. Anthony mine, Pinal County, Arizona.

BOLEITE   Philipsburg, Granite County, Montana.

BOLEITE (SEM photograph)   Philipsburg, Granite County, Montana.

BOLEITE on ancient slag   Laurium, Greece.

BOLEITE   Boleo, Baja California, Mexico.

BORACITE   Westeregeln, Saxony, Germany.

BORNITE   Cornwall, England.

BORNITE   Mamainse Point, Ontario, Canada.

BOTRYOGEN   Redington mine, Napa County, California.

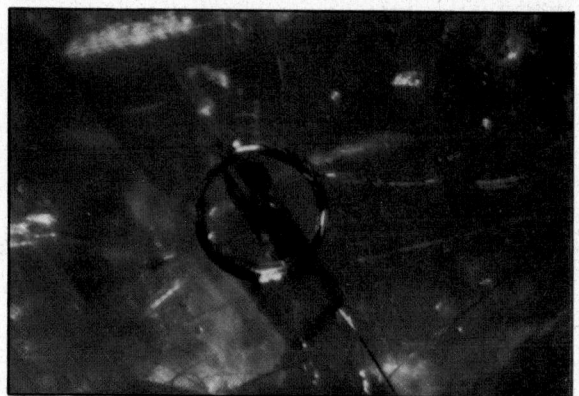

BOULANGERITE ring in fluorite   Madoc, Ontario, Canada.

BOULANGERITE   Noche Buena, Zacatecas, Mexico.

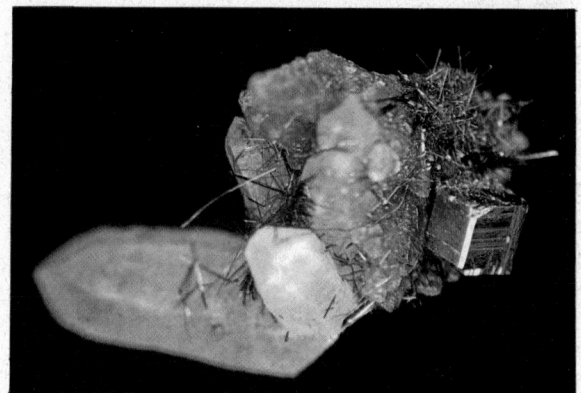

BOULANGERITE with pyrite and quartz   Noche Buena, Zacatecas, Mexico.

BOURNONITE   Julcani mine, Huancavelica, Peru.

BOURNONITE (columnar)   Julcani mine, Huancavelica, Peru.

BOURNONITE   Herodsfoot mine, Liskeard, Cornwall, England.

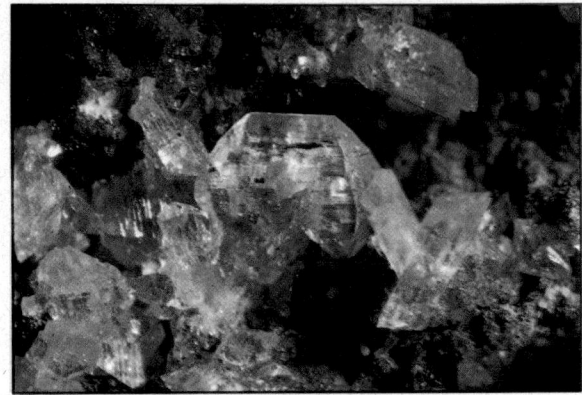

BRANDTITE   Harstig mine, Pajsberg, Vermland, Sweden.

BRANNOCKITE   Kings Mountain, Cleveland County, North Carolina.

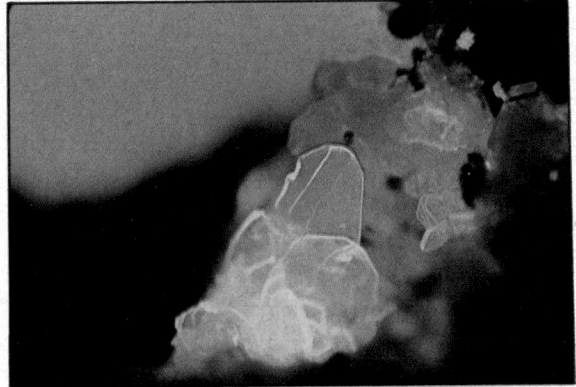

BRANNOCKITE (short-wave ultraviolet light)   Kings Mountain, Cleveland County, North Carolina.

BRAVOITE   Gey, Eifel district, Germany.

BRAVOITE with chalcopyrite   Gey, Eifel district, Germany.

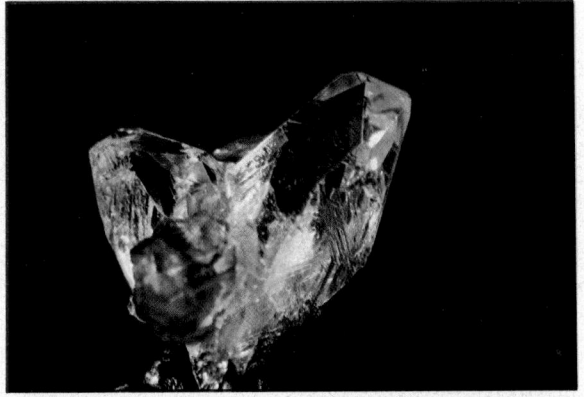

BRAZILIANITE   Conselheira Pena, Minas Geraes, Brazil.

BREITHAUPTITE   Sarrabus, Sardinia.

BROCHANTITE   Douglas Hill, Ludwig, Nevada.

BROCHANTITE on fluorite   Bingham, Socorro County, New Mexico.

BROMLITE   Fallowfield, Hexham, England.

BROOKITE with anatase Brook's Farm, Latimore, Cleveland County, North Carolina.

BUDDINGTONITE Sulphur Bank mine, Lake County, California.

CACOXENITE Rothlaufchen, near Waldgirmes, Wehlar, Germany.

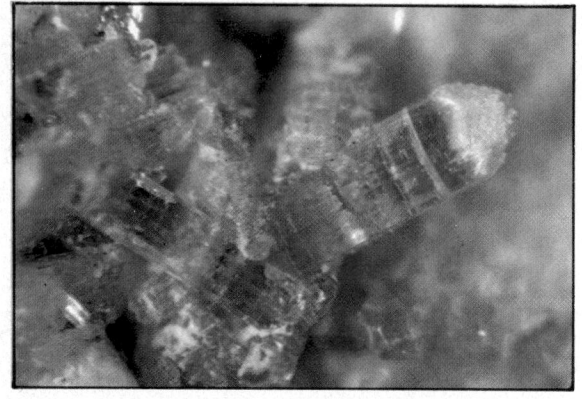

BURBANKITE Mont St. Hilaire, Quebec, Canada.

CACOXENITE spicules on beraunite Mt. Holly Springs, Cumberland County, Pennsylvania.

CACOXENITE Rothlaufchen, near Waldgirmes, Wehlar, Germany.

CAHNITE, Franklin, Sussex County, New Jersey.

CALAVERITE   Cripple Creek, Teller County, Colorado.

CALAVERITE with fluorite   Cripple Creek, Teller County, Colorado.

CALCITE on wulfenite   Mapimi, Durango, Mexico.

CALCITE   Cornwall, England.

CALCITE enclosing copper   Houghton, Michigan.

CALCITE on calcite   Warsaw, Hancock County, Illinois.

CALCITE on fluorite   Rosiclare, Hardin County, Illinois.

CALCITE on amethyst   Guanajuato, Mexico.

CALEDONITE    Leadhills, Lanarkshire, Scotland.

CALEDONITE    Mammoth-St. Anthony mine, Pinal County, Arizona.

CALEDONITE    Leadhills, Lanarkshire, Scotland.

CALLAGHANITE    Gabbs, Nye County, Nevada.

CALLAGHANITE    Gabbs, Nye County, Nevada.

CALOMEL    Near Belgrade, Yugoslavia.

CALUMETITE on copper    Centennial mine, near Calumet, Michigan.

CARMINITE   Eureka, Tintic, Utah.

CARMINITE   Tsumeb, South West Africa.

CARMINITE   Tsumeb, South West Africa.

CARNALLITE   Hattorf mine, Phillipstal, Hersfeld, Germany.

CARNOTITE   Anderson mine, Yavapai County, Arizona.

CARPHOLITE   Schlaggenwald, Bohemia, Czechoslovakia.

CASSITERITE   Camborne, Cornwall, England.

CASSITERITE (twin)   Malviento mine, Zacatecas, Mexico.

25

CATAPLEIITE   Mont St. Hilaire, Quebec, Canada.

CAVANSITE   Owyhee Dam, Malheur County, Oregon.

CELESTITE   Schneeberg, Saxony, Germany.

CELESTITE   Pugh quarry, Custar, Ohio.

CENOSITE   Nordmark, Vermland, Sweden.

CERARGYRITE on rosasite   Central Eureka mine, Eureka, Utah.

CERARGYRITE on silver   Broken Hill, New South Wales, Australia.

CERUSSITE (twinned) with brochantite and wulfenite
Tsumeb, South West Africa.

CERUSSITE (twin)   Tsumeb, South West Africa.

CERVANTITE with stibnite   Kapnik, Roumania.

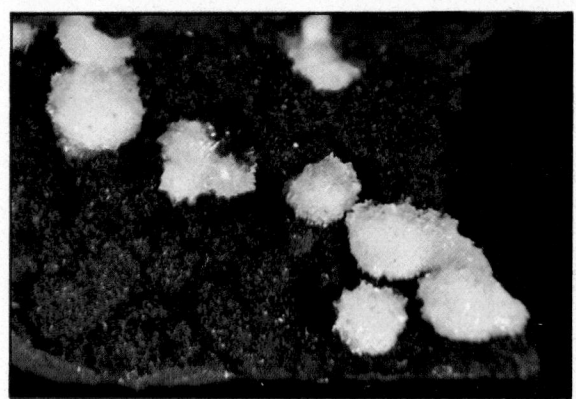

CHALCOALUMITE   Bisbee, Cochise County, Arizona.

CHALCOCITE   Carn Brea, Redruth, Cornwall, England.

CHALCOCITE (twin)   Butte, Silver Bow County, Montana.

CHALCOPHYLLITE   El Teniente mine, Rancagua, Chile.

CHALCOPHYLLITE   El Teniente mine, Rancagua, Chile.

CHALCOPYRITE   French Creek, Chester County, Pennsylvania.

CHALCOPYRITE inclusions in calcite   Faraday Uranium Mine, Bancroft, Ontario, Canada.

CHALCOPYRITE (iridescent)

CHAMBERSITE   Barber's Hill salt dome, Chambers County, Texas.

CHERVETITE with francevillite   Mounana, Gabon.

CHERVETITE with francevillite   Mounana, Gabon.

CHILDRENITE   George and Charlotte mine, Tavistock, Devon, England.

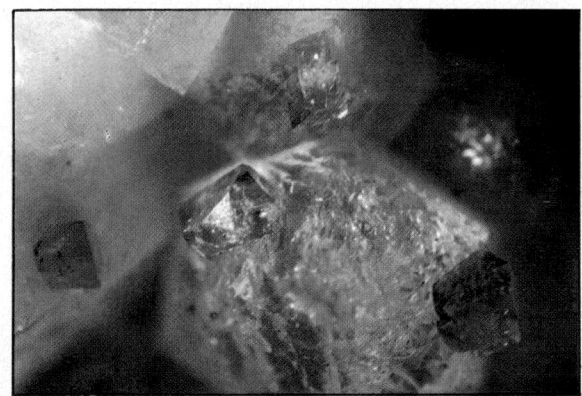

CHILDRENITE   George and Charlotte mine, Tavistock, Devon, England.

CHLOROPHOENICITE on zincite   Franklin, Sussex County, New Jersey.

CHRYSOBERYL (cyclic twin)   Collatina, Rio Doce, Espirito Santo, Brazil.

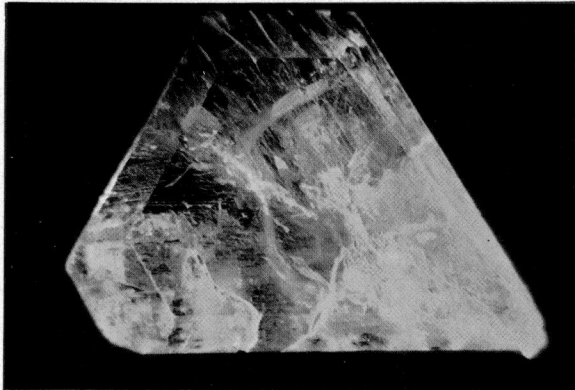

CHRYSOBERYL (twinned on (130))   Lac Alaotra, Madagasca Madagascar.

CINNABAR (acicular)   Aetna mine, Napa County, California.

CINNABAR (twin)   Nikitovska, USSR.

CINNABAR (twin)   Almaden, Spain.

CINNABAR (twin)   Almaden, Spain.

CLIFFORDITE   San Miguel mine, Moctezuma, Sonora, Mexico.

CLINOCLASE   Majuba Hill, Pershing County, Nevada.

CLINOHEDRITE with hancockite   Franklin, Sussex County, New Jersey.

CLINOHEDRITE   Franklin, Sussex County, New Jersey.

CLINOHUMITE   Rio Freddo, Naples, Italy

COBALTITE   Kibblehouse quarry, Perkiomenville, Pennsylvania.

COBALTITE   Espanola, Ontario, Canada.

COLEMANITE   Boron, Kern County, California.

COLLINSITE   Tip Top mine, Custer County, South Dakota.

COLUMBITE on albite   Strickland quarry, Portland, Connecticut.

CONICHALCITE with lavendulan   Gold Hill, Tooele County, Utah.

CONICHALCITE   Gold Hill, Tooele County, Utah.

CONNELLITE with aurichalcite   Czar mine, Bisbee, Cochise Co., Arizona.

CONNELLITE on cuprite   Bisbee, Cochise County, Arizona.

31

COPIAPITE   Calama, Chile.

COPPER   Mohawk mine, Keweenaw County, Michigan.

COPPER in gypsum   Mission mine, Pima County, Arizona.

COPPER   Tsumeb, South West Africa.

COPPER   Mohawk mine, Calumet, Keweenaw Peninsula, Michigan.

CORNETITE   Yerington, Lyon County, Nevada.

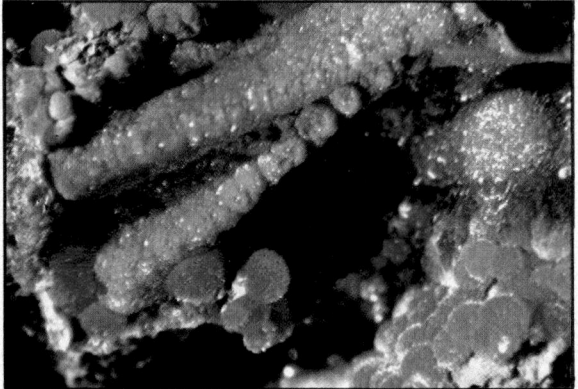
CORNWALLITE with cuprian austinite   Gold Hill, Tooele County, Utah.

BEST REF. IN ENGLISH: Agrell, S. O., Bown, M. G., and McKie, D., *Am. Min.*, **50**: 278 (1965).

**DEHRNITE** = Variety of carbonate-apatite
$(Ca, Na, K)_5 (PO_4, CO_3)_3 (OH)$

## DELAFOSSITE
$CuFeO_2$

CRYSTAL SYSTEM: Hexagonal
CLASS: $\bar{3}\ 2/m$
SPACE GROUP: $R\bar{3}m$
Z: 3
LATTICE CONSTANTS:
  $a = 3.04$
  $c = 17.12$
3 STRONGEST DIFFRACTION LINES:
  2.508 (100)
  1.512 ( 40)
  2.86 ( 35)
  HARDNESS: 5½
  DENSITY: 5.41 (Meas.)
          5.503 (Calc.)
  CLEAVAGE: $\{10\bar{1}0\}$ imperfect
Brittle.
  HABIT: Crystals tabular to equant. Usually massive; botryoidal, spherulitic. Twinning on $\{0001\}$, as contact twins.
  COLOR-LUSTER: Black. Opaque. Metallic. Streak black.
  MODE OF OCCURRENCE: Occurs chiefly as a secondary mineral, less commonly as a primary mineral, in copper ore deposits. Found at Eureka and Kimberly, Nevada; at the Pope-Shenon mine, near Salmon, Lemhi County, Idaho; as crystals and botryoidal crusts at Bisbee, Cochise County, Arizona; and at deposits in Mexico, Spain, Germany, and USSR.
  BEST REF. IN ENGLISH: *Am. Min.*, **53**: 1779 (1968). Palache, et al., "Dana's System of Mineralogy," 7th Ed., v. I, p. 674–675, New York, Wiley, 1944.

## DELESSITE (Melanolite)
$(Mg, Fe^{2+}, Fe^{3+}, Al)_6 (Si, Al)_4 O_{10} (O, OH)_8$
Chlorite Group

CRYSTAL SYSTEM: Monoclinic
CLASS: $2/m$
SPACE GROUP: $C2/m$
Z: 1
LATTICE CONSTANTS:
  $a \sim 5.3$
  $b \sim 9.2$
  $c \sim 14.3$
  $\beta \sim 97°$
3 STRONGEST DIFFRACTION LINES: Similar to thuringite with 14Å line stronger than 7Å.
OPTICAL CONSTANTS:
  $\alpha < 1.63$
  $\gamma = 1.650$
  $(-)2V$ = small to $0°$

HARDNESS: 2–3
DENSITY: 2.73 (Meas.)
CLEAVAGE: $\{001\}$ perfect
HABIT: Massive; foliated or granular. Twinning on $\{001\}$.
COLOR-LUSTER: Black with greenish tinge. Streak and powder olive green to greenish gray.
  MODE OF OCCURRENCE: Found at the long-abandoned Milk Row diabase quarry on Granite Street, Sommerville, Massachusetts.
  BEST REF. IN ENGLISH: Frondel, Clifford, *Am. Min.*, **40**: 1090–1094 (1955). Deer, Howie, and Zussman, "Rock Forming Minerals," v. 3, p. 131–163, New York, Wiley, 1962.

## DELHAYELITE
$(Na, K)_{10} Ca_5 Al_6 Si_{32} O_{80} (Cl_2, F_2, SO_4)_3 \cdot 18H_2O$

CRYSTAL SYSTEM: Orthorhombic
CLASS: $2/m\ 2/m\ 2/m$
SPACE GROUP: Pnmm
Z: 1
LATTICE CONSTANTS:
  $a = 13.06$ (or 6.53)
  $b = 24.65$
  $c = 7.04$
3 STRONGEST DIFFRACTION LINES:
  3.078 (100)
  12.30 ( 35)
  6.158 ( 25)
OPTICAL CONSTANTS:
  $\beta = 1.532$
  $2V_\alpha = 83°$
HARDNESS: Not determined.
DENSITY: $2.60 \pm 0.03$ (Meas.)
CLEAVAGE: $\{010\}$ distinct
HABIT: As platy crystals.
COLOR-LUSTER: Colorless.
  MODE OF OCCURRENCE: Occurs in kalsilite-melilite-nephelinite lava, Mt. Shaheru, Zaire.
  BEST REF. IN ENGLISH: Sahama, Th. G., and Hytonen, Kai, *Min. Mag.*, **32**: 6–9 (1959).

## DELLAITE (natural material inadequately described, data are for synthetic analog)
$Ca_6 Si_3 O_{11} (OH)_2$

CRYSTAL SYSTEM: Triclinic
Z: 2
LATTICE CONSTANTS:
  $a = 6.84$    $\alpha = 90°45'$
  $b = 6.94$    $\beta = 97°22'$
  $c = 12.89$   $\gamma = 98°16'$
3 STRONGEST DIFFRACTION LINES:
  (synthetic)
  2.290 (100)
  3.435 ( 70)
  3.067 ( 50)
  DENSITY: 2.94 (Calc. synthetic)
  HABIT: Crystals prismatic, elongated along $a$-axis. Twinning on $\{010\}$ (synthetic).

COLOR-LUSTER: Colorless, transparent, vitreous (synthetic).

MODE OF OCCURRENCE: Occurs in the late-stage veins which cut the metamorphic assemblage at Kilchoan, Ardnamurchan, Scotland.

BEST REF. IN ENGLISH: Agrell, S. O., *Min. Mag.*, **34**: 1–15 (1965).

## DELORENZITE = Tanteuxenite (tantalian euxenite)

## DELRIOITE
$CaSrV_2O_6(OH)_2 \cdot 3H_2O$

CRYSTAL SYSTEM: Monoclinic
CLASS: m or 2/m
SPACE GROUP: Ia or I2/a
Z: 8
LATTICE CONSTANTS:
  $a = 17.170$
  $b = 7.081$
  $c = 14.644$
  $\beta = 102°29'$
3 STRONGEST DIFFRACTION LINES:
  6.525 (100)
  3.539 ( 80)
  4.387 ( 60)
OPTICAL CONSTANTS:
  $\alpha = 1.783$
  $\beta = 1.834$
  $\gamma = 1.866$
(–)2V = medium to large
HARDNESS: ~ 2
DENSITY: 3.1 ± 0.1 (Meas.)
        3.16 (Calc.)
CLEAVAGE: Not determined.
HABIT: As intimate intergrowths with metadelrioite forming radial aggregates of fibrous acicular crystals. Twinning on {100} common.
COLOR-LUSTER: Pale yellowish green. Translucent. Vitreous to pearly.
MODE OF OCCURRENCE: Occurs as an efflorescence on sandstone on the dump at the Hummer portal of the Jo Dandy mine in Paradox Valley, Montrose County, Colorado.
BEST REF. IN ENGLISH: Thompson, Mary E., and Sherwood, Alexander, *Am. Min.*, **44**: 261–264 (1959). Smith, Marie Lindberg, *Am. Min.*, **55**: 185–200 (1970).

## DELTAITE = Crandallite plus hydroxylapatite

## DELVAUXITE (Inadequately described mineral)
$Fe_4(PO_4)_2(OH)_6 \cdot nH_2O$

CRYSTAL SYSTEM: Unknown.
OPTICAL CONSTANT:
  $N = 1.72$
  HARDNESS: 2½–4
  DENSITY: 1.85–2.83 (Meas.)

CLEAVAGE: Not determined. Fracture conchoidal.
HABIT: As concretionary masses, botryoidal crusts and coatings; stalactitic. Often gel-like.
COLOR-LUSTER: Yellowish brown to brown, reddish brown to brownish black. Vitreous, greasy, weak waxy. Streak yellow.
MODE OF OCCURRENCE: Occurs at Berneau, Liége, Belgium, and at localities in Austria, Germany, Czechoslovakia, and USSR.
BEST REF. IN ENGLISH: Palache, et al., "Dana's System of Mineralogy," 7th Ed., v. II, p. 935–936, New York, Wiley, 1951.

## DEMESMAEKERITE
$Pb_2Cu_5(UO_2)_2(SeO_3)_6(OH)_6 \cdot 2H_2O$

CRYSTAL SYSTEM: Triclinic
CLASS: 1 or $\bar{1}$
SPACE GROUP: P1 or P$\bar{1}$
Z: 1
LATTICE CONSTANTS:
  $a = 11.90$      $\alpha = 90°11'$
  $b = 10.02$      $\beta = 100°01'$
  $c = 5.63$       $\gamma = 91°49'$
3 STRONGEST DIFFRACTION LINES:
  2.97 (100)
  5.42 ( 80)
  5.89 ( 60)
OPTICAL CONSTANTS:
  $\alpha' = 1.835$
  $\gamma' = 1.910$
    (+)
HARDNESS: 3–4
DENSITY: 5.28 (Meas.)
        5.45 (Calc.)
CLEAVAGE: None observed.
HABIT: Crystals elongated [100] and flattened (100). Dominant faces {100} and {010}.
COLOR-LUSTER: Bottle green, turning brownish on dehydration.
MODE OF OCCURRENCE: Occurs associated with cuprosklodowskite, kasolite, malachite, guilleminite, chalcomenite and selenian digenite in the lower part of the oxidation zone of the Cu-Co deposit of Musonoi, near Kolwezi, Katanga, Zaire.
BEST REF. IN ENGLISH: Cesbron, F., Bachet, B., and Oosterbosche, R., *Am. Min.*, **51**: 1815–1816 (1966).

## DENNINGITE
$(Mn, Ca, Zn)Te_2O_5$

CRYSTAL SYSTEM: Tetragonal
CLASS: 4/m 2/m 2/m
SPACE GROUP: P4$_2$/nbc
Z: 8
LATTICE CONSTANTS:
  $a = 8.82$
  $c = 13.04$

3 STRONGEST DIFFRACTION LINES:
4.42 (100)
3.38 ( 80)
3.12 ( 70)
OPTICAL CONSTANTS:
$\omega$ = 1.89
$\epsilon$ = 2.00
(+)
HARDNESS: 4
DENSITY: 5.05 (Meas.)
5.07 (Calc.)
CLEAVAGE: {001} perfect
Fracture conchoidal.

HABIT: Small euhedral crystals, which are thin plates, octagonal in outline.

COLOR-LUSTER: Colorless to pale green; adamantine.

MODE OF OCCURRENCE: Occurs associated with tellurium, tellurite, paratellurite, and several other tellurites, especially spiroffite, at Moctezuma, Sonora, Mexico.

BEST REF. IN ENGLISH: Mandarino, J. A., Williams, S. J., and Mitchell, R. S., *Can. Min.*, 7: 340–341 (1961) (abst.). Mandarino, J. A., Williams, S. J., and Mitchell, R. S., *Can. Min.*, 7: 443–452 (1963).

# DERBYLITE
$Fe_6 Ti_6 Sb_2 O_{23}$

CRYSTAL SYSTEM: Orthorhombic
CLASS: 2/m 2/m 2/m
LATTICE CONSTANTS:
$a:b:c$ = 0.9661 : 1 : 0.5502
OPTICAL CONSTANTS:
(Li)
$\alpha$ = 2.45
$\beta$ = 2.45
$\gamma$ = 2.51
(+)2V ~ 0°
HARDNESS: 5
DENSITY: 4.53
CLEAVAGE: None.
Fracture conchoidal. Very brittle.

HABIT: As prismatic crystals, commonly twinned on {011} as cruciform twins, rarely as trillings.

COLOR-LUSTER: Black. Opaque. Resinous.

MODE OF OCCURRENCE: Occurs associated with tripuhyite, monazite, xenotime, zircon, lewisite, and other minerals in the cinnabar-bearing gravels of Tripuhy near Ouro Preto, Minas Geraes, Brazil.

BEST REF. IN ENGLISH: Palache, et al., "Dana's System of Mineralogy," 7th Ed., v. II, p. 1025–1026, New York, Wiley, 1951.

# DERRIKSITE
$Cu_4 (UO_2)(SeO_3)_2 (OH)_6 \cdot H_2O$

CRYSTAL SYSTEM: Orthorhombic
CLASS: 2/m 2/m 2/m or mm2
SPACE GROUP: Pnmm or Pnm2
Z: 2

LATTICE CONSTANTS:
$a$ = 5.57
$b$ = 19.07
$c$ = 5.96
3 STRONGEST DIFFRACTION LINES:
4.78 (100)
4.072 ( 80)
3.748 ( 80)
OPTICAL CONSTANTS:
$\alpha$ = 1.77
$\beta$ = 1.85
$\gamma$ = 1.89
Biaxial (–)
HARDNESS: Not determined.
DENSITY: 4.72 (Calc.)
CLEAVAGE: {010} very good

HABIT: Crystals elongated [001], flattened on (100). Forms: {010}, {100}, {110}, {121}. Euhedral; up to 0.7mm in size.

COLOR-LUSTER: Green to bottle green. Translucent.

MODE OF OCCURRENCE: The mineral occurs in association with chalcomenite and demesmaekerite as microcrystalline crusts on selenian digenite in the oxidation zone of the cobalt-copper deposit of Musonoi, Katanga, Zaire.

BEST REF. IN ENGLISH: Cesbron, Fabian, Pierrot, Roland, and Verbeek, Theodore, *Am. Min.*, 57: 1912–1913 (1972).

# DESCLOIZITE
$Pb(Zn, Cu)VO_4 OH$
Descloizite - Mottramite Series

CRYSTAL SYSTEM: Orthorhombic
CLASS: 2/m 2/m 2/m
SPACE GROUP: Pnam
Z: 4
LATTICE CONSTANTS:
$a$ = 7.607
$b$ = 9.446
$c$ = 6.074
3 STRONGEST DIFFRACTION LINES:
3.23 (100)
5.12 ( 80)
2.90 ( 80)
OPTICAL CONSTANTS:
$\alpha$ = 2.185
$\beta$ = 2.265
$\gamma$ = 2.35
(–)2V ~ 90°
HARDNESS: 3–3½
DENSITY: 6.24-6.26 (Meas.)
6.144 (Calc.)
CLEAVAGE: None.
Fracture uneven to conchoidal. Brittle.

HABIT: Crystals variable in form; usually pyramidal, prismatic, or tabular; also short prismatic. Crystal faces often rough or uneven. As plumose aggregates, drusy crusts, stalactitic, botryoidal with coarse fibrous structure, and massive granular.

COLOR-LUSTER: Orange-red to dark reddish brown to dark blackish brown; also dark shades of green. Transparent to opaque, vitreous to greasy.

MODE OF OCCURRENCE: Occurs widespread chiefly as a secondary mineral in the oxidation zone of ore deposits, often associated with pyromorphite, mimetite, vanadinite, calcite, and cerussite. Found in the United States at the Silver Queen mine, Galena, South Dakota, and at many places in Arizona, New Mexico, Nevada, Utah, and other western states. Large deposits providing superb crystal groups occur at Tsumeb, Grootfontein, and at many other localities in the Otavi district, South West Africa. Other occurrences are found in Mexico, Austria, Italy, Argentina, Germany, Algeria, Zaire, Rhodesia, and Tunisia.

BEST REF. IN ENGLISH: Palache, et al., "Dana's System of Mineralogy," 7th Ed., v. II, p. 811–815, New York, Wiley, 1951.

## DESMINE = Stilbite

## DESPUJOLSITE
$Ca_3Mn^{4+}(SO_4)_2(OH)_6 \cdot 3H_2O$

CRYSTAL SYSTEM: Hexagonal
CLASS: $\bar{6}$ m 2
SPACE GROUP: $P\bar{6}2c$
Z: 2
LATTICE CONSTANTS:
  $a$ = 8.56
  $c$ = 10.76
3 STRONGEST DIFFRACTION LINES:
  3.34 (100)
  4.26 ( 80)
  2.129 ( 80)
OPTICAL CONSTANTS:
  $\omega$ = 1.656
  $\epsilon$ = 1.682
    (+)
HARDNESS: ~ 2½
DENSITY: 2.46 (Meas.)
         2.54 (Calc.)
CLEAVAGE: Fracture conchoidal. Brittle.
HABIT: Aggregates of hexagonal prisms up to 0.5 mm long. Crystals show $(10\bar{1}0)$ dominant, $(10\bar{1}2)$, and $(0001)$.
COLOR-LUSTER: Lemon yellow; vitreous.
MODE OF OCCURRENCE: Occurs in cavities between crystals of gaudefroyite at Tachgagalt, Morocco.
BEST REF. IN ENGLISH: Gaudefroy, C., Granger, M. M., Permingeat, F., and Protas, J., *Am. Min.*, **54**: 326 (1969).

## DESTINEZITE = Diadochite

## DEVILLITE (Devilline)
$CaCu_4(SO_4)_2(OH)_6 \cdot 3H_2O$

CRYSTAL SYSTEM: Monoclinic
CLASS: 2/m
SPACE GROUP: $P2_1/c$
Z: 2

LATTICE CONSTANTS:
  $a$ = 10.97
  $b$ = 6.13
  $c$ = 10.44
  $\beta$ = 105°27′
3 STRONGEST DIFFRACTION LINES:
  10.2 (100)
  5.06 ( 80)
  3.38 ( 70)
OPTICAL CONSTANTS:
  $\alpha$ = 1.62
  $\beta$ = 1.652
  $\gamma$ = 1.656
(−)2V = 37°
HARDNESS: 2½
DENSITY: 3.13 (Meas.)
CLEAVAGE: {001} perfect
          {110} distinct
          {101} distinct
          {10$\bar{1}$} distinct
Not brittle.
HABIT: Crystals flattened six-sided plates striated [010]. As crusts or rosettes composed of thin crystals. Twinning on {010}.
COLOR-LUSTER: Dark emerald green to bluish green. Transparent. Vitreous; somewhat pearly on base. Streak pale green.
MODE OF OCCURRENCE: Occurs at the Ecton copper mine, Montgomery County, Pennsylvania; in Cornwall, England; at Finosa, Corsica; at Herrengrund, Czechoslovakia; and in the Uspensky mine, Kazakhstan, USSR.
BEST REF. IN ENGLISH: Palache, et al., "Dana's System of Mineralogy," 7th Ed., v. II, p. 590–592, New York, Wiley, 1951.

## DEWEYLITE = Mixture of clinochrysotile or lizardite and stevensite

## DEWINDTITE
$Pb(UO_2)_2(PO_4)_2 \cdot 3H_2O$

CRYSTAL SYSTEM: Orthorhombic
CLASS: mm2 or 2/m 2/m 2/m
SPACE GROUP: Bmb2 or Bmmb
Z: ≃ 12 (?)
LATTICE CONSTANTS:
  $a$ = 16.05
  $b$ = 17.50
  $c$ = 13.64
3 STRONGEST DIFFRACTION LINES:
  8.01 (100)
  5.89 (100)
  3.14 ( 90)
OPTICAL CONSTANTS:
  $\alpha$ = 1.760–1.762
  $\beta$ = 1.768–1.767
  $\gamma$ = 1.770–1.768
(−)2V = moderate
HARDNESS: Not determined.

DENSITY: 5.03 (Meas.)
5.01 (Calc.)
CLEAVAGE: {100} perfect
Brittle.

HABIT: Microscopic tablets flattened {100} and terminated by {001}. {100} striated parallel to c-axis. Also very fine grained and compact.

COLOR-LUSTER: Canary yellow; powder pale yellow. Translucent. Fluoresces green in ultraviolet light.

MODE OF OCCURRENCE: Occurs as a secondary mineral associated with torbernite, dumontite, and kasolite at Shinkolobwe, Katanga, Zaire, and at Wölsendorf, Bavaria, Germany.

BEST REF. IN ENGLISH: Frondel, Clifford USGS Bull. 1064, p. 230–232 (1958).

# DIABANTITE
$(Mg, Fe^{2+}, Al)_6 (Si, Al)_4 O_{10} (OH)_8$
Chlorite Group

CRYSTAL SYSTEM: Monoclinic
CLASS: 2/m
SPACE GROUP: C2/m
Z: 1
LATTICE CONSTANTS:
$a \sim 5.3$
$b \sim 9.2$
$c \sim 14.3$
$\beta \sim 97°$
3 STRONGEST DIFFRACTION LINES:
7.08  (100)
14.0  ( 80)
3.541 ( 80)
OPTICAL CONSTANTS:
$\beta = 1.61–1.63$
$\gamma - \alpha = .000–.004$
$(-)2V$
HARDNESS: 2–2½
DENSITY: 2.77–2.93 (Meas.)
3.035 (Calc.)
CLEAVAGE: {001} perfect

HABIT: Massive, compact, fibrous or with foliated, radiated, and concentric structure.

COLOR-LUSTER: Dark green to greenish black.

MODE OF OCCURRENCE: Occurs in diabase at Farmington Hills, Connecticut, and at Westfield, Massachusetts; also at localities in Germany and elsewhere.

BEST REF. IN ENGLISH: Deer, Howie, and Zussman, "Rock Forming Minerals," v. 3, p. 131–163, New York, Wiley, 1962.

# DIABOLEÏTE
$Pb_2 CuCl_2 (OH)_4$

CRYSTAL SYSTEM: Tetragonal
CLASS: 4mm
SPACE GROUP: P4mm
Z: 1
LATTICE CONSTANTS:
$a = 5.870$
$c = 5.494$

3 STRONGEST DIFFRACTION LINES:
5.507 (100)
3.314 (100)
2.292 (100)
OPTICAL CONSTANTS:
$\omega = 1.98$
$\epsilon = 1.85$
$(-)$
HARDNESS: 2½
DENSITY: 5.42 (Meas.)
5.41 (Calc.)
CLEAVAGE: {001} perfect
Fracture conchoidal. Brittle.

HABIT: Crystals tabular {001} with square outline. As aggregates of thin plates; also massive, finely crystalline.

COLOR-LUSTER: Deep blue; transparent to translucent. Vitreous. Streak pale blue.

MODE OF OCCURRENCE: Occurs as excellent small crystals associated with cerussite, boleite, linarite, and other secondary minerals in the Collins vein at the Mammoth mine, Tiger, Arizona; and also as tiny crystals associated with other secondary lead minerals at Higher Pitts farm, Mendip Hills, Somerset, England.

BEST REF. IN ENGLISH: Palache, et al., "Dana's System of Mineralogy," 7th Ed., v. II, p. 82–83, New York, Wiley, 1951.

# DIADOCHITE (Destinezite)
$Fe_2 PO_4 (SO_4)(OH) \cdot 5H_2O$

CRYSTAL SYSTEM: Triclinic
Z: 2
LATTICE CONSTANTS:
$a = 9.61$    $\alpha = 98°49'$
$b = 9.77$    $\beta = 108°01'$
$c = 7.36$    $\gamma = 63°59'$
3 STRONGEST DIFFRACTION LINES:
4.35 (100)
8.3 ( 90)
8.7 ( 80)
OPTICAL CONSTANTS:
$\alpha = 1.62$
$\beta = 1.64$
$\gamma = 1.67$
$(+)2V = small$
HARDNESS: $\sim$ 3–4
DENSITY: 2.0–2.4
CLEAVAGE: Fracture conchoidal to uneven or earthy. Brittle to pulverent.

HABIT: Usually gel-like and amorphous; as crusts or masses, botryoidal, stalactitic. Also microcrystalline in masses composed of minute six-sided plates of various habits.

COLOR-LUSTER: Deep yellowish brown, dark amber; also reddish brown, greenish yellow, pale green, yellowish white. Translucent to opaque. Resinous, waxy to dull.

MODE OF OCCURRENCE: Occurs as a recent deposit in mine workings, and as a near-surface secondary mineral. Found in the New Idria quicksilver mine, San Benito County, California; at Roberts Mountain, near Eureka, Nevada; in the Black Hills, South Dakota; at Shady, Polk County,

Arkansas; and at numerous other localities in the United States and elsewhere throughout the world.

BEST REF. IN ENGLISH: Palache, et al., "Dana's System of Mineralogy," 7th Ed., v. II, p. 1011–1013, New York, Wiley, 1951.

# DIAMOND

C

Polymorphous with graphite, chaoite, lonsdaleite

CRYSTAL SYSTEM: Cubic
CLASS: 4/m $\bar{3}$ 2/m
SPACE GROUP: Fd3m
Z: 8
LATTICE CONSTANT:
  $a = 3.567$
3 STRONGEST DIFFRACTION LINES:
  2.06   (100)
  1.261  ( 25)
  1.0754 ( 16)
OPTICAL CONSTANT:
  $N = 2.4175$ (Na)
  HARDNESS: 10
  DENSITY: 3.5142 (Meas.)
          3.5150 (Calc.)
  CLEAVAGE: {111} perfect
Fracture conchoidal. Brittle.

HABIT: Crystals usually octahedral; also cubic, dodecahedral, tetrahedral; often flattened and etched {111}. Faces commonly curved and often striated. Also spherical with radiated structure; rarely massive. Twinning on {111} and {001}, common.

COLOR-LUSTER: Colorless, white to blue-white, gray, various shades of yellow, brown, orange, pink, red, lavender blue, green, black. Transparent, translucent, rarely opaque. Adamantine to greasy. Sometimes fluorescent under ultraviolet light.

MODE OF OCCURRENCE: Occurs chiefly in kimberlite pipes or volcanic necks associated with olivine, pyrope, phlogopite, and other minerals; in conglomerates and alluvial deposits; and rarely in meteorites. Found in the United States in kimberlite near Murfreesboro, Pike County, Arkansas; and as isolated occurrences in gravels in California, Oregon, Idaho, Colorado, Wisconsin, Virginia, North Carolina, and Georgia. Kimberlite pipes and alluvial deposits in southern and central Africa are the principal sources for commercial diamonds; other significant deposits are found in Brazil, Venezuela, USSR, India, and Australia.

BEST REF. IN ENGLISH: Palache, et al., Dana's System of Mineralogy," 7th Ed., v. I, p. 146–151, New York, Wiley, 1944.

# DIAPHORITE

$Ag_3Pb_2Sb_3S_8$

CRYSTAL SYSTEM: Monoclinic
CLASS: 2/m
SPACE GROUP: $P2_1/c$
Z: 4

LATTICE CONSTANTS:
  $a = 17.87$
  $b = 5.90$
  $c = 15.85$
  $\beta = 116°09'$
3 STRONGEST DIFFRACTION LINES:
  3.296 (100)
  2.809 ( 80)
  2.945 ( 40)
  HARDNESS: 2½–3
  DENSITY: 5.97 (Meas.)
          5.97 (Calc.)
  CLEAVAGE: None.
Fracture subconchoidal to uneven. Brittle.

HABIT: Crystals prismatic, sometimes complex. Usually striated parallel to elongation. Twinning on {120} and {241}.

COLOR-LUSTER: Steel gray. Opaque. Metallic.

MODE OF OCCURRENCE: Occurs associated with pyrargyrite, galena, dolomite, and quartz in the Lake Chelan district, Okanogan County, Washington; at Catorce, San Luis Potosi, Mexico; at Zancudo, Columbia; at Pribram, Czechoslovakia; and at several mines in Saxony, Germany.

BEST REF. IN ENGLISH: Palache, et al., "Dana's System of Mineralogy," 7th Ed., v. I, p. 414–415, New York, Wiley, 1944.

# DIASPORE

$HAlO_2$

Dimorphous with boehmite
Var. Manganoan (mangandiaspore)

CRYSTAL SYSTEM: Orthorhombic
CLASS: 2/m 2/m 2/m
SPACE GROUP: Pbnm
Z: 4
LATTICE CONSTANTS:
  $a = 4.40$
  $b = 9.39$
  $c = 2.84$
3 STRONGEST DIFFRACTION LINES:
  3.99 (100)
  2.317 ( 56)
  2.131 ( 52)
OPTICAL CONSTANTS:
  $\alpha = 1.702$
  $\beta = 1.722$ (Na)
  $\gamma = 1.750$
  $(+)2V = 84°–85°$
  HARDNESS: 6½–7
  DENSITY: 3.3–3.5 (Meas.)
          3.380 (Calc.)
  CLEAVAGE: {010} perfect
           {110} distinct
           {100} in traces
Fracture conchoidal. Brittle.

HABIT: Crystals commonly thin elongated plates; sometimes acicular or tabular; striated. Also massive; foliated, scaly, stalactitic. Frequently disseminated. Twinning on {061}, or on {021}, rare, producing pseudohexagonal aggregates.

COLOR-LUSTER: Color various: shades of white to colorless, yellowish, greenish, lilac, pink, brownish. Mangandiaspore rose red to dark red. Transparent to subtranslucent. Vitreous, pearly on cleavages.

MODE OF OCCURRENCE: Occurs chiefly with corundum or emery, magnetite, spinel, margarodite, chlorite, dolomite, in metamorphic limestones, chloritic schists, and altered igneous rocks; also extensively in bauxite and aluminous clays. Found in California, Colorado, Arkansas, Missouri, Pennsylvania, Massachusetts, Connecticut, North Carolina, and elsewhere in the United States. Other notable occurrences are found in Greenland, England, Norway, Sweden, France, Switzerland, Germany, Hungary, Greece, USSR, Japan, China, and in the Postmasburg district, South Africa (mangandiaspore).

BEST REF. IN ENGLISH: Palache, et al., "Dana's System of Mineralogy," 7th Ed., v. I, p. 675-679, New York, Wiley, 1944.

## γ-DICALCIUM SILICATE
$Ca_2SiO_4$
Dimorph of larnite

CRYSTAL SYSTEM: Orthorhombic
CLASS: 2/m 2/m 2/m
SPACE GROUP: Pbnm
Z: 4
LATTICE CONSTANTS:
 (synthetic)
 $a = 5.06$
 $b = 11.28$
 $c = 6.78$
3 STRONGEST DIFFRACTION LINES:
 2.73 (100)
 1.909 ( 80)
 3.01 ( 70)
OPTICAL CONSTANTS:
 (synthetic)
 $\alpha = 1.640$
 $\beta = 1.645$
 $\gamma = 1.654$
 $(-)2V = 60°$
HARDNESS: 5-6
DENSITY: $\sim 3.0$ (Meas.)
 2.970 (Calc.)
HABIT: Fibrous.
COLOR-LUSTER: Luster dull to vitreous.
MODE OF OCCURRENCE: Occurs in limestones and marbles in contact zone at Marble Canyon, Texas.
BEST REF. IN ENGLISH: Bridge, Thomas E., *Am. Min.*, **51**: 1766-1774 (1966).

## DICKINSONITE
$H_2Na_6(Mn, Fe, Ca, Mg)_{14}(PO_4)_{12} \cdot H_2O$

CRYSTAL SYSTEM: Monoclinic
CLASS: 2/m
SPACE GROUP: C2/c
Z: 4

LATTICE CONSTANTS:
 $a = 16.70$ kX
 $b = 9.95$
 $c = 24.69$
 $\beta = 104°41'$
3 STRONGEST DIFFRACTION LINES:
 3.05 (100)
 2.72 ( 90)
 3.22 ( 70)
OPTICAL CONSTANTS:
 (+)2V
HARDNESS: 3½-4
DENSITY: 3.41 (Meas.)
 3.42 (Calc.)
CLEAVAGE: {001} perfect, easy.
Fracture uneven. Very brittle.
HABIT: Crystals tabular, often pseudorhombohedral; triangular striations on {001}. Commonly foliated to micaceous; also curved lamellar or radiated; as disseminated scales.
COLOR-LUSTER: Yellowish green to olive green to brownish green. Transparent to translucent. Vitreous, pearly on cleavage. Streak white.
MODE OF OCCURRENCE: Occurs in granite pegmatites associated with lithiophilite, rhodochrosite, and a wide variety of secondary phosphate minerals. Found sparingly at Branchville and at Portland, Connecticut, and at the Berry quarry near Poland, Maine.
BEST REF. IN ENGLISH: Palache, et al., "Dana's System of Mineralogy," 7th Ed., v. II, p. 717-719, New York, Wiley, 1951. Fisher, D. J., *Am. Min.*, **50**: 1647-1669 (1965).

## DICKITE
$Al_2Si_2O_5(OH)_4$
Kaolin Group (Kandite)
Trimorphous with kaolinite and nacrite

CRYSTAL SYSTEM: Monoclinic
CLASS: m
SPACE GROUP: Cc
Z: 4
LATTICE CONSTANTS:
 $a = 5.15$
 $b = 8.95$
 $c = 14.42$
 $\beta = 96.8°$
3 STRONGEST DIFFRACTION LINES:
 7.15 (100+)
 3.58 (100+)
 2.33 ( 90 )
OPTICAL CONSTANTS:
 $\alpha = 1.560-1.564$
 $\beta = 1.561-1.566$
 $\gamma = 1.566-1.570$
 (+)2V = 50°-80°
HARDNESS: 2-2½
DENSITY: $\sim 2.60$ (Meas.)
 2.599 (Calc.)
CLEAVAGE: {001} perfect
Scales flexible, inelastic; masses plastic when moist.

HABIT: Crystals thin tabular, pseudohexagonal, minute. Often in piles of stacked platelets. Usually massive; compact, friable, or mealy. Twinning rare.

COLOR-LUSTER: Colorless, white; sometimes tinted yellowish or brownish. Transparent to translucent. Satiny to dull earthy.

MODE OF OCCURRENCE: Occurs as a hydrothermal mineral in association with quartz and sulfides in ore deposits. Typical occurrences are found in Colorado, especially in the Nederland tungsten district, Boulder County; in the LaPlata district, LaPlata and Montezuma counties; abundantly in mines of the Red Mountain district, Ouray County; and with rhodochrosite, huebnerite, and other minerals at the Sweet Home mine, near Alma, Park County. Also found in Pike County, Arkansas; in Schuylkill, Pennsylvania; on the island of Anglesey, Wales; from the Karacheku massif, Karkaralinsk, Kazakh steppe, USSR; and at Shokozan, Japan.

BEST REF. IN ENGLISH: Deer, Howie, and Zussman, "Rock Forming Minerals," v. 3, p. 194–212, New York, Wiley, 1962.

# DIDYMOLITE = Plagioclase

# DIENERITE (Inadequately described mineral)
$Ni_3As$

CRYSTAL SYSTEM: Cubic

HABIT: Known only as a single cubic crystal 5 mm along its edge from Radstadt, Salzburg, Austria.

COLOR-LUSTER: Grayish white. Opaque. Bright metallic.

BEST REF. IN ENGLISH: Palache, et al., "Dana's System of Mineralogy," v. I, p. 175, New York, Wiley, 1944.

# DIETRICHITE
$(Zn, Fe, Mn)Al_2(SO_4)_4 \cdot 22H_2O$
Halotrichite Group

CRYSTAL SYSTEM: Monoclinic
CLASS: 2
SPACE GROUP: P2
Z: 2
LATTICE CONSTANTS: Similar to halotrichite.
3 STRONGEST DIFFRACTION LINES:
Similar to halotrichite.
OPTICAL CONSTANTS:
$\alpha = 1.475$
$\beta = 1.480$
$\gamma = 1.488$
(+)2V = large
HARDNESS: 2
DENSITY: Not determined.
CLEAVAGE: Not determined.
HABIT: As efflorescent fibrous crusts.
COLOR-LUSTER: Grayish white to brownish yellow. Silky.
MODE OF OCCURRENCE: Found at Felsőbánya, Roumania, as a recent efflorescence in mine workings.

BEST REF. IN ENGLISH: Palache, et al., "Dana's System of Mineralogy," 7th Ed., v. II, p. 528–529, New York, Wiley, 1951.

# DIETZEITE
$Ca_2(IO_3)_2CrO_4$

CRYSTAL SYSTEM: Monoclinic
CLASS: 2/m
SPACE GROUP: $P2_1/c$
Z: 4
LATTICE CONSTANTS:
$a = 10.16$
$b = 7.30$
$c = 14.03$
$\beta = 106°32'$
OPTICAL CONSTANTS:
$\alpha = 1.825$
$\beta = 1.842$
$\gamma = 1.857$
(−)2V = 86°
HARDNESS: 3½
DENSITY: 3.617 (Meas.)
3.61 (Calc.)
CLEAVAGE: {100} imperfect
Fracture conchoidal.
HABIT: As minute elongated tabular crystals. Usually columnar or as fibrous crusts.
COLOR-LUSTER: Deep golden yellow. Transparent.
MODE OF OCCURRENCE: Occurs in Chile at the Oficina Maria Elena, near Tocapilla, in vugs in massive nitrate rock associated with lopezite, tarapacaite, and ulexite; also in nitrate deposits in the Atacama desert, Antofagasta province.

BEST REF. IN ENGLISH: Palache, et al., "Dana's System of Mineralogy," 7th Ed., v. II, p. 318–319, New York, Wiley, 1951.

# DIGENITE
$Cu_{2-x}S$

CRYSTAL SYSTEM: Cubic
CLASS: $4/m\,\bar{3}\,2/m$
SPACE GROUP: Fm3m
Z: 4
LATTICE CONSTANT:
$a = 5.54$
3 STRONGEST DIFFRACTION LINES:
1.973 (100)
3.21 ( 40)
2.79 ( 40)
HARDNESS: 2½–3
DENSITY: 5.6 (Meas.)
5.715 (Calc.)
CLEAVAGE: {111} synthetic
Conchoidal fracture. Brittle.
HABIT: Distinct crystals rare. Usually massive.
COLOR-LUSTER: Deep blue to black. Opaque. Submetallic.
MODE OF OCCURRENCE: Occurs in copper ore deposits, often associated with pyrite, chalcopyrite, chalcocite, born-

ite, and other copper minerals. Found abundantly at Butte, Montana, and at Kennecott, Alaska; as crystals at the United Verde mine, Jerome, Arizona; and at localities in Mexico, Sweden, and South West Africa.

BEST REF. IN ENGLISH: Palache, et al., "Dana's System of Mineralogy," 7th Ed., v. I, p. 180–182, New York, Wiley, 1944.

**DILLNITE** = Variety of zunyite containing fluorine.

# DIMORPHITE (II)
$As_4S_3$

CRYSTAL SYSTEM: Orthorhombic
CLASS: $2/m\ 2/m\ 2/m$ or $mm2$
SPACE GROUP: Pnma or $Pn2_1a$
Z: 4
LATTICE CONSTANTS:
  $a = 11.24$
  $b = 9.90$
  $c = 6.56$
3 STRONGEST DIFFRACTION LINES:
  2.95 (100)
  3.92 ( 90)
  2.13 ( 80)
OPTICAL CONSTANTS:
  $\beta = 1.6654$
  Biaxial (+)
  HARDNESS: $1\frac{1}{2}$
  DENSITY: 3.58 (Meas.)
          3.60 (Meas. synthetic)
  CLEAVAGE: None.
Brittle.
HABIT: Crystals dipyramidal with forms $\{111\}$, $\{110\}$, and $\{101\}$, usually in groups of minute parallel individuals.
COLOR-LUSTER: Orange-yellow. Transparent. Adamantine.
MODE OF OCCURRENCE: Occurs with native sulfur, realgar, sal ammoniac, and other minerals at a fumarole of the Solfatara, Phlegraean fields, Italy.
BEST REF. IN ENGLISH: Palache, et al., "Dana's System of Mineralogy," 7th Ed., v. I, p. 197–198, New York, Wiley, 1944. Frankel, L. S., and Zoltai, T., *Zeitschrift für Kristal.*, **138**: 161–166 (1973).

# DIOPSIDE
$MgCaSi_2O_6$
Pyroxene Group
Var. Schefferite (manganoan); Salite

CRYSTAL SYSTEM: Monoclinic
CLASS: $2/m$
SPACE GROUP: C2/c
Z: 4
LATTICE CONSTANTS:
  $a = 9.73$
  $b = 8.91$
  $c = 5.25$
  $\beta = 105°50'$

3 STRONGEST DIFFRACTION LINES:
  2.99 (100)
  2.52 ( 65)
  2.89 ( 40)
OPTICAL CONSTANTS:
  $\alpha = 1.664$–$1.695$
  $\beta = 1.672$–$1.701$
  $\gamma = 1.695$–$1.721$
  $(+)2V_\gamma = 50°$–$60°$
HARDNESS: $5\frac{1}{2}$–$6\frac{1}{2}$
DENSITY: 3.22–3.38 (Meas.)
        3.273 (Calc.)
CLEAVAGE: $\{110\}$ good
          $\{100\}$ parting
          $\{010\}$ parting
Fracture uneven to conchoidal. Brittle.
HABIT: Crystals commonly short prismatic; also massive, lamellar, columnar, or granular. Twinning on $\{001\}$ or $\{100\}$ common, simple and multiple.
COLOR-LUSTER: Colorless, white, gray, pale green to dark greenish black, yellowish brown to reddish brown (schefferite), rarely blue. Transparent to nearly opaque. Vitreous, often dull. Streak white or grayish.
MODE OF OCCURRENCE: Occurs as a common and widespread mineral, chiefly in calcium-rich metamorphic rocks, in some basic and ultrabasic rocks, and less commonly in meteorites. Typical occurrences are found in California, Montana, South Dakota, Colorado, Pennsylvania, New York, New Jersey (schefferite), and Tennessee. Also in Canada, Sweden, Switzerland, Italy, Austria, Finland, USSR, South Africa, Madagascar, Ceylon, India, Burma, Korea, Japan, and elsewhere.
BEST REF. IN ENGLISH: Deer, Howie, and Zussman, "Rock Forming Minerals," v. 2, p. 42–74, New York, Wiley, 1963.

# DIOPTASE
$CuSiO_2(OH)_2$

CRYSTAL SYSTEM: Hexagonal
CLASS: $\bar{3}$
SPACE GROUP: $R\bar{3}$
Z: 18
LATTICE CONSTANTS:
  $a = 14.66$
  $c = 7.83$
3 STRONGEST DIFFRACTION LINES:
  2.60 (100)
  7.28 ( 50)
  2.12 ( 40)
OPTICAL CONSTANTS:
  $\omega = 1.644$–$1.658$
  $\epsilon = 1.697$–$1.7094$
HARDNESS: 5
DENSITY: 3.28–3.35 (Meas.)
        3.298 (Calc.)
CLEAVAGE: $\{10\bar{1}1\}$ perfect
Fracture uneven to conchoidal. Brittle.
HABIT: Crystals short to long prismatic, often terminated by rhombohedron. Also in crystalline aggregates; massive.
COLOR-LUSTER: Emerald green to deep bluish green.

Transparent to translucent. Vitreous; somewhat greasy on cleavage or fracture. Streak pale greenish blue.

MODE OF OCCURRENCE: Occurs in the oxidation zone of copper deposits, or in cavities in rocks associated with such deposits. Found in the Soda Lake Mountains, San Bernardino County, California; in Arizona as fine microcrystals associated with cerussite, murdochite, and orange wulfenite at the Mammoth-St. Anthony mine, Tiger, and elsewhere in Pinal County, and at localities in Gila, Greenlee, and Yuma counties; at Copiapo in Atacama and at other places in Chile; as small brilliant crystals on limestone near Altyn-Tube, Khirgiz Steppes, USSR; in Katanga, Zaire; as excellent large crystals at Mindouli and at other deposits in the Niari River basin in southern Congo; from Guchab near Otavi, and as superb crystals with calcite at Tsumeb, South West Africa.

BEST REF. IN ENGLISH: Winchell and Winchell, "Elements of Optical Mineralogy," 4th Ed., Pt. 2, p. 453, New York, Wiley, 1951.

# DIPYRE

$mNa_4(Al_3Si_9O_{24})Cl$ with $nCa_4(Al_6Si_6O_{24})CO_3$
$Ma_{80}Me_{20}$ to $Ma_{50}Me_{50}$
Scapolite Group

CRYSTAL SYSTEM: Tetragonal
CLASS: 4/m
SPACE GROUP: I4/m
Z: 2
LATTICE CONSTANTS:
  $a \simeq 12.15$
  $c \simeq 7.55$
OPTICAL CONSTANTS:
  $\omega = 1.545–1.610$
  $\epsilon = 1.540–1.570$
HARDNESS: 5½–6
DENSITY: 2.57–2.69 (Meas.)
CLEAVAGE: {100} distinct
          {110} distinct
Fracture uneven to conchoidal. Brittle.

HABIT: Crystals prismatic, often coarse. Also massive, granular; sometimes columnar.

COLOR-LUSTER: Colorless, white, gray, bluish, greenish, yellowish, pink, violet, and brownish. Transparent to translucent. Vitreous to somewhat pearly or resinous. Streak uncolored. Sometimes fluorescent under long-wave ultraviolet light.

MODE OF OCCURRENCE: Occurs mainly in regionally metamorphosed rocks, contact zones, altered basic igneous rocks, and ejected volcanic blocks. Typical occurrences are found in Ontario and Quebec, Canada; also in Norway and USSR.

BEST REF. IN ENGLISH: Shaw, D. M., *J. Petr.*, 1: 218; 261 (1960). Deer, Howie, and Zussman, "Rock Forming Minerals," v. 4, p. 321–337, New York, Wiley, 1963.

# DISTHENE = Kyanite

# DITTMARITE

$(NH_4)MgPO_4 \cdot H_2O$

CRYSTAL SYSTEM: Orthorhombic
CLASS: mm2
SPACE GROUP: $Pmn2_1$
Z: 2
LATTICE CONSTANTS:
  $a = 5.606$
  $b = 8.758$
  $c = 4.788$
3 STRONGEST DIFFRACTION LINES:
  8.77 (100)
  2.80 ( 50)
  2.92 ( 40)
OPTICAL CONSTANTS:
  $\alpha = 1.549$
  $\beta = 1.569$
  $\gamma = 1.571$
  $(-)2V = 40°$
HARDNESS: Low
DENSITY: 2.19 (Calc.)
HABIT: As minute crystals.
COLOR-LUSTER: Colorless. Transparent.
MODE OF OCCURRENCE: The mineral occurs in bat guano, associated with newberyite, struvite, schertelite, and hannayite, in the Skipton Caves, southwest of Ballarat, Victoria, Australia.

BEST REF. IN ENGLISH: Mrose, M. E., U.S. G. S. Prof. Pap. 750-A, p. A115 (1971).

# DIXENITE

$Mn_6(OH)_2SiO_4(AsO_3)_2$

CRYSTAL SYSTEM: Hexagonal
CLASS: $\bar{3}2/m$, 3m or 32
SPACE GROUP: $R\bar{3}m$, R3m or R32
Z: 9
LATTICE CONSTANTS:
  $a = 8.21$
  $c = 37.39$
3 STRONGEST DIFFRACTION LINES:
  2.92 (100)
  4.10 ( 90)
  2.96 ( 80)
OPTICAL CONSTANTS:
  $\omega = 1.96$
  $\epsilon = ?$
  (+)
HARDNESS: 3–4
DENSITY: 4.2 (Meas.)
CLEAVAGE: {0001} perfect, micaceous
HABIT: Massive; in aggregates of thin folia.
COLOR-LUSTER: Bronze to nearly black. Translucent.
MODE OF OCCURRENCE: Occurs in association with adelite at Långban, Vermland, Sweden.

BEST REF. IN ENGLISH: Wickman, F. E., *Geol. Fören. Stockh. Förhandl.*, 72: 64 (1950).

# DJALMAITE = Uranium-rich microlite

# DJERFISHERITE

$K_3Cu_3(Fe,Ni)_{11}S_{14}$ (Talnakh)
$K_3(Cu,Na)(Fe,Ni)_{12}S_{14}$ (Meteorite)

CRYSTAL SYSTEM: Cubic
CLASS: $2/m\,\overline{3}$ or 23
SPACE GROUP: Pm3 or P23
Z: 2
LATTICE CONSTANT:
 $a = 10.34$
3 STRONGEST DIFFRACTION LINES:
 (Meteorite)    (Talnakh)
 1.828 (100)    1.843 (100)
 2.985 ( 70)    3.33 ( 70)
 2.372 ( 60)    3.17 ( 70)
 HARDNESS: Microhardness 172 kg/mm$^2$.
 HABIT: Tiny grains; 0.02–0.4 mm in diameter.
 COLOR-LUSTER: Greenish brown; submetallic.
 MODE OF OCCURRENCE: Occurs associated with nickel-iron, troilite, schreibersite, clinoenstatite, tridymite, cristobalite, daubreelite, roedderite, and alabandite in the Kota Kota and St. Marks enstatite chondrites. Also found associated with talnakhite, chalcopyrite, pentlandite, magnetite, valleriite, sphalerite, and platinum minerals in the Talnakh copper-nickel deposit, Noril'sk, western Siberia.
 BEST REF. IN ENGLISH: Fuchs, L. H., *Science*, **153**: 166–167 (1966). (Meteorite.)   Genkin, A. D., Troneva, N. V., and Zhuravlev, M. N., *Am. Min.*, **55**: 1071 (1970). (Talnakh.)

# DJURLEITE

$Cu_{1.96}S$

CRYSTAL SYSTEM: Probably orthorhombic
Z: 128
LATTICE CONSTANTS:
 $a = 26.90$
 $b = 15.72$
 $c = 13.57$
3 STRONGEST DIFFRACTION LINES:
 1.871 (100)
 2.387 ( 90)
 1.964 ( 90)
 3.386 ( 50) diagnostic
 HARDNESS: 2½–3
 DENSITY: 5.63 (Meas.)
 CLEAVAGE: Fracture conchoidal. Brittle.
 HABIT: Crystals short prismatic or thick tabular. Also compact massive. Twinning on {110} common.
 COLOR-LUSTER: Blackish lead gray. Opaque. Metallic.
 MODE OF OCCURRENCE: Occurs principally as a supergene mineral in the enriched zone of sulfide deposits. Found at the Leonard mine, Butte, Montana; Salvadora mine, Milpillas, and Barrana de Cobre, Chihuahua, Mexico; Morococha, Peru; Bagacay, Samar Island, Philippine Islands; Tsumeb, South West Africa; and at the Ani mine, Akita Prefecture, and at other deposits in Japan.
 BEST REF. IN ENGLISH: Roseboom, Eugene H. Jr., *Am. Min.*, **47**: 1181–1184 (1962).

# DOLEROPHANE (Dolerophanite)

$Cu_2SO_5$

CRYSTAL SYSTEM: Monoclinic
CLASS: $2/m$
SPACE GROUP: C2/m
Z: 4
LATTICE CONSTANTS:
 $a = 9.355$
 $b = 6.312$
 $c = 7.628$
 $\beta = 122°17.5'$
3 STRONGEST DIFFRACTION LINES:
 3.623 (100)
 6.443 ( 50)
 2.615 ( 40)
OPTICAL CONSTANTS:
  $\alpha = 1.715$
  $\beta = 1.820$
  $\gamma = 1.880$
 $(+)2V = 85°$
 HARDNESS: 3
 DENSITY: 4.17 (Meas.)
        4.171 (Calc.)
 CLEAVAGE: $\{\overline{1}01\}$ perfect
 HABIT: As minute, elongated, somewhat tabular crystals, often striated parallel to elongation.
 COLOR-LUSTER: Brown to nearly black. Translucent to opaque. Streak yellowish brown.

 Soluble in water.
 MODE OF OCCURRENCE: Found as a sublimate on Mt. Vesuvius, Italy, associated with euchlorin, eriochalcite, and chalcocyanite.
 BEST REF. IN ENGLISH: Palache, et al., "Dana's System of Mineralogy," 7th Ed., v. II, p. 551–553, New York, Wiley, 1951.

# DOLOMITE

$CaMg(CO_3)_2$

CRYSTAL SYSTEM: Hexagonal
CLASS: $\overline{3}$
SPACE GROUP: $R\overline{3}$
Z: 3
LATTICE CONSTANTS:
 $a \simeq 4.83$
 $c \simeq 16.0$
3 STRONGEST DIFFRACTION LINES:
 2.883 (100)
 1.785 ( 60)
 2.191 ( 50)
OPTICAL CONSTANTS:
  (Na)
 $\omega = 1.679$
 $\epsilon = 1.501$
  $(-)$
 HARDNESS: 3½–4
 DENSITY: 2.85 (Meas.)
 CLEAVAGE: $\{10\overline{1}1\}$ perfect
Fracture subconchoidal. Brittle.

HABIT: Commonly simple rhombohedrons, often with curved faces. Also prismatic and terminated by rhombohedrons; rarely tabular or octahedral. As crystal aggregates composed of saddle-shaped forms. Massive, fine to coarse granular; rarely fibrous or pisolitic. Twinning on $\{0001\}$ common; also twinning on $\{10\bar{1}0\}$; $\{11\bar{2}0\}$; $\{10\bar{1}1\}$; and $\{02\bar{2}1\}$.

COLOR-LUSTER: Colorless, white, grayish, greenish, pale brown, pinkish. Transparent to subtranslucent. Vitreous to pearly.

MODE OF OCCURRENCE: Occurs widespread as extensive strata formed under various conditions and in many different ways; also in hydrothermal vein deposits; in cavities or veins in limestone or dolomitic rocks; in veins in serpentine; and in altered basic igneous rocks containing magnesium. In the United States it is found abundantly as fine saddle-shaped crystals in the Missouri-Oklahoma-Kansas lead-zinc deposits; in quartz geodes in, the vicinity of Keokuk, Iowa; as fine crystals in the Lockport dolomite in New York; and in Vermont, New Jersey, Michigan, and in many western mining districts. Fine specimens occur at St. Eustache, Quebec, Canada; at Guanajuato and other places in Mexico; in Bahia, Brazil; at Eugui, Navarra, Spain; in Germany, Austria, Switzerland, Italy, and at many other European localities.

BEST REF. IN ENGLISH: Deer, Howie, and Zussman, "Rock Forming Minerals," v. 5, p. 278-294, New York, Wiley, 1962.

# DOLORESITE
$H_8V_6O_{16}$

CRYSTAL SYSTEM: Monoclinic
CLASS: 2/m
SPACE GROUP: C2/m
Z: 1
LATTICE CONSTANTS:
  $a = 19.64$
  $b = 2.99$
  $c = 4.83$
  $\beta = 103°55'$
3 STRONGEST DIFFRACTION LINES:
  4.70 (100)
  3.83 ( 50)
  2.45 ( 50)
HARDNESS: Not determined.
DENSITY: 3.27-3.33 (Meas.)
         3.41 (Calc.)
CLEAVAGE: Fibrous.
HABIT: Crystals very minute; maximum size 0.1 mm. Usually massive with radiating botryoidal structure. Lamellar twinning practically universal on submicroscopic scale; twin plane (100).
COLOR-LUSTER: Dark brown when pure; usually nearly black. Opaque, except in small fragments. Submetallic, tarnishing dark bronze. Streak greenish black.
MODE OF OCCURRENCE: Occurs in relatively unoxidized vanadium-uranium ores in association with coffinite, haggite, montroseite, paramontroseite, and other vanadium oxides. Found in Colorado at the La Sal No. 2, Corvusite, Matchless, Arrowhead, Black Mama, and Lums-

den No. 2 mines in Mesa County, and at the Golden Cycle, J. J., and Peanut mines in Montrose County. Also at the Mi Vida mine, San Juan County, Utah; in Valencia County, New Mexico; at the Monument No. 2 mine, Apache County, Arizona; and at Carlile, Crook County, Wyoming.

BEST REF. IN ENGLISH: Stern, T. W., Stieff, L. R., Evans, H. T. Jr., and Sherwood, A. M., *Am. Min.*, **42**: 587-593 (1957). Evans, Howard T. Jr., and Mrose, Mary E., *Am. Min.*, **45**: 1144-1166 (1960).

# DOMEYKITE
$Cu_3As$

CRYSTAL SYSTEM: Cubic
CLASS: $\bar{4}$3m
SPACE GROUP: I$\bar{4}$3d
Z: 16
LATTICE CONSTANTS:
  $a = 9.611$
  $a = 19.241$ (pseudocubic)
3 STRONGEST DIFFRACTION LINES:
  2.05 (100)
  1.888 ( 70)
  1.965 ( 50)
HARDNESS: 3-3½
DENSITY: 7.92-8.10 (Meas.)
         7.95 (Calc.)
CLEAVAGE: Fracture uneven.
HABIT: Massive; botryoidal or reniform.
COLOR-LUSTER: Silver white to steel gray; tarnishes yellowish brown or iridescent, and alters rapidly on exposure to a brownish powdery coating. Opaque. Metallic.
MODE OF OCCURRENCE: Occurs at the Mohawk and other copper mines in Keweenaw County, and in Houghton County, Michigan; in Canada on Michipicoten Island in Lake Superior; and at localities in Mexico, Chile, England, Germany, and Sweden.
BEST REF. IN ENGLISH: Palache, et al., "Dana's System of Mineralogy," 7th Ed., v. I, p. 172, New York, Wiley, 1944.

# β-DOMEYKITE
$Cu_3As$

CRYSTAL SYSTEM: Hexagonal
CLASS: $\bar{3}$ 2/m
SPACE GROUP: R$\bar{3}$c
Z: 6
LATTICE CONSTANTS:
  $a = 7.10$ kX
  $c = 7.25$
3 STRONGEST DIFFRACTION LINES:
  2.08 (100)
  2.02 (100)
  1.445 ( 50)
HARDNESS: 3-3½
DENSITY: 7.2-7.9 (Meas.)
         8.13 (Calc. x-ray)
CLEAVAGE: Fracture uneven.
HABIT: Massive.

COLOR-LUSTER: Tin white to steel gray. Opaque. Metallic.

MODE OF OCCURRENCE: Occurs at Běloves, near Náchod, northeastern Bohemia, Czechoslovakia, and at Mesanki, Iran.

BEST REF. IN ENGLISH: *Min. Abs.*, **12**: 201.

## DONATHITE
$(Fe,Mg)(Cr,Fe)_2O_4$
Dimorph of chromite

CRYSTAL SYSTEM: Tetragonal
CLASS: 4/m 4/m 4/m
SPACE GROUP: P4/nnm
LATTICE CONSTANTS:
 $a = 8.342$
 $c = 8.305$
3 STRONGEST DIFFRACTION LINES:
 2.514 (100)
 2.086 ( 60)
 4.825 ( 50)
OPTICAL CONSTANTS: In reflected light shows distinct anisotropy.
 HARDNESS: 6½–7
 DENSITY: ~5.0 (Meas.)
          5.106, 5.060 (Calc.)
 CLEAVAGE: None.
Fracture uneven. Brittle.
 HABIT: Massive. Commonly twinned. Strongly magnetic.
 COLOR-LUSTER: Black. Opaque. Metallic. Streak blackish brown.
 MODE OF OCCURRENCE: Occurs at Hestmandö Island, Norway.
 BEST REF. IN ENGLISH: Seeliger, E., and Mücke, A., *Am. Min.*, **54**: 1218–1219 (1969).

## DOUGLASITE
$K_2FeCl_4 \cdot 2H_2O$

CRYSTAL SYSTEM: Monoclinic
LATTICE CONSTANTS:
 $a:b:c = 0.737:1:0.504$
  $\beta = 104°50'$
OPTICAL CONSTANTS:
 $\omega = 1.488$
 $\epsilon = 1.500$
  (+)
 HARDNESS: Not determined.
 DENSITY: 2.162 (Meas. synthetic)
 CLEAVAGE: $\{\overline{2}01\}$ (synthetic)
 HABIT: Massive, coarse granular.
 COLOR-LUSTER: Light green; alters to brownish red on exposure. Vitreous.
 MODE OF OCCURRENCE: Occurs associated with halite, sylvite, and carnallite at Douglashall northwest of Strassfurt, Saxony, Germany.
 BEST REF. IN ENGLISH: Palache, et al., "Dana's System of Mineralogy," 7th Ed., v. II, p. 100, New York, Wiley, 1951.

## DOVERITE = Synchysite-(Y)

## DRAVITE
$NaMg_3Al_6B_3Si_6O_{27}(OH,F)_4$
Tourmaline Group

CRYSTAL SYSTEM: Hexagonal
CLASS: 3m
SPACE GROUP: R3m
Z: 3
LATTICE CONSTANTS:
 $a = 15.947$
 $c = 7.194$
3 STRONGEST DIFFRACTION LINES:
 2.576 (100)
 3.99 ( 85)
 2.961 ( 85)
OPTICAL CONSTANTS:
 $\omega = 1.634$
 $\epsilon = 1.611$
  (–)
 HARDNESS: 7
 DENSITY: 3.03–3.15 (Meas.)
          3.018 (Calc.)
 CLEAVAGE: $\{11\overline{2}0\}$ very indistinct
            $\{10\overline{1}1\}$ very indistinct
Fracture uneven to conchoidal. Brittle.
 HABIT: Crystals short to long prismatic, sometimes nearly equant. Often striated vertically. Usually 3-, 6-, or 9-sided in cross section. As single crystals or as parallel or radiating groups. Also massive, compact, or as disseminated grains. Twinning on $\{10\overline{1}1\}$ or $\{40\overline{4}1\}$, rare.
 COLOR-LUSTER: Brown, brownish black, black. Transparent to nearly opaque. Vitreous to somewhat resinous. Streak uncolored.
 MODE OF OCCURRENCE: Occurs chiefly in metamorphic and metasomatic rocks; also in pegmatites, basic igneous rocks, and as a detrital mineral. Typical occurrences are found in California, South Dakota, and other western states; as fine crystals at Gouverneur, St. Lawrence County, New York; and in New Jersey, Connecticut, and New Hampshire. Also in England, Austria, Czechoslovakia, Yugoslavia, Tanzania, New Zealand, and as exceptional large euhedral crystals at Yinnietharra, Western Australia.
 BEST REF. IN ENGLISH: Deer, Howie, and Zussman, "Rock Forming Minerals," v. I, p. 300–319, New York, Wiley, 1962.

## DRESSERITE
$Ba_2Al_4(CO_3)_4(OH)_8 \cdot 3H_2O$

CRYSTAL SYSTEM: Orthorhombic
CLASS: 2/m 2/m 2/m
SPACE GROUP: Pbmm
Z: 2
LATTICE CONSTANTS:
 $a = 9.27$
 $b = 16.83$
 $c = 5.63$

3 STRONGEST DIFFRACTION LINES:
  8.09 (100)
  6.23 ( 60)
  3.66 ( 50)
OPTICAL CONSTANTS:
  $\alpha = 1.518$
  $\beta = (?)$
  $\gamma = 1.601$
  $(-)2V = 30–40°$
HARDNESS: 2½–3
DENSITY: 2.96 (Meas.)
        3.06 (Calc.)
CLEAVAGE: Not determined.
HABIT: Spherical aggregates of tapering fibers elongated parallel to $c$.
COLOR-LUSTER: White; transparent. Vitreous to silky. Streak white.
MODE OF OCCURRENCE: Occurs in cavities in an alkalic sill intruding limestone at St.-Michel, Montreal Island, Quebec, Canada as spheres and hemispheres ranging from 1 to 3 mm in diameter. It is found associated with weloganite, plagioclase, quartz, dawsonite, and a powdery hydrous aluminum oxide which gives an X-ray diffraction pattern similar to that of gibbsite.
BEST REF. IN ENGLISH: Jambor, J. L., Fong, D. G., and Sabina, Ann P., *Can. Min.*, **10**(Pt. 1): 84–89 (1969).

# DUFRENITE
$Fe^{2+}Fe_4^{3+}(PO_4)_3(OH)_5 \cdot 2H_2O$

CRYSTAL SYSTEM: Monoclinic
CLASS: 2/m
SPACE GROUP: C2/c
Z: 4
LATTICE CONSTANTS:
  $a = 25.84$
  $b = 5.126$
  $c = 13.78$
  $\beta = 111°12'$
3 STRONGEST DIFFRACTION LINES:
  3.151 (100)
  12.00 ( 90)
  5.002 ( 90)
OPTICAL CONSTANTS:
  $\alpha = 1.820–1.842$
  $\beta = 1.830–1.850$
  $\gamma = 1.875–1.925$
  $(+)2V = small$
HARDNESS: 3½–4½
DENSITY: 3.1–3.34 (Meas.)
CLEAVAGE: {100} perfect
Brittle.
HABIT: Crystals rare, indistinct, rounded, in aggregates. Usually as botryoidal crusts or masses with radial-fibrous structure.
COLOR-LUSTER: Dark green to greenish black, becoming greenish brown to reddish brown due to oxidation. Translucent to opaque. Vitreous to silky.
MODE OF OCCURRENCE: Occurs as a secondary mineral at Rock Run, Cherokee County, Alabama; at Wheal Phoenix, Cornwall, England; and at deposits in Saxony, Hesse, Westphalia, and Thuringia, Germany.
BEST REF. IN ENGLISH: Palache, et al., "Dana's System of Mineralogy," 7th Ed., v. II, p. 873–875, New York, Wiley, 1951. *Am. Min.*, **55**: 135–169 (1970).

# DUFRENOYSITE
$Pb_2As_2S_5$

CRYSTAL SYSTEM: Monoclinic
CLASS: 2/m
SPACE GROUP: P2₁/m
Z: 8
LATTICE CONSTANTS:
  $a = 8.41$
  $b = 25.85$
  $c = 7.88$
  $\beta = 90°30'$
3 STRONGEST DIFFRACTION LINES:
  3.74 (100)
  3.00 ( 90)
  2.70 ( 80)
OPTICAL CONSTANT:
  $n > 2.72$
HARDNESS: 3
DENSITY: 5.53 (Meas.)
        5.61 (Calc.)
CLEAVAGE: {010} perfect
Fracture conchoidal. Brittle.
HABIT: Crystals tabular and somewhat elongated; sometimes striated parallel to elongation. Twinning on {001}.
COLOR-LUSTER: Steel gray to lead gray. Opaque. Metallic. Streak reddish brown.
MODE OF OCCURRENCE: Occurs in the Cerro Gordo district, Inyo County, California; in the Banner and Wood River districts, Idaho; at Hall, Tirol, Austria; in crystalline dolomite at the Lengenbach quarry in the Binnental, Valais, Switzerland; and at Dundas, Tasmania.
BEST REF. IN ENGLISH: Palache, et al., "Dana's System of Mineralogy," 7th Ed., v. I, p. 442–445, New York, Wiley, 1944.

# DUFTITE
$CuPbAsO_4OH$

CRYSTAL SYSTEM: Orthorhombic
CLASS: 2/m 2/m 2/m
SPACE GROUP: Pnma
Z: 4
LATTICE CONSTANTS:
  $a = 7.81$
  $b = 9.19$
  $c = 6.08$
3 STRONGEST DIFFRACTION LINES:
  3.26 (100)
  2.85 ( 80)
  2.65 ( 80)

OPTICAL CONSTANTS:
$\alpha = 2.04$
$\beta = 2.08$
$\gamma = 2.10$
$(-)2V = $ large
HARDNESS: 3
DENSITY: 6.40 (Meas.)
6.49 (Calc.)
CLEAVAGE: Not determined.
HABIT: Crystals minute with rough rounded faces; as crusts and aggregates.
COLOR-LUSTER: Apple green to olive green to grayish green. Translucent. Vitreous to dull.
MODE OF OCCURRENCE: Occurs associated with wulfenite, malachite, azurite, and other minerals at Tsumeb, South West Africa.
BEST REF. IN ENGLISH: Palache, et al., "Dana's System of Mineralogy," 7th Ed., v. II, p. 810-811, New York, Wiley, 1951.

## DUMONTITE
$Pb_2(UO_2)_3(PO_4)_2(OH)_4 \cdot 3H_2O$

CRYSTAL SYSTEM: Monoclinic
CLASS: 2/m
SPACE GROUP: $P2_1/m$
Z: 2
LATTICE CONSTANTS:
$a = 8.16$
$b = 16.73$
$c = 7.02$
3 STRONGEST DIFFRACTION LINES:
4.27 (100)
3.00 ( 85)
2.95 ( 85)
OPTICAL CONSTANTS:
$\alpha = 1.85$
$\beta = 1.87$
$\gamma \simeq 1.89$
$(+)2V = $ large
HARDNESS: Not determined.
DENSITY: 5.65 (Meas.)
CLEAVAGE: Not determined.
HABIT: Minute crystals elongated on $c$-axis and flattened on {010}. Striated parallel to $c$-axis.
COLOR-LUSTER: Pale yellow to golden yellow. Translucent. Powder pale yellow. Fluoresces weak green in ultraviolet light.
MODE OF OCCURRENCE: Occurs as a secondary mineral with kasolite, autunite, and uranophane at the White Oak mine near Nogales, Arizona, and with kasolite at the Green Monster mine, Goodsprings, Nevada. It is also found associated with parsonsite and metatorbernite at Shinkolobwe, Katanga, Zaire.
BEST REF. IN ENGLISH: Frondel, Clifford, USGS Bull. 1064, p. 236-238 (1958).

## DUMORTIERITE
$Al_7O_3(BO_3)(SiO_4)_3$

CRYSTAL SYSTEM: Orthorhombic

CLASS: 2/m 2/m 2/m
SPACE GROUP: Pcmn
Z: 4
LATTICE CONSTANTS:
$a = 11.79$
$b = 20.209$
$c = 4.7015$
3 STRONGEST DIFFRACTION LINES:
2.549 (100)
5.89 ( 90)
5.09 ( 90)
OPTICAL CONSTANTS:
$\alpha = 1.6860$
$\beta = 1.722$
$\gamma = 1.7229$
$(-)2V = 13°$
HARDNESS: 8½
DENSITY: 3.41 (Meas.)
3.390 (Calc.)
CLEAVAGE: {100} good
{110} imperfect
HABIT: Crystals prismatic, rare. Usually massive, columnar, fibrous, or granular.
COLOR-LUSTER: Blue, violet, pinkish, brown. Transparent to translucent. Vitreous to dull. Streak white.
MODE OF OCCURRENCE: Occurs in aluminum-rich metamorphic rocks and rarely in pegmatites. Found in Imperial, Kern, Riverside, and San Diego counties, California; at Woodstock, Washington; in large quantities at Oreana, Humboldt County, Nevada; in the Ruby range, east of Dillon, Montana; in Fremont County, Colorado; in Yuma County, Arizona; and in New York. Other occurrences are found in Canada, Mexico, Brazil, Norway, France, Madagascar, and elsewhere.
BEST REF. IN ENGLISH: Winchell and Winchell," Elements of Optical Mineralogy," 4th Ed., pt. 2, p. 259, New York, Wiley, 1951.

## DUNDASITE
$Pb_2Al_4(CO_3)_4(OH)_8 \cdot 3H_2O$

CRYSTAL SYSTEM: Orthorhombic
CLASS: 2/m 2/m 2/m
SPACE GROUP: Pbmm
Z: 2
LATTICE CONSTANTS:
$a = 9.05$
$b = 16.35$
$c = 5.61$
3 STRONGEST DIFFRACTION LINES:
7.91 (100)
3.60 ( 80)
3.09 ( 60)
OPTICAL CONSTANTS:
$\alpha = 1.602$
$\beta \simeq 1.716$
$\gamma = 1.750$
$(-)2V = $ very large
HARDNESS: 2
DENSITY: 3.41, 3.55 (Meas.)
3.81 (Calc.)

CLEAVAGE: {010} perfect
HABIT: As small rounded aggregates of radiating crystals, and as matted crusts.
COLOR-LUSTER: White; transparent. Vitreous to silky.
MODE OF OCCURRENCE: Occurs as a secondary mineral in the oxidation zone of lead deposits. Found associated with crocoite in the Adelaide Proprietary mine, Dundas, and with cerussite at the Hercules mine, Mt. Read, Tasmania; at Gonnesa, Sardinia; and at localities in Ireland, Wales, and England.
BEST REF. IN ENGLISH: Palache, et al., "Dana's System of Mineralogy," 7th Ed., v. II, p. 279–280, New York, Wiley, 1951.

## DURANGITE
$NaAlAsO_4F$

CRYSTAL SYSTEM: Monoclinic
CLASS: 2/m
SPACE GROUP: C2/c
Z: 4
LATTICE CONSTANTS:
$a = 6.53$ kX
$b = 8.46$
$c = 7.00$
$\beta = 115°00'$
OPTICAL CONSTANTS:
$\alpha = 1.634$
$\beta = 1.673$
$\gamma = 1.685$
$(-)2V = 45°$
HARDNESS: 5
DENSITY: 3.94–4.07 (Meas.)
3.616 (Calc.)
CLEAVAGE: {110} distinct
Fracture uneven. Brittle.
HABIT: Crystals oblique pyramidal; faces commonly rough and dull. Synthetic crystals prismatic or tabular, often twinned on {001}.
COLOR-LUSTER: Light to dark orange-red. Translucent. Vitreous. Streak yellowish.
MODE OF OCCURRENCE: Occurs at the Barranca tin mine, Durango, Mexico, associated with cassiterite, topaz, and hematite, and in a pegmatite near Lake Ramsay, Lunenberg County, Nova Scotia, with cassiterite and amblygonite.
BEST REF. IN ENGLISH: Palache, et al., "Dana's System of Mineralogy," 7th Ed., v. II, p. 829–831, New York, Wiley, 1951.

## DUSSERTITE
$BaFe_3(AsO_4)_2(OH)_5$

CRYSTAL SYSTEM: Hexagonal
SPACE GROUP: R lattice
Z: 3
LATTICE CONSTANTS:
$a = 7.40$
$c = 17.48$

3 STRONGEST DIFFRACTION LINES:
3.13 (100)
2.32 ( 70)
2.00 ( 70)
OPTICAL CONSTANTS:
$\omega = 1.87$
$\epsilon = 1.85$
(-)
HARDNESS: 3½
DENSITY: 3.75 (Meas.)
4.12 (Calc.)
CLEAVAGE: Not determined.
HABIT: As crusts or as rosettes of minute flattened crystals.
COLOR-LUSTER: Green to yellowish green.
MODE OF OCCURRENCE: Occurs associated with carminite, arseniosiderite, and limonite at Mina Ojuela, Mapimi, Durango, Mexico, and on quartz at Djebel Debar, Constantine, Algeria.
BEST REF. IN ENGLISH: Palache, et al., "Dana's System of Mineralogy," 7th Ed., v. II, p. 839–840, New York, Wiley, 1951.

## DUTTONITE
$VO(OH)_2$

CRYSTAL SYSTEM: Monoclinic
CLASS: 2/m
SPACE GROUP: I2/c
Z: 4
LATTICE CONSTANTS:
$a = 8.80$
$b = 3.95$
$c = 5.96$
$\beta = 90°40'$
3 STRONGEST DIFFRACTION LINES:
4.40 (100)
3.61 ( 85)
1.838 ( 21)
HARDNESS: ~2½
DENSITY: 3.24 (Calc.)
CLEAVAGE: {100} distinct
HABIT: Six-sided platy crystals, flattened parallel to c{001}; pseudo-orthorhombic in aspect. As crusts and coatings. Crystals range from extremely minute to 0.5 mm in longest dimension.
COLOR-LUSTER: Pale brown.
MODE OF OCCURRENCE: Occurs associated with melanovanadite and native selenium at the Peanut mine, in the upper vanadium-uranium ore-bearing sandstone of the Salt Wash Sandstone member of the Morrison formation of Late Jurassic age, Montrose County, Colorado; also with native selenium and simplotite at the Sundown Claim, Slick Rock mining district, San Miguel County, Colorado.
BEST REF. IN ENGLISH: Thompson, M. E., Roach, Carl H., and Meyrowitz, Robert, *Am. Min.*, **42**: 455–460 (1957).

## DYPINGITE (Inadequately described mineral)
$Mg_5(CO_3)_4(OH)_2 \cdot 5H_2O$

CRYSTAL SYSTEM: Unknown

3 STRONGEST DIFFRACTION LINES:
 10.6 (100)
 5.86 ( 90)
 6.34 ( 60)
OPTICAL CONSTANTS:
 $\alpha$ = 1.508
 $\beta$ = 1.510
 $\gamma$ = 1.516
HARDNESS: Not determined.
DENSITY: 2.15 (Calc.)
CLEAVAGE: Not determined.

HABIT: Globular aggregates with radiating structure. Average size 0.3 mm.
COLOR-LUSTER: White; pearly. Fluorescent light blue and phosphorescent yellow-green under ultraviolet light.
MODE OF OCCURRENCE: Occurs as a thin cover on serpentine in the Dypingdal serpentine-magnesite deposit, Snarum, south Norway.
BEST REF. IN ENGLISH: Raade, Gunnar, *Am. Min.*, **55**: 1457–1465 (1970).

**DYSANALYTE** = Niobian perovskite

## DYSCRASITE
Ag$_3$Sb

CRYSTAL SYSTEM: Orthorhombic
CLASS: mm2
SPACE GROUP: Pm2m
Z: 1
LATTICE CONSTANTS:
 $a$ = 4.820
 $b$ = 5.225
 $c$ = 2.990
3 STRONGEST DIFFRACTION LINES:
 2.29 (100)
 2.42 ( 40)
 1.37 ( 40)
HARDNESS: 3½–4
DENSITY: 9.74 (Meas.)
         9.75 (Calc.)
CLEAVAGE:  {001} distinct
           {011} distinct
           {110} imperfect
Fracture uneven. Sectile.

HABIT: Crystals pyramidal. Commonly massive, granular or foliated. Twinning on {110} repeated, forming hexagonal aggregates.
COLOR-LUSTER: Silvery white; tarnishes yellowish, gray, or blackish. Opaque. Metallic.
MODE OF OCCURRENCE: Occurs as a primary mineral in silver ore deposits, commonly associated with other sulfides and calcite. Found in the ores from Poughkeepsie Gulch, near Silverton, San Juan County, Colorado; in the Reese River district, Nevada; at Cobalt, Ontario, Canada; as a minor ore mineral and as fine specimens at St. Andreasberg and elsewhere in Germany; at Ste. Marie aux Mines, Alsace, France; and as large masses at the Consols mine, Broken Hill, New South Wales, Australia.
BEST REF. IN ENGLISH: Palache, et al., "Dana's System of Mineralogy," 7th Ed., v. I, p. 173–175, New York, Wiley, 1944.

## DZHALINDITE
In(OH)$_3$

CRYSTAL SYSTEM: Cubic
CLASS: 2/m $\bar{3}$
SPACE GROUP: Im3
Z: 8
LATTICE CONSTANT:
 $a$ = 7.958
3 STRONGEST DIFFRACTION LINES:
 3.96 (100)
 2.80 ( 90)
 1.778 ( 90)
OPTICAL CONSTANT:
 $N$ = 1.716–1.725
HARDNESS: Not determined.
DENSITY: 4.34 (Calc.)
CLEAVAGE: Not determined.
HABIT: Small grains, as an alteration product of indite.
COLOR-LUSTER: Yellow-brown.
MODE OF OCCURRENCE: Occurs in cassiterite from the Dzhalindin deposit, Little Khingan Ridge, Far Eastern Siberia.
BEST REF. IN ENGLISH: Genkin, A. D., and Murav'eva, I. V., *Am. Min.*, **49**: 439–440 (1964).

**DZHULUKULITE** = Nickeloan cobaltite

# E

## EAKERITE
$Ca_2SnAl_2Si_6O_{16}(OH)_6$

CRYSTAL SYSTEM: Monoclinic
CLASS: 2/m
SPACE GROUP: $P2_1/a$
Z: 2
LATTICE CONSTANTS:
  $a = 15.829$
  $b = 7.721$
  $c = 7.438$
  $\beta = 101°19'$
3 STRONGEST DIFFRACTION LINES:
  4.81 (100)
  5.25 ( 90)
  7.31 ( 80)
OPTICAL CONSTANTS:
  Biaxial, positive, $2V \sim 35°$
  $n\alpha = 1.584$
  $n\beta = 1.586$  (Na light)
  $n\gamma = 1.600$
  HARDNESS: $\sim 5\frac{1}{2}$
  DENSITY: 2.93 (Meas.)
           2.931 (Calc.)
  CLEAVAGE: None.
Fracture conchoidal.
  HABIT: Elongated prismatic crystals up to 5.0 mm in length. Elongation [001]; forms {111}, {210}, {410}, {201}, {$\bar{2}$01}, {001}, and {100}. Prism zone striated parallel to [001].
  COLOR-LUSTER: Colorless to milky white, transparent, vitreous.
  MODE OF OCCURRENCE: Occurs in seams with quartz, albite, bavenite, and apatite in spodumene-rich pegmatite at the Foote Mineral Company spodumene mine, Kings Mountain, Cleveland County, North Carolina.
  BEST REF. IN ENGLISH: Leavens, Peter B., White, John S. Jr., and Hey, Max H., *Min. Record*, **1** (3): 92-96 (1970).

## EARLANDITE (Inadequately described mineral)
$Ca_3(C_6H_5O_7)_2 \cdot 4H_2O$

CRYSTAL SYSTEM: Unknown
OPTICAL CONSTANTS:
  $\alpha = 1.515$
  $\beta = 1.530$
  $\gamma = 1.580$
  $(+)2V = 60°$(Calc.)
HARDNESS: Not determined.
DENSITY: 1.95
CLEAVAGE: Not determined.
HABIT: As knobby, fine-grained nodules.
COLOR-LUSTER: White to pale yellow.
MODE OF OCCURRENCE: Found in oceanic bottom samples from the Weddell Sea, Antarctica.
  BEST REF. IN ENGLISH: Palache, et al., "Dana's System of Mineralogy," 7th Ed., v. II, p. 1105-1106, New York, Wiley, 1951.

## ECDEMITE (Ekdemite) (Inadequately described mineral)
$Pb_3Cl_2(AsO_4)$

CRYSTAL SYSTEM: Tetragonal
OPTICAL CONSTANTS:
  (Li)
  $\omega = 2.32$
  $\epsilon = 2.25$
  (-)
HARDNESS: 2½-3
DENSITY: 7.14
CLEAVAGE: {001} perfect
HABIT: As small tabular crystals, coarsely foliated masses, and thin incrustations.
COLOR-LUSTER: Greenish yellow to yellow, orange. Translucent. Vitreous to greasy.
MODE OF OCCURRENCE: Occurs as thin coatings on willemite and wulfenite at the Mammoth mine, Tiger, Pinal

County, Arizona; and as crystals and masses associated with heliophyllite at Långban, Jacobsberg, and Harstigen, Sweden.

BEST REF. IN ENGLISH: Palache, et al., "Dana's System of Mineralogy," 7th Ed., v. II, p. 1036–1037, New York, Wiley, 1951.

## ECKERMANNITE
$Na_3(Mg, Li)_4(Al, Fe)Si_8O_{22}(OH,F)_2$
Amphibole Group

CRYSTAL SYSTEM: Monoclinic
CLASS: 2/m
SPACE GROUP: C2/m
Z: 2
LATTICE CONSTANTS:
  $a$ = 9.762
  $b$ = 17.892
  $c$ = 5.284
  $\beta$ = 103.168°
3 STRONGEST DIFFRACTION LINES:
  3.10 (100)
  2.71 ( 80)
  3.40 ( 70)
OPTICAL CONSTANTS:
  $\alpha$ = 1.612–1.638
  $\beta$ = 1.625–1.652
  $\gamma$ = 1.630–1.654
  $(-)2V_\alpha$ = 80°–15°
HARDNESS: 5–6
DENSITY: 3.0–3.17 (Meas.)
      2.971 (Calc.)
CLEAVAGE: {110} perfect
      {010} parting
Fracture uneven. Brittle.

HABIT: Crystals long prismatic; also in prismatic aggregates. Twinning on {100} simple, lamellar.

COLOR-LUSTER: Dark bluish green. Translucent. Vitreous.

MODE OF OCCURRENCE: Occurs as a constituent of certain plutonic alkali igneous rocks such as nepheline-syenite. Typical occurrences are found at Camp Albion, Boulder County, Colorado; at Norra Kärr, southern Sweden; and at Tawmaw, Burma.

BEST REF. IN ENGLISH: Deer, Howie, and Zussman, "Rock Forming Minerals," v. 2, p. 364–374, New York, Wiley, 1963. *Am. Min.,* **29**: 455–456 (1944).

## ECKRITE = Variety of amphibole

## EDENITE (Ferroedenite)
$NaCa_2(Mg, Fe^{2+})_5Si_7AlO_{22}(OH, F)_2$
Amphibole Group
Hornblende Series

CRYSTAL SYSTEM: Monoclinic
CLASS: 2/m
SPACE GROUP: C2/m
Z: 2

LATTICE CONSTANTS:
  $a \simeq 9.9$
  $b \simeq 18.0$
  $c \simeq 5.3$
  $\beta \simeq 105°$
3 STRONGEST DIFFRACTION LINES: See hornblende.
OPTICAL CONSTANTS:
  $\alpha$ = 1.615–1.705
  $\beta$ = 1.618–1.714
  $\gamma$ = 1.632–1.730
  $(\pm)2V$ = 95°–27°
HARDNESS: 5–6
DENSITY: 3.0–3.06 (Meas.)
CLEAVAGE: {110} perfect
      {001} parting
      {100} parting
Fracture uneven to subconchoidal. Brittle.

HABIT: Crystals short prismatic. Also massive; resembles tremolite or anthophyllite. Twinning on {100} common, simple, lamellar.

COLOR-LUSTER: White, gray, pale to dark green, also colorless. Transparent to nearly opaque. Vitreous to silky.

MODE OF OCCURRENCE: Occurs as a constituent of both igneous and metamorphic rocks. Typical occurrences are found at Spanish Peak, Plumas County, California; Edenville, New York; Eganville, Ontario, Canada; and Kotaki, Niigata Prefecture, Japan.

BEST REF. IN ENGLISH: Deer, Howie, and Zussman, "Rock Forming Minerals," v. 2, p. 263–314, New York, Wiley, 1963.

## EDINGTONITE
$Ba Al_2 Si_3 O_{10} \cdot 4H_2O$
Zeolite Group

CRYSTAL SYSTEM: Orthorhombic (pseudotetragonal)
CLASS: 222
SPACE GROUP: $P2_1 2_1 2$
Z: 2
LATTICE CONSTANTS
  $a$ = 9.523
  $b$ = 9.644
  $c$ = 6.506
3 STRONGEST DIFFRACTION LINES:
  3.58 (100)
  2.749 (100)
  4.80 ( 90)
OPTICAL CONSTANTS:
  $\alpha$ = 1.5405
  $\beta$ = 1.5528
  $\gamma$ = 1.5569
  $(-)2V$ = 53°52'
HARDNESS: 4
DENSITY: 2.777 (Meas.)
      2.821 (Calc.)
CLEAVAGE: {110} perfect
HABIT: Crystals pyramidal, minute. Also massive.

COLOR-LUSTER: White, grayish, pink. Transparent to translucent. Vitreous.

MODE OF OCCURRENCE: Occurs in association with harmotome and other zeolites in basic igneous rock at several localities in the vicinity of Old Kilpatrick, Dunbartonshire, Scotland; also found at the Böhlet mine, Westergotland, Sweden.

BEST REF. IN ENGLISH: Deer, Howie, and Zussman, "Rock Forming Minerals," v. 4, p. 359-376, New York, Wiley, 1963.

## EGLESTONITE
$Hg_4Cl_2O$

CRYSTAL SYSTEM: Cubic
CLASS: $4/m\ \bar{3}\ 2/m$
SPACE GROUP: Im3m
Z: 3
LATTICE CONSTANT:
  $a = 8.02$
3 STRONGEST DIFFRACTION LINES:
  1.89 (100)
  3.28 ( 90)
  2.54 ( 60)
OPTICAL CONSTANT:
  $N = 2.49$ (Li)
  HARDNESS: 2½
  DENSITY: 8.327 (Meas.)
       8.56 (Calc.)
  CLEAVAGE: Fracture conchoidal to uneven. Brittle.
  HABIT: Crystals dodecahedral, striated parallel to the edges; also cubic or octahedral. Also massive, as crusts; sometimes hair-like.
  COLOR-LUSTER: Yellow, orange-yellow to brownish yellow. On exposure to light becomes dark brown then black. Translucent. Bright adamantine to resinous. Streak yellowish, rapidly becoming black.
  MODE OF OCCURRENCE: Occurs in California associated with montroydite, calomel, native mercury, and cinnabar south of the New Idria mine, San Benito County; with montroydite, calomel, and other mercury minerals in serpentine near Redwood City, and with cinnabar, mercury, and calomel near Palo Alto, San Mateo County. Found with calomel, mercury, montroydite, and terlinguaite in the mercury deposit at Terlingua, Brewster County, Texas, and with cinnabar in Pike County, Arkansas. It also occurs at the Monarch cinnabar mine, Transvaal, South Africa.
  BEST REF. IN ENGLISH: Palache, et al., "Dana's System of Mineralogy," 7th Ed., v. II, p. 51-52, New York, Wiley, 1951.

## EGUEIITE (Inadequately described mineral)
Near $CaFe_{14}^{3+}(PO_4)_{10}(OH)_{12} \cdot 21H_2O$

CRYSTAL SYSTEM: Monoclinic (?)
  HARDNESS: Not determined.
  DENSITY: 2.60
  CLEAVAGE: Very friable.
  HABIT: As small nodules with fibrous-lamellar structure.
  COLOR-LUSTER: Brownish yellow. Vitreous to slightly greasy. Streak yellow.
  MODE OF OCCURRENCE: Found associated with thenardite and trona in clay in the Eguéï region, Chad.

BEST REF. IN ENGLISH: Palache, et al., "Dana's System of Mineralogy," 7th Ed., v. II, p. 955, New York, Wiley, 1951.

## EITELITE
$Na_2Mg(CO_3)_2$

CRYSTAL SYSTEM: Hexagonal
CLASS: $\bar{3}$
SPACE GROUP: $R\bar{3}$
Z: 1
LATTICE CONSTANTS:
  $a = 6.168$
  $\alpha = 47°14'$
3 STRONGEST DIFFRACTION LINES:
  2.61 (100)
  1.89 ( 60) (synthetic)
  2.73 ( 50)
OPTICAL CONSTANTS:
    $\omega = 1.6052$
    $\epsilon = 1.4502$
  $(-)2V$ = small
  HARDNESS: 3½
  DENSITY: 2.737 (Meas.)
  CLEAVAGE: {0001} distinct
  HABIT: Crystals rhombohedral, up to 17 mm in size. Principal forms: {0001}, {10$\bar{1}$1}, and {01$\bar{1}$2}.
  COLOR-LUSTER: Colorless, white. Transparent. Vitreous.
  MODE OF OCCURRENCE: Eitelite occurs as the predominant mineral in a core section from the Mapco Shrine Hospital No. 1 well, Duchesne County, Utah, associated with trona, nahcolite, shortite, and magnesioriebeckite; also found in association with searlesite, reedmergnerite, shortite, leucosphenite, and crocidolite in the Green River (Eocene) formation from the Carter Oil Company, Kermit Poulson No. 1 well, Utah.
  BEST REF. IN ENGLISH: Milton, Charles, Axelrod, Joseph M., and Grimaldi, Frank S., *Am. Min.*, **40**: 326-327 (1955). Pabst, Adolf, *Am. Min.*, **58**: 211-217 (1973).

## EKANITE
$(Ca, Na, K, Th)_2Si_4O_{10}$

CRYSTAL SYSTEM: Tetragonal
CLASS: 4/mmm  SPACE GROUP: P4/mcc  Z: 4
LATTICE CONSTANTS: $a = 7.58, c = 14.77$
3 STRONGEST DIFFRACTION LINES:
  3.38 (100), 3.32 (55), 5.30 (45)
OPTICAL CONSTANTS: $\omega = 1.573, \epsilon = 1.572$ (-)
  HARDNESS: 5
  DENSITY: 3.32 (Calc.)
  CLEAVAGE: None.
  HABIT: Crystals elongated along $c$; forms {100} and {001}; often twinned. Also massive, as pebbles.
  COLOR-LUSTER: Dark brown, green. Transparent to translucent.
  MODE OF OCCURRENCE: Occurs as fine crystals at Mont St. Hilaire, Quebec, Canada. Also found in central Asia, and in the gem pits of Eheliyagoda, Raknapura district, Ceylon.

BEST REF. IN ENGLISH: Anderson, B. W., Claringbull, G. F., Davis, R. J., and Hill, D. K., *Nature*, **190** (4780): 997 (1961). *Can. Min.*, v. 11, p. 913-929 (1973).

## EKMANITE
$(Fe,Mn,Mg)_6 (Si,Al)_8 O_{20} (OH)_8 \cdot 2H_2O$

CRYSTAL SYSTEM: Orthorhombic (?)
Z: 1
LATTICE CONSTANTS:
  *a* = 5.54
  *b* = 9.60
  *c* = 12.08
3 STRONGEST DIFFRACTION LINES:
  7.26 (100)
  3.54 ( 70)
  3.61 ( 50)
OPTICAL CONSTANTS:
  $\alpha$ = 1.582
  $\beta = \gamma$ = 1.670
  $(-)2V \cong 0°$
HARDNESS: 2-2½
DENSITY: 2.79 (Meas.)
CLEAVAGE: {001} perfect
HABIT: Massive; foliated, foliated columnar; also scaly.
COLOR-LUSTER: Gray to black; also greenish. Translucent to nearly opaque. Pearly to submetallic.
MODE OF OCCURRENCE: Occurs in association with pyrochroite filling cavities in magnetite ore at Brunsjo, Sweden.
BEST REF. IN ENGLISH: Nagy, Bartholomew, *Am. Min.*, **39**: 946-956 (1954).

## ELBAITE
$Na(Li,Al)_3 Al_6 B_3 Si_6 O_{27} (OH,F)_4$
Tourmaline Group
Var: Rubellite
    Indicolite
    Achroite, etc.

CRYSTAL SYSTEM: Hexagonal
CLASS: 3m
SPACE GROUP: R3m
Z: 3
LATTICE CONSTANTS:
  *a* = 15.843
  *c* = 7.102
3 STRONGEST DIFFRACTION LINES:
  
  |  | (synthetic) |
  |---|---|
  | 2.576 (100) | 2.560 (100) |
  | 3.99 ( 85) | 2.878 ( 85) |
  | 2.961 ( 85) | 3.96 ( 60) |

OPTICAL CONSTANTS:
  $\omega$ = 1.640-1.655
  $\epsilon$ = 1.615-1.620
    (-)
HARDNESS: 7
DENSITY: 3.03-3.10 (Meas.)
CLEAVAGE: {11$\bar{2}$0} very indistinct
         {10$\bar{1}$1} very indistinct
Fracture uneven to conchoidal. Brittle.

HABIT: Crystals short to long prismatic, vertically striated; also acicular, rarely flattened as thin tablets. Usually 3-, 6-, or 9-sided. Commonly hemimorphic. As single crystals or as parallel or radiating groups. Also massive, compact; columnar to fibrous. Twinning on {10$\bar{1}$1} or {40$\bar{4}$1}, rare.
COLOR-LUSTER: Shades of green, blue, red, yellow; rarely white or colorless. Colors often in zonal arrangement parallel to basal plane or prism faces. Transparent to translucent. Vitreous. Streak uncolored.
MODE OF OCCURRENCE: Occurs chiefly in granite pegmatites in association with lepidolite, quartz, and feldspar. Notable occurrences are found at Pala and elsewhere in San Diego County, California; in Gunnison County, Colorado; in the Black Hills, South Dakota; at Newry and other localities in southwestern Maine; and in Middlesex County, Connecticut. Also in Ontario, Canada; as exceptional crystals of widely varied size and color in Minas Geraes, Brazil; also as fine crystals on the island of Elba, and in Germany, Switzerland, USSR, Ceylon, Nepal, Mozambique, South West Africa, and Madagascar.
BEST REF. IN ENGLISH: Deer, Howie, and Zussman, "Rock Forming Minerals," v. 1, p. 300-319, New York, Wiley, 1962.

**ELECTRUM** = Variety of gold containing more than 20% silver

## ELLESTADITE
$Ca_5 \{(Si, S, P, C) O_4\}_3 (Cl, F, OH)$
Apatite Group

CRYSTAL SYSTEM: Hexagonal
CLASS: 6/m
SPACE GROUP: $P6_3/m$
Z: 2
LATTICE CONSTANTS:
  *a* = 9.53
  *c* = 6.91
3 STRONGEST DIFFRACTION LINES:
  2.86 (100)
  2.76 ( 60)
  1.97 ( 60)
OPTICAL CONSTANTS:
  $\omega$ = 1.655
  $\epsilon$ = 1.650
    (-)
HARDNESS: ~5
DENSITY: 3.068 (Meas.)
CLEAVAGE: {0001} indistinct
         {10$\bar{1}$0} indistinct
HABIT: Massive, granular.
COLOR-LUSTER: Pale rose. Translucent. Vitreous.
MODE OF OCCURRENCE: Occurs as veinlets in blue calcite associated with wilkeïte, idocrase, diopside, and wollastonite in contact-metamorphosed marble at Crestmore, Riverside County, California.
BEST REF. IN ENGLISH: Palache, et al., "Dana's System of Mineralogy," 7th Ed., v. II, p. 906, New York, Wiley, 1951.

**ELLSWORTHITE** = Variety of pyrochlore containing uranium

## ELPASOLITE
$K_2NaAlF_6$

CRYSTAL SYSTEM: Cubic
CLASS: $2/m\bar{3}$
SPACE GROUP: Pa3
Z: 4
LATTICE CONSTANT:
  $a = 8.093$ kX
3 STRONGEST DIFFRACTION LINES:
  2.86 (100)
  2.02 (100)
  2.34 ( 80)
OPTICAL CONSTANT:
  $N = 1.376$
HARDNESS: 2½
DENSITY: 2.995 (Meas.)
         3.015 (Calc.)
CLEAVAGE: None.
Fracture uneven.
HABIT: Massive, rarely as indistinct crystals.
COLOR-LUSTER: Colorless; transparent. Vitreous to greasy.
MODE OF OCCURRENCE: Occurs as compact irregular masses in massive pachnolite, and as indistinct aggregates of crystals as much as 0.5 mm in size lining cavities in massive pachnolite, in a cryolite-bearing pegmatite at St. Peters Dome, El Paso County, Colorado.
BEST REF. IN ENGLISH: Palache, et al., "Dana's System of Mineralogy," 7th Ed., v. II, p. 114, New York, Wiley, 1951.

## ELPIDITE
$Na_2ZrSi_6O_{15} \cdot 3H_2O$

CRYSTAL SYSTEM: Orthorhombic
CLASS: mm2
SPACE GROUP: $Pb2_1m$
Z: 2
LATTICE CONSTANTS:
  $a = 7.4$ kX
  $b = 14.4$
  $c = 7.05$
3 STRONGEST DIFFRACTION LINES:
  3.25 (100)
  3.10 ( 80)
  2.94 ( 80)
OPTICAL CONSTANTS:
  $\alpha = 1.563$
  $\beta = 1.569$
  $\gamma = 1.577$
  (+)2V = 76–89°
HARDNESS: 5
DENSITY: 2.524–2.615 (Meas.)
CLEAVAGE: Prismatic; perfect, in two directions.
Fracture splintery.

HABIT: Crystals thin prismatic, vertically striated, up to 9 × 1.5 cm in size. Also fine-fibrous in fan-shaped aggregates.
COLOR-LUSTER: Colorless, yellow-white; also light brick red due to impurities. Transparent. Silky or vitreous.
MODE OF OCCURRENCE: Occurs chiefly in albitized nepheline-syenite pegmatites in association with albite, aegirine, and other minerals. Found at Mont St. Hilaire, Quebec, Canada; in the Julianehaab district, Greenland; on the Kola peninsula, USSR; and as exceptional crystals up to 30 cm long in quartz-microcline-aegirine pegmatites in Tarbagatai, Eastern Kazakhstan.
BEST REF. IN ENGLISH: Vlasov, K. A., "Mineralogy of Rare Elements," v. II, p. 365–367, Israel Program for Scientific Translations, 1966.

## ELYITE
$Pb_4Cu(SO_4)(OH)_8$

CRYSTAL SYSTEM: Monoclinic
CLASS: 2/m
SPACE GROUP: $P2_1/a$
Z: 2
LATTICE CONSTANTS:
  $a = 14.248$
  $b = 5.768$
  $c = 7.309$
  $\beta = 100°26'$
3 STRONGEST DIFFRACTION LINES:
  6.999 (100)
  7.189 ( 99)
  2.995 ( 73)
OPTICAL CONSTANTS:
  $a = 1.990$
  $\beta = 1.993$
  $\gamma = 1.994$
  $2V_x = 76°$ at 480; 68° at 530; 66° at 580; and 64° at 630 nm.
HARDNESS: 2
DENSITY: $\sim 6$ (Meas.)
         6.321 (Calc.)
CLEAVAGE: {001} good
Sectile.
HABIT: Crystals prismatic to fibrous; as radiating sprays or tufts, sometimes matted. Mirror twinning in {001} common, nonrepetitive.
COLOR-LUSTER: Violet. Transparent. Silky in fibrous crystal groups. Streak pale violet to white.
MODE OF OCCURRENCE: Occurs in cavities in massive sulfides associated with langite, serpierite, and supergene galena at Ward, Nevada.
BEST REF. IN ENGLISH: Williams, Sidney A., *Am. Min.,* **57**: 364–367 (1972).

## EMBOLITE
Ag(Cl,Br)

CRYSTAL SYSTEM: Cubic
CLASS: $4/m\bar{3}2/m$

SPACE GROUP: Fm3m
Z: 4
LATTICE CONSTANTS:
 (solid solution)
 $a = 5.55$–$5.78$
3 STRONGEST DIFFRACTION LINES:
 2.81 (100)
 1.99 ( 60)
 1.26 ( 40)
OPTICAL CONSTANTS: Isotropic.
 $n = 2.15$
 HARDNESS: 2½
 DENSITY: 5.7–5.8 (Meas.)
 CLEAVAGE: None.
Fracture subconchoidal to uneven. Sectile, ductile, very plastic.
 HABIT: Crystals commonly cubic, often with large {111} and {011}. Usually massive; in waxy coatings and crusts. Also as horn-like masses. Twinning on {111} common.
 COLOR-LUSTER: Usually gray, yellowish, greenish gray. Transparent to translucent. Waxy, resinous to adamantine.
 MODE OF OCCURRENCE: Embolite occurs as a secondary mineral in the oxidized zone of silver deposits frequently associated with native silver, iron and manganese oxides, and secondary lead and copper minerals. Found widespread in the western United States, especially in California, Nevada, Arizona, New Mexico, Colorado, and South Dakota. Notable amounts also occur at Chañarcillo and at other places in Chile; also at Broken Hill, New South Wales, and at Silver Reef, Victoria, Australia.
 BEST REF. IN ENGLISH: Palache, et al., "Dana's System of Mineralogy," 7th Ed., v. II, p. 11–15, New York, Wiley, 1951.

## EMBREYITE
$Pb_5 (CrO_4)_2 (PO_4)_2 \cdot H_2O$

CRYSTAL SYSTEM: Monoclinic
CLASS: 2/m
SPACE GROUP: $P2_1/m$
Z: 1
LATTICE CONSTANTS:
 $a = 9.755$
 $b = 5.636$
 $c = 7.135$
 $\beta = 103°05'$
3 STRONGEST DIFFRACTION LINES:
 3.167 (100)
 4.751 ( 60)
 2.818 ( 60)
OPTICAL CONSTANTS:
 $\alpha = 2.20$
 $\beta = \gamma = 2.36$
Birefringence high.
 $2V = 0°$ to $11°$
 HARDNESS: 3½
 DENSITY: 6.45 (Meas.)
      6.41 (Calc.)
 CLEAVAGE: None.
Fracture irregular. Brittle.

 HABIT: As drusy crystalline crusts composed of minute tabular crystals with the plane of flattening approximately normal to the surface of the crust. Crystals are composite and show sectored zoning and multiple twinning.
 COLOR-LUSTER: Various shades of orange. Transparent to translucent. Dull to sparkling and resinous. Streak yellow.
 MODE OF OCCURRENCE: Occurs in association with crocoite, cerussite, phoenicochroite, and vauquelinite at Beresov, in the Urals, USSR.
 BEST REF. IN ENGLISH: Williams, S. A., *Min. Mag.*, 38: 790–793 (1972).

## EMERALD = Gem variety of beryl

## EMMONSITE (Durdenite)
$Fe_2 (TeO_3)_3 \cdot 2H_2O$

CRYSTAL SYSTEM: Monoclinic
3 STRONGEST DIFFRACTION LINES:
 3.14 (100)
 2.87 ( 90)
 2.52 ( 90)
OPTICAL CONSTANTS:
 $\alpha = 1.962$
 $\beta = 2.09$
 $\gamma = 2.10$–$2.12$
 $(-)2V = $ small, $20°$
 HARDNESS: ~5
 DENSITY: 4.52–4.53 (Meas.)
 CLEAVAGE: {010} perfect
Two other cleavage planes form angles of 85 and 95° to the {010} plane.
 HABIT: As compact microcrystalline masses, fibrous crusts, patches of minute acicular crystals, and thin scaly coatings.
 COLOR-LUSTER: Yellowish green. Transparent to translucent. Vitreous to dull.
 MODE OF OCCURRENCE: Occurs as a secondary mineral in the oxidation zone of deposits containing native tellurium or tellurides. Found at the Moose, W.P.H., and Deadwood mines, Cripple Creek, Teller County, Colorado; at the Clinton mine, Lawrence County, South Dakota; near Tombstone, Arizona; at Carson Hill, Calaveras County, California; near Silver City, New Mexico; with mackayite at Goldfield, Nevada; and at the El Pomo mine, Tegucigalpa, Honduras.
 BEST REF. IN ENGLISH: Palache, et al., "Dana's System of Mineralogy," 7th Ed., v. II, p. 640–641, New York, Wiley, 1951.

## EMPLECTITE
$CuBiS_2$

CRYSTAL SYSTEM: Orthorhombic
CLASS: 2/m 2/m 2/m
SPACE GROUP: Pnam
Z: 4

LATTICE CONSTANTS:
$a = 6.125$
$b = 14.512$
$c = 3.890$
3 STRONGEST DIFFRACTION LINES:
3.05 (100)
3.23 ( 90)
3.13 ( 70)
HARDNESS: 2
DENSITY: 6.3-6.5 (Meas.)
6.429 (Calc.)
CLEAVAGE: {010} perfect
{001} imperfect
Fracture conchoidal to uneven. Brittle.
HABIT: Crystals short prismatic, flattened {010}, and striated vertically.
COLOR-LUSTER: Gray to tin white. Opaque. Metallic.
MODE OF OCCURRENCE: Occurs in vein deposits commonly associated with chalcopyrite and other sulfides. Found in Colorado in the Mike Rabbit mine, Chaffee County; at the Union Pacific mine, Jefferson County; and in the Missouri mine and neighboring veins in Hall Valley, Park County. It also occurs in Chile, Norway, Czechoslovakia, and at several deposits in Saxony and elsewhere in Germany.
BEST REF. IN ENGLISH: Palache, et al., "Dana's System of Mineralogy," 7th Ed., v. I, p. 435-437, New York, Wiley, 1944.

# EMPRESSITE
$Ag_{2-x}Te_{1+x}$ (x = 0.1-0.5)

CRYSTAL SYSTEM: Orthorhombic (hexagonal forms reported)
CLASS: 2/m 2/m 2/m, mm2
SPACE GROUP: Pmnm or Pmn2
Z: 16
LATTICE CONSTANTS:
$a = 8.90$
$b = 20.07$
$c = 4.62$
3 STRONGEST DIFFRACTION LINES:
2.70 (100)
2.23 ( 80)
3.81 ( 60)
HARDNESS: 3.5
DENSITY: 7.61
CLEAVAGE: None.
Uneven to subconchoidal fracture. Brittle.
HABIT: Compact granular masses.
COLOR-LUSTER: Pale bronze; tarnishes darker on exposure; metallic. Streak grayish black to black.
MODE OF OCCURRENCE: With galena and native tellurium at the Empress Josephine mine, Kerber Creek district, Colorado.
BEST REF. IN ENGLISH: Honea, Russell M., *Am. Min.*, 49: 325-338 (1964).

# ENALITE = Thorite (uranoan)

# ENARGITE
$Cu_3AsS_4$
Dimorphous with luzonite

CRYSTAL SYSTEM: Orthorhombic
CLASS: mm2
SPACE GROUP: $Pnm2_1$
Z: 2
LATTICE CONSTANTS:
$a = 6.46$
$b = 7.43$
$c = 6.18$
3 STRONGEST DIFFRACTION LINES:
3.22 (100)
1.859 ( 90)
2.87 ( 80)
OPTICAL CONSTANTS:
$\alpha = 3.081$
$\beta = 3.089$
$\gamma = 3.120$
(+)2V = 54°
HARDNESS: 3
DENSITY: 4.45 (Meas.)
4.40 (Calc.)
CLEAVAGE: {110} perfect
{100} distinct
{010} distinct
{001} indistinct
Fracture uneven. Brittle.
HABIT: Crystals commonly prismatic along c-axis and terminated by basal pinacoids; also tabular {001}. Prism zone striated parallel to c-axis. Usually massive, granular or prismatic. Twinning on {320} common; sometimes cyclically as sixlings.
COLOR-LUSTER: Grayish black to iron black. Opaque. Metallic, tarnishing dull. Streak grayish black.
MODE OF OCCURRENCE: Occurs chiefly in medium-temperature hydrothermal ore veins or replacement deposits associated with sulfides and quartz; less commonly in low-temperature deposits with chalcopyrite, galena, and sphalerite on dolomite, and in the anhydrite cap rock of salt domes. Found in the United States as exceptional crystals and as an ore mineral at Butte, Silver Bow County, Montana, and in small amounts in Colorado, Utah, Nevada, California, Louisiana, Arkansas, Missouri, and Alaska. Other occurrences are found in Mexico, Chile, Peru, Bolivia, Argentina, Yugoslavia, Hungary, Austria, Sardinia, Philippine Islands, Taiwan, and Tsumeb, South West Africa.
BEST REF. IN ENGLISH: Palache, et al., "Dana's System of Mineralogy," 7th Ed., v. I, p. 389-392, New York, Wiley, 1944.

# ENDELLITE (Hydrohalloysite)
$Al_2Si_2O_5(OH)_4 \cdot 4H_2O$
Kaolinite (Kandite) Group

CRYSTAL SYSTEM: Monoclinic
Z: 2

LATTICE CONSTANTS:
$a \simeq 5.2$
$b \simeq 8.9$
$c \simeq 10.1$
$\beta \simeq 92°18'$
3 STRONGEST DIFFRACTION LINES:
4.42 (100)
10.1 ( 90)
3.34 ( 90)
OPTICAL CONSTANT:
$N \simeq 1.537$
HARDNESS: 2-2½
DENSITY: 2.11-2.17 (Meas.)
CLEAVAGE: None.
Fracture earthy.
HABIT: Tubular, ultramicroscopic in size. As compact to mealy masses.
COLOR-LUSTER: Colorless, white; sometimes tinted various colors by impurities. Transparent to translucent. Pearly to dull earthy.
MODE OF OCCURRENCE: Typical occurrences are found at Anamosa, Iowa, and at Eureka, Utah.
BEST REF. IN ENGLISH: Grim, R., "Clay Mineralogy," 2nd Ed., p. 38, (1968).

## ENDLICHITE = Variety of vanadinite containing arsenic

## ENGLISHITE (Inadequately described mineral)
$K_2 Ca_4 Al_8 (PO_4)_8 (OH)_{10} \cdot 9H_2O$

CRYSTAL SYSTEM: Monoclinic (?)
3 STRONGEST DIFFRACTION LINES:
9.3 (100)
2.86 ( 70)
1.72 ( 60)
OPTICAL CONSTANTS:
$\alpha = 1.570$
$\beta = ?$
$\gamma = 1.572$
(-)2V = small
HARDNESS: ~3
DENSITY: ~2.65
CLEAVAGE: {001} perfect, micaceous
HABIT: As aggregates and layers of curved and composite plates.
COLOR-LUSTER: Colorless. Transparent. Vitreous; pearly on cleavage.
MODE OF OCCURRENCE: Occurs with crandallite, wardite, and other phosphate minerals in variscite nodules at Fairfield, Utah County, Utah.
BEST REF. IN ENGLISH: Palache, et al., "Dana's System of Mineralogy," 7th Ed., v. II, p. 957-958, New York, Wiley, 1951.

## ENSTATITE
$MgSiO_3$
Pyroxene Group
Var: Bronzite

CRYSTAL SYSTEM: Orthorhombic
CLASS: 2/m 2/m 2/m
SPACE GROUP: Pbca
Z: 16
LATTICE CONSTANTS:
$a = 18.228$
$b = 8.805$
$c = 5.185$
3 STRONGEST DIFFRACTION LINES:
3.167 (100)
2.872 ( 85)
2.494 ( 50)
OPTICAL CONSTANTS:
$\alpha = 1.650-1.662$
$\beta = 1.653-1.671$
$\gamma = 1.658-1.680$
(+)2V = 55°-90°
HARDNESS: 5-6
DENSITY: 3.209-3.431 (Meas.)
3.194 (Calc.)
CLEAVAGE: {210} good
{100} parting
{010} parting
Fracture uneven. Brittle.
HABIT: Crystals prismatic, rare. Usually massive, lamellar or fibrous. Twinning on {100}, simple and lamellar.
COLOR-LUSTER: Colorless, yellowish or greenish white, gray, olive green, brown. Transparent to nearly opaque. Vitreous to pearly. Streak uncolored, grayish.
MODE OF OCCURRENCE: Occurs widespread chiefly as a common constituent in basic or ultrabasic igneous rocks such as norites, pyroxenites, gabbros, and peridotites; in thermally and regionally metamorphosed rocks; and in stony and metallic meteorites. Typical occurrences are found in the serpentinized rocks of the Coast ranges and Sierra Nevada in California, and at localities in Colorado, Montana, Arizona, Pennsylvania, New York, Delaware, Maryland and North Carolina. Also in Ireland, Scotland, Norway, Germany, Austria, Czechoslovakia, Finland, South Africa, Japan, and elsewhere.
BEST REF. IN ENGLISH: Deer, Howie, and Zussman, "Rock Forming Minerals," v. 2, p. 8-41, New York, Wiley, 1963.

## EOSITE = Variety of wulfenite containing vanadium in substitution for molybdenum

## EOSPHORITE
$(Mn,Fe)AlPO_4 (OH)_2 \cdot H_2O$
Childrenite Series

CRYSTAL SYSTEM: Monoclinic (pseudo-orthorhombic)
SPACE GROUP: Bbam (pseudocell)
Z: 8
LATTICE CONSTANTS:
$a = 10.45$
$b = 13.49$
$c = 6.93$
$\beta = 90°$

3 STRONGEST DIFFRACTION LINES:
2.81 (100)
5.27 ( 40)
2.42 ( 40)
OPTICAL CONSTANTS:
$\alpha = 1.638-1.639$
$\beta = 1.660-1.664$
$\gamma = 1.667-1.671$
$(-)2V = 50°$
HARDNESS: 5
DENSITY: 3.05 (Pure Mn end-member.)
CLEAVAGE: {100} poor
Fracture uneven to subconchoidal.

HABIT: Crystals short to long prismatic; often flattened on {010}. Common forms a{100}, b{010}, m{110}, s{121}. Also as coarse radial aggregates. Twinning by reflection on {100}.

COLOR-LUSTER: Colorless, pale pink, pale yellow, light brown, reddish brown, black. Transparent to translucent. Vitreous to resinous.

MODE OF OCCURRENCE: Occurs in granite pegmatite associated with manganese phosphates at Branchville, Connecticut; at Red Hill, Black Mountain, Newry, Hebron, Poland, Mt. Mica, and Buckfield, Maine; at the Hugo mine, Keystone, South Dakota; Palermo mine, North Groton, New Hampshire; Hagendorf, Bavaria, Germany; and in Minas Geraes, Brazil as superb pink flattened long prismatic crystals as much as 1 cm wide and 4 cm in length.

BEST REF. IN ENGLISH: Hurlbut, Cornelius S. Jr., *Am. Min.*, **35**: 793–805 (1950).


# EPHESITE
$Na(LiAl_2)(Al_2Si_2)O_{10}(OH)_2$
Margarite Group

CRYSTAL SYSTEM: Monoclinic
CLASS: 2/m or m
SPACE GROUP: C2/c or Cc
Z: 2
LATTICE CONSTANTS:
$a = 5.120$
$b = 8.853$
$c = 19.303$
$\beta = 95°5'$
3 STRONGEST DIFFRACTION LINES:
3.20 (100)
9.59 ( 65)
1.92 ( 25)
OPTICAL CONSTANTS:
$\alpha = 1.592$
$\beta = 1.624$
$\gamma = 1.625$
HARDNESS: 3½–4½
DENSITY: 2.984 (Meas.)
2.965 (Calc.)
CLEAVAGE: {001} perfect
Laminae brittle.

HABIT: Crystals indistinct; up to 13 mm in diameter and 10 mm in length. Also as small flakes up to 5 mm across. Crystals often twinned by a 180° rotation about [310] or [3$\bar{1}$0].

COLOR-LUSTER: Pink. Translucent. Vitreous; pearly on cleavage.

MODE OF OCCURRENCE: Occurs at the emery deposits of Gumuch-Dagh near Ephesus, Asia Minor; also found in the Postmasburg district, South Africa, associated with brown mangan-diaspore, massive braunite, and bixbyite.

BEST REF. IN ENGLISH: Schaller, Waldemar T., Carron, Maxwell K., and Fleischer, Michael, *Am. Min.*, **52**: 1689–1696 (1967).


# EPIDIDYMITE
$NaBeSi_3O_7(OH)$
Dimorphous with eudidymite

CRYSTAL SYSTEM: Orthorhombic
CLASS: 2/m 2/m 2/m
SPACE GROUP: Pnam
Z: 8
LATTICE CONSTANTS:
$a = 12.66$
$b = 7.34$
$c = 13.48$
3 STRONGEST DIFFRACTION LINES:
3.40 (100)
3.09 (100)
2.99 (100)
OPTICAL CONSTANTS:
$\alpha = 1.536-1.542$
$\beta = 1.541-1.544$
$\gamma = 1.542-1.548$
$(+)2V = 30-32°$
HARDNESS: 6-7
DENSITY: 2.55 (Meas.)
2.579 (Calc.)
CLEAVAGE: {001} perfect
{100} distinct
Brittle.

HABIT: Crystals tabular, up to 2 × 2 × 0.5 cm. Also finely crystalline to cryptocrystalline; coarsely crystalline-micaceous; spherulitic with radial-fibrous structure. Twinning on {001} common, often as trillings.

COLOR-LUSTER: Usually white or colorless; also various shades of gray, yellow, blue, violet. Transparent to translucent. Vitreous; pearly on cleavages.

MODE OF OCCURRENCE: Occurs in nepheline-syenite pegmatites in association with albite, elpidite, analcime, natrolite, neptunite, fluorite, and other minerals. Found at Mont St. Hilaire, Quebec, Canada; in the Julianehaab district, Greenland; in the Langesundfjord district, Norway; and in the Lovozero and Khibiny alkali massifs, Kola peninsula, USSR.

BEST REF. IN ENGLISH: Vlasov, K. A., "Mineralogy of Rare Elements," v. II, p. 135–140, Israel Program for Scientific Translations, 1966. Robinson, Paul D., and Fang. J. H., *Am. Min.*, **55**: 1541-1549 (1970).


# EPIDOTE (Pistacite)
$Ca_2Al_2FeOSiO_4Si_2O_7(OH)$
Var. Tawmawite

CRYSTAL SYSTEM: Monoclinic

CLASS: 2/m
SPACE GROUP: P2$_1$/m
Z: 2
LATTICE CONSTANTS:
 $a$ = 8.90
 $b$ = 5.63
 $c$ = 10.20
 $\beta$ = 115°24′
3 STRONGEST DIFFRACTION LINES:
 2.90 (100)
 2.68 (100)
 2.69 ( 70)
OPTICAL CONSTANTS:
 $\alpha$ = 1.740
 $\beta$ = 1.768
 $\gamma$ = 1.787
 (−)2V = 74°
HARDNESS: 6-7
DENSITY: 3.35-3.5 (Meas.)
 3.208 (Calc.)
CLEAVAGE: {001} perfect
Fracture uneven. Brittle.
HABIT: Crystals short to long prismatic, often deeply striated; also thick tabular, or acicular. Commonly massive, coarse to fine granular; also fibrous, parallel or divergent. Twinning on {100}, lamellar; not common.
COLOR-LUSTER: Usually yellowish green to brownish green; also gray, grayish white, greenish black, black, deep green (tawmawite), rarely colorless. Transparent to nearly opaque. Vitreous to somewhat pearly or resinous. Streak uncolored or grayish.
MODE OF OCCURRENCE: Occurs chiefly in regionally metamorphosed igneous and sedimentary rocks; in acid igneous rocks contaminated with calc-silicate material; as an alteration of plagioclase feldspars; and in contact zones. Typical occurrences are found near Sulzer, Prince of Wales Island, Alaska; in Inyo, Kern, Riverside, and San Diego counties, California; in Adams County, Idaho; in Colorado at the Calumet iron mine in Chaffee County, near Ohio City and on Italian Mountain in Gunnison County, and on Epidote Hill, Park County; and at localities in Michigan, Connecticut, Massachusetts, and New Hampshire. Also in Baja California, Mexico; at Arendal, Norway; as fine crystals at Bourg d'Oisans, France; at many places in Switzerland; in the Ala Valley, Piedmont, Italy; as exceptional crystals in the Untersulzbachtal, Salzburg, Austria; and in Czechoslovakia, USSR, Japan, Korea, Australia, Madagascar, and elsewhere. The chromium variety, tawmawite, occurs at Tawmaw, Burma, and at Outokumpu, Finland.
BEST REF. IN ENGLISH: Deer, Howie, and Zussman, "Rock Forming Minerals," v. 1, p. 193-210, New York, Wiley, 1962.

## EPIGENITE (Inadequately described mineral)
(Cu, Fe)$_5$AsS$_6$ (?)

CRYSTAL SYSTEM: Orthorhombic (?)
 HARDNESS: 3½
 DENSITY: 4.5 (Meas.)
 CLEAVAGE: Fracture uneven.

HABIT: As short prismatic crystals resembling arsenopyrite, often in crust-like aggregates.
COLOR-LUSTER: Steel gray; tarnishes black, then blue. Opaque. Metallic. Streak black.
MODE OF OCCURRENCE: Found implanted on barite and associated with wittichenite, chalcopyrite, calcite, and fluorite, at the Neuglück mine, Wittichen, Germany.
BEST REF. IN ENGLISH: Palache, et al., "Dana's System of Mineralogy," 7th Ed., v. I, p. 361-362, New York, Wiley, 1944.

## EPIIANTHINITE = Schoepite

## EPISTILBITE
CaAl$_2$Si$_6$O$_{16}$ · 5H$_2$O
Zeolite Group

CRYSTAL SYSTEM: Monoclinic
CLASS: 2/m or m
SPACE GROUP: C2/m or Cm
Z: 3
LATTICE CONSTANTS:
 $a$ = 8.92
 $b$ = 17.73
 $c$ = 10.21
 $\beta$ = 124°20′
3 STRONGEST DIFFRACTION LINES:
 3.45 (100)
 8.89 ( 90)
 3.21 ( 90)
OPTICAL CONSTANTS:
 $\alpha$ = 1.485-1.505
 $\beta$ = 1.497-1.515
 $\gamma$ = 1.497-1.519
 (−)2V ≃ 44°
HARDNESS: 4
DENSITY: 2.25 (Meas.)
 2.266 (Calc.)
CLEAVAGE: {010} perfect
Fracture uneven. Brittle.
HABIT: Crystals prismatic, always twinned; in radiated spherical aggregates; also granular. Twinning on {100} or {010} common, often as cruciform interpenetrant twins.
COLOR-LUSTER: Colorless, white, pinkish. Transparent to translucent. Vitreous.
MODE OF OCCURRENCE: Occurs in trap rock at Bergen Hill, New Jersey; in pegmatite in association with beryl and bertrandite at Bedford, New York; in cavities in basalt in the Lanakai hills, Hawaii; and at localities in Nova Scotia, Iceland, Faroe Islands, Isle of Skye, Switzerland, and at Poona, India.
BEST REF. IN ENGLISH: Deer, Howie, and Zussman, "Rock Forming Minerals," v. 4, p. 377-385, New York, Wiley, 1963.

## EPISTOLITE
Na$_2$(Nb, Ti)$_2$Si$_2$O$_9$ · nH$_2$O

CRYSTAL SYSTEM: Monoclinic

LATTICE CONSTANT:
$\beta = 74°42'$

3 STRONGEST DIFFRACTION LINES:
6.11 (100)
4.33 (100)
2.72 ( 80)

OPTICAL CONSTANTS:
$\alpha = 1.610$
$\beta = 1.650\text{-}1.720$
$\gamma = 1.682\text{-}1.770$
$(-)2V = \sim 60\text{-}80°$

HARDNESS: 1-1½
DENSITY: 2.65-2.89 (Meas.)
CLEAVAGE: {001} perfect
{110} distinct
Very brittle. Friable.

HABIT: Crystals rectangular plates, up to 1 mm thick and 2 cm long. Also as thin lamellar aggregates.

COLOR-LUSTER: White, yellowish white, gray, light brown. Translucent; thin fragments transparent. Pearly.

MODE OF OCCURRENCE: Occurs in alkali pegmatites and albitites in association with microcline, albite, aegirine, eudialyte, sodalite, steenstrupine, and other minerals. Found in the Julianehaab district, Greenland, and in the Lovozero alkali massif, USSR.

BEST REF. IN ENGLISH: Vlasov, K. A., "Mineralogy of Rare Elements," v. II, p. 562-564, Israel Program for Scientific Translations, 1966.

## EPSOMITE

$MgSO_4 \cdot 7H_2O$

CRYSTAL SYSTEM: Orthorhombic
CLASS: 222
SPACE GROUP: $P2_1 2_1 2_1$
Z: 4
LATTICE CONSTANTS:
$a = 11.94$ kX
$b = 12.03$
$c = 6.865$

3 STRONGEST DIFFRACTION LINES:
4.21 (100)
5.35 ( 25)
2.677 ( 25)

OPTICAL CONSTANTS:
$\alpha = 1.4299\text{-}1.4400$
$\beta = 1.4523\text{-}1.4623$
$\gamma = 1.4572\text{-}1.4694$
$(-)2V \sim 50°$ (Calc.)

HARDNESS: 2-2½
DENSITY: 1.677 (Meas.)
1.678 (Calc.)
CLEAVAGE: {010} perfect
{101} distinct
Fracture conchoidal.

HABIT: Rarely as crystals; synthetic crystals short prismatic to equant. Twinning on {110} rare. Usually as fibrous to acicular crusts; stalactitic; as rounded masses; as efflorescences.

COLOR-LUSTER: Crystals colorless; transparent; vitreous. Aggregates white, pinkish, greenish; translucent; silky to earthy.

Very soluble in water. Effloresces in dry air. Taste saline, bitter.

MODE OF OCCURRENCE: Occurs widespread as a common efflorescence on the walls of mine workings, in limestone caverns, and in sheltered places on magnesian rock outcrops. It also is common in the waters of salt lakes and mineral springs; in oceanic and lacustrine salt deposits; and as a fumarolic deposit. Found in the United States as crystals up to several feet long on Kruger Mountain, near Oroville, Washington; at many places in Albany, Natrona, and Carbon counties, Wyoming; in cinnabar mines and many other environments in California; in Nevada, New Mexico, Montana, Utah, and other western states; and in limestone caverns in Indiana, Kentucky, and Tennessee. It also occurs at Epsom, Surrey, England; in France, Germany, Czechoslovakia, on Mt. Vesuvius, Italy, and at many other places throughout the world.

BEST REF. IN ENGLISH: Palache, et al., "Dana's System of Mineralogy," 7th Ed., v. II, p. 509-513, New York, Wiley, 1951.

## ERICAITE

$(Fe, Mg, Mn)_3 B_7 O_{13} Cl$
Iron analogue of boracite

CRYSTAL SYSTEM: Orthorhombic (?)
LATTICE CONSTANTS:
$a = 8.53$
$b = 8.60$
$c = 12.15$
HARDNESS: 7-7½
DENSITY: 3.17-3.27
CLEAVAGE: None.
Fracture conchoidal to uneven.

HABIT: Crystals pseudocubic, up to 4 mm edge length. Form (100) dominant, with (111) less prominent. Crystals often zoned.

COLOR-LUSTER: Light green to raspberry red and black; vitreous.

MODE OF OCCURRENCE: Occurs in halite-anhydrite and halite-anhydrite-kieserite rock in several potassium salt deposits in the southern Harz, Germany.

BEST REF. IN ENGLISH: Heide, F., Kühn, Robert, and Schaacke, Ingeburg, *Am. Min.*, **41**: 372 (1956).

## ERICSSONITE

$BaMn_2 Fe(Si_2 O_7)(O, OH)$
Dimorphous with orthoericssonite

CRYSTAL SYSTEM: Monoclinic
CLASS: 2/m or 2
SPACE GROUP: C2/m or C2
Z: 4
LATTICE CONSTANTS:
$a = 20.42$
$b = 7.03$
$c = 5.34$
$\beta = 95°30'$

3 STRONGEST DIFFRACTION LINES:
  3.510 (100)
  2.687 ( 70) broad
  2.780 ( 60) broad - This line distinguishes ericssonite from orthoericssonite.
  HARDNESS: 4½
  DENSITY: 4.21 (Meas.)
  CLEAVAGE: {100} perfect
              {011} distinct
Very brittle.
  HABIT: Massive; as embedded plates up to 2 cm in diameter.
  COLOR-LUSTER: Reddish black. Streak rich brown.

  Weakly magnetic.
  MODE OF OCCURRENCE: Occurs intergrown with ortho-ericssonite in schefferite-rhodonite-tephroite skarn at Lång-ban, Sweden.
  BEST REF. IN ENGLISH: Moore, Paul B., *Lithos*, **4**: 137–145 (1971).

## ERIKITE = Rhabdophane or monazite

## ERINITE = Cornwallite

## ERIOCHALCITE
$CuCl_2 \cdot 2H_2O$

CRYSTAL SYSTEM: Orthorhombic
CLASS: 2/m 2/m 2/m
SPACE GROUP: Pbmn
Z: 2
LATTICE CONSTANTS:
  $a = 7.38$
  $b = 8.04$
  $c = 3.72$
3 STRONGEST DIFFRACTION LINES:
  5.44 (100)
  4.02 ( 14)
  2.637 ( 10)
OPTICAL CONSTANTS:
    $\alpha = 1.646$
    $\beta = 1.685$ (Na)
    $\gamma = 1.745$
  (+)2V = 75°
  HARDNESS: 2½
  DENSITY: 2.47 (Meas. synthetic)
           2.55 (Calc.)
  CLEAVAGE: {110} perfect
              {001} good
Fracture conchoidal.
  HABIT: As wool-like aggregates; also as somewhat flat-tened aggregates of small crystals elongated along c-axis, with center of aggregates composed of bent, deeply grooved spire-like crystals.
  COLOR-LUSTER: Greenish blue to bluish green, some-times with yellowish tint. Transparent. Vitreous.
  MODE OF OCCURRENCE: Occurs as a sublimation prod-uct in fumaroles on Mt. Vesuvius, Italy, and as a secondary mineral associated with bandylite and atacamite at the Queténa mine, Antofagasta province, Chile.
  BEST REF. IN ENGLISH: Palache, et al., "Dana's Sys-tem of Mineralogy," 7th Ed., v. II, p. 44–46, New York, Wiley, 1951.

## ERIONITE
$(Ca, Na_2, K_2)_{1.5}Al_9Si_{27}O_{72} \cdot 27H_2O$
Zeolite Group

CRYSTAL SYSTEM: Hexagonal
CLASS: 6/m 2/m 2/m
SPACE GROUP: $P6_3/mmc$
Z: 1
LATTICE CONSTANTS:
  $a = 13.20$
  $c = 15.07$
3 STRONGEST DIFFRACTION LINES:
  11.41 (100)
  6.61 ( 73)
  4.322 ( 67)
OPTICAL CONSTANTS:
  (variable)
  $\omega = 1.4711$
  $\epsilon = 1.4740$
    (+)
  HARDNESS: Not determined.
  DENSITY: ~2.02
  CLEAVAGE: Not determined.
  HABIT: Crystals minute, prismatic, in radiating groups. Usually finely fibrous, wool-like.
  COLOR-LUSTER: White.
  MODE OF OCCURRENCE: Occurs in association with opal in fractures in rhyolitic tuff near Durkee, Baker County, Oregon; in association with paulingite in basalt dredged from the Columbia River, near Wenatchee, Wash-ington; and at localities in Nevada, Wyoming, South Dakota, and the Faroe Islands.
  BEST REF. IN ENGLISH: *Am. Min.*, **52**: 1785–1794 (1967). Bennett, J. M., and Gard, J. A., *Nature*, **214**: 1005–1006 (1967). Sheppard, Richard A., and Gude, Arthur J., 3rd, *Am. Min.*, **54**: 875–886 (1969).

## ERLICHMANITE
$OsS_2$

CRYSTAL SYSTEM: Cubic
CLASS: 2/m $\overline{3}$
SPACE GROUP: Pa3
Z: 4
LATTICE CONSTANT:
  $a \simeq 5.62$
3 STRONGEST DIFFRACTION LINES:
  3.24 (100)
  2.810 ( 85) (synthetic)
  1.694 ( 85)
  HARDNESS: Not determined.
  DENSITY: 9.59 (Calc.)
  CLEAVAGE: Not determined.
  HABIT: Massive; as minute grains ~ 20 μm in size.

COLOR-LUSTER: Gray. Opaque. Metallic.

MODE OF OCCURRENCE: Found in ferroplatinum from noble-metal placers collected at the MacIntosh Mine, Willow Creek, Trinity River, Humboldt County, California; also in a platinum-metal nugget from western Ethiopian laterites.

BEST REF. IN ENGLISH: Snetsinger, Kenneth G., *Am. Min.*, **56**: 1501–1506 (1971).

## ERNSTITE
$(Mn^{2+}_{1-x}Fe^{3+}_x)Al(PO_4)(OH)_{2-x}O_x$ (x = 0–1)

CRYSTAL SYSTEM: Monoclinic
CLASS: 2/m or m
SPACE GROUP: A2/a or Aa
Z: 8
LATTICE CONSTANTS:
  $a$ = 13.32
  $b$ = 10.497
  $c$ = 6.969
  $\beta$ = 90°22′
3 STRONGEST DIFFRACTION LINES:
  2.829 (100)
  2.836 ( 80)
  2.438 ( 50)
OPTICAL CONSTANTS:
  $\alpha$ = 1.678
  $\beta$ = 1.706  (Na)
  $\gamma$ = 1.721
  (−)2V = 74°
HARDNESS: 3–3½
DENSITY: 3.07 (Meas.)
     3.086 (Calc.)
CLEAVAGE: {010} good
      {100} good
HABIT: As radiating aggregates 10–15 mm long. Untwinned.

COLOR-LUSTER: Yellow-brown.

MODE OF OCCURRENCE: Occurs as an oxidation product of eosphorite (of which relicts remain) in granitic pegmatite near Karibib, South West Africa.

BEST REF. IN ENGLISH: Seeliger, E., and Mücke, A., *Am. Min.*, **56**: 637 (1971).

## ERYTHRITE
$Co_3(AsO_4)_2 \cdot 8H_2O$
Erythrite-Annabergite Series

CRYSTAL SYSTEM: Monoclinic
CLASS: 2/m
SPACE GROUP: C2/m
Z: 2
LATTICE CONSTANTS:
  $a$ = 10.184 kX
  $b$ = 13.340
  $c$ = 4.730
  $\beta$ = 105°01′
3 STRONGEST DIFFRACTION LINES:
  6.65  (100)
  1.677 ( 14)
  3.22  ( 12)

OPTICAL CONSTANTS:
  $\alpha$ = 1.622–1.629
  $\beta$ = 1.658–1.663
  $\gamma$ = 1.681–1.701
  (+)2V = large, 90° (also negative)
HARDNESS: 1½–2½
DENSITY: 3.18 (Meas.)
     3.18 (Calc.)
CLEAVAGE: {010} perfect
      {100} indistinct
      {$\bar{1}$02} indistinct
Flexible in thin laminae. Sectile.

HABIT: Crystals prismatic to acicular, flattened on {010}; commonly striated. As bladed aggregates or stellate groups; also globular tufts of minute crystals; fibrous, earthy.

COLOR-LUSTER: Deep purplish red to pale pink or nearly colorless. Transparent to translucent. Weakly adamantine, pearly on {010}; also dull. Streak lighter shade than color.

MODE OF OCCURRENCE: Occurs as a secondary mineral in the oxidation zone of cobalt-bearing ore deposits. Found in the United States in the Blackbird district, Lemhi County, Idaho, and at deposits in California, Nevada, Arizona, and New Mexico. Fine specimens occur at the Nipissing mine and at other places in the Cobalt area, Ontario, Canada; near Alamos, Sonora, Mexico; at several localities in Chile; in Germany, especially at Schneeberg; and in England, France, Sweden, Austria, Czechoslovakia, and Switzerland. Superb crystals up to one inch or more long, often associated with roselite, occur at Bou Azzer, Anti Atlas, Morocco.

BEST REF. IN ENGLISH: Palache, et al., "Dana's System of Mineralogy," 7th Ed., v. II, p. 746–750, New York, Wiley, 1951.

## ERYTHROSIDERITE
$K_2FeCl_5 \cdot H_2O$
Erythrosiderite Series

CRYSTAL SYSTEM: Orthorhombic
CLASS: 2/m 2/m 2/m
SPACE GROUP: Pnma
Z: 4
LATTICE CONSTANTS:
  $a$ = 13.75
  $b$ = 9.924
  $c$ = 6.93
OPTICAL CONSTANTS:
  $\alpha$ = 1.715
  $\beta$ = 1.75
  $\gamma$ = 1.80
Biaxial (+), 2V = 62°
HARDNESS: Not determined.
DENSITY: 2.372 (Synthetic)
CLEAVAGE: {210} perfect
      {011} perfect
HABIT: Crystals tabular, flattened {100}.
COLOR-LUSTER: Red to brownish red. Vitreous.

Deliquescent.

MODE OF OCCURRENCE: Occurs as an efflorescence on rinneite at Strassfurt, Saxony, Germany.

BEST REF. IN ENGLISH: Palache, et al., "Dana's System of Mineralogy," 7th Ed., v. II, p. 101–103, New York, Wiley, 1951.

**ESCHWEGEITE** = Synonym for tanteuxenite

**ESCHYNITE** = Aeschynite

## ESKEBORNITE
$CuFeSe_2$

CRYSTAL SYSTEM: Cubic
CLASS: $4/m\,\bar{3}\,2/m$
SPACE GROUP: Pm3m
Z: 2
LATTICE CONSTANT:
  $a = 5.53$
3 STRONGEST DIFFRACTION LINES:
  1.960 (100)
  3.19 ( 95)
  1.671 ( 80)
HARDNESS: 3–3½
DENSITY: Not determined.
CLEAVAGE: {001} perfect
HABIT: Crystals tabular, up to 1 mm in size. Usually as microscopic embedded grains.
COLOR-LUSTER: Brass yellow. Opaque. Metallic.
MODE OF OCCURRENCE: Occurs in association with chalcopyrite, clausthalite, tiemannite, berzelianite, and naumannite in dolomite veins in the Eskeborn adit, Tilkerode, Harz Mountains, Germany.
BEST REF. IN ENGLISH: *Am. Min.*, **46**: 467 (1961). Vlasov, K. A., "Mineralogy of Rare Elements," v. II, p. 654–655, Israel Program for Scientific Translations, 1966.

## ESKOLAITE
$Cr_2O_3$

CRYSTAL SYSTEM: Hexagonal
CLASS: $\bar{3}m$
SPACE GROUP: $R\bar{3}c$
Z: 6
LATTICE CONSTANTS:
  $a = 4.958$
  $c = 13.60$
3 STRONGEST DIFFRACTION LINES:
  1.67 (100)
  2.67 ( 97)
  3.63 ( 96)
HARDNESS: Microhardness (Vickers) ~3200 kg/mm².
DENSITY: 5.18 (Meas.)
         5.208 (Calc.)
CLEAVAGE: None.
Fracture uneven. Brittle.
HABIT: Crystals long prismatic to thick platy.
COLOR-LUSTER: Black; metallic. Opaque. Light green in finely divided material. Streak green.

MODE OF OCCURRENCE: Occurs in fresh, well-developed crystals, 1–12 mm in size, in chromium-bearing tremolite skarn at the Outokumpu mine in Finland, and as a major constituent of black pebbles found in the bed of the Merume River, Guyana.
BEST REF. IN ENGLISH: Kuovo, Olavi, and Vuorelainen, Yrjö, *Am. Min.*, **43**: 1098–1106 (1958). Tennyson, Christel, *Am. Min.*, **46**: 998–999 (1961).

## ESPERITE (Calcium-larsenite)
$(Ca, Pb)ZnSiO_4$

CRYSTAL SYSTEM: Monoclinic
CLASS: $2/m$
SPACE GROUP: $B2_1/m$
Z: 48
LATTICE CONSTANTS:
  $a = 17.63$
  $b = 8.27$
  $c = 30.52$
  $\beta = 90°$
3 STRONGEST DIFFRACTION LINES:
  3.017 (100)
  2.534 ( 75)
  7.62 ( 45)
OPTICAL CONSTANTS:
  $\alpha = 1.762$
  $\beta = 1.770$
  $\gamma = 1.774$
  $(-)2V = 40°$
HARDNESS: 5+
DENSITY: 4.28 (Meas.)
         4.25 (Calc.)
CLEAVAGE: {010} distinct
          {100} distinct
          {101} poor
Brittle.
HABIT: Massive, coarse granular.
COLOR-LUSTER: White. Transparent to translucent. Vitreous to somewhat greasy. Fluoresces brilliant lemon yellow under short-wave ultraviolet light.
MODE OF OCCURRENCE: Occurs in association with willemite, franklinite, and other minerals, at Franklin, Sussex County, New Jersey.
BEST REF. IN ENGLISH: Palache, Charles, USGS Prof. Paper 180, p. 81–82 (1935). Moore, Paul B., and Ribbe, Paul H., *Am. Min.*, **50**: 1170–1178 (1965).

## ETTRINGITE
$Ca_6Al_2(SO_4)_3(OH)_{12} \cdot 24H_2O$

CRYSTAL SYSTEM: Hexagonal
CLASS: $6/m\,2/m\,2/m$
SPACE GROUP: $P6_3/mmc$
Z: 8
LATTICE CONSTANTS:
  $a = 22.46$
  $c \simeq 21.44$

3 STRONGEST DIFFRACTION LINES:
9.65 (100)
5.58 ( 80)
3.21 ( 60)
OPTICAL CONSTANTS:
$\omega = 1.491$
$\epsilon = 1.470$
(−)
HARDNESS: 2–2½
DENSITY: 1.77 (Meas.)
1.79 (Calc.)
CLEAVAGE: $\{10\overline{1}0\}$ perfect
HABIT: Flattened hexagonal dipyramids $\{10\overline{1}2\}$; also prismatic {0001}, usually without terminal faces, but occasionally terminated by rhombohedral faces; also as thin fibers.
COLOR-LUSTER: Colorless and transparent; also milky white; vitreous.
MODE OF OCCURRENCE: Found in cavities of metamorphosed limestone inclusions in lava near Ettringen, Rhine Province, Germany; in excellent crystals up to 4 mm across associated with thomsonite and large crystals of clinohedrite at Franklin, New Jersey; also occurs at Scawt Hill, County Antrim, Ireland; Crestmore, California; and in the Lucky Cuss mine, Tombstone, Cochise County, Arizona. Also widespread in the Hatrurim and Ramleh areas, Israel.
BEST REF. IN ENGLISH: Hurlbut, Cornelius S. Jr., and Baum, John L., *Am. Min.*, **45**: 1137–1143 (1960).

# EUCAIRITE
AgCuSe

CRYSTAL SYSTEM: Orthorhombic
Z: 10
LATTICE CONSTANTS:
$a = 4.105$
$b = 20.35$
$c = 6.31$
3 STRONGEST DIFFRACTION LINES:
2.12 (100)
2.61 ( 70)
2.88 ( 50)
HARDNESS: 2½
DENSITY: 7.6–7.8 (Meas.)
7.91 (Calc.)
CLEAVAGE: None.
Fracture subconchoidal to uneven. Brittle.
HABIT: Massive and granular, also as thin metallic films staining calcite.
COLOR-LUSTER: Brilliant silver white and lead gray; tarnishes bright bronze; opaque; metallic. Streak shining.
MODE OF OCCURRENCE: Occurs associated with athabascaite and other selenides in hematite-stained carbonate vein material in basalt at the Martin Lake mine, Lake Athabasca area, northern Saskatchewan, Canada; also found disseminated in a serpentinized gangue of quartz and calcite at the copper mine of Skrikerum, Kalmar, Sweden; in the Harz mountains, Germany; in highly altered copper ore at Sierra de Umango, La Rioja, Argentina associated with

umangite, tiemannite, chalcomenite, and malachite; and sparingly at several copper deposits in Chile.
BEST REF. IN ENGLISH: Palache, et al., "Dana's System of Mineralogy," 7th Ed., v. I, p. 183–184, New York, Wiley, 1944. Early, J. W., *Am. Min.*, **35**: 337–364 (1950).

# EUCHLORIN (Inadequately described mineral)
$(K, Na)_8 Cu_9 (SO_4)_{10} (OH)_6$ (?)

CRYSTAL SYSTEM: Orthorhombic
OPTICAL CONSTANTS:
$\alpha = 1.580$
$\beta = 1.605$
$\gamma = 1.644$
(+)2V = moderately large
HARDNESS: Not determined.
DENSITY: Not determined.
CLEAVAGE: Not determined.
HABIT: Crystals slightly elongated rectangular tablets. As an incrustation.
COLOR-LUSTER: Emerald green.

Somewhat soluble in water.
MODE OF OCCURRENCE: Found as a fumarolic deposit on Mt. Vesuvius, Italy, associated with chalcocyanite, dolerophanite, melanothallite, and eriochalcite.
BEST REF. IN ENGLISH: Palache, et al., "Dana's System of Mineralogy," 7th Ed., v. II, p. 571, New York, Wiley, 1951.

# EUCHROITE
$Cu_2 AsO_4 OH \cdot 3H_2O$

CRYSTAL SYSTEM: Orthorhombic
CLASS: 2 2 2
SPACE GROUP: $P2_1 2_1 2_1$
Z: 4
LATTICE CONSTANTS:
$a = 10.05$ kX
$b = 10.50$
$c = 6.11$
3 STRONGEST DIFFRACTION LINES:
5.34 (100)
2.83 ( 90)
7.37 ( 80)
OPTICAL CONSTANTS:
$\alpha = 1.695$
$\beta = 1.698$
$\gamma = 1.733$
(+)2V = 29°
HARDNESS: 3½–4
DENSITY: 3.44 (Meas.)
3.45 (Calc.)
CLEAVAGE: {101} indistinct
{110} indistinct
Fracture subconchoidal to uneven. Brittle.
HABIT: Crystals short prismatic [010] to equant; rarely thick tabular {100}.
COLOR-LUSTER: Bright emerald green. Transparent to translucent. Vitreous.

MODE OF OCCURRENCE: Occurs as excellent crystals associated with olivenite lining crevices in sericite schist at Libethen, Hungary; and associated with strashimirite, azurite, malachite, and olivenite in the zone of oxidation of the Zapachitsa copper deposit, western Stara-Planina, Bulgaria.

BEST REF. IN ENGLISH: Berry, L. G., *Am. Min.*, **36**: 484–503 (1951). Palache, et al., "Dana's System of Mineralogy," 7th Ed., v. II, p. 934–935, New York, Wiley, 1951.

# EUCLASE
$BeAlSiO_4OH$

CRYSTAL SYSTEM: Monoclinic
CLASS: 2/m
SPACE GROUP: $P2_1/a$
Z: 4
LATTICE CONSTANTS:
 $a = 4.76$
 $b = 14.27$
 $c = 4.63$
 $\beta = 100°16'$
3 STRONGEST DIFFRACTION LINES:
 7.15 (100)
 3.219 ( 50)
 3.836 ( 35)
OPTICAL CONSTANTS:
 $\alpha = 1.651$
 $\beta = 1.655$
 $\gamma = 1.671$
 (+)2V = 50°
HARDNESS: 7½
DENSITY: 3.05–3.10 (Meas.)
 2.987 (Meas. Colorado)
 3.115 (Calc. Minas Geraes, Brazil)
CLEAVAGE: {010} perfect
 {110} imperfect
 {001} imperfect
Fracture conchoidal. Brittle.

HABIT: Crystals commonly long prismatic; also short and stout prismatic with {010} face dominant.

COLOR-LUSTER: Colorless, whitish, pale green, pale blue; transparent to translucent; vitreous.

MODE OF OCCURRENCE: Occurs as small crystals and crystal clusters in greisen deposits at the Boomer mine and in the Redskin Gulch area, Park County, Colorado; as superb gem crystals in clay-rich quartz lenses in phyllite and in pegmatites in Minas Geraes, Brazil; in vuggy quartz vein deposits in mica-schist at several localities in the Austrian Alps; in granite pegmatite in Bavaria, Ireland, Norway, and as fine crystals in Tanzania; also in placer deposits in the south Ural Mountains, Russia, and in Guyana.

BEST REF. IN ENGLISH: Vlasov, K. A., "Mineralogy of Rare Elements," v. II, p. 85–87, Israel Program for Scientific Translations, 1966.

# EUCOLITE = Variety of eudialyte rich in calcium

# EUCRASITE = Variety of thorite high in rare earths

# EUCRYPTITE
$LiAlSiO_4$

CRYSTAL SYSTEM: Hexagonal
CLASS: $\bar{3}$
SPACE GROUP: $R\bar{3}$
Z: 18
LATTICE CONSTANTS:
 $a = 13.48$
 $c = 9.001$
3 STRONGEST DIFFRACTION LINES:
 3.362 (100)
 3.965 ( 90)
 2.733 ( 70)
OPTICAL CONSTANTS:
 $\omega = 1.572$
 $\epsilon = 1.586$
 (+)
HARDNESS: 6½
DENSITY: 2.657 (Meas.)
 2.661 (Calc.)
CLEAVAGE: {10$\bar{1}$1} indistinct
Fracture conchoidal. Brittle.

HABIT: As single euhedral crystals up to 3 × 2 × 2 cm in size. Prominent forms {0001}, {10$\bar{1}$0}, and {11$\bar{2}$0}; others present as small truncating faces. Usually in massive granular aggregates.

COLOR-LUSTER: Colorless, white. Transparent. Vitreous. Fluoresces pink under ultraviolet light.

MODE OF OCCURRENCE: Occurs sparingly, intergrown with albite, at Center Stafford, New Hampshire and Branchville, Connecticut; in pegmatite at the Harding mine, Dixon, New Mexico; and as a major lithium ore mineral in pegmatite at Bikita, Rhodesia.

BEST REF. IN ENGLISH: Hurlbut, Cornelius S. Jr., *Am. Min.*, **47**: 557–561 (1962).

# EUDIALYTE
$Na_4(Ca, Fe, Ce, Mn)_2 ZrSi_6 O_{17}(OH, Cl)_2$
Var. Eucolite

CRYSTAL SYSTEM: Trigonal
CLASS: $\bar{3}$ 2/m
SPACE GROUP: $R\bar{3}m$
Z: 12
LATTICE CONSTANTS:
 $a = 14.34$
 $c = 30.21$
3 STRONGEST DIFFRACTION LINES:
 7.19 (100)
 5.74 ( 80)
 2.87 ( 80)
OPTICAL CONSTANTS:
 $\omega = 1.591–1.623$
 $\epsilon = 1.594–1.633$
 Optic sign variable.
HARDNESS: 5–5½
DENSITY: 2.74–2.98 (Meas.)
 2.82 (Calc. OH = Cl)
CLEAVAGE: {0001} indistinct
Fracture uneven. Brittle.

HABIT: Crystals thick to thin tabular of hexagonal or trigonal habit, short to long prismatic, rhombohedral.

COLOR-LUSTER: Various shades of yellowish brown, brownish red, pink, and red. Translucent; thin fragments transparent. Vitreous to greasy or dull. Streak colorless.

MODE OF OCCURRENCE: Occurs chiefly in nepheline-syenites and their pegmatites in association with microcline, nepheline, aegirine, and other minerals. Found with lamprophyllite in the Bear Paw Mountains, Montana; at Magnet Cove, Hot Spring County, Arkansas; at Mont St. Hilaire, Quebec, and other localities in Canada; as fine crystals in the Julianehaab district, Greenland; at Carlingford, Ireland; in the Langesundfjord district, Norway (eucolite); common in the alkali massifs of the Kola peninsula and elsewhere in the USSR; and at deposits in Madagascar.

BEST REF. IN ENGLISH: Vlasov, K. A., "Mineralogy of Rare Elements," v. II, p. 355-364, Israel Program for Scientific Translations, 1966.


# EUDIDYMITE

$NaBeSi_3O_7OH$
Dimorphous with epididymite

CRYSTAL SYSTEM: Monoclinic
CLASS: 2/m
SPACE GROUP: C2/c
Z: 8
LATTICE CONSTANTS:
 $a = 12.64$
 $b = 7.38$
 $c = 14.02$
 $\beta = 103°43'$
3 STRONGEST DIFFRACTION LINES:
 3.163 (100)
 3.398 ( 80)
 3.074 ( 80)
OPTICAL CONSTANTS:
 $\alpha = 1.536-1.542$
 $\beta = 1.541-1.544$
 $\gamma = 1.542-1.548$
 (+)2V = 30-32°
 HARDNESS: 6-7
 DENSITY: 2.551-2.554 (Meas.)
 2.577 (Calc.)
 CLEAVAGE: {001} perfect
 {100} distinct
Brittle.

HABIT: Crystals tabular, up to $2 \times 2 \times 0.5$ cm. Also finely crystalline to cryptocrystalline; coarsely crystalline-micaceous; spherulitic with radial-fibrous structure. Twinning on {001}, common, lamellar; sometimes as swallow-tail twins.

COLOR-LUSTER: Usually white or colorless; also various shades of gray, yellow, blue, violet. Transparent to translucent. Vitreous; pearly on cleavages.

MODE OF OCCURRENCE: Occurs in nepheline-syenite pegmatites in association with albite, elipidite, analcime, natrolite, neptunite, fluorite, and other minerals. Found in the Julianehaab district, Greenland; in the Langesundfjord

district, Norway; and in the Lovozero and Khibiny alkali massifs, Kola peninsula, USSR.

BEST REF. IN ENGLISH: Vlasov, K. A., "Mineralogy of Rare Elements," v. II, p. 135-140, Israel Program for Scientific Translations, 1966.


# EULITE = Variety of hypersthene


# EULYTITE (Eulytine)

$Bi_4Si_3O_{12}$

CRYSTAL SYSTEM: Cubic
CLASS: $\bar{4}$3m
SPACE GROUP: $I\bar{4}3d$
Z: 4
LATTICE CONSTANT:
 $a = 10.27$
3 STRONGEST DIFFRACTION LINES:
 3.20 (100)
 2.70 (100)
 4.13 ( 80)
OPTICAL CONSTANT:
 $N = 2.05$
 HARDNESS: 4½
 DENSITY: 6.6 (Meas.)
 6.76 (Calc.)
 CLEAVAGE: {110} indistinct
Brittle.

HABIT: Crystals tetrahedral, minute. Also in spherical forms.

COLOR-LUSTER: Colorless, yellowish, grayish, dark brown. Transparent to translucent. Vitreous.

MODE OF OCCURRENCE: Occurs as crystals implanted on quartz at Johanngeorgenstadt, and in association with native bismuth near Schneeberg, Saxony, Germany.

BEST REF. IN ENGLISH: Winchell and Winchell, "Elements of Optical Mineralogy," 4th Ed., pt. 2, p. 494, New York, Wiley, 1951.


# EUXENITE

$(Y, Er, Ce, La, U)(Nb, Ti, Ta)_2 (O, OH)_6$
Euxenite-Polycrase series
Var. Tanteuxenite-Ta predominant

CRYSTAL SYSTEM: Orthorhombic
CLASS: 2/m 2/m 2/m
SPACE GROUP: Pcan
Z: 4
LATTICE CONSTANTS:
 $a = 5.52$
 $b = 14.16-14.57$
 $c = 5.16$
3 STRONGEST DIFFRACTION LINES:
 2.99 (100)
 3.66 ( 40) Metamict
 2.95 ( 40)
OTPICAL CONSTANTS:
 Isotropic (metamict)
 $N = 2.06-2.24$

HARDNESS: 5½–6½
DENSITY: 4.30–5.87 depending on Ta content and degree of metamictization and hydration.
CLEAVAGE: None.
Fracture conchoidal. Brittle.
HABIT: Crystals short prismatic, sometimes flattened {010}. Commonly as parallel, subparallel, or divergent crystal aggregates. Also massive, compact. Twinning on {201}, common; twins flattened {010} and striated [100].
COLOR-LUSTER: Black, sometimes with faint brownish or greenish tinge. Opaque; translucent in thin fragments. Brilliant submetallic; sometimes vitreous or resinous. Streak grayish, yellowish, or brownish.
MODE OF OCCURRENCE: Occurs in granite pegmatites often in association with monazite and other rare-earth minerals, biotite, muscovite, magnetite, ilmenite, garnet, and beryl; also as a detrital mineral derived from pegmatites or granitic rocks. Found in Kern, San Bernardino, and Tulare counties, California; as large crystals near Encampment, Carbon County, Wyoming; in Chaffee, Fremont, Jefferson, and Park counties, Colorado; and at Morton, Delaware County, Pennsylvania. Other notable occurrences are in Canada, Greenland, Brazil, Norway, Finland, Zaire, Madagascar, and Australia.
BEST REF. IN ENGLISH: Palache, et al., "Dana's System of Mineralogy," 7th Ed., v. I, p. 787–792, New York, Wiley, 1944.

# EVANSITE (Inadequately described mineral)
$Al_3PO_4(OH)_6 \cdot 6H_2O(?)$

CRYSTAL SYSTEM: Unknown
OPTICAL CONSTANT:
$N \sim 1.445–1.485$
HARDNESS: 3–4
DENSITY: 1.8–2.2
CLEAVAGE: Fracture conchoidal. Very brittle.
HABIT: Massive; stalactitic, botryoidal, or reniform. Opal-like.
COLOR-LUSTER: Colorless, white, or tinted yellow, green, or blue; sometimes red to brown due to impurities. Transparent to translucent. Vitreous to resinous or waxy. Streak whitish.
MODE OF OCCURRENCE: Occurs as a secondary mineral often associated with allophane and limonite. Found at Goldburg, Custer County, Idaho; at Coalville and Columbiana in the Coosa coal field, Alabama; and in England, France, Spain, Czechoslovakia, Hungary, Tasmania, and Madagascar.
BEST REF. IN ENGLISH: Palache, et al., "Dana's System of Mineralogy," 7th Ed., v. II, p. 923–924, New York, Wiley, 1951.

# EVEITE
$Mn_2(OH)(AsO_4)$
Manganese analogue of adamite

CRYSTAL SYSTEM: Orthorhombic
CLASS: 2/m 2/m 2/m

SPACE GROUP: Pnnm
Z: 4
LATTICE CONSTANTS:
$a = 8.57$
$b = 8.77$
$c = 6.27$
3 STRONGEST DIFFRACTION LINES:
4.39 (100)
3.058 ( 90)
5.09 ( 80)
OPTICAL CONSTANTS:
$\alpha = 1.700$
$\beta = 1.715$
$\gamma = 1.732$
$(+)2V = 65°$
HARDNESS: 4
DENSITY: 3.76 (Meas.)
CLEAVAGE: {101} fair
HABIT: Tabular or sheaf-like crystals.
COLOR-LUSTER: Apple green. Streak white.
MODE OF OCCURRENCE: Occurs as an open-fissure mineral encrusting cavities and fractures in hausmannite, Fe-Mn oxides, and carbonates at Långban, Sweden.
BEST REF. IN ENGLISH: Moore, Paul B., $Am. Min.$, 55: 319–320 (1970).

# EVENKITE
$C_{24}H_{50}$ (or $C_{21}H_{44}$)

CRYSTAL SYSTEM: Monoclinic
CLASS: 2/m
SPACE GROUP: Probably P2₁/a
Z: 2
LATTICE CONSTANTS:
$a = 7.52$
$b = 4.98$
$c = 32.50$
$\beta \sim 90°$
3 STRONGEST DIFFRACTION LINES:
4.18 (100)
3.74 ( 90)
2.25 ( 80)
HARDNESS: 1
DENSITY: 0.920 (Meas.)
CLEAVAGE: Mica-like, good.
HABIT: Pseudohexagonal tabular crystals; flattened perpendicular to c{100}.
COLOR-LUSTER: Colorless or yellowish; wax-like.
MODE OF OCCURRENCE: Occurs as wax-like crystals in geodes associated with a vein cutting vesicular tuff in the Evenki district, Lower Tunguska River, Siberia.
BEST REF. IN ENGLISH: Strunz, H., and Contag, B., $Am. Min.$, 50: 2109 (1965).

# EWALDITE
$Ba(Ce,Y,Na,Ca)(CO_3)_2$
Dimorphous with mackelveyite

CRYSTAL SYSTEM: Hexagonal
CLASS: 6/m 2/m 2/m

SPACE GROUP: $P6_3mc$
Z: 2
LATTICE CONSTANTS:
  $a = 5.284$
  $c = 12.78$
OPTICAL CONSTANTS:
  $\omega = 1.646$ (Na)
  $\epsilon \leqslant 1.572$
HARDNESS: Not determined.
DENSITY: 3.25 (Meas.)
         3.37 (Calc.)
CLEAVAGE: None apparent.
HABIT: As small, less than 0.5 mm, hemimorphic poly-crystals consisting of two phases in syntactic intergrowth.
COLOR-LUSTER: Bluish green.
MODE OF OCCURRENCE: Occurs as minute crystals in drill core from four subsurface localities, some 20 miles apart, in the Green River formation in Sweetwater County, Wyoming, in association with mackelveyite, labuntsovite, leucosphenite, searlesite, and other typical minerals of the Green River formation.
BEST REF. IN ENGLISH: Donnay, Gabrielle, Donnay, J. D. H., and Preston, H., *Am. Min.*, **56**: 2156 (1971).

# EYLETTERSITE

$(Th, Pb)Al_3(PO_4, SiO_4)_2(OH)_6$ (?)
Crandallite Series

CRYSTAL SYSTEM: Hexagonal
CLASS: $\bar{3}\,2/m$
SPACE GROUP: $R\bar{3}m$
Z: 1
LATTICE CONSTANTS:
  $a = 6.98\text{–}6.99$
  $c = 16.66\text{–}16.72$
3 STRONGEST DIFFRACTION LINES:
  2.95 (100)
  3.51 ( 60)
  5.70 ( 55)
OPTICAL CONSTANTS: Weakly anisotropic.
                   $N = 1.61\text{–}1.66$
HARDNESS: Not determined.
DENSITY: 3.38–3.44 (Meas.)
         3.44, 3.50 (Calc.)
CLEAVAGE: None.
HABIT: Massive; as pulverulent nodules.
COLOR-LUSTER: Creamy white.
MODE OF OCCURRENCE: The mineral occurs in the Kobokobo pegmatite, Kivu, Zaire. It is derived by the alteration of thorian crandallite.

BEST REF. IN ENGLISH: Van Wambeke, L., *Am. Min.*, **56**: 1366-1384 (1971). Van Wambeke, L., *Min. Abs.*, **23**: 317 (72-3341) (1972).

# EZCURRITE

$Na_4B_{10}O_{17}\cdot 7H_2O$

CRYSTAL SYSTEM: Triclinic
CLASS: $\bar{1}$
SPACE GROUP: $P\bar{1}$
Z: 1
LATTICE CONSTANTS:
  $a = 8.598$     $\alpha = 102°45'$
  $b = 9.570$     $\beta = 107°30'$
  $c = 6.576$     $\gamma = 71°31'$
3 STRONGEST DIFFRACTION LINES:
  6.936 (100)
  3.074 ( 38)
  4.494 ( 29)
OPTICAL CONSTANTS:
  $\alpha = 1.468$
  $\beta = 1.507$ (Na)
  $\gamma = 1.529$
  $(-)2V = 73.5°$
HARDNESS: 3 parallel $c$
              3½ perpendicular $c$
DENSITY: 2.053 (Meas.)
         2.049 (Calc.)
CLEAVAGE: $\{110\}$ excellent
          $\{010\}$ good
          $\{100\}$ fair
          $\{1\bar{2}6\}$ fair
          $\{1\bar{1}0\}$ poor
          $\{101\}$ poor
HABIT: Crystals elongated on [001] and terminated by three major faces. Forms $\{001\}$, $\{010\}$, $\{100\}$, $\{110\}$, $\{1\bar{1}0\}$, $\{101\}$, $\{\bar{1}01\}$, and $\{\bar{1}\bar{2}6\}$. Crystals up to 0.5 mm in length. Also cleavable masses with bladed fibrous structure, sometimes radiating, up to 7 cm long and 1.5 cm wide.
COLOR-LUSTER: Colorless, transparent; vitreous to satiny.
MODE OF OCCURRENCE: Occurs associated with borax and kernite at the Tincalayu borax mine, Salta Province, Argentina.
BEST REF. IN ENGLISH: Muessig, Siegfried, and Allen, Robert D., *Econ. Geol.*, **52**: 426-437 (1957). Hurlbut, C. S. Jr., and Aristarian, L. F., *Am. Min.*, **52**: 1048-1059 (1967).

# F

## FABIANITE
$CaB_3O_5OH$

CRYSTAL SYSTEM: Monoclinic
CLASS: 2/m
SPACE GROUP: $P2_1/a$
Z: 1
LATTICE CONSTANTS:
  $a = 6.593$
  $b = 10.488$
  $c = 6.365$
  $\beta = 113°23'$
3 STRONGEST DIFFRACTION LINES:
  3.269 (100)
  2.920 ( 87)
  3.032 ( 83)
OPTICAL CONSTANTS:
  $\alpha = 1.6085$
  $\beta = 1.6375$
  $\gamma = 1.6500$
  (-)2V = 65°
HARDNESS: 6
DENSITY: 2.77 (Meas.)
        2.788 (Calc.)
CLEAVAGE: {011}
HABIT: Prismatic crystals, 0.3–25 mm in size. Forms {001}, {100}, {120}, {110}, {320}, {011}, {021}.
COLOR-LUSTER: Colorless; vitreous. Fluorescent brownish yellow under ultraviolet light.
MODE OF OCCURRENCE: Occurs in a rocksalt drill core associated with halite, anhydrite, howlite, and szaibelyite at Rehden near Diepholz, Germany.
BEST REF. IN ENGLISH: *Am. Min.*, 48: 212–213 (1963).   Erd, Richard C., Eberlein, G. Donald, and Christ, C. L., *Can. Min.*, **10** (Pt. 1): 108–112 (1969).


## FAHEYITE
$(Mn,Mg,Na)Be_2 Fe_2^{3+}(PO_4)_4 \cdot 6H_2O$

CRYSTAL SYSTEM: Hexagonal
CLASS: 622
SPACE GROUP: $P6_2 22$
Z: 3
LATTICE CONSTANTS:
  $a = 9.43$
  $c = 16.00$
3 STRONGEST DIFFRACTION LINES:
  5.72 (100)
  7.28 ( 90)
  3.244 ( 60)
OPTICAL CONSTANTS:
  $\omega = 1.631$
  $\epsilon = 1.652$
    (+)
HARDNESS: Not determined.
DENSITY: 2.660 (Meas.)
        2.670 (Calc.)
CLEAVAGE: Perfect, parallel to *c*-axis.
HABIT: Tufts, rosettes, and botryoidal masses of fibers. Fibers elongated parallel to *c*-axis. Individual fibers average about 0.08 mm in length and 0.01 mm in thickness.
COLOR-LUSTER: White, Bluish white, brownish white.
MODE OF OCCURRENCE: Occurs coating other minerals, such as muscovite, quartz, variscite, and frondelite at the Sapucaia pegmatite mine, Minas Geraes, Brazil.
BEST REF. IN ENGLISH: Lindberg, Mary Louise, and Murata, K. J., *Am. Min.*, 38: 263–270 (1953).


## FAIRCHILDITE
$K_2 Ca(CO_3)_2$

CRYSTAL SYSTEM: Hexagonal
CLASS: 6/m 2/m 2/m
SPACE GROUP: P6/mmc
Z: 2
LATTICE CONSTANTS:
  $a = 5.272$
  $c = 13.280$
3 STRONGEST DIFFRACTION LINES:
  3.19 (100)
  2.64 ( 80)
  6.64 ( 50)

OPTICAL CONSTANTS:
$\omega = 1.530$
$\epsilon = 1.48$
$(-)$
HARDNESS: Not determined.
DENSITY: 2.446 (Meas.)
CLEAVAGE: {0001} good
HABIT: As microscopic hexagonal plates flattened on {0001}.
COLOR-LUSTER: Colorless.
MODE OF OCCURRENCE: Occurs at many localities in the western United States as clinkers formed by the fusion of wood ash in partly burned fir, hemlock, and other trees.
BEST REF. IN ENGLISH: Milton, Charles, and Axelrod, Joseph M., *Am. Min.*, 32: 607 (1947). Mrose, Mary E., Rose, H. J. Jr., and Marinkenko, J. W., Geol. Soc. Amer. Program Ann. Meet., Nov. 14–16, 1966, p. 146 (1966) (abstr.).

# FAIRFIELDITE

$Ca_2(Mn^{2+},Fe^{2+})(PO_4)_2 \cdot 2H_2O$
Manganese analogue of messelite

CRYSTAL SYSTEM: Triclinic
CLASS: $\bar{1}$
SPACE GROUP: $P\bar{1}$
Z: 1
LATTICE CONSTANTS:
$a = 5.77$ kX      $\alpha = 102°05'$
$b = 6.56$            $\beta = 108°42\frac{1}{2}'$
$c = 5.47$            $\gamma = 90°05\frac{1}{2}'$
3 STRONGEST DIFFRACTION LINES:
3.23 (100)
6.40 ( 90)
3.03 ( 80)
OPTICAL CONSTANTS:
$\alpha = 1.636–1.640$
$\beta = 1.644–1.650$
$\gamma = 1.654–1.660$
$(+)2V \simeq 86°$
HARDNESS: $3\frac{1}{2}$
DENSITY: 3.08 (Meas.)
3.09 (Calc.)
CLEAVAGE: {001} perfect
{010} good
{1$\bar{1}$0} distinct
Fracture uneven. Brittle.
HABIT: Crystals prismatic to equant. Usually lamellar, foliated, or fibrous; in radiating masses.
COLOR-LUSTER: White; also yellowish, greenish white. Transparent. Vitreous to pearly. Streak white.
MODE OF OCCURRENCE: The mineral occurs in granite pegmatite at Branchville, Fairfield County, Connecticut; at Buckfield and Poland, Maine; and at Hühnerkobel, Bavaria, Germany.
BEST REF. IN ENGLISH: Palache, et al., "Dana's System of Mineralogy," 7th Ed., v. II, p. 720–722, New York, Wiley, 1951.

# FAMATINITE

$Cu_3SbS_4$

CRYSTAL SYSTEM: Tetragonal
CLASS: $\bar{4}$ 2m
SPACE GROUP: I$\bar{4}$2m
Z: 2
LATTICE CONSTANTS:
$a = 5.38$
$c = 10.76$
3 STRONGEST DIFFRACTION LINES:
3.08 (100)
1.89 ( 70)
1.61 ( 50)
HARDNESS: $\sim 3.5$
DENSITY: 4.635 (Meas.)
4.660 (Calc.)
CLEAVAGE: {101} good
{100} distinct
Fracture uneven to conchoidal. Brittle.
HABIT: Usually massive, fine-grained. More rarely coarse-grained. Crystals distinct equant, rare; usually rough with some or all faces curved and etched. Polished section exhibits polysynthetic twinning.
COLOR-LUSTER: Deep pinkish brown with grayish tinge. Dull metallic. Opaque. Streak black.
MODE OF OCCURRENCE: Found in low- to medium-intensity copper deposits often associated with enargite, tetrahedrite-tennantite, pyrite, chalcopyrite, covellite; and more rarely sphalerite, bismuthinite, ruby silvers, native silver, gold, and marcasite. The gangue minerals barite and drusy quartz are so common as to be almost characteristic. The most important localities are Mankayan, Luzon; Famatina, Argentina; Goldfield, Nevada; Hokuetsu, Japan; Kinkwaseki, Taiwan; Cerro de Pasco and Morococha, Peru; Butte, Montana; and Cananea, Sonora, Mexico.
BEST REF. IN ENGLISH: Gaines, Richard V., *Am. Min.*, 42: 766–779 (1957).

# FARALLONITE (Species status uncertain.)

$2MgO \cdot W_2O_5 \cdot SiO_2 \cdot nH_2O(?)$

CRYSTAL SYSTEM: Monoclinic (?)
HARDNESS: Not determined.
DENSITY: Not determined.
CLEAVAGE: {100} good
Fracture flat conchoidal.
HABIT: Massive, cryptocrystalline.
COLOR-LUSTER: Whitish to sky blue. Translucent. Waxy to pearly.
MODE OF OCCURRENCE: Occurs intimately mixed with anthoinite as an alteration product of wolframite. Found on outcrops during the rainy season at the Farallon mine, Tasna, Bolivia; also from the Zongo district, Bolivia.
BEST REF. IN ENGLISH: Kohanowski, N. N., *Mines Mag.* (Colo.), 43 (2): 17–24, 51, 59 (1953).

# FAROELITE = Variety of thompsonite

# FARRINGTONITE
$Mg_3(PO_4)_2$

CRYSTAL SYSTEM: Monoclinic
CLASS: 2/m
SPACE GROUP: $P2_1/n$
Z: 2
LATTICE CONSTANTS:
  $a = 7.60$
  $b = 8.23$
  $c = 5.08$
  $\beta = 94°05'$
3 STRONGEST DIFFRACTION LINES:
  3.44 (100)
  3.85 ( 90)
  4.08 ( 50)
OPTICAL CONSTANTS:
  $\alpha = 1.540$
  $\beta = 1.544$
  $\gamma = 1.559$
  $(+)2V = 54°-55°$
HARDNESS: Not determined.
DENSITY: 2.76 (Calc.)
CLEAVAGE: {100} fair to good
         {010} fair to good
HABIT: Massive.
COLOR-LUSTER: Wax-white to yellow.
MODE OF OCCURRENCE: Occurs peripheral to olivine nodules in the Springwater pallasite meteorite, from near Springwater, Saskatchewan, Canada.
BEST REF. IN ENGLISH: Dufresne, E. R., and Roy, S. K., *Am. Min.*, **46**: 1513 (1961).

# FASSAITE
$Ca(Mg, Fe^{3+}, Al)(Si, Al)_2O_6$
Pyroxene Group

CRYSTAL SYSTEM: Monoclinic
CLASS: 2/m
SPACE GROUP: C2/c
Z: 4
LATTICE CONSTANTS:
  $a \simeq 9.71$
  $b \simeq 8.86$
  $c \simeq 5.26$
  $\beta \simeq 106°$
OPTICAL CONSTANTS:
  $\alpha = 1.676-1.712$
  $\beta = 1.683-1.719$
  $\gamma = 1.702-1.736$
  $(+)2V = 51°-62°$
HARDNESS: 6
DENSITY: 2.96-3.34 (Meas.)
CLEAVAGE: {110} good
         {100} parting
Fracture uneven to conchoidal. Brittle.
HABIT: Crystals short prismatic resembling epidote. Also massive, as disseminated grains. Twinning on {100} simple, lamellar.

COLOR-LUSTER: Pale to dark green, black. Translucent to nearly opaque. Vitreous to dull.
MODE OF OCCURRENCE: Occurs chiefly in contact zones or in metamorphosed limestones and dolomites. Found in association with spinel, garnet, and clintonite, near Helena, Montana; in the Fassa valley, Trentino, Italy; and at localities in Scotland, Sweden, Ceylon, and elsewhere.
BEST REF. IN ENGLISH: Deer, Howie, and Zussman, "Rock Forming Minerals," v. 2, p. 161-166, New York, Wiley, 1963.    Knopf, A., and Lee, D. E., *Am. Min.*, **42**: 73-77 (1957).

# FAUJASITE
$(Na_2, Ca)Al_2Si_4O_{12} \cdot 6H_2O$
Zeolite Group

CRYSTAL SYSTEM: Cubic
CLASS: $4/m\ \overline{3}\ 2/m$
SPACE GROUP: Fd3m
Z: 40
LATTICE CONSTANT:
  $a \simeq 24.65$
3 STRONGEST DIFFRACTION LINES:
  14.3 (100)
  5.66 (100)
  3.76 (100)
OPTICAL CONSTANT:
  $N \simeq 1.48$
HARDNESS: 5
DENSITY: 1.92 (Meas.)
CLEAVAGE: {111} distinct
HABIT: Crystals octahedral. Twinning on {111} common.
COLOR-LUSTER: Colorless, white, or stained by impurities. Transparent in thin section. Vitreous.
MODE OF OCCURRENCE: Occurs with augite in a limburgite at Sasbach in the Kaiserstuhl, Baden, Germany; also found in association with other zeolites in the Aar and St. Gotthard massifs of Switzerland.
BEST REF. IN ENGLISH: Deer, Howie, and Zussman, "Rock Forming Minerals," v. 4, p. 392, 397, New York, Wiley, 1963.

# FAUSERITE = Manganoan epsomite

# FAUSTITE (Inadequately described mineral)
$(Zn, Cu)Al_6(PO_4)_4(OH)_8 \cdot 5H_2O$

CRYSTAL SYSTEM: Triclinic
3 STRONGEST DIFFRACTION LINES:
  3.68 (100)
  2.89 ( 80)
  6.70 ( 70)
HARDNESS: 5½
DENSITY: 2.92 (Meas.)
CLEAVAGE: Fracture slightly conchoidal to smooth. Brittle.

HABIT: Massive, compact.

COLOR-LUSTER: Apple green; waxy to dull. Opaque. Streak white to pale yellow-green.

MODE OF OCCURRENCE: Occurs as a vein filling and as nodules in altered shales at Copper King mine, Maggie Creek district, Eureka County, Nevada.

BEST REF. IN ENGLISH: Erd, Richard C., Foster, Margaret D., and Proctor, Paul D., *Am. Min.*, **38**: 964-972 (1953).

# FAYALITE
$Fe_2SiO_4$ (90-100 Atomic percent $Fe^{2+}$)
Olivine Group

CRYSTAL SYSTEM: Orthorhombic
CLASS: 2/m 2/m 2/m
SPACE GROUP: Pbnm
Z: 4
LATTICE CONSTANTS:
  $a = 4.817$
  $b = 10.477$
  $c = 6.105$
3 STRONGEST DIFFRACTION LINES:
  2.831 (100)
  2.501 ( 70)
  2.566 ( 50)
OPTICAL CONSTANTS:
  $\alpha = 1.827$
  $\beta = 1.869$
  $\gamma = 1.879$
  $(-)2V = 134°$
HARDNESS: 7
DENSITY: 4.32 (Meas.)
        4.317 (Calc.)
CLEAVAGE: {010} imperfect
          {100} imperfect
Fracture conchoidal. Brittle.

HABIT: Crystals usually thick tabular, often with wedge-shaped terminations; small. Usually massive, compact or granular. Twinning on {100}, not common.

COLOR-LUSTER: Greenish yellow, yellowish brown, brown. Transparent to translucent. Vitreous to greasy. Streak uncolored.

MODE OF OCCURRENCE: Occurs in small amounts in many acid and alkaline volcanic and plutonic rocks; in quartz syenites; in lithophysae in obsidian and in cavities in liparite; and in metamorphosed iron-rich sediments. Typical occurrences are found in Imperial, Inyo, Riverside, and Siskiyou counties, California; in the Cheyenne Mountain area, El Paso County, Colorado; at Obsidian Cliff, Yellowstone Park, Wyoming; at Rockport, Essex County, Massachusetts; and at Salt Lake Crater, Oahu, Hawaii. Also at localities in Sweden, France, Sicily, Germany, Finland, New Caledonia, and elsewhere.

BEST REF. IN ENGLISH: Deer, Howie, and Zussman, "Rock Forming Minerals," v. 1, p. 1-33, New York, Wiley, 1962.

# FEDORITE (Inadequately described mineral)
$(Na, K)Ca(Si, Al)_4(O, OH)_{10} \cdot 1.5H_2O$

CRYSTAL SYSTEM: Monoclinic (?)
3 STRONGEST DIFFRACTION LINES:
  2.93 (100b)
  2.97 ( 90 )
  1.826 ( 90b)
OPTICAL CONSTANTS:
  $\alpha = 1.522$
  $\beta = 1.530$
  $\gamma = 1.531$
  $(-)2V = 32°$
HARDNESS: Not determined.
DENSITY: 2.58 (Meas.)
CLEAVAGE: {001} micaceous
HABIT: Pseudohexagonal tabular crystals, resembling muscovite.

COLOR-LUSTER: Colorless to pale raspberry red; vitreous.

MODE OF OCCURRENCE: Occurs as fine veinlets, associated with narsarsukite, and partly replaced by quartz and apophyllite, in fenitized sandstone of the Tur'yii Peninsula, Kola, USSR.

BEST REF. IN ENGLISH: Kukharenko, A. A., et al., *Am. Min.*, **52**: 561-562 (1967).

# FEDOROVITE = Variety of diopside

# FEITKNECHTITE (Inadequately described mineral)
$\beta$-MnO(OH)
Trimorphous with manganite and groutite

CRYSTAL SYSTEM: Hexagonal
3 STRONGEST DIFFRACTION LINES:
  4.62 (100)
  2.64 ( 50)
  2.36 ( 20) broad
HARDNESS: Not determined.
DENSITY: 3.80 (Calc.)
CLEAVAGE: Not determined.
HABIT: Crystals hexagonal platelets, submicroscopic.

COLOR-LUSTER: Iron black to brownish black. Translucent to nearly opaque. Dull.

MODE OF OCCURRENCE: Occurs in intimate intergrowth with hausmannite as an oxidation product of pyrochroite at Franklin, Sussex County, New Jersey; at Långban and Pajsberg, Sweden; and at the Noda-Tamagawa mine, Japan.

BEST REF. IN ENGLISH: Bricker, Owen, *Am. Min.*, **50**: 1313-1318 (1965).

# FELDSPAR = Group name for aluminosilicates of Na, K, Ca, and Ba
See: Albite          Hyalophane
     Andesine        Labradorite
     Anorthite       Microcline
     Anorthoclase    Oligoclase
     Bytownite       Orthoclase
     Celsian         Paracelsian

## FELSÖBÁNYITE (Felsöbányaite)(Inadequately described mineral)
$Al_4SO_4(OH)_{10} \cdot 5H_2O$

CRYSTAL SYSTEM: Probably orthorhombic
3 STRONGEST DIFFRACTION LINES:
   4.78 (100)
   4.63 (100)
   2.27 ( 50)
HARDNESS: 1½
DENSITY: 2.35 (Meas.)
CLEAVAGE: {001} perfect,
            also {100}, {010} good
HABIT: Spherulitic aggregates of concentrically arranged thin tabular crystals.
COLOR-LUSTER: Colorless, yellow, white; vitreous, pearly on cleavage surfaces.
MODE OF OCCURRENCE: Occurs associated with marcasite, stibnite, barite, and quartz at Fëlsobánya, Roumania.
BEST REF. IN ENGLISH: Koch, S., and Sarudi, I., *Am. Min.*, **50**: 812 (1965).

## FENAKSITE
$(K, Na, Ca)_4(Fe^{2+}, Fe^{3+}, Mg, Mn)_2(Si_4O_{10})_2(OH, F)$

CRYSTAL SYSTEM: Triclinic
CLASS: $\bar{1}$
SPACE GROUP: $P\bar{1}$
Z: 1
LATTICE CONSTANTS:
   $a = 8.20$      $\alpha = 98°58'$
   $b = 9.97$      $\beta = 114°47'$
   $c = 6.97$      $\gamma = 105°02'$
3 STRONGEST DIFFRACTION LINES:
   3.03 (100)
   2.46 ( 70)
   3.55 ⎫
   3.44 ⎭ (70) doublet
OPTICAL CONSTANTS:
   $\alpha = 1.541$
   $\beta = 1.560$
   $\gamma = 1.567$
(+)2V = 84°
HARDNESS: 5-5½
DENSITY: 2.744 (Meas.)
CLEAVAGE: {100}, {010} good
HABIT: Grains up to 2-4 cm in diameter.
COLOR-LUSTER: Light rose; pearly on cleavages.
MODE OF OCCURRENCE: Occurs in pegmatite in the Khibina Tundra, USSR associated with canasite and lamprophyllite.
BEST REF. IN ENGLISH: Dorfman, M. D., Rogachev, D. D., Goroshchenko, Z. I., and Mokretsova, A. V., *Am. Min.*, **45**: 252-253 (1960).

## FENGHUANGLITE = Variety of britholite containing thorium.

## FERBERITE
$FeWO_4$
Wolframite Series

CRYSTAL SYSTEM: Monoclinic
CLASS: 2/m
SPACE GROUP: P2/c
Z: 2
LATTICE CONSTANTS:
   $a = 4.730$
   $b = 5.703$
   $c = 4.952$
   $\beta = 90°$
3 STRONGEST DIFFRACTION LINES:
   2.940 (100) broad
   4.736 ( 40)
   3.745 ( 35)
OPTICAL CONSTANTS:
   $\alpha = 2.255$
   $\beta = 2.305$ (Li)
   $\gamma = 2.414$
(+)2V = 68°
HARDNESS: 4-4½
DENSITY: 7.51 (Meas.)
         7.518 (Calc. x-ray)
CLEAVAGE: {010} perfect
Fracture uneven. Brittle.
HABIT: Crystals commonly wedge-shaped; also short prismatic and flattened {100}; usually striated. Also as groups of bladed crystals; massive. Twinning on {100} and {023} common.
COLOR-LUSTER: Black. Opaque. Submetallic. Streak brownish black to black.
MODE OF OCCURRENCE: Occurs chiefly in high-temperature hydrothermal ore veins and quartz veins in or near granitic rocks, and in medium-temperature vein deposits. It is found in large quantity, often as beautifully developed crystals, in the Nederland tungsten belt of Boulder County, Colorado, and less abundantly in Clear Creek and Gilpin counties. Other occurrences are found in Idaho, South Dakota, New Mexico, Arizona, Greenland, Bolivia, France, Spain, Germany, and Australia.
BEST REF. IN ENGLISH: Palache, et al., "Dana's System of Mineralogy," 7th Ed., v. II, p. 1064-1072, New York, Wiley, 1951.

## FERDISILICITE
$FeSi_2$

CRYSTAL SYSTEM: Tetragonal
LATTICE CONSTANTS:
   $a = 2.69$
   $c = 5.08$
3 STRONGEST DIFFRACTION LINES:
   1.846 (100)
   2.371 ( 90)
   1.775 ( 90)
HARDNESS: ~6.25
Microhardness 707-811 (759 av.) kg/mm².
DENSITY: 5.05 (Calc.)

CLEAVAGE: None.
Conchoidal fracture. Brittle.
HABIT: Grains 0.1–3 mm in size. Serrated, specular, dendritic with only traces of crystal form.
COLOR-LUSTER: Steel gray. Metallic. Opaque.
MODE OF OCCURRENCE: Found in placers and drill core samples of sandstones of the Poltava series near Zachativsk station, Donets region, USSR, associated with fersilicite. Also found in a drill core of epidote amphibolite in the Surskii region, USSR.
BEST REF. IN ENGLISH: Gevork'yan, V. Kh., Litvin, A. L., and Povarennykh, A. S., *Am. Min.*, **54**: 1737 (1969).

## FERGHANITE (Inadequately described mineral)
$LiH(UO_2/OH)_4(VO_4)_2 \cdot 2H_2O$

CRYSTAL SYSTEM: Probably orthorhombic
HARDNESS: ~2½
DENSITY: 3.31 (Meas.)
CLEAVAGE: {001} perfect
Also a second prismatic cleavage.
HABIT: Minute scales with six-sided outline and flattened on {001}.
COLOR-LUSTER: Sulfur yellow; translucent; waxy.
MODE OF OCCURRENCE: Occurs associated with other secondary uranium minerals, possibly as an alteration product of tyuyamunite, at Tyuya Muyun, Ferghana district, Turkestan, USSR.
BEST REF. IN ENGLISH: Frondel, Clifford, USGS Bull. 1064, p. 260–261 (1958).

## FERGUSONITE
$(Y, Er, Ce, Fe)(Nb, Ta, Ti)O_4$
Fergusonite Series

CRYSTAL SYSTEM: Tetragonal
CLASS: 4/m
SPACE GROUP: $I4_1/a$
Z: 4
LATTICE CONSTANTS:
$a$ = 5.16
$c$ = 10.89
3 STRONGEST DIFFRACTION LINES:
3.12 (100) Metamict
2.96 ( 90)
1.901 ( 50)
OPTICAL CONSTANTS:
Isotropic (metamict)
$N$ = 2.05–2.19
HARDNESS: 5½–6½
DENSITY: 5.38 (Calc. for $YNbO_4$)
CLEAVAGE: {111} in traces
Fracture subconchoidal. Brittle.
HABIT: Crystals short to long prismatic; also pyramidal. Usually as irregular masses or grains.
COLOR-LUSTER: Brownish black to black; crystal surfaces usually brown, gray, or yellow due to alteration. Opaque, except in thin fragments. Vitreous, submetallic; externally dull. Streak variable; greenish gray, yellowish brown, brown.

MODE OF OCCURRENCE: Occurs associated with euxenite, monazite, gadolinite, and other rare-earth minerals in granitic pegmatites; less commonly in placer deposits. Found in Riverside and San Diego counties, California; at Suppington, Gallatin County, Montana; in El Paso and Jefferson counties, Colorado; at Baringer Hill, Llano County, Texas; and at localities in Massachusetts, Connecticut, North Carolina, South Carolina, and Virginia. Other notable occurrences are found in Greenland, Norway, Sweden, Poland, Finland, USSR, Madagascar, Japan, and elsewhere.
BEST REF. IN ENGLISH: Palache, et al., "Dana's System of Mineralogy," 7th Ed., v. I, p. 757–763, New York, Wiley, 1944.

## β-FERGUSONITE
Polymorph of fergusonite

CRYSTAL SYSTEM: Monoclinic
CLASS: 2
SPACE GROUP: I2
Z: 4
LATTICE CONSTANTS:
$a$ = 5.12
$b$ = 10.89
$c$ = 5.20
$\beta$ = 88°10'
HABIT: Crystals 0.05–0.2 mm in size.
COLOR-LUSTER: Light yellow.
MODE OF OCCURRENCE: Occurs in the apical parts of microcline granite stocks, Central Asia.
BEST REF. IN ENGLISH: Gorshevskaya, S. A., Sidorenko, G. A., and Smorchkov, I. E., *Am. Min.*, **46**: 1516–1517 (1961).

## FERMORITE
$(Ca, Sr)_5\{(As, P)O_4\}_3(F, OH)$
Apatite Group

CRYSTAL SYSTEM: Hexagonal
CLASS: 6/m
SPACE GROUP: $P6_3/m$
Z: 2
LATTICE CONSTANTS:
$a$ = 9.55
$c$ = 6.98
3 STRONGEST DIFFRACTION LINES:
2.86 (100)
2.75 ( 60)
3.49 ( 50)
OPTICAL CONSTANTS:
Mean index = 1.660
uniaxial (–)
HARDNESS: 5
DENSITY: 3.518
CLEAVAGE: Fracture uneven.
HABIT: Massive, granular.
COLOR-LUSTER: Pale pinkish white. Translucent. Greasy. Streak white.

MODE OF OCCURRENCE: Occurs in veinlets in manganese ore at Sitipár, Chindwara district, Central Provinces, India.

BEST REF. IN ENGLISH: Palache, et al., "Dana's System of Mineralogy," 7th Ed., v. II, p. 904, New York, Wiley, 1951.

## FERNANDINITE (Inadequately described mineral)
$CaV_{12}O_{30} \cdot 14H_2O$

CRYSTAL SYSTEM: Unknown
LATTICE CONSTANTS:
  $a = 11.69$
  $b = 3.674$
HARDNESS: Not determined.
DENSITY: Not determined.
CLEAVAGE: Not determined.
HABIT: Massive, cryptocrystalline to fibrous; rarely as angular plates or flakes flattened {001}.
COLOR-LUSTER: Dull green.
MODE OF OCCURRENCE: Occurs in the oxidized zone of the vanadium deposit at Minasragra, near Cerro de Pasco, Peru.
BEST REF. IN ENGLISH: Schaller, W. T., *Washington Acad. Sci. J.*, **5**: 7 (1915).   Ross, Malcolm, *Am. Min.*, **44**: 322–341 (1959).

## FERRAZITE = A mixture

## FERRI-BEIDELLITE = Variety of beidellite

## FERRICOPIAPITE = Ferrian copiapite (oxidized copiapite)

## FERRIERITE
$(Na, K)_2 MgAl_3 Si_{15} O_{36} (OH) \cdot 9H_2O$
Zeolite Group

CRYSTAL SYSTEM: Orthorhombic
CLASS: 2/m 2/m 2/m
SPACE GROUP: Immm
Z: 2
LATTICE CONSTANTS:
  $a = 14.14$
  $b = 19.12$
  $c = 7.48$
3 STRONGEST DIFFRACTION LINES:
  3.537 (100)
  3.781 ( 65)
  9.51 ( 50)
OPTICAL CONSTANTS:
    $\alpha = 1.478$
    $\beta = 1.479$
    $\gamma = 1.482$
  (+)2V $\simeq$ 50°
HARDNESS: 3–3½
DENSITY: 2.136 (Meas.)
        2.11 (Calc.)

CLEAVAGE: {100} perfect
HABIT: Crystals thin tabular; in radiating groups.
COLOR-LUSTER: Colorless, white. Transparent to translucent. Vitreous to pearly.
MODE OF OCCURRENCE: Occurs at Leavitt Lake, Sonora Pass, Mono County, and associated with clinoptilolite in cavities in an andesitic breccia near Agoura, Los Angeles County, California; with chalcedony and occasionally opal in seams in basalt in a railroad cut on the north shore of Kamloops Lake, and with collinsite and carbonate-apatite at Francois Lake, British Columbia, Canada; and in lamellar masses near Albero Bassi, Vicenza, Italy.
BEST REF. IN ENGLISH: Deer, Howie, and Zussman, "Rock Forming Minerals," v. 4, p. 381–383, New York, Wiley, 1963.    Staples, Lloyd W., *Am. Min.*, **40**: 1095–1099 (1955).    Wise, W. S., Nokleberg, W. J., and Kokinos, M., *Am. Min.*, **54**: 887–895 (1969).

## FERRIHALLOYSITE = Variety of halloysite

## FERRIMOLYBDITE
$Fe_2 Mo_3 O_{12} \cdot 8H_2O$ with some $FeMoO_4 OH \cdot 3H_2O$

CRYSTAL SYSTEM: Probably orthorhombic
3 STRONGEST DIFFRACTION LINES:
  7.4  (100)
  2.29 ( 60)
  3.33 ( 50)
OPTICAL CONSTANTS:
  $\alpha = 1.791–1.806$
  $\beta = 1.808–1.827$
  $\gamma = 1.997–2.005$
  (+)2V small, 28°
HARDNESS: 1–2
DENSITY: 4.46 (Meas. synthetic-crystals)
        3.06 (Meas. synthetic-powder)
CLEAVAGE: Not determined.
HABIT: Microcrystalline; as bundles of fine fibers, crusts and aggregates of tufted or radial fibers, and as an earthy coating or powder.
COLOR-LUSTER: Yellow (Munsell-8.0 $Y\frac{5}{8}$).  Adamantine to silky when fibrous; earthy when encrusting. Streak pale yellow.
MODE OF OCCURRENCE: Occurs widespread as an oxidation product of sulfide ores of molybdenum. Typical occurrences include Mt. Mulgine, Western Australia; Kingsgate, New South Wales; Yeniseisk, Siberia; Bivongi, Calabria, Italy; Renfrew, Ontario, Canada; Climax, Telluride, and Hortense, Colorado; Santa Rita Mountains, Arizona; Little Cottonwood Canyon, Utah; at least 15 counties in California; Westmoreland, New Hampshire; and Chester, Pennsylvania.
BEST REF. IN ENGLISH: Kerr, Paul F., Thomas, Arthur W., and Langer, Arthur M., *Am. Min.*, **48**: 14–32 (1963).

## FERRINATRITE
$Na_3 Fe^{3+} (SO_4)_3 \cdot 3H_2O$

CRYSTAL SYSTEM: Hexagonal

CLASS: $\bar{3}$
SPACE GROUP: $P\bar{3}$
Z: 6
LATTICE CONSTANTS:
  $a = 15.57$
  $c = 8.67$
3 STRONGEST DIFFRACTION LINES:
  7.80 (100)
  2.91 ( 80)
  4.38 ( 60)
OPTICAL CONSTANTS:
  $\omega = 1.556$
  $\epsilon = 1.610$
  (+)
HARDNESS: 2½
DENSITY: 2.55-2.61 (Meas.)
          2.55 (Calc.)
CLEAVAGE: $\{10\bar{1}0\}$ perfect
          $\{11\bar{2}0\}$ less perfect
Fracture splintery. Brittle.

HABIT: Crystals short prismatic. As cleavable masses; cryptocrystalline; as fibrous aggregates; and as isolated crystals or radiated groups.

COLOR-LUSTER: Grayish white, pale green to bluish green, purplish; rarely colorless. Transparent. Vitreous.

MODE OF OCCURRENCE: Occurs as a secondary mineral with other sulfates in the Atacama Desert, northern Chile, especially at Chuquicamata, near Quetana, at Alcaparrosa, and at Mina la Compania, south of Sierra Gorda. It also is found as a fumarolic deposit at Mount Vesuvius, Italy, associated with alum and aphthitalite.

BEST REF. IN ENGLISH: Palache, et al., "Dana's System of Mineralogy," 7th Ed., v. II, p. 456-458, New York, Wiley, 1951.

# FERRIRICHTERITE (Chiklite) (Variety of richterite)

$(Na, Ca)_3(Mg, Fe^{3+}, Mn)_5(Si, Al)_8O_{22}(OH)_2$
Amphibole Group

CRYSTAL SYSTEM: Monoclinic
CLASS: 2/m
SPACE GROUP: C2/m
Z: 2
LATTICE CONSTANTS:
  $a \simeq 9.8$
  $b \simeq 17.9$
  $c \simeq 5.3$
  $\beta \simeq 104°$
3 STRONGEST DIFFRACTION LINES:
  2.70 (100)
  2.52 ( 50)
  3.38 ( 40)
OPTICAL CONSTANTS:
  $\alpha = 1.685$
  $\beta = 1.700$
  $\gamma = 1.712$
  $2V = 82°$
HARDNESS: 5-6
DENSITY: 3.44 (Meas.)

CLEAVAGE: $\{110\}$ perfect
         $\{001\}$ parting
         $\{100\}$ parting
Fracture uneven. Brittle.

HABIT: Crystals long prismatic with well-developed $\{100\}$ and $\{110\}$ forms.

COLOR-LUSTER: Deep violet. Translucent. Pearly on cleavage faces. Streak pale violet.

MODE OF OCCURRENCE: Occurs in pegmatite cutting manganese ore and muscovite schist at the Sitasaongi mine, Chikla area, Bhandara district, India.

BEST REF. IN ENGLISH: Bilgrami, S. A., *Min. Mag.*, **30**: 633-644 (1954).

# FERRI-SICKLERITE

$(Li, Fe^{3+}, Mn^{2+})PO_4$
Sicklerite-Ferri-sicklerite Series

CRYSTAL SYSTEM: Orthorhombic
CLASS: 2/m 2/m 2/m
SPACE GROUP: Pmnb
Z: 4
LATTICE CONSTANTS:
  $a = 5.939$ kX
  $b = 10.086$
  $c = 4.787$
3 STRONGEST DIFFRACTION LINES:
  3.01 (100)
  2.53 (100)
  4.32 ( 95)
OPTICAL CONSTANTS:
  $\alpha = 1.750$
  $\beta = 1.770$
  $\gamma = 1.780$
  $(-)2V =$ medium large
HARDNESS: $\sim 4$
DENSITY: 3.2-3.4 (Meas.)
CLEAVAGE: $\{100\}$ good
HABIT: Massive.

COLOR-LUSTER: Yellowish brown to dark brown. Opaque. Dull.

MODE OF OCCURRENCE: Occurs as an alteration of triphylite in the zone of weathering in granite pegmatites. Found at numerous places in the Keystone and Custer districts, Black Hills, South Dakota; in pegmatites in New Hampshire and Maine; at Varuträsk, Sweden; and in France, Germany, and Finland.

BEST REF. IN ENGLISH: Palache, et al., "Dana's System of Mineralogy," 7th Ed., v. II, p. 672-673, New York, Wiley, 1951.

# FERRISYMPLESITE (Inadequately described mineral)

$Fe_3(AsO_4)_2(OH)_3 \cdot 5H_2O$

CRYSTAL SYSTEM: Unknown
HARDNESS: Not determined.
DENSITY: 2.885 (Meas.)
CLEAVAGE: Not determined.
HABIT: As small irregular masses; fibrous.
COLOR-LUSTER: Deep amber brown. Resinous.

MODE OF OCCURRENCE: Occurs intimately admixed with erythrite and annabergite in the Hudson Bay mine, Cobalt, Ontario, Canada.

BEST REF. IN ENGLISH: Walker, T. L., and Parsons, A. L., Univ. Toronto Studies. Geol. Ser. no. 17, p. 16 (1924).

# FERRITUNGSTITE
$Ca_2 Fe_2^{2+} Fe_2^{3+} (WO_4)_7 \cdot 9H_2O$

CRYSTAL SYSTEM: Tetragonal
Z: 1
LATTICE CONSTANTS:
 $a = 10.28$
 $c = 7.28$
3 STRONGEST DIFFRACTION LINES:
 5.94 (100)
 2.966 (100)
 3.10 ( 90)
OPTICAL CONSTANTS:
 $\omega, \epsilon \sim 2.09-2.15$
HARDNESS: Not determined.
DENSITY: 5.2 (Meas.)
HABIT: Crystals dipyramidal, some exhibiting prismatic faces; also commonly in small fibers. As ocherous powder lining cavities. Single fragments do not exceed 0.025 mm in diameter.
COLOR-LUSTER: Bright yellow, translucent.
MODE OF OCCURRENCE: Occurs in cavities in a limonitic gossan at the Nevada Scheelite mine, Mineral County, Nevada; and also mixed with jarosite at the type locality, the Germania tungsten mine, Stevens County, Washington. It is also reported as an alteration of wolframite in the Cerro Liquinaste, Jujuy, Argentina.
BEST REF. IN ENGLISH: Richter, D. H., Reichen, Laura E., and Lemmon, D. M., Am. Min., 42: 83-90 (1957).

# FERROAUGITE = Variety of augite

# FERROAXINITE (See Axinite Group)
$Ca_2 (Fe, Mn) Al_2 BSi_4 O_{15} (OH)$ with Ca > 1.5; Fe > Mn

# FERROCARPHOLITE
$(Fe, Mg) Al_2 Si_2 O_6 (OH)_4$

CRYSTAL SYSTEM: Orthorhombic
CLASS: 2/m 2/m 2/m
SPACE GROUP: Ccca
Z: 8
LATTICE CONSTANTS:
 $a = 13.77$
 $b = 20.18$
 $c = 5.11$
OPTICAL CONSTANTS:
 $\alpha = 1.628$
 $\beta = 1.644$
 $\gamma = 1.647$
 $2V = 49°$

HARDNESS: 5½
DENSITY: 3.04 (Meas.)
CLEAVAGE: {010} perfect
 {110} indistinct
HABIT: Prismatic crystals, elongated parallel to c-axis. Observed crystal faces {010}, {100}, and {110}. Aggregates of parallel fibers up to 1 cm in length.
COLOR-LUSTER: Dark green.
MODE OF OCCURRENCE: Occurs associated with small quantities of rutile, leucoxene, zircon, and tourmaline, in a cobble of vein-quartz west of Tomata, eastern Central Celebes, Netherlands Indies; also found at Calabria in southern Italy.
BEST REF. IN ENGLISH: de Roever, W. P., Am. Min., 36: 736-745 (1951).

# FERROHEXAHYDRITE
$FeSO_4 \cdot 6H_2O$

CRYSTAL SYSTEM: Monoclinic
CLASS: 2/m
SPACE GROUP: C2/c
Z: 8
LATTICE CONSTANTS:
 $a = 10.08$
 $b = 7.28$
 $c = 24.59$
3 STRONGEST DIFFRACTION LINES:
 4.43 (100)
 2.97 ( 70)
 2.93 ( 70)
OPTICAL CONSTANTS:
 $\alpha = 1.468$
 $\gamma = 1.498$
HARDNESS: Not determined.
DENSITY: 1.934 (Calc.)
CLEAVAGE: Not determined.
HABIT: As fine acicular and capillary crystals, up to 5-6 mm in length; also fibrous. Unstable in the atmosphere.
COLOR-LUSTER: Colorless. Transparent. Vitreous.
MODE OF OCCURRENCE: Occurs at Vesuvius, Italy; also as an alteration product of melanterite in argillites of the Lower Carboniferous in northeastern Tatar, USSR.
BEST REF. IN ENGLISH: Vlasov, V. V., and Kuznetsov, A. V., Am. Min., 48: 433 (1963).

# FERROHORTONOLITE
$(Fe, Mg)_2 SiO_4$ (70-90 atomic percent $Fe^{2+}$)
Olivine Group

CRYSTAL SYSTEM: Orthorhombic
CLASS: 2/m 2/m 2/m
SPACE GROUP: Pbnm
Z: 4
LATTICE CONSTANTS:
 $a = 4.804$
 $b = 10.403$
 $c = 6.070$

3 STRONGEST DIFFRACTION LINES:
 2.83 (100)
 2.50 (100)
 2.57 (100)
OPTICAL CONSTANTS:
 $\alpha = 1.786$
 $\beta = 1.822$
 $\gamma = 1.833$
 $(-)2V = 58°$
HARDNESS: 7
DENSITY: 4.15 (Meas.)
CLEAVAGE: {010} imperfect
 {100} imperfect
Fracture conchoidal. Brittle.
 HABIT: Crystals thick tabular, rough. Usually massive, compact or granular.
 COLOR-LUSTER: Greenish yellow to brown. Transparent to translucent. Vitreous to greasy. Streak uncolored.
 MODE OF OCCURRENCE: Occurs chiefly in basalt, diabase, gabbro, and related rocks; also in meteorites. Typical occurrences are found at Beaver Bay, Minnesota; in the Julianehaab district, Greenland; and at localities in Scotland, Austria, Finland, Uganda, South Africa, Japan, and at Alice Springs, Australia.
 BEST REF. IN ENGLISH: Deer, Howie, and Zussman, "Rock Forming Minerals," v. 1, p. 1-33, New York, Wiley, 1961.

## FERROSALITE = Variety of hedenbergite

## FERROSELITE
$FeSe_2$

CRYSTAL SYSTEM: Orthorhombic
CLASS: 2/m 2/m 2/m
SPACE GROUP: Pnnm
Z: 2
LATTICE CONSTANTS:
 $a = 4.8001$
 $b = 5.776$
 $c = 3.5850$
3 STRONGEST DIFFRACTION LINES:
 2.59 (100)
 2.49 ( 85)
 1.89 ( 80)
HARDNESS: 6-6½
DENSITY: 7.21 (Meas.)
 7.07 (Calc.)
CLEAVAGE: Perfect parallel to elongation. Very brittle.
 HABIT: Crystals prismatic; 0.2-0.5 mm long and up to 0.1 mm in cross section. Form {110} dominant, commonly striated longitudinally. {011} noted. Penetration twinning marked.
 COLOR-LUSTER: Steel gray to tin white with rose tint; tarnishes bronze. Opaque. Metallic. Streak black.
 MODE OF OCCURRENCE: Occurs with minor chalcopyrite and pyrite cementing sandstones and pelites in the Middle Devonian deposits of the Tuvinsk Autonomous Territory; also found at Temple Mountain, Utah.

BEST REF. IN ENGLISH: Buryanova, E. Z., and Komkov, A. I., *Am. Min.*, **41**: 671 (1956). *Geochim. et Cosmochim. Acta*, **15**: 73 (1958).

## FERROSILITE = The component $FeSiO_3$ in the Pyroxene Group

## FERRUCCITE
$NaBF_4$

CRYSTAL SYSTEM: Orthorhombic
CLASS: 2/m 2/m 2/m
SPACE GROUP: Cmcm
Z: 4
LATTICE CONSTANTS:
 $a = 6.82$
 $b = 6.25$
 $c = 6.77$
3 STRONGEST DIFFRACTION LINES:
 3.39 (100)
 3.41 ( 85)
 2.31 ( 40)
OPTICAL CONSTANTS:
 $\alpha = 1.301$
 $\beta = 1.3012$ (Calc.)
 $\gamma = 1.3068$
 $(+)2V = 11°25'$
HARDNESS: ~3
DENSITY: 2.496 (Meas.)
 2.511 (Calc.)
CLEAVAGE: {100}
 {010}
 {001}
HABIT: Crystals minute, tabular. As dense crusts.
COLOR-LUSTER: Colorless, white.

Taste bitter.
 MODE OF OCCURRENCE: Occurs admixed with sassolite and other salts as a sublimation product around fumaroles on Mt. Vesuvius, Italy.
 BEST REF. IN ENGLISH: Palache, et al., "Dana's System of Mineralogy," 7th Ed., v. II, p. 98-99, New York, Wiley, 1951.

## FERSILICITE
$FeSi$

CRYSTAL SYSTEM: Cubic
LATTICE CONSTANT:
 $a = 4.48$
3 STRONGEST DIFFRACTION LINES:
 1.991 (100)
 1.817 ( 90)
 3.143 ( 50)
HARDNESS: ~6.5
Microhardness 776-838 (812 av.) $kg/mm^2$.
DENSITY: 6.18 (Calc.)
CLEAVAGE: None.
Conchoidal fracture. Brittle.

HABIT: Grains 0.1–3 mm in size. Serrated, specular, dendritic with only traces of crystal form.

COLOR-LUSTER: Tin white. Metallic. Opaque.

MODE OF OCCURRENCE: Found in placers and drill core samples of sandstones of the Poltava series near Zachativsk station, Donets region, USSR, associated with ferdisilicite.

BEST REF. IN ENGLISH: Gevork'yan, V. Kh., Litvin, A. L., and Povarennykh, A. S., *Am. Min.*, **54**: 1737 (1969).


# FERSMANITE
$(Na, Ca)_2 (Ti, Nb) Si(O, F)_6$

CRYSTAL SYSTEM: Monoclinic
LATTICE CONSTANTS:
$a:b:c = 0.99113 : 1 : 0.99618$
$\beta = 97°16'$
3 STRONGEST DIFFRACTION LINES:
3.03 (100)
2.80 ( 90)
1.90 ( 60)
OPTICAL CONSTANTS:
$\alpha = 1.873$
$\beta = 1.886$
$\gamma = 1.914–1.939$
$(-)2V = 0°–7°$
HARDNESS: 5–5½
DENSITY: 3.44–3.46 (Meas.)
CLEAVAGE: None.
{001} parting in traces.
Fracture uneven.

HABIT: Crystals usually pseudotetragonal; generally distorted; from 0.2 to 1.2 cm in diameter.

COLOR-LUSTER: Light brown to dark brown. Weakly translucent. Vitreous. Streak white with pale brownish tinge.

MODE OF OCCURRENCE: Occurs in association with pectolite, apatite, sulfides, and other minerals in nepheline-feldspathoid pegmatites. Found only from the Khibiny Massif, Kola peninsula, USSR.

BEST REF. IN ENGLISH: Vlasov, K. A., "Mineralogy of Rare Elements," v. II, p. 564–566, Israel Program for Scientific Translations, 1966.


# FERSMITE
$(Ca, Ce, Na)(Nb, Ti, Fe, Al)_2 (O, OH, F)_6$

CRYSTAL SYSTEM: Orthorhombic
CLASS: 2/m 2/m 2/m
SPACE GROUP: Pcan
Z: 4
LATTICE CONSTANTS:
$a = 5.718$
$b = 14.91$
$c = 5.221$
3 STRONGEST DIFFRACTION LINES:
3.049 (100)
3.762 ( 21)
1.527 ( 15)

OPTICAL CONSTANTS:
$\alpha = 2.07$
$\beta = 2.08$ (Calc.) (Li)
$\gamma = 2.19$
$(+)2V = 20°–25°$
HARDNESS: 4–4½
DENSITY: 4.69–4.79 (Meas.)
4.72 (Meas. synthetic)
CLEAVAGE: None. Fracture uneven to subconchoidal. Brittle.

HABIT: Euhedral prismatic crystals from 0.5 to 1.0 mm across, anhedral grains somewhat larger. Unit prism {110} predominating, less commonly tabular parallel to {100}. Prism faces striated.

COLOR-LUSTER: Dark brown to black; subvitreous to resinous. Streak grayish brown.

MODE OF OCCURRENCE: Occurs as inclusions and intergrowths with a tantalum-free columbite associated with monazite, ancylite, barite, quartz, and apatite in a fine-grained buff-colored marble at the Dark Star claim, Bitterroot Base Line, Ravalli County, Montana. Also from the pegmatites of the Vishnevye Mountains, Central Urals, USSR.

BEST REF. IN ENGLISH: Hess, H. D., and Trumpour, H. J., *Am. Min.*, **44**: 1–8 (1959).


# FERVANITE (Inadequately described mineral)
$Fe_4 V_4 O_{16} \cdot 5H_2O$

CRYSTAL SYSTEM: Probably Monoclinic
LATTICE CONSTANTS:
$a = 9.02$
$c = 6.65$
$\beta = 103°20'$
HARDNESS: Not determined.
DENSITY: Not determined.
CLEAVAGE: None apparent.
HABIT: Fibrous, elongated [001] and flattened {010}.
COLOR-LUSTER: Golden brown, brilliant luster.
MODE OF OCCURRENCE: Occurs in uranium-vanadium deposits at the Hummer mine, Montrose County, and at Gypsum Valley, San Miguel County, Colorado; in the LaSal Mountains, Grand County, Utah; and widespread at other localities in the uranium-vanadium districts of the Colorado Plateau.

BEST REF. IN ENGLISH: Hess, F. L., and Henderson, E. P., *Am. Min.*, **16**: 273–277 (1931). Ross, Malcolm, *Am. Min.*, **44**: 322–341 (1959).


# FIBROFERRITE
$Fe^{3+}SO_4 OH \cdot 5H_2O$

CRYSTAL SYSTEM: Hexagonal
CLASS: 3 or $\bar{3}$
SPACE GROUP: R3 or R$\bar{3}$
Z: 18
LATTICE CONSTANTS:
$a = 24.12$
$c = 7.63$

3 STRONGEST DIFFRACTION LINES:
12.1 (100)
2.98 ( 80)
6.96 ( 60)
OPTICAL CONSTANTS:
$\omega$ = 1.532
$\epsilon$ = 1.570
(+)
HARDNESS: ~2½
DENSITY: 1.92 (Meas.)
2.013 (Calc.)
CLEAVAGE: {0001} perfect

HABIT: As crusts and masses composed of fine fibers; also radial-fibrous; botryoidal. Commonly as trillings.

COLOR-LUSTER: White to pale yellow; also greenish gray, pale green to yellowish green. Silky, pearly, rarely sub-metallic. Decomposed by water.

MODE OF OCCURRENCE: Occurs as a secondary mineral formed by the oxidation of pyrite, commonly associated with melanterite, copiapite, jarosite, and other sulfates. Found as crystals in solution pits in melanterite at the Dexter No. 7 mine, Calf Mesa, San Rafael Swell, Utah; in the Black Iron Mine, Gilman district, Eagle County, Colorado; in Napa, San Bernardino, and Trinity counties, California; and from Genette Mountain, Arizona. It also occurs in veins up to ten feet in width at the Santa Elena mine, La Alcaparrosa, San Juan province, Argentina; and at localities in Canada, Chile, Italy, France, Austria, Czechoslovakia, Cyprus, and USSR.

BEST REF. IN ENGLISH: Palache, et al., "Dana's System of Mineralogy," 7th Ed., v. II, p. 614–616, New York, Wiley, 1951.

## FIBROLITE = Sillimanite

## FICHTELITE = A crystalline hydrocarbon
$C_{18}H_{32}$

## FIEDLERITE
$Pb_3Cl_4(OH)_2$

CRYSTAL SYSTEM: Monoclinic
CLASS: 2/m
SPACE GROUP: $P2_1/a$
Z: 4
LATTICE CONSTANTS:
$a$ = 16.59
$b$ = 8.00
$c$ = 7.19
$\beta$ = 102°12′
3 STRONGEST DIFFRACTION LINES:
3.89 (100)
2.55 (100)
2.81 ( 80)
OPTICAL CONSTANTS:
$\alpha$ = 1.98
$\beta$ = 2.04
$\gamma$ = 2.10
(−)2V = large

HARDNESS: ~3½
DENSITY: 5.88 (Meas.)
5.64 (Calc.)
CLEAVAGE: {100} good
Not brittle.

HABIT: Crystals small, lath-like; tabular {100} and elongated along $c$-axis. Twinning on {100} common.

COLOR-LUSTER: Colorless, white. Transparent. Adamantine.

MODE OF OCCURRENCE: Occurs associated with laurionite, paralaurionite, penfieldite, and other secondary lead minerals in vugs in ancient lead slags exposed to the action of sea water at Laurium, Greece.

BEST REF. IN ENGLISH: Palache, et al., "Dana's System of Mineralogy," 7th Ed., v. II, p. 67–69, New York, Wiley, 1951.

## FILLOWITE
$H_2Na_6(Mn, Fe, Ca)_{14}(PO_4)_{12} \cdot H_2O(?)$

CRYSTAL SYSTEM: Hexagonal
CLASS: $\bar{3}$
SPACE GROUP: $R\bar{3}$
Z: $\simeq 9$
LATTICE CONSTANTS:
$a$ = 15.25      $\alpha_{rh}$ = 16.91
$c$ = 43.32      $\alpha$ = 53°31′
3 STRONGEST DIFFRACTION LINES:
2.814 (100)
3.017 ( 70)
3.640 ( 65)
OPTICAL CONSTANTS:
$\omega$ = 1.671
$\epsilon$ = 1.676
(+)2V = small
HARDNESS: 4½
DENSITY: 3.43 (Meas.)
3.42 (Calc.)
CLEAVAGE: {001} perfect
Fracture uneven. Brittle.

HABIT: Crystals psuedorhombohedral. As granular crystalline masses.

COLOR-LUSTER: Yellowish to reddish brown. Transparent to translucent. Subresinous to greasy. Streak white.

MODE OF OCCURRENCE: Occurs in granite pegmatite associated with reddingite, fairfieldite, and triploidite at Branchville, Connecticut.

BEST REF. IN ENGLISH: Palache, et al., "Dana's System of Mineralogy," 7th Ed., v. II, p. 719–720, New York, Wiley, 1951. *Am. Min.*, **50**: 1647–1669 (1965).

## FINCHENITE = Variety of apatite

## FINNEMANITE
$Pb_5(AsO_3)_3Cl$

CRYSTAL SYSTEM: Hexagonal
CLASS: 6
SPACE GROUP: $P6_3$
Z: 2

LATTICE CONSTANTS:
$a = 10.23$
$c = 7.00$
3 STRONGEST DIFFRACTION LINES:
3.03 (100)
2.88 ( 90)
3.35 ( 60)
OPTICAL CONSTANTS:
$\omega = 2.2949$
$\epsilon = 2.2847$ (Na)
(–)
HARDNESS: 2½
DESNTIY: 7.265 (Meas.)
7.55 (Calc.)
CLEAVAGE: $\{10\bar{1}1\}$ distinct
HABIT: Crystals small, prismatic; as crusts.
COLOR-LUSTER: Gray to black. Translucent to opaque. Subadamantine.
MODE OF OCCURRENCE: Occurs lining crevices in granular hematite at Långban, Sweden.
BEST REF. IN ENGLISH: Palache, et al., "Dana's System of Mineralogy," 7th Ed., v. II, p. 1038–1039, New York, Wiley, 1951.

## FISCHERITE = Wavellite

## FISCHESSERITE
$Ag_3AuSe_2$
Selenium analogue of petzite

CRYSTAL SYSTEM: Cubic
CLASS: 432
SPACE GROUP: I432
Z: 8
LATTICE CONSTANT:
$a = 9.967$
3 STRONGEST DIFFRACTION LINES:
2.662 (100)
2.229 ( 80)
2.035 ( 80)
OPTICAL CONSTANTS: Isotropic. Reflectances (av. of 2) measured at 12 wavelengths showed a max. at 460 nm. $R_{max}$ and $R_{min}$ are: 420 nm, 29.8, 29.1; 460, 31.5, 30.1; 480, 29.9, 29.1; 540, 26.2, 25.9; 640, 25.0, 24.2%.
HARDNESS: 2
DENSITY: 9.05 (Calc.)
CLEAVAGE: Not observed.
HABIT: Massive; in xenomorphic grains.
COLOR-LUSTER: Light pink in reflected light. Opaque.
MODE OF OCCURRENCE: Occurs associated with clausthalite, native gold, and naumannite in carbonate veins at Predborice, Bohemia, Czechoslovakia.
BEST REF. IN ENGLISH: Johan, Zdenek, Picot, Paul, Pierrot, Roland, and Kvacek, M., *Am. Min.*, **57**: 1554 (1972).

## FIZÉLYITE
$Ag_2Pb_5Sb_8S_{18}$

CRYSTAL SYSTEM: Orthorhombic

CLASS: 2/m 2/m 2/m
SPACE GROUP: Pnmm
Z: 4 (?)
LATTICE CONSTANTS:
$a = 13.14$
$b = 19.23$
$c = 8.72$
3 STRONGEST DIFFRACTION LINES:
3.34 (100)
3.49 ( 50)
2.90 ( 50)
HARDNESS: 2
DENSITY: 5.56 (Meas.)
5.226 (Calc.)
CLEAVAGE: {010}
Brittle.
HABIT: Crystals prismatic, deeply striated, without terminal faces.
COLOR-LUSTER: Dark lead gray. Opaque. Metallic. Streak dark gray.
MODE OF OCCURRENCE: Occurs as small crystals associated with semseyite, pyrite, galena, sphalerite, pyrrhotite, dolomite, and quartz, at Kisbánya, Roumania.
BEST REF. IN ENGLISH: Palache, et al., "Dana's System of Mineralogy," 7th Ed., v. I, p. 450, New York, Wiley, 1944.

## FLAGSTAFFITE
$C_{10}H_{22}O_3$

CRYSTAL SYSTEM: Orthorhombic
CLASS: mm2
SPACE GROUP: Fdd2
Z: 16
LATTICE CONSTANTS:
$a = 18.60$
$b = 23.00$
$c = 10.86$
3 STRONGEST DIFFRACTION LINES:
4.93 (100)
8.76 ( 50)
7.19 ( 50)
OPTICAL CONSTANTS:
$\alpha = 1.505$
$\beta = 1.512$
$\gamma = 1.524$
(+)2V = 77°
HARDNESS: Not determined.
DENSITY: 1.09 (Meas.)
CLEAVAGE: {110} imperfect
HABIT: Minute prisms; common forms {110} and {111}, less common {010} and {011}.
COLOR-LUSTER: Colorless, transparent.
MODE OF OCCURRENCE: Occurs as well-formed crystals in radial cracks of fossil logs near the San Francisco Peaks, north of Flagstaff, Arizona.
BEST REF. IN ENGLISH: Strunz, H., and Contag, B., *Am. Min.*, **50**: 2109 (1965).

## FLAJOLOTITE = Tripuhyite

# FLEISCHERITE
$Pb_3Ge(SO_4)_2(OH)_6 \cdot 3H_2O$

CRYSTAL SYSTEM: Hexagonal
CLASS: 6/m 2/m 2/m
SPACE GROUP: Probably $P6_3/mmc$
Z: 2
LATTICE CONSTANTS:
 $a$ = 8.89
 $c$ = 10.86
3 STRONGEST DIFFRACTION LINES:
 3.619 (100)
 2.635 ( 80)
 3.437 ( 60)
OPTICAL CONSTANTS:
 $\omega$ = 1.747
 $\epsilon$ = 1.776
  (+)
HARDNESS: Low.
DENSITY: 4.2-4.4 (Meas.)
     4.59 (Calc.)
CLEAVAGE: None observed.
HABIT: Fibrous aggregates.
COLOR-LUSTER: White to pale rose; silky.
MODE OF OCCURRENCE: Occurs associated with cerussite, mimetite, and altered tennantite, also as a crust on plumbojarosite and mimetite on dolomite in the upper oxidation zone of the Tsumeb Mine, South West Africa.
BEST REF. IN ENGLISH: Frondel, C., and Strunz, H., *Am. Min.*, **45**: 1313 (1960).

# FLINKITE
$Mn_2^{2+}Mn^{3+}AsO_4(OH)_4$

CRYSTAL SYSTEM: Orthorhombic
CLASS: 2/m 2/m 2/m
SPACE GROUP: Pnma
Z: 4
LATTICE CONSTANTS:
 $a$ = 9.55
 $b$ = 13.11
 $c$ = 5.35
3 STRONGEST DIFFRACTION LINES:
 4.733 (100 )
 4.386 (100b)
 2.662 (100 )
OPTICAL CONSTANTS:
 $\alpha$ = 1.783
 $\beta$ = 1.801
 $\gamma$ = 1.834
 (+)2V = large
HARDNESS: 4½
DENSITY: 3.78 (Meas.)
     3.73 (Calc.)
CLEAVAGE: Brittle.
HABIT: Feathery aggregates of thin, tabular, slightly elongated crystals.
COLOR-LUSTER: Deep greenish brown. Transparent. Vitreous to greasy.
MODE OF OCCURRENCE: Occurs as an extremely rare mineral implanted upon caryopilite, barite, and other arsenates at the Harstigen Mine, Pajsberg, Värmland, Sweden.
BEST REF. IN ENGLISH: Palache, et al., "Dana's System of Mineralogy," 7th Ed., v. II, p. 793-794, New York, Wiley, 1951. Moore, Paul B., *Am. Min.*, **52**: 1603-1613 (1967).

**FLINT** = Variety of quartz

# FLORENCITE
$CeAl_3(PO_4)_2(OH)_6$

CRYSTAL SYSTEM: Hexagonal
CLASS: $\bar{3}$ 2/m
SPACE GROUP: $R\bar{3}m$
Z: 3
LATTICE CONSTANTS:
 $a$ = 6.974
 $c$ = 16.36
3 STRONGEST DIFFRACTION LINES:
 2.95 (100)
 5.71 ( 70)
 3.50 ( 70)
OPTICAL CONSTANTS:
 $\omega$ = 1.695, 1.680
 $\epsilon$ = 1.705
  (+)
HARDNESS: 5-6
DENSITY: 3.457-3.71 (Meas.)
CLEAVAGE: {0001} good
      {11$\bar{2}$0} in traces
Fracture subconchoidal to splintery.
HABIT: As small rhombohedral or pseudocubic crystals, and as rounded grains.
COLOR-LUSTER: Pink, pale yellow. Transparent to translucent. Resinous to greasy.
MODE OF OCCURRENCE: Occurs as an accessory constituent of mica schists and in sands in the vicinity of Ouro Preto and Diamantina, Minas Geraes, Brazil; in placer deposits in the Ural Mountains, USSR; in pegmatite at Klein Spitzkopje, South West Africa; and in the Kangankunde carbonatite, Malawi (Nyasaland).
BEST REF. IN ENGLISH: Palache, et al., "Dana's System of Mineralogy," 7th Ed., v. II, p. 838-839, New York, Wiley, 1951.

# FLUELLITE (Kreuzbergite)
$Al_2PO_4F_2OH \cdot 7H_2O$

CRYSTAL SYSTEM: Orthorhombic
CLASS: 2/m 2/m 2/m
SPACE GROUP: Fddd
Z: 8
LATTICE CONSTANTS:
 $a$ = 8.546
 $b$ = 11.222
 $c$ = 21.158

3 STRONGEST DIFFRACTION LINES:
6.48 (100)
3.24 ( 70)
3.09 ( 60)
OPTICAL CONSTANTS:
$\alpha$ = 1.473–1.490
$\beta$ = 1.490–1.496
$\gamma$ = 1.506–1.511
(+)2V = very large
HARDNESS: 3
DENSITY: 2.18 (Meas.)
2.16 (Calc.)
CLEAVAGE: {010} indistinct
{111} indistinct
HABIT: Crystals small, dipyramidal {111}, often modified by small {010}.
COLOR-LUSTER: Colorless, pale yellow, white. Transparent. Vitreous.
MODE OF OCCURRENCE: Found as minute crystals on quartz, associated with fluorite, apatite (tavistockite), arsenopyrite, and torbernite at Stenna Gwyn, near St. Austell, Cornwall, England. It also occurs with secondary iron phosphate minerals in pegmatite at Hagendorf and Pleystein, Bavaria, and as an alteration of triplite at Königswart, Bohemia, Czechoslovakia.
BEST REF. IN ENGLISH: Palache, et al., "Dana's System of Mineralogy," 7th Ed., v. II, p. 124–126, New York, Wiley, 1951. Gay, Brian B., and Jeffrey, G. A., *Am. Min.*, **51**: 1579–1592 (1966).

## FLUOBORITE (Nocerite)
$Mg_3F_3(BO_3)$

CRYSTAL SYSTEM: Hexagonal
CLASS: 6/m
SPACE GROUP: $P6_3/m$
Z: 2
LATTICE CONSTANTS:
$a$ = 8.924
$c$ = 3.115
3 STRONGEST DIFFRACTION LINES:
4.41 (100)
2.41 ( 95)
2.12 ( 90)
OPTICAL CONSTANTS:
$\omega$ = 1.570
$\epsilon$ = 1.534
(–)
HARDNESS: 3½
DENSITY: 2.98 (Meas.)
3.01 (Calc.)
CLEAVAGE: {0001} indistinct
HABIT: Crystals prismatic, up to 5.0 X 1.0 mm in size; as subparallel or divergent aggregates of prismatic crystals; also as compact fibrous masses or felted aggregates.
COLOR-LUSTER: Colorless, white. Transparent to translucent. Vitreous. Fibrous or felted aggregates silky.
MODE OF OCCURRENCE: Occurs as crystals in thermally metamorphosed impure limestones at Crestmore, Riverside County, and with calcite in a contact zone at the New Method mine, near Ludlow, San Bernardino County,

California; and as a hydrothermal mineral in veinlets in massive franklinite ore at Sterling Hill, Sussex County, New Jersey. It also is found at the Tallgruvan mine, Norberg, Sweden; at Nocera, Italy; at Broadford, Skye, Scotland; and at deposits in France, Finland, Malaya, Manchuria, and in the Hol Kol mine, Korea.
BEST REF. IN ENGLISH: Palache, et al., "Dana's System of Mineralogy," 7th Ed., v. II, p. 369–370, New York, Wiley, 1951.

## FLUOCERITE (Tysonite)
$(Ce, La)F_3$

CRYSTAL SYSTEM: Hexagonal
CLASS: $\bar{3}$ 2/m
SPACE GROUP: $P\bar{3}c$
Z: 6
LATTICE CONSTANTS:
$a$ = 7.124 kX
$c$ = 7.280
3 STRONGEST DIFFRACTION LINES:
3.19 (100)
2.05 ( 80)
2.00 ( 80)
OPTICAL CONSTANTS:
$\omega$ = 1.612–1.618
$\epsilon$ = 1.607–1.611
(–)
HARDNESS: 4–5
DENSITY: 5.93–6.14 (Meas.)
6.08 (Calc. Ce : La = 1 : 1)
CLEAVAGE: {0001} distinct
{11$\bar{2}$0} indistinct
Fracture conchoidal to uneven. Brittle.
HABIT: Crystals prismatic, elongated along c-axis, or tabular {0001}. Also massive, coarsely granular.
COLOR-LUSTER: Yellowish to reddish brown. Transparent to translucent. Vitreous to resinous; pearly on cleavage. Streak whitish.
MODE OF OCCURRENCE: Occurs in Colorado near the base of Stove Mountain, and at Crystal Park near Manitou, El Paso County; in the South Platte-Lake George area rare-earth pegmatites in Douglas, Teller, and Park counties; and associated with bastnaesite in pegmatite at the Black Cloud mine, Teller County. It is also found in pegmatite at Finbo, Österby, and Broddbo in Dalarne, Sweden.
BEST REF. IN ENGLISH: Palache, et al., "Dana's System of Mineralogy," 7th Ed., v. II, p. 48–50, New York, Wiley, 1951.

## FLUORAPATITE
$Ca_5(PO_4)_3F$
Apatite Series

CRYSTAL SYSTEM: Hexagonal
CLASS: 6/m
SPACE GROUP: $P6_3/m$
Z: 2

LATTICE CONSTANTS:
$a = 9.36$ kX
$c = 6.88$
3 STRONGEST DIFFRACTION LINES:
2.800 (100)
2.702 ( 60)
2.772 ( 55)
OPTICAL CONSTANTS:
$\omega = 1.6357$
$\epsilon = 1.6328$ (Na)
(−)
HARDNESS: 5
DENSITY: 3.1–3.2 (Meas.)
3.201 (Calc.)
CLEAVAGE: {0001} indistinct
{10$\bar{1}$0} trace
Fracture conchoidal to uneven. Brittle.

HABIT: Crystals short to long prismatic or thin to thick tabular, often complex. Also massive, compact to coarse granular; globular or reniform; stalactitic; fibrous; oolitic; earthy; as nodular concretions and as bedded deposits. Twinning on {11$\bar{2}$1} and {10$\bar{1}$3}, rare.

COLOR-LUSTER: Colorless, white, gray, yellow to yellowish green, pale to dark green, pale to dark bluish green, pale to very dark blue, violet-blue, violet, purple, various shades of red and brown. Transparent to opaque. Vitreous to subresinous, rarely silky. Often fluorescent, phosphorescent or thermoluminescent. Streak white.

MODE OF OCCURRENCE: Occurs chiefly as a common and widespread mineral in many igneous rocks; in pegmatites; in hydrothermal and Alpine-type veins; as bedded deposits of marine origin; in metamorphic rocks; and as detrital deposits. Found as deep blue anhedral crystals two feet or more in diameter and several feet in length at the Hugo mine Keystone, South Dakota; and as superb crystals in Maine, New York, New Hampshire, Connecticut, South Dakota, and California. Fine specimens also are found at many localities in Ontario and Quebec, Canada; in Durango, Mexico; and in Bolivia, Brazil, Spain, Germany, Czechoslovakia, Switzerland, Austria, Italy, France, England, USSR, Sweden, Burma, and Japan. Recently found as magnificent crystals at Panasqueira, Portugal.

BEST REF. IN ENGLISH: Palache, et al., "Dana's System of Mineralogy," 7th Ed., v. II, p. 879–889, New York, Wiley, 1951.

## FLUORITE
$CaF_2$

CRYSTAL SYSTEM: Cubic
CLASS: $4/m\bar{3}\,2/m$
SPACE GROUP: Fm3m
Z: 4
LATTICE CONSTANT:
$a = 5.462$
3 STRONGEST DIFFRACTION LINES:
1.931 (100)
3.153 ( 94)
1.647 ( 35)
OPTICAL CONSTANT:
$N = 1.4322$ (Li)
HARDNESS: 4

DENSITY: 3.180 (Meas.)
3.180 (Calc.)
CLEAVAGE: {111} perfect
Fracture subconchoidal to splintery. Brittle.

HABIT: Crystals usually cubic, less commonly in octahedrons, rarely in dodecahedrons. Modifying faces occasionally present. Twinning frequent on {111}, often as penetration twins of cubes. Also massive, coarse to fine granular, botryoidal, rarely fibrous or columnar.

COLOR-LUSTER: Colorless, commonly shades of purple, blue, green, yellow; also white, pink, crimson to brownish red, brown, bluish black. Color often unequally distributed or zoned. Vitreous. Transparent to translucent. Fluoresces blue, yellow, white, reddish, or pale violet under ultraviolet light. Some varieties thermoluminescent.

MODE OF OCCURRENCE: Occurs widespread as a common mineral in hydrothermal ore deposits; in cavities in sedimentary rocks; in pneumatolytic deposits; as a hot springs deposit; in miarolitic cavities in Alpine-type veins; in certain granites; and uncommonly in pegmatites. In the United States notable localities include Hardin and Pope counties in Illinois and adjacent areas of Kentucky; Clay Center, Ohio; the Hansonburg district, New Mexico; near Silverton, San Juan County, and other places in Colorado; at Muscalonge Lake, Jefferson County, New York; and at Westmoreland, Cheshire County, New Hampshire. In Canada exceptional crystals occur at Madoc and at the Faraday uranium mine, Bancroft, Ontario. Superb crystal groups occur in Cumberland, Durham, and at other places in England; at Wölsendorf, Bavaria and at many other localities in Germany; in France, Italy, and as superb rose-colored octahedrons in Alpine-type deposits in Switzerland. Yttrium varieties occur in pegmatite at the White Cloud mine, Jefferson County, and at other localities in Colorado; in northern Norway; and on the Khibina Tundra, Kola peninsula, USSR.

BEST REF. IN ENGLISH: Palache, et al., "Dana's System of Mineralogy," 7th Ed., v. II, p. 29–37, New York, Wiley, 1951.

## FLUOR-RICHTERITE = Variety of richterite
$Na(Ca, Na)Mg_5Si_8O_{22}F_2$
Amphibole Group

CRYSTAL SYSTEM: Monoclinic
CLASS: 2/m
SPACE GROUP: C2/m
Z: 2
LATTICE CONSTANTS:
$a = 9.823$
$b = 17.957$
$c = 5.268$
$\beta = 104°20'$
3 STRONGEST DIFFRACTION LINES:
3.13 (100)
8.42 (100)
2.80 ( 75)
OPTICAL CONSTANTS:
$\alpha = 1.603$
$\beta = 1.614$
$\gamma = 1.622$
$(-)2V = 72°$

HARDNESS: 5-6
DENSITY: 3.035 (Calc.)
CLEAVAGE: {110} perfect
{001} parting
{100} parting
Fracture uneven. Brittle.
HABIT: Crystals long prismatic, commonly double terminated. Twinning on {100} simple, lamellar.
COLOR-LUSTER: Dark grayish green to blackish green. Translucent; thin fragments transparent. Vitreous.
MODE OF OCCURRENCE: Occurs as magnificent single crystals up to 10 cm or more in length, embedded in white to pinkish calcite, at Wilberforce, Ontario, Canada.
BEST REF. IN ENGLISH: Kohn and Comeford, *Am. Min.*, **40**: 410 (1955).

## FLUOTARAMITE  = Magnesioarfvedsonite

## FORBESITE (Inadequately described mineral)
$H_2 (Ni, Co)_2 (AsO_4)_2 \cdot 7H_2O$

CRYSTAL SYSTEM: Unknown
HARDNESS: 2½
DENSITY: 3.134 (Meas.)
CLEAVAGE: Not determined.
HABIT: As radial-fibrous crusts.
COLOR-LUSTER: Grayish white. Dull to silky or resinous.
MODE OF OCCURRENCE: Occurs as a secondary mineral in decomposed basic igneous rock at a mine near Flamenco, in the Atacama desert, Chile.
BEST REF. IN ENGLISH: Palache, et al., "Dana's System of Mineralogy," 7th Ed., v. II, p. 711, New York, Wiley, 1951.

## FORMANITE
$Y(Ta, Nb)O_4$
Fergusonite Series

CRYSTAL SYSTEM: Tetragonal
CLASS: 4/m
SPACE GROUP: $I4_1/a$
Z: 4
LATTICE CONSTANTS:
$a$ = 7.76 (synthetic)
$c$ = 11.32
3 STRONGEST DIFFRACTION LINES: Metamict.
OPTICAL CONSTANTS:
Isotropic (metamict)
$N$ = 2.05-2.19
HARDNESS: 5½-6½
DENSITY: 7.03 (Calc. for $YTaO_4$)
CLEAVAGE: {111} in traces
Fracture subconchoidal. Brittle.
HABIT: As irregular masses or grains.
COLOR-LUSTER: Brownish black to black; surfaces commonly discolored and dull from alteration. Opaque, except in thin fragments. Vitreous, submetallic.
MODE OF OCCURRENCE: Occurs abundantly in placer deposits associated with cassiterite, euxenite, and monazite at Cooglegong, Western Australia.

BEST REF. IN ENGLISH: Palache, et al., "Dana's System of Mineralogy," 7th Ed., v. I, p. 757-763, New York, Wiley, 1944.

## FORNACITE
$(Pb, Cu)_3 \{(Cr, As)O_4\}_2 (OH)$

CRYSTAL SYSTEM: Monoclinic
CLASS: 2/m
SPACE GROUP: $P2_1/c$
Z: 4
LATTICE CONSTANTS:
$a$ = 7.91
$b$ = 5.91
$c$ = 17.46
3 STRONGEST DIFFRACTION LINES:
3.31 (100)
2.98 (100)
2.88 (100)
OPTICAL CONSTANTS:
$\alpha$ = 2.14
$\gamma$ = 2.24
(+)2V
HARDNESS: Not determined.
DENSITY: 6.27 (Meas.)
6.40 (Calc.)
CLEAVAGE: None
HABIT: As small prismatic crystals, up to 4 mm long.
COLOR-LUSTER: Olive green. Streak saffron yellow.
MODE OF OCCURRENCE: Occurs associated with dioptase at Renéville, Congo, and also found at Sébarz, Iran.
BEST REF. IN ENGLISH: Bariand, P., and Herpin, P., *Am. Min.*, **49**: 447 (1964).

## FORSTERITE
$Mg_2 SiO_4$  (0-10 atomic percent $Fe^{2+}$)
Olivine Group

CRYSTAL SYSTEM: Orthorhombic
CLASS: 2/m 2/m 2/m
SPACE GROUP: Pbnm
Z: 4
LATTICE CONSTANTS:
$a$ = 4.756
$b$ = 10.195
$c$ = 5.981
3 STRONGEST DIFFRACTION LINES:
2.458 (100)
3.883 ( 70)
2.512 ( 70)
OPTICAL CONSTANTS:
$\alpha$ = 1.635
$\beta$ = 1.651
$\gamma$ = 1.670
(+)2V = 82°
HARDNESS: 7
DENSITY: 3.275 (Meas.)
3.271 (Calc.)
3.213 (Calc. Synthetic)

CLEAVAGE: {010} imperfect
{100} imperfect
Fracture conchoidal. Brittle.

HABIT: Crystals usually thick tabular, vertically striated, often with wedge-shaped terminations. Usually massive, compact or granular; in embedded grains, irregular or rounded. Twinning {100}, {011}, not common.

COLOR-LUSTER: Green, lemon yellow, white. Transparent to translucent. Vitreous to greasy. Streak uncolored.

MODE OF OCCURRENCE: Occurs chiefly in basic and ultrabasic igneous rocks, and in thermally metamorphosed impure dolomitic limestones. Typical occurrences are found in the new City quarry at Riverside in Riverside County, and at Cascade Canyon in San Bernardino County, California; near Custer, Custer County, South Dakota; at Bolton and Rockport, Massachusetts; and at localities in Greenland, Norway, Sweden, Italy, Germany, Finland, USSR, and Burma.

BEST REF. IN ENGLISH: Deer, Howie, and Zussman, "Rock Forming Minerals," v. I, p. 1-33, New York, Wiley, 1962.

## FOSHAGITE
$Ca_4Si_3O_9(OH)_2$

CRYSTAL SYSTEM: Monoclinic
SPACE GROUP: C lattice
Z: 4
LATTICE CONSTANTS:
  $a = 10.32$
  $b = 7.36$
  $c = 14.07$
  $\beta = 106°24'$
3 STRONGEST DIFFRACTION LINES:
  2.92 (100)
  1.74 ( 90) broad
  6.8 ( 70)
OPTICAL CONSTANTS:
  $\beta = 1.594$
  $\gamma = 1.598$
  HARDNESS: Not determined.
  DENSITY: 2.36 (Meas.)
  CLEAVAGE: {001} distinct
  HABIT: Compact fibrous.
  COLOR-LUSTER: Snow white; silky.

MODE OF OCCURRENCE: Occurs in veins, with fibers at times several inches long, associated with idocrase, thaumasite, and blue calcite at Crestmore, Riverside County, California.

BEST REF. IN ENGLISH: Gard, J. A., and Taylor, H. F. W., *Am. Min.*, 43: 1-15 (1958).

## FOSHALLASITE (cf Afwillite)
## (Inadequately described mineral)
$Ca_3Si_2O_7 \cdot 3H_2O$ (?)

CRYSTAL SYSTEM: Probably orthorhombic
OPTICAL CONSTANTS:
  $\alpha = 1.535$
  $\beta = 1.542$
  $\gamma = 1.549$
  $(-)2V = 12°-18°$

HARDNESS: 2½-3
DENSITY: 2.5 (Meas.)
CLEAVAGE: {100} perfect
HABIT: Crystals tabular, elongated along {100} {010}; striated vertically. In radial aggregates.
COLOR-LUSTER: Snow white.
MODE OF OCCURRENCE: Occurs in association with calcite and mesolite in veins at the Lovtschorrite mine at Yukspor, Khibina Tundra, USSR.

BEST REF. IN ENGLISH: Winchell and Winchell, "Elements of Optical Mineralogy," 4th Ed., Pt. 2, p. 478, New York, Wiley, 1951.

## FOUCHERITE = Bořickite? (Inadequately described mineral)
Near $Ca(Fe, Al)_4(PO_4)_2(OH)_8 \cdot 7H_2O$

## FOURMARIERITE
$PbU_4O_{13} \cdot 4H_2O$

CRYSTAL SYSTEM: Orthorhombic
CLASS: 2/m 2/m 2/m or mm2
SPACE GROUP: Bbmm or Bbm2
Z: 8
LATTICE CONSTANTS:
  $a = 14.00$
  $b = 16.47$
  $c = 14.39$
3 STRONGEST DIFFRACTION LINES:
  7.20 (100)
  3.58 ( 50)
  3.18 ( 50)
OPTICAL CONSTANTS:
  $\alpha = 1.863$
  $\beta = 1.885$
  $\gamma = 1.890$
  $(-)2V = 50°$
HARDNESS: 3-4
DENSITY: 5.74 (Meas.)
       5.76 (Calc.)
CLEAVAGE: {001} perfect
       {100} imperfect
HABIT: Crystals tabular on {001}, elongated [010], pseudohexagonal; microscopic up to 2 mm in length. Also as compact crystalline masses.

COLOR-LUSTER: Reddish orange to carmine red; rarely reddish brown to brown. Transparent to translucent. Adamantine.

MODE OF OCCURRENCE: Occurs as alteration pseudomorphs after crystals of uraninite in pegmatites; as excellent crystals associated with torbernite, kasolite, and other secondary uranium minerals at Shinkolobwe, Katanga, Zaire; in the Colorado Plateau uranium province as an alteration product of uraninite; and in many other types of deposits containing altered uraninite.

BEST REF. IN ENGLISH: Christ, C. L., and Clark, Joan R., *Am. Min.*, 45: 1026-1061 (1960).

## FOWLERITE = Variety of rhodonite with up to 10% ZnO

## FRAIPONTITE (Inadequately described mineral)
$Zn_8Al_4(SiO_4)_5(OH)_8 \cdot 7H_2O$

CRYSTAL SYSTEM: Unknown
3 STRONGEST DIFFRACTION LINES:
  7.00 (100)
  3.52 ( 70)
  2.63 ( 30)
OPTICAL CONSTANTS:
  Biaxial (−)
HARDNESS: Not determined.
DENSITY: Not determined.
CLEAVAGE: Not determined.
HABIT: Massive; as thin fibrous crusts.
COLOR-LUSTER: Yellowish white. Silky.
MODE OF OCCURRENCE: Occurs on smithsonite probably from Vielle-Montagne, Belgium.
BEST REF. IN ENGLISH: Cesàro, G., *Am. Min.*, **13**: 492 (1928).

## FRANCEVILLEITE
$(Ba, Pb)(UO_2)_2(VO_4)_2 \cdot 5H_2O$
Barium analogue of meta-tyuyamunite

CRYSTAL SYSTEM: Orthorhombic
CLASS: 2/m 2/m 2/m
Z: 4
LATTICE CONSTANTS:
  $a = 10.3$
  $b = 8.3$
  $c = 16.6$
3 STRONGEST DIFFRACTION LINES:
  8.30 (100)
  2.98 ( 80)
  4.17 ( 60)
OPTICAL CONSTANTS:
  $\alpha = 1.750-1.785$
  $\beta = 1.910-1.952$
  $\gamma = 1.945-2.002$
  $(-)2V = 52°$
HARDNESS: 3
DENSITY: 4.55
CLEAVAGE: {001} perfect
HABIT: As crystal-plates several millimeters in thickness; as impregnations; and as cryptocrystalline veinlets.
COLOR-LUSTER: Yellow, yellowish green to brownish yellow. Adamantine to pearly.
MODE OF OCCURRENCE: Occurs in sandstones in the region of Franceville, Gabon.
BEST REF. IN ENGLISH: Branche, Georges; Ropert, Marie-Edith; Chantret, Francis; Morignat, Bernard; and Pouget, Robert, *Am. Min.*, **43**: 180 (1958).

## FRANCKEITE
$Pb_5Sn_3Sb_2S_{14}$

CRYSTAL SYSTEM: Triclinic
CLASS: $\bar{1}$
SPACE GROUP: $P\bar{1}$
Z: 8

LATTICE CONSTANTS:
  $a = 46.9$      $\alpha = 90°$
  $b = 5.82$      $\beta = 94°40'$
  $c = 17.3$      $\gamma = 90°$
3 STRONGEST DIFFRACTION LINES:
  3.44 (100)
  2.91 (100)
  2.86 (100)
HARDNESS: 2½-3
DENSITY: 5.88-5.92 (Meas.)
CLEAVAGE: {001} perfect
Flexible; inelastic. Somewhat malleable.
HABIT: Crystals thin tabular; somewhat elongated; vertically striated; often bent or warped. Often in spherical aggregates. Usually massive, foliated or radial. Twinning complex.
COLOR-LUSTER: Grayish black. Opaque. Metallic. Streak grayish black.
MODE OF OCCURRENCE: Occurs in the Thompson mine, Darwin district, Inyo County, and with meneghinite and stannite in the limestone contact rock of the Kalkar quarry, near Santa Cruz, Santa Cruz County, California. Found in the tin veins of Bolivia, sometimes in large amounts, especially at Huanuni, Poopó, Chocoya, Colquechaca, Oruro, and Llallagua.
BEST REF. IN ENGLISH: Palache, et al., "Dana's System of Mineralogy," 7th Ed., v. I, p. 448-450, New York, Wiley, 1944.

## FRANCOLITE = Variety of apatite
(carbonatian-fluorapatite)

## FRANKLINITE
$(Zn, Mn, Fe^{2+})(Fe^{3+}, Mn^{3+})_2O_4$
Magnetite Series
Spinel Group

CRYSTAL SYSTEM: Cubic
CLASS: $4/m \bar{3} 2/m$
SPACE GROUP: Fd3m
Z: 8
LATTICE CONSTANT:
  $a = 8.403$ (variable)
3 STRONGEST DIFFRACTION LINES:
  2.55  (100)
  1.499 ( 80)
  2.99  ( 70)
OPTICAL CONSTANT:
  $N = 2.36$ (Li)
HARDNESS: 5½-6½
DENSITY: 5.07-5.22 (Meas.)
        5.163 (Calc.)
CLEAVAGE: {111} parting, fair.
Fracture uneven to subconchoidal. Brittle.
HABIT: Crystals octahedral, up to 7 inches on edge, sometimes modified, edges often rounded. Also massive, compact, coarse to fine granular.
COLOR-LUSTER: Iron black. Opaque, except in thin fragments. Metallic to dull. Streak reddish brown to black.
Weakly magnetic.
MODE OF OCCURRENCE: Occurs associated with cal-

cite, willemite, zincite, tephroite, rhodonite, and other minerals, in the zinc deposits at Franklin and Sterling Hill, Sussex County, New Jersey.

BEST REF. IN ENGLISH: Palache, et al., "Dana's System of Mineralogy," 7th Ed., v. I, p. 698–707, New York, Wiley, 1944.

# FREBOLDITE
CoSe

CRYSTAL SYSTEM: Hexagonal
CLASS: 6/m 2/m 2/m
SPACE GROUP: P6/mmc
Z: 2
LATTICE CONSTANTS:
  $a = 3.61$
  $c = 5.28$
3 STRONGEST DIFFRACTION LINES:
  2.05 (100)
  1.066 (100)
  0.963 (100)
HARDNESS: Not determined.
DENSITY: Not determined.
CLEAVAGE: Not determined.
HABIT: Massive; as embedded grains. Observed only in thin sections.
COLOR-LUSTER: Similar to niccolite.
MODE OF OCCURRENCE: Occurs in association with hastite, trogtalite, bornhardtite, and other selenides in dolomite veinlets in the Trogtal quarries, Harz Mountains, Germany.
BEST REF. IN ENGLISH: Vlasov, K. A., "Mineralogy of Rare Elements," v. II, p. 663–664, Israel Program for Scientific Translations, 1966.

# FREIBERGITE = Variety of tetrahedrite containing silver

# FREIESLEBENITE
PbAgSbS$_3$

CRYSTAL SYSTEM: Monoclinic
CLASS: 2/m
SPACE GROUP: P2$_1$/a
Z: 4
LATTICE CONSTANTS:
  $a = 7.53$
  $b = 12.79$
  $c = 5.95$
  $\beta = 92°14'$
3 STRONGEST DIFFRACTION LINES:
  2.83 (100)
  3.48 ( 80)
  2.98 ( 70)
HARDNESS: 2–2½
DENSITY: 6.20–6.23 (Meas.)
        6.22 (Calc.)
CLEAVAGE: {110} imperfect
Fracture subconchoidal to uneven. Brittle.

HABIT: Crystals prismatic, vertically striated. Twinning on {100}.
COLOR-LUSTER: Lead gray to silver white. Opaque. Metallic. Streak same as color.
MODE OF OCCURRENCE: Occurs as excellent crystals associated with pyrargyrite, argentite, galena, and siderite at Hiendelaencina, Spain; with argentite at Freiberg, Saxony, Germany; and at localities in Czechoslovakia and Roumania.
BEST REF. IN ENGLISH: Palache, et al., "Dana's System of Mineralogy," 7th Ed., v. I, p. 416–418, New York, Wiley, 1944.

# FREIRINITE = Lavendulan

# FREMONTITE = Natromontebrasite

# FRESNOITE
Ba$_2$TiSi$_2$O$_8$

CRYSTAL SYSTEM: Tetragonal
CLASS: 4/m 2/m 2/m, 4mm, or $\overline{4}$2m
SPACE GROUP: P4/mbm, P4bm, or P$\overline{4}$b2
Z: 2
LATTICE CONSTANTS:
  $a = 8.52$
  $c = 5.210$
3 STRONGEST DIFFRACTION LINES:
  3.08 (100)
  3.305 ( 50)
  2.150 ( 30)
OPTICAL CONSTANTS:
  $\omega = 1.775$
  $\epsilon = 1.765$
    (–)
HARDNESS: 3–4
DENSITY: 4.43 (Meas.)
        4.45 (Calc.)
CLEAVAGE: {001} fair
HABIT: Subhedral to euhedral crystals elongated slightly in direction of c-axis. Grain size of crystals range from less than 0.1 mm up to 3 mm. Aggregates of grains range up to 5 mm.
COLOR-LUSTER: Lemon or canary yellow; vitreous. White streak. Fluoresces yellow under sw ultra-violet light.
MODE OF OCCURRENCE: Found in gneissic sanbornite-quartz rock and in quartzite that is poor or lacking in sanbornite associated with celsian, taramellite, diopside, and pyrrhotite in a narrow zone 2½ miles long near a granodiorite contact in eastern Fresno county, California.
BEST REF. IN ENGLISH: Alfors, John T., Stinson, Melvin C., Matthews, Robert A., and Pabst, Adolf, *Am. Min.*, **50**: 314–340 (1965).

# FREUDENBERGITE
Na$_2$(Ti, Nb)$_6$Fe$_2^{3+}$(O, OH)$_{18}$

CRYSTAL SYSTEM: Monoclinic
CLASS: 2, m or 2/m

SPACE GROUP: C2, Cm or C2/m
Z: 1
LATTICE CONSTANTS:
(monoclinic)
$a = 12.305$
$b = 3.822$
$c = 6.500$
$\beta = 107.3°$
3 STRONGEST DIFFRACTION LINES:
3.627 (100)
1.911 ( 90)
3.100 ( 80)
OPTICAL CONSTANTS:
$N \sim 2.37$–$2.42(+)$
HARDNESS: Abrasion hardness less than that of hematite and diopside.
DENSITY: 4.38 (Calc.)
CLEAVAGE: Basal and prismatic—good.
HABIT: Small xenomorphic grains averaging 0.15 mm long, 0.05 mm thick.
COLOR-LUSTER: Blackish; transparent under microscope. Powder olive gray to steel gray. Streak pale yellow-brown.
MODE OF OCCURRENCE: Occurs in an apatite-rich alkali syenite from Michelsberg, Katzenbuckel, Odenwald, Germany.
BEST REF. IN ENGLISH: McKie, Duncan, *Zeit. für Kristallog.*, **119**: 157–160 (1963).

# FREYALITE = Variety of thorite

# FRIEDELITE
$(Mn, Fe)_8 Si_6 O_{18} (OH, Cl)_4 \cdot 3H_2O$

CRYSTAL SYSTEM: Hexagonal
SPACE GROUP: R lattice
Z: 6
LATTICE CONSTANTS:
$a = 13.40$
$c = 21.43$
3 STRONGEST DIFFRACTION LINES:
2.56 (100)
7.17 ( 90)
3.60 ( 70)
OPTICAL CONSTANTS:
$\omega = 1.654$–$1.664$
$\epsilon = 1.625$–$1.629$
(–)
HARDNESS: 4–5
DENSITY: 3.041–3.059 (Meas.)
CLEAVAGE: {0001} perfect.
Fracture uneven. Brittle.
HABIT: Crystals commonly tabular, often markedly hemimorphic; rarely as slender needles. Also massive, cryptocrystalline, and as fibrous or lamellar aggregates.
COLOR-LUSTER: Pale pink to dark red, brownish red to brown; rarely light to dark yellow. Transparent to translucent. Vitreous to somewhat pearly.
MODE OF OCCURRENCE: Occurs in association with franklinite, willemite and other minerals, at Franklin and

Sterling Hill, Sussex County, New Jersey; at the Sjö mines near Örebro, and at the Harstig mine at Pajsberg, Sweden; and in the manganese mine at Adervielle, Hautes Pyrenees, France.
BEST REF. IN ENGLISH: Palache, Charles, USGS Prof. Paper 180, p. 88–90 (1935).

# FRITZSCHEITE (Species status uncertain)
$Mn(UO_2)_2 (VO_4)_2 \cdot 10H_2O$ (?)
Autunite Group

CRYSTAL SYSTEM: Tetragonal
HARDNESS: 2½–3½
DENSITY: 3.50 (Meas.)
CLEAVAGE: {001} perfect
{100} distinct
Brittle.
HABIT: Crystals platy.
COLOR-LUSTER: Reddish brown to hyacinth red. Translucent. Vitreous to pearly.
MODE OF OCCURRENCE: Occurs in association with torbernite at the Georg Wagsfort mine, Johanngeorgenstadt, Saxony, Germany, and in a hematite deposit as parallel borders on crystals of autunite at Neuhammer, Bohemia, Czechoslovakia.
BEST REF. IN ENGLISH: Frondel, Clifford, USGS Bull. 1064, p. 195–196 (1958).

# FROHBERGITE
$FeTe_2$

CRYSTAL SYSTEM: Orthorhombic
CLASS: 2/m 2/m 2/m
SPACE GROUP: Pnnm
Z: 2
LATTICE CONSTANTS:
$a = 5.28$
$b = 6.26$
$c = 3.85$
3 STRONGEST DIFFRACTION LINES:
2.81 (100)
2.71 ( 70)
2.07 ( 50)
HARDNESS: 3–4
DENSITY: 8.067 (Calc.)
CLEAVAGE: None.
Brittle.
HABIT: Massive, as minute embedded grains.
COLOR-LUSTER: Observed only in polished section.
MODE OF OCCURRENCE: Found in association with native gold, altaite, petzite, melonite, montbrayite, covellite, chalcopyrite, pyrite, and other minerals, at the Robb Montbray mine, Montbray Township, Abitibi County, Quebec, Canada.
BEST REF. IN ENGLISH: Vlasov, K. A., "Mineralogy of Rare Elements," v. II, p. 745–746, Israel Program for Scientific Translations, 1966.

## FROLOVITE (Inadequately described mineral)
$CaB_2O_4 \cdot 4H_2O$

CRYSTAL SYSTEM: Unknown
3 STRONGEST DIFFRACTION LINES:
  6.084 (100)
  3.858 ( 90)
  3.471 ( 80)
OPTICAL CONSTANTS:
  $\beta = 1.572$
  $\gamma = 1.586$
  $2V = 75°$
HARDNESS: 3.5
DENSITY: 2.14
CLEAVAGE: Not determined.
HABIT: Massive; fine-grained.
COLOR-LUSTER: White with grayish tint; translucent in fine splinters; dull.
MODE OF OCCURRENCE: Occurs associated with calciborite, calcite, garnet, and magnetite, and replaces calciborite, in limestones of Middle Devonian age in the Novo-Frolovsk contact-metasomatic copper deposits, Tur'insk region, northern Urals, USSR.
BEST REF. IN ENGLISH: Petrova, E. S., *Am. Min.*, **43**: 385–386 (1958).

## FRONDELITE
$(Mn^{2+}, Fe^{2+})Fe_4^{3+}(PO_4)_3(OH)_5$
Frondelite-Rockbridgeite Series

CRYSTAL SYSTEM: Orthorhombic
CLASS: 222
SPACE GROUP: $B22_12$
Z: 4
LATTICE CONSTANTS:
  $a = 13.89$
  $b = 17.01$
  $c = 5.21$
3 STRONGEST DIFFRACTION LINES:
  3.20 (100)
  3.38 ( 50)
  1.598 ( 50)
OPTICAL CONSTANTS:
  $\alpha = 1.860$
  $\beta = 1.880$
  $\gamma = 1.893$
  $(-)2V$ = moderate
HARDNESS: 4½
DENSITY: 3.476 (Meas.)
         3.473 (Calc.)
CLEAVAGE: {100} perfect
          {010} distinct
          {001} distinct
Fracture uneven. Brittle.
HABIT: As botryoidal crusts and masses with radial-fibrous or fine columnar structure.
COLOR-LUSTER: Dark olive green to greenish black, becoming brownish green to reddish brown on oxidation. Aggregates often exhibit concentric color banding. Subtranslucent. Vitreous to dull.

MODE OF OCCURRENCE: Occurs as a secondary mineral as an alteration product of triphylite or other manganese-iron phosphates in granite pegmatites. Found with metastrengite at the Fletcher mine, North Groton, New Hampshire, and in pegmatites in the Black Hills, South Dakota. It also occurs at the Sapucaia pegmatite, Minas Geraes, Brazil.
BEST REF. IN ENGLISH: Palache, et al., "Dana's System of Mineralogy," 7th Ed., v. II, p. 867–869, New York, Wiley, 1951.

## FROODITE
$\alpha\text{-}PdBi_2$

CRYSTAL SYSTEM: Monoclinic
CLASS: 2/m
SPACE GROUP: C2/m
Z: 4
LATTICE CONSTANTS:
  $a = 12.75$
  $b = 4.29$
  $c = 5.67$
  $\beta = 102°52'$
3 STRONGEST DIFFRACTION LINES:
  2.77 (100)
  1.556 ( 80)
  2.97 ( 70)
DENSITY: 12.5–12.6
CLEAVAGE: {001} very perfect
          {100} less perfect
Fracture uneven. Brittle.
HABIT: As tiny grains.
COLOR-LUSTER: Gray; metallic, splendent on fresh cleavage, tarnishes rapidly.
MODE OF OCCURRENCE: Found in mill concentrates of arsenic- and lead-copper-rich ores from the Frood Mine, Sudbury District, Ontario, Canada.
BEST REF. IN ENGLISH: Hawley, J. E., and Berry, L. G., *Can. Min.*, **6**: 200–209 (1958).

**FUCHSITE** = Variety of muscovite with up to 4.8% $Cr_2O_3$

## FUKUCHILITE
$Cu_3FeS_8$
Pyrite Group

CRYSTAL SYSTEM: Cubic
Z: 1
LATTICE CONSTANT:
  $a = 5.58$
3 STRONGEST DIFFRACTION LINES:
  2.789 (100)
  3.21 ( 80)
  1.685 ( 80)
HARDNESS: ~6
DENSITY: 4.80 (Calc.)
CLEAVAGE: Not determined.
HABIT: Minute grains.
COLOR-LUSTER: Dark brownish gray; submetallic.

MODE OF OCCURRENCE: Occurs in barite-bearing gypsum (and/or anhydrite ores) as minute grains in interstices of small masses consisting of barite, covellite, and pyrite at the Hanawa Mine, Akita Prefecture, Japan.

BEST REF. IN ENGLISH: Kajiwara, Yoshimichi, *Am. Min.*, **55**: 1811 (1970).

# FÜLÖPPITE
$Pb_3Sb_8S_{15}$

CRYSTAL SYSTEM: Monoclinic
CLASS: 2/m
SPACE GROUP: A2/a
Z: 4
LATTICE CONSTANTS:
  $a = 16.90$
  $b = 11.69$
  $c = 13.39$
  $\beta = 94°42'$

3 STRONGEST DIFFRACTION LINES:
  3.87 (100)
  3.22 ( 70)
  2.96 ( 40)
HARDNESS: 2½
DENSITY: 5.22–5.23 (Meas.)
        5.23 (Calc.)
CLEAVAGE: None.
Fracture uneven. Brittle.
HABIT: Crystals commonly short prismatic and pyramidal. Rarely thick tabular. Striated; often curved.
COLOR-LUSTER: Lead gray; often tarnished bronze or steel blue. Opaque. Metallic. Streak reddish gray.
MODE OF OCCURRENCE: Occurs associated with zinkenite, sphalerite, and dolomite in the Kereszthegy mine, Nagyág, Roumania.
BEST REF. IN ENGLISH: Palache, et al., "Dana's System of Mineralogy," 7th Ed., v. I, p. 463–464, New York, Wiley, 1944.

# FYNCHENITE = Thorium britholite

# G

## GABRIELSONITE
$PbFe(AsO_4)(OH)$

CRYSTAL SYSTEM: Orthorhombic
CLASS: mm2
SPACE GROUP: $P2_1ma$
Z: 4
LATTICE CONSTANTS:
  $a = 7.86$
  $b = 5.98$
  $c = 8.62$
3 STRONGEST DIFFRACTION LINES:
  3.192 (100)
  3.074 (100)
  2.706 ( 40)
OPTICAL CONSTANTS:
  $N > 2.00$
  $(-)2V = 80-90°$
HARDNESS: 3½
DENSITY: 6.67 (Meas.)
       6.69 (Calc.)
CLEAVAGE: None.
Very brittle.
  HABIT: As aggregates and masses of crude, rounded, crystals of indeterminate morphology.
  COLOR-LUSTER: Black; adamantine. Small fragments transparent greenish brown. Streak pale chocolate brown.
  MODE OF OCCURRENCE: Occurs in centimeter-size lumps associated with roméite, nadorite, calcite, and barite in specimens collected from the "Hindenberg" stope at Långban, Sweden.
  BEST REF. IN ENGLISH: Moore, P. B., *Am. Min.*, **53**: 1063-1064 (1968).

## GADOLINITE
$Be_2FeY_2Si_2O_{10}$

CRYSTAL SYSTEM: Monoclinic
CLASS: 2/m
SPACE GROUP: $P2_1/a$
Z: 2
LATTICE CONSTANTS:
  $a = 9.89$
  $b = 7.55$
  $c = 4.66$
  $\beta = 90°34'$
3 STRONGEST DIFFRACTION LINES:
  3.13 (100)
  2.83 (100)
  2.56 ( 90)
OPTICAL CONSTANTS: Usually metamict
  $\alpha = 1.78-1.77$
  $\gamma = 1.82-1.78$
  $(+)2V = 85°$
HARDNESS: 6½-7
DENSITY: 4.0-4.65 (Meas.)
       4.44 (Calc.)
CLEAVAGE: None.
Fracture conchoidal. Brittle.
  HABIT: Crystals often prismatic; sometimes flattened {001}; usually rough and coarse. Commonly massive, compact.
  COLOR-LUSTER: Black, greenish black, brown; rarely light green. Translucent to transparent. Vitreous to greasy. Streak greenish gray.
  MODE OF OCCURRENCE: Occurs chiefly in granites and granite pegmatites, often in association with fluorite and allanite. Found at several places in Colorado, especially in Clear Creek, Fremont, and Jefferson counties; in Llano County, Texas, and Mohave County, Arizona; and at localities in Greenland, Norway, Sweden, Switzerland, USSR, Japan, Australia, and elsewhere.
  BEST REF. IN ENGLISH: Vlasov, K. A., "Mineralogy of Rare Elements," v. II, p. 235-239, Israel Program for Scientific Translations, 1966.

## GAGARINITE
$NaCaY(F, Cl)_6$

CRYSTAL SYSTEM: Hexagonal
Z: 1
LATTICE CONSTANTS:
  $a = 5.99$
  $c = 3.53$

3 STRONGEST DIFFRACTION LINES:
    1.717 (100) broad
    2.086 ( 90)
    3.00  ( 50)
OPTICAL CONSTANTS:
    $\omega$ = 1.472
    $\epsilon$ = 1.492
      (+)
HARDNESS: 4½
DENSITY: 4.21 (Meas.)
CLEAVAGE: Prismatic–fair. Brittle.
HABIT: Massive; cryptocrystalline and crystalline.
COLOR-LUSTER: Cream, yellowish, rose; dull to vitreous. Streak white.
MODE OF OCCURRENCE: Occurs in albitized granites and associated quartz-microcline veins of "one of the granitic massifs of Kazakhstan" and in "analogous rocks of other regions of the USSR." The mineral alters easily and is replaced by aggregates of tengerite, synchisite, and yttrofluorite.
BEST REF. IN ENGLISH: Stepanov, A. V., and Severov, E. A., *Am. Min.*, **47**: 805 (1962).

# GAGEITE
$H_8(Mn, Mg)_7Si_2O_{15}$

CRYSTAL SYSTEM: Orthorhombic
CLASS: 2/m 2/m 2/m or mm2
SPACE GROUP: Pnnm or Pnn2
Z: 2
LATTICE CONSTANTS:
    $a$ = 13.79
    $b$ = 13.68
    $c$ = 3.279
3 STRONGEST DIFFRACTION LINES:
    6.87  (100)
    2.758 ( 80)
    2.707 ( 80)
HARDNESS: Not determined.
DENSITY: 3.584 (Meas.)
         3.554 (Calc.)
CLEAVAGE: Not determined.
HABIT: Acicular crystals up to 0.1 mm in thickness consisting of basal pinacoid and unit prism; as finely matted fibers or bundles, and in radial aggregates.
COLOR-LUSTER: Colorless to pink. Vitreous.
MODE OF OCCURRENCE: Occurs in open cavities associated with late-stage low-temperature hydrothermal veins at Franklin, New Jersey.
BEST REF. IN ENGLISH: Palache, Charles, USGS Prof. Paper 180, p. 111 (1935).  Moore, Paul B., *Am. Min.*, **53**: 309–315 (1968).  Moore, Paul B., *Am. Min.*, **54**: 1005–1017 (1969).

# GAHNITE
$ZnAl_2O_4$
Spinel Series

CRYSTAL SYSTEM: Cubic
CLASS: 4/m $\bar{3}$ 2/m

SPACE GROUP: Fd3m
Z: 8
LATTICE CONSTANTS:
    $a$ = 8.062 (synthetic)
3 STRONGEST DIFFRACTION LINES:
    2.438 (100)
    2.861 ( 84)
    1.429 ( 43)
OPTICAL CONSTANT:
    $N$ = 1.790
HARDNESS: 7½–8
DENSITY: 4.607 (Calc.)
CLEAVAGE: {111} parting, indistinct
Fracture conchoidal to uneven. Brittle.
HABIT: Crystals usually octahedral, sometimes modified; rarely cubic or dodecahedral. Also massive, coarse granular to compact, or as rounded grains. Twinning on {111}, sometimes as sixlings by repeated twinning.
COLOR-LUSTER: Greenish black, bluish black, dark bluish green, brown, yellowish. Translucent to nearly opaque. Vitreous to greasy. Streak grayish.
MODE OF OCCURRENCE: Occurs chiefly in crystalline schists, contact metamorphosed limestones, high-temperature replacement ore deposits, granite pegmatites, and placer deposits.  In the United States notable occurrences are found in California, Colorado, Maine, Connecticut, Massachusetts, Pennsylvania, New Jersey, Maryland, North Carolina, and Georgia.  It also occurs in Canada, Brazil, Sweden, Spain, Italy, Germany, Finland, India, Australia, and Madagascar.
BEST REF. IN ENGLISH: Palache, et al., "Dana's System of Mineralogy," 7th Ed., v. I, p. 689–697, New York, Wiley, 1944.

# GALAXITE
$MnAl_2O_4$
Spinel Series

CRYSTAL SYSTEM: Cubic
CLASS: 4/m $\bar{3}$ 2/m
SPACE GROUP: Fd3m
Z: 8
LATTICE CONSTANT:
    $a$ = 8.271 (synthetic)
3 STRONGEST DIFFRACTION LINES:
    2.492 (100)
    2.921 ( 60)
    1.460 ( 45)
OPTICAL CONSTANT:
    $N$ = 1.923
HARDNESS: 7½–8
DENSITY: 4.077 (Calc.)
CLEAVAGE: Fracture conchoidal to uneven.
HABIT: Massive; fine granular. As embedded grains.
COLOR-LUSTER: Black. Opaque. Streak reddish brown.
MODE OF OCCURRENCE: Occurs at Bald Knob, near Galax, Alleghany County, North Carolina, in a vein with alleghanyite, tephroite, rhodonite, spessartine, and calcite.
BEST REF. IN ENGLISH: Palache, et al., "Dana's System of Mineralogy," 7th Ed., v. I, p. 689–697, New York, Wiley, 1944.

## GALEITE
$Na_{15}(SO_4)_5F_4Cl$

CRYSTAL SYSTEM: Hexagonal
CLASS: $\bar{3}\,2/m$
SPACE GROUP: $P\bar{3}m$
Z: 3
LATTICE CONSTANTS:
  $a = 12.17$
  $c = 13.94$
3 STRONGEST DIFFRACTION LINES:
  2.79 (100)
  3.52 ( 80)
  3.68 ( 70)
OPTICAL CONSTANTS:
  $\omega = 1.447$
  $\epsilon = 1.449$
    (+)
  HARDNESS: Not determined.
  DENSITY: 2.605 (Meas.)
           2.610 (Calc.)
  HABIT: Small nodular aggregates of crystals up to 1 mm in diameter. Crystals hexagonal, barrel-shaped, rarely tabular, and a rhombohedron with rho 66°25′ is the most persistent form.
  COLOR-LUSTER: White.
  MODE OF OCCURRENCE: Occurs embedded in clay, often associated with gaylussite and northupite, in drill cores from Searles Lake, California.
  BEST REF. IN ENGLISH: Pabst, Adolph, Sawyer, D. L., and Switzer, George, GSA Bull., **66**: 1658-1659 (1955). *Am. Min.*, **48**: 485-510 (1963).

## GALENA
PbS

CRYSTAL SYSTEM: Cubic
CLASS: $4/m\,\bar{3}\,2/m$
SPACE GROUP: Fm3m
Z: 4
LATTICE CONSTANT:
  $a = 5.936$
3 STRONGEST DIFFRACTION LINES:
  2.969 (100)
  3.429 ( 84)
  2.099 ( 57)
OPTICAL CONSTANT:
  $N = 3.912$ (Na)
  HARDNESS: 2½
  DENSITY: 7.58 (Meas.)
           7.596 (Calc.)
  CLEAVAGE: {001} perfect, easy
Fracture subconchoidal.
  HABIT: Crystals usually cubic, octahedral, or cubo-octahedral; also tabular, skeletal, reticulated. Commonly massive, cleavable; coarse or fine granular; fibrous; plumose. Twinning on {111}, {114}, {144}.
  COLOR-LUSTER: Lead gray. Opaque. Bright metallic. Streak lead gray.
  MODE OF OCCURRENCE: Occurs chiefly in limestones, dolomites, and other sedimentary rocks; in hydrothermal ore veins; in contact metamorphic deposits; and rarely in pegmatites. Found widespread in the United States, especially in Missouri, Oklahoma, Kansas, Illinois, Iowa, Wisconsin, South Dakota, Colorado, Montana, Idaho, Utah, and California. Important deposits also occur in Canada, Mexico, Chile, Peru, Bolivia, England, Scotland, France, Belgium, Italy, Sardinia, Germany, Austria, Roumania, Czechoslovakia, USSR, and New South Wales, Australia.
  BEST REF. IN ENGLISH: Palache, et al., "Dana's System of Mineralogy," 7th Ed., v. I, p. 200-204, New York, Wiley, 1944.

## GALENOBISMUTITE (Bismutoplagionite, Cannizzarite)
$PbBi_2S_4$

CRYSTAL SYSTEM: Orthorhombic
CLASS: $2/m\,2/m\,2/m$
SPACE GROUP: Pnam
Z: 4
LATTICE CONSTANTS:
  $a = 11.65$
  $b = 14.49$
  $c = 4.08$
3 STRONGEST DIFFRACTION LINES:
  3.45 (100)
  1.961 ( 50)
  3.03 ( 40)
  HARDNESS: 2½-3½
  DENSITY: 7.04 (Meas.)
           7.15 (Calc.)
  CLEAVAGE: {110} good
Flexible.
  HABIT: Crystals lath-like, as extremely thin plates, or acicular; frequently twisted or bent; vertically striated. Usually massive; compact, fibrous to columnar.
  COLOR-LUSTER: Light gray to lead gray to tin white, sometimes tarnished yellowish or iridescent. Opaque. Metallic. Streak black.
  MODE OF OCCURRENCE: Occurs at the Greenback and Tucson mines in the Leadville district, Lake County, Colorado; with pyrite and other sulfides at Wickes, Jefferson County, Montana; at the Belzazzar mine, Quartzburg district, Idaho; with cosalite, native gold, and quartz, at the Cariboo mine, Barkerville, British Columbia, Canada; at Gladhammar and Nordmark, Sweden; and in fumaroles on Vulcano, Lipari Islands, Italy.
  BEST REF. IN ENGLISH: Palache, et al., "Dana's System of Mineralogy," 7th Ed., v. I, p. 471-472, New York, Wiley, 1944.

## GALLITE
$CuGaS_2$

CRYSTAL SYSTEM: Tetragonal
CLASS: $\bar{4}2m$
SPACE GROUP: $I\bar{4}2d$
Z: 4
LATTICE CONSTANTS:
  $a = 5.35$
  $c = 10.48$

3 STRONGEST DIFFRACTION LINES:
3.064 (100)
1.876 ( 70)
1.611 ( 60)
OPTICAL CONSTANTS: Anisotropy low.
HARDNESS: 3–3½
DENSITY: 4.40 (Calc.)
CLEAVAGE: Not determined.
HABIT: Massive; as minute embedded grains and as exsolution lamellae.
COLOR-LUSTER: Gray. Opaque. Metallic. Streak gray-black.
MODE OF OCCURRENCE: The mineral occurs in renierite and germanite and as exsolution lamellae along (100) and (111) of sphalerite at Tsumeb, South West Africa, and at the Prince Leopold Mine, Kipushi, Zaire.
BEST REF. IN ENGLISH: Strunz, H., Geier, B. H., and Seeliger, E., *Am. Min.*, **44**: 906 (1959).

## GAMAGARITE (Inadequately described mineral)
$Ba_4 (Fe, Mn)_2 V_4 O_{15} (OH)_2$

CRYSTAL SYSTEM: Monoclinic
LATTICE CONSTANTS:
$c/a = 1.15$
$\beta = 117°20'$
OPTICAL CONSTANTS:
$\alpha = 2.016$
$\beta = 2.040$
$\gamma = 2.130$
$(+)2V = 46°–62°$
HARDNESS: 4½–5
DENSITY: 4.62 (Meas.)
CLEAVAGE: {001} distinct
{100} distinct
{$\bar{1}$01} indistinct
Brittle.
HABIT: Crystals prismatic, somewhat flattened, indistinct. Also as aggregates of needles up to 1 cm or more in length.
COLOR-LUSTER: Dark brown to nearly black. Adamantine.
MODE OF OCCURRENCE: Occurs in association with diaspore, ephesite, and bixbyite (sitaparite), on the Gloucester farm, about 16 miles north of Postmasburg, South Africa.
BEST REF. IN ENGLISH: de Villiers, J. E., *Am. Min.*, **28**: 329–335 (1943).

## GANOMALITE (Inadequately described mineral)
$Pb_6 Ca_4 (OH)_2 (Si_2 O_7)_3$

CRYSTAL SYSTEM: Hexagonal
LATTICE CONSTANT:
$c/a = 1.317$
OPTICAL CONSTANTS:
$\omega = 1.910$
$\epsilon = 1.945$
(+)
HARDNESS: 3

DENSITY: 5.74
CLEAVAGE: Prismatic.
HABIT: Crystals prismatic; usually massive, granular, or as curved plates.
COLOR-LUSTER: Colorless, white to gray. Transparent to translucent. Vitreous to pearly.
MODE OF OCCURRENCE: Occurs at Långban, and as a common mineral in skarn at Jacobsberg, Sweden.
BEST REF. IN ENGLISH: Winchell and Winchell, "Elements of Optical Mineralogy," 4th Ed., Pt. 2, p. 478–479, New York, Wiley, 1951.

## GANOPHYLLITE
$NaMn_3 (OH)_4 (Si, Al)_4 O_{10}$
Dimorphous with bannisterite

CRYSTAL SYSTEM: Monoclinic
CLASS: 2/m
SPACE GROUP: A2/a
Z: 12
LATTICE CONSTANTS:
$a = 16.60$
$b = 27.04$
$c = 50.34$
$\beta = 94°10'$
3 STRONGEST DIFFRACTION LINES:
12.5 (100)
3.14 ( 25)
2.70 ( 14)
OPTICAL CONSTANTS:
$\alpha = 1.573$
$\beta = 1.611$
$\gamma = 1.613$
$(-)2V = small$
HARDNESS: 4–4½
DENSITY: 2.84–2.878 (Meas.)
2.923 (Calc.)
CLEAVAGE: {001} perfect, easy
Brittle.
HABIT: Crystals tabular or short prismatic; also foliated, micaceous; as rosettes.
COLOR-LUSTER: Brown to brownish yellow. Translucent.
MODE OF OCCURRENCE: Occurs in seams with barite in a manganese boulder found near Alum Rock Park, Santa Clara County, California; in association with rhodonite, willemite, axinite, and bustamite at Franklin, Sussex County, New Jersey; and at the Sjö mine near Grythyttan, and at Harstigen, near Persberg, Sweden.
BEST REF. IN ENGLISH: Palache, Charles, USGS Prof. Paper 180, p. 114 (1935).

## GARNET = Mineral Group of general formula:
$A_3 B_2 (SiO_4)_3$, A = Ca, Mg, $Fe^{2+}$, $Mn^{2+}$; B = Al, $Fe^{3+}$, Cr, V, Zr, Ti.

See:
| | |
|---|---|
| Almandine | Kimzeyite |
| Andradite | Knorringite |
| Goldmanite | Majorite |
| Grossular | Pyrope |
| Henritermierite | Spessartine |
| Hydrogrossular | Uvarovite |

**GARNIERITE** = General term for hydrous nickel silicates $(Ni, Mg)_3 Si_2 O_5 (OH)_4$

## GARRELSITE
$Ba_3 NaSi_2 B_6 O_{13} (OH)_7$

CRYSTAL SYSTEM: Monoclinic
CLASS: 2/m or m
SPACE GROUP: A2/a or Aa
Z: 8
LATTICE CONSTANTS:
   $a = 13.43$
   $b = 8.45$
   $c = 14.61$
   $\beta = 114°19'$
3 STRONGEST DIFFRACTION LINES:
   3.02 (100)
   2.07 ( 90)
   2.00 ( 90)
OPTICAL CONSTANTS:
   $\alpha = 1.620$
   $\beta = 1.633$
   $\gamma = 1.640$
   $(-)2V = 72°$
HARDNESS: Not determined.
DENSITY: 3.68 (Meas.)
          3.73 (Calc.)
CLEAVAGE: Not determined.
HABIT: Crystals have four-sided bipyramidal shape, formed by prismatic faces (110) and (1$\bar{1}$0), and pyramidal faces (21$\bar{1}$) and (2$\bar{1}\bar{1}$). Faces striated by alteration of prism and pyramid.
COLOR-LUSTER: Colorless.
MODE OF OCCURRENCE: Found in brown dolomitic shale containing nahcolite, shortite, searlesite, and microscopic wurtzite in a drill core from a depth of 2139 to 2370 feet at the Sun Oil Co. South Ouray No. 1 well, Uintah County, Utah.
BEST REF. IN ENGLISH: Milton, Charles, Axelrod, J. M., and Grimaldi, F. S., *Bull. Geol. Soc. Am.*, **66**: 1957 (1955).

## GARRONITE
$Na_2 Ca_5 Al_{12} Si_{20} O_{64} \cdot 27H_2 O$
Zeolite Group

CRYSTAL SYSTEM: Tetragonal
LATTICE CONSTANTS:
   $a = 9.85–10.01$
   $c = 9.87–10.32$
3 STRONGEST DIFFRACTION LINES:
   4.12 (100)
   3.14 (100)
   2.66 (100)
OPTICAL CONSTANTS:
   $\omega = 1.500–1.515$
   $\epsilon = 1.502–1.512$
   (+) or (−)
HARDNESS: Not given (probably ~4).
DENSITY: 2.13–2.17 (Meas.)
          2.201 (Calc.)

CLEAVAGE: Two cleavages at about 90° to each other and parallel to the lengths of the individuals.
HABIT: Radiating aggregates in amygdules.
COLOR-LUSTER: Not given (probably white).
MODE OF OCCURRENCE: Occurs as amygdule fillings in Tertiary basalts at 10 localities in eastern Antrim, Ireland and at 22 localities in eastern Iceland. Other zeolites are present such as: chabazite, thomsonite, levyne, natrolite, phillipsite, heulandite and mesolite.
BEST REF. IN ENGLISH: Walker, G. P. L., *Min. Mag.*, **33**: 173–186 (1962).

## GASPEITE
$(Ni, Mg, Fe)CO_3$
Calcite Group

CRYSTAL SYSTEM: Hexagonal
CLASS: $\bar{3}$ 2/m
SPACE GROUP: R$\bar{3}$m
Z: 3
LATTICE CONSTANTS:
   $a = 4.621$
   $c = 14.93$
3 STRONGEST DIFFRACTION LINES:
   2.741 (100)
   1.692 ( 45)
   3.543 ( 36)
OPTICAL CONSTANTS:
   $\omega = 1.83$
   $\epsilon = 1.61$
   (−)
HARDNESS: 4.5–5
DENSITY: 3.71 (Meas.)
          3.748 (Calc.)
CLEAVAGE: Rhombohedral $\{10\bar{1}1\}$ good. Uneven fracture.
HABIT: Rhombohedral crystals up to 0.5 mm in length.
COLOR-LUSTER: Light green; vitreous to dull. Yellow-green streak.
MODE OF OCCURRENCE: Found as a light green 2-foot wide vein enclosed in a varicolored siliceous dolomite in the Gaspé Peninsula, Quebec, Canada.
BEST REF. IN ENGLISH: Kohls, D. W., and Rodda, J. L., *Am. Min.*, **51**: 677–684 (1966).

## GASTUNITE = Weeksite
$(K, Na)_2 (UO_2)_3 (Si_2 O_5)_4 \cdot 8H_2 O$

CRYSTAL SYSTEM: Orthorhombic
CLASS: 2/m 2/m 2/m
SPACE GROUP: Pnna
LATTICE CONSTANTS:
   $a = 14.24$
   $b = 35.84$
   $c = 14.20$
3 STRONGEST DIFFRACTION LINES:
   7.12 (100)
   3.56 ( 90)
   9.10 ( 80)

CORUNDUM (ruby) Ampanihy, Madagascar.

CORUNDUM Theraka, Kenya.

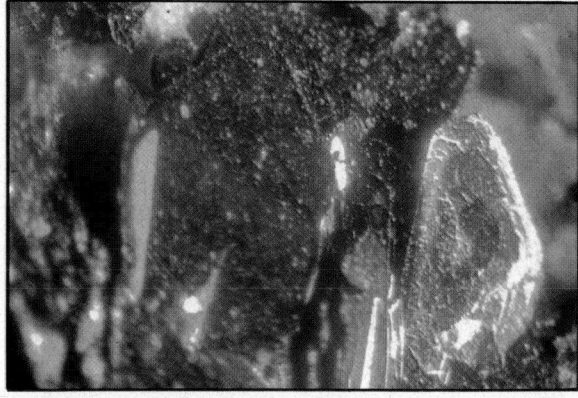

COUSINITE Musonoi, Katanga, Zaire.

COVELLITE Butte, Silver Bow County, Montana.

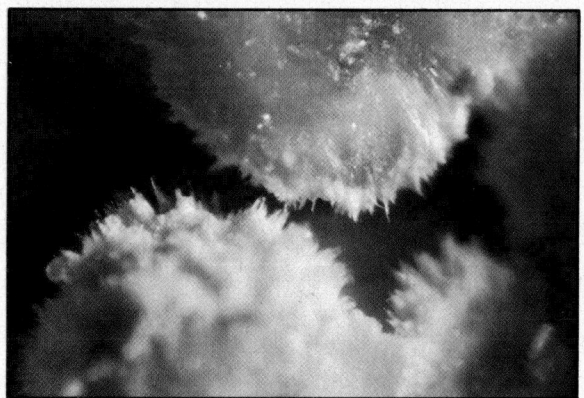

CRANDALLITE Clay Canyon, Fairfield, Utah County, Utah.

CREEDITE Santa Eulalia, Chihuahua, Mexico.

CREEDITE Santa Eulalia, Chihuahua, Mexico.

CRISTOBALITE in obsidian    Pachuca, Hidalgo, Mexico.

CROCIDOLITE (Variety of riebeckite)    Salzburg, Austria.

CROCOITE with pyromorphite    Beresov, Urals, USSR.

CROCOITE    Dundas, Tasmania.

CROCOITE    Dundas, Tasmania.

CRONSTEDTITE    Kisbanya, Roumania.

CRYOLITE   Ivigtut, Greenland.

CUBANITE with pyrrhotite   Morrò Velho, Minas Geraes, Brazil.

CUBANITE with pyrrhotite   Morro Velho, Minas Geraes, Brazil.

CUMENGITE   Boleo, Baja California, Mexico.

CUMENGITE   Boleo, Baja California, Mexico.

CUMENGITE in ancient slag   Laurium, Greece.

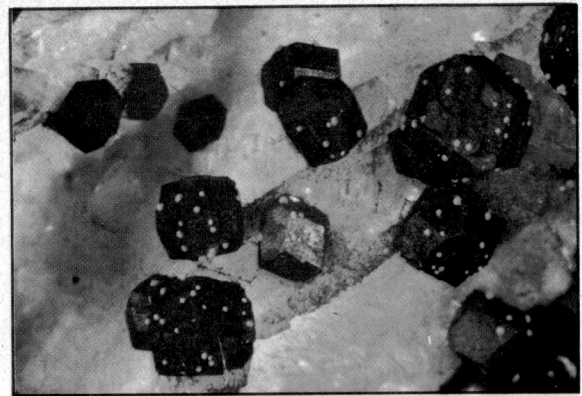

CUPRITE with malachite   Tsumeb, South West Africa.

CUPRITE, Wheal Gorland, St. Day, Cornwall, England.

CUPRITE with malachite   Tsumeb, South West Africa.

CUPRITE  (Variety chalcotrichite)  Morenci, Greenlee County, Arizona.

CUPRITE on and in calcite   Ajo, Pima County, Arizona.

CUPRITE (Variety chalcotrichite   SEM photograph)   Bisbee, Cochise County, Arizona.

CUPRITE   Tsumeb, South West Africa.

CUPRITE with calcite   Tsumeb, South West Africa.

CURIENITE   Mounana, Gabon.

CURITE with kasolite and torbernite   Shinkolobwe, Katanga, Zaire.

CURITE on soddyite   Shinkolobwe, Katanga, Zaire.

CURITE with torbernite   Shinkolobwe, Katanga, Zaire.

CYANOTRICHITE   Grandview mine, Coconino County, Arizona.

CYANOTRICHITE   Moldavia, Banat, Roumania.

CYLINDRITE   Poopo, Bolivia.

DANBURITE Charcas, San Luis Potosi, Mexico.

DACHIARDITE Speranza, Elba.

DATOLITE with epidote Lane's quarry, Westfield, Massachusetts.

DATOLITE with epidote Lane's quarry, Westfield, Massachusetts.

DAWSONITE on calcite Francon quarry, Montreal, Quebec, Canada.

DEMESMAEKERITE Musonoi, Katanga, Zaire.

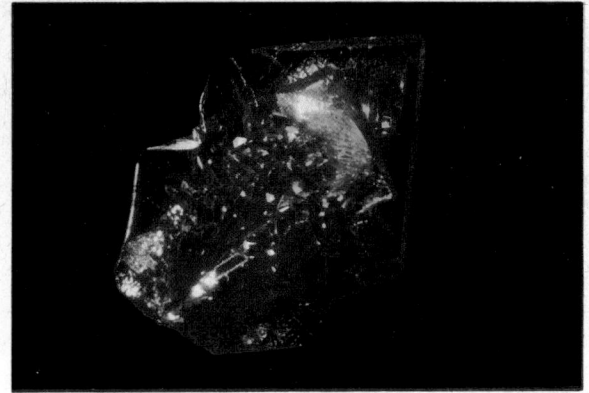

DESCLOIZITE   Grootfontein, South West Africa.

DESCLOIZITE   Near Georgetown, Grant County, New Mexico.

DESCLOIZITE   Lake Valley, Sierra County, New Mexico.

DEVILLITE   Ecton mine, Montgomery County, Pennsylvania.

DEVILLITE   Herrengrund, Czechoslovakia.

DIABOLEITE   Mammoth-St. Anthony mine, Pinal County, Arizona.

DIABOLEITE   Mammoth-St. Anthony mine, Pinal County, Arizona.

DIABOLEITE   Laurium, Greece.

DIAMOND in kimberlite   Kimberley, South Africa.

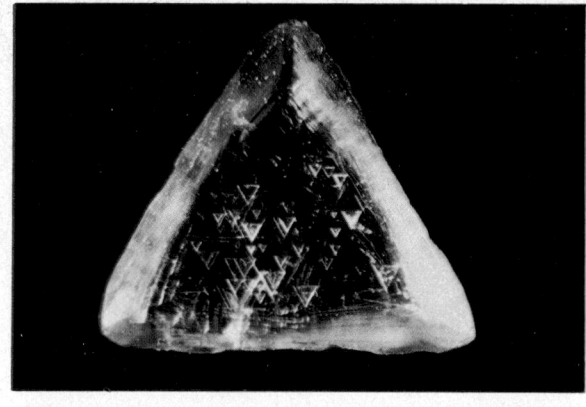

DIAMOND   Zaire, Katanga.

DIAMOND   Zaire, Katanga.

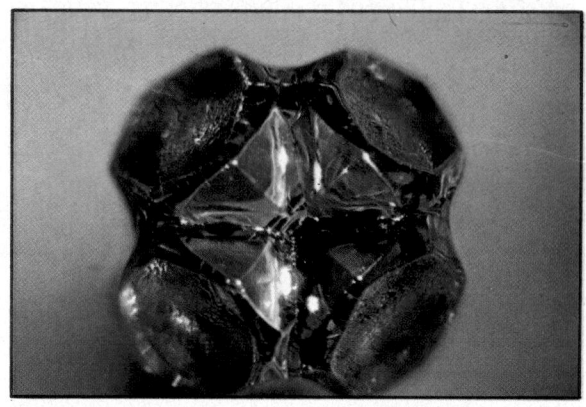

DIAMOND   Brazil.

DIAMOND   Kimberley, South Africa.

DIAMOND (twin)   Zaire, Katanga.

DIAMOND (parallel growth)   Zaire, Katanga.

DIAMOND (twins)   Kimberley, South Africa.

40

DIAMOND (trigon)   Kimberley, South Africa.

DIAMOND (trigons)   Kimberley, South Africa.

DIAMOND (trigons)   Kimberley, South Africa.

DIAMOND (trigons)   Kimberley, South Africa.

DIAMOND (etching on a cube face)   Zaire, Katanga.

DIAMOND (trigons)   Kimberley, South Africa.

DIAMOND (trigons)   Kimberley, South Africa.

DIAMOND   Zaire, Katanga.

DICKINSONITE   Branchville, Connecticut.

DIOPSIDE   Pizzo Cerbandome, Switzerland.

DIOPSIDE with grossular   Ala, Piedmont, Italy.

DIOPTASE   Tsumeb, South West Africa.

DIOPTASE   Tsumeb, South West Africa.

DIOPTASE   Tsumeb, South West Africa.

DOLOMITE   Lengenbach quarry, Binnental, Switzerland.

DOLOMITE with pyrite and marcasite Picher, Ottawa County, Oklahoma.

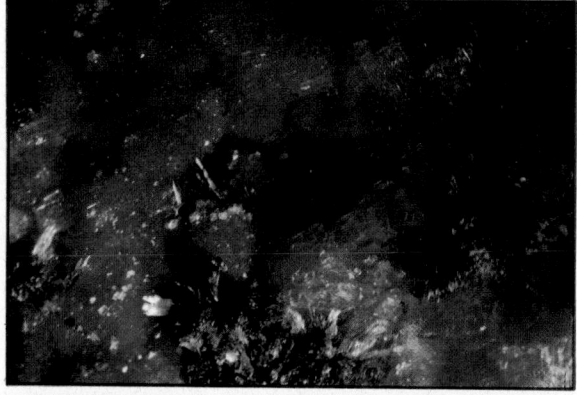

DOLOMITE enclosing cuprite, with malachite   Tsumeb, South West Africa.

DRESSERITE   Francon quarry, Montreal, Quebec, Canada.

DUFRENITE   Indian Mountain, Cherokee County, Alabama.

DUFTITE with mimetite and austinite   Mapimi, Durango, Mexico.

DUNDASITE with crocoite   Dundas, Tasmania.

43

DURANGITE   Durango, Mexico.

EAKERITE   Foote mine, Kings Mountain, North Carolina.

EAKERITE   Foote mine, Kings Mountain, North Carolina.

EGLESTONITE   Terlingua, Brewster County, Texas.

EGLESTONITE   Terlingua, Brewster County, Texas.

EKANITE   Mont St. Hilaire, Quebec, Canada.

EKANITE   Mont St. Hilaire, Quebec, Canada.

ELBAITE   Newry, Oxford County, Maine.

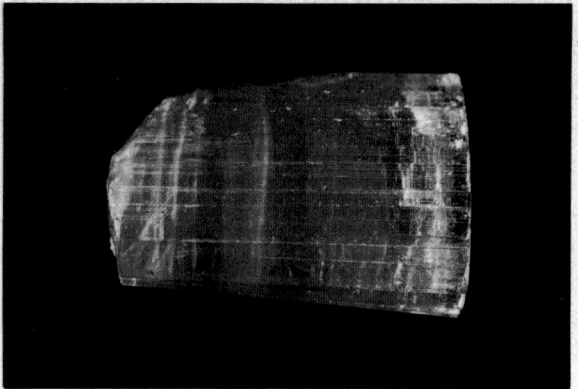

ELBAITE   Minas Geraes, Brazil.

ELBAITE   Dunton mine, Newry, Oxford County, Maine.

ELBAITE   Minas Geraes, Brazil.

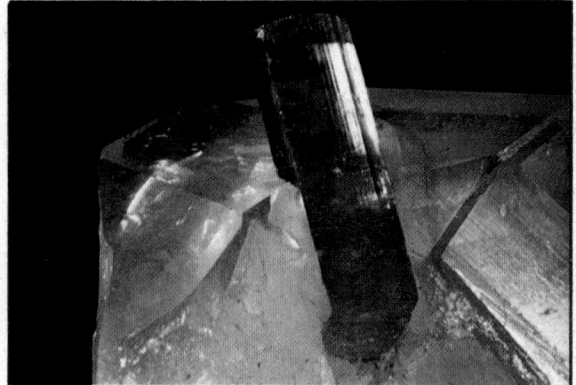

ELBAITE   Governador Valadares, Minas Geraes, Brazil.

ELBAITE   Minas Geraes, Brazil.

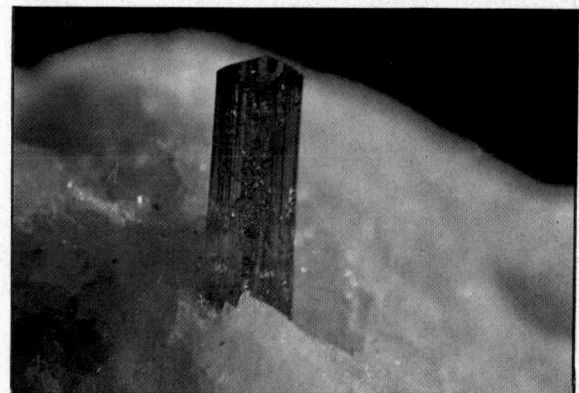

ELBAITE   Lengenbach quarry, Binnental, Switzerland.

45

EMBOLITE with malachite  Bisbee, Cochise County, Arizona.

EMMONSITE  Moctezuma, Sonora, Mexico.

EMMONSITE  Moctezuma, Sonora, Mexico.

EMPLECTITE with chalcopyrite  Near Schwarzenberg, Saxony, Germany.

ENARGITE  Butte, Silver Bow County, Montana.

ENARGITE  Longfellow mine, Red Mountain Pass, San Juan County, Colorado.

ENARGITE with anglesite  Tintic Standard mine, Tintic District, Utah.

ENARGITE  Tintic Standard mine, Tintic District, Utah.

ENGLISHITE   Clay Canyon, Fairfield, Utah County, Utah.

EOSPHORTIE   Taquaral, Minas Geraes, Brazil.

EOSPHORITE   Newry, Oxford County, Maine.

EOSPHORITE   Taquaral, Minas Geraes, Brazil.

EOSPHORITE   Taquaral, Minas Geraes, Brazil.

EPIDIDYMITE   Mont St. Hilaire, Quebec, Canada.

EPIDIDYMITE   Langesundfjord, Norway.

EPIDOTE   Untersulzbachtal, Salzburg, Austria.

EPIDOTE   Lane's quarry, Westfield, Massachusetts.

ERYTHRITE   Bou Azzer, Morocco.

ERYTHRITE   Bou Azzer, Morocco.

EUCHROITE   Libethen, Hungary.

EUDIALYTE   Mont St. Hilaire, Quebec, Canada.

OPTICAL CONSTANTS:
$\alpha$ = 1.604
$\beta$ = 1.610
$\gamma$ = 1.621
(+)2V = moderate
HARDNESS: 2
DENSITY: 3.96 (Meas.)
3.97 (Calc.)
CLEAVAGE: {010} perfect
HABIT: Radial aggregates of acicular to fibrous crystals; also dense microcrystalline aggregates pseudomorphous after uraninite.

COLOR-LUSTER: Yellow to greenish yellow (radial aggregates). Dull straw yellow (dense material). Fluoresces light yellow-green in ultraviolet light.

MODE OF OCCURRENCE: Occurs as a secondary mineral in radial aggregates coating chalcedony at the Red Knob mine, Muggins Mountains, Yuma County, Arizona; in vesicular cavities in welded tuff at the Mammoth prospect, Presidio County, Texas; and in dense fine-grained pseudomorphs after uraninite from the Easton area, Pennsylvania. It is also found associated with uranophane, beta-uranophane, schroeckingerite, calcite, fluorite, and stilbite in crevices in gneiss on the Radhausberg near Bad Gastein, Salzburg, Austria.

BEST REF. IN ENGLISH: Honea, Russell M., *Am. Min.*, **44**: 1047–1056 (1959).

# GAUDEFROYITE

$Ca_4Mn_{3-x}[(BO_3)_3/(CO_3)/(O, OH)_3]$ with x $\approx$ 0.17

CRYSTAL SYSTEM: Hexagonal
CLASS: 6
SPACE GROUP: $P6_3$
Z: 2
LATTICE CONSTANTS:
$a$ = 10 .6
$c$ = 5.90
3 STRONGEST DIFFRACTION LINES:
2.95 (100)
2.62 (100)
2.46 (100)
OPTICAL CONSTANTS:
$\omega$ = 1.81
$\epsilon$ = 2.02
(+)
HARDNESS: 6
DENSITY: 3.35–3.50 (Meas.)
3.44 (Calc.)
CLEAVAGE: Brittle. Conchoidal fracture.
HABIT: Acicular hexagonal prisms up to 5 cm long, with pyramidal terminations.
COLOR-LUSTER: Black; brilliant to dull.
MODE OF OCCURRENCE: Occurs associated with marokite, braunite, hausmannite, crednerite, polianite (as an alteration product along the gaudefroyite cleavage), and five as yet unidentified minerals, in a gangue of white, non-manganiferous calcite, quartz, and locally brucite, on the dumps of vein no. 2, Tachgagalt, 17 km SSW of Ouarzazate, Morocco.

BEST REF. IN ENGLISH: Jouravsky, G., and Permingeat, F., *Am. Min.*, **50**: 806–807 (1965).

# GAYLUSSITE (Gay-lussite)

$Na_2Ca(CO_3)_2 \cdot 5H_2O$

CRYSTAL SYSTEM: Monoclinic
CLASS: 2/m
SPACE GROUP: I2/a
Z: 4
LATTICE CONSTANTS:
$a$ = 11.589
$b$ = 7.779
$c$ = 11.207
$\beta$ = 101°58'
3 STRONGEST DIFFRACTION LINES:
3.205 (100)
6.407 ( 57)
2.635 ( 54)
OPTICAL CONSTANTS:
$\alpha$ = 1.445
$\beta$ = 1.516
$\gamma$ = 1.522
(−)2V = 34°
HARDNESS: 2½–3
DENSITY: 1.995 (Meas.)
1.989 (Calc.)
CLEAVAGE: {110} perfect
{001} indistinct
Fracture conchoidal. Brittle.
HABIT: Crystals flattened, wedge-shaped, often elongated.

COLOR-LUSTER: Colorless, white, grayish, yellowish. Transparent to translucent. Vitreous.

MODE OF OCCURRENCE: Occurs abundantly as euhedral crystals up to 2 inches or more in length at Searles Lake, San Bernardino County; in the muds of Deep Spring Lake and in the lake bed deposits of Owens Lake, Inyo County; in the China Lake deposits, Kern County; and with glauberite and northupite in the muds of Borax Lake, Lake County, California. It is also found at Independence Rock, Wyoming; in the Carson desert near Ragtown, Nevada; at Lagunillas, near Merida, Venezuela; and in the eastern Gobi desert, Mongolia.

BEST REF. IN ENGLISH: Palache, et al., "Dana's System of Mineralogy," 7th Ed., v. II, p. 234–235, New York, Wiley, 1951. Fang, J. H., Robinson, P. D., Cerven, J. F., and Wolf, L. A., *Am. Min.*, **52**: 1570–1572 (1967).

# GEARKSUTITE

$CaAlF_4OH \cdot H_2O$

CRYSTAL SYSTEM: Probably monoclinic
3 STRONGEST DIFFRACTION LINES:
4.55 (100)
3.15 ( 80)
2.28 ( 80)
OPTICAL CONSTANTS:
$\alpha$ = 1.448
$\beta$ = 1.454
$\gamma$ = 1.456
(−)2V = moderate
HARDNESS: 2 (aggregates)

DENSITY: 2.768 (Meas.)
CLEAVAGE: Fracture (masses) subconchoidal to uneven.
HABIT: Massive; chalk-like.
COLOR-LUSTER: White; dull.
MODE OF OCCURRENCE: Occurs in Colorado associated with fluorite at several mines in the Jamestown district, Boulder County; as a powdery cavity filling in pachnolite at the cryolite deposit at St. Peters Dome, El Paso County, and as chalk-like masses up to several inches in diameter in a fluorite-barite deposit at Wagon Wheel Gap, Mineral County. It also occurs in a deposit between Hot Springs and Warm Springs, Virginia; with thomsenolite in the cryolite deposit at Ivigtut, Greenland; at Miask, Urals, USSR; and as nodular masses in sandstone at Gingin, Western Australia.
BEST REF. IN ENGLISH: Palache, et al., "Dana's System of Mineralogy," 7th Ed., v. II, p. 119–120, New York, Wiley, 1951.

# GEDRITE (Aluminian anthophyllite)

$(Mg, Fe^{2+}, Al)_7 (Si, Al)_8 O_{22} (OH)_2$
Amphibole Group

CRYSTAL SYSTEM: Orthorhombic
CLASS: 2/m 2/m 2/m
SPACE GROUP: Pnma
Z: 4
LATTICE CONSTANTS:
$a = 18.594$
$b = 17.890$
$c = 5.304$
3 STRONGEST DIFFRACTION LINES:
3.06 (100)
8.27 ( 80)
3.23 ( 70)
OPTICAL CONSTANTS:
$\alpha = 1.671$
$\beta = 1.681$
$\gamma = 1.690$
$(-)2V = 75°$
HARDNESS: 5½–6
DENSITY: 3.15–3.57 (Meas.)
3.334 (Calc.)
CLEAVAGE: {110} perfect
{010} imperfect
{100} imperfect
HABIT: Crystals prismatic, rare. Usually massive, fibrous or lamellar.
COLOR-LUSTER: White, gray, greenish, brownish green, clove brown, yellowish brown, dark brown. Transparent to nearly opaque. Vitreous to silky, somewhat pearly on cleavage. Streak uncolored or grayish.
MODE OF OCCURRENCE: Occurs only in metamorphic rocks, such as schists and gneisses, or in metasomatic rocks. Found relatively widespread, at times as the principal constituent of the rock. Typical occurrences are found in Idaho, Montana, Arizona, Maine, and North Carolina. Also in Greenland, Scotland, Finland, USSR, India, Japan, and Australia.
BEST REF. IN ENGLISH: Deer, Howie, and Zussman, "Rock Forming Minerals," v. 2, p. 211–229, New York, Wiley, 1963.

# GEDROITZITE (Species status uncertain. Related to vermiculite).

$Na_2 Al_2 Si_3 O_{10} \cdot 2H_2O$ (?)

# GEHLENITE

$Ca_2 Al_2 SiO_7$
Melilite Group
Isomorphous with åkermanite

CRYSTAL SYSTEM: Tetragonal
CLASS: $\bar{4}$ 2m
SPACE GROUP: P$\bar{4}2_1$m
Z: 2
LATTICE CONSTANTS:
$a = 7.690$
$c = 5.067$
3 STRONGEST DIFFRACTION LINES:
2.85 (100)
1.75 (100)
2.43 ( 70)
OPTICAL CONSTANTS:
$\omega = 1.669$
$\epsilon = 1.658$
(-)
HARDNESS: 5–6
DENSITY: 3.038 (Meas.)
3.03 (Calc.)
CLEAVAGE: {001} distinct
{110} indistinct
Fracture uneven to conchoidal. Brittle.
HABIT: Crystals usually short prismatic. Commonly massive, granular. Twinning on {100}, {001}.
COLOR-LUSTER: Colorless, grayish green to brown, yellowish. Transparent to translucent. Vitreous to resinous.
MODE OF OCCURRENCE: Occurs in calcium-rich basic eruptive rocks; in thermally metamorphosed impure limestones; and in furnace slags and artificial melts. Found at Crestmore, Riverside County, and in Inyo and Tulare counties, California; in Luna County, New Mexico; at Iron Hill, Gunnison County, Colorado; at Velardeña, Durango. Mexico; in Quebec, Canada; and at localities in Ireland, Italy, Roumania, USSR, Java, and elsewhere.
BEST REF. IN ENGLISH: Deer, Howie, and Zussman, "Rock Forming Minerals," v. 1, p. 236–255, New York, Wiley, 1962.

# GEIKIELITE

$MgTiO_3$
Ilmenite Group

CRYSTAL SYSTEM: Hexagonal
CLASS: $\bar{3}$
SPACE GROUP: R$\bar{3}$
Z: 6
LATTICE CONSTANTS:
$a = 5.086$
$c = 14.093$
3 STRONGEST DIFFRACTION LINES:
2.722 (100)
2.218 ( 70)
2.527 ( 55)

OPTICAL CONSTANTS:
 $\omega$ = 2.31
 $\epsilon$ = 1.95
 (−)
HARDNESS: 5-6
DENSITY: 4.05 (Meas.)
 4.03 (Calc.)
CLEAVAGE: {$10\bar{1}1$} distinct
Fracture conchoidal.
 HABIT: Crystals short prismatic, up to 6-7 mm in diameter. Also as embedded grains.
 COLOR-LUSTER: Black. Opaque. Submetallic. Streak purple-brown.
 MODE OF OCCURRENCE: Occurs in highly metamorphosed magnesian marbles in the Santa Lucia Mountains, Monterey County, and associated with calcite, brucite, and spinel in metamorphosed magnesian limestones at Crestmore, Riverside County, California; as excellent large crystals with diopside and hydromagnesite at Wakefield, Quebec, Canada; with chlorites in the chrome-spinel deposits of the southern Ural Mountains, and in chromium-chlorites in a serpentinized body of dunite near Mount Jemorakly-Tube, North Caucasus, USSR; and in the gem gravels of Ceylon.
 BEST REF. IN ENGLISH: Palache, et al., "Dana's System of Mineralogy," 7th Ed., v. I, p. 535-541, New York, Wiley, 1944.

# GENTHELVITE
(Zn, Fe, Mn)$_4$Be$_3$Si$_3$O$_{12}$S
Helvite Group

CRYSTAL SYSTEM: Cubic
CLASS: $\bar{4}$3m
SPACE GROUP: P$\bar{4}$3n
Z: 2
LATTICE CONSTANT:
 $a$ = 8.11-8.15
3 STRONGEST DIFFRACTION LINES:
 3.33 ((100)
 2.174 (100)
 1.918 (100)
OPTICAL CONSTANT:
 $N$ = 1.742-1.745
 HARDNESS: 6-6½
 DENSITY: 3.62 (Meas.)
 3.70 (Calc.)
CLEAVAGE: None.
Fracture uneven. Brittle.
 HABIT: Crystals tetrahedral, up to 4.0 × 5.5 cm in size. Also as rounded aggregates up to 25 cm in diameter.
 COLOR-LUSTER: Colorless, white, yellowish white, bluish green, emerald green, purple-pink, pinkish red; also brown to black in weathered varieties. Transparent. Vitreous.
 MODE OF OCCURRENCE: Occurs at three localities on or near Stove Mountain, El Paso County, Colorado; as fine crystals with analcime in carbonatite at Mont St. Hilaire, Quebec, Canada; in alkaline pegmatites in the Lovozero massif, Kola Peninsula, USSR; and in albitized columbite-bearing members of the granitic suite that makes up the Jos-Bukuru complex in northern Nigeria.

 BEST REF. IN ENGLISH: Vlasov, K. A., "Mineralogy of Rare Elements," v. II, p. 119-127, Israel Program for Scientific Translations, 1966.

# GENTHITE = Garnierite

# GEOCRONITE
Pb$_5$SbAsS$_8$
Isostructural with jordanite.

CRYSTAL SYSTEM: Monoclinic
CLASS: 2/m
SPACE GROUP: P2$_1$/m
Z: 6
LATTICE CONSTANTS:
 $a$ = 9.0
 $b$ = 31.9
 $c$ = 8.5
 $\beta$ = 118°
3 STRONGEST DIFFRACTION LINES:
 3.54 (100)
 3.06 ( 90)
 2.89 ( 90)
 HARDNESS: 2½
Talmage hardness ~B.
 DENSITY: 6.4 (Meas.)
 6.51 (Calc.)
 CLEAVAGE: {100} indistinct
Fracture uneven.
 HABIT: Crystals tabular {010}. Usually massive, granular, and earthy. Twinning on {001} common, lamellar.
 COLOR-LUSTER: Galena white; sometimes with greenish, bluish green, or olive tinge. Opaque. Metallic. Streak same as color.
 MODE OF OCCURRENCE: Occurs in the Silver King mine, Park City, Utah; at the Livingston mine, Mackay, Custer County, Idaho; at the Defiance mine, Cerro Gordo, California; in a vein in limestone at Pietrasanta, Val di Castello, Italy; at Meredo, Asturias, Spain (schulzite); at Sala and other localities in Örebro, Sweden, and in the Kilbricken mine, County Clare, Ireland (kilbrickenite).
 BEST REF. IN ENGLISH: Douglass, Robert M., Murphy, Michael, J., and Pabst, A., *Am. Min.*, **39**: 908-928 (1954). Palache, et al., "Dana's System of Mineralogy," 7th Ed., v. I, p. 395-397, New York, Wiley, 1944.

# GEORGIADESITE
Pb$_3$AsO$_4$Cl$_3$

CRYSTAL SYSTEM: Monoclinic
LATTICE CONSTANTS:
 $a:b:c$ = 1.7675:1:1.5438
 $\beta$ = 102°33½'
OPTICAL CONSTANTS:
 $\alpha$ = 2.17
 $\beta$ = 2.17
 $\gamma$ = 2.18
 (+)2V = very large
 HARDNESS: 3½

DENSITY: 7.1 (Meas.)

CLEAVAGE: None.

HABIT: As stubby pseudohexagonal tablets with platy structure, apparently due to lamellar twinning on {100} and {$\bar{1}$04}. {001} grooved and striated.

COLOR-LUSTER: White, brownish yellow. Resinous.

MODE OF OCCURRENCE: Found as a single specimen at Laurium, Greece, in altered lead slag associated with laurionite, matlockite, and fiedlerite.

BEST REF. IN ENGLISH: Palache, et al., "Dana's System of Mineralogy," 7th Ed., v. II, p. 791–792, New York, Wiley, 1951.

## GERASIMOVSKITE (Inadequately described mineral)

$(Mn, Ca)_2 (Nb, Ti)_5 O_{12} \cdot 9H_2O$
Niobium analogue of belyankinite

CRYSTAL SYSTEM: Unknown

3 STRONGEST DIFFRACTION LINES: Amorphous (but has diffuse lines at 2.60, 1.85, 1.64).

OPTICAL CONSTANTS:

$\alpha \simeq 1.74$
$\beta \sim \gamma \simeq 1.81$
$(-)2V = 18°$

HARDNESS: 2

DENSITY: 2.52–2.58

CLEAVAGE: One perfect.

HABIT: Platy masses up to $1.5 \times 1.0 \times 0.3$ cm in size.

COLOR-LUSTER: Brown to gray or light gray; pearly.

MODE OF OCCURRENCE: Occurs in ussingite-bearing pegmatites of Punkarua Mt., Mt. Nepkha, and Mt. Allua, Lovozero massif, Kola Peninsula, USSR.

BEST REF. IN ENGLISH: Semenov, E. I., *Am. Min.*, **43**: 1220–1221 (1958).

## GERHARDTITE

$Cu_2 NO_3 (OH)_3$

CRYSTAL SYSTEM: Orthorhombic

CLASS: 222

SPACE GROUP: $P2_1 2_1 2_1$

Z: 4

LATTICE CONSTANTS:

$a = 5.592$
$b = 6.075$
$c = 13.812$

3 STRONGEST DIFFRACTION LINES:

6.91 (100)
2.624 ( 80)
2.310 ( 80)

OPTICAL CONSTANTS:

$\alpha = 1.703$
$\beta = 1.713$
$\gamma = 1.722$
$(+)2V = large$

HARDNESS: 2

DENSITY: 3.40–3.43 (Meas.)
3.40 (Calc.)

CLEAVAGE: {001} perfect
{100} good

Flexible.

HABIT: Crystals thick tabular, pyramid zone striated.

COLOR-LUSTER: Emerald green to dark green. Transparent. Streak light green.

MODE OF OCCURRENCE: Occurs as a secondary mineral in the oxidation zone of copper deposits. Found in vugs in massive cuprite at the United Verde mine, Jerome, Arizona, and at Likasi and Kalabi, Katanga, Zaire.

BEST REF. IN ENGLISH: Palache, et al., "Dana's System of Mineralogy," 7th Ed., v. II, p. 308–309, New York, Wiley, 1951.

## GERMANITE

$Cu_3 (Ge, Ga, Fe, Zn)(S, As)_4$

CRYSTAL SYSTEM: Cubic

CLASS: $\bar{4}$3m

SPACE GROUP: $P\bar{4}3n$

Z: 1

LATTICE CONSTANT:

$a = 5.299$

3 STRONGEST DIFFRACTION LINES:

3.054 (100)
1.87 ( 80)
1.596 ( 70)

HARDNESS: 4

DENSITY: 4.46–4.59 (Meas.)
4.30 (Calc.)

CLEAVAGE: None.

Brittle.

HABIT: Massive.

COLOR-LUSTER: Dark reddish gray; opaque. Metallic.

MODE OF OCCURRENCE: Occurs intimately associated with tennantite, enargite, pyrite, galena, sphalerite, azurite, and malachite at Tsumeb, South West Africa.

BEST REF. IN ENGLISH: "Dana's System of Mineralogy," 7th Ed., v. I, p. 385–386, New York, Wiley, 1944. Murdoch, Joseph, *Am. Min.*, **38**: 794–801 (1953).

## GERSDORFFITE

NiAsS

CRYSTAL SYSTEM: Cubic

CLASS: 23

SPACE GROUP: $P2_1 3$

Z: 4

LATTICE CONSTANT:

$a = 5.719$

3 STRONGEST DIFFRACTION LINES:

2.545 (100)
2.325 ( 90)
1.716 ( 80)

HARDNESS: 5½

DENSITY: 5.9 (Av.)
5.967 (Calc.)

CLEAVAGE: {001} perfect

Fracture uneven. Brittle.

HABIT: Crystals octahedral, cubo-octahedral, or pyritohedral. Faces commonly striated like pyrite. Also massive, granular and lamellar.

COLOR-LUSTER: Tin white to steel gray, often tarnished gray or grayish black. Opaque. Metallic. Streak grayish black.

MODE OF OCCURRENCE: Occurs as irregular veinlets in uraninite at the Caribou mine, Boulder County, and in galena-chalcopyrite-pyrite ore at the Homestake mine, near Tennessee Pass, Lake County, Colorado; at the Pine Tree mine, Mariposa County, California; and at Phoenixville, Pennsylvania. It is also found in Ontario, Canada, and at deposits in Germany, Czechoslovakia, Sweden, and Rhodesia. The antimonian variety (coryinite) occurs at Chatham, Connecticut, and at Olsa, Carinthia, Austria.

BEST REF. IN ENGLISH: Palache, et al., "Dana's System of Mineralogy," 7th Ed., v. I, p. 298–300, New York, Wiley, 1944.

# GERSTLEYITE
$(Na, Li)_4 As_2 Sb_8 S_{17} \cdot 6H_2O$

CRYSTAL SYSTEM: Probably monoclinic
3 STRONGEST DIFFRACTION LINES:
  11.85 (100)
  3.05 ( 90)
  5.64 ( 70)
OPTICAL CONSTANTS:
  $N > 2.01$
  HARDNESS: 2½
  DENSITY: 3.62 (Meas.)
  CLEAVAGE: {010} perfect
             {100} perfect
             {001} poor

HABIT: As groups of small, thick plates with rough surfaces; as fine-granular aggregates; or as platy-fibrous spherules up to 1 inch in diameter.

COLOR-LUSTER: Cinnabar red to blackish red. Weakly adamantine. Powder bright cinnabar red, darkening on exposure.

MODE OF OCCURRENCE: Occurs in clay at the Baker mine, Kramer district, Kern County, California.

BEST REF. IN ENGLISH: Frondel, Clifford, and Morgan, Vincent, *Am. Min.*, **41**: 839–843 (1956).

# GETCHELLITE
$AsSbS_3$

CRYSTAL SYSTEM: Monoclinic
CLASS: 2/m
SPACE GROUP: $P2_1/a$
Z: 8
LATTICE CONSTANTS:
  $a = 11.87$
  $b = 9.03$
  $c = 10.14$
  $\beta = 116°10'$
3 STRONGEST DIFFRACTION LINES:
  2.89 (100)
  4.44 ( 80)
  3.63 ( 70)

OPTICAL CONSTANTS:
  $\alpha > 2.11$ (white light)
  $\beta > 2.72$ (Li)
  (+)2V < 46°
  HARDNESS: 1½–2
  DENSITY: 3.92 (Meas.)
           4.01 (Calc.)
  CLEAVAGE: {001} perfect micaceous, yielding flexible but inelastic lamellae. Fracture splintery which suggests a second less perfect cleavage. Sectile.

HABIT: Equant subhedral crystals less than 0.5 mm in diameter; anhedral grains up to 4 mm in maximum dimension.

COLOR-LUSTER: Dark blood red; pearly to vitreous on fresh cleavage surfaces, elsewhere resinous. Cleavage surfaces and crystal faces often exhibit purple to green iridescent tarnish. Streak orange-red.

MODE OF OCCURRENCE: Occurs in an epithermal arsenical gold deposit intimately associated with orpiment, realgar, stibnite, cinnabar, and quartz, at the Getchell Mine, Humboldt County, Nevada.

BEST REF. IN ENGLISH: Weissberg, B. G., *Am. Min.*, **50**: 1817–1826 (1965).

# GEVERSITE
$PtSb_2$
Antimony analogue of sperrylite

CRYSTAL SYSTEM: Cubic
CLASS: m3
SPACE GROUP: Pa3
Z: 4
LATTICE CONSTANT:
  $a = 6.44$
3 STRONGEST DIFFRACTION LINES:
  1.94 (100)
  1.72 ( 60)
  2.92 ( 40)
HARDNESS: 4½–5
HABIT: Tiny grains.
COLOR-LUSTER: Light gray.
MODE OF OCCURRENCE: Occurs in concentrates of platinum minerals from the Driekop mine, eastern Transvaal.

BEST REF. IN ENGLISH: Stumpfl, E. F., *Min. Mag.*, **32**: 833–847 (1961).

# GIANNETTITE (Inadequately described mineral)
Near $Na_3 Ca_3 Mn(Zr, Fe) TiSi_6 O_{21} Cl$

CRYSTAL SYSTEM: Triclinic
OPTICAL CONSTANTS:
  $\alpha = 1.663$
  $\beta = 1.664$
  $\gamma = 1.675$
  (+)2V = 30°
  HARDNESS: Not determined.
  DENSITY: Not determined.
  CLEAVAGE: {100} perfect
             {010} imperfect
             {001} imperfect

HABIT: Crystals prismatic, small. Twinning on {100}, polysynthetic.

COLOR-LUSTER: Colorless, pale yellowish.

MODE OF OCCURRENCE: Occurs in alkali rocks in association with nepheline, aegirine, potash feldspar, catapleiite, eudialyte, and zeolites, on the Pocos de Caldas plateau, Brazil.

BEST REF. IN ENGLISH: Vlasov, K. A., "Mineralogy of Rare Elements," v. II, p. 385–386, Israel Program for Scientific Translations, 1966.

# GIBBSITE
$Al(OH)_3$
Trimorphous with bayerite and nordstrandite

CRYSTAL SYSTEM: Monoclinic
CLASS: 2/m
SPACE GROUP: $P2_1/n$
Z: 8
LATTICE CONSTANTS:
  $a = 8.624$
  $b = 5.060$
  $c = 9.700$
  $\beta = 94°34'$
3 STRONGEST DIFFRACTION LINES:
  4.82 (100)
  4.34 ( 40)
  4.30 ( 20)
OPTICAL CONSTANTS:
  $\alpha = 1.568$
  $\beta = 1.568$
  $\gamma = 1.587$
  $(+)2V \simeq 0°$
HARDNESS: 2½–3½
DENSITY: 2.40 (Meas.)
         2.435 (Calc.)
CLEAVAGE: {001} perfect
Tough.

HABIT: Crystals tabular, hexagonal in aspect, up to several inches long. Usually massive; as chalcedony-like coatings and crusts, stalactitic, concretionary, sometimes with indistinct fibrous structure; also compact, earthy. Twinning on {001} and about [130] as twin axis, common; on {100} and {110}, not common.

COLOR-LUSTER: White, grayish, greenish, reddish white. Translucent to transparent. Vitreous, pearly on cleavages.

MODE OF OCCURRENCE: Occurs as an alteration product of aluminum-bearing minerals in laterite and bauxite deposits; also as a low-temperature hydrothermal mineral in alkalic or other igneous rocks. Found in California, Arizona, Arkansas, Pennsylvania, Massachusetts, and New York. Also in Brazil, Guyana, Norway, France, Italy, Germany, Hungary, USSR, India, Madagascar, Tasmania, and elsewhere.

BEST REF. IN ENGLISH: Palache, et al., "Dana's System of Mineralogy," 7th Ed., v. I, p. 663–667, New York, Wiley, 1944.

# GIESSENITE
$Pb_8CuBi_6Sb_{1.5}S_{30}$ (?)

CRYSTAL SYSTEM: Orthorhombic
CLASS: 2 2 2
SPACE GROUP: $P2_12_12_1$
LATTICE CONSTANTS:
  $a = 34.5$
  $b = 38.3$
  $c = 4.08$
3 STRONGEST DIFFRACTION LINES:
  3.62 (100)
  3.32 ( 50)
  3.15 ( 50)
HARDNESS: ~2½
DENSITY: Not determined.
CLEAVAGE: Not determined.
HABIT: As fine needles.
COLOR-LUSTER: Grayish black; metallic. Opaque.

MODE OF OCCURRENCE: Occurs associated with galena, pyrite, rutile, and tennantite in the dolomite of Binnatal, Valais, Switzerland.

BEST REF. IN ENGLISH: St. Graeser, *Am. Min.*, **50**: 264 (1965).

# GILLESPITE
$BaFeSi_4O_{10}$

CRYSTAL SYSTEM: Tetragonal
CLASS: 4/m 2/m 2/m
SPACE GROUP: P4/ncc
Z: 4
LATTICE CONSTANTS:
  $a = 7.495$
  $c = 16.050$
3 STRONGEST DIFFRACTION LINES:
  3.39 (100)
  4.41 ( 70)
  3.22 ( 70)
OPTICAL CONSTANTS:
  $\omega = 1.621$
  $\epsilon = 1.619$
    $(-)$
HARDNESS: 3
DENSITY: 3.402 (Meas.)
CLEAVAGE: {001} distinct
          {100} indistinct
Brittle.
HABIT: Massive compact, and in embedded grains.
COLOR-LUSTER: Red. Translucent. Vitreous.

MODE OF OCCURRENCE: Occurs in association with sanbornite, celsian, taramellite, witherite, quartz, and other minerals. Found at Big Creek and at the Rush Creek sanbornite deposit in Fresno County, and in a vein in quartzite near Trumbull Peak, Mariposa County, California; also in a rock specimen from a moraine near the head of Dry Delta, Alaska Range, Alaska.

BEST REF. IN ENGLISH: Schaller, Waldemar T., *Am. Min.*, **14**: 319–322 (1929).

# GINORITE
$Ca_2B_{14}O_{23} \cdot 8H_2O$

CRYSTAL SYSTEM: Orthorhombic
3 STRONGEST DIFFRACTION LINES:
  7.14 (100)
  5.36 ( 30)
  3.27 ( 20)
OPTICAL CONSTANTS:
  $\alpha = 1.517$
  $\beta = 1.524$ (Calc.)
  $\gamma = 1.577$
  (+)2V = 42°
HARDNESS: 3½
DENSITY: 2.07-2.09 (Meas.)
CLEAVAGE: {010}
HABIT: As rhomb-shaped plates; also as dense masses and small pellets.
COLOR-LUSTER: White. Transparent to translucent.
MODE OF OCCURRENCE: Found as pellets, which average about 1-2 mm in diameter, embedded in a pale yellowish brown matrix of sassolite and clay in efflorescent masses in colemanite-veined basalt near the head of Twenty Mule Team Canyon, Death Valley, California. Also found associated with mirabilite at the Clinton Quarry, Windsor, Nova Scotia, Canada, and with calcite in veins in sandstone at Sasso Pisano, Tuscany, Italy.
BEST REF. IN ENGLISH: Allen, Robert D., and Kramer, Henry, *Am. Min.*, **42**: 56-61 (1957).

# GIORGIOSITE (Species status uncertain.)
A basic magnesium carbonate

3 STRONGEST DIFFRACTION LINES:
  3.40 (100)
  3.29 ( 70)
  2.92 ( 60)
MODE OF OCCURRENCE: Found as spherulitic growths in white crusts formed on lava of 1866 at Alphroëssa, Santorin island.
BEST REF. IN ENGLISH: Raade, Gunnar, *Am. Min.*, **55**: 1457-1465 (1970).

# GISMONDINE (Gismondite)
$CaAl_2Si_2O_8 \cdot 4H_2O$
Zeolite Group

CRYSTAL SYSTEM: Monoclinic
CLASS: 2/m
SPACE GROUP: $P2_1/c$
Z: 4
LATTICE CONSTANTS:
  $a = 10.02$
  $b = 10.63$
  $c = 9.83$
  $\beta = 92°42'$

3 STRONGEST DIFFRACTION LINES:
  2.71 (100)
  4.25 ( 70)
  3.19 ( 70)
OPTICAL CONSTANTS:
  $\alpha = 1.520-1.521$
  $\gamma = 1.522-1.525$
  2V = large
HARDNESS: 4½
DENSITY: 2.27 (Meas.)
        2.28 (Calc.)
CLEAVAGE: {101} distinct
HABIT: Crystals pseudotetragonal bipyramids produced by twinning on {110} and {001}.
COLOR-LUSTER: Colorless, white, bluish white, grayish, reddish. Transparent to translucent. Vitreous.
MODE OF OCCURRENCE: Occurs at Fritz's Island, Berks County, Pennsylvania; at Alexander Dam, Kauai, Hawaii; in basaltic lavas in Iceland and in County Antrim, Ireland; in leucitic lava at Capo di Bove, Rome, Italy; in cavities in leucite tephrite in Czechoslovakia; and with chlorite in highly altered granite in Queensland, Australia.
BEST REF. IN ENGLISH: Deer, Howie, and Zussman, "Rock Forming Minerals," v. 4, p. 401-407, New York, Wiley, 1963. Fischer, Karl, *Am. Min.*, **48**: 664-672 (1963).

# GLADITE (Inadequately described mineral)
$CuPbBi_5S_9$

CRYSTAL SYSTEM: Orthorhombic (?)
HARDNESS: 2-3
DENSITY: 6.96 (Meas.)
CLEAVAGE: {010} good
        {100} fair
HABIT: Crystals prismatic; up to 2 cm long and 2-6 mm thick.
COLOR-LUSTER: Lead gray. Opaque. Metallic. Streak black.
MODE OF OCCURRENCE: Occurs with lead bismuth sulfides and quartz at Gladhammar, Kalmar, Sweden.
BEST REF. IN ENGLISH: Palache, et al., "Dana's System of Mineralogy," 7th Ed., v. I, p. 483, New York, Wiley, 1944.

# GLASERITE = Variety of aphthitalite
Near $K_3Na(SO_4)_2$

# GLAUBERITE
$Na_2Ca(SO_4)_2$

CRYSTAL SYSTEM: Monoclinic
CLASS: 2/m
SPACE GROUP: C2/c
Z: 4

LATTICE CONSTANTS:
  $a = 10.129$
  $b = 8.306$
  $c = 8.533$
  $\beta = 112°11'$
3 STRONGEST DIFFRACTION LINES:
  3.13 (100)
  6.22 ( 80)
  2.66 ( 80)
OPTICAL CONSTANTS:
  $\alpha = 1.515$
  $\beta = 1.535$
  $\gamma = 1.536$
  $(-)2V = 7°$
HARDNESS: 2½-3
DENSITY: 2.80 (Meas.)
         2.78 (Calc.)
CLEAVAGE: {001} perfect
          {110} indistinct
Fracture conchoidal. Brittle.

HABIT: Crystals varied in habit; usually tabular with sharp beveled edges; also prismatic or dipyramidal. {001} and {111} often striated parallel to their intersection. As single crystals and groups.

COLOR-LUSTER: Colorless, gray, yellowish. Transparent to translucent. Vitreous; pearly on basal cleavage. Whitens in water. Streak white. Taste saline.

MODE OF OCCURRENCE: Occurs widespread in salt deposits, clastic sediments, and nitrate deposits, in cavities in basic extrusive rocks, and as a fumarolic deposit. Commonly found as platy crystals at Searles Lake, San Bernardino County; at Borax Lake, Lake County; and at localities in Imperial and Inyo counties, California; also from Camp Verde, Yavapai County, Arizona; and in New Mexico and Texas. Among many other places, it is found at deposits in Canada, Chile, Spain, France, Germany, Austria, USSR, and India.

BEST REF. IN ENGLISH: Palache, et al., "Dana's System of Mineralogy," 7th Ed., v. II, p. 431-433, New York, Wiley, 1951. Araki, Takaharu, and Zoltai, Tibor, *Am. Min.*, **52**: 1272-1277 (1967).


## GLAUBERS SALT = Mirabilite


## GLAUCOCERINITE (Glaucokerinite)
## (Inadequately described mineral)
$(Zn, Cu)_{10}Al_4SO_4(OH)_{30} \cdot 2H_2O$

CRYSTAL SYSTEM: Probably monoclinic
OPTICAL CONSTANT:
  $\gamma = 1.542$
  HARDNESS: 1
  DENSITY: 2.75
  CLEAVAGE: Not determined.
  HABIT: Massive with radial-fibrous structure and concentric color banding.
  COLOR-LUSTER: White to sky blue; also gray, greenish or brownish due to impurities.
  MODE OF OCCURRENCE: Found sparingly at Laurium, Greece, associated with malachite, smithsonite, gypsum, and adamite.

BEST REF. IN ENGLISH: Palache, et al., "Dana's System of Mineralogy," 7th Ed., v. II, p. 574, New York, Wiley, 1951.


## GLAUCOCHROITE
$CaMnSiO_4$

CRYSTAL SYSTEM: Orthorhombic
CLASS: 2/m 2/m 2/m
SPACE GROUP: Pbnm
Z: 4
LATTICE CONSTANTS:
  $a = 4.92$
  $b = 11.14$
  $c = 6.50$
3 STRONGEST DIFFRACTION LINES:
  1.85 (100)
  2.69 ( 80)
  2.63 ( 80)
OPTICAL CONSTANTS:
  $\alpha = 1.685$
  $\beta = 1.723$
  $\gamma = 1.736$
  $(-)2V = 61°$
HARDNESS: 6
DENSITY: 3.48 (Meas.)
         3.488 (Calc.)
CLEAVAGE: {001} indistinct
Very brittle.

HABIT: Crystals long prismatic, single or in compound aggregates. Rarely massive, coarse to fine granular. Twinning on {011}, penetration and contact.

COLOR-LUSTER: Bluish green, white, pinkish. Translucent. Vitreous.

MODE OF OCCURRENCE: Occurs in association with nasonite, willemite, garnet, axinite, hardystonite, tephroite, and franklinite at Franklin, Sussex County, New Jersey.

BEST REF. IN ENGLISH: Palache, Charles, USGS Prof. Paper 180, p. 79-80 (1935). O'Mara, J. H., *Am. Min.*, **36**: 918 (1951).


## GLAUCODOT
$(Co, Fe)AsS$

CRYSTAL SYSTEM: Orthorhombic
CLASS: 2/m 2/m 2/m
SPACE GROUP: Cmmm
Z: 24
LATTICE CONSTANTS:
  $a = 6.63$
  $b = 28.33$
  $c = 5.63$
3 STRONGEST DIFFRACTION LINES:
  2.72 ((100)
  1.827 ( 90)
  2.45 ( 80)
HARDNESS: 5
DENSITY: 6.055 (Meas.)
         6.155 (Calc.)

CLEAVAGE: {010} perfect
　　　　　　{101} distinct
Fracture uneven. Brittle.

HABIT: Crystals short prismatic, striated in the prism zone. Also massive. Twinning on {101}; also on {012} as trillings or cruciform twins.

COLOR-LUSTER: Grayish tin white to reddish silver white. Opaque. Metallic. Streak black.

MODE OF OCCURRENCE: Occurs associated with cobaltite and pyrite at the Standard Consolidated mine, Sumpter, Baker County, Oregon; in the Atacama district, Chile; as excellent single and twinned crystals at Hakansbo, Sweden; and at localities in Norway, Roumania, and Tasmania.

BEST REF. IN ENGLISH: Palache, et al., "Dana's System of Mineralogy," 7th Ed., v. I, p. 322–325, New York, Wiley, 1944.

# GLAUCOKERINITE = Glaucocerinite

# GLAUCONITE
$(K, Na)(Al, Fe^{3+}, Mg)_2 (Al, Si)_4 O_{10} (OH)_2$
(Mica Group)

CRYSTAL SYSTEM: Monoclinic
CLASS: m or 2/m
SPACE GROUP: Cm or C2/m
Z: 2
LATTICE CONSTANTS:
　$a = 5.25$
　$b = 9.09$
　$c = 10.03$
　$\beta \simeq 100°$
3 STRONGEST DIFFRACTION LINES:
　10.1 (100)
　2.59 (100)
　4.53 ( 80)
OPTICAL CONSTANTS:
　$\alpha = 1.592–1.610$
　$\beta = 1.614–1.641$
　$\gamma = 1.614–1.641$
　$(-)2V = 0°–20°$
HARDNESS: 2
DENSITY: 2.4–2.95 (Meas.)
　　　　　2.903 (Calc.)
CLEAVAGE: {001} perfect

HABIT: Crystals lath-like, extremely minute. Usually in minute rounded fine-grained aggregates of irregular platelets.

COLOR-LUSTER: Usually dull green, yellowish green, or blue-green. Translucent, nearly opaque. Dull or glistening.

MODE OF OCCURRENCE: Occurs in rocks of marine origin in deposits representing nearly all geological ages. Found abundantly in "greensands"; also in impure limestones, sandstones, and siltstones. Typical occurrences are found in California, Arizona, Colorado, South Dakota, Missouri, Mississippi, Minnesota, Pennsylvania, and New Jersey. Also in England, Belgium, France, USSR, India, and New Zealand.

BEST REF. IN ENGLISH: Deer, Howie, and Zussman, "Rock Forming Minerals," v. 3, p. 35–41, New York, Wiley, 1962.

# GLAUCOPARGASITE = Hornblende (?)

# GLAUCOPHANE
$Na_2 (Mg, Fe^{2+})_3 Al_2 Si_8 O_{22} (OH)_2$
Amphibole Group

CRYSTAL SYSTEM: Monoclinic
CLASS: 2/m
SPACE GROUP: C2/m
Z: 2
LATTICE CONSTANTS:
　$a = 9.541$
　$b = 17.740$
　$c = 5.295$
　$\beta = 103°40'$
3 STRONGEST DIFFRACTION LINES:
　2.714 (100)
　3.120 ( 90)
　2.502 ( 80)
OPTICAL CONSTANTS:
　$\alpha = 1.606–1.661$
　$\beta = 1.622–1.667$
　$\gamma = 1.627–1.670$
　$(-)2V = 0°–50°$
HARDNESS: 6
DENSITY: 3.08–3.15 (Meas.)
　　　　　2.908 (Calc.)
CLEAVAGE: {110} perfect
Fracture uneven to conchoidal. Brittle.

HABIT: Crystals usually slender prismatic. Commonly massive; fibrous, columnar, or granular. Twinning on {100} simple, lamellar.

COLOR-LUSTER: Grayish, bluish black, lavender blue, azure blue. Translucent. Vitreous to dull, sometimes pearly. Streak grayish blue.

MODE OF OCCURRENCE: Occurs in crystalline schists, commonly in association with chlorite, muscovite, stilpnomelane, lawsonite, epidote, pumpellyite, almandine, and jadeite. Found as a widespread constituent of the schists of the Coast Ranges of California; at Iron Hill, Gunnison County, Colorado; and at the Homestake mine, Lawrence County, South Dakota. Also found in Scotland, Switzerland, Italy, on the islands of Syra and Corsica, and at many places in Japan.

BEST REF. IN ENGLISH: Deer, Howie, and Zussman, "Rock Forming Minerals," v. 2, p. 333–351, New York, Wiley, 1963. Papike, J. J., and Clark, Joan R., *Am. Min.*, 53: 1156–1173 (1968).

# GLOCKERITE (Species status uncertain)
Near $Fe_4^{3+}(SO_4)(OH)_{10} \cdot 1–3H_2O$

CRYSTAL SYSTEM: Unknown
OPTICAL CONSTANTS:
　$\alpha = 1.76$
　$\gamma = 1.81$
HARDNESS: Not determined.
DENSITY: Not determined.
CLEAVAGE: Fracture conchoidal to powdery. Brittle.
HABIT: As crusts; stalactitic; earthy.

COLOR-LUSTER: Pale yellow to brown, brownish black to black; greenish. Translucent to opaque. Resinous or dull. Streak pale yellow to brown.

MODE OF OCCURRENCE: Occurs as a secondary mineral formed by the oxidation of iron sulfides. Found at mines in Clear Creek, Gilpin, and Teller counties, Colorado; at the Gap mine and at Germantown, Lancaster County, Pennsylvania; and at localities in England, Germany, Norway, Sweden, and the USSR.

BEST REF. IN ENGLISH: Palache, et al., "Dana's System of Mineralogy," 7th Ed., v. II, p. 587–588, New York, Wiley, 1951.

# GMELINITE
$(Na_2, Ca)Al_2Si_4O_{12} \cdot 6H_2O$
Zeolite Group

CRYSTAL SYSTEM: Hexagonal
CLASS: 6/m 2/m 2/m
SPACE GROUP: $P6_3/mmc$
Z: 4
LATTICE CONSTANTS:
$a \simeq 13.72$
$c \simeq 9.95$
3 STRONGEST DIFFRACTION LINES:
4.10 (100)
12.0 ( 90)
2.96 ( 80)
OPTICAL CONSTANTS:
$\omega$ = 1.476–1.494
$\epsilon$ = 1.474–1.480
(–)
HARDNESS: 4½
DENSITY: 2.04–2.17 (Meas.)
2.098 (Calc.)
CLEAVAGE: {10$\bar{1}$0} distinct
{0001} parting
Fracture uneven. Brittle.

HABIT: Crystals pyramidal, tabular, or rhombohedral in aspect; striated vertically. Penetration twins common.

COLOR-LUSTER: Colorless, white, yellowish, greenish white, reddish white, pink. Transparent to translucent. Vitreous.

MODE OF OCCURRENCE: Occurs chiefly in the amygdules of basaltic lavas and related igneous rocks in association with other zeolites. Found near Springfield, Lane County, Oregon; in the trap rock of northeastern New Jersey, especially at Bergen Hill and Great Notch; in the Bay of Fundy district, Nova Scotia; in County Antrim, Ireland; at Talisker on the Isle of Skye, Scotland; and at localities in Italy, Cyprus, Germany, and as fine crystals from Flinders Island, Victoria, Australia.

BEST REF. IN ENGLISH: Deer, Howie, and Zussman, "Rock Forming Minerals," v. 4, p. 387–400, New York, Wiley, 1963.

# GODLEVSKITE
$Ni_7S_6$

CRYSTAL SYSTEM: Orthorhombic
CLASS: 222, mm2, or 2/m 2/m 2/m
SPACE GROUP: C222, Cm2m, Cmm2, or Cmmm
LATTICE CONSTANTS:
$a$ = 9.180
$b$ = 11.263
$c$ = 9.457
3 STRONGEST DIFFRACTION LINES:
2.85 (100)
1.803 ( 90)
1.795 ( 80)
HARDNESS: Microhardness 383–418 (397 av.) kg/mm$^2$, 40–50 g load.

HABIT: Aggregates and single grains up to 1 mm, but rarely more than 0.3 mm in size. Twinning is common and often complex, sometimes forming "elbow" bends.

COLOR-LUSTER: Under reflecting microscope pale yellow with weak birefringence in shades of yellow. Anisotropy strong with color effects from bluish to reddish.

MODE OF OCCURRENCE: Occurs in bornite and bornite-chalcopyrite veins of the Noril'sk deposit and the Talnakh deposit, USSR. It is replaced in part by bornite and by millerite and has been observed replacing pentlandite.

BEST REF. IN ENGLISH: Kulagov, E. A., Evstigneeva, T. L., and Yushko-Zakharova, O. E., *Am. Min.*, **55**: 317–318 (1970).

# GOETHITE
$HFeO_2$
Trimorphous with lepidocrocite and akaganeite

CRYSTAL SYSTEM: Orthorhombic
CLASS: 2/m 2/m 2/m
SPACE GROUP: Pbnm
Z: 4
LATTICE CONSTANTS:
$a$ = 4.596
$b$ = 9.957
$c$ = 3.021
3 STRONGEST DIFFRACTION LINES:
4.18 (100)
2.69 ( 30)
2.452 ( 25)
OPTICAL CONSTANTS:
$\alpha$ = 2.260
$\beta$ = 2.393 (Na)
$\gamma$ = 2.398
(–)2V = small to medium
HARDNESS: 5–5½
DENSITY: 3.3–4.3 (Meas.)
4.264 (Calc.)
CLEAVAGE: {010} perfect
{100} distinct
Fracture uneven. Brittle.

HABIT: Crystals prismatic along c-axis and vertically striated, or as thin tablets flattened {010}. In tufts, druses, or radiating clusters of capillary, acicular, or thin prismatic crystals. Usually as colloform masses with radial fibrous or concentric structure; bladed or columnar; compact; concretionary; pisolitic; oolitic; earthy.

COLOR-LUSTER: Crystals blackish brown; massive varieties reddish or yellowish brown; earthy varieties brownish yellow to ocher yellow. Opaque; thin fragments translucent. Adamantine-metallic to dull. Silky when fibrous. Streak variable: shades of orange- to brownish yellow.

MODE OF OCCURRENCE: Occurs widespread and abundant as an alteration product of iron-bearing minerals such as pyrite, magnetite, chalcopyrite, and siderite. It is commonly associated with hematite, manganese oxides, calcite, clays, and other minerals, in such deposits as gossans, bogs, springs, laterites, and rarely in low-temperature hydrothermal veins. Notable deposits in the United States occur in Utah, Arizona, Colorado, South Dakota, Texas, Alabama, Virginia, West Virginia, Georgia, Tennessee, and the Lake Superior hematite region, especially at Marquette and Negaunee, Michigan. Other important occurrences are found in Cuba, Brazil, Chile, England, France, Germany, Czechoslovakia, USSR, and elsewhere.

BEST REF. IN ENGLISH: Palache, et al., "Dana's System of Mineralogy," 7th Ed., v. I, p. 680–685, New York, Wiley, 1944.

# GOLD
Au

CRYSTAL SYSTEM: Cubic
CLASS: $4/m\,\bar{3}\,2/m$
SPACE GROUP: Fm3m
Z: 4
LATTICE CONSTANT:
  $a = 4.0781$
3 STRONGEST DIFFRACTION LINES:
  2.355 (100)
  2.039 ( 52)
  1.230 ( 36)
OPTICAL CONSTANT:
  $N = 0.366$ (Na)
  HARDNESS: 2½–3
  DENSITY: 19.297 at 0° (Meas.)
       19.309 at 0° (Calc.)
  CLEAVAGE: None.
Fracture hackly. Very malleable and ductile.
  HABIT: Crystals octahedral, dodecahedral, or cubic; often flattened and elongated. Commonly dendritic, arborescent, reticulated, filiform, spongy. Also massive, as nuggets, flattened grains, or scales. Twinning on {111} common, often repeated.
  COLOR-LUSTER: Gold yellow, also silver white to orange-red due to impurities. Opaque. Metallic. Streak same as color.
  MODE OF OCCURRENCE: Occurs chiefly in hydrothermal veins, commonly associated with pyrite and other sulfides, in auriferous conglomerates, and in placer deposits. Found widespread throughout the world, especially in the western United States, Alaska, Canada, Mexico, Brazil, Chile, Colombia, USSR, Roumania, China, India, Tasmania, Australia, and the Witwatersrand district, Transvaal, Africa.
  BEST REF. IN ENGLISH: Palache, et al., "Dana's System of Mineralogy," 7th Ed., v. I, p. 90–95, New York, Wiley, 1944.

# GOLD AMALGAM
(Au, Hg)

CRYSTAL SYSTEM: Cubic
LATTICE CONSTANT:
  $a = 17.85$
  HARDNESS: Not determined.
  DENSITY: 15.47 (Meas.)
  CLEAVAGE: Fracture conchoidal. Plastic, or brittle.
  HABIT: Massive; as grains or rounded fragments.
  COLOR-LUSTER: White to yellowish. Opaque. Metallic.
  MODE OF OCCURRENCE: Occurs in California in the Odin drift mine near Nevada City, Nevada County, and in the region around Mariposa, Mariposa County; in Colorado as small pellets in quartz of the Gold and Silver Chief vein on Cornett Creek, San Miguel County; and at localities in Colombia and Borneo.
  BEST REF. IN ENGLISH: Palache, et al., "Dana's System of Mineralogy," 7th Ed., v. I, p. 105, New York, Wiley, 1944.

# GOLDFIELDITE (Inadequately described mineral)
$Cu_3(Te, Sb, As)S_4$  Te:Sb:As = 55:29:13
Tetrahedrite Group

CRYSTAL SYSTEM: Probably cubic
  HARDNESS: 3–3½
  DENSITY: Not determined.
  CLEAVAGE: Not determined. Brittle.
  HABIT: Massive; as crusts.
  COLOR-LUSTER: Dark lead gray. Opaque. Metallic.
  MODE OF OCCURRENCE: Occurs with marcasite and a number of unidentified minerals at the Mohawk mine, Goldfield, Nevada.
  BEST REF. IN ENGLISH: Levy, Claude, Am. Min., 53: 2105 (1968).

# GOLDICHITE
$KFe(SO_4)_2 \cdot 4H_2O$

CRYSTAL SYSTEM: Monoclinic
CLASS: 2/m
SPACE GROUP: $P2_1/c$
Z: 4
LATTICE CONSTANTS:
  $a = 10.45$
  $b = 10.53$
  $c = 9.15$
  $\beta = 101°49'$
3 STRONGEST DIFFRACTION LINES:
  3.068 (100)
  7.35 ( 90)
  10.29 ( 80)
OPTICAL CONSTANTS:
    $\alpha = 1.582$
    $\beta = 1.602$
    $\gamma = 1.629$
  (+)2V = 82°
  HARDNESS: ~2.5
  DENSITY: 2.43 (Meas.)
       2.419 (Calc.)
  CLEAVAGE: {100} perfect

HABIT: Crystals commonly singly terminated prismatic laths flattened parallel {100} with forms {100}, {110}, and {011}. Radiating clusters of crystals and fine-grained crystalline encrustations.

COLOR-LUSTER: Pale yellowish green; lavender tint under artificial light.

MODE OF OCCURRENCE: Occurs as crystals up to 4 mm in length associated with coquimbite, halotrichite, roemerite, alunogen, copiapite, and other sulfates, in a talus slope below a small, pyrite-rich uranium deposit in the Triassic Shinarump conglomerate at the Dexter No. 7 mine, Calf Mesa, San Rafael Swell, Utah.

BEST REF. IN ENGLISH: Rosenzweig, A., and Gross, Eugene B., *Am. Min.*, **40**: 469–480 (1955).

# GOLDMANITE
$Ca_3V_2Si_3O_{12}$
Garnet Group

CRYSTAL SYSTEM: Cubic
CLASS: $4/m\,\bar{3}\,2/m$
SPACE GROUP: Ia3d
Z: 8
LATTICE CONSTANT:
$a = 12.011$
3 STRONGEST DIFFRACTION LINES:
2.688 (100)
3.005 ( 65)
1.607 ( 49)
OPTICAL CONSTANT:
$N = 1.821$–1.855
DENSITY: 3.74 (Meas.)
3.737 (Calc.)
CLEAVAGE: None.
HABIT: Dodecahedral crystals rarely exceeding 0.1 mm and averaging 0.02 mm in diameter. Usually anhedral.
COLOR-LUSTER: Dark green to brownish green.
MODE OF OCCURRENCE: Occurs as minute grains and crystals embedded in vanadium clay and calcite in a metamorphosed uranium-vanadium deposit in the Entrada Sandstone in the Laguna uranium mining district, about 45 miles west of Albuquerque, New Mexico.
BEST REF. IN ENGLISH: Moench, Robert H., and Meyrowitz, Robert, *Am. Min.*, **49**: 644–655 (1964).

# GONNARDITE
$Na_2Ca[(Al, Si)_5O_{10}]_2 \cdot 6H_2O$
Zeolite Group

CRYSTAL SYSTEM: Orthorhombic (pseudotetragonal)
Z: 2
LATTICE CONSTANTS:
$a = 13.38$
$b = 13.38$
$c = 6.66$
3 STRONGEST DIFFRACTION LINES:
2.92 (100)
5.93 ( 80)
4.44 ( 60)

OPTICAL CONSTANTS:
$\alpha = 1.497$–1.506
$\beta = 1.499$–1.508
$(-)2V = 50°$
HARDNESS: 4½–5
DENSITY: 2.26 (Meas.)
2.252 (Calc.)
CLEAVAGE: Not determined.
HABIT: Massive, fibrous; in spherules with radiating structure.
COLOR-LUSTER: White. Translucent. Silky.
MODE OF OCCURRENCE: Occurs with wollastonite and pyrite in the Commercial quarry at Crestmore, Riverside County, California; in cavities in basalt at Chaux de Bergonne, Puy-de-Dôme, France; with thomsonite in vesicular basalt at Aci Castello and Aci Trezza, Sicily; in leucite tephrite at Capo di Bove, Rome, Italy; at Weilberg, Rhineland, Germany; at Kloch, Styria, Austria; and in the Langesundfjord district, Norway.
BEST REF. IN ENGLISH: Deer, Howie, and Zussman, "Rock Forming Minerals," v. 4, p. 358–376, New York, Wiley, 1963.

# GONYERITE
$(Mn, Mg, Fe)_6Si_4O_{10}(OH)_8$
Chlorite Group

CRYSTAL SYSTEM: Orthorhombic (pseudohexagonal)
Z: 4
LATTICE CONSTANTS:
$a = 9.46$
$b = 28.8$
$c = 5.47$
3 STRONGEST DIFFRACTION LINES:
7.23 (100)
3.61 ( 80)
4.79 ( 50)
OPTICAL CONSTANTS:
$\alpha = 1.646$
$\beta = 1.664$
$\gamma = 1.664$
$(-)2V = 0°$
HARDNESS: 2½
DENSITY: 3.01 (Meas.)
3.03 (Calc.)
CLEAVAGE: {001} perfect
Foliae flexible but inelastic.
HABIT: Rounded radial aggregates of laths and plates up to several millimeters in length. As tiny aggregates of pseudohexagonal plates.
COLOR-LUSTER: Deep brown. Powder chocolate brown.
MODE OF OCCURRENCE: Occurs associated with barite, berzeliite, bementite, and garnet in small hydrothermal veinlets cutting skarn at Långban, Sweden.
BEST REF. IN ENGLISH: Frondel, Clifford, *Am. Min.*, **40**: 1090–1094 (1955).

# GOONGARRITE = Mixture cosalite and galena

## GORCEIXITE
$BaAl_3(PO_4)_2(OH)_5 \cdot H_2O$

CRYSTAL SYSTEM: Hexagonal
CLASS: $\bar{3}2/m$
SPACE GROUP: $R\bar{3}m$
Z: 3
LATTICE CONSTANTS:
  $a = 7.02$
  $c = 16.87$
3 STRONGEST DIFFRACTION LINES:
  2.98 (100)
  5.73 ( 90)
  3.52 ( 80)
OPTICAL CONSTANTS:
  $N = 1.625$; uniaxial (+)
  HARDNESS: 6
  DENSITY: 3.323 (Meas.)
            3.297 (Calc.)
  CLEAVAGE: Fracture porcelaneous.
  HABIT: Botryoidal, radial-fibrous aggregates and as grains and pebbles.
  COLOR-LUSTER: Whitish, brown, sometimes mottled. Vitreous to dull.
  MODE OF OCCURRENCE: Occurs as spheroids in novaculite near Hot Springs, Garland County, Arkansas. It also is found at Felsobanya, Roumania; and as favas in the diamantiferous sands of Guyana, Brazil, and Africa.
  BEST REF. IN ENGLISH: Palache, et al., "Dana's System of Mineralogy," 7th Ed., v. II, p. 833, New York, Wiley, 1951.

## GORDONITE
$MgAl_2(PO_4)_2(OH)_2 \cdot 8H_2O$

CRYSTAL SYSTEM: Triclinic
CLASS: $\bar{1}$
SPACE GROUP: $P\bar{1}$
Z: 1
LATTICE CONSTANTS:
  $a = 5.24$      $\alpha = 107°25'$
  $b = 10.49$     $\beta = 111°04'$
  $c = 6.96$      $\gamma = 72°22'$
3 STRONGEST DIFFRACTION LINES:
  9.78 (100)
  3.17 ( 80)
  2.83 ( 70)
OPTICAL CONSTANTS:
      $\alpha = 1.534$
      $\beta = 1.543$
      $\gamma = 1.558$
  (+)2V = 73°
  HARDNESS: 3½
  DENSITY: 2.23 (Meas.)
            2.22 (Calc.)
  CLEAVAGE: {010} perfect
            {100} fair
            {001} poor
Fracture conchoidal. Brittle.
  HABIT: Crystals rare, minute, prismatic to platy; vertically striated. Usually as sheaf-like aggregates.

COLOR-LUSTER: Colorless, smoky white, pale pink, pale green. Transparent. Vitreous, pearly on {010}. Streak white.
  MODE OF OCCURRENCE: Occurs associated with crandallite and wardite in variscite nodules at Fairfield, Utah County, Utah.
  BEST REF. IN ENGLISH: Palache, et al., "Dana's System of Mineralogy," 7th Ed., v. II, p. 975-976, New York, Wiley, 1951.

## GÖRGEYITE
$K_2Ca_5(SO_4)_6 \cdot 1$ or $1½H_2O$

CRYSTAL SYSTEM: Monoclinic
CLASS: 2/m
SPACE GROUP: C2/c
Z: 4
LATTICE CONSTANTS:
  $a = 17.48$
  $b = 6.83$
  $c = 18.23$
  $\beta = 113.3°$
3 STRONGEST DIFFRACTION LINES:
  3.01  (100)
  3.16  ( 70)
  2.817 ( 40)
OPTICAL CONSTANTS:
      $\alpha = 1.560$
      $\beta = 1.569$ (Na)
      $\gamma = 1.584$
  (+)2V = 79°
  HARDNESS: ~3½
  DENSITY: 2.95 (Meas.)
            2.90 (Calc.)
  CLEAVAGE: {100} distinct
Fracture splintery.
  HABIT: Crystals thin tabular with c(100) dominant, also a(100), s(111), and m(110).
  COLOR-LUSTER: Colorless to yellowish; vitreous.
  MODE OF OCCURRENCE: Occurs with glauberite and minor halite and polyhalite in the Leopold horizon, Ischl salt deposit, Austria.
  BEST REF. IN ENGLISH: Mayrhofer, Heimo, *Am. Min.*, **39**: 403 (1954).

## GOSLARITE
$ZnSO_4 \cdot 7H_2O$

CRYSTAL SYSTEM: Orthorhombic
CLASS: 2 2 2
SPACE GROUP: $P2_12_12_1$
Z: 4
LATTICE CONSTANTS:
  $a = 11.85$ kX
  $b = 12.09$
  $c = 6.83$
3 STRONGEST DIFFRACTION LINES:
  4.21 (100)
  5.36 ( 80)
  4.18 ( 50)

OPTICAL CONSTANTS·
$\alpha$ = 1.447–1.463
$\beta$ = 1.475–1.480
$\gamma$ = 1.470–1.485
(-)2V = moderate, small
HARDNESS: 2–2½
DENSITY: 1.978 (Meas.)
1.972 (Calc.)
CLEAVAGE: {010} perfect
Brittle.

HABIT: Massive, granular or fibrous. Also stalactitic and as efflorescent crusts. Synthetic crystals stout prismatic.

COLOR-LUSTER: Crystals colorless; transparent; vitreous. Also white, greenish, bluish, brownish when massive; transparent to translucent; vitreous, silky, or dull.

Dehydrates readily. Soluble in water. Taste astringent, metallic.

MODE OF OCCURRENCE: Occurs widespread as a postmining efflorescence in mines that contain sphalerite or other zinc minerals. Found in the United States in Missouri, Kansas, Oklahoma, New Mexico, Arizona, California, Colorado, Utah, Montana, and in other states. Among many other localities, it is found in Mexico, Argentina, Peru, France, Spain, Sweden, and Germany.

BEST REF. IN ENGLISH: Palache, et al., "Dana's System of Mineralogy," 7th Ed., v. II, p. 513–516, New York, Wiley, 1951.

# GÖTZENITE
$(Ca, Na)_7 (Ti, Al)_2 Si_4 O_{15} (F, OH)_3$

CRYSTAL SYSTEM: Triclinic
Z: 1
LATTICE CONSTANTS:
$a$ = 9.65     $\alpha$ = 90°
$b$ = 7.32     $\beta$ = 101.3°
$c$ = 5.74     $\gamma$ = 101.1°
3 STRONGEST DIFFRACTION LINES:
3.100 (100)
2.986 (100)
1.911 ( 50)
OPTICAL CONSTANTS:
$\alpha$ = 1.660
$\beta$ = 1.662
$\gamma$ = 1.670
(+)2V = 52°
HARDNESS: Not determined.
DENSITY: 3.138 (Meas.)
CLEAVAGE: {100} perfect
{001} good
HABIT: Prismatic crystals up to 0.5 mm long.
COLOR-LUSTER: Colorless.
MODE OF OCCURRENCE: Occurs associated with melilite, nepheline, clinopyroxene, kalsilite, kirschsteinite, combeite, sodalite, magnetite, perovskite, apatite, hornblende, biotite, and an unidentified mineral, in melilite-nephelinite lava from Mt. Shaheru, Zaire.

BEST REF. IN ENGLISH: Sahama, Th. G., and Hytönen, Kai, *Min. Mag.*, **31**: 503–510 (1957).

# GOWERITE
$CaB_6 O_{10} \cdot 5H_2 O$

CRYSTAL SYSTEM: Monoclinic
CLASS: 2/m
SPACE GROUP: $P2_1/n$
Z: 4
LATTICE CONSTANTS:
$a$ = 11.03
$b$ = 16.40
$c$ = 6.577
$\beta$ = 90°56′
3 STRONGEST DIFFRACTION LINES:
8.2 (100)
3.19 ( 45)
9.2 ( 25)
OPTICAL CONSTANTS:
$\alpha$ = 1.484
$\beta$ = 1.501
$\gamma$ = 1.550
(+)2V = 63°
HARDNESS: 3
DENSITY: 2.00 ± 0.01 (Meas.)
1.982 (Calc.)
CLEAVAGE: {001?} distinct
{100?} imperfect
Fracture uneven. Brittle.
HABIT: Crystals long prismatic [001] to needle-like; grouped in radiating globular clusters. Average crystals are 0.8 mm in length by 0.06 mm and 0.02 mm in width and breadth. Occur up to 1.5 mm in length.

COLOR-LUSTER: Colorless, transparent; vitreous. Aggregates white. Streak white.

MODE OF OCCURRENCE: Occurs associated with ginorite, ulexite, meyerhofferite, sassolite, hydroboracite, and an undescribed magnesium borate in the Furnace Creek borate deposits of the Death Valley region, California. Minute globular clusters of gowerite form from the weathering of colemanite and priceite veins in some basaltic rocks in the Furnace Creek formation of late Tertiary age.

BEST REF. IN ENGLISH: Erd, Richard C., McAllister, James F., and Almond, Hy, *Am. Min.*, **44**: 911–919 (1959). Christ, C. L., and Clark, Joan R., *Am. Min.*, **45**: 230–234 (1960).

# GOYAZITE (Hamlinite)
$SrAl_3 (PO_4)_2 (OH)_5 \cdot H_2 O$

CRYSTAL SYSTEM: Hexagonal
CLASS: $\bar{3}$ 2/m
SPACE GROUP: $R\bar{3}m$
Z: 3
LATTICE CONSTANTS:
$a$ = 6.97
$c$ = 16.51
3 STRONGEST DIFFRACTION LINES:
5.73 (100)
2.96 (100)
3.49 ( 80)

OPTICAL CONSTANTS:
$\omega = 1.620-1.653$
$\epsilon = 1.630-1.661$
(+)
HARDNESS: 4½–5
DENSITY: 3.26 (Meas.)
3.29 (Calc.)
CLEAVAGE: {0001} perfect
HABIT: Crystals small, rhombohedral, pseudocubic, or tabular. Rhombohedral faces commonly striated horizontally. Also as rounded grains or pebbles.
COLOR-LUSTER: Colorless, yellowish, pink. Transparent. Resinous to greasy, pearly on base.
MODE OF OCCURRENCE: Occurs abundantly associated with ferberite throughout the Nederland tungsten belt of Boulder County, Colorado, especially at the Eagle Rock mine; found sparingly in pegmatites in Maine, New Hampshire, and South Dakota; near Diamantina, Minas Geraes, Brazil; at Lengenbach and in the Simplon tunnel, Switzerland; and in the USSR.
BEST REF. IN ENGLISH: Palache, et al., "Dana's System of Mineralogy," 7th Ed., v. II, p. 834–835, New York, Wiley, 1951.

# GRAFTONITE
$(Fe^{2+}, Mn^{2+}, Ca)_3 (PO_4)_2$

CRYSTAL SYSTEM: Monoclinic
CLASS: 2/m
SPACE GROUP: $P2_1/c$
Z: 4
LATTICE CONSTANTS:
$a = 8.87$
$b = 11.57$
$c = 6.17$
$\beta = 99°12'$
3 STRONGEST DIFFRACTION LINES:
2.86 (100)
3.50 ( 90)
2.715 ( 70)
OPTICAL CONSTANTS:
$\alpha = 1.695-1.709$
$\beta = 1.699-1.714$
$\gamma = 1.719-1.736$
(+)2V = small, 43°, 60°
HARDNESS: 5
DENSITY: 3.67–3.79 (Meas.)
3.72 (Calc.)
also 3.66 (Calc.)
CLEAVAGE: {010} distinct
{100} indistinct
Fracture subconchoidal to uneven.
HABIT: Massive, cleavable; rarely as rough composite crystals.
COLOR-LUSTER: Salmon pink to reddish brown; often dark brown from alteration. Vitreous to resinous. Translucent.
MODE OF OCCURRENCE: Occurs as a primary mineral in granite pegmatite as fresh masses as much as 3 feet in diameter associated with sarcopside and several secondary phosphates at the Bull Moose mine, near Custer, South Dakota. Also as clove brown masses as much as 18 inches in diameter at the Ross mine; in masses up to 6 inches across at the Victory mine near Custer; and in masses as much as 4 inches across associated with arrojadite at the Nickel Plate mine, Keystone, South Dakota. Also found in granite pegmatite at the Rice mine, Palermo mine, and Melvin Mountain, New Hampshire; at Greenwood, Maine; in the Kondakovo district, Eastern Siberia, USSR; at Olgiasca, Lake Como, Italy; and at Brissago, Tessin, Switzerland.
BEST REF. IN ENGLISH: "Dana's System of Mineralogy," 7th Ed., v. 2, p. 686–688, New York, Wiley, 1951. Lindberg, Mary Louise, *Am. Min.*, **35**: 59–76 (1950). Roberts, W. L., and Rapp, George Jr., *SDSM&T Bull.*, **18**, 98–99 (1965). Calvo, C., *Am. Min.*, **53**: 742–750 (1968).

# GRANDIDIERITE
$(Mg, Fe)Al_3 BSiO_9$

CRYSTAL SYSTEM: Orthorhombic
CLASS: 2/m 2/m 2/m
SPACE GROUP: Pbnm
Z: 4
LATTICE CONSTANTS:
$a = 10.34$
$b = 10.98$
$c = 5.76$
3 STRONGEST DIFFRACTION LINES:
5.17 (100)
5.04 (100)
5.48 ( 80)
OPTICAL CONSTANTS:
$\alpha = 1.590-1.602$
$\beta = 1.618-1.636$
$\gamma = 1.623-1.639$
(–)2V = 30°
HARDNESS: 7½
DENSITY: 2.976 (Meas.)
3.099 (Calc. Mg = Fe)
CLEAVAGE: {100} perfect
{010} distinct
Fracture uneven. Brittle.
HABIT: Crystals elongated, indistinct. As prismatic masses.
COLOR-LUSTER: Greenish blue. Translucent. Vitreous.
MODE OF OCCURRENCE: Occurs in pegmatite at Andrahomana and elsewhere in Madagascar.
BEST REF. IN ENGLISH: Winchell and Winchell, "Elements of Optical Mineralogy," 4th Ed., Pt. 2, p. 497, New York, Wiley, 1951.

# GRANGESITE = Mn-brunsvigite

# GRANTSITE
$Na_4 CaV_2^{4+} V_{10}^{5+} O_{32} \cdot 8H_2O$

CRYSTAL SYSTEM: Monoclinic

CLASS: 2/m, m, or 2
SPACE GROUP: C2/m (or Cm or C2)
Z: 1
LATTICE CONSTANTS:
  $a = 17.545$
  $b = 3.60$
  $c = 12.41$
  $\beta = 95°15'$
3 STRONGEST DIFFRACTION LINES:
  8.74 (100)
  3.61 ( 30)
  12.4 ( 18)
OPTICAL CONSTANTS:
  $\alpha = 1.82$
  $\beta > 2.0$
  $\gamma > 2.0$
  (−)2V
HARDNESS: Soft; smears easily.
DENSITY: 2.94 (Meas.)
          2.95 (Calc.)
CLEAVAGE: Not determined.
HABIT: Microscopic bladed crystals elongated parallel to b-axis.
COLOR-LUSTER: Dark olive green to greenish black. Silky or pearly to subadamantine. Streak olive green to brownish green.
MODE OF OCCURRENCE: Occurs as aggregates of microscopic fibers in thin seams and as interstitial filling in sandstone, in fracture fillings in coalified wood, and in botryoidal coatings pseudomorphous after the low-valent vanadium oxides paramontroseite and häggite. Found in the F-33 mine near Grants, New Mexico; in the Golden Cycle and the LaSalle mines in Montrose County, Colorado; and in the Parco No. 23 mine, Grand County, Utah.
BEST REF. IN ENGLISH: Weeks, A. D., Lindberg, M. L., Truesdell, A. H., and Meyrowitz, Robert, *Am. Min.*, **49**: 1511–1526 (1964).

# GRAPHITE
C
Polymorphous with diamond, chaoite, lonsdaleite

CRYSTAL SYSTEM: Hexagonal
CLASS: 6/m 2/m 2/m
SPACE GROUP: P6₃/mmc
Z: 4
LATTICE CONSTANTS:
  $a = 2.455$
  $c = 6.69$
3 STRONGEST DIFFRACTION LINES:
  3.35 (100)
  1.675 ( 80)
  1.541 ( 60)
OPTICAL CONSTANTS:
  $\omega = 1.93–2.07$ (red)
  uniaxial (−)
HARDNESS: 1–2
DENSITY: 2.09–2.23 (Meas.)
          2.2667 (Calc.)
CLEAVAGE: {0001} perfect, easy
Flexible, inelastic. Greasy feel. Sectile.

HABIT: As thin hexagonal tabular crystals, flattened {0001}, commonly exhibiting triangular striations. Usually in coarse to fine foliated masses; also in radiate aggregates, scaly, columnar, granular, or earthy.
COLOR-LUSTER: Iron black to steel gray. Opaque. Metallic; also dull, earthy.
MODE OF OCCURRENCE: Occurs chiefly in rocks which have suffered intense regional or igneous metamorphism. Found widespread in the United States, especially in the vicinity of Ticonderoga, New York; at Franklin and Sterling Hill, New Jersey; and at numerous deposits in Maine, Rhode Island, Pennsylvania, North Carolina, Alabama, South Dakota, Montana, New Mexico, California, and other western states. Among many other widespread occurrences throughout the world, it also is found in Canada, Mexico, Greenland, England, Germany and other places in central Europe, USSR, and as large commercial deposits in Ceylon.
BEST REF. IN ENGLISH: Palache, et al., "Dana's System of Mineralogy," 7th Ed., v. I, p. 152–154, New York, Wiley, 1944.

# GRATONITE
Pb₉As₄S₁₅

CRYSTAL SYSTEM: Hexagonal
CLASS: 3m
SPACE GROUP: R3m
Z: 3
LATTICE CONSTANTS:
  $a = 17.69$
  $c = 7.83$
3 STRONGEST DIFFRACTION LINES:
  3.43 (100)
  3.75 ( 60)
  2.734 ( 50)
HARDNESS: 2½
DENSITY: 6.22 (Meas.)
          6.17 (Calc.)
CLEAVAGE: None.
Brittle.
HABIT: Crystals prismatic [0001] with {11$\bar{2}$0} dominant; {02$\bar{2}$1} large, commonly the only terminal form present. Also massive.
COLOR-LUSTER: Dark lead gray. Opaque. Metallic. Streak black.
MODE OF OCCURRENCE: Occurs in small amounts in vugs in pyritic ore, associated with tetrahedrite, enargite, and realgar, at the Excelsior mine, Cerro de Pasco, Peru.
BEST REF. IN ENGLISH: Palache, et al., "Dana's System of Mineralogy," 7th Ed., v. I, p. 397–398, New York, Wiley, 1944.

# GRAYITE
(Th, Pb, Ca)PO₄ · H₂O

CRYSTAL SYSTEM: Hexagonal
CLASS: 622
SPACE GROUP: P6₂22
Z: 3

3 STRONGEST DIFFRACTION LINES:
3.04 (100)
2.82 ( 80)
2.14 ( 80)
HARDNESS: 3-4
DENSITY: 3.7-4.3
CLEAVAGE: None.
Fracture conchoidal.

HABIT: Massive; very fine grained; also powdery.

COLOR-LUSTER: Pale yellow; reddish brown. Resinous.

MODE OF OCCURRENCE: Occurs in a small vein in Precambrian rocks southwest of Gunnison in Gunnison County, and in the Wet Mountains of south-central Colorado; in the limestone member of the Pliocene Moonstone formation in Fremont County, Wyoming; and in the Mtoko district, Rhodesia.

BEST REF. IN ENGLISH: Dooley, J. R., Jr., and Hathaway, John C., USGS Prof. Paper 424-C, p. 339-341 (1961).


# GREENALITE
$(Fe^{2+}, Fe^{3+})_{5-6} Si_4 O_{10} (OH)_8$
Septechlorite Group

CRYSTAL SYSTEM: Orthorhombic
Z: 2
LATTICE CONSTANTS:
$a = 5.56$
$b = 9.60$
$c = 7.21$
3 STRONGEST DIFFRACTION LINES:
2.59 (100)
2.20 ( 90)
7.21 ( 70)
OPTICAL CONSTANTS: Nearly isotropic
$n \sim 1.674$ (Na)
HARDNESS: Not determined.
DENSITY: 2.85-3.15 (Meas.)
3.61 (Calc.)
CLEAVAGE: None.
HABIT: Massive; as minute rounded granules usually between 0.1 and 1.0 mm in diameter.

COLOR-LUSTER: Light yellowish green to bluish green in transmitted light; dark green to black in reflected light. Translucent to nearly opaque. Dull.

Somewhat magnetic.

MODE OF OCCURRENCE: Occurs as a primary mineral in the Biwabik formation of the Mesabi Range in Minnesota.

BEST REF. IN ENGLISH: Jolliffe, Fred, *Am. Min.*, **20**: 405-425 (1935). Gruner, J. W., *Am. Min.*, **21**: 449-455 (1936).


# GREENOCKITE
CdS
Dimorphous with hawleyite

CRYSTAL SYSTEM: Hexagonal
CLASS: 6 mm
SPACE GROUP: $C6_3mc$
Z: 2

LATTICE CONSTANTS:
$a = 4.142$
$c = 6.724$
3 STRONGEST DIFFRACTION LINES:
3.16 (100)
3.58 ( 75)
3.36 ( 60)
OPTICAL CONSTANTS:
$\omega = 2.529$
$\epsilon = 2.506$ (Na)
(+)
HARDNESS: 3-3½
DENSITY: 4.820 (Meas. synthetic)
4.772 (Calc.)
CLEAVAGE: $\{11\bar{2}2\}$ distinct
$\{0001\}$ imperfect
Fracture conchoidal. Brittle.

HABIT: Crystals hemimorphic pyramidal, thick tabular, and prismatic. Often striated horizontally and in oscillatory combination. Commonly as an earthy coating. Twin plane $\{0001\}$ and $\{11\bar{2}2\}$ not common; cyclic twins common (Llallagua).

COLOR-LUSTER: Various shades of yellow and orange to deep red. Strong adamantine to resinous. Streak orange-yellow to brick red.

MODE OF OCCURRENCE: Occurs as minute crystals with prehnite and zeolites at Paterson, New Jersey. Found as an earthy coating at Franklin, New Jersey; Friedensville, Pennsylvania; in the Joplin District, Missouri; at Eureka, Nevada; Hanover, New Mexico; at Topaz, Mono County, California; and as a bright yellow coating on smithsonite in Marion County, Arkansas. Originally found as superb crystals up to ½ inch in size on prehnite, with natrolite and calcite, in cavities in a labradorite porphyry at Bishopton, Renfrew, Scotland; also found widespread as a coating on sphalerite or smithsonite at many European localities. In the Bolivian tin deposits it is found at Llallagua as minute red crystals associated with quartz, marcasite, cassiterite, and wavellite; and at the Asunta tin-silver mine near San Vicente as elegant deep garnet red crystals up to 2 mm in size in fissures and small vugs.

BEST REF. IN ENGLISH: Palache, et al., "Dana's System of Mineralogy," 7th Ed., v. I, p. 228-230, New York, Wiley, 1944.


# GREIGITE (Melnikovite)
$Fe_3S_4$
Linnaeite Group

CRYSTAL SYSTEM: Cubic
CLASS: $4/m \bar{3} 2/m$
SPACE GROUP: Fd3m
Z: 8
LATTICE CONSTANT:
$a = 9.876$
3 STRONGEST DIFFRACTION LINES:
2.980 (100)
1.746 ( 77)
2.469 ( 55)
HARDNESS: 4-4½
DENSITY: 4.049 (Meas.)
4.079 (Calc.)

CLEAVAGE: None observed.
HABIT: Balls of intergrown octahedra with curved faces averaging 0.3 mm on an edge. Infrequently cubes. Also as tiny grains. Strongly magnetic.
COLOR-LUSTER: Pink, often tarnished a metallic blue; metallic. Sooty black (Skinner et al.)
MODE OF OCCURRENCE: Occurs as tiny grains and crystals in certain clay layers of the upper part of the Tropico Group, a Tertiary locustrine sequence in the Kramer-Four Corners area, San Bernardino County, California; in small quantities associated with stibnite throughout the ore bodies of the Lojane mine about 40 km north of Skopje, Macedonia, in southeast Yugoslavia; and also associated with pyrite, marcasite, and carbonates from a deposit in Zacatecas, Mexico.
BEST REF. IN ENGLISH: Skinner, Brian J., Erd, Richard C., and Grimaldi, Frank S., *Am. Min.*, **49**: 543–555 (1964).    Williams, Sidney A., *Am. Min.*, **53**: 2087–2088 (1968).

## GRIFFITHITE = Ferroan saponite

## GRIMALDIITE
CrO(OH)
Trimorphous with bracewellite and guyanaite

CRYSTAL SYSTEM: Hexagonal
HARDNESS: Not determined.
DENSITY: Not determined.
CLEAVAGE: Not determined.
HABIT: Intergrown with macconnellite forming composite platy hexagonal crystals.
COLOR-LUSTER: Red-brown.
MODE OF OCCURRENCE: Occurs in complex microscopic intergrowth with eskolaite, bracewellite, guyanaite, and macconnellite in "merumite" from alluvial gravels in Guyana.
BEST REF. IN ENGLISH: Milton, C., et al., Geol. Soc. Am., Program 1967 Annual Meetings, p. 151–152 (1967) (abst.).

## GRIMSELITE
$K_3Na(UO_2)(CO_3)_3 \cdot H_2O$

CRYSTAL SYSTEM: Hexagonal
CLASS: $\bar{6}m2$
SPACE GROUP: $P\bar{6}2c$
Z: 2
LATTICE CONSTANTS:
  $a$ = 9.30
  $c$ = 8.26
3 STRONGEST DIFFRACTION LINES:
  5.76 (100)
  8.09 ( 80)
  3.08 ( 80)
OPTICAL CONSTANTS:
  $\omega$ = 1.601
  $\epsilon$ = 1.480
  (−)

HARDNESS: 2–2½
DENSITY: 3.30 (Meas. Syn.)
          3.27 (Calc.)
CLEAVAGE: None.
Fracture conchoidal. Brittle.
HABIT: Crystals show forms m$\{10\bar{1}0\}$ (dominant), x$\{10\bar{1}1\}$, and c$\{0001\}$. As crusts of fine-grained aggregates, mostly of anhedral grains.
COLOR-LUSTER: Yellow. Streak pale yellow.
MODE OF OCCURRENCE: The mineral occurs associated with schroekingerite, monohydrocalcite, and two unnamed carbonate minerals, where aplite granite and granodiorite are cut by fissure veins and mineralized zones in the tunnel between Gerstenegg and Sommerlach, Grimsel area, Aar massif, Oberhasli, Canton Bern, Switzerland.
BEST REF. IN ENGLISH: Walenta, Kurt, *Am. Min.*, **58**: 139 (1973).

## GRIPHITE
$(Na, Al, Ca, Fe)_3 Mn_2 (PO_4)_{2.5} (OH)_2$

CRYSTAL SYSTEM: Cubic
CLASS: $2/m\bar{3}$
SPACE GROUP: $P2_1/a\bar{3}$
Z: 8
LATTICE CONSTANTS:
  $a$ = 12.222
3 STRONGEST DIFFRACTION LINES:
  (heated)      (unheated)
  2.73 (100)    2.75 (100)
  3.04 ( 30)    2.98 ( 30)
  2.95 ( 30)    2.00 ( 30)
OPTICAL CONSTANT:
  $N$ = 1.63–1.66
HARDNESS: 5½
DENSITY: 3.40 (Meas.)
          3.399 (Calc.)
CLEAVAGE: None.
Fracture uneven to conchoidal. Brittle.
HABIT: Massive, compact.
COLOR-LUSTER: Dark brown to brownish black. Translucent. Resinous to vitreous.
MODE OF OCCURRENCE: The mineral occurs as masses ranging upwards of 50 pounds in weight in granitic pegmatite at the Riverton lode (Everly mine), and as nodular masses as much as 6 feet in diameter at the Sitting Bull mine, near Keystone, Pennington County, South Dakota. It also is found at Mt. Ida, Northern Territory, Australia, and in Turkestan, USSR.
BEST REF. IN ENGLISH: Palache, et al., "Dana's System of Mineralogy," 7th Ed., v. II, p. 843–844, New York, Wiley, 1951.    McConnell, D., *Am. Min.*, **27**: 452–461 (1942).    Peacor, Donald R., and Simmons, Wm. B. Jr., *Am. Min.*, **57**: 269–272 (1972).

## GROCHAUITE = Variety of sheridanite

## GROSSULAR (Grossularite)
$Ca_3 Al_2 Si_3 O_{12}$
Garnet Group

Var. Rosolite
   Hessonite

CRYSTAL SYSTEM: Cubic
CLASS: $4/m\,\bar{3}\,2/m$
SPACE GROUP: Ia3d
Z: 8
LATTICE CONSTANT:
 $a = 11.851$
3 STRONGEST DIFFRACTION LINES:
 2.65 (100)
 1.58 ( 90)
 2.96 ( 80)
OPTICAL CONSTANT:
 $N = 1.734–1.75$
 HARDNESS: 6½–7
 DENSITY: 3.4–3.6 (Meas.)
 CLEAVAGE: None.
              {110} parting, sometimes distinct.
Fracture uneven to conchoidal. Brittle.
 HABIT: Crystals usually dodecahedrons or trapezohedrons; also in combination, or with hexoctahedron. Also massive, compact; fine or coarse granular; or as embedded grains.
 COLOR-LUSTER: Colorless, white, gray, yellow, yellowish green, green, yellowish brown, brown, pink, red, black. Transparent to nearly opaque. Vitreous to resinous. Streak white.
 MODE OF OCCURRENCE: Occurs chiefly in metamorphosed impure calcareous rocks, especially in contact zones, in association with wollastonite, idocrase, diopside, scapolite, and calcite; also in certain schists and in serpentinite. Found widespread in the western United States, especially in Eldorado, Inyo, Riverside, San Diego, Tulare, and Tuolumne counties, California, and in Chaffee County, Colorado. Fine crystals occur near Minot, Maine; at Warren, New Hampshire; and at Eden Mills, Vermont. Also in Gatineau and Megantic counties, Quebec, Canada; at Lake Jaco, Chihuahua, and near Xalostoc, Morelos, Mexico; in the Ala Valley, Piedmont, and at many other places in Italy; and at localities in Switzerland, USSR, Korea, Ceylon, and Australia.
 BEST REF. IN ENGLISH: Deer, Howie, and Zussman, "Rock Forming Minerals," v. 1, p. 77–112, New York, Wiley, 1962.

# GROUTITE
$HMnO_2$
Trimorphous with feitknechtite and manganite

CRYSTAL SYSTEM: Orthorhombic
CLASS: 2/m 2/m 2/m
SPACE GROUP: Pbnm
Z: 4
LATTICE CONSTANTS:
 $a = 4.56$
 $b = 10.70$
 $c = 2.85$
3 STRONGEST DIFFRACTION LINES:
 4.20 (100)
 2.81 ( 70)
 2.67 ( 70)

HARDNESS: ~5½
DENSITY: 4.144 (Meas.)
         4.172 (Calc.)
CLEAVAGE: {010} perfect
          {100} distinct
Fracture uneven. Brittle.
 HABIT: Crystals usually wedge or lens-shaped with rounded faces; [001] zone striated parallel to c-axis; in dense groups. Also as slender striated prisms.
 COLOR-LUSTER: Jet black. Opaque. Brilliant submetallic to adamantine. Streak dark brown.
 MODE OF OCCURRENCE: Occurs as fine specimens at the Sagamore, Mangan No. 2, and Mahnomen pits and elsewhere in the Cuyuna iron range in Minnesota; also as small crystals coating calcite in vugs in talc at the No. Two and One Half mine, Talcville, New York. An antimonian variety occurs in very small amounts at Franklin, Sussex County, New Jersey.
 BEST REF. IN ENGLISH: Gruner, John W., *Am. Min.*, 32: 654–659 (1947).

# GROVESITE (Inadequately described mineral)
$(Mn, Fe, Al)_{13}(Al, Si)_8 O_{22}(OH)_{14}$ (?)

CRYSTAL SYSTEM: Orthorhombic
LATTICE CONSTANTS:
 $a = 9.54$
 $b = 14.36$
 $c = 5.51$
OPTICAL CONSTANTS:
 $\alpha = 1.658$
 $\beta = \gamma = 1.667$ (Na)
DENSITY: 3.150 (Meas.)
HABIT: Small rosettes, up to 0.5 mm across.
COLOR-LUSTER: Blackish brown.
 MODE OF OCCURRENCE: Occurs at the No. 5 orebody of the Benallt mine, Rhiw, North Wales.
 BEST REF. IN ENGLISH: Bannister, F. A., Hey, M. M., and Smith, W. Campbell, *Min. Mag.*, 30: 645–647 (1955).

# GRUENLINGITE (Inadequately described mineral)
$Bi_4TeS_3$

CRYSTAL SYSTEM: Probably hexagonal
3 STRONGEST DIFFRACTION LINES:
 3.11 (100)
 2.13 ( 30)
 2.25 ( 20)
HARDNESS: 2
DENSITY: 8.08 (Meas.)
CLEAVAGE: Perfect, one direction. Flexible.
HABIT: Massive, lamellar.
COLOR-LUSTER: Steel gray; tarnishes darker or iridescent. Opaque. Metallic.
 MODE OF OCCURRENCE: Occurs in quartz with native bismuth and bismuthinite, at Carrock Fell, Cumberland, England, and mixed with native bismuth in dolomite in the Serrania de Ronda, Spain.
 BEST REF. IN ENGLISH: Palache, et al., "Dana's System of Mineralogy," 7th Ed., v. I, p. 164–165, New York,

Wiley, 1944. Vlasov, K. A., "Mineralogy of Rare Elements," p. 739, Israel Program for Scientific Translations, 1966.

## GRUNERITE
$(Fe, Mg)_7Si_8O_{22}(OH)_2$
Amphibole Group
Var. Amosite

CRYSTAL SYSTEM: Monoclinic
CLASS: 2/m
SPACE GROUP: C2/m
Z: 2
LATTICE CONSTANTS:
$a \simeq 9.6$
$b \simeq 18.3$
$c \simeq 5.3$
$\beta \simeq 101°50'$
3 STRONGEST DIFFRACTION LINES:
8.33 (100)
2.77 ( 90)
3.07 ( 80)
OPTICAL CONSTANTS:
$\alpha = 1.665-1.696$
$\beta = 1.675-1.709$
$\gamma = 1.698-1.729$
$(-)2V = 90°-96°$
HARDNESS: 5-6
DENSITY: 3.44-3.60 (Meas.)
3.54 (Calc.)
CLEAVAGE: {110} good
HABIT: Fibrous or fibrolamellar, often radiated. Twinning on {100} very common, simple, lamellar.
COLOR-LUSTER: Ash gray, dark green, brown. Translucent to nearly opaque. Silky.
MODE OF OCCURRENCE: Occurs chiefly in contact or regionally metamorphosed iron-rich rocks. Typical occurrences are found in Arizona, Colorado, South Dakota, Minnesota, Michigan, and Massachusetts. Also in the Wabash iron formation, Labrador, Canada, and at localities in Scotland, France, Finland, South Africa (amosite), and Australia.
BEST REF. IN ENGLISH: Deer, Howie, and Zussman, "Rock Forming Minerals," v. 2, p. 234-248, New York, Wiley, 1963.

## GUANAJUATITE
$Bi_2(Se, S)_3$

CRYSTAL SYSTEM: Orthorhombic
CLASS: 2/m 2/m 2/m
SPACE GROUP: Pbnm
Z: 4
LATTICE CONSTANTS:
$a = 11.32$
$b = 11.48$
$c = 4.17$
3 STRONGEST DIFFRACTION LINES:
3.19 (100)
3.65 ( 90)
1.989 ( 70)

HARDNESS: 2½-3½
DENSITY: 6.25-6.98 (Meas.)
CLEAVAGE: {010} distinct
{001} indistinct
Somewhat sectile.
HABIT: As semicompact masses of striated acicular crystals. Also massive, with fine granular or fibrous texture.
COLOR-LUSTER: Lead gray with bluish tint; opaque; metallic. Streak shiny gray.
MODE OF OCCURRENCE: Occurs as grains and small masses in quartz near the mouth of Kirtley Creek northeast of Salmon, Lemhi County, Idaho. Also found in association with native bismuth, bismuthinite, and pyrite at the La Industrial and Santa Catarina mines near Guanajuato, Mexico; at Fahlun, Sweden; and at Andreasberg, Harz Mountains, Germany.
BEST REF. IN ENGLISH: Palache, et al., "Dana's System of Mineralogy," 7th Ed., v. I, p. 278-279, New York, Wiley, 1944.

## GUDMUNDITE
FeSbS

CRYSTAL SYSTEM: Monoclinic
CLASS: 2/m
SPACE GROUP: $P2_1/d$
Z: 4
LATTICE CONSTANTS:
$a = 6.03$
$b = 5.93$
$c = 6.03$
$\beta = 112°44'$
3 STRONGEST DIFFRACTION LINES:
2.56 (100)
1.912 ( 80)
1.410 ( 70)
HARDNESS: ~6
DENSITY: 6.72 (Meas.)
6.91 (Calc.)
CLEAVAGE: None.
Fracture uneven. Brittle.
HABIT: Crystals prismatic, commonly twinned on {101} producing butterfly and cruciform twins.
COLOR-LUSTER: Silver white to steel gray. Opaque. Metallic.
MODE OF OCCURRENCE: Occurs in sulfide deposits, often associated with other sulfides and lead sulfantimonides. Found in the Yellowknife district, Northwest Territories, Canada; at several deposits in Sweden, especially at Gudmundstorp, Vastmanland; and at localities in Norway, Germany, and Turkey.
BEST REF. IN ENGLISH: Palache, et al., "Dana's System of Mineralogy," 7th Ed., v. I, p. 325-326, New York, Wiley, 1944.

## GUERINITE
$Ca_5H_2(AsO_4)_4 \cdot 9H_2O$

CRYSTAL SYSTEM: Monoclinic or triclinic
LATTICE CONSTANT:
$c = 6.69$

3 STRONGEST DIFFRACTION LINES:
14.00 (100)
3.89 ( 80)
3.01 ( 80)

OPTICAL CONSTANTS:
$\alpha$ = 1.576
$\beta$ = 1.582
$\gamma$ = 1.584
2V ~ 10°

HARDNESS: 1½

DENSITY: 2.68 (Meas.)

CLEAVAGE: Perfect parallel to plane of plates; medium parallel to elongation of crystals and nearly perpendicular to the first; and less perfect, nearly perpendicular to the first, and at 87° to the second.

HABIT: Spherulites and rosettes, rarely as single crystals 0.2–0.3 mm long, acicular to wedge-shaped.

COLOR-LUSTER: Colorless, white in aggregates; vitreous to pearly.

MODE OF OCCURRENCE: Found on two museum samples, one labeled "wapplerite" from the Daniel Mine, Schneeberg, Saxony, and the other labeled "pharmacolite," from Richelsdorf, Hessia

BEST REF. IN ENGLISH: Nefedov, E. I., *Am. Min.*, **47**: 416–417 (1962).    Pierrot, R., *Am. Min.*, **50**: 812 (1965).

# GUETTARDITE
$Pb_9(Sb, As)_{16}S_{33}$

CRYSTAL SYSTEM: Monoclinic
CLASS: 2/m
SPACE GROUP: $P2_1/a$
Z: 1
LATTICE CONSTANTS:
$a$ = 20.095
$b$ = 7.946
$c$ = 8.783
$\beta$ = 101°12'

3 STRONGEST DIFFRACTION LINES:
3.52  (100)
2.795 ( 90)
4.19  ( 50)

HARDNESS: Talmadge hardness B. Vickers microhardness 187 (180–197) kg/mm$^2$ with 50-g load.

DENSITY:  5.31 (Predicted)
          5.49 (Calc.)

CLEAVAGE: Perfect.
Conchoidal fracture. Very brittle.

HABIT: Isolated anhedral grains.    Polysynthetically twinned. Composition plane parallel to {100}.

COLOR-LUSTER: Grayish black; metallic. Streak black with brown tint.

MODE OF OCCURRENCE: Occurs sparingly in a small prospect pit in Precambrian marble at Madoc, Ontario, Canada.

BEST REF. IN ENGLISH: Jambor, J. L., *Can. Min.*, **9**, (Pt. 2): 191–213 (1967).

# GUGIAITE = Variety of meliphane

# GUILDITE
$CuFe(SO_4)_2(OH) \cdot 4H_2O$

CRYSTAL SYSTEM: Monoclinic
CLASS: 2/m or 2
SPACE GROUP: $P2_1/m$ or $P2_1$
Z: 2
LATTICE CONSTANTS:
$a$ = 9.786
$b$ = 7.134
$c$ = 7.263
$\beta$ = 105°17'

3 STRONGEST DIFFRACTION LINES:
3.144 (100)
9.46 ( 35)
4.998 ( 27)

OPTICAL CONSTANTS:
$\alpha$ = 1.622–1.623
$\beta$ = 1.628–1.630
$\gamma$ = 1.681–1.684
(+)2V $\approx$ 62°

HARDNESS: 2.5

DENSITY: 2.695 (Meas.)

CLEAVAGE:  {001} perfect
           {100} perfect
Fracture conchoidal. Brittle.

HABIT: Crystals short prismatic, pseudocubic, up to 5.0 mm in width.

COLOR-LUSTER: Honey yellow to deep chestnut brown; translucent, transparent on thin edges; vitreous. Streak pale canary yellow.

MODE OF OCCURRENCE: Occurs associated with coquimbite and other sulfates formed by a mine fire at the United Verde mine, Jerome, Arizona.

BEST REF. IN ENGLISH: Laughon, R. B., *Am. Min.*, **55**: 502–505 (1970).

# GUILLEMINITE
$Ba(UO_2)_3(OH)_4(SeO_3)_2 \cdot 3H_2O$

CRYSTAL SYSTEM: Orthorhombic
CLASS: 2/m 2/m 2/m
SPACE GROUP: Pncm
Z: 2
LATTICE CONSTANTS:
$a$ = 7.25
$b$ = 16.84
$c$ = 7.08

3 STRONGEST DIFFRACTION LINES:
8.39 (100)
7.29 (100)
3.55 ( 80)

OPTICAL CONSTANTS:
$\alpha$ = 1.720
$\beta$ = 1.798
$\gamma$ = 1.805
(-)2V $\simeq$ 35°

HARDNESS: Not determined.

DENSITY: 4.88 (Meas.)
         4.92 (Calc.)

CLEAVAGE: {100} perfect
{010} good
Brittle.

HABIT: As tabular crystals up to 0.4 × 0.2 mm in size in vugs; also as coatings and silky masses up to one centimeter in diameter.

COLOR-LUSTER: Canary yellow.

MODE OF OCCURRENCE: Occurs in the oxidized zone of the copper-cobalt deposit of Musonoi, Katanga, Zaire.

BEST REF. IN ENGLISH: Pierrot, R., Toussaint, J., and Verbeek, T., *Am. Min.*, **50**: 2103 (1965).

## GUITERMANITE (Species status uncertain)
$Pb_{10}As_6S_{19}$

CRYSTAL SYSTEM: Unknown
3 STRONGEST DIFFRACTION LINES:
3.19 (100)
2.22 ( 70)
1.82 ( 70)
HARDNESS: 3
DENSITY: 5.94
CLEAVAGE: Not determined.
Fracture uneven. Brittle.
HABIT: Massive, compact.
COLOR-LUSTER: Bluish gray. Opaque. Metallic.
MODE OF OCCURRENCE: Occurs associated with zunyite, enargite, pyrite, barite, and kaolin at the Zuni mine, San Juan County, Colorado.
BEST REF. IN ENGLISH: Palache, et al., "Dana's System of Mineralogy," 7th Ed., v. I, p. 401, New York, Wiley, 1944.

## GUMMITE = Generic name for gum-like uranium minerals

## GUMUCIONITE = Variety of sphalerite containing arsenic

## GUNNBJARNITE = A ferrian sepiolite
Near $(Mg, Ca, Fe^{2+})_3(Fe^{3+}, Al)_2Si_6O_{18} \cdot 3H_2O$

## GUNNINGITE
$ZnSO_4 \cdot H_2O$

CRYSTAL SYSTEM: Monoclinic
CLASS: 2/m
SPACE GROUP: A2/a
Z: 4
LATTICE CONSTANTS:
$a = 7.566$
$b = 7.586$ (synthetic)
$c = 6.954$
$\beta = 115°56'$
3 STRONGEST DIFFRACTION LINES:
3.42 (100)
4.77 ( 55)
3.07 ( 45)

OPTICAL CONSTANTS:
$\alpha' = 1.570$
$\beta' = 1.576$
$\gamma' = 1.630$
HARDNESS: ~2½
DENSITY: 3.195 (Meas. synthetic)
3.321 (Calc.)
CLEAVAGE: Not determined.
HABIT: Very fine-grained; as an efflorescence. Easily soluble in cold water.
COLOR-LUSTER: White; vitreous. Streak white.
MODE OF OCCURRENCE: Occurs sparsely on sphalerite, associated with "limonite", scorodite, gypsum, and other supergene minerals closely associated with the zinc sulfide, on dumps and in the relatively dry portions of eight mines in the Keno Hill–Galena Hill district, Central Yukon.
BEST REF. IN ENGLISH: Jambor, J. L., and Boyle, R. W., *Can. Min.*, 7: 209–218 (1962).

## GUSTAVITE
$Bi_{11}Pb_5Ag_3S_{24}$

CRYSTAL SYSTEM: Orthorhombic
CLASS: 2/m 2/m 2/m or mm2
SPACE GROUP: Bbmm, Bb2₁m, or Bbm2
Z: 1
LATTICE CONSTANTS:
$a = 13.548$
$b = 19.449$
$c = 4.105$
3 STRONGEST DIFFRACTION LINES:
3.363 (100)
2.996 (100)
2.895 (100)
HARDNESS: Polishing hardness less than cosalite, higher than galena.
DENSITY: 7.01 (Calc.)
CLEAVAGE: Rare cleavage parallel to the tabular plates.
HABIT: As tabular grains, sometimes slightly bent. Maximum grain size 2 × 2 × 0.5 mm.
COLOR-LUSTER: Steel gray. Metallic. Opaque.
MODE OF OCCURRENCE: Found associated with berryite and phase X $(Bi_{10}Pb_7Ag_2S_{24})$ in a suite of Bi-Pb-Ag-(Cu)-sulfosalt minerals in the cryolite deposit at Ivigtut, southwest Greenland.
BEST REF. IN ENGLISH: Karup-Møller, S., *Can. Min.*, **10** (pt. 2): 174–190 (1970).

## GUTSEVICHITE (Inadequately described mineral)
$(Al, Fe)_3(PO_4, VO_4)_2(OH)_3 \cdot 8H_2O$ (?)

CRYSTAL SYSTEM: Unknown
3 STRONGEST DIFFRACTION LINES:
4.082 (100)
2.506 ( 90)
1.820 ( 80)
OPTICAL CONSTANT:
$N = 1.560-1.575$
HARDNESS: 2½
DENSITY: 1.90-2.00 (Meas.)
CLEAVAGE: Not determined.
HABIT: Massive; as crusts, concretions, and cavity fillings.

COLOR-LUSTER: Yellowish green to greenish brown to dark brown. Translucent. Waxy to dull.

MODE OF OCCURRENCE: Occurs in the zone of oxidation of vanadium-containing shales of the Middle Cambrian in northwestern Kazakhstan.

BEST REF. IN ENGLISH: Ankinovich, E. A., *Am. Min.*, **46**: 1200 (1961).

# GUYANAITE
CrO(OH)
Trimorphous with bracewellite, grimaldiite
Isostructural with InOOH

CRYSTAL SYSTEM: Orthorhombic
HARDNESS: Not determined.
DENSITY: Not determined.
CLEAVAGE: Not determined.
HABIT: Massive, microcrystalline.
COLOR-LUSTER: Brown
MODE OF OCCURRENCE: Occurs in complex microscopic intergrowth with eskolaite, bracewellite, grimaldiite, and macconnellite, in "merumite" from alluvial gravels in Guyana.

BEST REF. IN ENGLISH: Milton, C., et al., Geol. Soc. Am., Program 1967 Annual Meetings, p. 151–152 (1967) (abstr.).

# GYPSUM
CaSO$_4$ · 2H$_2$O

CRYSTAL SYSTEM: Monoclinic
CLASS: 2/m
SPACE GROUP: A2/a
Z: 4
LATTICE CONSTANTS:
  $a$ = 5.67 kX
  $b$ = 15.15
  $c$ = 6.28
  $\beta$ = 113°50'
3 STRONGEST DIFFRACTION LINES:
  7.56 (100)
  3.059 ( 55)
  4.27 ( 50)
OPTICAL CONSTANTS:
  $\alpha$ = 1.5207
  $\beta$ = 1.5230 (Na)
  $\gamma$ = 1.5299
  (+)2V = 58°
HARDNESS: 2
DENSITY: 2.32 (Meas.)
       2.315 (Calc.)
CLEAVAGE: {010} perfect, easy
       {100} distinct
       {011} distinct
Flexible but not elastic. Fracture splintery.

HABIT: Crystals commonly thin to thick tabular, diamond-shaped; also short to long prismatic, up to 10 feet long; acicular; lenticular, often intergrown as rosettes; massive, fine to coarse granular (alabaster); fibrous (satin-spar); as distorted formations on cavern walls (helectites); pulverulent; concretionary. Twinning on {100} common as swallow-tail twins; also on {$\bar{1}$01} butterfly twins; rarely on {$\bar{1}$11}.

COLOR-LUSTER: Colorless and transparent (selenite); also white, gray, yellowish, greenish, reddish, or brownish when massive. Subvitreous, crystals pearly on cleavages. Streak white. Sometimes crystals fluorescent and phosphorescent greenish white under ultraviolet light.

MODE OF OCCURRENCE: Occurs abundantly and widespread chiefly in sedimentary deposits, especially in Permian and Triassic formations; in saline lakes and playas; as an efflorescence on certain soils; in the oxidized portions of ore deposits; and in deposits associated with volcanic activity. Large commercial deposits occur in California, Utah, Colorado, South Dakota, New Mexico, Iowa, Kansas, Michigan, New York, and in many other states; as fine crystals in Utah, Oklahoma, South Dakota, Ohio, and New York. It occurs in large deposits in eastern Canada; as crystals up to 5 feet long at Naica, Chihuahua, Mexico; as crystals up to 10 feet long at the Braden mine, Chile; and as large deposits in France, Sicily, Germany, Poland, Austria, USSR, and at many other places.

BEST REF. IN ENGLISH: Palache, et al., "Dana's System of Mineralogy," 7th Ed., v. II, p. 482–486, New York, Wiley, 1951. Deer, Howie, and Zussman, "Rock Forming Minerals," v. 5, p. 202–218, New York, Wiley, 1962.

# GYROLITE
Ca$_2$Si$_3$O$_7$(OH)$_2$ · H$_2$O

CRYSTAL SYSTEM: Hexagonal
CLASS: 6
SPACE GROUP: P6$_1$ or P6$_5$
Z: 48
LATTICE CONSTANTS:
  $a$ = 9.72
  $c$ = 132.8
3 STRONGEST DIFFRACTION LINES:
  22. (100)
  3.12 (100)
  11.0 ( 80)
OPTICAL CONSTANTS:
  $\omega$ = 1.549
  $\epsilon$ = 1.536
   (–)
HARDNESS: 3–4
DENSITY: 2.34–2.45 (Meas.)
       2.40 (Calc.)
CLEAVAGE: {0001} perfect
Fracture uneven. Brittle.

HABIT: Massive; lamellar-radiate structure; concretionary.
COLOR-LUSTER: Colorless, white. Transparent to translucent. Vitreous.

MODE OF OCCURRENCE: Occurs in crevices of rocks as a secondary mineral formed by the alteration of lime silicates. Found in association with apophyllite at Fort Point in San Francisco County, and at the New Almaden mine, Santa Clara County, California; also at localities in Nova Scotia, Brazil, Greenland, Ireland, Scotland, Czechoslovakia, India, and elsewhere.

BEST REF. IN ENGLISH: Mackay, A. L., and Taylor, H. F. W., *Min. Mag.*, **30**: 80–91 (1953). Meyer, J. W., and Jaunarajs, K. L., *Am. Min.*, **46**: 913–933 (1961).

**HACKMANITE** = Variety of sodalite containing sulfur

## HAGENDORFITE
$(Na, Ca)(Fe^{2+}, Mn^{2+})_2 (PO_4)_2$

CRYSTAL SYSTEM: Monoclinic
CLASS: 2/m
SPACE GROUP: $I2_1/a$
LATTICE CONSTANTS:
  $a$ = 10.93
  $b$ = 12.59
  $c$ = 6.52
  $\beta$ = 97°59′
3 STRONGEST DIFFRACTION LINES:
  2.686 (100)
  2.593 ( 75)
  3.42 ( 60)
OPTICAL CONSTANTS:
  $\alpha$ = 1.735
  $\beta$ = 1.742
  $\gamma$ = 1.745
  (−)2V
  HARDNESS: 3½
  DENSITY: 3.71
CLEAVAGE: Three; one good, one less good, and the third poor.
HABIT: Massive.
COLOR-LUSTER: Greenish black.
MODE OF OCCURRENCE: Occurs associated with triphylite, wolfeite, and hematite at the Hagendorf - South pegmatite, Bavaria; also from Norrö, Sweden.
BEST REF. IN ENGLISH: Fisher, D. J., *Bull. Geol. Soc. Am.*, **67**: 1694-95 (1956).

## HÄGGITE
$V_2O_2(OH)_3$

CRYSTAL SYSTEM: Monoclinic
CLASS: 2/m
SPACE GROUP: C2/m
Z: 2

LATTICE CONSTANTS:
  $a$ = 12.17
  $b$ = 2.99
  $c$ = 4.83
  $\beta$ = 98°15′
3 STRONGEST DIFFRACTION LINES:
  4.80 (100)
  4.05 ( 50)
  3.02 ( 25)
HARDNESS: Not determined.
DENSITY: Not determined.
CLEAVAGE: None.
HABIT: Crystals very minute; as a phase intergrown on a fine scale in parallel orientation with doloresite. Usually massive; finely crystalline.
COLOR-LUSTER: Black. Opaque. Submetallic.
MODE OF OCCURRENCE: Occurs in relatively unoxidized vanadium-uranium ores in association with doloresite, coffinite, montroseite, paramontroseite, and other vanadium oxides. Found in sandstone at Carlile, Crook County, Wyoming, and at the Runge mine, near Edgemont, Fall River County, South Dakota.
BEST REF. IN ENGLISH: Evans, Howard T. Jr., and Mrose, Mary E., *Acta Cryst.*, **11**: 56–58 (1958). Evans, Howard T. Jr., and Mrose, Mary E., *Am. Min.*, **45**: 1144–1166 (1960).

## HAIDINGERITE
$CaHAsO_4 \cdot H_2O$

CRYSTAL SYSTEM: Orthorhombic
CLASS: 2/m 2/m 2/m
SPACE GROUP: Pcnb
Z: 8
LATTICE CONSTANTS:
  $a$ = 6.94
  $b$ = 16.15
  $c$ = 7.94
3 STRONGEST DIFFRACTION LINES:
  5.22 (100)
  2.955 ( 80)
  8.06 ( 60)

OPTICAL CONSTANTS:
$\alpha = 1.590$
$\beta = 1.602$
$\gamma = 1.638$
$(+)2V \approx 58°$
HARDNESS: 2-2½
DENSITY: 2.95 (Meas.)
2.959 (Calc.)
CLEAVAGE: {010} perfect
Sectile.
HABIT: Crystals rare; short prismatic or equant. Usually as fibrous or fine grained botryoidal coatings.
COLOR-LUSTER: Colorless, white. Transparent to translucent. Vitreous to adamantine; pearly on cleavage.
MODE OF OCCURRENCE: Occurs as a secondary mineral in arsenic-bearing ore deposits. Found at the White Caps mine, Manhattan, Nye County, Nevada, associated with pharmacolite and other arsenates; at Joachimsthal (Jachymov), Bohemia, Czechoslovakia; and at Wittichen, Johanngeorgenstadt, and Schneeberg, Germany.
BEST REF. IN ENGLISH: Palache, et al., "Dana's System of Mineralogy," 7th Ed., v. II, p. 708-709, New York, Wiley, 1951.


# HAINITE (Species status uncertain)
Silicate of Na, Ca, Ti, and Zr

CRYSTAL SYSTEM: Unknown
OPTICAL CONSTANTS:
$\beta \sim 1.7$
$(+)2V$ = large
HARDNESS: $\sim 5$
DENSITY: 3.184 (Meas.)
CLEAVAGE: {010} perfect
{100} in traces
HABIT: Crystals prismatic. Twinning on {100}.
COLOR-LUSTER: Yellow. Translucent. Adamantine.
MODE OF OCCURRENCE: Occurs in phonolites and tinguaites in association with aegirine in the vicinity of Vrchni Hajn, Bohemia, Czechoslovakia.
BEST REF. IN ENGLISH: Vlasov, K. A., "Mineralogy of Rare Elements," v. II, p. 385, Israel Program for Scientific Translations, 1966.


# HAIWEEITE
$CaU_2^{4+}Si_6O_{19} \cdot 5H_2O$

CRYSTAL SYSTEM: Monoclinic
CLASS: 2/m
SPACE GROUP: Probably P2/c
LATTICE CONSTANTS:
$a = 15.4$
$b = 7.05$
$c = 7.10$
$\beta = 107°52'$
3 STRONGEST DIFFRACTION LINES:
9.14 (100)
4.556 ( 60)
4.42 ( 60)

OPTICAL CONSTANTS:
$\alpha = 1.571$
$\beta = 1.575$
$\gamma = 1.578$
$(-)2V \sim 15°$
HARDNESS: 3½
DENSITY: 3.35 (Meas.)
CLEAVAGE: {100} good
HABIT: Crystals needle-like, fractions of a millimeter in size. As spherulitic aggregates; also as single flake-like grains.
COLOR-LUSTER: Pale yellow to greenish yellow. Pearly. Fluoresces weak dull green under ultraviolet light.
MODE OF OCCURRENCE: Occurs on fracture surfaces in granite and in voids of the adjacent lake bed deposits near the Haiwee Reservoir in the Coso Mountains, California; and also along fractures of a tourmaline-bearing granite and correlated intruded pegmatites, associated with autunite, meta-autunite, uranophane, beta-uranophane, phosphuranylite, torbernite, meta-torbernite and uranium opal at Perus, 25 Km north of Sao Paulo, Brazil.
BEST REF. IN ENGLISH: McBurney, T. C., and Murdoch, Joseph, Am. Min., 44: 839-843 (1959).


# HAKITE
$(Cu, Hg)_{12}Sb_3(S, Se)_{13}$
Tetrahedrite Group
Selenium analogue of tetrahedrite

CRYSTAL SYSTEM: Cubic
CLASS: $\bar{4}3m$
SPACE GROUP: $I\bar{4}3m$ (probably)
Z: 1
LATTICE CONSTANT:
$a = 10.88$
3 STRONGEST DIFFRACTION LINES:
3.140 (100)
1.925 ( 90)
1.639 ( 80)
OPTICAL CONSTANTS: In reflected light isotropic, color creamy white to clear brown. Reflectances at 12 wavelengths (max. at 480nm) for two analyses: 420nm, 32.8, 31.5; 480, 34.0, 33.5; 580, 33.2, 33.6; 640, 33.6, 33.2%.
HARDNESS: Microhardness 352 kg/mm² for 20 g load, 306 for 40 g load.
DENSITY: 6.3 (Meas.)
CLEAVAGE: Not determined.
HABIT: Massive; in xenomorphic grains up to 0.3 mm in size.
COLOR-LUSTER: Gray-brown. Opaque. Metallic.
MODE OF OCCURRENCE: Occurs associated with clausthalite, permingeatite, umangite, and other selenides, in epithermal calcite veins at Predborice, Bohemia, Czechoslovakia.
BEST REF. IN ENGLISH: Johan, Zdenek, and Kvacek, Milan, Am. Min., 57: 1553-1554 (1972).


# HALITE
NaCl

CRYSTAL SYSTEM: Cubic
CLASS: $4/m \bar{3} 2/m$

SPACE GROUP: Fm3m
Z: 4
LATTICE CONSTANT:
  $a = 5.6387$
3 STRONGEST DIFFRACTION LINES:
  2.821 (100)
  1.994 ( 55)
  1.628 ( 15)
OPTICAL CONSTANT:
  $N = 1.5443$ (Na)
  HARDNESS: 2
  DENSITY: 2.168 (Meas.)
            2.1637(Calc.)
  CLEAVAGE: {001} perfect
Fracture conchoidal. Brittle.
  HABIT: Crystals cubic, rarely octahedral; crystals often cavernous or hopper-shaped. Also massive, compact to granular; and rarely columnar or stalactitic.
  COLOR-LUSTER: Colorless, white, yellow, orange, reddish, purple, and blue. Vitreous. Transparent to translucent. Soluble in water. Fluoresces sometimes orange, reddish, or greenish due to inclusions of organic or inorganic impurities.
  MODE OF OCCURRENCE: Occurs widespread chiefly as extensive sedimentary deposits ranging from a few inches to over a thousand feet in thickness; as an efflorescence in playa deposits and upon the walls of mine workings; and as a sublimation product in areas of volcanism. In the United States important deposits occur in New York, Michigan, Ohio, Kansas, and many other states. Large deposits are also found in Canada, Columbia, Peru, England, France, Germany, Austria, USSR, India, Algeria, and many other places.
  BEST REF. IN ENGLISH: Palache, et al., "Dana's System of Mineralogy," 7th Ed., v. II, p. 4-7, New York, Wiley, 1951.

# HALLIMONDITE
$Pb_2(UO_2)(AsO_4)_2$

CRYSTAL SYSTEM: Triclinic
CLASS: $\bar{1}$
SPACE GROUP: P$\bar{1}$
Z: 2
LATTICE CONSTANTS:
  $a = 7.123$    $\alpha = 100°34'$
  $b = 10.469$   $\beta = 94°48'$
  $c = 6.844$    $\gamma = 91°16'$
3 STRONGEST DIFFRACTION LINES:
  3.40 (100)
  2.84 ( 90)
  1.74 ( 60)
OPTICAL CONSTANTS:
  $\alpha = 1.882$
  $\beta = ?$
  $\gamma = 1.915$
  (+)2V $\simeq 80°$
  HARDNESS: 2½-3
  DENSITY: 6.39 (Meas. synthetic)
            6.40 (Calc.)
  CLEAVAGE: No distinct cleavage. Conchoidal fracture.
  HABIT: Crystals tabular; more or less elongated along the $c$-axis. Length does not exceed 0.4 mm.

COLOR-LUSTER: Yellow; transparent to translucent: subadamantine. Pale yellow streak.
  MODE OF OCCURRENCE: Found as small crystals and coatings in cavities and fractures of quartz, associated with hügelite and mimetite, at the Michael mine near Reichenbach in the Black Forest, Germany.
  BEST REF. IN ENGLISH: Walenta, Kurt; *Am. Min.*, **50**: 1143-1157 (1965).

# HALLOYSITE (metahalloysite)
$Al_2Si_2O_5(OH)_4 \cdot 2H_2O$
Kaolinite Group (Kandite)

CRYSTAL SYSTEM: Monoclinic
CLASS: m
SPACE GROUP: Am
Z: 2
LATTICE CONSTANTS:
  $a = 5.16$
  $b = 8.94$
  $c = 7.4$
  $\beta \sim 100°$
3 STRONGEST DIFFRACTION LINES:
  4.41 (100)
  7.4 ( 95)
  4.34 ( 70)
OPTICAL CONSTANT:
  $N = 1.555$
  HARDNESS: 2-2½
  DENSITY: 2-2.2
  CLEAVAGE: None. Fracture earthy.
  HABIT: Tubular; ultramicroscopic in size. As compact to mealy masses.
  COLOR-LUSTER: Colorless, white; sometimes tinted yellowish, brownish, reddish, bluish by impurities. Transparent to translucent. Pearly to dull earthy.
  MODE OF OCCURRENCE: Occurs widespread, often in association with kaolinite, as a common clay mineral formed by the weathering or hydrothermal alteration of feldspars and other aluminous silicate minerals. Typical occurrences are listed under kaolinite.
  BEST REF. IN ENGLISH: Faust, George T., *Am. Min.*, **40**: 1110-1118 (1955). Deer, Howie, and Zussman, "Rock Forming Minerals," v. 3, 194-212, New York, Wiley, 1962.

# HALOTRICHITE
$Fe^{2+}Al_2(SO_4)_4 \cdot 22H_2O$

CRYSTAL SYSTEM: Monoclinic
CLASS: 2/m
SPACE GROUP: P2/m
Z: 4
LATTICE CONSTANTS:
  $a = 20.47$ kX
  $b = 24.24$
  $c = 6.167$
  $\beta = 96°48'$

3 STRONGEST DIFFRACTION LINES:
4.77 (100)
3.48 (100)
4.29 ( 55)
OPTICAL CONSTANTS:
$\alpha = 1.480$
$\beta = 1.486$ (Na)
$\gamma = 1.490$
$(-)2V = 35°$
HARDNESS: 1½
DENSITY: 1.895 (Meas.)
1.95 (Calc.)
CLEAVAGE: {010} poor
Fracture conchoidal. Brittle.

HABIT: Crystals acicular, rarely terminated. As aggregates of fibrous or acicular crystals; as an incrustation or efflorescence; sometimes matted.

COLOR-LUSTER: Colorless, white, yellowish, greenish. Transparent to translucent. Vitreous.

Soluble in water. Taste astringent.

MODE OF OCCURRENCE: Occurs widespread chiefly as a weathering product of pyrite-bearing and aluminous rocks, in pyritic ore deposits and coal veins, and as a hot springs or fumarolic deposit. Found in the United States at the Dexter No. 7 mine, Calf Mesa, San Rafael Swell, Utah, associated with coquimbite and other sulfates; at many places in California; in the Alum Mountain district, Grant County, New Mexico; and in several other western states. Among other localities, it occurs in South America, especially in northern Chile; in Germany, France, Italy, Finland, and Sweden.

BEST REF. IN ENGLISH: Palache, et al., "Dana's System of Mineralogy," 7th Ed., v. II, p. 523–527, New York, Wiley, 1951.


# HALURGITE (Inadequately described mineral)
$Mg_2B_8O_{14} \cdot 5H_2O$

CRYSTAL SYSTEM: Probably orthorhombic or pseudo-orthorhombic
3 STRONGEST DIFFRACTION LINES:
3.87 (100)
3.29 (100)
4.81 ( 90)
OPTICAL CONSTANTS:
$\alpha = 1.532$
$\beta = 1.545$
$\gamma = 1.572$
$(+)2V = 70°$
HARDNESS: 2.5–3
DENSITY: 2.19 (Meas.)
CLEAVAGE: Not determined.
HABIT: As fine-grained masses; rarely as platy crystals 0.01 to 0.25 mm in size.
COLOR-LUSTER: White.
MODE OF OCCURRENCE: Occurs in rock salt with boracite, kaliborite, pinnoite, and anhydrite at a depth of about 400 meters in the Kungur saline rocks at an unspecified locality, probably in the Inder basin.
BEST REF. IN ENGLISH: Lovanova, V. V., *Am. Min.*, 47: 1217–1218 (1962).


# HAMBERGITE
$Be_2(OH, F)BO_3$

CRYSTAL SYSTEM: Orthorhombic
CLASS: 2/m 2/m 2/m
SPACE GROUP: Pbca
Z: 8
LATTICE CONSTANTS:
$a = 9.76$
$b = 12.23$
$c = 4.43$
3 STRONGEST DIFFRACTION LINES:
3.82 (100)
3.13 ( 90)
4.53 ( 80)
OPTICAL CONSTANTS:
$\alpha = 1.55$
$\beta = 1.59$
$\gamma = 1.63$
$(+)2V = 87°$
HARDNESS: 7½
DENSITY: 2.372 (Meas.)
2.365 (Calc.)
CLEAVAGE: {010} perfect
{100} good
Brittle.

HABIT: Crystals prismatic [100], often flattened {100}; sometimes twinned on {110}; occasionally doubly terminated. Individual crystals range in size up to 5.5 × 3.6 × 1.1 cm.

COLOR-LUSTER: Colorless, white, grayish white, yellowish white. Transparent to translucent or semi-opaque. Vitreous to dull.

MODE OF OCCURRENCE: Occurs in syenite pegmatite near Halgaråen, Langesundsfjord, Norway; in alkali-rich pegmatites at Anjanabanoana and other localities in Madagascar; at Sušice, Czechoslovakia; in gem gravels from Kashmir, India; and in pegmatite at the Little Three mine, near Ramona, and at the Himalaya mine, near Mesa Grande, San Diego County, California.

BEST REF. IN ENGLISH: Switzer, George, Clarke, Roy S. Jr., Sinkankas, John, and Worthing, Helen W., *Am. Min.*, 50: 85–95 (1965).


# HAMMARITE (Inadequately described mineral)
$Pb_2Cu_2Bi_4S_9$

CRYSTAL SYSTEM: Monoclinic (?)
LATTICE CONSTANTS:
$a:b = 1.048:1$
HARDNESS: 3–4
DENSITY: Not determined.
CLEAVAGE: {010} poor
Fracture conchoidal.
HABIT: Crystals short prismatic to needle-like, without terminal faces; faces curved.
COLOR-LUSTER: Steel-gray with red tint. Opaque. Metallic. Streak black.
MODE OF OCCURRENCE: Occurs as crystals implanted on drusy quartz at Gladhammer, Kalmar, Sweden.

BEST REF. IN ENGLISH: Palache, et al., "Dana's System of Mineralogy," 7th Ed., v. I, p. 442, New York, Wiley, 1944.

## HANCOCKITE

$(Pb, Ca, Sr)_2 (Al, Fe)_3 Si_3 O_{12} OH$

Epidote Group

CRYSTAL SYSTEM: Monoclinic
CLASS: 2/m
SPACE GROUP: $P2_1/m$
Z: 2
LATTICE CONSTANTS:
   $a = 9.03$
   $b = 5.62$
   $c = 10.29$
   $\beta = 115°56'$
3 STRONGEST DIFFRACTION LINES:
   2.91 (100)
   3.49 ( 50)
   2.60 ( 50)
OPTICAL CONSTANTS:
   $\alpha = 1.788$
   $\beta = 1.81$
   $\gamma = 1.830$
   $(-)2V \sim 50°$
HARDNESS: 6-7
DENSITY: 4.03 (Meas.)
CLEAVAGE: {001} perfect
Fracture uneven. Brittle.
   HABIT: Crystals lath-shaped with rounded faces showing characteristic epidote habit; striated vertically; very small. As drusy cellular aggregates and compact masses.
   COLOR-LUSTER: Crystals yellowish brown; masses dull brick red, brownish red, maroon. Translucent. Vitreous.
   MODE OF OCCURRENCE: Occurs in association with garnet, biotite, axinite, willemite, and other minerals, at Franklin, Sussex County, New Jersey.
   BEST REF. IN ENGLISH: Palache, Charles, USGS Prof. Paper 180, p. 98 (1935).

## HANKSITE

$Na_{22} K (SO_4)_9 (CO_3)_2 Cl$

CRYSTAL SYSTEM: Hexagonal
CLASS: 6/m
SPACE GROUP: $P6_3/m$
Z: 2
LATTICE CONSTANTS:
   $a = 10.46$ kX
   $c = 21.18$
3 STRONGEST DIFFRACTION LINES:
   3.78 (100)
   2.78 (100)
   2.61 (100)
OPTICAL CONSTANTS:
   $\omega = 1.481$
   $\epsilon = 1.461$
   $(-)$
HARDNESS: 3-3½

DENSITY: 2.56 (Meas.)
          2.57 (Calc.)
CLEAVAGE: {0001} good
Fracture uneven. Brittle.
   HABIT: As tabular to short prismatic hexagonal crystals up to 3 inches in size, terminated by large basal pinacoids and modified by the dipyramid {$10\bar{1}2$}; also as quartzoids with large {$10\bar{1}2$}. Crystals commonly interpenetrant.
   COLOR-LUSTER: Colorless; also yellowish, gray to grayish black due to inclusions. Occasionally distinctly zoned parallel to prism faces by clay inclusions. Transparent to translucent. Vitreous to dull. Streak white. Taste saline. Sometimes fluoresces weak yellow under long-wave ultraviolet light.
   MODE OF OCCURRENCE: Occurs abundantly, associated with trona, halite, borax, and other minerals, in the saline beds of Searles Lake, San Bernardino County; in minute crystals with trona at Mono Lake, Mono County; at Soda Lake, San Luis Obispo County; and in the borax fields of Death Valley, Inyo County, California.
   BEST REF. IN ENGLISH: Palache, et al., "Dana's System of Mineralogy," 7th Ed., v. II, p. 628-629, New York, Wiley, 1951.

## HANLÉITE = Uvarovite

## HANNAYITE

$(NH_4)_2 Mg_3 H_4 (PO_4)_4 \cdot 8H_2O$

CRYSTAL SYSTEM: Triclinic
CLASS: $\bar{1}$
SPACE GROUP: $P\bar{1}$
Z: 1
LATTICE CONSTANTS:
(synthetic)
   $a = 7.70$       $a = 76.0°$
   $b = 11.51$      $\beta = 99.8°$
   $c = 6.70$       $\gamma = 115.8°$
3 STRONGEST DIFFRACTION LINES:
   6.96 (100)
   3.46 ( 75)   (synthetic)
   5.15 ( 30)
OPTICAL CONSTANTS:
   $\alpha = 1.504-1.555$
   $\beta = 1.522-1.572$
   $\gamma = 1.539-1.579$
   $(-)2V = 45°-90°$
HARDNESS: Not determined. Soft.
DENSITY: 2.03 (Meas.)
          2.03 (Calc. synthetic)
CLEAVAGE: {001}}perfect
          {110} poor
          {$1\bar{1}0$} poor
          {130} poor
   HABIT: As small thin prismatic crystals elongated and striated parallel to c-axis. Synthetic crystals tabular {100}, elongate along c-axis, with prominent {$01\bar{1}$}, modified by narrow {110}.
   COLOR-LUSTER: Yellowish. Transparent. Vitreous.

MODE OF OCCURRENCE: Occurs in bat guano associated with struvite, newberyite, and brushite in the Skipton caves, Ballarat, Victoria, Australia.

BEST REF. IN ENGLISH: Palache, et al., "Dana's System of Mineralogy," 7th Ed., v. II, p. 699–700, New York, Wiley, 1951, Frazier, A. William, Lehr, James R., and Smith, James P., *Am. Min.*, 48: 635–641 (1963).

## HANUŠITE = Stevensite plus pectolite

## HARADAITE
$SrVSi_2O_7$

CRYSTAL SYSTEM: Orthorhombic
CLASS: 2/m 2/m 2/m
SPACE GROUP: Amam
Z: 4
LATTICE CONSTANTS:
 $a = 7.06$
 $b = 14.64$
 $c = 5.33$
3 STRONGEST DIFFRACTION LINES:
 3.20 (100)
 2.88 ( 90)
 3.65 ( 40)
OPTICAL CONSTANTS:
 $\alpha = 1.71$
 $\beta = 1.72$
 $\gamma = 1.73$
 2V = 90°
HARDNESS: Not determined.
DENSITY: 3.80 (Meas.)
 3.75 (Calc.)
CLEAVAGE: {010} perfect
HABIT: Massive.
COLOR-LUSTER: Light blue.
MODE OF OCCURRENCE: Occurs at the Yomato mine, Iwate, Japan.
BEST REF. IN ENGLISH: Takeuchi, Yoshio, and Joswil, Werner, *Am. Min.*, 56: 1123 (1971).

## HARDYSTONITE
$Ca_2ZnSi_2O_7$
Melilite Group

CRYSTAL SYSTEM: Tetragonal
CLASS: $\bar{4}$ 2m
SPACE GROUP: P$\bar{4}2_1$m
Z: 2
LATTICE CONSTANTS:
 $a = 7.823$
 $c = 5.013$
3 STRONGEST DIFFRACTION LINES:
 2.868 (100)
 3.085 ( 60)
 3.711 ( 50)
OPTICAL CONSTANTS:
 $\omega = 1.6691$
 $\epsilon = 1.6568$ (Na)
 (−)

HARDNESS: 3–4
DENSITY: 3.443 (Meas.)
 3.39 (Calc.)
CLEAVAGE: {001} distinct
 {100} indistinct
 {110} indistinct
Brittle.
HABIT: Massive, granular, or as disseminated grains.
COLOR-LUSTER: White, pinkish to light brown. Translucent. Vitreous.
MODE OF OCCURRENCE: Occurs in association with willemite, rhodonite, franklinite, idocrase, and apatite at Franklin, Sussex County, New Jersey.
BEST REF. IN ENGLISH: Palache, Charles, "The Minerals of Franklin and Sterling Hill, Sussex County, New Jersey"; USGS Prof. Paper 180, p. 93–94 (1935).

## HARKERITE
$Ca_{48}Mg_{16}Al_3(BO_3)_{15}(CO_3)_{18}(SiO_4)_{12}Cl_2(OH)_6 \cdot 3H_2O$

CRYSTAL SYSTEM: Cubic
CLASS: 4/m $\bar{3}$ 2/m
LATTICE CONSTANT:
 $a = 29.53$
3 STRONGEST DIFFRACTION LINES:
 2.61 (100)
 1.84 ( 90)
 2.13 ( 80)
OPTICAL CONSTANT:
 $N = 1.653$
DENSITY: 2.959
CLEAVAGE: None.
HABIT: As simple octahedral crystals.
COLOR-LUSTER: Colorless, vitreous.
MODE OF OCCURRENCE: Occurs in a skarn containing monticellite, calcite, and accessory bornite, chalcocite, magnetite, and diopside, at the contact of dolomitic limestones with Tertiary granite, in the Broadford area, Isle of Skye.
BEST REF. IN ENGLISH: Tilley, C. E., *Min. Mag.*, 29: 621–666 (1951). Ostrovskaya, I. V., Pertsev, N. N., and Nikitina, I. B., *Am. Min.*, 51: 1820 (1966).

## HARMOTOME
$BaAl_2Si_6O_{16} \cdot 6H_2O$
Zeolite Group

CRYSTAL SYSTEM: Monoclinic
CLASS: 2/m or 2
SPACE GROUP: P$2_1$/m or P$2_1$
Z: 2
LATTICE CONSTANTS:
 $a = 9.87$
 $b = 14.14$
 $c = 8.71$
 $\beta = 124°45'$
3 STRONGEST DIFFRACTION LINES:
 3.12 (100)
 2.70 (100)
 2.67 (100)

OPTICAL CONSTANTS:
$\alpha = 1.505$
$\beta = 1.508$
$\gamma = 1.512$
$(+)2V = 82°$
HARDNESS: 4½
DENSITY: 2.41–2.50 (Meas.)
CLEAVAGE: {010} distinct
{001} indistinct
Fracture uneven to subconchoidal. Brittle.

HABIT: Crystals usually complex penetration twins, often orthorhombic or tetragonal in aspect; as separate groups or radiating aggregates. Twinning on {001}, {021}, {110}.

COLOR-LUSTER: Colorless, white, gray, pink, yellow, brown. Transparent to translucent. Vitreous. Streak white.

MODE OF OCCURRENCE: Occurs chiefly in cavities in basalts and related igneous rocks; also in association with manganese mineralization, and in lenses in gneiss. Found near Ossining, Westchester County, New York; at Nisikkatch Lake, Saskatchewan, and at Rabbit Mountain, Ontario, Canada; as fine crystals at Strontian, Argyllshire, and elsewhere in Scotland; and at localities in North Wales, Norway, Germany, Finland, and USSR.

BEST REF. IN ENGLISH: Deer, Howie, and Zussman, "Rock Forming Minerals," v. 4, p. 386–400, New York, Wiley, 1963.


# HARSTIGITE

$MnCa_6(Be_2OOH)_2[Si_3O_{10}]_2$

CRYSTAL SYSTEM: Orthorhombic
CLASS: 2/m 2/m 2/m
SPACE GROUP: Pcmn
Z: 4
LATTICE CONSTANTS:
$a = 13.90$
$b = 13.62$
$c = 9.68$
3 STRONGEST DIFFRACTION LINES:
2.695 (100)
2.817 ( 50)
2.788 ( 50)
OPTICAL CONSTANTS:
$\alpha = 1.678$
$\beta = 1.68$
$\gamma = 1.683$
$(+)2V = 52°$
HARDNESS: 5½
DENSITY: 3.16
CLEAVAGE: None.
HABIT: Crystals stout prismatic, small.
COLOR-LUSTER: Colorless. Transparent. Vitreous.
MODE OF OCCURRENCE: Occurs as a very rare mineral associated with hausmannite, dolomite, reddish orange andradite, manganoan humite, and prismatic barite crystals at the Harstigen manganese mine near Pajsberg, Värmland, Sweden.

BEST REF. IN ENGLISH: Winchell and Winchell, "Elements of Optical Mineralogy," 4th Ed., Pt. 2, p. 479, New York, Wiley, 1951. Moore, Paul B., *Am. Min.*, 53: 309–315, 1418–1420 (1968).


# HARTTITE = Svanbergite


# HASTINGSITE (Ferrohastingsite)

$NaCa_2(Fe, Mg, Al)_5(Al_2Si_6)O_{22}(OH)_2$
Amphibole Group
Hornblende Series
Var. Dashkesanite

CRYSTAL SYSTEM: Monoclinic
CLASS: 2/m
SPACE GROUP: C2/m
Z: 2
LATTICE CONSTANTS:
$a = 9.912$
$b = 18.030$
$c = 5.296$
$\beta = 103.946°$
3 STRONGEST DIFFRACTION LINES:
8.43 (100)
3.13 ( 70)
2.71 ( 60)
OPTICAL CONSTANTS:
$\alpha = 1.67–1.71$
$\beta = 1.68–1.73$
$\gamma = 1.69–1.73$
$(-)2V \sim 10°–80°$
HARDNESS: 5–6
DENSITY: 3.17–3.59 (Meas.)
3.140 (Calc.)
CLEAVAGE: {110} perfect
{001} parting
{100} parting
Fracture uneven to subconchoidal. Brittle.

HABIT: Crystals short to long prismatic. Also massive, compact. Twinning on {100} common, simple, lamellar.

COLOR-LUSTER: Dark green, black. Translucent to nearly opaque. Vitreous.

MODE OF OCCURRENCE: Occurs as a constituent of both igneous and metamorphic rocks. Typical occurrences are found in the Highwood Mountains, Montana; at Iron Hill, Gunnison County, Colorado; in the Henry Mountains, Utah; and in Riverside County and elsewhere in California. Also in Hastings County, Ontario, Canada, and at localities in Ireland, Scotland, Norway, Sweden, Finland, USSR (dashkesanite), Ghana, and Japan.

BEST REF. IN ENGLISH: Deer, Howie, and Zussman, "Rock Forming Minerals," v. 2, p. 263–314, New York, Wiley, 1963.


# HASTITE

$CoSe_2$

CRYSTAL SYSTEM: Orthorhombic
CLASS: 2/m 2/m 2/m
SPACE GROUP: Pmnn
Z: 2

LATTICE CONSTANTS:
 $a = 3.60$
 $b = 4.84$
 $c = 5.72$
3 STRONGEST DIFFRACTION LINES:
 2.6 (100)
 2.5 (100)
 1.9 (100)
HARDNESS: Not determined. Hard.
DENSITY: 7.22 (Calc.)
CLEAVAGE: Not determined.
HABIT: Idiomorphic and radiating crystals. Twinning observed.
COLOR-LUSTER: Light brownish red to dark reddish violet.
MODE OF OCCURRENCE: Occurs associated with trogtalite, bornhardtite, and an unnamed cobalt selenide as intergrowths in clausthalite at Steinbruch Trogtal, near Lautenthal, Harz, Germany.
BEST REF. IN ENGLISH: Ramdohr, Paul, and Schmitt, Marg., *Am. Min.*, **41**: 164–165 (1956).

**HATCHETTOLITE** = Variety of pyrochlore (uranium-tantalum rich)

# HATCHITE
$(Pb, Tl)_2 AgAs_2 S_5$

CRYSTAL SYSTEM: Triclinic
CLASS: $\bar{1}$
SPACE GROUP: $P\bar{1}$
LATTICE CONSTANTS:
 $a = 7.92$    $\alpha = 105°40'$
 $b = 9.03$    $\beta = 112°57'$
 $c = 7.71$    $\gamma = 64°48'$
3 STRONGEST DIFFRACTION LINES:
 2.88 (100)
 3.35 ( 60)
 3.44 ( 40)
HARDNESS: Not determined.
DENSITY: Not determined.
CLEAVAGE: Not determined.
HABIT: Found as five small highly modified crystals deposited upon a crystal presumed to be rathite.
COLOR-LUSTER: Lead gray. Streak chocolate brown.
MODE OF OCCURRENCE: Found in crystalline dolomite at the Lengenbach quarry in the Binnental, Valais, Switzerland.
BEST REF. IN ENGLISH: Palache, et al., "Dana's System of Mineralogy," 7th Ed., v. I, p. 487, New York, Wiley, 1944. Nowacki, W., and Bahezre, C., *Am. Min.*, **49**: 446 (1964). *Am. Min.*, **56**: 361–362 (1971).

# HAUCHECORNITE
$Ni_9 (Bi, Sb)_2 S_8$

CRYSTAL SYSTEM: Tetragonal
CLASS: 4/m 2/m 2/m
SPACE GROUP: P4/mmm
Z: 1

LATTICE CONSTANTS:
 $a = 7.29$
 $c = 5.40$
3 STRONGEST DIFFRACTION LINES:
 2.79 (100)
 2.39 ( 60)
 2.30 ( 60)
HARDNESS: 5
DENSITY: 6.36 (Meas.)
         6.58 (Calc.)
CLEAVAGE: None. Fracture flat conchoidal.
HABIT: Crystals commonly tabular (001); also bipyramidal to short prismatic.
COLOR-LUSTER: Light bronze yellow, tarnishing darker. Metallic luster on fresh surface. Streak gray-black.
MODE OF OCCURRENCE: Found in a pocket opened in 1884, containing about 5 tons of ore in siderite gangue, in the Friedrich mine, near Wissen a.d. Sieg, Westphalia, Germany, associated with millerite, bismuthian ullmannite, siegenite, sphalerite, bismuthinite, and quartz. Recently found in the nickel deposits at Sudbury, Ontario, Canada.
BEST REF. IN ENGLISH: Peacock, M. A., *Am. Min.*, **35**: 440–446 (1950).

# HAUERITE
$MnS_2$
Pyrite Group

CRYSTAL SYSTEM: Cubic
CLASS: 2/m $\bar{3}$
SPACE GROUP: Pa3
Z: 4
LATTICE CONSTANT:
 $a = 6.10$
3 STRONGEST DIFFRACTION LINES:
 3.07  (100)
 1.843 ( 70)
 2.75  ( 50)
OPTICAL CONSTANT:
 $N = 2.69$ (Li)
HARDNESS: 4
DENSITY: 3.463 (Meas.)
         3.444 (Calc.)
CLEAVAGE: {001} perfect
Fracture subconchoidal to uneven. Brittle.
HABIT: Crystals octahedral or cubo-octahedral, up to 2.5 cm along the octahedral edge. Also as globular aggregates.
COLOR-LUSTER: Reddish brown to brownish black. Opaque. Metallic-adamantine to dull. Streak brownish red.
MODE OF OCCURRENCE: Occurs in the capping of salt domes in Galveston and Matagorda counties, Texas; as fine octahedral crystals associated with gypsum, calcite, and sulfur in clay at Destricello and Raddusa, Sicily; at Kalinka and Schemnitz, Czechoslovakia; and in crystalline schists of the Lake Wakatipu district, New Zealand.
BEST REF. IN ENGLISH: Palache, et al., "Dana's System of Mineralogy," 7th Ed., v. I, p. 293–294, New York, Wiley, 1944.

# HAUSMANNITE
$Mn_2^{2+}Mn^{4+}O_4$

CRYSTAL SYSTEM: Tetragonal
CLASS: 4/m 2/m 2/m
SPACE GROUP: $I4_1/amd$
Z: 4
LATTICE CONSTANTS:
  $a = 5.76$
  $c = 9.44$
3 STRONGEST DIFFRACTION LINES:
  2.49 (100)
  2.77 ( 90)
  1.544 ( 80)
OPTICAL CONSTANTS:
  $\omega = 2.46$
  $\epsilon = 2.15$ (Li)
  (−)
HARDNESS: 5½
DENSITY: 4.84 (Meas.)
        4.84 (Calc.)
CLEAVAGE: {001} nearly perfect
         {112} indistinct
         {011} indistinct
Fracture uneven. Brittle.
HABIT: Crystals pseudo-octahedral. Usually as coherent granular masses. Twinning on {112} common, often repeated as fivelings; also lamellar.
COLOR-LUSTER: Brownish black. Opaque, except in very thin fragments. Submetallic. Streak brown.
MODE OF OCCURRENCE: Occurs chiefly in high-temperature hydrothermal veins and in contact metamorphic deposits, commonly associated with pyrolusite, psilomelane, magnetite, hematite, barite, and braunite. Found in Mariposa, Nevada, Placer, Plumas, San Joaquin, San Luis Obispo, Santa Clara, Stanislaus, and Trinity counties, California; and in Washington, Colorado, Arkansas, and Franklin, New Jersey. Other notable deposits occur in Brazil, England, Scotland, Sweden, Italy, Switzerland, Germany, Bulgaria, India, and elsewhere.
BEST REF. IN ENGLISH: Palache, et al., "Dana's System of Mineralogy," 7th Ed., v. I, p. 712–715, New York, Wiley, 1944.

# HAÜYNE (Haüynite)
$(Na, Ca)_{4-8}Al_6Si_6O_{24}(SO_4)_{1-2}$
Sodalite Group

CRYSTAL SYSTEM: Cubic
CLASS: $\bar{4}$ 3m
SPACE GROUP: $P\bar{4}3n$
Z: 1
LATTICE CONSTANT:
  $a = 9.13$
3 STRONGEST DIFFRACTION LINES:
  3.72 (100)
  2.63 ( 50)
  6.45 ( 30)
OPTICAL CONSTANT:
  $N = 1.496–1.505$
HARDNESS: 5½–6

DENSITY: 2.44–2.50 (Meas.)
CLEAVAGE: {110} distinct
Fracture uneven to conchoidal. Brittle.
HABIT: Crystals usually dodecahedral or octahedral. Commonly in rounded grains. Twinning on {111} common, sometimes as penetration twins; also polysynthetic or contact twins.
COLOR-LUSTER: Usually blue; also white or shades of gray, green, yellow, red. Transparent to translucent. Vitreous to greasy.
MODE OF OCCURRENCE: Occurs in phonolites and related igneous rocks in association with leucite or nepheline. Found near Winnett, Montana; in Lawrence County, South Dakota; in the Cripple Creek district and elsewhere in Colorado; in the Monteregian area of Quebec, Canada; on the island of Tahiti; in the Auvergne region, France; at Laacher See, Rhineland, and at other places in Germany; at many localities in Italy; and at Jebel Tourguejid, Morocco.
BEST REF. IN ENGLISH: Deer, Howie, and Zussman, "Rock Forming Minerals," v. 4, p. 289–302, New York, Wiley, 1963.

# HAWLEYITE
$\beta$-CdS
Dimorph of greenockite

CRYSTAL SYSTEM: Cubic
CLASS: $\bar{4}$3m
SPACE GROUP: $F\bar{4}3m$
Z: 4
LATTICE CONSTANT:
  $a = 5.818$
3 STRONGEST DIFFRACTION LINES:
  3.36 (100)
  2.058 ( 80)
  1.753 ( 60)
HARDNESS: Not determined.
DENSITY: 4.87 (Calc.)
CLEAVAGE: Not observed.
HABIT: As fine-grained earthy coatings.
COLOR-LUSTER: Bright yellow.
MODE OF OCCURRENCE: Occurs as an earthy coating on sphalerite and siderite, in vugs and along late fractures in the Hector-Calumet mine, Galena Hill, Yukon Territory, Canada.
BEST REF. IN ENGLISH: Traill, R. J., and Boyle, R. W., *Am. Min.*, **40**: 555–559 (1955).

# HAXONITE
$(Fe, Ni)_{23}C_6$

CRYSTAL SYSTEM: Cubic
3 STRONGEST DIFFRACTION LINES:
  2.356 (100)
  2.151 (100)
  1.863 ( 80)
HARDNESS: 5½–6
DENSITY: Not determined.
CLEAVAGE: Not determined. Brittle.

HABIT: Massive; as spiky plates a few microns across and up to a millimeter long.

COLOR-LUSTER: Tin white. Opaque.

Strongly magnetic.

MODE OF OCCURRENCE: Occurs in association with cohenite and kamacite in all classes of iron meteorites. Typical occurrences are found in the Canyon Diablo, Edmonton, and Tazewell meteorites.

BEST REF. IN ENGLISH: Scott, E. R. D., *Nat. Phys. Sci.*, **229**: 61–62 (January 11, 1971).

## HAYCOCKITE
$Cu_4Fe_5S_8$

CRYSTAL SYSTEM: Orthorhombic, pseudotetragonal
SPACE GROUP: P lattice
Z: 12
LATTICE CONSTANTS:
  $a \simeq b = 10.71$
  $c = 31.56$
3 STRONGEST DIFFRACTION LINES:
  3.07 (100)
  1.88 ( 80)
  1.89 ( 60)
HARDNESS: Microhardness 263 kg/mm$^2$.
DENSITY: 4.35 (Calc.)
CLEAVAGE: Not determined.
HABIT: Massive. Observed in polished sections in areas up to about 500 $\mu$ in diameter. Twinning polysynthetic.
COLOR-LUSTER: Brass yellow. Opaque. Metallic.
MODE OF OCCURRENCE: Occurs in association with mooihoekite in the Duluth Gabbro complex, Minnesota. Also found in a pipe-shaped hortonolite-dunite pegmatite in the Norite Zone of the Bushveld Igneous Complex at Mooihoek Farm, Lydenburg District, Transvaal, South Africa.
BEST REF. IN ENGLISH: Cabri, Louis J., and Hall, Sydney R., *Am. Min.*, **57**: 689–708 (1972).

## HEAZLEWOODITE
$Ni_3S_2$

CRYSTAL SYSTEM: Hexagonal
CLASS: 32
SPACE GROUP: R32
Z: 3
LATTICE CONSTANTS:
  $a = 5.730$
  $c = 7.125$
3 STRONGEST DIFFRACTION LINES:
  1.83 (100)
  2.89 ( 90)
  1.66 ( 80)
HARDNESS: 4
DENSITY: 5.82 (Meas.)
       5.87 (Calc.)
CLEAVAGE: None.
HABIT: Massive, compact.

COLOR-LUSTER: Light bronze. Opaque. Metallic. Streak pale bronze.

Not magnetic.

MODE OF OCCURRENCE: Occurs in a short, one inch wide stringer of sulfides on the lower contact of a serpentinized peridotite dike, which cuts a series of silicified tuffs and limestones, near the top of Miles Ridge, Yukon Territory, Canada. Also found with magnetite in serpentine at Heazlewood, Tasmania.

BEST REF. IN ENGLISH: Peacock, M. A., Univ. of Toronto Studies, Geol. Ser., No. 51, p. 59–69 (1947).

## HECTORITE
$Na_{0.33}(Mg, Li)_3Si_4O_{10}(F, OH)_2$
Montmorillonite (Smectite) Group

CRYSTAL SYSTEM: Monoclinic
CLASS: 2/m
SPACE GROUP: Probably C2/m
LATTICE CONSTANTS:
  $a = 5.25$
  $b = 9.18$
  $c > 10.0$
  $\beta \simeq 99°$
3 STRONGEST DIFFRACTION LINES:
  4.58 (100)
  1.53 (100)
  15.8 ( 80)
OPTICAL CONSTANTS:
  $\alpha \simeq 1.49$
  $\beta = 1.50$
  $\gamma = 1.52$
  $(-)2V = $ small
HARDNESS: 1–2
DENSITY: Variable, 2–3
CLEAVAGE: {001} perfect
HABIT: Massive, very fine grained; clay-like.
COLOR-LUSTER: White. Dull.
MODE OF OCCURRENCE: Occurs as an alteration of clinoptilolite in a bentonite deposit near Hector, San Bernardino County, California.
BEST REF. IN ENGLISH: Deer, Howie, and Zussman, "Rock Forming Minerals," v. 3, p. 226–245, New York, Wiley, 1962.

## HEDENBERGITE
$CaFeSi_2O_6$
Pyroxene Group
Var. Jeffersonite (Mn-Zn rich)
    Ferrosalite

CRYSTAL SYSTEM: Monoclinic
CLASS: 2/m
SPACE GROUP: C2/c
Z: 4
LATTICE CONSTANTS:
  $a = 9.85$
  $b = 9.02$
  $c = 5.26$
  $\beta = 104°20'$

3 STRONGEST DIFFRACTION LINES:
2.97 (100)
2.53 ( 50)
2.56 ( 30)
OPTICAL CONSTANTS:
$\alpha$ = 1.716–1.726
$\beta$ = 1.723–1.730
$\gamma$ = 1.741–1.751
(+)2V = 52°–62°
HARDNESS: 6
DENSITY: 3.50–3.56 (Meas.)
3.64 (Calc.)
CLEAVAGE: {110} good
{100} parting
{010} parting
Fracture uneven to conchoidal. Brittle.

HABIT: Crystals short prismatic; usually massive, lamellar. Twinning on {001} or {100} common, simple and multiple.

COLOR-LUSTER: Brownish green, grayish green, dark green, grayish black, black. Translucent to nearly opaque. Vitreous to resinous or dull. Streak white or grayish.

MODE OF OCCURRENCE: Occurs chiefly in limestone contact zones; iron-rich metamorphic rocks; and in granites, prophyries, syenites, and other igneous rocks. Typical occurrences are found in California, Arizona, New Mexico, Utah, Colorado, South Dakota, New York, and New Jersey (jeffersonite). Also in Greenland, England, Sweden, Italy, Finland, USSR, Nigeria, and Australia.

BEST REF. IN ENGLISH: Deer, Howie, and Zussman, "Rock Forming Minerals," v. 2, p. 42–74, New York, Wiley, 1963.

# HEDLEYITE

$Bi_7Te_3$

CRYSTAL SYSTEM: Hexagonal
CLASS: $\bar{3}$ 2/m
SPACE GROUP: R$\bar{3}$m
LATTICE CONSTANTS:
$a$ = 39.68
$c$ = 118.8
3 STRONGEST DIFFRACTION LINES:
3.25 (100)
2.37 ( 50)
2.24 ( 40)
HARDNESS: 2
DENSITY: 8.91 (Meas.)
8.93 (Calc.)
CLEAVAGE: {0001} perfect
Thin lamellae ductile and elastic.

HABIT: Massive, lamellar; as plates up to 6 mm wide.

COLOR-LUSTER: Tin-white. Opaque. Metallic. Tarnishes iron black.

MODE OF OCCURRENCE: Occurs in association with native bismuth, gold, joseite B, arsenopyrite, molybdenite, and pyrrhotite in quartz veinlets in garnet-pyroxene-epidote skarn, at the Good Hope mine, near Hedley, British Columbia, and also with native gold, altaite, and hessite on Upper Baruesh Creek, Yukon, Canada.

BEST REF. IN ENGLISH: Vlasov, K. A., Mineralogy of Rare Elements," v. II, p. 732–733, Israel Program for Scientific Translations, 1966.

# HEDYPHANE

$(Ca,Pb)_5(AsO_4)_3Cl$
Svabite Series
Apatite Group

CRYSTAL SYSTEM: Hexagonal
CLASS: 6/m
SPACE GROUP: P6$_3$/m
Z: 2
LATTICE CONSTANTS:
$a$ = 10.2
$c$ = 7.31
3 STRONGEST DIFFRACTION LINES:
3.03 (100)
2.96 ( 60)
2.92 ( 50)
OPTICAL CONSTANTS:
$\omega$ = 1.958
$\epsilon$ = 1.948  (Na)
(+)
HARDNESS: 4½
DENSITY: 5.82 (Meas.)
CLEAVAGE: {10$\bar{1}$1}
Fracture subconchoidal. Brittle.

HABIT: Crystals prismatic; also pyramidal to thick tabular. Also massive, coarse granular.

COLOR-LUSTER: White, buff, yellowish white, bluish. Translucent. Bright, greasy to resinous.

MODE OF OCCURRENCE: Occurs associated with calcite, willemite, rhodonite, and native copper in veins an inch or more thick at Franklin, Sussex County, New Jersey. It also is found at the Harstig mine, Pajsberg, and at Långban, Sweden.

BEST REF. IN ENGLISH: Palache, et al., "Dana's System of Mineralogy," 7th Ed., v. II, p. 900–902, New York, Wiley, 1951.

# HEIDORNITE

$Na_2Ca_3B_5O_8(SO_4)_2Cl(OH)_2$

CRYSTAL SYSTEM: Monoclinic
CLASS: 2/m
SPACE GROUP: C2/c
Z: 4
LATTICE CONSTANTS:
$a$ = 10.21
$b$ = 7.84
$c$ = 18.79
$\beta$ = 93°30′
OPTICAL CONSTANTS:
$\alpha$ = 1.579
$\beta$ = 1.588  ($\lambda$ = 587 m$\mu$)
$\gamma$ = 1.604
(+)2V = 63°–77°
HARDNESS: 4–5
DENSITY: 2.753 (Meas.)
2.70 (Calc.)

CLEAVAGE: {001} perfect
HABIT: Crystals steep, spear-like, with forms {110}, {11$\bar{1}$}, and {11$\bar{2}$} dominant. Crystals terminated (some doubly), up to 7 cm long.
COLOR-LUSTER: Colorless, transparent.
MODE OF OCCURRENCE: Occurs, mixed with an equal amount of glauberite, in a cavity in a drill core from 1968 meters depth in the upper anhydrite of the Zechstein formation on the German-Dutch border northwest of Nordhorn.
BEST REF. IN ENGLISH: v. Engelhardt, Wolf, and Füchtbauer, Hans, *Am. Min.*, **42**: 120–121 (1957).

# HEIKOLITE (or Heikkolite) = Arfvedsonite

# HEINRICHITE
$Ba(UO_2)_2(AsO_4)_2 \cdot 10-12H_2O$

CRYSTAL SYSTEM: Tetragonal
LATTICE CONSTANTS:
$a$ = 7.13
$c$ = 20.56
3 STRONGEST DIFFRACTION LINES:
3.57 (100)
8.89 ( 80)
5.03 ( 80)
OPTICAL CONSTANTS:
$\omega$ = 1.605
$\epsilon$ = 1.573
(−)
HARDNESS: 2.5
DENSITY: 3.61 (Calc.)
CLEAVAGE: {001} perfect
        {100} distinct
HABIT: Crystals tabular {001}; up to 1 mm on a side; not more than 0.1 mm thick. Forms: {001}, {100}, and {110}.
COLOR-LUSTER: Yellow to green, transparent to translucent; vitreous to pearly. Fluoresces bright green to greenish yellow under short and long wave ultraviolet light.
MODE OF OCCURRENCE: Occurs coating fractures and lining vugs in light gray, altered, silicified rhyolite tuff in the White King mine, 14 miles northwest of Lakeview, Lake County, Oregon. Also as an alteration of pitchblende associated with zeunerite, novacekite, erythrite, arseniosiderite, and pitticite at Wittichen, Schiltach, and Reinerzau in the Black Forest, Germany.
BEST REF. IN ENGLISH: Gross, Eugene B., Corey, Alice S., Mitchell, Richard S., and Walenta, Kurt, *Am. Min.*, **43**: 1134–1143 (1958).

# HELIOPHYLLITE
$Pb_3AsO_{4-n}Cl_{2n+1}$

CRYSTAL SYSTEM: Orthorhombic
LATTICE CONSTANTS:
$a$ = 10.823
$b$ = 10.783
$c$ = 25.580

3 STRONGEST DIFFRACTION LINES:
2.84 (100)
2.70 ( 40)
3.66 ( 35)
OPTICAL CONSTANTS:
(−)2V = large
HARDNESS: ~2
DENSITY: 6.886 (Meas.)
        7.33 (Calc.)
CLEAVAGE: {011} nearly perfect
HABIT: As acute pyramidal crystals with inclined faces horizontally striated; also tabular. Usually massive, coarsely foliated or granular.
COLOR-LUSTER: Yellow to greenish yellow. Translucent. Vitreous to greasy.
MODE OF OCCURRENCE: Occurs intergrown with ecdemite at Harstigen and Jacobsberg, and as individual crystals at Långban, Sweden.
BEST REF. IN ENGLISH: Palache, et al., "Dana's System of Mineralogy," 7th Ed., v. II, p. 1037–1038, New York, Wiley, 1951.

# HELIOTROPE = Variety of quartz

# HELLANDITE
$Ca_3(Y,Yb \cdots)_4B_4Si_6O_{27} \cdot 3H_2O$

CRYSTAL SYSTEM: Monoclinic
LATTICE CONSTANT:
$\beta$ = 109°45′
3 STRONGEST DIFFRACTION LINES:
2.63 (100)
2.82 ( 60) Commonly metamict.
1.89 ( 60)
OPTICAL CONSTANTS:
$\beta \approx 1.65$
$\gamma - \alpha \approx 0.01$
(+)2V $\simeq$ 80°
HARDNESS: 5½
DENSITY: 3.35–3.60 (Meas.)
CLEAVAGE: None. Fracture conchoidal. Brittle.
HABIT: Crystals tabular or prismatic. Twinning on {001} and {100}.
COLOR-LUSTER: Red, brown, blackish. Translucent. Vitreous.
MODE OF OCCURRENCE: Occurs in granite pegmatites in association with tourmaline, allanite, thorite, apatite, zircon, phenacite, and other minerals, at Kragerø, Telemark, Norway.
BEST REF. IN ENGLISH: Vlasov, K. A., "Mineralogy of Rare Elements," v. II, p. 242–243 Israel Program for Scientific Translations, 1966. Oftedal, Ivar, *Am. Min.*, **51**: 534 (1966).

# HELLYERITE (Inadequately described mineral)
$NiCO_3 \cdot 6H_2O$

CRYSTAL SYSTEM: Unknown

3 STRONGEST DIFFRACTION LINES:
9.4 (100)
6.06 (100)
3.65 ( 70)
OPTICAL CONSTANTS:
$\alpha = 1.455$
$\beta = 1.503$
$\gamma = 1.549$
$(-)2V = 85°$
HARDNESS: 2½
DENSITY: 1.97 (Meas.)
CLEAVAGE: One perfect; two good but less perfect.
HABIT: As fine-grained coatings.
COLOR-LUSTER: Pale blue; vitreous.
MODE OF OCCURRENCE: Occurs as thin coatings, associated with zaratite, on shear planes within a body of serpentinite at the old Lord Brassey nickel mine, at Heazlewood, Tasmania.
BEST REF. IN ENGLISH: Williams, K. L., Threadgold, I. M., and Hounslow, A. W., *Am. Min.*, **44**: 533–538 (1959).


## HELVINE (Helvite)
$Mn_7FeBe_6Si_6O_{24}S_2$

CRYSTAL SYSTEM: Cubic
CLASS: $\bar{4}3m$
SPACE GROUP: $P\bar{4}3n$
Z: 1
LATTICE CONSTANT:
$a = 8.20–8.27$
3 STRONGEST DIFFRACTION LINES:
3.38 (100)
1.954 ( 80)
2.218 ( 50)
OPTICAL CONSTANT:
$N = 1.728–1.747$
HARDNESS: 6
DENSITY: 3.17–3.37 (Meas.)
CLEAVAGE: {111} distinct
Fracture conchoidal to uneven. Brittle.
HABIT: Crystals usually tedrahedral, sometimes octahedral in aspect, up to several centimeters in size. Also as rounded aggregates.
COLOR-LUSTER: Brown, reddish, gray, yellowish gray, yellow, yellowish green. Transparent to translucent. Vitreous to resinous.
MODE OF OCCURRENCE: Occurs chiefly in granite pegmatites; syenites and nepheline-syenite pegmatites; gneiss; magnetite-fluorite skarns; and hydrothermal veins. Found at Pala and Rincon, San Diego County, California; at Butte, Montana; at the Sunnyside mine, Silverton district, San Juan County, Colorado; in Sierra County, New Mexico; and at Amelia, Virginia. Also at Mont St. Hilaire, Quebec, Canada, and at localities in Brazil, Iceland, Norway, Germany, USSR, Rhodesia, and elsewhere.
BEST REF. IN ENGLISH: Vlasov, K. A., "Mineralogy of Rare Elements," v. II, p. 119–127, Israel Program for Scientific Translations, 1966.


## HEMAFIBRITE = Variety of synadelphite

## HEMATITE
$\alpha$-$Fe_2O_3$

CRYSTAL SYSTEM: Hexagonal
CLASS: $\bar{3}2/m$
SPACE GROUP: $R\bar{3}c$
Z: 6 (hexagonal), 2 (rhombohedral)
LATTICE CONSTANTS:
| (hexagonal) | (rhombohedral) |
|---|---|
| $a = 5.0317$ | $a = 5.420$ |
| $c = 13.737$ | $\alpha = 55°14'$ |

3 STRONGEST DIFFRACTION LINES:
2.69 (100)
1.690 ( 60)
2.51 ( 50)
OPTICAL CONSTANTS:
$\omega = 3.22$
$\epsilon = 2.94$ $(\lambda = 589\ m\mu)$
$(-)$
HARDNESS: 5–6
DENSITY: 5.26 (Meas.)
5.256 (Calc.)
CLEAVAGE: None.
{0001} parting, due to twinning.
{$10\bar{1}1$} parting, due to twinning.
Fracture uneven to subconchoidal. Brittle.
HABIT: Crystals usually thin to thick tabular, rhombohedral, pyramidal, or rarely prismatic. Tabular crystals sometimes arranged in rosettes called "iron roses." Basal pinacoid commonly striated. Usually massive, compact columnar, fibrous, reniform "kidney ore," botryoidal, stalactitic, micaceous to platy; also granular, concretionary, oolitic, or earthy. Twinning on {0001}, as penetration twins; also lamellar twinning on {$10\bar{1}1$}.
COLOR-LUSTER: Crystals steel-gray to iron-black, sometimes tarnished iridescent; thin fragments deep blood red. Massive and earthy material dull brownish red to bright red. Opaque, except in thin scales. Metallic, submetallic, or dull. Streak deep red or brownish red.
MODE OF OCCURRENCE: Occurs widespread as the most abundant iron ore, chiefly in extensive thick beds of sedimentary origin; also as an accessory mineral in igneous rocks; in vein deposits, often as a gossan; as a sublimation product in lavas; in many metamorphic rocks; and in contact metamorphic deposits. Found as enormous deposits, up to 1,000 feet in thickness, in the Lake Superior region of northern Minnesota, Michigan, and Wisconsin; other important deposits occur in New York, Alabama, Tennessee, Pennsylvania, Missouri, South Dakota, and Wyoming. Notable deposits also occur in Canada, Mexico, Cuba, Venezuela, Brazil, England, Norway, Sweden, France, Italy, Switzerland, Austria, Germany, Roumania, Czechoslovakia, USSR, and Australia.
BEST REF. IN ENGLISH: Palache, et al., "Dana's System of Mineralogy," 7th Ed., v. I, p. 527–534, New York, Wiley, 1944.


## HEMATOLITE
$Mn_4Al(OH)_2[AsO_4][AsO_3]_2$

CRYSTAL SYSTEM: Hexagonal

CLASS: $\bar{3}$
SPACE GROUP: $R\bar{3}$
Z: 2 (rhombohedral), 6 (hexagonal)
LATTICE CONSTANTS:

  (hexagonal)    (rhombohedral)
  $a = 8.27$      $a = 13.07$
  $c = 36.51$    $\alpha = 55°14'$

3 STRONGEST DIFFRACTION LINES:
  2.39 (100)
  1.562 ( 90)
  6.12 ( 80)
OPTICAL CONSTANTS:
  $\omega = 1.733$
  $\epsilon = 1.714$
    (–)
HARDNESS: 3½
DENSITY: 3.49 (Meas.)
         3.48 (Calc.)
CLEAVAGE: {0001} perfect
Fracture uneven. Brittle.

HABIT: Crystals thick tabular or rhombohedral. Rhombohedral faces striated horizontally.

COLOR-LUSTER: Brownish red to almost black. Translucent to nearly opaque. Vitreous; pearly to submetallic on cleavage.

MODE OF OCCURRENCE: Occurs in crystalline limestone associated with jacobsite, magnetite, barite, and fluorite at the Moss mine, Nordmark, Sweden.

BEST REF. IN ENGLISH: Palache, et al., "Dana's System of Mineralogy," 7th Ed., v. II, p. 777–778, New York, Wiley, 1951.

# HEMATOPHANITE
$Pb_4Fe_4O_9(OH,Cl)_2$

CRYSTAL SYSTEM: Tetragonal
CLASS: 4/m 2/m 2/m, 422, 4mm, or $\bar{4}m2$
SPACE GROUP: P lattice
Z: 3
LATTICE CONSTANTS:
  $a = 7.80$
  $c = 15.23$
3 STRONGEST DIFFRACTION LINES:
  2.18 (100)
  3.1 ( 50)
  1.55 ( 40)
OPTICAL CONSTANTS:
  uniaxial (–)
HARDNESS: 2–3
DENSITY: 7.70 (Meas.)
CLEAVAGE: {001} nearly perfect
HABIT: Crystals thin tabular, flattened on {001}; as lamellar aggregates.

COLOR-LUSTER: Dark reddish brown. Opaque, except in very thin fragments. Submetallic. Streak yellowish red.

MODE OF OCCURRENCE: Occurs in granular limestone associated with jacobsite, native copper, andradite, plumboferrite, and cuprite at the Jakobsberg manganese mine, Nordmark, Sweden.

BEST REF. IN ENGLISH: Palache, et al., "Dana's System of Mineralogy," 7th Ed., v. I, p. 728–729, New York, Wiley, 1944.

# HEMIHEDRITE
$ZnF_2[Pb_5(CrO_4)_3SiO_4]_2$

CRYSTAL SYSTEM: Triclinic
CLASS: 1
SPACE GROUP: P1
Z: 1
LATTICE CONSTANTS:
  $a = 9.497$    $\alpha = 120°30'$
  $b = 11.443$   $\beta = 92°06'$
  $c = 10.841$   $\gamma = 55°50'$
3 STRONGEST DIFFRACTION LINES:
  3.301 (100)
  4.872 ( 90)
  4.364 ( 80)
OPTICAL CONSTANTS:
  $\alpha = 2.105$
  $\beta = 2.32$ (25°C)
  $\gamma = 2.65$
  (+)2V = 88° (Calc.)
HARDNESS: 3
DENSITY: 6.42 (Meas.)
         6.50 (Calc.)
HABIT: Crystals exhibit triclinic hemihedral symmetry. Commonly twinned ($\bar{2}\bar{2}3$); ($0\bar{1}0$) and ($0\bar{1}2$) twinning less common. Crystals vary in length from 0.2 to 10 mm; average 0.5 mm.

COLOR-LUSTER: Bright orange to brown to almost black. Streak saffron yellow.

MODE OF OCCURRENCE: Found in the oxide zone of lead-bearing veins at the Florence Lead-Silver mine, Tortilla Mountains, Pinal County, Arizona, and at the Rat Tail claim near Wickenburg, Maricopa County, Arizona. Associated minerals may include cerussite, wulfenite, mimetite, willemite, minium, vauquelinite, and phoenicochroite.

BEST REF. IN ENGLISH: Williams, Sidney A., and Anthony, John W., *Am. Min.*, **55**: 1088–1102 (1970).

# HEMIMORPHITE
$Zn_4Si_2O_7(OH)_2 \cdot H_2O$

CRYSTAL SYSTEM: Orthorhombic
CLASS: mm2
SPACE GROUP: Imm2
Z: 2
LATTICE CONSTANTS:
  $a = 8.38$
  $b = 10.70$
  $c = 5.11$
3 STRONGEST DIFFRACTION LINES:
  3.104 (100)
  6.60 ( 86)
  3.288 ( 75)
OPTICAL CONSTANTS:
  $\alpha = 1.614$
  $\beta = 1.617$
  $\gamma = 1.636$
  (+)2V = 46°

HARDNESS: 4½–5
DENSITY: 3.4–3.5 (Meas.)
         3.366 (Calc.)
CLEAVAGE: {110} perfect
           {101} imperfect
           {001} in traces
Fracture uneven to subconchoidal. Brittle.

HABIT: Crystals commonly thin tabular, vertically striated; doubly terminated crystals show distinct hemimorphic development. Often in fan-shaped aggregates. Also massive, granular; stalactitic, or as compact mammillary masses with fibrous structure. Twinning on {001}, rare.

COLOR-LUSTER: Usually white or colorless; also pale blue, greenish, gray, yellowish, brown. Transparent to translucent. Vitreous; sometimes slightly silky or dull. Streak uncolored.

MODE OF OCCURRENCE: Occurs as a secondary mineral in the oxidized zone of ore deposits or in stratified calcareous rocks; rarely in granitic pegmatites. Common associated minerals are sphalerite, smithsonite, galena, cerussite, anglesite, calcite, and aurichalcite. Found widespread in the United States, especially in California, Nevada, Montana, Utah, Colorado, Arizona, New Mexico, South Dakota, Missouri, Pennsylvania, New Jersey, and Virginia. Also as fine crystals at Mapimi in Durango, and at Santa Eulalia, Chihuahua, Mexico; and at localities in England, Belgium, France, Spain, Sardinia, Germany, Austria, Roumania, USSR, Algeria, and elsewhere.

BEST REF. IN ENGLISH: Kostov, Ivan, "Mineralogy," p. 324–325, Edinburgh and London, Oliver and Boyd, 1968.

# HEMUSITE
$Cu_6SnMoS_8$

CRYSTAL SYSTEM: Cubic
LATTICE CONSTANT:
  $a = 10.82$
3 STRONGEST DIFFRACTION LINES:
  3.11 (100)
  1.919 ( 50)
  1.858 ( 30)
HARDNESS: ~4
DENSITY: Not determined.
CLEAVAGE: Not determined.
HABIT: As rounded isometric grains and aggregates usually about 0.05 mm in diameter.
COLOR-LUSTER: Gray. Opaque. Metallic.

Nonmagnetic.
MODE OF OCCURRENCE: Occurs in association with enargite, luzonite, colusite, stannoidite, renierite, tennantite, chalcopyrite, pyrite, and other minerals, in the copper ore deposit of Chelopech, Bulgaria.

BEST REF. IN ENGLISH: Terziev, G. I., *Am. Min.*, **56**: 1847–1854 (1971).

# HENDERSONITE
$Ca_2V_{1+x}^{4+}V_{8-x}^{5+}(O,OH)_{24} \cdot 8H_2O$

CRYSTAL SYSTEM: Orthorhombic
CLASS: 2/m 2/m 2/m or mm2

SPACE GROUP: Pnam or Pna2$_1$
Z: 4
LATTICE CONSTANTS:
  $a = 12.40$
  $b = 18.92$
  $c = 10.77$
3 STRONGEST DIFFRACTION LINES:
  9.45 (100)
  3.113 ( 36)
  4.70 ( 16)
OPTICAL CONSTANTS:
  $\alpha < 2.0$
  $\beta > 2.01$
  $\gamma > 2.01$
  (−)2V = medium
HARDNESS: ~2.5
DENSITY: 2.77–2.79 (Meas.)
         2.80$_5$ (Calc.)
CLEAVAGE: Not determined.
HABIT: Elongated, six-sided platy crystals. Usually aggregates of subparallel to parallel microscopic fibers or blades.

COLOR-LUSTER: Dark greenish black to black; pearly to subadamantine. Turns brownish on exposure. Streak dark brownish green.

MODE OF OCCURRENCE: Found in partly oxidized ore associated with paramontroseite, simplotite, melanovanadite, sherwoodite, and corvusite in thin seams and veinlets in vanadium-uranium deposits in the Salt Wash Sandstone Member of the Morrison Formation (Upper Jurassic) at the J. J. mine, Paradox Valley, Montrose County, Colorado, and in one of the Eastside mines, San Juan County, New Mexico.

BEST REF. IN ENGLISH: Lindberg, M. L., Weeks, A. D., Thompson, M. E., Elston, D. P., and Meyrowitz, Robert, *Am. Min.*, **47**: 1252–1272 (1962).

# HENDRICKSITE
$K(Zn,Mn)_3(Si_3Al)O_{10}(OH)_2$
Mica Group

CRYSTAL SYSTEM: Monoclinic
CLASS: m or 2/m
SPACE GROUP: Cm or C2/m
Z: 2
LATTICE CONSTANTS:
  $a = 5.37$
  $b = 9.32$
  $c = 10.30$
  $\beta = 99°$
3 STRONGEST DIFFRACTION LINES:
  10.20 (100)
  3.398 ( 60)
  5.094 ( 36)
OPTICAL CONSTANTS: Mean index = 1.686
HARDNESS: 2.5–3
DENSITY: 3.4
CLEAVAGE: {001} perfect
HABIT: Crystals tabular or short prismatic.
COLOR-LUSTER: Coppery red to reddish black.

MODE OF OCCURRENCE: Occurs abundantly as rough crystals and anhedral plates up to a foot in size at Franklin, New Jersey.

BEST REF. IN ENGLISH: Frondel, Clifford, and Ito, Jun, *Am. Min.*, **51**: 1107–1123 (1966).

## HENRITERMIERITE

$Ca_3 (Mn_{1.5}^{3+} Al_{0.5})(SiO_4)_2 (OH)_4$
Garnet Group

CRYSTAL SYSTEM: Tetragonal
CLASS: 4/m 2/m 2/m
SPACE GROUP: $I4_1/acd$
Z: 8
LATTICE CONSTANTS:
  $a$ = 12.39
  $c$ = 11.91
3 STRONGEST DIFFRACTION LINES:
  2.75  (100)
  2.516 ( 80)
  4.37  ( 60)
OPTICAL CONSTANTS:
  $\omega$ = 1.765
  $\epsilon$ = 1.800
  (+)
HARDNESS:  Not determined.
DENSITY:  3.34 (Meas.)
          3.40 (Calc.)
CLEAVAGE:  None. Fracture, conchoidal.
HABIT:  Aggregates of small grains up to 0.5 mm; mostly about 0.2 mm in diameter.
COLOR-LUSTER:  Clove to apricot brown; vitreous.
MODE OF OCCURRENCE: Occurs filling interstices between crystals of marokite, hausmannite, and rare gaudefroyite in the Tachgagalt manganese mine, Morocco.
BEST REF. IN ENGLISH: Gaudefroy, C., Orliac, M., Permingeat, F., and Parfenoff, A., *Am. Min.*, **54**: 1739 (1969).

## HERCYNITE

$FeAl_2O_4$
Spinel Series

CRYSTAL SYSTEM:  Cubic
CLASS:  $4/m\,\bar{3}\,2/m$
SPACE GROUP:  Fd3m
Z: 8
LATTICE CONSTANT:
  $a$ = 8.136
3 STRONGEST DIFFRACTION LINES:
  2.45  (100)    2.45 (100)
  1.562 ( 90)    2.02 ( 80)
  1.434 ( 90)    1.43 ( 80)
OPTICAL CONSTANT:
  $N$ = 1.83
HARDNESS:  7½–8
DENSITY:  4.323 (Calc.)
CLEAVAGE:  Fracture conchoidal to uneven.
HABIT:  Massive; fine granular. As embedded grains.
COLOR-LUSTER:  Black. Opaque. Streak dark green.

MODE OF OCCURRENCE: Occurs in emery deposits at Peekskill, Westchester County, New York, and near Whittles, Pittsylvania County, Virginia. Also at localities in Brazil, Germany, Switzerland, Czechoslovakia, India, Madagascar, and Tasmania.

BEST REF. IN ENGLISH: Palache, et al., "Dana's System of Mineralogy," 7th Ed., v. I, p. 689–697, New York, Wiley, 1944.

## HERDERITE

$CaBePO_4 (F, OH)$

CRYSTAL SYSTEM: Monoclinic
CLASS: 2/m
SPACE GROUP: $P2_1/c$
Z: 4
LATTICE CONSTANTS:
  $a$ = 4.80
  $b$ = 7.68
  $c$ = 9.80
  $\beta$ = 90°06′
3 STRONGEST DIFFRACTION LINES:
  3.14 (100)
  2.86 ( 80)
  2.20 ( 70)
OPTICAL CONSTANTS:
  $\alpha$ = 1.591–1.592
  $\beta$ = 1.611–1.612
  $\gamma$ = 1.619–1.621
  (–)2V = 67° ~ 75°
HARDNESS: 5–5½
DENSITY: 2.95–3.01 (Meas.)
          2.94 (Calc.)
CLEAVAGE: {110} interrupted
Fracture subconchoidal.
HABIT: Crystals stout prismatic or thick tabular. Commonly pseudo-orthorhombic in appearance. Also as botryoidal crusts or aggregates with radial-fibrous structure. Commonly twinned on {001} or {100}.
COLOR-LUSTER: Colorless to pale yellow or greenish white. Transparent to translucent. Vitreous.
MODE OF OCCURRENCE: Occurs as a late-stage hydrothermal mineral in granite pegmatites. Found at the Fletcher and Palermo mines, near North Groton, New Hampshire; as fine crystals at several places in Maine, especially at Stoneham, Newry, Hebron, Paris, Greenwood, and Buckfield in Oxford County, and at Poland, Topsham, and Auburn, in Androscoggin County. It also occurs as superb crystals up to 5 × 12 cm in size in Minas Geraes, Brazil; at localities in Bavaria and Saxony, Germany; and at Mursinsk, Ural Mountains, USSR.

BEST REF. IN ENGLISH: Palache, et al., "Dana's System of Mineralogy," 7th Ed., v. II, p. 820–822, New York, Wiley, 1951.

## HERRERITE = Variety of smithsonite containing copper

## HERSCHELITE

$(Na, Ca, K)AlSi_2O_6 \cdot 3H_2O$
Isostructural with chabazite
Zeolite Group

CRYSTAL SYSTEM: Hexagonal
CLASS: $\bar{3}$ 2/m
SPACE GROUP: R$\bar{3}$m
Z: 3
LATTICE CONSTANTS:
  $a$ = 13.799
  $c$ = 15.102
3 STRONGEST DIFFRACTION LINES:
  2.93 (100)
  4.32 ( 65)
  9.36 ( 50)
OPTICAL CONSTANTS:
  $\omega$ = 1.479
  $\epsilon$ = 1.481
  (+)
HARDNESS: 4-5
DENSITY: 2.08-2.16 (Meas.)
         2.050 (Calc.)
CLEAVAGE: $\{10\bar{1}1\}$ distinct
Fracture uneven. Brittle.
  HABIT: As aggregates of hexagonal plates.
  COLOR-LUSTER: White. Transparent to translucent. Vitreous.
  MODE OF OCCURRENCE: Occurs in cavities in lava at Aci Castello, on the flanks of Mt. Etna, Sicily; also found at Richmond, Victoria, Australia.
  BEST REF. IN ENGLISH: Mason, Brian, *Am. Min.*, 47: 985-987 (1962).

## HERZENBERGITE
SnS

CRYSTAL SYSTEM: Orthorhombic
CLASS: 2/m 2/m 2/m
SPACE GROUP: Pbnm
Z: 4
LATTICE CONSTANTS:
  $a$ = 4.33
  $b$ = 11.18
  $c$ = 3.98
3 STRONGEST DIFFRACTION LINES:
  2.793 (100)
  1.399 ( 70)
  2.831 ( 25)
HARDNESS: Not determined.
DENSITY: 5.197 (Calc. X-ray)
CLEAVAGE: $\{010\}$
HABIT: Massive, fine-grained.
  COLOR-LUSTER: Black. Opaque. Metallic. Streak black.
  MODE OF OCCURRENCE: Occurs associated with cassiterite, pyrite, and quartz in the Maria-Teresa mine, near Huari, Bolivia.
  BEST REF. IN ENGLISH: Palache, et al., "Dana's System of Mineralogy," 7th Ed., v. I, p. 259-260, New York, Wiley, 1944.

## HESSITE
Ag$_2$Te

CRYSTAL SYSTEM: Monoclinic

CLASS: 2/m
SPACE GROUP: P2$_1$/c
Z: 4
LATTICE CONSTANTS:
  $a$ = 8.09
  $b$ = 4.48
  $c$ = 8.96
  $\beta$ = 123°20'
3 STRONGEST DIFFRACTION LINES:
  2.31 (100)
  2.87 ( 80)
  2.25 ( 70)
HARDNESS: 2-3
DENSITY: 8.24-8.45 (Meas.)
         8.402 (Calc.)
CLEAVAGE: $\{100\}$ indistinct
Fracture even. Sectile.
  HABIT: Crystals pseudocubic, highly modified, and generally irregularly developed and distorted. Also massive, compact or fine-grained. Lamellar twinning under the microscope.
  COLOR-LUSTER: Lead gray to steel gray. Opaque. Metallic.
  MODE OF OCCURRENCE: Occurs in low- and medium-temperature hydrothermal vein deposits in association with other tellurides, native tellurium, and gold. Found in Calaveras, El Dorado, Mono, Nevada, Shasta, Siskiyou, and Tuolumne counties, California; in Boulder, Clear Creek, Eagle, Lake, La Plata, Montezuma, Ouray, Saguache, and Teller counties, Colorado; and at deposits in Mexico, Chile, Roumania, USSR, Western Australia, and elsewhere.
  BEST REF. IN ENGLISH: Vlasov, K. A., "Mineralogy of Rare Elements," v. II, p. 717-721, Israel Program for Scientific Translations, 1966. Sindeeva, N. D., "Mineralogy and Types of Deposits of Selenium and Tellurium" (1964).

**HESSONITE** = Variety of grossular containing iron.

## HETAEROLITE
ZnMn$_2$O$_4$

CRYSTAL SYSTEM: Tetragonal
CLASS: 4/m 2/m 2/m
SPACE GROUP: I4$_1$/amd
Z: 4
LATTICE CONSTANTS:
  $a$ = 5.74
  $c$ = 9.15
3 STRONGEST DIFFRACTION LINES:
  2.462 (100)
  2.707 ( 90)
  3.040 ( 80)
OPTICAL CONSTANTS:
  $\omega$ = 2.34
  $\epsilon$ = 2.14
  (−)
HARDNESS: 6
DENSITY: 5.18 (Meas.)
         5.23 (Calc.)
CLEAVAGE: $\{001\}$ indistinct
Fracture uneven. Brittle.

HABIT: Crystals octahedral. Commonly massive. Twinning on {112}, as fivelings.

COLOR-LUSTER: Black. Opaque, except in very thin fragments. Submetallic. Streak dark brown.

MODE OF OCCURRENCE: Occurs in the zinc deposits at Franklin and Sterling Hill, Sussex County, New Jersey, associated with willemite, franklinite, chalcophanite, and calcite.

BEST REF. IN ENGLISH: Palache, et al., "Dana's System of Mineralogy," 7th Ed., v. I, p. 715–717, New York, Wiley, 1944.

## HETEROGENITE (Stainierite) (Mindigite)
$CoO(OH)$

CRYSTAL SYSTEM: Hexagonal
CLASS: $\bar{3}\ 2/m$
SPACE GROUP: $R\bar{3}m$
Z: 3
LATTICE CONSTANTS:
  $a = 2.849$
  $c = 13.130$
3 STRONGEST DIFFRACTION LINES:
  4.40  (100)
  2.315 ( 80)
  1.804 ( 60)
HARDNESS: 4–5
DENSITY: 4.13–4.47 (Meas.)
CLEAVAGE: One parallel to elongation; one inclined. Fracture conchoidal to uneven (aggregates).

HABIT: Crystals minute, needle-like, with hexagonal or pseudohexagonal cross section. Usually massive, microcrystalline; as botryoidal and mammillary crusts.

COLOR-LUSTER: Black, steel gray. Opaque. Metallic to dull, earthy. Streak black, brownish black, dark brown.

MODE OF OCCURRENCE: Occurs as an alteration product of cobalt-bearing minerals, often associated with malachite or hematite. Found in the Goodsprings district, Nevada; also at deposits in Chile, Spain, Germany, Transvaal, and abundantly at Mindigi and elsewhere in Katanga, Zaire.

BEST REF. IN ENGLISH: Hey, M. H., *Min. Mag.*, 33: 253–259 (1962).

## HETEROMORPHITE
$Pb_7Sb_8S_{19}$

CRYSTAL SYSTEM: Monoclinic
LATTICE CONSTANTS:
  $a = 11.93$ kX
  $b = 8.31$
  $c = 14.18$
  $\beta = 106°30'$
3 STRONGEST DIFFRACTION LINES:
  2.95 (100)
  3.24 ( 90)
  2.14 ( 80)
HARDNESS: 2½–3
DENSITY: 5.73 (Meas.)
CLEAVAGE: {112} good
Fracture uneven. Brittle.

HABIT: Crystals pyramidal; somewhat elongated; often distorted and composed of subparallel individuals. Striated and rounded parallel [110]. Commonly massive.

COLOR-LUSTER: Black. Opaque. Metallic. Streak black.

MODE OF OCCURRENCE: Occurs associated with sphalerite at the antimony mines of Arnsberg, Westphalia, Germany; also found at Kara Kamar, Tadzhik, USSR.

BEST REF. IN ENGLISH: Palache, et al., "Dana's System of Mineralogy," 7th Ed., v. I, p. 465–466, New York, Wiley, 1944.

## HETEROSITE
$(Fe^{3+}, Mn^{3+})PO_4$
Heterosite-Pupurite Series

CRYSTAL SYSTEM: Orthorhombic
CLASS: $2/m\ 2/m\ 2/m$
SPACE GROUP: Pmnb
Z: 4
LATTICE CONSTANTS:
  $a = 5.82$
  $b = 9.68$
  $c = 4.76$
3 STRONGEST DIFFRACTION LINES:
  3.48 (100)
  4.29 ( 75)
  2.73 ( 75)
OPTICAL CONSTANTS:
  $\alpha = 1.86$
  $\beta = 1.89$
  $\gamma = 1.91$
  2V = large
HARDNESS: 4–4½
DENSITY: 3.409 (Meas.)
        3.702 (Calc.)
CLEAVAGE: {100} good
          {010} imperfect
Brittle. Fracture uneven.

HABIT: Massive.

COLOR-LUSTER: Deep rose to reddish purple. Usually externally dark brown to brownish black due to alteration. Subtranslucent to opaque. Dull to satiny. Streak reddish purple.

MODE OF OCCURRENCE: Occurs as an alteration of triphylite in the zone of weathering in granite pegmatites. Found at numerous places in the Keystone and Custer districts, Black Hills, South Dakota; in pegmatites in New Hampshire, Maine, and Massachusetts; at Varuträsk, Sweden; and in Portugal, France, Germany, Australia, and Erongo, South West Africa.

BEST REF. IN ENGLISH: Palache, et al., "Dana's System of Mineralogy," 7th Ed., v. II, p. 675–677, New York, Wiley, 1951.

## HEULANDITE
$(Ca, Na_2)Al_2Si_7O_{18} \cdot 6H_2O$
Zeolite Group

CRYSTAL SYSTEM: Monoclinic
CLASS: m

SPACE GROUP: Cm
Z: 4
LATTICE CONSTANTS:
$a = 17.73$
$b = 17.82$
$c = 7.43$
$\beta = 116°20'$
3 STRONGEST DIFFRACTION LINES:
3.917 (100)
2.959 ( 90)
8.845 ( 80)
OPTICAL CONSTANTS:
$\alpha = 1.496$
$\beta = 1.498$
$\gamma = 1.504$
$(+)2V = 35°$
HARDNESS: 3½–4
DENSITY: 2.1–2.2 (Meas.)
CLEAVAGE: {010} perfect
Fracture uneven. Brittle.

HABIT: Crystals trapezoidal, tabular parallel to {010}; often in subparallel aggregates. Also massive, granular.

COLOR-LUSTER: Colorless, white, gray, yellow, pink, red, brown. Transparent to translucent. Vitreous; pearly on {010}. Streak uncolored.

MODE OF OCCURRENCE: Occurs chiefly in cavities in basalts and andesites in association with other zeolites; also in skarns and mica gneiss; less commonly in bedded deposits or in sandstone. Found as excellent crystals in the trap rocks of northern New Jersey; near Baltimore, Maryland; at Ritter Hot Springs and elsewhere in Oregon; and in Hawaii. Other notable occurrences are found in Nova Scotia, Brazil, Iceland, Island of Skye, Scotland, Switzerland, Austria, Germany, Yugoslavia, USSR, India, Japan, New Zealand, and Australia.

BEST REF. IN ENGLISH: Deer, Howie, and Zussman, "Rock Forming Minerals," v. 4, p. 377–384, New York, Wiley, 1963. Merkle, A. B., and Slaughter, M., *Am. Min.*, **53**: 1120–1138 (1968).

## HEWETTITE
$CaV_6O_{16} \cdot 9H_2O$

CRYSTAL SYSTEM: Probably Orthorhombic
LATTICE CONSTANTS:
$a = 12.23$
$b = 3.605$
OPTICAL CONSTANTS:
$\alpha = 1.77$
$\beta = 2.18$ (Li)
$\gamma = 2.35–2.4$
$(-)2V$ medium
HARDNESS: Not determined.
DENSITY: 2.55 (with $9H_2O$)
2.618 (air-dried)
CLEAVAGE: Not determined.
HABIT: Nodular aggregates and coatings of fibers, elongate [010]. Crystals are well-developed laths elongated [010], and flattened {001}.

COLOR-LUSTER: Deep red; dull to silky. Powder maroon to brownish red.

MODE OF OCCURRENCE: Occurs in fractures below deposits containing corvusite at several localities in the Edgemont uranium-vanadium mining area, Fall River County, South Dakota; at the Hummer mine, Montrose County, Colorado; and also at the Cactus Rat mine, Grand County, Utah. Also found in the oxidized zone of the vanadium deposit at Minasragra, near Cerro de Pasco, Peru; and from Ferghana, USSR.

BEST REF. IN ENGLISH: Palache, et al., "Dana's System of Mineralogy," 7th Ed., v. II, p. 1060–1061, New York, Wiley, 1951. Ross, Malcolm, *Am. Min.*, **44**: 322–341 (1959).

## HEXAHYDRITE
$MgSO_4 \cdot 6H_2O$

CRYSTAL SYSTEM: Monoclinic
CLASS: 2/m
SPACE GROUP: C2/c
Z: 8
LATTICE CONSTANTS:
$a = 10.06$
$b = 7.16$
$c = 24.39$
$\beta = 98°34'$
3 STRONGEST DIFFRACTION LINES:
4.43 (100)
4.04 ( 90)
2.941 ( 80)
OPTICAL CONSTANTS:
$\alpha = 1.426$
$\beta = 1.453$
$\gamma = 1.456$
$(-)2V = 38°$
HARDNESS: Not determined.
DENSITY: 1.757 (Meas.)
1.745 (Calc.)
CLEAVAGE: {100} perfect
Fracture conchoidal.

HABIT: Crystals thick tabular, rare. Usually coarse columnar to fibrous. Twinning on {001} and {110}.

COLOR-LUSTER: Colorless, white, pale greenish. Transparent, usually opaque. Vitreous to pearly.

Soluble in water. Taste bitter, salty.

MODE OF OCCURRENCE: Occurs as a dehydration product of epsomite, sometimes as pseudomorphs. Found on the Bonaparte River, British Columbia, Canada; at Boleslaw, near Olkusz, Poland; at Kladno, Czechoslovakia; with halite in the Saki salt lakes, Crimea, USSR; and near Oroville, Okanogan County, Washington, as an efflorescence on epsomite.

BEST REF. IN ENGLISH: Palache, et al., "Dana's System of Mineralogy," 7th Ed., v. II, p. 494–495, New York, Wiley, 1951.

## HEXASTANNITE (Species status uncertain)
Near $Cu_{4.5}(Fe, Zn)_2SnS_7$

# HEYITE
$Pb_5 Fe_2 (VO_4)_2 O_4$

CRYSTAL SYSTEM: Monoclinic
CLASS: 2/m
SPACE GROUP: $P2_1/m$
Z: 1
LATTICE CONSTANTS:
  $a = 8.910$
  $b = 6.017$
  $c = 7.734$
  $\beta = 111°53'$
3 STRONGEST DIFFRACTION LINES:
  3.248 (100)
  2.970 ( 69)
  2.767 ( 61)
OPTICAL CONSTANTS:
        $\alpha = 2.185$
        $\beta = 2.219$
        $\gamma = 2.266$
  Biaxial. $2V = 89°$
  HARDNESS: 4
  DENSITY:  6.3 (Meas.)
            6.284 (Calc.)
  CLEAVAGE:  None. Fracture irregular. Brittle.
  HABIT: Crystals tabular $\{100\}$, elongate on $b$ axis, up to 0.4 mm in length. Forms: $a$ $\{100\}$, $c$ $\{001\}$, $d$ $\{\bar{1}01\}$, and $m$ $\{110\}$. Twinning on $\{110\}$ common.
  COLOR-LUSTER: Yellow-orange. Transparent. Streak amber yellow.
  MODE OF OCCURRENCE: Heyite is found in very minute amounts in silicified limestone at the Betty Joe claim nearly Ely, White Pine County, Nevada. It occurs on and replaces corroded tungstenian wulfenite in association with pyromorphite, cerussite, chrysocolla, and other oxidation zone minerals derived from galena, chalcopyrite, and pyrite.
  BEST REF. IN ENGLISH: Williams, Sidney A. *Min. Mag.*, **39**: 65–68 (1973).

# HEYROVSKÝITE
$Pb_6 Bi_2 S_9$

CRYSTAL SYSTEM: Orthorhombic
CLASS: 2/m 2/m 2/m or mm2
SPACE GROUP: Bbmm or Bbm2
Z: 4
LATTICE CONSTANTS:
  $a = 13.705$
  $b = 31.194$
  $c = 4.121$
3 STRONGEST DIFFRACTION LINES: Identical to phase II of Otto and Strunz (1968).
OPTICAL CONSTANTS: Strongly anistropic; reflectance strongly variable, roughly the same as of galena.
  HARDNESS: Microhardness (VHN) (50 g load) 166–234 kg/mm$^2$.
  DENSITY: 7.17 (Meas.)
           7.18 (Calc.)
  CLEAVAGE: Not determined.

HABIT: Crystals elongated parallel $c$-axis, flattened on $\{010\}$; dominant forms $\{010\}$ and $\{140\}$.
  COLOR-LUSTER: Lead gray. Opaque. Metallic.
  MODE OF OCCURRENCE: Occurs in association with galena and cosalite at Hůrky, Czechoslovakia.
  BEST REF. IN ENGLISH: Klominsky, J., Rieder, M., Kieft, C., and Mráz, L., *Min. Abs.*, **23**: 128 (1972).

# HIBONITE
$(Ca, Ce)(Al, Ti, Mg)_{12} O_{18}$

CRYSTAL SYSTEM: Hexagonal
CLASS: 6/m 2/m 2/m
SPACE GROUP:  $P6_3/mmc$
Z: 2
LATTICE CONSTANTS:
  $a = 5.61$
  $c = 22.16$
OPTICAL CONSTANTS:
  $\omega = 1.807$
  $\epsilon = 1.79$
  HARDNESS: 7½–8
  DENSITY: 3.84
  CLEAVAGE: $\{0001\}$ easy
Parting on $\{10\bar{1}0\}$. Fracture subconchoidal.
  HABIT:  As hexagonal prisms flattened parallel to $\{0001\}$ or in steep pyramids; the face $\{0001\}$ commonly divided into six sectors; crystals up to 4 cm in largest dimension.
  COLOR-LUSTER: Brownish black to black, reddish brown in thin fragments.
  MODE OF OCCURRENCE: Occurs in a metamorphosed limestone rich in calcic plagioclase, associated with corundum, spinel, and thorianite, and also in an alluvial deposit, at Elsiva, Fort Dauphin region, Madagascar.
  BEST REF. IN ENGLISH: Curien, Hubert, Guillemin, Claude, Orcel, Jean, and Sternberg, Micheline, *Am. Min.*, **42**: 119 (1957).

# HIBSCHITE = Hydrogrossular

# HIDALGOITE
$PbAl_3 (AsO_4)(SO_4)(OH)_6$

CRYSTAL SYSTEM:  Hexagonal
Z:  3
LATTICE CONSTANTS:
  $a = 7.04$
  $c = 16.99$
3 STRONGEST DIFFRACTION LINES:
  2.981 (100)
  5.73 ( 90)
  3.51 ( 90)
OPTICAL CONSTANT:
  $N = 1.713$  (Na)
  HARDNESS: 4½
  DENSITY: 3.96 (Meas.)
           4.27 (Calc.)
  CLEAVAGE: None.
Fracture conchoidal to irregular.  Brittle.

HABIT: Massive; dense, porcelain-like. Also porous or cavernous.

COLOR-LUSTER: White, light gray, light green. Translucent to nearly opaque. Dull, chalky.

MODE OF OCCURRENCE: Occurs in association with iron oxides filling fractures and cavities in oxidized ore at the Gold Hill mine, Tooele County, Utah. Also found in a meter-wide vein associated with limonite, hydromica, orthoclase, tourmaline, beaudantite, and a mineral tentatively referred to as ferrian hidalgoite, at the San Pascual mine, Zimapan mining district, Hidalgo, Mexico; and at a deposit at Cap Garrone, France.

BEST REF. IN ENGLISH: Smith, Robert, L., Simons, Frank S., and Vlisidis, Angelina C., *Am. Min.*, 38: 1218–1224 (1953). Clarkson, John F., Roberts, Willard L., and Lingard, Amos L., *Min. Rec.*, 2: 212–213 (1971).

# HIDDENITE = Emerald-green variety of spodumene containing Cr

# HIERATITE
$K_2SiF_6$

CRYSTAL SYSTEM: Cubic
CLASS: $4/m \bar{3} 2/m$
SPACE GROUP: Fm3m
Z: 4
LATTICE CONSTANT:
$a = 8.184$
3 STRONGEST DIFFRACTION LINES:
4.699 (100)
2.349 ( 70)
2.877 ( 65)
OPTICAL CONSTANT:
$N = 1.340$ (Na)
HARDNESS: ~2½
DENSITY: 2.665 (Meas. synthetic)
2.668 (Calc.)
CLEAVAGE: {111} perfect
HABIT: As stalactitic concretions. Dense to spongy in texture. Synthetic crystals cubo-octahedral and octahedral.

COLOR-LUSTER: Colorless, white, gray. Transparent. Vitreous.

MODE OF OCCURRENCE: Occurs associated with avogadrite and malladrite as a fumarolic deposit on Mt. Vesuvius, Italy, and with sulfur, realgar, sassolite, glauberite, mirabilite, and various alums on Vulcano Island.

BEST REF. IN ENGLISH: Palache, et al., "Dana's System of Mineralogy," 7th Ed., v. II, p. 103–104, New York, Wiley, 1951.

# HIGH-CHALCOCITE = High-temperature form of chalcocite, stable above 105°C.
$Cu_2S$

# HILGARDITE
$Ca_2B_5O_8(OH)_2Cl$
Dimorphous with parahilgardite

CRYSTAL SYSTEM: Monoclinic
CLASS: m
SPACE GROUP: Cc
Z: 4
LATTICE CONSTANTS:
$a = 6.31$
$b = 11.33$
$c = 11.44$
$\beta = 90°$
3 STRONGEST DIFFRACTION LINES:
2.85 (100)
2.11 ( 80)
1.99 ( 80)
OPTICAL CONSTANTS:
$\alpha = 1.630$
$\beta = 1.636$ (Na)
$\gamma = 1.664$
(+)2V = 35°
HARDNESS: 5
DENSITY: 2.71 (Meas.)
2.694 (Calc.)
CLEAVAGE: {010} perfect
{100} perfect, difficult
HABIT: Crystals tabular with pronounced hemimorphic aspect.

COLOR-LUSTER: Colorless, transparent. Vitreous.

MODE OF OCCURRENCE: Occurs in the water-insoluble residue of rock salt from the Choctaw salt dome, Iberville Parish, Louisiana.

BEST REF. IN ENGLISH: Palache, et al., "Dana's System of Mineralogy," 7th Ed., v. II, p. 382–383, New York, Wiley, 1951.

# HILLEBRANDITE
$Ca_2SiO_4 \cdot H_2O$

CRYSTAL SYSTEM: Monoclinic
CLASS: 2/m
SPACE GROUP: $P2_1/a$
Z: 12
LATTICE CONSTANTS:
$a = 16.60$
$b = 7.26$
$c = 11.85$
$\beta = 90°$
3 STRONGEST DIFFRACTION LINES:
2.92 (100)
4.76 ( 90)
3.33 ( 90)
OPTICAL CONSTANTS:
$\alpha = 1.605$
$\beta \simeq 1.61$
$\gamma = 1.612$
2V ≃ 60°
HARDNESS: 5½
DENSITY: 2.66 (Meas.)
2.626 (Calc.)
CLEAVAGE: Prismatic.
HABIT: Crystals fibrous, radiating.

COLOR-LUSTER: Porcelain white or greenish; colorless and transparent in section.

## HINSDALITE

$(Pb,Sr)Al_3PO_4SO_4(OH)_6$

CRYSTAL SYSTEM: Hexagonal
CLASS: $\bar{3}\,2/m$
SPACE GROUP: $R\bar{3}m$
Z: 3
LATTICE CONSTANTS:
  $a = 6.99$
  $c = 16.8$
3 STRONGEST DIFFRACTION LINES:
  2.78 (100)
  2.96 ( 80)
  5.59 ( 65)
OPTICAL CONSTANTS:
  $\omega = 1.671$
  $\epsilon = 1.689$
   (+)
HARDNESS: 4½
DENSITY: 3.65 (Meas.)
       4.072 (Calc.)
CLEAVAGE: {0001} perfect
HABIT: Crystals pseudocubic, rhombohedral, or tabular, sometimes with rough or dull faces. Also massive, granular.
COLOR-LUSTER: Nearly colorless, with yellowish or greenish tint. Transparent to translucent. Vitreous to greasy.
MODE OF OCCURRENCE: Occurs as excellent microcrystals implanted on enargite or covellite crystals at Butte, Montana; and abundantly in a vein cutting volcanic rocks in the Golden Fleece mine, near Lake City, Hinsdale County, Colorado.
BEST REF. IN ENGLISH: Palache, et al., "Dana's System of Mineralogy," 7th Ed., v. II, p. 1004, New York, Wiley, 1951.

## HIORTDAHLITE

$(Ca, Na)_{13}Zr_3Si_9(O, OH, F)_{33}$

CRYSTAL SYSTEM: Triclinic
LATTICE CONSTANTS:
  $a = 10.93$    $\alpha = 90°29'$
  $b = 10.31$    $\beta = 108°50'$
  $c = 7.33$     $\gamma = 90°08'$
3 STRONGEST DIFFRACTION LINES:
  2.979 (100)
  2.847 (100)
  3.25 ( 65)
OPTICAL CONSTANTS:
  $\alpha = 1.652$
  $\beta = 1.658$
  $\gamma = 1.665$
  (+)2V = 90°

MODE OF OCCURRENCE: Occurs in association with wollastonite, garnet, and carbonates in a contact zone between limestone and diorite in the Velardena mining district, Durango, Mexico.
BEST REF. IN ENGLISH: Winchell and Winchell, "Elements of Optical Mineralogy," 4th Ed., Pt. 2, p. 506, New York, Wiley, 1951.

HARDNESS: 5½
DENSITY: 3.267 (Meas.)
CLEAVAGE: {110} distinct
         {1$\bar{1}$0} distinct
Very brittle.
HABIT: Crystals usually tabular, flattened {100}; also pseudotetragonal showing equal development of $b$ {010} and $a$ {101}. Twinning on {100}, polysynthetic.
COLOR-LUSTER: Light yellow, honey yellow, yellowish brown. Translucent to transparent. Vitreous; greasy on fracture.
MODE OF OCCURRENCE: Occurs in alkali rocks and pegmatites in association with feldspars, nepheline, aegirine, meliphane, biotite, astrophyllite, sphene, and other minerals. Found in the Langesundfjord district, Norway; at Mte. Somma, Italy; on the Iles de Los, Guinea; and from the Korgeredab massif, Tuva, USSR.
BEST REF. IN ENGLISH: Vlasov, K. A., "Mineralogy of Rare Elements," v. II, p. 379–381, Israel Program for Scientific Translations, 1966.

## HISINGERITE (Inadequately described mineral)

$Fe_2^{3+}Si_2O_5(OH)_4 \cdot 2H_2O$

CRYSTAL SYSTEM: Hexagonal (?)
3 STRONGEST DIFFRACTION LINES:
  4.53 (100)
  2.58 (100)
  1.53 ( 70)
OPTICAL CONSTANTS:
  $N \sim 1.44$–1.73
  $\alpha = 1.715$
  $\gamma = 1.730$
(−)2V = 0° to small
HARDNESS: 2.5–3
DENSITY: 2.50–3.0
CLEAVAGE: None. Fracture conchoidal. Brittle.
HABIT: Massive; compact.
COLOR-LUSTER: Black to brownish black; resinous. Streak yellowish brown.
MODE OF OCCURRENCE: Occurs throughout the world in rocks of various ages and in many geological settings. Found as a common alteration product in the iron-rich gabbroic rocks of the Beaver Bay complex, and in veinlets up to 5 mm in width in the Bewabik formation in the Morris mine near Hibbing, in northern Minnesota; near Blaine, Idaho; associated with chalcopyrite and pyrite in the Wilcox mine, Parry Sound, Ontario; at Montauban mines, Quebec; in the Nicholson mine, Saskatchewan; also at Riddarhyttan and Långban, Sweden; Wheal Jane, Cornwall, England; and at the Kawayma mine, Japan.
BEST REF. IN ENGLISH: Whelan, J. A., and Goldich, S. S., *Am. Min.*, **46**: 1412–1423 (1961).

## HJELMITE = Pyrochlore plus tapiolite (?)

## HOCARTITE

$Ag_2FeSnS_4$

CRYSTAL SYSTEM: Tetragonal
Z: 2
LATTICE CONSTANTS:
 $a = 5.74$
 $c = 10.96$
3 STRONGEST DIFFRACTION LINES:
 3.26 (100)
 1.98 ( 80)
 1.72 ( 70)
HARDNESS: ~4
HABIT: Grains less than 1 mm in diameter.
COLOR-LUSTER: Brownish gray in polished section.
MODE OF OCCURRENCE: Occurs as inclusions in sphalerite and wurtzite and in oriented intergrowths with stannite at the tin mines of Tacama, Hocaya, and Chocaya, Bolivia, and at Fournial, Cantal, France.
BEST REF. IN ENGLISH: Caye, R., Laurent, Y., Picot, P., Pierrot, R., and Levy, C., *Am. Min.*, **54**: 573 (1969).

## HODGKINSONITE
$MnZn_2(OH)_2SiO_4$

CRYSTAL SYSTEM: Monoclinic
CLASS: 2/m
SPACE GROUP: $P2_1/c$
Z: 4
LATTICE CONSTANTS:
 $a = 11.76$
 $b = 5.31$
 $c = 8.17$
 $\beta = 95°28'$
3 STRONGEST DIFFRACTION LINES:
 2.864 (100)
 2.957 ( 90)
 1.547 ( 85)
OPTICAL CONSTANTS:
 $\alpha = 1.720$
 $\beta = 1.741$
 $\gamma = 1.746$
 $(-)2V = 52°$
HARDNESS: 4½-5
DENSITY: 3.91-3.99 (Meas.)
 4.08 (Calc.)
CLEAVAGE: {001} perfect
Brittle.
HABIT: Crystals varied in habit; usually acute pyramidal or stout prismatic. Also massive, granular.
COLOR-LUSTER: Bright pink to reddish brown. Transparent to translucent. Vitreous.
MODE OF OCCURRENCE: Occurs in association with willemite, franklinite, barite, calcite, tephroite, and other minerals, sometimes as crystals up to 2 cm across, at Franklin, Sussex County, New Jersey.
BEST REF. IN ENGLISH: Palache, Charles, USGS Prof. Paper 180, p. 108-111 (1935). Hardie, L. A., et al., *Am. Min.*, **49**: 415-420 (1964).

## HODRUSHITE
$Cu_8Bi_{12}S_{22}$

CRYSTAL SYSTEM: Monoclinic

CLASS: 2/m
SPACE GROUP: A2/m
LATTICE CONSTANTS:
 $a = 27.21$
 $b = 3.93$
 $c = 17.58$
 $\beta = 92°09'$
3 STRONGEST DIFFRACTION LINES:
 3.102 (100)
 3.62 ( 80)
 2.715 ( 80)
HARDNESS: Microhardness 187-213 $kg/mm^2$; average 200 $kg/mm^2$.
DENSITY: 6.35 (Meas.)
 6.451 (Calc.)
CLEAVAGE: None. Very brittle.
HABIT: As needle-shaped crystals less than 1 mm in length, irregular grains, or fine-grained aggregates. Rarely as vertically striated columnar or platy crystals up to 5 mm long.
COLOR-LUSTER: Steel gray with faint yellowish tint; tarnishes rapidly to brownish bronze. Opaque. Metallic.
MODE OF OCCURRENCE: Occurs associated with quartz, hematite, and locally with chalcopyrite and wittichenite (?) on the 150- and 250-meter mining levels of the Rosalia vein in Banská-Hodruše near Banska-Štiavnica, Czechoslovakia.
BEST REF. IN ENGLISH: Koděra, M., Kupčik, V., and Makovický, E., *Min. Mag.*, **37**: 641-648 (1970).

## HOEFERITE = Chapmanite

## HOEGBOMITE (Högbomite)
$Mg(Al,Fe,Ti)_4O_7$

CRYSTAL SYSTEM: Hexagonal (+ tetragonal)
CLASS: $\bar{3}\,2/m$ (18R), $\bar{6}m2$ (5H)
SPACE GROUP: $R\bar{3}m$ (18R), $P\bar{6}2m$, or $P\bar{6}m2$ (5H)
Z: 18 (18R), 1 (5H)
LATTICE CONSTANTS:

| (5H) | (18R) | (tetragonal) |
|---|---|---|
| $a = 5.718$ | $a = 5.738$ | $a = 8.34$ |
| $c = 23.02$ | $c = 83.36$ | $c = 7.96$ |

3 STRONGEST DIFFRACTION LINES:

| (5H) | (18R) | (tetragonal) |
|---|---|---|
| 2.43 (100) | 2.44 (100) | 2.87 (100) |
| 1.426 ( 80) | 2.87 ( 60) | 4.6 ( 70) |
| 2.49 ( 60) | 1.434 ( 60) | 4.17 ( 70) |

OPTICAL CONSTANTS:

| (5H) | (18R) | (tetragonal) |
|---|---|---|
| $\omega = 1.805$ | $\omega = 1.848$ | $\omega = 1.80 - 1.82$ |
| $\epsilon = 1.783$ | $\epsilon = 1.823$ | $\epsilon = 1.84 - 1.85$ |
| (-) | (-) | (+) |

HARDNESS: 6½
DENSITY: 3.81
CLEAVAGE: {0001} imperfect
Fracture conchoidal. Brittle.
HABIT: Crystals thin to thick tabular, rare. Usually as minute embedded grains. Twinning on {0001}, sometimes repeated.

COLOR-LUSTER: Black. Opaque. Metallic adamantine. Streak gray. Weakly magnetic.

MODE OF OCCURRENCE: Occurs in emery deposits near Peekskill, New York, and near Whittles, Pittsylvania County, Virginia. Also with corundum in magnetite ore at Rodstand, Norway, and in the Routevare district, Sweden.

BEST REF. IN ENGLISH: Palache, et al., "Dana's System of Mineralogy," 7th Ed., v. I, p. 723–724, New York, Wiley, 1944. *Am. Min.*, 37: 600–608 (1952).

# HOERNESITE
$Mg_3(AsO_4)_2 \cdot 8H_2O$

CRYSTAL SYSTEM: Monoclinic
CLASS: 2/m
SPACE GROUP: C2/m
Z: 2
LATTICE CONSTANTS:
 $a = 10.26$
 $b = 13.44$
 $c = 4.74$
 $\beta = 104.9°$
3 STRONGEST DIFFRACTION LINES:
 6.692 (100)
 3.005 ( 50)
 2.712 ( 50)
OPTICAL CONSTANTS:
 $\alpha = 1.563$
 $\beta = 1.571$
 $\gamma = 1.596$
 (+)2V = 60°
HARDNESS: 1
DENSITY: 2.57 (Meas.)
 2.57 (Calc.)
CLEAVAGE: {010} perfect
 {100} poor
Flexible.
HABIT: Crystals prismatic, elongated along *c*-axis and flattened {010}. Also columnar, foliated.
COLOR-LUSTER: White. Transparent. Pearly on cleavage.
MODE OF OCCURRENCE: Occurs in blocks of metamorphosed limestone in tuff at Fiano, near Naples, Italy; at Nagyág, Roumania; and Joachimsthal (Jackymov), Bohemia, Czechoslovakia.
BEST REF. IN ENGLISH: Palache, et al., "Dana's System of Mineralogy," 7th Ed., v. II, p. 755–756, New York, Wiley, 1951. Koritnig, S., and Süsse, P., *Am. Min.*, 52: 1588 (1967).

# HOHMANNITE
$Fe_2(OH)_2(SO_4)_2 \cdot 7H_2O$

CRYSTAL SYSTEM: Triclinic
CLASS: $\bar{1}$
SPACE GROUP: P$\bar{1}$
Z: 2
LATTICE CONSTANTS:
 $a = 9.05$     $\alpha = 90°09'$
 $b = 10.88$    $\beta = 90°35'$
 $c = 7.17$     $\gamma = 106°58'$

3 STRONGEST DIFFRACTION LINES:
 7.92 (100)
 8.69 ( 80)
 10.4 ( 60)
OPTICAL CONSTANTS:
 $\alpha = 1.559$
 $\beta = 1.643$   (Na)
 $\gamma = 1.655$
 (−)2V = 40°
HARDNESS: 3
DENSITY: 2.2 (Meas.)
 2.28 (Calc.)
CLEAVAGE: {010} perfect
 {110} distinct
 {1$\bar{1}$0} distinct
HABIT: Crystals short prismatic, often with rounded faces. Usually as granular aggregates of subhedral crystals.
COLOR-LUSTER: Brown, orangish brown, reddish brown. Transparent to translucent. Vitreous. Powder orange-yellow. Dehydrates rapidly and alters to metahohmannite.
MODE OF OCCURRENCE: Occurs as crystals in brecciated opalite at the Redington quicksilver mine, Knoxville, Napa County, California; and associated with other sulfates at Chuquicamata, Quetena, and Sierra Gorda, Chile.
BEST REF. IN ENGLISH: Palache, et al., "Dana's System of Mineralogy," 7th Ed., v. II, p. 613–614, New York, Wiley, 1951.

# HOKUTOLITE = Variety of barite containing about 18% PbO

# HOLDENITE
$(Mn, Zn)_6(AsO_4)(OH)_5O_2$

CRYSTAL SYSTEM: Orthorhombic
CLASS: 2/m 2/m 2/m
SPACE GROUP: Bmam
Z: 12
LATTICE CONSTANTS:
 $a = 11.97$ kX
 $b = 31.15$
 $c = 8.58$
3 STRONGEST DIFFRACTION LINES:
 2.99 (100)
 3.61 ( 90)
 2.47 ( 80)
OPTICAL CONSTANTS:
 $\alpha = 1.769$
 $\beta = 1.770$
 $\gamma = 1.785$
 (+)2V = 30°20'
HARDNESS: 4
DENSITY: 4.11 (Meas.)
 4.118 (Calc.)
CLEAVAGE: {010} poor
Fracture subconchoidal.
HABIT: Crystals thick tabular to equant, up to $1/3$ inch in size.
COLOR-LUSTER: Pink to yellowish red and deep red. Translucent. Vitreous.

MODE OF OCCURRENCE: Found as a single specimen at Franklin, Sussex County, New Jersey. The mineral occurs in a veinlet cutting massive franklinite ore.

BEST REF. IN ENGLISH: Palache, et al., "Dana's System of Mineralogy," 7th Ed., v. II, p. 775-777, New York, Wiley, 1951.

# HOLLANDITE

$Ba(Mn^{2+}, Mn^{4+}, Fe^{3+})_8O_{16}$
Cryptomelane Group

CRYSTAL SYSTEM: Monoclinic
CLASS: 2/m
SPACE GROUP: $P2_1/n$
Z: 2
LATTICE CONSTANTS:
  $a = 10.02$
  $b = 5.76$
  $c = 9.89$
  $\beta = 90°36'$
3 STRONGEST DIFFRACTION LINES:
  3.13 (100)
  2.40 ( 90)
  3.47 ( 80)
HARDNESS: 6
DENSITY: 4.95
CLEAVAGE: Prismatic, distinct. Brittle.
HABIT: Crystals short prismatic, terminated by a flat pyramid. Usually massive; fibrous. Sometimes concretionary.
COLOR-LUSTER: Black, grayish black to silvery gray. Opaque. Metallic to dull. Streak black.
MODE OF OCCURRENCE: Occurs as concretions in Virgin Valley, Humboldt County, Nevada; at Ultevis, Sweden; at Langenberg, Saxony, Germany; and in manganese deposits at Kajlidongri and elsewhere in India.
BEST REF. IN ENGLISH: Palache, et al., "Dana's System of Mineralogy," 7th Ed., v. I, p. 743-744, New York, Wiley, 1944.

# HOLLINGSWORTHITE (Inadequately described mineral)

(Rh, Pd)AsS

CRYSTAL SYSTEM: Unknown
  HARDNESS: >6
  HABIT: Small grains (maximum diameter about 40 $\mu$) closely intergrown with Rh-rich sperrylite, Rh-free sperrylite and geversite.
  COLOR-LUSTER: Medium grey. Opaque. Metallic.
  MODE OF OCCURRENCE: Found in platinum concentrates from the Driekop mine, Transvaal, South Africa.
  BEST REF. IN ENGLISH: Stumpfl, E. F., and Clark, A. M., Am. Min., 50: 1068-1074 (1965).

# HOLMQUISTITE

$Li_2(Mg, Fe^{2+})_3(Al, Fe^{3+})_2Si_8O_{22}(OH)_2$
Lithium analogue of anthophyllite
Amphibole Group

CRYSTAL SYSTEM: Orthorhombic
CLASS: 2/m 2/m 2/m
SPACE GROUP: Pnma
Z: 4
LATTICE CONSTANTS:
  $a = 18.30$
  $b = 17.69$
  $c = 5.30$
3 STRONGEST DIFFRACTION LINES:
  8.107 (100)
  3.000 ( 90)
  4.427 ( 70)
OPTICAL CONSTANTS:
  $\alpha = 1.622-1.642$
  $\beta = 1.642-1.660$
  $\gamma = 1.646-1.666$
  $(-)2V = 49°$
HARDNESS: 5-6
DENSITY: 3.13 (Meas.)
        3.09 (Calc.)
CLEAVAGE: {110} perfect
          {001}, {112}, {113} parting
HABIT: Crystals slender prismatic or acicular, often vertically striated, up to 10cm long. Also massive, as radiating fibrous aggregates.
COLOR-LUSTER: Dark violet (almost black) to light sky blue. Transparent to translucent. Vitreous.
MODE OF OCCURRENCE: Occurs at or near the contacts between lithium-rich pegmatites and country rocks of basic composition in association with quartz, plagioclase, biotite, tourmaline, and other minerals. Found in the wallrock at the Edison spodumene mine, Pennington County, South Dakota; at the Hiddenite mine, Alexander County, North Carolina; at Barraute, Quebec, Canada; at Utö, Södermanland, Sweden; in various regions of the USSR, and elsewhere.
BEST REF. IN ENGLISH: Vlasov, K. A., "Mineralogy of Rare Elements," v. II, p. 13-17, Israel Program for Scientific Translations, 1966.

# HOLTITE

$(Al, Sb, Ta)_7(B, Si)_4O_{18}$ (?)
Related to dumortierite

CRYSTAL SYSTEM: Orthorhombic
CLASS: 2/m 2/m 2/m
SPACE GROUP: Pmcn
Z: 4
LATTICE CONSTANTS:
  $a = 11.905$
  $b = 20.355$
  $c = 4.690$
3 STRONGEST DIFFRACTION LINES:
  10.28 (100)
  2.94 ( 40)
  5.89 ( 34)
OPTICAL CONSTANTS:
  $\alpha = 1.743-1.746$
  $\beta = 1.756-1.759$
  $\gamma = 1.758-1.761$
  $2V = 49°-55°$

HARDNESS: 8½
DENSITY: 3.90 ± 0.02 (meas.)
CLEAVAGE: {010}distinct (May be parting.)
HABIT: Crystals prismatic, pseudohexagonal; also acicular. As pebbles from 2 to 15 mm in size. Multiple twinning on {110}.
COLOR-LUSTER: Pale brownish white to brown, greenish. Vitreous to resinous or dull. Fluoresces dull orange under short-wave ultraviolet light; bright yellow under longwave.
MODE OF OCCURRENCE: Occurs in association with stibiotantalite and tantalite in an alluvial deposit near Greenbushes, Western Australia.
BEST REF. IN ENGLISH: Pryce, M. W., *Min. Mag.*, 38: 21–25 (1971).

# HOMILITE
$(Ca, Fe)_3 B_2 Si_2 O_{10}$

CRYSTAL SYSTEM: Monoclinic
CLASS: 2/m
SPACE GROUP: $P2_1/a$
Z: 2
LATTICE CONSTANTS:
$a = 9.67$
$b = 7.57$
$c = 4.74$
$\beta = 90°22'$
3 STRONGEST DIFFRACTION LINES:
3.10 (100)
2.52 (100)
2.83 ( 90)
OPTICAL CONSTANTS:
$\alpha = 1.715$
$\beta = 1.725$
$\gamma = 1.738$
$(+)2V = 80°$
HARDNESS: 5
DENSITY: 3.36–3.38 (Meas.)
3.58 (Calc.)
CLEAVAGE: None. Brittle.
HABIT: Crystals usually tabular {001}.
COLOR-LUSTER: Dark brown to black.
MODE OF OCCURRENCE: Occurs on Stokö and other islands in the Langesundfjord, Norway.
BEST REF. IN ENGLISH: Winchell and Winchell, "Elements of Optical Mineralogy," 4th Ed., Pt. 2, 356–357, New York, Wiley, 1951.

# HONESSITE (Inadequately described mineral)
Basic sulfate of Fe and Ni

HABIT: Extremely fine-grained and obscurely fibrous.
COLOR-LUSTER: Green to brown; dull.
MODE OF OCCURRENCE: Occurs as an alteration product of millerite near Linden, Iowa County, Wisconsin.
BEST REF. IN ENGLISH: Heyl, Allen, V., Milton, Charles, and Axelrod, Joseph M., *Bull Geol. Soc. Am.*, 67: 1706 (1956) (abst.).

# HOPEITE
$Zn_3 (PO_4)_2 \cdot 4H_2 O$
Dimorphous with parahopeite

CRYSTAL SYSTEM: Orthorhombic
CLASS: 2/m 2/m 2/m
SPACE GROUP: Pnma
Z: 4
LATTICE CONSTANTS:
$a = 10.66$
$b = 18.36$
$c = 5.04$
3 STRONGEST DIFFRACTION LINES:
9.04 (100)
4.57 (100)
2.857 (100)
OPTICAL CONSTANTS:
$\alpha = 1.574$
$\beta = 1.582$
$\gamma = 1.582$
$(-)2V = $ small
HARDNESS: 3¼
DENSITY: 3.05 (Meas.)
3.08 (Calc.)
CLEAVAGE: {010} perfect
{100} good
{001} poor
Fracture uneven. Brittle.
HABIT: Crystals short to long prismatic or tabular. Crystal faces commonly irregular. As isolated individuals, aggregates, or crusts. Also compact massive.
COLOR-LUSTER: Colorless, white, grayish white, pale yellow. Transparent to translucent. Vitreous, pearly on perfect cleavage. Streak white.
MODE OF OCCURRENCE: Occurs as a secondary mineral in zinc-bearing ore deposits. Found as superb crystal groups associated with tarbuttite and other secondary minerals at the Broken Hill mine, Zambia; with spencerite at the Hudson Bay mine, Salmo, British Columbia, Canada; and in small amounts at Altenberg, Belgium.
BEST REF. IN ENGLISH: Palache, et al., "Dana's System of Mineralogy," 7th Ed., v. II, p. 734–737, New York, Wiley, 1951.

# HORNBLENDE also see: Tschermakite, Pargasite, Hastingsite
$(Ca, Na, K)_{2-3} (Mg, Fe^{2+}, Fe^{3+}, Al)_5 (Si, Al)_8 O_{22} (OH)_2$
Amphibole Group

CRYSTAL SYSTEM: Monoclinic
CLASS: 2/m
SPACE GROUP: C2/m
Z: 2
LATTICE CONSTANTS:
$a = 9.88$
$b = 18.02$
$c = 5.33$
$\beta = 105°30'$
3 STRONGEST DIFFRACTION LINES:
2.70 (100)
3.09 ( 95)
3.38 ( 90)

OPTICAL CONSTANTS:
$\alpha = 1.615 - 1.705$
$\beta = 1.618 - 1.714$
$\gamma = 1.632 - 1.730$
$(-)2V \sim 65°$ (variable)
HARDNESS: 5–6
DENSITY: 3.02–3.27 (Meas.)
3.208 (Calc.)
CLEAVAGE: {110} perfect
{001} parting
{100} parting
Fracture eneven to subconchoidal. Brittle.

HABIT: Crystals long to short prismatic, often appearing nearly hexagonal in cross section and with rhombohedral-like terminations. Also massive; compact, granular, columnar, bladed or fibrous. Twinning on {100} common, simple, lamellar.

COLOR-LUSTER: Green to dark green, greenish brown, black. Translucent to nearly opaque. Vitreous to silky.

MODE OF OCCURRENCE: Occurs widespread, often as the predominant mineral, in many igneous and metamorphic rocks. Typical occurrences are found in California, Idaho, Utah, Colorado, Arizona, South Dakota, Pennsylvania, New York, and elsewhere in the United States. Also in Canada, Scotland, England, Norway, Sweden, Italy, Austria, Czechoslovakia, Finland, USSR, India, Korea, Japan, Australia, and New Zealand.

BEST REF. IN ENGLISH: Deer, Howie, and Zussman, "Rock Forming Minerals," v. 2, p. 263–314, New York, Wiley, 1963.

# HOROBETSUITE
$(Bi, Sb)_2 S_3$

CRYSTAL SYSTEM: Orthorhombic
LATTICE CONSTANTS:
$a = 11.24$
$b = 11.28$
$c = 3.90$
HARDNESS: $\sim 2$
DENSITY: 5.449 (Meas.)
HABIT: Prismatic crystals striated parallel to prism zone; 1.0–7.0 mm in length, and 0.2–2.0 mm in diameter.

COLOR-LUSTER: Lead gray. Opaque. Metallic. Streak black.

MODE OF OCCURRENCE: Occurs with free sulfur and iron sulfide minerals in the sulfur deposit at the Horobetsu mine, Hokkaido, Japan.

BEST REF. IN ENGLISH: Hayase, Kitaro, *Am. Min.*, 43: 623–624 (1958).

# HORSFORDITE (Inadequately described mineral)
$Cu_5 Sb$

CRYSTAL SYSTEM: Unknown
HARDNESS: 4–5
DENSITY: 8.812 (Meas.)
CLEAVAGE: Fracture uneven. Brittle.
HABIT: Massive.

COLOR-LUSTER: Silver white; tarnishes rapidly. Opaque. Metallic, brilliant.

MODE OF OCCURRENCE: Occurs abundantly near Mytilene in the eastern part of Asia Minor.

BEST REF. IN ENGLISH: Palache, et al., "Dana's System of Mineralogy," 7th Ed., v. I, p. 173, New York, Wiley, 1944.

# HORTONOLITE
$(Fe, Mg)_2 SiO_4$ (50–70 atomic percent $Fe^{2+}$)
Olivine Group

CRYSTAL SYSTEM: Orthorhombic
CLASS: 2/m 2/m 2/m
SPACE GROUP: Pbnm
Z: 4
LATTICE CONSTANTS:
$a = 4.799$
$b = 10.393$
$c = 6.063$
3 STRONGEST DIFFRACTION LINES:
2.810 (100)
2.489 ( 70)
2.549 ( 60)
OPTICAL CONSTANTS:
$\alpha = 1.752$
$\beta = 1.781$
$\gamma = 1.795$
$(-)2V = 65°$
HARDNESS: 7
DENSITY: 3.88 (Meas.)
3.908 (Calc.)
CLEAVAGE: {010} imperfect
{100} imperfect
Fracture conchoidal. Brittle.

HABIT: Crystals thick tabular, rough. Usually massive, compact or granular. Twinning on {100}, not common.

COLOR-LUSTER: Light green to brown. Transparent to translucent. Vitreous to greasy. Streak uncolored.

MODE OF OCCURRENCE: Occurs chiefly in diabase, gabbro, and related rocks. Typical occurrences are found at Monroe, Orange County, New York; at the Iron Hill mine, Cumberland, Rhode Island; and at localities in Greenland, Scotland, Czechoslovakia, Finland, Uganda, and Transvaal, South Africa.

BEST REF. IN ENGLISH: Deer, Howie, and Zussman, "Rock Forming Minerals," v. I, p. 1–33, New York, Wiley, 1962.

# HOWIEITE
$Na(Fe, Mn)_{10} (Fe, Al)_2 Si_{12} O_{31} (OH)_{13}$

CRYSTAL SYSTEM: Triclinic
Z: 1
LATTICE CONSTANTS:
$a = 10.17$   $\alpha = 91.3°$
$b = 9.72$   $\beta = 70.7°$
$c = 9.56$   $\gamma = 109.0°$

3 STRONGEST DIFFRACTION LINES:
  9.18 (100)
  7.91 ( 80)
  3.25 ( 65)
OPTICAL CONSTANTS:
  $\alpha$ = 1.701
  $\beta$ = 1.720
  $\gamma$ = 1.734
  (-)2V = 65°
DENSITY: 3.378 (Meas.)
CLEAVAGE: {010} good
           {100} fair
           {2$\overline{1}$0} weak
HABIT: Bladed crystals.
COLOR-LUSTER: Dark green to black.
MODE OF OCCURRENCE: Occurs in some of the metamorphosed shales, siliceous ironstones, and impure limestones of the Franciscan formation in the Laytonville district, Mendocino County, California.
BEST REF. IN ENGLISH: Agrell, S. O., Bown, M. G., and McKie, D., *Am. Min.*, **50**: 278 (1965).


# HOWLITE
$Ca_2 B_5 SiO_9 (OH)_5$

CRYSTAL SYSTEM: Monoclinic
CLASS: 2/m
SPACE GROUP: $P2_1/c$
Z: 2
LATTICE CONSTANTS:
  $a$ = 12.78-12.93
  $b$ = 9.33
  $c$ = 8.60
  $\beta$ = 104.83°
3 STRONGEST DIFFRACTION LINES:
  4.14 (100)
  6.24 ( 90)
  3.10 ( 75)
OPTICAL CONSTANTS:
  $\alpha$ = 1.583-1.586
  $\beta$ = 1.596-1.598
  $\gamma$ = 1.605
  (-)2V = large
HARDNESS: 3½; often less.
DENSITY: 2.45 (Meas.)
         2.432 (Calc.)
CLEAVAGE: Fracture of porcelaneous types nearly even and smooth.
HABIT: Crystals tabular, up to 1.0 mm across, usually attached by one end of symmetry axis. {100} dominant, {001} and {011} well developed. {104} or {102} sometimes present. Also as compact nodular masses, chalklike, earthy, scaly. Porcelaneous.
COLOR-LUSTER: White; subvitreous.
MODE OF OCCURRENCE: Occurs as microscopic crystals encrusting massive howlite at the Old Sterling Borax mine, Tick Canyon, California. Also found in California as nodules up to several hundred pounds in weight in the colemanite deposit at Lang, Los Angeles County; in veins in Gower Gulch, Inyo County; and with ulexite and bakerite in the Mohave desert near Daggett, San Bernardino County.

Also found as small nodules associated with ulexite near Windsor, Hants County, and at Wentworth, Nova Scotia.
BEST REF. IN ENGLISH: Palache, et al., "Dana's System of Mineralogy," 7th Ed., v. II, p. 362-363, New York, Wiley, 1951. Murdoch, Joseph, *Am. Min.*, **42**: 521-524 (1957). *Am. Min.*, **55**: 716-728 (1970).


# HSIANGHUALITE
$Ca_3 Li_2 Be_3 (SiO_4)_3 F_2$

CRYSTAL SYSTEM: Cubic
CLASS: 432
SPACE GROUP: $I4_1 32$
Z: 8
LATTICE CONSTANT:
  $a$ = 12.897
3 STRONGEST DIFFRACTION LINES:
  2.746 (100)
  2.209 (100)
  2.090 ( 90)
OPTICAL CONSTANT:
  $N$ = 1.6132
HARDNESS: 6½
DENSITY: 2.97-3.00 (Meas.)
CLEAVAGE: None.
HABIT: Trisoctahedral or dodecahedral crystals and granular masses.
COLOR-LUSTER: White, transparent to translucent; vitreous.
MODE OF OCCURRENCE: Occurs in phlogopite veins in a light-colored band of green and white banded metamorphosed Devonian limestone in Hunan Province, China.
BEST REF. IN ENGLISH: Huang, Wen Hui; Tu, Shao-Hua; Wang, K'ung-Hai; Chao, Chun-Lin; and Yu, Cheng-Chih, *Am. Min.*, **44**: 1327-1328 (1959). Beus, A. A., *Am. Min.*, **46**: 244 (1961).


# HUANGHOITE
$BaCe(CO_3)_2 F$

CRYSTAL SYSTEM: Hexagonal
Z: 3
LATTICE CONSTANTS:
  $a$ = 5.1
  $c$ = 19.6
3 STRONGEST DIFFRACTION LINES:
  3.21  (100)
  1.937 (100)
  2.01  ( 90)
OPTICAL CONSTANTS:
  $\omega$ = 1.765
  $\epsilon$ = 1.603
    (-)
HARDNESS: ~ 4.7
DENSITY: 4.67-4.51 (Meas.)
         4.49 (Calc.)
CLEAVAGE: {0001} distinct
Fracture irregular.
HABIT: Platy masses up to 10 × 5 × 1 cm.
COLOR-LUSTER: Honey yellow to yellowish green. Translucent; greasy.

MODE OF OCCURRENCE: Occurs widely distributed in hydrothermal deposits, including calcite veins, genetically associated with alkalic grano-syenites and the enclosing hydrothermally altered dolomites, near the Huang Ho River, China. Associated minerals are aegirine, fluorite, magnetite, hematite, monazite, bastnaesite, parisite, and aeschynite.

BEST REF. IN ENGLISH: Semenov, E. I., and Chang, P'ei-Shan, *Am. Min.*, **48**: 1179, (1963).

# HUEBNERITE
$MnWO_4$
Wolframite Series

CRYSTAL SYSTEM: Monoclinic
CLASS: 2/m
SPACE GROUP: P2/c
Z: 2
LATTICE CONSTANTS:
$a$ = 4.82 kX
$b$ = 5.76
$c$ = 4.97
$\beta$ = 90°53'
3 STRONGEST DIFFRACTION LINES:
2.996 (100)
2.954 ( 95)
4.84 ( 65)
OPTICAL CONSTANTS:
$\alpha$ = 2.17-2.20
$\beta$ = 2.22
$\gamma$ = 2.30-2.32
(+)2V = large (~ 73° Calc.)
HARDNESS: 4-4½
DENSITY: 7.18 (Meas.)
7.234 (Calc.)
CLEAVAGE: {010} perfect
Fracture uneven. Brittle.
HABIT: Crystals usually short to long prismatic, flattened or tabular {100}, and striated or furrowed in direction of elongation. Commonly as groups of parallel or subparallel crystals, or as radiating groups of crystals. Twinning on {100} common.
COLOR-LUSTER: Yellowish brown to reddish brown, rarely brownish black; sometimes tarnished iridescent. Transparent to translucent. Submetallic to resinous. Streak yellow to reddish brown or greenish gray.
MODE OF OCCURRENCE: Occurs in high-temperature hydrothermal ore veins and quartz veins in or near granitic rocks, in medium-temperature veins, and in small amounts in epithermal veins. Found at many localities in the western United States, especially as superb crystals with rhodochrosite and quartz crystals at the Sweet Home mine, Alma, Park County, and as beautiful sheaf-like clusters of crystals as much as 3 inches long near Silverton, San Juan County, Colorado. It also occurs in Idaho, Nevada, New Mexico, Arizona, and South Dakota. Among other localities, it is found in France, Czechoslovakia, Australia, and as excellent crystals associated with quartz crystals at Pasto Bueno, Peru.

BEST REF. IN ENGLISH: Palache, et al., "Dana's System of Mineralogy," 7th Ed., v. II, p. 1064-1072, New York, Wiley, 1951.

# HUEMULITE
$Na_4MgV_{10}O_{28} \cdot 24H_2O$

CRYSTAL SYSTEM: Triclinic
CLASS: 1 or $\bar{1}$
SPACE GROUP: P1 or P$\bar{1}$
Z: 1
LATTICE CONSTANTS:
$a$ = 11.770 $\quad$ $\alpha$ = 107°13'
$b$ = 11.838 $\quad$ $\beta$ = 112°10'
$c$ = 9.018 $\quad$ $\gamma$ = 101°30'
3 STRONGEST DIFFRACTION LINES:
7.62 (100)
10.59 ( 90)
9.12 ( 60)
OPTICAL CONSTANTS:
$\alpha$ = 1.679
$\beta$ = 1.734 (Na)
$\gamma$ = 1.742
(-)2V = 25°-30°
HARDNESS: 2.5-3 (Recrystallized)
DENSITY: 2.39 (Meas.)
2.404 (Calc.)
CLEAVAGE: {001} perfect
{010} good
HABIT: Aggregates of microscopic fibers; rounded masses forming thin films. Recrystallized and synthetic material is tabular, elongated parallel to the $c$-axis.
COLOR-LUSTER: Yellowish orange to orange; dull. Synthetic and recrystallized material yellowish orange to reddish orange; vitreous to subadamantine.
MODE OF OCCURRENCE: Found as botryoidal masses, thin films, or interstitial filling in sandstone in several uraniferous ore-bodies of the "sandstone-type deposits," in the southwestern part of Mendoza province, Argentina.

BEST REF. IN ENGLISH: Gordillo, C. E., Linares, E., Toubes, R. O., and Winchell, H., *Am. Min.*, **51**: 1-13 (1966).

# HUGELITE (Inadequately described mineral)
$Pb_2(UO_2)_3(AsO_4)_2(OH)_4 \cdot 3H_2O$

CRYSTAL SYSTEM: Monoclinic
OPTICAL CONSTANTS:
biaxial (+) to uniaxial (+)
$\alpha$ = 1.898
$\gamma$ = 1.915
2V varies greatly with wavelength and decreases to 0° (580-600 m$\mu$) and becomes uniaxial.
HARDNESS: Not determined.
DENSITY: ~ 5.1
CLEAVAGE: {100} good
HABIT: Crystals tabular on (100), elongated along $c$-axis, up to 2-3 mm long. Striated parallel to $c$-axis. Forms: m {110}, n{011}, c {001}, and {100}.
COLOR-LUSTER: Brown to orange-yellow; transparent to translucent. Greasy to adamantine.
MODE OF OCCURRENCE: Occurs rarely, filling cavities in hornstone breccia, associated with cerussite, mimetite, hallimondite, and zeunerite, probably from the Michael vein, Reichenbach, near Lahr, Baden, Germany.

BEST REF. IN ENGLISH: Walenta, Kurt, and Wimmenauer, Wolfhard, *Am. Min.*, **47**: 418–419 (1962).

# HÜHNERKOBELITE
$(Na_2, Ca)(Fe^{2+}, Mn^{2+})_2 (PO_4)_2$

CRYSTAL SYSTEM: Monoclinic
CLASS: 2/m
SPACE GROUP: $I2_1/a$
Z: 1
LATTICE CONSTANTS:
  $a = 10.89$
  $b = 12.54$
  $c = 6.46$
  $\beta = 97°33'$
3 STRONGEST DIFFRACTION LINES:
  2.703 (100)
  6.24 ( 70)
  3.085 ( 55)
OPTICAL CONSTANTS:
  $\alpha = 1.718$–$1.727$
  $\gamma = 1.731$–$1.738$
  (+)
HARDNESS: 5
DENSITY: 3.5–3.6 (Meas.)
CLEAVAGE: {001} good
         {010} distinct
HABIT: Crystals short prismatic, elongated along [010] or [001] with well-developed prism faces and w{211}, and poor f{132} predominating. Also as granular cleavable masses.
COLOR-LUSTER: Greenish black, dull olive green. Translucent. Vitreous.
MODE OF OCCURRENCE: The mineral occurs as crystals up to 3.0 mm across, associated with dickinsonite and numerous other phosphates, at the Palermo No. 1 pegmatite, North Groton, New Hampshire. Also found at Norrö on the island of Ranö, Sweden.
BEST REF. IN ENGLISH: Palache, et al., "Dana's System of Mineralogy," 7th Ed., v. II, p. 669–670, New York, Wiley, 1951.   Moore, Paul B., *Am. Min.*, **50**: 713–717 (1965).

# HULSITE (Paigeite)
$(Fe^{2+}, Mg^{2+}, Fe^{3+}, Sn^{4+})_3 BO_3 O_2$

CRYSTAL SYSTEM: Monoclinic
CLASS: 2 or 2/m
SPACE GROUP: P2 or P2/m
Z: 2
LATTICE CONSTANTS:
  $a = 10.684$
  $b = 3.099$
  $c = 5.438$
  $\beta = 94°09'$
3 STRONGEST DIFFRACTION LINES:
  2.664 (100)
  5.35 ( 70)
  2.585 ( 70)
HARDNESS: ~ 3

DENSITY: 4.5–4.62 (Meas.)
CLEAVAGE: {110} good
HABIT: Crystals small, rectangular, with uneven faces. Also as tabular masses.
COLOR-LUSTER: Black. Submetallic to vitreous.
MODE OF OCCURRENCE: Occurs in metamorphosed limestone at Brooks Mountain, Seward Peninsula, Alaska, associated with diopside, vesuvianite, magnetite, and garnet.
BEST REF. IN ENGLISH: Knopf and Schaller, *Am. J. Sci.*, **25**: 323 (1908).   Clark, Joan R., *Am. Min.*, **50**: 249–254 (1965).

# HUMBERSTONITE (Chile-Loweite)
$Na_7 K_3 Mg_2 (SO_4)_6 (NO_3)_2 \cdot 6H_2O$

CRYSTAL SYSTEM: Hexagonal
CLASS: $\bar{3}$
SPACE GROUP: $R\bar{3}$
Z: 3 (hexagonal), 1 (rhombohedral)
LATTICE CONSTANTS:
  (hexagonal cell)
  $a = 10.900$
  $c = 24.410$
3 STRONGEST DIFFRACTION LINES:
  3.393 (100)
  2.724 ( 70)
  8.137 ( 60)
OPTICAL CONSTANTS:
  $\omega = 1.474$
  $\epsilon = 1.436$
  (–)
HARDNESS: ~ 2½
DENSITY: 2.252 (Meas.)
        2.252 (Calc.)
CLEAVAGE: {0001} perfect
Fracture irregular. Brittle.
HABIT: Aggregates of thin, hexagonal-shaped crystals up to 0.2 mm in diameter. Crystals flattened [0001]; observed forms c {0001} and r {10$\bar{1}$1}.
COLOR-LUSTER: Colorless, transparent; vitreous.
MODE OF OCCURRENCE: Occurs associated with astrakhanite (bloedite) and soda-niter at several places in the nitrate fields in the Atacama Desert, Chile, especially in the Taltal district, being thickest (as much as 40 cm) and most widespread in the vicinity of Oficina Alemania.
BEST REF. IN ENGLISH: Kühn, Robert, and von Hodenberg, Renate, *Am. Min.*, **55**: 1072 (1970).   Ericksen, G. E., Fahey, J. J., and Mrose, M. E., *Am. Min.*, **53**: 507 (1968). Mrose, Mary E., Fahey, J. J., and Ericksen, G. E., *Am. Min.*, **55**: 1518–1533 (1970).

# HUMBOLDTINE
$FeC_2O_4 \cdot 2H_2O$

CRYSTAL SYSTEM: Orthorhombic
CLASS: mm 2
SPACE GROUP: Fmm 2
Z: 8

LATTICE CONSTANTS:
$a = 11.98$
$b = 15.46$ (synthetic)
$c = 5.56$
3 STRONGEST DIFFRACTION LINES:
4.75 (100)
2.98 ( 35)
2.61 ( 35)
OPTICAL CONSTANTS:
$\alpha \sim 1.494$
$\beta \sim 1.561$
$\gamma \sim 1.692$
$(+)2V$ = large
HARDNESS: 1½–2
DENSITY: 2.33 (Meas.)
2.32 (Calc.)
CLEAVAGE: {110} perfect
{100} imperfect
{010} imperfect
HABIT: As elongated prismatic or platy crystals; rare. Usually capillary or as crusts with fibrous structure; also fine-granular to earthy and compact.

COLOR-LUSTER: Yellow to brownish yellow. Transparent to translucent. Resinous to dull.

MODE OF OCCURRENCE: Occurs at Kettle Point, Bosanquet township, Ontario, Canada; and in coal deposits at localities in Germany and Czechoslovakia.

BEST REF. IN ENGLISH: Palache, et al., "Dana's System of Mineralogy," 7th Ed., v. II, p. 1102–1103, New York, Wiley, 1951.

# HUMITE
$Mg_7Si_3O_{12}(F, OH)_2$
Humite Group

CRYSTAL SYSTEM: Orthorhombic
CLASS: 2/m 2/m 2/m
SPACE GROUP: Pbnm
Z: 4
LATTICE CONSTANTS:
$a = 20.86$
$b = 4.74$
$c = 10.23$
3 STRONGEST DIFFRACTION LINES:
2.256 (100)
2.438 ( 70)
1.479 ( 70)
OPTICAL CONSTANTS:
$\alpha = 1.607–1.643$
$\beta = 1.619–1.653$
$\gamma = 1.639–1.675$
$(+)2V = 65°–84°$
HARDNESS: 6
DENSITY: 3.20–3.32 (Meas.)
3.201 (Calc.)
CLEAVAGE: {100} indistinct
Fracture uneven to subconchoidal. Brittle.
HABIT: Crystals small, varied in habit, usually highly modified.

COLOR-LUSTER: White, yellow, dark orange, brown. Transparent to translucent. Vitreous.

MODE OF OCCURRENCE: Occurs chiefly in contact zones in limestone or dolomite, also in veins. Typical occurrences are found at the Tilly Foster iron mine at Brewster, Putnam County, New York; in the Llanos de Juanar, Malaga, Spain; at Mte. Somma, Italy; near Filipstad, Vermland, Sweden; and at several places in Finland.

BEST REF. IN ENGLISH: Deer, Howie, and Zussman, "Rock Forming Minerals," v. 1, p. 47–58, New York, Wiley, 1967. Jones, Norris W., Ribbe, P. H., and Gibbs, G. V., *Am. Min.*, **54**: 391–411 (1969).

# HUMMERITE
$KMgV_5O_{14} \cdot 8H_2O$

CRYSTAL SYSTEM: Triclinic
CLASS: $\bar{1}$
SPACE GROUP: $P\bar{1}$
Z: 2
LATTICE CONSTANTS:
$a = 10.81$ $\alpha = 106°04'$
$b = 11.01$ $\beta = 107°49'$
$c = 8.85$ $\gamma = 65°40'$
3 STRONGEST DIFFRACTION LINES:
8.2 (100)
7.4 ( 70)
2.73 ( 60)
OPTICAL CONSTANTS: Mean index = 1.81. $(-)2V$
HARDNESS: Not determined.
DENSITY: 2.53 (Calc.)
CLEAVAGE: Not determined.
HABIT: Crystals elongated parallel to $c$-axis or tabular parallel to the $b$- or $c$-axes. As thin veinlets and crusts.

COLOR-LUSTER: Bright orange, yellow. Translucent.

MODE OF OCCURRENCE: Occurs in gray clay at the Hummer mine, and as an efflorescence on sandstone at the North Star mine, in Paradox Valley, Montrose County, Colorado; also as an efflorescence on mine walls in the Edgemont mining area, Fall River County, South Dakota.

BEST REF. IN ENGLISH: Weeks, Alice Dowse, Cisney, Evelyn A., and Sherwood, Alexander M., *Am. Min.*, **36**: 326–327 (1951).

# HUNGCHAOITE (Inadequately described mineral)
$MgB_4O_7 \cdot 9H_2O$

CRYSTAL SYSTEM: Monoclinic, pseudohexagonal
3 STRONGEST DIFFRACTION LINES:
4.07 (100)
2.902 (100)
2.734 ( 90)
OPTICAL CONSTANTS:
$\alpha = 1.442$
$\beta = 1.485$
$\gamma = 1.490$
$(-)2V = 36°$
COLOR-LUSTER: White.
MODE OF OCCURRENCE: Occurs in ulexite nodules in "one of the borate deposits of lacustrine origin of China."

BEST REF. IN ENGLISH: I-Hua, Chün; Hsien-Te, Hsieh; Tze-Chang, Chien; and Lai-Pao, Liu, *Am. Min.*, **50**: 262 (1965).

# HUNTITE
$CaMg_3(CO_3)_4$

CRYSTAL SYSTEM: Hexagonal
CLASS: 32
SPACE GROUP: R32
Z: 3 (hex.)
 1 (rhomb.)
LATTICE CONSTANTS:
 $a = 9.505$ $a_{rh} = 6.075$
 $c = 7.821$ $\alpha = 102°56'$
3 STRONGEST DIFFRACTION LINES:
 2.83 (100)
 1.973 ( 45)
 2.89 ( 43)
HARDNESS: Soft; smooth feel; adheres to tongue.
DENSITY: 2.696 (Meas.)
 2.70 (Calc.)
CLEAVAGE: Subconchoidal fracture (masses). Brittle.
HABIT: Compact, chalk-like masses; exceedingly fine-grained and fibrous; diameter of fibers generally less than 0.002 mm.
COLOR-LUSTER: White; earthy.
MODE OF OCCURRENCE: Occurs as a cave carbonate in areas where the country rocks are magnesium-rich; as a near-surface weathering product of magnesium-rich rocks such as brucite, marbles, dolomites, serpentinites, or magnesites; and as a diagenetic mineral in Recent sediment sequences. Found at numerous localities including Currant Creek and Gabbs, Nevada; Carlsbad Caverns, New Mexico; Crestmore, California; in la Grotte de la Clamouse in the Herault Region, France; Dorog, Hungary; Kurgashinkin, Uzbek SSR; Tea Tree Gully, Australia; and along the Trucial Coast, Persian Gulf. Aggregates weighing several kilograms occur in serpentinites in the vicinity of Yazd, Iran.
BEST REF. IN ENGLISH: Faust, George T., *Am. Min.*, **38**: 4–24 (1953).

# HUREAULITE
$H_2(Mn, Fe)_5(PO_4)_4 \cdot 4H_2O$

CRYSTAL SYSTEM: Monoclinic
CLASS: 2/m
SPACE GROUP: C2/c
Z: 4
LATTICE CONSTANTS:
 $a = 17.42$
 $b = 9.12$
 $c = 9.50$
 $\beta = 96°40'$
3 STRONGEST DIFFRACTION LINES:
 3.14 (100)
 8.01 ( 70)
 2.983 ( 70)
OPTICAL CONSTANTS:
 $\alpha = 1.637-1.652$
 $\beta = 1.645-1.658$
 $\gamma = 1.649-1.663$
 $(-)2V = 75°$
HARDNESS: 3½

DENSITY: 3.191 (Meas.)
 3.23 (Calc.)
CLEAVAGE: {100} good
Fracture uneven. Brittle.
HABIT: Crystals short to long prismatic, up to 3.0 cm long; also tabular or equant. Crystals isolated or grouped. Also massive, compact, scaly, or imperfectly fibrous.
COLOR-LUSTER: Pale rose, violet-rose, red, yellowish, orange, orange-red, brownish orange, yellowish to reddish brown, gray, colorless. Transparent to translucent. Bright vitreous to greasy. Streak nearly white.
MODE OF OCCURRENCE: Occurs as a secondary mineral in granite pegmatites associated with triphylite, lithiophilite, graftonite, and a wide variety of secondary phosphate minerals. Found widespread in the pegmatite districts of the Black Hills, South Dakota, especially at the Bull Moose and Tip Top mines in the Custer area, and at the Big Chief and Hesnard mines near Keystone. It also occurs at Branchville and at Portland, Connecticut; at the Palermo mine, North Groton, New Hampshire; and at Pala, San Diego County, California. Other localities include Huréaux and La Vilate, Haute Vienne, France; Mangualde, Portugal; Hagendorf, Bavaria, Germany; and Michelsdorf, Silesia, Poland.
BEST REF. IN ENGLISH: Palache, et al., "Dana's System of Mineralogy," 7th Ed., v. II, p. 700–702, New York, Wiley, 1951.

# HURLBUTITE
$CaBe_2(PO_4)_2$

CRYSTAL SYSTEM: Orthorhombic
CLASS: 2/m 2/m 2/m
SPACE GROUP: Pmmm
Z: 4
LATTICE CONSTANTS:
 $a = 8.29$
 $b = 8.80$
 $c = 7.81$
3 STRONGEST DIFFRACTION LINES:
 3.67 (100)
 3.03 ( 90)
 2.78 ( 90)
OPTICAL CONSTANTS:
 $\alpha = 1.595$
 $\beta = 1.601$
 $\gamma = 1.604$
 $(-)2V = 70°$
HARDNESS: 6
DENSITY: 2.877 (Meas.)
 2.88 (Calc.)
CLEAVAGE: Fracture conchoidal. Brittle.
HABIT: Crystals stout prismatic [001], relatively large development of c{001} and m{110}. Crystals range in size from 4 to 25 mm or more along [110]. Faces delicately etched and exhibit striations on {110}.
COLOR-LUSTER: Colorless to greenish white; when stained, yellow. Vitreous to greasy. Transparent to translucent. Streak white.
MODE OF OCCURRENCE: Occurs associated with muscovite, albite, triphylite, light-smoky quartz, and siderite in

pegmatite at the Smith mine, Chandler's Mill, Newport, New Hampshire.

BEST REF. IN ENGLISH: Mrose, Mary E., *Am. Min.*, **37**: 931-940 (1952).

## HUTCHINSONITE

$(Tl, Pb)_2 (Cu, Ag)As_5 S_{10}$

CRYSTAL SYSTEM: Orthorhombic
CLASS: 2/m 2/m 2/m
SPACE GROUP: Pbca
Z: 8
LATTICE CONSTANTS:
  $a = 10.78$
  $b = 35.28$
  $c = 8.14$
3 STRONGEST DIFFRACTION LINES:
  2.74 (100)
  3.78 ( 70)
  3.05 ( 60)
OPTICAL CONSTANTS:
  $\alpha = 3.078$
  $\beta = 3.176$ (Na)
  $\gamma = 3.188$
  $(-)2V = 37°24'$
HARDNESS: 1½-2
DENSITY: 4.6 (Meas.)
         4.58 (Calc.)
CLEAVAGE: {010} good
Fracture conchoidal. Brittle.
HABIT: Crystals prismatic to needle-like. As radiating tufts.
COLOR-LUSTER: Cherry red. Subtranslucent. Adamantine. Streak cherry red.
MODE OF OCCURRENCE: Occurs in crystalline dolomite associated with rathite, sartorite, pyrite, sphalerite, realgar, and orpiment, at the Lengenbach quarry, Binnental, Valais, Switzerland. Also found at Quiruvilca, Peru.
BEST REF. IN ENGLISH: Palache, et al., "Dana's System of Mineralogy," 7th Ed., v. I, p. 468-469, New York, Wiley, 1944.

## HUTTONITE

$ThSiO_4$
Dimorphous with thorite

CRYSTAL SYSTEM: Monoclinic
CLASS: 2/m
SPACE GROUP: $P2_1/n$
Z: 4
LATTICE CONSTANTS:
  $a = 6.80$
  $b = 6.96$
  $c = 6.54$
  $\beta = 104°55'$
3 STRONGEST DIFFRACTION LINES:
  3.09 (100)
  2.89 ( 90)
  4.23 ( 75)

OPTICAL CONSTANTS:
  $\alpha = 1.898$
  $\beta = 1.900$
  $\gamma = 1.922$
  $(+)2V = 25°$
HARDNESS: Not determined.
DENSITY: 7.1 (Meas.)
         7.18 (Calc.)
CLEAVAGE: {001} distinct
          {100} indistinct
Fracture conchoidal.
HABIT: Anhedral grains; 0.2 mm in maximum dimension.
COLOR-LUSTER: Colorless to very pale cream; adamantine. Fluoresces dull white with pink tinge under ultraviolet light.
MODE OF OCCURRENCE: Occurs as minute grains associated with scheelite, cassiterite, uranothorite, zircon, ilmenite, and gold in sands in the Gillespie's Beach area and and at several other localities in South Westland, New Zealand.
BEST REF. IN ENGLISH: Pabst, A., and Hutton, C. Osborne, *Am. Min.*, **36**: 60-69 (1951).

## HYACINTH = Variety of zircon

## HYALITE = Variety of opal

## HYALOPHANE

$(K, Ba)(Al, Si)_2 Si_2 O_8$
Barium Feldspar Group

CRYSTAL SYSTEM: Monoclinic
CLASS: 2/m
SPACE GROUP: C2/m
Z: 4
LATTICE CONSTANTS:
  $a = 8.52$
  $b = 12.95$
  $c = 7.14$
  $\beta \simeq 116°$
3 STRONGEST DIFFRACTION LINES:
  3.24 (100)
  3.31 ( 90)
  3.00 ( 70)
OPTICAL CONSTANTS:
  $\alpha = 1.520-1.542$
  $\beta = 1.524-1.545$
  $\gamma = 1.526-1.547$
  $2V = 101-132°$
HARDNESS: 6-6½
DENSITY: 2.58-2.88 (Meas.)
         2.945 (Calc.)
CLEAVAGE: {001} perfect
          {010} distinct
Fracture uneven to conchoidal. Brittle.
HABIT: Crystals prismatic, orthorhombic in aspect. Also massive. Carlsbad, Baveno, and Manebach twinning.
COLOR-LUSTER: Colorless, white, yellow. Transparent to translucent, vitreous.

MODE OF OCCURRENCE: Occurs as phenocrysts in analcite phonolite in the Highland Mountains, Montana; at Franklin, Sussex County, New Jersey; near Nisikkatch Lake, Saskatchewan, Canada; at Jakobsberg and near Örebro, Sweden; near Imfeld, Valais, Switzerland; at Slyudyanka, Transbaikalia, USSR; in gneiss at Broken Hill, New South Wales, Australia; and in manganese deposits at Otjosondu, South West Africa, and in the Kaso mine, Tochigi Prefecture, Japan.

BEST REF. IN ENGLISH: Deer, Howie, and Zussman, "Rock Forming Minerals," v. 4, p. 166–178, New York, Wiley, 1963.

# HYALOSIDERITE
$(Mg, Fe)_2 SiO_4$  (30–50 Atomic percent $Fe^{2+}$)
Olivine Group

CRYSTAL SYSTEM: Orthorhombic
CLASS: 2/m 2/m 2/m
SPACE GROUP: Pbnm
Z: 4
LATTICE CONSTANTS:
 $a = 4.783$
 $b = 10.335$
 $c = 6.031$
3 STRONGEST DIFFRACTION LINES:
 2.791 (100)
 2.533 ( 60)
 2.475 ( 60)
OPTICAL CONSTANTS:
 $\alpha = 1.721$
 $\beta = 1.748$
 $\gamma = 1.757$
 $(-)2V \sim 80°$
HARDNESS: 7
DENSITY: 3.69 (Meas.) $(Mg_{0.64}Fe_{0.36})_2 SiO_4$
        3.648 (Calc.)
CLEAVAGE: {010} imperfect
          {100} imperfect
Fracture conchoidal. Brittle.

HABIT: Crystals usually thick tabular, vertically striated, often with wedge-shaped terminations. Usually massive, compact or granular; in embedded grains, irregular or rounded. Twinning on {100}, {011}, not common.

COLOR-LUSTER: Green, lemon yellow. Transparent to translucent. Vitreous to greasy. Streak uncolored.

MODE OF OCCURRENCE: Occurs as a constituent of basic igneous rocks such as basalts, gabbros, and dolerites; in thermally metamorphosed impure dolomitic limestones; and in meteorites. Typical occurrences are found at Monroe, Orange County, New York; Tripyramid Mountain, New Hampshire; in the Julianehaab district, Greenland; at Sasbach, Baden, Germany; and at Moroto, Uganda.

BEST REF. IN ENGLISH: Deer, Howie, and Zussman, "Rock Forming Minerals," v. 1, p. 1–33, New York, Wiley, 1962.

# HYALOTEKITE (Inadequately described mineral)
$(Pb, Ca, Ba)_4 BSi_6 O_{17} (OH, F)$

CRYSTAL SYSTEM: Orthorhombic (?)
3 STRONGEST DIFFRACTION LINES:
 3.45 (100)
 3.53 ( 80)
 2.94 ( 80)
OPTICAL CONSTANTS:
 $\alpha = 1.963$
 $\beta = 1.963$
 $\gamma = 1.966$
 $(+)2V =$ small
HARDNESS: 5–5½
DENSITY: 3.81 (Meas.)
CLEAVAGE: Two, at 90°
HABIT: Massive, coarsely crystalline.
COLOR-LUSTER: White, gray. Translucent. Vitreous.
MODE OF OCCURRENCE: Occurs as a rare mineral in association with barylite at Långban, Sweden.

BEST REF. IN ENGLISH: Winchell and Winchell, "Elements of Optical Mineralogy," 4th Ed., Pt. 2, p. 401, New York, Wiley, 1951.

# HYDROBASALUMINITE (Inadequately described mineral)
$Al_4 SO_4 (OH)_{10} \cdot 36H_2O$

CRYSTAL SYSTEM: Unknown
3 STRONGEST DIFFRACTION LINES:
 12.6 (100)
 6.18 ( 70)
 5.29 ( 70)
HARDNESS: Not determined.
DENSITY: Not determined.
CLEAVAGE: Not determined. Plastic.
HABIT: As clay-like masses.
COLOR-LUSTER: White.
MODE OF OCCURRENCE: Occurs associated with basaluminite at the Lodge Pit siderite deposit at Irchester, Northamptonshire, England.

BEST REF. IN ENGLISH: Palache, et al., "Dana's System of Mineralogy," 7th Ed., v. II, p. 586, New York, Wiley, 1951.

# HYDROBIOTITE = Mixture of biotite and vermiculite

# HYDROBORACITE
$CaMgB_6 O_{11} \cdot 6H_2O$

CRYSTAL SYSTEM: Monoclinic
Z: 2
LATTICE CONSTANTS:
 $a = 11.76$
 $b = 6.68$
 $c = 8.20$
 $\beta = 102.8°$
3 STRONGEST DIFFRACTION LINES:
 5.77 (100)
 6.68 ( 85)
 3.32 ( 50)

OPTICAL CONSTANTS:
$\alpha = 1.520-1.523$
$\beta = 1.534-1.535$
$\gamma = 1.569-1.571$
$(+)2V = 60°-66°$
HARDNESS: 2-3
DENSITY: 2.167 (Meas.)
CLEAVAGE: {010} perfect

HABIT: Crystals long prismatic or acicular, elongated along c-axis. Also massive, compact and fine-grained; radiating or columnar; lamellar-fibrous.

COLOR-LUSTER: Colorless, white. Transparent. Vitreous; fibrous masses silky.

MODE OF OCCURRENCE: Occurs in California as needle-like crystals near Ryan, Death Valley, Inyo County; as fibrous crystalline masses in the open pit at Boron, Kern County; and at the Frazier borax mine, Ventura County. It is also found with halite at Strassfurt, Germany; at Yakal, Turkey; and in the Inder region, Kazakhstan, USSR.

BEST REF. IN ENGLISH: Palache, et al., "Dana's System of Mineralogy," 7th Ed., v. II, p. 353-355, New York, Wiley, 1951.

# HYDROCALUMITE
$Ca_2Al(OH)_7 \cdot 3H_2O$

CRYSTAL SYSTEM: Monoclinic
CLASS: 2
SPACE GROUP: $P2_1$
Z: 8
LATTICE CONSTANTS:
$a = 9.6$
$b = 11.4$
$c = 16.84$
$\beta = 111°$
3 STRONGEST DIFFRACTION LINES:
7.86 (100)
3.93 ( 60)
2.89 ( 50)
OPTICAL CONSTANTS:
$\alpha = 1.535$
$\beta = 1.553$
$\gamma = 1.557$
$(-)2V = 24°$
HARDNESS: 3
DENSITY: 2.15 (Meas.)
2.15 (Calc.)
CLEAVAGE: {001} perfect
HABIT: Massive.
COLOR-LUSTER: Colorless to pale green. Transparent. Vitreous, pearly on cleavages.

MODE OF OCCURRENCE: Occurs in vugs in a larnite rock in the contact zone at Scawt Hill, County Antrim, Ireland.

BEST REF. IN ENGLISH: Palache, et al., "Dana's System of Mineralogy," 7th Ed., v. I, p. 667-668, New York, Wiley, 1944.

# HYDROCERUSSITE
$Pb_3(CO_3)_2(OH)_2$

CRYSTAL SYSTEM: Hexagonal
Z: 9
LATTICE CONSTANTS:
$a = 8.97$
$c = 23.8$
3 STRONGEST DIFFRACTION LINES:
2.623 (100)
3.61 ( 90)
3.29 ( 90)
OPTICAL CONSTANTS:
$\omega = 2.09$
$\epsilon = 1.94$
$(-)$
HARDNESS: 3½
DENSITY: 6.80 (Meas.)
6.94 (Calc.)
CLEAVAGE: {0001} perfect
Brittle.

HABIT: Crystals thin to thick tabular, flattened {0001}, with hexagonal outline. Also steep pyramidal.

COLOR-LUSTER: Colorless, white, gray. Transparent to translucent. Adamantine, pearly on {0001}.

MODE OF OCCURRENCE: Occurs as a secondary mineral in oxidized ore as excellent crystals associated with boleite, cerussite, dioptase, diaboleite, and other secondary minerals in the Collins vein of the Mammoth mine, Tiger, Arizona; and sparingly at Kellogg, Idaho. It is also found in the Mendip Hills, Somerset, and at other places in England; at Långban, Sweden; at Laurium, Greece; and in the Altai Mountains, USSR.

BEST REF. IN ENGLISH: Palache, et al., "Dana's System of Mineralogy," 7th Ed., v. II, p. 270-271, New York, Wiley, 1951.

# HYDROCHLORBORITE (Inadequately described mineral)
$Ca_4B_8O_{15}Cl_2 \cdot 22H_2O$

CRYSTAL SYSTEM: Unknown
3 STRONGEST DIFFRACTION LINES:
8.40 (100)
5.98 ( 90)
2.605 ( 80)
OPTICAL CONSTANTS:
$\alpha = 1.5008$
$\beta = 1.5036$
$\gamma = 1.5199$
$(+)2V = 45°48'$
HARDNESS: 2½
DENSITY: 1.83 (Meas.)
CLEAVAGE: Not determined.
HABIT: Massive, dense.
COLOR-LUSTER: Colorless, transparent; vitreous.

MODE OF OCCURRENCE: Occurs associated with ulexite, halite, and gypsum in crusts overlying boron-containing clays in Tertiary sediments, China.

BEST REF. IN ENGLISH: Tzu-Chiang, Chien, and Shu-Chen, Chen, *Am. Min.*, **50**: 2099-2100 (1965).

# HYDROCYANITE = Chalcocyanite

**HYDROFLUORITE** = A natural gas (HF)

**HYDROGLAUBERITE (Inadequately described mineral)**
Na$_4$Ca(SO$_4$)$_3$ · 2H$_2$O

CRYSTAL SYSTEM: Probably orthorhombic
3 STRONGEST DIFFRACTION LINES:
  3.08 (100)
  9.2 ( 90)
  2.78 ( 90) broad
OPTICAL CONSTANTS:
  $\alpha'$ = 1.488
  $\gamma'$ = 1.500
  (−)2V
HARDNESS: Not determined.
DENSITY: Not determined.
CLEAVAGE: Not determined.
HABIT: Dense masses composed of fibers not more than 0.1 mm in size.
COLOR-LUSTER: Snow white; silky.

Taste salty, slightly bitter.
MODE OF OCCURRENCE: Occurs as an alteration product of glauberite, which is associated with halite, mirabilite, astrakhanite, and polyhalite in Tertiary sediments in the Karakalpakii ASSR.
BEST REF. IN ENGLISH: Slyusareva, M. N., *Am. Min.*, **55**: 321 (1970).

**HYDROGROSSULAR (Hibschite, Plazolite)**
Ca$_3$Al$_2$(SiO$_4$)$_{3-x}$(OH)$_{4x}$
Garnet Group

CRYSTAL SYSTEM: Cubic
CLASS: 4/m $\bar{3}$ 2/m
SPACE GROUP: Ia3d
Z: 8
LATTICE CONSTANT:
  $a$ = 11.85–12.16
3 STRONGEST DIFFRACTION LINES:
  2.71 (100)
  1.62 (100)
  3.03 ( 80)
OPTICAL CONSTANT:
  $N$ = 1.675–1.734
HARDNESS: 6–6½
DENSITY: 3.0–3.3 (Meas.)
           3.116 (Calc.)
CLEAVAGE: None.
HABIT: Crystals usually octahedral, minute; rarely tetrahedral. Also massive, compact.
COLOR-LUSTER: Colorless, white; also tinted pale gray, brown, pink, or green. Transparent to translucent. Vitreous to somewhat greasy
MODE OF OCCURRENCE: Occurs massive and as crystals up to 3/4 inch in size at Crestmore, Riverside County, and in association with prehnite and jadeite in San Benito County, California. Also found in metamorphosed marl at Marienberg, Bohemia, Czechoslovakia; in calc-silicate rock at Nikortzminda, Caucasus, Georgia, USSR; intergrown with albite and diopside at King Island, Tasmania; in jade-like masses in the Bushveld complex, Transvaal; at Tokatoka, New Zealand; and elsewhere.
BEST REF. IN ENGLISH: Deer, Howie, and Zussman, "Rock Forming Minerals," v. 1, p. 77–112, New York, Wiley, 1962.

**HYDROHALITE (Species status uncertain)**
NaCl · 2H$_2$O

CRYSTAL SYSTEM: Probably Monoclinic
MODE OF OCCURRENCE: Found as a product of crystallization of sea water or of saline spring waters at a temperature below 0°C.

**HYDROHAUSMANNITE** = Mixture of feitknechtite and Mn$_3$O$_4$

**HYDROHETAEROLITE**
Zn$_2$Mn$_4$O$_8$ · H$_2$O

CRYSTAL SYSTEM: Tetragonal
CLASS: 4/m 2/m 2/m
SPACE GROUP: I4$_1$/amd
Z: 2
LATTICE CONSTANTS:
  $a$ = 5.735
  $c$ = 9.005
3 STRONGEST DIFFRACTION LINES:
  2.47 (100)
  2.66 ( 80)
  3.02 ( 70)
OPTICAL CONSTANTS:
  $\omega$ = 2.26
  $\epsilon$ = 2.10
  (−)
HARDNESS: 5–6
DENSITY: 4.64 (Meas.)
CLEAVAGE: Parallel to elongation of fibers.
HABIT: Massive; fibrous crusts with botryoidal surface.
COLOR-LUSTER: Dark brown to brownish black. Submetallic. Streak dark brown.
MODE OF OCCURRENCE: Occurs in oxidized ore associated with chalcophanite, smithsonite, and hemimorphite at the Wolftone mine, Leadville, Colorado, and with chalcophanite at Sterling Hill, Sussex County, New Jersey.
BEST REF. IN ENGLISH: McAndrew, John, *Am. Min.*, **41**: 268–275 (1956). "Dana's System of Mineralogy," 7th Ed., v. I, p. 717–718, New York, Wiley, 1944.

**HYDROMAGNESITE**
Mg$_5$(CO$_3$)$_4$(OH)$_2$ · 4H$_2$O

CRYSTAL SYSTEM: Monoclinic
CLASS: 2/m
SPACE GROUP: P2$_1$/c
Z: 2

LATTICE CONSTANTS:
$a = 10.11$
$b = 8.94$
$c = 8.38$
$\beta = 114°35'$

3 STRONGEST DIFFRACTION LINES:
5.79 (100)
2.899 ( 82)
9.20 ( 39)

OPTICAL CONSTANTS:
$\alpha = 1.523$
$\beta = 1.527$ (Na)
$\gamma = 1.545$
(+)2V = moderate

HARDNESS: 3½
DENSITY: 2.253 (Meas.)
CLEAVAGE: {010} perfect
Brittle.

HABIT: Crystals acicular or bladed, elongated along c-axis and flattened on (100). Large crystals vertically striated on (100) by repetition of {110} and {100} forms. As sprays, rosettes, or crusts. Also massive, powdery, or chalk-like. Twinning on {100} very common.

COLOR-LUSTER: Colorless, white. Transparent. Crystals vitreous; aggregates pearly, silky, or dull.

MODE OF OCCURRENCE: Occurs widespread, sometimes in deposits having economic significance, as an alteration of serpentine or other magnesian rocks. Found in the United States, especially at numerous localities in California; at Luning, Nevada; as excellent crystals associated with artinite on Long Island, New York; at Hoboken, New Jersey; and at the chrome mines near Texas, Lancaster County, Pennsylvania. It also occurs abundantly in British Columbia, Canada, and in many European deposits. Exceptional crystals occur in the chromite deposits of Iran, notably as crystals up to 3 cm in the old open pit of Dovez, near the Shariar mine, Minab district, and as magnificent crystals up to 10 cm in groups of 40 × 30 cm in the Soghan mine, Esfandaque district, south of Kerman.

BEST REF. IN ENGLISH: Bariand, P., Cesbron, F., Vachey, H., and Sadrzadeh, M., *Min. Rec.*, **4**: 18–20 (1973).

# HYDROMARCHITE
$Sn_3O_2(OH)_2$

CRYSTAL SYSTEM: Triclinic
Z: 1
LATTICE CONSTANTS:
$a = 11.5$     $\alpha = 99°$
$b = 6.03$     $\beta = 60°30'$
$c = 19.8$     $\gamma = 88°30'$

3 STRONGEST DIFFRACTION LINES:
3.50 (100)
2.773 ( 90)
2.961 ( 80)

HARDNESS: Not determined.
DENSITY: Not determined.
CLEAVAGE: Not determined.
HABIT: As thin crusts composed of minute crystals.
COLOR-LUSTER: White.
MODE OF OCCURRENCE: Tin pannikins lost from the overturned canoe of a voyager between 1801 and 1821 were recovered 15 feet below the surface of the water at Boundary Falls, Winnipeg River, Ontario, Canada. Some of the surfaces have a thin crust composed of the alteration products hydromarchite and romarchite.

BEST REF. IN ENGLISH: Organ, R. M., and Mandarino, J. A., *Can. Min.*, **10**: 916 (1971) (abst.).

# HYDROMELANOTHALLITE (Species status uncertain.)
$Cu_2(OH)_2Cl_2 \cdot H_2O$

CRYSTAL SYSTEM: Probably orthorhombic
OPTICAL CONSTANTS:
$\alpha = 1.580$
$\beta = 1.605$ (variable)
$\gamma = 1.644$
(+)2V = moderately large

COLOR-LUSTER: Black; changes rapidly to green on exposure.

MODE OF OCCURRENCE: Occurs with eriochalcite, chalcocyanite, euchlorin, and dolerophanite as a sublimation product in the crater of Vesuvius, Italy.

BEST REF. IN ENGLISH: Frondel, Clifford, *Min. Mag.*, **29**: 42–43 (1950).

# HYDROMICA (See Illite)

# HYDROMOLYSITE (Inadequately described mineral)
$FeCl_3 \cdot 6H_2O$

CRYSTAL SYSTEM: Unknown
3 STRONGEST DIFFRACTION LINES:
6.0 (100)
3.14 ( 75) (synthetic)
2.76 ( 50)

HARDNESS: Not determined.
DENSITY: Not determined.
CLEAVAGE: Not determined.
HABIT: Massive; as crusts.
COLOR-LUSTER: Orange-red.
MODE OF OCCURRENCE: Occurs as an incrustation formed by the action of sea water driven by storms into pyritic deposits of Rio Marina, Elba.

BEST REF. IN ENGLISH: Garavelli, Carlo, *Am. Min.*, **44**: 908 (1959).

# HYDROMUSCOVITE = Illite

# HYDRONIUM JAROSITE
$(H_3O)Fe_3(SO_4)_2(OH)_6$
Alunite Group
Jarosite-Hydronium Jarosite Series

CRYSTAL SYSTEM: Hexagonal
CLASS: 3m

SPACE GROUP: R3m
Z: 3
LATTICE CONSTANTS:
 $a = 7.31$
 $c = 17.12$
3 STRONGEST DIFFRACTION LINES:
 5.11 (100)
 3.13 ( 80)
 3.09 ( 80)
 HARDNESS: 4-4½
 DENSITY: 2.496-2.905 (Meas.)
 CLEAVAGE: {0001} distinct
Fracture uneven to conchoidal. Brittle.

 HABIT: Crystals thin tabular or rhombohedral, up to 1.0 mm or more in size. Also as earthy or scaly aggregates and crusts; reniform.
 COLOR-LUSTER: Golden yellow, dark straw yellow, reddish brown to brown. Transparent to translucent. Vitreous; also resinous or glistening to dull.
 MODE OF OCCURRENCE: Occurs as superb crystals and fine-crystalline aggregates near Majuba Hill, Pershing County, Nevada; at the Eureka Hill mine, Tintic district, Juab County, and at Lion Hills, Tooele County, Utah; in abundance in well-weathered exposures of the Pierre shale, Yankton County, South Dakota; and at Upper Freemont, Pennsylvania. Also found in the oxidation zone of the iron ores at the "Staszic" pyrite mine at Rudki, Poland, and at localities in Greenland, Mexico, Chile, France, Finland, Czechoslovakia, USSR, and elsewhere.
 BEST REF. IN ENGLISH: Palache, et al., "Dana's System of Mineralogy," 7th Ed., v. II, p. 566-567, New York, Wiley, 1951. Kubisz, J., *Min. Abs.*, **15**: 364 (1962). Brophy, Gerald P., and Sheridan, Michael F., *Am. Min.*, **50**: 1595-1607 (1965).


## HYDROPARAGONITE = Brammallite


## HYDROPHILITE
$CaCl_2$ (?)

CRYSTAL SYSTEM: Orthorhombic
CLASS: 2/m 2/m 2/m
SPACE GROUP: Pnnm
Z: 2
LATTICE CONSTANTS:
 $a = 6.24$ kX
 $b = 6.43$
 $c = 4.20$ (synthetic)
3 STRONGEST DIFFRACTION LINES:
 4.49 (100)
 3.05 ( 80)
 2.33 ( 60)
OPTICAL CONSTANTS:
 $\alpha = 1.600$
 $\beta = 1.605$
 $\gamma = 1.613$
 (+)2V = moderate
 HARDNESS: Not determined.
 DENSITY: 2.22 (Meas. Synthetic)
       2.17 (Calc. Synthetic)
 CLEAVAGE: Prismatic, perfect.

 HABIT: Massive; as mealy crusts and disseminated crystalline particles. Twinning polysynthetic, complex.
 COLOR-LUSTER: White. Transparent to translucent.

Deliquescent.
 MODE OF OCCURRENCE: Reported from Lüneberg, Hanover, Germany; Etna on Sicily; Tarapacá, Chile; Chincha, Peru; Barren Island, Bay of Bengal; and Guy's Cliff, Warwickshire, England.
 BEST REF. IN ENGLISH: Palache, et al., "Dana's System of Mineralogy," 7th Ed., v. II, p. 41-42, New York, Wiley, 1951.


## HYDROSCARBROITE
$Al_{14}(CO_3)_3(OH)_{36} \cdot nH_2O$

CRYSTAL SYSTEM: Triclinic (See scarbroite)
3 STRONGEST DIFFRACTION LINES:
 9.0 (100)
 8.74 (100)
 4.47 ( 50)
 HARDNESS: Soft.
 DENSITY: Not determined
 CLEAVAGE: Not determined.
 HABIT: Massive; composed of thin plate-like crystallites about $1\mu$ across basal planes.
 COLOR-LUSTER: White.
 MODE OF OCCURRENCE: Occurs with scarbroite in vertical fissures in sandstone at South Bay, Scarborough, Yorkshire, England. Spontaneously dehydrates to scarbroite when permitted to come into equilibrium with the atmosphere at room temperature.
 BEST REF. IN ENGLISH: Duffin, W. J., and Goodyear, J., *Min. Mag.*, **32**: 353-362 (1960). Brindley, G. W., and Comer, J. J., *Min. Mag.*, **32**: 363-365 (1960).


## HYDROTALCITE
$Mg_6Al_2CO_3(OH)_{16} \cdot 4H_2O$
Dimorphous with manasseite

CRYSTAL SYSTEM: Hexagonal
SPACE GROUP: R lattice
Z: 3
LATTICE CONSTANTS:
 $a = 6.13$
 $c = 46.15$
3 STRONGEST DIFFRACTION LINES:
 7.69 (100)
 3.88 ( 70)
 2.58 ( 20)
OPTICAL CONSTANTS:
 $\omega = 1.511-1.531$
 $\epsilon = 1.495-1.529$
  (-)
 HARDNESS: 2
 DENSITY: 2.06 (Meas.)
       2.16 (Calc.)
 CLEAVAGE: {0001} perfect
Laminae flexible, inelastic. Greasy feel.
 HABIT: Massive, foliated, lamellar, or fibrous-lamellar.

COLOR-LUSTER: White. Transparent. Pearly to waxy. Streak white.

MODE OF OCCURRENCE: Occurs at Vernon, Sussex County, New Jersey; in Orange, Oxbow, and St. Lawrence counties, New York; with serpentine at Nordmark and Snarum, Norway; and at several localities in the USSR.

BEST REF. IN ENGLISH: Palache, et al., "Dana's System of Mineralogy," 7th Ed., v. I, p. 653–655, New York, Wiley, 1944.

## HYDROTHIONITE = A natural gas
$H_2S$

## HYDROTROILITE (Species status uncertain.)
Colloidal hydrous ferrous sulfide

COLOR-LUSTER: Black.

MODE OF OCCURRENCE: Occurs disseminated in many muds and clays.

BEST REF. IN ENGLISH: Palache, et al., "Dana's System of Mineralogy," 7th Ed., v. I, p. 236, New York, Wiley, 1944.

## HYDROTUNGSTITE
$H_2WO_4 \cdot H_2O$

CRYSTAL SYSTEM: Monoclinic (pseudo-orthorhombic)
CLASS: 2/m
SPACE GROUP: P2/m
Z: 2
LATTICE CONSTANTS:
 $a = 7.45$
 $b = 6.92$
 $c = 3.72$
 $\beta \sim 90°$
3 STRONGEST DIFFRACTION LINES:
 6.95 (100)
 3.27 ( 80)
 3.73 ( 70)
OPTICAL CONSTANTS:
  $\alpha = 1.70$
  $\beta = 1.95$
  $\gamma = 2.04$
 $(-)2V = 52°$
HARDNESS: $\sim 2$
DENSITY: 4.60 (Meas.)
      4.64 (Calc.)
CLEAVAGE: {010} imperfect
HABIT: Crystals tabular, ranging from 5 to 75 $\mu$ in size. As minute reticulated intergrowths. Twinning polysynthetic, parallel to prism faces.
COLOR-LUSTER: Dark green to yellowish green. Vitreous to dull.
MODE OF OCCURRENCE: Occurs as an alteration product of ferberite near Oruro, Bolivia.
BEST REF. IN ENGLISH: Kerr, Paul F., and Young, Ford, *Am. Min.*, **29**: 192–210 (1944). *Am. Min.*, **48**: 935 (1963).

## HYDROUGRANDITE
$(Ca, Mg, Fe)_3 (Fe, Al)_2 (SiO_4)_{3-x} (OH)_{4x}$

CRYSTAL SYSTEM: Cubic
LATTICE CONSTANT:
 $a = 12.063$
3 STRONGEST DIFFRACTION LINES:
 2.705 (100)
 1.614 (100)
 3.02 ( 80)
OPTICAL CONSTANT:
 $N = 1.825–1.830$
 HARDNESS: $>5$
 DENSITY: 3.45 (Meas.)
 HABIT: Small grains.
 COLOR-LUSTER: Green; vitreous.
 MODE OF OCCURRENCE: Occurs associated with olivine, augite, and plagioclase in peridotites from Hsiaosungshan, China.
 BEST REF. IN ENGLISH: Lung, Tsao Yung, *Am. Min.*, **50**: 2100 (1965).

## HYDROXYLAPATITE (Hydroxyapatite)
$Ca_5 (PO_4)_3 OH$
Apatite Series

CRYSTAL SYSTEM: Hexagonal
CLASS: 6/m
SPACE GROUP: $P6_3/m$
Z: 2
LATTICE CONSTANTS:
 $a = 9.418$
 $c = 6.884$
3 STRONGEST DIFFRACTION LINES:
 2.814 (100)
 2.778 ( 60)
 2.720 ( 60)
OPTICAL CONSTANTS:
  $\omega = 1.651$
  $\epsilon = 1.644$ (Na)
  $(-)$
 HARDNESS: 5
 DENSITY: 3.08 (Meas.)
       3.16 (Calc.)
 CLEAVAGE: {0001} indistinct
        {10$\bar{1}$0} trace
Fracture uneven to conchoidal. Brittle.
 HABIT: See fluorapatite.
 COLOR-LUSTER: See fluorapatite.
 MODE OF OCCURRENCE: Occurs in talc-schist near Holly Springs, Cherokee County, Georgia, in talc-schist in the Hospenthal, Uri, Switzerland; and in diallage-serpentine rock at Rossa in the Val Devero, Ossola, Italy.
 BEST REF. IN ENGLISH: Palache, et al., "Dana's System of Mineralogy," 7th Ed., v. II, p. 879–889, New York, Wiley, 1951.

## HYDROXYL-BASTNAESITE
$(Ce, La)(CO_3)(OH, F)$ with OH : F = 0.87 : 0.14

CRYSTAL SYSTEM: Hexagonal
CLASS: $\bar{6}$m2
SPACE GROUP: P$\bar{6}$2c
Z: 6
LATTICE CONSTANTS:
$a$ = 7.23
$c$ = 9.98
3 STRONGEST DIFFRACTION LINES:
2.92 (100)
3.59 ( 90)
2.09 ( 90)
OPTICAL CONSTANTS:
$\omega$ = 1.760
$\epsilon$ = 1.870
    (+)
HARDNESS: ~4
DENSITY: 4.745 (Meas.)
CLEAVAGE: {11$\bar{2}$0} imperfect
HABIT: Reniform aggregates.
COLOR-LUSTER: Wax yellow to dark brown, rarely colorless. Vitreous to greasy.
MODE OF OCCURRENCE: Occurs in cavities of late carbonatite veins, associated with barite, strontianite, monazite, ancylite, altered burbankite, pyrite, and other sulfides in an unspecified massif of alkalic-ultrabasic rocks.
BEST REF. IN ENGLISH: Kirillov, A. S., *Am. Min.*, **50**: 805 (1965).


# HYDROXYLELLESTADITE
$Ca_{10}(SiO_4)_3(SO_4)_3(OH, Cl, F)_2$ with OH > Cl, F
Apatite Group
OH analogue of ellestadite

CRYSTAL SYSTEM: Hexagonal
CLASS: 6/m
SPACE GROUP: P$6_3$/m
Z: 1
LATTICE CONSTANTS:
$a$ = 9.491
$c$ = 6.921
3 STRONGEST DIFFRACTION LINES:
2.839 (100)
2.739 ( 60)
2.655 ( 45)
OPTICAL CONSTANTS:
$\omega$ = 1.654
$\epsilon$ = 1.650
    (−)
HARDNESS: 4½
DENSITY: 3.018 (Meas.)
CLEAVAGE: Prismatic
HABIT: Massive; as cleavable aggregates up to 100 kg in weight.
COLOR-LUSTER: Pale purplish. Translucent. Vitreous.
MODE OF OCCURRENCE: Occurs in association with diopside, wollastonite, xanthophyllite, idocrase, and calcite, at the Chichibu mine, Saitama Prefecture, Japan.
BEST REF. IN ENGLISH: Harada, Kazuo; Nagashima, Kozo; Nakao, Kazuso; and Kato, Akira, *Am. Min.*, **56**: 1507–1518 (1971).

**HYDROXYL-HERDERITE** = Variety of herderite with (OH) > F


# HYDROZINCITE
$Zn_5(CO_3)_2(OH)_6$

CRYSTAL SYSTEM: Monoclinic
CLASS: 2/m
SPACE GROUP: C2/m
Z: 2
LATTICE CONSTANTS:
$a$ = 13.479
$b$ = 6.320
$c$ = 5.368
$\beta$ = 95°30′
3 STRONGEST DIFFRACTION LINES:
6.66 (100)
2.72 ( 70)
3.14 ( 50)
OPTICAL CONTSANTS:
$\alpha$ = 1.635–1.650
$\beta$ = 1.736
$\gamma$ = 1.740(?)–1.750
(−)2V = 40° (Calc.)
HARDNESS: 2–2½
DENSITY: Usually 3.5–3.8 (Meas.).
CLEAVAGE: {100} perfect.
Very brittle.
HABIT: Massive, compact to earthy. As crusts, stalactitic, reniform, and as dense masses. Crystals flattened, elongated laths, often tapering to a sharp point, and seldom more than 1.0 mm in length.
COLOR-LUSTER: White, gray and pale shades of yellow, pink, lavender, and brown. Crystals pearly; massive material dull to silky to shining. Sometimes fluoresces blue under ultraviolet light.
MODE OF OCCURRENCE: Occurs as a secondary mineral formed by the alteration of sphalerite in the oxidation zone of ore deposits, and rarely in pegmatites. Found in abundance at Goodsprings, Nevada; sparingly in pegmatite at the Tin Mountain mine, Custer County, South Dakota; in California, Colorado, Utah, New Mexico, the Missouri-Oklahoma-Kansas lead-zinc district, Arkansas, Wisconsin, Pennsylvania, and at Franklin, New Jersey. It also occurs abundantly as blade-like crystals at Mapimi, Mexico; in Santander province, Spain; near Constantine, Algeria; and at many other worldwide localities.
BEST REF. IN ENGLISH: Palache, et al., "Dana's System of Mineralogy," 7th Ed., v. II, p. 247–249, New York, Wiley, 1951.


# HYPERSTHENE
(Mg, Fe)SiO$_3$
Pyroxene Group
Var. Bronzite
    Ferrohypersthene
    Eulite

CRYSTAL SYSTEM: Orthorhombic
CLASS: 2/m 2/m 2/m

SPACE GROUP: Pbca

Z: 16

LATTICE CONSTANTS:

 $a = 18.20$

 $b = 8.86$

 $c = 5.20$

3 STRONGEST DIFFRACTION LINES:

 3.20 (100)

 2.89 ( 80)

 1.49 ( 80)

OPTICAL CONSTANTS:

 $\alpha = 1.692$

 $\beta = 1.702$

 $\gamma = 1.705$

 $(-)2V = 90°$

HARDNESS: 5-6

DENSITY: 3.42–3.84 (Meas.)

CLEAVAGE: {210} good

 {100} parting

 {010} parting

Fracture uneven. Brittle.

HABIT: Crystals short prismatic, rare. Usually massive, lamellar. Twinning on {100}, simple and lamellar.

COLOR-LUSTER: Brownish green, grayish black, greenish black, brown. Translucent to nearly opaque. Pearly, silky, vitreous, or bronzy. Streak grayish to brownish gray.

MODE OF OCCURRENCE: Occurs widespread as a common constituent in certain basic igneous rocks such as norites, gabbros, and andesites; in charnockite, gneiss, amphibolite, and other metamorphic rocks; and in certain meteorites. Typical occurrences are found in California, Colorado, Arizona, Delaware and New York. Also in Canada, Greenland, Scotland, Norway, Germany, Roumania, Finland, India, Japan, and elsewhere.

BEST REF. IN ENGLISH: Deer, Howie, and Zussman, "Rock Forming Minerals," v. 2, p. 8–41, New York, Wiley, 1963.

EULYTITE   Schneeberg, Saxony, Germany.

EULYTITE   Schneeberg, Saxony, Germany.

EVEITE   Franklin, Sussex County, New Jersey.

FAIRFIELDITE   Foote Mine, Kings Mountain, North Carolina.

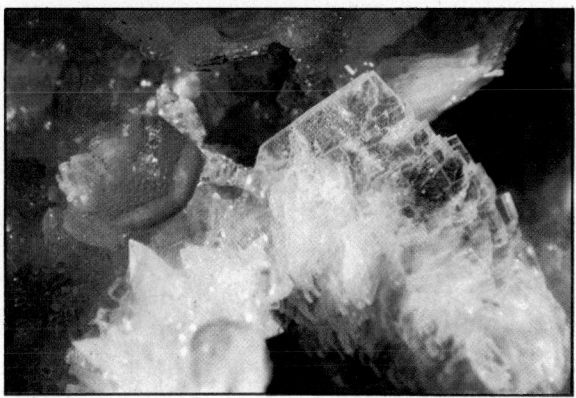

FAIRFIELDITE with fluorapatite   Foote Mine, Kings Mountain, North Carolina.

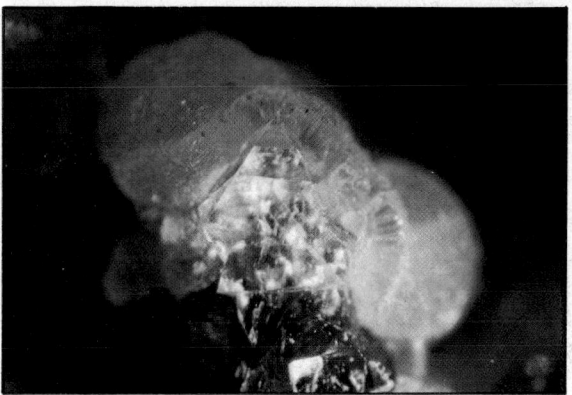

FAUJASITE   Sasbach, Baden, Germany.

FAUJASITE   Kaiserstuhl, Baden, Germany.

FAYALITE on cristobalite    Coso Hot Springs, Inyo County, California.

FELSOBANYITE pseudomorphous after stibnite    Felsobanya, Roumania.

FERBERITE    Nederland, Boulder County, Colorado.

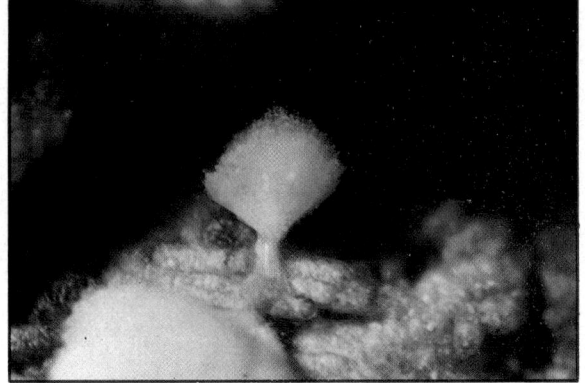

FERRIERITE    Agoura, Los Angeles County, California.

FIEDLERITE    Laurium, Greece.

FIZELYITE    Julcani mine, Huancavelica, Peru.

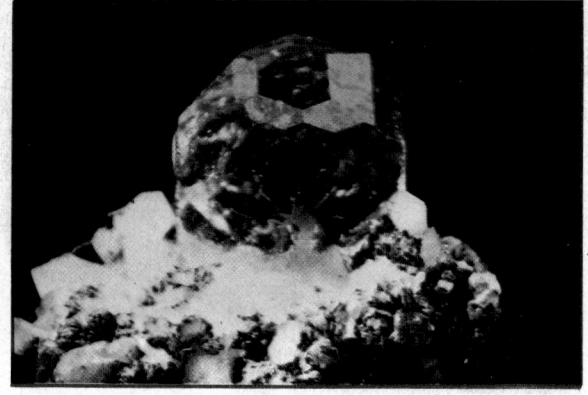

FLUORAPATITE    Auburn, Androscoggin County, Maine.

FLUORAPATITE    Hugo mine, Keystone, Pennington County, South Dakota.

FLUORAPATITE    Goschenalpen, Uri, Switzerland.

FLUORAPATITE    Foote Mine, Kings Mountain, North Carolina.

FLUORAPATITE with epidote and byssolite    Untersulzbachtal, Salzburg, Austria.

FLUORAPATITE    Auburn, Androscoggin County, Maine.

FLUORITE   Catron County, New Mexico.

FLUORITE on mimetite   Laurium, Greece.

FLUORITE with sphalerite   Rosiclare, Hardin County, Illinois.

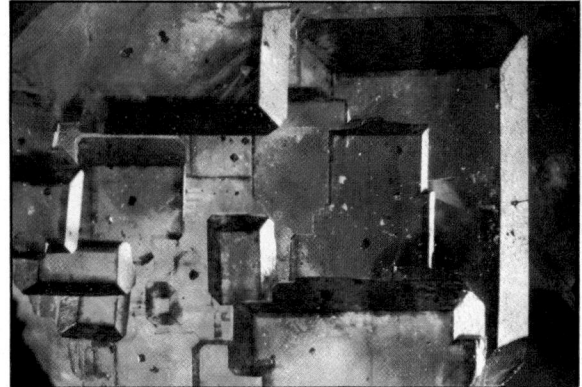

FLUORITE   Carabia, Asturias, Spain.

FLUORITE   Tyrol, Austria.

FLUORITE (zoned color)   Strickland quarry, Portland, Connecticut.

52

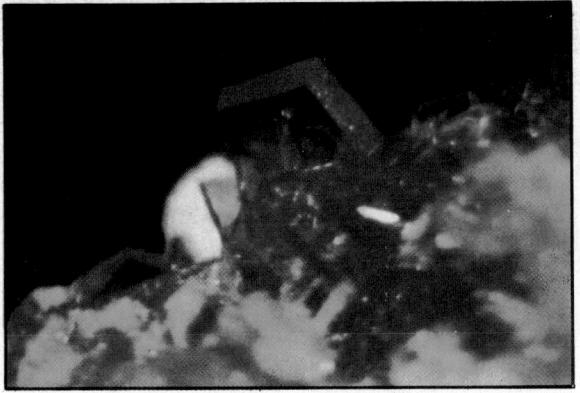

FOURMARIERITE   Shinkolobwe, Katanga, Zaire.

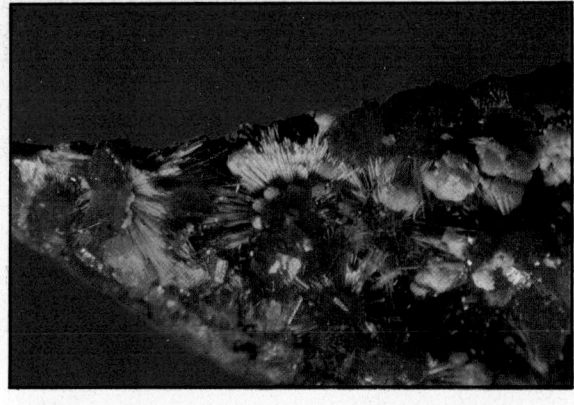

FOURMARIERITE with studtite   Shinkolobwe, Katanga, Zaire.

FRANCEVILLITE   Mounana, Gabon.

FRANCEVILLITE   Mounana, Gabon.

FRANCEVILLITE   Mounana, Gabon.

FRANCEVILLITE   Mounana, Gabon.

FRANCKEITE   Llallagua, Bolivia.

FRANKLINITE   Franklin, Sussex County, New Jersey.

FRANKLINITE   Franklin, Sussex County, New Jersey.

FRANKLINITE with hodgkinsonite   Franklin, Sussex County, New Jersey.

FREIESLEBENITE with stannite   Julcani mine, Huancavelica, Peru.

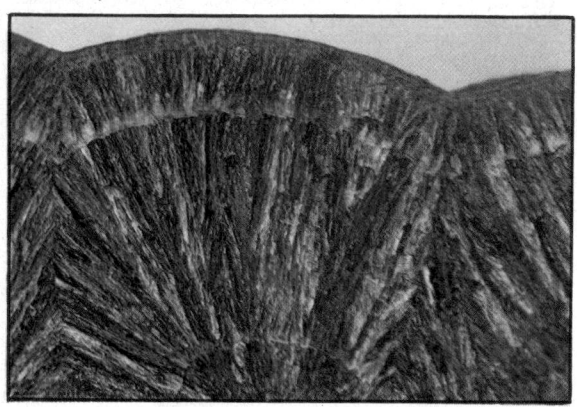

FRONDELITE   Sapucaia, Conselheira Pena, Minas Geraes, Brazil.

GAGEITE   Franklin, Sussex County, New Jersey.

GAHNITE   Chichester, Delaware County, Pennsylvania.

GALENA    Braubach, Nassau, Germany.

GALENA pseudomorphous after pyromorphite    Truro, Cornwall, England.

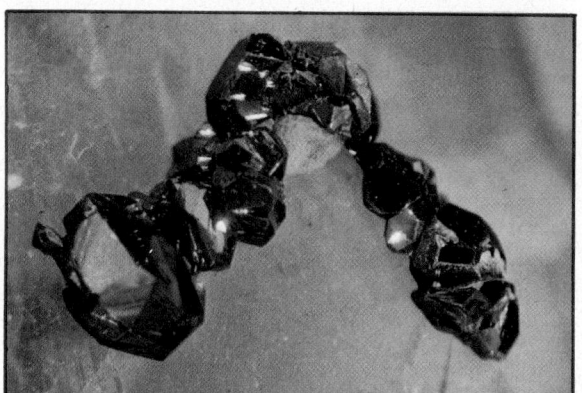

GALENA    Freiberg, Saxony, Germany.

GANOPHYLLITE    Franklin, Sussex County, New Jersey.

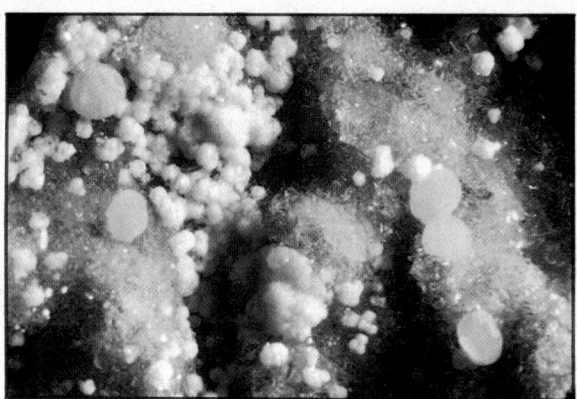

GARRONITE    Goble, Columbia County, Oregon.

GEHLENITE    Monzoni, Val di Fassa, Trentino, Italy.

GENTHELVITE with aegirine    Mont St. Hilaire, Quebec, Canada.

GERSDORFFITE   Hager, Hesse Nassau, Germany.

GERSDORFFITE   Goslar, Harz Mountains, Germany.

GERSTLEYITE   Boron, Kern County, California.

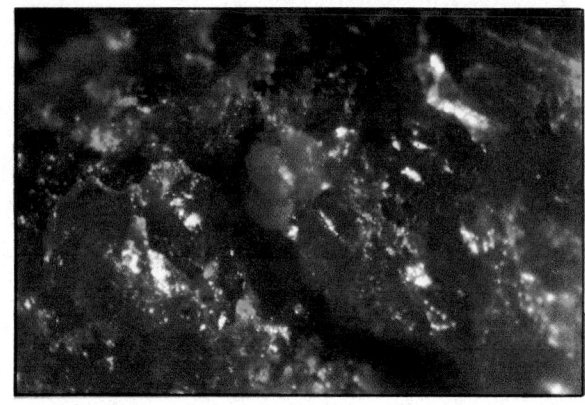

GETCHELLITE   Getchell mine, Humboldt County, Nevada.

GIBBSITE   Langesundfjord, Norway.

GISMONDITE with phillipsite   Vesuvius, Italy.

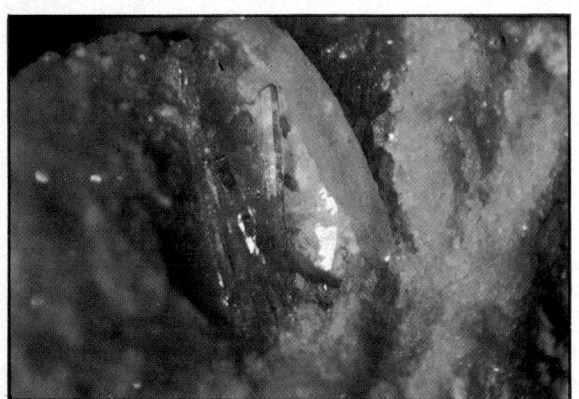

GLAUCOCHROITE   Franklin, Sussex County, New Jersey.

GLAUCODOT   Dixie Creek, Grant County, Oregon.

GMELINITE   Flinders, Victoria, Australia.

GOETHITE on quartz

GOETHITE   Imperial mine, Michigamme, Michigan.

GOETHITE (iridescent)   Zacatecas, Mexico.

GOETHITE   Botallack mine, Cornwall, England.

GOETHITE in quartz   Botallack mine, Cornwall, England.

GOETHITE spray in amethyst   Brazil.

GOLD    Placer County, California.

GOLD    Felsobanya, Roumania.

GOLD    Felsobanya, Roumania.

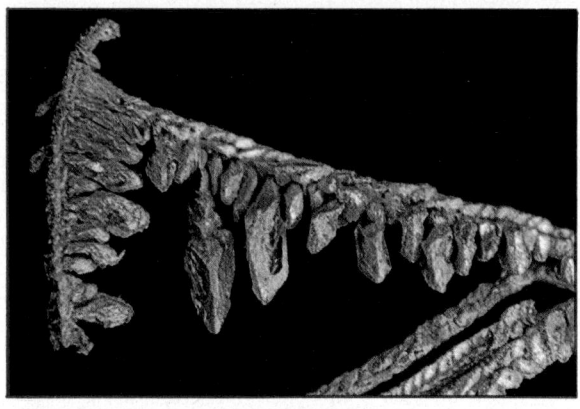

GOLD    Dugas mine, Cleveland, Georgia.

GOLD    California.

GOLD    Verespatak, Roumania.

GOLD    Cle Elum, Kittitas County, Washington.

GOLD    Morro Velho mine, Minas Geraes, Brazil.

GOLD (electrum)   Kalgoorlie, Western Australia.

GORDONITE   Clay Canyon, Fairfield, Utah County, Utah.

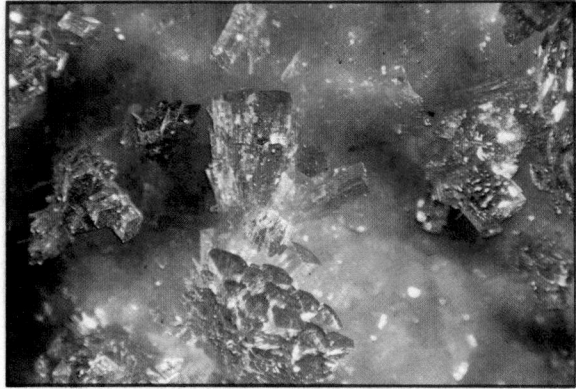

GORDONITE on variscite   Clay Canyon, Fairfield, Utah County, Utah.

GOWERITE with nobleite   Scramble mine, Death Valley, California.

GOYAZITE   Lengenbach quarry, Binnental, Switzerland.

GOYAZITE   Greenwood, Oxford County, Maine.

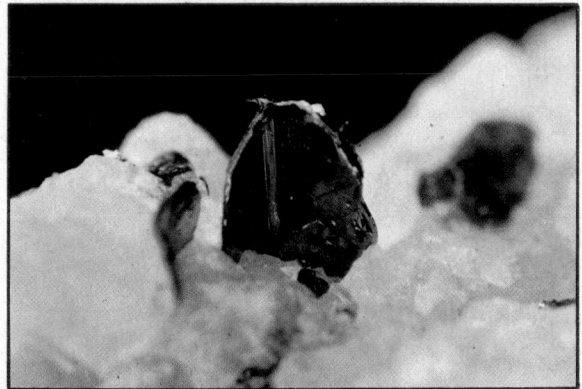

GRAPHITE   Kropfmuhl, Bavaria, Germany.

GRATONITE   Cerro de Pasco, Peru.

GREENOCKITE on anglesite in galena   Tintic Standard
mine, Tintic district, Utah.

GREENOCKITE   Bishopton, Renfrew, Scotland.

GREENOCKITE   Summit, Union County, New Jersey.

GROSSULAR (demantoid)   Val Malenco, Tyrol, Italy.

GROSSULAR with diopside   Ala, Piedmont, Italy.

GROSSULAR   Asbestos, Quebec, Canada.

GROSSULAR   Asbestos, Quebec, Canada.

GROSSULAR   Vasko, Roumania.

GUARINITE   Vesuvius, Italy.

GUILLEMINITE   Musonoi, Katanga, Zaire.

GYPSUM   Niccioleta, Tuscany, Italy.

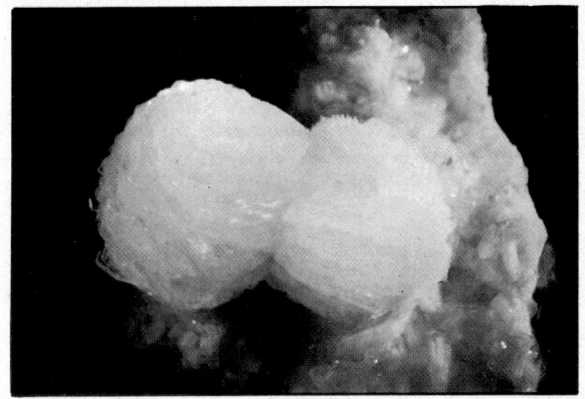

GYROLITE on apophyllite   Poona, India.

HALITE   Verde Valley, Yavapai County, Arizona.

HALITE, hopper crystal, Yachats, Oregon.

HANCOCKITE on axinite   Franklin, Sussex County, New
Jersey.

HARMOTOME   St. Andreasberg, Harz Mountains, Germany.

61

HARMOTOME   St. Andreasberg, Harz Mountains, Germany.

HAUERITE   Raddusa, Sicily.

HAUERITE   Jefferson County, Texas.

HAUYNE   Laacher See, Eifel district, Rhineland, Germany.

HEDENBERGITE   Moss mine, Nordmark, Sweden.

HEDYPHANE with rhodonite   Franklin, Sussex County, New Jersey.

HELVITE   Schwarzenberg, Saxony, Germany.

HEMATITE   Diamond Lake, Douglas County, Oregon.

HEMATITE   Puy de la Tache, France.

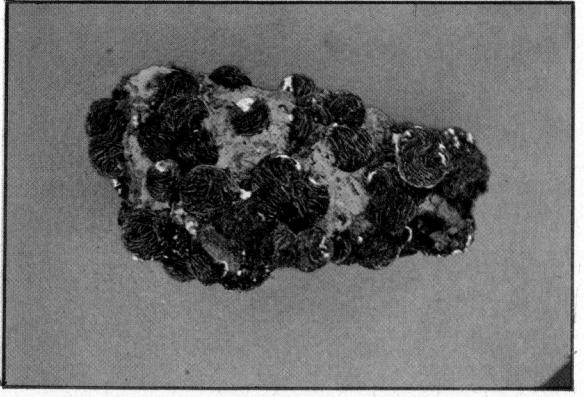

HEMATITE rosettes   Cleator Moor, Cumberland, England.

HEMATITE   Paterson, Passaic County, New Jersey.

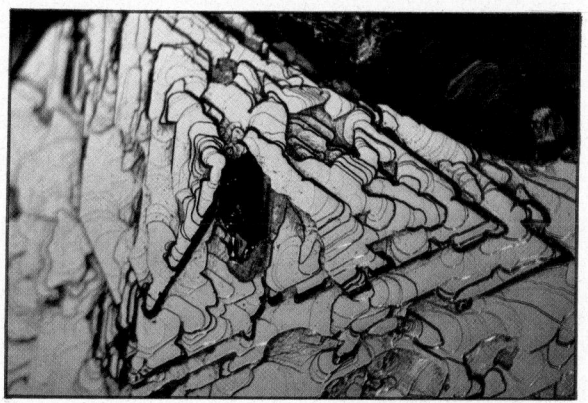

HEMATITE (surface figures)   Rio Oro, Costa Rica.

HEMIMORPHITE   Mapimi, Durango, Mexico.

HEMIMORPHITE with aurichalcite   Bingham, Socorro
County, New Mexico.

HERDERITE   Newry, Oxford County, Maine.

HESSITE   Botes, Roumania.

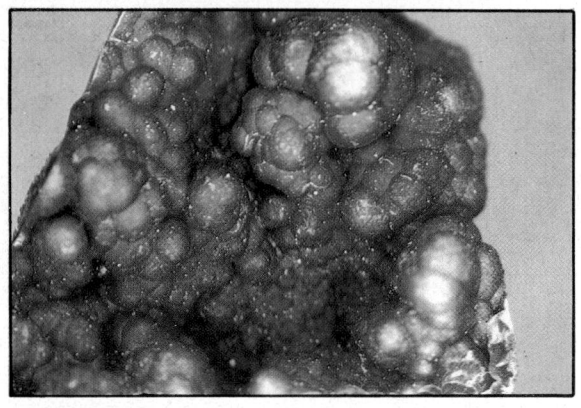

HETEROGENITE   Mindigi, Katanga, Zaire.

HEULANDITE   Poona, India.

HEULANDITE   Gunnahdah, New South Wales, Australia.

HIDALGOITE   Gold Hill, Tooele County, Utah.

HINSDALITE with covellite   Butte, Silver Bow County, Montana.

# I

## IANTHINITE
$UO_2 \cdot 5UO_3 \cdot 10H_2O$

CRYSTAL SYSTEM: Orthorhombic
Z: 4
LATTICE CONSTANTS:
 $a = 11.52$
 $b = 7.15$
 $c = 30.3$
3 STRONGEST DIFFRACTION LINES:
 7.63 (100)
 3.81 ( 80)
 3.24 ( 80)
OPTICAL CONSTANTS:
 $\alpha = 1.685$
 $\beta = 1.91$
 $\gamma = 1.93$
 $(-)2V = 58°$
HARDNESS: 2–3
DENSITY: 5.16 (Meas.)
CLEAVAGE: {001} perfect
 also {100}
HABIT: Crystals commonly minute plates or laths elongated along $b$-axis and flattened on {001}. Also thick tabular on {001}, or short prismatic along $b$-axis.

COLOR-LUSTER: Black to violet; submetallic; transparent. Streak brownish violet.

MODE OF OCCURRENCE: Occurs as minute crystals in pitchblende associated with schoepite, becquerelite, and other secondary uranium minerals from the Marshall Pass area, Saguache County, Colorado. Also found encrusting or veining altered uraninite at Shinkolobwe, Katanga, Zaire. at Woelsendorf, Bavaria; and at Bois Noirs, Bigay, and La Crouzille, Puy-de-Dôme, France.

BEST REF. IN ENGLISH: Frondel, Clifford, USGS Bull. 1064, p. 56–60 (1958). Guillemin, C., and Protas, J., *Am. Min.*, **44**: 1103–1104 (1959).

## ICE
$H_2O$ — Solid phase at or below $0°C$

CRYSTAL SYSTEM: Hexagonal
CLASS: 6/m 2/m 2/m
SPACE GROUP: $P6_3/mmc$
Z: 4
LATTICE CONSTANTS:
 $a = 4.51$
 $c = 7.35$
3 STRONGEST DIFFRACTION LINES:
 3.90 (100)
 3.66 (100)
 2.25 ( 90)
OPTICAL CONSTANTS:
 $\omega = 1.30907$
 $\epsilon = 1.31052$ (Na)
 (+)
HARDNESS: 1½
DENSITY: 0.9167
CLEAVAGE: None.
Fracture conchoidal. Brittle.

HABIT: Usually as very thin flattened six-rayed stellate snow crystals. Also as skeletal hexagonal prisms; compact aggregates of elongated lath-like crystals formed on the surface of water; in delicate forms as hoar frost. Sometimes as hailstones composed of hexagonal prisms or with concentric structure. Twinning on {0001} and {000$\bar{1}$} common.

COLOR-LUSTER: Colorless in thin layers; pale blue to greenish blue in thick layers. Often white due to flaws or included gas bubbles. Transparent. Vitreous. Streak colorless.

MODE OF OCCURRENCE: Occurs chiefly as a coating over lakes, ponds, rivers, and other bodies of water at low temperatures; as glaciers and icebergs; from atmospheric water vapor as snow, hail, sleet, and frost; and as stalactitic deposits in caverns and mine workings.

BEST REF. IN ENGLISH: Palache, et al., "Dana's System of Mineralogy," 7th Ed., v. I, p. 494–498, New York, Wiley, 1944.

## IDAITE
$Cu_3FeS_4$

CRYSTAL SYSTEM: Hexagonal

Z: 1
LATTICE CONSTANTS:
$a = 3.777$
$c = 11.18$
3 STRONGEST DIFFRACTION LINES:
3.134 (100)
2.818 ( 80)
1.887 ( 65)
HARDNESS: 2½
DENSITY: 4.2 (Meas. Synthetic)
        4.54 (Calc. Synthetic)
HABIT: Massive; lamellar.
COLOR-LUSTER: Between copper red and pinchbeck brown, opaque; metallic.
MODE OF OCCURRENCE: Occurs as a lamellar decomposition product of bornite, commonly associated with fine spindles of chalcopyrite, at the Ida mine, Khan, South West Africa, and widely distributed at many other copper deposits containing bornite.
BEST REF. IN ENGLISH: Frenzel, Gerhard, *Am. Min.,* **43**: 1219 (1958). Yund, Richard A., *Am. Min.,* **48**: 672–676 (1963).


**IDDINGSITE** = A mixture of silicates


## IDOCRASE (Vesuvianite)

$Ca_{10}Mg_2Al_4(SiO_4)_5(Si_2O_7)_2(OH)_4$
Var. Cyprine—blue
    Californite—compact, jade-like

CRYSTAL SYSTEM: Tetragonal
CLASS: 4/m 2/m 2/m
SPACE GROUP: P4/nnc
Z: 4
LATTICE CONSTANTS:
$a = 15.5\text{–}15.6$
$c \simeq 11.80$
3 STRONGEST DIFFRACTION LINES:
2.752 (100)
2.593 ( 80)
1.621 ( 60)
OPTICAL CONSTANTS:
$\omega = 1.703\text{–}1.752$
$\epsilon = 1.700\text{–}1.746$
    (-)
HARDNESS: 6–7
DENSITY: 3.33–3.45 (Meas.)
CLEAVAGE: {110} indistinct
              {100} very indistinct
              {001} very indistinct
Fracture uneven to conchoidal. Brittle.
HABIT: Crystals usually short prismatic; also pyramidal. Commonly massive, granular, cryptocrystalline, columnar.
COLOR-LUSTER: Various shades of green, brown, white, yellow, red, rarely blue. Transparent to translucent. Vitreous; sometimes resinous. Streak white.
MODE OF OCCURRENCE: Occurs in metamorphosed calcareous rocks, nepheline-syenites, and veins and pockets in ultrabasic rocks. Common associated minerals are grossular, epidote, diopside, phlogopite, calcite, and wollastonite.

Notable occurrences are found at Crestmore in Riverside County, in Siskiyou County (californite), and at many other places in California; as fine yellow crystals in Lyon County, Nevada; near Helena, Montana; at Magnet Cove, Arkansas; at Franklin, New Jersey (cyprine); and at localities in New York, Vermont, and Maine. Also in British Columbia, Ontario, and Quebec, Canada; at Lake Jaco, Chihuahua, Mexico; and at many places in Norway, Italy, Roumania, Czechoslovakia, Finland, USSR, Japan, Korea, and elsewhere.
BEST REF. IN ENGLISH: Deer, Howie, and Zussman, "Rock Forming Minerals," v. 1, p. 113–120, New York, Wiley, 1962.


## IDRIALITE (Curtisite)

$\sim C_{22}H_{14}$

CRYSTAL SYSTEM: Orthorhombic
Z: 4
LATTICE CONSTANTS:
$a = 8.07$
$b = 6.42$
$c = 27.75$
3 STRONGEST DIFFRACTION LINES:
4.94 (100)
3.39 ( 80)
4.04 ( 60)
HARDNESS: <2
DENSITY: 1.236 (Meas.)
CLEAVAGE: Perfect basal. Conchoidal fracture.
HABIT: Small irregular square or six-sided flakes; minute tabular crystals.
COLOR-LUSTER: Greenish yellow to light brown; vitreous to adamantine. Fluoresces bluish white in ultraviolet light.
MODE OF OCCURRENCE: Occurs with mercury ore at the type locality, Idria, Yugoslavia. Also occurs (curtisite) in a sandstone associated with opaline silica, realgar, and metacinnabarite, in the hot spring area at Skaggs Springs, Sonoma County, California.
BEST REF. IN ENGLISH: Strunz, H., and Contag, B., *Am. Min.,* **50**: 2109–2110 (1965).


**IGALIKITE** = Mixture of analcime and muscovite


## IIMORIITE

$Y_5(SiO_4)_3(OH)_3$

CRYSTAL SYSTEM: Triclinic
CLASS: $\bar{1}$
SPACE GROUP: $P\bar{1}$
Z: 1
LATTICE CONSTANTS:
$a = 11.6$
$b = 6.65 \times 2$
$c = 2 \times 13.1$
3 STRONGEST DIFFRACTION LINES:
2.96 (100)
3.02 ( 65)
2.78 ( 50)

OPTICAL CONSTANTS:
$\alpha = 1.786$
$\beta = 1.827$
Biaxial $(-)2V = 5-15°$
HARDNESS: 5½–6
DENSITY: 4.21 (Meas.)
4.22 (Calc.)
CLEAVAGE: {011} distinct
HABIT: Massive; up to 3 × 3 × 2 cm in size.
COLOR-LUSTER: Purplish gray. Vitreous. Streak white.
MODE OF OCCURRENCE: Occurs associated with biotite, fergusonite, monazite, uraninite, and an undetermined mineral, in quartz-microcline pegmatite at Fusamata, Kawamata-machi, Fukushima Prefecture, Japan. Also found as an alteration product of thalenite at the nearby Suishoyama (formerly Iisaka) pegmatite.
BEST REF. IN ENGLISH: Kato, Akira, and Nagashima, Kozo, *Am. Min.*, **58**: 140 (1973).

## IKAITE
$CaCO_3 \cdot 6H_2O$

CRYSTAL SYSTEM: Monoclinic
CLASS: 2/m or m
SPACE GROUP: C2/c or Cc
Z: 4
LATTICE CONSTANTS:
$a = 8.87$
$b = 8.23$
$c = 11.02$
$\beta = 110.2°$
3 STRONGEST DIFFRACTION LINES:
5.17 (100)
4.16 ( 80) Synthetic
5.85 ( 80)
OPTICAL CONSTANTS:
$\alpha \sim 1.455$
$\beta \sim 1.538$
$\gamma \sim 1.545$
$2V \sim 45°$
DENSITY: 1.80 (Meas.)
HABIT: Columnar-shaped skerries forming pillars reaching to within half a meter of the water surface.
COLOR-LUSTER: White, chalk-like.
MODE OF OCCURRENCE: Occurs in the inner part of Ika Fjord, 8 km south of Ivigtut, Greenland.
BEST REF. IN ENGLISH: Pauly, Hans, *Am. Min.*, **49**: 439 (1964).

## IKUNOLITE
$Bi_4 (S, Se)_3$

CRYSTAL SYSTEM: Hexagonal
CLASS: $\bar{3}$ 2/m
SPACE GROUP: R$\bar{3}$m
Z: 3 (hex.), 1 (rhomb.)
LATTICE CONSTANTS:
$a = 4.15$        $a_{rh} = 13.28$
$c = 39.19$        $\alpha = 18°00'$

3 STRONGEST DIFFRACTION LINES:
3.022 (100)
4.34 ( 50)
2.205 ( 30)
HARDNESS: 2
DENSITY: 7.8 (Meas.)
7.97 (Calc.)
CLEAVAGE: {0001} perfect
Very flexible.
HABIT: Massive; as minute grains.
COLOR-LUSTER: Lead gray. Metallic. Opaque. Streak dark gray.
MODE OF OCCURRENCE: Occurs with ferberite, bismuth, bismuthinite, and some cassiterite in a quartz vein at the Ikuno Mine, Hyôgo Prefecture, Japan.
BEST REF. IN ENGLISH: Kato, Akira, *Am. Min.*, **45**: 477–478 (1960).

## ILESITE
$(Mn, Zn, Fe)SO_4 \cdot 4H_2O$

CRYSTAL SYSTEM: Monoclinic
CLASS: 2/n
SPACE GROUP: P2$_1$/n
Z: 4
LATTICE CONSTANTS:
$a = 5.94$
$b = 13.76$
$c = 8.01$
$\beta = 90°48'$
3 STRONGEST DIFFRACTION LINES:
4.48 (100)
5.56 ( 65)
3.96 ( 65)
OPTICAL CONSTANTS:
$\alpha = 1.511$
$\beta = 1.519$ (synthetic)
$\gamma = 1.521$
$(-)2V$ = moderate
HARDNESS: Not determined.
DENSITY: 2.261 (Meas. snythetic)
2.263 (Calc. synthetic)
CLEAVAGE: Not determined.
HABIT: As crystalline aggregates; prismatic, often terminated by truncated pyramids.
COLOR-LUSTER: Clear green, becoming white on partial dehydration. Transparent.
MODE OF OCCURRENCE: Occurs as a secondary mineral in the oxidation zone of sulfide vein deposits at several mines in Hall Valley, Montezuma district, Park County, Colorado, especially on the McDonnell claim as streaks from 2 to 8 inches wide.
BEST REF. IN ENGLISH: Palache, et al., "Dana's System of Mineralogy," 7th Ed., v. II, p. 486–487, New York, Wiley, 1951.

## ILIMAUSSITE
$Ba_2 Na_4 CeFeNb_2 Si_8 O_{28} \cdot 5H_2O$

CRYSTAL SYSTEM: Hexagonal

CLASS: 6/m 2/m 2/m
SPACE GROUP: P6$_3$/mcm
Z: 3
LATTICE CONSTANTS:
  $a$ = 10.80
  $c$ = 20.31
3 STRONGEST DIFFRACTION LINES:
  2.67 (100)
  3.25 ( 60)
  3.12 ( 50)
OPTICAL CONSTANTS:
  $\omega$ = 1.689
  $\epsilon$ = 1.695
    (+)
HARDNESS: ~4
DENSITY: 3.6 (Meas.)
CLEAVAGE: Not determined.
HABIT: As lamellar aggregates, up to 15 × 10 × 3 mm.
COLOR-LUSTER: Brownish yellow; resinous. Alters easily, becoming dull and nontransparent.
MODE OF OCCURRENCE: Occurs associated with chkalovite and epistolite in a hydrothermal ussingite-analcime vein cutting sodalite syenite on Nakalaq, Ilimaussaq alkalic massif, South Greenland.
BEST REF. IN ENGLISH: Semenov, E. I., Kazakova, M. E., and Bukin, V. J., *Am. Min.*, **54**: 992-993 (1969).

## ILLITE
K$_{1-1.5}$Al$_4$(Si$_{7-6.5}$Al$_{1-1.5}$O$_{20}$)(OH)$_4$
Illite Group

CRYSTAL SYSTEM: Monoclinic
Z: 1
LATTICE CONSTANTS:
  $a$ = 5.2
  $b$ = 9.0
  $c$ = 9.95
  $\beta$ = 95°30'
3 STRONGEST DIFFRACTION LINES:
  10.0 (100)
  4.48 ( 90)
  3.33 ( 90)
OPTICAL CONSTANTS:
  $\alpha$ = 1.54-1.57
  $\beta$ = 1.57-1.61
  $\gamma$ = 1.57-1.61
(-)2V ⩽ 10°
HARDNESS: 1-2
DENSITY: 2.6-2.9
CLEAVAGE: {001} perfect
HABIT: Massive; extremely fine grained, often admixed with other clay minerals. Sometimes as micaceous hexagonal microflakes ranging from 200 to 2000 Å in size.
COLOR-LUSTER: Colorless in thin section; usually white; also various pale colors. Dull.
MODE OF OCCURRENCE: Illite, a clay mineral structurally related to the micas, is the dominant clay mineral in shales and many other sedimentary deposits throughout the world. It also occurs as a hydrothermal mineral in certain ore deposits and in alteration zones around hot springs.
BEST REF. IN ENGLISH: Deer, Howie, and Zussman, "Rock Forming Minerals," v. 3, p. 213-225, New York, Wiley, 1962.

## ILMAJOKITE
(Na, Ba, RE)$_{10}$Ti$_5$(Si, Al)$_{14}$O$_{22}$(OH)$_{44}$ · nH$_2$O

CRYSTAL SYSTEM: Monoclinic (?)
Z: 1
LATTICE CONSTANTS:
  $a$ ~ 23
  $b$ ~ 24.4
  $c$ ~ 37
3 STRONGEST DIFFRACTION LINES:
  11.5 (100)
  4.3 (100)
  2.44 (100)
OPTICAL CONSTANTS:
  $\alpha$ = 1.573
  $\beta$ = 1.576
  $\gamma$ = 1.579
(+)2V = 90°
HARDNESS: 1
DENSITY: 2.20 (Meas.)
CLEAVAGE: Perfect cleavages on the rhombic prism and pinacoid intersect at 72°. Brittle.
HABIT: As granular deposits, crusts, and brushes of crystals up to 2 mm long.
COLOR-LUSTER: Bright yellow. Vitreous.
MODE OF OCCURRENCE: Occurs on the walls of cavities in the central natrolitic zone of pegmatites in the Lovozero Tundra, near the valley of the Ilmajok River, Kola Peninsula, USSR. Associated minerals are aegirine, sphalerite, halite, and mountainite.
BEST REF. IN ENGLISH: Bussen, I. V., et al., *Am. Min.*, **58**: 139-140 (1973).

## ILMENITE
FeTiO$_3$

CRYSTAL SYSTEM: Hexagonal
CLASS: $\bar{3}$
SPACE GROUP: R$\bar{3}$
Z: 6
LATTICE CONSTANTS:
             (rhombohedral)
  $a$ = 5.083    $a_{rh}$ = 5.52
  $c$ = 14.04    $\alpha$ = 54°50'
3 STRONGEST DIFFRACTION LINES:
  2.74 (100)
  1.72 (100)
  2.54 ( 85)
OPTICAL CONSTANTS:
  $\omega$ = { birefringence
  $\epsilon$ = { very strong }
      (-)
HARDNESS: 5-6
DENSITY: 4.72 ± 0.04
         4.79 (Calc.)
CLEAVAGE: None.
          {0001} parting
          {10$\bar{1}$1} parting

Fracture conchoidal to uneven. Brittle.

HABIT: Crystals thick tabular; rarely acute rhombohedral. Also massive, lamellar to compact; granular; as sand. Twinning on {0001}; and on {10$\bar{1}$1}, lamellar.

COLOR-LUSTER: Iron black. Opaque. Metallic to dull. Streak black.

MODE OF OCCURRENCE: Occurs widespread as a common accessory mineral in many igneous and metamorphic rocks; as a major component in some black beach sand deposits; and in vein deposits and pegmatites. Found in California, Idaho, Colorado, Wyoming, South Dakota, Arkansas, Minnesota, Kentucky, Pennsylvania, Massachusetts, New York, Rhode Island, Connecticut, Virginia, Florida, and many other states. Notable deposits also occur in Canada, Mexico, England, Norway, Sweden, France, Italy, Switzerland, USSR, India, and Australia.

BEST REF. IN ENGLISH: Palache, et al., "Dana's System of Mineralogy," 7th Ed., v. I, p. 534–541, New York, Wiley, 1944.

# ILMENORUTILE = Niobian rutile

# ILSEMANNITE (Inadequately described mineral)
$Mo_3O_8 \cdot nH_2O$ (?)

CRYSTAL SYSTEM: Unknown
3 STRONGEST DIFFRACTION LINES:
3.36 (100)
4.23 ( 50) (Not reliable)
1.64 ( 50)
HARDNESS: Not determined.
DENSITY: Not determined.
CLEAVAGE: Not determined.
HABIT: Massive; as earthy crusts, stains or disseminated pigment.

COLOR-LUSTER: Black, blue-black, blue; becoming blue on exposure. Readily soluble in water, giving a deep-blue colored solution.

MODE OF OCCURRENCE: Occurs as an alteration product of molybdenite or other molybdenum minerals. Found in Inyo, Kern, and Shasta counties, California; at the Kiggins mine, Clackamas County, Oregon; at Ouray, Uintah County, and in the Marysvale area, Piute County, Utah; in Boulder, Clear Creek, Costilla, Jackson, Jefferson, Saguache, San Miguel, and Teller counties, Colorado; and at the Runge mine, Fall River County, and in Pennington and Shannon counties, South Dakota. It also occurs in Germany, South Africa, and Australia.

BEST REF. IN ENGLISH: Palache, et al., "Dana's System of Mineralogy," 7th Ed., v. I, p. 603–604, New York, Wiley, 1944.

# ILVAITE
$CaFe_2^{2+}Fe^{3+}Si_2O_8OH$

CRYSTAL SYSTEM: Orthorhombic
CLASS: 2/m 2/m 2/m
SPACE GROUP: Pbnm
Z: 4

LATTICE CONSTANTS:
$a$ = 8.78
$b$ = 12.99
$c$ = 5.85
3 STRONGEST DIFFRACTION LINES:
2.709 (100)
2.839 ( 85)
7.28 ( 70)
OPTICAL CONSTANTS:
$\alpha$ = 1.727
$\beta$ = 1.870
$\gamma$ = 1.883
(−)2V = 20°–30°
HARDNESS: 5½–6
DENSITY: 3.8–4.1 (Meas.)
4.064 (Calc.)
CLEAVAGE: {001} distinct
{010} distinct
Fracture uneven. Brittle.

HABIT: Crystals thick prismatic, diamond-shaped in cross-section, vertically striated. Also massive, compact or columnar.

COLOR-LUSTER: Black to grayish-black. Opaque. Glossy to dull submetallic. Streak black, often with brownish or greenish tint.

MODE OF OCCURRENCE: Occurs chiefly as a contact-metasomatic mineral; in iron, zinc and copper ore deposits; and in sodalite-syenite. Found in Fresno, Shasta, and Sonoma counties, California; as superb crystals up to three inches long at the Laxey mine, South Mountain, Owyhee County, Idaho; at North Clear Creek, Gilpin County, Colorado; at the Hanover-Empire zinc mine, Grant County, New Mexico; in the Julianehaab district, Greenland; as fine crystals at Rio Marina and Capo Calamita, Elba; on the island of Seriphos, Cyclades, Greece; and at localities in Germany, Italy, Algeria, Japan, and elsewhere.

BEST REF. IN ENGLISH: Winchell and Winchell, "Elements of Optical Mineralogy," 4th Ed., Pt. 2, p. 511, New York, Wiley, 1951.

# IMGREITE
NiTe

CRYSTAL SYSTEM: Hexagonal
LATTICE CONSTANTS:
(synthetic)
$a$ = 3.96
$c$ = 5.38
3 STRONGEST DIFFRACTION LINES:
2.84 (100)
2.10 ( 70)
1.56 ( 70)
HARDNESS: Microhardness 210–220 kg/mm$^2$.
DENSITY: 8.47 (Calc.)
CLEAVAGE: Not determined.
HABIT: Massive; as microscopic inclusions in hessite.
COLOR-LUSTER: Pale rose. Opaque. Metallic.
MODE OF OCCURRENCE: Occurs in association with sylvanite, calaverite, and other tellurides in the copper-nickel ores at the Nittis-Kumuzhya deposit, Monchegorsk region, USSR.

BEST REF. IN ENGLISH: Yushko-Zakharova, O. E., *Am. Min.*, **49**: 1151 (1964).

## IMHOFITE
$Tl_6CuAs_{16}S_{40}$

CRYSTAL SYSTEM: Monoclinic
CLASS: 2 or 2/m
SPACE GROUP: $P2_1$ or $P2_1/m$
LATTICE CONSTANTS:
  $a = 8.77$
  $b = 24.51$
  $c = 11.44$
  $\beta \simeq 107°$
3 STRONGEST DIFFRACTION LINES:
  2.675 (100)
  2.883 ( 90)
  3.960 ( 70)
HARDNESS: Very soft. Vickers hardness 38.
DENSITY: Not determined.
CLEAVAGE: Not determined.
HABIT: Thin plates, about 0.06 X 0.03 X 0.001 mm in size; also aggregates.
COLOR-LUSTER: Copper red; translucent.
MODE OF OCCURRENCE: Occurs sparingly at the Lengenbach quarry, Binnatal, Canton Wallis, Switzerland.
BEST REF. IN ENGLISH: Burri, G., Graeser, S., Morumo, F., and Nowacki, W., *Am. Min.*, **51**: 531 (1966). Nowacki, Werner, *Am. Min.*, **54**: 1498 (1969).

## INDERBORITE
$CaMgB_6O_{11} \cdot 11H_2O$

CRYSTAL SYSTEM: Monoclinic
CLASS: 2/m
SPACE GROUP: A2/a
Z: 4
LATTICE CONSTANTS:
  $a = 19.11$
  $b = 7.46$
  $c = 12.22$
  $\beta = 90°48'$
3 STRONGEST DIFFRACTION LINES:
  3.35 (100)
  3.26 ( 80)
  3.07 ( 80)
OPTICAL CONSTANTS:
    $\alpha = 1.482$
    $\beta = 1.512$ (Na)
    $\gamma = 1.530$
(−)2V = 77°
HARDNESS: 3½
DENSITY: 2.004 (Meas.)
CLEAVAGE: {100} good.
Fracture conchoidal.
HABIT: As well-developed prismatic crystals up to 2.0 cm. in size; also as coarsely crystalline aggregates.
COLOR-LUSTER: Colorless; also white. Transparent. Vitreous.
MODE OF OCCURRENCE: Occurs associated with cole-

manite, ulexite, and other borate minerals in the deposits of Inder Lake, western Kazakhstan, USSR.
BEST REF. IN ENGLISH: Palache, et al., "Dana's System of Mineralogy," 7th Ed., v. II, p. 355–356, New York, Wiley, 1951.

## INDERITE (Lesserite)
$Mg_2B_6O_{11} \cdot 15H_2O$
Dimorphous with kurnakovite

CRYSTAL SYSTEM: Monoclinic
CLASS: 2/m
SPACE GROUP: $P2_1/a$
Z: 2
LATTICE CONSTANTS:
  $a = 12.12$
  $b = 13.18$
  $c = 6.83$
  $\beta = 104°49'$ (X-ray), 104°32' (morph.)
3 STRONGEST DIFFRACTION LINES:
  5.66 (100)
  3.26 (100)
  5.01 ( 90)
OPTICAL CONSTANTS:
    $\alpha = 1.488$
    $\beta = 1.491$
    $\gamma = 1.505$
(+)2V = 37°
HARDNESS: 2½
DENSITY: 1.785 (Meas.)
        1.76 (Calc.)
CLEAVAGE: {110} good
        {001} indistinct
Fracture flat conchoidal to irregular. Brittle.
HABIT: Prismatic crystals up to 10 cm long and 1 cm in cross section. Forms: {110}, {120}, and {001}. Doubly terminated crystals common. Also as reniform nodular aggregates.
COLOR-LUSTER: Colorless, transparent; weak vitreous. Massive aggregates white to pink.
MODE OF OCCURRENCE: Occurs in the Jenifer mine and in large quantities in the Open Pit at Boron in the Kramer borate mining area, Kern County, California. Also found as small nodules in red clay at the Inder borate deposit in western Kazakhstan, USSR.
BEST REF. IN ENGLISH: *Am. Min.*, **41**: 927 (1956).

## INDIALITE (α-Cordierite)
$(Mg, Fe)_2Al_4Si_5O_{18}$
Isostructural with beryl

CRYSTAL SYSTEM: Hexagonal
CLASS: 6/m 2/m 2/m
SPACE GROUP: P6/mcc
Z: 2
LATTICE CONSTANTS:
  $a = 9.812$
  $c = 9.351$

3 STRONGEST DIFFRACTION LINES:
8.48 (100)
4.096 ( 55)
3.037 ( 55)
OPTICAL CONSTANTS:
$\omega$ = 1.539
$\epsilon$ = 1.534
(–)
HARDNESS: 7–7½
DENSITY: Not determined.
CLEAVAGE: Not determined.
HABIT: Crystals minute hexagonal prisms, single or in clusters. Also as small irregular grains.
COLOR-LUSTER: Pleochroic in thin section, colorless to pale mauve. Transparent. Vitreous.
MODE OF OCCURRENCE: Occurs in sedimentary rocks fused by the burning of a coal seam in the Bokaro coalfield, India.
BEST REF. IN ENGLISH: Miyashiro, Akiho, and Iiyama, Toshimichi, *Am. Min.*, **40**: 787 (1955). *Am. J. Sci.*, **253**: 193 (1955). Venkatesh, V., *Am. Min.*, **37**: 831–848 (1952).


**INDICOLITE** = Blue variety of tourmaline


**INDIGIRITE (Inadequately described mineral)**
$Mg_2Al_2(CO_3)_4(OH)_2 \cdot 15H_2O$

CRYSTAL SYSTEM: Unknown
LATTICE CONSTANTS:
$b(?)$ = 3.16 (perpendicular to elongation)
$c$ = 6.23 (parallel to elongation)
3 STRONGEST DIFFRACTION LINES:
5.80 (100)
7.62 ( 90)
5.24 ( 90)
OPTICAL CONSTANTS:
$\alpha$ = 1.472
$\gamma$ = 1.502
HARDNESS: ~2
DENSITY: 1.6 (Meas.)
CLEAVAGE: Not determined. Plates and needles flexible.
HABIT: As rosette-like, radiating fibrous aggregates of very fine fibers, needles or plates up to 1 mm in length.
COLOR-LUSTER: Snow white. Translucent. Vitreous to silky.
MODE OF OCCURRENCE: Occurs in association with gypsum, quartz, gibbsite, and amorphous iron oxide in the oxidation zone of the Sarylakh gold-antimony deposit in the basin of the upper part of the Indigirki River, northeastern Yakutia, USSR.
BEST REF. IN ENGLISH: Indolev, L. N., Zhdanov, Yu. Ya., Kashertseva, K. I., Soknev, Y. S., and Del'Yanidi, K. I., *Am. Min.*, **57**: 326–327 (1972).


**INDITE**
$FeIn_2S_4$

CRYSTAL SYSTEM: Cubic

CLASS: 4/m $\bar{3}$ 2/m
SPACE GROUP: Fd3m
Z: 8
LATTICE CONSTANT:
$a$ = 10.62
3 STRONGEST DIFFRACTION LINES:
3.20 (100)
1.877 ( 90)
2.05 ( 70)
HARDNESS: Microhardness 309 kg/mm². Scratched by a steel needle.
DENSITY: 4.67 (Calc.)
CLEAVAGE: Not determined.
HABIT: Small grains; usually less than 0.2 mm, rarely up to 0.5 mm.
COLOR-LUSTER: Iron black; metallic. Opaque.
MODE OF OCCURRENCE: Occurs in cassiterite from the Dzhalindin deposit, Little Khingan Ridge, Far Eastern Siberia.
BEST REF. IN ENGLISH: Genkin, A. D., and Murav'eva, I. V., *Am. Min.*, **49**: 439–440 (1964).


**INDIUM**
In

CRYSTAL SYSTEM: Tetragonal
CLASS: 4/m 2/m 2/m
SPACE GROUP: I4/mmm
Z: 2
LATTICE CONSTANTS:
$a$ = 3.25
$c$ = 4.95
3 STRONGEST DIFFRACTION LINES:
2.715 (100)
2.298 ( 36)
1.683 ( 24)
HARDNESS: Microhardness 130–159, av. 142 kg/mm².
DENSITY: 7.2–7.42 (Meas.)
7.286 (Calc.)
HABIT: Massive; as embedded grains up to 1 mm in size.
COLOR-LUSTER: Gray with yellowish tint. Opaque. Metallic.
MODE OF OCCURRENCE: Occurs in close association with native lead in greisenized and albitized granite in Eastern Transbaikal.
BEST REF. IN ENGLISH: Ivanov, V. V., *Am. Min.*, **52**: 299 (1967).


**INESITE**
$Ca_2Mn_7Si_{10}O_{28}(OH)_2 \cdot 5H_2O$

CRYSTAL SYSTEM: Triclinic
CLASS: $\bar{1}$
SPACE GROUP: P$\bar{1}$
Z: 1
LATTICE CONSTANTS:
$a$ = 8.927    $\alpha$ = 91°48′
$b$ = 9.245    $\beta$ = 132°35′
$c$ = 11.954    $\gamma$ = 94°22′

3 STRONGEST DIFFRACTION LINES:
9.16 (100)
2.92 ( 80)
2.84 ( 80)
OPTICAL CONSTANTS:
$\alpha = 1.6178$
$\beta = 1.6384$
$\gamma = 1.6519$
$(-)2V = 76°50'$
HARDNESS: 5½
DENSITY: 3.033 (Meas.)
3.041 (Calc.)
CLEAVAGE: {010} perfect
{100} good
Brittle.

HABIT: Single crystals prismatic or tabular; elongated on c-axis, from 1 to 15 mm in length and 1–8 mm in width; often exhibit a "chisel-shape"; rarely doubly terminated. Also as acicular radiating aggregates and veinlets.

COLOR-LUSTER: Various shades of pink and pinkish orange; darkens to brown on exposure to light and atmosphere. Translucent. Vitreous to somewhat silky.

MODE OF OCCURRENCE: Occurs as fine crystal aggregates and veinlets in association with rhodochrosite and bementite at the Hale Creek mine, Trinity County, California; as spectacular hemispherical masses, often encrusted with colorless apophyllite crystals, on the 19 level of the New Broken Hill Consolidated mine, Australia; at Jacobsberg, Långban, and at the Harstig mine near Persberg, Sweden; at Villa Carona, Durango, Mexico; at Nanzenbach, Germany; near Rendaiji, Japan; and at several other localities.

BEST REF. IN ENGLISH: Ryall, William R., and Threadgold, Ian M., *Am. Min.*, 53: 1614–1634 (1968).

# INNELITE
$Na_2(Ba, K)_4(Ca, Mg, Fe)Ti_3SiO_4O_{18}(OH, F)_{1.5}(SO_4)$

CRYSTAL SYSTEM: Triclinic
CLASS: 1
SPACE GROUP: P1
Z: 1
LATTICE CONSTANTS:
$a = 5.38$      $\alpha \sim 99°$
$b = 7.14$      $\beta = 95°$
$c = 14.76$    $\gamma \sim 90°$
3 STRONGEST DIFFRACTION LINES:
3.92 (100)
3.04 ( 60)
2.95 ( 60)
OPTICAL CONSTANTS:
$\alpha = 1.726$
$\beta = 1.737$
$\gamma = 1.766$
$(+)2V \simeq 82°$
HARDNESS: 4½–5
DENSITY: 3.96 (Meas.)
CLEAVAGE: {010} perfect
{110} perfect
{1$\bar{0}$1} perfect
{001} distinct
Brittle.

HABIT: As plates a few millimeters to some centimeters in size; crystal faces rare. Polysynthetic Manebach twinning.

COLOR-LUSTER: Pale yellow to brown. Vitreous on cleavages, oily on fracture.

MODE OF OCCURRENCE: Occurs in pulaskites and in aegirine-eckermannite-microcline pegmatites of the Inagli massif, and in shonkinite of the Yakokutsk massif, southern Yakutsk, USSR.

BEST REF. IN ENGLISH: Kravchenko, S. M., Vlasova, E. V., Kazakova, M. E., Ilokhin, V. V., and Abrashev, K. K., *Am. Min.*, 47: 805–806 (1962).

# INSIZWAITE
$Pt(Bi, Sb)_2$

CRYSTAL SYSTEM: Cubic
CLASS: $2/m \bar{3}$
SPACE GROUP: Pa3
Z: 1
LATTICE CONSTANT:
$a = 6.625$
3 STRONGEST DIFFRACTION LINES:
1.99 (100)
2.96 ( 80)
2.70 ( 80)
OPTICAL CONSTANTS: Isotropic.
HARDNESS: Microhardness 488–540 (av. 519) kg/mm² with 25 g load.
DENSITY: 12.8 (Calc.)
CLEAVAGE: Not determined.
HABIT: Massive; as small rounded grains.
COLOR-LUSTER: White under reflected light (in air and in oil). Opaque. Metallic.

MODE OF OCCURRENCE: The mineral occurs in a vein associated with pentlandite (containing exsolved chalcopyrite) and chalcopyrite (containing exsolved bornite) and with parkerite and niggliite at the Insizwa deposit, Waterfall Gorge, Pondoland, and East Griqualand areas (Transkei), South Africa.

BEST REF. IN ENGLISH: Cabri, L. J., and Harris, D. C., *Min. Mag.*, 38: 794–800 (1972).

# INYOITE
$Ca_2B_6O_{11} \cdot 13H_2O$

CRYSTAL SYSTEM: Monoclinic
CLASS: 2/m
SPACE GROUP: $P2_1/a$
Z: 2
LATTICE CONSTANTS:
$a = 10.63$
$b = 12.06$
$c = 8.405$
$\beta = 114°01'$
3 STRONGEST DIFFRACTION LINES:
3.03  (100)
7.59  ( 70)
2.286 ( 25)

OPTICAL CONSTANTS:
$\alpha = 1.492$-$1.495$
$\beta = 1.505$-$1.51$
$\gamma = 1.516$-$1.520$
$(-)2V = 70°$-$84°$
HARDNESS: 2
DENSITY: 1.875 (Meas.)
1.873 (Calc.)
CLEAVAGE: {001} good
Fracture irregular. Brittle.

HABIT: Crystals short prismatic [001] to tabular {001}, with {110} and {001} dominant. Massive granular; spherulitic aggregates.

COLOR-LUSTER: Colorless and transparent. Vitreous. White and cloudy when partially dehydrated.

MODE OF OCCURRENCE: Occurs abundantly in the borate deposits of the Inder region, western Kazakhstan, USSR; with meyerhofferite, colemanite, and priceite in the Mount Blanco borate district, near Death Valley, Inyo County, California; and lining cavities in gypsum at Hillsborough, New Brunswick, Canada.

BEST REF. IN ENGLISH: "Dana's System of Mineralogy," 7th Ed., v. 2, p. 358–360, New York, Wiley, 1951. Christ, C. L., *Am. Min.*, **38**: 912–918 (1953).

## IODARGYRITE = Iodyrite

## IODYRITE (Iodargyrite)
AgI

CRYSTAL SYSTEM: Hexagonal
CLASS: 6mm
SPACE GROUP: $P6_3mc$
Z: 2
LATTICE CONSTANTS:
$a = 4.58$ kX
$c = 7.49$
3 STRONGEST DIFFRACTION LINES:
3.73 (100)
2.296 ( 85)
3.98 ( 60)
OPTICAL CONSTANTS:
$\omega = 2.21$
$\epsilon = 2.22$
(+)
HARDNESS: 1½
DENSITY: 5.69 (Meas.)
5.70 (Calc.)
CLEAVAGE: {0001} perfect
Fracture conchoidal. Sectile; flexible.

HABIT: Crystals prismatic [0001], or tabular {0001}; distinct hemimorphic development. In rosettes and parallel groups. Also as scaly or lamellar masses. Twinned commonly as fourlings.

COLOR-LUSTER: Colorless when fresh, pale yellow on exposure to light; also yellow, greenish yellow, gray, brownish. Transparent to translucent. Resinous to adamantine, pearly on cleavage. Streak yellow.

MODE OF OCCURRENCE: Occurs as a secondary mineral in oxidized silver ore deposits. In the United States found as minute crystals in the Chrysotile mine, Leadville district, Lake County, and at the Red Cloud mine, Boulder County, Colorado; at Goldfield, and as fine crystals at Tonapah, Nevada; at Lake Valley, Sierra County, New Mexico, and at the Commonwealth mine, Cochise County, Arizona. It is also found in large amounts as excellent greenish yellow crystals with limonite at the Broken Hill mine, New South Wales, Australia; at Chanarcillo and other places in Chile; at several silver deposits in Mexico; and also in Spain, France, Germany, USSR, and in the Katanga district, Zaire.

BEST REF. IN ENGLISH: Palache, et al., "Dana's System of Mineralogy," 7th Ed., v. II, p. 22–25, New York, Wiley, 1951.

## IOLITE = Cordierite

## IOWAITE
$Mg_4Fe^{3+}(OH)_8OCl \cdot 2$-$4H_2O$

CRYSTAL SYSTEM: Hexagonal
LATTICE CONSTANTS:
$a = 3.119$
$c = 24.25$
3 STRONGEST DIFFRACTION LINES:
8.109 (100)
4.047 ( 40)
2.363 ( 27)
OPTICAL CONSTANTS:
$\omega = 1.543$
$\epsilon = 1.533$
(-)
HARDNESS: ~1.5
DENSITY: 2.11 (Meas.)
CLEAVAGE: {0001} perfect
HABIT: Platy crystals up to 2–3 mm in length.

COLOR-LUSTER: Bluish green; on exposure to the atmosphere, undergoes color changes from bluish green to light green, to whitish green, to whitish green with a rusty red overtone, due to loss of zeolitic water. Greasy luster; greasy or soapy feel.

MODE OF OCCURRENCE: Found in diamond drill core at a depth of 1000–1500 feet in Precambrian serpentinite from Sioux County, Iowa. It is associated with chrysotile, dolomite, brucite, calcite, magnesite, and pyrite, all of which post-date the serpentinization of the rock.

BEST REF. IN ENGLISH: Kohls, D. W., and Rodda, J. L., *Am. Min.*, **52**: 1261-1271 (1967). Allmann, R., and Donnay, J. D. H., *Am. Min.*, **54**: 296-299 (1969).

## IRANITE
$PbCrO_4 \cdot H_2O$

CRYSTAL SYSTEM: Triclinic
CLASS: 1
SPACE GROUP: P1
Z: 8

LATTICE CONSTANTS:
$a = 10.02$  $\alpha = 104°30'$
$b = 9.54$  $\beta = 66°$
$c = 9.89$  $\gamma = 108°30'$
3 STRONGEST DIFFRACTION LINES:
3.60 (100)
3.49 (100)
3.28 (100)
OPTICAL CONSTANTS:
$\alpha = 2.25–2.30$
$\gamma = 2.40–2.50$
$2V$ = large
HARDNESS:  Not determined.
DENSITY:  5.8 (Calc.)
HABIT:  Crystals less than 0.5 mm long.  Forms $\{\overline{1}00\}$, $\{010\}$, $\{011\}$ most prominent, and $\{\overline{1}0\overline{1}\}$, $\{00\overline{1}\}$, $\{10\overline{2}\}$.
COLOR-LUSTER:  Saffron yellow; vitreous.
MODE OF OCCURRENCE:  Occurs in very small amount associated with dioptase, fornacite, and another lead chromate at the ancient Sebarz Mine, northeast of Anarak, Central Iran.
BEST REF. IN ENGLISH:  Bariand, P., and Herpin, P., *Am. Min.*, **48**: 1417 (1963).


# IRARSITE
(Ir, Ru, Rh, Pt) AsS

CRYSTAL SYSTEM:  Cubic
Z:  1
LATTICE CONSTANT:
$a = 5.777$
3 STRONGEST DIFFRACTION LINES:
3.32 (100)
2.87 (100)
1.74 (100)
HARDNESS:  Microhardness 976 kg/mm$^2$.  Not scratched by a steel needle.
DENSITY:  11.92 (Meas.)
CLEAVAGE:  Not determined.  Brittle.
HABIT:  Massive.
COLOR-LUSTER:  Iron black.  Opaque.  Metallic.
MODE OF OCCURRENCE:  Occurs in chromite in hortonolite dunite from the Onverwacht deposit, South Africa, intergrown with native platinum and with ruthenian hollingworthite.
BEST REF. IN ENGLISH:  Genkin, A. D., Zhuravlev, N. N., Troneva, N. V., and Murav'eva, I. V., *Am. Min.*, **52**: 1580 (1967).


# IRIDOSMINE
Ir, Os
Var. Platinian
     Rhodian
     Ruthenian

CRYSTAL SYSTEM:  Hexagonal (cubic)
CLASS:  6/m 2/m 2/m (4/m $\overline{3}$ 2/m)
SPACE GROUP:  P6$_3$/mmc (Fm3m)
Z:  2 (4)

LATTICE CONSTANTS:
(hexagonal)  (cubic)
$a = 2.719$  $a = 3.88$
$c = 4.276$
3 STRONGEST DIFFRACTION LINES:
(cubic)
2.23 (100)
1.17 ( 70)
1.93 ( 50)
HARDNESS:  6–7
DENSITY:  19–21
CLEAVAGE:  {0001} perfect, difficult
Slightly malleable to brittle.
HABIT:  Crystals tabular, flattened {0001}; rarely short prismatic.
COLOR-LUSTER:  Tin white.  Opaque.  Metallic.  Streak gray.
MODE OF OCCURRENCE:  Occurs chiefly in placer deposits in association with native platinum.  Found in Butte, Del Norte, Humboldt, Placer, Shasta, and Trinity counties, California; in southern Oregon; in British Columbia and Quebec, Canada; in the gold-bearing conglomerates of the Witwatersrand, South Africa; and at deposits in Brazil, Columbia, USSR, Borneo, Australia, and Tasmania.
BEST REF. IN ENGLISH:  Palache, et al., "Dana's System of Mineralogy," 7th Ed., v. I, p. 111–113, New York, Wiley, 1944.


# IRIGINITE
U(MoO$_4$)$_2$ (OH)$_2$ · 2H$_2$O

CRYSTAL SYSTEM:  Monoclinic
3 STRONGEST DIFFRACTION LINES:
6.35  (100)
3.21  (100)
2.621 ( 55)
OPTICAL CONSTANTS:
$\alpha = 1.730$
$\beta \sim 1.82$
$\gamma \sim 1.93$
HARDNESS:  1–2
DENSITY:  3.84 (Meas.)
         3.975 (Calc.)
CLEAVAGE:  Not determined.
HABIT:  Fine-grained dense aggregates.
COLOR-LUSTER:  Canary yellow; vitreous to dull.
MODE OF OCCURRENCE:  Occurs associated with brannerite and with other uranium-molybdenum minerals in granulated albitite at an unspecified locality, presumably in the USSR; also disseminated in a channel sandstone in the Chadron formation on the south slope of Hart Table southwest of Scenic, South Dakota.
BEST REF. IN ENGLISH:  *Am. Min.*, **43**: 379 (1958). Epshtein, G. Yu., *Am. Min.*, **45**: 257–258 (1960).


# IRINITE = Thorian Loparite

# IRON
Fe
Var.  Kamacite, meteoric $\alpha$-Fe, Ni

CRYSTAL SYSTEM: Cubic
CLASS: 4/m $\bar{3}$ 2/m
Z: 2 (terrestrial; 54 (kamacite)
LATTICE CONSTANTS:
 (terrestrial)          (kamacite)
 $a = 2.866$    $a = 2.865$  (disordered phase)
               $a \approx 8.60$   (super lattice)
3 STRONGEST DIFFRACTION LINES:
 (terrestrial)      (kamacite)
 2.0268 (100)     2.031 (100)
 1.1702 ( 30)     1.967 ( 60)
 1.4332 ( 20)     1.435 ( 30)
OPTICAL CONSTANT:
 $N = 2.36, 1.73$
 HARDNESS: 4
 DENSITY: 7.3–7.87
          7.87 (Calc.)
 CLEAVAGE: {001}
  Parting on {112}
Fracture hackly. Malleable.
 HABIT: Crystals rare. Terrestrial iron usually as small grains up to masses 20 tons in weight. Meteoric iron in plates and lamellar masses, often intergrown with nickel-iron.
 COLOR-LUSTER: Steel gray to iron black. Opaque. Metallic. Magnetic.
 MODE OF OCCURRENCE: Occurs commonly in meteorites (kamacite); more rarely in igneous rocks, especially basalts, and in carbonaceous sedimentary deposits. Found in the United States in shale near New Brunswick, New Jersey, and in coal deposits at Cameron, Clinton County, Missouri. It also occurs as large masses in basalt on Disko Island, Greenland; as minute pellets in feldspar in the Nipissing district and elsewhere in Ontario, Canada; and at localities in Ireland, Scotland, France, Germany, Poland, Czechoslovakia, and USSR.
 BEST REF. IN ENGLISH: Palache, et al., "Dana's System of Mineralogy," 7th Ed., v. I, p. 114–116, New York, Wiley, 1944.


**IRON CORDIERITE** = Sekaninaite


**IRON-MONTICELLITE** = Kirchsteinite


**ISHIGANEITE** = Mixture of cryptomelane and birnessite


**ISHIKAWAITE (Uranium-rich variety of samarskite)**
(Fe, U, Y)$_2$ (Nb, Ta)$_2$O$_7$

CRYSTAL SYSTEM: Orthorhombic
 HARDNESS: 5–6
 DENSITY: 6.2–6.4 (Meas.)
 CLEAVAGE: None. Fracture conchoidal. Brittle.
 HABIT: Crystals tabular, up to 10 × 6 × 2 mm in size.
 COLOR-LUSTER: Black. Opaque. Resinous. Streak dark brown. Radioactive.
 MODE OF OCCURRENCE: Occurs in pegmatites con-

taining samarskite, fergusonite, zircon, monazite, xenotime, and other minerals, in the Ishikawa district, Iwaki Province, Japan.
 BEST REF. IN ENGLISH: Vlasov, K. A., "Mineralogy of Rare Elements," v. II, p. 533–544, Israel Program for Scientific Translations, 1966.


**ISOCLASITE (Inadequately described mineral)**
Ca$_2$PO$_4$OH · 2H$_2$O

CRYSTAL SYSTEM: Monoclinic
OPTICAL CONSTANTS:
  $\alpha = 1.565$
  $\beta = 1.568$
  $\gamma = 1.580$
 (+)2V = medium
 HARDNESS: 1½
 DENSITY: 2.92 (Meas.)
 CLEAVAGE: {010} distinct
 HABIT: As minute prismatic crystals and as cotton-like fibers.
 COLOR-LUSTER: Colorless, white. Transparent to translucent. Vitreous; pearly on cleavage.
 MODE OF OCCURRENCE: Occurs with chalcedony and dolomite at Joachimsthal, Czechoslovakia.
 BEST REF. IN ENGLISH: Palache, et al., "Dana's System of Mineralogy," 7th Ed., v. II, p. 933–934, New York, Wiley, 1951.


**ISOKITE**
CaMgPO$_4$F

CRYSTAL SYSTEM: Monoclinic
CLASS: 2/m
SPACE GROUP: C2/c
Z: 4
LATTICE CONSTANTS:
 $a = 6.52$
 $b = 8.75$
 $c = 7.51$
 $\beta = 121°28'$
3 STRONGEST DIFFRACTION LINES:
 3.19 (100)
 3.02 (100)
 2.63 (100)
OPTICAL CONSTANTS:
  $\alpha = 1.590$
  $\beta = 1.594$
  $\gamma = 1.614$
 (+)2V = 51°
 HARDNESS: ~5
 DENSITY: 3.27 (Meas.)
          3.29 (Calc.)
 CLEAVAGE: {010} very good
 HABIT: As fibrous spherulites up to 3.0 mm in diameter; also as crystalline plates showing forms {001}, {100}, {101}, {10$\bar{1}$}, and {10$\bar{2}$}.
 COLOR-LUSTER: Colorless to buff or pinkish; silky to slightly pearly. Fluoresces blue under long-wave ultraviolet light.

MODE OF OCCURRENCE: Occurs associated with strontian fluorapatite, ilmenite, monazite, barite, pyrite, sellaite, and pyrochlore in ankeritic rocks of a carbonatite plug at Nkumbwa Hill, 15 miles east of Isoka, Zambia.

BEST REF. IN ENGLISH: Deans, T., and McConnell, J. D. C., *Min. Mag.*, **30**: 681–690 (1955).

## ISOSTANNITE = Cubic phase of stannite
$Cu_2FeSnS_4$ (?)

## ISTISUITE (Inadequately described mineral.)
$(Ca, Na)_7(Si, Al)_8(O, OH)_{24}$

CRYSTAL SYSTEM: Monoclinic
HABIT: Elongated prisms and in fibers.
COLOR-LUSTER: Gray.
MODE OF OCCURRENCE: Occurs with wollastonite in the skarn zone near Istisu, Dalidag, Little Caucasus.

BEST REF. IN ENGLISH: Kashkai, M. A., and Mamedov, A. I., *Am. Min.*, **42**: 117 (1957).

## ITOÏTE
$Pb_3GeO_2(OH_2)(SO_4)_2$

CRYSTAL SYSTEM: Orthorhombic
CLASS: 2/m 2/m 2/m
SPACE GROUP: Pnma
Z: 4
LATTICE CONSTANTS:
$a = 8.47$
$b = 5.38$
$c = 6.94$
3 STRONGEST DIFFRACTION LINES:
2.065 (100)
3.326 ( 90)
3.003 ( 90)
OPTICAL CONSTANTS: Mean index = 1.84–1.85
HARDNESS: Not determined.
DENSITY: 6.67 (Calc.)
CLEAVAGE: Not determined.
HABIT: Fine-grained pseudomorphs after fleischerite.
COLOR-LUSTER: White; silky.
MODE OF OCCURRENCE: Occurs associated with cerussite, mimetite, and altered tennantite, also as a crust on plumbojarosite and mimetite on dolomite in the upper oxidation zone of the Tsumeb mine, South West Africa.

BEST REF. IN ENGLISH: Frondel, C., and Strunz, H., *Am. Min.*, **45**: 1313 (1960).

## IVANOVITE (Species status uncertain)
Hydrated borate-chloride of Ca, K

CRYSTAL SYSTEM: Monoclinic
LATTICE CONSTANTS:
$a = 14.72$
$b = 8.02$
$c = 8.60$
$\beta = 91°8'$
OPTICAL CONSTANTS:
$\alpha = 1.504$
$\beta = 1.523$
$\gamma = 1.531$
$(-)2V = 58°–72°$
CLEAVAGE: {010} good
HABIT: Crystals short prismatic or tabular.
MODE OF OCCURRENCE: Found in salts of the Inder borate deposits.

BEST REF. IN ENGLISH: Nefedov, E. I., *Am. Min.*, **40**: 552 (1955).

## IXIOLITE
$(Ta, Fe, Sn, Nb, Mn)_4O_8$

CRYSTAL SYSTEM: Orthorhombic
CLASS: 2/m 2/m 2/m
SPACE GROUP: Pcan
Z: 1
LATTICE CONSTANTS:
$a = 5.731$
$b = 4.742$
$c = 5.152$
3 STRONGEST DIFFRACTION LINES:
2.98 (100)
3.65 ( 32)
1.459 ( 29)
HARDNESS: 6–6½
DENSITY: 7.03–7.23 (Meas.)
7.392 (Calc.)
CLEAVAGE: Fracture uneven to subconchoidal. Brittle.
HABIT: Crystals prismatic, rectangular, with forms {100}, {001}, and {010}. Twinning on {103}, not common.
COLOR-LUSTER: Blackish gray to steel gray. Opaque. Metallic.
MODE OF OCCURRENCE: Found in pegmatite at Skogböle in Kimito, Finland.

BEST REF. IN ENGLISH: Nickel, E. H., Rowland, J. F., and McAdam, R. C., *Am. Min.*, **48**: 961–979 (1963).

# J

## JACOBSITE
(Mn, Fe, Mg)(Fe, Mn)$_2$O$_4$
Magnetite Series
Spinel Group

CRYSTAL SYSTEM: Cubic
CLASS: 4/m $\bar{3}$ 2/m
SPACE GROUP: Fd3m
Z: 8
LATTICE CONSTANT:
 $a$ = 8.42–8.51
3 STRONGEST DIFFRACTION LINES:
 2.563 (100)
 1.503 ( 40)
 3.005 ( 35)
OPTICAL CONSTANT:
 $N$ = 2.3
 HARDNESS: 5½–6½
 DENSITY: 4.76 (Meas.)
      5.03 (Calc.)
 CLEAVAGE: None.
Fracture uneven to subconchoidal. Brittle.
 HABIT: Crystals octahedral, usually distorted. Usually massive, coarse or fine granular.
 COLOR-LUSTER: Black. Opaque. Metallic to dull. Streak brown. Weakly magnetic.
 MODE OF OCCURRENCE: Occurs in Sweden at Örebro, Vester Silfberg, Långban, and in crystalline limestone at Jacobsberg. Also at localities in Hungary, Bulgaria, and India.
 BEST REF. IN ENGLISH: Palache, et al., "Dana's System of Mineralogy," 7th Ed., v. I, p. 698–707, New York, Wiley, 1944.

## JADEITE
NaAlSi$_2$O$_6$
Pyroxene Group

CRYSTAL SYSTEM: Monoclinic
CLASS: 2/m
SPACE GROUP: C2/c
Z: 4
LATTICE CONSTANTS:
 $a$ = 9.418
 $b$ = 8.562
 $c$ = 5.219
 $\beta$ = 107.58°
3 STRONGEST DIFFRACTION LINES:
 2.83 (100)
 2.42 ( 90)
 2.92 ( 80)
OPTICAL CONSTANTS:
 $\alpha$ = 1.640
 $\beta$ = 1.645
 $\gamma$ = 1.652
 (+)2V = 67°
 HARDNESS: 6 (crystals)
      6½–7 (massive)
 DENSITY: 3.245 (Meas.)
 CLEAVAGE: {110} good
Massive material extremely tough, fracture splintery.
 HABIT: Crystals rare in vugs in massive jadeite; as small elongated prismatic crystals with {110} dominant; also platy rectangular with {100} dominant. Zone [100] usually striated. Doubly terminated crystals very rare. Twinning on {100} common. Commonly fine granular to coarse granular, fibrous-foliated, and as alluvial boulders, sometimes tons in weight.
 COLOR-LUSTER: Crystals colorless with light to dark green tips; also dull white; vitreous. Transparent. Massive material pale green to emerald green, greenish white to white, gray, mauve, also stained by iron oxides to various shade of brown, red, orange or yellow; transparent to translucent; greasy to vitreous. Streak colorless.
 MODE OF OCCURRENCE: Occurs as lenses, stringers, and nodules in Franciscan cherts and graywackes in Contra Costa, Marin, Sonoma, and San Benito counties, and in crystals near Cloverdale, Mendocino County, California. Also found in abundance associated with serpentine derived from olivine rock and in alluvial deposits in Upper Burma. It also comes from Tibet, the province of Yünan, China, Guatemala, Mexico, and New Zealand.
 BEST REF. IN ENGLISH: Wolfe, C. W., *Am. Min.*, **40**: 248–260 (1955). *Am. Min.*, **51**: 956–975 (1966).

# JAGOITE
$(Pb, Ca)_3 Fe^{3+} Si_3 O_{10} (Cl, OH)$

CRYSTAL SYSTEM: Hexagonal
CLASS: 3 or $\bar{3}$
SPACE GROUP: P3 or P$\bar{3}$
LATTICE CONSTANTS:
  $a = 8.65$
  $c = 33.5$
3 STRONGEST DIFFRACTION LINES:
  3.40 (100)
  2.80 ( 80)
  4.16 ( 50)
OPTICAL CONSTANTS:
  Uniaxial (–), $n \sim 2.0$
  HARDNESS: 3
  DENSITY: 5.43
  CLEAVAGE: {0001} perfect
  HABIT: Fine-grained micaceous aggregates of plates.
  COLOR-LUSTER: Yellow-green; vitreous, shining on cleavage surfaces. Streak yellow.
  MODE OF OCCURRENCE: Occurs associated with quartz and an unidentified mineral, commonly surrounded by a zone of black melanotekite, as a rare mineral in hematite ore in the "Camberra" stope, Långban, Sweden.
  BEST REF. IN ENGLISH: Blix, Ragnar, Gabrielson, Olof, and Wickman, Frans E., *Am. Min.*, **43**: 387 (1958).

# JAHNSITE
$CaMn^{2+}Mg_2 (H_2O)_8 Fe_2^{3+}(OH)_2 (PO_4)_4$

CRYSTAL SYSTEM: Monoclinic
CLASS: 2/m
SPACE GROUP: P2/a
Z: 2
LATTICE CONSTANTS:
  $a = 14.94$
  $b = 7.14$
  $c = 9.93$
  $\beta = 110.168°$
3 STRONGEST DIFFRACTION LINES:
  9.27 (100)
  2.825 ( 80)
  4.91 ( 60)
OPTICAL CONSTANTS:
  $\alpha = 1.640$
  $\beta = 1.658$
  $\gamma = 1.670$
  Biaxial (–), 2V large.
  HARDNESS: 4
  DENSITY: 2.706, 2.718 (Meas.)
    2.715 (Calc.)
  CLEAVAGE: {001} good
Brittle.
  HABIT: Crystals commonly well developed and euhedral, striated parallel to [010], short to long prismatic parallel to [010], often tabular parallel to $a${100}; usually 0.1–0.5 mm, rarely up to 5 mm in length. Frequently as parallel aggregates, twinned by reflection on $c${001}.
  COLOR-LUSTER: Nut-brown, purplish-brown, yellow, yellow-orange and greenish yellow. Transparent to translucent. Vitreous to subadamantine.

MODE OF OCCURRENCE: Jahnsite occurs in moderate abundance in granite pegmatites as a late stage product in corroded triphylite-heterosite-ferrisicklerite-rockbridgeite masses, associated with leucophosphite, hureaulite, vivianite, laueite, collinsite, robertsite, and other secondary phosphate minerals. Found in South Dakota at the Tip Top, Bull Moose, White Elephant, and Linwood pegmatites in Custer County, and at the Big Chief, Gap Lode, and Hesnard mines in Pennington County. It also occurs with laueite and strunzite at the Palermo No. 1 mine, North Groton, New Hampshire.
  BEST REF. IN ENGLISH: Moore, Paul Brian, *Am. Min.*, **59**: 48–59 (1974).

# JALPAITE
$Ag_3 CuS_2$

CRYSTAL SYSTEM: Tetragonal
CLASS: 4/m 2/m 2/m
SPACE GROUP: I4$_1$/amd
Z: 8
LATTICE CONSTANTS:
  $a = 8.633$
  $c = 11.743$
3 STRONGEST DIFFRACTION LINES:
  2.345 (100)
  2.794 ( 90)
  2.740 ( 90)
  HARDNESS: 2–2½
Microhardness 22.8–29.7, av. 25.6 with 100 g load.
  DENSITY: 6.82 (Meas.)
    6.827 (Calc.)
  CLEAVAGE: Prismatic, good. Fracture subconchoidal. Sectile.
  HABIT: Foliated or irregularly grained masses.
  COLOR-LUSTER: Gray on fresh fracture; tarnishes lustrous dark gray or iridescent with prominent yellow and yellow-orange resembling chalcopyrite. Metallic. Streak black.
  MODE OF OCCURRENCE: Occurs admixed with mckinstryite intimately associated with chalcopyrite at Jalpa, and as rounded laths with chalcopyrite at Tepic, Mexico; found associated with chalcopyrite, tetrahedrite, galena, native silver, and intergrown with argentite in an ore vein at Bohutín, near Příbram, Bohemia, Czechoslovakia; in Colorado as large coarse masses associated with galena, sphalerite, pyrite, mckinstryite, and covellite at the Payrock mine, Silver Plume, Clear Creek County, and as large masses at the E. C. Hite mine, Boulder County, and at the Geyser mine, Silver Cliff, Custer County.
  BEST REF. IN ENGLISH: Johan, Z., *Am. Min.*, **53**: 1778 (1968). Grybeck, Donald, and Finney, J. J., *Am. Min.*, **53**: 1530–1543 (1968).

# JAMESONITE
$Pb_4 FeSb_6 S_{14}$

CRYSTAL SYSTEM: Monoclinic
CLASS: 2/m
SPACE GROUP: P2$_1$/a
Z: 2

LATTICE CONSTANTS:
$a = 15.65$
$b = 19.03$
$c = 4.03$
$\beta = 91°48'$
3 STRONGEST DIFFRACTION LINES:
3.43 (100)
3.70 ( 35)
2.813 ( 35)
HARDNESS: 2½
DENSITY: 5.63 (Meas.)
5.67 (Calc.)
CLEAVAGE: {001} good
Brittle.
HABIT: Crystals needle-like to fibrous, striated in direction of elongation. As felt-like, fibrous aggregates; also massive, as columnar aggregates of prismatic crystals, radial or plumose. Twinning on {100}.
COLOR-LUSTER: Gray-black, sometimes tarnished iridescent. Opaque. Metallic. Streak gray-black.
MODE OF OCCURRENCE: Occurs in low- to medium-temperature hydrothermal vein deposits, commonly associated with other sulfosalts, sulfides, carbonates, and quartz. Found in the United States in California, Utah, Idaho, Nevada, Colorado, and Arkansas. It also occurs in British Columbia, Ontario, and New Brunswick, Canada; as fine felted masses on pyrite at Mina Noche Buena near Mazapil, Zacatecas, Mexico; at Oruro and elsewhere in Bolivia; and at deposits in Argentina, England, Spain, Germany, Czechoslovakia, Roumania, Yugoslavia, and Sardinia.
BEST REF. IN ENGLISH: Palache, et al., "Dana's System of Mineralogy," 7th Ed., v. I, p. 451-455, New York, Wiley, 1944.

**JARGON** = Zircon

**JARLITE**
$NaSr_3Al_3F_{16}$

CRYSTAL SYSTEM: Monoclinic
CLASS: 2/m, 2, or m
SPACE GROUP: C2/m, C2, or Cm
Z: 4
LATTICE CONSTANTS:
$a = 15.99$
$b = 10.82$
$c = 7.24$
$\beta = 101°49'$
3 STRONGEST DIFFRACTION LINES:
2.98 (100)
3.19 ( 90)
2.15 ( 70)
OPTICAL CONSTANTS:
$\alpha = 1.429$
$\beta = 1.433$ (Na)
$\gamma = 1.436$
$2V \approx 90°$ (variable sign and number)
HARDNESS: 4-4½
DENSITY: 3.87 (Meas.)
3.61 (Calc.)

CLEAVAGE: Not determined.
HABIT: Crystals tabular, minute, elongated along c-axis. Also massive, sometimes with radial columnar structure.
COLOR-LUSTER: Colorless, white, gray. Transparent to translucent. Vitreous.
MODE OF OCCURRENCE: Occurs at the cryolite deposit at Ivigtut, Greenland, associated with cryolite, thomsenolite, chiolite, topaz, and fluorite.
BEST REF. IN ENGLISH: Palache, et al., "Dana's System of Mineralogy," 7th Ed., v. II, p. 118-119, New York, Wiley, 1951.

**JAROSITE**
$KFe_3^{3+}(SO_4)_2(OH)_6$
Alunite Group

CRYSTAL SYSTEM: Hexagonal
CLASS: 3
SPACE GROUP: R3
Z: 3 (hex.) 1 (rhomb.)
LATTICE CONSTANTS:
(hexagonal)    (rhombohedral)
$a = 7.20$     $a_{rh} = 7.34$
$c = 17.00$    $\alpha = 61°38'$
3 STRONGEST DIFFRACTION LINES:
3.08 (100)
3.11 ( 60)
2.292 ( 50)
OPTICAL CONSTANTS:
$\omega = 1.820$
$\epsilon = 1.715$
(-)
HARDNESS: 2½-3½
DENSITY: 2.9-3.26 (Meas.)
3.25 (Calc.)
CLEAVAGE: {0001} distinct
Fracture conchoidal to uneven. Brittle.
HABIT: Crystals usually minute, tabular or pseudocubic; sometimes rounded or distorted. As crusts or coatings of tiny crystals; massive, granular; fibrous; as rounded masses; earthy.
COLOR-LUSTER: Pale yellow to yellowish brown to brown. Translucent. Vitreous to resinous. Earthy masses dull, ocherous. Streak pale yellow.
MODE OF OCCURRENCE: Occurs widespread as a secondary mineral in cracks and seams in ferruginous ores and rocks. Found at many localities in the United States, especially in the northern Black Hills, South Dakota, and in Virginia, Colorado, California, Utah, Nevada, Idaho, and Arizona. Among many other localities, it also is found in Bolivia, Chile, France, Germany, Spain, Czechoslovakia, USSR, and Australia.
BEST REF. IN ENGLISH: Palache, et al., "Dana's System of Mineralogy," 7th Ed., v. II, p. 560-562, New York, Wiley, 1951.

**JASPER** = Variety of quartz

**JEFFERISITE** = Variety of vermiculite

**JEFFERSONITE** = Variety of hedenbergite-rich in manganese and zinc
Pyroxene Group

## JENNITE
$Na_2 Ca_8 (SiO_3)_3 Si_2 O_7 (OH)_6 \cdot 8H_2O$

CRYSTAL SYSTEM: Triclinic
Z: 1
LATTICE CONSTANTS:
  $a = 10.56$    $\alpha = 99°42'$
  $b = 7.25$     $\beta = 97°40'$
  $c = 10.81$    $\gamma = 110°04'$
3 STRONGEST DIFFRACTION LINES:
  10.5 (100)
  2.92 ( 80)
  3.04 ( 60)
OPTICAL CONSTANTS:
    $\alpha = 1.552$
    $\beta = 1.564$
    $\gamma = 1.571$
  (−)2V = 74° (Calc.)
  HARDNESS: Not determined.
  DENSITY: 2.32 (Meas.)
           2.31 (Calc.)
  CLEAVAGE: {001} distinct
  HABIT: Crystals blade-shaped, small.  Also as fibrous aggregates.
  COLOR-LUSTER: White.
  MODE OF OCCURRENCE: Occurs as a late-stage mineral partially filling open spaces in fractured calcite-monticellite-hercynite and vesuvianite-wollastonite contact rock at Crestmore, California.
  BEST REF. IN ENGLISH: Carpenter, A. B., Chalmers, R. A., Gard, J. A., Speakman, K., and Taylor, H. F. W., *Am. Min.*, **51**: 56–74 (1966).

## JEREMEJEVITE (Eremeyevite)
$Al_6 B_5 O_{15} (OH)_3$

CRYSTAL SYSTEM: Hexagonal
CLASS: 6/m
SPACE GROUP: C6₃/m
Z: 12
LATTICE CONSTANTS:
  $a = 8.57$ kX
  $c = 8.17$
3 STRONGEST DIFFRACTION LINES:
  4.27  (100)
  1.388 (100)
  2.19  ( 90)
OPTICAL CONSTANTS:
    $\omega = 1.653$
    $\epsilon = 1.640$
  (sometimes biaxial)
  HARDNESS: 6½
  DENSITY: 3.28 (Meas.)
           3.27 (Calc.)
  CLEAVAGE: None. Fracture conchoidal.
  HABIT: As elongated hexagonal prisms, up to several

inches in length, with irregular terminations.  Crystals are composite with uniaxial outer zone, and biaxial core which is apparently monoclinic.
  COLOR-LUSTER: Colorless to pale yellowish brown. Vitreous.
  MODE OF OCCURRENCE: Found as a few single crystals loose in debris on Mt. Soktuj, Nerchinsk district, eastern Siberia, USSR.
  BEST REF. IN ENGLISH: "Dana's System of Mineralogy," 7th Ed., v. II, p. 330–332, New York, Wiley, 1951.

**JEROMITE** = A mixture

**JEŽEKITE** = Morinite

## JIMBOITE
$Mn_3 B_2 O_6$
Manganese analogue of kotoite

CRYSTAL SYSTEM: Orthorhombic
CLASS: 2/m 2/m 2/m
SPACE GROUP: Pnmn
Z: 2
LATTICE CONSTANTS:
  $a = 5.640$
  $b = 8.715$
  $c = 4.637$
3 STRONGEST DIFFRACTION LINES:
  2.78 (100)
  2.59 ( 50)
  2.33 ( 40)
OPTICAL CONSTANTS:
    $\alpha = 1.792$
    $\beta = 1.794$
    $\gamma = 1.821$
  (+)2V = 30°
  HARDNESS: 5½
  DENSITY: 3.98 (Meas.)
           4.09 (Calc.)
  CLEAVAGE: {110} perfect
            {101} parting
  HABIT: Massive.
  COLOR-LUSTER: Light purplish brown; vitreous.
  MODE OF OCCURRENCE: Occurs associated with rhodochrosite, galaxite, jacobsite, tephroite, alabandite, galena, pyrrhotite, and chalcopyrite in banded carbonate ores on the 18th level of the Kaso mine, Kanuma City, Tochigi Prefecture, Japan.
  BEST REF. IN ENGLISH: Watanabe, T., Kato, A., Matsumoto, T., and Ito, J., *Proc. Japan. Acad.* **39**: 170–175 (1963).

## JOAQUINITE
$Ba_2 NaCe_2 Fe(Ti, Nb)_2 Si_8 O_{26} (OH, F)_2$

CRYSTAL SYSTEM: Orthorhombic
CLASS: 2/m 2/m 2/m

SPACE GROUP: Cmcm
Z: 2
LATTICE CONSTANTS:
$a$ = 9.680
$b$ = 10.539
$c$ = 22.345
3 STRONGEST DIFFRACTION LINES:
2.80 (100)
5.58 ( 70)
2.95 ( 18)
OPTICAL CONSTANTS:
$\alpha$ = 1.748-1.754
$\beta$ = 1.760-1.767
$\gamma$ = 1.762-1.823
(+)2V = 40°
HARDNESS: 5½
DENSITY: 3.89 (Meas.)
3.93 (Calc. Ti = Fe)
CLEAVAGE: Brittle.
HABIT: Crystals tabular to equant; minute.
COLOR-LUSTER: Honey yellow to brown. Transparent to translucent. Vitreous.
MODE OF OCCURRENCE: Occurs in association with neptunite, benitoite, and natrolite near the headwaters of the San Benito River in San Benito County, California.
BEST REF. IN ENGLISH: Semenov, E. I., et al., *Am. Min.* **52**: 1762-1769 (1967).

# JOESMITHITE
(Ca, Pb)Ca$_2$ (Mg, Fe$^{2+}$, Fe$^{3+}$)$_5$ [Si$_6$Be$_2$O$_{22}$] (OH)$_2$

CRYSTAL SYSTEM: Monoclinic
CLASS: 2/m
SPACE GROUP: P2/a
Z: 2
LATTICE CONSTANTS:
$a$ = 9.88
$b$ = 17.87
$c$ = 5.227
$\beta$ = 105°40′
3 STRONGEST DIFFRACTION LINES:
3.33 (100)
2.564 ( 60)
2.530 ( 60)
OPTICAL CONSTANTS:
$\alpha$ = 1.747
$\beta$ = 1.765
$\gamma$ = 1.78
(+)2V = large
HARDNESS: 5½
DENSITY: 3.83 (Meas.)
CLEAVAGE: {110} perfect
HABIT: Prismatic crystals, some fully terminated. Forms {010}, {100}, {110}, {011}, {$\bar{1}$12}, and {$\bar{1}$13}.
COLOR-LUSTER: Black. Streak pale brown.
MODE OF OCCURRENCE: Occurs in a hematite-magnetite-schefferite skarn assemblage, lining cavities and wholly enclosed in younger calcite, at Långban, Sweden.
BEST REF. IN ENGLISH: Moore, Paul B., *Am. Min.*, **54**: 577-578 (1969). Moore, Paul B., *Min. Mag.*, **36**: 876-879 (1968); *Mineral. Soc. Am. Spec. Pap.* **2**, 111-115 (1969).

# JOHACHIDOLITE
CaAlB$_3$O$_7$

CRYSTAL SYSTEM: Orthorhombic
CLASS: 2/m 2/m 2/m
SPACE GROUP: Cmma
Z: 4
LATTICE CONSTANTS:
$a$ = 7.970
$b$ = 11.722
$c$ = 4.374
OPTICAL CONSTANTS:
$\alpha$ = 1.715
$\beta$ = 1.720
$\gamma$ = 1.726
(-)2V = 72° (Calc.)
HARDNESS: 6½-7
DENSITY: 3.44 (Calc.)
CLEAVAGE: None.
HABIT: As grains and lamellar masses.
COLOR-LUSTER: Colorless. Transparent. Vitreous. Fluoresces intense blue under ultraviolet light.
MODE OF OCCURRENCE: Johachidolite occurs in nepheline dikes cutting limestone in the Johachido district, Kenkyohokudo prefecture, Korea.
BEST REF. IN ENGLISH: Palache, et al., "Dana's System of Mineralogy," 7th Ed., v. II, p. 384, New York, Wiley, 1951. Moore, P. B. and Araki, T., *Nature Physical Science*, **240**: 63-65 (Nov. 20, 1972).

# JOHANNITE
Cu(UO$_2$)$_2$ (SO$_4$)$_2$ (OH)$_2$ · 6H$_2$O

CRYSTAL SYSTEM: Triclinic
CLASS: $\bar{1}$
SPACE GROUP: P$\bar{1}$
Z: 1
LATTICE CONSTANTS:
$a$ = 8.92   $\alpha$ = 110°
$b$ = 9.59   $\beta$ = 111°59′
$c$ = 6.84   $\gamma$ = 100°18′
3 STRONGEST DIFFRACTION LINES:
7.73 (100)
6.16 ( 90)
3.41 ( 80)
OPTICAL CONSTANTS:
$\alpha$ = 1.572-1.577
$\beta$ = 1.592-1.597
$\gamma$ = 1.612-1.616
2V ≈ 90° (+), (-)
HARDNESS: 2-2½
DENSITY: 3.32 (Meas.)
3.27 (Calc.)
CLEAVAGE: {100} good
Not brittle.
HABIT: Crystals thick tabular on {100} and elongated along $c$-axis; also prismatic [001]. As subparallel or drusy aggregates, and as small spheroidal aggregates and coatings composed of scales or lath-like fibers. Twins common with twin axis [001] and composition plane {010}; both simple twins and lamellar.

COLOR-LUSTER: Emerald green to greenish yellow, transparent to translucent. Vitreous. Streak pale green. Not fluorescent.

MODE OF OCCURRENCE: Occurs associated with zippeite in the oxidized zone of uraninite veins in the Central City district, Gilpin County, Colorado; in Utah at the Happy Jack Mine, White Canyon, San Juan County, at the Oyler mine, Henry Mountains district, at Marysvale, and at the Frey No. 4 mine, White Canyon district. Found very sparingly on altered uraninite at Great Bear Lake, Canada; as crystals at Joachimsthal, Bohemia and at Johanngeorgenstadt, Germany; and associated with chalcanthite as an efflorescence on mine walls in the Taboshar uranium deposit in the Kara-Mazar Mountains, North Tadjikistan, USSR.

BEST REF. IN ENGLISH: Frondel, Clifford, USGS Bull. 1064, p. 130–135 (1958).

# JOHANNSENITE

$Ca(Mn, Fe^{2+})Si_2O_6$
Pyroxene Group
Manganese analogue of diopside and hedenbergite

CRYSTAL SYSTEM: Monoclinic
CLASS: 2/m
SPACE GROUP: C2/c
Z: 4
LATTICE CONSTANTS:
  $a = 9.83$
  $b = 9.04$
  $c = 5.27$
  $\beta = 105°$
3 STRONGEST DIFFRACTION LINES:
  3.02 (100)
  2.55 ( 80)
  2.60 ( 60)
OPTICAL CONSTANTS:
  $\alpha = 1.703-1.716$
  $\beta = 1.711-1.728$
  $\gamma = 1.732-1.745$
  $(+)2V = 68°-70°$
HARDNESS: 6
DENSITY: 3.44–3.55 (Meas.)
         3.561 (Calc.)
CLEAVAGE: {110} good
          {001} parting
          {010} parting
          {100} parting
Fracture uneven to conchoidal. Brittle.

HABIT: Crystals short prismatic. Usually massive, columnar, or as radiating and spherulitic aggregates of fibers and prisms. Twinning on {100} common, simple and lamellar.

COLOR-LUSTER: Gray, greenish, brown, black. Translucent to nearly opaque. Surface commonly stained black by manganese oxide.

MODE OF OCCURRENCE: Occurs chiefly in association with rhodonite and bustamite in metasomatized limestones; less frequently in quartz and calcite veins in rhyolite. Found at the Star mine, Vanadium, New Mexico; at Pueblo, Mexico; in limestone at Monte Civillina, Recoaro, Italy; and at Broken Hill, New South Wales, Australia.

BEST REF. IN ENGLISH: Deer, Howie, and Zussman, "Rock Forming Minerals," v. 2, p. 75–78, New York, Wiley, 1963.

# JOHNSTRUPITE = Mosandrite

# JORDANITE

$(Pb, Tl)_{13}As_7S_{23}$

CRYSTAL SYSTEM: Monoclinic
CLASS: 2/m
SPACE GROUP: $P2_1/m$
Z: 2
LATTICE CONSTANTS:
  $a = 8.87$
  $b = 31.65$
  $c = 8.40$
  $\beta = 118°6'$
3 STRONGEST DIFFRACTION LINES:
  2.238 (100)
  2.97 ( 90)
  3.15 ( 80)
HARDNESS: 3
DENSITY: 6.44 (Meas.)
         6.49 (Calc.)
CLEAVAGE: {010} very perfect
          {001} parting
Fracture conchoidal. Brittle.

HABIT: Crystals tabular with pseudohexagonal aspect. Rarely reniform. Twinning on {001} common, often lamellar, on {$\bar{2}$01} common; on {$\bar{1}$01} and {101} rare.

COLOR-LUSTER: Lead gray, commonly tarnished iridescent. Metallic. Opaque. Streak black.

MODE OF OCCURRENCE: Occurs associated with lengenbachite and other sulfosalts in cavities in dolomite at the Lengenbach quarry, in the Binnental, Valais, Switzerland; with galena and sphalerite near Beuthen, Upper Silesia, Poland, and at Nagyág, Roumania; and with barite at Aomori, Japan.

BEST REF. IN ENGLISH: Palache, et al., "Dana's System of Mineralogy," 7th Ed., v. I, p. 398–401, New York, Wiley, 1944.

# JORDISITE (Species status uncertain)

$MoS_2$

CRYSTAL SYSTEM: Unknown
3 STRONGEST DIFFRACTION LINES: Amorphous to X-rays.
HARDNESS: Soft.
DENSITY: Not determined.
CLEAVAGE: Fracture of aggregates earthy or irregular.

HABIT: Massive; as colloidal aggregates or disseminations of extremely small particles.

COLOR-LUSTER: Black. Opaque. Dull.

MODE OF OCCURRENCE: Occurs in association with ilsemannite in uranium ore deposits in the Kern River uranium area, Kern County, California; at the Bullion Monarch mine, Marysvale, Utah; at the Black King No. 5 claim at Placerville, San Miguel County, Colorado; and at the

Runge mine, near Edgemont, Fall River County, South Dakota. Also found at the Himmelsfürst mine, Freiberg, Saxony, Germany.

BEST REF. IN ENGLISH: Palache, et al., "Dana's System of Mineralogy," 7th Ed., v. I, p. 331, New York, Wiley, 1944.

## JOSEITE
$Bi_3Te(Se, S)$

CRYSTAL SYSTEM: Hexagonal
CLASS: $\bar{3}$
SPACE GROUP: $R\bar{3}m$
Z: 3
LATTICE CONSTANTS:

| (joseite A) | (joseite B) |
|---|---|
| $a = 4.25$ | $a = 4.34$ |
| $c = 39.77$ | $c = 40.83$ |

3 STRONGEST DIFFRACTION LINES:

| (joseite A) | (joseite B) |
|---|---|
| 3.08 (100) | 3.16 (100) |
| 2.24 ( 50) | 2.17 ( 50) |
| 2.11 ( 50) | 2.30 ( 40) |
| $Bi_{4+x}Te_{1-x}S_2$ | $Bi_{4+x}Te_{2-x}S$ |

HARDNESS: 2
DENSITY: 8.10 (Meas.) A
8.3 (Meas.) B
CLEAVAGE: Perfect, one direction. Flexible.
HABIT: As irregular laminated masses.
COLOR-LUSTER: Steel gray to grayish black; tarnishes darker or iridescent. Opaque. Metallic.
MODE OF OCCURRENCE: Joseite A occurs at Glacier Gulch, Smithers, British Columbia, Canada; joseite B is found in granular limestone at San Jose, Minas Geraes, Brazil.

BEST REF. IN ENGLISH: Palache, et al., "Dana's System of Mineralogy," 7th Ed., v. I, p. 166-167, New York, Wiley, 1944. Sindeeva, N. D., "Mineralogy and Types of Deposits of Selenium and Tellurium" (1964).

## JOSEITE C (Inadequately described mineral)
$Bi_{16}(TeS_3)_3$ with Te:S ~ 1:3

CRYSTAL SYSTEM: Unknown
3 STRONGEST DIFFRACTION LINES:
3.03 (100)
2.23 ( 50) doublet
2.10 ( 50)
HARDNESS: Microhardness 71.2 perpendicular, 89.9 parallel to basal cleavage.
DENSITY: Not determined.
CLEAVAGE: Perfect basal.
HABIT: Massive; as very minute grains.
COLOR-LUSTER: Steel gray. Opaque. Metallic.
MODE OF OCCURRENCE: Occurs at the contact of quartz and native bismuth in the Sokhondo deposit, eastern Transbaikal, USSR.

BEST REF. IN ENGLISH: Godovikov, A. A., Kochetkova, K. V., and Lavrent'ev, Yu. G., *Am. Min.*, **56**: 1839-1840 (1971).

## JOSEPHINITE = Nickel-iron

## JOURAVSKITE
$Ca_6Mn_2(SO_4)_2(CO_3)_2(OH)_{12} \cdot 24H_2O$

CRYSTAL SYSTEM: Hexagonal
CLASS: 6
SPACE GROUP: $P6_3$
Z: 1
LATTICE CONSTANTS:
$a = 11.06$
$c = 10.50$
3 STRONGEST DIFFRACTION LINES:
9.6 (100)
5.53 ( 60)
3.81 ( 60)
OPTICAL CONSTANTS:
$\omega = 1.556$
$\epsilon = 1.540$
(-)
HARDNESS: 2½
DENSITY: 1.95 ± 0.01 (Meas.)
1.93 ± 0.02 (Calc.)
CLEAVAGE: {100} good
HABIT: Tiny embedded grains up to several tenths of a millimeter in diameter; rarely show crystal faces, prism {100} and dipyramid {102}.
COLOR-LUSTER: Greenish yellow to greenish orange.
MODE OF OCCURRENCE: Occurs in masses 1-5 mm across in a dark mass of manganese minerals, intimately associated with calcite, on the dumps of the Tachgagalt No. 2 vein, Anti-Atlas, Morocco.

BEST REF. IN ENGLISH: Gaudefroy, C., and Permingeat, F., *Am. Min.*, **50**: 2102-2103 (1965).

## JULGOLDITE
$Ca_2Fe^{2+}(Fe^{3+}, Al)_2(SiO_4)(Si_2O_7)(OH)_2 \cdot H_2O$

CRYSTAL SYSTEM: Monoclinic
CLASS: 2/m
SPACE GROUP: A2/m
Z: 4
LATTICE CONSTANTS:
$a = 8.92$
$b = 6.09$
$c = 19.37$
$\beta = 97°30'$
3 STRONGEST DIFFRACTION LINES:
2.950 (100)
3.84 ( 80)
4.80 ( 70)
OPTICAL CONSTANTS:
$\alpha = 1.776$
$\beta = 1.814$
$\gamma = 1.836$
(-)2V = 50°-70°
HARDNESS: 4½
DENSITY: 3.602 (Meas.)
3.605 (Calc.)
CLEAVAGE: {100} good

HABIT: Crystals flat prismatic to bladed, up to 2 mm long. Twinning on {001} common, often repeated.

COLOR-LUSTER: Deep lustrous black. Almost sub-metallic. Small fragments greenish black to green. Streak greenish olive with bluish tinge.

MODE OF OCCURRENCE: Occurs filling cavities and embedded in large plates of apophyllite and minor barite, which constitute a fissure filling in granular hematite-magnetite ore from the "Amerika" stope, Långban, Sweden.

BEST REF. IN ENGLISH: Moore, Paul B., *Am. Min.*, **56**: 2157–2158 (1971).

# JULIËNITE
$Na_2Co(SCN)_4 \cdot 8H_2O$

CRYSTAL SYSTEM: Tetragonal
CLASS: 4/m
SPACE GROUP: $P4_2/n$
LATTICE CONSTANTS:
 $a = 19.0$
 $c = 5.47$

3 STRONGEST DIFFRACTION LINES:
 3.55 (100)
 3.23 (100)
 1.375 (100)
OPTICAL CONSTANTS:
 $\omega = 1.556$
 $\epsilon = 1.645$
 (+)
HARDNESS: Not determined.
DENSITY: 1.648 (Meas. synthetic)
 1.645 (Calc.)
CLEAVAGE: [001]
HABIT: As thin crusts of minute needle-like crystals elongated along $c$-axis.
COLOR-LUSTER: Blue.

Soluble in water.

MODE OF OCCURRENCE: Occurs at Chamibumba, near Kambove, Katanga, Zaire.

BEST REF. IN ENGLISH: Palache, et al., "Dana's System of Mineralogy," 7th Ed., v. II, p. 1106–1107, New York, Wiley, 1951.

# K

## KAERSUTITE (Titanian Hornblende)

$Ca_2(Na, K)(Mg, Fe^{2+}, Fe^{3+})_4 TiSi_6 Al_2 O_{22}(O, OH, F)_2$
Amphibole Group

CRYSTAL SYSTEM: Monoclinic
CLASS: 2/m
SPACE GROUP: C2/m
Z: 2
LATTICE CONSTANTS:
  $a \simeq 9.9$
  $b \simeq 18.21$
  $c \simeq 5.4$
  $\beta \simeq 106°$
3 STRONGEST DIFFRACTION LINES:
  2.69 (100)
  3.11 ( 80)
  8.38 ( 65)
OPTICAL CONSTANTS:
  $\alpha$ = 1.670-1.689
  $\beta$ = 1.690-1.741
  $\gamma$ = 1.700-1.772
  (-)2V = 66°-82°
HARDNESS: 5-6
DENSITY: 3.2-3.28
CLEAVAGE: {110} perfect
          {001} parting
          {100} parting
Fracture uneven to subconchoidal. Brittle.
  HABIT: Crystals short prismatic. Also massive, compact, as embedded grains. Twinning on {100} common, simple, lamellar.
  COLOR-LUSTER: Dark brown to black. Translucent to nearly opaque. Vitreous to somewhat resinous.
  MODE OF OCCURRENCE: Occurs mainly in volcanic rocks, especially as an abundant constituent in camptonites. Found as phenocrysts up to 4 inches in diameter in camptonite dikes near Boulder Dam, Mohave County, Arizona; at Kaersut and Skaergaard, Greenland; at Wart Holm, Copinshay, Orkneys; and at localities in USSR, Japan, and New Zealand.
  BEST REF. IN ENGLISH: Deer, Howie, and Zussman, "Rock Forming Minerals," v. 2, p. 321-327, New York, Wiley, 1963.

## KAHLERITE

$Fe^{2+}(UO_2)_2(AsO_4)_2 \cdot 12H_2O$

CRYSTAL SYSTEM: Tetragonal
CLASS: 4/m
SPACE GROUP: $P4_2/n$
Z: 8
LATTICE CONSTANTS:
  $a$ = 14.30
  $c$ = 21.97 (synthetic)
3 STRONGEST DIFFRACTION LINES:
  3.53 (100)
  11.1 ( 80)
  5.55 ( 50)
OPTICAL CONSTANTS:
  $\epsilon$ = 1.632
  $\omega$ = 1.634
  (-)
HARDNESS: Not determined.
DENSITY: Not determined.
CLEAVAGE: {001} excellent
HABIT: Tabular crystals 2.0 mm in size. Forms: $t$ {111}, $i$ {021}, $y$ {012}, $P$ {011}, and $n$ {010}.
  COLOR-LUSTER: Lemon yellow to yellowish green. Not fluorescent.
  MODE OF OCCURRENCE: Occurs associated with scorodite, symplesite, and pitticite on altered loellingite at Huttenberg, Austria.
  BEST REF. IN ENGLISH: Meixner, Heinz, *Am. Min.*, **39**: 1038 (1954). Frondel, Clifford, USGS Bull. 1064, p. 204-205 (1958).

## KAINITE

$KMgSO_4Cl \cdot 3H_2O$

CRYSTAL SYSTEM: Monoclinic
CLASS: 2/m
SPACE GROUP: C2/m
Z: 16

LATTICE CONSTANTS:
$a = 19.75$
$b = 16.24$
$c = 9.86$
$\beta = 94°55'$
3 STRONGEST DIFFRACTION LINES:
3.157 (100)
3.017 ( 70)
2.725 ( 30)
OPTICAL CONSTANTS:
$\alpha = 1.494$
$\beta = 1.505$
$\gamma = 1.516$
$(-)2V \sim 90°$
HARDNESS: 2½–3
DENSITY: 2.15 (Meas.)
2.24 (Calc.)
CLEAVAGE: {001} perfect
Fracture splintery to smooth. Brittle.
HABIT: Crystals thick tabular to equant, often highly modified, up to several centimeters in size. As aggregates and coatings in cracks or vugs. Also massive, granular.
COLOR-LUSTER: Colorless, also discolored gray, blue, violet, reddish or yellowish by impurities. Transparent to translucent. Vitreous.

Soluble in water. Taste saline, bitter.
MODE OF OCCURRENCE: Occurs in bedded salt deposits of oceanic origin often associated with halite, sylvite, polyhalite, carnallite, and other salt minerals. Found in the Carlsbad area, Eddy County, New Mexico, and in adjacent parts of Texas; in thick beds in the Permian salt deposits in central Germany, north and south of the Harz mountains; at Kalusz, Poland; and at the Stebniksk mine, USSR.
BEST REF. IN ENGLISH: Palache, et al., "Dana's System of Mineralogy," 7th Ed., v. II, p. 594–596, New York, Wiley, 1951.

# KAINOSITE = Cenosite

# KALIBORITE (Paternoite)
$HKMg_2B_{12}O_{21} \cdot 9H_2O$

CRYSTAL SYSTEM: Monoclinic
CLASS: 2/m
SPACE GROUP: C2/c
LATTICE CONSTANTS:
$a = 18.93$
$b = 8.62$
$c = 14.97$
$\beta = 99°54'$
3 STRONGEST DIFFRACTION LINES:
7.22 (100)
6.215 (100)
3.104 ( 70)
OPTICAL CONSTANTS:
$\alpha = 1.5081$
$\beta = 1.5255$ (Na)
$\gamma = 1.5500$
$(+)2V = 80°38'$
HARDNESS: 4–4½

DENSITY: 2.116 (Meas.)
CLEAVAGE: {001} perfect
{$\bar{1}$01} perfect
{100} good
HABIT: As minute crystals, often in aggregates. Also as fine-granular nodular masses.
COLOR-LUSTER: Colorless, white, or reddish brown. Transparent. Vitreous.
MODE OF OCCURRENCE: Occurs in salt deposits at Leopoldshall, at Neustrassfurt, and near Aschersleben, Saxony, Germany; as small nodules in bloedite beds in the salt deposits at Mte. Sambuco, Calascibetta, Sicily; at Sallent Spain; and in the Inder Lake deposits, Kazakhstan, USSR.
BEST REF. IN ENGLISH: Palache, et al., "Dana's System of Mineralogy," 7th Ed., v. II, p. 367–368, New York, Wiley, 1951. *Am. Min.*, **50**: 1079–1083 (1965).

# KALICINE (Kalicinite)
$KHCO_3$

CRYSTAL SYSTEM: Monoclinic
CLASS: 2/m
SPACE GROUP: $P2_1/a$
Z: 4
LATTICE CONSTANTS:
$a = 15.01$ kX
$b = 5.69$
$c = 3.68$
$\beta = 104°30'$
3 STRONGEST DIFFRACTION LINES:
(Chypis)   (synthetic)
2.84 (100)   3.67 (100)
3.68 ( 32)   2.63 ( 90)
2.62 ( 32)   2.86 ( 85)
OPTICAL CONSTANTS:
$\alpha = 1.380$
$\beta = 1.482$ (Na)
$\gamma = 1.578$
$(-)2V = 81.5°$
HARDNESS: Soft.
DENSITY: 2.168 (Meas. synthetic)
2.17 (Calc.)
CLEAVAGE: {100}
{001}
{101}
HABIT: Crystals short prismatic, elongated along c-axis. As fine-crystalline aggregates.
COLOR-LUSTER: Colorless, white, yellowish. Transparent.
MODE OF OCCURRENCE: Found at Chypis, Canton Wallis, Switzerland.
BEST REF. IN ENGLISH: Palache, et al., "Dana's System of Mineralogy," 7th Ed., v. II, p. 136–137, New York, Wiley, 1951.

# KALINITE (Inadequately described mineral)
$KAl(SO_4)_2 \cdot 11H_2O$

CRYSTAL SYSTEM: Monoclinic

3 STRONGEST DIFFRACTION LINES:
  4.83 (100)
  4.32 (100)
  4.11 (100)
OPTICAL CONSTANTS:
  $\alpha = 1.429\text{-}1.430$
  $\beta = 1.452$
  $\gamma = 1.456\text{-}1.458$
  $(-)2V = 52°$
HARDNESS: ~2-2½
DENSITY: Not determined.
CLEAVAGE: Not determined.
HABIT: As an efflorescence; fibrous.
COLOR-LUSTER: Colorless, white. Transparent.

Soluble in water. Taste sweetish and astringent.
MODE OF OCCURRENCE: Occurs at Turkey Creek, Jefferson County, Colorado, and probably at many other localities in that state; also in California, Chile, and Australia.
BEST REF. IN ENGLISH: Palache, et al., "Dana's System of Mineralogy," 7th Ed., v. II, p. 471, New York, Wiley, 1951.

## KALIOPHILITE
$KAlSiO_4$
Dimorphous with kalsilite

CRYSTAL SYSTEM: Hexagonal
CLASS: 622
SPACE GROUP: $P6_3 22$
Z: 54
LATTICE CONSTANTS:
  $a = 26.930$
  $c = 8.522$
3 STRONGEST DIFFRACTION LINES:
  3.09 (100)
  2.593 ( 30)
  2.131 ( 25)
OPTICAL CONSTANTS:
  $\omega = 1.537$
  $\epsilon = 1.533$
    $(-)$
HARDNESS: 6
DENSITY: 2.49-2.67 (Meas.)
CLEAVAGE: {0001} indistinct
          {10$\bar{1}$0} indistinct
HABIT: Crystals thick prismatic or acicular.
COLOR-LUSTER: Colorless. Transparent. Vitreous.
MODE OF OCCURRENCE: Occurs in ejected blocks of biotite pyroxenite and augite-melilite-calcite rock at Mte. Somma, and also in ejected blocks in association with leucite, hauyne, and other minerals, at Albano, Latium, Italy.
BEST REF. IN ENGLISH: Deer, Howie, and Zussman, "Rock Forming Minerals," v. 4, p. 231-270, New York, Wiley, 1963.

## KALISTRONTITE
$K_2Sr(SO_4)_2$

CRYSTAL SYSTEM: Hexagonal
Z: 3

LATTICE CONSTANTS:
  $a = 5.45$
  $c = 20.7$
3 STRONGEST DIFFRACTION LINES:
  3.12 (100)
  1.904 ( 90)
  2.73 ( 80)
OPTICAL CONSTANTS:
  $\omega = 1.569$
  $\epsilon = 1.549$
    $(-)$
HARDNESS: 2
DENSITY: 3.20 (Meas.)
         3.32 (Calc.)
CLEAVAGE: {0001} perfect
Brittle.
HABIT: Single crystals, elongated along the prism, up to 15-20 mm long; also as flattened hexagonal tablets up to 22 mm in diameter.
COLOR-LUSTER: Colorless, transparent, vitreous.
MODE OF OCCURRENCE: Found in drill core of saline anhydrite, containing clay and dolomite and a little sylvite, from a depth of 447 meters near Alshtan, Bachkir Autonomous S.S.R.
BEST REF. IN ENGLISH: Voronova, M. L., *Am. Min.*, **48**: 708-709 (1963).

## KALKOWSKYN (Kalkowskite) = Pseudorutile

## KALLILITE = Variety of ullmannite containing bismuth

## KALSILITE
$KAlSiO_4$
Dimorphous with kaliophilite

CRYSTAL SYSTEM: Hexagonal
CLASS: 622
SPACE GROUP: $P6_3 22$
Z: 2
LATTICE CONSTANTS:
  $a \simeq 5.2$
  $c \simeq 8.7$
3 STRONGEST DIFFRACTION LINES:
  3.12 (100)
  2.58 ( 50)
  3.97 ( 45)
OPTICAL CONSTANTS:
  $\omega = 1.538\text{-}1.543$
  $\epsilon = 1.532\text{-}1.537$
    $(-)$
HARDNESS: 6
DENSITY: 2.59-2.625 (Meas.)
CLEAVAGE: {10$\bar{1}$0} poor
          {0001} poor
Fracture subconchoidal. Brittle.
HABIT: Massive, compact; as embedded grains.
COLOR-LUSTER: Colorless, white, gray. Transparent to translucent. Vitreous to greasy.

MODE OF OCCURRENCE: Occurs as complex phenocrysts and as a constituent of the groundmass of certain lavas. Found at San Venanzo, Umbria, Italy; at the Baruta crater, Nyiragongo area, Zaire; and at the Kyambobo crater, Bunyaruguru field, southwestern Uganda.

BEST REF. IN ENGLISH: Deer, Howie, and Zussman, "Rock Forming Minerals," v. 4, p. 231–270, New York, Wiley, 1963.

**KAMACITE** = Meteoric iron containing nickel

**KAMAREZITE** = Brochantite

## KÄMMERERITE
$(Mg, Cr)_6(AlSi_3)O_{10}(OH)_8$
Chlorite Group

CRYSTAL SYSTEM: Triclinic
CLASS: 1
SPACE GROUP: C1
Z: 4
LATTICE CONSTANTS:
 $a = 5.335$
 $b = 9.240$
 $c = 28.735$
 $\beta = 97°5'$
3 STRONGEST DIFFRACTION LINES:
 7.21 (100)
 2.42 (100)
 14.5 ( 90)
OPTICAL CONSTANTS:
 $\alpha = 1.5970$
 $\beta = 1.5980$
 $\gamma = 1.5996$
HARDNESS: 2–2½
DENSITY: 2.645 (Meas.)
 2.64 (Calc.)
CLEAVAGE: {001} perfect, micaceous
Laminae flexible.
HABIT: Crystals hexagonal in shape, bounded by steep six-sided pyramids.
COLOR-LUSTER: Red to purplish red. Vitreous. Transparent to translucent.
MODE OF OCCURRENCE: Occurs in chromite deposits, often in association with uvarovite. Found widespread in the chromite deposits of northern California, and at Texas, Lancaster County, Pennsylvania. Also at Lake Itkul near Miask and at Bisersk, USSR; at Erzincan, Turkey; and elsewhere.
BEST REF. IN ENGLISH: Deer, Howie, and Zussman, "Rock Forming Minerals," v. 3, p. 145, 146, New York, Wiley, 1962.

**KANDITE** = Synonym for Kaolinite Clay Mineral Group

## KAOLINITE
$Al_2Si_2O_5(OH)_4$
Kaolin Group (Kandite)
Trimorphous with dickite and nacrite

CRYSTAL SYSTEM: Triclinic
CLASS: 1
SPACE GROUP: P1
Z: 2
LATTICE CONSTANTS:
 $a = 5.155$ $\alpha = 91.68°$
 $b = 8.959$ $\beta = 104.87°$
 $c = 7.407$ $\gamma = 89.94°$
3 STRONGEST DIFFRACTION LINES:
 7.17 (100)
 1.49 ( 90)
 3.58 ( 80)
OPTICAL CONSTANTS:
 $\alpha = 1.553$–1.565
 $\beta = 1.559$–1.569
 $\gamma = 1.560$–1.570
(–)2V = 24°–50°
HARDNESS: 2–2½
DENSITY: 2.6–2.63 (Meas.)
 2.594 (Calc.)
CLEAVAGE: {001} perfect
Scales flexible, inelastic; often plastic when moist.
HABIT: Crystals commonly thin hexagonal platelets or scales up to 2 mm in size; also as elongated plates or curved laths. Usually massive; compact, friable or mealy. Twinning rare.
COLOR-LUSTER: Colorless, white; sometimes tinted yellowish, brownish, reddish, or bluish. Transparent to translucent. Pearly to dull earthy.
MODE OF OCCURRENCE: Occurs widespread as a very common clay mineral formed mainly by the weathering or hydrothermal alteration of feldspars and other aluminous silicate minerals. Typical occurrences are found in California, Arizona, New Mexico, Colorado, Utah, Montana, South Dakota, Arkansas, Pennsylvania, Vermont, Delaware, Maryland, Virginia, North Carolina, South Carolina, and Georgia. Also in England, France, Germany, Czechoslovakia, and elsewhere. Minute crystals occur in a kaolinite horizon beneath the Alloway clay of Salem County, New Jersey.
BEST REF. IN ENGLISH: Deer, Howie, and Zussman, "Rock Forming Minerals," v. 3, p. 194–212, New York, Wiley, 1962.

**KAPNITE** = Ferroan variety of smithsonite

**KARACHAITE** = Fibrous variety of serpentine

## KARELIANITE
$V_2O_3$

CRYSTAL SYSTEM: Hexagonal
CLASS: $\bar{3}$ 2/m
SPACE GROUP: R$\bar{3}$c
Z: 2
LATTICE CONSTANTS:
 $a = 4.99$
 $c = 13.98$

3 STRONGEST DIFFRACTION LINES:
    1.69 (100)
    2.70 ( 80)
    3.65 ( 60)
HARDNESS: 8–9
DENSITY: 4.87 (Meas.)
CLEAVAGE: Conchoidal fracture.
HABIT: Embedded grains up to 0.3 mm in size. Prismatic.
COLOR-LUSTER: Black.
MODE OF OCCURRENCE: Found in ore boulders derived from the Outokumpu orebody southeast of the Outokumpu ore field, Finland, associated with pyrrhotite, chalcopyrite, and pyrite.
BEST REF. IN ENGLISH: Long, J. V. P., Vuorelainen, Yrjö, and Kouvo, Olavi, *Am. Min.*, **48**: 33–41 (1963).

## KARNASURTITE (Kozhanovite)

$(La, Ce, Th)(Ti, Nb)(Al, Fe)(Si, P)_2 O_7 (OH)_4 \cdot 3H_2 O(?)$

CRYSTAL SYSTEM: Hexagonal (?)
5 STRONGEST DIFFRACTION LINES:
    3.10 (70)
    2.88 (70)
    3.29 (60)
    3.49 (50)
    1.723 (50)
Amorphous to x-rays; after being heated at 900°, gives pattern close to huttonite.
OPTICAL CONSTANTS:
    $\omega = 1.617$
    $\epsilon = 1.595$
       (−)
HARDNESS: 2
DENSITY: 2.89–2.95
CLEAVAGE: One good. Brittle.
HABIT: Grains up to 1 cm in diameter; also aggregates of platy crystals up to 10 × 6 × 2 cm.
COLOR-LUSTER: Honey yellow when fresh to pale yellow when altered. Greasy. Streak yellowish.
MODE OF OCCURRENCE: Occurs associated with schizolite, natrolite, and epididymite, and is replaced by polylithionite, in a zoned pegmatitic stock at Mt. Karnasurt, Lovozero alkaline massif, Kola Peninsula.
BEST REF. IN ENGLISH: Kuz'menko, M. V., Kozhanov, S. I. *Am. Min.*, **45**: 1133–1134 (1960).

## KARPATITE = Carpathite

## KARPINSKITE

$(Mg, Ni)_2 Si_2 O_5 (OH)_2 (?)$

CRYSTAL SYSTEM: Monoclinic (?)
STRONGEST DIFFRACTION LINES:
    11.00, 7.71, 4.76, 3.75, 1.555
OPTICAL CONSTANTS:
    (deeply colored)        (pale colored)
    $\alpha = 1.570$            $\alpha = 1.553$
    $\gamma = 1.594$            $\gamma = 1.569$
       (−)2V

HARDNESS: 2½–3
DENSITY: 2.53–2.63 (Meas.)
CLEAVAGE: Not determined.
HABIT: Massive, cryptocrystalline; in plates and monoclinic prisms 0.3–0.8 mm long.
COLOR-LUSTER: Colorless to light blue to deep greenish blue. Translucent. Dull to weakly greasy.
MODE OF OCCURRENCE: Occurs as veinlets in "kerolitized" serpentinites of the Nizhne-Tagilsk serpentine massif, USSR.
BEST REF. IN ENGLISH: Rukavishnikova, I. A., *Am. Min.*, **42**: 584 (1957).

## KARPINSKYITE = A mixture of leifite and a zinc-bearing clay

## KASOITE = A variety of celsian

## KASOLITE

$PbUO_2 SiO_4 \cdot H_2 O$

CRYSTAL SYSTEM: Monoclinic
CLASS: 2/m
SPACE GROUP: $P2_1/a$
Z: 4
LATTICE CONSTANTS:
    $a = 13.24$
    $b = 6.94$
    $c = 6.70$
    $\beta = 104°20'$
3 STRONGEST DIFFRACTION LINES:
    2.92 (100)
    3.06 ( 80)
    4.21 ( 60)
OPTICAL CONSTANTS:
    $\alpha = 1.89–1.90$
    $\beta = 1.90–1.91$
    $\gamma = 1.95–1.97$
(+)2V = 43°
HARDNESS: ~4½
DENSITY: 5.83–6.5 (Meas.)
          6.256 (Calc.)
CLEAVAGE: {001} perfect
          {100} indistinct
          {010} indistinct
Brittle.
HABIT: Crystals stout prismatic, up to 4 mm long, elongated along b-axis; also flattened on {001} and lath-like by extension along b-axis. {100} commonly striated parallel to edge with {001}. Forms: {001}, {100}, {110}, and {111}. As subparallel crystal aggregates, rosettes, and radial-fibrous aggregates. Also as compact masses and dense crusts.
COLOR-LUSTER: Ocher yellow to yellowish brown to reddish orange; rarely bottle green, grayish green, greenish black. Transparent to translucent, also opaque. Crystals subadamantine to greasy. Massive material dull to earthy. Not fluorescent.
MODE OF OCCURRENCE: Occurs in the United States as a component of "gummite" pseudomorphs after uraninite

in pegmatites at the Ruggles mine near Grafton Center, New Hampshire; at the Ingersoll mine at Keystone, and at other localities in the Black Hills, South Dakota. It is also found in Nevada in the East Walker River area, Lyon County, and in the Goodsprings district, Clark County. In Canada found as crusts on altered uraninite at the Nicholson mines, Lake Athabaska, Saskatchewan, and at Great Bear Lake. It occurs at several localities in France including Kersegalec, Lignol, Morbihan; Grury, Saône-et-Loire; and in the Puy-de-Dôme district. Superb crystals are found associated with metatorbernite, curite, soddyite, and other secondary uranium minerals at Shinkolobwe, Kalongwe, Kambove, and Luiswishi, Katanga, Zaire.

BEST REF. IN ENGLISH: Frondel, Clifford, USGS Bull. 1064, p. 315–319 (1958).

## KASPARITE = Cobaltoan pickeringite

## KASSITE
$CaTi_2O_4(OH)_2$

CRYSTAL SYSTEM: Orthorhombic
Z: 4
LATTICE CONSTANTS:
  $a = 9.02$
  $b = 9.56$
  $c = 5.26$
3 STRONGEST DIFFRACTION LINES:
  3.30 (100)
  1.761 (100)
  4.77 ( 50)
OPTICAL CONSTANTS:
  $\alpha = 1.95$
  $\beta = 2.13$
  $\gamma = 2.21$
  $(-)2V = 58°$
HARDNESS: 5
DENSITY: 3.42 (Meas.)
       3.418 (Calc.)
CLEAVAGE: {010} perfect
         {101} distinct
Very brittle.
HABIT: Crystals flattened on {010}; twinning on (101) and (181) very common.
COLOR-LUSTER: Yellowish to pale yellow; adamantine.
MODE OF OCCURRENCE: Occurs as crystals on the walls of miarolitic cavities and also forms pseudomorphs after perovskite, replaces ilmenite, and is covered by sphene in alkalic pegmatites, Afrikanda massif, Kola Peninsula.
BEST REF. IN ENGLISH: Kukharenko, A. A., et al., *Am. Min.*, **52**: 559–560 (1967).

## KATOPHORITE (Kataphorite, Taramite)
$Na_2Ca(Fe^{3+}, Al)_5(AlSi_7)O_{22}(OH)_2$
Amphibole Group

CRYSTAL SYSTEM: Monoclinic
(See arfvedsonite for analogous crystal data)

OPTICAL CONSTANTS:
  $\alpha = 1.639–1.681$
  $\beta = 1.658–1.688$
  $\gamma = 1.660–1.690$
  $(-)2V = 0°–50°$
HARDNESS: 5
DENSITY: 3.33–3.48 (Meas.)
CLEAVAGE: {110} perfect
         {010} parting
Brittle.
HABIT: Crystals prismatic. Twinning on {100}.
COLOR-LUSTER: Rose red, dark red-brown, bluish black. Translucent. Vitreous.
MODE OF OCCURRENCE: Occurs chiefly in basic alkali rocks such as theralite and shonkinite. Found at Mariupol, Ukraine, USSR, and in the Oslo district, Norway.
BEST REF. IN ENGLISH: Deer, Howie, and Zussman, "Rock Forming Minerals," v. 2, p. 359–363, New York, Wiley, 1963.

## KATOPTRITE = Catoptrite

## KAWAZULITE
$BiTe_2Se$
Selenium analogue of tetradymite

CRYSTAL SYSTEM: Hexagonal
CLASS: $\bar{3}2/m$, 3m, or 32
SPACE GROUP: $R\bar{3}m$, R3m, or R32
Z: 3
LATTICE CONSTANTS:
  $a = 4.24$
  $c = 29.66$
3 STRONGEST DIFFRACTION LINES:
  3.12 (100)
  2.31 ( 50)
  4.92 ( 40)
OPTICAL CONSTANTS: Optically anisotropic, brownish gray to gray. Pleochroism pinkish creamy white to white with a creamy tint. Reflectance 40–50%.
HARDNESS: 1½
DENSITY: 8.08 (Calc.)
CLEAVAGE: {0001} perfect
Flexible.
HABIT: As very thin foils up to 4 mm across, with maximum thickness of 50 $\mu$.
COLOR-LUSTER: Silver to tin white. Opaque. Metallic. Streak light steel gray.
MODE OF OCCURRENCE: The mineral occurs with selenium-bearing tellurium in a quartz vein at the Kawazu mine, Shizuoka Prefecture, Japan.
BEST REF. IN ENGLISH: Kato, Akira, *Am. Min.*, **57**: 1312 (1972).

## KEHOEITE = A mixture

## KEILHAUITE = Yttrian titanite

## KELDYSHITE
$(Na, H)_2 ZrSi_2 O_7$

CRYSTAL SYSTEM: Triclinic
CLASS: $\bar{1}$
SPACE GROUP: Probably $P\bar{1}$
LATTICE CONSTANTS:
  $a = 6.66$  $\alpha = 92°45'$
  $b = 8.33$  $\beta = 94°13'$
  $c = 5.42$  $\gamma = 72°20'$
3 STRONGEST DIFFRACTION LINES:
  3.99 (100)
  4.18 ( 80)
  2.97 ( 52)
OPTICAL CONSTANTS:
  $\alpha = 1.670$
  $\beta = ?$
  $\gamma = 1.710$
  $2V = 60°$
HARDNESS: 3.8–4.3
DENSITY: ~3.30
CLEAVAGE: Two poor, which intersect at about 90°. Fracture irregular. Very brittle.
HABIT: Irregular grains up to 4 mm in size and as aggregates up to 6 mm. Under microscope shows very fine polysynthetic twinning.
COLOR-LUSTER: White; vitreous to greasy. Transparent in thin fragments.
MODE OF OCCURRENCE: Occurs as a primary mineral in foyaites consisting of microcline (partly albitized), nepheline, sodalite, aegirine, alkali amphibole, and accessory eudialyite and ramsayite in drill cores from the western and northwestern slopes of the Lovozero massif, Kola Peninsula, USSR.
BEST REF. IN ENGLISH: *Am. Min.*, **55**: 1072 (1970). Gerasimovskii, V. I., *Am. Min.*, **47**: 1216 (1962).

## KEMMLITZITE
$(Sr, Ce)Al_3 (AsO_4)[(P, S)O_4](OH)_6$

CRYSTAL SYSTEM: Hexagonal
Z: 3
LATTICE CONSTANTS:
  (hex.)        (rhomb.)
  $a = 7.072$   $a_{rh} = 6.837$
  $c = 16.51$   $\alpha = 61°51'$
3 STRONGEST DIFFRACTION LINES:
  2.959 (100)
  3.514 ( 90)
  1.903 ( 90)
OPTICAL CONSTANTS:
  $\omega = 1.701$
  $\epsilon = 1.707$ (Na)
    (+)
HARDNESS: 5.5
DENSITY: 3.63 (Meas.)
          3.601 (Calc.)
CLEAVAGE: Basal—indistinct.
HABIT: Pseudocubic rhombohedral crystals with average rhombohedron edge 0.1–0.15 mm. Basal pinacoid rarely present.

COLOR-LUSTER: Light grayish brown, partly colorless, and partly cloudy brownish with zonal structure.
MODE OF OCCURRENCE: Found associated with apatite, anatase, and zircon in the heavy fraction separated from kaolinized quartz porphyry from the Kemmlitz deposit, Saxony, East Germany.
BEST REF. IN ENGLISH: Hak, J., Johan, Z., Kuacek, M., and Liebscher, W., *Am. Min.*, **55**: 320–321 (1970).

## KEMPITE
$Mn_2 Cl(OH)_3$

CRYSTAL SYSTEM: Orthorhombic
CLASS: 2/m 2/m 2/m
SPACE GROUP: Pnam
LATTICE CONSTANTS:
  $a = 6.47$
  $b = 9.52$
  $c = 7.12$
OPTICAL CONSTANTS:
  $\alpha = 1.684$
  $\beta = 1.695$ (with orange filter)
  $\gamma = 1.698$
  $(-)2V = 54°52'$ (Calc.)
HARDNESS: ~3½
DENSITY: ~2.94 (Meas.)
CLEAVAGE: None reported.
HABIT: Crystals prismatic, elongated along $c$-axis. Very minute.
COLOR-LUSTER: Emerald green.
MODE OF OCCURRENCE: Occurs associated with pyrochroite, hausmannite, and other manganese minerals in a manganese boulder from Alum Rock Park, near San Jose, Santa Clara County, California.
BEST REF. IN ENGLISH: Palache, et al., "Dana's System of Mineralogy," 7th Ed., v. II, p. 73–74, New York, Wiley, 1951.

## KENNEDYITE
$Fe_2^{3+}MgTi_3O_{10}$
Pseudobrookite Group

CRYSTAL SYSTEM: Orthorhombic
Z: 2
LATTICE CONSTANTS:
  $a = 9.77$
  $b = 9.95$
  $c = 3.73$
3 STRONGEST DIFFRACTION LINES:
  3.485 (100)
  4.88 ( 80)
  2.743 ( 80)
HARDNESS: Not determined.
DENSITY: 4.07 (Meas.)
HABIT: Crystals lath-shaped, up to 2.0 mm in length.
COLOR-LUSTER: Black, in small grains dark brown, translucent. Metallic, brilliant.
MODE OF OCCURRENCE: Occurs in a ground-mass of alkali feldspar in a sill at the base of the Karroo volcanic succession, Mateke Hills, Rhodesia.

BEST REF. IN ENGLISH: Von Knorring, O., and Cox, K. G., *Min. Mag.*, **32**: 676–682 (1961).

# KENTROLITE
$Pb_2 Mn_2^{3+} Si_2 O_9$
Isostructural with melanotekite

CRYSTAL SYSTEM: Orthorhombic
CLASS: 222
SPACE GROUP: $C222_1$
Z: 4
LATTICE CONSTANTS:
  $a = 7.006$
  $b = 11.071$
  $c = 9.980$
3 STRONGEST DIFFRACTION LINES:
  2.90 (100)
  2.86 (100)
  2.74 (100)
OPTICAL CONSTANTS:
  $\alpha = 2.10$
  $\beta = 2.20$
  $\gamma = 2.31$
  $(+)2V = 88°$
HARDNESS: 5
DENSITY: 6.2 (Meas.)
         6.237 (Calc.)
CLEAVAGE: {110} distinct
Brittle.
HABIT: Crystals prismatic, minute, often in sheaf-like forms. Usually massive.
COLOR-LUSTER: Dark reddish brown; black on surface.
MODE OF OCCURRENCE: Occurs sparingly as minute crystals in association with calcite and willemite at Franklin, Sussex County, New Jersey. Also in manganese deposits at Långban, Pajsberg, Harstigen, and Jakobsberg, Sweden; near Ozieri, Sardinia; and in southern Chile.
BEST REF. IN ENGLISH: Glasser, F. P., *Am. Min.*, **52**: 1085–1093 (1967). Winchell and Winchell, "Elements of Optical Mineralogy," 4th Ed., Pt. 2, p. 523, New York, Wiley, 1951.

# KENYAITE
$Na_2 Si_{22} O_{41} (OH)_8 \cdot 6H_2O$

CRYSTAL SYSTEM: Monoclinic
LATTICE CONSTANTS:
  $a = 7.79$
  $b = 19.72$
  $c = 6.91$
  $\beta = 95°54'$
3 STRONGEST DIFFRACTION LINES:
  19.68 (100)
  3.428 ( 85)
  9.925 ( 50)
HARDNESS: Not determined.
DENSITY: 3.18 (Calc.) (Probably too high)
CLEAVAGE: Not determined.
HABIT: Nodular concretions, sometimes with cores of chert.

COLOR-LUSTER: White.
MODE OF OCCURRENCE: Occurs with magadiite in brown silts of the High Magadi beds deposited over a large area from saline brines of the precursor of Lake Magada, Kenya.
BEST REF. IN ENGLISH: Eugster, H. P., *Science*, **157**: 1177–1180 (1967). *Am. Min.*, **53**: 2061 (1968).

# KERCHENITE = Oxidized vivianite
$Fe^{2+} Fe_2^{3+} (PO_4)_2 (OH)_2 \cdot 6H_2O$

# KERMESITE
$Sb_2 S_2 O$

CRYSTAL SYSTEM: Monoclinic (triclinic?)
CLASS: 2/m
SPACE GROUP: C2/m
Z: 8
LATTICE CONSTANTS:
  $a = 10.97$
  $b = 8.19$
  $c = 10.36$
  $\beta = 101°45'$
OPTICAL CONSTANTS:
  $\alpha > 2.72$
  $\beta = 2.74$
  $\gamma = ?$
  $(+)2V = small$
HARDNESS: 1–1½
DENSITY: 4.68 (Meas.)
         4.69 (Calc.)
CLEAVAGE: {001} perfect
          {100} distinct
Sectile. Thin laminae flexible.
HABIT: Crystals lath-shaped; as radiating aggregates or hairlike.
COLOR-LUSTER: Cherry red. Translucent to opaque. Adamantine. Streak brownish red.
MODE OF OCCURRENCE: Occurs as an alteration of stibnite or native antimony, commonly associated with other secondary antimony minerals. Found at the Mojave antimony mine, Kern County, and in the Blind Spring mining district, Mono County, California; at Burke, Shoshone County, and in Valley County, Idaho; at deposits in Quebec, Nova Scotia, and New Brunswick, Canada; with native antimony at Nuevo Tepache, Sonora, Mexico; and at localities in France, Germany, Czechoslovakia, Sardinia, Algeria, and Zambia.
BEST REF. IN ENGLISH: Palache, et al., "Dana's System of Mineralogy," 7th Ed., v. I, p. 279-280, New York, Wiley, 1944.

# KERNITE
$Na_2 B_4 O_7 \cdot 4H_2O$

CRYSTAL SYSTEM: Monoclinic
CLASS: 2/m
SPACE GROUP: $P2_1/n$
Z: 4

LATTICE CONSTANTS:
$a = 15.52$ kX
$b = 9.14$
$c = 6.96$
$\beta = 97°29'$
3 STRONGEST DIFFRACTION LINES:
7.41 (100)
6.63 ( 80)
2.47 ( 40)
OPTICAL CONSTANTS:
$\alpha = 1.454$
$\beta = 1.472$
$\gamma = 1.488$
$(-)2V = 80°$
HARDNESS: 2½–3
DENSITY: 1.908 (Meas.)
1.93 (Calc.)
CLEAVAGE: {100} perfect
{001} perfect
{$\bar{2}$01} distinct
Very brittle; splintery.

HABIT: Crystals nearly equant; rare. Massive; usually as cleavable masses with fibrous structure. Twinning on {110}.

COLOR-LUSTER: Colorless and transparent when fresh; usually white and opaque due to surface alteration coating of tincalconite. Dull to vitreous; silky or pearly on cleavage.

MODE OF OCCURRENCE: Occurs as veins, irregular masses, and as crystals up to 3 X 8 feet in size in large playa deposits near Boron, Kramer district, Kern County, California; it is also found associated with borax and ezcurrite at the Tincalayu borax mine, Salta Province, Argentina.

BEST REF. IN ENGLISH: Palache, et al., "Dana's System of Mineralogy," 7th Ed., v. II, p. 335–337, New York, Wiley, 1951.

## KERSTENITE
$PbSeO_4$

CRYSTAL SYSTEM: Orthorhombic
CLASS: 2/m 2/m 2/m
SPACE GROUP: Pnma
Z: 4
LATTICE CONSTANTS:
$a = 5.50$
$b = 7.04$ (synthetic)
$c = 8.48$
3 STRONGEST DIFFRACTION LINES:
3.03 (100)
3.36 ( 75)
3.25 ( 75)
OPTICAL CONSTANTS:
$\alpha = 1.96$
$\gamma = 1.98$
$2V = 50°$
HARDNESS: 3½
DENSITY: 7.08 (Calc.)
CLEAVAGE: Imperfect perpendicular to elongation.
HABIT: As prismatic and acicular crystals up to 1.0 mm long. Also botryoidal.
COLOR-LUSTER: Colorless to greenish yellow, sulfur yellow. Greasy to vitreous.

MODE OF OCCURRENCE: Occurs as an oxidation product of primary lead and copper selenides in veins in the Cerro de Cacheuta, Mendoza, Argentina; at Pacajake, Bolivia; and at the Friedrichsglück mine, near Hildburghausen, Germany.

BEST REF. IN ENGLISH: Goni, J., and Guillemin, C., Am. Min., 39: 850 (1954).

## KËSTERITE (Kosterite)
$Cu_2(Zn, Fe)SnS_4$
Stannite Group
Zinc analogue of stannite

CRYSTAL SYSTEM: Tetragonal (Pseudocubic)
CLASS: $\bar{4}$ 2m
SPACE GROUP: I$\bar{4}$2m
Z: 2
LATTICE CONSTANTS:
$a = 5.44$
$c = 10.88$
3 STRONGEST DIFFRACTION LINES:
3.15 (100)
1.93 ( 70)
1.65 ( 50)
HARDNESS: 4½
DENSITY: 4.54–4.59 (Meas.)
4.50 (Calc. Zn:Fe = 2:1)
CLEAVAGE: None.
HABIT: Massive.
COLOR-LUSTER: Greenish black. Opaque. Streak black.
MODE OF OCCURRENCE: Occurs in quartz-sulfide veinlets in the Kêster deposit in the Yano-Adychansk region, USSR. Also found at the Snowflake mine, Revelstoke mining division, British Columbia, Canada.

BEST REF. IN ENGLISH: Am. Min., 43: 1222–1223 (1958). Ivanov, V. V., and Pyatenko, Yu. A., Am. Min., 44: 1329 (1959).

## KETTNERITE
$CaBi(CO_3)OF$

CRYSTAL SYSTEM: Tetragonal
CLASS: 4/m 2/m 2/m
SPACE GROUP: P4/nmm
Z: 2
LATTICE CONSTANTS:
$a = 3.79$
$c = 13.59$
DENSITY: 5.80
HABIT: Small plates, usually 0.2–0.3 mm in size.
COLOR-LUSTER: Brown, brownish yellow to lemon yellow.
MODE OF OCCURRENCE: Occurs associated with fluorite, bismuth, bismuthinite, hematite, and altered topaz in cavities in a quartz vein cutting pegmatitic potassium feldspar near Krupka (Graupen), Krušné Hory (Erzgebirge), N.W. Bohemia, Czechoslovakia.

BEST REF. IN ENGLISH: Zak, Lubor, and Syneček, Vladimir, Am. Min., 42: 121 (1957).

**KHLOPINITE** = Variety of samarskite

**KHUNIITE** = Variety of hemihedrite.

## KIESERITE
$MgSO_4 \cdot H_2O$

CRYSTAL SYSTEM: Monoclinic
CLASS: 2/m
SPACE GROUP: C2/c
Z: 4
LATTICE CONSTANTS:
$a = 6.90$
$b = 7.71$
$c = 7.54$
$\beta = 116°5.5'$
3 STRONGEST DIFFRACTION LINES:
3.409 (100)
4.84 ( 90)
3.331 ( 90)
OPTICAL CONSTANTS:
$\alpha = 1.520$
$\beta = 1.533$
$\gamma = 1.584$
(+)2V = 55° (Na)
HARDNESS: 3½
DENSITY: 2.571 (Meas.)
2.571 (Calc.)
CLEAVAGE: {110} perfect
{111} perfect
{$\bar{1}$11} imperfect
{$\bar{1}$01} imperfect
{011} imperfect
Friable.
HABIT: Usually massive, coarse to fine granular. Crystals bipyramidal, often highly modified, rare. Twinning on {001} rare.
COLOR-LUSTER: Colorless, white, grayish, white, yellowish. Transparent to translucent. Vitreous.
MODE OF OCCURRENCE: Occurs chiefly in sedimentary deposits of oceanic origin associated with halite and other salts, and rarely as a product of volcanic activity. In the United States it is found in Grand County, Utah, and in deposits in the Permian basin of southeastern New Mexico, especially in the vicinity of Carlsbad, and also in adjacent parts of Texas. It also occurs at many localities in Germany, Poland, Austria, Sicily, USSR, and the Punjab Salt Range, India.
BEST REF. IN ENGLISH: Palache, et al., "Dana's System of Mineralogy," 7th Ed., v. II, p. 477-479, New York, Wiley, 1951.

## KILCHOANITE
$Ca_3Si_2O_7$
Dimorphous with rankinite

CRYSTAL SYSTEM: Orthorhombic
CLASS: 2/m 2/m 2/m or mm2
SPACE GROUP: Imam or Ima2
Z: 8

LATTICE CONSTANTS:
$a = 11.42$
$b = 5.09$
$c = 21.95$
3 STRONGEST DIFFRACTION LINES:
2.89 (100)
2.68 ( 95)
3.07 ( 90)
OPTICAL CONSTANTS:
$\alpha = 1.647$
$\gamma = 1.650$
(+)2V = 60°
HARDNESS: Not determined.
DENSITY: 2.992
CLEAVAGE: None.
HABIT: Massive; as a replacement of rankinite.
COLOR-LUSTER: Colorless.
MODE OF OCCURRENCE: Occurs associated with rankinite, which it replaces, spurrite, melilite, cuspidine, grossular, wollastonite, and vesuvianite in limestones thermally metamorphosed by gabbro at Kilchoan, Ardnamurchan, Scotland, and Carlingford, Ireland.
BEST REF. IN ENGLISH: Agrell, S. O., and Gay, P., *Nature*, **189** (4766): 743 (1961).

## KIMZEYITE
$Ca_3(Zr, Ti)_2(Al, Si)_3O_{12}$
Garnet Group

CRYSTAL SYSTEM: Cubic
CLASS: 4/m $\bar{3}$ 2/m
SPACE GROUP: Ia3d
Z: 8
LATTICE CONSTANT:
$a = 12.46$
3 STRONGEST DIFFRACTION LINES:
1.667 (100)
2.539 ( 90)
2.79 ( 80)
OPTICAL CONSTANT:
$N = 1.94$
HARDNESS: ~7
DENSITY: 4.0 (Meas.)
4.03 (Calc.)
CLEAVAGE: None.
HABIT: Dodecahedrons modified by trapezohedron. Crystals usually less than 1 mm in diameter, rarely up to 5 mm.
COLOR-LUSTER: Dark brown; vitreous. Streak light brown.
MODE OF OCCURRENCE: Occurs in a light-colored phase of carbonatite which is associated with ijolite in the Kimzey calcite quarry, Magnet Cove, Arkansas. Associated minerals include abundant apatite, monticellite, calcite, perovskite (dysanalyte), magnetite, and minor biotite, pyrite, and vesuvianite.
BEST REF. IN ENGLISH: Milton, Charles, Ingram, Blanche L., and Blade, Lawrence V., *Am. Min.*, **46**: 533-548 (1961).

## KINGITE
$Al_3(PO_4)_2(OH, F)_3 \cdot 9H_2O$

CRYSTAL SYSTEM:  Triclinic
Z:  2
LATTICE CONSTANTS:
$a = 9.15$    $\alpha = 98.6°$
$b = 10.00$    $\beta = 93.6°$
$c = 7.24$    $\gamma = 93.2°$
3 STRONGEST DIFFRACTION LINES:
  9.1  (100)
  3.45 ( 80)
  3.48 ( 65)
OPTICAL CONSTANTS:  Mean index ~ 1.514.
  HARDNESS:  Not determined.
  DENSITY:  2.21–2.30 (Meas.)
        2.465 (Calc.)
  HABIT:  Nodular masses up to 2 inches in diameter; platelets or irregular thick fragments about 0.5 $\mu$m–1.0 $\mu$m in size.
  COLOR-LUSTER:  White.
  MODE OF OCCURRENCE:  Occurs coated and veined by brown powdery material consisting of talc, quartz, and halite, with possibly a little gypsum and hematite or goethite, at Robertstown, South Australia, and also in Precambrian sediments near the Clinton phosphate workings, 64 miles west–southwest of the first deposit.
  BEST REF. IN ENGLISH:  Norrish, K., Rogers, Lillian E. R., and Shapter, R. E., *Min. Mag.*, **31**: 351–357 (1957). Kato, Toshio, *Am. Min.*, **55**: 515–517 (1970).

## KINOITE
$Cu_2Ca_2Si_3O_{10} \cdot 2H_2O$

CRYSTAL SYSTEM:  Monoclinic
CLASS:  2/m
SPACE GROUP:  P2$_1$/m
Z:  2
LATTICE CONSTANTS:
  $a = 6.990$
  $b = 12.890$
  $c = 5.654$
  $\beta = 96°05'$
3 STRONGEST DIFFRACTION LINES:
  4.72  (100)
  3.052 ( 80)
  6.441 ( 70)
OPTICAL CONSTANTS:
    $\alpha = 1.638$
    $\beta = 1.665$
    $\gamma = 1.676$
  $(-)2V = 68°$
  HARDNESS:  ~5
  DENSITY:  3.16 (Meas.)
        3.193 (Calc.)
  CLEAVAGE:  {010} excellent
        {100} {001} distinct
  HABIT:  Crystals tabular in the plane of the $b$ and $c$-axis and are slightly elongated by extension on $c$.  [$hkO$] zone striated and presents curved aspect because of development of vicinal forms.

COLOR-LUSTER:  Deep azure blue; transparent to translucent.
  MODE OF OCCURRENCE:  Occurs as single crystals nearly 1 mm in length and in 1 mm thick veinlets associated with apophyllite, native copper, and copper sulfides in the northern Santa Rita Mountains, Pima County, Arizona.  It was found in drill cores which cut skarn developed in a Paleozoic limestone sequence.
  BEST REF. IN ENGLISH:  Anthony, John W., and Laughon, Robert B., *Am. Min.*, **55**: 709–715 (1970).

## KIPUSHITE = Veszelyite

## KIROVITE = Magnesian variety of melanterite

## KIRSCHSTEINITE
$CaFeSiO_4$
Iron analogue of monticellite

CRYSTAL SYSTEM:  Orthorhombic
CLASS:  2/m 2/m 2/m
SPACE GROUP:  Pbnm
Z:  4
LATTICE CONSTANTS:
  $a = 4.886$
  $b = 11.146$
  $c = 6.434$
3 STRONGEST DIFFRACTION LINES:
  2.949 (100)
  2.680 ( 85)
  2.604 ( 80)
OPTICAL CONSTANTS:
    $\alpha = 1.696$
    $\beta = 1.734$
    $\gamma = 1.743$
  $(-)2V = 50°$
  DENSITY:  3.434 (Meas.)
  HABIT:  Massive.
  COLOR-LUSTER:  Slightly greenish; colorless in thin section.
  MODE OF OCCURRENCE:  Occurs associated with melilite, nepheline, clinopyroxene, kalsilite, götzenite, combeite, sodalite, magnetite, perovskite apatite, hornblende, biotite, and an unidentified mineral, in melilite-nephelinite lava from Mt. Shaheru, Zaire.
  BEST REF. IN ENGLISH:  Sahama, Th. G., and Hytönen, Kai, *Min. Mag.*, **31**: 698–699 (1957).

## KITKAITE
NiTeSe

CRYSTAL SYSTEM:  Hexagonal
CLASS:  $\bar{3}$ 2/m
SPACE GROUP:  P$\bar{3}$m1
Z:  1
LATTICE CONSTANTS:
  $a = 3.716$
  $c = 5.126$

3 STRONGEST DIFFRACTION LINES:
2.729 (100)
2.007 ( 45)
1.510 ( 35)
HARDNESS: 110 kg/mm$^2$
DENSITY: 7.22 (Meas.)
7.19 (Calc.)
CLEAVAGE: {001} good
HABIT: Massive.
COLOR-LUSTER: Pale yellow. Opaque. Metallic.
MODE OF OCCURRENCE: Occurs in albitite in narrow carbonate-bearing fissure veinlets associated with clausthalite, hematite, selenian polydymite, selenian linnaeite, and locally with Ni-, Co- and Se-bearing pyrite in Kuusamo, northeast Finland.
BEST REF. IN ENGLISH: Häkli, T. A., Vourelainen, Y., and Sahama, Th. G., *Am. Min.*, **50**: 581–586 (1965).


**KLAPROTHITE** = A mixture of wittichenite and emplectite


**KLEBELSBERGITE** (Inadequately described mineral)
Sulfate of Sb

CRYSTAL SYSTEM: Monoclinic
LATTICE CONSTANT:
$\beta = 91°48'$
OPTICAL CONSTANTS:
$\alpha > 1.74$
(–)2V
HARDNESS: Not determined.
DENSITY: Not determined.
CLEAVAGE: Not determined. Brittle.
HABIT: Crystals acicular, in tufts.
COLOR-LUSTER: Dark yellow. Transparent.
MODE OF OCCURRENCE: Found sparingly at Felsöbánya, Roumania, associated with stibnite.
BEST REF. IN ENGLISH: Palache, et al., "Dana's System of Mineralogy," 7th Ed., v. II, p. 583, New York, Wiley, 1951.


**KLEINITE**
$Hg_2 N(Cl, SO_4) \cdot nH_2O$

CRYSTAL SYSTEM: Hexagonal (above 130°C), triclinic (below 130°C)
CLASS: 6/m 2/m 2/m
SPACE GROUP: Probably C6$_3$/mmc
Z: 18
LATTICE CONSTANTS:
(hex.)
$a = 13.56$
$c = 11.13$
Metastable at room temperature.
3 STRONGEST DIFFRACTION LINES:
2.91 (100)
2.62 (100)
3.88 ( 60)

OPTICAL CONSTANTS:
(triclinic) (hexagonal)
$\alpha = 2.16$ $\omega = 2.19$
$\beta = 2.18$ $\epsilon = 2.21$
$\gamma = 2.18$ (+)
(–)2V = small to medium
HARDNESS: 3½–4
DENSITY: ~8.0 (Meas.)
CLEAVAGE: {0001} imperfect but easy
{10$\bar{1}$0} imperfect
Brittle.
HABIT: Crystals short prismatic, elongated along c-axis.
COLOR-LUSTER: Pale yellow to orange. Transparent to translucent. Adamantine to greasy. Streak yellow.
MODE OF OCCURRENCE: Occurs associated with terlinguaite, calomel, cinnabar, eglestonite, mercury, and montroydite at the mercury deposits in Cretaceous limestone in the Terlingua district, Brewster County, Texas.
BEST REF. IN ENGLISH: Palache, et al., "Dana's System of Mineralogy," 7th Ed., v. II, p. 87–89, New York, Wiley, 1951.


**KLEMENTITE** = Variety of thuringite


**KLOCKMANNITE**
CuSe

CRYSTAL SYSTEM: Hexagonal
CLASS: 6/m 2/m 2/m
SPACE GROUP: P6$_3$/mmc
Z: 6
LATTICE CONSTANTS:
$a = 3.938$
$c = 17.25$
3 STRONGEST DIFFRACTION LINES:
2.88 (100)
3.18 ( 90)
1.967 ( 80)
HARDNESS: 1.9–2.7
Microhardness 57–86 kg/mm$^2$.
DENSITY: 5.99 (Meas.)
6.12 (Calc.)
CLEAVAGE: {0001} perfect
HABIT: As dense granular or flaky aggregates.
COLOR-LUSTER: Dark gray to bluish black. Opaque. Metallic on fresh fracture, rapidly becoming dull.
MODE OF OCCURRENCE: Occurs in Canada at the Ransom Property, Ontario, and in substantial amounts in uranium ores from the Lake Athabasca area, northern Saskatchewan. It also is found in hydrothermal veins at Sierra de Umango and Cerro de Cacheuta, Argentina; at Lehrbach and Tilkerode, Harz Mountains, Germany; and at Skrikerum, Sweden.
BEST REF. IN ENGLISH: Vlasov, K. A., "Mineralogy of Rare Elements," v. II, p. 643–644, Israel Program for Scientific Translations, 1966. Sindeeva, N. D., "Mineralogy and Types of Selenium and Tellurium," (1964).

# KNEBELITE

$(Fe, Mn)_2 SiO_4$
Olivine Group

CRYSTAL SYSTEM: Orthorhombic
CLASS: 2/m 2/m 2/m
SPACE GROUP: Pnma
Z: 4
LATTICE CONSTANTS:
  $a = 4.82-4.85$
  $b = 10.49-10.60$
  $c = 6.09-6.22$
3 STRONGEST DIFFRACTION LINES:
  2.86 (100)
  2.53 ( 80)
  5.31 ( 30)
OPTICAL CONSTANTS:
  $\alpha = 1.775-1.815$
  $\beta = 1.810-1.853$
  $\gamma = 1.826-1.867$
  $(-)2V = 44°-61°$
HARDNESS: 6½
DENSITY: 3.96-4.25 (Meas.)
CLEAVAGE: {010} good
          {001} imperfect
          {110} imperfect
Fracture conchoidal. Brittle.
  HABIT: Crystals short prismatic. Usually massive, compact or granular.
  COLOR-LUSTER: Gray-black, brownish black. Translucent. Colorless or pale yellow in thin section. Vitreous to greasy. Streak uncolored.
  MODE OF OCCURRENCE: Occurs mainly as a metamorphic product in iron-manganese ore deposits. Found at the Blue Bell mine, Kootenay Lake, British Columbia, Canada; at the Treburland manganese mine, Cornwall, England; at Dannemora and elsewhere in Sweden; and at the Kaso and Kyurazawa mines, Tochigi prefecture, Japan.
  BEST REF. IN ENGLISH: Deer, Howie, and Zussman, "Rock Forming Minerals," v. 1, p. 34-40, New York, Wiley, 1962.

# KNIPOVICHITE (Species status uncertain.)

Hydrated carbonate of Cr, Al, and Ca

OPTICAL CONSTANTS:
  $\alpha = 1.540$
  $\beta = ?$
  $\gamma = 1.592$
  $(-)2V = small$
HARDNESS: 2-3
DENSITY: 2.229
CLEAVAGE: Not determined.
HABIT: Fibrous radial aggregates.
COLOR-LUSTER: Pink.
MODE OF OCCURRENCE: Occurs in dolomite and quartz veins.
  BEST REF. IN ENGLISH: Nefedov, E. I., *Am. Min.*, **40**: 551 (1955).

# KNOPITE = Cerian perovskite

# KNORRINGITE (Chromium Pyrope)

$Mg_3 Cr_2 (SiO_4)_3$
Garnet Group

CRYSTAL SYSTEM: Cubic
CLASS: $4/m \overline{3} 2/m$
SPACE GROUP: Ia3d
Z: 8
LATTICE CONSTANT:
  $a = 11.65$
OPTICAL CONSTANT:
  $N = 1.803$
  HARDNESS: Not determined. Probably ~7.
  DENSITY: 3.756 (Meas.)
           3.852 (Calc.)
  CLEAVAGE: None.
  HABIT: Massive; as very minute grains.
  COLOR-LUSTER: Green; vitreous.
  MODE OF OCCURRENCE: Found sparingly at the Kao kimberlite pipe, Lesotho.
  BEST REF. IN ENGLISH: Nixon, Peter H., and Hornung, George, *Am. Min.*, **53**: 1833-1840 (1968).

# KOBEÏTE

$(Y, Fe, U)(Ti, Nb, Ta)_2 (O, OH)_6$

CRYSTAL SYSTEM: Cubic (?)
Z: 1
LATTICE CONSTANT:
  $a = 5.03$ (620°C)
3 STRONGEST DIFFRACTION LINES:
  2.89 (100)      4.16 (100)
  1.772 ( 80)     3.74 ( 60)
  1.513 ( 70)     3.13 ( 60)
  Metamict. Heated    Pattern from
  in vacuo for 1      single uncrushed
  hour at 620°C.      fragment.
DENSITY: $5^+$ (Meas.)
CLEAVAGE: Conchoidal fracture. Brittle.
  HABIT: Radiated crystals and parallel growths; crystals long prismatic to stumpy and irregular, coarsely striated parallel to length.
  COLOR-LUSTER: Deep coffee brown to black; vitreous to resinous.
  MODE OF OCCURRENCE: Occurs in pegmatite associated with parisite, zircon, monazite, tscheffkinite, xenotime, biotite, albite, quartz, and muscovite at Shiraishi, Kobe-mura, Kyoto Prefecture, Japan; also as slender prismatic crystals in a cobble of graphic-granite in gravels of the Paringa River, South Westland, New Zealand.
  BEST REF. IN ENGLISH: Hutton, C. Osborne, *Am. Min.*, **42**: 342-353 (1957).

# KOBELLITE

$Pb_2 (Bi, Sb)_2 S_5$
(Kobellite-Tintinaite Series)

CRYSTAL SYSTEM: Orthorhombic
CLASS: 2/m 2/m 2/m

SPACE GROUP: Pnmm
Z: 12
LATTICE CONSTANTS:
$a = 22.62$
$b = 34.08$
$c = 4.02$
3 STRONGEST DIFFRACTION LINES:
3.54 (100)
3.41 ( 90)
2.72 ( 50)
HARDNESS: 2½–3
DENSITY: 6.48 (Meas. Hvena, Sweden)
6.51 (Calc.)
CLEAVAGE: {010} distinct
HABIT: Massive; fibrous or radiated; also fine granular.
COLOR-LUSTER: Lead gray. Metallic. Opaque. Streak black.
MODE OF OCCURRENCE: Occurs in richly mineralized granite pegmatite veins associated with bismuthinite, jamesonite, and tetrahedrite, at the Superior Stone Quarry, Wake County, North Carolina; at the Silver Bell mine, Ouray, and elsewhere in Colorado; at the Dodger Tungsten mine, Salmo, British Columbia; with cobaltite at Hvena, Sweden; and at several other reported localities.
BEST REF. IN ENGLISH: Harris, D. C., Jambor, J. L., Lachance, G. R., and Thorpe, R. I., *Can. Min.*, **9** (Pt. 3): 371–382 (1968).

## KOCHUBEITE (Kotschubeite) = Variety of clinochlore with more than 4% $Cr_2O_3$

## KOECHLINITE
$(BiO)_2 MoO_4$

CRYSTAL SYSTEM: Orthorhombic
CLASS: 2/m 2/m 2/m
SPACE GROUP: Cmc(a?)
Z: 4
LATTICE CONSTANTS:
$a = 5.502$
$b = 16.213$
$c = 5.483$
3 STRONGEST DIFFRACTION LINES:
3.15 (100)
1.652 ( 20)
2.741 ( 18)
OPTICAL CONSTANTS:
$\alpha = 2.52$
$\beta = 2.61$
$\gamma = 2.67$
(–)2V = large
HARDNESS: Not determined.
DENSITY: 8.26 (Meas.)
8.281 (Calc.)
CLEAVAGE: {010} perfect
{0kl} imperfect
Very brittle.
HABIT: Crystals thin, square to rectangular laths and plates striated parallel [010]. Also massive, earthy. Twinning on {101}, contact or penetration.

COLOR-LUSTER: Greenish yellow; yellow, white. Transparent. Streak pale greenish yellow.
MODE OF OCCURRENCE: Occurs associated with quartz, native bismuth, and smaltite at the Daniel mine, Schneeberg, Saxony, Germany; as earthy masses with bismoclite at Bygoo, New South Wales; and at the Dunallan mine, Coolgardie, Western Australia.
BEST REF. IN ENGLISH: Palache, et al., "Dana's System of Mineralogy," 7th Ed., v. II, p. 1092–1093, New York, Wiley, 1951.

## KOENENITE
$Na_4 Mg_9 Al_4 Cl_{12} (OH)_{22}$

CRYSTAL SYSTEM: Hexagonal
3 STRONGEST DIFFRACTION LINES:
3.50 (100)
2.85 ( 50)
2.34 ( 50)
OPTICAL CONSTANTS:
$\omega = 1.52$
$\epsilon = 1.55$
(+)
HARDNESS: ~1½
DENSITY: 1.98 (Meas.)
CLEAVAGE: {0001} perfect
Thin fragments flexible.
HABIT: As crusts of indistinct crystals; also as rosettes of intergrown tablets.
COLOR-LUSTER: Colorless; often pale yellow to dark red due to inclusions of hematite scales. Transparent to translucent. Pearly on cleavages.
MODE OF OCCURRENCE: Occurs in potash mines in the vicinity of Wathlingen, Hanover, Germany, associated with halite, carnallite, sylvite, and anhydrite.
BEST REF. IN ENGLISH: Palache, et al., "Dana's System of Mineralogy," 7th Ed., v. II, p. 86–87, New York, Wiley, 1951.

## KOETTIGITE (Köttigite)
$Zn_3 (AsO_4)_2 \cdot 8H_2O$

CRYSTAL SYSTEM: Monoclinic
CLASS: 2/m
SPACE GROUP: C2/m
Z: 2
LATTICE CONSTANTS:
$a = 10.11$ kX
$b = 13.31$
$c = 4.70$
$\beta = 103°50'$
3 STRONGEST DIFFRACTION LINES:
3.20 (100)
3.00 (100)
2.72 ( 50)
OPTICAL CONSTANTS:
$\alpha = 1.622$
$\beta = 1.638$ (Na)
$\gamma = 1.671$
(+)2V = 74°

HARDNESS: 2½–3
DENSITY: 3.33 (Meas.)
          3.32 (Calc.)
CLEAVAGE: {010} perfect
HABIT: Crystals prismatic, flattened on {010}. Massive; as crusts with fibrous structure.
COLOR-LUSTER: Carmine red, brownish. Translucent. Silky on fracture. Streak reddish white.
MODE OF OCCURRENCE: Occurs sparingly as a secondary mineral at the Daniel mine, Schneeberg, Saxony, Germany. Also found at Mapimi, Durango, Mexico.
BEST REF. IN ENGLISH: Palache, et al., "Dana's System of Mineralogy," 7th Ed., v. II, p. 751–752, New York, Wiley, 1951.

## KOGARKOITE (= Schairerite?) (Species status uncertain)
$Na_3(SO_4)F$

## KOIVINITE = Florencite

## KOKTAITE
$(NH_4)_2 Ca(SO_4)_2 \cdot H_2O$

CRYSTAL SYSTEM: Monoclinic
CLASS: 2/m or 2
SPACE GROUP: $P2_1/m$ or $P2_1$
Z: 2
LATTICE CONSTANTS:
  $a = 10.17$
  $b = 7.15$
  $c = 6.34$
  $\beta = 102°45'$
3 STRONGEST DIFFRACTION LINES:
  9.83 (100)
  3.30 ( 65)
  4.96 ( 40)
OPTICAL CONSTANTS:
  $\alpha = 1.524$
  $\beta = 1.532$
  $\gamma = 1.536$
  $(-)2V = 72°$
HARDNESS: Not determined.
DENSITY: 2.09 (Meas.)
CLEAVAGE: None.
HABIT: Crystals acicular or fibrous. Twinning on {100} common.
COLOR-LUSTER: Colorless, white.
MODE OF OCCURRENCE: Occurs associated with ammonia-alum and mascagnite as pseudomorphs after gypsum on tailing-piles of a lignite mine at Žeravice, near Kyjov, Czechoslovakia.
BEST REF. IN ENGLISH: Palache, et al., "Dana's System of Mineralogy," 7th Ed., v. II, p. 444, New York, Wiley, 1951.

## KOLBECKITE (Sterrettite, Eggonite)
$ScPO_4 \cdot 2H_2O$

CRYSTAL SYSTEM: Orthorhombic
CLASS: 2 2 2
SPACE GROUP: $P2_1 2_1 2_1$
Z: 2
LATTICE CONSTANTS:
  $a = 8.90$ kX
  $b = 10.20$
  $c = 5.43$
3 STRONGEST DIFFRACTION LINES:
  4.88 (100)
  4.51 ( 80)
  2.90 ( 80)
OPTICAL CONSTANTS:
  $\alpha = 1.572$
  $\beta = 1.590$
  $\gamma = 1.601$
  $(-)2V = 60°$
HARDNESS: 5
DENSITY: 2.44 (Meas.)
          2.47 (Calc.)
CLEAVAGE: {110} fair
          {100} poor
          {001} poor
Fracture conchoidal. Brittle.
HABIT: Crystals short prismatic, minute. Apparently twinned on {001} and perhaps {031} as a subordinate composition plane.
COLOR-LUSTER: Colorless, yellowish, bright blue to blue-gray. Transparent. Vitreous to pearly.
MODE OF OCCURRENCE: Occurs with wardite and crandallite in altered variscite nodules at Fairfield, Utah County, Utah; on specimens of silver ore from Felsöbánya, Roumania; and very sparingly in a quartz-wolframite vein in the Sadisdorf copper mine near Schmiedeberg, Saxony, Germany.
BEST REF. IN ENGLISH: Mrose, M. E., and Wappner, Blanca, *Bull. Geol. Soc. Am.*, **70**: 1648–1649 (1959) (abs.).

## KOLOVRATITE (Species status uncertain)
Vanadate of Ni and Zn

CRYSTAL SYSTEM: Unknown
3 STRONGEST DIFFRACTION LINES:
  11.6  (100)
  5.83 (100)
  3.88 (100)
OPTICAL CONSTANTS: Mean index = 1.577.
HARDNESS: Not determined.
DENSITY: Not determined.
CLEAVAGE: Not determined.
HABIT: As botryoidal crusts and incrustations.
COLOR-LUSTER: Yellow to greenish yellow.
MODE OF OCCURRENCE: Occurs widespread in carbonaceous slates and quartz schists in the Ferghana district, Turkestan, USSR.
BEST REF. IN ENGLISH: Palache, et al., "Dana's System of Mineralogy," 7th Ed., v. II, p. 1048–1049, New York, Wiley, 1951.

# KOMAROVITE
$(Ca, Mn)Nb_2Si_2O_9(O, F) \cdot 3.5 H_2O$
Ca-Mn-Nb analogue of labuntsovite

CRYSTAL SYSTEM: Orthorhombic
Z: 18
LATTICE CONSTANTS:
  $a = 21.30$
  $b = 14.00$
  $c = 17.19$
3 STRONGEST DIFFRACTION LINES:
  3.16 (100)
  12.2 ( 70)
  1.783 ( 47)
OPTICAL CONSTANTS: Optically biaxial, positive, $\eta$s
$\alpha$ 1.750, $\beta$ 1.766, $\gamma$ 1.85, 2V = 48°.
HARDNESS: 1½–2
DENSITY: 3.0 (Meas.)
          2.96 (Calc.)
CLEAVAGE: {001} fair
HABIT: Massive; foliated, platy.
COLOR-LUSTER: Pale rose. Dull. Streak white.
MODE OF OCCURRENCE: The mineral occurs associated with late albite and redeposited fine-grained natrolite in alkalic rocks of Mt. Karnasurt, Kola Peninsula, USSR. It fills fractures in natrolite and is coated by hydrous manganese oxides.
BEST REF. IN ENGLISH: Portnov, A. M., Krivokoneva, G. K., and Stolyarova, T. I., *Am. Min.*, **57**: 1315–1316 (1972).

# KONGSBERGITE = Mercurian silver

# KONINCKITE (Strengite or Metastrengite ?)
$FePO_4 \cdot 3H_2O$

CRYSTAL SYSTEM: Orthorhombic (?)
OPTICAL CONSTANTS:
  $\alpha = 1.645$
  $\beta = 1.65$
  $\gamma = 1.656$
  2V = not very large
HARDNESS: 3½
DENSITY: 2.3
CLEAVAGE: One; transverse to elongation.
HABIT: As small spherical aggregates of radiating needles.
COLOR-LUSTER: Yellow. Transparent. Vitreous.
MODE OF OCCURRENCE: Occurs associated with richellite, halloysite, and allophane at Richelle near Visé, Belgium.
BEST REF. IN ENGLISH: Palache, et al., "Dana's System of Mineralogy," 7th Ed., v. II, p. 763, New York, Wiley, 1951.

# KOPPITE = Pyrochlore

# KORNELITE
$Fe_2(SO_4)_3 \cdot 7H_2O$

CRYSTAL SYSTEM: Monoclinic
CLASS: 2/m
SPACE GROUP: $P2_1/n$
Z: 4
LATTICE CONSTANTS:
  $a = 14.29$
  $b = 20.10$
  $c = 5.45$
  $\beta = 97°01'$
3 STRONGEST DIFFRACTION LINES:
  10.0 (100)
  6.64 ( 80)
  4.67 ( 80)
OPTICAL CONSTANTS:
  $\alpha = 1.567$
  $\beta = 1.581$
  $\gamma = 1.638$
  (+)2V = 49°–62°
HARDNESS: Not determined.
DENSITY: 2.30 (Meas.)
          2.29 (Calc.)
CLEAVAGE: {010}
HABIT: Crystals lath-like or acicular. As globular masses with radial-fibrous structure, and as crusts and tufted aggregates.
COLOR-LUSTER: Pale pink to violet. Fibrous masses silky.

Soluble in water.
MODE OF OCCURRENCE: Occurs as a secondary mineral in the Tintic Standard mine, Juab County, Utah; at the Copper Queen mine, Bisbee, Cochise County, Arizona, associated with coquimbite and other sulfates; and with coquimbite and voltaite at Szomolnok, Hungary.
BEST REF. IN ENGLISH: Palache, et al., "Dana's System of Mineralogy," 7th Ed., v. II, p. 530–532, New York, Wiley, 1951.

# KORNERUPINE
$Mg_3Al_6(Si, Al, B)_5O_{21}(OH)$

CRYSTAL SYSTEM: Orthorhombic
CLASS: 2/m 2/m 2/m
SPACE GROUP: Ccmm or Cmmm
Z: 4
LATTICE CONSTANTS:
  $a = 13.65–13.69$ kX
  $b = 15.91–15.93$
  $c = 6.68$
3 STRONGEST DIFFRACTION LINES:
  2.64 (100)
  3.03 ( 80)
  2.12 ( 60)
OPTICAL CONSTANTS:
  $\alpha = 1.6650–1.682$
  $\beta = 1.6766–1.696$
  $\gamma = 1.6770–1.699$
  2V = 3°–48°
HARDNESS: 6–7
DENSITY: 3.273–3.449 (Meas.)
          3.315 (Calc.)
CLEAVAGE: {110} imperfect to distinct

HABIT: Fibrous to columnar aggregates; crystals prismatic with forms {010}, {100}, and {110}.

COLOR-LUSTER: Colorless, white, pink, greenish yellow, sea green, dark green, brown, black. Transparent to translucent. Vitreous.

MODE OF OCCURRENCE: Occurs as well-defined crystals, as much as 2 inches long and ½ inch in diameter, with quartz, orthoclase, biotite, tourmaline, andalusite, almandite, rutile, and other minerals near Lac Ste-Marie, Gatineau County, Quebec, Canada. Also in clear large crystals of gem quality and sea-green color at Itrongay, Madagascar; together with zircon, garnet, and spinel in gem gravels from Ceylon; in an albitic facies of a "granulite" at Waldheim, Saxony, Germany; near Port Shepstone, Natal, South Africa; Fiskernaes, Greenland; and Antarctica.

BEST REF. IN ENGLISH: Girault, J. P., *Am. Min.*, 37: 531–541 (1952).

## KORZHINSKITE (Inadequately described mineral)
$CaB_2O_4 \cdot H_2O$

CRYSTAL SYSTEM: Unknown
3 STRONGEST DIFFRACTION LINES:
  2.02 (100)
  3.11 ( 70)
  2.81 ( 70)
OPTICAL CONSTANTS:
  $\alpha$ = 1.642
  $\beta$ = 1.647
  $\gamma$ = 1.672
  (+)2V = 43.6°
CLEAVAGE: One in direction of elongation.
HABIT: Lamellar aggregates of prismatic grains.
COLOR-LUSTER: Colorless, transparent.
MODE OF OCCURRENCE: Occurs associated with sibirskite, calciborite, calcite, and a little andradite-grossular garnet and dolomite in limestones of Middle Devonian age in the Novo-Frolovsk contact-metasomatic copper deposits, Tur'insk region, northern Urals.

BEST REF. IN ENGLISH: Malinko, S. V., *Am. Min.*, 49: 441 (1964).

## KOSTOVITE (Inadequately described mineral)
$AuCuTe_4$

CRYSTAL SYSTEM: Monoclinic (?)
3 STRONGEST DIFFRACTION LINES:
  3.03 (100)
  2.10 ( 90)
  2.93 ( 60)
HARDNESS: ~2-2.5
DENSITY: Not determined.
CLEAVAGE: Distinct in one direction. Brittle.
HABIT: Isometric to slightly elongated grains, 0.02–0.05 mm in diameter, enclosed in other ore-forming minerals, or as aggregates.
COLOR-LUSTER: Grayish white; metallic. Opaque.
MODE OF OCCURRENCE: Found in the copper ore deposit of Chelopech, Bulgaria, in association with tennantite, chalcopyrite, pyrite, native tellurium, native gold, tellurides, and other minerals.

BEST REF. IN ENGLISH: Terziev, G., *Am. Min.*, 51: 29–36 (1966).

## KOTOITE
$Mg_3B_2O_6$

CRYSTAL SYSTEM: Orthorhombic
CLASS: 2/m 2/m 2/m
SPACE GROUP: Pnmn
Z: 2
LATTICE CONSTANTS:
  $a$ = 5.398
  $b$ = 8.416
  $c$ = 4.497
3 STRONGEST DIFFRACTION LINES:
  2.67 (100)
  2.23 (100)
  2.18 (100)
OPTICAL CONSTANTS:
  $\alpha$ = 1.652
  $\beta$ = 1.653      (Na)
  $\gamma$ = 1.673–1.674
  (+)2V = 21°
HARDNESS: 6½
DENSITY: 3.04 (Meas.)
         3.099 (Calc.)
CLEAVAGE: {110} perfect
          {101} parting
HABIT: Massive granular, and as disseminated grains. Twinning plane {101}.
COLOR-LUSTER: Colorless. Transparent. Vitreous.
MODE OF OCCURRENCE: Occurs in the contact zone of a granite intrusion into dolomite at the Hol Kol mine, Suan, Korea, and in the interior of nodules of szaibelyite in marble at Rézbánya, Roumania.

BEST REF. IN ENGLISH: Palache, et al., "Dana's System of Mineralogy," 7th Ed., v. II, p. 328–329, New York, Wiley, 1951.

## KOTSCHUBEITE = Kochubeite

## KÖTTIGITE = Koettigite

## KOTULSKITE
Pd(Te, Bi)

CRYSTAL SYSTEM: Hexagonal
CLASS: 6/m 2/m 2/m
SPACE GROUP: $P6_3$/mmc
Z: 2
LATTICE CONSTANTS:
  $a$ = 4.19
  $c$ = 5.67
3 STRONGEST DIFFRACTION LINES:
  3.05 (100)
  2.24 ( 90)
  2.09 ( 90)

HARDNESS: Not determined.
CLEAVAGE: None.
HABIT: Very minute grains.
COLOR-LUSTER: Cream color in reflected light. Reflecting power 66%. Strongly anisotropic with color effects from brownish blue to gray-blue.
MODE OF OCCURRENCE: Occurs as inclusions in chalcopyrite, often intergrown with moncheite and michenerite, from the Monchegorsk deposits, Monche Tundra, USSR.
BEST REF. IN ENGLISH: Genkin, A. D., Zhuravlev, N. N., and Smirnova, E. M., *Am. Min.*, **48**: 1181 (1963). Kingston, G. A., *Min. Mag.*, **35**: 815-834 (1966).

## KOUTEKITE
Cu₂As

CRYSTAL SYSTEM: Hexagonal
Z: 42
LATTICE CONSTANTS:
$a = 11.51$
$c = 14.54$
3 STRONGEST DIFFRACTION LINES:
2.078 (100)
2.024 (100)
1.994 (100)
HARDNESS: 4-4½
DENSITY: 8.48 (Meas.)
CLEAVAGE: Not determined.
HABIT: Fine grains.
COLOR-LUSTER: Bluish gray under microscope; metallic. Streak black.
MODE OF OCCURRENCE: Occurs intergrown with an unknown copper arsenide associated with arsenic, silver, smaltite, loellingite, and calcite "in the bicarbonate gangue" of specimens from Černy Dul, Giant Mountains, Bohemia.
BEST REF. IN ENGLISH: Johan, Zdeněk, *Nature*, **181**: (4622): 1553-1554 (1958). *Am. Min.*, **46**: 467 (1961).

## KÔZULITE
(Na, K, Ca)₃ (Mn, Mg, Fe, Al)₅ Si₈O₂₂ (OH, F)₂
Amphibole Group

CRYSTAL SYSTEM: Monoclinic
CLASS: 2/m
SPACE GROUP: C2/m
Z: 2
LATTICE CONSTANTS:
$a = 9.91$
$b = 18.11$
$c = 5.30$
$\beta = 104.6°$
3 STRONGEST DIFFRACTION LINES:
8.512 (100)
3.153 ( 67)
2.827 ( 31)
OPTICAL CONSTANTS:
$\alpha = 1.685$ (Na)
$\beta = 1.717$ (mean)
$\gamma = 1.720$
$(-)2V = 34°-36°$ (Na)

HARDNESS: 5
DENSITY: 3.30 (Meas.)
3.36 (Calc.)
CLEAVAGE: {110} perfect
HABIT: Banded aggregates of short prismatic crystals up to 3.5 × 2.0 × 1.5 mm in size.
COLOR-LUSTER: Reddish black to black; vitreous. Streak light purplish brown.
MODE OF OCCURRENCE: Occurs associated with braunite, rhodonite, manganiferous alkali pyroxene and amphibole, and quartz, where sedimentary beds have been metamorphosed by the intrusion of granodiorite, in the manganese deposit of the Tanahota mine, Iwate Prefecture, Japan.
BEST REF. IN ENGLISH: Nambu, Matsuo, Tanida, Katsutoshi, and Kitamura, Tsuyoshi, *Am. Min.*, **55**: 1815-1816 (1970).

## KRAUSITE
KFe³⁺(SO₄)₂ · H₂O

CRYSTAL SYSTEM: Monoclinic
CLASS: 2/m
SPACE GROUP: P2₁/m
Z: 2
LATTICE CONSTANTS:
$a = 7.908$
$b = 5.152$
$c = 8.988$
$\beta = 102°45'$
3 STRONGEST DIFFRACTION LINES:
3.09 (100)
4.40 ( 74)
6.59 ( 66)
OPTICAL CONSTANTS:
$\alpha = 1.588$
$\beta = 1.650$
$\gamma = 1.722$
$(+)2V = $ large
HARDNESS: 2½
DENSITY: 2.840 (Meas.)
2.839 (Calc.)
CLEAVAGE: {001} perfect
{100} good
HABIT: Crystals usually short prismatic; also tabular to equant. As crusts.
COLOR-LUSTER: Pale lemon yellow. Transparent. Vitreous.
MODE OF OCCURRENCE: Occurs associated with alunite, voltaite, coquimbite, and other minerals at the "Sulfur hole" east of the borax mines in the Calico Mountains, San Bernardino County, California. It is also found associated with halotrichite and other iron sulfates in limestone at the Santa Maria mine, Velardeña, Durango, Mexico.
BEST REF. IN ENGLISH: Palache, et al., "Dana's System of Mineralogy," 7th Ed., v. II, p. 462-463, New York, Wiley, 1951. Graeber, Edward J., Morosin, Bruno, and Rosenzweig, Abraham, *Am. Min.*, **50**: 1929-1936 (1965).

# KRAUSKOPFITE
$BaSi_2O_5 \cdot 3H_2O$

CRYSTAL SYSTEM: Monoclinic
CLASS: 2/m
SPACE GROUP: $P2_1/a$
Z: 4
LATTICE CONSTANTS:
  $a = 8.460$
  $b = 10.622$
  $c = 7.837$
  $\beta = 94°32'$
3 STRONGEST DIFFRACTION LINES:
  3.84 (100)
  6.36 ( 45)
  5.34 ( 45)
OPTICAL CONSTANTS:
    $\alpha = 1.574$
    $\beta = 1.587$
    $\gamma = 1.599$
  $(-)2V = 88°$
HARDNESS: $\sim 4$
DENSITY: 3.14 (Meas.)
         3.10 (Calc.)
CLEAVAGE: {010} perfect
         {001} perfect
A third poor cleavage or fracture at a high angle to the other two.
HABIT: Equidimensional to short prismatic grains up to 6 mm long; commonly grains are 3 mm or less in length.
COLOR-LUSTER: White to colorless; subvitreous; pearly on cleavage plates. White streak.
MODE OF OCCURRENCE: Found as a vein-forming mineral commonly associated with macdonaldite, opal, and witherite in sanbornite-bearing metamorphic rocks in eastern Fresno County, California.
BEST REF. IN ENGLISH: Alfors, John T., Stinson, Melvin C., Matthews, Robert A., and Pabst, Adolf, *Am. Min.*, **50**: 314–340 (1965).

# KREMERSITE
$(NH_4,K)_2FeCl_5 \cdot H_2O$
Erythrosiderite Series

CRYSTAL SYSTEM: Orthorhombic
CLASS: 2/m 2/m 2/m
SPACE GROUP: Pnma
Z: 4
LATTICE CONSTANTS:
  $a = 13.78$
  $b = 9.85$
  $c = 7.09$
OPTICAL CONSTANTS:
    $\alpha = 1.715$
    $\beta = 1.75$
    $\gamma = 1.80$
  $(+)2V = 62°$
HARDNESS: Not determined.
DENSITY: 2.00 (Synthetic)
CLEAVAGE: {210} perfect
         {011} perfect

HABIT: Crystals pseudo-octahedral.
COLOR-LUSTER: Red to brownish red. Vitreous.

Deliquescent.
MODE OF OCCURRENCE: Occurs in fumaroles on Mt. Etna, and in association with hematite and molysite in fumaroles on Mt. Vesuvius, Italy.
BEST REF. IN ENGLISH: Palache, et al., "Dana's System of Mineralogy," 7th Ed., v. II, p. 101–103, New York, Wiley, 1951.

# KRENNERITE
$AuTe_2$

CRYSTAL SYSTEM: Orthorhombic
CLASS: mm2
SPACE GROUP: Pma2
Z: 8
LATTICE CONSTANTS:
  $a = 16.54$
  $b = 8.82$
  $c = 4.46$
3 STRONGEST DIFFRACTION LINES:
  3.03 (100)
  2.11 ( 70)
  2.94 ( 60)
HARDNESS: 2-3
DENSITY: 8.63 (Meas.)
         8.86 (Calc.)
CLEAVAGE: {001} perfect
Fracture subconchoidal to uneven. Brittle.
HABIT: Crystals commonly short prismatic [001]; vertical zone striated [001].
COLOR-LUSTER: Silver white to pale yellowish white, opaque; metallic.
MODE OF OCCURRENCE: Occurs usually in low-temperature veins; also in moderate- to high-temperature veins, but deposited toward the low-temperature limit for the deposit in which it occurs. Found with calaverite at several mines in the Cripple Creek district, Teller County, Colorado; in Montbray Township, Rouyn district, Quebec, Canada; at Nagyág, Transylvania; and associated with other tellurides at Kalgoorlie and Mulgabbie, Western Australia.
BEST REF. IN ENGLISH: Palache, et al., "Dana's System of Mineralogy," 7th Ed., v. I, p. 333–335, New York, Wiley, 1944. Tunell, George, and Murata, K. J., *Am. Min.*, **35**: 959–984 (1950).

# KRIBERGITE (Species status uncertain)
Near $Al_5(PO_4)_3SO_4(OH)_4 \cdot 4H_2O$

3 STRONGEST DIFFRACTION LINES:
  11.6 (100)
  5.02 ( 25)
  6.62 ( 20)
OPTICAL CONSTANTS:
  mean index = 1.484
HARDNESS: Not determined.
DENSITY: 1.92
CLEAVAGE: Not determined.

HABIT: Massive, compact, chalk-like.
COLOR-LUSTER: White.
MODE OF OCCURRENCE: Occurs as a cavity filling in copper-rich pyrite at the Kristineberg mine, Västerbotten, Sweden.
BEST REF. IN ENLGISH: Palache, et al., "Dana's System of Mineralogy," 7th Ed., v. II, p. 1011, New York, Wiley, 1951.

## KRINOVITE
$NaMg_2CrSi_3O_{10}$

CRYSTAL SYSTEM: Monoclinic
Z: 32
LATTICE CONSTANTS:
  $a$ = 19.48
  $b$ = 29.18
  $c$ = 10.25
  $\beta$ = 103°
3 STRONGEST DIFFRACTION LINES:
  2.501 (100)
  2.655 ( 90)
  2.893 ( 80)
OPTICAL CONSTANTS:
  $\alpha$ = 1.712
  $\beta$ = 1.725
  $\gamma$ = 1.760
  (+)2V = 61°
HARDNESS: 5.5-7
DENSITY: 3.38 (Meas.)
         3.44 (Calc.)
CLEAVAGE: None observed.
HABIT: Minute subhedral grains (largest approximately 200 $\mu$).
COLOR-LUSTER: Deep emerald green.
MODE OF OCCURRENCE: Occurs disseminated within graphite nodules in the octahedrite meteorites Canyon Diablo, Wichita County, and Youndegin.
BEST REF. IN ENGLISH: Olsen, Edward, and Fuchs, Louis, *Science*, **161**: 786-787 (1968).

## KROEHNKITE (Kröhnkite)
$Na_2Cu(SO_4)_2 \cdot 2H_2O$

CRYSTAL SYSTEM: Monoclinic
CLASS: 2/m
SPACE GROUP: $P2_1/c$
Z: 2
LATTICE CONSTANTS:
  $a$ = 5.78 kX
  $b$ = 12.58
  $c$ = 5.48
  $\beta$ = 108°30'
3 STRONGEST DIFFRACTION LINES:
  6.33 (100)
  3.278 ( 90)
  2.757 ( 90)

OPTICAL CONSTANTS:
  $\alpha$ = 1.544
  $\beta$ = 1.578 (Na)
  $\gamma$ = 1.601
  (−)2V = 78°42'
HARDNESS: 2½-3
DENSITY: 2.90 (Meas.)
         2.95 (Calc.)
CLEAVAGE: {010} perfect
          {$\bar{1}$01} imperfect
Fracture conchoidal.
HABIT: Crystals short prismatic; also octahedral and somewhat elongated. Massive, granular; and as columnar or fibrous crusts or aggregates. Twinning on {101} common.
COLOR-LUSTER: Light blue to dark sky blue; greenish blue to yellowish green, on exposure. Transparent to translucent. Vitreous to greasy.

Easily soluble in water.
MODE OF OCCURRENCE: Occurs as a secondary mineral at several copper ore deposits in northern Chile, especially at Chuquicamata where it is found in abundance associated with antlerite, leightonite, atacamite, chalcanthite, and natrochalcite.
BEST REF. IN ENGLISH: Palache, et al., "Dana's System of Mineralogy," 7th Ed., v. II, p. 444-446, New York, Wiley, 1951.

## KRYZHANOVSKITE (Kruzhanovskite)
$Fe_3^{3+}(OH)_3(PO_4)_2$

CRYSTAL SYSTEM: Orthorhombic
CLASS: 2/m 2/m 2/m
SPACE GROUP: Pbna
Z: 4
LATTICE CONSTANTS:
  $a$ = 9.404
  $b$ = 9.973
  $c$ = 8.536
3 STRONGEST DIFFRACTION LINES:
  3.156 (100)
  4.996 ( 70)
  4.701 ( 50)
OPTICAL CONSTANTS:
  2V = 40°-45°
HARDNESS: 3½-4
DENSITY: 3.31 (Meas.)
         3.35 (Calc.)
CLEAVAGE: {001} parting
HABIT: Crystals prismatic, poorly formed, up to 2-3 cm. in diameter.
COLOR-LUSTER: Deep red-brown to greenish brown, bronze on parting surface. Translucent. Vitreous to dull. Streak yellowish brown.
MODE OF OCCURRENCE: Kryzhanovskite occurs intimately associated with sicklerite as the outer part of large nodules of altered triphylite in the Kalbinsk pegmatites, U.S.S.R.; also as an alteration of phosphoferrite at the Bull Moose, Big Chief, and Dan Patch pegmatites, Black Hills, South Dakota.
BEST REF. IN ENGLISH: Ginzburg, A. I., *Am. Min.*,

36: 382 (1951). Moore, Paul Brian, *Am. Min.*, 56: 1–17 (1971).

## KTENASITE
$(Cu,Zn)_3(SO_4)(OH)_4 \cdot 2H_2O$

CRYSTAL SYSTEM: Monoclinic
CLASS: 2/m
SPACE GROUP: $P2_1/c$
Z: 8
LATTICE CONSTANTS:
  $a = 11.16$
  $b = 6.11$
  $c = 23.74$
  $\beta = 95°24'$
3 STRONGEST DIFFRACTION LINES:
  12.9 (100)
  2.74 (100)
  5.86 ( 90)
OPTICAL CONSTANTS:
  $\alpha = 1.571$
  $\beta = 1.613$
  $\gamma = 1.623$
  $(-)2V = 51°$
HARDNESS: 2–2½
DENSITY: 2.969
  HABIT: Groups of tabular crystals up to 1.0 mm in size. Forms {001}, {100}, {013}, {101}, and {201}.
  COLOR-LUSTER: Blue-green, transparent; vitreous.
  MODE OF OCCURRENCE: Occurs with blue serpierite, glaucocerinite, and porous smithsonite in a specimen from the Kamaresa mine, Laurium, Greece, in the collection of the University of Thessaloniki.
  BEST REF. IN ENGLISH: Kokkoros, P., *Am. Min.*, 36: 381 (1951).

## KULLERUDITE
$NiSe_2$

CRYSTAL SYSTEM: Orthorhombic
CLASS: 2/m 2/m 2/m
SPACE GROUP: Pnnm
Z: 2
LATTICE CONSTANTS:
  $a = 4.89$
  $b = 5.96$
  $c = 3.67$
3 STRONGEST DIFFRACTION LINES:
  2.64 (100)
  2.545 (100)
  2.935 ( 80)
HARDNESS: Rather soft, polishes well.
DENSITY: 6.72 (Calc.)
CLEAVAGE: Not determined.
HABIT: Massive; very fine grained.
COLOR-LUSTER: Lead gray. Opaque. Metallic.
MODE OF OCCURRENCE: Occurs as an alteration product of wilkmanite in sills of albite diabase in a schist formation, associated with low-grade uranium mineralization at Kuusamo, northeast Finland.

BEST REF. IN ENGLISH: Vuorelainen, Y., Huhma, A., and Häkli, A., *Am. Min.*, 50: 520 (1965).

## KUNZITE = Transparent lilac variety of spodumene

## KUPFFERITE = Theoretical end-member of cummingtonite–grunerite series.

## KUPLETSKITE
$(K,Na)_2(Mn,Fe)_4(Ti,Nb)Si_4O_{14}(OH)_2$
Manganese analogue of astrophyllite
Var. Cesium-kupletskite

CRYSTAL SYSTEM: Monoclinic
OPTICAL CONSTANTS:
  $\alpha = 1.656$
  $\beta = 1.699$
  $\gamma = 1.731$
  $(-)2V = 79°$
HARDNESS: ~3
DENSITY: 3.201–3.229
CLEAVAGE: [100] perfect
  HABIT: Lamellar masses up to $5.0 \times 3.0 \times 1.0$ cm in size, consisting of individual plates; also as fine plates.
  COLOR-LUSTER: Dark brown to black. Streak brown.
  MODE OF OCCURRENCE: Occurs associated with schizolite, neptunite, microcline, eudialyte, nepheline, and aegirine in four pegmatites of the Lovozero massif of alkalic rocks, Kola Peninsula, USSR.
  BEST REF. IN ENGLISH: Semenov, E. I., *Am. Min.*, 42: 118–119 (1957).

## KURCHATOVITE
$Ca(Mg,Mn)B_2O_5$

CRYSTAL SYSTEM: Orthorhombic
Z: 24
LATTICE CONSTANTS:
  $a = 11.15$
  $b = 36.4$
  $c = 5.55$
3 STRONGEST DIFFRACTION LINES:
  2.78 (100)
  1.922 ( 90)
  2.67 ( 80)
OPTICAL CONSTANTS:
  $\alpha = 1.635$
  $\beta = 1.681$
  $\gamma = 1.698$
  $(-)2V = 66°$
HARDNESS: 4½
DENSITY: 3.02 (Meas.)
         3.027 (Calc.)
CLEAVAGE: {010} perfect
HABIT: Massive.
  COLOR-LUSTER: Pale gray; vitreous. Fluoresces bright violet in long-wave ultraviolet light.
  MODE OF OCCURRENCE: Occurs associated with vesu-

vianite, garnet, svabite, magnetite, and sphalerite in a skarn in Siberia.

BEST REF. IN ENGLISH: Malinko, S. V., Lisitsyn, A. E., Dorofeeva, K. A., Ostravskaya, I. V., and Shashkin, D. P., *Am. Min.*, **51**: 1817–1818 (1966).

# KURGANTAITE (Inadequately described mineral)
$(Sr,Ca)_2 B_4 O_8 \cdot H_2 O(?)$

CRYSTAL SYSTEM: Unknown
3 STRONGEST DIFFRACTION LINES:
  3.17 (100)
  2.82 ( 90)
  2.51 ( 80)
OPTICAL CONSTANTS:
   $\alpha = 1.641$
   $\gamma = 1.682$
  2V = very small
HARDNESS: Hard, scratches glass.
DENSITY: $\sim 3.0$
CLEAVAGE: Not determined.
HABIT: Massive; as fine-grained nodules up to 4 cm in diameter.
COLOR-LUSTER: White.
MODE OF OCCURRENCE: Occurs in gypsum-anhydrite beds in the western Kurgan-Tau (western Kazakhstan), USSR, in association with sylvite, carnallite, halite, and anhydrite.
BEST REF. IN ENGLISH: Yarzhemskii, Ya. Ya., *Am. Min.*, **40**: 941 (1955).

# KURNAKOVITE
$Mg_2 B_6 O_{11} \cdot 15H_2 O$
Dimorphous with inderite

CRYSTAL SYSTEM: Triclinic
CLASS: 2/m
SPACE GROUP: $P2_1/a$
Z: 2
LATTICE CONSTANTS:
  $a = 12.12$
  $b = 13.18$
  $c = 6.83$
  $\beta = 104°49'$
3 STRONGEST DIFFRACTION LINES:
  5.66 (100)
  3.26 (100)
  5.01 ( 90)
OPTICAL CONSTANTS:
   $\alpha = 1.488$
   $\beta = 1.491$
   $\gamma = 1.505$
  (+)2V = 37°
HARDNESS: 3
DENSITY: 1.785 (Meas.)
      1.76 (Calc.)
CLEAVAGE: {110} good
       {001} indistinct
Fracture conchoidal. Brittle.

HABIT: As large blocky crystals, often in groups. Also as large cleavable masses, and as dense granular aggregates.
COLOR-LUSTER: Colorless, often with white surface alteration coating. Transparent. Vitreous; pearly on cleavage.
MODE OF OCCURRENCE: The mineral occurs abundantly as large masses and rough to well-formed crystals up to 24 inches across at Boron, Kern County, California. Also found at the Inder borate deposit in western Kazakhstan, USSR.

# KURSKITE = Variety of apatite

# KURUMSAKITE (Inadequately described mineral)
$(Zn,Ni,Ca)_8 Al_8 V_2 Si_5 O_{35} \cdot 27H_2 O(?)$

CRYSTAL SYSTEM: Orthorhombic (?)
3 STRONGEST DIFFRACTION LINES:
  1.53 ⎱
  3.91 ⎰ Intensities not reported.
  2.61 ⎰
OPTICAL CONSTANTS:
   $\alpha = 1.616$
   $\beta = ?$
   $\gamma = 1.622–1.623$
  (+)2V $\approx 35°$
HARDNESS: Not determined.
DENSITY: 4.03 (Meas.)
CLEAVAGE: Not determined.
HABIT: As radiating to finely felted fibers.
COLOR-LUSTER: Greenish yellow to bright yellow. Vitreous to silky.
MODE OF OCCURRENCE: Occurs on the walls of cavities and open fissures in bituminous schists of Kara-Tau, USSR.
BEST REF. IN ENGLISH: Ankinovich, E. A., *Am. Min.*, **42**: 583–584 (1957).

# KÜSTELITE = Variety of silver

# KUTINAITE
$Cu_2 AgAs$

CRYSTAL SYSTEM: Cubic
Z: 28
LATTICE CONSTANT:
  $a = 11.76$
3 STRONGEST DIFFRACTION LINES:
  2.259 (100)
  2.078 (100)
  2.702 ( 90)
HARDNESS: Vickers microhardness; 7.6 g load, 387 $kg/mm^2$; 15.2 g load, 398 $kg/mm^2$.
DENSITY: 8.38 (Meas. synthetic)
CLEAVAGE: Malleable.
HABIT: Massive; fine-grained.

COLOR-LUSTER: Silvery gray. Metallic. Opaque.

MODE OF OCCURRENCE: Found very sparingly associated with koutekite, nóvakite, and paxite in carbonate-rich veins of the Černý Důl ore deposit in the Giant Mountains of northern Bohemia.

BEST REF. IN ENGLISH: Hak, J., Johan, Z., and Skinner, Brian J., *Am. Min.*, 55: 1083–1087 (1970).

## KUTNAHORITE
$Ca(Mn, Mg, Fe)(CO_3)_2$

CRYSTAL SYSTEM: Hexagonal
CLASS: $\bar{3}$
SPACE GROUP: $R\bar{3}$
Z: 1
LATTICE CONSTANTS:
$a = 4.85$
$c = 16.34$
3 STRONGEST DIFFRACTION LINES:
2.94 (100)
1.814 ( 30)
1.837 ( 25)
OPTICAL CONSTANTS:
$\omega = 1.727$
$\epsilon = 1.535$
  (−)
HARDNESS: 3½–4
DENSITY: 3.12 (Meas.)
CLEAVAGE: $\{10\bar{1}1\}$ perfect
Fracture subconchoidal. Brittle.

HABIT: Anhedral masses and cleavable aggregates.

COLOR-LUSTER: White, pale pink, pale rose. Translucent.

MODE OF OCCURRENCE: Occurs at Franklin, New Jersey, at Chvaletice and Kutná Hora, Czechoslovakia, at Providencia, Mexico, and at the Ryûjima mine, Nagano Prefecture, Japan.

BEST REF. IN ENGLISH: Palache, et al., "Dana's System of Mineralogy," 7th Ed., v. II, p. 217, New York, Wiley, 1951. Frondel, Clifford, and Bauer, L. H., *Am. Min.*, 40: 748–760 (1955). Tsusue, Akio, *Am. Min.*, 52: 1751–1761 (1967).

## KYANITE (Disthene)
$Al_2SiO_5$
Trimorphous with andalusite, sillimanite

CRYSTAL SYSTEM: Triclinic
CLASS: $\bar{1}$
SPACE GROUP: $P\bar{1}$
Z: 4
LATTICE CONSTANTS:
$a = 7.10$    $\alpha = 90°5½'$
$b = 7.74$    $\beta = 101°2'$
$c = 5.57$    $\gamma = 105°44½'$

3 STRONGEST DIFFRACTION LINES:
3.18 (100)
1.38 ( 75)
3.35 ( 65)
OPTICAL CONSTANTS:
$\alpha = 1.712–1.718$
$\beta = 1.721–1.723$
$\gamma = 1.727–1.734$
$(−)2V = 82°–83°$
HARDNESS: 4–7½, variable face to face, and dependant on crystallographic direction.
DENSITY: 3.53–3.67 (Meas.)
CLEAVAGE: $\{100\}$ perfect
           $\{010\}$ distinct
           $\{001\}$ parting
Somewhat flexible.

HABIT: Crystals long-bladed, flattened $\{100\}$, and elongated parallel to *c*-axis; often bent or twisted. Also massive, bladed to somewhat fibrous. Twinning on $\{100\}$, lamellar; also multiple twinning on $\{001\}$, by pressure.

COLOR-LUSTER: Usually blue; also white, gray, green, yellow, pink, or nearly black. Color often varying in a single crystal. Transparent to translucent. Vitreous to pearly. Streak uncolored.

MODE OF OCCURRENCE: Occurs widespread, chiefly in paragonite schist, muscovite schist, gneiss, and in granite and granitic pegmatites. Found in California, Idaho, Arizona, Colorado, and other western states; at many places in the New England states, especially in Vermont, Connecticut, Massachusetts, Virginia, North Carolina, and Georgia. Other notable occurrences are found in Canada, Brazil, Ireland, Sweden, France, Italy, Switzerland, Austria, Czechoslovakia, Roumania, USSR, Kenya, Zaire, India, Korea, and Australia.

BEST REF. IN ENGLISH: Deer, Howie, and Zussman, "Rock Forming Minerals," v. 1, p. 137–143, New York, Wiley, 1962.

## KYANOPHILITE (Species status uncertain)
$(K,Na)Al_2Si_2O_7(OH)$

CRYSTAL SYSTEM: Triclinic (?)
HARDNESS: 2½–4
DENSITY: 2.89–2.90 (Meas.)
CLEAVAGE: Not determined.
HABIT: Massive; as nodular plumose aggregates.
COLOR-LUSTER: Apple green.
MODE OF OCCURRENCE: Occurs as an alteration product of kyanite and sillimanite in kyanite-graphite schists and sillimanite-biotite gneisses at Mavinhalli, Mysore, India.

BEST REF. IN ENGLISH: Rao, B. Rama, *Min. Abs.*, 10: 7 (1949).

# L

**LABITE** = Chrysotile (?)


## LABRADORITE
$mCaAl_2Si_2O_8$ with $nNaAlSi_3O_8$
$Ab_{50}An_{50}$ to $Ab_{30}An_{70}$
Plagioclase Feldspar Group

CRYSTAL SYSTEM: Triclinic
CLASS: $\bar{1}$
SPACE GROUP: $C\bar{1}$
Z: 4
LATTICE CONSTANTS:
  $a = 8.176$    $\alpha = 93°27.1'$
  $b = 12.865$   $\beta = 116°2.9'$
  $c = 7.102$    $\gamma = 90°30.6'$
3 STRONGEST DIFFRACTION LINES:
  3.18 (100)
  3.76 ( 70)
  3.21 ( 70)
OPTICAL CONSTANTS:
    $\alpha = 1.5625$
    $\beta = 1.5668$
    $\gamma = 1.5718$
  (+)2V = 85°
HARDNESS: 6–6½
DENSITY: 2.69–2.72 (Meas.)
         2.710 (Calc.)
CLEAVAGE:  {001} perfect
            {010} nearly perfect
            {110} imperfect
Fracture uneven to conchoidal. Brittle.

HABIT: Crystals tabular, flattened along *b*-axis; not common. Usually massive, cleavable, granular, or compact. Twinning common. Principal twin laws: Carlsbad, albite, and pericline.

COLOR-LUSTER: Colorless, white, gray; frequently shows play of bright colors—blue, green, red, yellow, and pearl gray. Transparent to translucent. Vitreous. Streak white.

MODE OF OCCURRENCE: Occurs widespread as a constituent of anorthosite, andesite, norite, basalt, gabbro, diorite, amphibolite, and other igneous and metamorphic rocks, especially the more basic types. Typical occurrences are found in California, Utah, South Dakota, Minnesota, Wisconsin, Pennsylvania, New York, and North Carolina. Also at localities in Greenland, Ireland, Norway, Sweden, Sicily, Roumania, South Africa, and as fine large masses displaying vivid play of colors in eastern Labrador, Canada, and near Lammenpaa, Finland.

BEST REF. IN ENGLISH: Deer, Howie, and Zussman, "Rock Forming Minerals," v. 4, p. 94–165, New York, Wiley, 1963. Barth, T. F. W., "Feldspars," New York, Wiley, 1969.


## LABUNTSOVITE
$(K,Ba,Na)(Ti,Nb)(Si,Al)_2(O,OH)_7 \cdot H_2O$

CRYSTAL SYSTEM: Monoclinic
CLASS: 2/m
SPACE GROUP: I2/m
LATTICE CONSTANTS:
  $a = 15.57$
  $b = 13.75$
  $c = 14.27$
  $\beta = 116°55'$
3 STRONGEST DIFFRACTION LINES:
  3.15  (100)
  2.56  ( 90)
  1.543 ( 90)
OPTICAL CONSTANTS:
    $\alpha = 1.689$–$1.695$
    $\beta = 1.698$–$1.702$
    $\gamma = 1.795$–$1.825$
  (+)2V ~ 20°–44°
HARDNESS: ~6
DENSITY: 2.901–2.96 (Meas.)
CLEAVAGE:  {100} perfect (ortho.)
            $\{\bar{1}02\}$ (mono.)
Brittle.

HABIT: Simple prismatic crystals up to 12 X 3 X 2 mm. Forms: c{001}, b{010}, m{110}, d{011}, and s{111}.

COLOR-LUSTER: Rose to brownish yellow.

MODE OF OCCURRENCE: Occurs in druses with albite and natrolite, and associated with aegirine, nepheline, microcline, eudialyte, ramsayite, and murmanite, widely distributed in the Lovozero and Khibina alkalic massifs, USSR. It is a hydrothermal alteration product of murmanite and is sometimes altered to anatase (leucoxene).

BEST REF. IN ENGLISH: Semenov, E. I., and Burova, T. A., *Am. Min.*, **41**: 163 (1956).

## LACROIXITE
NaAl(PO$_4$)(OH, F)

CRYSTAL SYSTEM: Monoclinic
CLASS: 2/m
SPACE GROUP: A2/a
Z: 4
LATTICE CONSTANTS:
  $a$ = 6.89
  $b$ = 8.22
  $c$ = 6.425
  $\beta$ = 115°30′
3 STRONGEST DIFFRACTION LINES:
  2.900 (100)
  2.470 ( 85)
  3.159 ( 71)
OPTICAL CONSTANTS:
  $\alpha$ = 1.545
  $\beta$ = 1.554
  $\gamma$ = 1.565
  (−)2V ≈ 90°
HARDNESS: 4½
DENSITY: 3.126 (Meas.)
CLEAVAGE: Parallel to {111} and {1$\bar{1}$1} indistinct.
HABIT: As fragmentary crystals.
COLOR-LUSTER Pale yellow to pale green. Vitreous to resinous.
MODE OF OCCURRENCE: Found with morinite, apatite, childrenite, roscherite, and tourmaline in druses in granite at Greifenstein, Saxony, Germany; at Konigswart in Marienbad, Bohemia, and Ječlov, near Jihlava, Czechoslovakia, in association with natromontebrasite; and at Strickland quarry, Portland, Connecticut, in association with pink apatite, pink brazilianite, augelite, and natromontebrasite.
BEST REF. IN ENGLISH: Palache, et al., "Dana's System of Mineralogy," 7th Ed., v. II, p. 783, New York, Wiley, 1951. Mrose, Mary E., Program and Abstracts, 20th Clay Min. Conf., p. 10 (1971).

## LAITAKARITE
Bi$_4$Se$_2$S

CRYSTAL SYSTEM: Hexagonal
CLASS: $\bar{3}$ 2/m
SPACE GROUP: R$\bar{3}$m
Z: 1 (rhomb.), 3 (hex.)
LATTICE CONSTANTS:
                    (rhomb.)
  $a$ = 4.225    $a_{rh}$ = 13.53
  $c$ = 39.93    $\alpha$ = 17°58′

3 STRONGEST DIFFRACTION LINES:
  3.072 (100)
  2.246 (100)
  2.112 ( 80)
HARDNESS: Soft. Flexible, inelastic.
DENSITY: 8.12 (Meas.)
CLEAVAGE: Basal–excellent.
HABIT: As foliated plates and sheets. Grains 0.5–2.0 mm in diameter.
COLOR-LUSTER: Galena white. Opaque. Metallic.
MODE OF OCCURRENCE: Occurs associated with chalcopyrite, native bismuth, sphalerite, molybdenite, silver, pyrite, and galena in veinlets in quartz-anthophyllite-cordierite-biotite rocks at Orijärvi, Finland.
BEST REF. IN ENGLISH: Vorma, Atso, *Am. Min.*, **47**: 806–807 (1962).

**LAMPADITE** = Variety of wad containing copper

## LAMPROPHYLLITE
Na$_2$(Sr,Ba)$_2$Ti$_3$(SiO$_4$)$_4$(OH,F)$_2$

CRYSTAL SYSTEM: Monoclinic (and orthorhombic)
CLASS: 2/m and 2/m 2/m 2/m
SPACE GROUP: C2/m and Pnmn
Z: 2 (mono), 2 (ortho)
LATTICE CONSTANTS:
  (mono.)         (ortho.)
  $a$ = 19.76    $a$ = 19.09  kX
  $b$ = 7.06     $b$ = 7.06
  $c$ = 5.40     $c$ = 5.36
  $\beta$ = 96.5
3 STRONGEST DIFFRACTION LINES:
  (mono.)         (ortho)
  2.773 (100)    3.17 (100)
  3.43 ( 55)     2.75 ( 90)
  2.130 ( 45)    2.12 ( 50)
OPTICAL CONSTANTS:
  $\alpha$ = 1.746 (variable)
  $\beta$ = 1.754
  $\gamma$ = 1.779
  (+)2V = 21°–43°
HARDNESS: 2–3
DENSITY: 3.44–3.53 (Meas.)
CLEAVAGE: {100} perfect, micaceous Fracture uneven. Brittle.
HABIT: Crystals tabular, elongated. Also as stellate aggregates and fine needle-like segregations. Twinning on {100}, sometimes polysynthetic.
COLOR-LUSTER: Golden brown to dark brown. Translucent. Vitreous and semimetallic in weathered varieties.
MODE OF OCCURRENCE: Occurs chiefly in nepheline-syenites and their pegmatites in association with microcline, nepheline, aegirine, eudialyte, and other minerals. Found in the Bear Paw Mountains, Montana; in the Langesundfjord district, Norway; on the Kola Peninsula and elsewhere in the USSR; and in South Africa.
BEST REF. IN ENGLISH: Vlasov, K. A., "Mineralogy of Rare Elements," v. II, p. 207–209, Israel Program for Scientific Translations, 1966.

# LANARKITE
$Pb_2SO_5$

CRYSTAL SYSTEM: Monoclinic
CLASS: 2/m
SPACE GROUP: C2/m
Z: 4
LATTICE CONSTANTS:
 $a$ = 13.73 kX
 $b$ = 5.68
 $c$ = 7.07
 $\beta$ = 116°13'
3 STRONGEST DIFFRACTION LINES:
 2.95 (100)
 3.33 ( 80)
 2.85 ( 30)
OPTICAL CONSTANTS:
 $\alpha$ = 1.928
 $\beta$ = 2.007 (Na)
 $\gamma$ = 2.036
 (–)2V $\approx$ 60°
HARDNESS: 2-2½
DENSITY: 6.92 (Meas.)
 7.08 (Calc.)
CLEAVAGE: {$\overline{2}01$} perfect
 {$\overline{4}01$} imperfect
 {201} imperfect
Thin laminae flexible.
 HABIT: Crystals prismatic; also massive. Polysynthetic twinning, rare.
 COLOR-LUSTER: Gray, greenish white, yellow. Transparent to translucent. Adamantine to resinous, pearly on cleavage. Streak white. Fluoresces yellow under ultraviolet light.
 MODE OF OCCURRENCE: Occurs as a secondary mineral in the oxidation zone of lead ore deposits. Found associated with caledonite, leadhillite, and cerussite at the Susanna mine, Leadhills, Lanarkshire, Scotland; at Tanne, Harz Mountains, and in the Herrensegen mine, Black Forest, Germany; and at localities in France, Austria, and Chile.
 BEST REF. IN ENGLISH: Palache, et al., "Dana's System of Mineralogy," 7th Ed., v. II, p. 550-551, New York, Wiley, 1951.

# LANDAUITE
$(Mn,Zn,Fe)(Fe,Ti)_3O_7$

CRYSTAL SYSTEM: Monoclinic
CLASS: m, or 2/m
SPACE GROUP: Aa or A2/a
Z: 4
LATTICE CONSTANTS:
 $a$ = 5.22
 $b$ = 8.95
 $c$ = 9.53
 $\beta$ = 107°35'
3 STRONGEST DIFFRACTION LINES:
 2.83 (100)
 2.11 ( 90)
 1.780 ( 80)

OPTICAL CONSTANTS:
 $\alpha$ = 2.373
 $\beta$ = ?
 $\gamma$ = 2.388
 (–)2V ~ 60°
HARDNESS: 7.5
DENSITY: 4.42 (Meas.)
 4.70 (Calc.)
CLEAVAGE: None. Fracture conchoidal.
 HABIT: Fine-grained aggregates of irregular form up to 1 mm in size.
 COLOR-LUSTER: Black; opaque, except in thin splinters translucent brownish green. Semimetallic. Streak gray.
 MODE OF OCCURRENCE: Occurs associated with albite, polylithionite, brookite, strontian chabazite, monazite, and bastnaesite in albite veinlets along the contact of the intrusive in massive quartz syenites, and superposed on pegmatites of syenitic composition, in the Burpala alkalic massif, northern Baikal.
 BEST REF. IN ENGLISH: Portnov, A. M., Nikolaeva, L. E., and Stolyarova, T. I., *Am. Min.*, **51**: 1546 (1966).

# LANDESITE
$Mn_{10}Fe_3^{3+}(PO_4)_8(OH)_5 \cdot 11H_2O$?

CRYSTAL SYSTEM: Orthorhombic
CLASS: 2/m 2/m 2/m
SPACE GROUP: Pbcn
Z: 4
LATTICE CONSTANTS:
 $a$ = 8.47
 $b$ = 9.43
 $c$ = 10.17
3 STRONGEST DIFFRACTION LINES:
 3.207 (100)
 5.096 ( 80)
 4.284 ( 80)
OPTICAL CONSTANTS:
 $\alpha$ = 1.720
 $\beta$ = 1.728
 $\gamma$ = 1.735
 (–)2V = large
HARDNESS: 3-3½
DENSITY: 3.026 (Meas.)
CLEAVAGE: {010} distinct
 HABIT: As octahedral-like crystals pseudomorphous after reddingite.
 COLOR-LUSTER: Brown.
 MODE OF OCCURRENCE: Occurs in granite pegmatite associated with lithiophilite, apatite, eosphorite, fairfieldite, and rhodochrosite at the Berry quarry, Poland, Maine.
 BEST REF. IN ENGLISH: Palache, et al., "Dana's System of Mineralogy," 7th Ed., v. II, p. 729-730, New York, Wiley, 1951. Moore, Paul B., *Am. Min.*, **49**: 1122-1125 (1964).

# LÅNGBANITE
$Mn_4^{2+}Mn_9^{3+}Sb^{5+}Si_2O_{24}$
Var. Ferrostibian

CRYSTAL SYSTEM: Hexagonal
CLASS: 6/m 2/m 2/m
SPACE GROUP: P6/mmm
Z: 3
LATTICE CONSTANTS:
 $a = 11.56$
 $c = 11.11$
3 STRONGEST DIFFRACTION LINES:
 2.55 (100)
 2.74 ( 90)
 2.80 ( 80)
OPTICAL CONSTANTS:
 $\omega = 2.36$
 $\epsilon = 2.31$
 (−)
HARDNESS: 6½
DENSITY: 4.6–4.8 (Meas.)
      4.684 (Calc.)
CLEAVAGE: None.
HABIT: Crystals prismatic, complex.
COLOR-LUSTER: Black. Opaque. Metallic.
MODE OF OCCURRENCE: Occurs in ore deposits at Långban, and at the Sjö mine, near Grythyttan, Sweden.
BEST REF. IN ENGLISH: Winchell and Winchell, "Elements of Optical Mineralogy," 4th Ed., Pt. 2, p. 529, New York, Wiley, 1951.


# LANGBEINITE
$K_2Mg_2(SO_4)_3$

CRYSTAL SYSTEM: Cubic
CLASS: 23
SPACE GROUP: $P2_13$
Z: 4
LATTICE CONSTANT:
 $a = 9.9211$
3 STRONGEST DIFFRACTION LINES:
 3.14 (100)
 2.65 ( 35)
 4.05 ( 25)
OPTICAL CONSTANT:
 $N = 1.536$
HARDNESS: 3½–4
DENSITY: 2.83 (Meas.)
      2.849 (Calc.)
CLEAVAGE: None. Fracture conchoidal. Brittle.
HABIT: Crystals rare. Usually massive; bedded; as nodular masses; and as disseminated grains.
COLOR-LUSTER: Colorless; sometimes whitish, gray, yellowish, greenish, pinkish to reddish, or violet. Transparent to translucent. Vitreous.
MODE OF OCCURRENCE: Occurs only in salt deposits of oceanic origin. Large quantities are mined from beds up to 7 feet in thickness near Carlsbad, Eddy County, New Mexico, and from deposits in Saskatchewan, Canada. It is also found in the potash deposits of northern Germany; in Austria; and in the Punjab Salt Range, India.
BEST REF. IN ENGLISH: Palache, et al., "Dana's System of Mineralogy," 7th Ed., v. II, p. 434–435, New York, Wiley, 1951.


# LANGISITE
(Co, Ni)As

CRYSTAL SYSTEM: Hexagonal
LATTICE CONSTANTS:
 $a = 3.538$
 $c = 5.127$
3 STRONGEST DIFFRACTION LINES:
 2.631 (100)
 1.966 ( 90)
 1.770 ( 80)
HARDNESS: Microhardness 780–857 kg/mm$^2$ 50 g load.
HABIT: Irregular grains and lamellae.
COLOR-LUSTER: In reflected light very weakly bireflecting, pinkish-buff. Opaque. Metallic.
MODE OF OCCURRENCE: Found as minute grains in safflorite associated with parkerite, cobalt pentlandite, siegenite, bismuthinite, native bismuth, maucherite, bravoite, and pyrite-marcasite in an ore-bearing fault vein in the Langis mine, Casey township in the Cobalt-Gowganda area, Ontario, Canada.
BEST REF. IN ENGLISH: Petruk, W., Harris, D. C., and Stewart, J. M., *Can. Min.*, **9** (Pt. 5): 597–616 (1969).


# LANGITE
$Cu_4SO_4(OH)_6 \cdot 2H_2O$

CRYSTAL SYSTEM: Orthorhombic
CLASS: mm2
SPACE GROUP: $Pmc2_1$
Z: 2
LATTICE CONSTANTS:
 $a = 6.02$
 $b = 11.2$
 $c = 7.12$
3 STRONGEST DIFFRACTION LINES:
 7.12 (100)
 3.56 ( 80)
 2.49 ( 60)
OPTICAL CONSTANTS:
 $\alpha = 1.654$
 $\beta = 1.713$
 $\gamma = 1.722$
(−)2V = 70°
HARDNESS: 2½–3
DENSITY: 3.31 (Meas.)
      3.26 (Calc.)
CLEAVAGE: {001} perfect
      {010} distinct
HABIT: Crystals small (up to 5.0 mm) elongate on {100} or equant. Often twinned repeatedly on {110}; twins tabular or scaly on {001}. Also as crusts or earthy.
COLOR-LUSTER: Sky blue to bluish green. Translucent. Vitreous, crusts silky.
MODE OF OCCURRENCE: Occurs as a secondary mineral, associated with gypsum and basic copper sulfates, at St. Blazey and St. Just, Cornwall, England; associated with chalcopyrite at Mollau, Haut Rhin, France; as small crystals coating superficially altered chalcopyrite in a quartz gangue in the Caroline tunnel of the Ward mine, near Ely, Nevada; and reported from several other localities.

BEST REF. IN ENGLISH: Palache, et al., "Dana's System of Mineralogy," 7th Ed., v. II, p. 583–585, New York, Wiley, 1951.

## LANSFORDITE
$MgCO_3 \cdot 5H_2O$

CRYSTAL SYSTEM: Monoclinic
CLASS: 2/m
SPACE GROUP: $P2_1/m$
Z: 4
LATTICE CONSTANTS:
  $a = 12.48$
  $b = 7.55$
  $c = 7.34$
  $\beta = 101°46'$
3 STRONGEST DIFFRACTION LINES:
  4.16 (100)
  3.85 (100)
  5.80 ( 80)
OPTICAL CONSTANTS:
  $\alpha = 1.465$
  $\beta = 1.468$
  $\gamma = 1.507$
  $(+)2V = 59°30'$
HARDNESS: 2½
DENSITY: 1.692 (Meas.)
         1.700 (Calc.)
CLEAVAGE:  {001} perfect
           {100} distinct
HABIT:  As minute short prismatic crystals; stalactitic.
COLOR-LUSTER: Colorless, white. Transparent to translucent. Vitreous. Becomes dull and opaque on exposure.
MODE OF OCCURRENCE: Found associated with nesquehonite in the anthracite mine at Nesquehoning near Lansford, Carbon County, Pennsylvania, as a recent product formed under normal atmospheric conditions of temperature and pressure. Also found associated with hydromagnesite at Atlin, British Columbia, Canada, and with nesquehonite to Cogne, Val d'Aosta, Piedmont, Italy.
BEST REF. IN ENGLISH: Palache, et al., "Dana's System of Mineralogy," 7th Ed., v. II, p. 228–230, New York, Wiley, 1951.

## LANTHANITE
$(La,Ce)_2 (CO_3)_3 \cdot 8H_2O$

CRYSTAL SYSTEM: Orthorhombic
CLASS: 2/m 2/m 2/m
Z: 4
LATTICE CONSTANTS:
  $a = 9.50$ kX
  $b = 17.1$
  $c = 9.00$
3 STRONGEST DIFFRACTION LINES:
  3.12 (100)
  9.1 ( 90)
  4.6 ( 70)

OPTICAL CONSTANTS:
  $\alpha = 1.52$
  $\beta = 1.587$  (Na)
  $\gamma = 1.613$
  $(-)2V \sim 63°$
HARDNESS: 2½–3
DENSITY: 2.69–2.74 (Meas.)
         2.73 (Calc.)
CLEAVAGE: {010} micaceous
Not brittle.
HABIT: Crystals lath-like, platy to thick tabular. As scales; also granular to earthy. Twinning on {101}.
COLOR-LUSTER: Colorless, white, yellowish, pinkish. Transparent. Pearly.
MODE OF OCCURRENCE: Occurs in oxidized zinc ores at Bethlehem, Lehigh County, Pennsylvania; also found at Moriah, Essex County, New York; near Manitou, El Paso County, Colorado; in pegmatite at Barringer Hill, Llano County, Texas; and as a coating on cerite at Bastnaes, Sweden.
BEST REF. IN ENGLISH: Palache, et al., "Dana's System of Mineralogy," 7th Ed., v. II, p. 241–243, New York, Wiley, 1951.

## LAPIS LAZULI = Lazurite

## LAPPARENTITE (Inadequately described mineral)
$Al_2 (SO_4)_2 (OH)_2 \cdot 9H_2O$

CRYSTAL SYSTEM: Orthorhombic (?)
OPTICAL CONSTANTS:
  $\alpha = 1.460$
  $\beta = 1.470$
  $\gamma = 1.484$
  $(+)2V = 80°$ (Calc.)
HARDNESS: $\sim 3$
DENSITY: 1.892 (Meas.)
CLEAVAGE: Not determined.
HABIT: Crystals rhomboidal tablets. Massive, chalk-like.
COLOR-LUSTER: White.
MODE OF OCCURRENCE: Occurs at Wackerdorf, Germany, and in association with ammonia alum and copiapite on burning coal waste-heaps at Libušín, Bohemia, Czechoslovakia.
BEST REF. IN ENGLISH: Palache, et al., "Dana's System of Mineralogy," 7th Ed., v. II, p. 601, New York, Wiley, 1951.

## LARDERELLITE
$NH_4 B_5 O_6 (OH)_4$

CRYSTAL SYSTEM: Monoclinic
CLASS: 2/m
SPACE GROUP: $P2_1/a$
Z: 4
LATTICE CONSTANTS:
  $a = 11.63$
  $b = 7.615$
  $c = 9.447$
  $\beta = 96°45'$

**3 STRONGEST DIFFRACTION LINES:**
4.70  (100)
2.921 (100)
2.887 (100)
**OPTICAL CONSTANTS:**
$\alpha = 1.509$
$\beta \simeq 1.52$
$\gamma = 1.561$
$(+)2V = 64°$
**HARDNESS:** Not determined.
**DENSITY:** 1.905 (Meas.)
1.887 (Calc.)
**CLEAVAGE:** {100} perfect
**HABIT:** Microscopic rhomboidal tablets flattened {100}.
**COLOR-LUSTER:** White; sometimes yellowish due to admixed impurities.
**MODE OF OCCURRENCE:** Occurs associated with ammonioborite and sassolite in the boric acid lagoon at Larderello, Tuscany, Italy.
**BEST REF. IN ENGLISH:** Clark, Joan R., *Am. Min.*, **45**: 1087–1093 (1960).

## LARNITE
$\beta$-$Ca_2SiO_4$
Dimorph of bredigite
Unstable at atmospheric temperatures and pressures.

**CRYSTAL SYSTEM:** Monoclinic
**CLASS:** 2/m
**SPACE GROUP:** $P2_1/n$
**Z:** 4
**LATTICE CONSTANTS:**
$a = 5.48$
$b = 6.76$ (Synthetic)
$c = 9.28$
$\beta = 94°33'$
**3 STRONGEST DIFFRACTION LINES:**
2.795 (100)
2.744 ( 95)
2.780 ( 90)
**OPTICAL CONSTANTS:**

| (synthetic) | (natural) |
|---|---|
| $\alpha = 1.715$ | $\alpha = 1.707$ |
| $\gamma = 1.740$ | $\beta = 1.715$ |
| $(+)2V = 63°$ | $\gamma = 1.730$ |

**HARDNESS:** ~6
**DENSITY:** 3.28 (Meas.)
3.31 (Calc.)
**CLEAVAGE:** {100} distinct
**HABIT:** Crystals tabular, imperfectly developed; also massive, granular. Twinning on {100} common, polysynthetic.
**COLOR-LUSTER:** Colorless to gray.  Transparent to translucent. Vitreous.
**MODE OF OCCURRENCE:** Occurs in association with bredigite and $\gamma$-$Ca_2SiO_4$ in a contact zone around a syenite-monzonite intrusion in Marble Canyon, Culberson County, Texas.  Also in association with spurrite and other minerals in the contact zone of limestone and Tertiary dolerite at Scawt Hill, near Larne, County Antrim, Ireland; and in a contact zone at Tokatoka, New Zealand.

**BEST REF. IN ENGLISH:** Tilley, C. E., *Min. Mag.*, **22**: 77–86, (1929).    Bridge, Thomas E., *Am. Min.*, **51**: 1766–1774 (1966).

## LARSENITE
$PbZnSiO_4$

**CRYSTAL SYSTEM:**  Orthorhombic
**CLASS:**  mm2
**SPACE GROUP:**  $Pna2_1$
**Z:**  8
**LATTICE CONSTANTS:**
$a = 8.244$
$b = 18.963$
$c = 5.06$
**3 STRONGEST DIFFRACTION LINES:**
3.17 (100)
3.01 ( 60)
2.83 ( 50)
**OPTICAL CONSTANTS:**
$\alpha = 1.92$
$\beta = 1.95$
$\gamma = 1.96$
$(-)2V = 80°$
**HARDNESS:** 3
**DENSITY:** 5.90 (Meas.)
6.141 (Calc.)
**CLEAVAGE:** {120} distinct
Brittle.
**HABIT:** Crystals slender prismatic, vertically striated; sometimes tabular, flattened {010}.
**COLOR-LUSTER:** White. Transparent. Adamantine.
**MODE OF OCCURRENCE:** Occurs in association with willemite, clinohedrite, esperite, hardystonite, hodgkinsonite and other minerals in veinlets at Franklin, Sussex County, New Jersey.
**BEST REF. IN ENGLISH:** Palache, Charles, USGS Prof. Paper 180, p. 80–81 (1935).  *Am. Min.*, **51**: 269 (1966).

## LATIUMITE
$K(Ca, Na)_3(Al, Si)_5O_{11}(SO_4, CO_3)$

**CRYSTAL SYSTEM:**  Monoclinic
**CLASS:** 2
**SPACE GROUP:** $P2_1$
**Z:** 2
**LATTICE CONSTANTS:**
$a = 12.06$
$b = 5.08$
$c = 10.81$
$\beta = 106°$
**3 STRONGEST DIFFRACTION LINES:**
2.86 (100)
3.06 ( 90)
2.96 ( 90)
**OPTICAL CONSTANTS:**
$\alpha = 1.600$
$\beta = 1.606$
$\gamma = 1.614$
$2V$ = large.  Sign (+) or (−).

HARDNESS: 5½-6
DENSITY: 2.93 (Meas.)
2.92 (Calc.)
CLEAVAGE: {100} perfect
HABIT: Crystals tabular, elongated; also massive. Twinning on {100} common.
COLOR-LUSTER: White. Vitreous.
MODE OF OCCURRENCE: Latiumite occurs with hedenbergitic pyroxene, grossular-andradite, melilite, leucite, haüyne, and rarely with kaliophilite, in two ejected blocks of the Alban Hills, Albano, Latium, Italy; also in metamorphic ejected blocks found in a pumice deposit near Pitigliano (Grosseto, Tuscany).
BEST REF. IN ENGLISH: Tilley, C. E. and Henry, N.F.M., *Min. Mag.*, **30**: 39-45 (1953). Cannillo, Elio; Negro, A. Dal; and Rossi, Giuseppe, *Am. Min.*, **58**: 466-470 (1973).

## LATRAPPITE
(Ca, Na)(Nb, Ti, Fe)$O_3$
Niobium analogue of perovskite

CRYSTAL SYSTEM: Orthorhombic
CLASS: 2/m 2/m 2/m
SPACE GROUP: Pcmn
Z: 4
LATTICE CONSTANTS:
$a$ = 5.448
$b$ = 7.777
$c$ = 5.553
3 STRONGEST DIFFRACTION LINES:
2.744 (100)
3.887 ( 79)
1.942 ( 57)
HARDNESS: 5½
DENSITY: 4.40 (Meas.)
4.457 (Calc.)
HABIT: Small cubic crystals (~0.2 mm). Complex twinning common.
COLOR-LUSTER: Black. Submetallic. Opaque.
MODE OF OCCURRENCE: Occurs disseminated in calcite at the carbonatite deposit at Oka, Quebec, Canada.
BEST REF. IN ENGLISH: Nickel, E. H., and McAdam, R. C., *Can. Min.*, **7**: 683-697 (1963). Nickel, E. H., *Can. Min.*, **8** (pt 1): 121-122 (1964).

## LAUBMANNITE
$Fe_3^{2+}Fe_6^{3+}(PO_4)_4(OH)_{12}$

CRYSTAL SYSTEM: Orthorhombic
CLASS: 2/m 2/m 2/m
SPACE GROUP: Pbma
Z: 4
LATTICE CONSTANTS:
$a$ = 13.95
$b$ = 30.77
$c$ = 5.16
3 STRONGEST DIFFRACTION LINES:
5.150 (100)
15.30 ( 90)
3.446 ( 70)

OPTICAL CONSTANTS:
$\alpha$ = 1.840
$\beta$ = 1.847
$\gamma$ = 1.892
(+)2V = moderate, variable
HARDNESS: 3½-4
DENSITY: 3.33 (Meas.)
CLEAVAGE: Two probable cleavages parallel to fiber length. Brittle.
HABIT: As botryoidal and mammillary aggregates with radial-fibrous or parallel-fibrous structure.
COLOR-LUSTER: Bright yellowish green, grayish green to greenish brown to brown. Vitreous to silky. Subtranslucent to opaque.
MODE OF OCCURRENCE: Occurs filling crevices in novaculite at Buckeye Mountain, near Shady, Polk County, Arkansas; at the Nitzelbuch mine, Amberg-Auerbach district, Bavaria, Germany; and in vugs in a gossan zone surrounding magnetite ore at Leveäniemi, Norbotten Province, Sweden.
BEST REF. IN ENGLISH: Moore, Paul B., *Am. Min.*, **55**: 135-169 (1970). Palache, et al., "Dana's System of Mineralogy," 7th Ed., v. II, p. 803, New York, Wiley, Wiley, 1951.

## LAUEÏTE
$MnFe_2(PO_4)_2(OH)_2 \cdot 8H_2O$
Dimorphous with strunzite

CRYSTAL SYSTEM: Triclinic
CLASS: $\bar{1}$
SPACE GROUP: P$\bar{1}$
Z: 1
LATTICE CONSTANTS:
$a$ = 5.28    $\alpha$ = 107°55'
$b$ = 10.66   $\beta$ = 110°59'
$c$ = 7.14    $\gamma$ = 71°07'
3 STRONGEST DIFFRACTION LINES:
9.91 (100)
3.28 ( 90)
4.95 ( 80)
OPTICAL CONSTANTS:
$\alpha$ < 1.612
$\beta$ = 1.658
$\gamma$ = 1.682
(-)2V = 50°
HARDNESS: 3
DENSITY: 2.44-2.49 (Meas.)
2.56 (Calc.)
CLEAVAGE: {010} perfect
HABIT: Wedge-shaped crystals up to 2.0 mm in size. Most common forms: {100}, {010}, {110}, {1$\bar{1}$0}, {0$\bar{1}$1}, and {011}.
COLOR-LUSTER: Honey-brown, yellowish-orange. Transparent to translucent. Vitreous.
MODE OF OCCURRENCE: Occurs on rockbridgeite at the Hagendorf pegmatite, eastern Bavaria; associated with strunzite, stewartite, and pseudolaueite, all implanted upon siderite and ludlamite at the Palermo No. 1 quarry, North Groton, New Hampshire; and at several pegmatites in the Keystone and Custer districts, Black Hills, South Dakota,

especially at the Tip-Top mine near Custer, and at the Big Chief mine, Glendale.

BEST REF. IN ENGLISH: Strunz, H., *Am. Min.*, **39**: 1038 (1954).   Moore, Paul Brian, *Am. Min.*, **50**: 1884–1892 (1965).   Bauer, Werner H., *Am. Min.*, **54**: 1312–1323 (1969).

# LAUMONTITE
$CaAl_2Si_4O_{12} \cdot 4H_2O$
Var. Leonhardite—partially dehydrated
$$CaAl_2Si_4O_{12} \cdot 3\frac{1}{2}H_2O$$
Zeolite Group

CRYSTAL SYSTEM: Monoclinic
CLASS: 2, or m
SPACE GROUP: C2 or Cm
Z: 4
LATTICE CONSTANTS:
 $a = 14.90$
 $b = 13.17$
 $c = 7.55$
 $\beta = 111°30'$
3 STRONGEST DIFFRACTION LINES:
 9.49 (100)
 4.156 ( 60)
 6.86 ( 35)
OPTICAL CONSTANTS:
 $\alpha = 1.502–1.514$
 $\beta = 1.512–1.522$
 $\gamma = 1.514–1.525$
 $(-)2V = 26°–47°$
HARDNESS: 3–4
DENSITY: 2.20–2.41 (Meas.)
 2.405 (Calc.)
CLEAVAGE: {010} perfect
 {110} perfect
Fracture uneven.
HABIT: Crystals usually square prisms with steep oblique terminations.   Also fibrous, columnar, radiating and divergent.  Twinning on {100}, sometimes as "swallow-tail" twins.
COLOR-LUSTER: White, gray, yellowish, pink, brownish. Transparent to translucent, becoming opaque and often powdery on exposure.  Vitreous to pearly.  Streak uncolored.
MODE OF OCCURRENCE: Occurs in veins and cavities of many rock types, chiefly in basalt and related rocks; in decomposed granite and pegmatite; in metamorphic and sedimentary rocks; and in metallic vein deposits. Found as exceptional crystals up to 6 inches in length at the Pine Creek tungsten mine, Bishop, California; as fine crystals in Oregon and Washington; and at localities in Arizona, Michigan, Pennsylvania, and in the trap rocks of northern New Jersey. Other notable occurrences are found in Nova Scotia, Brazil, Scotland, England, Norway, Italy, Switzerland, Germany, Austria, Roumania, India, and New Zealand.
BEST REF. IN ENGLISH: Deer, Howie, and Zussman, "Rock Forming Minerals," v. 4, p. 401–407, New York, Wiley, 1963.

# LAUNAYITE
$Pb_{22}Sb_{26}S_{61}$

CRYSTAL SYSTEM: Monoclinic
CLASS: 2/m
SPACE GROUP: C 2/m (?)
Z: 2
LATTICE CONSTANTS:
 $a = 42.58$
 $b = 4.015$
 $c = 32.16$
 $\beta = 102°01'$
3 STRONGEST DIFFRACTION LINES:
 3.45 (100)
 4.17 ( 80)
 2.92 ( 80)
HARDNESS: Vickers microhardness 179 (171–197) kg/mm$^2$ with 50 g load.
DENSITY: 5.75 (predicted)
 5.83 (Calc.)
CLEAVAGE: {100} perfect
 {001} perfect
HABIT: Fibrous; elongation [010].   Observed only in polished section.
COLOR-LUSTER: Lead gray. Metallic. Opaque. Streak black.
MODE OF OCCURRENCE: Occurs associated with veenite and boulangerite in a small prospect pit in Precambrian marble at Madoc, Ontario, Canada.
BEST REF. IN ENGLISH: Jambor, J. L., *Can. Min.*, **9** (Pt. 2): 191–213 (1967).

# LAURIONITE
PbClOH
Dimorphous with Paralaurionite

CRYSTAL SYSTEM: Orthorhombic
CLASS: 2/m 2/m 2/m
SPACE GROUP: Pnam
Z: 4
LATTICE CONSTANTS:
 $a = 7.1$
 $b = 9.7$
 $c = 4.05$
3 STRONGEST DIFFRACTION LINES:
 3.30 (100)
 4.01 ( 90)
 2.52 ( 70)
OPTICAL CONSTANTS:
 $\alpha = 2.077$
 $\beta = 2.116$ (Na)
 $\gamma = 2.158$
 $(-)2V = $ large
HARDNESS: 3–3½
DENSITY: 6.241 (Meas.)
 6.14 (Calc.)
CLEAVAGE: {101} distinct
Not brittle.
HABIT: Crystals thin to thick tabular {100}; elongated along c-axis. {100} sometimes striated.

COLOR-LUSTER: Colorless, white. Transparent. Adamantine; pearly on {100}.

MODE OF OCCURRENCE: Occurs associated with paralaurionite, fiedlerite, phosgenite, and other secondary lead minerals as excellent small crystals in vugs in lead slags exposed to the action of sea water at Laurium, Greece; and also in oxidized ore associated with paralaurionite, phosgenite, and anglesite, at the Wheal Rose mine near Sithney, Cornwall, England.

BEST REF. IN ENGLISH: Palache, et al., "Dana's System of Mineralogy," 7th Ed., v. II, p. 62–64, New York, Wiley, 1951.

# LAURITE
$RuS_2$

CRYSTAL SYSTEM: Cubic
CLASS: $2/m\bar{3}$
SPACE GROUP: Pa3
Z: 4
LATTICE CONSTANT:
$a = 5.60$
3 STRONGEST DIFFRACTION LINES:
1.695 (100)
3.25 ( 80)
2.81 ( 70)
HARDNESS: 7½
DENSITY: 6.–6.99 (Meas.)
6.211 (Calc.)
CLEAVAGE: {111} perfect
Fracture subconchoidal. Brittle.

HABIT: Crystals minute; cubic, octahedral, or pyritohedral. Usually as small rounded grains.

COLOR-LUSTER: Iron black. Opaque. Bright metallic. Powder dark gray.

MODE OF OCCURRENCE: Occurs in platinum concentrates from the Bushveld igneous complex, Transvaal, South Africa; in the platinum sands of Borneo; and reported from Oregon and Colombia.

BEST REF. IN ENGLISH: Palache, et al., "Dana's System of Mineralogy," 7th Ed., v. I, p. 291–292, New York, Wiley, 1944.

# LAUSENITE (Inadequately described mineral)
$Fe_2(SO_4)_3 \cdot 6H_2O$

CRYSTAL SYSTEM: Monoclinic
OPTICAL CONSTANTS:
$\alpha = 1.598$
$\beta = 1.628$
$\gamma = 1.654$
(–)2V = large
HARDNESS: Not determined.
DENSITY: Not determined.
CLEAVAGE: Not determined.
HABIT: As aggregates of minute fibers, elongated parallel to c-axis.
COLOR-LUSTER: White. Silky.
MODE OF OCCURRENCE: Occurs in the fire zone of the United Verde mine, Jerome, Arizona, associated with copiapite and other sulfates.

BEST REF. IN ENGLISH: Palache, et al., "Dana's System of Mineralogy," 7th Ed., v. II, p. 530, New York, Wiley, 1951.

# LAUTARITE
$Ca(IO_3)_2$

CRYSTAL SYSTEM: Monoclinic
CLASS: 2/m
SPACE GROUP: $P2_1/a$
Z: 4
LATTICE CONSTANTS:
$a = 7.18$ kX
$b = 11.38$
$c = 7.32$
$\beta = 106°22'$
3 STRONGEST DIFFRACTION LINES:
4.27 (100)
3.04 (100)
3.24 ( 67)
OPTICAL CONSTANTS:
$\alpha = 1.792$
$\beta = 1.840$
$\gamma = 1.888$
(+)2V $\simeq 90°$
HARDNESS: 3½–4
DENSITY: 4.519 (Meas.)
4.48 (Calc.)
CLEAVAGE: {011} good
{100} trace
{110} trace
HABIT: Crystals short prismatic; striated and rounded in direction of elongation. As radial aggregates.

COLOR-LUSTER: Colorless, yellowish. Transparent.

MODE OF OCCURRENCE: Occurs associated with gypsum in nitrate deposits in the Atacama desert, Antofagasta Province, Chile.

BEST REF. IN ENGLISH: Palache, et al., "Dana's System of Mineralogy," 7th Ed., v. II, p. 312–313, New York, Wiley, 1951.

# LAUTITE
CuAsS

CRYSTAL SYSTEM: Orthorhombic
CLASS: mm2
SPACE GROUP: $Pbn2_1$
Z: 4
LATTICE CONSTANTS:
$a = 5.47$
$b = 11.47$
$c = 3.78$
3 STRONGEST DIFFRACTION LINES:
3.10 (100)
1.903 ( 80)
1.610 ( 60)
HARDNESS: 3–3½

DENSITY: 4.9 (Meas.)
  4.818 (Calc.)
CLEAVAGE: {001}
Brittle.
HABIT: Crystals tabular or short prismatic, striated on {001}. Also massive, compact to fine granular; radiated; columnar to fibrous. Twinning on {110}.
COLOR-LUSTER: Black to steel gray with reddish tint. Opaque. Metallic.
MODE OF OCCURRENCE: Occurs in mesothermal vein deposits commonly associated with other sulfides, native arsenic, tennantite, proustite, and pyrargyrite. Found at Lauta, Saxony, Germany, and at the Gabe Gottes mine, near Ste. Marie aux Mines, Alsace-Lorraine, France.
BEST REF. IN ENGLISH: Palache, et al., "Dana's System of Mineralogy," 7th Ed., v. I, p. 327–328, New York, Wiley, 1944.

# LAVENDULAN (Freirinite)
$(Ca, Na)_2 Cu_5 (AsO_4)_4 Cl \cdot 4-5H_2O$

CRYSTAL SYSTEM: Orthorhombic
Z: 8
LATTICE CONSTANTS:
  $a = 9.73$
  $b = 41.0$
  $c = 9.85$
3 STRONGEST DIFFRACTION LINES:
  9.77 (100)
  3.11 ( 70)
  4.87 ( 50)
OPTICAL CONSTANTS:
  Freirinite      (biaxial)
  (uniaxial)      $\alpha = 1.66$
  $\omega = 1.748$    $\beta = 1.715$
  $\epsilon = 1.645$    $\gamma = 1.734$
     (-)         $2V = 33°$
HARDNESS: 2½
DENSITY: 3.54 (Meas.)
CLEAVAGE: {001} good
  {110} imperfect
HABIT: As aggregates of fine flakes and as thin botryoidal crusts.
COLOR-LUSTER: Blue to lavender blue. Translucent. Vitreous to waxy, also silky.
MODE OF OCCURRENCE: Occurs associated with olivenite and conichalcite at Gold Hill, Tooele County, Utah; with erythrite, malachite, cuprite, and cobaltian wad at the Blanca mine, San Juan, Chile; and with erythrite at Joachimsthal (Jachymov), Bohemia, Czechoslovakia.
BEST REF. IN ENGLISH: Guillemin, C., *Am. Min.*, **42**: 123–124 (1957).

# LÅVENITE
$(Na, Ca, Mn)_3 (Zr, Ti, Fe)(SiO_4)_2 F$
Var. Titan-Låvenite

CRYSTAL SYSTEM: Monoclinic
CLASS: 2/m
SPACE GROUP: P2₁/a
Z: 4

LATTICE CONSTANTS:
  $a = 10.54$
  $b = 9.90$
  $c = 7.14$
  $\beta = 108°12'$
3 STRONGEST DIFFRACTION LINES:
  2.89 (100)
  2.81 ( 90)
  3.22 ( 50)
OPTICAL CONSTANTS:
  $\alpha = 1.67$
  $\beta = 1.69$
  $\gamma = 1.72$
  $(-)2V = 40-70°$
HARDNESS: 6
DENSITY: 3.51–3.547 (Meas.)
CLEAVAGE: {100} perfect
Fracture uneven. Brittle.
HABIT: Crystals prismatic. Also massive or as acicular radial aggregates.
COLOR-LUSTER: Colorless, light to dark yellow, brown, brownish red to blackish brown. Translucent; thin fragments transparent. Vitreous to dull and greasy.
MODE OF OCCURRENCE: Occurs in alkali rocks and their pegmatites in association with feldspars, nepheline, sphene, mosandrite, eudialyte, catapleiite, zircon, and other minerals. Found on the Pocos de Caldas plateau, Brazil; on the island of Låven and elsewhere in the Lagensundfjord district, Norway, on the Kola Peninsula (titan-låvenite); and at localities in France, Portugal, Azores, Madagascar, and Iles de Los, Guinea.
BEST REF. IN ENGLISH: Vlasov, K. A., "Mineralogy of Rare Elements," v. II, p. 381–384, Israel Program for Scientific Translations, 1966.

# LAVROVITE = Variety of diopside with small amounts of Cr and V

# LAWRENCITE
$FeCl_2$

CRYSTAL SYSTEM: Hexagonal
CLASS: $\bar{3}$ 2/m
SPACE GROUP: R$\bar{3}$m
Z: 1 (rhomb.), 3 (hex.)
LATTICE CONSTANTS:
  $a = 3.5775$      $a_{rh} = 6.19$
  $c = 17.52$      $\alpha = 33°33½'$
3 STRONGEST DIFFRACTION LINES:
  2.54 (100)
  5.9 ( 63)
  1.80 ( 63)
OPTICAL CONSTANTS:
  $\omega = 1.567$
  $\omega - \epsilon$ = weak
     (-)
HARDNESS: Not determined. Soft.
DENSITY: 3.162 (Meas.)
  3.22 (Calc.)

CLEAVAGE: Perfect $\{0001\}$ cleavage inferred from crystal structure.
HABIT: Massive.
COLOR-LUSTER: Green to brown.

Deliquescent.
MODE OF OCCURRENCE: Occurs as inclusions in native iron at Ovifak, Greenland, and as a fissure filling in many iron meteorites.
BEST REF. IN ENGLISH: Palache, et al., "Dana's System of Mineralogy," 7th Ed., v. II, p. 40, New York, Wiley, 1951.

# LAWSONITE
$CaAl_2(OH)_2(Si_2O_7) \cdot H_2O$

CRYSTAL SYSTEM: Orthorhombic
CLASS: 2/m 2/m 2/m
SPACE GROUP: Ccmm
Z: 4
LATTICE CONSTANTS:
$a = 8.75-8.90$
$b = 5.75-5.84$
$c = 13.09-13.33$
3 STRONGEST DIFFRACTION LINES:
2.624 (100)
1.550 ( 80)
2.726 ( 70)
OPTICAL CONSTANTS:
$\alpha = 1.665$
$\beta = 1.674-1.675$
$\gamma = 1.684-1.686$
$(+)2V = 76°-87°$
HARDNESS: 6
DENSITY: 3.05-3.12 (Meas.)
3.094 (Calc.)
CLEAVAGE: $\{100\}$ perfect
$\{010\}$ perfect
$\{101\}$ imperfect
HABIT: Crystals prismatic, sometimes tabular parallel to $c$. As subhedral tablets; also massive, granular.
COLOR-LUSTER: Colorless, white, gray, blue, pinkish. Translucent. Vitreous to greasy. Streak white.
MODE OF OCCURRENCE: Occurs as a low-temperature metamorphic mineral chiefly in glaucophane schists. Found widespread in the metamorphic rocks of the Coast Ranges in California, notably on the Tiburon Peninsula, Marin County, and as fine euhedral crystals up to 2 inches in length near Covelo, Mendocino County. Also at Santa Clara, Cuba, and at localities in France, Italy, Corsica, Celebes, New Caledonia, and in the Kantô Mountains, Japan.
BEST REF. IN ENGLISH: Deer, Howie, and Zussman, "Rock Forming Minerals," v. 1, p. 221-226, New York, Wiley, 1962.

# LAZAREVIĆITE = Arsenosulvanite

# LAZULITE
$(Mg, Fe^{2+})Al_2(PO_4)_2(OH)_2$
Lazulite-Scorzalite Series

CRYSTAL SYSTEM: Monoclinic
CLASS: 2/m
SPACE GROUP: $P2_1/c$
Z: 2
LATTICE CONSTANTS:
$a = 7.15$
$b = 7.28$
$c = 7.25$
$\beta = 120°35'$
3 STRONGEST DIFFRACTION LINES:
3.072 (100)
3.136 ( 95)
6.15 ( 75)
OPTICAL CONSTANTS:
$\alpha = 1.604-1.635$
$\beta = 1.633-1.663$
$\gamma = 1.642-1.673$
$(-)2V = 68.9°$
HARDNESS: 5½-6
DENSITY: 3.08 (Extrapolated for Mg end member)
3.14 (Calc.)
CLEAVAGE: $\{110\}$ indistinct to good
$\{101\}$ indistinct
Fracture uneven to splintery. Brittle.
HABIT: Crystals usually acute pyramidal, with large $\{111\}$ and $\{\bar{1}11\}$ and small $\{101\}$; also tabular on $\{101\}$ or $\{\bar{1}11\}$. Massive, compact to granular. Commonly twinned on composition plane $a[100]$.
COLOR-LUSTER: Deep azure blue to light blue; also bluish green. Translucent to opaque; rarely transparent. Vitreous to dull.
MODE OF OCCURRENCE: Occurs in quartz veins, granitic pegmatites, and highly metamorphosed rocks, particularly quartzites. Often found associated with quartz, pyrophyllite, kyanite, muscovite, andalusite, sillimanite, rutile, corundum, and garnet. Found as crystals up to 6 inches in length at Hörrsjoberg, Sweden, and as fine crystals at many European localities, also in Bolivia, Brazil, and in many deposits in Madagascar. Superb crystals are found in the United States especially at Graves Mountain, Lincoln County, Georgia, and large masses to 6 inches across occur at the Champion mine, Mono County, California.
BEST REF. IN ENGLISH: Palache, et al., "Dana's System of Mineralogy, "7th Ed., v. 2, p. 908-911, New York, Wiley, 1951. Pecora, W. T., and Fahey, J. J., *Am. Min.*, **35**: 1-18 (1950).

# LAZURITE (Lapis Lazuli)
$(Na, Ca)_8(Al, Si)_{12}O_{24}(S, SO_4)$
Sodalite Group

CRYSTAL SYSTEM: Cubic
CLASS: $\bar{4}$3m
SPACE GROUP: $P\bar{4}3m$
Z: 1
LATTICE CONSTANT:
$a = 9.09$ (Varies with substitution)
3 STRONGEST DIFFRACTION LINES:
3.71 (100)
2.62 ( 80)
2.87 ( 45)

OPTICAL CONSTANT:

$N \approx 1.50$

HARDNESS: 5-5½

DENSITY: 2.38-2.45 (Meas.)

CLEAVAGE: {110} imperfect

Fracture uneven. Brittle.

HABIT: Crystals rare; usually coarse dodecahedrons up to one inch in diameter. Commonly massive, compact.

COLOR-LUSTER: Deep blue, azure blue, violet blue, greenish blue. Translucent. Dull. Streak bright blue.

MODE OF OCCURRENCE: Occurs as a contact metamorphic mineral in limestones, often in association with calcite and enclosing small grains of pyrite. Found in Cascade Canyon, San Bernardino County, California; in association with diopside, amphibole, mica, calcite, and pyrite on North Italian Mountain, Gunnison County, Colorado; in the Andes of Ovalle, Chile; at Vesuvius and in the Albani Mountains, Italy; near Lake Baikal, USSR; and as euhedral crystals and irregular masses in marble in the Kokcha River valley, Badakhstan, northeast Afghanistan.

BEST REF. IN ENGLISH: Winchell and Winchell, "Elements of Optical Mineralogy," 4th Ed., Pt 2, p. 350–351, New York, Wiley, 1951.

# LEAD

Pb

CRYSTAL SYSTEM: Cubic

CLASS: $4/m \bar{3} 2/m$

SPACE GROUP: Fm3m

Z: 4

LATTICE CONSTANT:

$a = 4.9496$

3 STRONGEST DIFFRACTION LINES:

2.855 (100)

2.475 ( 50)

1.493 ( 32)

HARDNESS: 1½

DENSITY: 11.347 at 20° (Meas.)

11.339 at 20° (Calc.)

CLEAVAGE: None. Very malleable; ductile.

HABIT: Crystals rare, usually octahedral, also cubic or dodecahedral. Commonly massive, as thin plates, wires, dendritic, or as rounded masses. Twinning on {111}.

COLOR-LUSTER: Gray white. Opaque. Metallic. Usually tarnished lead gray; dull. Streak lead gray.

MODE OF OCCURRENCE: Occurs very sparingly as scales, globules, and irregular branching masses among the rare minerals of the Parker shaft, Franklin, Sussex County, New Jersey. It also is found at the Jay Gould mine, Blaine County, Idaho, and has been reported from other western States. Excellent crystal groups exceeding 100 pounds in weight have been found at Långban, and small crystals at the Harstig mine, Pajsberg, Sweden. Other occurrences are found on Madeira island; at Jalapa, Mexico; and at several places in the USSR.

BEST REF. IN ENGLISH: Palache, et al., "Dana's System of Mineralogy," 7th Ed., v. I, p. 102–103, New York, Wiley, 1944.

# LEADHILLITE

$Pb_4 SO_4 (CO_3)_2 (OH)_2$

Low-temperature polymorph of susannite

CRYSTAL SYSTEM: Monoclinic

CLASS: 2/m

SPACE GROUP: $P2_1/a$

Z: 8

LATTICE CONSTANTS:

$a = 9.08$

$b = 20.76$

$c = 11.56$

$\beta = 90°27.5'$

3 STRONGEST DIFFRACTION LINES:

3.53 (100)

2.917 ( 60)

2.605 ( 50)

OPTICAL CONSTANTS:

$\alpha = 1.87$

$\beta = 2.00$

$\gamma = 2.01$

$(-)2V \simeq 10°$

HARDNESS: 2½-3

DENSITY: 6.55 (Meas.)

6.57 (Calc.)

CLEAVAGE: {001} perfect, easy.

Fracture conchoidal.

HABIT: Crystals often pseudohexagonal; thin to thick tabular {001}; occasionally prismatic or equant. Also massive, granular. Commonly contact twins of the aragonite type.

COLOR-LUSTER: Colorless, white, gray, yellowish, pale green to pale blue. Transparent to translucent. Resinous to adamantine. Occasionally fluoresces yellow under ultraviolet light.

MODE OF OCCURRENCE: Occurs in the oxidized zone of lead deposits as a secondary mineral associated with cerussite, anglesite, diaboleite, lanarkite, matlockite, linarite, pyromorphite, and galena. Among the many known localities may be mentioned Leadhills and Wanlockhead, Scotland; Matlock, Derbyshire, England; Tsumeb, South West Africa; Tintic district, Utah; near Granby and Joplin, Missouri; and in superb specimens often associated with diaboleite at the Mammoth mine, Tiger, Pinal County, Arizona.

BEST REF. IN ENGLISH: Mrose, M. E., and Christian, R. P., Can. Min., 10: 141 (1969) (abstr.). Palache, et al., "Dana's System of Mineralogy," 7th Ed., v. 2, p. 295–298, New York, Wiley, 1951.

# LEAD HYDROXYAPATITE (Species status uncertain)

$Pb_5 (PO_4)_3 OH(?)$

CRYSTAL SYSTEM: Hexagonal

Z: 2

3 STRONGEST DIFFRACTION LINES:

2.965 (100)

4.11 ( 40) (synthetic)

2.852 ( 40)

MODE OF OCCURRENCE: Occurs rarely with pyromorphite, pseudomorphous after galena, at Whyte's Cleuch in the Leadhills-Wanlockhead lead-zinc district, Scotland.

BEST REF. IN ENGLISH: Temple, A. K., *Trans. Roy. Soc. Edinburgh*, **63**: 85-114 (1955-56) (Publ. 1957).


## LECHATELIERITE = Natural fused silica
$SiO_2$


## LECONTITE
$Na(NH_4, K)SO_4 \cdot 2H_2O$

CRYSTAL SYSTEM: Orthorhombic
CLASS: 222
SPACE GROUP: $P2_1 2_1 2_1$
Z: 4
LATTICE CONSTANTS:
$a$ = 8.23-8.24
$b$ = 12.85-12.88
$c$ = 6.24-6.25
3 STRONGEST DIFFRACTION LINES:
5.068 (100)
3.802 ( 94)
3.033 ( 89)
OPTICAL CONSTANTS:
$\alpha$ = 1.440
$\beta$ = 1.454
$\gamma$ = 1.455
(-)2V = 29°44' (Calc.)
HARDNESS: 2-2½
DENSITY: 1.745 (Meas. synthetic)
1.737 (Calc.)
CLEAVAGE: {011} distinct
HABIT: Crystals long prismatic, narrow; also stubby; up to one inch in length; also fine-grained masses.
COLOR-LUSTER: Colorless. Transparent. Vitreous.

Taste bitter, saline.
MODE OF OCCURRENCE: Found embedded in bat guano at Taylor's Cave of Las Piedras, Comayagua, Honduras.
BEST REF. IN ENGLISH: Faust, Robert J., and Bloss, F. Donald, *Am. Min.*, **48**: 180-188 (1963).


## LEDIKITE = Trioctahedral analogue of illite


## LEGRANDITE
$Zn_2(OH)AsO_4 \cdot H_2O$

CRYSTAL SYSTEM: Monoclinic
CLASS: 2/m
SPACE GROUP: $P2_1/c$
Z: 8
LATTICE CONSTANTS:
$a$ = 12.80
$b$ = 7.94
$c$ = 10.22
$\beta$ = 104°12'

3 STRONGEST DIFFRACTION LINES:
4.08 (100)
6.68 ( 71)
5.93 ( 71)
OPTICAL CONSTANTS:
$\alpha$ = 1.675-1.702
$\beta$ = 1.690-1.709
$\gamma$ = 1.735-1.740
(+)2V = 50°
HARDNESS: 4½
DENSITY: 3.98 (Meas.)
4.015 (Calc.)
CLEAVAGE: {100} fair to poor
Fracture uneven. Brittle.
HABIT: Crystals long prismatic, elongated parallel to [001]. Aggregates of crystals in sprays or fans common. Crystals vary in length from 1 mm to 6 cm; maximum diameter 7.5 mm.
COLOR-LUSTER: Colorless to wax yellow. Transparent to translucent. Vitreous.
MODE OF OCCURRENCE: Occurs in cavities in a hard, compact limonite matrix, usually associated with crystals of adamite, at the Ojuela mine, Mapimi, Durango, Mexico. Described as a new species from the Flor de Peña mine, Lampazos, Nuevo Leon, Mexico by Drugman and Hey (1932).
BEST REF. IN ENGLISH: Desautels, Paul E., and Clarke, Roy S. Jr., *Am. Min.*, **48**: 1258-1265 (1963).


## LEHIITE (inadequately described mineral)
$(K, Na)_2 Ca_5 Al_8 (PO_4)_8 (OH)_{12} \cdot 6H_2O$ (?)

CRYSTAL SYSTEM: Unknown
OPTICAL CONSTANTS:
$\alpha$ = 1.600
$\beta$ = 1.615
$\gamma$ = 1.629
(-)2V = very large
HARDNESS: 5½
DENSITY: 2.89
CLEAVAGE: Not determined.
HABIT: As layers and crusts composed of subparallel fibrous aggregates.
COLOR-LUSTER: White to gray.
MODE OF OCCURRENCE: Occurs associated with wardite, crandallite, and other phosphate minerals in the outer shell of altered variscite nodules at Fairfield, near Lehi, Utah County, Utah.
BEST REF. IN ENGLISH: Palache, et al., "Dana's System of Mineralogy," 7th Ed., v. II, p. 942-943, New York, Wiley, 1951.


## LEIFITE
$(Na, H_3O)_2 (Si, Al, Be, B)_7 (O, OH, F)_{14}$

CRYSTAL SYSTEM: Hexagonal
CLASS: 32
SPACE GROUP: P321 or P312
Z: 3

LATTICE CONSTANTS:
  $a = 14.351$
  $c = 4.854$
3 STRONGEST DIFFRACTION LINES:
  3.13 (100)
  3.34 ( 70)
  2.45 ( 70)
OPTICAL CONSTANTS:
  $\omega = 1.516$
  $\epsilon = 1.520$
  (+)
HARDNESS: 6
DENSITY: 2.57 (Meas.)
        2.49 (Calc.)
CLEAVAGE: $\{10\bar{1}0\}$ distinct
Brittle.
  HABIT: Crystals acicular, deeply striated.
  COLOR-LUSTER: Colorless. Transparent. Vitreous.
  MODE OF OCCURRENCE: Occurs in cavities in alkali-pegmatite in association with microcline, calcite, zinnwaldite, and acmite at Narsarsuk, Greenland.
  BEST REF. IN ENGLISH: Winchell and Winchell, "Elements of Optical Mineralogy," 4th Ed., Pt. 2, p. 355, New York, Wiley, 1951.


# LEIGHTONITE

$K_2 Ca_2 Cu(SO_4)_4 \cdot 4H_2O$

CRYSTAL SYSTEM: Orthorhombic
CLASS: 2/m 2/m 2/m
SPACE GROUP: Fmmm
Z: 4
LATTICE CONSTANTS:
  $a = 11.67$
  $b = 16.52$
  $c = 7.49$
3 STRONGEST DIFFRACTION LINES:
  2.90  (100)
  3.18  ( 60)
  1.781 ( 30)
OPTICAL CONSTANTS:
  $\alpha = 1.578$
  $\beta = 1.587$ (Na)
  $\gamma = 1.595$
  $(-)2V = 86°09'$ (Calc.)
HARDNESS: 3
DENSITY: 2.95 (Meas.)
         2.95 (Calc.)
CLEAVAGE: None.
  HABIT: Crystals lath-like; rarely equant; as crusts or crossfiber veinlets.
  COLOR-LUSTER: Pale bluish green, pale blue. Transparent to translucent. Vitreous.
  MODE OF OCCURRENCE: Occurs associated with kroehnkite and atacamite at Chuquicamata, Chile.
  BEST REF. IN ENGLISH: Palache, et al., "Dana's System of Mineralogy," 7th Ed., v. II, p. 461–462, New York, Wiley, 1951.


# LEMNÄSITE = Alluaudite.


# LEMOYNITE

$(Na, Ca)_3 Zr_2 Si_8 O_{22} \cdot 8H_2O$

CRYSTAL SYSTEM: Monoclinic
CLASS: Probably 2/m
SPACE GROUP: Probably C2/m
Z: 2
LATTICE CONSTANTS:
  $a = 10.48$
  $b = 16.20$
  $c = 9.07$
  $\beta = 105°20'$
3 STRONGEST DIFFRACTION LINES:
  8.01  (100)
  3.562 ( 49)
  2.807 ( 48)
OPTICAL CONSTANTS:
  $\alpha = 1.540$
  $\beta = 1.553$
  $\gamma = 1.570$
  $(+)2V = 80°$
HARDNESS: ~4
DENSITY: 2.29 (Meas.)
         2.26 (Calc.)
CLEAVAGE: $\{001\}$ distinct
          $\{010\}$ distinct
  HABIT: Crystals prismatic, very minute (0.05 mm in diameter). In spherulites about 0.5 cm in diameter.
  COLOR-LUSTER: Whitish with very light yellow tint.
  MODE OF OCCURRENCE: Occurs in small amounts in pegmatites of the St. Hilaire alkaline massif, Quebec, Canada.
  BEST REF. IN ENGLISH: Perrault, Guy, et al., *Can. Min.*, 9(Pt. 5): 585 (1969).


# LENGENBACHITE

$(Ag, Cu)_2 Pb_6 As_4 S_{13}$

CRYSTAL SYSTEM: Monoclinic
CLASS: 2/m
SPACE GROUP: $P2_1/m$
LATTICE CONSTANTS:
  $a = 34.82$
  $b = 11.49$
  $c = 36.72$
  $\beta = 94°19'$
3 STRONGEST DIFFRACTION LINES:
  3.06 (100)
  2.84 ( 90)
  4.60 ( 30)
HARDNESS: Soft (marks paper).
DENSITY: 5.80–5.85 (Meas.)
         5.781 (Calc.)
  CLEAVAGE: One perfect parallel to flat face.  Also two other fair cleavages across the flat face.
Flexible but not elastic. Malleable.
  HABIT: Crystals very thin, blade-shaped, sometimes curled. Flat face striated in direction of elongation.
  COLOR-LUSTER: Steel gray, sometimes with iridescent tarnish. Opaque. Metallic. Streak black.

MODE OF OCCURRENCE: Found in the cavities of a crystalline dolomite, associated with pyrite and often deposited upon jordanite, at Lengenbach, in the Binnental, Valais, Switzerland.

BEST REF. IN ENGLISH: Palache, et al., "Dana's System of Mineralogy," 7th Ed., v. I, p. 398, New York, Wiley, 1944.

## LENOBLITE (inadequately described mineral)
$V_2O_4 \cdot 2H_2O$

CRYSTAL SYSTEM: Unknown
3 STRONGEST DIFFRACTION LINES:
   4.59 (100)
   3.722 ( 90)
   2.528 ( 80)
HARDNESS: Not determined.
DENSITY: Not determined.
CLEAVAGE: Not determined.
HABIT: Massive.
COLOR-LUSTER: Azure blue; becomes greenish due to oxidation.
MODE OF OCCURRENCE: Occurs with duttonite in the uranium deposit at Mounana, Gabon.
BEST REF. IN ENGLISH: Cesbron, Fabien, and Vachey, Hélène, *Am. Min.*, **56**: 635–636 (1971).

## LEONHARDITE = Variety of laumontite

## LEONHARDTITE = Starkeyite

## LEONITE
$K_2Mg(SO_4)_2 \cdot 4H_2O$

CRYSTAL SYSTEM: Monoclinic
CLASS: 2/m
SPACE GROUP: C2/m
Z: 4
LATTICE CONSTANTS:
   $a = 11.78$
   $b = 9.52$
   $c = 9.88$
   $\beta = 95°17.1'$
3 STRONGEST DIFFRACTION LINES:
   (synthetic)
   5.87 (100)     3.42 (100)
   2.371 ( 50)    3.04 ( 50)
   1.946 ( 45)    2.38 ( 45)
OPTICAL CONSTANTS:
   $\alpha = 1.479$
   $\beta = 1.482$
   $\gamma = 1.487$
   (+)2V near 90°
HARDNESS: 2½-3
DENSITY: 2.201 (Meas.)
           2.205 (Calc.)
CLEAVAGE: No apparent cleavage.
Fracture conchoidal.

HABIT: Crystals tabular and elongated along c-axis, usually as crude crystals and irregular grains intergrown with other salt minerals.
COLOR-LUSTER: Colorless, also yellowish. Transparent. Vitreous to waxy.

Taste bitter.

MODE OF OCCURRENCE: Occurs in oceanic salt deposits in Eddy and Lea Counties, New Mexico, associated with kainite, sylvite, halite, and polyhalite; and sparingly in the potash deposits of Germany.
BEST REF. IN ENGLISH: Palache, et al., "Dana's System of Mineralogy," 7th Ed., v. II, p. 450–451, New York, Wiley, 1951.

## LEPIDOCROCITE
$\gamma$-FeO · OH
Trimorphous with goethite and akaganeite

CRYSTAL SYSTEM: Orthorhombic
CLASS: 2/m 2/m 2/m
SPACE GROUP: Amam
Z: 4
LATTICE CONSTANTS:
   $a = 3.86$
   $b = 12.50$
   $c = 3.06$
3 STRONGEST DIFFRACTION LINES:
   6.26 (100)
   3.29 ( 90)
   2.47 ( 80)
OPTICAL CONSTANTS:
   $\alpha = 1.94$
   $\beta = 2.20$
   $\gamma = 2.51$
   (-)2V = 83°
HARDNESS: 5
DENSITY: 3.854 (Meas.)
           3.97 (Calc.)
CLEAVAGE: {010} perfect
           {100} less perfect
           {001} good
Brittle.
HABIT: Crystals thin, scale-like, flattened and somewhat elongated. As isolated crystals attached by an edge to matrix, or as druses. Usually massive, bladed to micaceous or fibrous.
COLOR-LUSTER: Deep red to reddish brown. Transparent. Submetallic. Streak dull orange.
MODE OF OCCURRENCE: Occurs as a secondary mineral commonly associated with goethite. Found at Iron Mountain, Shasta County, California; in the Black Hills, South Dakota; in the Lake Superior iron region; and at Easton, Pennsylvania. It also occurs at Mapimi, Durango, Mexico, and at localities in France, Germany, USSR, and elsewhere.
BEST REF. IN ENGLISH: Palache, et al., "Dana's System of Mineralogy," 7th Ed., v. I, p. 642–645, New York, Wiley, 1944.

## LEPIDOLITE
$K(Li, Al)_3(Si, Al)_4O_{10}(F, OH)_2$
Mica Group

CRYSTAL SYSTEM: Monoclinic (and hexagonal)
CLASS: 2/m or m
SPACE GROUP: C2/m or Cm
LATTICE CONSTANTS:

| C2/m or Cm | C2/c | $P3_112$ or $P3_212$ |
|---|---|---|
| $Z = 1$ | $Z = 2$ | $Z = 3$ |
| $a = 5.3$ | $a = 9.2$ | $a = 5.3$ |
| $b = 9.2$ | $b = 5.3$ | $c = 30.0$ |
| $c = 10.2$ | $c = 20.0$ | |
| $\beta = 100°$ | $\beta = 98°$ | |

3 STRONGEST DIFFRACTION LINES:
   3M: 9.93 (100), 3.33 (100), 2.61 (80)
   3T: 3.32 (100), 2.58 (75), 10.0 (50)
   1M: 3.34 (100), 10.0 (75), 4.99 (75)
   $2M_2$: 2.58 (100), 1.99 (80), 10.0 (60)
   Ortho. $Ccm2_1$: 3.31 (100), 2.59 (100), 1.98 (100)
OPTICAL CONSTANTS:
   $\alpha = 1.525-1.548$
   $\beta = 1.551-1.585$
   $\gamma = 1.554-1.587$
   $(-)2V = 0°-58°$
HARDNESS: 2½-3
DENSITY: 2.8-3.3 (Meas.)
CLEAVAGE: {001} perfect, micaceous
                   {110} imperfect
                   {010} imperfect
Flexible, elastic
   HABIT: Crystals tabular, pseudohexagonal. Also as thick cleavable masses, coarse to fine scaly aggregates, and aggregates of short cylindrical books. Twinning on {001}, rare.
   COLOR-LUSTER: Various shades of pink and purple; also colorless, white, grayish, yellowish. Transparent to translucent. Pearly. Streak uncolored.
   MODE OF OCCURRENCE: Occurs almost exclusively in granite pegmatites, often in association with elbaite, spodumene, amblygonite, and other minerals; less commonly in granites, aplites, and high-temperature tin-bearing veins. Notable occurrences are found in the Pala district, San Diego County, California; at the Brown Derby mine, Gunnison County, Colorado; at the Ingersoll mine, Pennington County, and elsewhere in the Black Hills, South Dakota; and at localities in Wyoming, Arizona, and New Mexico. Also in Maine, Connecticut, and other New England states. Other important occurrences are found at Bernic Lake, Manitoba, Canada, and in Brazil, Sweden, Germany, Czechoslovakia, Finland, USSR, Mozambique, Madagascar, and Japan.
   BEST REF. IN ENGLISH: Deer, Howie, and Zussman, "Rock Forming Minerals," v. 3, p. 85-91, New York, Wiley, 1962.

## LEPIDOMELANE = Ferrian biotite

## LERMONTOVITE (Inadequately described mineral)
$(U, Ca, Ce)_3 (PO_4)_4 \cdot 6H_2O$

CRYSTAL SYSTEM: Unknown
OPTICAL CONSTANTS:
   $\alpha = 1.562-1.574; 1562-1.702$
   $\beta = ?$
   $\gamma = 1.702-1.726; 1.574-1.726$

HARDNESS: Not determined.
DENSITY: 4.50
CLEAVAGE: Not determined. Very brittle.
HABIT: As botryoidal aggregates of radial fibrous needles.
COLOR-LUSTER: Grayish green. Dull; silky in fractures.
MODE OF OCCURRENCE: Occurs under sharply reducing conditions in the zone of cementation of hydrothermal deposits, associated with molybdenum sulfate, marcasite, hydrous silicates, and "thallium ocher" at an unspecified locality, presumably in the USSR.
   BEST REF. IN ENGLISH: Am. Min., 43: 379-380 (1958).

## LESSERITE = Inderite

## LESSINGITE = Variety of britholite

## LETOVICITE
$(NH_4)_3 H(SO_4)_2$

CRYSTAL SYSTEM: Monoclinic
CLASS: 2/m
3 STRONGEST DIFFRACTION LINES:
   4.95 (100)
   4.98 ( 85) (synthetic)
   3.77 ( 80)
OPTICAL CONSTANTS:
   $\alpha = 1.501$
   $\beta = 1.516$ (Na)
   $\gamma = 1.525$
   $(-)2V = 75°$ (Calc.)
HARDNESS: Not determined.
DENSITY: 1.83 (Meas. synthetic)
CLEAVAGE: {001} distinct
Fracture uneven.
   HABIT: Crystals minute pseudohexagonal plates, flattened on {001}; as granular masses. Lamellar twinning common.
   COLOR-LUSTER: Colorless, white, transparent.
   MODE OF OCCURRENCE: Occurs as a product formed during the burning of coal mine waste-heaps at Letovice, and near Kladno, Czechoslovakia.
   BEST REF. IN ENGLISH: Palache, et al., "Dana's System of Mineralogy," 7th Ed., v. II, p. 397, New York, Wiley, 1951.

## LEUCHTENBERGITE = Variety of clinochlore

## LEUCITE
$KAISi_2 O_6$

CRYSTAL SYSTEM: Tetragonal (pseudocubic)
CLASS: 4/m
SPACE GROUP: $I4_1/a$
Z: 16
LATTICE CONSTANTS:
   $a \simeq 13.0$
   $c \simeq 13.8$

3 STRONGEST DIFFRACTION LINES:
3.266 (100)
3.438 ( 85)
5.39 ( 80)
OPTICAL CONSTANTS:
$n \sim 1.510$
$\epsilon - \omega$ = very small
(+) uniaxial
HARDNESS: 5½–6
DENSITY: 2.47–2.50 (Meas.)
2.472 (Calc.)
CLEAVAGE: {110} very imperfect
Fracture conchoidal. Brittle.
HABIT: Crystals usually trapezohedral, often with {100} or {110}. Faces sometimes finely striated due to twinning. Also as disseminated grains or massive granular. Twinning on {110} common, repeated.
COLOR-LUSTER: Colorless, white, gray. Transparent to translucent. Vitreous. Streak uncolored.
MODE OF OCCURRENCE: Occurs in potassium-rich basic lavas. Found in the Leucite Hills and the Absaroka Range, Wyoming; in the Bearpaw and Highland Mountains, Montana; at Magnet Cove, Hot Spring County, Arkansas; and at Hamburg, Sussex County, New Jersey. Other notable occurrences are found on the shores of Vancouver Island, British Columbia, Canada; at numerous places in Italy; and at localities in France, Germany, Zaire, Uganda, Australia, and elsewhere.
BEST REF. IN ENGLISH: Deer, Howie, and Zussman, "Rock Forming Minerals," v. 4, p. 276–288, New York, Wiley, 1963.


# LEUCOAUGITE = Variety of augite low in iron


# LEUCOCHALCITE = Olivenite


# LEUCOGLAUCITE = Ferrinatrite


# LEUCOPHANE (Leucophanite)
$(Ca, Na)_2 BeSi_2 (O, F, OH)_7$

CRYSTAL SYSTEM: Orthorhombic (pseudotetragonal)
CLASS: 222
SPACE GROUP: $P2_1 2_1 2$
Z: 4
LATTICE CONSTANTS:
$a = 7.38$
$b = 9.96$
$c = 7.38$
3 STRONGEST DIFFRACTION LINES:
2.75 (100)
3.60 ( 50)
2.97 ( 50)
OPTICAL CONSTANTS:
$\alpha = 1.570 - 1.571$
$\beta = 1.595 - 1.594$
$\gamma = 1.596 - 1.598$
$(-)2V = 39° - 40°$

HARDNESS: 4
DENSITY: 2.96–2.98 (Meas.)
CLEAVAGE: {001} perfect
{100} distinct
{010} distinct
Fracture uneven. Brittle.
HABIT: Crystals tabular, pseudotetragonal; also spherulitic. Twinning on {001} or {110} common, sometimes as penetration twins.
COLOR-LUSTER: White, greenish yellow, yellow. Transparent to translucent. Vitreous.
MODE OF OCCURRENCE: Occurs in nepheline-syenite pegmatites at Mont St. Hilaire, Quebec, Canada; in the Julianehaab district, Greenland; in the Langesundfjord district, Norway; and associated with apatite, fluorite, polylithionite, and other minerals, in the Lovozero Massif, Kola Peninsula, USSR.
BEST REF. IN ENGLISH: Vlasov, K. A., "Mineralogy of Rare Elements," v. II, p. 129–134, Israel Program for Scientific Translations, 1966.


# LEUCOPHOENICITE
$Mn_7 Si_3 O_{12} (OH)_2$

CRYSTAL SYSTEM: Monoclinic
CLASS: 2/m
SPACE GROUP: $P2_1/a$
Z: 2
LATTICE CONSTANTS:
$a = 10.842$
$b = 4.826$
$c = 11.324$
$\beta = 103.93°$
3 STRONGEST DIFFRACTION LINES:
1.806 (100)
2.877 ( 90)
2.684 ( 80)
OPTICAL CONSTANTS:
$\alpha = 1.751$
$\beta = 1.771$
$\gamma = 1.782$
HARDNESS: 5½–6
DENSITY: 3.848 (Meas.)
CLEAVAGE: {001} imperfect
Brittle.
HABIT: Crystals varied in habit, small, rare; usually slender prismatic or thin tabular; often striated. Usually massive granular or as isolated grains. Twinning on {001} common; contact or interpenetrant.
COLOR-LUSTER: Light pink to deep pink, purplish red, brown. Transparent to translucent. Vitreous.
MODE OF OCCURRENCE: Leucophoenicite occurs as crystals in late stage open hydrothermal veins and as granular masses in ore and skarn at Franklin, Sussex County, New Jersey, in association with willemite, franklinite, calcite, idocrase, garnet, sussexite and other minerals.
BEST REF. IN ENGLISH: Palache, Charles, USGS Prof. Paper 180, p. 103–105 (1935). Cook, David, *Am. Min.*, **54**: 1392–1398 (1969). Moore, Paul B., *Am. Min.*, **55**: 1146–1166 (1970).

# LEUCOPHOSPHITE

$KFe_2^{3+}(OH)(H_2O)(PO_4)_2 \cdot H_2O$

CRYSTAL SYSTEM: Monoclinic (pseudo-orthorhombic)
CLASS: 2/m
SPACE GROUP: $P2_1/n$
Z: 4
LATTICE CONSTANTS:
  $a = 9.782$
  $b = 9.658$
  $c = 9.751$
  $\beta = 102.24°$
3 STRONGEST DIFFRACTION LINES:
  6.79  (100)
  5.99  ( 70)
  3.061 ( 70)
OPTICAL CONSTANTS:
  $\alpha = 1.707$
  $\beta = 1.721$
  $\gamma = 1.739$
  (+)2V
HARDNESS: 3½
DENSITY: 2.948 (Meas.)
        2.957 (Calc.)
CLEAVAGE: {100} excellent
Very brittle.
  HABIT: Crystals usually short prismatic with forms {$\bar{1}01$}, {$\bar{1}11$}, {011}, and {110} dominant. Also lamellar or as fine-grained chalk-like masses.
  COLOR-LUSTER: Pegmatites: Buff to chocolate brown greenish, reddish, reddish-brown. Vitreous. Transparent to translucent. Sedimentary origin: White to greenish. Dull.
  MODE OF OCCURRENCE: Occurs abundantly in superb crystals as much as 1 cm in length, and as lamellar masses several centimeters across, associated with rockbridgeite in altered triphylite nodules at the Tip-Top mine, Custer, South Dakota. Also in fine crystals at the White Elephant, Bull Moose, and Rock Ridge mines near Custer, and at the Gap Lode near Hill City, South Dakota. Found as minute crystals in vugs in frondelite at the Sapucaia pegmatite mine, Minas Geraes, Brazil; also as fine-grained chalk-like masses from two deposits, both of sedimentary origin, at Bomi Hill, western Liberia, and from Weelhamby Lake, Nighanboun Hills, Western Australia.
  BEST REF. IN ENGLISH: Lindberg, Mary Louise, *Am. Min.*, **42**: 214–221 (1957). Moore, Paul Brian, *Am. Min.*, **57**: 397–410 (1972).

# LEUCOSPHENITE

$BaNa_4Ti_2B_2Si_{10}O_{30}$

CRYSTAL SYSTEM: Monoclinic
CLASS: 2/m
SPACE GROUP: C2/m
Z: 2
LATTICE CONSTANTS:
  $a = 9.76–9.80$
  $b = 16.69–16.85$
  $c = 7.10–7.21$
  $\beta = 93°23'$
3 STRONGEST DIFFRACTION LINES:
  4.22 (100)
  8.45 ( 90)
  3.37 ( 70)
OPTICAL CONSTANTS:
  $\alpha = 1.645$
  $\beta = 1.661$
  $\gamma = 1.688$
  (+)2V = 77°
HARDNESS: 6½
DENSITY:  3.05–3.089 (Meas.)
        3.103 (Calc.)
CLEAVAGE: {010} distinct, {001} fair
Brittle.
  HABIT: Crystals tabular or wedge-shaped. Twinning on {001} common.
  COLOR-LUSTER: Colorless, white, pale blue. Transparent to translucent. Vitreous.
  MODE OF OCCURRENCE: Occurs widespread in the Green River formation of Tertiary age in Wyoming and Utah as well-formed crystals several millimeters in length, commonly associated with shortite or analcime. Also found at Mont St. Hilaire, Quebec, Canada; Narsarsuk, Greenland; South Yakutia, USSR; and the Alaiski massif, Turkestan.
  BEST REF. IN ENGLISH: Pabst, A., and Milton, Charles, *Am. Min.*, **57**: 1801–1822 (1972).

# LEUCOXENE = Synonym for sphene. Much "leucoxene" is anatase.

# LEVYNE (Levynite)

$(Na, Ca)_2(Al, Si)_9O_{18} \cdot 8H_2O$
Zeolite Group

CRYSTAL SYSTEM: Hexagonal
CLASS: $\bar{3}$ 2/m
SPACE GROUP: $R\bar{3}m$
Z: 6
LATTICE CONSTANTS:
  $a \simeq 13.3$
  $c \simeq 22.5$
3 STRONGEST DIFFRACTION LINES:
  4.10  (100)
  2.815 ( 80)
  8.19  ( 65)
OPTICAL CONSTANTS:
  $\omega = 1.496–1.505$
  $\epsilon = 1.491–1.500$
    (–)
HARDNESS: 4–4½
DENSITY: 2.09–2.16 (Meas.)
        2.121 (Calc.)
CLEAVAGE: Fracture uneven. Brittle.
  HABIT: Crystals thin tabular {0001}, or in sheaf-like aggregates.
  COLOR-LUSTER: Colorless, white, grayish, or tinted yellowish or reddish by impurities. Transparent to translucent. Vitreous.

MODE OF OCCURRENCE: Occurs as superb crystals in cavities in basalt in Grant County, Oregon; at North Table Mountain, Jefferson County, Colorado; and at localities in Greenland, Iceland, Faroe Islands, and Ireland.

BEST REF. IN ENGLISH: Deer, Howie, and Zussman, "Rock Forming Minerals," v. 4, p. 387–400, New York, Wiley, 1963.


**LEWISITE** = Titanian romeite


**LEWISTONITE** = Variety of hydroxylapatite


## LIBERITE
$Li_2BeSiO_4$

CRYSTAL SYSTEM: Monoclinic
CLASS: m
SPACE GROUP: Pn
Z: 2
LATTICE CONSTANTS:
  $a = 6.111$
  $b = 4.946$
  $c = 4.706$
3 STRONGEST DIFFRACTION LINES:
  3.799 (100)
  3.709 (100)
  2.581 (100)
OPTICAL CONSTANTS:
  $\alpha = 1.6220$
  $\beta = 1.6332$
  $\gamma = 1.6380$
  $(-)2V = 66°18'$
HARDNESS: ~ 7
DENSITY: 2.688 (Meas.)
CLEAVAGE: {010} perfect
                {100} distinct
                {001} distinct
Brittle.
HABIT: Minute aggregates occasionally showing pinacoidal faces.
COLOR-LUSTER: Pale-yellow to brown. Vitreous on cleavages, slightly oily on fractures.
MODE OF OCCURRENCE: Occurs in association with lepidolite, natrolite, eucryptite, hsianghualite, cassiterite, and scheelite in veins cutting ribbon rock (lepidolite-fluorite-magnetite) in tactites from the Nanling Ranges, south China.
BEST REF. IN ENGLISH: Chao, Ch'un-Lin, *Am. Min.*, **50**: 519 (1965).


## LIBETHENITE
$Cu_2PO_4OH$

CRYSTAL SYSTEM: Orthorhombic
CLASS: 2/m 2/m 2/m
SPACE GROUP: Pnnm
Z: 4
LATTICE CONSTANTS:
  $a = 8.10$
  $b = 8.45$
  $c = 5.91$
3 STRONGEST DIFFRACTION LINES:
  4.81 (100)
  2.63 (100)
  5.85 ( 90)
OPTICAL CONSTANTS:
  $\alpha = 1.701$
  $\beta = 1.743$
  $\gamma = 1.787$
  $(-)2V = $ nearly 90°
HARDNESS: 4
DENSITY: 3.97 (Meas.)
                3.93 (Calc.)
CLEAVAGE: {100} indistinct
                {010} indistinct
Fracture conchoidal to uneven.
HABIT: Crystals equant or short prismatic. As isolated crystals or as crusts.
COLOR-LUSTER: Light to dark green and blackish green; also light to dark olive green. Transparent to translucent. Vitreous to greasy.
MODE OF OCCURRENCE: Occurs chiefly as a secondary mineral in the oxidation zone of copper-bearing ore deposits. Found as excellent microcrystals at Inspiration and Castle Dome, Gila County, and in the Coronado mine, Greenlee County, Arizona; in the Santa Rita district, Grant County, New Mexico; at Yerington, Lyon County, Nevada; in glaucophane schist near Llanada, San Benito County, California; and at the Perkiomen mine, Montgomery County, Pennsylvania. It also occurs as crystals up to 6 mm in size at Libethen, near Neusohl, Roumania; and at localities in Chile, England, France, USSR, and Zaire.
BEST REF. IN ENGLISH: Palache, et al., "Dana's System of Mineralogy," 7th Ed., v. II, p. 862–864, New York, Wiley, 1951.


## LIEBIGITE
$Ca_2U(CO_3)_4 \cdot 10H_2O$

CRYSTAL SYSTEM: Orthorhombic
CLASS: mm2
SPACE GROUP: Bba2
Z: 8
LATTICE CONSTANTS:
  $a = 16.71$
  $b = 17.55$
  $c = 13.79$
3 STRONGEST DIFFRACTION LINES:
  6.81 (100)
  8.68 ( 90)
  5.40 ( 90)
OPTICAL CONSTANTS:
  $\alpha = 1.497$
  $\beta = 1.502$
  $\gamma = 1.539$
  $(+)2V = 40°$
HARDNESS: 2½–3

DENSITY: 2.41 (Meas.)
2.43 (Calc.)
CLEAVAGE: {100} distinct
HABIT: Crystals equant or short prismatic [001], generally indistinct with rounded edges and convex or vicinal faces. Usually as granular or scaly crusts and films, sometimes with small-botryoidal surface.
COLOR-LUSTER: Green to yellowish green, transparent to translucent. Vitreous, pearly on cleavage surfaces. Fluorescent green in ultraviolet light.
MODE OF OCCURRENCE: Occurs in Wyoming as a secondary mineral at the Lucky Mc mine in the Wind River Basin, at the Silver Cliff mine at Lusk, and at Pumpkin Buttes; in Utah at the Mi Vida mine, San Juan County, and at the Black Ape mine in the Yellow Cat district; in New Mexico at the Hanosh mines near Grants; and at Jim Thorpe, Pennsylvania. It also occurs at the Nicholson mine, Goldfields, Saskatchewan, Canada; as a coating on uraninite from near Adrianople, Turkey; in abundance in the oxidized zone of uraninite veins and as an efflorescence on mine walls at Joachimsthal, Bohemia; at Schmiedeberg, Eisleben, and Schneeberg, Germany; at Stripa, Sweden; and at Wheal Basset, Cornwall, England.
BEST REF. IN ENGLISH: Frondel, Clifford, USGS Bull. 1064, p. 108–112 (1958).

## LIKASITE
$Cu_6PO_4(NO_3)_2(OH)_7$

CRYSTAL SYSTEM: Orthorhombic
CLASS: 2/m 2/m 2/m
SPACE GROUP: Pmcb
Z: 2
LATTICE CONSTANTS:
 $a = 6.72$
 $b = 21.65$
 $c = 5.79$
OPTICAL CONSTANTS:
 $\alpha = ?$
 $\beta = 1.61$
 $\gamma = 1.69$
HARDNESS: Not determined.
DENSITY: 2.96–2.98 (Meas.)
CLEAVAGE: {010} perfect and easy
HABIT: Crystals tabular with $c${001} dominant; also $r${101}, $k${105}, $g${108}, $b${010}, $d${012}, $e${014}, and $f${018}.
COLOR-LUSTER: Blue.
MODE OF OCCURRENCE: Occurs as crystals implanted on cuprite or as blue masses in cuprite, surrounded by a crust of malachite and by malachite pseudomorphs after likasite, at the Likasi copper mine, Zaire.
BEST REF. IN ENGLISH: Schoep, A., Borchert, W., and Kohler, K., Am. Min., 40; 942 (1955).

## LILLIANITE
$Pb_3Bi_2S_6$

CRYSTAL SYSTEM: Orthorhombic
CLASS: 2/m 2/m 2/m

SPACE GROUP: Bbmm
Z: 4
LATTIC CONSTANTS:
 (synthetic)
 $a = 13.5$ kX
 $b = 20.7$
 $c = 4.15$
3 STRONGEST DIFFRACTION LINES:
 (natural)   (synthetic)
 3.51 (100)   3.52 (100)
 2.92 ( 90)   2.06 (100)
 2.14 ( 90)   2.92 ( 90)
HARDNESS: 2–3
DENSITY: 7.06–7.14 (Meas. synthetic)
7.09 (Calc.)
CLEAVAGE: None. Fracture conchoidal.
HABIT: Elongated platy crystals; fibrous.
COLOR-LUSTER: Steel gray. Opaque. Metallic. Streak black.
MODE OF OCCURRENCE: Occurs at the Spokoinoe wolframite deposit, Eastern Transbaikal, and at the Bukuinsk deposit, Czechoslovakia. Also reported from the Monticello and other mines near the mouth of Cleveland Gulch, Hinsdale County, and in the Silverton district, San Juan County, Colorado; also at deposits in Finland and Sweden.
BEST REF. IN ENGLISH: Syritso, L. F., and Senderova, V. M., Am. Min., 50: 811 (1965). Am. Min., 54: 579 (1969).

## LIMAITE = Stannian gahnite

## LIME
CaO

CRYSTAL SYSTEM: Cubic
CLASS: 4/m $\bar{3}$ 2/m
SPACE GROUP: Fm3m
Z: 4
LATTICE CONSTANT:
 $a = 4.797$ (synthetic)
3 STRONGEST DIFFRACTION LINES:
 2.405 (100)
 1.701 ( 45)
 2.778 ( 34)
OPTICAL CONSTANT:
 $N = 1.838$
HARDNESS: 3½
DENSITY: 3.345
CLEAVAGE: {001} perfect
{011} parting
HABIT: Massive.
COLOR-LUSTER: White.
MODE OF OCCURRENCE: Occurs in blocks of calcareous rocks enclosed in the lava at Mt. Vesuvius, Italy.
BEST REF. IN ENGLISH: Palache, et al., "Dana's System of Mineralogy," 7th Ed., v. I, p. 503, New York, Wiley, 1944.

## LIMONITE = General term for hydrous ferric iron oxides

# LINARITE
$PbCuSO_4(OH)_2$

CRYSTAL SYSTEM: Monoclinic
CLASS: 2/m
SPACE GROUP: $P2_1/m$
Z: 2
LATTICE CONSTANTS:
$a = 9.66$ kX
$b = 5.64$
$c = 4.67$
$\beta = 102°48'$
3 STRONGEST DIFFRACTION LINES:
3.12 (100)
3.53 ( 70)
1.79 ( 60)
OPTICAL CONSTANTS:
$\alpha = 1.809$
$\beta = 1.838$ (Na)
$\gamma = 1.859$
$(-)2V = 80°$
HARDNESS: 2½
DENSITY: 5.30 (Meas.)
5.319 (Calc.)
CLEAVAGE: {100} perfect
{001} imperfect
Fracture conchoidal. Brittle.
HABIT: Crystals elongated, usually thin tabular, ranging up to 10 cm in length; often in groups or isolated single crystals; also as crusts and aggregates of prismatic crystals in random orientation. Twinning on {100} common.
COLOR-LUSTER: Dark azure blue. Transparent to translucent. Vitreous to subadamantine. Streak pale blue.
MODE OF OCCURRENCE: Occurs widespread as a secondary mineral in the oxidation zone of lead and copper ore deposits. Found in the United States as superb large crystals at the Mammoth mine, Tiger, Pinal County, and other places in Arizona; as fine specimens at the Blanchard mine, Socorro County, New Mexico; at numerous localities in Inyo County and other deposits in California; in Montana, Utah, Idaho, and Nevada. It also occurs as fine specimens at many localities in England, Scotland, Spain, Germany, Sardinia, USSR, Canada, Argentina, Peru, Chile, Japan, Australia, and at Tsumeb, South West Africa.
BEST REF. IN ENGLISH: Palache, et al., "Dana's System of Mineralogy," 7th Ed., v. II, p. 553-555, New York, Wiley, 1951.

# LINDACKERITE (Inadequately described mineral)
$H_2Cu(AsO_4)_4 \cdot 8-9H_2O$

CRYSTAL SYSTEM: Monoclinic
3 STRONGEST DIFFRACTION LINES:
10.2 (100)
3.19 ( 80)
3.95 ( 60)
OPTICAL CONSTANTS:
$\alpha = 1.629$
$\beta = 1.662$
$\gamma = 1.727$
$(+)2V \sim 73°$ (Calc.)

HARDNESS: 2-2½
DENSITY: 3.20 (Meas.)
CLEAVAGE: {010} perfect
HABIT: As tiny lath-like crystals grouped in rosettes or crusts.
COLOR-LUSTER: Apple green. Vitreous. Streak white to pale green.
MODE OF OCCURRENCE: Occurs as a secondary mineral associated with annabergite, erythrite, and other secondary minerals at Joachimsthal (Jachymov), Bohemia, Czechoslovakia.
BEST REF. IN ENGLISH: Palache, et al., "Dana's System of Mineralogy," 7th Ed., v. II, p. 1007-1008, New York, Wiley, 1951.

# LINDGRENITE
$Cu_3(MoO_4)_2(OH)_2$

CRYSTAL SYSTEM: Monoclinic
CLASS: 2/m
SPACE GROUP: $P2_1/n$
Z: 2
LATTICE CONSTANTS:
$a = 5.613$
$b = 14.03$
$c = 5.405$
$\beta = 98°23'$
3 STRONGEST DIFFRACTION LINES:
3.50 (100)
4.15 ( 55)
2.67 ( 40)
OPTICAL CONSTANTS:
$\alpha = 1.930$
$\beta = 2.002$
$\gamma = 2.020$
$(-)2V = 71°$
HARDNESS: 4½
DENSITY: 4.26 (Meas.)
4.29 (Calc.)
CLEAVAGE: {010} perfect
{101} very poor
{100} very poor
HABIT: Crystals usually tabular or platy; rarely acicular; sometimes doubly terminated. {010} striated parallel [001]. Also massive, platy.
COLOR-LUSTER: Green; yellowish green in thin plates. Transparent to translucent. Greasy.
MODE OF OCCURRENCE: Occurs on specimens of tactite from the dump of the upper tunnel of the Helena mine, about four miles northeast of Cuprum, Adams County, Idaho; in porphyry copper deposits at Inspiration, Gila County, Arizona; and with antlerite, quartz, limonite, and hematite at Chuquicamata, Chile.
BEST REF. IN ENGLISH: Palache, et al., "Dana's System of Mineralogy," 7th Ed., v. II, p. 1094-1095, New York, Wiley, 1951.

# LINDSTRÖMITE (Inadequately described mineral)
$CuPbBi_3S_6$ (?)

HODGKINSONITE on zincite   Franklin, Sussex County, New Jersey.

HONESSITE   Hall's Gap, Lincoln County, Kentucky.

HOPEITE   Broken Hill, Zambia.

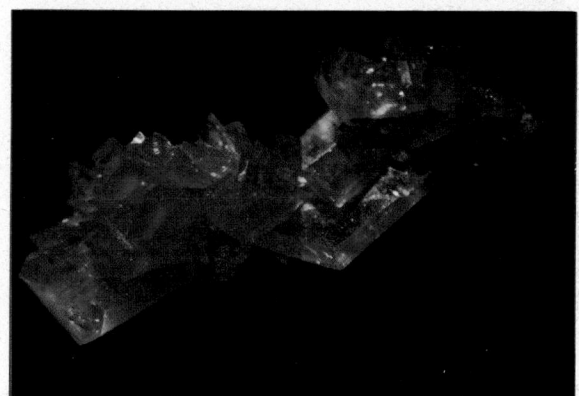

HOPEITE   Broken Hill, Zambia.

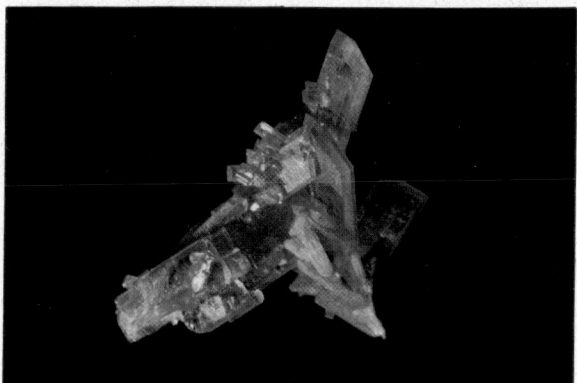

HOPEITE   Broken Hill, Zambia.

HORNBLENDE   Monte Somma, Italy.

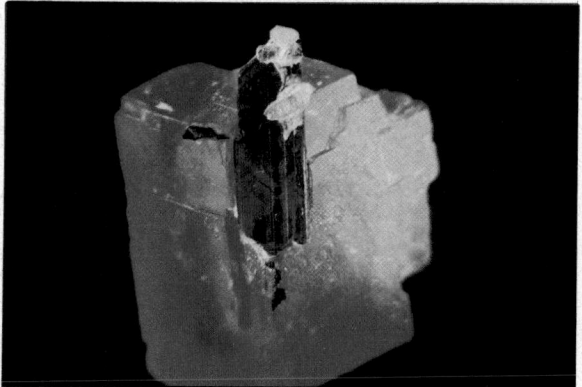

HUEBNERITE on rhodochrosite   Sweet Home mine, Alma, Park County, Colorado.

HUEBNERITE   Silverton, San Juan County, Colorado.

HUMITE   Vesuvius, Italy.

HUREAULITE   Tip Top mine, Custer County, South Dakota.

HUTCHINSONITE with orpiment   Quiruvilca, Peru.

HUTCHINSONITE   Lengenbach quarry, Binnental, Switzerland.

HYPERSTHENE   Diamond Lake, Douglas County, Oregon.

IANTHINITE   Shinkolobwe, Katanga, Zaire.

IDOCRASE    Asbestos, Quebec, Canada.

IDOCRASE    Asbestos, Quebec, Canada.

IDOCRASE    Asbestos, Quebec, Canada.

ILMENITE    Francon quarry, Montreal, Quebec, Canada.

ILSEMANNITE    Quinn Draw, Badlands, South Dakota.

ILVAITE    Laxey mine, South Mountain, Owyhee County, Idaho.

IODYRITE    Broken Hill, New South Wales, Australia.

INESITE    Hayfork, Trinity County, California.

67

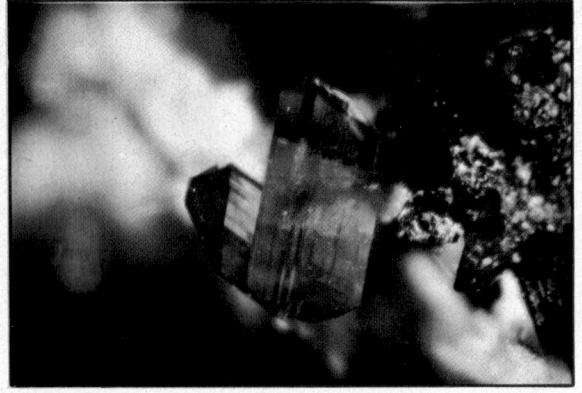

JAHNSITE   Tip Top mine, Custer County, South Dakota.

JAHNSITE   Tip Top mine, Custer County, South Dakota.

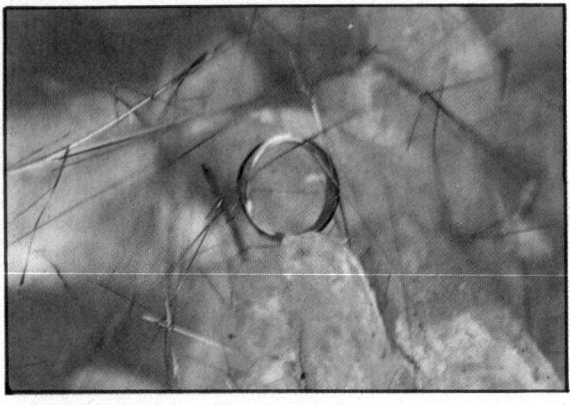

JAMESONITE   Felsobanya, Roumania.

JAROSITE   Laurium, Greece.

JOAQUINITE   San Benito County, California.

JOAQUINITE   San Benito County, California.

JOHANNSENITE on axinite   Franklin, Sussex County, New Jersey.

JORDANITE   Lengenbach quarry, Binnental, Switzerland.

KAMMERERITE   Kop Duglari, near Erzerum, Turkey.

KAMMERERITE   Texas, Lancaster County, Pennsylvania.

KASOLITE   Musonoi, Katanga, Zaire.

KASOLITE   Musonoi, Katanga, Zaire.

KASOLITE   Musonoi, Katanga, Zaire.

KASOLITE on torbernite   Shinkolobwe, Katanga, Zaire.

KERMESITE with stibnite   Braunsdorf, Saxony, Germany.

KINOITE   Santa Rita Mountains, Pima County, Arizona.

KLEBELSBERGITE with stibnite   Felsobanya, Roumania.

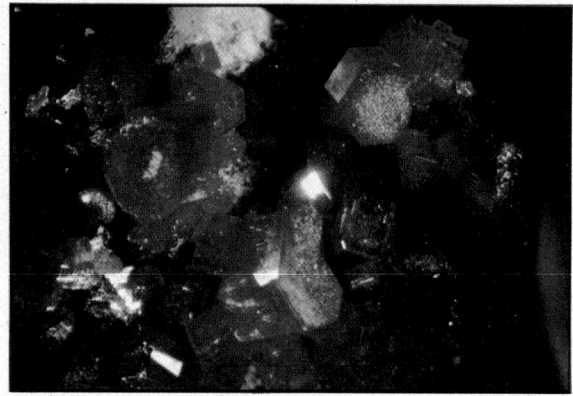

KLEINITE   Terlingua, Brewster County, Texas.

KOLBECKITE   Clay Canyon, Fairfield, Utah County, Utah.

KÖTTIGITE   Mapimi, Durango, Mexico.

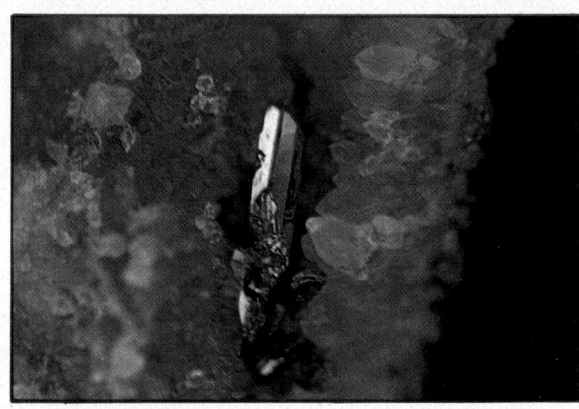

KRENNERITE   Cripple Creek, Teller County, Colorado.

KROEHNKITE   Chuquicamata, Chile.

KROEHNKITE   Chuquicamata, Chile.

KRYZHANOVSKITE with siderite   Big Chief Mine,
Pennington County, South Dakota.

KRYZHANOVSKITE with siderite   Big Chief Mine,
Pennington County, South Dakota.

KYANITE   Helderman's Mountain, Lincoln County, North
Carolina.

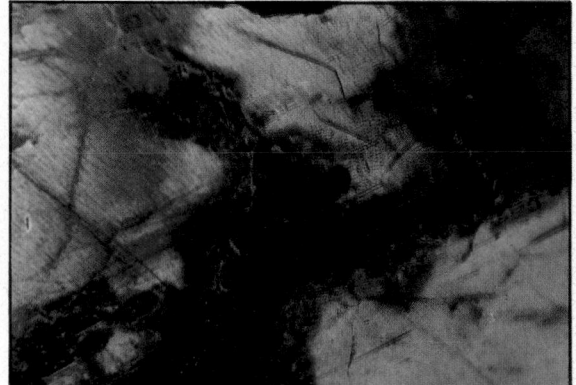

LABRADORITE   Labrador, Canada.

LABUNTSOVITE   Mont St. Hilaire, Quebec, Canada.

LANARKITE   Leadhills, Lanarkshire, Scotland.

LANGBANITE   Langban, Sweden.

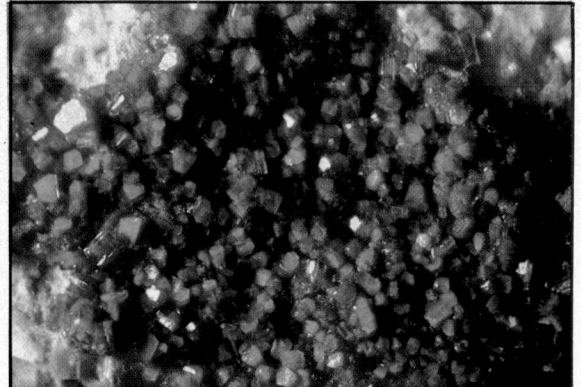

LANGITE   St. Just, Cornwall, England.

LARSENITE on clinohedrite   Franklin, Sussex County, New Jersey.

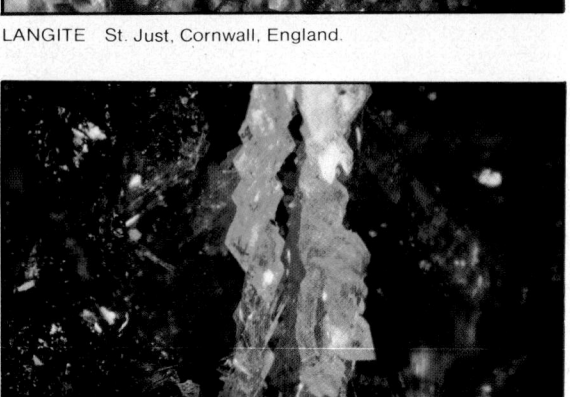

LAUEITE (parallel growth) with strunzite   Hagendorf, Bavaria, Germany.

LAUEITE   White Elephant mine, Custer County, South Dakota.

LAUEITE   Fletcher mine, North Groton, New Hampshire.

LAURIONITE   Laurium, Greece.

LAVENDULAN   Gold Hill, Tooele County, Utah.

LAWSONITE   Cazadero, Sonoma County, California.

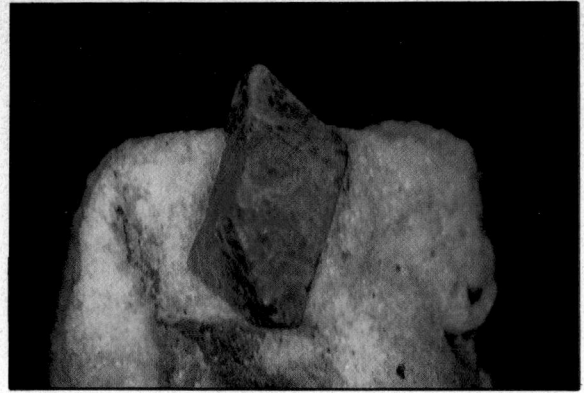

LAZULITE   Graves Mountain, Lincoln County, Georgia.

LEAD   Langban, Sweden.

LEADHILLITE   Campbell Shaft, Bisbee, Cochise County, Arizona.

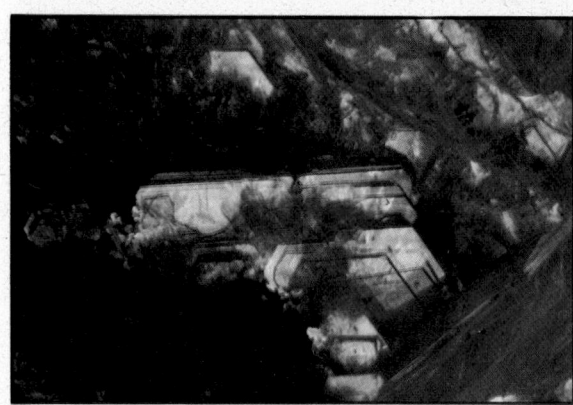

LEADHILLITE   Campbell Shaft, Bisbee, Cochise County, Arizona.

LEADHILLITE   Campbell Shaft, Bisbee, Cochise County, Arizona.

LEGRANDITE   Mapimi, Durango, Mexico.

LEGRANDITE   Mapimi, Durango, Mexico.

LEGRANDITE   Mapimi, Durango, Mexico.

LEMOYNITE   Mont St. Hilaire, Quebec, Canada.

LEPIDOLITE on elbaite   Newry, Oxford County, Maine.

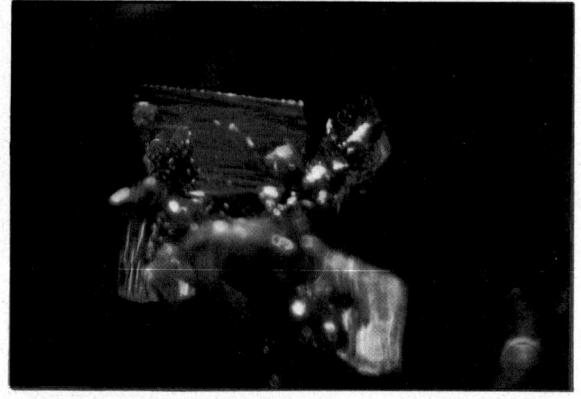

LEPIDOCROCITE   Siegen, Westphalia, Germany.

LEUCOPHANITE   Mont St. Hilaire, Quebec, Canada.

LEUCOPHOENICITE   Franklin, Sussex County, New Jersey.

LEUCOPHOSPHITE with laueite   Fletcher mine, North Groton, New Hampshire.

LEUCOPHOSPHITE   White Elephant mine, Custer County, South Dakota.

LEUCOPHOSPHITE   Tip Top mine, Custer County, South Dakota.

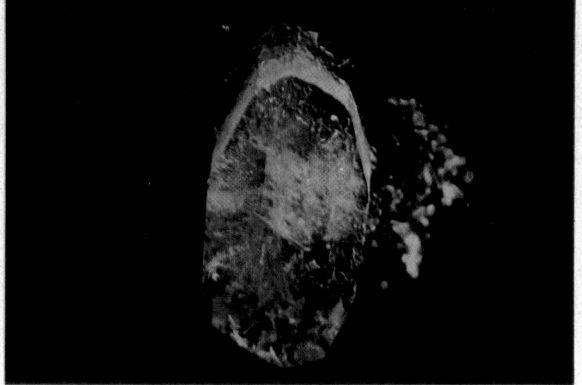

LEUCOSPHENITE   Mont St. Hilaire, Quebec, Canada.

LEVYNITE   Near Mt. Vernon, Grant County, Oregon.

LEVYNITE   Beech Creek, Grant County, Oregon.

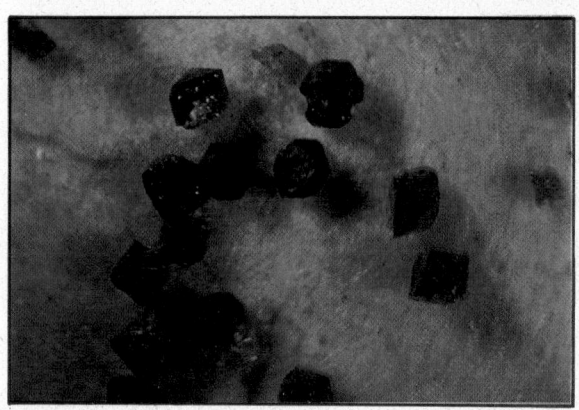

LEWISITE   Tripuhy, Minas Geraes, Brazil.

LIBETHENITE   Libethen, Roumania.

LIBETHENITE on hallyosite   Silver Bell, Pinal County, Arizona.

LINARITE   Mammoth-St. Anthony mine, Pinal County, Arizona.

LINARITE   Bingham, Socorro County, New Mexico.

75

LIROCONITE   Wheal Gorland, St. Day, Cornwall, England.

LIROCONITE   Wheal Gorland, St. Day, Cornwall, England.

LIROCONITE   Wheal Gorland, St. Day, Cornwall, England.

LORANDITE   Allchar, Macedonia, Yugoslavia.

LUDLAMITE   Blackbird mine, Lemhi County, Idaho.

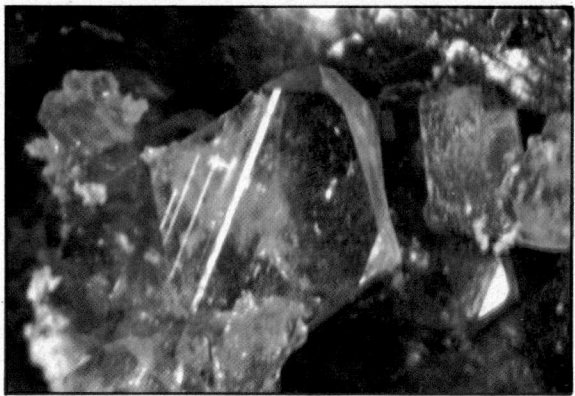

LUDLAMITE   Big Chief mine, Glendale, Pennington County, South Dakota.

LUDLOCKITE with calcite   Tsumeb, South West Africa.

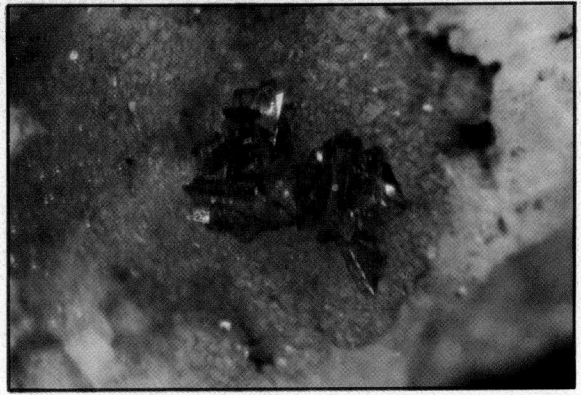

MACKAYITE   Goldfield, Esmeralda County, Nevada.

MAGNETITE with biotite   Vesuvius, Italy.

MAGNETITE   Nordmark, Vermland, Sweden.

MALACHITE   Mapimi, Durango, Mexico.

MALACHITE   Bisbee, Cochise County, Arizona.

MALACHITE   Mapimi, Durango, Mexico.

MALACHITE   Horhausen, Rhineland, Germany.

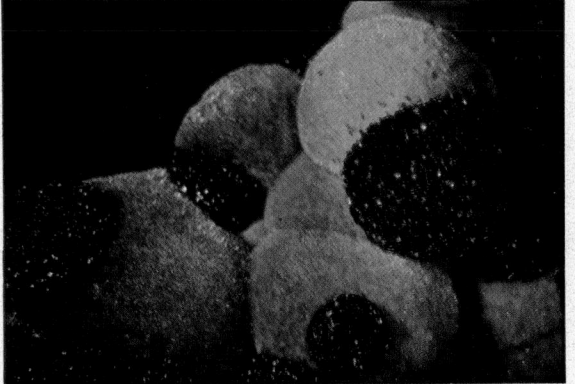

MALACHITE with azurite   Tsumeb, South West Africa.

MANGANITE   Pribram, Bohemia, Czechoslovakia.

MANGANITE· Lerida, Spain.

MARCASITE   Racine, Racine County, Wisconsin.

MARCASITE   Dundas, Ontario, Canada.

MARSHITE on cuprite, coated by aurichalcite   Broken Hill,
New South Wales, Australia.

MARTHOZITE   Musonoi, Kolwezi, Katanga, Zaire.

MATLOCKITE   Laurium, Greece.

MAUCHERITE   Eisleben, Thuringia, Germany.

MASUYITE with ruthertordine   Shinkolobwe mine, Katanga, Zaire.

MEIONITE   Vesuvius, Italy.

MEIONITE   Vesuvius, Italy.

MEIONITE   Vesuvius, Italy.

MELANOPHLOGITE   Racalmuto, Sicily.

MELILITE with nepheline   Eifel district, Rhineland, Germany.

79

MENEGHINITE   Bottino, Tuscany, Italy.

MERCURY   Socrates mine, Sonoma County, California.

MESOLITE with heulandite   Rock Creek, Stevenson, Washington.

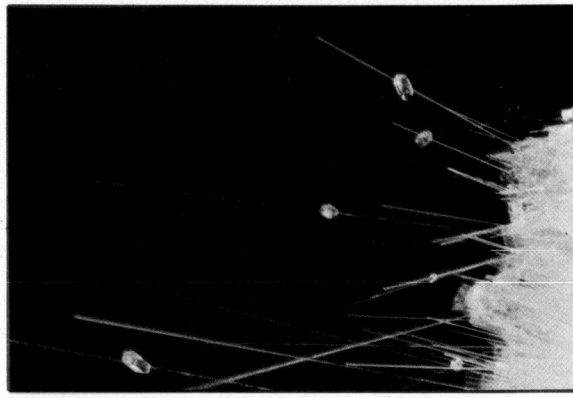

MESOLITE with calcite   Bear Creek quarry, Drain, Oregon.

MESSELITE   Big Chief mine, Glendale, Pennington County, South Dakota.

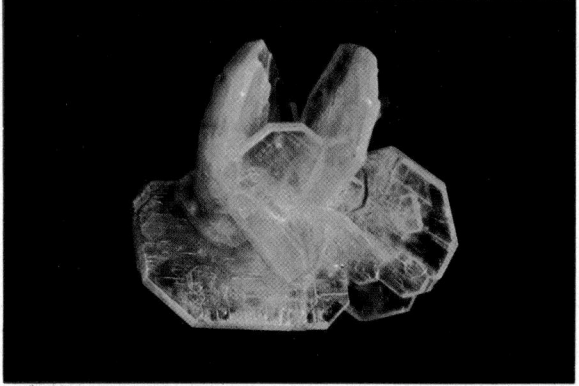

META-AUTUNITE   Carretona, Spain.

META-AUTUNITE (short-wave ultraviolet light)   Carretona, Spain.

CRYSTAL SYSTEM: Orthorhombic (?)
HARDNESS: 3–3½
DENSITY: 7.01 (Meas.)
CLEAVAGE: {100} good
{010} good
{230} good
Fracture conchoidal to uneven.
HABIT: Crystals prismatic, without terminal faces; vertically striated.
COLOR-LUSTER: Lead gray. Opaque. Metallic. Streak black.
MODE OF OCCURRENCE: Occurs as crystals up to 1 cm long on quartz at Gladhammar, Kalmar, Sweden.
BEST REF. IN ENGLISH: Palache, et al., "Dana's System of Mineralogy," 7th Ed., v. I, p. 459–460, New York, Wiley, 1944.

# LINNAEITE
$Co_3S_4$

CRYSTAL SYSTEM: Cubic
CLASS: 4/m $\bar{3}$ 2/m
SPACE GROUP: Fd3m
Z: 8
LATTICE CONSTANT:
$a = 9.401$
3 STRONGEST DIFFRACTION LINES:
2.83 (100)
1.67 ( 80)
2.36 ( 70)
HARDNESS: 4½–5½
DENSITY: 4.8–5.0 (Meas.)
4.88 (Calc.)
CLEAVAGE: {001} imperfect
Fracture subconchoidal to uneven.
HABIT: Crystals usually octahedral. Commonly massive, compact to granular. Twinning on {111}.
COLOR-LUSTER: Light gray to steel gray; rapidly tarnishes copper red or violet-gray. Opaque. Metallic.
MODE OF OCCURRENCE: Occurs in hydrothermal vein deposits commonly associated with other sulfide minerals. Found as microscopic crystals at the Klau quicksilver mine, Santa Lucia Range, San Luis Obispo County, California, and at the Springfield and Mineral Hill mines, Carroll County, Maryland. It also occurs in the Siegen district of Westphalia, Germany; at Gladhammar and elsewhere in Sweden; and as fine crystals at Katanga, Zaire, and at N'Kana, Zambia.
BEST REF. IN ENGLISH: Palache, et al., "Dana's System of Mineralogy," 7th Ed., v. I, p. 262–265, New York, Wiley, 1944.

# LIPSCOMBITE
$(Fe^{2+}, Mn^{2+})Fe_2^{3+}(PO_4)_2(OH)_2$

CRYSTAL SYSTEM: Tetragonal
CLASS: 422
SPACE GROUP: $P4_12_12$
Z: 4

LATTICE CONSTANTS:
$a = 7.40$
$c = 12.81$
3 STRONGEST DIFFRACTION LINES:
3.314 (100)
3.206 ( 60)
1.656 ( 40)
DENSITY: 3.66 (Meas.)
3.706 (Calc.)
HABIT: Aggregates of microscopic crystals.
COLOR-LUSTER: Olive green to black. Splendent. Opaque.
MODE OF OCCURRENCE: Found intimately intergrown with cyrilovite, leucophosphite, and metastrengite in vugs in the alkali-altered zone in frondelite at the Sapucaia pegmatite mine, Municipio of Counselheiro Pena, Minas Geraes, Brazil, and has been identified from other granitic pegmatites.
BEST REF. IN ENGLISH: Lindberg, M. L., *Am. Min.*, 47: 353–359 (1962). *Am. Min.*, 55: 135–169 (1970).

# LIROCONITE
$Cu_2Al(As,P)O_4(OH)_4 \cdot 4H_2O$

CRYSTAL SYSTEM: Monoclinic
CLASS: 2/m
SPACE GROUP: I2/a
Z: 4
LATTICE CONSTANTS:
$a = 12.67$ kX
$b = 7.55$
$c = 9.86$
$\beta = 91°23'$
3 STRONGEST DIFFRACTION LINES:
6.46 (100)
3.01 (100)
5.95 ( 90)
OPTICAL CONSTANTS:
$\alpha = 1.612$
$\beta = 1.652$
$\gamma = 1.675$
(–)2V ~ 72° (Calc.)
HARDNESS: 2–2½
DENSITY: 2.926–3.01 (Meas.)
2.95 (Calc. As:P = 3:1)
CLEAVAGE: {110} indistinct
{011} indistinct
Fracture conchoidal to uneven. Brittle.
HABIT: Crystals flattened and wedge-shaped with forms $m$ {110} and $e$ {011} nearly equally developed; faces of both striated parallel to intersection edge. Crystals occur singly, up to 1 cm in length, and as subparallel groups; also coarsely granular.
COLOR-LUSTER: Sky blue to verdigris green. Transparent to translucent. Vitreous to resinous.
MODE OF OCCURRENCE: Occurs very sparingly in the oxidized zone of copper deposits associated with limonite, azurite, malachite, cuprite, clinoclasite, olivenite, and chalcophyllite. Found at several mines in Cornwall and Devonshire, England; at several localities in Germany; Urals,

USSR; Herrengrund, Czechoslovakia; and with linarite and caledonite at the Cerro Gordo mine, Inyo County, California.

BEST REF. IN ENGLISH: Palache, et al., "Dana's System of Mineralogy," 7th Ed., v. 2, p. 921–922, New York, Wiley, 1951. Berry, L. G., *Am. Min.*, **36**: 484–503 (1951).


## LISKEARDITE (Inadequately described mineral)
(Al, Fe)$_3$AsO$_4$(OH)$_6$ · 5H$_2$O

CRYSTAL SYSTEM: Monoclinic or orthorhombic
3 STRONGEST DIFFRACTION LINES:
  17.6 (100)
  3.33 (100)
  8.65 ( 80)
OPTICAL CONSTANTS:
  $\alpha$ = 1.661
  $\beta$ = 1.675
  $\gamma$ = 1.689
  (+)2V = nearly 90°
HARDNESS: Not determined. Soft.
DENSITY: 3.011 (Meas.)
CLEAVAGE: One, parallel to elongation.
HABIT: As fibrolamellar aggregates of crystals, and as thin crusts.
COLOR-LUSTER: White; often tinted greenish, bluish, or brownish white.
MODE OF OCCURRENCE: Occurs coating quartz, pyrite, and other minerals at Liskeard, Cornwall, England. It also is found at Cap Garonne, Var, France.
BEST REF. IN ENGLISH: Palache, et al., "Dana's System of Mineralogy," 7th Ed., v. II, p. 924, New York, Wiley, 1951.


## LITHARGE
PbO
Dimorphous with massicot

CRYSTAL SYSTEM: Tetragonal
CLASS: 4/m 2/m 2/m
SPACE GROUP: P4/nmm
Z: 2
LATTICE CONSTANTS:
  $a$ = 3.986
  $c$ = 5.011
3 STRONGEST DIFFRACTION LINES:
  3.115 (100)
  2.809 ( 62)
  1.872 ( 37)
OPTICAL CONSTANTS:
  $\omega$ = 2.665
  $\epsilon$ = 2.535 (Li)
  (−)
HARDNESS: 2
DENSITY: 9.355
CLEAVAGE: {110} perfect
HABIT: As alteration borders on massicot, and as crusts. Synthetic crystals tabular.
COLOR-LUSTER: Red. Transparent. Oily to dull.
MODE OF OCCURRENCE: Occurs in Inyo, Kern, and

San Bernardino counties, California; in the Mineral Hill district, Blaine County, Idaho; and probably in many other oxidized lead deposits throughout the world.

BEST REF. IN ENGLISH: Palache, et al., "Dana's System of Mineralogy," 7th Ed., v. I, p. 514–515, New York, Wiley, 1944.


## LITHIDIONITE = A mixture of quartz and carbonates


## LITHIONITE = Synonym for lepidolite


## LITHIOPHILITE
Li(Mn$^{2+}$, Fe$^{2+}$)PO$_4$
Lithiophilite-Triphylite Series

CRYSTAL SYSTEM: Orthorhombic
CLASS: 2/m 2/m 2/m
SPACE GROUP: Pmnb
Z: 4
LATTICE CONSTANTS:
  $a$ = 6.05
  $b$ = 10.32
  $c$ = 4.71
3 STRONGEST DIFFRACTION LINES:
  2.531 (100)
  3.01 ( 90)
  3.469 ( 50)
OPTICAL CONSTANTS:
  $\alpha$ = 1.669
  $\beta$ = 1.673
  $\gamma$ = 1.682
  (+)2V ~ 65°
HARDNESS: 4–5
DENSITY: 3.34 (Meas. Mn:Fe = 9:1)
        3.54 (Calc. LiMnPO$_4$)
CLEAVAGE: {100} nearly perfect
         {010} imperfect
         {011} interrupted
Fracture subconchoidal to uneven.
HABIT: As large anhedral or subhedral crystals with rough faces. Usually massive, cleavable.
COLOR-LUSTER: Light reddish brown, yellowish brown, salmon; often externally brownish to blackish due to alteration. Translucent to transparent. Resinous to vitreous. Streak colorless to dirty white.
MODE OF OCCURRENCE: Occurs as a primary mineral in granite pegmatites, often partly or completely altered to a wide variety of secondary phosphate minerals. Found at several localities in the Black Hills, South Dakota, especially at the Custer Mountain Lode, Custer district. It also occurs at Branchville and Portland, Connecticut; at several places in Oxford County, Maine; at Pala, San Diego County, California; in Canada; in the Karibib area, and in Namaqualand, South Africa; at Mangualde, Portugal; and in Argentina.
BEST REF. IN ENGLISH: Palache, et al., "Dana's System of Mineralogy," 7th Ed., v. II, p. 665–669, New York, Wiley, 1951.

# LITHIOPHORITE

$(Al, Li)MnO_2(OH)_2$

CRYSTAL SYSTEM: Monoclinic
CLASS: 2/m
SPACE GROUP: C2/m
Z: 2
LATTICE CONSTANTS:
  $a = 5.06$
  $b = 2.91$
  $c = 9.55$
  $\beta = 100°30'$
3 STRONGEST DIFFRACTION LINES:
  4.71 (100)
  2.37 ( 70)
  1.88 ( 70)
  HARDNESS: 3
  DENSITY: 3.14–3.4 (Meas.)
          3.33 (Calc.)
  CLEAVAGE: {001} perfect
  HABIT: Massive, compact, botryoidal; also as fine scales and dendritic.
  COLOR-LUSTER: Bluish black to black. Opaque. Dull to metallic. Streak blackish gray to black.
  MODE OF OCCURRENCE: Occurs in the Artillery Mountains area, Arizona; at White Oak Mountain, Tennessee; at Charlottesville, Virginia; in the Schneeberg district, Saxony, Germany; at Salm Chateau, Belgium; in Silesia, Poland; and at Gloucester Farm, Postmasburg, South Africa.
  BEST REF. IN ENGLISH: Ramsdell, L. S., *Am. Min.*, **17**: 143–149 (1932). *Am. Min.*, **52**: 1545–1549 (1967).

# LITHIOPHOSPHATE

$Li_3PO_4$

CRYSTAL SYSTEM: Orthorhombic
CLASS: mm2
SPACE GROUP: Pmn2₁
Z: 2
LATTICE CONSTANTS:
  $a = 6.1155$
  $b = 5.2340$ (synthetic)
  $c = 4.8452$
3 STRONGEST DIFFRACTION LINES:
  3.965 (100)
  2.635 (100)
  3.794 ( 90)
OPTICAL CONSTANTS:
  $\alpha = 1.550-1.553$
  $\beta = 1.557-1.558$
  $\gamma = 1.566-1.567$
  (+)2V = 69°–80°
HARDNESS: 4
DENSITY: 2.478 (Meas.)
        2.479 (Calc.)
CLEAVAGE: {010} perfect
          {110} distinct
  HABIT: Crystals elongated parallel to *b* and superficially resemble topaz, up to 1 cm in length. Forms:

{110}, {101}, {010}, {013}, and {001}. As masses up to 5.0 × 9.0 cm in size.
  COLOR-LUSTER: Colorless, white, light rose. Vitreous.
  MODE OF OCCURRENCE: Occurs as a hydrothermal replacement of montebrasite in the central zone of a pegmatite in amphibolite on the Kola Peninsula, USSR; also found as crystals and masses on quartz and albite crystal druses which line open fissures in massive albite-microcline-quartz-spodumene-muscovite pegmatite at Kings Mountain, Cleveland County, North Carolina.
  BEST REF. IN ENGLISH: Matias, V. V., and Bondareva, A. M., *Am. Min.*, **42**: 585 (1957). White, John S. Jr., *Am. Min.*, **54**: 1467–1469 (1969).

# LIVEINGITE (Rathite II, see rathite)

$Pb_9As_{13}S_{28}$

CRYSTAL SYSTEM: Monoclinic
CLASS: 2
SPACE GROUP: P2₁
Z: 4
LATTICE CONSTANTS:
  $a = 8.43$
  $b = 70.70$
  $c = 7.91$
  $\beta \simeq 90°$
HARDNESS: 3
DENSITY: 5.3 (Meas.)
CLEAVAGE: Not determined.
HABIT: As minute crystals. Twinning on {100}.
COLOR-LUSTER: Lead gray. Opaque. Metallic.
MODE OF OCCURRENCE: Found as two small crystals at the Lengenbach quarry in the Binnental, Valais, Switzerland.
  BEST REF. IN ENGLISH: Palache, et al., "Dana's System of Mineralogy," 7th Ed., v. I, p. 462, New York, Wiley, 1944.

# LIVINGSTONITE

$HgSb_4S_8$

CRYSTAL SYSTEM: Triclinic or monoclinic
CLASS: $\bar{1}$ or 2/m
SPACE GROUP: P$\bar{1}$ or A2/a
Z: 1 (triclinic), 8 (monoclinic)
LATTICE CONSTANTS:

| | |
|---|---|
| $a = 7.65$ | $a = 30.25$ |
| $b = 10.82$ or | $b = 3.98$ |
| $c = 3.99$ | $c = 21.60$ |
| $\alpha = 99°12\frac{1}{2}'$ | $\beta = 104°10'$ |
| $\beta = 102°01'$ | |
| $\gamma = 73°48'$ | |

3 STRONGEST DIFFRACTION LINES:
  2.99 (100)
  3.73 ( 70)
  3.47 ( 70)
OPTICAL CONSTANTS:
  $\alpha > 2.72$
  $\beta \approx 3$
  $\gamma = ?$

Biaxial; probably negative
HARDNESS: 2
DENSITY: 5.00 (Meas.)
    4.88 (Calc.)
CLEAVAGE: {010} perfect
        {100} perfect
        {001} poor
Flexible.

HABIT: Minute elongated needles; also columnar to fibrous massive; in globular masses of interlaced needles.

COLOR-LUSTER: Blackish gray. Nearly opaque. Metallic to adamantine. Streak red.

MODE OF OCCURRENCE: Occurs in a matrix of calcite and gypsum associated with stibnite, cinnabar, sulfur, and valentinite at Huitzuco, Guerrero, and with gypsum and sulfur at Guadalcázar, San Luis Potosi, Mexico.

BEST REF. IN ENGLISH: Palache, et al., "Dana's System of Mineralogy," 7th Ed., v. I, p. 485–486, New York, Wiley, 1944.    Gorman, D. H., *Am. Min.*, **36**: 480–483 (1951).

## LIZARDITE
$Mg_3Si_2O_5(OH)_4$
Serpentine Group

CRYSTAL SYSTEM: Monoclinic
CLASS: m
SPACE GROUP: Cm
Z: 2
LATTICE CONSTANTS:
    $a = 5.31$
    $b = 9.2$
    $c = 7.31$
    $\beta \simeq 90°$
3 STRONGEST DIFFRACTION LINES:
    7.4 (100)
    2.51 (100)
    4.6 ( 80)
OPTICAL CONSTANTS:
    $\alpha = 1.538$–$1.554$
    $\beta = ?$
    $\gamma = 1.546$–$1.560$
    $(-)2V = ?$
HARDNESS: 2½
DENSITY: ~ 2.55 (Meas.)
    2.58 (Calc.)
CLEAVAGE: {001} perfect
HABIT: Massive; fine to coarse grained; compact. Also in small scales.

COLOR-LUSTER: Green, white. Translucent.

MODE OF OCCURRENCE: Occurs chiefly in association with chrysotile as a component of massive green serpentine. Typical occurrences are found at Aboutville, New York, and at Kennack Cove, Lizard, Cornwall, England.

BEST REF. IN ENGLISH: Deer, Howie, and Zussman, "Rock Forming Minerals, v. 3, p. 170-190, New York, Wiley, 1962.

## LODOCHNIKITE = Brannerite

## LODOCHNIKOVITE
## (Species status uncertain)
Oxide of Fe, Al, Mg, and Ca

CRYSTAL SYSTEM: Orthorhombic
LATTICE CONSTANTS:
    $a = 17.1$
    $b = 23.6$
    $c = 5.71$
HARDNESS: 7½
MODE OF OCCURRENCE: Occurs in skarn with calcite, spinel, titanomagnetite, idocrase, and corundum, presumably in USSR.

BEST REF. IN ENGLISH: Nefedov, E. I., *Am. Min.*, **40**: 551 (1955); **41**: 672 (1956).

## LOELLINGITE
$Fe_{1+x}As_{2-x}$
Var. Geyerite, sulfurian
    Glaucopyrite, cobaltian

CRYSTAL SYSTEM: Orthorhombic
CLASS: 2/m 2/m 2/m
SPACE GROUP: Pnnm
Z: 2
LATTICE CONSTANTS:
    $a = 5.25$
    $b = 5.92$
    $c = 2.85$
3 STRONGEST DIFFRACTION LINES:
    2.61 (100)
    2.33 ( 90)
    1.638 ( 80)
HARDNESS: 5-5½
DENSITY: 7.43 (Meas.)
    7.48 (Calc.)
CLEAVAGE: {010} sometimes distinct
        {101} sometimes distinct
Fracture uneven. Brittle.

HABIT: Crystals commonly prismatic, up to 5 × 30 cm in size; also pyramidal and doubly terminated. Often massive. Twinning on {011}, sometimes as trillings.

COLOR-LUSTER: Silver white to steel gray. Opaque. Metallic. Streak grayish black.

MODE OF OCCURRENCE: Occurs chiefly in mesothermal vein deposits associated with other sulfides and calcite, or in pegmatites. Found widespread in the pegmatites of the Black Hills, South Dakota, especially at the Ingersoll mine, Keystone, as exceptional crystals and as masses exceeding 600 pounds in weight. It also occurs in Gunnison and San Juan counties, Colorado; in Amador, Inyo, Riverside, and San Diego counties, California; at Franklin, New Jersey; in Maine, New Hampshire, and New York; abundantly in the Cobalt district, Ontario, Canada; at deposits in Bolivia, Chile, Spain, Germany, and Norway; and at Lölling, Carinthia, Austria.

BEST REF. IN ENGLISH: Palache, et al., "Dana's System of Mineralogy," 7th Ed., v. I, p. 303–307, New York, Wiley, 1944.

## LOEWEITE (Löweite)
$Na_{12}Mg_7(SO_4)_{13} \cdot 15H_2O$

CRYSTAL SYSTEM: Hexagonal
CLASS: $\bar{3}$
SPACE GROUP: $R\bar{3}$
Z: 3
LATTICE CONSTANTS:
   (rhomb. cell)
   $a_{rh} = 11.769$
   $\alpha = 106.5°$
3 STRONGEST DIFFRACTION LINES:
   3.17 (100)
   4.29 ( 95)
   4.04 ( 95)
OPTICAL CONSTANTS:
   $\omega = 1.490$
   $\epsilon = 1.471$ (Na)
   $(-)$
HARDNESS: 2½–3
DENSITY: 2.36–2.38 (Meas.)
         2.36 (Calc.)
CLEAVAGE: Not determined. Fracture conchoidal.
HABIT: Massive; as embedded anhedral grains.
COLOR-LUSTER: Colorless to gray or pale yellow; also reddish yellow due to iron oxide inclusions. Transparent. Vitreous.

Soluble in water. Taste bitter.
MODE OF OCCURRENCE: Occurs as a constituent of oceanic salt deposits. Found in the Permian age Salada formation, Carlsbad, New Mexico; in the Strassfurt district, Germany; and as crystalline masses associated with anhydrite at Hallstatt and Bad Ischl, Upper Austria.
BEST REF. IN ENGLISH: Palache, et al., "Dana's System of Mineralogy," 7th Ed., v. II, p. 446–447, New York, Wiley, 1951. Fang, J. H., and Robinson, P. D., *Am. Min.*, 55: 378–386 (1970).

## LOKKAITE
$(Y, Ca)_2(CO_3)_3 \cdot 2H_2O$

CRYSTAL SYSTEM: Orthorhombic
CLASS: mm2, or 2/m 2/m 2/m
SPACE GROUP: Possibly Pb2m, Pbm2, Pbmm
LATTICE CONSTANTS:
   $a = 39.07$
   $b = 6.079$
   $c = 9.19$
3 STRONGEST DIFFRACTION LINES:
   3.808 (100)
   4.594 ( 75)
   3.902 ( 60)
OPTICAL CONSTANTS:
   $\alpha = 1.569$
   $\beta = 1.592$
   $\gamma = 1.620$
HARDNESS: Not determined
DENSITY: Not determined.
CLEAVAGE: Not determined.
Brittle.

HABIT: Massive; as fibrous radial aggregates 0.5–1.0 mm in diameter.
COLOR-LUSTER: White.
MODE OF OCCURRENCE: Occurs as an alteration product of rare-earth minerals at the Pyörönmaa pegmatite in Kangasala, southwest Finland.
BEST REF. IN ENGLISH: Perttunen, Vesa, *Am. Min.*, 56: 1838 (1971).

## LOMBAARDITE = Allanite

## LOMONOSOVITE
$Na_8(Mn, Fe, Ca, Mg)Ti_3Si_4P_2O_{24}$

CRYSTAL SYSTEM: Triclinic
CLASS: 1
SPACE GROUP: P1
Z: 1
LATTICE CONSTANTS:
   $a = 5.40$     $\alpha = 100°$
   $b = 7.03$     $\beta = 96°$
   $c = 14.3$     $\gamma = 90°$
3 STRONGEST DIFFRACTION LINES:
   2.83 (100)
   1.778 ( 90)
   1.840 ( 80)
OPTICAL CONSTANTS:
   $\alpha = 1.670$
   $\beta = 1.750$
   $\gamma = 1.778$
   $(-)2V = 56°$
HARDNESS: 3–4
DENSITY: 3.15 (Meas.)
CLEAVAGE: {100} perfect.
Fracture uneven. Brittle.
HABIT: Laminated tabular crystals up to 7 X 5 X 0.6 cm in size. Fine polysynthetic twinning.
COLOR-LUSTER: Dark cinnamon brown to black, also rose-violet; vitreous to adamantine on cleavage, vitreous to greasy on fracture. Transparent in thin fragments.
MODE OF OCCURRENCE: Occurs associated with hackmanite, ussingite, lamprophyllite, eudialyte, arfvedsonite, microcline, ramsayite, and aegirine, in pegmatites in syenite, Lovozero Range, Kola Peninsula, USSR.
BEST REF. IN ENGLISH: Gerasimovsky, V. I., *Am. Min.*, 35: 1092–1093 (1950).

## LONSDALEITE
C
Dimorphous with diamond, chaoite, graphite

CRYSTAL SYSTEM: Hexagonal
CLASS: 6/m 2/m 2/m
SPACE GROUP: $P6_3/mmc$
Z: 4
LATTICE CONSTANTS:
   $a = 2.52$
   $c = 4.12$

3 STRONGEST DIFFRACTION LINES:
2.19 (100)
2.06 (100)
1.26 ( 75)
DENSITY: > 3.3 (Meas.)
3.51 (Calc.)
HABIT: Cubes and cubo-octahedrons up to 0.7 mm in size.
COLOR-LUSTER: Pale brownish yellow under microscope.
MODE OF OCCURRENCE: Found in the Canyon Diablo and Goalpara meteorites.
BEST REF. IN ENGLISH: Frondel, Clifford, and Marvin, Ursula B., *Science*, **155**: 995–997 (1967).

## LOPARITE = Niobian perovskite

## LOPEZITE
$K_2Cr_2O_7$

CRYSTAL SYSTEM: Triclinic
CLASS: $\bar{1}$
SPACE GROUP: $P\bar{1}$
Z: 4
LATTICE CONSTANTS:
$a = 7.50$    $\alpha = 98°00'$
$b = 13.40$   $\beta = 90°51'$
$c = 7.38$    $\gamma = 96°13'$
3 STRONGEST DIFFRACTION LINES:
3.32 (100)
3.49 ( 30)
3.68 ( 20)
OPTICAL CONSTANTS:
$\alpha = 1.714$
$\beta = 1.732$
$\gamma = 1.805$
$(+)2V = 50°$
HARDNESS: 2½
DENSITY: 2.69 (Meas. synthetic)
2.660 (Calc. synthetic)
CLEAVAGE: {010} perfect ⎫
{100} distinct ⎬ (synthetic)
{001} distinct ⎭
HABIT: As minute spherical aggregates. Synthetic crystals short prismatic.
COLOR-LUSTER: Orange, orange-red, red. Transparent. Vitreous.

Soluble in water.
MODE OF OCCURRENCE: Occurs as aggregates up to 1 mm in diameter with dietzeite, tarapacaite, and ulexite in cavities in nitrate rock at Oficina Rosario and Maria Elena, Antofagasta province, Chile.
BEST REF. IN ENGLISH: Palache, et al., "Dana's System of Mineralogy," 7th Ed., v. II, p. 645–646, New York, Wiley, 1951.

## LORANDITE
$TlAsS_2$
Isomorphous with miargyrite

CRYSTAL SYSTEM: Monoclinic
CLASS: 2/m
SPACE GROUP: $P2_1/a$
Z: 8
LATTICE CONSTANTS:
$a = 12.27$
$b = 11.33$
$c = 6.11$
$\beta = 104°02'$
3 STRONGEST DIFFRACTION LINES:
3.59 (100)
2.88 ( 80)
2.97 ( 70)
OPTICAL CONSTANTS:
$a > 2.72$ (Li)
$\gamma - \alpha$ = extreme
2V = large, probably positive
HARDNESS: 2–2½
DENSITY: 5.5288–5.5362 (Meas.)
5.53 (Calc.)
CLEAVAGE: {100} perfect
{201} distinct
{001} distinct
Flexible, readily separates into cleavage lamellae and fibers.
HABIT: Crystals tabular or short prismatic with striations parallel to *b*- or *c*-axis. Also as thin veinlets or fine granular.
COLOR-LUSTER: Cochineal to carmine red, surface often blackish lead gray, sometimes coated with ocher yellow powder. Translucent to transparent. Metallic adamantine. Streak dark cerise red.
MODE OF OCCURRENCE: Occurs associated with realgar, orpiment, and barite, incrusting massive fine-grained pyrite, at the Rambler mine, Encampment, Wyoming; as veinlets in realgar and as crystals up to 1 cm in size at the arsenic-antimony deposit at Allchar, Macedonia, Yugoslavia; and at the Dzhizhikrut antimony-mercury deposit, Tadzhikistan, USSR.
BEST REF. IN ENGLISH: Vlasov, K. A., "Mineralogy of Rare Elements," v. II, p. 592–598, Israel Program for Scientific Translations, 1966.

## LORANSKITE = Variety of euxenite

## LORENZENITE (Ramsayite)
$Na_2Ti_2Si_2O_9$

CRYSTAL SYSTEM: Orthorhombic
CLASS: 2/m 2/m 2/m
SPACE GROUP: Pcnb
Z: 4
LATTICE CONSTANTS:
$a = 8.66$
$b = 14.42$
$c = 5.18$
3 STRONGEST DIFFRACTION LINES:
2.74 (100)
3.33 ( 70)
1.60 ( 50)

OPTICAL CONSTANTS:
$\alpha = 1.91$
$\beta = 2.01$
$\gamma = 2.03$
$(-)2V = 38°-40°$
HARDNESS: 6
DENSITY: 3.43 (Meas.)
            3.413 (Calc.)
CLEAVAGE: {010} distinct
HABIT: Crystals acicular, minute.
COLOR-LUSTER: Brown to black.
MODE OF OCCURRENCE: Occurs in nepheline syenite at Mont St. Hilaire, Quebec, Canada; at Narsarsuk, Greenland; and on the Kola Peninsula, USSR.
BEST REF. IN ENGLISH: Sahama, Th. G., *Am. Min.*, **32**: 59–63 (1947).  Winchell and Winchell, "Elements of Optical Mineralogy," 4th Ed., Pt. 2, p. 401, New York, Wiley, 1951.

# LORETTOITE
$Pb_7Cl_2O_6$

CRYSTAL SYSTEM: Orthorhombic
CLASS: 2/m 2/m 2/m
SPACE GROUP: Fmmm
Z: 2
LATTICE CONSTANTS:
$a = 5.50$
$b = 22.9$
$c = 5.49$
3 STRONGEST DIFFRACTION LINES:
2.98 (100)
2.77 ( 80)
1.993 ( 70)
OPTICAL CONSTANTS:
$\omega = 2.35-2.40$
$\epsilon = 2.33-2.37$  (Li)
    $(-)$
HARDNESS: 2½–3
DENSITY: 7.39 (Loretto)
            7.95 (Argentina)
CLEAVAGE: {001} perfect
HABIT: Compact masses of coarse fibers or blades.
COLOR-LUSTER: Yellow to reddish yellow. Translucent. Adamantine. Streak yellow.
MODE OF OCCURRENCE: Occurs with other lead minerals at Loretto, Lawrence County, Tennessee, and at the Cerro de las Coronas, Chubut, Argentina.
BEST REF. IN ENGLISH: Palache, et al., "Dana's System of Mineralogy," 7th Ed., v. II, p. 56, New York, Wiley, 1951.

# LOSEYITE
$(Zn, Mn)_7(CO_3)_2(OH)_{10}$

CRYSTAL SYSTEM: Monoclinic
CLASS: 2/m
SPACE GROUP: A2/a
Z: 4

LATTICE CONSTANTS:
$a = 16.2$ kX
$b = 5.55$
$c = 14.92$
$\beta = 95°24'$
3 STRONGEST DIFFRACTION LINES:
3.68 (100)
2.63 (100)
3.80 ( 90)
OPTICAL CONSTANTS:
$\alpha = 1.637$
$\beta = 1.648$
$\gamma = 1.676$
$(+)2V = 64°$
HARDNESS: ~ 3
DENSITY: 3.27 (Meas.)
            3.37 (Calc.)
CLEAVAGE: No apparent cleavage.
HABIT: Crystals lath-like, elongated [010]. As aggregates or radiating bundles of crystals.
COLOR-LUSTER: Bluish white, brownish. Transparent.
MODE OF OCCURRENCE: Occurs in small veinlets in massive ore, associated with calcite, altered pyrochroite, and other minerals, at Franklin, Sussex County, New Jersey.
BEST REF. IN ENGLISH: Palache, et al., "Dana's System of Mineralogy," 7th Ed., v. II, p. 244–245, New York, Wiley, 1951.

# LOUGHLINITE (Inadequately described mineral)
$Na_2Mg_3Si_6O_{16} \cdot 8H_2O$

CRYSTAL SYSTEM: Uncertain
LATTICE CONSTANT:
$c = 5.3$ (only datum known)
3 STRONGEST DIFFRACTION LINES:
12.9 (100)
4.34 ( 18)
4.51 (  7)
OPTICAL CONSTANTS:
$\alpha = 1.500$
$\beta = 1.505$
$\gamma = 1.525$
$(+)2V = 60°$ (Calc.)
HARDNESS: Not determined.
DENSITY: 2.165 (Meas.)
CLEAVAGE: Not determined.
HABIT: Massive, fibrous.
COLOR-LUSTER: Pearly white. Silky.
MODE OF OCCURRENCE: Found in veins, usually less than 3 cm long, that vary in thickness up to 1 cm, which is the maximum length of the fibers, in dolomitic oil shale at three locations between 18 and 20 miles west of Green River, Sweetwater County, Wyoming.
BEST REF. IN ENGLISH: Fahey, Joseph J., Ross, Malcolm, and Axelrod, Joseph M., *Am. Min.*, **45**: 270–281 (1960).

# LOVOZERITE
$(Na, Ca)_2(Zr, Ti)(Si_6O_{13})(OH)_6 \cdot 3H_2O$

CRYSTAL SYSTEM: Monoclinic
CLASS: 2
SPACE GROUP: C2
Z: 2
LATTICE CONSTANTS:
 $a = 10.48$
 $b = 10.20$
 $c = 7.33$
 $\beta = 92°30'$
3 STRONGEST DIFFRACTION LINES:
 3.90 (100)
 4.36 ( 50)
 5.22 ( 20)
OPTICAL CONSTANTS:
 $\omega = 1.561$
 $\epsilon = 1.549$
 $(-)$
HARDNESS: 5
DENSITY: 2.3–2.7 (Meas.)
CLEAVAGE: Not determined; in two directions. Brittle. Fracture conchoidal. Altered varieties friable.
HABIT: Massive; as irregular grains from 0.3 to 30 mm in size. Also as pseudomorphs after eudialyte. Twinning common, polysynthetic.
COLOR-LUSTER: Reddish brown to dark brown to black. Opaque. Resinous.
MODE OF OCCURRENCE: Occurs in nepheline-syenites, in association with microcline, aegirine, nepheline, eudialyte, and other minerals. Found in the Julianehaab district, Greenland, and in the Lovozero alkali massif and elsewhere on the Kola Peninsula, USSR.
BEST REF. IN ENGLISH: Vlasov, K. A., "Mineralogy of Rare Elements," v. II, p. 364–365, Israel Program for Scientific Translations, 1966.

# LÖWIGITE = Synonym of alunite

# LUCKITE = Manganoan variety of melanterite

# LUDLAMITE
$Fe_3(PO_4)_2 \cdot 4H_2O$

CRYSTAL SYSTEM: Monoclinic
CLASS: 2/m
SPACE GROUP: $P2_1/a$
Z: 2
LATTICE CONSTANTS:
 $a = 10.48$ kX
 $b = 4.63$
 $c = 9.16$
 $\beta = 100°36'$
3 STRONGEST DIFFRACTION LINES:
 3.96 (100)
 2.765 (100)
 2.543 (100)
OPTICAL CONSTANTS:
 $\alpha = 1.650$–1.653
 $\beta = 1.667$–1.675
 $\gamma = 1.688$–1.697
 $(+)2V = 82°$

HARDNESS: 3½
DENSITY: 3.19 (Meas.)
 3.21 (Calc.)
CLEAVAGE: {001} perfect
 {100} indistinct
HABIT: Crystals thin to thick tabular, often wedge-shaped. Also massive, granular.
COLOR-LUSTER: Bright green, apple green, pale green to greenish white to colorless. Transparent to translucent. Vitreous.
MODE OF OCCURRENCE: Occurs as a secondary mineral in the oxidation zone of ore deposits, and as an alteration product of primary iron phosphate minerals in granite pegmatites. Found as superb crystals up to ½ inch in size, often with vivianite, in the Blackbird district, Lemhi County, Idaho. In South Dakota as crystalline masses up to 1 foot in diameter, with vugs containing crystals up to 7 mm in length, at the Hesnard mine, Keystone, and as fine crystals in pegmatite at the Big Chief, Ferguson, and Dan Patch mines in the same district; and as excellent crystals in pegmatite at the Bull Moose and Tip Top mines, Custer district. It is also found at the Palermo pegmatite, North Groton, New Hampshire; at the Wheal Jane mine, Truro, Cornwall, England; and at Hagendorf, Bavaria, Germany.
BEST REF. IN ENGLISH: Palache, et al., "Dana's System of Mineralogy," 7th Ed., v. II, p. 952–953, New York, Wiley, 1951.

# LUDLOCKITE
$(Fe, Pb)As_2O_6$

CRYSTAL SYSTEM: Triclinic
Z: 9
LATTICE CONSTANTS:
 $a = 10.41$    $\alpha = 113.9°$
 $b = 11.95$    $\beta = 99.7°$
 $c = 9.86$     $\gamma = 82.7°$
3 STRONGEST DIFFRACTION LINES:
 8.81  (100)
 2.935 ( 80)
 3.33  ( 40)
OPTICAL CONSTANTS: Biaxial, positive, $\alpha$ 1.96, $\beta$ 2.055, $\gamma > 2.11$, Z near $\alpha$, optic axial plane perpendicular to $\{0\bar{1}1\}$. Pleochroic, X yellow, Y deep yellow, Z orange-yellow.
HARDNESS: 1½–2
DENSITY: 4.40 (Meas.)
 4.35 (Calc.)
CLEAVAGE: $\{0\bar{1}1\}$ micaceous, combines with other easy $\{0kl\}$ cleavages to give ready fraying of crystals into fibers. Sectile and flexible.
HABIT: Crystals elongated [100], flattened $\{0\bar{1}1\}$; form $\{021\}$ prominent; no terminal faces. Twinning on $\{0\bar{1}1\}$, lamellar.
COLOR-LUSTER: Red. Subadamantine. Streak light brown.
MODE OF OCCURRENCE: Occurs associated with zincian siderite in a cavity of ore from the "germanite" section at Tsumeb, South West Africa.
BEST REF. IN ENGLISH: Davis, R. J., Embrey, P. G., and Hey, M. H., *Am. Min.*, **57**: 1003–1004 (1972).

## LUDWIGITE
$(Mg, Fe^{2+})_2 Fe^{3+} BO_5$

CRYSTAL SYSTEM: Orthorhombic
CLASS: 2/m 2/m 2/m
SPACE GROUP: Pbam
Z: 4
LATTICE CONSTANTS:
 $a = 9.26$
 $b = 12.26$
 $c = 3.05$
3 STRONGEST DIFFRACTION LINES:
 5.12 (100)
 2.547 ( 70)
 2.515 ( 70)
OPTICAL CONSTANTS:
 $\alpha = 1.83-1.85$
 $\beta = 1.83-1.85$
 $\gamma = 1.94-2.02$
 (+)2V = 20°-45°
HARDNESS: 5
DENSITY: 3.86 (Meas.)
 3.92 (Calc.)
CLEAVAGE: {001} perfect
HABIT: Crystals rare, prismatic with rhombic cross section. Commonly as fibrous masses, granular, or as embedded aggregates of needle-like crystals.
COLOR-LUSTER: Dark green, greenish black to black. Opaque. Silky, when fibrous.
MODE OF OCCURRENCE: Occurs as a high-temperature mineral in contact metamorphic deposits. Found in the United States in some of the cassiterite ores at Gorman, Kern County, and in iron deposits in San Bernardino County, California; at several mines in the Big and Little Cottonwood districts, Salt Lake County, Utah; in Colorado Gulch near Helena, and at Philipsburg, Montana; and at localities in Nevada and Idaho. It also occurs at Tallgruvan, Sweden; in magnetite skarn deposits in Norway; at Moravicza in the Banat, Rumania; in the Hol Kol mine, Suan, Korea; and at the Hanayama mine, Fukushima Prefecture, Japan.
BEST REF. IN ENGLISH: Palache, et al., Dana's System of Mineralogy," 7th Ed., v. II, p. 321-324, New York, Wiley, 1951. GSA Bull., 66: 1540-1541 (1955).

## LUESHITE
$NaNbO_3$
Perovskite Group

CRYSTAL SYSTEM: Orthorhombic (pseudocubic)
CLASS: 222
SPACE GROUP: $P222_1$
Z: 1
LATTICE CONSTANTS:
 (orthorhombic)  (pseudocubic)
 $a = 5.51$   $a = 3.91$
 $b = 5.53$
 $c = 15.50$
3 STRONGEST DIFFRACTION LINES:
 2.76 (100)
 3.90 ( 85)
 1.95 ( 70)

OPTICAL CONSTANTS:
 $\eta \sim 2.30$
 (+)2V ~ 45°
HARDNESS: 5½
DENSITY: 4.44 (Meas.)
 4.55 (Calc.)
CLEAVAGE: {001} imperfect, easy
HABIT: Cubes up to 1.5 cm on an edge, but mostly 0.5-1.0 cm. Faces lightly striated.
COLOR-LUSTER: Black; thin fragments reddish brown. Streak gray.
MODE OF OCCURRENCE: Occurs as incrustations on a compact yellow mica at the contact of cancrinite syenite with carbonatite containing pyrochlore at Lueshe, 150 km north of Goma, Zaire. Also found at a vermiculite mine, Gem Park, Fremont, Colorado.
BEST REF. IN ENGLISH: Safiannikoff, A., *Am. Min.*, 46: 1004 (1961). USGS Prof. Paper 450-C, p. 4-6 (1962).

## LUNAR MINERALS
Species identified in samples collected from the Moon by Apollo astronauts:

| | | |
|---|---|---|
| Apatite | Perovskite | Rutile |
| *Armalcolite | Plagioclase | Schreibersite |
| Baddeleyite | Anorthite | Spinel |
| Chromite | Bytownite | *Tranquillityite |
| Cohenite | Labradorite | Tridymite |
| Cristobalite | Pyroxene | Troilite |
| Ilmenite | Augite | Ulvospinel |
| Iron | Pigeonite | Whitlockite |
| Nickel-iron | *Pyroxferroite | Zircon |
| Olivine | Quartz | |

*New species.

Many additional minerals also have been tentatively identified.
BEST REF. IN ENGLISH: *Science*, 167: (3918), (1970). Mason, Brian, *Min. Rec.*, 1: 16-22 (1970).

## LÜNEBURGITE (Lueneburgite)
$Mg_3 (PO_4)_2 B_2 O_3 \cdot 8H_2O$

CRYSTAL SYSTEM: Monoclinic
CLASS: Probably 2/m
SPACE GROUP: Probably C2/m
LATTICE CONSTANTS:
 $a = 10.10$
 $b = 7.62$
 $c = 9.81$
 $\beta = 97°24'$
Z: 2
OPTICAL CONSTANTS:
 $\alpha = 1.520-1.522$
 $\beta = 1.54-1.541$
 $\gamma = 1.545-1.548$
 (-)2V = moderate
HARDNESS: ~ 2
DENSITY: 2.05 (Meas.)
CLEAVAGE: {110} and {1$\bar{1}$0} fair

HABIT: Crystals minute pseudohexagonal tablets. As flattened masses and nodules with fibrous to earthy structure.

COLOR-LUSTER: White to brownish white; also green.

MODE OF OCCURRENCE: Occurs in clay associated with halite, sylvite, and polyhalite in the Permian salt basin in the vicinity of Carlsbad, New Mexico and in adjacent parts of Texas. It also is found in marl at Lüneburg, Hannover, Germany; and in a guano deposit at Mejillones, Peru.

BEST REF. IN ENGLISH: Palache, et al., "Dana's System of Mineralogy," 7th Ed., v. II, p. 385, New York, Wiley, 1951.

**LUSAKITE** = Variety of staurolite containing up to 8% CoO

**LUSSATITE, LUSSATINE** = Fibrous varieties of Cristobalite

## LUSUNGITE
$(Sr, Pb)Fe_3(PO_4)_2(OH)_5 \cdot H_2O$

CRYSTAL SYSTEM: Hexagonal
CLASS: $\bar{3}$ 2/m
SPACE GROUP: $R\bar{3}m$
Z: 3 (hex.)
LATTICE CONSTANTS:
  $a = 7.04$     $a_{rh} = 6.92$
  $c = 16.80$     $\alpha = 61°12'$
3 STRONGEST DIFFRACTION LINES:
  2.98 (100)
  5.77 ( 90)
  3.53 ( 61)
OPTICAL CONSTANT:
  $n \sim 1.77–1.855$
  uniaxial, probably positive
HARDNESS: Not determined.
DENSITY: Not determined.
CLEAVAGE: Not determined.
HABIT: Massive, pulverulent.
COLOR-LUSTER: Dark Brown.
MODE OF OCCURRENCE: Occurs associated with goethite (limonite) and quartz in the phosphate zone of the Kobokobo pegmatite, Zaire.

BEST REF. IN ENGLISH: Van Wambeke, L., *Am. Min.*, 44: 906–907 (1959).

## LUZONITE
$Cu_3AsS_4$
Dimorphous with enargite

CRYSTAL SYSTEM: Tetragonal
CLASS: $\bar{4}2m$
SPACE GROUP: $I\bar{4}2m$
Z: 2
LATTICE CONSTANTS:
  (synthetic)
  $a = 5.290$
  $c = 10.465$
3 STRONGEST DIFFRACTION LINES:
  3.046 (100)
  1.855 ( 90)
  1.592 ( 70)
HARDNESS: $\sim 3.5$
DENSITY: 4.380 (Meas.)
        4.438 (Calc.)
CLEAVAGE: {101} good
         {100} distinct
Fracture uneven to conchoidal. Brittle.

HABIT: Usually massive, fine-grained. More rarely coarse-grained. Crystals distinct equant, rare; usually rough, with some or all faces curved and etched. Polished section exhibits polysynthetic twinning.

COLOR-LUSTER: Deep pinkish brown with grayish tinge. Dull, metallic. Opaque. Streak black.

MODE OF OCCURRENCE: Found in low- to medium-intensity copper deposits often associated with enargite, tetrahedrite-tennantite, pyrite, chalcopyrite, covellite; and more rarely sphalerite, bismuthinite, ruby silvers, native silver, gold, and marcasite. The gangue minerals barite and drusy quartz are so common as to be almost characteristic. The most important localities are Mankayan, Luzon; Famatina, Argentina; Goldfield, Nevada; Hokuetsu, Japan; Kinkwaseki, Taiwan; Cerro de Pasco and Morococha, Peru; Butte, Montana; and Cananea, Mexico.

BEST REF. IN ENGLISH: Gaines, Richard V. *Am. Min.*, 42: 766–779 (1957).

**LYNDOCHITE** = Euxenite

# M

## MACALLISTERITE
$Mg_2B_{12}O_{20} \cdot 15H_2O$

CRYSTAL SYSTEM: Hexagonal
CLASS: $\bar{3}2/m$
SPACE GROUP: $R\bar{3}c$
Z: 6 (hex.)
LATTICE CONSTANTS:
  $a = 11.543$    $a_{rh} = 13.60$
  $c = 35.556$    $\alpha = 50°14'$
3 STRONGEST DIFFRACTION LINES:
  8.715 (100)
  5.772 ( 50)
  4.063 ( 50)
OPTICAL CONSTANTS:
  $\omega = 1.504$
  $\epsilon = 1.459$
   (-)
HARDNESS: 2½
DENSITY: 1.867 (Meas.)
         1.867 (Calc.)
CLEAVAGE: {0001} good
         {01$\bar{1}$2} good
Brittle.
  HABIT: Crystals rhombohedral, with {01$\bar{1}$2} and {0001} always present; {01$\bar{1}$2} dominant; {10$\bar{1}$4} present on smaller crystals. As aggregates of minute crystals and as random intergrowths of crystals up to 12 mm in maximum dimension in nodular masses.
  COLOR-LUSTER: Colorless, light amber; transparent.
  MODE OF OCCURERNCE: Occurs as small pellets of crystal clusters intimately intermixed with ginorite and sassolite in the Furnace Creek Wash area, Death Valley region, Inyo County, California; also in nodules with riva-davite in a plug-like deposit of massive borax at the Tincalayu Mine, Salta, Argentina; and associated with hydroboracite and hungchaoite as an efflorescence from a continental saline lake in China.
  BEST REF. IN ENGLISH: Schaller, Waldemar T., Vlisidis, Angelina C., and Mrose, Mary E., *Am. Min.*, **50**: 629-640 (1965). Aristarain, L. F., and Hurlbut, C. S. Jr., *Am. Min.*, **52**: 1776-1784 (1967).

## MACCONNELLITE (Inadequately described mineral)
$CuCrO_2$

CRYSTAL SYSTEM: Hexagonal-R
  HARDNESS: Not determined.
  DENSITY: Not determined.
  CLEAVAGE: Not determined.
  HABIT: Intergrown with grimaldiite forming composite platy hexagonal crystals.
  COLOR-LUSTER: Red-brown.
  MODE OF OCCURRENCE: Occurs in complex microscopic intergrowth with eskolaite, bracewellite, guyanaite, and grimaldiite in "merumite" from alluvial gravels in Guyana.
  BEST REF. IN ENGLISH: Milton, C., et al., Geol. Soc. Am., Program 1967 Annual Meetings, p. 151-152 (1967). (abstr.)

## MACDONALDITE
$BaCa_4Si_{15}O_{35} \cdot 11H_2O$

CRYSTAL SYSTEM: Orthorhombic
CLASS: 2/m 2/m 2/m or mm2
SPACE GROUP: Bmmb, B2mb, or Bm2₁b
Z: 4
LATTICE CONSTANTS:
  $a = 14.06$
  $b = 23.52$
  $c = 13.08$
3 STRONGEST DIFFRACTION LINES:
  6.5  (100)
  4.36 ( 80)
  6.3  ( 50)
OPTICAL CONSTANTS:
  $\alpha = 1.518$
  $\beta = 1.524$
  $\gamma = 1.530$
  2V = 90° (-) or (+)
HARDNESS: 3½-4
DENSITY: 2.27 (Meas.)
         2.27 (Calc.)

CLEAVAGE: {010} perfect
{001} good
{100} poor, or a fracture

HABIT: Tiny needle-like crystals exhibit the forms {010}, {001}, and {100} and are elongate parallel to the $a$-axis.

COLOR-LUSTER: White to colorless; satiny to vitreous. White streak.

MODE OF OCCURRENCE: Occurs in grains that range from less than 1 mm in length to 6 mm in sanbornite-bearing metamorphic rocks which outcrop in a narrow zone 2½ miles long near a granodiorite contact in eastern Fresno County, California.

BEST REF. IN ENGLISH: Alfors, John T., Stinson, Melvin C., Matthews, Robert A., and Pabst, Adolf, *Am. Min.*, **50**: 314-340 (1965).

# MACEDONITE
$PbTiO_3$

CRYSTAL SYSTEM: Tetragonal
LATTICE CONSTANTS:
$a = 3.889$
$c = 4.209$
3 STRONGEST DIFFRACTION LINES:
2.843 (100)
1.603 ( 90)
2.728 ( 80)
HARDNESS: 5½
DENSITY: 7.82 (Meas.)
8.09 (Calc. x-ray)
CLEAVAGE: None.

HABIT: Crystals rare; prismatic with forms {100}, {001}, {110}, {101}, and {111}. Usually as irregular grains ranging from a few tenths of a micron up to 0.2 mm in size.

COLOR-LUSTER: Black with weak brownish tint. Opaque. Vitreous.

MODE OF OCCURRENCE: Occurs sparingly in amazonite quartz syenite veins in Crni Kamen near Prilep, Macedonia, Yugoslavia; also as an inclusion in hematite and magnetoplumbite at Långban, Sweden.

BEST REF. IN ENGLISH: Radusinović, Dušan, and Markov, Cvetko, *Am. Min.*, **56**: 387-394 (1971).

# MACGOVERNITE (McGovernite)
$Mn_9 Mg_4 Zn_2 As_2 Si_2 O_{17} (OH)_{14}$

CRYSTAL SYSTEM: Hexagonal
CLASS: $\bar{3}$ 2/m or 3m
SPACE GROUP: R$\bar{3}$c or R3c
Z: 18 (hex.), 6 (rhomb.)
LATTICE CONSTANTS:
$a = 8.22$        $a_{rh} = 68.7$
$c = 205.5$        $\alpha = 6°52'$
OPTICAL CONSTANT:
$\omega = 1.761$
(+)
HARDNESS: Not determined.
DENSITY: 3.719 (Meas.)
CLEAVAGE: {0001} perfect, micaceous
Brittle.

HABIT: Massive, granular; also as slightly deformed mica-like "books."

COLOR-LUSTER: Light brown to deep reddish brown. Translucent. Pearly.

MODE OF OCCURRENCE: Occurs in association with franklinite, willemite, zincite, calcite, and other minerals, at Sterling Hill, Sussex County, New Jersey.

BEST REF. IN ENGLISH: Wuensch, Bernhardt J., *Am. Min.*, **45**: 937-945 (1960).    Palache, Charles, USGS Prof. Paper 180, p. 91 (1935).

# MACKAYITE
$FeTe_2 O_5 (OH)$ (?)

CRYSTAL SYSTEM: Tetragonal
CLASS: 4/m 2/m 2/m
SPACE GROUP: I4/acd
Z: 16
LATTICE CONSTANTS:
$a = 11.704$
$c = 14.95$ (variable)
3 STRONGEST DIFFRACTION LINES:
3.17 (100)
1.61 ( 90)
3.32 ( 80)
OPTICAL CONSTANTS:
$\omega = 2.19$
$\epsilon = 2.21$   (Li)
(+)
HARDNESS: 4½
DENSITY: 4.86 (Meas.)
4.79 (Calc.)
CLEAVAGE: Fracture subconchoidal. Brittle.

HABIT: Crystals pyramidal to short prismatic.

COLOR-LUSTER: Yellowish green to brownish green. Transparent. Vitreous. Streak pale green.

MODE OF OCCURRENCE: Occurs as minute well-developed crystals, associated with emmonsite and other minerals, in the oxidation zone of a vein containing pyrite and native tellurium at Goldfield, Nevada.

BEST REF. IN ENGLISH: *Am. Min.*, **55**: 1072 (1970). Palache, et al., "Dana's System of Mineralogy," 7th Ed., v. II, p. 642-643, New York, Wiley, 1951.

# MACKELVEYITE (Mckelveyite)
$(Na, Ca) (Ba, Y, U)_2 (CO_3)_3 \cdot 1-2H_2O$
Dimorphous with ewaldite

CRYSTAL SYSTEM: Hexagonal
CLASS: 3m
SPACE GROUP: P31m
Z: 6
LATTICE CONSTANTS:
$a = 9.174$
$c = 19.154$
3 STRONGEST DIFFRACTION LINES:
2.942 (100)
4.47 ( 85)
6.40 ( 35)

OPTICAL CONSTANTS:
$\omega = 1.66$
$\epsilon = 1.57$
$(-)$
HARDNESS: Not determined.
DENSITY: 3.58 (Meas.) aggregates
3.47 (Meas.) crystals
3.62 (Calc.)
CLEAVAGE: No good cleavages.
HABIT: Small, less than 0.5 mm, hemimorphic crystals and aggregates of crystal plates. Crystals are simple in development with maximum of six measurable forms; $\{0001\}$, $\{000\bar{1}\}$, $\{10\bar{1}2\}$, and $\{01\bar{1}2\}$ dominant, also $\{10\bar{1}1\}$, and $\{01\bar{1}1\}$.
COLOR-LUSTER: Lime yellow; usually black due to amorphous carbonaceous inclusions.
MODE OF OCCURRENCE: Found as minute crystals in drill core from four subsurface localities, some 20 miles apart, in the Green River formation in Sweetwater County, Wyoming, associated with labuntsovite, ewaldite, leucosphenite, searlesite, and other typical minerals of the Green River formation.
BEST REF. IN ENGLISH: Milton, Charles, Ingram, Blanche, Clark, Joan R., and Dwornik, Edward J., *Am. Min.* **50**: 593-612 (1965). Desautels, Paul E., *Am. Min.*, **52**: 860-864 (1967). Donnay, Gabrielle, and Donnay, J. D. H., *Am. Min.*, **56**: 2156 (1971).

# MACKINAWITE
$(Fe, Ni)_9 S_8$

CRYSTAL SYSTEM: Tetragonal
CLASS: $4/m\ 2/m\ 2/m$
SPACE GROUP: P4/nmm
Z: 2
LATTICE CONSTANTS:
$a = 3.676$
$c = 5.032$
3 STRONGEST DIFFRACTION LINES:
5.03 (100)
2.97 ( 80)
2.31 ( 80)
HARDNESS: $\sim 2\frac{1}{2}$
Vickers hardness 58 with 50 g load.
DENSITY: 4.30 (Calc.)
CLEAVAGE: $\{001\}$ perfect
HABIT: Massive, as extremely minute feathery lamellae or irregular grains up to 1.0 mm in diameter; also as idiomorphic crystals up to $0.2 \times 10$-15 mm in size. Crystals usually tetragonal basal plates or simple pyramidal combinations.
COLOR-LUSTER: Bronze, fresh fracture surfaces white-gray. Powder black.
MODE OF OCCURRENCE: Occurs at the Mackinaw mine, Snohomish County, Washington; in dunites and pyroxenites of the Muskox intrusion, Northwest Territories, Canada; in the carbonatite and pegmatite of Loolekop, Phalaborwa complex, in the Bushveld igneous complex, and from Insizwa, Cape Province, South Africa; also in ore at the Outokumpo mine, Finland, and in sulfide ores of the Singhbhum copper belt, India.

BEST REF. IN ENGLISH: Evans, H. T. Jr., Milton, Charles, Chao, E. C. T., Adler, Isidore, Mead, Cynthia, Ingram, Blanche, and Berner, R. A., USGS Prof. Paper 475-D p. 64-69 (1964).

# MACKINSTRYITE (McKinstryite)
$(Ag, Cu)_2 S$

CRYSTAL SYSTEM: Orthorhombic
CLASS: $2/m\ 2/m\ 2/m$ or mm2
SPACE GROUP: Pnam or Pna2$_1$
Z: 32
LATTICE CONSTANTS:
$a = 14.043$
$b = 15.677$
$c = 7.803$
3 STRONGEST DIFFRACTION LINES:
2.606 (100)
2.070 ( 70)
3.508 ( 60)
HARDNESS: Talmadge hardness B.
DENSITY: 6.61 (Meas. av. of 6)
CLEAVAGE: One, poorly developed. Fracture subconchoidal.
HABIT: As masses of intergrown crystals 0.2-3mm in diameter.
COLOR-LUSTER: Steel gray, tarnishing to dark gray or black. Opaque. Metallic. Streak dark gray.
MODE OF OCCURRENCE: Occurs associated with chalcopyrite, stromeyerite, gangue calcite and actinolite, and minor silver and arsenopyrite in a specimen collected in 1907 at the Foster mine, Cobalt, Ontario, Canada.
BEST REF. IN ENGLISH: Skinner, Brian J., Jambor, John L., and Ross, Malcolm, *Econ. Geol.*, **61**: 1383-1389 (1966).

# MADOCITE
$Pb_{17}(Sb, As)_{16} S_{41}$

CRYSTAL SYSTEM: Orthorhombic
CLASS: $2/m\ 2/m\ 2/m$ or mm2
SPACE GROUP: Pbam or Pba2
Z: 4
LATTICE CONSTANTS:
$a = 27.2$
$b = 34.1$
$c = 2 \times 4.06$
3 STRONGEST DIFFRACTION LINES:
3.396 (100)
3.355 ( 90)
2.720 ( 80)
HARDNESS: Talmadge hardness B+. Vickers microhardness 155 (141-171) kg/mm$^2$ with 50 g load.
DENSITY: 5.99 (Predicted)
5.98 (Calc.)
CLEAVAGE: $\{010\}$ perfect
Fracture conchoidal.
HABIT: Small grains, elongated and striated along [001].
COLOR-LUSTER: Gray-black. Opaque. Metallic. Streak gray-black.

MODE OF OCCURRENCE: Occurs associated with boulangerite and jamesonite in a small prospect pit in Precambrian marble at Madoc, Ontario, Canada.

BEST REF. IN ENGLISH: Jambor, J. L., *Can. Min.*, **9** (Pt. 2) 191–213 (1967).

# MAGADIITE
$NaSi_7O_{13}(OH)_3 \cdot 3H_2O$

CRYSTAL SYSTEM: Monoclinic (?) (or tetragonal)
LATTICE CONSTANTS:
  (tetragonal)
  $a = 12.620$
  $c = 15.573$
3 STRONGEST DIFFRACTION LINES:
  15.41 (100)
  3.435 ( 80)
  3.146 ( 50)
  HABIT: Putty-like laminae, varying in thickness from a few inches to 2 feet.
  COLOR-LUSTER: White.
MODE OF OCCURRENCE: Occurs in brown silts of the High Magadi beds deposited over a large area from saline brines of the precursor of Lake Magadi, Kenya; in veins cutting playa sediments and in muds associated with alkali salts and brines at Alkali Lake playa, Oregon; also found in altered volcanic rocks about 3 miles north of Trinity Lake, Trinity County, California; and reported from Lac Tchad, Africa.

BEST REF. IN ENGLISH: Eugster, H. P., *Science*, **157**: 1177–1180 (1967). Rooney, Thomas P., Jones, Blair F., and McNeal, James T., *Am. Min.*, **54**: 1034–1043 (1969).

# MAGBASITE (Inadequately described mineral)
$KBa(Al,Sc)(Mg,Fe^{2+})_6Si_6O_{20}F_2$

CRYSTAL SYSTEM: Unknown
3 STRONGEST DIFFRACTION LINES:
  3.63 (100)
  3.2 (100)
  2.59 ( 80)
OPTICAL CONSTANTS:
  $\alpha = 1.597$
  $\beta = 1.609$
  $\gamma = 1.615$
  $(-)2V = 70°$
HARDNESS: ~5
DENSITY: 3.41 (Meas.)
  HABIT: Fan-shaped fine acicular and felt-like deposits up to 0.5 cm in size, somewhat resembling tremolite.
  COLOR-LUSTER: Colorless or rose-violet; vitreous.
MODE OF OCCURRENCE: Occurs associated with fluorite, barite, and parisite, genetically associated with alkalic barkevitic granosyenites occurring in dolomites in "one of the Asiatic hydrothermal formations."

BEST REF. IN ENGLISH: Semenov, E. I., Khomyakov, A. P., and Bykova, A. V., *Am. Min.*, **51**: 530–531 (1966).

# MAGHEMITE
$\gamma\text{-}Fe_2O_3$

CRYSTAL SYSTEM: Cubic
CLASS: 23
SPACE GROUP: $P2_13$
LATTICE CONSTANTS:
  (synthetic)
  $a = 8.339$
3 STRONGEST DIFFRACTION LINES:
  2.52 (100)
  2.95 (100)
  1.61 (100)
  HARDNESS: 5
  DENSITY: 5.0–5.2 (Meas.)
  CLEAVAGE: Not determined.
  HABIT: Massive.
  COLOR-LUSTER: Brown. Nearly opaque. Dull. Streak brown.

Strongly magnetic.
MODE OF OCCURRENCE: Occurs as a secondary mineral formed at low temperatures in gossans. Found widespread in small amounts, especially in California, Oklahoma, Canada, Germany, South Africa, and elsewhere.

BEST REF. IN ENGLISH: Palache, et al., "Dana's System of Mineralogy," 7th Ed., v. I, p. 708–709, New York, Wiley, 1944.

# MAGNESIOCHROMITE (Picrochromite)
$(Mg,Fe)(Cr,Al)_2O_4$
Spinel Group

CRYSTAL SYSTEM: Cubic
CLASS: $4/m\ \bar{3}\ 2/m$
SPACE GROUP: Fd3m
Z: 8
LATTICE CONSTANT:
  $a = 8.277$
3 STRONGEST DIFFRACTION LINES:
  2.49 (100)
  1.466 ( 70)
  1.593 ( 60)
  HARDNESS: 5½
  DENSITY: 4.2 ± 0.1
       4.43 (Calc. $MgCr_2O_4$)
  CLEAVAGE: None. Fracture uneven. Brittle.
  HABIT: Crystals octahedral, rare. Usually massive, compact to fine granular.
  COLOR-LUSTER: Black. Opaque. Metallic.

Sometimes weakly magnetic.
MODE OF OCCURRENCE: Occurs in olivine-rich (commonly serpentinized) igneous rocks, often associated with pyroxene, magnetite, and pyrrhotite. Found at the Caribou pit, Coleraine Township and elsewhere in Quebec; in British Columbia; and at deposits in Germany, Bulgaria, New Zealand, and New Caledonia.

BEST REF. IN ENGLISH: Palache, et al., "Dana's System of Mineralogy," 7th Ed., v. I, p. 709–712, New York, Wiley, 1944.

## MAGNESIOCOPIAPITE (Knoxvillite)
$MgFe_4(SO_4)_6(OH)_2 \cdot 20H_2O$
Copiapite Group

CRYSTAL SYSTEM: Triclinic
CLASS: $\bar{1}$
SPACE GROUP: $P\bar{1}$
Z: 1
LATTICE CONSTANTS:
  $a = 7.33$ kX    $\alpha = 93°51'$
  $b = 18.15$    $\beta = 101°30'$
  $c = 7.27$    $\gamma = 99°23'$
OPTICAL CONSTANTS:
  $\alpha = 1.507$
  $\beta = 1.529$
  $\gamma = 1.576$
  (+)2V = 67° (Calc.)
HARDNESS: 2½–3
DENSITY: 2.08–2.17
CLEAVAGE: {010} perfect
       {$\bar{1}01$} imperfect
HABIT: Crystals tabular. Usually as loose scaly aggregates; also granular or as crusts.
COLOR-LUSTER: Various shades of yellow and greenish yellow. Transparent to translucent. Pearly.
MODE OF OCCURRENCE: Occurs as a secondary mineral formed by the oxidation of sulfides. Found in association with redingtonite and mercury minerals at the old Redington mine at Knoxville, Napa County, California; also with botryogen at Falun, Sweden.
BEST REF. IN ENGLISH: Palache, et al., "Dana's System of Mineralogy," 7th Ed., v. II, p. 623–627, New York, Wiley, 1951.

## MAGNESIOFERRITE
$(Mg,Fe)Fe_2O_4$
Magnetite Series
Spinel Group

CRYSTAL SYSTEM: Cubic
CLASS: $4/m\,\bar{3}\,2/m$
SPACE GROUP: Fd3m
Z: 8
LATTICE CONSTANT:
  $a = 8.366$
3 STRONGEST DIFFRACTION LINES:
  2.525 (100)
  2.96 ( 40)
  1.481 ( 35)
OPTICAL CONSTANT:
  $N = 2.38$ (synthetic)
HARDNESS: 5½–6½
DENSITY: 4.436 (Meas.)
       4.55 (Calc.)
CLEAVAGE: Not determined.
Fracture uneven to subconchoidal. Brittle.
HABIT: Crystals minute, octahedral. Usually massive, fine granular. Twinning on {111} common.
COLOR-LUSTER: Black to brownish black. Opaque. Metallic, splendent to dull. Streak black.

Strongly magnetic.

MODE OF OCCURRENCE: Occurs in the crystalline limestone of the Crestmore quarry, Riverside County, and in chlorite schist from near the benitoite mine, San Benito County, California; at Magnet Cove, Arkansas; in the Puy-de-Dôme region, France; and as a deposit in volcanic fumaroles in Italy, Sicily, and the Lipari Islands.
BEST REF. IN ENGLISH: Palache, et al., "Dana's System of Mineralogy," 7th Ed., v. I, p. 698–707, New York, Wiley, 1944.

## MAGNESIOKATOPHORITE
$Na_2CaMg_4(Fe^{+3},Al)Si_7AlO_{22}(OH,F)_2$
Amphibole Group

CRYSTAL SYSTEM: Monoclinic (See kataphorite)
OPTICAL CONSTANTS:
  $\alpha = 1.639$–1.681
  $\beta = 1.658$–1.688
  $\gamma = 1.660$–1.690
  (–)2V = 0°–50°
HARDNESS: 5
DENSITY: 3.20 (Meas.)
CLEAVAGE: {110} perfect
       {010} parting
Brittle.
HABIT: Crystals prismatic. Twinning on {100}.
COLOR-LUSTER: Rose-red, dark red-brown, bluish black. Translucent. Vitreous.
MODE OF OCCURRENCE: Occurs chiefly in basic alkali rocks such as theralite and shonkinite. Found in the Shields River basin, Montana, and Khibina, Kola Peninsula, USSR.
BEST REF. IN ENGLISH: Deer, Howie, and Zussman, "Rock Forming Minerals," v. 2, p. 359–363, New York, Wiley, 1963.

## MAGNESIORIEBECKITE (Rhodusite, crocidolite)
$Na_2Mg_3Fe_2^{+3}Si_8O_{22}(OH)_2$
Amphibole Group

CRYSTAL SYSTEM: Monoclinic
CLASS: 2/m
SPACE GROUP: C2/m
Z: 2
LATTICE CONSTANTS:
  $a = 9.79$
  $b = 18.12$
  $c = 5.28$
  $\beta = 102°51'$
3 STRONGEST DIFFRACTION LINES:
  8.45 (100)
  3.135 ( 90)
  2.724 ( 80)
HARDNESS: 5
DENSITY: 3.02–3.192 (Meas.)
       3.075 (Calc.)
CLEAVAGE: {110} perfect
Fracture uneven. Brittle.
HABIT: Crystals long prismatic, striated parallel to

elongation. Also massive; fibrous, columnar, or granular. Twinning on {100} simple, lamellar.

COLOR-LUSTER: Dark blue to black. Translucent to nearly opaque. Vitreous, silky.

MODE OF OCCURRENCE: Occurs as an authigenic mineral in rocks of the Green River formation in Utah, Colorado, and Wyoming; in granulite at Glen Lui, Aberdeenshire, Scotland; in amphibole schist in the Krivoi Rog basin, Ukraine, USSR (rhodusite); in hematite ore at Zesfontein, South West Africa; and at localities in Bolivia (crocidolite), Japan, and South Australia (crocidolite).

BEST REF. IN ENGLISH: Deer, Howie, and Zussman, "Rock Forming Minerals," v. 2, p. 333–351, New York, Wiley, 1963.

## MAGNESITE
$MgCO_3$

CRYSTAL SYSTEM: Hexagonal
CLASS: $\bar{3}$ 2/m
SPACE GROUP: R$\bar{3}$c
Z: 2 (rhomb.), 6 (hex.)
LATTICE CONSTANTS:
  $a = 4.579$      $a_{rh} = 5.61$
  $c = 14.845$      $\alpha = 48°10'$
3 STRONGEST DIFFRACTION LINES:
  2.742 (100)
  2.102 ( 45)
  1.700 ( 35)
OPTICAL CONSTANTS:
  $\omega = 1.700$
  $\epsilon = 1.509$
    (–)
HARDNESS: 3¾–4¼
DENSITY: 3.0–3.1 (Meas.)
        3.08 (Calc.)
CLEAVAGE: {10$\bar{1}$1} perfect
Fracture conchoidal. Brittle.

HABIT: Crystals not common; usually rhombohedral; rarely prismatic, tabular, or scalenohedral. Commonly massive, compact, coarse- to fine-grained, chalky to porcelaneous. Also lamellar or fibrous.

COLOR-LUSTER: Colorless, white, gray, yellowish to brown. Transparent to translucent. Vitreous; also dull.

MODE OF OCCURRENCE: Occurs widespread as an alteration product of rocks rich in magnesium; as beds in metamorphic rocks; in sedimentary deposits; as a gangue mineral in hydrothermal ore veins; and rarely as a primary mineral in igneous rock. It occurs widespread in the western United States, especially in serpentinous rocks extending along the Coast Range in California, and in large deposits in Clark County and in the Paradise Range, Nevada. In Europe it is found as excellent crystals at Oberdorf and at other places in Austria; with serpentine at Snarum, Norway; and at many other localities. Superb crystals occur near Brumado in the Serra das Eguas, Bahia, Brazil. Other important deposits are found in Manchuria, Korea, India, Algeria, Zaire, and South Africa.

BEST REF. IN ENGLISH: Palache, et al., "Dana's System of Mineralogy," 7th Ed., v. II, p. 162–166, New York, Wiley, 1951.

## MAGNESIUM-CHLOROPHOENICITE
$(Mg,Mn)_5 (AsO_4)(OH)_7$

CRYSTAL SYSTEM: Monoclinic
CLASS: 2/m
SPACE GROUP: C2/m
OPTICAL CONSTANTS:
  $\alpha = 1.669$
  $\beta = 1.672$
  $\gamma = 1.677$
HARDNESS: Not determined.
DENSITY: 3.37 (Meas.)
CLEAVAGE: One perfect in direction of elongation.
HABIT: As rosettes and radial-fibrous aggregates.
COLOR-LUSTER: Colorless to white.
MODE OF OCCURRENCE: Found as radial aggregates up to 0.4 inch in diameter implanted on the surface of a narrow open vein composed of zincite and carbonates at Franklin, Sussex County, New Jersey.

BEST REF. IN ENGLISH: Palache, et al., "Dana's System of Mineralogy," 7th Ed., v. II, p. 780, New York, Wiley, 1951.

## MAGNETITE
$Fe_3O_4$
Magnetite Series
Spinel Group

CRYSTAL SYSTEM: Cubic
CLASS: 4/m $\bar{3}$ 2/m
SPACE GROUP: Fd3m
Z: 8
LATTICE CONSTANT:
  $a = 8.374$
3 STRONGEST DIFFRACTION LINES:
  2.530 (100)
  1.614 ( 85)
  1.483 ( 85)
OPTICAL CONSTANT:
  $N = 2.42$ (Na)
HARDNESS: 5½–6½
DENSITY: 5.175 (Meas.)
        5.197 (Calc.)
CLEAVAGE: None. {111} parting, good.
Fracture uneven to subconchoidal. Brittle.

HABIT: Crystals usually octahedral, sometimes highly modified; also dodecahedral, striated. Commonly massive, compact, fine to coarse granular. Twinning on {111}, common, as lamellar twins or spinel twins.

COLOR-LUSTER: Iron black, grayish black. Opaque. Splendent metallic to dull. Streak black.

Strongly magnetic.

MODE OF OCCURRENCE: Occurs abundantly and widespread chiefly as magmatic segregations; in sulfide vein deposits; as an accessory mineral in metamorphic rocks, pegmatites, and igneous rocks; as replacement deposits; and as a detrital mineral. Notable United States occurrences are found in California, Utah, New Mexico, Colorado, Wyoming, South Dakota, Arkansas, Texas, New York, New Jersey, and North Carolina. Other major deposits occur in Canada, Mexico, Norway, Sweden, Italy, Switzerland,

Germany, Austria, Hungary, USSR, South Africa, and elsewhere.

BEST REF. IN ENGLISH: Palache, et al., "Dana's System of Mineralogy," 7th Ed., v. I, p. 698–707, New York, Wiley, 1944.

# MAGNETOPLUMBITE
$(Pb,Mn)_2 Fe_6 O_{11}$

CRYSTAL SYSTEM: Hexagonal
CLASS: 6/m 2/m 2/m
SPACE GROUP: $P6_3/mmc$
Z: 2
LATTICE CONSTANTS:
   $a = 5.88$
   $c = 23.02$
3 STRONGEST DIFFRACTION LINES:
   2.77 (100)
   2.63 (100)
   1.62 ( 70)
HARDNESS: 6
DENSITY: 5.517 (Meas.)
CLEAVAGE: {0001} perfect
HABIT: Crystals steep pyramidal, doubly terminated.
COLOR-LUSTER: Grayish black. Opaque. Submetallic. Streak dark brown.

Strongly magnetic.
MODE OF OCCURRENCE: Occurs associated with kentrolite and manganophyllite at Långban, Sweden.
BEST REF. IN ENGLISH: Palache, et al., "Dana's System of Mineralogy," 7th Ed., v. I, p. 728, New York, Wiley, 1944.   Rouse, Roland C., and Peacor, Donald R., *Am. Min.*, 53: 869–879 (1968).

# MAGNÏOPHILITE = Magnesian graftonite

# MAGNIOTRIPLITE = Magnesian triplite

# MAGNOCOLUMBITE
$(Mg,Fe,Mn)(Nb,Ta)_2 O_6$
Magnesium analogue of columbite

CRYSTAL SYSTEM: Orthorhombic
CLASS: 2/m 2/m 2/m
SPACE GROUP: Pnca
Z: 4
LATTICE CONSTANTS:
   $a = 5.02$
   $b = 14.17$
   $c = 5.65$
3 STRONGEST DIFFRACTION LINES:
   2.955 (100)
   1.723 ( 90)
   1.535 ( 90)
OPTICAL CONSTANTS:
   $\alpha = 2.33$
   $\beta = ?$
   $\gamma = 2.40$
   $(+)2V \sim 80°$

HARDNESS: ~6
DENSITY: 5.17 (Meas.)
        5.23 (Calc.)
CLEAVAGE: {010}
          {100}
Fracture uneven.
HABIT: Acicular and tabular crystals 0.1–0.2 cm with rough surfaces. Twinning commonly observed.
COLOR-LUSTER: Black to brownish black; semimetallic. Streak dark brown. Thin splinters translucent brownish red.
MODE OF OCCURRENCE: Occurs commonly intergrown with ilmenorutile in a pegmatite in dolomitic marbles at Kugi-Lyal, southwest Pamir, USSR.
BEST REF. IN ENGLISH: Mathias, V. V., Rossovskii, L. N., Shostatskii, A. N., and Kumskova, N. M., *Am. Min.*, 48: 1182–1183 (1963).

# MAGNOPHORITE = Variety of richterite

# MAGNUSSONITE
$(Mn,Mg,Cu)_5 (AsO_3)_3 (OH,Cl)$

CRYSTAL SYSTEM: Cubic (possibly tetragonal)
CLASS: 4/m $\bar{3}$ 2/m
SPACE GROUP: Ia3d
Z: 32
LATTICE CONSTANTS:
                (tetrag.)
   $a = 19.70$    $a = 19.58$
                 $c = 19.72$
                 $I4_1/amd$
3 STRONGEST DIFFRACTION LINES:
   2.85 (100)
   3.12 ( 30)
   2.47 ( 30)
HARDNESS: 3½–4
DENSITY: 4.30 (Meas.)
         4.23 (Calc.)
HABIT: Fine-grained incrustations.
COLOR-LUSTER: Grass green to emerald green, sometimes blue green. Vitreous. Streak white.
MODE OF OCCURRENCE: Occurs in fissures, usually in dolomite impregnated by hausmannite, or in fine-grained hematite with calcite, trigonite, dixenite, brown manganiferous serpentine, and an unidentified black mineral that contains iron, magnesium, copper, and silicon, at Långban, Sweden; a tetragonal form from the Brattfors mine, Nordmarks Odalfält, Sweden, is in space group $I4_1/amd$, $a$ 19.58, $c$ 19.72 Å.
BEST REF. IN ENGLISH: Gabrielson, O., *Am. Min.*, 42: 581 (1957).   Moore, Paul B., *Am. Min.*, 56: 639 (1971).

# MAJORITE
$Mg_3 (Fe,Al,Si)_2 Si_3 O_{12}$
Garnet Group

CRYSTAL SYSTEM: Cubic
CLASS: 4/m $\bar{3}$ 2/m

SPACE GROUP: Ia3d
Z: 8
LATTICE CONSTANT:
  $a = 11.524$
3 STRONGEST DIFFRACTION LINES: Similar to pyrope.
  HARDNESS: $\sim 7-7\frac{1}{2}$
  DENSITY: $\sim 4$
  CLEAVAGE: None.
  HABIT: As minute embedded grains.
  COLOR-LUSTER: Purple.
  MODE OF OCCURRENCE: Occurs associated with ringwoodite, olivine, kamacite, and goethite, often coarsely interwoven with pyroxene, in the Coorara meteorite.
  BEST REF. IN ENGLISH: Smith, J. V., and Mason, Brian, *Science*, **168**: 832–833 (1970).

## MAKATITE
$Na_2 Si_4 O_9 \cdot 5H_2O$

CRYSTAL SYSTEM: Orthorhombic
Z: 1
LATTICE CONSTANTS:
  $a = 16.840$
  $b = 10.256$
  $c = 19.146$
3 STRONGEST DIFFRACTION LINES:
  5.09 (100)
  2.97 ( 57)
  9.04 ( 53)
OPTICAL CONSTANTS:
  $n_{min} = 1.472$
  $n_{max} = 1.487$
  HARDNESS: Not determined.
  DENSITY: 2.072 (Meas.)
  CLEAVAGE: Not determined.
  HABIT: Crystals acicular, 0.05–2.0 $\mu$m wide and commonly 5–30 $\mu$m long; in radiating aggregates. Also as spherulites 0.05–0.3 mm in diameter.
  COLOR-LUSTER: White.
  MODE OF OCCURRENCE: Found in cores of trona from three drill holes in the Evaporite Series of Lake Magadi, Kenya. Makatite commonly replaces trona and is locally pseudomorphous after bladed trona. The pseudomorphs are spongy in appearance because of the spherulitic character of the makatite.
  BEST REF. IN ENGLISH: Sheppard, Richard A., Gude, Arthur J. III, and Hay, R. L., *Am. Min.*, **55**: 358–366 (1970).

## MÄKINENITE
$\gamma$-NiSe

CRYSTAL SYSTEM: Hexagonal
CLASS: 3m
SPACE GROUP: Probably R3m
Z: 9
LATTICE CONSTANTS:
  $a = 10.01$
  $c = 3.28$

3 STRONGEST DIFFRACTION LINES:
  2.88 (100)
  2.63 (100)
  2.325 (100)
  HARDNESS: Almost as soft as clausthalite.
  DENSITY: 7.22 (Calc.)
  CLEAVAGE: Not determined.
  HABIT: Massive.
  COLOR-LUSTER: Pure yellow in oil; orange-yellow in air. Pleochroism strong, yellow to greenish yellow. Anisotropism strong, in oil pale green to pale orange-yellow, in air glowing cinder red to blue-green or green.
  MODE OF OCCURRENCE: Occurs with clausthalite and selenian melonite in sills of albite diabase in a schist formation, associated with low-grade uranium mineralization at Kuusamo, northeast Finland.
  BEST REF. IN ENGLISH: Vourelainen, Y., Huhma, A., and Häkli, A., *Am. Min.*, **50**: 520 (1965).

## MALACHITE
$Cu_2 CO_3 (OH)_2$

CRYSTAL SYSTEM: Monoclinic
CLASS: 2/m
SPACE GROUP: $P2_1/a$
Z: 4
LATTICE CONSTANTS:
  $a = 9.48$
  $b = 12.03$
  $c = 3.21$
  $\beta = 98°$
3 STRONGEST DIFFRACTION LINES:
  2.857 (100)
  3.693 ( 85)
  5.055 ( 75)
OPTICAL CONSTANTS:
  $\alpha = 1.655$
  $\beta = 1.875$
  $\gamma = 1.909$
  $(-)2V \simeq 43°$ (Na)
  HARDNESS: $3\frac{1}{2}-4$
  DENSITY: 4.05 (Meas.)
         4.00 (Calc.)
(Massive material ranges down to $\sim 3.6$)
  CLEAVAGE: $\{\bar{2}01\}$ perfect
            $\{010\}$ fair
Fracture subconchoidal to uneven (massive). Crystals brittle. Massive material tough.
  HABIT: Crystals small, usually acicular or short to long prismatic with wedge-shaped terminations. Very commonly twinned on $\{100\}$. Usually massive, often as thick compact crusts with botryoidal or mammillary surface and fibrous, banded structure. Also stalactitic, as loosely coherent aggregates, and as thin films and stains.
  COLOR-LUSTER: Bright green to dark or blackish green. Translucent to opaque. Crystals vitreous to adamantine. Fibrous varieties silky or velvety; often dull or earthy. Streak pale green.
  MODE OF OCCURRENCE: Occurs widespread as a secondary mineral in the oxidation zone of copper deposits, closely associated with azurite and other secondary minerals.

Found widespread throughout the mining districts of the western United States, especially in Arizona at Bisbee, Cochise County, and in Gila, Greenlee, and Pima counties. It also occurs abundantly in Grant and Socorro counties, New Mexico; the Tintic district, Utah; and in the eastern United States at Ducktown, Tennessee, and at the Jones mine, Berks County, Pennsylvania. Other famous deposits include those of Goumeshevsk and Mednorudyansk in the Ural Mountains, USSR, where compact masses up to 50 tons have been found; Luishwishi, Katanga, Zaire; Tsumeb, South West Africa; Bwana Mkubwa, Zambia; the Burra district, South Australia, and Broken Hill, New South Wales, Australia.

BEST REF. IN ENGLISH: Palache, et al., "Dana's System of Mineralogy," 7th Ed., v. II, p. 252–256, New York, Wiley, 1951.

## MALACOLITE = Synonym for diopside

## MALACON = Variety of zircon

## MALAYITE

$CaSnSiO_5$
Isostructural with sphene

CRYSTAL SYSTEM: Monoclinic (or triclinic)
3 STRONGEST DIFFRACTION LINES:
  3.28 (100)
  5.03 ( 80)
  2.66 ( 80)
OPTICAL CONSTANTS:
  $\alpha = 1.765$
  $\beta = 1.784$
  $\gamma = 1.799$
  $(-)2V \simeq 85°$
HARDNESS: 3½–4
DENSITY: 4.3 (Meas.)
CLEAVAGE: Not determined.
HABIT: Massive.
COLOR-LUSTER: Pale yellow to colorless. Translucent. Fluoresces yellowish green under short-wave ultraviolet light.
MODE OF OCCURRENCE: Occurs in association with varlamoffite as a coating on cassiterite crystals from Sungei Lah section, Chenderiang, Malaya.
BEST REF. IN ENGLISH: Ingham, F. T., and Bradford, E. F., *Am. Min.*, **46**: 768–769 (1961).

## MALDONITE

$Au_2Bi$

CRYSTAL SYSTEM: Cubic
CLASS: $4/m\,\bar{3}\,2/m$
SPACE GROUP: Fd3m
Z: 8
LATTICE CONSTANT:
  $a = 7.958$

3 STRONGEST DIFFRACTION LINES:
  2.41 (100)
  1.537 ( 60)
  2.30 ( 50)
HARDNESS: 1½–2
DENSITY: 15.46 (Meas.)
        15.70 (Calc.)
CLEAVAGE: {001} distinct
Malleable and sectile.
HABIT: Massive granular and as thin coatings. Synthetic crystals octahedral.
COLOR-LUSTER: Silver white with pink tint, tarnishes copper red to black. Metallic.
MODE OF OCCURRENCE: Occurs in gold mines at Nuggety Reef and Union Reef, Maldon, Victoria, Australia.
BEST REF. IN ENGLISH: Palache, et al., "Dana's System of Mineralogy," 7th Ed., v. I, p. 95–96, New York, Wiley, 1944.

## MALLADRITE

$Na_2SiF_6$

CRYSTAL SYSTEM: Hexagonal
CLASS: 32
SPACE GROUP: P321
Z: 3
LATTICE CONSTANTS:
  $a = 8.86$
  $c = 5.04$
3 STRONGEST DIFFRACTION LINES:
  2.287 (100)
  1.801 ( 80)
  3.34 ( 70)
OPTICAL CONSTANTS:
  $\omega = 1.3125$
  $\epsilon = 1.3089$ (Na)
  $(-)$
HARDNESS: Not determined.
DENSITY: 2.714 (Meas. synthetic)
CLEAVAGE: Not determined
HABIT: As crusts. Synthetic crystals prismatic with basal plane and pyramid.
COLOR-LUSTER: Pale rose.
MODE OF OCCURRENCE: Occurs associated with sal-ammoniac, feruccite, avogadrite, and hieratite as an incrustation on lava at Mt. Vesuvius, Italy.
BEST REF. IN ENGLISH: Palache, et al., "Dana's System of Mineralogy," 7th Ed., v. II, p. 105–106, New York, Wiley, 1951.

## MALLARDITE (Inadequately described mineral)

$MnSO_4 \cdot 7H_2O$

CRYSTAL SYSTEM: Monoclinic
CLASS: 2/m
OPTICAL CONSTANT:
  $(+)2V$
HARDNESS: $\sim 2$
DENSITY: 1.846 (Meas. Synthetic)
CLEAVAGE: {001} good

HABIT: As fibrous masses and crusts. Synthetic crystals tabular, flattened on {001}.

COLOR-LUSTER: Pale rose. Vitreous. Dehydrates rapidly becoming opaque and mealy. Soluble in water.

MODE OF OCCURRENCE: Occurs as an oxidation product associated with manganoan melanterite at the Lucky Boy mine, Butterfield Canyon, Salt Lake County, Utah; and in the Bayard area of the Central district, Grant County, New Mexico.

BEST REF. IN ENGLISH: Palache, et al., "Dana's System of Mineralogy," 7th Ed., v. II, p. 507-508, New York, Wiley, 1951.

# MANANDONITE
$LiAl_4(AlBSi_2O_{10})(OH)_8$
Chlorite Group

CRYSTAL SYSTEM: Monoclinic
Z: 2
LATTICE CONSTANTS:
$a = 5.23$
$b = 8.92$
$c = 14.11$
$\beta = 97°45'$
OPTICAL CONSTANTS:
$\beta = 1.6$
$\gamma - \alpha = 0.014$
$(+)2V \approx 25°-30°$
HARDNESS: $\sim 2\frac{1}{2}$
DENSITY: 2.89 (Meas.)
CLEAVAGE: {001} perfect, micaceous
HABIT: As lamellar aggregates and crusts of hexagonal plates.
COLOR-LUSTER: Colorless. Transparent. Pearly.

Easily fusible.

MODE OF OCCURRENCE: Occurs in pegmatite in association with rubellite tourmaline, lepidolite, and quartz, at Antandrokomby, near Mt. Bity on the Manandona River, Madagascar.

BEST REF. IN ENGLISH: Vlasov, K. A., "Mineralogy of Rare Elements," v. II, p. 35, Israel Program for Scientific Translations, 1966.

# MANASSEITE
$Mg_6Al_2CO_3(OH)_{16} \cdot 4H_2O$
Dimorphous with hydrotalcite

CRYSTAL SYSTEM: Hexagonal
Z: 1
LATTICE CONSTANTS:
$a = 6.12$
$c = 15.34$
3 STRONGEST DIFFRACTION LINES:
7.67 (100)
1.84 ( 60)
2.60 ( 50)
OPTICAL CONSTANTS:
$\omega = 1.524$
$\epsilon = 1.510$
(-)

HARDNESS: 2
DENSITY: 2.05 (Meas.)
2.00 (Calc.)
CLEAVAGE: {0001} perfect
Laminae flexible, inelastic. Greasy feel.
HABIT: Massive, foliated; lamellar on {0001}.
COLOR-LUSTER: White, pale bluish, grayish, brownish white. Transparent. Pearly, waxy.
MODE OF OCCURRENCE: Occurs associated with serpentine at Amity, Orange County, New York; also at Nordmark and Snarum, Norway.
BEST REF. IN ENGLISH: Palache, et al., "Dana's System of Mineralogy," 7th Ed., v. I, p. 658-659, New York, Wiley, 1944.

# MANGAN-ALLUAUDITE = Variety of alluaudite

# MANGANAXINITE (See Axinite Group)
$Ca_2(Mn,Fe)Al_2BSi_4O_{15}(OH)$ with $Ca > 1.5$ and $Mn > Fe$

# MANGANBERZELIITE
$(Ca,Na)_3(Mn,Mg)_2(AsO_4)_3$
Berzeliite-Manganberzeliite Series

CRYSTAL SYSTEM: Cubic
CLASS: $4/m\bar{3}2/m$
SPACE GROUP: Ia3d
Z: 8
LATTICE CONSTANT:
$a \simeq 12.5$ kX
OPTICAL CONSTANT:
$N = 1.790$ ($\lambda = 535$ m$\mu$)
HARDNESS: $4\frac{1}{2}-5$
DENSITY: 4.21 (Meas. Franklin)
$\sim 4.46$ (Mn end-member)
CLEAVAGE: None.
Fracture uneven to subconchoidal. Brittle.
HABIT: Crystals rare; as minute trapezohedrons with small modifying faces. Usually massive or as rounded grains.
COLOR-LUSTER: Honey yellow to orange-yellow to yellowish red. Transparent to translucent. Resinous.
MODE OF OCCURRENCE: Occurs as granular veinlets up to ¾ inch thick cutting franklinite-willemite ore at Franklin, Sussex County, New Jersey. It also is found in Sweden in limestone skarn associated with caryinite, hausmannite, tephroite, rhodonite, and other minerals, at Långban; at the Sjö mine, Grythytte parish; and at the Moss mine, Nordmark.
BEST REF. IN ENGLISH: Palache, et al., "Dana's System of Mineralogy," 7th Ed., v. II, p. 681-683, New York, Wiley, 1951.

# MANGANESE-HÖRNESITE
$(Mn, Mg)_3(AsO_4)_2 \cdot 8H_2O$

CRYSTAL SYSTEM: Monoclinic
CLASS: $2/m$

SPACE GROUP: Probably $P2_1/c$
Z: 4
LATTICE CONSTANTS:
$a = 10.38$
$b = 28.09$
$c = 4.774$
$\beta = 105°40'$
3 STRONGEST DIFFRACTION LINES:
7.01 (100)
8.19 ( 80)
3.02 ( 70)
OPTICAL CONSTANTS:
$\alpha = 1.579$
$\beta = 1.589$
$\gamma = 1.609$
$(+)2V = 65°-70°$
HARDNESS: 1
DENSITY: 2.64 (Meas.)
2.76 (Calc.)
CLEAVAGE: {010} perfect
{100} poor
HABIT: Massive.
COLOR-LUSTER: White.
MODE OF OCCURRENCE: Occurs as crusts in fissures at Långban, Sweden.
BEST REF. IN ENGLISH: Gabrielson, O., *Am. Min.*, **39**: 159 (1954).

# MANGANITE
MnO(OH)
Trimorphous with feitknechtite, groutite

CRYSTAL SYSTEM: Monoclinic
CLASS: 2/m
SPACE GROUP: $B2_1/d$
Z: 8
LATTICE CONSTANTS:
$a = 8.94$
$b = 5.28$
$c = 5.74$
$\beta = 90°$
3 STRONGEST DIFFRACTION LINES:
3.40 (100)
2.64 ( 60)
2.28 ( 50)
OPTICAL CONSTANTS:
$\alpha = 2.25$
$\beta = 2.25$ (Li)
$\gamma = 2.53$
$(+)2V = $ small
HARDNESS: 4
DENSITY: 4.33 (Meas.)
4.30 (Calc.)
CLEAVAGE: {010} perfect
{110} imperfect
{001} imperfect
Fracture uneven. Brittle.
HABIT: Crystals short to long prismatic, vertically striated, with flat or wedge-shaped terminations; sometimes highly modified. Often grouped in bundles. Also massive, fibrous to columnar; granular; concretionary; stalactitic. Twinning on {011} as contact or penetration twins; on {100} lamellar.
COLOR-LUSTER: Black to dark steel gray. Opaque. Submetallic to dull. Streak reddish brown to black.
MODE OF OCCURRENCE: Occurs chiefly as a low-temperature hydrothermal vein mineral; as deposits formed by circulating meteoric waters; or as a bog, lacustrine, or shallow marine deposit. Found widespread in small amounts in the United States, especially in California, Arizona, New Mexico, Utah, Colorado, South Dakota, Minnesota, Michigan, New Jersey, Virginia, and Georgia. Other notable occurrences are found in Germany, especially as superb crystals at Ilfeld, Harz Mountains; and at localities in Canada, England, Scotland, Sweden, France, Roumania, China, and elsewhere.
BEST REF. IN ENGLISH: Palache, et al., "Dana's System of Mineralogy," 7th Ed., v. I, p. 646–650, New York, Wiley, 1944.

# MANGAN-NEPTUNITE  See Neptunite
$(Na, K)_2(Mn, Fe)TiSi_4O_{12}$

# MANGANOCALCITE = Variety of calcite containing manganese

# MANGANOLANGBEINITE
$K_2Mn_2(SO_4)_3$

CRYSTAL SYSTEM: Cubic
CLASS: 23
SPACE GROUP: $P2_13$
Z: 4
LATTICE CONSTANT:
$a = 10.014$
3 STRONGEST DIFFRACTION LINES:
3.20 (100)
2.70 ( 50)
3.05 ( 18)
OPTICAL CONSTANT:
$N = 1.572$
HARDNESS: Not determined.
DENSITY: 3.02 (Meas.)
3.057 (Calc.)
CLEAVAGE: Not determined.
HABIT: Crystals small tetrahedra.
COLOR-LUSTER: Rose red.
MODE OF OCCURRENCE: Occurs in a cavern in lava on Mt. Vesuvius, Italy, associated with halite, sylvite, and other minerals.
BEST REF. IN ENGLISH: Palache, et al., "Dana's System of Mineralogy," 7th Ed., v. II, p. 435, New York, Wiley, 1951.

# MANGANOPHYLLITE = Variety of biotite containing up to 18% MnO

# MANGANOSIDERITE = Variety of rhodochrosite

## MANGANOSITE
MnO

CRYSTAL SYSTEM: Cubic
CLASS: $4/m\ \bar{3}\ 2/m$
SPACE GROUP: Fm3m
Z: 4
LATTICE CONSTANT:
$a = 4.436$
3 STRONGEST DIFFRACTION LINES:
2.223 (100)
2.568 ( 60)
1.571 ( 60)
OPTICAL CONSTANT:
$N_{green} = 2.19$
HARDNESS: 5½
DENSITY: 5.364 (Meas.)
5.365 (Calc. x-ray)
CLEAVAGE: {001} fair
Fracture somewhat fibrous.

HABIT: Crystals octahedral, sometimes modified by dodecahedron or cube. Commonly as masses or irregular grains.

COLOR-LUSTER: Emerald green, becoming black on exposure. Transparent. Vitreous. Streak brown.

MODE OF OCCURRENCE: Occurs with franklinite, willemite, and zincite, at Franklin, Sussex County, New Jersey; and in Sweden at Långban and Nordmark associated with pyrochroite in dolomite.

BEST REF. IN ENGLISH: Palache, et al., "Dana's System of Mineralogy," 7th Ed., v. I, p. 501–502, New York, Wiley, 1944.

## MANGANOSTEENSTRUPINE
(Ce, La, Th, Ca)Mn(SiO$_3$)(OH)$_2$ · 2H$_2$O
Manganese analogue of steenstrupine

For crystal data see steenstrupine
OPTICAL CONSTANT:
$N > 1.80$
HARDNESS: 5½–6
DENSITY: 3.288
CLEAVAGE: Fracture conchoidal. Brittle.
HABIT: Aggregates up to 10 cm in diameter. Amorphous.
COLOR-LUSTER: Black, opaque, brownish red in thin fragments. Resinous. Streak dark brown.

MODE OF OCCURRENCE: Occurs in hackmanite syenites, associated with microcline, natrolite, hackmanite, aegirine, schizolite, and nenadkevichite, in pegmatites on the northeast slope of Mt. Karnasurt, Lovozero massif, Kola Peninsula, USSR.

BEST REF. IN ENGLISH: Vlasov, K. A., Kuz'menko, M. V., and Es'kova, E. M., *Am. Min.*, **45**: 1132 (1960).

## MANGANOSTIBITE
Mn$_7^{2+}$Sb$^{5+}$As$^{5+}$O$_{12}$

CRYSTAL SYSTEM: Orthorhombic
CLASS: $2/m\ 2/m\ 2/m$
SPACE GROUP: Ibmm
Z: 4

LATTICE CONSTANTS:
$a = 8.72$
$b = 18.85$
$c = 6.06$
3 STRONGEST DIFFRACTION LINES:
2.651 (100)
4.965 ( 60)
4.380 ( 60)
OPTICAL CONSTANTS:
$\alpha = 1.92$
$\beta = 1.95$
$\gamma = 1.96$
$(-)2V = $ small
HARDNESS: Not determined.
DENSITY: 4.949 (Meas.)
5.00 (Calc.)
CLEAVAGE: Fracture uneven.
HABIT: Embedded grains, sometimes with crystal outlines.
COLOR-LUSTER: Black, opaque; greasy. Streak chocolate brown.

MODE OF OCCURRENCE: Found associated with sonolite and catoptrite in marble at the Brattfors mine, Nordmarks Odalfält, Värmland, Sweden.

BEST REF. IN ENGLISH: Moore, Paul B., *Am. Min.*, **53**: 1779 (1968). Moore, Paul B., *Am. Min.*, **55**: 1489–1499 (1970).

## MANGANOTANTALITE
(Mn, Fe)(Ta, Nb)$_2$O$_6$
Columbite-Tantalite Series

CRYSTAL SYSTEM: Orthorhombic
CLASS: $2/m\ 2/m\ 2/m$
SPACE GROUP: Pcan
Z: 4
LATTICE CONSTANTS:
$a = 5.09$
$b = 14.39$ (variable)
$c = 5.76$
3 STRONGEST DIFFRACTION LINES:
2.99 (100)
3.69 ( 90)
2.41 ( 70)
OPTICAL CONSTANTS:
$\alpha = 2.19$
$\beta = 2.25$
$\gamma = 2.34$
$(+)2V = $ large
HARDNESS: 6–6½
DENSITY: 8.003 (Calc.)
CLEAVAGE: {010} distinct
{100} indistinct
Fracture subconchoidal to uneven. Brittle.

HABIT: Crystals short prismatic, equant, or thick tabular. Also massive. Twinning on {201}, common.

COLOR-LUSTER: Brownish black with reddish brown or red internal reflections. Translucent; thin fragments transparent. Vitreous to resinous. Streak dark red.

MODE OF OCCURRENCE: Occurs chiefly in granite pegmatites in association with albite, microcline, lepidolite, and other minerals; less commonly in placer deposits. Found near Pala, San Diego County, California; at the Walden quarry, Portland, Connecticut; and in the vicinity of Amelia, Virginia. Also found in Minas Geraes, Brazil, and at localities in Sweden, USSR, Australia, and Mozambique.

BEST REF. IN ENGLISH: Palache, et al., "Dana's System of Mineralogy," 7th Ed., v. I, p. 780–787, New York, Wiley, 1944.

# MANGANPYROSMALITE
$(Mn, Fe)_8 (Si_6 O_{15})(OH, Cl)_{10}$

CRYSTAL SYSTEM: Hexagonal
CLASS: $\bar{3}$ 2/m
SPACE GROUP: P$\bar{3}$m1
Z: 2
LATTICE CONSTANTS:
 $a$ = 13.36
 $c$ = 7.16
3 STRONGEST DIFFRACTION LINES:
 7.16 (100)
 2.683 ( 90)
 3.583 ( 80)
OPTICAL CONSTANTS:
 $\omega$ = 1.669
 $\epsilon$ = 1.631
 (–)
HARDNESS: 4½
DENSITY: 3.13 (Meas.)
 3.14 (Calc.)
CLEAVAGE: {0001} perfect
HABIT: Massive granular, grains range up to 0.3 mm in size.
COLOR-LUSTER: Pure brown.
MODE OF OCCURRENCE: Occurs associated with friedelite, bementite, and willemite as veinlets in franklinite ore at Sterling Hill, Sussex County, New Jersey.
BEST REF. IN ENGLISH: Frondel, Clifford, and Bauer, L. H., *Am. Min.*, **38**: 755–760 (1953).

# MANJIROITE
$NaMn_8 O_{16} \cdot nH_2O$
Na analogue of cryptomelane

CRYSTAL SYSTEM: Tetragonal
Z: 1
LATTICE CONSTANTS:
 $a$ = 9.916
 $c$ = 2.864
3 STRONGEST DIFFRACTION LINES:
 2.406 (100)
 7.017 ( 98)
 3.136 ( 92)
HARDNESS: 7
DENSITY: 4.29 (Meas.)
CLEAVAGE: None. Conchoidal fracture.
HABIT: Massive; compact.
COLOR-LUSTER: Dark brownish gray; dull. Streak brownish black.

MODE OF OCCURRENCE: Occurs associated with pyrolusite, nsutite, birnessite, cryptomelane, and goethite, in masses up to 10 × 8 × 5 cm in size, in the oxidation zone of rhodonite-tephroite-rhodochrosite bedded ore deposits in the Kohare mine and at five other deposits, Iwate Prefecture, Japan.
BEST REF. IN ENGLISH: Nabu, Matsuo, and Tanida, Katsutoshi, *Am. Min.*, **53**: 2103 (1968).

# MANSFIELDITE
$AlAsO_4 \cdot 2H_2O$
Scorodite-Mansfieldite Series.

CRYSTAL SYSTEM: Orthorhombic
CLASS: 2/m 2/m 2/m
SPACE GROUP: Pcab
Z: 8
LATTICE CONSTANTS:
 $a$ = 10.08
 $b$ = 9.76
 $c$ = 8.72
3 STRONGEST DIFFRACTION LINES:
 5.45 (100)
 4.36 (100)
 3.09 ( 90)
OPTICAL CONSTANTS:
 $\alpha$ = 1.622 –1.631
 $\beta$ = 1.624–1.649
 $\gamma$ = 1.642–1.663
 (+)2V = 30°–68°
HARDNESS: 3½–4
DENSITY: 3.031 (Meas.)
 3.126 (Calc.)
CLEAVAGE: {201} imperfect
 {001} trace
 {100} trace
HABIT: As fibrous crusts and porous masses.
COLOR-LUSTER: White to pale gray. Transparent in crystals. Vitreous.
MODE OF OCCURRENCE: Occurs at Hobart Butte, Lane County, Oregon, associated with scorodite, realgar, and kaolinite.
BEST REF. IN ENGLISH: Palache, et al., "Dana's System of Mineralogy," 7th Ed., v. II, p. 763–767, New York, Wiley, 1951.

# MANSJÖITE = Variety of diopside

# MARCASITE
$FeS_2$
Dimorphous with Pyrite

CRYSTAL SYSTEM: Orthorhombic
CLASS: 2/m 2/m 2/m
SPACE GROUP: Pnnm
Z: 2
LATTICE CONSTANTS:
 $a$ = 4.436
 $b$ = 5.414
 $c$ = 3.381

3 STRONGEST DIFFRACTION LINES:
2.71 (100)
1.76 ( 63)
3.44 ( 40)
HARDNESS: 6–6½
DENSITY: 4.92 (Meas.)
4.875 (Calc.)
CLEAVAGE: {101} distinct
{110} in traces
Fracture uneven. Brittle.

HABIT: Crystals usually tabular, flattened on {010}; also pyramidal, prismatic, or capillary. Faces often curved. Commonly massive; fine granular, stalactitic, reniform, and globular; often with radial structure. Twinning on {101} common, producing cockscomb and spear shapes or stellate fivelings. Sometimes twinned on {011}.

COLOR-LUSTER: Pale brass yellow to tin white, darkening on exposure; iridescent when tarnished. Opaque. Metallic. Streak greenish black.

MODE OF OCCURRENCE: Occurs chiefly as a low-temperature mineral formed from highly acid solutions. Found frequently in clay, shale, chalk, limestone, and coal beds, and as a supergene mineral in vein deposits. It occurs widespread in the United States, especially in the lead-zinc deposits of Missouri, Oklahoma, Kansas, Illinois, and Wisconsin. Other notable occurrences are found in Mexico, Bolivia, England, France, Germany, and Czechoslovakia.

BEST REF. IN ENGLISH: Palache, et al., "Dana's System of Mineralogy," 7th Ed., v. I, p. 311–315, New York, Wiley, 1944.

## MARGARITE

$CaAl_4Si_2O_{10}(OH)_2$
Mica Group

CRYSTAL SYSTEM: Monoclinic
CLASS: 2/m
SPACE GROUP: C2/c
Z: 4
LATTICE CONSTANTS:
$a = 5.13$
$b = 8.92$
$c = 19.50$
$\beta = 95°$
3 STRONGEST DIFFRACTION LINES:
3.180 (100)
1.908 ( 35)
2.517 ( 25)
OPTICAL CONSTANTS:
$\alpha = 1.630-1.638$
$\beta = 1.642-1.648$
$\gamma = 1.644-1.650$
$(-)2V = 40°-67°$
HARDNESS: 3½–4½
DENSITY: 3.0–3.1 (Meas.)
3.077 (Calc.)
CLEAVAGE: {001} perfect
Laminae brittle.

HABIT: Crystals tabular, pseudohexagonal, rare. Usually in confused platy aggregates or scaly masses. Twinning on {001}, twin axis [310].

COLOR-LUSTER: Grayish pink, pink, pale yellow, pale green. Translucent. Pearly. Streak uncolored.

MODE OF OCCURRENCE: Occurs in association with corundum and diaspore in metamorphic emery deposits; also associated with staurolite and tourmaline in chlorite and mica schists. Found with corundum in the plumasite near Meadow Valley, Plumas County, and in glaucophane-bearing rocks at several localities in California; at Chester, Hampden County, Massachusetts; with corundum in Chester and Delaware counties, Pennsylvania; and near Buck Creek, Clay County, North Carolina. Also at localities in Italy, Austria, Greece, USSR, Turkey, South Africa, Japan, and Australia.

BEST REF. IN ENGLISH: Deer, Howie, and Zussman, "Rock Forming Minerals," v. 3, p. 95–98, New York, Wiley, 1962.

## MARGAROSANITE

$(Ca, Mn)_2PbSi_3O_9$

CRYSTAL SYSTEM: Triclinic
CLASS: 1, or $\bar{1}$
SPACE GROUP: P1 or P$\bar{1}$
Z: 2
LATTICE CONSTANTS:
$a = 6.77$ $\alpha = 110°35'$
$b = 9.64$ $\beta = 102°0'$
$c = 6.75$ $\gamma = 88°30'$
3 STRONGEST DIFFRACTION LINES:
2.978 (100)
3.03 ( 50)
2.673 ( 40)
OPTICAL CONSTANTS:
$\alpha = 1.727$
$\beta = 1.771$
$\gamma = 1.798$
$(-)2V = 78°$
HARDNESS: 2.5–3
DENSITY: 4.33 (Meas.)
4.30 (Calc.)
CLEAVAGE: {010} perfect
{100} good
{001} fair

HABIT: As lamellar masses of thin plates. Plates show rhombic outlines of the two lesser cleavages.

COLOR-LUSTER: Colorless, transparent; pearly on cleavage faces. Streak white.

MODE OF OCCURRENCE: Occurs associated with almandite, hancockite, roeblingite, nasonite, franklinite, willemite, yellow axinite, datolite, manganophyllite (?), and barite, at Franklin, New Jersey; and with nasonite, schefferite, apophyllite, calcite, and thaumasite at Långban, Sweden.

BEST REF. IN ENGLISH: Armstrong, Richard Lee, Am. Min., 48: 698–703 (1963). Am. Min., 49: 781–782 (1964).

## MARIALITE

$mNa_4(Al_3Si_9O_{24})Cl$ with $nCa_4(Al_6Si_6O_{24})CO_3$
$Ma_{100}Me_0$ to $Ma_{80}Me_{20}$
Scapolite Group

CRYSTAL SYSTEM: Tetragonal
CLASS: 4/m
SPACE GROUP: P4/m
Z: 2
LATTICE CONSTANTS:
  $a$ = 12.075
  $c$ = 7.516
3 STRONGEST DIFFRACTION LINES:
  3.44 (100)
  3.03 (100)
  3.78 ( 90)
OPTICAL CONSTANTS:
  $\omega$ = 1.546–1.550
  $\epsilon$ = 1.540–1.541
  (–)
HARDNESS: 5½–6
DENSITY: 2.50–2.62 (Meas.)
CLEAVAGE: {100} distinct
           {110} distinct
Fracture uneven to conchoidal. Brittle.
HABIT: Crystals prismatic, often coarse. Also massive, granular; sometimes columnar.
COLOR-LUSTER: Colorless, white, gray, bluish, greenish, yellowish, pink, violet, and brownish. Transparent to translucent. Vitreous to somewhat pearly or resinous. Streak uncolored.
MODE OF OCCURRENCE: Occurs mainly in regionally metamorphosed rocks, contact zones, altered igneous rocks, and ejected volcanic blocks. Typical occurrences are found in Ontario and Quebec, Canada, and north Karelia, USSR.
BEST REF. IN ENGLISH: Shaw, D. M., *Jr. Petr.*, 1: 218; 261 (1960). Deer, Howie, and Zussman, "Rock Forming Minerals," v. 4, p. 321–337, New York, Wiley, 1963.

## MARIGNACITE = Cerian pyrochlore

## MARIPOSITE = Variety of phengite containing chromium

## MARMATITE = Variety of sphalerite containing more than 10% Fe

## MAROKITE
$CaMn_2O_4$

CRYSTAL SYSTEM: Orthorhombic
CLASS: 2/m 2/m 2/m
SPACE GROUP: Pmab
Z: 4
LATTICE CONSTANTS:
  $a$ = 9.71
  $b$ = 10.03
  $c$ = 3.16
3 STRONGEST DIFFRACTION LINES:
  2.71 (100)
  2.22 (100)
  2.29 ( 80)

OPTICAL CONSTANTS:
  $\alpha \simeq 2.10$
  $\beta \simeq \gamma \simeq 2.42$
  $(-)2V = 20°-25°$
HARDNESS: Not determined.
DENSITY: 4.64 (Meas.)
          4.63 (Calc.)
CLEAVAGE: {100} perfect
           {001} good
HABIT: As large untwinned crystals flattened on {010}. Maximum size 5 × 1.5 × 0.5 cm.
COLOR-LUSTER: Black, opaque with dark red internal reflection; bright luster. Streak reddish brown.
MODE OF OCCURRENCE: Occurs in a gangue of manganese-free calcite and barite, associated with hausmannite, braunite, rare crednerite and polianite, and as yet unidentified minerals, on the dump of no. 2 vein at Tachgagalt, 17 km south-southeast of Ouarzazate, Morocco; also as a rare mineral in the ore from Black Rock Mine, N.W. Cape Province, Republic of South Africa.
BEST REF. IN ENGLISH: Gaudefroy, C., Jouravsky, C., and Permingeat, F., *Am. Min.*, **49**: 817 (1964).

## MARRITE
$PbAgAsS_3$

CRYSTAL SYSTEM: Monoclinic
CLASS: 2/m
SPACE GROUP: P2$_1$/a
Z: 4
LATTICE CONSTANTS:
  $a$ = 7.291
  $b$ = 12.68
  $c$ = 5.998
  $\beta$ = 91°13′
3 STRONGEST DIFFRACTION LINES:
  3.45 (100)
  2.75 (100)
  3.00 ( 70)
HARDNESS: 3
DENSITY: 6.23
          5.822 (Calc.)
CLEAVAGE: None. Fracture conchoidal. Brittle.
HABIT: Crystals equant to tabular {010}; striated [100].
COLOR-LUSTER: Lead gray to steel gray; metallic, often with iridescent tarnish. Streak black with brown tinge.
MODE OF OCCURRENCE: Occurs associated with lengenbachite and rathite in dolomite at the Lengenbach quarry, Binnatal, Valais, Switzerland.
BEST REF. IN ENGLISH: Wuensch, B. J., and Nowacki, W., *Am. Min.*, **50**: 812 (1965).

## MARSHITE
$CuI$

CRYSTAL SYSTEM: Cubic
CLASS: $\bar{4}$ 3m
SPACE GROUP: F$\bar{4}$3m
Z: 4

LATTICE CONSTANT:
$a = 6.047$ kX
3 STRONGEST DIFFRACTION LINES:
3.49 (100)
2.139 ( 55)
1.824 ( 30)
OPTICAL CONSTANT:
$N = 2.346$ (Na)
HARDNESS: 2½
DENSITY: 5.68 (Meas.)
5.60 (Calc.)
CLEAVAGE: {011} perfect
Fracture conchoidal. Brittle.
HABIT: Crystals tetrahedral, often modified by {001} or other forms; also cubo-octahedral. As isolated crystals and crusts. Twinning on {111}, sometimes repeated.
COLOR-LUSTER: Colorless to pale yellow when fresh; flesh pink to dark brownish red on exposure. Transparent. Adamantine. Streak yellow. Fluoresces dark red in ultraviolet light.
MODE OF OCCURRENCE: Occurs associated with native copper, cuprite, cerussite, and iron and manganese oxides at Broken Hill, New South Wales, Australia, and with leightonite, atacamite, and gypsum at Chuquicamata, Chile.
BEST REF. IN ENGLISH: Palache, et al., "Dana's System of Mineralogy, 7th Ed., v. II, p. 20–22, New York, Wiley, 1951.

## MARTHOZITE
$Cu(UO_2)_3(SeO_3)_3(OH)_2 \cdot 7H_2O$

CRYSTAL SYSTEM: Orthorhombic
CLASS: 2/m 2/m 2/m or mm2
SPACE GROUP: Pnma or Pn2$_1$a
Z: 4
LATTICE CONSTANTS:
$a = 16.40$
$b = 17.20$
$c = 6.98$
3 STRONGEST DIFFRACTION LINES:
8.23 (100)
3.09 (100)
3.22 ( 80)
OPTICAL CONSTANTS:
$\alpha$ (not measured)
$\beta = 1.780-1.785$
$\gamma = 1.780-1.785$
$(-)2V = 39°$ (Na)
HARDNESS: Not determined.
DENSITY: 4.4 (Meas.)
4.7 (Calc.)
CLEAVAGE: {100} perfect
{010} imperfect
HABIT: Millimeter-size cyrstals, flattened on (100); this face is striated parallel to [001]. Faces noted {100}, {111}, {011}, {2$\bar{3}$0} large; {101}, {201}, {301}, {211}, {210}, and {230} vicinal.
COLOR-LUSTER: Yellowish green to greenish brown.
MODE OF OCCURRENCE: Occurs associated with guilleminite, demesmaekerite, kasolite, cuprosklodowskite, ura-
notile, malachite, chalcomenite, and sengierite in the zone of oxidation at the Musonoi deposit, Katanga, Zaire.
BEST REF. IN ENGLISH: Cesbron, Fabian, Oosterbosch, R., and Pierrot, Roland, *Am. Min.*, **55**: 533 (1970).

**MARTITE** = Pseudomorph of hematite after magnetite

## MASCAGNITE
$(NH_4)_2 SO_4$

CRYSTAL SYSTEM: Orthorhombic
CLASS: 2/m 2/m 2/m
SPACE GROUP: Pmcn
Z: 4
LATTICE CONSTANTS:
$a = 5.98$ kX
$b = 10.62$
$c = 7.78$
3 STRONGEST DIFFRACTION LINES:
4.33 (100)
4.39 ( 65)
3.06 ( 55)
OPTICAL CONSTANTS:
$\alpha = 1.520$
$\beta = 1.523$ (Na)
$\gamma = 1.533$
$(+)2V = 52°12'$
HARDNESS: 2-2½
DENSITY: 1.768 (Meas.)
1.769 (Calc.)
CLEAVAGE: {001} good
Fracture uneven. Somewhat sectile.
HABIT: Crystals rare, prismatic. Usually as crusts and stalactitic. Twinning on {110}; also polysynthetic.
COLOR-LUSTER: Colorless, gray, yellowish gray, yellow. Crystals transparent; aggregates translucent to opaque. Vitreous to dull.

Taste bitter.
MODE OF OCCURRENCE: Occurs associated with boussingaultite and epsomite near The Geysers, Sonoma County, California. It is also found as a sublimation product in fumaroles at Mt. Etna, Sicily; at Mt. Vesuvius, Italy; and on the Nyamlagira volcano, Zaire; in guano deposits of the Chincha and Guañape Islands; and as a sublimation product from burning coal seams at many localities.
BEST REF. IN ENGLISH: Palache, et al., "Dana's System of Mineralogy," 7th Ed., v. II, p. 398–399, New York, Wiley, 1951.

## MASSICOT
PbO
Dimorphous with litharge

CRYSTAL SYSTEM: Orthorhombic
CLASS: 2/m 2/m 2/m
SPACE GROUP: Pbma
Z: 4

LATTICE CONSTANTS:
 $a = 5.459$
 $b = 4.723$
 $c = 5.859$
3 STRONGEST DIFFRACTION LINES:
 3.067 (100)
 2.946 ( 31)
 2.744 ( 28)
OPTICAL CONSTANTS:
 $\alpha = 2.51$
 $\beta = 2.61$  (Li)
 $\gamma = 2.71$
 (+)2V = 90° [(−) for blue]
HARDNESS: 2
DENSITY: 9.642
CLEAVAGE: {100} distinct
 {010} distinct
 {110} in traces
Flexible, inelastic.
HABIT: Massive, earthy to scaly. Synthetic crystals tabular.
COLOR-LUSTER: Yellow, reddish yellow. Translucent; thin fragments transparent. Oily to dull. Streak lighter than color.
MODE OF OCCURRENCE: Occurs as an oxidation product of primary lead minerals, commonly associated with galena, secondary lead, copper, or antimony minerals, and limonite. Found in Wythe County, Virginia, and widespread in the mining districts of the western United States, especially in California, Nevada, and Colorado. It also occurs in Mexico, Bolivia, France, Italy, Germany, Czechoslovakia, Roumania, Greece, Transvaal, and New South Wales, Australia.
BEST REF. IN ENGLISH: Palache, et al., "Dana's System of Mineralogy," 7th Ed., v. I, p. 516–517, New York, Wiley, 1944.

# MASUYITE
$UO_3 \cdot 2H_2O$

CRYSTAL SYSTEM: Orthorhombic
CLASS: 2/m 2/m 2/m
SPACE GROUP: Pban
LATTICE CONSTANTS:
 $a = 13.98$
 $b = 14.20$
 $c = 12.11$
3 STRONGEST DIFFRACTION LINES:
 7.08 (100)
 3.52 ( 70)
 3.12 ( 50)
OPTICAL CONSTANTS:
 $\alpha = 1.785$
 $\beta = 1.895$
 $\gamma = 1.915$
 (−)2V ~ 50°
HARDNESS: Not determined.
DENSITY: 5.08 (Meas.)
 5.028 (Calc.)
CLEAVAGE: {001} perfect
 {010} imperfect

HABIT: Crystals tabular on {001}; pseudohexagonal. Often extensively twinned, both on {110} and {1̄30}. Crystals very minute in size.
COLOR-LUSTER: Reddish orange to carmine red. Transparent.
MODE OF OCCURRENCE: Occurs in small cavities in uraninite associated with uranophane, sklodowskite, and sharpite, at Shinkolobwe, Katanga, Zaire.
BEST REF. IN ENGLISH: Christ, C. L., and Clark, Joan R., *Am. Min.*, 45: 1026–1061 (1960).

# MATILDITE
$AgBiS_2$

CRYSTAL SYSTEM: Hexagonal
Z: 12
LATTICE CONSTANTS:
 $a = 8.12$
 $c = 19.02$
3 STRONGEST DIFFRACTION LINES:
 2.83 (100)
 3.30 ( 80)
 2.03 ( 60)
HARDNESS: 2½
Microhardness (25 g load) range from VHN 59 to 76.
DENSITY: 6.99 (Calc.)
CLEAVAGE: None. Fracture uneven. Very brittle.
HABIT: Crystals rare; indistinct acicular or prismatic; vertically striated; up to 25mm long. Usually massive, granular.
COLOR-LUSTER: Gray to iron black. Opaque. Metallic. Streak light gray.
MODE OF OCCURRENCE: Occurs in medium-to high-temperature vein deposits and pegmatites, commonly associated with tetrahedrite and sulfide minerals. Found associated with granular quartz in the Lake City district, Hinsdale County, and elsewhere in Colorado; at the Mayflower mine, Boise County, Idaho; and at the Essex mine, Inyo County, California. It also occurs at several places in Canada; in Greenland, Bolivia, Peru, Spain, Germany, Switzerland, and Japan.
BEST REF. IN ENGLISH: Harris, D. C., and Thorpe, R. I., *Can. Min.*, 9: (pt. 5): 655–662 (1969). Palache, et al., "Dana's System of Mineralogy," 7th Ed., v. I, p. 429–430, New York, Wiley, 1944.

# MATLOCKITE
PbFCl

CRYSTAL SYSTEM: Tetragonal
CLASS: 4/m 2/m 2/m
SPACE GROUP: P4/nmm
Z: 2
LATTICE CONSTANTS:
 $a = 4.09$ kX
 $c = 7.21$
3 STRONGEST DIFFRACTION LINES:
 3.56 (100)
 3.61 ( 70)
 2.904 ( 47)

OPTICAL CONSTANTS:
 ω = 2.145
 ε = 2.006   (Na)
 (−)
HARDNESS: 2½–3
DENSITY: 7.12 (Meas.)
      7.13 (Calc.)
CLEAVAGE: {001} perfect
Fracture subconchoidal to uneven. Brittle.

HABIT: Crystals tabular {001}; also stout pyramidal with small prism faces. As rosettes or aggregates of platy crystals. Also massive, lamellar.

COLOR-LUSTER: Colorless, yellow, pale yellowish brown, greenish. Transparent. Adamantine; somewhat pearly on {001}.

MODE OF OCCURRENCE: Found associated with cerussite, diaboleite, boleite, and leadhillite at the Mammoth mine, Tiger, Arizona; also found near Matlock, Derbyshire, England; Laurium, Greece; and at Challacolla, Tarapacá, Chile.

BEST REF. IN ENGLISH: Palache, et al., "Dana's System of Mineralogy," 7th Ed., v. II, p. 59–60, New York, Wiley, 1951.

# MATRAITE
γ - ZnS
Trimorphous with sphalerite, wurtzite

CRYSTAL SYSTEM: Hexagonal
CLASS: 3m
SPACE GROUP: R3m
Z: 3
LATTICE CONSTANTS:
 $a$ = 3.8
 $c$ = 9.4
HARDNESS: 3½–4
DENSITY: Not determined.
CLEAVAGE: {10$\bar{1}$0} distinct
Brittle.

HABIT: Crystals pyramidal; in scepter-shaped aggregates which are oriented intergrowths of sphalerite with γ-ZnS.

COLOR-LUSTER: Brownish yellow. Transparent. Vitreous.

MODE OF OCCURRENCE: Occurs in association with sphalerite, wurtzite, galena, chalcopyrite, and pyrite at a deposit in the Matra Mountains, Hungary.

BEST REF. IN ENGLISH: Koch, S., and Sasvari, Kalman, Am. Min., 45: 1131 (1960).

# MATTAGAMITE
CoTe$_2$

CRYSTAL SYSTEM: Orthorhombic
Z: 1
LATTICE CONSTANTS:
 $a$ = 5.31
 $b$ = 6.31
 $c$ = 3.89

3 STRONGEST DIFFRACTION LINES:
 2.805 (100)
 2.703 ( 80)
 2.066 ( 60)
OPTICAL CONSTANTS: Weakly anisotropic.
HARDNESS: Microhardness 630 kg/mm$^2$ with 15 g load.
DENSITY: Not determined.
HABIT: Massive; as minute equidimensional or blade-like grains isolated in altaite and as thin irregular rims on pyrrhotite and chalcopyrite in contact with altaite.

COLOR-LUSTER: Violet under reflected light.

MODE OF OCCURRENCE: Ferroan mattagamite was found in only one specimen, which contained fairly abundant chalcopyrite and sphalerite, from a small zone of telluride ore within the Mattagami Lake mine near the town of Matagami, Quebec, Canada.

BEST REF. IN ENGLISH: Thorpe, R. I., and Harris, D. C., Can. Min., 12: 55–60 (1973).

# MATTEUCCITE
NaHSO$_4$ · H$_2$O

CRYSTAL SYSTEM: Monoclinic
CLASS: m
SPACE GROUP: Aa
Z: 4
LATTICE CONSTANTS:
 $a$ = 8.22
 $b$ = 7.79
 $c$ = 7.81
 β = 119°56′
3 STRONGEST DIFFRACTION LINES:
 3.55 (100)
 3.43 (100) (synthetic)
 5.2 ( 60)
OPTICAL CONSTANTS:
 (+)2V = 86° (synthetic)
HARDNESS: Not determined.
DENSITY: 2.117 (synthetic)
CLEAVAGE: Not determined.
HABIT: Massive, stalactitic.
COLOR-LUSTER: Colorless.

MODE OF OCCURRENCE: Occurs intimately mixed with mercallite and ralstonite in stalactites from the 1933 eruption of Vesuvius.

BEST REF. IN ENGLISH: Carobbi, Guido, and Cipriani, Curcio, Am. Min., 39: 848 (1954).

# MAUCHERITE
Ni$_{11}$As$_8$

CRYSTAL SYSTEM: Tetragonal
CLASS: 422
SPACE GROUP: P4$_{1,3}$2$_1$2
Z: 4
LATTICE CONSTANTS:
 $a$ = 6.868
 $c$ = 21.80

3 STRONGEST DIFFRACTION LINES:
 2.01 (100)
 1.713 (100)
 2.69 ( 90)
HARDNESS: 5
DENSITY: 7.95 (Meas.)
        8.03 (Calc.)
CLEAVAGE: None. Fracture uneven. Brittle.
HABIT: Crystals usually tabular, flattened on {001}; also pyramidal. Usually massive, radial-fibrous or granular. Twinning reported on {106} and {203}.
COLOR-LUSTER: Silvery gray with reddish tint; tarnishes copper red. Opaque. Metallic. Streak blackish gray.
MODE OF OCCURRENCE: Occurs in Canada associated with nickel minerals in the Cobalt, Sudbury, and Timiskaming districts, Ontario; and with millerite, pyroxene, and calcite at Orford, Quebec. It is also found in Hesse, Saxony, and Thuringia, Germany; in Styria, Austria; and at Los Jarales, Malaga, Spain.
BEST REF. IN ENGLISH: Palache, et al., "Dana's System of Mineralogy," 7th Ed., v. I, p. 192–194, New York, Wiley, 1944.

## MAUSITE = Metavoltine

## MAUZELIITE = Plumbian romeite

## MAVINITE = Chloritoid

## MAWSONITE
$Cu_7Fe_2SnS_{10}$

CRYSTAL SYSTEM: Cubic (or pseudocubic)
SPACE GROUP: I lattice
LATTICE CONSTANT:
 $a = 10.74$
3 STRONGEST DIFFRACTION LINES:
 3.09 (100)
 1.895 ( 80)
 1.618 ( 60)
HARDNESS: $\approx 3\frac{1}{2}$–4
DENSITY: Not determined.
CLEAVAGE: No cleavage visible on polished surfaces; fractures resulting from hardness indentation tests suggest two imperfect cleavages at approximately right angles.
HABIT: Massive.
COLOR-LUSTER: Brownish orange in polished section. Opaque.
MODE OF OCCURRENCE: Occurs typically as rounded to irregular inclusions in bornite in the size range 0.05–1.3 mm associated with chalcocite, chalcopyrite, tetrahedrite, pyrite, galena, tennantite, and an enargite-type phase, at Mt. Lyell, Tasmania. Also found at Tingha, New South Wales, Australia.
BEST REF. IN ENGLISH: Markham, N. L., and Lawrence, L. J., *Am. Min.*, **50**: 900–908 (1965).

## MAYENITE
$Ca_{12}Al_{14}O_{33}$

CRYSTAL SYSTEM: Cubic
CLASS: $\bar{4}3m$
SPACE GROUP: $I\bar{4}3d$
Z: 2
LATTICE CONSTANT:
 $a = 12.02$ (Calc.)
3 STRONGEST DIFFRACTION LINES:
 2.68 (100)
 4.89 ( 95)
 2.45 ( 50)
OPTICAL CONSTANT:
 $N = 1.643$
HARDNESS: Not determined.
DENSITY: 2.85 (Meas. av.)
HABIT: Rounded grains with maximum diameter of 60 $\mu$.
COLOR-LUSTER: Colorless, transparent.
MODE OF OCCURRENCE: Occurs associated with ettringite, calcite, wollastonite, brownmillerite, gehlenite, larnite, diopside, grossular, pyrrhotite, spinel, afwillite, hydrocalumite, and portlandite. It is found in several thermally metamorphosed marly limestone inclusions of Tertiary (?) age in effusive volcanic rocks of the Ettringer Bellerberg (leucite-tephrite), related to the Quaternary extrusives of the Laacher. See area, near Mayen, Eifel district (Rhineland-Palatinate), West Germany.
BEST REF. IN ENGLISH: Hentschel, Gerhard, *Am. Min.*, **50**: 2106–2107 (1965).

## MBOZIITE
$Na_2CaFe_3^{2+}Fe_2^{3+}Al_2Si_6O_{22}(OH)_2$
Amphibole Group

CRYSTAL SYSTEM: Monoclinic
Z: 2
LATTICE CONSTANTS:
 $a = 9.952$
 $b = 18.101$
 $c = 5.322$
 $\beta = 105.499°$
3 STRONGEST DIFFRACTION LINES:
 3.15 (100)
 8.53 ( 70)
 2.732 ( 50)
OPTICAL CONSTANTS:
 $\alpha = 1.705$
 $\beta = 1.713$
 $\gamma = 1.715$
HARDNESS: 5–6
DENSITY: Not determined.
CLEAVAGE: {110} perfect
HABIT: Crystals up to 10 cm long and 4 cm thick; also as prismatic grains up to 3 mm long and 0.5–2 mm thick.
COLOR-LUSTER: Black; deep blue-green.
MODE OF OCCURRENCE: Occurs with oligoclase, potash feldspar, nepheline, sodalite, aegirine-augite, biotite, sphene, and apatite in a 15-foot foyaite dike which cuts nepheline-bearing gneisses in the Mbozi syenite-gabbro com-

plex in southwest Tanzania. Also from the Darkainle nepheline-syenite complex, Borama district, Somali Republic.

BEST REF. IN ENGLISH: Brock, P. W. G., Gellatly, D. C., and von Knorring, O., *Min. Mag.*, **33**: 1057–1065 (1964).

## MCGOVERNITE = Macgovernite

## MCKELVEYITE = Mackelveyite

## MEDMONTITE = Chrysocolla plus mica

## MEERSCHAUM = Sepiolite

## MEIONITE

$mCa_4(Al_6Si_6O_{24})CO_3$ with $nNa_4(Al_3Si_9O_{24})Cl$
$Ma_{20}Me_{80}$ to $Ma_0Me_{100}$
Scapolite Group

CRYSTAL SYSTEM: Tetragonal
CLASS: 4/m
SPACE GROUP: P4/m
Z: 2
LATTICE CONSTANTS:
 $a = 12.13$
 $c = 7.69$
OPTICAL CONSTANTS:
 $\omega = 1.590–1.600$
 $\epsilon = 1.556–1.562$
 $(-)$
HARDNESS: 5½–6
DENSITY: 2.72–2.78 (Meas.)
CLEAVAGE: {100} distinct
 {110} distinct
Fracture uneven to conchoidal. Brittle.

HABIT: Crystals prismatic, often coarse. Also massive, granular; sometimes columnar.

COLOR-LUSTER: Colorless, white, gray, bluish, greenish, yellowish, pink, violet and brownish. Transparent to translucent. Vitreous to somewhat pearly or resinous. Streak uncolored.

MODE OF OCCURRENCE: Occurs mainly in regionally metamorphosed rocks, contact zones, altered basic igneous rocks, and ejected volcanic blocks. Typical occurrences are found in Italy, Finland, and USSR.

BEST REF. IN ENGLISH: Shaw, D. M., *J. Petr.*, **1**: 218; 261 (1960). Deer, Howie, and Zussman, "Rock Forming Minerals," v. 4, p. 321–337, New York, Wiley, 1963.

## MELACONITE = Synonym for tenorite

## MELANITE = Variety of andradite containing titanium

## MELANOCERITE

$Ce_4CaBSi_2O_{12}(OH)$
Var. Caryocerite (Th-rich)

CRYSTAL SYSTEM: Hexagonal
LATTICE CONSTANTS:
 $c/a = 1.2554$
  $a = 9.35$ } calcined
  $c = 6.88$ } at 600°
3 STRONGEST DIFFRACTION LINES:
 2.81 (100)  Caryocerite (metamict)
 3.44 ( 40)  (calcined at 600–1000°)
 1.84 ( 40)
OPTICAL CONSTANTS:
 $n \simeq 1.733–1.76$, uniaxial $(-)$, often isotropic
HARDNESS: 6
DENSITY: 4.13–4.29 (Meas.)
CLEAVAGE: None. Fracture conchoidal. Brittle.
HABIT: Crystals rhombohedral. Usually massive, compact.
COLOR-LUSTER: Brown, black. Resinous. Translucent.
MODE OF OCCURRENCE: Occurs in alkali pegmatites in association with pyroxene, astrophyllite, lepidomelane, gadolinite, and other rare-earth minerals. Found at the Cardiff uranium mine, Wilberforce, Ontario, Canada; in the Langesundfjord district, Norway; and at localities in Siberia.

BEST REF. IN ENGLISH: Vlasov, K. A., "Mineralogy of Rare Elements," v. II, p. 301–302, Israel Program for Scientific Translations, 1966.

## MELANOPHLOGITE

$SiO_2$

CRYSTAL SYSTEM: Supercell, tetragonal; subcell, cubic.
SPACE GROUP: Supercell, $P4_2/nbc$; Subcell, $P4_232$.
Z: 48 (cubic)
LATTICE CONSTANTS:
 (supercell)  (subcell)
 $a = 26.82$  13.4
 $c = 13.37$
 (Supercell transforms at 1050°C to subcell.)
3 STRONGEST DIFFRACTION LINES:
 3.579 (100)
 6.000 ( 80)
 3.870 ( 70)
OPTICAL CONSTANT:
 $N = 1.425$
HARDNESS: 6½–7
DENSITY: 2.005; 2.052 (Meas.)
 2.001; 2.06 (Calc.)
CLEAVAGE: None. Brittle.
HABIT: Crystals simple cubes up to 4mm along edge, rare. Usually as complex intergrown rounded aggregates forming thin crusts.

COLOR-LUSTER: Colorless, pale yellow to deep reddish brown; turbid-white when altered. Transparent to translucent. Vitreous.

MODE OF OCCURRENCE: Occurs as single crystals and interlocking intergrowths on sulfur crystals from Agrigento (formerly Girghenti) and Palermo, Sicily. Also found as fine crystals on lussatite in a vein cavity in a metamorphosed sedimentary pyrite-rhodochrosite deposit at Chvaletice, Bohemia, Czechoslovakia.

BEST REF. IN ENGLISH: Skinner, Brian J., and Appleman, Daniel E., *Am. Min.*, **48**: 854–867 (1963). Žák, Lubor, *Am. Min.*, **57**: 779–796 (1972).

## MELANOSTIBITE (Melanostibian, Lamprostibian)
$Mn(Sb, Fe)O_3$

CRYSTAL SYSTEM: Hexagonal
CLASS: $\bar{3}$
SPACE GROUP: $R\bar{3}$
Z: 2
LATTICE CONSTANTS:
  $a = 5.226$
  $c = 14.325$
3 STRONGEST DIFFRACTION LINES:
  2.806 (100)
  2.613 ( 80)
  1.766 ( 80)
OPTICAL CONSTANT:
  $N = 2.12$
  Uniaxial
HARDNESS: $4^+$
DENSITY: 5.63 (Calc.)
CLEAVAGE: {0001} perfect
Brittle.
HABIT: As porous sponge-like aggregates of small striated crystals in parallel growth, and as thin plates. Crystals often somewhat elongated.
COLOR-LUSTER: Coal black. Small fragments deep blood red and transparent. Streak and powder light rouge red to cherry red, more vivid than hematite.
MODE OF OCCURRENCE: Occurs in association with calcite and tephroite in dolomite rock at the Sjö mine, Grythyttan, Örebro, Sweden.
BEST REF. IN ENGLISH: Moore, Paul B., *Ark. Min. Geol.*, **4**: 449–458 (1968).

## MELANOTEKITE
$Pb_2 Fe_2^{3+} Si_2 O_9$

CRYSTAL SYSTEM: Orthorhombic
CLASS: 222
SPACE GROUP: $C222_1$
Z: 4
LATTICE CONSTANTS:
  $a = 6.97$
  $b = 11.0$
  $c = 9.83$
3 STRONGEST DIFFRACTION LINES:
  2.914 (100)
  2.732 (100)
  5.523 ( 80)
OPTICAL CONSTANTS:
  $\alpha = 2.12$
  $\beta = 2.17$
  $\gamma = 2.31$
  $(+)2V = 67°$
HARDNESS: $6\frac{1}{2}$

DENSITY: 5.7–6.19 (Meas.)
        6.225 (Calc.)
CLEAVAGE: Two—unequal.
HABIT: Crystals prismatic; also massive, cleavable.
COLOR-LUSTER: Black to blackish gray. Nearly opaque. Metallic to greasy.
MODE OF OCCURRENCE: Occurs as crystals at Hillsboro, Sierra County, New Mexico; with native lead at Långban, and as a common mineral at Pajsberg, near Harstigen, Sweden.
BEST REF. IN ENGLISH: Winchell and Winchell, "Elements of Optical Mineralogy," 4th Ed., Pt. 2, p. 525, New York, Wiley, 1951. *Am. Min.*, **52**: 1085–1093 (1967).

## MELANOTHALLITE (Species status uncertain, see hydromelanothallite)
CuClOH

HABIT: As scales.
COLOR-LUSTER: Black.
MODE OF OCCURRENCE: Occurs in association with eriochalcite, chalcocyanite, euchlorin, and dolerophanite as a sublimation product in fumaroles at Mt. Vesuvius, Italy.
BEST REF. IN ENGLISH: Palache et al., "Dana's System of Mineralogy," 7th Ed., v. II, p. 44, New York, Wiley, 1951.

## MELANOVANADITE
$Ca_2 V_4^{4+} V_6^{5+} O_{25} \cdot nH_2O$

CRYSTAL SYSTEM: Triclinic
CLASS: 1
SPACE GROUP: P1
Z: 1
LATTICE CONSTANTS:

|  |  | (B-cell) |
|---|---|---|
| $a = 6.36$ | | $a = 7.97$ |
| $b = 16.86$ | | $b = 16.86$ |
| $c = 6.27$ | | $c = 9.81$ |
| $\alpha = 90°0'$ | | $\alpha = 87°55'$ |
| $\beta = 101°50'$ | | $\beta = 90°45'$ |
| $\gamma = 93°10'$ | | $\gamma = 92°30'$ |

OPTICAL CONSTANTS:
  $\alpha = 1.73$
  $\beta = 1.96$
  $\gamma = 1.98$
  $(-)2V = $ medium
HARDNESS: $2\frac{1}{2}$
DENSITY: 3.477
CLEAVAGE: {010} perfect
Brittle.
HABIT: As velvety, divergent aggregates of elongated prismatic crystals with striated or rounded prism faces.
COLOR-LUSTER: Black. Opaque. Submetallic. Streak dark reddish brown.
MODE OF OCCURRENCE: Occurs in crevices in black shale associated with pascoite and other minerals at Minasragra, near Cerro de Pasco, Peru.
BEST REF. IN ENGLISH: Palache, et al., "Dana's System of Mineralogy," 7th Ed., v. II, p. 1058–1060, New York, Wiley, 1951.

# MELANTERITE
$FeSO_4 \cdot 7H_2O$
Var. Pisanite $(Fe, Cu)SO_4 \cdot 7H_2O$
Kirovite $(Fe, Mg)SO_4 \cdot 7H_2O$

CRYSTAL SYSTEM: Monoclinic
CLASS: 2/m
SPACE GROUP: $P2_1/c$
Z: 4
LATTICE CONSTANTS:
  $a = 14.07$
  $b = 6.50$
  $c = 11.04$
  $\beta = 105°34'$
3 STRONGEST DIFFRACTION LINES:
  4.90 (100)
  3.78 ( 64)
  3.23 ( 20)
OPTICAL CONSTANTS:
  $\alpha = 1.4713$
  $\beta = 1.478$ (Na-synthetic)
  $\gamma = 1.4856$
  $(+)2V = 85°27'$
HARDNESS: 2
DENSITY: 1.90 (Meas.)
CLEAVAGE: {001} perfect
  {110} distinct
Fracture conchoidal. Brittle.
HABIT: Crystals equant to short prismatic, also thick tabular or octahedral. Cuprian variety pseudorhombohedral or short to long prismatic. Usually in stalactitic, fibrous, or pulverulent aggregates or crusts.
COLOR-LUSTER: Various shades of green, greenish blue, blue, white. Translucent. Vitreous, also silky. Becomes yellowish white and opaque on exposure. Streak colorless.

Soluble in water. Taste sweetish, metallic, astringent.
MODE OF OCCURRENCE: Occurs as a typical weathering product of pyrite, marcasite, and cupriferous pyrite ores, and is invariably found as efflorescences in pyritic ore bodies. It is found widespread in the United States, especially at Ducktown, Tennessee (pisanite); and in California, Utah, Nevada, Montana, South Dakota, and Colorado. Other important localities include Szomolnok, Czechoslovakia; Idria, Italy; Rammelsberg and Bodenmais, Germany; Rio Tinto, Spain; Falun, Sweden; and the Kalata mine, Kirovgrad, USSR (kirovite).
BEST REF. IN ENGLISH: Palache, et al., "Dana's System of Mineralogy," 7th Ed., v. II, p. 499–504, New York, Wiley, 1951.

# MELILITE = Synonym for Åkermanite-Gehlenite Group

# MELIPHANITE (Melinophane)
$(Ca, Na)_2 Be(Si, Al)_2 (O, F)_7$
Var. Gugiaite

CRYSTAL SYSTEM: Tetragonal
CLASS: $\bar{4}$ or $\bar{4}2m$
SPACE GROUP: $P\bar{4}2_1m$ or $P\bar{4}$
Z: 8

LATTICE CONSTANTS:
  $a = 10.60$
  $c = 9.90$
3 STRONGEST DIFFRACTION LINES:
  (meliphane)    (gugiaite)
  2.75 (100)     2.765 (100)
  2.96 ( 50)     1.709 ( 70)
  1.702 ( 50)    1.485 ( 70)
OPTICAL CONSTANTS:
  $\omega = 1.612$
  $\epsilon = 1.593$
  (−)
HARDNESS: 5–5½
DENSITY: 3.0–3.034 (Meas.)
  3.03 (Calc. Gugiaite)
CLEAVAGE: {010} perfect
  {001} distinct
  {110} indistinct
Fracture uneven. Brittle.
HABIT: Crystals thin tabular or as short square pyramids. Also as tabular or lamellar aggregates up to several centimeters in diameter.
COLOR-LUSTER: Yellow, also colorless. Transparent. Vitreous.
MODE OF OCCURRENCE: Occurs in nepheline-syenite pegmatites in the Julianehaab district, Greenland, and in the Langesundfjord district, Norway. Also found in cavities in skarn rocks and disseminated in melanite near the village of Gugiya, China (gugiaite).
BEST REF. IN ENGLISH: Vlasov, K. A., "Mineralogy of Rare Elements," v. II, p. 129–134, Israel Program for Scientific Translations, 1966.

# MELKOVITE (Inadequately described mineral)
$CaFeH_6 (MoO_4)_4 (PO_4) \cdot 6H_2O$

3 STRONGEST DIFFRACTION LINES:
  2.92 (100)
  3.54 ( 90)
  8.42 ( 80)
OPTICAL CONSTANTS: Mean index = 1.838.
HARDNESS: ~3
DENSITY: 2.971 (Meas.)
CLEAVAGE: One perfect. Brittle.
HABIT: Finely crystalline powdery films and veinlets consisting of platy crystals usually 0.001–0.002 mm in diameter.
COLOR-LUSTER: Lemon yellow to brownish yellow; dull to waxy.
MODE OF OCCURRENCE: Occurs associated with hydrous iron oxides, powellite, iriginite, ferrimolybdite, jarosite, and autunite in the oxidation zone of a molybdenite-fluorite deposit in the Shunak Mountains, central Kazakhstan.
BEST REF. IN ENGLISH: Egorov, B. L., Dara, A. D., Senderova, V. M., *Am. Min.*, 55: 320 (1970).

# MELLITE
$C_{12} Al_2 O_{12} \cdot 18H_2O$

CRYSTAL SYSTEM: Tetragonal

CLASS: 422(?)
SPACE GROUP: $P4_122$ or $P4_322$
Z: 16
LATTICE CONSTANTS:
 $a = 22.0$ kX
 $c = 23.3$
3 STRONGEST DIFFRACTION LINES:
 7.99 (100)
 4.23 ( 70)
 5.80 ( 55)
OPTICAL CONSTANTS:
 $\omega = 1.539$
 $\epsilon = 1.511$ (Na)
  (−)
HARDNESS: 2-2½
DENSITY: 1.64 (Meas.)
 1.704 (Calc. x-ray)
CLEAVAGE: {011} indistinct
Fracture conchoidal. Somewhat sectile.
 HABIT: Crystals prismatic or pyramidal; small and rare. Also massive, fine granular, nodular, or as coatings.
 COLOR-LUSTER: Golden brown to brownish or reddish; rarely white. Transparent to translucent. Resinous to vitreous. Fluoresces blue under short-wave ultraviolet light.
 MODE OF OCCURRENCE: Occurs as a secondary deposit in cracks and cavities in lignite and brown coal. Found in the Paris basin, Seine, France; at Artern, Thuringia, and near Bitterfeld, Saxony, Germany; and at localities in Czechoslovakia and USSR.
 BEST REF. IN ENGLISH: Palache, et al., "Dana's System of Mineralogy," 7th Ed., v. II, p. 1104–1105, New York, Wiley, 1951.

## MELNIKOVITE = Greigite

## MELONITE
$NiTe_2$

CRYSTAL SYSTEM: Hexagonal
CLASS: $\bar{3}2/m$
SPACE GROUP: $P\bar{3}m1$
Z: 1
LATTICE CONSTANTS:
 $a = 3.842$
 $c = 5.266$
3 STRONGEST DIFFRACTION LINES:
 2.82 (100)
 1.549 ( 60)
 2.06 ( 50)
HARDNESS: 1-1½
DENSITY: 7.72 (Meas.)
 7.73 (Calc.)
CLEAVAGE: {0001} perfect
Brittle.
 HABIT: As hexagonal lamellae, and indistinct laminated particles.
 COLOR-LUSTER: Reddish white, tarnishing brown. Opaque. Metallic. Streak dark gray.
 MODE OF OCCURRENCE: Occurs chiefly in hydrothermal vein deposits associated with other tellurides,

sulfides, gold, and quartz; less commonly found in Sudbury type copper-nickel deposits. Occurs at the Melones and Stanislaus mines, Carson Hill, Calaveras County, California; and at deposits in Boulder and Teller Counties, Colorado. It also is found at localities in Canada, USSR, and South Australia.
 BEST REF. IN ENGLISH: Vlasov, K. A., "Mineralogy of Rare Elements," v. II, p. 690–696, Israel Program for Scientific Translations, 1966.

## MELONJOSEPHITE
$Ca(Fe^{2+},Mg)Fe^{3+}(PO_4)_2(OH)$

CRYSTAL SYSTEM: Orthorhombic
CLASS: 2/m 2/m 2/m or mm2
SPACE GROUP: Pbam or Pba2
Z: 4
LATTICE CONSTANTS:
 $a = 9.548$
 $b = 10.847$
 $c = 6.380$
3 STRONGEST DIFFRACTION LINES:
 3.049 (100)
 5.418 ( 92)
 2.709 ( 89)
OPTICAL CONSTANTS:
 $\alpha = 1.720$
 $\beta = 1.770$
 $\gamma = 1.800$
Biaxial (−)2V = 80-85°
HARDNESS: ~5
DENSITY: 3.65 (Meas.)
 3.61 (Calc.)
CLEAVAGE: {110} perfect
 {010} imperfect
HABIT: Massive.
 COLOR-LUSTER: Dark green, nearly black. Brilliant, somewhat resinous.
 MODE OF OCCURRENCE: The mineral occurs near the quartz core of the Angarf-Sud pegmatite, Zenaga Plain, Morocco, in a mass of a mineral of the alluaudite group surrounding a core of triphylite.
 BEST REF. IN ENGLISH: Fransolet, Andre-Mathiew, IMA Abstract (73-12).

## MENDELEEVITE =Variety of betafite

## MENDIPITE
$Pb_3Cl_2O_2$

CRYSTAL SYSTEM: Orthorhombic
CLASS: 222
SPACE GROUP: $P2_12_12_1$
Z: 4
LATTICE CONSTANTS:
 $a = 9.52$
 $b = 11.95$
 $c = 5.87$

3 STRONGEST DIFFRACTION LINES:
3.09 (100)
3.04 ( 70)
2.78 ( 70)
OPTICAL CONSTANTS:
$\alpha = 2.24$
$\beta = 2.27$
$\gamma = 2.31$
(+)2V = nearly 90°
HARDNESS: 2½
DENSITY: 7.24 (Meas.)
7.22 (Calc.)
CLEAVAGE: {110} perfect
{100} distinct
{010} distinct
Fracture conchoidal to uneven.

HABIT: As columnar or fibrous masses, often radiated.

COLOR-LUSTER: Colorless, white to gray, often with yellowish, bluish, or reddish tint. Transparent to translucent. Adamantine to resinous on fractures, pearly to silky on cleavages. Streak white.

MODE OF OCCURRENCE: Occurs as nodular masses associated with cerussite, hydrocerussite, pyromorphite, diaboleite, malachite, calcite, and manganese oxide minerals at Higher Pitts farm and near Churchill, Mendip Hills, Somerset, England. It also has been found in the Altai Mountains, USSR, and at the Kunibert mine near Brilon, Westphalia, Germany.

BEST REF. IN ENGLISH: Palache, et al., "Dana's System of Mineralogy," 7th Ed., v. II, p. 56–58, New York, Wiley, 1951.

# MENDOZITE
$NaAl(SO_4)_2 \cdot 11H_2O$

CRYSTAL SYSTEM: Monoclinic
CLASS: 2/m
LATTICE CONSTANTS:
$a:b:c = 2.5060:1:0.9125$
$\beta = 109°1'$
OPTICAL CONSTANTS:
$\alpha = 1.449$
$\beta = 1.461$ (synthetic)
$\gamma = 1.463$
(−)2V = 56°
HARDNESS: ~3
DENSITY: 1.730–1.765 (Meas.)
CLEAVAGE: {100} good
{001} indistinct
{010} indistinct
HABIT: As crusts and fibrous masses. Synthetic crystals prismatic or pseudorhombohedral.

COLOR-LUSTER: Colorless, white, Transparent.

MODE OF OCCURRENCE: Occurs in California as crusts on the walls of an old tunnel near Volcano, Amador County; on the Pritchard Ranch near St. Helena, Napa County; and near Hidden Springs, San Bernardino County. It is also found near Red Cliff, Eagle County, Colorado; in Box Elder County, Utah; in Callaway and St. Louis counties, Missouri; at Chuquicamata and at other places in northern Chile; near Mendoza, Argentina; and at Shimané, Idzumo, Japan.

BEST REF. IN ENGLISH: Palache, et al., "Dana's System of Mineralogy," 7th Ed., v. II, p. 469–471, New York, Wiley, 1951.

# MENEGHINITE
$CuPb_{13}Sb_7S_{24}$ or $Pb_{13}Sb_7S_{23}$

CRYSTAL SYSTEM: Orthorhombic
CLASS: mm2
SPACE GROUP: Pnm2₁
Z: 2
LATTICE CONSTANTS:
$a = 11.36$
$b = 24.04$
$c = 8.26$
3 STRONGEST DIFFRACTION LINES:
3.29 (100)
3.70 ( 90)
2.91 ( 80)
HARDNESS: 2½
DENSITY: 6.40–6.42 (Meas.)
6.41 (Calc.)
CLEAVAGE: {010} perfect, interrupted
{001} difficult
Fracture conchoidal. Brittle.

HABIT: Crystals slender prismatic, striated parallel to elongation. Also massive, fibrous to compact.

COLOR-LUSTER: Blackish lead gray. Opaque. Bright metallic. Streak black, shining.

MODE OF OCCURRENCE: Occurs associated with franckeite and stannite in the contact rock of the Kalkar limestone quarry, near Santa Cruz, Santa Cruz County, California; in gneiss at Marble Lake, Frontenac County, Ontario, Canada; as fine crystals at Bottino, Tuscany, Italy; and at localities in Sweden and Germany.

BEST REF. IN ENGLISH: Palache, et al., "Dana's System of Mineralogy," 7th Ed., v. I, p. 402–404, New York, Wiley, 1944.

# MERCALLITE
$KHSO_4$

CRYSTAL SYSTEM: Orthorhombic
CLASS: 2/m 2/m 2/m
SPACE GROUP: Pbca
Z: 16
LATTICE CONSTANTS:
$a \simeq 9.79$
$b \simeq 18.93$
$c \simeq 8.40$
3 STRONGEST DIFFRACTION LINES:
3.84 (100)
3.52 ( 85)
3.41 ( 85)
OPTICAL CONSTANTS:
$\alpha = 1.445$
$\beta = 1.460$ (Calc.)
$\gamma = 1.491$
(+)2V = 56°

HARDNESS: Not determined.
DENSITY: 2.318 (Meas. syn.)
2.317 (Calc. syn.)
CLEAVAGE: None.
HABIT: As stalactites composed of minute tabular crystals.
COLOR-LUSTER: Colorless, also sky blue. Transparent. Vitreous.

Taste acid.
MODE OF OCCURRENCE: Occurs admixed with other sulfates and fluorides in fumaroles in the crater of Mt. Vesuvius, Italy.
BEST REF. IN ENGLISH: Palache, et al., "Dana's System of Mineralogy," 7th Ed., v. II, p. 395, New York, Wiley, 1951.

# MERCURY
Hg

CRYSTAL SYSTEM: Hexagonal
CLASS: $\bar{3}$m
SPACE GROUP: R$\bar{3}$m
Z: 3 (hex.), 1 (rhomb.)
LATTICE CONSTANTS:
$a = 3.457$    $a_{rh} = 2.9863$
$c = 6.664$    $\alpha = 70°44.6'$
DENSITY: 14.382 (Meas.)
14.492 (Calc.)
HABIT: Liquid; forms rhombohedral crystals at $-39°C$.
COLOR-LUSTER: Tin white. Opaque. Brilliant metallic.
MODE OF OCCURRENCE: Occurs associated with cinnabar near recent volcanic rocks and hot springs deposits. Found widespread in California, especially in the Coast Ranges, extending from Del Norte County to Santa Barbara County; the metal was so abundant at the Rattlesnake mine, in Sonoma County, that it would spurt out of the ore when a pick was sunk in. It also occurs at Terlingua, Texas; at Almaden, Spain; at Idria, Italy; and at deposits in Yugoslavia and Germany.
BEST REF. IN ENGLISH: Palache, et al., "Dana's System of Mineralogy," 7th Ed., v. I, p. 103, New York, Wiley, 1944.

# MERENSKYITE
(Pd, Pt)(Te, Bi)$_2$

CRYSTAL SYSTEM: Hexagonal
LATTICE CONSTANTS:
$a = 3.978$
$c = 5.125$
3 STRONGEST DIFFRACTION LINES:
2.92 (100)
2.10 ( 60)
3.07 ( 30)
HARDNESS: Polishing hardness less than pentlandite, greater than chalcopyrite, and just greater than kotulskite.
HABIT: Minute grains.
COLOR-LUSTER: In reflected light white with weak pleochroism from white to grayish white. In oil pleochroism more distinct, white with slight creamy tint to light grayish white. Distinctly to strongly anisotropic with dark brown to light greenish gray polarization colors.
MODE OF OCCURRENCE: Occurs with other platinoid bismuthotellurides in the Merensky Reef at Rustenburg platinum mine in the western Bushveld, Transvaal, Africa.
BEST REF. IN ENGLISH: Kingston, G. A., Min. Mag., 35: 815-834 (1966).

# MERRIHUEITE
(K, Na)$_2$(Fe, Mg)$_5$Si$_{12}$O$_{30}$

CRYSTAL SYSTEM: Hexagonal
CLASS: 6/m 2/m 2/m
SPACE GROUP: P6/mcc
LATTICE CONSTANTS:
$a = 10.16$
$c = 14.32$
3 STRONGEST DIFFRACTION LINES:
3.73 (100)
2.774 (100)
3.23 ( 90)
OPTICAL CONSTANT:
$n \sim 1.559-1.592$
HARDNESS: Not determined.
DENSITY: 2.87 (Calc.)
CLEAVAGE: Not determined.
HABIT: Grains up to 150 $\mu$ in diameter.
COLOR-LUSTER: Greenish blue.
MODE OF OCCURRENCE: Occurs as rare inclusions in twinned clinoenstatite and clinobronzite in the Mezö-Madaras chondritic meteorite.
BEST REF. IN ENGLISH: Dodd, Robert T. Jr., van Schmus, W. Randall, and Marvin, Ursula B., Science, 149: 972-974 (1965).

# MERRILLITE = Meteoritic whitlockite

# MERTIEITE
Pd$_5$(Sb, As)$_2$

CRYSTAL SYSTEM: Pseudohexagonal, possibly monoclinic
Z: 1
LATTICE CONSTANTS:
$a = 15.04$
$c = 22.41$
3 STRONGEST DIFFRACTION LINES:
2.28 (100)
2.17 (100)
2.23 ( 60)
OPTICAL CONSTANTS: Distinctly anisotropic.
HARDNESS: Vickers hardness 570-593 with 50 g load.
DENSITY: Not determined.
CLEAVAGE: Not determined.
HABIT: Massive; as discrete grains 0.10-0.50 mm in maximum dimensions.
COLOR-LUSTER: Brassy yellow. Opaque. Metallic.
MODE OF OCCURRENCE: Occurs very sparingly in

precious-metal placer concentrates at Goodnews Bay, Alaska.

BEST REF. IN ENGLISH: Desborough, George A., Finney, J. J., and Leonard, B. F., *Am. Min.*, **58**: 1-10 (1973).

**MERUMITE** = A complex intergrowth of at least three major and several minor phases
Hydrated oxide of Cr (with some Al)

## MERWINITE
$Ca_3 MgSi_2 O_8$

CRYSTAL SYSTEM: Monoclinic
CLASS: 2/m
SPACE GROUP: $P2_1/a$
Z: 4
LATTICE CONSTANTS:
  $a = 13.254$
  $b = 5.293$
  $c = 9.328$
  $\beta = 91.90°$
3 STRONGEST DIFFRACTION LINES:
  2.672 (100)
  2.749 ( 40) synthetic
  2.213 ( 40)
OPTICAL CONSTANTS:
  $\alpha = 1.706$
  $\beta = 1.711-1.712$
  $\gamma = 1.724$
  (+)2V = 71°
HARDNESS: 6
DENSITY: 3.15 (Meas.)
        3.33; 3.31 (Calc.)
CLEAVAGE: {100} perfect
Brittle.
HABIT: Massive; compact, granular. Natural crystals rare, rounded and pitted, and often show polysynthetic twinning. Synthetic crystals prismatic or thick tabular.
COLOR-LUSTER: Colorless to pale greenish. Transparent to translucent. Vitreous to greasy.
MODE OF OCCURRENCE: Merwinite occurs in contact zones in association with spurrite, gehlenite, larnite, monticellite, and spinel. Found in the limestone quarries at Crestmore, Riverside County, California; near Neihart, Montana; at Velardena, Durango, Mexico; and at Scawt Hill, County Antrim, Ireland.
BEST REF. IN ENGLISH: Larsen, Esper S. and Foshag, William F., *Am. Min.*, **6**: 143-148 (1921). Moore, Paul Brian and Araki, Takaharu, *Am. Min.*, **57**: 1355-1374 (1972).

**MESITITE** = Ferrian magnesite

## MESOLITE
$Na_2 Ca_2 Al_6 Si_9 O_{30} \cdot 8H_2O$
Zeolite Group

CRYSTAL SYSTEM: Monoclinic
CLASS: 2

SPACE GROUP: C2
Z: 8
LATTICE CONSTANTS:
  $a = 56.7$
  $b = 6.55$
  $c = 18.48$
  $\beta = 90°$
3 STRONGEST DIFFRACTION LINES:
  2.86 (100)
  5.79 ( 70)
  4.35 ( 50) broad
OPTICAL CONSTANTS:
  $\alpha \simeq 1.504$
  $\beta \simeq 1.505$
  $\gamma \simeq 1.5053$
  (+)2V ≃ 80°
HARDNESS: 5
DENSITY: 2.259 (Meas.)
        2.250 (Calc.)
CLEAVAGE: {101}, {10$\bar{1}$} perfect
Fracture uneven. Brittle. Compact masses tough.
HABIT: Crystals acicular or fibrous; as divergent tufts or compact masses of radiating crystals. Always twinned {100}.
COLOR-LUSTER: Colorless, white. Transparent. Vitreous; or silky when fibrous.
MODE OF OCCURRENCE: Occurs in cavities in volcanic rocks, usually in association with other zeolites. Found in Baker, Douglas, Klackamas, and Lane counties, Oregon; in Kern County and other places in California; at Table Mountain, Jefferson County, Colorado; on Fritz Island, Berks County, Pennsylvania; and in the trap rocks of northern New Jersey. Also found in the Bay of Fundy district, Nova Scotia; and at localities in Greenland, Iceland, Faroe Islands, Ireland, Scotland, Sicily, India, Australia, and elsewhere.
BEST REF. IN ENGLISH: Deer, Howie, and Zussman, "Rock Forming Minerals," v. 4, p. 358-376, New York, Wiley, 1963.

## MESSELITE (Neomesselite)
$Ca_2 (Fe^{2+}, Mn^{2+})(PO_4)_2 \cdot 2H_2O$
Iron analogue of fairfieldite

CRYSTAL SYSTEM: Triclinic
CLASS: $\bar{1}$
SPACE GROUP: $P\bar{1}$
Z: 1
LATTICE CONSTANTS:
  $a \simeq 5.8$    $\alpha \simeq 102°$
  $b \simeq 6.6$    $\beta \simeq 109°$
  $c \simeq 5.5$    $\gamma \simeq 90°$
3 STRONGEST DIFFRACTION LINES:
  6.34 (100)
  3.17 (100)
  3.02 ( 80)
OPTICAL CONSTANTS:
  $\alpha = 1.653$
  $\beta = 1.659$
  $\gamma = 1.676$
  (+)2V = 20° to 35°

HARDNESS: 3½
DENSITY: 3.16 (Meas.)
CLEAVAGE: {001} perfect
{010} good
{1 $\bar{1}$ 0} distinct
Fracture uneven. Brittle.

HABIT: Crystals prismatic to equant. Usually lamellar, foliated, or fibrous; in radiating aggregates.

COLOR-LUSTER: Colorless, white to pale greenish white and greenish gray. Transparent. Vitreous to pearly. Streak white.

MODE OF OCCURRENCE: Messelite occurs as a late hydrothermal mineral in granite pegmatites. Found associated with graftonite, phosphoferrite, ludlamite, vivianite, and hureaulite at the Bull Moose mine, and with hureaulite in altered triphylite at the Tip Top mine, Custer County; as large crystals with siderite and ludlamite at the Big Chief mine, and with triphylite at the Ingersoll and Dan Patch mines near Keystone, Pennington County, South Dakota. It also occurs at the Palermo mine, North Groton, New Hampshire, and abundantly at King's Mountain, North Carolina.

BEST REF. IN ENGLISH: Frondel, Clifford, *Am. Min.*, **40**: 828–833 (1955).

## META-ALUMINITE
$Al_2(SO_4)(OH)_4 \cdot 5H_2O$

CRYSTAL SYSTEM: Monoclinic
3 STRONGEST DIFFRACTION LINES:
4.48 (100)
8.35 ( 79)
4.36 ( 67)
OPTICAL CONSTANTS:
$\alpha = 1.497$
$\beta = 1.512$
$\gamma \simeq 1.513$
(-)2V = small
DENSITY: 1.85 (Meas.)
CLEAVAGE: Not determined.
HABIT: Crystals minute laths; also as microcrystalline masses.
COLOR-LUSTER: White; silky.
MODE OF OCCURRENCE: Occurs abundantly associated with basaluminite and gypsum in veinlets in sandstone at the Fuemrole mine, Temple mountain, Emery County, Utah.
BEST REF. IN ENGLISH; Frondel, Clifford, *Am. Min.*, **53**: 717–721 (1968).

## META-ALUNOGEN (Inadequately described mineral)
$Al_4(SO_4)_6 \cdot 27H_2O$

CRYSTAL SYSTEM: Monoclinic
OPTICAL CONSTANTS:
$\alpha = 1.469$
$\beta = 1.473$
$\gamma = 1.491$
(+)2V = moderate

HARDNESS: Not determined.
DENSITY: Not determined.
CLEAVAGE: {010} perfect
HABIT: Massive.
COLOR-LUSTER: White. Waxy or pearly.
MODE OF OCCURRENCE: Occurs as a natural dehydration product of alunogen in veins in altered andesite at Francisco de Vergara, Antofagasta, Chile.
BEST REF. IN ENGLISH: Gordon, Samuel G., "Notulae Naturae," *Acad. Nat. Sci., Philadelphia*, No. 1, 9 pp (1942).

## META-ANKOLEITE
$K_2(UO_2)_2(PO_4)_2 \cdot 6H_2O$

CRYSTAL SYSTEM: Tetragonal
CLASS: 4/m 2/m 2/m
SPACE GROUP: P4/nmm
Z: 1
LATTICE CONSTANTS:
$a = 6.993$
$c = 8.891$
3 STRONGEST DIFFRACTION LINES:
8.92 (100)
3.73 ( 65)
3.25 ( 55)
OPTICAL CONSTANTS:
$n \sim 1.580 (-)$
HARDNESS: Not determined.
DENSITY: 3.54 (Calc.)
CLEAVAGE: {001} perfect, micaceous
{100} distinct
HABIT: As plates, usually about 0.5 mm and up to 1 mm in diameter.
COLOR-LUSTER: Yellow. Fluoresces yellow-green in both short-wave and long-wave ultraviolet light.
MODE OF OCCURRENCE: Occurs intergrown with a phosphuranylite-type mineral associated with muscovite, quartz, and albite in the Mungenyi pegmatite, Ankole district, southwest Uganda. It also occurs associated with quartz, sericitized microcline, and accessory monazite, spinel, barite, and zircon in poorly cemented feldspathic quartzite, Sebungive district, Rhodesia.
BEST REF. IN ENGLISH: Gallather, M. J., and Atkin, D., *Bull. Geol. Surv. Great Brit.*, **25**: 49 (1966).

## META-AUTUNITE
$Ca(UO_2)_2(PO_4)_2 \cdot 2-6H_2O$

CRYSTAL SYSTEM: Tetragonal
CLASS: 4/mmm or 422
SPACE GROUP: P4/nmm or $P4_2 22$
Z: 1, 16
LATTICE CONSTANTS:
(P4/nmm) ($P4_2 22$)
$a = 6.99$ (variable) $a = 19.78$
$c = 8.44$ $c = 16.92$
Z = 1 Z = 16
3 STRONGEST DIFFRACTION LINES:
8.47 (100)
3.61 ( 85)
2.11 ( 70)

OPTICAL CONSTANTS:
$\alpha = 1.585-1.600$
$\beta = 1.595-1.610$
$\gamma = 1.595-1.613$
$(-)2V$ = small, variable
HARDNESS: 2-2½
DENSITY: 3.45-3.55 (Meas.)
3.53 (Calc.)
CLEAVAGE: {001} perfect
{100} indistinct
HABIT: As dehydration pseudomorphs after autunite; morphology same as that species.

COLOR-LUSTER: Lemon yellow, greenish yellow to yellowish green. Translucent to opaque. Pearly to dull. Fluoresces pale yellowish green in ultraviolet light.

MODE OF OCCURRENCE: Occurs as a dehydration product of autunite that has been exposed in the open air; localities are given under autunite.

BEST REF. IN ENGLISH: Frondel, Clifford, USGS Bull. 1064, p. 205-207 (1958). Ross, Malcolm, *Am. Min.*, **48**: 1389-1393 (1963).

# METABORITE
$HBO_2$

CRYSTAL SYSTEM: Cubic
CLASS: $\bar{4}3m$
SPACE GROUP: $P\bar{4}3n$
Z: 24
LATTICE CONSTANT:
$a = 8.89$
3 STRONGEST DIFFRACTION LINES:
4.442 (100)
1.939 ( 95)
3.627 ( 50)
OPTICAL CONSTANT:
$N = 1.618$
HARDNESS: 5
DENSITY: 2.47 (Meas.)
2.489 (Calc.)
CLEAVAGE: Brittle. Fracture conchoidal.
HABIT: Aggregates up to 1 cm in size, showing single faces of the tetrahedron and trigonal tristetrahedron. Slowly soluble in water.

COLOR-LUSTER: Colorless to brownish; vitreous.

MODE OF OCCURRENCE: Occurs in rock salt with small amounts of anhydrite, boracite, aksaite, ginorite, hilgardite, strontiborite, and halurgite at an unspecified locality, presumably in Kazakhstan.

BEST REF. IN ENGLISH: Lobanova, V. V., and Avrova, N. P., *Am. Min.*, **50**: 261-262 (1965).

# METACHALCOPHYLLITE = Artificially dehydrated chalcophyllite
$Cu_{18}Al_2(AsO_4)_3(SO_4)_3(OH)_{27}$

# METACINNABARITE (Metacinnabar)
HgS
Dimorphous with cinnabar

CRYSTAL SYSTEM: Cubic
CLASS: $\bar{4}3m$
SPACE GROUP: $F\bar{4}3m$
Z: 4
LATTICE CONSTANT:
$a = 5.853$
3 STRONGEST DIFFRACTION LINES:
3.38 (100)
2.068 ( 55)
1.7644 ( 45)
HARDNESS: 3
DENSITY: 7.65 (Meas.)
7.65 (Calc.)
CLEAVAGE: Fracture uneven to subconchoidal. Brittle.
HABIT: Usually massive; crystals rare, tetrahedral. Twinning on {111} common.

COLOR-LUSTER: Grayish black. Opaque. Metallic. Streak black.

MODE OF OCCURRENCE: Occurs chiefly in the upper portions of mercury deposits, often associated with cinnabar and other sulfides. Found widespread at the quicksilver deposits in the Coast Ranges of California, extending from Del Norte County to Santa Barbara County; at Marysvale, Utah; and in Alaska. It also occurs in Canada, Mexico, Spain, Italy, Czechoslovakia, Roumania, and China.

BEST REF. IN ENGLISH: Palache, et al., "Dana's System of Mineralogy," 7th Ed., v. I, p. 215-217, New York, Wiley, 1944.

# METADELRIOITE
$CaSrV_2O_6(OH)_2$

CRYSTAL SYSTEM: Triclinic
CLASS: $\bar{1}$ or 1
SPACE GROUP: $P\bar{1}$ or P1
Z: 2
LATTICE CONSTANTS:
$a = 7.343$    $\alpha = 119°39'$
$b = 8.382$    $\beta = 90°16'$
$c = 5.117$    $\gamma = 102°49'$
3 STRONGEST DIFFRACTION LINES:
4.944 (100)
3.457 (100)
4.732 ( 80)
HARDNESS: ~2
DENSITY: 4.3 (Meas.)
4.21 (Calc.)
CLEAVAGE: Not determined.
HABIT: As aggregates of fibrous acicular crystals; often as intergrowths with delrioite.

COLOR-LUSTER: Light yellow-green; vitreous to pearly.

MODE OF OCCURRENCE: Occurs as an efflorescence on sandstone on the dump at the Hummer portal of the Jo Dandy mine in Paradox Valley, Montrose County, Colorado.

BEST REF. IN ENGLISH: Smith, Marie Lindberg, *Am. Min.*, **55**: 185-200 (1970).

# METAHAIWEEITE
$Ca(UO_2)_2Si_6O_{15} \cdot nH_2O (n < 5)$

CRYSTAL SYSTEM: Monoclinic (?)
CLASS: 2/m
SPACE GROUP: Probably P2/c
LATTICE CONSTANTS: Related to those of haiweeite.
3 STRONGEST DIFFRACTION LINES:
6.98 (100)
8.81 ( 50)
5.85 ( 50)
OPTICAL CONSTANTS:
$\alpha = 1.611$
$\beta = 1.620$
$\gamma = 1.645$
HARDNESS: 3½
DENSITY: 3.35 (Meas.)
CLEAVAGE: {100} distinct
HABIT: Crystals poorly developed; as single flake-like grains or spherical aggregates with radial structure.
COLOR-LUSTER: Pale yellow to greenish yellow. Translucent. Pearly. Fluoresces weak dull green under ultraviolet light.
MODE OF OCCURRENCE: Occurs on fracture surfaces in granite and in voids of the adjacent lake bed deposits near the Haiwee Reservoir in the Coso Mountains, California, intimately associated with haiweeite.
BEST REF. IN ENGLISH: McBurney, T. C., and Murdoch, Joseph, *Am. Min.*, 44: 839–843 (1959).


# METAHALLOYSITE = Halloysite


# METAHEINRICHITE
$Ba(UO_2)_2(AsO_4)_2 \cdot 8H_2O$

CRYSTAL SYSTEM: Tetragonal
CLASS: 4/m, 422, or 4
SPACE GROUP: $P4_2 22$, $P4_2/m$, or $P4_2$
Z: 2
LATTICE CONSTANTS:
$a = 7.07$
$c = 17.74$
3 STRONGEST DIFFRACTION LINES:
8.55 (100)
3.71 (100)
5.41 ( 90)
OPTICAL CONSTANTS:
$\omega = 1.637$
$\epsilon = 1.609$
(–)
HARDNESS: 2.5
DENSITY: 4.04 (Meas.)
4.09 (Calc.)
CLEAVAGE: {001} perfect
{100} distinct
HABIT: Crystals tabular {001}; up to 1 mm on a side; not more than 0.1 mm thick. Forms: {001}, {100}, and {110}.
COLOR-LUSTER: Yellow to green, transparent to translucent; vitreous to pearly. Fluoresces bright green to greenish yellow under short- and long-wave ultraviolet light.
MODE OF OCCURRENCE: Occurs coating fractures and lining vugs in light gray, altered, silicified rhyolite tuff in the White King mine, 14 miles northwest of Lakeview, Lake County, Oregon. Also occurs as an alteration of pitchblende associated with zeunerite, novacekite, erythrite, arseniosiderite, and pitticite at Wittichen, Schiltach, and Reinerzau in the Black Forest, Germany.
BEST REF. IN ENGLISH: Gross, Eugene B., Corey, Alice S., Mitchell, Richard S., and Walenta, Kurt, *Am. Min.*, 43: 1134–1143 (1958).


# METAHEWETTITE
$CaV_6O_{16} \cdot 9H_2O$

CRYSTAL SYSTEM: Monoclinic
CLASS: 2/m
SPACE GROUP: $P2_1/m$
Z: 1
LATTICE CONSTANTS:
$a = 12.18$
$b = 3.614$
$c = 7.80$
$\beta = 95°15'$
OPTICAL CONSTANTS:
$\alpha = 1.70$
$\beta = 2.10$ (Li)
$\gamma \sim 2.23$
(–)2V = 52° (Calc.)
HARDNESS: Not determined.
DENSITY: 2.94 (Meas. with $3H_2O$)
CLEAVAGE: Not determined.
HABIT: As nodular aggregates and coatings of fibers, elongate [010]. Crystals are well-developed laths elongated [010], and flattened {001}.
COLOR-LUSTER: Deep red; dull to silky. Powder maroon to brownish red.
MODE OF OCCURRENCE: Occurs as an impregnation in sandstone in the lower Fall River flood plain sediments at the Get Me Rich mine near Edgemont, Fall River County, South Dakota; in Paradox Valley, Montrose County, Colorado; at Thompson's, Grand County, Utah; and at several other localities in the Colorado Plateau uranium-vanadium district.
BEST REF. IN ENGLISH: Palache, et al., "Dana's System of Mineralogy," 7th Ed., v. II, p. 1061–1062, New York, Wiley, 1951. Ross, Malcolm, *Am. Min.*, 44: 322–341, (1959).


# METAHOHMANNITE (Inadequately described mineral)
$Fe_2(SO_4)_2(OH)_2 \cdot 3H_2O$

CRYSTAL SYSTEM: Unknown
OPTICAL CONSTANTS:
$\alpha = 1.709$
$\beta = 1.718$
$\gamma = 1.734$
(+)2V $\approx$ 70°
HARDNESS: Not determined.
DENSITY: Not determined.
CLEAVAGE: Not determined.
HABIT: Massive, pulverulent.

COLOR-LUSTER: Orange.

MODE OF OCCURRENCE: Occurs at Alcaparossa and Chuquicamata, Chile, as an alteration product of hohmannite.

BEST REF. IN ENGLISH: Palache, et al., "Dana's System of Mineralogy," 7th Ed., v. II, p. 608, New York, Wiley, 1951.

## METAKAHLERITE
$Fe(UO_2)_2(AsO_4)_2 \cdot 8H_2O$

CRYSTAL SYSTEM: Tetragonal
Z: 1
LATTICE CONSTANTS:
  $a = 7.18$
  $c = 8.58$
3 STRONGEST DIFFRACTION LINES:
  (Sophia Shaft)   (synthetic)
  3.59 (100)   8.86 (100)
  8.55 ( 90)   3.59 (100)
  4.29 ( 60)   1.608 ( 70)
OPTICAL CONSTANTS:
  $\omega = 1.642$
  $\epsilon = 1.608$
Uniaxial to biaxial, negative, with 2V up to 22°.
HARDNESS: Not determined
DENSITY: 3.84 (Calc.)
CLEAVAGE: {001} excellent
  {100} good
HABIT: As scaly aggregates.
COLOR-LUSTER: Sulfur yellow; pearly on {001}.
MODE OF OCCURRENCE: Occurs in the Sophia Shaft, Wittichen, Baden, Germany.
BEST REF. IN ENGLISH: Walenta, Kurt, Am. Min., 45: 254 (1960).

## METAKIRCHHEIMERITE
$Co(UO_2)_2(AsO_4)_2 \cdot 8H_2O$

CRYSTAL SYSTEM: Tetragonal
LATTICE CONSTANTS: Cell probably similar to metakahlerite.
3 STRONGEST DIFFRACTION LINES:
  8.55 (100)
  3.56 (100)
  5.07 ( 60)
OPTICAL CONSTANTS:
  $\omega = 1.644$
  $\epsilon = 1.617$
Uniaxial to biaxial (−) with 2V = 0°-20°
HARDNESS: 2-2½
DENSITY: < 3.33
CLEAVAGE: {001} excellent
HABIT: As tabular crystals, and crusts.
COLOR-LUSTER: Pale rose; pearly on cleavage.
MODE OF OCCURRENCE: Occurs associated with metakahlerite, novacekite, meta-heinrichite, and erythrite in the Sophia shaft, Wittichen, Baden, Germany.
BEST REF. IN ENGLISH: Walenta, K., Am. Min., 44: 466 (1959).

## METANOVACEKITE
$Mg(UO_2)_2(AsO_4)_2 \cdot 4\text{-}8H_2O$
Autunite Group

CRYSTAL SYSTEM: Tetragonal
CLASS: 4/m
SPACE GROUP: P4/n
Z: 1
LATTICE CONSTANTS:
  $a = 7.16$
  $c = 8.58$
3 STRONGEST DIFFRACTION LINES:
  8.52 (100)
  3.57 ( 90)
  2.14 ( 60)
OPTICAL CONSTANTS:
  $\omega = 1.632$
  $\epsilon = 1.595$
  (−)
HARDNESS: 2½
DENSITY: 3.51 (Meas.)
  3.716 (Calc.)
CLEAVAGE: {001} perfect
  {010} indistinct
  {110} indistinct
HABIT: Crystals rectangular plates flattened on {001}. As crusts or porous interlocking aggregates of thin plates and scales; also as lamellar aggregates.
COLOR-LUSTER: Dull pale yellow, yellow. Nearly opaque. Fluoresces dull green in ultraviolet light.
MODE OF OCCURRENCE: Occurs at the Anton mine, Black Forest, Germany.
BEST REF. IN ENGLISH: Walenta, Kurt, J. Geol. Landes. Baden-Wurtemburg, 3: 17 (1958). Walenta, Kurt, Tscherm. min. pet. Mitt., 9: 111-174 (1964).

## METAROSSITE
$CaV_2O_6 \cdot 2H_2O$

CRYSTAL SYSTEM: Triclinic
Z: 2
LATTICE CONSTANTS:
  $a = 7.065$  $\alpha = 96°39'$
  $b = 7.769$  $\beta = 105°47'$
  $c = 6.215$  $\gamma = 92°58'$
OPTICAL CONSTANTS:
  $\alpha = 1.840$
  $\beta > 1.85$
  $\gamma > 1.85$
(+)2V = large
HARDNESS: Not determined. Soft.
DENSITY: 2.45 (Meas.)
CLEAVAGE: Not determined. Friable.
HABIT: As platy to flaky masses and veinlets. Twinned on $(0\bar{1}1)$.
COLOR-LUSTER: Light yellow. Pearly to dull.

Soluble in water.
MODE OF OCCURRENCE: Occurs as a dehydration product of rossite in carnotite-bearing sandstone in Bull Pen Canyon, and in ore from the Buckhorn claim, Slick Rock district, San Miguel County, Colorado.

BEST REF. IN ENGLISH: Palache, et al., "Dana's System of Mineralogy," 7th Ed., v. II, p. 1054–1055, New York, Wiley, 1951.

# METASCHODERITE
$Al_2 (PO_4)(VO_4) \cdot 6H_2O$

CRYSTAL SYSTEM: Monoclinic
CLASS: 2/m
SPACE GROUP: P2/m
Z: 4
LATTICE CONSTANTS:
  $a = 11.4$
  $b = 14.9$
  $c = 9.2$
  $\beta = 79°$
3 STRONGEST DIFFRACTION LINES:
  7.5 (100)
  14.9 ( 60)
  11.1 ( 40)
OPTICAL CONSTANTS:
  $\alpha = 1.598$
  $\beta = 1.604$
  $\gamma = 1.626$
  (+)2V = 59° (Calc.)
HARDNESS: ~ 2
DENSITY: 1.610 (Calc.)
CLEAVAGE: Not determined.
HABIT: Microscopic bladed tabular crystals, elongated parallel to (001).
COLOR-LUSTER: Yellowish orange.
MODE OF OCCURRENCE: Occurs sparsely as microcrystalline coatings associated with wavellite and vashegyite along fractures in phosphatic cherts of lower Paleozoic age near Eureka, Nevada.
BEST REF. IN ENGLISH: Hausen, Donald M., *Am. Min.*, **47**: 637–648 (1962).

# METASCHOEPITE (Schoepite II)
$UO_3 \cdot 1\text{-}2H_2O$
A natural dehydration product of schoepite

CRYSTAL SYSTEM: Orthorhombic
CLASS: 2/m 2/m 2/m
SPACE GROUP: Pbna
Z: 32
LATTICE CONSTANTS:
  $a = 13.99$
  $b = 16.72$
  $c = 14.73$
HARDNESS: ~ 2½
DENSITY: Not determined.
CLEAVAGE: {001} perfect
              {010} distinct
Brittle.
HABIT: Crystals tabular on {001}, elongated [010], pseudohexagonal aspect.
COLOR-LUSTER: Yellow.
MODE OF OCCURRENCE: Occurs as an alteration product of schoepite I at the Shinkolobwe mine, Katanga, Zaire.

BEST REF. IN ENGLISH: Christ, C. L., and Clark, Joan R., *Am. Min.*, **45**: 1026–1061 (1960).

# METASCOLECITE = High temperature polymorph of scolecite.

# METASIDERONATRITE
$Na_4 Fe_2 (OH)_2 (SO_4)_4 \cdot 3H_2O$

CRYSTAL SYSTEM: Orthorhombic
CLASS: 2/m 2/m 2/m
Z: 2
LATTICE CONSTANTS:
  $a:b:c = 0.4571:1:0.1187$
3 STRONGEST DIFFRACTION LINES:
  3.66 (100)
  2.73 ( 80)
  7.93 ( 60)
OPTICAL CONSTANTS:
  $\alpha = 1.543$
  $\beta = 1.575$
  $\gamma = 1.634$
  (+)2V = 60°
HARDNESS: 2½
DENSITY: 2.46 (Meas.)
CLEAVAGE: {100} perfect
              {010} perfect
              {001} distinct
Fracture fibrous.
HABIT: Usually as coarse to fine crystalline aggregates. Crystals prismatic, rare.
COLOR-LUSTER: Golden yellow to straw yellow. Silky.

Soluble in boiling water.
MODE OF OCCURRENCE: Occurs associated with metavoltine, ferrinatrite, and other sulfates at Chuquicamata, Antofagasta province, Chile.
BEST REF. IN ENGLISH: Palache, et al., "Dana's System of Mineralogy," 7th Ed., v. II, p. 603–604, New York, Wiley, 1951.

# METASTIBNITE (Inadequately described mineral)
$Sb_2 S_3$

CRYSTAL SYSTEM: Amorphous
HARDNESS: 2–3
DENSITY: Not determined.
CLEAVAGE: None.
HABIT: Massive; as mammillary coatings and stains.
COLOR-LUSTER: Red, purple, purplish gray. Transparent in thin splinters. Submetallic. Streak red.
MODE OF OCCURRENCE: Occurs in California with stibnite and oxides of antimony at the Bishop antimony mine, Inyo County; at Sulphur Bank, Lake County; and in siliceous sinter at The Geysers in Sonoma County. Also found at Steamboat Springs, Washoe County, Nevada, and at Oruro and elsewhere in Bolivia.
BEST REF. IN ENGLISH: Palache, et al., "Dana's System of Mineralogy," 7th Ed., v. I, p. 275, New York, Wiley, 1944.

## METASTRENGITE (Phosphosiderite)
$FePO_4 \cdot 2H_2O$
Dimorphous with strengite

CRYSTAL SYSTEM: Monoclinic
CLASS: 2/m
SPACE GROUP: $P2_1/n$
Z: 4
LATTICE CONSTANTS:
  $a = 5.32$
  $b = 9.87$
  $c = 8.69$
  $\beta = 90°36'$
3 STRONGEST DIFFRACTION LINES:
  2.78 (100)
  4.69 ( 80)
  4.37 ( 80)
OPTICAL CONSTANTS:
  $\alpha = 1.692$
  $\beta = 1.725$ (Na)
  $\gamma = 1.738$
  $(-)2V = 62°$
HARDNESS: 3½–4
DENSITY: 2.76 (Meas.)
         2.76 (Calc.)
CLEAVAGE: {010} good
          {001} indistinct
Fracture uneven. Brittle.

HABIT: Crystals tabular or stout prismatic. Also as botryoidal crusts and masses with radial-fibrous structure. Twinning on {101}, common as interpenetration twins.

COLOR-LUSTER: Reddish violet, peach-blossom red, purple, wine yellow, nearly colorless. Transparent to translucent. Vitreous to subresinous.

MODE OF OCCURRENCE: Occurs as superb crystals up to 1 cm or more in length associated with strengite crystals in vugs in barbosalite at the Bull Moose mine, Custer County, and in other pegmatites in the Black Hills, South Dakota; in rockbridgeite with bermanite at the Williams prospect, Coosa County, Alabama; in pegmatite at Pala, San Diego County, California, and at the Fletcher and Palermo quarries near North Groton, New Hampshire; and with strengite at Manhattan, Nevada. It also is found near Chanteloube, France; in the Kalterborn mine near Eiserfeld, at Hagendorf, and at other places in Germany; in Sweden, Sardinia, and on Malpelo Island, Pacific Ocean.

BEST REF. IN ENGLISH: Palache, et al., "Dana's System of Mineralogy," 7th Ed., v. II, p. 769–771, New York, Wiley, 1951.

## METATORBERNITE
$Cu(UO_2)_2(PO_4)_2 \cdot 8H_2O$

CRYSTAL SYSTEM: Tetragonal
CLASS: 4/m
SPACE GROUP: P4/n
Z: 1 or 2
LATTICE CONSTANTS:

| (Z = 1) | (Z = 2) |
|---|---|
| $a = 6.96$ | $a = 6.969$ |
| $c = 8.62$ | $c = 17.306$ |

3 STRONGEST DIFFRACTION LINES:
  8.71 (100)
  3.68 (100)
  3.48 ( 80)
OPTICAL CONSTANTS:
  $\omega \simeq 16.26$
  $\epsilon \simeq 1.624$
    (+)
HARDNESS: 2½
DENSITY: 3.7–3.8 (Meas.)
         3.70 (Calc.)
CLEAVAGE: {001} perfect
Brittle.

HABIT: Crystals thin tablets flattened on {001}. Dominant forms {001} and {100}. As lamellar aggregates, rosettes, and sheaflike or subparallel aggregates. Also as dehydration pseudomorphs after torbernite.

COLOR-LUSTER: Pale green to dark green; transparent to translucent. Vitreous to subadamantine, pearly on {001}. Also dull in pseudomorphs after torbernite. Not fluorescent.

MODE OF OCCURRENCE: Occurs chiefly as a secondary mineral in the oxidized portion of veins or other deposits containing uraninite together with copper-containing minerals, and possibly as a low-temperature hydrothermal mineral in vein deposits. Found widespread in small amounts in in the copper-uranium deposits of the Colorado Plateau; in uraniferous lignite in the North Cave Hills, Harding County, South Dakota; and in quartzite in the Wilson Creek area, Gila County, Arizona. It is also found at the Nicholson mine, Lake Athabasca, Saskatchewan, Canada; in the Redruth area, Cornwall, England; in the Puy-de-Dôme district, France; at several localities in the Erzgebirge of Saxony and Bohemia; and as excellent specimens in the Katanga district, Zaire.

BEST REF. IN ENGLISH: Frondel, Clifford, USGS Bull. 1064, p. 208–211 (1958). Ross, Malcolm, Evans, H. T. Jr., and Appleman, D. E., *Am. Min.*, **49**: 1603–1621 (1964).

## METATYUYAMUNITE
$Ca(UO_2)_2(VO_4)_2 \cdot 3\text{-}5H_2O$

CRYSTAL SYSTEM: Orthorhombic
CLASS: 2/m 2/m 2/m
SPACE GROUP: Pnam
Z: 4
LATTICE CONSTANTS:
  $a = 10.54$
  $b = 8.49$
  $c = 17.34$
3 STRONGEST DIFFRACTION LINES:
  8.51 (100)
  4.22 ( 80)
  3.26 ( 60)
OPTICAL CONSTANTS:
  $\alpha = 1.68$ (Calc.)
  $\beta = 1.835$
  $\gamma = 1.865$
  $(-)2V = 45°$
HARDNESS: $\sim 2$
DENSITY: 3.8–3.9 (Meas.)

CLEAVAGE: {001} perfect
{010} distinct
{100} distinct

HABIT: Commonly compact to microcrystalline masses; also pulverulent, as thin coatings and films, and as an impregnation in sandstone or limestone. Crystals are tiny scales and laths flattened on {001} and elongated along $a$-axis.

COLOR-LUSTER: Canary yellow to greenish yellow. On exposure to sunlight color changes to dark green or yellow green. Adamantine to waxy. Not fluorescent.

MODE OF OCCURRENCE: Found with tyuyamunite and carnotite as coatings or disseminations in sandstone or limestone at many localities on the Colorado Plateau; in the Shiprock and Grants districts, New Mexico; at several localities in western South Dakota, Wyoming, and from Tyuya Muyun, Ferghana, Turkestan, U.S.S.R.

BEST REF. IN ENGLISH: Frondel, Clifford, USGS Bull. 1064, p. 254–257 (1958).

## META-URANOCIRCITE I AND II

I—$Ba(UO_2)_2(PO_4)_2 \cdot 8H_2O$
II—$Ba(UO_2)_2(PO_4)_2 \cdot 6H_2O$

CRYSTAL SYSTEM: Tetragonal
CLASS: 4/m 2/m 2/m or 4/m
SPACE GROUP: $P4_2/n$ (I), P4/nmm (II)
Z: 2
LATTICE CONSTANTS:
$a = 6.96$
$c = 16.90$
3 STRONGEST DIFFRACTION LINES:

| I | II |
|---|----|
| 8.93 (100) | 8.55 (100) |
| 3.73 ( 80) | 3.61 ( 90) |
| 5.48 ( 70) | 5.39 ( 70) |

OPTICAL CONSTANTS:
$\epsilon$ or $\alpha = 1.610$
$\beta = 1.623$
$\omega$ or $\gamma = 1.623$
HARDNESS: 2–2½
DENSITY: 3.95 (Meas. I)
4.00 (Meas. II)
CLEAVAGE: {001} perfect
{100} distinct
Not Brittle. Thin flakes flexible.

HABIT: Crystals thin plates flattened on {001} with rectangular outline. As subparallel aggregates or fan-like groups.

COLOR-LUSTER: Yellow green; transparent to translucent. Pearly on {001}. Fluoresces green in ultraviolet light.

MODE OF OCCURRENCE: Found chiefly as a secondary mineral, but may be a primary low-temperature deposit in some vein occurrences. In the United States found at the Davier mine, Lawrence County, disseminated in uraniferous lignite in the Slim Buttes area, Harding County, and in a channel sandstone in the Chadron formation, Badlands, South Dakota; and with fluorite in the Honeycomb Hills, Utah. It also occurs in Germany at Menzenschevand (both I and II); at Schneeberg and at Bergen, Saxony; and at Wölsendorf, Bavaria. Also found in the Banat, Rumania;

at Rosmaneira, Spain; and in an alluvial deposit near Antsirabe, Madagascar.

BEST REF. IN ENGLISH: Frondel, Clifford, USGS Bull. 1064, p. 211–215 (1958). Walenta, K., *Min. Abs.* **17**: 695 (1966).

## META-URANOPILITE (Inadequately described mineral)

$(UO_2)_6SO_4(OH)_{10} \cdot 5H_2O$

CRYSTAL SYSTEM: Orthorhombic
OPTICAL CONSTANTS:
$\alpha = 1.72$
$\beta = 1.76$
$\gamma = 1.76$
$(-)2V$
HARDNESS: Not determined.
DENSITY: Not determined.
CLEAVAGE: Not determined.
HABIT: Fibrous, as needles and laths.
COLOR-LUSTER: Gray, grayish green, brownish, yellow. Fluoresces yellowish green in ultraviolet light.

MODE OF OCCURRENCE: Occurs at Joachimsthal, Bohemia, and is also reported from Cornwall, England.

BEST REF. IN ENGLISH: Frondel, Clifford, *Am. Min.*, **37**: 950–959 (1952).

## META-URANOSPINITE

$Ca(UO_2)_2(AsO_4)_2 \cdot 6\text{-}8H_2O$
Meta-autunite Group

CRYSTAL SYSTEM: Tetragonal
CLASS: 4/m 2/m 2/m
SPACE GROUP: P4/nmm
LATTICE CONSTANTS: Similar to meta-autunite.
3 STRONGEST DIFFRACTION LINES:
8.65 (100)
3.57 (100)
3.31 ( 90)
OPTICAL CONSTANT:
$\gamma = 1.618$
HARDNESS: 2–3
DENSITY: Not determined. Presumably higher than uranospinite.
CLEAVAGE: {001} perfect
{100} distinct
Brittle.

HABIT: Crystals small thin rectangular plates flattened on {001}.

COLOR-LUSTER: Lemon yellow to siskin green. Pearly on {001}. Fluoresces bright lemon yellow in ultraviolet light.

MODE OF OCCURRENCE: Name given to the dehydration product of uranospinite from the Clara Shaft, Wittichen, Baden, Germany. Readily rehydrates to uranospinite.

BEST REF. IN ENGLISH: Walenta, Kurt, *Am. Min.*, **45**: 254 (1960).

## METAVANDENDRIESSCHEITE
$PbU_7O_{22} \cdot nH_2O(n<12)$
A natural dehydration product of vandendriesscheite

CRYSTAL SYSTEM: Orthorhombic
Z: 240 units of $UO_2(OH)_2$ per cell
LATTICE CONSTANTS:
 $a = 14.07$
 $b = 41.31$
 $c = 43.33$
OPTICAL CONSTANTS:
  $\alpha = 1.780$
  $\beta = 1.850$
  $\gamma = 1.860$
  $(-)2V \simeq 60°$
PHYSICAL PROPERTIES: See vandendriesscheite.
MODE OF OCCURRENCE: See vandendriesscheite.
BEST REF. IN ENGLISH: *Am. Min.*, **45**: 1026-1061 (1960).

## METAVANURALITE
$Al(UO_2)_2(VO_4)_2(OH) \cdot 8H_2O$

CRYSTAL SYSTEM: Triclinic
CLASS: $\bar{1}$ or 1
SPACE GROUP: $P\bar{1}$ or P1
Z: 2
LATTICE CONSTANTS:
 $a = 10.46$    $\alpha = 75°53'$
 $b = 8.44$     $\beta = 102°50'$
 $c = 10.43$    $\gamma = 90°$
3 STRONGEST DIFFRACTION LINES:
 9.92 (100)
 4.174 ( 90)
 3.155 ( 90)
HARDNESS: ~2
CLEAVAGE: {001} perfect, easy.
HABIT: Crystals plates on {001}. Forms: {001} dominant; also {101}, {$\bar{1}$01}
COLOR-LUSTER: Citron yellow.
MODE OF OCCURRENCE: Occurs with plumboan francevillite, chervetite, and brackebuschite in a supergene deposit at Mounana, Gabon.
BEST REF. IN ENGLISH: Cesbron, Fabien, *Am. Min.*, **56**: 637 (1971).

## METAVARISCITE
$AlPO_4 \cdot 2H_2O$
Dimorphous with variscite

CRYSTAL SYSTEM: Monoclinic
CLASS: 2/m
SPACE GROUP: $P2_1/n$
Z: 4
LATTICE CONSTANTS:
 $a = 8.47$
 $b = 9.47$
 $c = 5.16$
 $\beta \sim 90°$

3 STRONGEST DIFFRACTION LINES:
 2.71 (100)
 4.25 ( 90)
 4.575 ( 80)
OPTICAL CONSTANTS:
  $\alpha = 1.551$
  $\beta = 1.558$
  $\gamma = 1.582$
  $(+)2V = 55°$
HARDNESS: ~3½
DENSITY: 2.54 (Meas.)
        2.535 (Calc.)
CLEAVAGE: {010}
Brittle.
HABIT: Crystals minute, thin to thick tabular, slightly elongated or equant. Rarely long prismatic. Also granular, massive. Twinning on {102} as contact twins.
COLOR-LUSTER: Pale green. Transparent to translucent. Vitreous. Streak white.
MODE OF OCCURRENCE: Occurs as fine microcrystals in cavities in massive variscite near Lucin, Box Elder County, Utah; at Candelaria, Nevada; and on Malpelo Island, Pacific Ocean.
BEST REF. IN ENGLISH: Palache, et al., "Dana's System of Mineralogy," 7th Ed., v. II, p. 767-769, New York, Wiley, 1951.

## METAVAUXITE
$Fe^{2+}Al_2(PO_4)_2(OH)_2 \cdot 8H_2O$

CRYSTAL SYSTEM: Monoclinic
CLASS: 2/m
SPACE GROUP: $P2_1/c$
Z: 2
LATTICE CONSTANTS:
 $a = 10.21$ kX
 $b = 9.57$
 $c = 6.93$
 $\beta = 98°02'$
3 STRONGEST DIFFRACTION LINES:
 2.75 (100)
 4.67 ( 90)
 4.32 ( 85)
OPTICAL CONSTANTS:
  $\alpha = 1.550$
  $\beta = 1.561$ (Na)
  $\gamma = 1.577$
  $(+)2V = $ large
HARDNESS: 3
DENSITY: 2.345 (Meas.)
        2.35 (Calc.)
CLEAVAGE: Brittle.
HABIT: Crystals long prismatic to acicular; as subparallel to radial aggregates.
COLOR-LUSTER: Colorless, white, pale green. Transparent to translucent. Vitreous; fibrous aggregates silky.
MODE OF OCCURRENCE: Occurs with paravauxite, vauxite, and wavellite, often encrusting quartz crystals, in the tin mines at Llallagua and Tasna, Bolivia.
BEST REF. IN ENGLISH: Palache, et al., "Dana's System of Mineralogy," 7th Ed., v. II, p. 971-972, New York, Wiley, 1951.

## METAVOLTINE
$(K, Na)_5 Fe_3 (SO_4)_6 (OH)_2 \cdot 8H_2O$

CRYSTAL SYSTEM: Hexagonal
Z: 8
LATTICE CONSTANTS:
  $a = 19.470$
  $c = 18.622$
3 STRONGEST DIFFRACTION LINES:
  9.06 (100)
  8.26 ( 90)
  3.28 ( 75)
OPTICAL CONSTANTS:
  $\omega = 1.588-1.595$
  $\epsilon = 1.573-1.581$
    (−)
HARDNESS: 2½
DENSITY: 2.40–2.51 (Meas.)
       2.437 (Calc.)
CLEAVAGE: {0001} perfect
HABIT: As minute tabular crystals with hexagonal outline; usually as scaly or granular aggregates.
COLOR-LUSTER: Yellowish brown to greenish brown; also orange-brown. Translucent. Resinous.

Partially water-soluble.
MODE OF OCCURRENCE: Occurs with krausite, coquimbite, alunite, and other sulfates near Borate, in the Calico Mountains, San Bernardino County, California. It also is found at Chuquicamata, and at other places in northern Chile; in Italy; and near Madeni Zakh, Iran, as an alteration of pyritic trachyte.
BEST REF. IN ENGLISH: Palache, et al., "Dana's System of Mineralogy," 7th Ed., v. II, p. 619–621, New York, Wiley, 1951.


## METAZELLERITE
$Ca(UO_2)(CO_3)_2 \cdot 3H_2O$

CRYSTAL SYSTEM: Orthorhombic
CLASS: mm2 or 2/m 2/m 2/m
SPACE GROUP: $Pbn2_1$ or Pbnm
Z: 4
LATTICE CONSTANTS:
  $a = 9.718$
  $b = 18.226$
  $c = 4.965$
3 STRONGEST DIFFRACTION LINES:
  9.10  (100)
  3.794 ( 50)
  4.695 ( 36)
OPTICAL CONSTANTS:
  Mean $n \sim 1.626$
HARDNESS: ~2
DENSITY: 3.414 (Calc.)
CLEAVAGE: Aggregated clumps fracture in direction of the elongation.
HABIT: Fibrous and hair-like, aggregating into pincushion clumps.
COLOR-LUSTER: Chalky yellow; dull luster.
MODE OF OCCURRENCE: Occurs as a partial dehydra-tion product of zellerite (q.v.). Described from the Lucky Mc uranium mine, Fremont County, Wyoming.
BEST REF. IN ENGLISH: Coleman, R. G., Ross, D. R., and Meyrowitz, R., *Am. Min.*, **51**: 1567–1578 (1966).


## METAZEUNERITE
$Cu(UO_2)_2 (AsO_4)_2 \cdot 8H_2O$

CRYSTAL SYSTEM: Tetragonal
CLASS: 4/m
SPACE GROUP: $P4_2/n$
Z: 2
LATTICE CONSTANTS:
  $a = 7.10$
  $c = 17.42$
3 STRONGEST DIFFRACTION LINES:
  8.86 (100)
  3.73 (100)
  5.57 ( 80)
OPTICAL CONSTANTS:
  $\omega = 1.650$
  $\epsilon = 1.626$
    (−)
HARDNESS: 2–2½
DENSITY: 3.64 (Meas.)
       3.79 (Calc.)
CLEAVAGE: {001} perfect
        {100} distinct
HABIT: Crystals rectangular tablets flattened on {001}; sometimes pyramidal with small {001} terminal faces. Also as micaceous, foliated, or subparallel aggregates of platy crystals. Observed as parallel growths with uranospinite and troegerite.
COLOR-LUSTER: Pale green, grass green to emerald green. Transparent to translucent. Vitreous; pearly on {001}. Fluoresces dull yellow green in ultraviolet light.
MODE OF OCCURRENCE: Occurs as a secondary mineral, chiefly in the oxidized zone of deposits containing uraninite together with arsenic-bearing minerals. In the United States it is found as well-developed crystals at the Myler mine, Majuba Hill, Pershing County, Nevada; in the Tintic district, Utah; at several localities in Montana; disseminated in uraniferous lignite in the Cave Hills and Slim Buttes areas, Harding County, South Dakota; and sparingly in many sandstone-type uranium deposits on the Colorado Plateau. It is also found at the Nicholson mines, Lake Athabasca, Saskatchewan, Canada; at Joachimsthal and Zinnwald, Bohemia; at the Anton mine, Black Forest, Baden, and in the Weisser Hirsch mine at Neustädtl near Schneeberg, Saxony, Germany; at Wheal Gorland, Cornwall, England; at Cap Garonne, France; and on the island of Monte Cristo, Italy.
BEST REF. IN ENGLISH: Frondel, Clifford, USGS Bull. 1064, p. 215–220 (1958).


## MEYERHOFFERITE
$Ca_2B_6O_{11} \cdot 7H_2O$

CRYSTAL SYSTEM: Triclinic
CLASS: $\bar{1}$

SPACE GROUP: P$\bar{1}$
Z: 1
LATTICE CONSTANTS:
  $a = 6.61$    $\alpha = 91°0'$ (or $99°46'$)
  $b = 8.35$    $\beta = 101°31'$
  $c = 6.49$    $\gamma = 86°55'$
3 STRONGEST DIFFRACTION LINES:
  8.31 (100)
  6.47 (100)
  4.97 ( 25)
OPTICAL CONSTANTS:
  $\alpha = 1.500$
  $\beta = 1.535$
  $\gamma = 1.560$
  $(-)2V = 78°$
HARDNESS: 2
DENSITY: 2.120 (Meas.)
      2.125 (Calc.)
CLEAVAGE: {010} perfect
      {100} imperfect
      {1$\bar{1}$0} imperfect
HABIT: Crystals prismatic, elongated [001] with {100} dominant and {110} less dominant. Also fibrous.

COLOR-LUSTER: Colorless to white, transparent to translucent. Crystals vitreous; fibrous masses silky.

MODE OF OCCURRENCE: Occurs as an alteration product of inyoite in the colemanite deposits of the Mt. Blanco district, near Death Valley, Inyo County, California.

BEST REF. IN ENGLISH: Palache, et al., "Dana's System of Mineralogy," 7th Ed., v. II, p. 356–358, New York, Wiley, 1951.    Christ, C. L., *Am. Min.*, **38**: 912–918 (1953).

# MEYMACITE
$WO_3 \cdot 2H_2O$

CRYSTAL SYSTEM: (Amorphous)
STRONGEST DIFFRACTION LINES:
  X-ray pattern shows only one diffuse line (3.81).
OPTICAL CONSTANTS: Isotropic, with $n > 2$.
  HARDNESS: Not determined.
  DENSITY: 3.94–4.10 (Meas.)
  CLEAVAGE: Conchoidal fracture.
  HABIT: Massive.
  COLOR-LUSTER: Yellow-brown; resinous to vitreous.
  MODE OF OCCURRENCE: Occurs as an alteration product of scheelite at Meymac, Corrèze, France.
  BEST REF. IN ENGLISH: Pierrot, R., and Van Tassel, R., *Am. Min.*, **53**: 1065 (1968).

# MIARGYRITE
$AgSbS_2$

CRYSTAL SYSTEM: Monoclinic
CLASS: m
SPACE GROUP: Aa
Z: 8
LATTICE CONSTANTS:
  $a = 13.20$
  $b = 4.40$
  $c = 12.86$
  $\beta = 98°37.5'$

3 STRONGEST DIFFRACTION LINES:
  2.89 (100)
  3.45 ( 90)
  2.75 ( 80)
OPTICAL CONSTANTS:
  $\beta > 2.72$ (Li)
  $(+)2V =$ medium
HARDNESS: 2½
DENSITY: 5.25 (Meas.)
      5.29 (Calc.)
CLEAVAGE: {010} imperfect
      {100} trace
      {101} trace
Fracture subconchoidal to uneven. Brittle.
HABIT: Crystals thick tabular; often deeply striated; sometimes complex. Also massive.

COLOR-LUSTER: Steel gray to iron black. Nearly opaque; thin fragments translucent and dark red in color. Metallic adamantine. Streak red.

MODE OF OCCURRENCE: Occurs in low-temperature hydrothermal vein deposits, commonly associated with other sulfosalts, sulfides, barite, calcite, and quartz. Found in Boulder, La Plata, Montezuma, Park, and Summit counties, Colorado; in the Flint and Silver City districts, Owyhee County, Idaho; and as well-formed crystals in the Randsburg district, San Bernardino County, California. It also occurs in Mexico, Bolivia, Chile, Spain, Germany, Czechoslovakia, and Roumania.

BEST REF. IN ENGLISH: Palache, et al., "Dana's System of Mineralogy," 7th Ed., v. I, p. 424–427, New York, Wiley, 1944.

# MICA
= Mineral group of general formula: $(K, Na, Ca)(Mg, Fe, Li, Al)_{2-3}(Si, Al)_4 O_{10}(OH, F)_2$

See:
| | |
|---|---|
| Anandite | Margarite |
| Biotite | Muscovite |
| Celadonite | Paragonite |
| Clintonite | Phlogopite |
| Ephesite | Polylithionite |
| Glauconite | Roscoelite |
| Hendricksite | Taeniolite |
| Illite | Zinnwaldite |
| Lepidolite | |

# MICHENERITE
PdBiTe

CRYSTAL SYSTEM: Cubic
CLASS: 23
SPACE GROUP: P$2_1$3
Z: 4
LATTICE CONSTANT:
  $a = 6.646(5)$
3 STRONGEST DIFFRACTION LINES:
  2.97 (100)
  2.003 (100)
  1.778 ( 80)
OPTICAL CONSTANTS: Under reflected light creamish white with tinge of gray in air and in oil. Isotropic.
  HARDNESS: 2½

DENSITY: ~10.0

CLEAVAGE: None. Fracture conchoidal. Brittle.

HABIT: Massive; as minute grains.

COLOR-LUSTER: Grayish white (like galena). Opaque. Metallic. Streak black.

MODE OF OCCURRENCE: The mineral occurs in samples from the Vermilion and Frood mines, Sudbury area, Ontario, Canada, as well as from a sample labeled "Copper Cliff concentrates." At the Vermilion mine michenerite is found associated with gold, froodite, and bismuthinite; at the Frood mine with cubanite, chalcopyrite, hessite, altaite, native bismuth, galena, and sperrylite. Also found in copper-nickel ores of the Oktyabrsko ore deposit, USSR.

BEST REF. IN ENGLISH: Hawley, J. E., and Berry, L. G., *Can. Min.*, **6**: 200-209 (1958). Cabri, L. J., and Harris, D. C., *Can. Min.*, **11**: 903-912 (1973).

# MICROCLINE

$KAlSi_3O_8$
Alkali Feldspar Group
Dimorphous with Orthoclase
var. Amazonite

CRYSTAL SYSTEM: Triclinic
CLASS: $\bar{1}$
SPACE GROUP: $C\bar{1}$
Z: 4
LATTICE CONSTANTS:

| | |
|---|---|
| $a = 8.58$ | $\alpha = 90°38.5'$ |
| $b = 12.97$ | $\beta = 115°56'$ |
| $c = 7.22$ | $\gamma = 87°41'$ |

3 STRONGEST DIFFRACTION LINES:
3.244 (100)
4.21 ( 60)
3.83 ( 50)
OPTICAL CONSTANTS:
$\alpha = 1.514-1.529$
$\beta = 1.518-1.533$
$\gamma = 1.521-1.539$
$(-)2V = 66°-103°$
HARDNESS: 6-6½
DENSITY: 2.55-2.63 (Meas.)
2.56 (Calc.)
CLEAVAGE: {001} perfect
{010} perfect
{100}, {110}, {$\bar{1}$10}, {$\bar{2}$01} partings.
Fracture uneven. Brittle.

HABIT: Crystals usually short prismatic, blocky, sometimes several feet across; also tabular, flattened along *b* axis. Commonly massive, cleavable to granular compact. Twinning very common; polysynthetic, also Carlsbad, Manebach, Baveno, and other twin laws.

COLOR-LUSTER: White, gray, yellowish, reddish, green. Transparent to translucent. Vitreous; often pearly on cleavages.

MODE OF OCCURRENCE: Occurs widespread as a common constituent of plutonic rocks such as pegmatites, granite, and syenite. Also widely distributed in crystalline schists; less commonly in contact zones and hydrothermal veins. Typical occurrences are found in the Pala district, California; Pikes Peak area, Colorado (amazonite); Black Hills,

South Dakota (perthite); Amelia, Virginia (amazonite); and at localities in Pennsylvania, Maine, New Hampshire, New York, Connecticut, and North Carolina. Also in Canada (amazonite, perthite), Mexico, Brazil (amazonite), Norway, Sweden, Germany, Italy, USSR (amazonite), Madagascar, India, Japan, and elsewhere.

BEST REF. IN ENGLISH: Deer, Howie, and Zussman, "Rock Forming Minerals," v. 4, p. 6-93, New York, Wiley, 1963.

# MICROLITE

$(Ca, Na, Fe)_2 (Ta, Nb)_2 (O, OH, F)_7$
Pyrochlore Group

CRYSTAL SYSTEM: Cubic
CLASS: $4/m\ \bar{3}\ 2/m$
SPACE GROUP: Fd3m
Z: 8
LATTICE CONSTANT:
$a = 10.41$
3 STRONGEST DIFFRACTION LINES:
2.98 (100)
1.836 ( 80)
1.563 ( 80)
OPTICAL CONSTANTS:
$N = 1.93-1.94$;
1.98-2.02
HARDNESS: 5-5½ (Lower when metamict.)
DENSITY: Usually 4.3-5.7 (Meas.)
6.31 (Calc.)
CLEAVAGE: {111} sometimes distinct, may be parting. Fracture subchonchoidal to uneven. Brittle.

HABIT: Crystals octahedral, up to 6 cm or more in size, often with modifying faces of {011}, {113}, or {001}. Also as embedded grains or irregular masses. Twinning on {111} not common.

COLOR-LUSTER: Pale yellow to brown, reddish, rarely emerald green. Translucent to opaque, rarely transparent. Vitreous to resinous. Not fluorescent.

MODE OF OCCURRENCE: Occurs chiefly as a primary mineral in pegmatites, often associated with lepidolite or spodumene. Found in the United States as superb large crystals at the Rutherford mines near Amelia, Virginia; at several localities in Connecticut; in Maine and Massachusetts; at the Tin Mountain and Ingersoll mines, Black Hills, South Dakota; at the Brown Derby mine, Gunnison County, and other places in Colorado; at the Harding mine, Taos County, New Mexico; and in the pegmatites of southern California. It is also found in small amounts in Greenland, Norway, Sweden, Finland, France, Madagascar, and Western Australia.

BEST REF. IN ENGLISH: Frondel, Clifford, USGS Bull. 1064, p. 326-333 (1958).

# MICROSOMMITE = Davyne

# MIERSITE

$(Ag, Cu)I$

CRYSTAL SYSTEM: Cubic

CLASS: $\bar{4}$ 3m
SPACE GROUP: F$\bar{4}$3m
Z: 4
LATTICE CONSTANT:
$a$ = 6.491 kX
3 STRONGEST DIFFRACTION LINES:
3.23 (100)
2.28 ( 80)
1.95 ( 80)
OPTICAL CONSTANT:
$N$ = 2.20
HARDNESS: 2½
DENSITY: 5.64 (Meas.)
5.68 (Meas. synthetic)
5.67 (Calc. synthetic)
CLEAVAGE: {011} perfect
Fracture conchoidal. Brittle.

HABIT: Crystals tetrahedral, rarely cubo-octahedral. As crusts and aggregates of distinct crystals. Twinning on {111} common, sometimes repeated.

COLOR-LUSTER: Canary yellow; transparent. Adamantine. Streak canary yellow.

MODE OF OCCURRENCE: Occurs with malachite, cuprite, cerussite, and iron manganese oxides at Broken Hill, New South Wales, Australia.

BEST REF. IN ENGLISH: Palache, et al., "Dana's System of Mineralogy," 7th Ed., v. II, p. 19-20, New York, Wiley, 1951.

## MIKHEEVITE = Görgeyite

## MILARITE
$KCa_2 AlBe_2 (Si_{12}O_{30}) \cdot H_2O$

CRYSTAL SYSTEM: Hexagonal
CLASS: 6/m 2/m 2/m
SPACE GROUP: P6/mcc
Z: 2
LATTICE CONSTANTS:
$a$ = 10.45
$c$ = 13.88
3 STRONGEST DIFFRACTION LINES:
3.307 (100)
2.880 ( 90)
4.160 ( 65)
OPTICAL CONSTANTS:
$\omega$ = 1.532-1.551
$\epsilon$ = 1.529-1.548
(–)
HARDNESS: 5½-6
DENSITY: 2.46-2.61 (Meas.)
2.524 (Calc.)
CLEAVAGE: None
Fracture conchoidal to uneven. Brittle.

HABIT: Crystals prismatic, up to 4 cm in length.

COLOR-LUSTER: Colorless, pale green, yellowish green. Transparent to translucent. Vitreous.

MODE OF OCCURRENCE: Occurs as fine crystals in association with adularia (valencianite) at the Valencia mine, Guanajuato, Mexico; in fissures and Alpine veins, aplites and pegmatites occurring in granites and syenites, in the St. Gotthard district and elsewhere in Switzerland; in microcline-perthite-quartz pegmatite at Vezna, Czechoslovakia; in low-temperature hydrothermal veins at Henneberg, Thuringia, East Germany; on the Kola Peninsula, USSR; and as excellent gemmy crystals at Swakopmund, S.W. Africa.

BEST REF. IN ENGLISH: Vlasov, K. A., "Mineralogy of Rare Elements," v. II, p. 108-111, Israel Program for Scientific Translations, 1966.

## MILLERITE
NiS

CRYSTAL SYSTEM: Hexagonal
CLASS: 3m
SPACE GROUP: R$\bar{3}$m
Z: 9 (hex.), 3 (rhomb.)
LATTICE CONSTANTS:
$a$ = 9.591
$c$ = 3.145
3 STRONGEST DIFFRACTION LINES:
2.777 (100)
1.863 ( 95)
2.513 ( 65)
HARDNESS: 3-3½
DENSITY: 5.41-5.42 (Calc.)
CLEAVAGE: {01$\bar{1}$2} perfect
{10$\bar{1}$1} perfect
Fracture uneven. Brittle.

HABIT: Usually as very slender to capillary crystals elongated in direction of $c$-axis; often in tufted to compact radiating groups, or as fibrous, loosely coherent interwoven masses. Also massive, cleavable.

COLOR-LUSTER: Brass yellow, bronze yellow; tarnishes greenish gray. Opaque. Metallic. Streak greenish black.

MODE OF OCCURRENCE: Occurs as a low-temperature mineral in limestone, dolomite, hematite, serpentines, and carbonate ore veins. Found at the Gap mine, Lancaster County, Pennsylvania; at the Sterling mine, Antwerp, New York; in geodes in limestone in Wisconsin, Illinois, Iowa, and Missouri; and with cinnabar at numerous deposits in California. It also occurs in Canada, especially as large cleavable masses at Timagami, Ontario; in cavities in siderite at Glamorgan, Wales; at Joachimstal, Czechoslovakia; and as fine specimens at several localities in Germany.

BEST REF. IN ENGLISH: Palache, et al., "Dana's System of Mineralogy," 7th Ed., v. I, p. 239-241, New York, Wiley, 1944.

## MILLISITE
$(Na, K)CaAl_6(PO_4)_4(OH)_9 \cdot 3H_2O$

CRYSTAL SYSTEM: Probably tetragonal
LATTICE CONSTANTS:
$a$ = 7.00
$c$ = 19.07
3 STRONGEST DIFFRACTION LINES:
4.84 (100vb)
2.98 ( 80b )
2.81 ( 80b )

OPTICAL CONSTANTS:
$\alpha$ = 1.584
$\beta$ = 1.598
$\gamma$ = 1.602
(+)2V = moderate
HARDNESS: 5½
DENSITY: 2.83 (Meas.)
2.87 (Calc.)
CLEAVAGE: {001} (?)
HABIT: As chalcedonic crusts or spherules with finely fibrous structure.
COLOR-LUSTER: White, light gray, greenish.
MODE OF OCCURRENCE: Occurs interlayered with wardite in variscite nodules at Fairfield, Utah County, Utah; as a microcrystalline intergrowth with crandallite as a major component of the aluminum phosphate zone of the Bone Valley formation of west-central Florida; and in the phosphate deposits of Thiès, Senegal, Africa.
BEST REF. IN ENGLISH: Palache, et al., "Dana's System of Mineralogy," 7th Ed., v. II, p. 941–942, New York, Wiley, 1951.

**MILOSCHITE** = Variety of allophane with 4% $Cr_2O_3$

## MIMETITE
$Pb_5(AsO_4)_3Cl$
Pyromorphite Series
Apatite Group

CRYSTAL SYSTEM: Monoclinic
CLASS: 2/m
SPACE GROUP: $P2_1/b$
Z: 4
LATTICE CONSTANTS:
$a$ = 10.24
$b$ = 20.48
$c$ = 7.45
$\beta$ = 120°
3 STRONGEST DIFFRACTION LINES:
3.06 (100)
3.01 ( 95)
2.96 ( 65)
HARDNESS: 3½–4
DENSITY: 7.28 (Calc.)
CLEAVAGE: Fracture subconchoidal to uneven. Brittle.
HABIT: Crystals usually acicular. Usually globular, botryoidal, reniform, or granular.
COLOR-LUSTER: Pale yellow to bright sulfur yellow to yellowish brown, orange-yellow, orange, white, colorless. Transparent to translucent. Subadamantine to resinous. Streak white.
MODE OF OCCURRENCE: Occurs as a secondary mineral in the oxidation zone of lead-bearing ore deposits. Found in the United States in California, Nevada, Utah, Arizona, Colorado, South Dakota, and Pennsylvania. Fine specimens occur at San Pedro Corallitos and at Santa Eulalia, Chihuahua, and at Mapimi, Durango, Mexico; in Cornwall and Cumberland, England; at Leadhills and Wanlockhead, Scotland; in Sweden, France, Germany, Czechoslovakia, USSR, Australia, and Algeria; and as magnificent crystal groups at Tsumeb, South West Africa.

BEST REF. IN ENGLISH: Palache, et al., "Dana's System of Mineralogy," 7th Ed., v. II, p. 889–895, New York, Wiley, 1951. Keppler, Ulrich, *Am. Min.*, **54**: 993 (1969).

## MINASRAGRITE
$VOSO_4 \cdot 5H_2O$

CRYSTAL SYSTEM: Monoclinic
CLASS: 2/m
SPACE GROUP: $P2_1/a$
Z: 4
LATTICE CONSTANTS:
$a$ = 12.947
$b$ = 9.748
$c$ = 7.005
$\beta$ = 110.93°
3 STRONGEST DIFFRACTION LINES:
5.135 (100)
3.907 ( 70)
5.431 ( 60)
OPTICAL CONSTANTS:
$\alpha$ = 1.513
$\beta$ = 1.536
$\gamma$ = 1.545
Biaxial (–), 2V medium large
HARDNESS: Soft
DENSITY: 2.03 (Meas.)
2.036 (Calc.)
CLEAVAGE: None.
HABIT: As an efflorescence of minute crystals; also as granular aggregates, spherulites, or mammillary masses.
COLOR-LUSTER: Blue. Vitreous.
MODE OF OCCURRENCE: The mineral occurs at Minasragra, near Cerro de Pasco, Peru, as an efflorescence on patronite.
BEST REF. IN ENGLISH: Palache, et al., "Dana's System of Mineralogy," 7th Ed., v. II, p. 437–438, New York, Wiley, 1951. Smith, Marie Lindberg, and Marinenko, John, *Am. Min.*, **58**: 531–534 (1973).

## MINGUZZITE
$K_3Fe(C_2O_4)_3 \cdot 3H_2O$

CRYSTAL SYSTEM: Monoclinic
LATTICE CONSTANTS:
$a:b:c$ = 0.9916:1:0.3895
$\beta$ = 94°13.5'
3 STRONGEST DIFFRACTION LINES:
6.9 (100)
3.61 ( 65)
2.18 ( 65)
OPTICAL CONSTANTS:
$\alpha$ = 1.501, 1.498
$\beta$ = 1.555, 1.554
$\gamma$ = 1.597, 1.594
(–)2V = 78° (Calc.)
HARDNESS: Not determined.
DENSITY: 2.080 (Meas.)
2.092 (Meas. Synthetic)
CLEAVAGE: {010} perfect

HABIT: Crystals tabular, tenths to hundredths of a millimeter in size. Forms: b {010}, e {11$\bar{1}$}, σ {111}, and m {110}.

COLOR-LUSTER: Green to yellow-green; vitreous.

MODE OF OCCURRENCE: Occurs associated with humboldtine in limonite at Cape Calamita, Elba.

BEST REF. IN ENGLISH: Garavelli, Carlo L., *Am. Min.*, **41**: 370 (1956).

## MINIUM
$Pb_3O_4$

CRYSTAL SYSTEM: Tetragonal
CLASS: $\bar{4}$m2
SPACE GROUP: P$\bar{4}$b2
Z: 4
LATTICE CONSTANTS:
  $a$ = 8.824
  $c$ = 6.654
3 STRONGEST DIFFRACTION LINES:
  3.38 (100)
  2.903 ( 50)
  2.787 ( 45)
HARDNESS: 2½
DENSITY: 8.9–9.2 (Meas. synthetic)
         8.925 (Calc.)
CLEAVAGE: {110} perfect
HABIT: Massive, earthy or powdery.
COLOR-LUSTER: Scarlet red to brownish red. Opaque. Dull to somewhat greasy. Streak orange-yellow.

MODE OF OCCURRENCE: Occurs as an oxidation product of galena and other lead minerals, commonly associated with galena, cerussite, massicot, and limonite. Found in Wythe County, Virginia; Mineral Point, Wisconsin; and widespread in the mining districts of the western United States, especially in California, Idaho, Utah, and Colorado. It also occurs in Mexico, England, Scotland, Sweden, Italy, Germany, and USSR.

BEST REF. IN ENGLISH: Palache, et al., "Dana's System of Mineralogy," 7th Ed., v. I, p. 517–519, New York, Wiley, 1944.

## MINNESOTAITE
$(Fe, Mg)_3 Si_4 O_{10} (OH)_2$
Iron analogue of talc

CRYSTAL SYSTEM: Monoclinic
Z: 4
LATTICE CONSTANTS:
  $a$ = 5.5
  $b$ = 9.38
  $c$ = 19.3
  $\beta$ = 99°30'
3 STRONGEST DIFFRACTION LINES:
  9.6 (100)
  2.52 ( 70)
  3.17 ( 50)
OPTICAL CONSTANTS:
  $\alpha$ = 1.592
  $\beta$ = ?    (Na)
  $\gamma$ = 1.623

(–)2V = small
HARDNESS: <3
DENSITY: 3.01 (Meas.)
         3.21 (Calc.)
CLEAVAGE: {001} perfect
HABIT: As microscopic plates or needles; disseminated or in compact confused aggregates.
COLOR-LUSTER: Greenish gray. Greasy to waxy or dull.
MODE OF OCCURRENCE: Occurs in association with quartz (chert), siderite, magnetite, stilpnomelane, and greenalite, in the Mesabi and Cuyuna iron ranges, Minnesota.

BEST REF. IN ENGLISH: Gruner, John W., *Am. Min.*, **29**: 363–372 (1944).    Blake, R., *Am. Min.* **50**: 148–169 (1965).

## MINYULITE
$KAl_2 (PO_4)_2 (OH,F) \cdot 4H_2O$

CRYSTAL SYSTEM: Orthorhombic
CLASS: mm2
SPACE GROUP: Pmm2
Z: 2
LATTICE CONSTANTS:
  $a$ = 9.35
  $b$ = 9.74
  $c$ = 5.52
3 STRONGEST DIFFRACTION LINES:
  5.6 (100)
  3.38 (100)
  6.8 ( 60)
OPTICAL CONSTANTS:
  $\alpha$ = 1.531
  $\beta$ = 1.534
  $\gamma$ = 1.538
  (+)2V large
HARDNESS: 3½
DENSITY: 2.45
CLEAVAGE: {001} perfect. Brittle.
HABIT: As acicular radiating aggregates.
COLOR-LUSTER: Colorless, white. Transparent. Silky.
MODE OF OCCURRENCE: Occurs at Noarlunga, South Australia, and in glauconitic phosphate deposits near Minyulo Well, Dandaragan, Western Australia.

BEST REF. IN ENGLISH: Palache, et al., "Dana's System of Mineralogy," 7th Ed., v. II, p. 970, New York, Wiley, 1951.    Spencer, Bannister, Hey, and Bennett, *Min. Mag.*, **26**: 313 (1943).

## MIRABILITE
$Na_2 SO_4 \cdot 10H_2O$

CRYSTAL SYSTEM: Monoclinic
CLASS: 2/m
SPACE GROUP: P2$_1$/a
Z: 4
LATTICE CONSTANTS:
  $a$ = 12.82
  $b$ = 10.35
  $c$ = 11.48
  $\beta$ = 107°40'

3 STRONGEST DIFFRACTION LINES:
  5.49 (100)
  3.21 ( 75)
  3.26 ( 60)
OPTICAL CONSTANTS:
  $\alpha$ = 1.396
  $\beta$ = 1.4103  (Na)
  $\gamma$ = 1.419 (Calc.)
  $(-)2V = 75°56'$
HARDNESS: 1½–2
DENSITY: 1.464 (Meas.)
          1.465 (Calc.)
CLEAVAGE: {100} perfect
  also {001}, {010}, and {011} in traces
Fracture conchoidal. Brittle.
  HABIT: As crystals resembling those of borax with chief forms a {100}, m {110}, b {010}, and c {001}. Often long prismatic to acicular [010], or granular to fibrous aggregates and efflorescent crusts. Twinning rare.
  COLOR-LUSTER: Colorless, white. Transparent to opaque. Vitreous.

  Taste bitter.
  MODE OF OCCURRENCE: Occurs widespread as a saline lake deposit, in playas, as a hot springs deposit, and as an efflorescence upon soil. Found in the United States in Imperial, Kern, Napa, San Bernardino, and San Luis Obispo Counties, California; at many places in Nevada; abundantly in Albany, Natrona, and other counties in Wyoming; along the south shore of Great Salt Lake, Utah; and in several other western states. It also occurs in Spain, Italy, Sicily, Austria, Czechoslovakia, and the USSR.
  BEST REF. IN ENGLISH: Palache, et al., "Dana's System of Mineralogy," 7th Ed., v. II, p. 439–442, New York, Wiley, 1951.

# MISENITE
KHSO$_4$

CRYSTAL SYSTEM: Monoclinic
LATTICE CONSTANTS:
  $a:b:c$ = 3.2196 : 1 : 2.1842
          $\beta$ = 102°05'
OPTICAL CONSTANTS:
  $\alpha$ = 1.475
  $\beta$ = 1.480
  $\gamma$ = 1.487
  (+)2V = large
HARDNESS: Not determined.
DENSITY: 2.32 (Meas.)
CLEAVAGE: {010} distinct
  HABIT: Crystals acicular or lath-like; as fibrous masses and aggregates.
  COLOR-LUSTER: Grayish white. Transparent to translucent. Pearly to silky.

  Taste bitter.
  MODE OF OCCURRENCE: Occurs as an efflorescence in a fumarole on Cape Miseno near Naples, Italy.
  BEST REF. IN ENGLISH: Palache, et al., "Dana's System of Mineralogy," 7th Ed., v. II, p. 396–397, New York, Wiley, 1951.

# MISERITE
(Ca, K, Na, Al, Y, etc.)SiO$_3$

CRYSTAL SYSTEM: Triclinic
Z: 16
LATTICE CONSTANTS:
  $a$ = 10.099   $\alpha$ = 96°24'
  $b$ = 15.996   $\beta$ = 111°04'
  $c$ = 7.367    $\gamma$ = 76°39'
3 STRONGEST DIFFRACTION LINES:
  3.14  (100)
  1.641 ( 80)
  2.377 ( 65)
OPTICAL CONSTANTS:
  $\alpha$ = 1.587
  $\beta$ = 1.589
  $\gamma$ = 1.594
  (+)2V = 65°(Meas.) 64°48'(Calc.).
HARDNESS: 5½–6
DENSITY: 2.84–2.926 (Meas.)
CLEAVAGE: {100} perfect
          {010} imperfect
  HABIT: Massive, cleavable. Also fine-grained fibrous; compact aggregates form slightly curved, scaly masses several inches in diameter, but rarely more than ¼ inch thick.
  COLOR-LUSTER: Pinkish to lavender; also red-brown. Translucent. Vitreous or pearly.
  MODE OF OCCURRENCE: Occurs in association with wollastonite in contact metamorphosed shale at Wilson Mineral Springs and at the Union Carbide Vanadium plant in the Magnet Cove area, Arkansas. Also found in association with hornblende, eudialyte, scapolite, and an unknown light gray mineral, in a carbonatite vein in the Kipawa Lake area, Temiscamingue County, Quebec, Canada.
  BEST REF. IN ENGLISH: Berry, L. G., Lin, Hsi-Che, and Davis, G. C., *Can. Min.*, **11**: 569 (1972).

# MISPICKEL = Arsenopyrite

# MITRIDATITE
Ca$_3$Fe$_4^{3+}$(OH)$_6$(H$_2$O)$_3$[PO$_4$]$_4$
Robertsite-Mitridatite Series

CRYSTAL SYSTEM: Monoclinic
CLASS: 2/m
SPACE GROUP: A2/a
Z: 8
LATTICE CONSTANTS:
  $a$ = 17.52
  $b$ = 19.35
  $c$ = 11.25
  $\beta$ = 95.92°
3 STRONGEST DIFFRACTION LINES:
  8.64  (100)
  2.721 ( 70)
  5.55  ( 60)
OPTICAL CONSTANTS:
  $\alpha$ = 1.785
  $\beta$ = 1.85
  $\gamma$ = 1.85
  Biaxial (−), 2V 5–10°

HARDNESS: 3½
DENSITY: 3.24 (Meas.)
  3.08 (Calc.)
CLEAVAGE: {100} good
Fracture uneven. Brittle

HABIT: Crystals thin tabular, pseudo-rhombohedral in aspect, minute. Forms: $a$ {100}, $c$ {001} (small), and $v$ {$\bar{4}$23}. Also as colloidal or fine grained nodules, crusts, veinlets, and coatings.

COLOR-LUSTER: Crystals deep red to bronzy-red; translucent to nearly opaque; resinous to adamantine. Colloidal or fine-grained material greenish yellow to green and dark green; also brownish green, and yellowish brown to brownish black; dull and earthy to resinous. Streak dull olive green.

MODE OF OCCURRENCE: Mitridatite crystals are found in association with jahnsite, collinsite, hureaulite, and hydroxylapatite lining cavities in altered triphylite at the White Elephant pegmatite, Custer County, and commonly as a stain or thin incrustation in numerous triphylite-lithiophilite-bearing pegmatites elsewhere in the Black Hills of South Dakota. It also occurs at the Yanysk-Tokila deposit and other localities on the Kerch and Taman peninsulas in the Soviet Union, and in several pegmatites in Central Africa.

BEST REF. IN ENGLISH: Moore, Paul Brian, *Am. Min.*, **59**: 48–59 (1974).

# MITSCHERLICHITE
$K_2 CuCl_4 \cdot 2H_2O$

CRYSTAL SYSTEM: Tetragonal
CLASS: 4/m 2/m 2/m
SPACE GROUP: $P4_2/mnm$
Z: 2
LATTICE CONSTANTS:
  $a$ = 7.45
  $c$ = 7.88
3 STRONGEST DIFFRACTION LINES:
  2.64 (100)
  2.71 ( 95)
  5.42 ( 70)
OPTICAL CONSTANTS:
  $\omega$ = 1.6438
  $\epsilon$ = 1.6097
    (–)
HARDNESS: 2½
DENSITY: 2.418 (Meas.)
  2.41 (Calc.)
CLEAVAGE: Not reported.
HABIT: Crystals short prismatic or pyramidal. Stalactitic.
COLOR-LUSTER: Green, greenish blue. Transparent. Vitreous.

Deliquescent.

MODE OF OCCURRENCE: Found as stalactites on the floor of the crater of Mt. Vesuvius, Italy, associated with metavoltine, sylvite, and gypsum.

BEST REF. IN ENGLISH: Palache, et al., "Dana's System of Mineralogy," 7th Ed., v. II, p. 100–101, New York, Wiley, 1951.

# MIXITE
$Cu_{12} Bi_2 (AsO_4)_6 (OH)_{12} \cdot 6H_2O$

CRYSTAL SYSTEM: Hexagonal
CLASS: 6/m
SPACE GROUP: $P6_3/m$
Z: 1
LATTICE CONSTANTS:
  $a$ = 13.63
  $c$ = 5.90
3 STRONGEST DIFFRACTION LINES:
  12.0 (100)
  2.46 ( 90)
  3.57 ( 80)
OPTICAL CONSTANTS:
  $\omega$ = 1.748–1.750
  $\epsilon > 1.81$
HARDNESS: 3–4
DENSITY: 3.79 (Meas.)
  3.77 (Calc.)
CLEAVAGE: Not determined.
HABIT: As needle-like crystals elongated along $c$-axis and striated vertically. As tufted or radiated aggregates or crusts; also compact, in rounded masses with concentric fibrous structure. Also as cross-fiber veinlets.

COLOR-LUSTER: Emerald green to bluish green, pale blue, pale green, whitish. Aggregates dull; single crystals brilliant. Streak lighter than color.

MODE OF OCCURRENCE: Occurs at the Ajax, Boss Tweed, and Carissa mines, Tintic district, Juab County, Utah; at the Cerro Gordo mine, Inyo County, California; at the El Carmen mine, Durango, Mexico; and at localities in Germany and Czechoslovakia.

BEST REF. IN ENGLISH: Palache, et al., "Dana's System of Mineralogy," 7th Ed., v. II, p. 943–944, New York, Wiley, 1951. Walenta, *Neues Jahrb. Min. Monatsh.* **223**: (1960).

# MIZZONITE
$mCa_4 (Al_6 Si_6 O_{24}) CO_3$ with $nNa_4 (Al_3 Si_9 O_{24}) Cl$
$Ma_{50} Me_{50}$ to $Ma_{20} Me_{80}$
Scapolite Group

CRYSTAL SYSTEM: Tetragonal
CLASS: 4/m
SPACE GROUP: I4/m
Z: 2
LATTICE CONSTANTS:
  $a$ = 12.169
  $c$ = 7.569
3 STRONGEST DIFFRACTION LINES:
  2.70 (100)
  3.47 ( 95)
  3.09 ( 75)
HARDNESS: 5½–6
DENSITY: 2.67–2.722 (Meas.)
CLEAVAGE: {100} distinct
  {110} distinct
Fracture uneven to conchoidal. Brittle.
HABIT: Crystals prismatic, often coarse. Also massive, granular; sometimes columnar.
COLOR-LUSTER: Colorless, white, gray, bluish, greenish, yellowish, pink, violet, and brownish. Transparent to translucent. Vitreous to somewhat pearly or resinous. Streak uncolored. Sometimes fluoresces bright yellow or dull red under long-wave ultraviolet light.

MODE OF OCCURRENCE: Occurs mainly in regionally metamorphosed rocks, contact zones, altered basic igneous rocks, and ejected volcanic blocks. Typical occurrences are found in Chaffee, Clear Creek, Jefferson, and Gunnison counties, Colorado; Custer County, South Dakota; Bucks County, Pennsylvania. Also in Ontario and Quebec, Canada, and at localities in Sweden, Finland, Ghana, Madagascar, and South Australia.

BEST REF. IN ENGLISH: Shaw, D. M., *J. Petr.*, **1**: 218; 261 (1960). Deer, Howie, and Zussman, "Rock Forming Minerals," v. 4, p. 321–337, New York, Wiley, 1963. *Am. Min.*, **51**: 1014–1027 (1966).

# MOCTEZUMITE
$Pb(UO_2)(TeO_3)_2$

CRYSTAL SYSTEM: Monoclinic
CLASS: 2/m
SPACE GROUP: $P2_1/c$
Z: 3
LATTICE CONSTANTS:
  $a = 7.82$
  $b = 7.07$
  $c = 13.84$
  $\beta = 93°37.5'$
3 STRONGEST DIFFRACTION LINES:
  3.156 (100)
  3.492 ( 90)
  2.997 ( 80)
OPTICAL CONSTANTS:
  All indices $> 2.11$
    $(-)2V = 5-10°$
  HARDNESS: ~3
  DENSITY: 5.73 (Meas.)
  CLEAVAGE: {100} perfect
HABIT: Crystals elongated parallel to $b$, and flattened parallel to (201). Most crystal faces are curved. Maximum observed size of crystals 3 mm long, 1 mm wide, and 0.2 mm thick.

COLOR-LUSTER: Bright orange; altered crystals dark orange to brownish orange.

MODE OF OCCURRENCE: Found as minute blades and rosettes in cracks and vugs in the oxidized zone of a tellurium-gold deposit at La Moctezuma mine, Moctezuma, Sonora, Mexico.

BEST REF. IN ENGLISH: Gaines, Richard V., *Am. Min.*, **50**: 1158–1163 (1965).

# MOFETTITE = A natural gas
$CO_2$

# MOHRITE
$(NH_4)_2 Fe(SO_4)_2 \cdot 6H_2O$

CRYSTAL SYSTEM: Monoclinic
CLASS: 2/m
SPACE GROUP: $P2_1/c$
Z: 2

LATTICE CONSTANTS:
  $a = 6.237$
  $b = 12.613$
  $c = 9.292$
  $\beta = 106°53'$
3 STRONGEST DIFFRACTION LINES:
  4.20 (100)
  4.13 ( 55)
  3.79 ( 55)
OPTICAL CONSTANTS:
  $\alpha = 1.486, 1.480$
  $\gamma = 1.497, 1.486$
  $(+)2V \approx 75°$
  DENSITY: 1.862, 1.800 (Meas.)
        1.870, 1.805 (Calc.)
  CLEAVAGE: {$\bar{1}02$} perfect
        {010} distinct
HABIT: As irregular lamina and minute quasi-euhedral crystals.

COLOR-LUSTER: Pale green; vitreous.

MODE OF OCCURRENCE: Occurs as incrustations from the boriferous soffioni of Travale, Val di Cecina, Tuscany, Italy.

BEST REF. IN ENGLISH: Garavelli, Carlo L., *Am. Min.*, **50**: 805 (1965).

# MOISSANITE
SiC

CRYSTAL SYSTEM: Hexagonal
CLASS: 6 mm
SPACE GROUP: $P6_3 mc$
Z: 6
LATTICE CONSTANTS:
  $a = 3.076$
  $c = 15.07$
3 STRONGEST DIFFRACTION LINES:
  1.54 (100)
  1.31 (100)
  2.51 ( 85)
OPTICAL CONSTANTS:
  $\omega = 2.654$
  $\epsilon = 2.697$ ($\lambda = 589$ m$\mu$)
    (+)
  HARDNESS: 9½
  DENSITY: 3.218 (Meas.)
        3.217 (Calc.)
  CLEAVAGE: Not determined. Fracture conchoidal.
  HABIT: As small hexagonal plates.
  COLOR-LUSTER: Green to emerald green, bluish, black. Metallic.

MODE OF OCCURRENCE: Occurs associated with minute diamonds in the meteoric iron of Canyon Diablo, Arizona.

BEST REF. IN ENGLISH: Palache, et al., "Dana's System of Mineralogy," 7th Ed., v. I, p. 123–124, New York, Wiley, 1944.

# MOLURANITE
$UO_2 \cdot 3UO_3 \cdot 7MoO_3 \cdot 2OH_2O$

CRYSTAL SYSTEM: Unknown

3 STRONGEST DIFFRACTION LINES: Amorphous to x-rays.

OPTICAL CONSTANT:

$N = 1.97-1.98$

HARDNESS: 3-4

DENSITY: $\sim 4$. (Meas.)

CLEAVAGE: Brittle.

HABIT: Massive; colloform

COLOR-LUSTER: Black, translucent, brown in thin fragments. Resinous.

MODE OF OCCURRENCE: Occurs in fine fissures in granulated albitite, associated with molybdenite, chalcopyrite, and galena, which are incrusted on fissures, moluranite forming accumulations in the central part with brannerite and with other uranium-molybdenum compounds at an unspecified deposit, presumably in the USSR.

BEST REF. IN ENGLISH: *Am. Min.*, **43**: 380 (1958). Epshtein, G. Yu., *Am. Min.*, **45**: 258 (1960).

# MOLYBDENITE
$MoS_2$

CRYSTAL SYSTEM: Hexagonal

CLASS: $6/m\ 2/m\ 2/m$

SPACE GROUP: $P6_3/mmc$ (2-H), R3m (3-R)

Z: 2 (2-H), 3 (3-R)

LATTICE CONSTANTS:

| (2-H) | (3-R) |
|-------|-------|
| $a = 3.15$ | $a = 3.16$ |
| $c = 12.30$ | $c = 18.33$ |

3 STRONGEST DIFFRACTION LINES:

| (2-H) | (3-R) |
|-------|-------|
| 6.15 (100) | 6.09 (100) |
| 2.277 ( 45) | 2.71 ( 70) |
| 1.830 ( 25) | 1.581 ( 70) |

OPTICAL CONSTANTS:

(2-H)

$\omega = 4.336$
$\epsilon = 2.035$ ($\gamma = 852$ nm)

(−)

HARDNESS: 1-1½

DENSITY: 4.62-5.06 (Meas.)
5.05 (Calc.)

CLEAVAGE: $\{0001\}$ perfect, easy Laminae very flexible; inelastic. Sectile.

HABIT: Crystals usually thin to thick tabular, with poorly developed faces and rough hexagonal aspect, or as short barrel-shaped prisms. Commonly as foliate or radiate masses, scales, or disseminated grains.

COLOR-LUSTER: Lead gray. Opaque. Metallic. Feel greasy. Streak greenish.

MODE OF OCCURRENCE: Occurs widespread as the most abundant molybdenum mineral; found in high-temperature veins, contact metamorphic deposits, granites, pegmatites, and aplites. Notable localities occur throughout the United States, especially at the Climax mine, Lake County, Colorado. Other major occurrences are found in Canada, Mexico, Peru, England, Norway, Sweden, Portugal, Germany, Czechoslovakia, USSR, Morocco and elsewhere in Africa, China, Japan, and Australia.

BEST REF. IN ENGLISH: Palache, et al., "Dana's System of Mineralogy," 7th Ed., v. I, p. 328–331, New York, Wiley, 1944.

# MOLYBDITE
$MoO_3$

CRYSTAL SYSTEM: Orthorhombic

CLASS; $2/m\ 2/m\ 2/m$

SPACE GROUP: Pbnm

Z: 4

LATTICE CONSTANTS:

$a = 3.954$
$b = 13.808$
$c = 3.690$

3 STRONGEST DIFFRACTION LINES:

3.80 (100)
3.25 (100)
3.45 ( 90)

OPTICAL CONSTANTS:

$\gamma$ and $\beta > 2.0$

(+)2V = very large

HARDNESS: Not determined.

DENSITY: 4.72 (Calc.)

CLEAVAGE: $\{100\}$ perfect
$\{001\}$ distinct
$\{010\}$-Perfect cleavage given in the literature for synthetic $MoO_3$ not observed due to thinness of crystals.

HABIT: As flat needles or thin plates, up to $5 \times 1 \times 0.0X$ mm in size, forming irregular aggregates. Crystals elongated and striated parallel to c-axis, flattened on $\{010\}$.

COLOR-LUSTER: Light greenish yellow to nearly colorless; transparent; adamantine.

MODE OF OCCURRENCE: Found in a quartz vein with molybdenite, near the contact with fine-grained topaz-quartz greisen, in Krupka, northwestern Bohemia, Czechoslovakia. It occurs in coarse-grained quartz in the cavities formed after total or partial "leaching" of molybdenite, or in their close vicinity.

BEST REF. IN ENGLISH: Čech, F., and Povondra, P., *Am. Min.*, **49**: 1497-1498 (1964).

# MOLYBDOMENITE
$PbSeO_3$

CRYSTAL SYSTEM: Monoclinic

CLASS: $2/m$

SPACE GROUP: $P2_1/m$

Z: 2

LATTICE CONSTANTS:

$a = 6.86$
$b = 5.48$
$c = 4.50$
$\beta = 112°45'$

3 STRONGEST DIFFRACTION LINES:

2.756 (100)
3.32 ( 90)
3.43 ( 70)

OPTICAL CONSTANTS:

$\alpha = 2.12$

$\gamma = 2.14$

$(-)2V = 80°$

HARDNESS: 3½

DENSITY: 7.07 (Meas. synthetic)

7.12 (Calc.)

CLEAVAGE: One perfect, another less distinct perpendicular to the first.

HABIT: As very thin scales.

COLOR-LUSTER: Colorless to yellowish white; pearly to greasy.

MODE OF OCCURRENCE: Occurs with chalcomenite in the Cerro de Cacheuta, Mendoza, Argentina; also found at Trogtal, Harz, Germany.

BEST REF. IN ENGLISH: Mandarino, J. A., *Can. Min.*, 8: 149-158 (1965). Goni, J., and Guillemin, C., *Am. Min.*, 39: 850 (1954).


# MOLYBDOPHYLLITE (Inadequately described mineral)

$Pb_2 Mg_2 Si_2 O_7 (OH)_2$

CRYSTAL SYSTEM: Hexagonal

SPACE GROUP: R cell

LATTICE CONSTANTS:

$c/a = 0.549$

3 STRONGEST DIFFRACTION LINES:

13.8 (100)

4.60 ( 90)

2.67 ( 70)

OPTICAL CONSTANTS:

$\omega = 1.815$

$\epsilon = 1.761$

(-)

HARDNESS: 3-4

DENSITY: 4.72 (Meas.)

CLEAVAGE: {0001} perfect

HABIT: Crystals platy, in foliated aggregates.

COLOR-LUSTER: Colorless, pale green. Transparent. Vitreous.

MODE OF OCCURRENCE: Occurs in association with hausmannite at Långban, Sweden.

BEST REF. IN ENGLISH: Winchell and Winchell, "Elements of Optical Mineralogy," 4th Ed., Pt. 2, p. 479, New York, Wiley, 1951.


# MOLYBDOSCHEELITE = Variety of scheelite containing up to 24% $MoO_3$


# MOLYSITE

$FeCl_3$

CRYSTAL SYSTEM: Hexagonal

CLASS: $\bar{3}$

SPACE GROUP: $R\bar{3}$

Z: 2

LATTICE CONSTANTS:

(hexag.)     (rhomb.)

$a = 5.92$ kX    $a_{rh} = 6.69$

$c = 17.26$    $\alpha = 52°30'$

3 STRONGEST DIFFRACTION LINES:

2.68 (100)

2.08 ( 40)

5.9 ( 32)

OPTICAL CONSTANTS:

Mean index 1.6, uniaxial (-)

HARDNESS: Not determined. Soft.

DENSITY: 2.904 (Meas.)

CLEAVAGE: {0001} perfect

HABIT: Massive, as coatings.

COLOR-LUSTER: Yellow to brownish red or purplish red.

Deliquescent.

MODE OF OCCURRENCE: Occurs as a rapidly decomposing sublimation product at Mt. Etna, Sicily; Mt. Vesuvius, Italy; and in other areas of volcanic activity.

BEST REF. IN ENGLISH: Palache, et al., "Dana's System of Mineralogy," 7th Ed., v. II, p. 47-48, New York, Wiley, 1951.


# MONAZITE

$(Ce, La, Y, Th)PO_4$

CRYSTAL SYSTEM: Monoclinic

CLASS: 2/m

SPACE GROUP: $P2_1/n$

Z: 4

LATTICE CONSTANTS:

$a = 6.77$

$b = 7.01$

$c = 6.43$

$\beta = 103°10'$

3 STRONGEST DIFFRACTION LINES:

3.09 (100)

2.87 ( 70)

3.30 ( 50)

OPTICAL CONSTANTS:

$\alpha = 1.785$

$\beta = 1.787$

$\gamma = 1.840$

$(+)2V \sim 12°$

HARDNESS: 5-5½

DENSITY: 4.6-5.4

CLEAVAGE: {100} distinct

{010} less distinct

{110} indistinct

{101} indistinct

{011} indistinct

Distinct parting often present on {001}; parting on {$\bar{1}$11} not common. Fracture conchoidal to uneven. Brittle.

HABIT: Crystals commonly small, thin to thick tabular on {100} or elongated along c-axis; sometimes equant, wedge-shaped or prismatic; also as large euhedral crystals weighing several pounds. Crystal faces often uneven, rough, or striated. Also massive granular; as a detrital mineral. Twinning on {100} common.

COLOR-LUSTER: Reddish brown, brown, yellowish brown, pink, yellow, greenish, grayish white, nearly white. Transparent to subtranslucent. Resinous, waxy, vitreous to subadamantine. Streak white or slightly colored.

MODE OF OCCURRENCE: Occurs widespread throughout the world as a detrital mineral in placer deposits and beach and river sands; in pegmatites, metamorphic rocks, and vein deposits. In the United States extensive detrital deposits are found in Idaho, Montana, Florida, North and South Carolina, and elsewhere in the Appalachian region. In Colorado it is produced in commercial amounts from the Climax molybdenum mine, and also occurs as fine crystals in many pegmatites throughout the state. Exceptional crystals weighing several pounds are found associated with euxenite near Encampment, Wyoming. It is also found in the Petaca district, New Mexico, and in pegmatite at Amelia, Virginia. Large commercial detrital deposits are found in Australia, Ceylon, India, Malaya, Nigeria, and Brazil. Fine crystals have been recovered from pegmatites at several localities in Norway, Finland, and Madagascar; in Switzerland fine crystals occur at several localities in Alpine vein deposits; and excellent twinned crystals are found associated with cassiterite at Callipampa, Bolivia.

BEST REF. IN ENGLISH: Palache, et al., "Dana's System of Mineralogy," 7th Ed., v. II, p. 691–696, New York, Wiley, 1951. Frondel, Clifford, USGS Bull. 1064, p. 150–160 (1958).

# MONCHEÏTE
(Pt, Pd)(Te, Bi)$_2$

CRYSTAL SYSTEM: Hexagonal
CLASS: $\bar{3}$ 2/m
SPACE GROUP: P$\bar{3}$m
Z: 1
LATTICE CONSTANTS:
  $a$ = 4.049
  $c$ = 5.288
3 STRONGEST DIFFRACTION LINES:
  2.93 (100)
  2.11 ( 80)
  2.02 ( 70)
HARDNESS: Not determined.
DENSITY: $\sim$ 10
CLEAVAGE: {0001}
HABIT: Massive; as minute grains, rarely up to 0.2 mm in size.
COLOR-LUSTER: Steel gray. Opaque. Metallic.
MODE OF OCCURRENCE: Occurs as inclusions in chalcopyrite from the Monchegorsk deposits, Monche Tundra, USSR.
BEST REF. IN ENGLISH: Genkin, A. D., Zhuravlev, N. N., and Smirnova, E. M., *Am. Min.*, 48: 1181 (1963).

# MONETITE
CaHPO$_4$

CRYSTAL SYSTEM: Triclinic
CLASS: $\bar{1}$

SPACE GROUP: P$\bar{1}$
Z: 4
LATTICE CONSTANTS:
  $a$ = 6.90    $\alpha$ = 96°21′
  $b$ = 7.00    $\beta$ = 91°16′
  $c$ = 6.65    $\gamma$ = 76°6′
3 STRONGEST DIFFRACTION LINES:
  2.96 (100)
  3.35 ( 75)
  3.37 ( 70)
OPTICAL CONSTANTS:
  $\alpha$ = 1.600, 1.587
  $\beta$ = 1.614
  $\gamma$ = 1.631, 1.640
  (+)2V = moderate to large
HARDNESS: 3½
DENSITY: 2.929 (Meas. synthetic)
         2.92 (Calc.)
CLEAVAGE: Indistinct in three directions. Fracture uneven. Brittle.
HABIT: As massive aggregates composed of minute flattened crystals with rough faces and rhombohedral outline; also stalactitic or as crusts.
COLOR-LUSTER: Colorless, white, pale yellowish. Transparent to translucent. Vitreous.
MODE OF OCCURRENCE: Occurs on the Islands of Moneta, Mona, and Los Monges in the Caribbean Sea; with newberyite and apatite on Ascension Island in the South Atlantic ocean; and in a cave deposit at Gunong Jerneh, Malaya.
BEST REF. IN ENGLISH: Palache, et al., "Dana's System of Mineralogy," 7th Ed., v. II, p. 660–661, New York, Wiley, 1951.

# MONHEIMITE = Ferroan variety of smithsonite

# MONIMOLITE
(Pb, Ca, Fe)$_3$(SbO$_4$)$_2$

CRYSTAL SYSTEM: Cubic
CLASS: 4/m $\bar{3}$ 2/m
LATTICE CONSTANT:
  $a$ = 10.47
3 STRONGEST DIFFRACTION LINES:
  3.48 (100)
  2.65 ( 40)
  5.8 ( 30)
HARDNESS: 4½–6
DENSITY: 5.94–7.29 (Meas.)
         6.58 (Calc.)
CLEAVAGE: {111} indistinct
Fracture conchoidal to splintery.
HABIT: Usually as octahedral crystals; rarely cubic.
COLOR-LUSTER: Yellow, orange-yellow, grayish green, dark brown. Translucent to opaque. Greasy to adamantine. Streak pale yellow to reddish brown.
MODE OF OCCURRENCE: Occurs associated with tephroite in calcite veins at the Harstig mine Pajsberg, and at Långban, Sweden.

BEST REF. IN ENGLISH: Palache, et al., "Dana's System of Mineralogy," 7th Ed., v. II, p. 1023, New York, Wiley, 1951.

## MONOHYDROCALCITE
$CaCO_3 \cdot H_2O$

CRYSTAL SYSTEM: Hexagonal
SPACE GROUP: $P3_1 12$
Z: 9
LATTICE CONSTANTS:
  (synthetic)
  $a = 10.62$
  $c = 7.54$
3 STRONGEST DIFFRACTION LINES:

| (natural) | (synthetic) |
|-----------|-------------|
| 2.17 (100) | 4.30 (100) |
| 1.926 (100) | 3.05 (100) |
| 4.49 ( 90) | 1.93 (100) |

OPTICAL CONSTANTS:
  $\omega = 1.545$
  $\epsilon = 1.590$
  (+) (natural), (−) (synthetic)
DENSITY: 2.38 (Meas.)
             2.376 (Calc.)
HABIT: Massive; fine-grained.
COLOR-LUSTER: White to gray.
MODE OF OCCURRENCE: Occurs with calcite in bottom sediments of Lake Issyk-Kul, Kirgizia, USSR.
BEST REF. IN ENGLISH: Semenov, E. I., *Am. Min.*, **49**: 1151 (1964).

## MONREPITE = Biotite

## MONSMEDITE
$K_2O \cdot Tl_2O_3 \cdot 8SO_3 \cdot 15H_2O$

CRYSTAL SYSTEM: Cubic
3 STRONGEST DIFFRACTION LINES:
  3.403 (100)
  3.560 ( 80)
  2.860 ( 80)
OPTICAL CONSTANT:
  $n \simeq 1.6081$
HARDNESS: 2
DENSITY: 3.00 (Meas.)
CLEAVAGE: Cubic, rarely octahedral.
HABIT: Cubes; from tenths of a millimeter to 1 cm, mostly 0.25–0.50 cm. Forms (111) and (110) also present.
COLOR-LUSTER: Dark green to black; nearly opaque in large crystals; pitchy.
MODE OF OCCURRENCE: Occurs as geodes with marcasite, barite, and iron sulfates in the oxidation zone of a hydrothermal deposit containing pyrite, sphalerite, galena, and accessory pyrargyrite, stephanite, chalcopyrite, and stibnite at Baia Sprie, Roumania.
BEST REF. IN ENGLISH: Götz, A., Mihalka, St., Ionita, I., and Toth, Z., *Am. Min.*, **54**: 1496 (1969).

## MONTANITE (Inadequately described mineral)
$(BiO)_2 TeO_4 \cdot 2H_2O$

CRYSTAL SYSTEM: Monoclinic (?)
OPTICAL CONSTANTS:
  $\beta = 2.09$
  $\gamma - \alpha = 0.01$
  $(-)2V = $ small
HARDNESS: Not determined. Soft.
DENSITY: 3.79 (Meas.)
CLEAVAGE: Not determined.
HABIT: Massive, compact to earthy.
COLOR-LUSTER: Yellowish, also greenish to white. Opaque. Waxy to dull.
MODE OF OCCURRENCE: Occurs at Highland, Montana, as an alteration of tetradymite. It is also reported from Colorado, New Mexico, North Carolina, New South Wales, Australia, and the USSR.
BEST REF. IN ENGLISH: Palache, et al., "Dana's System of Mineralogy," 7th Ed., v. II, p. 636–637, New York, Wiley, 1951.

## MONTBRAYITE
$Au_2 Te_3$

CRYSTAL SYSTEM: Triclinic
CLASS: $\bar{1}$
SPACE GROUP: $P\bar{1}$
Z: 12
LATTICE CONSTANTS:
  $a = 12.10$    $\alpha = 104°30.5'$
  $b = 13.46$    $\beta = 97°34.5'$
  $c = 10.80$    $\gamma = 107°53.5'$
3 STRONGEST DIFFRACTION LINES:
  2.09 (100)
  2.98 ( 80)
  2.93 ( 80)
HARDNESS: 2½
DENSITY: 9.94 (Meas.)
CLEAVAGE: $\{1\bar{1}0\}$ distinct
           $\{0\bar{1}1\}$ distinct
           $\{1\bar{1}1\}$ distinct
Fracture flat conchoidal. Very brittle.
HABIT: Massive, as irregular grains.
COLOR-LUSTER: Tin white to pale yellow. Opaque. Metallic.
MODE OF OCCURRENCE: Found in association with native gold, altaite, petzite, melonite, frohbergite, chalcopyrite, pyrite, and other minerals, at the Robb Montbray mine, Montbray Township, Abitibi County, Quebec, Canada.
BEST REF. IN ENGLISH: Vlasov, K. A., "Mineralogy of Rare Elements," v. II, p. 696–698, Israel Program for Scientific Translations, 1966.

## MONTEBRASITE
$(Li, Na)Al(PO_4)(OH, F)$
Amblygonite Series

CRYSTAL SYSTEM: Triclinic
CLASS: $\bar{1}$

SPACE GROUP: P$\bar{1}$
Z: 2
LATTICE CONSTANTS:
$a = 5.19$      $\alpha = 112°2.5'$
$b = 7.12$      $\beta = 97°49.5'$
$c = 5.04$      $\gamma = 68°7.5'$
3 STRONGEST DIFFRACTION LINES:
2.968 (100)
3.164 ( 90)
4.672 ( 70)
OPTICAL CONSTANTS:
$\alpha = 1.57–1.60$
$\beta = 1.61$
$\gamma = 1.62$
$(-)2V = 81°30'$
HARDNESS: 5½–6
DENSITY:  3.027 (Meas.)
          2.98 (Calc.)
CLEAVAGE: {100} perfect
          {110} good
          {0$\bar{1}$1} distinct
          {001} imperfect
Fracture conchoidal to uneven.  Brittle.

HABIT: Crystals equant to short prismatic, commonly with rough faces. Usually as large cleavable masses. Twinning on {$\bar{1}\bar{1}$1} common.

COLOR-LUSTER: Usually white to grayish white; also colorless, yellowish, pinkish, tan, greenish, bluish. Transparent to translucent.  Vitreous to greasy; pearly on cleavages.

MODE OF OCCURRENCE: Occurs in granite pegmatites often in masses and crystals of very large size. Found widespread in the Black Hills, South Dakota, and at localities in Maine and Connecticut. It also occurs in Nova Scotia, Brazil, Western Australia, South West Africa, France, Spain, and Sweden. Many reported occurrences of amblygonite probably pertain to montebrasite.

BEST REF. IN ENGLISH: Palache, et al., "Dana's System of Mineralogy," 7th Ed., v. II, p. 823–827, New York, Wiley, 1951.

## MONTEPONITE = Cadmium oxide

## MONTESITE (Species status uncertain)
PbSn$_4$S$_5$

CRYSTAL SYSTEM: Unknown
HARDNESS: >2½
DENSITY: 5.66
HABIT: Massive.
COLOR-LUSTER: Silvery white.
MODE OF OCCURRENCE: Occurs at López Huayco-Ocurí, Bolivia, associated with cassiterite.
BEST REF. IN ENGLISH: Herzenberg, Roberto, *Am. Min.*, **35**: 334–335 (1950).

## MONTGOMERYITE
Ca$_2$Al$_2$(PO$_4$)$_3$(OH) · 7H$_2$O

CRYSTAL SYSTEM: Monoclinic

CLASS: 2/m
SPACE GROUP: C2/c
Z: 2
LATTICE CONSTANTS:
$a = 10.01$
$b = 24.15$
$c = 6.26$
$\beta = 91°28'$
3 STRONGEST DIFFRACTION LINES:
5.09  (100)
12.0  ( 90)
2.882 ( 50)
OPTICAL CONSTANTS:
$\alpha = 1.572$
$\beta = 1.578$
$\gamma = 1.582$
$(-)2V = 75°$
HARDNESS: 4
DENSITY: 2.46 (Meas. Etta mine)
         2.53 (Meas. Fairfield, Utah)
         2.51 (Calc.)
CLEAVAGE: {010} perfect
          {100} poor
HABIT: Crystals minute, lath-like, often in subparallel growths. Also massive, as subparallel aggregates of coarse plates.

COLOR-LUSTER: Deep green, rarely pale green to white to colorless.  Transparent to translucent.  Vitreous.

MODE OF OCCURRENCE: Occurs associated with crandallite and other phosphate minerals in variscite nodules at Fairfield, Utah County, Utah; and with mitridatite and frondelite-rockbridgeite in altered triphylite nodules at the Etta mine, Keystone, Pennington County, South Dakota.

BEST REF. IN ENGLISH: Mead, Cynthia W., and Mrose, Mary E., U.S. Geol. Surv. Prof. Pap. 600-D, p. D204–D206 (1968).

## MONTICELLITE
MgCaSiO$_4$

CRYSTAL SYSTEM: Orthorhombic
CLASS: 2/m 2/m 2/m
SPACE GROUP: Pnma
Z: 4
LATTICE CONSTANTS:
$a = 4.815$
$b = 11.08$
$c = 6.37$
3 STRONGEST DIFFRACTION LINES:
2.666 (100)
3.637 ( 40)
2.586 ( 40)
OPTICAL CONSTANTS:
$\alpha = 1.639–1.654$
$\beta = 1.646–1.664$  (synthetic)
$\gamma = 1.653–1.674$
$(-)2V = 72°–82°$
HARDNESS: 5½
DENSITY: 3.08–3.27 (Meas.)
         3.054 (Calc.)
CLEAVAGE: {010} indistinct
Brittle.

HABIT: Crystals prismatic, rarely up to 8 cm in length. Usually massive, granular, or as disseminated grains. Twinning on {031}.

COLOR-LUSTER: Colorless, gray, greenish. Transparent to translucent. Vitreous.

MODE OF OCCURRENCE: Occurs mainly as a metamorphic or metasomatic mineral; also as a constituent of some ultrabasic rocks. Typical occurrences are found at Crestmore, Riverside County, and at the Dewey mine near Valley Wells, San Bernardino County, California; in carbonatite at Magnet Cove, Arkansas, and at Oka, Quebec, Canada; and at localities in Scotland, Italy, USSR, Zaire, and near Hobart, Tasmania.

BEST REF. IN ENGLISH: Deer, Howie, and Zussman, "Rock Forming Minerals," v. 1, p. 41–46, New York, Wiley, 1962.

## MONTMORILLONITE

$(Na, Ca)_{0.33}(Al, Mg)_2 Si_4 O_{10}(OH)_2 \cdot nH_2O$
Montmorillonite (Smectite) Group

CRYSTAL SYSTEM: Monoclinic
CLASS: 2/m
SPACE GROUP: C2/m
Z: 1
LATTICE CONSTANTS:
$a$ = 5.17
$b$ = 8.94
$c$ = 9.95
$\beta$ = 99°54′
3 STRONGEST DIFFRACTION LINES: $d_{100}$ spacings continuous between about 10 and 21Å, varying with humidity and nature of cation. Basal spacing expands with glycerol treatment.
OPTICAL CONSTANTS:
$\alpha$ = 1.48–1.57
$\beta$ = 1.50–1.60
$\gamma$ = 1.50–1.60
(–)2V = mostly small
HARDNESS: 1–2
DENSITY: Variable, 2–3
CLEAVAGE: {001} perfect
HABIT: Massive, very fine grained; clay-like.
COLOR-LUSTER: White, gray, yellowish, greenish, pink. Dull.

MODE OF OCCURRENCE: Occurs as the principal constituent of bentonite clay deposits; also found widespread in soils, sedimentary and metamorphic rocks, and deposits of hydrothermal origin. Typical occurrences are found in California, Arizona, New Mexico, Colorado, Wyoming, South Dakota, Mississippi, Maine, and Connecticut. Also in France, Italy, Czechoslovakia, Roumania, and elsewhere.

BEST REF. IN ENGLISH: Deer, Howie, and Zussman, "Rock Forming Minerals," v. 3, p. 226–245, New York, Wiley, 1962.

## MONTROSEITE

$(V, Fe)O \cdot OH$

CRYSTAL SYSTEM: Orthorhombic
CLASS: 2/m 2/m 2/m

SPACE GROUP: Pbnm
Z: 4
LATTICE CONSTANTS:
$a$ = 4.82
$b$ = 9.48
$c$ = 2.93
3 STRONGEST DIFFRACTION LINES:
4.31 (100)
2.644 (100)
3.38 ( 80)
HARDNESS: Not determined.
DENSITY: 4.0 (Meas.)
4.11 (Calc. for 8% FeO)
CLEAVAGE: {010} good
{110} good
Brittle.

HABIT: Microscopic bladed crystals; common forms are b {010}, m {110}, p {121}, and a large vicinal form is approximately {0, 10, 1}.

COLOR-LUSTER: Black; opaque; submetallic. Streak black.

MODE OF OCCURRENCE: Occurs at several mines in the Salt Wash sandstone member of the Morrison formation (late Jurassic) in the Uravan mineral belt in western Colorado; at several localities in Emery and Grand Counties, Utah; and at the Runge mine, Edgemont, Fall River County, South Dakota. It is often associated with corvusite, hewettite, haggite, and paramontroseite.

BEST REF. IN ENGLISH: Weeks, Alice D., Cisney, Evelyn A., and Sherwood, Alexander M., *Am. Min.*, **38**: 1235–1241 (1953).

## MONTROYDITE

HgO

CRYSTAL SYSTEM: Orthorhombic
CLASS: 2/m 2/m 2/m
SPACE GROUP: Pnma
Z: 4 (synthetic)
LATTICE CONSTANTS:
$a$ = 6.613
$b$ = 3.513
$c$ = 5.504
3 STRONGEST DIFFRACTION LINES:
2.967 (100)
2.834 ( 80) (synthetic)
2.408 ( 65)
OPTICAL CONSTANTS:
$\alpha$ = 2.37
$\beta$ = 2.5 (Li)
$\gamma$ = 2.65
(+)2V = large
HARDNESS: 2½
DENSITY: 11.2 (synthetic)
11.209 (Calc.)
CLEAVAGE: {010} perfect
Sectile. Flexible, inelastic.

HABIT: Crystals slender prismatic, often twisted or bent; also equant or somewhat flattened; terminal forms commonly striated. As vermicular, tubular, or rounded aggregates of tiny crystals. Also massive, pulverulent.

COLOR-LUSTER: Dark red, brownish red to brown. Translucent to transparent. Vitreous to adamantine. Streak yellowish brown.

MODE OF OCCURRENCE: Occurs in California at the Red Elephant mine, Lake County; associated with calomel, mercury, cinnabar, and eglestonite in a silicate-carbonate rock, near the New Idria mine, and as bladed crystals at Clear Creek, San Benito County; as long prismatic and bent crystals near Redwood City, San Mateo County; and at the Esperanza and Socrates mines, Sonoma County. Also found at Terlingua, Brewster County, Texas, associated with mercury, terlinguaite, and calcite.

BEST REF. IN ENGLISH: Palache, et al., "Dana's System of Mineralogy," 7th Ed., v. I, p. 511–514, New York, Wiley, 1944.

# MOOIHOEKITE
$Cu_9Fe_9S_{16}$

CRYSTAL SYSTEM: Tetragonal
CLASS: $\bar{4}2m$
SPACE GROUP: $P\bar{4}2m$
Z: 1
LATTICE CONSTANTS:
   $a$ = 10.58
   $c$ = 5.37
3 STRONGEST DIFFRACTION LINES:
   3.07 (100)
   1.89 ( 80)
   1.60 ( 60)
HARDNESS: Microhardness 261 kg/mm$^2$.
DENSITY: 4.36 (Meas.)
            4.37 (Calc.)
CLEAVAGE: Not determined.
HABIT: Massive. Observed in polished sections in areas up to 1 mm$^2$ and as an intergrowth with haycockite.
COLOR-LUSTER: Brass yellow. Tarnishes various hues of pinkish brown to dark purple. Opaque. Metallic.
MODE OF OCCURRENCE: Occurs in association with haycockite in the Duluth Gabbro complex, Minnesota. Also found in a pipe-shaped hortonolite-dunite pegmatite in the Norite Zone of the Bushveld Igneous Complex at Mooihoek Farm, Lydenburg District, Transvaal, South Africa.
BEST REF. IN ENGLISH: Cabri, Louis J., and Hall, Sydney R., *Am. Min.*, **57**: 689–708 (1972).

# MOOREITE
$(Mg, Zn, Mn)_8 SO_4 (OH)_{14} \cdot 3-4H_2O$

CRYSTAL SYSTEM: Monoclinic
CLASS: 2/m
SPACE GROUP: $P2_1/a$
Z: 4
LATTICE CONSTANTS:
   $a$ = 11.18
   $b$ = 20.28
   $c$ = 8.23
   $\beta$ = 92°55′

3 STRONGEST DIFFRACTION LINES:
   10.37 (100)
   5.14 (100)
   8.29 ( 90)
OPTICAL CONSTANTS:
   $\alpha$ = 1.533
   $\beta$ = 1.545
   $\gamma$ = 1.547
   $(-)2V \sim 50°$
HARDNESS: 3
DENSITY: 2.47 (Meas.)
            2.52 (Calc.)
CLEAVAGE: {010} perfect
HABIT: Crystals tabular to platy. Usually as subparallel aggregates of minute crystals. Sometimes as well-developed doubly terminated crystals.
COLOR-LUSTER: Colorless. Transparent. Vitreous.
MODE OF OCCURRENCE: Occurs in cavities and crevices in pyrochroite in a vein in calcite-franklinite-willemite ore at Sterling Hill, Sussex County, New Jersey.
BEST REF. IN ENGLISH: Palache, et al., "Dana's System of Mineralogy," 7th Ed., v. II, p. 574–575, New York, Wiley, 1951.    Finney, J. J., *Am. Min.*, **54**: 973–975 (1969).

# MOORHOUSEITE
$CoSO_4 \cdot 6H_2O$

CRYSTAL SYSTEM: Monoclinic
CLASS: 2/m
SPACE GROUP: C2/c
Z: 8
LATTICE CONSTANTS:
   $a$ = 10.0
   $b$ = 7.2
   $c$ = 24.3
   $\beta$ = 98°22′
3 STRONGEST DIFFRACTION LINES:
   4.39 (100)
   4.01 ( 60)
   2.91 ( 50)
OPTICAL CONSTANTS:
   $\alpha$ = 1.470
   $\beta$ = 1.497
   $\gamma$ = 1.497
   $(-)2V$ = very small
HARDNESS: 2½
DENSITY: 1.97 (Meas.)
            2.04 (Calc.)
CLEAVAGE: Conchoidal fracture.
HABIT: Granular to fine-grained columnar, as water-soluble efflorescences.
COLOR-LUSTER: Pink; vitreous. Streak white.
MODE OF OCCURRENCE: Occurs on a 2 × 4 × 8 inch specimen of sulfides in a barite-siderite matrix from the Magnet Cove Barium Corporation mine about 2½ miles southwest of Walton, Nova Scotia.
BEST REF. IN ENGLISH: Jambor, J. L., and Boyle, R. W., *Can. Min.*, **8**: 166–171 (1965).

# MORAESITE
$Be_2 PO_4 OH \cdot 4H_2O$

CRYSTAL SYSTEM: Monoclinic
CLASS: m or 2/m
SPACE GROUP: Cc or C2/c
Z: 12
LATTICE CONSTANTS:
$a = 8.55$
$b = 36.90$
$c = 7.13$
$\beta = 97°41'$
3 STRONGEST DIFFRACTION LINES:
7.00 (100)
3.278 ( 90)
4.24 ( 60)
OPTICAL CONSTANTS:
$\alpha = 1.462$
$\beta = 1.482$
$\gamma = 1.490$
$(-)2V = 65°$
HARDNESS: Not determined.
DENSITY: 1.805 (Meas.)
1.806 (Calc.)
CLEAVAGE: Two excellent cleavages, the traces of which on {100} are parallel to the b- and c-axes.
HABIT: Spherulitic masses, as distinct crystals, and as crusts with a coarse fibrous structure. Crystals are singly terminated needles, acicular [001] with forms {100} and {130} and faces (131) and (1$\bar{3}$1).
COLOR-LUSTER: White. Streak white.
MODE OF OCCURRENCE: Occurs on the walls of vugs that have developed in or adjacent to beryl, and on surfaces of albite, quartz, and muscovite, often associated with frondelite and other secondary phosphates at the Sapucaia pegmatite mine, Minas Geraes, Brazil.
BEST REF. IN ENGLISH: Lindberg, M. L., Pecora, W. T., and Barbosa, A. L. de M., *Am. Min.*, **38**: 1126–1133 (1953).

# MORDENITE (Ptilolite)
$(Ca, Na_2, K_2)Al_2Si_{10}O_{24} \cdot 7H_2O$
Zeolite Group

CRYSTAL SYSTEM: Orthorhombic
CLASS: mm2 or 2/m 2/m 2/m
SPACE GROUP: Cmc2 or Cmcm
Z: 4
LATTICE CONSTANTS:
$a = 18.13$
$b = 20.49$
$c = 7.52$
3 STRONGEST DIFFRACTION LINES:
3.48 (100)
3.22 (100)
9.10 ( 90)
OPTICAL CONSTANTS:
$\alpha = 1.472-1.483$
$\beta = 1.475-1.485$
$\gamma = 1.477-1.487$
$(+)2V = 76°-104°$; also $(-)$
HARDNESS: 4-5
DENSITY: 2.12-2.15 (Meas.)
CLEAVAGE: {100} perfect
{010} distinct

HABIT: Crystals prismatic, vertically striated, minute. Usually acicular to fine fibrous, often in cottony aggregates. Also compact, porcelaneous.
COLOR-LUSTER: Colorless, white, or stained yellowish or pinkish. Transparent to translucent. Vitreous to silky.
MODE OF OCCURRENCE: Occurs in veins and cavities in igneous rocks; as a hydration product of glasses; and as an authigenic deposit in sediments. Found in Los Angeles and Riverside counties, California; at Challis, Custer County, Idaho; in the Tintic district, Utah; from North and South Table Mountains in Jefferson County, and at other places in Colorado; and in the Hoodoo Mountains, Wyoming. Also at Morden, Kings County, Nova Scotia, Canada, and at localities in Scotland, Yugoslavia, USSR, Japan, New Zealand, and elsewhere.
BEST REF. IN ENGLISH: Deer, Howie, and Zussman, "Rock Forming Minerals," v. 4, p. 401–407, New York, Wiley, 1963.

# MORENOSITE
$NiSO_4 \cdot 7H_2O$

CRYSTAL SYSTEM: Orthorhombic
CLASS: 222
SPACE GROUP: $P2_12_12_1$
Z: 4
LATTICE CONSTANTS:
$a = 11.86$
$b = 12.08$
$c = 6.81$
3 STRONGEST DIFFRACTION LINES:
4.20 (100)
5.3 ( 60)
2.85 ( 25)
OPTICAL CONSTANTS:
$\alpha = 1.47$
$\beta = 1.49$ (Na)
$\gamma = 1.49$
$(-)2V = 41°54'$
HARDNESS: 2-2½
DENSITY: 1.976 (Meas.)
1.912 (Calc.)
CLEAVAGE: {010} distinct
Fracture conchoidal.
HABIT: Stalactitic and as efflorescent crusts. Synthetic crystals short prismatic.
COLOR-LUSTER: Greenish white to green. Transparent to translucent. Vitreous. Streak pale greenish white.

Soluble in water. Taste astringent, metallic
MODE OF OCCURRENCE: Occurs as an oxidation product of nickel-bearing sulfides. Found in California associated with marcasite at the Klau quicksilver mine, San Luis Obispo County; at the Phoenix cinnabar mine, Napa County; with erythrite in the Friday copper mine, near Julian, San Diego County; and at the Island Mountain copper mine, Trinity County. It also is found at the Copper King mine, Boulder County, Colorado; at the Gap Nickel mine, Lancaster County, Pennsylvania; in the Sudbury and Algoma districts, Ontario, Canada; and in Peru, Spain, France, Italy, Germany, and Algeria.

BEST REF. IN ENGLISH: Palache, et al., "Dana's System of Mineralogy," 7th Ed., v. II, p. 516–519, New York, Wiley, 1951.

**MORGANITE** = Gem variety of beryl (Cs-bearing)

**MORINITE (Ježekite)**
$Ca_4 Na_2 Al_4 (PO_4)_4 O_2 F_6 \cdot 5H_2O$

CRYSTAL SYSTEM: Monoclinic
CLASS: 2/m
SPACE GROUP: $P2_1/m$
Z: 1
LATTICE CONSTANTS:
 $a$ = 9.456
 $b$ = 10.690
 $c$ = 5.445
 $\beta$ = 105°27.5'
3 STRONGEST DIFFRACTION LINES:
 2.94 (100)
 3.47 ( 80)
 1.783 ( 80)
OPTICAL CONSTANTS:
 $\alpha$ = 1.551
 $\beta$ = 1.563
 $\gamma$ = 1.565
 (−)2V = 43°
HARDNESS: 4–4½
DENSITY: 2.962 (Meas.)
 2.911 (Calc.)
CLEAVAGE: {100} perfect
 {001} imperfect
Fracture subconchoidal to uneven. Brittle.
 HABIT: Crystals long prismatic or tabular, vertically striated. Also columnar, radial-fibrous, or coarsely crystalline.
 COLOR-LUSTER: Colorless, white, pale pink. Translucent. Vitreous to slightly oily.
 MODE OF OCCURRENCE: Occurs as large crystalline masses and as prismatic crystals up to 1 cm in length, associated with apatite, augelite, wardite, and montebrasite, at the Hugo mine, Keystone, South Dakota. It also is found in pegmatite at Montebras, Creuze, France; at Greifenstein, Saxony, Germany; and at Viitaniemi, Finland.
 BEST REF. IN ENGLISH: Fisher, D. Jerome, and Runner, J. J., *Am. Min.*, **43**: 585–594 (1958).

**MORION** = Smoky quartz

**MOSANDRITE (Rinkolite, Johnstrupite, Lovchorrite, Rinkite)**
$(Na, Ca, Ce)_3 Ti(SiO_4)_2 F$

CRYSTAL SYSTEM: Monoclinic
Z: 4
LATTICE CONSTANTS:
 $a$ = 18.41
 $b$ = 5.64
 $c$ = 7.43
 $\beta$ = 93°

3 STRONGEST DIFFRACTION LINES:
 3.053 (100)
 2.687 ( 20)
 3.561 ( 15)
OPTICAL CONSTANTS:
 $\alpha$ = 1.662–1.643
 $\beta$ = 1.667–1.645
 $\gamma$ = 1.681–1.651
 (+)2V = 43°–87°
HARDNESS: 5
DENSITY: 2.93–3.5 (Meas.)
 3.46 (Calc.)
CLEAVAGE: {100} distinct
Fracture uneven to conchoidal. Brittle.
 HABIT: Crystals long prismatic or tabular. Prism faces vertically striated. Also massive; colloidal. Twinning on {100}, lamellar.
 COLOR-LUSTER: Yellow, yellowish green to reddish brown or brown. Translucent. Vitreous on cleavage, greasy on fracture.
 MODE OF OCCURRENCE: Occurs in nepheline-syenites and nepheline-syenite pegmatites. Found at Mont St. Hilaire, Quebec, Canada (rinkite); in the Julianehaab district, Greenland; on the islands of Låven and Stokö, and elsewhere in the Langesundfjord district, Norway; from the Khibiny and Lovozero alkali massifs on the Kola Peninsula, USSR; and in Transvaal, South Africa.
 BEST REF. IN ENGLISH: Vlasov, K. A., "Mineralogy of Rare Elements," v. II, p. 312–316, Israel Program for Scientific Translations, 1966.

**MOSCHELLANDSBERGITE**
$\gamma\text{-}Ag_2 Hg_3$

CRYSTAL SYSTEM: Cubic
CLASS: $4/m\ \bar{3}\ 2/m$
SPACE GROUP: Im3m
Z: 10
LATTICE CONSTANT:
 $a$ = 10.1
3 STRONGEST DIFFRACTION LINES:
 2.36 (100)
 1.365 ( 70)
 1.236 ( 60)
HARDNESS: 3½
DENSITY: 13.71 (Meas.)
 13.49 (Calc.)
CLEAVAGE: {011} distinct
 {001} distinct
Fracture conchoidal. Brittle.
 HABIT: Crystals dodecahedral, often modified by cube and/or trapezohedron. Also massive, granular.
 COLOR-LUSTER: Silver white. Opaque. Bright metallic.
 MODE OF OCCURRENCE: Occurs as fine crystals at Moschellandsberg, Bavaria, Germany; in the Chalanches mine, near Allemont, France; and at Sala, Sweden.
 BEST REF. IN ENGLISH: Palache, et al., "Dana's System of Mineralogy," 7th Ed., v. I, p. 103–104, New York, Wiley, 1944.

## MOSESITE
$Hg_2N(Cl, SO_4, MoO_4, CO_3) \cdot H_2O$

CRYSTAL SYSTEM: Cubic
CLASS: $\bar{4}3m$
SPACE GROUP: $F\bar{4}3m$
Z: 8
LATTICE CONSTANT:
 $a = 9.524$
3 STRONGEST DIFFRACTION LINES:
 2.74 (100)
 2.86 ( 80)
 1.68 ( 70)
OPTICAL CONSTANT:
 $N > 1.95$
 HARDNESS: 3½
 DENSITY: 7.72 (Meas.)
    7.53 (Calc.)
 CLEAVAGE: Octahedral {111} imperfect
Fracture uneven. Brittle.
 HABIT: Crystals usually octahedral; also cubo-octahedral or cubical. Often spinel twins of the simple octahedron.
 COLOR-LUSTER: Lemon yellow to canary yellow, becoming light olive green on long exposure to light. Adamantine. Streak very pale yellow.
 MODE OF OCCURRENCE: Found very sparingly associated with montroydite, calcite, and gypsum at Terlingua, Brewster County, Texas; with native mercury and cinnabar at the Clack quicksilver mine, Fitting district, Pershing County, Nevada; at El Doctor, Querétaro, Mexico; and at Huahuaxtla, Guerrero, Mexico.
 BEST REF. IN ENGLISH: Switzer, George, Foshag, W. F., Murata, K. J., and Fahey, J. J., *Am. Min.*, 38: 1225–1234 (1953).

## MOSSITE
$Fe(Nb, Ta)_2O_6$
Tapiolite Series

CRYSTAL SYSTEM: Tetragonal
CLASS: $4/m\ 2/m\ 2/m$
SPACE GROUP: $P4_2/mnm$
Z: 2
LATTICE CONSTANTS:
 $a = 4.711$
 $c = 9.12$
OPTICAL CONSTANTS:
 $\omega = 2.27$
 $\epsilon = 2.42$ (Li)
  (+)
HARDNESS: 6–6½
DENSITY: 6.93 (Calc. for Nb:Ta = 1:1)
CLEAVAGE: None.
Fracture uneven to subconchoidal.
 HABIT: Crystals short prismatic or equant, often appearing monoclinic or orthorhombic due to distortion. Twinning on {013} common, simple and polysynthetic.
 COLOR-LUSTER: Black. Opaque. Submetallic to subadamantine. Streak blackish brown to brownish gray.
 MODE OF OCCURRENCE: Occurs in granite pegmatites in Råde parish, near Moss, Norway.

BEST REF. IN ENGLISH: Palache, et al., "Dana's System of Mineralogy," 7th Ed., v. I, p. 775–778, New York, Wiley, 1944.

## MOTTRAMITE (Psittacinite)
$Pb(Cu, Zn)VO_4OH$
Descloizite-Mottramite Series

CRYSTAL SYSTEM: Orthorhombic
CLASS: $2/m\ 2/m\ 2/m$
SPACE GROUP: Pnam
Z: 4
LATTICE CONSTANTS:
 $a = 7.685$
 $b = 9.301$
 $c = 6.02$
3 STRONGEST DIFFRACTION LINES:
 3.24 (100)
 5.07 ( 80)
 2.87 ( 80)
OPTICAL CONSTANTS:
 $\alpha = 2.17$
 $\beta = 2.26$
 $\gamma = 2.32$
 $(-)2V \sim 73°$
 HARDNESS: 3–3½
 DENSITY: $\sim 5.9$
 CLEAVAGE: None.
Fracture uneven to conchoidal. Brittle.
 HABIT: Crystals variable in form; usually pyramidal, prismatic, or tabular. Crystal faces often rough or uneven. As plumose aggregates, drusy crusts, stalactitic, botryoidal with coarse fibrous structure, and massive granular.
 COLOR-LUSTER: Grass green to dark green, also brown to nearly black. Transparent to opaque. Vitreous to greasy.
 MODE OF OCCURRENCE: Occurs chiefly in the oxidation zone of ore deposits often associated with descloizite, pyromorphite, mimetite, vanadinite, calcite, and cerussite. Found in the United States at the Mammoth mine, Tiger, Pinal County, and at other places in Arizona; in Sierra County, New Mexico; at Crestmore, Riverside County, California, and in other western states. It also occurs at Mottram St. Andrew, Cheshire, England; at Tsumeb, Grootfontein, and at many other localities in the Otavi District, South West Africa; and in Sardinia, Bolivia, and Chile.
 BEST REF. IN ENGLISH: Palache, et al., "Dana's System of Mineralogy," 7th Ed., v. II, p. 811–815, New York, Wiley, 1951. Kingsbury and Hartley, *Min. Mag.*, 31: 289 (1956).

## MOUNANAITE
$PbFe_2(VO_4)_2(OH)_2$

CRYSTAL SYSTEM: Triclinic
CLASS: $\bar{1}$
SPACE GROUP: $P\bar{1}$
Z: 1
3 STRONGEST DIFFRACTION LINES:
 4.66 (100)
 4.56 ( 80)
 2.77 ( 80)

LATTICE CONSTANTS:

| | |
|---|---|
| $a = 5.55$ | $\alpha = 111°01'$ |
| $b = 7.66$ | $\beta = 112°07'$ |
| $c = 5.56$ | $\gamma = 94°09'$ |

OPTICAL CONSTANTS: All indices above 2.09.

HARDNESS: Not determined.

DENSITY: 4.85 (Meas.)
4.89 (Calc.)

HABIT: Crystals elongated on $c$ and platy on (010). Forms {010} dominant, {100}, {110}, {011}, {11$\bar{1}$}, {12$\bar{1}$}, {02$\bar{1}$}, {01$\bar{1}$}, {021}. Most crystals twinned by rotation around [001] or on (1$\bar{1}$1).

COLOR-LUSTER: Brownish red.

MODE OF OCCURRENCE: Occurs in very small amounts associated with goethite, francevillite, and curienite in the uranium-vanadium deposit of Mounana, Haut-Ogoue, Gabon, Africa.

BEST REF. IN ENGLISH: Cesbron, Fabien, and Fritsche, Jean, *Am. Min.*, **54**: 1738–1739 (1969).


## MOUNTAINITE

$(Ca, Na_2, K_2)_2 Si_4 O_{10} \cdot 3H_2O$

CRYSTAL SYSTEM: Monoclinic

CLASS: 2/m

SPACE GROUP: $P2_1/a$, $P2_1/c$, or $P2_1/n$

Z: 8

LATTICE CONSTANTS:

$a = 13.51$
$b = 13.10$
$c = 13.51$
$\beta = 104°$

OPTICAL CONSTANTS:

$\alpha = 1.504$
$\beta = 1.510$
$\gamma = 1.519$

HARDNESS: Not determined.

DENSITY: 2.36 (Meas.)

HABIT: As matted silky fibers in rosettes up to 2.0 mm in size.

COLOR-LUSTER: White; silky.

MODE OF OCCURRENCE: Occurs associated with rhodesite at the Bultfontein mine, Kimberley, Republic of South Africa.

BEST REF. IN ENGLISH: Gard, J. A., Taylor, H. F. W., and Chalmers, R. A., *Min. Mag.*, **31**: 611–623 (1957).


## MOURITE (Inadequately described mineral)

$U^{4+}Mo_5^{6+}O_{12}(OH)_{10}$

CRYSTAL SYSTEM: Unknown

3 STRONGEST DIFFRACTION LINES:

5.897 (100)
12.77 ( 90)
2.871 ( 80)

OPTICAL CONSTANT:

$n > 1.780$

HARDNESS: ~3

DENSITY: > 4.2

HABIT: As fan-shaped, radiating-fibrous, or vermiform

aggregates of plates with rectangular outlines of a few microns size.

COLOR-LUSTER: Violet. Streak violet-blue.

MODE OF OCCURRENCE: Occurs as nodules, crusts, and plates in the supergene zone of a deposit, the hypogene ore of which contains pitchblende, molybdenite, jordisite, pyrite, and other sulfides at an unspecified locality presumably in the USSR.

BEST REF. IN ENGLISH: Kopchenova, E. V., Skvortsova, K. V., Silantieva, N. I., Sidorenko, G. A., and Mikhailova, L. V., *Am. Min.*, **47**: 1217 (1962).


## MUIRITE

$Ba_{10}Ca_2MnTiSi_{10}O_{30}(OH, Cl, F)_{10}$

CRYSTAL SYSTEM: Tetragonal

CLASS: 4/m 2/m 2/m, 4mm, 422, or $\bar{4}$ 2m

SPACE GROUP: P4/mmm, P4mm, P422, P$\bar{4}$m2, or P$\bar{4}$2m

Z: 1

LATTICE CONSTANTS:

$a = 13.942$
$c = 5.590$

3 STRONGEST DIFFRACTION LINES:

2.91 (100)
4.42 ( 75)
3.73 ( 60)

OPTICAL CONSTANTS:

$\omega = 1.697$
$\epsilon = 1.704$
(+)

HARDNESS: ~2½

DENSITY: 3.86 (Meas.)
3.88 (Calc.)

CLEAVAGE: {001} indistinct
{100} indistinct

HABIT: Subhedral to euhedral tetragonal crystals; anhedral grains. Grain size ranges from a fraction of a millimeter to aggregates up to 3 mm.

COLOR-LUSTER: Orange; subvitreous. Pale orange streak.

MODE OF OCCURRENCE: Found sparsely distributed in sanbornite-bearing metamorphic rocks which outcrop in a narrow zone 2½ miles long near a granodiorite contact in eastern Fresno County, California.

BEST REF. IN ENGLISH: Alfors, John T., Stinson, Melvin C., Matthews, Robert A., and Pabst, Adolf, *Am. Min.*, **50**: 314–340 (1965). *Am. Min.*, **50**: 1500–1503, (1965).


## MUKHINITE

$Ca_2(Al_2V)(SiO_4)_3(OH)$

A clinozoisite with one-third the Al replaced by V

CRYSTAL SYSTEM: Monoclinic

(For crystal data see clinozoisite)

3 STRONGEST DIFFRACTION LINES:

2.892 (100)
2.682 ( 80)
2.600 ( 80)

METATORBERNITE   Chalk Mt. mine, Mitchell County, North Carolina.

METATORBERNITE   Chalk Mt. mine, Mitchell County, North Carolina.

METASTRENGITE   Bull Moose mine, Custer County, South Dakota.

METASTRENGITE (twin)   Bull Moose mine, Custer County, South Dakota.

METASTRENGITE (twin)   Bull Moose mine, Custer County, South Dakota.

META-URANOCIRCITE   Bergen, Voigtland, Saxony, Germany.

METAVARISCITE   Lucin, Box Elder County, Utah.

MIARGYRITE   Braunsdorf, Saxony, Germany.

MICROCLINE   Piedmont, Italy.

MICROLITE   Topsham, Maine.

MILARITE   Swakopmund, South West Africa.

MILARITE   Swakopmund, South West Africa.

MILARITE   Val Giuf, Grisons, Switzerland.

MILLERITE   Rhondda, Glamorgan, Wales.

MILLERITE (ring)   Hall's Gap, Kentucky.

MIMETITE   Tsumeb, South West Africa.

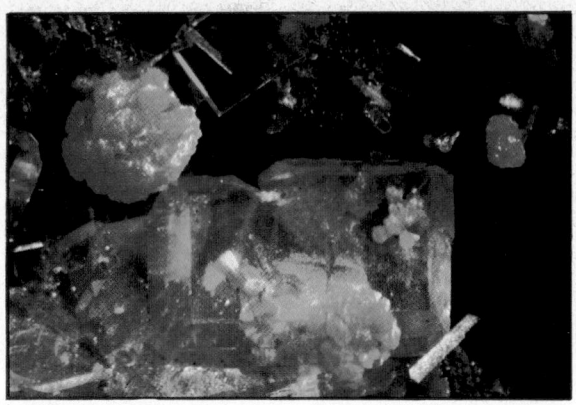

MIMETITE with wulfenite   San Francisco mine, Sonora, Mexico.

MIMETITE   Rawley mine, near Theba, Maricopa County, Arizona.

MIMETITE   Tsumeb, South West Africa.

MIMETITE (campylite)   Caldbeck Fells, Cumberland, England.

MIMETITE   Tsumeb, South West Africa.

MIXITE   Laurium, Greece.

MOLYBDENITE   Haddam, Connecticut.

MONTROYDITE with mercury   Socrates mine, Sonoma County, California.

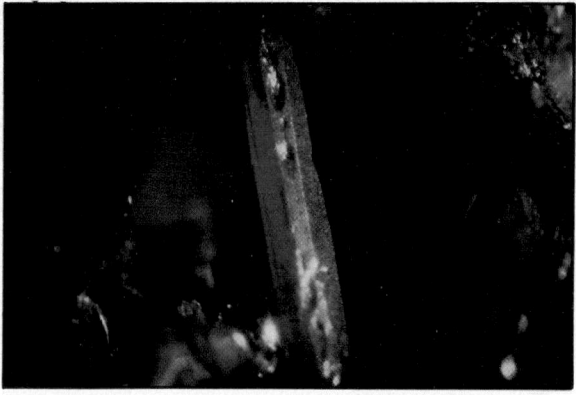

MONTROYDITE   Terlingua, Brewster County, Texas.

MONTGOMERYITE   Clay Canyon, Fairfield, Utah County, Utah.

MONTGOMERYITE   Clay Canyon, Fairfield, Utah County, Utah.

MONTEBRASITE   Palermo mine, North Groton, New Hampshire.

MONAZITE   Graubunden, Val Tavetsch, Grisons, Switzerland.

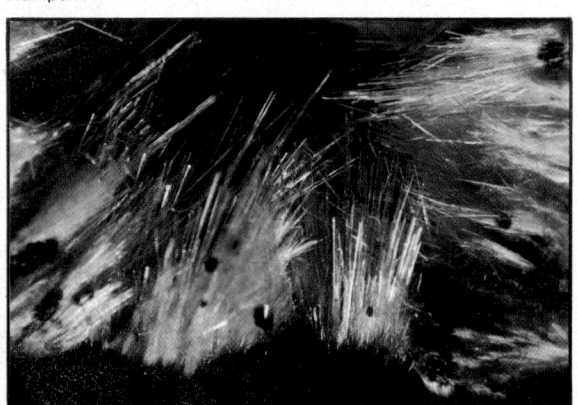

MORDENITE   Monte Lake, British Columbia, Canada.

MOTTRAMITE   Tsumeb, South West Africa.

84

MOUNANAITE with chervetite   Mounana, Gabon.

MUSCOVITE   New Hampshire.

MUSCOVITE (twin)   Nello Teer quarry, Raleigh, North Carolina.

MUSCOVITE, chromian (fuchsite)   Lengenbach quarry, Binnental, Switzerland.

NAGYAGITE   Nagyag, Roumania.

NANTOKITE   Nova Scotia, Canada.

NARSARSUKITE   Mont St. Hilaire, Quebec, Canada.

85

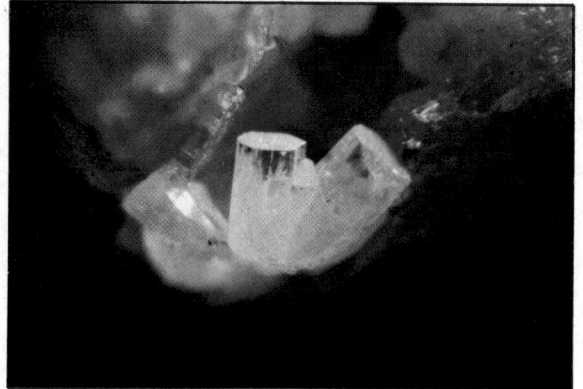

NASONITE   Franklin, Sussex County, New Jersey.

NASONITE   Franklin, Sussex County, New Jersey.

NATROCHALCITE with kroehnkite   Chuquicamata, Chile.

NATROJAROSITE on jarosite   Laurium, Greece.

NATROLITE   Snake Hill, Secaucus, New Jersey.

NATROLITE   Springfield, Lane County, Oregon.

NATROLITE   Tick Canyon, near Lang, California.

NATROLITE   Oregon.

NENADKEVICHITE   Mont St. Hilaire, Quebec, Canada.

NEPHELINE with melilite   Vesuvius, Italy.

NEPTUNITE with benitoite   San Benito County, California.

NEPTUNITE (twin)   San Benito County, California.

NICCOLITE   Mansfield, Thuringia, Germany.

NICKEL-SKUTTERUDITE (chloanthite)   Schneeberg, Saxony, Germany.

OFFRETITE   Rock Island Dam, Columbia River, Washington.

OLIVENITE   Majuba Hill, Pershing County, Nevada.

OLIVENITE   Tsumeb, South West Africa.

OLIVENITE (leucochalcite)   Majuba Hill, Pershing County, Nevada.

OLIVENITE   Majuba Hill, Pershing County, Nevada.

OPAL   Queensland, Australia.

ORDONEZITE   Santin mine, Cerro de las Fajas, Santa Caterina, Guanajuato, Mexico.

ORPIMENT   Getchell mine, Golconda, Humboldt County, Nevada.

ORTHOCLASE with quartz   Switzerland.

88

OSMIRIDIUM   Ural Mountains, USSR.

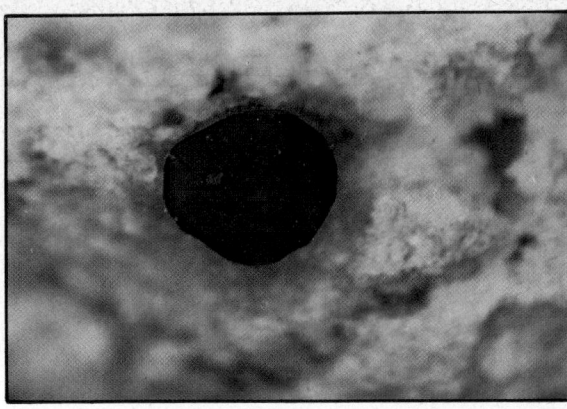

OSUMILITE   Sakkabira, Kagoshima Prefecture, Japan.

OTAVITE on tsumebite   Tsumeb, South West Africa.

OVERITE   Fairfield, Utah County, Utah.

PACHNOLITE   Ivigtut, Greenland.

PALERMOITE with whitlockite   Palermo mine, North Groton, New Hampshire.

PARACELCIAN   Benallt mine, Rhiw, Caernarvonshire, Wales.

PARADAMITE   Mapimi, Durango, Mexico.

PARALAURIONITE   Laurium, Greece.

PARATACAMITE   Mason Pass, near Yerington, Lyon County, Nevada.

PARAVAUXITE   Llallagua, Bolivia.

PARAVAUXITE with childrenite   Llallagua, Bolivia.

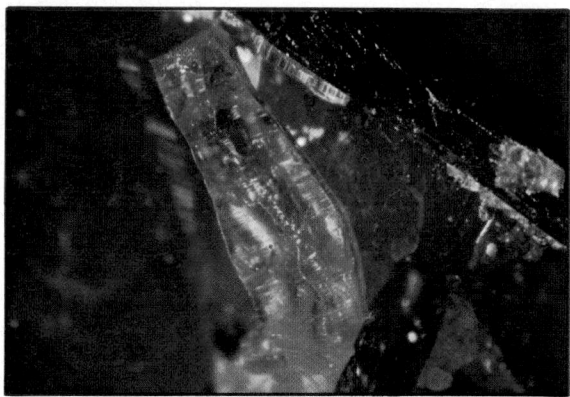

PARISITE   West Quincy, Massachusetts.

PARISITE   Ivigtut, Greenland.

PASCOITE   Minasragra, Peru.

PECTOLITE   De Sourdy quarry, Mont St. Hilaire, Quebec, Canada.

PECTOLITE   De Sourdy quarry, Mont St. Hilaire, Quebec, Canada.

PEROVSKITE   San Benito County, California.

PEROVSKITE   San Benito County, California.

PEROVSKITE, dendritic   Hannebacher, Eifel, West Germany.

PEROVSKITE   San Benito County, California.

PETZITE on gold   Bald Mountain, near Sonora, Tuolumne County, California.

PETZITE on gold   Bald Mountain, near Sonora, Tuolumne County, California.

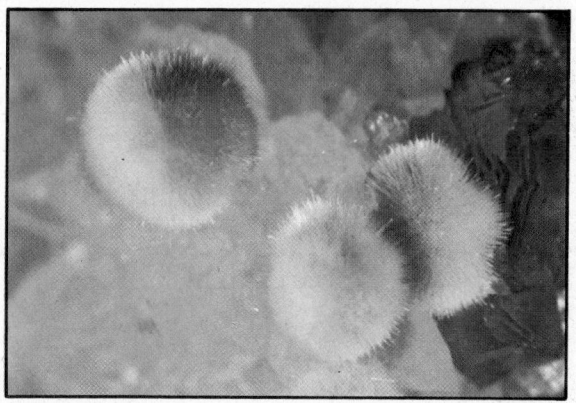

PHARMACOLITE with siderite   Schneeberg, Saxony, Germany.

PHARMACOSIDERITE   Richmond-Sitting Bull mine, Galena, Lawrence County, South Dakota.

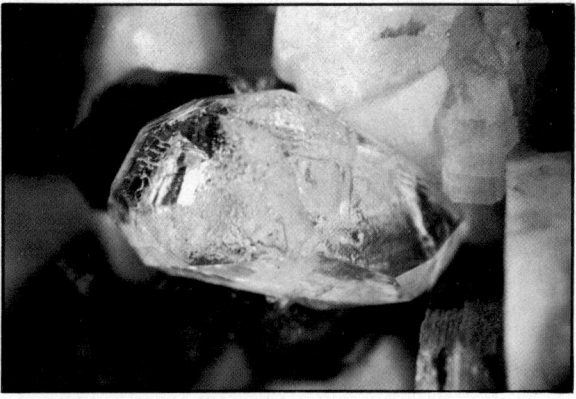

PHENAKITE   Striegau, Austria.

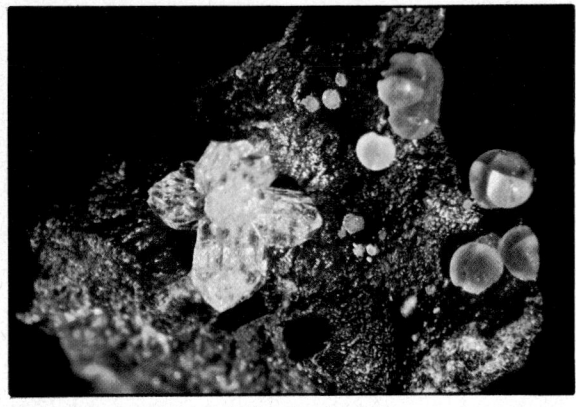

PHILLIPSITE   Richmond, Victoria, Australia.

PHLOGOPITE   Langban, Sweden.

PHLOGOPITE   Crestmore, Riverside County, California.

PHOENICOCHROITE with hemihedrite   Near Wickenburg, Maricopa County, Arizona.

PHOSGENITE in ancient slag   Laurium, Greece.

PHOSPHURANYLITE (barian)   Bergen, Vogtland, Germany.

PLANCHEITE with conichalcite   Tsumeb, South West Africa.

PLANCHEITE   Kambove, Katanga, Zaire.

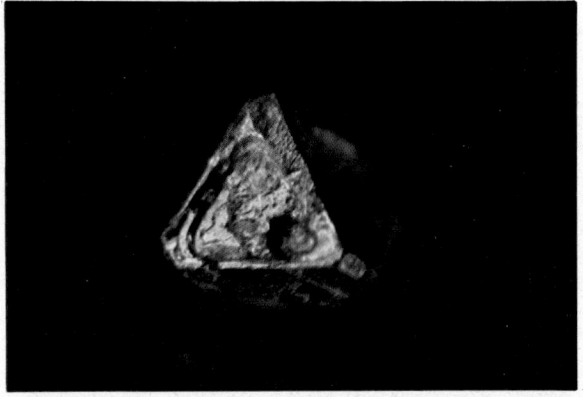

PLATINUM   Broken Hill, New South Wales, Australia.

PLATTNERITE on calcite   Mapimi, Durango, Mexico.

POLYBASITE   Guanajuato, Mexico.

POLYBASITE   Las Chispas mine, Arizpe, Sonora, Mexico.

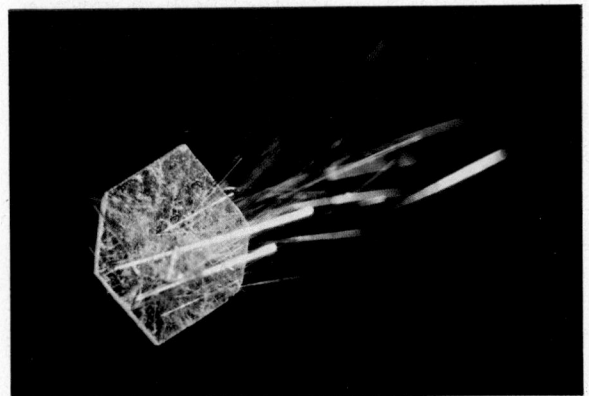

PREHNITE with actinolite   Cornog, Chester County, Pennsylvania.

PROSOPITE   St. Peter's Dome, El Paso County, Colorado.

PROUSTITE   Aue, Saxony, Germany.

PROUSTITE   Chanarcillo, Chile.

PROUSTITE   O'Brien mine, near Cobalt, Ontario, Canada.

PROUSTITE   Niederschlema, Saxony, Germany.

PROUSTITE   Erzgebirge, Saxony, Germany.

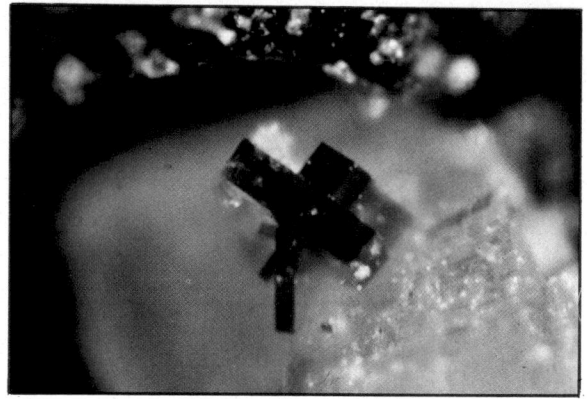

PSEUDOBOLEITE in ancient slag   Laurium, Greece.

PSEUDOBROOKITE   Thomas Range, Juab County, Utah.

PSEUDOBROOKITE   Thomas Range, Juab County, Utah.

PSEUDOBROOKITE   Taylor Creek, Sierra County, New Mexico.

PSEUDOMALACHITE   Virneberg, Rhineland, Germany.

PUCHERITE   Schneeberg, Saxony, Germany.

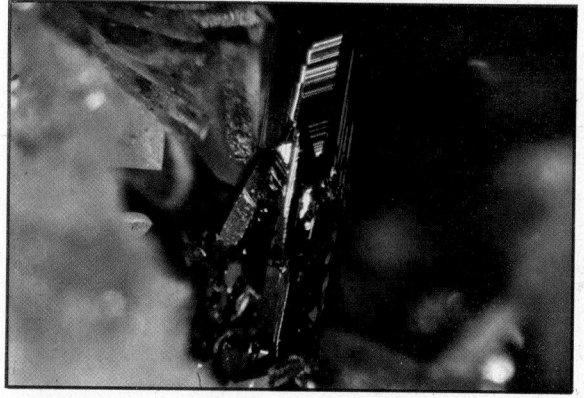

PYRARGYRITE   Guadelupe mine, Zacatecas, Mexico.

PYRARGYRITE   St. Andreasberg, Harz, Germany.

PYRITE   Pribram, Bohemia.

PYRITE (iridescent)   Naica, Chihuahua, Mexico.

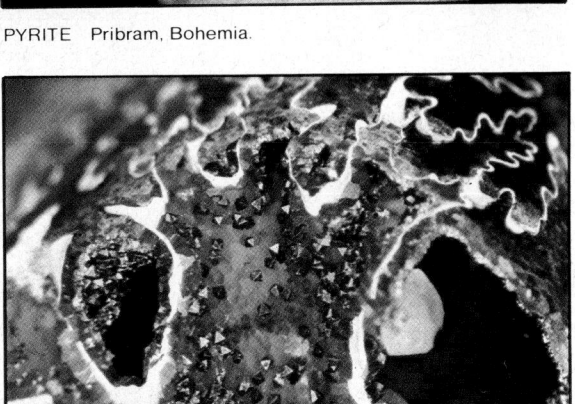

PYRITE lining chamber of Baculite   Pierre Shale, Cretaceous, Elk Creek, Meade County, South Dakota.

PYRITE   Logrono, Spain.

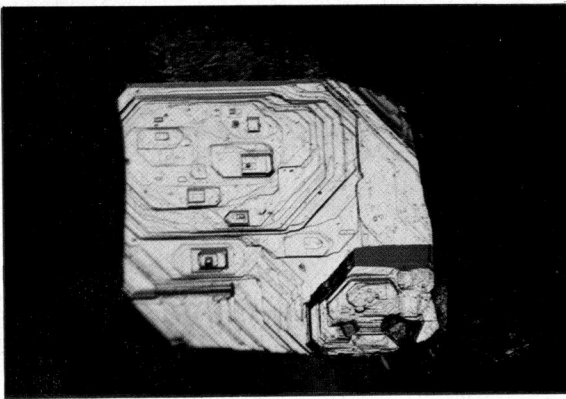

PYRITE   Yavapai County, Arizona.

PYRITE   Cerro de Pasco, Peru.

PYRITE with quartz   Kansas mine, Arizona.

PYRITE   Naica, Chihuahua, Mexico.

OPTICAL CONSTANTS:

$\alpha = 1.723$
$\beta = 1.733$
$\gamma = 1.755$
$(+)2V = 88°$

HARDNESS: 8
CLEAVAGE: {001} very perfect
{100} less perfect than {001}
Brittle.

HABIT: Irregular, short prismatic crystals up to $1 \times 2.5$ mm; also as irregular grains up to 5 mm in diameter. Some crystals twinned on {100}.

COLOR-LUSTER: Black with brownish tint; vitreous. Streak pale brownish gray.

MODE OF OCCURRENCE: Occurs in marble from the roof of the Tashelginsk iron ore deposit, Gornaya Shoriya, western Siberia, associated with goldmanite in areas of marble enriched in pale green muscovite and disseminated pyrite, pyrrhotite, chalcopyrite, and sphalerite.

BEST REF. IN ENGLISH: Shepel, A. B., and Karpenko, M. V., *Am. Min.*, 55: 321–322 (1970).


# MULLITE
$Al_6Si_2O_{13}$

CRYSTAL SYSTEM: Orthorhombic
CLASS: 2/m 2/m 2/m
SPACE GROUP: Pbam
Z: 3/4
LATTICE CONSTANTS:

$a = 7.5456$
$b = 7.6898$ (synthetic)
$c = 2.8842$

3 STRONGEST DIFFRACTION LINES:

3.390 (100)
3.428 ( 95)
2.206 ( 60)

OPTICAL CONSTANTS:

$\alpha = 1.637$
$\beta = 1.641$
$\gamma = 1.652$

HARDNESS: 6–7
DENSITY: 3.03–3.16 (Meas.)
3.170 (Calc.)
CLEAVAGE: {010} distinct
HABIT: Crystals prismatic.
COLOR-LUSTER: Colorless to pale pink. Transparent to translucent. Vitreous.

MODE OF OCCURRENCE: Occurs in fused argillaceous inclusions in Tertiary eruptive rocks on the island of Mull, Scotland; also found in slags and firebricks. Upon intense heating, the minerals andalusite, sillimanite, and kyanite are transformed to mullite.

BEST REF. IN ENGLISH: Bowen, N. L., Greig, J. W., and Zies, E. G., *J. Wash. Acad. Sci.*, 14: 183 (1924).   NBS Monograph 25, Sec. 3, 1964.


# MURDOCHITE
$Cu_6PbO_8$

CRYSTAL SYSTEM: Cubic
CLASS: 4/m $\bar{3}$ 2/m
SPACE GROUP: Fm3m
Z: 4
LATTICE CONSTANT:

$a = 9.210$

3 STRONGEST DIFFRACTION LINES:

5.30 (100)
2.659 (100)
1.629 (100)

HARDNESS: 4
DENSITY: 6.47 (Meas.)
6.1 (Calc.)
HABIT: Minute cubes and octahedrons.
COLOR-LUSTER: Black, lustrous; opaque.

MODE OF OCCURRENCE: Found as tiny octahedra, less than 1 mm in the longest dimension, on the surface of and embedded within plates of wulfenite and on the surface of crystals of fluorite, associated with hemimorphite, willemite, and quartz at the Mammoth mine, Tiger, Arizona. It also occurs as superb cubic crystals on barite or quartz and associated with linarite and brochantite at the Blanchard mine, near Bingham, New Mexico; at the Higgins mine, Bisbee, Arizona; and at Mapimi, Mexico.

BEST REF. IN ENGLISH: Fahey, Joseph J., Christ, C. L., and Clark, Joan R., *Am. Min.*, 40: 905–916 (1955).


# MURMANITE
$Na_2(Ti, Nb)_2Si_2O_9 \cdot H_2O$

CRYSTAL SYSTEM: Monoclinic
3 STRONGEST DIFFRACTION LINES:

2.88 (100)
5.77 ( 60)
3.82 ( 50)

OPTICAL CONSTANTS:

$\alpha = 1.735–1.682$
$\beta = 1.765–1.770$
$\gamma = 1.807–1.839$
$(-)2V = 50°–75°$

HARDNESS: 2–3
DENSITY: 2.76–2.84 (Meas. fresh)
2.64–2.74 (Meas. altered)
CLEAVAGE: {001} perfect
Fracture uneven. Very brittle.

HABIT: Euhedral crystals rare; as aggregates of platy crystals up to $8 \times 8 \times 3$ cm, or as small flaky and lamellar masses.

COLOR-LUSTER: Lilac to bright pink; altered varieties silvery, yellow, brown, brownish black to black. Opaque. Vitreous on cleavage; greasy on fracture.

MODE OF OCCURRENCE: Occurs in alkali pegmatites and nepheline-syenites in association with aegirine, microcline, eudialyte, sodalite, lamprophyllite, and other minerals. Found in the Khibiny and Lovozero massifs, Kola Peninsula, USSR.

BEST REF. IN ENGLISH: Vlasov, K. A., "Mineralogy of Rare Elements," v. II, p. 558–562, Israel Program for Scientific Translations, 1966.


**MUROMONTITE** = Variety of allanite containing up to 4% BeO

# MUSCOVITE
$KAl_3Si_3O_{10}(OH)_2$
Mica Group
Var. Fuchsite
    Mariposite
    Phengite
    Sericite

CRYSTAL SYSTEM: Monoclinic
CLASS: 2/m
SPACE GROUP: C2/c
Z: 4
LATTICE CONSTANTS:
  (most common polytype)
      $a = 5.19$
      $b = 9.04$
      $c = 20.08$
      $\beta = 95°30'$
3 STRONGEST DIFFRACTION LINES:
   (2M$_1$)       (3T)
  3.32 (100)    9.97 (100)
  9.95 ( 95)    3.33 (100)
  2.57 ( 55)    4.99 ( 55)
OPTICAL CONSTANTS:
     $\alpha = 1.552-1.574$
     $\beta = 1.582-1.610$
     $\gamma = 1.587-1.616$
  $(-)2V = 30°-47°$
  HARDNESS: 2½ parallel (001)
                4 perpendicular (001)
  DENSITY: 2.77-2.88 (Meas.)
  CLEAVAGE: {001} perfect
Thin laminae flexible and elastic; tough.

HABIT: Crystals usually tabular; hexagonal or diamond-shaped in cross section. Commonly lamellar or scaly massive; cryptocrystalline or compact massive (sericite); in plumose, stellate, or rounded aggregates; and as disseminated scales. Twinning on {001} common.

COLOR-LUSTER: Colorless, various shades of gray, green, yellow, brown, violet; also rose red, deep ruby red. Transparent to translucent. Vitreous to pearly or silky. Streak colorless.

MODE OF OCCURRENCE: Occurs abundantly and widespread in many geological environments, especially in granites and granitic pegmatites; in phyllites, schists, and gneisses; and in detrital or authigenic sediments. Notable occurrences are found in the Black Hills, South Dakota, and at localities in California, Colorado, New Mexico, Maine, New Hampshire, Massachusetts, Connecticut, Pennsylvania, Virginia, and North Carolina. Also in Canada, Brazil, Greenland, Norway, Sweden, Austria, Switzerland, USSR, India, and elsewhere. Crystals 3 × 4 feet in size have been mined from pegmatites in the Custer district, South Dakota; and a single crystal 10 feet in diameter and 15 feet in length, weighing 85 tons, was removed from the Inikurti mine, Nellore, India.

BEST REF. IN ENGLISH: Deer, Howie, and Zussman, "Rock Forming Minerals," v. 3, p. 11-30, New York, Wiley, 1962.

# MUSKOXITE
$Mg_7Fe_4^{3+}O_{13} \cdot 10H_2O$

CRYSTAL SYSTEM: Hexagonal (?)
LATTICE CONSTANTS:
  $a = 3.1$
  $c = $ uncertain
3 STRONGEST DIFFRACTION LINES:
  2.31 (100)
  4.6 ( 80)
  1.74 ( 50)
OPTICAL CONSTANTS:
  mean index = 1.80-1.81
     $(-)2V = 15°-40°$
  HARDNESS: $\sim 3$
  DENSITY: 3.20 Meas.
  CLEAVAGE: Perfect basal.

HABIT: Aggregates of intergrown minute crystals, less commonly as individual plates with hexagonal outline, paper-thin {001}.

COLOR-LUSTER: Dark reddish brown; vitreous; streak light orange-brown.

MODE OF OCCURRENCE: Occurs with serpentine in small fractures in drill cores from the Muskox Intrusion, Northwest Territories, Canada.

BEST REF. IN ENGLISH: Jambor, J. L., *Am. Min.*, 54: 684-696 (1969).

# MUTHMANNITE (Species status doubtful)
(Ag, Au)Te
May be mixture of Calaverite and Petzite

CRYSTAL SYSTEM: Unknown
HARDNESS: 2½
DENSITY: Unknown
CLEAVAGE: One perfect in zone of elongation.
HABIT: Crystals tabular, usually elongated in one direction.

COLOR-LUSTER: Bright brass yellow; gray-white on fresh fracture. Opaque. Metallic. Powder iron gray.

MODE OF OCCURRENCE: Occurs in association with other tellurides at Nagyag, Transylvania, Roumania.

BEST REF. IN ENGLISH: Palache, et al., "Dana's System of Mineralogy," 7th Ed., v. I, p. 260-261, New York, Wiley, 1944. Sindeeva, N. D., "Mineralogy and Types of Deposits of Selenium and Tellurium," Wiley-Interscience, 1964.

# N

## NACRITE

$Al_2Si_2O_5(OH)_4$
Kaolinite (Kandite) Group
Trimorphous with Kaolinite and Dickite

CRYSTAL SYSTEM: Monoclinic
CLASS: m
SPACE GROUP: Cc
Z: 4
LATTICE CONSTANTS:
 $a = 8.909$
 $b = 5.146$
 $c = 15.697$
 $\beta = 113°42'$
3 STRONGEST DIFFRACTION LINES:
 7.18 (100)
 4.36 ( 80)
 3.59 ( 80)
OPTICAL CONSTANTS:
 $\alpha = 1.557$
 $\beta = 1.562$
 $\gamma = 1.563$
 $(-)2V = 40°$, also $(+)2V \approx 90°$
HARDNESS: 2-2½
DENSITY: ~ 2.60 (Meas.)
 2.582 (Calc.)
CLEAVAGE: {001} perfect
Scales flexible, inelastic; masses plastic when moist.

HABIT: Crystals thin tabular, pseudohexagonal, minute. Usually massive; compact, friable or mealy. Twinning rare.

COLOR-LUSTER: Colorless, white; sometimes tinted yellowish or brownish. Satiny, pearly to dull earthy.

MODE OF OCCURRENCE: Occurs predominantly as a hydrothermal mineral in ore deposits. Found at the Eureka shaft, St. Peters Dome, El Paso County, Colorado; at the Tracey mine, Michigan; and near Freiberg, Saxony, Germany.

BEST REF. IN ENGLISH: Deer, Howie, and Zussman, "Rock Forming Minerals," v. 3, p. 194-212, New York, Wiley, 1962. Bailey, W., *Am. Min.*, **48**: 1196 (1963).

## NADORITE

$PbSbO_2Cl$

CRYSTAL SYSTEM: Orthorhombic
CLASS: 2/m 2/m 2/m
SPACE GROUP: Bmmb
Z: 4
LATTICE CONSTANTS:
 $a = 5.60$
 $b = 5.44$
 $c = 12.22$
3 STRONGEST DIFFRACTION LINES:
 2.800 (100)
 3.71 ( 30)
 1.945 ( 30)
OPTICAL CONSTANTS:
 $\alpha = 2.30$
 $\beta = 2.35$ (Li)
 $\gamma = 2.40$
 $(+)2V = $ large
HARDNESS: 3½-4
DENSITY: 7.02 (Meas.)
 7.072 (Calc. X-ray)
CLEAVAGE: {010} perfect
Brittle.

HABIT: Crystals tabular or prismatic; also lenticular with octagonal or square outline. Often in divergent or sub-parallel groups. Twinning on {101} rather common.

COLOR-LUSTER: Grayish brown, brownish yellow to yellow. Translucent. Resinous to adamantine.

MODE OF OCCURRENCE: Occurs as excellent specimens composed of thin tabular crystals, often associated with bindheimite and smithsonite, at Djebel Nador, Constantine, Algeria. It also is found at Långban, Jacobsberg, and Harstigen, Sweden; and as an alteration of jamesonite at the Bodannon mine, St. Endellion, Cornwall, England.

BEST REF. IN ENGLISH: Palache, et al., "Dana's System of Mineralogy," 7th Ed., v. II, p. 1039-1040, New York, Wiley, 1951.

**NAGATELITE** = Variety of allanite

**NAGYAGITE**
$Pb_5Au(Te, Sb)_4S_{5-8}$

CRYSTAL SYSTEM: Tetragonal
CLASS: $\bar{4}$
SPACE GROUP: $P\bar{4}$
Z: 1
LATTICE CONSTANTS:
　$a = 4.14$
　$c = 30.15$
3 STRONGEST DIFFRACTION LINES:
　3.02　(100)
　2.81　( 60)
　1.506 ( 60)
HARDNESS: 1–1½
DENSITY: 7.45–7.55 (Meas.)
　　　　　7.55 (Calc.)
CLEAVAGE: {010} perfect
Somewhat malleable; laminae flexible.
　HABIT: As thin tabular striated crystals. Usually massive; granular or foliated. Twinning on {010}.
　COLOR-LUSTER: Blackish lead gray. Opaque. Metallic, bright on fresh cleavage surfaces. Streak blackish lead gray.
　MODE OF OCCURRENCE: Occurs in hydrothermal vein deposits commonly associated with native gold, sulfides, tellurides, and carbonate minerals. Found as exceptional specimens, often associated with rhodochrosite, at Nagyag, Roumania; at Kalgoorlie, Western Australia, associated with tetrahedrite and tellurides; in the Tavua district, Vitu Levu, Fiji Islands; and at the Sylvia mine, Tararu Creek, New Zealand. Other occurrences have been reported from Japan, Hungary, Canada, North Carolina, Colorado, and California.
　BEST REF. IN ENGLISH: Palache, et al., "Dana's System of Mineralogy," 7th Ed., v. I, p. 168–170, New York, Wiley, 1944.

**NAHCOLITE**
$NaHCO_3$

CRYSTAL SYSTEM: Monoclinic
CLASS: 2/m
SPACE GROUP: $P2_1/n$
Z: 4
LATTICE CONSTANTS:
　$a = 7.51$ kX
　$b = 9.70$
　$c = 3.53$
　$\beta = 93°19'$
3 STRONGEST DIFFRACTION LINES:
　2.97 (100)
　2.60 ( 90)
　3.08 ( 25)
OPTICAL CONSTANTS:
　　$\alpha = 1.377$
　　$\beta = 1.503$
　　$\gamma = 1.583$
　$(-)2V \sim 75°$

HARDNESS: 2½
DENSITY: 2.21 (Meas.)
　　　　　2.165 (Calc.)
CLEAVAGE: {101} perfect
　　　　　{111} good
　　　　　{100} distinct
Fracture conchoidal.
　HABIT: Crystals prismatic, elongated along $c$-axis. As friable crystal aggregates; as fibrous veinlets; and as concretionary crystalline masses up to 5 feet in diameter.
　COLOR-LUSTER: Colorless, white; sometimes discolored gray to brown and black by impurities. Transparent to translucent. Vitreous to resinous.
　MODE OF OCCURRENCE: Occurs as large concretionary masses and layers in unweathered oil-shale of the Green River formation at the Bureau of Mines oil-shale mine near Rifle, Garfield County and in other parts of northwestern Colorado. It also occurs as beds up to 2 feet in thickness at various horizons in the beds at Searles Lake, San Bernardino County, and in small amounts in the muds of Deep Spring Lake, Inyo County, California. Other minor occurrences have been reported from Italy, Egypt, and near Lake Magadi, Kenya.
　BEST REF. IN ENGLISH: Palache, et al., "Dana's System of Mineralogy," 7th Ed., v. II, p. 134–135, New York, Wiley, 1951.

**NAKASÉITE** = Andorite

**NAMAQUALITE** = Cyanotrichite

**NANTOKITE**
CuCl

CRYSTAL SYSTEM: Cubic
CLASS: $\bar{4}$3m
SPACE GROUP: $F\bar{4}3m$
Z: 4
LATTICE CONSTANT:
　$a = 5.418$
3 STRONGEST DIFFRACTION LINES:
　3.12　(100)
　1.915 ( 55)
　1.633 ( 30)
OPTICAL CONSTANT:
　$N = 1.930$ (synthetic)
HARDNESS: 2½
DENSITY: 4.136 (Meas. synthetic)
　　　　　4.22 (Calc.)
CLEAVAGE: {011}
Fracture conchoidal.
　HABIT: Massive, granular.
　COLOR-LUSTER: Colorless to white; also grayish to greenish. Transparent. Adamantine.
　MODE OF OCCURRENCE: Occurs at copper mines near Nantoko, Copiapó, Chile, associated with native copper, cuprite, atacamite, and hematite, and with a similar mineral assemblage at Broken Hill, New South Wales, Australia.

BEST REF. IN ENGLISH: Palache, et al., "Dana's System of Mineralogy," 7th Ed., v. II, p. 18-19, New York, Wiley, 1951.

# NARSARSUKITE
$Na_2(Ti, Fe)Si_4(O, F)_{11}$

CRYSTAL SYSTEM: Tetragonal
CLASS: 4/m
SPACE GROUP: I4/m
Z: 4
LATTICE CONSTANTS:
  $a = 10.74$
  $c = 7.90$
3 STRONGEST DIFFRACTION LINES:
  5.37  (100)
  3.394 ( 80)
  3.260 ( 80)
OPTICAL CONSTANTS:
  $\omega = 1.604-1.609$
  $\epsilon = 1.625-1.653$
     (+)
HARDNESS: 6-7
DENSITY: 2.783 (Meas.)
CLEAVAGE:  {100} perfect
              {110} perfect
Fracture uneven. Brittle.
  HABIT: Crystals short prismatic or tabular, flattened {001}.
  COLOR-LUSTER: Colorless, yellow, green, brown. Transparent to translucent. Vitreous.
  MODE OF OCCURRENCE: Occurs as small crystals in association with aegirine and quartz at Sage Creek and Halfbreed Creek in the Sweetgrass Hills of north-central Montana. Also found in carbonatite at Mont St. Hilaire, Quebec, Canada; as fine crystals at Igdlutalik and Narsarsuk, Greenland; and on the Kola Peninsula, USSR.
  BEST REF. IN ENGLISH: Winchell and Winchell, "Elements of Optical Mineralogy," 4th Ed., Pt. 2, p. 358-359, New York, Wiley, 1951.    Peacor, Donald R., and Buerger, M. J., *Am. Min.*, **47**: 539-556 (1962).

# NASINITE (Inadequately described mineral)
$Na_4B_{10}O_{17} \cdot 7H_2O$

CRYSTAL SYSTEM: Monoclinic
3 STRONGEST DIFFRACTION LINES:
  5.27 (100)
  5.99 ( 95)
  2.90 ( 70)
OPTICAL CONSTANTS:
  $\alpha = 1.494$
  $\beta = 1.512$
  $\gamma = 1.524$
  $(-)2V = 66.8°$
HARDNESS: Not determined.
DENSITY: Not determined.
CLEAVAGE: Not determined.
HABIT: Tiny crystals and microcrystalline clusters.
MODE OF OCCURRENCE: Occurs as incrustations of a mechanically inseparable mixture of nasinite, biringuccite, thenardite, orpiment, and minor silicates on tubing of the "Hole of the Storehouse" drilled in 1927 at Lardarello, Tuscany, Italy.
  BEST REF. IN ENGLISH: Cipriani, Curzio, and Vannuccini, Piero, *Am. Min.*, **48**: 709-711 (1963).

# NASLEDOVITE (Inadequately described mineral)
$PbMn_3Al_4(CO_3)_4(SO_4)O_5 \cdot 5H_2O$

CRYSTAL SYSTEM: Unknown
3 STRONGEST DIFFRACTION LINES:
  3.261 (100)
  2.028 ( 60)
  2.019 ( 60)
OPTICAL CONSTANT:
  $\gamma = 1.591$
HARDNESS: 2
DENSITY: 3.069
  HABIT: Radial-fibrous oölites, 2.0-3.0 mm in size.
  COLOR-LUSTER: Snow white; silky. Covered by a thin film of dark brown to reddish material.
  MODE OF OCCURRENCE: Occurs in sooty pyrolusite and ocherous iron oxide with crystals of cerussite as a fissure filling in altered granodiorite porphyries that enclose polymetallic ores at Sardob, eastern part of Altyn-Topkansk ore field, Kuraminsk Mountains.
  BEST REF. IN ENGLISH: Enikeev, M. R., *Am. Min.*, **44**: 1325 (1959).

# NASONITE
$Ca_4Pb_6Si_6O_{21}Cl_2$

CRYSTAL SYSTEM: Hexagonal
CLASS: 6/m, or 6
SPACE GROUP: $P6_3/m$ or $P6_3$
Z: 2
LATTICE CONSTANTS:
  $a = 10.06$ kX
  $c = 13.24$
3 STRONGEST DIFFRACTION LINES:
  3.27 (100)
  1.81 (100)
  3.16 ( 90)
OPTICAL CONSTANTS:
  $\omega = 1.9453$
  $\epsilon = 1.9710$
     (+)
HARDNESS: 4
DENSITY: 5.55 (Meas.)
         5.63 (Calc.)
  CLEAVAGE: Basal, good. Prismatic, indistinct.
  HABIT: Crystals prismatic $\{10\bar{1}0\}$, $\{11\bar{2}0\}$, with $\{10\bar{1}1\}$ and/or {0001} as terminal forms; minute, very rare. Usually massive granular.
  COLOR-LUSTER: White; greasy to adamantine.
  MODE OF OCCURRENCE: Occurs associated with barysilite, granular datolite, white fibrous prehnite, willemite, axinite, hancockite, garnet, clinohedrite, and manganophyllite below the 800 level in pillar 910 south of the Pal-

mer shaft, and associated with glaucochroite, garnet, axinite, and barite from the Parker shaft, Franklin, New Jersey. Also found in fissures filled with calcite associated with schefferite, native lead, apophyllite, margarosanite, and thaumasite at Långban, Sweden.

BEST REF. IN ENGLISH: Palache, Charles, U.S.G.S. Prof. Paper 180, p. 92-93 (1935)    Frondel, Clifford, and Bauer, L. H., *Am. Min.*, **36**: 534-537 (1951).

# NATROALUNITE
$NaAl_3(SO_4)_2(OH)_6$
Alunite Group

CRYSTAL SYSTEM: Hexagonal
CLASS: 3m
SPACE GROUP: R3m
Z: 3
LATTICE CONSTANTS:
  $a = 6.977$
  $c = 16.74$
3 STRONGEST DIFFRACTION LINES:
  2.96 (100)
  4.90 ( 75)
  2.97 ( 70)
OPTICAL CONSTANTS:
  $\omega = 1.568$
  $\epsilon = 1.590$
  (+)
HARDNESS: 3½-4
DENSITY: 2.6-2.9 (Meas.)
CLEAVAGE: {0001} distinct
              {01$\bar{1}$2} in traces
Fracture conchoidal. Brittle.

HABIT: Crystals rhombohedral, often pseudocubic; also tabular or lenticular; as druses or aggregates. Usually massive, granular to dense compact; also columnar or fibrous.

COLOR-LUSTER: White; often discolored grayish, yellowish, reddish, brownish. Transparent to nearly opaque. Vitreous, pearly on base. Streak white.

MODE OF OCCURRENCE: Occurs widespread chiefly as a mineral formed by solfotaric processes on volcanic rocks, and less commonly in metamorphic rocks and as a vein mineral in ore deposits. Found at many localities in the western United States, especially at the Big Star deposit, Marysvale, Utah, and in California, Colorado, Arizona, and Hawaii. Other deposits are found in Chile, Western Australia, and Japan.

BEST REF. IN ENGLISH: Palache, et al., "Dana's System of Mineralogy," 7th Ed., v. II, p. 556-560, New York, Wiley, 1951.

# NATROCHALCITE
$NaCu_2(SO_4)_2OH \cdot H_2O$

CRYSTAL SYSTEM: Monoclinic
CLASS: 2/m
SPACE GROUP: I2/m
Z: 2

LATTICE CONSTANTS:
  $a = 8.33$
  $b = 6.16$
  $c = 7.44$
  $\beta = 112°54'$
3 STRONGEST DIFFRACTION LINES:
  2.80 (100)
  6.57 ( 80)
  3.44 ( 80)
OPTICAL CONSTANTS:
  $\alpha = 1.649$
  $\beta = 1.655$
  $\gamma = 1.714$
  (+)2V = 36°48'
HARDNESS: 4½
DENSITY: 3.49 (Meas.)
          3.54 (Calc.)
CLEAVAGE: {001} perfect
Fracture uneven. Brittle.

HABIT: Crystals pyramidal; also as cross-fiber veinlets and crystalline crusts.

COLOR-LUSTER: Emerald green. Transparent. Vitreous. Streak greenish white.

MODE OF OCCURRENCE: Occurs associated with gypsum, antlerite, atacamite, bloedite, brochantite, and chalcanthite, at Chuquicamata, Antofagasta province, Chile.

BEST REF. IN ENGLISH: Palache, et al., "Dana's System of Mineralogy," 7th Ed., v. II, p. 602-603, New York, Wiley, 1951.

# NATRODAVYNE = Davyne

# NATROJAROSITE
$NaFe_3^{3+}(SO_4)_2(OH)_6$
Alunite Group

CRYSTAL SYSTEM: Hexagonal
CLASS: 3m
SPACE GROUP: R3m
Z: 3 (hexag.), 1 (rhomb.)
LATTICE CONSTANTS:
  (hexag.)          (rhomb.)
  $a = 7.18$ kX     $a_{rh} = 6.835$
  $c = 16.30$       $\alpha = 63°23'$
3 STRONGEST DIFFRACTION LINES:
  5.06 (100)
  3.06 ( 80)
  3.12 ( 70)
OPTICAL CONSTANTS:
  $\omega = 1.832$
  $\epsilon = 1.750$
  (-)
HARDNESS: 3
DENSITY: 3.18 (Meas.)
          2.878 (Calc.)
CLEAVAGE: {0001} perfect
Fracture conchoidal. Brittle.

HABIT: Crystals usually minute, tabular with hexagonal outline or pseudocubic. As crusts or coatings of tiny crystals; also earthy.

COLOR-LUSTER: Ochre yellow to brown. Transparent to translucent. Vitreous; earthy material dull. Streak pale yellow.

MODE OF OCCURRENCE: Occurs as a secondary mineral in cracks and seams in ferruginous ores and rocks. Found in the United States, especially at the Buxton mine, Lawrence County, and at other places in the northern Black Hills, South Dakota; in sediments of the Gypsum Springs anticline, Carbon County, Montana; at the Georgia Sunset claim, near Kingman, Arizona; in Esmeralda County, Nevada; and at the Gold Hill mine, Tooele County, Utah. It also is found in Mexico, Chile, Denmark, Norway, USSR, and Czechoslovakia.

BEST REF. IN ENGLISH: Palache, et al., "Dana's System of Mineralogy," 7th Ed., v. II, p. 563–565, New York, Wiley, 1951.

# NATROLITE
$Na_2 Al_2 Si_3 O_{10} \cdot 2H_2 O$
Zeolite Group

CRYSTAL SYSTEM: Orthorhombic
CLASS: mm2
SPACE GROUP: Fdd2
Z: 8
LATTICE CONSTANTS:
  $a = 18.295$
  $b = 18.615$
  $c = 6.603$
3 STRONGEST DIFFRACTION LINES:
  2.85 (100)
  5.89 ( 85)
  2.87 ( 80)
OPTICAL CONSTANTS:
  $\alpha = 1.473-1.483$
  $\beta = 1.476-1.486$
  $\gamma = 1.485-1.496$
  $(+)2V = 58°-64°$
HARDNESS: 5-5½
DENSITY: 2.20-2.26 (Meas.)
         2.215 (Calc.)
CLEAVAGE: {110} perfect
          {010} parting
Fracture uneven. Brittle.

HABIT: Crystals prismatic, slender to acicular, vertically striated; also short prismatic. Commonly fibrous, radiating; also massive, granular or compact. Twinning on {110}, {011}, {031} rare.

COLOR-LUSTER: Colorless, white, gray, yellowish, reddish. Transparent to translucent. Vitreous to somewhat pearly.

MODE OF OCCURRENCE: Occurs chiefly in cavities in basalt and related rocks; also as an alteration product of nepheline or sodalite in nepheline-syenites, phonolites, and related rocks; and as an alteration product of plagioclase in some aplites and dolerites. Found as fine crystals in San Benito County and at other places in California; near Springfield, Lane County, and near Medford, Jackson County, Oregon; on the Olympic Peninsula, Washington; near Livingston, Montana; at Table Mountain, Jefferson County, Colorado; in Montgomery County, Pennsylvania; and in the trap rocks of

northern New Jersey. Also found in Canada as fine crystals at Mont St. Hilaire, and as crystals up to 3 feet long and 4 inches across at an asbestos mine in Quebec; at many localities in the Bay of Fundy district, Nova Scotia; and in the Ice Valley region of British Columbia. Other notable occurrences are found in Greenland, Iceland, Faroe Islands, Ireland, Scotland, Norway, France, Germany, Czechoslovakia, and elsewhere.

BEST REF. IN ENGLISH: Deer, Howie, and Zussman, "Rock Forming Minerals," v. 4, p. 358–376, New York, Wiley, 1963.

# NATROMONTEBRASITE (Fremontite)
$(Na, Li)Al(PO_4)(OH, F)$
Amblygonite Series

CRYSTAL SYSTEM: Triclinic
CLASS: $\bar{1}$
SPACE GROUP: $P\bar{1}$
Z: 2
LATTICE CONSTANTS:
  $a = 5.18$ kX      $\alpha = 112°02.5'$
  $b = 7.11$         $\beta = 97°49.5'$
  $c = 5.03$         $\gamma = 68°07.5'$
OPTICAL CONSTANTS:
  $\alpha = 1.594$
  $\beta = 1.603$
  $\gamma = 1.615$
  $(+)2V$ = very large
HARDNESS: 5½-6
DENSITY: 3.04-3.1 (Meas.)
CLEAVAGE: {100} perfect
          {110} good
          {0$\bar{1}$1} distinct
          {001} imperfect
Fracture subconchoidal to uneven. Brittle.

HABIT: Crystals short prismatic, rough; also massive.

COLOR-LUSTER: Grayish white to white. Translucent. Vitreous.

MODE OF OCCURRENCE: Occurs in granite pegmatite associated with pink tourmaline and lepidolite at Eight Mile Park, Fremont County, Colorado. Also found at the Strickland quarry, Portland, Connecticut, and at localities in Czechoslovakia and Madagascar.

BEST REF. IN ENGLISH: Palache, et al., "Dana's System of Mineralogy," 7th Ed., v. II, p. 823–827, New York, Wiley, 1951.

# NATRON
$Na_2 CO_3 \cdot 10H_2 O$

CRYSTAL SYSTEM: Monoclinic
CLASS: 2/m
SPACE GROUP: I2/a
Z: 4
LATTICE CONSTANTS:
  $a = 12.57$
  $b = 9.01$
  $c = 13.47$
  $\beta = 121°26'$

3 STRONGEST DIFFRACTION LINES:
3.036 (100)
3.015 ( 70)
2.894 ( 60)
OPTICAL CONSTANTS:
$\alpha$ = 1.405
$\beta$ = 1.425
$\gamma$ = 1.440
(−)2V = large
HARDNESS: 1–1½
DENSITY: 1.44 (Meas.)
1.46 (Calc.)
CLEAVAGE: {001} distinct
{010} imperfect
{110} in traces
Fracture conchoidal. Brittle.
HABIT: As an efflorescence, and as crusts and coatings. Synthetic crystals tabular.
COLOR-LUSTER: Colorless, white; often yellowish or grayish due to impurities. Crystals vitreous.

Taste alkaline.
MODE OF OCCURRENCE: Occurs most commonly in solution in soda lakes, as in the water of Owens Lake and other soda lakes in Inyo County, and at Searles Lake, San Bernardino County, California. It is also found at Ragtown, Nevada; in the soda lakes of Wady Natrum, Egypt; as an efflorescence on lava at Mt. Etna, Sicily and on Mt. Vesuvius, Italy; in the Gobi Desert, Mongolia; and at other localities.
BEST REF. IN ENGLISH: Palache, et al., "Dana's System of Mineralogy," 7th Ed., v. II, p. 230-231, New York, Wiley, 1951.

## NATRONBIOTITE = Variety of biotite containing sodium

## NATRONIOBITE (Inadequately described mineral)
$NaNbO_3$
Dimorphous with lueshite

CRYSTAL SYSTEM: Probably monoclinic
3 STRONGEST DIFFRACTION LINES:
2.97 (100)
3.06 ( 90)
1.60 ( 80)
OPTICAL CONSTANTS:
$\alpha$ = 2.10-2.13
$\beta$ = 2.19-2.21
$\gamma$ = 2.21-2.24
(+)2V = 10°-30°
HARDNESS: 5½-6
DENSITY: 4.40 (Meas.)
CLEAVAGE: Not determined.
HABIT: As irregular grains and fine-grained aggregates.
COLOR-LUSTER: Yellowish, brownish, blackish.
MODE OF OCCURRENCE: Occurs in dolomite carbonatites containing apatite, phlogopite, dysanalyte, and pyrochlore, in the Lesnaya Baraka and Sallanlatvi massifs, Kola Peninsula, USSR.
BEST REF. IN ENGLISH: Bulakh, A. G., Kukharenko, A. A., Knipovich, Yu. N., Kondrat'eva, V. V., Baklanova, K. A., and Baranova, E. N., *Am. Min.*, 47: 1483 (1962).

## NATROPHILITE
$NaMnPO_4$

CRYSTAL SYSTEM: Orthorhombic
CLASS: 2/m 2/m 2/m
SPACE GROUP: Pnam
Z: 4
LATTICE CONSTANTS:
$a$ = 10.53
$b$ = 5.00
$c$ = 6.29
3 STRONGEST DIFFRACTION LINES:
(high natrophilite)
2.72 (100)
2.60 ( 80)
3.72 ( 70)
OPTICAL CONSTANTS:
$\alpha$ = 1.671
$\beta$ = 1.674
$\gamma$ = 1.684
(+)2V ~ 75°
HARDNESS: 4½-5
DENSITY: 3.41 (Meas.)
3.47 (Calc.)
CLEAVAGE: {100} good
{010} indistinct
{021} interrupted
Fracture conchoidal.
HABIT: Crystals rare, stout prismatic. Usually granular or as cleavable masses.
COLOR-LUSTER: Deep wine yellow. Transparent to translucent. Bright resinous, pearly on good cleavage.
MODE OF OCCURRENCE: Occurs sparingly in granite pegmatite at Branchville, Connecticut, associated with lithiophilite and secondary phosphate minerals.
BEST REF. IN ENGLISH: Palache, et al., "Dana's System of Mineralogy," 7th Ed., v. II, p. 670-671, New York, Wiley, 1951. Fisher, Jerome, *Am. Min.*, 50: 1096-1097 (1965).

## NATROPHOSPHATE
$Na_6H(PO_4)_2F \cdot 17H_2O$

CRYSTAL SYSTEM: Cubic
SPACE GROUP: Fd3c
Z: 56
LATTICE CONSTANT:
$a$ = 27.79
3 STRONGEST DIFFRACTION LINES:
2.67 (100)
2.42 ( 90)
8.18 ( 70)
OPTICAL CONSTANTS: Constantly isotropic with $n$1.460-1.462.
HARDNESS: ~2½
DENSITY: 1.71-1.722 (Meas.)
CLEAVAGE: {111} imperfect
Fracture conchoidal.

HABIT: Massive; as dense monomineralic aggregates of irregular form, up to 5 × 3 cm in size.

COLOR-LUSTER: Colorless. Transparent. Vitreous to greasy. On exposure to air, surface alters to a secondary powdery coating. Fluoresces weak orange under ultraviolet radiation.

MODE OF OCCURRENCE: Occurs in the central cavernous zone of a pegmatite of Yukspore Mt., Khibina massif, USSR. It is found in close association with fibrous aegirine, delhayelite, strontian apatite, natrolite, pectolite, and villiaumite in cavities, the walls of which are covered by druses of small crystals of albite, aegirine, and rarely villiaumite.

BEST REF. IN ENGLISH: Kapustin, Yu. L., Bykova, A. V., and Bukin, V. I., *Int. Geol. Rev.*, **14**: 984–989 (1972).

# NAUJAKASITE

$(Na, K)_6 (Fe, Mn, Ca)(Al, Fe)_4 Si_8 O_{26} \cdot H_2 O$

CRYSTAL SYSTEM: Monoclinic
CLASS: 2/m
SPACE GROUP: C2/m (?)
Z: 2
LATTICE CONSTANTS:
$a = 15.039$
$b = 7.991$
$c = 10.487$
$\beta = 113°40'$
3 STRONGEST DIFFRACTION LINES:
3.99 (100)
3.56 ( 70)
2.26 ( 70)
OPTICAL CONSTANTS:
$\alpha = 1.537$
$\beta = 1.550$
$\gamma = 1.556$
$(-)2V = 52°–71°$
HARDNESS: 2–3
DENSITY: 2.622 (Meas.)
2.661 (Calc.)
CLEAVAGE: {001} micaceous
{$\bar{4}$01} distinct
{010} distinct
Brittle.

HABIT: Crystals platy; diameter 1–5 mm, thickness less than 1 mm.

COLOR-LUSTER: Gray to silver white; pearly on basal plane.

MODE OF OCCURRENCE: Found in a single specimen as an aggregate of mica-like plates, intimately associated with arfvedsonite, from Naujakasik, Tunugdliarfik Fjord, southwestern Greenland.

BEST REF. IN ENGLISH: Petersen, Ole V., *Am. Min.*, **53**: 1780 (1968). Bøggild, O. B., *Am. Min.*, **20**: 138 (1935).

# NAUMANNITE

$\beta\text{-}Ag_2 Se$

CRYSTAL SYSTEM: Orthorhombic (cubic above 133°)
Z: 4

LATTICE CONSTANTS:
$a = 4.336$
$b = 7.067$
$c = 7.753$
3 STRONGEST DIFFRACTION LINES:
2.66 (100)
2.56 (100)
2.23 ( 60)
HARDNESS: 2½
DENSITY: 7.69–7.79 (Meas.)
7.866 (Calc.)
CLEAVAGE: None observed.
Poor hackly fracture. Sectile and malleable.

HABIT: In cubes; also massive granular and as thin plates.

COLOR-LUSTER: Grayish iron black tarnishing in time iridescent brownish; opaque; metallic. Streak iron black.

MODE OF OCCURRENCE: Occurs associated with argentian tetrahedrite, gold, electrum, pyrite, and chalcopyrite as thin bands in milky quartz at Republic, Washington; formerly found in abundance at the De Lamar mine, Silver City district, Owyhee County, Idaho. Also found with chalcopyrite, clausthalite and other selenides in carbonate gangue at Tilkerode, Harz mountains, Germany; and at Cerro de Cacheuta, Mendoza, Argentina.

BEST REF. IN ENGLISH: Palache, et al., "Dana's System of Mineralogy," 7th Ed., v. I, p. 179–180, New York, Wiley, 1944. Earley, J. W., *Am. Min.*, **35**: 337–364 (1950). Sindeeva, N. D., "Mineralogy and Types of Deposits of Selenium and Tellurium," Wiley-Interscience, 1964.

# NAVAJOITE

$V_2 O_5 \cdot 3H_2 O$

CRYSTAL SYSTEM: Probably monoclinic
Z: 6
LATTICE CONSTANTS:
$a = 17.43$
$b = 3.65$
$c = 12.25$
$\beta = 97°30'$
3 STRONGEST DIFFRACTION LINES:
12.1 (100)
10.6 ( 60)
2.90 ( 50)
OPTICAL CONSTANTS:
$\alpha = 1.905$
$\beta = 2.02$
$\gamma > 2.02$
HARDNESS: <2
DENSITY: 2.56 (Meas.)
CLEAVAGE: Not observed.

HABIT: Fibrous; as a coating; and as thin well-developed laths up to several microns in length. Laths and fibers are elongated [010] and flattened {001}.

COLOR-LUSTER: Dark brown; adamantine on freshly broken surfaces. Streak brown.

MODE OF OCCURRENCE: Occurs as a coating around pebbles and sand grains and as thin seams in sandstone and clay lenses in a vanadium-uranium deposit in the Shinarump conglomerate (Triassic) at the Monument No. 2 mine, Navajo Indian Reservation, Apache County, Arizona.

BEST REF. IN ENGLISH: Weeks, Alice D., Thompson, Mary E., and Sherwood, Alexander M., *Am. Min.*, **40**: 207–212 (1955). Ross, Malcolm, *Am. Min.*, **44**: 322–341 (1959).

## NEIGHBORITE
$NaMgF_3$

CRYSTAL SYSTEM: Orthorhombic
CLASS: 2/m 2/m 2/m
SPACE GROUP: Pcmn
Z: 4
LATTICE CONSTANTS:
 $a$ = 5.363
 $b$ = 7.676
 $c$ = 5.503
3 STRONGEST DIFFRACTION LINES:
 1.918 (100)
 2.71 ( 50)
 3.83 ( 35)
OPTICAL CONSTANTS:
 Mean index = 1.364 (Na)
 HARDNESS: 4.5
 DENSITY: 3.03 (Meas.)
      3.06 (Calc.)
 CLEAVAGE: No distinct cleavage. Fracture uneven.
 HABIT: As twinned octahedral crystals 0.1–0.5 mm in diameter; also as rounded grains.
 COLOR-LUSTER: Colorless to pink; transparent to translucent. Vitreous.
 MODE OF OCCURRENCE: Found associated with dolomite and quartz, and the accessory minerals burbankite, nahcolite, barytocalcite, garrelsite, wurtzite, calcite, and pyrite in the dolomitic oil shale of the Eocene Green River formation of South Ouray, Uintah County, Utah.
 BEST REF. IN ENGLISH: Chao, E. C. T., Evans, Howard T. Jr., Skinner, Brian J., and Milton, Charles, *Am. Min.*, **46**: 379–393 (1961).

## NEKOITE
$CaSi_2O_5 \cdot 2H_2O$

CRYSTAL SYSTEM: Triclinic
Z: 3
LATTICE CONSTANTS:
 $a$ = 7.60   $\alpha$ = 111°48′
 $b$ = 7.32   $\beta$ = 86°12′
 $c$ = 9.86   $\gamma$ = 103°54′
3 STRONGEST DIFFRACTION LINES:
 9.25 ( 100)
 3.36 (100b)
 2.82 ( 100)
OPTICAL CONSTANTS:
 Mean index = 1.535
      (+)2V ≈ 70°
HARDNESS: Not determined.
DENSITY: 2.206 (Meas.)
      2.28 (Calc.)

CLEAVAGE: {100} distinct
HABIT: Crystals needle-like showing repeated twinning with lamellae parallel to cleavage. Also finely fibrous.
COLOR-LUSTER: White. Pearly.
MODE OF OCCURRENCE: Occurs in association with wilkeite in the limestone at Crestmore, Riverside County, California.
BEST REF. IN ENGLISH: Gard, J. A., and Taylor, H. F. W. *Min. Mag.*, **31**: 5–20 (1956).

## NEMALITE = Ferroan brucite

## NENADKEVICHITE
$(Na, Ca)(Nb, Ti)Si_2O_7 \cdot 2H_2O$

CRYSTAL SYSTEM: Probably orthorhombic
3 STRONGEST DIFFRACTION LINES:
 3.20 (100)
 3.10 (100)
 1.427 (100)
OPTICAL CONSTANTS:
 $\alpha$ = 1.659
 $\beta$ = 1.686
 $\gamma$ = 1.785
 2V = 46°
 HARDNESS: ~5
 DENSITY: 2.838 (Meas. brown)
      2.885 (Meas. rose)
 CLEAVAGE: {001} poor
Fracture uneven.
 HABIT: As foliated masses from a few millimeters to 4 × 2.5 × 0.4 cm in size.
 COLOR-LUSTER: Dark brown, brown to rose; dull. Opaque. Streak pale rose.
 MODE OF OCCURRENCE: Occurs in alkali pegmatites of the Lovozero massif, Kola Peninsula, USSR, in association with microcline, aegirine, and eudialyte.
 BEST REF. IN ENGLISH: Kuz'menko, M. V., and Kazakova, M. E., *Am. Min.*, **40**: 1154 (1955).

## NEOMESSELITE = Messelite

## NEOTOCITE (Inadequately described mineral)
Near $Mn_2Fe_2Si_4O_{13} \cdot 6H_2O(?)$

CRYSTAL SYSTEM: Unknown
3 STRONGEST DIFFRACTION LINES:
 4.36 (100b)
 3.59 (100b)
 1.54 (100b)
HARDNESS: 3–4
DENSITY: 2.43 (Meas.)
CLEAVAGE: None. Fracture conchoidal. Brittle.
HABIT: Massive; compact.
COLOR-LUSTER: Black, dark brown. Nearly opaque. Resinous.
MODE OF OCCURRENCE: Occurs as an alteration product of rhodonite and other manganese silicates. Typical

occurrences are found at the Montreal mine, Gogebic range, Iron County, Wisconsin, and on Salt Spring Island, British Columbia, Canada.

BEST REF. IN ENGLISH: Whelan, J. A., and Goldich, S. S., *Am. Min.*, **46**: 1412–1423 (1961).

## NEPHELINE (Nephelite)

(Na, K)AlSiO$_4$

CRYSTAL SYSTEM: Hexagonal
CLASS: 6
SPACE GROUP: P6$_3$
Z: 8
LATTICE CONSTANTS:
 $a \simeq 10.0$
 $c \simeq 8.4$
3 STRONGEST DIFFRACTION LINES:
 3.027 (100)
 3.870 ( 60)
 3.294 ( 40)
OPTICAL CONSTANTS:
 $\omega$ = 1.529–1.546
 $\epsilon$ = 1.526–1.542
 (–)
HARDNESS: 5½–6
DENSITY: 2.55–2.665 (Meas.)
 2.653 (Calc.)
CLEAVAGE: {10$\bar{1}$0} indistinct
 {0001} indistinct
Fracture subconchoidal. Brittle.

HABIT: Crystals usually simple hexagonal prisms, often with rough faces. Commonly massive, compact; or as embedded grains. Twinning on {10$\bar{1}$0}, {33$\bar{6}$5}, {11$\bar{2}$2}.

COLOR-LUSTER: Colorless, white, gray, yellowish, greenish or bluish gray, dark green, brownish red, brick red. Transparent to nearly opaque. Vitreous to greasy. Streak white.

MODE OF OCCURRENCE: Occurs chiefly in plutonic and volcanic rocks, and in pegmatites associated with nepheline syenites. Found in Lawrence County, South Dakota; in the Highwood, Bearpaw, and Judith Mountains, Montana; and at localities in Colorado, Texas, Arkansas, Massachusetts, New Hampshire, Maine, and elsewhere. Notable occurrences also are found in Ontario and other provinces of Canada; in the Julianehaab district, Greenland; in the Langesundfjord area, Norway; and at localities in Scotland, Italy, Roumania, Finland, USSR, Burma, Korea, New Zealand, Cameroon, Zaire, Kenya, Transvaal, and many other places.

BEST REF. IN ENGLISH: Deer, Howie, and Zussman, "Rock Forming Minerals," v. 4, p. 231–270, New York, Wiley, 1963.

## NEPHRITE = Compact variety of actinolite or tremolite

## NEPOUITE = Nickel-antigorite

Ni$_6$Si$_4$O$_{10}$(OH)$_8$

## NEPTUNITE

(Na, K)$_2$(Fe$^{2+}$, Mn)TiSi$_4$O$_{12}$
Mangan-neptunite (Mn > Fe$^{2+}$)

CRYSTAL SYSTEM: Monoclinic
CLASS: m
SPACE GROUP: Cc
Z: 8
LATTICE CONSTANTS:
 $a$ = 16.46
 $b$ = 12.50
 $c$ = 10.01
 $\beta$ = 115°26′
3 STRONGEST DIFFRACTION LINES:
 3.186 (100)
 9.6 ( 60)
 3.517 ( 45)
OPTICAL CONSTANTS:
 $\alpha$ = 1.690–1.6908
 $\beta$ = 1.6927–1.700
 $\gamma$ = 1.7194–1.736
 (+)2V = 49°, 36°
HARDNESS: 5–6
DENSITY: 3.19–3.23 (Meas.)
 3.24 (Calc.)
CLEAVAGE: {110} perfect
Fracture conchoidal. Brittle.

HABIT: Crystals prismatic, usually square in cross section.

COLOR-LUSTER: Black; often with deep reddish brown internal reflections. Nearly opaque. Vitreous. Streak reddish brown.

MODE OF OCCURRENCE: Occurs as superb crystals up to 6 cm long in association with benitoite, joaquinite and natrolite at the benitoite gem mine near the headwaters of the San Benito River, San Benito County, California. Also found in association with eudialyte in nepheline-syenite in the Julianehaab district, Greenland. Mangan-neptunite occurs in carbonatite at Mont St. Hilaire, Quebec, Canada, and on the Kola Peninsula, USSR.

BEST REF. IN ENGLISH: Winchell and Winchell, "Elements of Optical Mineralogy," 4th Ed., Pt. 2, p. 463, New York, Wiley, 1951. Berry, L. G., *Can. Min.*, **7**: 679 (1963).

## NESQUEHONITE

MgCO$_3$ · 3H$_2$O

CRYSTAL SYSTEM: Monoclinic
CLASS: 2/m
SPACE GROUP: P2$_1$/n
Z: 4
LATTICE CONSTANTS:
 $a$ = 12.00
 $b$ = 5.39
 $c$ = 7.68
 $\beta$ = 90.45°
3 STRONGEST DIFFRACTION LINES:
 6.5 (100)
 3.86 ( 90)
 2.61 ( 70)
OPTICAL CONSTANTS:
 $\alpha$ = 1.417 (Calc.)
 $\beta$ = 1.503
 $\gamma$ = 1.527 (Na)
 (–)2V = 53°

HARDNESS: 2½
DENSITY: 1.852 (Meas.)
1.856 (Calc.)
CLEAVAGE: {110} perfect
{001} distinct
Fracture splintery to fibrous.

HABIT: Crystals prismatic, deeply striated parallel to *c*-axis. As tufts of radiating needle-like crystals, and as flattened radiated coatings; also as felted aggregates and botryoidal.

COLOR-LUSTER: Colorless, white. Transparent to translucent. Vitreous.

MODE OF OCCURRENCE: Occurs as a recent product formed under normal atmospheric conditions of temperature and pressure. It is found at Nesquehoning near Lansford, Pennsylvania; with hydromagnesite near the Florence Mack mercury mine, south of New Idria, San Benito County, California; and in Italy, France, and Austria.

BEST REF. IN ENGLISH: Palache, et al., "Dana's System of Mineralogy," 7th Ed., v. II, p. 225–227, New York, Wiley, 1951.

# NEWBERYITE
$MgHPO_4 \cdot 3H_2O$

CRYSTAL SYSTEM: Orthorhombic
CLASS: 2/m 2/m 2/m
SPACE GROUP: Pbca
Z: 8
LATTICE CONSTANTS:
$a = 10.215$
$b = 10.681$
$c = 10.014$
3 STRONGEST DIFFRACTION LINES:

| (synthetic) | (natural) |
|---|---|
| 3.45 (100) | 4.71 (100) |
| 3.05 ( 80) | 5.94 ( 99) |
| 5.9 ( 40) | 3.46 ( 94) |

OPTICAL CONSTANTS:
$\alpha = 1.514$
$\beta = 1.517$
$\gamma = 1.533$
$(+)2V = 44°46'$
HARDNESS: 3–3½
DENSITY: 2.10 (Meas.)
CLEAVAGE: {010} perfect
{001} imperfect
Fracture uneven to subconchoidal. Brittle.

HABIT: Crystals short prismatic, tabular, or equant.

COLOR-LUSTER: Colorless. Often grayish or brownish due to impurities. Transparent. Vitreous.

MODE OF OCCURRENCE: Occurs as crystals up to one inch in size associated with struvite and hannayite in bat guano in the Skipton Caves, Ballarat, Victoria, Australia; at Mejillones, Chile; on Ascension Island in the South Atlantic; and on La Reunion Island, Indian Ocean.

BEST REF. IN ENGLISH: Palache, et al., "Dana's System of Mineralogy," 7th Ed., v. II, p. 709–711, New York, Wiley, 1951.

# NEYITE
$Pb_7(Cu, Ag)_2 Bi_6 S_{17}$

CRYSTAL SYSTEM: Monoclinic
CLASS: 2/m
SPACE GROUP: C2/m
Z: 8
LATTICE CONSTANTS:
$a = 37.5$
$b = 4.07$
$c = 41.6$
$\beta = 96.8°$
3 STRONGEST DIFFRACTION LINES:
3.72 (100)
3.51 (100)
2.92 (100)
HARDNESS: 2.5
DENSITY: 7.02 (Meas.)
7.16 (Calc.)
CLEAVAGE: None.
Fracture flat conchoidal. Very brittle.

HABIT: As aggregates of prismatic to bladed crystals with stepped surfaces on the broad faces. Individual crystals have dimensions of 0.1 × 0.1 × 3.0 mm.

COLOR-LUSTER: Lead gray. Bright metallic. Opaque. Tarnishes yellow, then Prussian blue.

MODE OF OCCURRENCE: Occurs in late vuggy quartz veins with pyrite, galena, sphalerite, chalcopyrite, aikinite, cosalite, and tetrahedrite at the Lime Creek molybdenum deposit near Alice Arm, British Columbia, Canada.

BEST REF. IN ENGLISH: Drummond, A. D., Trotter, J., Thompson, R. M., and Gower, J. A., *Can. Min.*, **10**(pt. 1): 90–96 (1969).

# NICCOLITE (Nickeline, Niccoline)
NiAs

CRYSTAL SYSTEM: Hexagonal
CLASS: 6/m 2/m 2/m
SPACE GROUP: $P6_3/mmc$
Z: 2
LATTICE CONSTANTS:
$a = 3.602$
$c = 5.009$
3 STRONGEST DIFFRACTION LINES:
2.66 (100)
1.961 ( 90)
1.811 ( 80)
OPTICAL CONSTANTS:
$\omega = 2.11$
$\epsilon = 1.80$ $(\lambda = 589 \text{ m}\mu)$
(–)
HARDNESS: 5–5½
DENSITY: 7.784 (Meas.)
7.834 (Calc.)
CLEAVAGE: None. Fracture uneven. Brittle.

HABIT: Crystals rare, small, pyramidal. Usually massive, reniform, with columnar structure, or disseminated. Twinning on $\{10\bar{1}1\}$, as fourlings.

COLOR-LUSTER: Pale copper red; tarnishes grayish to blackish. Opaque. Metallic. Streak pale brownish black.

MODE OF OCCURRENCE: Occurs in vein deposits, and in norites or deposits derived from them, commonly associated with silver, cobalt, or nickel ores. Found sparingly in the United States at the Stanislaus mine, Calaveras County, and at Long Lake, Inyo County, California; in the Copper King mine, Boulder County, and at the Gem Mine, Fremont County, Colorado; and at Franklin, New Jersey. It occurs abundantly in the Cobalt, Gowganda, Sudbury, and Thunder Bay districts, Ontario, Canada; in Sonora, Mexico; massive and as small crystals at Eisleben and elsewhere in Germany; and at localities in France, Austria, Czechoslovakia, and Japan.

BEST REF. IN ENGLISH: Palache, et al., "Dana's System of Mineralogy," 7th Ed., v. I, p. 236–238, New York, Wiley, 1944.

# NICKEL
Ni

CRYSTAL SYSTEM: Cubic
CLASS: $4/m\ \bar{3}\ 2/m$
SPACE GROUP: Fm3m
Z: 4
LATTICE CONSTANT:
  $a = 3.524$
3 STRONGEST DIFFRACTION LINES:
  2.034 (100)
  1.762 ( 42)
  1.246 ( 21)
DENSITY: 8.91 (Calc.)
HABIT: As euhedral cubes or intergrown cubes up to 0.1 mm in size, sometimes twinned on (111).
COLOR-LUSTER: Grayish white. Opaque. Metallic.
MODE OF OCCURRENCE: Occurs enclosed in heazlewoodite from Bogota, near Canala, New Caledonia.
BEST REF. IN ENGLISH: Ramdohr, Paul, Am. Min., 53: 348 (1968).

# NICKELHEXAHYDRITE
$NiSO_4 \cdot 6H_2O$

CRYSTAL SYSTEM: Monoclinic
CLASS: 2/m
SPACE GROUP: C2/c
Z: 8
LATTICE CONSTANTS:
  $a = 9.84$
  $b = 7.17$  (Synthetic)
  $c = 24.0$
  $\beta = 97.5°$
3 STRONGEST DIFFRACTION LINES:
  4.35 (100)
  3.98 ( 90)
  2.89 ( 90)
OPTICAL CONSTANTS:
  $\alpha = 1.469$–$1.470$
  $\gamma = 1.493$–$1.494$

HARDNESS: Not determined.
DENSITY: 2.07 (Calc.)
CLEAVAGE: {010} perfect
          {100} good
HABIT: As crusts and coatings.
COLOR-LUSTER: Bluish green; vitreous.
MODE OF OCCURRENCE: Occurs in the mine pit of the Severnaya mine, Noril'sk deposits, USSR.
BEST REF. IN ENGLISH: Oleinikov, B. V., Shvartsev, S. L., Mandrikova, N. T., and Oleinikova, N. N., Am. Min., 51: 529–530 (1966).

# NICKELINE = Niccolite

# NICKEL-IRON (Josephinite, Awaruite)
Fe, Ni

CRYSTAL SYSTEM: Cubic
CLASS: $4/m\ \bar{3}\ 2/m$
SPACE GROUP: Fm3m
Z: 4
LATTICE CONSTANT:
  $a = 3.560$
3 STRONGEST DIFFRACTION LINES:
  2.06 (100)
  1.78 ( 30)
  1.26 ( 20)
HARDNESS: 5
DENSITY: 7.8–8.22 (Meas.)
CLEAVAGE: None. Malleable; flexible.
HABIT: As fine scales, grains, and rounded masses sometimes exceeding 100 pounds in weight. Also in meteorites as intergrowths with kamacite.
COLOR-LUSTER: Silver white to grayish white. Opaque. Metallic.

Strongly magnetic.
MODE OF OCCURRENCE: Occurs in large masses in Jackson and Josephine counties, Oregon; in Del Norte County, California; on the Fraser River, British Columbia, Canada; and at localities in Italy, USSR, and New Zealand.
BEST REF. IN ENGLISH: Palache, et al., "Dana's System of Mineralogy," 7th Ed., v. I, p. 117–119, New York, Wiley, 1944.

# NICKEL SAPONITE = Pimelite

# NICKEL-SKUTTERUDITE (Chloanthite)
$(Ni, Co)As_3$

CRYSTAL SYSTEM: Cubic
CLASS: $2/m\ \bar{3}$
SPACE GROUP: Im3
Z: 8
LATTICE CONSTANT:
  $a \simeq 8.2$
HARDNESS: 5½–6

DENSITY: 6.5 ± 0.4
CLEAVAGE: {001} distinct
{111} distinct
{011} in traces
Fracture uneven to conchoidal. Brittle.

HABIT: Crystals cubic, octahedral, or cubo-octahedral; sometimes with dodecahedral or pyritohedral modifications. In reticulated shapes or distorted aggregates of crystals. Commonly massive, fine granular.

COLOR-LUSTER: Silver gray to tin white; occasionally tarnished iridescent or gray. Opaque. Metallic. Streak black.

MODE OF OCCURRENCE: Occurs in medium- to high-temperature hydrothermal veins, commonly associated with other cobalt and nickel minerals, arsenopyrite, native silver and bismuth, and silver sulfosalts. Found in the Bullards Peak district near Silver City, Grant County, New Mexico; in the Cobalt and other districts in Canada; near Ste. Marie aux Mines, Alsace-Lorraine, France; and at Schneeberg and elsewhere in Saxony, Germany.

BEST REF. IN ENGLISH: Palache, et al., "Dana's System of Mineralogy," 7th Ed., v. I, p. 342–346, New York, Wiley, 1944.

## NIFONTOVITE (Inadequately described mineral)
$CaB_2O_4 \cdot 3H_2O$

CRYSTAL SYSTEM: Monoclinic or triclinic
3 STRONGEST DIFFRACTION LINES:
2.41 (100)
7.04 ( 80)
2.21 ( 80)
OPTICAL CONSTANTS:
$\alpha = 1.575$
$\beta = 1.578$
$\gamma = 1.584$
(+)2V = 76°
HARDNESS: 3.5
DENSITY: 2.36
CLEAVAGE: Poor parallel to elongation.
HABIT: Massive.
COLOR-LUSTER: Colorless; vitreous. Fluoresces violet under long-wave ultraviolet light.

MODE OF OCCURRENCE: Occurs associated with andradite-grossularite garnet and szaibelyite in a skarn formed by the intrusion of quartz diorite into limestone in the Tur'insk region, northern Urals.

BEST REF. IN ENGLISH: Malinko, S. V., and Lisitsyn, A. E., *Am. Min.*, 47: 172 (1962).

## NIGERITE
$(Zn, Fe^{2+}, Mg)(Sn, Zn)_2(Al, Fe^{3+})_{12}O_{22}(OH)_2$

CRYSTAL SYSTEM: Hexagonal
CLASS: $\bar{3}$ 2/m or 3m
SPACE GROUP: P$\bar{3}$m or P31m
Z: 3
LATTICE CONSTANTS:
$a = 5.72$
$c = 13.86$

3 STRONGEST DIFFRACTION LINES:
2.42 (100)
2.85 ( 90)
1.55 ( 60)
OPTICAL CONSTANTS:
$\omega = 1.80$
$\epsilon = 1.81$
(+)
HARDNESS: 8½
DENSITY: 4.51 (Meas.)
CLEAVAGE: None. Brittle.
HABIT: Crystals hexagonal plates up to 5mm or more in diameter.
COLOR-LUSTER: Lustrous brown. Nearly opaque.

Weakly magnetic.

MODE OF OCCURRENCE: Occurs in association with cassiterite, columbite, gahnite, andalusite, sillimanite, quartz, and chrysoberyl in quartz-sillimanite rocks associated with tin-bearing pegmatites in the Egba district, Kabba Province, central Nigeria.

BEST REF. IN ENGLISH: *Min. Mag.*, 28: 118–136 (1947). *Am. Min.*, 52: 864–866 (1967).

## NIGGLIITE
PtSn

CRYSTAL SYSTEM: Hexagonal
CLASS: 6/m 2/m 2/m
SPACE GROUP: P6$_3$/mmc
Z: 2
LATTICE CONSTANTS:
$a = 4.111$
$c = 5.446$
3 STRONGEST DIFFRACTION LINES:
2.15 (100)
2.05 (100)
1.486 (100)
HARDNESS: 3
DENSITY: 13.44 (Calc.)
CLEAVAGE: None. Brittle.
HABIT: As minute grains.
COLOR-LUSTER: Silver white. Opaque. Metallic.

MODE OF OCCURRENCE: Found associated with cubanite, galena, chalcopyrite, and parkerite in concentrates of the oxidized ore from the mine dumps at Waterfall Gorge, Insizwa, South Africa. Also in association with stannopalladinite, hessite, and tellurides of platinum and palladium at Monchegorsk, USSR.

BEST REF. IN ENGLISH: Palache, et al., "Dana's System of Mineralogy," 7th Ed., v. I, p. 347, New York, Wiley, 1944. Groeneveld Meijer, W. O. J., *Am. Min.*, 40: 693–696 (1955).

## NIGRIN or NIGRINE = Variety of rutile containing up to ~ 11% Fe in substitution for Ti

## NIMITE
$(Ni, Mg, Fe, Al)_6(AlSi_3)O_{10}(OH)_8$
Chlorite Group

CRYSTAL SYSTEM: Monoclinic
CLASS: 2/m
SPACE GROUP: C2/m
LATTICE CONSTANTS:
  $a = 5.320$
  $b = 9.214$
  $c = 14.302$
  $\beta = 97.10°$
3 STRONGEST DIFFRACTION LINES:
  7.10 (100)
  3.55 ( 45)
  14.2 ( 25)
OPTICAL CONSTANTS:
  $\alpha = 1.637$
  $\beta = 1.647$
  $\gamma \simeq \beta$
  $2V \simeq 15°$
HARDNESS: 3
DENSITY: 3.19 (Meas.)
        3.20 (Calc.)
CLEAVAGE: {001} pronounced
HABIT: Massive; as minute irregular veins in talc.
COLOR-LUSTER: Yellowish green.
MODE OF OCCURRENCE: Occurs associated with ferroan trevorite, violarite, millerite, willemseite, and secondary reevesite and goethite in a small body of nickeliferous rocks at the contact between quartzite and ultramafic rocks, about two miles west of the Scotia talc mine, Barberton Mountain Land, Transvaal.
BEST REF. IN ENGLISH: de Waal, S. A., *Am. Min.*, **55**: 18–30 (1970).   Heimstra, S. A., and de Waal, S. A., *Am. Min.*, **54**: 1739–1740 (1969).

# NINGYOITE
$(Ca, U, Ce)_2 (PO_4)_2 \cdot 1\text{-}2H_2O$

CRYSTAL SYSTEM: Orthorhombic
CLASS: 222
SPACE GROUP: P222
Z: 3
LATTICE CONSTANTS:
  $a = 6.78$
  $b = 12.10$
  $c = 6.38$
3 STRONGEST DIFFRACTION LINES:
  3.02 (100)
  2.81 ( 80)
  2.13 ( 80)
OPTICAL CONSTANTS:
  $N \sim 1.64\text{-}1.71$
HARDNESS: Not determined.
DENSITY: 4.75 (Calc.)
CLEAVAGE: Not determined.
HABIT: Crystals acicular or elongated lozenge-shaped; extremely small, about $5\mu$.
COLOR-LUSTER: Brownish green to brown in transmitted light.
MODE OF OCCURRENCE: Occurs as a coating on pyrite and other minerals or filling cracks and cavities in the ore in an oxidized zone of the Ningyo-toge mine, Tottori Prefecture, Japan.

BEST REF. IN ENGLISH: Muto, Tadashi, Meyrowitz, Robert, Pommer, Alfred M., and Murano, Toru, *Am. Min.*, **44**: 633–650 (1959).

# NININGERITE
(Mg, Fe, Mn)S

CRYSTAL SYSTEM: Cubic
LATTICE CONSTANT:
  $a \simeq 5.17$
STRONGEST DIFFRACTION LINES:
  2.584 } Abee enstatite chondrite.
  1.829 }
HARDNESS: Not determined.
DENSITY: 3.21–3.68 (Calc.)
CLEAVAGE: Not determined.
HABIT: Massive.
COLOR-LUSTER: Opaque in thin section, gray in reflected light, isotropic.
MODE OF OCCURRENCE: Occurs intimately intergrown with nickel-iron and troilite in the Abee, Saint Sauveur, Adhi-kot, Indarch, St. Mark's, and Kota-Kota enstatite chondrites.
BEST REF. IN ENGLISH: Keil, Klaus, and Snetsinger, K. G., *Science*, **155**: 451–543 (1967).

# NIOBOLOPARITE (Niobian loparite) = Variety of perovskite containing thorium

# NIOBOPHYLLITE
$(K, Na)_3 (Fe, Mn)_6 (Nb, Ti)_2 Si_8 (O, OH, F)_{31}$
Niobium analogue of astrophyllite

CRYSTAL SYSTEM: Triclinic
CLASS: 1 or $\bar{1}$
SPACE GROUP: P1 or P$\bar{1}$
Z: 1
LATTICE CONSTANTS:
  $a = 5.391$    $\alpha = 113.1°$
  $b = 11.88$    $\beta = 94.5°$
  $c = 11.66$    $\gamma = 103.1°$
3 STRONGEST DIFFRACTION LINES:
  3.506 (100)
  10.52 ( 90)
  2.778 ( 80)
OPTICAL CONSTANTS:
  $\alpha = 1.724$
  $\beta = 1.760$
  $\gamma = 1.772$
  $(-)2V = 60°$
HARDNESS: Not determined.
DENSITY: 3.42 (Meas.)
        3.406 (Calc.)
CLEAVAGE: {001} perfect
Secondary cleavage parallel to a (0kl) plane ill-defined.
HABIT: As polycrystalline aggregates of small flakes. Some flakes twinned by reflection on (001) plane.
COLOR-LUSTER: Brown.
MODE OF OCCURRENCE: Occurs with aegirine-augite,

barylite, eudidymite, neptunite, manganoan pectolite, pyrochlore, joaquinite, sphalerite, and galena in a band of paragneiss, consisting chiefly of albite and arfvedsonite, in the Seal Lake area, Labrador.

BEST REF. IN ENGLISH: Nickel, E. H., Rowland, J. F., and Charette, D. J., *Can. Min.*, **8**(pt. 1): 40–52 (1964).

## NIOCALITE
$Ca_4 NbSi_2 O_{10}(O, F)$

CRYSTAL SYSTEM: Monoclinic
CLASS: 2/m or m
SPACE GROUP: Pa or P2/a
Z: 4
LATTICE CONSTANTS:
  $a = 10.83$
  $b = 10.42$
  $c = 7.38$
  $\beta = 109°40'$
3 STRONGEST DIFFRACTION LINES:
  3.01 (100)
  2.89 ( 60)
  2.85 ( 60)
OPTICAL CONSTANTS:
  $\alpha = 1.701$
  $\beta = 1.714$
  $\gamma = 1.720$
  $(-)2V = 56°$
HARDNESS: ~6
DENSITY: 3.32 (Meas.)
CLEAVAGE: None. Fracture uneven. Very Brittle.
HABIT: Crystals prismatic, up to 10 mm in length and 1 mm in width. Crystals normally twinned on *b*-axis.
COLOR-LUSTER: Light yellow; vitreous. Transparent.
MODE OF OCCURRENCE: Occurs associated with apatite, diopside, biotite, pyrochlore, and niobian perovskite in calcite in carbonatite at Oka, Quebec, Canada.
BEST REF. IN ENGLISH: Nickel, E. H., *Am. Min.*, **41**: 785–786 (1956). Nickel, Rowland, and Maxwell, *Can. Min.*, **6**: 264 (1958).

## NISBITE
$NiSb_2$

CRYSTAL SYSTEM: Orthorhombic
CLASS: Probably 2/m 2/m 2/m
SPACE GROUP: Probably Pnnm
Z: 2
LATTICE CONSTANTS:
  $a = 5.16$
  $b = 6.30$
  $c = 3.84$
3 STRONGEST DIFFRACTION LINES:
  2.76 (100)
  1.84 ( 80)
  2.70 ( 60)
HARDNESS: Microhardness 479 kg/mm$^2$ with 5 g weight.
DENSITY: 8.0 (Calc.)
CLEAVAGE: Not determined.
HABIT: Irregular grains, up to 20$\mu$ in size.

COLOR-LUSTER: Weakly bireflecting in reflected light; appears white in air and oil. Opaque. Metallic.
MODE OF OCCURRENCE: Found in drill core in a massive base-metal sulfide ore associated with chalcopyrite, breithauptite, pyrargyrite, galena, pyrrhotite, and tetrahedrite from Mulcahy Township, District of Kenora, Red Lake Mining Division, Ontario, Canada.
BEST REF. IN ENGLISH: Cabri, L. J., Harris, D. C., and Stewart, J. M., *Can. Min.*, **10**(Pt. 2): 232–246 (1970).

## NISSONITE
$Cu_2 Mg_2 (PO_4)_2 (OH)_2 \cdot 5H_2 O$

CRYSTAL SYSTEM: Monoclinic
CLASS: 2/m or m
SPACE GROUP: C2/c or Cc
Z: 4
LATTICE CONSTANTS:
  $a = 22.58$
  $b = 5.027$
  $c = 10.514$
  $\beta = 99°20'$
3 STRONGEST DIFFRACTION LINES:
  11.14  (100)
  2.785 ( 25)
  4.374 ( 21)
OPTICAL CONSTANTS:
  $\alpha = 1.584$
  $\beta = 1.620$
  $\gamma = 1.621$
  $(-)2V = 19°$ (Calc.)
HARDNESS: 2½
DENSITY: 2.73 (Meas.)
      2.74 (Calc.)
CLEAVAGE: {100} fair
HABIT: Crystals tabular {100} and elongated [001]. Forms c {001}, a {100}, and q {$\bar{1}11$}. As minute diamond-shaped crystals; more abundantly as thin crusts of crystal aggregates.
COLOR-LUSTER: Bluish green.
MODE OF OCCURRENCE: Occurs associated with malachite, azurite, libethenite, turquoise, chrysocolla, cuprite, barite, calcite, gypsum, riebeckite, and crossite in a small copper prospect in metamorphic rocks of the Franciscan Formation, Panoche Valley, California.
BEST REF. IN ENGLISH: Mrose, Mary E., Meyrowitz, Robert, Alfors, J. T., and Chesterman, C. W., (abstr.) Geol. Soc. Amer., Program Ann. Meet. Nov. 14-16, 1966, p. 145-146 (1966).

## NITER (Nitre)
$KNO_3$

CRYSTAL SYSTEM: Orthorhombic
CLASS: 2/m 2/m 2/m
SPACE GROUP: Pmcn
Z: 4
LATTICE CONSTANTS:
  $a = 5.42$ kX
  $b = 9.17$
  $c = 6.45$

3 STRONGEST DIFFRACTION LINES:
3.78   (100)
3.73   ( 56)  (synthetic)
3.033 ( 55)
OPTICAL CONSTANTS:
$\alpha = 1.332$
$\beta = 1.504$
$\gamma = 1.504$
$(-)2V = 7°$
HARDNESS: 2
DENSITY: 2.109 (Meas.)
         2.104 (Calc.)
CLEAVAGE: {011} Nearly perfect
          {010} Good
          {110} Imperfect
Fracture subconchoidal to uneven. Brittle.

HABIT: Commonly as thin crusts or small tufts; also massive, earthy to granular. Synthetic crystals prismatic, often exhibiting twinning on {110}.

COLOR-LUSTER: Colorless, white; often discolored by impurities. Transparent. Vitreous.

Taste saline.

MODE OF OCCURRENCE: Occurs widespread, usually in small amounts, as an efflorescence upon soil or in caverns. Found in the United States in caverns in Kentucky, Tennessee, Ohio, Alabama, New Mexico, and Idaho. It is also found at several localities in California, especially along the old shorelines of Death Valley and along the Amargosa River in Inyo County. Extensive deposits occur in the desert regions of northern Chile, and it is also found in Bolivia, Italy, Spain, USSR, and at many other places.

BEST REF. IN ENGLISH: Palache, et al., "Dana's System of Mineralogy," 7th Ed., v. II, p. 303–305, New York, Wiley, 1951.


# NITRATINE = Soda-niter


# NITROBARITE
$Ba(NO_3)_2$

CRYSTAL SYSTEM: Cubic
CLASS: $2/m\overline{3}$
SPACE GROUP: Pa3
Z: 4
LATTICE CONSTANT:
$a = 8.11$ kX
3 STRONGEST DIFFRACTION LINES:
2.448 (100)
4.68   ( 95) (synthetic)
2.343 ( 55)
OPTICAL CONSTANT:
$N = 1.5714$
HARDNESS: Not determined.
DENSITY: 3.250 (Meas.)
         3.244 (Calc.)
CLEAVAGE: None.
HABIT: Crystals octahedral. Twinning on {111}.
COLOR-LUSTER: Colorless. Transparent.
MODE OF OCCURRENCE: Known only from one specimen, presumably from the nitrate deposits of northern Chile.

BEST REF. IN ENGLISH: Palache, et al., "Dana's System of Mineralogy," 7th Ed., v. II, p. 305–306, New York, Wiley, 1951.


# NITROCALCITE
$Ca(NO_3)_2 \cdot 4H_2O$

CRYSTAL SYSTEM: Monoclinic
CLASS: 2/m
Z: 4
LATTICE CONSTANTS:
$a = 14.46$
$b = 9.17$
$c = 6.30$
$\beta = 98°00'$
3 STRONGEST DIFFRACTION LINES:
5.2   (100)
7.8   ( 45) (synthetic)
2.81 ( 40)
OPTICAL CONSTANTS:
$\alpha = 1.465$
$\beta = 1.498$ (synthetic)
$\gamma = 1.504$
$(-)2V \simeq 50°$
HARDNESS: Soft.
DENSITY: 1.90 (Meas.)
CLEAVAGE: Not determined.
HABIT: As an efflorescence. Synthetic crystals long prismatic.
COLOR-LUSTER: White, gray. Transparent.

Very soluble in water. Taste bitter.

MODE OF OCCURRENCE: Occurs as an efflorescence upon soil, in limestone caverns, and upon calcareous rocks. Found in the United States in California, Arizona, New Mexico, and in limestone caverns in Indiana and Kentucky. It also occurs at localities in France, Spain, and Hungary.

BEST REF. IN ENGLISH: Palache, et al., "Dana's System of Mineralogy," 7th Ed., v. II, p. 306–307, New York, Wiley, 1951.


# NITROMAGNESITE
$Mg(NO_3)_2 \cdot 6H_2O$

CRYSTAL SYSTEM: Monoclinic
CLASS: 2/m
SPACE GROUP: $P2_1/c$
Z: 2
LATTICE CONSTANTS:
$a = 6.194$
$b = 12.71$
$c = 6.600$
$\beta = 93°$
3 STRONGEST DIFFRACTION LINES:
3.295 (100)
2.925 ( 75)
4.43   ( 55)
OPTICAL CONSTANTS:
$\alpha = 1.34$
$\beta = 1.506$
$\gamma = 1.506$

$(-)2V \simeq 5°$
HARDNESS: Not determined.
DENSITY: 1.58 (Meas.)
1.64 (Calc.)
CLEAVAGE: {110} perfect
HABIT: As an efflorescence. Synthetic crystals long prismatic.
COLOR-LUSTER: Colorless, white. Transparent. Vitreous.
MODE OF OCCURRENCE: Occurs as an efflorescence in a limestone cavern in Madison County, Kentucky, and in marl from Baume and Salina, Jura, France.
BEST REF. IN ENGLISH: Palache, et al., "Dana's System of Mineralogy," 7th Ed., v. II, p. 307, New York, Wiley, 1951.

## NOBLEITE
$CaB_6O_{10} \cdot 4H_2O$

CRYSTAL SYSTEM: Monoclinic
CLASS: 2/m
SPACE GROUP: $P2_1/a$
Z: 4
LATTICE CONSTANTS:
$a = 14.56$
$b = 8.016$
$c = 9.838$
$\beta = 111°45'$
3 STRONGEST DIFFRACTION LINES:
6.79 (100)
3.39 ( 31)
5.18 ( 9)
OPTICAL CONSTANTS:
$\alpha = 1.500$
$\beta = 1.520$
$\gamma = 1.554$
$(+)2V = 76°$
HARDNESS: 3
DENSITY: 2.09 (Meas.)
2.09 (Calc.)
CLEAVAGE: {100} perfect
{001} indistinct
Fracture uneven; sectile; flexible in thin crystals or cleavage flakes; inelastic.
HABIT: As platy euhedral crystals; also as mammillary coatings formed by plates arranged subparallel to {100}. {100} plate form commonly elongated [010] or [001]. Large crystals have hexagonal aspect; smaller crystals commonly rhomb-shaped. Contact twinning, with (100) as twin plane and composition surface common.
COLOR-LUSTER: Colorless and transparent; subvitreous, pearly on cleavage plates. Fine-grained aggregates white. Streak white.
MODE OF OCCURRENCE: Occurs as platy crystals up to 3 mm in length, produced by the weathering of colemanite and priceite veins in altered basaltic rocks of the Furnace Creek formation of late Tertiary age, in the Furnace Creek borate deposits of the Death Valley region, California.
BEST REF. IN ENGLISH: Erd, Richard C., McAllister, James F., and Vlisidis, Angelina C., *Am. Min.*, **46**: 560-571 (1961).

## NOCERITE = Synonym for fluoborite

## NOGIZAWALITE = A mixture

## NOHLITE = Synonym for samarskite

## NOLANITE
$Fe_3V_7O_{16}$

CRYSTAL SYSTEM: Hexagonal
CLASS: 6mm
SPACE GROUP: $P6_3mc$
Z: 1
LATTICE CONSTANTS:
$a = 5.846$
$c = 9.254$
3 STRONGEST DIFFRACTION LINES:
3.44 (100)
2.66 ( 90)
2.49 ( 90)
HARDNESS: 5-5½
DENSITY: 4.69 (Meas.)
4.60 (Calc.)
CLEAVAGE: Fracture uneven.
HABIT: Subhedral to euhedral hexagonal plates up to 1 mm in diameter; as radiating crusts.
COLOR-LUSTER: Black. Opaque. Submetallic. Powder brownish black.
MODE OF OCCURRENCE: Occurs intimately associated with dolomite, quartz, calcite, pyrite, hematite, pitchblende, chalcopyrite, ilmenite, and galena in several uranium mines in the Beaverlodge region, Saskatchewan, Canada; also found in metamorphosed and altered greenstones closely associated with native gold and numerous tellurides in the Kalgoorlie mining district, Western Australia.
BEST REF. IN ENGLISH: Robinson, S. C., Evans, H. T. Jr., Schaller, W. T., and Fahey, J. J., *Am. Min.*, **42**: 619-628 (1957). Radtke, Arthur S., *Am. Min.*, **52**: 734-743 (1967).

## NONTRONITE (Chloropal, Faratsihite)
$Na_{0.33}Fe_2^{3+}(Al, Si)_4O_{10}(OH)_2 \cdot nH_2O$
Montmorillonite (Smectite) Group

CRYSTAL SYSTEM: Monoclinic
CLASS: 2/m
SPACE GROUP: C2/m
Z: 1
LATTICE CONSTANTS:
$a = 5.23$
$b = 9.11$
$c = 15-15.5$
3 STRONGEST DIFFRACTION LINES:

| (nontronite) | (faratsihite) |
|---|---|
| 15.4 (100) | 13.9 (100) |
| 4.56 (100) | 4.44 ( 80) |
| 1.52 (100) | 3.54 ( 50) |

OPTICAL CONSTANTS:
  (−)2V = moderate
  HARDNESS: 1–2
  DENSITY: Variable, 2–3
  CLEAVAGE: {001} perfect.
Fracture conchoidal and splintery to earthy.
  HABIT: Massive, very fine grained; clay-like.
  COLOR-LUSTER: Pale yellow to olive green or green. Dull to somewhat resinous or waxy. Nearly opaque.
  MODE OF OCCURRENCE: Occurs as an alteration product of volcanic glasses; also in mineral veins, often in association with opal and quartz. Typical occurrences are found in California, Washington, Nevada, Utah, Colorado, Texas, and Pennsylvania. Also in France, Germany, Czechoslovakia, and Madagascar (faratsihite).
  BEST REF. IN ENGLISH: Deer, Howie, and Zussman, "Rock Forming Minerals," v. 3, p. 226–245, New York, Wiley, 1962.


## NORALITE = Hornblende


## NORBERGITE
$Mg_3SiO_4(F, OH)_2$
Humite Group

CRYSTAL SYSTEM: Orthorhombic
CLASS: 2/m 2/m 2/m
SPACE GROUP: Pbnm
Z: 4
LATTICE CONSTANTS:
  $a$ = 8.72
  $b$ = 4.70
  $c$ = 10.20
  $\beta$ = 90°
3 STRONGEST DIFFRACTION LINES:
  3.058 (100)
  2.230 ( 80)
  2.639 ( 75)
OPTICAL CONSTANTS:
  $\alpha$ = 1.563–1.567
  $\beta$ = 1.567–1.579
  $\gamma$ = 1.590–1.593
  (+)2V = 44°–50°
HARDNESS: 6–6½
DENSITY: 3.177 (Meas.)
         3.186 (Calc.)
CLEAVAGE: None.
Fracture uneven to subconchoidal. Brittle.
  HABIT: Crystals varied in habit, usually highly modified. Also as rounded grains.
  COLOR-LUSTER: Tan, yellow, yellow-orange, orange-brown. Transparent to translucent. Vitreous to somewhat resinous.
  MODE OF OCCURRENCE: Occurs in contact zones in limestone or dolomite; also in ore deposits. Found in the Sparta-Franklin district, Sussex County, New Jersey; in Orange County, New York; and at the Ostanmosoa iron mine, Norberg, Vastmanland, Sweden.
  BEST REF. IN ENGLISH: Deer, Howie, and Zussman, "Rock Forming Minerals," v. I, p. 47–58, New York, Wiley,

1962. Gibbs, G. V., and Ribbe, P. H., *Am. Min.*, **54**: 376–390 (1969).


## NORDENSKIÖLDINE
$CaSnB_2O_6$

CRYSTAL SYSTEM: Hexagonal
CLASS: $\bar{3}$
SPACE GROUP: $R\bar{3}$
Z: 3 (hexag.), 1 (rhomb.)
LATTICE CONSTANTS:
  (hexag.)              (rhomb.)
  $a$ = 4.853          $a_{rh}$ = 6.001
  $c$ = 15.920         $\alpha$ = 47°41½′
3 STRONGEST DIFFRACTION LINES:
  2.892 (100)
  1.804 (100)
  3.707 ( 60)
OPTICAL CONSTANTS:
  $\omega$ = 1.778
  $\epsilon$ = 1.660
  (−)
HARDNESS: 5½–6
DENSITY: 4.20 (Meas.)
         4.189 (Calc.)
CLEAVAGE: {0001} perfect
          {10$\bar{1}$1} indistinct
Fracture conchoidal. Brittle.
  HABIT: Crystals thin to thick tabular, flattened on {0001}. Also as thick, lens-like crystals and as parallel growths with cassiterite, calcite, and siderite.
  COLOR-LUSTER: Colorless, various shades of yellow. Transparent. Vitreous; pearly on {0001}.
  MODE OF OCCURRENCE: Occurs in very small amounts in an alkaline pegmatite on Arö island, Langesundfjord, Norway, associated with feldspar, meliphane, zircon, and other minerals. It also is found associated with stannite and other sulfides, tourmaline, calcite, siderite, and cassiterite in an ore pipe in marble at Arandis, South West Africa.
  BEST REF. IN ENGLISH: Palache, et al., "Dana's System of Mineralogy," 7th Ed., v. II, p. 332–333, New York, Wiley, 1951.


## NORDITE
$Na_3Ce(Sr, Ca)(Mn, Mg, Fe, Zn)_2Si_6O_{18}$

CRYSTAL SYSTEM: Orthorhombic
CLASS: 2/m 2/m 2/m
SPACE GROUP: Pcaa
Z: 4
LATTICE CONSTANTS:
  $a$ = 14.27
  $b$ = 5.16
  $c$ = 19.45
3 STRONGEST DIFFRACTION LINES:
  2.95 (100)
  2.86 (100)
  1.764 ( 80)

OPTICAL CONSTANTS:
$\alpha$ = 1.619–1.621
$\beta$ = 1.63–1.64
$\gamma$ = 1.642–1.655
(–)2V = 30°
HARDNESS: 5
DENSITY: 3.43–3.49 (Meas.)
3.48 (Calc.)
CLEAVAGE: {100} distinct
Brittle.

HABIT: Crystals tabular, flattened {100}, up to 1 cm long; sometimes forms radial-fibrous aggregates of curved and lamellar crystals up to 1.5 cm in diameter.

COLOR-LUSTER: Light brown to dark brown. Translucent. Vitreous, greasy on fracture.

MODE OF OCCURRENCE: Occurs in sodalite-syenites and related ussingite pegmatites in association with sodalite, steenstrupine, and other minerals, in the Lovozero massif, Kola Peninsula, USSR.

BEST REF. IN ENGLISH: Vlasov, K. A., "Mineralogy of Rare Elements," v. II, p. 316–317, Israel Program for Scientific Translations, 1966. Bakakin, V. V., Belov, N. V., Borisov, S. V., and Solovyeva, L. P., *Am. Min.*, **55**: 1167–1181 (1970).

## NORDMARKITE = Variety of staurolite rich in manganese

## NORDSTRANDITE

Al(OH)$_3$
Trimorphous with gibbsite and bayerite

CRYSTAL SYSTEM: Monoclinic
Z: 8
LATTICE CONSTANTS:
$a$ = 8.89    $\alpha$ = 92°56′
$b$ = 5.00    $\beta$ = 110°23′
$c$ = 10.24   $\gamma$ = 90°32′
3 STRONGEST DIFFRACTION LINES:
4.789 (100)
2.263 ( 15)
4.322 ( 12)
OPTICAL CONSTANTS:
$\alpha$ = 1.580
$\beta$ = 1.580
$\gamma$ = 1.596
(+)2V = low
HARDNESS: 3
DENSITY: 2.436 (Meas.)
2.510 (Calc.)
CLEAVAGE: {001} perfect
HABIT: As thick tabular crystals (<10–75 $\mu$ and up to 250 $\mu$), which lie on basal plane and are rhombic in plan. Also aggregates up to 1 mm in size.

COLOR-LUSTER: Colorless to coral pink and reddish brown (due to goethite).

MODE OF OCCURRENCE: Occurs in secondary solution cavities in limestone, mainly in the basal part near the contact with deeply weathered basalt flows on the island of Guam. Also found in soil from the edge of a sinkhole in a limestone cliff at Gunony Kapor near Bau, western Sarawak.

BEST REF. IN ENGLISH: Wall, J. R. D., Wolfenden, E. B., Beard, E. H., and Deams, T., *Nature*, **196**(4851): 264–265 (1962). Hathaway, J. C., and Schlanger, S. O., *Nature*, **196**(4851): 265–266 (1962).

## NORILSKITE = Intergrowths of several platinum minerals

## NORSETHITE

BaMg(CO$_3$)$_2$

CRYSTAL SYSTEM: Hexagonal
CLASS: 32
SPACE GROUP: R32
Z: 3
LATTICE CONSTANTS:
$a$ = 5.02
$c$ = 16.75
3 STRONGEST DIFFRACTION LINES:
3.015 (100)
3.860 ( 35)
2.656 ( 35)
OPTICAL CONSTANTS:
$\omega$ = 1.694
$\epsilon$ = 1.519
(–)
HARDNESS: ~3.5
DENSITY: 3.837 (Meas.)
3.840 (Calc.)
CLEAVAGE: Rhombohedral {10$\bar{1}$1} good. Hackly fracture.

HABIT: As circular plates or flattened rhombohedral crystals 0.2–2.0 mm in diameter. Some crystals and irregular grains show corrugated or flattened edges; thick corrugated grains are pumpkin-shaped. Also as coarse-grained masses.

COLOR-LUSTER: Clear to milky white; vitreous to pearly.

MODE OF OCCURRENCE: Found associated with shortite, labuntsovite, searlesite, loughlinite, pyrite, and quartz in dolomitic black oil shale below the main trona bed in the Westvaco trona mine about 18 miles west of Green River, Wyoming; also occurs in large quantities in a zinc-lead-copper deposit at Rosh Pinah, situated about 100 miles southeast of Aus, South West Africa.

BEST REF. IN ENGLISH: Mrose, Mary E., Chao, E. C. T., Fahey, Joseph J., and Milton, Charles, *Am. Min.*, **46**: 420–429 (1961).

## NORTHUPITE

Na$_3$Mg(CO$_3$)$_2$Cl

CRYSTAL SYSTEM: Cubic
CLASS: 2/m $\bar{3}$ or 4/m $\bar{3}$ 2/m
SPACE GROUP: Fd3 or Fd3m
Z: 16

LATTICE CONSTANT:
$a \cong 13.98$
3 STRONGEST DIFFRACTION LINES:
2.69 (100)
2.48 (100)
8.00 ( 80)
OPTICAL CONSTANT:
$N = 1.5144$ (Na)
HARDNESS: 3½–4
DENSITY: 2.380 (Meas.)
            2.40 (Calc.)
CLEAVAGE: None. Fracture conchoidal. Brittle.
HABIT: Crystals octahedral.
COLOR-LUSTER: Colorless; often yellowish, gray, or brownish due to impurities. Transparent. Vitreous.
MODE OF OCCURRENCE: Occurs as fine octahedral crystals associated with galeite, tychite, and trona at Searles Lake, San Bernardino County, and with gaylussite and pirssonite in trona at Borax Lake, Lake County, California. It is also found in drill cores from oil shale near Green River, Sweetwater County, Wyoming, associated with gaylussite, shortite, trona, pirssonite, and bradleyite.
BEST REF. IN ENGLISH: Palache, et al., "Dana's System of Mineralogy," 7th Ed., v. II, p. 278–279, New York, Wiley, 1951.

# NOSEAN (Noselite)
$Na_8 Al_6 Si_6 O_{24} SO_4$
Sodalite Group

CRYSTAL SYSTEM: Cubic
CLASS: $\bar{4}$ 3m
SPACE GROUP: P$\bar{4}$3m
Z: 1
LATTICE CONSTANT:
$a = 9.05$
3 STRONGEST DIFFRACTION LINES:
3.71  (100)
2.628 ( 75)
6.45  ( 70)
OPTICAL CONSTANT:
$N = 1.495$
HARDNESS: 5½
DENSITY: 2.30–2.40 (Meas.)
            2.205 (Calc.)
CLEAVAGE: {110} indistinct
HABIT: Crystals dodecahedral. Usually massive, granular. Twinning on {111}.
COLOR-LUSTER: Colorless, white, gray, bluish, brown, reddish, or black. Transparent to translucent. Vitreous.
MODE OF OCCURRENCE: Occurs exclusively in phonolites, similar igneous rocks, and volcanic bombs. Found in phonolite and tinguaite in Lawrence County, South Dakota; as phenocrysts in the phonolite of Wolf Rock, Cornwall, England; in the Laacher See, Lower Rhine, Germany; in the Cape Verde Islands; in northern China; and in the Cripple Creek district, Colorado.
BEST REF. IN ENGLISH: Deer, Howie, and Zussman, "Rock Forming Minerals," v. 4, p. 289–302, New York, Wiley, 1963.

# NOVACEKITE
$Mg(UO_2)_2 (AsO_4)_2 \cdot 12H_2O$
Autunite Group

CRYSTAL SYSTEM: Tetragonal
CLASS: 4/m
SPACE GROUP: P4$_2$/n
Z: 2
LATTICE CONSTANTS:
$a = 7.11$
$c = 20.06$ (variable)
3 STRONGEST DIFFRACTION LINES:
10.15 (100)
 3.58 ( 90)
 5.06 ( 80)
OPTICAL CONSTANTS:
$\epsilon$ or $\alpha = 1.620–1.625$
$\beta = 1.620–1.641$
$\omega$ or $\gamma = 1.620–1.641$
$2V = 0°–20°$
HARDNESS: 2½
DENSITY: ~3.7
CLEAVAGE: {001} perfect
            {010} indistinct
            {110} indistinct
HABIT: Crystals rectangular plates flattened on {001} ranging in size up to 0.5 mm on an edge. As crusts or porous interlocking aggregates of thin plates and scales; also as lamellar aggregates filling tiny veinlets and cavities.
COLOR-LUSTER: Pale yellow, yellow. Crystals lose luster and become opaque upon dehydration. Transparent to translucent. Fluoresces dull green in ultraviolet light.
MODE OF OCCURRENCE: Occurs as a coating on sandstone in the Woodrow area, Laguna Reservation, Valencia County, New Mexico; also found near Aldama, Chihuahua, Mexico, and at Schneeberg, Saxony, Germany.
BEST REF. IN ENGLISH: Frondel, Clifford, USGS Bull. 1064, p. 177–183 (1958).

# NOVACULITE = Variety of quartz

# NOVÁKITE
$(Cu, Ag)_4 As_3$

CRYSTAL SYSTEM: Tetragonal, pseudocubic
Z: 12
LATTICE CONSTANTS:
$a = 9.99$
$c = 14.03$
3 STRONGEST DIFFRACTION LINES:
1.877 (100)
1.959 ( 90)
1.998 ( 80)
HARDNESS: 3–3.5
DENSITY: 6.7 (Meas.)
CLEAVAGE: None.
HABIT: As irregular aggregates; grains up to 3 cm in size.
COLOR-LUSTER: Steel gray; rapidly tarnishes to iridescent, then black. Opaque. Metallic. Streak black.

MODE OF OCCURRENCE: Found as granular aggregates in carbonate veins near Cerný Důl, in the Giant Mountains, northern Bohemia, associated with arsenic, arsenolamprite, koutekite, silver, löllingite, chalcocite, smaltite, chalcopyrite, bornite, pitchblende, and a new undescribed copper arsenide mineral.

BEST REF. IN ENGLISH: Johan, Z., and Hak, J., *Am. Min.*, **46**: 885–891 (1961).


# NOWACKIITE
$Cu_6Zn_3As_4S_{12}$

CRYSTAL SYSTEM: Hexagonal
CLASS: 3
SPACE GROUP: R3
Z: 3(hex.), 1(rhomb.)
LATTICE CONSTANTS:
$a = 13.44$   $a_{rh} = 8.34$
$c = 9.17$    $\alpha = 107°20'$
3 STRONGEST DIFFRACTION LINES:
  1.019 (100b)
  1.081 ( 75b)
  1.887 ( 63 )
HARDNESS: Not determined.
DENSITY: 4.3 (Calc.)
CLEAVAGE: Not determined.
HABIT: Crystals up to 0.3 mm in size.
COLOR-LUSTER: Lead gray to black. Opaque.
MODE OF OCCURRENCE: Found as about 10 minute crystals on a honey-yellow sphalerite crystal on dolomite at the Lengenbach quarry, Binnatal, Canton Wallis, Switzerland.

BEST REF. IN ENGLISH: Marumo, F., *Am. Min.*, **54**: 1497–1498 (1969).   Marumo, F., and Burri, G., *Am. Min.*, **51**: 532 (1966).


# NSUTITE ($\gamma$-MnO$_2$)
$Mn^{4+}_{1-x}Mn^{2+}_xO_{2-2x}(OH)_{2x}$
(x = 0.06–0.07)

CRYSTAL SYSTEM: Hexagonal
Z: 12
LATTICE CONSTANTS:
  $a = 9.65$
  $c = 4.43$
3 STRONGEST DIFFRACTION LINES:
  1.635 (100)
  4.00 ( 98)
  2.33 ( 69)

HARDNESS: 8.5
          6.5–7 (Manganoan)
DENSITY: 4.86 (Calc.)
         4.24–4.67 (Meas.)
         3.86 (Manganoan)
HABIT: Platy to wedge-like crystals from less than 0.1 $\mu$ up to 50 $\mu$. Porous to dense aggregates.
COLOR-LUSTER: Dark gray to black. Opaque.
MODE OF OCCURRENCE: Occurs as a major constituent in manganese ores from Nsuta, Ghana, and has been detected from numerous manganese deposits throughout the world.

BEST REF. IN ENGLISH: Zwicker, W. K., Groeneveld Meijer, W. O. J., and Jaffe, H. W., *Am. Min.*, **47**: 246–266 (1962).   Faulring, G. M., *Am. Min.*, **50**: 170–179 (1965).


# NUFFIELDITE
$Pb_{10}Cu_4Bi_{10}S_{27}$

CRYSTAL SYSTEM: Orthorhombic
CLASS: 2/m 2/m 2/m or mm2
SPACE GROUP: Pnam or Pna2$_1$
Z: 1
LATTICE CONSTANTS:
  $a = 14.61$
  $b = 21.38$
  $c = 4.03$
3 STRONGEST DIFFRACTION LINES:
  3.66 (100)
  3.54 (100)
  4.00 ( 90)
HARDNESS: Microhardness 149–178 kg/mm$^2$ with 15 g load.
DENSITY: 7.01 (Meas.)
         7.006 (Calc.)
CLEAVAGE: Excellent parallel to [001]. Indistinct normal to [001]. Fracture uneven to flat conchoidal. Brittle.
HABIT: Crystals prismatic to acicular, deeply striated parallel to [001]; as bundles of parallel unterminated crystals. Crystals up to 3 mm in length. Also as stubby unterminated crystals in compact subparallel aggregates.
COLOR-LUSTER: Lead gray to steel gray; opaque. Bright metallic; tarnishes pale iridescent grayish green to reddish brown. Streak dark greenish gray to black.
MODE OF OCCURRENCE: Found in small vugs in narrow quartz veins associated with pyrite and molybdenite in the Lime Creek quartz diorite stock near Alice Arm, British Columbia, Canada.

BEST REF. IN ENGLISH: Kingston, P. W., *Can. Min.*, **9**(pt. 4): 439–452 (1968).

## OBRUCHEVITE

$(Y, Na, Ca, U)(Nb, Ta, Ti, Fe)_2 (O, OH)_7$
Pyrochlore Group

CRYSTAL SYSTEM: Cubic
LATTICE CONSTANT:
  $a = 10.17$
3 STRONGEST DIFFRACTION LINES:
  2.975 (100)
  1.695 ( 90) Metamict
  1.488 ( 70)
OPTICAL CONSTANT:
  $N = 1.830-1.835$
  HARDNESS: 4½-5
  DENSITY: 3.60-3.80
  CLEAVAGE: None. Fracture conchoidal.
  HABIT: As dense masses without crystal form.
  COLOR-LUSTER: Various shades of brown, usually chocolate brown; vitreous to adamantine.
  MODE OF OCCURRENCE: Occurs associated with garnet, fergusonite, and columbite in albite-muscovite replacement zones in a pegmatite in the Alakurtt region, northwest Karelia, USSR.
  BEST REF. IN ENGLISH: Kalita, A. P., *Am. Min.*, **43**: 797 (1958).


## OCTAHEDRITE = Anatase


## OELLACHERITE = Variety of muscovite with up to 10% BaO


## OFFRETITE

$(K, Ca)_3 (Al_5 Si_{13})O_{36} \cdot 14H_2 O$
Zeolite Group

CRYSTAL SYSTEM: Hexagonal
CLASS: $\bar{6}$ m2
SPACE GROUP: P$\bar{6}$m2
Z: 2

LATTICE CONSTANTS:
  $a = 13.263$
  $c = 15.111$
3 STRONGEST DIFFRACTION LINES:
  11.50 (100)
   2.88 ( 65)
   4.35 ( 60)
OPTICAL CONSTANTS:
  $\omega = 1.489$
  $\epsilon = 1.486$
    (-)
  HARDNESS: Not determined.
  DENSITY: 2.13 (Meas.)
  CLEAVAGE: {0001} distinct
Fracture uneven. Brittle.
  HABIT: Crystals prismatic, minute, vertically striated, often rounded. Also in hemispherical forms with radiate structure.
  COLOR-LUSTER: Colorless to white. Transparent to translucent. Vitreous.
  MODE OF OCCURRENCE: Occurs in amygdaloidal basalt at Mt. Simiouse near Montbrison, Loire, France, and in basalt from Palau Island, Caroline Islands. Recently found as cross-fiber coatings on levynite crystals in amygdules in basalt from Beech Creek, Oregon.
  BEST REF. IN ENGLISH: Bennett, J. M., and Gard, J. A., *Nature*, **214**: 1005-1006 (1967). Sheppard, Richard A., and Gude, Arthur J., 3rd, *Am. Min.*, **54**: 875-886 (1969).


## OKENITE

$CaSi_2 O_4 (OH)_2 \cdot H_2 O$

CRYSTAL SYSTEM: Triclinic
Z: 9
LATTICE CONSTANTS:
  $a = 9.84$          $\alpha = 90°$
  $b = 7.20$          $\beta = 103.9°$
  $c = 21.33$         $\gamma = 111.5°$
3 STRONGEST DIFFRACTION LINES:
  (India)        (Greenland)
  21.0  (100)    8.91 (100)
   8.8  ( 80)    2.96 ( 80)
   3.56 ( 80)    2.83 ( 80)

OPTICAL CONSTANTS:
$\alpha = 1.530$
$\beta = ?$
$\gamma = 1.541$ (variable)
$(-)2V = $ large
HARDNESS: 4½-5
DENSITY: 2.28-2.332 (Meas.)
2.33 (Calc.)
CLEAVAGE: {010} perfect
HABIT: Crystals blade-shaped, flattened {010}. Commonly as fibrous interlaced masses; also compact. Twinning on {010}.
COLOR-LUSTER: White; sometimes with yellowish or bluish tint. Transparent to translucent. Vitreous to pearly.
MODE OF OCCURRENCE: Occurs in amygdules in basalt or related eruptive rocks. Found on Disco Island, Greenland; Faroe Islands; Iceland; Chile; and in the Syhadree Mountains, Bombay, India.
BEST REF. IN ENGLISH: Winchell and Winchell, "Elements of Optical Mineralogy," 4th Ed., Pt. 2, p. 358, New York, Wiley, 1951.

## OLDHAMITE

(Ca, Mn)S

CRYSTAL SYSTEM: Cubic
CLASS: $4/m \bar{3} 2/m$
SPACE GROUP: Fm3m
Z: 4
LATTICE CONSTANT:
$a = 5.686$
3 STRONGEST DIFFRACTION LINES:
2.846 (100)
2.013 ( 70) (synthetic)
1.6439 ( 21)
OPTICAL CONSTANT:
$N = 2.137$
HARDNESS: 4
DENSITY: 2.58 (Meas.)
2.594 (Calc.)
CLEAVAGE: {001}
HABIT: As small spherules.
COLOR-LUSTER: Light brown. Transparent.
MODE OF OCCURRENCE: Occurs embedded in the Busti meteorite, India, and in other meteorites.
BEST REF. IN ENGLISH: Palache, et al., "Dana's System of Mineralogy," 7th Ed., v. I, p. 208-209, New York, Wiley, 1944.

## OLIGOCLASE

$mNaAlSi_3O_8$ with $nCaAl_2Si_2O_8$
$Ab_{90}An_{10}$ to $Ab_{70}An_{30}$
Plagioclase Feldspar Group

CRYSTAL SYSTEM: Triclinic
CLASS: $\bar{1}$
SPACE GROUP: $C\bar{1}$
Z: 2

LATTICE CONSTANTS:
$a = 8.163$  $\alpha = 93.39°$
$b = 12.875$  $\beta = 116.27°$
$c = 7.107$  $\gamma = 90.29°$
3 STRONGEST DIFFRACTION LINES:
3.20 (100)
4.02 ( 80)
3.74 ( 80)
OPTICAL CONSTANTS:
$\alpha = 1.542$
$\beta = 1.546$
$\gamma = 1.549$
$(-)2V = 82°$
HARDNESS: 6-6½
DENSITY: 2.63-2.67 (Meas.)
CLEAVAGE: {001} perfect
{010} nearly perfect
{110} imperfect
Fracture uneven to conchoidal. Brittle.
HABIT: Crystals tabular, flattened along b-axis; not common. Usually massive, cleavable, granular, or compact. Twinning common. Principal twin laws: Carlsbad, albite, and pericline.
COLOR-LUSTER: Colorless, white, gray, greenish, yellowish, brown, reddish; occasionally shows brilliant reflections from inclusions (sunstone). Transparent to translucent. Vitreous. Streak white.
MODE OF OCCURRENCE: Occurs widespread in pegmatites; also as a common rock-forming mineral in granite, syenite, porphyries, andesite, trachyte, gneiss, and crystalline schists. Typical occurrences are found in California, New Mexico, South Dakota, Pennsylvania, Maine, New York, and North Carolina. Also in Canada, Norway (sunstone), Sweden, USSR, Kenya, and Australia.
BEST REF. IN ENGLISH: Deer, Howie, and Zussman, "Rock Forming Minerals," v. 4, p. 94-165, New York, Wiley, 1963.

**OLIGONITE** = Variety of siderite. Manganese-bearing.

## OLIVEIRAITE (Inadequately described mineral. Doubtful.)

$Zr_3Ti_2O_{10} \cdot 2H_2O$

CRYSTAL SYSTEM: Unknown (metamict)
3 STRONGEST DIFFRACTION LINES: Metamict.
HARDNESS: Not determined.
DENSITY: Not determined.
CLEAVAGE: None.
HABIT: Massive.
COLOR-LUSTER: Yellowish green in transmitted light.
MODE OF OCCURRENCE: Occurs as an alteration product of euxenite near Pomba, Minas Geraes, Brazil.
BEST REF. IN ENGLISH: Palache, et al., "Dana's System of Mineralogy," 7th Ed., v. I, p. 774, New York, Wiley, 1944.

## OLIVENITE (Leucochalcite)

$Cu_2AsO_4OH$

CRYSTAL SYSTEM: Orthorhombic
CLASS: 2/m 2/m 2/m or 222
SPACE GROUP: Pmnm or P2$_1$2$_1$2$_1$
Z: 4
LATTICE CONSTANTS:
  $a$ = 8.16 kX
  $b$ = 8.54
  $c$ = 5.86
3 STRONGEST DIFFRACTION LINES:
  2.98 (100)
  4.82 ( 90)
  5.91 ( 70)
OPTICAL CONSTANTS:
  $\alpha$ = 1.772–1.780
  $\beta$ = 1.810–1.820
  $\gamma$ = 1.863–1.865
  (+), sometimes (−), 2V ~90°
HARDNESS: 3
DENSITY: 4.378 (Meas.)
            4.45 (Calc.)
CLEAVAGE: {011} indistinct
          {110} indistinct
Fracture uneven to conchoidal.
  HABIT: Short prismatic to acicular [001], often elongated [100]; sometimes tabular on {011}, {001}, or {100}. Also globular or reniform with fibrous structure; massive, granular to earthy; nodular.
  COLOR-LUSTER: Olive green; also greenish brown and brown; less commonly straw yellow, grayish green, grayish white, and white. Translucent to opaque. Vitreous to adamantine. Fibrous varieties pearly to silky.
  MODE OF OCCURRENCE: Occurs as a secondary mineral in the oxidized zone of ore deposits often associated with azurite, malachite, limonite, scorodite, dioptase, and calcite. Found as superb crystals at Tsumeb, South West Africa; at numerous mines in Cornwall and other localities in England; commonly at the Mammoth, American Eagle, and other mines in the Tintic district, Utah; in fine crystals at the Myler mine, Majuba Hill, Pershing County, Nevada; and at numerous additional localities in many parts of the world.
  BEST REF. IN ENGLISH: Palache, et al., "Dana's System of Mineralogy," 7th Ed., v. 2, p. 859–861, New York, Wiley, 1951. Berry, L. G., *Am. Min.*, **36**: 484–503 (1951).

## OLIVINE = Group name for orthorhombic Mg$_2$SiO$_4$·Fe$_2$SiO$_4$ (forsterite-fayalite) Series, dimorphous with ringwoodite

## OLSACHERITE
Pb$_2$(SO$_4$)(SeO$_4$)

CRYSTAL SYSTEM: Orthorhombic
CLASS: 222
SPACE GROUP: P22$_1$2
Z: 4
LATTICE CONSTANTS:
  $a$ = 8.42
  $b$ = 10.96
  $c$ = 7.00

3 STRONGEST DIFFRACTION LINES:
  3.338 (100)
  3.234 (100)
  3.015 (100)
OPTICAL CONSTANTS:
  $\alpha$ = 1.945
  $\beta$ = 1.966
  $\gamma$ = 1.983
  (−)2V = 80°
HARDNESS: 3–3½
DENSITY: 6.55 (Calc.)
CLEAVAGE: {101} good
          {010} fair
  HABIT: Crystals well-formed, 0.1–1.5 mm long, elongated on [010]. Length 10–15 times the width.
  COLOR-LUSTER: Colorless, partially transparent with colorless streak and vitreous luster; some crystals present a very thin whitish coating with a dull luster.
  MODE OF OCCURRENCE: Olsacherite is an alteration product of penroseite and occurs very sparingly in well-formed crystals coating the walls of small cracks or cavities in the Pacajake mine, near Hiaco, Bolivia. In this deposit the primary minerals are: penroseite, clausthalite, naumannite, tiemannite, pyrite, siderite, barite, hematite, calcite; and the secondary minerals are: cerussite, anglesite, wulfenite, limonite, chalcomenite, ahlfeldite, and native selenium.
  BEST REF. IN ENGLISH: Hurlbut, C. S. Jr., and Aristarain, L. F., *Am. Min.*, **54**: 1519–1527 (1969).

## OLSHANSKYITE (Inadequately described mineral)
Ca$_3$[B(OH)$_4$]$_4$(OH)$_2$

CRYSTAL SYSTEM: Apparently monoclinic, might be triclinic
3 STRONGEST DIFFRACTION LINES:
  2.81 (100)
  3.05 ( 80)
  7.61 ( 50)
OPTICAL CONSTANTS:
  $\alpha$ = 1.557
  $\beta$ = 1.568 (Calc.)
  $\gamma$ = 1.570
  (−)2V ≃ 54°
HARDNESS: 4
DENSITY: 2.23 (Meas.)
CLEAVAGE:
  HABIT: As tr... tically twinned cr...
  COLOR-LUSTE...
  MODE OF OCC... in "magnesian ska...
  BEST REF. IN... I. B., and Pertsev...

## OMPHACITE
(Ca, Na)(Mg, Fe$^{2+}$...
Pyroxene Group

CRYSTAL SYST...

CLASS: 2/m
SPACE GROUP: C2/c
Z: 4
3 STRONGEST DIFFRACTION LINES:
2.98 (100)
1.40 ( 80)
2.13 ( 70)
OPTICAL CONSTANTS:
$\alpha$ = 1.662–1.691
$\beta$ = 1.670–1.700
$\gamma$ = 1.688–1.718
(+)2V = 58°–83°
HARDNESS: 5–6
DENSITY: 3.29–3.39 (Meas.)
3.362 (Calc.)
CLEAVAGE: {110} good
{100} parting
Fracture uneven to conchoidal. Brittle.

HABIT: Massive, granular to foliated. Often interlaminated with amphibole. Twinning on {100} common, simple, lamellar.

COLOR-LUSTER: Grass green, dark green. Translucent. Vitreous to silky.

MODE OF OCCURRENCE: Occurs in association with garnet as a constituent of eclogites and related rocks. Typical occurrences are found at Healdsburg, Sonoma County, California, and at localities in Greenalnd, Norway, France, Germany, Austria, Ghana, and Japan.

BEST REF. IN ENGLISH: Deer, Howie, and Zussman, "Rock Forming Minerals," v. 2, p. 154–160, New York, Wiley, 1963.

## ONOFRITE = Variety of metacinnabar
Hg(S, Se)

## ONORATOITE
$Sb_8O_{11}Cl_2$

CRYSTAL SYSTEM: Triclinic
CLASS: $\bar{1}$
SPACE GROUP: $P\bar{1}$
Z: 1
LATTICE CONSTANTS:
$a$ = 18.92     $\alpha \simeq \gamma \simeq 90°$
$b$ = 4.03      $\beta$ = 110°
$c$ = 10.31
3 STRONGEST DIFFRACTION LINES:
3.190 (100)
4.39 ( 65)
2.677 ( 60)
TICAL CONSTANTS:
$> \beta$ > 2.18
$\gamma$ > 2.23

: Not determined.
(Meas.)
(Calc.)
etermined.
ular, elongated [010], flattened
{201}, {401}, {412}.

COLOR-LUSTER: White.

MODE OF OCCURRENCE: Occurs as an oxidation product of stibnite in the old antimony mines of Cetine di Cotoriano, near Rosia, Siena, Tuscany, Italy.

BEST REF. IN ENGLISH: Belluomini, G., Fornaseri, M., and Nicoletti, M., *Min. Mag.*, **36**: 1037–1044 (1968).

## OOSTERBOSCHITE
$(Pd, Cu)_7Se_5$

CRYSTAL SYSTEM: Orthorhombic
Z: 8
LATTICE CONSTANTS:
$a$ = 10.42
$b$ = 10.60
$c$ = 14.43
3 STRONGEST DIFFRACTION LINES:
2.647 (100)
2.600 ( 80)
1.847 ( 80)
OPTICAL CONSTANTS: Anisotropy medium strong with bluish gray to brownish gray colors. Reflectances at 12 wavelengths show maxima at 560 nm; with min. and max. values of R: 460 nm, 40.9, 44.5; 560, 45.5, 49.8; 640, 43.0, 48.1%.
HARDNESS: Microhardness 340 kg/mm$^2$.
DENSITY: 8.48 (Calc.)
CLEAVAGE: Not determined.

HABIT: Massive; as grains up to 0.4 mm. Twinning polysynthetic.

COLOR-LUSTER: Light white-yellow with creamy tint in reflected light. Opaque.

MODE OF OCCURRENCE: The mineral occurs associated with trogtalite, covellite, and selenian digenite in the oxidation zone of the Musonoi copper-cobalt deposit, Katanga, Zaire.

BEST REF. IN ENGLISH: Johan, Zdenek, Picot, Paul, Pierrot, Roland, and Verbeek, Theodore, *Am. Min.*, **57**: 1553 (1972).

## OPAL (hydrated silica containing cristobalite)
$SiO_2 \cdot nH_2O$

CRYSTAL SYSTEM: Noncrystalline
OPTICAL CONSTANT:
$N$ = 1.435–1.455
HARDNESS: 5½–6½
DENSITY: 1.99–2.25 (Meas.)
CLEAVAGE: None.
Fracture conchoidal; also irregular or splintery. Brittle to very brittle.

HABIT: Massive; compact, porous or earthy; as microcrystalline aggregates of cristobalite crystallites containing abundant nonessential water. Often rock-forming or as veinlets, cavity fillings, concretions, or pseudomorphous after other minerals. Commonly botryoidal, reniform, globular, coralloidal, stalactitic.

COLOR-LUSTER: Colorless, white, bluish white; also yellow to yellowish or reddish brown, brown, orange, greenish, bluish, gray to black. Often exhibits striking internal play of colors by reflected light. Transparent to nearly opaque. Vitreous, resinous, pearly, waxy. Streak white. Sometimes fluoresces green to yellowish green under ultra-violet light.

MODE OF OCCURRENCE: The synonomy and varietal nomenclature of opal is extensive; principal names in contemporary usage are: geyserite, siliceous sinter, diatomaceous earth, tripolite, fire opal, precious opal, noble opal, girasol opal, common opal, hyalite, moss opal, milk opal, wood opal, cacholong, hydrophane, and tabasheer.

Occurs widespread in a great variety of geological environments as a low-temperature mineral usually confined to surface or near-surface deposits. Typical occurrences are found in California (diatomaceous earth); Virgin Valley, Nevada (precious opal); and in Washington, Oregon, Idaho, Wyoming, South Dakota, Nebraska, Colorado, Arizona, New Mexico, North Carolina, and Georgia. Fine gem opal is found in Mexico, Honduras, Hungary, and at numerous localities in New South Wales and Queensland, Australia.

BEST REF. IN ENGLISH: Frondel, Clifford, "Dana's System of Mineralogy," 7th Ed., v. III, p. 287–306, New York, Wiley, 1962.


## ORCELITE (Inadequately described mineral)
$Ni_2As$

CRYSTAL SYSTEM: Hexagonal
3 STRONGEST DIFFRACTION LINES:
  1.977 (100)
  1.918 (100)
  2.110 ( 40)
HARDNESS: Not determined.
DENSITY: 6.5 (Meas.)
CLEAVAGE: Not determined.
HABIT: Massive; lamellar structure.
COLOR-LUSTER: Rose bronze. Opaque. Metallic.
MODE OF OCCURRENCE: Found in drill cores in serpentinized harzburgite, Tiebaghi massif, New Caledonia.
BEST REF. IN ENGLISH: Caillère, Simonne, Avias, Jacques, and Falgueirettes, Jean, *Am. Min.*, **45**: 753–754 (1960).


## ORDOÑEZITE
$ZnSb_2O_6$

CRYSTAL SYSTEM: Tetragonal
CLASS: 4/m 2/m 2/m
SPACE GROUP: P4/mnm
Z: 2
LATTICE CONSTANTS:
  $a = 4.67$
  $c = 9.24$
3 STRONGEST DIFFRACTION LINES:
  1.72 (100)
  3.26 ( 90)
  2.55 ( 80)
OPTICAL CONSTANTS:
  $\epsilon > \omega > 1.95$
    (+)

HARDNESS: 6½
DENSITY: 6.635 (Meas.)
CLEAVAGE: None. Fracture conchoidal.
HABIT: As small (to 2 mm) tetragonal crystals with forms {001}, {011} and {110}; usually twinned on {013}. Also as drusy or stalactitic masses.
COLOR-LUSTER: Very light to very dark brown, variable from colorless to pearl gray, olive buff to dark olive, light yellowish olive, bone brown (Ridgeway), transparent. Adamantine.
MODE OF OCCURRENCE: Occurs in small veinlets in rhyolitic rocks associated with cassiterite, tridymite, cristobalite, sanidine, topaz, hematite, and fluorite at the Santín mine, Cerro de las Fajas, Santa Caterina, Guanajuato, Mexico.
BEST REF. IN ENGLISH: Switzer, George, and Foshag, W. F., *Am. Min.*, **40**: 64–69 (1955).


## OREGONITE
$Ni_2FeAs_2$

CRYSTAL SYSTEM: Hexagonal
Z: 3
LATTICE CONSTANTS:
  $a = 6.083$
  $c = 7.130$
3 STRONGEST DIFFRACTION LINES:
  2.314 (100)
  2.119 (100)
  1.991 ( 70)
HARDNESS: ~5
DENSITY: 6.92 (Calc.)
HABIT: As fine-grained water-rolled pebbles with smooth brown crust.
COLOR-LUSTER: Metallic white with high reflectivity under the microscope. Opaque.
MODE OF OCCURRENCE: Occurs associated with small amounts of native copper, bornite, chalcopyrite, molybdenite, chromite, and perhaps niccolite in a gangue of penninite and serpentine in Josephine Creek, Oregon.
BEST REF. IN ENGLISH: Ramdohr, Paul, and Schmitt, Margaret, *Am. Min.*, **45**: 1130 (1960).


## ORIENTITE
$Ca_2Mn_3^{3+}Si_3O_{12}(OH)$

CRYSTAL SYSTEM: Orthorhombic
CLASS: 2/m 2/m 2/m
SPACE GROUP: Bbmm
Z: 4
LATTICE CONSTANTS:
  $a = 9.04$
  $b = 19.14$
  $c = 6.08$
3 STRONGEST DIFFRACTION LINES:
  2.704 (100)
  5.06 ( 90)
  4.394 ( 90)

OPTICAL CONSTANTS:
$\alpha = 1.756$
$\beta = 1.777$
$\gamma = 1.794$
$(-)2V = 68°-83°$
HARDNESS: 4½–5
DENSITY: 3.05 (Meas.)
3.09 (Calc.)
CLEAVAGE: {001} imperfect
{120} imperfect
HABIT: Crystals prismatic to tabular, up to 1 mm in length. Dominant forms {110}, {001}, and {010}.
COLOR-LUSTER: Deep red-brown to almost black. Resinous to subvitreous. Streak brown.
MODE OF OCCURRENCE: Occurs as a minor constituent of oxide manganese ores at several localities in Oriente Province, Cuba.
BEST REF. IN ENGLISH: Sclar, Charles B., *Am. Min.*, **46**: 226–232 (1961).

## ORLITE (= Kasolite ?) (species status uncertain)
$Pb_3(UO_2)_3(Si_2O_7)_2 \cdot 6H_2O$

3 STRONGEST DIFFRACTION LINES:
3.226 (100)
1.678 ( 70)
6.356 ( 50)
OPTICAL CONSTANTS:
$\beta = 1.788$
$\gamma = 1.793$
DENSITY: 5.307
HABIT: As radiating aggregates of fine acicular crystallites, measuring tenths of a millimeter in size.
COLOR-LUSTER: Light creamy yellow; waxy.
MODE OF OCCURRENCE: Occurs associated with uranophane and kasolite in the middle horizon of the oxidation zone of an uranium deposit in liparite at an unspecified locality in USSR.
BEST REF. IN ENGLISH: *Am. Min.*, **43**: 381 (1958).

## ORPIMENT
$As_2S_3$

CRYSTAL SYSTEM: Monoclinic
CLASS: 2/m
SPACE GROUP: $P2_1/n$
Z: 4
LATTICE CONSTANTS:
$a = 11.46$
$b = 9.59$
$c = 4.24$
$\beta = 90.45°$
3 STRONGEST DIFFRACTION LINES:
4.82 (100)
2.70 ( 53)
4.00 ( 47)
OPTICAL CONSTANTS:
$\alpha = 2.4$
$\beta = 2.81$ (Li)
$\gamma = 3.02$
$(-)2V = 76°$

HARDNESS: 1½–2
DENSITY: 3.49 (Meas.)
3.52 (Calc.)
CLEAVAGE: {010} perfect
{100} in traces
Sectile. Cleavage lamellae flexible, inelastic.
HABIT: Crystals short prismatic, elongated along *c*-axis, or monoclinic in aspect. Usually as foliated, columnar, reniform, granular, or powdery aggregates. Twinning on {100}.
COLOR-LUSTER: Bright yellow-orange to lemon yellow or brownish yellow. Transparent to translucent. Resinous, pearly on cleavage. Streak pale yellow.
MODE OF OCCURRENCE: Occurs in low-temperature hydrothermal veins and as a hot springs deposit, commonly associated with realgar and stibnite. Found in Nevada as fine crystals and foliated masses at the Getchell mine near Golconda, Humboldt County, with realgar at Manhattan, Nye County, and at Steamboat Springs, Washoe County; as large crystals with calcite at Mercur, Tooele County, Utah; in Kern, Lake, Siskiyou, Sonoma, and Trinity counties, California; and in Yellowstone Park, Wyoming. It also occurs in Peru, France, Italy, Germany, Hungary, Greece, Yugoslavia, Turkey, China, Japan, and as exceptional crystals up to 5 cm in size at Lukhumis, Georgia, USSR.
BEST REF. IN ENGLISH: Palache, et al., "Dana's System of Mineralogy," 7th Ed., v. I, p. 266–269, New York, Wiley, 1944.

## ORTHITE = Synonym for allanite

## ORTHO-ANTIGORITE = Structural variety of antigorite

## ORTHOCHAMOSITE = Orthorhombic chamosite
$(Mg, Fe^{2+})_3 Fe^{3+}(AlSi_3)O_{10}(OH)_8$
Septechlorite Group

## ORTHO-CHRYSOTILE = Structural variety of chrysotile
$Mg_3Si_2O_5(OH)_4$
Serpentine Group

## ORTHOCLASE
$KAlSi_3O_8$
Alkali Feldspar Group
Dimorphous with microcline
Var. Adularia
Sanidine
Valencianite

CRYSTAL SYSTEM: Monoclinic
CLASS: 2/m
SPACE GROUP: C2/m
Z: 4

LATTICE CONSTANTS:
$a = 8.56$
$b = 13.00$
$c = 7.19$
$\beta = 116°$

3 STRONGEST DIFFRACTION LINES:
3.18 (100)
4.02 ( 90)
3.80 ( 80)

OPTICAL CONSTANTS:
$\alpha = 1.518-1.529$
$\beta = 1.522-1.533$
$\gamma = 1.522-1.539$
$(-)2V = 33°-103°$

HARDNESS: 6-6½

DENSITY: 2.55-2.63 (Meas.)
2.570 (Calc.)

CLEAVAGE: {001} perfect
{010} perfect
{100}, {110}, {$\bar{1}$10}, {$\bar{2}$01} partings. Fracture uneven to conchoidal. Brittle.

HABIT: Crystals usually short prismatic, blocky, sometimes orthorhombic or tetragonal in aspect; also tabular, flattened along $b$-axis. Commonly massive, cleavable to granular, lamellar or cryptocrystalline. Twinning very common; simple, multiple, and repeated. Principal twin laws: Carlsbad, Baveno, Manebach.

COLOR-LUSTER: Colorless, white, gray, yellow, reddish, greenish. Transparent to translucent. Vitreous to pearly. Streak white.

MODE OF OCCURRENCE: Occurs widespread as a common constituent of igneous rocks such as pegmatites, granites, syenites, rhyolites, and trachytes; in crystalline schists and other metamorphic rocks; in ore veins; as a constituent of arkosic sedimentary rocks; and as an authigenic mineral in sedimentary deposits. Typical occurrences are found in Alaska, California, Oregon (sanidine), Nevada, Idaho (valencianite), Colorado, Arizona, New Mexico, Texas, South Dakota, Arkansas, Pennsylvania, New York, and other New England states. Also in Canada, Mexico (valencianite), England, Norway, France, Italy, Switzerland (adularia), Austria, Germany (sanidine), Czechoslovakia, Yugoslavia, USSR, Madagascar, Taiwan, Thailand, Korea, Japan, and elsewhere.

BEST REF. IN ENGLISH: Deer, Howie, and Zussman, "Rock Forming Minerals," v. 4, p. 6-93, New York, Wiley, 1963. Barth, T. F. W., "Feldspars," Wiley-Interscience, 1969.

## ORTHOERICSSONITE
$BaMn_2 Fe(Si_2 O_7)(O, OH)$
Dimorphous with ericssonite

CRYSTAL SYSTEM: Orthorhombic
CLASS: 2/m 2/m 2/m or mm2
SPACE GROUP: Pnmn or Pn2n
Z: 4
LATTICE CONSTANTS:
$a = 20.32$
$b = 7.03$
$c = 5.34$

3 STRONGEST DIFFRACTION LINES:
3.510 (100)
2.687 ( 70)
2.132 ( 70)*
*This line distinguishes orthoericssonite from ericssonite.

OPTICAL CONSTANTS:
$\alpha = 1.807$
$\beta = 1.833$
$\gamma = 1.89$
$(+)2V = 43°$

HARDNESS: 4½

DENSITY: 4.21 (Meas.)

CLEAVAGE: {100} perfect
{011} distinct
Very brittle.

HABIT: Massive; as embedded plates up to 2 cm in diameter.

COLOR-LUSTER: Reddish black. Streak rich brown.

Weakly magnetic.

MODE OF OCCURRENCE: Occurs intergrown with ericssonite in schefferite-rhodonite-tephroite skarn at Långban, Sweden.

BEST REF. IN ENGLISH: Moore, Paul B., *Lithos*, **4**: 137-145 (1971).

## ORTHOFERROSILITE
$FeSiO_3$
Pyroxene Group

CRYSTAL SYSTEM: Orthorhombic
CLASS: 2/m 2/m 2/m
SPACE GROUP: Pbca
Z: 16
LATTICE CONSTANTS:
$a = 18.433$
$b = 9.060$
$c = 5.258$

3 STRONGEST DIFFRACTION LINES:
4.62 (100)
2.911 ( 90)
6.48 ( 70)

OPTICAL CONSTANTS:
$\alpha = 1.755-1.768$
$\beta = 1.763-1.770$
$\gamma = 1.772-1.788$
$(+)2V = 55°-90°$

HARDNESS: 5-6

DENSITY: 3.88 (Meas.)
4.043 (Calc.)

CLEAVAGE: {210} good
{100} parting
{010} parting
Fracture uneven. Brittle.

HABIT: Crystals short prismatic, rare. Usually massive. Twinning on {100}.

COLOR-LUSTER: Green, dark brown. Vitreous. Translucent to nearly opaque.

MODE OF OCCURRENCE: Occurs in thermally metamorphosed iron-rich rock, Yu-hsi-kou district, Manchuria.

BEST REF. IN ENGLISH: Deer, Howie, and Zussman, "Rock Forming Minerals," v. 2, p. 8-41, New York, Wiley, 1963.

# ORTHOPINAKIOLITE
$Mg_3MnMn_2B_2O_{10}$
Dimorph of pinakiolite

CRYSTAL SYSTEM: Orthorhombic
CLASS: mm2 or 2/m 2/m 2/m
SPACE GROUP: Pnn2 or Pnnm
Z: 8
LATTICE CONSTANTS:
  $a = 18.45$
  $b = 12.70$
  $c = 6.07$
3 STRONGEST DIFFRACTION LINES:
  2.59 (100)
  5.17 ( 90)
  2.52 ( 90)
HARDNESS: 6
DENSITY: 4.03 (Meas.)
CLEAVAGE: {010} good
Very brittle.
HABIT: As thin tablets {010} with rectangular outline. Rarely short prismatic. Crystals often bent or broken.
COLOR-LUSTER: Black; metallic. Opaque. Streak brownish gray.
MODE OF OCCURRENCE: Occurs in granular dolomite associated with hausmannite and manganophyllite at Långban, Sweden.
BEST REF. IN ENGLISH: Randmets, Rein, *Am. Min.*, **46**: 768 (1961).

# OSANNITE = Synonym for riebeckite

# OSARIZAWAITE
$PbCuAl_2(SO_4)_2(OH)_6$
Aluminum analogue of beaverite

CRYSTAL SYSTEM: Hexagonal
CLASS: 3m
SPACE GROUP: R3m
Z: 3 (hexag.), 1 (rhomb.)
LATTICE CONSTANTS:
  (hexag.)        (rhomb.)
  $a = 7.05$    $a_{rh} = 7.045$
  $c = 17.25$    $\alpha = 60°03'$
3 STRONGEST DIFFRACTION LINES:
  3.00 (100)
  5.75 ( 70)
  3.52 ( 60)
OPTICAL CONSTANTS:
  $\omega = 1.71$
  $\epsilon = 1.74$
  (+)
HARDNESS: Not determined.
DENSITY: 4.037 (Meas.)
        4.114 (Calc.)
CLEAVAGE: Not determined.
HABIT: As friable aggregates of minute crystals commonly showing hexagonal outline and rarely the development of the rhombohedron.

COLOR-LUSTER: Close to Ridgway's "Veronese Green." Earthy.
MODE OF OCCURRENCE: Occurs as earthy friable encrustations on barite and quartz together with minor clay, hematite, and goethite in the Marble Bar area of Western Australia. Also in the oxidized zone of lead-zinc-copper veins associated with anglesite, kaolin, limonite, linarite, azurite, brochantite, malachite, chalcocite, covellite, sulfur, chalcedony, and hydrous manganese oxides at the type locality, the Ozarizawa Mine, Akita Prefecture, Japan.
BEST REF. IN ENGLISH: Morris, R. C., *Am. Min.*, **47**: 1079-1093 (1962). Taguchi, Yasuro, *Am. Min.*, **47**: 1216-1217 (1962). Morris, R. C., *Am. Min.*, **48**: 947 (1963).

# OSARSITE
(Os, Ru)AsS

CRYSTAL SYSTEM: Monoclinic
Z: 4
LATTICE CONSTANTS:
  $a = 5.933$
  $b = 5.916$
  $c = 6.009$
  $\beta = 112°21'$
3 STRONGEST DIFFRACTION LINES:
  3.79 (100)
  1.892 (100)
  1.870 ( 80)
OPTICAL CONSTANTS: Anisotropy in air weak but distinct.
HARDNESS: Not determined.
DENSITY: 8.44 (Calc.)
CLEAVAGE: Not determined.
HABIT: Known only as one single discrete 150 $\mu$ grain; in polycrystalline intergrowth with irarsite.
COLOR-LUSTER: Gray in reflected light. Opaque. Metallic.
MODE OF OCCURRENCE: Found in platinum placer sands from Gold Bluff, Humboldt County, California.
BEST REF. IN ENGLISH: Snetsinger, K. G., *Am. Min.*, **57**: 1029-1036 (1972).

# OSBORNITE
TiN

CRYSTAL SYSTEM: Cubic
CLASS: 4/m $\bar{3}$ 2/m
SPACE GROUP: Fm3m
Z: 4
LATTICE CONSTANT:
  $a = 4.235$
3 STRONGEST DIFFRACTION LINES:
  2.12 (100)
  2.44 ( 75)
  1.496 ( 55)
HARDNESS: Not determined.
DENSITY: 5.40 (Calc.)
CLEAVAGE: Not determined.

HABIT: As minute octahedral crystals.

COLOR-LUSTER: Golden yellow.

MODE OF OCCURRENCE: Occurs in oldhamite and pyroxene in the Busti, India, meteorite.

BEST REF. IN ENGLISH: Palache, et al., "Dana's System of Mineralogy," 7th Ed., v. I, p. 124, New York, Wiley, 1944.

## OSMIRIDIUM = Iridosmine

## OSMIUM = Siserskite

## OSUMILITE
$(K, Na)(Mg, Fe^{2+})_2 (Al, Fe^{3+})_3 (Si, Al)_{12} O_{30} \cdot H_2O$

CRYSTAL SYSTEM: Hexagonal
CLASS: 6/m 2/m 2/m
SPACE GROUP: P6/mcc
Z: 2
LATTICE CONSTANTS:
  $a = 10.155$
  $c = 14.284$
3 STRONGEST DIFFRACTION LINES:
  3.24 (100)
  7.17 ( 80)
  5.08 ( 80)
OPTICAL CONSTANTS:
  $\omega = 1.550$   or   $\omega = 1.545-1.547$
  $\epsilon = 1.546$        $\epsilon = 1.549-1.551$ (Na)
                              (+)
DENSITY: 2.64 (Meas.)
CLEAVAGE: None. Brittle.

HABIT: Crystals short prismatic, elongated parallel to $c$-axis, or tabular {0001}. Well-developed forms: c{0001}, m{10$\bar{1}$0}, and a{11$\bar{2}$0}; faces of j{21$\bar{3}$0} common, usually very small. Crystals range from less than 1 mm to 5 mm in length.

COLOR-LUSTER: Dark blue to black; light blue in thin section. Translucent. Vitreous.

MODE OF OCCURRENCE: Occurs as fine crystals in association with cristobalite, hortonolite, feldspar, and quartz in vugs in andesite at MacKenzie Pass, Lane County, Oregon. Also found in hypersthene plagioliparite (rhyodacite) volcanic rock which contains andesine, tridymite, quartz, hypersthene, biotite, and magnetite at Sakkabira, Kagosima Prefecture, Kyushu, Japan.

BEST REF. IN ENGLISH: Miyashiro, Akiho, *Am. Min.*, **41**: 104-116 (1956). Brown, G. E., and Gibbs, G. V., *Am. Min.*, **54**: 101-116 (1969).

## OTAVITE
$CdCO_3$

CRYSTAL SYSTEM: Hexagonal
CLASS: $\bar{3}$ 2/m
SPACE GROUP: R$\bar{3}$c
Z: 6 (hexag.), 2 (rhomb.)

LATTICE CONSTANTS:
  (hexag.)          (rhomb.)
  $a = 4.912$       $a_{rh} = 6.11$
  $c = 16.199$      $\alpha = 47°24'$
3 STRONGEST DIFFRACTION LINES:
  2.95 (100)
  3.78 ( 80)
  2.46 ( 35)
HARDNESS: Not determined.
DENSITY: 4.96 (Meas. synthetic)
         5.03 (Calc.)
CLEAVAGE: Not brittle.
HABIT: As crusts of minute rhombohedral crystals.
COLOR-LUSTER: White to yellow-brown and reddish. Brilliant, adamantine.
MODE OF OCCURRENCE: Occurs as a secondary mineral associated with smithsonite, azurite, malachite, cerussite, pyromorphite, and olivenite at Tsumeb, South West Africa.

BEST REF. IN ENGLISH: Palache, et al., "Dana's System of Mineralogy," 7th Ed., v. II, p. 181, New York, Wiley, 1951.

## OTTEMANITE
$Sn_2S_3$

CRYSTAL SYSTEM: Orthorhombic
CLASS: 2/m 2/m 2/m
SPACE GROUP: Pnam
Z: 1
LATTICE CONSTANTS:
  $a = 8.79$
  $b = 14.02$  (synthetic)
  $c = 3.74$
3 STRONGEST DIFFRACTION LINES:
  4.131 (100)
  5.495 ( 75) (synthetic)
  2.670 ( 45)
HARDNESS: Less than for herzenbergite (SnS).
DENSITY: Not determined.
CLEAVAGE: Not determined.
HABIT: Observed in polished sections as small laths, commonly twinned, replacing stannite and replaced by cassiterite.
COLOR-LUSTER: Gray by reflected light, weakly pleochroic, strongly anisotropic; internal reflections orange-brown. Opaque.
MODE OF OCCURRENCE: Occurs in tin sulfide ore from a zone of secondary enrichment or oxidation at Cerro de Potosi, Bolivia.

BEST REF. IN ENGLISH: Möh, Gunter H., and Berndt, Fritz, *Am. Min.*, **50**: 2107 (1965).

## OTTRELITE = Variety of chloritoid rich in manganese

## OVERITE
$CaMg(H_2O)_4 Al(OH) [PO_4]_2$
Aluminum analogue of segelerite

CRYSTAL SYSTEM: Orthorhombic
CLASS: 2/m 2/m 2/m
SPACE GROUP: Pcaa
Z: 8
LATTICE CONSTANTS:
  $a = 14.78$
  $b = 18.78$
  $c = 7.14$
3 STRONGEST DIFFRACTION LINES:
  2.832 (100)
  9.4 ( 80)
  5.29 ( 60)
OPTICAL CONSTANTS:
  $\alpha = 1.568$
  $\beta = 1.574$
  $\gamma = 1.580$
  $(-)2V = 75°$
HARDNESS: 3½–4
DENSITY: 2.53 (Meas.)
        2.48 (Calc.)
CLEAVAGE: {010} perfect
          {100} poor
Brittle.
HABIT: Crystals minute, platy to lath-like, often as subparallel growths on {010}. Also massive, as subparallel aggregates of coarse plates.
COLOR-LUSTER: Light green to colorless. Transparent to translucent. Vitreous.
MODE OF OCCURRENCE: Overite occurs associated with crandallite and other phosphate minerals in variscite nodules at Fairfield, Utah County, Utah.
BEST REF. IN ENGLISH: Palache, et al., "Dana's System of Mineralogy," 7th Ed., v. II, p. 979–980, New York, Wiley, 1951. Moore, Paul Brian, *Am. Min.*, **59**: 48–59 (1974).

## OWYHEEITE
$Ag_2 Pb_5 Sb_6 S_{15}$

CRYSTAL SYSTEM: Orthorhombic
CLASS: 2/m 2/m 2/m
SPACE GROUP: Pnam
Z: 4
LATTICE CONSTANTS:
  $a = 22.82$
  $b = 27.20$
  $c = 8.19$
3 STRONGEST DIFFRACTION LINES:
  3.25 (100)
  3.49 ( 70)
  2.84 ( 60)
HARDNESS: 2½
DENSITY: 6.22–6.51 (Meas.)
        6.41 (Calc.)

CLEAVAGE: {001}. Brittle; thin fibers flexible.
HABIT: Crystals acicular; also massive with distinct to somewhat indistinct fibrous structure.
COLOR-LUSTER: Lead gray to silver white, tarnishing yellowish. Opaque. Metallic. Streak reddish brown.
MODE OF OCCURRENCE: Occurs in Idaho at the Poorman mine, Owyhee County, associated with pyrargyrite, sphalerite, and quartz, and in the Banner district, Boise County; in Nevada at Morey, Nye County, associated with sphalerite; and in Colorado at the Domingo and Garfield mines, Gunnison County.
BEST REF. IN ENGLISH: Palache, et al., "Dana's System of Mineralogy," 7th Ed., v. I, p. 423, New York, Wiley, 1944. Robinson, *Am. Min.*, **34**: 401 (1949).

## OXAMMITE
$(NH_4)_2 C_2 O_4 \cdot H_2 O$

CRYSTAL SYSTEM: Orthorhombic
CLASS: 222
SPACE GROUP: $P2_1 2_1 2$
Z: 2
LATTICE CONSTANTS:
  $a = 8.035$
  $b = 10.31$ (synthetic)
  $c = 3.801$
3 STRONGEST DIFFRACTION LINES:
  2.666 (100)
  6.32 ( 95) (synthetic)
  3.057 ( 95)
OPTICAL CONSTANTS:
  $\alpha = 1.438$
  $\beta = 1.547$ (Na) (synthetic)
  $\gamma = 1.595$
  $(-)2V = 62°$
HARDNESS: 2½
DENSITY: ~1.5
        1.541 (Calc.)
CLEAVAGE: {001} distinct
HABIT: Usually as lamellar masses, pulverulent; rarely as distinct crystals.
COLOR-LUSTER: Colorless (synthetic) to yellowish white. Transparent.
MODE OF OCCURRENCE: Occurs with mascagnite in guano deposits on the Guañape Islands, Peru.
BEST REF. IN ENGLISH: Palache, et al., "Dana's System of Mineralogy," 7th Ed., v. II, p. 1103–1104, New York, Wiley, 1951.

## OXYKERCHENITE = Oxidized vivianite
$Fe_8 (OH)_8 (PO_4)_6 \cdot 17H_2 O$

# P

## PABSTITE
$BaSnSi_3O_9$

CRYSTAL SYSTEM: Hexagonal
CLASS: $\bar{6}m2$
SPACE GROUP: $P\bar{6}c2$
Z: 2
LATTICE CONSTANTS:
  $a = 6.706$
  $c = 9.829$
3 STRONGEST DIFFRACTION LINES:
  3.77 (100)
  2.78 ( 90)
  3.37 ( 50)
OPTICAL CONSTANTS:
  $\omega = 1.685$
  $\epsilon = 1.674$
    (−)
HARDNESS: ~6
DENSITY: 4.03 (Meas.)
         4.07 (Calc.)
CLEAVAGE: Not determined.
HABIT: As small anhedral grains; rarely in crystal form showing trigonal outline. Grains commonly less than 2 mm in diameter; aggregates do not exceed 10 mm.
COLOR-LUSTER: Colorless to white with a pink tinge on fresh surface. Fluoresces bluish white in short-wave ultraviolet light.
MODE OF OCCURRENCE: Occurs in small fractures and as scattered grains in association with quartz, calcite, tremolite, witherite, phlogopite, diopside, forsterite, and taramellite, at the Pacific Limestone Products quarry at Santa Cruz, California.
BEST REF. IN ENGLISH: Gross, Eugene B., Wainwright, John E. N., Evans, Bernard W., *Am. Min.*, **50**: 1164–1169 (1965).

## PACHNOLITE
$NaCaAlF_6 \cdot H_2O$

CRYSTAL SYSTEM: Monoclinic
CLASS: 2/m or m

SPACE GROUP: C2/c or Cc
Z: 16
LATTICE CONSTANTS:
  $a = 12.12$ kX
  $b = 10.39$
  $c = 15.68$
3 STRONGEST DIFFRACTION LINES:
  3.95 (100)
  1.971 ( 90)
  2.79 ( 70)
OPTICAL CONSTANTS:
  $\alpha = 1.411$
  $\beta = 1.420$
  $\gamma = 1.413$
(+)2V = 76°
HARDNESS: 3
DENSITY: 2.983 (Meas.)
         2.97 (Calc.)
CLEAVAGE: {001} indistinct
Fracture uneven. Brittle.
HABIT: Crystals slender prismatic, elongated along c-axis. Commonly acutely terminated. Prism faces striated parallel to intersection with {001}. Also massive, cleavable to granular, and as chalcedony-like masses. Twinning on {100} common.
COLOR-LUSTER: Colorless to white. Transparent to translucent. Vitreous.
MODE OF OCCURRENCE: Occurs as an alteration product of cryolite, associated with elpasolite and other secondary fluorides, in pegmatite at St. Peters Dome, El Paso County, Colorado. It is also found as aggregates of excellent microcrystals at the cryolite deposit at Ivigtut, Greenland.
BEST REF. IN ENGLISH: Palache, et al., "Dana's System of Mineralogy," 7th Ed., v. II, p. 114–116, New York, Wiley, 1951.

## PAIGEITE = Vonsenite

## PAINITE
$Al_{20}Ca_4BSiO_{38}$

CRYSTAL SYSTEM: Hexagonal

CLASS: 6, 6/m, or 622
SPACE GROUP: $P6_3/m$ (if holohedral), $P6_3$, $P6_322$
Z: 2
LATTICE CONSTANTS:
$a = 8.725$
$c = 8.46$
3 STRONGEST DIFFRACTION LINES:
5.76 (100)
2.52 (100)
3.70 ( 80)
OPTICAL CONSTANTS:
$\omega = 1.8159$
$\epsilon = 1.7875$
(-)
HARDNESS: ~8
DENSITY: 4.00 (Meas.)
3.985 (Calc.)
HABIT: Only one terminated gem crystal weighing 1.7 g known. Pseudo-orthorhombic appearance; faces $a\{10\bar{1}0\}$, $m\{11\bar{2}0\}$ well-developed and large; $c\{0001\}$, $o\{11\bar{2}1\}$, $p\{11\bar{2}2\}$, $q\{20\bar{2}1\}$, $r\{10\bar{1}1\}$, $s\{10\bar{1}2\}$, $k\{12\bar{3}0\}$, and $l\{13\bar{4}0\}$ also present.
COLOR-LUSTER: Garnet red, transparent; vitreous.
MODE OF OCCURRENCE: Found near Mogok, Burma.
BEST REF. IN ENGLISH: Claringbull, G. F., Hey, Max H., and Payne, C. J., *Min. Mag.*, 31: 420–425 (1957).

# PALERMOITE
$(Li, Na)_2 (Sr, Ca)Al_4(PO_4)_4(OH)_4$

CRYSTAL SYSTEM: Orthorhombic
CLASS: 2/m 2/m 2/m
SPACE GROUP: Immm
Z: 4
LATTICE CONSTANTS:
$a = 7.315$
$b = 15.849$
$c = 11.556$
3 STRONGEST DIFFRACTION LINES:
3.089 (100)
4.360 ( 64)
3.129 ( 60)
OPTICAL CONSTANTS:
$\alpha = 1.627$
$\beta = 1.642$
$\gamma = 1.644$
$(-)2V \sim 20°$
HARDNESS: 5½
DENSITY: 3.22 (Meas.)
3.24 (Calc.)
CLEAVAGE: {100} perfect
{001} fair
Fracture fibrous to subconchoidal. Brittle.
HABIT: Crystals long prismatic, vertically striated.
COLOR-LUSTER: Colorless to white. Transparent. Vitreous to subadamantine. Streak white. Fluoresces white in direct X-ray beam.
MODE OF OCCURRENCE: Occurs in association with siderite, childrenite-eosphorite, green fibrous beraunite, whitlockite, brazilianite, apatite, and quartz as a late hydrothermal mineral in open cavities at the Palermo pegmatite, North Groton, New Hampshire.
BEST REF. IN ENGLISH: Mrose, Mary E., *Am. Min.*, 38: 354 (1953). Frondel, Clifford, and Ito, Jun, *Am. Min.*, 50: 777–779 (1965).

# PALLADIUM
Pd

CRYSTAL SYSTEM: Cubic
CLASS: $4/m\ \bar{3}\ 2/m$
SPACE GROUP: Fm3m
Z: 4
LATTICE CONSTANT:
$a = 3.8824$
3 STRONGEST DIFFRACTION LINES:
2.246 (100)
1.945 ( 42)
1.376 ( 25)
HARDNESS: 4½–5
DENSITY: 11.9 (Meas.)
12.04 (Calc.)
CLEAVAGE: None. Malleable and ductile.
HABIT: Usually as small grains, sometimes with radial-fibrous texture.
COLOR-LUSTER: Whitish steel gray. Opaque. Metallic.
MODE OF OCCURRENCE: Occurs very sparingly at Itabira, Minas Geraes, Brazil; in the alluvial deposits of the river Pinto, department of Cauca, Colombia; in the Ural Mountains, USSR; and in the platinum deposits in the Transvaal, Africa.
BEST REF. IN ENGLISH: Palache, et al., "Dana's System of Mineralogy," 7th Ed., v. I, p. 109–110, New York, Wiley, 1944.

# PALMIERITE
$(K, Na)_2 Pb(SO_4)_2$

CRYSTAL SYSTEM: Hexagonal
CLASS: $\bar{3}\ 2/m$
SPACE GROUP: $R\bar{3}m$
Z: 3
LATTICE CONSTANTS:
$a = 5.500$
$c = 20.86$ (synthetic)
3 STRONGEST DIFFRACTION LINES:
3.14 (100)
1.76 ( 90)
1.24 ( 90)
OPTICAL CONSTANTS:
$\omega = 1.712$
$\omega - \epsilon = $ strong
(-)
HARDNESS: Not determined.
DENSITY: 4.33 (Meas.)
4.352 (Calc.)
CLEAVAGE: Not determined.
HABIT: As microscopic hexagonal micaceous plates, flattened on {0001}.

COLOR-LUSTER: Colorless, white. Vitreous, pearly on {0001}.

MODE OF OCCURRENCE: Found in fumarole deposits formed after various eruptions at Mt. Vesuvius, Italy.

BEST REF. IN ENGLISH: Palache, et al., "Dana's System of Mineralogy," 7th Ed., v. II, p. 403–404, New York, Wiley, 1951.

## PALYGORSKITE (Attapulgite)
$(Mg, Al)_2 Si_4 O_{10}(OH) \cdot 4H_2O$

CRYSTAL SYSTEM: Monoclinic (and orthorhombic)
Palygorskite has many polytypes.
CLASS: 2/m
SPACE GROUP: C2/m
Z: 4
LATTICE CONSTANTS:
 $a \simeq 12.7$
 $b \simeq 17.9$
 $c \simeq 5.2$
 $\beta \approx 95°$
3 STRONGEST DIFFRACTION LINES:
 10.50 (100)
 3.23 (100)
 4.49 ( 80)
HARDNESS: Soft.
DENSITY: 2.217 (Meas.)
CLEAVAGE: {110} easy
Tough.
HABIT: Crystals lath-shaped, extremely elongated, in bundles. Usually as thin flexible sheets, composed of minute interlaced fibers, resembling leather or parchment.
COLOR-LUSTER: White, gray. Translucent. Dull.
MODE OF OCCURRENCE: Occurs mainly in hydrothermal veins or in altered serpentine or granitic rocks. Found as large sheets in association with calcite and barite crystals at Metaline Falls, Washington; at Attapulgas, Georgia; and at localities in Scotland, England, Shetland Islands, France, USSR, Morocco, and elsewhere.
BEST REF. IN ENGLISH: Christ, C. L., Hathaway, J. C., Hostetler, P. B., and Shepard, Anna O., *Am. Min.*, **54**: 198–205 (1969).

## PANDAITE
$(Ba, Sr)_2 (Nb, Ti)_2 (O, OH)_7$
Pyrochlore Group
Analogue of rijkeboerite

CRYSTAL SYSTEM: Cubic
CLASS: 4/m $\bar{3}$ 2/m
SPACE GROUP: Fd3m
Z: 8
LATTICE CONSTANT:
 $a = 10.562$
3 STRONGEST DIFFRACTION LINES:
 3.045 (100)
 6.10 ( 80)
 1.865 ( 70)
OPTICAL CONSTANT:
 $N = 2.85$

HARDNESS: 4½–5
DENSITY: 4.00 (Meas.)
 4.01 (Calc.)
CLEAVAGE: {111} very poor
Fracture conchoidal.
HABIT: As small euhedral crystals showing an octahedron with small cube faces.
COLOR-LUSTER: Yellowish gray to light olive gray.
MODE OF OCCURRENCE: Occurs in a highly weathered biotite-rich rock occurring in a roof pendant of the Mbeya (also called the Panda Hill) carbonatite, near Mbeya, Tanzania.
BEST REF. IN ENGLISH: Jager, E., Niggli, E., and Van der Veen, A. H., *Min. Mag.*, **32**: 10-25 (1959).

## PANDERMITE = Priceite

## PANETHITE
$(Na, Ca, K)_2 (Mg, Fe, Mn)_2 (PO_4)_2$

CRYSTAL SYSTEM: Monoclinic
CLASS: 2/m
SPACE GROUP: $P2_1/n$
Z: 32
LATTICE CONSTANTS:
 $a = 10.224$
 $b = 14.715$
 $c = 26.258$
 $\beta = 91°32'$
3 STRONGEST DIFFRACTION LINES:
 3.007 (100)
 2.710 ( 70)
 5.10 ( 60)
OPTICAL CONSTANTS:
 $\alpha = 1.567$
 $\beta = 1.576$
 $\gamma = 1.579$
 $(-)2V = 51°$
HARDNESS: Not determined.
DENSITY: 2.9-3.0 (Meas.)
 2.99 (Calc.)
CLEAVAGE: Not determined.
HABIT: Massive; as minute grains.
COLOR-LUSTER: Pale amber, transparent.
MODE OF OCCURRENCE: Occurs associated with whitlockite, brianite, albite, and enstatite in small pockets in the metallic phase of the Dayton octahedrite meteorite.
BEST REF. IN ENGLISH: Fuchs, L. H., Olson, E., and Henderson, E. P., *Am. Min.*, **53**: 509 (1968).

## PAPAGOITE
$CaCuAlSi_2 O_6(OH)_3$

CRYSTAL SYSTEM: Monoclinic
CLASS: 2/m
SPACE GROUP: C2/m
Z: 4

LATTICE CONSTANTS:
$a = 12.91$
$b = 11.48$
$c = 4.69$
$\beta = 100°38'$
3 STRONGEST DIFFRACTION LINES:
2.874 (100)
4.29 ( 90)
2.204 ( 90)
OPTICAL CONSTANTS:
$\alpha = 1.607$
$\beta = 1.641$
$\gamma = 1.672$
$(-)2V = 78°$
HARDNESS: 5–5½
DENSITY: 3.25 (Meas.)
          3.25 (Calc.)
CLEAVAGE: {100} distinct
Brittle.
HABIT: Crystals approximately equidimensional, usually less than 1 mm in length; slightly flattened parallel to {001} with faces in the zone [010] well developed. Commonly as microcrystalline aggregates and coatings.
COLOR-LUSTER: Cerulean blue. Transparent to translucent. Vitreous.
MODE OF OCCURRENCE: Found with fibrous radiating aggregates of ajoite in narrow veinlets and as coatings in metasomatically altered rocks at Ajo, Pima County, Arizona.
BEST REF. IN ENGLISH: Hutton, C. Osborne, and Vlisidis, Angelina C., *Am. Min.*, **45**: 599–611 (1960).

# PARABUTLERITE
$Fe^{3+}SO_4OH \cdot 2H_2O$
Dimorphous with butlerite

CRYSTAL SYSTEM: Orthorhombic
CLASS: 2/m 2/m 2/m
SPACE GROUP: Pmnb
Z: 8
LATTICE CONSTANTS:
$a = 7.38$
$b = 20.13$
$c = 7.22$
3 STRONGEST DIFFRACTION LINES:
4.99 (100)
3.11 (100)
5.85 ( 60)
OPTICAL CONSTANTS:
$\alpha = 1.598$
$\beta = 1.663$
$\gamma = 1.737$
$(+)2V = 87°$
HARDNESS: 2½
DENSITY: 2.55 (Meas.)
          2.538 (Calc.)
CLEAVAGE: {110} poor
Fracture conchoidal. Brittle.
HABIT: Crystals prismatic, striated in direction of elongation. Twinning on {142} rare.
COLOR-LUSTER: Light orange to light brownish orange. Transparent. Vitreous.

MODE OF OCCURRENCE: Occurs associated with butlerite, copiapite, and melanterite, at the Dexter No. 7 mine, Calf Mesa, San Rafael Swell, Utah; at Chuquicamata, Alcaparrosa, and Quetena, Chile; and with butlerite and other sulfates at the Santa Elena mine, La Alcaparrosa, Argentina.
BEST REF. IN ENGLISH: Palache, et al., "Dana's System of Mineralogy," 7th Ed., v. II, p. 610–611, New York, Wiley, 1951.

# PARACELSIAN
$BaAl_2Si_2O_8$
Barium Feldspar Group
Dimorphous with celsian
Isostructural with danburite

CRYSTAL SYSTEM: Monoclinic, strongly pseudo-orthorhombic
CLASS: 2/m
SPACE GROUP: $P2_1/c$
Z: 4
LATTICE CONSTANTS:
$a = 8.58$
$b = 9.583$
$c = 9.08$
$\beta \simeq 90°$
3 STRONGEST DIFFRACTION LINES:
4.00 (100)
3.80 ( 70)
2.99 ( 50)
OPTICAL CONSTANTS:
$\alpha = 1.5702$
$\beta = 1.5824$
$\gamma = 1.5869$
$(-)2V = 52.7°$
HARDNESS: 6
DENSITY: 3.31–3.32 (Meas.)
          3.342 (Calc.)
CLEAVAGE: {110} poor
Fracture uneven. Brittle.
HABIT: Crystals prismatic, often large and well-formed, resembling topaz. Twinning complex, multiple.
COLOR-LUSTER: Colorless. Transparent. Vitreous.
MODE OF OCCURRENCE: Occurs in a band in shale and sandstone at the Benallt manganese mine, Rhiw, Lleyn Peninsula, Caernarvonshire, Wales.
BEST REF. IN ENGLISH: Deer, Howie, and Zussman, "Rock Forming Minerals," v. 4, p. 166–178, New York, Wiley, 1963.

# PARACOQUIMBITE
$Fe_2(SO_4)_3 \cdot 9H_2O$
Dimorphous with coquimbite

CRYSTAL SYSTEM: Hexagonal
CLASS: $\bar{3}$
SPACE GROUP: $R\bar{3}$
Z: 12
LATTICE CONSTANTS:
$a = 10.93$
$c = 51.30$

3 STRONGEST DIFFRACTION LINES:
    8.26 (100)
    2.76 ( 75)
    5.45 ( 65)
OPTICAL CONSTANTS:
    $\omega = 1.550$
    $\epsilon = 1.555$
HARDNESS: 2½
DENSITY: 2.11 (Meas.)
                2.115 (Calc.)
CLEAVAGE: $\{01\bar{1}2\}$ imperfect
                $\{10\bar{1}4\}$ imperfect
HABIT: Crystals rhombohedral; also equant, pseudo-cubic, or prismatic. Massive, granular. Twinning on $\{0001\}$ common.
COLOR-LUSTER: Pale violet. Transparent. Vitreous.
MODE OF OCCURRENCE: The mineral occurs associated with coquimbite, often as scepter-like overgrowths or in parallel position, at Alcaparrosa, Tierra Amarilla, and Quetena, Chile.
BEST REF. IN ENGLISH: Palache, et al., "Dana's System of Mineralogy," 7th Ed., v. II, p. 534–535, New York, Wiley, 1951. Robinson, Paul D., and Fang, J. H., *Am. Min.*, **56**: 1567–1572 (1971).

## PARACOSTIBITE
CoSbS

CRYSTAL SYSTEM: Orthorhombic
CLASS: 2/m 2/m 2/m
SPACE GROUP: Pbca
Z: 8
LATTICE CONSTANTS:
    $a = 5.764$
    $b = 5.952$
    $c = 11.635$
3 STRONGEST DIFFRACTION LINES:
    2.555 (100)
    5.813 ( 80)
    2.035 ( 80)
HARDNESS: Microhardness 1009 kg/mm$^2$ with 15 g wt.
DENSITY: 7.1 (Calc.)
HABIT: As irregular and subhedral grains up to 130 $\mu$ in size.
COLOR-LUSTER: In reflected light weakly bireflecting, white with faint grayish tinge, in both air and oil immersion. Opaque. Metallic.
MODE OF OCCURRENCE: Found in drill core in a massive base-metal sulfide ore associated with pyrargyrite, pyrrhotite, antimonial silver, galena, and sphalerite from Mulcahy Township, District of Kenora, Red Lake Mining Division, Ontario, Canada.
BEST REF. IN ENGLISH: Cabri, L. J., Harris, D. C., and Stewart, J. M., *Can. Min.*, **10**(Pt. 2): 232–246 (1970).

## PARADAMITE
Zn$_2$AsO$_4$OH
Dimorphous with adamite

CRYSTAL SYSTEM: Triclinic

CLASS: $\bar{1}$
SPACE GROUP: P$\bar{1}$
Z: 2
LATTICE CONSTANTS:
    $a = 5.807$    $\alpha = 104°15'$
    $b = 6.666$    $\beta = 87°52'$
    $c = 5.627$    $\gamma = 103°12'$
3 STRONGEST DIFFRACTION LINES:
    6.33 (100)
    3.71 (100)
    2.99 ( 90)
OPTICAL CONSTANTS:
    $\alpha = 1.726$
    $\beta = 1.771$
    $\gamma = 1.780$
    $(-)2V = 50°$
HARDNESS: 3½
DENSITY: 4.55 (Meas.)
                4.67 (Calc.)
CLEAVAGE: $\{010\}$ perfect
HABIT: As sheaf-like aggregates of crystals; also as somewhat rounded and striated equant crystals up to 5 mm in size.
COLOR-LUSTER: Pale yellow, transparent; vitreous.
MODE OF OCCURRENCE: Occurs associated with mimetite and adamite on a matrix of limonite at the Ojuela Mine, Mapimi, Durango, Mexico.
BEST REF. IN ENGLISH: Switzer, George, *Science*, **123**(3206): 1039 (1956). Finney, J. J., *Am. Min.*, **51**: 1218–1220 (1966).

## PARADOCRASITE
Sb$_2$(Sb, As)$_2$

CRYSTAL SYSTEM: Monoclinic
CLASS: 2
SPACE GROUP: C2
Z: 1
LATTICE CONSTANTS:
    $a = 7.252$
    $b = 4.172$
    $c = 4.431$
    $\beta = 123°08.4'$
3 STRONGEST DIFFRACTION LINES:
    3.05 (100)
    2.219 ( 80)
    2.093 ( 80)
HARDNESS: Microhardness at 100 g load = 104–134, mean 118.
DENSITY: 6.52 (Meas.)
                6.44 (Calc.)
CLEAVAGE: None. $\{010\}$ parting, perfect; several less perfect partings. Brittle.
HABIT: Crystals stubby prismatic, sometimes striated parallel to long axis, up to 0.5 long; in diversely oriented aggregates. Also as grooved pseudohexagonal plates. Twinning on $\{010\}$, $\{\bar{2}01\}$, $\{001\}$, and $\{\bar{1}01\}$, polysynthetic.
COLOR-LUSTER: Brilliant silver white. Opaque. Metallic. Streak black.
MODE OF OCCURRENCE: Occurs replacing calcite in a

specimen of supposed "dyscrasite" from Broken Hill, New South Wales, Australia.

BEST REF. IN ENGLISH: Leonard, B. F., Mead, Cynthia W., and Finney, J. J., *Am. Min.*, **56**: 1127-1146 (1971).

# PARAGONITE
$NaAl_3Si_3O_{10}(OH)_2$
Mica Group

CRYSTAL SYSTEM:  Monoclinic
CLASS:  2/m or m
SPACE GROUP:  C2/c or Cc
Z: 4
LATTICE CONSTANTS:
    $a = 5.13$
    $b = 8.89$
    $c = 19.32$
    $\beta = 95°$
3 STRONGEST DIFFRACTION LINES:
    2.522 (100)
    4.39 ( 90)
    3.203 ( 80)
OPTICAL CONSTANTS:
    $\alpha = 1.564-1.580$
    $\beta = 1.594-1.609$
    $\gamma = 1.600-1.609$
    $(-)2V = 0°-40°$
HARDNESS: 2½
DENSITY:  2.78-2.90 (Meas.)
          2.907 (Calc.)
CLEAVAGE:  {001} perfect
HABIT:  Massive, compact; also as fine scales.
COLOR-LUSTER:  Colorless, pale yellow.  Transparent to translucent. Pearly.
MODE OF OCCURRENCE:  Occurs chiefly in schists, phyllites, gneisses, quartz veins, and fine-grained sediments. Typical occurrences are found in the Leadville district, Lake County, Colorado; in the schist of Glebe Mountain, southern Vermont; and in Campbell and Franklin counties, Virginia. Also at Monte Campione and elsewhere in Switzerland, and in Piedmont, Italy.
BEST REF. IN ENGLISH:  Deer, Howie, and Zussman, "Rock Forming Minerals," v. 3, p. 31-34, New York, Wiley, 1962.  *Am. Min.*, **49**: 183-190 (1964).

# PARAGUANAJUATITE
$Bi_2(Se, S)_3$

CRYSTAL SYSTEM: Hexagonal
CLASS:  $\bar{3}$ 2/m
SPACE GROUP:  $R\bar{3}m$
Z: 3
LATTICE CONSTANTS:
    $a = 4.125$
    $c = 28.56$
3 STRONGEST DIFFRACTION LINES:
    3.03 (100)
    2.23 ( 60)
    1.404 ( 40)
HARDNESS:  2½-3

DENSITY:  7.704 (Calc.)
CLEAVAGE:  {0001} perfect
Somewhat sectile.
HABIT:  Massive, lamellar or granular.
COLOR-LUSTER:  Lead gray. Opaque. Metallic. Streak gray, shiny.
MODE OF OCCURRENCE:  Occurs intergrown with guanajuatite at the Santa Catarina mine, Sierra de Santa Rosa, near Guanajuato, Mexico; at Fahlun, Sweden; and at one deposit in Japan.
BEST REF. IN ENGLISH:  Vlasov, K. A., "Mineralogy of Rare Elements," v. II, p. 639-640, Israel Program for Scientific Translations, 1966.

# PARAHILGARDITE
$Ca_2B_5O_8(OH)_2Cl$
Dimorphous with Hilgardite

CRYSTAL SYSTEM:  Triclinic
CLASS:  1
SPACE GROUP:  P1
Z:  3
LATTICE CONSTANTS:
    $a = 6.31$      $\alpha = 84.0°$
    $b = 6.484$    $\beta = 79.6°$
    $c = 17.50$    $\gamma = 60.9°$
3 STRONGEST DIFFRACTION LINES:
    2.87 (100)
    2.83 (100)
    2.755 ( 80)
OPTICAL CONSTANTS:
    $\alpha = 1.630$
    $\beta = 1.636$ (Na)
    $\gamma = 1.664$
    $(+)2V = 35°$
HARDNESS:  5
DENSITY:  2.71 (Meas.)
          2.720 (Calc.)
CLEAVAGE:  {010} perfect
            {100} perfect
HABIT:  As oriented intergrowths with hilgardite.
COLOR-LUSTER:  Colorless, transparent. Vitreous.
MODE OF OCCURRENCE:  Occurs in the water-insoluble residue of rock salt from the Choctaw salt dome, Iberville Parish, Louisiana.
BEST REF. IN ENGLISH:  Palache, et al., "Dana's System of Mineralogy," 7th Ed., v. II, p. 383-384, New York, Wiley, 1951.

# PARAHOPEITE
$Zn_3(PO_4)_2 \cdot 4H_2O$
Dimorphous with hopeite

CRYSTAL SYSTEM:  Triclinic
CLASS:  $\bar{1}$
SPACE GROUP:  $P\bar{1}$
Z:  1
LATTICE CONSTANTS:
    $a = 5.755$ kX    $\alpha = 93°17.5'$
    $b = 7.535$        $\beta = 91°55'$
    $c = 5.292$        $\gamma = 91°19'$

3 STRONGEST DIFFRACTION LINES:
7.56 (100)
2.99 ( 90)
4.48 ( 70)
OPTICAL CONSTANTS:
$\alpha$ = 1.614
$\beta$ = 1.625
$\gamma$ = 1.637
(+)2V = nearly 90°
HARDNESS: 3¾
DENSITY: 3.31 (Meas.)
3.304 (Calc.)
CLEAVAGE: {010} perfect
HABIT: Crystals tabular, elongated parallel to $c$-axis; often as subparallel aggregates or radial groups. Twinning on {100} common, polysynthetic.
COLOR-LUSTER: Colorless. Transparent. Vitreous, pearly on cleavage.
MODE OF OCCURRENCE: Occurs as a secondary mineral in zinc-bearing ore deposits. Found as excellent crystal groups associated with tarbuttite, hemimorphite, pyromorphite, and limonite at the Broken Hill mine, Zambia; it also occurs associated with hopeite and other secondary zinc minerals at Salmo, British Columbia, Canada.
BEST REF. IN ENGLISH: Palache, et al., "Dana's System of Mineralogy," 7th Ed., v. II, p. 733–734, New York, Wiley, 1951.

## PARAJAMESONITE (Inadequately described mineral)
$Pb_4 FeSb_6 S_{14}$
Dimorphous with jamesonite

CRYSTAL SYSTEM: Orthorhombic
3 STRONGEST DIFFRACTION LINES:
4.203 (100)
3.777 ( 60)
2.483 ( 30)
HARDNESS: 2½–3
DENSITY: 5.479–5.485 (Meas.)
CLEAVAGE: None.
HABIT: Crystals columnar with imperfect, rounded faces and without good terminations, up to 8 × 2 mm in size.
COLOR-LUSTER: Gray. Opaque. Metallic.
MODE OF OCCURRENCE: Occurs in association with sphalerite, galena, pyrrhotite, and chalcopyrite at the Herzsa mine, Kisbanya, Roumania.
BEST REF. IN ENGLISH: Zsivny, Viktor, and Naray-Szabo, Istvan V., Am. Min., 34: 133 (1949).

## PARALAURIONITE
PbCl(OH)
Dimorphous with laurionite

CRYSTAL SYSTEM: Monoclinic
CLASS: 2/m
SPACE GROUP: C2/m
Z: 4

LATTICE CONSTANTS:
$a$ = 10.79
$b$ = 3.98
$c$ = 7.19
$\beta$ = 117°13'
3 STRONGEST DIFFRACTION LINES:
5.14 (100)
3.21 (100)
2.51 ( 90)
OPTICAL CONSTANTS:
$\alpha$ = 2.05
$\beta$ = 2.15
$\gamma$ = 2.20
(−)2V = medium to large
HARDNESS: Not determined. Soft.
DENSITY: 6.15 (Meas.)
6.28 (Calc.)
CLEAVAGE: {001} perfect, easy
Not brittle.
HABIT: Crystals lath-like, also thin tabular {100}. Terminations often show pyramid faces at the corners. Twinning on {100} very common.
COLOR-LUSTER: Colorless, white, rarely violet or greenish. Transparent. Subadamantine.
MODE OF OCCURRENCE: Occurs rarely associated with diaboleite, matlockite, leadhillite, and other secondary lead minerals at the Mammoth mine, Tiger, Arizona. It is also found as excellent small crystals associated with laurionite in vugs in lead slags, altered by the action of sea water, at Laurium Greece; with phosgenite at the Wheal Rose mine near Sithney, Cornwall, England; and at the San Rafael mine in the Sierra Gorda, Chile.
BEST REF. IN ENGLISH: Palache, et al., "Dana's System of Mineralogy," 7th Ed., v. II, p. 64–66, New York, Wiley, 1951.

## PARALUMINITE (Species status in doubt)
$Al_4 SO_4 (OH)_{10} \cdot 10H_2 O$ (?)

3 STRONGEST DIFFRACTION LINES:
4.50 (100)
2.27 ( 80)
8.7 ( 70)
OPTICAL CONSTANTS:
$\alpha$ = 1.462–1.463
$\beta$ = 1.470
$\gamma$ = 1.471
(−)2V = 0° to small
HARDNESS: Soft
DENSITY: 1.85–2.13 (Meas.)
CLEAVAGE: Not determined.
HABIT: Massive, fibrous; nodular.
COLOR-LUSTER: White to pale yellow, chalky.
MODE OF OCCURRENCE: Occurs at Huelgoat, Brittany, France; near Halle, Saxony, and at Garnsdorf, Thuringia, Germany; in yellow clay at Gorelyi Spring, southwest border of Viliui Basin, and at Mt. Sokolova, near Saratov on the Danube, USSR.
BEST REF. IN ENGLISH: Palache, et al., "Dana's System of Mineralogy," 7th Ed., v. II, p. 586–587, New York, Wiley, 1951.

# PARAMELACONITE
$Cu^{2+}_{1-2x}Cu^{1+}_{2x}O_{1-x}$

CRYSTAL SYSTEM: Tetragonal
CLASS: 4/m 2/m 2/m
SPACE GROUP: $I4_1/amd$
Z: 16
LATTICE CONSTANTS:
  $a = 5.83$
  $c = 9.88$
3 STRONGEST DIFFRACTION LINES:
  2.50 (100)
  1.58 ( 80)
  1.25 ( 80)
  HARDNESS: 4½
  DENSITY: 6.106 (Calc.)
  CLEAVAGE: None. Fracture conchoidal.
  HABIT: Crystals stout prismatic, striated.
  COLOR-LUSTER: Black. Opaque. Brilliant adamantine. Streak brownish black.
  MODE OF OCCURRENCE: Occurs as crystals up to 8 mm in length, associated with malachite, dioptase, cuprite, tenorite, and chrysocolla, at the Algomah mine, Ontonagon County, Michigan; and associated with connellite, malachite, tenorite, cuprite, and goethite, at the Copper Queen mine, Bisbee, Arizona.
  BEST REF. IN ENGLISH: Palache, et al., "Dana's System of Mineralogy," 7th Ed., v. I, p. 510–511, New York, Wiley, 1944.

# PARAMONTROSEITE
$VO_2$

CRYSTAL SYSTEM: Orthorhombic
CLASS: 2/m 2/m 2/m
SPACE GROUP: Pbnm
Z: 4
LATTICE CONSTANTS:
  $a = 4.89$
  $b = 9.39$
  $c = 2.93$
3 STRONGEST DIFFRACTION LINES: See montroseite.
  HARDNESS: Not determined.
  DENSITY: 4.0 (Meas.)
  CLEAVAGE: {010} good
Brittle.
  HABIT: As pseudomorphs after microscopic bladed crystals of montroseite.
  COLOR-LUSTER: Black; opaque. Submetallic. Streak black.
  MODE OF OCCURRENCE: Occurs as a metastable form of $VO_2$ resulting from the oxidation of montroseite.
  BEST REF. IN ENGLISH: Evans, Howard T., Jr., and Mrose, Mary E., *Am. Min.*, **40**: 861–875 (1955).

# PARARAMMELSBERGITE
$NiAs_2$
Dimorphous with rammelsbergite

CRYSTAL SYSTEM: Orthorhombic
CLASS: 2/m 2/m 2/m
SPACE GROUP: Pbma
Z: 8
LATTICE CONSTANTS:
  $a = 5.770$
  $b = 5.838$
  $c = 11.419$
3 STRONGEST DIFFRACTION LINES:
  2.559 (100)
  2.521 ( 95)
  2.371 ( 65)
  HARDNESS: ~5
  DENSITY: 7.25 (Meas.)
           7.20 (Calc.)
  CLEAVAGE: {001} perfect
Fracture uneven.
  HABIT: Crystals poorly developed rectangular tablets. Usually massive with reniform structure.
  COLOR-LUSTER: Tin-white. Opaque. Metallic. Streak grayish black.
  MODE OF OCCURRENCE: Occurs in mesothermal vein deposits associated with other nickel and cobalt minerals, loellingite, and calcite. Found at Cobalt, Gowganda, and South Lorrain, Ontario, Canada.
  BEST REF. IN ENGLISH: Palache, et al., "Dana's System of Mineralogy," 7th Ed., v. I, p. 310–311, New York, Wiley, 1944.

# PARA-SCHACHNERITE
$Ag_3Hg_2$

CRYSTAL SYSTEM: Orthorhombic
SPACE GROUP: Probably Cmcm
Z: 2
LATTICE CONSTANTS:
  $a = 2.961$
  $b = 5.13$
  $c = 4.83$
3 STRONGEST DIFFRACTION LINES:
  2.267 (100)
  2.404 ( 60)
  1.263 ( 60)
OPTICAL CONSTANTS: Birefringent with colors brownish-rose-white on $c$, creamy white on $a$ and $b$.
  HARDNESS: Polishing hardness distinctly higher than that of moschellandesbergite.
  DENSITY: 12.98 (Calc.)
  HABIT: As crystals up to 1 cm long; usually much smaller. Always twinned, complex, trillings common. Twin plane (110).
  COLOR-LUSTER: Creamy white in reflected light. Opaque. Metallic.
  MODE OF OCCURRENCE: Occurs as an alteration of moschellandesbergite in the zone of oxidation of the old mercury mine at Landsberg near Obermoschel, Pfalz, Germany. Associated minerals are schachnerite, mercurian silver, limonite, ankerite, argentite, and cinnabar.
  BEST REF. IN ENGLISH: Seeliger, E., and Mücke, A., *Am. Min.*, **58**: 347 (1973).

## PARASCHOEPITE (Schoepite III)

$UO_3 \cdot 2H_2O(?)$
A natural dehydration product of schoepite

CRYSTAL SYSTEM: Orthorhombic
CLASS: 2/m 2/m 2/m
SPACE GROUP: Pbca
Z: 32
LATTICE CONSTANTS:
  $a$ = 14.12
  $b$ = 16.83
  $c$ = 15.22
OPTICAL CONSTANTS:
  $\alpha$ = 1.700
  $\beta$ = 1.750
  $\gamma$ = 1.770
  2V = 40°
HARDNESS: 2-3
DENSITY: Not determined.
CLEAVAGE:, {001} perfect
            {010} distinct
Brittle.
  HABIT: Crystals tabular on {001}, elongated [010], pseudohexagonal aspect.
  COLOR-LUSTER: Yellow.
  MODE OF OCCURRENCE: Occurs as an alteration product of Schoepite I at the Shinkolobwe mine, Katanga, Zaire.
  BEST REF. IN ENGLISH: Schoep, Alfred, and Stradiot, Sadi, *Am. Min.*, 32: 344-350 (1947).  Christ, C. L., and Clark, Joan R., *Am. Min.*, 45: 1026-1061 (1960).


## PARASYMPLESITE

$Fe_3(AsO_4)_2 \cdot 8H_2O$
Dimorphous with symplesite

CRYSTAL SYSTEM: Monoclinic
CLASS: 2/m
SPACE GROUP: C2/m
Z: 2
LATTICE CONSTANTS:
  $a$ = 10.25
  $b$ = 13.48
  $c$ = 4.71
  $\beta$ = 103°50'
3 STRONGEST DIFFRACTION LINES:
  6.83  (100)
  7.063 ( 40)
  9.006 ( 18)
OPTICAL CONSTANTS:
  $\alpha$ = 1.628
  $\beta$ = 1.660
  $\gamma$ = 1.705
(-)2V = large
HARDNESS: 2
DENSITY: 3.07 (Meas.)
            3.097 (Calc.)
CLEAVAGE: {010} very perfect
  HABIT: Crystals up to 3.0 X 1.5 X 1.0 mm in size. Prominent faces: b{010}, w{$\bar{2}$01}, m{110}, E{$\bar{5}$02}, and t{$\bar{4}$01}.
  COLOR-LUSTER: Light greenish blue.

MODE OF OCCURRENCE: Occurs at Kiura, Ohita, Japan.
  BEST REF. IN ENGLISH: Ito, Tei-Ichi; Minato, Hideo; and Sakurai, Kin'ichi, *Am. Min.*, 40: 368 (1955).


## PARATACAMITE

$Cu_2(OH)_3Cl$
Dimorphous with atacamite

CRYSTAL SYSTEM: Hexagonal
CLASS: $\bar{3}$ 2/m
SPACE GROUP: R$\bar{3}$m
Z: 24
LATTICE CONSTANTS:
  $a$ = 13.68
  $c$ = 13.98
3 STRONGEST DIFFRACTION LINES:
  5.44 (100)
  2.27 ( 85)
  2.78 ( 80)
OPTICAL CONSTANTS:
  $\omega$ = 1.842
  $\epsilon$ = 1.848
    (+)
HARDNESS: 3
DENSITY: 3.74 (Meas.)
            3.75 (Calc.)
CLEAVAGE: {10$\bar{1}$1} good
Fracture conchoidal to uneven.
  HABIT: Crystals rhombohedral, commonly twinned on {10$\bar{1}$1}.  Also massive, granular, and as powdery incrustations.
  COLOR-LUSTER: Green to greenish black. Translucent to nearly opaque. Vitreous; also earthy to dull.
  MODE OF OCCURRENCE: Occurs as minute crystals coating fractures in rock at a copper prospect at Mason Pass near Yerington, Lyon County, Nevada. It is also found in the Sierra Gorda and at other localities in Chile; in the Botallack mine, St. Just, Cornwall, England; at Capo Calamita, Italy; and at Broken Hill, New South Wales, Australia.
  BEST REF. IN ENGLISH: Palache, et al., "Dana's System of Mineralogy," 7th Ed., v. II, p. 74-76, New York, Wiley, 1951.


## PARATELLURITE

$TeO_2$
Dimorphous with tellurite

CRYSTAL SYSTEM: Tetragonal
CLASS: 422
SPACE GROUP: $P4_12_12$ or $P4_32_12$
Z: 4
LATTICE CONSTANTS:
  $a$ = 4.810
  $c$ = 7.613
3 STRONGEST DIFFRACTION LINES:
  2.988 (100)
  3.404 ( 86)
  1.873 ( 55)
HARDNESS: 1

DENSITY: 5.60 (Meas.)
6.017 (Calc.)
CLEAVAGE: Not determined.
HABIT: Massive; fine-grained.
COLOR-LUSTER: Grayish white; resinous to waxy.
MODE OF OCCURRENCE: Occurs very sparingly with tellurite in thin seams in native tellurium as an alteration product of tellurium and tellurite at Cananea, Sonora, Mexico.
BEST REF. IN ENGLISH: Switzer, George, and Swanson, Howard E., *Am. Min.*, **45**: 1272–1274 (1960).

# PARAVAUXITE
$Fe^{2+}Al(PO_4)_2(OH)_2 \cdot 10H_2O$

CRYSTAL SYSTEM: Triclinic
CLASS: $\bar{1}$
SPACE GROUP: $P\bar{1}$
Z: 1
LATTICE CONSTANTS:
$a = 5.23$    $\alpha = 107°17'$
$b = 10.52$    $\beta = 111°24'$
$c = 6.96$    $\gamma = 72°29'$
3 STRONGEST DIFFRACTION LINES:
9.82 (100)
6.38 ( 90)
4.20 ( 90)
OPTICAL CONSTANTS:
$\alpha = 1.552$
$\beta = 1.559$
$\gamma = 1.572$
$(+)2V = 72°$
HARDNESS: 3
DENSITY: 2.36 (Meas.)
2.38 (Calc.)
CLEAVAGE: {010} perfect
Fracture conchoidal. Brittle.
HABIT: Crystals short prismatic to thick tabular. As randomly intergrown, subparallel, or radial aggregates.
COLOR-LUSTER: Colorless to pale greenish white. Transparent to translucent. Vitreous; pearly on cleavage. Streak white.
MODE OF OCCURRENCE: Occurs as fine specimens in the tin deposits at Llallagua, Bolivia, often associated with vauxite, metavauxite, wavellite, and quartz.
BEST REF. IN ENGLISH: Palache, et al., "Dana's System of Mineralogy," 7th Ed., v. II, p. 972–973, New York, Wiley, 1951. Hurlbut, Cornelius S. Jr., and Honea, Russell M., *Am. Min.*, **47**: 1–8 (1962).

# PARAVIVIANITE (= oxidized vivianite?)

# PARAWOLLASTONITE
$\alpha\text{-}CaSiO_3$

CRYSTAL SYSTEM: Monoclinic
CLASS: 2

SPACE GROUP: $P2_1$
Z: 12
LATTICE CONSTANTS:
$a = 15.42$
$b = 7.32$
$c = 7.07$
$\beta = 95°24'$
3 STRONGEST DIFFRACTION LINES:
2.97 (100)
3.83 ( 80)
3.52 ( 80)
OPTICAL CONSTANTS:
$\alpha = 1.618$
$\beta = 1.630$
$\gamma = 1.632$
$(-)2V = 38°–60°$
HARDNESS: 4½–5
DENSITY: 2.913 (Meas.)
2.93 (Calc.)
CLEAVAGE: {100} perfect
{001} good
{$\bar{1}02$} good
Fracture splintery.
HABIT: Crystals commonly tabular; usually massive, cleavable to fibrous; also granular and compact. Twinning on {100} common.
COLOR-LUSTER: White or gray; rarely yellowish. Transparent to translucent. Vitreous to pearly; somewhat silky when fibrous.
MODE OF OCCURRENCE: Occurs chiefly in contacts of igneous intrusions into limestone, and in limestone blocks ejected from volcanoes. Found at Crestmore, Riverside County, California; at Mt. Somma, Italy; and at Csiklova, Roumania.
BEST REF. IN ENGLISH: Deer, Howie, and Zussman, "Rock Forming Minerals," v. II, p. 167–175, New York, Wiley, 1963.

# PARGASITE
$NaCaMg_4Al(Si, Al)_8O_{22}(OH)_2$
Amphibole Group
Hornblende Series

CRYSTAL SYSTEM: Monoclinic
CLASS: 2/m
SPACE GROUP: C2/m
Z: 2
LATTICE CONSTANTS:
$a \simeq 9.9$
$b \simeq 18.0$
$c \simeq 5.3$
$\beta \simeq 105°30'$
3 STRONGEST DIFFRACTION LINES:
3.12 (100)
8.43 ( 40)
3.27 ( 35)
OPTICAL CONSTANTS:
$\alpha = 1.613$
$\beta = 1.618$
$\gamma = 1.635$
$(+)2V = 120°$

HARDNESS: 5-6
DENSITY: 3.069-3.181 (Meas.)
CLEAVAGE: {110} perfect
{001} parting
{100} parting
Fracture uneven to subconchoidal. Brittle.
HABIT: Crystals short to long prismatic. Also massive, compact. Twinning on {100} common, simple, lamellar.
COLOR-LUSTER: Light brown, brown, bluish green, grayish black. Translucent to nearly opaque. Vitreous.
MODE OF OCCURRENCE: Occurs as a constituent of both igneous and metamorphic rocks. Typical occurrences are found in the Twin Lakes region, Fresno County, California; in Custer County, South Dakota; and at Burlington, Pennsylvania. Also at Pargas, Finland, and at localities in Venezuela, Scotland, Sweden, Austria, USSR, and Japan.
BEST REF. IN ENGLISH: Deer, Howie and Zussman, "Rock Forming Minerals," v. 2, p. 263-314, New York, Wiley, 1963.

# PARISITE
$Ca(Ce, La)_2(CO_3)_3F_2$

CRYSTAL SYSTEM: Hexagonal
CLASS: 3
SPACE GROUP: R3
Z: 6
LATTICE CONSTANTS:
$a$ = 7.091 kX
$c$ = 27.93
3 STRONGEST DIFFRACTION LINES:
2.04 (100)
1.28 (100)
2.83 (100)
OPTICAL CONSTANTS:
$\omega$ = 1.676 (Na)
$\epsilon$ = 1.757
(+)
HARDNESS: 4½
DENSITY: 4.36 (Meas.)
4.38 (Calc.)
CLEAVAGE: {0001} distinct (parting or cleavage)
Fracture subconchoidal to splintery. Brittle.
HABIT: Crystals small acute double hexagonal pyramids; occasionally prismatic in appearance. Also rhombohedral. Lateral faces commonly striated. {0001} often present as small to large faces.
COLOR-LUSTER: Brownish yellow, brown, wax yellow to grayish yellow; transparent to translucent. Vitreous to resinous; {0001} parting pearly.
MODE OF OCCURRENCE: Occurs in veinlets and pockets in carbonaceous shale beds in the emerald deposits in the Muzo district, Colombia; in pegmatite pipes in riebeckite-aegirite-granite at Quincy, Massachusetts; near Pyrites, Ravalli County, Montana in an altered trachyte; and from several localities in Norway, Italy, Madagascar, and Manchuria.
BEST REF. IN ENGLISH: Palache, et al., "Dana's System of Mineralogy," 7th Ed., v. 2, p. 282-285, New York, Wiley, 1951. Donnay, Gabrielle, and Donnay, J. D. H., Am. Min., 38: 932-963 (1953).

# PARKERITE
$Ni_3(Bi, Pb)_2S_2$

CRYSTAL SYSTEM: Orthorhombic
CLASS: 2/m 2/m 2/m
SPACE GROUP: Pmam
Z: 1
LATTICE CONSTANTS:
$a$ = 5.545
$b$ = 5.731
$c$ = 4.052
3 STRONGEST DIFFRACTION LINES:
2.86 (100)
2.34 ( 90)
4.02 ( 70)
HARDNESS: 2+
DENSITY: 8.4 (Meas.)
8.5 (Calc.)
CLEAVAGE: {010} excellent
HABIT: Massive; as minute grains.
COLOR-LUSTER: Creamy-white in polished section. Opaque. Metallic.
MODE OF OCCURRENCE: Parkerite occurs as an uncommon mineral in certain arsenide-sulfide assemblages, frequently associated with niccolite, bismuth, bismuthinite and maucherite. Found in Canada at Sudbury and in the Cobalt-Gowganda area in Ontario; also at Great Slave Lake, Northwest Territories. A variety with a significant amount of lead in substitution for bismuth occurs at Insizwa, South Africa.
BEST REF. IN ENGLISH: Michener, C. E. and Peacock, M. A., Am. Min., 28: 343-355 (1943). Fleet, Michael E., Am. Min., 58: 435-439 (1973).

# PARSETTENSITE
$KMn_{10}Si_{12}O_{30}(OH)_{12}$

CRYSTAL SYSTEM: Hexagonal
LATTICE CONSTANTS:
$a$ = 22.5
$c$ = 38.0
OPTICAL CONSTANTS:
$\omega$ = 1.576
$\epsilon$ = 1.546
(-)
HARDNESS: ~ 1.5
DENSITY: 2.59 (Meas.)
CLEAVAGE: {001} perfect
HABIT: Massive, somewhat micaceous.
COLOR-LUSTER: Copper red. Somewhat metallic.
MODE OF OCCURRENCE: Occurs in the manganese deposits of Val d'Err, Canton Graubünden, Switzerland.
BEST REF. IN ENGLISH: Jakob, J., Am. Min., 10: 107 (1925).

# PARSONSITE
$Pb_2UO_2(PO_4)_2 \cdot 2H_2O$

CRYSTAL SYSTEM: Triclinic
CLASS: $\bar{1}$

SPACE GROUP: P$\bar{1}$
Z: 2
LATTICE CONSTANTS:

| | |
|---|---|
| $a = 6.862$ | $\alpha = 101°26'$ |
| $b = 10.425$ | $\beta = 98°15'$ |
| $c = 6.684$ | $\gamma = 86°17'$ |

3 STRONGEST DIFFRACTION LINES:
3.283 (100)
3.253 (100)
4.233 ( 65)
OPTICAL CONSTANTS:
$\alpha = 1.85$
$\gamma = 1.86$
$(-)2V \sim 11°-26°$
HARDNESS: 2½-3
DENSITY: 5.72 (Meas.)
6.29 (Meas. Synthetic)
CLEAVAGE: {010} indistinct
Fracture conchoidal.

HABIT: Crystals prismatic, up to 4 mm long, elongated [001] and flattened on {010}; as acicular crystals without distinct form; as rudely radial clusters or tufts of crystals; also as radial fibrous masses and earthy crusts.

COLOR-LUSTER: Very pale yellow to yellowish amber, greenish brown, chocolate brown, rarely pale rose. Transparent to translucent. Subadamantine to greasy. Not fluorescent.

MODE OF OCCURRENCE: Occurs as a secondary mineral associated with autunite and phosphuranylite in pegmatite at the Ruggles mine, Grafton Center, New Hampshire. In France it is found associated with torbernite at Lachaux, Puy-de-Dôme, and with renardite and torbernite at Grury, Saône et Loire; at Wölsendorf, Bavaria, Germany, with uranocircite; and with metatorbernite at Shinkolobwe, Katanga, Zaire.

BEST REF. IN ENGLISH: Frondel, Clifford, USGS Bull. 1064, p. 233-236 (1958).

## PARTRIDGEITE = Variety of bixbyite

## PARTZITE
$Cu_2 Sb_2 (O, OH)_7$ (?)
Copper analogue of bindheimite

CRYSTAL SYSTEM: Cubic
CLASS: 4/m $\bar{3}$ 2/m
SPACE GROUP: Fd3m
Z: 8
LATTICE CONSTANT:
$a = 10.25$
3 STRONGEST DIFFRACTION LINES:
2.95 (100)
5.91 ( 90)
1.81 ( 80)
OPTICAL CONSTANT:
$N = 1.61-1.82$
HARDNESS: 3-4
DENSITY: 2.96-3.96 (Meas.)
(5.95 $Cu_2 Sb_2 O_7$)

CLEAVAGE: None. Fracture conchoidal to uneven.
HABIT: Massive.
COLOR-LUSTER: Olive green; black tarnish.
MODE OF OCCURRENCE: Occurs as an alteration of antimonial sulfide ores at various mines in the Blind Springs mining district, Mono County, California.

BEST REF. IN ENGLISH: Palache, et al., "Dana's System of Mineralogy," 7th Ed., v. I, p. 599, New York, Wiley, 1944. Mason, Brian, and Vitaliano, C. J., *Min. Mag.*, **30**: 100-112 (1953).

## PARWELITE
$(Mn, Mg)_5 Sb(Si, As)_2 O_{10-11}$

CRYSTAL SYSTEM: Monoclinic
CLASS: 2/m or 2
SPACE GROUP: I 2/m or I2
Z: 8
LATTICE CONSTANTS:
$a = 9.76$
$b = 19.32$
$c = 10.06$
$\beta = 95°54'$
3 STRONGEST DIFFRACTION LINES:
2.734 (100)
2.915 ( 95)
2.314 ( 65)
OPTICAL CONSTANTS:
$\alpha, \beta = 1.85$
$\gamma = 1.88$
$(+)2V \simeq 27°$
HARDNESS: 5½
DENSITY: 4.62 (Meas.)
CLEAVAGE: {010} poor to fair
Subconchoidal fracture.
HABIT: Crystals prismatic; stubby.
COLOR-LUSTER: Yellowish brown, transparent to cloudy.

MODE OF OCCURRENCE: Occurs associated with långbanite, spessartine, hausmannite, berzeliite, and caryinite in manganoan carbonate at Långban, Sweden.

BEST REF. IN ENGLISH: Moore, Paul B., *Am. Min.*, **55**: 323 (1970).

## PASCOITE
$Ca_2 V_6 O_{17} \cdot 11H_2O$

CRYSTAL SYSTEM: Monoclinic
CLASS: 2 or 2/m
SPACE GROUP: I2 or I2/m
Z: 1
LATTICE CONSTANTS:
$a = 16.834$
$b = 10.156$
$c = 10.921$
$\beta = 93°08'$
3 STRONGEST DIFFRACTION LINES:
5.5 (100)
4.67 (100)
5.1 ( 80)

OPTICAL CONSTANTS:
$\alpha = 1.775$
$\beta = 1.815$
$\gamma = 1.825$
$(-)2V = 50\frac{1}{2}°$ (Na)
HARDNESS: ~2½
DENSITY: 1.87 (Meas.)
1.88 (Calc.)
CLEAVAGE: {010} distinct
Fracture conchoidal.
HABIT: As granular crusts, rarely exhibiting minute lath-like crystals with oblique terminations.
COLOR-LUSTER: Yellowish orange to dark reddish orange; yellowish brown when partially dehydrated. Translucent. Vitreous to subadamantine to dull. Streak yellow.

Soluble in water.
MODE OF OCCURRENCE: Occurs as colorful granular crusts on sandstone in the Ambrosia Lakes district, McKinley County, New Mexico; in ore from the Henry Clay claim and the Bitter Creek and Mill No. 1 mines, Montrose County, and at the LaSal No. 2 mine, Mesa County, Colorado; with carnotite in sandstone in Garfield and Grand counties, Utah; and in the vanadium deposit at Minas Ragra, near Cerro de Pasco, Peru.
BEST REF. IN ENGLISH: Palache, et al., "Dana's System of Mineralogy," 7th Ed., v. II, p. 1055-1056, New York, Wiley, 1951.

## PATERNOITE = Kaliborite

## PATRONITE
A vanadium sulfide near $VS_4$

CRYSTAL SYSTEM: Monoclinic
CLASS: 2/m
SPACE GROUP: I2/c
Z: 8
LATTICE CONSTANTS:
$a = 6.78$
$b = 10.42$
$c = 12.11$
$\beta = 100.8°$
3 STRONGEST DIFFRACTION LINES:
5.15 (100)
5.47 ( 60)
3.92 ( 50)
HABIT: Massive, very fine grained. Admixed with other sulfides.
COLOR-LUSTER: Greenish black to black.
MODE OF OCCURRENCE: Occurs in graphite-like masses as an important vanadium ore in association with a hydrocarbon, bravoite, and many secondary vanadium minerals at Minas Ragra, near Cerro de Pasco, Peru.
BEST REF. IN ENGLISH: Palache, et al., "Dana's System of Mineralogy," 7th Ed., v. I, p. 347, New York, Wiley, 1944.

## PAULINGITE
$\sim K_2 (Ca, Ba)_{1.3} (Si, Al)_{12} O_{24} \cdot 14H_2O$
Zeolite Group

CRYSTAL SYSTEM: Cubic
CLASS: $4/m\ \overline{3}\ 2/m$
SPACE GROUP: Im3m
LATTICE CONSTANT:
$a = 35.10$
3 STRONGEST DIFFRACTION LINES:
8.29 (100)
6.88 (100)
4.78 ( 90)
OPTICAL CONSTANT:
$N = 1.473$ (Na)
HARDNESS: ~5
DENSITY: Not determined.
CLEAVAGE: No distinct cleavage.
HABIT: As rhombic dodecahedra 0.1-1.0 mm in diameter.
COLOR-LUSTER: Colorless, transparent; vitreous.
MODE OF OCCURRENCE: Occurs in vesicles in basalt from the Columbia River near Wenatchee, Washington; also near Riggins, Idaho.
BEST REF. IN ENGLISH: Kamb, W. Barclay, and Oke, William C., Am. Min., 45: 79-91 (1960).

## PAVONITE
$AgBi_3 S_5$

CRYSTAL SYSTEM: Monoclinic
CLASS: 2/m
SPACE GROUP: A 2/m
Z: 4
LATTICE CONSTANTS:
$a = 16.34$
$b = 4.03$
$c = 13.35$
$\beta = 94°30'$
3 STRONGEST DIFFRACTION LINES:
2.84 (100)
2.01 ( 70)
3.58 ( 60)
HARDNESS: ~2
DENSITY: 6.54 (Meas.)
6.79 (Calc.)
CLEAVAGE: Not observed.
HABIT: As tiny bladed crystals elongated parallel b[010]; massive.
COLOR-LUSTER: Lead gray to tin white. Metallic. Opaque. Streak lead gray.
MODE OF OCCURRENCE: Found associated with bismuthinite and ankerite, Porvenir mine, Cerro Bonete (Aceroduyok), Province Sur-Lipez, Bolivia.
BEST REF. IN ENGLISH: Nuffield, E. W., Am. Min., 39: 409-415 (1954).

## PAXITE
$Cu_2 As_3$

CRYSTAL SYSTEM: Orthorhombic
Z: 10

LATTICE CONSTANTS:
$a = 12.84$
$b = 11.50$
$c = 7.654$
3 STRONGEST DIFFRACTION LINES:
3.164 (100)
3.633 ( 80)
2.772 ( 70)
HARDNESS: 3½–4
DENSITY: Approx. 5.3 (Meas.)
5.14 (Calc.)
CLEAVAGE: One perfect
HABIT: Massive, lamellar structure.
COLOR-LUSTER: Light steel gray on fresh surface; tarnishes rapidly and finally becomes black. Metallic on fresh surface. Opaque. Streak black.
MODE OF OCCURRENCE: Occurs with arsenic, arsenolamprite, silver, loellingite, niccolite, koutekite, nováite, chalcocite, skutterudite, bornite, chalcopyrite, tiemannite, clausthalite, uraninite, hematite, and fluorite as local inclusions in calcite veins which are in pyroxenic gneisses and muscovitic mica schists at Cerný Důl in Krkonoše (Giant Mts.), Bohemia. It is intergrown with nováite, koutekite, and especially arsenic.
BEST REF. IN ENGLISH: Johan, Z., *Am. Min.*, **47**: 1484–1485 (1962).

# PEARCEITE

$Ag_{16}As_2S_{11}$
Pearceite-Antimonpearceite Series
Dimorphous with arsenpolybasite

CRYSTAL SYSTEM: Monoclinic (pseudohexagonal)
CLASS: 2/m
SPACE GROUP: C2/m
Z: 2
LATTICE CONSTANTS:
$a = 12.61$
$b = 7.28$
$c = 11.88$
$\beta = 90°0'$
3 STRONGEST DIFFRACTION LINES:
3.00 (100)
2.84 ( 90)
3.11 ( 50)
HARDNESS: 3
DENSITY: 6.13 (Meas)
6.07 (Calc.)
CLEAVAGE: None. Fracture irregular to conchoidal.
HABIT: Crystals thin tabular, flattened and striated on {001}. Also massive compact. Twinning on {110}.
COLOR-LUSTER: Black. Nearly opaque. Metallic. Streak black.
MODE OF OCCURRENCE: Occurs in low- to medium-temperature silver vein deposits, commonly associated with native silver, argentite, silver and lead sulfosalts, carbonates, fluorite, barite, and quartz. Found widespread in Colorado especially at the Mollie Gibson mine, Pitkin County, where large quantities of nearly pure pearceite that yielded 10,000–16,000 ounces of silver to the ton were mined. It

also occurs in Idaho, Utah, Montana, Mexico, Chile, Spain, and Czechoslovakia.
BEST REF. IN ENGLISH: Palache, et al., "Dana's System of Mineralogy," 7th Ed., v. I, p. 353, 354, New York, Wiley, 1944.

# PECORAITE (Inadequately described mineral)

$Ni_3Si_2O_5(OH)_4$
Serpentine Group

CRYSTAL SYSTEM: Monoclinic (?)
3 STRONGEST DIFFRACTION LINES:
7.43 (100)
3.66 ( 80)
1.529 ( 80)
OPTICAL CONSTANT:
$n \sim 1.565–1.603$
HARDNESS: Not determined.
DENSITY: 3.084 (Meas.) on material containing adsorbed water.
HABIT: As small grains 0.1–5 mm in diameter.
COLOR-LUSTER: Oriental green.
MODE OF OCCURRENCE: Occurs associated with maghemite and quartz, and less cassidyite and reevesite filling cracks in the Wolf Creek meteorite, Western Australia.
BEST REF. IN ENGLISH: Faust, G. T., Fahey, J. J., Mason, Brian, and Dwornik, E. J., *Science*, **165**: 59–60 (1969).

# PECTOLITE

$NaCa_2Si_3O_8OH$

CRYSTAL SYSTEM: Triclinic
CLASS: $\bar{1}$
SPACE GROUP: $P\bar{1}$
Z: 2
LATTICE CONSTANTS:
$a = 7.99$    $\alpha = 90°23'$
$b = 7.04$    $\beta = 95°17'$
$c = 7.02$    $\gamma = 102°28'$
3 STRONGEST DIFFRACTION LINES:
2.921 (100)
3.10 ( 80)
3.90 ( 60)
OPTICAL CONSTANTS:
$\alpha = 1.595–1.610$
$\beta = 1.605–1.615$
$\gamma = 1.632–1.645$
$(+)2V = 50°–63°$
HARDNESS: 4½–5
DENSITY: 2.74–2.88 (Meas.)
2.876 (Calc.)
CLEAVAGE: {001} perfect
{100} perfect
HABIT: As aggregates of needle-like crystals elongated parallel to b-axis, usually radiating and forming globular masses.
COLOR-LUSTER: Colorless, white. Transparent to translucent. Vitreous or silky.

MODE OF OCCURRENCE: Occurs chiefly in cavities in basaltic rocks, often in association with zeolites, and less commonly in lime-rich metamorphic rocks, in alkaline igneous rocks, and in mica peridotite. Found in basalts at many localities in the vicinity of Paterson, and in the zinc deposits at Franklin and Sterling Hill, New Jersey; in mica peridotite in Woodson County, Kansas; at Thetford Mines, Quebec, and at Lake Nipigon, Ontario, Canada; and at deposits in Greenland, Scotland, Sweden, Czechoslovakia, USSR, Morocco, South Africa, Japan, and on the Island of Los, Guinea.

BEST REF. IN ENGLISH: Deer, Howie, and Zussman, "Rock Forming Minerals," v. II, p. 176–181, New York, Wiley, 1963.

## PELIGOTITE = Johannite

## PELLYITE
$Ba_2Ca(Fe, Mg)_2Si_6O_{17}$

CRYSTAL SYSTEM: Orthorhombic
CLASS: 2/m 2/m 2/m or mm2
SPACE GROUP: Cmcm, Cmc2$_1$, or C2cm
Z: 4
LATTICE CONSTANTS:
  $a = 15.677$
  $b = 7.151$
  $c = 14.209$
3 STRONGEST DIFFRACTION LINES:
  3.43 (100)
  3.19 ( 65)
  2.308 ( 60)
OPTICAL CONSTANTS:
  $\alpha = 1.643$
  $\beta = 1.645$ } Na
  $\gamma = 1.649$
  (+)2V = 47° (est.)
HARDNESS: 6
DENSITY: 3.51 (Meas.)
         3.48 (Calc.)
CLEAVAGE: Prismatic, poor. Fracture conchoidal.
HABIT: Massive, crystalline; average grain size 2 mm.
COLOR-LUSTER: Colorless to pale yellow. Translucent. Vitreous.
MODE OF OCCURRENCE: Occurs as a constituent of skarns which have developed in original limestone bodies adjacent to igneous contacts. Found in association with gillespite, sanbornite, taramellite, barite, witherite, chalcopyrite, andradite, hedenbergite, and quartz near the headwaters of the Pelly River, Yukon Territory. A similar occurrence is found in Fresno County, California.

BEST REF. IN ENGLISH: Montgomery, J. H., Thompson, R. M., and Meagher, E. P., Can. Min., 11: 444–447 (1972).

## PENDLETONITE = Carpathite

## PENFIELDITE
$Pb_2Cl_3OH$

CRYSTAL SYSTEM: Hexagonal
CLASS: 6/m 2/m 2/m
SPACE GROUP: P6/m
Z: 36
LATTICE CONSTANTS:
  $a = 11.28$
  $c = 48.65$
3 STRONGEST DIFFRACTION LINES:
  3.73 (100)
  3.14 ( 80)
  3.31 ( 60)
OPTICAL CONSTANTS:
  $\omega = 2.13$
  $\epsilon = 2.21$
  (+)
HARDNESS: Not determined.
DENSITY: 5.82 (Meas.)
         6.00 (Calc.)
CLEAVAGE: {0001} distinct
Not brittle.
HABIT: Crystals prismatic, elongated along c-axis, or steeply pyramidal. Also tabular {0001}. As very minute crystals usually grouped in parallel position. Twinning on {0001} common.
COLOR-LUSTER: Colorless, white; also with yellowish or bluish tinge from impurities. Transparent. Adamantine to greasy.
MODE OF OCCURRENCE: Occurs associated with laurionite, fiedlerite, and other secondary lead minerals as minute crystals in vugs in lead slags exposed to the action of sea water at Laurium, Greece. It is also found with percylite in the Sierra Gorda, Antofagasta, Chile.

BEST REF. IN ENGLISH: Palache, et al., "Dana's System of Mineralogy," 7th Ed., v. II, p. 66–67, New York, Wiley, 1951.

## PENNAITE (Species status in doubt)
Silicate and chloride of Na, Ca, $Fe^{2+}$, Mn, Ti, and Zr

CRYSTAL SYSTEM: Triclinic ?
OPTICAL CONSTANTS:
  $\beta = 1.70$
  $\gamma - \alpha = 0.044$
  (+)2V = 25°
HABIT: Crystals prismatic. Twinning polysynthetic.
COLOR-LUSTER: Yellow to light brown.
MODE OF OCCURRENCE: Occurs in foyaites, phonolites, tinguaites, and lujaurites, in association with nepheline, eudialyte, and other minerals in zirconium deposits on the Pocos de Caldas plateau, Brazil.

BEST REF. IN ENGLISH: Vlasov, K. A., "Mineralogy of Rare Elements," v. II, p. 386, Israel Program for Scientific Translations 1966.

## PENNANTITE
$(Mn, Al)_6(Si, Al)_4O_{10}(OH)_8$
Chlorite Group

CRYSTAL SYSTEM: Monoclinic
CLASS: 2/m

SPACE GROUP: C2/m
Z: 4
LATTICE CONSTANTS:
 $a = 5.43$
 $b = 9.4$
 $c = 28.5$
 $\beta = 97°25'$
3 STRONGEST DIFFRACTION LINES: Similar to thuringite.
OPTICAL CONSTANTS:
 $\alpha = 1.646$
 $\beta = \gamma = 1.661$
 $(-)2V \approx 0°$
HARDNESS: 2-3
DENSITY: 3.06 (Meas.)
CLEAVAGE: {001} perfect
HABIT: Massive; as minute flakes. Twinning on {001}, twin plane [310].
COLOR-LUSTER: Orange-red. Translucent.
MODE OF OCCURRENCE: Occurs in association with analcime, paragonite, banalsite, ganophyllite, and spessartine in the manganese ores of the Benallt mine, Caernarvonshire, Wales.
BEST REF. IN ENGLISH: Deer, Howie, and Zussman, "Rock Forming Minerals," v. 3, p. 131-163, New York, Wiley, 1962.

## PENNINITE (Pennine)
$(Mg, Fe, Al)_6 (Si, Al)_4 O_{10} (OH)_8$
Chlorite Group

CRYSTAL SYSTEM: Monoclinic
CLASS: 2/m
SPACE GROUP: C2/m
Z: 2
LATTICE CONSTANTS:
 $a = 5.34$
 $b = 9.24$
 $c = 14.38$
 $\beta = 97°06'$
3 STRONGEST DIFFRACTION LINES:
 7.19 (100)
 4.80 (100)
 3.60 (100)
OPTICAL CONSTANTS:
 $\beta = 1.56-1.59$
 $\gamma - \alpha = 0.000-0.004$
 $(\pm)2V = $ small
HARDNESS: 2-2½
DENSITY: 2.6-2.85 (Meas.)
CLEAVAGE: {001} perfect
Laminae flexible, inelastic.
HABIT: Crystals thick tabular, steep rhombohedral, or in tapering six-sided pyramids. Rhombohedral faces often horizontally striated. Often in crested groups. Also massive, compact or scaly. Twinning on {001}, twin plane [310].
COLOR-LUSTER: Emerald green to olive green; sometimes silver white, yellowish. Transparent to translucent. Vitreous; pearly on cleavage.
MODE OF OCCURRENCE: Occurs mainly with serpentine or in crystalline schists. Found in the Ala valley,

Piedmont, and in the Pfitschtal, Trentino, Italy; from the Binnatal and in the Zermatt region, Switzerland; and from the Zillertal, Tyrol, Austria.
BEST REF. IN ENGLISH: Deer, Howie, and Zussman, "Rock Forming Minerals," v. 3, p. 131-163, New York, Wiley, 1962.

## PENROSEITE (Blockite)
$(Ni, Cu)Se_2$

CRYSTAL SYSTEM: Cubic
CLASS: $2/m \bar{3}$
SPACE GROUP: Pa3
Z: 4
LATTICE CONSTANT:
 $a = 5.991$
3 STRONGEST DIFFRACTION LINES:
 2.68 (100)
 2.45 (100)
 1.806 ( 90)
HARDNESS: 2½-3
DENSITY: 6.58-6.74 (Meas.)
CLEAVAGE: {001} perfect
 {011} distinct
Brittle.
HABIT: Botryoidal or reniform with radiating columnar structure; also granular to massive.
COLOR-LUSTER: Steel gray, tarnishing rapidly to dull lead gray; opaque; metallic. Streak black.
MODE OF OCCURRENCE: Occurs associated with clausthalite at Colquechaca, and associated with naumannite, pyrite, and chalcopyrite in a siderite vein in quartzite at the Hiaca silver mine, approximately 30 km east-northeast of Colquechaca, Bolivia.
BEST REF. IN ENGLISH: Palache, et al., "Dana's System of Mineralogy," 7th Ed., v. I, p. 294-296, New York, Wiley, 1944. Early, J. W., Am. Min., 35: 337-364 (1950). Sindeeva, N. D., "Mineralogy and Types of Deposits of Selenium and Tellurium," Wiley-Interscience, 1964.

## PENTAGONITE
$Ca(VO)(Si_4 O_{10}) \cdot 4H_2 O$
Dimorph of Cavansite

CRYSTAL SYSTEM: Orthorhombic
CLASS: mm2
SPACE GROUP: Ccm2$_1$
Z: 4
LATTICE CONSTANTS:
 $a = 10.298$
 $b = 13.999$
 $c = 8.89$
3 STRONGEST DIFFRACTION LINES:
 6.071 (100)
 3.920 (100)
 3.755 (100)
OPTICAL CONSTANTS:
 $\alpha = 1.533$
 $\beta = 1.544$
 $\gamma = 1.547$
 $(-)2V = 50°$

HARDNESS: ~3–4
DENSITY: 2.33 (Calc.)
CLEAVAGE: {010} good
Brittle.

HABIT: Crystals prismatic, elongated parallel to $c$, with dome terminations; minute. Often as spherulitic rosettes up to 5 mm in diameter. Commonly twinned to form fivelings with star-shaped cross sections.

COLOR-LUSTER: Greenish-blue. Transparent. Vitreous.

MODE OF OCCURRENCE: Pentagonite is found in cavities and veinlets in a brown tuff near Owyhee Dam, Malheur County, Oregon. It occurs associated with cavansite, calcite, stilbite, heulandite, analcime, and apophyllite.

BEST REF. IN ENGLISH: Lloyd W. Staples, Howard T. Evans, Jr., and James R. Lindsay, *Am. Min.*, **58**: 405–411 (1973). Howard T. Evans, Jr., *Am. Min.*, **58**: 412–424 (1973).


# PENTAHYDRITE
$MgSO_4 \cdot 5H_2O$

CRYSTAL SYSTEM: Triclinic
LATTICE CONSTANTS:
$a:b:c = 0.621:1:0.5605$
$\alpha = 98°30'$
$\beta = 109°00'$ (synthetic)
$\gamma = 75°05'$
OPTICAL CONSTANTS:
$\beta = 1.491$ (synthetic)
$(-)2V = 45°08'$
HARDNESS: Not determined.
DENSITY: 1.718 (Meas. synthetic)
CLEAVAGE: None.
HABIT: Massive, fine granular or slightly platy. Synthetic crystals elongated with prominent a$\{100\}$ and w$\{\bar{1}11\}$.
COLOR-LUSTER: White, light blue, light greenish blue.
MODE OF OCCURRENCE: Occurs as a constituent of efflorescent salts at The Geysers, Sonoma County, California; at Cripple Creek, Teller County, Colorado; in the Comstock Lode, Virginia City, Nevada; and with chalcanthite at Copaquire, Tarapacá Province, Chile.

BEST REF. IN ENGLISH: Palache, et al., "Dana's System of Mineralogy," 7th Ed., v. II, p. 492–493, New York, Wiley, 1951.


# PENTAHYDROBORITE (Inadequately described mineral)
$CaB_2O_4 \cdot 5H_2O$

CRYSTAL SYSTEM: Unknown
3 STRONGEST DIFFRACTION LINES:
7.04 (100)
2.99 ( 90)
3.54 ( 80)
OPTICAL CONSTANTS:
$\alpha = 1.531$
$\beta = 1.536$
$\gamma = 1.544$
$(+)2V = 73°$
HARDNESS: 2.5

DENSITY: 2.00
CLEAVAGE: None.
HABIT: Small grains, showing no crystal forms.
COLOR-LUSTER: Colorless. Fluoresces violet under short-wave ultraviolet light.

MODE OF OCCURRENCE: Occurs associated with garnet, magnetite, and szaibelyite at the contact of Middle Devonian limestone with quartz diorite in a copper deposit of the skarn type at an unspecified locality in the Urals.

BEST REF. IN ENGLISH: Malinko, S. V., *Am. Min.*, **47**: 1482 (1962).


# PENTAHYDROCALCITE (Species status in doubt)
$CaCO_3 \cdot 5H_2O(?)$

CRYSTAL SYSTEM: Monoclinic
OPTICAL CONSTANTS:
$\alpha = 1.460$
$\beta = 1.535$ (Na)
$\gamma = 1.545$
$(-)2V = 38°$
HARDNESS: Not determined.
DENSITY: 1.75 (Meas.)
CLEAVAGE: Not determined.
HABIT: As acute rhombohedrons.
COLOR-LUSTER: Colorless, transparent.
MODE OF OCCURRENCE: Observed as a recent formation in a water pipe, and deposited upon wood in a brook near Christiana, Norway.

BEST REF. IN ENGLISH: Palache, et al., "Dana's System of Mineralogy," 7th Ed., v. II, p. 228, New York, Wiley, 1951.


# PENTLANDITE
$(Fe, Ni)_9S_8$

CRYSTAL SYSTEM: Cubic
CLASS: $4/m\,\bar{3}\,2/m$
SPACE GROUP: Fm3m
Z: 4
LATTICE CONSTANT:
$a = 9.93–10.19$
3 STRONGEST DIFFRACTION LINES:
1.775 (100)
3.03 ( 80)
1.931 ( 50)
HARDNESS: 3½–4
DENSITY: 4.6–5.0 (Meas.)
4.956 (Calc. Fe:Ni = 2:1)
CLEAVAGE: None. Parting {111}.
Fracture conchoidal. Brittle.
HABIT: Massive, granular.
COLOR-LUSTER: Light bronze yellow. Opaque. Metallic. Streak bronze brown.

MODE OF OCCURRENCE: Occurs chiefly in highly basic rocks associated with pyrrhotite. Found sparingly in the United States in San Diego and Ventura Counties, California; in Clark County, Nevada; and in southeastern Alaska. It is mined as a major nickel ore at Sudbury, Ontario, Canada, and also occurs in Norway and Transvaal, South Africa.

BEST REF. IN ENGLISH: Palache, et al., "Dana's System of Mineralogy," 7th Ed., v. I, p. 242–243, New York, Wiley, 1944.

## PERCYLITE (Inadequately described mineral)
$PbCuCl_2 (OH)_2$

CRYSTAL SYSTEM: Cubic (?)
OPTICAL CONSTANT:
  $N = 2.05$
  HARDNESS: 2½
  DENSITY: Not determined.
  CLEAVAGE: Not reported.
  HABIT: Crystals small, often cubical or dodecahedral. Also massive.
  COLOR-LUSTER: Sky blue; transparent. Vitreous. Streak sky blue.
  MODE OF OCCURRENCE: Occurs as a secondary mineral at the Beatriz mine in the Sierra Gorda, Antofagasta, and in the Cerro de Challocolla, Tarapacá, Chile; in Sonora, Mexico; and in Namaqualand, South Africa.
  BEST REF. IN ENGLISH: Palache, et al., "Dana's System of Mineralogy," 7th Ed., v. II, p. 81–82, New York, Wiley, 1951.

## PERICLASE
MgO

CRYSTAL SYSTEM: Cubic
CLASS: $4/m\,\bar{3}\,2/m$
SPACE GROUP: Fm3m
Z: 4
LATTICE CONSTANT:
  $a = 4.203$
3 STRONGEST DIFFRACTION LINES:
  2.106 (100)
  1.489 ( 52)
  1.216 ( 12)
OPTICAL CONSTANT:
  $N = 1.736$ (Na)
  HARDNESS: 5½
  DENSITY: 3.56 (Meas. synthetic)
             3.581 (Calc.)
  CLEAVAGE: {001} perfect
             {111} imperfect
  HABIT: Crystals octahedral, cubic, cubo-octahedral, dodecahedral. Commonly as rounded grains.
  COLOR-LUSTER: Colorless, white, gray, yellow, brownish yellow, green, black. Transparent. Vitreous. Streak white.
  MODE OF OCCURRENCE: Occurs in marbles as a high-temperature metamorphic mineral. Found in Riverside County, California, at the old and new City quarries, the Wet Weather quarry, and at the Jensen quarry; in the dolomitic xenoliths of the Organ Mountains batholith of New Mexico; and at localities in Spain, Sweden, Czechoslovakia, Italy, and Sardinia.

BEST REF. IN ENGLISH: Palache, et al., "Dana's System of Mineralogy," 7th Ed., v. I, p. 499–500, New York, Wiley, 1944.

**PERIDOT** = Gem variety of olivine

**PERISTERITE** = Variety of sodic plagioclase

## PERITE
$PbBiO_2 Cl$
Bismuth analogue of nadorite

CRYSTAL SYSTEM: Orthorhombic
CLASS: $2/m\,2/m\,2/m$
SPACE GROUP: Bmmb
Z: 4
LATTICE CONSTANTS:
  $a = 5.627$
  $b = 5.575$
  $c = 12.425$
3 STRONGEST DIFFRACTION LINES:
  2.86 (100)
  1.620 ( 90)
  3.77 ( 80)
OPTICAL CONSTANT:
  $n > 2.4$
  HARDNESS: 3
  DENSITY: 8.16 (Meas.)
  CLEAVAGE: Distinct, perpendicular to $c$-axis.
  HABIT: As tabular plates about 0.5 mm in size.
  COLOR-LUSTER: Sulfur yellow; adamantine.
  MODE OF OCCURRENCE: Occurs in fissures in a skarn of hausmannite, calcite, a ludwigite-like mineral, and a few crystals of a red unidentified mineral in the Råman drift (130 m level), Långban, Sweden.
  BEST REF. IN ENGLISH: Gillberg, Marianne, *Am. Min.*, **46**: 765 (1961).

## PERMINGEATITE
$Cu_3 (Sb, As)Se_4$
Luzonite-Germanite Group
Selenium analogue of famatinite

CRYSTAL SYSTEM: Tetragonal
CLASS: 42m
SPACE GROUP: I42m
Z: 2
LATTICE CONSTANTS:
  $a = 5.63$
  $c = 11.23$
3 STRONGEST DIFFRACTION LINES:
  3.251 (100)
  1.986 ( 90)
  1.697 ( 80)
  HARDNESS: 4–4½
  DENSITY: 5.82 (Calc.)
  CLEAVAGE: Not determined.
  HABIT: Massive; as small grains.

COLOR-LUSTER: Light brownish pink in polished section. Opaque. Metallic.

MODE OF OCCURRENCE: Occurs in association with berzelianite, hakite, eskebornite, umangite, hematite, naumannite, clausthalite, ferroselite, klockmannite, uraninite, native gold, pyrite, chalcopyrite, and goethite in a carbonate vein at Předořice, Bohemia, Czechoslovakia.

BEST REF. IN ENGLISH: Johan, Z., Picot, P., Pierrot, R., and Kvacek, M., *Am. Min.*, **57**: 1554-1555 (1972).

## PEROVSKITE
(Ca, Na, Fe$^{2+}$, Ce)(Ti, Nb)O$_3$
Var. Dysanalyte (niobian)
     Knopite (cerian)

CRYSTAL SYSTEM: Orthorhombic
CLASS: 2/m 2/m 2/m
SPACE GROUP: Pnma
Z: 4
LATTICE CONSTANTS:
  $a$ = 5.44
  $b$ = 7.64
  $c$ = 5.38
3 STRONGEST DIFFRACTION LINES:
  2.70 (100)
  1.91 ( 50)
  2.72 ( 40)
OPTICAL CONSTANTS:
  $N$ = 2.34 (variable)
HARDNESS: 5½
DENSITY: 4.01 (Meas.)
        4.04 (Calc.)
CLEAVAGE: {001} imperfect
Fracture subconchoidal to uneven. Brittle.

HABIT: Crystals commonly cubic, often highly modified, striated parallel to cube edges. Also octahedral (especially in Ce and Nb varieties). Rarely massive granular or reniform masses. Twinning on {111} as penetration twins, also complex lamellar.

COLOR-LUSTER: Black, dark brown, amber to yellow. Transparent to opaque. Adamantine, metallic adamantine, metallic, or dull. Streak colorless to pale gray.

MODE OF OCCURRENCE: Occurs chiefly as an accessory mineral in ultrabasic and basic alkaline rocks, in chlorite or talc schists, and in contact metamorphosed limestones. Found as small octahedral crystals at the Commercial quarry, Crestmore, Riverside County, and as lustrous crystals up to 5 mm across near the benitoite mine, San Benito County, California; in the Bearpaw Mountains, Montana; in Gunnison County, Colorado; as fine crystals (dysanalyte) at Magnet Cove, Arkansas; and at localities in Kentucky and New York. It also occurs abundantly in the Oka carbonatite, Quebec, and elsewhere in Canada; and at deposits in Brazil, Sweden, Italy, Germany, and especially as fine large crystals near Zermatt, Switzerland, and at Achmatowsk and near Slatoust, Urals. USSR.

BEST REF. IN ENGLISH: Deer, Howie, and Zussman, "Rock Forming Minerals," v. 5, p. 48-55, New York, Wiley, 1962.

## PERRIERITE
(Ca, Ce, Th)$_4$(Mg, Fe)$_2$(Ti, Fe)$_3$Si$_4$O$_{22}$
Dimorph of chevkinite

CRYSTAL SYSTEM: Monoclinic
CLASS: 2/m
SPACE GROUP: C2/m
Z: 2
LATTICE CONSTANTS:
  $a$ = 13.61
  $b$ = 5.68
  $c$ = 11.73
  $\beta$ = 113.5°
3 STRONGEST DIFFRACTION LINES:
  2.96 (100)
  5.34 ( 65)
  2.82 ( 65)
OPTICAL CONSTANTS:
  $\alpha$ = 1.90-1.95
  $\beta$ = 2.01 (Calc.)
  $\gamma$ = 2.02-2.06
  (–)2V = 60°
HARDNESS: 5½
DENSITY: 4.3
CLEAVAGE: None apparent.
Fracture uneven to conchoidal.

HABIT: Crystals elongate, slightly flattened prisms resembling epidote; usually less than 0.2 mm in size. Frequently twinned on {100}.

COLOR-LUSTER: Black to brownish; resinous. Streak brown.

MODE OF OCCURRENCE: Occurs associated with pyroxene, ilmenite, garnet, and magnetite in sands derived from the weathering of tuffs at Nettuno, Italy; also found at Kobe-mura, Japan; in silicic ash beds of early Miocene age along Goshen Hole Rim in southeastern Wyoming and near Chimney Rock in southwestern Nebraska; in Amherst, Bedford, and Nelson counties, Virginia; and reported from several other localities in many parts of the world.

BEST REF. IN ENGLISH: Bonatti, Stefano, and Gottardi, Glauco, *Am. Min.*, **36**: 926 (1951). Bonatti, Stefano, *Am. Min.*, **44**: 115-137 (1959). Gottardi, Glauco, *Am. Min.*, **45**: 1-15 (1960). Mitchell, Richard S., *Am. Min.*, **51**: 1394-1405 (1966).

## PERRYITE (Inadequately described mineral)
(Ni, Fe)$_5$ (Si, P)$_2$ with deficiency in Si + P
Formula determined by microprobe analysis.

CRYSTAL SYSTEM: Unknown
MODE OF OCCURRENCE: Found in the Kota-Kota, South Oman, Horse Creek, Mount Egerton, and Norton County enstatite chondrites.

BEST REF. IN ENGLISH: Reed, S. J. B., *Min. Mag.*, **36**: 850-854 (1968). Wai, C. M., *Min. Mag.*, **37**: 905-908 (1970).

## PETALITE
LiAlSi$_4$O$_{10}$

CRYSTAL SYSTEM: Monoclinic
CLASS: 2/m
SPACE GROUP: P2/a
Z: 2
LATTICE CONSTANTS:
  $a = 11.76$
  $b = 5.14$
  $c = 7.62$
  $\beta = 112°24'$
3 STRONGEST DIFFRACTION LINES:
  3.731 (100)
  3.670 ( 90)
  3.515 ( 30)
OPTICAL CONSTANTS:
  $\alpha = 1.504$
  $\beta = 1.510$
  $\gamma = 1.516$
  $(+)2V = 83°24'$
HARDNESS: 6-6½
DENSITY: 2.3-2.5 (Meas.)
        2.454 (Calc.)
CLEAVAGE: {001} perfect
        {201} distinct
Fracture subconchoidal. Brittle.

HABIT: Crystals euhedral; small, rare. Usually massive, as large cleavable blocky segregations. Twinning on {001}, polysynthetic.

COLOR-LUSTER: Colorless, white, gray, yellow. Transparent to translucent. Vitreous to pearly.

MODE OF OCCURRENCE: Occurs in granite pegmatites in association with cleavelandite, quartz, and lepidolite. Found at Pala and Rincon, San Diego County, California; at Bolton, Worcester County, Massachusetts; at Peru, Oxford County, Maine; on the Island of Utö, Sweden; Elba, Italy; USSR; and elsewhere.

BEST REF. IN ENGLISH: Vlasov, K. A., "Mineralogy of Rare Elements," v. II, p. 35-38, Israel Program for Scientific Translations, 1966.

# PETZITE
$Ag_3AuTe_2$

CRYSTAL SYSTEM: Cubic
CLASS: 432
SPACE GROUP: I4₁32
Z: 8
LATTICE CONSTANT:
  $a = 10.38$
3 STRONGEST DIFFRACTION LINES:
  2.77 (100)
  2.12 ( 80)
  2.03 ( 70)
HARDNESS: 2½-3
DENSITY: 8.7-9.4 (Meas.)
        8.74 (Calc.)
CLEAVAGE: {001} ?
Fracture subconchoidal. Slightly sectile to brittle.

HABIT: Commonly massive, fine granular to compact; rarely as small single crystals intimately intergrown with hessite.

COLOR-LUSTER: Steel gray to iron black. often tarnished; metallic. Opaque.

MODE OF OCCURRENCE: Occurs in vein deposits with other tellurides at Bótes and Nagyág, Transylvania, Roumania; Timmins, Ontario, Canada; Kalgoorlie, Western Australia; and at several localities in Colorado and California.

BEST REF. IN ENGLISH: Frueh, Alfred J. Jr., *Am. Min.*, **44**: 693-701 (1959). Vlasov, K. A., "Mineralogy of Rare Elements," v. II, p. 712-717, Israel Program for Scientific Translations, 1966.

# PHACOLITE = Variety of Chabazite

# PHARMACOLITE
$CaHAsO_4 \cdot 2H_2O$

CRYSTAL SYSTEM: Monoclinic
CLASS: m or 2/m
SPACE GROUP: A2/a or Aa
Z: 4
LATTICE CONSTANTS:
  $a = 6.00$ kX
  $b = 15.40$
  $c = 6.29$
  $\beta = 114°47'$
3 STRONGEST DIFFRACTION LINES:
  4.32 (100)
  3.09 (100)
  2.71 ( 90)
OPTICAL CONSTANTS:
  $\alpha = 1.583$
  $\beta = 1.589$ (Na)
  $\gamma = 1.594$
  $(-)2V = 79°24'$
HARDNESS: 2-2½
DENSITY: 2.68 (Meas.)
        2.674 (Calc.)
CLEAVAGE: {010} perfect
Fracture uneven. Thin laminae flexible.

HABIT: Crystals flattened on {010} or acicular; small and rare. Usually botryoidal, stalactitic, or as clusters of acicular crystals or thin fibers.

COLOR-LUSTER: Colorless, white, grayish. Transparent to translucent. Vitreous, pearly on cleavage. Fibrous material silky.

MODE OF OCCURRENCE: Occurs with erythrite and smaltite at the O.K. mine, San Gabriel Canyon, Los Angeles County, California; and with haidingerite at the White Cap mine, Manhattan, Nye County, Nevada. It also is found at Wittichen in the Black Forest and at several other localities in Germany; at Joachimsthal (Jachymov), Bohemia, Czechoslovakia; and at Markirch, Alsace, France.

BEST REF. IN ENGLISH: Palache, et al., "Dana's System of Mineralogy," 7th Ed., v. II, p. 706-708, New York, Wiley, 1951.

# PHARMACOSIDERITE
$Fe_3(AsO_4)_2(OH)_3 \cdot 5H_2O$

CRYSTAL SYSTEM: Cubic
CLASS: $\bar{4}$3m
SPACE GROUP: P$\bar{4}$3m
Z: 3/2
LATTICE CONSTANT:
$a = 7.91$
3 STRONGEST DIFFRACTION LINES:
7.98 (100)
3.25 ( 50)
2.816 ( 50)
OPTICAL CONSTANT:
$N = 1.693$
HARDNESS: 2½
DENSITY: 2.798 (Meas.)
2.869 (Calc. 6H$_2$O)
CLEAVAGE: {100} imperfect to good
Fracture uneven. Somewhat sectile.

HABIT: Crystals usually cubes with diagonally striated faces; also tetrahedral. Occasionally granular or earthy.

COLOR-LUSTER: Olive green to emerald green, amber to dark brown or reddish brown. Transparent to translucent. Adamantine to greasy.

MODE OF OCCURRENCE: Occurs as an oxidation product of arsenic-rich minerals, and less commonly as a hydrothermal deposit. Found sparingly in granite pegmatite at the White Elephant mine, Custer County, and associated with beudantite and scorodite in a lead-silver ore deposit at the Richmond-Sitting Bull mine, Galena, Lawrence County, South Dakota; as fine crystals at the Myler mine, Majuba Hill, Pershing County, Nevada; and at the Gold Hill mine and elsewhere in the Tintic district, Juab County, Utah. It also occurs at many localities in Cornwall, England, and in France, Germany, Czechoslovakia, Italy, and Algeria.

BEST REF. IN ENGLISH: Palache, et al., "Dana's System of Mineralogy," 7th Ed., v. II, p. 995–997, New York, Wiley, 1951.

# PHENACITE (Phenakite)
Be$_2$SiO$_4$

CRYSTAL SYSTEM: Hexagonal
CLASS: $\bar{3}$
SPACE GROUP: R$\bar{3}$
Z: 18
LATTICE CONSTANTS:
$a = 12.45$    $a_{rh} = 7.70$
$c = 8.23$    $\alpha = 108°01'$
3 STRONGEST DIFFRACTION LINES:
3.119 (100)
3.66 ( 80)
2.518 ( 75)
OPTICAL CONSTANTS:
$\omega = 1.654$
$\epsilon = 1.670$
(+)
HARDNESS: 7½–8
DENSITY: 2.93–3.00 (Meas.)
2.960 (Calc.)

CLEAVAGE: {11$\bar{2}$0} distinct
{10$\bar{1}$1} imperfect
Fracture conchoidal. Brittle.

HABIT: Crystals rhombohedral, prismatic, occasionally long-prismatic and acicular. Crystals range in size from microscopic up to 5 × 10 × 18 cm. Also granular, as acicular columnar aggregates, and as radiating-fibrous spherulites. Twinning on {10$\bar{1}$0} common.

COLOR-LUSTER: Colorless; also wine yellow, yellow, pink, pinkish red, brown. Transparent. Vitreous.

MODE OF OCCURRENCE: Occurs in granite pegmatites, greisens, hydrothermal and Alpine veins. Found in the Vandenberg Catherina mine, Pala, San Diego County, California; at Mount Antero in Chaffee County, Crystal Park in El Paso County, and on Crystal Peak in Teller County, Colorado; on Bald Face Mountain, Carroll County, New Hampshire; at Lords Hill, Oxford County, Maine; and at the Morefield mine, Amelia County, Virginia. Other notable occurrences are found at San Miguel di Piracicaba, Minas Geraes, Brazil; Kragerö, Norway; on the Takowaja River, near Sverdlovsk, Urals, USSR; and in France, Switzerland, Czechoslovakia, and elsewhere.

BEST REF. IN ENGLISH: Vlasov, K. A., "Mineralogy of Rare Elements," v. II, p. 81–85, Israel Program for Scientific Translations, 1966.

# PHENGITE = High-silica variety of muscovite

# PHILIPSTADITE = Hornblende

# PHILLIPSITE
Near (K$_2$, Na$_2$, Ca)Al$_2$Si$_4$O$_{12}$ · 4½H$_2$O
Zeolite Group

CRYSTAL SYSTEM: Monoclinic
CLASS: 2/m or 2
SPACE GROUP: P2$_1$/m or P2$_1$
Z: 2
LATTICE CONSTANTS:
$a = 10.02$
$b = 14.28$
$c = 8.64$
$\beta = 125°40'$
3 STRONGEST DIFFRACTION LINES:
7.15 (100)
3.20 (100)
4.12 ( 80)
OPTICAL CONSTANTS:
$\alpha = 1.483–1.504$
$\beta = 1.484–1.509$
$\gamma = 1.486–1.514$
(+)2V = 60°–80°
HARDNESS: 4–4½
DENSITY: 2.20 (Meas.)
CLEAVAGE: {010} distinct
{100} distinct
Fracture uneven. Brittle.

HABIT: Crystals penetration twins, often simulating orthorhombic or tetragonal forms. Crystals isolated or in druses. Also in spherulites. Twinning on {001}, {021}, {110}.

COLOR-LUSTER: Colorless, white, sometimes reddish. Transparent to translucent. Vitreous.

MODE OF OCCURRENCE: Occurs in cavities in basalt, phonolite, and related rocks; in saline lake deposits; in calcareous deep sea sediments; and as a hot springs deposit. Found in Inyo, Kern, Plumas, Riverside, and San Bernardino counties, California; at the Moiliili quarry, Honolulu, Hawaii; as fine crystals at Capo di Bove and elsewhere in Italy; and at localities in Sicily, Northern Ireland, Germany, Czechoslovakia, Uganda, and Australia.

BEST REF. IN ENGLISH: Deer, Howie, and Zussman, "Rock Forming Minerals," v. 4, p. 386-399, New York, Wiley, 1963. *Am. Min.*, **52**: 1785-1794 (1967).

## PHLOGOPITE
$KMg_3AlSi_3O_{10}(OH)_2$
Mica Group

CRYSTAL SYSTEM: Monoclinic, hexagonal
CLASS: 2/m and 32
SPACE GROUP: C2/m (mono.), $P3_112$ (hex.)
Z: 2 (1M), 6 (3T)
LATTICE CONSTANTS:

| | 1M | 3T |
|---|---|---|
| $a$ = | 5.36 | $a$ = 5.3 (?) |
| $b$ = | 9.29 | |
| $c$ = | 10.41 | $c$ = 30.06 |
| $\beta$ = | 100°0' | |

3 STRONGEST DIFFRACTION LINES:

| 1M | 3T |
|---|---|
| 9.94 (>100) | 10.13 (>100) |
| 3.35 (>100) | 3.35 (>100) |
| 2.61 ( 30) | 2.01 (>100) |

OPTICAL CONSTANTS:
  $\alpha$ = 1.530-1.590
  $\beta$ = 1.557-1.637
  $\gamma$ = 1.558-1.637
(-)2V = 0°-15°
HARDNESS: 2-2½
DENSITY: 2.76-2.90 (Meas.)
CLEAVAGE: {001} perfect
Thin laminae flexible and elastic, tough.

HABIT: Crystals prismatic, usually tapered; often large and coarse. Also as plates and scales. Twinning on {001}, twin axis [310].

COLOR-LUSTER: Usually yellowish brown to brownish red; also colorless, white, greenish. Transparent to translucent. Pearly, often submetallic on cleavage surface. Streak colorless. Often exhibits asterism in transmitted light.

MODE OF OCCURRENCE: Occurs chiefly in metamorphic limestones and ultrabasic rocks. Typical occurrences are found in California, Colorado, South Dakota, New York, and New Jersey. Also as fine large pseudohexagonal crystals in Quebec and south central and eastern portions of Ontario, Canada, and at localities in Sweden, Italy, Switzerland, Roumania, Finland, USSR, South Africa, Madagascar, Ceylon, India, Australia, and New Zealand.

BEST REF. IN ENGLISH: Deer, Howie, and Zussman, "Rock Forming Minerals," v. 3, p. 42-54, New York, Wiley, 1962. Steinfink, Hugo, *Am. Min.*, **47**: 886-896 (1962).

## PHOENICOCHROITE
$Pb_2O(CrO_4)$

CRYSTAL SYSTEM: Monoclinic
CLASS: 2/m
SPACE GROUP: C2/m
Z: 4
LATTICE CONSTANTS:
  $a$ = 14.001
  $b$ = 5.675
  $c$ = 7.137
  $\beta$ = 115°13'
3 STRONGEST DIFFRACTION LINES:
  3.380 (100)
  2.979 (100)
  6.43 ( 50)
OPTICAL CONSTANTS:
  $\alpha$ = 2.38
  $\beta$ = 2.44 (Na)
  $\gamma$ = 2.65
(+)2V = 58°
HARDNESS: 2½
DENSITY: 7.01 (Meas.)
         7.075 (Calc.)
CLEAVAGE: {$\bar{2}$01} good
          {001} fair
          {010} fair
          {011} fair

HABIT: As large polycrystalline masses, rarely exhibiting crystal faces. Forms: {010}, {100}, {$\bar{2}$01}, and {$\bar{2}$11}. Also as anhedral cleavable masses.

COLOR-LUSTER: Dark cochineal red, translucent to transparent on thin edges. Resinous or adamantine. Streak yellowish orange.

MODE OF OCCURRENCE: Occurs associated with crocoite, vauquelinite, and pyromorphite in altered quartz-galena veins in limestone at Beresov, Urals, USSR; at the Adelaide Proprietary mine, Dundas, Tasmania; as cleavage plates up to 25 mm across associated with hemihedrite, crocoite, cerussite, mimetite, and vauquelinite in the oxidized zone of galena-bearing veins at the Potter-Cramer property and at several other deposits in the Wickenburg area, Arizona.

BEST REF. IN ENGLISH: Williams, Sidney A., McLean, John, and Anthony, John W., *Am. Min.*, **55**: 784-792 (1970).

## PHOSGENITE
$Pb_2CO_3Cl_2$

CRYSTAL SYSTEM: Tetragonal
CLASS: 4/m 2/m 2/m
SPACE GROUP: P4/mbm
Z: 4
LATTICE CONSTANTS:
  $a$ = 8.112
  $c$ = 8.814

3 STRONGEST DIFFRACTION LINES:
2.79 (100)
3.61 ( 80)
2.56 ( 80)
OPTICAL CONSTANTS:
$\omega = 2.1181$
$\epsilon = 2.1446$ (Na)
(+)
HARDNESS: 2–3
DENSITY: 6.133 (Meas.)
6.136 (Calc.)
CLEAVAGE: {001} distinct
{110} distinct
{100} indistinct
Fracture conchoidal.

HABIT: Crystals short to long prismatic; also thick tabular. Sometimes massive, granular.

COLOR-LUSTER: Colorless, white, yellowish white, gray, various shades of brown, also greenish or pinkish. Transparent to translucent. Adamantine. Streak white. Sometimes fluoresces in shades of yellow under ultraviolet light.

MODE OF OCCURRENCE: Occurs as a secondary mineral in the oxidation zone of lead deposits, often associated with other secondary lead minerals. Found in the United States as acicular crystals in quartz at the Silver Sprout mine, Inyo County, California; in abundance as large masses at the Terrible mine at Oak Creek, Custer County, Colorado; with diaboleite at the Mammoth mine, Pinal County, Arizona; and at localities in New Mexico and Massachusetts. It also occurs as superb crystals up to 5 inches across at Monte Poni and at other places in Sicily; as fine crystals at Matlock and elsewhere in Derbyshire, England; at Tarnow, Poland; and in the USSR, Tasmania, Australia, Tunisia, and at Tsumeb, South West Africa.

BEST REF. IN ENGLISH: Palache, et al., "Dana's System of Mineralogy," 7th Ed., v. II, p. 256–259, New York, Wiley, 1951.

# PHOSPHOFERRITE
(Fe, Mn)$_3$(PO$_4$)$_2$ · 3H$_2$O
Reddingite-Phosphoferrite Series

CRYSTAL SYSTEM: Orthorhombic
CLASS: mm2
SPACE GROUP: P2na
Z: 4
LATTICE CONSTANTS:
$a = 9.41$
$b = 10.02$
$c = 8.66$
3 STRONGEST DIFFRACTION LINES:
3.18 (100)
2.724 ( 80)
4.25 ( 70)
OPTICAL CONSTANTS:
$\alpha = 1.672$
$\beta = 1.680$
$\gamma = 1.700$
(+)2V = 68°
HARDNESS: 3–3½

DENSITY: 3.29 (Meas.)
3.340 (Calc. Fe$_2$Mn)
CLEAVAGE: {010} poor
Fracture uneven. Brittle.

HABIT: Crystals octahedral with {111}, or tabular {010}; often in parallel grouping. Also massive, granular, and coarse fibrous.

COLOR-LUSTER: Colorless, pale green; often light to dark reddish brown from alteration. Transparent to translucent. Vitreous to resinous.

MODE OF OCCURRENCE: The mineral occurs as a hydrothermal alteration of triphylite or graftonite in granite pegmatites. Found in South Dakota as fine crystals with ludlamite, vivianite, messelite, and graftonite, at the Bull Moose mine, Custer County, and in association with ludlamite, vivianite, and triphylite at the Dan Patch and Big Chief mines, Pennington County. Originally found at Hagendorf, Bavaria, Germany, with ludlamite, vivianite, and triploidite.

BEST REF. IN ENGLISH: Palache, et al., "Dana's System of Mineralogy," 7th Ed., v. II, p. 727–729, New York, Wiley, 1951.

# PHOSPHOPHYLLITE
Zn$_2$(Fe, Mn)(PO$_4$)$_2$ · 4H$_2$O

CRYSTAL SYSTEM: Monoclinic
CLASS: 2/m
SPACE GROUP: P2$_1$/c
Z: 2
LATTICE CONSTANTS:
$a = 10.25$
$b = 5.09$
$c = 10.51$
$\beta = 120°15'$
3 STRONGEST DIFFRACTION LINES:
2.825 (100)
4.40 ( 95)
8.84 ( 80)
OPTICAL CONSTANTS:
$\alpha = 1.595$
$\beta = 1.614$ (Na)
$\gamma = 1.616$
(−)2V ~ 45°
HARDNESS: 3–3½
DENSITY: 3.13 (Meas.)
3.145 (Calc.)
CLEAVAGE: {100} perfect
{010} distinct
{$\bar{1}$02} distinct
Brittle. Fracture uneven.

HABIT: Crystals short to long prismatic or thick tabular. As isolated individuals or in groups. Twinning on {100} common; on {$\bar{1}$02} rare.

COLOR-LUSTER: Colorless to vivid bluish green. Transparent. Vitreous.

MODE OF OCCURRENCE: Occurs in vugs in massive sulfides at Potosi, Bolivia, as magnificent transparent single crystals up to 3 × 2 inches in size. It also is found in granite pegmatite as small crystals associated with triplite, triphylite, sphalerite, apatite, and a wide variety of secondary phosphate minerals at Hagendorf, Bavaria, Germany.

BEST REF. IN ENGLISH: Palache, et al., "Dana's System of Mineralogy," 7th Ed., v. II, p. 738–739, New York, Wiley, 1951.    Kleber, et al., *Acta. Cryst.*, **14**: 795 (1961).


# PHOSPHORROESSLERITE
$MgH(PO_4) \cdot 7H_2O$

CRYSTAL SYSTEM: Monoclinic
CLASS: 2/m
SPACE GROUP: A2/a
Z: 8
LATTICE CONSTANTS:
$a = 11.35$ kX
$b = 25.36$
$c = 6.60$
$\beta = 95°$
OPTICAL CONSTANTS:
$\alpha = 1.477$
$\beta = 1.485$
$\gamma = 1.486$
$(-)2V = 38°10'$
HARDNESS: 2½
DENSITY: 1.725 (Meas.)
1.717 (Calc.)
CLEAVAGE: Fracture conchoidal.
HABIT: Crystals short prismatic or equant, also skeletal or crusted.
COLOR-LUSTER: Colorless; usually yellowish due to impurities.  Transparent to translucent.  Vitreous when fresh; usually dull.
MODE OF OCCURRENCE: Found in abandoned mine workings near Schellgaden, Salzburg province, Austria.
BEST REF. IN ENGLISH: Palache, et al., "Dana's System of Mineralogy," 7th Ed., v. II, p. 713–714, New York, Wiley, 1951.


# PHOSPHOSIDERITE = Metastrengite


# PHOSPHURANYLITE
$Ca(UO_2)_4(PO_4)_2(OH)_4 \cdot 7H_2O$

CRYSTAL SYSTEM: Orthorhombic
CLASS: 2/m 2/m 2/m
SPACE GROUP: Bmmb
Z: 6
LATTICE CONSTANTS:
$a = 15.85$
$b = 17.42$
$c = 13.76$
3 STRONGEST DIFFRACTION LINES:
7.91 (100)
3.96 ( 60)
3.15 ( 60)
OPTICAL CONSTANTS:
$\alpha = 1.658 - 1.690$
$\beta = 1.699 - 1.724$
$\gamma = 1.699 - 1.724$
$(-)2V = 0° - 25°, 51°$

HARDNESS: ~2½
DENSITY: ~4.1 (Meas.)
CLEAVAGE: {100} perfect
{010} indistinct
Not brittle.
HABIT: Tiny scales, plates, or laths with rectangular outline.  As dense, earthy aggregates or thin coatings and crusts.
COLOR-LUSTER: Light golden yellow to deep yellow.  Translucent.  Not fluorescent.
MODE OF OCCURRENCE: Occurs widespread in small amounts as a secondary mineral in the weathered zone of pegmatites that carry uraninite, and also found sparingly in several sandstone-type uranium deposits on the Colorado Plateau.  In the United States it is found in pegmatite at the Flat Rock and Buchanan mines, Mitchell County, North Carolina; in several pegmatites in the New England area, including the Ruggles and Palermo mines, Grafton County, New Hampshire; and at the Ferguson Lode claim near Keystone, Pennington County, South Dakota, associated with uraninite, meta-autunite, and beta-uranophane.  Notable foreign localities include LaCrouzille and Margnac, Haute-Vienne, France; Carrasca and Urgeirica, Portugal; Wolsendorf, Bavaria, Germany; and Memoes, Rio Grande do Norte, Brazil.
BEST REF. IN ENGLISH: Frondel, Clifford, USGS Bull. 1064, p. 222–227 (1958).


# PICKERINGITE
$MgAl_2(SO_4)_4 \cdot 22H_2O$

CRYSTAL SYSTEM: Monoclinic
CLASS: 2
SPACE GROUP: P2
Z: 4
LATTICE CONSTANTS:
$a = 20.8$
$b = 24.2$
$c = 6.18$
$\beta = 95°$
3 STRONGEST DIFFRACTION LINES:
4.82 (100)
3.51 ( 90)
4.32 ( 35)
OPTICAL CONSTANTS:
$\alpha = 1.475$
$\beta = 1.480$
$\gamma = 1.483$
$(-)2V = 60°$
HARDNESS: 1½
DENSITY: 1.79 (Meas.)
1.84 (Calc.)
CLEAVAGE: {010} poor
Fracture conchoidal.  Brittle.
HABIT: Crystals acicular, rarely terminated.  As aggregates of fibrous or acicular crystals; as an incrustation or efflorescence; sometimes matted.
COLOR-LUSTER: Colorless, white, or with yellowish or reddish tinge.  Transparent to translucent.  Vitreous.

Soluble in water.  Taste astringent.
MODE OF OCCURRENCE: Occurs widespread chiefly as a weathering product of pyrite-bearing and aluminous rocks,

in pyritic ore deposits and coal veins, and in caverns. Found in the United States, especially in Inyo, San Bernardino, Shasta, and Sonoma counties, California; in Quay and San Miguel counties, New Mexico; and in Colorado, Utah, Nevada, and South Dakota. It also is found in Canada, northern Chile, Germany, Austria, Czechoslovakia, France, Italy, and at many other localities.

BEST REF. IN ENGLISH: Palache, et al., "Dana's System of Mineralogy," 7th Ed., v. II, p. 523–527, New York, Wiley, 1951.


## PICOTITE = Chromian spinel


## PICOTPAULITE
$TiFe_2S_3$

CRYSTAL SYSTEM: Orthorhombic
CLASS: 222, mm2, or 2/m 2/m 2/m
SPACE GROUP: C222, Cmm2, Amm2, or Cmmm
Z: 4
LATTICE CONSTANTS:
 $a = 4.50$
 $b = 10.72$
 $c = 9.04$
3 STRONGEST DIFFRACTION LINES:
 2.912 (100)
 4.26 ( 90)
 3.80 ( 70)
HARDNESS: Microhardness 41 kg/mm$^2$.
DENSITY: 5.20 (Calc.)
CLEAVAGE: Not determined.
HABIT: Crystals pseudohexagonal. Twinning on {120} common, as penetration twins.
COLOR-LUSTER: Bronze. Creamy white in reflected light.
MODE OF OCCURRENCE: Occurs in association with raguinite and pyrite in the realgar-lorandite association from the Allchar deposit, Macedonia, Yugoslavia.
BEST REF. IN ENGLISH: Johan, Z., Pierrot, R., Schubnel, H. J., and Permingeat, F., *Min. Abs.*, **22**: 228 (1971).


## PICROCHROMITE = Magnesiochromite
$MgCr_2O_4$


## PICROEPIDOTE = Magnesium epidote


## PICROMERITE
$K_2Mg(SO_4)_2 \cdot 6H_2O$

CRYSTAL SYSTEM: Monoclinic
CLASS: 2/m
SPACE GROUP: P2$_1$/a
Z: 2
LATTICE CONSTANTS:
 $a = 9.064$
 $b = 12.256$
 $c = 6.113$
 $\beta = 104.791°$

3 STRONGEST DIFFRACTION LINES:
 3.70 (100)
 4.15 ( 75)
 4.05 ( 75)
OPTICAL CONSTANTS:
 $\alpha = 1.4607$
 $\beta = 1.4629$  (Na)
 $\gamma = 1.4755$
 (+)2V = 47°54'
HARDNESS: 2½
DENSITY: 2.028 (Meas.)
 2.025 (Calc.)
CLEAVAGE: {$\bar{2}$01} perfect
HABIT: Usually massive or as crusts on other salts. Crystals short prismatic.
COLOR-LUSTER: Colorless, white; also gray, yellowish, or reddish due to admixed impurities. Transparent. Vitreous.

Taste bitter. Easily soluble in water.
MODE OF OCCURRENCE: Occurs in beds up to 3 feet thick in the kainite zone of an oceanic salt deposit at Aschersleben, and with anhydrite and halite at Leopoldshall, Germany; at Kalusz, Poland, associated with epsomite; and as an uncommon fumarolic deposit on Mt. Vesuvius, Italy.
BEST REF. IN ENGLISH: Palache, et al., "Dana's System of Mineralogy," 7th Ed., v. II, p. 453–454, New York, Wiley, 1951.


## PICROPHARMACOLITE
$H_2Ca_4Mg(AsO_4)_4 \cdot 12H_2O$

CRYSTAL SYSTEM: Monoclinic
Z: 4
LATTICE CONSTANTS:
 $a \simeq 21$ (or $a = 13.7$?)
 $b \simeq 13.5$
 $c = 6.74$
 $\beta \simeq 140°$
3 STRONGEST DIFFRACTION LINES:
 13.50 (100)
 3.18 ( 80)
 9.20 ( 60)
OPTICAL CONSTANTS:
 $\alpha = 1.566$ (Calc.)
 $\beta = 1.571$
 $\gamma = 1.578$
 2V $\simeq 50°$
HARDNESS: Not determined.
DENSITY: 2.62 (Meas.)
 2.606 (Calc. Ca$_{1.5}$Mg$_{1.5}$)
CLEAVAGE: {100} perfect
 {010} perfect
HABIT: As tiny needles, mammillary incrustations, or concretions and radial-structured nodules up to 1 cm in diameter.
COLOR-LUSTER: Colorless, white, slightly pearly; transparent to translucent.
MODE OF OCCURRENCE: Occurs at Sainte-Marie-aux-Mines, France; at Riechelsdorf and Freiberg, Saxony, Germany; and coating dolomite at Joplin, Missouri.

BEST REF. IN ENGLISH: Pierrot, R., *Am. Min.*, 47: 1222 (1962).

## PICROTEPHROITE = Variety of tephroite
$(Mn, Mg_2)SiO_4$

## PIEDMONTITE = Piemontite

## PIEMONTITE (Piedmontite)
$Ca_2(Al, Fe, Mn^{3+})_3Si_3O_{12}OH$
Epidote Group
Var. Withamite

CRYSTAL SYSTEM: Monoclinic
CLASS: 2/m
SPACE GROUP: $P2_1/m$
Z: 2
LATTICE CONSTANTS:
$a = 8.95$
$b = 5.70$
$c = 9.41$
$\beta = 115°42'$
3 STRONGEST DIFFRACTION LINES:
2.91 (100)
3.50 ( 80)
2.84 ( 80)
OPTICAL CONSTANTS:
$\alpha = 1.732-1.794$
$\beta = 1.750-1.807$
$\gamma = 1.762-1.829$
$(+)2V = 64°-85°$
HARDNESS: 6
DENSITY: 3.45-3.52 (Meas.)
　　　　　3.50 (Calc.)
CLEAVAGE: {001} perfect
Fracture uneven. Brittle.
HABIT: Crystals prismatic or acicular, similar to epidote. Commonly massive, coarse to fine granular; also fibrous, parallel or divergent. Twinning {100}, lamellar; not common.
COLOR-LUSTER: Reddish brown to reddish black; also carmine red to straw yellow. Translucent to nearly opaque.
MODE OF OCCURRENCE: Occurs in schists, andesites (withamite), rhyolites, and felsite; also in metasomatic manganese ore deposits; rarely in pegmatites. Found at many places in California, especially in Los Angeles, Madera, Plumas, Riverside, and San Bernardino counties; in the Tucson Mountains, Pat Hills, and Santa Rita Mountains, Arizona; near Annapolis, Iron County, Missouri; in Adams County, Pennsylvania; at Glen Coe, Argyllshire, Scotland (withamite); in the Nordmark district, Vermland, Sweden; on Ile de Groix, Morbihan, France; in sericite schist at the type locality, Piemonte, Italy; in porphyry at Djebel Dokhan, Egypt; in glaucophane schists on the island of Shikoku, Japan; and in quartz schist at Black Peak, northwestern Otago, New Zealand.
BEST REF. IN ENGLISH: Deer, Howie, and Zussman, "Rock Forming Minerals," v. 1, p. 194-210, New York, Wiley, 1962.

## PIERROTITE
$Tl_2(Sb, As)_{10}S_{17}$

CRYSTAL SYSTEM: Orthorhombic (?)
LATTICE CONSTANTS:
$a = 8.77$
$b = 38.8$
$c = 8.03$
3 STRONGEST DIFFRACTION LINES:
3.59 (100)
3.49 ( 90)
2.70 ( 90)
HARDNESS: ~3½
DENSITY: 4.97
CLEAVAGE: Not determined.
HABIT: Massive; as crystalline aggregates.
COLOR-LUSTER: Grayish black. Metallic. Transparent and dark red in section.
MODE OF OCCURRENCE: Occurs in association with stibnite, realgar, pyrite, and an amorphous thallium mineral in quartz veins from Jas-Roux, Hautes-Alpes, France.
BEST REF. IN ENGLISH: Guillemin, C., Johan, Z., Lanforêt, C., and Picot, P., *Min. Abs.*, 21: 70-3428 (1970).

## PIGEONITE
$(Mg, Fe, Ca)(Mg, Fe)Si_2O_6$
Pyroxene Group

CRYSTAL SYSTEM: Monoclinic
CLASS: 2/m
SPACE GROUP: $P2_1/c$
Z: 4
LATTICE CONSTANTS:
$a = 9.73$
$b = 8.95$
$c = 5.26$
$\beta = 108°33'$
3 STRONGEST DIFFRACTION LINES:
3.02 (100)
2.90 (100)
3.21 ( 80)
OPTICAL CONSTANTS:
$\alpha = 1.682-1.722$
$\beta = 1.684-1.722$
$\gamma = 1.705-1.751$
$(+)2V = 0°-30°$
HARDNESS: 6
DENSITY: 3.30-3.46 (Meas.)
　　　　　3.540 (Calc.)
CLEAVAGE: {110} good
　　　　　{001} parting
　　　　　{010} parting
　　　　　{100} parting
Fracture uneven to conchoidal. Brittle.
HABIT: Crystals short prismatic. Usually as disseminated microphenocrysts and small grains. Twinning on {001} or {100}, simple or lamellar.
COLOR-LUSTER: Brown, light purplish brown, greenish brown, black. Translucent to nearly opaque. Vitreous to dull.
MODE OF OCCURRENCE: Occurs chiefly as small

grains in the groundmass of volcanic rocks, and is particularly characteristic in andesites and dacite; also found in some meteorites. Occurs in the Beaver Bay diabase at Pigeon Point, Minnesota, and at localities in Arizona, Colorado, Montana, New Jersey, and Virginia. Also in Greenland, Scotland, Germany, Pakistan, Japan, and elsewhere.

BEST REF. IN ENGLISH: Deer, Howie, and Zussman, "Rock Forming Minerals," v. 2, p. 143–153, New York, Wiley, 1963.  *Am. Min.*, **55**: 1195–1209 (1970).

**PILBARITE** = Mixture of thorogummite and kasolite

**PILOLITE** = Aluminum-bearing sepiolite

## PIMELITE
$(Ni, Mg)_3 Si_4 O_{10}(OH)_2 \cdot 4H_2O$
Montmorillonite (Smectite) Group

CRYSTAL SYSTEM:  Monoclinic
  HARDNESS:  2½
  DENSITY:  2.71–2.76 (Meas.)
  CLEAVAGE:  {001} perfect
  HABIT:  Massive, very fine grained; as compact aggregates or earthy.
  COLOR-LUSTER:  Apple green to white. Translucent. Dull, greasy. Streak greenish white.
  MODE OF OCCURRENCE:  Occurs at Křemže, southern Bohemia, Czechoslovakia.
  BEST REF. IN ENGLISH: "Dana's System of Mineralogy," 6th Ed., p. 677, 1892.

## PINAKIOLITE
$Mg_3 Mn^{2+} Mn_2^{3+} B_2 O_{10}$
Dimorph of orthopinakiolite

CRYSTAL SYSTEM:  Monoclinic
CLASS:  2/m
SPACE GROUP:  P 2$_1$/m
Z:  2
LATTICE CONSTANTS:
  $a = 11.02$  $a = 5.36$
  $b = 5.98$   $b = 5.98$   or
  $c = 5.36$   $c = 12.73$
  $\beta = 95°48'$  $\beta = 120°34'$
3 STRONGEST DIFFRACTION LINES:
  2.51 (100)
  2.70 ( 90)
  5.42 ( 80)
OPTICAL CONSTANTS:
  $\alpha = 1.908$
  $\beta = 2.05$
  $\gamma = 2.065$
  (–)2V = 32°
HARDNESS:  6
DENSITY:  3.88 (Meas.)
          3.96 (Calc.)

CLEAVAGE:  {010} good
Very brittle.
HABIT:  Crystals thin tabular with rectangular outline; rarely short prismatic. Crystals often broken or bent. Twinning on {011} common.
COLOR-LUSTER:  Black. Opaque. Brilliant metallic; prism faces dull. Streak brownish gray.
MODE OF OCCURRENCE:  Occurs in granular dolomite associated with hausmannite and manganophyllite at Långban, Sweden.
BEST REF. IN ENGLISH: Palache, et al., "Dana's System of Mineralogy," 7th Ed., v. II, p. 324–325, New York, Wiley, 1951.

**PINITE** = Mixture of muscovite and chlorite

## PINNOITE
$MgB_2O_4 \cdot 3H_2O$

CRYSTAL SYSTEM:  Tetragonal
CLASS:  4 or 4/m
SPACE GROUP:  P4$_2$ or P4$_2$/m
Z:  4
LATTICE CONSTANTS:
  $a = 7.617$
  $c = 8.190$
OPTICAL CONSTANTS:
  $\omega = 1.565$
  $\epsilon = 1.575$  (Na)
    (+)
HARDNESS:  3½
DENSITY:  2.29 (Meas.)
          2.292 (Calc.)
CLEAVAGE:  Fracture uneven.
HABIT:  Crystals short prismatic, elongated along c-axis, rare. Commonly crystalline or fine granular. Also nodular with crystalline surface and radial-fibrous structure.
COLOR-LUSTER:  Pale yellow, yellow, yellowish green. Translucent. Vitreous.
MODE OF OCCURRENCE:  Occurs in salt deposits at Aschersleben and Strassfurt, Saxony, and at Leopoldshall, Anhalt, Germany.
BEST REF. IN ENGLISH: Palache, et al., "Dana's System of Mineralogy," 7th Ed., v. II, p. 334–335, New York, Wiley, 1951.

## PINTADOITE (Species status in doubt)
$Ca_2 V_2 O_7 \cdot 9H_2O$

HARDNESS:  Not determined
DENSITY:  Not determined.
CLEAVAGE:  Not determined.
HABIT:  As a thin efflorescence.
COLOR-LUSTER:  Light to dark green.

Slowly soluble in water.
MODE OF OCCURRENCE:  Occurs on outcroppings of the McElmo sandstone in Pintado Canyon, San Juan County, and elsewhere in southeastern Utah.

BEST REF. IN ENGLISH: Hess and Schaller, *J. Washington Ac. Sc.*, 4: 576 (1914).

## PIRSSONITE
$Na_2Ca(CO_3)_2 \cdot 2H_2O$

CRYSTAL SYSTEM: Orthorhombic
CLASS: mm2
SPACE GROUP: Fdd2
Z: 8
LATTICE CONSTANTS:
  $a$ = 11.32
  $b$ = 20.06
  $c$ = 6.00
3 STRONGEST DIFFRACTION LINES:
  2.51 (100)
  2.65 ( 90)
  2.57 ( 80)
OPTICAL CONSTANTS:
  $\alpha$ = 1.5043
  $\beta$ = 1.5095 (Na)
  $\gamma$ = 1.5751
  (+)2V = 31°26'
HARDNESS: 3-3½
DENSITY: 2.367 (Meas.)
        2.352 (Calc.)
CLEAVAGE: Fracture conchoidal. Brittle.
HABIT: Crystals short prismatic, elongated along $c$-axis; tabular {010}; or pyramidal.
COLOR-LUSTER: Colorless, white, grayish. Transparent. Vitreous.
MODE OF OCCURRENCE: Occurs abundantly as euhedral crystals in the mud layers, and as consolidated crystalline strata at Searles Lake, San Bernardino County; with gaylussite and northupite in trona at Borax Lake, Lake County; and as an efflorescence from Deep Spring Lake, Inyo County, California. It also has been found in drill cores from oil shale near Green River, Sweetwater County, Wyoming.
BEST REF. IN ENGLISH: Palache, et al., "Dana's System of Mineralogy," 7th Ed., v. II, p. 232-233, New York, Wiley, 1951.

**PISANITE** = Cuprian variety of melanterite
$(Fe, Cu, Zn)SO_4 \cdot 7H_2O$

**PISEKITE** = Inadequately described metamict niobate-tantalate near samarskite in composition

**PISTACITE** = Variety of epidote

**PISTOMESITE** = Variety of siderite

**PITCHBLENDE** = Variety of uraninite

## PITTICITE (Inadequately described mineral)
Near $Fe_2^{3+}AsO_4SO_4OH \cdot nH_2O$

CRYSTAL SYSTEM: Unknown
OPTICAL CONSTANT:
  $N \simeq 1.616-1.635$
  HARDNESS: ~2-3
  DENSITY: ~2.2-2.5
  CLEAVAGE: Fracture conchoidal, splintery, or earthy. Opaline masses brittle.
  HABIT: As opaline crusts; also stalactitic, reniform, botryoidal, earthy, or as coatings.
  COLOR-LUSTER: Various shades of light to dark brown, yellow, gray, whitish. Transparent to opaque. Vitreous, greasy, dull. Streak yellow to white.
  MODE OF OCCURRENCE: Occurs as a near-surface secondary mineral and as a recent deposit from mine or spring waters. Found with scorodite as an alteration product of arsenopyrite near the mouth of Devils Gulch, Mariposa County, and at the Carlin mine, near Jamestown, Tuolumne County, California; at Manhattan, Nevada; and at localities in England, Germany, Czechoslovakia, and Roumania.
  BEST REF. IN ENGLISH: Palache, et al., "Dana's System of Mineralogy," 7th Ed., v. II, p. 1014-1015, New York, Wiley, 1951.

**PLAGIOCLASE** = Group name for triclinic feldspars of general formula: $(Na, Ca)Al(Al, Si)Si_2O_8$. End members Albite (Ab)$(NaAlSi_3O_8)$, Anorthite (An)$(CaAl_2Si_2O_8)$. See Albite, Oligoclase, Andesine, Labradorite, Bytownite, Anorthite.

## PLAGIONITE
$Pb_5Sb_8S_{17}$

CRYSTAL SYSTEM: Monoclinic
CLASS: 2/m
SPACE GROUP: C2/c
Z: 4
LATTICE CONSTANTS:
  $a$ = 13.45
  $b$ = 11.81
  $c$ = 19.94
  $\beta$ = 107°11'
3 STRONGEST DIFFRACTION LINES:
  3.21 (100)
  3.26 ( 90)
  2.91 ( 90)
HARDNESS: 2½
DENSITY: 5.54 (Meas.)
        5.57 (Calc.)
CLEAVAGE: {112} distinct
Fracture uneven to conchoidal. Brittle.
HABIT: Crystals commonly thick tabular, sometimes short prismatic. Striated [$\bar{1}10$]. Usually massive, compact to granular.
COLOR-LUSTER: Blackish gray. Opaque. Metallic. Streak blackish gray.
MODE OF OCCURRENCE: Occurs with cassiterite and franckeite at Oruro, Bolivia; at Leyraux, Cantal, France; with other sulfosalts at Wolfsberg, Harz Mountains, and at Arnsberg, Westphalia, Germany.

BEST REF. IN ENGLISH: Palache, et al., "Dana's System of Mineralogy," 7th Ed., v. I, p. 464-465, New York, Wiley, 1944.    Jambor, J. L., *Min. Mag.*, **37**: 442-446 (1969).

## PLANCHEITE
$Cu_8(Si_4O_{11})_2(OH)_4 \cdot H_2O$

CRYSTAL SYSTEM:  Orthorhombic
CLASS:  2/m 2/m 2/m
SPACE GROUP:  Pcnb
Z:  4
LATTICE CONSTANTS:
  $a = 19.04$
  $b = 20.01$
  $c = 5.27$
3 STRONGEST DIFFRACTION LINES:
  4.43 (100)
  3.50 ( 70)
  3.32 ( 70)
OPTICAL CONSTANTS:
  $\alpha = 1.743$      $\alpha = 1.697$
  $\beta = ?$    or    $\beta = 1.718$
  $\gamma = 1.795$      $\gamma = 1.741$
  $(+)2V = 80°$      $2V = 88\frac{1}{2}°$(Calc.)
HARDNESS: 5½
DENSITY: 3.65-3.80 (Meas.)
         3.85 (Calc.)
CLEAVAGE:  Not determined.
HABIT:  Fibrous; often in compact radial aggregates.
COLOR-LUSTER:  Pale to dark blue. Translucent. Satiny.
MODE OF OCCURRENCE:  Occurs as a secondary mineral in the oxidation zone of copper deposits in association with malachite and other secondary copper minerals. Found at Mindigi and Tantara, Katanga, Zaire. Also as crystalline grains and masses disseminated in compact green conichalcite at the Table Mountain mine, near Klondyke, in the Galiuro Mountains, Pinal County, Arizona.
BEST REF. IN ENGLISH: Evans, H. T. Jr., and Mrose, Mary E., *Science*, **154**: 506-507 (1966).

## PLANERITE = Variety intermediate between turquoise and coeruleolactite

## PLASMA = Variety of quartz

## PLATINIRIDIUM
Ir, Pt

CRYSTAL SYSTEM:  Cubic
CLASS: 4/m $\bar{3}$ 2/m
SPACE GROUP:  Fm3m for nearly pure Ir
Z:  4
LATTICE CONSTANT:
  $a = 3.8312$
  HARDNESS: 6-7
  DENSITY: 22.65-22.84 (Meas.)
           22.66 (Calc., 0°)

CLEAVAGE:  {001} indistinct.
Fracture hackly.  Slightly malleable.
HABIT:  Crystals usually cubic, commonly as nuggets or as angular or rounded grains.  Twinning on {111}, polysynthetic.
COLOR-LUSTER:  Silver white with yellow tint; gray on fracture. Opaque. Metallic.
MODE OF OCCURRENCE:  Occurs as nuggets on the Enright claim, 3 miles above Trinity Center, Trinity County, California; in Brazil; in the Ural Mountains, USSR; and near Mandalay, Burma.
BEST REF. IN ENGLISH: Palache, et al., "Dana's System of Mineralogy," 7th Ed., v. I, p. 110-111, New York, Wiley, 1944.

## PLATINUM
Pt

CRYSTAL SYSTEM:  Cubic
CLASS:  4/m $\bar{3}$ 2/m
SPACE GROUP:  Fm3m
Z:  4
LATTICE CONSTANT:
  $a = 3.9237$
3 STRONGEST DIFFRACTION LINES:
  2.265  (100)
  1.962  ( 53)
  1.1826 ( 33)
  HARDNESS:  4-4½
  DENSITY:  21.440 (Meas.)
            21.472 (Calc. X-ray)
CLEAVAGE:  None.
Fracture hackly.  Malleable and ductile.
HABIT:  Crystals cubic, small and often distorted. Commonly as minute grains or scales; also as nuggets up to 26 pounds in weight.
COLOR-LUSTER:  Whitish steel gray to dark gray. Opaque. Metallic.

Sometimes magnetic.
MODE OF OCCURRENCE:  Occurs chiefly in placer deposits or in basic and ultrabasic rocks; rarely in contact metamorphic deposits and quartz veins. Found widespread in the gold sands of California, especially in Trinity County where nuggets up to 2½ ounces in weight have been reported; in Curry County and elsewhere in Oregon; at Platinum, Alaska; and in Burke and Rutherford counties, North Carolina.  It also occurs in Canada, Honduras, Haiti, Brazil, Colombia, Peru, Ireland, Germany, Finland, USSR, New Zealand, Australia, Borneo, Madagascar, and Transvaal, Africa.
BEST REF. IN ENGLISH: Palache, et al., "Dana's System of Mineralogy," 7th Ed., v. I, p. 106-109, New York, Wiley, 1944.

## PLATTNERITE
$PbO_2$

CRYSTAL SYSTEM:  Tetragonal
CLASS:  4/m 2/m 2/m

SPACE GROUP: $P4_2/mnm$
Z: 2
LATTICE CONSTANTS:
  $a = 4.931$
  $c = 3.367$
3 STRONGEST DIFFRACTION LINES:
  3.50 (100)
  2.80 (100)
  1.856 (100)
HARDNESS: 5½
DENSITY: 9.42 (Meas.)
        9.63 (Calc.)
CLEAVAGE: None. Brittle.
HABIT: Crystals slender prismatic, elongated parallel to $c$-axis. Often in botryoidal or nodular masses; dense to fibrous. Twinning on {011}, as penetration twins.
COLOR-LUSTER: Crystals black; massive material brownish black to iron black. Opaque. Brilliant metallic adamantine. Sometimes becomes dull on exposure. Streak brown.
MODE OF OCCURRENCE: Occurs as a secondary mineral resulting from the alteration of other lead-bearing minerals. Found as masses weighing as much as 200 pounds in the Morning mine, Mullan, Shoshone County, Idaho, and elsewhere in the state; at Carbonate, Lawrence County, South Dakota; as fine crystals with murdochite at the Blanchard mine, Socorro County, New Mexico; abundantly as tiny crystals associated with hemimorphite, aurichalcite, and other secondary minerals at Mapimi, Durango, Mexico; at Leadhills and Wanlockhead, Scotland; and with other lead oxides at Tsumeb, South West Africa.
BEST REF. IN ENGLISH: Palache, et al., "Dana's System of Mineralogy," 7th Ed., v. I, p. 581-583, New York, Wiley, 1944.

## PLATYNITE
$PbBi_2(Se, S)_3$ with Se:S = 2:1

CRYSTAL SYSTEM: Hexagonal
SPACE GROUP: R lattice
Z: 8
LATTICE CONSTANTS:
  $a = 8.49$
  $c = 20.80$
HARDNESS: 2-3
DENSITY: 7.98 (Meas.)
CLEAVAGE: {0001} good
         {10$\bar{1}$1} fair
HABIT: As thin plates or foliae.
COLOR-LUSTER: Dark gray to black. Opaque. Metallic. Streak black, shining.
MODE OF OCCURRENCE: Occurs associated with chalcopyrite in quartz at Fahlun, Sweden.
BEST REF. IN ENGLISH: Palache, et al., "Dana's System of Mineralogy," 7th Ed., v. I, p. 474, New York, Wiley, 1944. Sindeeva, N. D., "Mineralogy and Types of Deposits of Selenium and Tellurium," Wiley-Interscience, 1964.

## PLAYFAIRITE
$Pb_{16}Sb_{18}S_{43}$

CRYSTAL SYSTEM: Monoclinic
CLASS: 2/m (?)
SPACE GROUP: $P2_1/m$ (?)
Z: 4
LATTICE CONSTANTS:
  $a = 45.12$
  $b = 4.139$
  $c = 21.45$
  $\beta = 92°32'$
3 STRONGEST DIFFRACTION LINES:
  3.39 (100)
  3.32 (100)
  2.785 ( 70)
HARDNESS: Vickers microhardness 154 (150-171) kg/mm$^2$ with 50 g load.
DENSITY: 5.80 (Predicted)
        5.72 (Calc.)
CLEAVAGE: {100} perfect
HABIT: Tabular; heavily striated parallel to elongation. Observed in very small amounts in polished section.
COLOR-LUSTER: Lead gray to black. Metallic. Opaque. Streak black.
MODE OF OCCURRENCE: Occurs in irregular microscopic veinlets associated with boulangerite in a small prospect pit in Precambrian marble at Madoc, Ontario, Canada.
BEST REF. IN ENGLISH: Jambor, J. L., *Can. Min.*, 9(Pt. 2): 191-213 (1967).

## PLAZOLITE = Hydrogrossular

## PLENARGYRITE = Matildite

## PLEONASTE = Ceylonite, ferroan spinel

## PLESSITE = Intergrowth of kamacite and taenite

## PLOMBIERITE
$Ca_5H_2Si_6O_{18} \cdot 6H_2O$ (?)

CRYSTAL SYSTEM: Orthorhombic
CLASS: 222
SPACE GROUP: $C222_1$
Z: 4
LATTICE CONSTANTS:
  $a = 11.24$
  $b = 28.0$
  $c = 7.30$
3 STRONGEST DIFFRACTION LINES:
  3.09 (100)
  2.81 (100)
  1.83 (100)
DENSITY: 2.018
HABIT: Massive; compact.
COLOR-LUSTER: White; translucent.
MODE OF OCCURRENCE: Reported from Crestmore, Riverside County, California; Ballycraigy, County Antrim, Ireland; and Kloech, Styria, Austria.

BEST REF. IN ENGLISH: Heller, L., and Taylor, H. F. W., "Crystallographic Data for the Calcium Silicates," H. M. Stationery Office, 79 pp. (1956).

## PLUMALSITE
$Pb_4 Al_2 (SiO_3)_7$

CRYSTAL SYSTEM: Orthorhombic
LATTICE CONSTANTS:
$a = 4.90$ ⎫
$b = 4.90$ ⎬ ?
$c = 4.08$ ⎭
3 STRONGEST DIFFRACTION LINES:
4.06 (100)
3.62 ( 80)
2.42 ( 80)
OPTICAL CONSTANT:
$n > 1.782$
HARDNESS: 5–6
DENSITY: 4.35–4.38 (Meas.)
CLEAVAGE: None.
HABIT: Massive; as sharply angular platy fragments.
COLOR-LUSTER: Various shades of yellow or green, sometimes colorless or black (from inclusions). Vitreous.
MODE OF OCCURRENCE: Occurs in continental deposits and the upper part of the weathered crust of crystalline rocks over a large area of the Ukrainian Shield.
BEST REF. IN ENGLISH: Gornyi, G. Ya., Dyadchenko, M. G., and Kudykina, T. A., *Am. Min.*, **53**: 349–350 (1968).

**PLUMANGITE** = Name proposed for mineral corresponding to $(Cu_{0.8}Zn_{0.15})O \cdot PbMn_4O_{10}$ occurring with murdochite at the T. Khuni mine, Anarak, Iran.

**PLUMBAGO** = Graphite

**PLUMBOBETAFITE** = Name proposed for a new lead-rich variety of the pyrochlore group which occurs as an accessory in a microcline-quartz-albite-aegirine-riebeckite dike cutting nepheline syenite of the Burpala alkalic massif, northern Baikal, USSR.

CRYSTAL SYSTEM: Cubic
LATTICE CONSTANT:
$a = 10.33$ (800°C)
3 STRONGEST DIFFRACTION LINES: Metamict.
DENSITY: 4.64 (Meas.)
CLEAVAGE: Fracture irregular.
HABIT: Crystals octahedral with curved faces, rare. Usually as rounded isometric grains up to 2–3 mm.
COLOR-LUSTER: Yellowish, occasionally with brownish black cores. Luster adamantine.
BEST REF. IN ENGLISH: *Am. Min.*, **55**: 1068 (1970).

## PLUMBOFERRITE
$PbFe_4O_7$

CRYSTAL SYSTEM: Hexagonal

CLASS: 32
SPACE GROUP: P312
Z: 42
LATTICE CONSTANTS:
$a = 11.86$
$c = 47.14$
3 STRONGEST DIFFRACTION LINES:
2.64 (100)
2.81 ( 90)
2.96 ( 80)
HARDNESS: 5
DENSITY: 6.07 (Meas.)
        6.55 (Calc.)
CLEAVAGE: {0001} distinct
HABIT: Crystals thick tabular, flattened on {0001}. Usually massive, cleavable.
COLOR-LUSTER: Black. Opaque. Submetallic. Streak red.
MODE OF OCCURRENCE: Occurs with jacobsite, native copper, and andradite in granular limestone at the Jakobsberg mine, Nordmark, and at the Sjö mine, Örebro, Sweden.
BEST REF. IN ENGLISH: Palache, et al., "Dana's System of Mineralogy," 7th Ed., v. I, p. 726–727, New York, Wiley, 1944.

## PLUMBOGUMMITE
$PbAl_3 (PO_4)_2 (OH)_5 \cdot H_2O$

CRYSTAL SYSTEM: Hexagonal
CLASS: $\bar{3}$ 2/m
SPACE GROUP: R$\bar{3}$m
Z: 3
LATTICE CONSTANTS:
$a = 7.018$
$b = 16.784$
3 STRONGEST DIFFRACTION LINES:
2.97 (100)
5.73 ( 80)
3.51 ( 60)
OPTICAL CONSTANTS:
$\omega = 1.653$
$\epsilon = 1.675$
   (+)
HARDNESS: 4½–5
DENSITY: 4.014 (Meas.)
        4.08 (Calc.)
CLEAVAGE: Not determined.
Fracture of masses uneven to subconchoidal. Brittle.
HABIT: Massive, compact; also as botryoidal or stalactitic crusts or masses with concentric structure. As minute crystals with hexagonal outline.
COLOR-LUSTER: Grayish white, yellowish gray to yellowish or reddish brown; also yellow, greenish or bluish. Translucent. Dull to resinous, gum like. Streak colorless to white.
MODE OF OCCURRENCE: Occurs as a secondary mineral in the oxidation zone of lead-bearing ore deposits, and as rolled fragments in diamantiferous alluvial deposits. Found at the Cerro Gordo mine, Inyo County, California; at the Dallas mine, Gilpin County, Colorado; with marcasite at the Canton mine, Georgia; and at Mine la Motte, Missouri.

It also occurs in the Diamantina district, Minas Geraes, Brazil; and at deposits in France and Cumberland, England.

BEST REF. IN ENGLISH: Palache, et al., "Dana's System of Mineralogy," 7th Ed., v. II, p. 831–832, New York, Wiley, 1951.

## PLUMBOJAROSITE
$PbFe_6(SO_4)_4(OH)_6$
Alunite Group

CRYSTAL SYSTEM: Hexagonal
CLASS: $\bar{3}\,2/m$
SPACE GROUP: $R\bar{3}m$
Z: 3 (hexag.), 1 (rhomb.)
LATTICE CONSTANTS:

| (hexag.) | (rhomb.) |
|---|---|
| $a = 7.20$ kX | $a_{rh} = 11.95$ |
| $c = 33.60$ | $\alpha = 35°05'$ |

3 STRONGEST DIFFRACTION LINES:
   3.066 (100)
   5.933 ( 95)
   1.829 ( 70)
OPTICAL CONSTANTS:
   $\omega = 1.875$
   $\epsilon = 1.786$
      (−)
HARDNESS: Not determined. Soft.
DENSITY: 3.64 (Meas.)
            3.60 (Calc.)
CLEAVAGE: $\{10\bar{1}4\}$ fair.
HABIT: As crusts or compact masses composed of microscopic hexagonal plates. Also as ocherous masses and crusts.
COLOR-LUSTER: Yellowish brown to dark brown. Dull to shining or silky.
MODE OF OCCURRENCE: Occurs widespread as a secondary mineral in the oxidation zone of lead deposits. Found at many localities in the western United States, especially in the Darwin district, Inyo County, and at the Cactus Queen mine, Mojave County, California; at San Jose and Three Brothers mines, Grant County, and in the Cooks Peak district, Luna County, New Mexico; at the Boss mine, Yellow Pine district, Nevada; in Beaver, Juab, Salt Lake, Tooele, Utah, and Washington Counties, Utah; in Arizona; and in the Picher district, Oklahoma. It also is found in Bolivia, at Laurium, Greece, and as a significant ore mineral in the lead mines at Bolkardag, Anatolia, Turkey.

BEST REF. IN ENGLISH: Palache, et al., "Dana's System of Mineralogy," 7th Ed., v. II, p. 568–570, New York, Wiley, 1951.

## PLUMBONACRITE = Hydrocerussite

## PLUMBOPALLADINITE
$Pd_3Pb_2$

CRYSTAL SYSTEM: Hexagonal
LATTICE CONSTANTS:
   $a = 4.470$
   $c = 5.719$

3 STRONGEST DIFFRACTION LINES:
   2.30 (100)
   2.23 (100)
   1.207 ( 60)
HARDNESS: ~5
Microhardness 407–441 kg/mm².
   DENSITY: Not determined.
   CLEAVAGE: Not determined.
   HABIT: Massive; as very minute grains in aggregates up to 0.7 mm in diameter.
   COLOR-LUSTER: In reflected light bright white with a faint rose tint, grayish white next to native silver.
   MODE OF OCCURRENCE: Occurs in veinlets of cubanite in talnakhite, usually closely associated with polarite, stannopalladinite, native silver, galena, and sphalerite in the Talnakh nickel-copper deposits in the Polar Urals, USSR.
   BEST REF. IN ENGLISH: Genkin, A. D., Evstigneeva, T. L., Vyal'sov, L. N., Laputina, I. P., and Troneva, N. V., *Am. Min.*, **56**: 1121 (1971).

## PLUMBOPYROCHLORE
$(Pb, Y, U, Ca)_{2-x}Nb_2O_6(OH)$
Pyrochlore Group

CRYSTAL SYSTEM: Cubic
CLASS: $4/m\,\bar{3}\,2/m$
SPACE GROUP: $Fd3m$
Z: 8
LATTICE CONSTANT:
   $a = 10.534$
3 STRONGEST DIFFRACTION LINES:
   3.02 (100)
   1.861 ( 90)
   1.581 ( 90)
OPTICAL CONSTANT:
   $N = 2.08$
HARDNESS: Not determined.
DENSITY: Not determined. Yttrian variety 5.04 (Meas.).
CLEAVAGE: None.
HABIT: Crystals octahedral; also as small grains.
COLOR-LUSTER: Crystals greenish yellow to red on periphery, dark brown at the center.
MODE OF OCCURRENCE: Occurs in metasomatically altered granitic rocks, Urals, associated with quartz, microcline, albite, and aegirine.
BEST REF. IN ENGLISH: Skorobogatova, N. V., Sidorenko, G. A., Dorofeeva, K. A., and Stolyarova, T. I., *Am. Min.*, **55**: 1068 (1970).

## PLUMOSITE = Boulangerite, also jamesonite

## PODOLITE = Synonym for dahllite

## POITEVINITE
$(Cu, Fe, Zn)SO_4 \cdot H_2O$

CRYSTAL SYSTEM: Monoclinic
CLASS: $2/m$

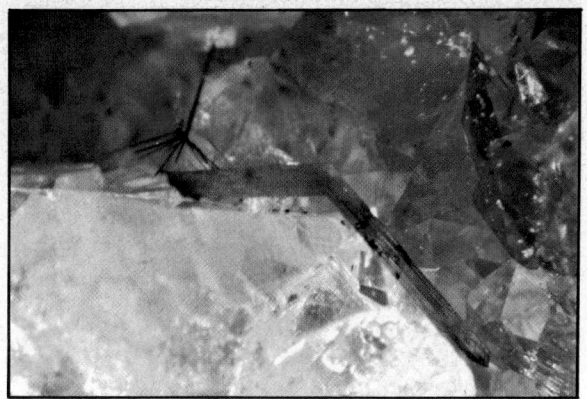

PYRITE   Jackson County, Indiana.

PYRITE (coils)   Hall's Gap, Kentucky.

PYRITE (acicular)   Rock Island Dam, Columbia River, Washington.

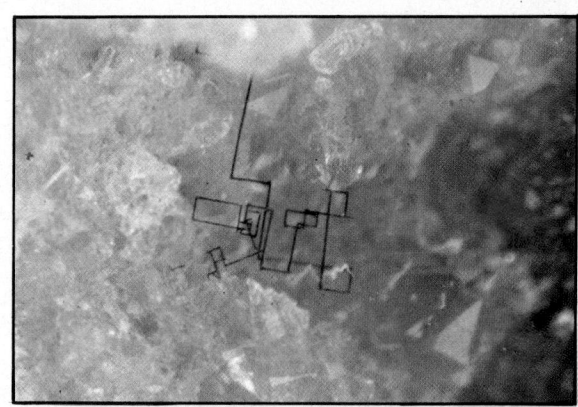

PYRITE   Hall's Gap, Kentucky.

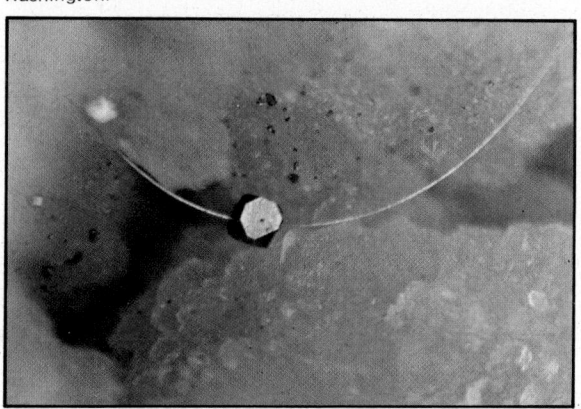

PYRITE on millerite   Hall's Gap, Kentucky.

PYRITE   Panasqueira, Portugal.

PYROCHLORE with catapleite on aegirine   Mont St. Hilaire, Quebec, Canada.

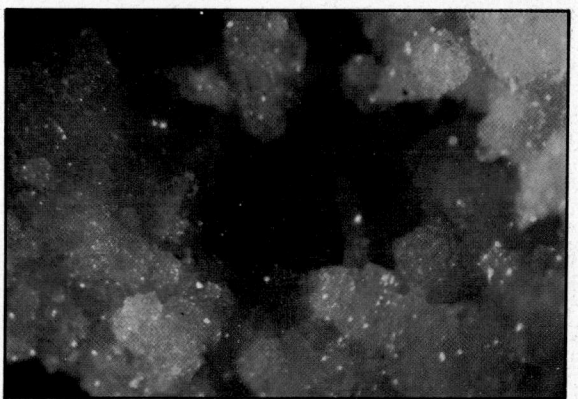

PYROCHROITE   Franklin, Sussex County, New Jersey.

PYROLUSITE   Holly Grove, Monroe County, Arkansas.

PYROLUSITE, dendritic   Malpais Hills, Arizona.

PYROMORPHITE   Broken Hill, Zambia.

PYROSMALITE   Nordmark, Vermland, Sweden.

PYROSTILPNITE   St. Andreasberg, Harz Mountains,
Germany.

PYROSTILPNITE   Guadelupe mine, Zacatecas, Mexico.

PYRRHOTITE   Morro Velho mine, Minas Geraes, Brazil.

PYRRHOTITE   Morro Velho mine, Minas Geraes, Brazil.

98

QUARTZ (rose)   Newry, Oxford County, Maine.

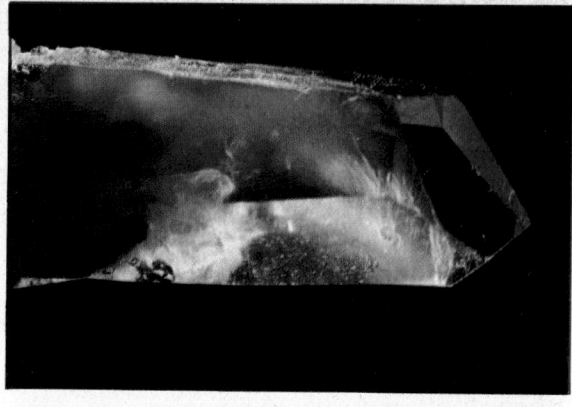

QUARTZ (blue)   Switzerland.

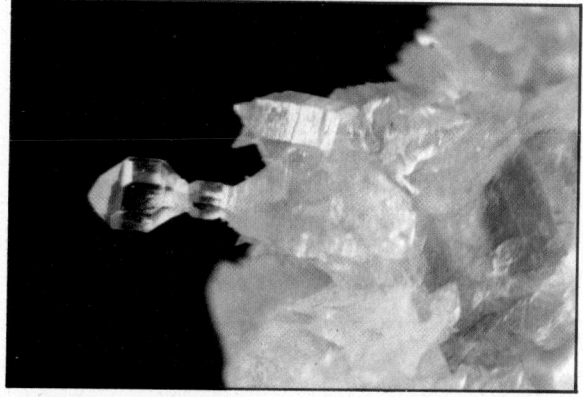

QUARTZ (sceptre)   Quartzite, Yuma County, Arizona.

QUARTZ (rose)   Taquaral, Minas Geraes, Brazil.

QUARTZ (smoky)   St. Gotthard, Switzerland.

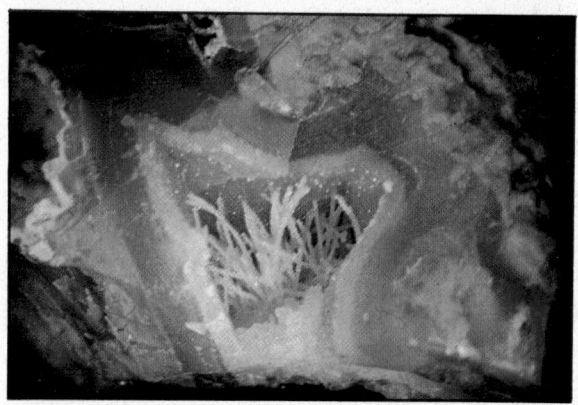

QUARTZ with chrysocolla   Globe, Gila County, Arizona.

QUARTZ (Japanese Twin)  San Antonio mine, Chihuahua, Mexico

QUARTZ with rutile inclusions   Brazil.

QUARTZ (amethyst)  Guanajuato, Mexico.

QUARTZ  Poona, India.

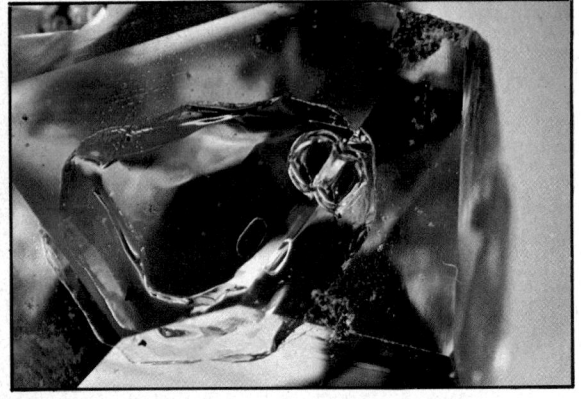

QUARTZ (moveable water bubbles)  Brazil.

QUARTZ (SEM photograph)  Quartzsite, Arizona.

RAMSAYITE  Mont St. Hilaire, Quebec, Canada.

RAMSDELLITE  Chihuahua, Mexico.

RAMSDELLITE  Chihuahua, Mexico.

RAMSDELLITE after groutite  Lake Valley, Sierra County, New Mexico.

RASPITE    Broken Hill, New South Wales, Australia.

RASPITE    Broken Hill, New South Wales, Australia.

RASPITE with stolzite    Broken Hill, New South Wales, Australia.

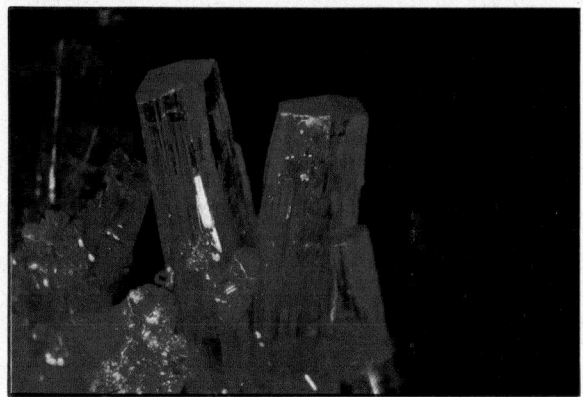

REALGAR    Getchell mine, Golconda, Humboldt County, Nevada.

REALGAR    Getchell mine, Golconda, Humboldt County, Nevada.

REDDINGITE    Palermo mine, North Groton, New Hampshire.

REALGAR    Getchell mine, Golconda, Humboldt County, Nevada.

101

RENARDITE on torbernite   Shinkolobwe. Katanga, Zaire.

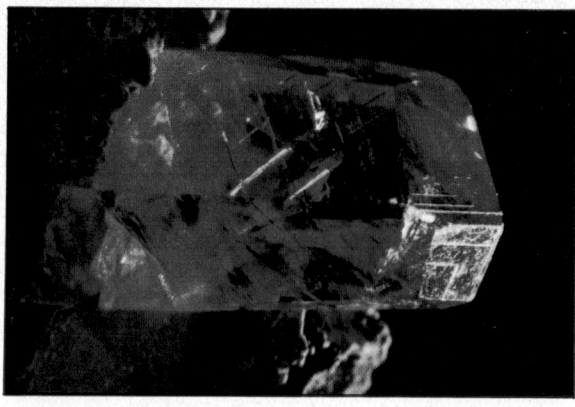

RHODOCHROSITE   Santa Eulalia, Chihuahua, Mexico.

RHODOCHROSITE   Silverton. San Juan County. Colorado.

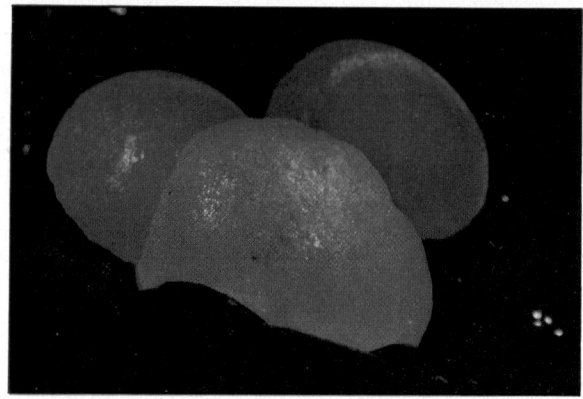

RHODOCHROSITE   Wolf Mine, Herdorf, Germany.

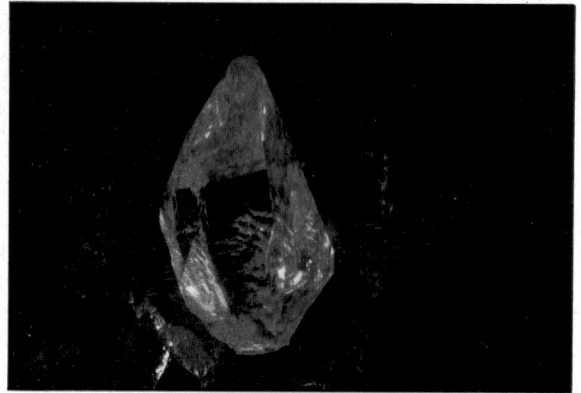

RHODOCHROSITE   Hotazel, South Africa.

RHODONITE with andradite   Franklin, Sussex County, New Jersey.

RHODONITE   Pajsberg, Sweden.

102

ROBERTSITE with leucophosphite  Tip Top Mine, Custer County, South Dakota.

ROBERTSITE on jahnsite  Gap Lode, Pennington County, South Dakota.

ROCKBRIDGEITE inclusions in leucophosphite  Tip Top mine, Custer County, South Dakota.

ROCKBRIDGEITE with leucophosphite  Tip Top mine, Custer County, South Dakota.

ROCKBRIDGEITE on hureaulite  Bull Moose Mine, Custer County, South Dakota.

ROSCHERITE  Rio Joquitehorha, Minas Geraes, Brazil.

ROSCHERITE  Newry. Oxford County, Maine.

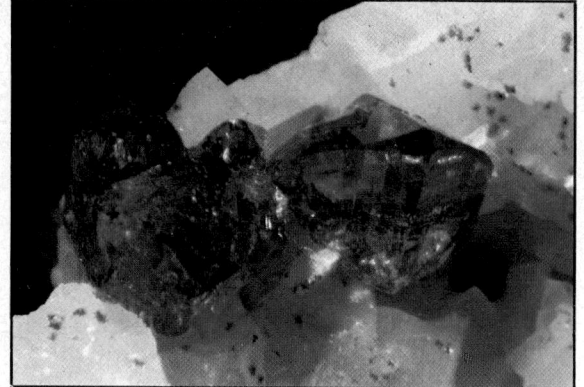

ROSELITE   Bou Azzer, Morocco.

ROSELITE   Bou Azzer, Morocco.

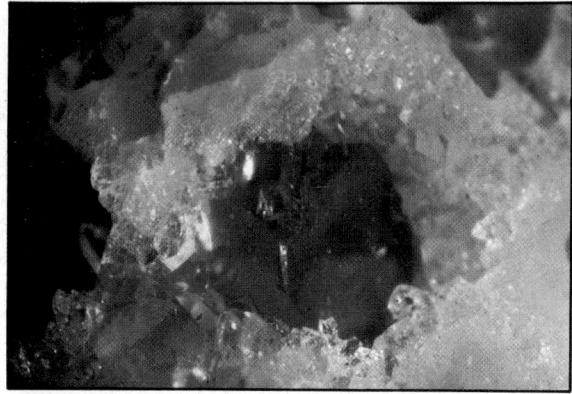

ROSELITE   Daniel mine, Schneeberg, Saxony, Germany.

RUTHERFORDINE   Shinkolobwe, Katanga, Zaire.

RUTILE in quartz   Minas Geraes, Brazil.

RUTILE   (cyclic twin) Magnet Cove, Hot Spring County, Arkansas.

RUTILE on beryl   Hiddenite, near Stony Point, Alexander County, North Carolina.

RUTILE on chlorite   Cavradi, Tavetsch, Switzerland.

104

RUTILE (reticulated)   Alexander County, North Carolina.

RUTILE in quartz   Brazil.

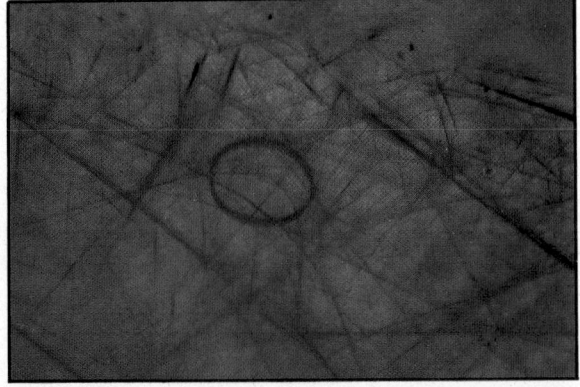

RUTILE ring in topaz   Tepetate, San Luis Potosi, Mexico.

RUTILE   Hiddenite, near Stony Point, Alexander County, North Carolina.

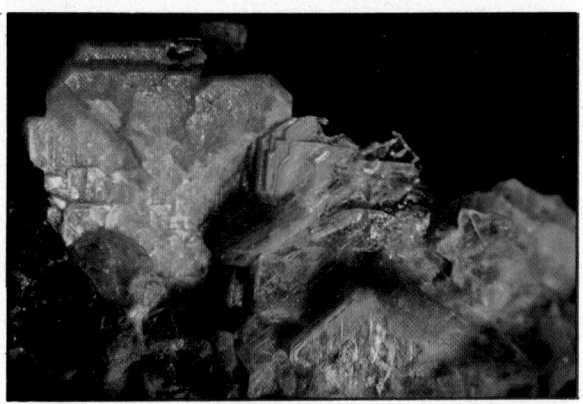

SABUGALITE   Haute Vienne, France.

SAL AMMONIAC   Mt. Etna, Sicily.

SALEEITE (short-wave ultraviolet light)   Rum Jungle, Northern Territory, Australia.

SALEEITE   Rum Jungle, Northern Territory, Australia.

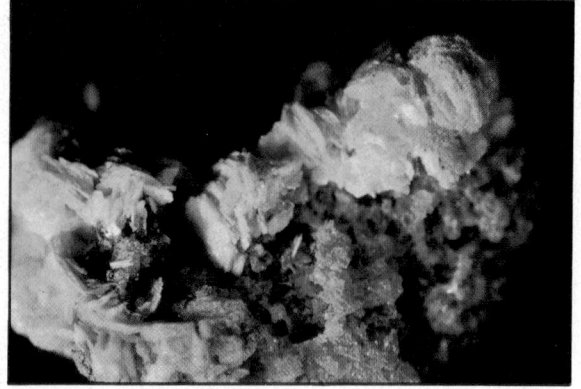

SALEEITE   Shinkolobwe, Katanga, Zaire.

SALESITE   Chuquicamata, Chile.

SARTORITE   Lengenbach quarry, Binnental, Switzerland.

SCHEELITE   Schneeberg, Saxony, Germany.

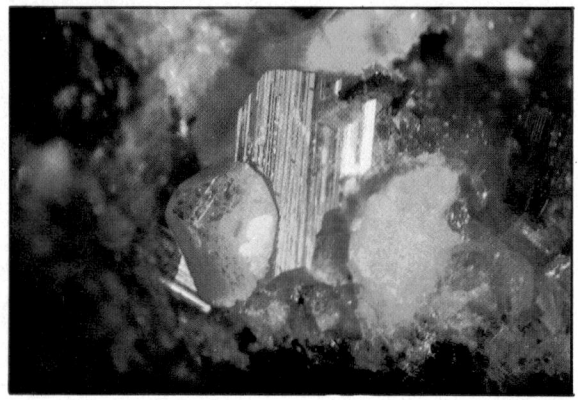

SCHOEPITE on becquerelite   Shinkolobwe, Katanga, Zaire.

SCHOEPITE   Shinkolobwe, Katanga, Zaire.

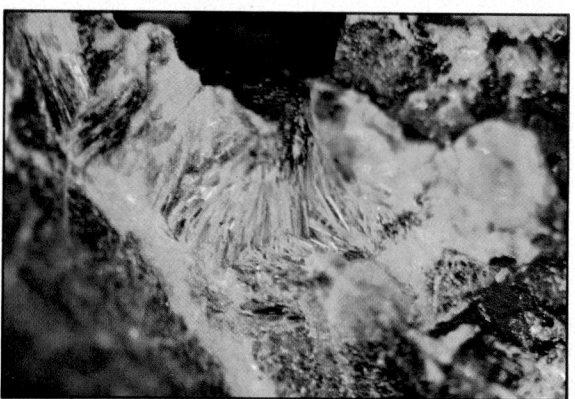

SCHOEPITE as an alteration of ianthinite   Shinkolobwe,
Katanga, Zaire.

SCHOLZITE   Near Blinman, South Australia.

SCHOLZITE   Near Blinman, South Australia.

SCHROECKINGERITE   Wamsutter, Sweetwater County, Wyoming.

SCHROECKINGERITE   (short-wave ultraviolet light) Wamsutter, Sweetwater County, Wyoming.

SCOLECITE   Fontana, Tahiti.

SCORODITE   Richmond-Sitting Bull mine, Galena, Lawrence County, South Dakota.

SCORODITE   Richmond-Sitting Bull mine, Galena, Lawrence County, South Dakota.

SCORODITE   Mammoth mine, Tintic district, Utah.

SEARLESITE   Boron, Kern County, California.

SEGELERITE   Tip Top mine, Custer County, South Dakota.

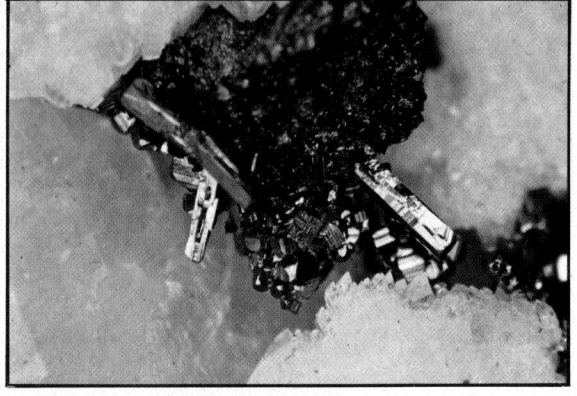

SELIGMANNITE   Lengenbach quarry, Binnental, Switzerland.

SELLAITE   Vesuvius, Italy.

SEMSEYITE   Kisbanya, Roumania.

SEMSEYITE   Julcani mine, Huancavelica, Peru.

SENGIERITE   Shinkolobwe, Katanga, Zaire.

SERANDITE    Mont St. Hilaire, Quebec, Canada.

SERPIERITE    Laurium, Greece.

SHATTUCKITE    Ajo, Pima County, Arizona.

SIDERITE with whitlockite    Palermo mine, North Groton, New Hampshire.

SIDERITE    Frostburg, Allegany County, Maryland.

SIDERITE with sphalerite    Gilman, Eagle County, Colorado.

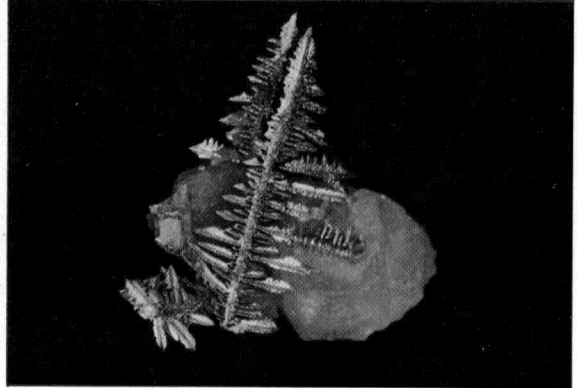

SILVER   Hope Bay mines, Northwest Territories, Canada.

SILVER   Hope Bay mines, Northwest Territories, Canada.

SILVER   Frontier mine, South Lorrain, Ontario, Canada.

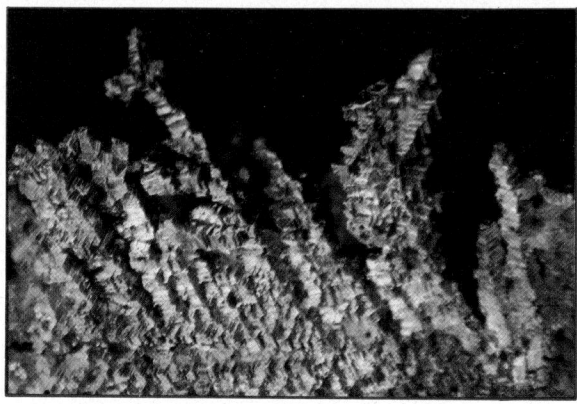

SILVER   Wurtemburg, Bavaria, Germany.

SILVER on cerussite   Bunker Hill and Sullivan mine, Kellogg, Idaho.

SINCOSITE   Ross Hannibal mine, near Lead, Lawrence County, South Dakota.

SKLODOWSKITE   Santa Eulalia, Chihuahua, Mexico.

SKUTTERUDITE   Bou Azzer, Morocco.

SKUTTERUDITE   Bou Azzer, Morocco.

SMALTITE (arsenic-deficient skutterudite)   Cobalt, Ontario, Canada.

SMITHSONITE   Tsumeb, South West Africa.

SMITHSONITE   Tsumeb, South West Africa.

SMITHSONITE   Tsumeb, South West Africa.

SMITHSONITE   Broken Hill, New South Wales, Australia.

SODDYITE with curite and torbernite   Shinkolobwe, Katanga, Zaire.

SODALITE   Vesuvius, Italy.

SPANGOLITE   Blanchard mine, Bingham, Socorro County, New Mexico.

SPERRYLITE in chalcopyrite   Sudbury, Ontario, Canada.

SPERRYLITE   Sudbury, Ontario, Canada.

SPESSARTINE on rhyolite   Thomas Range, Utah.

SPHAEROCOBALTITE on quartz   Bou Azzer, Morocco.

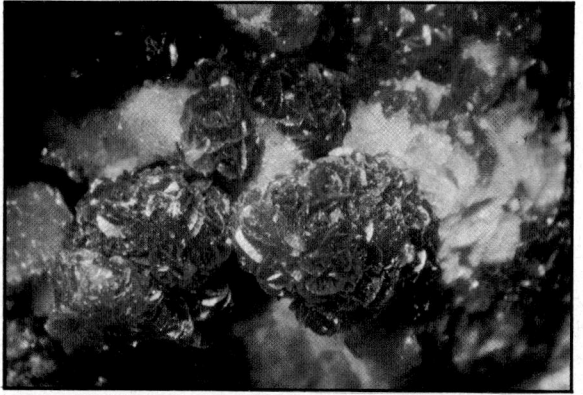

SPHAEROCOBALTITE   Musonoi mine, Kolwezi, Katanga, Zaire.

SPACE GROUP: A2/a
Z: 4
LATTICE CONSTANTS:
  $a = 7.480$
  $b = 7.424$
  $c = 7.053$
  $\beta = 114°40'$
3 STRONGEST DIFFRACTION LINES:
  3.46 (100)
  4.72 ( 50)
  3.08 ( 50)
OPTICAL CONSTANTS:
  $\alpha = 1.626$
  $\beta = 1.671$
  $\gamma = 1.699$
  $2V = 75°$
HARDNESS: 3–3½
DENSITY: 3.30 (Meas.)
         3.300 (Calc.)
CLEAVAGE: None.
HABIT: Massive, very fine grained, powdery to vermiform.
COLOR-LUSTER: Salmon.
MODE OF OCCURRENCE: Occurs in association with bonattite at Hat Creek, Bonaparte River area, Lillooet district, British Columbia, Canada.
BEST REF. IN ENGLISH: Jambor, J. L., Lachance, G. R., and Courville, S., *Can. Min.*, 8: 109–110 (1964).

# POLARITE
Pd(Pb, Bi)

CRYSTAL SYSTEM: Orthorhombic
LATTICE CONSTANTS:
  $a = 7.191$
  $b = 8.693$
  $c = 10.681$
3 STRONGEST DIFFRACTION LINES:
  2.65 (100)
  2.16 ( 90)
  2.25 ( 50)
HARDNESS: Microhardness 168–232 kg/mm².
DENSITY: Not determined.
CLEAVAGE: Not determined.
HABIT: Massive; as minute grains up to 0.3 mm in size.
COLOR-LUSTER: In polished section, white with yellowish tint. Opaque. Metallic.
MODE OF OCCURRENCE: Occurs intergrown with stannopalladinite, plumbopalladinite, nickeloan platinum, sphalerite, and native silver, associated with chalcopyrite, talnakhite, and cubanite in vein ores of the Talnakh deposit in the Polar Urals.
BEST REF. IN ENGLISH: Genkin, A. D., Evstigneeva, T. L., Troneva, N. V., and Vyal'sov, L. N., *Am. Min.*, 55: 1810 (1970).

# POLIANITE = Synonym for pyrolusite

# POLLUCITE
$Cs_{1-x}Na_x AlSi_2 O_6 \cdot xH_2O$  x~0.3
Zeolite Group

CRYSTAL SYSTEM: Cubic
CLASS: $4/m\,\overline{3}\,2/m$
SPACE GROUP: Ia3d
Z: 16
LATTICE CONSTANT:
  $a = 13.682$
3 STRONGEST DIFFRACTION LINES:
  3.421 (100)
  2.907 (100)
  3.652 ( 80)
OPTICAL CONSTANT:
  $N = 1.520$
HARDNESS: 6½–7
DENSITY: 2.936 (Meas.)
         2.94 (Calc.)
CLEAVAGE: None.
Fracture conchoidal to uneven. Brittle.
HABIT: Crystals cubic, sometimes dodecahedral; rare. Usually massive, fine-grained.
COLOR-LUSTER: Colorless, white, gray; sometimes tinted pale pink, blue, or violet. Transparent. Vitreous, slightly greasy.
MODE OF OCCURRENCE: Occurs in granite pegmatites, sometimes as large segregations, in association with microcline, cleavelandite, quartz, amblygonite, spodumene, lepidolite, and other minerals. Found as masses 3–4 feet thick at the Tin Mountain mine, Custer County, South Dakota; at Pala, Mesa Grande, and Vulcan Mountain, San Diego County, California; at Hebron, Greenwood, Mt. Mica, and Buckfield, Maine; and at localities in Connecticut and Massachusetts. Extensive deposits occur at Bernic Lake, Manitoba, Canada, and Karibib, South West Africa. Other occurrences are found at Varutrask, Sweden; Elba, Italy; Finland; and Kazakhstan, USSR.
BEST REF. IN ENGLISH: Vlasov, K. A., "Mineralogy of Rare Elements," v. II, p. 61–65, Israel Program for Scientific Translations, 1966.    Newnham, R. E., *Am. Min.*, 52: 1515–1518 (1967).

# POLYARGYRITE (Species status uncertain)
$Ag_{24}Sb_2 S_{15}$

CRYSTAL SYSTEM: Cubic
HARDNESS: 2½
DENSITY: 6.974 (Meas.)
CLEAVAGE: {001}
Fracture uneven. Sectile and malleable.
HABIT: As modified cubo-octahedral crystals, often indistinct and distorted.
COLOR-LUSTER: Blackish gray to iron black. Opaque. Metallic. Streak black.
MODE OF OCCURRENCE: Found associated with argentite, tetrahedrite, and dolomite at the Wenzel mine, Wolfach, Germany.
BEST REF. IN ENGLISH: Palache, et al., "Dana's System of Mineralogy," 7th Ed., v. I, p. 355, New York, Wiley, 1944.

# POLYBASITE

$Ag_{16}Sb_2S_{11}$

Polybasite-Arsenpolybasite Series

Dimorphous with antimonpearceite

CRYSTAL SYSTEM: Monoclinic (pseudo hexagonal)

CLASS: 2/m

SPACE GROUP: C2/m

Z: 16

LATTICE CONSTANTS:

$a = 26.12$

$b = 15.08$

$c = 23.89$

$\beta = 90°0'$

4 STRONGEST DIFFRACTION LINES:

3.00    (100)

3.19    ( 90)

2.88    ( 80)

2.69-2.64 ( 60)*

*Best diagnostic line. Absent in Pearceite Series.

OPTICAL CONSTANTS:

$N > 2.72$ (Li)

(+)2V = 22°

HARDNESS: 2-3

DENSITY: 6.30 (Meas.)

6.33 (Calc.)

CLEAVAGE: {001} imperfect

Fracture uneven. Brittle.

HABIT: Crystals thin tabular, flattened on {001}; usually pseudohexagonal, also pseudorhombohedral. Triangular striations on {001}. Also massive, compact. Twinning on {110}, repeated.

COLOR-LUSTER: Iron black; dark ruby red in thin splinters. Nearly opaque. Metallic. Streak black.

MODE OF OCCURRENCE: Occurs in low- to medium-temperature silver vein deposits, commonly associated with native silver, argentite, galena, and other silver and lead minerals. Found widespread in small amounts in the western United States, especially in Colorado, South Dakota, Montana, Idaho, Nevada, Arizona, and California. It also occurs in Canada; at many places in Mexico, especially as splendid large crystals at Arizpe and Las Chiapas, Sonora; and in Chile, Peru, Sardinia, Czechoslovakia, Germany, and New South Wales, Australia.

BEST REF. IN ENGLISH: Palache, et al., "Dana's System of Mineralogy," 7th Ed., v. I, p. 351-353, New York, Wiley, 1944.

# POLYCRASE

(Y, Er, Ce, La, U)(Ti, Nb, Ta)$_2$(O, OH)$_6$

Euxenite-Polycrase Series

CRYSTAL SYSTEM: Orthorhombic

CLASS: 2/m 2/m 2/m (?)

SPACE GROUP: Pcan (?)

Z: 4 (?)

LATTICE CONSTANTS:

$a = 5.55$kX

$b = 14.62$

$c = 5.19$

3 STRONGEST DIFFRACTION LINES: Metamict.

OPTICAL CONSTANTS:

Isotropic (metamict)

$N = 2.248$

HARDNESS: 5½-6½

DENSITY: 4.30-5.87 depending on Ta content and degree of metamictization and hydration.

CLEAVAGE: None. Fracture conchoidal. Brittle.

HABIT: Crystals short prismatic, often flattened {010}. Commonly as parallel, subparallel, or divergent crystal aggregates. Also massive, compact. Twinning on {201}, common; twins flattened {010} and striated [100].

COLOR-LUSTER: Black, sometimes with faint brownish or greenish tinge. Opaque; translucent in thin fragments. Brilliant submetallic; sometimes vitreous or resinous. Streak grayish, yellowish, or brownish.

MODE OF OCCURRENCE: Occurs in granite pegmatites often in association with monazite, and other rare earth minerals, biotite, muscovite, magnetite, ilmenite, garnet, and beryl; also as a detrital mineral derived from pegmatites or granitic rocks. Found in placer deposits in Boise County, Idaho; at Baringer Hill, Llano County, Texas; in Henderson County, North Carolina; and near Marietta, Greenville County, South Carolina. Other notable occurrences are found in Canada, Brazil, Norway, Sweden, USSR, and elsewhere.

BEST REF. IN ENGLISH: Palache, et al., "Dana's System of Mineralogy," 7th Ed., v. I, p. 787-792, New York, Wiley, 1944.

# POLYDYMITE

$Ni_3S_4$

Linnaeite Group

CRYSTAL SYSTEM: Cubic

CLASS: 4/m $\bar{3}$ 2/m

SPACE GROUP: Fd3m

Z: 8

LATTICE CONSTANT:

$a = 9.405$

3 STRONGEST DIFFRACTION LINES:

1.674 (100)

2.85 ( 90)

2.36 ( 90)

HARDNESS: 4½-5½

DENSITY: 4.5-4.8 (Meas.)

4.83 (Calc.)

CLEAVAGE: {001} imperfect

Fracture subconchoidal to uneven.

HABIT: Crystals usually octahedral. Commonly massive; compact to granular. Twinning on {111}.

COLOR-LUSTER: Light gray to steel gray; tarnishes copper red or violet-gray. Opaque. Metallic.

MODE OF OCCURRENCE: Occurs in hydrothermal vein deposits, commonly associated with other sulfide minerals, in the Siegen district of Westphalia, Germany, especially at Müsen.

BEST REF. IN ENGLISH: Palache, et al., "Dana's System of Mineralogy," 7th Ed., v. I, p. 262-265, New York, Wiley, 1944.

# POLYHALITE
$K_2MgCa_2(SO_4)_4 \cdot 2H_2O$

CRYSTAL SYSTEM: Triclinic
CLASS: $\bar{1}$ or 1
SPACE GROUP: $P\bar{1}$ or P1
Z: 1
LATTICE CONSTANTS:

| | |
|---|---|
| $a = 6.952$ | $\alpha = 104.06°$ |
| $b = 8.886$ | $\beta = 113.94°$ |
| $c = 6.954$ | $\gamma = 101.15°$ |

3 STRONGEST DIFFRACTION LINES:

| | |
|---|---|
| 2.90 (100) | 3.18 (100) |
| 3.18 ( 70) | 2.91 ( 25) |
| 2.85 ( 16) | 2.89 ( 25) |

OPTICAL CONSTANTS:
$\alpha = 1.547$
$\beta = 1.560$
$\gamma = 1.567$
$(-)2V = 62°$
HARDNESS: 3½
DENSITY: 2.78 (Meas.)
2.683 (Calc.)
CLEAVAGE: $\{10\bar{1}\}$ perfect
HABIT: Usually massive or fibrous to foliated. Rarely as small highly modified tabular or elongated crystals, commonly twinned on $\{010\}$, $\{100\}$, or polysynthetic.
COLOR-LUSTER: Colorless, white, gray; usually flesh pink to brick red due to iron oxide inclusions. Transparent to translucent. Vitreous to resinous.
MODE OF OCCURRENCE: Occurs widespread, chiefly as extensive sedimentary deposits associated with anhydrite, halite, sylvite, carnallite, kieserite, and other salts, and rarely as a product of volcanic activity. In the United States it is found in Grand County, Utah, and in large deposits in the Permian basin of southeastern New Mexico, especially in the vicinity of Carlsbad, and also in adjacent parts of Texas. It also is found at deposits in Germany, Poland, Austria, France, England, USSR, Iran, and encrusting lava at Mt. Vesuvius, Italy.
BEST REF. IN ENGLISH: Palache, et al., "Dana's System of Mineralogy," 7th Ed., v. II, p. 458–460, New York, Wiley, 1951.

# POLYLITHIONITE
$KLi_2Al(Si_4O_{10})(F, OH)_2$
Mica Group

CRYSTAL SYSTEM: Monoclinic
CLASS: 2/m
SPACE GROUP: C2/m
Z: 2
LATTICE CONSTANTS:
(synthetic)
$a = 5.186$
$b = 8.968$
$c = 10.029$
$\beta = 100°24'$

3 STRONGEST DIFFRACTION LINES:
(synthetic)
3.59 (100)
3.31 (100)
3.07 (100)
OPTICAL CONSTANTS:
(synthetic)
$\alpha = 1.529$
$\beta = 1.552$
$\gamma = 1.555$
$(-)2V \sim 30°$
HARDNESS: 2-3
DENSITY: 2.583-2.757 (Meas. cryptocrystalline)
2.817 (Meas. crystals)
CLEAVAGE: $\{001\}$ perfect
HABIT: Crystals tabular, hexahedral, up to 5 cm in diameter. Also as scaly aggregates and cryptocrystalline masses. As pseudomorphs after natrolite.
COLOR-LUSTER: White, pink, cream-colored, sometimes greenish or bluish. Transparent. Pearly; cryptocrystalline varieties waxy.
MODE OF OCCURRENCE: Occurs in carbonatites in association with natrolite, microcline, tainiolite, and other minerals. Found at Mont St. Hilaire, Quebec, Canada; in the Julianehaab district, Greenland; and on the Kola Peninsula, USSR.
BEST REF. IN ENGLISH: Vlasov, K. A., "Mineralogy of Rare Elements," v. II, p. 25–29, Israel Program for Scientific Translations, 1966.

# POLYMIGNITE
$(Ca, Fe, Y, Th)(Nb, Ti, Ta)O_4$

CRYSTAL SYSTEM: Orthorhombic
CLASS: 2/m 2/m 2/m
3 STRONGEST DIFFRACTION LINES:
2.94 (100) Metamict
1.82 ( 80) (ignited at 1000°C for 1 hr)
1.75 ( 50)
OPTICAL CONSTANT:
Isotropic—metamict
$N = 2.215$
HARDNESS: 6½
DENSITY: 4.77-4.85 (Meas.)
CLEAVAGE: $\{100\}$ in traces ?
$\{010\}$ in traces ?
Fracture conchoidal. Brittle.
HABIT: Crystals prismatic, sometimes flattened; vertically striated; up to 5 cm long.
COLOR-LUSTER: Black. Opaque; translucent in very thin fragments. Brilliant submetallic. Streak dark brown.
MODE OF OCCURRENCE: Occurs at Beverly, Massachusetts; and in pegmatite at Fredriksvärn and on the Island of Svenor, Norway.
BEST REF. IN ENGLISH: Vlasov, K. A., "Mineralogy of Rare Elements," v. II, p. 496–499, Israel Program for Scientific Translations, 1966.

# PORPEZITE = Variety of gold with up to 10% Pd

## PORTLANDITE
$Ca(OH)_2$

CRYSTAL SYSTEM: Hexagonal
CLASS: $\bar{3}\ 2/m$
SPACE GROUP: $P\bar{3}m1$
Z: 1
LATTICE CONSTANTS:
  $a = 3.585$
  $c = 4.895$
3 STRONGEST DIFFRACTION LINES:
  2.628 (100)
  4.90 ( 74)
  1.927 ( 42)
OPTICAL CONSTANTS:
  $\omega = 1.574$
  $\epsilon = 1.547$
    (−)
HARDNESS: 2
DENSITY: 2.230 (Meas.)
        2.241 (Calc.)
CLEAVAGE: {0001} perfect
Flexible. Sectile.
  HABIT: As minute hexagonal plates.
  COLOR-LUSTER: Colorless. Transparent. Pearly on cleavages.
  MODE OF OCCURRENCE: Occurs in larnite-spurrite contact rocks at Scawt Hill, county Antrim, Ireland, associated with calcite and afwillite. Also reported from Vesuvius, Italy.
  BEST REF. IN ENGLISH: Palache, et al., "Dana's System of Mineralogy," 7th Ed., v. I, p. 641–642, New York, Wiley, 1944.

## POSNJAKITE
$Cu_4(SO_4)(OH)_6 \cdot H_2O$

CRYSTAL SYSTEM: Monoclinic
CLASS: $2/m$
SPACE GROUP: C2/c
Z: 4
LATTICE CONSTANTS:
  $a = 14.236$
  $b = 6.340$
  $c = 10.571$
  $\beta = 102°55'$
3 STRONGEST DIFFRACTION LINES:
  6.94 (100)
  3.47 ( 30)
  2.70 ( 25)
OPTICAL CONSTANTS:
  $\alpha = 1.625$
  $\beta = 1.680$
  $\gamma = 1.706$
  (−)2V = 57°
HARDNESS: 2-3
DENSITY: 3.35 (Meas.)
        3.36 (Calc.)
CLEAVAGE: Not determined.
  HABIT: Crystals tabular, small, up to 0.2-0.5 mm; in small grains; as thin films.

COLOR-LUSTER: Blue to dark blue; vitreous. Streak bluish.
  MODE OF OCCURRENCE: Occurs associated with aurichalcite and other secondary minerals near oxidized chalcopyrite in quartz veins of the Nura-Talkinsk tungsten deposits, central Kazakhstan.
  BEST REF. IN ENGLISH: Komkov, A. I., and Nefedov, E. I., *Am. Min.*, **52**: 1582–1583 (1967).

## POTARITE
PdHg

CRYSTAL SYSTEM: Tetragonal, cubic
CLASS: 4/m 2/m 2/m, 23
SPACE GROUP: P4/mmm, $P2_13$
Z: 1 (tetrag.), 4 (cubic)
LATTICE CONSTANTS:
  (tetrag.)        (cubic)
  $a = 3.020$      $a = 5.21$
  $c = 3.706$
3 STRONGEST DIFFRACTION LINES:
  2.34  (100)
  1.274 ( 80)
  1.402 ( 55)
HARDNESS: 3½
DENSITY: 14.88 (Meas.)
        15.09 (Calc.)
CLEAVAGE: None. Brittle
  HABIT: As small grains or nuggets with slightly divergent fibrous or columnar structure.
  COLOR-LUSTER: Silver white. Opaque. Bright metallic. Streak silver white.
  MODE OF OCCURRENCE: Occurs in the diamond-bearing placer deposits in the Potaro River region, Guyana.
  BEST REF. IN ENGLISH: Palache, et al., "Dana's System of Mineralogy," 7th Ed., v. I, p. 105, New York, Wiley, 1944. *Am. Min.*, **45**: 1093–1097 (1960).

## POTASH ALUM
$KAl(SO_4)_2 \cdot 12H_2O$

CRYSTAL SYSTEM: Cubic
CLASS: $2/m\ \bar{3}$
SPACE GROUP: Pa3
Z: 4
LATTICE CONSTANT:
  $a = 12.133$ kX
3 STRONGEST DIFFRACTION LINES:
  4.298 (100)
  3.250 ( 55)
  4.053 ( 45)
OPTICAL CONSTANT:
  $N = 1.4562$ (Na)
HARDNESS: 2-2½
DENSITY: 1.757 (Meas.)
        1.754 (Calc.)
CLEAVAGE: {111} in traces
Fracture conchoidal.
  HABIT: Usually massive, granular or columnar; also

stalactitic, and as powdery crusts and coatings. Crystals cubic or octahedral.

COLOR-LUSTER: Colorless, also white. Transparent. Vitreous.

Soluble in water. Taste sweetish and astringent.

MODE OF OCCURRENCE: Occurs as an efflorescence or filling cracks in argillaceous rocks; in brown coals containing disseminated iron sulfides; and in fumaroles and solfotaras. Found at numerous localities in California, Colorado, South Dakota, Nevada, and other western states; in Alum Cave, Sevier County, Tennessee; in Chile, England, Scotland, France, Spain, Germany, and Italy.

BEST REF. IN ENGLISH: Palache, et al., "Dana's System of Mineralogy," 7th Ed., v. II, p. 472–474, New York, Wiley, 1951.

# POUGHITE
$Fe_2(TeO_3)_2(SO_4) \cdot 3H_2O$

CRYSTAL SYSTEM: Orthorhombic
CLASS: 2/m 2/m 2/m
SPACE GROUP: Pmnb
Z: 4
LATTICE CONSTANTS:
 $a = 9.66$
 $b = 14.20$
 $c = 7.86$
3 STRONGEST DIFFRACTION LINES:
 7.10 (100)
 5.74 (100)
 3.239 ( 70)
OPTICAL CONSTANTS:
 $\alpha = 1.72$
 $\beta = 1.985$
 $\gamma = 1.990$
 $(-)2V = 15°-20°$
HARDNESS: 2½
DENSITY: 3.75 (Meas.)
CLEAVAGE: {010} perfect
 {101} good
HABIT: Crystals tabular, diamond-shaped, flattened parallel to (010).
COLOR-LUSTER: Dark yellow to brownish yellow and greenish yellow. Synthetic poughite is pale yellow to bright sulfur yellow.
MODE OF OCCURRENCE: Found sparingly as thin drusy crusts and small botryoidal rosettes at the Moctezuma mine, Sonora, Mexico. Also occurs at the El Plomo mine, Ojojona District, Tegucigalpa, Honduras.
BEST REF. IN ENGLISH: Gaines, Richard V., *Am. Min.*, **53**: 1075–1080 (1968).

# POWELLITE
$Ca(Mo, W)O_4$
Isostructural with scheelite

CRYSTAL SYSTEM: Tetragonal
CLASS: 4/m
SPACE GROUP: I $4_1$/a
Z: 4
LATTICE CONSTANTS:
 $a = 5.23$ kX
 $c = 11.44$
3 STRONGEST DIFFRACTION LINES:
 3.10 (100)
 1.929 ( 30) (synthetic)
 4.76 ( 25)
OPTICAL CONSTANTS:
 $\omega = 1.967$
 $\epsilon = 1.978$
 (+)
HARDNESS: 3½-4
DENSITY: 4.23 (Meas.)
 4.26 (Calc. X-ray)
CLEAVAGE: {112} indistinct
 {011} indistinct
 {001} indistinct
HABIT: Crystals usually pyramidal; less commonly thin tabular. Pyramid faces sometimes striated. As crusts of intergrown crystals. Also massive, foliated, pulverulent to ocherous.
COLOR-LUSTER: Straw yellow, greenish yellow, brown, grayish white to gray, pale greenish blue to blue, blackish blue. Transparent. Subadamantine to greasy. Foliated material often pearly. Fluoresces yellowish white to golden yellow.
MODE OF OCCURRENCE: Occurs as a secondary mineral in the oxidation zone of ore deposits. Found at the Isle Royale and South Hecla mines, Houghton County, Michigan; at several deposits in the western United States, especially in California, Nevada, Utah, Arizona, New Mexico, and Colorado; and in Turkey, USSR, and Morocco.
BEST REF. IN ENGLISH: Palache, et al., "Dana's System of Mineralogy," 7th Ed., v. II, p. 1079–1081, New York, Wiley, 1951.

**PRASE** = Variety of quartz

# PREHNITE
$Ca_2 Al_2 Si_3 O_{10}(OH)_2$

CRYSTAL SYSTEM: Orthorhombic
CLASS: mm2
SPACE GROUP: P2cm
Z: 2
LATTICE CONSTANTS:
 $a = 4.61$
 $b = 5.47$
 $c = 18.48$
3 STRONGEST DIFFRACTION LINES:
 3.08 (100)
 2.55 (100)
 3.48 ( 90)
OPTICAL CONSTANTS:
 $\alpha = 1.611-1.632$
 $\beta = 1.615-1.642$
 $\gamma = 1.632-1.665$
 $(+)2V = 65°-69°$
HARDNESS: 6-6½
DENSITY: 2.90-2.95 (Meas.)
 2.94 (Calc.)

CLEAVAGE:  {001} distinct
                     {110} indistinct
Fracture uneven. Brittle.

HABIT: Crystals usually tabular, also prismatic or steep pyramidal. Individual crystals rare, commonly in tabular groups or barrel-shaped aggregates. Usually as compact granular masses or in botryoidal, stalactitic or reniform forms with columnar, radiated or lamellar structure.

COLOR-LUSTER: Pale to dark green, yellow, gray, white, colorless. Transparent to translucent. Vitreous to somewhat pearly. Streak uncolored.

MODE OF OCCURRENCE: Occurs chiefly as a secondary or hydrothermal mineral in cavities in basic igneous rocks, commonly associated with calcite, datolite, pectolite, and zeolites; also in granite, gneiss, diorite, metamorphosed limestones, and veins. Found in small amounts at several localities in California, especially at Crestmore in Riverside County; on Mt. Sneffels, Ouray County, Colorado; with native copper in amygdaloidal rocks in Keweenaw County, Michigan; at the Lane trap rock quarry, near Westfield, Hampden County, Massachusetts; at Farmington, Hartford County, Connecticut; abundantly in the trap rocks of northern New Jersey, especially at Paterson, Bergen Hill, and Great Notch; and at the Fairfax quarry, near Centreville, Fairfax County, Virginia. Also found in Canada, Scotland, France, as fine crystals from Switzerland, Italy, Austria, Germany, Czechoslovakia, USSR, South Africa, Pakistan, Australia, New Zealand, and elsewhere.

BEST REF. IN ENGLISH: Deer, Howie, and Zussman, "Rock Forming Minerals," v. 3, p. 263-266, New York, Wiley, 1962.

## PREOBRAZHENSKITE
$Mg_3B_{11}O_{15}(OH)_9$

CRYSTAL SYSTEM: Orthorhombic, pseudohexagonal
Z: 4
LATTICE CONSTANTS:
  $a = 16.30$
  $b = 9.14$
  $c = 10.15$
OPTICAL CONSTANTS:
  $\beta \simeq \alpha = 1.573-1.576$
  $\gamma = 1.594-1.596$
  Nearly uniaxial, (+)2V = small
HARDNESS: 4½-5
DENSITY: Not given
HABIT: Massive; fine-grained, nodular.
COLOR-LUSTER: Colorless, lemon yellow, dark gray.
MODE OF OCCURRENCE: The mineral occurs in fine-grained halite-polyhalite rock and encloses kaliborite and boracite. In places it has been partially replaced by inyoite. Found widespread in small amounts in several parts of the saliferous strata of the Inder uplift, western Kazakhstan, USSR.
BEST REF. IN ENGLISH: Yarzhemskii, Ya. Ya., *Am. Min.*, 42: 704 (1957). Ostrovskaya, I. V., and Nikitina, I. B., *Am. Min.*, 55: 1071 (1970).

## PRIBRAMITE = Cadmium-bearing sphalerite

## PRICEITE (Pandermite)
$Ca_4B_{10}O_{19} \cdot 7H_2O$

CRYSTAL SYSTEM: Triclinic (?)
LATTICE CONSTANTS:
  $a:b:c = 0.551:1:?$
    $\beta \simeq 110°$
3 STRONGEST DIFFRACTION LINES:
  10.92 (100)
  3.63 (100)
  5.46 ( 80)
OPTICAL CONSTANTS:
  $\alpha = 1.571-1.573$
  $\beta = 1.590-1.591$
  $\gamma = 1.593-1.594$
(-)2V $\simeq$ 32°
HARDNESS: 3-3½
DENSITY: 2.42 (Meas.)
CLEAVAGE: {001} perfect. Fracture earthy to conchoidal.
HABIT: As nodules or irregular masses; soft and chalky to hard and compact. Rhombic crystal outline observed under microscopic.
COLOR-LUSTER: White; earthy.
MODE OF OCCURRENCE: Found as compact rounded masses associated with aragonite near Chetco, Curry County, Oregon in an apparent hot spring deposit; also occurs as nodules in clay, associated with gypsum and colemanite in Furnace Creek wash, Death Valley, California; and as nodules and masses up to a ton in weight underlying beds of gypsum and clay at Sultan Tschair, Turkey.
BEST REF. IN ENGLISH: Palache, et al., "Dana's System of Mineralogy," 7th Ed., v. 2, p. 341-343, New York, Wiley, 1951.

## PRIDERITE
$(K, Ba)(Ti, Fe^{3+})_8O_{16}$

CRYSTAL SYSTEM: Tetragonal
CLASS: 4/m
SPACE GROUP: I4/m
Z: 1
LATTICE CONSTANTS:
  $a = 10.11$
  $c = 2.964$
3 STRONGEST DIFFRACTION LINES:
  3.20 (100)
  7.14 ( 70)
  5.05 ( 70)
OPTICAL CONSTANT:
  $\omega > 2.10$ (+)
DENSITY: 3.86 (Meas.)
CLEAVAGE: Basal, perfect. Parallel to prism, fair.
HABIT: As stout rectangular prisms with a lamellation parallel to base up to 1.0 X 0.5 mm in size; also as rods about 0.05 mm long.
COLOR-LUSTER: Black, reddish; adamantine. Streak gray.
MODE OF OCCURRENCE: Occurs as a characteristic accessory in the leucite-bearing rocks of the west Kimberley area, Western Australia.

BEST REF. IN ENGLISH: Norrish, K., *Min. Mag.*, **29**: 496–501 (1951).


**PRIORITE** = Aeschynite-(Y)


**PRISMATINE** = Variety of kornerupine


## PROBERTITE
$NaCaB_5O_9 \cdot 5H_2O$

CRYSTAL SYSTEM: Monoclinic
CLASS: 2/m
SPACE GROUP: $P2_1/a$
Z: 4
LATTICE CONSTANTS:
 $a = 13.43$
 $b = 12.57$
 $c = 6.589$
 $\beta = 100°15'$
3 STRONGEST DIFFRACTION LINES:
 9.12 (100)
 2.807 ( 35)
 6.62 ( 20)
OPTICAL CONSTANTS:
 $\alpha = 1.514$
 $\beta = 1.524$
 $\gamma = 1.543$
 $(+)2V = 73°$
HARDNESS: 3½
DENSITY: 2.141 (Meas.)
 2.131 (Calc.)
CLEAVAGE: {110} perfect
Brittle.
HABIT: Single crystals rare; acicular [001]; often flattened {100}; rarely flattened {110}. Commonly as radial aggregates composed of acicular crystals up to 3 cm in length; spherulitic; as compact aggregates.
COLOR-LUSTER: Colorless, transparent; vitreous.
MODE OF OCCURRENCE: Occurs associated with colemanite and ulexite in the Ryan district, Inyo County, in crystals at the Baker mine, Boron, Kern County, and with ulexite at Lang, Los Angeles County, California.
BEST REF. IN ENGLISH: Clark, Joan R., and Christ, C. L., *Am. Min.*, **44**: 712–719 (1959).


**PROCHLORITE** = Ripidolite


**PROIDONITE** = A natural gas
$SiF_4$


## PROSOPITE
$CaAl_2(F, OH)_8$ with F:OH ≈ 5:3

CRYSTAL SYSTEM: Monoclinic
CLASS: 2/m
SPACE GROUP: A2/a
Z: 4
LATTICE CONSTANTS:
 $a = 7.32$ kX
 $b = 11.11$
 $c = 6.69$
 $\beta = 95°00'$
3 STRONGEST DIFFRACTION LINES:
 4.35 (100)
 2.13 ( 60)
 1.837 ( 60)
OPTICAL CONSTANTS:
 $\alpha = 1.501$
 $\beta = 1.503$
 $\gamma = 1.510$
 $(+)2V = 62°45'$ (Na)
HARDNESS: 4½
DENSITY: 2.894 (Meas.)
 2.898 (Calc.)
CLEAVAGE: {111} perfect
Fracture conchoidal to uneven. Brittle.
HABIT: As minute tabular crystals; and as embedded blades or bundles of blades. Also massive, fine granular to powdery.
COLOR-LUSTER: Colorless, white, gray. Transparent to translucent. Vitreous.
MODE OF OCCURRENCE: Occurs associated with pachnolite, thomsenolite, and gearksutite at the cryolite deposit at St. Peters Dome, El Paso County, Colorado. It also is found in the Dugway district, Tooele County, Utah, with fluorite, pyrite, and quartz; at Altenberg, Saxony, Germany; in tin veins and greisen at Schlaggenwald, Bohemia; and at Mt. Bischoff, Tasmania.
BEST REF. IN ENGLISH: Palache, et al., "Dana's System of Mineralogy," 7th Ed., v. II, p. 121–123, New York, Wiley, 1951.


## PROUSTITE
$Ag_3AsS_3$
Dimorphous with xanthoconite

CRYSTAL SYSTEM: Hexagonal
CLASS: $\bar{3}$ 2/m
SPACE GROUP: R3c
Z: 6 (hex.) 2 (rhomb.)
LATTICE CONSTANTS:
 (hexag. cell) (rhomb.)
 $a = 10.77$ $a_{rh} = 6.86$
 $c = 8.67$ $\alpha = 103°30.5'$
3 STRONGEST DIFFRACTION LINES:
 2.74 (100)
 2.48 ( 90)
 3.27 ( 80)
OPTICAL CONSTANTS:
 $\omega = 3.0877$
 $\epsilon = 2.7924$ (Na)
 (−)
HARDNESS: 2–2½
DENSITY: 5.55–5.64 (Meas.)
 5.62 (Calc.)

CLEAVAGE: $\{10\bar{1}1\}$ distinct
Fracture conchoidal to uneven. Brittle.

HABIT: Crystals prismatic, rhombohedral, or scalenohedral; often highly modified. Usually massive; compact, disseminated, in crusts or bands. Twinning on $\{10\bar{1}4\}$ and $\{10\bar{1}1\}$ common; on $\{0001\}$ and $\{01\bar{1}2\}$ less common.

COLOR-LUSTER: Scarlet to vermilion red. Transparent to translucent. Adamantine to submetallic. Streak bright red.

MODE OF OCCURRENCE: Occurs in low-temperature hydrothermal vein deposits, commonly associated with other sulfosalts, silver, galena, pyrite, calcite, dolomite, and quartz. Found in small amounts in the western United States, especially in the Silver City district, Owyhee County, Idaho, and in Colorado, Nevada, and California. It also occurs in the Cobalt district, Ontario, Canada, often as fine crystals associated with pyrargyrite; at Batopilas, Chihuahua, and elsewhere in Mexico; as superb crystals up to 3 inches long and 1 inch thick at the Dolores mine, Chanarcillo, Chili; at Sarrabus, Sardinia; as fine crystals at Freiberg and other places in Saxony, Germany; and at silver deposits in France, Czechoslovakia, and Roumania.

BEST REF. IN ENGLISH: Palache, et al., "Dana's System of Mineralogy," 7th Ed., v. I, p. 366–369, New York, Wiley, 1944.

## PRZHEVALSKITE
$Pb(UO_2)_2(PO_4)_2 \cdot 2H_2O$

CRYSTAL SYSTEM: Orthorhombic
3 STRONGEST DIFFRACTION LINES:
  3.610 (100)
  9.080 ( 90)
  1.619 ( 60)
OPTICAL CONSTANTS:
  $\alpha$ = 1.739
  $\beta$ = 1.749
  $\gamma$ = 1.752
  $(-)2V \sim 30°$
CLEAVAGE: $\{001\}$ distinct
HABIT: As foliated aggregates of tabular crystals, ranging in size from 0.1 to 1.0 mm.
COLOR-LUSTER: Bright yellow with faint greenish tint; vitreous to adamantine.
MODE OF OCCURRENCE: Occurs in very small amounts associated with torbernite, autunite, dumontite, renardite, uranophane, metahalloysite, hydrous oxides of iron and manganese, and wulfenite in the oxidation zone of a pitchblende-sulfide deposit at an unspecified locality, presumably in the USSR.
BEST REF. IN ENGLISH: Am. Min., 43: 381–382 (1958).

## PSEUDOAUTUNITE
$(H_3O)_4Ca_2(UO_2)_2(PO_4)_4 \cdot 5H_2O$ (?)

CRYSTAL SYSTEM: Orthorhombic (?)
LATTICE CONSTANTS:
  $a = b = 6.94$ kX
  $c = 12.85$

3 STRONGEST DIFFRACTION LINES:
  6.2  (100)
  3.25 (100)
  1.92 ( 90)
OPTICAL CONSTANTS:
  $\alpha$ = 1.541
  $\beta$ = 1.568 (Calc.)
  $\gamma$ = 1.570
  $(-)2V = 32°$
HARDNESS: Not determined.
DENSITY: 3.28 (Meas.)
         3.29 (Calc.)
CLEAVAGE: $\{001\}$ perfect, micaceous
HABIT: Crystals hexagonal in aspect, platy, up to 0.1 mm in size; as crusts.
COLOR-LUSTER: Pale yellow to white. Fluoresces intense greenish yellow in short-wave ultraviolet light; weakly in long-wave.
MODE OF OCCURRENCE: Occurs in fissures and cavities of albite-acmite veins in association with calcite, pyrochlore, oxidized sulfides, and sometimes apatite, in northern Karelia, USSR.
BEST REF. IN ENGLISH: Sergeev, A. S., Am. Min., 50: 1505–1506 (1965).

## PSEUDOBOLÉITE
$Pb_5Cu_4Cl_{10}(OH)_8 \cdot 2H_2O$

CRYSTAL SYSTEM: Tetragonal
Z: 2
LATTICE CONSTANTS:
  $a = 15.294$
  $c = 30.588$
3 STRONGEST DIFFRACTION LINES:
  4.43 (100)
  3.83 (100)
  2.71 ( 95)
OPTICAL CONSTANTS:
  $\omega$ = 2.03
  $\epsilon$ = 2.00
  $(-)$
HARDNESS: 2½
DENSITY: 4.85 (Meas.)
         4.89 (Calc.)
CLEAVAGE: $\{001\}$ perfect
          $\{101\}$ nearly perfect
HABIT: As a parallel growth upon boleite.
COLOR-LUSTER: Indigo blue. Translucent. Pearly on cleavages.
MODE OF OCCURRENCE: Occurs as a secondary mineral associated with boleite and cumengite at Boleo, Baja California, Mexico. Also found at Chancay, Peru, and Laurium, Greece.
BEST REF. IN ENGLISH: Palache, et al., "Dana's System of Mineralogy," 7th Ed., v. II, p. 80–81, 1951.

## PSEUDOBROOKITE
$Fe_2TiO_5$

CRYSTAL SYSTEM: Orthorhombic
CLASS: 2/m 2/m 2/m

SPACE GROUP: Bbmm
Z: 4
LATTICE CONSTANTS:
  $a = 9.79$
  $b = 9.93$
  $c = 3.725$
3 STRONGEST DIFFRACTION LINES:
  3.483 (100)
  2.748 ( 80)
  4.902 ( 45)
OPTICAL CONSTANTS:
    $\alpha = 2.38$
    $\beta = 2.39$  (Li)
    $\gamma = 2.42$
  (+)2V = 50°
HARDNESS: 6
DENSITY: 4.39 (Meas.)
         4.39 (Calc.)
CLEAVAGE: {010} distinct
Fracture subconchoidal to uneven. Brittle.
  HABIT: Crystals commonly tabular, elongated, and verti-cally striated; also long prismatic or acicular.
  COLOR-LUSTER: Black, brownish black to reddish brown. Opaque, except in thin fragments. Metallic adamantine. Streak reddish brown to yellowish brown.
  MODE OF OCCURRENCE: Occurs in cavities in basalt on Red Cone, Crater Lake, Oregon; as fine crystals asso-ciated with small topaz crystals and minute plates of hematite in lithophysae in rhyolite in the Thomas Moun-tains, Utah; and in volcanic igneous rocks at localities in Costa Rica, Norway, France, Spain, Germany, Rumania, and elsewhere.
  BEST REF. IN ENGLISH: Palache, et al., "Dana's Sys-tem of Mineralogy," 7th Ed., v. I, p. 736–738, New York, Wiley, 1944.

# PSEUDOCOTUNNITE
$K_2 PbCl_4$ (?)

CRYSTAL SYSTEM: Probably orthorhombic
Z: 4
LATTICE CONSTANTS:
  $a = 9.82$
  $b = 11.80$ (synthetic)
  $c = 5.77$
HARDNESS: Not determined.
DENSITY: Not determined.
CLEAVAGE: Not determined.
  HABIT: Crystals needle-like or lath-like; also as warty or dendritic crusts.
  COLOR-LUSTER: Colorless, white, yellowish. Dull.
  MODE OF OCCURRENCE: Found as a sublimate in fumaroles of the eruptions of Mt. Vesuvius, Italy in 1872 and 1906.
  BEST REF. IN ENGLISH: Palache, et al., "Dana's Sys-tem of Mineralogy," 7th Ed., v. II, p. 96–97, New York, Wiley, 1951.

# PSEUDOLAUEÏTE
$MnFe_2 (PO_4)_2 (OH)_2 \cdot 8H_2 O$

CRYSTAL SYSTEM: Monoclinic
CLASS: 2/m
SPACE GROUP: $P2_1/a$
Z: 2
LATTICE CONSTANTS:
  $a = 9.647$
  $b = 7.428$
  $c = 10.194$
  $\beta = 104.63°$
3 STRONGEST DIFFRACTION LINES:
  9.926 (100)
  5.869 ( 70)
  3.472 ( 40)
OPTICAL CONSTANTS:
    $\alpha = 1.626$
    $\beta = 1.650$
    $\gamma = 1.686$
  (+)2V = 80°
HARDNESS: 3
DENSITY: 2.463 (Meas.)
         2.51 (Calc.)
CLEAVAGE: Not determined. Brittle.
  HABIT: Crystals prismatic to thick tabular with forms a{100}, c{001}, m{110}, and rare {011} and {$\bar{2}$01}.
  COLOR-LUSTER: Orange-yellow; vitreous.
  MODE OF OCCURRENCE: Occurs as incrustations and as the core of stewartite crystals associated with oxides of manganese and iron in pegmatite at Hagendorf, Bavaria.
  BEST REF. IN ENGLISH: Strunz, H., Am. Min., 41: 815 (1956). Baur, Werner H., Am. Min., 54: 1312–1323 (1969).

# PSEUDOMALACHITE (Tagilite)
$Cu_5 (PO_4)_2 (OH)_4 \cdot H_2 O$

CRYSTAL SYSTEM: Monoclinic
CLASS: 2/m
SPACE GROUP: $P2_1/a$
Z: 2
LATTICE CONSTANTS:
  $a = 17.06$
  $b = 5.76$
  $c = 4.49$
  $\beta = 91°02'$
3 STRONGEST DIFFRACTION LINES:
  4.49  (100)
  2.386 ( 70)
  2.443 ( 60)
OPTICAL CONSTANTS:
    $\alpha = 1.791$
    $\beta = 1.856$
    $\gamma = 1.867$
  (-)2V = 48°
HARDNESS: 4½-5
DENSITY: 4.30-4.35 (Meas.—coarse crystalline material)
         4.08-4.21 (Meas.—radiating and cryptocrys-talline)
         4.34 (Calc.)
CLEAVAGE: {100} perfect and difficult
Fracture splintery.
  HABIT: Crystals short prismatic [001]; often in subpar-

allel aggregates; also reniform, botryoidal, massive with radial fibrous structure and concentric banding, foliated, microcrystalline or dense, colloform. Twinning on {100}.

COLOR-LUSTER: Dark green to blackish green (crystals); Fibrous material bluish green to green. Translucent. Vitreous.

MODE OF OCCURRENCE: Occurs as a secondary mineral associated with quartz, chalcedony, malachite, chrysocolla, tenorite, and limonite at numerous localities in France, Germany, Czechoslovakia, Roumania, USSR, Chile, England, Belgium, Zambia, Zaire, Western Australia, and United States.

BEST REF. IN ENGLISH: Palache, et al., "Dana's System of Mineralogy," 7th Ed., v. 2, p. 799–801, New York, Wiley, 1951.    Berry, L. G., *Am. Min.*, **35**: 365–385 (1950).

# PSEUDONATROLITE = Mordenite

# PSEUDORUTILE (Transition phase between the atomic arrangements in ilmenite and rutile ?)
$Fe_2 Ti_3 O_9$ (?)

CRYSTAL SYSTEM: Hexagonal
LATTICE CONSTANTS:
   $a = 2.872$
   $c = 4.594$
3 STRONGEST DIFFRACTION LINES:
   1.69 (100)
   2.49 ( 60)
   2.19 ( 50)
OPTICAL CONSTANT:
   $N > 1.769$
   HARDNESS: 3½
   DENSITY: 4.13 (Calc.)
   CLEAVAGE: Fracture conchoidal.
   HABIT: As thin irregular plates with fibrous structure.
   COLOR-LUSTER: Light brown to black. Transparent in thin fragments, submetallic to waxy. Streak reddish brown.
   MODE OF OCCURRENCE: Occurs as an alteration product of ilmenite. Observed in concentrates from Florida, New Jersey, India, and Brazil..
   BEST REF. IN ENGLISH: Teufer, G., and Temple, A. K., *Nature*, **211**: 179–181 (1966).

# PSEUDOTHURINGITE = Variety of thuringite with $Si = 2.0$–$2.5$ and $Mg < Fe^{2+}$

# PSEUDOWAVELLITE = Crandallite

# PSEUDOWOLLASTONITE
$\beta$-$CaSiO_3$

CRYSTAL SYSTEM: Triclinic
CLASS: $\bar{1}$
SPACE GROUP: P$\bar{1}$

Z: 24
LATTICE CONSTANTS:
   $a = 6.90$        $\alpha = 90°$
   $b = 11.78$      $\beta = 90°48'$
   $c = 19.65$      $\gamma = 119°18'$
3 STRONGEST DIFFRACTION LINES:
   3.20 (100)
   2.79 ( 80)
   1.96 ( 80)
OPTICAL CONSTANTS:
   $\alpha = 1.610$
   $\beta = 1.611$
   $\gamma = 1.654$
   (+)2V
   HARDNESS: 5
   DENSITY: 2.912 (Meas.)
                2.90 (Calc.)
   CLEAVAGE: {001}
   HABIT: As equant grains. Twinning on {001} lamellar.
   COLOR-LUSTER: Colorless. Transparent. Vitreous.
   MODE OF OCCURRENCE: Found sparingly in pyrometamorphosed Tertiary rocks in southwest Iran.
   BEST REF. IN ENGLISH: Deer, Howie, and Zussman, "Rock Forming Minerals," v. II, p. 167–175, New York, Wiley, 1963.

# PSILOMELANE (Romanechite)
$BaMn^{2+}Mn_8^{4+}O_{16}(OH)_4$

CRYSTAL SYSTEM: Monoclinic
CLASS: 2/m
SPACE GROUP: C2/n
Z: 2
LATTICE CONSTANTS:
   $a = 13.94$
   $b = 2.846$
   $c = 9.683$
   $\beta = 92.32°$
3 STRONGEST DIFFRACTION LINES:
   2.41 (100)
   2.19 ( 85)
   3.48 ( 60)
   HARDNESS: 5–6, decreasing in earthy varieties.
   DENSITY: 6.45 (Calc.)
   CLEAVAGE: Not determined.
   HABIT: Massive; botryoidal, reniform, stalactitic, sometimes with concentric banding; also earthy and pulverulent.
   COLOR-LUSTER: Black to steel gray. Opaque. Submetallic, dull. Streak brownish black to black, shining.
   MODE OF OCCURRENCE: Occurs as a secondary mineral formed by the weathering of manganese carbonates and silicates. Found chiefly as large residual deposits; replacement deposits in calcareous or dolomitic rocks; and as concretions in lake deposits, swamp deposits, and clays. Found widespread, commonly associated with pyrolusite, in the United States, Scotland, Sweden, Belgium, France, Germany, India, and elsewhere.
   BEST REF. IN ENGLISH: Mukherjee, *Min. Mag.*, **35**: 643–655 (1965).    *Am. Min.*, **45**: 176–187 (1960).

# PUCHERITE

$BiVO_4$

CRYSTAL SYSTEM: Orthorhombic
CLASS: 2/m 2/m 2/m
SPACE GROUP: Pnca
Z: 4
LATTICE CONSTANTS:
  $a = 5.332$
  $b = 5.060$
  $c = 12.02$
3 STRONGEST DIFFRACTION LINES:
  3.499 (100)
  2.702 (100)
  4.644 ( 55)
OPTICAL CONSTANTS:
  $\alpha = 2.41$
  $\beta = 2.50$
  $\gamma = 2.51$
  $(-)2V = 19°$
HARDNESS: 4
DENSITY: 6.25 (Meas.)
         6.63 (Calc.)
CLEAVAGE: {001} perfect
Fracture subconchoidal. Brittle.

HABIT: Crystals small, often with curved faces; usually tabular {001}, also acicular; {111} striated parallel edge {001} {111}. Also massive, ocherous.

COLOR-LUSTER: Yellowish brown and reddish brown to dark brownish red. Transparent to opaque. Vitreous to adamantine. Streak yellow.

MODE OF OCCURRENCE: Occurs with native bismuth in pegmatite at the Stewart and Pala Chief mines near Pala, and with bismutite at the Victor mine, Rincon, San Diego County, California. Also found in the oxidized portion of a bismuth-silver-uranium-copper vein in the Pucher shaft of the Wolfgang mine at Schneeberg, and from Sosa near Eibenstock, Saxony, Germany; in pegmatites at Mt. Bity and Ampangabe, Madagascar; and at Brejauba, Minas Geraes, Brazil.

BEST REF. IN ENGLISH: Palache, et al., "Dana's System of Mineralogy," 7th Ed., v. II, p. 1050–1052, New York, Wiley, 1951.    Qurashi, M. M., and Barnes, W. H., *Am. Min.*, **37**: 423–426 (1952).

# PUMPELLYITE

$Ca_2 MgAl_2 (SiO_4)(Si_2 O_7)(OH)_2 \cdot H_2 O$

CRYSTAL SYSTEM: Monoclinic
CLASS: 2/m
SPACE GROUP: A2/m
Z: 2
LATTICE CONSTANTS:
  $a = 8.81$
  $b = 5.94$
  $c = 19.14$
  $\beta = 97.6°$
3 STRONGEST DIFFRACTION LINES:
  2.90 (100)
  3.79 ( 50)
  2.74 ( 50)

OPTICAL CONSTANTS:
  $\alpha = 1.674-1.702$
  $\beta = 1.675-1.715$
  $\gamma = 1.688-1.722$
  $(+)2V = 26°-85°$
HARDNESS: 6
DENSITY: 3.18–3.23 (Meas.)
CLEAVAGE: {001} distinct
          {100} distinct

HABIT: Crystals usually fibrous, or as narrow plates flattened {001}. As stellate needle-like clusters, or in dense mats of randomly oriented fibers. Twinning on {001}, {100}, common.

COLOR-LUSTER: Green, bluish green, brown. Translucent.

MODE OF OCCURRENCE: Occurs widespread as a relatively common mineral in rocks of widely different composition and geological environment. Found in amygdaloidal cavities in copper-bearing ores at Calument, Houghton County, Michigan; with lawsonite in glaucophane schist at many localities in California; in the trap rocks of northern New Jersey; and at localities in Scotland, France, Netherlands, Austria, Finland, USSR, Haiti, Japan, Borneo, New Zealand, South Africa, and elsewhere.

BEST REF. IN ENGLISH: Deer, Howie, and Zussman, "Rock Forming Minerals" v. I, p. 227–235, New York, Wiley, 1962.

# PURPURITE

$(Mn^{3+}, Fe^{3+})PO_4$
Heterosite-Purpurite Series

CRYSTAL SYSTEM: Orthorhombic
CLASS: 2/m 2/m 2/m
SPACE GROUP: Pmnb
Z: 4
LATTICE CONSTANTS:
  $a = 5.83$
  $b = 9.70$
  $c = 4.77$
3 STRONGEST DIFFRACTION LINES:
  2.952 (100)
  2.448 (100)
  4.37 ( 70)
OPTICAL CONSTANTS:
  $\alpha = 1.85$
  $\beta = 1.86$
  $\gamma = 1.92$
  2V = moderate
HARDNESS: 4–4½
DENSITY: 3.69 (Calc.)
CLEAVAGE: {100} good
          {010} imperfect
Fracture uneven. Brittle.

HABIT: Massive.

COLOR-LUSTER: Deep rose to reddish purple. Usually externally dark brown to brownish black, due to alteration. Subtranslucent to opaque. Dull to satiny. Streak reddish purple.

MODE OF OCCURRENCE: Occurs as an alteration of lithiophilite in the zone of weathering in granite pegmatites. Found at the Custer Mountain lode, Custer County, and at other places in the Black Hills, South Dakota; at Pala and Rincon, San Diego County, California; at Kings Mountain, Gaston County, North Carolina; at Chanteloube, France, at Feiteira, Portugal; and at Wodgina, Western Australia.

BEST REF. IN ENGLISH: Palache, et al., "Dana's System of Mineralogy," 7th Ed., v. II, p. 675-677, New York, Wiley, 1951.

**PYCNOCHLORITE** = Variety of chamosite with Si = 2.8-3.1 and $Fe^{2+}/Fe^{2+} + Mg = 0.2$-0.5

## PYRARGYRITE
$Ag_3SbS_3$

CRYSTAL SYSTEM: Hexagonal
CLASS: $\bar{3}$ 2/m
SPACE GROUP: R3c
Z: 6 (hexag.), 2 (rhomb.)
LATTICE CONSTANTS:
(hexag.)     (rhomb.)
$a = 11.03$   $a_{rh} = 7.0$
$c = 8.72$    $\alpha = 103°58'$
3 STRONGEST DIFFRACTION LINES:
2.779 (100)
2.566 ( 90)
3.22 ( 90)
OPTICAL CONSTANTS:
$\omega = 3.084$
$\epsilon = 2.881$ (Li)
(–)
HARDNESS: 2½
DENSITY: 5.85 (Meas.)
          5.82 (Calc.)
CLEAVAGE: $\{10\bar{1}1\}$ distinct
            $\{01\bar{1}2\}$ indistinct
Fracture conchoidal to uneven. Brittle.
HABIT: Crystals commonly prismatic; hemimorphic development sometimes distinct. Also steep scalenohedral. Usually massive; compact, disseminated, or as crusts. Twinning on $\{10\bar{1}4\}$ common; on $\{10\bar{1}1\}$, $\{01\bar{1}2\}$, and about $[11\bar{2}0]$ less common.
COLOR-LUSTER: Deep red. Translucent. Adamantine to submetallic. Streak dark red.
MODE OF OCCURRENCE: Occurs in low-temperature hydrothermal vein deposits, commonly associated with other sulfosalts, silver, galena, pyrite, calcite, dolomite, and quartz. Found widespread in the western United States, especially in Colorado, Idaho, Nevada, and California. It also occurs in British Columbia and Ontario, Canada; at numerous localities in Mexico, Chile, and Bolivia; in Spain and Czechoslovakia; and as exceptional crystals in the Harz Mountains and in Saxony, Germany.
BEST REF. IN ENGLISH: Palache, et al., "Dana's System of Mineralogy," 7th Ed., v. I, p. 362-366, New York, Wiley, 1944.

## PYRITE
$FeS_2$
Dimorphous with marcasite

CRYSTAL SYSTEM: Cubic
CLASS: 2/m $\bar{3}$
SPACE GROUP: Pa3
Z: 4
LATTICE CONSTANT:
$a = 5.405$
3 STRONGEST DIFFRACTION LINES:
1.6332 (100)
2.709 ( 85)
2.423 ( 65)
OPTICAL CONSTANTS: Opaque.
HARDNESS: 6-6½
DENSITY: 5.00-5.028 (Meas.)
            5.013 (Calc.)
CLEAVAGE: $\{100\}$ indistinct
           $\{311\}$ very indistinct
           $\{110\}$ occasional parting
Fracture conchoidal to uneven. Brittle.
HABIT: Crystals usually cubic, pyritohedral, octahedral, or combinations of these forms; frequently abnormally developed; sometimes acicular. Parallel striations common on cubes and pyritohedrons. Frequently massive or disseminated; granular, reniform, botryoidal, stalactitic. Pyritohedrons sometimes twinned on $\{011\}$ producing "iron cross" twins.
COLOR-LUSTER: Pale brass yellow; often with iridescent tarnish. Opaque. Metallic.
MODE OF OCCURRENCE: Occurs as the most abundant and widespread sulfide mineral in rocks of all ages. Found in low- to high-temperature vein deposits; in igneous rocks and pegmatites; as magmatic segregations; in contact metamorphic ore deposits and metamorphosed sedimentary rocks; in coal beds, shale, limestone, and other sedimentary deposits; and rarely as a sublimation product. Notable localities occur throughout the United States, Canada, Mexico, Chile, Bolivia, Peru, England, Italy (especially on the island of Elba), France, Germany, Czechoslovakia, Spain, Norway, Sweden, Japan, Tasmania, and elsewhere.
BEST REF. IN ENGLISH: Palache, et al., "Dana's System of Mineralogy," 7th Ed., v. I, p. 282-290, New York, Wiley, 1944.

## PYROAURITE
$Mg_6Fe_2^{3+}CO_3(OH)_{16} \cdot 4H_2O$
Dimorphous with sjogrenite

CRYSTAL SYSTEM: Hexagonal
CLASS: $\bar{3}$ 2/m
SPACE GROUP: R$\bar{3}$m
Z: 3
LATTICE CONSTANTS:
$a = 6.19$
$c = 46.54$
3 STRONGEST DIFFRACTION LINES:
7.77 (100)
3.89 ( 80)
2.62 ( 50)

OPTICAL CONSTANTS:
$\omega = 1.564$
$\epsilon = 1.543$
$(-)$
HARDNESS: 2½
DENSITY: 2.12 (Meas.)
     2.12 (Calc.)
CLEAVAGE: $\{0001\}$ perfect
Laminae flexible, inelastic.
HABIT: Crystals thin to thick tabular, flattened on $\{0001\}$. Also fibrous.
COLOR-LUSTER: Yellowish white, brownish white, silvery white, greenish, colorless. Transparent. Vitreous to waxy, pearly.
MODE OF OCCURRENCE: Occurs chiefly in serpentine or as a low-temperature hydrothermal vein mineral. Found at Blue Mont, Maryland; in a dolomitic rock at Rutherglen, Ontario, Canada; and at localities in Scotland (igelstromite), Sweden, Italy, Austria, Greece, and USSR.
BEST REF. IN ENGLISH: Palache, et al., "Dana's System of Mineralogy," 7th Ed., v. I, p. 656-658, New York, Wiley, 1944.  Allmann, R., *Am. Min.*, **54**: 296-299 (1969).

# PYROBELONITE
$PbMnVO_4OH$

CRYSTAL SYSTEM: Orthorhombic
CLASS: 2/m 2/m 2/m
SPACE GROUP: Pnam
Z: 4
LATTICE CONSTANTS:
$a = 7.644$
$b = 9.508$
$c = 6.182$
3 STRONGEST DIFFRACTION LINES:
3.252 (100)
5.172 ( 70)
2.643 ( 70)
OPTICAL CONSTANTS:
$\alpha = 2.32$
$\beta = 2.36$
$\gamma = 2.37$
$(-)2V = 29°$
HARDNESS: 3½
DENSITY: 5.377 (Meas.)
     5.39 (Calc.)
CLEAVAGE: Not observed.
Fracture conchoidal. Brittle.
HABIT: Crystals acicular [001].
COLOR-LUSTER: Scarlet red. Transparent. Adamantine. Streak orangish or reddish.
MODE OF OCCURRENCE: Occurs as minute grains and crystals associated with hausmannite, pyrochroite, manganite, calcite, and barite at Långban, Sweden.
BEST REF. IN ENGLISH: Palache, et al., "Dana's System of Mineralogy," 7th Ed., v. 2, p. 815-816, New York, Wiley, 1951.  Barnes, W. H., and Ahmed, F. R., *Can. Min.*, **10**(Pt. 1): 117-123 (1969).

# PYROCHLORE
$(Na, Ca, U)_2 (Nb, Ta, Ti)_2 O_6 (OH, F)$
Pyrochlore-Microlite Series
Var. Ellsworthite
Hatchettolite

CRYSTAL SYSTEM: Cubic
CLASS: $4/m \bar{3} 2/m$
SPACE GROUP: Fd3m
Z: 8
LATTICE CONSTANT:
$a = 10.35\text{-}10.43$
3 STRONGEST DIFFRACTION LINES:
3.00 (100)
1.838 ( 60)
1.568 ( 50)
OPTICAL CONSTANT:
$N = 1.96\text{-}2.01$
HARDNESS: 5-5½ (Lower when metamict.)
DENSITY: 4.48 (Meas.)
     4.31 (Calc.)
CLEAVAGE: $\{111\}$ sometimes distinct, may be parting Fracture subconchoidal to uneven. Brittle.
HABIT: Crystals octahedral, up to 1 cm or more in size, often with modifying faces of $\{011\}$, $\{113\}$, or $\{001\}$. Also as embedded grains or irregular masses. Twinning on $\{111\}$ not common.
COLOR-LUSTER: Yellowish brown, reddish brown, brown to black. Translucent to opaque. Vitreous to resinous. Not fluorescent.
MODE OF OCCURRENCE: Occurs as a primary mineral in carbonatites and pegmatites, and as an accessory mineral in nepheline syenites and other alkalic rocks. Found in the United States associated with zircon, astrophyllite, and fluorite in a riebeckite syenite at St. Peters Dome, El Paso County, and in carbonatite deposits in the Powderhorn area, Gunnison County, Colorado; also in Maine, North Carolina, California, and Wisconsin. In Canada it occurs at the Verity Property, Blue River, British Columbia; as superb crystals in carbonatite at Oka, Quebec; as large masses of uranoan variety (ellsworthite) at the McDonald mine, Hybla, and as the variety hatchettolite at the Woodcox mine, Hastings County, Ontario. It is also found in Greenland, France, Germany, Italy, USSR, Sweden, at many places in Norway, and as excellent crystals, sometimes twinned, at Mbeya, Tanzania.
BEST REF. IN ENGLISH: Frondel, Clifford, USGS Bull. 1064, p. 326-333 (1958).

# PYROCHROITE
$Mn(OH)_2$

CRYSTAL SYSTEM: Hexagonal
CLASS: $\bar{3} 2/m$
SPACE GROUP: P$\bar{3}$m1
Z: 1
LATTICE CONSTANTS:
$a = 3.34$
$c = 4.68$

3 STRONGEST DIFFRACTION LINES:
2.45 (100)
4.72 ( 90)
1.826 ( 60)

OPTICAL CONSTANTS:
$\omega = 1.723$
$\epsilon = 1.681$
(−)

HARDNESS: 2½

DENSITY: 3.25 (Meas.)
3.274 (Calc.)

CLEAVAGE: {0001} perfect
Thin foliae flexible.

HABIT: Crystals tabular, rhombohedral, or rarely prismatic. Commonly massive, foliated, or as thin veinlets.

COLOR-LUSTER: Colorless, white, pale green, or blue; on exposure becomes dark brown and then black. Transparent in thin fragments, becoming opaque with alteration. Pearly on cleavages when fresh, becoming dull when altered.

MODE OF OCCURRENCE: Occurs chiefly as a low-temperature hydrothermal mineral, commonly associated with hausmannite, calcite, dolomite, rhodochrosite. Found as a major component of a boulder of manganese ore near Alum Rock Park, Santa Clara County, California; at Franklin and Sterling Hill, Sussex County, New Jersey; at Långban and elsewhere in Sweden; near Prijedor, Bosnia, Yugoslavia; and at Gonzen, St. Gall, Switzerland.

BEST REF. IN ENGLISH: Palache, et al., "Dana's System of Mineralogy," 7th Ed., v. I, p. 639–641, New York, Wiley, 1944.

# PYROLUSITE

$MnO_2$
Dimorphous with ramsdellite

CRYSTAL SYSTEM: Tetragonal
CLASS: 4/m 2/m 2/m
SPACE GROUP: $P4_2/mnm$
Z: 2

LATTICE CONSTANTS:
$a = 4.42$
$c = 2.87$ (variable)

3 STRONGEST DIFFRACTION LINES:
3.14 (100)
2.41 ( 50)
1.63 ( 50)

OPTICAL CONSTANTS: Opaque.

HARDNESS: 6-6½ (crystals)
2-6 (massive)

DENSITY: 5.06 (Meas.)
5.148 (Calc.)

CLEAVAGE: {110} perfect
Fracture uneven. Brittle.

HABIT: Crystals (polianite) short to long prismatic, elongated parallel to c-axis; also equant. Commonly massive, compact, columnar or fibrous; concretionary; stalactitic; granular to pulverulent; dendritic. Frequently pseudomorphous after manganite. Twinning on {031} and {032}, rare.

COLOR-LUSTER: Black or dark steel gray. Opaque. Metallic to dull. Streak black or bluish black.

MODE OF OCCURRENCE: Occurs as a secondary mineral resulting from the alteration of manganite or other manganese-bearing minerals. Found widespread throughout the United States, especially in California, Nevada, Arizona, New Mexico, Montana, Colorado, South Dakota, Texas, Arkansas, Tennessee, Georgia, Virginia, Connecticut, Massachusetts, New Hampshire, and Vermont. Other notable deposits occur in Canada, Mexico, Cuba, Brazil, France, Spain, Germany, Czechoslovakia, Roumania, USSR, India, Ghana, and elsewhere.

BEST REF. IN ENGLISH: Palache, et al., "Dana's System of Mineralogy," 7th Ed., v. I, p. 562–566, New York, Wiley, 1944.

# PYROMORPHITE

$Pb_5(PO_4)_3Cl$
Pyromorphite Series

CRYSTAL SYSTEM: Hexagonal
CLASS: 6/m
SPACE GROUP: $P6_3/m$
Z: 2

LATTICE CONSTANTS:
$a = 10.00$
$c = 7.33$

3 STRONGEST DIFFRACTION LINES:
2.957 (100)
2.990 ( 80)
2.065 ( 70)

OPTICAL CONSTANTS:
$\omega = 2.058$
$\epsilon = 2.048$ (Na)
(−)

HARDNESS: 3½-4

DENSITY: 7.04 (Meas.)
7.10 (Calc.)

CLEAVAGE: {10$\bar{1}$1} trace
Fracture uneven to subconchoidal. Brittle.

HABIT: Crystals short hexagonal prisms, commonly barrel-shaped and often cavernous; also equant, tabular, or pyramidal. As branching groups of prismatic crystals in nearly parallel position, tapering to slender point. Commonly globular, reniform, and botryoidal with subcolumnar structure; also granular, fibrous, earthy.

COLOR-LUSTER: Various shades of green, yellow, orange, brown, gray; rarely colorless or white. Transparent to translucent. Subadamantine to resinous. Streak white.

MODE OF OCCURRENCE: Occurs widespread as a secondary mineral in the oxidation zone of lead-bearing ore deposits. Found as fine specimens in the Coeur d'Alene district, Shoshone County, Idaho; at the Wheatley mines, Phoenixville, Chester County, and the Ecton and Perkiomen mines, Montgomery County, Pennsylvania; at Galena and Carbonate, Lawrence County, South Dakota; and at deposits in Colorado, New Mexico, Utah, California, and other western states. Exceptional specimens also are found at the Society Girl claim, Moyie, British Columbia, Canada; at Bad Ems and other places in Germany; and at localities in Mexico; England, Scotland, France, Italy, Czechoslovakia, USSR, Zaire, Algeria, Burma, and at Broken Hill, New South Wales, Australia.

BEST REF. IN ENGLISH: Palache, et al., "Dana's System of Mineralogy," 7th Ed., v. II, p. 889–895, New York, Wiley, 1951.

## PYROPE
$Mg_3 Al_2 Si_3 O_{12}$
Garnet Group
Var. Rhodolite

CRYSTAL SYSTEM: Cubic
CLASS: $4/m\,\bar{3}\,2/m$
SPACE GROUP: Ia3d
Z: 8
LATTICE CONSTANT:
  $a = 11.459$
3 STRONGEST DIFFRACTION LINES:
  2.58 (100)
  1.54 (100)
  2.88 ( 60)
  2.46 ( 30)*
*This line distinguishes pyrope from other garnets.
OPTICAL CONSTANT:
  $N = 1.714$
HARDNESS: 7–7½
DENSITY: 3.5–3.8 (Meas.)
CLEAVAGE: None.
          {110} parting sometimes distinct.
Fracture conchoidal. Brittle.
  HABIT: Crystals rare, usually dodecahedrons or trapezohedrons; often as rounded pebbles or embedded grains.
  COLOR-LUSTER: Pinkish red, purplish red, orange-red, crimson, to nearly black. Transparent to translucent. Vitreous.
  MODE OF OCCURRENCE: Occurs in peridotites and associated serpentinites, and in sands and gravels derived from them; also in eclogites, hornblende-garnet-plagioclase rocks, and in certain Precambrian anorthosites. Found in San Juan County, Utah; Apache County, Arizona; McKinley County, New Mexico; near Murfreesboro, Pike County, Arkansas; and in Macon County, North Carolina (rhodolite). Other notable occurrences are found in Bohemia, Czechoslovakia; Madras, India; in the diamond-bearing peridotites of South Africa; and in Queensland and New South Wales, Australia.
  BEST REF. IN ENGLISH: Deer, Howie, and Zussman, "Rock Forming Minerals," v. 1, p. 77–112, New York, Wiley, 1962.

## PYROPHANITE
$MnTiO_3$
Ilmenite Group

CRYSTAL SYSTEM: Hexagonal
CLASS: $\bar{3}$
SPACE GROUP: $R\bar{3}$
Z: 2 (rhomb.), 6 (hexag.)
LATTICE CONSTANTS:
  (hexag.)          (rhomb.)
  $a = 5.137$       $a_{rh} = 5.62$
  $c = 14.29$       $\alpha = 54°16'$

3 STRONGEST DIFFRACTION LINES:
  2.78  (100)
  2.57  ( 90)
  1.745 ( 80)
OPTICAL CONSTANTS:
  $\omega = 2.481$
  $\epsilon = 2.21$  (Na)
    (-)
HARDNESS: 5–6
DENSITY: 4.54 (Meas.)
         4.58 (Calc.)
CLEAVAGE: {02$\bar{2}$1} perfect
          {10$\bar{1}$2} distinct
  HABIT: As small tabular crystals and embedded grains; also fine scaly.
  COLOR-LUSTER: Deep blood red to greenish yellow. Metallic to submetallic.
  MODE OF OCCURRENCE: Occurs as minute grains embedded in garnet and quartz at the Piquery mine, Queluz district, Minas Geraes, Brazil; associated with pennantite at the Benallt mine, Rhiw, Caernarvonshire, Wales; in drusy cavities in manganese ore associated with garnet, ganophyllite, manganophyllite, and calcite at the Harstig mine, Pajsberg, Sweden; and widely distributed in Japanese manganese ores.
  BEST REF. IN ENGLISH: Palache, et al., "Dana's System of Mineralogy," 7th Ed., v. I, p. 535–541, New York, Wiley, 1944.

## PYROPHYLLITE
$Al_2 Si_4 O_{10}(OH)_2$

CRYSTAL SYSTEM: Monoclinic
CLASS: 2/m
SPACE GROUP: C2/c
Z: 4
LATTICE CONSTANTS:
  $a = 5.16$
  $b = 8.90$
  $c = 18.64$
  $\beta = 99°55'$
3 STRONGEST DIFFRACTION LINES:
  3.08 (100)
  9.21 ( 60)
  4.58 ( 50)
OPTICAL CONSTANTS:
  $\alpha = 1.534–1.556$
  $\beta = 1.586–1.589$
  $\gamma = 1.596–1.601$
  $(-)2V = 53°–62°$
HARDNESS: 1–2
DENSITY: 2.65–2.90 (Meas.)
         2.844 (Calc.)
CLEAVAGE: {001} perfect
Laminae flexible, inelastic. Feel greasy.
  HABIT: Crystals subhedral, tabular parallel to {010} and much elongated; often curved and distorted. Usually foliated, radiated lamellar or fibrous; also granular to compact.
  COLOR-LUSTER: White, grayish white, yellowish, pale

blue, greenish, grayish or brownish green. Transparent to translucent. Pearly to dull and glistening.

MODE OF OCCURRENCE: Occurs chiefly in schistose rocks, often in association with andalusite, kyanite, sillimanite, and lazulite, or in hydrothermal veins with quartz and micas. Typical occurrences are found in Madera, Mariposa, Mono, San Bernardino, and San Diego counties, California; in Mohave and Yuma counties, Arizona; in Schuylkill County, Pennsylvania; as extensive beds at Deep River, Guilford County, and at other places in North Carolina; in the Chesterfield district, Chesterfield County, South Carolina; and at Graves Mountain, Lincoln County, Georgia. Also found in Minas Geraes, Brazil, and at localities in Mexico, Sweden, Belgium, Switzerland, USSR, South Africa, Korea, Japan, and elsewhere.

BEST REF. IN ENGLISH: Deer, Howie, and Zussman, "Rock Forming Minerals," v. 3, p. 115–120, New York, Wiley, 1962.

# PYROSMALITE
$(Mn, Fe)_{14}Si_3O_7(OH, Cl)_6$

CRYSTAL SYSTEM: Hexagonal
CLASS: $\bar{3}\,2/m$
SPACE GROUP: $P\bar{3}m1$
LATTICE CONSTANTS:
$a = 13.36$
$c = 7.16$
3 STRONGEST DIFFRACTION LINES:
2.69 (100)
7.16 ( 80)
2.25 ( 70)
OPTICAL CONSTANTS:
$\omega = 1.675-1.682$
$\epsilon = 1.636-1.647$
HARDNESS: 4-4½
DENSITY: 3.06-3.19 (Meas.)
3.121 (Calc.)
CLEAVAGE: {0001} perfect
Fracture uneven. Brittle.
HABIT: Crystals thick hexagonal prisms or tabular; also massive, foliated.
COLOR-LUSTER: Dark green to pale brown; also externally gray, pale green, brown due to alteration. Translucent. Somewhat pearly.
MODE OF OCCURRENCE: Occurs in association with calcite, pyroxene, apophyllite, and magnetite at iron mines at Dannemora in Upsala, and in Nordmark, Sweden. Also found at Broken Hill, New South Wales, Australia.
BEST REF. IN ENGLISH: Winchell and Winchell, "Elements of Optical Mineralogy," 4th Ed., Pt. 2, p. 359, New York, Wiley, 1951.

# PYROSTILPNITE
$Ag_3SbS_3$

CRYSTAL SYSTEM: Monoclinic
CLASS: 2/m
SPACE GROUP: $P2_1/c$
Z: 4

LATTICE CONSTANTS:
$a = 6.83$ kX
$b = 15.81$
$c = 6.23$
$\beta = 117°9'$
3 STRONGEST DIFFRACTION LINES:
2.85 (100)
2.65 ( 50)
2.42 ( 50)
OPTICAL CONSTANTS: Indices very high.
HARDNESS: 2
DENSITY: 5.94 (Meas.)
5.97 (Calc.)
CLEAVAGE: {010} perfect
Fracture conchoidal. Thin plates somewhat flexible.
HABIT: Crystals tabular, commonly forming flat rhombs; also lath-like. Striated on {010}. Often in sheaf-like aggregates resembling stilbite. Twinning on {100} common.
COLOR-LUSTER: Hyacinth red. Translucent. Adamantine. Streak yellowish orange.
MODE OF OCCURRENCE: Occurs in low-temperature hydrothermal vein deposits, commonly associated with ruby silvers and other silver minerals. Found sparingly at the California Rand mine, San Bernardino County, California; in the Silver City district, Owyhee County, Idaho; at Chanarcillo, Chile; with pyrargyrite and proustite at Andreasberg, Harz, Germany; at Příbram, Czechoslovakia; and at the Long Tunnel mine, Heazlewood, Tasmania.
BEST REF. IN ENGLISH: Palache, et al., "Dana's System of Mineralogy," 7th Ed., v. I, p. 369–370, New York, Wiley, 1944.

# PYROXENE = Mineral group of general formula: $ABSi_2O_6$, A = Na, Ca, Mg, $Fe^{2+}$, B = Mg, $Fe^{2+}$, Al

See: Aegirine (Acmite)
Augite
Bustamite
Clinoenstatite
Clinoferrosilite
Clinohypersthene
Diopside
Enstatite
Fassaite
Hedenbergite
Hypersthene
Johannsenite
Omphacite
Orthoferrosilite
Pigeonite
Rhodonite
Spodumene
Ureyite

# PYROXFERROITE
$(Fe, Ca, Mg, Mn)SiO_3$

CRYSTAL SYSTEM: Triclinic
CLASS: 1 or $\bar{1}$
SPACE GROUP: P1 or $P\bar{1}$
Z: 14
LATTICE CONSTANTS:
$a = 6.62$ $\alpha = 114.4°$
$b = 7.54$ $\beta = 82.7°$
$c = 17.35$ $\gamma = 94.5°$
3 STRONGEST DIFFRACTION LINES:
2.934 (100)
2.674 ( 60)
3.09 ( 45)

OPTICAL CONSTANTS:
$\alpha = 1.753$
$\beta = 1.755$
$\gamma = 1.766$
$(+)2V = 35°$
HARDNESS: Not determined.
DENSITY: 3.68–3.76 (Meas.)
3.82–3.83 (Calc.)
CLEAVAGE: Not determined.
HABIT: As discrete embedded grains.
COLOR-LUSTER: Yellow.
MODE OF OCCURRENCE: Occurs in rocks collected by the Apollo XI mission to the Sea of Tranquillity, Moon. The rocks are microgabbros or diabases with major clinopyroxene, plagioclase, and ilmenite.
BEST REF. IN ENGLISH: *Geochim. Cosmochim. Acta. Suppl. Proc. Apollo XI Lunar Sci. Conf.* **1**: 65–79 (1970).

# PYROXMANGITE
(Mn, Fe)SiO$_3$

CRYSTAL SYSTEM: Triclinic
CLASS: $\bar{1}$
SPACE GROUP: P$\bar{1}$
Z: 14
LATTICE CONSTANTS:
$a = 7.56$ $\alpha = 84.0°$
$b = 17.45$ $\beta = 94.3°$
$c = 6.67$ $\gamma = 113.7°$
3 STRONGEST DIFFRACTION LINES:
2.95 (100)
2.18 ( 70)
3.12 ( 60)
OPTICAL CONSTANTS:
$\alpha = 1.726$–1.748
$\beta = 1.728$–1.750
$\gamma = 1.744$–1.764
$(+)2V = 35°$–46°
HARDNESS: 5½–6
DENSITY: 3.61–3.80 (Meas.)
3.91 (Calc.)
CLEAVAGE: {110} perfect
{1$\bar{1}$0} perfect
{010} poor
{001} poor
Brittle.
HABIT: Crystals commonly tabular. Usually massive, cleavable to compact, also in embedded grains. Twinning on {110} lamellar; on {001}, simple; not common.
COLOR-LUSTER: Pale pink to rose pink and purplish pink, also yellowish red-brown to brown to dark brown due to alteration. Transparent to translucent. Vitreous to pearly on fresh surfaces.
MODE OF OCCURRENCE: Occurs in metamorphic or metasomatic rocks in association with rhodochrosite, spessartine, alabandite, tephroite, alleghanyite, and hausmannite; also in impure quartzites and in grunerite-garnet schist. Found at the Big Indian manganese deposit near Randsburg, Kern County, California; near Boise, Idaho; at Iva, South Carolina; and at localities in Scotland, Sweden, Finland, Japan, and at Broken Hill, New South Wales, Australia.
BEST REF. IN ENGLISH: Deer, Howie, and Zussman, "Rock Forming Minerals," v. 2, p. 196–202, New York, Wiley, 1963.

# PYRRHOTITE
Fe$_{1-x}$S

CRYSTAL SYSTEM: Orthorhombic, hexagonal, monoclinic
Z: 64
LATTICE CONSTANTS:
(mono.)
$a = 11.880$
$b = 6.852$
$c = 22.738$
$\beta = 90°33'$
3 STRONGEST DIFFRACTION LINES:
(Mono.)
2.057 (100)
2.966 ( 90)
2.635 ( 90)
OPTICAL CONSTANTS: Opaque.
HARDNESS: 3½–4½
DENSITY: 4.53–4.77 (Meas.)
CLEAVAGE: None.
Basal parting sometimes distinct. Fracture subconchoidal to uneven. Brittle.
HABIT: As well-formed platy to tabular or bipyramidal crystals, single or twinned on {10$\bar{1}$2}. Usually massive, as granular aggregates.
COLOR-LUSTER: Bronze yellow to bronze red; often tarnishes dark brown. Opaque. Metallic. Streak grayish black.

Often highly magnetic.
MODE OF OCCURRENCE: Occurs chiefly as magmatic segregations in basic igneous rocks such as gabbros, peridotites, and norites, commonly associated with pyrite and other sulfides. It is found also in veins, metamorphic deposits, and rarely in pegmatites. Important localities include Ducktown, Tennessee; Lancaster County, Pennsylvania; Standish, Maine; Homestake mine, Lead, South Dakota; Sudbury, Ontario, Canada; Chihuahua, Mexico; Llallagua, Bolivia; Morro Velho mine, Minas Geraes, Brazil; Kongsberg, Norway; Trepča, Yugoslavia; Kisbánya, Roumania; Trentino, Italy; Andreasberg, Bodenmais, Freiberg, and Schneeberg, Germany.
BEST REF. IN ENGLISH: Palache, et al., "Dana's System of Mineralogy," 7th Ed., v. I, p. 231–235, New York, Wiley, 1944.

# α-QUARTZ
$SiO_2$

### CRYSTALLINE VARIETIES:

*Rock Crystal:* Colorless, transparent. Excellent crystals common.

*Amethyst:* Various shades of purple or violet. Transparent to translucent.

*Citrine:* Pale yellow, yellow, yellow-brown, reddish brown. Transparent to translucent. Often erroneously called "topaz."

*Smoky Quartz:* Pale smoky brown to almost black. Transparent to nearly opaque. Also called cairngorm or morion.

*Rose Quartz:* Pale pink to deep rose red, occasionally with purplish tinge. Sometimes exhibits distinct asterism. Rarely in euhedral crystals.

*Milky Quartz:* Milky white to grayish white. Translucent to nearly opaque. Luster often greasy.

*Blue Quartz:* Pale blue, grayish blue to lavender blue. Translucent. Somewhat chatoyant.

### CRYSTALLINE VARIETIES WITH INCLUSIONS:

*Rutilated Quartz:* Contains distinct acicular rutile crystals in sprays or random orientation.

*Aventurine:* Spangled with glistening scales of micaceous minerals such as mica or hematite.

*Tiger-eye:* Finely fibrous, chatoyant. Formed by complete or partial replacement of asbestos, especially crocidolite. Yellow, yellow-brown, brown, reddish brown, bluish; also grayish green, green.

*Ferruginous Quartz:* Contains inclusions of goethite or hematite. Brown to red.

A variety of other minerals are found as inclusions in quartz crystals, including tourmaline, various amphiboles, stibnite, pyrite, scheelite, gold, copper, and epidote.

### CRYPTOCRYSTALLINE VARIETIES:

*Chalcedony:* Composed of microscopic fibers. Often mammillary, botryoidal, stalactitic; concretionary; as geodes; in veinlets; as pseudomorphs after other minerals; as a replacement of organic materials. Color white, grayish, blue, brown, black. Luster waxy. Transparent to translucent. Often fluorescent in various shades of green.

*Agate:* Distinctly banded chalcedony with successive layers differing in color.

*Onyx:* Banded agate with layers in parallel straight lines. Often used for cameos.

*Moss Agate:* Gray, bluish, or milky transparent to translucent chalcedony containing dendritic inclusions.

*Sard:* Uniformly colored light brown to dark brown translucent chalcedony.

*Carnelian:* Uniformly colored blood red, flesh red to reddish brown translucent chalcedony.

*Chrysoprase:* Apple green translucent subvariety of chalcedony.

*Plasma:* Microgranular or microfibrous variety of quartz colored various shades of green.

*Prase:* Leek green translucent variety of quartz.

*Heliotrope:* Greenish variety of chalcedony or plasma containing red spots of iron oxide.

*Bloodstone:* Greenish variety of chalcedony or plasma containing red spots of iron oxide.

*Flint:* Compact microcrystalline variety of quartz commonly found as nodules in chalk or marly limestone.

*Chert:* Compact microcrystalline variety of quartz commonly found in bedded deposits or as lenses.

*Novaculite:* Compact microcrystalline variety of quartz of uniform texture frequently used for whetstones.

CRYSTAL SYSTEM: Hexagonal

CLASS: 3 2

SPACE GROUP: $P3_1 2$ (Right-handed),
$P3_2 2$ (Left-handed)

Z: 3

LATTICE CONSTANTS:
$a = 4.90$
$c = 5.39$

3 STRONGEST DIFFRACTION LINES:
3.343 (100)
4.26 ( 35)
1.817 ( 17)

OPTICAL CONSTANTS:
$\omega = 1.544$
$\epsilon = 1.553$
(+)

HARDNESS: 7

DENSITY: 2.65 (Meas.)
2.647 (Calc.)

CLEAVAGE: Seldom distinct; observed on $r\{10\bar{1}1\}$,

$z\{01\bar{1}1\}$, $m\{10\bar{1}0\}$, $c\{0001\}$, $a\{11\bar{2}0\}$, $s\{11\bar{2}1\}$, and $x\{5\bar{1}\bar{6}1\}$.

FRACTURE: Conchoidal to subconchoidal (crystals); flat conchoidal to uneven or splintery (massive); uneven, hackly or splintery (fibrous). Brittle; cryptocrystalline varieties tough.

HABIT: Crystals hexagonal; enantiomorphic. Short to long prismatic, usually elongated parallel to $c$-axis with $m\{10\bar{1}0\}$ faces horizontally striated; often terminated by rhombohedrons $r\{10\bar{1}1\}$ and $z\{01\bar{1}1\}$. Often double 6-sided pyramids (quartzoids). Crystals frequently distorted, twisted, bent, skeletal. Often in druses; also isolated. Commonly massive, coarse granular to cryptocrystalline; stalactitic; concretionary; as sand. Twinning very common; principal twin laws: Dauphiné, Brazilian, Combined, and Japanese.

COLOR-LUSTER: Usually colorless, white; also shades of gray, yellow, purple, violet, pink, brown, nearly black. (See Varieties.) Transparent to translucent. Vitreous; sometimes greasy or waxy. Streak white; faintly tinted in colored varieties.

MODE OF OCCURRENCE: Quartz, found in rocks of all ages and in many ore deposits, is the most common of all minerals. It is a very important rock-forming mineral, being an essential constituent of many igneous, metamorphic, and sedimentary rocks. Also found abundantly in unconsolidated sand and gravel deposits. Notable occurrences are found in California, Arizona, Montana, Wyoming, South Dakota, Colorado, Arkansas, Iowa, Michigan, Maine, New York, Pennsylvania, Virginia, and North Carolina. Also in Canada, Mexico, Brazil, Uruguay, Scotland, England, France, Italy, Switzerland, Germany, Austria, Czechoslovakia, Roumania, USSR, South Africa, Madagascar, Ceylon, India, Japan, Australia, and elsewhere.

BEST REF. IN ENGLISH: Frondel, Clifford, "Dana's System of Mineralogy," 7th Ed., v. III, p. 9–250, (1962).

## β-Quartz
SiO$_2$
Stable between 573° and 870°C

CRYSTAL SYSTEM: Hexagonal
CLASS: 6 2 2
SPACE GROUP: C6$_2$2 or C6$_4$2
Z: 3
LATTICE CONSTANTS:
(at 575°C)
$a = 4.999$
$c = 5.457$
3 STRONGEST DIFFRACTION LINES:
3.40 (100)
4.34 ( 20)
1.837 ( 10)
OPTICAL CONSTANTS:
$\omega = 1.5329$ (Na) at 580°C
$\epsilon = 1.5405$
(+)
HARDNESS: 7
DENSITY: ~2.53 at 600°C
CLEAVAGE: $\{10\bar{1}1\}$ good
$\{10\bar{1}0\}$ parting
Fracture conchoidal. Brittle.

HABIT: Crystals always α-quartz paramorphs; typically dipyramidal; often rounded and rough. Enantiomorphous. Twinning on $\{10\bar{1}1\}$, Esterel law, very common; on $\{11\bar{2}2\}$, Verespatak law, less common; other twin laws rare.

COLOR-LUSTER: White, milky white, gray, colorless. Transparent to nearly opaque. Vitreous. Usually fractured internally.

MODE OF OCCURRENCE: Occurs mainly in cavities or as phenocrysts in rhyolite and other acidic rocks. Typical occurrences are found in the western United States, England, France, Germany, Roumania, USSR, Japan, and elsewhere.

BEST REF. IN ENGLISH: Frondel, Clifford, "Dana's System of Mineralogy," 7th Ed., v. III, p. 251–258, New York, Wiley, 1962.

## QUENSELITE
PbMnO$_2$ · OH

CRYSTAL SYSTEM: Monoclinic
CLASS: 2
SPACE GROUP: P2
Z: 4
LATTICE CONSTANTS:
$a = 5.61$
$b = 5.68$
$c = 9.16$
$\beta = 93°18'$
3 STRONGEST DIFFRACTION LINES:
3.04 (100)
2.72 ( 80)
3.68 ( 70)
OPTICAL CONSTANTS:
$\alpha \simeq 2.30$
(+)2V (?)
HARDNESS: 2½
DENSITY: 6.84 (Meas.)
7.091 (Calc.)
CLEAVAGE: $\{001\}$ perfect
Thin laminae, flexible.

HABIT: Crystals tabular, sometimes slightly elongated. Striated on $\{011\}$.

COLOR-LUSTER: Black. Opaque. Metallic to adamantine. Streak dark brownish gray.

MODE OF OCCURRENCE: Occurs associated with barite and calcite at Långban, Sweden.

BEST REF. IN ENGLISH: Palache, et al., "Dana's System of Mineralogy," 7th Ed., v. I, p. 729–730, New York, Wiley, 1944.

## QUENSTEDTITE
Fe$_2$(SO$_4$)$_3$ · 10H$_2$O

CRYSTAL SYSTEM: Triclinic
CLASS: $\bar{1}$
SPACE GROUP: P$\bar{1}$
Z: 2

LATTICE CONSTANTS:
$a = 6.15$     $\alpha = 94°10'$
$b = 23.77$   $\beta = 101°44.5'$
$c = 6.56$     $\gamma = 96°18.5'$

3 STRONGEST DIFFRACTION LINES:
4.08 (100)
5.78 ( 80)
4.19 ( 80)

OPTICAL CONSTANTS:
$\alpha = 1.547$
$\beta = 1.566$ (Na)
$\gamma = 1.594$
$(+)2V = 70°$

HARDNESS: 2½

DENSITY: 2.147 (Meas.)
2.14 (Calc.)

CLEAVAGE: {010} perfect
{100} good

HABIT: Usually as minute highly modified tabular or short prismatic crystals in aggregates. Twinning on {010} common.

COLOR-LUSTER: Pale violet to reddish violet. Transparent.

Soluble in water.

MODE OF OCCURRENCE: Occurs with copiapite and coquimbite at Alcaparrosa, and at Tierra Amarilla, Chile.

BEST REF. IN ENGLISH: Palache, et al., "Dana's System of Mineralogy," 7th Ed., v. II, p. 535–536, New York, Wiley, 1951.

**QUERCYITE** = Carbonate-apatite

**QUETENITE** = Botryogen

## RABBITTITE

$Ca_3 Mg_3 (UO_2)_2 (CO_3)_6 (OH)_4 \cdot 18H_2O$

CRYSTAL SYSTEM: Monoclinic
Z: 8
LATTICE CONSTANTS:
  $a = 32.6$
  $b = 23.8$
  $c = 9.45$
  $\beta = 90°$
3 STRONGEST DIFFRACTION LINES:
  8.24 (100)
  7.79 ( 80)
  4.37 ( 80)
OPTICAL CONSTANTS:
  $\alpha = 1.502$
  $\beta = 1.508$
  $\gamma = 1.525$
  (+?)2V
HARDNESS: ~2½
DENSITY: 2.57 (Meas.)
  CLEAVAGE: 2 perfect prismatic; {001} good
Not brittle.
  HABIT: As bundles of finely acicular crystals elongated parallel to c-axis; fibrous.
  COLOR-LUSTER: Pale green; silky. Fluoresces cream-yellow in short-wave ultraviolet light.
  MODE OF OCCURRENCE: Occurs as an efflorescence on a pillar of high-grade ore near the portal of the Lucky Strike No. 2 mine, San Rafael district, Emery County, Utah.
  BEST REF. IN ENGLISH: Thompson, Mary E., Weeks, Alice D., and Sherwood, Alexander M., *Am. Min.*, **40**: 201-212 (1955).

## RAGUINITE

$TlFeS_2$

CRYSTAL SYSTEM: Orthorhombic
Z: 8
LATTICE CONSTANTS:
  $a = 12.40$
  $b = 10.44$
  $c = 5.26$
3 STRONGEST DIFFRACTION LINES:
  2.89 (100)
  4.17 ( 60)
  3.35 ( 60)
  HARDNESS: Not determined.
  DENSITY: 6.4 (Meas.)
          6.29 (Calc.)
  CLEAVAGE: None.
  HABIT: As pseudohexagonal plates, pseudomorphs after an unknown mineral, consisting of bundles of fibers intimately mixed with pyrite.
  COLOR-LUSTER: Brilliant bronze.
  MODE OF OCCURRENCE: Occurs associated with lorandite, orpiment, realgar; and vrbaite at Allchar, Macedonia, Yugoslavia.
  BEST REF. IN ENGLISH: Johan, Zdenek, Picot, Paul, and Pierrot, Roland, *Am. Min.*, **54**: 1741 (1969). Laurent, Yvette; Picot, Paul; Pierrot, Roland; Permingeat, Francois, and Ivanov, T., *Am. Min.*, **54**: 1495 (1969).

## RALSTONITE

$NaMgAl(F, OH)_6 \cdot H_2O$

CRYSTAL SYSTEM: Cubic
CLASS: $4/m\ \bar{3}\ 2/m$
SPACE GROUP: Fd3m
Z: 8
LATTICE CONSTANT:
  $a = 9.89$
3 STRONGEST DIFFRACTION LINES:
  5.74 (100)
  1.765 ( 85)
  2.88 ( 75)
OPTICAL CONSTANT:
  $N = 1.399$ (Na)
  HARDNESS: 4½

DENSITY: 2.56–2.62 (Meas.)
2.56 (Calc.)
CLEAVAGE: {111} imperfect
Fracture uneven. Brittle.
HABIT: Crystals octahedral, cubo-octahedral, or cubic with small octahedral faces.
COLOR-LUSTER: Colorless, white. Transparent to translucent. Vitreous
MODE OF OCCURRENCE: Occurs as crystals, which range from microscopic to 1.0 mm in diameter, coating crystals of pachnolite, or lining cavities in thomsenolite and pachnolite, at the cryolite deposit at St. Peters Dome, El Paso County, Colorado. It is also found with thomsenolite and cryolite in pegmatite at Ivigtut, Greenland.
BEST REF. IN ENGLISH: Palache, et al., "Dana's System of Mineralogy," 7th Ed., v. II, p. 126–127, New York, Wiley, 1951.  *Am. Min.*, 50: 1851–1864 (1965).

# RAMDOHRITE (Andorite VI)
$Ag_2Pb_3Sb_6S_{13}$

CRYSTAL SYSTEM: Orthorhombic
CLASS: 2/m 2/m 2/m or mm2
SPACE GROUP: Pmcm or P2cn
Z: 12
LATTICE CONSTANTS:
$a = 13.04$
$b = 19.15$
$c = 25.74$
3 STRONGEST DIFFRACTION LINES:
3.33 (100)
2.91 ( 60)
3.45 ( 40)
HARDNESS: ~3
DENSITY: 5.44 (Meas.)
CLEAVAGE: Not determined.
Fracture uneven. Brittle.
HABIT: Crystals long prismatic or thick lance-shaped. Twinning lamellar.
COLOR-LUSTER: Gray-black. Opaque. Metallic. Streak gray-black.
MODE OF OCCURRENCE: Occurs associated with stannite, tetrahedrite, jamesonite, cassiterite, pyrite, and sphalerite, at Chocaya, Potosí, Bolivia.
BEST REF. IN ENGLISH: Palache, et al., "Dana's System of Mineralogy," 7th Ed., v. I, p. 450–451, New York, Wiley, 1944.

# RAMEAUITE
$K_2CaU_6O_{20} \cdot 9H_2O$

CRYSTAL SYSTEM: Monoclinic
CLASS: 2/m, m
SPACE GROUP: C 2/c or Cc
Z: 4
LATTICE CONSTANTS:
$a = 13.97$
$b = 14.26$
$c = 14.22$
$\beta = 121°1'$

3 STRONGEST DIFFRACTION LINES:
7.12 (100)
3.495 (100)
3.139 (100)
OPTICAL CONSTANTS:
$2V = 32°$
HARDNESS: Not determined.
DENSITY: 5.60 (Meas.)
5.55 (calc.)
CLEAVAGE: {010} good
HABIT: Crystals small (up to 1 mm long), always twinned on {100}, flattened on {010} and elongate parallel to [001].
COLOR-LUSTER: Orange.
MODE OF OCCURRENCE: The mineral occurs with calcite and α-uranophane on pitchblende at Margnac, in the Massif Central, France.
BEST REF. IN ENGLISH: Cesbron, F., Brown, W. L., Bariand, P., and Geffroy, J., *Min. Mag.*, 38: 781–789 (1972).

# RAMMELSBERGITE
$NiAs_2$
Dimorphous with pararammelsbergite

CRYSTAL SYSTEM: Orthorhombic
CLASS: 2/m 2/m 2/m
SPACE GROUP: Pnnm
Z: 2
LATTICE CONSTANTS:
$a = 4.78$
$b = 5.78$
$c = 3.53$
3 STRONGEST DIFFRACTION LINES:
2.552 (100)
2.476 ( 90)
2.843 ( 65)
HARDNESS: 5½–6
DENSITY: 6.97 (Meas.)
7.05 (Calc.)
CLEAVAGE: Not determined. Fracture uneven.
HABIT: Crystals short prismatic, up to 1.0 mm in size, showing forms {001}, {100}, {230}, {011}, and {101}. Usually massive; fine-granular to prismatic or radial-fibrous in structure.
COLOR-LUSTER: Tin white with pinkish tint. Opaque. Metallic.
MODE OF OCCURRENCE: Occurs chiefly in mesothermal vein deposits associated with other nickel and cobalt minerals, domeykite, loellingite, uraninite, native silver, and bismuth. Found massive and as minute crystals at Mohawk, Michigan; in the Cobalt district, Ontario, and at the Eldorado mine, Great Bear Lake, Canada; and at deposits in Chile, France, Italy, Sardinia, Switzerland, Austria, and Germany.
BEST REF. IN ENGLISH: Palache, et al., "Dana's System of Mineralogy," 7th Ed., v. I, p. 309–310, New York, Wiley, 1944.

# RAMSAYITE = Lorenzenite

# RAMSDELLITE
$\gamma\text{-}MnO_2$
Dimorphous with pyrolusite

CRYSTAL SYSTEM: Orthorhombic
CLASS: 2/m 2/m 2/m
SPACE GROUP: Pbnm
Z: 4
LATTICE CONSTANTS:
  $a = 4.52$
  $b = 9.27$
  $c = 2.87$
3 STRONGEST DIFFRACTION LINES:
  4.06 (100)
  2.32 ( 65)
  2.55 ( 60)
  HARDNESS: 2–4; usually ~3
  DENSITY: 4.37–4.83 (Meas.)
           4.84 (Calc.)
  CLEAVAGES: On three pinacoids and a prism. Brittle.
  HABIT: As pseudomorphs after groutite crystals. Also massive; platy or finely crystalline.
  COLOR-LUSTER: Steel gray to iron black. Opaque. Brilliant metallic. Streak dull black.
  MODE OF OCCURRENCE: Occurs in fissures and faults in limestone in the manganese deposits at Lake Valley, Sierra County, New Mexico; in manganese veins at Butte, Montana; at the Tolbard mine, Riverside County, California; in Bannock County, Idaho; and at the Monroe-Tener mine near Chisholm, Minnesota. Also at localities in Canada, Mexico, Czechoslovakia, Turkey, India, and elsewhere.
  BEST REF. IN ENGLISH: Fleischer, Michael, Richmond, W. E., and Evans, Howard T. Jr., *Am. Min.*, **47**: 47–58 (1962).

# RANCIÉITE (Inadequately described mineral)
$(Ca, Mn^{2+})Mn_4^{4+}O_9 \cdot 3H_2O$

CRYSTAL SYSTEM: Unknown
3 STRONGEST DIFFRACTION LINES:
  7.52 (100)
  2.70 ( 90)
  3.74 ( 80)
  HARDNESS: Not determined.
  DENSITY: 3.2–3.3 (Meas.)
  CLEAVAGE: Not determined.
  HABIT: Massive, fine lamellar.
  COLOR-LUSTER: Black, brownish, silver gray. Transparent in thin fragments with brown color. Bright metallic.
  MODE OF OCCURRENCE: Occurs in Oriente Province, Cuba, and in cavities in limonite at Rancié, Ariege, and elsewhere in France.
  BEST REF. IN ENGLISH: Richmond, W. E., Fleischer, Michael, and Mrose, Mary E., *Am. Min.*, **54**: 1741–1742 (1969). Palache, et al., "Dana's System of Mineralogy," 7th Ed., v. I, p. 572, New York, Wiley, 1944.

# RANKAMAITE
$(Na, K, Pb, Li)_3(Ta, Nb, Al)_{11}(O, OH)_{30}$

CRYSTAL SYSTEM: Orthorhombic
CLASS: 222, mm2, or 2/m 2/m 2/m
SPACE GROUP: C222, Cmm2, Cm2m, or Cmmn
Z: 2
LATTICE CONSTANTS:
  $a = 17.19$
  $b = 17.70$
  $c = 3.933$
3 STRONGEST DIFFRACTION LINES:
  2.970 (100)
  3.011 ( 80)
  3.375 ( 60)
OPTICAL CONSTANT:
  $N > 2.1$
  HARDNESS: 3–4
  DENSITY: 5.5 (Meas.)
           5.84 (Calc.)
  CLEAVAGE: None.
  HABIT: Massive; as water-worn pebbles.
  COLOR-LUSTER: White to creamy white, felted, resembling sillimanite.
  MODE OF OCCURRENCE: Occurs associated with simpsonite, manganotantalite, cassiterite, and muscovite in heavy mineral concentrates from alluvial deposits, Mumba area, Kivu, eastern Zaire.
  BEST REF. IN ENGLISH: Von Knorring, Oleg; Vorma, Atso, and Nixon, P. H., *Am. Min.*, **55**: 1814 (1970).

# RANKINITE
$Ca_3Si_2O_7$

CRYSTAL SYSTEM: Monoclinic
CLASS: 2/m
SPACE GROUP: $P2_1/a$
Z: 4
LATTICE CONSTANTS:
  $a = 10.55$
  $b = 8.88$
  $c = 7.85$
  $\beta = 120.1°$
3 STRONGEST DIFFRACTION LINES:
  2.72 (100)
  3.18 ( 80)
  4.48 ( 70)
OPTICAL CONSTANTS:
  $\alpha = 1.641$
  $\beta = 1.644$
  $\gamma = 1.650$
  $(+)2V = 64°$
  HARDNESS: 5½
  DENSITY: 2.96–3.00 (Meas.)
           3.00 (Calc.)
  CLEAVAGE: None.
  HABIT: Massive; as rounded or irregular grains.
  COLOR-LUSTER: Colorless, white. Transparent. Vitreous.
  MODE OF OCCURRENCE: Occurs in association with larnite, wollastonite, and melilite at Scawt Hill, County Antrim, Ireland; also found at Takatoka, New Zealand.
  BEST REF. IN ENGLISH: Tilley, C. E., *Am. Min.*, **27**: 720 (1942).

# RANQUILITE
$Ca(UO_2)_2Si_6O_{15} \cdot 12H_2O$ (?)

CRYSTAL SYSTEM: Orthorhombic
Z: 8
LATTICE CONSTANTS:
 $a = 17.64$
 $b = 14.28$
 $c = 18.48$
3 STRONGEST DIFFRACTION LINES:
 9.30 (100)
 4.62 ( 43)
 4.47 ( 40)
OPTICAL CONSTANTS: Mean index = 1.564
 DENSITY: 2.89–3.32
 HABIT: As aggregates of microcrystals in random orientation.
 COLOR-LUSTER: Pale yellow.
 MODE OF OCCURRENCE: Occurs in fissures in gypsum associated with calcite and some limonite in the Ranquil-C6 area, Portezuelo Hill, Malargüe Department, Mendoza Province, Argentina; also reported from the San Sebastian mine, Sañogasta area, Chilecito, La Rioja Province.
 BEST REF. IN ENGLISH: de Abeldo, M. Jiménez, et al., *Am. Min.*, **45**: 1078–1086 (1960).

# RANSOMITE
$CuFe_2^{3+}(SO_4)_4 \cdot 6H_2O$

CRYSTAL SYSTEM: Monoclinic
CLASS: 2/m
SPACE GROUP: $P2_1/c$
Z: 2
LATTICE CONSTANTS:
 $a = 4.811$
 $b = 16.217$
 $c = 10.403$
 $\beta = 93°01'$
OPTICAL CONSTANTS:
 $\alpha = 1.631$
 $\beta = 1.643$
 $\gamma = 1.695$
 (+)2V = small
HARDNESS: 2½
DENSITY: 2.632 (Meas.)
CLEAVAGE: {010} perfect
HABIT: As crusts and radiating tufts of acicular crystals.
COLOR-LUSTER: Bright sky blue. Transparent. Vitreous, pearly on cleavage.
MODE OF OCCURRENCE: Occurs associated with other sulfates in the fire zone in the United Verde mine, Jerome, Arizona.
BEST REF. IN ENGLISH: Palache, et al., "Dana's System of Mineralogy," 7th Ed., v. II, p. 519–520, New York, Wiley, 1951.   Wood, M., *Am. Min.* **55**: 729–734 (1970).

# RASHLEIGHITE = Ferrian turquoise

# RASPITE
$PbWO_4$
Dimorphous with stolzite

CRYSTAL SYSTEM: Monoclinic
CLASS: 2/m
SPACE GROUP: $P2_1/n$
Z: 4
LATTICE CONSTANTS:
 $a = 13.07$
 $b = 5.00$
 $c = 5.58$
 $\beta = 96°18'$
3 STRONGEST DIFFRACTION LINES:
 3.224 (100)
 2.760 ( 60)
 3.619 ( 55)
OPTICAL CONSTANTS:
 $\alpha = 2.27$
 $\beta = 2.27$
 $\gamma = 2.30$
 (+)2V ~ 0°
HARDNESS: 2½–3
DENSITY: 8.465 (Meas.)
 8.517 (Calc.)
CLEAVAGE: {100} perfect
HABIT: Crystals commonly thick to thin tabular, slightly elongated and striated parallel [010]. Twinning on {100} common; less common on {$\bar{1}$02}.
 COLOR-LUSTER: Light yellow to yellowish brown, gray. Adamantine.
 MODE OF OCCURRENCE: Occurs associated with stolzite in oxidized ore at the Proprietary mine, Broken Hill, New South Wales, Australia; in gold placer deposits near Sumidouro, Minas Geraes, Brazil; and in tin veins on the Cerro Estano, near Guanajuato, Mexico.
 BEST REF. IN ENGLISH: Palache, et al., "Dana's System of Mineralogy," 7th Ed., v. II, p. 1089–1090, New York, Wiley, 1951.

# RASVUMITE
$K_3Fe_9S_{14}$

CRYSTAL SYSTEM: Orthorhombic
Z: 1
LATTICE CONSTANTS:
 $a = 9.12$
 $b = 11.08$
 $c = 5.47$
3 STRONGEST DIFFRACTION LINES:
 2.99  (100)
 1.799 ( 90)
 5.54  ( 80)
HARDNESS: Microhardness 243–433, av. 333.9 kg/mm².
 DENSITY: 3.1 (Meas.)
 3.18 (Calc.)
CLEAVAGE: {110} perfect
HABIT: As aggregates of very fine, often curved needles, in grains up to 2 mm in size.

COLOR-LUSTER: Steel gray, becoming dull black on oxidation. Opaque. Metallic.

MODE OF OCCURRENCE: Occurs in association with K feldspar, nepheline, aegirine, djerfisherite, pyrrhotite, cubanite, and two unknown copper-bearing minerals, in pegmatites of the Rasvumchorr and Kukisvumchorr apatite deposits, Khibina massif, Kola Peninsula, USSR.

BEST REF. IN ENGLISH: Sokolova, M. N., Dobrovol'-skaya, M. G., Organova, N. I., and Dmitrik, A. L., *Am. Min.*, **56**: 1121–1122 (1971).

# RATHITE
$Pb_7As_9S_{20}$
Found in three forms, I, II, III

CRYSTAL SYSTEM: Monoclinic
CLASS: 2 or 2/m (I)
SPACE GROUP: $P2_1/n$ (I), $P2_1$ (II, III)
Z: 2 (I)
LATTICE CONSTANTS:

| (I) | (II) | (III) |
|---|---|---|
| $a = 25.1$ | $a = 8.43$ | $a = 24$ |
| $b = 7.91$ | $b = 70.9$ | $b = 7.91$ |
| $c = 8.46$ | $c = 7.91$ | $c = 8.43$ |
| $\beta = 99°$ | $\beta = 90°$ | $\beta = 90°$ |

3 STRONGEST DIFFRACTION LINES:
(I)
2.75 (100)
3.60 ( 80)
3.39 ( 70)
HARDNESS: 3
DENSITY:

| | (I) | (III) |
|---|---|---|
| | 5.41 (Meas.) | 5.38 (Meas.) |
| | 5.34 (Calc.) | 5.35 (Calc.) |

CLEAVAGE: {100} perfect, parting on {010}. Fracture conchoidal.

HABIT: Crystals prismatic; striated parallel to elongation. Twinning on {100} producing polysynthetic pseudoorthorhombic crystals. Also twinning on {$\overline{3}01$}.

COLOR-LUSTER: Lead gray; tarnishes iridescent. Opaque. Metallic. Streak dark brown.

MODE OF OCCURRENCE: Occurs in crystalline dolomite associated with sartorite and other rare sulfosalts at the Lengenbach quarry in the Binnental, Valais, Switzerland.

BEST REF. IN ENGLISH: Palache, et al., "Dana's System of Mineralogy," 7th Ed., v. I, p. 455–457 New York, Wiley, 1944.

# RAUENTHALITE (Inadequately described mineral)
$Ca_3(AsO_4)_2 \cdot 10H_2O$

CRYSTAL SYSTEM: Probably monoclinic or triclinic
3 STRONGEST DIFFRACTION LINES:
10.8 (100)
3.36 ( 80)
2.44 ( 60)

OPTICAL CONSTANTS:
$\alpha = 1.540$ (Calc.)
$\beta = 1.552$
$\gamma = 1.570$
$(+)2V \simeq 85°$
HARDNESS: Not determined.
DENSITY: 2.36 (Meas.)
CLEAVAGE: Not observed.
HABIT: Small rounded, irregular, polycrystalline spherules; also as tiny crystals.
COLOR-LUSTER: Snow white to colorless.
MODE OF OCCURRENCE: Occurs at the 40 meter level of the Gabe Gottes vein at Sainte-Marie-aux-Mines, Alsace, France.
BEST REF. IN ENGLISH: Pierrot, R., *Am. Min.*, **50**: 805–806 (1965).

# RAUVITE
$Ca(UO_2)_2V_{10}O_{28} \cdot 16H_2O$

CRYSTAL SYSTEM: Unknown
3 STRONGEST DIFFRACTION LINES:
10.70 (100)
2.95 ( 50)
3.49 ( 40)
HARDNESS: Not determined.
DENSITY: 2.92 (Meas.)
CLEAVAGE: Waxy and sectile to brittle and glassy.
HABIT: Very fine grained; as dense, slickensided masses, botryoidal crusts, and filmy coatings; also as an interstitial filling in sandstone.
COLOR-LUSTER: Purplish or bluish black to brownish red or dirty orange yellow. Adamantine to waxy. Streak yellowish brown to olive. Not fluorescent.
MODE OF OCCURRENCE: Occurs in Utah filling cracks and impregnating sandstone of the Shinarump member of the Chinle formation, often associated with carnotite, uvanite, hewettite, metatorbernite, and gypsum, at Flat Top, Temple Rock, South Temple Wash, and at the Temple Mountain mine, Emery County; at the Cactus Rat and Corvusite mines, Thompson district, Grand County; and at the Monument No. 2 mine, San Juan County. In Colorado it is found at the Small Spot and Arrowhead mines, Mesa County, and at the Hummer mines, Montrose County. It also occurs in the uranium-vanadium ores of the Edgemont area, Fall River County, South Dakota, especially at the Road Hog No. 3A, Get Me Rich, Ridgerunner, Virginia C, and Hot Point mines.
BEST REF. IN ENGLISH: Frondel, Clifford, USGS Bull. 1064, p. 263–264 (1958).

# REALGAR
AsS

CRYSTAL SYSTEM: Monoclinic
CLASS: 2/m
SPACE GROUP: $P2_1/n$
Z: 16

LATTICE CONSTANTS:
 $a = 9.27$
 $b = 13.50$
 $c = 6.56$
 $\beta = 106°33'$
3 STRONGEST DIFFRACTION LINES:
 5.40 (100)
 3.19 ( 90)
 2.94 ( 80)
OPTICAL CONSTANTS:
 $\alpha = 2.538$
 $\beta = 2.684$  ($\lambda = 590$ m$\mu$)
 $\gamma = 2.704$
 $(-)2V = 40°34'$ ($\lambda = 648$ m$\mu$)
 HARDNESS: 1½–2
 DENSITY: 3.56 (Meas.)
  3.59 (Calc.)
 CLEAVAGE: {010} distinct
  {$\bar{1}$01} indistinct
  {100} indistinct
  {120} indistinct
Fracture conchoidal; sectile.

HABIT: Crystals short prismatic, striated along $c$-axis. Usually massive, compact, coarse to fine grained. Also as powdery crusts or coatings. Twinning on {100}; as contact twins.

COLOR-LUSTER: Dark red to orange-red. Transparent to translucent. Resinous to greasy. Alters to yellowish orange powder on long exposure to light. Streak orange-yellow.

MODE OF OCCURRENCE: Occurs chiefly in low-temperature hydrothermal vein deposits associated with orpiment and stibnite; less commonly as a sublimation product or as a hot springs deposit. Found in United States as excellent crystals and crystal groups at the Getchell mine, Humboldt County, and as large masses with orpiment at Manhattan, Nye County, Nevada; as fine crystals at Mercur, Tooele County, Utah; with ulexite at Boron, Kern County, California; in Snohomish County, Washington; and sparingly in the Homestake gold mine, Lead, South Dakota. It also occurs widespread at numerous European localities, especially as fine crystals at the Legenbach quarry, Binnental, Switzerland, and at Felsobanya and Nagyag, Roumania.

BEST REF. IN ENGLISH: Palache, et al., "Dana's System of Mineralogy," 7th Ed., v. I, p. 255–258, New York, Wiley, 1944.

## RECTORITE (Allevardite)
Interstratified pyrophyllite-vermiculite

## REDDINGITE
$(Mn, Fe)_3 (PO_4)_2 \cdot 3H_2O$
Reddingite-Phosphoferrite Series

CRYSTAL SYSTEM: Orthorhombic
CLASS: 2/m 2/m 2/m
SPACE GROUP: Pbna
Z: 4

LATTICE CONSTANTS:
 $a = 9.49$
 $b = 10.08$
 $c = 8.70$
3 STRONGEST DIFFRACTION LINES:
 3.20 (100)
 2.737 ( 80)
 4.28 ( 70)
OPTICAL CONSTANTS:
 $\alpha = 1.643–1.658$
 $\beta = 1.648–1.664$
 $\gamma = 1.674–1.685$
 $(+)2V = 41°–65°$
 HARDNESS: 3–3½
 DENSITY: 3.23 (Meas.)
  3.24 (Calc. Mn:Fe = 3:1)
 CLEAVAGE: {010} poor
Fracture uneven. Brittle.

HABIT: Crystals octahedral with large {111}, or tabular {010}; often in parallel grouping. Also massive, granular, and coarse fibrous.

COLOR-LUSTER: Pinkish to yellowish white, colorless; often reddish brown to dark brown from alteration. Transparent to translucent. Vitreous to resinous.

MODE OF OCCURRENCE: The mineral occurs as a hydrothermal alteration of lithiophilite in granite pegmatites. Found with fairfieldite and other secondary phosphate minerals at Branchville, Fairfield County, Connecticut; also at Buckfield and Poland, Maine.

BEST REF. IN ENGLISH: Palache, et al., "Dana's System of Mineralogy," 7th Ed., v. II, p. 727–729, New York, Wiley, 1951.

## REDINGTONITE (Inadequately described mineral)
$(Fe, Mg, Ni)(Cr, Al)_2 (SO_4)_4 \cdot 22H_2O$
Halotrichite Group

CRYSTAL SYSTEM: Probably monoclinic
HARDNESS: Not determined.
DENSITY: 1.761 (Meas.)
CLEAVAGE: Not determined.
HABIT: Massive, with fine parallel-fibrous structure.
COLOR-LUSTER: Pale purple. Silky.

Soluble in water.

MODE OF OCCURRENCE: Found mixed with magnesiocopiapite in the Redington mercury mine, Knoxville, Napa County, California.

BEST REF. IN ENGLISH: Palache, et al., "Dana's System of Mineralogy," 7th Ed., v. II, p. 529, New York, Wiley, 1951.

## REDLEDGEITE (Chromrutile)
$Mg_4 Cr_6 Ti_{23} Si_2 O_{61} (OH)_4$ (?)

CRYSTAL SYSTEM: Tetragonal
CLASS: 4/m
SPACE GROUP: I4$_1$/a
Z: 8

LATTICE CONSTANTS:
$a = 20.23$
$c = 5.88$
3 STRONGEST DIFFRACTION LINES:
3.19 (100)
2.46 ( 90)
2.22 ( 90)
DENSITY: 3.72 (Meas.)
3.66 (Calc.)
HABIT: Crystals prismatic or equant.
COLOR-LUSTER: Black, brilliant.
MODE OF OCCURRENCE: Occurs with kammererite in crevices in chromite at the Red Ledge mine, Washington district, Nevada County, California.
BEST REF. IN ENGLISH: Gordon and Shannon, *Am. Min.*, **13**: 69 (1928). Strunz, H., *Am. Min.*, **46**: 1201 (1961).


# REEDMERGNERITE
$NaBSi_3O_8$

CRYSTAL SYSTEM: Triclinic
CLASS: $\bar{1}$
SPACE GROUP: $C\bar{1}$
Z: 4
LATTICE CONSTANTS:
$a = 7.833$        $\alpha = 93°18.5'$
$b = 12.360$      $\beta = 116°21.1'$
$c = 6.803$        $\gamma = 92°03.3'$
3 STRONGEST DIFFRACTION LINES:
3.037 (100)
3.076 ( 90)
3.561 ( 90)
HARDNESS: 6-6½
DENSITY: 2.776 (Meas.)
2.779 (Calc.)
CLEAVAGE: {001} perfect
HABIT: Crystals stubby prisms with characteristic wedge-shaped ends; always single, never twinned or composite; largest about 2 mm long.
COLOR-LUSTER: Colorless. Transparent. Vitreous.
MODE OF OCCURRENCE: Occurs with eitelite, shortite, nahcolite, searlesite, leucosphenite, acmite, analcime, and magnesioriebeckite in unmetamorphosed dolomitic oil shales of the Green River formation in several oil wells in Duchesne County, Utah.
BEST REF. IN ENGLISH: Milton, Charles, Chao, E. T. C., Axelrod, Joseph M., and Grimaldi, Frank S., *Am. Min.*, **45**: 188-199 (1960). Appleman, Daniel E., and Clark, Joan R., *Am. Min.*, **50**: 1827-1850 (1965).


# REEVESITE
$Ni_6Fe_2(OH)_{16}CO_3 \cdot 4H_2O$

CRYSTAL SYSTEM: Hexagonal
Z: 3
LATTICE CONSTANTS:
$a = 6.15$
$c = 45.61$

3 STRONGEST DIFFRACTION LINES:
7.63 (100)
2.60 ( 81)
3.80 ( 73)
OPTICAL CONSTANTS:
$\omega = 1.735$
$\epsilon = 1.65$
(-)
HARDNESS: Not determined.
DENSITY: 2.80 (Meas.) and 2.88 (Meas.)
2.78 (Calc.)        2.87 (Calc.)
CLEAVAGE: Not determined.
HABIT: As fine-grained aggregates; individual crystallites are hexagonal platelets up to 0.1 mm in diameter and 0.02 mm thick.
COLOR-LUSTER: Bright yellow, greenish yellow.
MODE OF OCCURRENCE: Occurs lining cavities and cracks in weathered meteorites from the Wolf Creek crater in Western Australia; also found in the nickel ore from the Bon Accord area in the Barberton Mountain Land, South Africa.
BEST REF. IN ENGLISH: White, John S. Jr., Henderson, E. P., and Mason, Brian, *Am. Min.*, **52**: 1190-1197 (1967).


# REFIKITE
$C_{20}H_{32}O_2$ (?)

CRYSTAL SYSTEM: Orthorhombic
CLASS: 222
SPACE GROUP: $P2_12_12$
Z: 4
LATTICE CONSTANTS:
$a = 10.43$
$b = 22.35$
$c = 7.98$
3 STRONGEST DIFFRACTION LINES:
5.50 (100)
6.09 ( 90)
5.20 ( 70)
HABIT: Crystals acicular.
MODE OF OCCURRENCE: Occurs in fossil roots of spruce in a swamp in southern Bavaria, Germany.
BEST REF. IN ENGLISH: Strunz, H., and Contag B., *Am. Min.*, **50**: 2110 (1965).


# REINERITE
$Zn_3(AsO_3)_2$

CRYSTAL SYSTEM: Orthorhombic
CLASS: 2/m 2/m 2/m
SPACE GROUP: Pmma
Z: 4
LATTICE CONSTANTS:
$a = 6.091$
$b = 14.397$
$c = 7.804$
3 STRONGEST DIFFRACTION LINES:
4.00  (100)
3.20  ( 80)
2.644 ( 80)

OPTICAL CONSTANTS:
  $\alpha = 1.74$
  $\beta = 1.79$ (Na)
  $\gamma = 1.82$
  (-)2V
HARDNESS: 5-5½
DENSITY: 4.270 (Meas.)
         4.278 (Calc.)
CLEAVAGE:    {110} good
             also {011}
             and {111}
HABIT: As rough crystals up to 4.7 X 2 cm with the faces {010}, {110}, and {001} dominant, and {012} present.

COLOR-LUSTER: Ocean blue to light yellow-green; vitreous to adamantine on fracture surfaces.

MODE OF OCCURRENCE: Occurs rarely, associated with chalcocite and a little bornite in cavities in the "second oxidation zone" at a depth of about 900 meters at the Tsumeb Mine, South West Africa.

BEST REF. IN ENGLISH: Geier, B. H., and Weber, K., *Am. Min.*, **44**: 207–208 (1959).

# RENARDITE

$Pb(UO_2)_4 (PO_4)_2 (OH)_4 \cdot 7H_2O$

CRYSTAL SYSTEM: Orthorhombic
CLASS: 2/m 2/m 2/m
SPACE GROUP: Bmmb
Z: 6
LATTICE CONSTANTS:
  $a = 16.01$
  $b = 17.5$
  $c = 13.7$
3 STRONGEST DIFFRACTION LINES:
  7.97 (100)
  3.99 ( 90)
  5.83 ( 80)
OPTICAL CONSTANTS:
  $\alpha = 1.715-1.721$
  $\beta = 1.736-1.741$
  $\gamma = 1.739-1.745$
  $(-)2V \approx 40-45°$
HARDNESS: 3½
DENSITY: 4.35 (Meas.)
         4.34 (Calc.)
CLEAVAGE: {100} perfect
Brittle.
HABIT: Crystals, up to 2.0 mm long, usually rectangular plates flattened on {100} with {010}, {101}, and rarely {001}. Also lath-like by elongation along c-axis. As microcrystalline crusts, small crystal aggregates, fibrous radiated nodules, and as lamellar masses.

COLOR-LUSTER: Lemon yellow, yellow, brownish yellow. Translucent to transparent; vitreous to adamantine. Not fluorescent.

MODE OF OCCURRENCE: Occurs as a secondary mineral associated with torbernite, kasolite, and dumontite at Shinkolobwe, Katanga, Zaire; as an alteration of uraninite at Grabo, Ivory Coast, Africa; and at a number of localities in France including Kersegalec near Lignol, Morbihan; in the vicinity of Lachaux, Puy-de-Dôme; and at LaFaye, Grury, Saone et Loire.

BEST REF. IN ENGLISH: Frondel, Clifford, USGS Bull. 1064, p. 227–230 (1958).

# RENIÉRITE

$Cu_3 (Fe, Ge, Zn)(As, S)_4$

CRYSTAL SYSTEM: Cubic or pseudocubic
CLASS: $\bar{4}$3m
SPACE GROUP: P$\bar{4}$3n
Z: 8
LATTICE CONSTANTS:
  (pseudocubic)
  $a = 10.583$
3 STRONGEST DIFFRACTION LINES:
  3.06 (100)
  1.87 ( 80)
  1.595 ( 60)
HARDNESS: 4.5
DENSITY: 4.3 (Meas.)
CLEAVAGE: None.
HABIT: Idiomorphic grains and in crystals up to 1.5 mm in vugs. Minute crystals have positive forms {111}, {hll} and {hhl} of good quality, and the negative form {111} curved and imperfect. Massive.

COLOR-LUSTER: Orange-bronze; opaque. Metallic.

Magnetic—has polarity.

MODE OF OCCURRENCE: Occurs as inclusions in chalcopyrite, sphalerite, galena, and tennantite at the Prince Leopold mine, Kipushi, Zaire; also as disseminated grains intergrown with galena, pyrite, tennantite, germanite, and bornite, in primary ore from the 2390 foot level of the Tsumeb mine, Tsumeb, Southwest Africa.

BEST REF. IN ENGLISH: Murdoch, Joseph, *Am. Min.*, **38**: 794–801 (1953). *Econ. Geol.*, **52**: 612–631 (1957).

# RETGERSITE

$NiSO_4 \cdot 6H_2O$

CRYSTAL SYSTEM: Tetragonal
CLASS: 4 2 2
SPACE GROUP: P$4_1 2_1 2$ or P$4_3 2_1 2$
Z: 4
LATTICE CONSTANTS:
  $a = 6.776$ kX
  $c = 18.249$
3 STRONGEST DIFFRACTION LINES:
  4.25 (100)
  4.57 ( 40)
  2.964 ( 20)
OPTICAL CONSTANTS:
  $\omega = 1.5109$
  $\epsilon = 1.4873$ (Na)
  (-)
HARDNESS: 2½
DENSITY: 2.07 (Meas. Synthetic)
         2.075 (Calc.)

CLEAVAGE: {001} perfect
{110} in traces
Fracture uneven to subconchoidal. Brittle.

HABIT: Crystals thick tabular; also prismatic, elongated along $c$-axis. As fibrous veinlets and crusts, and as twisted tufts.

COLOR-LUSTER: Dark emerald green with blue tint. Vitreous. Streak greenish white.

Soluble in water. Taste bitter.

MODE OF OCCURRENCE: Occurs as a secondary mineral at the Gap Nickel mine, Lancaster County, Pennsylvania; as an alteration product of niccolite in Cottonwood Canyon, Churchill County, Nevada; as minute crystals on patronite at Minas Ragra, Peru; at Lobenstein, Thuringia, and at Lichtenberg, Germany.

BEST REF. IN ENGLISH: Palache, et al., "Dana's System of Mineralogy," 7th Ed., v. II, p. 497–498, New York, Wiley, 1951.

# RETZIAN
$Mn_2 Y(AsO_4)(OH)_4$

CRYSTAL SYSTEM: Orthorhombic
CLASS: 2/m 2/m 2/m
SPACE GROUP: Pban
Z: 4
LATTICE CONSTANTS:
$a = 5.670$
$b = 12.03$
$c = 4.863$
3 STRONGEST DIFFRACTION LINES:
2.717 (100)
3.53 ( 80)
1.848 ( 50)
OPTICAL CONSTANTS:
$\alpha = 1.777$
$\beta = 1.788$
$\gamma = 1.800$
(+)2V = large
HARDNESS: 4
DENSITY: 4.15 (Meas.)
4.14 (Calc.)
CLEAVAGE: Fracture conchoidal to uneven.
HABIT: Crystals prismatic or tabular.
COLOR-LUSTER: Dark brown. Subtranslucent. Vitreous to greasy. Streak light brown.
MODE OF OCCURRENCE: Occurs sparingly associated with jacobsite in dolomite at the Moss mine, Nordmark, Sweden.
BEST REF. IN ENGLISH: Moore, Paul B., *Am. Min.*, **53**: 1779 (1968). Moore, Paul B., *Am. Min.*, **52**: 1603–1613 (1967). Palache, et al., "Dana's System of Mineralogy," 7th Ed., v. II, p. 794–795, New York, Wiley, 1951.

# REYERITE
$(Na, K)_4 Ca_{14} Si_{24} O_{60}(OH)_5 \cdot 5H_2O$

CRYSTAL SYSTEM: Hexagonal

CLASS: $\bar{3}$ or 3
SPACE GROUP: P$\bar{3}$ or P3
Z: 1
LATTICE CONSTANTS:
$a = 9.764$
$c = 19.070$
3 STRONGEST DIFFRACTION LINES:
3.17 (100)
2.855 ( 80)
2.659 ( 80)
OPTICAL CONSTANTS:
$\omega = 1.563$
$\epsilon = 1.558$
(-)
HARDNESS: 3-4
DENSITY: 2.54–2.578 (Meas.)
2.59 (Calc.)
CLEAVAGE: {0001} perfect
Fracture uneven. Brittle.
HABIT: Massive; as micaceous spherical aggregates.
COLOR-LUSTER: Colorless, white. Transparent to translucent. Vitreous; pearly on cleavage.
MODE OF OCCURRENCE: Reyerite occurs in chlorite-containing amygdules in a diabase dike at Rawlins quarry, Brunswick County, Virginia. It is also found within a tuff at Niokornak, Greenland, and within the plateau basalts at 'S Airde Beinn, northern Mull, Scotland.
BEST REF. IN ENGLISH: Clement, Stephen C. and Ribbe, Paul H., *Am. Min.*, **58**: 517–522 (1973).

# RÉZBÁNYITE (Inadequately described mineral)
$Pb_3 Cu_2 Bi_{10} S_{19}$

CRYSTAL SYSTEM: Unknown
HARDNESS: 2½
DENSITY: 6.24 (Rezbanya)
6.89 (Vaskö)
CLEAVAGE: Indistinct. Fracture uneven.
HABIT: Massive; compact to fine granular.
COLOR-LUSTER: Light lead gray, tarnishing dull. Opaque. Metallic. Streak black.
MODE OF OCCURRENCE: Occurs associated with chalcopyrite, calcite, and quartz, at Rézbánya, Roumania, and with dolomite and quartz at Vaskö.
BEST REF. IN ENGLISH: Palache, et al., "Dana's System of Mineralogy," 7th Ed., v. I, p. 470–471, New York, Wiley, 1944.

# RHABDITE = Schreibersite

# RHABDOPHANE
$(Ce, Y, La, Di)(PO_4) \cdot H_2O$

CRYSTAL SYSTEM: Hexagonal
CLASS: 6 2 2
SPACE GROUP: P$6_2$22
Z: 3
LATTICE CONSTANTS:
$a = 6.98$
$c = 6.39$

3 STRONGEST DIFFRACTION LINES:
3.02 (100)
4.40 ( 80)
2.83 ( 80)
OPTICAL CONSTANTS:
$\omega = 1.654$
$\epsilon = 1.703$
(+)
HARDNESS: 3½
DENSITY: 3.94-4.01 (Meas.)
CLEAVAGE: Fracture uneven.
HABIT: As stalactitic or botryoidal incrustations with radial-fibrous structure.
COLOR-LUSTER: Brown, pinkish, or yellowish white. Translucent. Greasy.
MODE OF OCCURRENCE: Occurs as thin crusts in the limonite deposit at Salisbury, Connecticut, and at a deposit in Cornwall, England.
BEST REF. IN ENGLISH: Palache, et al., "Dana's System of Mineralogy," 7th Ed., v. II, p. 774, New York, Wiley, 1951.

# RHODESITE
$(Ca, Na_2, K_2)_8 Si_{16}O_{40} \cdot 11H_2O$

CRYSTAL SYSTEM: Orthorhombic
CLASS: mm2 or 2/m 2/m 2/m
SPACE GROUP: $Pmn2_1$ (or Pmmn)
Z: 1
LATTICE CONSTANTS:
$a = 23.636$
$b = 6.549$
$c = 7.037$
3 STRONGEST DIFFRACTION LINES:
6.55 (100)
4.39 ( 45)
5.90 ( 35)
OPTICAL CONSTANTS:
$\alpha = 1.501$-1.502
$\beta = 1.505$
$\gamma = 1.513$-1.515
(+)2V = low
HARDNESS: Not determined.
DENSITY: ~2.36 (Meas.)
CLEAVAGE: {100} good
HABIT: Matted silky fibers in rosettes up to 2.0 mm in size.
COLOR-LUSTER: White; silky.
MODE OF OCCURRENCE: Occurs associated with mountainite at the Bultfontein mine, Kimberley, Republic of South Africa; also about 35 miles northwest of Redding, Trinity County, California, associated with magadiite and trioctahedral montmorillonite in an altered silicic lava.
BEST REF. IN ENGLISH: Mountain, E. D., Min. Mag., 31: 607-610 (1957). Sheppard, Richard A., and Gude, Arthur J. 3rd, Am. Min., 54: 251-255 (1969).

# RHODIZITE
$CsB_{12}Be_4Al_4O_{28}$

CRYSTAL SYSTEM: Cubic
CLASS: $\bar{4}3m$
SPACE GROUP: $P\bar{4}3m$
Z: 1
LATTICE CONSTANT:
$a = 7.317$
3 STRONGEST DIFFRACTION LINES:
2.983 (100)
3.27 ( 50)
2.440 ( 50)
OPTICAL CONSTANT:
$N = 1.6935$ (Na)
HARDNESS: 8½
DENSITY: 3.44 (Meas.)
3.478 (Calc.)
CLEAVAGE: {111} difficult
{$\bar{1}11$} difficult
Fracture conchoidal.
HABIT: Crystals dodecahedral or tetrahedral, up to 2.0 cm in size.
COLOR-LUSTER: Colorless, white, yellowish white, yellow, grayish. Vitreous to adamantine. Transparent to translucent.
MODE OF OCCURRENCE: Occurs only as crystals in pegmatite associated with deep-red rubellite tourmaline. Found in Madagascar at Antandrokomby near Mt. Bity, at Manjaka, and at Ambalanilaifotsy, Ambatofinandrohama. It also occurs as minute crystals on red tourmaline at localities near Mursinsk, Ural Mountains, USSR.
BEST REF. IN ENGLISH: Palache, et al., "Dana's System of Mineralogy," 7th Ed., v. II, p. 329-330, New York, Wiley, 1951. Frondel, Clifford, and Ito, Jun, Am. Min., 51: 533-534 (1966). Buerger, Martin J., and Taxer, Karlheinz, Science, 152 (3721): 500-502 (1966).

# RHODOCHROSITE
$MnCO_3$

CRYSTAL SYSTEM: Hexagonal
CLASS: $\bar{3} 2/m$
SPACE GROUP: $R\bar{3}c$
Z: 2 (rhomb.), 6 (hexag.)
LATTICE CONSTANTS:

| (hexag.) | (rhomb.) |
|---|---|
| $a = 4.73$ kX | $a_{rh} = 5.84$ |
| $c = 15.51$ | $\alpha = 47°46'$ |

3 STRONGEST DIFFRACTION LINES:
2.84 (100)
3.66 ( 35)
1.763 ( 35)
OPTICAL CONSTANTS:
$\omega = 1.816$
$\epsilon = 1.597$
(-)
HARDNESS: 3½-4
DENSITY: 3.70 (Meas. pure $MnCO_3$). Usually 3.4-3.6
CLEAVAGE: {$10\bar{1}1$} perfect
Sometimes parting {$01\bar{1}2$}. Fracture conchoidal to uneven. Brittle.
HABIT: Crystals rhombohedral; rarely prismatic, scalenohedral, or thick tabular. Commonly massive, compact to coarsely granular. Also stalactitic, botryoidal, or globular.

COLOR-LUSTER: Pale pink to deep red; orangish red, yellowish gray, tan, brown. Vitreous; sometimes pearly. Transparent to translucent.

MODE OF OCCURRENCE: Occurs as a gangue mineral in hydrothermal ore veins; in high-temperature contact metasomatic deposits; as a secondary mineral in manganese deposits; and rarely as a late-stage mineral in pegmatites. Found in the United States as spectacular vivid red gemmy rhombohedrons, as much as 3 inches on an edge, associated with sphalerite, huebnerite, quartz, and other minerals at the Sweet Home mine, Park County, and also as superb specimens in Chaffee, Gilpin, Lake, Saguache, and San Juan Counties, Colorado; also as fine pink crystal groups and large masses in several mines at Butte, Montana. It also occurs as fine pink cleavable masses near Magdalena, Sonora, Mexico; as exceptional stalactitic aggregates in Catamarca province, Argentina; and at numerous localities in Europe, especially at Kapnik, Roumania, in Saxony, Westphalia, Harz Mountains, and other places in Germany. Recently found as superb gemmy scalenohedral crystal groups at Hotazel, near Kuruman, South Africa.

BEST REF. IN ENGLISH: Palache, et al., "Dana's System of Mineralogy," 7th Ed., v. II, p. 171-175, New York, Wiley, 1951.

# RHODONITE
MnSiO$_3$
(Pyroxene Group)

CRYSTAL SYSTEM: Triclinic
CLASS: $\bar{1}$
SPACE GROUP: P$\bar{1}$
Z: 10
LATTICE CONSTANTS:
 $a$ = 7.66   $\alpha$ = 86.0°
 $b$ = 12.27   $\beta$ = 93.2°
 $c$ = 6.68   $\gamma$ = 111.1°
3 STRONGEST DIFFRACTION LINES:
 2.772 (100)
 2.980 ( 65)
 2.924 ( 65)
OPTICAL CONSTANTS:
  $\alpha$ = 1.711-1.738
  $\beta$ = 1.716-1.741
  $\gamma$ = 1.724-1.751
 (+)2V = 63°-76°
HARDNESS: 5½-6½
DENSITY: 3.57-3.76 (Meas.)
     3.726 (Calc.)
CLEAVAGE: {110} perfect
     {1$\bar{1}$0} perfect
     {001} good
Fracture conchoidal to uneven; very tough when compact.
HABIT: Crystals commonly tabular parallel to {001}, usually rough with rounded edges. Commonly massive, cleavable to compact; also fine to coarse granular.
COLOR-LUSTER: Pink to rose red to brownish red, often veined by black alteration products; rarely yellow, gray. Transparent to translucent. Vitreous; somewhat pearly on cleavages.

MODE OF OCCURRENCE: Occurs in manganese-bearing ore bodies, usually formed by metasomatic or hydrothermal processes or by metamorphism of sedimentary deposits. Found as excellent large crystals, often associated with calcite, willemite, and franklinite, at Franklin, New Jersey; at many deposits in California and Colorado; in Montana, Massachusetts, Maine, and elsewhere in the United States. Other important deposits occur in Brazil, England, Sweden, Italy, Hungary, Finland, USSR, South Africa, India, Japan, and as fine crystals in galena at Broken Hill, New South Wales, Australia.

BEST REF. IN ENGLISH: Deer, Howie, and Zussman, "Rock Forming Minerals," v. II, p. 182-190, New York, Wiley, 1963.

# RHODOSTANNITE
Cu$_2$FeSn$_3$S$_8$

CRYSTAL SYSTEM: Hexagonal
Z: 3
LATTICE CONSTANTS:
 $a$ = 7.27
 $c$ = 18.07
3 STRONGEST DIFFRACTION LINES:
 3.12 (100)
 5.93 ( 60)
 2.58 ( 50)
HARDNESS: Vickers microhardness 243-266 kg/mm$^2$.
DENSITY: Not determined.
HABIT: Massive.
COLOR-LUSTER: Reddish. Opaque. Metallic.
MODE OF OCCURRENCE: Occurs associated with stannite, of which it appears to be an alteration product, in a sample from Vila Apacheta, Bolivia.
BEST REF. IN ENGLISH: Springer, G., *Min. Mag.*, **36**: 1045-1051 (1968).

# RHODUSITE =Magnesioriebeckite

# RHOMBOCLASE
HFe$^{3+}$(SO$_4$)$_2$ · 4H$_2$O

CRYSTAL SYSTEM: Orthorhombic
CLASS: 2/m 2/m 2/m
LATTICE CONSTANTS:
 $a:b:c$ = 0.558:1:0.937
OPTICAL CONSTANTS:
  $\alpha$ = 1.534
  $\beta$ = 1.553
  $\gamma$ = 1.638 (Na)
 (+)2V = 27°
HARDNESS: 2
DENSITY: 2.23 (Meas.)
CLEAVAGE: {001} perfect
     {110} good
Fracture conchoidal to fibrous. Thin cleavages flexible.
HABIT: Crystals thin flattened plates; also stalactitic with bladed radiating structure.

COLOR-LUSTER: Colorless, white, gray, yellowish, greenish, bluish. Transparent. Vitreous to pearly.

MODE OF OCCURRENCE: Occurs with roemerite and other sulfates at Alcaparrosa, near Cerritos Bay, Chile; as incrustations in abandoned workings at the Esperanza mine, Cerro de Pasco, Peru; and as an alteration of pyrite, associated with szomolnokite and other sulfates, at Szomolnok, Czechoslovakia.

BEST REF. IN ENGLISH: Palache, et al., "Dana's System of Mineralogy," 7th Ed., v. II, p. 436–437, New York, Wiley, 1951.

# RHÖNITE = Variety of aenigmatite

# RICHELLITE
$Ca_3 Fe_{10} (PO_4)_8 (OH, F)_{12} \cdot nH_2O$

CRYSTAL SYSTEM: Tetragonal
CLASS: 4 (?)
SPACE GROUP: P4 (?)
LATTICE CONSTANTS:
$a = 5.18$
$c = 12.61$ heated (500°C, 30 min.)
3 STRONGEST DIFFRACTION LINES:
3.24 (100)
1.590 ( 80) (heated)
3.58 ( 65)
HARDNESS: 2–3
DENSITY: ~2.0
CLEAVAGE: Not determined.
HABIT: Massive; compact or foliated, also as globules with radial-fibrous structure.
COLOR-LUSTER: Yellowish to reddish brown. Greasy to horn-like.
MODE OF OCCURRENCE: Occurs associated with halloysite and allophane at Richelle near Visé, Belgium.
BEST REF. IN ENGLISH: Palache, et al., "Dana's System of Mineralogy," 7th Ed., v. II, p. 956–957, New York, Wiley, 1951.

# RICHETITE (Species status uncertain)
Hydrous oxide of Pb and U

CRYSTAL SYSTEM: Monoclinic
3 STRONGEST DIFFRACTION LINES: Metamict.
OPTICAL CONSTANTS:
$\sim 2.0 < \beta$ and $\gamma < \sim 2.7$
$(-)2V$ = very large
HARDNESS: Not determined.
DENSITY: Not determined.
CLEAVAGE: {010} perfect
Also a second less well developed cleavage in the zone [010].
HABIT: Crystals minute pseudohexagonal tablets flattened on {010}; as subparallel, lamellar aggregates on {010}. Crystal corners truncated by tiny oblique faces. Commonly twinned [010].
COLOR-LUSTER: Black, adamantine.
MODE OF OCCURRENCE: Occurs as a secondary mineral associated with uranophane at Shinkolobwe, Katanga, Zaire.

BEST REF. IN ENGLISH: Frondel, Clifford, USGS Bull. 1064, p. 91–92 (1958).

# RICHTERITE
$(Na, K)_2 (Mg, Mn, Ca)_6 Si_8 O_{22} (OH)_2$
Amphibole Group

CRYSTAL SYSTEM: Monoclinic
CLASS: 2/m
SPACE GROUP: C2/m
Z: 2
LATTICE CONSTANTS:
$a = 9.82$
$b = 17.96$
$c = 5.27$
$\beta = 104°20'$
3 STRONGEST DIFFRACTION LINES:
8.55 (100)
2.71 ( 80) broad
3.38 ( 70)
OPTICAL CONSTANTS:
$\alpha = 1.605–1.685$
$\beta = 1.618–1.700$
$\gamma = 1.627–1.712$
$(-)2V = 66°–87°$
HARDNESS: 5–6
DENSITY: 2.97–3.132 (Meas.)
          3.035 (Calc.)
CLEAVAGE: {110} perfect
          {001} parting
          {100} parting
Fracture uneven. Brittle.
HABIT: Crystals long prismatic, rarely terminated. Twinning on {100} simple, lamellar.
COLOR-LUSTER: Brown, yellow, brownish red, rose red, pale to dark green. Transparent to translucent. Vitreous.
MODE OF OCCURRENCE: Occurs in contact metasomatic deposits; in certain igneous rocks; and in thermally metamorphosed limestones. Typical occurrences are found near Libby, Montana; in the Leucite Hills, Wyoming; at Iron Hill, Gunnison County, Colorado; and at localities in Sweden, Madagascar, Burma, and Australia.
BEST REF. IN ENGLISH: Deer, Howie, and Zussman, "Rock Forming Minerals," v. 2, p. 352–358, New York, Wiley, 1963.

# RICKARDITE
$Cu_{2-x} Te$ (x ~ 0.6)

CRYSTAL SYSTEM: Tetragonal
CLASS: 4/m 2/m 2/m
SPACE GROUP: P4/nmm
Z: 2
LATTICE CONSTANTS:
$a = 3.97$
$c = 6.11$

3 STRONGEST DIFFRACTION LINES:
2.07 (100)
3.35 ( 60)
2.55 ( 40)
HARDNESS: 3½
DENSITY: 7.54 (Meas.)
          7.42 (Calc.)
CLEAVAGE: None. Fracture uneven. Brittle.
HABIT: Massive, compact.
COLOR-LUSTER: Purple-red; rapidly dulls on exposure. Opaque. Metallic. Streak reddish.
MODE OF OCCURRENCE: Occurs with other tellurides in gold-quartz veins and in copper sulfide ores. Found as small lenticular masses, commonly associated with native tellurium and weissite, at the Good Hope mine at Vulcan, Gunnison County, and sparingly at the Empress Josephine mine, Saguache County, Colorado. It is also reported from Warren, Arizona; the San Sebastian mine, San Salvador; in the Pionerskoe gold ore deposit, USSR; at Kalgoorlie, Western Australia; at the Tein mine, Hokkaido, and the Hinokizawa mine, Rendansu, Japan; and at the Ookiep copper mine, South Africa.
BEST REF. IN ENGLISH: Vlasov, K. A., "Mineralogy of Rare Elements," v. II, p. 687–688, Israel Program for Scientific Translations, 1966. Sindeeva, N. D., "Mineralogy and Types of Deposits of Selenium and Tellurium," Wiley-Interscience, 1964.

# RIEBECKITE (Crocidolite)
$Na_2 Fe_3^{2+} Fe_2^{3+} Si_8 O_{22} (OH)_2$
Amphibole Group

CRYSTAL SYSTEM: Monoclinic
CLASS: 2/m
SPACE GROUP: C2/m
Z: 2
LATTICE CONSTANTS:
 $a \simeq 9.75$
 $b \simeq 18.0$
 $c \simeq 5.3$
 $\beta \simeq 103°$
3 STRONGEST DIFFRACTION LINES:
8.42 (100)
2.72 (100)
3.09 ( 80)
OPTICAL CONSTANTS:
 $\alpha = 1.654–1.701$
 $\beta = 1.662–1.711$
 $\gamma = 1.668–1.717$
 $2V = 40°–90° (-), (+)$
HARDNESS: 5
DENSITY: 3.32–3.382 (Meas.)
          3.43 (Calc.)
CLEAVAGE: {110} perfect
Fracture uneven. Brittle.
HABIT: Crystals long prismatic, striated parallel to elongation. Also massive; fibrous, columnar, or granular. Twinning on {100} simple, lamellar.
COLOR-LUSTER: Dark blue to black. Translucent to nearly opaque. Vitreous, silky.

MODE OF OCCURRENCE: Occurs mainly in granites, syenites, rhyolites, trachytes, bedded ironstones (crocidolite), and regionally metamorphosed schists. Typical occurrences are found in Mendocino County and elsewhere in California; at St. Peters Dome, El Paso County, Colorado; in the San Francisco Mountains, Coconino County, Arizona; at Cumberland Hill, Providence, Rhode Island; and near Quincy, Norfolk County, Massachusetts. Also found in Canada, Greenland, Scotland, Portugal, France, Austria, Corsica, Nigeria, South Africa (crocidolite), Zambia, Korea, Madagascar, and Australia.
BEST REF. IN ENGLISH: Deer, Howie, and Zussman, "Rock Forming Minerals," v. 2, p. 333–351, New York, Wiley, 1963.

# RIJKEBOERITE
$Ba(Ta, Nb)_2 (O, OH)_7$
Pyrochlore Group
Analogue of pandaite

CRYSTAL SYSTEM: Cubic
CLASS: $4/m \bar{3} 2/m$
SPACE GROUP: Fd3m
Z: 8
LATTICE CONSTANT:
 $a = 10.570$
3 STRONGEST DIFFRACTION LINES:
3.034 (100)
6.04 ( 80)
3.18 ( 65)
HARDNESS: 4½–5
DENSITY: 5.71 (Meas.)
          5.60 (Calc.)
CLEAVAGE: Not determined.
HABIT: Massive.
COLOR-LUSTER: Pink, reddish, yellowish brown, white, colorless.
MODE OF OCCURRENCE: Occurs in cassiterite-rich alluvials from a pegmatite at Chi-Chico, São João del Rei, Minas Geraes, Brazil.
BEST REF. IN ENGLISH: van der Veen, A. H., *Am. Min.*, **48**: 1415 (1963).

# RINGWOODITE
$(Mg, Fe)_2 SiO_4$
Dimorphous with olivine

CRYSTAL SYSTEM: Cubic
Z: 8
LATTICE CONSTANT:
 $a = 8.113$
3 STRONGEST DIFFRACTION LINES:
2.447 (100)
2.028 ( 58)
1.434 ( 58)
OPTICAL CONSTANT:
 $N = 1.768$
HARDNESS: Not determined.
DENSITY: 3.90 (Calc.)
CLEAVAGE: Not determined.

HABIT: As rounded grains up to 100 $\mu$m in diameter.
COLOR-LUSTER: Purple; bluish gray.
MODE OF OCCURRENCE: Occurs in veins and as pseudomorphs after olivine in the Tenham chondrite meteorite.
BEST REF. IN ENGLISH: Binns, R. A., Davis, R. J., and Reed, S. J. B., *Nature*, **221**: 943–944 (1969).

## RINKITE = Mosandrite

## RINKOLITE = Mosandrite

## RINNEITE
$K_3NaFeCl_6$

CRYSTAL SYSTEM: Hexagonal
CLASS: $\bar{3}\ 2/m$
SPACE GROUP: $R\bar{3}c$
Z: 2 (rhomb.), 6 (hexag.)
LATTICE CONSTANTS:

|  (hexag.) | (rhomb.) |
|---|---|
| $a = 11.90$ | $a_{rh} = 8.289$ |
| $c = 13.91$ | $\alpha = 91°45'$ |

3 STRONGEST DIFFRACTION LINES:
2.51 (100)
6.00 ( 55)
2.59 ( 45)
OPTICAL CONSTANTS:
$\omega = 1.5886$
$\epsilon = 1.5894$ (Na)
(+)
HARDNESS: 3
DENSITY: 2.347 (Meas.)
2.406 (Calc.)
CLEAVAGE: $\{11\bar{2}0\}$ good
Fracture conchoidal, splintery.
HABIT: Massive granular. Synthetic crystals short prismatic or thick tabular.
COLOR-LUSTER: Colorless; commonly yellow, rose, or violet, becoming brown on exposure. Brilliant, often silky.

Taste astringent.
MODE OF OCCURRENCE: Occurs in saline deposits in Germany near Norhausen, Saxony, and at several places in Hannover, associated with halite, anhydrite, sylvite, and kieserite.
BEST REF. IN ENGLISH: Palache, et al., "Dana's System of Mineralogy," 7th Ed., v. II, p. 107–108, New York, Wiley, 1951.

## RIONITE = Bismuthian tennantite

## RIPIDOLITE (Prochlorite)
$(Mg, Fe, Al)_6(Si, Al)_4O_{10}(OH)_8$
Chlorite Group

CRYSTAL SYSTEM: Monoclinic
CLASS: $2/m$
SPACE GROUP: C2/m (?)
Z: 4
LATTICE CONSTANTS:
$a = 5.360$
$b = 9.283$
$c = 14.044$
$\beta = 97°15'$
3 STRONGEST DIFFRACTION LINES:
7.07 (100)
14.1 ( 80)
3.54 ( 50)
OPTICAL CONSTANTS:
$\alpha = 1.616$
$\beta = 1.616$
$\gamma = 1.620$
(+)2V = very small
HARDNESS: 2–3
DENSITY: 2.88–3.08 (Meas.)
2.899 (Calc.)
CLEAVAGE: $\{001\}$ perfect
Laminae flexible, inelastic.
HABIT: Crystals commonly tabular, hexagonal in outline. Usually massive, foliated, scaly, or granular. Twinning on $\{001\}$, twin plane [310].
COLOR-LUSTER: Green to dark green or grayish green, brownish. Transparent to translucent. Pearly. Streak uncolored or pale greenish.
MODE OF OCCURRENCE: Occurs relatively widespread in schists and other metamorphic rocks; less commonly in ore veins. Typical occurrences are found in the southern Appalachian region of the eastern United States, and at localities in England, Madagascar, and New Zealand.
BEST REF. IN ENGLISH: Deer, Howie, and Zussman, "Rock Forming Minerals," v. 3, p. 131–163, New York, Wiley, 1962.

## RIVADAVITE
$Na_6MgB_{24}O_{40} \cdot 22H_2O$

CRYSTAL SYSTEM: Monoclinic
CLASS: $2/m$
SPACE GROUP: $P2_1/m$
Z: 1
LATTICE CONSTANTS:
$a = 14.779$
$b = 8.010$
$c = 11.128$
$\beta = 105°57'$
3 STRONGEST DIFFRACTION LINES:
14.2 (100)
7.59 (100)
2.95 ( 66)
OPTICAL CONSTANTS:
$\alpha = 1.470$
$\beta = 1.481$
$\gamma = 1.497$
(+)2V = 80°
HARDNESS: 3½

DENSITY: 1.905 (Meas.)
1.910 (Calc.)
CLEAVAGE: {100} excellent
{$\bar{1}$01} excellent
{010} poor
Extremely brittle; tends to break in splintery fragments elongated on the *b*-axis.

HABIT: Crystals prismatic, elongated on *b*-axis and flattened {100}; up to 3 mm long. As nodular masses from 1 to 9 cm in diameter composed of aggregates of single crystals.

COLOR-LUSTER: Colorless. Transparent. Vitreous. Aggregates white.

MODE OF OCCURRENCE: Occurs embedded in borax at the Tincalayu borax deposit, Province of Salta, Argentina.

BEST REF. IN ENGLISH: Hurlbut, C. S. Jr., and Aristarain, L. F., *Am. Min.*, **52**: 326–335 (1967).


# RIVERSIDEITE
$Ca_5 Si_6 O_{16} (OH)_2 \cdot 2H_2O$

CRYSTAL SYSTEM: Orthorhombic
CLASS: 222
SPACE GROUP: $B22_1 2$
Z: 4
LATTICE CONSTANTS:
*a* = 11.12
*b* = 18.6
*c* = 7.32
3 STRONGEST DIFFRACTION LINES:
3.04 (100)
9.67 ( 80)
4.83 ( 80)
HARDNESS: 3
DENSITY: 2.64 (Meas.)
CLEAVAGE: Not determined.
HABIT: Massive, fibrous.
COLOR-LUSTER: White. Silky.
MODE OF OCCURRENCE: Occurs in admixture with wilkeite at Crestmore, Riverside County, California.

BEST REF. IN ENGLISH: Eakle, Arthur S., *Univ. Calif., Bull. Dept. Geol.*, **10** (19): 344–346 (1917). Taylor, H. F. W., *Min. Mag.*, **30**: 155–165 (1953).


# ROBERTSITE
$Ca_3 Mn_4^{3+} (OH)_6 (H_2O)_3 [PO_4]_4$
Robertsite-Mitridatite Series

CRYSTAL SYSTEM: Monoclinic
CLASS: 2/m
SPACE GROUP: A2/a
Z: 8
LATTICE CONSTANTS:
*a* = 17.36
*b* = 19.53
*c* = 11.30
β = 96°

3 STRONGEST DIFFRACTION LINES:
8.63 (100)
2.75 ( 60)
5.61 ( 50)
OPTICAL CONSTANTS:
α = 1.775
β = 1.82
γ = 1.82
Biaxial (–), 2V ~ 8°
HARDNESS: 3½
DENSITY: 3.13, 3.17 (Meas.)
3.05 (Calc.)
CLEAVAGE: {100} good
Fracture uneven. Brittle.

HABIT: Crystals thick tabular, wedge-shaped, small; very thick crystals pseudo-rhombohedral in aspect. Forms: *a*{100}, *c*{001}, and *p*{031}. As bunched radial aggregates composed of plates up to 5 mm across; as botryoidal aggregates with fibrous structure; and as feathery aggregates. Frequently twinned by rotation normal to {100}.

COLOR-LUSTER: Light to dark reddish-brown, deep red to bronzy, shiny black. Translucent to nearly opaque. Silky; resinous to adamantine. Streak chocolate-brown.

MODE OF OCCURRENCE: Robertsite occurs in moderate abundance in granite pegmatites as a late stage product in corroded triphylite-heterosite-ferrisicklerite-rockbridgeite masses, usually associated with whitlockite, collinsite, hydroxylapatite, jahnsite, leucophosphite and other secondary phosphate minerals. Found at the Tip Top, Linwood, and White Elephant mines in Custer County, and at the Gap Lode mine, Pennington County, South Dakota.

BEST REF. IN ENGLISH: Moore, Paul Brian, *Am. Min.*, **59**: 48–59 (1974).


# ROBINSONITE
$Pb_7 Sb_{12} S_{25}$

CRYSTAL SYSTEM: Triclinic
CLASS: $\bar{1}$
SPACE GROUP: P$\bar{1}$
Z: 1
LATTICE CONSTANTS:
*a* = 16.51      α = 96°4′
*b* = 17.62      β = 96°22′
*c* = 3.97       γ = 91°12′
3 STRONGEST DIFFRACTION LINES:
3.39 (100)
4.04 ( 80)
3.92 ( 80)
HARDNESS: 2½–3
DENSITY: 5.20–5.34 (Meas.)
5.40 (Calc.)
CLEAVAGE: Not observed.
Fracture irregular. Brittle.

HABIT: Crystals slender prismatic [001], striated [001]; also massive, fibrous to compact.

COLOR-LUSTER: Bluish lead gray. Opaque. Metallic.

MODE OF OCCURRENCE: Occurs as a primary mineral with pyrite, sphalerite, stibnite, and boulangerite in small pieces in oxidized ore bodies at the Red Bird mercury mine, Pershing County, Nevada.

BEST REF. IN ENGLISH: Berry, L. G., Fahey, Joseph J., and Bailey, Edgar H., *Am. Min.*, **37**: 438–446 (1952).

## ROCKBRIDGEITE
$(Fe^{2+}, Mn)Fe_4^{3+}(PO_4)_3(OH)_5$
Frondelite-Rockbridgeite Series

CRYSTAL SYSTEM: Orthorhombic
CLASS: 2/m 2/m 2/m
SPACE GROUP: Bbmm
Z: 4
LATTICE CONSTANTS:
 $a = 13.783$
 $b = 16.805$
 $c = 5.172$
3 STRONGEST DIFFRACTION LINES:
 3.196 (100)
 4.842 ( 50)
 3.573 ( 50)
OPTICAL CONSTANTS:
 $\alpha = 1.875$
 $\beta = 1.880$
 $\gamma = 1.897$
 (+)2V = moderate
HARDNESS: 4½
DENSITY: 3.3–3.49 (Meas.)
CLEAVAGE: {100} perfect
 {010} distinct
 {001} distinct
Fracture uneven. Brittle.
HABIT: As minute elongated slender prismatic crystals, often as isolated individuals or as radial aggregates. Usually as botryoidal crusts or as masses with radial-fibrous or fine columnar structure.
COLOR-LUSTER: Light to dark olive green to nearly black, often becoming brownish green to reddish brown on oxidation. Aggregates often exhibit concentric color banding. Subtranslucent. Vitreous to dull.
MODE OF OCCURRENCE: Occurs in "limonite" beds, in novaculite deposits, and as an alteration product of triphylite or other iron-manganese phosphate minerals in granite pegmatites. Found widespread in the Black Hills, South Dakota, especially at the Bull Moose and Tip Top mines near Custer, and at the Big Chief mine near Keystone, as excellent microcrystals and large fibrous masses; in prospect pits in the Ouachita Mountains, Arkansas; in Cherokee and Coosa Counties, Alabama; at Greenbelt, Maryland; in Rockbridge County, Virginia; and at the Fletcher and Palermo mines, near North Groton, New Hampshire. It also is found in France, Germany, USSR, and Brazil.
BEST REF. IN ENGLISH: Palache, et al., "Dana's System of Mineralogy," 7th Ed., v. II, p. 867–869, New York, Wiley, 1951. Moore, Paul, *Am. Min.*, **55**: 135–169 (1970).

## ROCK-CRYSTAL = Quartz

## RODALQUILARITE
$H_3Fe_2(TeO_3)_4Cl$

CRYSTAL SYSTEM: Triclinic
Z: 1
LATTICE CONSTANTS:
 $a = 8.89$    $\alpha = 103°10'$
 $b = 5.08$    $\beta = 107°5'$
 $c = 6.63$    $\gamma = 77°52'$
3 STRONGEST DIFFRACTION LINES:
 4.24 (100)
 2.62 ( 80)
 3.31 ( 60)
OPTICAL CONSTANTS:
 $N \sim 2.2$
 $(-)2V \approx 38°$
HARDNESS: 2–3
DENSITY: 5.05–5.15 (Meas.)
 5.14 (Calc.)
CLEAVAGE: One good cleavage.
Very brittle, easily crushed to a greenish yellow powder.
HABIT: Tiny stout crystals, usually less than 0.1 mm; also as crusts.
COLOR-LUSTER: Grass green to emerald green; greasy.
MODE OF OCCURRENCE: Occurs associated with native gold, jarosite, and rare emmonsite, probably as an alteration product of gold tellurides, in quartz vugs in the zone of oxidation of the Rodalquilar gold deposit, Almeria Province, Spain.
BEST REF. IN ENGLISH: Lopez, J. Sierra, Leal, G., Pierrot, R., Laurent, Y., Protas J., and DuSausoy, Y., *Am. Min.*, **53**: 2104–2105 (1968).

## ROEBLINGITE
$Pb_2Ca_7Si_6O_{14}(OH)_{10}(SO_4)_2$

CRYSTAL SYSTEM: Monoclinic
CLASS: m or 2/m
SPACE GROUP: Cc or C2/c
Z: 2
LATTICE CONSTANTS:
 $a = 13.27$
 $b = 8.38$
 $c = 13.09$
 $\beta = 103.86°$
3 STRONGEST DIFFRACTION LINES:
 2.94 (100)
 3.00 ( 90)
 3.10 ( 80) or 3.17 (80)
OPTICAL CONSTANTS:
 $\alpha = 1.64$
 $\beta = 1.64$
 $\gamma = 1.66$
 (+)2V = small
HARDNESS: 3
DENSITY: 3.433 (Meas.)
CLEAVAGE: {001} perfect
 {100} imperfect
HABIT: As compact masses composed of minute prismatic crystals; also fibrous or platy.
COLOR-LUSTER: White, pink. Translucent. Dull to somewhat resinous.
MODE OF OCCURRENCE: Occurs in association with garnet, axinite, esperite, clinohedrite, and other minerals,

at Franklin, Sussex County, New Jersey; also in fracture fillings at Långban, Sweden.

BEST REF. IN ENGLISH: Palache, Charles, USGS Prof. Paper No. 180, p. 113 (1935). Foit, Franklin F. Jr., *Am. Min.*, **51**: 504-508 (1966).

# ROEDDERITE
$(Na, K)_2 Mg_5 Si_{12} O_{30}$

CRYSTAL SYSTEM: Hexagonal
Z: 2
LATTICE CONSTANTS:
 $a = 10.139$
 $c = 14.275$
3 STRONGEST DIFFRACTION LINES:
 3.570 (100)
 3.239 ( 77)
 2.922 ( 67)
OPTICAL CONSTANTS:
 $\omega = 1.537$
 $\epsilon = 1.542$
HARDNESS: Not determined.
DENSITY: 2.63 (Calc.)
 ~2.6 (Meas.)
CLEAVAGE: None.
HABIT: Flat plates partly bounded laterally by faces intersecting at 120°.
COLOR-LUSTER: Colorless.
MODE OF OCCURRENCE: Found as an accessory in the black enstatite chondrite Indarch that fell on April 7, 1891, in Transcaucasia. (Meunier, S., *Compt. rendus*, **125**: 894 [1897]); also in the octahedrite iron meteorite Wichita County (USA).
BEST REF. IN ENGLISH: Fuchs, Louis H., Frondel, Clifford, and Klein, Cornelis Jr., *Am. Min.*, **51**: 949-955 (1966).

# ROEMERITE
$Fe^{2+}Fe_2^{3+}(SO_4)_4 \cdot 14H_2O$

CRYSTAL SYSTEM: Triclinic
CLASS: $\bar{1}$
SPACE GROUP: $P\bar{1}$
Z: 1
LATTICE CONSTANTS:
 $a = 6.463$  $\alpha = 90°32'$
 $b = 15.309$  $\beta = 101°5'$
 $c = 6.341$  $\gamma = 85°44'$
3 STRONGEST DIFFRACTION LINES:
 4.79 (100)
 4.03 ( 90)
 5.05 ( 50)
OPTICAL CONSTANTS:
 $\alpha = 1.519-1.524$
 $\beta = 1.570-1.571$
 $\gamma = 1.578-1.583$
 $(-)2V \simeq 45°-51°$
HARDNESS: 3-3½
DENSITY: 2.18 (Meas.)
 2.173 (Calc.)

CLEAVAGE: {010} perfect
 {001} distinct
Fracture uneven. Brittle.
HABIT: Crystals thick tabular; cuboidal. As crystal aggregates and crusts; granular; stalactitic.
COLOR-LUSTER: Clove brown to yellow, also violet-brown. Translucent. Vitreous, resinous, or greasy.

Soluble in water. Taste saline, astringent.
MODE OF OCCURRENCE: Occurs abundantly as crystalline masses and minute crystals, associated with coquimbite and other sulfates, at the Dexter No. 7 mine, Calf Mesa, San Rafael Swell, Utah; as an alteration of pyrrhotite at Island Mountain, Trinity County, and in Alpine, Contra Costa, Napa, San Benito, and San Bernardino counties, California; and at the United Verde mine, Jerome, Arizona. It also is found in Chile, Germany, Czechoslovakia, USSR, and with voltaite at Madeni Zakh, Iran.
BEST REF. IN ENGLISH: Palache, et al., "Dana's System of Mineralogy," 7th Ed., v. II, p. 520-522, New York, Wiley, 1951. *Am. Min.*, **55**: p. 78-89 (1970).

# ROENTGENITE (Röntgenite)
$Ca_2 (Ce, La)_3 (CO_3)_5 F_3$

CRYSTAL SYSTEM: Hexagonal
CLASS: 3
SPACE GROUP: R3
Z: 9
LATTICE CONSTANTS:
 $a = 7.131$
 $c = 69.41$
OPTICAL CONSTANTS:
 $\omega = 1.662$ (Na)
 $\epsilon = 1.756$
HARDNESS: Not determined. Probably ~4½.
DENSITY: 4.19 (Calc.)
CLEAVAGE: None observed. Fracture conchoidal.
HABIT: Similar to parisite and synchysite; most crystals singly terminated, with $q\{10\bar{1}1\}$ and $p\{01\bar{1}2\}$ dominant, truncated by a small $c\{0003\}$. The two pyramids are striated horizontally.
COLOR-LUSTER: Wax yellow to brown; transparent to translucent.
MODE OF OCCURRENCE: Occurs as minute crystals intimately associated with synchysite, parisite, and bastnaesite at Narsarsuk, Greenland.
BEST REF. IN ENGLISH: Donnay, Gabrielle, *Am. Min.*, **38**: 868-870 (1953). Donnay, Gabrielle, and Donnay, J. D. H., *Am. Min.*, **38**: 932-963 (1953).

# ROEPPERITE = Variety of tephroite

# ROESSLERITE
$MgH(AsO_4) \cdot 7H_2O$

CRYSTAL SYSTEM: Monoclinic
CLASS: 2/m or m
SPACE GROUP: A2/a or Aa
Z: 8

LATTICE CONSTANTS:
$a = 11.61$
$b = 25.92$
$c = 6.728$
$\beta = 95°23'$

3 STRONGEST DIFFRACTION LINES:
4.08 (100)
6.49 ( 80)
4.72 ( 80)

OPTICAL CONSTANTS:
$\alpha = 1.525$
$\beta = 1.53$
$\gamma = 1.550$
(+)2V = small

HARDNESS: 2–3

DENSITY: 1.930 (Meas.)
1.913 (Calc.)

CLEAVAGE: {111} imperfect

HABIT: Crystals short prismatic, small and rare. As fine-grained or fibrous crusts.

COLOR-LUSTER: Colorless, white. Transparent. Vitreous to dull.

Slightly soluble in water.

MODE OF OCCURRENCE: Occurs as a secondary mineral in arsenic-bearing ore deposits. Found in Germany at Bieber, near Hanau; at the Sophia mine, near Wittichen, and at the Wolfgang mine, near Alpirsbach, in the Black Forest; and also at Joachimsthal (Jachymov), Bohemia, Czechoslovakia.

BEST REF. IN ENGLISH: Palache, et al., "Dana's System of Mineralogy," 7th Ed., v. II, p. 712–713, New York, Wiley, 1951.

# ROGGIANITE

$NaCa_6Al_9Si_{13}O_{46} \cdot 20H_2O$

CRYSTAL SYSTEM: Tetragonal
CLASS: 4mm, $\overline{4}$2m, or 4/m 2/m 2/m
SPACE GROUP: I4cm, I$\overline{4}$c2, or I4/mcm
LATTICE CONSTANTS:
$a = 18.37$
$c = 9.14$

3 STRONGEST DIFFRACTION LINES:
13.08 (100)
9.27 ( 80)
6.13 ( 80)

OPTICAL CONSTANTS:
$\omega = 1.527$
$\epsilon = 1.535$
(+)

HARDNESS: Not determined.

DENSITY: 2.02 (Meas.–certainly too low.)

CLEAVAGE: {110} perfect

HABIT: As fibrous aggregates; elongated along c-axis.

COLOR-LUSTER: Whitish yellow; white.

MODE OF OCCURRENCE: Occurs coating fractures in a sodium feldspar dike cutting gneiss at Alpa Rosso, Val Vigezzo, Novara Prov., Italy.

BEST REF. IN ENGLISH: Passaglia, E., and Gard, J. A., *Am. Min.*, **55**: 322–323 (1970).

# ROMANECHITE = Psilomelane

# ROMARCHITE

SnO

CRYSTAL SYSTEM: Tetragonal
CLASS: 4/m 2/m 2/m
SPACE GROUP: P4/nmm
Z: 1
LATTICE CONSTANTS:
$a = 3.79$
$c = 4.83$

3 STRONGEST DIFFRACTION LINES:
2.98 (100)
1.601 ( 90)
1.799 ( 70)

HARDNESS: Not determined.

DENSITY: Not determined.

CLEAVAGE: Not determined.

HABIT: As thin crusts composed of minute crystals.

COLOR-LUSTER: Black.

MODE OF OCCURRENCE: Tin pannikins lost from the overturned canoe of a voyager between 1801 and 1821 were recovered 15 feet below the surface of the water at Boundary Falls, Winnipeg River, Ontario, Canada. Some of the surfaces have a thin crust composed of the alteration products romarchite and hydromarchite.

BEST REF. IN ENGLISH: Organ, R. M., and Mandarino, J. A., *Can. Min.*, **10**: 916 (1971) (abs.)

# ROMÉITE

$(Ca, Fe, Na)_2(Sb, Ti)_2(O, OH)_7$

CRYSTAL SYSTEM: Cubic
CLASS: 4/m $\overline{3}$ 2/m
SPACE GROUP: Fd3m
Z: 8
LATTICE CONSTANT:
$a = 10.29$ kX (variable)

3 STRONGEST DIFFRACTION LINES:
2.94 (100)
1.813 (100)
1.548 (100)

OPTICAL CONSTANT:
$N = 1.817–1.87$

HARDNESS: 5½–6½

DENSITY: 4.7–5.4 (Meas.)
5.31 (Calc.)

CLEAVAGE: {111} imperfect
Fracture uneven to splintery.

HABIT: Crystals small, octahedral, embedded or in groups. Also massive. Twinning on {111}, not common.

COLOR-LUSTER: Pale yellow to dark brown. Transparent to nearly opaque. Vitreous to greasy. Streak pale yellow to nearly colorless.

MODE OF OCCURRENCE: Occurs in manganese ores at Miguel Burnier, and with cinnabar in placer deposits at Tripuhy near Ouro Preto, Minas Geraes, Brazil; at Långban and other places in Sweden; and at deposits in Italy, Austria, and Algeria.

BEST REF. IN ENGLISH: Palache, et al., "Dana's System of Mineralogy," 7th Ed., v. II, p. 1020–1022, New York, Wiley, 1951.

## ROOSEVELTITE
$BiAsO_4$
Isostructural with monazite

CRYSTAL SYSTEM: Monoclinic
OPTICAL CONSTANTS:
$N > 1.74$
HARDNESS: 4–4½
DENSITY: 6.86 (Meas.)
CLEAVAGE: Not determined.
Brittle.
HABIT: As thin botryoidal crusts.
COLOR-LUSTER: Gray. Adamantine.
MODE OF OCCURRENCE: Occurs in cassiterite veinlets in dacitic and rhyolitic lava flows at Santiaguilo, Potosi, Bolivia.
BEST REF. IN ENGLISH: Palache, et al., "Dana's System of Mineralogy," 7th Ed., v. II, p. 697, New York, Wiley, 1951.

## ROQUESITE
$CuInS_2$

CRYSTAL SYSTEM: Tetragonal
CLASS: $\bar{4}2m$
SPACE GROUP: $I\bar{4}2d$
Z: 4
LATTICE CONSTANTS:
$a = 5.51$
$c = 11.05$
3 STRONGEST DIFFRACTION LINES:
3.16 (100)
1.94 ( 80)
1.66 ( 60)
OPTICAL CONSTANTS:
uniaxial (+)
HARDNESS: 3½–4
DENSITY: Not determined.
4.80 (Calc.)
CLEAVAGE: Not determined.
HABIT: As subhedral crystals to rounded grains up to 10 μ in size.
COLOR-LUSTER: Gray with slight bluish tint. Opaque.
MODE OF OCCURRENCE: Occurs associated with chalcopyrite, sphalerite, calcite, or fluorite at Mt. Pleasant, about 35 miles south of Fredricton, New Brunswick, Canada; as inclusions in bornite, associated with chalcopyrite, wittichenite, chalcocite, covellite, and a little sphalerite at the copper-tin-iron deposit of Charrier, Allier, France; and with chalcopyrite at Akenobe, Japan.
BEST REF. IN ENGLISH: Picot, P., and Pierrot, R., *Am. Min.*, **48**: 1178–1179 (1963).

## ROSASITE
$(Cu, Zn)_2 CO_3 (OH)_2$

CRYSTAL SYSTEM: Monoclinic
Z: 4
LATTICE CONSTANTS:
$a = 9.4$
$b = 12.3$
$c = 3.4$
$\beta \sim 90°$
3 STRONGEST DIFFRACTION LINES:
3.68 (100)
2.59 (100)
5.04 ( 80)
OPTICAL CONSTANTS:
$\alpha = 1.672–1.708$
$\beta \sim 1.83$
$\gamma = 1.832$
$(-)2V$ = very small
HARDNESS: ~4½
DENSITY: 4.0–4.2 (Meas.)
CLEAVAGE: In two directions at right angles.
Brittle.
HABIT: Botryoidal or mammillary crusts with fibrous structure; and as felt-like crusts of microscopic crystals.
COLOR-LUSTER: Bluish green to green; also sky blue.
MODE OF OCCURRENCE: Occurs as a secondary mineral in the oxidation zone of zinc-copper-lead ore deposits. Found associated with smithsonite and hemimorphite at Cerro Gordo, Inyo County, California; at the Myler mine, Majuba Hill, Pershing County, and near Wellington, Nevada; with aurichalcite at Kelly, Socorro County, New Mexico; and at several localities in Cochise County and elsewhere in Arizona. It also occurs at the Rosas mine, Sulcis, Sardinia; at Tsumeb, South West Africa; and as fine specimens at Mapimi, Durango, Mexico.
BEST REF. IN ENGLISH: Palache, et al., "Dana's System of Mineralogy," 7th Ed., v. II, p. 251–252, New York, Wiley, 1951.

## ROSCHERITE
$(Ca, Mn, Fe)_3 Be_3 (PO_4)_3 (OH)_3 \cdot 2H_2O$

CRYSTAL SYSTEM: Monoclinic
CLASS: 2/m
SPACE GROUP: C2/c
Z: 4
LATTICE CONSTANTS:
$a = 15.95$
$b = 11.95$
$c = 6.62$
$\beta = 94°50'$
3 STRONGEST DIFFRACTION LINES:
5.96 (100)
9.58 ( 90)
3.18 ( 70)
OPTICAL CONSTANTS:
$\alpha = 1.636$    $\alpha = 1.624$
$\beta = 1.641$    $\beta = 1.639$
$\gamma = 1.651$    $\gamma = 1.643$
$(+)2V$    $(-)2V$ = large
HARDNESS: 4½
DENSITY: 2.934 (Meas.)
2.93 (Calc.)

CLEAVAGE:  {001} good
{010} distinct

HABIT:  Crystals tabular parallel to (100); also short prismatic [001], with six- or eight-sided cross section; and as aggregates of thin plates or radiating fibrous masses.

COLOR-LUSTER:  Olive green, light brown, dark brown, reddish brown.

MODE OF OCCURRENCE:  Occurs in pegmatite at the Nevel quarry, Newry and at Black Mountain, Maine; as single crystals, crystal aggregates, and granular crusts in vugs in muscovite, associated with faheyite, variscite, frondelite, beryl, and euhedral crystals of quartz at the Sapucaia pegmatite mine, near Conselheiro Pena, Minas Geraes, Brazil. Also found with morinite, apatite, lacroixite, childrenite, and tourmaline in drusy cavities of a granite at Greifenstein, near Ehrenfriedersdorf, Saxony, Germany.

BEST REF. IN ENGLISH: Palache, et al., "Dana's System of Mineralogy," 7th Ed., v. II, p. 968–969, New York, Wiley, 1951.    Lindberg, Mary Louise, *Am. Min.*, **43**: 824–838 (1958).

# ROSCOELITE

K(V, Al, Mg)$_3$(Al, Si$_3$)O$_{10}$(OH)$_2$
Mica Group

CRYSTAL SYSTEM:  Monoclinic
CLASS:  2/m
SPACE GROUP:  C2/m
Z:  2
LATTICE CONSTANTS:
$a$ = 5.26
$b$ = 9.09      (synthetic)
$c$ = 10.25
$\beta$ = 101.0°
3 STRONGEST DIFFRACTION LINES:
10.0 (100)
4.54 ( 80)  (1M)
3.35 ( 80)
OPTICAL CONSTANTS:
$\alpha$ = 1.610
$\beta$ = 1.685
$\gamma$ = 1.704
(–)2V = 24.5°–39.5°
HARDNESS:  2½
DENSITY:  2.97 (Meas.)
CLEAVAGE:  {001} perfect
HABIT:  As minute scales, often in stellate groups.

COLOR-LUSTER:  Clove brown to greenish brown, also dark green. Translucent. Pearly.

MODE OF OCCURRENCE:  Found interlaminated with native gold at the Stuckslager or Sam Sims mine on Granite Creek, near Coloma, and at several other localities in El Dorado County, California; as an associate of telluride minerals from widespread localities in Colorado; at numerous uranium-vanadium mines in Utah, Arizona, and Colorado on the Colorado Plateau; and at Kalgoorlie, Western Australia.

BEST REF. IN ENGLISH: Deer, Howie, and Zussman, "Rock Forming Minerals," v. 3, p. 11–30, New York, Wiley, 1962.

# ROSELITE

Ca$_2$(Co, Mg)(AsO$_4$)$_2$ · 2H$_2$O
Dimorphous with beta-roselite

CRYSTAL SYSTEM:  Monoclinic
CLASS:  2/m
SPACE GROUP:  P2$_1$/c
Z:  2
LATTICE CONSTANTS:
$a$ = 5.60 kX
$b$ = 12.80
$c$ = 5.60
OPTICAL CONSTANTS:
$\alpha$ = 1.694          $\alpha$ = 1.725
$\beta$ = 1.704  and   $\beta$ = 1.728
$\gamma$ = 1.719          $\gamma$ = 1.735
(+)2V = 75°          (+)2V = 60°
HARDNESS:  3½
DENSITY:  3.50–3.74 (Meas.)
3.65 (Calc.)
CLEAVAGE:  {010} perfect, easy
HABIT:  Crystals short prismatic or thick tabular; terminations usually simple.  As crystal druses and as spherical aggregates. Twinning on {100} common, often as fourlings.

COLOR-LUSTER:  Dark rose red to pink. Transparent to translucent. Vitreous.

MODE OF OCCURRENCE:  Occurs as superb crystals associated with erythrite at Bou Azzer, Morocco; at the Daniel and Rappold mines at Schneeberg, Saxony, and at Shapbach, Bavaria, Germany.

BEST REF. IN ENGLISH: Palache, et al., "Dana's System of Mineralogy," 7th Ed., v. II, p. 723–725, New York, Wiley, 1951.

# ROSENBUSCHITE

(Ca, Na)$_3$(Zr, Ti)Si$_2$O$_8$F

CRYSTAL SYSTEM:  Triclinic
CLASS:  $\bar{1}$
SPACE GROUP:  P$\bar{1}$
Z:  4
LATTICE CONSTANTS:
$a$ = 10.12 kX      $\alpha$ = 91°21′
$b$ = 11.39          $\beta$ = 99°38.5′
$c$ = 7.27            $\gamma$ = 111°54.5′
3 STRONGEST DIFFRACTION LINES:
2.94 (100)
3.06 ( 80)
1.89 ( 60)
OPTICAL CONSTANTS:
$\alpha$ = 1.678
$\beta$ = 1.687
$\gamma$ = 1.705
(+)2V = 78°
HARDNESS:  5–6
DENSITY:  3.30–3.315 (Meas.)
CLEAVAGE:  {100} perfect.
Fracture uneven. Brittle.

HABIT:  Crystals prismatic or acicular, up to 2mm long. Usually massive with radial-fibrous structure.

COLOR-LUSTER: Gray to orange to brownish. Transparent in thin fragments. Silky to subvitreous. Sometimes fluorescent yellowish to greenish white under ultraviolet light.

MODE OF OCCURRENCE: Occurs in nepheline-syenites in association with aegirine, woehlerite, mosandrite, eucolite, astrophyllite, and other minerals. Found at Red Hill, Moultonboro, Carroll County, New Hampshire; in Brazil; in the Langesundfjord district, Norway; and from the Lovozero massif, Kola Peninsula, USSR.

BEST REF. IN ENGLISH: Vlasov, K. A., "Mineralogy of Rare Elements," v. II, p. 388–390, Israel Program for Scientific Translations, 1966.

## ROSENHAHNITE
(CaSiO$_3$)$_3$ · H$_2$O with minor substitution of Ba for Ca

CRYSTAL SYSTEM: Triclinic
CLASS: $\bar{1}$ or 1
SPACE GROUP: P$\bar{1}$ or P1
Z: 2
LATTICE CONSTANTS:
  $a = 6.946$     $\alpha = 108°39'$
  $b = 9.474$     $\beta = 94°49'$
  $c = 6.809$     $\gamma = 95°43'$
3 STRONGEST DIFFRACTION LINES:
  3.201 (100)
  2.965 ( 90)
  3.043 ( 60)
OPTICAL CONSTANTS:
  $\alpha = 1.625$
  $\beta = 1.640$
  $\gamma = 1.646$
  $(-)2V \approx 64°$
HARDNESS: 4.5–5
DENSITY: 2.89 (Meas.)
         2.905 (Calc.)
CLEAVAGE: {001} perfect
          {100} good
          {010} good
HABIT: Crystals tabular to lath-like with {010} the most prominent form. The crystals occur in irregular clusters and often are dull and etched.

COLOR-LUSTER: Colorless to buff.

MODE OF OCCURRENCE: Found as crystals up to a maximum size of 5 X 2 X 1 mm in narrow veins in brecciated, fine-grained diopside-garnet metasedimentary rock along the east bank of the Russian River, north of Cloverdale, Mendocino County, California.

BEST REF. IN ENGLISH: Pabst, A., Gross, E. B., and Alfors, J. T., *Am. Min.*, **52**: 336–351 (1967).

## ROSICKÝITE (γ-sulfur)
S
Trimorphous with α-sulfur and β-sulfur

CRYSTAL SYSTEM: Monoclinic
CLASS: 2/m
SPACE GROUP: P2/n
Z: 32

LATTICE CONSTANTS:
  $a = 8.52$
  $b = 13.05$
  $c = 8.23$
  $\beta = 112°53'$
OPTICAL CONSTANTS:
  $(-)2V = $ large
HARDNESS: Soft
DENSITY: < 2.075
CLEAVAGE: None.
Gliding on {001}, {101}, {$\bar{1}$01}.
HABIT: Crystals equidimensional to platy or acicular; usually less than 0.5 mm in diameter. Sometimes skeletal. Inverts to α-sulfur at room temperature. Twinning on {101}.

COLOR-LUSTER: Colorless to pale yellow with greenish tinge. Adamantine.

MODE OF OCCURRENCE: Occurs in the center of walnut-sized limonite nodules found in a thin stratum of clay at Havérna, near Letovíce, Mähren, Czechoslovakia; also reported from a fumarole at Vulcano, Lipari Islands.

BEST REF. IN ENGLISH: Sekanina, J., *Am. Min.*, **17**: 251–252 (1932).

## ROSIÉRÉSITE (Species status in doubt)
Hydrous phosphate of Pb, Cu, Al

CRYSTAL SYSTEM: Unknown
OPTICAL CONSTANT:
  $N \sim 1.50$
HARDNESS: Not determined.
DENSITY: 2.2 (Meas.)
CLEAVAGE: None.
HABIT: As opaline stalactitic masses with concentric structure.

COLOR-LUSTER: Greenish yellow, yellow, pale brown.

MODE OF OCCURRENCE: Occurs as a recent deposit on mine walls at the copper mine at Rosiérés, Tarn, France.

BEST REF. IN ENGLISH: Palache, Charles, et al., "Dana's System of Mineralogy," 7th Ed., v. II, p. 924, New York, Wiley, 1951.

## ROSSITE
CaV$_2$O$_6$ · 4H$_2$O

CRYSTAL SYSTEM: Triclinic
CLASS: $\bar{1}$
SPACE GROUP: P$\bar{1}$
Z: 2
LATTICE CONSTANTS:
  $a = 8.53$     $\alpha = 101°32'$
  $b = 8.56$     $\beta = 114°58'$
  $c = 7.02$     $\gamma = 103°23'$
OPTICAL CONSTANTS:
  $\alpha = 1.710$
  $\beta = 1.770$   (synthetic)
  $\gamma = 1.840$
  $(-?)2V = $ large
HARDNESS: 2–3
DENSITY: 2.45

CLEAVAGE: {010} good

Brittle.

HABIT: As small glassy kernels embedded in metarossite. Synthetic crystals usually elongated [001] and flattened on {010}; rarely tabular, lath-like, or needle-like. Twinning on {100}.

COLOR-LUSTER: Yellow. Transparent. Vitreous to pearly.

Slowly soluble in water.

MODE OF OCCURRENCE: Occurs with metarossite filling small veinlets in carnotite-bearing sandstone at Bull Pen Canyon, San Miguel County, Colorado, about 5 miles southeast of Summit Point post office, Utah.

BEST REF. IN ENGLISH: Palache, et al., "Dana's System of Mineralogy," 7th Ed., v. II, p. 1053–1054, New York, Wiley, 1951.

## ROUBAULTITE
$Cu_2(UO_2)_3(OH)_{10} \cdot 5H_2O$

CRYSTAL SYSTEM: Triclinic
CLASS: 1 or $\bar{1}$
SPACE GROUP: P1 or P$\bar{1}$
Z: 1
LATTICE CONSTANTS:
  $a = 7.73$    $\alpha = 86°29'$
  $b = 6.87$    $\beta = 134°12'$
  $c = 10.87$   $\gamma = 93°10'$
3 STRONGEST DIFFRACTION LINES:
  5.55 (100)
  7.74 ( 90)
  6.88 ( 80)
OPTICAL CONSTANTS:
  $\alpha = 1.700$
  $\beta = 1.800$
  $\gamma = 1.84$
  HARDNESS: ~3
  DENSITY: 5.02 (Calc.)
  CLEAVAGE: {100} perfect
            {010} distinct
HABIT: Crystals platy; in rosettes up to 3.0 mm in diameter.
COLOR-LUSTER: Green.
MODE OF OCCURRENCE: Occurs in association with becquerelite, vandenbrandeite, soddyite, and cuprosklodowskite, in the oxidation zone of the uranium deposit at Shinkolobwe, Katanga, Zaire.
BEST REF. IN ENGLISH: Cesbron, F., Pierrot, R., and Verbeek, T., *Min. Abs.*, **22**: 229 (1971).

## ROWEITE
$Ca_2Mn_2^{2+}(OH)_4[B_4O_7(OH)_2]$

CRYSTAL SYSTEM: Orthorhombic
CLASS: 2/m 2/m 2/m
SPACE GROUP: Pbam
Z: 4
LATTICE CONSTANTS:
  $a = 9.057$
  $b = 13.357$
  $c = 8.289$

OPTICAL CONSTANTS:
  $\alpha = 1.648$
  $\beta = 1.660$
  $\gamma = 1.663$
  $(-)2V = 15°$
HARDNESS: ~5
DENSITY: 2.935 (Meas.)
         2.93 (Calc.)
CLEAVAGE: {101} poor
Brittle.
HABIT: Crystals lath-like, elongated along $c$-axis and flattened {010}.
COLOR-LUSTER: Light brown. Transparent.
MODE OF OCCURRENCE: Found associated with willemite and thomsonite in veinlets cutting franklinite-zincite ore at Franklin, Sussex County, New Jersey. Only one specimen is known.
BEST REF. IN ENGLISH: Palache, et al., "Dana's System of Mineralogy," 7th Ed., v. II, p. 377–378, New York, Wiley, 1951.

## ROWLANDITE = Thalenite

## ROZENITE
$FeSO_4 \cdot 4H_2O$

CRYSTAL SYSTEM: Monoclinic
CLASS: 2/m
SPACE GROUP: P2$_1$/n
Z: 4
LATTICE CONSTANTS:
  $a = 5.945$
  $b = 13.59$
  $c = 7.94$
  $\beta = 90°30'$
3 STRONGEST DIFFRACTION LINES:
  4.47 (100)
  5.46 ( 90)
  3.97 ( 70)
OPTICAL CONSTANTS:
  $\alpha = 1.526–1.528$
  $\beta = 1.536–1.537$
  $\gamma = 1.541–1.545$
  $(-)2V$ near 90°
HARDNESS: Not determined.
DENSITY: 2.195 (Meas.)
         2.29 (Calc.)
CLEAVAGE: Not determined.
HABIT: Crystals tabular, very minute, about 0.05 X 0.15 $\mu$ in size. Usually as microcrystalline crusts.
COLOR-LUSTER: Colorless, white. Dull.
MODE OF OCCURRENCE: Occurs in Township 44, Range 28, west of 1st meridian, Manitoba, Canada; from the Thames River, England; and as an efflorescence on weathered gneisses containing pyrite on the slopes of Ornak, Western High Tatara, and in an old gallery of the "Staszic" pyrite mine at Rudki, Poland.
BEST REF. IN ENGLISH: Kubisz, J., *Min. Abs.*, **15**: 364 (1962). Jambor, J. L., and Traill, R. J., *Can. Min.*, **7**: 751–763 (1963).

**RUBELLITE** = Pink or red variety of elbaite

**RUBY** = Gem variety of corundum

## RUSAKOVITE (Inadequately described mineral)
(Fe, Al)$_5$ {(V, P)O$_4$}$_2$ (OH)$_9$ · 3H$_2$O

CRYSTAL SYSTEM: Unknown
3 STRONGEST DIFFRACTION LINES:
  3.21 (100)
  2.945 ( 90)
  2.441 ( 80)
OPTICAL CONSTANT:
  $n$ = 1.833
  HARDNESS: 1.5-2
  DENSITY: 2.73-2.80 (Meas.)
  HABIT: As crusts, veinlets, and reniform concretions.
  COLOR-LUSTER: Yellow-orange to reddish yellow; dull. Streak ocher yellow.
  MODE OF OCCURRENCE: Occurs in a surface, partially oxidized layer of carbonaceous shale, which contains apatite, colophane, vanadium-bearing mica, and sulfides of copper, zinc, lead, and vanadium in northwestern Kara-Tau, Kazakhstan.
  BEST REF. IN ENGLISH: Ankinovich, E. A., *Am. Min.*, **45**: 1316 (1960).

## RUSSELLITE
Bi$_2$WO$_6$

CRYSTAL SYSTEM: Tetragonal
CLASS: $\bar{4}$2m or 4/m 2/m 2/m
SPACE GROUP: I$\bar{4}$2d or I4/amd
Z: 4
LATTICE CONSTANTS:
  $a$ = 5.42
  $c$ = 11.3
3 STRONGEST DIFFRACTION LINES:
  3.08 (100)
  1.64 (100)
  1.91 ( 90)
OPTICAL CONSTANTS:
  Mean index = 2.2
  Uniaxial (+)
  HARDNESS: 3½
  DENSITY: 7.35 (Meas.)
           7.40 (Calc.)
  CLEAVAGE: Not determined.
  HABIT: Massive, compact, fine-grained.
  COLOR-LUSTER: Pale yellow to greenish.
  MODE OF OCCURRENCE: Occurs associated with native bismuth, bismuthinite, wolframite, cassiterite, and other minerals, at the Castle-an-Dinas mine, St. Columb Major, Cornwall, England.
  BEST REF. IN ENGLISH: Palache, et al., "Dana's System of Mineralogy," 7th Ed., v. I, p. 604-605, New York, Wiley, 1944.

## RUSTUMITE
Ca$_4$Si$_2$O$_7$(OH)$_2$

CRYSTAL SYSTEM: Monoclinic
CLASS: m or 2/m
SPACE GROUP: Cc or C2/c
Z: 10
LATTICE CONSTANTS:
  $a$ = 7.62
  $b$ = 18.55
  $c$ = 15.51
  $\beta$ = 104°20'
3 STRONGEST DIFFRACTION LINES:
  3.03 (100)
  2.89 ( 90)
  3.19 ( 80)
  HARDNESS: Not determined.
  DENSITY: 2.835 (Calc.)
  CLEAVAGE: {100} poor
                {010} very poor
                {001} very poor
  HABIT: As crude tabular crystals up to 2 mm in length. Lamellar twinning on (100) common.
  COLOR-LUSTER: Colorless with faint hazy turbidity.
  MODE OF OCCURRENCE: Occurs associated with åkermanite, merwinite, larnite, spurrite, rankinite, and kilchoanite in a metamorphic assemblage at Kilchoan, Ardnamurchan, Scotland.
  BEST REF. IN ENGLISH: Agrell, S. O., *Min. Mag.*, **34** (Tilley Volume): 1-15 (1965).

## RUTHERFORDINE
UO$_2$CO$_3$

CRYSTAL SYSTEM: Orthorhombic
CLASS: 2/m 2/m 2/m
SPACE GROUP: Pmmm
Z: 2
LATTICE CONSTANTS:
  $a$ = 4.845
  $b$ = 9.205
  $c$ = 4.296
3 STRONGEST DIFFRACTION LINES:
  4.60 (100)
  3.21 ( 90)
  4.29 ( 80)
OPTICAL CONSTANTS:
      $\alpha$ = 1.700-1.723
      $\beta$ = 1.716-1.730
      $\gamma$ = 1.755-1.795
  (+)2V = 46° (variable)
  HARDNESS: Not determined.
  DENSITY: 5.7 (Meas.)
           5.724 (Calc.)
  CLEAVAGE: {010} perfect
                {001} distinct
  HABIT: Thin lath-like single crystals up to 3 mm in length; elongated [001], large {100}, and less dominant {010}. As microscopic fibrous aggregates. Crystals often striated on {100} parallel to [001] and often form radiating clusters.

COLOR-LUSTER: White, pale yellow to amber-orange and brownish. Also yellowish green. Aggregates dull to earthy. Fibrous crusts silky.

MODE OF OCCURRENCE: Occurs associated with becquerelite and masuyite crystals at Shinkolobwe, Katanga, Zaire; in pegmatite resulting from the alteration of uraninite at Morogoro, Tanzania; the Beryl Mountain pegmatite, New Hampshire; and Newry, Maine. Also at several localities in the Colorado Plateau uranium district, at the Lucky Mc No. 20 mine, Fremont County, Wyoming, and from Peak Smelters east of Edwards Creek, South Australia.

BEST REF. IN ENGLISH: Clark, Joan R., and Christ, C. L., *Am. Min.*, **41**: 844–850 (1956).    Frondel, Clifford, USGS Bull. 1064, p. 104–106 (1958).

# RUTILE

$TiO_2$
Trimorphous with anatase, brookite
Var. Struverite-tantalian
      Ilmenorutile-niobian

CRYSTAL SYSTEM: Tetragonal
CLASS: $4/m\ 2/m\ 2/m$
SPACE GROUP: $P4/mnm$
Z: 2
LATTICE CONSTANTS:
  $a = 4.58$
  $c = 2.95$
3 STRONGEST DIFFRACTION LINES:
  3.245 (100)
  1.687 ( 50)
  2.489 ( 41)
OPTICAL CONSTANTS:
  $\omega = 2.6124$  (Na)
  $\epsilon = 2.8993$
     (+)

HARDNESS: 6–6½
DENSITY: 4.23 (Meas.)
       4.250 (Calc.)
       4.2–5.6 (Meas: Nb and Ta varieties)
CLEAVAGE: {110} distinct
         {100} indistinct
         {111} in traces
Parting on {092} and {011}. Fracture conchoidal to uneven. Brittle.

HABIT: Crystals short prismatic, commonly striated parallel to *c*-axis; also slender prismatic to acicular and vertically striated; rarely pyramidal. Also massive, compact granular. Twinning on {011} common, forming knee-shaped twins, sixlings, and eightlings; also twinning on {031} as "arrowhead" twins; and on {092}, as twin gliding plane.

COLOR-LUSTER: Usually reddish brown to red; also yellow, orange-yellow, bluish, grayish black to black; rarely greenish. Transparent to translucent, Nb, Ta, and Fe varieties opaque. Splendent adamantine or submetallic. Streak pale brown to yellowish. Nb-Ta varieties grayish or greenish black.

MODE OF OCCURRENCE: Occurs widespread chiefly in gneiss, schist, quartzite, crystalline limestone and dolomite, alluvial deposits, and as an accessory mineral in many types of igneous rocks. Notable occurrences are found in California, South Dakota (struverite), Arkansas, Pennsylvania, Virginia, North Carolina, Florida, and Graves Mountain, Georgia. Also Canada, Brazil, Norway, Sweden, France, Italy, Switzerland, Germany, Austria, Czechoslovakia, Roumania, USSR, Madagascar, and elsewhere. Fine specimens of quartz, containing inclusions of needle-like crystals of rutile, occur in Switzerland and Minas Geraes, Brazil.

BEST REF. IN ENGLISH: Palache, et al., "Dana's System of Mineralogy," 7th Ed., v. I, p. 554–562, New York, Wiley, 1944.

# S

## SABUGALITE

$HAl(UO_2)_4(PO_4)_4 \cdot 16H_2O$

CRYSTAL SYSTEM: Tetragonal
CLASS: 4/m 2/m 2/m
SPACE GROUP: I4/mmm
Z: 1
LATTICE CONSTANTS:
  $a = 6.96$
  $c = 19.3$
3 STRONGEST DIFFRACTION LINES:
  9.69 (100)
  4.86 ( 90)
  3.47 ( 80)
OPTICAL CONSTANTS:
  $\alpha$ or $\epsilon = 1.564-1.565$
      $\beta = 1.581-1.583$
  $\gamma$ or $\omega = 1.582-1.584$
      $2V = 0°$ to moderate
  HARDNESS: 2½
  DENSITY: 3.2 (Meas.)
              3.15 (Calc.)
  CLEAVAGE: {001} perfect
  HABIT: Crystals thin plates up to 1.0 mm on edge, square or rectangular lath-like in shape. Plane of flattening uneven, warped or striated. As densely aggregated crusts.
  COLOR-LUSTER: Bright yellow to lemon yellow; transparent to translucent. Weakly vitreous. Fluoresces lemon yellow in ultraviolet light.
  MODE OF OCCURRENCE: Occurs in sandstone-type uranium deposits at the Huskon and Arrow Head claims near Cameron, Arizona, and at the Happy Jack mine, White Canyon, San Juan County, Utah. In Wyoming it is found at the Lucky Mc No. 20 mine, the Blue Buck claim, and in the Blarco Group, Fremont County; at the Del Linch claim, Johnson County; and at the Poison Creek claim, Crook County. It occurs in Portugal as a secondary mineral in the weathered outcrops of uraninite veins at Mina da Quarta Feira, Sabugal County, Beira Alta Province; at the Mina de Coitos, Bendada Parish; and at Kariz in Minko Province. It is also found at Margnac II, Haute-Vienne, France.
  BEST REF. IN ENGLISH: Frondel, Clifford, USGS Bull. 1064, p. 196-200 (1958).

## SAFFLORITE

$(Co, Fe)As_2$

CRYSTAL SYSTEM: Orthorhombic and monoclinic
CLASS: 2/m, 2/m 2/m 2/m
SPACE GROUP: Pnnm, P2/m
Z: 2
LATTICE CONSTANTS:
  (Pnnm)        (P2/m)
  $a = 5.81$      $a = 5.051$
  $b = 6.36$      $b = 5.873$
  $c = 4.87$      $c = 3.127$    (synthetic)
                  $\beta = 90°27'$
3 STRONGEST DIFFRACTION LINES:
  2.586 (100)
  2.573 ( 95)
  2.361 ( 90)
OPTICAL CONSTANT: Opaque.
  HARDNESS: 4½-5
  DENSITY: 7.2 (Meas.)
              7.471 (Calc.)
  CLEAVAGE: {100} distinct
Fracture uneven to conchoidal. Brittle.
  HABIT: Crystals short to long prismatic, resembling arsenopyrite. Usually massive with radial-fibrous structure. Twinning on {101} producing cruciform twins or star-shaped trillings; on {011} producing fivelings.
  COLOR-LUSTER: Tin-white, tarnishing to dark gray. Opaque. Metallic. Streak grayish black.
  MODE OF OCCURRENCE: Occurs chiefly in mesothermal vein deposits associated with niccolite, smaltite, and other cobalt-nickel minerals. Found at the Eldorado mine, Great Bear Lake, and at South Lorrain, Ontario, Canada; with native silver and rammelsbergite at Batopolis, Mexico; and at deposits in Germany, Sweden, and Sardinia.
  BEST REF. IN ENGLISH: Palache, et al., "Dana's System of Mineralogy," 7th Ed., v. I, p. 307-309, New York, Wiley, 1944. *Am. Min.*, **48**: 279 (1963).

## SAHAMALITE

$(Mg, Fe^{2+})(Ce, La)_2(CO_3)_4$

CRYSTAL SYSTEM: Monoclinic
CLASS: 2/m
SPACE GROUP: $P2_1/a$
Z: 2
LATTICE CONSTANTS:
$a = 5.92$
$b = 16.21$
$c = 4.63$
$\beta = 106°45'$
3 STRONGEST DIFFRACTION LINES:
3.90 (100)
3.65 (100)
2.87 (100)
OPTICAL CONSTANTS:
$\alpha = 1.679$
$\beta = 1.776$
$\gamma = 1.807$
(−)2V = 57°
HARDNESS: Not determined.
DENSITY: 4.30 (Meas.)
4.30 (Calc.)
CLEAVAGE: {010} poor
HABIT: Crystals tabular, parallel to {$\overline{2}$01}; dominant forms are {$\overline{2}$01}, {$\overline{1}$31}, {120}, {110}, {010}, and {011}. Crystal size 0.01–0.2 mm.
COLOR-LUSTER: Colorless.
MODE OF OCCURRENCE: Occurs associated with parisite and altered bastnaesite in barite-dolomite rock near Mountain Pass, San Bernardino County, California.
BEST REF. IN ENGLISH: Jaffe, Howard W., Meyrowitz, Robert, Evans, Howard T. Jr., *Am. Min.*, **38**: 741–754 (1953).

## SAHLINITE
$Pb_{14}(AsO_4)_2O_9Cl_4$

CRYSTAL SYSTEM: Monoclinic
3 STRONGEST DIFFRACTION LINES:
3.01 (100)
2.94 (100)
2.81 (100)
OPTICAL CONSTANTS: Indices of refraction are very high.
(−)2V
HARDNESS: 2–3
DENSITY: 7.95 (Meas.)
CLEAVAGE: {010} perfect
HABIT: As aggregates of minute thin scales.
COLOR-LUSTER: Pale yellow.
MODE OF OCCURRENCE: Occurs sparingly associated with hausmannite in dolomite at Långban, Sweden.
BEST REF. IN ENGLISH: Palache, et al., "Dana's System of Mineralogy," 7th Ed., v. II, p. 775, New York, Wiley, 1951.

## SAINFELDITE
$H_2Ca_5(AsO_4)_4 \cdot 4H_2O$

CRYSTAL SYSTEM: Monoclinic
CLASS: 2/m
SPACE GROUP: C2/c

Z: 4
LATTICE CONSTANTS:
$a = 18.64$
$b = 9.81$
$c = 10.12$
$\beta \simeq 97°$
3 STRONGEST DIFFRACTION LINES:
3.37 (100)
3.18 ( 80)
8.7 ( 60)
OPTICAL CONSTANTS:
$\alpha = 1.600$
$\beta = 1.610$
$\gamma = 1.616$
$2V \simeq 80°$ (−)
HARDNESS: Not determined.
DENSITY: 3.04 (Meas.)
3.00 (Calc.)
HABIT: Minute flattened rosettes or radiating crystals. Crystals elongate parallel to $c$, flattened on {100}.
COLOR-LUSTER: Colorless to light pink; transparent.
MODE OF OCCURRENCE: Occurs at the 40 meter level of the Gabe Gottes vein, Sainte-Marie-aux-Mines, Alsace, France, and at Jachymov, Bohemia.
BEST REF. IN ENGLISH: Pierrot, R., *Am. Min.*, **50**: 806 (1965).

## SAKHAITE
$Ca_{12}Mg_4(CO_3)_4(BO_3)_7Cl(OH)_2 \cdot H_2O$

CRYSTAL SYSTEM: Cubic
CLASS: 4/m $\overline{3}$ 2/m, 432, $\overline{4}$3m, 2/m$\overline{3}$ or 23
SPACE GROUP: Fm3m, F43, F$\overline{4}$3m, Fm3, or F23
Z: 4
LATTICE CONSTANT:
$a = 14.64$
3 STRONGEST DIFFRACTION LINES:
2.58 (100)
2.108 ( 56)
5.16 ( 22)
OPTICAL CONSTANT:
$N = 1.638–1.641$
HARDNESS: 5
DENSITY: 2.78–2.83 (Meas.)
CLEAVAGE: None observed.
HABIT: Massive; in lens-like bodies.
COLOR-LUSTER: Gray to grayish white; colorless and transparent in fine particles. Vitreous to slightly greasy.
MODE OF OCCURRENCE: Occurs associated with kotoite, clinohumite, forsterite, suanite, spinel, ludwigite, szaibelyite, and sphalerite in several areas separated by 25 km of magnesian skarn at the contact of a Mesozoic granitic massif with Middle and Lower Paleozoic dolomites in Siberia.
BEST REF. IN ENGLISH: Ostrovskaya, I. V., Pertsev, N. N., and Nikitina, I. B., *Am. Min.*, **51**: 1817 (1966).

## SAKHAROVAITE (bismuth jamesonite)
(Pb, Fe)(Bi, Sb)$_2$S$_4$

CRYSTAL SYSTEM: Monoclinic (?)
Z: 1
3 STRONGEST DIFFRACTION LINES: X-ray pattern corresponds closely to jamesonite with additional strong lines at 4.03 and 3.76.
HARDNESS: Not determined (low).
DENSITY: Not determined.
CLEAVAGE: One perfect.
HABIT: Crystals fine capillary, up to 1 cm long; in radiating aggregates.
COLOR-LUSTER: Lead gray. Opaque. Metallic.
MODE OF OCCURRENCE: Occurs in association with realgar, cinnabar, and quartz cyrstals in cavities in carbonate veinlets cutting arsenopyrite ore in the Ustarasaisk deposits, western Tyan-Shan.
BEST REF. IN ENGLISH: Sakharova, M. S., *Am. Min.*, **41**: 814 (1956).

# SAKURAIITE
$(Cu, Fe, Zn)_3(In, Sn)S_4$
Indium analogue of kesterite

CRYSTAL SYSTEM: Tetragonal (pseudocubic)
Z: 2
LATTICE CONSTANTS:
  $a = 5.455$
  $c = 10.9$
3 STRONGEST DIFFRACTION LINES:
  3.15 (100)
  1.927 ( 40)
  1.650 ( 20)
HARDNESS: 4
DENSITY: 4.45 (Calc.)
CLEAVAGE: None.
HABIT: Massive, forming an exsolution-like texture with stannite up to 0.5 × 0.03 mm in size in the polished section.
COLOR-LUSTER: Greenish steel gray; metallic. Streak lead gray with olive tint.
MODE OF OCCURRENCE: Occurs associated with stannite, sphalerite, chalcopyrite, cassiterite, matildite, danaite, quartz, and calcite in a banded-structured vein of the Ikuno mine, Hyogo Prefecture, Japan.
BEST REF. IN ENGLISH: Kato, Akira, *Am. Min.*, **53**: 1421 (1968).

# SAL-AMMONIAC
$NH_4Cl$

CRYSTAL SYSTEM: Cubic
CLASS: $4/m\,\bar{3}\,2/m$
SPACE GROUP: Pm3m
Z: 1
LATTICE CONSTANT:
  $a = 3.8758$
3 STRONGEST DIFFRACTION LINES:
  2.740 (100)
  3.87 ( 25)
  1.582 ( 25)
OPTICAL CONSTANT:
  $N = 1.639$

HARDNESS: 1–2
DENSITY: 1.519
CLEAVAGE: {111} imperfect
Fracture conchoidal. Somewhat sectile, not easily powdered.
HABIT: Crystals commonly trapezohedral, also dodecahedral or cubic. Crystal faces often curved or stepped. Usually as skeletal or dendritic aggregates; also as crusts or fibrous masses; stalactitic; earthy. Twinning on {111} common.
COLOR-LUSTER: Colorless, white; also grayish, yellow, or brown from impurities. Transparent. Vitreous.

Taste saline, stinging.
MODE OF OCCURRENCE: Occurs widespread chiefly as a sublimation product in burned oil-shales or coal seams, and about volcanic fumaroles. Found in the United States in the Monterey shale of Burning Mountain, Los Angeles County, and associated with sulfur from burned oil-shales on the Hope Ranch, Santa Barbara County, California; also around fumaroles at Kilauea, Hawaii. Thick crusts of sal-ammoniac are found at Mt. Vesuvius, Italy; on the island of Stromboli; on Mt. Etna, Sicily; and at many other places.
BEST REF. IN ENGLISH: Palache, et al., "Dana's System of Mineralogy," 7th Ed., v. II, p. 15–18, New York, Wiley, 1951.

# SALÉEITE
$Mg(UO_2)_2(PO_4)_2 \cdot 8{-}10H_2O$
Autunite Group

CRYSTAL SYSTEM: Tetragonal
CLASS: $4/m\,2/m\,2/m$
SPACE GROUP: I4/mmm
Z: 2
LATTICE CONSTANTS:
  $a = 7.01$
  $c = 19.84$
3 STRONGEST DIFFRACTION LINES:
  9.85 (100)
  3.49 ( 90)
  4.95 ( 80)
OPTICAL CONSTANTS:
  $\epsilon$ or $\alpha = 1.544{-}1.565$
  $\beta = 1.570{-}1.582$
  $\omega$ or $\gamma = 1.571{-}1.585$
  $2V = 0°{-}10°, 61°{-}65°$
HARDNESS: 2½
DENSITY: 3.27 (Meas.)
CLEAVAGE: {001} perfect
  {010} indistinct
  {110} indistinct
HABIT: Crystals rectangular plates flattened on {001} ranging in size up to 3.0 mm on edge, as flattened bipyramids. Forms are {001}, {010}, {012}, and {120}. In fan-shaped groups, subparallel aggregates with torbernite, and as aggregates of thin plates and scales.
COLOR-LUSTER: Pale yellow, yellow, yellowish green. Crystals lose their luster and become opaque upon dehydration. Translucent to transparent. Fluoresces brilliant yellow-green in ultraviolet light.
MODE OF OCCURRENCE: Occurs sparsely disseminated in carnotite-bearing sandstone at the Gull mine, Fall River

County, and as a constituent of uraniferous lignite in the Cave Hills and Slim Buttes areas, Harding County, South Dakota. In Europe it occurs with autunite and limonite in quartz at Plessis, near Mortagne, Deux-Sèvres, France; with meta-autunite, phosphuranylite, and sabugalite in altered granite at Mina da Quarta Seira, Sabugal, Portugal; an arsenatian variety is found at Schneeberg, Saxony, Germany, associated with uranophane and zeunerite. Also found as superb crystals at Rum Jungle, Australia, and at Shinkolobwe, Katanga, Zaire.

BEST REF. IN ENGLISH: Frondel, Clifford, USGS Bull. 1064, p. 177–183 (1958).

## SALESITE
$CuIO_3OH$

CRYSTAL SYSTEM: Orthorhombic
CLASS: 2/m 2/m 2/m
SPACE GROUP: Pnma
Z: 4
LATTICE CONSTANTS:
  $a = 10.773$
  $b = 6.696$
  $c = 4.780$
3 STRONGEST DIFFRACTION LINES:
  4.37 (100)
  3.66 ( 65)
  2.39 ( 60)
OPTICAL CONSTANTS:
  $\alpha = 1.786$
  $\beta = 2.070$
  $\gamma = 2.075$
  $(-)2V = 0°–5°$
HARDNESS: 3
DENSITY: 4.77 (Meas.)
        4.888 (Calc.)
CLEAVAGE: {110} perfect
Brittle.
HABIT: As tiny stout prismatic crystals with pyramidal terminations.
COLOR-LUSTER: Bluish green. Transparent. Vitreous.
MODE OF OCCURRENCE: Occurs as a secondary mineral in oxidized ore at Chuquicamata, Chile.
BEST REF. IN ENGLISH: Palache, et al., "Dana's System of Mineralogy," 7th Ed., v. II, p. 315–316, New York, Wiley, 1951.

## SALITE = Intermediate member of Diopside-Hedenbergite series

## SALMONSITE (Inadequately described mineral)
$Mn_9^{2+}Fe_2^{3+}(PO_4)_8 \cdot 14H_2O$

CRYSTAL SYSTEM: Probably orthorhombic
3 STRONGEST DIFFRACTION LINES:
  9.42 (100 )
  3.16 (100b)
  2.853 (100 )

OPTICAL CONSTANTS:
  $\alpha = 1.655$
  $\beta = 1.66$
  $\gamma = 1.670$
  $(+)2V =$ very large
HARDNESS: 4
DENSITY: 2.88 (Meas.)
CLEAVAGE: Two pinacoidal.
HABIT: As fibrous masses.
COLOR-LUSTER: Pale tan.
MODE OF OCCURRENCE: Occurs in granite pegmatite as an alteration product of hureaulite at the Stewart mine, Pala, San Diego County, California, associated with blue strengite.
BEST REF. IN ENGLISH: Palache, et al., "Dana's System of Mineralogy," 7th Ed., v. II, p. 730–731, New York, Wiley, 1951.

## SALTPETER = Niter

## SAMARSKITE
$(Fe, Y, U)_2(Nb, Ti, Ta)_2O_7$

CRYSTAL SYSTEM: Orthorhombic
3 STRONGEST DIFFRACTION LINES:
  2.98 (100)
  2.92 ( 90)  heated (1050°C)
  3.13 ( 40)
OPTICAL CONSTANTS:
  Metamict and isotropic
    N = 220  (variable)
HARDNESS: 5–6
DENSITY: Widely variable, depending upon chemical composition and degree of hydration. Usually ~5.25–5.69.
CLEAVAGE: {010} indistinct
Fracture conchoidal. Brittle.
HABIT: Crystals prismatic with rectangular cross section; sometimes elongated, with large {101}; also tabular, flattened on {010} or {100}. Crystals rough. Commonly massive, compact.
COLOR-LUSTER: Velvety black, becoming dull on alteration. Exterior commonly brownish or yellowish brown. Opaque; translucent in thin fragments. Resinous, vitreous, or submetallic on fresh fracture; crystal faces greasy to dull. Streak black to reddish brown on unaltered material.
MODE OF OCCURRENCE: Occurs in granite pegmatites often associated with columbite; rarely as a detrital mineral in placer deposits. Found at the Southern Pacific silica quarry near Nuevo, Riverside County, California; near Idaho City, Boise County, Idaho; in Douglas, Fremont, Gunnison, and Jefferson counties, Colorado; at Topsham, Maine; and at localities in Connecticut, Maryland, and in abundance at many places in North Carolina. It also occurs in Canada; as fine crystals in Brazil and Madagascar; and in Norway, Sweden, USSR, Zaire, India, Borneo, Japan, and elsewhere.
BEST REF. IN ENGLISH: Palache, et al., "Dana's System of Mineralogy," 7th Ed., v. I, p. 797–800, New York, Wiley, 1944.

## SAMIRESITE = Uranium and lead pyrochlore

## SAMPLEITE
$NaCaCu_5(PO_4)_4Cl \cdot 5H_2O$

CRYSTAL SYSTEM: Orthorhombic
CLASS: 2/m 2/m 2/m
Z: 8
LATTICE CONSTANTS:
  $a = 9.70$ kX
  $b = 38.40$
  $c = 9.65$
3 STRONGEST DIFFRACTION LINES:
  9.60 (100)
  3.04 (100)
  4.30 ( 80)
OPTICAL CONSTANTS:
  $\alpha = 1.629$
  $\beta = 1.677$ (Na)
  $\gamma = 1.679$
  $(-)2V = 22°34'$ (Calc.)
HARDNESS: ~4
DENSITY: 3.20 (Meas.)
         3.274 (Calc.)
CLEAVAGE: {010} perfect
          {100} good
          {001} good
HABIT: Crystals very thin, lath-like, elongated along $c$-axis and flattened {010}.
COLOR-LUSTER: Light blue to bluish green. Pearly on {010}.
MODE OF OCCURRENCE: Occurs associated with gypsum, limonite, jarosite, and other minerals in highly oxidized ore at Chuquicamata, Chile.
BEST REF. IN ENGLISH: Palache, et al., "Dana's System of Mineralogy," 7th Ed., v. II, p. 945–946, New York, Wiley, 1951.

## SAMSONITE
$Ag_4MnSb_2S_6$

CRYSTAL SYSTEM: Monoclinic
CLASS: 2/m
SPACE GROUP: $P2_1/n$
Z: 2
LATTICE CONSTANTS:
  $a = 10.29$
  $b = 8.05$
  $c = 6.61$
  $\beta = 92°41'$
3 STRONGEST DIFFRACTION LINES:
  3.20 (100)
  3.01 ( 90)
  2.59 ( 60)
OPTICAL CONSTANTS: Nearly opaque.
HARDNESS: 2½
DENSITY: 5.51 (Meas.)
         5.56 (Calc.)
CLEAVAGE: None.
Fracture conchoidal. Brittle.
HABIT: Crystals prismatic.
COLOR-LUSTER: Steel black. Nearly opaque; thin fragments translucent dark red to brown. Metallic. Streak dark red.
MODE OF OCCURRENCE: Occurs in vugs associated with galena, tetrahedrite, pyrargyrite, dyscrasite, pyrolusite, calcite, and quartz in the Samson vein at Andreasberg, Harz Mountains, Germany.
BEST REF. IN ENGLISH: Palache, et al., "Dana's System of Mineralogy," 7th Ed., v. I, p. 393–395, New York, Wiley, 1944.

## SANBORNITE
$BaSi_2O_5$

CRYSTAL SYSTEM: Orthorhombic
CLASS: 2/m 2/m 2/m
SPACE GROUP: Pcmn
Z: 4
LATTICE CONSTANTS:
  $a = 4.63$
  $b = 7.69$
  $c = 13.53$
3 STRONGEST DIFFRACTION LINES:
  3.97 (100)
  3.092 ( 75)
  3.342 ( 70)
OPTICAL CONSTANTS:
  $\alpha = 1.598$
  $\beta = 1.617$
  $\gamma = 1.625$
  $(-)2V = 67°$
HARDNESS: ~5
DENSITY: 3.712 (Meas.)
         3.77 (Calc.)
CLEAVAGE: {001} perfect
          {100} distinct
          {010} indistinct
HABIT: As anhedral plates up to 2 or 3 cm across and up to 5 mm in thickness.
COLOR-LUSTER: White to colorless; translucent to transparent. Cleavage lamellae pearly.
MODE OF OCCURRENCE: Occurs associated with gillespite, quartz, witherite, and other minerals near Incline, Mariposa County, California.
BEST REF. IN ENGLISH: Douglass, Robert M., *Am. Min.*, **43**: 517–536 (1958).

## SANDERITE (Inadequately described mineral)
$\alpha$-$MgSO_4 \cdot 2H_2O$

CRYSTAL SYSTEM: Unknown
3 STRONGEST DIFFRACTION LINES:
  3.10 (100)
  4.25 ( 90)
  4.42 ( 80)
DENSITY: 2.37
MODE OF OCCURRENCE: Occurs mixed with leonhardtite, pentahydrite, and hexahydrite as an efflorescence on kieserite in the 601 meter level of the "Hansa I" shaft, Empelde, and the 675 meter level of the "Niedersachsen" shaft near Wathlingen, Germany.
BEST REF. IN ENGLISH: None.

**SANIDINE** = Monoclinic disordered orthoclase
$KAlSi_3O_8$

## SANJUANITE
$Al_2(PO_4)(SO_4)OH \cdot 9H_2O$

CRYSTAL SYSTEM: Monoclinic (?)
3 STRONGEST DIFFRACTION LINES:
  10.77 (100)
  4.13 ( 55)
  5.28 ( 38)
OPTICAL CONSTANTS:
  $\alpha = 1.484$
  $\gamma = 1.499$
  HARDNESS: 3
  DENSITY: 1.94 (Meas.)
  CLEAVAGE: Fracture earthy or uneven.
  HABIT: As aggregates of microscopic fibers, arranged in parallel or divergent bundles, forming chalk-like compact masses.
  COLOR-LUSTER: White; dull to silky.
  MODE OF OCCURRENCE: Found sparingly in veinlets 2–5 cm thick in dark carboniferous slates on the eastern slope of Sierra Chica de Zonda, Department of Pocito, San Juan Province, Argentina.
  BEST REF. IN ENGLISH: de Abeledo, M. E. J., Angelelli, V., de Benyacar, M. A. R., and Gordillo, C., *Am. Min.*, **53**: 1–8 (1968).

## SANMARTINITE
$(Zn, Fe, Ca, Mn)WO_4$
Wolframite Group

CRYSTAL SYSTEM: Monoclinic
CLASS: 2/m
SPACE GROUP: P2/c
Z: 2
LATTICE CONSTANTS:
  $a = 4.712$ kX
  $b = 5.738$
  $c = 4.958$
  $\beta = 90°28'$
3 STRONGEST DIFFRACTION LINES:
  2.926 (100)
  1.700 ( 60)
  2.467 ( 50)
  HARDNESS: Not determined.
  DENSITY: 6.70 (Meas.)
      6.62 (Calc.)
  CLEAVAGE: {010} perfect
  HABIT: As microscopic tabular crystals, reticular aggregates, and fine-granular masses.
  COLOR-LUSTER: Dark brown to brownish black. Resinous.
  MODE OF OCCURRENCE: Occurs associated with willemite as an alteration of scheelite in a quartz vein in Los Cerrillos, 7 km south of San Martin, San Luis province, Argentina.
  BEST REF. IN ENGLISH: Palache, et al., "Dana's System of Mineralogy, 7th Ed., v. II, p. 1072–1073, New York, Wiley, 1951.

## SANTAFEITE
$Na_2Mn_3^{4+}(Mn, Ca, Sr)_6(V, As)_6O_{28} \cdot 8H_2O$

CRYSTAL SYSTEM: Orthorhombic
CLASS: 222
SPACE GROUP: $B22_12$
Z: 2
LATTICE CONSTANTS:
  $a = 9.25$
  $b = 30.00$
  $c = 6.33$
3 STRONGEST DIFFRACTION LINES:
  14.9 (100)
  7.47 ( 50)
  2.702 ( 50)
OPTICAL CONSTANTS:
  $\alpha = 2.01$
  DENSITY: 3.379 (Meas.)
  CLEAVAGE: {010} easy and perfect
        {110} distinct
Very brittle.
  HABIT: As rosettes of radially oriented acicular crystals. Rosettes range from about 1.5–4 mm in diameter. A typical rosette 2.8 mm in diameter consists of crystals approximately 0.3 mm along $\alpha$-axis, 0.07 mm along $b$-axis, and 1.4 mm along $c$-axis.
  COLOR-LUSTER: Black, subadamantine. Streak brown.
  MODE OF OCCURRENCE: Occurs as an encrustation of small rosettes of acicular crystals on an outcrop joint surface of Todilto limestone in the Grants uranium district, McKinley County, New Mexico.
  BEST REF. IN ENGLISH: Sun, Ming-Shan, and Weber, Robert H., *Am. Min.*, **43**: 677–687 (1958).

## SANTITE
$KB_5O_8 \cdot 4H_2O$

CRYSTAL SYSTEM: Orthorhombic
CLASS: mm2
SPACE GROUP: Aba2
Z: 4
LATTICE CONSTANTS:
  $a = 11.10$
  $b = 11.18$
  $c = 9.08$
3 STRONGEST DIFFRACTION LINES:
  3.36 (100)
  3.52 ( 85)
  5.60 ( 70)
OPTICAL CONSTANTS:
  $\alpha = 1.422$
  $\beta = 1.435$
  $\gamma = 1.480$
  (+)2V
  HARDNESS: Not determined.
  DENSITY: 1.735 (Meas.)
      1.740 (Calc.)

CLEAVAGE: Not determined.
HABIT: As irregular grains up to 0.06 mm long.
COLOR-LUSTER: Colorless.
MODE OF OCCURRENCE: Occurs mixed with larderellite and sassolite at Larderello, Tuscany, Italy.
BEST REF. IN ENGLISH: Merlino, Stefano, and Sartori, Franco, *Am. Min.*, **56**: 636 (1971).

## SAPONITE
$(Mg, Al, Fe)_3 (Al, Si)_4 O_{10} (OH)_2$
Montmorillonite (Smectite) Group

CRYSTAL SYSTEM: Monoclinic
LATTICE CONSTANTS:

| (Glycollated) | (Dried) |
|---|---|
| $a = 5.3$ | $a = 5.3$ |
| $b = 9.14$ | $b = 9.16$ |
| $c = 16.9$ | $c = 12.4$ |

3 STRONGEST DIFFRACTION LINES:

| (Glycollated) | (Dried) |
|---|---|
| 17.0 (100) | 12.3 (100) |
| 3.37 ( 80) | 1.53 ( 70) |
| 1.54 ( 70) | 3.10 ( 50) |

OPTICAL CONSTANTS:
$\alpha = 1.48-1.53$
$\beta = 1.486-1.493$
$\gamma = 1.50-1.59$
$(-)2V =$ moderate
HARDNESS: 1-2
DENSITY: 2.24-2.30 (Meas.)
2.10 (Calc.)
CLEAVAGE: {001} perfect
Plastic when moist; brittle on drying.
HABIT: Massive, very fine grained; nodular, in veins, or as a cavity filling.
COLOR-LUSTER: White; also tinted yellowish, grayish green, bluish, reddish. Greasy.
MODE OF OCCURRENCE: Occurs chiefly in association with mineral veins or in amygdaloidal cavities in basic eruptive rocks. Typical occurrences are found near Milford, Beaver County, Utah; in the copper deposits of Keweenaw Peninsula, Michigan; in amygdaloid on the north shore of Lake Superior, Minnesota; and at localities in Canada, Scotland, England, Sweden, Czechoslovakia, Transvaal, Japan, and elsewhere.
BEST REF. IN ENGLISH: Deer, Howie, and Zussman, "Rock Forming Minerals," v. 3, p. 226-245, New York, Wiley, 1962.

## SAPPHIRE = Gem variety of corundum

## SAPPHIRINE
$Mg_{3.5} Al_{9.0} Si_{1.5} O_{20}$

CRYSTAL SYSTEM: Monoclinic
CLASS: 2/m
SPACE GROUP: $P2_1/a$
Z: 4

LATTICE CONSTANTS:
$a = 11.266$
$b = 14.401$
$c = 9.929$
$\beta = 125.46°$
3 STRONGEST DIFFRACTION LINES:
2.447 (100)
2.015 ( 90)
2.990 ( 65)
OPTICAL CONSTANTS:
$\alpha = 1.716$
$\beta = 1.721$
$\gamma = 1.723$
$(-)2V = 47.5°$
HARDNESS: 7½
DENSITY: 3.4-3.5 (Meas.)
3.486 (Calc.)
CLEAVAGE: {100} indistinct
{001} indistinct
{010} indistinct
Fracture uneven to subconchoidal.
HABIT: Crystals tabular. Usually granular, disseminated.
COLOR-LUSTER: Pale blue to bluish or greenish gray, green. Transparent. Vitreous.
MODE OF OCCURRENCE: Occurs chiefly as a high-temperature metamorphic mineral in aluminum- and magnesium-rich rocks which are low in silicon. Found in the emery district near Peekskill, Westchester County, New York; at St. Urbain, Quebec, Canada; and at localities in Greenland, Italy, South Africa, Madagascar, India, and elsewhere.
BEST REF. IN ENGLISH: Palache, et al., "Dana's System of Mineralogy," 7th Ed., v. I, p. 724-726, New York, Wiley, 1944. Moore, Paul B., *Am. Min.*, **54**: 31-49 (1969).

## SARCOLITE
$NaCa_4 Al_3 Si_5 O_{19}$

CRYSTAL SYSTEM: Tetragonal
CLASS: 4/m
SPACE GROUP: I4/m(?)
Z: 6
LATTICE CONSTANTS:
$a = 12.43$
$c = 15.59$
3 STRONGEST DIFFRACTION LINES:
2.75 (100)
3.34 ( 80)
2.85 ( 70)
OPTICAL CONSTANTS:
$\omega = 1.604-1.640$
$\epsilon = 1.615-1.657$
(+)
HARDNESS: 6
DENSITY: 2.92 (Meas.)
2.961 (Calc.)
HABIT: Crystals nearly equant.
COLOR-LUSTER: Flesh red. Translucent. Vitreous.
MODE OF OCCURRENCE: Occurs in volcanic rock at Mte. Somma, Vesuvius, Italy.
BEST REF. IN ENGLISH: Winchell and Winchell, "Ele-

ments of Optical Mineralogy," 4th Ed., pt. 2, p. 496, New York, Wiley, 1951.

## SARCOPSIDE
$(Fe, Mn, Mg)_3(PO_4)_2$

CRYSTAL SYSTEM: Monoclinic
CLASS: 2/m
SPACE GROUP: $P2_1/a$
Z: 2
LATTICE CONSTANTS:
  $a = 10.47$
  $b = 4.80$
  $c = 6.06$
  $\beta \sim 90°$
3 STRONGEST DIFFRACTION LINES:
  3.03 (100)
  3.54 ( 80)
  6.06 ( 50)
OPTICAL CONSTANTS:
  $\alpha = 1.670$
  $\beta = 1.728$
  $\gamma = 1.732$
  $(-)2V = 28°$
HARDNESS: 4
DENSITY: 3.79 (Meas.)
         3.798 (Calc.)
CLEAVAGE: {100} good
          {001} good
          {010} poor
Fracture uneven to splintery.
  HABIT: As irregular masses with fibrous structure, commonly as intergrowths with graftonite. Rarely in distorted six-sided plates. Twinning on {001}, polysynthetic.
  COLOR-LUSTER: Colorless when fresh; usually gray to reddish brown to brown due to slight alteration. Transparent to subtranslucent. Vitreous to silky and glistening when slightly altered.
  MODE OF OCCURRENCE: Occurs in granite pegmatite as large masses intergrown with graftonite at the Bull Moose mine, Custer County, South Dakota; at the French King No. 2 quarry, East Alstead, and at Deering, New Hampshire; at Otav and at Domažlice, Bohemia, Czechoslovakia; near Tirschenreuth, Germany; and near Michelsdorf, Silesia, Poland.
  BEST REF. IN ENGLISH: Hurlbut, C. S. Jr., *Am. Min.*, **50**: 1698–1707 (1965). Peacor, Donald R., *Am. Min.*, **54**: 969–972 (1969). Peacor, D. R., and Garske, David, *Am. Min.*, **40**: 1149–1150 (1964).

**SARD** = Variety of Quartz

## SARKINITE
$Mn_2AsO_4OH$

CRYSTAL SYSTEM: Monoclinic
CLASS: 2/m
SPACE GROUP: $P2_1/c$
Z: 16

LATTICE CONSTANTS:
  $a = 12.68$
  $b = 13.54$
  $c = 10.17$
  $\beta = 108°44'$
3 STRONGEST DIFFRACTION LINES:
  3.17 (100)
  3.04 (100)
  3.29 ( 70)
OPTICAL CONSTANTS:
  $\alpha = 1.793$
  $\beta = 1.807$
  $\gamma = 1.809$
  $(-)2V = 83°$
HARDNESS: 4–5
DENSITY: 4.08–4.18 (Meas.)
         4.04 (Calc.)
CLEAVAGE: {100} distinct
Fracture subconchoidal to uneven.
  HABIT: Crystals commonly thick tabular and slightly elongated; also short prismatic. {100} and {001} usually uneven. Also granular; as spherical forms.
  COLOR-LUSTER: Dark red to reddish yellow to yellow. Greasy. Streak red to yellow.
  MODE OF OCCURRENCE: Occurs in Sweden at Långban associated with hausmannite and hedyphane; with jacobsite, magnetite, tephroite, and other minerals at the Sjö mine, Grythytte parish; and with calcite, barite, native lead, and bementite at the Harstig mine, Pajsberg.
  BEST REF. IN ENGLISH: Palache, et al., "Dana's System of Mineralogy," 7th Ed., v. II, p. 855–857, New York, Wiley, 1951.

## SARMIENTITE
$Fe_2^{3+}AsO_4SO_4OH \cdot 5H_2O$

CRYSTAL SYSTEM: Monoclinic
CLASS: 2/m
SPACE GROUP: $P2_1/c$
Z: 4
LATTICE CONSTANTS:
  $a = 6.55$
  $b = 18.55$
  $c = 9.70$
  $\beta = 97°39'$
3 STRONGEST DIFFRACTION LINES:
  9.29 (100)
  4.64 ( 90)
  4.26 ( 80)
OPTICAL CONSTANTS:
  $\alpha = 1.628$
  $\beta = 1.635$
  $\gamma = 1.698$
  $(+)2V = 38°$(Calc.)
HARDNESS: Not determined.
DENSITY: 2.58 (Meas.)
         2.58 (Calc.)
CLEAVAGE: Not determined.
  HABIT: Crystals microscopic, prismatic, elongated along $c$-axis and flattened {010}. As nodular masses.
  COLOR-LUSTER: Pale yellowish orange.

MODE OF OCCURRENCE: Occurs as nodules associated with fibroferrite and other sulfates in the gossan of pyritic sulfide veins in diabase at the Santa Elena mine, La Alcaparrosa, San Juan, Argentina.

BEST REF. IN ENGLISH: Palache, et al., "Dana's System of Mineralogy," 7th Ed., v. II, p. 1013-1014, New York, Wiley, 1951. de Abeledo, M. E. J., and de Benyacar, M. A. R., *Am. Min.*, **53**: 2077-2082 (1968).


# SARTORITE
$PbAs_2S_4$

CRYSTAL SYSTEM: Orthorhombic
CLASS: 222
SPACE GROUP: $P2_12_12_1$
Z: 4
LATTICE CONSTANTS:
(pseudocell)
$a = 19.52$
$b = 4.16$
$c = 7.88$
3 STRONGEST DIFFRACTION LINES:
2.752 (100)
3.52 ( 90)
2.947 ( 90)
OPTICAL CONSTANTS: Opaque.
HARDNESS: 3
DENSITY: 5.10 (Meas.)
5.07 (Calc.)
CLEAVAGE: {100} fair
Fracture conchoidal. Very brittle.
HABIT: Crystals prismatic and deeply striated. Terminations often rounded and irregular, occasionally cavernous. Often as parallel groupings. Twinning on {100}, often repeated.
COLOR-LUSTER: Lead gray. Opaque. Metallic. Streak brown.
MODE OF OCCURRENCE: Occurs as crystals, up to several centimeters long, associated with other rare sulfosalts, realgar, and pyrite, in crystalline dolomite, at the Lengenbach quarry in the Binnental, Valais, Switzerland.
BEST REF. IN ENGLISH: Palache, et al., "Dana's System of Mineralogy," 7th Ed., v. I, p. 478-481, New York, Wiley, 1944.


# SARYARKITE
$(Ca, Y, Th)_2 Al_4 (SiO_4, PO_4)_4 (OH) \cdot 9H_2O$

CRYSTAL SYSTEM: Tetragonal
CLASS: 422
SPACE GROUP: $P42_12$ or $P4_22_12$
Z: 4
LATTICE CONSTANTS:
$a = 8.213$
$c = 6.55$
3 STRONGEST DIFFRACTION LINES:
3.014 (100)
2.827 (100)
1.854 (100)

OPTICAL CONSTANTS:
$\omega = 1.606$
$\epsilon = 1.620$ (Na)
(+)
HARDNESS: 3½-4
DENSITY: 3.07-3.15 (Meas.)
3.35 (Calc.)
CLEAVAGE: None.
HABIT: Massive.
COLOR-LUSTER: White; dull to greasy.
MODE OF OCCURRENCE: Occurs associated with barite, molybdenite, pyrite, hematite, hydrous iron oxides, and a thorium mineral in Devonian propylitized acid effusives and in altered granitic rocks, associated with secondary quartzites containing sericite, alunite, and andalusite from an unspecified locality, presumably in Kazakhstan, USSR.
BEST REF. IN ENGLISH: Korl, O. F., Chernov, V. I., Shipovalov, Yu. V., and Khan, G. A., *Am. Min.*, **49**: 1775-1776 (1964).


# SASSOLITE
$B(OH)_3$

CRYSTAL SYSTEM: Triclinic
CLASS: $\bar{1}$
SPACE GROUP: $P\bar{1}$
Z: 4
LATTICE CONSTANTS:
$a = 7.039$    $\alpha = 92°35'$
$b = 7.053$    $\beta = 101°10'$
$c = 6.578$    $\gamma = 119°50'$
3 STRONGEST DIFFRACTION LINES:
3.18 (100)
6.04 ( 20)
1.591 ( 20)
OPTICAL CONSTANTS:
$\alpha = 1.340$
$\beta = 1.456$
$\gamma = 1.459$
(−)2V = very small
HARDNESS: 1
DENSITY: 1.46-1.52 (Meas.)
1.48 (Calc.)
CLEAVAGE: {001} perfect, micaceous
Crystals flexible.
HABIT: Crystals tabular {001}, pseudohexagonal; as plates less than 1 mm in diameter; rarely needle-like [110]. Commonly as coatings and small scales.
COLOR-LUSTER: White to gray; often discolored to shades of yellow or brown by impurities. Transparent. Pearly luster.
MODE OF OCCURRENCE: Occurs as small plates associated with ginorite in efflorescent masses in colemanite-veined basalt near the head of Twenty Mule Team Canyon, Death Valley, California. Found abundantly in the waters of the Tuscan lagoons, Italy, and in other natural waters at Wiesbaden and Aachen, Germany, and in Lake County, California. Also found abundantly as a sublimation product at Mt. Vesuvius, Italy and at Vulcano in the Lipari Islands. Also found at Steamboat Springs, Nevada; The Geysers,

Sonoma County, California; Norris Geyser Basin, Yellowstone Park, Wyoming; Kramer borate district, California.

BEST REF. IN ENGLISH: "Dana's System of Mineralogy," 7th Ed., v. I, p. 662–663, New York, Wiley, 1944.

## SATIMOLITE
$KNa_2Al_4B_6O_{15}Cl_3 \cdot 13H_2O$

CRYSTAL SYSTEM: Orthorhombic
LATTICE CONSTANTS:
  $a = 12.62$
  $b = 18.64$
  $c = 6.97$
3 STRONGEST DIFFRACTION LINES:
  3.20 (100)
  9.5 ( 90)
  6.3 ( 90)
OPTICAL CONSTANTS:
  $\alpha = 1.535$
  $\beta = 1.552$
  $\gamma = 1.552$
  $(-)2V$ = very large
CLEAVAGE: Aggregates crumble under light pressure into a fine chalk-like powder.
HABIT: As minute irregular grains forming dense rounded aggregates up to 6–8 mm in diameter; rarely as tabular rhombic forms up to 0.02 mm in size.
COLOR-LUSTER: White; chalk-like.
MODE OF OCCURRENCE: The mineral occurs in clay-polyhalite-halite, clay-boracite-polyhalite, rarely clay-kieserite-polyhalite in the Kungar strata of the near Caspian region; also within crystals of kaliborite, and with halite filling fissures in clay-polyhalite rocks in the Kazakh SSR.
BEST REF. IN ENGLISH: *Am. Min.*, **55**: 1069 (1970).

## SATPAEVITE (Inadequately described mineral)
$Al_{12}(V^{4+}O)_2V_6O_{35} \cdot 30H_2O$

CRYSTAL SYSTEM: Unknown.
3 STRONGEST DIFFRACTION LINES:
  1.926 (100)
  2.366 ( 90)
  1.469 ( 90)
OPTICAL CONSTANTS:
  $\alpha' = 1.676$
  $\gamma' = 1.690$
  $(+)2V \simeq 70°$
HARDNESS: 1½
DENSITY: 2.4
CLEAVAGE: Perfect pinacoidal.
HABIT: Floury aggregates of fine grains, occasionally foliated. As veinlets and crusts.
COLOR-LUSTER: Canary yellow to saffron yellow; dull, pearly on cleavage.
MODE OF OCCURRENCE: Occurs associated with gypsum, steigerite, hewettite, and delvauxite at shallow depth in a decomposed carbonaceous shale at several localities in the Kurumsak and Balasauskandyk ore fields, northwestern Kara-Tau, USSR.
BEST REF. IN ENGLISH: Ankinovich, E. A., *Am. Min.*, **44**: 1325–1326 (1959).

## SAUCONITE
$Na_{0.33}Zn_3(Si, Al)_4O_{10}(OH)_2 \cdot 4H_2O$
Montmorillonite (Smectite) Group

CRYSTAL SYSTEM: Monoclinic
3 STRONGEST DIFFRACTION LINES:
  15.4 (100)
  2.67 (100)
  1.54 (100)
OPTICAL CONSTANTS:
  $\alpha = 1.564$
  $\beta = ?$
  $\gamma = 1.605$
  $(-)2V$ = small
HARDNESS: 1–2
DENSITY: Not determined.
CLEAVAGE: {001} perfect
HABIT: Massive, very fine grained; clay-like. Also as minute micaceous plates in parallel arrangement forming laminated masses.
COLOR-LUSTER: Reddish brown, mottled brownish yellow. Dull.
MODE OF OCCURRENCE: Occurs at the New Discovery, Yankee Doodle, and other mines in the Leadville district, Lake County, Colorado; at the Coon Hollow mine, Boone County, Arkansas; at the Liberty mine, Meekers Grove, Wisconsin; in the Saucon Valley, Pennsylvania; and elsewhere.
BEST REF. IN ENGLISH: Ross, C. S., *Am. Min.*, **31**: 411 (1946).

## SBORGITE
$NaB_5O_8 \cdot 5H_2O$

CRYSTAL SYSTEM: Triclinic
CLASS: 2/m
SPACE GROUP: C2/c
Z: 8
LATTICE CONSTANTS:
  $a = 11.10$
  $b = 16.35$
  $c = 13.59$  (Synthetic)
  $\beta = 113°10'$
3 STRONGEST DIFFRACTION LINES:
  4.60 (100)
  3.30 ( 77)
  3.20 ( 76)
OPTICAL CONSTANTS:
  $\alpha = 1.431$
  $\beta = 1.438$
  $\gamma = 1.507$ (Calc.)
  $(+)2V = 35°$ (synthetic)
HARDNESS: Not determined.
DENSITY: 1.713 (Meas. Synthetic)
CLEAVAGE: Not determined.
HABIT: Crystals anhedral to euhedral, up to 1 mm in diameter. In sugary-textured crystal aggregates. Also stalactitic and as fine-grained crusts.
COLOR-LUSTER: Colorless. Transparent. Vitreous.
MODE OF OCCURRENCE: Occurs as a surface or near-surface deposit in association with halite, ulexite, thenardite,

colemanite, and other borate minerals, at three separate places near the Twenty Mule Team Canyon road, Furnace Creek area, Death Valley, California. Also found with borax and thenardite as fine-grained mixtures encrusting artificial conduits for natural steam at Larderello, Italy.

BEST REF. IN ENGLISH: Cipriani, Curzio, *Am. Min.*, **43**: 378 (1958). McAllister, James F., USGS Prof. Paper 424B, Article 129, p. B299–B301 (1961).

# SCACCHITE
$MnCl_2$

CRYSTAL SYSTEM: Hexagonal
CLASS: $\bar{3}2/m$
SPACE GROUP: $R\bar{3}m$
Z: 3 (hexag.), 1 (rhomb.)
LATTICE CONSTANTS:

| (hexag.) | (rhomb.) |
|---|---|
| $a = 3.68$ kX | $a_{rh} = 6.20$ |
| $c = 17.48$ | $\alpha = 34°32'$ |

3 STRONGEST DIFFRACTION LINES:
  5.85 (100)
  2.59 ( 80)
  3.16 ( 25)
HARDNESS: Not determined. Soft.
DENSITY: 2.977 (Meas.)
          3.04 (Calc.)
CLEAVAGE: {0001} perfect?
HABIT: As deliquescent crystalline crusts.
COLOR-LUSTER: Colorless, commonly rose red, darkening to brown or brownish red on exposure.
MODE OF OCCURRENCE: Occurs in fumaroles on Mt. Vesuvius, Italy, associated with sylvite, halite, and other salts.
BEST REF. IN ENGLISH: Palache, et al., "Dana's System of Mineralogy," 7th Ed., v. II, p. 40–41, New York, Wiley, 1951.

# SCANDIAN IXIOLITE = Ixiolite

# SCAPOLITE (Wernerite) = Group Name
See Marialite
     Dipyre
     Mizzonite
     Meionite
General Formula:
  $(Na, Ca, K)_4 Al_3 (Al, Si)_3 Si_6 O_{24} (Cl, F, OH, CO_3, SO_4)$

# SCARBROITE
$Al_{14}(CO_3)_3(OH)_{36}$

CRYSTAL SYSTEM: Triclinic
Z: 4
LATTICE CONSTANTS:

| | |
|---|---|
| $a = 9.94$ | $\alpha = 98.7°$ |
| $b = 14.88$ | $\beta = 96.5°$ |
| $c = 26.47$ | $\gamma = 89.0°$ |

3 STRONGEST DIFFRACTION LINES:
  8.64 (100) broad
  4.32 ( 80)
  1.430 ( 80) broad
OPTICAL CONSTANTS:
Mean $n = 1.509$
HARDNESS: Soft
DENSITY: 2.17 (Meas. after heating to 100°)
          2.03 (Calc.)
CLEAVAGE: Not determined.
HABIT: Massive, compact; composed of thin plate-like crystallites about 1 $\mu$ across basal planes.
COLOR–LUSTER: White.
MODE OF OCCURRENCE: Occurs with hydroscarbroite in vertical fissures in sandstone at South Bay, Scarborough, Yorkshire, England.
BEST REF. IN ENGLISH: Duffin, W. J., and Goodyear, J., *Nature*, **180** (No. 4593): 977 (1957). Duffin, W. J., and Goodyear, J., *Min. Mag.*, **32**: 353–362 (1960).

# SCAWTITE
$Ca_7 Si_6 O_{18}(CO_3) \cdot 2H_2 O$

CRYSTAL SYSTEM: Monoclinic
CLASS: 2/m
SPACE GROUP: I2/m
Z: 2
LATTICE CONSTANTS:
  $a = 10.18$
  $b = 15.22$
  $c = 6.64$
  $\beta = 100°48'$
3 STRONGEST DIFFRACTION LINES:
  3.03 (100)
  2.99 (100)
  1.89 ( 80)
OPTICAL CONSTANTS:
  $\alpha = 1.595–1.603$
  $\beta = 1.605–1.609$
  $\gamma = 1.622–1.618$
(+)2V = 74°–78°
HARDNESS: 4½–5
DENSITY: 2.77 (Meas.)
          2.74 (Calc.)
CLEAVAGE: {010} fair
          {001} perfect (?)
HABIT: As bundles of thin tabular crystals in subparallel or slightly divergent groups. Forms: {100} dominant, also poorly developed {110}, {120}, {130}, {010}, and {101}. Also as microscopic lath-like grains.
COLOR-LUSTER: Colorless; vitreous.
MODE OF OCCURRENCE: Occurs in thin veins in a matrix of massive diopside-wollastonite-spurrite rock on the 910 foot level of the commercial quarry, Crestmore, California; in contact zone material from south of Neihart, Montana; at Scawt Hill, County Antrim, Ireland; at Ballycraigy, Larne, Northern Ireland; and at Tokatoka, North Auckland, New Zealand.
BEST REF. IN ENGLISH: Winchell and Winchell, "Elements of Optical Mineralogy," 4th Ed., Pt. 2, p. 394, New

York, Wiley, 1951.    Murdoch, Joseph, *Am. Min.*, **40**: 505–514 (1955).

## SCHACHNERITE

$Ag_{1.1}Hg_{0.9}$

CRYSTAL SYSTEM: Hexagonal
CLASS: 6/m 2/m 2/m
SPACE GROUP: $P6_3/mmc$
Z: 2
LATTICE CONSTANTS:
  $a$ = 2.978
  $c$ = 4.842
3 STRONGEST DIFFRACTION LINES:
  2.273 (100)
  0.859 ( 60)
  2.420 ( 50)
OPTICAL CONSTANTS: Anisotropy weak.
  HARDNESS: Polishing hardness low.
  DENSITY: 13.52 (Calc.)
  HABIT: As crystals up to 1 cm long; usually much smaller.
  COLOR-LUSTER: Gray in reflected light.   Opaque. Metallic.
  MODE OF OCCURRENCE: Occurs as an alteration of moschellandesbergite in the zone of oxidation of the old mercury mine at Landsberg near Obermoschel, Pfalz, Germany. Associated minerals are para-schachnerite, mercurian silver, limonite, ankerite, argentite, and cinnabar.
  BEST REF. IN ENGLISH: Seeliger, E., and Mücke, A., *Am. Min.*, **58**: 347 (1973).

## SCHAFARZIKITE

$FeSb_2O_4$

CRYSTAL SYSTEM:  Tetragonal
CLASS:  4/m 2/m 2/m
SPACE GROUP: $P4_2/mbc$
Z: 4
LATTICE CONSTANTS:
  $a$ = 8.59 kX
  $c$ = 5.92
OPTICAL CONSTANTS:
  $\omega > 1.74$
  $\epsilon$ = ?
    (+)
  HARDNESS: 3½
  DENSITY: 4.3
  CLEAVAGE:  {110} perfect
             {100} very good
             {001} trace
  HABIT: Crystals prismatic. {100} striated parallel [001].
  COLOR-LUSTER: Red to reddish brown; thin splinters yellow. Opaque. Metallic.
  MODE OF OCCURRENCE:  Occurs in very small amounts on stibnite, associated with senarmontite, valentinite, and kermesite, at Pernek, Czechoslovakia.
  BEST REF. IN ENGLISH: Palache, et al., "Dana's System of Mineralogy," 7th Ed., v. II, p. 1035–1036, New York, Wiley, 1951.    Zeemann, J., *Am. Min.*, **37**: 136 (1952).

## SCHAIRERITE

$Na_3SO_4(F, Cl)$

CRYSTAL SYSTEM:  Hexagonal
CLASS: 3 2 or $\bar{3}$ 2/m
SPACE GROUP:  C312 or $P\bar{3}1m$
Z:  21
LATTICE CONSTANTS:
  $a$ = 12.17
  $c$ = 19.29
3 STRONGEST DIFFRACTION LINES:
  2.76 (100)
  3.52 ( 80)
  3.79 ( 70)
OPTICAL CONSTANTS:
  $\omega$ = 1.440
  $\epsilon$ = 1.445
    (+)
  HARDNESS: 3½
  DENSITY: 2.612, 2.63 (Meas.)
           2.60 (Calc.)
  CLEAVAGE: Fracture conchoidal. Brittle.
  HABIT:  As minute crystals with pronounced trigonal habit with the suggestion of rhombohedral symmetry. Twinning on {0001} common.
  COLOR-LUSTER: Colorless, transparent. Vitreous.
  MODE OF OCCURRENCE: Occurs associated with tychite, pirssonite, gaylussite, trona, thenardite, hanksite, and calcite at Searles Lake, San Bernardino County, California.
  BEST REF. IN ENGLISH: Palache, et al., "Dana's System of Mineralogy," 7th Ed., v. II, p. 547–548, New York, Wiley, 1951.    Wolfe, C. W., and Caras, Alice, *Am. Min.*, **36**: 912–915 (1951).

## SCHALLERITE

$(Mn, Fe)_8Si_6As(O, OH, Cl)_{26}$

CRYSTAL SYSTEM:  Hexagonal
Z:  4
LATTICE CONSTANTS:
  $a$ = 13.43
  $c$ = 14.31
3 STRONGEST DIFFRACTION LINES:
  1.688 (100)
  2.673 ( 60)
  1.511 ( 60)
OPTICAL CONSTANTS:
  $\omega$ = 1.643, 1.679
  $\epsilon$ = 1.681, 1.704
    (−)
  HARDNESS: ~5
  DENSITY: 3.368 (Meas.)
  CLEAVAGE:  {0001} distinct
  HABIT: Massive, granular.
  COLOR-LUSTER: Reddish brown.   Waxy on fracture; pearly on cleavage.
  MODE OF OCCURRENCE: Occurs in association with calcite and rhodonite at Franklin, Sussex County, New Jersey.

BEST REF. IN ENGLISH: Palache, Charles, USGS Prof. Paper 180, p. 90 (1935).

## SCHAPBACHITE = Mixture of galena and matildite

## SCHAURTEITE
$Ca_3 Ge^{4+}[(OH)_6|(SO_4)_2] \cdot 3H_2O$

CRYSTAL SYSTEM: Hexagonal
CLASS: $6/m\ 2/m\ 2/m$, $6mm$, or $\bar{6}m2$
SPACE GROUP: $P6_3/mmc$, $P6_3mc$, or $P\bar{6}2c$
Z: 2
LATTICE CONSTANTS:
  $a = 8.525$
  $c = 10.803$
3 STRONGEST DIFFRACTION LINES:
  3.34  (100)
  4.26  ( 70)
  2.139 ( 60)
OPTICAL CONSTANTS:
  $\omega = 1.569$
  $\epsilon = 1.581$
    (+)
HARDNESS: Not determined.
DENSITY: ~2.65 (Meas.)
          2.64 (Calc.)
HABIT: Crystals fine fibrous to needle-like up to 2 mm in length.
COLOR-LUSTER: White; silky.
MODE OF OCCURRENCE: Occurs as an alteration product of germanium-bearing ores at Tsumeb, South West Africa.
BEST REF. IN ENGLISH: Strunz, H., and Tennyson, Ch., *Am. Min.*, **53**: 507 (1968).

## SCHEELITE
$CaWO_4$

CRYSTAL SYSTEM: Tetragonal
CLASS: $4/m$
SPACE GROUP: $I4_1/a$
Z: 4
LATTICE CONSTANTS:
  $a = 5.246$ kX
  $c = 11.349$
3 STRONGEST DIFFRACTION LINES:
  3.10  (100)
  4.76  ( 55)
  3.072 ( 30)
OPTICAL CONSTANTS:
  $\omega = 1.9375$  $(\lambda = 570\ m\mu)$
  $\epsilon = 1.9208$
    (+)
HARDNESS: 4½–5
DENSITY: 6.10 (Meas.)
         6.12 (Calc. X-ray)
CLEAVAGE: {101} distinct
        {112} interrupted
        {001} indistinct
Fracture subconchoidal to uneven.

HABIT: Crystals octahedral or tabular; often diagonally striated and rough. Also massive, granular; columnar. Twinning on {110} common, as penetration or contact twins.
COLOR-LUSTER: Colorless, white, gray, pale yellowish white to brownish, orange-yellow, greenish, purplish, reddish. Transparent to translucent. Vitreous to adamantine. Fluoresces bright bluish white to white or yellowish white under short-wave ultraviolet light. Streak white.
MODE OF OCCURRENCE: Occurs widespread in contact metamorphic deposits, in hydrothermal veins and pegmatites, and in placer deposits, commonly associated with wolframite. Found in the United States at Trumbull, Fairfield County, Connecticut; in Custer, Lawrence, and Pennington counties, South Dakota; near Mill City and at other places in Nevada; at numerous localities in California, especially near Darwin, Inyo County; as fine crystals in Cochise and Mohave counties, Arizona; and in New Mexico, Colorado, Idaho, Montana, and Utah. It also occurs in Canada, Mexico, Bolivia, Brazil, Peru, England, France, Austria, Germany, Czechoslovakia, Italy, Switzerland, Sardinia, USSR, Malaysia, Burma, Australia, and as exceptional crystals from 6 to 13 inches in length at localities in Japan and Korea.
BEST REF. IN ENGLISH: Palache, et al., "Dana's System of Mineralogy," 7th Ed., v. II, p. 1074–1079, New York, Wiley, 1951.

## SCHEFFERITE = Manganoan diopside

## SCHERTELITE
$(NH_4)_2 MgH_2 (PO_4)_2 \cdot 4H_2O$

CRYSTAL SYSTEM: Orthorhombic
CLASS: $2/m\ 2/m\ 2/m$
SPACE GROUP: Pbca
Z: 8
LATTICE CONSTANTS:
  $a = 11.47$
  $b = 23.63$
  $c = 8.62$
3 STRONGEST DIFFRACTION LINES:
  5.94 (100)
  2.97 ( 43)
  5.21 ( 37)
OPTICAL CONSTANTS:
    $\alpha = 1.508$
    $\beta = 1.515$
    $\gamma = 1.523$
(+)2V = nearly 90°
HARDNESS: Not determined.
DENSITY: 1.83 (Calc.)
CLEAVAGE: None. Brittle.
HABIT: As small indistinct flat crystals. Synthetic crystals tabular {100}, with nearly equally developed {010}, {001}, {111}, and {110}; also as single stout rods elongated along c-axis.
COLOR-LUSTER: Colorless, transparent. Vitreous.

Soluble in water.
MODE OF OCCURRENCE: Occurs in a bat guano de-

posit associated with struvite and newberyite in the Skipton Caves, Ballarat, Victoria, Australia.

BEST REF. IN ENGLISH: Palache, et al., "Dana's System of Mineralogy," 7th Ed., v. II, p. 699, New York, Wiley, 1951.   Frazier, A. William, Lehr, James R., and Smith, James P., *Am. Min.*, **48**: 635–641 (1963).

## SCHETELIGITE (Probably a member of the pyrochlore group)

$(Ca, Mn, Sb, Bi)_2 (Ti, Ta, Nb, W)_2 (O, OH)_7$
Sometimes contains Yt

CRYSTAL SYSTEM:  Probably cubic
OPTICAL CONSTANTS:  Isotropic
  HARDNESS: 5½
  DENSITY:  4.74
  CLEAVAGE: None.  Fracture conchoidal.
  HABIT: Crystals rough; indistinct.  Metamict.
  COLOR-LUSTER:  Brilliant black.  Opaque, except in very thin fragments.  Streak pale yellow to grayish.
  MODE OF OCCURRENCE:  Occurs in pegmatite at Torvelona, Iveland, Norway.
  BEST REF. IN ENGLISH: Palache, et al., "Dana's System of Mineralogy," 7th Ed., v. I, p. 757, New York, Wiley, 1944.

## SCHIRMERITE

$AgPb_2 Bi_3 S_7$

CRYSTAL SYSTEM: Cubic
CLASS: $4/m \bar{3} 2/m$
SPACE GROUP: Fm3m
LATTICE CONSTANTS:
  $a = 10.38$ (5.69?) variable.
3 STRONGEST DIFFRACTION LINES:
  2.82  (100)
  3.27 (( 70)
  1.961 ( 60)
  HARDNESS:  2
  DENSITY:  6.737 (Meas.)
  CLEAVAGE:  None.
Fracture uneven.  Brittle.
  HABIT:  Massive, fine granular.
  COLOR-LUSTER:  Lead gray to iron black.  Opaque. Metallic.
  MODE OF OCCURRENCE: Occurs disseminated in quartz in the Treasure Vault lode, and in other veins that parallel the Treasure Vault, in Park and Summit counties, Colorado.  It is also reported from the Lake City district, Hinsdale County, and from the Santa Cruz mine, San Miguel County, Colorado.
  BEST REF. IN ENGLISH: Palache, et al., "Dana's System of Mineralogy," 7th Ed., v. I, p. 424, New York, Wiley, 1944.

## SCHIZOLITE = Manganoan Pectolite

## SCHMITTERITE

$UO_2 TeO_3$

CRYSTAL SYSTEM:  Orthorhombic
CLASS:  2/m 2/m 2/m
SPACE GROUP:  Pmab
Z:  4
LATTICE CONSTANTS:
  $a = 7.860$
  $b = 10.089$
  $c = 5.363$
3 STRONGEST DIFFRACTION LINES:
  3.682 (100)
  5.35  ( 90)
  3.099 ( 90)
OPTICAL CONSTANTS:
  $N > 2.00$
  (–)2V
  HARDNESS:  ~1
  DENSITY:  6.878 (Meas.)
            6.916 (Calc.)
  CLEAVAGE:  {100}
  HABIT:  Crystals blade-like, elongated parallel to *a*-axis; as small micaceous rosettes up to 1 mm in diameter.
  COLOR-LUSTER:  Colorless to light yellow.  Pearly.
  MODE OF OCCURRENCE:  Occurs very sparingly, usually associated with emmonsite, at the Moctezuma mine, Moctezuma, Sonora, Mexico.
  BEST REF. IN ENGLISH:  Gaines, Richard V., *Am. Min.*, **56**: 411–415 (1971).

## SCHODERITE

$Al_2 (PO_4)(VO_4) \cdot 8H_2O$

CRYSTAL SYSTEM:  Monoclinic
CLASS:  2/m
SPACE GROUP:  P2/m
Z:  4
LATTICE CONSTANTS:
  $a = 11.4$
  $b = 15.8$
  $c = 9.2$
  $\beta = 79°$
3 STRONGEST DIFFRACTION LINES:
  7.9  (100)
  15.8 ( 40)
  11.1 ( 20)
OPTICAL CONSTANTS:
  $\alpha = 1.542$
  $\beta = 1.548$
  $\gamma = 1.566$
  (+)2V = 61° (Calc.)
  HARDNESS:  ~2
  DENSITY:  1.88 (Meas.)
  HABIT:  Microscopic bladed tabular crystals, elongated parallel to (001).
  COLOR-LUSTER:  Yellowish orange.
  MODE OF OCCURRENCE:  Occurs sparsely as yellowish orange microcrystalline coatings associated with wavellite and vashegyite along fractures in phosphatic cherts of lower Paleozoic age near Eureka, Nevada.

BEST REF. IN ENGLISH: Hausen, Donald M., *Am. Min.*, **47**: 637–648 (1962).

# SCHOENFLIESITE
$MgSn(OH)_6$

CRYSTAL SYSTEM: Cubic
CLASS: $2/m\,\bar{3}$
SPACE GROUP: Pn3
Z: 4
LATTICE CONSTANT:
 $a = 7.759$
3 STRONGEST DIFFRACTION LINES:
 3.88 (100)
 4.48 ( 50)
 2.74 ( 25)
OPTICAL CONSTANTS: Mean $n$ for synthetic material 1.590.
 HARDNESS: Not determined.
 DENSITY: 3.483 (Calc.)
 CLEAVAGE: Not determined.
 HABIT: Massive, very fine grained.
 COLOR-LUSTER: Not reported.
 MODE OF OCCURRENCE: Found in association with hematite, maghemite, goethite, hulsite, and fluorite as a component of altered hulsite from Brooks Mountain, Alaska.
 BEST REF. IN ENGLISH: Faust, George T., and Schaller, W. T., *Am. Min.*, **57**: 1557–1558 (1972).

# SCHOEPITE
$UO_3 \cdot 2H_2O$

CRYSTAL SYSTEM: Orthorhombic
CLASS: $2/m\ 2/m\ 2/m$
SPACE GROUP: Pbca
Z: 32
LATTICE CONSTANTS:
 $a = 14.33$
 $b = 16.79$
 $c = 14.73$
3 STRONGEST DIFFRACTION LINES:
 7.28 (100)
 5.08 ( 70)
 3.44 ( 25)
OPTICAL CONSTANTS:
  $\alpha = 1.70$ (Calc.), 1.690
  $\beta = 1.720, 1.714$
  $\gamma = 1.735$
 $(-)2V \simeq 75°$
 HARDNESS: ~2½
 DENSITY: 4.8 (Meas.)
     4.83 (Calc.)
 CLEAVAGE: {001} perfect
      {010} imperfect
Brittle.
 HABIT: Crystals tabular on {001}, elongated [010], psuedohexagonal; also prismatic elongated [001]. Rarely as dense microcrystalline aggregates.
 COLOR-LUSTER: Sulfur yellow to lemon yellow, occa-sionally brownish yellow to amber. Transparent to translu-cent. Adamantine; pearly on cleavage. Powder pale yellow.
 MODE OF OCCURRENCE: Found at Shinkolobwe, Katanga, Zaire, with becquerelite, curite, and vandendreis-scheite. Occurs at Wölsendorf, Bavaria with billietite; at Margnac II, Haute-Vienne, France, with uraninite and "gum-mite"; in the Colorado Plateau uranium district in small amounts as an alteration of uraninite; and in several other uranium districts containing altered uraninite.
 BEST REF. IN ENGLISH: Christ, C. L., and Clark, Joan R., *Am. Min.*, **45**: 1026–1061 (1960).

# SCHOLZITE
$CaZn_2(PO_4)_2 \cdot 2H_2O$

CRYSTAL SYSTEM: Orthorhombic
CLASS: $2/m\ 2/m\ 2/m$
SPACE GROUP: Pbmm
Z: 12
LATTICE CONSTANTS:
 $a = 17.14$
 $b = 22.19$
 $c = 6.61$
3 STRONGEST DIFFRACTION LINES:
 8.588 (100)
 2.788 ( 90)
 4.230 ( 70)
OPTICAL CONSTANTS:
  $\alpha = 1.581$
  $\beta = 1.586$
  $\gamma = 1.596$
 $(+)2V = 70°$
 HARDNESS: 3–3½
 DENSITY: 3.11 (Meas.)
      3.14 (Calc.)
 CLEAVAGE: {100} fair
Brittle.
 HABIT: Crystals prismatic; platy; up to 5 mm long.
 COLOR-LUSTER: Colorless to white. Transparent to translucent. Vitreous.
 MODE OF OCCURRENCE: Occurs associated with sphalerite, triplite, feldspar, and quartz in pegmatite at Hagendorf near Pleystein, Oberpfalz, Bavaria, Germany. Also found as crystals encrusting gossan about 40 miles southeast of Blinman, South Australia.
 BEST REF. IN ENGLISH: Strunz, H., *Am. Min.*, **36**: 382 (1951). Strunz, H., and Tennyson, Ch., *Am. Min.*, **46**: 1519 (1961).

# SCHORL
$Na(Fe, Mn)_3Al_6B_3Si_6O_{27}(OH, F)_4$
Tourmaline Group

CRYSTAL SYSTEM: Hexagonal
CLASS: 3m
SPACE GROUP: R3m
Z: 3
LATTICE CONSTANTS:
 $a = 15.93–16.03$
 $c = 7.12–7.19$

3 STRONGEST DIFFRACTION LINES:
2.576 (100)
3.99 ( 85)
2.961 ( 85)
OPTICAL CONSTANTS:
$\omega$ = 1.655–1.675
$\epsilon$ = 1.625–1.650
(–)
HARDNESS: 7
DENSITY: 3.10–3.25 (Meas.)
CLEAVAGE: {11$\bar{2}$0} very indistinct
{10$\bar{1}$1} very indistinct
Fracture uneven to conchoidal. Brittle.

HABIT: Crystals short to long prismatic, vertically striated; also acicular; rarely flattened as thin tablets. Usually 3, 6, or 9-sided. Commonly hemimorphic. As single crystals or as parallel or radiating groups. Also massive, compact; columnar to fibrous. Twinning on {10$\bar{1}$1} or {40$\bar{4}$1} rare.

COLOR-LUSTER: Black, brownish black, bluish black. Translucent to opaque. Vitreous to resinous. Streak uncolored.

MODE OF OCCURRENCE: Occurs widespread as a characteristic mineral in granite pegmatites, some granites, pneumatolytic veins, in some metamorphic and metasomatic rocks, and as a detrital mineral. Common associated minerals are quartz, muscovite, and feldspar. Typical occurrences are found in California, Colorado, South Dakota, Pennsylvania, Maine, New Hampshire, Massachusetts, New York, Connecticut, New Jersey, and North Carolina. Also in Canada, Mexico, Brazil, Greenland, Norway, England, Czechoslovakia, USSR, Madagascar, Japan, and as exceptional euhedral crystals at Yinnietharra, Western Australia.

BEST REF. IN ENGLISH: Deer, Howie, and Zussman, "Rock Forming Minerals," v. 1, p. 300–319, New York, Wiley, 1962.

## SCHORLOMITE = Titanium-rich andradite garnet

## SCHREIBERSITE (Rhabdite)
(Fe, Ni)$_3$P

CRYSTAL SYSTEM: Tetragonal
CLASS: $\bar{4}$
SPACE GROUP: I$\bar{4}$
Z: 8
LATTICE CONSTANTS:
$a$ = 9.013
$c$ = 4.424
3 STRONGEST DIFFRACTION LINES:
2.19 (100)
2.11 ( 70)
1.972 ( 70)
OPTICAL CONSTANTS: Opaque.
HARDNESS: 6½–7
DENSITY: 7.0–7.8 (Meas.)
7.24 (Calc.)
CLEAVAGE: {001} perfect
{010} or {110} imperfect
Very brittle.

HABIT: Crystals rare, often rounded. As plates, tablets, rods, or needles.
COLOR-LUSTER: Silver white to tin white; tarnishes to brass yellow or brown. Opaque. Metallic.

Strongly magnetic.

MODE OF OCCURRENCE: Occurs as a combustion product in the coal mines of Commentry and Cranzac, France. It is also found as a component in all iron meteorites.

BEST REF. IN ENGLISH: Palache, et al., "Dana's System of Mineralogy," 7th Ed., v. I, p. 124–126, New York, Wiley, 1944.

## SCHROECKINGERITE (Schröckingerite) (Dakeite)
NaCa$_3$UO$_2$SO$_4$(CO$_3$)$_3$F · 10H$_2$O

CRYSTAL SYSTEM: Orthorhombic
CLASS: 2/m 2/m 2/m
SPACE GROUP: Cmmm
Z: 4
LATTICE CONSTANTS:
$a$ = 9.69
$b$ = 16.83
$c$ = 14.26
3 STRONGEST DIFFRACTION LINES:
7.26 (100)
4.796 ( 80)
8.48 ( 70)
OPTICAL CONSTANTS:
$\alpha$ = 1.495
$\beta$ = 1.543
$\gamma$ = 1.544 (synthetic)
(–)2V = 16°
HARDNESS: 2½
DENSITY: 2.55 (Meas.)
CLEAVAGE: {001} perfect, micaceous
Thin crystals and cleavage lamelae flexible.

HABIT: As pseudohexagonal plates up to 1.0 mm in diameter and 0.02 mm thick. Also as crusts, clusters, rosettes, or globular aggregates of scales flattened on {001}.

COLOR-LUSTER: Greenish yellow, transparent. Weakly vitreous, pearly on cleavage surfaces. Fluorescent bright yellowish green under ultraviolet light.

MODE OF OCCURRENCE: Occurs widespread as a secondary mineral, often of very recent formation. Found abundantly (dakeite) as spherical aggregates in gypsum-bearing clays along Lost Creek about 40 miles north-northwest of Wamsutter, Sweetwater County, Wyoming; as an efflorescence on tunnel walls at the Hillside mine, Yavapai County, Arizona; as excellent crystals at the Shinarump No. 3 mine, Seven Mile Canyon, near Moab, and at many other localities in Utah. Also found at Joachimsthal, Bohemia; Johanngeorgenstadt, Saxony; from the Radhausberg near Bad Gastein, Austria; and at the La Soberania mine, San Isidro, Mendoza Province, Argentina.

BEST REF. IN ENGLISH: Frondel, Clifford, USGS Bull. 1064, p. 121–126 (1958). Smith, Deane K., *Am. Min.*, **44**: 1020–1025 (1959).

## SCHUBNELITE
$Fe_2(VO_4)_2 \cdot 2H_2O$

CRYSTAL SYSTEM: Triclinic
CLASS: $\bar{1}$
SPACE GROUP: $P\bar{1}$
Z: 1
LATTICE CONSTANTS:
  $a = 6.59$    $\alpha = 125°$
  $b = 5.43$    $\beta = 104°$
  $c = 6.62$    $\gamma = 84°43'$
3 STRONGEST DIFFRACTION LINES:
  4.47 (100)
  3.21 (100)
  5.15 ( 80)
OPTICAL CONSTANTS:
  Indices > 2
  HARDNESS: Not determined.
  DENSITY: 3.28 (Meas.)
  CLEAVAGE: Not determined.
  HABIT: Crystals minute, up to 0.5 mm, with forms {010}, {100}, and {001}. Twinning on $\{2\bar{1}2\}$.
  COLOR-LUSTER: Black. Transparent yellowish to greenish brown in thin fragments.
  MODE OF OCCURRENCE: Occurs in association with fervanite at the base of the oxidation zone of the uranium deposit at Mounana, Gabon.
  BEST REF. IN ENGLISH: Cesbron, F., Am. Min., 57: 1556-1557 (1972).

## SCHUETTEÏTE
$Hg_3(SO_4)O_2$

CRYSTAL SYSTEM: Hexagonal
CLASS: 32
SPACE GROUP: $P3_121$
Z: 3
LATTICE CONSTANTS:
  $a = 7.07$
  $c = 10.9$
3 STRONGEST DIFFRACTION LINES:
  2.91 (100)
  3.34 ( 70)
  2.60 ( 55)
OPTICAL CONSTANTS:
  Indices > 2.10
  Uniaxial (-)
  HARDNESS: ~3
  DENSITY: 8.18 (Meas.)
             8.36 (Calc.)
  CLEAVAGE: Not determined.
  HABIT: Crystals thin hexagonal plates, minute. As thin films or crusts.
  COLOR-LUSTER: Canary yellow, greenish yellow, orange-yellow.
  MODE OF OCCURRENCE: Occurs as a supergene mercury mineral in several quicksilver deposits in the more arid parts of the western United States, and is a common product on dumps of burnt ore and bricks from old mercury furnaces. Found in California in the Mt. Diablo mine, Contra Costa County, and as coatings on cinnabar-bearing

basalt at the Sulphur Bank mine in Lake County; with opalite at the Idaho-Almaden mine east of Weiser, Idaho; at the Opalite mine in southeastern Oregon; and at several deposits in Nevada, especially at the Silver Cloud mine north of Battle Mountain.
  BEST REF. IN ENGLISH: Bailey, E. H., Hildebrand, F. A., Christ, C. L., and Fahey, J. J., Am. Min., 44: 1026-1038 (1959).

## SCHUILINGITE
$Pb_3Ca_6Cu_2(CO_3)_8(OH)_6 \cdot 6H_2O$

CRYSTAL SYSTEM: Monoclinic
3 STRONGEST DIFFRACTION LINES:
  4.78 (100)
  3.85 ( 80)
  9.56 ( 60)
OPTICAL CONSTANTS:
  $\alpha = 1.710$ (Calc.)
  $\beta = 1.755$
  $\gamma = 1.775$
  $(-)2V = 66°$
  HARDNESS: 3-4
  DENSITY: 5.2 (Meas.)
  CLEAVAGE: {110} perfect
  HABIT: Crystals acicular, up to 0.3 mm long; in crusts.
  COLOR-LUSTER: Turquoise to azure blue. Transparent. Adamantine. Powder pale blue.
  MODE OF OCCURRENCE: Found at Kalompe, Katanga, Zaire, associated with cerussite, calcite, pyrite, and other minerals.
  BEST REF. IN ENGLISH: Palache, et al., "Dana's System of Mineralogy," v. II, p. 252, New York, Wiley, 1951. Guillemin, C., and Pierrot, R., Am. Min., 43: 796 (1958).

## SCHULTENITE
$PbHAsO_4$

CRYSTAL SYSTEM: Monoclinic
CLASS: m, or 2/m
SPACE GROUP: Pa or P2/a
Z: 2
LATTICE CONSTANTS:
  $a = 5.83$
  $b = 6.76$
  $c = 4.85$
  $\beta = 95.5°$
3 STRONGEST DIFFRACTION LINES:
  3.35 (100)
  3.13 ( 90)
  1.95 ( 80)
OPTICAL CONSTANTS:
  $\alpha = 1.8903$ (Calc.)
  $\beta = 1.9097$
  $\gamma = 1.9765$ (Na)
  $(+)2V = 58°14'$
  HARDNESS: 2½
  DENSITY: 5.943 (Meas.)
             6.06 (Calc.)

CLEAVAGE: {010} good
Brittle.

HABIT: Crystals flattened, usually with rhombohedral outline resembling gypsum. Faces commonly striated.

COLOR-LUSTER: ·Colorless. Transparent. Vitreous, brilliant. Streak white.

MODE OF OCCURRENCE: Occurs as a secondary mineral with anglesite, bayldonite, and other secondary minerals at Tsumeb, South West Africa.

BEST REF. IN ENGLISH: Palache, et al., "Dana's System of Mineralogy," 7th Ed., v. II, p. 661–663, New York, Wiley, 1951.


**SCHULZITE** = Variety of geocronite


**SCHWARTZEMBERGITE**
$Pb_6(IO_3)_2Cl_4O_2(OH)_2$

CRYSTAL SYSTEM: Tetragonal
CLASS: 4/m 2/m 2/m
SPACE GROUP: Probably $P4_2/nnm$
Z: 1
LATTICE CONSTANTS:
 $a$ = 5.614
 $c$ = 12.459
OPTICAL CONSTANTS:
 $\alpha$ = 2.25
 $\beta$ = 2.35
 $\gamma$ = 2.36 (Li)
 (−)2V = small
HARDNESS: 2–2½
DENSITY: 7.39 (Meas.)
 7.09 (Calc.)
CLEAVAGE: {001} distinct

HABIT: Crystals flat-pyramidal, rounded. As crusts and masses, compact to earthy.

COLOR-LUSTER: Pale yellow to yellowish brown to reddish brown. Adamantine. Streak pale yellow.

MODE OF OCCURRENCE: Occurs as a secondary mineral in oxidized ore at several localities in northern Chile, as at Cachinal in the Atacama desert, and the San Rafael mine, Sierra Gorda, Caracoles district.

BEST REF. IN ENGLISH: Palache, et al., "Dana's System of Mineralogy," 7th Ed., v. II, p. 317–318, New York, Wiley, 1951. Mücke, A., *Am Min.*, **55**: 1814 (1970).


**SCHWATZITE** = Mercurian tetrahedrite


**SCOLECITE**
$CaAl_2Si_3O_{10} \cdot 3H_2O$
Zeolite Group

CRYSTAL SYSTEM: Monoclinic
CLASS: m
SPACE GROUP: Cc
Z: 8

LATTICE CONSTANTS:
 $a$ = 18.5
 $b$ = 18.99
 $c$ = 6.55
 $\beta$ = 90°39′
3 STRONGEST DIFFRACTION LINES:
 6.64 (100)
 4.44 (100)
 2.89 ( 80)
OPTICAL CONSTANTS:
 $\alpha$ = 1.507–1.513
 $\beta$ = 1.516–1.520
 $\gamma$ = 1.517–1.521
 (−)2V = 36°–56°
HARDNESS: 5
DENSITY: 2.27 (Meas.)
 2.274 (Calc.)
CLEAVAGE: {110} perfect
Fracture uneven. Brittle.

HABIT: Crystals slender prismatic, vertically striated. Often in fibrous radiated masses. Twinning on {100} common; on {001} or {110} rare.

COLOR-LUSTER: Colorless, white. Transparent to translucent. Vitreous, silky when fibrous.

MODE OF OCCURRENCE: Occurs chiefly in cavities in basalt and related rocks; also in schists and contact zones. Found at the Engels mine, Plumas County, and at Crestmore, Riverside County, California; at Table Mountain, Jefferson County, Colorado; in the trap rocks of northern New Jersey; in the Bay of Fundy district, Nova Scotia; as fine crystals in Rio Grande do Sul, Brazil; and at localities in Greenland, Iceland, Faroe Islands, Scotland, Switzerland, Germany, India, and elsewhere.

BEST REF. IN ENGLISH: Deer, Howie, and Zussman, "Rock Forming Minerals," v. 4, p. 358–376, New York, Wiley, 1963.


**SCORODITE**
$FeAsO_4 \cdot 2H_2O$
Scorodite-Mansfieldite Series

CRYSTAL SYSTEM: Orthorhombic
CLASS: 2/m 2/m 2/m
SPACE GROUP: Pcab
Z: 8
LATTICE CONSTANTS:
 $a$ = 10.30 kX
 $b$ = 10.01
 $c$ = 8.92
3 STRONGEST DIFFRACTION LINES:
 4.50 (100)
 5.65 ( 80)
 3.20 ( 80)
OPTICAL CONSTANTS:
 $\alpha$ = 1.784
 $\beta$ = 1.795
 $\gamma$ = 1.814
 (+)2V ~ 75°
HARDNESS: 3½–4
DENSITY: 3.278 (Meas.)
 3.31 (Calc.)

CLEAVAGE: {201} imperfect
{001} trace
{100} trace
Fracture subconchoidal. Brittle.

HABIT: Crystals pyramidal; also tabular or short prismatic. Usually as drusy crusts or irregular groups. Also massive, crystalline, or porous; dense to earthy.

COLOR-LUSTER: Pale grayish green, yellowish brown to brown, colorless, bluish green, blue, violet; earthy material usually pale gray, pale green to brownish green. Vitreous to resinous; also dull. Transparent to translucent.

MODE OF OCCURRENCE: Occurs chiefly as a secondary mineral formed by the oxidation of arsenic-bearing minerals. Found widespread in the United States, especially in Beaver, Juab, and Tooele counties, Utah; at Galena, and in pegmatites in Custer and Pennington counties, South Dakota; and in California, Washington, Idaho, Nevada, Wyoming, Georgia, and New York. It also occurs in Ontario, Canada; as large blue crystals from Mexico; near Ouro Preto, Minas Geraes, Brazil; and at numerous deposits in Europe, Africa, and Asia.

BEST REF. IN ENGLISH: Palache, et al., "Dana's System of Mineralogy," 7th Ed., v. II, p. 763–767, New York, Wiley, 1951.

## SCORZALITE

$(Fe^{2+}, Mg)Al_2(PO_4)_2(OH)_2$
Lazulite-Scorzalite Series

CRYSTAL SYSTEM: Monoclinic
CLASS: 2/m
SPACE GROUP: $P2_1/n$
Z: 2
LATTICE CONSTANTS:
$a = 7.15$
$b = 7.32$
$c = 7.14$
$\beta = 119°00'$
3 STRONGEST DIFFRACTION LINES:
3.24 (100)
3.20 (100)
3.14 ( 80)
OPTICAL CONSTANTS:
$\alpha = 1.633$
$\beta = 1.663$
$\gamma = 1.673$
$(-)2V = 62°$
HARDNESS: 5½–6
DENSITY: 3.38 (extrapolated for Fe end member)
3.39 (Calc.)
CLEAVAGE: {110} indistinct to good
{101} indistinct
Fracture uneven. Brittle.

HABIT: Massive, compact to granular.

COLOR-LUSTER: Deep azure blue to bluish green. Subtranslucent to opaque. Vitreous to dull. Streak white.

MODE OF OCCURRENCE: Occurs as irregular grains in the border zone of granitic pegmatite associated with quartz, muscovite, albite, apatite, brazilianite, and souzalite at the Corrego Frio mine, Minas Geraes, Brazil; also in granitic pegmatite in masses up to 2 inches in diameter, associated with quartz, tourmaline, wyllieite, muscovite, and albite at the Victory mine, Custer, South Dakota; and as irregular masses in a quartz dike at the White Mountain andalusite mine, Mono County, California.

BEST REF. IN ENGLISH: "Dana's System of Mineralogy," 7th Ed., v. 2, p. 908–911, New York, Wiley, 1951. Pecora, W. T., and Fahey, J. J., *Am. Min.*, **35**: 1–18 (1950).

## SEAMANITE

$Mn_3^{2+}(OH)_2[B(OH)_4][PO_4]$

CRYSTAL SYSTEM: Orthorhombic
CLASS: 2/m 2/m 2/m
SPACE GROUP: Pbnm
Z: 4
LATTICE CONSTANTS:
$a = 7.811$
$b = 15.114$
$c = 6.691$
3 STRONGEST DIFFRACTION LINES:
6.917 (100)
2.843 ( 80)
3.777 ( 70)
OPTICAL CONSTANTS:
$\alpha = 1.640$
$\beta = 1.663$
$\gamma = 1.665$
$(+)2V = 40°$
HARDNESS: 4
DENSITY: 3.128 (Meas.)
3.132 (Calc.)
CLEAVAGE: {001} distinct
Brittle.

HABIT: Crystals acicular, elongated along c-axis.

COLOR-LUSTER: Pale pink to yellow or yellowish brown. Transparent. Vitreous.

MODE OF OCCURRENCE: Seamanite occurs very sparingly as fine micro crystals associated with sussexite, calcite, and manganic oxyhydroxides in fractures cutting siliceous rock at the Chicagon mine, Iron River, Iron County, Michigan.

BEST REF. IN ENGLISH: Palache, et al., "Dana's System of Mineralogy," 7th Ed., v. II, p. 388–389, New York, Wiley, 1951. Moore, Paul B., and Ghose, Subrata, *Am. Min.*, **56**: 1527–1538 (1971).

## SEARLESITE

$NaBSi_2O_6 \cdot H_2O$

CRYSTAL SYSTEM: Monoclinic
CLASS: 2
SPACE GROUP: $P2_1$
Z: 2
LATTICE CONSTANTS:
$a = 7.972$
$b = 7.052$
$c = 4.900$
$\beta = 93°57'$

3 STRONGEST DIFFRACTION LINES:
  8.01 (100)
  4.06 ( 50)
  3.48 ( 40)
OPTICAL CONSTANTS:
  $\alpha$ = 1.516
  $\beta$ = 1.531
  $\gamma$ = 1.535
  $(-)2V = 55°$
HARDNESS: Soft
DENSITY: 2.460 (Meas.)
         2.46 (Calc.)
CLEAVAGE: {100} perfect
          {$\bar{1}$02} imperfect
          {010} imperfect
HABIT: As small prismatic crystals, large subhedral crystals, or spherulites composed of radiating fibers.
COLOR-LUSTER: White; also light brown (due to presence of minute quantities of organic matter).
MODE OF OCCURRENCE: Occurs as white spherulites about 1mm in diameter in clay at a depth of 540 feet at Searles Lake, San Bernardino County, California; as prismatic crystals up to 3mm in length in the Silver Peak Range, Esmeralda County, Nevada; and as subhedral crystals up to 7 inches long, 4 inches wide, and 1 inch thick in a stratum of trona at a depth of 1500 feet in the Green River formation in Sweetwater County, Wyoming.
BEST REF. IN ENGLISH: Fahey, Joseph J., and Axelrod, Joseph M., *Am. Min.*, **35**: 1014–1020 (1950).

## SEDERHOLMITE
$\beta$-NiSe

CRYSTAL SYSTEM: Hexagonal
CLASS: 6/m 2/m 2/m
SPACE GROUP: $P6_3/mmc$
Z: 2
LATTICE CONSTANTS:
  (Ni rich)      (Ni deficient)
  $a$ = 3.65      $a$ = 3.624
  $c$ = 5.34      $c$ = 5.288
3 STRONGEST DIFFRACTION LINES:
  2.70  (100)
  2.015 ( 80)
  1.806 ( 60)
DENSITY: 7.06 (Calc.)
HABIT: Massive; as embedded grains.
COLOR-LUSTER: Yellow to orange-yellow.
MODE OF OCCURRENCE: Occurs as grains in clausthalite, and with wilkmanite in penroseite clusters, in calcite veins in sills of albite diabase in a schist formation, associated with low-grade uranium mineralization, at Kuusamo, northeast Finland.
BEST REF. IN ENGLISH: Vuorelainen, Y., Huhma, A., and Häkli, A., *Am. Min.*, **50**: 519 (1965).

## SEDOVITE
$U(MoO_4)_2$

CRYSTAL SYSTEM: Orthorhombic

Z: 1
LATTICE CONSTANTS:
  $a$ = 3.36
  $b$ = 11.08
  $c$ = 6.42
3 STRONGEST DIFFRACTION LINES:
  3.193 (100)
  11.04 ( 90)
  3.370 ( 90)
OPTICAL CONSTANTS:
  Indices > 1.789
HARDNESS: ~ 3
DENSITY: 4.2 (Meas.)
CLEAVAGE: One parallel to elongation.
HABIT: Radiating-fibrous bundles of crystals, tenths to hundredths of a millimeter long. Also powdery.
COLOR-LUSTER: Brown to reddish brown. Nearly opaque.
MODE OF OCCURRENCE: Occurs associated with wulfenite, large platy crystals of gypsum, and mourite in the supergene zone of a uranium-molybdenum deposit. No locality given.
BEST REF. IN ENGLISH: Skvortsova, K. V., and Sidorenko, G. A., *Am. Min.*, **51**: 530 (1966).

## SEELIGERITE
$Pb_3(IO_3)Cl_3O$

CRYSTAL SYSTEM: Orthorhombic, pseudotetragonal
CLASS: 2 2 2
SPACE GROUP: $C222_1$
Z: 8
LATTICE CONSTANTS:
  $a = b$ = 7.964
       $c$ = 27.288
3 STRONGEST DIFFRACTION LINES:
  3.219 (100)
  3.649 ( 90)
  2.785 ( 80)
OPTICAL CONSTANTS:
  $\alpha$ = 2.12
  $\beta = \gamma$ = 2.32
  $(-)2V = -4°$
HARDNESS: Not determined.
DENSITY: 6.83 (Meas. synthetic)
         7.052 (Calc.)
CLEAVAGE: {001} perfect
          {110} good
          {100} rare
          {010} rare
HABIT: As thin plates.
COLOR-LUSTER: Bright yellow.
MODE OF OCCURRENCE: Occurs in small amounts associated with boleite, paralaurionite, and schwartzembergite, at Mina Santa Ana, near Caracoles, Sierra Gorda, Chile.
BEST REF. IN ENGLISH: Mücke, Arno, *Am. Min.*, **57**: 327–328 (1972).

## SEGELERITE
$CaMg(H_2O)_4Fe^{3+}(OH)[PO_4]_2$
$Fe^{3+}$ analogue of Overite

SPHALERITE on dolomite. Lengenbach quarry, Binnental, Switzerland.

SPHALERITE Nordmark, Vermland, Sweden.

SPHALERITE Naica, Chihuahua, Mexico.

SPHALERITE Joplin, Jasper County, Missouri.

SPHENE (twin) Zillertal, Tyrol, Austria.

SPHENE (twin) Binnental, Switzerland.

SPINEL Mogok district, Burma.

SPINEL Vesuvius, Italy.

113

SPODUMENE (hiddenite)  Hiddenite, near Stony Point, Alexander County, North Carolina.

STEENSTRUPINE  Kangerdluarsuk, Greenland.

STEPHANITE  Eisleben, Thuringia, Germany.

STEPHANITE  Pribram, Bohemiá, Czechoslovakia.

STEWARTITE with metastrengite  Stewart mine, Pala, San Diego County, California.

STIBNITE  Felsobanya, Roumania.

STIBNITE  Felsobanya, Roumania.

STIBNITE  Ambrose mine, near Hollister, San Benito County, California.

STICHTITE   Kaapsche Hoop, Transvaal, South Africa.

STILBITE with heulandite   Bernardsville, New Jersey.

STILPNOMELANE with hematite   Antwerp, Jefferson County, New York.

STOLZITE with pyromorphite   Broken Hill, New South Wales, Australia.

STOLZITE with raspite   Broken Hill, New South Wales, Australia.

STRENGITE   Indian Mountain, Cherokee County, Alabama.

STRONTIANITE   Mt. Union, Mifflin County, Pennsylvania.

STRUNZITE   Big Chief mine, Glendale, Pennington County, South Dakota.

STRUNZITE   Hagendorf, Bavaria, Germany.

STRUNZITE   Hagendorf, Bavaria, Germany.

SULPHOHALITE   Searles Lake, San Bernardino County, California.

SULFUR   Girgenti, Sicily.

SVABITE   Harstig mine, Pajsberg, Sweden.

SYLVANITE   Nagyag, Roumania.

TARBUTTITE   Broken Hill, Zambia.

TARBUTTITE   Broken Hill, Zambia.

116

TAVORITE   Tip Top mine, Custer County, South Dakota.

TANTALITE (twin)   Greenbushes, Western Australia.

TELLURITE   Moctezuma, Sonora, Mexico.

TELLURIUM   Kawazu mine, near Simoda, Izu Peninsula, Japan.

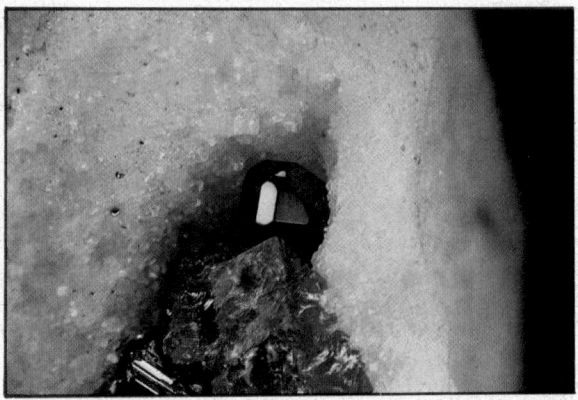

TENNANTITE (binnite)   Lengenbach quarry, Binnental, Switzerland.

TENNANTITE   Dresser quarry, Walton, Nova Scotia, Canada.

TENNANTITE (binnite)   Lengenbach quarry, Binnental, Switzerland.

TEPHROITE with willemite   Franklin, Sussex County, New Jersey.

117

TERLINGUAITE with montroydite   Terlingua, Brewster
County, Texas.

TETRAHEDRITE with pyrite   Carlotta mine, Chanarcillo,
Chile.

TETRAHEDRITE   Siegen, Westphalia, Germany.

THOMSENOLITE   Ivigtut, Greenland.

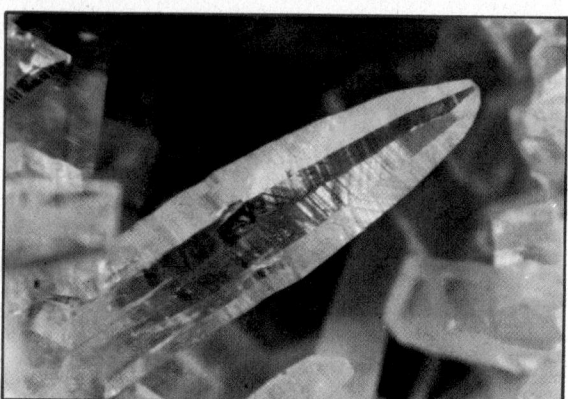

THOMSENOLITE   Ivigtut, Greenland.

THOMSONITE   Whipple quarry, Drain, Douglas County,
Oregon.

THORIANITE   Oka, Quebec, Canada.

TIEMANNITE   Marysvale, Piute County, Utah.

TOPAZ    Tepetate, San Luis Potosi, Mexico.

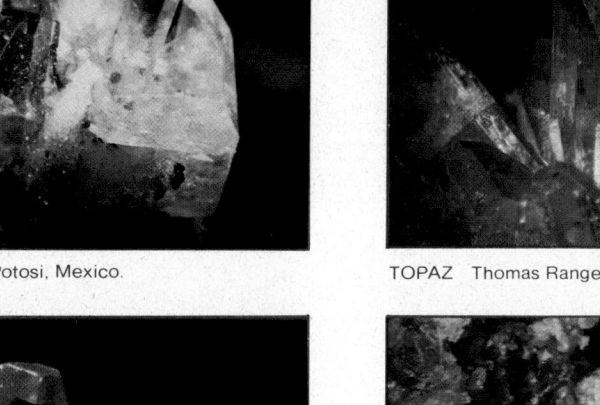

TOPAZ    Thomas Range, Utah.

TOPAZ    Tepetate, San Luis Potosi, Mexico.

TORBERNITE    Wheal Basset, Illogan, Cornwall, England.

TORBERNITE    Shinkolobwe, Katanga, Zaire.

TRIDYMITE    Diamond Lake, Douglas County, Oregon.

TSUMEBITE on azurite    Tsumeb, South West Africa.

TUHUALITE    Opo Bay, Mayor Island, New Zealand.

119

TUNELLITE   Boron, Kern County, California.

TURQUOISE   Lynch Station, Campbell County, Virginia.

TYROLITE   Eureka, Juab County, Utah.

URANINITE   Bryson City, North Carolina.

URANOPHANE   Haystack Mountain, near Grants, Valencia County, New Mexico.

URANOPHANE   Santa Eulalia, Chihuahua, Mexico.

UVAROVITE   Bysserok, Ural Mountains, USSR.

UVAROVITE with chrome diopside   Outokumpu, Finland.

UVAROVITE (vicinal hexoctahedron) Jacksonville, Tuolumne Co., California.

VALENTINITE with kermesite Lac Nicollet, Sherbrooke, Quebec, Canada.

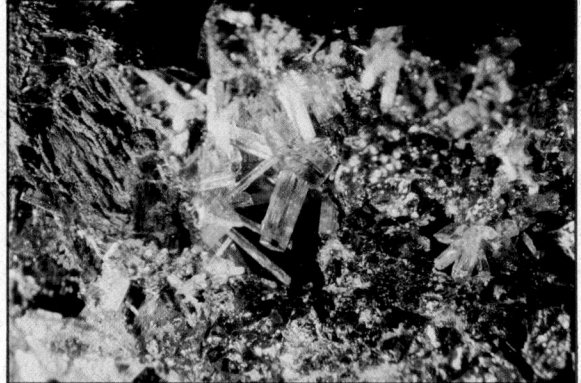

VALENTINITE on sphalerite and galena Ouro Preto, Minas Geraes, Brazil.

VALENTINITE Pribram, Bohemia, Czechoslovakia.

VANADINITE Olsen mine, Superior, Pinal County, Arizona.

VANADINITE Apache mine Gila County, Arizona.

VANADINITE Old Yuma mine, Pima County, Arizona.

VANADINITE Lake Valley, Sierra County, New Mexico.

VANDENBRANDEITE   Musonoi mine, Kolwezi, Katanga, Zaire.

VANURALITE   Mounana, Gabon.

VAUQUELINITE   Beresov, Urals, USSR.

VAUXITE   Llallagua, Bolivia.

VILLIAUMITE   Mont St. Hilaire, Quebec, Canada.

VIVIANITE on ludlamite   Big Chief mine, Glendale, Pennington County, South Dakota.

VIVIANITE on ludlamite   Palermo mine, North Groton, New Hampshire.

VOLBORTHITE    Monument Valley, Navajo County, Arizona.

VOLBORTHITE    Near Thompson, Grand County, Utah.

WAKABAYASHILITE    Manhattan, Nye County, Nevada.

WARDITE with lewistonite    Fairfield, Utah County, Utah.

WARDITE with eosphorite    Taquaral, Minas Geraes, Brazil.

WARDITE    Wolfsberg, Carinthia, Austria.

WAVELLITE with beraunite    Mt. Holly Springs, Cumberland
County, Pennsylvania.

WAVELLITE on pyrophyllite    Near Staley, Randolph County,
North Carolina.

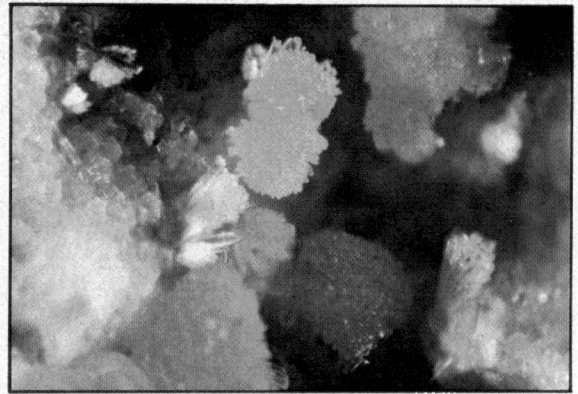

WEEKSITE   Anderson mine, Yavapai County, Arizona.

WEINSCHENKITE (churchite)   Vesuvius. Rockbridge County. Virginia.

WELOGANITE   Francon quarry, Montreal, Quebec, Canada.

WERMLANDITE   Långban, Sweden.

WHEWELLITE (twin)   Burgk near Dresden, Saxony, Germany.

WHERRYITE   Mammoth-St. Anthony mine, Pinal County, Arizona.

WHEWELLITE (twin)   Burgk near Dresden, Saxony, Germany.

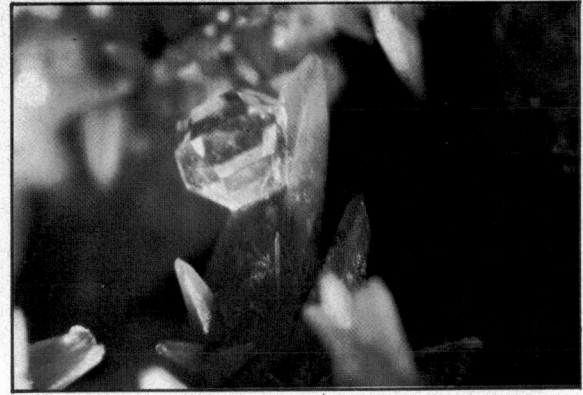

WHITLOCKITE on siderite    Palermo mine, North Groton, New Hampshire.

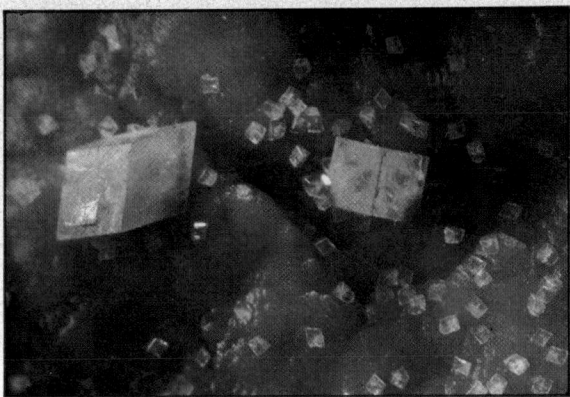

WHITLOCKITE on carbonate-apatite    Tip Top mine, Custer, County, South Dakota.

WILLEMITE    Franklin, Sussex County, New Jersey.

WILLEMITE (short-wave ultraviolet light)    Franklin, Sussex County, New Jersey.

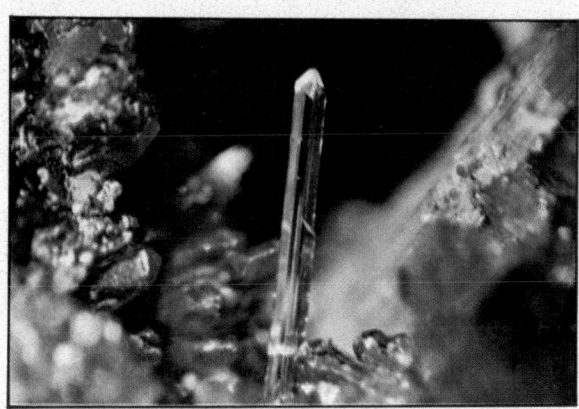

WILLEMITE    Franklin, Sussex County, New Jersey.

WITHERITE    Fallowfield, Hexham, England.

WOLLASTONITE    Vesuvius, Italy.

WOODHOUSEITE    White Mountains, Mono County, California.

WOODHOUSEITE    White Mountains, Mono County, California.

WULFENITE    Red Cloud mine, Yuma County, Arizona.

WULFENITE with torbernite    Shinkolobwe, Katanga, Zaire.

WULFENITE    Los Lamentos, Chihuahua, Mexico.

WULFENITE    Los Lamentos, Chihuahua, Mexico.

WULFENITE    Lost Soldier mine, Lyon County, Nevada.

WULFENITE    (Theba, Arizona)

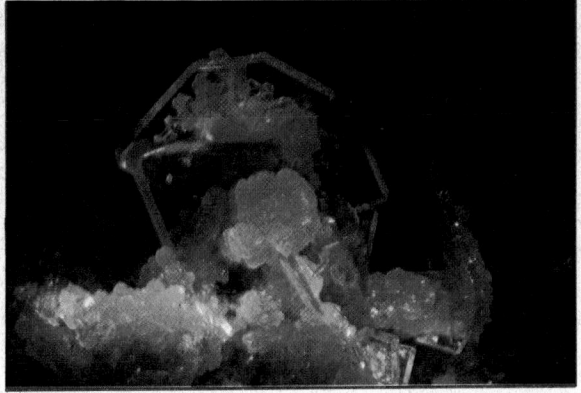

WULFENITE with mimetite   San Francisco mine, Sonora, Mexico.

WULFENITE   Red Cloud mine, Yuma County, Arizona.

WULFENITE   Arizona.

WURTZITE   Thomaston Dam, Connecticut.

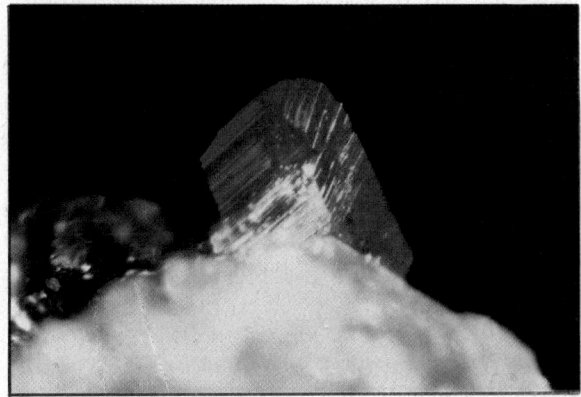

XANTHOCONITE   Joachimstal (Jachymov), Bohemia, Czechoslovakia.

XANTHOCONITE   Keeley mine, near Cobalt, Ontario, Canada.

XENOTIME   Rauris, Salzburg, Austria.

YEDLINITE  Mammoth-St. Anthony mine, Pinal County, Arizona.

ZINCITE  Franklin, Sussex County, New Jersey.

ZINCITE  Franklin, Sussex County, New Jersey.

ZIRCON  Laacher See, Rhineland, Germany.

ZIRCON  Sarance mine, Bancroft, Ontario, Canada.

ZIRCON  Schelgaden, Tyrol, Austria.

ZUNYITE  Zuni mine, Silverton, San Juan County, Colorado.

CRYSTAL SYSTEM: Orthorhombic
CLASS: 2/m 2/m 2/m
SPACE GROUP: Pcca
Z: 8
LATTICE CONSTANTS:
$a$ = 14.826
$b$ = 18.751
$c$ = 7.307
3 STRONGEST DIFFRACTION LINES:
2.868 (100)
9.31 ( 90)
5.34 ( 60)
OPTICAL CONSTANTS:
$\alpha$ = 1.618
$\beta$ = 1.635
$\gamma$ = 1.650
Biaxial (−), 2V large.
HARDNESS: 4⁻
DENSITY: 2.67 (Meas.)
         2.610 (Calc.)
CLEAVAGE: {010} perfect
HABIT: Crystals long prismatic, vertically striated, nearly square in cross-section, and up to 0.5 mm in length. Forms: $a\{100\}$, $b\{010\}$, $m\{110\}$, and $q\{121\}$.
COLOR-LUSTER: Pale yellow-green, chartreuse, inclining to colorless. Transparent to translucent. Vitreous.
MODE OF OCCURRENCE: Segelerite occurs sparsely along joint fractures in dense ferrisicklerite-rockbridgeite, usually in close association with collinsite, minor hydroxylapatite and globules of red-brown robertsite, at the Tip Top pegmatite, Custer County, South Dakota.
BEST REF. IN ENGLISH: Moore, Paul Brian, *Am. Min.*, **59**: 48–59 (1974).

## SEIDOZERITE
$Na_2 (Zr, Ti, Mn)_2 Si_2 O_8 F_2$

CRYSTAL SYSTEDM: Monoclinic
CLASS: 2/m
SPACE GROUP: P2/c
Z: 4
LATTICE CONSTANTS:
$a$ = 5.53
$b$ = 7.10
$c$ = 18.30
$\beta$ = 102°43′
3 STRONGEST DIFFRACTION LINES:
2.97 (100)
2.87 ( 70)
1.830 ( 70)
OPTICAL CONSTANTS:
$\alpha$ = 1.725
$\beta$ = 1.758
$\gamma$ = 1.830
(+)2V = 68°
HARDNESS: 4–5
DENSITY: 3.472 (Meas.)
         3.59 (Calc.)
CLEAVAGE: {001} perfect
Brittle.
HABIT: Radiating crystals up to 5 × 1 cm in size; prismatic, elongated parallel $b$-axis. Forms: $a\{100\}$, $c\{001\}$, $e\{203\}$, $b\{010\}$, $d\{011\}$, and $p\{111\}$. Also fibrous or spherulitic.
COLOR-LUSTER: Brownish red to reddish yellow, translucent with red color. Vitreous.
MODE OF OCCURRENCE: Occurs with microcline, aegirine, nepheline, apatite, pyrochlore, magnetite, ilmenite, titanolavenite, and eudialyte in four nepheline syenite pegmatites of the Murai and Uel'kuai Rivers, Seidozero region, Lovozero massif, Kola Peninsula, USSR.
BEST REF. IN ENGLISH: Semenov, E. I., Kazakova, M. E., and Simonov, V. I., *Am. Min.*, **44**: 467–468 (1959). Vlasov, K. A., "Mineralogy of Rare Elements," v. II, p. 386–388, Israel Program for Scientific Translations, 1966.

## SEKANINAITE (Iron-Cordierite)
$(Fe, Mg)_2 Al_4 Si_5 O_{18}$

CRYSTAL SYSTEM: Orthorhombic
CLASS: 2/m 2/m 2/m
SPACE GROUP: Cccm
Z: 4
LATTICE CONSTANTS:
$a \simeq 9.7$
$b \simeq 17.1$
$c \simeq 9.4$
3 STRONGEST DIFFRACTION LINES:
3.39 (100)
8.63 ( 70)
8.55 ( 70)
OPTICAL CONSTANTS:
$\alpha$ = 1.559
$\beta$ = 1.569
$\gamma$ = 1.573
(−)2V = large
HARDNESS: 7–7½
DENSITY: 2.772 (Meas.)
CLEAVAGE: {010} distinct
          {001} indistinct
          {100} indistinct
Fracture conchoidal. Brittle.
HABIT: Crystals usually conical in form, imperfectly developed, up to several decimeters in size.
COLOR-LUSTER: Light blue, bluish violet. Transparent to translucent. Vitreous.
MODE OF OCCURRENCE: Occurs as giant crystals in a zoned pegmatite body near Dolni Bory, western Moravia, Czechoslovakia.
BEST REF. IN ENGLISH: Stanek, J., and Miskovsky, J., *Am. Min.*, **50**: 264–265 (1965) [Named in 1968.]

## SELENITE = Synonym for transparent gypsum

## SELENIUM
Se

CRYSTAL SYSTEM: Hexagonal
CLASS: 32
SPACE GROUP: P3$_1$21 or P3$_2$21

Z: 3
LATTICE CONSTANTS:
  $a = 4.366$
  $c = 4.954$
3 STRONGEST DIFFRACTION LINES:
  3.00  (100)
  3.78  ( 55)
  2.072 ( 35)
OPTICAL CONSTANTS:
  $\omega = 3.0$
    $\epsilon = 4.04$  (Na)
      (+)
HARDNESS: 2
DENSITY: 4.80 (Meas.)
         4.809 (Calc.)
CLEAVAGE: $\{01\bar{1}2\}$ distinct
Crystals very flexible.

HABIT: Crystals acicular [0001]; often hollow and tube-like; as felty aggregates or clustered in sheets; also as small glossy globules.

COLOR-LUSTER: Gray, purple-gray, metallic. Also reddish, vitreous, when globular.

MODE OF OCCURRENCE: Occurs on fracture surfaces and disseminated throughout limited areas of the Lakota sandstone at the Road Hog No. 1A mine near Edgemont, Fall River County, South Dakota; in high-grade vanadium-uranium ore from the Peanut mine, Bull Canyon, Montrose County, Colorado; associated with zippeite, metatyuyamunite, metarossite, montroseite, and corvusite, at the Parco No. 23 mine, Thompsons district, Grand County, Utah; as acicular crystals and reddish glassy coatings on sandstone in the Ambrosia Lakes district, McKinley County, New Mexico; as needles up to 2 cm long encrusting quartzite and fritted sandstone in the fire zone of the United Verde mine, Jerome, Yavapai County, Arizona; and at Kladno, Bohemia.

BEST REF. IN ENGLISH: Palache, et al., "Dana's System of Mineralogy," 7th Ed., v. I, p. 136–137, New York, Wiley, 1944.

## SELENOLITE
$SeO_2$

CRYSTAL SYSTEM: Tetragonal
CLASS: 4/m 2/m 2/m or 4mm
SPACE GROUP: $P4_2/mbc$ or $P4_2bc$
Z: 8
LATTICE CONSTANTS:
  $a = 8.37$
  $c = 5.06$
3 STRONGEST DIFFRACTION LINES:
  3.01 (100)
  3.74 ( 60)
  4.18 ( 55)
OPTICAL CONSTANTS:
  Indices ~ 1.9–2.0
HARDNESS: Not determined. Soft.
DENSITY: 4.16 (Calc.)
CLEAVAGE: Not determined.
HABIT: Crystals minute, acicular.
COLOR-LUSTER: Colorless, white. Vitreous.
MODE OF OCCURRENCE: Occurs as minute needles

associated with molybdomenite and cerussite at Cacheuta, Mendosa Province, Argentina; in lava from the crater of Vesuvius, Italy; and in a tin-ore deposit of the Magadan region, USSR.

BEST REF. IN ENGLISH: Vlasov, K. A., "Mineralogy of Rare Elements," v. II, p. 673, Israel Program for Scientific Translations, 1966.

## SELENOVAESITE = Selenian vaesite

## SELENSULPHUR = Selenian sulfur

## SELEN-TELLURIUM (Inadequately described mineral)
(Se, Te)

CRYSTAL SYSTEM: Unknown
HARDNESS: 2–2½
DENSITY: Not determined.
CLEAVAGE: Hexagonal prismatic, perfect.
Brittle.
HABIT: Massive.
COLOR-LUSTER: Blackish gray. Opaque. Metallic. Streak black.

MODE OF OCCURRENCE: Occurs associated with quartz and barite at the El Plomo silver mine, Ojojona district, Tegucigalpa, Honduras.

BEST REF. IN ENGLISH: Palache, et al., "Dana's System of Mineralogy," 7th Ed., v. I, p. 137–138, New York, Wiley, 1944.

## SELIGMANNITE
$CuPbAsS_3$

CRYSTAL SYSTEM: Orthorhombic
CLASS: 2/m 2/m 2/m, or mm2
SPACE GROUP: $Pn2_1m$, Pnmm
Z: 4
LATTICE CONSTANTS:
  $(Pn2_1m)$      (Pnmm)
  $a = 8.08$      $a = 8.11$
  $b = 8.75$      $b = 8.74$
  $c = 7.65$      $c = 7.63$
3 STRONGEST DIFFRACTION LINES:
  2.72  (100)
  3.85  ( 80)
  1.749 ( 70)
HARDNESS: 3
DENSITY: 5.38–5.44 (Meas.)
         5.42 (Calc.)
CLEAVAGE: $\{001\}$ very poor
          $\{100\}$ very poor
          $\{010\}$ very poor
Fracture conchoidal. Brittle.

HABIT: Crystals equant; short prismatic to tabular. Often striated. Twinning on $\{110\}$, very common.

COLOR-LUSTER: Dark lead gray to black. Opaque. Metallic. Streak dark brown to purple-black.

MODE OF OCCURRENCE: Occurs associated with tennantite, pyrite, and sphalerite at Bingham Canyon and Carrs Fork, Salt Lake County, Utah; at Butte and Emery, Montana; in cavities in crystalline dolomite associated with tennantite and other sulfosalts at the Lengenbach quarry in the Binnental, Valais, Switzerland; and at Kalgoorlie, Western Australia.

BEST REF. IN ENGLISH: Palache, et al., "Dana's System of Mineralogy," 7th Ed., v. I, p. 411–412, New York, Wiley, 1944.


# SELLAITE
$MgF_2$

CRYSTAL SYSTEM: Tetragonal
CLASS: 4/m 2/m 2/m
SPACE GROUP: $P4_2/mnm$
Z: 2
LATTICE CONSTANTS:
 $a = 4.660$
 $c = 3.078$
3 STRONGEST DIFFRACTION LINES:
 3.27 (100)
 2.231 ( 95) (synthetic)
 1.711 ( 75)
OPTICAL CONSTANTS:
 $\omega = 1.378$
 $\epsilon = 1.390$
 (+)
HARDNESS: 5
DENSITY: 3.148 (Meas.)
 3.14 (Calc.)
CLEAVAGE: {010} perfect
 {110} perfect
Fracture conchoidal. Brittle.

HABIT: Crystals stubby prismatic to acicular. Also as fibrous aggregates.

COLOR-LUSTER: Colorless, white. Transparent. Vitreous.

MODE OF OCCURRENCE: Occurs in bituminous dolomite-anhydrite rock at Bleicherode, Harz Mountains, Germany, and near Moutiers, Savoy, France. It is also found in Italy at Mt. Vesuvius associated with anhydrite, and at Carrara associated with gypsum in cavities in marble.

BEST REF. IN ENGLISH: Palache, et al., "Dana's System of Mineralogy," 7th Ed., v. II, p. 37–39, New York, Wiley, 1951.


# SEMENOVITE
$(Na, Ce, Ca)_{12}(Be, Si)Si_{12}O_{40}(F, OH)_8 \cdot H_2O$

CRYSTAL SYSTEM: Tetragonal
CLASS: 4/m
SPACE GROUP: $P4_2/n$
Z: 2
LATTICE CONSTANTS:
 $a = 13.87$
 $c = 9.89$

3 STRONGEST DIFFRACTION LINES:
 3.28 (100)
 2.84 (100)
 2.73 (100)
HARDNESS: 3½–4
DENSITY: 3.140 (Meas.)
CLEAVAGE: None. Fracture uneven.

HABIT: Crystals tetragonal bipyramidal, with {111} and rarely minor {110}. Usually from 0.1 to 1.0 mm in size; rarely 1 cm. Crystal faces rather uneven. Always forms intricate mimetic (tetragonal) interpenetration twins; twinning plane (110).

COLOR-LUSTER: Brown, reddish brown, gray. Transparent to translucent. Vitreous.

MODE OF OCCURRENCE: Occurs in cavities in one epi-eudidymite bearing albitite on the Taseq Slope, Ilimaussaq, Greenland.

BEST REF. IN ENGLISH: Petersen, Ole V., and Rönsbo, Jörn G., *Lithos*, 5: 163–173 (1972).


# SEMSEYITE
$Pb_9Sb_8S_{21}$

CRYSTAL SYSTEM: Monoclinic
CLASS: 2/m
SPACE GROUP: C2/c
Z: 4
LATTICE CONSTANTS:
 $a = 13.48$
 $b = 11.87$
 $c = 24.48$
 $\beta = 105°45'$
3 STRONGEST DIFFRACTION LINES:
 3.26 (100)
 3.81 ( 90)
 2.95 ( 90)
HARDNESS: 2½
DENSITY: 6.08 (Meas.)
 6.15 (Calc.)
CLEAVAGE: {112} perfect.
Brittle.

HABIT: Crystals tabular or prismatic; often twisted. As aggregates and groups composed of subparallel crystals.

COLOR-LUSTER: Lead gray to black. Opaque. Metallic, tarnishing dull. Streak black.

MODE OF OCCURRENCE: Occurs associated with pyrite at Oruro, Bolivia; with stibnite at Glendinning, Dumfriesshire, Scotland; at Wolfsberg, Harz Mountains, Germany; as fine specimens at Kisbanya, Felsobanya, and Rodna, Roumania; and at the Julcani mine, Huancavelica, Peru.

BEST REF. IN ENGLISH: Palache, et al., "Dana's System of Mineralogy," 7th Ed., v. I, p. 466–468, New York, Wiley, 1944.


# SENAITE
$Pb(Ti, Fe, Mn, Mg)_{24}O_{38}$
Ilmenite Group

CRYSTAL SYSTEM: Hexagonal
CLASS: 3 or $\bar{3}$

SPACE GROUP: R3 or R$\overline{3}$
Z: 3
LATTICE CONSTANTS:
$a$ = 10.42
$c$ = 20.86
3 STRONGEST DIFFRACTION LINES:
2.894 (100)
3.43 ( 90)
1.981 ( 75)
HARDNESS: 6+
DENSITY: 5.301 (Meas.)
CLEAVAGE: Not determined.
Fracture conchoidal.
HABIT: As rough crystals and rounded fragments.
COLOR-LUSTER: Black. Opaque, except in very thin fragments. Submetallic. Streak brownish black.
MODE OF OCCURRENCE: Occurs in the diamond-bearing placer deposits near Diamantina, Minas Geraes, Brazil.
BEST REF. IN ENGLISH: Palache, et al., "Dana's System of Mineralogy," 7th Ed., v. I, p. 541–542, New York, Wiley, 1944. Rouse, Roland C., and Peacor, Donald R., *Am. Min.*, **53**: 869–879 (1968).

## SENARMONTITE
$Sb_2O_3$
Dimorphous with valentinite

CRYSTAL SYSTEM: Cubic
CLASS: 4/m $\overline{3}$ 2/m
SPACE GROUP: Fd3m
Z: 16
LATTICE CONSTANT:
$a$ = 11.14
3 STRONGEST DIFFRACTION LINES:
3.218 (100)
1.972 ( 42)
2.788 ( 40)
OPTICAL CONSTANT:
$N$ = 2.087 (Na)
HARDNESS: 2–2½
DENSITY: 5.50 (Meas.)
5.583 (Calc.)
CLEAVAGE: {111} in traces.
Fracture uneven. Brittle.
HABIT: Crystals octahedral, up to 3 cm along an axis; also massive, granular, or as crusts.
COLOR-LUSTER: Colorless to grayish white. Transparent. Resinous. Streak white.
MODE OF OCCURRENCE: Occurs as a secondary mineral formed by the alteration of stibnite or other antimony minerals. Found with cervantite and stibnite at the property of the Skidoo Mining Company, Inyo County, California; associated with boulangerite at the Tuttle mine and elsewhere in the Silver City district, Pennington County, South Dakota; at South Ham, Wolfe County, Quebec, Canada; as superb large crystals at Hamimat, Constantine, Algeria; and at localities in France, Italy, Germany, and Sardinia.
BEST REF. IN ENGLISH: Palache, et al., "Dana's System of Mineralogy," 7th Ed., v. I, p. 544–545, New York, Wiley, 1944.

## SENGIERITE
$Cu(UO_2)_2(VO_4)_2 \cdot 6H_2O$

CRYSTAL SYSTEM: Monoclinic
CLASS: 2/m
SPACE GROUP: P2$_1$/a
Z: 2
LATTICE CONSTANTS:
$a$ = 10.62
$b$ = 8.10
$c$ = 10.11
$\beta$ = 103°36′
3 STRONGEST DIFFRACTION LINES:
9.82 (100)
4.91 ( 80)
3.735 ( 60)
OPTICAL CONSTANTS:
$\alpha$ = 1.76–1.77
$\beta$ = 1.92–1.94
$\gamma$ = 1.94–1.97
(–)2V = 37°–39°
HARDNESS: 2½
DENSITY: 4.05 (Meas.)
CLEAVAGE: {001} perfect
Brittle.
HABIT: Crystals thin plates with pseudohexagonal outline and range up to 2.0 mm in size. Tabular on {001}; forms observed are {001}, {111}, {010}, and rarely {110} and {011}. As coatings in open fractures.
COLOR-LUSTER: Green to slightly yellowish green. Transparent. Vitreous to adamantine.
MODE OF OCCURRENCE: Occurs associated with fibrous malachite, chrysocolla, vandenbrandeite, and secondary oxides of iron, cobalt, and nickel at Luishwishi, Haut-Katanga, Zaire; also found associated with volborthite and cuprosklodowskite at Amelal, Aragana-Bigoudine region, Morocco; and as efflorescent patches up to 20 mm in diameter on chalcocite from the Cole Shaft, Bisbee, Arizona.
BEST REF. IN ENGLISH: Frondel, Clifford, USGS Bull. 1064, p. 258–260 (1958).

## SEPIOLITE (Meerschaum)
$Mg_2Si_3O_8 \cdot 2H_2O$

CRYSTAL SYSTEM: Orthorhombic
Z: 8
LATTICE CONSTANTS:
$a$ = 13.50
$b$ = 26.97
$c$ = 5.25
3 STRONGEST DIFFRACTION LINES:
12.3 (100)
2.58 ( 60)
4.4 ( 35)

OPTICAL CONSTANTS:

$\alpha = 1.515-1.520, 1.506$

$\beta = ?$

$\gamma = 1.525-1.529, 1.526$

$(-)2V = 0°-60°$

HARDNESS: 2-2½

Impressible by finger nail. Adheres to tongue.

DENSITY: ~2.

Dry porous masses float on water.

CLEAVAGE: Not determined.

HABIT: Massive; fine fibrous. Usually compact nodular, earthy, or clay-like.

COLOR-LUSTER: White, grayish, yellowish, or tinted bluish green or reddish. Nearly opaque. Dull.

MODE OF OCCURRENCE: Occurs mainly as an alteration product of serpentine or magnesite. Found as an authigenic mineral in lacustrine sediments of Mound Lake in Lynn and Terry counties, Texas; in small veins in calcite and as pellets in limestone at Crestmore, Riverside County, California; in Yavapai County, Arizona; in Grant County and near Sapella, San Miguel County, New Mexico; in Duchesne and Tooele counties, Utah; and in Chester and Delaware counties, Pennsylvania. Also as fine nodular masses on the plains of Eskisehir, Turkey, and at localities in Spain, Czechoslovakia, Greece, and Morocco.

BEST REF. IN ENGLISH: Winchell and Winchell, "Elements of Optical Mineralogy," 4th Ed., Pt. 2, p. 444, New York, Wiley, 1951. Brindley, G. W., *Am. Min.*, **44**: 495-500 (1959).

# SERANDITE

$Mn_2 NaSi_3 O_8 (OH)$

CRYSTAL SYSTEM: Triclinic (pseudomonoclinic)

CLASS: $\bar{1}$

SPACE GROUP: $P\bar{1}$

Z: 2

LATTICE CONSTANT:

$\beta = 94°30'$

3 STRONGEST DIFFRACTION LINES:

3.19 (100)

3.00 (100)

2.85 (100)

OPTICAL CONSTANTS:

$\alpha = 1.660$

$\beta = 1.664$

$\gamma = 1.688$

$(+)2V = 35.5°$

HARDNESS: 4½-5

DENSITY: 3.32 (Meas.)

CLEAVAGE: {001} perfect
{100} perfect

Fracture uneven. Brittle.

HABIT: Crystals thick tabular or prismatic; often in aggregates of randomly intergrown crystals.

COLOR-LUSTER: Rose red, pink. Transparent. Vitreous, pearly on cleavages.

MODE OF OCCURRENCE: Occurs as superb large crystals and crystal aggregates in association with analcime, aegirine, and other minerals in carbonatite at Mont St.

Hilaire, Quebec, Canada; also in nepheline syenite on the island of Rouma, Los Islands, Guinea.

BEST REF. IN ENGLISH: Winchell and Winchell, "Elements of Optical Mineralogy," 4th Ed., Pt. 2, p. 443, New York, Wiley, 1951.

# SERENDIBITE (Inadequately described mineral)

$Ca_4 (Mg, Fe, Al)_6 (Al, Fe)_9 (Si, Al)_6 B_3 O_{40}$

CRYSTAL SYSTEM: Triclinic

OPTICAL CONSTANTS:

$\alpha = 1.701$

$\beta = 1.703$

$\gamma = 1.706$

2V (+) and (−)

HARDNESS: 6½-7

DENSITY: 3.42 (Meas.)

CLEAVAGE: None.

HABIT: Massive; as aggregates of irregular grains. Polysynthetic twinning.

COLOR-LUSTER: Grayish blue-green, blue, dark blue. Vitreous.

MODE OF OCCURRENCE: Occurs as a contact mineral in limestone at the new City quarry, and in calcite and colorless idocrase at the Commercial quarry, Crestmore, Riverside County, California; in association with diopside, phlogopite, spinel, scapolite, orthoclase, plagioclase, calcite, and other minerals in Grenville limestone near its contact with an intrusive granite a few miles west of Johnsburg, Warren County, New York; at Gangapitija, Ceylon; and in Tanzania.

BEST REF. IN ENGLISH: Larsen, E. S., and Schaller, W. T., *Am. Min.*, **17**: 457-465 (1932).

# SERICITE = Hydrous muscovite

# SERPENTINE = Mineral group of general formula:

$(Mg, Fe)_3 Si_2 O_5 (OH)_4$

See: Antigorite

Chrysotile

Clinochrysotile

Lizardite

Pecoraite

# SERPIERITE

$Ca(Cu, Zn)_4 (SO_4)_2 (OH)_6 \cdot 3H_2O$

CRYSTAL SYSTEM: Monoclinic

CLASS: 2/m

SPACE GROUP: C2/c

Z: 1

LATTICE CONSTANTS:

$a = 22.186$

$b = 6.250$

$c = 21.853$

$\beta = 113°22'$

3 STRONGEST DIFFRACTION LINES:
10.22 (100)
5.09 ( 90)
3.386 ( 90)
OPTICAL CONSTANTS:
$\alpha = 1.583$
$\beta = 1.641$
$\gamma = 1.648$
(−)2V = 37° (Calc.)
HARDNESS: Not determined.
DENSITY: 3.07 (Meas.)
3.08 (Calc.)
CLEAVAGE: {001} perfect
HABIT: Crystals minute, lath-like, elongated along *c*-axis and flattened {001}. As tufts, crusts, and rounded masses with satiny surface.
COLOR-LUSTER: Sky blue. Transparent. Vitreous; pearly on cleavage.
MODE OF OCCURRENCE: Occurs with smithsonite at Ross Island, Killarney, Ireland, and at Laurium, Greece; and associated with cyanotrichite and linarite at Akchagyl, Kazakhstan, USSR.
BEST REF. IN ENGLISH: Faraone, Domenico, Sabelli, Cesare, and Zanazzi, Pier F., *Am. Min.*, **54**: 328–329 (1969).

## SEYBERTITE = Clintonite

## SEYRIGITE = Variety of scheelite containing up to 24% $MoO_3$

## SHANDITE
$Ni_3Pb_2S_2$

CRYSTAL SYSTEM: Hexagonal
CLASS: $\bar{3}$ 2/m
SPACE GROUP: $R\bar{3}m$
Z: 3 (hexag.), 1 (rhomb.)
LATTICE CONSTANTS:

| (hexag.) | (rhomb.) |
|---|---|
| $a = 5.565$ | $a_{rh} = 5.565$ |
| $c = 13.631$ | $\alpha = 60°$ |

3 STRONGEST DIFFRACTION LINES:
2.80 (100)
2.27 ( 80)
3.97 ( 70)
HARDNESS: 4
DENSITY: 8.72 (Meas.)
8.86 (Calc.)
CLEAVAGE: $\{10\bar{1}1\}$ perfect
HABIT: Massive; as embedded grains.
COLOR-LUSTER: Brass yellow. Opaque. Metallic.
MODE OF OCCURRENCE: Occurs in association with pentlandite, heazlewoodite, and other sulfides, at Trial Harbour, Tasmania.
BEST REF. IN ENGLISH: Peacock, M. A., and McAndrew, John, *Am. Min.*, **35**: 425–439 (1950).

## SHANNONITE = Monticellite

## SHARPITE
$(UO_2)(CO_3) \cdot H_2O$ ?

CRYSTAL SYSTEM: Probably orthorhombic
3 STRONGEST DIFFRACTION LINES:
4.49 (100)
3.93 ( 60)
2.99 ( 60)
OPTICAL CONSTANTS:
$\alpha = 1.633$
$\beta = ?$
$\gamma \approx 1.72$
(+ ?)2V
HARDNESS: ~2½
DENSITY: >3.33
CLEAVAGE: Not determined.
HABIT: In crusts of thin radiating fibers; as rosettes.
COLOR-LUSTER: Greenish yellow to olive green.
MODE OF OCCURRENCE: Occurs associated with uranophane, curite, becquerelite, and masuyite at Shinkolobwe, Katanga, Zaire. Also found in France at Kruth, Haut-Rhin, with uraninite and billietite, and at Brugeaud near Bessines, Haute-Vienne.
BEST REF. IN ENGLISH: Frondel, Clifford, USGS Bull. 1064, p. 106–108 (1958).

## SHATTUCKITE
$Cu_5(SiO_3)_4(OH)_2$

CRYSTAL SYSTEM: Orthorhombic
CLASS: 2/m 2/m 2/m
SPACE GROUP: Pcab
Z: 4
LATTICE CONSTANTS:
$a = 9.876$
$b = 19.812$
$c = 5.381$
3 STRONGEST DIFFRACTION LINES:
4.42 (100)
4.96 ( 90)
3.50 ( 80)
OPTICAL CONSTANTS:
$\alpha = 1.752–1.753$
$\beta = 1.782$
$\gamma = 1.815$
2V = large
HARDNESS: Not determined.
DENSITY: 4.11 (Meas.)
4.138 (Calc.)
CLEAVAGE: {010} excellent
{100} excellent
HABIT: Crystals slender prismatic, 1–2 mm long; prominent forms {100}, {010}, {110}; in radiated spherulitic aggregates. Also compact, granular, fibrous.
COLOR-LUSTER: Blue to dark blue. Translucent. Vitreous to silky.
MODE OF OCCURRENCE: Occurs as an alteration product of other secondary copper minerals. Found in Arizona at the Shattuck mine, Bisbee district, Cochise County, and in association with ajoite in veins at the New Cornelia mine, Ajo, Pima County.

BEST REF. IN ENGLISH: Evans, H. T. Jr., and Mrose, Mary E., *Science*, **154**: 506, 507 (1966). Mrose, M. E., and Vlisidis, A. C., *Am. Min.*, **51**: 266–267 (1966) (abstr.).

## SHCHERBAKOVITE
$Na(K, Ba)_2 (Ti, Nb)_2 (Si_2 O_7)_2$

CRYSTAL SYSTEM: Orthorhombic
CLASS: 2/m 2/m 2/m, mm2, or 222
SPACE GROUP: I222, $I2_1 2_1 2_1$, Immm, $Imm2_1$
Z: 4
LATTICE CONSTANTS:
  $a$ = 10.55
  $b$ = 13.92
  $c$ = 8.10
3 STRONGEST DIFFRACTION LINES:
  2.90 (100)
  2.64 ( 70)
  1.69 ( 70)
OPTICAL CONSTANTS:
  $\alpha$ = 1.707
  $\beta$ = 1.745
  $\gamma$ = 1.776
  (–)2V = 82°
HARDNESS: 6½
DENSITY: 2.968 (Meas.)
CLEAVAGE: Two; observed under microscope. Fracture uneven. Brittle.
HABIT: Crystals prismatic elongated, up to 5.0 × 1.5 × 1.0 cm in size. Terminated crystals not common.
COLOR-LUSTER: Dark brown; vitreous, greasy on fracture surface. Translucent on thin edges.
MODE OF OCCURRENCE: Occurs in a pegmatite in alkalic rocks (rischorrite), associated with and including natrolite-pectolite aggregates, Khibina Tundra, USSR.
BEST REF. IN ENGLISH: Es'kova, E. M., and Kazakova, M. E., *Am. Min.*, **40**: 788 (1955).

## SHCHERBINAITE
$V_2 O_5$

CRYSTAL SYSTEM: Orthorhombic
Z: 1
LATTICE CONSTANTS:
  $a$ = 4.35
  $b$ = 11.53
  $c$ = 3.57
3 STRONGEST DIFFRACTION LINES:
  4.339 (100)
  2.883 ( 50)
  4.067 ( 28)
OPTICAL CONSTANTS:
  $N$ = 2.42
HARDNESS: Not determined.
DENSITY: ~3.2 (Meas.)
       3.36 (Synthetic)
CLEAVAGE: Brittle, easily split along the length.
HABIT: As finely fibrous aggregates of acicular crystals less than 1.5 mm long and less than 0.1 mm thick.
COLOR-LUSTER: Yellow-green. Translucent. Vitreous.

MODE OF OCCURRENCE: Shcherbinaite occurs on the walls of fissures in the "New" cupola of Bezymyanny Volcano, Kamchatka, where fumarolic gases rich in HCl and HF are escaping.
BEST REF. IN ENGLISH: Borisenko, L. F., *Am. Min.*, **58**: 560 (1973).

## SHERIDANITE
$(Mg, Al)_6 (Si, Al)_4 O_{10} (OH)_8$
Chlorite Group

CRYSTAL SYSTEM: Monoclinic
CLASS: 2/m
SPACE GROUP: C2/m (?)
Z: 4
LATTICE CONSTANTS:
  $a$ = 5.326
  $b$ = 9.226
  $c$ = 14.127
  $\beta$ = 97°10′
3 STRONGEST DIFFRACTION LINES:
  2.543 (100)
  7.114 ( 70)
  14.1 ( 60)
OPTICAL CONSTANTS:
  $\alpha$ = 1.578
  $\beta$ = 1.580
  $\gamma$ = 1.586
  (+)2V = small
HARDNESS: 2–3
DENSITY: 2.68–2.80 (Meas.)
       2.700 (Calc.)
CLEAVAGE: {001} perfect
HABIT: Massive; in scaly aggregates. Sometimes spherulitic. Twinning on {001}, twin plane [310].
COLOR-LUSTER: Colorless, greenish white, yellowish green, olive-green. Transparent to translucent. Pearly.
MODE OF OCCURRENCE: Occurs chiefly in schists or other metamorphic rocks. Typical occurrences are found in Sheridan County, Wyoming; at Ducktown, Tennessee; at Camberousse, Savoy, France; and at Aorere, Nelson, New Zealand.
BEST REF. IN ENGLISH: Deer, Howie, and Zussman, "Rock Forming Minerals," v. 3, p. 131–163, New York, Wiley, 1962.

## SHERWOODITE
$Ca_3 (V^{4+}O)_2 V_6^{5+} O_{20} \cdot 15H_2O$

CRYSTAL SYSTEM: Tetragonal
CLASS: 4/m 2/m 2/m
SPACE GROUP: I4/amd
Z: 16
LATTICE CONSTANTS:
  $a$ = 27.8
  $c$ = 13.8
3 STRONGEST DIFFRACTION LINES:
  12.3 (100)
  10.0 (100)
   9.3 ( 80)

OPTICAL CONSTANTS:
$\omega = 1.765$
$\epsilon = 1.735$
(−)
HARDNESS: ~2
DENSITY: 2.8 (Meas.)
2.86 (Calc.)
CLEAVAGE: Fracture subconchoidal to uneven.
HABIT: Crystals equant or slightly flattened; forms m{100} and d{011}. Also as polycrystalline aggregates.
COLOR-LUSTER: Dark blue-black; vitreous to earthy. Blue-green to yellow-green when altered.
MODE OF OCCURRENCE: Occurs in small amounts in the Matchless mine, Mesa County, Colorado, and in many other vanadium-uranium mines on the Colorado Plateau.
BEST REF. IN ENGLISH: Thompson, Mary E., Roach, Carl H., and Meyrowitz, Robert, *Am. Min.*, **43**: 749–755 (1958).

## SHILKINITE = Illite

## SHORTITE
$Na_2 Ca_2 (CO_3)_3$

CRYSTAL SYSTEM: Orthorhombic
CLASS: mm2
SPACE GROUP: C2mm
Z: 2
LATTICE CONSTANTS:
$a = 7.10$ kX
$b = 10.97$
$c = 4.98$
3 STRONGEST DIFFRACTION LINES:
2.56 (100)
5.52 ( 70)
4.96 ( 70)
OPTICAL CONSTANTS:
$\alpha = 1.531$
$\beta = 1.555$
$\gamma = 1.570$
(−)2V = 75° (Calc.)
HARDNESS: 3
DENSITY: 2.60 (Meas.)
2.605 (Calc.)
CLEAVAGE: {010} distinct
Fracture conchoidal. Brittle.
HABIT: Crystals wedge-shaped, up to 15 mm in maximum dimension, either tabular {100} or equant to short prismatic [100], {011} striated [100] and in oscillatory combination with {00$\bar{1}$}.
COLOR-LUSTER: Colorless to pale yellow, transparent; vitreous.
MODE OF OCCURRENCE: Found as individual crystals concentrated in bands varying from a fraction of an inch to several inches in thickness in the drill core of the Sun Oil Company's Ouray well in the Green River formation, Uintah County, Utah; it also occurs with calcite and pyrite embedded in an Eocene montmorillonite-type clay from an oil well west of Green River, Sweetwater County, Wyoming.
BEST REF. IN ENGLISH: Palache, et al., "Dana's System of Mineralogy," 7th Ed., v. II, p. 222–223, New York, Wiley, 1951.

## SHUNGITE = Variety of graphite. Black, vitreous with some water.

## SIBERITE = Variety of elbaite. Purplish to red.

## SIBIRSKITE (Inadequately described mineral)
$CaHBO_3$

CRYSTAL SYSTEM: Probably orthorhombic
3 STRONGEST DIFFRACTION LINES:
2.93 (100)
2.58 (100)
1.878 ( 60)
OPTICAL CONSTANTS:
$\alpha = 1.555$
$\beta = 1.643$
$\gamma = 1.658$
(−)2V = 43°
HABIT: As aggregates of irregular grains; also in diamond-shaped forms 1.0–1.5 mm in size.
COLOR-LUSTER: Colorless; often colored dark gray by chlorite.
MODE OF OCCURRENCE: Occurs associated with korzhinskite, calciborite, calcite, and a little andradite-grossular garnet and dolomite in limestones of Middle Devonian age in the Novo-Frolovsk contact metasomatic copper deposits, Tur'insk region, northern Urals. Also occurs mixed with calcite, a chlorite, and a little pyrite in skarns near the contact of Middle Cambrian limestones with granites at an unspecified locality, presumably in Siberia.
BEST REF. IN ENGLISH: Vasilkova, N. N., *Am. Min.*, **48**: 433 (1963).

## SICKLERITE
$Li(Mn^{2+}, Fe^{3+})PO_4$

CRYSTAL SYSTEM: Orthorhombic
CLASS: 2/m 2/m 2/m
SPACE GROUP: Pmnb
Z: 4
LATTICE CONSTANTS:
$a = 5.95$
$b = 10.10$
$c = 4.80$
3 STRONGEST DIFFRACTION LINES:
3.013 (100)
2.531 (100)
4.32 ( 95)
OPTICAL CONSTANTS:
$\alpha = 1.715$
$\beta = 1.735$
$\gamma = 1.745$
(−)2V = medium large
HARDNESS: ~4
DENSITY: 3.45 (Meas.)

CLEAVAGE: {100} good

HABIT: Massive.

COLOR-LUSTER: Yellowish brown to dark brown. Opaque. Dull.

MODE OF OCCURRENCE: Occurs as an alteration of lithiophilite in the zone of weathering in granite pegmatites. Found at the Custer Mountain Lode, Custer County, South Dakota, and at other places in the Black Hills; at Pala, San Diego County, California; at Wodgina, Western Australia; at Varuträsk, Sweden; and in Finland.

BEST REF. IN ENGLISH: Palache, et al., "Dana's System of Mineralogy," 7th Ed., v. II, p. 672–673, New York, Wiley, 1951.

## SIDERAZOT (Inadequately described mineral)
$Fe_5 N_2$

CRYSTAL SYSTEM: Hexagonal

CLASS: $\bar{3}$ 2/m

SPACE GROUP: $P\bar{3}/m$

LATTICE CONSTANTS:

$a = 2.74$

$c = 4.39$

HARDNESS: Not determined.

DENSITY: 3.147 (Meas.)

CLEAVAGE: Not determined.

HABIT: As thin coatings.

COLOR-LUSTER: Silvery white. Metallic.

MODE OF OCCURRENCE: Found as a coating on lava on Mt. Etna, Sicily, and at Mt. Vesuvius, Italy.

BEST REF. IN ENGLISH: Palache, et al., "Dana's System of Mineralogy," 7th Ed., v. I, p. 126, New York, Wiley, 1944.

## SIDERITE
$FeCO_3$
Calcite Group

CRYSTAL SYSTEM: Hexagonal

CLASS: $\bar{3}$ 2/m

SPACE GROUP: $R\bar{3}c$

Z: 2 (rhomb.), 6 (hexag.)

LATTICE CONSTANTS:

(hexag.)          (rhomb.)

$a = 4.688$      $a_{rh} = 5.796$

$c = 15.375$     $\alpha = 47°43'$

3 STRONGEST DIFFRACTION LINES:

2.79 (100)

1.734 ( 80)

3.59 ( 60)

OPTICAL CONSTANTS:

$\omega = 1.8728$

$\epsilon = 1.6331$ (Na)

(−)

HARDNESS: 3¾–4¼

DENSITY: 3.96 (Meas.)

CLEAVAGE: {10$\bar{1}$1} perfect

Fracture uneven to conchoidal. Brittle.

HABIT: Crystals commonly rhombohedral; also tabular, prismatic, scalenohedral. Crystal faces often curved. Commonly massive, coarse to fine granular. Rarely botryoidal or globular; oolitic; also admixed with silica or clay forming stony or earthy masses and concretions. Twinning not common.

COLOR-LUSTER: Pale yellowish, yellowish gray, pale green, greenish gray, ash gray, yellowish brown, grayish brown, brown, reddish brown, blackish brown. Rarely white to almost colorless, sometimes tarnished iridescent. Translucent to subtranslucent. Vitreous; also pearly or silky. Stony or earthy masses dull. Streak white.

MODE OF OCCURRENCE: Occurs widespread in bedded sedimentary deposits; as a gangue mineral in hydrothermal ore veins; in basaltic igneous rocks and rarely in pegmatites; and in certain metamorphic rocks. In the United States, fine cleavable masses and crystals are found at Roxbury, Litchfield County, Connecticut; at many localities in Colorado, especially in the Gilman district, Eagle County and in the Crystal Peak area, Teller County; and in numerous mining districts throughout the western part of the country. In Canada exceptional large crystals and crystal groups occur in carbonatite at Mont St. Hilaire, Quebec. Large cleavage rhombs are found in cryolite at Ivigtut, Greenland. Fine crystal specimens occur at the Morro Velho gold mine, Minas Geraes, Brazil; and at many places in Cornwall, England, Germany, Austria, France, and Italy.

BEST REF. IN ENGLISH: Palache, et al., "Dana's System of Mineralogy," 7th Ed., v. II, p. 166–171, New York, Wiley, 1951.

## SIDERONATRITE
$Na_2 Fe^{3+}(SO_4)_2 OH \cdot 3H_2O$

CRYSTAL SYSTEM: Orthorhombic

CLASS: 2/m 2/m 2/m

SPACE GROUP: Pbnm

Z: 4

LATTICE CONSTANTS:

$a = 7.27$

$b = 20.50$

$c = 7.15$

3 STRONGEST DIFFRACTION LINES:

10.2 (100)

3.01 ( 80)

3.38 ( 60)

OPTICAL CONSTANTS:

$\alpha = 1.508$

$\beta = 1.525$

$\gamma = 1.586$

(+)2V ~ 58° (Calc.)

HARDNESS: 1½–2½

DENSITY: 2.28 (Meas.)

2.276 (Calc.)

CLEAVAGE: {100} perfect

HABIT: Minute needles. As fibrous crusts or nodular masses, and earthy.

COLOR-LUSTER: Straw yellow to yellow to yellow-brown, also pale orange. Streak yellowish white to pale yellow.

Soluble in boiling water.

MODE OF OCCURRENCE: Occurs in Chile with other secondary sulfates at Mina La Compania, south of Sierra

Gorda, and at the San Simon mine, Tarapacá province; at Potosi, Bolivia, with voltaite; and on the Island of Cheleken, Caspian Sea.

BEST REF. IN ENGLISH: Palache, et al., "Dana's System of Mineralogy," 7th Ed., v. II, p. 604–605, New York, Wiley, 1951.

**SIDEROPHYLLITE** = Ferroan variety of biotite

**SIDEROPLESITE** = Variety of siderite containing magnesium

## SIDEROTIL
$FeSO_4 \cdot 5H_2O$

CRYSTAL SYSTEM: Triclinic
CLASS: $\bar{1}$
SPACE GROUP: $P\bar{1}$
Z: 2
LATTICE CONSTANTS:
  $a = 6.26$      $\alpha = 97°15'$
  $b = 10.63$    $\beta = 109°40'$
  $c = 6.06$      $\gamma = 75°0'$
3 STRONGEST DIFFRACTION LINES:
  4.89 (100)
  3.73 ( 80)
  5.57 ( 60)
OPTICAL CONSTANTS:
  $\alpha = 1.513–1.515$
  $\beta = 1.525–1.526$
  $\gamma = 1.534–1.536$
  $(-)2V = 50°–60°$
HARDNESS: Not determined.
DENSITY: 2.1–2.2 (Meas.)
          2.212 (Calc.)
CLEAVAGE: Not determined.
HABIT: As divergent groups of acicular crystals, and as fibrous crusts.
COLOR-LUSTER: White to yellowish, pale greenish white.
MODE OF OCCURRENCE: Occurs as a dehydration product of melanterite in the Mount Diablo quicksilver mine, Contra Costa County, California; at the Alvarado mine, Clifton district, Tooele County, Utah; at Yerington, Lyon County, Nevada; and elsewhere.
BEST REF. IN ENGLISH: Palache, et al., "Dana's System of Mineralogy," 7th Ed., v. II, p. 491–492, New York, Wiley, 1951. Jambor, J. L., and Traill, R. J., *Can. Min.*, 7: 751–763 (1963).

## SIEGENITE
$(Co, Ni)_3S_4$
Linnaeite Group

CRYSTAL SYSTEM: Cubic
CLASS: $4/m\,\bar{3}\,2/m$
SPACE GROUP: Fd3m
Z: 8

LATTICE CONSTANT:
  $a = 9.41$
3 STRONGEST DIFFRACTION LINES:
  2.86 (100)
  1.67 ( 80)
  2.36 ( 70)
OPTICAL CONSTANTS: Opaque.
  HARDNESS: 4½–5½
  DENSITY: 4.5–4.8 (Meas.)
          4.826 (Calc.) Co:Ni = 1:1
  CLEAVAGE: {001} imperfect
Fracture subconchoidal to uneven.
HABIT: Crystals usually octahedral. Commonly massive, compact to granular. Twinning on {111}.
COLOR-LUSTER: Light gray to steel gray; rapidly tarnishes copper red or violet gray. Opaque. Metallic.
MODE OF OCCURRENCE: Occurs in hydrothermal vein deposits commonly associated with other sulfide minerals. Found at Mine LaMotte, Missouri, in the Siegen district of Westphalia, Germany, especially at Müsen; and at Kladno, Czechoslovakia.
BEST REF. IN ENGLISH: Palache, et al., "Dana's System of Mineralogy," 7th Ed., v. I, p. 262–265, New York, Wiley, 1944.

## SIGLOITE
$(Fe^{2+}, Fe^{3+})Al_2(PO_4)_2(O, OH) \cdot 8H_2O$

CRYSTAL SYSTEM: Triclinic
CLASS: $\bar{1}$
SPACE GROUP: $P\bar{1}$
Z: 1
LATTICE CONSTANTS:
  $a = 5.26$      $\alpha = 106°58'$
  $b = 10.52$    $\beta = 111°30'$
  $c = 7.06$      $\gamma = 69°30'$
3 STRONGEST DIFFRACTION LINES:
  9.69 (100)
  6.46 ( 90)
  4.86 ( 90)
OPTICAL CONSTANTS:
  $\alpha = 1.563$
  $\beta = 1.586$
  $\gamma = 1.619$
  $(+)2V = 76°$
HARDNESS: 3
DENSITY: 2.35 (Meas.)
          2.36 (Calc.)
CLEAVAGE: {010} perfect
          {001} good
HABIT: Crystals short prismatic [001] to thick tabular {010}. An oxidation pseudomorph after paravauxite.
COLOR-LUSTER: Pale straw yellow to light brown.
MODE OF OCCURRENCE: Found on the 305 meter level of the Contacto vein, Siglo XX mine, Llallagua, Bolivia, in cavities filled with intergrowths of crystals and in other cavities with crystals perched on wavellite overgrowths on quartz.
BEST REF. IN ENGLISH: Hurlbut, Cornelius S. Jr., and Honea, Russell, *Am. Min.*, 47: 1–8 (1962).

# SILHYDRITE
$3SiO_2 \cdot H_2O$

CRYSTAL SYSTEM: Orthorhombic
Z: 1
LATTICE CONSTANTS:
  $a = 14.519$
  $b = 18.80$
  $c = 15.938$
3 STRONGEST DIFFRACTION LINES:
  14.50  (100)
   3.424 ( 82)
   3.143 ( 35)
OPTICAL CONSTANTS: Optically similar to magadiite.
Mean $N$ 1.466; birefringence low, about 0.006.
  HARDNESS: No mineral hardness can be measured; dry specimens have bulk hardness ∼1.
  DENSITY:  2.141 (Meas.)
                  2.116 (Calc.)
(Bulk density of aggregates 0.65.)
  CLEAVAGE: Fracture surfaces of aggregates rough and subconchoidal.
  HABIT: As monomineralic masses of crystals that are individually less than 4 $\mu$m long. Distinctively punky, light in heft, and very porous when dry—a small piece up to 1 cm across can be moved by a light air puff.
  COLOR-LUSTER: White; often strained by hydrous iron oxides.

  Insoluble in water.
  MODE OF OCCURRENCE: The mineral occurs at the magadiite deposit in the Bonanza King quadrangle, Trinity County, California. Evidence indicates that near-surface water leached sodium from magadiite forming silhydrite.
  BEST REF. IN ENGLISH: Gude, Arthur J. 3d, and Sheppard, Richard A., *Am. Min.*, **57**: 1053-1065 (1972).


# SILLÉNITE
$\gamma$-$Bi_2 O_3$
Dimorphous with bismite

CRYSTAL SYSTEM: Cubic
CLASS: 23
SPACE GROUP: I23
Z: 12
LATTICE CONSTANTS:
   $a = 10.08$
  or $a = 10.76$
3 STRONGEST DIFFRACTION LINES:
  3.22 (100)
  1.74 ( 90)
  2.73 ( 80)
OPTICAL CONSTANT:
  $N > 2.42$
  HARDNESS: Not determined. Soft.
  DENSITY: 8.80 (Calc.)
  CLEAVAGE: Not determined.
  HABIT: Massive, fine-grained; earthy.
  COLOR-LUSTER: Green, yellowish green, grayish green, brownish green. Translucent; thin fragments transparent. Waxy to dull.

MODE OF OCCURRENCE: Occurs as a secondary mineral associated with native bismuth and bismutite at Durango, Mexico.
  BEST REF. IN ENGLISH; Palache, et al., "Dana's System of Mineralogy," 7th Ed., v. I, p. 601, New York, Wiley, 1944.


# SILLIMANITE
$Al_2 SiO_5$
Trimorphous with andalusite, kyanite

CRYSTAL SYSTEM: Orthorhombic
CLASS: 2/m 2/m 2/m
SPACE GROUP: Pbnm
Z: 4
LATTICE CONSTANTS:
  $a = 7.44$
  $b = 7.59$
  $c = 5.75$
3 STRONGEST DIFFRACTION LINES:
  3.42 (100)
  3.37 ( 65)
  2.20 ( 60)
OPTICAL CONSTANTS:
      $\alpha = 1.654$-$1.661$
      $\beta = 1.658$-$1.662$
      $\gamma = 1.673$-$1.683$
  $(+)2V = 21°$-$30°$
  HARDNESS: 6½-7½
  DENSITY: 3.23-3.27 (Meas.)
                  3.26 (Calc.)
  CLEAVAGE: {010} perfect
Fracture uneven.
  HABIT: Crystals long prismatic, nearly square in cross section, and vertically striated. Usually massive, fibrous to somewhat columnar; also radiating.
  COLOR-LUSTER: Colorless, white, gray, yellowish, brownish, greenish, bluish. Transparent to translucent. Vitreous to silky. Streak uncolored.
  MODE OF OCCURRENCE: Occurs mainly in schists, gneisses, and granites, often in association with andalusite, cordierite, or corundum. Found in the Boehls Butte quadrangle, Idaho; in Custer and Pennington counties, South Dakota; and at localities in Oklahoma, Pennsylvania, New York, Connecticut, Delaware, North Carolina, and South Carolina. Also in Canada, Ireland, Scotland, France, Germany, Czechoslovakia, India, Ceylon, Burma, Korea, South Africa, and Madagascar.
  BEST REF. IN ENGLISH: Deer, Howie, and Zussman, "Rock Forming Minerals," v. 1, p. 121-128, New York, Wiley, 1962.


# SILVER
Ag

CRYSTAL SYSTEM: Cubic
CLASS: 4/m $\bar{3}$ 2/m
SPACE GROUP: Fm3m
Z: 4

LATTICE CONSTANT:
 $a = 4.085$
3 STRONGEST DIFFRACTION LINES:
 2.359 (100)
 2.044 ( 40)
 1.231 ( 26)
OPTICAL CONSTANTS: Opaque, isotropic.
 HARDNESS: 2½-3
 DENSITY: 10.50
 CLEAVAGE: None.
Malleable and ductile. Fracture hackly.

HABIT: Crystals cubic, octahedral, or dodecahedral; often in parallel groups; also elongated, arborescent, reticulated, or as thin to thick wires. Also massive, sometimes in thick sheets, as a coating, or as scales. Twinning on {111} common, often repeated.

COLOR-LUSTER: Silver white; often tarnished gray to black. Opaque. Metallic.

MODE OF OCCURRENCE: Occurs chiefly in the oxidized zone of ore deposits, commonly associated with other silver minerals, and also as a primary mineral of hydrothermal origin. Found widespread throughout the world, especially in the western United States, Canada, Mexico, Bolivia, Chile, France, Germany, Czechoslovakia, Norway, USSR, Sardinia, and Australia.

BEST REF. IN ENGLISH: Palache, et al., "Dana's System of Mineralogy," 7th Ed., v. I, p. 96-99, New York, Wiley, 1944.

## SILVIALITE (Sylvialite) = Hypothetical end-member of Scapolite Group

## SIMPLOTITE
$CaV_4O_9 \cdot 5H_2O$

CRYSTAL SYSTEM: Monoclinic
CLASS: 2/m, m, or 2
SPACE GROUP: A2/m, Am, or A2
Z: 4
LATTICE CONSTANTS:
 $a = 8.39$
 $b = 17.02$
 $c = 8.37$
 $\beta = 90°25'$
3 STRONGEST DIFFRACTION LINES:
 8.51 (100)
 2.62 ( 25)
 3.14 ( 18)
OPTICAL CONSTANTS:
 $\alpha = 1.705$
 $\beta = 1.767$
 $\gamma = 1.769$
 $(-)2V \sim 25°$
 HARDNESS: $\sim 1$
 DENSITY: 2.64 (Meas.)
 CLEAVAGE: {010} micaceous
HABIT: Micaceous plates and hemispherical aggregates of plates.

COLOR-LUSTER: Thin flakes yellow-green; aggregates black or greenish black. Vitreous. Powder brownish black.

MODE OF OCCURRENCE: Occurs coating fracture surfaces in sandstone associated with duttonite, melanovanadite, and native selenium at the Peanut mine, Montrose County, Colorado, and at several other vanadium-uranium mines on the Colorado Plateau.

BEST REF. IN ENGLISH: Thompson, M. E., Roach, C. H., and Meyrowitz, Robert, *Am. Min.*, 43: 16-24 (1958).

## SIMPSONITE
$Al_4Ta_3O_{13}OH$

CRYSTAL SYSTEM: Hexagonal
CLASS: 6/m
SPACE GROUP: P6/m
Z: 1
LATTICE CONSTANTS:
 $a = 7.377$
 $c = 4.514$
3 STRONGEST DIFFRACTION LINES:
 1.64 (100)
 1.39 ( 95)
 2.84 ( 85)
OPTICAL CONSTANTS:
 $\omega = 2.034$
 $\epsilon = 1.976$
 $(-)$
 HARDNESS: 7-7½
 DENSITY: 5.92-6.84 (Meas.)
  6.826 (Calc.)
 CLEAVAGE: None. Fracture conchoidal. Brittle.

HABIT: Crystals tabular or short prismatic parallel to [0001]; also bipyramidal. As individual crystals and crystalline masses up to 5 × 5 × 4 cm in size.

COLOR-LUSTER: Colorless, pale yellow, cream-colored to light brown. Transparent. Vitreous, adamantine. Fluoresces pale yellow, bluish white, or yellowish blue under ultraviolet light.

MODE OF OCCURRENCE: Occurs in granite pegmatites in association with tantalite, microlite, tapiolite, beryl, quartz, muscovite, and other minerals. Found at Alto do Giz, near Equador, Rio Grande do Norte, Brazil; on the Kola Peninsula, USSR; at Tabba-Tabba, Western Australia; and at the Mdara mine, Bikita, Rhodesia.

BEST REF. IN ENGLISH: Palache, et al., "Dana's System of Mineralogy," 7th Ed., v. I, p. 771, New York, Wiley, 1944. Vlasov, K. A., "Mineralogy of Rare Elements," v. II, p. 446-449, Israel Program for Scientific Translations, 1966.

## SINCOSITE
$Ca(VO)_2(PO_4)_2 \cdot 5H_2O$

CRYSTAL SYSTEM: Tetragonal
3 STRONGEST DIFFRACTION LINES:
 6.28 (100)
 3.14 ( 60)
 3.35 ( 20)
OPTICAL CONSTANTS:
 $\omega \sim 1.680$
 $\epsilon \sim 1.655$
 $(-)$ (sometimes biaxial)

HARDNESS: Not determined. Soft.
DENSITY: ~2.84
CLEAVAGE: {001} perfect
{100} distinct
{110} distinct
Brittle.

HABIT: Crystals thin to thick tabular, usually square in outline; often in compact, irregular or scaly masses. Also as flattened, nearly circular rosettes, or as radial aggregates and small nodules admixed with carbonaceous material and gypsum. Twinning on {110}, rare.

COLOR-LUSTER: Grass green, yellowish green to olive green or brownish green. Vitreous; altered crystals submetallic. Transparent to translucent. Streak green.

MODE OF OCCURRENCE: Occurs in vugs, small cavities, and incrusting minute fissures in siliceous gold ore at the Ross Hannibal mine, near Lead, Lawrence County, South Dakota. Also found near Bloomington, Bear Lake County, Idaho, and as thin veinlets and irregular masses in carbonaceous shale near Sincos, Junin, Peru.

BEST REF. IN ENGLISH: Palache, et al., "Dana's System of Mineralogy," 7th Ed., v. II, p. 1057–1058, New York, Wiley, 1951.

## SINHALITE
$MgAlBO_4$

CRYSTAL SYSTEM: Orthorhombic
CLASS: 2/m 2/m 2/m
SPACE GROUP: Pbmm
Z: 4
LATTICE CONSTANTS:
$a = 4.328$
$b = 9.878$
$c = 5.675$
3 STRONGEST DIFFRACTION LINES:
2.14 (100)
1.62 ( 90)
3.24 ( 70)
OPTICAL CONSTANTS:
$\beta = 1.6969$
$\gamma = 1.7065$
$2V = 56°$
HARDNESS: ~6½–7
DENSITY: 3.475–3.50 (Meas.)
3.446 (Calc.)
HABIT: Massive; as rounded grains.
COLOR-LUSTER: Yellow to dark brown, greenish brown, light pink to brownish pink. Transparent. Vitreous.
MODE OF OCCURRENCE: The mineral occurs associated with serendibite in thin layers of a contact zone between limestone and an intrusive granite in Warren County, New York, and in a skarn mineral association in marble in the Handeni area, northeast Tanzania. Fine gemstones have been cut from rolled pebbles presumably from alluvial deposits of Ceylon.
BEST REF. IN ENGLISH: Claringbull, G. F., and Hey, Max H., Min. Mag., 29: 841–849 (1952).

**SINICITE** = Uranium-rich variety of aeschynite

## SINNERITE
$Cu_6As_4S_9$

CRYSTAL SYSTEM: Triclinic
CLASS: 1
SPACE GROUP: P1
Z: 2
LATTICE CONSTANTS:
$a = 9.06$ $\alpha = 90°00'$
$b = 9.83$ $\beta = 109°30'$
$c = 9.08$ $\gamma = 107°48'$
3 STRONGEST DIFFRACTION LINES:
3.024 (100)
1.852 ( 80)
1.581 ( 70)
HARDNESS: Vickers hardness 373.
DENSITY: 5.2 (Meas.)
CLEAVAGE: Irregular. Brittle.
HABIT: Found as a single highly twinned crystal, about $3mm^3$, on a crystal of binnite. Twinned on planes (010), (110), and (410).
COLOR-LUSTER: Steel gray. Opaque. Metallic. Streak gray-black.
MODE OF OCCURRENCE: Found in the Lengenbach quarry, Binnental, Valais, Switzerland.
BEST REF. IN ENGLISH: Marumo, F., and Nowacki, W., Min. Abs., 17: 255 (1965). Nowacki, W., Marumo, F., and Takeuchi, Y., Min. Abs., 17: 74 (1965). Makovicky, Emil, and Skinner, Brian J., Am. Min., 57: 824–834 (1972).

## SINOITE
$Si_2N_2O$

CRYSTAL SYSTEM: Orthorhombic
Z: 4
LATTICE CONSTANTS:
$a = 5.473$
$b = 8.843$
$c = 4.835$
3 STRONGEST DIFFRACTION LINES:
3.36 (100)
4.43 (100)
4.66 ( 80)
OPTICAL CONSTANTS:
$\alpha = 1.740$
$\gamma = 1.855$
$(-)2V = small$
HARDNESS: Not determined.
DENSITY: 2.84 (Calc.)
CLEAVAGE: None.
HABIT: Crystals lath-like, up to 0.2 mm long, and as irregular grains.
COLOR-LUSTER: Colorless in thin section. Light gray in polished section.
MODE OF OCCURRENCE: Occurs as an accessory mineral in the Jajh deh Kot Lalu enstatite chondrite meteorite, which fell in the Sind Province, Pakistan, on May 2, 1926.
BEST REF. IN ENGLISH: Andersen, C. A., Keil, Klaus, and Mason, Brian, Science, 146 (3641): 256–257 (1964).

## SISERSKITE (Osmiridium)
Os, Ir
Var. Platinian
     Rhodian
     Ruthenian

CRYSTAL SYSTEM: Hexagonal
CLASS: 6/m 2/m 2/m
SPACE GROUP: $P6_3/mmc$
Z: 2
LATTICE CONSTANTS:
  $a = 2.714$
  $c = 4.314$
3 STRONGEST DIFFRACTION LINES:
  2.13 (100)
  1.22 (100)
  1.35 ( 60)
OPTICAL CONSTANTS: Opaque.
  HARDNESS: 6-7
  DENSITY: 19-21
  CLEAVAGE: {0001} perfect, difficult
Slightly malleable to brittle.
  HABIT: Crystals tabular, flattened {0001}; rarely short prismatic. Usually as cleavage flakes or irregular flattened grains.
  COLOR-LUSTER: Light steel gray. Metallic. Opaque. Streak gray.
  MODE OF OCCURRENCE: Occurs in association with platinum near Stanislaus, Tuolumne County, California; in the Ural Mountains, USSR; and in Borneo.
  BEST REF. IN ENGLISH: Palache, et al., "Dana's System of Mineralogy," 7th Ed., v. I, p. 111-113, New York, Wiley, 1944.

**SISMONDINE** = Variety of chloritoid. Magnesium-rich.

**SITAPARITE** = Variety of bixbyite containing 0-30% $Fe_2O_3$. Formed at temperatures below 650°C.

## SJÖGRENITE
$Mg_6Fe_2^{3+}CO_3(OH)_{16} \cdot 4H_2O$
Dimorphous with pyroaurite

CRYSTAL SYSTEM: Hexagonal
CLASS: 6/m 2/m 2/m
SPACE GROUP: $P6_3/mmc$
Z: 1
LATTICE CONSTANTS:
  $a = 6.21$
  $c = 15.60$
3 STRONGEST DIFFRACTION LINES:
  7.79 (100)
  3.89 ( 80)
  1.86 ( 40)
OPTICAL CONSTANTS:
  $\omega = 1.573$
  $\epsilon = 1.550$
  (-)
HARDNESS: 2½

DENSITY: 2.11 (Meas.)
         2.11 (Calc.)
CLEAVAGE: {0001} perfect
Laminae flexible; inelastic.
  HABIT: Crystals thin plates with hexagonal outline; flattened on {0001}.
  COLOR-LUSTER: White, yellowish, brownish. Transparent. Vitreous to waxy, pearly on cleavages.
  MODE OF OCCURRENCE: Occurs as a low-temperature hydrothermal mineral, associated with calcite and pyroaurite, at Långban, Sweden.
  BEST REF. IN ENGLISH: Palache, et al., "Dana's System of Mineralogy," 7th Ed., v. I, p. 659-660, New York, Wiley, 1944.

## SKLODOWSKITE
$Mg(UO_2)_2Si_2O_7 \cdot 6H_2O$

CRYSTAL SYSTEM: Monoclinic
CLASS: 2/m
SPACE GROUP: I 2/m
Z: 2
LATTICE CONSTANTS:
  $a = 16.74$
  $b = 7.01$
  $c = 6.59$
  $\beta = 96°$
3 STRONGEST DIFFRACTION LINES:
  8.42 (100)
  4.19 ( 80)
  3.27 ( 70)
OPTICAL CONSTANTS:
  $\alpha = 1.613$
  $\beta = 1.635$
  $\gamma = 1.657$
  $(-)2V$ = very large
HARDNESS: 2-3
DENSITY: 3.64 (Meas.)
        3.638 (Calc.)
CLEAVAGE: {100} perfect
Brittle.
  HABIT: As prismatic acicular crystals elongated [010]. Forms: {001}, {100}, {301}, and {011}. Often as rosettes, as mammillary crusts, radial-fibrous aggregates, and as granular or earthy masses.
  COLOR-LUSTER: Pale yellow to greenish yellow. Subvitreous. Fibrous aggregates silky. Not fluorescent.
  MODE OF OCCURRENCE: Occurs as a secondary mineral at the New Haven mine, Crook County, Wyoming; in Utah at the Oyler Tunnel claim near Fruita, Wayne County, and at Honeycomb Hills, Juab County; and also in the Grants area, New Mexico. In Katanga, Zaire, associated with cuprosklodowskite at Kalongwe, and with uraninite, becquerelite, curite, and other secondary uranium minerals at Shinkolobwe.
  BEST REF. IN ENGLISH: Frondel, Clifford, USGS Bull. 1064, p. 300-304 (1958).

**SKOLITE** = Aluminum-rich variety of glauconite

# SKUTTERUDITE
$(Co, Ni)As_3$
Var. Smaltite

CRYSTAL SYSTEM: Cubic
CLASS: $2/m \bar{3}$
SPACE GROUP: Im3
Z: 8
LATTICE CONSTANT:
 $a = 8.19$
3 STRONGEST DIFFRACTION LINES:
 2.592 (100)
 2.193 ( 35)
 1.835 ( 35)
OPTICAL CONSTANTS: Opaque, isotropic.
 HARDNESS: 5½–6
 DENSITY: 6.1–6.9 (Meas.)
  6.821 (Calc.)
 CLEAVAGE: {001} distinct
  {111} distinct
  {011} in traces
Fracture uneven to conchoidal. Brittle.
 HABIT: Crystals cubic, octahedral, cubo-octahedral, and rarely pyritohedral, sometimes with dodecahedral modification; in reticulated shapes or distorted aggregates of crystals. Commonly massive, fine granular.
 COLOR-LUSTER: Tin white, occasionally tarnished iridescent or gray. Opaque. Metallic. Streak black.
 MODE OF OCCURRENCE: Occurs in medium- to high-temperature hydrothermal veins, commonly associated with other cobalt and nickel minerals, arsenopyrite, native silver, native bismuth, and silver sulfosalts. Found at the Horace Porter mine, Elk Mountains, Gunnison County, Colorado; in the Cobalt and other districts in Canada; as crystals at Skutterud, near Modum, Norway; at Schneeberg and elsewhere in Saxony, Germany; and as superb crystals associated with calcite at Irhtem, Morocco.
 BEST REF. IN ENGLISH: Palache, et al., "Dana's System of Mineralogy," 7th Ed., v. I, p. 342–346, New York, Wiley, 1944.

# SLAVIKITE
$MgFe_3^{3+}(SO_4)_4(OH)_3 \cdot 18H_2O$

CRYSTAL SYSTEM: Hexagonal
SPACE GROUP: R (?)
LATTICE CONSTANTS:
 $c/a = 1.389$
3 STRONGEST DIFFRACTION LINES:
 9.04 (100)
 11.7 ( 80)
 5.83 ( 80)
OPTICAL CONSTANTS:
 $\omega = 1.530–1.537$
 $\epsilon = 1.495–1.506$
  (–)
HARDNESS: Not determined.
DENSITY: 1.89–1.99 (Meas.)
CLEAVAGE: Not determined.
HABIT: Crystals minute, tabular, flattened on {0001} with prominent {10$\bar{1}$1}. As crusts, scales, and fine granular.

COLOR-LUSTER: Greenish yellow. Vitreous.
 MODE OF OCCURRENCE: Occurs as a secondary mineral formed by the alteration of pyrite. Found at the Santa Elena mine, La Alcaparrosa, San Juan province, Argentina; at Saint-Cyprien-sur-Durdou, Aveyron, France; and in Austria and Czechoslovakia.
 BEST REF. IN ENGLISH: Palache, et al., "Dana's System of Mineralogy," 7th Ed., v. II, p. 621–622, New York, Wiley, 1951.

# SMALTITE = Arsenic-deficient variety of skutterudite
$(Co, Ni)As_{3-x}$ where $x \approx 0.5–1.0$

CRYSTAL SYSTEM: Cubic
CLASS: $2/m \bar{3}$
SPACE GROUP: Im3
Z: 8
LATTICE CONSTANT:
 $a = 8.24$
OPTICAL CONSTANTS: Opaque, isotropic.
 HARDNESS: 5½–6
 DENSITY: ~6.1
 CLEAVAGE: {001} distinct
  {111} distinct
  {011} in traces
Fracture uneven to conchoidal. Brittle.
 HABIT: Crystals cubic, octahedral, or cubo-octahedral; sometimes with dodecahedral or pyritohedral modifications. In reticulated shapes or distorted aggregates of crystals. Commonly massive, fine granular. Twinning on {112} as sixlings.
 COLOR-LUSTER: Silver gray to tin white; occasionally tarnished iridescent or gray. Opaque. Metallic. Streak black.
 MODE OF OCCURRENCE: Occurs in medium- to high-temperature hydrothermal veins, commonly associated with other cobalt and nickel minerals, arsenopyrite, native silver and bismuth, and silver sulfosalts. Found sparingly in Calaveras, Los Angeles, Inyo, Lassen, Napa, Nevada, and Siskiyou counties, California; at the Horace Porter and American Eagle mines on White Rock Mountain, and at the Luona mine on Teocalli Mountain, Gunnison County, Colorado; in the Cobalt and other districts in Canada; and at deposits in Chile, Switzerland, and Germany, especially in the vicinity of Schneeberg.
 BEST REF. IN ENGLISH: Palache, et al., "Dana's System of Mineralogy," 7th Ed., v. I, p. 342–346, New York, Wiley, 1944.

# SMARAGDITE = Variety of actinolite

# SMITHITE
$AgAsS_2$

CRYSTAL SYSTEM: Monoclinic
CLASS: 2/m
SPACE GROUP: A2/a
Z: 24

LATTICE CONSTANTS:
 $a = 17.20$
 $b = 7.76$
 $c = 15.16$ (synthetic)
 $\beta = 101°12'$
3 STRONGEST DIFFRACTION LINES:
 2.82 (100)
 3.21 ( 80)
 2.72 ( 60)
OPTICAL CONSTANTS:
  $\beta = 3.27$ (Na)
 $(-)2V = 65°$
HARDNESS: 1½-2
DENSITY: 4.88 (Meas.)
    4.93 (Calc.)
CLEAVAGE: {100} perfect
Fracture conchoidal. Brittle.

HABIT: Crystals equant; tabular {100}, with hexagonal aspect.

COLOR-LUSTER: Light red to brownish orange. Opaque; thin fragments translucent. Adamantine. Streak red.

MODE OF OCCURRENCE: Occurs in crystalline dolomite associated with other sulfosalts, realgar, orpiment, sphalerite, and pyrite, at the Lengenbach quarry in the Binnental, Valais, Switzerland.

BEST REF. IN ENGLISH: Palache, et al., "Dana's System of Mineralogy," 7th Ed., v. I, p. 430-432, New York, Wiley, 1944.

# SMITHSONITE
$ZnCO_3$
Calcite Group

CRYSTAL SYSTEM: Hexagonal
CLASS: $\bar{3}$ 2/m
SPACE GROUP: $R\bar{3}c$
Z: 2 (rhomb.), 6 (hexag.)
LATTICE CONSTANTS:
  (hexag.)        (rhomb.)
 $a = 4.651$      $a_{rh} = 5.669$
 $c = 14.977$     $\alpha = 48°26'$
3 STRONGEST DIFFRACTION LINES:
 2.75 (100)
 3.55 ( 50)
 1.703 ( 45)
OPTICAL CONSTANTS:
 $\omega = 1.848$
 $\epsilon = 1.621$ (Na)
  $(-)$
HARDNESS: 4-4½
DENSITY: 4.30-4.45 (Meas.)
    4.43 (Calc.)
CLEAVAGE: {10$\bar{1}$1} near perfect
Fracture subconchoidal to uneven. Brittle.

HABIT: Crystals rhombohedral, faces often curved and rough; rarely scalenohedral. Commonly botryoidal, stalactitic, or reniform; massive, granular to compact; as crystalline crusts; also earthy and as cellular masses.

COLOR-LUSTER: White, grayish, pale to deep yellow, yellowish brown to brown, pale green to bright apple green, bluish green, blue, pale to deep pink, purplish; rarely color-less. Transparent to translucent. Vitreous to pearly. Earthy and cellular masses dull. Streak white.

MODE OF OCCURRENCE: Occurs as a secondary mineral in the oxidized zone of ore deposits; as a replacement of calcareous rocks adjacent to ore deposits; and rarely as a secondary mineral derived from sphalerite in pegmatites. Found widespread in the United States, especially as superb colorful botryoidal encrustations and masses at Kelly, Socorro County, New Mexico; as thick yellow banded crusts in lead mines at Yellville, Marion County, Arkansas; in pegmatite associated with hemimorphite, aurichalcite, hydrozincite, columbite, and cassiterite, at the Tin Mountain mine, Custer County, South Dakota; and in large deposits in Colorado, Montana, and Utah. Colorful specimens occur at Laurium, Greece, and in the lead-zinc deposits of Sardinia, Italy. Superb colorless and transparent crystals up to 1 cm in size are found at the Broken Hill mine, Zambia, and magnificent pink, white, and grayish crystal specimens at Tsumeb, South West Africa. It also occurs in Germany, Austria, Belgium, Italy, France, Spain, Algeria, Tunisia, and Australia.

BEST REF. IN ENGLISH: Palache, et al., "Dana's System of Mineralogy," 7th Ed., v. II, p. 176-181, New York, Wiley, 1951.

# SMOLIANINOVITE (Smolyaninovite)
(Inadequately described mineral)
$(Co, Ni, Mg, Cu)_4 (Fe, Al)_2 (AsO_4)_4 (OH) \cdot 10H_2$

CRYSTAL SYSTEM: Unknown
3 STRONGEST DIFFRACTION LINES:
 10.87 (100)
 3.16 ( 70)
 2.88 ( 60)
OPTICAL CONSTANTS:
 Mean index = 1.625
 HARDNESS: ~2
 DENSITY: 2.43-2.49
HABIT: As earthy to dense aggregates of finely fibrous material.

COLOR-LUSTER: Yellow; silky.

MODE OF OCCURRENCE: Occurs widely distributed as an oxidation product of smaltite, safflorite, and rammelsbergite, especially at Bou-Azzer, Morocco.

BEST REF. IN ENGLISH: Yakhontova, L. K., *Am. Min.*, **42**: 307-308 (1957).

# SMYTHITE
$(Fe, Ni)_9 S_{11}$

CRYSTAL SYSTEM: Possibly hexagonal or monoclinic (pseudorhombohedral)
Z: 1
LATTICE CONSTANTS:
 $a = 3.47$ (pseudorhombohedral)
 $c = 34.4$
3 STRONGEST DIFFRACTION LINES:
 1.732 (100)
 1.897 ( 80)
 1.979 ( 70)

OPTICAL CONSTANTS: Opaque.
HARDNESS: Soft; readily scratched with a steel needle.
DENSITY: ~4.32 (Meas.)
~4.32 (Calc.)
CLEAVAGE: Basal, perfect.
Fracture subconchoidal to smooth. Laminae flexible and elastic.
HABIT: Crystals hexagonal in outline, platy, ranging from 0.05 to 1.25 mm in diameter.
COLOR-LUSTER: Jet black with tinge of brown; light bronze yellow if basal plane is used as a mirror. Luster on basal surface metallic and splendent. Streak dark gray.
MODE OF OCCURRENCE: Occurs mainly as a low-temperature oxidation product of monoclinic pyrrhotite. Found as inclusions in calcite crystals in quartz geodes, and infrequently embedded in dolomite, barite, or rarely in quartz, at several localities in Monroe County, and near Medora, Jackson County, Indiana; also found at Boron, California; in low-grade nickel deposits in Manitoba, Ontario, and Quebec; in the Kimmerian sedimentary iron ores of the Kerch Peninsula, USSR; and in many nickel ores, particularly supergene ores, in Australia.
BEST REF. IN ENGLISH: Erd, Richard C., Evans, Howard T. Jr., and Richter, Donald H., *Am. Min.*, **42**: 309–333 (1957). Taylor, Lawrence A., *Am. Min.*, **55**: 1650–1658 (1970). Taylor, Lawrence A., and Williams, K. L., *Am. Min.*, **57**: 1571–1577 (1972).

# SODA ALUM
$NaAl(SO_4)_2 \cdot 12H_2O$

CRYSTAL SYSTEM: Cubic
CLASS: $2/m\ \overline{3}$
SPACE GROUP: Pa3
Z: 4
LATTICE CONSTANT:
$a = 12.19$ kX
3 STRONGEST DIFFRACTION LINES:
4.23 (100)
3.65 ( 50) (Not reliable)
3.98 ( 40)
OPTICAL CONSTANT:
$N = 1.4388$ (Na)
HARDNESS: ~3
DENSITY: 1.675 (Meas.)
1.68 (Calc.)
CLEAVAGE: Fracture conchoidal.
HABIT: Crystals octahedral (synthetic).
COLOR-LUSTER: Colorless. Transparent. Vitreous.

Soluble in water.
MODE OF OCCURRENCE: A number of occurrences reported but not authenticated. Possible occurrences listed under mendozite (q.v.).
BEST REF. IN ENGLISH: Palache, et al., "Dana's System of Mineralogy," 7th Ed., v. II, p. 474, New York, Wiley, 1951.

# SODA-DRAVITE = Variety of dravite
$NaMg_3B_3Al_6Si_6O_{27}(O, OH)_4$

# SODALITE
$Na_4Al_3Si_3O_{12}Cl$
Var. Hackmanite

CRYSTAL SYSTEM: Cubic
CLASS: $\overline{4}$ 3m
SPACE GROUP: P$\overline{4}$3n
Z: 2
LATTICE CONSTANT:
$a = 8.878$
3 STRONGEST DIFFRACTION LINES:
3.63 (100)
6.30 ( 80)
2.10 ( 80)
OPTICAL CONSTANT:
$N = 1.483–1.487$
HARDNESS: 5½–6
DENSITY: 2.14–2.4 (Meas.)
2.31 (Calc.)
CLEAVAGE: {110} poor
Fracture uneven to conchoidal. Brittle.
HABIT: Crystals usually dodecahedral. Also massive, as embedded grains; sometimes nodular with concentric structure. Twinning on {111} common.
COLOR-LUSTER: Colorless, white, yellowish, greenish, light to dark blue, reddish. Transparent to translucent. Vitreous to greasy. Streak uncolored.
MODE OF OCCURRENCE: Occurs chiefly in nepheline-syenites and related rock types, in metasomatized calcareous rocks, and in cavities in ejected volcanic blocks. Found in the Bearpaw Mountains, Montana; in phonolite in Lawrence County, South Dakota, and at Cripple Creek, Colorado; at Magnet Cove, Arkansas (hackmanite); at Litchfield and other places in Maine; at Red Hill, New Hampshire; near Salem, Essex County, Massachusetts; and at Beemerville, New Jersey. Other notable occurrences are found at Bancroft, Ontario, and elsewhere in Canada; and at localities in Brazil, Bolivia, Greenland, Scotland, Norway, Italy, Germany, Roumania, USSR, Angola, Transvaal, Zambia, Burma, and Korea.
BEST REF. IN ENGLISH: Deer, Howie, and Zussman, "Rock Forming Minerals," v. 4, p. 289–302, New York, Wiley, 1963.

# SODA-NITER (Soda-nitre)
$NaNO_3$

CRYSTAL SYSTEM: Hexagonal
CLASS: $\overline{3}$ 2/m
SPACE GROUP: R$\overline{3}$c
Z: 2 (rhomb.), 6 (hexag.)
LATTICE CONSTANTS:
$a = 5.07$     $a_{rh} = 6.32$
$c = 16.81$     $\alpha = 47°15'$
3 STRONGEST DIFFRACTION LINES:
3.03 (100)
2.311 ( 25)
1.898 ( 16)
OPTICAL CONSTANTS:
$\omega = 1.5874$
$\epsilon = 1.3361$ (Na)
(–)

HARDNESS: 1½–2
DENSITY: 2.266 (Meas.)
       2.25 (Calc.)
CLEAVAGE: {10$\bar{1}$1} perfect
        {01$\bar{1}$2} imperfect
        {0001} imperfect
Fracture conchoidal. Slightly sectile.

HABIT: Crystals rhombohedral. Commonly massive granular; as crusts. Twinning common; usually on {01$\bar{1}$2}, {02$\bar{2}$1}, or {0001}.

COLOR-LUSTER: Colorless, white; often discolored gray, yellowish, or brownish by impurities. Transparent. Vitreous.

Easily soluble in water.

MODE OF OCCURRENCE: Occurs widespread in arid regions, usually as an efflorescence upon soil. Found in the United States in California, Nevada, Utah, New Mexico, and Texas. Large deposits occur in a region approximately 450 miles long and from 10 to 50 miles wide in the deserts of northern Chile. It also is found in Peru, Bolivia, North Africa, India, and the USSR.

BEST REF. IN ENGLISH: Palache, et al., "Dana's System of Mineralogy," 7th Ed., v. II, p. 300–302, New York, Wiley, 1951.

## SODA-TREMOLITE = Richterite

## SODDYITE
$(UO_2)_5(SiO_4)_2(OH)_2 \cdot 5H_2O$

CRYSTAL SYSTEM: Orthorhombic
CLASS: 2/m 2/m 2/m
SPACE GROUP: Fddd
Z: 3
LATTICE CONSTANTS:
 $a$ = 8.32
 $b$ = 11.21
 $c$ = 18.71
3 STRONGEST DIFFRACTION LINES:
 4.57 (100)
 6.28 ( 90)
 3.36 ( 90)
OPTICAL CONSTANTS:
 $\alpha$ = 1.650–1.654
 $\beta$ = 1.685
 $\gamma$ = 1.699–1.715
 (–)2V = large
HARDNESS: 3½
DENSITY: 4.70 (Meas.)
       4.75 (Calc.)
CLEAVAGE: {001} perfect
        {111} good
HABIT: Crystals bipyramidal {111}, up to 3 mm long, with horizontal striations and sometimes with small base {001}. Rarely thick tabular on {001}. Crystals often distorted by elongation parallel to [110], giving platy or prismatic appearance. As subparallel aggregates, divergent clusters, or as radial fibrous aggregates; also massive to earthy and as fibrous veinlets.

COLOR-LUSTER: Small crystals amber yellow and transparent. Large crystals canary yellow and opaque. Fibrous and massive material yellowish green to dull yellow. Vitreous to resinous, dependent upon transparency of crystals; also dull or earthy. Powder pale yellow. Not fluorescent.

MODE OF OCCURRENCE: Occurs in the United States associated with uranophane and kasolite as pseudomorphs after uraninite at the Ruggles pegmatite, Grafton Center, New Hampshire; also found at the Lucky Mc mine, Fremont County, Wyoming; at the Jack Pile mine, Laguna, New Mexico; at the Steel City mine, Yavapai County, Arizona, and in the Honeycomb Hills area, Utah. It also occurs in the pegmatites of Norrabees south of the Orange river in Namaqualand, South Africa; and at Shinkolobwe, Katanga, Zaire it is found massive with curite and as excellent crystals with metatorbernite, curite, sklodowskite, and kasolite.

BEST REF. IN ENGLISH: Frondel, Clifford, USGS Bull. 1064, p. 312–315 (1958).

## SODIUM AUTUNITE
$Na(UO_2)(PO_4) \cdot 4H_2O$
Autunite Group

CRYSTAL SYSTEM: Tetragonal
CLASS: 4/m 2/m 2/m
SPACE GROUP: P4/nmm
Z: 2
LATTICE CONSTANTS:
 $a$ = 6.97
 $c$ = 8.69
3 STRONGEST DIFFRACTION LINES:
 3.67  (100)
 2.675 ( 80)
 1.566 ( 80)
OPTICAL CONSTANTS:
 $\omega$ = 1.578
 $\epsilon$ = 1.559
   (–)
HARDNESS: 2–2½
DENSITY: 3.584 (Meas.)
       3.89 (Calc. for $8H_2O$)
CLEAVAGE: {001} perfect
        {100} less perfect
Brittle.
HABIT: As thin plates; also in foliated and radiating masses.

COLOR-LUSTER: Lemon yellow to lettuce yellow; vitreous, pearly on basal plane. Fluoresces strong yellow-green under ultraviolet light.

MODE OF OCCURRENCE: Occurs in several lignite deposits in the Cave Hills and Slim Buttes areas in Harding County, South Dakota; also "in one of the granodiorite massifs of the USSR."

BEST REF. IN ENGLISH: Chernikov, A. A., Krutetskaya, O. V., and Organov, N. I., *Am. Min.*, **43**: 383 (1958).

## SODIUM BETPAKDALITE
$(Na, Ca)_3Fe_2[(As_2O_4)(MoO_4)_6] \cdot 15H_2O$

CRYSTAL SYSTEM: Monoclinic
Z: 4

LATTICE CONSTANTS:
$a = 11.28$
$b = 19.30$
$c = 17.67$
$\beta = 94°30'$
3 STRONGEST DIFFRACTION LINES:
8.73 (100)
3.63 ( 90)
1.84 ( 80)
OPTICAL CONSTANTS:
$\alpha = 1.792$
$\gamma = 1.810$
HARDNESS: Not determined.
DENSITY: 2.02 (Meas.)
CLEAVAGE: Not determined.
HABIT: Crystals pseudohexagonal in outline, a few microns in size. As microcrystalline or powdery films and crusts.
COLOR-LUSTER: Lemon yellow. Dull.
MODE OF OCCURRENCE: The mineral occurs associated with goethite, gypsum, opal, natrojarosite, and ferrimolybdite in the oxidation zone of a molybdenum deposit, presumably in the USSR.
BEST REF. IN ENGLISH: Skvortsova, K. V., Nestrova, Yu. S., Sidorenko, G. A., Arapova, G. A., Dara, A. D., and Rybakova, L. I., *Am. Min.*, **57**: 1312–1313 (1972).

## SODIUM-URANOSPINITE
$(Na_2, Ca)(UO_2)_2(AsO_4)_2 \cdot 5H_2O$

CRYSTAL SYSTEM: Tetragonal
CLASS: 4/m 2/m 2/m
SPACE GROUP: P4/nmm
Z: 1
LATTICE CONSTANTS:
$a = 7.12$
$c = 8.61$
3 STRONGEST DIFFRACTION LINES:
8.42 (100)
3.63 ( 90) (synthetic)
3.27 ( 80)

OPTICAL CONSTANTS:
$\alpha = 1.585$
$\gamma = 1.612$
$(-)2V$ = very small
HARDNESS: 2½
DENSITY: 3.846 (Meas.)
CLEAVAGE: {001} perfect
{010} distinct
{100} distinct
HABIT: In tabular to elongated crystals up to 2 cm in length; as radial fibrous aggregates; and as square crystals pseudomorphous after metazeunerite (sometimes with a core of metazeunerite).
COLOR-LUSTER: Yellow-green to lemon and straw yellow; vitreous, pearly on basal plane. Fluoresces strong yellow-green under ultraviolet light.
MODE OF OCCURRENCE: Occurs in the oxidation zone of a primary hydrothermal deposit containing pitchblende, arsenopyrite, pyrite, and galena in carbonatized Devonian felsite-porphyry and tuffaceous breccia at an unspecified locality, presumably in USSR.
BEST REF. IN ENGLISH: Kopchenova, E. V., and Skvortsova, K. V., *Am. Min.*, **43**: 383–384 (1958).

## SOGDIANITE
$(K, Na)_2 Li_2 (Li, Fe, Al, Ti)_2 Zr_2 (Si_2O_5)_6$

CRYSTAL SYSTEM: Hexagonal
CLASS: 6/m 2/m 2/m
SPACE GROUP: P6/mmm
Z: 2
LATTICE CONSTANTS:
$a = 10.059$
$c = 14.052$
3 STRONGEST DIFFRACTION LINES:
3.19 (100)
2.89 (100)
4.08 ( 90)
OPTICAL CONSTANTS:
$\omega = 1.606$
$\epsilon = 1.608$
(–)
HARDNESS: 7
DENSITY: 2.90 (Meas.)
CLEAVAGE: {0001} perfect
HABIT: Massive; platy.
COLOR-LUSTER: Violet (like kunzite); transparent; vitreous.
MODE OF OCCURRENCE: Found as platy deposits up to 10 X 7 X 4 cm, associated with rare earth minerals such as thorite and stillwellite, in a vein pegmatoidal body corresponding roughly to an alkalic granite consisting of microcline, quartz, and aegirine in the Alai Range, Tadzhik SSR.
BEST REF. IN ENGLISH: Dusmatov, V. D., Efimov, A. F., Kataeva, Z. T., Khoroshilova, L. A., and Yanulov, K. P., *Am. Min.*, **54**: 1221–1222 (1969).

## SÖHNGEITE
$Ga(OH)_3$

CRYSTAL SYSTEM: Cubic
CLASS: $2/m\,\bar{3}$
SPACE GROUP: Probably Im3
Z: 8
LATTICE CONSTANT:
$a = 7.47$
3 STRONGEST DIFFRACTION LINES:
3.74 (100)
1.669 ( 70)
2.63 ( 60)
OPTICAL CONSTANT:
$N = 1.736$
HARDNESS: 4–4½
DENSITY: 3.84 (Meas.)
3.847 (Calc.)
HABIT: As crystalline aggregates up to 5 mm in size.
COLOR-LUSTER: Light brown.
MODE OF OCCURRENCE: Occurs on corroded german-

ite containing gallite at the Tsumeb mine, South West Africa.

BEST REF. IN ENGLISH: Strunz, H., *Am. Min.*, **51**: 1815 (1966).


## SOMBRERITE = Synonym for carbonate-apatite


## SOMMAIRITE = Zincian melanterite


## SONOLITE
$Mn_9(SiO_4)_4(OH, F)_2$
Manganese analogue of clinohumite

CRYSTAL SYSTEM: Monoclinic
CLASS: 2/m
SPACE GROUP: $P2_1/a$
Z: 2
LATTICE CONSTANTS:
  a = 10.66
  b = 4.88
  c = 14.33
  β = 100°34′
3 STRONGEST DIFFRACTION LINES:
  2.869 (100)
  1.743 (100)
  2.651 ( 70)
OPTICAL CONSTANTS:
  α = 1.765, 1.763
  β = 1.778, 1.779
  γ = 1.787, 1.793
  (–)2V = 75.5°–82°
HARDNESS: 5.5
DENSITY: 3.82 (Meas.)
        3.97 (Calc.)
CLEAVAGE: None.
HABIT: Crystals prismatic, often anhedral. Also massive, fine-grained.
COLOR-LUSTER: Dull reddish orange, pinkish brown to dark reddish brown to dark brown. Dull to vitreous.
MODE OF OCCURRENCE: Occurs associated with rhodochrosite, galaxite, and pyrochroite at the Sono, Hanawa, and other manganese mines in Japan; also as grains up to one inch in size in coarse franklinite-willemite-zincite ore, and in veinlets associated with green willemite and chlorite at Franklin and Sterling Hill, New Jersey.
BEST REF. IN ENGLISH: Yoshinaga, Mayum, *Am. Min.*, **48**: 1413 (1963).  Cook, David, *Am. Min.*, **54**: 1392–1398 (1969).


## SONORAITE
$FeTeO_3(OH) \cdot H_2O$

CRYSTAL SYSTEM: Monoclinic
CLASS: 2/m
SPACE GROUP: $P2_1/c$
Z: 4

LATTICE CONSTANTS:
  a = 10.984
  b = 10.268
  c = 7.917
  β = 108°29′
3 STRONGEST DIFFRACTION LINES:
  10.40 (100)
  4.66 ( 80)
  3.11 ( 80)
OPTICAL CONSTANTS:
  α = 2.018
  β = 2.023
  γ = 2.025
  (–)2V = 20°–25°
HARDNESS: ~3
DENSITY: 3.95 (Meas.)
        4.179 (Calc.)
CLEAVAGE: Not determined.
HABIT: Crystals platy, thin; often in sheaves and rosettes forming subparallel and radiating aggregates from 0.5 to 1.0 mm across.
COLOR-LUSTER: Yellowish green, transparent; vitreous.
MODE OF OCCURRENCE: Found at the Moctezuma tellurium-gold mine in Sonora, Mexico, associated with emmonsite, anglesite, limonite, and quartz. The total amount of material found probably amounted to less than 20 mg.
BEST REF. IN ENGLISH: Gaines, Richard V., Donnay, Gabrielle, and Hey, Max H., Am. Min., **53**: 1828–1832 (1968).


## SORBYITE
$Pb_{17}(Sb, As)_{22}S_{50}$

CRYSTAL SYSTEM: Monoclinic
CLASS: 2/m
SPACE GROUP: C2/m
Z: 2
LATTICE CONSTANTS:
  a = 45.15
  b = 4.146 (true cell x2)
  c = 26.53
  β = 113°25′
3 STRONGEST DIFFRACTION LINES:
  3.44 (100)
  3.38 ( 90)
  4.13 ( 60)
HARDNESS: Vickers microhardness 175 (172–186) kg/mm² with 50 g load.
DENSITY: 5.59 (Predicted)
        5.52 (Calc.)
CLEAVAGE: [001] perfect
HABIT: Crystals equant to thin tabular {100}, always elongate [010], and commonly heavily striated [010].
COLOR-LUSTER: Black; metallic.  Massive material probably lead gray. Streak black.
MODE OF OCCURRENCE: Occurs very sparingly in a small prospect pit in Precambrian marble at Madoc, Ontario, Canada.
BEST REF. IN ENGLISH: Jambor, J. L., *Can. Min.*, **9** (Pt. 2): 191–213 (1967).

## SORENSENITE
$Na_4SnBe_2Si_6O_{16}(OH)_4$

CRYSTAL SYSTEM: Monoclinic
CLASS: 2/m or m
SPACE GROUP: C2/c or Cc
Z: 4
LATTICE CONSTANTS:
  $a = 18.58$
  $b = 7.45$
  $c = 12.05$
  $\beta = 98°09'$
3 STRONGEST DIFFRACTION LINES:
  (Nakalaq)      (Kvanefjeld)
  2.918 (100)    2.92 (100)
  3.41  ( 85)    2.87 (100)
  2.960 ( 85)    3.36 ( 90)
OPTICAL CONSTANTS:
  (Nakalaq)      (Kvanefjeld)
  $\alpha = 1.579$      $\alpha = 1.576$
  $\beta = 1.585$      $\beta = 1.581$ ($\lambda = 5893$ Å)
  $\gamma = 1.586$      $\gamma = 1.584$
  (−)2V small to 75°
  HARDNESS: 5.5
  DENSITY: 2.9 (Meas.)
           2.90 (Calc.)
CLEAVAGE: Distinct in two directions at 63 ± 2°.
Brittle.
HABIT: Crystals acicular or elongated tabular, up to 10 ×
1 × 1 cm in size.
COLOR-LUSTER: Colorless to pinkish, altering to milky
white. Fine silky.
MODE OF OCCURRENCE: Occurs in the Ilimaussaq
alkalic intrusive, south Greenland, at Nakalaq in hydrothermal veins associated with analcime, microcline, sodalite,
neptunite, and aegirine, and at Kvanefjeld in coarse-grained
rock containing microcline, nepheline, analcime, and arfvedsonite. Other minerals present include neptunite, apatite,
sphalerite, and the Be minerals beryllite and chkalovite.
BEST REF. IN ENGLISH: Semenov, E. I., Gerassimov-
sky, V. I., Maksimova, N. V., Andersen, S., and Petersen,
O. V., *Am. Min.*, **51**: 1547–1548 (1966).

## SOUXITE = Varlamoffite

## SOUZALITE
$(Mg, Fe^{2+})_3(Al, Fe^{3+})_4(PO_4)_4(OH)_6 \cdot 2H_2O$

CRYSTAL SYSTEM: Probably monoclinic
3 STRONGEST DIFFRACTION LINES:
  2.690 (100)
  3.79  ( 90)
  5.35  ( 80)
OPTICAL CONSTANTS:
    $\alpha = 1.618$
    $\beta = 1.642$
    $\gamma = 1.652$
  (−)2V = 68°
  HARDNESS: 5½–6
  DENSITY: 3.087 (Meas.)
CLEAVAGE: One good, one fair, approximately at right
angles.

HABIT: As coarse fibrous masses. Twinning polysyn-
thetic.
COLOR-LUSTER: Dark green.
MODE OF OCCURRENCE: Occurs as an alteration prod-
uct of scorzalite at the Corrego Frio pegmatite, Minas
Geraes, Brazil.
BEST REF. IN ENGLISH: Pecora, W. T., and Fahey,
J. J., *Am. Min.*, **34**: 83 (1949).

## SPADAITE (Inadequately described mineral)
$MgSiO_2(OH)_2 \cdot H_2O$ (?)

CRYSTAL SYSTEM: Unknown
OPTICAL CONSTANTS:
  $\alpha = 1.521$
  $\beta = 1.525$
  $\gamma = 1.545$
  (+)2V small to medium
HARDNESS: 2½
DENSITY: 2.2
CLEAVAGE: Not determined.
HABIT: Massive, dense. Under microscope indistinctly
platy or columnar.
COLOR-LUSTER: Cream to pinkish. Pearly or greasy.
Colorless in section.
MODE OF OCCURRENCE: Occurs in association with
quartz, chalcopyrite, bornite, native gold, garnet, diopside,
wollastonite, and other silicates, in gold ore at the Cane
Springs, Alvaredo, and Midas mines, Gold Hill, Tooele
County, Utah. Also found in association with wollastonite
at Capo di Bove, near Rome, Italy, and in dolerite at Sas-
bach, Kaiserstuhl, Germany.
BEST REF. IN ENGLISH: Schaller, W. T., and Nolan,
T. B., *Am. Min.*, **16**: 231–236 (1931).

## SPANGOLITE
$Cu_6AlSO_4(OH)_{12}Cl \cdot 3H_2O$

CRYSTAL SYSTEM: Hexagonal
CLASS: 3m
SPACE GROUP: P3c1
Z: 2
LATTICE CONSTANTS:
  $a = 8.245$
  $c = 14.34$
3 STRONGEST DIFFRACTION LINES:
  7.1  (100)
  3.59 ( 80)
  2.54 ( 70)
OPTICAL CONSTANTS:
  $\omega = 1.681–1.686$
  $\epsilon = 1.627–1.638$
     (−)
  HARDNESS: ~3
  DENSITY: 3.14 (Meas.)
           3.14 (Calc.)
CLEAVAGE: {0001} perfect
          {10$\bar{1}$1} distinct
Fracture conchoidal. Brittle.

HABIT: Crystals short prismatic or tabular, usually holohedral, less commonly hemimorphic. Crystals hexagonal in aspect. Prism and trigonal pyramid faces striated horizontally. Twinning on {0001} rare.

COLOR-LUSTER: Dark green, emerald green, bluish green. Vitreous. Transparent to translucent. Streak pale green.

MODE OF OCCURRENCE: Occurs as a secondary mineral in the oxidized zone of copper-bearing ore deposits, often associated with malachite, brochantite, azurite, linarite, chrysocolla, cuprite, and other minerals. Found in Arizona at the Czar mine and as superb large crystals at the Copper Queen mine, Bisbee; in the Tombstone area, Cochise County; and at other places. As fine crystals in seams and cavities at the Blanchard mine, Socorro County, New Mexico; at the Myler mine, Majuba Hill, Pershing County, Nevada; in the Grand Central mine, Juab County, Utah; at St. Day, Cornwall, England; and from Arenas, Sardinia.

BEST REF. IN ENGLISH: Palache, et al., "Dana's System of Mineralogy," 7th Ed., v. II, p. 576–578, New York, Wiley, 1951.

# SPECULARITE = Flaky, micaceous variety of hematite

# SPENCERITE
$Zn_4(PO_4)_2(OH)_2 \cdot 3H_2O$

CRYSTAL SYSTEM: Monoclinic
CLASS: 2/m
SPACE GROUP: P2/c
Z: 2
LATTICE CONSTANTS:
 $a = 10.441$
 $b = 5.280$
 $c = 11.196$
 $\beta = 116°50'$
3 STRONGEST DIFFRACTION LINES:
 9.4 (100)
 3.501 ( 55)
 2.332 ( 30)
OPTICAL CONSTANTS:
  $\alpha = 1.586$
  $\beta = 1.602$
  $\gamma = 1.606$
 (−)2V = 49°
HARDNESS: 3
DENSITY: 3.145 (Meas.)
    3.253 (Calc.)
CLEAVAGE: {100} perfect
    {010} good
    {001} distinct
HABIT: Crystals tabular, elongated [001]. Terminations blunt or lance-like. Commonly striated. Also massive, stalactitic, with platy to columnar structure. Twinning on {100}, polysynthetic.

COLOR-LUSTER: White. Vitreous; pearly on cleavages.

MODE OF OCCURRENCE: Occurs in oxidized zinc ore associated with hemimorphite and hopeite at the Hudson Bay mine, Salmo, British Columbia, Canada.

BEST REF. IN ENGLISH: Palache, et al., "Dana's System of Mineralogy," 7th Ed., v. II, p. 931–933, New York, Wiley, 1951.

# SPENCITE = Tritomite (Y)

# SPERRYLITE
$PtAs_2$

CRYSTAL SYSTEM: Cubic
CLASS: $2/m\,\bar{3}$
SPACE GROUP: Pa3
Z: 4
LATTICE CONSTANT:
 $a = 5.941$
3 STRONGEST DIFFRACTION LINES:
 1.801 (100)
 1.148 ( 70)
 2.98 ( 60)
OPTICAL CONSTANTS: Opaque, isotropic.
 HARDNESS: 6–7
 DENSITY: 10.46–10.6 (Meas.)
     10.59 (Calc.)
 CLEAVAGE: {001} indistinct
Fracture conchoidal. Brittle.

HABIT: Crystals cubic or cubo-octahedral; sometimes highly modified.

COLOR-LUSTER: Tin white. Opaque. Bright metallic. Streak black.

MODE OF OCCURRENCE: Occurs associated with covellite at the Rambler mine, near Encampment, Wyoming; in gravels near Franklin, Macon County, North Carolina; at the Frood, Vermilion, and Victoria mines, Sudbury district, Ontario, Canada; in the Nikolayevskiy placer gold deposits on the Tok River, in Amur, eastern Siberia; and as exceptional crystals up to 2 cm on edge in the Waterburg district, Transvaal, South Africa.

BEST REF. IN ENGLISH: Palache, et al., "Dana's System of Mineralogy," 7th Ed., v. I, p. 292–293, New York, Wiley, 1944.

# SPESSARTINE (Spessartite)
$Mn_3Al_2Si_3O_{12}$
Garnet Group

CRYSTAL SYSTEM: Cubic
CLASS: $4/m\,\bar{3}\,2/m$
SPACE GROUP: Ia3d
Z: 8
LATTICE CONSTANT:
 $a = 11.621$
3 STRONGEST DIFFRACTION LINES:
 2.60 (100)
 1.56 ( 40) (synthetic)
 1.61 ( 30)
OPTICAL CONSTANTS:
 $N = 1.800$
 HARDNESS: 7–7½
 DENSITY: 3.8–4.25 (Meas.)

CLEAVAGE: None.
{110} parting sometimes distinct.
Fracture uneven to conchoidal. Brittle.

HABIT: Crystals usually dodecahedrons or trapezohedrons; also in combination, or with hexoctahedron. Crystal faces often striated. Also massive, compact; fine or coarse granular; or as embedded grains.

COLOR-LUSTER: Various shades of red, reddish orange, yellowish brown, reddish brown to brown. Transparent to translucent. Vitreous. Streak white.

MODE OF OCCURRENCE: Occurs principally in granite pegmatites, gneiss, quartzite, schist, and lithophysae in rhyolite; less commonly in skarn deposits. Found as superb crystals in pegmatite in San Diego County and elsewhere in California; as fine crystals in rhyolite near Ely, White Pine County, Nevada, and near Nathrop, Chaffee County, Colorado; as excellent specimens at the Rutherford mine, Amelia County, Virginia; and at localities in New Mexico, Pennsylvania, and North Carolina. Other notable occurrences are found in Brazil, England, Wales, Finland, Czechoslovakia, New Zealand, and Madagascar.

BEST REF. IN ENGLISH: Deer, Howie, and Zussman, "Rock Forming Minerals," v. 1, p. 77–112, New York, Wiley, 1962.

# SPHAEROCOBALTITE (Spherocobaltite) =
Cobaltocalcite

# SPHAEROSIDERITE = Variety of siderite exhibiting globular or spheroidal forms

# SPHALERITE
ZnS
Trimorphous with wurtzite, matraite
Var. Cleiophane (colorless, white)
    Marmatite (black; Fe-rich)

CRYSTAL SYSTEM: Cubic
CLASS: $\bar{4}$3m
SPACE GROUP: F$\bar{4}$3m
Z: 4
LATTICE CONSTANT:
    $a = 5.43$
3 STRONGEST DIFFRACTION LINES:
    3.123 (100)
    1.912 ( 51)
    1.633 ( 30)
OPTICAL CONSTANT:
    $N = 2.369$ (Na)
HARDNESS: 3½–4
DENSITY: 3.9–4.1 (Meas.)
        4.096 (Calc. X-ray)
CLEAVAGE: {011} perfect
Fracture conchoidal. Brittle.

HABIT: Crystals usually tetrahedral, often developed as to appear octahedral; also dodecahedral; rounded faces common. Also massive, cleavable; granular; stalactitic; botryoidal; concretionary. Twinning on {111} very common.

COLOR-LUSTER: Brown, black, yellow, green, red, gray, white, colorless. Transparent to translucent. Resinous to adamantine. Streak pale brown to colorless.

MODE OF OCCURRENCE: Found widespread as the most abundant zinc mineral, commonly associated with galena and other sulfides in limestones, dolomites, and other sedimentary rocks; in hydrothermal ore veins; in contact metamorphic deposits; and rarely in pegmatites. Important deposits occur in Missouri, Oklahoma, Kansas, Illinois, Wisconsin, Montana, Colorado, Idaho, and in many other states. Also in Canada, Mexico, England, Scotland, Sweden, France, Spain, Germany, Czechoslovakia, Roumania, Japan, and New South Wales, Australia.

BEST REF. IN ENGLISH: Palache, et al., "Dana's System of Mineralogy," 7th Ed., v. I, p. 210–215, New York, Wiley, 1944.

# SPHENE (Titanite)
CaTiSiO$_5$

CRYSTAL SYSTEM: Monoclinic
CLASS: 2/m
SPACE GROUP: C2/c
Z: 4
LATTICE CONSTANTS:
    $a = 6.56$
    $b = 8.72$
    $c = 7.44$
    $\beta = 119°43'$
3 STRONGEST DIFFRACTION LINES:
    3.233 (100)
    2.989 ( 90)
    2.595 ( 90)
OPTICAL CONSTANTS:
    $\alpha = 1.843$–1.950
    $\beta = 1.870$–2.034
    $\gamma = 1.943$–2.110
    (+)2V = 17°–40°
HARDNESS: 5–5½
DENSITY: 3.45–3.55 (Meas.)
        3.53 (Calc.)
CLEAVAGE: {110} distinct
        {221} parting, common, due to twinning lamellae.
Fracture uneven to conchoidal. Brittle.

HABIT: Crystals varied in habit; usually wedge-shaped and flattened {001}; also prismatic. Also massive, compact; sometimes lamellar. Twinning on {100} common; also lamellar twinning on {221}.

COLOR-LUSTER: Colorless, yellow, green, gray, brown, rose red, and black, often varying in a single crystal. Adamantine to resinous. Transparent to nearly opaque. Streak white.

MODE OF OCCURRENCE: Occurs widespread chiefly as an accessory mineral in igneous rocks; also in schists, gneisses, and other metamorphic rocks; in beds of iron ore; and as a detrital mineral in sedimentary deposits. It is especially common in nepheline-syenite rocks. Notable occurrences are found at Crestmore, Riverside County, California; at Magnet Cove, Arkansas; at Franklin, New Jersey; in the Tilly Foster mine at Brewster, and at other places in New

York; in Quebec and Ontario, Canada; and at localities in Greenland, Brazil, Norway, France, Italy, Austria, Switzerland, Czechoslovakia, Finland, USSR, Pakistan, Madagascar, New Zealand, and elsewhere.

BEST REF. IN ENGLISH: Deer, Howie, and Zussman, "Rock Forming Minerals," v. 1, p. 69–76, New York, Wiley, 1962.

## SPINEL GROUP

See:  Chromite     Jacobsite
      Coulsonite     Magnesiochromite
      Cuprospinel     Magnetite
      Franklinite     Spinel
      Gahnite     Trevorite
      Galaxite     Ulvöspinel
      Hercynite

## SPINEL

$MgAl_2O_4$
Spinel Series

CRYSTAL SYSTEM: Cubic
CLASS: $4/m\,\bar{3}\,2/m$
SPACE GROUP: Fd3m
Z: 8
LATTICE CONSTANT:
    $a = 8.086$
3 STRONGEST DIFFRACTION LINES:
    2.436 (100)
    2.021 ( 70)  (synthetic)
    1.429 ( 60)
OPTICAL CONSTANT:
    $N \simeq 1.720$
    HARDNESS: 7½–8
    DENSITY: 3.581 (Calc.)
    CLEAVAGE: {111} parting, indistinct
Fracture conchoidal; also uneven. Brittle.
    HABIT: Crystals usually octahedral, sometimes modified; rarely cubic or dodecahedral. Also massive, coarse granular to compact, or as rounded grains. Twinning on {111}, sometimes as sixlings by repeated twinning.
    COLOR-LUSTER: Various shades of red, blue, green, brown, black. Transparent to opaque. Vitreous to nearly dull. Streak white.
    MODE OF OCCURRENCE: Occurs as a common metamorphic mineral found chiefly in crystalline limestones, gneiss, and serpentine; also as an accessory mineral in basic igneous rocks, and in placer deposits. In the United States notable occurrences are found in California, Montana, Colorado, New York, New Jersey, Massachusetts, North Carolina, Virginia, and Alabama. It also occurs in Canada, France, Italy, Sweden, Germany, Finland, USSR, Madagascar, India, and as fine gem crystals in Burma and Ceylon.
    BEST REF. IN ENGLISH: Palache, et al., "Dana's System of Mineralogy," 7th Ed., v. I, p. 689–697, New York, Wiley, 1944.

## SPIROFFITE

$(Mn, Zn)_2 Te_3 O_8$

CRYSTAL SYSTEM: Monoclinic
CLASS: m or 2/m
SPACE GROUP: Cc or C2/c
Z: 4
LATTICE CONSTANTS:
    $a = 13.00$
    $b = 5.38$
    $c = 12.12$
    $\beta = 98°$
3 STRONGEST DIFFRACTION LINES:
    4.98 (100)
    3.00 (100)
    4.06 ( 80)
OPTICAL CONSTANTS:
    $\alpha = 1.85$
    $\beta = 1.91$
    $\gamma > 2.10$
(+)2V = 55°
HARDNESS: ~3½
DENSITY: 5.01 (Meas.)
HABIT: As small cleavable masses.
COLOR-LUSTER: Red to purple; adamantine.
MODE OF OCCURRENCE: Occurs associated with tellurite, paratellurite, native tellurium, and several other tellurites near Moctezuma, Sonora, Mexico.
    BEST REF. IN ENGLISH: Mandarino, J. A., Williams, S. J., and Mitchell, R. S., MSA Special Paper 1, p. 305–309 (1963). *Am. Min.*, **47**: 196 (1962).

## SPODIOPHYLLITE (Species status uncertain)

$(Na, K)_4 (Mg, Fe)_3 (Fe, Al)_2 Si_8 O_{24}$

CRYSTAL SYSTEM: Monoclinic (?)
OPTICAL CONSTANTS:
    Uniaxial (–)
    HARDNESS: 3–3.25
    DENSITY: 2.633 (Meas.)
    CLEAVAGE: {001} perfect, micaceous
Brittle, laminae not flexible.
    HABIT: Crystals, prismatic, pseudohexagonal. Rough, elongated.
    COLOR-LUSTER: Ash gray. Transparent to translucent. Dull; pearly on cleavage.
    MODE OF OCCURRENCE: Occurs very sparingly at Narsarsuk, Greenland, in association with aegirine, albite, rhodochrosite, zircon, and ancylite.
    BEST REF. IN ENGLISH: Dana and Ford, "Dana's System of Mineralogy," 6th Ed., App. II, p. 96, New York, Wiley, 1914.

## SPODUMENE (Triphane)

$LiAlSi_2O_6$
Pyroxene Group
Var. Hiddenite (green)
      Kunzite (pink)

CRYSTAL SYSTEM: Monoclinic
CLASS: m
SPACE GROUP: C2

Z: 4
LATTICE CONSTANTS:
 *a* = 9.52
 *b*= 8.32
 *c* = 5.25
 $\beta$ = 110°28′
3 STRONGEST DIFFRACTION LINES:
 2.93 (100)
 2.80 ( 80)
 1.57 ( 70)
OPTICAL CONSTANTS:
  $\alpha$ = 1.653–1.670
  $\beta$ = 1.660–1.669
  $\gamma$ = 1.665–1.682
 (+)2V = 55°–66°
HARDNESS: 6½–7½
DENSITY: 3.0–3.2 (Meas.)
    3.19 (Calc.)
CLEAVAGE: {110} perfect
    {010} parting, common
    {100} parting, less common
Fracture uneven—hackly to subconchoidal. Brittle.

HABIT: Crystals prismatic, often flattened and vertically striated. Up to 47 feet long and 3 X 6 feet in cross section. Also as cleavable masses. Twinning on {100} common.

COLOR-LUSTER: Colorless, white, gray, yellowish, greenish, emerald green, pink, violet. Transparent to translucent. Vitreous to dull. Streak white. Sometimes fluorescent under ultraviolet light.

MODE OF OCCURRENCE: Occurs in granite pegmatites in association with quartz, feldspar, muscovite, columbite-tantalite, and other minerals. Mined chiefly for its lithium content; also for gemstones (kunzite, hiddenite). Found widespread in the Black Hills of South Dakota, especially as gigantic crystals at the Etta mine, Pennington County, and the Tin Mountain mine, Custer County. Other major occurrences are found in the Pala district, California (kunzite); at Hiddenite, Alexander County (hiddenite), and at Kings Mountain, North Carolina; and at deposits in Maine, Massachusetts and Connecticut. Other notable localities include Bernic Lake, Manitoba; Madagascar (gem varieties); Minas Geraes, Brazil (gem varieites); Sweden; USSR; and elsewhere.

BEST REF. IN ENGLISH: Vlasov, K. A., "Mineralogy of Rare Elements," v. II, p. 6–11, Israel Program for Scientific Translations, 1966.

## SPURRITE
$Ca_5 Si_2 O_8 CO_3$

CRYSTAL SYSTEM: Monoclinic
CLASS: 2/m
SPACE GROUP: $P2_1/a$
Z: 4
LATTICE CONSTANTS:
 *a* = 10.49
 *b* = 6.705
 *c* = 14.15
 $\beta$ = 101°19′
3 STRONGEST DIFFRACTION LINES:
 2.701 (100)
 2.646 ( 65)
 2.667 ( 60)

OPTICAL CONSTANTS:
  $\alpha$ = 1.640
  $\beta$ = 1.674
  $\gamma$ = 1.679
 (–)2V = 39.5°
HARDNESS: 5
DENSITY: 3.0 (Meas.)
    3.025 (Calc.)
CLEAVAGE: {001} distinct
    {100} poor
HABIT: Crystals anhedral; usually massive, granular. Twinning on {20$\bar{5}$} simple; on {001} polysynthetic.

COLOR-LUSTER: Gray, lavender-gray. Translucent. Vitreous.

MODE OF OCCURRENCE: Occurs in association with gehlenite and merwinite in limestone at Crestmore, Riverside County, California. Also found as a contact mineral at Tres Hermanes, Luna County, New Mexico; in the Velardena mining district, Durango, Mexico; at Scawt Hill, County Antrim, Ireland; and elsewhere.

BEST REF. IN ENGLISH: Winchell and Winchell, "Elements of Optical Mineralogy," 4th Ed., Pt. 2, p. 516, New York, Wiley, 1951.

## STAFFELITE = Francolite

## STAINIERITE = Synonym for heterogenite

## STANFIELDITE
$Ca_4 (Mg, Mn, Fe)_5 (PO_4)_6$

CRYSTAL SYSTEM: Monoclinic
CLASS: 2/m
SPACE GROUP: P2/c (?)
Z: 8
LATTICE CONSTANTS:
 *a* = 17.267
 *b* = 10.002
 *c* = 22.917
 $\beta$ = 100°31′
3 STRONGEST DIFFRACTION LINES:
 2.817 (100)
 3.747 ( 80)
 2.505 ( 80)
OPTICAL CONSTANTS:
  $\alpha$ = 1.619
  $\beta$ = 1.622
  $\gamma$ = 1.631
 (+)2V = 50°
HARDNESS: 4–5, nearer to 5.
DENSITY: 3.15 (Meas.)
    3.15 (Calc. for Mg/Fe = 3:2)
CLEAVAGE: Not determined.
HABIT: Massive; as irregular grains and thin veinlets.
COLOR-LUSTER: Reddish to amber; under the microscope clear and transparent.
MODE OF OCCURRENCE: Occurs as grains up to 1 mm in diameter along the walls of cracks in the Estherville meso-siderite meteorite, and as veinlets less than 1 mm in width

cutting olivine in the Santa Rosalia, Albin, Finnmarken, Imilac, Mt. Vernon, and Newport pallasites.

BEST REF. IN ENGLISH: Fuchs, L. H., *Science*, **158**: 910-911 (1967).

# STANNITE
$Cu_2FeSnS_4$

CRYSTAL SYSTEM: Tetragonal
CLASS: $\bar{4}2m$
SPACE GROUP: $I\bar{4}2m$
Z: 2
LATTICE CONSTANTS:
  $a = 5.46$
  $c = 10.725$
3 STRONGEST DIFFRACTION LINES:
  3.12 (100)
  1.922 ( 70)
  1.642 ( 40)
OPTICAL CONSTANTS: Opaque.
  HARDNESS: 4
  DENSITY: 4.3-4.5 (Meas.)
      4.437 (Calc.)
  CLEAVAGE: {110} indistinct
      {001} indistinct
Fracture uneven. Brittle.
  HABIT: Crystals commonly pseudotetrahedral or pseudo-dodecahedral due to twinning; often striated. Usually massive, granular. Twinning on {102} and {112}.
  COLOR-LUSTER: Steel gray to grayish black; often tarnishes bluish. Opaque. Metallic. Streak blackish.
  MODE OF OCCURRENCE: Occurs chiefly in hydrothermal vein deposits, and rarely in pegmatites, commonly associated with cassiterite, wolframite, and sulfide minerals. Found sparingly in pegmatite at the Peerless and Etta mines, Keystone, Pennington County, South Dakota; on the Seward Peninsula, Alaska; at several places in Bolivia; abundantly in the tin deposits of Cornwall, England; at Zinnwald, Czechoslovakia; and in the vicinity of Zeehan, Tasmania.
  BEST REF. IN ENGLISH: Palache, et al., "Dana's System of Mineralogy," 7th Ed., v. I, p. 224-226, New York, Wiley, 1944.

# STANNOIDITE
$Cu_5(Fe, Zn)_2SnS_8$

CRYSTAL SYSTEM: Orthorhombic
CLASS: 2 2 2, 2/m 2/m 2/m or mm2
SPACE GROUP: I222, $I2_12_12_1$, Immm, or Imm2
LATTICE CONSTANTS:
  $a = 10.76$
  $b = 5.40$
  $c = 16.09$
3 STRONGEST DIFFRACTION LINES:
  3.11 (100 )
  1.906 ( 70 )
  1.621 ( 20b)
  HARDNESS: ~4
  DENSITY: 4.29 (Calc.)

CLEAVAGE: None.
Fracture uneven to subconchoidal.
  HABIT: Massive.
  COLOR-LUSTER: Brass brown; metallic. Streak dark brown gray.
  MODE OF OCCURRENCE: Occurs with chalcopyrite and quartz, surrounding stannite, which replaces and surrounds stannite, at the Konjo Mine, Okayama Prefecture, Japan.
  BEST REF. IN ENGLISH: Kato, Akira, *Am. Min.*, **54**: 1495-1496 (1969).

# STANNOPALLADINITE (Inadequately described mineral)
$Pd_5Sn_3$

CRYSTAL SYSTEM: Cubic
3 STRONGEST DIFFRACTION LINES:
  2.27 (100)
  2.20 (100)
  1.58 ( 70)
  HARDNESS: Microhardness 387-452 kg/mm$^2$.
  DENSITY: 10.2 (Meas.)
  CLEAVAGE: Not determined.
  HABIT: Crystals elongated and rounded.
  COLOR-LUSTER: Brown-rose. Opaque.
  MODE OF OCCURRENCE: Occurs intergrown with niggliite in association with hessite and tellurides of platinum and palladium at Monchegorsk, USSR.
  BEST REF. IN ENGLISH: Maslenitzky, I. N., Faleev, P. V., and Iskyul, E. V., *Min. Abs.*, **10**: 453 (1949). *Am. Min.*, **56**: 360-361 (1971).

# STARINGITE
$(Fe, Mn)_x(Sn, Ti)_{6-3x}(Ta, Nb)_{2x}O_{12}$

CRYSTAL SYSTEM: Tetragonal
CLASS: 4/m 2/m 2/m
SPACE GROUP: P4/mnm
Z: 1
LATTICE CONSTANTS:
  $a = 4.742$
  $c = 9.535$
3 STRONGEST DIFFRACTION LINES:
  3.36 (100)
  1.76 ( 90)
  2.644 ( 80)
OPTICAL CONSTANTS: Uniaxial (+)
  HARDNESS: 6-6½
  DENSITY: 7.17 (Calc.)
  CLEAVAGE: None.
  HABIT: Massive, compact.
  COLOR-LUSTER: Black.
  MODE OF OCCURRENCE: Occurs as small exsolution bodies (20-50 $\mu$) in tapiolite from granite pegmatites at Seridozenho and Pedra Lavreda, Paraiba State, Brazil.
  BEST REF. IN ENGLISH: *Min. Mag.*, **37**: 447-454 (1969).

## STARKEYITE (Leonhardtite)
$MgSO_4 \cdot 4H_2O$

CRYSTAL SYSTEM: Monoclinic
CLASS: 2/m
SPACE GROUP: $P2_1/n$
Z: 4
LATTICE CONSTANTS:
 $a = 5.922$
 $b = 13.604$
 $c = 7.905$
 $\beta = 90°51'$
3 STRONGEST DIFFRACTION LINES:
 4.48 (100)
 2.95 (100)
 5.5 ( 60)
OPTICAL CONSTANTS:
 $\alpha = 1.490$
 $\beta = 1.491$
 $\gamma = 1.497$
 $(+)2V = 50°$
HARDNESS: Not determined.
DENSITY: 2.007 (Calc.)
CLEAVAGE: Not determined.
HABIT: Massive; as a powdery efflorescence.
COLOR-LUSTER: White; dull.
MODE OF OCCURRENCE: Occurs mixed with sanderite, pentahydrite, and hexahydrite as an efflorescence on kieserite in the 601 meter level of the "Hansa I" shaft, Empelde, and the 675 meter level of the "Niedersachsen" shaft near Wathlingen, Hanover, Germany; also as a powdery efflorescence on an altered mixture of pyrite and marcasite at the Starkey mine, Madison County, Missouri.
BEST REF. IN ENGLISH: Grawe, Oliver R., Missouri Geol. Survey and Water Resources, p. 209–210 (1945). Grawe, Oliver R., *Am. Min.*, **41**: 662 (1956).


## STAUROLITE
$Fe_2Al_9Si_4O_{22}(OH)_2$

CRYSTAL SYSTEM: Monoclinic, pseudo-orthorhombic
CLASS: 2/m
SPACE GROUP: C2/m
Z: 2
LATTICE CONSTANTS:
 $a = 7.87$
 $b = 16.62$ (variable)
 $c = 5.66$
 $\beta \simeq 90.0°$
3 STRONGEST DIFFRACTION LINES:
 3.012 (100)
 2.693 (100)
 2.372 ( 80)
OPTICAL CONSTANTS:
 $\alpha = 1.739-1.747$
 $\beta = 1.745-1.753$
 $\gamma = 1.752-1.761$
 $(+)2V = 82°-90°$
HARDNESS: 7-7½

DENSITY: 3.65-3.83 (Meas.)
 3.686 (Calc.)
CLEAVAGE: {010} distinct
Fracture uneven to subconchoidal. Brittle.
HABIT: Crystals short prismatic, often with rough surfaces. Twinning common, as cruciform twins; on {031} giving a cross near 90°, and on {231} giving an oblique cross at 60°31'.
COLOR-LUSTER: Dark brown, reddish brown, yellowish brown, brownish black. Translucent to nearly opaque. Vitreous to resinous. Streak uncolored to grayish.
MODE OF OCCURRENCE: Occurs widespread as a product of regional metamorphism, particularly in mica schists and gneisses, in association with garnet, quartz, muscovite, and kyanite. Typical occurrences are found in Rio Arriba County, New Mexico; Black Hills, South Dakota; Patrick County, Virginia; Fannin County, Georgia; and North Carolina, Massachusetts, New Hampshire, and Maine. Also in Canada, Brazil, Greenland, Ireland, Scotland, France, Switzerland, Finland, and Zambia.
BEST REF. IN ENGLISH: Deer, Howie, and Zussman, "Rock Forming Minerals," v. 1, p. 151–160, New York, Wiley, 1962. Smith, J. V., *Am. Min.*, **53**: 1139–1155 (1968).


## STEATITE = Synonym for talc


## STEENSTRUPINE
$CeNaMnSi_3O_9$

CRYSTAL SYSTEM: Hexagonal
Z: 3
LATTICE CONSTANTS:
 $a = 9.47$
 $c = 15.39$
3 STRONGEST DIFFRACTION LINES:
 3.09 (100)
 2.87 (100) (Often metamict)
 4.2 ( 70)
OPTICAL CONSTANTS:
 $\omega = 1.665$
 $\epsilon = 1.663$
 (–)
HARDNESS: 5
DENSITY: 3.1-3.6 (Meas.)
CLEAVAGE: None. Fracture conchoidal.
HABIT: Crystals rhombohedral; usually as irregular masses up to 10 cm in size.
COLOR-LUSTER: Brownish red to black. Opaque.
MODE OF OCCURRENCE: Occurs in nepheline-sodalite syenite pegmatites and in syenite in the Julianehaab district, Greenland, and from the Lovozero massif, Kola Peninsula, USSR.
BEST REF. IN ENGLISH: Vlasov, K. A., "Mineralogy of Rare Elements," v. II, p. 321–324, Israel Program for Scientific Translations, 1966.


## STEIGERITE (Inadequately described mineral)
$AlVO_4 \cdot 3H_2O$

CRYSTAL SYSTEM: Unknown
LATTICE CONSTANTS:
$a$ = 12.92
$c$ = 10.98
$\beta$ = 121°23′
OPTICAL CONSTANTS: Mean index = 1.71
HARDNESS: Not determined.
DENSITY: Not determined.
CLEAVAGE: Not determined.
HABIT: As compact aggregates composed of thin, poorly developed laths and angular plates, usually less than 1 $\mu$ in size. Laths and flakes flattened {010}.
COLOR-LUSTER: Canary yellow, waxy.
MODE OF OCCURRENCE: Occurs associated with fervanite and gypsum coating fractures in and around nodular concretions of corvusite in sandstone along the north wall of Gypsum Valley, San Miguel County, Colorado.
BEST REF. IN ENGLISH: Henderson, E. P., *Am. Min.*, **20**: 769-772 (1935).    Ross, Malcolm, *Am. Min.*, **44**: 322–341 (1959).


**STELLERITE** = Morphological variety of stilbite exhibiting rectangular habit


## STENHUGGARITE
$Ca_2 Fe_2^{2+} Sb_2^{5+} O_3 (As^{3+}O_3)_4$

CRYSTAL SYSTEM: Tetragonal
CLASS: 4/m 2/m 2/m
SPACE GROUP: I4₁/amd
Z: 8
LATTICE CONSTANTS:
$a$ = 16.12
$c$ = 10.70
3 STRONGEST DIFFRACTION LINES:
2.985 (100)
1.845 ( 40)
2.548 ( 35)
HARDNESS: 4⁻
DENSITY: 4.63 (Meas.)
CLEAVAGE: None.
HABIT: Crystals minute, nearly equant, with forms {100}, {211}, and {301}.
COLOR-LUSTER: Orange. Streak bright yellow.
MODE OF OCCURRENCE: Occurs very rarely in pockets in granular hematite ore as an "early fissure" mineral at Långban, Sweden.
BEST REF. IN ENGLISH: Moore, Paul B., *Am. Min.*, **56**: 636-637 (1971).


## STENONITE
$Sr_2 AlCO_3 F_5$

CRYSTAL SYSTEM: Monoclinic
CLASS: probably 2/m
SPACE GROUP: Probably P2₁/m
Z: 4

LATTICE CONSTANTS:
$a$ = 5.447
$b$ = 8.688
$c$ = 13.14
$\beta$ = 98°20′
3 STRONGEST DIFFRACTION LINES:
3.387 (100)
2.166 (100)
1.925 ( 80)
OPTICAL CONSTANTS:
$\alpha$ = 1.452
$\beta$ = 1.527
$\gamma$ = 1.538
(−)2V = 43°
HARDNESS: ~3.5
DENSITY: 3.86 (Meas.)
3.86 (Calc.)
CLEAVAGE: One basal {001}
Two prismatic {120}.
HABIT: Massive; as grains up to 4 × 1 cm in size.
COLOR-LUSTER: Colorless to white; vitreous.
MODE OF OCCURRENCE: Occurs associated with jarlite, cryolite, weberite, fluorite, pyrite, sphalerite, galena, chalcopyrite, and hydrous fluorides including ralstonite, pachnolite, prosopite, and topaz, mainly in the contact zone between siderite-cryolite ore and masses rich in fluorite, at the cryolite mine, Ivigtut, Greenland.
BEST REF. IN ENGLISH: Pauly, Hans, *Am. Min.*, **48**: 1178 (1963).


## STEPANOVITE
$NaMg\{Fe(C_2O_4)_3\} \cdot 8\text{-}9H_2O$

CRYSTAL SYSTEM: Hexagonal
Z: 6
LATTICE CONSTANTS:
$a$ = 9.84
$c$ = 36.67
OPTICAL CONSTANTS:
$\omega$ = 1.515
$\epsilon$ = 1.417
(−)
HARDNESS: 2
DENSITY: 1.69 (Meas.)
1.69 (Calc.)
CLEAVAGE: None observed. Fracture irregular.
HABIT: As xenomorphic grains. Synthetic crystals rhombohedral or hexagonal prismatic, with twinning on {0001}.
COLOR-LUSTER: Greenish; vitreous.

Easily soluble in water.
MODE OF OCCURRENCE: Occurs in veinlets in coal in the Tyllakh coal deposit in the estuary of the Lena River, USSR.
BEST REF. IN ENGLISH: Knipovich, Yu. N., Komkov, A. I., and Nefedov, E. I., *Am. Min.*, **49**: 442-443 (1964).


## STEPHANITE
$Ag_5 SbS_4$

CRYSTAL SYSTEM: Orthorhombic
CLASS: mm2
SPACE GROUP: $Cmc2_1$
Z: 4
LATTICE CONSTANTS:
  $a = 7.70$
  $b = 12.32$
  $c = 8.48$
3 STRONGEST DIFFRACTION LINES:
  3.08 (100)
  2.58 ( 90)
  2.89 ( 60)
OPTICAL CONSTANTS: Opaque.
  HARDNESS: 2–2½
  DENSITY: 6.25 (Meas.)
              6.47 (Calc.)
  CLEAVAGE: {010} imperfect
              {021} imperfect
Fracture uneven to subconchoidal. Brittle.
  HABIT: Crystals usually short prismatic to tabular, often with oblique striations on prism faces. Also massive, compact or disseminated. Twinning on {110}, often repeated, pseudohexagonal; rarely on {010}, {100}, or {130}.
  COLOR-LUSTER: Iron black. Opaque. Metallic. Streak iron black.
  MODE OF OCCURRENCE: Occurs in silver vein deposits typically associated with native silver, other silver minerals, tetrahedrite, and sulfides. Found in the western United States especially at the Comstock Lode, Virginia City, Nevada, in California, and in Colorado. It also occurs in Ontario, Canada, and at mines in Mexico, Bolivia, Chile, England, Spain, Sardinia, Czechoslovakia, Hungary, Norway, and as fine crystals at Andreasberg and other German localities.
  BEST REF. IN ENGLISH: Palache, et al., "Dana's System of Mineralogy," 7th Ed., v. I, p. 358–361, New York, Wiley, 1944.

## STERCORITE
$Na(NH_4)H(PO_4) \cdot 4H_2O$

CRYSTAL SYSTEM: Triclinic (pseudomonoclinic)
LATTICE CONSTANTS:
  $a:b:c = 2.908:1:1.859$
      $\beta = 98°30'$
3 STRONGEST DIFFRACTION LINES:
  6.6  (100)
  2.89 (100)
  4.6  ( 50)
OPTICAL CONSTANTS:
  $\alpha = 1.439$
  $\beta = 1.442$
  $\gamma = 1.469$
  $(+)2V = 35°34'$
  HARDNESS: 2
  DENSITY: 1.554 (Meas.)
  CLEAVAGE: None.
  HABIT: As crystalline masses and nodules. Synthetic crystals short prismatic. Twinning on {010} repeated, simulating monoclinic symmetry.

COLOR-LUSTER: Colorless, white, yellowish, brownish. Transparent. Vitreous.

  Soluble in water.
  MODE OF OCCURRENCE: Occurs in guano deposits on the Guañape Islands, off the coast of Peru, and on Ichaboe Island, South West Africa.
  BEST REF. IN ENGLISH: Palache, et al., "Dana's System of Mineralogy," 7th Ed., v. II, p. 698–699, New York, Wiley, 1951.

## STERNBERGITE
$AgFe_2S_3$
Dimorphous with argentopyrite

CRYSTAL SYSTEM: Orthorhombic
CLASS: 2/m 2/m 2/m
SPACE GROUP: Ccmm
Z: 8
LATTICE CONSTANTS:
  $a = 6.61$
  $b = 11.64$
  $c = 12.67$
3 STRONGEST DIFFRACTION LINES:
  4.29 (100)
  2.79 ( 70)
  3.22 ( 50)
OPTICAL CONSTANTS: Opaque.
  HARDNESS: 1–1½
  DENSITY: 4.25 (Meas.)
              4.29 (Calc.)
  CLEAVAGE: {010} perfect, easy
Thin laminae flexible.
  HABIT: Crystals thin pseudohexagonal plates; as fan-shaped aggregates or rosettes. Crystals faintly striated on {001}. Twinning on {130} common.
  COLOR-LUSTER: Golden brown; occasionally tarnishes violet-blue. Opaque. Metallic.
  MODE OF OCCURRENCE: Occurs in the silver mines of the Georgetown district, Clear Creek County, Colorado; sparingly at Crestmore, Riverside County, California; with silver ores at Joachimstal and at Pribram, Czechoslovakia; and at Schneeberg and elsewhere in Germany.
  BEST REF. IN ENGLISH: Palache, et al., "Dana's System of Mineralogy," 7th Ed., v. I, p. 246–248, New York, Wiley, 1944.

## STERRETTITE = Kolbeckite

## STERRYITE
$Pb_{12}(Sb, As)_{10}S_{27}$

CRYSTAL SYSTEM: Orthorhombic
CLASS: 2/m 2/m 2/m
SPACE GROUP: Pbam
Z: 4
LATTICE CONSTANTS:
  $a = 28.48$
  $b = 42.62$
  $c = 4.090$

3 STRONGEST DIFFRACTION LINES:
  3.26 (100)
  3.68 ( 90)
  2.836 ( 70)
HARDNESS: Talmadge hardness B.
DENSITY: 6.00 (Predicted)
  5.91 (Calc.)
CLEAVAGE: Perfect cleavage parallel to elongation.
HABIT: Plumose; bundles of fibers elongated parallel to [001]. Observed in polished section as needle-like laths and anhedral grains.
COLOR-LUSTER: Black. Streak black.
MODE OF OCCURRENCE: Occurs associated with veenite in a small prospect pit in Precambrian marble at Madoc, Ontario, Canada.
BEST REF. IN ENGLISH: Jambor, J. L., *Can. Min.*, **9** (Pt. 2) 191–213 (1967).

## STETEFELDTITE (Stetefeldite)
$Ag_{2-y}Sb_{2-x}(O, OH, H_2O)_{6-7}$
Silver analogue of bindheimite

CRYSTAL SYSTEM: Cubic
CLASS: $4/m\,\bar{3}\,2/m$
SPACE GROUP: Fd3m
Z: 8
LATTICE CONSTANT:
  $a = 10.46$
3 STRONGEST DIFFRACTION LINES:
  3.02 (100)
  2.61 ( 70)
  1.85 ( 70)
OPTICAL CONSTANT:
  $N = 1.95$
HARDNESS: 3½–4½
DENSITY: 4.6 (Meas.)
  5.38 (Calc. for $Ag_1Sb_2O_7$)
CLEAVAGE: Not determined.
HABIT: Massive.
COLOR-LUSTER: Black to brown. Streak shining.
MODE OF OCCURRENCE: Occurs in association with chalcocite and pyrite at Belmont, Nye County, Nevada. A similar mineral is found at the Potosi mine, near Huancavelica, Peru.
BEST REF. IN ENGLISH: Mason, Brian, and Vitaliano, C. J., *Min. Mag.*, **30**: 100–112 (1953).

## STEVENSITE
$Mg_3Si_4O_{10}(OH)_2$
Montmorillonite (Smectite) Group

CRYSTAL SYSTEM: Monoclinic
Z: 1
LATTICE CONSTANTS:
  $a = b/\sqrt{3}$
  $b = 9.12$
  $c$ = variable
3 STRONGEST DIFFRACTION LINES:
  ~12. (100)
  4.54 (100)
  2.62 ( 90)

HARDNESS: 2.5
DENSITY: 2.15–2.565 (Meas.)
CLEAVAGE: {001} perfect
HABIT: Massive; fine-grained; typically pseudomorphous after pectolite.
COLOR-LUSTER: Pink, buff, amber. Translucent. Dull to somewhat resinous.
MODE OF OCCURRENCE: Occurs in trap rock at several localities in northern New Jersey, notably in the vicinity of Paterson and Springfield. Also found as a hydrothermal alteration product of dunite at Mine Creek, Bakersville, Mitchell County, North Carolina.
BEST REF. IN ENGLISH: Faust, George T., and Murata, K. J., *Am. Min.*, **38**: 973–987 (1953). Faust, George T., Hathaway, John C., and Millot, Georges, *Am. Min.*, **44**: 342–370 (1959).

## STEWARTITE
$MnFe_2(OH)_2(PO_4)_2 \cdot 8H_2O$

CRYSTAL SYSTEM: Triclinic
CLASS: $\bar{1}$ or 1
SPACE GROUP: $P\bar{1}$ or P1
Z: 2
LATTICE CONSTANTS:
  $a = 2 \times 5.23$    $\alpha = 90°35'$
  $b = 10.77$    $\beta = 109°58'$
  $c = 7.25$    $\gamma = 71°21'$
3 STRONGEST DIFFRACTION LINES:
  9.98 (100)
  3.93 ( 40)
  2.99 ( 40)
OPTICAL CONSTANTS:
  $\alpha = 1.612–1.63$
  $\beta = 1.653–1.658$
  $\gamma = 1.660–1.681$
(–)2V = large
HARDNESS: Not determined.
DENSITY: 2.94 (Meas.)
CLEAVAGE: One distinct.
HABIT: As minute crystals and tufts of fibers.
COLOR-LUSTER: Yellow to brownish yellow. Transparent to translucent. Vitreous to somewhat silky.
MODE OF OCCURRENCE: Occurs in granite pegmatite as minute yellowish crystals associated with rockbridgeite, metastrengite, and other secondary phosphate minerals in altered triphylite nodules at the Hesnard mine, Keystone, Pennington County, South Dakota; at the Stewart mine, Pala, San Diego County, California; at the Fletcher and Palermo pegmatites, North Groton, New Hampshire; and at Newry, Maine. It also is found at Alto Boqueiro, Rio Grande do Norte, Brazil, and at Chanteloube, Haute-Vienne, France.
BEST REF. IN ENGLISH: Palache, et al., "Dana's System of Mineralogy," 7th Ed., v. II, p. 730, New York, Wiley, 1951. Peacor, Donald R., Am. Min., **48**: 913–914 (1963).

## STIBICONITE
$Sb_3O_6(OH)$

CRYSTAL SYSTEM: Cubic
CLASS: 4/m $\bar{3}$ 2/m
SPACE GROUP: Fd3m
Z: 8
LATTICE CONSTANT:
  $a = 10.27$
3 STRONGEST DIFFRACTION LINES:
  2.96 (100)
  5.93 ( 90)
  1.81 ( 80)
OPTICAL CONSTANT:
  $N = 1.605-2.05$
  HARDNESS: 3–7 increasing with increase in density.
  DENSITY: 3.3–5.5 (Meas.)
  CLEAVAGE: Not determined.
  HABIT: Massive, fine-grained, compact. As crusts, botryoidal, pulverulent.
  COLOR-LUSTER: Chalky white to pale yellow; also orange, brown, gray, black; darker colors due to impurities. Transparent to translucent. Pearly to earthy; vitreous, opaline.
  MODE OF OCCURRENCE: Occurs as an alteration product of stibnite, commonly associated with valentinite. Found in Inyo, Kern, San Benito, San Bernardino, and San Luis Obispo counties, California; and at localities in Nevada, Idaho, Utah, Arizona, South Dakota, and Arkansas. It also occurs widespread in Mexico, especially as superb pseudomorphs after stibnite at Catorce, San Luis Potosi; and at deposits in Bolivia, Peru, England, Spain, Italy, Rumania, Algeria, Borneo, Australia, and China.
  BEST REF. IN ENGLISH: Palache, et al., "Dana's System of Mineralogy," 7th Ed., v. I, p. 597–598, New York, Wiley, 1944. Vitaliano, Charles J., and Mason, Brian, *Am. Min.*, **37**: 982–999 (1952).

## STIBIOCOLUMBITE
SbNbO$_4$
Stibiotantalite Series

CRYSTAL SYSTEM: Orthorhombic
CLASS: mm2
SPACE GROUP: Pc2$_1$n
Z: 4
LATTICE CONSTANTS:
  $a = 4.929$
  $b = 11.797$ (synthetic)
  $c = 5.559$
3 STRONGEST DIFFRACTION LINES:
  3.13  (100)
  2.947 ( 30)  (synthetic)
  1.735 ( 20)
OPTICAL CONSTANTS:
  $\alpha = 2.3977$
  $\beta = 2.4190$  (Na)
  $\gamma = 2.4588$
  $(+)2V = 73°25'$ (Calc.)
  HARDNESS: 5½
  DENSITY: 5.98 (Meas. for Ta:Nb = 1:4.8)
           5.68 (Calc. for SbNbO$_4$)
  CLEAVAGE: {010} distinct
            {100} indistinct
Fracture subconchoidal. Brittle.

HABIT: Crystals prismatic, sometimes flattened parallel to {100}. Prism faces vertically striated. Twinning on {010}, polysynthetic.
  COLOR-LUSTER: Brown, reddish brown, brownish yellow. Translucent; transparent in very thin fragments. Adamantine to resinous. Streak yellowish brown to light yellow.

  Pyroelectric.
  MODE OF OCCURRENCE: Occurs in pegmatite associated with pink cesium-rich beryl, pink tourmaline, and lepidolite, at Mesa Grande, San Diego County, California.
  BEST REF. IN ENGLISH: Palache, et al., "Dana's System of Mineralogy," 7th Ed., v. I, p. 767–769, New York, Wiley, 1944.

## STIBIOLUZONITE = Famatinite

## STIBIOPALLADINITE
Pd$_5$Sb$_2$

CRYSTAL SYSTEM: Orthorhombic (?)
CLASS: 2 2 2
SPACE GROUP: P22$_1$2
Z: 1
LATTICE CONSTANTS:
  $a = 12.80$
  $b = 15.04$
  $c = 11.36$
3 STRONGEST DIFFRACTION LINES:
  2.270 (100)
  2.194 (100)
  1.577 ( 50)
OPTICAL CONSTANTS: Anisotropism distinct in air and in oil.
  HARDNESS: 4–5
  DENSITY: 9.5 (Meas.)
  CLEAVAGE: None. Fracture uneven.
  HABIT: Massive; as minute grains.
  COLOR-LUSTER: Silvery white to steel gray. Opaque. Metallic.
  MODE OF OCCURRENCE: Occurs in platinum deposits in the vicinity of Potgietersrust, Transvaal, South Africa, and in the copper-nickel ores of the Noril'sk deposit, USSR.
  BEST REF. IN ENGLISH: Palache, et al., "Dana's System of Mineralogy," 7th Ed., v. I, p. 175, New York, Wiley, 1944. Desborough, George A., Finney, J. J., and Leonard, B. F., *Am. Min.*, **58**: 1–10 (1973).

## STIBIOTANTALITE
Sb(Ta, Nb)O$_4$
Stibiotantalite Series

CRYSTAL SYSTEM: Orthorhombic
CLASS: mm2
SPACE GROUP: Pc2$_1$n
Z: 4
LATTICE CONSTANTS:
  $a = 4.90$
  $b = 11.79$
  $c = 5.57$

3 STRONGEST DIFFRACTION LINES:
3.13 (100)
3.55 ( 60)
2.96 ( 60)
OPTICAL CONSTANTS:
$\alpha$ = 2.3742
$\beta$ = 2.4039
$\gamma$ = 2.4568
(+)2V = 75°5′ (Calc.)
HARDNESS: 5–5½
DENSITY: 7.345 (Meas. for Ta:Nb = 19:1)
7.53 (Calc. SbTaO$_4$)
CLEAVAGE: {010} distinct
{100} indistinct
Fracture subconchoidal. Brittle.

HABIT: Crystals prismatic [001]; thin tabular, to thick nearly equant. {010} and {110} striated parallel [001]. Twinning on {010}, polysynthetic.

COLOR-LUSTER: Dark brown to light yellowish brown; also yellowish steel gray, reddish yellow, reddish brown, greenish yellow. Transparent to translucent. Vitreous to resinous.

MODE OF OCCURRENCE: Occurs as excellent tabular crystals associated with pink tourmaline, lepidolite, pink beryl, quartz, and, rarely, cassiterite in pegmatite at Mesa Grande and Pala, San Diego County, California; in pegmatite at Topsham, Maine, and as pure crystal fragments up to 5 X 3½ X 2½ cm in size at the Brown Derby No. 1 pegmatite, Gunnison County, Colorado; also in pegmatite at Varutrask, Sweden, and in the northern part of Kola Peninsula, USSR. Found as brilliantly faced crystals weighing as much as 10 pounds at the Muiane pegmatite, Ribaue-Alto Ligonha pegmatite district, Mozambique.

BEST REF. IN ENGLISH: Palache, et al., "Dana's System of Mineralogy," 7th Ed., v. I, p. 767–769, New York, Wiley, 1944.

# STIBNITE
Sb$_2$S$_3$

CRYSTAL SYSTEM: Orthorhombic
CLASS: 2/m 2/m 2/m
SPACE GROUP: Pbnm
Z: 4
LATTICE CONSTANTS:
$a$ = 11.20
$b$ = 11.28
$c$ = 3.83
3 STRONGEST DIFFRACTION LINES:
2.764 (100)
3.053 ( 95)
3.556 ( 70)
OPTICAL CONSTANTS:
$\alpha$ = 3.194
$\beta$ = 4.046
$\gamma$ = 4.303
(–)2V = 25°45′ (Calc.)
HARDNESS: 2
DENSITY: 4.63–4.66 (Meas.)
4.63 (Calc.)

CLEAVAGE: {010} perfect, easy
{100} imperfect
{110} imperfect
Very flexible; not elastic. Fracture uneven to subconchoidal.

HABIT: Crystals usually slender prismatic, often vertically striated, bent or twisted. As radiated groups or jumbled aggregates of acicular crystals; also as bladed, columnar, granular, or compact masses. Twinning on {130} or {120}, rare.

COLOR-LUSTER: Pale to dark lead gray; sometimes tarnished iridescent, bluish, or blackish. Opaque. Metallic, often brilliant. Streak gray to dark gray.

MODE OF OCCURRENCE: Occurs chiefly in low-temperature hydrothermal vein or replacement deposits, commonly associated with realgar, orpiment, galena, barite, cinnabar, pyrite, sphalerite, lead sulfantimonides, calcite, and gold. Found widespread in small amounts in the western United States, especially in the Manhattan district, Nevada; the Coeur d'Alene district, Idaho; and at numerous localities in California, Oregon, Washington, Alaska, Montana, South Dakota, Colorado, and Sevier County, Arkansas. Superb crystals and crystal groups occur at Ichinokawa, Shikoku, Japan; at Bau, Sarawak, Borneo; at Felsobanya, Roumania; and at Kremnitz and Schemnitz, Czechoslovakia. Significant deposits also occur in Canada, Mexico, Peru, Germany, Italy, France, Algeria, and China.

BEST REF. IN ENGLISH: Palache, et al., "Dana's System of Mineralogy," 7th Ed., v. I, p. 270–275, New York, Wiley, 1944.

# STICHTITE
Mg$_6$Cr$_2$CO$_3$(OH)$_{16}$ · 4H$_2$O
Dimorphous with barbertonite

CRYSTAL SYSTEM: Hexagonal
SPACE GROUP: R lattice
Z: 3
LATTICE CONSTANTS:
$a$ = 6.18
$c$ = 46.38
3 STRONGEST DIFFRACTION LINES:
7.8 (100)
3.91 ( 90)
2.60 ( 40)
OPTICAL CONSTANTS:
$\omega$ = 1.545
$\epsilon$ = 1.518
(–)
HARDNESS: 1½–2
DENSITY: 2.16 (Meas.)
2.11 (Calc.)
CLEAVAGE: {0001} perfect
Laminae flexible, inelastic. Greasy feel.

HABIT: Massive, foliated, lamellar, fibrous; in micaceous scales.

COLOR-LUSTER: Bright lilac to rose-pink. Translucent; thin flakes transparent. Pearly, waxy, greasy. Streak white to pale lilac.

MODE OF OCCURRENCE: Occurs in serpentine rock, commonly associated with chromite. Found at the Megantic mine, Black Lake, Quebec, Canada; at Dundas, Tasmania;

and at Kaapsche Hoop, Barberton district, Transvaal, South Africa.

BEST REF. IN ENGLISH: Palache, et al., "Dana's System of Mineralogy," 7th Ed., v. I, p. 655–656, New York, Wiley, 1944.

## STILBITE (Desmine)
$NaCa_2Al_5Si_{13}O_{36} \cdot 16H_2O$
Zeolite Group

CRYSTAL SYSTEM: Monoclinic
CLASS: 2/m
SPACE GROUP: C2/m
Z: 2
LATTICE CONSTANTS:
  $a = 13.63$
  $b = 18.17$
  $c = 11.31$
  $\beta = 129°10'$
3 STRONGEST DIFFRACTION LINES:
  9.04 (100)
  4.07 ( 95)
  3.04 ( 70)
OPTICAL CONSTANTS:
  $\alpha = 1.484-1.500$
  $\beta = 1.492-1.507$
  $\gamma = 1.494-1.513$
  $(-)2V = 30°-49°$
HARDNESS: 3½–4
DENSITY: 2.09–2.20 (Meas.)
        2.186 (Calc.)
CLEAVAGE: {010} perfect
          {100} indistinct
Fracture uneven. Brittle.
HABIT: Crystals usually cruciform penetration twins simulating rhombic forms. Twinned crystals often in sub-parallel position forming sheaf-like aggregates; rarely as isolated individuals. Also divergent or radiated, massive bladed, or globular. Twinning on {001}.
COLOR-LUSTER: White, gray, yellowish, pink, reddish, orange, light to dark brown. Transparent to translucent. Vitreous; pearly on cleavage. Streak uncolored.
MODE OF OCCURRENCE: Occurs chiefly in amygdules and cavities in basalt, andesite, and related volcanic rocks; also as a hydrothermal mineral in crevices in metamorphic rocks, in cavities in granitic pegmatites, and in certain hot springs deposits. Notable occurrences are found in Cowlitz County, Washington; at many localities in Oregon; in San Diego County and elsewhere in California; in Montgomery County, Pennsylvania; in the trap rock of northern New Jersey, especially at Paterson and Great Notch; and at localities in Nova Scotia, Mexico, Brazil, Iceland, Faroe Islands, Scotland, Norway, Switzerland, Italy, Germany, Austria, Czechoslovakia, Poland, Australia, and the Deccan trap area of India.
BEST REF. IN ENGLISH: Deer, Howie, and Zussman, "Rock Forming Minerals," v. 4, p. 377–385, New York, Wiley, 1963. *Am. Min.*, **55**: 387–397 (1970).

## STILLEITE
ZnSe

CRYSTAL SYSTEM: Cubic
CLASS: $\bar{4}3m$
SPACE GROUP: $F\bar{4}3m$
Z: 4
LATTICE CONSTANT:
  $a = 5.67$
3 STRONGEST DIFFRACTION LINES:
  3.273 (100)
  2.003 ( 70)
  1.707 ( 44)
HARDNESS: ~5
DENSITY: 5.29 (Calc.)
CLEAVAGE: Not determined.
HABIT: Massive; observed only in thin sections.
COLOR-LUSTER: Gray. Metallic. Opaque.
MODE OF OCCURRENCE: Occurs associated with pyrite, linnaeite, clausthalite, two unidentified minerals and dolomite in an ore sample from Shinkolobwe, Katanga, Zaire.
BEST REF. IN ENGLISH: Ramdohr, Paul, *Am. Min.*, **42**: 584 (1957).

## STILLWELLITE
(Ce, La, Ca)BSiO$_5$

CRYSTAL SYSTEM: Hexagonal
CLASS: $\bar{3}2/m$
SPACE GROUP: $P\bar{3}m$ (P3$_1$12?)
Z: 3
LATTICE CONSTANTS:
  $a = 6.907$
  $c = 6.768$
3 STRONGEST DIFFRACTIONS LINES:
  (Norway)
  2.943 (100)
  4.48 ( 90)
  3.45 ( 90)
OPTICAL CONSTANTS:
  $\omega = 1.765$
  $\epsilon = 1.780$
    (+)
HARDNESS: Not determined.
DENSITY: 4.612 (Meas.)
CLEAVAGE: Not determined.
HABIT: Crystals rhombohedral; up to 5.0 mm in diameter. Usually massive, compact.
COLOR-LUSTER: Lavender-gray or brownish yellow. Translucent. Resinous. Colorless in thin section.
MODE OF OCCURRENCE: Occurs associated with much allanite and with garnet and uraninite as a metasomatic replacement of metamorphosed calcareous sediments at the Mary Kathleen Lease, 34 miles east of Mt. Isa, Queensland, Australia. Also found in the Langesundfjord district, Norway.
BEST REF. IN ENGLISH: McAndrew, J., and Scott, T. R., *Nature*, **176** (4480): 509–510 (1955).

## STILPNOMELANE
$K(Fe, Mg, Al)_3Si_4O_{10}(OH)_2 \cdot nH_2O$

CRYSTAL SYSTEM: Triclinic
Z: 12
LATTICE CONSTANTS:
  $a = 21.72$
  $b = 21.72$
  $c = 17.74$
  $\beta = 95.86°$
3 STRONGEST DIFFRACTION LINES:
  12.3 (100)
  4.16 (100)
  2.55 (100)
OPTICAL CONSTANTS:
  $\alpha = 1.543-1.634$
  $\beta = \gamma = 1.576-1.745$
  $(-)2V \simeq 0°$
HARDNESS: 3-4
DENSITY: 2.59-2.96 (Meas.)
CLEAVAGE: {001} perfect
          {010} imperfect
Brittle.
HABIT: As thin foliated plates; also fibrous or as velvety coatings.
COLOR-LUSTER: Black, deep reddish brown, golden brown, dark green. Translucent to nearly opaque. Pearly to subvitreous, sometimes bronze-toned submetallic.
MODE OF OCCURRENCE: Occurs mainly as a widespread and abundant constituent of schists in association with chlorite, epidote, lawsonite, and other minerals; also in iron ore deposits, commonly associated with magnetite, goethite, hematite, and pyrite. Typical occurrences are found near Longvale, Mendocino County, and elsewhere in California; in the Animikian iron formation, Minnesota; at Crystal Falls, Michigan; at the Sterling mine, Antwerp, New York; and near Rockyhill, Somerset County, New Jersey. Also in Canada, Greenland, Sweden, Switzerland, Germany, Czechoslovakia, Poland, Finland, USSR, Japan, and New Zealand.
BEST REF. IN ENGLISH: Deer, Howie, and Zussman, "Rock Forming Minerals," v. 3, p. 103-114, New York, Wiley, 1962.

# STISHOVITE

$SiO_2$
Polymorphous with quartz, tridymite, cristobalite, coesite

CRYSTAL SYSTEM: Tetragonal
CLASS: 4/m 2/m 2/m
SPACE GROUP: P4/mnm
Z: 2
LATTICE CONSTANTS:
  $a = 4.1790$
  $c = 2.6649$
3 STRONGEST DIFFRACTION LINES:
  2.959 (100)
  1.530 ( 50)
  1.981 ( 35)
OPTICAL CONSTANTS:
  $\omega = 1.799$
  $\epsilon = 1.826$ (synthetic)
    (+)

HARDNESS: Microhardness 2080 kg/mm$^2$ parallel to elongation; 1700 kg/mm$^2$ perpendicular to elongation.
DENSITY: 4.35 (Meas. synthetic)
         4.28 (Calc.)
CLEAVAGE: Not determined.
HABIT: Massive, fine granular, usually of submicron size.
COLOR-LUSTER: Colorless. Transparent. Vitreous.
MODE OF OCCURRENCE: Occurs in association with coesite and silica glass in sandstone at Meteor Crater, Arizona.
BEST REF. IN ENGLISH: Chao, E. C. T., Fahey, J. J., Littler, Janet, and Milton, D. J., *J. Geophys. Res.*, **67**: 419-421 (1962). Frondel, C., "Dana's System of Mineralogy," 7th Ed., v. III, p. 317-318, New York, Wiley, 1962.

# STISTAITE

SnSb

CRYSTAL SYSTEM: Cubic
LATTICE CONSTANT:
  $a = 4.15$
3 STRONGEST DIFFRACTION LINES:
  3.09 ((100)
  2.19 (100)
  1.374 ( 80)
HARDNESS: Microhardness 103-127, av. 115 kg/mm$^2$.
DENSITY: Not determined.
CLEAVAGE: None. Malleable.
HABIT: Crystals cubic, 0.02-0.15 mm in size.
COLOR-LUSTER: Light gray. Opaque. Metallic.
MODE OF OCCURRENCE: Occurs in placer samples collected from a region of Silurian shaly-sandy sediments near the right effluent of the Elkiaidan River, eastern North Nuratin Range, Uzbekistan, in association with unnamed Cu (Sn, Sb).
BEST REF. IN ENGLISH: Nikolaeva, E. P., Grigorenko, V. A., Galarkina, S. D., and Tsypinka, P. E., *Am. Min.*, **56**: 358 (1971).

# STOKESITE

$CaSnSi_3O_9 \cdot 2H_2O$

CRYSTAL SYSTEM: Orthorhombic
CLASS: 2/m 2/m 2/m
SPACE GROUP: Pnna
Z: 4
LATTICE CONSTANTS:
  $a = 14.41$
  $b = 11.61$
  $c = 5.23$
3 STRONGEST DIFFRACTION LINES:
  3.99 (100 )
  2.89 (100b)
  7.25 ( 80 )
OPTICAL CONSTANTS:
  $\alpha = 1.609$
  $\beta = 1.6125$
  $\gamma = 1.619$
  $(+)2V = 69.5°$
HARDNESS: 6

DENSITY: 3.185 (Meas.)
3.211 (Calc.)
CLEAVAGE: {110} perfect
{010} poor
HABIT: Crystals pyramidal with large {010} faces.
COLOR-LUSTER: Colorless.
MODE OF OCCURRENCE: Occurs in association with axinite at Roscommon Cliff, St. Just, Cornwall, England.
BEST REF. IN ENGLISH: Winchell and Winchell, "Elements of Optical Mineralogy," 4th Ed., pt. 2, p. 454, New York, Wiley, 1951. *Min. Mag.*, **32**: 433 (1960).


## STOLZITE
PbWO$_4$
Dimorphous with raspite

CRYSTAL SYSTEM: Tetragonal
CLASS: 4/m
SPACE GROUP: I4$_1$/a
Z: 4
LATTICE CONSTANTS:
$a$ = 5.46
$c$ = 12.05
3 STRONGEST DIFFRACTION LINES:
3.25 (100)
2.024 ( 35)
1.660 ( 35)
OPTICAL CONSTANTS:
$\omega$ = 2.27
$\epsilon$ = 2.19
(−)
HARDNESS: 2½–3
DENSITY: 7.9–8.34 (Meas.)
8.40 (Calc.)
CLEAVAGE: {001} imperfect
{011} indistinct
Fracture conchoidal to uneven. Brittle.
HABIT: Crystals usually dipyramidal or thick tabular; rarely prismatic. Prism faces horizontally striated; pyramidal faces also commonly striated.
COLOR-LUSTER: Dull yellowish, yellowish gray, tan to brown or reddish brown, yellowish red to red, green. Transparent to translucent. Subadamantine, resinous. Streak uncolored.
MODE OF OCCURRENCE: Occurs as a secondary mineral in the oxidation zone of tungsten-bearing ore deposits. Found in the Grouse Creek Mountains, Box Elder County, at the Emma mine, Salt Lake County, and in the Clifton district, Tooele County, Utah; at several places in Arizona; at the Wheatley mines, Chester County, Pennsylvania; and at the Manhan lead mine, Southampton, Massachusetts. It also occurs in Brazil, England, Sardinia, Germany, Nigeria, and as superb crystals up to one inch in size at the Proprietary mine, Broken Hill, New South Wales, Australia.
BEST REF. IN ENGLISH: Palache, et al., "Dana's System of Mineralogy," 7th Ed., v. II, p. 1087–1089, New York, Wiley, 1951.


## STOTTITE
Fe$^{2+}$Ge(OH)$_6$

CRYSTAL SYSTEM: Tetragonal
CLASS: 4/m
SPACE GROUP: P4$_2$/n
Z: 4
LATTICE CONSTANTS:
$a$ = 7.55
$c$ = 7.47
3 STRONGEST DIFFRACTION LINES:
3.767 (100)
2.655 ( 80)
1.687 ( 70)
OPTICAL CONSTANTS:
$\omega$ = 1.738
$\epsilon$ = 1.728
(−)
HARDNESS: 4½
DENSITY: 3.596 (Meas.)
CLEAVAGE: {100} good
{010} good
{001} fair
HABIT: Pseudo-octahedral crystals up to 1 cm in size. Forms: p{111} dominant, and small faces of a{100}, c{001}, m{110}, d{011}, and e{012}.
COLOR-LUSTER: Brown; greasy. Streak gray-white.
MODE OF OCCURRENCE: Occurs at a depth of about 1000 meters in Level 30 at the Tsumeb Mine, South West Africa.
BEST REF. IN ENGLISH: Strunz, H., Söhnge, G., and Geier, B. H., *Am. Min.*, **43**: 1006 (1958).


## STRANSKIITE
Zn$_2$Cu(AsO$_4$)$_2$

CRYSTAL SYSTEM: Triclinic
CLASS: $\bar{1}$
SPACE GROUP: P$\bar{1}$
Z: 2
LATTICE CONSTANTS:
$a$ = 5.07    $\alpha$ = 111°
$b$ = 6.77    $\beta$ = 113.5°
$c$ = 5.28    $\gamma$ = 86°
OPTICAL CONSTANTS:
$\alpha$ = 1.795
$\beta$ = 1.842
$\gamma$ = 1.874
(−)2V = 80°
HARDNESS: 4
DENSITY: 5.23
CLEAVAGE: {010} perfect
{100} good
also {001}
{$\bar{1}$01}
HABIT: As radiating aggregates.
COLOR-LUSTER: Cyan blue.
MODE OF OCCURRENCE: Occurs on chalcocite at the 1000 meter level of the Tsumeb mine, South West Africa.
BEST REF. IN ENGLISH: Strunz, H., *Am. Min.*, **45**: 1315 (1960).


## STRASHIMIRITE
Cu$_8$(AsO$_4$)$_4$(OH)$_4$ · 5H$_2$O

CRYSTAL SYSTEM: Monoclinic
CLASS: 2/m, 2, or m
SPACE GROUP: P2/m, P2, or Pm
Z: 6
LATTICE CONSTANTS:
 $a = 9.71$
 $b = 18.85$
 $c = 8.94$
 $\beta = 97°12'$
3 STRONGEST DIFFRACTION LINES:
 18.74 (100)
 2.86 (100)
 8.97 ( 90)
OPTICAL CONSTANTS:
 $\alpha = 1.726$
 $\beta = ?$
 $\gamma = 1.747$ (Na)
 $(-)2V = 70°$
HARDNESS: Not determined.
DENSITY: 3.81 (Calc.)
CLEAVAGE: Not determined.
HABIT: As platy to fibrous aggregates in spherulites up to 0.5 mm in size.
COLOR-LUSTER: White to pale green; pearly to greasy.
MODE OF OCCURRENCE: Occurs replacing tyrolite and cornwallite associated with azurite, olivenite, malachite, and euchroite in the zone of oxidation of the Zapachitsa copper deposit, western Stara-Planina, Bulgaria.
BEST REF. IN ENGLISH: Mincheva-Stefanova, I., *Am. Min.*, **54**: 1221 (1969).


# STRENGITE

$FePO_4 \cdot 2H_2O$
Variscite-Strengite Series
Dimorphous with metastrengite

CRYSTAL SYSTEM: Orthorhombic
CLASS: 2/m 2/m 2/m
SPACE GROUP: Pcab
Z: 8
LATTICE CONSTANTS:
 $a = 10.05$
 $b = 9.92$
 $c = 8.74$
3 STRONGEST DIFFRACTION LINES:
 4.38 (100)
 5.50 ( 80)
 3.11 ( 80)
OPTICAL CONSTANTS:
 $\alpha = 1.707$
 $\beta = 1.719$
 $\gamma = 1.741$
 (+)2V = moderate, sometimes small
HARDNESS: 3½–4½
DENSITY: 2.87 (Meas.)
 2.848 (Calc.) X-ray
CLEAVAGE: {010} good
 {001} poor
Fracture conchoidal. Brittle.
HABIT: Crystals octahedral, thin to thick tabular, stout prismatic. As small spherical aggregates with radial-fibrous structure and drusy surface, and as crusts. Twinning on {201} not common.
COLOR-LUSTER: Colorless, pale to deep violet, red, carmine. Transparent to translucent. Vitreous. Streak white.
MODE OF OCCURRENCE: Occurs as a secondary mineral formed by the alteration of iron-rich phosphate minerals. Found as superb crystals up to 4 mm in size associated with metastrengite in vugs in barbosalite at the Bull Moose mine, Custer County, South Dakota; as spherical aggregates and fine crystals with cacoxenite on limonite at Indian Mountain, Cherokee County, Alabama; in pegmatites in the Pala district, San Diego County, California; and at localities in Nevada, Virginia, Pennsylvania, and New Hampshire. It also occurs at several deposits in Germany, especially at Hagendorf, Bavaria; in Sweden; and in altered triphylite at Mangualde, Portugal.
BEST REF. IN ENGLISH: Palache, et al., "Dana's System of Mineralogy," 7th Ed., v. II, p. 756–761, New York, Wiley, 1951.


# STROMEYERITE

AgCuS

CRYSTAL SYSTEM: Orthorhombic
CLASS: 2/m 2/m 2/m
SPACE GROUP: Bbmm
Z: 4
LATTICE CONSTANTS:
 $a = 6.66$
 $b = 7.99$
 $c = 4.06$
3 STRONGEST DIFFRACTION LINES:
 2.61 (100)
 3.33 ( 80)
 2.07 ( 80)
HARDNESS: 2½–3
DENSITY: 6.2–6.3 (Meas.)
 6.26 (Calc.)
CLEAVAGE: None. Fracture conchoidal. Brittle.
HABIT: Crystals pseudohexagonal prisms formed by twinning on {110}. Usually massive, compact.
COLOR-LUSTER: Dark steel gray; rapidly tarnishes dark blue. Opaque. Metallic. Streak steel gray.
MODE OF OCCURRENCE: Occurs widespread in Colorado, often as the chief silver ore mineral, at deposits in Boulder, Clear Creek, Custer, Eagle, Park, Pitkin, Saguache, San Juan, and Summit counties; and at localities in California, Arizona, and Montana. It also is found in Ontario and British Columbia, Canada; in Mexico, Chile, Peru, Germany, USSR, and Tasmania.
BEST REF. IN ENGLISH: Palache, et al., "Dana's System of Mineralogy," 7th Ed., v. I, p. 190–191, New York, Wiley, 1944.


# STRONTIANITE

$SrCO_3$
Aragonite Group

CRYSTAL SYSTEM: Orthorhombic

CLASS: 2/m 2/m 2/m
SPACE GROUP: Pmcn
Z: 4
LATTICE CONSTANTS:
$a = 5.118$ kX
$b = 8.404$
$c = 6.082$
3 STRONGEST DIFFRACTION LINES:
3.535 (100)
3.450 ( 70)
2.053 ( 50)
OPTICAL CONSTANTS:
$\alpha = 1.5199$
$\beta = 1.6666$ (Na)
$\gamma = 1.6685$
$(-)2V = 7°07'$(Calc.)
HARDNESS: 3½
DENSITY: 3.785 (Meas.)
CLEAVAGE: {110} nearly perfect
{021} poor
{010} in traces
Fracture subconchoidal to uneven. Brittle.
HABIT: Crystals short to long prismatic, elongated along c-axis; often acicular or spear-shaped. Commonly massive, bladed to fibrous; granular; concretionary. Twinning very common, often repeated.
COLOR-LUSTER: Colorless, white, gray, yellowish, yellowish brown, greenish, reddish. Transparent to translucent. Vitreous to resinous.
MODE OF OCCURRENCE: Occurs as a low-temperature mineral as veins and geodes in limestones and marls, and as a near-surface gangue mineral in sulfide vein deposits. In the United States it is found as large deposits in limestone in the Strontium Hills, San Bernardino County, California; as small crystals at Woodville, Sandusky County, Ohio, and in the vicinity of Schoharie, New York; and in Pennsylvania, New Mexico, Texas, Louisiana, South Dakota, and Washington. In Canada it is found in Carleton County, Ontario, and in the Caribou district, British Columbia. Excellent crystals occur at several places in the Tyrol, Austria, and in Westphalia, Germany. It is also found at Strontian, Argyllshire, Scotland; in Mexico; and in India.
BEST REF. IN ENGLISH: Palache, et al., "Dana's System of Mineralogy," 7th Ed., v. II, p. 196–200, New York, Wiley, 1951.

## STRONTIOBORITE
$SrB_8O_{13} \cdot 2H_2O$?

CRYSTAL SYSTEM: Monoclinic
CLASS: 2
SPACE GROUP: $P2_1$
LATTICE CONSTANTS:
$a = 9.83$
$b = 8.22$
$c = 7.55$
$\beta = 107°46'$
3 STRONGEST DIFFRACTION LINES:
7.20 (100)
4.09 ( 90)
3.29 ( 80)

OPTICAL CONSTANTS:
$\alpha = 1.470$
$\beta = 1.510$
$\gamma = 1.579$
$2V \sim 85°$ (+), sometimes (−)
HARDNESS: Not determined.
DENSITY: 2.806
CLEAVAGE: Not determined. Very brittle.
HABIT: Small plates, mostly 0.10–0.15 mm, but up to 2.0 mm in size.
COLOR-LUSTER: Colorless. Transparent.
MODE OF OCCURRENCE: Occurs in rock salt with fine-lamellar structure in the Kungur strata of the near-Caspian region.
BEST REF. IN ENGLISH: Lobanova, V. V., *Am. Min.*, **46**: 768 (1961). Kondrat'eva, V. V., *Am. Min.*, **50**: 1508 (1965).

## STRONTIOGINORITE (Volkolvite)
$(Sr, Ca)_2 B_{14}O_{23} \cdot 8H_2O$

CRYSTAL SYSTEM: Monoclinic
CLASS: 2/m
SPACE GROUP: $P2_1/a$
Z: 4
LATTICE CONSTANTS:
$a = 12.85$
$b = 14.48$
$c = 12.84$
$\beta = 101°35'$
3 STRONGEST DIFFRACTION LINES:
7.25 (100)
2.10 ( 80)
5.40 ( 60)
OPTICAL CONSTANTS:
$\alpha = 1.512$
$\beta = 1.524$
$\gamma = 1.577$ (Na)
$(+)2V = 53°$
HARDNESS: 2–3
DENSITY: 2.25 (Meas.)
CLEAVAGE: {010} perfect, easy
{001} very good
HABIT: Tabular crystals up to 3.0 × 1.0 × 0.5 mm in size. Dominant forms: {110}, {010}, and {111}; also {001}, {100}, {250}, {131}, {141}.
COLOR-LUSTER: Colorless, transparent; greasy.
MODE OF OCCURRENCE: Found in the water-insoluble residues of the "Old Halite" bed from the Königshall-Hindenburg potassium mine at Reyershausen, northern Germany.
BEST REF. IN ENGLISH: Braitsch, Otto, *Am. Min.*, **45**: 478 (1960).

## STRONTIOHILGARDITE
$(Sr, Ca)_2 Ba_5 O_8 (OH)_2 Cl$

CRYSTAL SYSTEM: Triclinic
CLASS: 1
SPACE GROUP: P1

Z: 1
LATTICE CONSTANTS:
$a = 6.38$   $\alpha = 75.4°$
$b = 6.480$   $\beta = 61.2°$
$c = 6.608$   $\gamma = 60.5°$
3 STRONGEST DIFFRACTION LINES:
2.89 (100)
2.82 (100)
2.775 (100)
OPTICAL CONSTANTS:
$\alpha = 1.638$
$\beta = 1.639$ (Calc.)($\lambda = 500$ nm)
$\gamma = 1.670$
(+)2V $\approx 19°$
HARDNESS: 5–7
DENSITY: 2.993 (Meas.)
CLEAVAGE: {001} good
{211} distinct
HABIT: Crystals up to 2.0 × 1.0 × 0.5 mm in size. Dominant forms: {010}, and {0$\bar{1}$0}; also {100}, {$\bar{1}$00}, {001}, {00$\bar{1}$}, {$\bar{1}$01}, {0$\bar{1}\bar{1}$}, {120}, and {111}.
COLOR-LUSTER: Colorless to pale yellow; vitreous to adamantine.
MODE OF OCCURRENCE: Found in insoluble residues from "Hartsalz" of the 820 meter level of the Königshall-Hindenburg IIK mine, Reyershausen, Germany.
BEST REF. IN ENGLISH: Braitsch, Otto, *Am. Min.*, **44**: 1102–1103 (1959).   Davies, W. O., and Machin, M. P., *Am. Min.*, **53**: 2084–2087 (1968).

## STRONTIUM-APATITE

$(Sr, Ca)_5 (PO_4)_3 (OH, F)$
Apatite Group

CRYSTAL SYSTEM: Hexagonal
CLASS: 6/m
SPACE GROUP: $P6_3/m$
Z: 2
LATTICE CONSTANTS:
$a = 9.66$ kX
$c = 7.19$
3 STRONGEST DIFFRACTION LINES:
2.89 (100)
3.167 ( 70)
2.78 ( 70)
OPTICAL CONSTANTS:
$\omega = 1.651$
$\epsilon = 1.637$
(–)
HARDNESS: 5
DENSITY: 3.84 (Meas.)
HABIT: Poorly formed oval crystals 0.2 × 1.0 cm in longitudinal section and as small crystals with corroded faces. Forms: prism {10$\bar{1}$0} and dypyramid {10$\bar{1}$1}.
COLOR-LUSTER: Pale green to yellowish green, colorless and transparent in small crystals; vitreous, on fracture greasy.
MODE OF OCCURRENCE: Occurs associated with batisite, innelite, ramsayite and rare eudialyte in sugary albite filling interstices between crystals of aegirine and eckermannite in veins in alkalic pegmatites that cut the dunite core of the concentrically zoned Inagil massif of ultrabasic-alkalic rocks, Southern Yakutia.
BEST REF. IN ENGLISH: Efimov, A. S., Kravchenko, S. M., and Vasil'eva, Z. V., *Am. Min.*, **47**: 808 (1962).

## STRUNZITE

$MnFe_2 (PO_4)_2 (OH)_2 \cdot 8H_2O$
Dimorphous with laueite

CRYSTAL SYSTEM: Monoclinic
CLASS: 2/m
SPACE GROUP: C2/c
Z: 4
LATTICE CONSTANTS:
$a = 9.80$
$b = 18.06$
$c = 7.34$
$\beta = 100°10'$
3 STRONGEST DIFFRACTION LINES:
9.02 (100)
5.32 ( 80)
4.35 ( 60)
OPTICAL CONSTANTS:
$\alpha = 1.619$
$\beta = 1.670$
$\gamma = 1.720$
(–)2V = moderate
HARDNESS: Not determined.
DENSITY: 2.47–2.56 (Meas.)
CLEAVAGE: Not determined.
HABIT: As divergent tufts and felted coatings of tiny hair-like or lath-like crystals. Laths flattened on (010). Twinning on (100).
COLOR-LUSTER: Straw yellow to brownish yellow.
MODE OF OCCURRENCE: Occurs as a near-surface weathering product in pegmatites that contain triphylite at Hagendorf and other localities in Bavaria; Palermo and Fletcher, New Hampshire; Norway, Newry, Rumford, and Unity, Maine; at numerous localities in the Keystone and Custer districts, South Dakota; and as films in weathered outcrops of phosphate rock of the Phosphoria formation at Rasmussen Valley, Idaho.
BEST REF. IN ENGLISH: Frondel, Clifford, *Am. Min.*, **43**: 793–794 (1958).

## STRÜVERITE = Tantalian rutile
$(Ti, Ta, Fe)_3O_6$

## STRUVITE
$NH_4MgPO_4 \cdot 6H_2O$

CRYSTAL SYSTEM: Orthorhombic
CLASS: mm2
SPACE GROUP: $Pm2_1n$
Z: 2
LATTICE CONSTANTS:
$a = 6.945$
$b = 11.208$ (synthetic)
$c = 6.1355$

3 STRONGEST DIFFRACTION LINES:
4.257 (100)
5.601 ( 60) (synthetic)
2.919 ( 55)
OPTICAL CONSTANTS:
$\alpha = 1.495$
$\beta = 1.496$
$\gamma = 1.504$ (Na)
(+)2V = 37°22'
HARDNESS: 2
DENSITY: 1.711 (Meas.)
1.706 (Calc.)
CLEAVAGE: {001} good
{100} poor
Fracture uneven to subconchoidal. Brittle.

HABIT: Crystals widely varied in habit; short prismatic, thick tabular, equant, coffin-shaped, wedge-shaped; distinctly hemimorphic. Crystals range up to 2.0 cm or more in size. Twinning on {001} common.

COLOR-LUSTER: Colorless, becoming white on dehydration. Transparent to translucent. Vitreous; dehydrated material dull.

MODE OF OCCURRENCE: Occurs in bat guano associated with newberyite and hannayite in the Skipton Caves, Ballarat, Victoria, Australia; at Saldanha Bay, Cape Province, South Africa; on La Reunion Island, Indian Ocean; and, among other localities, in Denmark and in Germany.

BEST REF. IN ENGLISH: Palache, et al., "Dana's System of Mineralogy," 7th Ed., v. II, p. 715–717, New York, Wiley, 1951.

# STUDTITE
$UO_4 \cdot 4H_2O$

CRYSTAL SYSTEM: Monoclinic
CLASS: 2, m, or 2/m
SPACE GROUP: C2, Cm, or C2/m
Z: 2
LATTICE CONSTANTS:
$a = 11.85$
$b = 6.80$
$c = 4.25$
$\beta = 93°51'$
3 STRONGEST DIFFRACTION LINES:
5.93 (100)
3.40 ( 80)
2.96 ( 60)
OPTICAL CONSTANTS:
$\alpha = 1.545$
$\beta = 1.555$
$\gamma = 1.68$
(+)2V = large
HARDNESS: Not determined.
DENSITY: 3.64 (Calc.)
CLEAVAGE: Fibers flexible.

HABIT: Crystals acicular, up to 4.0 mm long and 0.01 mm in diameter; as radial aggregates and thin crusts.

COLOR-LUSTER: Whitish yellow to yellow.

MODE OF OCCURRENCE: Occurs sparingly as a secondary mineral associated with uranophane and rutherford-

ine at Shinkolobwe, Katanga, Zaire, and at Menzenschwand, Black Forest, Germany.

BEST REF. IN ENGLISH: Walenta, Kurt, *Am. Min.*, **59**: 166–171 (1974).

# STUETZITE
$Ag_5Te_3$

CRYSTAL SYSTEM: Hexagonal
CLASS: 6/m 2/m 2/m
SPACE GROUP: P6/mmm
Z: 7
LATTICE CONSTANTS:
$a = 13.38$
$c = 8.45$
3 STRONGEST DIFFRACTON LINES:
2.16 (100)
2.55 ( 80)
3.03 ( 70)
HARDNESS: 3½
DENSITY: 8.00 (Meas.)
8.18 (Calc.)
CLEAVAGE: None.
Subconchoidal to uneven fracture. Brittle.

HABIT: Massive, compact, granular aggregates. Rarely as hexagonal equant highly modified crystals. Synthetic crystals short prismatic with well-developed pyramidal terminations.

COLOR-LUSTER: Dark lead gray; rapidly develops dark bronze to iridescent tarnish. Metallic.

MODE OF OCCURRENCE: Occurs associated with empressite at the Empress Josephine mine, Kerber Creek district, Saguache County, Colorado; with altaite at the Red Cloud mine, Gold Hill district, Boulder County, Colorado; with sylvanite in the Kalgoorlie district, Western Australia; also from Vatukoula, island of Viti Levu, Fijian group; the May Day mine, La Plata district, La Plata County, Colorado; and the Golden Fleece mine, Lake City, Hinsdale County, Colorado.

BEST REF. IN ENGLISH: Honea, Russell M., *Am. Min.*, **49**: 325–338 (1964).

# STURTITE
A doubtful hydrous Mn silicate

CRYSTAL SYSTEM: Unknown
OPTICAL CONSTANTS: Isotropic.
HARDNESS: >3
DENSITY: 2.054 (Meas.)
CLEAVAGE: None.
Fracture subconchoidal to uneven. Very brittle.

HABIT: Massive; compact.

COLOR-LUSTER: Black. Vitreous inclining to greasy. Streak yellowish brown.

MODE OF OCCURRENCE: Found associated with quartz, spessartine, rhodochrosite, calcite, galena, sphalerite, rhodonite, and manganhedenbergite, at the Zinc Corporation mine, Broken Hill, New South Wales, Australia.

BEST REF. IN ENGLISH: Hodge-Smith, T., *Am. Min.*, **15**: 537 (1930).

**STYLOTYPITE** = Tetrahedrite

## SUANITE
$Mg_2B_2O_5$

CRYSTAL SYSTEM: Monoclinic
CLASS: 2/m
SPACE GROUP: $P2_1/a$
Z: 4
LATTICE CONSTANTS:
  $a = 12.31$
  $b = 3.120$
  $c = 9.20$
  $\beta = 104°20'$
3 STRONGEST DIFFRACTION LINES:
  2.556 (100)
  2.010 ( 60)
  2.823 ( 50)
OPTICAL CONSTANTS:
  $\alpha = 1.596$
  $\beta = 1.639$
  $\gamma = 1.670$
  $(-)2V = 70°$
HARDNESS: 5½
DENSITY: 2.91 (Meas.)
          2.91 (Calc.)
CLEAVAGE: Perfect prismatic.
HABIT: As fibrous aggregates.
COLOR-LUSTER: White; silky to pearly.
MODE OF OCCURRENCE: The mineral occurs in kotoite marble associated with szaibelyite, calcite, kotoite, spinel, and clinohumite at the Hol Kol mine, Suan district, North Korea.
BEST REF. IN ENGLISH: *Am. Min.*, **39**: 692(1954). *Am. Min.*, **40**: 941 (1955).   *Am. Min.*, **48**: 915 (1963).

**SUCCINITE** = Variety of amber

## SUDOITE
$(Al, Fe, Mg)_{4-5}(Si, Al)_4O_{10}(OH)_8$
Chlorite Group

CRYSTAL SYSTEM: Monoclinic
CLASS: 2/m
SPACE GROUP: C2/m
Z: 2
LATTICE CONSTANTS:
  $a = 5.237$
  $b = 9.070$
  $c = 14.285$
  $\beta = 97°02'$
3 STRONGEST DIFFRACTION LINES:
  2.501 (100)
  4.52 ( 85)
  14.2 ( 80)
HARDNESS: Not determined.
DENSITY: 2.653 (Calc.)
CLEAVAGE: {001} perfect
HABIT: Massive, fine-grained.

COLOR-LUSTER: Whitish. Translucent.
MODE OF OCCURRENCE: Occurs disseminated in hematite ore at the Tracy mine, Negaunee, Michigan; also as a hydrothermal mineral at Kaiserbachtal, Südpfalz, Germany, and at the Kamakita mine, Japan.
BEST REF. IN ENGLISH: Eggleston, R. A., and Bailey, S. W., *Am. Min.*, **52**: 673–689 (1967).

## SUKULAITE
$Ta_2Sn_2O_7$

CRYSTAL SYSTEM: Cubic
CLASS: $4/m\,\bar{3}\,2/m$
SPACE GROUP: Fd3m
Z: 8
LATTICE CONSTANT:
  $a = 10.57$
3 STRONGEST DIFFRACTION LINES:
  3.046 (100)
  1.866 ( 80)
  1.589 ( 80)
HARDNESS: Polishing hardness greater than that of cassiterite.
DENSITY: 8.33 (Meas. synthetic)
HABIT: As minute inclusions in cassiterite.
COLOR-LUSTER: Yellowish brown; translucent.
MODE OF OCCURRENCE: Occurs forming narrow rims around wodginite, and in small grains replacing wodginite almost completely, in cassiterite samples collected about 100 years ago from granite pegmatite in Sukula, Tammela, southwest Finland.
BEST REF. IN ENGLISH: Vorma, Atso, and Siivola, Jaakko, *Am. Min.*, **53**: 2103–2104 (1968).

## SULFOBORITE
$Mg_3SO_4B_2O_4(OH)_2 \cdot 4H_2O$

CRYSTAL SYSTEM: Orthorhombic
CLASS: 2/m 2/m 2/m
SPACE GROUP: Pcmn
Z: 4
LATTICE CONSTANTS:
  $a = 7.79$
  $b = 12.54$
  $c = 10.14$
3 STRONGEST DIFFRACTION LINES:
  3.47 (100)
  3.09 (100)
  2.05 (100)
OPTICAL CONSTANTS:
  $\alpha = 1.527$
  $\beta = 1.536$   (Na)
  $\gamma = 1.551$
  $(-)2V = 79°36'$
HARDNESS: 4–4½
DENSITY: 2.440 (Meas.)
CLEAVAGE: {110} good
          {001} fair
Brittle.

HABIT: Crystals short to long prismatic.
COLOR-LUSTER: Colorless, transparent.
MODE OF OCCURRENCE: Occurs in small amounts in the salt deposits at Westeregeln, Saxony, and near Wittmar, Brunswick, Germany; also in the Inder Lake region, western Kazakhstan, USSR.
BEST REF. IN ENGLISH: Palache, et al., "Dana's System of Mineralogy," 7th Ed., v. II, p. 387–388, New York, Wiley, 1951.

## SULFOHALITE = Sulphohalite

## SULFUR ($\alpha$-sulfur)
S
Trimorphous with $\beta$-sulfur and rosickyite
Var. Selenian; Selensulphur

CRYSTAL SYSTEM: Orthorhombic
CLASS: 2/m 2/m 2/m
SPACE GROUP: Fddd
Z: 128
LATTICE CONSTANTS:
  $a = 10.45$
  $b = 12.84$
  $c = 24.46$
3 STRONGEST DIFFRACTION LINES:
  3.85 (100)
  3.21 ( 60)
  3.44 ( 40)
OPTICAL CONSTANTS:
  $\alpha = 1.9579$
  $\beta = 2.0377$
  $\gamma = 2.2452$
  (+)2V = 68°58'
HARDNESS: 1½–2½
DENSITY: 2.07 (Meas.)
  2.076 (Calc.)
CLEAVAGE: {001} imperfect
  {110} imperfect
  {111} imperfect
  {111} parting
Fracture conchoidal to uneven. Brittle to somewhat sectile.
HABIT: Crystals usually bipyramidal; also thick tabular {100}, or disphenoidal {111}. Faces usually exhibit uneven development. Also massive; as crusts; stalactitic; as a powder. Twinning on {011}, {101}, {110}, rare.
COLOR-LUSTER: Sulfur yellow to yellowish brown or yellowish gray; also reddish, greenish. Transparent to translucent. Resinous to greasy. Streak white.
MODE OF OCCURRENCE: Occurs chiefly in deposits connected with volcanic processes or hot springs; in sedimentary rocks; in salt domes; and less commonly as a secondary mineral in ore deposits. Found as large deposits in Louisiana and Texas, and widespread in the western United States, especially in California, Utah, Nevada, Colorado, Wyoming, and Hawaii. It also occurs in Baja California and elsewhere in Mexico; in the volcanoes of the Chilean and Argentine Andes; as superb crystals at numerous places in Sicily and Italy; and at many other deposits throughout the world.

BEST REF. IN ENGLISH: Palache, et al., "Dana's System of Mineralogy," 7th Ed., v. I, p. 140–144, New York, Wiley, 1944.

## $\beta$-SULFUR
S
Trimorphous with $\alpha$-sulfur and rosickyite

CRYSTAL SYSTEM: Monoclinic
CLASS: 2/m
SPACE GROUP: P2$_1$/c
Z: 48
LATTICE CONSTANTS:
  $a = 11.02$
  $b = 10.96$
  $c = 10.90$
  $\beta = 96°44'$
3 STRONGEST DIFFRACTION LINES:
  3.29 (100)
  6.65 ( 25)
  3.74 ( 20)
OPTICAL CONSTANT:
  (–)2V = 58°
HARDNESS: 2½
DENSITY: 1.958–1.982 (Meas.)
  1.96 (Calc.)
CLEAVAGE: {001} imperfect
  {110} imperfect
Fracture conchoidal to uneven. Brittle.
HABIT: Crystals thick tabular; pseudo-isometric. Often skeletal. Twinning on {011}, {012}, {100}.
COLOR-LUSTER: Pale yellow, nearly colorless. Commonly brownish due to organic inclusions. Transparent to translucent. Adamantine, resinous.
MODE OF OCCURRENCE: Occurs in fumaroles at Vulcano, Lipari Islands, and at Mt. Vesuvius, Italy.
BEST REF. IN ENGLISH: Palache, et al., "Dana's System of Mineralogy," 7th Ed., v. I, p. 144–145, New York, Wiley, 1944.

## $\gamma$-SULFUR = Rosickýite

## SULFURITE = $\beta$-sulfur

## SULFUROSITE = A natural gas
SO$_2$

## SULPHOHALITE (Sulfohalite)
Na$_6$(SO$_4$)$_2$FCl

CRYSTAL SYSTEM: Cubic
CLASS: 4/m $\bar{3}$ 2/m
SPACE GROUP: Fm3m
Z: 4
LATTICE CONSTANT:
  $a = 10.065$

3 STRONGEST DIFFRACTION LINES:
  3.56  (100)
  2.91  ( 80)
  1.779 ( 65)
OPTICAL CONSTANT:
  $N$ = 1.455
  HARDNESS: 3½
  DENSITY: 2.505 (Meas.)
             2.505 (Calc.)
  CLEAVAGE: Fracture conchoidal.
  HABIT: Crystals usually dodecahedral, cubo-octahedral, or octahedral, ranging up to 3 cm in size.
  COLOR-LUSTER: Colorless, gray, pale greenish yellow. Transparent. Vitreous to greasy.

  Soluble in water. Taste slightly saline.
  MODE OF OCCURRENCE: Occurs associated with hanksite in the salt deposits of Searles Lake, San Bernardino County, California, and in the Otjiwalundo salt pan, South West Africa.
  BEST REF. IN ENGLISH: Palache, et al., "Dana's System of Mineralogy," 7th Ed., v. II, p. 548–549, New York, Wiley, 1951.

# SULVANITE
$Cu_3VS_4$

CRYSTAL SYSTEM: Cubic
CLASS: $\bar{4}3m$
SPACE GROUP: $P\bar{4}3m$
Z: 1
LATTICE CONSTANT:
  $a$ = 5.3912
3 STRONGEST DIFFRACTION LINES:
  5.4   (100)
  1.910 ( 90)
  3.12  ( 50)
OPTICAL CONSTANTS: Opaque.
  HARDNESS: 3½
  DENSITY: 3.86 (Meas.)
             3.92 (Calc.)
  CLEAVAGE: {001} perfect
  HABIT: Crystals cubic; commonly massive.
  COLOR-LUSTER: Bronze gold. Opaque. Metallic; dull tarnish. Streak black.
  MODE OF OCCURRENCE: Occurs as altered crystals and massive in limestone and calcite breccia in the Mercur district, Tooele County, Utah; in the Sierra de Cordoba, Argentina; and associated with gypsum, malachite, azurite, cuprodescloizite, and atacamite in the Burra-Burra district, South Australia.
  BEST REF. IN ENGLISH: Palache, et al., "Dana's System of Mineralogy," 7th Ed., v. I, p. 384–385, New York, Wiley, 1944. Trojer, Felix J., *Am. Min.*, **51**: 890–894 (1966).

# SUOLUNITE
$Ca_2H_2Si_2O_7 \cdot H_2O$

CRYSTAL SYSTEM: Orthorhombic
CLASS: mm2

SPACE GROUP: Fdd2
Z: 8
LATTICE CONSTANTS:
  $a$ = 11.15
  $b$ = 19.67
  $c$ = 6.08
3 STRONGEST DIFFRACTION LINES:
  4.03
  3.11 (Intensities not reported)
  2.80
OPTICAL CONSTANTS:
  $\alpha$ = 1.610
  $\beta$ = 1.620
  $\gamma$ = 1.623
  $(-)2V \sim 30$–$50°$
  HARDNESS: Not determined.
  DENSITY: 2.683 (Meas.)
  CLEAVAGE: None.
  HABIT: Granular; fine-grained.
  COLOR-LUSTER: Pure white; vitreous to resinous.
  MODE OF OCCURRENCE: Occurs in a vein 2–3 cm wide cutting harzburgite rocks in the center of an ultrabasic rock mass at a depth of about 100 meters in Inner Mongolia.
  BEST REF. IN ENGLISH: Huang, Yung-Hwei, *Am. Min.*, **53**: 349 (1968).

# SURSASSITE
$Mn_5Al_4Si_5O_{21} \cdot 3H_2O$

CRYSTAL SYSTEM: Monoclinic
CLASS: 2/m or 2
SPACE GROUP: $P2_1/m$ or $P2_1$
Z: 1
LATTICE CONSTANTS:
  $a$ = 8.82
  $b$ = 5.84
  $c$ = 9.71
  $\beta$ = 108°24′
3 STRONGEST DIFFRACTION LINES:
  2.82 (100)
  2.88 ( 70)
  2.57 ( 70)
OPTICAL CONSTANTS:
  $\alpha$ = 1.735–1.736
  $\beta$ = 1.753–1.755
  $\gamma$ = 1.766–1.767
  $(-)2V \sim 65°$
  HARDNESS: Not determined.
  DENSITY: 3.256 (Meas.)
             3.194 (Calc.)
  CLEAVAGE: {101} distinct
  HABIT: Massive, fibrous.
  COLOR-LUSTER: Deep reddish brown to copper red. Translucent. Silky to somewhat dull.
  MODE OF OCCURRENCE: Occurs in association with barite, calcite, and quartz in veinlets cutting Middle Silurian iron-manganese ores about 5 miles west of Woodstock, New Brunswick, Canada; also as small dense radiated botryoidal masses in cracks in the radiolarian cherts of the manganese deposits of Val d'Err, Graubünden, Switzerland.

BEST REF. IN ENGLISH: Heinrich, E. Wm., *Can. Min.*, **7**: 291-300 (1962). Freed, R. L., *Am. Min.*, **49**: 168-173 (1964).

## SUSANNITE
$Pb_4(SO_4)(CO_3)_2(OH)_2$
Dimorphous with leadhillite

CRYSTAL SYSTEM: Hexagonal
CLASS: $\bar{3}$
SPACE GROUP: $R\bar{3}$
LATTICE CONSTANTS:
 $a = 9.05$
 $c = 11.54$
3 STRONGEST DIFFRACTION LINES:
 3.53 (100)
 2.917 ( 60)
 2.605 ( 50)
OPTICAL CONSTANTS:
 Uniaxial, also $2V = 0°-3°$
 HARDNESS: 2½-3
 DENSITY: 6.55 (Meas.)
 CLEAVAGE: {0001} perfect
Fracture conchoidal. Somewhat sectile.
 HABIT: As acute rhombohedral crystals with modifying faces of the basal pinacoid or a hexagonal prism.
 COLOR-LUSTER: Colorless to greenish or yellowish. Transparent to translucent. Adamantine to resinous.
 MODE OF OCCURRENCE: Occurs as a secondary mineral in the oxidation zone of lead deposits. Found at the Mammoth mine, Tiger, Pinal County, Arizona; at the Susanna mine, Leadhills, Lanarkshire, Scotland; in Moldavia, Rumania; and at Nerchinsk, USSR.
 BEST REF. IN ENGLISH: Mrose, M. E., and Christian, R. P., *Can. Min.*, **10**: 141 (1969) (abstr.). Palache, et al., "Dana's System of Mineralogy," 7th Ed., v. II, p. 298-299 New York, Wiley, 1951.

## SUSSEXITE
$MnBO_3H$

CRYSTAL SYSTEM: Orthorhombic
Z: 8
LATTICE CONSTANTS:
 $a = 10.70$
 $b = 12.77$
 $c = 3.25$
3 STRONGEST DIFFRACTION LINES:
 6.32 (100)
 2.744 ( 80)
 2.478 ( 80)
OPTICAL CONSTANTS:
 $\alpha = 1.670$
 $\beta = 1.728$
 $\gamma = 1.732$
 $(-)2V \sim 25°$
 HARDNESS: 3-3½
 DENSITY: 3.30 (Meas.)
     3.432 (Calc.)
 CLEAVAGE: Not determined.

 HABIT: As veinlets or masses with fibrous structure; also as embedded masses, sometimes dense porcelaneous or chalky.
 COLOR-LUSTER: White, pinkish, straw yellow. Silky to dull or earthy. Streak white.
 MODE OF OCCURRENCE: Occurs associated with leucophoenicite, pyrochroite, and other minerals in hydrothermal veinlets in massive franklinite ore at Franklin, Sussex County, New Jersey; and as veinlets in chert and hematite at the Chicagon mine, Iron County, Michigan. It also is found at the Gonzen mine, Switzerland.
 BEST REF. IN ENGLISH: Palache, et al., "Dana's System of Mineralogy," 7th Ed., v. II, p. 375-377, New York, Wiley, 1951.

## SVABITE
$Ca_5(AsO_4)_3(F, Cl, OH)$
Svabite Series
Apatite Group

CRYSTAL SYSTEM: Hexagonal
CLASS: 6/m
SPACE GROUP: $P6_3/m$
Z: 2
LATTICE CONSTANTS:
 $a = 9.95$    $a = 9.75$
 $c = 6.75$ or $c = 6.92$
3 STRONGEST DIFFRACTION LINES:
 2.91 (100)
 2.84 ( 90)
 2.82 ( 90)
OPTICAL CONSTANTS:
 $\omega = 1.706$
 $\epsilon = 1.698$
  (-)
 HARDNESS: 4-5
 DENSITY: 3.5-3.8 (Meas.)
      3.67 (Calc.)
 CLEAVAGE: {10$\bar{1}$0} indistinct
Brittle.
 HABIT: Crystals short prismatic. Also massive.
 COLOR-LUSTER: Colorless, yellowish white to gray and grayish green. Transparent to translucent. Vitreous to resinous.
 MODE OF OCCURRENCE: Occurs widely disseminated in small amounts in the zinc ores at Franklin, Sussex County, New Jersey; and at Långban, Jacobsberg, and the Harstig mine, Pajsberg, Sweden.
 BEST REF. IN ENGLISH: Palache, et al., "Dana's System of Mineralogy," 7th Ed., v. II, p. 899-900, New York, Wiley, 1951.

## SVANBERGITE
$SrAl_3PO_4SO_4(OH)_6$

CRYSTAL SYSTEM: Hexagonal
CLASS: $\bar{3}$ 2/m
SPACE GROUP: $R\bar{3}m$
Z: 1 (rhomb.), 3 (hexag.)

LATTICE CONSTANTS:
$a = 6.99$      $a_{rh} = 6.91$
$c = 16.75$     $\alpha = 60°58.5'$
3 STRONGEST DIFFRACTION LINES:
2.98 (100)
2.22 (100)
5.74 ( 90)
OPTICAL CONSTANTS:
$\omega = 1.631$ (Na)
$\epsilon = 1.646$
(+)
HARDNESS: 5
DENSITY: 3.22 (Meas.)
          3.24 (Calc.)
CLEAVAGE: {0001} distinct
HABIT: Crystals rhombohedral or pseudocubic. Also granular.
COLOR-LUSTER: Colorless to yellow, rose, reddish brown. Translucent. Vitreous to adamantine.
MODE OF OCCURRENCE: Occurs associated with pyrophyllite and diopside at Thorne, Mineral County, Nevada; at the Westanå iron mine in Skåne, and at Horrsjöberg, Sweden; with pyrophyllite near Chalmoux, Saône et Loire, France; and in the diamond gravels of Bahia, Brazil.
BEST REF. IN ENGLISH: Palache, et al., "Dana's System of Mineralogy," 7th Ed., v. II, p. 1005–1006, New York, Wiley, 1951.

**SVIDNEITE** = A transition member in the kataphorite-arfvedsonite series

## SWARTZITE
$CaMgUO_2(CO_3)_3 \cdot 12H_2O$

CRYSTAL SYSTEM: Monoclinic
CLASS: 2/m or 2
SPACE GROUP: $P2_1/m$ or $P2_1$
Z: 2
LATTICE CONSTANTS:
$a = 11.12$
$b = 14.72$
$c = 6.47$
$\beta = 99°26'$
3 STRONGEST DIFFRACTION LINES:
8.76 (100)
5.50 (100)
7.31 ( 90)
OPTICAL CONSTANTS:
$\alpha = 1.465$
$\beta = 1.51$
$\gamma = 1.540$
(−)2V = 40° (Calc.)
HARDNESS: Not determined.
DENSITY: 2.3 (Meas.)
CLEAVAGE: Not determined.
HABIT: Crystals minute, prismatic along [001]. As crusts.
COLOR-LUSTER: Bright green; turns dull yellowish white on dehydration. Fluoresces bright yellowish green in ultraviolet light.
MODE OF OCCURRENCE: Occurs as an efflorescence on mine walls in the Hillside mine, Yavapai County, Arizona, associated with bayleyite, andersonite, and schroeckingerite. Also found at the Coral claim, Elk Ridge, San Juan County, Utah.
BEST REF. IN ENGLISH: Frondel, Clifford, USGS Bull. 1064, p. 117–119 (1958).

## SWEDENBORGITE
$NaBe_4SbO_7$

CRYSTAL SYSTEM: Hexagonal
CLASS: 6mm
SPACE GROUP: $P6_3mc$
Z: 2
LATTICE CONSTANTS:
$a = 5.42$ kX
$c = 8.80$
3 STRONGEST DIFFRACTION LINES:
4.20 (100)
2.72 ( 90)
2.51 ( 90)
OPTICAL CONSTANTS:
$\omega = 1.772$
$\epsilon = 1.770$   ($\lambda = 589$ nm)
(−)
HARDNESS: ~8
DENSITY: 4.285 (Meas.)
          4.316 (Calc.)
CLEAVAGE: {0001} distinct
Fracture subconchoidal.
HABIT: As short prismatic crystals up to 8 mm in size.
COLOR-LUSTER: Colorless, pale wine yellow to honey yellow. Transparent. Vitreous.
MODE OF OCCURRENCE: Occurs in skarn associated with calcite, manganophyllite, bromelite, hematite, and richterite at Långban, Sweden.
BEST REF. IN ENGLISH: Palache, et al., "Dana's System of Mineralogy," 7th Ed., v. II, p. 1027–1029, New York, Wiley, 1951.

## SWITZERITE
$(Mn, Fe)_3(PO_4)_2 \cdot 4H_2O$

CRYSTAL SYSTEM: Monoclinic
CLASS: 2/m
SPACE GROUP: P2/a
Z: 8
LATTICE CONSTANTS:
$a = 17.099$
$b = 12.694$
$c = 8.282$
$\beta = 95°55'$
3 STRONGEST DIFFRACTION LINES:
8.550 (100)
2.585 ( 60)
7.128 ( 40)
OPTICAL CONSTANTS:
$\alpha = 1.602$ (Calc.)
$\beta = 1.628$
$\gamma = 1.632$
(−)2V = 42°

HARDNESS: ~2½
DENSITY: 2.95 (Meas.)
3.18 (Calc.)
CLEAVAGE: {100} perfect
{010} fair
HABIT: As micaceous flakes or bladed crystals, up to 3 mm long, flattened on {100} and elongated [001].
COLOR-LUSTER: Pale pink or light golden brown. Medium brown to chocolate brown when oxidized. Luster pearly to adamantine on the cleavage.
MODE OF OCCURRENCE: Found at the Foote Mineral Company spodumene mine, Kings Mountain, North Carolina, in association with vivianite in seams in spodumene-rich pegmatite.
BEST REF. IN ENGLISH: Leavens, Peter B., and White, John S. Jr., *Am. Min.*, **52**: 1595-1602 (1967).

# SYLVANITE
AgAuTe₄

CRYSTAL SYSTEM: Monoclinic
CLASS: 2/m̄
SPACE GROUP: P2/n, P2/c
Z: 2
LATTICE CONSTANTS:
| (P2/n) | (P2/c) |
|---|---|
| a = 8.94 | a = 8.94 |
| b = 4.48 | b = 4.48 |
| c = 8.83 | c = 14.59 |
| β = 110°22′ | β = 145°26′ |

3 STRONGEST DIFFRACTION LINES:
3.05 (100)
2.14 ( 50)
2.25 ( 30)
HARDNESS: 1½-2
DENSITY: 8.16 (Meas.)
8.17 (Calc.)
CLEAVAGE: {010} perfect
Fracture uneven. Brittle.
HABIT: Crystals varied in habit; usually thick tabular or short prismatic. Also bladed, imperfectly columnar, or granular. Twinning on {100} common, often as penetration twins producing arborescent forms resembling written characters (graphic ore).
COLOR-LUSTER: Silver white, sometimes with yellowish tint. Opaque. Metallic, brilliant. Streak silver white.
MODE OF OCCURRENCE: Occurs in low- to high-temperature vein deposits, commonly associated with other tellurides, sulfides, fluorite, carbonates, gold, tellurium, and quartz. Found widespread in the western United States, especially in the Cripple Creek district, Teller County, and elsewhere in Colorado; in Calaveras, Trinity, Tuolumne, and Yuba counties, California; and at deposits in Oregon, Idaho, Montana, and South Dakota. It also occurs in Canada, Roumania, USSR, and Western Australia.
BEST REF. IN ENGLISH: Palache, et al., "Dana's System of Mineralogy," 7th Ed., v. I, p. 338-341, New York, Wiley, 1944.

**SYLVINITE** = Mixture of sylvite and halite

# SYLVITE
KCl

CRYSTAL SYSTEM: Cubic
CLASS: 4/m 3̄ 2/m
SPACE GROUP: Fm3m
Z: 4
LATTICE CONSTANT:
a = 6.2910
3 STRONGEST DIFFRACTION LINES:
3.15 (100)
2.22 ( 59)
1.82 ( 23)
OPTICAL CONSTANT:
N = 1.49031 (Na)
HARDNESS: 2
DENSITY: 1.993 (Meas.)
1.99 (Calc.)
CLEAVAGE: {001} perfect
Fracture uneven. Brittle.
HABIT: Crystals cubic; rarely octahedral or cubo-octahedral. Also massive, compact to granular, as crusts; and columnar.
COLOR-LUSTER: White, gray, bluish; also reddish due to minute inclusions of hematite. Transparent. Vitreous.

Soluble in water.
MODE OF OCCURRENCE: Occurs chiefly as extensive thick sedimentary deposits, often associated with halite, gypsum, anhydrite, carnallite, polyhalite, kieserite, and kainite. In the United States it is found in large deposits in the Permian basin of southeastern New Mexico, especially in the vicinity of Carlsbad, and also in adjacent parts of Texas. Extensive deposits are also found in Saskatchewan, Canada; at Strassfurt and other places in Germany; and at Kaluszyn, Poland; it also occurs as a sublimation product on Mt. Etna, Sicily and Mt. Vesuvius, Italy; and in Peru and Chile it is found in several nitrate deposits.
BEST REF. IN ENGLISH: Palache, et al., "Dana's System of Mineralogy," 7th Ed., v. II, p. 7-9, New York, Wiley, 1951.

# SYMPLESITE
Fe₃(AsO₄)₂ · 8H₂O
Dimorphous with parasymplesite

CRYSTAL SYSTEM: Triclinic
CLASS: 1̄
SPACE GROUP: P1̄
Z: 1
LATTICE CONSTANTS:
| a = 7.87 | α = 99°55′ |
| b = 9.41 | β = 97°22.5′ |
| c = 4.72 | γ = 105°57.5′ |

3 STRONGEST DIFFRACTION LINES:
6.79 (100)
7.50 ( 16)
8.97 ( 16)

OPTICAL CONSTANTS:
  $\alpha = 1.635$
  $\beta = 1.668$
  $\gamma = 1.702$
  $(-)2V = 86.5°$
HARDNESS: ~2½
DENSITY: 3.01 (Meas.)
        3.02 (Calc.)
CLEAVAGE: {1$\bar{1}$0} perfect
Fracture uneven. Brittle.
HABIT: As small imperfect crystals, and as spherical aggregates with radial-fibrous structure. Earthy.
COLOR-LUSTER: Light blue, light green to greenish black, deep indigo blue. Transparent to translucent. Vitreous; pearly on cleavage; also dull, earthy.
MODE OF OCCURRENCE: Occurs as a light blue powder filling cavities in loellingite in pegmatite at the Custer Mountain Lode, Custer County, South Dakota. It also is found at Tagish Lake, Yukon, Canada; at Lobenstein and at other places in Germany; and at localities in Italy, Austria, Roumania, Czechoslovakia, and Tasmania.
BEST REF. IN ENGLISH: Palache, et al., "Dana's System of Mineralogy," 7th Ed., v. II, p. 752-753, New York, Wiley, 1951.

# SYNADELPHITE
$Mn_9^{2+}(OH)_9(H_2O)_2(AsO_3)(AsO_4)_2$

CRYSTAL SYSTEM: Orthorhombic
CLASS: 2/m 2/m 2/m
SPACE GROUP: Pnma
Z: 4
LATTICE CONSTANTS:
  $a = 10.754$
  $b = 18.865$
  $c = 9.884$
3 STRONGEST DIFFRACTION LINES:
  8.72 (100)
  2.64 ( 90)
  9.36 ( 70)
OPTICAL CONSTANTS:
  $\alpha = 1.750$
  $\beta = 1.751$
  $\gamma = 1.761$
  $(+)2V = 37°$
HARDNESS: 4½
DENSITY: 3.57 (Meas.) to 3.79 for high Pb varieties.
        3.59 (Calc.)
CLEAVAGE: {010} imperfect
Fracture conchoidal to uneven. Brittle.
HABIT: Crystals short prismatic; prism faces commonly striated. Single individuals are fourlings twinned by reflection on (100) and (010). Also massive, as crusts.
COLOR-LUSTER: Crystals commonly zoned in color; center parts colorless, exterior red, reddish brown, dark brown to black. Centers vitreous; exteriors adamantine to sub-metallic. Transparent in colorless material; translucent to opaque in plumbian varieties.
MODE OF OCCURRENCE: Synadelphite occurs in cavities in low-temperature hydrothermal veins cutting hausmannite ores in dolomitic marbles at several small abandoned ore-bodies in the province of Värmland, Sweden, in particular Östra Mossgruven in Nordmarks Odalfält and at Långban.
BEST REF. IN ENGLISH: Palache, et al., "Dana's System of Mineralogy," 7th Ed., v. II, p. 780-782, New York, Wiley, 1951. Moore, Paul B., *Am. Min.*, **55**: 2023-2037 (1970).

# SYNCHYSITE (Synchisite)
$(Ce, La)Ca(CO_3)_2F$

CRYSTAL SYSTEM: Orthorhombic
SPACE GROUP: C lattice
Z: 2
LATTICE CONSTANTS:
  $a = 4.10$
  $b = 7.10$
  $c = 9.12$
  $\beta = 90°$
3 STRONGEST DIFFRACTION LINES:
  3.55 (100)
  2.80 (100)
  9.1 ( 60)
OPTICAL CONSTANTS:
  $\omega = 1.641-1.650$
  $\epsilon = 1.74-1.745$ (+)
HARDNESS: 4½
DENSITY: 3.902-4.15 (Meas.)
        3.99 (Calc.)
CLEAVAGE: {0001} (May be parting)
Fracture subconchoidal to splintery. Brittle.
HABIT: Crystals thin to thick tabular, pseudohexagonal, minute. Twinning very common.
COLOR-LUSTER: Gray to grayish yellow to brown. Translucent. Vitreous to greasy.
MODE OF OCCURRENCE: Occurs associated with thorite in the Powderhorn district, Gunnison County, Colorado; with parisite near Pyrites, Ravalli County, Montana, and at the Ballou and Fallon Brothers quarries, Quincy, Massachusetts; in abundance as fine platy crystals associated with large siderite crystals, albite, and zircon in carbonatite at Mont St. Hilaire, Quebec, Canada; in pegmatite veins in nepheline-syenite associated with cordylite, ancylite, neptunite, and aegirine at Narsarsuk, Greenland; at Predazzo, Italy; in an Alpine-type vein at Piz Blas, Graubunden, Switzerland; and in the Potgietersrust district, Transvaal, South Africa.
BEST REF. IN ENGLISH: Donnay, Gabrielle, and Donnay, J. D. H., *Am. Min.*, **38**: 932-963 (1953).

# SYNCHYSITE-(Y) syn. Doverite
$(Y, Ce)Ca(CO_3)_2F$

CRYSTAL SYSTEM: Orthorhombic
SPACE GROUP: C lattice
Z: 2
LATTICE CONSTANTS:
  $a = 4.00$
  $b = 6.92$
  $c = 9.00$

3 STRONGEST DIFFRACTION LINES:
1.89 (100)
2.00 ( 90)
1.83 ( 90)
OPTICAL CONSTANTS:
$\omega = 1.643$
$\epsilon = 1.73$
(+)
HARDNESS: 6½ (?)
DENSITY: 3.89–4.1+ (Meas.)
CLEAVAGE: Not determined.
Fracture uneven to subconchoidal. Brittle.
HABIT: Massive; as fine-grained compact masses.
COLOR-LUSTER: Reddish brown. Dull; nonmetallic. Streak brownish.
MODE OF OCCURRENCE: Occurs associated with xenotime, hematite, and quartz, at the Scrub Oaks iron mine at Dover, Morris County, New Jersey. Also found in Colorado at the White Cloud pegmatite near South Platte, and associated with cenosite at the Henry pegmatite, Cotapaxi.
BEST REF. IN ENGLISH: Smith, William E., et al., *Am. Min.*, **45**: 92–98 (1960). Butler, J. R., and Hall, R., *Econ. Geol.*, **55**: 1541–1550 (1960). Levinson, A. A., and Borup, R. A., *Am. Min.*, **47**: 337–343 (1962).

# SYNGENITE
$K_2Ca(SO_4)_2 \cdot H_2O$

CRYSTAL SYSTEM: Monoclinic
CLASS: 2/m
SPACE GROUP: $P2_1/m$
Z: 2
LATTICE CONSTANTS:
$a = 9.55$
$b = 7.13$
$c = 6.00$
$\beta = 105°$
3 STRONGEST DIFFRACTION LINES:
2.86 (100)
3.16 ( 75)
9.52 ( 60)
OPTICAL CONSTANTS:
$\alpha = 1.5010$
$\beta = 1.5166$ (Na)
$\gamma = 1.5176$
$(-)2V = 28°18'$
HARDNESS: 2½
DENSITY: 2.603 (Meas.)
2.597 (Calc.)
CLEAVAGE: {110} perfect
{100} perfect
{010} distinct
Fracture conchoidal.
HABIT: Crystals tabular or prismatic, often highly modified and sometimes several centimeters in size. Striated parallel to *c*-axis. Twinning on {100} common.
COLOR-LUSTER: Colorless, white. Transparent to translucent. Vitreous.
MODE OF OCCURRENCE: Occurs sparingly in salt deposits of oceanic origin, and as a volcanic product. Found as crusts on lava in Heleakala crater, Maui, Hawaiian Islands;

on Mount Vesuvius, Italy; at the Glück Auf potash mine near Sondershausen, Germany; and in the salt deposits of Galicia, Poland.
BEST REF. IN ENGLISH: Palache, et al., "Dana's System of Mineralogy," 7th Ed., v. II, p. 442–444, New York, Wiley, 1951.

# SZAIBELYITE (Camsellite)
$(Mg, Mn)BO_3H$

CRYSTAL SYSTEM: Orthorhombic
CLASS: 222
SPACE GROUP: $P2_12_12$
Z: 8
LATTICE CONSTANTS:
$a = 10.34$
$b = 12.45$
$c = 3.21$
3 STRONGEST DIFFRACTION LINES:
6.20 (100)
2.202 ( 80)
2.657 ( 75)
OPTICAL CONSTANTS:
$\alpha = 1.575$
$\beta = 1.646$
$\gamma = 1.650$
$(-)2V \sim 25°$
HARDNESS: 3–3½
DENSITY: 2.60 (Meas.)
CLEAVAGE: Not determined.
HABIT: As fibrous veinlets or masses; also as embedded nodules, sometimes dense or chalk-like.
COLOR-LUSTER: White to straw yellow. Silky to dull or earthy. Streak white.
MODE OF OCCURRENCE: Occurs as impregnations and coatings on serpentine near Stinson Beach, Marin County, California; at Blind Mountain, Lincoln County, Nevada; as veinlets in iron ore at the Eureka mine, Gogebic range, Michigan; near Douglas Lake, British Columbia, Canada; and at localities in Sweden, Germany, Hungary, USSR, and at the Hol Kol mine, Suan, Korea.
BEST REF. IN ENGLISH: Palache, et al., "Dana's System of Mineralogy," 7th Ed., v. II, p. 375–377, New York, Wiley, 1951.

# SZMIKITE
$MnSO_4 \cdot H_2O$

CRYSTAL SYSTEM: Monoclinic
CLASS: 2/m
SPACE GROUP: A2/a
Z: 4
LATTICE CONSTANTS:
$a = 7.758$
$b = 7.612$ (synthetic)
$c = 7.126$
$\beta = 115°42.5'$

3 STRONGEST DIFFRACTION LINES:
3.499 (100)
3.135 ( 40) (synthetic)
4.90 ( 35)

OPTICAL CONSTANTS:
$\alpha = 1.562$
$\beta = 1.595$
$\gamma = 1.632$
(+)2V near 90°

HARDNESS: 1½

DENSITY: 2.845–3.21 (Meas. synthetic)
2.960 (Calc.)

CLEAVAGE: Fracture splintery.

HABIT: As stalactitic masses.

COLOR-LUSTER: Grayish white, brownish white, reddish, rose red.

MODE OF OCCURRENCE: Found as an efflorescence at Felsöbánya, Roumania.

BEST REF. IN ENGLISH: Palache, et al., "Dana's System of Mineralogy," 7th Ed., v. II, p. 481, New York, Wiley, 1951.

# SZOMOLNOKITE
$FeSO_4 \cdot H_2O$

CRYSTAL SYSTEM: Monoclinic
CLASS: 2/m
SPACE GROUP: A2/a
Z: 4

LATTICE CONSTANTS:
$a = 7.624$
$b = 7.468$ (synthetic)
$c = 7.123$
$\beta = 115°52'$

3 STRONGEST DIFFRACTION LINES:
3.44 (100)
3.12 ( 40) (synthetic)
2.52 ( 35)

OPTICAL CONSTANTS:
$\alpha = 1.591$
$\beta = 1.623$
$\gamma = 1.663$
(+)2V = 80°

HARDNESS: 2½

DENSITY: 3.03–3.07 (Meas.)
3.092 (Calc. synthetic)

CLEAVAGE: Fracture conchoidal to uneven. Brittle.

HABIT: Crystals bipyramidal, often distorted. Also globular, stalactitic.

COLOR-LUSTER: Colorless, yellow, yellowish to reddish brown, bluish. Transparent to translucent. Vitreous.

Soluble in water.

MODE OF OCCURRENCE: Occurs at the Tintic Standard mine, Juab County, Utah; at the Santa Elena mine, San Juan Province, Argentina, with copiapite and other sulfates; at Alcaparrosa and Quetana, Chile; at Szomolnok (Smolnik) and near Skrivan, Czechoslovakia.

BEST REF. IN ENGLISH: Palache, et al., "Dana's System of Mineralogy," 7th Ed., v. II, p. 479–480, New York, Wiley, 1951.

# T

## TAAFFEITE (4H and 9R)
$BeMgAl_4O_8$

CRYSTAL SYSTEM: Hexagonal
CLASS: 622 and $\bar{3}2/m$
SPACE GROUP: $P6_322$, $R\bar{3}m$
Z: 4, 9
LATTICE CONSTANTS:

| 4H (hexag.) | 9R (trig.) | 9R (rhomb.) |
|---|---|---|
| $a = 5.72$ | $a = 5.675$ | $a_{rh} = 14.085$ |
| $c = 18.38$ | $c = 41.096$ | $\alpha = 23.24$ |

3 STRONGEST DIFFRACTION LINES:

| 4H | 9R |
|---|---|
| 2.43 (100) | 2.41 (100) |
| 2.05 ( 80) | 1.42 ( 80) |
| 1.428 ( 80) | 2.05 ( 70) |

OPTICAL CONSTANTS:

| 4H | 9R |
|---|---|
| $\omega = 1.722$ | $\omega = 1.739$ |
| $\epsilon = 1.777$ | $\epsilon = 1.735$ |
| (-) (?) | (-) |

HARDNESS: 8–8½
DENSITY: 3.60–3.613 (Meas.)
CLEAVAGE: Not determined.
HABIT: Crystals hexagonal or lenticular, small. Also as fine-grained aggregates.
COLOR-LUSTER: Colorless, greenish, pinkish lilac. Transparent to translucent. Vitreous.
MODE OF OCCURRENCE: Occurs in skarns formed at the contact of Devonian dolomites and dolomitized limestones with an intrusion of beryllium-bearing granite in China, in association with calcite, fluorite, cancrinite, chrysoberyl, and green spinel. Also, faceted gemstones weighing 0.87 and 1.419 carats have been cut from taaffeite presumably found as rolled pebbles in alluvial deposits of Ceylon.
BEST REF. IN ENGLISH: Anderson, B. W., and Claringbull, G. F., *Min. Mag.*, **29**: 765–772 (1951). Vlasov, K. A., "Mineralogy of Rare Elements," v. II, p. 77–79, Israel Program for Scientific Translations, 1966.

## TACHARANITE (Inadequately described mineral)
$(Ca, Mg, Al)(Si, Al)O_2 \cdot H_2O$

CRYSTAL SYSTEM: Unknown
3 STRONGEST DIFFRACTION LINES:
12.7 (100)
3.05 ( 80)
2.89 ( 70)
HARDNESS: Not determined.
DENSITY: 2.36 (Meas.)
CLEAVAGE: Not determined.
HABIT: Massive.
COLOR-LUSTER: White.
MODE OF OCCURRENCE: Occurs with a variety of associations of hydrous calcium silicates and zeolites, including mesolite, gyrolite, thomsonite, tobermorite, xonotlite, as well as calcite and saponite in amygdules from olivine dolerite at Portree, Isle of Skye.
BEST REF. IN ENGLISH: Sweet, Jessie M., Bothwell, D. I., and Williams, D. L., *Min. Mag.*, **32**: 745–753 (1961).

## TACHHYDRITE (Tachyhydrite)
$CaMg_2Cl_6 \cdot 12H_2O$

CRYSTAL SYSTEM: Hexagonal
LATTICE CONSTANTS:
$a:c = 1:1.76$
3 STRONGEST DIFFRACTION LINES:
2.60 (100)
3.09 ( 50)
3.80 ( 30)
OPTICAL CONSTANTS:
$\omega = 1.520$
$\epsilon = 1.512$ (Na)
(-)
HARDNESS: 2
DENSITY: 1.667 (Meas.)
CLEAVAGE: $\{10\bar{1}1\}$ perfect
HABIT: Massive; as rounded masses.

COLOR-LUSTER: Colorless, yellow. Transparent. Vitreous.

Deliquescent, taste bitter.

MODE OF OCCURRENCE: Occurs in potash-rich saline deposits in the Strassfurt district, Saxony, Germany, associated with kainite, carnallite, halite, sylvite, and other salts.

BEST REF. IN ENGLISH: Palache, et al., "Dana's System of Mineralogy," 7th Ed., v. II, p. 95–96, New York, Wiley, 1951.

## TADZHIKITE
$Ca_3(Ce,Y)_2(Ti,Al,Fe)B_4Si_4O_{22}$
Hellandite Group

CRYSTAL SYSTEM: Triclinic
Z: 2(?)
LATTICE CONSTANTS:
  $a = 17.93$
  $b = 4.71$
  $c = 10.39$
  $\beta = 100°45'$
  "$\alpha$ and $\gamma$ close to 90°"
3 STRONGEST DIFFRACTION LINES:
  2.65 (100)
  1.913 ( 55)
  4.97 ( 30)
OPTICAL CONSTANTS:
  $\alpha = 1.761, 1.750$
  $\beta = ?$
  $\gamma = 1.772, 1.763$
  2V = 80°–92°
HARDNESS: 6
DENSITY: 3.73; 3.86 (Meas.)
         3.732 (Calc.)
CLEAVAGE: {010} distinct
HABIT: Crystals prismatic, flattened along {010}; also as curved plates. Twinning polysynthetic.
COLOR-LUSTER: Pale grayish brown, dark brown. Translucent. Vitreous.
MODE OF OCCURRENCE: Occurs in zoned pegmatites in the Turkestan-Alai alkalic province of the Tadzhikaya Republic, Russian central Asia, in albitized portions of an outer quartz-aegirine-microcline zone, and with polylithionite and riebeckite-arfvedsonite in the core.
BEST REF. IN ENGLISH: Efimov, A. F., Dusmatov, V. D., Alkhazov, V. Yu., Pudovkina, Z. G., and Kazakova, M. E., *Am. Min.*, **56**: 1838–1839 (1971).

## TAENIOLITE (Tainiolite)
$KLiMg_2Si_4O_{10}F_2$
Mica Group

CRYSTAL SYSTEM: Monoclinic
CLASS: 2/m
SPACE GROUP: C2/m
Z: 2
LATTICE CONSTANTS:
  $a = 5.27$
  $b = 9.13$   (synthetic)
  $c = 10.12$
  $\beta = 100°$

3 STRONGEST DIFFRACTION LINES:
  3.34 (100)
  10.04 ( 75)
  5.01 ( 45)
OPTICAL CONSTANTS:
  $\alpha = 1.524–1.522$
  $\gamma = \beta = 1.554–1.553$
HARDNESS: 2½–3
DENSITY: 2.82–2.90 (Meas.)
CLEAVAGE: {001} perfect, micaceous
Foliae flexible, somewhat elastic.
HABIT: Crystals tabular, pseudohexagonal, up to 3.5 × 5 × 0.5 cm. Also as scales and cryptocrystalline aggregates. Twinning as penetration trillings rare.
COLOR-LUSTER: Colorless to brownish. Transparent. Vitreous.
MODE OF OCCURRENCE: Occurs in nepheline-syenite pegmatites (carbonatites) in association with natrolite, apatite, neptunite, and other minerals. Found at Magnet Cove, Hot Spring County, Arkansas; at Narsarsuk, Greenland; and in the Lovozero massif, Kola Peninsula, USSR.
BEST REF. IN ENGLISH: Vlasov, K. A., "Mineralogy of Rare Elements," v. II, p. 29–31, Israel Program for Scientific Translations, 1966.

## TAENITE (Nickel-iron)
$\gamma$-Fe, Ni (~32% Ni)

CRYSTAL SYSTEM: Cubic
CLASS: 4/m $\bar{3}$ 2/m
SPACE GROUP: Fm3m
Z: 32
LATTICE CONSTANT:
  $a = 7.168$
3 STRONGEST DIFFRACTION LINES:
  3.34 (100)
  2.88 ( 80)
  2.53 ( 80)
HARDNESS: 5
DENSITY: 7.8–8.22 (Meas.)
CLEAVAGE: None. Malleable; flexible.
HABIT: Massive; in intergrowths or as narrow borders around kamacite in meteorites. Also as large flexible tablets.
COLOR-LUSTER: Silver white to grayish white. Opaque. Metallic.

Strongly magnetic.
MODE OF OCCURRENCE: Occurs in all octahedrites which exhibit Widmannstätten structure; also in some ataxites.
BEST REF. IN ENGLISH: Palache, et al., "Dana's System of Mineralogy," 7th Ed., v. I, p. 117–119, New York, Wiley, 1944. *Am. Min.*, **51**: 37–55 (1966).

## TAGILITE = Pseudomalachite

## TAKANELITE
$(Mn^{2+},Ca)Mn_4^{4+}O_9 \cdot nH_2O$ (Idealized)
Manganous manganese analogue of rancieite

CRYSTAL SYSTEM: Hexagonal
Z: 3
LATTICE CONSTANTS:
  $a = 8.68$
  $c = 9.00$
3 STRONGEST DIFFRACTION LINES:
  7.57 (100)
  3.765 ( 25)
  2.349 ( 20)
  HARDNESS: Vickers hardness (100 g load) 480 kg/mm$^2$.
  DENSITY: 3.41 (Meas. on impure material)
              3.78 (Calc.)
  CLEAVAGE: None.
  HABIT: Massive; in irregular-shaped nodules 1–15 cm across.
  COLOR-LUSTER: Steel gray to black. Opaque. Submetallic to dull. Streak brownish black.
  MODE OF OCCURRENCE: The mineral occurs as microscopic intergrowths with braunite, halloysite, goethite, and quartz, in the oxidation zone of the braunite-rhodochrosite-caryopilite bedded deposit at the Nomura mine, Ehime Prefecture, Japan.
  BEST REF. IN ENGLISH: Nambu, Matsuo, and Tanida, Katsutoshi, *Am. Min.*, **56**: 1487–1488 (1971).


## TAKOVITE (Inadequately described mineral)
$Ni_5Al_4O_2(OH)_{18} \cdot 6H_2O$

3 STRONGEST DIFFRACTION LINES:
  7.566 (100)
  3.767 ( 90)
  2.552 ( 90)
OPTICAL CONSTANTS: Extinction parallel, elongation positive, ns $\alpha'$ 1.598, $\gamma'$ 1.605.
  HABIT: Massive.
  COLOR-LUSTER: Bluish green.
  MODE OF OCCURRENCE: Occurs associated with allophane, gibbsite, and "aidyrlite," at the contact of limestone and metamorphosed serpentinite, at Takova, Serbia. Also reported to occur in altered peridotite at Mueo, New Caledonia, and as a component of bauxite from an unspecified deposit in Greece.
  BEST REF. IN ENGLISH: Maksimovak, Z., *Am. Min.*, **57**: 1559 (1972).


## TALASSKITE = Variety of fayalite


## TALC (Steatite)
$Mg_3Si_4O_{10}(OH)_2$

CRYSTAL SYSTEM: Monoclinic
CLASS: 2/m
SPACE GROUP: C2/c. Also Triclinic Cl or C$\bar{1}$.
Z: 4 (Monoclinic)

LATTICE CONSTANTS:
| (Monoclinic) | (Triclinic) |
|---|---|
| $a = 5.28$ | $a = 5.25$ |
| $b = 9.15$ | $b = 9.13$ |
| $c = 18.9$ | $c = 9.44$ |
| $\beta = 100°15'$ | $\alpha = 90°46'$ |
| | $\beta = 98°55'$ |
| | $\gamma = 90°0'$ |

3 STRONGEST DIFFRACTION LINES:
| (Monoclinic) | (Triclinic) |
|---|---|
| 9.35 (100) | 9.34 (100) |
| 1.53 ( 55) | 4.56 ( 80) |
| 4.59 ( 45) | 3.12 ( 80) |

OPTICAL CONSTANTS:
| (Monoclinic) | (Triclinic) |
|---|---|
| $\alpha = 1.539–1.550$ | $\alpha = 1.545$ |
| $\beta = 1.589–1.594$ | $\beta = 1.584$ |
| $\gamma = 1.589–1.600$ | $\gamma = 1.584$ |
| $(-)2V = 0°–30°$ | $(+)2V = 0°–10°$ |

HARDNESS: 1
Greasy feel.
  DENSITY: 2.58–2.83 (Meas.)
              2.777 (Calc.)
  CLEAVAGE: {001} perfect
Laminae flexible, inelastic.
  HABIT: Crystals commonly thin tabular, up to 1 cm in width. Usually massive, fine-grained compact; also as foliated or fibrous masses or in globular stellate groups.
  COLOR-LUSTER: Pale green to dark green or greenish gray; also white, silvery white, gray, brownish. Translucent. Pearly, greasy, or dull. Streak white.
  MODE OF OCCURRENCE: Occurs widespread as a common mineral of secondary origin formed chiefly by the hydrothermal alteration of ultrabasic rocks or the thermal metamorphism of siliceous dolomites. Typical occurrences are found in California, Arizona, Colorado, South Dakota, Pennsylvania, Vermont, New Hampshire, Massachusetts, New York, Virginia, North Carolina, and Georgia. Also in Canada, England, Norway, Sweden, France, Italy, Austria, Switzerland, Germany, USSR, South Africa, India, China, and Australia.
  BEST REF. IN ENGLISH: Deer, Howie, and Zussman, "Rock Forming Minerals," v. 3, p. 121–130, New York, Wiley, 1962. Ross, Malcolm, Smith, William L., and Ashton, William H., *Am. Min.*, **53**: 751–769 (1968).


## TALMESSITE (Arsenate-belovite; belovite of Nefedov)
$Ca_2Mg(AsO_4)_2 \cdot 2H_2O$
Isomorphous with beta-roselite

CRYSTAL SYSTEM: Triclinic
CLASS: $\bar{1}$
SPACE GROUP: P$\bar{1}$
Z: 1
LATTICE CONSTANTS:
| | |
|---|---|
| $a = 5.89$ | $\alpha = 112°38'$ |
| $b = 7.69$ | $\beta = 70°49'$ |
| $c = 5.56$ | $\gamma = 119°25'$ |

3 STRONGEST DIFFRACTION LINES:
3.07 (100)
2.77 (100)
3.21 ( 80)
OPTICAL CONSTANTS:
$\alpha$ = 1.672
$\beta$ = 1.685
$\gamma$ = 1.698
(-)2V ~ 90°
HARDNESS: 5
DENSITY: 3.2-3.5 (Meas.)
3.483 (Calc.)
HABIT: As druses of minute crystals and as fine crystalline aggregates.
COLOR-LUSTER: Colorless, white, pale green.
MODE OF OCCURRENCE: Occurs in the zone of oxidation at the Talmessi mine, 35 km west of Anarak, Central Iran, associated with aragonite and dolomite; also at an unspecified locality in the USSR. A cobaltian variety has been reported from Bou Azzer, Morocco.
BEST REF. IN ENGLISH: Bariand, P., and Herpin, P., *Am. Min.*, **45**: 1315 (1960). *Am. Min.*, **50**: 813 (1965).

# TALNAKHITE
$Cu_9(Fe, Ni)_8S_{16}$

CRYSTAL SYSTEM: Cubic
CLASS: Possibly $\bar{4}$ 3m
SPACE GROUP: Possibly $\bar{I4}$ 3m
Z: 2
LATTICE CONSTANT:
$a$ = 10.64
3 STRONGEST DIFFRACTION LINES:
3.043 (100)
1.879 ( 90)
1.598 ( 70)
HARDNESS: Not determined.
DENSITY: 4.24 (Meas.)
4.36 (Calc.)
CLEAVAGE: Not determined.
HABIT: Massive.
COLOR-LUSTER: Brass yellow; tarnishes rapidly to hues of pink and brown, then iridescent. Opaque. Metallic.
MODE OF OCCURRENCE: Occurs in association with chalcopyrite, cubanite, djerfisherite, and pentlandite at the Talnakh ore deposit, Noril'sk, western Siberia.
BEST REF. IN ENGLISH: Cabri, L. J., *Econ. Geol.*, **62**: 910-925 (1967).

# TAMARUGITE
$NaAl(SO_4)_2 \cdot 6H_2O$

CRYSTAL SYSTEM: Monoclinic
CLASS: 2/m
SPACE GROUP: $P2_1/a$
Z: 4
LATTICE CONSTANTS:
$a$ = 7.353
$b$ = 25.225
$c$ = 6.097
$\beta$ = 95.2°

3 STRONGEST DIFFRACTION LINES:
4.223 (100)
4.207 ( 80)
3.647 ( 59)
OPTICAL CONSTANTS:
$\alpha$ = 1.484
$\beta$ = 1.486 (Na)
$\gamma$ = 1.497
(+)2V ~ 60°
HARDNESS: ~3
DENSITY: 2.06 (Meas.)
2.066 (Calc.)
CLEAVAGE: {010} perfect
HABIT: Crystals thin tabular or short prismatic. Usually as fine-grained or fibrous masses.
COLOR-LUSTER: Colorless. Transparent. Vitreous.

Soluble in water. Taste sweetish and astringent.

MODE OF OCCURRENCE: Occurs as an alteration of mendozite at Fulton, Calloway County, and near Eureka, St. Louis County, Missouri. It is also found as a secondary mineral associated with sideronatrite at Mina de la Compania, Sierra Gorda, and at several other places in northern Chile; in the sulfur cave at Miseno, Italy; on the island of Cyprus; and at other localities.
BEST REF. IN ENGLISH: Palache, et al., "Dana's System of Mineralogy," 7th Ed., v. II, p. 466-468, New York, Wiley, 1951. Robinson, P. D., and Fang, J. H., *Am. Min.*, **54**: 19-30 (1969).

# TANGEÏTE = Calciovolborthite

# TANTALITE
$(Fe, Mn)(Ta, Nb)_2O_6$
Columbite-Tantalite Series
Dimorphous with tapiolite

CRYSTAL SYSTEM: Orthorhombic
CLASS: 2/m 2/m 2/m
SPACE GROUP: Pcan
Z: 4
LATTICE CONSTANTS:
$a$ = 5.73
$b$ = 14.24
$c$ = 5.08
3 STRONGEST DIFFRACTION LINES:
2.96 (100)
3.66 ( 50)
1.721 ( 20)
OPTICAL CONSTANTS:
$\alpha$ = 2.26
$\beta$ = 2.30-2.40
$\gamma$ = 2.43
(+)2V = large
HARDNESS: 6-6½
DENSITY: 8.2 decreasing linearly with increase in $Nb_2O_5$ content.
CLEAVAGE: {010} distinct
{100} less distinct.
Fracture uneven to subconchoidal. Brittle.

HABIT: Crystals thin to thick tabular, short prismatic, equant; less commonly pyramidal. As large aggregates of parallel to divergent crystals; also massive, compact. Twinning on {201} common, often heart-shaped with pinnate striations parallel to {010}, or as pseudohexagonal trillings. Twinning on {203} or {501} uncommon.

COLOR-LUSTER: Black to brownish black; manganoan varieties often reddish brown. Often tarnished iridescent. Opaque, except in very thin fragments. Submetallic to weakly vitreous. Streak black, brownish black to reddish brown.

MODE OF OCCURRENCE: Occurs in granite pegmatites in association with albite, microcline, quartz, spodumene, beryl, lepidolite, montebrasite, tourmaline, muscovite, cassiterite, and other minerals; also in placer deposits in areas of pegmatites and granitic rocks. Found abundantly, often as large well-formed crystals, at numerous deposits in Custer, Lawrence, and Pennington counties, South Dakota; also at localities in California, Colorado, Wyoming, and in many New England pegmatites. It also occurs in Canada, Brazil, Sweden, France, Finland, USSR, Rhodesia, Western Australia, and elsewhere.

BEST REF. IN ENGLISH: Palache, et al., "Dana's System of Mineralogy," 7th Ed., v. I, p. 780-787, New York, Wiley, 1944.

# TANTALUM
Ta

CRYSTAL SYSTEM: Cubic
CLASS: $4/m\,\bar{3}\,2/m$
SPACE GROUP: Im3m
Z: 2
LATTICE CONSTANT:
$a = 3.2959$
3 STRONGEST DIFFRACTION LINES:
2.338 (100)
2.024 ( 47)
1.221 ( 24)
HARDNESS: 6-7
DENSITY: 11.2 (Meas.) (impure?)
16.411 (Calc. X-ray)
CLEAVAGE: Not determined.
HABIT: As minute cubic crystals and grains.
COLOR-LUSTER: Grayish yellow. Bright luster.
MODE OF OCCURRENCE: Occurs sparingly in the alluvial gold deposits in the Altai and Ural Mountains, USSR.
BEST REF. IN ENGLISH: Palache, et al., "Dana's System of Mineralogy," 7th Ed., v. I, p. 126, New York, Wiley, 1944.

# TANTEUXENITE = Tantalian euxenite

# TANZANITE = Gem variety of zoisite

# TAPIOLITE
(Fe, Mn)(Ta, Nb)$_2$O$_6$
Tapiolite Series
Dimorphous with tantalite

CRYSTAL SYSTEM: Tetragonal
CLASS: 4/m 2/m 2/m
SPACE GROUP: P4$_2$/mnm
Z: 2
LATTICE CONSTANTS:
$a = 4.750-4.753$
$c = 9.208-9.278$
3 STRONGEST DIFFRACTION LINES:
1.740 (100)
3.33 ( 80)
2.56 ( 80)
OPTICAL CONSTANTS:
$\omega = 2.27$
$\epsilon = 2.42$ (Li)
(+)
HARDNESS: 6-6½
DENSITY: 7.82 (Meas.)
8.13 (Calc. for FeTa$_2$O$_6$)
CLEAVAGE: None.
Fracture uneven to subconchoidal.
HABIT: Crystals short prismatic or equant, often appearing monoclinic or orthorhombic due to distortion. Twinning on {013} common, simple and polysynthetic.
COLOR-LUSTER: Black. Opaque. Submetallic to subadamantine. Streak blackish brown to brownish gray.
MODE OF OCCURRENCE: Occurs chiefly in albitized granite pegmatites or in placer deposits in the vicinity of granite pegmatites. Found in South Dakota at the Old Mike mine, Custer County, and at the Hardesty Homestead mine, near Keystone, Pennington County; as fine crystals with quartz and albite in Fremont County, Wyoming; at Paris and Topsham, Maine; in Ceara, Brazil; at Chanteloube, France; at several places in Kimito and Tammela parishes, Finland; and at localities in USSR, Morocco, and Western Australia.
BEST REF. IN ENGLISH: Vlasov, K. A., "Mineralogy of Rare Elements," v. II, p. 449-452, Israel Program for Scientific Translations, 1966.

# TARAMELLITE
Ba$_4$(Fe, Mg)Fe$_2^{3+}$TiSi$_8$O$_{24}$(OH)$_2$

CRYSTAL SYSTEM: Orthorhombic
CLASS: mm2 or 2/m 2/m 2/m
SPACE GROUP: Pnm2$_1$ or Pnmm
Z: 2
LATTICE CONSTANTS:
$a = 13.95$
$b = 7.05$
$c = 12.01$
3 STRONGEST DIFFRACTION LINES:
3.01 (100)
2.584 ( 55)
3.83 ( 50)
OPTICAL CONSTANTS:
$\alpha = 1.77$
$\beta = 1.774$
$\gamma = 1.83$
HARDNESS: 5½
DENSITY: 3.9 (Meas.)
CLEAVAGE: {100} perfect

HABIT: Massive; fibrous.

COLOR-LUSTER: Reddish brown.

MODE OF OCCURRENCE: Occurs at several localities in California, notably in association with sanbornite, quartz, witherite, pyrrhotite, and diopside near Rush Creek, and with gillespite and sanbornite at Big Creek, Fresno County; at the sanbornite locality near Incline, Mariposa County; and at the Kalkar quarry, near Santa Cruz, Santa Cruz County. Also found in limestone at Candoglia, Piedmont, Italy.

BEST REF. IN ENGLISH: Winchell and Winchell, "Elements of Optical Mineralogy," 4th Ed., Pt. 2, p. 401, New York, Wiley, 1951.

## TARAMITE = Katophorite

## TARANAKITE
$H_6 K_3 Al_5 (PO_4)_8 \cdot 18H_2O$

CRYSTAL SYSTEM: Hexagonal

CLASS: 3m or $\bar{3}$ 2/m

SPACE GROUP: R3c or R$\bar{3}$c

Z: 6

LATTICE CONSTANTS:

$a = 8.71$

$c = 96.1$

3 STRONGEST DIFFRACTION LINES:

15.5 (100)

7.6 ( 80)

3.83 ( 80)

HARDNESS: Very soft.

DENSITY: 2.09 (Meas.)

2.11 (Calc.)

HABIT: As minute lath-like crystals; also massive, clay-like, pulverulent to compact.

COLOR-LUSTER: White; also gray or yellowish white.

MODE OF OCCURRENCE: Occurs as flour-like masses near contacts between guano and clay and along fractures within brecciated clay in Pig Hole Cave, Giles County, Virginia. Also found at widespread localities as a product of the reaction of phosphatic solutions derived from bat or sea-bird guano on clays or aluminous rocks. Described from the Sugarloaves, Taranaki, New Zealand; also occurs at Mt. Alburno, Salerno, Italy; in caves at Grotte de Minerve, Hérault, France; and on the island of Rèunion, Indian Ocean.

BEST REF. IN ENGLISH: Palache, et al., "Dana's System of Mineralogy," 7th Ed., v. II, p. 999–1000, New York, Wiley, 1951. *Am. Min.*, **44**: 138–142 (1959).

## TARAPACAITE
$K_2 CrO_4$

CRYSTAL SYSTEM: Orthorhombic

CLASS: 2/m 2/m 2/m

SPACE GROUP: Pmcn

Z: 4

LATTICE CONSTANTS:

$a = 5.92$ kX

$b = 10.40$

$c = 7.61$

3 STRONGEST DIFFRACTION LINES:

3.078 (100)

2.988 ( 75)

2.960 ( 40)

OPTICAL CONSTANTS:

$\alpha = 1.687$

$\beta = 1.722$

$\gamma = 1.731$

$(-)2V = 52°$

HARDNESS: Not determined.

DENSITY: 2.74 (Meas. synthetic)

2.735 (Calc.)

CLEAVAGE: {001} distinct

{010} distinct

HABIT: Crystals thick tabular, flattened on {001}. Twinning on {110}, pseudohexagonal.

COLOR-LUSTER: Bright yellow. Transparent.

Soluble in water.

MODE OF OCCURRENCE: Occurs associated with dietzeite and lopezite in the nitrate deposits in Antofagasta, Atacama, and Tarapaca provinces, Chile.

BEST REF. IN ENGLISH: Palache, et al., "Dana's System of Mineralogy," 7th Ed., v. II, p. 644–645, New York, Wiley, 1951.

## TARASOVITE (an interlayered mica-clay)
$NaK(H_3O)Al_8(Si, Al)_{16}O_{40}(OH)_8 \cdot 2H_2O \cdot 0.42 (Ca, Na)$

CRYSTAL SYSTEM: Monoclinic or triclinic

LATTICE CONSTANTS:

$a = 5.13$

$b = 8.88$

$c = 19.7$

$\beta = 95.0°$

3 STRONGEST DIFFRACTION LINES:

43.75 (100)

10.55 ( 93)

21.8 ( 70)

OPTICAL CONSTANTS:

$\alpha = 1.544$

$\beta = 1.578$

$\gamma = 1.586$

$(-)2V = 23°$

DENSITY: 2.36 (Meas.)

HABIT: As tangled platy aggregates; individual plates up to 0.3 cm.

MODE OF OCCURRENCE: Occurs in the selvage of quartz veins in Middle Carboniferous sandy slates, at Nagolnaya Tarasovka, Donbas region, USSR.

BEST REF. IN ENGLISH: Lazarenko, E. K., and Korolev, Yu. M., *Am. Min.*, **56**: 1123 (1971).

## TARBUTTITE
$Zn_2 PO_4 (OH)$

CRYSTAL SYSTEM: Triclinic

CLASS: $\bar{1}$

SPACE GROUP: P$\bar{1}$

Z: 2

LATTICE CONSTANTS:

| | |
|---|---|
| $a = 5.657$ | $\alpha = 102°27'$ |
| $b = 6.432$ | $\beta = 87°42'$ |
| $c = 5.521$ | $\gamma = 102°34'$ |

3 STRONGEST DIFFRACTION LINES:
   2.78 (100)
   6.12 ( 90)
   3.70 ( 90)
OPTICAL CONSTANTS:
   $\alpha = 1.660$
   $\beta = 1.705$
   $\gamma = 1.713$ (Na)
   $(-)2V = 50°$
HARDNESS: 3¾
DENSITY: 4.140 (Meas.)
         4.21 (Calc.)
CLEAVAGE: {010} perfect
Fracture uneven. Brittle.

HABIT: Crystals short prismatic or equant, commonly rounded and deeply striated. As aggregates or crusts.

COLOR-LUSTER: Colorless, white, or pale shades of yellow, brown, red, or green. Transparent to translucent. Vitreous; pearly on cleavage. Streak white.

MODE OF OCCURRENCE: Occurs as a secondary mineral associated with limonite, hopeite, hemimorphite, pyromorphite, vanadinite, and other secondary minerals, in the zone of oxidation at the zinc mines of Broken Hill, Zambia.

BEST REF. IN ENGLISH: Palache, et al., "Dana's System of Mineralogy," 7th Ed., v. II, p. 869–871, New York, Wiley, 1951. Finney, J. J., *Am. Min.*, **51**: 1218–1220 (1966).

## TARNOWITZITE = Plumbian aragonite

## TATARSKITE (Inadequately described mineral)
$Ca_6Mg_2(SO_4)_2(CO_3)_2Cl_4(OH)_4 \cdot 7H_2O$

CRYSTAL SYSTEM: Probably orthorhombic
3 STRONGEST DIFFRACTION LINES:
   2.967 (100)
   2.625 ( 90)
   5.34 ( 80)
OPTICAL CONSTANTS:
   $\alpha = 1.567$
   $\beta = 1.654$
   $\gamma = 1.7222$
   $(-)2V = 83°$ (Calc.)
HARDNESS: 2½
DENSITY: 2.341 (Meas.)
CLEAVAGE: Two average cleavages on pinacoids.
HABIT: Massive, coarsely crystalline.
COLOR-LUSTER: Colorless to slightly yellowish; transparent; vitreous, pearly on cleavage.
MODE OF OCCURRENCE: Found associated with halite, bischofite, magnesite, hilgardite, and strontio-hilgardite in a drill core in anhydrite rock, of the Caspian depression, at depths of 850–900 meters. It is in part present as pseudomorphs after an unknown prismatic mineral.

BEST REF. IN ENGLISH: Lobanova, V. V., *Am. Min.*, **49**: 1151–1152 (1964).

## TAURISCITE
$FeSO_4 \cdot 7H_2O$
Isostructural with epsomite

CRYSTAL SYSTEM: Orthorhombic
CLASS: 2 2 2
SPACE GROUP: $P2_12_12_1$
   HARDNESS: ~2
   DENSITY: 1.875 (Meas.)
   CLEAVAGE: Not determined.
   HABIT: As well-formed short prismatic to equant crystals up to 1 cm in size.
   COLOR-LUSTER: Green; transparent; vitreous.
   MODE OF OCCURRENCE: Occurs associated with potash alum and melanterite at Wingdälle, Canton Uri, Switzerland, and also reported from Szomolnok, Czechoslovakia.

   BEST REF. IN ENGLISH: Palache, et al., "Dana's System of Mineralogy," 7th Ed., v. II, p. 519, New York, Wiley, 1951.

## TAVISTOCKITE = Fluorapatite

## TAVORITE
$LiFe^{3+}PO_4OH$

CRYSTAL SYSTEM: Triclinic
3 STRONGEST DIFFRACTION LINES:
   3.045 (100)
   3.285 ( 90)
   4.99 ( 50)
OPTICAL CONSTANTS: Mean index = 1.807
   HARDNESS: Not determined.
   DENSITY: 3.288 (Meas.)
   CLEAVAGE: 3 distinct cleavages.
   HABIT: Crystals equant or short prismatic; often in druses of interlocking crystals. Also in fine granular crystalline masses and coatings.
   COLOR-LUSTER: Grass green to yellowish green; transparent to translucent. Vitreous. Coatings earthy.
   MODE OF OCCURRENCE: Occurs as minute yellow-green to grass green crystals in sheaf-like aggregates lining cavities in altered triphylite at the Bull Moose mine; and as superb grass green crystals up to 1 mm in size, associated with leucophosphite and hureaulite crystals in altered triphylite, at the Tip-Top mine, Custer County, South Dakota, and in several other pegmatites in the Custer and Keystone districts in the same state. Found originally sparsely disseminated in altered triphylite at the Sapucaia pegmatite mine, Minas Geraes, Brazil, and subsequently identified from pegmatites in Germany, Czechoslovakia, Madagascar, and the New England area.

   BEST REF. IN ENGLISH: Lindberg, M. L., and Pecora, W. T., *Am. Min.*, **40**: 952–966 (1955).

## TAWMAWITE = Chromium variety of epidote

## TAYLORITE (ammonian arcanite)
$K_{2-x}(NH_4)_xSO_4$ with $x \sim 0.35$

CRYSTAL SYSTEM: Orthorhombic
SPACE GROUP: Pmcn
Z: 4
LATTICE CONSTANTS:
$a \simeq 5.8$
$b \simeq 10.1$
$c \simeq 7.5$
3 STRONGEST DIFFRACTION LINES:
2.92 (100)
4.21 ( 80)
3.04 ( 70)
OPTICAL CONSTANTS:
$\alpha = 1.5008$
$\beta = 1.5037$
$\gamma = 1.5062$
(+)2V = large
HARDNESS: 2
DENSITY: Not determined.
CLEAVAGE: Not determined.
HABIT: Massive; as compact lumps or concretions with a crystalline structure.
COLOR-LUSTER: White or creamy.

Taste pungent and bitter.
MODE OF OCCURRENCE: Occurs in guano from Guanapé, Peru, and from the Chincha Islands off the coast of Peru.
BEST REF. IN ENGLISH: Palache, et al., "Dana's System of Mineralogy," 7th Ed., v. II, p. 400, New York, Wiley, 1951. Winchell, Horace, and Benoit, Richard J., *Am. Min.*, **36**: 590–602 (1951).

## TAZHERANITE
(Zr, Ca, Ti)O$_2$

CRYSTAL SYSTEM: Cubic
CLASS: 4/m $\bar{3}$ 2/m, 432, or $\bar{4}$ 3m
SPACE GROUP: Fm3m, F432, or F$\bar{4}$3c
Z: 3
LATTICE CONSTANT:
$a = 5.108$
3 STRONGEST DIFFRACTION LINES:
2.94 (100)
1.804 (100)
1.539 (100)
OPTICAL CONSTANT:
$N = 2.35$
HARDNESS: 7.5
DENSITY: 5.01 (Meas.)
CLEAVAGE: None. Brittle.
HABIT: Irregular grains; also as rounded or thick tabular isometric crystals 0.01–0.2, rarely up to 1.5 mm, some with perfect faces.
COLOR-LUSTER: Yellowish orange to reddish orange, rarely cherry red. Adamantine to greasy.
MODE OF OCCURRENCE: Occurs in calciphyres which form bands in periclase-brucite marbles within the alkalic and nepheline syenites of the Tazheran alkalic massif, west of Lake Baikal, Siberia. Associated minerals are calcite, dolomite, spinel, forsterite, pyrrhotite, melilite, ludwigite, clinohumite, magnesioferrite, calzirtite, baddeleyite, giekielite, perovskite, rutile, and zircon.
BEST REF. IN ENGLISH: Konev, A. A., Ushchapovskaya, Z. F., Kashaev, A. A., and Lebedeva, V. S., *Am. Min.*, **55**: 318 (1970).

## TEALLITE
PbSnS$_2$

CRYSTAL SYSTEM: Orthorhombic
CLASS: 2/m 2/m 2/m
SPACE GROUP: Pbnm
Z: 2
LATTICE CONSTANTS:
(synthetic)
$a = 4.266$
$b = 11.419$
$c = 4.090$
3 STRONGEST DIFFRACTION LINES:
2.856 (100)
3.327 ( 16)    (synthetic)
3.416 ( 10)
HARDNESS: 1½
DENSITY: 6.36 (Meas.)
6.57 (Calc.)
CLEAVAGE: {001} perfect
Flexible; inelastic. Somewhat malleable.
HABIT: Crystals tabular; thin; nearly square outline. Commonly striated. Usually massive aggregates of thin molybdenite-like lamellae with irregular edges.
COLOR-LUSTER: Lead gray to iron-gray to grayish black. Opaque. Metallic; sometimes tarnished dull or iridescent. Streak black.
MODE OF OCCURRENCE: Occurs associated with cassiterite and various sulfide minerals in the tin veins of Ocuri, Huanuni, Colquiri, Monserrat, Colquechaca, and elsewhere in Bolivia. Also reported to occur at Freiberg, Saxony, Germany.
BEST REF. IN ENGLISH: Palache, et al., "Dana's System of Mineralogy," 7th Ed., v. I, p. 439–441, New York, Wiley, 1944.

## TEEPLEITE
Na$_2$B(OH)$_4$Cl

CRYSTAL SYSTEM: Tetragonal
CLASS: 4/m 2/m 2/m
SPACE GROUP: P4/nmm
Z: 2
LATTICE CONSTANTS:
$a = 7.26$
$c = 4.85$
3 STRONGEST DIFFRACTION LINES:
2.697 (100)
2.015 ( 90)
2.90 ( 80)
OPTICAL CONSTANTS:
$\omega = 1.519$
$\epsilon = 1.503$
(−)

HARDNESS: 3–3½

DENSITY: 2.076 (Meas.)

2.07 (Calc.)

CLEAVAGE: Fracture subconchoidal to irregular. Very brittle.

HABIT: Crystals tabular, flattened on {001}. As interpenetrating groups of crystals.

COLOR-LUSTER: White; transparent. Vitreous to greasy, becoming dull on exposure.

MODE OF OCCURRENCE: Occurs with trona and halite in Borax Lake, near Clear Lake, Lake County, and in the lower layers of the beds at Searles Lake, San Bernardino County, California.

BEST REF. IN ENGLISH: Palache, et al., "Dana's System of Mineralogy," 7th Ed., v. II, p. 372–373, New York, Wiley, 1951. Ross and Edwards, *Am. Min.*, **44**: 875 (1959).

# TEINEITE

Cu(Te, S)O$_3$ · 2H$_2$O

Isostructural with chalcomenite

CRYSTAL SYSTEM: Orthorhombic

CLASS: 2 2 2

SPACE GROUP: P2$_1$ 2$_1$ 2$_1$

Z: 4

LATTICE CONSTANTS:

$a$ = 6.63

$b$ = 9.61

$c$ = 7.43

3 STRONGEST DIFFRACTION LINES:

3.45 (100)

5.45 ( 90) broad

3.06 ( 80)

OPTICAL CONSTANTS:

$\alpha$ = 1.767

$\beta$ = 1.782

$\gamma$ = 1.791

(–)2V = 36°

HARDNESS: 2½

DENSITY: 3.80 (Meas.)

3.85 (Calc.)

CLEAVAGE: {001} distinct

{100} indistinct

Brittle.

HABIT: Crystals prismatic, up to 1 cm long. Faces deeply etched. As aggregates of intergrown crystals and as crusts.

COLOR-LUSTER: Sky blue to cobalt blue or bluish gray. Streak bluish white.

MODE OF OCCURRENCE: Occurs in association with native tellurium, sylvanite, tetrahedrite, and pyrite in quartz-chalcedony and barite veins, at the Teine gold-ore deposit, Hokkaido, Japan.

BEST REF. IN ENGLISH: Vlasov, K. A., "Mineralogy of Rare Elements," v. II, P. 759–760, Israel Program for Scientific Translations, 1966.

# TELLURANTIMONY

Sb$_2$Te$_3$

Isostructural with tellurbismuth

CRYSTAL SYSTEM: Hexagonal

CLASS: $\bar{3}$2/m

SPACE GROUP: R$\bar{3}$m

Z: 1

LATTICE CONSTANTS:

$a$ = 4.258

$c$ = 30.516

3 STRONGEST DIFFRACTION LINES:

3.156 (100)

2.129 ( 80)

2.348 ( 70)

OPTICAL CONSTANTS: Anisotropism moderate.

HARDNESS: Microhardness 39.6-61.3 kg/mm$^2$ with 25 gram load.

DENSITY: Not determined.

HABIT: Crystals lath-shaped up to 175 microns wide and 350 microns long. Twinning perpendicular to the elongation axis common.

COLOR-LUSTER: Weakly pleochroic under reflected light with colors from pink to cream.

MODE OF OCCURRENCE: Tellurantimony occurs as scattered inclusions in altaite in specimens from a small zone of telluride ore within the Mattagami Lake Mine near the town of Matagami, Quebec, Canada.

BEST REF. IN ENGLISH: Thorpe, R. I., and Harris, D. C., *Can. Min.*, **12**: 55–60 (1973).

# TELLURITE

TeO$_2$

Dimorphous with paratellurite

CRYSTAL SYSTEM: Orthorhombic

CLASS: 2/m 2/m 2/m

SPACE GROUP: Pbca

Z: 8

LATTICE CONSTANTS:

$a$ = 5.607

$b$ = 12.034

$c$ = 5.463

3 STRONGEST DIFFRACTION LINES:

3.280 (100)

3.723 ( 95)

3.008 ( 50)

OPTICAL CONSTANTS:

$\alpha$ = 2.00

$\beta$ = 2.18 (Li)

$\gamma$ = 2.35

(+)2V = nearly 90°

HARDNESS: 2

DENSITY: 5.90 (Meas.)

5.83 (Calc.)

CLEAVAGE: {010} perfect

Flexible.

HABIT: Crystals acicular, or as thin striated plates. Also as spherical masses with radial structure, and as pulverulent coatings.

COLOR-LUSTER: White, straw yellow to yellowish orange. Transparent. Adamantine.

MODE OF OCCURRENCE: Occurs as an oxidation product of tellurium minerals, often intimately associated with native tellurium. Found in Boulder, Gunnison, and Teller counties, Colorado; in Jefferson Canyon, Nye County, Nevada; as fine specimens at Moctezuma, Sonora, Mexico; and at localities in Roumania, USSR, and Japan.

BEST REF. IN ENGLISH: Palache, et al., "Dana's System of Mineralogy," 7th Ed., v. I, p. 593–595, New York, Wiley, 1944.

# TELLURIUM
Te

CRYSTAL SYSTEM: Hexagonal
CLASS: 32
SPACE GROUP: $P3_1 21$ or $P3_2 21$
Z: 3
LATTICE CONSTANTS:
$a = 4.447$
$c = 5.915$
3 STRONGEST DIFFRACTION LINES:
3.230 (100)
2.351 ( 37)
2.228 ( 31)
OPTICAL CONSTANTS: Opaque.
HARDNESS: 2–2½
DENSITY: 6.25 (Meas.)
6.2365 (Calc.)
CLEAVAGE: $\{10\bar{1}0\}$ perfect
$\{0001\}$ imperfect
Brittle.
HABIT: Crystals usually small, prismatic to acicular. Usually massive, columnar to fine granular.
COLOR-LUSTER: Tin white. Opaque. Metallic. Streak gray.
MODE OF OCCURRENCE: Occurs in hydrothermal vein deposits often associated with gold and silver tellurides, native gold, and galena. Found at numerous mines in Boulder, Gunnison, La Plata, Montezuma, Saguache, and Teller counties, Colorado; in Butte, Calaveras, Mariposa, Shasta, and Tuolumne counties, California; and at Delamar, Lincoln County, Nevada. It also occurs as large crystalline masses at Moctezuma, Sonora, Mexico; at localities in Roumania, Japan, Western Australia, and as large crystals in the mines of Balia, Asia Minor.
BEST REF. IN ENGLISH: Palache, et al., "Dana's System of Mineralogy," 7th Ed., v. I, p. 138–139, New York, Wiley, 1944.

# TELLUROBISMUTHITE (Wehrlite)
$Bi_2 Te_3$

CRYSTAL SYSTEM: Hexagonal
CLASS: $\bar{3} 2/m$
SPACE GROUP: $R\bar{3}m$
Z: 3 (hexag.), 1 (rhomb.)
LATTICE CONSTANTS:
(hexag.)     (rhomb.)
$a = 4.375$    $a_{rh} = 10.44$
$c = 30.39$    $\alpha = 24°11.5'$

3 STRONGEST DIFFRACTION LINES:
3.222 (100)
2.376 ( 25)
2.192 ( 25)
OPTICAL CONSTANTS: Opaque.
HARDNESS: 1½–2
DENSITY: 7.815 (Meas.)
8.05 (Calc.)
CLEAVAGE: $\{0001\}$ perfect
Laminae flexible, inelastic. Somewhat sectile.
HABIT: As foliated masses or irregular plates.
COLOR-LUSTER: Pale lead-gray. Opaque. Metallic. Streak same as color.
MODE OF OCCURRENCE: Occurs in Colorado at the Treasure Vault mine, Clear Creek County; on Mount Chipeta and near Whitehorn, Fremont County; and at the Hamilton and Little Gerald mines on the slopes of Sierra Blanca, which marks the junction of Huerfano, Costilla, and Alamosa counties. It also is found in a gold placer at Highland, Montana; at the Little Mildred mine, Hachita, New Mexico; at deposits in Georgia and Virginia; and at localities in Canada, Sweden, and Japan.
BEST REF. IN ENGLISH: Palache, et al., "Dana's System of Mineralogy," 7th Ed., v. I, p. 160–161, New York, Wiley, 1944.

# TENGERITE (Inadequately described mineral)
$CaY_3 (CO_3)_4 (OH)_3 \cdot 3H_2O$

CRYSTAL SYSTEM: Unknown
3 STRONGEST DIFFRACTION LINES:
3.86 (100)
4.55 ( 70)
2.95 ( 70)
OPTICAL CONSTANTS:
$\alpha = 1.555$    $\alpha = 1.622$
$\beta = 1.57$   and $\beta = ?$
$\gamma = 1.585$    $\gamma = 1.642$
(+)2V = large
HARDNESS: Not determined.
DENSITY: 3.12 (Meas.)
CLEAVAGE: Not determined.
HABIT: As thin crystalline to powdery coatings; also as thin fibrous mammillary crusts.
COLOR-LUSTER: White, dull.
MODE OF OCCURRENCE: Occurs as a coating on gadolinite in the Roscoe dike, Clear Creek Canyon, Jefferson County, Colorado; as a coating on melanocerite at the Cardiff uranium mine, Wilberforce, Ontario, Canada; at Ytterby, Kragerö, and Åskagen, Sweden; at Hundholmen, Norway; and at Iisaka, Japan.
BEST REF. IN ENGLISH: Palache, et al., "Dana's System of Mineralogy," 7th Ed., v. II, p. 275–276, New York, Wiley, 1951.

# TENNANTITE
$(Cu, Fe)_{12} As_4 S_{13}$
Tetrahedrite Series

CRYSTAL SYSTEM: Cubic

CLASS: $\bar{4}$3m
SPACE GROUP: I$\bar{4}$3m
Z: 2
LATTICE CONSTANT:
 $a$ = 10.19
3 STRONGEST DIFFRACTION LINES:
 2.94 (100)
 1.801 ( 80)
 1.535 ( 50)
OPTICAL CONSTANTS: Opaque except in very thin splinters. $N > 2.72$
 HARDNESS: 3–4½
 DENSITY: 4.59–4.75 (Meas.)
 CLEAVAGE: None.
Fracture uneven to subconchoidal. Brittle.
 HABIT: Crystals tetrahedral, often modified by other forms. Also massive, coarse granular to compact. Twin axis [111]; contact or penetration twins common. Often repeated.
 COLOR-LUSTER: Steel gray to iron black. Opaque. Metallic, sometimes splendent. Streak black to brown to dark red.
 MODE OF OCCURRENCE: Occurs chiefly in low- to high-temperature hydrothermal ore veins associated with sulfides, carbonates, barite, fluorite, and quartz; less frequently in high-temperature veins or contact metasomatic deposits. Found in the United States in Virginia, North Carolina, Colorado, Montana, Utah, and Idaho. It also occurs as exceptional crystals in Zacatecas, Mexico; and at localities in Peru, England, Norway, Sweden, Switzerland, Germany, Poland, Korea, and Tsumeb, South West Africa.
 BEST REF. IN ENGLISH: Palache, et al., "Dana's System of Mineralogy," 7th Ed., v. I, p. 374–384, New York, Wiley, 1944.

# TENORITE
CuO

CRYSTAL SYSTEM: Monoclinic
CLASS: 2/m
SPACE GROUP: C2/c
Z: 4
LATTICE CONSTANTS:
 $a$ = 4.653
 $b$ = 3.425
 $c$ = 5.129
 $\beta$ = 99°28′
3 STRONGEST DIFFRACTION LINES:
 2.52 (100)
 2.32 ( 96)
 2.53 ( 49)
OPTICAL CONSTANTS:
 $\beta$ = 2.63 (red)
 $\beta$ = 3.17 (blue)
 2V = large
 HARDNESS: 3½
 DENSITY: 6.45 (Meas. Synthetic)
     6.51 (Calc.)
 CLEAVAGE: In zones [011] and [0$\bar{1}$1]. Fracture uneven to conchoidal. Brittle. Thin fragments flexible and elastic.

 HABIT: Crystals very thin, elongated, striated laths. Also as thin scales or curved plates. Usually massive—compact to powdery. Twinning on {011} common, producing feather-like, dendritic, or stellate forms.
 COLOR-LUSTER: Steel gray to black. Opaque; thin scales translucent. Metallic.
 MODE OF OCCURRENCE: Occurs chiefly in the oxidation zone of copper deposits, often associated with malachite, azurite, limonite, manganese oxides, cuprite, and other secondary minerals; also as a sublimation product deposited upon lavas in volcanic regions. In the United States notable deposits occur in California, Oregon, Nevada, Arizona, New Mexico, Montana, Utah, Michigan, and Tennessee. Other major occurrences are found in Bolivia, Chile, England, Scotland, Spain, France, Italy, Germany, Czechoslovakia, Roumania, USSR, and Tsumeb, South West Africa.
 BEST REF. IN ENGLISH: Palache, et al., "Dana's System of Mineralogy," 7th Ed., v. I, p. 507–510, New York, Wiley, 1944.

# TEPHROITE
$Mn_2SiO_4$
Olivine Group

CRYSTAL SYSTEM: Orthorhombic
CLASS: 2/m 2/m 2/m
SPACE GROUP: Pnma
Z: 4
LATTICE CONSTANTS:
 $a$ = 4.86–4.90
 $b$ = 10.60–10.62
 $c$ = 6.22–6.25
3 STRONGEST DIFFRACTION LINES:
 2.56 (100)
 3.61 ( 85)
 2.86 ( 85)
OPTICAL CONSTANTS:
  $\alpha$ = 1.770–1.788
  $\beta$ = 1.807–1.810
  $\gamma$ = 1.817–1.825
 (−)2V = 60°–70°
 HARDNESS: 6
 DENSITY: 4.113 (Meas.)
     4.15 (Calc.)
 CLEAVAGE: {010} distinct
     {001} imperfect
Fracture uneven to conchoidal. Brittle.
 HABIT: Crystals generally short prismatic, highly modified, usually elongated parallel to $c$-axis. Also massive, compact, or as disseminated grains. Twinning on {011}, not common.
 COLOR-LUSTER: Gray, olive green, bluish green, flesh red, reddish brown. Transparent to translucent. Vitreous to greasy.
 MODE OF OCCURRENCE: Occurs chiefly in iron-manganese ore deposits and their associated skarns; rarely in metamorphic rocks derived from manganese-rich sediments. Found in Santa Clara County and other places in California; at the Sunnyside mine, San Juan County, Colorado; and in abundance, often as fine crystals, at Franklin and Sterling Hill, New Jersey. Other notable occurrences are found at

the Benallt mine, Wales; in the Treburland mine, Cornwall, England; at Långban and other localities in Sweden; and at deposits in France, Australia, Japan, Antarctica, and elsewhere.

BEST REF. IN ENGLISH: Deer, Howie, and Zussman, "Rock Forming Minerals," v. I, p. 34-40, New York, Wiley, 1961.

## TEREMKOVITE = Owyheeite

## TERLINGUAITE
$Hg_2ClO$

CRYSTAL SYSTEM: Monoclinic
CLASS: 2/m
SPACE GROUP: C 2/c
Z: 4
LATTICE CONSTANTS:
  $a$ = 12.01
  $b$ = 5.91
  $c$ = 9.49
  $\beta$ = 106°
3 STRONGEST DIFFRACTION LINES:
  3.26 (100)
  2.51 (100)
  2.81 ( 60)
OPTICAL CONSTANTS:
      $\alpha$ = 2.35
      $\beta$ = 2.64      (Li)
      $\gamma$ = 2.66
  (-)2V $\simeq$ 20° (red)
  HARDNESS: 2½
  DENSITY: 8.7 (Meas.)
          8.73 (Calc.)
  CLEAVAGE: $\{\bar{1}01\}$ perfect
Brittle.
  HABIT: Crystals prismatic [010] often flattened on $\{001\}$. Other forms rare. As massive aggregates of distorted crystals, also as fine powder.
  COLOR-LUSTER: Yellow to greenish yellow, also brown. Yellowish crystals and powder become olive green on exposure to light. Transparent to translucent. Brilliant adamantine.
  MODE OF OCCURRENCE: Found with calomel, cinnabar, eglestonite, kleinite, mercury, and montroydite at the mercury deposits in Cretaceous limestone in the Terlingua district, Brewster County, Texas.
  BEST REF. IN ENGLISH: Palache, et al., "Dana's System of Mineralogy, 7th Ed., v. II, p. 52-56, New York, Wiley, 1951.

## TERNOVSKITE = Riebeckite

## TERTSCHITE (Inadequately described mineral)
$Ca_4B_{10}O_{19} \cdot 20H_2O$

CRYSTAL SYSTEM: Monoclinic (?)

3 STRONGEST DIFFRACTION LINES:
  2.83 ⎫
  2.35 ⎬ Intensities not given.
  2.02 ⎭
OPTICAL CONSTANTS:
  $\alpha'$ = 1.502
  $\gamma'$ = 1.517
  HARDNESS: Not determined.
  DENSITY: Not determined.
  CLEAVAGE: Not determined.
  HABIT: Massive, fibrous; resembling ulexite.
  COLOR-LUSTER: Snow white; silky. Fluoresces light to deep blue-violet under ultraviolet light.
  MODE OF OCCURRENCE: Occurs at the Kurtpinari mine, Faras, Turkey.
  BEST REF. IN ENGLISH: Meixner, Heinz, *Am. Min.*, **39**: 849 (1954).

## TERUGGITE
$Ca_4MgB_{12}As_2O_{28} \cdot 18H_2O$

CRYSTAL SYSTEM: Monoclinic
CLASS: 2/m
SPACE GROUP: $P2_1/a$
Z: 2
LATTICE CONSTANTS:
  $a$ = 15.68
  $b$ = 19.90
  $c$ = 6.25
  $\beta$ = 100°05'
3 STRONGEST DIFFRACTION LINES:
  12.13  (100)
  2.785 ( 30)
  9.98  ( 22)
OPTICAL CONSTANTS:
      $\alpha$ = 1.526
      $\beta$ = 1.528
      $\gamma$ = 1.551
  (+)2V = 33°
  HARDNESS: 2½
  DENSITY: 2.149 (Meas.)
          2.139 (Calc.)
  CLEAVAGE: $\{001\}$ good
          $\{110\}$ fair
  HABIT: Crystals acicular, greatly elongated on the $c$-axis with rhomboidal cross section perpendicular to this direction. Crystals measure between 30 and 110 $\mu$m in length.
  COLOR-LUSTER: Colorless and transparent; vitreous. Aggregates present a powdery white appearance.
  MODE OF OCCURRENCE: Occurs in white cauliflower-shaped nodules, ranging from 2 to 6 cm in diameter, in a spring deposit at the Loma Blanca borate deposit, province of Jujuy, Argentina. It is associated with inyoite, calcite, ulexite, aragonite, and realgar.
  BEST REF. IN ENGLISH: Aristarain, L. F., and Hurlbut, C. S. Jr., *Am. Min.*, **53**: 1815-1827 (1968).

## TESCHEMACHERITE
$(NH_4)HCO_3$

CRYSTAL SYSTEM: Orthorhombic
CLASS: 2/m 2/m 2/m
SPACE GROUP: Pnaa
Z: 8
LATTICE CONSTANTS:
 $a$ = 8.76 kX
 $b$ = 10.79
 $c$ = 7.29
3 STRONGEST DIFFRACTION LINES:
 3.00 (100)
 5.34 ( 60)
 3.62 ( 55)
OPTICAL CONSTANTS:
  $\alpha$ = 1.4227
  $\beta$ = 1.5358   (Na)
  $\gamma$ = 1.5545
 $(-)2V$ = 41°38′ (Calc.)
 HARDNESS: 1½
 DENSITY: 1.58 (Meas.)
 CLEAVAGE: {110} perfect
Brittle.
 HABIT: Massive, compact crystalline.
 COLOR-LUSTER: Colorless, white, yellowish. Transparent.
 MODE OF OCCURRENCE: Occurs in guano deposits on Chincha and Guañape Islands, Peru, and at Saldanha Bay, South Africa.
 BEST REF. IN ENGLISH: Palache, et al., "Dana's System of Mineralogy," 7th Ed., v. II, p. 137-138, New York, Wiley, 1951.


## TETRADYMITE
$Bi_2 Te_2 S$

CRYSTAL SYSTEM: Hexagonal
CLASS: $\bar{3}$ 2/m
SPACE GROUP: R$\bar{3}$ m
Z: 3 (hexag.), 1 (rhomb.)
LATTICE CONSTANTS:
 (hexag.)        (rhomb.)
 $a$ = 4.32        $a_{rh}$ = 10.31
 $c$ = 30.01       $\alpha$ = 24°10′
3 STRONGEST DIFFRACTION LINES:
 3.11  (100)
 2.29  ( 40)
 1.645 ( 40)
OPTICAL CONSTANTS: Opaque.
 HARDNESS: 1½–2
 DENSITY: 7.1–7.5 (Meas.)
        7.21 (Calc.)
 CLEAVAGE: {0001} perfect
Laminae flexible, inelastic.
 HABIT: Crystals usually indistinct; steep pyramidal with pyramid faces striated horizontally. Commonly massive, granular to foliated; also bladed. Twinning on {01$\bar{1}$8} or {01$\bar{1}$5} as fourlings.
 COLOR-LUSTER: Pale steel-gray; tarnishes iridescent and dull. Opaque. Metallic. Streak pale steel gray.
 MODE OF OCCURRENCE: Occurs chiefly in quartz veins and in contact metamorphic deposits, commonly associated with native gold, tellurides, and sulfides. Found widespread

in small amounts in the United States, especially in California, Arizona, New Mexico, Colorado, Idaho, Montana, South Dakota, North and South Carolina, Georgia, and Virginia. It also occurs at many gold deposits in Canada; near Sorata, Bolivia; as excellent crystals near Schemnitz, Czechoslovakia; and at localities in Norway, Sweden, Roumania, Rhodesia, Australia, and Japan.
 BEST REF. IN ENGLISH: Palache, et al., "Dana's System of Mineralogy," 7th Ed., v. I, p. 161-164, New York, Wiley, 1944.


## TETRAHEDRITE
$Cu_{12} Sb_4 S_{13}$
Tetrahedrite Series

CRYSTAL SYSTEM: Cubic
CLASS: $\bar{4}$ 3m
SPACE GROUP: I$\bar{4}$3m
Z: 2
LATTICE CONSTANT:
 $a$ = 10.33
3 STRONGEST DIFFRACTION LINES:
 3.00  (100)
 1.831 ( 60)
 1.563 ( 30)
OPTICAL CONSTANTS: Opaque except in thin splinters.
        $N > 2.72$ (Li)
 HARDNESS: 3–4½
 DENSITY: 4.6–5.1 increasing with Sb or Ag
        4.99 (Calc.)
 CLEAVAGE: None.
Fracture uneven to subconchoidal. Brittle.
 HABIT: Crystals tetrahedral, often modified by other forms. Also massive, coarse granular to compact. Twin axis [111]; contact or penetration twins common. Often repeated.
 COLOR-LUSTER: Steel gray to iron black. Opaque. Metallic, sometimes splendent. Streak black to brown to dark red.
 MODE OF OCCURRENCE: Occurs chiefly in low- to medium-temperature hydrothermal ore veins associated with sulfides, carbonates, barite, fluorite, and quartz; less frequently in high-temperature veins, contact metasomatic deposits, and pegmatites. Found widespread throughout the western United States: as fine crystals at Bingham, Salt Lake County, Utah; and in large amounts in California, Idaho, Montana, Colorado, Nevada, Arizona, and New Mexico. It also occurs in Canada, Mexico, Bolivia, Chile, Peru, England, France, Germany, Roumania, Czechoslovakia, Switzerland, Sweden, Algeria, and Australia.
 BEST REF. IN ENGLISH: Palache, et al., "Dana's System of Mineralogy," 7th Ed., v. I, p. 374-384, New York, Wiley, 1944.


## TETRAWICKMANITE
$MnSn(OH)_6$
Dimorphous with wickmanite

CRYSTAL SYSTEM: Tetragonal
CLASS: 4/m

SPACE GROUP: $P4_2/n$
Z: 1
LATTICE CONSTANTS:
  $a = 7.7870$
  $c = 7.797$
3 STRONGEST DIFFRACTION LINES:
  3.94 (100)
  2.77 ( 90)
  1.76 ( 50)
OPTICAL CONSTANTS:
  $\omega = 1.724$
  $\epsilon = 1.720$
    (−)
HARDNESS: Not determined.
DENSITY: 3.65 (Meas.)
          3.79 (Calc.)
CLEAVAGE: None.
HABIT: Crystals resemble minute wulfenite crystals; prominent pyramid {112}, the pinacoid {001}, and very minor prism {100}. Also as minute, rounded, rough-surfaced, globule-like masses.
COLOR-LUSTER: Honey yellow, brownish orange. Transparent to translucent. Vitreous.
MODE OF OCCURRENCE: Occurs very sparsely in pegmatite at the Foote Mineral Company spodumene mine, Kings Mountain, North Carolina. It is found associated with bavenite, eakerite, siderite-rhodochrosite, quartz, and albite.
BEST REF. IN ENGLISH: White, John S. Jr., and Nelen, Joseph A., *Min. Rec.*, 4: 24–29 (1973).


# THALENITE
$Y_2Si_2O_7$
Var. Yttrialite
     Rowlandite

CRYSTAL SYSTEM: Monoclinic
LATTICE CONSTANTS:
  $a:b:c = 0.919:1:0.648$
    $\beta = 97°05'$
3 STRONGEST DIFFRACTION LINES:
  3.10 (100)
  2.81 ( 40)
  2.75 ( 30)
OPTICAL CONSTANTS: Generally metamict.
      $\alpha = 1.731$
      $\beta = 1.738$
      $\gamma = 1.744$
  (−)2V = 68°
HARDNESS: 6
DENSITY: 4.3–4.6 (Meas.)
CLEAVAGE: None. Fracture conchoidal. Brittle.
HABIT: Crystals tabular or prismatic. Usually compact massive.
COLOR-LUSTER: Flesh pink, brownish green, brown. Translucent. Greasy.
MODE OF OCCURRENCE: Occurs in granite pegmatites in association with fergusonite, allanite, gadolinite, cyrtolite, and magnetite. Found at Baringer Hill, Llano County, Texas; at Åskagen and at Österby, Sweden; at Hundholmen,

Norway; at Iizaka, Fukushima Prefecture, Japan; in European USSR and Siberia; and at the Snowflake Feldspar mine, Teller County, Colorado.
BEST REF. IN ENGLISH: Vlasov, K. A., "Mineralogy of Rare Elements," v. II, p. 243–246, Israel Program for Scientific Translations, 1966.


# THAUMASITE
$Ca_3Si(OH)_6(CO_3)(SO_4) \cdot 12H_2O$

CRYSTAL SYSTEM: Hexagonal
CLASS: 6
SPACE GROUP: $P6_3$
Z: 2
LATTICE CONSTANTS:
  $a = 10.90$
  $c = 10.29$
3 STRONGEST DIFFRACTION LINES:
  9.66  (100)
  3.792 ( 75)
  4.582 ( 65)
OPTICAL CONSTANTS:
  $\omega = 1.500-1.507$
  $\epsilon = 1.464-1.468$
    (−)
HARDNESS: 3½
DENSITY: 1.91 (Meas.)
          1.916 (Calc.)
CLEAVAGE: {$10\bar{1}1$} in traces. Brittle.
HABIT: Crystals acicular to filiform. Usually massive, compact.
COLOR-LUSTER: Colorless, white. Transparent to translucent. Vitreous to somewhat silky or greasy.
MODE OF OCCURRENCE: Occurs in association with spurrite or ettringite at Crestmore, Riverside County, California; at the Old Hickory mine, Beaver County, Utah; at the Lucky Cuss mine, Cochise County, Arizona; as fine crystalline masses at Paterson and West Paterson, New Jersey; as minute crystals and large crystalline aggregates associated with prehnite near Centreville, Fairfax County, Virginia; at Långban, Sweden; and elsewhere.
BEST REF. IN ENGLISH: Winchell and Winchell, "Elements of Optical Mineralogy," 4th Ed., pt. 2, p. 179, New York, Wiley, 1951.


# THENARDITE
$Na_2SO_4$

CRYSTAL SYSTEM: Orthorhombic
CLASS: 2/m 2/m 2/m
SPACE GROUP: Fddd
Z: 8
LATTICE CONSTANTS:
  $a = 9.75$ kX
  $b = 12.29$
  $c = 5.85$
3 STRONGEST DIFFRACTION LINES:
  2.783 (100)
  4.66 ( 73)
  3.178 ( 51)

OPTICAL CONSTANTS:
$\alpha = 1.464 – 1.471$
$\beta = 1.473 – 1.477$
$\gamma = 1.4812 – 1.485$
$(+)2V = 82°35'$
HARDNESS: 2½–3
DENSITY: 2.664 (Meas.)
2.663 (Calc.)
CLEAVAGE: {010} perfect
{101} fair
{100} incomplete
Fracture uneven to hackly. Not brittle.

HABIT: Crystals tabular, dipyramidal, rarely prismatic. Crystals up to several inches in length common. Also as crusts and efflorescences. Twinning on {110} common; also on {011}.

COLOR-LUSTER: Colorless, grayish white, yellowish, brownish, reddish. Transparent to translucent. Vitreous to resinous.

Taste salty.

MODE OF OCCURRENCE: Occurs widespread throughout the world in salt lake and playa deposits and as an efflorescence upon the soil in arid regions; also as a deposit formed around fumaroles and upon recent lavas. Important deposits are found at Searles Lake, San Bernardino County and at other places in California; and in Nevada, Arizona, Chile, Spain, Italy, Sicily, USSR, and Egypt.

BEST REF. IN ENGLISH: Palache, et al., "Dana's System of Mineralogy," 7th Ed., v. II, p. 404–407, New York, Wiley, 1951.

## THERMONATRITE
$Na_2CO_3 \cdot H_2O$

CRYSTAL SYSTEM: Orthorhombic
CLASS: mm2
SPACE GROUP: $P2_1ab$
Z: 4
LATTICE CONSTANTS:
$a = 6.44$ kX
$b = 10.72$ (synthetic)
$c = 5.24$
3 STRONGEST DIFFRACTION LINES:
2.768 (100)
2.372 ( 60) (synthetic)
2.753 ( 55)
OPTICAL CONSTANTS:
$\alpha = 1.420$
$\beta = 1.506 – 1.509$ (synthetic)
$\gamma = 1.524 – 1.525$
$(-)2V = 48°$
HARDNESS: 1–1½
DENSITY: 2.255 (Meas. synthetic)
2.259 (Calc.)
CLEAVAGE: {100}, difficult
Somewhat sectile.

HABIT: As a crust or efflorescence; also as clusters of platy needles.

COLOR-LUSTER: Colorless, white, grayish, yellowish. Transparent. Vitreous.

Taste alkaline.

MODE OF OCCURRENCE: Occurs as a deposit from saline lakes, as an efflorescence on the soil in arid regions, and as a fumarole product. It is found in California associated with sborgite near Twenty Mule Team Canyon, as efflorescences from Deep Spring Lake, and as efflorescent coatings in Death Valley, Inyo County; and associated with halite crystals from Borax Lake, Lake County. It also occurs on Mt. Vesuvius, Italy; as a bedded deposit in the Gorodki oil field in the Kama region, USSR; at Szegedin and Debreczin, Hungary; and in the desert region of the Sudan, Egypt.

BEST REF. IN ENGLISH: Palache, et al., "Dana's System of Mineralogy," 7th Ed., v. II, p. 224–225, New York, Wiley, 1951.

## THOMSENOLITE
$NaCaAlF_6 \cdot H_2O$

CRYSTAL SYSTEM: Monoclinic
CLASS: 2/m
SPACE GROUP: $P2_1/c$
Z: 4
LATTICE CONSTANTS:
$a = 5.57$
$b = 5.50$
$c = 16.10$
$\beta = 96°23'$
3 STRONGEST DIFFRACTION LINES:
4.02 (100)
1.963 ( 90)
1.996 ( 80)
OPTICAL CONSTANTS:
$\alpha = 1.4072$
$\beta = 1.4136$
$\gamma = 1.4150$
$(-)2V = 50°$
HARDNESS: 2
DENSITY: 2.981 (Meas.)
2.99 (Calc.)
CLEAVAGE: {001} perfect
{110} distinct
Fracture uneven. Brittle.

HABIT: Crystals often resemble cubes. Commonly prismatic, also tabular {001}. Prism faces striated parallel to {001}. Also as stalactitic masses and chalcedonic crusts.

COLOR-LUSTER: Colorless, white; sometimes reddish or brownish due to included hematite. Transparent to translucent. Vitreous. Somewhat pearly on perfect cleavage.

MODE OF OCCURRENCE: Occurs as an alteration product of cryolite in pegmatite as microscopic crystals mixed with pachnolite at St. Peters Dome, El Paso County, Colorado. It also occurs associated with ralstonite, pachnolite, and fluorite in the cryolite deposit at Ivigtut, Greenland; and at Miask, Ilmen Mountains, USSR.

BEST REF. IN ENGLISH: Palache, et al., "Dana's System of Mineralogy," 7th Ed., v. II, p. 116–118, New York, Wiley, 1951.

## THOMSONITE
$NaCa_2Al_5Si_5O_{20} \cdot 6H_2O$
Zeolite Group

CRYSTAL SYSTEM: Orthorhombic
CLASS: 2/m 2/m 2/m
SPACE GROUP: Pnma
Z: 4
LATTICE CONSTANTS:
  $a = 13.07$
  $b = 13.09$
  $c = 2 \times 6.63$
3 STRONGEST DIFFRACTION LINES:
  2.86 (100)
  2.95 ( 80)
  2.68 ( 80)
OPTICAL CONSTANTS:
  $\alpha = 1.497 - 1.530$
  $\beta = 1.513 - 1.533$
  $\gamma = 1.518 - 1.544$
  $(+)2V = 42° - 75°$
HARDNESS: 5-5½
DENSITY: 2.25-2.40 (Meas.)
         2.366 (Calc.)
CLEAVAGE: {010} perfect
          {100} distinct
Fracture uneven to subconchoidal. Brittle.

HABIT: Crystals prismatic, acicular, or blade-like; vertically striated; rare. Usually as radiating or lamellar aggregates; compact. Twinning on {110}.

COLOR-LUSTER: Colorless, white, yellowish, pink; also greenish (lintonite). Transparent to translucent. Vitreous to somewhat pearly. Streak uncolored.

MODE OF OCCURRENCE: Occurs chiefly in amygdules or crevices in basalt and related igneous rocks, often in association with other zeolites, and also in schists or contact rocks. Found at Springfield, Lane County, and at other places in Oregon; in Colusa, Kern, Los Angeles, Plumas, Riverside, and San Bernardino counties, California; at Table Mountain, Jefferson County, Colorado; at Magnet Cove, Arkansas; as agate-like water-worn pebbles near Grand Marais, Cook County, Minnesota; in the trap rock of northern New Jersey; and at localities in Nova Scotia, Greenland, Ireland, Faroe Islands, Scotland, Italy, Germany, Czechoslovakia, India, and elsewhere

BEST REF. IN ENGLISH: Deer, Howie, and Zussman, "Rock Forming Minerals," v. 4, p. 359-376, New York, Wiley, 1963.

## THORBASTNAESITE (Bastnaesite-Th)
Th(Ca, Ce)(CO₃)₂F₂ · 3H₂O

$Th(Ca, Ce)(CO_3)_2F_2 \cdot 3H_2O$

CRYSTAL SYSTEM: Hexagonal
CLASS: $\bar{6}$ 2m
SPACE GROUP: P$\bar{6}$2c
Z: 3
LATTICE CONSTANTS:
  $a = 6.99$
  $c = 9.71$
3 STRONGEST DIFFRACTION LINES:
  2.85 (100)
  2.03 (100)
  3.54 ( 80)
OPTICAL CONSTANT:
  $n \sim 1.670 - 1.678$

HARDNESS: Not determined.
DENSITY: 4.04 (Meas.)
         5.70 (Calc.)
(Discrepancy ascribed to molecular water.)
CLEAVAGE: Not determined.
HABIT: Massive; cryptocrystalline.
COLOR-LUSTER: Brown.
MODE OF OCCURRENCE: Occurs as an accessory in iron-rich saccharoidal albitites in exocontact alkaline metasomatic rocks of "one of the nepheline syenite intrusives of Eastern Siberia."

BEST REF. IN ENGLISH: Pavlenko, A. S., Orlova, L. P., Akhmanova, M. V., and Tobelko, K. I., *Am. Min.*, **50**: 1505 (1965).

## THOREAULITE
SnTa₂O₇

$SnTa_2O_7$

CRYSTAL SYSTEM: Monoclinic
CLASS: 2/m
SPACE GROUP: C2/c
Z: 4
LATTICE CONSTANTS:
  $a = 17.140$
  $b = 4.865$
  $c = 5.548$
  $\beta = 91.0°$
3 STRONGEST DIFFRACTION LINES:
  3.10 (100)
  3.07 (100)
  2.86 ( 80)
OPTICAL CONSTANTS:
  $\alpha = 2.39$
  $\beta = 2.42$
  $\gamma = 2.52$
  $(+)2V = 25° - 35°$
HARDNESS: 5½-6
DENSITY: 7.6-7.9 (Meas.)
         7.5 (Calc.)
CLEAVAGE: {100} perfect
          {011} imperfect

HABIT: Crystals short prismatic parallel to [001]; subhedral. As irregular masses up to 5 × 10 cm. Twinning on {010}, lamellar.

COLOR-LUSTER: Brown to yellowish. Adamantine to resinous. Opaque, translucent in thin fragments. Streak light yellow, sometimes with faint greenish tinge.

MODE OF OCCURRENCE: Occurs in granite pegmatites in association with cassiterite, microlite, tantalite, albite, lepidolite, amblygonite, and pollucite. Found at Monono, Katanga, and at Punia, Maniema district, Kivu, Zaire; at Sebeia, Rwanda; and in the USSR.

BEST REF. IN ENGLISH: Vlasov, K. A., "Mineralogy of Rare Elements," v. II, p. 539-541, Israel Program for Scientific Translations, 1966. *Am. Min.*, **55**: 367 (1970).

## THORIANITE
ThO₂

$ThO_2$

CRYSTAL SYSTEM: Cubic

CLASS: 4/m $\overline{3}$ 2/m
SPACE GROUP: Fm3m
Z: 4
LATTICE CONSTANT:
 $a = 5.595$ (variable)
3 STRONGEST DIFFRACTION LINES:
 3.234 (100)
 1.689 ( 64)
 1.980 ( 58)
OPTICAL CONSTANTS: Opaque.
 HARDNESS: 6½–7
 DENSITY: 9.991 (Meas.)
  10.0 (Calc.)
 CLEAVAGE: {001} poor
Fracture conchoidal to uneven. Brittle.
 HABIT: Crystals commonly cubic or cubo-octahedral. Found as embedded grains or crystals; and as crystals, often water-worn, in alluvial deposits. Interpenetration twins on {111} very common.
 COLOR-LUSTER: Black, brownish black to dark gray. Opaque submetallic to resinous.
 MODE OF OCCURRENCE: Occurs chiefly as a primary mineral in pegmatites and as a detrital mineral derived from pegmatites. Found as crystals up to one inch in size in serpentine at Easton, Pennsylvania; in black sands in the Missouri River near Helena, Montana; the Scott River, Siskiyou County, California; and in the Nixon Fork and Wiseman districts, Alaska. In Canada it is found in pegmatites, veins transitional to pegmatites, and metasomatized zones in crystalline limestones at many localities in Quebec and Ontario. It is also found in a carbonatite at Phalabora, eastern Transvaal; and occurs widespread in stream gravels in Ceylon; also in alluvial deposits in Madagascar, Travancore, and Transbaikalia, Siberia.
 BEST REF. IN ENGLISH: Frondel, Clifford, USGS Bull. 1064, p. 47–53 (1958).


# THORITE
$ThSiO_4$
Var. Uranothorite

CRYSTAL SYSTEM: Tetragonal
CLASS: 4/m 2/m 2/m
SPACE GROUP: I4/amd
Z: 4
LATTICE CONSTANTS:
 $a = 7.12$
 $c = 6.32$
3 STRONGEST DIFFRACTION LINES:
 3.53 (100)
 4.69 ( 90) (Usually metamict)
 2.65 ( 80)
OPTICAL CONSTANTS:
Usually isotropic, metamict
  $N = 1.664–1.87$
rare: $\omega = 1.818–1.825, 1.78$
  $\epsilon = 1.839–1.840, 1.79$
   (+)
 HARDNESS: ~4½
 DENSITY: 6.7–4.1 (Meas., decreasing with increase of water content).

 CLEAVAGE: {100} distinct
Cleavage not apparent in metamict crystals. Fracture conchoidal to subconchoidal or splintery. Brittle.
 HABIT: Crystals usually short prismatic, up to 8 cm in diameter, with {100} and the pyramid {101}; also pyramidal {101} with prism faces small or absent. Also massive, as small reniform masses, and as embedded anhedral grains and masses.
 COLOR-LUSTER: Brownish yellow, yellow to orange yellow; also brownish black to dark brown and reddish brown; less commonly greenish black to green. Color often mottled or irregular within a single crystal. Transparent in thin section or small grains; translucent to opaque in large masses. Vitreous, resinous to greasy when fresh. Dull, when partly altered. Not fluorescent.
 MODE OF OCCURRENCE: Occurs widespread as a primary mineral chiefly in pegmatites, metasomatized zones in impure limestones, hydrothermal veins, and in detrital deposits. Prominent localities in the United States include Henderson County, North Carolina; as crystals from Baringer Hill, Llano County, Texas; and at numerous deposits in Colorado, Idaho, Montana, California, New Mexico, and Alaska. In Canada it is found at many localities in the Grenville rocks of Quebec and Ontario; as large crystals in pegmatite at the Macdonald mine, Hybla, and abundantly as large altered crystals in a pyroxene skarn at the Kemp uranium mines near Wilberforce, Ontario. It is also found at many places on the Langesundfjord, and at other localities in Norway; in Scotland, Italy, Sicily, and the Batum Region, Caucasus, USSR; as detrital grains in Ceylon; and as fine crystals from several deposits in Madagascar.
 BEST REF. IN ENGLISH: Frondel, Clifford, USGS Bull. 1064, p. 265–276 (1958).


# THOROGUMMITE
$(Th, U)(SiO_4)_{1-x}(OH)_{4x}$

CRYSTAL SYSTEM: Tetragonal
Z: 4
LATTICE CONSTANTS:
 $a = 7.068$
 $c = 6.260$
3 STRONGEST DIFFRACTION LINES:
 3.537 (100)
 4.695 ( 90)
 2.653 ( 60)
OPTICAL CONSTANTS:
 $N = 1.54–1.64$
 HARDNESS: Not determined.
 DENSITY: 3.2–5.4 (Meas.) Variation due primarily to content of nonessential water.
 CLEAVAGE: Not determined.
 HABIT: As dense fine-grained aggregates pseudomorphous after thorite or other thorium minerals; as opaline to microcrystalline crusts; and as firm to earthy nodular masses.
 COLOR-LUSTER: Yellowish brown, greenish gray, buff to nearly white. Dull to subvitreous. Varieties containing quadrivalent uranium are black, subvitreous. Not fluorescent.
 MODE OF OCCURRENCE: Occurs as a secondary mineral formed by the alteration of thorite, thorianite, or yttri-

alite. Found in the United States as pseudomorphs after thorite at Baringer Hill, Llano County, Texas; and as an alteration of thorianite at Easton, Pennsylvania. It also occurs at Hybla, Ontario, Canada; at Wodgina, Western Australia; as an alteration of thorite at Haicheng Prefecture, South Manchuria; and as an alteration product of yttrialite associated with biotite and feldspar at Iisaka-mura, Dategun, Fukushima Prefecture, Japan.

BEST REF. IN ENGLISH: Frondel, Clifford, USGS Bull. 1064, p. 280-285 (1958).

## THOROSTEENSTRUPINE
$(Ca, Th, Mn)_3 Si_4 O_{11} F \cdot 6H_2 O$

CRYSTAL SYSTEM: Unknown
3 STRONGEST DIFFRACTION LINES:
4.08 (100)⎫ Metamict. Heated at
3.25 (100)⎬ 900°C for 30 minutes
2.61 (100)⎭
OPTICAL CONSTANTS:
$N$ = 1.63-1.66
HARDNESS: ~ 4
DENSITY: 3.02 (Meas.)
CLEAVAGE: Not determined.
Fracture conchoidal. Brittle.
HABIT: As platy crystals, usually 2-5 mm, sometimes up to 1 cm in length, with rough faces. Weakly magnetic.
COLOR-LUSTER: Dark brown, nearly black; translucent reddish brown in thin splinters; greasy to vitreous. Streak dark brown.
MODE OF OCCURRENCE: Occurs associated with microcline, albite, aegirine-augite, quartz, fluorite, thorite, and miserite that contains rare earths, in metasomatic veins at an unspecified locality in "eastern Siberia."
BEST REF. IN ENGLISH: Kupriyanova, I. I., Stolyarova, T. I., and Sidorenko, G. A., *Am. Min.*, **48**: 433-434 (1963).

## THOROTUNGSTITE = Yttrotungstite

## THORTVEITITE
$(Sc, Y)_2 Si_2 O_7$

CRYSTAL SYSTEM: Monoclinic
CLASS: 2/m
SPACE GROUP: C2/m
Z: 2
LATTICE CONSTANTS:
(synthetic)
$a$ = 6.508
$b$ = 8.506
$c$ = 4.677
$\beta$ = 102°43.4′
3 STRONGEST DIFFRACTION LINES:
(synthetic)
3.13 (100)
3.11 (100)
2.93 ( 45)

OPTICAL CONSTANTS:
$\alpha$ = 1.750-1.756
$\beta$ = 1.789-1.793
$\gamma$ = 1.800-1.809
$(-)2V$ = 60°-65°
HARDNESS: 6-7
DENSITY: 3.58 (Meas.)
3.394 (Calc.)
CLEAVAGE: {110} distinct
{001} parting
Fracture fine-conchoidal to uneven. Brittle.
HABIT: Crystals prismatic, usually distorted and somewhat pointed. Most crystals only exhibit faces of {110} prism. Up to 35 cm long and 4 cm thick. Twinning on {110} common.
COLOR-LUSTER: Grayish green. Translucent to transparent. Vitreous. Streak pale grayish green.
MODE OF OCCURRENCE: Occurs in granite pegmatites rich in rare earth elements in association with euxenite, monazite, fergusonite, beryl, and other minerals. Found in Iveland parish and elsewhere in Norway; in aplites of the Shilovo-Koneva massif, Urals, USSR; at Befanamo, Madagascar; and in Japan.
BEST REF. IN ENGLISH: Vlasov, K. A., "Mineralogy of Rare Elements," v. II, p. 212-217, Israel Program for Scientific Translations, 1966.

## THORUTITE (Smirnovite)
$(Th, U, Ca)Ti_2 (O, OH)_6$

CRYSTAL SYSTEM: Unknown
3 STRONGEST DIFFRACTION LINES:
3.17 (100) ⎫
1.728 ( 85) ⎬ Metamict. After heating
1.695 ( 85) ⎭ to 1000°C
OPTICAL CONSTANTS:
Isotropic $N$ > 2.1
DENSITY: 5.82 (Meas.)
CLEAVAGE: None. Fracture conchoidal.
HABIT: Crystals short prismatic, up to 2 cm in length and 0.5-1.0 cm in diameter.
COLOR-LUSTER: Black; resinous. Translucent, dark brown, on thin edges. Streak pale brown.
MODE OF OCCURRENCE: Occurs associated with thorite, zircon, calcite, and a little barite and galena in a syenite massif in veins of microcline and sericitized nepheline at an unspecified locality, presumably in the USSR.
BEST REF. IN ENGLISH: Gotman, Ya. D., and Khapaev, I. A., *Am. Min.*, **43**: 1007 (1958).

## THUCHOLITE = Mixture of uraninite and carbonaceous matter

## THULITE = Variety of zoisite rich in manganese

## THURINGITE
$(Fe^{2+}, Fe^{3+}, Mg)_6 (Al, Si)_4 O_{10} (OH)_8$
Chlorite Group

CRYSTAL SYSTEM: Monoclinic
CLASS: 2/m (?)
SPACE GROUP: C2/m (?)
Z: 4
LATTICE CONSTANTS:
 $a = 5.307$
 $b = 9.192$
 $c = 14.069$
3 STRONGEST DIFFRACTION LINES:
 7.07 (100)
 14.1 ( 90)
 3.54 ( 60)
OPTICAL CONSTANTS:
 $\alpha = 1.670$
 $\beta = 1.685$
 $\gamma = 1.685$
 (–)2V
 HARDNESS: 2–3
 DENSITY: 2.96–3.31 (Meas.)
          2.985 (Calc.)
 CLEAVAGE: {001} perfect
 HABIT: Massive; in scaly aggregates. Twinning on {001}, twin plane (310).
 COLOR-LUSTER: Olive green, greenish brown to dark brown. Translucent. Pearly.
 MODE OF OCCURRENCE: Occurs chiefly as a vein mineral or with iron ores in various metamorphic rocks. Typical occurrences are found in the Creede district, Mineral County, Colorado; at Hot Springs, Garland County, Arkansas; in the Lake Superior iron region, Michigan; and in Jefferson County, West Virginia. Also in Thuringia and elsewhere in Germany; and at localities in Austria, Czechoslovakia, and South Africa.
 BEST REF. IN ENGLISH: Deer, Howie, and Zussman, "Rock Forming Minerals," v. 3, p. 131–163, New York, Wiley, 1962.

# TIEMANNITE
HgSe

CRYSTAL SYSTEM: Cubic
CLASS: $\bar{4}$3m
SPACE GROUP: F$\bar{4}$3m
Z: 4
LATTICE CONSTANT:
 $a = 6.084$
3 STRONGEST DIFFRACTION LINES:
 3.51 (100)
 2.151 ( 50)
 1.835 ( 30)
OPTICAL CONSTANTS: Opaque.
 HARDNESS: 2½
 DENSITY: 8.24–8.27 (Meas.)
          8.24 (Calc.)
 CLEAVAGE: None.
Fracture conchoidal to uneven. Brittle.
 HABIT: Crystals tetrahedral. {111} usually dull, {$\bar{1}$11} bright; frequently striated [1$\bar{1}$0]. Commonly massive, compact granular.
 COLOR-LUSTER: Lead gray with purplish blue tint.

Weathered surfaces dull reddish brown to black. Opaque; metallic. Streak nearly black.
 MODE OF OCCURRENCE: Occurs associated with barite, calcite, and manganese oxide in a vein in limestone near Marysvale, Piute County, Utah; and at Clausthal, Tilkerode, and Zorge in the Harz Mountains, Germany.
 BEST REF. IN ENGLISH: Palache, et al., "Dana's System of Mineralogy," 7th Ed., v. I, p. 217–218, New York, Wiley, 1944. Early, J. W., *Am. Min.*, **35**: 337–364 (1950).

# TIENSHANITE
BaNa$_2$MnTiB$_2$Si$_6$O$_{20}$

CRYSTAL SYSTEM: Hexagonal
CLASS: 6/m, 6, or $\bar{6}$
SPACE GROUP: P6/m, P6, or P$\bar{6}$
Z: 1
LATTICE CONSTANTS:
 $a = 16.755$
 $c = 10.435$
3 STRONGEST DIFFRACTION LINES:
 4.19 (100)
 3.177 ( 90)
 3.474 ( 80)
OPTICAL CONSTANTS:
 $\omega = 1.666$
 $\epsilon = 1.653$
  (–)
 HARDNESS: 6–6½
 DENSITY: 3.29 (Meas.)
 CLEAVAGE: {0001} distinct
Brittle.
 HABIT: Massive; as finely crystalline aggregates up to 5 × 6 cm in size.
 COLOR-LUSTER: Pistachio green. Vitreous.
 MODE OF OCCURRENCE: Occurs in association with pyrochlore, astrophyllite, bafertisite, stillwellite, danburite, datolite, galena, sphene, and calcite in a narrow vein of quartz-aegirine-microcline pegmatite, cutting alkali syenite of a massif of the Turkestan-Alai alkalic province, southern Tienshan.
 BEST REF. IN ENGLISH: Dusmatov, V. D., Efimov, A. F., Alkhazov, V. Yu., Kazakova, M. E., and Mumyatskaya, N. G., *Am. Min.*, **53**: 1426 (1968).

# TIKHONENKOVITE
SrAlF$_4$(OH) · H$_2$O

CRYSTAL SYSTEM: Monoclinic
Z: 4
LATTICE CONSTANTS:
 $a = 8.73$
 $b = 10.62$
 $c = 5.02$
 $\beta = 102°43'$
3 STRONGEST DIFFRACTION LINES:
 4.89 (100)
 3.64 ( 90)
 3.27 ( 80)

OPTICAL CONSTANTS:

$\alpha = 1.452$

$\beta = 1.456$

$\gamma = 1.458$

$(-)2V = 70°$

HARDNESS: 3½

DENSITY: 3.26 (Meas.)

CLEAVAGE: {001} perfect

Fracture conchoidal to uneven.

HABIT: As small crystals, up to 5 mm in diameter; combinations of several prisms with {100} and {001}.

COLOR-LUSTER: Colorless to slightly rosy. Vitreous.

MODE OF OCCURRENCE: Occurs in fissures and druses in limonite-hematite ores, with gearksutite, fluorite, celestite, strontianite, and barite in the oxidation zone of siderite-rich ores of Karasug in the western part of the Tannu-Ola Range, Tuva.

BEST REF. IN ENGLISH: Khomyakov, A. P., Stepanov, V. I., Moleva, V. A., and Pudovkina, Z. V., *Am. Min.*, **49**: 1774-1775 (1964).

# TILASITE
$CaMgAsO_4F$

CRYSTAL SYSTEM: Monoclinic

CLASS: m

SPACE GROUP: Aa

Z: 4

LATTICE CONSTANTS:

$a = 7.553$

$b = 8.951$

$c = 6.701$

$\beta = 120°58'$

3 STRONGEST DIFFRACTION LINES:

3.246 (100)

2.682 ( 90)

3.073 ( 80)

OPTICAL CONSTANTS:

$\alpha = 1.640$

$\beta = 1.660$

$\gamma = 1.675$

$(-)2V = 82°44'$

HARDNESS: 5

DENSITY: 3.77 (Meas.)

3.78 (Calc.)

CLEAVAGE: {10$\bar{1}$} good

Parting on {1$\bar{3}$3}, {10$\bar{2}$}, and {0$\bar{1}$1}. Brittle.

HABIT: Crystals often complex, up to 6.0 mm in greatest dimension; sometimes elongated, and flattened on {010}. Also massive. Twinning on {001} common.

COLOR-LUSTER: Gray to violet-gray, apple green to olive green. Translucent. Vitreous to resinous.

MODE OF OCCURRENCE: Occurs in veinlets in crystalline limestone with calcite, braunite, and conichalcite in an outcrop near the White Tail Deer mine, Bisbee district, Arizona. It also is found in dolomitic limestone at Långban, Sweden, and associated with braunite, barite, and other minerals at Kajlidongri, Jhabua State, India.

BEST REF. IN ENGLISH: Williams, Sidney A., *Min. Record*, **1**: 68-69 (1970). Palache, et al., "Dana's System of Mineralogy," 7th Ed., v. II, p. 827-829, New York, Wiley, 1951.

# TILLEYITE
$Ca_5Si_2O_7(CO_3)_2$

CRYSTAL SYSTEM: Monoclinic

CLASS: 2/m

SPACE GROUP: $P2_1/a$

Z: 4

LATTICE CONSTANTS:

$a = 15.025$

$b = 10.269$

$c = 7.628$

$\beta = 105°50'$

3 STRONGEST DIFFRACTION LINES:

3.01 (100)

3.10 ( 60)

1.90 ( 45)

OPTICAL CONSTANTS:

$\alpha = 1.612$

$\beta = 1.632$

$\gamma = 1.653$

$(+)2V = 85°-89°$

HARDNESS: Not determined.

DENSITY: 2.84 (Meas.)

2.88 (Calc.)

CLEAVAGE: {100 } perfect

{101?} good

{$\bar{1}$01?} poor

HABIT: Massive; also as rounded grains or irregular plates. Twinning on {101}, common.

COLOR-LUSTER: Colorless, white.

MODE OF OCCURRENCE: Occurs in a contact zone at Crestmore, Riverside County, California; also in thermally metamorphosed limestones at Carlingford, Ireland.

BEST REF. IN ENGLISH: Larsen, Esper S., and Dunham, Kingsley C., *Am. Min.*, **18**: 469-473 (1933). Nockolds, S. R., *Min. Mag.*, **28**, 151-158 (1947).

# TIN
Sn

CRYSTAL SYSTEM: Cubic ($\alpha$), tetragonal ($\beta$)

CLASS: 4/m $\bar{3}$ 2/m ($\alpha$), 4/m 2/m 2/m ($\beta$)

SPACE GROUP: Fd3m ($\alpha$), I4$_1$/amd ($\beta$)

Z: 8($\alpha$), 4($\beta$)

LATTICE CONSTANTS:

| ($\alpha$) | ($\beta$) |
|---|---|
| $a = 6.47$ | $a = 5.820$ |
| | $c = 3.175$ |

3 STRONGEST DIFFRACTION LINES:

| ($\alpha$) | ($\beta$) |
|---|---|
| 3.75 (100) | 2.92 (100) |
| 2.29 ( 83) | 2.79 ( 90) |
| 1.96 ( 53) | 2.02 ( 74) |

OPTICAL CONSTANTS: Opaque.

HARDNESS: 2

DENSITY: 5.765 (α Calc.)
7.31 (β Meas.)
7.278 (β Calc.)

CLEAVAGE: No distinct cleavage. Fracture hackly. Ductile, malleable.

HABIT: As minute embedded grains; also small rounded grains in placer deposits.

COLOR-LUSTER: Tin white, opaque. Metallic.

MODE OF OCCURRENCE: Occurs in discrete grains up to 1.5 mm in diameter in calcite associated with pitchblende, pyrite, chalcopyrite, bornite, chalcocite, covellite, sphalerite, and galena at Nesbitt La Bine uranium mines, Beaverlodge, Saskatchewan, Canada; also as small rounded grains up to 1 mm in diameter near Oban, New South Wales, Australia. Reported in volcanic gases on the islands of Volcano and Stromboli.

BEST REF. IN ENGLISH: Palache, et al., "Dana's System of Mineralogy," 7th Ed., v. I, p. 126, 127, New York, Wiley, 1944.


## TINAKSITE
$NaK_2Ca_2TiSi_7O_{19}(OH)$

CRYSTAL SYSTEM: Triclinic
CLASS: 1 or $\bar{1}$
SPACE GROUP: P1 or P$\bar{1}$
Z: 2
LATTICE CONSTANTS:
$a = 10.35$ $\alpha = 91°00'$
$b = 12.17$ $\beta = 99°20'$
$c = 7.05$ $\gamma = 92°30'$
3 STRONGEST DIFFRACTION LINES:
3.03 (100)
3.25 ( 80)
2.331 ( 56)
OPTICAL CONSTANTS:
$\alpha = 1.593$
$\beta = 1.621$
$\gamma = 1.666$
(+)2V = 74°-78°
HARDNESS: 6
DENSITY: 2.82 (Meas.)
2.85 (Calc.)
CLEAVAGE: {010} perfect
{110} imperfect

HABIT: Crystals prismatic, from tenths of a millimeter to several centimeters; often in radiating fibrous aggregates and rosettes up to 3-5 cm in diameter.

COLOR-LUSTER: Pale yellow; vitreous on cleavage planes.

MODE OF OCCURRENCE: Occurs as an accessory in potassium feldspar metasomatites in the border part of the Murunsk massif, northwestern Aldan, at the contact of the massif with limestones.

BEST REF. IN ENGLISH: Rogov, Yu. G., Rogova, V. P., Voronkou, A. A., and Moleva, V. A., *Am. Min.*, **50**: 2098-2099 (1965).


## TINCAL = Borax


## TINCALCONITE
$Na_2B_4O_5(OH)_4 \cdot 3H_2O$

CRYSTAL SYSTEM: Hexagonal
CLASS: 32
SPACE GROUP: R32
Z: 9
LATTICE CONSTANTS:
(synthetic)
$a = 11.09$
$c = 21.07$
3 STRONGEST DIFFRACTION LINES:
(synthetic)
2.92 (100)
4.38 ( 90)
3.44 ( 55)
OPTICAL CONSTANTS:
$\omega = 1.461$
$\epsilon = 1.477$ (synthetic)
(+)
DENSITY: 1.880 (Meas.)
1.94 (Calc.)
HABIT: As a fine-grained powder.
COLOR-LUSTER: White. Opaque. Dull.

MODE OF OCCURRENCE: Tincalconite occurs as a hydration or dehydration product of other borate minerals. Typical occurrences: Searles Lake, San Bernardino County, and Kramer District, Kern County, California.

BEST REF. IN ENGLISH: Palache, et al., "Dana's System of Mineralogy," 7th Ed., v. II, p. 337-339, New York, Wiley, 1951. Giacovazzo, Carmelo; Menchetti, Silvio; and Scordari, Fernando, *Am. Min.*, **58**: 523-530 (1973).


## TINTICITE (Inadequately described mineral)
$Fe_6(PO_4)_4(OH)_6 \cdot 7H_2O$

CRYSTAL SYSTEM: Orthorhombic (?)
3 STRONGEST DIFFRACTION LINES:
3.91 (100)
3.28 ( 95)
3.01 ( 85)
OPTICAL CONSTANTS:
Mean index ≈ 1.745
HARDNESS: ~2½
DENSITY: ~2.8
CLEAVAGE: Not determined.
HABIT: As dense, porcelaneous to earthy masses.
COLOR-LUSTER: White with yellowish green tint.

MODE OF OCCURRENCE: Occurs in a limestone cavern near the Tintic Standard mine, Juab County, Utah, associated with jarosite and limonite.

BEST REF. IN ENGLISH: Palache, et al., "Dana's System of Mineralogy," 7th Ed., v. II, p. 970-971, New York, Wiley, 1951.


## TINTINAITE
$Pb_5(Sb, Bi)_8S_{17}$
Kobellite-Tintinaite Series

CRYSTAL SYSTEM: Orthorhombic

CLASS: 2/m 2/m 2/m
SPACE GROUP: Pnnm
Z: 4
LATTICE CONSTANTS:
$a = 22.30$
$b = 34.00$
$c = 4.04$
3 STRONGEST DIFFRACTION LINES:
3.40 (100)
3.52 ( 80)
2.71 ( 70)
HARDNESS: Microhardness C- (Talmadge).
DENSITY: 5.51 (Predicted)
5.48 (Calc.)
CLEAVAGE: {010} distinct
HABIT: Massive; bladed, in parallel aggregates and radiating fibrous clusters.
COLOR-LUSTER: Lead gray; metallic. Black streak.
MODE OF OCCURRENCE: Occurs as small masses measuring up to 2 mm in diameter and as veinlets filling interstices and fractures in malachite-stained weathered quartz, principally associated with jamesonite, in ore from the Tintina silver mines, Tintina, Yukon. Bismuthian tintinaite occurs as blades forming parallel aggregates up to 2 mm × 0.5 mm in quartz, associated with pyrite, arsenopyrite, pyrrhotite, chalcopyrite, and magnetite, at the Deer Park mine, Rossland, British Columbia, Canada.
BEST REF. IN ENGLISH: Harris, D. C., Jambor, J. L., Lachance, G. R., and Thorpe, R. I., *Can. Min.*, **9** (Pt. 3); 371-382 (1968).

## TINZENITE (See Axinite Group)
$(Ca, Mn, Fe)_3 Al_2 BSi_4 O_{15} (OH)$ with $Ca < 1.5$ and $Mn > Fe$

## TIRODITE (Manganoan Cummingtonite)
$(Mg, Mn, Fe)_7 Si_8 O_{22} (OH)_2$
Amphibole Group

CRYSTAL SYSTEM: Monoclinic
CLASS: 2/m
SPACE GROUP: C2/m
Z: 2
LATTICE CONSTANTS:
$a = 9.799$
$b = 17.99$
$c = 5.289$
$\beta = 103.89°$
3 STRONGEST DIFFRACTION LINES:
8.42 (100)
3.13 ( 90)
3.27 ( 30)
OPTICAL CONSTANTS:
(impure samples)
$\alpha = 1.630-1.638$
$\beta = 1.644-1.651$
$\gamma = 1.652-1.665$
$2V = 89°-107°$
HARDNESS: 6-6½
DENSITY: 3.07-3.13 (Meas.)

CLEAVAGE: {110} perfect
{100} imperfect
{001} parting
Brittle.
HABIT: Crystals long prismatic to acicular. Usually in compact aggregates. Twinning on {100} common, multiple.
COLOR-LUSTER: Pale greenish white, light green, pink to rose red, tan. Transparent to translucent. Vitreous.
MODE OF OCCURRENCE: Occurs as bladed crystals up to 1 × 3 inches in size at Talcville, New York; in the Wabush iron formation, Labrador, Canada; at Nsuta, Ghana; and in manganese ore deposits at Tirodi and elsewhere in India.
BEST REF. IN ENGLISH: Dunn, J. A., and Roy, O. C., *Misc. Notes, Geol. Sur. India Rec.*, **73**: 295 (1938). *Am. Min.*, **46**: 642-653; 637-641 (1961). *Am. Min.*, **49**: 963-982 (1964).

## TITANAUGITE = Variety of augite containing up to 5% $TiO_2$

## TITANITE = Synonym for sphene

## TOBERMORITE
$Ca_5 Si_6 O_{17} \cdot 5H_2 O$

CRYSTAL SYSTEM: Orthorhombic
CLASS: 222
SPACE GROUP: C222₁
Z: 4
LATTICE CONSTANTS:
$a = 11.3$
$b = 7.33$
$c = 22.6$
3 STRONGEST DIFFRACTION LINES:
11.3 (100)
2.83 (100)
5.55 ( 90)
OPTICAL CONSTANTS:
$\alpha = 1.570$
$\beta = 1.571$
$\gamma = 1.575$
$(+)2V = small$
HARDNESS: 2½
DENSITY: 2.423-2.44 (Meas.)
CLEAVAGE: {001} distinct
{100} distinct
HABIT: Massive; finely fibrous, granular, or as {001} plates.
COLOR-LUSTER: White, pale pinkish white. Translucent. Silky.
MODE OF OCCURRENCE: Occurs pseudomorphic after wilkeite in association with blue calcite, monticellite, and vesuvianite at Crestmore, Riverside County, California. Also found at Goldfield, Nevada; filling cavities in rock near Tobermory, Island of Mull; and at localities in Italy, Austria, Israel, Taiwan, and elsewhere.
BEST REF. IN ENGLISH: *Min. Mag.*, **31**: 361-370 (1956). *Min. Mag.*, **30**: 293-305 (1954).

## TOCHILINITE

I 2(FeS) $\cdot$ 1.67 (Mg, Fe)(OH)$_2$
II 2 (FeS)$_2$ $\cdot$ 3(Mg, Fe)(OH)$_2$
Iron analogue of valleriite

CRYSTAL SYSTEM: Monoclinic
Z: 6
LATTICE CONSTANTS:

|  | (I) | (II) |
|---|---|---|
| $a =$ | 5.37 | 5.42 |
| $b =$ | 15.65 | 15.77 |
| $c =$ | 10.72 | 10.74 |
| $\beta =$ | 95° | 95° |

3 STRONGEST DIFFRACTION LINES:
   5.34 (100)
  10.68 ( 90)
   1.85 ( 70)
OPTICAL CONSTANTS: Strongly birefringent, anisotropy strong, from pinkish cream to gray. Reflectances (Rg' and Rp') at 12 wavelengths from 450 to 950 nm: 450, 15.4 9.1; 550, 18.1, 8.5; 589, 18.9, 8.4; 650, 20.2, 8.0; 750, 21.4, 7.7; 950, 19.3, 6.8%.
   HARDNESS: Microhardness 15.0–48.8 kg/mm$^2$ with 5 g load.
   DENSITY: 2.96, 2.99 (Meas.)
           3.03 (Calc.)
   CLEAVAGE: {001} distinct
   HABIT: Crystals acicular, in radiating-fibrous aggregates up to 5–6 mm in size; also as aggregates of small grains.
   COLOR-LUSTER: Bronze-black. Opaque.
   MODE OF OCCURRENCE: The mineral occurs in serpentinite (lizardite) from drill cores of the copper-nickel deposits of the Lower Mamon intrusive complex and in basic rocks of the Staromelovatskii intrusive, Voronezh region, USSR.
   BEST REF. IN ENGLISH: Organova, N. I., Genkin, A. D., Drits, V. A., Molotkov, S. P., Kuz'mina, O. V., and Dmitrik, A. L., *Am. Min.*, **57**: 1552 (1972).

## TOCORNALITE (Inadequately described mineral)

(Ag, Hg)I

3 STRONGEST DIFFRACTION LINES:
   2.644 (100)
   3.76 ( 90)
   3.61 ( 90)
   HARDNESS: Soft.
   HABIT: Granular massive.
   COLOR-LUSTER: Pale yellow to bright yellow; rapidly darkens or turns gray-green, then black, on exposure to light.
   MODE OF OCCURRENCE: Occurs in association with embolite at Broken Hill, New South Wales, Australia. Also found at Chanarcillo, Chile.
   BEST REF. IN ENGLISH: Palache, et al., "Dana's System of Mineralogy," 7th Ed., v. II, p. 25, New York, Wiley, 1951. Mason, Brian, *Smithsonian Contrib. Earth Sci.*, **9**: 79–80 (1972).

## TODOROKITE

(Mn, Ca, Mg)Mn$_3^{4+}$O$_7$ $\cdot$ H$_2$O

CRYSTAL SYSTEM: Monoclinic, pseudo-orthorhombic
LATTICE CONSTANTS:
   $a = 9.75$
   $b = 2.84$
   $c = 9.59$
   $\beta = 90°$
3 STRONGEST DIFFRACTION LINES:
   9.68 (100)
   4.80 ( 80) (variable)
   2.39 ( 40)
OPTICAL CONSTANTS:
   All $n > 2.00$
   HARDNESS: Not determined. Aggregates soft.
   DENSITY: 3.49 (Average - Berman balance)
           3.66–3.82 (Pycnometer)
   CLEAVAGE: {001} perfect
            {100} perfect
Fracture fibrous.
   HABIT: Crystals very minute thin plates flattened on {001}. As nodular masses with concentric layering; stalactitic; columnar to fine fibrous aggregates; and as thin coatings.
   COLOR-LUSTER: Dusky purplish gray, dark brown to dark brownish black and sooty black. Opaque. Metallic to dull. Streak brownish black to dark brown.
   MODE OF OCCURRENCE: Occurs as a secondary mineral resulting from the alteration of other manganese-bearing minerals. Found sparingly in California, Montana (zincian), South Dakota, Tennessee, and Sterling Hill, New Jersey. It also occurs in Cuba, Mexico, Brazil, France, Portugal, Austria, Ghana, Philippines, Bikini Atoll, Saipan Island, and at the Todoroki mine, Hokkaido, Japan.
   BEST REF. IN ENGLISH: Straczek, J. A., et al., *Am. Min.*, **45**: 1174–1184 (1960). Frondel, C., et al., *Am. Min.*, **45**: 1167–1173 (1960).

## TOMBARTHITE

Y$_4$(Si, H$_4$)$_4$O$_{12-x}$(OH)$_{4+2x}$

CRYSTAL SYSTEM: Monoclinic
CLASS: 2/m
SPACE GROUP: P2$_1$/n
Z: 1
LATTICE CONSTANTS:

| (I) | (II) |
|---|---|
| $a = 7.12$ | $a = 7.12$ |
| $b = 7.29$ | $b = 7.24$ |
| $c = 6.71$ | $c = 6.69$ |
| $\beta = 102°41'$ | $\beta = 102°29'$ |

3 STRONGEST DIFFRACTION LINES:

| (I) | (II) | |
|---|---|---|
| 6.55 (100) | 6.95 (100) | (slightly |
| 3.42 ( 80) | 6.53 (100) | metamict) |
| 3.23 ( 70) | 3.60 (100) | |

OPTICAL CONSTANT:
   $N = 1.639$
   HARDNESS: 5–6

DENSITY: 3.51 (Meas. I)
3.68 (Calc. I)
3.65 (Meas. II)
3.64 (Calc. II)
CLEAVAGE: Fracture conchoidal.
HABIT: Massive.
COLOR-LUSTER: Brownish black; dull.
MODE OF OCCURRENCE: Occurs in masses up to more than 1 cm wide, surrounded by feldspar, and in contact with or intergrown with thalenite, in a feldspar quarry in the Högetveit pegmatite dike, Evje parish, Norway.
BEST REF. IN ENGLISH: Neumann, Henrich, and Nilssen, Borghild, *Am. Min.*, **54**: 327–328 (1969).

# TOPAZ
$Al_2SiO_4(F,OH)_2$
Var. Pycnite—columnar, very compact.

CRYSTAL SYSTEM: Orthorhombic
CLASS: 2/m 2/m 2/m
SPACE GROUP: Pmnb
Z: 4
LATTICE CONSTANTS:
$a = 8.394$
$b = 8.800$
$c = 4.650$
3 STRONGEST DIFFRACTION LINES:
2.937 (100)
3.195 ( 65)
3.693 ( 60)
OPTICAL CONSTANTS:
$\alpha = 1.606–1.629$
$\beta = 1.609–1.631$
$\gamma = 1.616–1.638$
$(+)2V = 48°–68°$
HARDNESS: 8
DENSITY: 3.49–3.57 (Meas.)
3.558 (Calc.)
CLEAVAGE: {001} perfect
Fracture subconchoidal to uneven. Brittle.
HABIT: As well-developed short to long prismatic crystals, often highly modified, occasionally up to several hundred pounds in weight. Also massive, coarse to fine granular, and columnar (pycnite).
COLOR-LUSTER: Colorless, white, gray, bluish, greenish, yellowish, yellow-brown to orange, pinkish to reddish. Transparent to translucent. Vitreous. Streak uncolored.
MODE OF OCCURRENCE: Occurs widespread in pegmatites and high-temperature quartz veins; in cavities in granites and rhyolites; in contact zones; and in alluvial deposits. Common associated minerals are quartz, microcline, albite, tourmaline, muscovite, beryl, fluorite, and cassiterite. Found as fine crystals in pegmatite in San Diego County, California; in rhyolite in the Thomas Mountains, Juab County, Utah; as superb crystals in the Pikes Peak region in Colorado; in Mason County, Texas; at Lords Hill and other places in Maine; and at localities in New Hampshire and Connecticut. Notable occurrences also are found in Mexico; Minas Geraes, Brazil; Mourne Mountains, Ireland; Norway and Sweden; Schneckenstein, Saxony, Germany; in the Adun-Chilon, Ilmen, and Ural Mountains, USSR; at Jos,

Nigeria; at Kleine Spitzkopje, South West Africa; and at localities in Ceylon, Burma, Japan, Australia, Tasmania, and elsewhere.
BEST REF. IN ENGLISH: Deer, Howie, and Zussman, "Rock Forming Minerals," v. 1, p. 145–150, New York, Wiley, 1962.

# TOPAZOLITE = Transparent yellowish green variety of andradite

# TORBERNITE
$Cu(UO_2)_2(PO_4)_2 \cdot 8–12H_2O$
Autunite Group

CRYSTAL SYSTEM: Tetragonal
CLASS: 4/m 2/m 2/m
SPACE GROUP: I4/mmm
Z: 2
LATTICE CONSTANTS:
| (synthetic) | (natural) |
|---|---|
| $a = 7.025$ | $a = 7.06$ |
| $c = 20.63$ | $c = 20.5$ |
3 STRONGEST DIFFRACTION LINES:
10.3 (100)
4.94 ( 90) (synthetic)
3.58 ( 90)
OPTICAL CONSTANTS:
$\omega = 1.590–1.592$
$\epsilon = 1.581–1.582$
(–)
HARDNESS: 2–2.5
DENSITY: 3.22 (Meas.)
3.28 (Calc. for 12 $H_2O$)
CLEAVAGE: {001} perfect
{100} indistinct
Brittle.
HABIT: Crystals thin to thick tabular on {001}, with rectangular or octagonal shape; rarely pyramidal and terminated by small faces of {001}. As scaly or lamellar aggregates; rarely as granular to almost earthy masses. Twinning on {110}, rare.
COLOR-LUSTER: Emerald green to grass green; sometimes leek green, apple green, or siskin green. Vitreous to subadamantine; pearly on {001}. Transparent to translucent. Streak paler than the color.
MODE OF OCCURRENCE: In the United States torbernite is widespread in small amounts in pegmatites, vein deposits, and several sedimentary rock types. Excellent crystal specimens occur at Chalk Mountain, near Spruce Pine, Mitchell County, North Carolina. It also is found in Sonora, Mexico; in large amounts in the oxidized uranium ores in the Katanga district, Zaire; at Mt. Painter, South Australia; associated with pitchblende in vein deposits in Czechoslovakia and Saxony, Germany; finely crystallized at several localities in Cornwall, England; in France at Cap Garonne and various places in the Puy-de-Dôme district; and at numerous other localities throughout the world.
BEST REF. IN ENGLISH: Berman, Robert, *Am. Min.*, **42**: 905–908 (1957). Frondel, Clifford, USGS Bull. 1064, p. 170–177 (1958).

## TORENDRIKITE = Riebeckite

## TORNEBOHMITE
$(Ce, La)_3 Si_2 O_8 OH$

CRYSTAL SYSTEM:  Hexagonal or pseudohexagonal
Z:  2
LATTICE CONSTANTS:
  $a = 7.74$
  $c = 8.58$
3 STRONGEST DIFFRACTION LINES:
  3.08 (100)
  2.82 ( 90)
  2.01 ( 60)
OPTICAL CONSTANTS:
    $\alpha = 1.845$
    $\beta = 1.852$
    $\gamma = 1.878$
  (+)2V = small
  HARDNESS: 4½
  DENSITY: 4.9 (Meas.)
  CLEAVAGE:  Not determined.
  HABIT:  Massive, fine-grained.
  COLOR-LUSTER:  Olive green to green.
  MODE OF OCCURRENCE: Occurs invariably as inclusions in cerite.  Found in the Jamestown district, Boulder County, Colorado; at the Bastnaes mine, Vastmanland, Sweden; and in alkali granite pegmatite in the Urals, USSR.
  BEST REF. IN ENGLISH: Vlasov, K. A., "Mineralogy of Rare Elements," v. II, p. 319–320, Israel Program for Scientific Translations, 1966.

## TORREYITE
$(Mg, Zn, Mn)_7 SO_4 (OH)_{12} \cdot 4H_2O$

CRYSTAL SYSTEM:  Monoclinic
3 STRONGEST DIFFRACTION LINES:
  13.06 (100)
  5.35 ( 90)
  3.85 ( 80)
OPTICAL CONSTANTS:
    $\alpha = 1.570$
    $\beta = 1.584$
    $\gamma = 1.585$
  (–)2V ~ 40°
  HARDNESS: 3
  DENSITY: 2.665 (Meas.)
  CLEAVAGE:  {010} good
  HABIT:  Massive, granular.
  COLOR-LUSTER:  Bluish white. Translucent. Vitreous to pearly.
  MODE OF OCCURRENCE: Occurs in cavities and crevices in pyrochroite in a vein in calcite-franklinite-willemite ore at Sterling Hill, Sussex County, New Jersey.
  BEST REF. IN ENGLISH:  Palache, et al., "Dana's System of Mineralogy," 7th Ed., v. II, p. 575–576, New York, Wiley, 1951.

## TOSUDITE = Interlayered dioctahedral chlorite and montmorillonite

## TOURMALINE = Group Name
  See:  Buergerite
      Dravite
      Elbaite
      Schorl

## TRANQUILLITYITE
$Fe_8 (Zr, Y)_2 Ti_3 Si_3 O_{24}$

CRYSTAL SYSTEM:  Hexagonal
Z:  3
LATTICE CONSTANTS:
  $a = 11.69$
  $c = 22.25$
3 STRONGEST DIFFRACTION LINES:
  3.23  (100)
  1.781 ( 70)
  2.155 ( 60)
OPTICAL CONSTANTS: Isotropic to weakly anisotropic, nonpleochroic. 2V = 40°
  HARDNESS:  Not determined.
  DENSITY: 4.7 (Calc.)
  CLEAVAGE:  Not determined.
  HABIT: As thin laths from a few microns to approximately $65 \times 15 \mu$ in size.
  COLOR-LUSTER:  Dark foxy red. Nearly opaque.
  MODE OF OCCURRENCE: Occurs in association with late-stage interstitial phases, such as troilite plus metal, pyroxferroite, cristobalite, and alkali feldspar, in both Apollo 11 and Apollo 12 basaltic rocks. The high uranium (56–93 ppm) and rare earth content makes the mineral particularly important in rare earth geochemistry and lead isotopic evolution of the lunar basalts.
  BEST REF. IN ENGLISH: Lovering, J. F. et al., *Proc. Second Lunar Sci. Conf.* (Suppl. to Geochim. Cosmochim. Acta 35), 1: 39–45 (1971).

## TRASKITE
$Ba_9 Fe_2 Ti_2 Si_{12} O_{36} (OH, Cl, F)_6 \cdot 6H_2O$

CRYSTAL SYSTEM:  Hexagonal
CLASS:  6/m 2/m 2/m, $\bar{6}$m2, or 6mn, or 622
SPACE GROUP: P6/mmm, P$\bar{6}$2m, P$\bar{6}$m2, P6mm, or P622
Z:  3
LATTICE CONSTANTS:
  $a = 17.88$
  $c = 12.30$
3 STRONGEST DIFFRACTION LINES:
  2.96 (100)
  15.4  ( 50)
  3.51 ( 40)
OPTICAL CONSTANTS:
  $\omega = 1.714$
    $\epsilon = 1.702$
      (–)
  HARDNESS:  ~5
  DENSITY: 3.71 (Meas.)
          3.75 (Calc.)
  CLEAVAGE: None. Conchoidal fracture.

HABIT: Crystals euhedral, well-formed; {0001} and {10$\bar{1}$0} are most prominent, but many minute pyramid faces occur on some crystals. Grain size ranges from 0.1 to 3 mm. Also as equant anhedral grains.

COLOR-LUSTER: Brownish red; vitreous. Pale reddish brown streak.

MODE OF OCCURRENCE: Occurs as tiny grains, generally less than 1 mm in diameter, in sanbornite-bearing metamorphic rocks which outcrop in a narrow zone 2½ miles long near a granodiorite contact in eastern Fresno County, California.

BEST REF. IN ENGLISH: Alfors, John T., Stinson, Melvin C., Matthews, Robert A., and Pabst, Adolph, *Am. Min.*, **50**: 314–340 (1965). *Am. Min.*, **50**: 1500–1503 (1965).


# TRECHMANNITE
AgAsS$_2$

CRYSTAL SYSTEM: Hexagonal
CLASS: $\bar{3}$ or 3
SPACE GROUP: R$\bar{3}$ or R3
LATTICE CONSTANTS:
$a$ = 13.98       $a_{rh}$ = 8.65
$c$ = 9.12        $\alpha$ = 108°17′
3 STRONGEST DIFFRACTION LINES:
2.702 (100)
3.15 ( 80)
1.887 ( 80)
OPTICAL CONSTANTS:
$\omega$ = 2.6  (Li)
$\epsilon$ = ?
(–)
HARDNESS: 1½–2
DENSITY: Not determined.
CLEAVAGE: {10$\bar{1}$1} good
{0001} distinct
Fracture conchoidal. Brittle.

HABIT: Crystals equant or short prismatic; sometimes irregular.

COLOR-LUSTER: Scarlet-vermilion. Opaque. Adamantine. Streak same as color.

MODE OF OCCURRENCE: Occurs in crystalline dolomite associated with tennantite, pyrite, seligmannite, and chrome-muscovite, at the Lengenbach quarry in the Binnental, Valais, Switzerland. [Trechmannite-$\alpha$, supposedly isomorphous with trechmannite, lead gray, same locality.]

BEST REF. IN ENGLISH: Palache, et al., "Dana's System of Mineralogy," 7th Ed., v. I, p. 432, New York, Wiley, 1944.


# TREMOLITE
Ca$_2$Mg$_5$Si$_8$O$_{22}$(OH)$_2$
0–20 mol. % (OH)$_2$Ca$_2$Fe$_5$Si$_8$O$_{22}$
Amphibole Group
var. Hexagonite

CRYSTAL SYSTEM: Monoclinic
CLASS: 2/m
SPACE GROUP: C2/m
Z: 2

LATTICE CONSTANTS:
$a$ = 9.84
$b$ = 18.02
$c$ = 5.27
$\beta$ = 104°57′
3 STRONGEST DIFFRACTION LINES:
8.38 (100)
3.12 (100)
2.71 ( 90)
OPTICAL CONSTANTS:
$\alpha$ = 1.560
$\beta$ = 1.613
$\gamma$ = 1.624
(–)2V $\simeq$ 81°
HARDNESS: 5–6
DENSITY: 2.9–3.2 (Meas.)
CLEAVAGE: {110} good
{100} parting
Fracture uneven to subconchoidal. Brittle; some compact varieties tough.

HABIT: Crystals commonly long-bladed; less frequently short and stout. Usually in fibrous or thin columnar aggregates, often radiated. Also massive, fibrous or granular. Twinning on {100} common, simple, lamellar.

COLOR-LUSTER: Colorless, white, gray, pale greenish, pink (hexagonite), brown. Transparent to translucent. Vitreous.

MODE OF OCCURRENCE: Occurs mainly in contact and regionally metamorphosed dolomites, magnesian limestones, and low-grade ultrabasic rocks. Typical occurrences are found throughout the western United States, especially in California, Arizona, Utah, Colorado, and South Dakota; at Canaan, Connecticut; at Lee, Massachusetts; and at numerous places in New York, notably at Edwards (hexagonite), and at Balmat. Also found in abundance in Ontario and Quebec, Canada, and at localities in Italy, Switzerland, Austria, and elsewhere.

BEST REF. IN ENGLISH: Deer, Howie, and Zussman, "Rock Forming Minerals," v. 2, p. 249–262, New York, Wiley, 1963. Ross, Malcolm, Smith, William L., and Ashton, William H., *Am. Min.*, **53**: 751–769 (1968).


# TREVORITE
NiFe$_2$O$_4$
Magnetite Series
Spinel Group

CRYSTAL SYSTEM: Cubic
CLASS: 4/m $\bar{3}$ 2/m
SPACE GROUP: Fd3m
Z: 8
LATTICE CONSTANT:
$a$ = 8.41 (varies)
3 STRONGEST DIFFRACTION LINES:
2.513 (100)
1.476 ( 40)
1.605 ( 30)
OPTICAL CONSTANTS: Only thin splinters will transmit light. $N$ = 2.3 (?)
HARDNESS: 5+

DENSITY: 5.165 (Meas.)
5.24 (Calc.)
CLEAVAGE: None.
Fracture uneven to subconchoidal. Brittle.
HABIT: Crystals octahedral, distorted. Commonly massive.
COLOR-LUSTER: Black; metallic. Opaque. Streak brown.

Strongly magnetic.
MODE OF OCCURRENCE: Occurs in a green talc rock on the Bon Accord farm, near Sheba Siding, Transvaal, South Africa.
BEST REF. IN ENGLISH: Palache, et al., "Dana's System of Mineralogy," 7th Ed., v. I, p. 698–707, New York, Wiley, 1944.

## TRICHALCITE = Tyrolite

## α-TRIDYMITE
$SiO_2$

CRYSTAL SYSTEM: Orthorhombic
CLASS: 222
SPACE GROUP: $C222_1$
Z: 160
LATTICE CONSTANTS:
$a = 9.91$
$b = 17.18$
$c = 40.78$
Variable. $c$ is reported as being many multiples of 4.07.
3 STRONGEST DIFFRACTION LINES:
4.30 (100)
4.09 ( 90)
3.80 ( 60)
OPTICAL CONSTANTS:
$\alpha = 1.468-1.479$
$\beta = 1.469-1.480$
$\gamma = 1.473-1.483$
$2V = 30°-76°$
HARDNESS: 7
DENSITY: 2.26 (Meas.)
CLEAVAGE: None.
Fracture conchoidal. Brittle to very brittle.
HABIT: Crystals tabular, pseudohexagonal, as inversion pseudomorphs after β-tridymite.
COLOR-LUSTER: Colorless to white. Transparent. Vitreous, sometimes pearly on {0001}.
MODE OF OCCURRENCE: α-Tridymite, originally formed as β-tridymite, occurs mainly in cavities and in the groundmass of volcanic rocks of Tertiary or younger age. Common associated minerals include sanidine, quartz, cristobalite, fayalite, hematite, augite, and hornblende. It is also a frequent constituent in stony meteorites. Typical occurrences are found in California, Oregon, Washington, Nevada, Utah, Wyoming, Colorado, and Arizona. Also found in andesite in the Cerro San Cristóbal, near Pachuca, Mexico, and at localities in Scotland, France, Italy, Germany, USSR, and Japan.
BEST REF. IN ENGLISH: Frondel, Clifford, "Dana's System of Mineralogy," 7th Ed., v. III, p. 259–272, New York, Wiley, 1962. Grant, Raymond W., *Am. Min.*, **52**: 536–541 (1967).

## β-TRIDYMITE
$SiO_2$

CRYSTAL SYSTEM: Hexagonal
CLASS: 6/m 2/m 2/m
SPACE GROUP: C6/mmc
Z: 4
LATTICE CONSTANTS:
$a = 5.03$
$c = 8.22$
3 STRONGEST DIFFRACTION LINES:
4.12 (100)
4.37 ( 80)
3.86 ( 60)
OPTICAL CONSTANTS: Uniaxial positive.
DENSITY: 2.197 (Calc.)
HABIT: Crystals usually minute, thin tabular, flattened on {0001}. The pyramid o{10$\bar{1}$3} commonly present. Also thick tabular. Sometimes as phenocrysts up to several centimeters across. Often as rosettes or in spherical and fan-shaped aggregates. Twinning on {10$\bar{1}$6} very common, contact and penetration; wedge-shaped, sheaf-like, or as trillings. Twinning on {30$\bar{3}$4}, uncommon.
MODE OF OCCURRENCE: Formed normally between 870° and 1470°C. Below 117°C it inverts to the orthorhombic modification (α- tridymite).
BEST REF. IN ENGLISH: Frondel, Clifford, "Dana's System of Mineralogy," 7th Ed., v. III, p. 259–272, New York, Wiley, 1962. *Am. Min.*, **52**: 536–541 (1967).

## TRIGONITE
$Pb_3Mn[AsO_3]_2[AsO_2(OH)]$

CRYSTAL SYSTEM: Monoclinic
CLASS: m
SPACE GROUP: Pn
Z: 2
LATTICE CONSTANTS:
$a = 7.25$ kX
$b = 6.80$
$c = 11.07$
$\beta = 91°49'$
3 STRONGEST DIFFRACTION LINES:
2.99 (100)
3.08 ( 90)
3.24 ( 70)
OPTICAL CONSTANTS:
$\alpha = 2.07$
$\beta = 2.10$
$\gamma = 2.12$
$(-)2V =$ very large
HARDNESS: 2-3
DENSITY: Variable, 6.1-7.1
CLEAVAGE: {010} perfect
{101} good
Fracture uneven.

HABIT: Crystals thick tabular with triangular outline, often highly modified. As subparallel growths.

COLOR-LUSTER: Yellow to brownish yellow to dark brown. Translucent. Vitreous to adamantine.

MODE OF OCCURRENCE: Occurs associated with native lead, magnussonite, and dixenite in crevices in dolomite-hausmannite ore at Långban, Sweden.

BEST REF. IN ENGLISH: Palache, et al., "Dana's System of Mineralogy," 7th Ed., v. II, p. 1032-1033, New York, Wiley, 1951.

## TRIHYDROCALCITE (Species status doubtful, probably a mixture of the monohydrate and the hexahydrate)

$CaCO_3 \cdot 3H_2O$

CRYSTAL SYSTEM: Unknown.
HARDNESS: Soft.
DENSITY: ~2.63 (?)
CLEAVAGE: Not determined.
HABIT: As microscopic fibers; in mold-like or felty coatings.
COLOR-LUSTER: White.
MODE OF OCCURRENCE: Occurs in a limestone cavern at Wolmsdorff, Glatz, Silesia, and in marl in the Lublin region, Poland.
BEST REF. IN ENGLISH: Palache, et al., "Dana's System of Mineralogy," 7th Ed., v. II, p. 227-228, New York, Wiley, 1951.

## TRIKALSILITE

(K, Na)AlSiO$_4$

CRYSTAL SYSTEM: Hexagonal
CLASS: 6 (probably)
SPACE GROUP: P6$_3$ (probably)
Z: 18
LATTICE CONSTANTS:
  $a = 15.4$
  $c = 8.6$
3 STRONGEST DIFFRACTION LINES:
  3.08 (100)
  3.05 ( 95) (synthetic)
  2.56 ( 45)
HARDNESS: 6
DENSITY: 2.630 (Calc.)
CLEAVAGE: Fracture subconchoidal. Brittle.
HABIT: Massive.
COLOR-LUSTER: Colorless. Transparent. Vitreous.
MODE OF OCCURRENCE: Occurs in parallel growth with nepheline in the lava from Kabfumu, North Kivu, Zaire.
BEST REF. IN ENGLISH: Sahama, Th. G., and Smith, J. V., *Am. Min.*, **42**: 286 (1957).

## TRIMERITE

CaMn$_2$(BeSiO$_4$)$_3$

CRYSTAL SYSTEM: Monoclinic

CLASS: 2/m
SPACE GROUP: P2$_1$/c
Z: 16
LATTICE CONSTANTS:
  $a = 16.4$
  $b = 7.62$
  $c = 27.92$
  $\beta = 90°09'$
3 STRONGEST DIFFRACTION LINES:
  2.764 (100)
  3.56 ( 40)
  2.229 ( 35)
OPTICAL CONSTANTS:
  $\alpha = 1.715$
  $\beta = 1.720$
  $\gamma = 1.725$
  $(-)2V = 83°$
HARDNESS: 6-7
DENSITY: 3.474 (Meas.)
        3.507 (Calc.)
CLEAVAGE: {001} distinct
Fracture conchoidal. Brittle.
HABIT: Crystals prismatic, pseudohexagonal, up to 8 X 12 mm. Twinning on {100}, {101}, and {10$\bar{1}$}, polysynthetic.
COLOR-LUSTER: Colorless, pink, yellowish red. Transparent. Vitreous.
MODE OF OCCURRENCE: Occurs sparingly at three manganese deposits in Sweden. Found as intergrowths with calcite, forming inclusions in fine-grained rock consisting of magnetite, pyroxene, garnet, and other minerals at Harstigen, and at Jacobsberg; also in segregations of calcite within fine-grained hematite at Långban.
BEST REF. IN ENGLISH: Vlasov, K. A., "Mineralogy of Rare Elements," v. II, p. 117-118, Israel Program for Scientific Translations, 1966.

## TRIPHANE = Synonym for spodumene

## TRIPHYLITE

Li(Fe$^{2+}$, Mn$^{2+}$)PO$_4$
Triphylite-Lithiophilite Series

CRYSTAL SYSTEM: Orthorhombic
CLASS: 2/m 2/m 2/m
SPACE GROUP: Pmnb
Z: 4
LATTICE CONSTANTS:
  $a = 6.0285$
  $b = 10.3586$
  $c = 4.7031$
3 STRONGEST DIFFRACTION LINES:
  2.54 (100)
  4.29 ( 90)
  3.51 ( 90)
OPTICAL CONSTANTS:
  $\alpha = 1.689-1.694$
  $\beta = 1.689-1.695$
  $\gamma = 1.695-1.702$
  $(+)2V = 0°-55°$; also $(-)$
HARDNESS: 4-5

DENSITY: 3.423 (Meas.)
 3.527 (Calc. Fe:Mn = 1:1)
CLEAVAGE: {100} nearly perfect
 {010} imperfect
 {011} interrupted
Fracture subconchoidal to uneven.

HABIT: Crystals rare; stout prismatic, often with rounded or rough faces. Usually massive, cleavable.

COLOR-LUSTER: Greenish gray, bluish gray; often externally brownish to blackish due to alteration. Translucent to transparent. Resinous to vitreous. Streak colorless to dirty white.

MODE OF OCCURRENCE: Occurs as a primary mineral in granite pegmatites, often partly or completely altered to a wide variety of secondary phosphate minerals. Found widespread in the Black Hills, South Dakota, often as anhedral or subhedral crystals up to 6 or more feet in size, especially at the Dan Patch, Big Chief, and Hesnard mines, Keystone district, and at the Bull Moose and Tip Top mines, Custer district. It also occurs as fine crystals at the Smith mine, Chandler's Mill, at Palermo, and at other places in New Hampshire; in Maine; and in the Pala district, California. Other occurrences are found in Canada, Brazil, France, Sweden, Germany, and Finland.

BEST REF. IN ENGLISH: Palache, et al., "Dana's System of Mineralogy," 7th Ed., v. II, p. 665–669, New York, Wiley, 1951. Finger, L. W. and Rapp, G. R. Jr., Carnegie Inst. Wash. Year Book 69: 290–292 (1970).


# TRIPLITE
$(Mn^{2+}, Fe^{2+}, Mg, Ca)_2 (PO_4)(F, OH)$
Triplite-Zwieselite Series
Var. Magnesian (Talktriplite)

CRYSTAL SYSTEM: Monoclinic
CLASS: 2/m
SPACE GROUP: I 2/m
Z: 8
LATTICE CONSTANTS:
 $a$ = 12.085
 $b$ = 6.536
 $c$ = 9.910
 $\beta$ = 105°38'
3 STRONGEST DIFFRACTION LINES:
 2.87 (100)
 3.05 ( 90)
 3.26 ( 70)
OPTICAL CONSTANTS:
 $\alpha$ = 1.643–1.696
 $\beta$ = 1.647–1.704
 $\gamma$ = 1.665–1.713
 (+)2V = 25°–76°
HARDNESS: 5–5½
DENSITY: 3.55–3.87 (Meas.)
CLEAVAGE: {001} good
 {010} fair
 {100} poor
Fracture subconchoidal to uneven.

HABIT: As rough, poorly developed crystals. Commonly massive.

COLOR-LUSTER: Reddish brown to dark brown; also salmon pink; often brownish black to black on alteration. Translucent to opaque. Vitreous to resinous. Streak brown to white.

MODE OF OCCURRENCE: Occurs widespread as a primary mineral in granite pegmatites, and sparingly in high-temperature vein deposits. Found in the United States in California, Nevada, Arizona, Colorado, South Dakota, Virginia, Connecticut, and Maine. It also occurs in Argentina, France, Germany, Czechoslovakia, Portugal, Sweden, Finland, South West Africa, Rhodesia, and South Australia.

BEST REF. IN ENGLISH: Palache, et al., "Dana's System of Mineralogy," 7th Ed., v. II, p. 849–852, New York, Wiley, 1951.


# TRIPLOIDITE
$(Mn^{2+}, Fe^{2+})_2 PO_4 OH$
Triploidite-Wolfeite Series

CRYSTAL SYSTEM: Monoclinic
CLASS: 2/m
SPACE GROUP: $P2_1/a$
Z: 16
LATTICE CONSTANTS:
 $a$ = 12.26
 $b$ = 13.38
 $c$ = 9.90
 $\beta$ = 108°04'
3 STRONGEST DIFFRACTION LINES:
 2.94 (100)
 3.10 ( 90)
 3.19 ( 80)
OPTICAL CONSTANTS:
 $\alpha$ = 1.725
 $\beta$ = 1.726
 $\gamma$ = 1.730
 (+)2V = moderate
HARDNESS: 4½–5
DENSITY: 3.697 (Meas.)
CLEAVAGE: {010} good
 {120} fair
Fracture subconchoidal to uneven.

HABIT: Crystals prismatic, vertically striated. Usually as columnar to parallel-fibrous aggregates; also divergent fibrous; granular.

COLOR-LUSTER: Pinkish, yellowish to yellowish brown or reddish brown; rarely greenish. Transparent to translucent. Vitreous to greasy. Streak whitish.

MODE OF OCCURRENCE: Occurs as pinkish masses in granite pegmatite, associated with alluaudite and loellingite, at the Ross mine, Custer County, and sparingly at the Peerless mine, Pennington County, South Dakota; with lithiophilite, rhodochrosite, and secondary phosphate minerals at Branchville, Fairfield County, Connecticut; and as an alteration of triplite at Wein, Czechoslovakia.

BEST REF. IN ENGLISH: Palache, et al., "Dana's System of Mineralogy," 7th Ed., v. II, p. 853–855, New York, Wiley, 1951.


# TRIPPKEÏTE
$CuAs_2 O_4$

CRYSTAL SYSTEM: Tetragonal
CLASS: 4/m 2/m 2/m
SPACE GROUP: $P4_2/mbc$
Z: 4
LATTICE CONSTANTS:
  $a = 8.55$ kX
  $c = 5.65$
OPTICAL CONSTANTS:
  $\omega = 1.90$
  $\epsilon = 2.12$
   (+)
HARDNESS: Not determined. Soft.
DENSITY: 4.8 (Meas.)
CLEAVAGE: {100} perfect
         {110} good
HABIT: Crystals short prismatic, often slightly bent.
COLOR-LUSTER: Greenish blue. Brilliant.
MODE OF OCCURRENCE: Occurs associated with olivenite in copper deposits near Copiapó, Chile.
BEST REF. IN ENGLISH: Palache, et al., "Dana's System of Mineralogy," 7th Ed., v. II, p. 1034, New York, Wiley, 1951.

# TRIPUHYITE
$FeSb_2O_6$

CRYSTAL SYSTEM: Tetragonal
Z: 3
LATTICE CONSTANTS:
  $a = 4.63$
  $c = 9.14$
3 STRONGEST DIFFRACTION LINES:
  3.28 (100)
  2.56 ( 90)
  1.72 ( 90)
OPTICAL CONSTANTS:
  $\alpha = 2.19$
  $\beta = 2.20$
  $\gamma = 2.33$
  (+)2V = small
HARDNESS: ~ 7
DENSITY: 5.82 (Meas.)
        6.14 (Calc. $3FeSbO_4$)
CLEAVAGE: Not determined.
HABIT: As microcrystalline aggregates; compact or earthy, nodular.
COLOR-LUSTER: Lemon yellow to brownish black. Translucent. Streak yellowish.
MODE OF OCCURRENCE: Occurs at El Antimonio, Sonora, Mexico; in placer deposits associated with cinnabar at Tripuhy near Ouro Preto, Minas Geraes, Brazil; as small veins in dacite lava near Doncellas, Jujuy, Argentina; and with nadorite, calcite, and celestite near Djebel Nador, Constantine, Algeria.
BEST REF. IN ENGLISH: Palache, et al., "Dana's System of Mineralogy," 7th Ed., v. II, p. 1024, New York, Wiley, 1951.

# TRITOMITE
$(Ce, La, Y, Th)_5 (Si, B)_3 (O, OH, F)_{13}$

CRYSTAL SYSTEM: Hexagonal
Z: 2
LATTICE CONSTANTS:
  $a = 9.35$
  $c = 6.88$
3 STRONGEST DIFFRACTION LINES:
  2.81 (100)
  3.44 ( 40)
  1.84 ( 40)
OPTICAL CONSTANTS:
  Isotropic; metamict. $N = 1.650–1.757$
HARDNESS: 5½
DENSITY: 4.2 (Meas.)
CLEAVAGE: None. Fracture conchoidal. Brittle.
HABIT: Crystals acute triangular pyramidal in form.
COLOR-LUSTER: Dark brown. Translucent in thin splinters. Vitreous to resinous.
MODE OF OCCURRENCE: The mineral occurs in nepheline syenite pegmatites at Låven, Brevik, and Barkevik, in the Langesundsfjord district of southern Norway.
BEST REF. IN ENGLISH: Jaffe, Howard W., and Molinski, Victor J., *Am. Min.*, **47**: 9–25 (1962).

# TRITOMITE-(Y) (Syn. Spencite)
$(Y, Ca, La, Fe)_5 (Si, B, Al)_3 (O, OH, F)_{13}$

CRYSTAL SYSTEM: Hexagonal
Z: 2
LATTICE CONSTANTS:
  (hexagonal)
  $a = 9.32$
  $c = 6.84$
3 STRONGEST DIFFRACTION LINES
  2.78 (100)
  4.02 ( 50)
  3.13 ( 50)
Heated to 1000°C. for 3 hours in air.
OPTICAL CONSTANTS:
  Isotropic; metamict. $N \sim 1.630–1.770$
HARDNESS: 3½ (Ontario)
         6½ (New Jersey)
DENSITY: 3.05 (Meas. Ontario)
        3.40 (Meas. New Jersey)
CLEAVAGE: None. Fracture conchoidal. Brittle.
HABIT: Massive, compact. Also as anhedral grains.
COLOR-LUSTER: Dark greenish black, reddish brown to brownish black. Translucent in thin splinters. Vitreous to resinous.
MODE OF OCCURRENCE: The mineral occurs intimately associated with magnetite, black zircon, apatite, and fergusonite in a granite pegmatite near Cranberry Lake, Sussex County, New Jersey; also as masses in a narrow pegmatite stringer in a vuggy pyroxenite, associated with calcite, red apatite crystals, diopside, purple fluorite, and wernerite at a prospect pit in Cardiff Township, Haliburton County, Ontario, Canada.
BEST REF. IN ENGLISH: Frondel, Clifford, *Can. Min.*, **6**: 576–581 (1961). Jaffe, Howard W., and Molinski, Victor J., *Am. Min.*, **47**: 9–25 (1962).

# TROEGERITE
$H_2(UO_2)_2(AsO_4)_2 \cdot 8H_2O$

CRYSTAL SYSTEM: Tetragonal
CLASS: 4/m 2/m 2/m
SPACE GROUP: P4/nmm
Z: 1
LATTICE CONSTANTS:
 $a = 7.16$
 $c = 8.80$ (synthetic)
3 STRONGEST DIFFRACTION LINES:
 8.59 (100)
 3.79 ( 90) (synthetic)
 3.30 ( 80)
OPTICAL CONSTANTS:
 (synthetic)
 $\omega = 1.612$
 $\epsilon = 1.584$
 (-)
HARDNESS: 2-3
DENSITY: 3.55 (Meas. synthetic)
CLEAVAGE: {001} perfect, micaceous
 {100} good
HABIT: Crystals thin tabular on {001}. Crystals usually composite with uneven faces; as subparallel aggregates. Inclined faces small and striated horizontally. Parallel growths with zeunerite common.
COLOR-LUSTER: Lemon yellow, transparent. Subvitreous, pearly on {001}. Fluoresces lemon yellow in ultraviolet light.
MODE OF OCCURRENCE: Occurs as a secondary mineral with metazeunerite and walpurgite in the Walpurgis vein of the Weisser Hirsch mine at Neustädtl and at the Daniel mine, Schneeberg, Saxony, Germany.
BEST REF. IN ENGLISH: Frondel, Clifford, USGS Bull. 1064, p. 187-191 (1958).

# TROGTALITE
$CoSe_2$

CRYSTAL SYSTEM: Cubic
CLASS: 2/m $\bar{3}$
SPACE GROUP: Pa3
Z: 4
LATTICE CONSTANT:
 $a = 5.87$
3 STRONGEST DIFFRACTION LINES:
 2.62 (100)
 2.39 (100)
 1.765 ( 90)
OPTICAL CONSTANTS: Isotropic.
HARDNESS: Hard.
DENSITY: 7.120 (Calc.)
CLEAVAGE: Not determined.
HABIT: Massive. Observed only in thin sections.
COLOR-LUSTER: Rose-violet.
MODE OF OCCURRENCE: Occurs associated with hastite, bornhardtite, and freboldite as intergrowths in clausthalite at Steinbruch Trogtal, near Lautenthal, Harz, Germany.

BEST REF. IN ENGLISH: Ramdohr, Paul, and Schmidt, Marg., *Am. Min.*, **41**: 164-165 (1956).

# TROILITE
FeS

CRYSTAL SYSTEM: Hexagonal
CLASS: 6/m 2/m 2/m
SPACE GROUP: P6₃/mmc
Z: 12
LATTICE CONSTANTS:
 $a = 5.958$
 $c = 11.74$ (synthetic)
3 STRONGEST DIFFRACTION LINES:
 2.09 (100)
 2.66 ( 60)
 1.72 ( 50)
OPTICAL CONSTANTS: Opaque.
HARDNESS: 3½-4½
DENSITY: 4.67-4.82 (Meas.)
 4.85 (Calc.)
CLEAVAGE: None.
HABIT: Massive, compact granular.
COLOR-LUSTER: Light grayish brown; rapidly tarnishes to bronze-brown. Opaque. Metallic. Streak black.
MODE OF OCCURRENCE: Occurs in a sheared zone of serpentine at a copper claim northeast of Crescent City, Del Norte County, California. Also frequently found as nodules in iron meteorites.
BEST REF. IN ENGLISH: Palache, et al., "Dana's System of Mineralogy," 7th Ed., v. I, p. 233, New York, Wiley, 1944.

# TROLLEITE
$Al_4(PO_4)_3(OH)_3$ (?)

CRYSTAL SYSTEM: Monoclinic or triclinic
3 STRONGEST DIFFRACTION LINES:
 3.20 (100)
 3.09 ( 80)
 2.51 ( 45)
OPTICAL CONSTANTS:
 $\alpha = 1.619$
 $\beta = 1.639$
 $\gamma = 1.643$ (Na)
 (-)2V = 49°
HARDNESS: ~8½ (synthetic)
 5½-6 (natural)
DENSITY: 3.09 (Meas.)
CLEAVAGE: Indistinct in two directions at an angle of 110°52'.
Fracture conchoidal to even.
HABIT: Massive, lamellar.
COLOR-LUSTER: Pale green. Vitreous.
MODE OF OCCURRENCE: Occurs in association with berlinite and other phosphates at the iron mine of Westanå in Kristianstad, Sweden.
BEST REF. IN ENGLISH: Palache, et al., "Dana's System of Mineralogy," 7th Ed., v. II, p. 911, New York,

Wiley, 1951.    Sclar, C. B., Carrison, L. C., and Schwartz, C. M., *Am. Min.*, **50**: 267 (1965).

## TRONA
$Na_3H(CO_3)_2 \cdot 2H_2O$

CRYSTAL SYSTEM:  Monoclinic
CLASS:  2/m
SPACE GROUP:  I2/a
Z:  4
LATTICE CONSTANTS:
$a = 20.11$
$b = 3.49$
$c = 10.31$
$\beta = 103°8'$
3 STRONGEST DIFFRACTION LINES:
2.659 (100)
3.08  ( 80)
9.88  ( 60)
OPTICAL CONSTANTS:
$\alpha = 1.412, 1.418$
$\beta = 1.492$
$\gamma = 1.540, 1.543$
$(-)2V \simeq 76°$  (Na)
HARDNESS:  2½–3
DENSITY:  2.11 (Meas.)
         2.13 (Calc.)
CLEAVAGE:  {100} perfect
           {$\bar{1}11$} indistinct
           {001} indistinct
HABIT:  Crystals elongated [010] and flattened {001}. Commonly massive, fibrous or columnar; also as thick plates.
COLOR-LUSTER:  Colorless, grayish, yellowish white, pale brownish.  Transparent to translucent.  Vitreous, glistening.
MODE OF OCCURRENCE:  Occurs as a saline lake deposit or as an efflorescence on soil in arid regions.  In California thick layers of solid trona occur with borax, thenardite, glauberite, hanksite, and other salts at Searles Lake, San Bernardino County; in the borax deposits of Death Valley and along the shores of Owens Lake, Inyo County; at Borax Lake, Lake County; and on the east edge of Mono Lake, Mono County.  It also occurs widespread in other western states, especially in Nevada and in western Wyoming.  Among other localities, it is found at Lake Magadi, Kenya, and in Mongolia, Tibet, Armenia, and Iran.
BEST REF. IN ENGLISH:  Palache, et al., "Dana's System of Mineralogy," 7th Ed., v. II, p. 138–140, New York, Wiley, 1951.    Pabst, A., *Am. Min.*, **44**: 274–281 (1959).

## TROOSTITE = Manganoan willemite

## TRUDELLITE = A mixture of chloraluminite and natroalunite

## TRUSCOTTITE
$(Ca, Mn)_2Si_4O_9(OH)_2$

CRYSTAL SYSTEM:  Hexagonal
CLASS:  3 or $\bar{3}$
SPACE GROUP:  P3 or P$\bar{3}$
Z:  6
LATTICE CONSTANTS:
$a = 9.752$
$c = 18.80$
3 STRONGEST DIFFRACTION LINES:
19.0  (100)
9.48 (100)
3.14 (100)
OPTICAL CONSTANTS:
$\omega = 1.552$
$\epsilon = 1.530$
$(-)2V$ small
DENSITY:  2.35 (Meas.)
CLEAVAGE:  {0001} perfect
HABIT:  Massive; as spherical aggregates.
COLOR-LUSTER:  White.  Translucent.  Pearly on cleavage.
MODE OF OCCURRENCE:  Truscottite occurs at the Toi mine, Shizuoka Prefecture, Japan, and at the Lelong Donok gold mine, southern Sumatra.
BEST REF. IN ENGLISH:  Mackay, A. L. and Taylor, H. F. W., *Min. Mag.*, **30**: 450–457 (1954).

## TRÜSTEDITE
$Ni_3Se_4$

CRYSTAL SYSTEM:  Cubic
CLASS:  Probably 4/m $\bar{3}$ 2/m
SPACE GROUP:  Probably Fd3m
Z:  8
LATTICE CONSTANT:
$a = 9.94$
3 STRONGEST DIFFRACTION LINES:
2.48  (100)
1.755 (100)
3.00  ( 80)
HARDNESS:  ~2½
Slightly softer than penroseite.
DENSITY:  6.62 (Calc.)
CLEAVAGE:  Not determined.
HABIT:  Crystals euhedral, minute.
COLOR-LUSTER:  Yellow, completely isotropic.  Reflectivity higher than those of sederholmite and penroseite.
MODE OF OCCURRENCE:  Occurs in clausthalite, associated with penroseite and sederholmite, in calcite-uranium-bearing veins in sills of albite diabase in a schist formation, at Kuusamo, northeast Finland.
BEST REF. IN ENGLISH:  Vourelainen, Y., Huhma, A., Häkli, A., *Am. Min.*, **50**: 520 (1965).

## TSCHEFFKINITE = Chevkinite

## TSCHERMAKITE (Ferrotschermakite)
$Ca_2Mg_3(Al, Fe)_2(Al_2Si_6)O_{22}(OH, F)_2$
Amphibole Group
Hornblende Series

CRYSTAL SYSTEM: Monoclinic
CLASS: 2/m
SPACE GROUP: C2/m
Z: 2
LATTICE CONSTANTS:
$a \simeq 9.9$
$b \simeq 18.0$
$c \simeq 5.3$
$\beta \simeq 105°$
3 STRONGEST DIFFRACTION LINES: See hornblende.
OPTICAL CONSTANTS:
$\alpha = 1.680$
$\beta = 1.695$
$\gamma = 1.698$
$(-)2V = 45°$
HARDNESS: 5–6
DENSITY: 3.14–3.35 (Meas.)
CLEAVAGE: {110} perfect
{001} parting
{100} parting
Fracture uneven to subconchoidal. Brittle.
HABIT: Crystals short prismatic. Also massive; compact. Twinning on {100} common, simple, lamellar.
COLOR-LUSTER: Green, dark green, black. Translucent to nearly opaque.
MODE OF OCCURRENCE: Occurs as a constituent of both igneous and metamorphic rocks. Typical occurrences are found in Greenland, Scotland, Germany, Finland, and Japan.
BEST REF. IN ENGLISH: Deer, Howie, and Zussman, "Rock Forming Minerals," v. 2, p. 263–314, New York, Wiley, 1963.

## TSCHERMIGITE = Ammonia alum

## TSUMCORITE
$PbZnFe(AsO_4)_2 \cdot H_2O$

CRYSTAL SYSTEM: Monoclinic
CLASS: 2/m
SPACE GROUP: C2/m
Z: 2
LATTICE CONSTANTS:
$a = 9.131$
$b = 6.326$
$c = 7.583$
$\beta = 115.3°$
3 STRONGEST DIFFRACTION LINES:
3.244 (100)
4.663 ( 90)
2.863 ( 90)
OPTICAL CONSTANTS:
$n \sim 1.9$. $2V \sim 90°$
Dichroism from yellow to yellowish green.
HARDNESS: 4½
DENSITY: ~5.2 (Meas.)
CLEAVAGE: Not determined.
HABIT: As spherulitic crusts or intergrown sub-parallel crystals.

COLOR-LUSTER: Yellow-brown or red-brown. Translucent.
MODE OF OCCURRENCE: Occurs in association with lead-iron arsenates and sulfates in the oxidation zone of the ore deposits at Tsumeb, South West Africa; also reported from Thasos, Greece.
BEST REF. IN ENGLISH: Geier, B. H., Kautz, K., and Müller, G., *Min. Abs.*, **23**: 129 (1972).

## TSUMEBITE
$Pb_2Cu(PO_4)(SO_4)(OH)$

CRYSTAL SYSTEM: Monoclinic
CLASS: 2/m
SPACE GROUP: $P2_1/m$
Z: 2
LATTICE CONSTANTS:
$a = 8.70$
$b = 5.80$
$c = 7.85$
$\beta = 111°30'$
OPTICAL CONSTANTS:
$\alpha = 1.885$
$\beta = 1.920$
$\gamma = 1.956$
$(+)2V = $ near 90°
HARDNESS: 3½
DENSITY: 6.133 (Meas.)
CLEAVAGE: Fracture uneven. Brittle.
HABIT: Crystals thick tabular, always twinned as trillings or more complex groups. As crusts of intergrown crystals. Twin plane {$\bar{1}22$}.
COLOR-LUSTER: Emerald green. Transparent. Brilliant, vitreous.
MODE OF OCCURRENCE: Occurs as a secondary mineral at Morenci, Greenlee County, Arizona; and with cerussite, smithsonite, and azurite at Tsumeb, South West Africa.
BEST REF. IN ENGLISH: Palache, et al., "Dana's System of Mineralogy," 7th Ed., v. II, p. 918–919, New York, Wiley, 1951.

## TUCANITE = Scarbroite

## TUGTUPITE
$Na_4BeAlSi_4O_{12}Cl$

CRYSTAL SYSTEM: Tetragonal
CLASS: $\bar{4}$
SPACE GROUP: $I\bar{4}$
Z: 2
LATTICE CONSTANTS:
$a = 8.583$
$c = 8.817$
3 STRONGEST DIFFRACTION LINES:
3.52 (100)
6.13 ( 80)
3.57 ( 60)

OPTICAL CONSTANTS:

$\omega = 1.496$

$\epsilon = 1.502$

Uniaxial (+) or (+)2V = 0°-10°

HARDNESS: ~ 4

DENSITY: 2.3 (Meas.)

   2.36 (Calc.)

CLEAVAGE: Distinct bipyramidal. Fracture conchoidal. Brittle.

HABIT: Massive, compact. Twinning on {101}.

COLOR-LUSTER: White to light pink or rose red; also bluish, greenish. Transparent to translucent. Vitreous to greasy. Fluoresces bright rose red under ultraviolet light.

MODE OF OCCURRENCE: Occurs as veins cutting chkalovite in an albite-analcime vein in nepheline syenite pegmatite, Tugtup agtakôrfia, Ilimaussaq, Greenland; also as an alteration product of chkalovite in pegmatites of Mt. Sengischorr and Mt. Punkaruaiv, Lovozero massif, Kola Peninsula, USSR.

BEST REF. IN ENGLISH: Sørensen, Henning, *Am. Min.*, **46**: 241 (1961); **48**: 1178 (1963).


# TUHUALITE

$(Na, K)_2 (Fe^{2+}, Fe^{3+}, Al)_3 Si_7 O_{18} (OH)_2$

CRYSTAL SYSTEM: Orthorhombic

CLASS: 2/m 2/m 2/m or mm2

SPACE GROUP: Cmca or C2ca

Z: 1

LATTICE CONSTANTS:

$a = 14.31$

$b = 17.28$

$c = 10.11$

3 STRONGEST DIFFRACTION LINES:

7.16 (100)

2.77 ( 90)

3.18 ( 80)

OPTICAL CONSTANTS:

   $\alpha = 1.608$

   $\beta = 1.612$

   $\gamma = 1.621$

(+)2V = 61°-70°

HARDNESS: 3-4

DENSITY: 2.89 (Meas.)

   2.86 (Calc.)

CLEAVAGE: {100} good

   {010} good

   {001} good

Very brittle.

HABIT: Crystals prismatic.

COLOR-LUSTER: Black to dark blue.

MODE OF OCCURRENCE: Occurs in volcanic rock (commendite) at Opo Bay, Mayor (Tuhua) Island, New Zealand.

BEST REF. IN ENGLISH: Hutton, C. Osborne, *Min. Mag.*, **31**: 96-106 (1956).


# TULAMEENITE

$Pt_2 FeCu$

CRYSTAL SYSTEM: Tetragonal

Z: 2

LATTICE CONSTANTS:

$a = 3.891$

$c = 3.577$

3 STRONGEST DIFFRACTION LINES:

2.179 (100)

1.163 ( 80)

1.094 ( 80)

OPTICAL CONSTANTS: Very weakly anisotropic.

HARDNESS: Microhardness 420-456 kg/mm$^2$ with 50 gm load.

DENSITY: 15.6 (Calc.)

HABIT: Massive; as minute grains.

COLOR-LUSTER: White in reflected light in oil and air.

MODE OF OCCURRENCE: Tulameenite occurs as a minor constituent of placer deposits along the valleys of the Tulameen and Similkameen Rivers in south-central British Columbia, Canada. It is found associated with cubic iron-bearing platinum as rounded to irregular areas up to about 400 $\mu m$ in diameter, as free grains, or as grains with complex inclusions.

BEST REF. IN ENGLISH: Cabri, Louis J., Owens, Dalton R., and Laflamme, J. H. Gilles, *Can. Min.*, **12**: 21-25 (1973).


# TUMITE = Axinite


# TUNDRITE

$Na_2 Ce_2 (Ti, Nb) SiO_8 \cdot 4H_2O$

CRYSTAL SYSTEM: Triclinic

CLASS: 1

SPACE GROUP: P1

Z: 2

LATTICE CONSTANTS:

$a = 4.94$   $\alpha = 105°20'$

$b = 7.57$   $\beta = 102°45'$

$c = 14.05$   $\gamma = 70°48'$

3 STRONGEST DIFFRACTION LINES:

2.78 (100)

2.72 (100)

3.06 ( 80)

OPTICAL CONSTANTS:

   $\alpha = 1.743$

   $\beta = 1.80$

   $\gamma = 1.88$

   (+)2V

HARDNESS: ~3

DENSITY: 3.70 (Meas.)

CLEAVAGE: Not observed. Brittle.

HABIT: Crystals acicular, up to 5 mm long; also as spherulites up to 15 mm in diameter.

COLOR-LUSTER: Brownish yellow to greenish yellow; vitreous.

MODE OF OCCURRENCE: Occurs associated with aegirine, lamprophyllite, and ramsayite in three nepheline syenite pegmatites at Mt. Nepkha, Lovozero tundra, Kola Peninsula, USSR.

BEST REF. IN ENGLISH: *Am. Min.*, **53**: 1780 (1968). Semenov, E. I., *Am. Min.*, **50**: 2097-2098 (1965).

## TUNELLITE
$SrB_6O_{10} \cdot 4H_2O$

CRYSTAL SYSTEM: Monoclinic
CLASS: 2/m
SPACE GROUP: $P2_1/a$
Z: 4
LATTICE CONSTANTS:
  $a = 14.390$
  $b = 8.213$
  $c = 9.934$
  $\beta = 114°02'$
3 STRONGEST DIFFRACTION LINES:
  6.57 (100)
  5.135 ( 30)
  6.97 ( 25)
OPTICAL CONSTANTS:
    $\alpha = 1.519$
    $\beta = 1.534$ (Na)
    $\gamma = 1.569$
  (+)2V = 68°
HARDNESS: 2½
DENSITY: 2.40 (Meas.)
        2.381 (Calc.)
CLEAVAGE: {100} perfect
        {001} distinct
HABIT: Prismatic and tabular crystals up to 4 cm in length; also as fine-grained nodules. Faces: {100}, {001}, {110}, {011}, {111}; faces of form {Okl} striated parallel to [001].
COLOR-LUSTER: Colorless; subvitreous to pearly on {100}.
MODE OF OCCURRENCE: Occurs as a secondary mineral in the open pit, Kramer, and also in the Furnace Creek area, Death Valley, California.
BEST REF. IN ENGLISH: Erd, R. C., Morgan, Vincent, and Clark, Joan R., U.S.G.S. Prof. Paper 424-C, 294–297 (1961). *Am. Min.*, **49**: 1549–1568 (1964).

## TUNGSTENITE
$WS_2$

CRYSTAL SYSTEM: Hexagonal
CLASS: 6/m 2/m 2/m
SPACE GROUP: P 6/mmc
Z: 2
LATTICE CONSTANTS:
  $a = 3.154$
  $c = 12.362$
3 STRONGEST DIFFRACTION LINES:
  6.18 (100)
  2.277 ( 35) (synthetic)
  2.731 ( 25)
OPTICAL CONSTANTS: Opaque.
HARDNESS: 2½
DENSITY: 7.75 (Meas.)
        7.732 (Calc.)
CLEAVAGE: {0001}
Sectile.
HABIT: Massive; as fine scaly or feathery aggregates.

COLOR-LUSTER: Lead gray. Opaque. Streak lead gray.
MODE OF OCCURRENCE: Occurs as a replacement deposit in limestone, associated with wolframite, pyrite, galena, sphalerite, argentite, tetrahedrite, and quartz, at the Emma mine, Little Cottonwood district, Salt Lake County, Utah.
BEST REF. IN ENGLISH: Palache, et al., "Dana's System of Mineralogy," 7th Ed., v. I, p. 331–332, New York, Wiley, 1944.

## TUNGSTITE
$WO_3 \cdot H_2O$

CRYSTAL SYSTEM: Orthorhombic
3 STRONGEST DIFFRACTION LINES:
  3.49 (100)
  2.68 ( 90)
  2.56 ( 90)
OPTICAL CONSTANTS:
    $\alpha = 2.09$
    $\beta = 2.24$
    $\gamma = 2.26$
  (−)2V = 26°
HARDNESS: 1–2.5
DENSITY: 5.5
CLEAVAGE: {001} perfect
        {110} imperfect
HABIT: Crystals microscopic, platy. Usually massive, earthy to powdery.
COLOR-LUSTER: Bright yellow, golden yellow, yellowish green. Transparent. Resinous.
MODE OF OCCURRENCE: Occurs as an alteration product of scheelite, wolframite, or other tungsten minerals. Found at the Margarita claim and at the Darwin mine, Inyo County, California; and at localities in Idaho, Nevada, Colorado, South Dakota, Connecticut, and North Carolina. Also at deposits in Bolivia, England, France, Sardinia, Tasmania, and Canada.
BEST REF. IN ENGLISH: Palache, et al., "Dana's System of Mineralogy," 7th Ed., v. I, p. 605–606, New York, Wiley, 1944. *Am. Min.*, **29**: 192–210 (1944).

## TUNGUSITE (Inadequately described mineral)
$Ca_4Fe_2Si_6O_{15}(OH)_6$

CRYSTAL SYSTEM: Unknown
3 STRONGEST DIFFRACTION LINES:
  1.818 (100)
  4.17 ( 80)
  3.12 ( 80)
OPTICAL CONSTANTS:
  $\omega$ (or $\beta = \gamma$) = 1.568
    (−)2V = 0°
HARDNESS: ~2
DENSITY: 2.59 (Meas.)
CLEAVAGE: Easily separable into very fine platelets. Platelets elastic, chlorite-like.
HABIT: As plates up to 0.5 cm, forming crusts of radiating-fibrous structure.

COLOR-LUSTER: Grass green to yellowish green; weak pearly.

MODE OF OCCURRENCE: Occurs associated with analcime, apophyllite, gyrolite, "zeolites," calcite, and quartz on the walls of amygdules and druses in spherulitic lava from the right bank of the Lower Tunguska River, 2 km above Tura, Siberia.

BEST REF. IN ENGLISH: Kudryasheva, V. I., *Am. Min.*, **52**: 927–928 (1967).

## TUNISITE
$NaHCa_2Al_4(CO_3)_4(OH)_{10}$

CRYSTAL SYSTEM: Tetragonal
CLASS: 4/m 2/m 2/m
SPACE GROUP: P4/nmm
Z: 2
LATTICE CONSTANTS:
  $a = 11.22$
  $c = 6.582$
3 STRONGEST DIFFRACTION LINES:
  5.615 (100)
  2.592 ( 90)
  3.551 ( 80)
OPTICAL CONSTANTS:
  $\omega = 1.573$
  $\epsilon = 1.599$
   (+)
HARDNESS: 4.5 (crystals)
          3.5 (aggregates)
DENSITY: 2.51 (Meas.)
         2.48 (Calc.)
CLEAVAGE: Basal and prismatic, very good.
HABIT: As fine-grained aggregates or in small tabular crystals. Rarely in macroscopic tabular crystals which form parallel groups.
COLOR-LUSTER: Colorless to white.
MODE OF OCCURRENCE: Found at the Pb-Zn deposit at Sakiet Sidi Youssef, Tunisia, associated with pyrite, marcasite, sphalerite, and calcite in hydrothermal veins.
BEST REF. IN ENGLISH: Johan, Zdenek; Povondra, Pavel, and Slánsky, Ervin, *Am. Min.*, **54**: 1–13 (1969).

## TURANITE (Inadequately described mineral)
$Cu_5(VO_4)_2(OH)_4$

CRYSTAL SYSTEM: Probably orthorhombic
OPTICAL CONSTANTS:
  $\alpha = 2.00$
  $\beta = 2.01$
  $\gamma = 2.02$
  (−)2V = medium
HARDNESS: 5
DENSITY: Not determined.
CLEAVAGE: Not determined.
HABIT: As reniform crusts and spherical concretions with radial-fibrous structure.
COLOR-LUSTER: Olive green.
MODE OF OCCURRENCE: Occurs with uranium and vanadium minerals in cavities in limestone at Tyuya Muyun, Ferghana, Turkestan, USSR.

BEST REF. IN ENGLISH: Palache, et al., "Dana's System of Mineralogy," 7th Ed., v. II, p. 818, New York, Wiley, 1951.

## TURGITE = Hematite with adsorbed water

## TURQUOIS(E)
$CuAl_6(PO_4)_4(OH)_8 \cdot 4-5H_2O$

CRYSTAL SYSTEM: Triclinic
CLASS: $\bar{1}$
SPACE GROUP: P$\bar{1}$
Z: 1
LATTICE CONSTANTS:
  $a = 7.424$     $\alpha = 68.61°$
  $b = 7.629$     $\beta = 69.71°$
  $c = 9.910$     $\gamma = 65.08°$
3 STRONGEST DIFFRACTION LINES:
  3.68 (100)
  2.91 ( 80)
  6.17 ( 70)
OPTICAL CONSTANTS:
  $\alpha = 1.61$
  $\beta = 1.62$ (Na)
  $\gamma = 1.65$
  (+)2V $\simeq 40°$
HARDNESS: 5–6
DENSITY: 2.6–2.8 (massive)
         2.84 (crystals)
         2.91 (calc.)
CLEAVAGE: {001} perfect
          {010} good
Fracture conchoidal to smooth (massive).
HABIT: Crystals rare, small, short prismatic. Usually massive, dense, cryptocrystalline to fine-granular; also concretionary, stalactitic, and as veinlets or crusts.
COLOR-LUSTER: Crystals: Bright blue. Transparent. Vitreous. Massive: Pale blue to sky blue, bluish green to apple green to greenish gray. Waxy or dull. Subtranslucent to opaque.
MODE OF OCCURRENCE: Occurs as a secondary mineral formed by the action of surface waters on aluminous rocks. Found as excellent microcrystals implanted on quartz or schist near Lynch Station, Campbell County, Virginia; as large deposits in the Los Cerillos Mountains, New Mexico, and elsewhere in the state; and in Arizona, Colorado, Nevada, and California. It also occurs in France, Germany, USSR, Egypt, and as fine gem-quality material in Iran.
BEST REF. IN ENGLISH: Cid-Dresdner, Hilda, *Am. Min.*, **50**: 283 (1965). Palache, et al., "Dana's System of Mineralogy," 7th Ed., v. II, p. 946–951, New York, Wiley, 1951.

## TUXTLITE = Pyroxene, diopside-jadeite

## TWINNITE
$Pb(Sb, As)_2S_4$

CRYSTAL SYSTEM:  Triclinic (pseudo-orthorhombic)
Z:  8
LATTICE CONSTANTS:
  $a$ = 19.583      $\alpha$ = 89°46′
  $b$ = 7.981       $\beta$ = 89°37′
  $c$ = 8.564       $\gamma$ = 90°16′
3 STRONGEST DIFFRACTION LINES:
  3.51  (100)
  2.344 ( 80)
  2.78  ( 70)
  HARDNESS:  Talmadge hardness B.    Vickers micro-hardness 147 (131–152) kg/mm$^2$ with 50 gm load.
  DENSITY:  5.26 (Predicted)
              5.323 (Calc.)
  CLEAVAGE:  {100} perfect
Brittle.
  HABIT:  Only two megascopic grains, the larger about 1.5 mm in diameter, and four microscopic grains have been found.  The grains are polysynthetically twinned, the trace of the composition plane being parallel to {100}.
  COLOR-LUSTER:  Black; metallic.    Black streak with slight brownish tint.
  MODE OF OCCURRENCE:  Found as an extremely rare mineral in a small prospect pit in Precambrian marble at Madoc, Ontario, Canada.
  BEST REF. IN ENGLISH: Jambor, J. L., *Can. Min.*, **9** (Pt. 2): 191–213 (1967).


# TYCHITE
$Na_6 Mg_2 SO_4 (CO_3)_4$

CRYSTAL SYSTEM:  Cubic
CLASS:  2/m $\bar{3}$
SPACE GROUP:  Fd3
Z:  8
LATTICE CONSTANT:
  $a$ = 13.898
3 STRONGEST DIFFRACTION LINES:
  2.674 (100)
  4.187 ( 76)
  2.459 ( 40)
OPTICAL CONSTANT:
  $N$ = 1.510  (Na)
  HARDNESS:  3½–4
  DENSITY:  2.549 (Meas.)
              2.586 (Calc.)
  CLEAVAGE: None.  Fracture conchoidal.  Brittle.
  HABIT:  Crystals minute unmodified octahedrons.
  COLOR-LUSTER:  Colorless; transparent.  Vitreous.
  MODE OF OCCURRENCE:  Occurs associated with northupite at several horizons in the Bottom Mud and the Mixed Layer at Searles Lake, San Bernardino County, California.
  BEST REF. IN ENGLISH: Palache, et al., "Dana's System of Mineralogy," 7th Ed., v. II, p. 294–295, New York, Wiley, 1951.    Keester, Kenneth L., Johnson, Gerald G. Jr., and Vand, Vladimir, *Am. Min.*, **54**: 302–305 (1969).


# TYRETSKITE
$(Ca, Sr)_2 B_5 O_8 (OH, Cl)_3$

CRYSTAL SYSTEM:  Triclinic
CLASS:  1 or $\bar{1}$
SPACE GROUP:  P1 or P$\bar{1}$
Z:  1
LATTICE CONSTANTS:
  $a$ = 6.44      $\alpha$ = 61°46′
  $b$ = 6.45      $\beta$ = 60°15′
  $c$ = 6.41      $\gamma$ = 73°30′
3 STRONGEST DIFFRACTION LINES:
  2.93 (100)
  2.86 (100)
  2.14 ( 90)
OPTICAL CONSTANTS:
  $\alpha$ = 1.637
  $\gamma$ = 1.670
$2V_\gamma \sim 46°$
HARDNESS:  Not determined.
DENSITY:  2.189 (Meas.)
CLEAVAGE:  Not determined.
HABIT:  Massive; as radiating fibrous aggregates.
COLOR-LUSTER:  White or brownish.
MODE OF OCCURRENCE:  Occurs in a dolomitic rock in association with sylvite, carnallite, halite, and anhydrite near the Tyret' station of the East Siberian railway, in the Lower Cambrian saline strata of the southeastern part of the Irkutsk amphitheatre, USSR.
BEST REF. IN ENGLISH: Ivanov, A. A., Yarzhemskii, Ya. Ya., and Kondrat'eva, V. V., *Min. Abs.*, **17**: 500–501 (1966).    Davies, W. O., and Machin, M. P., *Am. Min.*, **53**: 2084–2087 (1968).


# TYROLITE (Trichalcite)
$Ca_2 Cu_9 (AsO_4)_4 (OH)_{10} \cdot 10H_2O$

CRYSTAL SYSTEM:  Orthorhombic
CLASS:  2/m 2/m 2/m
SPACE GROUP:  Pmma
Z:  4
LATTICE CONSTANTS:
  $a$ = 10.50
  $b$ = 54.71
  $c$ = 5.59
3 STRONGEST DIFFRACTION LINES:
  28.0  (100)
  14.1  ( 80)
  2.98  ( 80)
OPTICAL CONSTANTS:
  $\alpha$ = 1.694
  $\beta$ = 1.726
  $\gamma$ = 1.730
$(-)2V \approx 36°$
HARDNESS:  ~2
DENSITY:  3.18 (Meas.)
            3.27 (Calc.)
CLEAVAGE:  {001} perfect
Thin laminae flexible.  Sectile.
HABIT:  Crystals rare, lath-like; as flattened scales, fan-shaped aggregates, drusy crusts, or reniform masses with foliated structure; also divergent fibrous.  Twinning on {101}, repeated, resulting in pseudohexagonal aggregates.

COLOR-LUSTER: Pale apple green to greenish blue to sky blue. Translucent. Vitreous; pearly on cleavage.

MODE OF OCCURRENCE: Occurs as a secondary mineral in the oxidation zone of copper-bearing ore deposits. Found as fine specimens associated with chalcophyllite at the Myler mine, Majuba Hill, Pershing County, Nevada; at the Eureka, Gold Chain, and Mammoth mines, Tintic district, Juab County, at the Centennial Alta mine, Salt Lake County, and at the Liberal King mine, Shoshone County, Utah. It also is found in England, France, Spain, Italy, Germany, Austria, Czechoslovakia, Rumania, and USSR.

BEST REF. IN ENGLISH: Palache, et al., "Dana's System of Mineralogy," 7th Ed., v. II, p. 925-926, New York, Wiley, 1951.

# TYRRELLITE

(Cu, Co, Ni)$_3$Se$_4$
Linnaeite Group

CRYSTAL SYSTEM: Cubic
CLASS: $4/m\ \bar{3}\ 2/m$
SPACE GROUP: Fm3m
Z: 9
LATTICE CONSTANT:
 $a = 10.005$
3 STRONGEST DIFFRACTION LINES:
 1.769 (100)
 2.501 ( 90)
 2.886 ( 70)
OPTICAL CONSTANTS: Isotropic, opaque.
 HARDNESS: 3½
 DENSITY: 6.6 (Meas.)
 CLEAVAGE: {001} distinct
 HABIT: Crystals cubic. Also as rounded embedded grains.
 COLOR-LUSTER: Bronze. Opaque. Metallic.
 MODE OF OCCURRENCE: Occurs in association with umangite, klockmannite, clausthalite, berzelianite, chalcopyrite, pyrite, and hematite in the Goldfields district, Saskatchewan, Canada.
 BEST REF. IN ENGLISH: Vlasov, K. A., "Mineralogy of Rare Elements," v. II, p. 656, Israel Program for Scientific Translations, 1966.

# TYSONITE = Fluocerite

# TYUYAMUNITE

Ca(UO$_2$)$_2$(VO$_4$)$_2$ · 5-8H$_2$O

CRYSTAL SYSTEM: Orthorhombic
CLASS: 2/m 2/m 2/m
SPACE GROUP: Pnan
Z: 4
LATTICE CONSTANTS:
 $a = 10.36$
 $b = 8.36$
 $c = 20.40$
3 STRONGEST DIFFRACTION LINES:
 10.18 (100)
 5.02 ( 90)
 3.20 ( 50)
OPTICAL CONSTANTS:
 $\alpha = 1.670-1.77$
 $\beta = 1.870-1.93$
 $\gamma = 1.895-1.97$
 $(-)2V = 40-55°$
HARDNESS: ~2
DENSITY: 3.3-3.6 (Meas.)
CLEAVAGE: {001} perfect, micaceous
 {010} distinct
 {100} distinct
Not brittle.
 HABIT: Crystals tiny scales and laths elongated along $b$-axis and flattened on {001}. Crystal faces dull and somewhat curved. Commonly massive, compact to microcrystalline; in fan-shaped or radial aggregates; also powdery, and as thin coatings; as an impregnation of limestone or sandstone.
 COLOR-LUSTER: Greenish yellow to yellow. Translucent to opaque. Crystals adamantine to waxy, pearly on {001}. Massive material dull to waxy. Occasionally fluoresces weak yellowish green in ultraviolet light.
 MODE OF OCCURRENCE: Occurs widespread as a secondary mineral often associated with metatyuyamunite, carnotite, and other secondary uranium and vanadium minerals. In the United States it is found abundantly as thick crusts and coatings on limestone in the Grants district, New Mexico; as coatings or disseminations in sandstone, limestone, and shale at numerous localities in western South Dakota; and at several deposits in Wyoming, Colorado, Utah, Nevada, Arizona, and Texas. It is also found associated with vanadinite, malachite, and a number of vanadates at Tyuya Muyun, Ferghana district, Turkestan, USSR.
 BEST REF. IN ENGLISH: Frondel, Clifford, USGS Bull. 1064, p. 248-253 (1958). Stern, et al., *Am. Min.*, **41**: 187 (1956).

# U

## UHLIGITE (Inadequately described mineral)
$Ca_3(Ti, Al, Zr)_9O_{20}$ (?)

CRYSTAL SYSTEM: Pseudocubic
LATTICE CONSTANT:
  $a = 7.639$
  HARDNESS: 5½
  DENSITY: 4.15 (Meas.)
  CLEAVAGE: {001} imperfect
Fracture conchoidal.
  HABIT: Crystals octahedral with modifying cube. Octahedral faces striated. Twinning on {111}.
  COLOR-LUSTER: Black. Opaque, except in thin fragments. Metallic.
  MODE OF OCCURRENCE: Occurs in nepheline-syenite at Lake Magadi, Kenya.
  BEST REF. IN ENGLISH: Palache, et al., "Dana's System of Mineralogy," 7th Ed., v. I, p. 735–736, New York, Wiley, 1944.

## UKLONSKOVITE
$NaMg(SO_4)(OH) \cdot 2H_2O$

CRYSTAL SYSTEM: Monoclinic
CLASS: 2/m
SPACE GROUP: $P2_1/m$
Z: 4
LATTICE CONSTANTS:
  $a = 13.15$
  $b = 7.19$
  $c = 5.72$
  $\beta = 90°37'$
3 STRONGEST DIFFRACTION LINES:
  3.56 (100)
  3.20 (100)
  3.018 (100)
OPTICAL CONSTANTS:
  $\alpha = 1.476$
  $\beta = ?$
  $\gamma = 1.500$
  (+)2V
HARDNESS: Not determined.

DENSITY: 2.42–2.5 (Meas.)
          2.404 (Calc.)
HABIT: Crystals prismatic, flattened, up to 2 mm in size.
COLOR-LUSTER: Transparent, colorless; vitreous.
MODE OF OCCURRENCE: Occurs associated with glauberite and polyhalite, at a depth of 80 meters, in cavities in clays covering a saline layer in Neogene rocks of Kara-Kalpakii, lower Amu-Darya River, USSR.
  BEST REF. IN ENGLISH: Slyusareva, M. N., *Am. Min.*, **50**: 520–521 (1965).

## ULEXITE
$NaCaB_5O_9 \cdot 8H_2O$

CRYSTAL SYSTEM: Triclinic
CLASS: $\bar{1}$
SPACE GROUP: $P\bar{1}$
Z: 2
LATTICE CONSTANTS:
  $a = 8.80$      $\alpha = 90°15'$
  $b = 12.86$    $\beta = 109°10'$
  $c = 6.67$      $\gamma = 105°5'$
3 STRONGEST DIFFRACTION LINES:
  12.2 (100)
  7.75 ( 80)
  6.00 ( 30)
OPTICAL CONSTANTS:
  $\alpha = 1.496$
  $\beta = 1.505$
  $\gamma = 1.519$
(+)2V = 78°
HARDNESS: 2½
DENSITY: 1.955 (Meas.)
          1.959 (Calc.)
CLEAVAGE: {010} perfect
          {1$\bar{1}$0} good
          {110} poor
Brittle.
  HABIT: Crystals usually acicular, greatly elongated [001]. Commonly as nodules or lense-like masses of radially oriented or randomly arranged acicular crystals; as tufted "cotton ball" masses; as crusts and compact veins composed of parallel fibers.

COLOR-LUSTER: Crystals colorless; vitreous. Aggregates white; silky or satiny, often chatoyant.

MODE OF OCCURRENCE: Occurs abundantly in playa deposits and desicated saline deposits in arid regions in California, Nevada, Argentina, Peru, Chile, and the USSR. Also in gypsum deposits associated with anhydrite, glauberite, and borates at several localities in the Maritime Provinces, Canada.

BEST REF. IN ENGLISH: Clark, J. R., and Christ, C. L., *Am. Min.*, **44**: 712–719 (1959).


# ULLMANNITE
NiSbS
Var. Willyamite (Co)
    Kallilite (Bi)

CRYSTAL SYSTEM: Cubic
CLASS: 23
SPACE GROUP: $P2_13$
Z: 4
LATTICE CONSTANT:
  $a = 5.881$
3 STRONGEST DIFFRACTION LINES:
  2.66 (100)
  2.41 ( 40)
  1.78 ( 40)
OPTICAL CONSTANTS: Opaque.
  HARDNESS: 5–5½
  DENSITY: 6.65 (Meas.)
       6.793 (Calc.)
  CLEAVAGE: {001} perfect
Fracture uneven. Brittle.
  HABIT: Crystals usually cubic; also octahedral, pyritohedral, or tetrahedral. Striated on cube faces. Usually massive, granular. Twinning on [110].
  COLOR-LUSTER: Tin white to steel gray. Opaque. Metallic. Streak grayish black.
  MODE OF OCCURRENCE: Occurs in vein deposits commonly associated with other nickel minerals and siderite. Found in the North Pole silver mine, Gunnison County, Colorado (kallilite). It also occurs in England, France, Sardinia, Austria, Germany, and New South Wales, Australia (willyamite).
  BEST REF. IN ENGLISH: Palache, et al., "Dana's System of Mineralogy," 7th Ed., v. I, p. 301–302, New York, Wiley, 1944.


# ULVÖSPINEL
$Fe_2TiO_4$
Spinel Group

CRYSTAL SYSTEM: Cubic
CLASS: $4/m\ \bar{3}\ 2/m$
SPACE GROUP: Fd3m
Z: 8
LATTICE CONSTANT:
  $a = 8.47$

3 STRONGEST DIFFRACTION LINES:
  2.574 (100)
  1.509 ( 45)
  1.643 ( 35)
HARDNESS: Not determined.
DENSITY: 4.771 (Calc.)
CLEAVAGE: Not determined.
HABIT: Massive; as very fine microscopic exsolution lamellae.
COLOR-LUSTER: Iron black. Reflectivety less than ilmenite.
MODE OF OCCURRENCE: Occurs as a common constituent of titaniferous magnetites, generally as very fine exsolution lamellae parallel to {100} of magnetite. Typical occurrences are found in Essex County, New York, and on Ulvö island, Ångermanland archipelago, northern Sweden. The mineral also occurs as a rare accessory in rock samples from the Moon.
BEST REF. IN ENGLISH: Ramdohr, Paul, *Econ. Geol.*, **48**: 677–688 (1953). *Am. Min.*, **40**: 138 (1955).


# UMANGITE
$Cu_3Se_2$

CRYSTAL SYSTEM: Orthorhombic
CLASS: 222
SPACE GROUP: $P2_12_12$
Z: 4
LATTICE CONSTANTS:
  $a = 6.40$
  $b = 12.46$
  $c = 4.28$
3 STRONGEST DIFFRACTION LINES:
  3.56  (100)
  1.829 ( 90)
  1.778 ( 80)
OPTICAL CONSTANTS: Opaque.
  HARDNESS: 3
  DENSITY: 6.44 (Meas.)
        6.49 (Calc.)
  CLEAVAGE: Two very poor rectangular cleavages.
Fracture conchoidal to uneven. Brittle.
  HABIT: Massive, in small grains or fine granular aggregates; sometimes exhibits lamellar twinning.
  COLOR-LUSTER: Bright bluish black with reddish tint; weathered surfaces dull iridescent purple. Opaque. Metallic. Streak black.
  MODE OF OCCURRENCE: Occurs as blebs and veinlets in grayish pink calcite gangue associated with berzelianite, athabascaite, and chalcocite at the Martin Lake mine, Lake Athabasca area, northern Saskatchewan, Canada. In Argentina found with clausthalite, chalcomenite, malachite, and calcite at Sierra de Umango, La Rioja, and with chalcomenite and malachite at Sierra de Cacheuta, Mendoza Province. Also found at Clausthal, Andreasberg, Lehrbach, Tilkerode, and Zorge in the Harz Mountains, Germany; and with berzelianite and chalcopyrite at the copper mine of Skrikerum, Kalmar, Sweden.
  BEST REF. IN ENGLISH: Palache, et al., "Dana's System of Mineralogy," 7th Ed., v. I, p. 194–195, New York,

Wiley, 1944. Early, J. W., *Am. Min.*, **35**: 337–364 (1950).

## UMOHOITE
$UO_2MoO_4 \cdot 4H_2O$

CRYSTAL SYSTEM: Monoclinic
CLASS: 2/m or 2
SPACE GROUP: $P2_1/m$ or $P2_1$
Z: 4
LATTICE CONSTANTS:
  $a = 14.30$
  $b = 7.50$
  $c = 6.38$
  $\beta = 99°5'$
3 STRONGEST DIFFRACTION LINES:
  8.42 (100)
  5.604 (100)
  3.349 ( 75)
OPTICAL CONSTANTS:
    $\alpha = 1.66$
    $\beta = 1.83$
    $\gamma = 1.91$
  $(-)2V = 65°$
HARDNESS: ~2
DENSITY: 4.55–4.93 (Meas.)
CLEAVAGE: None.
HABIT: Crystals plate-like; also fine-grained crystalline, platy, or foliated aggregates.
COLOR-LUSTER: Black to bluish black; opaque. Bright submetallic.
MODE OF OCCURRENCE: Occurs as thin veinlets or disseminated specks in an alteration zone of clay minerals along a uranium-bearing vein at the Freedom No. 2 mine, Marysvale, Utah; it is also found at the Alyce Tolino mine, Cameron, Arizona; at the Lucky Mc mine, Gas Hills, Wyoming; and at an unspecified locality in the USSR.
BEST REF. IN ENGLISH: Frondel, Clifford, USGS Bull. 1064, p. 148–149 (1958). *Am. Min.*, **44**: 1248 (1959).

## UNGEMACHITE
$Na_8K_3Fe^{3+}(SO_4)_6(OH)_2 \cdot 10H_2O$

CRYSTAL SYSTEM: Hexagonal
CLASS: $\bar{3}$
SPACE GROUP: $R\bar{3}$
Z: 1 (rhomb.), 3 (hexag.)
LATTICE CONSTANTS:
  $a = 10.84$ kX    $a_{rh} = 10.37$
  $c = 24.82$        $\alpha = 62°59.5'$
3 STRONGEST DIFFRACTION LINES:
  3.43 (100)
  8.33 ( 60)
  2.72 ( 50)
OPTICAL CONSTANTS:
  $\omega = 1.502$
  $\epsilon = 1.449$
    $(-)$
HARDNESS: 2½

DENSITY: 2.287 (Meas.)
         2.29 (Calc.)
CLEAVAGE: {0001} perfect
Fracture uneven. Brittle.
HABIT: Crystals thick tabular, flattened on {0001}. Pseudorhombohedral.
COLOR-LUSTER: Colorless to pale yellow. Transparent. Vitreous.
MODE OF OCCURRENCE: Occurs at Chuquicamata, Chile, associated with jarosite, sideronatrite, metasideronatrite, and clino-ungemachite.
BEST REF. IN ENGLISH: Palache, et al., "Dana's System of Mineralogy," 7th Ed., v. II, p. 596–597, New York, Wiley, 1951.

## URALBORITE
$CaB_2O_4 \cdot 2H_2O$

CRYSTAL SYSTEM: Monoclinic
CLASS: 2/m
SPACE GROUP: $P2_1/n$
3 STRONGEST DIFFRACTION LINES:
  7.61 (100)
  2.13 (100)
  2.97 ( 90)
OPTICAL CONSTANTS:
    $\alpha = 1.604$
    $\beta = 1.609$
    $\gamma = 1.615$
  $(+)2V = 85°$
HARDNESS: ~4
DENSITY: 2.60
CLEAVAGE: Parallel to elongation—distinct.
HABIT: As radiating-fibrous aggregates of prismatic crystals up to 0.5–0.7 cm in size.
COLOR-LUSTER: Colorless, transparent. Fluoresces violet in long-wave ultraviolet light.
MODE OF OCCURRENCE: Occurs associated with garnet, magnetite, szaibelyite, and frolovite in deep drill cores from the contact of Middle Devonian limestone with quartz diorite at a copper deposit of the skarn type at an unspecified locality in the Urals.
BEST REF. IN ENGLISH: Malinko, S. V., *Am. Min.*, **47**: 1482 (1962).

**URALITE** = Actinolite pseudomorphous after pyroxene

## URALOLITE
$CaBe_3(PO_4)_2(OH)_2 \cdot 4H_2O$

CRYSTAL SYSTEM: Monoclinic
Z: 8
LATTICE CONSTANTS:
  $a = 8.43$
  $b = 39.50$
  $c = 7.12$
  $\beta = 94°58'$

3 STRONGEST DIFFRACTION LINES:
  3.56 (100)
  3.04 ( 80)
  3.20 ( 70)
OPTICAL CONSTANTS:
  $\alpha$ = 1.510
  $\beta$ = 1.525
  $\gamma$ = 1.536
  (-)2V
HARDNESS: 2½
DENSITY: Probably 2.05–2.14
             2.042 (Calc.)
CLEAVAGE: Probable cleavage perpendicular to elongation and others along elongation.

HABIT: As concretions composed of radiating fibrous spherulites 2–3 mm in diameter, and in sheaf-like growths.

COLOR-LUSTER: Colorless to white, sometimes stained brown by iron oxides. Fibrous aggregates silky; needles vitreous.

MODE OF OCCURRENCE: Occurs in kaolin-hydromuscovite rocks (from the Urals) containing fluorite, beryl, apatite, crandallite, moraesite, and herderite.

BEST REF. IN ENGLISH: Grigor'ev, N. A., *Am. Min.*, **49**: 1776 (1964).


## URAMPHITE (Inadequately described mineral)
$NH_4 UO_2 PO_4 \cdot 3H_2O$

CRYSTAL SYSTEM: Unknown
3 STRONGEST DIFFRACTION LINES:
  3.78  (100)
  2.22  ( 90)
  1.694 ( 90)
OPTICAL CONSTANTS:
    $\alpha$ = 1.564
    $\beta = \gamma$ = 1.585
    (-)2V = 0°–3°
  HARDNESS: Not determined.
  DENSITY: 3.7
  CLEAVAGE: Distinct in two directions.
  HABIT: As square tablets up to 0.2 X 0.2 mm in size, in small rosettes, and as lichen-like deposits.
  COLOR-LUSTER: Bottle green to pale green; vitreous. Fluoresces medium yellow-green under ultraviolet light.
  MODE OF OCCURRENCE: Occurs in the oxidized zone of a uranium-coal deposit, in fractures in the coal 20–50 meters below the surface, at an unspecified locality in USSR.
  BEST REF. IN ENGLISH: Nekrasova, Z. A., *Am. Min.*, **44**: 464 (1959).


## URANINITE
$UO_2$

CRYSTAL SYSTEM: Cubic
CLASS: $4/m \bar{3} 2/m$
SPACE GROUP: Fm3m
Z: 4
LATTICE CONSTANT:
  $a$ = 5.368–5.555

3 STRONGEST DIFFRACTION LINES:
  3.157 (100)
  1.934 ( 49)
  2.735 ( 48)
OPTICAL CONSTANTS: Opaque.
HARDNESS: 5–6
DENSITY: 7.5–10.0 (Meas. most natural crystals)
             6.5–9.0 (Meas. pitchblende)
             10.95 (Meas. synthetic $UO_2$).
CLEAVAGE: Octahedral, observed in polished sections. Fracture conchoidal to uneven. Brittle.

HABIT: Crystals commonly cubic or cubo-octahedral, less commonly as unmodified octahedra or dodecahedra. Also as dendritic aggregates of small crystals; massive, dense to granular; occasionally pulverulent; and as botryoidal crusts and masses, often with banded, radial-fibrous to columnar structure (pitchblende). Twins on {111} rare.

COLOR-LUSTER: Black to brownish or grayish black; usually opaque. Submetallic, also pitchy or greasy. Streak black to brownish black, grayish or olive green.

MODE OF OCCURRENCE: Occurs mainly in hydrothermal vein deposits, in bedded sedimentary rocks, in pegmatites, and widespread in smaller amounts in a large variety of environments. In the United States it is found in vein deposits in Gilpin County, Colorado and the Marysvale district, Utah; in many sandstone-type deposits in Colorado, Utah, New Mexico, and Arizona; and in pegmatites in New England, especially at the Ruggles pegmatite, Grafton Center, New Hampshire; in South Dakota at the Ingersoll Mine, Keystone, and in several other Black Hills pegmatites; and in North Carolina. Other major deposits include the Bancroft district, Ontario, where crystals up to 5 pounds in weight have been recovered, the Blind River area, Great Bear Lake, and many other districts in Canada; the Katanga district, Zaire; Rum Jungle, Northern Territory, Australia; and at Joachimsthal, Schneeberg, and nearby localities in the Erzegebirge of Saxony, Germany, and Bohemia, Czechoslovakia.

BEST REF. IN ENGLISH: Frondel, Clifford, USGS Bull. 1064, p. 11–47 (1958).


## URANITE = General term for uranyl phosphates and arsenates of the autunite and meta-autunite groups


## URANOCIRCITE
Uranocircite I: $Ba(UO_2)_2 (PO_4)_2 \cdot 12H_2O$
Uranocircite II: $Ba(UO_2)_2 (PO_4)_2 \cdot 10H_2O$
Autunite Group

CRYSTAL SYSTEM: Probably tetragonal
CLASS: 4/m 2/m 2/m
SPACE GROUP: P4/nnc (II)
Z: 2 (II)
LATTICE CONSTANTS:

|         | (I)        | (II)       |
|---------|------------|------------|
| $a$     | $\sim 7$   | $a$ = 7.01 |
| $c$     | = 22.59    | $c$ = 20.46 |

3 STRONGEST DIFFRACTION LINES:
  5.10 (100)
  2.04 (100)  (uranocircite II)
  10.1 ( 60)

OPTICAL CONSTANTS:
 (I)      (II)
 $\alpha = 1.562$    $\alpha = 1.574$   $2V = 70°$
          $\beta = 1.583$
          $\gamma = 1.588$
HARDNESS: 2-2½
DENSITY: 3.46 (Calc. 10 $H_2O$)
CLEAVAGE: {001} perfect
          {100} distinct
Not brittle. Thin plates flexible.
 HABIT: Crystals thin tabular with rectangular outline. Crystals usually as subparallel, composite, or fan-like aggregates.
 COLOR-LUSTER: Yellow. Transparent. Pearly on {001}. Fluoresces green under ultraviolet light.
 MODE OF OCCURRENCE: Uranocircite I and II occur with other hydrated phases of $Ba(UO_2)_2(PO_4)_2$ at Farnwitte near Menzenschwand, Black Forest, Germany.
 BEST REF. IN ENGLISH: Frondel, Clifford, USGS Bull. 1064, p. 211-215 and 177 (1958). Walenta, K., *Min. Abs.*, **17**: 695 (1966).

## URANOPHANE
$Ca(UO_2)_2Si_2O_7 \cdot 6H_2O$

CRYSTAL SYSTEM: Monoclinic
CLASS: 2
SPACE GROUP: $P2_1$
Z: 2
LATTICE CONSTANTS:
 $a = 15.87$
 $b = 7.05$
 $c = 6.66$
 $\beta = 97°15'$
3 STRONGEST DIFFRACTION LINES:
 7.88 (100)
 3.94 ( 90)
 2.99 ( 80)
OPTICAL CONSTANTS:
 $\alpha = 1.642-1.648$
 $\beta = 1.661-1.667$ (Na)
 $\gamma = 1.667-1.675$
 $(-)2V = 32°, 37°-45°$
HARDNESS: ~2½
DENSITY: 3.83 (Meas.)
          3.85 (Calc.)
CLEAVAGE: {100} perfect
Brittle.
 HABIT: Crystals prismatic, up to 8 mm long, nearly square in cross section. Acicular to hair-like. As radiated or tufted aggregates or crusts. Also as dense microcrystalline masses and pulverulent or felt-like coatings.
 COLOR-LUSTER: Pale yellow, lemon yellow to yellowish green, also yellowish orange to amber brown. Transparent to translucent. Vitreous to greasy, pearly on cleavage. Massive material usually dull, earthy, to waxy. Crystals fluoresce weak yellowish green in ultraviolet light. Massive material usually does not respond.
 MODE OF OCCURRENCE: Occurs in pegmatites as pseudomorphs after uraninite crystals; as coatings along joint planes or crevices in igneous or other rocks, pre-

sumably deposited by meteoric waters; and in the oxidized zone of vein deposits of uraninite. In the United States it is found in sedimentary rocks and several pegmatites in the Black Hills of South Dakota; in pegmatites in the Spruce Pine district, Mitchell County, North Carolina, and at the Ruggles mine and other places in New Hampshire. Superb crystals and radiated masses have been recovered from the Hanosh mine, Grants district, New Mexico. It is also found at Lusk and Pumpkin Buttes, Wyoming; and at several localities in Utah, Arizona, Colorado, California, and other western states. In Canada it is found as beautiful radiated fibrous aggregates in vugs at the Faraday uranium mine, Bancroft, Ontario. Also found abundantly at Wölsendorf and other localities in Saxony, Germany; as exceptional crystals in the Katanga district, Zaire; and widespread in small amounts at many other uranium deposits throughout the world.
 BEST REF. IN ENGLISH: Frondel, Clifford, USGS Bull. 1064, p. 294-300 (1958).

## URANOPILITE
$(UO_2)_6SO_4(OH)_{10} \cdot 12H_2O$

CRYSTAL SYSTEM: Probably monoclinic
LATTICE CONSTANT:
 $c = 8.91$
3 STRONGEST DIFFRACTION LINES:
 7.12 (100)
 9.18 ( 80)
 4.28 ( 80)
OPTICAL CONSTANTS:
 $\alpha = 1.621-1.623$
 $\beta = 1.623-1.625$
 $\gamma = 1.632-1.634$
     (+)2V
HARDNESS: Not determined. Soft.
DENSITY: 3.96 (Meas.)
CLEAVAGE: {010} perfect
 HABIT: As thin felted crusts, films, and as small irregular to globular aggregates and rosettes of tiny lathlike crystals, elongated [001] and flattened {010}.
 COLOR-LUSTER: Bright lemon yellow to straw yellow; silky. Fluoresces strong yellowish green in ultraviolet light.
 MODE OF OCCURRENCE: Occurs as a secondary mineral associated with johannite and zippeite-like minerals at the Happy Jack mine, San Juan County and with gypsum at Capitol Wash, Wayne County, Utah; also found associated with uraninite in the Thornburg mine, Los Ochos district, Saguache County, Colorado; and at the Snowball claim, Fremont County, Wyoming. In Canada it is found at Goldfields, Saskatchewan and at Great Bear Lake and Hottah Lake, Northwest Territory. Among other localities, it is found in Europe at Wheal Owles, Cornwall, England; at LaCrouzille, Haute-Vienne and at Grury, Saône-et-Loire, France; in the Vizeu district Portugal; at Johanngeorgenstadt, Saxony; and at Joachimsthal and Příbram, Bohemia.
 BEST REF. IN ENGLISH: Frondel, Clifford, USGS Bull. 1064, p. 135-140 (1958).

## URANOSPATHITE (Species status uncertain)
$Cu(UO_2)_2 \{(As, P)O_4\}_2 \cdot 12(?)H_2O$

CRYSTAL SYSTEM: Probably tetragonal
OPTICAL CONSTANTS:
$\alpha \simeq 1.49$
$\beta = 1.510$
$\gamma = 1.521$
$(-)2V = 69°$ (Na)
HARDNESS: Not determined.
DENSITY: 2.50 (Meas.)
CLEAVAGE: {001} perfect
{100} good
{010} fibrous
HABIT: As fan-like groups of thin rectangular plates and laths flattened on {001}, elongated along b-axis and striated parallel to elongation. Sometimes twinned on {110}.
COLOR-LUSTER: Yellow to pale green and bluish green.
MODE OF OCCURRENCE: Occurs in parallel growths with bassetite at Redruth, Cornwall, England.
BEST REF. IN ENGLISH: Hallimond, A. F., *Min. Mag.*, 17: 221 (1915). Frondel, Clifford, USGS Bull. 1064, p. 194–195 (1958).

## URANOSPHERITE (Uranosphaerite)
$Bi_2U_2O_9 \cdot 3H_2O$

CRYSTAL SYSTEM: Orthorhombic or monoclinic
3 STRONGEST DIFFRACTION LINES:
3.16 (100)
1.83 ( 80)
3.87 ( 70)
OPTICAL CONSTANTS:
$\alpha = 1.955-1.959$
$\beta = 1.985-1.981$
$\gamma = 2.05-2.060$
$(+)2V = $ large
HARDNESS: 2–3
DENSITY: 6.36 (Meas.)
CLEAVAGE: {100} distinct
HABIT: As half-globular aggregates, sometimes rough and drusy, composed of minute acutely terminated crystals elongated along the c-axis. Aggregates do not exceed 0.5 mm in diameter.
COLOR-LUSTER: Reddish orange to orange yellow and golden. Dull luster; greasy on fracture surfaces. Streak paler than the color.
MODE OF OCCURRENCE: Occurs associated with walpurgite, uranospinite, troegerite, zeunerite, erythrite, and black cobalt oxide in a vein carrying uraninite, native bismuth, and cobalt-nickel arsenides at the Weisser Hirsch mine at Neustädtl, near Schneeberg, Saxony, Germany.
BEST REF. IN ENGLISH: Berman, Robert, *Am. Min.*, 42: 905–908 (1957). Frondel, Clifford, USGS Bull. 1064, p. 98–99 (1958).

## URANOSPINITE
$Ca(UO_2)_2 (AsO_4)_4 \cdot 10H_2O$
Autunite Group

CRYSTAL SYSTEM: Tetragonal
CLASS: 4/m 2/m 2/m
SPACE GROUP: I4/mmm
Z: 2
3 STRONGEST DIFFRACTION LINES:
10.16 (100)
3.56 ( 90)
3.39 ( 80)
OPTICAL CONSTANTS:
$\epsilon$ or $\alpha = 1.55-1.596$
$\beta = 1.567-1.619$
$\omega$ or $\gamma = 1.572-1.621$
Uniaxial (–) or $(-)2V \sim 0°-62°$
HARDNESS: 2–3
DENSITY: 3.45 (Meas.)
3.65 (Calc. Synthetic)
CLEAVAGE: {001} perfect
{100} distinct
Brittle.
HABIT: Crystals small thin rectangular plates flattened on {001}.
COLOR-LUSTER: Lemon yellow to siskin green. Pearly on {001}. Fluoresces bright lemon yellow in ultraviolet light.
MODE OF OCCURRENCE: Occurs as a secondary mineral at Honeycomb Hills, Fish Springs mining district, Juab County, and also 9 miles south of Pahreah, Kane County, Utah. It is found as a secondary mineral derived from the alteration of uraninite and primary arsenides in the Weisser Hirsch mine at Neustädtl, near Schneeberg, Saxony, associated with walpurgite, metauranocircite, troegerite, and metazeunerite. It has also been reported as an alteration product of uraninite associated with niccolite at the Solitaria mine, Argentina, and as a secondary mineral at Radium Hill, Mount Painter, South Australia.
BEST REF. IN ENGLISH: Frondel, Clifford, USGS Bull. 1064, p. 183–187 (1958).

## URANOTHORITE = Variety of thorite
$(Th, U) SiO_4$

## URANOTILE = Uranophane

## URBANITE = Aegirine-augite

## UREYITE
$NaCrSi_2O_6$
Pyroxene Group

CRYSTAL SYSTEM: Monoclinic
CLASS: 2/m
SPACE GROUP: C2/c
Z: 4
LATTICE CONSTANTS:
$a = 9.560$
$b = 8.746$
$c = 5.270$
$\beta = 107.38°$

3 STRONGEST DIFFRACTION LINES:
  2.961 (100)
  2.877 ( 70)
  2.181 ( 70)

OPTICAL CONSTANTS:
  $\alpha$ = 1.740–1.748
  $\beta$ = 1.756
  $\gamma$ = 1.762–1.765     (Na)
  (−)2V = 60°–70°

HARDNESS: Not determined.

DENSITY: 3.60 (Calc.)

CLEAVAGE: {110} distinct
           {001} parting pronounced

HABIT: As polycrystalline aggregates.

COLOR-LUSTER: Emerald-green.

MODE OF OCCURRENCE: Occurs in daubreelite in the Coahuila and Hex River Mountains meteorites, and embedded in cliftonite rims of troilite nodules in the Toluca meteorite.

BEST REF. IN ENGLISH: Frondel, Clifford, and Klein, Cornelis Jr., *Science*, **149**: 742–744 (1965).

## URSILITE (Inadequately described mineral)
$(Ca, Mg)_2 (UO_2)_2 Si_5 O_{14} \cdot 9H_2O$ (?)

CRYSTAL SYSTEM: Unknown

3 STRONGEST DIFFRACTION LINES:

| (Ca-ursilite) | (Mg-ursilite) |
|---|---|
| 3.37 (100) | 4.98 (100) |
| 3.02 (100) | 3.06 (100) |
| 4.56 ( 90) | 2.30 ( 90) |

OPTICAL CONSTANTS:

| (Ca-ursilite) | (Mg-ursilite) |
|---|---|
| $\alpha$ = 1.548 | $\alpha$ = 1.543 |
| $\gamma$ = 1.556 | $\gamma$ = 1.550 |
| (−)2V | |

HARDNESS: 3

DENSITY: 3.034 (Meas. Ca-ursilite)
          3.254 (Meas. Mg-ursilite)

CLEAVAGE: Brittle.

HABIT: As earthy or nodular incrustations, rarely as radiating spherulites.

COLOR-LUSTER: Lemon yellow.  Fluoresces greenish yellow under ultraviolet light.

MODE OF OCCURRENCE: Occurs along joints in quartz porphyries, associated with kaolinite and calcite, and rarely with uranophane and sklodowskite at an unspecified locality in USSR.

BEST REF. IN ENGLISH: Chernikov, A. A., Krumetskaya, O. V., and Sidel'nikova, V. D., *Am. Min.*, **44**: 464–465 (1959).

## USOVITE
$Ba_2 MgAl_2 F_{12}$

CRYSTAL SYSTEM: Probably orthorhombic

3 STRONGEST DIFFRACTION LINES:
  3.41 (100)
  2.04 ( 70)
  1.75 ( 70)

OPTICAL CONSTANTS:
  $\alpha$ = 1.441
  $\beta$ = 1.442
  $\gamma$ = 1.444
  (+)2V = 70°

HARDNESS: 3½

CLEAVAGE: One perfect. Fracture irregular.

HABIT: In irregular grains 0.5–3.0 mm, also as indistinct elongated platy forms up to 1.5 × 0.5 cm.

COLOR-LUSTER: Brown to dark brown, color uneven and spotty. Translucent yellowish brown in fine fragments. Streak white. Vitreous to greasy.

MODE OF OCCURRENCE: Occurs intergrown with green and colorless fluorite associated with greenish muscovite, zeolites, and halloysite in a fluorite vein in the Upper Noiby River area, Yenisei region, Siberia.

BEST REF. IN ENGLISH: Nozhkin, A. D., Gavrilenko, V. A., and Moleva, V. A., *Am. Min.*, **52**: 1582 (1967).

## USSINGITE
$Na_4 Al_2 Si_6 O_{17} \cdot H_2O$

CRYSTAL SYSTEM: Triclinic

CLASS: 1

SPACE GROUP: P1

Z: 1

LATTICE CONSTANTS:
  $a$ = 7.256     $\alpha$ = 102°35′
  $b$ = 8.72      $\beta$ = 115°40′
  $c$ = 7.15      $\gamma$ = 79°13′

3 STRONGEST DIFFRACTION LINES:
  2.95 (100)
  2.69 (100)
  6.35 ( 90)

OPTICAL CONSTANTS:
  $\alpha$ = 1.504
  $\beta$ = 1.509
  $\gamma$ = 1.545
  (+)2V = 36°

HARDNESS: 6½

DENSITY: 2.46–2.5 (Meas.)
          2.53 (Calc.)

CLEAVAGE: {110} perfect
           {1$\bar{1}$0} perfect

HABIT: Massive. Twinning on {010}, common.

COLOR-LUSTER: Pale to dark violet-red.

MODE OF OCCURRENCE: Occurs in pegmatite at Kangerdluarsuk, Greenland; also found on the Kola Peninsula, USSR.

BEST REF. IN ENGLISH: Winchell and Winchell "Elements of Optical Mineralogy," 4th Ed., pt. 2, p. 443, New York, Wiley, 1951.

## USTARASITE (Inadequately described mineral)
$Pb(Bi, Sb)_6 S_{10}$

3 STRONGEST DIFFRACTION LINES:
  3.527 (100) broad
  1.483 (100)
  1.057 (100)

HARDNESS: 2½
CLEAVAGE: One perfect. One indistinct.
HABIT: Prismatic crystals; sometimes bent or twisted.
COLOR-LUSTER: Silvery gray to gray; metallic.
MODE OF OCCURRENCE: Occurs in quartz-bismuthinite veins in the deposits of Ustarasaisk, western Tyan-Shan.
BEST REF. IN ENGLISH: Sakharova, M. S., *Am. Min.*, **41**: 814 (1956).

# UVANITE
$(UO_2)_2V_6O_{17} \cdot 15H_2O$

CRYSTAL SYSTEM: Orthorhombic
3 STRONGEST DIFFRACTION LINES:
2.94 (100)
1.71 ( 90)
2.24 ( 80)
OPTICAL CONSTANTS:
$\alpha = 1.817$
$\beta = 1.879$
$\gamma = 2.057$
$(+)2V = 52°$
HARDNESS: Not determined.
DENSITY: Not determined.
CLEAVAGE: Two pinacoidal.
HABIT: As microcrystalline masses and coatings.
COLOR-LUSTER: Brownish yellow. Not fluorescent.
MODE OF OCCURRENCE: Occurs associated with gypsum, hyalite opal, and several secondary uranium and vanadium minerals in asphaltic sandstone of the Shinarump member of the Chinle formation at Temple Rock, Emery County, Utah.
BEST REF. IN ENGLISH: Frondel, Clifford, USGS Bull. 1064, p. 261-263 (1958).

# UVAROVITE
$Ca_3Cr_2Si_3O_{12}$
Garnet Group

CRYSTAL SYSTEM: Cubic
CLASS: $4/m\ \bar{3}\ 2/m$
SPACE GROUP: Ia3d
Z: 8
LATTICE CONSTANT:
$a = 12.00$
3 STRONGEST DIFFRACTION LINES:
2.684 (100)
2.999 ( 70)
1.603 ( 60)
OPTICAL CONSTANT:
$N = 1.86$
HARDNESS: 6½-7
DENSITY: 3.4-3.8 (Meas.)
3.848 (Calc.)
CLEAVAGE: None.
Fracture uneven to conchoidal. Brittle.
HABIT: Crystals usually dodecahedrons or trapezohedrons; also in combination, or with hexoctahedron. Dodecahedral faces often striated. Also massive, coarse granular, or as embedded grains.
COLOR-LUSTER: Emerald green. Transparent to translucent. Vitreous. Streak white.
MODE OF OCCURRENCE: Occurs in association with chromite in serpentine, in skarn deposits, and in metamorphosed limestones. Found sparingly at numerous chromite deposits in northern California, and near Riddle, Grant County, Oregon; with diopside at Orford, and in serpentine at Thetford, Quebec, Canada; in serpentinite near Röros, Norway; as fine crystals at Outokumpu, Finland; in chromite near Bisersk, Urals, USSR; in the Bushveld complex of South Africa; and elsewhere.
BEST REF. IN ENGLISH: Deer, Howie, and Zussman, "Rock Forming Minerals," v. I, p. 77-112, New York, Wiley, 1962.

**UVITE** = A hypothetical end-member of tourmaline $CaMg_4Al_5B_3Si_6O_{27}(OH)_4$

**UZBEKITE** = Volborthite

# V

## VAESITE

$NiS_2$
Pyrite Group

CRYSTAL SYSTEM: Cubic
CLASS: $2/m \, \overline{3}$
SPACE GROUP: Pa3
Z: 4
LATTICE CONSTANT:
$a = 5.67932$
3 STRONGEST DIFFRACTION LINES:
2.83 (100)
1.71 ( 80)
1.09 ( 60)
HARDNESS: Not determined.
DENSITY: 4.45 (Calc.)
CLEAVAGE: {001} perfect
HABIT: Crystals octahedral or cubic.
COLOR-LUSTER: Gray. Opaque. Metallic.
MODE OF OCCURRENCE: Occurs in association with pyrite, randomly disseminated in dolomite, at the Kasompi mine, about 70 km west-southwest of Kambove, Katanga, Zaire.
BEST REF. IN ENGLISH: Kerr, Paul F., *Am. Min.*, **30**: 483–497 (1945).

## VALENTINITE

$Sb_2O_3$
Dimorphous with senarmontite

CRYSTAL SYSTEM: Orthorhombic
CLASS: $2/m \, 2/m \, 2/m$
SPACE GROUP: Pccn
Z: 4
LATTICE CONSTANTS:
$a = 4.92$
$b = 12.46$
$c = 5.42$
3 STRONGEST DIFFRACTION LINES:
3.142 (100)
3.118 ( 75)
3.494 ( 25)

OPTICAL CONSTANTS:
$\alpha = 2.18$
$\beta = 2.35$ (Na)
$\gamma = 2.35$
$(-)2V$ = very small
HARDNESS: 2½–3
DENSITY: 5.76 (Meas.)
5.828 (Calc.)
CLEAVAGE: {110} perfect
{010} imperfect
Brittle.
HABIT: Crystals usually prismatic: also tabular. Prism faces often rounded by striations. As aggregates. Also massive, with granular, lamellar, or columnar structure.
COLOR-LUSTER: Colorless, white, grayish, yellowish, brownish, or reddish. Adamantine, pearly on cleavage. Transparent. Streak white.
MODE OF OCCURRENCE: Occurs as a secondary mineral formed by the alteration of stibnite or other antimony minerals. Found in Contra Costa, Inyo, Kern, San Benito, San Bernardino, and Santa Clara counties, California; in the Ochoco district, Crook County, Oregon; at the Stanley Stibnite mine, Shoshone County, Idaho; in the Silver City district, Pennington County, South Dakota; and at localities in Canada, Mexico, Bolivia, Brazil, France, Italy, Germany, Czechoslovakia, Roumania, and Algeria.
BEST REF. IN ENGLISH: Palache, et al., "Dana's System of Mineralogy," 7th Ed., v. I, p. 547–550, New York, Wiley, 1944.

## VALLERIITE

$2(Fe, Cu)_2S_2 \cdot 3(Mg, Al)(OH)_2$

CRYSTAL SYSTEM: Hexagonal
CLASS: $\overline{3} \, 2/m$
SPACE GROUP: $R\overline{3}m$
Z: 3/2
LATTICE CONSTANTS:
$a = 3.792$
$c = 34.10$

3 STRONGEST DIFFRACTION LINES:
11.4 (100)
5.71 (100)
3.27 ( 60)
HARDNESS: Very soft; sooty.
DENSITY: 3.09–3.14 (Meas.)
4.26 (X-ray)
CLEAVAGE: {0001} perfect
HABIT: Massive.
COLOR-LUSTER: Bronze-yellow. Opaque. Metallic.
MODE OF OCCURRENCE: Occurs in high temperature copper deposits at Loolekop, eastern Transvaal, South Africa, and at the Aurora mine, Nya-Kopparberg, Sweden. Also found at the Elizabeth mine, South Stratford, Vermont; Muskox intrusion, Northwest Territory, Canada; and Vihaute, Finland.
BEST REF. IN ENGLISH: Evans, et al., USGS Prof. Paper 475-D, D64–69 (1963).

# VANADINITE

$Pb_5(VO_4)_3Cl$
Pyromorphite Series
Apatite Group

CRYSTAL SYSTEM: Hexagonal
CLASS: 6/m
SPACE GROUP: $P6_3/m$
Z: 2
LATTICE CONSTANTS:
$a = 10.323$
$c = 7.346$
3 STRONGEST DIFFRACTION LINES:
2.988 (100)
3.068 ( 85)
3.384 ( 60)
OPTICAL CONSTANTS:
$\omega = 2.416$
$\epsilon = 2.350$ (Na)
(–)
HARDNESS: 2¾–3
DENSITY: 6.88 (Meas.)
6.929 (Calc.)
CLEAVAGE: Fracture conchoidal to uneven. Brittle.
HABIT: Crystals short to long prismatic, also acicular to hair-like. Sometimes skeletal or cavernous. As isolated crystals, druses, or aggregates. Also globular.
COLOR-LUSTER: Bright red, orange-red, brownish red to rich brown, pale yellowish to yellow to brownish yellow; rarely white or colorless. Transparent to nearly opaque. Resinous to subadamantine. Streak white or yellowish.
MODE OF OCCURRENCE: Occurs chiefly as a secondary mineral in the oxidation zone of lead-bearing ore deposits. Found as fine specimens at many localities in Arizona, especially at the Apache mine, Gila County; at the Mammoth mine, Tiger, Pinal County; and in Pima and Yuma counties. It also occurs in New Mexico, California, Utah, South Dakota, and Colorado. Excellent specimens also are found at several places in Chihuahua and elsewhere in Mexico; in Argentina, Scotland, Sardinia, Austria, USSR, Algeria, Tunisia, and as magnificent crystal groups at Mibladen, Morocco.

BEST REF. IN ENGLISH: Palache, et al., "Dana's System of Mineralogy," 7th Ed., v. II, p. 895–898, New York, Wiley, 1951.

# VANALITE (Inadequately described mineral)

$NaAl_8V_{10}O_{38} \cdot 30H_2O$

CRYSTAL SYSTEM: Unknown
3 STRONGEST DIFFRACTION LINES:
3.313 (100) kX
2.262 ( 80)
1.518 ( 60)
OPTICAL CONSTANTS:
$\alpha = 1.710$
$\beta = ?$
$\gamma = 1.735$
HARDNESS: Not determined.
DENSITY: 2.3–2.4
CLEAVAGE: Not determined.
HABIT: Crystals wedge-shaped averaging 0.025 mm in size.
COLOR-LUSTER: Bright yellow with orange tint; waxy to vitreous, in friable material dull.
MODE OF OCCURRENCE: Occurs in weathered shales as incrustations on joint planes and as veinlets and cavity fillings associated with halloysite, montmorillonite, and other clay minerals in northwestern Kara-Tau, Kazakhstan.
BEST REF. IN ENGLISH: Ankinovich, E. A., *Am. Min.*, **48**: 1180 (1963).

# VANDENBRANDEITE

$CuUO_4 \cdot 2H_2O$

CRYSTAL SYSTEM: Triclinic
CLASS $\bar{1}$
SPACE GROUP: $P\bar{1}$
Z: 2
LATTICE CONSTANTS:
$a = 7.86$    $\alpha = 91°52'$
$b = 5.44$    $\beta = 102°00'$
$c = 6.10$    $\gamma = 89°37'$
3 STRONGEST DIFFRACTION LINES:
4.44 (100)
5.26 ( 90)
2.97 ( 80)
OPTICAL CONSTANTS:
$\alpha = 1.76–1.77$
$\beta = 1.78–1.792$
$\gamma = 1.80$
(–?)2V = large, near 90°
HARDNESS: 4
DENSITY: 5.08 (Meas.)
5.26 (Calc.)
CLEAVAGE: {110} perfect
HABIT: Rarely as very small crystals tabular {001} with the forms {110}, {1$\bar{1}$0}, {10$\bar{1}$}, and {01$\bar{1}$}. As radial fibrous rosettes, and as lamellar masses.
COLOR-LUSTER: Dark green to blackish green, vitreous; translucent, transparent in thin flakes.

MODE OF OCCURRENCE: Occurs as a secondary mineral, formed by the oxidation of uraninite and associated copper minerals. In Katanga, Zaire, it is found at Luiswishi with cuprosklodowskite; at Shinkolobwe with curite, uranophane, and wad; at Kambove with pitchblende, kasolite, and cuprosklodowskite; and at Kalongwe associated with kasolite, sklodowskite, cuprosklodowskite, malachite, and pitchblende.

BEST REF. IN ENGLISH: Frondel, Clifford, USGS Bull. 1064, 100–103 (1958).

# VANDENDREISSCHEITE
$PbU_7O_{22} \cdot 12H_2O$

CRYSTAL SYSTEM: Orthorhombic
CLASS: 2/m 2/m 2/m or mm2
SPACE GROUP: Pmma, $P2_1$ma, or Pm2a
Z: 36?
LATTICE CONSTANTS:
 $a$ = 14.07
 $b$ = 40.85
 $c$ = 43.33
3 STRONGEST DIFFRACTION LINES:
 7.25 (100)
 3.61 (100)
 3.17 ( 75)
OPTICAL CONSTANTS:
 $\alpha$ = 1.780
 $\beta$ = 1.850
 $\gamma$ = 1.860
 2V $\simeq$ 60°
 HARDNESS: $\sim$ 3
 DENSITY: 5.45 (Meas.)
 CLEAVAGE: {001} perfect
HABIT: Crystals prismatic, elongated [100], (010) striated parallel to [100]; also tabular on {001}, elongated [100], pseudohexagonal. Commonly as dense microcrystalline aggregates.
COLOR-LUSTER: Orange to amber, adamantine. Transparent to translucent.
MODE OF OCCURRENCE: Found as minute crystals associated with fourmarierite and rutherfordine at Shinkolobwe, Katanga, Zaire; as tiny crystals associated with fourmarierite on altered uraninite at Great Bear Lake, Canada; as a component of the so-called "gummite" pseudomorphs after uraninite found in numerous granitic pegmatites in the New England pegmatite area, North Carolina, and in the Black Hills, South Dakota.
BEST REF. IN ENGLISH: Christ, C. L., and Clark, Joan R., Am. Min., 45: 1026–1061 (1960).

# VANOXITE (Inadequately described mineral)
$V_4^{4+}V_2^{5+}O_{13} \cdot 8H_2O$ (?)

CRYSTAL SYSTEM: Unknown
OPTICAL CONSTANTS: Opaque.
 HARDNESS: Not determined.
 DENSITY: Not determined.
 CLEAVAGE: Not determined.

HABIT: Crystals microscopic with rhombohedral aspect. Usually massive, as a cement in sandstone.
COLOR-LUSTER: Black. Opaque.
MODE OF OCCURRENCE: Occurs associated with gypsum as a replacement of wood at the Jo Dandy mine, on the southwest wall of Paradox Valley; in ore from the Henry Clay claim at Long Park; and from the Bill Bryan claim in Wild Steer Canyon, 6 miles south of Gateway, Montrose County, Colorado.
BEST REF. IN ENGLISH: Palache, et al., "Dana's System of Mineralogy," 7th Ed., v. I, p. 601–602, New York, Wiley, 1944.

# VANTHOFFITE
$Na_6Mg(SO_4)_4$

CRYSTAL SYSTEM: Monoclinic
CLASS: 2/m
SPACE GROUP: $P2_1$/c
Z: 2
LATTICE CONSTANTS:
 $a$ = 9.80
 $b$ = 9.22
 $c$ = 8.20
 $\beta$ = 113°30′
3 STRONGEST DIFFRACTION LINES:
 2.91 (100)
 3.44 ( 95)
 3.43 ( 95)
OPTICAL CONSTANTS:
 $\alpha$ = 1.486
 $\beta$ = 1.488
 $\gamma$ = 1.489
 (−)2V = 83°
 HARDNESS: 3½
 DENSITY: 2.694 (Meas.)
 CLEAVAGE: Not determined.
Fracture conchoidal to uneven. Friable.
 HABIT: Massive; as bedded aggregates or anhedral grains.
 COLOR-LUSTER: Colorless, transparent. Vitreous to pearly.
MODE OF OCCURRENCE: Occurs associated with massive bloedite at Bertram siding, Imperial County, California; at the Berlepsch mine near Strassfurt, at the Wilhelmshall mine near Halberstadt, and at other potash deposits in Germany. Also found in the Permain age Salado formation at Carlsbad, New Mexico.
BEST REF. IN ENGLISH: Palache, et al., "Dana's System of Mineralogy," 7th Ed., v. II, p. 430, New York, Wiley, 1951.

# VANURALITE
$Al(UO_2)_2(VO_4)_2(OH) \cdot 11H_2O$

CRYSTAL SYSTEM: Monoclinic
CLASS: 2/m
SPACE GROUP: A2/a
Z: 4

LATTICE CONSTANTS:
$a = 10.55$
$b = 8.44$
$c = 24.52$
$\beta = 103°$
3 STRONGEST DIFFRACTION LINES:
11.98 (100)
5.98 ( 90)
3.795 ( 80)
OPTICAL CONSTANTS:
$\alpha = 1.65$
$\beta = 1.85$
$\gamma = 1.90$
$(-)2V = 44°$
HARDNESS: ~2
DENSITY: 3.62 (Meas.)
CLEAVAGE: {001} perfect, easy
HABIT: As crystal plates flattened on {001}. Forms: {001} dominant; {101}, {$\bar{1}$01}, {111}, {$\bar{1}$11}, and {011}.
COLOR-LUSTER: Citron yellow.
MODE OF OCCURRENCE: Occurs with plumboan francevillite, chervetite, and brackebuschite in a supergene deposit at Mounana, Gabon.
BEST REF. IN ENGLISH: Branche, Georges; Bariand, Pierre; Chantret, Francis; Pouget, Robert; and Rimsky, Alexandre, *Am. Min.*, **48**: 1415-1416 (1963). Cesbron, Fabien, *Am. Min.*, **56**: 639-640 (1971).

# VANURANYLITE
$(H_3O)_2(UO_2)_2(VO_4)_2 \cdot 3.6H_2O$ ?

CRYSTAL SYSTEM: Monoclinic (?)
Z: 4 (?)
LATTICE CONSTANTS:
$a = 10.49$
$b = 8.37$
$c = 20.30$
$\beta = 90°$
3 STRONGEST DIFFRACTION LINES:
5.00 (100)
3.23 (100)
2.11 ( 80)
OPTICAL CONSTANTS:
$\alpha < 1.710$
$1.92 < \beta < 1.95$
$\gamma > 1.95$
$(-)2V = $ small
HARDNESS: ~2
DENSITY: 3.644 (Meas.)
3.252 (Calc.)
CLEAVAGE: {001} perfect
HABIT: As platy pseudohexagonal crystals; very minute in size. As fine crusts.
COLOR-LUSTER: Intense yellow. Does not fluoresce under ultraviolet light.
MODE OF OCCURRENCE: Occurs associated with uranophane and soddyite along fractures in sandstones in the oxidation zone. Locality not given.
BEST REF. IN ENGLISH: Buryanova, E. Z., Strokova, G. S., and Shitov, V. A., *Am. Min.*, **51**: 1548 (1966).

# VARISCITE
$AIPO_4 \cdot 2H_2O$
Variscite-Strengite Series
Dimorphous with metavariscite

CRYSTAL SYSTEM: Orthorhombic
CLASS: 2/m 2/m 2/m
SPACE GROUP: Pcab
Z: 8
LATTICE CONSTANTS:
$a = 9.87$
$b = 9.57$
$c = 8.52$
3 STRONGEST DIFFRACTION LINES:
5.365 (100)
4.257 (100)
3.039 (100)
OPTICAL CONSTANTS:
$\alpha = 1.563$
$\beta = 1.588$
$\gamma = 1.594$
$(-)2V = $ moderate
HARDNESS: 3½-4½
DENSITY: 2.57 (Meas.)
2.61 (Calc.)
CLEAVAGE: {010} good
{001} poor
Fracture splintery to uneven (massive), conchoidal (crystals).
HABIT: Crystals octahedral (rare). Usually massive, as crusts, veinlets, or nodules. Also opal-like. Twinning on {201} not common.
COLOR-LUSTER: Pale green to emerald green, also bluish green to colorless. Transparent to translucent. Crystals vitreous; massive material waxy to dull.
MODE OF OCCURRENCE: Occurs chiefly in deposits formed by the action of phosphatic meteoric waters on aluminous rocks. Found in Utah as nodules up to one foot in diameter near Fairfield, Utah County, associated with crandallite and other phosphate minerals; at Lucin, Box Elder County; and at several other places in the state. It also occurs as incrustations on rock at Bear, Montgomery County, Arkansas, and in California, Nevada, Arizona, and Pennsylvania. Other occurrences include localities in Czechoslovakia, Germany, Austria, Redondo Island in the Antilles, Western Australia, and Rio Grande do Norte, Brazil.
BEST REF. IN ENGLISH: Palache, et al., "Dana's System of Mineralogy," 7th Ed., v. II, p. 756-761, New York, Wiley, 1951.

# VARLAMOFFITE (Souxite)
$(Sn, Fe)(O, OH)_2$

CRYSTAL SYSTEM: Tetragonal
CLASS: 4/m 2/m 2/m
SPACE GROUP: $P4_2/mnm$
Z: 2
LATTICE CONSTANTS:
$a = 4.71$
$c = 3.13$

# 3 STRONGEST DIFFRACTION LINES:
1.75 (100)
3.33 ( 80)
2.60 ( 80)
HARDNESS: Not determined.
DENSITY: 2.52–2.61 (Meas.)
CLEAVAGE: None.
HABIT: Massive; earthy, porous.
COLOR-LUSTER: Yellow. Waxy.
MODE OF OCCURRENCE: Occurs in tin-bearing veins and granite pegmatites. Found near Keystone, Pennington County, South Dakota; at the Cligga mine, Cornwall, England; near Atondo, Kalima, Zaire; at Chenderiang Ridge, Malaya; and at the Sardine tin mine, North Queensland, Australia.
BEST REF. IN ENGLISH: Buttgenbach, H., *Am. Min.*, **34**: 618 (1949). Taylor, R. G., Morgan, W. R., and Phillips, D. N., *Min. Mag.*, **37**: 624–628 (1970).

# VARULITE
$(Na_2, Ca)(Mn, Fe)_2(PO_4)_2$

CRYSTAL SYSTEM: Monoclinic
CLASS: 2/m
SPACE GROUP: $I2_1/a$
3 STRONGEST DIFFRACTION LINES:
2.74 (100)
3.50 ( 40)
2.56 ( 40)
OPTICAL CONSTANTS:
$\alpha$ = 1.708, 1.720
$\gamma$ = 1.722, 1.732
(+)2V = large, 70° (Na)
HARDNESS: 5
DENSITY: 3.581 (Meas.)
CLEAVAGE: {001} good
{010} distinct
HABIT: Massive, granular.
COLOR-LUSTER: Dull yellowish green. Vitreous.
MODE OF OCCURRENCE: Occurs in granite pegmatites at Varuträsk and at Skrumpetorp, Sweden.
BEST REF. IN ENGLISH: Palache, et al., "Dana's System of Mineralogy," 7th Ed., v. II, p. 669–670, New York, Wiley, 1951.

# VASHEGYITE (Inadequately described mineral)
$Al_4(PO_4)_3(OH)_3 \cdot 13H_2O$ (?)

CRYSTAL SYSTEM: Unknown
3 STRONGEST DIFFRACTION LINES:
10.5 (100)
7.2 ( 70)
2.90 ( 70)
OPTICAL CONSTANTS:
Mean index = 1.48
HARDNESS: 2–3
DENSITY: 1.96 (Meas.)
CLEAVAGE: Not determined.
HABIT: Massive, compact to porous.

COLOR-LUSTER: White, pale green to yellow and brownish. Subtranslucent to opaque. Dull.
MODE OF OCCURRENCE: Occurs with variscite in altered slate near Manhattan, Nye County, Nevada, and at localities in Germany, Czechoslovakia, and Hungary.
BEST REF. IN ENGLISH: Palache, et al., "Dana's System of Mineralogy," 7th Ed., v. II, p. 999, New York, Wiley, 1951.

# VATERITE
$\mu$ $CaCO_3$
Trimorphous with calcite, aragonite

CRYSTAL SYSTEM: Hexagonal
CLASS: 6/m 2/m 2/m
SPACE GROUP: $P6_3/mmc$
Z: 12
LATTICE CONSTANTS:
(synthetic)
$a$ = 7.135
$c$ = 16.98
3 STRONGEST DIFFRACTION LINES:
3.58 (100)
3.30 (100) (synthetic)
2.73 (100)
OPTICAL CONSTANTS:
(synthetic)
$\omega$ = 1.550
$\epsilon$ = 1.650
(+)
HARDNESS: ~3
DENSITY: 2.54 (Meas.)
2.65 (Calc.)
HABIT: Finely fibrous, platy.
COLOR-LUSTER: Colorless.
MODE OF OCCURRENCE: Occurs in hydrogel pseudomorphs after larnite at Ballycraigy, Larne, Northern Ireland.
BEST REF. IN ENGLISH: Palache, et al., "Dana's System of Mineralogy, 7th Ed., v. II, p. 181–182, New York, Wiley, 1951. McConnell, J. D. C., *Min. Mag.*, **32**: 535–544 (1960).

# VAUQUELINITE
$Pb_2Cu(CrO_4)(PO_4)(OH)$

CRYSTAL SYSTEM: Monoclinic
CLASS: 2/m
SPACE GROUP: $P2_1/n$
Z: 4
LATTICE CONSTANTS:
$a$ = 13.68
$b$ = 5.83
$c$ = 9.53
$\beta$ = 93°58'
3 STRONGEST DIFFRACTION LINES:
3.31 (100)
4.73 ( 70)
2.89 ( 60)

OPTICAL CONSTANTS:

$\alpha = 2.11$
$\beta = 2.22$
$\gamma = 2.22$
$(-)2V \sim 0°$

HARDNESS: 2½-3

DENSITY: 5.986-6.06 (Meas.)
6.03 (Calc. for $Pb_2Cu[CrO_4][PO_4]$)

CLEAVAGE: Fracture uneven. Brittle.

HABIT: Crystals usually wedge-shaped, minute. As irregular aggregates; fibrous; compact; granular. Twinning on {102}.

COLOR-LUSTER: Green to brown to nearly black. Transparent to translucent. Adamantine to resinous. Streak greenish or brownish.

MODE OF OCCURRENCE: Occurs as a secondary mineral associated with crocoite, pyromorphite, and mimetite at Beresov, Urals, USSR. Also reported from numerous unauthenticated localities.

BEST REF. IN ENGLISH: Palache, et al., "Dana's System of Mineralogy," 7th Ed., v. II, p. 650-652, New York, Wiley, 1951.


# VAUXITE
$Fe^{2+}Al_2(PO_4)_2(OH)_2 \cdot 6H_2O$

CRYSTAL SYSTEM: Triclinic
CLASS: $\bar{1}$
SPACE GROUP: $P\bar{1}$
Z: 2
LATTICE CONSTANTS:

$a = 9.13$     $\alpha = 98.3°$
$b = 11.59$    $\beta = 92.0°$
$c = 6.14$     $\gamma = 108.4°$

3 STRONGEST DIFFRACTION LINES:
5.45 (100)
5.97 ( 80)
4.94 ( 60)

OPTICAL CONSTANTS:
$\alpha = 1.551$
$\beta = 1.555$ (Na)
$\gamma = 1.562$
$(+)2V = 32°$

HARDNESS: 3½

DENSITY: 2.40 (Meas.)
2.41 (Calc.)

CLEAVAGE: None. Brittle.

HABIT: Crystals minute, tabular, and somewhat elongated. As radial or subparallel aggregates. Twinning on {010}.

COLOR-LUSTER: Pale sky blue to dark blue. Transparent to translucent. Vitreous. Streak white.

MODE OF OCCURRENCE: Occurs as excellent microcrystals associated with paravauxite, quartz, and wavellite in the tin deposits at Llallagua, Bolivia.

BEST REF. IN ENGLISH: Palache, et al., "Dana's System of Mineralogy," 7th Ed., v. II, p. 974-975, New York, Wiley, 1951. Baur, W. H., and Rao, B. Rama, *Am. Min.*, **53**: 1025-1028 (1968).


# VÄYRYNENITE
$BeMnPO_4(OH, F)$

CRYSTAL SYSTEM: Monoclinic
CLASS: 2/m
SPACE GROUP: $P2_1/a$
Z: 4
LATTICE CONSTANTS:

$a = 5.411$
$b = 14.49$
$c = 4.73$
$\beta = 102°45'$

3 STRONGEST DIFFRACTION LINES:
3.45 (100)
7.25 ( 85)
2.89 ( 85)

OPTICAL CONSTANTS:
$\alpha = 1.638$
$\beta = 1.658$
$\gamma = 1.664$
$(-)2V = 54°08'$

HARDNESS: 5

DENSITY: 3.183-3.215 (Meas.)
3.23 (Calc.)

CLEAVAGE: {001} distinct
Brittle.

HABIT: Crystals prismatic; in drusy aggregates. Usually massive, granular.

COLOR-LUSTER: Pale pink to rose red; transparent; vitreous.

MODE OF OCCURRENCE: Occurs associated with herderite, hurlbutite, beryllonite, microcline, morinite, and muscovite in lithium pegmatite at Viitaniemi, Eräjärvi, Finland.

BEST REF. IN ENGLISH: Volborth, A., *Am. Min.*, **39**: 848 (1954). Volborth, A., *Am. Min.*, **41**: 371 (1956).


# VEATCHITE
$Sr_2B_{11}O_{16}(OH)_5 \cdot H_2O$
Dimorphous with p-veatchite

CRYSTAL SYSTEM: Monoclinic
CLASS: m
SPACE GROUP: Aa
Z: 4
LATTICE CONSTANTS:

$a = 20.81$
$b = 11.74$
$c = 6.637$
$\beta = 92°2'$

3 STRONGEST DIFFRACTION LINES:
10.5 (100)
3.34 (100)
2.61 ( 70)

OPTICAL CONSTANTS:
$\alpha = 1.551$
$\beta = 1.553$ (Na)
$\gamma = 1.62$
$(+)2V = 37°$

HARDNESS: 2

DENSITY: 2.62 (Meas.)

CLEAVAGE: {010} perfect
{001} imperfect

HABIT: Crystals platy, flattened on {100}, elongated [001] or [042]; slender prismatic to fibrous. As groups of divergent plates and cross-fiber veins.

COLOR-LUSTER: Colorless, transparent; also white. Vitreous, pearly on cleavage. Fibrous masses silky.

MODE OF OCCURRENCE: Found as thin cross-fiber veins with limestone and howlite and as minute crystals on colemanite in a borate deposit at Lang, Los Angeles County, and as crystals in drill core from the Four Corners area, Kramer district, San Bernardino County, California.

BEST REF. IN ENGLISH: Clark, Joan R., and Mrose, Mary E., *Am. Min.*, **45**: 1221–1229 (1960).    Jager and Lehmann, *Z. anorg. allgem. chem.*, **323**: 286–291 (1963).


# p-VEATCHITE

$Sr_2B_{11}O_{16}(OH)_5 \cdot H_2O$
A polymorph of veatchite

CRYSTAL SYSTEM: Monoclinic
CLASS: 2/m or 2
SPACE GROUP: $P2_1/m$ or $P2_1$
Z: 4
LATTICE CONSTANTS:
  $a = 6.72$
  $b = 20.81$
  $c = 6.64$
  $\beta = 119°04'$
3 STRONGEST DIFFRACTION LINES:
  10.4   (100)
  3.31  ( 60)
  2.592 ( 30)
OPTICAL CONSTANTS:
  $\alpha = 1.550$
  $\beta = 1.553$
  $\gamma = 1.620$
  (+)2V = 25°
HARDNESS: 2
DENSITY: 2.60–2.65 (Meas.)
CLEAVAGE: {010} perfect and easy
{001} imperfect
HABIT: Like veatchite.
COLOR-LUSTER: Transparent to white; vitreous, pearly on cleavage.
MODE OF OCCURRENCE: Occurs at the Königshall-Hindenburg Mine, Reyershausen, Germany.
BEST REF. IN ENGLISH: Braitsch, Otto, *Am. Min.*, **44**: 1323–1324 (1959).    Clark and Mrose, *Am. Min.*, **45**: 1221 (1960).


# VEENITE

$Pb_2(Sb, As)_2S_5$

CRYSTAL SYSTEM: Orthorhombic
CLASS: 2/m 2/m 2/m or mm2
SPACE GROUP: Pmcn or $P2_1cn$
Z: 8

LATTICE CONSTANTS:
  $a = 4.22 \times 2$
  $b = 26.2$
  $c = 7.90$
3 STRONGEST DIFFRACTION LINES:
  3.81 (100)
  3.03 ( 90)
  3.42 ( 80)
HARDNESS: Talmadge hardness C-.    Vickers micro-hardness 164 (156–172) $kg/mm^2$ with 50 g load.
DENSITY: 5.92 (Meas.)
5.96 (Calc.)
CLEAVAGE: None observed.
Fracture conchoidal. Very brittle.
HABIT: Commonly massive or as disseminated anhedral grains. Crystals elongated [100] with prism zone almost equant, but slightly flattened on {010}. Faces roughened and grooved.
COLOR-LUSTER: Steel gray; metallic.    Streak black with faint brownish tint.
MODE OF OCCURRENCE: Occurs as minute crystals and in masses up to 2 cm in diameter associated with calcite and boulangerite in a small prospect pit in Precambrian marble at Madoc, Ontario, Canada.
BEST REF. IN ENGLISH: Jambor, J. L., *Can. Min.*, **9** (Pt. 1): 7–24 (1967).


**VERDELITE** = Green variety of tourmaline


**VERMICULITE** = Group name with general formula $(Mg, Fe, Al)_3(Al, Si)_4O_{10}(OH)_2 \cdot 4H_2O$


# VERPLANCKITE

$Ba_2(Mn, Fe, Ti)Si_2O_6(O, OH, Cl, F)_2 \cdot 3H_2O$

CRYSTAL SYSTEM: Hexagonal
CLASS: 6/m 2/m 2/m, $\bar{6}$m2, 6mm, or 622
SPACE GROUP: P6/mmm, P$\bar{6}$2m, P$\bar{6}$m2, P6mm, or P622
Z: 6
LATTICE CONSTANTS:
  $a = 16.35$
  $c = 7.17$
3 STRONGEST DIFFRACTION LINES:
  3.95 (100)
  2.97 ( 70)
  2.738 ( 70)
OPTICAL CONSTANTS:
  $\omega = 1.683$
  $\epsilon = 1.672$
  (–)
HARDNESS: 2½–3
DENSITY: 3.52 (Meas.)
3.46 (Calc.)
CLEAVAGE: {11$\bar{2}$0} good
{0001} poor
HABIT: As prismatic subhedral to euhedral hexagonal crystals elongated parallel to the c-axis. Crystals range in size from a fraction of a millimeter to 3 mm in length.

COLOR-LUSTER: Brownish orange to light brownish yellow; vitreous. Pale orange streak.

MODE OF OCCURRENCE: Occurs as radial masses of prismatic crystals and as disseminated grains associated with quartz, sanbornite, celsian, diopside, taramellite, pyrrhotite, pyrite, fresnoite, muirite, and traskite in gneissic sanbornite-quartz rock in a narrow zone 2½ miles long near a granodiorite contact in eastern Fresno County, California.

BEST REF. IN ENGLISH: Alfors, John T., Stinson, Melvin C., Matthews, Robert A., and Pabst, Adolf, *Am. Min.*, **50**: 314–340 (1965). Alfors and Putman, *Am. Min.*, **50**: 1500–1503 (1965).

# VÉSIGNIÉITE (Inadequately described mineral)
$BaCu_3(VO_4)_2(OH)_2$

CRYSTAL SYSTEM: Unknown
3 STRONGEST DIFFRACTION LINES:
  3.21 (100)
  2.72 ( 75)
  2.29 ( 75)
OPTICAL CONSTANTS:
  $\alpha = 2.01$
  $\beta = 2.04$
  $\gamma = 2.08$
  $(-)2V \sim 60°$
HARDNESS: 3–4
DENSITY: 4.05 (Meas.)
CLEAVAGE: Basal, good.
HABIT: Lamellar aggregates and as polysynthetic twins with pseudohexagonal outline up to more than 0.5 mm in diameter.
COLOR-LUSTER: Yellow-green to dark olive green; vitreous.
MODE OF OCCURRENCE: Occurs with barite and calcite in vugs in manganese ores at Friedrichsroda, Thuringia, Germany; also on quartzite at Perm, Urals, and at Agalik, Uzbekistan.
BEST REF. IN ENGLISH: Guillemin, Claude, *Am. Min.*, **40**: 942–943 (1955).

# VESUVIANITE = Synonym for idocrase

# VESZELYITE (Arakawaite, Kipushite)
$(Cu, Zn)_3(PO_4)(OH)_3 \cdot 2H_2O$

CRYSTAL SYSTEM: Monoclinic
CLASS: 2/m
SPACE GROUP: $P2_1/a$
Z: 4
LATTICE CONSTANTS:
  $a = 9.84$
  $b = 10.17$
  $c = 7.48$
  $\beta = 103°25'$
3 STRONGEST DIFFRACTION LINES:
  3.642 (100)
  6.96 ( 40)
  4.489 ( 30)

OPTICAL CONSTANTS:
  $\alpha = 1.640, 1.618$
  $\beta = 1.658, 1.622$
  $\gamma = 1.695, 1.658$
  $(+)2V \sim 71°-38½°$
HARDNESS: 3½–4
DENSITY: 3.4 (Meas.)
       3.42 (Calc. Cu:Zn = 3:2)
CLEAVAGE:  {001}
          {110}
HABIT: Crystals short prismatic, thick tabular, equant, or octahedral. As granular aggregates.
COLOR-LUSTER: Green, greenish blue to dark blue. Translucent. Vitreous.
MODE OF OCCURRENCE: Occurs at Moravicza (Vaskö), Banat, Roumania; at Kipushi, Katanga, Zaire (kipushite); at the Broken Hill mine, Kabwe, Zambia; and at the Arakawa mine, Ugo province, Japan (arakawaite).
BEST REF. IN ENGLISH: Palache, et al., "Dana's System of Mineralogy," 7th Ed., v. II, p. 916–918, New York, Wiley, 1951.

# VIBERTITE = Bassanite

# VILATÉITE = Metastrengite

# VILLAMANINITE
$(Cu, Ni, Co, Fe)(S, Se)_2$
Pyrite Group

CRYSTAL SYSTEM: Cubic
CLASS: $2/m\,\overline{3}$
SPACE GROUP: Pa3
Z: 4
LATTICE CONSTANT:
  $a = 5.693$
  HARDNESS: 4½
  DENSITY: 4.523 (Meas.)
  CLEAVAGE: None. Brittle.
HABIT: Crystals cubo-octahedral, in irregular groups. Also as radiating nodular masses.
COLOR-LUSTER: Iron black. Opaque. Metallic.
MODE OF OCCURRENCE: Occurs in dolomite in the Cármenes district, near Villamanín, León, Spain.
BEST REF. IN ENGLISH: Hey, M. H., *Min. Mag.*, **33**: 169–170 (1962).

# VILLIAUMITE
NaF

CRYSTAL SYSTEM: Cubic
CLASS: $4/m\,\overline{3}\,2/m$
SPACE GROUP: Fm3m
Z: 4
LATTICE CONSTANTS:
  (annealed)        (natural)
  $a = 4.628$       $a = 4.630-4.633$

3 STRONGEST DIFFRACTION LINES:
2.319 (100)
1.639 ( 60)
1.338 ( 17)
OPTICAL CONSTANT:
$N$ = 1.3270
HARDNESS: 2-2½
DENSITY: 2.79 (Meas.)
2.81 (Calc.)
CLEAVAGE: {001} perfect
Brittle.
HABIT: Crystals minute. Usually massive, granular.
COLOR-LUSTER: Deep carmine red, lavender pink to light orange, transparent to translucent. Vitreous. Streak white.
MODE OF OCCURRENCE: Occurs filling small miarolitic cavities in the nepheline syenites of Rouma, Isles de Los, Guinea, West Africa; in the alkalic rocks in the Lovozero and Khibina massifs, Kola Peninsula, USSR; in the Ilimaussaq intrusion, Greenland; in saline lake deposits in Magadi area, Kenya; at the Desourdy quarry, Mont St. Hilaire, Quebec, Canada; and in the phonolite sill which forms Peck's Mesa, eastern Colfax County, New Mexico.
BEST REF. IN ENGLISH: Stormer, J. C. Jr., and Carmichael, I.S.E., *Am. Min.*, **55**: 126-134 (1970).

## VIMSITE
$CaB_2O_2(OH)_4$

CRYSTAL SYSTEM: Monoclinic
CLASS: 2/m
SPACE GROUP: C2/c
Z: 4
LATTICE CONSTANTS:
$a$ = 10.02
$b$ = 9.71
$c$ = 4.44
$\beta$ = 92°
3 STRONGEST DIFFRACTION LINES:
3.48 (100)
2.222 ( 80)
3.72 ( 70)
OPTICAL CONSTANTS:
$\alpha$ = 1.585
$\beta$ = 1.614
$\gamma$ = 1.614
$(-)2V$ = 28°
HARDNESS: 4
DENSITY: 2.54 (Meas.)
2.56 (Calc. for 2.2 $H_2O$)
CLEAVAGE: Perfect along elongation.
HABIT: Crystals up to 2 mm in size.
COLOR-LUSTER: Colorless; transparent; vitreous.
MODE OF OCCURRENCE: Occurs with uralborite in skarn from the Urals, USSR.
BEST REF. IN ENGLISH: Shashkin, D. P., Simonov, M. A., and Belov, N. V., *Am. Min.*, **54**: 1219-1220 (1969).

## VINOGRADOVITE
$(Na, Ca, K)_4 Ti_4 AlSi_6 O_{23} \cdot 2H_2O$

CRYSTAL SYSTEM: Monoclinic, pseudo-orthorhombic
LATTICE CONSTANTS:
(goniometric data)
$a:b:c$ = 1.18:1:0.76
$\beta$ = 91°58'
3 STRONGEST DIFFRACTION LINES:
3.21 (100)
3.07 (100)
1.614 ( 80)
OPTICAL CONSTANTS:
$\alpha$ = 1.745
$\beta$ = 1.770
$\gamma$ = 1.775
$(-)2V$ = 41°
HARDNESS: ~4
DENSITY: 2.878
CLEAVAGE: {010} perfect
Fracture uneven. Brittle.
HABIT: As spherulites and in aggregates of fibrous crystals; less commonly as fine prismatic crystals showing faces $b\{010\}$, $m\{110\}$, $n\{410\}$, $d\{101\}$, and $D\{\bar{1}01\}$.
COLOR-LUSTER: Colorless to white; vitreous.
MODE OF OCCURRENCE: Occurs in cavities of natrolite or analcite in the central part, or replacing ramsayite and lamprophyllite in the contact zone, at 12 localities in nepheline syenite pegmatites of the Khibina and Lovozero massifs, Kola Peninsula, USSR.
BEST REF. IN ENGLISH: Semenov, E. I., Bohnshtedt-Kupletskaya, E. M., Moleva, V. A., and Sludskaya, N. N., *Am. Min.*, **42**: 308 (1957).

## VIOLAITE = Diopside

## VIOLARITE
$FeNi_2S_4$
Linnaeite Group

CRYSTAL SYSTEM: Cubic
CLASS: m3m
SPACE GROUP: Fd3m
Z: 8
LATTICE CONSTANTS:
$a$ = 9.445 (variable)
3 STRONGEST DIFFRACTION LINES:
2.85 (100)
1.674 ( 80)
1.820 ( 60)
OPTICAL CONSTANTS: Opaque.
HARDNESS: 4½-5½
DENSITY: 4.5-4.8 (Meas.)
4.75 (Calc.)
CLEAVAGE: {001} imperfect
Fracture subconchoidal to uneven.
HABIT: Massive; granular to compact.
COLOR-LUSTER: Light gray to steel gray; rapidly tarnishes copper red or violet-gray. Opaque. Metallic.
MODE OF OCCURRENCE: Occurs with pyrrhotite and chalcopyrite in the nickel ores of the Friday mine at Julian, San Diego County, California; near Linden, Iowa County, and elsewhere in Wisconsin; with millerite in the vicinity of

St. Louis, Missouri; and rarely in the nickel ores at the Vermilion mine, Sudbury, Ontario, Canada.

BEST REF. IN ENGLISH: Palache, et al., "Dana's System of Mineralogy," 7th Ed., v. I, p. 262–265, New York, Wiley, 1944.

**VIRIDINE** = Manganian andalusite

## VISÉITE
$NaCa_5Al_{10}Si_3P_5O_{30}(OH, F)_{18} \cdot 16H_2O$
Zeolite Group

CRYSTAL SYSTEM: Cubic or pseudocubic
Z: 2
LATTICE CONSTANT:
  $a = 13.65$
3 STRONGEST DIFFRACTION LINES:
  2.92 (100)
  1.74 ( 60)
  3.46 ( 50)
OPTICAL CONSTANT:
  $N = 1.53$
  HARDNESS: 3–4
  DENSITY: 2.2 (Meas.)
          2.17 (Calc.)
  CLEAVAGE: None. Brittle.
  HABIT: Massive; as minute nodes.
  COLOR-LUSTER: White, bluish white, yellowish white. Translucent; transparent in thin fragments. Vitreous.
  MODE OF OCCURRENCE: Occurs in association with delvauxite at Visé, Belgium.
  BEST REF. IN ENGLISH: Mélon, J., *Am. Min.*, **30**: 548 (1945). Deer, Howie, and Zussman, "Rock Forming Minerals," v. 4, p. 342, 345, 408, New York, Wiley, 1963. McConnell, Duncan, *Am. Min.*, **37**: 609–617 (1952).

## VISHNEVITE (Sulfate Cancrinite)
$(Na, Ca, K)_{6-7}Al_6Si_6O_{24}(SO_4, CO_3, Cl)_{1.0-1.5} \cdot 1–5H_2O$
Cancrinite Group

CRYSTAL SYSTEM: Hexagonal
CLASS: 6 2 2
SPACE GROUP: $P6_32$
Z: 1
LATTICE CONSTANTS:
  $a = 12.58–12.76$
  $c = 5.11–5.20$
OPTICAL CONSTANTS:
  $\omega = 1.507–1.490$
  $\epsilon = 1.495–1.488$
    (–)
  HARDNESS: 5–6
  DENSITY: 2.32–2.42 (Meas.)
  CLEAVAGE: $\{10\bar{1}0\}$ perfect
              $\{0001\}$ poor
Fracture uneven. Brittle.
  HABIT: Crystals prismatic, rare. Usually massive. Lamellar twinning rare.

COLOR-LUSTER: Colorless, white, yellow, orange, pink to reddish, pale blue to pale bluish gray. Transparent to translucent. Vitreous, pearly or greasy. Streak colorless.
  MODE OF OCCURRENCE: Occurs as a primary mineral in certain alkali rocks and as a common alteration product of nepheline. Found at Iron Hill, Gunnison County, Colorado; Loch Borolan, Assynt, Scotland; Iivaara, Kuusamo, Finland; and at Vishnevy Gory, Urals, and elsewhere in the USSR.
  BEST REF. IN ENGLISH: Deer, Howie, and Zussman, "Rock Forming Minerals," v. 4, p. 310–310, New York, Wiley, 1963.

## VIVIANITE
$Fe_3(PO_4)_2 \cdot 8H_2O$

CRYSTAL SYSTEM: Monoclinic
CLASS: 2/m
SPACE GROUP: C2/m
Z: 2
LATTICE CONSTANTS:
  $a = 10.039$ kX
  $b = 13.388$
  $c = 4.687$
  $\beta = 104°18'$
3 STRONGEST DIFFRACTION LINES:
  6.80 (100)
  2.97 ( 67) (synthetic)
  2.71 ( 67)
OPTICAL CONSTANTS:
  $\alpha = 1.5788–1.616$
  $\beta = 1.6024–1.656$
  $\gamma = 1.6294–1.675$
  $(+)2V = 83\frac{1}{2}°–63\frac{1}{2}°$
  HARDNESS: 1½–2
  DENSITY: 2.68 (Meas.)
          2.71 (Calc.)
  CLEAVAGE: $\{010\}$ perfect, easy
              $\{\bar{1}06\}$ trace
              $\{100\}$ trace
Fracture fibrous. Thin lamellae flexible. Sectile.
  HABIT: Crystals short to long prismatic, elongated along c-axis, often flattened; also tabular or equant. Crystals often rounded. Commonly grouped in clusters or as radial and stellate groups. Also massive bladed or fibrous, as crusts, concretionary, and earthy.
  COLOR-LUSTER: Colorless when fresh; darkens rapidly to shades of green or blue; on further exposure to light becoming dark green, dark bluish green, dark purplish, or bluish black. Streak colorless to bluish white, soon changing to dark blue or brown.
  MODE OF OCCURRENCE: Occurs as a secondary mineral in some metallic ore veins; as an alteration product of primary phosphate minerals in pegmatites; and as concretions or associated with organic matter in clays and other sedimentary deposits. Found as superb crystals with ludlamite in the Blackbird District, Lemhi County, Idaho; as fine crystals up to 5 inches long in Bingham Canyon, Salt Lake County, Utah; as magnificent crystal groups at Richmond, Virginia; in pegmatites in the Black Hills, South Dakota; and at localities in Colorado, California, New Jersey, Delaware,

Maryland, Washington, D.C., and Florida. Excellent dark green crystals up to 6 inches or more in length occur at Llallagua and Poopo, Bolivia; spectacular crystals up to 4 feet long near N'gaoundere, Cameroons; and at places in Canada, England, Yugoslavia, Czechoslovakia, Roumania, France, Germany, USSR, Australia, and Japan.

BEST REF. IN ENGLISH: Palache, et al., "Dana's System of Mineralogy," 7th Ed., v. II, p. 742-746, New York, Wiley, 1951.

## VLADIMIRITE
$Ca_5H_2(AsO_4)_4 \cdot 5H_2O$

CRYSTAL SYSTEM: Monoclinic
CLASS: 2/m
SPACE GROUP: $P2_1/c$
Z: 3
LATTICE CONSTANTS:
 $a = 5.81$
 $b = 10.19$
 $c = 22.7$
 $\beta = 97°19'$
3 STRONGEST DIFFRACTION LINES:
 2.79 (100)
 4.15 ( 80)
 4.00 ( 80)
OPTICAL CONSTANTS:
 $\alpha = 1.650$
 $\beta = 1.654-1.656$
 $\gamma = 1.661$
 $(-)2V = 70°$
HARDNESS: 3½
DENSITY: 3.14 (Meas.)
 3.17 (Calc.)
CLEAVAGE: One, good.
HABIT: As acicular radial aggregates.
COLOR-LUSTER: Pale rose.
MODE OF OCCURRENCE: Occurs in the zone of oxidation in an ore deposit at Kovouaksi, Touva, USSR.
BEST REF. IN ENGLISH: Pierrot, R., Am. Min., 50: 813 (1965).

## VLASOVITE
$Na_2ZrSi_4O_{11}$

CRYSTAL SYSTEM: Monoclinic
CLASS: 2/m
SPACE GROUP: C2/c
Z: 4
LATTICE CONSTANTS:
 $a = 10.98$
 $b = 10.00$
 $c = 8.52$
 $\beta = 100°24'$
3 STRONGEST DIFFRACTION LINES:
 3.26 (100)
 2.966 ( 94)
 5.02 ( 72)

OPTICAL CONSTANTS:
 $\alpha = 1.706$
 $\beta = 1.623$
 $\gamma = 1.628$
 $(-)2V = 50°-56°$
HARDNESS: 6
DENSITY: 2.97 (Meas.)
CLEAVAGE: {010} distinct
Also a second imperfect cleavage at an angle of 88° to the first. Fracture irregular to conchoidal. Brittle.
HABIT: Grains up to 0.5 X 1 X 1.5 cm in size.
COLOR-LUSTER: Colorless; greasy, vitreous to pearly on cleavage. Transparent. Altered material fluoresces strong orange-yellow under ultraviolet light.
MODE OF OCCURRENCE: Occurs associated with microcline, albite, arfvedsonite, aegirine, eudialyte, apatite, and fluorite in the contact zone between pegmatites and fenites near Mt. Vavnbed, Lovozero massif, Kola Peninsula, USSR.
BEST REF. IN ENGLISH: Tikhonenkova, R. P., and Kazakova, M. E., Am. Min., 46: 1202-1203 (1961). Vlasov, K. A., "Mineralogy of Rare Elements," v. II, p. 373-375, Israel Program for Scientific Translations, 1966.

## VOGLITE (Inadequately described mineral)
$\sim Ca_2Cu(UO_2)(CO_3)_4 \cdot 6H_2O$

CRYSTAL SYSTEM: Triclinic or monoclinic
3 STRONGEST DIFFRACTION LINES: No X-ray data.
OPTICAL CONSTANTS:
 $\alpha = 1.541, 1.513$
 $\beta = 1.547, 1.525$
 $\gamma = 1.564, 1.542$
 $(+)2V = 60°$ (Na), 90°
HARDNESS: Not determined.
DENSITY: Not determined.
CLEAVAGE: {010} perfect
HABIT: As coatings and aggregates of tiny scales. Scales are rhomboidal in outline, resembling gypsum, and are striated on the flat side parallel to an edge.
COLOR-LUSTER: Emerald green to bright grass green. Pearly on cleavages. Streak pale green.
MODE OF OCCURRENCE: Voglite occurs as a secondary mineral incrusting sandstone at Frey Point, Utah; also found very sparingly with liebigite as an alteration product of uraninite from the Elias mine, Joachimsthal, Bohemia, Czechoslovakia.
BEST REF. IN ENGLISH: Frondel, Clifford, USGS Bull. 1064, p. 126-128 (1958).

## VOLBORTHITE
$Cu_3(VO_4)_2 \cdot 3H_2O$

CRYSTAL SYSTEM: Monoclinic
3 STRONGEST DIFFRACTION LINES:
 7.26 (100) or 7.18 (100)
 2.87 ( 80)
 2.56 ( 80)

OPTICAL CONSTANTS:
$\alpha = 1.793$
$\beta = 1.801$
$\gamma = 1.816$
$(-)2V = $ large
HARDNESS: $3\frac{1}{2}$
DENSITY: 3.5–3.8
CLEAVAGE: {001} perfect

HABIT: Massive, compact; scaly crusts and scales with triangular or hexagonal outline; rosette-like, honeycombed, or boxwork-like aggregates; radiating fibrous and circular in outline; spherulitic. Lamellar twinning common.

COLOR-LUSTER: Dark olive green to green and yellowish green, bright yellow, brownish yellow, light brown, blackish brown. Translucent to subtranslucent. Vitreous to pearly.

MODE OF OCCURRENCE: Occurs as a secondary mineral in sandstone at Richardson, Grand County, Utah; coating cracks in conglomerate and shale in Monument Valley, Navajo County, Arizona; as a weathering product of a thin, interlava sedimentary rock of Upper Triassic age west of Menzies Bay on Vancouver Island, and north of Gowland Harbour on Quadra Island, British Columbia. Also found in the Uzbekistana district in Ferghana, at Syssersk and Nizhne Tagilsk in the Ural Mountains, and in Permian sandstone at Woskressensk in Perm, USSR.

BEST REF. IN ENGLISH: Jambor, J. L., *Am. Min.*, **45**: 1307–1309 (1960). Palache, et al., "Dana's System of Mineralogy," 7th Ed., v. II, p. 818–819, New York, Wiley, 1951.

## VOLKONSKOITE = Chromian Nontronite

## VOLKOLVITE = Strontioginorite

## VOLKOVSKITE
$(CaSr)B_6O_{10} \cdot 3H_2O$

CRYSTAL SYSTEM: Monoclinic
CLASS: 2
SPACE GROUP: $P2_1$
Z: 8
LATTICE CONSTANTS:
$a = 6.57$
$b = 48.30$
$c = 6.51$
$\beta = 119°05'$
3 STRONGEST DIFFRACTION LINES:
8.1 (100)
3.28 ( 90)
2.63 ( 80)
OPTICAL CONSTANTS:
$\alpha = 1.536$
$\beta = 1.539$
$\gamma = 1.603$
$(+)2V = 24°$ (Calc.)
HARDNESS: Not determined.
DENSITY: 2.29–2.34 (Meas.)
2.39 (Calc.)

CLEAVAGE: {010} perfect
{001} good
Brittle; crushes to small splinters.

HABIT: Plates of elongated outline with angles near 60° and 120°, sometimes rhomb-like. Plates up to 1.5 × 0.5 × 0.05 mm in size.

COLOR-LUSTER: Colorless; vitreous.

MODE OF OCCURRENCE: Occurs associated with anhydrite, sylvite, hilgardite, and boracite, dispersed in mass of hard salt. No locality given; presumably Inder borate deposits.

BEST REF. IN ENGLISH: Kondrat'eva, V. V., Ostrovskaya, I. V., and Yarzhemskii, Ya. Ya., *Am. Min.*, **51**: 1550 (1966).

## VOLTAITE
$K_2 Fe_5^{2+} Fe_4^{3+} (SO_4)_{12} \cdot 18H_2O(?)$

CRYSTAL SYSTEM: Cubic
CLASS: $4/m\,\overline{3}\,2/m$
SPACE GROUP: Fd3c
Z: 16
LATTICE CONSTANT:
$a = 27.36$
3 STRONGEST DIFFRACTION LINES:
3.54 (100)
3.41 ( 90)
5.57 ( 70)
OPTICAL CONSTANT:
$N = 1.593$–1.608
HARDNESS: 3
DENSITY: 2.7 (Meas.)
CLEAVAGE: None. Fracture conchoidal.

HABIT: Crystals octahedral, dodecahedral, and in combinations of these. Also massive, granular.

COLOR-LUSTER: Dark olive green to black. Opaque. Resinous. Streak grayish green.

MODE OF OCCURRENCE: Occurs in Arizona as fine crystals up to 1 cm in size in the fire zone at the United Verde mine, Jerome, Yavapai County, and with coquimbite and other sulfates in the Copper Queen mine, Bisbee, Cochise County. Also found at the Redington mine, Napa County, and at localities in Contra Costa, San Benito, San Bernardino, Shasta, and Sonoma Counties, California; and at the Dexter No. 7 mine, Calf Mesa, San Rafael Swell, Utah. It also occurs in Bolivia, Chile, Spain, Germany, Czechoslovakia, and Cyprus; on Mt. Vesuvius and at other places in Italy; and at the Konomai mine and Yanahara mines, Japan.

BEST REF. IN ENGLISH: Palache, et al., "Dana's System of Mineralogy," 7th Ed., v. II, p. 464–465, New York, Wiley, 1951.

## VOLTZITE = A mixture of wurtzite plus an organometallic compound of zinc.

## VOLYNSKITE
$AgBiTe_2$

CRYSTAL SYSTEM: Orthorhombic

3 STRONGEST DIFFRACTION LINES:
 3.09 (100)
 3.21 ( 80)
 2.21 ( 50)
HARDNESS: Microhardness 55–99 kg/mm$^2$.
DENSITY: Not determined.
CLEAVAGE: One perfect. Two imperfect. Brittle.
HABIT: Massive; as embedded grains less than 0.3 mm in size.
COLOR-LUSTER: Bright lead gray; metallic. Opaque. Pale purplish in reflected light.
MODE OF OCCURRENCE: Occurs as inclusions in tellurobismuthite, in association with altaite, hessite, and other tellurides, in a gold-bearing ore from Armenia, USSR.
BEST REF. IN ENGLISH: Bezsmertnaya, M. S., and Soboleva, L. N., *Am. Min.*, **51**: 531 (1966).

## VONSENITE (Paigeite)
$(Fe^{2+}, Mg)_2 Fe^{3+}BO_5$

CRYSTAL SYSTEM: Orthorhombic
CLASS: 2/m 2/m 2/m
SPACE GROUP: Pbam
Z: 4
LATTICE CONSTANTS:
 $a = 9.37$
 $b = 12.35$
 $c = 3.05$
3 STRONGEST DIFFRACTION LINES:
 2.580 (100)
 5.16 ( 50)
 2.372 ( 25)
OPTICAL CONSTANTS: $\alpha < 1.914$
 HARDNESS: 5
 DENSITY: 4.21–4.77 (Meas.)
 CLEAVAGE: None apparent.
HABIT: As granular polygonal aggregates and occasionally stubby prisms having diamond-shaped cross section.
COLOR-LUSTER: Black. Opaque. Submetallic to adamantine.
MODE OF OCCURRENCE: Occurs in skarn in the pyrometasomatic magnetite deposit at Jayville, and at the Clifton mine, near Degrasse, New York; in the old City quarry at Riverside, Riverside County, and associated with magnetite and hedenbergite in the Gabilan Range, Monterey County, California; and in the Chersky Range in Yakutia, Siberia, USSR.
BEST REF. IN ENGLISH: Clark, Joan R., *Am. Min.*, **50**: 249–254 (1965).

## VOROBYEVITE = Variety of beryl rich in cesium and lithium

## VRBAITE
$Tl_4 Hg_3 Sb_2 As_8 S_{20}$

CRYSTAL SYSTEM: Orthorhombic
CLASS: 2/m 2/m 2/m
SPACE GROUP: Cmca
Z: 4

LATTICE CONSTANTS:
 $a = 13.189$
 $b = 23.235$
 $c = 11.223$
3 STRONGEST DIFFRACTION LINES:
 4.04 (100)
 3.33 ( 80)
 2.57 ( 80)
HARDNESS: 3½
DENSITY: 5.3 (Meas.)
     5.29 (Calc.)
CLEAVAGE: {010} good
Fracture conchoidal to uneven. Brittle.
HABIT: Crystals small, thick tabular or pyramidal. Usually in groups.
COLOR-LUSTER: Dark grayish black with bluish tinge. Thin fragments translucent and dark red. Submetallic to metallic. Streak pale yellowish red.
MODE OF OCCURRENCE: Occurs associated with realgar and orpiment at Allchar, Macedonia, Yugoslavia.
BEST REF. IN ENGLISH: *Am. Min.*, **53**: 351 (1968). Palache, et al., "Dana's System of Mineralogy," 7th Ed., v. I, p. 484–485, New York, Wiley, 1944.

## VREDENBURGITE = Mixture of jacobsite and hausmannite

## VUDYAVRITE (Inadequately described mineral)
$Ce_2 Ti_2 Si_4 O_{15} \cdot 14H_2O$ (?)

CRYSTAL SYSTEM: Not determined.
OPTICAL CONSTANT:
 $N = 1.64–1.58$
 HARDNESS: ~3
 DENSITY: 2.40–2.52 (Meas.)
CLEAVAGE: Fracture conchoidal; exterior powdery, friable.
HABIT: Massive; as an ocherous weathering product of mosandrite, occasionally as pseudomorphs.
COLOR-LUSTER: Reddish-brown, outer part yellowish. Greasy.
MODE OF OCCURRENCE: Occurs in the Langesundfjord district, Norway; also in the Khibiny and Lovozero massifs, Kola Peninsula, USSR.
BEST REF. IN ENGLISH: Vlasov, K. A., "Mineralogy of Rare Elements," v. II, p. 325–328, Israel Program for Scientific Translations, 1966.

## VULCANITE
CuTe

CRYSTAL SYSTEM: Orthorhombic
CLASS: Probably 2/m 2/m 2/m
SPACE GROUP: Probably Pmnm
Z: 2
LATTICE CONSTANTS:
 $a = 4.09$
 $b = 6.95$
 $c = 3.15$

3 STRONGEST DIFFRACTION LINES:
2.03 (100)
2.86 ( 70)
3.52 ( 60)
HARDNESS: 1–2
DENSITY: 7.1 (Calc.)
CLEAVAGE: One prominent cleavage and one less prominent are probably pinacoidal.
HABIT: As aggregates of prismatic or blade-like to irregular grains.
COLOR-LUSTER: Light bronze to yellow-bronze. Metallic.
MODE OF OCCURRENCE: Occurs intergrown with rickardite forming layers 0.5–3.0 mm thick encrusting rock fragments at the Good Hope Mine, Vulcan, Gunnison County, Colorado.
BEST REF. IN ENGLISH: Cameron, Eugene N., and Threadgold, Ian M., *Am. Min.*, **46**: 258–268 (1961).

# VYSOTSKITE
(Pd, Ni)S

CRYSTAL SYSTEM: Tetragonal
CLASS: Probably 4/m
SPACE GROUP: Probably $P4_2/m$
Z: 8
LATTICE CONSTANTS:
$a = 6.371$
$c = 6.540$
3 STRONGEST DIFFRACTION LINES:
2.91 (100)
2.86 (100)
2.61 ( 80)
HARDNESS: Microhardness 32 kg/mm$^2$. Scratched by steel needle.
DENSITY: 6.69 (Meas. synthetic)
6.73 (Calc. synthetic)
HABIT: As minute irregular masses, rarely in well-formed prismatic crystals up to 0.07 mm.
COLOR-LUSTER: Silvery; metallic.
MODE OF OCCURRENCE: Occurs associated with chalcopyrite, millerite, nickelian pyrite, linnaeite, and cooperite in disseminated ores in andesine diabases in the Nor'ilsk deposits, USSR.
BEST REF. IN ENGLISH: Genkin, A. D., Zvyzgintsev, O. E., *Am. Min.*, **48**: 708 (1963).

WAD = Generic term for hydrated oxides of Mn


## WADEITE
$K_2CaZrSi_4O_{12}$

CRYSTAL SYSTEM: Hexagonal
CLASS: 6/m
SPACE GROUP: $P6_3/m$
Z: 2
LATTICE CONSTANTS:
 $a$ = 6.893
 $c$ = 10.172
3 STRONGEST DIFFRACTION LINES:
 2.85 (100)
 3.85 ( 80)
 5.97 ( 60)
OPTICAL CONSTANTS:
 $\omega$ = 1.625
 $\epsilon$ = 1.655
 (+)
HARDNESS: 5½-6
DENSITY: 3.10-3.13 (Meas.)
 3.16 (Calc.)
CLEAVAGE: None. Fracture conchoidal. Brittle.
HABIT: Crystals prismatic, hexagonal in cross section; usually terminated by basal pinacoids, sometimes with modifying pyramid faces.
COLOR-LUSTER: Colorless, light pink, lilac. Transparent. Adamantine.
MODE OF OCCURRENCE: Occurs in potassium-rich rocks in association with leucite, zeolites, apatite, and other minerals. Found in western Kimberley, Australia, and from the Khibiny massif, Kola peninsula, USSR.
BEST REF. IN ENGLISH: Vlasov, K. A., "Mineralogy of Rare Elements," v. II, p. 371-372, Israel Program for Scientific Translations, 1966.


## WAGNERITE
$(Mg, Fe, Mn, Ca)_2PO_4F$

CRYSTAL SYSTEM: Monoclinic
CLASS: 2/m
SPACE GROUP: $P2_1/a$
Z: 16
LATTICE CONSTANTS:
 $a$ = 11.90 kX
 $b$ = 12.51
 $c$ = 9.63
 $\beta$ = 108°07′
3 STRONGEST DIFFRACTION LINES:
 2.993 (100)
 2.839 (100)
 3.150 ( 90)
OPTICAL CONSTANTS:
 $\alpha$ = 1.5678
 $\beta$ = 1.5719 (Na)
 $\gamma$ = 1.5824
 (+)2V = 28°24.5′
HARDNESS: 5-5½
DENSITY: 3.15 (Meas.)
 3.15 (Calc.)
CLEAVAGE: {100} imperfect
 {120} imperfect
 {001} in traces
Fracture splintery to subconchoidal.
HABIT: Crystals prismatic, vertically striated or rounded, sometimes large and rough. Also massive.
COLOR-LUSTER: Various shades of yellow, grayish, greenish, yellowish red to brick brown. Translucent to opaque. Vitreous to resinous.
MODE OF OCCURRENCE: Occurs in quartz veins associated with chlorite, magnesite, and lazulite at Höllgraben and Radelgraben in Salzburg, Austria; at several places near Bamle, Norway; at Hallsjoberget, Sweden; in pegmatite near Mangualde, Portugal; and in lava on Mt. Vesuvius, Italy.
BEST REF. IN ENGLISH: Palache, et al., "Dana's System of Mineralogy," 7th Ed., v. II, p. 845-848, New York, Wiley, 1951.


## WAIRAKITE
$CaAl_2Si_4O_{12} \cdot 2H_2O$
Zeolite Group
Calcium analogue of analcime

CRYSTAL SYSTEM: Monoclinic
CLASS: m or 2/m
SPACE GROUP: Ia or I2/a
Z: 8
LATTICE CONSTANTS:
$a = 13.69$
$b = 13.68$
$c = 13.56$
$\beta = 90.5°$
3 STRONGEST DIFFRACTION LINES:
(synthetic)
3.41 (100)
5.57 ( 90)
2.90 ( 80)
OPTICAL CONSTANTS:
$\alpha = 1.498$
$\gamma = 1.502$ (Na)
$2V = 70°–105°$
HARDNESS: 5½–6
DENSITY: 2.26
CLEAVAGE: {100} distinct
Brittle.

HABIT: Crystals subhedral, up to 15 mm, sometimes showing octahedral-like faces. Twinning on {110}, lamellar.

COLOR-LUSTER: Colorless to white; vitreous to dull.

MODE OF OCCURRENCE: Occurs in tuffaceous sandstones and breccias, vitric tuffs and welded tuffs in drill cores from levels 600–2890 feet, and in rocks ejected by steam from some of the drill holes at Wairakei, about 4 miles from Lake Taupo, in the central part of the North Island, New Zealand. Also found in a graywacke fragment from The Geysers, California.

BEST REF. IN ENGLISH: Steiner, A., *Min. Mag.*, **30**: 690–698 (1955). Coombs, D. S., *Min. Mag.*, **30**: 699–708 (1955).

## WAIRAUITE
CoFe

CRYSTAL SYSTEM: Cubic
LATTICE CONSTANT:
$a = 2.856$
HARDNESS: Polishing hardness similar to awaruite.
DENSITY: 8.23 (Calc.)

HABIT: As euhedral grains; cube and octahedron common forms. Grains usually less than $2\mu$ in diameter. Some grains zoned, consisting of a core of awaruite surrounded by wairauite.

COLOR-LUSTER: Steel gray. Opaque. Metallic.

Highly magnetic.

MODE OF OCCURRENCE: Occurs with awaruite in the Red Hills serpentinites of the Wairau Valley, South Island, New Zealand. Also from Muskox intrusion, Northwest Territory, Canada.

BEST REF. IN ENGLISH: Challis, G. A., and Long, J. V. P., *Min. Mag.*, **33**: 942–948 (1964).

## WAKABAYASHILITE
$(As, Sb)_2 S_3$ with As:Sb approx. 10:1

CRYSTAL SYSTEM: Monoclinic
SPACE GROUP: $P2_1$ or $P2_1/m$
Z: 6
LATTICE CONSTANTS:
$a = 25.17$
$b = 6.48$
$c = 25.24$
$\beta = 120.0°$
3 STRONGEST DIFFRACTION LINES:
6.33 (100)
4.78 ( 50)
3.50 ( 50)
OPTICAL CONSTANTS: Optically similar to orpiment, weakly anisotropic, strongly birefringent, $n$'s close to those of orpiment. Internal reflections golden yellow.
HARDNESS: 1½
DENSITY: 3.96 (Meas.)
4.06 (Calc.)
CLEAVAGE: {100} perfect
{010} perfect
{10$\bar{1}$} perfect
Flexible.

HABIT: Crystals prismatic or fibrous, elongated along $b$, up to 2 cm long.

COLOR-LUSTER: Golden to lemon yellow. Silky (for aggregates of fibers) to resinous. Streak orange-yellow, turns orange on grinding.

MODE OF OCCURRENCE: The mineral occurs associated with realgar, orpiment, and calcite at the White Caps mine, Manhattan, Nevada. Also found with realgar, orpiment, pyrite, and stibnite in druses of quartz at the Nishinomaki mine, Gumma Prefecture, Japan.

BEST REF. IN ENGLISH: Kato, Akira; Sakurai, Kin Ichi; and Ohsumi, Kazumasa, *Am. Min.*, **57**: 1311–1312 (1972).

## WAKEFIELDITE
$YVO_4$
Xenotime Group

CRYSTAL SYSTEM: Tetragonal
CLASS: 4/m 2/m 2/m
SPACE GROUP: $I4_1/amd$
Z: 4
LATTICE CONSTANTS:
$a = 7.105$
$c = 6.29$
3 STRONGEST DIFFRACTION LINES:
3.56 (100)
1.824 ( 75)
2.66 ( 50)
OPTICAL CONSTANTS:
(synthetic)
$\omega = 2.00$
$\epsilon = 2.14$
(+)
HARDNESS: ~5
DENSITY: 4.21 (Meas. Synthetic)
4.25 (Calc.)
CLEAVAGE: Not determined.
HABIT: Massive; pulverulent.
COLOR-LUSTER: Pale tan to yellow.

MODE OF OCCURRENCE: Occurs very sparingly associated with quartz, allanite, biotite, chlorite, bismutite, and unidentified rare earth minerals in Precambrian pegmatite at the Evans-Lou feldspar mine near Wakefield, Quebec, Canada.

BEST REF. IN ENGLISH: Miles, Norman M., Hogarth, Donald D., and Russell, Douglas S., *Am. Min.*, **56**: 395–410 (1971).

# WALLISITE
PbTlCuAs$_2$S$_5$

CRYSTAL SYSTEM: Triclinic
CLASS: $\bar{1}$
SPACE GROUP: P$\bar{1}$
LATTICE CONSTANTS:
  $a = 7.98$     $\alpha = 106°05'$
  $b = 8.98$     $\beta = 114°27'$
  $c = 7.76$     $\gamma = 65°30'$
3 STRONGEST DIFFRACTION LINES:
  3.339 (100)
  2.834 ( 40)
  2.879 ( 30)
  HARDNESS: Not determined.
  DENSITY: 5.71 (Calc.)
  CLEAVAGE: $\{\bar{1}\bar{1}1\}$ good
  HABIT: Massive.
  COLOR-LUSTER: Lead gray. Opaque. Metallic.
  MODE OF OCCURRENCE: Occurs as an overgrowth on rathite-I at the Lengenbach quarry, Binnatal, Canton Wallis, Switzerland.

BEST REF. IN ENGLISH: Takeuchi, Y., Ohmasa, M., and Nowacki, Werner, *Am. Min.*, **54**: 1497 (1969). Nowacki, Werner, *Am. Min.*, **51**: 532 (1966).

# WALPURGITE
(BiO)$_4$UO$_2$(AsO$_4$)$_2$ · 3H$_2$O

CRYSTAL SYSTEM: Triclinic
CLASS: $\bar{1}$
SPACE GROUP: P$\bar{1}$
Z: 1
LATTICE CONSTANTS:
  $a = 7.13$     $\alpha = 101°40'$
  $b = 10.44$    $\beta = 110°49'$
  $c = 5.49$     $\gamma = 88°17'$
3 STRONGEST DIFFRACTION LINES:
  3.11 (100)
  3.25 ( 50)
  3.05 ( 50)
OPTICAL CONSTANTS:
  $\alpha = 1.871$–1.91
  $\beta = 1.975$–2.00
  $\gamma = 2.005$–2.06
  $(-)2V = 50°$–60°
  HARDNESS: 3½
  DENSITY: 5.95 (Meas.)
         6.69 (Calc.)
  CLEAVAGE: {010} perfect
Brittle.

HABIT: Crystals lath-like, up to 2 mm long, elongated along *c*-axis and flattened on {010}. As subparallel aggregates and radial groupings. Commonly twinned on {010}, the twinned crystals appearing like gypsum.
  COLOR-LUSTER: Colorless, wax yellow to straw yellow. Transparent to translucent. Adamantine to greasy. Not fluorescent.
  MODE OF OCCURRENCE: Occurs as a secondary mineral associated with torbernite, zeunerite, troegerite, uranospinite, uranospherite, erythrite, and black cobalt oxide in the oxidized zone of a vein carrying uraninite, native bismuth, and cobalt-nickel arsenides in the Walpurgis vein in the Weisser Hirsch mine at Neustädtel near Schneeberg, Saxony, Germany. Also found at Joachimsthal, Bohemia.

BEST REF. IN ENGLISH: Frondel, Clifford, USGS Bull. 1064, p. 239–242 (1958).

# WALSTROMITE
BaCa$_2$Si$_3$O$_9$

CRYSTAL SYSTEM: Triclinic
CLASS: $\bar{1}$ or 1
SPACE GROUP: P$\bar{1}$ or P1
Z: 2
LATTICE CONSTANTS:
  $a = 6.743$    $\alpha = 69°51'$
  $b = 9.607$    $\beta = 102°14'$
  $c = 6.687$    $\gamma = 97°6.5'$
3 STRONGEST DIFFRACTION LINES:
  2.99 (100)
  6.58 ( 20)
  2.70 ( 20)
OPTICAL CONSTANTS:
  $\alpha = 1.668$
  $\beta = 1.684$
  $\gamma = 1.685$
  $(-)2V = 30°$
  HARDNESS: ~3½
  DENSITY: 3.67 (Meas.)
  CLEAVAGE: {011} nearly perfect
           {010} nearly perfect
           {100} nearly perfect
  HABIT: Crystals nearly equant to short prismatic, anhedral to subhedral. Grain size ranges from a fraction of a millimeter to 15 mm in length.
  COLOR-LUSTER: White to colorless, subvitreous, pearly on cleavage plates. Fluoresces dull pink under short-wave ultraviolet light; bright pink under long-wave ultraviolet light.
  MODE OF OCCURRENCE: Occurs as individual grains disseminated in sanbornite-quartz rock in eastern Fresno County, California, associated with wollastonite, celsian, taramellite, pyrrhotite, pyrite, witherite, fresnoite, and an unidentified blue iron-barium silicate mineral.

BEST REF. IN ENGLISH: Alfors, John T., Stinson, Melvin C., Matthews, Robert A., and Pabst, Adolf, *Am. Min.*, **50**: 314–340 (1965).

# WALTHERITE = Walpurgite

# WARDITE
$NaAl_3(PO_4)_2(OH)_4 \cdot 2H_2O$

CRYSTAL SYSTEM: Tetragonal
CLASS: 422
SPACE GROUP: $P4_12_12$ and $P4_32_12$
Z: 4
LATTICE CONSTANTS:
　$a$ = 7.04
　$c$ = 18.88
3 STRONGEST DIFFRACTION LINES:
　4.77 (100)
　4.73 (100)
　3.09 ( 80)
OPTICAL CONSTANTS:
　$\omega$ = 1.586–1.594
　$\epsilon$ = 1.595–1.604
　　(+)
HARDNESS: 5
DENSITY: 2.81 (Meas.)
　　　　　2.81 (Calc.)
CLEAVAGE: {001} perfect
HABIT: Crystals pyramidal, symmetrically developed with the same forms at opposite ends of the $c$-axis. Also as crusts and granular aggregates, as subparallel aggregates of coarse fibers, and as radially fibrous and concentrically banded spherulites. Crystals often striated horizontally.
COLOR-LUSTER: Colorless, white, pale green to bluish green; transparent to translucent. Vitreous.
MODE OF OCCURRENCE: Occurs associated with millisite and crandallite in variscite nodules at Fairfield, Utah County, and in the variscite deposits at Amatrice Hill and Lucin, Utah; as crystals up to 10 mm in maximum dimension coating quartz in pegmatite at Beryl Mountain near West Andover, New Hampshire; in narrow veins in massive amblygonite in the Stewart mine, Pala, San Diego County, California; and as small, well-formed crystals associated with silky-white apatite, morinite, augelite, and montebrasite at the Hugo mine, Keystone, South Dakota. Also found as an alteration product of amblygonite at Montebras, Creuse, France; and in the pegmatite of the alto Patrimonio, at Piedras Lavradas, Paraíba, Brazil.
BEST REF. IN ENGLISH: Palache, et al., "Dana's System of Mineralogy," 7th Ed., v. II, p. 940–941, New York, Wiley, 1951.　Hurlbut, Cornelius S. Jr., *Am. Min.*, **37**: 849–852 (1952).

# WARDSMITHITE
$Ca_5MgB_{24}O_{42} \cdot 30H_2O$

CRYSTAL SYSTEM: Hexagonal or pseudohexagonal
3 STRONGEST DIFFRACTION LINES:
　13.5　(100)
　12.3　( 62)
　6.12 ( 55)
OPTICAL CONSTANTS:
　$\omega$ = 1.490
　$\epsilon$ = 1.476
　　(–)
HARDNESS: 2½

DENSITY: 1.88 (Meas.)
CLEAVAGE: {0001?}
HABIT: Platy; occurs as aggregates of subparallel plates to 75 $\mu$m in diameter, uncommonly as single crystals to 15 $\mu$m in diameter.
COLOR-LUSTER: Colorless.
MODE OF OCCURRENCE: Found at two localities in the Death Valley region, Inyo County, California, on weathered veins of colemanite or priceite in the Furnace Creek Formation (Pliocene). It occurs intimately associated with gowerite, ulexite, and colemanite.
BEST REF. IN ENGLISH: Erd, Richard C., McAllister, James F., and Vlisidis, Angelina C., *Am. Min.*, **55**: 349–357 (1970).

**WARTHAITE** = A mixture of cosalite and galena

# WARWICKITE
$(Mg, Fe)_3TiB_2O_8$

CRYSTAL SYSTEM: Orthorhombic
CLASS: 2/m 2/m 2/m
SPACE GROUP: Pnam
Z: 2
LATTICE CONSTANTS:
　$a$ = 9.20
　$b$ = 9.45
　$c$ = 3.01
3 STRONGEST DIFFRACTION LINES:
　2.58 (100)
　1.60 ( 50)
　1.50 ( 50)
OPTICAL CONSTANTS:
　$\alpha$ = 1.806
　$\beta$ = 1.809
　$\gamma$ = 1.830
　(+)2V = small, variable
HARDNESS: 3½–4
DENSITY: 3.35 (Meas.)
　　　　　3.432 (Calc.)
CLEAVAGE: {100} perfect
Fracture uneven. Brittle.
HABIT: As embedded slender prismatic crystals with rounded terminations.
COLOR-LUSTER: Dark brown to dull black. Subtranslucent; transparent in minute grains. Subvitreous, pearly, to dull. Cleavage surface submetallic. Streak bluish black.
MODE OF OCCURRENCE: Occurs in crystalline limestone associated with spinel, ilmenite, magnetite, diopside, and chondrodite near Warwick, Orange County, New York.
BEST REF. IN ENGLISH: Palache, et al., "Dana's System of Mineralogy," 7th Ed., v. II, p. 326–327, New York, Wiley, 1951.

**WATER** = Liquid phase of $H_2O$ existing between 0° and 100°C (see Ice).

## WATTEVILLEITE (Inadequately described mineral)

$Na_2Ca(SO_4)_2 \cdot 4H_2O$ (?)

CRYSTAL SYSTEM: Orthorhombic or monoclinic
OPTICAL CONSTANTS:
$\alpha = 1.435$
$\beta = 1.455$
$\gamma = 1.459$
$(-)2V \simeq 48°$
HARDNESS: Not determined.
DENSITY: 1.81 (Meas.)
CLEAVAGE: Not determined.
HABIT: Crystals minute, acicular or fibrous, in aggregates.
COLOR-LUSTER: White; silky.

Soluble in water. Taste sweetish, astringent.
MODE OF OCCURRENCE: Occurs in pyrite-rich lignite in the Einigkeit mine, near Bischofsheim, Germany.
BEST REF. IN ENGLISH: Palache, et al., "Dana's System of Mineralogy," 7th Ed., v. II, p. 452, New York, Wiley, 1951.

## WAVELLITE

$Al_3(OH)_3(PO_4)_2 \cdot 5H_2O$

CRYSTAL SYSTEM: Orthorhombic
CLASS: 2/m 2/m 2/m
SPACE GROUP: Pcmn
Z: 4
LATTICE CONSTANTS:
$a = 9.60$ kX
$b = 17.31$
$c = 6.98$
3 STRONGEST DIFFRACTION LINES:
3.437 (100)
3.263 (100)
8.58 ( 80)
OPTICAL CONSTANTS:
$\alpha = 1.520-1.535$
$\beta = 1.526-1.543$
$\gamma = 1.545-1.561$
$(+)2V \sim 71°$
HARDNESS: 3¼-4
DENSITY: 2.36 (Meas.)
2.37 (Calc.)
CLEAVAGE: {110} perfect
{101} good
{010} distinct
Fracture subconchoidal to uneven. Brittle.
HABIT: Crystals minute, rare, stout to long prismatic. Usually as acicular radiating aggregates, often distinctly spherical; also as crusts, stalactitic, or rarely opaline.
COLOR-LUSTER: White to greenish white to green, also yellowish green to yellow and yellowish brown, brown to brownish black, rarely bluish or colorless. Transparent to translucent. Vitreous to resinous or pearly. Streak white.
MODE OF OCCURRENCE: Occurs as a secondary mineral in hydrothermal veins, in certain aluminous metamorphic rocks, and in phosphate-rock and limonite deposits. Found widespread as fine specimens in Garland, Hot Spring, and Montgomery counties, Arkansas; at East Whiteland, Chester County, and elsewhere in Pennsylvania; in St. Clair County, Alabama; at Dunellen, Marion County, Florida; in the King Turquois mine, Saguache County, and in the Cripple Creek district, Teller County, Colorado; and near Slate Mountain, El Dorado County, California. It also is found in Bolivia, England, Ireland, France, Portugal, Germany, Czechoslovakia, Bulgaria, Roumania, and Tasmania.
BEST REF. IN ENGLISH: Palache, et al., "Dana's System of Mineralogy," 7th Ed., v. II, p. 962-964, New York, Wiley, 1951.

## WAYLANDITE

$(Bi, Ca)Al_3(PO_4, SiO_4)_2(OH)_6$

CRYSTAL SYSTEM: Hexagonal
CLASS: $\bar{3}$ 2/m
SPACE GROUP: $R\bar{3}m$
Z: 3
LATTICE CONSTANTS:
$a = 6.9649$
$c = 16.256$
3 STRONGEST DIFFRACTION LINES:
2.93 (100)
5.66 ( 71)
3.483 ( 50)
OPTICAL CONSTANTS:
Uniaxial (-)
HARDNESS: 4-5
DENSITY: 3.86 (Calc.)
CLEAVAGE: Not determined. Fracture uneven.
HABIT: Massive; fine-grained, compact.
COLOR-LUSTER: White; vitreous to dull.
MODE OF OCCURRENCE: Occurs as veinlets and marginal crusts as a replacement of bismutotantalite in the lithium pegmatite at Wampewo Hill, Busiro County of Buganda, Uganda.
BEST REF. IN ENGLISH: Von Knorring, Oleg, and Mrose, Mary E., Geol. Soc. Am. Special Paper 73: 256 (1962).

## WEBERITE

$Na_2MgAlF_7$

CRYSTAL SYSTEM: Orthorhombic
CLASS: 2/m 2/m 2/m, 2 2 2, or mm2
SPACE GROUP: Immm, I222, $I2_12_12_1$, or I2mm
Z: 4
LATTICE CONSTANTS:
$a = 7.30$
$b = 9.99$
$c = 7.06$
3 STRONGEST DIFFRACTION LINES:
1.779 (100)
2.95 ( 90)
2.89 ( 90)
OPTICAL CONSTANTS:
$\alpha = 1.346$
$\beta = 1.348$ (Na)
$\gamma = 1.350$
$(+)2V \approx 83°$

HARDNESS: 3½
DENSITY: 2.96 (Meas.)
2.97 (Calc.)
CLEAVAGE: {110} poor
Fracture uneven.
HABIT: Irregular grains and fine-grained masses.
COLOR-LUSTER: Grayish white, translucent; vitreous. Streak white.
MODE OF OCCURRENCE: Occurs associated with rutile, thomsenolite, pachnolite, fluorite, and prosopite at St. Peters Dome, El Paso County, Colorado; and found with cryolite, chiolite, fluorite, "ivigtite," topaz, and pyrite at the cryolite deposit at Ivigtut, Greenland.
BEST REF. IN ENGLISH: Palache, et al., "Dana's System of Mineralogy," 7th Ed., v. II, p. 127–128, New York, Wiley, 1951.

# WEDDELLITE
$CaC_2O_4 \cdot 2-3H_2O$

CRYSTAL SYSTEM: Tetragonal
CLASS: 4/m
SPACE GROUP: I4/m
Z: 8
LATTICE CONSTANTS:
$a$ = 12.40 kX
$c$ = 7.37
3 STRONGEST DIFFRACTION LINES:
6.18 (100)
2.775 ( 65)
4.42 ( 30)
OPTICAL CONSTANTS:
$\omega$ = 1.523
$\epsilon$ = 1.544
(+)
HARDNESS: ~4
DENSITY: 1.94 (Meas.)
1.940 (Calc.)
CLEAVAGE: None. Fracture subconchoidal.
HABIT: As minute isolated pyramidal crystals; occasionally in small aggregates.
COLOR-LUSTER: Colorless, white; also yellowish brown to brown due to organic impurities. Transparent.
MODE OF OCCURRENCE: Occurs in oceanic bottom samples from the Weddell Sea, Antarctica.
BEST REF. IN ENGLISH: Palache, et al., "Dana's System of Mineralogy," 7th Ed., v. II, p. 1101–1102, New York, Wiley, 1951.

# WEEKSITE (Gastunite)
$K_2(UO_2)_2(Si_2O_5)_3 \cdot 4H_2O$

CRYSTAL SYSTEM: Orthorhombic
CLASS: 2/m 2/m 2/m
SPACE GROUP: Pnnb
Z: 16
LATTICE CONSTANTS:
$a$ = 14.26
$b$ = 35.88
$c$ = 14.20

3 STRONGEST DIFFRACTION LINES:
7.11 (100)
5.57 ( 90)
8.98 ( 80)
OPTICAL CONSTANTS:
$\alpha$ = 1.596
$\beta$ = 1.603
$\gamma$ = 1.606
$(-)2V \sim 60°$
HARDNESS: Soft—crushes more easily than gypsum.
DENSITY: ~4.1 (Meas.)
4.02 (Calc.)
CLEAVAGE: Not determined.
HABIT: As small spherulites of radiating crystals.
COLOR-LUSTER: Yellow; waxy to silky.
MODE OF OCCURRENCE: Occurs in opal veinlets in rhyolite and as replacements of pebbles in a tuffaceous conglomerate in the Thomas Range in central Juab County, Utah. Also identified from eight other localities in Pennsylvania, California, Wyoming, New Mexico, Mexico, Arizona, and Texas.
BEST REF. IN ENGLISH: Outerbridge, W. F., Staatz, M. H., Meyrowitz, R., and Pommer, A. M., *Am. Min.*, 45: 39–52 (1960).

# WEGSCHEIDERITE
$Na_5H_3(CO_3)_4$

CRYSTAL SYSTEM: Triclinic
CLASS: $\bar{1}$
SPACE GROUP: P$\bar{1}$
Z: 2
LATTICE CONSTANTS:
$a$ = 10.04    $\alpha$ = 91°55'
$b$ = 15.56    $\beta$ = 95°49'
$c$ = 3.466    $\gamma$ = 108°40'
3 STRONGEST DIFFRACTION LINES:
2.95 (100)
3.68 ( 60)
2.64 ( 60)
OPTICAL CONSTANTS:
$\alpha$ = 1.433
$\beta$ = 1.519
$\gamma$ = 1.528
$(-)2V = 32.4°$ (Na)
HARDNESS: 2½-3
DENSITY: 2.341 (Meas.)
2.334 (Calc.)
CLEAVAGE: Prismatic, distinct. Uneven to subconchoidal fracture.
HABIT: Crystals acicular to bladed; up to 5 cm long.
COLOR-LUSTER: Colorless; vitreous.
MODE OF OCCURRENCE: Found in drill cores in a lacustrine deposit of middle Eocene age, associated with trona, nahcolite, and halite, in Sweetwater County, Wyoming.
BEST REF. IN ENGLISH: Fahey, Joseph J., and Yorks, K. P., *Am. Min.*, 48: 400–403 (1963). Appelman, D. E., *Am. Min.*, 48: 404–406 (1963).

# WEHRLITE = Tellurobismuthite

## WEIBULLITE
$Bi_6Pb_4S_9Se_4$

CRYSTAL SYSTEM: Monoclinic
CLASS: 2/m
SPACE GROUP: P2/m
Z: 2
LATTICE CONSTANTS:
 $a = 18.07$
 $b = 4.05$
 $c = 17.57$
 $\beta = 94°29'$
3 STRONGEST DIFFRACTION LINES:
 3.82 (100)
 3.05 ( 90)
 2.04 ( 60)
 HARDNESS: 2-3
 DENSITY: 6.97; 7.145 (Meas.)
       6.96 (Calc.)
 CLEAVAGE: One good, parallel the elongation.
Flexible; inelastic. Very brittle.
 HABIT: Crystals prismatic, indistinct. Usually massive, fibrous to prismatic, or foliated.
 COLOR-LUSTER: Steel gray. Opaque. Metallic. Streak dark steel gray.
 MODE OF OCCURRENCE: Occurs mixed with laitakarite and bismuthinite at Fahlun, Sweden.
 BEST REF. IN ENGLISH: Karup-Møller, S., *Am. Min.*, **56**: 639 (1971).

## WEILERITE
Probably $BaAl_3(AsO_4)(SO_4)(OH)_6$

CRYSTAL SYSTEM: Hexagonal
Z: 1 (rhomb.), 3 (hexag.)
LATTICE CONSTANTS:
 $a = 7.05$     $a_{rh} = 7.02$
 $c = 15.16$    $\alpha = 60°18'$
3 STRONGEST DIFFRACTION LINES:
 3.01 (100)
 5.80 ( 80)
 3.54 ( 70)
OPTICAL CONSTANTS:
 $n$ min. 1.688, max 1.698
Uniaxial, sign not determined.
 HARDNESS: Not determined.
 DENSITY: 3.75 (Calc.)
 CLEAVAGE: Not determined.
 HABIT: Idiomorphic crystals less than 10 $\mu$ in size, apparently cubes, but actually rhombohedrons. Crystals are zoned. As earthy crusts.
 COLOR-LUSTER: White.
 MODE OF OCCURRENCE: Occurs associated with adamite and mimetite on hornstone and barite at Weiler bei Lahr, Schwarzwald, Germany.
 BEST REF. IN ENGLISH: Walenta, Kurt, and Wimmenauer, Wolfhard, *Am. Min.*, **47**: 415 (1962). Walenta, Kurt, *Am. Min.*, **52**: 1588-1589 (1967).

## WEILITE
$CaHAsO_4$

CRYSTAL SYSTEM: Triclinic
CLASS: $\bar{1}$
SPACE GROUP: $P\bar{1}$
Z: 4
LATTICE CONSTANTS:
 $a = 7.11$     $\alpha = 94°19'$
 $b = 6.94$     $\beta = 101°35'$
 $c = 7.15$     $\gamma = 87°22'$
3 STRONGEST DIFFRACTION LINES:
 3.43 (100)
 3.05 (100)
 2.79 ( 70)
OPTICAL CONSTANTS:
 $\alpha' = 1.664$
 $\gamma' = 1.688$
 $(-)2V \simeq 82°$
 HARDNESS: Not determined.
 DENSITY: 3.48 (Meas.)
       3.45 (Calc.)
 CLEAVAGE: Not determined.
 HABIT: Powdery incrustations or pseudomorphs after pharmacolite or haidingerite.
 COLOR-LUSTER: White, porcelaneous; greasy to slightly pearly.
 MODE OF OCCURRENCE: Occurs as a secondary mineral in the oxidized zone of arsenic-rich veins at unspecified deposits in France and Germany.
 BEST REF. IN ENGLISH: Herpin, P., and Pierrot, R., *Am. Min.*, **49**: 816 (1964).

## WEINSCHENKITE (Churchite)
$(Y, Er)PO_4 \cdot 2H_2O$

CRYSTAL SYSTEM: Monoclinic
CLASS: 2/m or m
SPACE GROUP: A2/a or Aa
Z: 4
LATTICE CONSTANTS:
 $a = 5.61$
 $b = 15.14$
 $c = 6.19$
 $\beta = 115.3°$
3 STRONGEST DIFFRACTION LINES:
 4.21 (100)
 7.50 ( 90)
 3.02 ( 90)
OPTICAL CONSTANTS:
 $\alpha = 1.600-1.605$
 $\beta = 1.608-1.612$
 $\gamma = 1.645$
 $(+)2V$ = medium small
 HARDNESS: 3
 DENSITY: 3.265 (Meas.)
 CLEAVAGE: $\{\bar{1}01\}$ perfect
Fracture conchoidal.
 HABIT: Crystals minute, lath-like, elongated along $c$-axis and flattened on $\{010\}$. As fan-shaped aggregates or rosettes, as crusts, and as spherulites with radial-fibrous structure.
 COLOR-LUSTER: Colorless, white, smoke gray. Transparent. Vitreous; pearly on cleavage. Also silky.

MODE OF OCCURRENCE: Occurs in a manganese-rich limonite deposit at the Kelly Bank mine, Rockbridge County, Virginia. It also is found associated with wavellite, cacoxenite, and other phosphate minerals in a limonite deposit at Auerbach, Oberpfalz, Bavaria, Germany; and at a copper deposit in Cornwall, England.

BEST REF. IN ENGLISH: Palache, et al., "Dana's System of Mineralogy," 7th Ed., v. II, p. 771–773, New York, Wiley, 1951.

## WEISSITE
$Cu_{2-x}Te$ (x = 0-0.33)

CRYSTAL SYSTEM: Hexagonal
CLASS: 6/m 2/m 2/m
SPACE GROUP: P6/mmm
Z: 54
LATTICE CONSTANTS:

| $(Cu_{2-x}Te)$ | $(Cu_2Te)$ |
|---|---|
| $a = 12.45$ | $a = 12.54$ |
| $c = 21.56$ | $c = 21.71$ |

3 STRONGEST DIFFRACTION LINES:
2.09 (100)
3.61 ( 70)
2.01 ( 70)
OPTICAL CONSTANTS: Opaque.
HARDNESS: 3
DENSITY: 6.44 (Calc. X-ray)
CLEAVAGE: None. Fracture uneven.
HABIT: Massive, compact.
COLOR-LUSTER: Dark bluish black, gradually tarnishing to black. Opaque. Metallic, shiny. Streak black.
MODE OF OCCURRENCE: Occurs chiefly in deposits of supergene origin, commonly associated with rickardite and other tellurium minerals. Found at the Good Hope and Mammoth Chimney mines at Vulcan, Gunnison County, Colorado; at Glava, Värmland, Sweden; in the Tein mine, Hokkaido, Japan; and at Kalgoorlie, Western Australia.
BEST REF. IN ENGLISH: Vlasov, K. A., "Mineralogy of the Rare Elements," v. II, p. 689–690, Israel Program for Scientific Translations, 1966.

## WELINITE
$Mn^{4+}Mn_3^{2+}O_3(SiO_4)$

CRYSTAL SYSTEM: Hexagonal
CLASS: 6
SPACE GROUP: Probably P6₃
Z: 2
LATTICE CONSTANTS:
$a = 8.155$
$c = 4.785$
3 STRONGEST DIFFRACTION LINES:
1.782 (100)
3.102 ( 80)
2.332 ( 80)
OPTICAL CONSTANTS:
$\omega = 1.864$
$\epsilon = 1.88$
(+)

HARDNESS: 4
DENSITY: 4.47 (Meas.)
4.41 (Calc.)
CLEAVAGE: {001} poor to distinct
Brittle.
HABIT: Crystals of indeterminate morphology; up to 2 cm in size.
COLOR-LUSTER: Deep red-brown to reddish black; resinous.
MODE OF OCCURRENCE: Occurs in small amounts embedded in calcite, barite, sarkinite, and adelite filling fissures in compact hausmannite ore at Långban, Sweden. It also occurs associated with hausmannite and dolomite at the Sjö mine, Grythyttan, Sweden.
BEST REF. IN ENGLISH: Moore, Paul B., *Am. Min.*, **53**: 1064 (1968). Moore, Paul B., *Arkiv. Min. Geol.*, **4**: 459–466 (1968).

## WELLSITE
$(Ba, Ca, K_2)(Al_2Si_3)O_{10} \cdot 3H_2O$
Zeolite Group

CRYSTAL SYSTEM: Monoclinic
CLASS: 2/m or 2
SPACE GROUP: P2₁/m or P2₁
Z: 2
LATTICE CONSTANTS:
$a \simeq 9.95$
$b \simeq 14.21$
$c \simeq 8.68$
$\beta = 126°33'$
3 STRONGEST DIFFRACTION LINES:
3.15 (100)
4.07 ( 80)
2.69 ( 80)
OPTICAL CONSTANTS:
$\beta = 1.50$
$\gamma = 1.503$
(+)2V = 39°
HARDNESS: 4-4½
DENSITY: 2.248 (Meas.)
CLEAVAGE: {010} good
{100} good
Fracture uneven. Brittle.
HABIT: Crystals uniformly penetration twins; twinning on {001}, {021}, {110}.
COLOR-LUSTER: Colorless to white. Transparent to translucent. Vitreous.
MODE OF OCCURRENCE: Occurs as small crystals on feldspar or corundum at the Buck Creek corundum mine, Clay County, North Carolina; at Vezna, Czechoslovakia; and in association with analcime, leonhardite, and calcite lining fissures in an igneous rock at the village of Kurtzy, near Simpferopol, Crimea, USSR.
BEST REF. IN ENGLISH: Deer, Howie, and Zussman, "Rock Forming Minerals," v. 4, p. 387–389, New York, Wiley, 1963. Winchell and Winchell, "Elements of Optical Mineralogy," 4th Ed., Pt. 2, p. 344, New York, Wiley, 1951.

# WELOGANITE
$Sr_5 Zr_2 C_9 H_8 O_{31}$

CRYSTAL SYSTEM: Hexagonal
CLASS: 32
SPACE GROUP: $P3_1$ or $P3_2$
Z: 2
LATTICE CONSTANTS:
  $a = 8.96$
  $c = 18.06$
3 STRONGEST DIFFRACTION LINES:
  2.809 (100)
  4.35 ( 90)
  2.590 ( 70)
OPTICAL CONSTANTS:
  $\alpha = 1.558$
  $\beta = 1.646$
  $\gamma = 1.648$
  $(-)2V \sim 15°$
  HARDNESS: 3½
  DENSITY: 3.20 (Meas.)
  CLEAVAGE: $\{0001\}$ perfect
Fracture conchoidal.

HABIT: Crystals roughly hexagonal in outline elongate parallel to $c$; prism faces heavily striated and grooved parallel to base due to oscillatory growth. Termination of crystals roughly pyramidal, often with flat pedion. Crystals range from 2 mm to more than 3 cm in length. Also massive.

COLOR-LUSTER: White, lemon yellow, to amber. Thin fragments colorless and transparent. Crystals frequently zoned from white to various shades of yellow on basal sections. Vitreous. Streak white.

MODE OF OCCURRENCE: Occurs in an alkalic sill which has intruded Ordovician limestone at St-Michel, Montreal Island, Quebec, Canada. It is found associated with calcite, quartz, dawsonite, dresserite, plagioclase, siderite, dolomite, strontianite, barite, celestite, fluorite, cryolite, zircon, and anatase.

BEST REF. IN ENGLISH: Sabina, Ann P., Jambor, J. L., and Plant, A. G., *Can. Min.*, **9** (Pt. 4): 468–477 (1968).

# WELSHITE
$Ca_2 Mg_4 Fe^{3+} Sb^{5+} O_2 [Si_4 Be_2 O_{18}]$ or
$Ca_2 Mg_4 Fe^{3+} Sb^{5+} O_2 [Si_3 Be_3 O_{16} (OH)_2]$

CRYSTAL SYSTEM: Triclinic, pseudo-monoclinic
CLASS: $\bar{1}$
SPACE GROUP: $C\bar{1}$
Z: 1
LATTICE CONSTANTS:
  $a = 9.68$
  $b = 14.77$
  $c = 5.14$
  $\beta = 101°30'(\alpha = \gamma \approx 90.0°)$
3 STRONGEST DIFFRACTION LINES:
  2.530 (100)
  7.324 ( 70)
  2.910 ( 70)

OPTICAL CONSTANTS:
  $N = 1.79$ (Calc.)
Biaxial (+) $2E \sim 45°$
  HARDNESS: $5^+$
  DENSITY: 3.77 (Meas.)
  CLEAVAGE: None. Fracture conchoidal.
  HABIT: Crystals thick prismatic, up to 3 mm in length; resemble monoclinic pyroxenes in development. Invariably polysynthetically twinned parallel to $\{010\}m$.

COLOR-LUSTER: Deep reddish-brown to reddish-black (orange-brown to deep tan in very thin fragments). Subadamantine. Streak pale brown.

MODE OF OCCURRENCE: Welshite occurs embedded in fine-grained hematite-dolomite lean ore at Långban, Sweden. Associated minerals include berzeliite, manganoan phlogopite, richterite, hausmannite, tilasite, roméite (var. weslienite), and very rare swedenborgite.

BEST REF. IN ENGLISH: Moore, Paul Brian, *Min. Rec.*, **1**: 161 (1970).

# WENKITE
$(Ba, Ca)_9 Al_9 Si_{12} O_{42} (SO_4)_2 (OH)_5$

CRYSTAL SYSTEM: Hexagonal
CLASS: Probably 6/m 2/m 2/m
SPACE GROUP: Probably P6/mmm
Z: 1
LATTICE CONSTANTS:
  $a = 13.528$
  $c = 7.471$
3 STRONGEST DIFFRACTION LINES:
  3.46 (100)
  2.69 ( 90)
  3.38 ( 80)
OPTICAL CONSTANTS:
  $\omega = 1.596$
  $\epsilon = 1.589$
  $(-)$
Occasionally anomalously biaxial $2V \leqslant 10°$.
  HARDNESS: 6
  DENSITY: 3.13 (Meas.)
       3.276 (Calc.)
  CLEAVAGE: Prismatic, very poor.
  HABIT: Crystals prismatic, up to 5 cm in length.
  COLOR-LUSTER: Light gray.

MODE OF OCCURRENCE: Occurs between barite layers and calc-silicate rock as product of strong metamorphism in the marbles of Candoglia, northern Italy.

BEST REF. IN ENGLISH: Papageorgakis, J., *Am. Min.*, **48**: 213 (1963).

# WENZELITE = Hureaulite

# WERMLANDITE
$Ca_2 Mg_{14} Al_4 (CO_3)(OH)_{42} \cdot 3OH_2 O$

CRYSTAL SYSTEM: Hexagonal
CLASS: 3m
SPACE GROUP: P3c1

Z: 2
LATTICE CONSTANTS:
$a = 9.260$
$c = 22.52$
3 STRONGEST DIFFRACTION LINES:
7.98 (100)
11.16 ( 70)
4.63 ( 50)
OPTICAL CONSTANTS:
$\omega = 1.493$
$\epsilon = 1.482$
HARDNESS: 1½
DENSITY: 1.932 (Meas.)
CLEAVAGE: {0001} perfect
HABIT: Crystals thin hexagonal plates.
COLOR-LUSTER: Greenish gray.
MODE OF OCCURRENCE: The mineral occurs in association with calcite and magnetite at Långban, Sweden.
BEST REF. IN ENGLISH: Moore, Paul B., *Lithos*, 4: 213–217 (1971).


## WERNERITE = Scapolite


## WESTERVELDITE
(Fe, Ni, Co) As

CRYSTAL SYSTEM: Orthorhombic
CLASS: 2/m 2/m 2/m
SPACE GROUP: Pmcn
Z: 4
LATTICE CONSTANTS:
$a = 3.457$
$b = 5.971$
$c = 5.332$
3 STRONGEST DIFFRACTION LINES:
2.604 (100)
1.988 ( 50)
2.083 ( 25)
HARDNESS: Polishing hardness greater than maucherite or niccolite.
DENSITY: 8.13 (Calc.)
CLEAVAGE: Not determined.
HABIT: Massive; as minute irregularly shaped blebs and stringers disseminated in maucherite.
COLOR-LUSTER: Brownish white to gray against maucherite in air. Opaque. Metallic.
MODE OF OCCURRENCE: Occurs in chromite-niccolite ore at the LaGallega mine, about 3 km east of Ojén, Málaga, Spain.
BEST REF. IN ENGLISH: Oen, I. S., Burke, E. A. J., Kieft, C., and Westerhof, A. B., *Am. Min.*, 57: 354–363 (1972).


## WESTGRENITE
(Bi, Ca) (Ta, Nb)$_2$ O$_6$ OH

CRYSTAL SYSTEM: Cubic
Z: 8

LATTICE CONSTANT:
$a = 10.485$
3 STRONGEST DIFFRACTION LINES:
3.03 (100)
6.06 ( 71)
3.16 ( 50)
OPTICAL CONSTANT:
$N > 2.00$
HARDNESS: 5
DENSITY: 6.5 (Meas.)
6.83 (Calc.)
CLEAVAGE: Fracture uneven.
HABIT: Massive; as veinlets, formed by replacement of bismutotantalite.
COLOR-LUSTER: Yellow, pink, brown; dull resinous.
MODE OF OCCURRENCE: Occurs as a late hydrothermal crystallization product in a lithium pegmatite at Wampewo Hill, Busiro County of Buganda, Uganda.
BEST REF. IN ENGLISH: Von Knorring, Oleg, and Mrose, Mary E., Geol. Soc. Am. Special Paper 73: 256 (1962).


## WHARTONITE = Nickeliferous pyrite


## WHERRYITE
Pb$_4$ Cu(SO$_4$)$_2$ CO$_3$ (Cl, OH)$_2$ O

CRYSTAL SYSTEM: Monoclinic
CLASS: 2, m, or 2/m
SPACE GROUP: C2, Cm, or C2/m
Z: 4
LATTICE CONSTANTS:
$a = 20.82$
$b = 5.79$
$c = 9.17$
$\beta = 91°17'$
3 STRONGEST DIFFRACTION LINES:
3.16 (100)
2.75 ( 50)
3.06 ( 45)
OPTICAL CONSTANTS:
$\alpha = 1.942$
$\beta = 2.010$ (Na)
$\gamma = 2.024$
$(-)2V = 50°$ (Calc.)
HARDNESS: Not determined.
DENSITY: 6.45 (Meas.)
7.22 (Calc.)
HABIT: Massive, fine granular; also fibrous through acicular to aggregates of relatively equidimensional crystals up to 0.4 mm in size.
COLOR-LUSTER: Light green, pale yellow, bright yellowish green.
MODE OF OCCURRENCE: Found in a vug just above the 760 foot level at the Mammoth mine, Tiger, Arizona, associated with chrysocolla, diaboléite, greenish paralaurionite, and leadhillite.
BEST REF. IN ENGLISH: Fahey, Joseph J., Daggett, E. B., and Gordon, Samuel G., *Am. Min.*, 35: 93–98 (1950). McLean, W. John, *Am. Min.*, 55: 505–508 (1970).

# WHEWELLITE
$CaC_2O_4 \cdot H_2O$

CRYSTAL SYSTEM: Monoclinic
CLASS: 2/m
SPACE GROUP: $P2_1/n$
Z: 8
LATTICE CONSTANTS:
  $a = 6.276$
  $b = 14.561$
  $c = 10.012$
  $\beta = 107°05'$
3 STRONGEST DIFFRACTION LINES:
  5.93  (100)
  3.65  (100)
  2.968 ( 80)
OPTICAL CONSTANTS:
     $\alpha = 1.489$
     $\beta = 1.553$
     $\gamma = 1.649$
  (+)2V ~ 80°
HARDNESS: 2½–3
DENSITY: 2.21–2.23 (Meas.)
         2.22 (Calc.)
CLEAVAGE: $\{\bar{1}01\}$ very good
           $\{001\}$ distinct
           $\{010\}$ distinct
           $\{110\}$ indistinct
Fracture conchoidal. Brittle.
  HABIT: As coarsely crystalline masses up to 8.0 cm across. Crystals equant or short prismatic [001], faces often irregularly developed. Twins very common, heart-shaped or prismatic and of pseudo-orthorhombic appearance.
  COLOR-LUSTER: Colorless, white, yellowish or brownish; transparent to translucent; vitreous to pearly.
  MODE OF OCCURRENCE: Occurs as crystals up to several inches in size in a coal seam at Burgk near Dresden, Saxony, Germany; also found in Germany at Zwickau, at Freiberg, and in septaria from near Hoheneggelsen. Found with lignite or coal at various localities in Czechoslovakia; in veins at the Sainte-Sylvestre mine, Urbeis, Alsace, France; from veins at Recsk and Kapnik in Hungary; and at several localities in the USSR. In the United States found in septarian concretions near Havre, Montana, and in the bed of the Huron River between Milan and Monroeville, Erie County, Ohio; also found sparingly in a fault opening in San Juan County, Utah, and as excellent crystals implanted on yellow calcite in septarian concretions along the Cheyenne River, Meade County, South Dakota.
  BEST REF. IN ENGLISH: Leavens, Peter B., *Am. Min.*, **53**: 455–463 (1968).

# WHITLOCKITE
$(Ca, Mg)_3(PO_4)_2$

CRYSTAL SYSTEM: Hexagonal
CLASS: $\bar{3}2/m$
SPACE GROUP: $R\bar{3}c$
Z: 21 (hexag.), 7 (rhomb.)

LATTICE CONSTANTS:
  (hexag.)        (rhomb.)
  $a = 10.32$ kX   $a_{rh} = 13.67$
  $c = 36.9$        $\alpha = 44°21'$
3 STRONGEST DIFFRACTION LINES:
  2.837 (100)
  2.572 ( 80)
  1.701 ( 80)
OPTICAL CONSTANTS:
  $\omega = 1.629$
  $\epsilon = 1.626$
    (–)
HARDNESS: 5
DENSITY: 3.12 (Meas.)
         3.123 (Calc.)
CLEAVAGE: None.
Fracture subconchoidal to uneven. Brittle.
  HABIT: Crystals rhombohedral, rarely tabular. Also coarse granular to earthy.
  COLOR-LUSTER: Colorless, white, gray, yellowish, pinkish. Transparent to translucent. Vitreous to resinous.
  MODE OF OCCURRENCE: Occurs as a secondary mineral in granite pegmatites. Found in Custer County, South Dakota, as fine crystals associated with quartz, carbonate-apatite, and a wide variety of secondary phosphate minerals at the Tip Top mine, and sparingly at the Bull Moose mine; also at the Palermo mine, North Groton, New Hampshire, with siderite, quartz, triphylite, and secondary phosphate minerals. It also occurs in phosphate-rock deposits on several islands in the West Indies; as a cave deposit at Sebdou, Oran, Algeria; as a common accessory mineral in meteorites; and has been identified in lunar rocks.
  BEST REF. IN ENGLISH: Palache, et al., "Dana's System of Mineralogy," 7th Ed., v. II, p. 684–686, New York, Wiley, 1951.

# WICKENBURGITE
$CaPb_3Al_2Si_{10}O_{24}(OH)_6$

CRYSTAL SYSTEM: Hexagonal
CLASS: 6/m 2/m 2/m
SPACE GROUP: $P6_3/mmc$
Z: 2
LATTICE CONSTANTS:
  $a = 8.53$
  $c = 20.16$
3 STRONGEST DIFFRACTION LINES:
  10.1  (100)
   3.26 ( 80)
   3.93 ( 60)
OPTICAL CONSTANTS:
  $\omega = 1.6918$
  $\epsilon = 1.6480$  (Na)
    (–)
HARDNESS: 5
DENSITY: 3.85 (Meas.)
         3.88 (Calc.)
CLEAVAGE: $\{0001\}$ indistinct
  HABIT: Crystals tabular, from 0.2 to 1.5 mm in size; dominant forms $\{0001\}$ and $\{10\bar{1}1\}$.

COLOR-LUSTER: White, colorless; rarely pink. Vitreous. Fluoresces dull orange under short-wave ultraviolet light.

MODE OF OCCURRENCE: Found in abundance associated with phoenicochroite, mimetite, cerussite, and willemite as an oxide zone mineral at several prospects near Wickenburg, Maricopa County, Arizona.

BEST REF. IN ENGLISH: Williams, Sidney A., *Am. Min.*, **53**: 1433–1438 (1968).

# WICKMANITE
$(Mn, Ca)Sn(OH)_6$

CRYSTAL SYSTEM: Cubic
CLASS: $4/m\,\bar{3}\,2/m$
SPACE GROUP: Pn3m
Z: 4
LATTICE CONSTANTS:
  $a = 7.873$
3 STRONGEST DIFFRACTION LINES:
  3.931 (100)
  1.762 ( 70)
  2.778 ( 60)
OPTICAL CONSTANT:
  $N = 1.705$ (White light)
  DENSITY: 3.89 (Meas.)
       3.82 (Calc.)
HABIT: Crystals octahedral.
COLOR-LUSTER: Brownish yellow.
MODE OF OCCURRENCE: Occurs as a late-stage, low-temperature mineral in three stopes in the western part of the workings at Långban, Sweden.

BEST REF. IN ENGLISH: Moore, Paul B., and Smith, J. V., *Am. Min.*, **53**: 1063 (1968).

# WIGHTMANITE
$Mg_9B_2O_{12} \cdot 8H_2O$

CRYSTAL SYSTEM: Triclinic
CLASS: $\bar{1}$
SPACE GROUP: P$\bar{1}$
Z: 1
LATTICE CONSTANTS:
  $a = 11.73$   $\alpha = 96°09'$
  $b = 11.44$   $\beta = 97°45.5'$
  $c = 3.089$   $\gamma = 105°52.5'$
3 STRONGEST DIFFRACTION LINES:
  10.70 (100)
  9.07 (100)
  3.03 ( 30)
OPTICAL CONSTANTS:
  $\alpha = 1.585$
  $\beta = 1.603$
  $\gamma = 1.604$
  $(-)2V = 33°$
HARDNESS: 5.5
DENSITY: 2.60 (Meas.)
     2.58 (Calc.)
CLEAVAGE: {010} perfect
      {100} fair

HABIT: Crystals prismatic, pseudohexagonal; single or in radiating clusters.

COLOR-LUSTER: Colorless with tinge of green. Vitreous.

MODE OF OCCURRENCE: Occurs in a matrix of coarsely crystalline dolomite-calcite rock at the Commercial Quarry, Crestmore, California, associated with fluoborite and ludwigite.

BEST REF. IN ENGLISH: Murdoch, Joseph, *Am. Min.*, **47**: 718–722 (1962).

# WIIKITE = Euxenite or obruchevite

# WILKEÏTE
$Ca_5(SiO_4, PO_4, SO_4)_3(O, OH, F)$
Apatite Group

CRYSTAL SYSTEM: Hexagonal
CLASS: $6/m$
SPACE GROUP: P$6_3$/m
Z: 2
LATTICE CONSTANTS:
  $a = 9.48$ kX
  $c = 6.91$
3 STRONGEST DIFFRACTION LINES:
  2.80 (100)
  2.70 ( 90)
  2.24 ( 80)
OPTICAL CONSTANTS:
  $\omega = 1.650$–$1.640$
  $\epsilon = 1.646$–$1.636$
     $(-)$
HARDNESS: $\sim 5$
DENSITY: 3.12–3.23 (Meas.)
CLEAVAGE: {0001} imperfect
Brittle.
HABIT: As indistinct rounded crystals; also massive granular.

COLOR-LUSTER: Pale pinkish, yellowish. Translucent. Vitreous to resinous.

MODE OF OCCURRENCE: Occurs in blue calcite with idocrase, diopside, garnet, and tobermorite, in contact metamorphosed marble at Crestmore, Riverside County, California. It also is found at Kyshtym, Urals, USSR, and probably in the sanidinites of the Laacher See, Rhineland, Germany.

BEST REF. IN ENGLISH: Palache, et al., "Dana's System of Mineralogy," 7th Ed., v. II, p. 905, New York, Wiley, 1951.

# WILKMANITE
$Ni_3Se_4$

CRYSTAL SYSTEM: Monoclinic
CLASS: $2/m$
SPACE GROUP: I2/m
Z: 2
LATTICE CONSTANTS:
  $a = 6.22$
  $b = 3.63$
  $c = 10.52$
  $\beta = 90.53°$

3 STRONGEST DIFFRACTION LINES:
2.70  (100)
2.02  (100)
1.800 (100)
HARDNESS: ~ 2½
DENSITY: 6.96 (Calc.)
CLEAVAGE: {001} good
HABIT: Massive.
COLOR-LUSTER: Pale grayish yellow, with high reflectivity.
MODE OF OCCURRENCE: Occurs as a primary mineral and as an alteration product of sederholmite in sills of albite diabase in a schist formation, associated with low-grade uranium mineralization at Kuusamo, northeast Finland.
BEST REF. IN ENGLISH: Vuorelainen, Y., Huhma, A., and Häkli, A., *Am. Min.*, **50**: 519 (1965).


# WILLEMITE
$Zn_2 SiO_4$

CRYSTAL SYSTEM: Hexagonal
CLASS: $\bar{3}$
SPACE GROUP: $R\bar{3}$
Z: 6 (rhomb.), 18 (hexag.)
LATTICE CONSTANTS:
$a = 13.94$        $a_{rh} = 8.63$
$c = 9.34$          $\alpha = 107°45'$
3 STRONGEST DIFFRACTION LINES:
2.634 (100)
2.834 ( 95)
3.486 ( 80)
OPTICAL CONSTANTS:
$\omega = 1.691$
$\epsilon = 1.719$
   (+)
HARDNESS: 5½
DENSITY: 3.89-4.19 (Meas.)
              4.250 (Calc.)
CLEAVAGE: {0001} poor
                 {11$\bar{2}$0} poor
Fracture conchoidal to uneven. Brittle.
HABIT: Crystals prismatic, hexagonal, short and stout to long and slender; often terminated by basal pinacoid and several rhombohedrons. Also massive, compact or fibrous; and as disseminated grains.
COLOR-LUSTER: Colorless, white, various shades of green, yellow, red, brown, gray. Transparent to translucent. Vitreous to resinous. Streak uncolored. Commonly fluoresces intense yellowish green under ultraviolet light; often strongly phosphorescent.
MODE OF OCCURRENCE: Occurs as an ore mineral and as superb crystals and fluorescent specimens at Franklin and Sterling Hill, Sussex County, New Jersey. Also found as a secondary mineral at the Cerro Gordo and Ygnacio mines, Inyo County, California; at the Talisman mine, Beaver County, Utah; as fine microcrystals at the Mammoth mine, Pinal County, and elsewhere in Arizona; and at localities in New Mexico and Colorado. Other notable occurrences are found in Greenland, Belgium, Algeria, Zaire, Zambia, and at Tsumeb, South West Africa.
BEST REF. IN ENGLISH: Winchell and Winchell, "Elements of Optical Mineralogy," 4th Ed., Pt. 2, p. 497, New York, Wiley, 1951.


# WILLEMSEITE (A nickel-rich talc)
$(Ni, Mg)_3 Si_4 O_{10} (OH)_2$

CRYSTAL SYSTEM: Monoclinic
CLASS: 2/m or m
SPACE GROUP: C2/c or Cc
Z: 4
LATTICE CONSTANTS:
$a = 5.316$ (or 5.136)
$b = 9.149$
$c = 18.994$
$\beta = 99.96°$
3 STRONGEST DIFFRACTION LINES:
9.40  (100)
3.12  ( 28)
2.503 ( 23)
OPTICAL CONSTANTS:
   $\alpha = 1.600$
   $\beta = 1.652$
   $\gamma = 1.655$ (Calc.)
$(-)2V \simeq 27°$
HARDNESS: 2
DENSITY: 3.31 (Meas.)
              3.348 (Calc.)
CLEAVAGE: {001} perfect
HABIT: Massive; fine-grained.
COLOR-LUSTER: Light green.
MODE OF OCCURRENCE: Occurs associated with ferroan trevorite, nimite, violarite, millerite, and secondary reevesite and goethite in a small tabular body of nickeliferous rocks at the contact between quartzite and ultramafic rocks, about 2 miles west of the Scotia talc mine, Barberton Mountain Land, Transvaal.
BEST REF. IN ENGLISH: de Waal, S. A., *Am. Min.*, **55**: 31-42 1970.   Heimstra, S. A., and deWaal, S. A., *Am. Min.*, **54**: 1740 (1969).


# WILLYAMITE
(Co, Ni)SbS with Co > Ni

CRYSTAL SYSTEM: Pseudocubic; symmetry is lower than orthorhombic.
LATTICE CONSTANTS:
$a = 5.918$ (pseudocubic cell)
$\alpha \cong \beta \cong \gamma \cong 90°$
OPTICAL CONSTANTS: Opaque.
HARDNESS: 5½
DENSITY: 6.76 (Meas.)
CLEAVAGE: {001} perfect
Fracture uneven. Brittle.
HABIT: Crystals pseudocubic, several millimeters to a side.
COLOR-LUSTER: Steel gray. Opaque. Metallic. Streak grayish black.
MODE OF OCCURRENCE: Occurs at the Consols lode, Broken Hill, Willyama Township, New South Wales, Australia.
BEST REF. IN ENGLISH: *Am. Min.*, **56**: 361 (1971).

**WILUITE** = Variety of idocrase. Anomalously biaxial (+).

**WINCHITE** = Variety of richterite

**WISERITE**
$Mn_4B_2O_5(OH, Cl)_4$

CRYSTAL SYSTEM: Tetragonal
Z: 4
LATTICE CONSTANTS:
  $a$ = 14.27 kX
  $c$ = 3.31
3 STRONGEST DIFFRACTION LINES:
  14.2 (100)
  2.53 ( 80)
  6.40 ( 60)
OPTICAL CONSTANTS:
  $\alpha$ = 1.700–1.717
  $\gamma$ = 1.753–1.76
Uniaxial to slightly biaxial (−)
HARDNESS: ~2½
DENSITY: 3.42
CLEAVAGE: {001} perfect
HABIT: As fibrous masses.
COLOR-LUSTER: White to brownish or reddish. Silky.
MODE OF OCCURRENCE: Occurs intimately intergrown with pyrochroite, sussexite, and an unidentified mineral (probably a manganese borate) at Gonzen, near Sargans, Switzerland.
BEST REF. IN ENGLISH: Eprecht, W. T., Schaller, W. T., and Vlisidis, A. C., *Am. Min.*, 45: 258 (1960).

**WITHAMITE** = Variety of piemontite

**WITHERITE**
$BaCO_3$
Aragonite Group

CRYSTAL SYSTEM: Orthorhombic
CLASS: 2/m 2/m 2/m
SPACE GROUP: Pmcn
Z: 4
LATTICE CONSTANTS:
  $a$ = 5.252 kX
  $b$ = 8.828
  $c$ = 6.544
3 STRONGEST DIFFRACTION LINES:
  3.72 (100)
  3.68 ( 53)
  2.15 ( 28)
OPTICAL CONSTANTS:
  $\alpha$ = 1.529
  $\beta$ = 1.676
  $\gamma$ = 1.677 (Na)
(−)2V = 16°
HARDNESS: 3–3½
DENSITY: 4.291 (Meas.)
  4.29 (Calc.)

CLEAVAGE: {010} distinct
  {110} imperfect
Fracture uneven.
HABIT: Crystals always twinned on {110}, forming pseudohexagonal dipyramids; also tabular or short prismatic with convex base. Faces commonly striated horizontally and are irregular or rough. Also massive; granular, coarse fibrous, or columnar.
COLOR-LUSTER: Colorless, white, gray; sometimes with pale yellow, greenish, or brown tinge. Transparent to translucent. Vitreous to resinous. Fluoresces and phosphoresces bluish white under short-wave ultraviolet light.
MODE OF OCCURRENCE: Occurs as a low-temperature mineral in hydrothermal vein deposits. It occurs in the United States as large crystals associated with fluorite and calcite at the Minerva mine at Rosiclare, Illinois, and is also found in Kentucky, Montana, Arizona, and California. Commercial deposits are found in northern England, and superb crystal groups occur near Hexham, Northumberland, and in Cumberland and Durham. Other deposits are found in Austria, Germany, Czechoslovakia, France, USSR, and Japan.
BEST REF. IN ENGLISH: Palache, et al., "Dana's System of Mineralogy," 7th Ed., v. II, p. 194–196, New York, Wiley, 1951.

**WITTICHENITE (Klaprothite)**
$Cu_3BiS_3$

CRYSTAL SYSTEM: Orthorhombic
CLASS: 222
SPACE GROUP: $P2_12_12_1$
Z: 4
LATTICE CONSTANTS:
  $a$ = 7.66
  $b$ = 10.31
  $c$ = 6.69
3 STRONGEST DIFFRACTION LINES:
  2.85 (100)
  3.08 ( 80)
  4.55 ( 40)
OPTICAL CONSTANTS: Opaque.
HARDNESS: 2–3
DENSITY: 6.19 (Meas.)
  6.19 (Calc.)
CLEAVAGE: {100} distinct.
Fracture uneven to conchoidal. Brittle.
HABIT: Crystals prismatic to thick tabular; also columnar or acicular. Massive.
COLOR-LUSTER: Steel gray to tin white, tarnishing pale lead gray, brass yellow, or iridescent. Opaque. Metallic. Streak black.
MODE OF OCCURRENCE: Occurs associated with bornite, chalcocite, and enargite at Butte, Montana; in the silver ores of the Silverton district, San Juan County, Colorado; at the Daniel, King David, Neugluck, and other mines near Wittichen, Baden, and at other localities in Germany; and at deposits in Peru and Japan.
BEST REF. IN ENGLISH: Nuffield, *Econ. Geol.*, 42: 147–160 (1947).

## WITTITE (Species status uncertain)
Pb$_5$Bi$_6$(S, Se)$_{14}$ (?)

CRYSTAL SYSTEM: Orthorhombic or monoclinic
  HARDNESS: 2-2½
  DENSITY: 7.12
  CLEAVAGE: One good.
  HABIT: Massive; resembles molybdenite.
  COLOR-LUSTER: Light lead gray. Opaque. Metallic. Streak black.
  MODE OF OCCURRENCE: Occurs in a cordierite-bearing amphibole rock associated with magnetite and quartz at Fahlun, Kopparberg, Sweden.
  BEST REF. IN ENGLISH: Palache, et al., "Dana's System of Mineralogy," 7th Ed., v. I, p. 451, New York, Wiley, 1944.

## WODANITE = Variety of biotite containing up to 12% TiO$_2$

## WODGINITE
(Ta, Nb, Sn, Mn, Fe)$_{16}$O$_{32}$

CRYSTAL SYSTEM: Monoclinic
CLASS: 2/m or m
SPACE GROUP: C2/c or Cc
Z: 1
LATTICE CONSTANTS:
  $a$ = 9.52
  $b$ = 11.47
  $c$ = 5.10
  $\beta$ = 91°18'
3 STRONGEST DIFFRACTION LINES:
  3.00 (100)
  3.67 ( 70)
  2.95 ( 70)
  HARDNESS: ~6
  DENSITY: 7.36 (Meas. Bernic Lake)
           7.81 (Calc. Bernic Lake)
           7.19 (Meas. Wodgina)
           7.69 (Calc. Wodgina)
  HABIT: Sphenoidal and irregular grains from less than 1 mm to about 10 mm in diameter; also as roughly tabular grains, some of which form partially radiating groups.
  COLOR-LUSTER: Reddish brown through dark brown to almost black; submetallic.
  MODE OF OCCURRENCE: Occurs in a matrix of granular albite, quartz, and muscovite at Wodgina, Western Australia; also in coarse, partially sericitized perthitic microcline and in fine-grained bluish white aplitic albite at Bernic Lake, Manitoba, Canada.
  BEST REF. IN ENGLISH: Nickel, E. H., Rowland, J. F., McAdam, R. C., Can. Min., 7: 390-402 (1963).

## WOEHLERITE (Wöhlerite)
NaCa$_2$(Zr, Nb)Si$_2$O$_8$(O, OH, F)

CRYSTAL SYSTEM: Monoclinic
CLASS: 2/m

SPACE GROUP: P2$_1$/m
Z: 4
LATTICE CONSTANTS:
  $a$ = 10.80
  $b$ = 10.26
  $c$ = 7.26
  $\beta$ = 108°57'
3 STRONGEST DIFFRACTION LINES:
  2.839 (100)
  2.998 ( 70)
  3.25  ( 60)
OPTICAL CONSTANTS:
  $\alpha$ = 1.700
  $\beta$ = 1.716
  $\gamma$ = 1.726
  (-)2V = 72°-77°
  HARDNESS: 5½-6
  DENSITY: 3.41-3.44 (Meas.)
  CLEAVAGE: {010} distinct
            {100} indistinct
            {110} indistinct
Fracture conchoidal to splintery. Brittle.
  HABIT: Crystals prismatic; also tabular, flattened {100}, and often rectangular in outline. Also as irregular grains up to 3 cm in size. Twinning on {100}, common; sometimes complex.
  COLOR-LUSTER: Light yellow to dark yellow, gray, brown. Transparent to translucent. Vitreous to greasy or dull. Streak yellowish white.
  MODE OF OCCURRENCE: Occurs in alkali rocks and pegmatites in association with feldspars, nepheline, aegirine, biotite, mosandrite, astrophyllite, rosenbuschite, and other minerals. Found associated with amphibole and zircon in syenite at Red Hill, Moultonboro, Carroll County, New Hampshire; in the Langesundfjord district, Norway; on the Kola Peninsula and elsewhere in the USSR; on the Islands of Los, Guinea, and in Tasmania.
  BEST REF. IN ENGLISH: Vlasov, K. A., "Mineralogy of Rare Elements," v. II, p. 377-379, Israel Program for Scientific Translations, 1966.

## WÖLCHITE = Synonym for bournonite

## WOLFACHITE (Species status uncertain)
Ni(As, Sb)S

CRYSTAL SYSTEM: Orthorhombic (?)
  HARDNESS: 4½-5
  DENSITY: 6.372 (Meas.)
  CLEAVAGE: Not determined. Fracture uneven.
  HABIT: Crystals small, resembling arsenopyrite; also as columnar radiating aggregates.
  COLOR-LUSTER: Silver white to tin white. Opaque. Metallic. Streak black.
  MODE OF OCCURRENCE: Occurs associated with dyscrasite, galena, and niccolite in calcite at Wolfach, Baden, Germany.
  BEST REF. IN ENGLISH: Palache, et al., "Dana's System of Mineralogy," 7th Ed., v. I, p. 326-327, New York, Wiley, 1944.

# WOLFEITE
$(Fe^{2+}, Mn^{2+})_2 PO_4 OH$
Triploidite-Wolfeite Series

CRYSTAL SYSTEM: Monoclinic
CLASS: 2/m
SPACE GROUP: $P2_1/a$
Z: 16
LATTICE CONSTANTS:
  $a = 12.20$
  $b = 13.17$
  $c = 9.79$
  $\beta = 108°00'$
3 STRONGEST DIFFRACTION LINES:
  2.93 (100)
  3.09 ( 90)
  3.18 ( 80)
OPTICAL CONSTANTS:
  $\alpha = 1.741$
  $\beta = 1.742$
  $\gamma = 1.746$
  (+)2V = moderate
HARDNESS: 4½–5
DENSITY: 3.79 (Meas.)
CLEAVAGE: {010} good
           {120} fair
Fracture subconchoidal to uneven.
  HABIT: Crystals prismatic, vertically striated. Usually as columnar to parallel fibrous aggregates; also divergent fibrous; granular.
  COLOR-LUSTER: Reddish brown to dark clove-brown; rarely greenish. Transparent to translucent. Vitreous to greasy. Streak whitish.
  MODE OF OCCURRENCE: Occurs in granite pegmatite at the Bull Moose mine, Custer County, South Dakota; as an alteration of triphylite at the Palermo quarry, North Groton, New Hampshire; at Hagendorf, Bavaria, Germany; at Cyrillhof, Czechoslovakia; and with triplite at Skrumpetorp, Sweden.
  BEST REF. IN ENGLISH: Palache, et al., "Dana's System of Mineralogy," 7th Ed., v. II, p. 853–855, New York, Wiley, 1951.

# WOLFRAMITE
(Fe, Mn)WO$_4$
Wolframite Series

CRYSTAL SYSTEM: Monoclinic
CLASS: 2/m
SPACE GROUP: P2/c
Z: 2
LATTICE CONSTANTS:
  $a = 4.78$ kX
  $b = 5.73$
  $c = 4.98$
  $\beta = 89°34'$
3 STRONGEST DIFFRACTION LINES:
  2.968 (100)
  2.946 ( 90)
  4.78 ( 60)
OPTICAL CONSTANTS:
  $\alpha = 2.26, 2.31$
  $\beta = 2.32$       (Li)
  $\gamma = 2.42, 2.46$
  (+)2V = 78°36'   (Calc.)
HARDNESS: 4–4½
DENSITY: 7.371 (Calc. X-ray)
CLEAVAGE: {010} perfect
Fracture uneven. Brittle.
  HABIT: Crystals usually short prismatic; also long prismatic and somewhat tabular {100}; sometimes tabular or equant. Usually striated in direction of elongation. As groups of subparallel crystals; also lamellar or massive granular; rarely as intergrowths of acicular crystals. Twinning on {100} or {023} common.
  COLOR-LUSTER: Dark grayish black, brownish black to iron black. Sometimes tarnished iridescent. Translucent to opaque. Submetallic. Streak reddish brown to brownish black or black.
  MODE OF OCCURRENCE: Occurs chiefly in high-temperature ore veins and quartz veins in or near granitic rocks, in medium-temperature ore deposits, in small amounts in epithermal veins, and in placer deposits. Found at numerous localities in Lawrence and Pennington counties, South Dakota; in Taos County, New Mexico; and at several deposits in Arizona. It also occurs widespread throughout the world, especially in Canada, Argentina, Bolivia, England, France, Spain, Portugal, Germany, Czechoslovakia, USSR, Malaysia, Burma, Iran, Korea, Japan, Australia, Tasmania, and Transvaal, Africa. Exceptional specimens occur at Schlaggenwald and Zinnwald, Bohemia, Czechoslovakia; at several places in Bolivia; and as large spectacular tabular crystals associated with purple apatite crystals at Panesqueira, Portugal.
  BEST REF. IN ENGLISH: Palache, et al., "Dana's System of Mineralogy," 7th Ed., v. II, p. 1064–1072, New York, Wiley, 1951.

# WOLFRAMIXIOLITE
(Nb, Ta, W, Ti, Fe, Zr, Mn, U, Ca, Mg)$_3$O$_6$

CRYSTAL SYSTEM: Monoclinic
CLASS: 2/m
SPACE GROUP: P2/c
Z: 2
LATTICE CONSTANTS:
  $a = 4.750$
  $b = 5.72$
  $c = 5.06$
  $\beta$ not given, but presumably 90°
3 STRONGEST DIFFRACTION LINES:
  2.963 (100)
  3.641 ( 70)
  1.716 ( 70)
OPTICAL CONSTANTS: Opaque.
  HARDNESS: Microhardness 412 kg/mm$^2$ with 50 g load.
DENSITY: 6.55 (Meas.)
          6.712 (Calc.)
CLEAVAGE: Prismatic—well-developed, and a poorer one nearly perpendicular to the first.

HABIT: Prismatic grains up to 0.22 mm in diameter.

COLOR-LUSTER: Black with gray tint; opaque. Occasionally exhibits brownish red tint under binocular microscope.

MODE OF OCCURRENCE: Occurs as prismatic grains intergrown with microcline, quartz, ilmenite, and fluorite in specimens from an unknown locality, now in the Fersman Mineralogical Museum of the Academy of Sciences, USSR.

BEST REF. IN ENGLISH: Ginsburg, A. L., Gorzhevskaya, S. A., Sidorenko, G. A., and Ukhina, T. A., *Am. Min.*, **55**: 318–319 (1970).

## WOLLASTONITE
$\alpha$-CaSiO$_3$

CRYSTAL SYSTEM: Triclinic
CLASS: $\bar{1}$
SPACE GROUP: P$\bar{1}$
Z: 6
LATTICE CONSTANTS:
$a = 7.94$       $\alpha = 90°03'$
$b = 7.32$       $\beta = 95°17'$
$c = 7.07$       $\gamma = 102°28'$
3 STRONGEST DIFFRACTION LINES:
2.97 (100)
3.83 ( 80)
3.52 ( 80)
OPTICAL CONSTANTS:
$\alpha = 1.616$–1.640
$\beta = 1.628$–1.650
$\gamma = 1.631$–1.653
$(-)2V = 38°$–60°
HARDNESS: 4½–5
DENSITY: 2.87–3.09 (Meas.)
          2.96 (Calc.)
CLEAVAGE: {100} perfect
          {001} good
          {$\bar{1}$02} good
Fracture splintery.

HABIT: Crystals commonly tabular; usually massive, cleavable to fibrous; also granular and compact. Twinning on {100} common.

COLOR-LUSTER: White to grayish; rarely colorless or very pale green. Transparent to translucent. Vitreous to pearly; somewhat silky when fibrous. Sometimes fluoresces under ultraviolet light (Franklin, N.J.).

MODE OF OCCURRENCE: Occurs chiefly as a common mineral of metamorphosed limestones, and less commonly in certain alkaline igneous rocks. Notable occurrences are found in Inyo, Kern, and Riverside counties and elsewhere in California; at many places in Colorado and other western states; at Willsboro, Essex County, New York; in Ontario and Quebec, Canada; in Chiapis, Mexico; and at localities in Norway, Italy, Roumania, and Finland.

BEST REF. IN ENGLISH: Deer, Howie, and Zussman, "Rock Forming Minerals," v. II, p. 167–175, New York, Wiley, 1963.

## WÖLSENDORFITE
(Pb, Ca)U$_2$O$_7$ · 2H$_2$O

CRYSTAL SYSTEM: Orthorhombic
Z: 6
LATTICE CONSTANTS:
$a = 11.95$
$b = 13.99$
$c = 7.02$
3 STRONGEST DIFFRACTION LINES:
3.09 (100)
3.47 ( 80)
6.93 ( 60)
OPTICAL CONSTANTS:
$n_1 = 2.09$
$n_2 = 2.05$
HARDNESS: Not determined.
DENSITY: 6.8 (Meas.)
CLEAVAGE: {001} good
HABIT: As crystalline masses, spherulites, and nodules.
COLOR-LUSTER: Orange-red to carmine red.

MODE OF OCCURRENCE: Occurs in fissures in fluorite at Wölsendorf, Bavaria; as incrustations on pitchblende at Great Bear Lake, North West Territory, Canada; with the pitchblende of Kerségalec, Lignol, France; and with other secondary uranium minerals at Shinkolobwe, Katanga, Zaire.

BEST REF. IN ENGLISH: Protas, Jean, *Am. Min.*, **42**: 919 (1957).

## WOODFORDITE = Ettringite

## WOODHOUSEITE
CaAl$_3$PO$_4$SO$_4$(OH)$_6$

CRYSTAL SYSTEM: Hexagonal
CLASS: $\bar{3}$2/m
SPACE GROUP: R$\bar{3}$m
Z: 1 (rhomb.), 3 (hexag.)
LATTICE CONSTANTS:
$a = 6.961$ kX       $a_{rh} = 6.75$
$c = 16.27$              $\alpha = 62°04'$
3 STRONGEST DIFFRACTION LINES:
2.94 (100)
1.89 ( 94)
2.19 ( 86)
OPTICAL CONSTANTS:
$\omega = 1.636$
$\epsilon = 1.647$
  (+)
HARDNESS: 4½
DENSITY: 3.012 (Meas.)
          3.001 (Calc.)
CLEAVAGE: {0001} perfect
HABIT: As small crystals, usually pseudocubic, also tabular. Crystal faces commonly rough, curved, or striated.
COLOR-LUSTER: White to flesh pink. Transparent to translucent. Vitreous; pearly on {0001}.

MODE OF OCCURRENCE: Occurs in veins with quartz, lazulite, augelite, and topaz in the andalusite deposit on the western slope of the White Mountains, 7 miles east of Mocalno, north of Bishop, Mono County, California.

BEST REF. IN ENGLISH: Palache, et al., "Dana's System of Mineralogy," 7th Ed., v. II, p. 1006–1007, New York, Wiley, 1951.

# WOODRUFFITE
$(Zn, Mn^{2+})Mn_3^{4+}O_7 \cdot 1–2H_2O$

CRYSTAL SYSTEM: Tetragonal
Z: 2
LATTICE CONSTANTS:
  $a = 8.42$
  $c = 9.28$
3 STRONGEST DIFFRACTION LINES:
  4.66 (100)
  2.66 ( 60)
  2.48 ( 60)
HARDNESS: $\sim 4\frac{1}{2}$
DENSITY: 4.01 (Meas.)
        3.98 (Calc.)
CLEAVAGE: Fracture smooth—conchoidal.
HABIT: Botryoidal masses, and soft almost pulverulent coatings.
COLOR-LUSTER: Iron black to chocolate brown. Dull.
MODE OF OCCURRENCE: Occurs associated with chalcophanite and zincian cryptomelane in secondary zinc ores at Sterling Hill, Sussex County, New Jersey; also found at Sandur, Mysore, India.
BEST REF. IN ENGLISH: Frondel, Clifford, *Am. Min.*, **38**: 761–769 (1953).

# WOODWARDITE (Inadequately described mineral)
$Cu_4 Al_2 SO_4 (OH)_{12} \cdot 4–6H_2O(?)$

3 STRONGEST DIFFRACTION LINES:
  10.9  (100)
   5.46 ( 60)
   3.66 ( 50)
OPTICAL CONSTANTS:
  $\alpha = 1.552, 1.571$
  $\beta = 1.555$
  $\gamma = 1.565, 1.576$
Variance probably due to variable water content.
     (+)2V
HARDNESS: Not determined.
DENSITY: 2.38 (Meas.)
CLEAVAGE: Not determined.
HABIT: As tiny rounded concretions with fine-fibrous structure; spherulitic.
COLOR-LUSTER: Turquoise blue to greenish blue. Translucent.
MODE OF OCCURRENCE: Occurs at Simmde Dallhuan, Drws-y-coed, Nantlle, Caernarvonshire, Wales; in Cornwall, England; and at Klausen, Trentino, Italy.
BEST REF. IN ENGLISH: Palache, et al., "Dana's System of Mineralogy," 7th Ed., v. II, p. 580, New York, Wiley, 1951.

# WULFENITE
$PbMoO_4$

CRYSTAL SYSTEM: Tetragonal
CLASS: 4/m
SPACE GROUP: $I4_1/a$
Z: 4
LATTICE CONSTANTS:
  $a = 5.435$
  $c = 12.11$
3 STRONGEST DIFFRACTION LINES:
  3.24  (100)
  2.021 ( 30)  (Synthetic)
  1.653 ( 25)
OPTICAL CONSTANTS:
  $\omega = 2.4053$
   $\epsilon = 2.2826$ (Na)
     (–)
HARDNESS: $2\frac{3}{4}$–3
DENSITY: 6.5–7.0 (Meas.)
        6.815 (Calc. X-ray)
CLEAVAGE: {011} distinct
          {001} indistinct
          {013} indistinct
Fracture uneven to subconchoidal. Brittle.
HABIT: Crystals usually square tabular, sometimes very thin, less commonly octahedral, prismatic, or cuboidal. Also massive, coarse to fine granular. Twinning on {001}, not common.
COLOR-LUSTER: Various shades of orange and yellow, orangish brown to brown; yellowish gray, tan, greenish brown. Transparent to translucent. Resinous to adamantine. Streak white.
MODE OF OCCURRENCE: Occurs as a secondary mineral in the oxidation zone of ore deposits, often associated with calcite, limonite, cerussite, vanadinite, mimetite, pyromorphite, galena, malachite, and other minerals. Found at the Wheatley mines, Chester County, Pennsylvania; at Loudville, Hampshire County, Massachusetts; and widespread in the western and southwestern United States. Notable localities in Arizona include the Glove mine at Amado; the Rawley mine, Theba, Maricopa County; the Hamburg and Red Cloud mines, Yuma County; and the Mammoth mine, Tiger, Pinal County. Fine specimens also occur at localities in New Mexico, Nevada, Utah, and South Dakota; at Los Lamentos, Mapimi, and at other places in Mexico; in Germany, Austria, Czechoslovakia, Sardinia, Algeria, Morocco, Australia, and at Tsumeb, South West Africa.
BEST REF. IN ENGLISH: Palache, et al., "Dana's System of Mineralogy," 7th Ed., v. II, p. 1081–1086, New York, Wiley, 1951.

# WURTZITE
ZnS or (Zn, Fe)S
Trimorphous with Sphalerite, Matraite

CRYSTAL SYSTEM: Hexagonal
CLASS: 6mm
SPACE GROUP: $P6_3mc$
Z: 2(2H), 4(4H), 6(6H), 8(8H), 10(10H)
LATTICE CONSTANTS:

|  (2H) | (4H) | (6H) |
|-------|------|------|
| $a = 3.820$ | $a = 3.814$ | $a = 3.821$ |
| $c = 6.260$ | $c = 12.46$ | $c = 18.73$ |

(8H)      (10H)
$a = 3.82$    $a = 3.824$
$c = 24.96$   $c = 31.20$

3 STRONGEST DIFFRACTION LINES:
3.31 (100)
3.13 ( 86) (Synthetic 2H)
2.93 ( 84)

OPTICAL CONSTANTS:
$\omega = 2.356$
$\epsilon = 2.378$ (Na)
(+)

HARDNESS: 3½–4

DENSITY: 3.98–4.08 (Meas. 2H)
      4.089 (Calc. 2H)

CLEAVAGE: $\{11\bar{2}0\}$ distinct
      $\{0001\}$ imperfect
Brittle.

HABIT: Crystals usually hemimorphic pyramidal; also short prismatic to tabular; often striated horizontally. Also as banded crusts; columnar; or fibrous.

COLOR-LUSTER: Brownish black to orange-brown. Translucent. Resinous. Streak brown. Often fluoresces orange under long-wave ultraviolet light.

MODE OF OCCURRENCE: Occurs chiefly in hydrothermal vein deposits frequently associated with or disseminated in sphalerite; found less commonly in sedimentary deposits. Occurs in the United States in clay-ironstone concretions in eastern Ohio and western Pennsylvania; in Connecticut; at ore deposits in Missouri, Montana, Nevada, Idaho, and Utah; and in contact limestone in Mono County, California. Other occurrences are found in Bolivia, Peru, England, Roumania, and Czechoslovakia.

BEST REF. IN ENGLISH: Palache, et al., "Dana's System of Mineralogy," 7th Ed., v. I, p. 227–228, New York, Wiley, 1944.

## WYARTITE
$Ca_3(UO_2)_6U(CO_3)_2(OH)_{16} \cdot 3-5H_2O$

CRYSTAL SYSTEM: Orthorhombic
CLASS: 222 (I), 2/m 2/m 2/m (II), or mm2 (II)
SPACE GROUP: $P2_12_12_1$(I), Pmcn(II), or $P2_1$cn(II)
Z: 2
LATTICE CONSTANTS:
    (I)      (II)
$a = 11.25$  $a = 11.25$
$b = 7.09$   $b = 7.09$
$c = 20.80$  $c = 16.83$

3 STRONGEST DIFFRACTION LINES:
Wyartite I    Wyartite II
10.3 (100)  8.46 (100)
5.19 ( 30)  3.47 ( 20)
3.47 ( 10)  4.22 ( 18)

OPTICAL CONSTANTS:
    (I)
$\beta = 1.89$
$\gamma = 1.91$
$(-)2V = 48°$

HARDNESS: 3–4
DENSITY: 4.69
CLEAVAGE: $\{001\}$ perfect
    also $\{010\}$
HABIT: Crystals small, striated, with $\{001\}$ predominant.
COLOR-LUSTER: Black to violet-black; greenish on alteration. Dull, on cleavage vitreous to submetallic. Opaque to translucent. Streak brownish violet.
MODE OF OCCURRENCE: Occurs as an alteration product of a red alteration product of uraninite (wölsendorfite?) at Shinkolobwe, Katanga, Zaire.
BEST REF. IN ENGLISH: Guillemin, C., and Protas, J., Am. Min., 44: 908 (1959). Clark, Joan R., Am. Min., 45: 200–208 (1960).

## WYLLIEITE
$Na_2 Fe_2^{2+} Al(PO_4)_3$

CRYSTAL SYSTEM: Monoclinic
CLASS: 2/m
SPACE GROUP: $P2_1/n$
Z: 4
LATTICE CONSTANTS:
$a = 11.868$
$b = 12.382$
$c = 6.354$
$\beta = 114.52°$

3 STRONGEST DIFFRACTION LINES:
2.693 (100)
2.674 (100)
6.15 ( 60)

OPTICAL CONSTANTS:
$\alpha = 1.688$
$\beta = 1.691$
$\gamma = 1.696$
Biaxial (+), $2V \sim 50°$
HARDNESS: $4^+$
DENSITY: 3.601 (Meas.)
    3.60 (Calc.)
CLEAVAGE: $\{010\}$ perfect
    $\{\bar{1}01\}$ distinct
Very brittle.

HABIT: Crystals euhedral or subhedral, coarse with uneven surfaces, up to six inches across. Often as large masses of interlocking crude crystals weighing up to fifty pounds.

COLOR-LUSTER: Deep bluish-green to deep oily green, greyish-green to greenish-black. Thin fragments pale green and transparent. Vitreous to submetallic. Streak and powder dirty olive-green.

MODE OF OCCURRENCE: Wyllieite occurs as a primary mineral in granite pegmatites. It is found in considerable abundance embedded in bull quartz, perthite and albitized perthite, frequently associated with muscovite, schorl, arrojadite, scorzalite, graftonite and sarcopside, at the Victory mine, Custer County, South Dakota; also with abundant arrojadite at the Nickel Plate mine near Keystone, Pennington County.

BEST REF. IN ENGLISH: Moore, Paul B. and Ito, Jun, Min. Rec., 4: 131–136 (1973).

# XANTHIOSITE
$Ni_3(AsO_4)_2$

CRYSTAL SYSTEM: Monoclinic
CLASS: 2/m
SPACE GROUP: $P2_1/a$
Z: 4
LATTICE CONSTANTS:
  $a = 10.174$
  $b = 9.548$
  $c = 5.766$
  $\beta = 92°58.5'$
3 STRONGEST DIFFRACTION LINES:
  2.53 (100)
  3.46 ( 80)
  2.75 ( 80)
HARDNESS: 4
DENSITY: 5.37 and 5.47 (Meas.)
          5.388 (Calc.)
CLEAVAGE: Not determined.
HABIT: Massive.
COLOR-LUSTER: Golden yellow.
MODE OF OCCURRENCE: Occurs with aerugite, various nickel-cobalt-arsenic minerals, traces of decomposing pitchblende, and various alteration products, in comby quartz vein-material at the South Terras mine, St. Stephen-in-Brannel, Cornwall, England; also found at Johanngeorgenstadt, Saxony, Germany.
BEST REF. IN ENGLISH: Davis, R. J., Hey, M. H., and Kingsbury, A. W. G., *Min. Mag.*, 35: 72-83 (1965).


# XANTHOARSENITE = Sarkinite


# XANTHOCONITE
$Ag_3AsS_3$
Dimorphous with proustite

CRYSTAL SYSTEM: Monoclinic
CLASS: 2/m
SPACE GROUP: A2/a, C2/c
Z: 8

LATTICE CONSTANTS:

| (A2/a) | (C2/c) |
|--------|--------|
| $a = 16.98$ | $a = 12.06$ |
| $b = 6.21$ | $b = 6.26$ |
| $c = 11.99$ or | $c = 17.08$ |
| $\beta = 110°10'$ | $\beta = 110°0'$ |

3 STRONGEST DIFFRACTION LINES:

| 3.00 (100) | 2.974 (100) |
|-----------|-------------|
| 2.82 ( 60) or | 3.01 ( 82) |
| 3.14 ( 30) | 2.819 ( 70) |

OPTICAL CONSTANTS:
  $N \sim 3.00$
  $(-)2V$
HARDNESS: 2–3
DENSITY: 5.54 (Meas.)
          5.53 (Calc.)
CLEAVAGE: {001} distinct
Fracture subconchoidal. Brittle.
HABIT: Crystals commonly tabular, pseudohexagonal, flattened on {001}. Rarely pyramidal or lath-like. Also massive, reniform. Twinning on {001} common.
COLOR-LUSTER: Yellowish orange to reddish orange to reddish brown. Transparent to translucent. Adamantine. Streak orange-yellow.
MODE OF OCCURRENCE: Occurs in low-temperature hydrothermal vein deposits, commonly associated with ruby silvers and other silver minerals. Found at the General Petite mine, Atlanta district, Idaho; as aggregates of minute crystals associated with proustite at several mines in the Cobalt district, Ontario, Canada; at Chanarcillo, Chile; as crystals on dolomite at Ste. Marie aux Mines, Alsace-Lorraine, France; at Freiberg and Schneeberg, Saxony, Germany; and in the silver-mining districts of Czechoslovakia.
BEST REF. IN ENGLISH: Palache, et al., "Dana's System of Mineralogy," 7th Ed., v. I, p. 371-372, New York, Wiley, 1944.


# XANTHOPHYLLITE = Variety of clintonite with optic axial plane parallel to (010)


# XANTHOXENITE
$Ca_4Fe_2(PO_4)_4(OH)_2 \cdot 3H_2O$

CRYSTAL SYSTEM: Triclinic
3 STRONGEST DIFFRACTION LINES:
3.05 (100)
2.73 ( 90)
3.22 ( 80)
OPTICAL CONSTANTS:
$\alpha = 1.704$
$\beta = 1.715$
$\gamma = 1.724$
$(-)2V = $ large
HARDNESS: $\sim 2\frac{1}{2}$
DENSITY: 2.8; 2.97 (Meas.)
CLEAVAGE: One perfect.
HABIT: Rarely, as distinct crystals; usually as masses and crusts composed of indistinct platy or lath-like crystals, or as fibrous, radiating, sheaf-like aggregates.
COLOR-LUSTER: Pale yellow to brownish yellow. Dull to waxy.
MODE OF OCCURRENCE: Occurs as an alteration product of triphylite at the Palermo mine, North Groton, New Hampshire, and at Hühnerkobel, Bavaria, Germany.
BEST REF. IN ENGLISH: Palache, et al., "Dana's System of Mineralogy," 7th Ed., v. II, p. 977–978, New York, Wiley, 1951.

# XENOTIME
$YPO_4$
Isostructural with zircon

CRYSTAL SYSTEM: Tetragonal
CLASS: 4/m 2/m 2/m
SPACE GROUP: I4/amd
Z: 4
LATTICE CONSTANTS:
$a = 6.88$ kX
$c = 6.03$
3 STRONGEST DIFFRACTION LINES:
3.45 (100)
2.56 ( 50)
1.768 ( 50)
OPTICAL CONSTANTS:
$\omega = 1.721, 1.720$
$\epsilon = 1.816, 1.827$ (Na)
(+)
HARDNESS: 4–5
DENSITY: 4.4–5.1 (Meas.)
CLEAVAGE: {100} perfect.
Fracture splintery to uneven. Brittle.
HABIT: Crystals short to long prismatic; also pyramidal or equant. As aggregates of rough crystals; as rosettes. Twinning on {111} rare.
COLOR-LUSTER: Yellowish brown to reddish brown; also pale gray, pale yellow, greenish, reddish. Vitreous to resinous. Translucent to opaque.

MODE OF OCCURRENCE: Occurs widespread in small amounts in acidic and alkalic igneous rocks; in pegmatites, metamorphic rocks, Alpine-type vein deposits, and as a detrital mineral. Found in the United States in California, Colorado, North Carolina, Alabama, Georgia, and New York. It also occurs in Brazil, Norway, Sweden, Germany, Poland, Switzerland, Madagascar, South Africa, New Zealand, India, and Japan.
BEST REF. IN ENGLISH: Palache, et al., "Dana's System of Mineralogy," 7th Ed., v. II, p. 688–691, New York, Wiley, 1951.

# XONOTLITE (Jurupaite)
$Ca_3 Si_3 O_8 (OH)_2$

CRYSTAL SYSTEM: Monoclinic
CLASS: 2/m
SPACE GROUP: P2/a
Z: 2
LATTICE CONSTANTS:
$a = 16.95$
$b = 7.34$
$c = 7.03$
$\beta \simeq 90°$
3 STRONGEST DIFFRACTION LINES:
3.08 (100)
3.24 ( 50)
2.82 ( 50)
OPTICAL CONSTANTS:
$\alpha = \beta = 1.583$
$\gamma = 1.593$
$(+)2V = $ Very small
HARDNESS: 6
DENSITY: 2.71 (Meas.)
2.709 (Calc.)
CLEAVAGE: One good. Tough.
HABIT: Crystals acicular. Usually massive, compact; matted, parallel fibrous.
COLOR-LUSTER: Chalky white, colorless, pink, light gray, medium gray. Translucent to transparent. Greasy, pearly, dull vitreous.
MODE OF OCCURRENCE: Occurs mainly as veinlets in serpentine or in contact zones. Found in Mendocino, Monterey, Riverside, San Francisco, and Santa Barbara counties, California; at Isle Royal, Michigan; and near Leesburg, Virginia. Also at Tetela de Xonotla, Mexico; near Yauco, Puerto Rico; and elsewhere.
BEST REF. IN ENGLISH: Winchell and Winchell, "Elements of Optical Mineralogy," 4th Ed., Pt. 2, p. 455, New York, Wiley, 1951. Kaye, C. A., *Am. Min.*, 38: 860–862 (1953).

# XYLOTILE = Probably an altered asbestos (chrysotile)

# Y

## YAGIITE

$(Na, K)_3 Mg_4 Al_6 (Si, Al)_{12} O_{60}$

CRYSTAL SYSTEM: Hexagonal
CLASS: 6/m 2/m 2/m
SPACE GROUP: P6/mcc
Z: 2
LATTICE CONSTANTS:
  $a = 10.09$
  $c = 14.29$
3 STRONGEST DIFFRACTION LINES:
  3.228 (100)
  5.059 ( 65)
  3.726 ( 50)
OPTICAL CONSTANTS:
  $\omega = 1.536$
  $\epsilon = 1.544$
   (+)
DENSITY: 2.70 (Calc.)
HABIT: Massive.
COLOR-LUSTER: Colorless.
MODE OF OCCURRENCE: Occurs interstitially to small equant grains of aluminous titanian diopside in a small silicate inclusion (0.8 mm in diameter) surrounded by nickel-iron in the Colomera iron meteorite.
BEST REF. IN ENGLISH: Bunch, Theodore E., and Fuchs, Louis H., *Am. Min.*, **54**: 14-18 (1969).


## YAMATOITE = Hypothetical end member
$Mn_3 V_2 (SiO_4)_3$
Garnet Group


## YAROSLAVITE
$Ca_3 Al_2 F_{10} (OH)_2 \cdot H_2O$

CRYSTAL SYSTEM: Orthorhombic
Z: 4
LATTICE CONSTANTS:
  $a = 8.74$
  $b = 5.53$
  $c = 4.51$

3 STRONGEST DIFFRACTION LINES:
  3.445 (100)
  2.222 ( 80)
  3.653 ( 70)
OPTICAL CONSTANTS:
  $\alpha = 1.413$
  $\gamma = 1.423$
  $(-)2V = 74°$
HARDNESS: Microhardness 264 kg/mm².
DENSITY: 3.09 (Meas.)
CLEAVAGE: Pinacoidal.  Fracture irregular.
HABIT: As oval or spherical aggregates with radiated fibrous structure.
COLOR-LUSTER: White, transparent; vitreous.
MODE OF OCCURRENCE: Occurs in small amounts in the zone of oxidation of banded sellaite-tourmaline-fluorite ores in "one of the hydrothermal fluorite-rare metal deposits of Siberia."
BEST REF. IN ENGLISH: Novikova, M. I., Sidorenko, G. A., and Kuznetsova, M. N., *Am. Min.*, **51**: 1546-1547 (1966).


## YAVAPAIITE
$KFe(SO_4)_2$

CRYSTAL SYSTEM: Monoclinic
CLASS: 2, 2/m, or m
SPACE GROUP: C2, C2/m, or Cm
Z: 2
LATTICE CONSTANTS:
  $a = 8.12$
  $b = 5.14$
  $c = 7.82$
  $\beta = 94°24'$
3 STRONGEST DIFFRACTION LINES:
  2.97 (100)
  7.85 ( 90)
  3.87 ( 80)
OPTICAL CONSTANTS:
  $\alpha = 1.593$
  $\beta = 1.684$
  $\gamma = 1.698$

$(-)2V = 30.5°$ (Na)
HARDNESS: 2½–3
DENSITY: 2.88 (Meas.)
2.92 (Calc.)
CLEAVAGE: {001} perfect
{100} perfect
{110} distinct
Fracture subconchoidal to uneven. Brittle

HABIT: Crystals short, stubby, with longer dimension parallel to [010]. {001} is dominant, with {101}, {100}, {201}, and unit prism {110} less well developed. Crystals do not exceed 0.6 mm in length.

COLOR-LUSTER: Pale pink, transparent; vitreous. Streak white; powder pale yellow.

MODE OF OCCURRENCE: Occurs very sparsely, associated with sulfur, voltaite, and other sulfates, at the United Verde copper mine, Jerome, Arizona.

BEST REF. IN ENGLISH: Hutton, C. Osborne, *Am. Min.*, 44: 1105–1114 (1959).

# YEATMANITE
$(Mn, Zn)_{13}Sb_2Si_4Zn_2O_{28}$

CRYSTAL SYSTEM: Triclinic
CLASS: $\bar{1}$ or 1
SPACE GROUP: $P\bar{1}$ or P1
Z: 1
LATTICE CONSTANTS:
$a = 5.52$ $\alpha = 92°48'$
$b = 11.56$ $\beta = 101°45'$
$c = 9.029$ $\gamma = 76°11'$
OPTICAL CONSTANTS:
$\alpha = 1.873$
$\beta = 1.905$
$\gamma = 1.910$
$(-)2V \simeq 49°$
HARDNESS: 4
DENSITY: 5.02 (Meas.)
5.37 (Calc.)
CLEAVAGE: {100} perfect
Brittle.

HABIT: Cyrstals pseudo-orthorhombic with twinning on {023} and multiple twinning on {010}.

COLOR-LUSTER: Clove brown. Streak light brown.

MODE OF OCCURRENCE: Occurs as thin crystalline plates embedded in willemite at Franklin, Sussex County, New Jersey.

BEST REF. IN ENGLISH: Palache, Charles, Bauer, L. H., and Berman, Harry, *Am. Min.*, 23: 527–530 (1938), Moore, Paul B., *Am. Min.*, 51: 1494–1500 (1966).

# YEDLINITE
$Pb_6CrCl_6O_6 \cdot 2H_2O$

CRYSTAL SYSTEM: Hexagonal
CLASS: $\bar{3}$
SPACE GROUP: $R\bar{3}$
Z: 3

LATTICE CONSTANTS:
$a = 12.868$
$c = 9.821$
3 STRONGEST DIFFRACTION LINES:
2.952 (100)
2.622 ( 68)
4.506 ( 65)
OPTICAL CONSTANTS:
$\omega = 2.125$
$\epsilon = 2.059$ $(-)$
HARDNESS: ~2½
DENSITY: 5.85 (Meas.)
CLEAVAGE: {110} distinct; somewhat sectile.
HABIT: Crystals prismatic, sometimes doubly terminated, up to 1 mm long. Forms: {110}, {1$\bar{1}$1}, {001}, {100} and {201}.

COLOR-LUSTER: Red-violet. Transparent to translucent. Vitreous. Streak white.

MODE OF OCCURRENCE: Occurs in association with diaboleite, quartz, wulfenite, dioptase, phosgenite, and wherryite from the Mammoth-St. Anthony mine, Tiger, Pinal County, Arizona.

BEST REF. IN ENGLISH: McLean, W. John, Bideaux, Richard A., and Thomssen, Richard W., *Am. Min.*, in press.

# YODERITE
$(Al, Mg, Fe)_2Si(O, OH)_5$

CRYSTAL SYSTEM: Monoclinic
CLASS: 2
SPACE GROUP: $P2_1$
Z: 4
LATTICE CONSTANTS:
$a = 8.10$
$b = 5.78$
$c = 7.28$
$\beta = 106°$
3 STRONGEST DIFFRACTION LINES:
3.50 (100)
3.03 ( 80)
2.61 ( 60)
OPTICAL CONSTANTS:
$\alpha = 1.689$
$\beta = 1.691$
$\gamma = 1.715$
$(+)2V = 25°$
HARDNESS: 6
DENSITY: 3.39
CLEAVAGE: Parting [001] moderately good. {100} poor.

HABIT: Anhedral grains, up to $12 \times 6 \times 3$ mm in size.

COLOR-LUSTER: Purple.

MODE OF OCCURRENCE: Occurs in quartz-yoderite-kyanite-talc schist at Mautia Hill, central Tanzania.

BEST REF. IN ENGLISH: McKie, Duncan, and Radford, A. J., *Min. Mag.*, 32: 282–307 (1959).

# YOKOSUKAITE = Nsutite

## YOSHIMURAITE
$(Ba, Sr)_2 TiMn_2 (SiO_4)_2 (PO_4, SO_4)(OH, Cl)$

CRYSTAL SYSTEM: Triclinic
CLASS: $\bar{1}$
SPACE GROUP: $P\bar{1}$
Z: 2
LATTICE CONSTANTS:
 $a = 7.00$     $\alpha = 93.5°$
 $b = 14.71$    $\beta = 90.2°$
 $c = 5.39$     $\gamma = 95.3°$
3 STRONGEST DIFFRACTION LINES:
 2.91 (100)
 4.85 ( 50)
 3.21 ( 50)
OPTICAL CONSTANTS:
  $\alpha = 1.763$
  $\beta = 1.777$
  $\gamma = 1.785$
 $(+)2V = 85°-90°$
 HARDNESS: 4½
 DENSITY: 4.13 (Meas.)
       4.21 (Calc.)
 CLEAVAGE: {010} perfect
       {10$\bar{1}$} distinct
       {101} distinct
Brittle.
 HABIT: Bladed, or in mica-like forms up to 5 cm long. Polysynthetic twinning on (010).
 COLOR-LUSTER: Orange-brown.
 MODE OF OCCURRENCE: Occurs in coarse-grained pegmatite along the boundary between massive manganese ore (tephroite with rhodonite, braunite, and hausmannite) and massive hornfels containing manganophyllite, richterite, and urbanite at the Noda-Tamagawa Mine, Iwate Prefecture, Japan.
 BEST REF. IN ENGLISH: Watanabe, Takeo, Takeuchi, Yoshio, and Ito, Jun, *Min. J.* (Japan), 3: 156–167 (1961) (in English).

## YTTRIALITE = Variety of thalenite

## YTTROCERITE = Cerian fluorite

## YTTROCOLUMBITE = Samarskite

## YTTROCRASITE
$(Y, Th)Ti_2 (O, OH)_6$ (?)

CRYSTAL SYSTEM: Probably orthorhombic
OPTICAL CONSTANTS: Partly metamict and isotropic, and partly weakly anisotropic. $N = 2.12-2.15$
 HARDNESS: 5½–6
 DENSITY: 4.80
 CLEAVAGE: Not determined.   Fracture uneven or slightly conchoidal.
 HABIT: Crystals prismatic, similar to polycrase crystals; also thin tabular.

COLOR-LUSTER: Black. Opaque; translucent in thin fragments. Crystal faces dull; resinous on fracture surfaces.
 Radioactive.
 MODE OF OCCURRENCE: Occurs in pegmatites at the Southern Pacific silica quarry at Nuevo, Riverside County, California, and in Burnet County, Texas, about 3 miles east of Baringer Hill.
 BEST REF. IN ENGLISH: Palache, et al., "Dana's System of Mineralogy," 7th Ed., v. I, p. 793, New York, Wiley, 1944.

## YTTROFLUORITE = Yttrian fluorite

## YTTROTANTALITE
$(Y, U, Fe)(Ta, Nb)O_4$

CRYSTAL SYSTEM: Monoclinic
CLASS: 2/m
SPACE GROUP: I2/a
LATTICE CONSTANTS:
 $a = 5.34$
 $b = 10.94$
 $c = 5.07$
 $\beta = 95°18'$
LATTICE CONSTANT:
 $a = 10.30$ (heated at 700°C for 3 hours–cubic phase)
3 STRONGEST DIFFRACTION LINES: Metamict.
OPTICAL CONSTANT:
 $N = 2.15$
 HARDNESS: 5–5½
 DENSITY: 5.7 (Meas.)
After ignition 6.15.
 CLEAVAGE: {010} indistinct
Fracture conchoidal. Brittle.
 HABIT: Crystals short prismatic parallel to [001] or tabular parallel to {010}, with {110} and {010} faces predominant.   Also massive, compact, and as irregular grains.
 COLOR-LUSTER: Black, blackish brown. Opaque; translucent in thin fragments. Submetallic, greasy. Streak gray.
 MODE OF OCCURRENCE: Occurs in pegmatites associated with monazite, samarskite, columbite, xenotime, gadolinite, and mica at Hattevick in Iveland, and near Berg in Råde, Norway; and associated with fergusonite in granite pegmatites of Ytterby, Sweden; also with garnet, albite, and altered topaz at localities near Fahlun, Sweden.
 BEST REF. IN ENGLISH: Palache, et al., "Dana's System of Mineralogy," 7th Ed., v. I, p. 763, 764, New York, Wiley, 1944.

## YTTROTITANITE = Variety of sphene containing yttrium

## YTTROTUNGSTITE (Thorotungstite)
$YW_2 O_6 (OH)_3$

CRYSTAL SYSTEM: Probably orthorhombic

3 STRONGEST DIFFRACTION LINES:
 3.36 (100)
 3.25 (100)
 2.01 ( 90)
OPTICAL CONSTANTS:
 $N > 1.74 (-)$
DENSITY: 5.55; 5.82 (Meas.)
CLEAVAGE: One good transverse cleavage. One fair longitudinal cleavage.
HABIT: As minute lath-like crystals in irregular masses.
COLOR-LUSTER: Yellowish.
MODE OF OCCURRENCE: Occurs as an alteration product of wolframite or scheelite in an eluvial cassiterite deposit at Kramat Pulai, Perak, Malaya.
BEST REF. IN ENGLISH: Palache, et al., "Dana's System of Mineralogy," 7th Ed., v. II, p. 1097, New York, Wiley, 1951.


# YUGAWARALITE
$CaAl_2Si_6O_{16} \cdot 4H_2O$
Zeolite Group

CRYSTAL SYSTEM: Monoclinic
CLASS: m
SPACE GROUP: Pc
Z: 2
LATTICE CONSTANTS:
 $a = 6.72$
 $b = 13.98$
 $c = 10.05$
 $\beta = 111°31'$

3 STRONGEST DIFFRACTION LINES:
 3.056 (100)
 5.82 ( 90)
 4.68 ( 85)
OPTICAL CONSTANTS:
 $\alpha = 1.495$
 $\beta = 1.497$
 $\gamma = 1.504$
 $(+)2V = 78°$
HARDNESS: 4½
DENSITY: 2.23 (Meas.)
 2.26 (Calc.)
CLEAVAGE: {010} imperfect
Very brittle.
HABIT: Crystals flat, with {010} dominant, also {100}, {001}, {110}, {120}, {011}, and {$\bar{1}$11}. Up to 5 × 2 × 10 mm in size.
COLOR-LUSTER: Colorless to white; vitreous; b{010} commonly iridescent.
MODE OF OCCURRENCE: Occurs in networks and veins and as crystals in cavities in andesite tuffs that have been altered by waters of hot springs at Yugawara hot spring, Kanagawa Prefecture, Japan; also found at Heina-bergsjokull, Iceland.
BEST REF. IN ENGLISH: Sakurai, K., and Hayashi, A., *Sci. Repts. Yokohama National Univ.*, Sec. **II**, No. 1: 69–77 (1952) (in English). Harada, Kazuo, Nagashima, Kozo, and Sakurai, Kin-Ichi, *Am. Min.*, **54**: 306–309 (1969).

**YUKONITE** = Arseniosiderite

**YUKSPORITE** = Pectolite

# Z

## ZAPATALITE
$Cu_3Al_4(PO_4)_3(OH)_9 \cdot 4H_2O$

CRYSTAL SYSTEM: Tetragonal
Z: 6
LATTICE CONSTANTS:
 $a = 15.22$
 $c = 11.52$
3 STRONGEST DIFFRACTION LINES:
 11.60 (100)
 7.62 (100)
 6.82 ( 70)
OPTICAL CONSTANTS:
 $\omega = 1.646$
 $\epsilon = 1.635$
 (−) occasionally biaxial (−) with variable 2V
HARDNESS: 1½
DENSITY: 3.016 (Meas.)
 3.017 (Calc.)
CLEAVAGE: {001} good
Sectile.
 HABIT: Massive; as poorly crystalline cavity fillings.
 COLOR-LUSTER: Pale blue (faience blue). Translucent. Streak pale blue.
 MODE OF OCCURRENCE: Occurs in association with libethenite, chenevixite, beaverite, alunite, and pseudomalachite in silicified and brecciated limestone at a small prospect northwest of Cerro Morita, 27 km southwest of Agua Prieta, Sonora, Mexico.
 BEST REF. IN ENGLISH: Williams, Sidney A., *Min. Mag.*, **38**: 541–544 (1972).

## ZARATITE
$Ni_3CO_3(OH)_4 \cdot 4H_2O$

CRYSTAL SYSTEM: Cubic
LATTICE CONSTANT:
 $a = 6.16$
3 STRONGEST DIFFRACTION LINES:
 5.07 (100)
 8.93 ( 70)
 2.45 ( 70)

OPTICAL CONSTANT:
 $N = 1.56–1.61$
HARDNESS: 3½
DENSITY: 2.57–2.69 (Meas.)
 2.67 (Calc.)
CLEAVAGE: Fracture conchoidal. Brittle.
HABIT: Massive, compact; also as mammillary incrustations.
 COLOR-LUSTER: Emerald green. Transparent to translucent. Vitreous to greasy.
 MODE OF OCCURRENCE: Occurs as a secondary mineral in serpentine and basic igneous rocks, often associated with chromite, and less commonly as an alteration of millerite or meteoric iron. Found in fractures in chromite in the Coast Range counties of California from Contra Costa County southward; at Low's mine and Wood's mine, Lancaster County, and at West Nottingham, Chester County, Pennsylvania; and at localities in Greenland, Austria, Spain, Czechoslovakia, Italy, USSR, Shetland Islands, India, Tasmania, New Zealand, and Australia.
 BEST REF. IN ENGLISH: Palache, et al., "Dana's System of Mineralogy," 7th Ed., v. II, p. 245–246, New York, Wiley, 1951.

## ZAVARITSKITE
BiOF

CRYSTAL SYSTEM: Tetragonal
CLASS: 4/m 2/m 2/m
SPACE GROUP: P4/nmm
Z: 2
LATTICE CONSTANTS:
 $a = 3.75$
 $c = 6.23$
3 STRONGEST DIFFRACTION LINES:
 (synthetic)
 2.65 (100)
 1.87 (100)
 1.62 (100)
OPTICAL CONSTANT:
 $n = 2.213$
HARDNESS: Not determined.

DENSITY: 7.88; 8.34 (Meas.)
 9.00 (Meas. synthetic)
 9.21 (Calc.)
CLEAVAGE: Not determined.
HABIT: Massive; fine-grained.
COLOR-LUSTER: Gray, translucent in very thin layers; semimetallic to greasy. Colorless under the microscope.
MODE OF OCCURRENCE: Occurs with bismutite as pseudomorphs after bismuthinite in quartz-topaz-siderophyllite greisens at the Sherlova Gory deposits, E. Transbaikal.
BEST REF. IN ENGLISH: Dolomanova, E. I., Senderova, V. M., and Yanchenko, M. T., *Am. Min.*, **48**: 210 (1963).

## ZEIRINGITE (Zeyringite) = Aragonite plus aurichalcite

## ZELLERITE
$Ca(UO_2)(CO_3)_2 \cdot 5H_2O$

CRYSTAL SYSTEM: Orthorhombic
CLASS: 2/m 2/m 2/m or mm2
SPACE GROUP: Pmnm or Pmn2$_1$
Z: 4
LATTICE CONSTANTS:
 $a = 11.220$
 $b = 19.252$
 $c = 4.933$
3 STRONGEST DIFFRACTION LINES:
 9.66 (100)
 4.85 ( 50)
 5.59 ( 35)
OPTICAL CONSTANTS:
 $\alpha = 1.536$
 $\beta = 1.559$
 $\gamma = 1.697$
 $(+)2V = 30°-40°$
HARDNESS: ~2
DENSITY: 3.25 (Meas.)
 3.242 (Calc.)
CLEAVAGE: Aggregated clumps fracture in direction of the elongation.
HABIT: Crystals fibrous, hair-like or needle-like, aggregating into pincushion clumps.
COLOR-LUSTER: Light lemon yellow; on dehydration assumes a chalky yellow color. Fluoresces weak mottled green under ultraviolet light. Dull luster.
MODE OF OCCURRENCE: Occurs intimately associated with gypsum and secondary iron oxides at the Lucky Mc uranium mine, Fremont County, Wyoming. Also found at the Pat No. 8 mine, Powder River Basin, Wyoming, and in the Grants uranium district in New Mexico.
BEST REF. IN ENGLISH: Coleman, R. G., Ross, D. R., and Meyrowitz, R., *Am. Min.*, **51**: 1567-1578 (1966).

## ZEMANNITE
$(Zn, Fe)_2 (TeO_3)_3 Na_x H_{2-x} \cdot nH_2O$

CRYSTAL SYSTEM: Hexagonal
CLASS: 6/m
SPACE GROUP: P6$_3$/m

Z: 2
LATTICE CONSTANTS:
 $a = 9.41$
 $c = 7.64$
3 STRONGEST DIFFRACTION LINES:
 8.15 (100)
 2.778 ( 90)
 4.07 ( 80)
OPTICAL CONSTANTS:
 $\omega = 1.85$
 $\epsilon = 1.93$
 (+)
DENSITY: >4.05 (Meas.)
CLEAVAGE: Very brittle.
HABIT: Crystals prismatic, hexagonal, terminated by a dipyramid.
COLOR-LUSTER: Light to dark brown; adamantine.
MODE OF OCCURRENCE: Occurs as minute crystals, less than 1 mm long, at Moctezuma, Sonora, Mexico.
BEST REF. IN ENGLISH: Mandarino, J. A., Matzat, E., and Williams, S. J., *Can. Min.*, **10**: 139-140 (1969) (Abstr.).

## ZEOLITE GROUP
Hydrated aluminosilicates of alkalis and alkaline earths with $(Al + Si):O = 1:2$; characterized by reversible loss of $H_2O$ without significant structural deformation.

See:
| | |
|---|---|
| Analcime | Laumontite |
| Brewsterite | Levyne (Levynite) |
| Chabazite | Mesolite |
| Clinoptilolite | Mordenite |
| Dachiardite | Natrolite |
| Edingtonite | Offretite |
| Epistilbite | Paulingite |
| Erionite | Phillipsite |
| Faujasite | Pollucite |
| Ferrierite | Scolecite |
| Garronite | Stilbite |
| Gismondine | Thomsonite |
| Gmelinite | Viséite |
| Gonnardite | Wairakite |
| Harmotome | Wellsite |
| Herschelite | Yugawaralite |
| Heulandite | |

## ZEOPHYLLITE
$Ca_4 (Si_3O_7)(OH)_4 F_2$

CRYSTAL SYSTEM: Triclinic
Z: 3
LATTICE CONSTANTS:
 $a = 9.34$   $\alpha = 90.0°$
 $b = 9.34$   $\beta = 110°$
 $c = 13.2$   $\gamma = 120°$
3 STRONGEST DIFFRACTION LINES:
 12.0 (100)
 3.03 ( 80)
 6.09 ( 60)
OPTICAL CONSTANTS: For light vibrating in cleavage plane: $n = 1.568-1.5698$.

$(-)2V = 0°-10°$
HARDNESS: 3
DENSITY: 2.747-2.764 (Meas.)
2.613 (Calc.)
CLEAVAGE: {001} perfect
HABIT: Crystals platy; in spherical forms with radiating foliated structure.
COLOR-LUSTER: White. Translucent; tips of crystals often opaque.
MODE OF OCCURRENCE: Occurs in association with apophyllite and zeolites in cavities in basalt at Alten Berg, Leitmeritz, Bohemia, Czechoslovakia.
BEST REF. IN ENGLISH: Chalmers, R. A., Dent, Lesley S., and Taylor, H. F. W., *Min. Mag.*, **31**: 726-735 (1958).

## ZEUNERITE
$Cu(UO_2)_2(AsO_4)_2 \cdot 12H_2O$
Autunite Group

CRYSTAL SYSTEM: Tetragonal
CLASS: 4/m 2/m 2/m
SPACE GROUP: P4/nnc
Z: 2
LATTICE CONSTANTS:
$a = 7.18$
$c = 21.06$
3 STRONGEST DIFFRACTION LINES:
10.3 (100)
3.59 (100)
5.20 ( 70)
OPTICAL CONSTANTS:
$\omega = 1.610$
$\epsilon = 1.582-1.583$
$(-)$
HARDNESS: 2½
DENSITY: 3.39 (Meas.)
CLEAVAGE: {001} perfect
{100} distinct
Brittle.
HABIT: Crystals tabular on {001} and resemble torbernite.
COLOR-LUSTER: Yellowish green, green to emerald green. Transparent when fully hydrated. Weakly vitreous. Not fluorescent.
MODE OF OCCURRENCE: Occurs as a secondary mineral in the oxidized zone of deposits containing uraninite together with arsenic-bearing minerals. Fully hydrated zeunerite has been found at the Dexter mine on Calf Mesa, San Rafael Swell, Utah, at the Anton mine, Black Forest, Germany, and at the Weisser Hirsch mine, Schneeberg, Saxony. Most localities reported for metazeunerite probably also refer to zeunerite.
BEST REF. IN ENGLISH: Berman, Robert, *Am. Min.*, **42**: 905-908 (1957). Frondel, Clifford, USGS Bull. 1064, p. 191-194 (1958).

## ZHEMCHUZHNIKOVITE
$NaMg(Al, Fe^{3+})(C_2O_4)_3 \cdot 8H_2O$

CRYSTAL SYSTEM: Hexagonal

Z: 6
LATTICE CONSTANTS:
$a = 16.67$ kX
$c = 12.51$
OPTICAL CONSTANTS:
$\omega = 1.479$
$\epsilon = 1.408$
$(-)$
HARDNESS: 2
DENSITY: 1.69 (Meas.)
1.66 (Calc.)
CLEAVAGE: Basal, fair.
HABIT: Natural crystals acicular to fibrous; synthetic crystals hexagonal, prismatic.
COLOR-LUSTER: Smoky green; vitreous.

Easily soluble in water.
MODE OF OCCURRENCE: Occurs in veinlets in coal in the Chaitumussk deposits, 200 km from the estuary of the Lena River, USSR.
BEST REF. IN ENGLISH: Knipovich, Yu. N., Komkov, A. I., and Nefedov, E. I., *Am. Min.*, **49**: 442-443 (1964).

## ZINALSITE (Inadequately described mineral)
$Zn_7Al_4(SiO_4)_6(OH)_2 \cdot 9H_2O$

CRYSTAL SYSTEM: Unknown
OPTICAL CONSTANTS:
Mean index = 1.56-1.58. Biaxial $(-)$
HARDNESS: 2½-3
DENSITY: 3.007 (Meas.)
CLEAVAGE: Not determined.
HABIT: As dense cryptocrystalline aggregates.
COLOR-LUSTER: White, rose, to reddish brown.
MODE OF OCCURRENCE: Occurs associated with hydromicas and other zinc clays in the zone of oxidation of the Akdzhal and Achisai deposits, Kazakhstan, and also found at Sterling Hill, New Jersey.
BEST REF. IN ENGLISH: Chukhrov, F. V., *Am. Min.*, **44**: 208-209 (1959).

## ZINC
Zn

CRYSTAL SYSTEM: Hexagonal
CLASS: 6/m 2/m 2/m
SPACE GROUP: $P6_3/mmc$
Z: 2
LATTICE CONSTANTS:
$a = 2.6591$
$c = 4.9353$
3 STRONGEST DIFFRACTION LINES:
2.091 (100)
2.473 ( 53)
2.308 ( 40)
OPTICAL CONSTANTS: Isotropic with $n$ 1.80; reflectivity 57%.
HARDNESS: 2
DENSITY: 6.9-7.2 (Meas.)
7.135 (Calc.)

CLEAVAGE: {0001} perfect
Brittle.
HABIT: As aggregates of ovoid or platy grains, not exceeding 15 μm in diameter.
COLOR-LUSTER: Bluish gray. Opaque. Metallic. Streak grayish white.
MODE OF OCCURRENCE: The mineral occurs as an oxidation product of sphalerite at Keno Hill, Yukon Territory, Canada; in the bedrock of the Shamshadinsk and Alaverdi areas and in the northern Armenian alluvial deposits; in spherulitic and irregular forms from the heavy fractions of Kazakhstan Upper Devonian biotite granites; and as an oxidation product of sphalerite and djurleite at the Dulcinea de Llampos copper mine near Copiapó, northern Chile.
BEST REF. IN ENGLISH: Boyle, R. W., *Can. Min.*, **6**: 692–694 (1961). Bartikyan, P. M., *Min. Abs.*, **18**: 200 (1967).

## ZINCALUMINITE (Inadequately described mineral)
$Zn_6 Al_6 (SO_4)_2 (OH)_{26} \cdot 5H_2O$

CRYSTAL SYSTEM: Probably orthorhombic
OPTICAL CONSTANTS:
$\omega = 1.534$
$\epsilon = 1.514$
(–)
HARDNESS: 2½–3
DENSITY: 2.26 (Meas.)
CLEAVAGE: Not determined.
HABIT: Crystals minute thin hexagonal plates. As crusts and tufted aggregates.
COLOR-LUSTER: White, bluish white, pale blue.
MODE OF OCCURRENCE: Occurs as a secondary mineral at the zinc mines at Laurium, Greece, associated with smithsonite.
BEST REF. IN ENGLISH: Palache, et al., "Dana's System of Mineralogy," 7th Ed., v. II, p. 579–580, New York, Wiley, 1951.

## ZINC BLENDE = Sphalerite

## ZINC-FAUSERITE
$(Mn, Mg, Zn)SO_4 \cdot 7H_2O$

CRYSTAL SYSTEM: Orthorhombic
LATTICE CONSTANTS:
$a:b:c = 0.9821:1:0.5615$
OPTICAL CONSTANTS:
Mean index = 1.465
(–)2V = large
HARDNESS: 2½
DENSITY: 1.9971
CLEAVAGE: {010} good
HABIT: As an efflorescent crust. Forms: {100}, {010}, {101}, {110}, and {111}.
COLOR-LUSTER: Pale rose. Loses water rapidly in air and is quickly covered by a white crust.

MODE OF OCCURRENCE: Occurs on Level XII at the Felsöbánya mine, Roumania.
BEST REF. IN ENGLISH: Tokody, Lázló, *Am. Min.*, **35**: 333 (1950).

## ZINCITE
(Zn, Mn)O

CRYSTAL SYSTEM: Hexagonal
CLASS: 6mm
SPACE GROUP: $P6_3mc$
Z: 2
LATTICE CONSTANTS:
$a = 3.242$
$c = 5.176$
3 STRONGEST DIFFRACTION LINES:
2.476 (100)
2.816 ( 71)
2.602 ( 56)
OPTICAL CONSTANTS:
$\omega = 2.013$
$\epsilon = 2.029$ (Na)
(+)
HARDNESS: 4
DENSITY: 5.684 (Meas.)
5.680 (Calc.)
CLEAVAGE: $\{10\bar{1}0\}$ perfect, difficult
$\{000\bar{1}\}$ parting-sometimes distinct
Fracture conchoidal. Brittle.
HABIT: Crystals hemimorphic pyramidal with large negative base, commonly rounded and corroded. Often massive, platy; also compact or granular. Twinning on {0001}.
COLOR-LUSTER: Deep yellow to orange-yellow to dark red. Translucent to transparent. Subadamantine. Streak orange-yellow.
MODE OF OCCURRENCE: Occurs associated with franklinite, willemite, and calcite in the zinc deposits at Franklin and Sterling Hill, Sussex County, New Jersey; at the Dick Weber mine, Saguache County, Colorado; and at localities in Spain, Italy, Poland, and Tasmania.
BEST REF. IN ENGLISH: Palache, et al., "Dana's System of Mineralogy," 7th Ed., v. I, p. 504–506, New York, Wiley, 1944.

## ZINC-MELANTERITE
$(Zn, Cu, Fe)SO_4 \cdot 7H_2O$

CRYSTAL SYSTEM: Monoclinic
CLASS: 2/m
OPTICAL CONSTANTS:
$\alpha = 1.479$
$\beta = 1.483$
$\gamma = 1.488$
(+)2V ~ 90°
HARDNESS: ~2
DENSITY: 2.02
CLEAVAGE: Not determined.
HABIT: Massive, columnar.
COLOR-LUSTER: Pale greenish blue. Vitreous.

Dehydrates readily at room temperature.

MODE OF OCCURRENCE: Occurs in large amounts as an alteration product of pyrite-chalcopyrite-sphalerite ore at the Good Hope and Vulcan mines, Gunnison County, Colorado.

BEST REF. IN ENGLISH: Palache, et al., "Dana's System of Mineralogy," 7th Ed., v. II, p. 508, New York, Wiley, 1951.

## ZINCOBOTRYOGEN
$(Zn, Mg, Mn^{2+})Fe^{3+}(SO_4)_2(OH) \cdot 7H_2O$

CRYSTAL SYSTEM: Monoclinic
CLASS: 2/m
SPACE GROUP: $P2_1/n$
Z: 4
LATTICE CONSTANTS:
 $a = 10.488$
 $b = 17.819$
 $c = 7.185$
 $\beta = 100°50'$
3 STRONGEST DIFFRACTION LINES:
 9.01 (100)
 5.24 ( 80)
 3.24 ( 60)
OPTICAL CONSTANTS:
  $\alpha = 1.542$
  $\beta = 1.551$
  $\gamma = 1.587$
 $(+)2V = 54°$  (Calc.)
HARDNESS: ~2.5
DENSITY: 2.201
HABIT: Radiating crystalline aggregates and single prismatic crystals, showing the forms b{010}, $\sigma${101}, l{120}, m{110}.

COLOR-LUSTER: Bright orange-red; vitreous to greasy.

MODE OF OCCURRENCE: Occurs associated with pickeringite in the oxidation zone of a lead-zinc deposit on the northern border of the extremely arid Tsaidam Basin, China.

BEST REF. IN ENGLISH: Tu-Chih, Kuang; Hsieh, Hsien-Te; Li, Hsi-Lin; and Yin, Shu-Shen, *Am. Min.*, **49**: 1776–1777 (1964).

## ZINCOCOPIAPITE
$ZnFe_4(SO_4)_6(OH)_2 \cdot 18H_2O$

CRYSTAL SYSTEM: Triclinic
CLASS: $\bar{1}$
SPACE GROUP: $P\bar{1}$
Z: 1
LATTICE CONSTANTS:
 $a = 7.35$    $\alpha = 93°50'$
 $b = 18.16$   $\beta = 101°30'$
 $c = 7.28$    $\gamma = 99°22'$
3 STRONGEST DIFFRACTION LINES:
 9.25 (100)
 6.12 ( 80)
 3.56 ( 60)

OPTICAL CONSTANTS:
  $\alpha = 1.534$
  $\beta = 1.554$
  $\gamma = 1.586$
 $(+)2V = 78°$  (Calc.)
HARDNESS: ~2
DENSITY: 2.181
HABIT: Compact, massive aggregates. Crystal forms b{010}, a{100}, e{0$\bar{1}$1}.

COLOR-LUSTER: Yellowish green; vitreous.

MODE OF OCCURRENCE: Occurs closely associated with copiapite, halotrichite, coquimbite, roemerite, sideronatrite, and melanterite in the oxidation zone of a lead-zinc deposit on the northern border of the extremely arid Tsaidam Basin, China.

BEST REF. IN ENGLISH: Tu-Chih, Kuang; Li, Hsi-Lin; Hsieh, Hsien-Te; and Yin, Shu-Shen, *Am. Min.*, **49**: 1777 (1964).

## ZINCROSASITE = Variety of rosasite with Zn > Cu

## ZINCSILITE (Inadequately described mineral)
near $Zn_3Si_4O_{10}(OH)_2 \cdot nH_2O$
Montmorillonite Group

CRYSTAL SYSTEM: Unknown
3 STRONGEST DIFFRACTION LINES:
 15.3  (100)
  4.09 ( 70)
  1.528 ( 60)
OPTICAL CONSTANTS:
  $\alpha = 1.514$
  $\beta = 1.559$
  $\gamma = 1.562$
 $(-)2V = 0°-22°$
HARDNESS: 1½–2
DENSITY: 2.61–2.71 (Meas.)
CLEAVAGE: {001} perfect
HABIT: Fine foliae or lamellae up to $2 \times 1.5 \times 0.5$ mm in size.

COLOR-LUSTER: White to bluish; pearly on cleavage face.

MODE OF OCCURRENCE: Occurs as a homo-axial pseudomorph after diopside associated with chrysocolla, supergene fluorite, opal, and manganese oxides in the zone of oxidation of the skarn galena-sphalerite-chalcopyrite deposits of Batystau, central Kazakhstan.

BEST REF. IN ENGLISH: Smol'yaninova, N. N., Moleva, V. A., and Organova, N. I., *Am. Min.*, **46**: 241 (1961).

## ZINKENITE (Zinckenite)
$Pb_6Sb_{14}S_{27}$

CRYSTAL SYSTEM: Hexagonal
CLASS: 6 or 6/m
SPACE GROUP: $P6_3$ or $P6_3/m$
Z: 12

LATTICE CONSTANTS:

$a = 44.14$
$c = 8.62$

3 STRONGEST DIFFRACTION LINES:

3.45 (100)
2.80 ( 40)
1.99 ( 30)

OPTICAL CONSTANTS: Opaque.

HARDNESS: 3-3½
DENSITY: 5.36 (Meas.)
5.35 (Calc.)
CLEAVAGE: $\{11\bar{2}0\}$ indistinct
Fracture uneven.

HABIT: Crystals slender prismatic, vertically striated. Usually massive, or in radial fibrous to columnar aggregates.

COLOR-LUSTER: Steel gray, sometimes tarnished iridescent. Opaque. Metallic. Streak steel gray.

MODE OF OCCURRENCE: Occurs in low- to medium-temperature vein deposits, commonly associated with boulangerite and other sulfosalts, sulfides, carbonates, and quartz. Found at the Sweet Home mine, Alma, Park County, and at the Brobdignag claim, San Juan County, Colorado; in Eureka and Nye counties, Nevada; at the Tuttle mine, Silver City, Pennington County, South Dakota; and in Sevier County, Arkansas. It also occurs in British Columbia, Canada, and at deposits in Bolivia, France, Germany, Czechoslovakia, Roumania, and Tasmania.

BEST REF. IN ENGLISH: Palache, et al., "Dana's System of Mineralogy," 7th Ed., v. I, p. 476-478, New York, Wiley, 1944.

## ZINNWALDITE

K(Li, Al, Fe)$_3$(Al, Si)$_4$O$_{10}$(OH, F)$_2$
Mica Group

CRYSTAL SYSTEM: Monoclinic
CLASS: 2/m
SPACE GROUP: C2/m or C2/c
Z: 4
LATTICE CONSTANTS:

$a = 5.27$
$b = 9.09$
$c = 20.14$
$\beta = 100°00'$

3 STRONGEST DIFFRACTION LINES:

3.29 (100)
9.8 ( 80)
1.98 ( 55)

OPTICAL CONSTANTS:

$\alpha \sim 1.55$
$\beta \sim 1.58$
$\gamma \sim 1.58$
$(-)2V \sim 30°$

HARDNESS: 2½-4
DENSITY: 2.9-3.3 (Meas.)
CLEAVAGE: $\{001\}$ perfect, micaceous
Laminae flexible, elastic.

HABIT: Crystals short prismatic or tabular. Also as disseminated scales or scaly aggregates. Twinning on $\{001\}$, twin axis (310).

COLOR-LUSTER: Gray, brown, sometimes dark green. Transparent. Vitreous; often pearly on cleavage.

MODE OF OCCURRENCE: Occurs chiefly in greisens, high-temperature quartz veins, and in granite pegmatites. Found in Riverside and San Diego counties, California; at Amelia, Virginia; in the York district, Seward Peninsula, Alaska; at Narsarsuk, Greenland; in tin veins of Cornwall, England, and at Zinnwald and elsewhere in Saxony, Germany; also at several places in the USSR.

BEST REF. IN ENGLISH: Vlasov, K. A., "Mineralogy of Rare Elements," v. II, p. 19-22, Israel Program for Scientific Translations, 1966.

## ZIPPEITE

(UO$_2$)$_2$(SO$_4$)(OH)$_2$ · 4H$_2$O

CRYSTAL SYSTEM: Orthorhombic
LATTICE CONSTANTS:

$a = 7.14$
$b = 17.34$
$c = 4.43$

3 STRONGEST DIFFRACTION LINES:

7.06 (100)
3.12 ( 90)
3.51 ( 80)

OPTICAL CONSTANTS:

$\alpha \simeq 1.64-1.68$
$\beta \simeq 1.72-1.715$
$\gamma \simeq 1.75-1.77$

HARDNESS: Not determined.
DENSITY: 3.66 (Meas. synthetic)
3.68 (Calc. synthetic)
CLEAVAGE: Not determined.
Probably $\{010\}$ perfect.

HABIT: As thin coatings or compact aggregates composed of microscopic crystals. Twinned crystals common.

COLOR-LUSTER: Orange yellow, golden yellow through orange-red and reddish brown. Transparent in crushed grains. Dull to earthy. Fluorescence widely variable and not dependable for diagnostic purposes.

MODE OF OCCURRENCE: Occurs as a secondary mineral, often of recent formation, as an efflorescence on mine walls or dump heaps. In the United States found in the pitchblende ores of several deposits in the Central City and adjoining district of Gilpin and Clear Creek counties, at the Thornburg mine, Los Ochos district, Saguache County, and coating mine walls at the Schwartzwalder mine, Jefferson County, Colorado. It is also found at the Happy Jack mine, White Canyon, the Lucky Strike No. 2 mine, Emery County, and at other places in Utah; at the Hillside mine, Yavapai County, Arizona; at the Woodrow mine, Valencia County, New Mexico; at Great Bear Lake, Canada; in Cornwall, England; and at Příbram and Schlaggenwald, Bohemia.

BEST REF. IN ENGLISH: Frondel, Clifford, USGS Bull. 1064, p. 141-147 (1958).

## ZIRCON

ZrSiO$_4$
Var. Cyrtolite

CRYSTAL SYSTEM: Tetragonal
CLASS: 4/m 2/m 2/m
SPACE GROUP: I4$_1$/amd
Z: 4
LATTICE CONSTANTS:
  $a = 6.60$
  $c = 5.98$
3 STRONGEST DIFFRACTION LINES:
  3.30 (100)
  4.43 ( 45)
  2.518 ( 45)
OPTICAL CONSTANTS:
  $\omega = 1.923$-1.960
  $\epsilon = 1.968$-2.015
    (+)
HARDNESS: 7½
DENSITY: 4.6-4.7 (Meas.)
        4.668 (Calc.)
        3.6-4.0 (Meas. metamict)
CLEAVAGE: {110} imperfect
          {111} poor
Fracture uneven; metamict varieties conchoidal. Brittle.

HABIT: Crystals short to long prismatic; dipyramidal. As single crystals, sheaf-like and radial-fibrous aggregates, and as irregular grains. Metamict varieties often as parallel or subparallel aggregates with curved faces and edges. Twinning on {111}, {100}, and {221} rare.

COLOR-LUSTER: Colorless; usually shades of brown, green, gray, yellow, and red. Transparent. Metamict varieties transparent to opaque. Vitreous to adamantine. Metamict varieties vitreous to greasy.

MODE OF OCCURRENCE: Occurs widespread as an accessory mineral in igneous rocks, in some metamorphic rocks, and as a common detrital mineral in some sedimentary deposits. Notable United States occurrences are found at St. Peter's Dome, El Paso County, Colorado; Tin Mountain mine, Custer County, South Dakota; Ouichita Mountains, Oklahoma; Llano County, Texas; Henderson County, North Carolina; and at localities in Maine, Massachusetts, New York, and New Jersey. In Quebec and Ontario, Canada, as fine crystals up to 12 X 4 inches in size and 15 pounds in weight. Other notable occurrences are found in Brazil, Norway, France, Italy, Germany, USSR, Rhodesia, Madagascar, Ceylon, Thailand, Burma, Korea, Australia, and elsewhere.

BEST REF. IN ENGLISH: Deer, Howie, and Zussman, "Rock Forming Minerals," v. 1, p. 59-68, New York, Wiley, 1962.

## ZIRCONOLITE

CaZrTi$_2$O$_7$
Var. Niobozirconolite

CRYSTAL SYSTEM: Cubic
Z: 1
LATTICE CONSTANTS: Metamict.
  $a = 5.02$ after heating to 700°-800°C
3 STRONGEST DIFFRACTION LINES:
  (after ignition)
    2.98 (100)
    1.82 ( 50)
    2.53 ( 30)

OPTICAL CONSTANT:
  $N = 2.06$-2.15
HARDNESS: 5½-6
DENSITY: 4-4.5 (Meas.)
CLEAVAGE: None. Fracture uneven to conchoidal.
HABIT: Crystals imperfect flattened octahedra. Usually as irregular masses up to 2-3 cm. Twinning on {111} (?)
COLOR-LUSTER: Brown to black. Opaque; translucent in thin fragments. Streak brownish yellow.

MODE OF OCCURRENCE: Occurs in highly altered pyroxenities in association with apatite, calcite, ilmenite, perovskite, and sphene. Found in the Afrikanda massif, Kola Peninsula, and in the Arbarastakh massif, Aldan, USSR. The niobium variety is reported from carbonatites.

BEST REF. IN ENGLISH: Vlasov, K. A., "Mineralogy of Rare Elements," v. II, p. 336-340, Israel Program for Scientific Translations, 1966.

## ZIRCOSULFATE

Zr(SO$_4$)$_2$ · 4H$_2$O

CRYSTAL SYSTEM: Orthorhombic
CLASS: 2/m 2/m 2/m
SPACE GROUP: Fddd
Z: 8
LATTICE CONSTANTS:
  $a = 25.92$
  $b = 11.62$ (synthetic)
  $c = 5.532$
3 STRONGEST DIFFRACTION LINES:
  4.33 (100)
  2.98 ( 90)
  2.33 ( 60)
OPTICAL CONSTANTS:
  $\alpha = 1.620$
  $\beta = 1.644$ (Calc.)
  $\gamma = 1.674$
(+)2V = 75°
HARDNESS: 2½-3
DENSITY: 2.85 (Meas.)
        2.83 (Meas. Synthetic)
        2.833 (Calc.)
CLEAVAGE: Not observed.
HABIT: As compact, powdery masses; particles are 0.01-0.03 mm in size, have rounded forms or rhombic sections.
COLOR-LUSTER: Colorless or white; dull.
MODE OF OCCURRENCE: Occurs associated with hisingerite, smithsonite, limonite, and alteration products of eudialyte filling a cavity 2 cm in diameter in nepheline syenite pegmatite of the Korgeredabin alkalic massif, southeastern Tuva.

BEST REF. IN ENGLISH: Kapustin, Yu. L., *Am. Min.*, 51: 529 (1966).

## ZIRFESITE (Inadequately described mineral)

Near ZrO$_2$Fe$_2$O$_3$SiO$_2$ · nH$_2$O

CRYSTAL SYSTEM: Unknown.
LATTICE CONSTANTS: Amorphous.

3 STRONGEST DIFFRACTION LINES:
(after ignition)
2.92 (100)
1.787 ( 90)
1.520 ( 90)
OPTICAL CONSTANT:
$N = 1.60-1.65$
HARDNESS: Not determined.
DENSITY: 2.70 (Meas.)
CLEAVAGE: None.
HABIT: Massive; as crusts and pseudomorphs after eudialyte, up to 3 cm in size. Pulverulent.
COLOR-LUSTER: Light yellow. Dull; isolated flakes pearly.
MODE OF OCCURRENCE: Occurs in nepheline-syenites as an alteration of eudialyte in the Khibiny and Lovozero massifs, Kola Peninsula, USSR.
BEST REF. IN ENGLISH: Vlasov, K. A., "Mineralogy of Rare Elements," v. II, p. 392-393, Israel Program for Scientific Translations, 1966.

## ZIRKELITE (Inadequately described mineral)
$(Ca, Fe, Th)_2 (Zr, Ti, Nb)_2 O_7$ (?)

CRYSTAL SYSTEM: Cubic (?)
OPTICAL CONSTANT:
$N = 2.19$
HARDNESS: 5½
DENSITY: 4.706; 4.741 (Meas.)
CLEAVAGE: None. Fracture conchoidal. Brittle.
HABIT: Crystals flattened octahedra. Twinning on {111} common, polysynthetic and as fourlings.
COLOR-LUSTER: Black. Opaque, except in very thin fragments. Resinous.
MODE OF OCCURRENCE: Occurs with baddeleyite and perovskite in magnetite pyroxenites (jacupirangites) of Jacupiranga, São Paulo, Brazil.
BEST REF. IN ENGLISH: Palache, et al., "Dana's System of Mineralogy," 7th Ed., v. I, p. 740, New York, Wiley, 1944.

## ZIRKLERITE (Inadequately described mineral)
$(Fe, Mg, Ca)_9 Al_4 Cl_{18} (OH)_{12} \cdot 14H_2O$ (?)

CRYSTAL SYSTEM: Probably hexagonal
OPTICAL CONSTANTS:
Mean index = 1.552. Uniaxial (+)
HARDNESS: ~3½
DENSITY: ~2.6 (Meas.)
CLEAVAGE: Rhombohedral.
HABIT: Massive, granular to fibrous.
COLOR-LUSTER: Light gray.
MODE OF OCCURRENCE: Occurs as the chief constituent of a rock occurring in breccia-like layers in halite and potash salts in the Adolfsglück mine at Hope, Hannover, and other potash deposits in northern Germany.
BEST REF. IN ENGLISH: Palache, et al., "Dana's System of Mineralogy," 7th Ed., v. II, p. 87, New York, Wiley, 1951.

## ZOISITE (Epidote Group)
$Ca_2 Al_3 Si_3 O_{12} OH$
Dimorphous with clinozoisite
Var. Thulite
Tanzanite

CRYSTAL SYSTEM: Orthorhombic
CLASS: 2/m 2/m 2/m
SPACE GROUP: Pnmc
Z: 4
LATTICE CONSTANTS:
$a = 16.2-16.3$
$b = 5.45-5.63$
$c = 10.04-10.21$
3 STRONGEST DIFFRACTION LINES:
2.69 (100)
2.87 ( 65)
4.03 ( 50)
OPTICAL CONSTANTS:
$\alpha = 1.685-1.705$
$\beta = 1.688-1.710$
$\gamma = 1.697-1.725$
$(+)2V = 0°-60°$
HARDNESS: 6½-7
DENSITY: 3.355 (Meas.)
3.358 (Calc.)
CLEAVAGE: {100} perfect
{010} imperfect
Fracture uneven to conchoidal. Brittle.
HABIT: Crystals usually prismatic, often deeply striated vertically. Commonly massive, compact to columnar.
COLOR-LUSTER: Gray, white, greenish gray, greenish brown, green; also pink (thulite), or colorless to blue or purple (tanzanite). Transparent to translucent. Vitreous; sometimes pearly on cleavage. Streak uncolored.
MODE OF OCCURRENCE: Occurs chiefly in regionally metamorphosed rocks such as calcareous shales and schists, basic igneous rocks, and argillaceous sandstones; also in altered granitic rocks and gneisses, metamorphosed impure dolomites and limestones, and quartz veins and pegmatites. Typical occurrences are found in Okanogan County, Washington (thulite); at numerous localities in California; in Douglas, Lyon, and Mineral counties, Nevada (thulite); in the Black Hills, South Dakota; at Ducktown, Tennessee; in Chester and Delaware counties, Pennsylvania; in Franklin and Hampshire counties, Massachusetts; and in Mitchell and Yancey counties, North Carolina (thulite). Also in Baja California, Mexico; in east Greenland (thulite); at Telemark and elsewhere in Norway; and at localities in Scotland, Italy, Germany, Austria, Finland, USSR, Japan, and as excellent crystals exhibiting a large number of forms in Tanzania (tanzanite).
BEST REF. IN ENGLISH: Deer, Howie, and Zussman, "Rock Forming Minerals," v. 1, p. 185-192, New York, Wiley, 1962. Dollase, W. A., *Am. Min.*, 53: 1883-1898 (1968). Hurlbut, C. S. Jr., *Am. Min.*, 54: 702-709 (1969).

## ZUNYITE
$Al_{13} Si_5 O_{20} (OH, F)_{18} Cl$

CRYSTAL SYSTEM: Cubic

CLASS: $\bar{4}$3m
SPACE GROUP: F$\bar{4}$3m
Z: 4
LATTICE CONSTANT:
 $a$ = 13.925
3 STRONGEST DIFFRACTION LINES:
 8.07 (100)
 4.21 (100)
 2.68 ( 90)
OPTICAL CONSTANT:
 $N$ = 1.600 (Na)
 HARDNESS: 7
 DENSITY: 2.873 (Meas.)
  2.855 (Calc.)
 CLEAVAGE: {111} easy
Brittle.
 HABIT: Crystals euhedral tetrahedrons, minute; also octahedrons, frequently twinned; less than 1 mm to about 5 mm in size.
 COLOR-LUSTER: Colorless, grayish white to flesh color. Transparent to translucent. Vitreous.
 MODE OF OCCURRENCE: Occurs at three localities in Colorado; as a vein mineral and disseminated through an altered porphyry at the Charter Oak mine in Ouray County; as an alteration product of feldspar in a silicified dike rock in the Bonanza district, Saguache County; and enclosed in guitermanite at the Zuni mine (type locality), San Juan County. Also found in highly aluminous shales and flagstones at Postmasburg, South Africa; in transparent tetrahedrons at Kuni Village, Gumma Prefecture, Japan; and at Beni-Embarek, Algeria.
 BEST REF. IN ENGLISH: Vermaas, F. H. S., *Am. Min.*, **37**: 960–965 (1952).

# ZUSSMANITE
K(Fe, Mg, Mn)$_{13}$(Si, Al)$_{18}$O$_{42}$(OH)$_{14}$

CRYSTAL SYSTEM: Hexagonal
CLASS: 3 or $\bar{3}$
SPACE GROUP: R3 or R$\bar{3}$
Z: 1 (rhomb.), 3 (hexag.)
LATTICE CONSTANTS:
 $a$ = 11.66   $a_{rh}$ = 11.69
 $c$ = 28.69   $\alpha$ = 59.8°
3 STRONGEST DIFFRACTION LINES:
 9.60 (100)
 4.78 ( 45)
 3.19 ( 25)
OPTICAL CONSTANTS:
 $\omega$ = 1.643
 $\epsilon$ = 1.623
  (−)
DENSITY: 3.146 (Meas.)
CLEAVAGE: {0001} perfect
HABIT: Crystals tabular.
COLOR-LUSTER: Pale green. Translucent. Vitreous.
MODE OF OCCURRENCE: Occurs in some of the metamorphosed shales, siliceous ironstones, and impure limestones of the Franciscan formation in the Laytonville district, Mendocino County, California.

BEST REF. IN ENGLISH: Agrell, S. O., Bown, M. G., and McKie, D., *Am. Min.*, **50**: 278 (1965).

# ZVYAGINTSEVITE
(Pd, Pt)$_3$Pb
Type material is a high tin-bearing variety, Pd$_3$ (Pb, Sn)

CRYSTAL SYSTEM: Cubic
CLASS: 4/m $\bar{3}$ 2/m
SPACE GROUP: Pm3m
Z: 1
LATTICE CONSTANT:
 $a$ = 4.02
3 STRONGEST DIFFRACTION LINES:
 2.32 (100)
 1.22 ( 90)
 2.01 ( 80)
OPTICAL CONSTANTS: Isotropic.
 HARDNESS: Vickers hardness 279 kg/mm$^2$ with 15 g load.
 DENSITY: 13.32 (Meas.)
  13.41 (Calc.)
 HABIT: Irregular grains from less than 4 $\mu$ to 250 $\mu$, also as veinlets up to 120 $\mu$ long.
 COLOR-LUSTER: White in reflected light; metallic. Powder black.
 MODE OF OCCURRENCE: Occurs associated with cubic chalcopyrite, cubanite, pentlandite, magnetite, valleriite, a silver-gold alloy, and a new mineral Pd(Bi, Pb), in an ore sample from Noril'sk, Siberia.
 BEST REF. IN ENGLISH: Genkin, A. D., Murav'eva, I. V., and Troneva, N. V., *Am. Min.*, **52**: 299 (1967).   Cabri, L. J., and Traill, R. J., *Can. Min.*, **8**: 541–550 (1966).

# ZWIESELITE
(Fe$^{2+}$, Mn$^{2+}$, Mg, Ca)$_2$(PO$_4$)(F, OH)
Triplite-Zwieselite Series

CRYSTAL SYSTEM: Monoclinic
CLASS: 2/m
SPACE GROUP: I2/m
Z: 8
LATTICE CONSTANTS:
 $a$ = 12.03 kX
 $b$ = 6.46
 $c$ = 10.03
 $\beta$ = 105°42′
OPTICAL CONSTANTS:
 $\alpha$ = 1.696
 $\beta$ = 1.704 (variable)
 $\gamma$ = 1.713
(+)2V = 87° (Calc.)
HARDNESS: 5-5½
DENSITY: 3.927-3.97 (Meas.)
CLEAVAGE: {001} good
 {010} fair
 {100} poor
Fracture subconchoidal to uneven.

HABIT: Crystals poorly developed; rough. Usually massive.

COLOR-LUSTER: Clove brown; brownish black to black on alteration. Translucent. Vitreous to resinous.

MODE OF OCCURRENCE: Occurs as a primary mineral in granite pegmatites. Found at Rabenstein, near Zwiesel, Bavaria, Germany; also at Kimoto, Finland.

BEST REF. IN ENGLISH: Palache, et al., "Dana's System of Mineralogy," 7th Ed., v. 2, p. 849–852, New York, Wiley, 1951.